DISCRETE MATHEMATICS AND ITS APPLICATIONS
Series Editor KENNETH H. ROSEN

HANDBOOK OF LINEAR ALGEBRA

DISCRETE MATHEMATICS AND ITS APPLICATIONS

Series Editor

Kenneth H. Rosen, Ph.D.

Juergen Bierbrauer, Introduction to Coding Theory

Kun-Mao Chao and Bang Ye Wu, Spanning Trees and Optimization Problems

Charalambos A. Charalambides, Enumerative Combinatorics

Henri Cohen, Gerhard Frey, et al., Handbook of Elliptic and Hyperelliptic Curve Cryptography

Charles J. Colbourn and Jeffrey H. Dinitz, Handbook of Combinatorial Designs, Second Edition

Steven Furino, Ying Miao, and Jianxing Yin, Frames and Resolvable Designs: Uses, Constructions, and Existence

Randy Goldberg and Lance Riek, A Practical Handbook of Speech Coders

Jacob E. Goodman and Joseph O'Rourke, Handbook of Discrete and Computational Geometry, Second Edition

Jonathan L. Gross and Jay Yellen, Graph Theory and Its Applications, Second Edition

Jonathan L. Gross and Jay Yellen, Handbook of Graph Theory

Darrel R. Hankerson, Greg A. Harris, and Peter D. Johnson, Introduction to Information Theory and Data Compression, Second Edition

Daryl D. Harms, Miroslav Kraetzl, Charles J. Colbourn, and John S. Devitt, Network Reliability: Experiments with a Symbolic Algebra Environment

Leslie Hogben, Handbook of Linear Algebra

Derek F. Holt with Bettina Eick and Eamonn A. O'Brien, Handbook of Computational Group Theory

David M. Jackson and Terry I. Visentin, An Atlas of Smaller Maps in Orientable and Nonorientable Surfaces

Richard E. Klima, Neil P. Sigmon, and Ernest L. Stitzinger, Applications of Abstract Algebra with Maple™ and MATLAB®, Second Edition

Patrick Knupp and Kambiz Salari, Verification of Computer Codes in Computational Science and Engineering

William Kocay and Donald L. Kreher, Graphs, Algorithms, and Optimization

Donald L. Kreher and Douglas R. Stinson, Combinatorial Algorithms: Generation Enumeration and Search

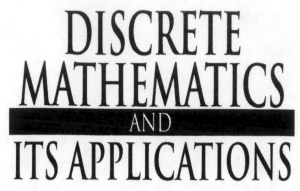

Continued Titles

Charles C. Lindner and Christopher A. Rodgers, Design Theory

Hang T. Lau, A Java Library of Graph Algorithms and Optimization

Alfred J. Menezes, Paul C. van Oorschot, and Scott A. Vanstone, Handbook of Applied Cryptography

Richard A. Mollin, Algebraic Number Theory

Richard A. Mollin, Codes: The Guide to Secrecy from Ancient to Modern Times

Richard A. Mollin, Fundamental Number Theory with Applications

Richard A. Mollin, An Introduction to Cryptography, Second Edition

Richard A. Mollin, Quadratics

Richard A. Mollin, RSA and Public-Key Cryptography

Carlos J. Moreno and Samuel S. Wagstaff, Jr., Sums of Squares of Integers

Dingyi Pei, Authentication Codes and Combinatorial Designs

Kenneth H. Rosen, Handbook of Discrete and Combinatorial Mathematics

Douglas R. Shier and K.T. Wallenius, Applied Mathematical Modeling: A Multidisciplinary Approach

Jörn Steuding, Diophantine Analysis

Douglas R. Stinson, Cryptography: Theory and Practice, Third Edition

Roberto Togneri and Christopher J. deSilva, Fundamentals of Information Theory and Coding Design

Lawrence C. Washington, Elliptic Curves: Number Theory and Cryptography

DISCRETE MATHEMATICS AND ITS APPLICATIONS
Series Editor KENNETH H. ROSEN

HANDBOOK OF LINEAR ALGEBRA

EDITED BY
LESLIE HOGBEN
IOWA STATE UNIVERSITY
AMES, IOWA, U.S.A.

ASSOCIATE EDITORS
RICHARD BRUALDI
ANNE GREENBAUM
ROY MATHIAS

Chapman & Hall/CRC
Taylor & Francis Group
Boca Raton London New York

Chapman & Hall/CRC is an imprint of the
Taylor & Francis Group, an informa business

Chapman & Hall/CRC
Taylor & Francis Group
6000 Broken Sound Parkway NW, Suite 300
Boca Raton, FL 33487-2742

Printed in the United States of America on acid-free paper
10 9 8 7 6 5 4 3 2

International Standard Book Number-10: 1-58488-510-6 (Hardcover)
International Standard Book Number-13: 978-1-58488-510-8 (Hardcover)

Visit the Taylor & Francis Web site at
http://www.taylorandfrancis.com

and the CRC Press Web site at
http://www.crcpress.com

Dedication

I dedicate this book to my husband, Mark Hunacek, with gratitude both for his support throughout this project and for our wonderful life together.

Acknowledgments

I would like to thank Executive Editor Bob Stern of Taylor & Francis Group, who envisioned this project and whose enthusiasm and support has helped carry it to completion. I also want to thank Yolanda Croasdale, Suzanne Lassandro, Jim McGovern, Jessica Vakili and Mimi Williams, for their expert guidance of this book through the production process.

I would like to thank the many authors whose work appears in this volume for the contributions of their time and expertise to this project, and for their patience with the revisions necessary to produce a unified whole from many parts.

Without the help of the associate editors, Richard Brualdi, Anne Greenbaum, and Roy Mathias, this book would not have been possible. They gave freely of their time, expertise, friendship, and moral support, and I cannot thank them enough.

I thank Iowa State University for providing a collegial and supportive environment in which to work, not only during the preparation of this book, but for more than 25 years.

Leslie Hogben

The Editor

Leslie Hogben, Ph.D., is a professor of mathematics at Iowa State University. She received her B.A. from Swarthmore College in 1974 and her Ph.D. in 1978 from Yale University under the direction of Nathan Jacobson. Although originally working in nonassociative algebra, she changed her focus to linear algebra in the mid-1990s.

Dr. Hogben is a frequent organizer of meetings, workshops, and special sessions in combinatorial linear algebra, including the workshop, "Spectra of Families of Matrices Described by Graphs, Digraphs, and Sign Patterns," hosted by American Institute of Mathematics in 2006 and the Topics in Linear Algebra Conference hosted by Iowa State University in 2002. She is the Assistant Secretary/Treasurer of the International Linear Algebra Society.

An active researcher herself, Dr. Hogben particularly enjoys introducing graduate and undergraduate students to mathematical research. She has three current or former doctoral students and nine master's students, and has worked with many additional graduate students in the Iowa State University Combinatorial Matrix Theory Research Group, which she founded. Dr. Hogben is the co-director of the NSF-sponsored REU "Mathematics and Computing Research Experiences for Undergraduates at Iowa State University" and has served as a research mentor to ten undergraduates.

Contributors

Marianne Akian
INRIA, France

Zhaojun Bai
University of California-Davis

Ravindra Bapat
Indian Statistical Institute

Francesco Barioli
University of Tennessee-Chattanooga

Wayne Barrett
Brigham Young University, UT

Christopher Beattie
Virginia Polytechnic Institute and State University

Peter Benner
Technische Universität Chemnitz, Germany

Dario A. Bini
Università di Pisa, Italy

Alberto Borobia
U. N. E. D, Spain

Murray R. Bremner
University of Saskatchewan, Canada

Richard A. Brualdi
University of Wisconsin-Madison

Ralph Byers
University of Kansas

Peter J. Cameron
Queen Mary, University of London, England

Alan Kaylor Cline
University of Texas

Fritz Colonius
Universität Augsburg, Germany

Robert M. Corless
University of Western Ontario, Canada

Biswa Nath Datta
Northern Illinois University

Jane Day
San Jose State University, CA

Luz M. DeAlba
Drake University, IA

James Demmel
University of California-Berkeley

Inderjit S. Dhillon
University of Texas

Zijian Diao
Ohio University Eastern

J. A. Dias da Silva
Universidade de Lisboa, Portugal

Jack Dongarra
University of Tennessee and Oakridge National Laboratory

Zlatko Drmač
University of Zagreb, Croatia

Victor Eijkhout
University of Tennessee

Mark Embree
Rice University, TX

Shaun M. Fallat
University of Regina, Canada

Miroslav Fiedler
Academy of Sciences of the Czech Republic

Roland W. Freund
University of California-Davis

Shmuel Friedland
University of Illinois-Chicago

Stéphane Gaubert
INRIA, France

Anne Greenbaum
University of Washington

Willem H. Haemers
Tilburg University, Netherlands

Frank J. Hall
Georgia State University

Lixing Han
University of Michigan-Flint

Per Christian Hansen
Technical University of Denmark

Daniel Hershkowitz
Technion, Israel

Nicholas J. Higham
University of Manchester, England

Leslie Hogben
Iowa State University

Randall Holmes
Auburn University, AL

Kenneth Howell
University of Alabama in Huntsville

Mark Hunacek
Iowa State University, Ames

David J. Jeffrey
University of Western Ontario, Canada

Charles R. Johnson
College of William and Mary, VA

Steve Kirkland
University of Regina, Canada

Wolfgang Kliemann
Iowa State University

Julien Langou
University of Tennessee

Amy N. Langville
The College of Charleston, SC

António Leal Duarte
Universidade de Coimbra, Portugal

Steven J. Leon
University of Massachusetts-Dartmouth

Chi-Kwong Li
College of William and Mary, VA

Ren-Cang Li
University of Texas-Arlington

Zhongshan Li
Georgia State University

Raphael Loewy
Technion, Israel

Armando Machado
Universidade de Lisboa, Portugal

Roy Mathias
University of Birmingham, England

Volker Mehrmann
Technical University Berlin, Germany

Beatrice Meini
Università di Pisa, Italy

Carl D. Meyer
North Carolina State University

Mark Mills
Central College, Iowa

Lucia I. Murakami
Universidade de São Paulo, Brazil

Michael G. Neubauer
California State University-Northridge

Michael Neumann
University of Connecticut

Esmond G. Ng
Lawrence Berkeley National Laboratory, CA

Michael Ng
Hong Kong Baptist University

Hans Bruun Nielsen
Technical University of Denmark

Simo Puntanen
University of Tampere, Finland

Robert Reams
Virginia Commonwealth University

Joachim Rosenthal
University of Zurich, Switzerland

Uriel G. Rothblum
Technion, Israel

Heikki Ruskeepää
University of Turku, Finland

Carlos M. Saiago
Universidade Nova de Lisboa, Portugal

Lorenzo Sadun
University of Texas

Hans Schneider
University of Wisconsin-Madison

George A. F. Seber
University of Auckland, NZ

Peter Šemrl
University of Ljubljana, Slovenia

Bryan L. Shader
University of Wyoming

Helene Shapiro
Swarthmore College, PA

Ivan P. Shestakov
Universidad de São Paulo, Brazil

Ivan Slapničar
University of Spilt, Croatia

Danny C. Sorensen
Rice University, TX

Michael Stewart
Georgia State University

Jeffrey L. Stuart
Pacific Lutheran University, WA

George P. H. Styan
McGill University, Canada

Tatjana Stykel
Technical University Berlin, Germany

Bit-Shun Tam
Tamkang University, Taiwan

T. Y. Tam
Auburn University, AL

Michael Tsatsomeros
Washington State University

Leonid N. Vaserstein
Pennsylvania State University

Amy Wangsness
Fitchburg State College, MA

Ian M. Wanless
Monash University, Australia

Jenny Wang
University of California-Davis

David S. Watkins
Washington State University

William Watkins
California State University-Northridge

Paul Weiner
St. Mary's University of Minnesota

Robert Wilson
Rutgers University, NJ

Henry Wolkowicz
University of Waterloo, Canada

Zhijun Wu
Iowa State University

Contents

Advanced Linear Algebra

Topics in Advanced Linear Algebra

Part II Combinatorial Matrix Theory and Graphs

Matrices and Graphs

Topics in Combinatorial Matrix Theory

Part III Numerical Methods

Numerical Methods for Linear Systems

Applications to Algebra

Part V Computational Software

Interactive Software for Linear Algebra

Packages of Subroutines for Linear Algebra

Preface

It is no exaggeration to say that linear algebra is a subject of central importance in both mathematics and a variety of other disciplines. It is used by virtually all mathematicians and by statisticians, physicists, biologists, computer scientists, engineers, and social scientists. Just as the basic idea of first semester differential calculus (approximating the graph of a function by its tangent line) provides information about the function, the process of linearization often allows difficult problems to be approximated by more manageable linear ones. This can provide insight into, and, thanks to ever-more-powerful computers, approximate solutions of the original problem. For this reason, people working in all the disciplines referred to above should find the *Handbook of Linear Algebra* an invaluable resource.

The *Handbook* is the first resource that presents complete coverage of linear algebra, combinatorial linear algebra, and numerical linear algebra, combined with extensive applications to a variety of fields and information on software packages for linear algebra in an easy to use handbook format.

Content

The *Handbook* covers the major topics of linear algebra at both the graduate and undergraduate level as well as its offshoots (numerical linear algebra and combinatorial linear algebra), its applications, and software packages for linear algebra computations. The *Handbook* takes the reader from the very elementary aspects of the subject to the frontiers of current research, and its format (consisting of a number of independent chapters each organized in the same standard way) should make this book accessible to readers with divergent backgrounds.

Format

There are five main parts in this book. The first part (Chapters 1 through Chapter 26) covers linear algebra; the second (Chapter 27 through Chapter 36) and third (Chapter 37 through Chapter 49) cover, respectively, combinatorial and numerical linear algebra, two important branches of the subject. Applications of linear algebra to other disciplines, both inside and outside of mathematics, comprise the fourth part of the book (Chapter 50 through Chapter 70). Part five (Chapter 71 through Chapter 77) addresses software packages useful for linear algebra computations.

Each chapter is written by a different author or team of authors, who are experts in the area covered. Each chapter is divided into sections, which are organized into the following uniform format:

- Definitions
- Facts
- Examples

Most relevant definitions appear within the Definitions segment of each chapter, but some terms that are used throughout linear algebra are not redefined in each chapter. The Glossary, covering the terminology of linear algebra, combinatorial linear algebra, and numerical linear algebra, is available at the end of the book to provide definitions of terms that appear in different chapters. In addition to the definition, the Glossary also provides the number of the chapter (and section, thereof) where the term is defined. The Notation Index serves the same purpose for symbols.

The Facts (which elsewhere might be called theorems, lemmas, etc.) are presented in list format, which allows the reader to locate desired information quickly. In lieu of proofs, references are provided for all facts. The references will also, of course, supply a source of additional information about the subject of the chapter. In this spirit, we have encouraged the authors to use texts or survey articles on the subject as references, where available.

The Examples illustrate the definitions and facts. Each section is short enough that it is easy to go back and forth between the Definitions/Facts and the Examples to see the illustration of a fact or definition. Some sections also contain brief applications following the Examples (major applications are treated in their own chapters).

Feedback

To see updates and provide feedback and errata reports, please consult the web page for this book: http://www.public.iastate.edu/~lhogben/HLA.html or contact the editor via email, LHogben@iastate.edu, with *HLA* in the subject heading.

Preliminaries

This chapter contains a variety of definitions of terms that are used throughout the rest of the book, but are not part of linear algebra and/or do not fit naturally into another chapter. Since these definitions have little connection with each other, a different organization is followed; the definitions are (loosely) alphabetized and each definition is followed by an example.

Algebra

An **(associative) algebra** is a vector space A over a field F together with a **multiplication** $(\mathbf{x}, \mathbf{y}) \mapsto \mathbf{xy}$ from $A \times A$ to A satisfying two *distributive* properties and *associativity*, i.e., for all $a, b \in F$ and all $\mathbf{x}, \mathbf{y}, \mathbf{z} \in A$:

$$(a\mathbf{x} + b\mathbf{y})\mathbf{z} = a(\mathbf{xz}) + b(\mathbf{yz}), \qquad \mathbf{x}(a\mathbf{y} + b\mathbf{z}) = a(\mathbf{xy}) + b(\mathbf{xz}) \qquad (\mathbf{xy})\mathbf{z} = \mathbf{x}(\mathbf{yz}).$$

Except in Chapter 69 and Chapter 70 the term *algebra* means associative algebra. In these two chapters, associativity is not assumed.

Examples:

The vector space of $n \times n$ matrices over a field F with matrix multiplication is an (associative) algebra.

Boundary

The **boundary** ∂S of a subset S of the real numbers or the complex numbers is the intersection of the closure of S and the closure of the complement of S.

Examples:

The boundary of $S = \{x \in \mathbb{C} : |z| \leq 1\}$ is $\partial S = \{x \in \mathbb{C} : |z| = 1\}$.

Complement

The **complement** of the set X in universe S, denoted $S \setminus X$, is all elements of S that are not in X. When the universe is clear (frequently the universe is $\{1, \ldots, n\}$) then this can be denoted X^c.

Examples:

For $S = \{1, 2, 3, 4, 5\}$ and $X = \{1, 3\}$, $S \setminus X = \{2, 4, 5\}$.

Complex Numbers

Let $a, b \in \mathbb{R}$. The symbol i denotes $\sqrt{-1}$.

The **complex conjugate** of a complex number $c = a + bi$ is $\overline{c} = a - bi$.
The **imaginary part** of $a + bi$ is $\text{im}(a + bi) = b$ and the **real part** is $\text{re}(a + bi) = a$.
The **absolute value** of $c = a + bi$ is $|c| = \sqrt{a^2 + b^2}$.

The **argument** of the nonzero complex number $re^{i\theta}$ is θ (with $r, \theta \in \mathbb{R}$ and $0 < r$ and $0 \leq \theta < 2\pi$).
The **open right half plane** $\mathbb{C}^+$ is $\{z \in \mathbb{C} : \text{re}(z) > 0\}$.
The **closed right half plane** $\mathbb{C}_0^+$ is $\{z \in \mathbb{C} : \text{re}(z) \geq 0\}$.
The **open left half plane** $\mathbb{C}^-$ is $\{z \in \mathbb{C} : \text{re}(z) < 0\}$.
The **closed left half plane** $\mathbb{C}^-$ is $\{z \in \mathbb{C} : \text{re}(z) \leq 0\}$.

Facts:

1. $|c| = c\bar{c}$
2. $|re^{i\theta}| = r$
3. $re^{i\theta} = r\cos\theta + r\sin\theta i$
4. $\overline{re^{i\theta}} = re^{-i\theta}$

Examples:

$\overline{2+3i} = 2 - 3i$, $\overline{1.4} = 1.4$, $1 + i = \sqrt{2}e^{i\pi/4}$.

Conjugate Partition

Let $\upsilon = (u_1, u_2, \ldots, u_n)$ be a sequence of integers such that $u_1 \geq u_2 \geq \cdots \geq u_n \geq 0$. The **conjugate partition** of υ is $\upsilon^* = (u_1^*, \ldots, u_t^*)$, where u_i^* is the number of js such that $u_j \geq i$. t is sometimes taken to be u_1, but is sometimes greater (obtained by extending with 0s).

Facts: If t is chosen to be the minimum, and $u_n > 0$, $\upsilon^{**} = \upsilon$.

Examples:

$(4, 3, 2, 2, 1)^* = (5, 4, 2, 1)$.

Convexity

Let V be a real or complex vector space.

Let $\{\mathbf{v}_1, \mathbf{v}_2, \ldots, \mathbf{v}_k\} \in V$. A vector of the form $a_1\mathbf{v}_1 + a_2\mathbf{v}_2 + \cdots + a_k\mathbf{v}_k$ with all the coefficients a_i nonnegative and $\sum a_i = 1$ is a **convex combination** of $\{\mathbf{v}_1, \mathbf{v}_2, \ldots, \mathbf{v}_k\}$.

A set $S \subseteq V$ is **convex** if any convex combination of vectors in S is in S.

The **convex hull** of S is the set of all convex combinations of S and is denoted by Con(S).

An **extreme point** of a closed convex set S is a point $\mathbf{v} \in S$ that is not a nontrivial convex combination of other points in S, i.e., $a\mathbf{x} + (1-a)\mathbf{y} = \mathbf{v}$ and $0 \leq a \leq 1$ implies $\mathbf{x} = \mathbf{y} = \mathbf{v}$.

A **convex polytope** is the convex hull of a finite set of vectors in $\mathbb{R}^n$.

Let $S \subseteq V$ be convex. A function $f : S \to \mathbb{R}$ is **convex** if for all $a \in \mathbb{R}$, $0 < a < 1$, $\mathbf{x}, \mathbf{y} \in S$, $f(a\mathbf{x} + (1-a)\mathbf{y}) \leq af(\mathbf{x}) + (1-a)f(\mathbf{y})$.

Facts:

1. A set $S \subseteq V$ is convex if and only if Con$(S) = S$.
2. The extreme points of Con(S) are contained in S.
3. [HJ85] *Krein-Milman Theorem*: A compact convex set is the convex hull of its extreme points.

Examples:

1. $[1.9, 0.8]^T$ is a convex combination of $[1, -1]^T$ and $[2, 1]^T$, since $[1.9, 0.8]^T = 0.1[1, -1]^T + 0.9[2, 1]^T$.
2. The set K of all $\mathbf{v} \in \mathbb{R}^3$ such that $v_i \geq 0, i = 1, 2, 3$ is a convex set. Its only extreme point is the zero vector.

Elementary Symmetric Function

The k**th elementary symmetric function of** $\alpha_i, i = 1, \ldots, n$ is

$$S_k(\alpha_1, \ldots, \alpha_n) = \sum_{1<i_1<i_2<\cdots<i_k<n} \alpha_{i_1}\alpha_{i_2}\ldots\alpha_{i_k}.$$

Examples:

$S_2(\alpha_1, \alpha_2, \alpha_3) = \alpha_1\alpha_2 + \alpha_1\alpha_3 + \alpha_2\alpha_3$,
$S_1(\alpha_1, \ldots, \alpha_n) = \alpha_1 + \alpha_2 + \cdots + \alpha_n$, $S_n(\alpha_1, \ldots, \alpha_n) = \alpha_1\alpha_2\ldots\alpha_n$.

Equivalence Relation

A binary relation $\equiv$ in a nonempty set S is an **equivalence relation** if it satisfies the following conditions:

1. (Reflexive) For all $a \in S, a \equiv a$.
2. (Symmetric) For all $a, b \in S, a \equiv b$ implies $b \equiv a$.
3. (Transitive) For all $a, b, c \in S, a \equiv b$ and $a \equiv b$ imply $a \equiv c$.

Examples:

Congruence mod n is an equivalence relation on the integers.

Field

A **field** is a set F with at least two elements together with a function $F \times F \to F$ called addition, denoted $(a, b) \to a + b$, and a function $F \times F \to F$ called multiplication, denoted $(a, b) \to ab$, which satisfy the following axioms:

1. (Commutativity) For each $a, b \in F, a + b = b + a$ and $ab = ba$.
2. (Associativity) For each $a, b, c \in F, (a + b) + c = a + (b + c)$ and $(ab)c = a(bc)$.
3. (Identities) There exist two elements 0 and 1 in F such that $0 + a = a$ and $1a = a$ for each $a \in F$.
4. (Inverses) For each $a \in F$, there exists an element $-a \in F$ such that $(-a) + a = 0$. For each nonzero $a \in F$, there exists an element $a^{-1} \in F$ such that $a^{-1}a = 1$.
5. (Distributivity) For each $a, b, c \in F, a(b + c) = ab + ac$.

Examples:

The real numbers, $\mathbb{R}$, the complex numbers, $\mathbb{C}$, and the rational numbers, $\mathbb{Q}$, are all fields. The set of integers, $\mathbb{Z}$, is not a field.

Greatest Integer Function

The **greatest integer** or **floor** function $\lfloor x \rfloor$ (defined on the real numbers) is the greatest integer less than or equal to x.

Examples:

$\lfloor 1.5 \rfloor = 1$, $\lfloor 1 \rfloor = 1$, $\lfloor -1.5 \rfloor = -2$.

Group

(See also Chapter 67 and Chapter 68.)

A **group** is a nonempty set G with a function $G \times G \to G$ denoted $(a, b) \to ab$, which satisfies the following axioms:

1. (Associativity) For each $a, b, c \in G$, $(ab)c = a(bc)$.
2. (Identity) There exists an element $e \in G$ such that $ea = a = ae$ for each $a \in G$.
3. (Inverses) For each $a \in G$, there exists an element $a^{-1} \in G$ such that $a^{-1}a = e = aa^{-1}$.

A group is **abelian** if $ab = ba$ for all $a, b \in G$.

Examples:

1. Any vector space is an abelian group under $+$.
2. The set of invertible $n \times n$ real matrices is a group under matrix multiplication.
3. The set of all permutations of a set is a group under composition.

Interlaces

Let $a_1 \geq a_2 \geq \cdots \geq a_n$ and $b_1 \geq b_2 \geq \cdots \geq b_{n-1}$, two sequences of real numbers arranged in decreasing order. Then the sequence $\{b_i\}$ **interlaces** the sequence $\{a_i\}$ if $a_n \leq b_{n-1} \leq a_{n-1} \cdots \leq b_1 \leq a_1$. Further, if all of the above inequalities can be taken to be strict, the sequence $\{b_i\}$ **strictly interlaces** the sequence $\{a_i\}$. Analogous definitions are given when the numbers are in increasing order.

Examples:

$7 \geq 2.2 \geq -1$ strictly interlaces $11 \geq \pi \geq 0 \geq -2.6$.

Majorization

Let $\alpha = (a_1, a_2, \ldots, a_n)$, $\beta = (b_1, b_2, \ldots, b_n)$ be sequences of real numbers.

$\alpha^{\downarrow} = (a_1^{\downarrow}, a_2^{\downarrow}, \ldots, a_n^{\downarrow})$ is the permutation of α with entries in nonincreasing order, i.e., $a_1^{\downarrow} \geq a_2^{\downarrow} \geq \ldots \geq a_n^{\downarrow}$.

$\alpha^{\uparrow} = (a_1^{\uparrow}, a_2^{\uparrow}, \ldots, a_n^{\uparrow})$ is the permutation of α with entries in nondecreasing order, i.e., $a_1^{\uparrow} \leq a_2^{\uparrow} \leq \ldots \leq a_n^{\uparrow}$.

α **weakly majorizes** β, written $\alpha \succeq_w \beta$ or $\beta \preceq_w \alpha$, if:

$$\sum_{i=1}^{k} a_i^{\downarrow} \geq \sum_{i=1}^{k} b_i^{\downarrow} \quad \text{for all } k = 1, \ldots n.$$

α **majorizes** β, written $\alpha \succeq \beta$ or $\beta \preceq \alpha$, if $\alpha \succeq_w \beta$ and $\sum_{i=1}^{n} a_i = \sum_{i=1}^{n} b_i$.

Examples:

1. If $\alpha = (2, 2, -1.3, 8, 7.7)$, then $\alpha^{\downarrow} = (8, 7.7, 2, 2, -1.3)$ and $\alpha^{\uparrow} = (-1.3, 2, 2, 7.7, 8)$.
2. $(5,3,1.5,1.5,1) \succeq (4,3,2,2,1)$ and $(6,5,0) \succeq_w (4,3,2)$.

Metric

A **metric** on a set S is a real-valued function $f : S \times S \to \mathbb{R}$ satisfying the following conditions:

1. For all $x, y \in S$, $f(x, y) \geq 0$.
2. For all $x \in S$, $f(x, x) = 0$.
3. For all $x, y \in S$, $f(x, y) = 0$ implies $x = y$.
4. For all $x, y \in S$, $f(x, y) = f(y, x)$.
5. For all $x, y, z \in S$, $f(x, y) + f(y, z) \geq f(x, z)$.

A metric is intended as a measure of distance between elements of the set.

Examples:

If $\|\cdot\|$ is a norm on a vector space, then $f(x, y) = \|\mathbf{x} - \mathbf{y}\|$ is a metric.

Multiset

A **multiset** is an unordered list of elements that allows repetition.

Examples:

Any set is a multiset, but $\{1, 1, 3, -2, -2, -2\}$ is a multiset that is not a set.

O and o

Let, f, g be real valued functions of $\mathbb{N}$ or $\mathbb{R}$, i.e., $f, g : \mathbb{N} \to \mathbb{R}$ or $f, g : \mathbb{R} \to \mathbb{R}$.

f is $O(g)$ (**big-oh** of g) if there exist constants C, k such that $|f(x)| \leq C|g(x)|$ for all $x \geq k$.

f is $o(g)$ (**little-oh** of g) if $\lim_{x\to\infty} \left|\frac{f(n)}{g(n)}\right| = 0$.

Examples:

$x^2 + x$ is $O(x^2)$ and $\ln x$ is $o(x)$.

Path-connected

A subset S of the complex numbers is **path-connected** if for any $x, y \in S$ there exists a continuous function $p : [0, 1] \to S$ with $p(0) = x$ and $p(1) = y$.

Examples:

$S = \{z \in \mathbb{C} : 1 \leq |z| \leq 2\}$ and the line $\{a + bi : a = 2b + 3\}$ are path-connected.

Permutations

A **permutation** is a one-to-one onto function from a set to itself.

The set of permutations of $\{1, \ldots, n\}$ is denoted S_n. The identity permutation is denoted ε_n. In this book, permutations are generally assumed to be elements of S_n for some n.

A **cycle** or k-**cycle** is a permutation τ such that there is a subset $\{a_1, \ldots, a_k\}$ of $\{1, \ldots, n\}$ satisfying $\tau(a_i) = a_{i+1}$ and $\tau(a_k) = a_1$; this is denoted $\tau = (a_1, a_2, \ldots, a_k)$. The **length** of this cycle is k.

A **transposition** is a 2-cycle.

A permutation is **even** (respectively, **odd**) if it can be written as the product of an even (odd) number of transpositions.

The **sign** of a permutation τ, denoted sgn τ, is +1 if τ is even and -1 if τ is odd.

Note: Permutations are functions and act from the left (see Examples).

Facts:

1. Every permutation can be expressed as a product of disjoint cycles. This expression is unique up to the order of the cycles in the decomposition and cyclic permutation within a cycle.
2. Every permutation can be written as a product of transpositions. If some such expression includes an even number of transpositions, then every such expression includes an even number of transpositions.
3. S_n with the operation of composition is a group.

Examples:

1. If $\tau = (1523) \in S_6$, then $\tau(1) = 5, \tau(2) = 3, \tau(3) = 1, \tau(4) = 4, \tau(5) = 2, \tau(6) = 6$.
2. (123)(12)=(13).
3. $\text{sgn}(1234) = -1$, because $(1234) = (14)(13)(12)$.

Ring

(See also Section 23.1)

A **ring** is a set R together with a function $R \times R \to R$ called addition, denoted $(a, b) \to a + b$, and a function $R \times R \to R$ called multiplication, denoted $(a, b) \to ab$, which satisfy the following axioms:

1. (Commutativity of +) For each $a, b \in R$, $a + b = b + a$.
2. (Associativity) For each $a, b, c \in R$, $(a + b) + c = a + (b + c)$ and $(ab)c = a(bc)$.
3. (+ identity) There exists an element 0 in R such that $0 + a = a$.
4. (+ inverse) For each $a \in R$, there exists an element $-a \in R$ such that $(-a) + a = 0$.
5. (Distributivity) For each $a, b, c \in R, a(b + c) = ab + ac$ and $(a + b)c = ac + bc$.

A **zero divisor** in a ring R is a nonzero element $a \in R$ such that there exists a nonzero $b \in R$ with $ab = 0$ or $ba = 0$.

Examples:

- The set of integers, $\mathbb{Z}$, is a ring.
- Any field is a ring.
- Let F be a field. Then $F^{n\times n}$, with matrix addition and matrix multiplication as the operations, is a ring. $E_{11} = \begin{bmatrix} 1 & 0 \\ 0 & 0 \end{bmatrix}$ and $E_{22} = \begin{bmatrix} 0 & 0 \\ 0 & 1 \end{bmatrix}$ are zero divisors since $E_{11}E_{22} = 0_2$.

Sign

(For sign of a permutation, see *permutation.*)

The **sign** of a complex number is defined by:

$$\text{sign}(z) = \begin{cases} z/|z|, & \text{if } z \neq 0; \\ 1, & \text{if } z = 0. \end{cases}$$

If z is a real number, this sign function yields 1 or -1.
This sign function is used in numerical linear algebra.
The **sign** of a real number (as used in sign patterns) is defined by:

$$\text{sgn}(a) = \begin{cases} +, & \text{if } a > 0; \\ 0, & \text{if } a = 0; \\ -, & \text{if } a < 0. \end{cases}$$

This sign function is used in combinatorial linear algebra, and the product of a sign and a real number is interpreted in the obvious way as a real number.

Warning: The two sign functions disagree on the sign of 0.

Examples:

$\text{sgn}(-1.3) = -$, $\text{sign}(-1.3) = -1$, $\text{sgn}(0) = 0$, $\text{sign}(0) = 1$,
$\text{sign}(1 + i) = \dfrac{(1 + i)}{\sqrt{2}}$.

References

[HJ85] [HJ85] R. Horn and C. R. Johnson. *Matrix Analysis*. Cambridge University Press, Cambridge, 1985.

I

Linear Algebra

Basic Linear Algebra

Matrices with Special Properties

Advanced Linear Algebra

Topics in Advanced Linear Algebra

Basic Linear Algebra

1

Vectors, Matrices, and Systems of Linear Equations

Jane Day
San Jose State University

Throughout this chapter, F will denote a field. The references [Lay03], [Leo02], and [SIF00] are good sources for more detail about much of the material in this chapter. They discuss primarily the field of real numbers, but the proofs are usually valid for any field.

1.1 Vector Spaces

Vectors are used in many applications. They often represent quantities that have both direction and magnitude, such as velocity or position, and can appear as functions, as n-tuples of scalars, or in other disguises. Whenever objects can be added and multiplied by scalars, they may be elements of some vector space. In this section, we formulate a general definition of vector space and establish its basic properties. An element of a field, such as the real numbers or the complex numbers, is called a scalar to distinguish it from a vector.

Definitions:

A **vector space over** F is a set V together with a function $V \times V \to V$ called **addition**, denoted $(\mathbf{x},\mathbf{y}) \to \mathbf{x}+\mathbf{y}$, and a function $F \times V \to V$ called **scalar multiplication** and denoted $(c,\mathbf{x}) \to c\mathbf{x}$, which satisfy the following axioms:

1. (Commutativity) For each $\mathbf{x}, \mathbf{y} \in V$, $\mathbf{x}+\mathbf{y}=\mathbf{y}+\mathbf{x}$.
2. (Associativity) For each $\mathbf{x}, \mathbf{y}, \mathbf{z} \in V$, $(\mathbf{x}+\mathbf{y})+\mathbf{z}=\mathbf{x}+(\mathbf{y}+\mathbf{z})$.
3. (Additive identity) There exists a **zero vector** in V, denoted $\mathbf{0}$, such that $\mathbf{0}+\mathbf{x}=\mathbf{x}$ for each $\mathbf{x} \in V$.
4. (Additive inverse) For each $\mathbf{x} \in V$, there exists $-\mathbf{x} \in V$ such that $(-\mathbf{x})+\mathbf{x}=\mathbf{0}$.
5. (Distributivity) For each $a \in F$ and $\mathbf{x}, \mathbf{y} \in V$, $a(\mathbf{x}+\mathbf{y})=a\mathbf{x}+a\mathbf{y}$.
6. (Distributivity) For each $a, b \in F$ and $\mathbf{x} \in V$, $(a+b)\,\mathbf{x}=a\mathbf{x}+b\mathbf{x}$.

7. (Associativity) For each $a, b \in F$ and $\mathbf{x} \in V$, $(ab)\,\mathbf{x} = a(b\mathbf{x})$.
8. For each $\mathbf{x} \in V$, $1\mathbf{x} = \mathbf{x}$.

The properties that for all $\mathbf{x}, \mathbf{y} \in V$, and $a \in F, \mathbf{x}+\mathbf{y} \in V$ and $a\mathbf{x} \in V$, are called **closure under addition** and **closure under scalar multiplication**, respectively. The elements of a vector space V are called **vectors**. A vector space is called **real** if $F = \mathbb{R}$, **complex** if $F = \mathbb{C}$.

If n is a positive integer, F^n denotes the set of all ordered n-tuples (written as columns). These are sometimes written instead as rows $[x_1 \ \cdots \ x_n]$ or $(x_1, \ldots, x_n)$. For $\mathbf{x} = \begin{bmatrix} x_1 \\ \vdots \\ x_n \end{bmatrix}, \mathbf{y} = \begin{bmatrix} y_1 \\ \vdots \\ y_n \end{bmatrix} \in F^n$ and $c \in F$, define addition and scalar multiplication coordinate-wise: $\mathbf{x} + \mathbf{y} = \begin{bmatrix} x_1 + y_1 \\ \vdots \\ x_n + y_n \end{bmatrix}$ and $c\mathbf{x} = \begin{bmatrix} cx_1 \\ \vdots \\ cx_n \end{bmatrix}$. Let $\mathbf{0}$ denote the n-tuple of zeros. For $\mathbf{x} \in F^n$, x_j is called the jth **coordinate** of $\mathbf{x}$.

A **subspace** of vector space V over field F is a subset of V, which is itself a vector space over F when the addition and scalar multiplication of V are used. If S_1 and S_2 are subsets of vector space V, define $S_1 + S_2 = \{\mathbf{x}+\mathbf{y} : \mathbf{x} \in S_1 \text{ and } \mathbf{y} \in S_2\}$.

Facts:

Let V be a vector space over F.

1. F^n is a vector space over F.
2. [FIS03, pp. 11–12] (Basic properties of a vector space):
   - The vector $\mathbf{0}$ is the only additive identity in V.
   - For each $\mathbf{x} \in V$, $-\mathbf{x}$ is the only additive inverse for $\mathbf{x}$ in V.
   - For each $\mathbf{x} \in V$, $-\mathbf{x} = (-1)\mathbf{x}$.
   - If $a \in F$ and $\mathbf{x} \in V$, then $a\mathbf{x} = \mathbf{0}$ if and only if $a = 0$ or $\mathbf{x} = \mathbf{0}$.
   - (Cancellation) If $\mathbf{x}, \mathbf{y}, \mathbf{z} \in V$ and $\mathbf{x}+\mathbf{y} = \mathbf{x}+\mathbf{z}$, then $\mathbf{y} = \mathbf{z}$.
3. [FIS03, pp. 16–17] Let W be a subset of V. The following are equivalent:
   - W is a subspace of V.
   - W is nonempty and closed under addition and scalar multiplication.
   - $\mathbf{0} \in W$ and for any $\mathbf{x}, \mathbf{y} \in W$ and $a, b \in F$, $a\mathbf{x} + b\mathbf{y} \in W$.
4. For any vector space V, $\{\mathbf{0}\}$ and V itself are subspaces of V.
5. [FIS03, p. 19] The intersection of any nonempty collection of subspaces of V is a subspace of V.
6. [FIS03, p. 22] Let W_1 and W_2 be subspaces of V. Then $W_1 + W_2$ is a subspace of V containing W_1 and W_2. It is the smallest subspace that contains them in the sense that any subspace that contains both W_1 and W_2 must contain $W_1 + W_2$.

Examples:

1. The set $\mathbb{R}^n$ of all ordered n-tuples of real numbers is a vector space over $\mathbb{R}$, and the set $\mathbb{C}^n$ of all ordered n-tuples of complex numbers is a vector space over $\mathbb{C}$. For instance, $\mathbf{x} = \begin{bmatrix} 3 \\ 0 \\ -1 \end{bmatrix}$ and $\mathbf{y} = \begin{bmatrix} 2i \\ 4 \\ 2-3i \end{bmatrix}$ are elements of $\mathbb{C}^3$; $\mathbf{x}+\mathbf{y} = \begin{bmatrix} 3+2i \\ 4 \\ 1-3i \end{bmatrix}$, $-\mathbf{y} = \begin{bmatrix} -2i \\ -4 \\ -2+3i \end{bmatrix}$, and $i\mathbf{y} = \begin{bmatrix} -2 \\ 4i \\ 3+2i \end{bmatrix}$.
2. Notice $\mathbb{R}^n$ is a subset of $\mathbb{C}^n$ but not a subspace of $\mathbb{C}^n$, since $\mathbb{R}^n$ is not closed under multiplication by nonreal numbers.

3. The vector spaces $\mathbb{R}$, $\mathbb{R}^2$, and $\mathbb{R}^3$ are the usual Euclidean spaces of analytic geometry. There are three types of subspaces of $\mathbb{R}^2$: $\{\mathbf{0}\}$, a line through the origin, and $\mathbb{R}^2$ itself. There are four types of subspaces of $\mathbb{R}^3$: $\{\mathbf{0}\}$, a line through the origin, a plane through the origin, and $\mathbb{R}^3$ itself. For instance, let $\mathbf{v} = (5, -1, -1)$ and $\mathbf{w} = (0, 3, -2)$. The lines $W_1 = \{s\mathbf{v} : s \in \mathbb{R}\}$ and $W_2 = \{s\mathbf{w} : s \in \mathbb{R}\}$ are subspaces of $\mathbb{R}^3$. The subspace $W_1 + W_2 = \{s\mathbf{v} + t\mathbf{w} : s, t \in \mathbb{R}\}$ is a plane. The set $\{s\mathbf{v} + \mathbf{w} : s \in \mathbb{R}\}$ is a line parallel to W_1, but is not a subspace. (For more information on geometry, see Chapter 65.)
4. Let $F[x]$ be the set of all polynomials in the single variable x, with coefficients from F. To add polynomials, add coefficients of like powers; to multiply a polynomial by an element of F, multiply each coefficient by that scalar. With these operations, $F[x]$ is a vector space over F. The zero polynomial z, with all coefficients 0, is the additive identity of $F[x]$. For $f \in F[x]$, the function $-f$ defined by $-f(x) = (-1)f(x)$ is the additive inverse of f.
5. In $F[x]$, the constant polynomials have degree 0. For $n > 0$, the polynomials with highest power term x^n are said to have degree n. For a nonnegative integer n, let $F[x; n]$ be the subset of $F[x]$ consisting of all polynomials of degree n or less. Then $F[x; n]$ is a subspace of $F[x]$.
6. When $n > 0$, the set of all polynomials of degree exactly n is not a subspace of $F[x]$ because it is not closed under addition or scalar multiplication. The set of all polynomials in $\mathbb{R}[x]$ with rational coefficients is not a subspace of $\mathbb{R}[x]$ because it is not closed under scalar multiplication.
7. Let V be the set of all infinite sequences $(a_1, a_2, a_3, \ldots)$, where each $a_j \in F$. Define addition and scalar multiplication coordinate-wise. Then V is a vector space over F.
8. Let X be a nonempty set and let $\mathcal{F}(X, F)$ be the set of all functions $f: X \to F$. Let $f, g \in \mathcal{F}(X, F)$ and define $f + g$ and cf pointwise, as $(f + g)(x) = f(x) + g(x)$ and $(cf)(x) = cf(x)$ for all $x \in X$. With these operations, $\mathcal{F}(X, F)$ is a vector space over F. The zero function is the additive identity and $(-1)f = -f$, the additive inverse of f.
9. Let X be a nonempty subset of $\mathbb{R}^n$. The set $C(X)$ of all continuous functions $f: X \to \mathbb{R}$ is a subspace of $\mathcal{F}(X, \mathbb{R})$. The set $\mathcal{D}(X)$ of all differentiable functions $f: X \to \mathbb{R}$ is a subspace of $C(X)$ and also of $\mathcal{F}(X, \mathbb{R})$.

1.2 Matrices

Matrices are rectangular arrays of scalars that are used in a great variety of ways, such as to solve linear systems, model linear behavior, and approximate nonlinear behavior. They are standard tools in almost every discipline, from sociology to physics and engineering.

Definitions:

An $m \times p$ **matrix** over F is an $m \times p$ rectangular array $A = \begin{bmatrix} a_{11} & \cdots & a_{1p} \\ \vdots & \cdots & \vdots \\ a_{m1} & \cdots & a_{mp} \end{bmatrix}$, with entries from F. The notation $A = [a_{ij}]$ that displays a typical entry is also used. The element a_{ij} of the matrix A is called the (i, j) **entry** of A and can also be denoted $(A)_{ij}$. The **shape** (or **size**) of A is $m \times p$, and A is **square** if $m = p$; in this case, m is also called the size of A. Two matrices $A = [a_{ij}]$ and $B = [b_{ij}]$ are said to be **equal** if they have the same shape and $a_{ij} = b_{ij}$ for all i, j. Let $A = [a_{ij}]$ and $B = [b_{ij}]$ be $m \times p$ matrices, and let c be a scalar. Define **addition** and **scalar multiplication** on the set of all $m \times p$ matrices over F entrywise, as $A + B = [a_{ij} + b_{ij}]$ and $cA = [ca_{ij}]$. The set of all $m \times p$ matrices over F with these operations is denoted $F^{m \times p}$.

If A is $m \times p$, **row** i is $[a_{i1}, \quad \ldots, \quad a_{ip}]$ and **column** j is $\begin{bmatrix} a_{1j} \\ \vdots \\ a_{mj} \end{bmatrix}$. These are called a **row vector** and a **column vector** respectively, and they belong to $F^{n \times 1}$ and $F^{1 \times n}$, respectively. The elements of F^n are identified with the elements of $F^{n \times 1}$ (or sometimes with the elements of $F^{1 \times n}$). Let $\mathbf{0}_{mp}$ denote the $m \times p$ matrix of zeros, often shortened to $\mathbf{0}$ when the size is clear. Define $-A = (-1)A$.

Let $A = [\mathbf{a}_1 \quad \dots \quad \mathbf{a}_p] \in F^{m \times p}$, where $\mathbf{a}_j$ is the jth column of A, and let $\mathbf{b} = \begin{bmatrix} b_1 \\ \vdots \\ b_p \end{bmatrix} \in F^{p \times 1}$. The **matrix–vector product** of A and $\mathbf{b}$ is $A\mathbf{b} = b_1\mathbf{a}_1 + \cdots + b_p\mathbf{a}_p$. Notice $A\mathbf{b}$ is $m \times 1$.

If $A \in F^{m \times p}$ and $C = [\mathbf{c}_1 \quad \dots \quad \mathbf{c}_n] \in F^{p \times n}$, define the **matrix product** of A and C as $AC = [A\mathbf{c}_1 \quad \dots \quad A\mathbf{c}_n]$. Notice AC is $m \times n$.

Square matrices A and B **commute** if $AB = BA$. When $i = j$, a_{ii} is a **diagonal entry** of A and the set of all its diagonal entries is the **main diagonal** of A. When $i \neq j$, a_{ij} is an **off-diagonal entry**.

The **trace** of A is the sum of all the diagonal entries of A, tr $A = \sum_{i=1}^{n} a_{ii}$.

A matrix $A = [a_{ij}]$ is **diagonal** if $a_{ij} = 0$ whenever $i \neq j$, **lower triangular** if $a_{ij} = 0$ whenever $i < j$, and **upper triangular** if $a_{ij} = 0$ whenever $i > j$. A **unit triangular** matrix is a lower or upper triangular matrix in which each diagonal entry is 1.

The **identity matrix** I_n, often shortened to I when the size is clear, is the $n \times n$ matrix with main diagonal entries 1 and other entries 0.

A **scalar matrix** is a scalar multiple of the identity matrix.

A **permutation matrix** is one whose rows are some rearrangement of the rows of an identity matrix.

Let $A \in F^{m \times p}$. The **transpose** of A, denoted A^T, is the $p \times m$ matrix whose (i, j) entry is the (j, i) entry of A.

The square matrix A is **symmetric** if $A^T = A$ and **skew-symmetric** if $A^T = -A$.

When $F = \mathbb{C}$, that is, when A has complex entries, the **Hermitian adjoint** of A is its conjugate transpose, $A^* = \bar{A}^T$; that is, the (i, j) entry of A^* is $\overline{a_{ji}}$. Some authors, such as [Leo02], write A^H instead of A^*.

The square matrix A is **Hermitian** if $A^* = A$ and **skew-Hermitian** if $A^* = -A$.

Let α be a nonempty set of row indices and β a nonempty set of column indices. A **submatrix** of A is a matrix $A[\alpha, \beta]$ obtained by choosing the entries of A, which lie in rows α and columns β. A **principal submatrix** of A is a submatrix of the form $A[\alpha, \alpha]$. A **leading principal submatrix** of A is one of the form $A[\{1, \dots, k\}, \{1, \dots, k\}]$.

Facts:

1. [SIF00, p. 5] $F^{m \times p}$ is a vector space over F. That is, if $\mathbf{0}, A, B, C \in F^{m \times p}$, and $c, d \in F$, then:
   - $A + B = B + A$
   - $(A + B) + C = A + (B + C)$
   - $A + \mathbf{0} = \mathbf{0} + A = A$
   - $A + (-A) = (-A) + A = \mathbf{0}$
   - $c(A + B) = cA + cB$
   - $(c + d)A = cA + dA$
   - $(cd)\, A = c(dA)$
   - $1A = A$
2. If $A \in F^{m \times p}$ and $C \in F^{p \times n}$, the (i, j) entry of AC is $(AC)_{ij} = \sum_{k=1}^{p} a_{ik}a_{kj}$. This is the matrix product of row i of A and column j of C.
3. [SIF00, p. 88] Let $c \in F$, let A and B be matrices over F, let I denote an identity matrix, and assume the shapes allow the following sums and products to be calculated. Then:
   - $AI = IA = A$
   - $A\mathbf{0} = \mathbf{0}$ and $\mathbf{0}A = \mathbf{0}$
   - $A(BC) = (AB)C$
   - $A(B + C) = AB + AC$
   - $(A + B)C = AC + BC$
   - $c(AB) = A(cB) = (cA)B$ for any scalar c

4. [SIF00, p. 5 and p. 20] Let $c \in F$, let A and B be matrices over F, and assume the shapes allow the following sums and products to be calculated. Then:
   - $(A^T)^T = A$
   - $(A+B)^T = A^T + B^T$
   - $(cA)^T = cA^T$
   - $(AB)^T = B^T A^T$
5. [Leo02, pp. 321–323] Let $c \in \mathbb{C}$, let A and B be matrices over $\mathbb{C}$, and assume the shapes allow the following sums and products to be calculated. Then:
   - $(A^*)^* = A$
   - $(A+B)^* = A^* + B^*$
   - $(cA)^* = \bar{c}A^*$
   - $(AB)^* = B^* A^*$
6. If A and B are $n \times n$ and upper (lower) triangular, then AB is upper (lower) triangular.

Examples:

1. Let $A = \begin{bmatrix} 1 & 2 & 3 \\ 4 & 5 & 6 \end{bmatrix}$ and $\mathbf{b} = \begin{bmatrix} 7 \\ 8 \\ -9 \end{bmatrix}$. By definition, $A\mathbf{b} = 7\begin{bmatrix} 1 \\ 4 \end{bmatrix} + 8\begin{bmatrix} 2 \\ 5 \end{bmatrix} - 9\begin{bmatrix} 3 \\ 6 \end{bmatrix} = \begin{bmatrix} -4 \\ 14 \end{bmatrix}$. Hand calculation of $A\mathbf{b}$ can be done more quickly using Fact 2: $A\mathbf{b} = \begin{bmatrix} 1\cdot 7 + 2\cdot 8 - 3\cdot 9 \\ 4\cdot 7 + 5\cdot 8 - 6\cdot 9 \end{bmatrix} = \begin{bmatrix} -4 \\ 14 \end{bmatrix}$.

2. Let $A = \begin{bmatrix} 1 & -3 & 4 \\ 2 & 0 & 8 \end{bmatrix}$, $B = \begin{bmatrix} -3 & 8 & 0 \\ 1 & 2 & -5 \end{bmatrix}$, and $C = \begin{bmatrix} 1 & -1 & 8 \\ 1 & 3 & 0 \\ 1 & 2 & -2 \end{bmatrix}$. Then $A + B = \begin{bmatrix} -2 & 5 & 4 \\ 3 & 2 & 3 \end{bmatrix}$ and $2A = \begin{bmatrix} 2 & -6 & 8 \\ 4 & 0 & 16 \end{bmatrix}$. The matrices $A + C$, BA, and AB are not defined, but

$$AC = \left[A\begin{bmatrix} 1 \\ 1 \\ 1 \end{bmatrix} \quad A\begin{bmatrix} -1 \\ 3 \\ 2 \end{bmatrix} \quad A\begin{bmatrix} 8 \\ 0 \\ -2 \end{bmatrix} \right] = \begin{bmatrix} 2 & -2 & 0 \\ 10 & 14 & 0 \end{bmatrix}.$$

3. Even when the shapes of A and B allow both AB and BA to be calculated, AB and BA are not usually equal. For instance, let $A = \begin{bmatrix} 1 & 0 \\ 0 & 2 \end{bmatrix}$ and $B = \begin{bmatrix} a & b \\ c & d \end{bmatrix}$; then $AB = \begin{bmatrix} a & b \\ 2c & 2d \end{bmatrix}$ and $BA = \begin{bmatrix} a & 2b \\ c & 2d \end{bmatrix}$, which will be equal only if $b = c = 0$.
4. The product of matrices can be a zero matrix even if neither has any zero entries. For example, if $A = \begin{bmatrix} 1 & -1 \\ 2 & -2 \end{bmatrix}$ and $B = \begin{bmatrix} 1 & 1 \\ 1 & 1 \end{bmatrix}$, then $AB = \begin{bmatrix} 0 & 0 \\ 0 & 0 \end{bmatrix}$. Notice that BA is also defined but has no zero entries: $BA = \begin{bmatrix} 3 & -3 \\ 3 & -3 \end{bmatrix}$.

5. The matrices $\begin{bmatrix} 1 & 0 & 0 \\ 0 & 0 & 0 \\ 0 & 0 & -9 \end{bmatrix}$ and $\begin{bmatrix} 1 & 0 & 0 \\ 0 & -3 & 0 \end{bmatrix}$ are diagonal, $\begin{bmatrix} 1 & 0 & 0 \\ 2 & 0 & 0 \\ 1 & 5 & -9 \end{bmatrix}$ and $\begin{bmatrix} 1 & 0 & 0 \\ 2 & -3 & 0 \end{bmatrix}$ are lower triangular, and $\begin{bmatrix} 1 & -4 & 7 \\ 0 & 1 & 2 \\ 0 & 0 & -9 \end{bmatrix}$ and $\begin{bmatrix} 1 & 2 & 3 \\ 0 & 4 & 5 \\ 0 & 0 & 0 \\ 0 & 0 & 0 \end{bmatrix}$ are upper triangular. The matrix $\begin{bmatrix} 1 & 0 & 0 \\ 2 & 1 & 0 \\ 1 & 5 & 1 \end{bmatrix}$ is unit lower triangular, and its transpose is unit upper triangular.

6. Examples of permutation matrices include every identity matrix, $\begin{bmatrix} 0 & 1 \\ 1 & 0 \end{bmatrix}$, $\begin{bmatrix} 0 & 0 & 1 \\ 0 & 1 & 0 \\ 1 & 0 & 0 \end{bmatrix}$, and $\begin{bmatrix} 0 & 0 & 1 \\ 1 & 0 & 0 \\ 0 & 1 & 0 \end{bmatrix}$.

7. Let $A = \begin{bmatrix} 1+i & -3i & 4 \\ 1+2i & 5i & 0 \end{bmatrix}$. Then $A^T = \begin{bmatrix} 1+i & 1+2i \\ -3i & 5i \\ 4 & 0 \end{bmatrix}$ and $A^* = \begin{bmatrix} 1-i & 1-2i \\ 3i & -5i \\ 4 & 0 \end{bmatrix}$.

8. The matrices $\begin{bmatrix} 1 & 2 & 3 \\ 2 & 4 & 5 \\ 3 & 5 & 6 \end{bmatrix}$ and $\begin{bmatrix} i & 2 & 3+2i \\ 2 & 4-i & 5i \\ 3+2i & 5i & 6 \end{bmatrix}$ are symmetric.

9. The matrices $\begin{bmatrix} 0 & 2 & 3 \\ -2 & 0 & 5 \\ -3 & -5 & 0 \end{bmatrix}$ and $\begin{bmatrix} 0 & 2 & 3+2i \\ -2 & 0 & -5 \\ -3-2i & 5 & 0 \end{bmatrix}$ are skew-symmetric.

10. The matrix $\begin{bmatrix} 1 & 2+i & 1-3i \\ 2-i & 0 & 1 \\ 1+3i & 1 & 6 \end{bmatrix}$ is Hermitian, and any real symmetric matrix, such as $\begin{bmatrix} 4 & 2 & 3 \\ 2 & 0 & 5 \\ 3 & 5 & -1 \end{bmatrix}$, is also Hermitian.

11. The matrix $\begin{bmatrix} i & 2 & -3+2i \\ -2 & 4i & 5 \\ 3+2i & -5 & 0 \end{bmatrix}$ is skew-Hermitian, and any real skew-symmetric matrix, such as $\begin{bmatrix} 0 & 2 & -3 \\ -2 & 0 & 5 \\ 3 & -5 & 0 \end{bmatrix}$, is also skew-Hermitian.

12. Let $A = \begin{bmatrix} 1 & 2 & 3 & 4 \\ 5 & 6 & 7 & 8 \\ 9 & 10 & 11 & 12 \\ 13 & 14 & 15 & 16 \end{bmatrix}$. Row 1 of A is $[1 \quad 2 \quad 3 \quad 4]$, column 3 is $\begin{bmatrix} 3 \\ 7 \\ 11 \\ 15 \end{bmatrix}$, and the submatrix in rows $\{1, 2, 4\}$ and columns $\{2, 3, 4\}$ is $A[\{1,2,4\},\{2,3,4\}] = \begin{bmatrix} 2 & 3 & 4 \\ 6 & 7 & 8 \\ 14 & 15 & 16 \end{bmatrix}$. A principal submatrix of A is $A[\{1, 2, 4\}, \{1, 2, 4\}] = \begin{bmatrix} 1 & 2 & 4 \\ 5 & 6 & 8 \\ 13 & 14 & 16 \end{bmatrix}$. The leading principal submatrices of A are $[1]$, $\begin{bmatrix} 1 & 2 \\ 5 & 6 \end{bmatrix}$, $\begin{bmatrix} 1 & 2 & 3 \\ 5 & 6 & 7 \\ 9 & 10 & 11 \end{bmatrix}$, and A itself.

1.3 Gaussian and Gauss–Jordan Elimination

Definitions:

Let A be a matrix with m rows.

When a row of A is not zero, its first nonzero entry is the **leading entry** of the row. The matrix A is in **row echelon form** (REF) when the following two conditions are met:

1. Any zero rows are below all nonzero rows.
2. For each nonzero row $i, i \leq m - 1$, either row $i + 1$ is zero or the leading entry of row $i + 1$ is in a column to the right of the column of the leading entry in row i.

The matrix A is in **reduced row echelon form** (RREF) if it is in row echelon form and the following third condition is also met:

3. If a_{ik} is the leading entry in row i, then $a_{ik} = 1$, and every entry of column k other than a_{ik} is zero.

Elementary row operations on a matrix are operations of the following types:

1. Add a multiple of one row to a different row.
2. Exchange two different rows.
3. Multiply one row by a nonzero scalar.

The matrix A is **row equivalent** to the matrix B if there is a sequence of elementary row operations that transforms A into B. The **reduced row echelon form** of A, RREF(A), is the matrix in reduced row echelon form that is row equivalent to A. A **row echelon form of** A is any matrix in row echelon form that is row equivalent to A. The **rank** of A, denoted rank A or rank(A), is the number of leading entries in RREF(A). If A is in row echelon form, the positions of the leading entries in its nonzero rows are called **pivot positions** and the entries in those positions are called **pivots.** A column (row) that contains a pivot position is a **pivot column (pivot row)**.

Gaussian Elimination is a process that uses elementary row operations in a particular way to change, or reduce, a matrix to row echelon form. **Gauss–Jordan Elimination** is a process that uses elementary row operations in a particular way to reduce a matrix to RREF. See Algorithm 1 below.

Facts:

Let $A \in F^{m \times p}$.

1. [Lay03, p. 15] The reduced row echelon form of A, RREF(A), exists and is unique.
2. A matrix in REF or RREF is upper triangular.
3. Every elementary row operation is reversible by an elementary row operation of the same type.
4. If A is row equivalent to B, then B is row equivalent to A.
5. If A is row equivalent to B, then RREF(A) = RREF(B) and rank A = rank B.
6. The number of nonzero rows in any row echelon form of A equals rank A.
7. If B is any row echelon form of A, the positions of the leading entries in B are the same as the positions of the leading entries of RREF(A).
8. [Lay03, pp. 17–20] (Gaussian and Gauss–Jordan Elimination Algorithms) When one or more pivots are relatively small, using the algorithms below in floating point arithmetic can yield inaccurate results. (See Chapter 38 for more accurate variations of them, and Chapter 75 for information on professional software implementations of such variations.)

Algorithm 1. Gaussian and Gauss-Jordan Elimination

Let $A \in F^{m\times p}$. Steps 1 to 4 below do Gaussian Elimination, reducing A to a matrix that is in row echelon form. Steps 1 to 6 do Gauss–Jordan Elimination, reducing A to RREF(A).

1. Let $U = A$ and $r = 1$. If $U = \mathbf{0}$, U is in RREF.
2. If $U \neq \mathbf{0}$, search the submatrix of U in rows r to m to find its first nonzero column, k, and the first nonzero entry, a_{ik}, in this column. If $i > r$, exchange rows r and i in U, thus getting a nonzero entry in position (r, k). Let U be the matrix created by this row exchange.
3. Add multiples of row r to the rows below it, to create zeros in column k below row r. Let U denote the new matrix.
4. If either $r = m - 1$ or rows $r + 1, \ldots, m$ are all zero, U is now in REF. Otherwise, let $r = r + 1$ and repeat steps 2, 3, and 4.
5. Let $k_1, \ldots, k_s$ be the pivot columns of U, so $(1, k_1), \ldots, (s, k_s)$ are the pivot positions. For $i = s$, $s - 1, \ldots, 2$, add multiples of row i to the rows above it to create zeros in column k_i above row i.
6. For $i = 1, \ldots, s$, divide row s by its leading entry. The resulting matrix is RREF(A).

Examples:

1. The RREF of a zero matrix is itself, and its rank is zero.

2. Let $A = \begin{bmatrix} 1 & 3 & 4 & -8 \\ 0 & 0 & 2 & 4 \\ 0 & 0 & 0 & 0 \end{bmatrix}$ and $B = \begin{bmatrix} 1 & 3 & 4 & -8 \\ 0 & 0 & 0 & 4 \\ 0 & 0 & 1 & 0 \end{bmatrix}$. Both are upper triangular, but A is in REF and B is not. Use Gauss–Jordan Elimination to calculate RREF(A) and RREF(B).

   For A, add (-2)(row 2) to row 1 and multiply row 2 by $\frac{1}{2}$. This yields RREF$(A) = \begin{bmatrix} 1 & 3 & 0 & -16 \\ 0 & 0 & 1 & 2 \\ 0 & 0 & 0 & 0 \end{bmatrix}$.

   For B, exchange rows 2 and 3 to get $\begin{bmatrix} 1 & 3 & 4 & -8 \\ 0 & 0 & 1 & 0 \\ 0 & 0 & 0 & 4 \end{bmatrix}$, which is in REF. Then add 2(row 3) to row 1 to get a new matrix. In this new matrix, add (-4)(row 2) to row 1, and multiply row 3 by $\frac{1}{4}$. This yields RREF$(B) = \begin{bmatrix} 1 & 3 & 0 & 0 \\ 0 & 0 & 1 & 0 \\ 0 & 0 & 0 & 1 \end{bmatrix}$.

   Observe that rank $(A) = 2$ and rank $(B) = 3$.

3. Apply Gauss–Jordan Elimination to $A = \begin{bmatrix} 2 & 6 & 4 & 4 \\ -4 & -12 & -8 & -7 \\ 0 & 0 & -1 & -4 \\ 1 & 3 & 1 & -2 \end{bmatrix}$.

   Step 1. Let $U^{(1)} = A$ and $r = 1$.
   Step 2. No row exchange is needed since $a_{11} \neq 0$.

   Step 3. Add (2)(row 1) to row 2, and $(-\frac{1}{2})$(row 1) to row 4 to get $U^{(2)} = \begin{bmatrix} 2 & 6 & 4 & 4 \\ 0 & 0 & 0 & 1 \\ 0 & 0 & 1 & 4 \\ 0 & 0 & -1 & -4 \end{bmatrix}$.

   Step 4. The submatrix in rows 2, 3, 4 is not zero, so let $r = 2$ and return to Step 2.

Step 2. Search the submatrix in rows 2 to 4 of $U^{(2)}$ to see that its first nonzero column is column 3 and the first nonzero entry in this column is in row 3 of $U^{(2)}$. Exchange rows 2 and 3 in $U^{(2)}$ to get

$$U^{(3)} = \begin{bmatrix} 2 & 6 & 4 & 4 \\ 0 & 0 & 1 & 4 \\ 0 & 0 & 0 & 1 \\ 0 & 0 & -1 & -4 \end{bmatrix}.$$

Step 3. Add row 2 to row 4 in $U^{(3)}$ to get $U^{(4)} = \begin{bmatrix} 2 & 6 & 4 & 4 \\ 0 & 0 & 1 & 4 \\ 0 & 0 & 0 & 1 \\ 0 & 0 & 0 & 0 \end{bmatrix}$.

Step 4. Now $U^{(4)}$ is in REF, so Gaussian Elimination is finished.

Step 5. The pivot positions are $(1, 1)$, $(2, 3)$, and $(3, 4)$. Add –4(row 3) to rows 1 and 2 of $U^{(4)}$ to get

$U^{(5)} = \begin{bmatrix} 2 & 6 & 4 & 0 \\ 0 & 0 & 1 & 0 \\ 0 & 0 & 0 & 1 \\ 0 & 0 & 0 & 0 \end{bmatrix}$. Add –4(row 2) of $U^{(5)}$ to row 1 of $U^{(5)}$ to get $U^{(6)} = \begin{bmatrix} 2 & 6 & 0 & 0 \\ 0 & 0 & 1 & 0 \\ 0 & 0 & 0 & 1 \\ 0 & 0 & 0 & 0 \end{bmatrix}$.

Step 6. Multiply row 1 of $U^{(6)}$ by $\frac{1}{2}$, obtaining $U^{(7)} = \begin{bmatrix} 1 & 3 & 0 & 0 \\ 0 & 0 & 1 & 0 \\ 0 & 0 & 0 & 1 \\ 0 & 0 & 0 & 0 \end{bmatrix}$, which is RREF$(A)$.

1.4 Systems of Linear Equations

Definitions:

A **linear equation** is an equation of the form $a_1x_1 + \cdots + a_px_p = b$ where $a_1, \ldots, a_p, b \in F$ and $x_1, \ldots, x_p$ are **variables.** The scalars a_j are **coefficients** and the scalar b is the **constant term.**

A **system of linear equations**, or **linear system**, is a set of one or more linear equations in the same variables, such as
$$\begin{array}{l} a_{11}x_1 + \cdots + a_{1p}x_p = b_1 \\ a_{21}x_2 + \cdots + a_{2p}x_p = b_2 \\ \cdots \\ a_{m1}x_1 + \cdots + a_{mp}x_p = b_m \end{array}.$$
A **solution** of the system is a p-tuple $(c_1, \ldots, c_p)$ such that letting $x_j = c_j$ for each j satisfies every equation. The **solution set** of the system is the set of all solutions. A system is **consistent** if there exists at least one solution; otherwise it is **inconsistent**. Systems are **equivalent** if they have the same solution set. If $b_j = 0$ for all j, the system is **homogeneous.** A formula that describes a general vector in the solution set is called the **general solution.**

For the system
$$\begin{array}{l} a_{11}x_1 + \cdots + a_{1p}x_p = b_1 \\ a_{21}x_2 + \cdots + a_{2p}x_p = b_2 \\ \cdots \\ a_{m1}x_1 + \cdots + a_{mp}x_p = b_m \end{array},$$
the $m \times p$ matrix $A = \begin{bmatrix} a_{11} & \cdots & a_{1p} \\ \vdots & \cdots & \vdots \\ a_{m1} & \cdots & a_{mp} \end{bmatrix}$ is the **coefficient matrix,** $\mathbf{b} = \begin{bmatrix} b_1 \\ \vdots \\ b_m \end{bmatrix}$ is the **constant vector,** and $\mathbf{x} = \begin{bmatrix} x_1 \\ \vdots \\ x_p \end{bmatrix}$ is the **unknown vector.** The $m \times (p+1)$ matrix $[A\ \mathbf{b}]$ is the **augmented matrix** of the system. It is customary to identify the system of linear equations with the matrix-vector equation $A\mathbf{x} = \mathbf{b}$. This is valid because a column vector $\mathbf{x} = \begin{bmatrix} c_1 \\ \vdots \\ c_p \end{bmatrix}$ satisfies $A\mathbf{x} = \mathbf{b}$ if and only if $(c_1, \ldots, c_p)$ is a solution of the linear system.

Observe that the coefficients of x_k are stored in column k of A. If $A\mathbf{x} = \mathbf{b}$ is equivalent to $C\mathbf{x} = \mathbf{d}$ and column k of C is a pivot column, then x_k is a **basic variable**; otherwise, x_k is a **free variable**.

Facts:

Let $A\mathbf{x} = \mathbf{b}$ be a linear system, where A is an $m \times p$ matrix.

1. [SIF00, pp. 27, 118] If elementary row operations are done to the augmented matrix $[A\ \mathbf{b}]$, obtaining a new matrix $[C\ \mathbf{d}]$, the new system $C\mathbf{x} = \mathbf{d}$ is equivalent to $A\mathbf{x} = \mathbf{b}$.
2. [SIF00, p. 24] There are three possibilities for the solution set of $A\mathbf{x} = \mathbf{b}$: either there are no solutions or there is exactly one solution or there is more than one solution. If there is more than one solution and F is infinite (such as the real numbers or complex numbers), then there are infinitely many solutions. If there is more than one solution and F is finite, then there are at least $|F|$ solutions.
3. A homogeneous system is always consistent (the zero vector $\mathbf{0}$ is always a solution).
4. The set of solutions to the homogeneous system $A\mathbf{x} = \mathbf{0}$ is a subspace of the vector space F^p.
5. [SIF00, p. 44] The system $A\mathbf{x} = \mathbf{b}$ is consistent if and only if $\mathbf{b}$ is not a pivot column of $[A\ \mathbf{b}]$, that is, if and only if $\text{rank}([A\ \mathbf{b}]) = \text{rank}\ A$.
6. [SIF00, pp. 29–32] Suppose $A\mathbf{x} = \mathbf{b}$ is consistent. It has a unique solution if and only there is a pivot position in each column of A, that is, if and only if there are no free variables in the equation $A\mathbf{x} = \mathbf{b}$. Suppose there are $t \geq 1$ nonpivot columns in A. Then there are t free variables in the system. If $\text{RREF}([A\ \mathbf{b}]) = [C\ \mathbf{d}]$, then the general solution of $C\mathbf{x} = \mathbf{d}$, hence of $A\mathbf{x} = \mathbf{b}$, can be written in the form $\mathbf{x} = s_1\mathbf{v}_1 + \cdots + s_t\mathbf{v}_t + \mathbf{w}$ where $\mathbf{v}_1, \ldots, \mathbf{v}_t, \mathbf{w}$ are column vectors and $s_1, \ldots, s_t$ are parameters, each representing one of the free variables. Thus $\mathbf{x} = \mathbf{w}$ is one solution of $A\mathbf{x} = \mathbf{b}$. Also, the general solution of $A\mathbf{x} = \mathbf{0}$ is $\mathbf{x} = s_1\mathbf{v}_1 + \cdots + s_t\mathbf{v}_t$.
7. [SIF00, pp. 29–32] (General solution of a linear system algorithm)

Algorithm 2: General Solution of a Linear System $A\mathbf{x} = \mathbf{b}$

This algorithm is intended for small systems using rational arithmetic. It is not the most efficient and when some pivots are relatively small, using this algorithm in floating point arithmetic can yield inaccurate results. (For more accurate and efficient algorithms, see Chapter 38.) Let $A \in F^{m\times p}$ and $\mathbf{b} \in F^{p\times 1}$.

1. Calculate $\text{RREF}([A\ \mathbf{b}])$, obtaining $[C\ \mathbf{d}]$.
2. If there is a pivot in the last column of $[C\ \mathbf{d}]$, stop. There is no solution.
3. Assume the last column of $[C\ \mathbf{d}]$ is not a pivot column, and let $\mathbf{d} = [d_1, \ldots, d_m]^T$.
   a. If $\text{rank}(C) = p$, so there exists a pivot in each column of C, then $\mathbf{x} = \mathbf{d}$ is the unique solution of the system.
   b. Suppose $\text{rank}\ C = r < p$.
      i. Write the system of linear equations represented by the nonzero rows of $[C\ \mathbf{d}]$. In each equation, the first nonzero term will be a basic variable, and each basic variable appears in only one of these equations.
      ii. Solve each equation for its basic variable and substitute parameter names for the $p - r$ free variables, say $s_1, \ldots, s_{p-r}$. This is the general solution of $C\mathbf{x} = \mathbf{d}$ and, thus, the general solution of $A\mathbf{x} = \mathbf{b}$.
      iii. To write the general solution in vector form, as $\mathbf{x} = s_1\mathbf{v}^{(1)} + \cdots + s_{p-r}\mathbf{v}^{(p-r)} + \mathbf{w}$, let (i, k_i) be the ith pivot position of C. Define $\mathbf{w} \in F^p$ by $w_{k_i} = d_i$ for $i = 1, \ldots, r$, and all other entries of $\mathbf{w}$ are 0. Let x_{u_j} be the jth free variable, and define the vectors $\mathbf{v}^{(j)} \in F^p$ as follows:

         For $j = 1, \ldots, p - r$,
         the u_j-entry of $\mathbf{v}^{(j)}$ is 1,
         for $i = 1, \ldots, r$, the k_i-entry of $\mathbf{v}^{(j)}$ is $-c_{iu_j}$,
         and all other entries of $\mathbf{v}^{(j)}$ are 0.

Examples:

1. The linear system $\begin{array}{r} x_1 + x_2 = 0 \\ -x_1 + x_2 = 0 \end{array}$ has augmented matrix $\begin{bmatrix} 1 & 1 & 0 \\ -1 & 1 & 0 \end{bmatrix}$. The RREF of this is $\begin{bmatrix} 1 & 0 & 0 \\ 0 & 1 & 0 \end{bmatrix}$, which is the augmented matrix for the equivalent system $\begin{array}{l} x_1 = 0 \\ x_2 = 0 \end{array}$. Thus, the original system has a unique solution in $\mathbb{R}^2$, (0,0). In vector form the solution is $\mathbf{x} = \begin{bmatrix} x_1 \\ x_2 \end{bmatrix} = \begin{bmatrix} 0 \\ 0 \end{bmatrix}$.

2. The system $\begin{array}{r} x_1 + x_2 = 2 \\ x_1 - x_2 = 0 \end{array}$ has a unique solution in $\mathbb{R}^2$, $(1, 1)$, or $\mathbf{x} = \begin{bmatrix} x_1 \\ x_2 \end{bmatrix} = \begin{bmatrix} 1 \\ 1 \end{bmatrix}$.

3. The system $\begin{array}{r} x_1 + x_2 + x_3 = 2 \\ x_2 + x_3 = 2 \\ x_3 = 0 \end{array}$ has a unique solution in $\mathbb{R}^3$, $(0, 2, 0)$, or $\mathbf{x} = \begin{bmatrix} 0 \\ 2 \\ 0 \end{bmatrix}$.

4. The system $\begin{array}{r} x_1 + x_2 = 2 \\ 2x_1 + 2x_2 = 4 \end{array}$ has infinitely many solutions in $\mathbb{R}^2$. The augmented matrix reduces to $\begin{bmatrix} 1 & 1 & 2 \\ 0 & 0 & 0 \end{bmatrix}$, so the only equation left is $x_1 + x_2 = 2$. Thus x_1 is basic and x_2 is free. Solving for x_1 and letting $x_2 = s$ gives $x_1 = -s + 2$. Then the general solution is $\begin{array}{l} x_1 = -s + 2 \\ x_2 = s \end{array}$, or all vectors of the form $(-s + 2, s)$. Letting $\mathbf{x} = \begin{bmatrix} x_1 \\ x_2 \end{bmatrix}$, the vector form of the general solution is $\mathbf{x} = \begin{bmatrix} -s + 2 \\ s \end{bmatrix} = s \begin{bmatrix} -1 \\ 1 \end{bmatrix} + \begin{bmatrix} 2 \\ 0 \end{bmatrix}$.

5. The system $\begin{array}{r} x_1 + x_2 + x_3 + x_4 = 1 \\ x_2 + x_3 - x_4 = 3 \end{array}$ has infinitely many solutions in $\mathbb{R}^4$. Its augmented matrix $\begin{bmatrix} 1 & 1 & 1 & 1 & 1 \\ 0 & 1 & 1 & -1 & 3 \end{bmatrix}$ reduces to $\begin{bmatrix} 1 & 0 & 0 & 2 & -2 \\ 0 & 1 & 1 & -1 & 3 \end{bmatrix}$. Thus, x_1 and x_2 are the basic variables, and x_3 and x_4 are free. Write each of the new equations and solve it for its basic variable to see $\begin{array}{l} x_1 = -2x_4 - 2 \\ x_2 = -x_3 + x_4 + 3 \end{array}$. Let $x_3 = s_1$ and $x_4 = s_2$ to get the general solution $\begin{array}{l} x_1 = -2s_2 - 2 \\ x_2 = -s_1 + s_2 + 3 \\ x_3 = s_1 \\ x_4 = s_2 \end{array}$, or $\mathbf{x} = s_1 \mathbf{v}^{(1)} + s_2 \mathbf{v}^{(2)} + \mathbf{w} = s_1 \begin{bmatrix} 0 \\ -1 \\ 1 \\ 0 \end{bmatrix} + s_2 \begin{bmatrix} -2 \\ 1 \\ 0 \\ 1 \end{bmatrix} + \begin{bmatrix} -2 \\ 3 \\ 0 \\ 0 \end{bmatrix}$.

6. These systems have no solutions: $\begin{array}{r} x_1 + x_2 = 0 \\ x_1 + x_2 = 1 \end{array}$ and $\begin{array}{r} x_1 + x_2 + x_3 = 0 \\ x_1 - x_2 - x_3 = 0 \\ x_2 + x_3 = 1 \end{array}$. This can be verified by inspection, or by calculating the RREF of the augmented matrix of each and observing that each has a pivot in its last column.

1.5 Matrix Inverses and Elementary Matrices

Invertibility is a strong and useful property. For example, when a linear system $A\mathbf{x} = \mathbf{b}$ has an invertible coefficient matrix A, it has a unique solution. The various characterizations of invertibility in Fact 10 below are also quite useful. Throughout this section, F will denote a field.

Definitions:

An $n \times n$ matrix A is **invertible**, or **nonsingular**, if there exists another $n \times n$ matrix B, called the **inverse** of A, such that $AB = BA = I_n$. The inverse of A is denoted A^{-1} (cf. Fact 1). If no such B exists, A is **not invertible**, or **singular**.

For an $n \times n$ matrix and a positive integer m, the ***m*th power** of A is $A^m = \underbrace{AA\dots A}_{m \text{ copies of } A}$. It is also convenient to define $A^0 = I_n$. If A is invertible, then $A^{-m} = (A^{-1})^m$.

An **elementary matrix** is a square matrix obtained by doing one elementary row operation to an identity matrix. Thus, there are three types:

1. A multiple of one row of I_n has been added to a different row.
2. Two different rows of I_n have been exchanged.
3. One row of I_n has been multiplied by a nonzero scalar.

Facts:

1. [SIF00, pp. 114–116] If $A \in F^{n\times n}$ is invertible, then its inverse is unique.
2. [SIF00, p. 128] (Method to compute A^{-1}) Suppose $A \in F^{n\times n}$. Create the matrix $[A\ I_n]$ and calculate its RREF, which will be of the form $[\text{RREF}(A)X]$. If $\text{RREF}(A) = I_n$, then A is invertible and $X = A^{-1}$. If $\text{RREF}(A) \neq I_n$, then A is not invertible. As with the Gaussian algorithm, this method is theoretically correct, but more accurate and efficient methods for calculating inverses are used in professional computer software. (See Chapter 75.)
3. [SIF00, pp. 114–116] If $A \in F^{n\times n}$ is invertible, then A^{-1} is invertible and $(A^{-1})^{-1} = A$.
4. [SIF00, pp. 114–116] If $A, B \in F^{n\times n}$ are invertible, then AB is invertible and $(AB)^{-1} = B^{-1}A^{-1}$.
5. [SIF00, pp. 114–116] If $A \in F^{n\times n}$ is invertible, then A^T is invertible and $(A^T)^{-1} = (A^{-1})^T$.
6. If $A \in F^{n\times n}$ is invertible, then for each $\mathbf{b} \in F^{n\times 1}$, $A\mathbf{x} = \mathbf{b}$ has a unique solution, and it is $\mathbf{x} = A^{-1}\mathbf{b}$.
7. [SIF00, p. 124] If $A \in F^{n\times n}$ and there exists $C \in F^{n\times n}$ such that either $AC = I_n$ or $CA = I_n$, then A is invertible and $A^{-1} = C$. That is, a left or right inverse for a square matrix is actually its unique two-sided inverse.
8. [SIF00, p. 117] Let E be an elementary matrix obtained by doing one elementary row operation to I_n. If that same row operation is done to an $n \times p$ matrix A, the result equals EA.
9. [SIF00, p. 117] An elementary matrix is invertible and its inverse is another elementary matrix of the same type.
10. [SIF00, pp. 126] (Invertible Matrix Theorem) (See Section 2.5.) When $A \in F^{n\times n}$, the following are equivalent:
    - A is invertible.
    - $\text{RREF}(A) = I_n$.
    - $\text{Rank}(A) = n$.
    - The only solution of $A\mathbf{x} = \mathbf{0}$ is $\mathbf{x} = \mathbf{0}$.
    - For every $\mathbf{b} \in F^{n\times 1}$, $A\mathbf{x} = \mathbf{b}$ has a unique solution.
    - For every $\mathbf{b} \in F^{n\times 1}$, $A\mathbf{x} = \mathbf{b}$ has a solution.
    - There exists $B \in F^{n\times n}$ such that $AB = I_n$.
    - There exists $C \in F^{n\times n}$ such that $CA = I_n$.
    - A^T is invertible.
    - There exist elementary matrices whose product equals A.
11. [SIF00, p. 148] and [Lay03, p.132] Let $A \in F^{n\times n}$ be upper (lower) triangular. Then A is invertible if and only if each diagonal entry is nonzero. If A is invertible, then A^{-1} is also upper (lower) triangular, and the diagonal entries of A^{-1} are the reciprocals of those of A. In particular, if L is a unit upper (lower) triangular matrix, then L^{-1} is also a unit upper (lower) triangular matrix.

12. Matrix powers obey the usual rules of exponents, i.e., when A^s and A^t are defined for integers s and t, then $A^sA^t = A^{s+t}, (A^s)^t = A^{st}$.

Examples:

1. For any n, the identity matrix I_n is invertible and is its own inverse. If P is a permutation matrix, it is invertible and $P^{-1} = P^T$.
2. If $A = \begin{bmatrix} 7 & 3 \\ 2 & 1 \end{bmatrix}$ and $B = \begin{bmatrix} 1 & -3 \\ -2 & 7 \end{bmatrix}$, then calculation shows $AB = BA = I_2$, so A is invertible and $A^{-1} = B$.
3. If $A = \begin{bmatrix} 0.2 & 4 & 1 \\ 0 & 2 & 1 \\ 0 & 0 & -1 \end{bmatrix}$, then $A^{-1} = \begin{bmatrix} 5 & -10 & -5 \\ 0 & 0.5 & 0.5 \\ 0 & 0 & -1 \end{bmatrix}$, as can be verified by multiplication.
4. The matrix $A = \begin{bmatrix} 1 & 2 \\ 2 & 4 \end{bmatrix}$ is not invertible since $\text{RREF}(A) \neq I_2$. Alternatively, if B is any 2×2 matrix, AB is of the form $\begin{bmatrix} r & s \\ 2r & 2s \end{bmatrix}$, which cannot equal I_2.
5. Let A be an $n \times n$ matrix A with a zero row (zero column). Then A is not invertible since $\text{RREF}(A) \neq I_n$. Alternatively, if B is any $n \times n$ matrix, AB has a zero row (BA has a zero column), so B is not an inverse for A.
6. If $A = \begin{bmatrix} a & b \\ c & d \end{bmatrix}$ is any 2×2 matrix, then A is invertible if and only if $ad - bc \neq 0$; further, when $ad - bc \neq 0$, $A^{-1} = \dfrac{1}{ad-bc}\begin{bmatrix} d & -b \\ -c & a \end{bmatrix}$. The scalar $ad - bc$ is called the determinant of A. (The determinant is defined for any $n \times n$ matrix in Section 4.1.) Using this formula, the matrix $A = \begin{bmatrix} 7 & 3 \\ 2 & 1 \end{bmatrix}$ from Example 2 (above) has determinant 1, so A is invertible and $A^{-1} = \begin{bmatrix} 1 & -3 \\ -2 & 7 \end{bmatrix}$, as noted above. The matrix $\begin{bmatrix} 1 & 2 \\ 2 & 4 \end{bmatrix}$ from Example 3 (above) is not invertible since its determinant is 0.
7. Let $A = \begin{bmatrix} 1 & 3 & 0 \\ 2 & 7 & 0 \\ 1 & 1 & 1 \end{bmatrix}$. Then $\text{RREF}([A \quad I_n]) = \begin{bmatrix} 1 & 0 & 0 & 7 & -3 & 0 \\ 0 & 1 & 0 & -2 & 1 & 0 \\ 0 & 0 & 1 & -5 & 2 & 1 \end{bmatrix}$, so A^{-1} exists and equals $\begin{bmatrix} 7 & -3 & 0 \\ -2 & 1 & 0 \\ -5 & 2 & 1 \end{bmatrix}$.

1.6 LU Factorization

This section discusses the *LU* and *PLU* factorizations of a matrix that arise naturally when Gaussian Elimination is done. Several other factorizations are widely used for real and complex matrices, such as the QR, Singular Value, and Cholesky Factorizations. (See Chapter 5 and Chapter 38.) Throughout this section, F will denote a field and A will denote a matrix over F. The material in this section and additional background can be found in [GV96, Sec. 3.2].

Definitions:

Let A be a matrix of any shape.

An *LU* **factorization**, or **triangular factorization**, of A is a factorization $A = LU$ where L is a square unit lower triangular matrix and U is upper triangular. A ***PLU* factorization** of A is a factorization of the form $PA = LU$ where P is a permutation matrix, L is square unit lower triangular, and U is upper triangular. An ***LDU* factorization** of A is a factorization $A = LDU$ where L is a square unit lower triangular matrix, D is a square diagonal matrix, and U is a unit upper triangular matrix.

A ***PLDU* factorization** of A is a factorization $PA = LDU$ where P is a permutation matrix, L is a square unit lower triangular matrix, D is a square diagonal matrix, and U is a unit upper triangular matrix.

Facts: [GV96, Sec. 3.2]

1. Let A be square. If each leading principal submatrix of A, except possibly A itself, is invertible, then A has an *LU* factorization. When A is invertible, A has an *LU* factorization if and only if each leading principal submatrix of A is invertible; in this case, the *LU* factorization is unique and there is also a unique *LDU* factorization of A.
2. Any matrix A has a *PLU* factorization. Algorithm 1 (Section 1.3) performs the addition of multiples of pivot rows to lower rows and perhaps row exchanges to obtain an REF matrix U. If instead, the same series of row exchanges are done to A before any pivoting, this creates PA where P is a permutation matrix, and then PA can be reduced to U without row exchanges. That is, there exist unit lower triangular matrices E_j such that $E_k \ldots E_1(PA) = U$. It follows that $PA = LU$, where $L = (E_k \ldots E_1)^{-1}$ is unit lower triangular and U is upper triangular.
3. In most professional software packages, the standard method for solving a square linear system $A\mathbf{x} = \mathbf{b}$, for which A is invertible, is to reduce A to an REF matrix U as in Fact 2 above, choosing row exchanges by a strategy to reduce pivot size. By keeping track of the exchanges and pivot operations done, this produces a *PLU* factorization of A. Then $A = P^T LU$ and $P^T LU\mathbf{x} = \mathbf{b}$ is the equation to be solved. Using forward substitution, $P^T L\mathbf{y} = \mathbf{b}$ can be solved quickly for $\mathbf{y}$, and then $U\mathbf{x} = \mathbf{y}$ can either be solved quickly for $\mathbf{x}$ by back substitutution, or be seen to be inconsistent. This method gives accurate results for most problems. There are other types of solution methods that can work more accurately or efficiently for special types of matrices. (See Chapter 7.)

Examples:

1. Calculate a *PLU* factorization for $A = \begin{bmatrix} 1 & 1 & 2 & 3 \\ -1 & -1 & -3 & 1 \\ 0 & 1 & 1 & 1 \\ -1 & 0 & -1 & 1 \end{bmatrix}$. If Gaussian Elimination is performed on A, after adding row 1 to rows 2 and 4, rows 2 and 3 must be exchanged and the final result is
$$U = E_3 P E_2 E_1 A = \begin{bmatrix} 1 & 1 & 2 & 3 \\ 0 & 1 & 1 & 1 \\ 0 & 0 & -1 & 4 \\ 0 & 0 & 0 & 3 \end{bmatrix}$$
where E_1, E_2, and E_3 are lower triangular unit matrices and P is a permutation matrix. This will not yield an *LU* factorization of A. But if the row exchange is done to A first, by multiplying A by $P = \begin{bmatrix} 1 & 0 & 0 & 0 \\ 0 & 0 & 1 & 0 \\ 0 & 1 & 0 & 0 \\ 0 & 0 & 0 & 1 \end{bmatrix}$, one gets $PA = \begin{bmatrix} 1 & 1 & 2 & 3 \\ 0 & 1 & 1 & 1 \\ -1 & -1 & -3 & 1 \\ -1 & 0 & -1 & 1 \end{bmatrix}$; then Gaussian Elimination can proceed without any row exchanges. Add row 1 to rows 3 and 4 to get
$$F_2 F_1 PA = \begin{bmatrix} 1 & 1 & 2 & 3 \\ 0 & 1 & 1 & 1 \\ 0 & 0 & -1 & 4 \\ 0 & 1 & 1 & 4 \end{bmatrix} \text{ where } F_1 = \begin{bmatrix} 1 & 0 & 0 & 0 \\ 0 & 1 & 0 & 0 \\ 1 & 0 & 1 & 0 \\ 0 & 0 & 0 & 1 \end{bmatrix} \text{ and } F_2 = \begin{bmatrix} 1 & 0 & 0 & 0 \\ 0 & 1 & 0 & 0 \\ 0 & 0 & 1 & 0 \\ 1 & 0 & 0 & 1 \end{bmatrix}.$$
Then add $(-1)(\text{row } 2)$ to row 4 to get $U = F_3\, F_2\, F_1\, PA = \begin{bmatrix} 1 & 1 & 2 & 3 \\ 0 & 1 & 1 & 1 \\ 0 & 0 & -1 & 4 \\ 0 & 0 & 0 & 3 \end{bmatrix}$, where $F_3 = \begin{bmatrix} 1 & 0 & 0 & 0 \\ 0 & 1 & 0 & 0 \\ 0 & 0 & 1 & 0 \\ 0 & -1 & 0 & 1 \end{bmatrix}$.

Note that U is the same upper triangular matrix as before. Finally, $L = (F_3 F_2 F_1)^{-1}$ is unit lower triangular and $PA = LU$ is true, so this is a PLU factorization of A. To get a $PLDU$ factorization, use the same P and L, and define $D = \begin{bmatrix} 1 & 0 & 0 & 0 \\ 0 & 1 & 0 & 0 \\ 0 & 0 & -1 & 0 \\ 0 & 0 & 0 & 3 \end{bmatrix}$ and $U = \begin{bmatrix} 1 & 1 & 2 & 3 \\ 0 & 1 & 1 & 1 \\ 0 & 0 & 1 & -4 \\ 0 & 0 & 0 & 1 \end{bmatrix}$.

2. Let $A = LU = \begin{bmatrix} 1 & 3 & 4 \\ -1 & -1 & -5 \\ 2 & 12 & 3 \end{bmatrix}$. Each leading principal submatrix of A is invertible so A has both LU and LDU factorizations:

$A = LU = \begin{bmatrix} 1 & 0 & 0 \\ -1 & 1 & 0 \\ 2 & 3 & 1 \end{bmatrix} \begin{bmatrix} 1 & 3 & 4 \\ 0 & 2 & -1 \\ 0 & 0 & -2 \end{bmatrix}$. This yields an LDU factorization of A, $\begin{bmatrix} 1 & 0 & 0 \\ -1 & 1 & 0 \\ 2 & 3 & 1 \end{bmatrix} \begin{bmatrix} 1 & 0 & 0 \\ 0 & 2 & 0 \\ 0 & 0 & -2 \end{bmatrix}$ $\begin{bmatrix} 1 & 3 & 4 \\ 0 & 1 & -0.5 \\ 0 & 0 & 1 \end{bmatrix}$. With the LU factorization, an equation such as $A\mathbf{x} = \begin{bmatrix} 1 \\ 1 \\ 0 \end{bmatrix}$ can be solved efficiently as follows. Use forward substitution to solve $L\mathbf{y} = \begin{bmatrix} 1 \\ 1 \\ 0 \end{bmatrix}$, getting $\mathbf{y} = \begin{bmatrix} 1 \\ 2 \\ -8 \end{bmatrix}$, and then backward substitution to solve $U\mathbf{x} = \mathbf{y}$, getting $\mathbf{x} = \begin{bmatrix} -24 \\ 3 \\ 4 \end{bmatrix}$.

3. Any invertible matrix whose $(1, 1)$ entry is zero, such as $\begin{bmatrix} 0 & 1 \\ 1 & 0 \end{bmatrix}$ or $\begin{bmatrix} 0 & -1 & 5 \\ 1 & 1 & 1 \\ 1 & 0 & 3 \end{bmatrix}$, does not have an LU factorization.

4. The matrix $A = \begin{bmatrix} 1 & 3 & 4 \\ -1 & -3 & -5 \\ 2 & 6 & 6 \end{bmatrix}$ is not invertible, nor is its leading principal 2×2 submatrix, but it does have an LU factorization: $A = LU = \begin{bmatrix} 1 & 0 & 0 \\ -1 & 1 & 0 \\ 2 & 3 & 1 \end{bmatrix} \begin{bmatrix} 1 & 3 & 4 \\ 0 & 0 & -1 \\ 0 & 0 & 1 \end{bmatrix}$. To find out if an equation such as $A\mathbf{x} = \begin{bmatrix} 1 \\ 1 \\ 0 \end{bmatrix}$ is consistent, notice $L\mathbf{y} = \begin{bmatrix} 1 \\ 1 \\ 0 \end{bmatrix}$ yields $\mathbf{y} = \begin{bmatrix} 1 \\ 2 \\ -8 \end{bmatrix}$, but $U\mathbf{x} = \mathbf{y}$ is inconsistent, hence $A\mathbf{x} = \begin{bmatrix} 1 \\ 1 \\ 0 \end{bmatrix}$ has no solution.

5. The matrix $A = \begin{bmatrix} 0 & -1 & 5 \\ 1 & 1 & 1 \\ 1 & 0 & 2 \end{bmatrix}$ has no LU factorization, but does have a PLU factorization with $P = \begin{bmatrix} 0 & 1 & 0 \\ 1 & 0 & 0 \\ 0 & 0 & 1 \end{bmatrix}$, $L = \begin{bmatrix} 1 & 0 & 0 \\ 0 & 1 & 0 \\ 1 & 1 & 1 \end{bmatrix}$, and $U = \begin{bmatrix} 1 & 1 & 1 \\ 0 & -1 & 5 \\ 0 & 0 & -4 \end{bmatrix}$.

References

[FIS03] S.H. Friedberg, A.J. Insel, and L.E. Spence. *Linear Algebra*, 3rd ed. Pearson Education, Upper Saddle River, NJ, 2003.

[GV96] G.H. Golub and C.F. Van Loan. *Matrix Computations*, 3rd ed. Johns Hopkins Press, Baltimore, MD, 1996.

[Lay03] David C. Lay. *Linear Algebra and Its Applications*, 3rd ed. Addison Wesley, Boston, 2003.

[Leo02] Steven J. Leon. *Linear Algebra with Applications*, 6th ed. Prentice Hall, Upper Saddle River, NJ, 2003.

[SIF00] L.E. Spence, A.J. Insel, and S.H. Friedberg. *Elementary Linear Algebra*. Prentice Hall, Upper Saddle River, NJ, 2000.

2

Linear Independence, Span, and Bases

Mark Mills
Central College

2.1 Span and Linear Independence

Let V be a vector space over a field F.

Definitions:

A **linear combination** of the vectors $\mathbf{v}_1, \mathbf{v}_2, \ldots, \mathbf{v}_k \in V$ is a sum of scalar multiples of these vectors; that is, $c_1\mathbf{v}_1 + c_2\mathbf{v}_2 + \cdots + c_k\mathbf{v}_k$, for some scalar coefficients $c_1, c_2, \ldots, c_k \in F$. If S is a set of vectors in V, a linear combination of vectors in S is a vector of the form $c_1\mathbf{v}_1 + c_2\mathbf{v}_2 + \cdots + c_k\mathbf{v}_k$ with $k \in \mathbb{N}, \mathbf{v}_i \in S, c_i \in F$. Note that S may be finite or infinite, but a linear combination is, by definition, a finite sum. The zero vector is defined to be a linear combination of the empty set.

When all the scalar coefficients in a linear combination are 0, it is a **trivial linear combination.** A sum over the empty set is also a trivial linear combination.

The **span** of the vectors $\mathbf{v}_1, \mathbf{v}_2, \ldots, \mathbf{v}_k \in V$ is the set of all linear combinations of these vectors, denoted by $\text{Span}(\mathbf{v}_1, \mathbf{v}_2, \ldots, \mathbf{v}_k)$. If S is a (finite or infinite) set of vectors in V, then the span of S, denoted by $\text{Span}(S)$, is the set of all linear combinations of vectors in S.

If $V = \text{Span}(S)$, then S **spans** the vector space V.

A (finite or infinite) set of vectors S in V is **linearly independent** if the only linear combination of distinct vectors in S that produces the zero vector is a trivial linear combination. That is, if $\mathbf{v}_i$ are distinct vectors in S and $c_1\mathbf{v}_1 + c_2\mathbf{v}_2 + \cdots + c_k\mathbf{v}_k = \mathbf{0}$, then $c_1 = c_2 = \cdots = c_k = 0$. Vectors that are not linearly independent are **linearly dependent**. That is, there exist distinct vectors $\mathbf{v}_1, \mathbf{v}_2, \ldots, \mathbf{v}_k \in S$ and $c_1, c_2, \ldots, c_k$ not all 0 such that $c_1\mathbf{v}_1 + c_2\mathbf{v}_2 + \cdots + c_k\mathbf{v}_k = \mathbf{0}$.

Facts: The following facts can be found in [Lay03, Sections 4.1 and 4.3].

1. $\text{Span}(\emptyset) = \{\mathbf{0}\}$.
2. A linear combination of a single vector $\mathbf{v}$ is simply a scalar multiple of $\mathbf{v}$.
3. In a vector space V, $\text{Span}(\mathbf{v}_1, \mathbf{v}_2, \ldots, \mathbf{v}_k)$ is a subspace of V.
4. Suppose the set of vectors $S = \{\mathbf{v}_1, \mathbf{v}_2, \ldots, \mathbf{v}_k\}$ spans the vector space V. If one of the vectors, say $\mathbf{v}_i$, is a linear combination of the remaining vectors, then the set formed from S by removing $\mathbf{v}_i$ still spans V.
5. Any single nonzero vector is linearly independent.
6. Two nonzero vectors are linearly independent if and only if neither is a scalar multiple of the other.
7. If S spans V and $S \subseteq T$, then T spans V.
8. If T is a linearly independent subset of V and $S \subseteq T$, then S is linearly independent.
9. Vectors $\mathbf{v}_1, \mathbf{v}_2, \ldots, \mathbf{v}_k$ are linearly dependent if and only if $\mathbf{v}_i = c_1\mathbf{v}_1 + \cdots + c_{i-1}\mathbf{v}_{i-1} + c_{i+1}\mathbf{v}_{i+1} + \cdots + c_k\mathbf{v}_k$, for some $1 \leq i \leq k$ and some scalars $c_1, \ldots, c_{i-1}, c_{i+1}, \ldots, c_k$. A set S of vectors in V is linearly dependent if and only if there exists $\mathbf{v} \in S$ such that $\mathbf{v}$ is a linear combination of other vectors in S.
10. Any set of vectors that includes the zero vector is linearly dependent.

Examples:

1. Linear combinations of $\begin{bmatrix} 1 \\ -1 \end{bmatrix}, \begin{bmatrix} 0 \\ 3 \end{bmatrix} \in \mathbb{R}^2$ are vectors of the form $c_1 \begin{bmatrix} 1 \\ -1 \end{bmatrix} + c_2 \begin{bmatrix} 0 \\ 3 \end{bmatrix} = \begin{bmatrix} c_1 \\ -c_1 + 3c_2 \end{bmatrix}$, for any scalars $c_1, c_2 \in \mathbb{R}$. Any vector of this form is in $\text{Span}\left(\begin{bmatrix} 1 \\ -1 \end{bmatrix}, \begin{bmatrix} 0 \\ 3 \end{bmatrix}\right)$. In fact, $\text{Span}\left(\begin{bmatrix} 1 \\ -1 \end{bmatrix}, \begin{bmatrix} 0 \\ 3 \end{bmatrix}\right) = \mathbb{R}^2$ and these vectors are linearly independent.
2. If $\mathbf{v} \in \mathbb{R}^n$ and $\mathbf{v} \neq \mathbf{0}$, then geometrically $\text{Span}(\mathbf{v})$ is a line in $\mathbb{R}^n$ through the origin.
3. Suppose $n \geq 2$ and $\mathbf{v}_1, \mathbf{v}_2 \in \mathbb{R}^n$ are linearly independent vectors. Then geometrically $\text{Span}(\mathbf{v}_1, \mathbf{v}_2)$ is a plane in $\mathbb{R}^n$ through the origin.
4. Any polynomial $p(x) \in \mathbb{R}[x]$ of degree less than or equal to 2 can easily be seen to be a linear combination of 1, x, and x^2. However, $p(x)$ is also a linear combination of 1, $1 + x$, and $1 + x^2$. So $\text{Span}(1, x, x^2) = \text{Span}(1, 1 + x, 1 + x^2) = \mathbb{R}[x; 2]$.
5. The n vectors $\mathbf{e}_1 = \begin{bmatrix} 1 \\ 0 \\ 0 \\ \vdots \\ 0 \end{bmatrix}, \mathbf{e}_2 = \begin{bmatrix} 0 \\ 1 \\ 0 \\ \vdots \\ 0 \end{bmatrix}, \ldots, \mathbf{e}_n = \begin{bmatrix} 0 \\ 0 \\ \vdots \\ 0 \\ 1 \end{bmatrix}$ span F^n, for any field F. These vectors are also linearly independent.
6. In $\mathbb{R}^2$, $\begin{bmatrix} 1 \\ -1 \end{bmatrix}$ and $\begin{bmatrix} 0 \\ 3 \end{bmatrix}$ are linearly independent. However, $\begin{bmatrix} 1 \\ -1 \end{bmatrix}, \begin{bmatrix} 0 \\ 3 \end{bmatrix}$, and $\begin{bmatrix} 1 \\ 5 \end{bmatrix}$ are linearly dependent, because $\begin{bmatrix} 1 \\ 5 \end{bmatrix} = \begin{bmatrix} 1 \\ -1 \end{bmatrix} + 2 \begin{bmatrix} 0 \\ 3 \end{bmatrix}$.
7. The infinite set $\{1, x, x^2, \ldots, x^n, \ldots\}$ is linearly independent in $F[x]$, for any field F.
8. In the vector space of continuous real-valued functions on the real line, $C(\mathbb{R})$, the set $\{\sin(x), \sin(2x), \ldots, \sin(nx), \cos(x), \cos(2x), \ldots, \cos(nx)\}$ is linearly independent for any $n \in \mathbb{N}$. The infinite set $\{\sin(x), \sin(2x), \ldots, \sin(nx), \ldots, \cos(x), \cos(2x), \ldots, \cos(nx), \ldots\}$ is also linearly independent in $C(\mathbb{R})$.

Applications:

1. The homogeneous differential equation $\frac{d^2y}{dx^2} - 3\frac{dy}{dx} + 2y = 0$ has as solutions $y_1(x) = e^{2x}$ and $y_2(x) = e^x$. Any linear combination $y(x) = c_1y_1(x) + c_2y_2(x)$ is a solution of the differential equation, and so $\text{Span}(e^{2x}, e^x)$ is contained in the set of solutions of the differential equation (called the solution space for the differential equation). In fact, the solution space is spanned by e^{2x} and e^x, and so is a subspace of the vector space of functions. In general, the solution space for a homogeneous differential equation is a vector space, meaning that any linear combination of solutions is again a solution.

2.2 Basis and Dimension of a Vector Space

Let V be a vector space over a field F.

Definitions:

A set of vectors $\mathcal{B}$ in a vector space V is a **basis** for V if

- $\mathcal{B}$ is a linearly independent set, and
- $\text{Span}(\mathcal{B}) = V$.

The set $\mathcal{E}_n = \left\{ \mathbf{e}_1 = \begin{bmatrix} 1 \\ 0 \\ 0 \\ \vdots \\ 0 \end{bmatrix}, \mathbf{e}_2 = \begin{bmatrix} 0 \\ 1 \\ 0 \\ \vdots \\ 0 \end{bmatrix}, \ldots, \mathbf{e}_n = \begin{bmatrix} 0 \\ 0 \\ \vdots \\ 0 \\ 1 \end{bmatrix} \right\}$ is the **standard basis** for F^n.

The number of vectors in a basis for a vector space V is the **dimension** of V, denoted by $\dim(V)$. If a basis for V contains a finite number of vectors, then V is **finite dimensional**. Otherwise, V is **infinite dimensional**, and we write $\dim(V) = \infty$.

Facts: All the following facts, except those with a specific reference, can be found in [Lay03, Sections 4.3 and 4.5].

1. Every vector space has a basis.
2. The standard basis for F^n is a basis for F^n, and so $\dim F^n = n$.
3. A basis $\mathcal{B}$ in a vector space V is the largest set of linearly independent vectors in V that contains $\mathcal{B}$, and it is the smallest set of vectors in V that contains $\mathcal{B}$ and spans V.
4. The empty set is a basis for the trivial vector space $\{\mathbf{0}\}$, and $\dim(\{\mathbf{0}\}) = 0$.
5. If the set $S = \{\mathbf{v}_1, \ldots, \mathbf{v}_p\}$ spans a vector space V, then some subset of S forms a basis for V. In particular, if one of the vectors, say $\mathbf{v}_i$, is a linear combination of the remaining vectors, then the set formed from S by removing $\mathbf{v}_i$ will be "closer" to a basis for V. This process can be continued until the remaining vectors form a basis for V.
6. If S is a linearly independent set in a vector space V, then S can be expanded, if necessary, to a basis for V.
7. No nontrivial vector space over a field with more than two elements has a unique basis.
8. If a vector space V has a basis containing n vectors, then every basis of V must contain n vectors. Similarly, if V has an infinite basis, then every basis of V must be infinite. So the dimension of V is unique.
9. Let $\dim(V) = n$ and let S be a set containing n vectors. The following are equivalent:
   - S is a basis for V.
   - S spans V.
   - S is linearly independent.

10. If $\dim(V) = n$, then any subset of V containing more than n vectors is linearly dependent.
11. If $\dim(V) = n$, then any subset of V containing fewer than n vectors does not span V.
12. [Lay03, Section 4.4] If $\mathcal{B} = \{\mathbf{b}_1, \ldots, \mathbf{b}_p\}$ is a basis for a vector space V, then each $\mathbf{x} \in V$ can be expressed as a unique linear combination of the vectors in $\mathcal{B}$. That is, for each $\mathbf{x} \in V$ there is a unique set of scalars $c_1, c_2, \ldots, c_p$ such that $\mathbf{x} = c_1\mathbf{b}_1 + c_2\mathbf{b}_2 + \cdots + c_p\mathbf{b}_p$.

Examples:

1. In $\mathbb{R}^2$, $\begin{bmatrix} 1 \\ -1 \end{bmatrix}$ and $\begin{bmatrix} 0 \\ 3 \end{bmatrix}$ are linearly independent, and they span $\mathbb{R}^2$. So they form a basis for $\mathbb{R}^2$ and $\dim(\mathbb{R}^2) = 2$.
2. In $F[x]$, the set $\{1, x, x^2, \ldots, x^n\}$ is a basis for $F[x; n]$ for any $n \in \mathbb{N}$. The infinite set $\{1, x, x^2, x^3, \ldots\}$ is a basis for $F[x]$, meaning $\dim(F[x]) = \infty$.
3. The set of $m \times n$ matrices E_{ij} having a 1 in the i, j-entry and zeros everywhere else forms a basis for $F^{m\times n}$. Since there are mn such matrices, $\dim(F^{m\times n}) = mn$.
4. The set $S = \left\{ \begin{bmatrix} 1 \\ 0 \end{bmatrix}, \begin{bmatrix} 0 \\ 1 \end{bmatrix}, \begin{bmatrix} 1 \\ 2 \end{bmatrix} \right\}$ clearly spans $\mathbb{R}^2$, but it is not a linearly independent set. However, removing any single vector from S will cause the remaining vectors to be a basis for $\mathbb{R}^2$, because any pair of vectors is linearly independent and still spans $\mathbb{R}^2$.
5. The set $S = \left\{ \begin{bmatrix} 1 \\ 1 \\ 0 \\ 0 \end{bmatrix}, \begin{bmatrix} 0 \\ 0 \\ 1 \\ 1 \end{bmatrix} \right\}$ is linearly independent, but it cannot be a basis for $\mathbb{R}^4$ since it does not span $\mathbb{R}^4$. However, we can start expanding it to a basis for $\mathbb{R}^4$ by first adding a vector that is not in the span of S, such as $\begin{bmatrix} 1 \\ 0 \\ 0 \\ 0 \end{bmatrix}$. Then since these three vectors still do not span $\mathbb{R}^4$, we can add a vector that is not in their span, such as $\begin{bmatrix} 0 \\ 0 \\ 1 \\ 0 \end{bmatrix}$. These four vectors now span $\mathbb{R}^4$ and they are linearly independent, so they form a basis for $\mathbb{R}^4$.
6. Additional techniques for determining whether a given finite set of vectors is linearly independent or spans a given subspace can be found in Sections 2.5 and 2.6.

Applications:

1. Because $y_1(x) = e^{2x}$ and $y_2(x) = e^x$ are linearly independent and span the solution space for the homogeneous differential equation $\frac{d^2y}{dx^2} - 3\frac{dy}{dx} + 2y = 0$, they form a basis for the solution space and the solution space has dimension 2.

2.3 Direct Sum Decompositions

Throughout this section, V will be a vector space over a field F, and W_i, for $i = 1, \ldots, k$, will be subspaces of V. For facts and general reading for this section, see [HK71].

Definitions:

The **sum** of subspaces W_i, for $i = 1, \ldots, k$, is $\sum_{i=1}^k W_i = W_1 + \cdots + W_k = \{\mathbf{w}_1 + \cdots + \mathbf{w}_k \mid \mathbf{w}_i \in W_i\}$. The sum $W_1 + \cdots + W_k$ is a **direct sum** if for all $i = 1, \ldots, k$, we have $W_i \cap \sum_{j \neq i} W_j = \{\mathbf{0}\}$. $W = W_1 \oplus \cdots \oplus W_k$ denotes that $W = W_1 + \cdots + W_k$ and the sum is direct. The subspaces W_i, for $i = i, \ldots, k$, are **independent** if for $\mathbf{w}_i \in W_i$, $\mathbf{w}_1 + \cdots + \mathbf{w}_k = \mathbf{0}$ implies $\mathbf{w}_i = \mathbf{0}$ for all $i = 1, \ldots, k$. Let V_i, for $i = 1, \ldots, k$, be vector spaces over F. The **external direct sum** of the V_i, denoted $V_1 \times \cdots \times V_k$, is the cartesian product of V_i, for $i = 1, \ldots, k$, with coordinate-wise operations. Let W be a subspace of V. An **additive coset** of W is a subset of the form $v + W = \{v + w \mid w \in W\}$ with $v \in V$. The **quotient** of V by W, denoted V/W, is the set of additive cosets of W with operations $(v_1 + W) + (v_2 + W) = (v_1 + v_2) + W$ and $c(v + W) = (cv) + W$, for any $c \in F$. Let $V = W \oplus U$, let $\mathcal{B}_W$ and $\mathcal{B}_U$ be bases for W and U respectively, and let $\mathcal{B} = \mathcal{B}_W \cup \mathcal{B}_U$. The **induced basis** of $\mathcal{B}$ in V/W is the set of vectors $\{u + W \mid u \in \mathcal{B}_U\}$.

Facts:

1. $W = W_1 \oplus W_2$ if and only if $W = W_1 + W_2$ and $W_1 \cap W_2 = \{\mathbf{0}\}$.
2. If W is a subspace of V, then there exists a subspace U of V such that $V = W \oplus U$. Note that U is not usually unique.
3. Let $W = W_1 + \cdots + W_k$. The following are equivalent:
   - $W = W_1 \oplus \cdots \oplus W_k$. That is, for all $i = 1, \ldots, k$, we have $W_i \cap \sum_{j \neq i} W_j = \{\mathbf{0}\}$.
   - $W_i \cap \sum_{j=1}^{i-1} W_j = \{\mathbf{0}\}$, for all $i = 2, \ldots, k$.
   - For each $\mathbf{w} \in W$, $\mathbf{w}$ can be expressed in exactly one way as a sum of vectors in $W_1, \ldots, W_k$. That is, there exist unique $\mathbf{w}_i \in W_i$, such that $\mathbf{w} = \mathbf{w}_1 + \cdots + \mathbf{w}_k$.
   - The subspaces W_i, for $i = 1, \ldots, k$, are independent.
   - If $\mathcal{B}_i$ is an (ordered) basis for W_i, then $\mathcal{B} = \bigcup_{i=1}^k \mathcal{B}_i$ is an (ordered) basis for W.
4. If $\mathcal{B}$ is a basis for V and $\mathcal{B}$ is partitioned into disjoint subsets $\mathcal{B}_i$, for $i = 1, \ldots, k$, then $V = \operatorname{Span}(\mathcal{B}_1) \oplus \cdots \oplus \operatorname{Span}(\mathcal{B}_k)$.
5. If S is a linearly independent subset of V and S is partitioned into disjoint subsets S_i, for $i = 1, \ldots, k$, then the subspaces $\operatorname{Span}(S_1), \ldots, \operatorname{Span}(S_k)$ are independent.
6. If V is finite dimensional and $V = W_1 + \cdots + W_k$, then $\dim(V) = \dim(W_1) + \cdots + \dim(W_k)$ if and only if $V = W_1 \oplus \cdots \oplus W_k$.
7. Let V_i, for $i = 1, \ldots, k$, be vector spaces over F.
   - $V_1 \times \cdots \times V_k$ is a vector space over F.
   - $\widehat{V_i} = \{(0, \ldots, 0, v_i, 0, \ldots, 0) \mid v_i \in V_i\}$ (where v_i is the ith coordinate) is a subspace of $V_1 \times \cdots \times V_k$.
   - $V_1 \times \cdots \times V_k = \widehat{V_1} \oplus \cdots \oplus \widehat{V_k}$.
   - If V_i, for $i = 1, \ldots, k$, are finite dimensional, then $\dim \widehat{V_i} = \dim V_i$ and $\dim(V_1 \times \cdots \times V_k) = \dim V_1 + \cdots + \dim V_k$.
8. If W is a subspace of V, then the quotient V/W is a vector space over F.
9. Let $V = W \oplus U$, let $\mathcal{B}_W$ and $\mathcal{B}_U$ be bases for W and U respectively, and let $\mathcal{B} = \mathcal{B}_W \cup \mathcal{B}_U$. The induced basis of $\mathcal{B}$ in V/W is a basis for V/W and $\dim(V/W) = \dim U$.

Examples:

1. Let $\mathcal{B} = \{\mathbf{v}_1, \ldots, \mathbf{v}_n\}$ be a basis for V. Then $V = \operatorname{Span}(\mathbf{v}_1) \oplus \cdots \oplus \operatorname{Span}(\mathbf{v}_n)$.
2. Let $X = \left\{ \begin{bmatrix} x \\ 0 \end{bmatrix} \mid x \in \mathbb{R} \right\}$, $Y = \left\{ \begin{bmatrix} 0 \\ y \end{bmatrix} \mid y \in \mathbb{R} \right\}$, and $Z = \left\{ \begin{bmatrix} z \\ z \end{bmatrix} \mid z \in \mathbb{R} \right\}$. Then $\mathbb{R}^2 = X \oplus Y = Y \oplus Z = X \oplus Z$.

3. In $F^{n\times n}$, let W_1 be the subspace of symmetric matrices and W_2 be the subspace of skew-symmetric matrices. Clearly, $W_1 \cap W_2 = \{\mathbf{0}\}$. For any $A \in F^{n\times n}$, $A = \frac{A+A^T}{2} + \frac{A-A^T}{2}$, where $\frac{A+A^T}{2} \in W_1$ and $\frac{A-A^T}{2} \in W_2$. Therefore, $F^{n\times n} = W_1 \oplus W_2$.
4. Recall that the function $f \in C(\mathbb{R})$ is even if $f(-x) = f(x)$ for all x, and f is odd if $f(-x) = -f(x)$ for all x. Let W_1 be the subspace of even functions and W_2 be the subspace of odd functions. Clearly, $W_1 \cap W_2 = \{\mathbf{0}\}$. For any $f \in C(\mathbb{R})$, $f = f_1 + f_2$, where $f_1(x) = \frac{f(x)+f(-x)}{2} \in W_1$ and $f_1(x) = \frac{f(x)-f(-x)}{2} \in W_2$. Therefore, $C(\mathbb{R}) = W_1 \oplus W_2$.
5. Given a subspace W of V, we can find a subspace U such that $V = W \oplus U$ by choosing a basis for W, extending this linearly independent set to a basis for V, and setting U equal to the span of the basis vectors not in W. For example, in $\mathbb{R}^3$, Let $W = \left\{ \begin{bmatrix} a \\ -2a \\ a \end{bmatrix} \mid a \in \mathbb{R} \right\}$. If $\mathbf{w} = \begin{bmatrix} 1 \\ -2 \\ 1 \end{bmatrix}$, then $\{\mathbf{w}\}$ is a basis for W. Extend this to a basis for $\mathbb{R}^3$, for example by adjoining $\mathbf{e}_1$ and $\mathbf{e}_2$. Thus, $V = W \oplus U$, where $U = \text{Span}(\mathbf{e}_1, \mathbf{e}_2)$. Note: there are many other ways to extend the basis, and many other possible U.
6. In the external direct sum $\mathbb{R}[x;2] \times \mathbb{R}^{2\times 2}$, $\left(2x^2+7, \begin{bmatrix} 1 & 2 \\ 3 & 4 \end{bmatrix}\right) + 3\left(x^2+4x-2, \begin{bmatrix} 0 & 1 \\ -1 & 0 \end{bmatrix}\right) = \left(5x^2+12x+1, \begin{bmatrix} 1 & 5 \\ 0 & 4 \end{bmatrix}\right)$.
7. The subspaces X, Y, Z of $\mathbb{R}^2$ in Example 2 have bases $\mathcal{B}_X = \left\{\begin{bmatrix} 1 \\ 0 \end{bmatrix}\right\}, \mathcal{B}_Y = \left\{\begin{bmatrix} 0 \\ 1 \end{bmatrix}\right\}, \mathcal{B}_Z = \left\{\begin{bmatrix} 1 \\ 1 \end{bmatrix}\right\}$, respectively. Then $\mathcal{B}_{XY} = \mathcal{B}_X \cup \mathcal{B}_Y$ and $\mathcal{B}_{XZ} = \mathcal{B}_X \cup \mathcal{B}_Z$ are bases for $\mathbb{R}^2$. In $\mathbb{R}^2/X$, the induced bases of $\mathcal{B}_{XY}$ and $\mathcal{B}_{XZ}$ are $\left\{\begin{bmatrix} 0 \\ 1 \end{bmatrix} + X\right\}$ and $\left\{\begin{bmatrix} 1 \\ 1 \end{bmatrix} + X\right\}$, respectively. These are equal because $\begin{bmatrix} 1 \\ 1 \end{bmatrix} + X = \begin{bmatrix} 0 \\ 1 \end{bmatrix} + \begin{bmatrix} 1 \\ 0 \end{bmatrix} + X = \begin{bmatrix} 0 \\ 1 \end{bmatrix} + X$.

2.4 Matrix Range, Null Space, Rank, and the Dimension Theorem

Definitions:

For any matrix $A \in F^{m\times n}$, the **range** of A, denoted by range(A), is the set of all linear combinations of the columns of A. If $A = [\mathbf{m}_1\ \mathbf{m}_2\ \ldots\ \mathbf{m}_n]$, then range$(A) = \text{Span}(\mathbf{m}_1, \mathbf{m}_2, \ldots, \mathbf{m}_n)$. The range of A is also called the **column space** of A.

The **row space** of A, denoted by RS(A), is the set of all linear combinations of the rows of A. If $A = [\mathbf{v}_1\ \mathbf{v}_2\ \ldots\ \mathbf{v}_m]^T$, then RS$(A) = \text{Span}(\mathbf{v}_1, \mathbf{v}_2, \ldots, \mathbf{v}_m)$.

The **kernel** of A, denoted by ker(A), is the set of all solutions to the homogeneous equation $A\mathbf{x} = \mathbf{0}$. The kernel of A is also called the **null space** of A, and its dimension is called the **nullity** of A, denoted by null(A).

The **rank** of A, denoted by rank(A), is the number of leading entries in the reduced row echelon form of A (or any row echelon form of A). (See Section 1.3 for more information.)

$A, B \in F^{m\times n}$ are **equivalent** if $B = C_1^{-1}AC_2$ for some invertible matrices $C_1 \in F^{m\times m}$ and $C_2 \in F^{n\times n}$. $A, B \in F^{n\times n}$ are **similar** if $B = C^{-1}AC$ for some invertible matrix $C \in F^{n\times n}$. For square matrices $A_1 \in F^{n_1\times n_1}, \ldots, A_k \in F^{n_k\times n_k}$, the **matrix direct sum** $A = A_1 \oplus \cdots \oplus A_k$ is the block diagonal matrix with the matrices A_i down the diagonal. That is, $A = \begin{bmatrix} A_1 & & \mathbf{0} \\ & \ddots & \\ \mathbf{0} & & A_k \end{bmatrix}$, where $A \in F^{n\times n}$ with $n = \sum_{i=1}^{k} n_i$.

Facts: Unless specified otherwise, the following facts can be found in [Lay03, Sections 2.8, 4.2, 4.5, and 4.6].

1. The range of an $m \times n$ matrix A is a subspace of F^m.
2. The columns of A corresponding to the pivot columns in the reduced row echelon form of A (or any row echelon form of A) give a basis for range(A). Let $\mathbf{v}_1, \mathbf{v}_2, \ldots, \mathbf{v}_k \in F^m$. If matrix $A = [\mathbf{v}_1\ \mathbf{v}_2\ \ldots\ \mathbf{v}_k]$, then a basis for range$(A)$ will be a linearly independent subset of $\mathbf{v}_1, \mathbf{v}_2, \ldots, \mathbf{v}_k$ having the same span.
3. $\dim(\text{range}(A)) = \text{rank}(A)$.
4. The kernel of an $m \times n$ matrix A is a subspace of F^n.
5. If the reduced row echelon form of A (or any row echelon form of A) has k pivot columns, then $\text{null}(A) = n - k$.
6. If two matrices A and B are row equivalent, then $\text{RS}(A) = \text{RS}(B)$.
7. The row space of an $m \times n$ matrix A is a subspace of F^n.
8. The pivot rows in the reduced row echelon form of A (or any row echelon form of A) give a basis for $\text{RS}(A)$.
9. $\dim(\text{RS}(A)) = \text{rank}(A)$.
10. $\text{rank}(A) = \text{rank}(A^T)$.
11. (Dimension Theorem) For any $A \in F^{m\times n}$, $n = \text{rank}(A) + \text{null}(A)$. Similarly, $m = \dim(\text{RS}(A)) + \text{null}(A^T)$.
12. A vector $\mathbf{b} \in F^m$ is in range(A) if and only if the equation $A\mathbf{x} = \mathbf{b}$ has a solution. So range$(A) = F^m$ if and only if the equation $A\mathbf{x} = \mathbf{b}$ has a solution for every $\mathbf{b} \in F^m$.
13. A vector $\mathbf{a} \in F^n$ is in $\text{RS}(A)$ if and only if the equation $A^T\mathbf{y} = \mathbf{a}$ has a solution. So $\text{RS}(A) = F^n$ if and only if the equation $A^T\mathbf{y} = \mathbf{a}$ has a solution for every $\mathbf{a} \in F^n$.
14. If $\mathbf{a}$ is a solution to the equation $A\mathbf{x} = \mathbf{b}$, then $\mathbf{a} + \mathbf{v}$ is also a solution for any $\mathbf{v} \in \ker(A)$.
15. [HJ85, p. 14] If $A \in F^{m\times n}$ is rank 1, then there are vectors $\mathbf{v} \in F^m$ and $\mathbf{u} \in F^n$ so that $A = \mathbf{v}\mathbf{u}^T$.
16. If $A \in F^{m\times n}$ is rank k, then A is a sum of k rank 1 matrices. That is, there exist $A_1, \ldots, A_k$ with $A = A_1 + \cdots + A_k$ and $\text{rank}(A_i) = 1$, for $i = 1, \ldots, k$.
17. [HJ85, p. 13] The following are all equivalent statements about a matrix $A \in F^{m\times n}$.
    (a) The rank of A is k.
    (b) $\dim(\text{range}(A)) = k$.
    (c) The reduced row echelon form of A has k pivot columns.
    (d) A row echelon form of A has k pivot columns.
    (e) The largest number of linearly independent columns of A is k.
    (f) The largest number of linearly independent rows of A is k.
18. [HJ85, p. 13] (Rank Inequalities) (Unless specified otherwise, assume that $A, B \in F^{m\times n}$.)
    (a) $\text{rank}(A) \leq \min(m, n)$.
    (b) If a new matrix B is created by deleting rows and/or columns of matrix A, then $\text{rank}(B) \leq \text{rank}(A)$.
    (c) $\text{rank}(A + B) \leq \text{rank}(A) + \text{rank}(B)$.
    (d) If A has a $p \times q$ submatrix of 0s, then $\text{rank}(A) \leq (m - p) + (n - q)$.

(e) If $A \in F^{m\times k}$ and $B \in F^{k\times n}$, then

$$\text{rank}(A) + \text{rank}(B) - k \leq \text{rank}(AB) \leq \min\{\text{rank}(A), \text{rank}(B)\}.$$

19. [HJ85, pp. 13–14] (Rank Equalities)
    (a) If $A \in \mathbb{C}^{m\times n}$, then $\text{rank}(A^*) = \text{rank}(A^T) = \text{rank}(\overline{A}) = \text{rank}(A)$.
    (b) If $A \in \mathbb{C}^{m\times n}$, then $\text{rank}(A^*A) = \text{rank}(A)$. If $A \in \mathbb{R}^{m\times n}$, then $\text{rank}(A^TA) = \text{rank}(A)$.
    (c) Rank is unchanged by left or right multiplication by a nonsingular matrix. That is, if $A \in F^{n\times n}$ and $B \in F^{m\times m}$ are nonsingular, and $M \in F^{m\times n}$, then

    $$\text{rank}(AM) = \text{rank}(M) = \text{rank}(MB) = \text{rank}(AMB).$$

    (d) If $A, B \in F^{m\times n}$, then $\text{rank}(A) = \text{rank}(B)$ if and only if there exist nonsingular matrices $X \in F^{m\times m}$ and $Y \in F^{n\times n}$ such that $A = XBY$ (i.e., if and only if A is equivalent to B).
    (e) If $A \in F^{m\times n}$ has rank k, then $A = XBY$, for some $X \in F^{m\times k}$, $Y \in F^{k\times n}$, and nonsingular $B \in F^{k\times k}$.
    (f) If $A_1 \in F^{n_1\times n_1}, \ldots, A_k \in F^{n_k\times n_k}$, then $\text{rank}(A_1 \oplus \cdots \oplus A_k) = \text{rank}(A_1) + \cdots + \text{rank}(A_k)$.
20. Let $A, B \in F^{n\times n}$ with A similar to B.
    (a) A is equivalent to B.
    (b) $\text{rank}(A) = \text{rank}(B)$.
    (c) $\text{tr}\, A = \text{tr}\, B$.
21. Equivalence of matrices is an equivalence relation on $F^{m\times n}$.
22. Similarity of matrices is an equivalence relation on $F^{n\times n}$.
23. If $A \in F^{m\times n}$ and $\text{rank}(A) = k$, then A is equivalent to $\begin{bmatrix} I_k & 0 \\ 0 & 0 \end{bmatrix}$, and so any two matrices of the same size and rank are equivalent.
24. (For information on the determination of whether two matrices are similar, see Chapter 6.)
25. [Lay03, Sec. 6.1] If $A \in \mathbb{R}^{n\times n}$, then for any $\mathbf{x} \in \text{RS}(A)$ and any $\mathbf{y} \in \ker(A)$, $\mathbf{x}^T\mathbf{y} = 0$. So the row space and kernel of a real matrix are orthogonal to one another. (See Chapter 5 for more on orthogonality.)

Examples:

1. If $A = \begin{bmatrix} 1 & 7 & -2 \\ 0 & -1 & 1 \\ 2 & 13 & -3 \end{bmatrix} \in \mathbb{R}^{3\times 3}$, then any vector of the form $\begin{bmatrix} a + 7b - 2c \\ -b + c \\ 2a + 13b - 3c \end{bmatrix} \left(= \begin{bmatrix} 1 & 7 & -2 \\ 0 & -1 & 1 \\ 2 & 13 & -3 \end{bmatrix} \begin{bmatrix} a \\ b \\ c \end{bmatrix} \right)$ is in range(A), for any $a, b, c \in \mathbb{R}$. Since a row echelon form of A is $\begin{bmatrix} 1 & 7 & -2 \\ 0 & 1 & -1 \\ 0 & 0 & 0 \end{bmatrix}$, we know that the set $\left\{ \begin{bmatrix} 1 \\ 0 \\ 2 \end{bmatrix}, \begin{bmatrix} 7 \\ -1 \\ 13 \end{bmatrix} \right\}$ is a basis for range(A), and the set $\left\{ \begin{bmatrix} 1 \\ 7 \\ -2 \end{bmatrix}, \begin{bmatrix} 0 \\ 1 \\ -1 \end{bmatrix} \right\}$ is a basis for RS(A). Since its reduced row echelon form is $\begin{bmatrix} 1 & 0 & 5 \\ 0 & 1 & -1 \\ 0 & 0 & 0 \end{bmatrix}$, the set $\left\{ \begin{bmatrix} 1 \\ 0 \\ 5 \end{bmatrix}, \begin{bmatrix} 0 \\ 1 \\ -1 \end{bmatrix} \right\}$ is another basis for RS(A).

2. If $A = \begin{bmatrix} 1 & 7 & -2 \\ 0 & -1 & 1 \\ 2 & 13 & -3 \end{bmatrix} \in \mathbb{R}^{3\times 3}$, then using the reduced row echelon form given in the previous example, solutions to $A\mathbf{x} = \mathbf{0}$ have the form $\mathbf{x} = c\begin{bmatrix} -5 \\ 1 \\ 1 \end{bmatrix}$, for any $c \in \mathbb{R}$. So $\ker(A) = \text{Span}\left(\begin{bmatrix} -5 \\ 1 \\ 1 \end{bmatrix}\right)$.

3. If $A \in \mathbb{R}^{3\times 5}$ has the reduced row echelon form $\begin{bmatrix} 1 & 0 & 3 & 0 & 2 \\ 0 & 1 & -2 & 0 & 7 \\ 0 & 0 & 0 & 1 & -1 \end{bmatrix}$, then any solution to $A\mathbf{x} = \mathbf{0}$ has the form

$$\mathbf{x} = c_1\begin{bmatrix} -3 \\ 2 \\ 1 \\ 0 \\ 0 \end{bmatrix} + c_2\begin{bmatrix} -2 \\ -7 \\ 0 \\ 1 \\ 1 \end{bmatrix}$$

for some $c_1, c_2 \in \mathbb{R}$. So,

$$\ker(A) = \text{Span}\left(\begin{bmatrix} -3 \\ 2 \\ 1 \\ 0 \\ 0 \end{bmatrix}, \begin{bmatrix} -2 \\ -7 \\ 0 \\ 1 \\ 1 \end{bmatrix}\right).$$

4. Example 1 above shows that $\left\{\begin{bmatrix} 1 \\ 0 \\ 2 \end{bmatrix}, \begin{bmatrix} 7 \\ -1 \\ 13 \end{bmatrix}\right\}$ is a linearly independent set having the same span as the set $\left\{\begin{bmatrix} 1 \\ 0 \\ 2 \end{bmatrix}, \begin{bmatrix} 7 \\ -1 \\ 13 \end{bmatrix}, \begin{bmatrix} -2 \\ 1 \\ -3 \end{bmatrix}\right\}$.

5. $\begin{bmatrix} 1 & 7 \\ 2 & -3 \end{bmatrix}$ is similar to $\begin{bmatrix} 37 & -46 \\ 31 & -39 \end{bmatrix}$ because $\begin{bmatrix} 37 & -46 \\ 31 & -39 \end{bmatrix} = \begin{bmatrix} -2 & 3 \\ 3 & -4 \end{bmatrix}^{-1} \begin{bmatrix} 1 & 7 \\ 2 & -3 \end{bmatrix} \begin{bmatrix} -2 & 3 \\ 3 & -4 \end{bmatrix}$.

2.5 Nonsingularity Characterizations

From the previous discussion, we can add to the list of nonsingularity characterizations of a square matrix that was started in the previous chapter.

Facts: The following facts can be found in [HJ85, p. 14] or [Lay03, Sections 2.3 and 4.6].

1. If $A \in F^{n\times n}$, then the following are equivalent.
   (a) A is nonsingular.
   (b) The columns of A are linearly independent.
   (c) The dimension of range(A) is n.

(d) The range of A is F^n.

(e) The equation $A\mathbf{x} = \mathbf{b}$ is consistent for each $\mathbf{b} \in F^n$.

(f) If the equation $A\mathbf{x} = \mathbf{b}$ is consistent, then the solution is unique.

(g) The equation $A\mathbf{x} = \mathbf{b}$ has a unique solution for each $\mathbf{b} \in F^n$.

(h) The rows of A are linearly independent.

(i) The dimension of $\text{RS}(A)$ is n.

(j) The row space of A is F^n.

(k) The dimension of $\ker(A)$ is 0.

(l) The only solution to $A\mathbf{x} = \mathbf{0}$ is $\mathbf{x} = \mathbf{0}$.

(m) The rank of A is n.

(n) The determinant of A is nonzero. (See Section 4.1 for the definition of the determinant.)

2.6 Coordinates and Change of Basis

Coordinates are used to transform a problem in a more abstract vector space (e.g., the vector space of polynomials of degree less than or equal to 3) to a problem in F^n.

Definitions:

Suppose that $\mathcal{B} = (\mathbf{b}_1, \mathbf{b}_2, \ldots, \mathbf{b}_n)$ is an ordered basis for a vector space V over a field F and $\mathbf{x} \in V$. The **coordinates of x relative to the ordered basis** $\mathcal{B}$ (or the **$\mathcal{B}$-coordinates of x**) are the scalar coefficients $c_1, c_2, \ldots, c_n \in F$ such that $\mathbf{x} = c_1\mathbf{x}_1 + c_2\mathbf{x}_2 + \cdots + c_n\mathbf{x}_n$. Whenever coordinates are involved, the vector space is assumed to be nonzero and finite dimensional.

If $c_1, c_2, \ldots, c_n$ are the $\mathcal{B}$-coordinates of $\mathbf{x}$, then the vector in F^n,

$$[\mathbf{x}]_{\mathcal{B}} = \begin{bmatrix} c_1 \\ c_2 \\ \vdots \\ c_n \end{bmatrix},$$

is the **coordinate vector of x relative to $\mathcal{B}$** or the **$\mathcal{B}$-coordinate vector of x.**

The mapping $\mathbf{x} \to [\mathbf{x}]_{\mathcal{B}}$ is the **coordinate mapping determined by $\mathcal{B}$.**

If $\mathcal{B}$ and $\mathcal{B}'$ are ordered bases for the vector space F^n, then the **change-of-basis matrix** from $\mathcal{B}$ to $\mathcal{B}'$ is the matrix whose columns are the $\mathcal{B}'$-coordinate vectors of the vectors in $\mathcal{B}$ and is denoted by ${}_{\mathcal{B}'}[I]_{\mathcal{B}}$. Such a matrix is also called a **transition matrix.**

Facts: The following facts can be found in [Lay03, Sections 4.4 and 4.7] or [HJ85, Section 0.10]:

1. For any vector $\mathbf{x} \in F^n$ with the standard ordered basis $\mathcal{E}_n = (\mathbf{e}_1, \mathbf{e}_2, \ldots, \mathbf{e}_n)$, we have $\mathbf{x} = [\mathbf{x}]_{\mathcal{E}_n}$.
2. For any ordered basis $\mathcal{B} = (\mathbf{b}_1, \ldots, \mathbf{b}_n)$ of a vector space V, we have $[\mathbf{b}_i]_{\mathcal{B}} = \mathbf{e}_i$.
3. If $\dim(V) = n$, then the coordinate mapping is a one-to-one linear transformation from V onto F^n. (See Chapter 3 for the definition of linear transformation.)
4. If $\mathcal{B}$ is an ordered basis for a vector space V and $\mathbf{v}_1, \mathbf{v}_2 \in V$, then $\mathbf{v}_1 = \mathbf{v}_2$ if and only if $[\mathbf{v}_1]_{\mathcal{B}} = [\mathbf{v}_2]_{\mathcal{B}}$.
5. Let V be a vector space over a field F, and suppose $\mathcal{B}$ is an ordered basis for V. Then for any $\mathbf{x}, \mathbf{v}_1, \ldots, \mathbf{v}_k \in V$ and $c_1, \ldots, c_k \in F$, $\mathbf{x} = c_1\mathbf{v}_1 + \cdots + c_k\mathbf{v}_k$ if and only if $[\mathbf{x}]_{\mathcal{B}} = c_1[\mathbf{v}_1]_{\mathcal{B}} + \cdots + c_k[\mathbf{v}_k]_{\mathcal{B}}$. So, for any $\mathbf{x}, \mathbf{v}_1, \ldots, \mathbf{v}_k \in V$, $\mathbf{x} \in \text{Span}(\mathbf{v}_1, \ldots, \mathbf{v}_k)$ if and only if $[\mathbf{x}]_{\mathcal{B}} \in \text{Span}([\mathbf{v}_1]_{\mathcal{B}}, \ldots, [\mathbf{v}_k]_{\mathcal{B}})$.
6. Suppose $\mathcal{B}$ is an ordered basis for an n-dimensional vector space V over a field F and $\mathbf{v}_1, \ldots, \mathbf{v}_k \in V$. The set $S = \{\mathbf{v}_1, \ldots, \mathbf{v}_k\}$ is linearly independent in V if and only if the set $S' = \{[\mathbf{v}_1]_{\mathcal{B}}, \ldots, [\mathbf{v}_k]_{\mathcal{B}}\}$ is linearly independent in F^n.

7. Let V be a vector space over a field F with $\dim(V) = n$, and suppose $\mathcal{B}$ is an ordered basis for V. Then $\text{Span}(\mathbf{v}_1, \mathbf{v}_2, \ldots, \mathbf{v}_k) = V$ for some $\mathbf{v}_1, \mathbf{v}_2, \ldots, \mathbf{v}_k \in V$ if and only if $\text{Span}([\mathbf{v}_1]_\mathcal{B}, [\mathbf{v}_2]_\mathcal{B}, \ldots, [\mathbf{v}_k]_\mathcal{B}) = F^n$.
8. Suppose $\mathcal{B}$ is an ordered basis for a vector space V over a field F with $\dim(V) = n$, and let $S = \{\mathbf{v}_1, \ldots, \mathbf{v}_n\}$ be a subset of V. Then S is a basis for V if and only if $\{[\mathbf{v}_1]_\mathcal{B}, \ldots, [\mathbf{v}_n]_\mathcal{B}\}$ is a basis for F^n if and only if the matrix $[[\mathbf{v}_1]_\mathcal{B}, \ldots, [\mathbf{v}_n]_\mathcal{B}]$ is invertible.
9. If $\mathcal{B}$ and $\mathcal{B}'$ are ordered bases for a vector space V, then $[\mathbf{x}]_{\mathcal{B}'} = {}_{\mathcal{B}'}[I]_\mathcal{B}\, [\mathbf{x}]_\mathcal{B}$ for any $\mathbf{x} \in V$. Furthermore, ${}_{\mathcal{B}'}[I]_\mathcal{B}$ is the only matrix such that for any $\mathbf{x} \in V$, $[\mathbf{x}]_{\mathcal{B}'} = {}_{\mathcal{B}'}[I]_\mathcal{B}\, [\mathbf{x}]_\mathcal{B}$.
10. Any change-of-basis matrix is invertible.
11. If B is invertible, then B is a change-of-basis matrix. Specifically, if $B = [\mathbf{b}_1 \cdots \mathbf{b}_n] \in F^{n\times n}$, then $B = {}_{\mathcal{E}_n}[I]_\mathcal{B}$, where $\mathcal{B} = (\mathbf{b}_1, \ldots, \mathbf{b}_n)$ is an ordered basis for F^n.
12. If $\mathcal{B} = (\mathbf{b}_1, \ldots, \mathbf{b}_n)$ is an ordered basis for F^n, then ${}_{\mathcal{E}_n}[I]_\mathcal{B} = [\mathbf{b}_1 \cdots \mathbf{b}_n]$.
13. If $\mathcal{B}$ and $\mathcal{B}'$ are ordered bases for a vector space V, then ${}_\mathcal{B}[I]_{\mathcal{B}'} = ({}_{\mathcal{B}'}[I]_\mathcal{B})^{-1}$.
14. If $\mathcal{B}$ and $\mathcal{B}'$ are ordered bases for F^n, then ${}_{\mathcal{B}'}[I]_\mathcal{B} = ({}_{\mathcal{B}'}[I]_{\mathcal{E}_n})({}_{\mathcal{E}_n}[I]_\mathcal{B})$.

Examples:

1. If $p(x) = a_n x^n + a_{n-1}x^{n-1} + \cdots + a_1 x + a_0 \in F[x; n]$ with the standard ordered basis $\mathcal{B} = (1, x, x^2, \ldots, x^n)$, then $[p(x)]_\mathcal{B} = \begin{bmatrix} a_0 \\ a_1 \\ \vdots \\ a_n \end{bmatrix}$.

2. The set $\mathcal{B} = \left(\begin{bmatrix} 1 \\ -1 \end{bmatrix}, \begin{bmatrix} 0 \\ 3 \end{bmatrix} \right)$ forms an ordered basis for $\mathbb{R}^2$. If $\mathcal{E}_2$ is the standard ordered basis for $\mathbb{R}^2$, then the change-of-basis matrix from $\mathcal{B}$ to $\mathcal{E}_2$ is ${}_{\mathcal{E}_2}[T]_\mathcal{B} = \begin{bmatrix} 1 & 0 \\ -1 & 3 \end{bmatrix}$, and $({}_{\mathcal{E}_2}[T]_\mathcal{B})^{-1} = \begin{bmatrix} 1 & 0 \\ \frac{1}{3} & \frac{1}{3} \end{bmatrix}$. So for $\mathbf{v} = \begin{bmatrix} 3 \\ 1 \end{bmatrix}$ in the standard ordered basis, we find that $[\mathbf{v}]_\mathcal{B} = ({}_{\mathcal{E}_2}[T]_\mathcal{B})^{-1}\mathbf{v} = \begin{bmatrix} 3 \\ \frac{4}{3} \end{bmatrix}$. To check this, we can easily see that $\mathbf{v} = \begin{bmatrix} 3 \\ 1 \end{bmatrix} = 3\begin{bmatrix} 1 \\ -1 \end{bmatrix} + \frac{4}{3}\begin{bmatrix} 0 \\ 3 \end{bmatrix}$.

3. The set $\mathcal{B}' = (1, 1+x, 1+x^2)$ is an ordered basis for $\mathbb{R}[x; 2]$, and using the standard ordered basis $\mathcal{B} = (1, x, x^2)$ for $\mathbb{R}[x; 2]$ we have ${}_\mathcal{B}[P]_{\mathcal{B}'} = \begin{bmatrix} 1 & 1 & 1 \\ 0 & 1 & 0 \\ 0 & 0 & 1 \end{bmatrix}$. So, $({}_\mathcal{B}[P]_{\mathcal{B}'})^{-1} = \begin{bmatrix} 1 & -1 & -1 \\ 0 & 1 & 0 \\ 0 & 0 & 1 \end{bmatrix}$ and $[5 - 2x + 3x^2]_{\mathcal{B}'} = ({}_\mathcal{B}[P]_{\mathcal{B}'})^{-1} \begin{bmatrix} 5 \\ -2 \\ 3 \end{bmatrix} = \begin{bmatrix} 4 \\ -2 \\ 3 \end{bmatrix}$. Of course, we can see $5 - 2x + 3x^2 = 4(1) - 2(1+x) + 3(1+x^2)$.

4. If we want to change from the ordered basis $\mathcal{B}_1 = \left(\begin{bmatrix} 1 \\ -1 \end{bmatrix}, \begin{bmatrix} 0 \\ 3 \end{bmatrix} \right)$ in $\mathbb{R}^2$ to the ordered basis $\mathcal{B}_2 = \left(\begin{bmatrix} 2 \\ 1 \end{bmatrix}, \begin{bmatrix} 5 \\ 0 \end{bmatrix} \right)$, then the resulting change-of-basis matrix is ${}_{\mathcal{B}_2}[T]_{\mathcal{B}_1} = ({}_{\mathcal{E}_2}[T]_{\mathcal{B}_2})^{-1}({}_{\mathcal{E}_2}[T]_{\mathcal{B}_1}) = \begin{bmatrix} 2 & 5 \\ 1 & 0 \end{bmatrix}^{-1} \begin{bmatrix} 1 & 0 \\ -1 & 3 \end{bmatrix} = \begin{bmatrix} -1 & 3 \\ \frac{3}{5} & -\frac{6}{5} \end{bmatrix}$.

5. Let $S = \{5 - 2x + 3x^2, 3 - x + 2x^2, 8 + 3x\}$ in $\mathbb{R}[x;2]$ with the standard ordered basis $\mathcal{B} = (1, x, x^2)$. The matrix $A = \begin{bmatrix} 5 & 3 & 8 \\ -2 & -1 & 3 \\ 3 & 2 & 0 \end{bmatrix}$ contains the $\mathcal{B}$-coordinate vectors for the polynomials in S and it has row echelon form $\begin{bmatrix} 5 & 3 & 8 \\ 0 & 1 & 31 \\ 0 & 0 & 1 \end{bmatrix}$. Since this row echelon form shows that A is nonsingular, we know by Fact 8 above that S is a basis for $\mathbb{R}[x;2]$.

2.7 Idempotence and Nilpotence

Definitions:

A is an **idempotent** if $A^2 = A$.

A is **nilpotent** if, for some $k \geq 0$, $A^k = 0$.

Facts: All of the following facts except those with a specific reference are immediate from the definitions.

1. Every idempotent except the identity matrix is singular.
2. Let $A \in F^{n \times n}$. The following statements are equivalent.
   (a) A is an idempotent.
   (b) $I - A$ is an idempotent.
   (c) If $\mathbf{v} \in \text{range}(A)$, then $A\mathbf{v} = \mathbf{v}$.
   (d) $F^n = \ker A \oplus \text{range} A$.
   (e) [HJ85, p. 37 and p. 148] A is similar to $\begin{bmatrix} I_k & 0 \\ 0 & 0 \end{bmatrix}$, for some $k \leq n$.
3. If A_1 and A_2 are idempotents of the same size and commute, then $A_1 A_2$ is an idempotent.
4. If A_1 and A_2 are idempotents of the same size and $A_1 A_2 = A_2 A_1 = 0$, then $A_1 + A_2$ is an idempotent.
5. If $A \in F^{n \times n}$ is nilpotent, then $A^n = 0$.
6. If A is nilpotent and B is of the same size and commutes with A, then AB is nilpotent.
7. If A_1 and A_2 are nilpotent matrices of the same size and $A_1 A_2 = A_2 A_1 = 0$, then $A_1 + A_2$ is nilpotent.

Examples:

1. $\begin{bmatrix} -8 & 12 \\ -6 & 9 \end{bmatrix}$ is an idempotent. $\begin{bmatrix} 1 & -1 \\ 1 & -1 \end{bmatrix}$ is nilpotent.

References

[Lay03] D. C. Lay. *Linear Algebra and Its Applications*, 3rd ed. Addison-Wesley, Reading, MA, 2003.
[HK71] K. H. Hoffman and R. Kunze. *Linear Algebra*, 2nd ed. Prentice-Hall, Upper Saddle River, NJ, 1971.
[HJ85] R. A. Horn and C. R. Johnson. *Matrix Analysis*. Cambridge University Press, Cambridge, 1985.

3

Linear Transformations

Francesco Barioli
University of Tennessee at Chattanooga

3.1 Basic Concepts

Let V, W be vector spaces over a field F.

Definitions:

A **linear transformation** (or **linear mapping**) is a mapping $T: V \rightarrow W$ such that, for each $\mathbf{u}, \mathbf{v} \in V$, and for each $c \in F$, $T(\mathbf{u}+\mathbf{v}) = T(\mathbf{u}) + T(\mathbf{v})$, and $T(c\mathbf{u}) = cT(\mathbf{u})$.

V is called the **domain** of the linear transformation $T: V \rightarrow W$.

W is called the **codomain** of the linear transformation $T: V \rightarrow W$.

The **identity** transformation $I_V: V \rightarrow V$ is defined by $I_V(\mathbf{v}) = \mathbf{v}$ for each $\mathbf{v} \in V$. I_V is also denoted by I.

The **zero** transformation $0: V \rightarrow W$ is defined by $0(\mathbf{v}) = \mathbf{0}_W$ for each $\mathbf{v} \in V$.

A **linear operator** is a linear transformation $T: V \rightarrow V$.

Facts:

Let $T: V \rightarrow W$ be a linear transformation. The following facts can be found in almost any elementary linear algebra text, including [Lan70, IV§1], [Sta69, §3.1], [Goo03, Chapter 4], and [Lay03, §1.8].

1. $T(\sum_1^n a_i\mathbf{v}_i) = \sum_1^n a_i T(\mathbf{v}_i)$, for any $a_i \in F$, $\mathbf{v}_i \in V$, $i = 1, \ldots, n$.
2. $T(\mathbf{0}_V) = \mathbf{0}_W$.
3. $T(-\mathbf{v}) = -T(\mathbf{v})$, for each $\mathbf{v} \in V$.
4. The identity transformation is a linear transformation.
5. The zero transformation is a linear transformation.

6. If $\mathcal{B} = \{\mathbf{v}_1, \ldots, \mathbf{v}_n\}$ is a basis for V, and $\mathbf{w}_1, \ldots, \mathbf{w}_n \in W$, then there exists a unique $T: V \to W$ such that $T(\mathbf{v}_i) = \mathbf{w}_i$ for each i.

Examples:

Examples 1 to 9 are linear transformations.

1. $T: \mathbb{R}^3 \to \mathbb{R}^2$ where $T\left(\begin{bmatrix} x \\ y \\ z \end{bmatrix}\right) = \begin{bmatrix} x + y \\ 2x - z \end{bmatrix}$.
2. $T: V \to V$, defined by $T(\mathbf{v}) = -\mathbf{v}$ for each $\mathbf{v} \in V$.
3. If $A \in F^{m\times n}$, $T: F^n \to F^m$, where $T(\mathbf{v}) = A\mathbf{v}$.
4. $T: F^{m\times n} \to F$, where $T(A) = \operatorname{tr} A$.
5. Let $\mathcal{C}([0, 1])$ be the vector space of all continuous functions on $[0, 1]$ into $\mathbb{R}$, and let $T: \mathcal{C}([0, 1]) \to \mathbb{R}$ be defined by $T(f) = \int_0^1 f(t)dt$.
6. Let V be the vector space of all functions $f: \mathbb{R} \to \mathbb{R}$ that have derivatives of all orders, and $D: V \to V$ be defined by $D(f) = f'$.
7. The transformation, which rotates every vector in the plane $\mathbb{R}^2$ through an angle θ.
8. The projection T onto the xy-plane of $\mathbb{R}^3$, i.e., $T\left(\begin{bmatrix} x \\ y \\ z \end{bmatrix}\right) = \begin{bmatrix} x \\ y \\ 0 \end{bmatrix}$.
9. $T: \mathbb{R}^3 \to \mathbb{R}^3$, where $T(\mathbf{v}) = \mathbf{b} \times \mathbf{v}$, for some $\mathbf{b} \in \mathbb{R}^3$.

   Examples 10 and 11 are not linear transformations.
10. $f: \mathbb{R}^2 \to \mathbb{R}^2$, where $f\left(\begin{bmatrix} x \\ y \end{bmatrix}\right) = \begin{bmatrix} y + 1 \\ x - y - 2 \end{bmatrix}$ is not a linear transformation because $f(\mathbf{0}) \neq \mathbf{0}$.
11. $f: \mathbb{R}^2 \to \mathbb{R}$, where $f\left(\begin{bmatrix} x \\ y \end{bmatrix}\right) = x^2$ is not a linear transformation because $f\left(2\begin{bmatrix} 1 \\ 0 \end{bmatrix}\right) = 4 \neq 2 = 2f\left(\begin{bmatrix} 1 \\ 0 \end{bmatrix}\right)$.

3.2 The Spaces $L(V,W)$ and $L(V,V)$

Let V, W be vector spaces over F.

Definitions:

$L(V, W)$ denotes the set of all linear transformations of V into W.

For each $T_1, T_2 \in L(V, W)$ the **sum** $T_1 + T_2$ is defined by $(T_1 + T_2)(\mathbf{v}) = T_1(\mathbf{v}) + T_2(\mathbf{v})$.

For each $c \in F$, $T \in L(V, W)$ the **scalar multiple** cT is defined by $(cT)(\mathbf{v}) = cT(\mathbf{v})$.

For each $T_1, T_2 \in L(V, V)$ the **product** $T_1 T_2$ is the composite mapping defined by $(T_1 T_2)(\mathbf{v}) = T_1(T_2(\mathbf{v}))$.

$T_1, T_2 \in L(V, V)$ **commute** if $T_1 T_2 = T_2 T_1$.

$T \in L(V, V)$ is a **scalar transformation** if, for some $c \in F$, $T(\mathbf{v}) = c\mathbf{v}$ for each $\mathbf{v} \in V$.

Facts:

Let $T, T_1, T_2 \in L(V, W)$. The following facts can be found in almost any elementary linear algebra text, including [Fin60, §3.2], [Lan70, IV §4], [Sta69, §3.6], [SW68, §4.3], and [Goo03, Chap. 4].

1. $T_1 + T_2 \in L(V, W)$.
2. $cT \in L(V, W)$.

3. If $T_1, T_2 \in L(V, V)$, then $T_1 T_2 \in L(V, V)$.
4. $L(V, W)$, with sum and scalar multiplication, is a vector space over F.
5. $L(V, V)$, with sum, scalar multiplication, and composition, is a linear algebra over F.
6. Let dim $V = n$ and dim $W = m$. Then dim $L(V, W) = mn$.
7. If dim $V > 1$, then there exist $T_1, T_2 \in L(V, V)$, which do not commute.
8. $T_0 \in L(V, V)$ commutes with all $T \in L(V, V)$ if and only if T_0 is a scalar transformation.

Examples:

1. For each $j = 1, \ldots, n$ let $T_j \in L(F^n, F^n)$ be defined by $T_j(\mathbf{x}) = x_j \mathbf{e}_j$. Then $\sum_{i=1}^{n} T_j$ is the identity transformation in V.
2. Let T_1 and T_2 be the transformations that rotates every vector in $\mathbb{R}^2$ through an angle θ_1 and θ_2 respectively. Then $T_1 T_2$ is the rotation through the angle $\theta_1 + \theta_2$.
3. Let T_1 be the rotation through an angle θ in $\mathbb{R}^2$ and let T_2 be the reflection on the horizontal axis, that is, $T_2(x, y) = (x, -y)$. Then T_1 and T_2 do not commute.

3.3 Matrix of a Linear Transformation

Let V, W be nonzero finite dimensional vector spaces over F.

Definitions:

The **linear transformation associated to a matrix** $A \in F^{m\times n}$ is $T_A \colon F^n \to F^m$ defined by $T_A(\mathbf{v}) = A\mathbf{v}$.

The **matrix associated to a linear transformation** $T \in L(V, W)$ and relative to the ordered bases $\mathcal{B} = (\mathbf{b}_1, \ldots, \mathbf{b}_n)$ of V, and $\mathcal{C}$ of W, is the matrix ${}_{\mathcal{C}}[T]_{\mathcal{B}} = [[T(\mathbf{b}_1)]_{\mathcal{C}} \cdots [T(\mathbf{b}_n)]_{\mathcal{C}}]$.

If $T \in L(F^n, F^m)$, then the **standard matrix** of T is $[T] = {}_{\mathcal{E}_m}[T]_{\mathcal{E}_n}$, where $\mathcal{E}_n$ is the standard basis for F^n.

Note: If $V = W$ and $\mathcal{B} = \mathcal{C}$, the matrix ${}_{\mathcal{B}}[T]_{\mathcal{B}}$ will be denoted by $[T]_{\mathcal{B}}$.

If $T \in L(V, V)$ and $\mathcal{B}$ is an ordered basis for V, then the **trace** of T is tr T = tr $[T]_{\mathcal{B}}$.

Facts:

Let $\mathcal{B}$ and $\mathcal{C}$ be ordered bases V and W, respectively. The following facts can be found in almost any elementary linear algebra text, including [Lan70, V §2], [Sta69, §3.4–3.6], [SW68, §4.3], and [Goo03, Chap. 4].

1. The trace of $T \in L(V, V)$ is independent of the ordered basis of V used to define it.
2. For $A, B \in F^{m\times n}$, $T_A = T_B$ if and only if $A = B$.
3. For any $T_1, T_2 \in L(V, W)$, ${}_{\mathcal{C}}[T_1]_{\mathcal{B}} = {}_{\mathcal{C}}[T_2]_{\mathcal{B}}$ if and only if $T_1 = T_2$.
4. If $T \in L(F^n, F^m)$, then $[T] = [T(\mathbf{e}_1) \cdots T(\mathbf{e}_n)]$.
5. The change-of-basis matrix from basis $\mathcal{B}$ to $\mathcal{C}$, ${}_{\mathcal{C}}[I]_{\mathcal{B}}$, as defined in Chapter 2.6, is the same matrix as the matrix of the identity transformation with respect to $\mathcal{B}$ and $\mathcal{C}$.
6. Let $A \in F^{m\times n}$ and let T_A be the linear transformation associated to A. Then $[T_A] = A$.
7. If $T \in L(F^n, F^m)$, then $T_{[T]} = T$.
8. For any $T_1, T_2 \in L(V, W)$, ${}_{\mathcal{C}}[T_1 + T_2]_{\mathcal{B}} = {}_{\mathcal{C}}[T_1]_{\mathcal{B}} + {}_{\mathcal{C}}[T_2]_{\mathcal{B}}$.
9. For any $T \in L(V, W)$, and $c \in F$, ${}_{\mathcal{C}}[cT]_{\mathcal{B}} = c\,{}_{\mathcal{C}}[T]_{\mathcal{B}}$.
10. For any $T_1, T_2 \in L(V, V)$, $[T_1 T_2]_{\mathcal{B}} = [T_1]_{\mathcal{B}} [T_2]_{\mathcal{B}}$.
11. If $T \in L(V, W)$, then, for each $\mathbf{v} \in V$, $[T(\mathbf{v})]_{\mathcal{C}} = {}_{\mathcal{C}}[T]_{\mathcal{B}} [\mathbf{v}]_{\mathcal{B}}$. Furthermore ${}_{\mathcal{C}}[T]_{\mathcal{B}}$ is the only matrix A such that, for each $\mathbf{v} \in V$, $[T(\mathbf{v})]_{\mathcal{C}} = A[\mathbf{v}]_{\mathcal{B}}$.

Examples:

1. Let T be the projection of $\mathbb{R}^3$ onto the xy-plane of $\mathbb{R}^3$. Then

$$[T] = \begin{bmatrix} 1 & 0 & 0 \\ 0 & 1 & 0 \\ 0 & 0 & 0 \end{bmatrix}.$$

2. Let T be the identity in F^n. Then $[T]_{\mathcal{B}} = I_n$.
3. Let T be the rotation by θ in $\mathbb{R}^2$. Then

$$[T] = \begin{bmatrix} \cos\theta & -\sin\theta \\ \sin\theta & \cos\theta \end{bmatrix}.$$

4. Let $D: \mathbb{R}[x; n] \to \mathbb{R}[x; n-1]$ be the derivative transformation, and let $\mathcal{B} = \{1, x, \ldots, x^n\}$, $\mathcal{C} = \{1, x, \ldots, x^{n-1}\}$. Then

$$_{\mathcal{C}}[T]_{\mathcal{B}} = \begin{bmatrix} 0 & 1 & 0 & \cdots & 0 \\ 0 & 0 & 2 & \cdots & 0 \\ \vdots & \vdots & \vdots & \ddots & \vdots \\ 0 & 0 & 0 & \cdots & n-1 \end{bmatrix}.$$

3.4 Change of Basis and Similarity

Let V, W be nonzero finite dimensional vector spaces over F.

Facts:

The following facts can be found in [Gan60, III §5–6] and [Goo03, Chap. 4].

1. Let $T \in L(V, W)$ and let $\mathcal{B}, \mathcal{B}'$ be bases of V, $\mathcal{C}, \mathcal{C}'$ be bases of W. Then

$$_{\mathcal{C}'}[T]_{\mathcal{B}'} = {}_{\mathcal{C}'}[I]_{\mathcal{C}}\, {}_{\mathcal{C}}[T]_{\mathcal{B}}\, {}_{\mathcal{B}}[I]_{\mathcal{B}'}.$$

2. Two $m \times n$ matrices are equivalent if and only if they represent the same linear transformation $T \in L(V, W)$, but possibly in different bases, as in Fact 1.
3. Any $m \times n$ matrix A of rank r is equivalent to the $m \times n$ matrix

$$\tilde{I}_r = \begin{bmatrix} I_r & 0 \\ 0 & 0 \end{bmatrix}.$$

4. Two $m \times n$ matrices are equivalent if and only if they have the same rank.
5. Two $n \times n$ matrices are similar if and only if they represent the same linear transformation $T \in L(V, V)$, but possibly in different bases, i.e., if A_1 is similar to A_2, then there is $T \in L(V, V)$ and ordered bases $\mathcal{B}_1, \mathcal{B}_2$ of V such that $A_i = [T]_{\mathcal{B}_i}$ and conversely.

Examples:

1. Let T be the projection on the x-axis of $\mathbb{R}^2$, i.e., $T(x, y) = (x, 0)$. If $\mathcal{B} = \{\mathbf{e}_1, \mathbf{e}_2\}$ and $\mathcal{C} = \{\mathbf{e}_1 + \mathbf{e}_2, \mathbf{e}_1 - \mathbf{e}_2\}$, then $[T]_{\mathcal{B}} = \begin{bmatrix} 1 & 0 \\ 0 & 0 \end{bmatrix}$, $[T]_{\mathcal{C}} = \begin{bmatrix} 1/2 & 1/2 \\ 1/2 & 1/2 \end{bmatrix}$, and $[T]_{\mathcal{C}} = Q^{-1}[T]_{\mathcal{B}} Q$ with $Q = \begin{bmatrix} 1 & 1 \\ 1 & -1 \end{bmatrix}$.

3.5 Kernel and Range

Let V, W be vector spaces over F and let $T \in L(V, W)$.

Definitions:

T is **one-to-one** (or **injective**) if $\mathbf{v}_1 \neq \mathbf{v}_2$ implies $T(\mathbf{v}_1) \neq T(\mathbf{v}_2)$.

The **kernel** (or **null space**) of T is the set $\ker T = \{\mathbf{v} \in V \mid T(\mathbf{v}) = \mathbf{0}\}$.

The **nullity** of T, denoted by null T, is the dimension of $\ker T$.

T is **onto** (or **surjective**) if, for each $\mathbf{w} \in W$, there exists $\mathbf{v} \in V$ such that $T(\mathbf{v}) = \mathbf{w}$.

The **range** (or **image**) of T is the set range $T = \{\mathbf{w} \in W \mid \exists \mathbf{v}, \mathbf{w} = T(\mathbf{v})\}$.

The **rank** of T, denoted by rank T, is the dimension of range T.

Facts:

The following facts can be found in [Fin60, §3.3], [Lan70, IV §3], [Sta69, §3.1–3.2], and [Goo03, Chap. 4].

1. $\ker T$ is a subspace of V.
2. The following statements are equivalent.
   (a) T is one-to-one.
   (b) $\ker T = \{\mathbf{0}\}$.
   (c) Each linearly independent set is mapped to a linearly independent set.
   (d) Each basis is mapped to a linearly independent set.
   (e) Some basis is mapped to a linearly independent set.
3. range T is a subspace of W.
4. rank $T = \operatorname{rank}_{\mathcal{C}}[T]_{\mathcal{B}}$ for any finite nonempty ordered bases $\mathcal{B}, \mathcal{C}$.
5. For $A \in F^{m\times n}$, $\ker T_A = \ker A$ and range $T_A =$ range A.
6. (Dimension Theorem) Let $T \in L(V, W)$ where V has finite dimension. Then null $T +$ rank $T = \dim V$.
7. Let $T \in L(V, V)$, where V has finite dimension, then T is one-to-one if and only if T is onto.
8. Let $T(\mathbf{v}) = \mathbf{w}$. Then $\{\mathbf{u} \in V \mid T(\mathbf{u}) = \mathbf{w}\} = \mathbf{v} + \ker T$.
9. Let $V = \operatorname{Span}\{\mathbf{v}_1, \ldots, \mathbf{v}_n\}$. Then range $T = \operatorname{Span}\{T(\mathbf{v}_1), \ldots, T(\mathbf{v}_n)\}$.
10. Let $T_1, T_2 \in L(V, V)$. Then $\ker T_1T_2 \supseteq \ker T_2$ and range $T_1T_2 \subseteq$ range T_1.
11. Let $T \in L(V, V)$. Then

$$\{\mathbf{0}\} \subseteq \ker T \subseteq \ker T^2 \subseteq \cdots \subseteq \ker T^k \subseteq \cdots$$

$$V \supseteq \operatorname{range} T \supseteq \operatorname{range} T^2 \supseteq \cdots \supseteq \operatorname{range} T^k \supseteq \cdots.$$

Furthermore, if, for some k, range $T^{k+1} =$ range T^k, then, for each $i \geqslant 1$, range $T^{k+i} =$ range T^k. If, for some k, $\ker T^{k+1} = \ker T^k$, then, for each $i \geqslant 1$, $\ker T^{k+i} = \ker T^k$.

Examples:

1. Let T be the projection of $\mathbb{R}^3$ onto the xy-plane of $\mathbb{R}^3$. Then $\ker T = \{(0, 0, z) : z \in \mathbb{R}\}$; range $T = \{(x, y, 0) : x, y \in \mathbb{R}\}$; null $T = 1$; and rank $T = 2$.
2. Let T be the linear transformation in Example 1 of Section 3.1. Then $\ker T = \operatorname{Span}\{[1 \ -1 \ 2]^T\}$, while range $T = \mathbb{R}^2$.
3. Let $D \in L(\mathbb{R}[x], \mathbb{R}[x])$ be the derivative transformation, then $\ker D$ consists of all constant polynomials, while range $D = \mathbb{R}[x]$. In particular, D is onto but is not one-to-one. Note that $\mathbb{R}[x]$ is not finite dimensional.

4. Let $T_1, T_2 \in L(F^{n\times n}, F^{n\times n})$ where $T_1(A) = \frac{1}{2}(A - A^T)$, $T_2(A) = \frac{1}{2}(A + A^T)$, then

$$\ker T_1 = \text{range } T_2 = \{n \times n \text{ symmetric matrices}\};$$
$$\ker T_2 = \text{range } T_1 = \{n \times n \text{ skew-symmetric matrices}\};$$
$$\text{null } T_1 = \text{rank } T_2 = \frac{n(n+1)}{2}; \qquad \text{null } T_2 = \text{rank } T_1 = \frac{n(n-1)}{2}.$$

5. Let $T(\mathbf{v}) = \mathbf{b} \times \mathbf{v}$ as in Example 9 of Section 3.1. Then $\ker T = \text{Span}\{\mathbf{b}\}$.

3.6 Invariant Subspaces and Projections

Let V be a vector space over F, and let $V = V_1 \oplus V_2$ for some V_1, V_2 subspaces of V. For each $\mathbf{v} \in V$, let $\mathbf{v}_i \in V_i$ denote the (unique) vector such that $\mathbf{v} = \mathbf{v}_1 + \mathbf{v}_2$ (see Section 2.3). Finally, let $T \in L(V, V)$.

Definitions:

For $i, j \in \{1, 2\}$, $i \neq j$, the **projection** onto V_i along V_j is the operator $\text{proj}_{V_i,V_j} : V \to V$ defined by $\text{proj}_{V_i,V_j}(\mathbf{v}) = \mathbf{v}_i$ for each $\mathbf{v} \in V$ (see also Chapter 5).

The **complementary projection** of the projection proj_{V_i,V_j} is the projection proj_{V_j,V_i}.

T is an **idempotent** if $T^2 = T$.

A subspace V_0 of V is **invariant** under T or T**-invariant** if $T(V_0) \subseteq V_0$.

The **fixed space** of T is $\text{fix } T = \{\mathbf{v} \in V \mid T(\mathbf{v}) = \mathbf{v}\}$.

T is **nilpotent** if, for some $k \geqslant 0$, $T^k = 0$.

Facts:

The following facts can be found in [Mal63, §43–44].

1. $\text{proj}_{V_i,V_j} \in L(V, V)$.
2. $\text{proj}_{V_1,V_2} + \text{proj}_{V_2,V_1} = I$, the identity linear operator in V.
3. $\text{range}\,(\text{proj}_{V_i,V_j}) = \ker(\text{proj}_{V_j,V_i}) = V_i$.
4. Sum and intersection of invariant subspaces are invariant subspaces.
5. If V has a nonzero subspace different from V that is invariant under T, then there exists a suitable ordered basis $\mathcal{B}$ of V such that $[T]_{\mathcal{B}} = \begin{bmatrix} A_{11} & A_{12} \\ 0 & A_{22} \end{bmatrix}$. Conversely, if $[T]_{\mathcal{B}} = \begin{bmatrix} A_{11} & A_{12} \\ 0 & A_{22} \end{bmatrix}$, where A_{11} is an m-by-m block, then the subspace spanned by the first m vectors in $\mathcal{B}$ is a T-invariant subspace.
6. Let T have two nonzero finite dimensional invariant subspaces V_1 and V_2, with ordered bases $\mathcal{B}_1$ and $\mathcal{B}_2$, respectively, such that $V_1 \oplus V_2 = V$. Let $T_1 \in L(V_1, V_1)$, $T_2 \in L(V_2, V_2)$ be the restrictions of T on V_1 and V_2, respectively, and let $\mathcal{B} = \mathcal{B}_1 \cup \mathcal{B}_2$. Then $[T]_{\mathcal{B}} = [T_1]_{\mathcal{B}_1} \oplus [T_2]_{\mathcal{B}_2}$.

The following facts can be found in [Hoh64, §6.15; §6.20].

7. Every idempotent except the identity is singular.
8. The statements 8a through 8e are equivalent. If V is finite dimensional, statement 8f is also equivalent to these statements.

   (a) T is an idempotent.

   (b) $I - T$ is an idempotent.

   (c) $\text{fix } T = \text{range } T$.

   (d) $V = \ker T \oplus \text{fix } T$.

(e) T is the projection onto V_1 along V_2 for some V_1, V_2, with $V = V_1 \oplus V_2$.

(f) There exists a basis $\mathcal{B}$ of V such that $[T]_{\mathcal{B}} = \begin{bmatrix} I & 0 \\ 0 & 0 \end{bmatrix}$.

9. If T_1 and T_2 are idempotents on V and commute, then $T_1 T_2$ is an idempotent.
10. If T_1 and T_2 are idempotents on V and $T_1 T_2 = T_2 T_1 = 0$, then $T_1 + T_2$ is an idempotent.
11. If $\dim V = n$ and $T \in L(V, V)$ is nilpotent, then $T^n = 0$.

Examples:

1. Example 8 of Section 3.1, $T : \mathbb{R}^3 \to \mathbb{R}^3$, where $T\left(\begin{bmatrix} x \\ y \\ z \end{bmatrix}\right) = \begin{bmatrix} x \\ y \\ 0 \end{bmatrix}$ is the projection onto Span$\{\mathbf{e}_1, \mathbf{e}_2\}$ along Span$\{\mathbf{e}_3\}$.
2. The zero subspace is T-invariant for any T.
3. T_1 and T_2, defined in Example 4 of Section 3.5, are the projection of $F^{n\times n}$ onto the subspace of n-by-n
   symmetric matrices along the subspace of n-by-n skew-symmetric matrices, and the projection of $F^{n\times n}$ onto the skew-symmetric matrices along the symmetric matrices, respectively.
4. Let T be a nilpotent linear transformation on V. Let $T^p = 0$ and $T^{p-1}(\mathbf{v}) \neq 0$. Then $S =$ Span$\{\mathbf{v}, T(\mathbf{v}), T^2(\mathbf{v}), \ldots, T^{p-1}(\mathbf{v})\}$ is a T-invariant subspace.

3.7 Isomorphism and Nonsingularity Characterization

Let U, V, W be vector spaces over F and let $T \in L(V, W)$.

Definitions:

T is **invertible** (or an **isomorphism**) if there exists a function $S: W \to V$ such that $ST = I_V$ and $TS = I_W$. S is called the **inverse** of T and is denoted by T^{-1}.

V and W are **isomorphic** if there exists an isomorphism of V onto W.

T is **nonsingular** if $\ker T = \{\mathbf{0}\}$; otherwise T is **singular**.

Facts:

The following facts can be found in [Fin60, §3.4], [Hoh64, §6.11], and [Lan70, IV §4]:

1. The inverse is unique.
2. T^{-1} is a linear transformation, invertible, and $(T^{-1})^{-1} = T$.
3. If $T_1 \in L(V, W)$ and $T_2 \in L(U, V)$, then $T_1 T_2$ is invertible if and only if T_1 and T_2 are invertible.
4. If $T_1 \in L(V, W)$ and $T_2 \in L(U, V)$, then $(T_1 T_2)^{-1} = T_2^{-1} T_1^{-1}$.
5. Let $T \in L(V, W)$, and let $\dim V = \dim W = n$. The following statements are equivalent:

   (a) T is invertible.

   (b) T is nonsingular.

   (c) T is one-to-one.

   (d) $\ker T = \{\mathbf{0}\}$.

   (e) null $T = 0$.

   (f) T is onto.

   (g) range $T = W$.

(h) rank $T = n$.

(i) T maps some bases of V to bases of W.

6. If V and W are isomorphic, then $\dim V = \dim W$.
7. If $\dim V = n > 0$, then V is isomorphic to F^n through φ defined by $\varphi(\mathbf{v}) = [\mathbf{v}]_{\mathcal{B}}$ for any ordered basis $\mathcal{B}$ of V.
8. Let $\dim V = n > 0$, $\dim W = m > 0$, and let $\mathcal{B}$ and $\mathcal{C}$ be ordered bases of V and W, respectively. Then $L(V, W)$ and $F^{m \times n}$ are isomorphic through φ defined by $\varphi(T) = {}_{\mathcal{C}}[T]_{\mathcal{B}}$.

Examples:

1. $V = F[x; n]$ and $W = F^{n+1}$ are isomorphic through $T \in L(V, W)$ defined by $T(\sum_0^n a_i x^i) = [a_0 \ldots a_n]^T$.
2. If V is an infinite dimensional vector space, a nonsingular linear operator $T \in L(V, V)$ need not be invertible. For example, let $T \in L(\mathbb{R}[x], \mathbb{R}[x])$ be defined by $T(p(x)) = xp(x)$. Then T is nonsingular but not invertible since T is not onto. For matrices, nonsingular and invertible are equivalent, since an $n \times n$ matrix over F is an operator on the finite dimensional vector F^n.

3.8 Linear Functionals and Annihilator

Let V, W be vector spaces over F.

Definitions:

A **linear functional** (or **linear form**) on V is a linear transformation from V to F.

The **dual space** of V is the vector space $V^* = L(V, F)$ of all linear functionals on V.

If V is nonzero and finite dimensional, the **dual basis** of a basis $\mathcal{B} = \{\mathbf{v}_1, \ldots, \mathbf{v}_n\}$ of V is the set $\mathcal{B}^* = \{f_1, \ldots, f_n\} \subseteq V^*$, such that $f_i(\mathbf{v}_j) = \delta_{ij}$ for each i, j.

The **bidual space** is the vector space $V^{**} = (V^*)^* = L(V^*, F)$.

The **annihilator** of a set $S \subseteq V$ is $S^a = \{f \in V^* \mid f(\mathbf{v}) = 0, \forall \mathbf{v} \in S\}$.

The **transpose** of $T \in L(V, W)$ is the mapping $T^T \in L(W^*, V^*)$ defined by setting, for each $g \in W^*$,

$$\begin{aligned} T^T(g) : V &\to F \\ \mathbf{v} &\mapsto g(T(\mathbf{v})). \end{aligned}$$

Facts:

The following facts can be found in [Hoh64, §6.19] and [SW68, §4.4].

1. For each $\mathbf{v} \in V$, $\mathbf{v} \neq 0$, there exists $f \in V^*$ such that $f(\mathbf{v}) \neq 0$.
2. For each $\mathbf{v} \in V$ define $h_{\mathbf{v}} \in L(V^*, F)$ by setting $h_{\mathbf{v}}(f) = f(\mathbf{v})$. Then the mapping

$$\begin{aligned} \varphi : V &\to V^{**} \\ \mathbf{v} &\mapsto h_{\mathbf{v}} \end{aligned}$$

is a one-to-one linear transformation. If V is finite dimensional, φ is an isomorphism of V onto V^{**}.

3. S^a is a subspace of V^*.
4. $\{0\}^a = V^*$; $V^a = \{0\}$.
5. $S^a = (\text{Span}\{S\})^a$.

The following facts hold for finite dimensional vector spaces.

6. If V is nonzero, for each basis $\mathcal{B}$ of V, the dual basis exists, is uniquely determined, and is a basis for V^*.
7. $\dim V = \dim V^*$.
8. If V is nonzero, each basis of V^* is the dual basis of some basis of V.

9. Let $\mathcal{B}$ be a basis for the nonzero vector space V. For each $\mathbf{v} \in V$, $f \in V^*$, $f(\mathbf{v}) = [f]_{\mathcal{B}^*}^T [\mathbf{v}]_{\mathcal{B}}$.
10. If S is a subspace of V, then $\dim S + \dim S^a = \dim V$.
11. If S is a subspace of V, then, by identifying V and V^{**}, $S = (S^a)^a$.
12. Let S_1, S_2 be subspaces of V such that $S_1^a = S_2^a$. Then $S_1 = S_2$.
13. Any subspace of V^* is the annihilator of some subspace S of V.
14. Let S_1, S_2 be subspaces of V. Then $(S_1 \cap S_2)^a = S_1^a + S_2^a$ and $(S_1 + S_2)^a = S_1^a \cap S_2^a$.
15. $\ker T^T = (\text{range } T)^a$.
16. $\text{rank } T = \text{rank } T^T$.
17. If $\mathcal{B}$ and $\mathcal{C}$ are nonempty bases of V and W, respectively, then ${}_{\mathcal{B}^*}[T^T]_{\mathcal{C}^*} = ({}_{\mathcal{C}}[T]_{\mathcal{B}})^T$.

Examples:

1. Let $V = C[a, b]$ be the vector space of continuous functions $\varphi : [a, b] \to \mathbb{R}$, and let $c \in [a, b]$. Then $f(\varphi) = \varphi(c)$ is a linear functional on V.
2. Let $V = C[a, b]$, $\psi \in V$, and $f(\varphi) = \int_a^b \varphi(t)\psi(t)dt$. Then f is a linear functional.
3. The trace is a linear functional on $F^{n \times n}$.
4. let $V = F^{m \times n}$. $\mathcal{B} = \{E_{ij} : 1 \leqslant i \leqslant m, 1 \leqslant j \leqslant n\}$ is a basis for V. The dual basis $\mathcal{B}^*$ consists of the linear functionals f_{ij}, $1 \leqslant i \leqslant m, 1 \leqslant j \leqslant n$, defined by $f_{ij}(A) = a_{ij}$.

References

[Fin60] D.T. Finkbeiner, *Introduction to Matrices and Linear Transformations*. San Francisco: W.H. Freeman, 1960.

[Gan60] F.R. Gantmacher, *The Theory of Matrices*. New York: Chelsea Publishing, 1960.

[Goo03] E.G. Goodaire. *Linear Algebra: a Pure and Applied First Course*. Upper Saddle River, NJ: Prentice Hall, 2003.

[Hoh64] F.E. Hohn, *Elementary Matrix Algebra*. New York: Macmillan, 1964.

[Lan70] S. Lang, *Introduction to Linear Algebra*. Reading, MA: Addison-Wesley, 1970.

[Lay03] D.C. Lay, *Linear Algebra and Its Applications, 3rd ed.* Boston: Addison-Wesley, 2003.

[Mal63] A.I. Maltsev, *Foundations of Linear Algebra*. San Francisco: W.H. Freeman, 1963.

[Sta69] J.H. Staib, *An Introduction to Matrices and Linear Transformations*. Reading, MA: Addison-Wesley, 1969.

[SW68] R.R. Stoll and E.T. Wong, *Linear Algebra*. New York: Academic Press, 1968.

4

Determinants and Eigenvalues

Luz M. DeAlba
Drake University

4.1 Determinants

Definitions:

The **determinant**, det A, of a matrix $A = [a_{ij}] \in F^{n\times n}$ is an element in F defined inductively:

- $\det [a] = a$.
- For $i, j \in \{1, 2, \ldots, n\}$, the ij**th minor** of A corresponding to a_{ij} is defined by $m_{ij} = \det A(\{i\}, \{j\})$.
- The ij**th cofactor** of a_{ij} is $c_{ij} = (-1)^{i+j} m_{ij}$.
- $\det A = \sum_{j=1}^{n} (-1)^{i+j} a_{ij} m_{ij} = \sum_{j=1}^{n} a_{ij} c_{ij}$ for $i \in \{1, 2, \ldots, n\}$.

This method of computing the determinant of a matrix is called **Laplace expansion of the determinant by minors along the** i**th row.**

The **determinant of a linear operator** $T : V \to V$ on a finite dimensional vector space, V, is defined as $\det(T) = \det([T]_{\mathcal{B}})$, where $\mathcal{B}$ is a basis for V.

Facts:

All matrices are assumed to be in $F^{n\times n}$, unless otherwise stated. All the following facts except those with a specific reference can be found in [Lay03, pp. 185–213] or [Goo03, pp. 167–193].

1. $\det \begin{bmatrix} a_{11} & a_{12} \\ a_{21} & a_{22} \end{bmatrix} = a_{11}a_{22} - a_{12}a_{21}$.
2. $\det A = \det \begin{bmatrix} a_{11} & a_{12} & a_{13} \\ a_{21} & a_{22} & a_{23} \\ a_{31} & a_{32} & a_{33} \end{bmatrix} = a_{11}a_{22}a_{33} + a_{21}a_{13}a_{32} + a_{31}a_{12}a_{23} - a_{31}a_{13}a_{22} - a_{21}a_{12}a_{33}$ $-a_{11}a_{32}a_{23}$.
3. The determinant is independent of the row i used to evaluate it.
4. (Expansion of the determinant by minors along the jth column) Let $j \in \{1, 2, \ldots, n\}$. Then $\det A = \sum_{i=1}^{n} (-1)^{i+j} a_{ij} m_{ij} = \sum_{i=1}^{n} a_{ij} c_{ij}$.
5. $\det I_n = 1$.
6. If A is a triangular matrix, then $\det A = a_{11}a_{22} \cdots a_{nn}$.

7. If B is a matrix obtained from A by interchanging two rows (or columns), then $\det B = -\det A$.
8. If B is a matrix obtained from A by multiplying one row (or column) by a nonzero constant r, then $\det B = r \det A$.
9. If B is a matrix obtained from A by adding to a row (or column) a multiple of another row (or column), then $\det B = \det A$.
10. If A, B, and C differ only in the rth row (or column), and the rth row (or column) of C is the sum of the rth rows (or columns) of A and B, then $\det C = \det A + \det B$.
11. If A is a matrix with a row (or column) of zeros, then $\det A = 0$.
12. If A is a matrix with two identical rows (or columns), then $\det A = 0$.
13. Let B be a row echelon form of A obtained by Gaussian elimination, using k row interchange operations and adding multiples of one row to another (see Algorithm 1 in Section 1.3). Then $\det A = (-1)^k \det B = (-1)^k b_{11} b_{22} \cdots b_{nn}$.
14. $\det A^T = \det A$.
15. If $A \in \mathbb{C}^{n\times n}$, then $\det A^* = \overline{\det A}$.
16. $\det AB = \det A \det B$.
17. If $c \in F$, then $\det(cA) = c^n \det A$.
18. A is nonsingular, that is A^{-1} exists, if and only if $\det A \neq 0$.
19. If A is nonsingular, then $\det\left(A^{-1}\right) = \frac{1}{\det A}$.
20. If S is nonsingular, then $\det\left(S^{-1}AS\right) = \det A$.
21. [HJ85] $\det A = \sum_\sigma \operatorname{sgn}\sigma\, a_{1\sigma(1)} a_{2\sigma(2)} \cdots a_{n\sigma(n)}$, where summation is over the $n!$ permutations, σ, of the n indices $\{1, 2, \ldots, n\}$. The weight "sgnσ" is 1 when σ is even and -1 when σ is odd. (See Preliminaries for more information on permutations.)
22. If $\mathbf{x}, \mathbf{y} \in F^n$, then $\det(I + \mathbf{x}\mathbf{y}^T) = 1 + \mathbf{y}^T\mathbf{x}$.
23. [FIS89] Let T be a linear operator on a finite dimensional vector space V. Let $\mathcal{B}$ and $\mathcal{B}'$ be bases for V. Then $\det(T) = \det\left([T]_{\mathcal{B}}\right) = \det\left([T]_{\mathcal{B}'}\right)$.
24. [FIS89] Let T be a linear operator on a finite dimensional vector space V. Then T is invertible if and only if $\det(T) \neq 0$.
25. [FIS89] Let T be an invertible linear operator on a finite dimensional vector space V. Then $\det(T^{-1}) = \frac{1}{\det(T)}$.
26. [FIS89] Let T and U be linear operators on a finite dimensional vector space V. Then $\det(TU) = \det(T) \cdot \det(U)$.

Examples:

1. Let $A = \begin{bmatrix} 3 & -2 & 4 \\ 2 & 5 & -6 \\ -3 & 1 & 5 \end{bmatrix}$. Expanding the determinant of A along the second column: $\det A = 2 \cdot \det \begin{bmatrix} 2 & -6 \\ -3 & 5 \end{bmatrix} + 5 \cdot \det \begin{bmatrix} 3 & 4 \\ -3 & 5 \end{bmatrix} - \det \begin{bmatrix} 3 & 4 \\ 2 & -6 \end{bmatrix} = 2 \cdot (-8) + 5 \cdot 27 + 26 = 145$.

2. Let $A = \begin{bmatrix} -1 & 3 & -2 & 4 \\ 2 & 5 & 8 & 1 \\ 7 & -4 & 0 & -6 \\ 0 & 3 & 1 & 5 \end{bmatrix}$. Expanding the determinant of A along the third row: $\det A = 7 \cdot \det \begin{bmatrix} 3 & -2 & 4 \\ 5 & 8 & 1 \\ 3 & 1 & 5 \end{bmatrix} + 4 \cdot \det \begin{bmatrix} -1 & -2 & 4 \\ 2 & 8 & 1 \\ 0 & 1 & 5 \end{bmatrix} + 6 \cdot \det \begin{bmatrix} -1 & 3 & -2 \\ 2 & 5 & 8 \\ 0 & 3 & 1 \end{bmatrix} = 557$.

3. Let $T : \mathbb{R}^2 \to \mathbb{R}^2$ defined by $T\left(\begin{bmatrix} x_1 \\ x_2 \end{bmatrix}\right) = \begin{bmatrix} 2x_1 - 3x_2 \\ x_1 + 6x_2 \end{bmatrix}$. With $\mathcal{B} = \left\{\begin{bmatrix} 1 \\ 0 \end{bmatrix}, \begin{bmatrix} 0 \\ 1 \end{bmatrix}\right\}$, then $\det\left([T]_{\mathcal{B}}\right) = \det \begin{bmatrix} 2 & -3 \\ 1 & 6 \end{bmatrix} = 15$. Now let $\mathcal{B}' = \left\{\begin{bmatrix} 1 \\ 1 \end{bmatrix}, \begin{bmatrix} 1 \\ 0 \end{bmatrix}\right\}$. Then $\det\left([T]_{\mathcal{B}'}\right) = \det \begin{bmatrix} 7 & 1 \\ -8 & 1 \end{bmatrix} = 15$.

Applications:

1. (Cramer's Rule) If $A \in F^{n\times n}$ is nonsingular, then the equation $A\mathbf{x} = \mathbf{b}$, where $\mathbf{x}, \mathbf{b} \in F^n$, has the unique solution $\mathbf{s} = \begin{bmatrix} s_1 \\ s_2 \\ \vdots \\ s_n \end{bmatrix}$, where $s_i = \frac{\det A_i}{\det A}$ and A_i is the matrix obtained from A by replacing the ith column with $\mathbf{b}$.
2. [Mey00, p. 486] (Vandermonde Determinant) $\det \begin{bmatrix} 1 & 1 & \cdots & 1 \\ x_1 & x_2 & \cdots & x_n \\ x_1^2 & x_2^2 & \cdots & x_n^2 \\ \vdots & \vdots & \ddots & \vdots \\ x_1^{n-1} & x_2^{n-1} & \cdots & x_n^{n-1} \end{bmatrix} = \prod_{1\leq i<j\leq n}(x_i - x_j)$.
3. [FB90, pp. 220–235] (Volume) Let $\mathbf{a}_1, \mathbf{a}_2, \ldots, \mathbf{a}_n$ be linearly independent vectors in $\mathbb{R}^m$. The volume, V, of the n-dimensional solid in $\mathbb{R}^m$, defined by $S = \{\sum_{i=1}^n t_i\mathbf{a}_i, 0 \leq t_i \leq 1, i = 1, 2, \ldots, n\}$, is given by $V = \sqrt{\det\left(A^T A\right)}$, where A is the matrix whose ith column is the vector $\mathbf{a}_i$.
   Let $m \geq n$ and $T : \mathbb{R}^n \to \mathbb{R}^m$ be a linear transformation whose standard matrix representation is the $m \times n$ matrix A. Let S be a region in $\mathbb{R}^n$ of volume V_S. Then the volume of the image of S under the transformation T is $V_{T(S)} = \sqrt{\det\left(A^T A\right)} \cdot V_S$.
4. [Uhl02, pp. 247–248] (Wronskian) Let $f_1, f_2, \ldots, f_n$ be $n-1$ times differentiable functions of the real variable x. The determinant

$$W(f_1, f_2, \ldots, f_n)(x) = \det \begin{bmatrix} f_1(x) & f_2(x) & \cdots & f_n(x) \\ f_1'(x) & f_2'(x) & \cdots & f_n'(x) \\ \vdots & \vdots & \ddots & \vdots \\ f_1^{(n-1)}(x) & f_2^{(n-1)}(x) & \cdots & f_n^{(n-1)}(x) \end{bmatrix}$$

   is called the **Wronskian** of $f_1, f_2, \ldots, f_n$. If $W(f_1, f_2, \ldots, f_n)(x) \neq 0$ for some $x \in \mathbb{R}$, then the functions $f_1, f_2, \ldots, f_n$ are linearly independent.

4.2 Determinants: Advanced Results

Definitions:

A **principal minor** is the determinant of a principal submatrix. (See Section 1.2.)

A **leading principal minor** is the determinant of a leading principal submatrix.

The sum of all the $k \times k$ principal minors of A is denoted $S_k(A)$.

The k**th compound matrix** of $A \in F^{m\times n}$ is the $\binom{m}{k} \times \binom{n}{k}$ matrix $C_k(A)$ whose entries are the $k \times k$ minors of A, usually in lexicographical order.

The **adjugate** of $A \in F^{n\times n}$ is the matrix $\operatorname{adj} A = [c_{ji}] = [c_{ij}]^T$, where c_{ij} is the ijth-cofactor.

The k**th adjugate** of $A \in F^{n\times n}$ is the $\binom{n}{k} \times \binom{n}{k}$ matrix $\operatorname{adj}^{(k)} A$, whose a_{ji} entry is the cofactor, in A, of the $(n-k)$th minor, of A, in the ijth position of the compound.

Let $\alpha \subseteq \{1, 2, \ldots, n\}$ and $A \in F^{n\times n}$ with $A[\alpha]$ nonsingular. The matrix

$$A/A[\alpha] = A[\alpha^c] - A[\alpha^c, \alpha]A[\alpha]^{-1}A[\alpha, \alpha^c]$$

is called the **Schur complement** of $A[\alpha]$.

Facts:

All matrices are assumed to be in $F^{n\times n}$, unless otherwise stated. All the following facts except those with a specific reference can be found in [Lay03, pp. 185–213] or [Goo03, pp. 167–193].

1. $A\,(\text{adj } A) = (\text{adj } A)\,A = (\det A)\,I_n$.
2. $\det(\text{adj } A) = (\det A)^{n-1}$.
3. If $\det A \neq 0$, then adj A is nonsingular, and $(\text{adj } A)^{-1} = (\det A)^{-1} A$.
4. [Ait56] (Method of Condensation) Let $A = \begin{bmatrix} a_{11} & a_{12} & a_{13} & \cdots & a_{1n} \\ a_{21} & a_{22} & a_{23} & \cdots & a_{2n} \\ a_{31} & a_{32} & a_{33} & \cdots & a_{3n} \\ \vdots & \vdots & \vdots & \ddots & \vdots \\ a_{n1} & a_{n2} & a_{n3} & \cdots & a_{nn} \end{bmatrix}$, and assume without loss of generality that $a_{11} \neq 0$, otherwise a nonzero element can be brought to the $(1,1)$ position by interchanging two rows, which will change the sign of the determinant. Multiply all the rows of A except the first by a_{11}. For $i = 2, 3, \ldots, n$, perform the row operations: replace row i with row $i - a_{i1} \cdot$ row 1. Thus $a_{11}^{n-1} \det A = \det \begin{bmatrix} a_{11} & a_{12} & a_{13} & \cdots & a_{1n} \\ 0 & a_{11}a_{22} - a_{21}a_{12} & a_{11}a_{23} - a_{21}a_{13} & \cdots & a_{11}a_{2n} - a_{21}a_{1n} \\ 0 & a_{11}a_{32} - a_{31}a_{12} & a_{11}a_{33} - a_{31}a_{13} & \cdots & a_{11}a_{3n} - a_{31}a_{1n} \\ \vdots & \vdots & \vdots & \ddots & \vdots \\ 0 & a_{11}a_{n2} - a_{n1}a_{12} & a_{11}a_{n3} - a_{n1}a_{12} & \cdots & a_{11}a_{nn} - a_{n1}a_{1n} \end{bmatrix}$.

   So, $\det A = \frac{1}{a_{11}^{n-2}} \cdot \det \begin{bmatrix} \det\begin{bmatrix} a_{11} & a_{12} \\ a_{21} & a_{22} \end{bmatrix} & \det\begin{bmatrix} a_{11} & a_{13} \\ a_{21} & a_{23} \end{bmatrix} & \cdots & \det\begin{bmatrix} a_{11} & a_{1n} \\ a_{21} & a_{2n} \end{bmatrix} \\ \det\begin{bmatrix} a_{11} & a_{12} \\ a_{31} & a_{32} \end{bmatrix} & \det\begin{bmatrix} a_{11} & a_{13} \\ a_{31} & a_{33} \end{bmatrix} & \cdots & \det\begin{bmatrix} a_{11} & a_{1n} \\ a_{31} & a_{32} \end{bmatrix} \\ \vdots & \vdots & \ddots & \vdots \\ \det\begin{bmatrix} a_{11} & a_{12} \\ a_{n1} & a_{n2} \end{bmatrix} & \det\begin{bmatrix} a_{11} & a_{12} \\ a_{n1} & a_{n3} \end{bmatrix} & \cdots & \det\begin{bmatrix} a_{11} & a_{1n} \\ a_{n1} & a_{nn} \end{bmatrix} \end{bmatrix}$.
5. [Ait56] $A^{(k)}(\text{adj}^{(k)} A) = (\text{adj}^{(k)} A)A^{(k)} = (\det A)I_n$.
6. [Ait56] $\det\left(A^{(k)}\right) = \det(A^r)$, where $r = \begin{pmatrix} n-1 \\ k-1 \end{pmatrix}$.
7. [Ait56] $\det\left(A^{(n-k)}\right) = \det(\text{adj}^{(k)} A)$.
8. [HJ85] If $A \in F^{n\times n}$, $B \in F^{m\times m}$, then $\det(A \otimes B) = (\det A)^m (\det B)^n$. (See Section 10.5 for the definition of $A \otimes B$.)
9. [Uhl02] For $A \in F^{n\times n}$, $\det A$ is the unique normalized, alternating, and multilinear function $d : F^{n\times n} \to F$. That is, $d(I_n) = 1$, $d(A) = -d(A')$, where A' denotes the matrix obtained from A, by interchanging two rows, and d is linear in each row of A, if the remaining rows of A are held fixed.
10. [HJ85] (Cauchy–Binet) Let $A \in F^{n\times k}$, $B \in F^{k\times n}$, and $C = AB$. Then

$$\det C[\alpha,\beta] = \sum_{\gamma} \det A[\alpha,\gamma] \det B[\gamma,\beta],$$

   where $\alpha \subseteq \{1,2,\ldots,m\}$, $\beta \subseteq \{1,2,\ldots,n\}$, with $|\alpha| = |\beta| = r$, $1 \leq r \leq \min\{m,k,n\}$, and the sum is taken over all sets $\gamma \subseteq \{1,2,\ldots,k\}$ with $|\gamma| = r$.
11. [HJ85] (Schur Complement). Let $A[\alpha]$ be nonsingular. Then

$$\det A = \det A[\alpha] \det\left(A[\alpha^c] - A[\alpha^c,\alpha]\, A[\alpha]^{-1} A[\alpha,\alpha^c]\right).$$

12. [HJ85] (Jacobi's Theorem) Let A be nonsingular and let $\alpha, \beta \subseteq \{1, 2, \ldots, n\}$, with $|\alpha| = |\beta|$. Then

$$\det A^{-1}[\alpha^c, \beta^c] = (-1)^{(\sum_{i\in\alpha} i + \sum_{j\in\beta} j)} \frac{\det A[\beta, \alpha]}{\det A}.$$

In particular, if $\alpha = \beta$. Then $\det A^{-1}[\alpha^c] = \frac{\det A[\alpha]}{\det A}$.

13. [HJ85] (Sylvester's Identity) Let $\alpha \subseteq \{1, 2, \ldots, n\}$ with $|\alpha| = k$, and $i, j \in \{1, 2, \ldots, n\}$, with $i, j \notin \alpha$. For $A \in F^{n\times n}$, let $B = [b_{ij}] \in F^{(n-k)\times(n-k)}$ be defined by $b_{ij} = \det A[\alpha \cup \{i\}, \alpha \cup \{j\}]$. Then

$$\det B = (\det A[\alpha])^{n-k-1} \det A.$$

Examples:

1. Let $A = \begin{bmatrix} 1 & 1 & 0 & 0 \\ 0 & 1 & 0 & 1 \\ 0 & 0 & -3 & 2 \\ 0 & -1 & -2 & -4 \end{bmatrix}$. $S_3(A) = 23$ because $\det A[\{1,2,3\}] = \det \begin{bmatrix} 1 & 1 & 0 \\ 0 & 1 & 0 \\ 0 & 0 & -3 \end{bmatrix} = -3$, $\det A[\{1,2,4\}] = \det \begin{bmatrix} 1 & 1 & 0 \\ 0 & 1 & 1 \\ 0 & -1 & -4 \end{bmatrix} = -3$, $\det A[\{1,3,4\}] = \det \begin{bmatrix} 1 & 0 & 0 \\ 0 & -3 & 2 \\ 0 & -2 & -4 \end{bmatrix} = 16$, and $\det A[\{2,3,4\}] = \det \begin{bmatrix} 1 & 0 & 1 \\ 0 & -3 & 2 \\ -1 & -2 & -4 \end{bmatrix} = 13$. From the Laplace expansion on the first column and $\det A[\{2,3,4\}] = 13$, it follows that $S_4(A) = \det A = 13$. Clearly, $S_1(A) = \operatorname{tr} A = -5$.

2. (kth compound) Let $A = \begin{bmatrix} 1 & 1 & 1 & 1 \\ 1 & 0 & 1 & 4 \\ 1 & 1 & 0 & 1 \\ 1 & 4 & 0 & 1 \end{bmatrix}$. Then $\det A = 9$, $C_2(A) = \begin{bmatrix} -1 & 0 & 3 & 1 & 4 & 3 \\ 0 & -1 & 0 & -1 & 0 & 1 \\ 3 & -1 & 0 & -4 & -3 & 1 \\ 1 & -1 & -3 & -1 & -4 & 1 \\ 4 & -1 & -3 & -4 & -16 & 1 \\ 3 & 0 & 0 & 0 & -3 & 0 \end{bmatrix}$, and $\det(C_2(A)) = 729$.

3. (Cauchy–Binet) Let $A = \begin{bmatrix} -1 & 3 & -1 \\ 2 & 0 & 0 \\ i & -4 & 0 \\ 0 & 1+i & 1 \end{bmatrix}$, $B = \begin{bmatrix} 3 & 2 & -3 & 0 \\ 0 & -4 & 4 & 3i \\ 7 & -6i & 5 & 4 \end{bmatrix}$, and $C = AB$.

   Then $\det C[\{2,4\},\{2,3\}] =$
   $\det A[\{2,4\},\{1,2\}] \det B[\{1,2\},\{2,3\}] +$
   $\det A[\{2,4\},\{1,3\}] \det B[\{1,3\},\{2,3\}] +$
   $\det A[\{2,4\},\{2,3\}] \det B[\{2,3\},\{2,3\}] = 12 - 44i$.

4. (Schur Complement) Let $A = \begin{bmatrix} a & \mathbf{b}^* \\ \mathbf{b} & C \end{bmatrix}$, where $a \in \mathbb{C}$, $\mathbf{b} \in \mathbb{C}^{n-1}$, and $C \in \mathbb{C}^{(n-1)\times(n-1)}$. If C is nonsingular, then $\det A = (a - \mathbf{b}^* C^{-1} \mathbf{b}) \det C$. If $a \neq 0$, then $\det A = a \det\left(C - \frac{1}{a}\mathbf{b}\mathbf{b}^*\right)$.

5. (Jacobi's Theorem) Let $A = \begin{bmatrix} 1 & -2 & 0 \\ 3 & 4 & 0 \\ -1 & 0 & 5 \end{bmatrix}$ and $\alpha = \{2\} = \beta$. By Jacobi's formula, $\det A^{-1}(2) = \frac{\det A[2]}{\det A} = \frac{4}{50} = \frac{2}{25}$. This can be readily verified by computing $A^{-1} = \begin{bmatrix} \frac{2}{5} & \frac{1}{5} & 0 \\ -\frac{3}{10} & \frac{1}{10} & 0 \\ \frac{2}{25} & \frac{1}{25} & \frac{1}{5} \end{bmatrix}$, and verifying $\det A^{-1}[\{1,3\}] = \frac{2}{25}$.

6. (Sylvester's Identity) Let $A = \begin{bmatrix} -7 & i & -3 \\ -i & -2 & 1+4i \\ -3 & 1-4i & 5 \end{bmatrix}$ and $\alpha = \{1\}$. Define $B \in \mathbb{C}^{2\times 2}$, with entries $b_{11} = \det A[\{1,2\}] = 13$, $b_{12} = \det A[\{1,2\},\{1,3\}] = -7 - 31i$, $b_{21} = \det A[\{1,3\},\{1,2\}] = -7 + 31i$, $b_{22} = \det A[\{1,3\}] = -44$. Then $-1582 = \det B = (\det A[\{1\}]) \det A = (-7) \det A$, so $\det A = 226$.

4.3 Eigenvalues and Eigenvectors

Definitions:

An element $\lambda \in F$ is an **eigenvalue** of a matrix $A \in F^{n\times n}$ if there exists a nonzero vector $\mathbf{x} \in F^n$ such that $A\mathbf{x} = \lambda\mathbf{x}$. The vector $\mathbf{x}$ is said to be an **eigenvector** of A corresponding to the eigenvalue λ. A nonzero row vector $\mathbf{y}$ is a **left eigenvector** of A, corresponding to the eigenvalue λ, if $\mathbf{y}A = \lambda\mathbf{y}$.

For $A \in F^{n\times n}$, the **characteristic polynomial** of A is given by $p_A(x) = \det(xI - A)$.

The **algebraic multiplicity**, $\alpha(\lambda)$, of $\lambda \in \sigma(A)$ is the number of times the eigenvalue occurs as a root in the characteristic polynomial of A.

The **spectrum** of $A \in F^{n\times n}$, $\sigma(A)$, is the multiset of all eigenvalues of A, with eigenvalue λ appearing $\alpha(\lambda)$ times in $\sigma(A)$.

The **spectral radius** of $A \in \mathbb{C}^{n\times n}$ is $\rho(A) = \max\{|\lambda| : \lambda \in \sigma(A)\}$.

Let $p(x) = c_n x^n + c_{n-1}x^{n-1} + \cdots + c_2 x^2 + c_1 x + c_0$ be a polynomial with coefficients in F. Then $p(A) = c_n A^n + c_{n-1}A^{n-1} + \cdots + c_2 A^2 + c_1 A + c_0 I$.

For $A \in F^{n\times n}$, the **minimal polynomial** of A, $q_A(x)$, is the unique monic polynomial of least degree for which $q_A(A) = 0$.

The vector space $\ker(A - \lambda I)$, for $\lambda \in \sigma(A)$, is the **eigenspace** of $A \in F^{n\times n}$ corresponding to λ, and is denoted by $E_\lambda(A)$.

The **geometric multiplicity**, $\gamma(\lambda)$, of an eigenvalue λ is the dimension of the eigenspace $E_\lambda(A)$.

An eigenvalue λ is **simple** if $\alpha(\lambda) = 1$.

An eigenvalue λ is **semisimple** if $\alpha(\lambda) = \gamma(\lambda)$.

For $K = \mathbb{C}$ or any other algebraically closed field, a matrix $A \in K^{n\times n}$ is **nonderogatory** if $\gamma(\lambda) = 1$ for all $\lambda \in \sigma(A)$, otherwise A is **derogatory**. Over an arbitrary field F, a matrix is nonderogatory (derogatory) if it is nonderogatory (derogatory) over the algebraic closure of F.

For $K = \mathbb{C}$ or any other algebraically closed field, a matrix $A \in K^{n\times n}$ is **nondefective** if every eigenvalue of A is semisimple, otherwise A is **defective.** Over an arbitrary field F, a matrix is nondefective (defective) if it is nondefective (defective) over the algebraic closure of F.

A matrix $A \in F^{n\times n}$ is **diagonalizable** if there exists a nonsingular matrix $B \in F^{n\times n}$, such that $A = BDB^{-1}$ for some diagonal matrix $D \in F^{n\times n}$.

For a monic polynomial $p(x) = x^n + c_{n-1}x^{n-1} + \cdots + c_2 x^2 + c_1 x + c_0$ with coefficients in F, the $n \times n$ matrix $C(p) = \begin{bmatrix} 0 & 0 & 0 & \cdots & 0 & -c_0 \\ 1 & 0 & 0 & \cdots & 0 & -c_1 \\ 0 & 1 & 0 & \cdots & 0 & -c_2 \\ \vdots & \vdots & \vdots & \ddots & \vdots & \vdots \\ 0 & 0 & 0 & \cdots & 1 & -c_{n-1} \end{bmatrix}$ is called the **companion matrix** of $p(x)$.

Let T be a linear operator on a finite dimensional vector space, V, over a field F. An element $\lambda \in F$ is an **eigenvalue** of T if there exists a nonzero vector $\mathbf{v} \in V$ such that $T(\mathbf{v}) = \lambda\mathbf{v}$. The vector $\mathbf{v}$ is said to be an **eigenvector** of T corresponding to the eigenvalue λ.

For a linear operator, T, on a finite dimensional vector space, V, with a basis, $\mathcal{B}$, the **characteristic polynomial** of T is given by $p_T(x) = \det([T])_{\mathcal{B}}$.

A linear operator T on a finite dimensional vector space, V, is **diagonalizable** if there exists a basis, $\mathcal{B}$, for V such that $[T]_{\mathcal{B}}$ is diagonalizable.

Facts:

These facts are grouped into the following categories: Eigenvalues and Eigenvectors, Diagonalization, Polynomials, Other Facts. All matrices are assumed to be in $F^{n\times n}$ unless otherwise stated. All the following facts, except those with a specific reference, can be found in [Mey00, pp. 489–660] or [Lay03, pp. 301–342].

Eigenvalues and Eigenvectors

1. $\lambda \in \sigma(A)$ if and only if $p_A(\lambda) = 0$.
2. For each eigenvalue λ of a matrix A, $1 \leq \gamma(\lambda) \leq \alpha(\lambda)$.
3. A simple eigenvalue is semisimple.
4. For any F, $|\sigma(A)| \leq n$. If $F = \mathbb{C}$ or any algebraically closed field, then $|\sigma(A)| = n$.
5. If $F = \mathbb{C}$ or any algebraically closed field, then $\det A = \prod_{i=1}^{n} \lambda_i, \lambda_i \in \sigma(A)$.
6. If $F = \mathbb{C}$ or any algebraically closed field, then $\operatorname{tr} A = \sum_{i=1}^{n} \lambda_i, \lambda_i \in \sigma(A)$.
7. For $A \in \mathbb{C}^{n\times n}$, $\lambda \in \sigma(A)$ if and only if $\bar{\lambda} \in \sigma(A^*)$.
8. For $A \in \mathbb{R}^{n\times n}$, viewing $A \in \mathbb{C}^{n\times n}$, $\lambda \in \sigma(A)$ if and only if $\bar{\lambda} \in \sigma(A)$.
9. If $A \in \mathbb{C}^{n\times n}$ is Hermitian (e.g., $A \in \mathbb{R}^{n\times n}$ is symmetric), then A has real eigenvalues and A can be diagonalized. (See also Section 7.2.)
10. A and A^T have the same eigenvalues with same algebraic multiplicities.
11. If $A = [a_{ij}]$ is triangular, then $\sigma(A) = \{a_{11}, a_{22}, \ldots, a_{nn}\}$.
12. If A has all row (column) sums equal to r, then r is an eigenvalue of A.
13. A is singular if and only if $\det A = 0$, if and only if $0 \in \sigma(A)$.
14. If A is nonsingular and λ is an eigenvalue of A of algebraic multiplicity $\alpha(\lambda)$, with corresponding eigenvector $\mathbf{x}$, then λ^{-1} is an eigenvalue of A^{-1} with algebraic multiplicity $\alpha(\lambda)$ and corresponding eigenvector $\mathbf{x}$.
15. Let $\lambda_1, \lambda_2, \ldots, \lambda_s$ be distinct eigenvalues of A. For each $i = 1, 2, \ldots, s$ let $\mathbf{x}_{i1}, \mathbf{x}_{i2}, \ldots, \mathbf{x}_{ir_i}$ be linearly independent eigenvectors corresponding to λ_i. Then the vectors $\mathbf{x}_{11}, \ldots, \mathbf{x}_{1r_1}, \mathbf{x}_{21}, \ldots, \mathbf{x}_{2r_2}, \ldots, \mathbf{x}_{s1}, \ldots, \mathbf{x}_{sr_s}$ are linearly independent.
16. [FIS89] Let T be a a linear operator on a finite dimensional vector space over a field F, with basis $\mathcal{B}$. Then $\lambda \in F$ is an eigenvalue of T if and only if λ is an eigenvalue of $[T]_{\mathcal{B}}$.
17. [FIS89] Let $\lambda_1, \lambda_2, \ldots, \lambda_s$ be distinct eigenvalues of the linear operator T, on a finite dimensional space V. For each $i = 1, 2, \ldots, s$ let $\mathbf{x}_{i1}, \mathbf{x}_{i2}, \ldots, \mathbf{x}_{ir_i}$ be linearly independent eigenvectors corresponding to λ_i. Then the vectors $\mathbf{x}_{11}, \ldots, \mathbf{x}_{1r_1}, \mathbf{x}_{21}, \ldots, \mathbf{x}_{2r_2}, \ldots, \mathbf{x}_{s1}, \ldots, \mathbf{x}_{sr_s}$ are linearly independent.
18. Let T be linear operator on a finite dimensional vector space V over a field F. Then $\lambda \in F$ is an eigenvalue of T if and only if $p_T(\lambda) = 0$.

Diagonalization

19. [Lew91, pp. 135–136] Let $\lambda_1, \lambda_2, \ldots, \lambda_s$ be distinct eigenvalues of A. If $A \in \mathbb{C}^{n\times n}$, then A is diagonalizable if and only if $\alpha(\lambda_i) = \gamma(\lambda_i)$ for $i = 1, 2, \ldots, s$. If $A \in \mathbb{R}^{n\times n}$, then A is diagonalizable by a nonsingular matrix $B \in \mathbb{R}^{n\times n}$ if and only if all the eigenvalues of A are real and $\alpha(\lambda_i) = \gamma(\lambda_i)$ for $i = 1, 2, \ldots, s$.
20. Method for Diagonalization of A over $\mathbb{C}$: This is a theoretical method using exact arithmetic and is undesirable in decimal arithmetic with rounding errors. See Chapter 43 for information on appropriate numerical methods.

    - Find the eigenvalues of A.
    - Find a basis $\mathbf{x}_{i1}, \ldots, \mathbf{x}_{ir_i}$ for $E_{\lambda_i}(A)$ for each of the distinct eigenvalues $\lambda_1, \ldots, \lambda_k$ of A.
    - If $r_1 + \cdots + r_k = n$, then let $B = [\mathbf{x}_{11} \ldots \mathbf{x}_{1r_1} \ldots \mathbf{x}_{k1} \ldots \mathbf{x}_{kr_k}]$. B is invertible and $D = B^{-1}AB$ is a diagonal matrix, whose diagonal entries are the eigenvalues of A, in the order that corresponds to the order of the columns of B. Else A is not diagonalizable.

21. A is diagonalizable if and only if A has n linearly independent eigenvectors.

22. A is diagonalizable if and only if $|\sigma(A)| = n$ and A is nondefective.
23. If A has n distinct eigenvalues, then A is diagonalizable.
24. A is diagonalizable if and only if $q_A(x)$ can be factored into distinct linear factors.
25. If A is diagonalizable, then so are $A^T, A^k, k \in \mathbb{N}$.
26. If A is nonsingular and diagonalizable, then A^{-1} is diagonalizable.
27. If A is an idempotent, then A is diagonalizable and $\sigma(A) \subseteq \{0, 1\}$.
28. If A is nilpotent, then $\sigma(A) = \{0\}$. If A is nilpotent and is not the zero matrix, then A is not diagonalizable.
29. [FIS89] Let T be a linear operator on a finite dimensional vector space V with a basis $\mathcal{B}$. Then T is diagonalizable if and only if $[T]_{\mathcal{B}}$ is diagonalizable.
30. [FIS89] A linear operator, T, on a finite dimensional vector space V is diagonalizable if and only if there exists a basis $\mathcal{B} = \{\mathbf{v_1}, \ldots, \mathbf{v_n}\}$ for V, and scalars $\lambda_1, \ldots, \lambda_n$, such that $T(\mathbf{v_i}) = \lambda_i \mathbf{v_i}$, for $1 \leq i \leq n$.
31. [FIS89] If a linear operator, T, on a vector space, V, of dimension n, has n distinct eigenvalues, then it is diagonalizable.
32. [FIS89] The characteristic polynomial of a diagonalizable linear operator on a finite dimensional vector space can be factored into linear terms.

Polynomials

33. [HJ85] (Cayley–Hamilton Theorem) Let $p_A(x) = x^n + a_{n-1}x^{n-1} + \cdots + a_1 x + a_0$ be the characteristic polynomial of A. Then $p_A(A) = A^n + a_{n-1}A^{n-1} + \cdots + a_1 A + a_0 I_n = 0$.
34. [FIS89] (Cayley–Hamilton Theorem for a Linear Operator) Let $p_T(x) = x^n + a_{n-1}x^{n-1} + \cdots + a_1 x + a_0$ be the characteristic polynomial of a linear operator, T, on a finite dimensional vector space, V. Then $p_T(T) = T^n + a_{n-1}T^{n-1} + \cdots + a_1 T + a_0 I_n = T_0$, where T_0 is the zero linear operator on V.
35. $p_{A^T}(x) = p_A(x)$.
36. The minimal polynomial $q_A(x)$ of a matrix A is a factor of the characteristic polynomial $p_A(x)$ of A.
37. If λ is an eigenvalue of A associated with the eigenvector $\mathbf{x}$, then $p(\lambda)$ is an eigenvalue of the matrix $p(A)$ associated with the eigenvector $\mathbf{x}$, where $p(x)$ is a polynomial with coefficients in F.
38. If B is nonsingular, $p_A(x) = p_{B^{-1}AB}(x)$, therefore, A and $B^{-1}AB$ have the same eigenvalues.
39. Let $p_A(x) = x^n + a_{n-1}x^{n-1} + \cdots + a_1 x + a_0$ be the characteristic polynomial of A. If $|\sigma(A)| = n$, then $a_k = (-1)^{n-k} S_{n-k}(\lambda_1, \ldots, \lambda_n), k = 0, 1, \ldots, n-1$, where $S_k(\lambda_1, \ldots, \lambda_n)$ is the kth symmetric function of the eigenvalues of A.
40. Let $p_A(x) = x^n + a_{n-1}x^{n-1} + \cdots + a_1 x + a_0$ be the characteristic polynomial of A. Then $a_k = (-1)^{n-k} S_{n-k}(A), k = 0, 1, \ldots, n-1$.
41. If $|\sigma(A)| = n$, then $S_k(A) = S_k(\lambda_1, \ldots, \lambda_n)$.
42. If $C(p)$ is the companion matrix of the polynomial $p(x)$, then $p(x) = p_{C(p)}(x) = q_{C(p)}(x)$.
43. [HJ85, p. 135] If $|\sigma(A)| = n$, A is nonderogatory, and B commutes with A, then there exists a polynomial $f(x)$ of degree less than n such that $B = f(A)$.

Other Facts:

44. If A is nonsingular and λ is an eigenvalue of A of algebraic multiplicity $\alpha(\lambda)$, with corresponding eigenvector $\mathbf{x}$, then $\det(A)\lambda^{-1}$ is an eigenvalue of adj A with algebraic multiplicity $\alpha(\lambda)$ and corresponding eigenvector $\mathbf{x}$.
45. [Lew91] If $\lambda \in \sigma(A)$, then any nonzero column of adj $(A - \lambda I)$ is an eigenvector of A corresponding to λ.
46. If $AB = BA$, then A and B have a common eigenvector.
47. If $A \in F^{m \times n}$ and $B \in F^{n \times m}$, then $\sigma(AB) = \sigma(BA)$ except for the zero eigenvalues.

48. If $A \in F^{m\times m}$ and $B \in F^{n\times n}$, $\lambda \in \sigma(A)$, $\mu \in \sigma(B)$, with corresponding eigenvectors $\mathbf{u}$ and $\mathbf{v}$, respectively, then $\lambda\mu \in \sigma(A \otimes B)$, with corresponding eigenvector $\mathbf{u} \otimes \mathbf{v}$. (See Section 10.5 for the definition of $A \otimes B$.)

Examples:

1. Let $A = \begin{bmatrix} 0 & -1 \\ 1 & 0 \end{bmatrix}$. Then, viewing $A \in \mathbb{C}^{n\times n}$, $\sigma(A) = \{-i, i\}$. That is, A has no eigenvalues over the reals.

2. Let $A = \begin{bmatrix} -3 & 7 & -1 \\ 6 & 8 & -2 \\ 72 & -28 & 19 \end{bmatrix}$. Then $p_A(x) = (x+6)(x-15)^2 = q_A(x)$, $\lambda_1 = -6$, $\alpha(\lambda_1) = 1$, $\gamma(\lambda_1) = 1$, $\lambda_2 = 15$, $\alpha(\lambda_2) = 2$, $\gamma(\lambda_2) = 1$. Also, a set of linearly independent eigenvectors is $\left\{ \begin{bmatrix} -1 \\ -2 \\ 4 \end{bmatrix}, \begin{bmatrix} -1 \\ 1 \\ 4 \end{bmatrix} \right\}$. So, A is not diagonalizable.

3. Let $A = \begin{bmatrix} 57 & -21 & 21 \\ -14 & 22 & -7 \\ -140 & 70 & -55 \end{bmatrix}$. Then $p_A(x) = (x+6)(x-15)^2$, $q_A(x) = (x+6)(x-15)$, $\lambda_1 = -6$, $\alpha(\lambda_1) = 1$, $\gamma(\lambda_1) = 1$, $\lambda_2 = 15$, $\alpha(\lambda_2) = 2$, $\gamma(\lambda_2) = 2$. Also, a set of linearly independent eigenvectors is $\left\{ \begin{bmatrix} -1 \\ 0 \\ 2 \end{bmatrix}, \begin{bmatrix} 1 \\ 2 \\ 0 \end{bmatrix}, \begin{bmatrix} -3 \\ 1 \\ 10 \end{bmatrix} \right\}$. So, A is diagonalizable.

4. Let $A = \begin{bmatrix} -5+4i & 1 & i \\ 2+8i & -4 & 2i \\ 20-4i & -4 & -i \end{bmatrix}$. Then $\sigma(A) = \{-6, -3, 3i\}$. If $B = \frac{1}{9}A^2 + 2A - 4I$, then $\sigma(B) = \{-12, -9, -5+6i\}$.

5. Let $A = \begin{bmatrix} -2 & 1 & 0 \\ 0 & -3i & 0 \\ 0 & 0 & 1 \end{bmatrix}$ and $B = \begin{bmatrix} -2 & 1 & 0 \\ 3 & -1 & 1 \\ 4 & 0 & 1 \end{bmatrix}$. B is nonsingular, so A and $B^{-1}AB = \begin{bmatrix} -1+3i & 1-i & i \\ 5+6i & -1-2i & 1+2i \\ 8-12i & -4+4i & 1-4i \end{bmatrix}$ have the same eigenvalues, which are given in the diagonal of A.

6. Let $A = \begin{bmatrix} 2 & -1 & 0 \\ 0 & 3 & 1 \end{bmatrix}$ and $B = \begin{bmatrix} 3 & 0 \\ 2 & 1 \\ 1 & 0 \end{bmatrix}$. Then $AB = \begin{bmatrix} 4 & -1 \\ 7 & 3 \end{bmatrix}$, $\sigma(AB) = \left\{\frac{1}{2}(7+3\sqrt{3}i), \frac{1}{2}(7-3\sqrt{3}i)\right\}$, $BA = \begin{bmatrix} 6 & -3 & 0 \\ 4 & 1 & 1 \\ 2 & -1 & 0 \end{bmatrix}$, and $\sigma(BA) = \left\{\frac{1}{2}(7+3\sqrt{3}i), \frac{1}{2}(7-3\sqrt{3}i), 0\right\}$.

7. Let $A = \begin{bmatrix} 1 & 1 & 0 \\ 0 & 1 & 0 \\ 0 & 0 & -3i \end{bmatrix}$ and $B = \begin{bmatrix} -2 & 7 & 0 \\ 0 & -2 & 0 \\ 0 & 0 & i \end{bmatrix}$. Then $AB = BA = \begin{bmatrix} -2 & 5 & 0 \\ 0 & -2 & 0 \\ 0 & 0 & 3 \end{bmatrix}$. A and B share the eigenvector $\mathbf{x} = \begin{bmatrix} 1 \\ 0 \\ 0 \end{bmatrix}$, corresponding to the eigenvalues $\lambda = 1$ of A and $\mu = -2$ of B.

8. Let $A = \begin{bmatrix} 1 & 1 & 0 & 0 \\ 0 & 1 & 0 & 1 \\ 0 & 0 & -3 & 2 \\ 0 & -1 & -2 & -4 \end{bmatrix}$. Then $p_A(x) = x^4 + 5x^3 + 4x^2 - 23x + 13$. $S_4(A), S_3(A)$, and $S_1(A)$ were computed in Example 1 of Section 4.2, and it is straightforward to verify that $S_2(A) = 4$. Comparing these values to the characteristic polynomial, $S_4(A) = 13 = (-1)^4 13$, $S_3(A) = 23 = (-1)^3(-23)$, $S_2(A) = (-1)^2 4$, and $S_1(A) = (-1)(5)$. It follows that $S_4(\lambda_1, \lambda_2, \lambda_3, \lambda_4) = \lambda_1\lambda_2\lambda_3\lambda_4 = 13$, $S_3(\lambda_1, \lambda_2, \lambda_3, \lambda_4) = 23$, $S_2(\lambda_1, \lambda_2, \lambda_3, \lambda_4) = 4$, and $S_1(\lambda_1, \lambda_2, \lambda_3, \lambda_4) = \lambda_1 + \lambda_2 + \lambda_3 + \lambda_4 = -5$ (these values can also be verified with a computer algebra system or numerical software).

9. Let $p(x) = x^3 - 7x^2 - 3x + 2$, $C = C(p) = \begin{bmatrix} 0 & 0 & -2 \\ 1 & 0 & 3 \\ 0 & 1 & 7 \end{bmatrix}$. Then $p_C(x) = x^3 - 7x^2 - 3x + 2 = p(x)$. Also, $p_C(C) = -C^3 + 7C^2 + 3C - 2I = -\begin{bmatrix} -2 & -14 & -104 \\ 3 & 19 & 142 \\ 7 & 52 & 383 \end{bmatrix} + 7\begin{bmatrix} 0 & -2 & -14 \\ 0 & 3 & 19 \\ 1 & 7 & 52 \end{bmatrix}$

$$+3\begin{bmatrix} 0 & 0 & -2 \\ 1 & 0 & 3 \\ 0 & 1 & 7 \end{bmatrix} - 2\begin{bmatrix} 1 & 0 & 0 \\ 0 & 1 & 0 \\ 0 & 0 & 1 \end{bmatrix} = \begin{bmatrix} 0 & 0 & 0 \\ 0 & 0 & 0 \\ 0 & 0 & 0 \end{bmatrix}.$$

Applications:

1. (Markov Chains) (See also Chapter 54 for more information.) A Markov Chain describes a process in which a system can be in any one of n states: $s_1, s_2, \ldots, s_n$. The probability of entering state s_i depends only on the state previously occupied by the system. The **transition probability** of entering state j, given that the system is in state i, is denoted by p_{ij}. The **transition matrix** is the matrix $P = [p_{ij}]$; its rows have sum 1. A (row or column) vector is a **probability vector** if its entries are nonnegative and sum to 1. The probabilty row vector $\pi^{(k)} = (\pi_1^{(k)}, \pi_2^{(k)}, \ldots \pi_n^{(k)})$, $k \geq 0$, is called the **state vector** of the system at time k if its ith entry is the probability that the system is in state s_i at time k. In particular, when $k = 0$, the state vector is called the **initial state vector** and its ith entry is the probability that the system begins at state s_i. It follows from probability theory that $\pi^{(k+1)} = \pi^{(k)} P$, and thus inductively that $\pi^{(k)} = \pi^{(0)} P^k$. If the entries of some power of P are all positive, then P is said to be **regular**. If P is a regular transition matrix, then as $n \to \infty$, $P^n \to \begin{bmatrix} \pi_1 & \pi_2 & \cdots & \pi_n \\ \pi_1 & \pi_2 & \cdots & \pi_n \\ \vdots & \vdots & \ddots & \vdots \\ \pi_1 & \pi_2 & \cdots & \pi_n \end{bmatrix}$. The row vector $\pi = (\pi_1, \pi_2, \ldots, \pi_n)$ is called the **steady state vector**, π is a probability vector, and as $n \to \infty$, $\pi^{(n)} \to \pi$. The vector π is the unique probability row vector with the property that $\pi P = \pi$. That is, π is the unique probability row vector that is a left eigenvector of P for eigenvalue 1.
2. (Differential Equations) [Mey00, pp. 541–546] Consider the system of linear differential equations $\begin{cases} x_1' = a_{11}x_1 + a_{12}x_2 + \ldots + a_{1n}x_n \\ x_2' = a_{21}x_1 + a_{22}x_2 + \ldots + a_{2n}x_n \\ \vdots \\ x_n' = a_{n1}x_1 + a_{n2}x_2 + \ldots + a_{nn}x_n \end{cases}$, where each of the unknowns $x_1, x_2, \ldots, x_n$

is a differentiable function of the real variable t. This system of linear differential equations can be written in matrix form as $\mathbf{x}' = A\mathbf{x}$, where $A = [a_{ij}]$, $\mathbf{x} = \begin{bmatrix} x_1(t) \\ x_2(t) \\ \vdots \\ x_n(t) \end{bmatrix}$, and $\mathbf{x}' = \begin{bmatrix} x_1'(t) \\ x_2'(t) \\ \vdots \\ x_n'(t) \end{bmatrix}$. If A is diagonalizable, there exists a nonsingular matrix B (the columns of the matrix B are linearly independent eigenvectors of A), such that $B^{-1}AB = D$ is a diagonal matrix, so $\mathbf{x}' = BDB^{-1}\mathbf{x}$, or $B^{-1}\mathbf{x}' = DB^{-1}\mathbf{x}$. Let $\mathbf{u} = B^{-1}\mathbf{x}$. The linear system of differential equations $\mathbf{u}' = D\mathbf{u}$ has solution $\mathbf{u} = \begin{bmatrix} k_1 e^{\lambda_1 t} \\ k_2 e^{\lambda_2 t} \\ \vdots \\ k_n e^{\lambda_n t} \end{bmatrix}$, where $\lambda_1, \lambda_2, \ldots, \lambda_n$ are the eigenvalues of A. It follows that $\mathbf{x} = B\mathbf{u}$. (See also Chapter 55.)

3. (Dynamical Systems) [Lay03, pp. 315–316] Consider the dynamical system given by $\mathbf{u}_{k+1} = A\mathbf{u}_k$, where $A = [a_{ij}]$, $\mathbf{u}_0 = \begin{bmatrix} a_1 \\ a_2 \\ \vdots \\ a_n \end{bmatrix}$. If A is diagonalizable, there exist n linearly independent eigenvectors, $\mathbf{x}_1, \mathbf{x}_2, \ldots, \mathbf{x}_n$, of A. The vector $\mathbf{u}_0$ can then be written as a linear combination of the eigenvectors, that is, $\mathbf{u}_0 = c_1\mathbf{x}_1 + c_2\mathbf{x}_2 + \cdots + c_n\mathbf{x}_n$. Then $\mathbf{u}_1 = A\mathbf{u}_0 = A(c_1\mathbf{x}_1 + c_2\mathbf{x}_2 + \cdots + c_n\mathbf{x}_n) = c_1\lambda_1\mathbf{x}_1 + c_2\lambda_2\mathbf{x}_2 + \cdots + c_n\lambda_n\mathbf{x}_n$. Inductively, $\mathbf{u}_{k+1} = A\mathbf{u}_k = c_1\lambda_1^{k+1}\mathbf{x}_1 + c_2\lambda_2^{k+1}\mathbf{x}_2 + \cdots + c_n\lambda_n^{k+1}\mathbf{x}_n$. Thus, the long-term behavior of the dynamical system can be studied using the eigenvalues of the matrix A. (See also Chapter 56.)

References

[Ait56] A. C. Aitken. *Determinants and Matrices*, 9th ed. Oliver and Boyd, Edinburgh, 1956.

[FB90] J. B. Fraleigh and R. A. Beauregard. *Linear Algebra*, 2nd ed. Addison-Wesley, Reading, PA, 1990.

[FIS89] S. H. Friedberg, A. J. Insel, and L. E. Spence. *Linear Algebra*, 2nd ed. Prentice Hall, Upper Saddle River, NJ, 1989.

[Goo03] E. G. Goodaire. *Linear Algebra a Pure and Applied First Course.* Prentice Hall, Upper Saddle River, NJ, 2003.

[HJ85] R. Horn and C. R. Johnson. *Matrix Analysis.* Cambridge University Press, Cambridge, 1985.

[Lay03] D. C. Lay. *Linear Algebra and Its Applications*, 3rd ed. Addison-Wesley, Boston, 2003.

[Lew91] D. W. Lewis. *Matrix Theory.* Word Scientific, Singapore, 1991.

[Mey00] C. D. Meyer. *Matrix Analysis and Applied Linear Algebra.* SIAM, Philadelphia, 2000.

[Uhl02] F. Uhlig. *Transform Linear Algebra.* Prentice Hall, Upper Saddle River, NJ, 2002.

5

Inner Product Spaces, Orthogonal Projection, Least Squares, and Singular Value Decomposition

Lixing Han
University of Michigan-Flint

Michael Neumann
University of Connecticut

5.1 Inner Product Spaces

Definitions:

Let V be a vector space over the field F, where $F = \mathbb{R}$ or $F = \mathbb{C}$. An **inner product** on V is a function $\langle \cdot, \cdot \rangle\colon V \times V \to F$ such that for all $\mathbf{u}, \mathbf{v}, \mathbf{w} \in V$ and $a, b \in F$, the following hold:

- $\langle \mathbf{v}, \mathbf{v} \rangle \geq 0$ and $\langle \mathbf{v}, \mathbf{v} \rangle = 0$ if and only if $\mathbf{v} = \mathbf{0}$.
- $\langle a\mathbf{u} + b\mathbf{v}, \mathbf{w} \rangle = a\langle \mathbf{u}, \mathbf{w} \rangle + b\langle \mathbf{v}, \mathbf{w} \rangle$.
- For $F = \mathbb{R}$: $\langle \mathbf{u}, \mathbf{v} \rangle = \langle \mathbf{v}, \mathbf{u} \rangle$; For $F = \mathbb{C}$: $\langle \mathbf{u}, \mathbf{v} \rangle = \overline{\langle \mathbf{v}, \mathbf{u} \rangle}$ (where bar denotes complex conjugation).

A real (or complex) **inner product space** is a vector space V over $\mathbb{R}$ (or $\mathbb{C}$), together with an inner product defined on it.

In an inner product space V, the **norm**, or **length**, of a vector $\mathbf{v} \in V$ is $\|\mathbf{v}\| = \sqrt{\langle \mathbf{v}, \mathbf{v} \rangle}$.

A vector $\mathbf{v} \in V$ is a **unit vector** if $\|\mathbf{v}\| = 1$.

The **angle** between two nonzero vectors $\mathbf{u}$ and $\mathbf{v}$ in a real inner product space is the real number θ, $0 \leq \theta \leq \pi$, such that $\langle \mathbf{u}, \mathbf{v} \rangle = \|\mathbf{u}\| \|\mathbf{v}\| \cos\theta$. See the Cauchy–Schwarz inequality (Fact 9 below).

Let V be an inner product space. The **distance** between two vectors $\mathbf{u}$ and $\mathbf{v}$ is $d(\mathbf{u}, \mathbf{v}) = \|\mathbf{u} - \mathbf{v}\|$.

A Hermitian matrix A is **positive definite** if $\mathbf{x}^* A\mathbf{x} > 0$ for all nonzero $\mathbf{x} \in \mathbb{C}^n$. (See Chapter 8 for more information on positive definite matrices.)

Facts:

All the following facts except those with a specific reference can be found in [Rom92, pp. 157–164].

1. The vector space $\mathbb{R}^n$ is an inner product space under the **standard inner product**, or **dot product**, defined by

$$\langle \mathbf{u}, \mathbf{v}\rangle = \mathbf{u}^T\mathbf{v} = \sum_{i=1}^{n} u_i v_i.$$

This inner product space is often called n–dimensional **Euclidean space.**

2. The vector space $\mathbb{C}^n$ is an inner product space under the **standard inner product**, defined by

$$\langle \mathbf{u}, \mathbf{v}\rangle = \mathbf{v}^*\mathbf{u} = \sum_{i=1}^{n} u_i \bar{v}_i.$$

This inner product space is often called n-dimensional **unitary space.**

3. [HJ85, p. 410] In $\mathbb{R}^n$, a function $\langle\cdot,\cdot\rangle\colon \mathbb{R}^n \times \mathbb{R}^n \to \mathbb{R}$ is an inner product if and only if there exists a real symmetric positive definite matrix G such that $\langle \mathbf{u}, \mathbf{v}\rangle = \mathbf{u}^T G\mathbf{v}$, for all $\mathbf{u}, \mathbf{v} \in \mathbb{R}^n$.
4. [HJ85, p. 410] In $\mathbb{C}^n$, a function $\langle\cdot,\cdot\rangle\colon \mathbb{C}^n \times \mathbb{C}^n \to \mathbb{C}$ is an inner product if and only if there exists a Hermitian positive definite matrix H such that $\langle \mathbf{u}, \mathbf{v}\rangle = \mathbf{v}^* H\mathbf{u}$, for all $\mathbf{u}, \mathbf{v} \in \mathbb{C}^n$.
5. Let l^2 be the vector space of all infinite complex sequences $\mathbf{v} = (v_n)$ with the property that $\sum_{n=1}^{\infty} |v_n|^2 < \infty$. Then l^2 is an inner product space under the inner product

$$\langle \mathbf{u}, \mathbf{v}\rangle = \sum_{n=1}^{\infty} u_n \bar{v}_n.$$

6. The vector space $C[a, b]$ of all continuous real-valued functions on the closed interval $[a, b]$ is an inner product space under the inner product

$$\langle f, g\rangle = \int_a^b f(x)g(x)dx.$$

7. If V is an inner product space and $\langle \mathbf{u}, \mathbf{w}\rangle = \langle \mathbf{v}, \mathbf{w}\rangle$ for all $\mathbf{w} \in V$, then $\mathbf{u} = \mathbf{v}$.
8. The inner product on an inner product space V, when restricted to vectors in a subspace S of V, is an inner product on S.
9. Let V be an inner product space. Then the norm function $\|\cdot\|$ on V has the following basic properties for all $\mathbf{u}, \mathbf{v} \in V$:
   - $\|\mathbf{v}\| \geq 0$ and $\|\mathbf{v}\| = 0$ if and only if $\mathbf{v} = \mathbf{0}$.
   - $\|a\mathbf{v}\| = |a|\|\mathbf{v}\|$, for all $a \in F$.
   - (The triangle inequality) $\|\mathbf{u} + \mathbf{v}\| \leq \|\mathbf{u}\| + \|\mathbf{v}\|$ with equality if and only if $\mathbf{v} = a\mathbf{u}$, for some $a \in F$.
   - (The Cauchy–Schwarz inequality) $|\langle \mathbf{u}, \mathbf{v}\rangle| \leq \|\mathbf{u}\|\|\mathbf{v}\|$ with equality if and only if $\mathbf{v} = a\mathbf{u}$, for some $a \in F$.
   - $|\|\mathbf{u}\| - \|\mathbf{v}\|| \leq \|\mathbf{u} - \mathbf{v}\|$.

- (The parallelogram law) $\|\mathbf{u}+\mathbf{v}\|^2 + \|\mathbf{u}-\mathbf{v}\|^2 = 2\|\mathbf{u}\|^2 + 2\|\mathbf{v}\|^2$.
- (Polarization identities)

$$4\langle \mathbf{u}, \mathbf{v}\rangle = \begin{cases} \|\mathbf{u}+\mathbf{v}\|^2 - \|\mathbf{u}-\mathbf{v}\|^2, \text{ if } F = \mathbb{R}. \\ \|\mathbf{u}+\mathbf{v}\|^2 - \|\mathbf{u}-\mathbf{v}\|^2 + i\|\mathbf{u}+i\mathbf{v}\|^2 - i\|\mathbf{u}-i\mathbf{v}\|^2, \text{ if } F = \mathbb{C}. \end{cases}$$

Examples:

1. Let $\mathbb{R}^4$ be the Euclidean space with the inner product $\langle \mathbf{u}, \mathbf{v}\rangle = \mathbf{u}^T\mathbf{v}$. Let $\mathbf{x} = [1, 2, 3, 4]^T \in \mathbb{R}^4$ and $\mathbf{y} = [3, -1, 0, 2]^T \in \mathbb{R}^4$ be two vectors. Then
   - $\langle \mathbf{x}, \mathbf{y}\rangle = 9$, $\|\mathbf{x}\| = \sqrt{30}$, and $\|\mathbf{y}\| = \sqrt{14}$.
   - The distance between $\mathbf{x}$ and $\mathbf{y}$ is $d(\mathbf{x}, \mathbf{y}) = \|\mathbf{x} - \mathbf{y}\| = \sqrt{26}$.
   - The angle between $\mathbf{x}$ and $\mathbf{y}$ is $\theta = \arccos \dfrac{9}{\sqrt{30}\sqrt{14}} = \arccos \dfrac{9}{2\sqrt{105}} \approx 1.116$ radians.
2. $\langle \mathbf{u}, \mathbf{v}\rangle = u_1v_1 + 2u_1v_2 + 2u_2v_1 + 6u_2v_2 = \mathbf{u}^T \begin{bmatrix} 1 & 2 \\ 2 & 6 \end{bmatrix} \mathbf{v}$ is an inner product on $\mathbb{R}^2$, as the matrix $G = \begin{bmatrix} 1 & 2 \\ 2 & 6 \end{bmatrix}$ is symmetric positive definite.
3. Let $C[-1, 1]$ be the vector space with the inner product $\langle f, g\rangle = \int_{-1}^{1} f(x)g(x)dx$ and let $f(x) = 1$ and $g(x) = x^2$ be two functions in $C[-1, 1]$. Then $\langle f, g\rangle = \int_{-1}^{1} x^2 dx = 2/3$, $\langle f, f\rangle = \int_{-1}^{1} 1 dx = 2$, and $\langle g, g\rangle = \int_{-1}^{1} x^4 dx = 2/5$. The angle between f and g is $\arccos(\sqrt{5}/3) \approx 0.730$ radians.
4. [Mey00, p. 286] $\langle A, B\rangle = \operatorname{tr}(AB^*)$ is an inner product on $\mathbb{C}^{m\times n}$.

5.2 Orthogonality

Definitions:

Let V be an inner product space. Two vectors $\mathbf{u}, \mathbf{v} \in V$ are **orthogonal** if $\langle \mathbf{u}, \mathbf{v}\rangle = 0$, and this is denoted by $\mathbf{u} \perp \mathbf{v}$.

A subset S of an inner product space V is an **orthogonal set** if $\mathbf{u} \perp \mathbf{v}$, for all $\mathbf{u}, \mathbf{v} \in S$ such that $\mathbf{u} \neq \mathbf{v}$.

A subset S of an inner product space V is an **orthonormal set** if S is an orthogonal set and each $\mathbf{v} \in S$ is a unit vector.

Two subsets S and W of an inner product space V are **orthogonal** if $\mathbf{u} \perp \mathbf{v}$, for all $\mathbf{u} \in S$ and $\mathbf{v} \in W$, and this is denoted by $S \perp W$.

The **orthogonal complement** of a subset S of an inner product space V is $S^\perp = \{\mathbf{w} \in V | \langle \mathbf{w}, \mathbf{v}\rangle = 0 \text{ for all } \mathbf{v} \in S\}$.

A **complete orthonormal set** M in an inner product space V is an orthonormal set of vectors in V such that for $\mathbf{v} \in V$, $\mathbf{v} \perp M$ implies that $\mathbf{v} = \mathbf{0}$.

An **orthogonal basis** for an inner product space V is an orthogonal set that is also a basis for V.

An **orthonormal basis** for V is an orthonormal set that is also a basis for V.

A matrix U is **unitary** if $U^*U = I$.

A real matrix Q is **orthogonal** if $Q^TQ = I$.

Facts:

1. [Mey00, p. 298] An orthogonal set of nonzero vectors is linearly independent. An orthonormal set of vectors is linearly independent.
2. [Rom92, p. 164] If S is a subset of an inner product space V, then $S^\perp$ is a subspace of V. Moreover, if S is a subspace of V, then $S \cap S^\perp = \{\mathbf{0}\}$.
3. [Mey00, p. 409] In an inner product space V, $\{\mathbf{0}\}^\perp = V$ and $V^\perp = \{\mathbf{0}\}$.
4. [Rom92, p. 168] If S is a finite dimensional subspace of an inner product space V, then for any $\mathbf{v} \in V$,
   - There are unique vectors $\mathbf{s} \in S$ and $\mathbf{t} \in S^\perp$ such that $\mathbf{v} = \mathbf{s} + \mathbf{t}$. This implies $V = S \oplus S^\perp$.
   - There is a unique linear operator P such that $P(\mathbf{v}) = \mathbf{s}$.
5. [Mey00, p. 404] If S is a subspace of an $n-$dimensional inner product space V, then
   - $(S^\perp)^\perp = S$.
   - $\dim(S^\perp) = n - \dim(S)$.
6. [Rom92, p. 174] If S is a subspace of an infinite dimensional inner product space, then $S \subseteq (S^\perp)^\perp$, but the two sets need not be equal.
7. [Rom92, p. 166] An orthonormal basis is a complete orthonormal set.
8. [Rom92, p. 166] In a finite-dimensional inner product space, a complete orthonormal set is a basis.
9. [Rom92, p. 165] In an infinite-dimensional inner product space, a complete orthonormal set may not be a basis.
10. [Rom92, p. 166] Every finite-dimensional inner product space has an orthonormal basis.
11. [Mey00, p. 299] Let $\mathcal{B} = \{\mathbf{u}_1, \mathbf{u}_2, \ldots, \mathbf{u}_n\}$ be an orthonormal basis for V. Every vector $\mathbf{v} \in V$ can be uniquely expressed as

$$\mathbf{v} = \sum_{i=1}^{n} \langle \mathbf{v}, \mathbf{u}_i \rangle \mathbf{u}_i.$$

    The expression on the right is called the **Fourier expansion** of $\mathbf{v}$ with respect to $\mathcal{B}$ and the scalars $\langle \mathbf{v}, \mathbf{u}_i \rangle$ are called the **Fourier coefficients.**
12. [Mey00, p. 305] (Pythagorean Theorem) If $\{\mathbf{v}_i\}_{i=1}^k$ is an orthogonal set of vectors in V, then $\| \sum_{i=1}^k \mathbf{v}_i \|^2 = \sum_{i=1}^k \|\mathbf{v}_i\|^2$.
13. [Rom92, p. 167] (Bessel's Inequality) If $\{\mathbf{u}_i\}_{i=1}^k$ is an orthonormal set of vectors in V, then $\|\mathbf{v}\|^2 \geq \sum_{i=1}^k |\langle \mathbf{v}, \mathbf{u}_i \rangle|^2$.
14. [Mey00, p. 305] (Parseval's Identity) Let $\mathcal{B} = \{\mathbf{u}_1, \mathbf{u}_2, \ldots, \mathbf{u}_n\}$ be an orthonormal basis for V. Then for each $\mathbf{v} \in V$, $\|\mathbf{v}\|^2 = \sum_{i=1}^{n} |\langle \mathbf{v}, \mathbf{u}_i \rangle|^2$.
15. [Mey00, p. 405] Let $A \in F^{m\times n}$, where $F = \mathbb{R}$ or $\mathbb{C}$. Then
   - $\ker(A)^\perp = \text{range}(A^*)$, $\text{range}(A)^\perp = \ker(A^*)$.
   - $F^m = \text{range}(A) \oplus \text{range}(A)^\perp = \text{range}(A) \oplus \ker(A^*)$.
   - $F^n = \ker(A) \oplus \ker(A)^\perp = \ker(A) \oplus \text{range}(A^*)$.
16. [Mey00, p. 321] (See also Section 7.1.) The following statements for a real matrix $Q \in \mathbb{R}^{n\times n}$ are equivalent:
   - Q is orthogonal.
   - Q has orthonormal columns.
   - Q has orthonormal rows.
   - $QQ^T = I$, where I is the identity matrix of order n.
   - For all $\mathbf{v} \in \mathbb{R}^n$, $\|Q\mathbf{v}\| = \|\mathbf{v}\|$.

17. [Mey00, p. 321] (See also Section 7.1.) The following statements for a complex matrix $U \in \mathbb{C}^{n\times n}$ are equivalent:
    - U is unitary.
    - U has orthonormal columns.
    - U has orthonormal rows.
    - $UU^* = I$, where I is the identity matrix of order n.
    - For all $\mathbf{v} \in \mathbb{C}^n$, $\|U\mathbf{v}\| = \|\mathbf{v}\|$.

Examples:

1. Let $C[-1,1]$ be the vector space with the inner product $\langle f, g\rangle = \int_{-1}^{1} f(x)g(x)dx$ and let $f(x) = 1$ and $g(x) = x$ be two functions in $C[-1,1]$. Then $\langle f, g\rangle = \int_{-1}^{1} xdx = 0$. Thus, $f \perp g$.
2. The standard basis $\{\mathbf{e}_1, \mathbf{e}_2, \ldots, \mathbf{e}_n\}$ is an orthonormal basis for the unitary space $\mathbb{C}^n$.
3. If $\{\mathbf{v}_1, \mathbf{v}_2, \cdots, \mathbf{v}_n\}$ is an orthogonal basis for $\mathbb{C}^n$ and $S = \text{span}\,\{\mathbf{v}_1, \mathbf{v}_2, \cdots, \mathbf{v}_k\}$ $(1 \le k \le n-1)$, then $S^\perp = \text{span}\,\{\mathbf{v}_{k+1}, \cdots, \mathbf{v}_n\}$.
4. The vectors $\mathbf{v}_1 = [2,2,1]^T, \mathbf{v}_2 = [1,-1,0]^T$, and $\mathbf{v}_3 = [-1,-1,4]^T$ are mutually orthogonal. They can be normalized to $\mathbf{u}_1 = \mathbf{v}_1/\|\mathbf{v}_1\| = [2/3, 2/3, 1/3]^T$, $\mathbf{u}_2 = \mathbf{v}_2/\|\mathbf{v}_2\| = [1/\sqrt{2}, -1/\sqrt{2}, 0]^T$, and $\mathbf{u}_3 = \mathbf{v}_3/\|\mathbf{v}_3\| = [-\sqrt{2}/6, -\sqrt{2}/6, 2\sqrt{2}/3]^T$. The set $\mathbb{B} = \{\mathbf{u}_1, \mathbf{u}_2, \mathbf{u}_3\}$ forms an orthonormal basis for the Euclidean space $\mathbb{R}^3$.
    - If $\mathbf{v} = [v_1, v_2, v_3]^T \in \mathbb{R}^3$, then $\mathbf{v} = \langle \mathbf{v}, \mathbf{u}_1\rangle\mathbf{u}_1 + \langle \mathbf{v}, \mathbf{u}_2\rangle\mathbf{u}_2 + \langle \mathbf{v}, \mathbf{u}_3\rangle\mathbf{u}_3$, that is,
    $$\mathbf{v} = \frac{2v_1 + 2v_2 + v_3}{3}\mathbf{u}_1 + \frac{v_1 - v_2}{\sqrt{2}}\mathbf{u}_2 + \frac{-v_1 - v_2 + 4v_3}{3\sqrt{2}}\mathbf{u}_3.$$
    - The matrix $Q = [\mathbf{u}_1, \mathbf{u}_2, \mathbf{u}_3] \in \mathbb{R}^{3\times 3}$ is an orthogonal matrix.
5. Let S be the subspace of $\mathbb{C}^3$ spanned by the vectors $\mathbf{u} = [i, 1, 1]^T$ and $\mathbf{v} = [1, i, 1]^T$. Then the orthogonal complement of S is
$$S^\perp = \{\mathbf{w} | \mathbf{w} = \alpha[1, 1, -1+i]^T, \text{ where } \alpha \in \mathbb{C}\}.$$
6. Consider the inner product space l^2 from Fact 5 in Section 5.1. Let $\mathcal{E} = \{\mathbf{e}_i | i = 1, 2, \ldots\}$, where $\mathbf{e}_i$ has a 1 on ith place and 0s elsewhere. It is clear that $\mathcal{E}$ is an orthonormal set. If $\mathbf{v} = (v_n) \perp \mathcal{E}$, then for each n, $v_n = \langle \mathbf{v}, \mathbf{e}_n\rangle = 0$. This implies $\mathbf{v} = \mathbf{0}$. Therefore, $\mathcal{E}$ is a complete orthonormal set. However, $\mathcal{E}$ is not a basis for l^2 as $S = \text{span}\{\mathcal{E}\} \neq l^2$. Further, $S^\perp = \{\mathbf{0}\}$. Thus, $(S^\perp)^\perp = l^2 \not\subseteq S$ and $l^2 \neq S \oplus S^\perp$.

5.3 Adjoints of Linear Operators on Inner Product Spaces

Let V be a finite dimensional (real or complex) inner product space and let T be a linear operator on V.

Definitions:

A linear operator T^* on V is called the **adjoint** of T if $\langle T(\mathbf{u}), \mathbf{v}\rangle = \langle \mathbf{u}, T^*(\mathbf{v})\rangle$ for all $\mathbf{u}, \mathbf{v} \in V$.

The linear operator T is **self-adjoint**, or **Hermitian**, if $T = T^*$; T is **unitary** if $T^*T = I_V$.

Facts:

The following facts can be found in [HK71].

1. Let f be a linear functional on V. Then there exists a unique $\mathbf{v} \in V$ such that $f(\mathbf{w}) = \langle \mathbf{w}, \mathbf{v} \rangle$ for all $\mathbf{w} \in V$.
2. The adjoint T^* of T exists and is unique.
3. Let $\mathcal{B} = (\mathbf{u_1}, \mathbf{u_2}, \ldots, \mathbf{u_n})$ be an ordered, orthonormal basis of V. Let $A = [T]_{\mathcal{B}}$. Then

$$a_{ij} = \langle T(\mathbf{u_j}), \mathbf{u_i} \rangle, \quad i, j = 1, 2, \ldots, n.$$

   Moreover, $[T^*]_{\mathcal{B}} = A^*$, the Hermitian adjoint of A.
4. (Properties of the adjoint operator)
   (a) $(T^*)^* = T$ for every linear operator T on V.
   (b) $(aT)^* = \bar{a}T^*$ for every linear operator T on V and every $a \in F$.
   (c) $(T + T_1)^* = T^* + T_1{}^*$ for every linear operators T, T_1 on V.
   (d) $(TT_1)^* = T_1{}^* T^*$ for every linear operators T, T_1 on V.
5. Let $\mathcal{B}$ be an ordered orthonormal basis of V and let $A = [T]_{\mathcal{B}}$. Then
   (a) T is self-adjoint if and only if A is a Hermitian matrix.
   (b) T is unitary if and only if A is a unitary matrix.

Examples:

1. Consider the space $\mathbb{R}^3$ equipped with the standard inner product and let $f(\mathbf{w}) = 3w_1 - 2w_3$. Then with $\mathbf{v} = [3, 0, -2]^T$, $f(\mathbf{w}) = \langle \mathbf{w}, \mathbf{v} \rangle$.
2. Consider the space $\mathbb{R}^3$ equipped with the standard inner product. Let $\mathbf{v} = \begin{bmatrix} x \\ y \\ z \end{bmatrix}$ and $T(\mathbf{v}) = \begin{bmatrix} 2x + y \\ y - 3z \\ x + y + z \end{bmatrix}$. Then $[T] = \begin{bmatrix} 2 & 1 & 0 \\ 0 & 1 & -3 \\ 1 & 1 & 1 \end{bmatrix}$, so $[T]^* = \begin{bmatrix} 2 & 0 & 1 \\ 1 & 1 & 1 \\ 0 & -3 & 1 \end{bmatrix}$, and $T^*(\mathbf{v}) = \begin{bmatrix} 2x + z \\ x + y + z \\ -3y + z \end{bmatrix}$.
3. Consider the space $\mathbb{C}^{n \times n}$ equipped with the inner product in Example 4 of section 5.1. Let $A, B \in \mathbb{C}^{n \times n}$ and let T be the linear operator on $\mathbb{C}^{n \times n}$ defined by $T(X) = AX + XB,\ X \in \mathbb{C}^{n \times n}$. Then $T^*(X) = A^*X + XB^*,\ X \in \mathbb{C}^{n \times n}$.
4. Let V be an inner product space and let T be a linear operator on V. For a fixed $\mathbf{u} \in V$, $f(\mathbf{w}) = \langle T(\mathbf{w}), \mathbf{u} \rangle$ is a linear functional. By Fact 1, there is a unique vector $\mathbf{v}$ such that $f(\mathbf{w}) = \langle \mathbf{w}, \mathbf{v} \rangle$. Then $T^*(\mathbf{u}) = \mathbf{v}$.

5.4 Orthogonal Projection

Definitions:

Let S be a finite-dimensional subspace of an inner product space V. Then according to Fact 4 in Section 5.2, each $\mathbf{v} \in V$ can be written uniquely as $\mathbf{v} = \mathbf{s} + \mathbf{t}$, where $\mathbf{s} \in S$ and $\mathbf{t} \in S^\perp$. The vector $\mathbf{s}$ is called the **orthogonal projection** of $\mathbf{v}$ onto S and is often written as $\mathrm{Proj}_S \mathbf{v}$, where the linear operator Proj_S is called the **orthogonal projection** onto S along $S^\perp$. When $V = \mathbb{C}^n$ or $V = \mathbb{R}^n$ with the standard inner product, the linear operator Proj_S is often identified with its standard matrix $[\mathrm{Proj}_S]$ and Proj_S is used to denote both the operator and the matrix.

Facts:

1. An orthogonal projection is a projection (as defined in Section 3.6).
2. [Mey00, p. 433] Suppose that P is a projection. The following statements are equivalent:

   - P is an orthogonal projection.
   - $P^* = P$.
   - $\text{range}(P) \perp \ker(P)$.

3. [Mey00, p. 430] If S is a subspace of a finite dimensional inner product space V, then

$$\text{Proj}_{S^\perp} = I - \text{Proj}_S.$$

4. [Mey00, p. 430] Let S be a p–dimensional subspace of the standard inner product space $\mathbb{C}^n$, and let the columns of matrices $M \in \mathbb{C}^{n\times p}$ and $N \in \mathbb{C}^{n\times(n-p)}$ be bases for S and $S^\perp$, respectively. Then the orthogonal projections onto S and $S^\perp$ are

$$\text{Proj}_S = M(M^*M)^{-1}M^* \quad \text{and} \quad \text{Proj}_{S^\perp} = N(N^*N)^{-1}N^*.$$

   If M and N contain orthonormal bases for S and $S^\perp$, then $\text{Proj}_S = MM^*$ and $\text{Proj}_{S^\perp} = NN^*$.
5. [Lay03, p. 399] If $\{\mathbf{u}_1, \ldots, \mathbf{u}_p\}$ is an orthonormal basis for a subspace S of $\mathbb{C}^n$, then for any $\mathbf{v} \in \mathbb{C}^n$,

$$\text{Proj}_S\mathbf{v} = (\mathbf{u}_1^*\mathbf{v})\mathbf{u}_1 + \cdots + (\mathbf{u}_p^*\mathbf{v})\mathbf{u}_p.$$

6. [TB97, p. 46] Let $\mathbf{v} \in \mathbb{C}^n$ be a nonzero vector. Then

   - $\text{Proj}_\mathbf{v} = \dfrac{\mathbf{v}\mathbf{v}^*}{\mathbf{v}^*\mathbf{v}}$ is the orthogonal projection onto the line $L = \text{span}\{\mathbf{v}\}$.
   - $\text{Proj}_{\perp\mathbf{v}} = I - \dfrac{\mathbf{v}\mathbf{v}^*}{\mathbf{v}^*\mathbf{v}}$ is the orthogonal projection onto $L^\perp$.

7. [Mey00, p. 435] (The Best Approximation Theorem) Let S be a finite dimensional subspace of an inner product space V and let $\mathbf{b}$ be a vector in V. Then $\text{Proj}_S\mathbf{b}$ is the unique vector in S that is closest to $\mathbf{b}$ in the sense that

$$\min_{\mathbf{s}\in S} \|\mathbf{b} - \mathbf{s}\| = \|\mathbf{b} - \text{Proj}_S\mathbf{b}\|.$$

   The vector $\text{Proj}_S\mathbf{b}$ is called **the best approximation to b by the elements of** S.

Examples:

1. Generally, an orthogonal projection $P \in \mathbb{C}^{n\times n}$ is not a unitary matrix.
2. Let $\{\mathbf{v}_1, \mathbf{v}_2, \cdots, \mathbf{v}_n\}$ be an orthogonal basis for $\mathbb{R}^n$ and let S the subspace of $\mathbb{R}^n$ spanned by $\{\mathbf{v}_1, \cdots, \mathbf{v}_k\}$, where $1 \le k \le n-1$. Then $\mathbf{w} = c_1\mathbf{v}_1 + c_2\mathbf{v}_2 + \cdots + c_n\mathbf{v}_n \in \mathbb{R}^n$ can be written as $\mathbf{w} = \mathbf{s} + \mathbf{t}$, where $\mathbf{s} = c_1\mathbf{v}_1 + \cdots + c_k\mathbf{v}_k \in S$ and $\mathbf{t} = c_{k+1}\mathbf{v}_{k+1} + \cdots + c_n\mathbf{v}_n \in S^\perp$.
3. Let $\mathbf{u}_1 = [2/3, 2/3, 1/3]^T$, $\mathbf{u}_2 = [1/3, -2/3, 2/3]^T$, and $\mathbf{x} = [2, 3, 5]^T$. Then $\{\mathbf{u}_1, \mathbf{u}_2\}$ is an orthonormal basis for the subspace $S = \text{span}\,\{\mathbf{u}_1, \mathbf{u}_2\}$ of $\mathbb{R}^3$.

   - The orthogonal projection of $\mathbf{x}$ onto S is

$$\text{Proj}_S\mathbf{x} = \left(\mathbf{u}_1^T\right)\mathbf{x}\mathbf{u}_1 + \left(\mathbf{u}_2^T\mathbf{x}\right)\mathbf{u}_2 = [4, 2, 3]^T.$$

   - The orthogonal projection of $\mathbf{x}$ onto $S^\perp$ is $\mathbf{y} = \mathbf{x} - \text{Proj}_S\mathbf{x} = [-2, 1, 2]^T$.
   - The vector in S that is closest to $\mathbf{x}$ is $\text{Proj}_S\mathbf{x} = [4, 2, 3]^T$.

- Let $M = [\mathbf{u}_1, \mathbf{u}_2]$. Then the orthogonal projection onto S is

$$\text{Proj}_S = MM^T = \frac{1}{9}\begin{bmatrix} 5 & 2 & 4 \\ 2 & 8 & -2 \\ 4 & -2 & 5 \end{bmatrix}.$$

- The orthogonal projection of any $\mathbf{v} \in \mathbb{R}^3$ onto S can be computed by $\text{Proj}_S\mathbf{v} = MM^T\mathbf{v}$. In particular, $MM^T\mathbf{x} = [4, 2, 3]^T$.

4. Let $\mathbf{w}_1 = [1, 1, 0]^T$ and $\mathbf{w}_2 = [1, 0, 1]^T$. Consider the subspace $W = \text{span}\{\mathbf{w}_1, \mathbf{w}_2\}$ of $\mathbb{R}^3$. Define the matrix $M = [\mathbf{w}_1, \mathbf{w}_2] = \begin{bmatrix} 1 & 1 \\ 1 & 0 \\ 0 & 1 \end{bmatrix}$. Then $M^T M = \begin{bmatrix} 2 & 1 \\ 1 & 2 \end{bmatrix}$.

- The orthogonal projection onto W is $\text{Proj}_W = M(M^T M)^{-1} M^T =$

$$\begin{bmatrix} 1 & 1 \\ 1 & 0 \\ 0 & 1 \end{bmatrix} \begin{bmatrix} 2 & 1 \\ 1 & 2 \end{bmatrix}^{-1} \begin{bmatrix} 1 & 1 & 0 \\ 1 & 0 & 1 \end{bmatrix} = \frac{1}{3}\begin{bmatrix} 2 & 1 & 1 \\ 1 & 2 & -1 \\ 1 & -1 & 2 \end{bmatrix}.$$

- The orthogonal projection of any $\mathbf{v} \in \mathbb{R}^3$ onto W can be computed by $\text{Proj}_W\mathbf{v}$. For $\mathbf{v} = [1, 2, 3]^T$, $\text{Proj}_W\mathbf{v} = \text{Proj}_W[1, 2, 3]^T = [7/3, 2/3, 5/3]^T$.

5.5 Gram–Schmidt Orthogonalization and QR Factorization

Definitions:

Let $\{\mathbf{a}_1, \mathbf{a}_2, \ldots, \mathbf{a}_n\}$ be a basis for a subspace S of an inner product space V. An orthonormal basis $\{\mathbf{u}_1, \mathbf{u}_2, \ldots, \mathbf{u}_n\}$ for S can be constructed using the following Gram–Schmidt orthogonalization process:

$$\mathbf{u}_1 = \frac{\mathbf{a}_1}{\|\mathbf{a}_1\|} \quad \text{and} \quad \mathbf{u}_k = \frac{\mathbf{a}_k - \sum_{i=1}^{k-1} \langle \mathbf{a}_k, \mathbf{u}_i \rangle \mathbf{u}_i}{\|\mathbf{a}_k - \sum_{i=1}^{k-1} \langle \mathbf{a}_k, \mathbf{u}_i \rangle \mathbf{u}_i\|}, \quad \text{for} \quad k = 2, \ldots, n.$$

A **reduced QR factorization** of $A \in \mathbb{C}^{m\times n}$ $(m \geq n)$ is a factorization $A = \hat{Q}\hat{R}$, where $\hat{Q} \in \mathbb{C}^{m\times n}$ has orthonormal columns and $\hat{R} \in \mathbb{C}^{n\times n}$ is an upper triangular matrix.
A **QR factorization** of $A \in \mathbb{C}^{m\times n}$ $(m \geq n)$ is a factorization $A = QR$, where $Q \in \mathbb{C}^{m\times m}$ is a unitary matrix and $R \in \mathbb{C}^{m\times n}$ is an upper triangular matrix with the last $m - n$ rows of R being zero.

Facts:

1. [TB97, p. 51] Each $A \in \mathbb{C}^{m\times n}$ $(m \geq n)$ has a full QR factorization $A = QR$. If $A \in \mathbb{R}^{m\times n}$, then both Q and R may be taken to be real.
2. [TB97, p. 52] Each $A \in \mathbb{C}^{m\times n}$ $(m \geq n)$ has a reduced QR factorization $A = \hat{Q}\hat{R}$. If $A \in \mathbb{R}^{m\times n}$, then both $\hat{Q}$ and $\hat{R}$ may be taken to be real.
3. [TB97, p. 52] Each $A \in \mathbb{C}^{m\times n}$ $(m \geq n)$ of full rank has a unique reduced QR factorization $A = \hat{Q}\hat{R}$, where $\hat{Q} \in \mathbb{C}^{m\times n}$ and $\hat{R} \in \mathbb{C}^{n\times n}$ with real $r_{ii} > 0$.
4. [TB97, p. 48] The orthonormal basis $\{\mathbf{u}_1, \mathbf{u}_2, \ldots, \mathbf{u}_n\}$ generated via the Gram–Schmidt orthogonalization process has the property

$$\text{Span}(\{\mathbf{u}_1, \mathbf{u}_2, \ldots, \mathbf{u}_k\}) = \text{Span}(\{\mathbf{a}_1, \mathbf{a}_2, \ldots, \mathbf{a}_k\}),$$

for $k = 1, 2, \ldots, n$.

5. [TB97, p. 51]

```
Algorithm 1: Classical Gram–Schmidt Orthogonalization:
input: a basis {a_1, a_2, ..., a_n} for a subspace S
output: an orthonormal basis {u_1, u_2, ..., u_n} for S
for j = 1 : n
        u_j := a_j
        for i = 1 : j − 1
                r_ij := ⟨a_j, u_i⟩
                u_j := u_j − r_ij u_i
        end
        r_jj := ||u_j||
        u_j := u_j / r_jj
end
```

6. [TB97, p. 58]

```
Algorithm 2: Modified Gram–Schmidt Orthogonalization
input: a basis {a_1, a_2, ..., a_n} for a subspace S
output: an orthonormal basis {u_1, u_2, ..., u_n} for S
w_i := a_i, i = 1 : n
for i = 1 : n
        r_ii := ||w_i||
        u_i := w_i / r_ii
        for j = i + 1 : n
                r_ij := ⟨w_j, u_i⟩
                w_j := w_j − r_ij u_i
        end
end
```

7. [Mey00, p. 315] If exact arithmetic is used, then Algorithms 1 and 2 generate the same orthonormal basis $\{\mathbf{u}_1, \mathbf{u}_2, \ldots, \mathbf{u}_n\}$ and the same r_{ij}, for $j \geq i$.
8. [GV96, pp. 230–232] If $A = [\mathbf{a}_1, \mathbf{a}_2, \ldots, \mathbf{a}_n] \in \mathbb{C}^{m\times n}$ $(m \geq n)$ is of full rank n, then the classic or modified Gram–Schmidt process leads to a reduced QR factorization $A = \hat{Q}\hat{R}$, with $\hat{Q} = [\mathbf{u}_1, \mathbf{u}_2, \ldots, \mathbf{u}_n]$ and $\hat{R}_{ij} = r_{ij}$, for $j \geq i$, and $\hat{R}_{ij} = 0$, for $j < i$.
9. [GV96, p. 232] The costs of Algorithm 1 and Algorithm 2 are both $2mn^2$ flops when applied to compute a reduced QR factorization of a matrix $A \in \mathbb{R}^{m\times n}$.
10. [Mey00, p. 317 and p. 349] For the QR factorization, Algorithm 1 and Algorithm 2 are not numerically stable. However, Algorithm 2 often yields better numerical results than Algorithm 1.
11. [Mey00, p. 349] Algorithm 2 is numerically stable when it is used to solve least squares problems.
12. (Numerically stable algorithms for computing the QR factorization using Householder reflections and Givens rotations are given in Chapter 38.)
13. [TB97, p. 54] (See also Chapter 38.) If $A = QR$ is a QR factorization of the rank n matrix $A \in \mathbb{C}^{n\times n}$, then the linear system $A\mathbf{x} = \mathbf{b}$ can be solved as follows:

- Compute the factorization $A = QR$.
- Compute the vector $\mathbf{c} = Q^*\mathbf{b}$.
- Solve $R\mathbf{x} = \mathbf{c}$ by performing back substitution.

Examples:

1. Consider the matrix $A = \begin{bmatrix} 1 & 2 \\ 2 & 0 \\ 0 & 2 \end{bmatrix}$.

   - A has a (full) QR factorization $A = QR$:

$$\begin{bmatrix} 1 & 2 \\ 2 & 0 \\ 0 & 2 \end{bmatrix} = \begin{bmatrix} \frac{1}{\sqrt{5}} & \frac{4}{3\sqrt{5}} & -\frac{2}{3} \\ \frac{2}{\sqrt{5}} & -\frac{2}{3\sqrt{5}} & \frac{1}{3} \\ 0 & \frac{\sqrt{5}}{3} & \frac{2}{3} \end{bmatrix} \begin{bmatrix} \sqrt{5} & \frac{2}{\sqrt{5}} \\ 0 & \frac{6}{\sqrt{5}} \\ 0 & 0 \end{bmatrix}.$$

   - A has a reduced QR factorization $A = \hat{Q}\hat{R}$:

$$\begin{bmatrix} 1 & 2 \\ 2 & 0 \\ 0 & 2 \end{bmatrix} = \begin{bmatrix} \frac{1}{\sqrt{5}} & \frac{4}{3\sqrt{5}} \\ \frac{2}{\sqrt{5}} & -\frac{2}{3\sqrt{5}} \\ 0 & \frac{\sqrt{5}}{3} \end{bmatrix} \begin{bmatrix} \sqrt{5} & \frac{2}{\sqrt{5}} \\ 0 & \frac{6}{\sqrt{5}} \end{bmatrix}.$$

2. Consider the matrix $A = \begin{bmatrix} 3 & 1 & -2 \\ 3 & -4 & 1 \\ 3 & -4 & -1 \\ 3 & 1 & 0 \end{bmatrix}$. Using the classic or modified Gram–Schmidt process gives the following reduced QR factorization:

$$\begin{bmatrix} 3 & 1 & -2 \\ 3 & -4 & 1 \\ 3 & -4 & -1 \\ 3 & 1 & 0 \end{bmatrix} = \begin{bmatrix} \frac{1}{2} & \frac{1}{2} & -\frac{1}{2} \\ \frac{1}{2} & -\frac{1}{2} & \frac{1}{2} \\ \frac{1}{2} & -\frac{1}{2} & -\frac{1}{2} \\ \frac{1}{2} & \frac{1}{2} & \frac{1}{2} \end{bmatrix} \begin{bmatrix} 6 & -3 & -1 \\ 0 & 5 & -1 \\ 0 & 0 & 2 \end{bmatrix}.$$

5.6 Singular Value Decomposition

Definitions:

A **singular value decomposition (SVD)** of a matrix $A \in \mathbb{C}^{m\times n}$ is a factorization

$$A = U\Sigma V^*, \quad \Sigma = \operatorname{diag}(\sigma_1, \sigma_2, \ldots, \sigma_p) \in \mathbb{R}^{m\times n}, \quad p = \min\{m, n\},$$

where $\sigma_1 \geq \sigma_2 \geq \ldots \geq \sigma_p \geq 0$ and both $U = [\mathbf{u}_1, \mathbf{u}_2, \ldots, \mathbf{u}_m] \in \mathbb{C}^{m\times m}$ and $V = [\mathbf{v}_1, \mathbf{v}_2, \ldots, \mathbf{v}_n] \in \mathbb{C}^{n\times n}$ are unitary. The diagonal entries of Σ are called the **singular values** of A. The columns of U are called **left singular vectors** of A and the columns of V are called **right singular vectors** of A.

Let $A \in \mathbb{C}^{m\times n}$ with rank $r \leq p = \min\{m, n\}$. A **reduced singular value decomposition (reduced SVD)** of A is a factorization

$$A = \hat{U}\hat{\Sigma}\hat{V}^*, \quad \hat{\Sigma} = \operatorname{diag}(\sigma_1, \sigma_2, \ldots, \sigma_r) \in \mathbb{R}^{r\times r},$$

where $\sigma_1 \geq \sigma_2 \geq \ldots \geq \sigma_r > 0$ and the columns of $\hat{U} = [\mathbf{u}_1, \mathbf{u}_2, \ldots, \mathbf{u}_r] \in \mathbb{C}^{m\times r}$ and the columns of $\hat{V} = [\mathbf{v}_1, \mathbf{v}_2, \ldots, \mathbf{v}_r] \in \mathbb{C}^{n\times r}$ are both orthonormal.

(See §8.4 and §3.7 for more information on singular value decomposition.)

Facts:

All the following facts except those with a specific reference can be found in [TB97, pp. 25–37].

1. Every $A \in \mathbb{C}^{m\times n}$ has a singular value decomposition $A = U\Sigma V^*$. If $A \in \mathbb{R}^{m\times n}$, then U and V may be taken to be real.
2. The singular values of a matrix are uniquely determined.
3. If $A \in \mathbb{C}^{m\times n}$ has a singular value decomposition $A = U\Sigma V^*$, then
$$A\mathbf{v}_j = \sigma_j \mathbf{u}_j, \quad A^*\mathbf{u}_j = \sigma_j \mathbf{v}_j, \quad \mathbf{u}_j^* A\mathbf{v}_j = \sigma_j,$$
for $j = 1, 2, \ldots, p = \min\{m, n\}$.
4. If $U\Sigma V^*$ is a singular value decomposition of A, then $V\Sigma^T U^*$ is a singular value decomposition of A^*.
5. If $A \in \mathbb{C}^{m\times n}$ has r nonzero singular values, then
   - $\text{rank}(A) = r$.
   - $A = \sum_{j=1}^{r} \sigma_j \mathbf{u}_j \mathbf{v}_j^*$.
   - $\ker(A) = \text{span}\{\mathbf{v}_{r+1}, \ldots, \mathbf{v}_n\}$.
   - $\text{range}(A) = \text{span}\{\mathbf{u}_1, \ldots, \mathbf{u}_r\}$.
6. Any $A \in \mathbb{C}^{m\times n}$ of rank $r \le p = \min\{m, n\}$ has a **reduced singular value decomposition,**
$$A = \hat{U}\hat{\Sigma}\hat{V}^*, \quad \hat{\Sigma} = \text{diag}(\sigma_1, \sigma_2, \ldots, \sigma_r) \in \mathbb{R}^{r\times r},$$
where $\sigma_1 \ge \sigma_2 \ge \cdots \ge \sigma_r > 0$ and the columns of $\hat{U} = [\mathbf{u}_1, \mathbf{u}_2, \ldots, \mathbf{u}_r] \in \mathbb{C}^{m\times r}$ and the columns of $\hat{V} = [\mathbf{v}_1, \mathbf{v}_2, \ldots, \mathbf{v}_r] \in \mathbb{C}^{n\times r}$ are both orthonormal. If $A \in \mathbb{R}^{m\times n}$, then $\hat{U}$ and $\hat{V}$ may be taken to be real.
7. If $\text{rank}(A) = r$, then A has r nonzero singular values.
8. The nonzero singular values of A are the square roots of the nonzero eigenvalues of A^*A or AA^*.
9. [HJ85, p. 414] If $U\Sigma V^*$ is a singular value decomposition of A, then the columns of V are eigenvectors of A^*A; the columns of U are eigenvectors of AA^*.
10. [HJ85, p. 418] Let $A \in \mathbb{C}^{m\times n}$ and $p = \min\{m, n\}$. Define
$$G = \begin{bmatrix} 0 & A \\ A^* & 0 \end{bmatrix} \in \mathbb{C}^{(m+n)\times(m+n)}.$$
If the singular values of A are $\sigma_1, \ldots, \sigma_p$, then the eigenvalues of G are $\sigma_1, \ldots, \sigma_p, -\sigma_1, \ldots, -\sigma_p$ and additional $|n - m|$ zeros.
11. If $A \in \mathbb{C}^{n\times n}$ is Hermitian with eigenvalues $\lambda_1, \lambda_2, \cdots, \lambda_n$, then the singular values of A are $|\lambda_1|, |\lambda_2|, \cdots, |\lambda_n|$.
12. For $A \in \mathbb{C}^{n\times n}$, $|\det A| = \sigma_1\sigma_2\cdots\sigma_n$.
13. [Aut15; Sch07] (**Eckart–Young Low Rank Approximation Theorem**)
Let $A = U\Sigma V^*$ be an SVD of $A \in \mathbb{C}^{m\times n}$ and $r = \text{rank}(A)$. For $k < r$, define $A_k = \sum_{j=1}^{k} \sigma_j \mathbf{u}_j \mathbf{v}_j^*$. Then
   - $\|A - A_k\|_2 = \min_{\text{rank}(B)\le k} \|A - B\|_2 = \sigma_{k+1}$;
   - $\|A - A_k\|_F = \min_{\text{rank}(B)\le k} \|A - B\|_F = \sqrt{\sum_{j=k+1}^{r} \sigma_j^2}$,

   where $\|M\|_2 = \max_{\|x\|_2=1} \|Mx\|_2$ and $\|M\|_F = \sqrt{\sum_{i=1}^{m}\sum_{j=1}^{n} m_{ij}^2}$ are the 2-norm and Frobenius norm of matrix M, respectively. (See Chapter 37 for more information on matrix norms.)

Examples:

Consider the matrices $A = \begin{bmatrix} 1 & 2 \\ 2 & 0 \\ 0 & 2 \end{bmatrix}$ and $B = A^T = \begin{bmatrix} 1 & 2 & 0 \\ 2 & 0 & 2 \end{bmatrix}$.

1. The eigenvalues of $A^T A = \begin{bmatrix} 5 & 2 \\ 2 & 8 \end{bmatrix}$ are 9 and 4. So, the singular values of A are 3 and 2.
2. Normalized eigenvectors for $A^T A$ are $\mathbf{v}_1 = \begin{bmatrix} \frac{1}{\sqrt{5}} \\ \frac{2}{\sqrt{5}} \end{bmatrix}$ and $\mathbf{v}_2 = \begin{bmatrix} \frac{2}{\sqrt{5}} \\ -\frac{1}{\sqrt{5}} \end{bmatrix}$.
3. $\mathbf{u}_1 = \frac{1}{3} A\mathbf{v}_1 = \begin{bmatrix} \frac{\sqrt{5}}{3} \\ \frac{2}{3\sqrt{5}} \\ \frac{4}{3\sqrt{5}} \end{bmatrix}$ and $\mathbf{u}_2 = \frac{1}{2} A\mathbf{v}_2 = \begin{bmatrix} 0 \\ \frac{2}{\sqrt{5}} \\ -\frac{1}{\sqrt{5}} \end{bmatrix}$. Application of the Gram–Schmidt process to $\mathbf{u}_1$, $\mathbf{u}_2$, and $\mathbf{e}_1$ produces $\mathbf{u}_3 = \begin{bmatrix} \frac{2}{3} \\ -\frac{1}{3} \\ -\frac{2}{3} \end{bmatrix}$.
4. A has the singular value decomposition $A = U \Sigma V^T$, where

$$U = \frac{1}{3\sqrt{5}} \begin{bmatrix} 5 & 0 & 2\sqrt{5} \\ 2 & 6 & -\sqrt{5} \\ 4 & -3 & -2\sqrt{5} \end{bmatrix}, \quad \Sigma = \begin{bmatrix} 3 & 0 \\ 0 & 2 \\ 0 & 0 \end{bmatrix}, \quad V = \frac{1}{\sqrt{5}} \begin{bmatrix} 1 & 2 \\ 2 & -1 \end{bmatrix}.$$

5. A has the reduced singular value decomposition $A = \hat{U} \hat{\Sigma} \hat{V}^T$, where

$$\hat{U} = \frac{1}{3\sqrt{5}} \begin{bmatrix} 5 & 0 \\ 2 & 6 \\ 4 & -3 \end{bmatrix}, \quad \hat{\Sigma} = \begin{bmatrix} 3 & 0 \\ 0 & 2 \end{bmatrix}, \quad \hat{V} = \frac{1}{\sqrt{5}} \begin{bmatrix} 1 & 2 \\ 2 & -1 \end{bmatrix}.$$

6. B has the singular value decomposition $B = U_B \Sigma_B V_B^T$, where

$$U_B = V_A = \frac{1}{\sqrt{5}} \begin{bmatrix} 1 & 2 \\ 2 & -1 \end{bmatrix}, \quad \Sigma_B = \begin{bmatrix} 3 & 0 & 0 \\ 0 & 2 & 0 \end{bmatrix}, \quad V_B = U_A = \frac{1}{3\sqrt{5}} \begin{bmatrix} 5 & 0 & 2\sqrt{5} \\ 2 & 6 & -\sqrt{5} \\ 4 & -3 & -2\sqrt{5} \end{bmatrix}.$$

($U_A = U$ and $V_A = V$ for A were given in Example 4.)

5.7 Pseudo-Inverse

Definitions:

A **Moore–Penrose pseudo-inverse** of a matrix $A \in \mathbb{C}^{m\times n}$ is a matrix $A^\dagger \in \mathbb{C}^{n\times m}$ that satisfies the following four **Penrose** conditions:

$$AA^\dagger A = A; \quad A^\dagger A A^\dagger = A^\dagger; \quad (AA^\dagger)^* = AA^\dagger; \quad (A^\dagger A)^* = A^\dagger A.$$

Facts:

All the following facts except those with a specific reference can be found in [Gra83, pp. 105–141].

1. Every $A \in \mathbb{C}^{m\times n}$ has a unique pseudo-inverse $A^\dagger$. If $A \in \mathbb{R}^{m\times n}$, then $A^\dagger$ is real.

2. [LH95, p. 38] If $A \in \mathbb{C}^{m\times n}$ of rank $r \leq \min\{m, n\}$ has an SVD $A = U\Sigma V^*$, then its pseudo-inverse is $A^\dagger = V\Sigma^\dagger U^*$, where

$$\Sigma^\dagger = \text{diag}(1/\sigma_1, \ldots, 1/\sigma_r, 0, \ldots, 0) \in \mathbb{R}^{n\times m}.$$

3. $\mathbf{0}_{mn}^\dagger = \mathbf{0}_{nm}$ and $J_{mn}^\dagger = \frac{1}{mn}J_{nm}$, where $\mathbf{0}_{mn} \in \mathbb{C}^{m\times n}$ is the all 0s matrix and $J_{mn} \in \mathbb{C}^{m\times n}$ is the all 1s matrix.
4. $(A^\dagger)^* = (A^*)^\dagger$; $(A^\dagger)^\dagger = A$.
5. If A is a nonsingular square matrix, then $A^\dagger = A^{-1}$.
6. If U has orthonormal columns or orthonormal rows, then $U^\dagger = U^*$.
7. If $A = A^*$ and $A = A^2$, then $A^\dagger = A$.
8. $A^\dagger = A^*$ if and only if A^*A is idempotent.
9. If $A = A^*$, then $AA^\dagger = A^\dagger A$.
10. If $U \in \mathbb{C}^{m\times n}$ is of rank n and satisfies $U^\dagger = U^*$, then U has orthonormal columns.
11. If $U \in \mathbb{C}^{m\times m}$ and $V \in \mathbb{C}^{n\times n}$ are unitary matrices, then $(UAV)^\dagger = V^*A^\dagger U^*$.
12. If $A \in \mathbb{C}^{m\times n}$ $(m \geq n)$ has full rank n, then $A^\dagger = (A^*A)^{-1}A^*$.
13. If $A \in \mathbb{C}^{m\times n}$ $(m \leq n)$ has full rank m, then $A^\dagger = A^*(AA^*)^{-1}$.
14. Let $A \in \mathbb{C}^{m\times n}$. Then
    - $A^\dagger A$, $AA^\dagger$, $I_n - A^\dagger A$, and $I_m - AA^\dagger$ are orthogonal projections.
    - $\text{rank}(A) = \text{rank}(A^\dagger) = \text{rank}(AA^\dagger) = \text{rank}(A^\dagger A)$.
    - $\text{rank}(I_n - A^\dagger A) = n - \text{rank}(A)$.
    - $\text{rank}(I_m - AA^\dagger) = m - \text{rank}(A)$.
15. If $A = A_1 + A_2 + \cdots + A_k$, $A_i^*A_j = 0$, and $A_iA_j^* = 0$, for all $i, j = 1, \cdots, k$, $i \neq j$, then $A^\dagger = A_1^\dagger + A_2^\dagger + \cdots + A_k^\dagger$.
16. If A is an $m \times r$ matrix of rank r and B is an $r \times n$ matrix of rank r, then $(AB)^\dagger = B^\dagger A^\dagger$.
17. $(A^*A)^\dagger = A^\dagger(A^*)^\dagger$; $(AA^*)^\dagger = (A^*)^\dagger A^\dagger$.
18. [Gre66] Each one of the following conditions is necessary and sufficient for $(AB)^\dagger = B^\dagger A^\dagger$:
    - $\text{range}(BB^*A^*) \subseteq \text{range}(A^*)$ and $\text{range}(A^*AB) \subseteq \text{range}(B)$.
    - $A^\dagger ABB^*$ and $A^*ABB^\dagger$ are both Hermitian matrices.
    - $A^\dagger ABB^*A^* = BB^*A^*$ and $BB^\dagger A^*AB = A^*AB$.
    - $A^\dagger ABB^*A^*ABB^\dagger = BB^*A^*A$.
    - $A^\dagger AB = B(AB)^\dagger AB$ and $BB^\dagger A^* = A^*AB(AB)^\dagger$.
19. Let $A \in \mathbb{C}^{m\times n}$ and $\mathbf{b} \in \mathbb{C}^m$. Then the system of equations $A\mathbf{x} = \mathbf{b}$ is consistent if and only if $AA^\dagger\mathbf{b} = \mathbf{b}$. Moreover, if $A\mathbf{x} = \mathbf{b}$ is consistent, then any solution to the system can be expressed as $\mathbf{x} = A^\dagger\mathbf{b} + (I_n - A^\dagger A)\mathbf{y}$ for some $\mathbf{y} \in \mathbb{C}^n$.

Examples:

1. The pseudo-inverse of the matrix $A = \begin{bmatrix} 1 & 2 \\ 2 & 0 \\ 0 & 2 \end{bmatrix}$ is $A^\dagger = \frac{1}{18}\begin{bmatrix} 2 & 8 & -2 \\ 4 & -2 & 5 \end{bmatrix}$.

2. $(AB)^\dagger = B^\dagger A^\dagger$ generally does not hold. For example, if

$$A = \begin{bmatrix} 1 & 0 \\ 0 & 0 \end{bmatrix} \quad \text{and} \quad B = \begin{bmatrix} 1 & 1 \\ 0 & 1 \end{bmatrix},$$

then

$$(AB)^\dagger = \begin{bmatrix} 1 & 1 \\ 0 & 0 \end{bmatrix}^\dagger = \frac{1}{2}\begin{bmatrix} 1 & 0 \\ 1 & 0 \end{bmatrix}.$$

However,

$$B^\dagger A^\dagger = \begin{bmatrix} 1 & 0 \\ 0 & 0 \end{bmatrix}.$$

5.8 Least Squares Problems

Definitions:

Given $A \in F^{m\times n}$ ($F = \mathbb{R}$ or $\mathbb{C}$), $m \geq n$, and $\mathbf{b} \in F^m$, the **least squares problem** is to find an $\mathbf{x}_0 \in F^n$ such that $\|\mathbf{b} - A\mathbf{x}\|$ is minimized: $\|\mathbf{b} - A\mathbf{x}_0\| = \min_{\mathbf{x}\in F^n} \|\mathbf{b} - A\mathbf{x}\|$.

- Such an $\mathbf{x}_0$ is called a **solution** to the least squares problem or a **least squares solution** to the linear system $A\mathbf{x} = \mathbf{b}$.
- The vector $\mathbf{r} = \mathbf{b} - A\mathbf{x} \in F^m$ is called the **residual.**
- If $\text{rank}(A) = n$, then the least squares problem is called the **full rank least squares problem.**
- If $\text{rank}(A) < n$, then the least squares problem is called the **rank–deficient least squares problem.**

The system $A^*A\mathbf{x} = A^*\mathbf{b}$ is called the **normal equation** for the least squares problem.
(See Chapter 39 for more information on least squares problems.)

Facts:

1. [Mey00, p. 439] Let $A \in F^{m\times n}$ ($F = \mathbb{R}$ or $\mathbb{C}$, $m \geq n$) and $\mathbf{b} \in F^m$ be given. Then the following statements are equivalent:
   - $\mathbf{x}_0$ is a solution for the least squares problem.
   - $\min_{\mathbf{x}\in F^n} \|\mathbf{b} - A\mathbf{x}\| = \|\mathbf{b} - A\mathbf{x}_0\|$.
   - $A\mathbf{x}_0 = P\mathbf{b}$, where P is the orthogonal projection onto range(A).
   - $A^*\mathbf{r}_0 = \mathbf{0}$, where $\mathbf{r}_0 = \mathbf{b} - A\mathbf{x}_0$.
   - $A^*A\mathbf{x}_0 = A^*\mathbf{b}$.
   - $\mathbf{x}_0 = A^\dagger\mathbf{b} + \mathbf{y}_0$ for some $\mathbf{y}_0 \in \ker(A)$.
2. [LH95, p. 36] If $A \in F^{m\times n}$ ($F = \mathbb{R}$ or $\mathbb{C}$, $m \geq n$) and $\text{rank}(A) = r \leq n$, then $\mathbf{x}_0 = A^\dagger\mathbf{b}$ is the unique solution of minimum length for the least squares problem.
3. [TB97, p. 81] If $A \in F^{m\times n}$ ($F = \mathbb{R}$ or $\mathbb{C}$, $m \geq n$) has full rank, then $\mathbf{x}_0 = A^\dagger\mathbf{b} = (A^*A)^{-1}A^*\mathbf{b}$ is the unique solution for the least squares problem.
4. [TB97, p. 83]

> **Algorithm 3**: Solving Full Rank Least Squares via QR Factorization
> input: matrix $A \in F^{m\times n}$ ($F = \mathbb{R}$ or $\mathbb{C}$, $m \geq n$) with full rank n and vector $\mathbf{b} \in F^m$
> output : solution $\mathbf{x}_0$ for $\min_{\mathbf{x}\in F^n} \|\mathbf{b} - A\mathbf{x}\|$
> compute the reduced QR factorization $A = \hat{Q}\hat{R}$;
> compute the vector $\mathbf{c} = \hat{Q}^*\mathbf{b}$;
> solve $\hat{R}\mathbf{x}_0 = \mathbf{c}$ using back substitution.

5. [TB97, p. 84]

Algorithm 4: Solving Full Rank Least Squares via SVD
input: matrix $A \in F^{m\times n}$ ($F = \mathbb{R}$ or $\mathbb{C}$, $m \geq n$) with full rank n and vector $\mathbf{b} \in F^m$
output : solution $\mathbf{x}_0$ for $\min_{\mathbf{x}\in F^n} \|\mathbf{b} - A\mathbf{x}\|$
compute the reduced SVD $A = \hat{U}\hat{\Sigma}\hat{V}^*$ with $\hat{\Sigma} = \text{diag}(\sigma_1, \sigma_2, \cdots, \sigma_n)$;
compute the vector $\mathbf{c} = \hat{U}^*\mathbf{b}$;
compute the vector $\mathbf{y}$: $y_i = c_i/\sigma_i$, $i = 1, 2, \cdots, n$;
compute $\mathbf{x}_0 = \hat{V}\mathbf{y}$.

6. [TB97, p. 82]

Algorithm 5: Solving Full Rank Least Squares via Normal Equations
input: matrix $A \in F^{m\times n}$ ($F = \mathbb{R}$ or $\mathbb{C}$, $m \geq n$) with full rank n and vector $\mathbf{b} \in F^m$
output : solution $\mathbf{x}_0$ for $\min_{\mathbf{x}\in F^n} \|\mathbf{b} - A\mathbf{x}\|$
compute the matrix A^*A and the vector $\mathbf{c} = A^*\mathbf{b}$;
solve the system $A^*A\mathbf{x}_0 = \mathbf{c}$ via the Cholesky factorization.

Examples:

1. Consider the inconsistent linear system $A\mathbf{x} = \mathbf{b}$, where

$$A = \begin{bmatrix} 1 & 2 \\ 2 & 0 \\ 0 & 2 \end{bmatrix}, \quad \mathbf{b} = \begin{bmatrix} 1 \\ 2 \\ 3 \end{bmatrix}.$$

Then the normal equations are given by $A^TA\mathbf{x} = A^T\mathbf{b}$, where

$$A^TA = \begin{bmatrix} 5 & 2 \\ 2 & 8 \end{bmatrix} \quad \text{and} \quad A^T\mathbf{b} = \begin{bmatrix} 5 \\ 8 \end{bmatrix}.$$

A least squares solution to the system $A\mathbf{x} = \mathbf{b}$ can be obtained via solving the normal equations:

$$\mathbf{x}_0 = (A^TA)^{-1}A^T\mathbf{b} = A^\dagger\mathbf{b} = \begin{bmatrix} 2/3 \\ 5/6 \end{bmatrix}.$$

2. We use Algorithm 3 to find a least squares solution of the system $A\mathbf{x} = \mathbf{b}$ given in Example 1. The reduced QR factorization $A = \hat{Q}\hat{R}$ found in Example 1 in Section 5.5 gives

$$\hat{Q}^T\mathbf{b} = \begin{bmatrix} \frac{1}{\sqrt{5}} & \frac{4}{3\sqrt{5}} \\ \frac{2}{\sqrt{5}} & -\frac{2}{3\sqrt{5}} \\ 0 & \frac{\sqrt{5}}{3} \end{bmatrix}^T \begin{bmatrix} 1 \\ 2 \\ 3 \end{bmatrix} = \begin{bmatrix} \sqrt{5} \\ \sqrt{5} \end{bmatrix}.$$

Now solving $\hat{R}\mathbf{x} = [\sqrt{5}, \sqrt{5}]^T$ gives the least squares solution $\mathbf{x}_0 = [2/3, 5/6]^T$.

3. We use Algorithm 4 to solve the same problem given in Example 1. Using the reduced singular value decomposition $A = \hat{U}\hat{\Sigma}\hat{V}^T$ obtained in Example 5, Section 5.6, we have

$$\mathbf{c} = \hat{U}^T\mathbf{b} = \frac{1}{3\sqrt{5}} \begin{bmatrix} 5 & 0 \\ 2 & 6 \\ 4 & -3 \end{bmatrix}^T \begin{bmatrix} 1 \\ 2 \\ 3 \end{bmatrix} = \begin{bmatrix} \frac{7}{\sqrt{5}} \\ \frac{1}{\sqrt{5}} \end{bmatrix}.$$

Now we compute $\mathbf{y} = [y_1, y_2]^T$:

$$y_1 = c_1/\sigma_1 = \frac{7}{3\sqrt{5}} \quad \text{and} \quad y_2 = c_2/\sigma_2 = \frac{1}{2\sqrt{5}}.$$

Finally, the least squares solution is obtained via

$$\mathbf{x}_0 = \hat{V}\mathbf{y} = \frac{1}{\sqrt{5}} \begin{bmatrix} 1 & 2 \\ 2 & -1 \end{bmatrix} \begin{bmatrix} \frac{7}{3\sqrt{5}} \\ \frac{1}{2\sqrt{5}} \end{bmatrix} = \begin{bmatrix} 2/3 \\ 5/6 \end{bmatrix}.$$

References

[Aut15] L. Auttone. Sur les Matrices Hypohermitiennes et sur les Matrices Unitaires. Ann. Univ. Lyon, Nouvelle Série I, Facs. 38:1–77, 1915.

[Gra83] F. A. Graybill. *Matrices with Applications in Statistics*. 2nd ed., Wadsworth Intl. Belmont, CA, 1983.

[Gre66] T. N. E. Greville. Notes on the generalized inverse of a matrix product. *SIAM Review*, 8:518–521, 1966.

[GV96] G. H. Golub and C. F. Van Loan. *Matrix Computations*. 3rd ed., Johns Hopkins University Press, Baltimore, 1996.

[Hal58] P. Halmos. *Finite-Dimensional Vector Spaces*. Van Nostrand, New York, 1958.

[HK71] K. H. Hoffman and R. Kunze. *Linear Algebra*. 2nd ed., Prentice Hall, Upper Saddle River, NJ, 1971.

[HJ85] R. A. Horn and C. R. Johnson. *Matrix Analysis*. Cambridge University Press, Cambridge, 1985.

[Lay03] D. Lay. *Linear Algebra and Its Applications*. 3rd ed., Addison Wesley, Boston, 2003.

[LH95] C. L. Lawson and R. J. Hanson. *Solving Least Squares Problems*. SIAM, Philadelphia, 1995.

[Mey00] C. Meyer. *Matrix Analysis and Applied Linear Algebra*. SIAM, Philadelphia, 2000.

[Rom92] S. Roman. *Advanced Linear Algebra*. Springer-Verlag, New York, 1992.

[Sch07] E. Schmidt. Zur Theorie der linearen und nichtliniearen Integralgleichungen. *Math Annal*, 63:433-476, 1907.

[TB97] L. N. Trefethen and D. Bau. *Numerical Linear Algebra*. SIAM, Philadelphia, 1997.

Matrices with Special Properties

6

Canonical Forms

Leslie Hogben
Iowa State University

A *canonical form* of a matrix is a special form with the properties that every matrix is associated to a matrix in that form (the canonical form of the matrix), it is unique or essentially unique (typically up to some type of permutation), and it has a particularly simple form (or a form well suited to a specific purpose). A canonical form partitions the set matrices in $F^{m\times n}$ into sets of matrices each having the same canonical form, and that canonical form matrix serves as the representative. The canonical form of a given matrix can provide important information about the matrix. For example, reduced row echelon form (RREF) is a canonical form that is useful in solving systems of linear equations; RREF partitions $F^{m\times n}$ into sets of row equivalent matrices.

The previous definition of a canonical form is far more general than the canonical forms discussed in this chapter. Here all matrices are square, and every matrix is similar to its canonical form. This chapter discusses the two most important canonical forms for square matrices over fields, the Jordan canonical form (and its real version) and (two versions of) the rational canonical form. These canonical forms capture the eigenstructure of a matrix and play important roles in many areas, for example, in matrix functions, Chapter 11, and in differential equations, Chapter 55. These canonical forms partition $F^{n\times n}$ into similarity classes.

The Jordan canonical form is most often used when all eigenvalues of the matrix $A \in F^{n\times n}$ lie in the field F, such as when the field is algebraically closed (e.g., $\mathbb{C}$), or when the field is $\mathbb{R}$; otherwise the rational canonical form is used (e.g., for $\mathbb{Q}$). The Smith normal form is a canonical form for square matrices over principal ideal domains (see Chapter 23); it is discussed here only as it pertains to the computation of the rational canonical form. If any one of these canonical forms is known, it is straightforward to determine the others (perhaps in the algebraic closure of the field F). Details are given in the sections on rational canonical form.

Results about each type of canonical form are presented in the section on that canonical form, which facilitates locating a result, but obscures the connections underlying the derivations of the results. The facts about all of the canonical forms discussed in this section can be derived from results about modules over a principal ideal domain; such a module-theoretic treatment is typically presented in abstract algebra texts, such as [DF04, Chap. 12].

None of the canonical forms discussed in this chapter is a continuous function of the entries of a matrix and, thus, the computation of such a canonical form is inherently unstable in finite precision arithmetic. (For information about perturbation theory of eigenvalues see Chapter 15; for information specifically about numerical computation of the Jordan canonical form, see [GV96, Chapter 7.6.5].)

6.1 Generalized Eigenvectors

The reader is advised to consult Section 4.3 for information about eigenvalues and eigenvectors. In this section and the next, F is taken to be an algebraically closed field to ensure that an $n \times n$ matrix has n eigenvalues, but many of the results could be rephrased for a matrix that has all its eigenvalues in F, without the assumption that F is algebraically closed. The real versions of the definitions and results are presented in Section 6.3.

Definitions:

Let F be an algebraically closed field (e.g., $\mathbb{C}$), let $A \in F^{n\times n}$, let $\mu_1, \ldots, \mu_r$ be the distinct eigenvalues of A, and let λ be any eigenvalue of A.

For k a nonnegative integer, the **k-eigenspace** of A at λ, denoted $N_\lambda^k(A)$, is $\ker(A - \lambda I)^k$.

The **index** of A at λ, denoted $\nu_\lambda(A)$, is the smallest integer k such that $N_\lambda^k(A) = N_\lambda^{k+1}(A)$. When λ and A are clear from the context, $\nu_\lambda(A)$ will be abbreviated to ν, and $\nu_{\mu_i}(A)$ to ν_i.

The **generalized eigenspace** of A at λ is the set $N_\lambda^\nu(A)$, where ν is the index of A at λ.

The vector $\mathbf{x} \in F^n$ is a **generalized eigenvector** of A for λ if $\mathbf{x} \neq \mathbf{0}$ and $\mathbf{x} \in N_\lambda^\nu(A)$.

Let V be a finite dimensional vector space over F, and let T be a linear operator on V. The definitions of **k-eigenspace** of T, **index**, and **generalized eigenspace** of T are analogous.

Facts:

Facts requiring proof for which no specific reference is given can be found in [HJ85, Chapter 3] or [Mey00, Chapter 7.8].

Notation: F is an algebraically closed field, $A \in F^{n\times n}$, V is an n dimensional vector space over F, $T \in L(V, V)$, $\mu_1, \ldots, \mu_r$ are the distinct eigenvalues of A or T, and $\lambda = \mu_i$ for some $i \in \{1, \ldots, r\}$.

1. An eigenvector for eigenvalue λ is a generalized eigenvector for λ, but the converse is not necessarily true.
2. The eigenspace for λ is the 1-eigenspace, i.e., $E_\lambda(A) = N_\lambda^1(A)$.
3. Every k-eigenspace is invariant under multiplication by A.
4. The dimension of the generalized eigenspace of A at λ is the algebraic multiplicity of λ, i.e., $\dim N_{\mu_i}^{\nu_i}(A) = \alpha_A(\mu_i)$.
5. A is diagonalizable if and only if $\nu_i = 1$ for $i = 1, \ldots, r$.
6. F^n is the vector space direct sum of the generalized eigenspaces, i.e.,

$$F^n = N_{\mu_1}^{\nu_1}(A) \oplus \cdots \oplus N_{\mu_r}^{\nu_r}(A).$$

   This is a special case of the Primary Decomposition Theorem (Fact 12 in Section 6.4).
7. Facts 1 to 6 remain true when the matrix A is replaced by the linear operator T.
8. If $\hat{T}$ denotes T restricted to $N_{\mu_i}^{\nu_i}(T)$, then the characteristic polynomial of $\hat{T}$ is $p_{\hat{T}}(x) = (x-\mu_i)^{\alpha(\mu_i)}$. In particular, $\hat{T} - \mu_i I$ is nilpotent.

Examples:

1. Let $A = \begin{bmatrix} 65 & 18 & -21 & 4 \\ -201 & -56 & 63 & -12 \\ 67 & 18 & -23 & 4 \\ 134 & 36 & -42 & 6 \end{bmatrix} \in \mathbb{C}^{4\times 4}$. $p_A(x) = x^4 + 8x^3 + 24x^2 + 32x + 16 = (x+2)^4$, so the only eigenvalue of A is -2 with algebraic multiplicity 4. The reduced row echelon form of $A + 2I$ is $\begin{bmatrix} 1 & \frac{18}{67} & -\frac{21}{67} & \frac{4}{67} \\ 0 & 0 & 0 & 0 \\ 0 & 0 & 0 & 0 \\ 0 & 0 & 0 & 0 \end{bmatrix}$, so $N^1_{-2}(A) = \text{Span}\left(\begin{bmatrix} -18 \\ 67 \\ 0 \\ 0 \end{bmatrix}, \begin{bmatrix} 21 \\ 0 \\ 67 \\ 0 \end{bmatrix}, \begin{bmatrix} -4 \\ 0 \\ 0 \\ 67 \end{bmatrix}\right)$. $(A+2I)^2 = 0$, so $N^2_{-2}(A) = \mathbb{C}^4$. Any vector not in $N^1_{-2}(A)$, e.g., $\mathbf{e}_1 = [1, 0, 0, 0]^T$, is a generalized eigenvector for -2 that is not an eigenvector for -2.

6.2 Jordan Canonical Form

The Jordan canonical form is perhaps the single most important and widely used similarity-based canonical form for (square) matrices.

Definitions:

Let F be an algebraically closed field (e.g., $\mathbb{C}$), and let $A \in F^{n\times n}$. (The real versions of the definitions and results are presented in Section 6.3.)

For $\lambda \in F$ and positive integer k, the **Jordan block** of size k with eigenvalue λ is the $k \times k$ matrix having every diagonal entry equal to λ, every first superdiagonal entry equal to 1, and every other entry equal to 0, i.e.,

$$J_k(\lambda) = \begin{bmatrix} \lambda & 1 & 0 & \cdots & 0 \\ 0 & \lambda & 1 & & 0 \\ \vdots & \ddots & \ddots & \ddots & \\ 0 & \cdots & 0 & \lambda & 1 \\ 0 & \cdots & 0 & 0 & \lambda \end{bmatrix}.$$

A **Jordan matrix** (or a **matrix in Jordan canonical form**) is a block diagonal matrix having Jordan blocks as the diagonal blocks, i.e., a matrix of the form $J_{k_1}(\lambda_1) \oplus \cdots \oplus J_{k_t}(\lambda_t)$ for some positive integers $t, k_1, \ldots, k_t$ and some $\lambda_1, \ldots, \lambda_t \in F$. (Note: the λ_i need not be distinct.)

A **Jordan canonical form** of matrix A, denoted J_A or JCF(A), is a Jordan matrix that is similar to A. It is conventional to group the blocks for the same eigenvalue together and to order the Jordan blocks with the same eigenvalue in nonincreasing size order.

The **Jordan invariants** of A are the following parameters:

- The set of distinct eigenvalues of A.
- For each eigenvalue λ, the number b_λ and sizes $p_1, \ldots, p_{b_\lambda}$ of the Jordan blocks with eigenvalue λ in a Jordan canonical form of A.

The **total number of Jordan blocks** in a Jordan canonical form of A is $\sum b_\mu$, where the sum is taken over all distinct eigenvalues μ.

If $J_A = C^{-1}AC$, then the ordered set of columns of C is called a **Jordan basis** for A.

Let $\mathbf{x}$ be an eigenvector for eigenvalue λ of A. If $\mathbf{x} \in \text{range}(A - \lambda I)^h - \text{range}(A - \lambda I)^{h+1}$. Then h is called the **depth** of $\mathbf{x}$.

Let $\mathbf{x}$ be an eigenvector of depth h for eigenvalue λ of A. A **Jordan chain** above $\mathbf{x}$ is a sequence of vectors $\mathbf{x}_0 = \mathbf{x}, \mathbf{x}_1, \ldots, \mathbf{x}_h$ satisfying $\mathbf{x}_i = (A - \lambda I)\mathbf{x}_{i+1}$ for $i = 0, \ldots, h-1$.

Let V be a finite dimensional vector space over F, and let T be a linear operator on V.

A **Jordan basis** for T is an ordered basis $\mathcal{B}$ of V, with respect to which the matrix ${}_{\mathcal{B}}[T]_{\mathcal{B}}$ of T is a Jordan matrix. In this case, ${}_{\mathcal{B}}[T]_{\mathcal{B}}$ is a Jordan canonical form of T, denoted JCF(T) or J_T, and the **Jordan invariants** of T are the Jordan invariants of JCF$(T) = {}_{\mathcal{B}}[T]_{\mathcal{B}}$.

Facts:

Facts requiring proof for which no specific reference is given can be found in [HJ85, Chapter 3] or [Mey00, Chapter 7.8].

Notation: F is an algebraically closed field, $A, B \in F^{n\times n}$, and λ is an eigenvalue of A.

1. A has a Jordan canonical form J_A, and J_A is unique up to permutation of the Jordan blocks. In particular, the Jordan invariants of A are uniquely determined by A.
2. A, B are similar if and only if they have the same Jordan invariants.
3. The Jordan invariants and, hence, the Jordan canonical form of A can be found from the eigenvalues and the ranks of powers of $A - \lambda I$. Specifically, the number of Jordan blocks of size k in J_A with eigenvalue λ is

$$\operatorname{rank}(A - \lambda I)^{k-1} + \operatorname{rank}(A - \lambda I)^{k+1} - 2\operatorname{rank}(A - \lambda I)^{k}.$$

4. The total number of Jordan blocks in a Jordan canonical form of A is the maximal number of linearly independent eigenvectors of A.
5. The number b_λ of Jordan blocks with eigenvalue λ in J_A equals the geometric multiplicity $\gamma_A(\lambda)$ of λ. A is nonderogatory if and only if for each eigenvalue λ of A, J_A has exactly one block with λ.
6. The size of the largest Jordan block with eigenvalue λ equals the multiplicity of λ as a root of the minimal polynomial $q_A(x)$ of A.
7. The size of the largest Jordan block with eigenvalue λ equals the size of the index $\nu_\lambda(A)$ of A at λ.
8. The sum of the sizes of all the Jordan blocks with eigenvalue λ in J_A (i.e., the number of times λ appears on the diagonal of the Jordan canonical form) equals the algebraic multiplicity $\alpha_A(\lambda)$ of λ.
9. Knowledge of both the characteristic and minimal polynomials suffices to determine the Jordan block sizes for any eigenvalue having algebraic multiplicity at most 3 and, hence, to determine the Jordan canonical form of A if no eigenvalue of A has algebraic multiplicity exceeding 3. This is not necessarily true when the algebraic multiplicity of an eigenvalue is 4 or greater (cf. Example 3 below).
10. Knowledge of the the algebraic multiplicity, geometric multiplicity, and index of an eigenvalue λ suffices to determine the Jordan block sizes for λ if the algebraic multiplicity of λ is at most 6. This is not necessarily true when the algebraic multiplicity of an eigenvalue is 7 or greater (cf. Example 4 below).
11. The following are equivalent:

    (a) A is similar to a diagonal matrix.

    (b) The total number of Jordan blocks of A equals n.

    (c) The size of every Jordan block in a Jordan canonical form J_A of A is 1.

12. If A is real, then nonreal eigenvalues of A occur in conjugate pairs; furthermore, if λ is a nonreal eigenvalue, then each size k Jordan block with eigenvalue λ can be paired with a size k Jordan block for $\overline{\lambda}$.
13. If $A = A_1 \oplus \cdots \oplus A_m$, then $J_{A_1} \oplus \cdots \oplus J_{A_m}$ is a Jordan canonical form of A.
14. [Mey00, Chapter 7.8] A Jordan basis and Jordan canonical form of A can be constructed by using Algorithm 1.

Algorithm 1: Jordan Basis and Jordan Canonical Form
Input: $A \in F^{n\times n}$, the distinct eigenvalues $\mu_1, \ldots, \mu_r$, the indices $\nu_1, \ldots, \nu_r$.
Output: $C \in F^{n\times n}$ such that $C^{-1}AC = J_A$.
Initially C has no columns.
FOR $i = 1, \ldots, r$ % working on eigenvalue μ_i
Step 1: Find a special basis $\mathcal{B}_{\mu_i}$ for $E_{\mu_i}(A)$.
(a) Initially $\mathcal{B}_{\mu_i}$ has no vectors.
(b) FOR $k = \nu_i - 1$ down to 0
Extend the set of vectors already found to a basis for range$(A - \mu_i I)^k \cap E_{\mu_i}(A)$.
(c) Denote the vectors of $\mathcal{B}_{\mu_i}$ by $\mathbf{b}_j$ (ordered as found in step (**b**)).
Step 2: For each vector $\mathbf{b}_j$ found in Step 1, build a Jordan chain above $\mathbf{b}_j$.
FOR $j = 1, \ldots, \dim\ker(A - \mu_i I)$ % working on $\mathbf{b}_j$
(a) Solve $(A - \mu_i I)^{h_j}\mathbf{u}_j = \mathbf{b}_j$ for $\mathbf{u}_j$ where h_j is the depth of $\mathbf{b}_j$.
(b) Insert $(A - \mu_i I)^{h_j}\mathbf{u}_j, (A - \mu_i I)^{h_j-1}\mathbf{u}_j, \ldots, (A - \mu_i I)\mathbf{u}_j, \mathbf{u}_j$
as the next $h + 1$ columns of C.

15. A and its transpose A^T have the same Jordan canonical form (and are, therefore, similar).
16. For a nilpotent matrix, the list of block sizes determines the Jordan canonical form or, equivalently, determines the similarity class. The number of similarity classes of nilpotent matrices of size n is the number of partitions of n.
17. Let J_A be a Jordan matrix, let D be the diagonal matrix having the same diagonal as J_A, and let $N = J_A - D$. Then N is nilpotent.
18. A can be expressed as the sum of a diagonalizable matrix A_D and a nilpotent matrix A_N, where A_D and A_N are polynomials in A (and A_D and A_N commute).
19. Let V be an n-dimensional vector space over F and T be a linear operator on V. Facts 1, 3 to 10, 16, and 18 remain true when matrix A is replaced by linear operator T; in particular, JCF(T) exists and is independent (up to permutation of the diagonal Jordan blocks) of the ordered basis of V used to compute it, and the Jordan invariants of T are independent of basis.

Examples:

1. $J_4(3) = \begin{bmatrix} 3 & 1 & 0 & 0 \\ 0 & 3 & 1 & 0 \\ 0 & 0 & 3 & 1 \\ 0 & 0 & 0 & 3 \end{bmatrix}.$
2. Let A be the matrix in Example 1 in Section 6.1. $p_A(x) = x^4 + 8x^3 + 24x^2 + 32x + 16 = (x+2)^4$, so the only eigenvalue of A is -2 with algebraic multiplicity 4. From Example 1 in section 6.1, A has 3 linearly independent eigenvectors for eigenvalue -2, so J_A has 3 Jordan blocks with eigenvalue -2. In this case, this is enough information to completely determine that $J_A = \begin{bmatrix} -2 & 1 & 0 & 0 \\ 0 & -2 & 0 & 0 \\ 0 & 0 & -2 & 0 \\ 0 & 0 & 0 & -2 \end{bmatrix}.$
3. The Jordan canonical form of A is not necessarily determined by the characteristic and minimal polynmials of A. For example, the Jordan matrices $A = J_2(0) \oplus J_1(0) \oplus J_1(0)$ and $B = J_2(0) \oplus J_2(0)$ are not similar to each other, but have $p_A(x) = p_B(x) = x^4$, $q_A(x) = q_B(x) = x^2$.
4. The Jordan canonical form of A is not necessarily determined by the eigenvalues and the algebraic multiplicity, geometric multiplicity, and index of each eigenvalue. For example, the Jordan matrices $A = J_3(0) \oplus J_3(0) \oplus J_1(0)$ and $B = J_3(0) \oplus J_2(0) \oplus J_2(0)$ are not similar to each other, but have $\alpha_A(0) = \alpha_B(0) = 7, \gamma_A(0) = \gamma_B(0) = 3, \nu_0(A) = \nu_0(B) = 3$ (and $p_A(x) = p_B(x) = x^7, q_A(x) = q_B(x) = x^3$).

TABLE 6.1 rank$(A-\lambda I)^k$

$k=$	1	2	3	4	5
$\lambda=1$	11	10	9	9	9
$\lambda=2$	12	10	10	10	10
$\lambda=3$	12	11	10	9	9

5. We use Algorithm 1 to find a matrix C such that $C^{-1}AC = J_A$ for
$A = \begin{bmatrix} -2 & 3 & 0 & 1 & -1 \\ 4 & 0 & 3 & 0 & -2 \\ 6 & -3 & 3 & -1 & -1 \\ -8 & 6 & -3 & 2 & 0 \\ 2 & 3 & 3 & 1 & -3 \end{bmatrix}$. Computations show that $p_A(x) = x^5$ and $\ker A = \text{Span}(\mathbf{z}_1, \mathbf{z}_2, \mathbf{z}_3)$, where $\mathbf{z}_1 = [3, 2, -4, 0, 0]^T, \mathbf{z}_2 = [0, 1, 0, -3, 0]^T, \mathbf{z}_3 = [3, 4, 0, 0, 6]^T$.
For Step 1, $A^3 = \mathbf{0}$, and range$(A^2) = \text{Span}(\mathbf{b}_1)$ where $\mathbf{b}_1 = [-1, -1, 0, -1, -2]^T$.
Then $\mathcal{B} = \{\mathbf{b}_1, \mathbf{z}_1, \mathbf{z}_2\}$ is a suitable basis (any 2 of $\{\mathbf{z}_1, \mathbf{z}_2, \mathbf{z}_3\}$ will work in this case).
For Step 2, construct a Jordan chain above $\mathbf{b}_1$ by solving $A^2\mathbf{u}_1 = \mathbf{b}_1$. There are many possible solutions; we choose $\mathbf{u}_1 = [0, 0, 0, 0, 1]^T$. Then $A\mathbf{u}_1 = [-1, -2, -1, 0, -3]^T$,
$$C = \begin{bmatrix} -1 & -1 & 0 & 3 & 0 \\ -1 & -2 & 0 & 2 & 1 \\ 0 & -1 & 0 & -4 & 0 \\ -1 & 0 & 0 & 0 & -3 \\ -2 & -3 & 1 & 0 & 0 \end{bmatrix}, \quad \text{and} \quad J_A = \begin{bmatrix} 0 & 1 & 0 \\ 0 & 0 & 1 \\ 0 & 0 & 0 \end{bmatrix} \oplus [0] \oplus [0].$$
6. We compute the Jordan canonical form of a 14×14 matrix A by the method in Fact 3, where the necessary data about the eigenvalues of A and ranks is given in Table 6.1.

$\lambda = 1$
- The number of blocks of size 1 is $14 + 10 - 2 \cdot 11 = 2$.
- The number of blocks of size 2 is $11 + 9 - 2 \cdot 10 = 0$.
- The number of blocks of size 3 is $10 + 9 - 2 \cdot 9 = 1$.

So $\nu_1 = 3$ and $b_1 = 3$.

$\lambda = 2$
- The number of blocks of size 1 is $14 + 10 - 2 \cdot 12 = 0$.
- The number of blocks of size 2 is $12 + 10 - 2 \cdot 10 = 2$.

So, $\nu_2 = 2$ and $b_2 = 2$.

$\lambda = 3$
- The number of blocks of size 1 is $14 + 11 - 2 \cdot 12 = 1$.
- The number of blocks of size 2 is $12 + 10 - 2 \cdot 11 = 0$.
- The number of blocks of size 3 is $11 + 9 - 2 \cdot 10 = 0$.
- The number of blocks of size 4 is $10 + 9 - 2 \cdot 9 = 1$.

So, $\nu_3 = 4$ and $b_3 = 2$.

From this information,
$$J_A = J_3(1) \oplus J_1(1) \oplus J_1(1) \oplus J_2(2) \oplus J_2(2) \oplus J_4(3) \oplus J_1(3).$$

6.3 Real-Jordan Canonical Form

The real-Jordan canonical form is used in applications to differential equations, dynamical systems, and control theory (see Chapter 56). The real-Jordan canonical form is discussed here only for matrices and with limited discussion of generalized eigenspaces; more generality is possible, and is readily derivable from the corresponding results for the Jordan canonical form.

Definitions:

Let $A \in \mathbb{R}^{n\times n}, \alpha, \beta, \alpha_j, \beta_j \in \mathbb{R}$.

The **real generalized eigenspace** of A at eigenvalue $\alpha + \beta i$ is

$$E(A, \alpha + \beta i) = \begin{cases} \ker((A^2 - 2\alpha A + (\alpha^2 + \beta^2)I)^\nu) & \text{if } \beta \neq 0 \\ N_\alpha^\nu(A) = \ker((A - \alpha I)^\nu) & \text{if } \beta = 0. \end{cases}$$

The vector $\mathbf{x} \in \mathbb{R}^n$ is a **real generalized eigenvector** of A for $\alpha + \beta i$ if $\mathbf{x} \neq \mathbf{0}$ and $\mathbf{x} \in E(A, \alpha + \beta i)$.

For $\alpha, \beta \in \mathbb{R}$ with $\beta \neq 0$, and even positive integer $2k$, the **real-Jordan block** of size $2k$ with eigenvalue $\alpha + \beta i$ is the $2k \times 2k$ matrix having k copies of $M_2(\alpha, \beta) = \begin{bmatrix} \alpha & \beta \\ -\beta & \alpha \end{bmatrix}$ on the (block matrix) diagonal, $k - 1$ copies of $I_2 = \begin{bmatrix} 1 & 0 \\ 0 & 1 \end{bmatrix}$ on the first (block matrix) superdiagonal, and copies of $0_2 = \begin{bmatrix} 0 & 0 \\ 0 & 0 \end{bmatrix}$ everywhere else, i.e.,

$$J_{2k}^{\mathbb{R}}(\alpha + \beta i) = \begin{bmatrix} M_2(\alpha,\beta) & I_2 & 0_2 & \cdots & 0_2 \\ 0_2 & M_2(\alpha,\beta) & I_2 & \cdots & 0_2 \\ \vdots & \ddots & \ddots & & \vdots \\ 0_2 & \cdots & 0_2 & M_2(\alpha,\beta) & I_2 \\ 0_2 & \cdots & 0_2 & 0_2 & M_2(\alpha,\beta) \end{bmatrix}.$$

A **real-Jordan matrix** (or a **matrix in real-Jordan canonical form**) is a block diagonal matrix having diagonal blocks that are Jordan blocks or real-Jordan blocks, i.e., a matrix of the form $J_{m_1}(\alpha_1) \oplus \cdots \oplus J_{m_t}(\alpha_t) \oplus J_{2k_{t+1}}^{\mathbb{R}}(\alpha_{t+1} + \beta_{t+1} i) \oplus \cdots \oplus J_{2k_s}^{\mathbb{R}}(\alpha_s + \beta_s i)$ (or a permutation of the direct summands).

A **real-Jordan canonical form** of matrix A, denoted $J_A^{\mathbb{R}}$ or $\mathrm{JCF}^{\mathbb{R}}(A)$, is a real-Jordan matrix that is similar to A. It is conventional to use $\beta_j > 0$, to group the blocks for the same eigenvalue together, and to order the Jordan blocks with the same eigenvalue in nonincreasing size order.

The **total number of Jordan blocks** in a real-Jordan canonical form of A is the number of blocks (Jordan or real-Jordan) in $J_A^{\mathbb{R}}$.

Facts:

Facts requiring proof for which no specific reference is given can be found in [HJ85, Chapter 3].

Notation: $A, B \in \mathbb{R}^{n\times n}, \alpha, \beta, \alpha_j, \beta_j \in \mathbb{R}$.

1. The real generalized eigenspace of a complex number $\lambda = \alpha + \beta i$ and its conjugate $\bar{\lambda} = \alpha - \beta i$ are equal, i.e., $E(A, \alpha + \beta i) = E(A, \alpha - \beta i)$.
2. The real-Jordan blocks with a nonreal complex number and its conjugate are similar to each other.
3. A has a real-Jordan canonical form $J_A^{\mathbb{R}}$, and $J_A^{\mathbb{R}}$ is unique up to permutation of the diagonal (real-) Jordan blocks.
4. A, B are similar if and only if their real-Jordan canonical forms have the same set of Jordan and real-Jordan blocks (although the order may vary).
5. If all the eigenvalues of A are real, then $J_A^{\mathbb{R}}$ is the same as J_A (up to the order of the Jordan blocks).
6. The real-Jordan canonical form of A can be computed from the Jordan canonical form of A. The nonreal eigenvalues occur in conjugate pairs, and if $\beta > 0$, then each size k Jordan block with eigenvalue $\alpha + \beta i$ can be paired with a size k Jordan block for $\alpha - \beta i$. Then $J_k(\alpha + \beta i) \oplus J_k(\alpha - \beta i)$ is replaced by $J_{2k}^{\mathbb{R}}(\alpha + \beta i)$. The Jordan blocks of $J_A^{\mathbb{R}}$ with real eigenvalues are the same as the those of J_A.
7. The total number of Jordan and real-Jordan blocks in a real-Jordan canonical form of A is the number of Jordan blocks with a real eigenvalue plus half the number of Jordan blocks with a nonreal eigenvalue in a Jordan canonical form of A.

8. If $\beta \neq 0$, the size of the largest real-Jordan block with eigenvalue $\alpha + \beta i$ is twice the multiplicity of $x^2 - 2\alpha x + (\alpha^2 + \beta^2)$ as a factor of the minimal polynomial $q_A(x)$ of A.
9. If $\beta \neq 0$, the sum of the sizes of all the real-Jordan blocks with eigenvalue $\alpha + \beta i$ in J_A equals twice the algebraic multiplicity $\alpha_A(\alpha + \beta i)$.
10. If $\beta \neq 0$, $\dim E(A, \alpha + \beta i) = \alpha_A(\alpha + \beta i)$.
11. If $A = A_1 \oplus \cdots \oplus A_m$, then $J^{\mathbb{R}}_{A_1} \oplus \cdots \oplus J^{\mathbb{R}}_{A_m}$ is a real-Jordan canonical form of A.

Examples:

1. Let $a = \begin{bmatrix} -10 & 6 & -4 & 4 & 0 \\ -17 & 10 & -4 & 6 & -1 \\ -4 & 2 & -3 & 2 & 1 \\ -11 & 6 & -11 & 6 & 3 \\ -4 & 2 & -4 & 2 & 2 \end{bmatrix}$. Since the characteristic and minimal polynomials of A are both $x^5 - 5x^4 + 12x^3 - 16x^2 + 12x - 4 = (x - 1)\left(x^2 - 2x + 2\right)^2$,

$$J^{\mathbb{R}}_A = \begin{bmatrix} 1 & 1 & 1 & 0 & 0 \\ -1 & 1 & 0 & 1 & 0 \\ 0 & 0 & 1 & 1 & 0 \\ 0 & 0 & -1 & 1 & 0 \\ 0 & 0 & 0 & 0 & 1 \end{bmatrix}.$$

2. The requirement that $\beta \neq 0$ is important. $A = \begin{bmatrix} 0 & 0 & 1 & 0 \\ 0 & 0 & 0 & 1 \\ 0 & 0 & 0 & 0 \\ 0 & 0 & 0 & 0 \end{bmatrix}$ is *not* a real-Jordan matrix;

$$J_A = \begin{bmatrix} 0 & 1 & 0 & 0 \\ 0 & 0 & 0 & 0 \\ 0 & 0 & 0 & 1 \\ 0 & 0 & 0 & 0 \end{bmatrix}.$$

6.4 Rational Canonical Form: Elementary Divisors

The elementary divisors rational canonical form is closely related to the Jordan canonical form (see Fact 7 below). A rational canonical form (either the elementary divisors or the invariant factors version, cf. Section 6.6) is used when it is desirable to stay within a field that is not algebraically closed, such as the rational numbers.

Definitions:

Let F be a field.

For a monic polynomial $p(x) = x^n + c_{n-1}x^{n-1} + \cdots + c_2x^2 + c_1x + c_0 \in F[x]$ (with $n \geq 1$), the **companion matrix** of $p(x)$ is the $n \times n$ matrix $C(p) = \begin{bmatrix} 0 & 0 & \cdots & -c_0 \\ 1 & 0 & \cdots & -c_1 \\ \vdots & \ddots & & \vdots \\ 0 & \cdots & 1 & -c_{n-1} \end{bmatrix}$.

An **elementary divisors rational canonical form matrix (ED-RCF matrix)** (over F) is a block diagonal matrix of the form $C(h_1^{m_1}) \oplus \cdots \oplus C(h_t^{m_t})$ where each $h_i(x)$ is a monic polynomial that is irreducible over F.

The **elementary divisors** of the ED-RCF matrix $C(h_1^{m_1}) \oplus \cdots \oplus C(h_t^{m_t})$ are the polynomials $h_i(x)^{m_i}$, $i = 1, \ldots t$.

An **elementary divisors rational canonical form** of matrix $A \in F^{n\times n}$, denoted $\mathrm{RCF}_{ED}(A)$, is an ED-RCF matrix that is similar to A. It is conventional to group the companion matrices associated with powers of the same irreducible polynomial together, and within such a group to order the blocks in size order.

The **elementary divisors** of A are the elementary divisors of $\mathrm{RCF}_{ED}(A)$.

Let V be a finite dimensional vector space over F, and let T be a linear operator on V.

An **ED-RCF basis** for T is an ordered basis $\mathcal{B}$ of V, with respect to which the matrix ${}_{\mathcal{B}}[T]_{\mathcal{B}}$ of T is an ED-RCF matrix. In this case, ${}_{\mathcal{B}}[T]_{\mathcal{B}}$ is an elementary divisors rational canonical form of T, denoted $\mathrm{RCF}_{ED}(T)$, and the **elementary divisors** of T are the elementary divisors of $\mathrm{RCF}_{ED}(T) = {}_{\mathcal{B}}[T]_{\mathcal{B}}$.

Let $q(x)$ be a monic polynomial over F.

A **primary decomposition** of a nonconstant monic polynomial $q(x)$ over F is a factorization $q(x) = (h_1(x))^{m_1} \cdots (h_r(x))^{m_r}$, where the $h_i(x), i = 1, \ldots, r$ are distinct monic irreducible polynomials over F.

The factors $(h_i(x))^{m_i}$ in a primary decomposition of $q(x)$ are the **primary factors** of $q(x)$.

Facts:

Facts requiring proof for which no specific reference is given can be found in [HK71, Chapter 7] or [DF04, Chapter 12].

1. The characteristic and minimal polynomials of the companion matrix $C(p)$ are both equal to $p(x)$.
2. Whether or not a matrix is an ED-RCF matrix depends on polynomial irreducibility, which depends on the field. See Example 1 below.
3. Every matrix $A \in F^{n\times n}$ is similar to an ED-RCF matrix, $\mathrm{RCF}_{ED}(A)$, and $\mathrm{RCF}_{ED}(A)$ is unique up to permutation of the companion matrix blocks on the diagonal. In particular, the elementary divisors of A are uniquely determined by A.
4. $A, B \in F^{n\times n}$ are similar if and only if they have the same elementary divisors.
5. (See Fact 3 in Section 6.2) For $A \in F^{n\times n}$, the elementary divisors and, hence, $\mathrm{RCF}_{ED}(A)$ can be found from the irreducible factors $h_i(x)$ of the characteristic polynomial of A and the ranks of powers of $h_i(A)$. Specifically, the number of times $h_i(x)^k$ appears as an elementary divisor of A is
$$\frac{1}{\deg h_i(x)}(\mathrm{rank}(h_i(A))^{k-1} + \mathrm{rank}(h_i(A))^{k+1} - 2\,\mathrm{rank}(h_i(A))^k).$$
6. If $A \in F^{n\times n}$ has n eigenvalues in F, then the elementary divisors of A are the polynomials $(x-\lambda)^k$, where the $J_k(\lambda)$ are the Jordan blocks of J_A.
7. There is a natural association between the diagonal blocks in the elementary divisors rational canonical form of A and the Jordan canonical form of A in $\hat{F}^{n\times n}$, where $\hat{F}$ is the algebraic closure of F. Let $h(x)^m$ be an elementary divisor of A, and factor $h(x)$ into monic linear factors over $\hat{F}$, $h(x) = (x-\lambda_1)\cdots(x-\lambda_t)$. If the roots of $h(x)$ are distinct (e.g., if the characteristic of F is 0 or F is a finite field), then the ED-RCF diagonal block $C(h^m)$ is associated with the Jordan blocks $J_m(\lambda_i), i = 1, \ldots, t$. If the characteristic of F is p and $h(x)$ has repeated roots, then all roots have the same multiplicity p^k (for some positive integer k) and the ED-RCF diagonal block $C(h^m)$ is associated with the Jordan blocks $J_{p^k m}(\lambda_i), i = 1, \ldots, t$.
8. [HK71, Chapter 4.5] Every monic polynomial $q(x)$ over F has a primary decomposition. The primary decomposition is unique up to the order of the monic irreducible polynomials, i.e., the set of primary factors of $q(x)$ is unique.
9. [HK71, Chapter 6.8] Let $q(x) \in F[x]$, let $h_i(x)^{m_i}, i = 1, \ldots, r$ be the primary factors, and define $f_i(x) = \frac{q(x)}{h_i(x)^{m_i}}$. Then there exist polynomials $g_i(x)$ such that $f_1(x)g_1(x) + \cdots + f_r(x)g_r(x) = 1$.

Let $A \in F^{n\times n}$ and let $q_A(x) = (h_1(x))^{m_1} \cdots (h_r(x))^{m_r}$ be a primary decomposition of its minimal polynomial.

10. Every primary factor $h_i(x)^{m_i}$ of $q_A(x)$ is an elementary divisor of A.
11. Every elementary divisor of A is of the form $(h_i(x))^m$ with $m \le m_i$ for some $i \in \{1, \ldots, r\}$.

12. [HK71, Chapter 6.8] *Primary Decomposition Theorem*
    (a) $F^n = \ker(h_1(A)^{m_1}) \oplus \cdots \oplus \ker(h_r(A)^{m_r})$.
    (b) Let f_i and g_i be as defined in Fact 9. Then for $i = 1, \ldots, r$, $E_i = f_i(A)g_i(A)$ is the projection onto $\ker(h_i(A)^{m_i})$ along $\ker(h_1(A)^{m_1}) \oplus \cdots \oplus \ker(h_{i-1}(A)^{m_{i-1}}) \oplus \ker(h_{i+1}(A)^{m_{i+1}}) \oplus \cdots \oplus \ker(h_r(A)^{m_r})$.
    (c) The $E_i = f_i(A)g_i(A)$ are mutually orthogonal idempotents (i.e., $E_i^2 = E_i$ and $E_iE_j = 0$ if $i \neq j$) and $I = E_1 + \cdots + E_r$.
13. [HK71, Chapter 6.7] If $A \in F^{n\times n}$ is diagonalizable, then $A = \mu_1 E_1 + \cdots + \mu_r E_R$ where the E_i are the projections defined in Fact 12 with primary factors $h_i(x)^{m_i} = (x - \mu_i)$ of $q_A(x)$.

Let V be an n-dimensional vector space over F, and let T be a linear operator on V.

14. Facts 3, 5 to 7, and 10 to 13 remain true when matrix A is replaced by linear operator T; in particular, $\mathrm{RCF}_{ED}(T)$ exists and is independent (up to permutation of the companion matrix diagonal blocks) of the ordered basis of V used to compute it, and the elementary divisors of T are independent of basis.
15. If $\hat{T}$ denotes T restricted to $\ker(h_i(T)^{m_i})$, then the minimal polynomial of $\hat{T}$ is $h_i(T)^{m_i}$.

Examples:

1. Let $A = [-1] \oplus [-1] \oplus \begin{bmatrix} 0 & 0 & -1 \\ 1 & 0 & -3 \\ 0 & 1 & -3 \end{bmatrix} \oplus \begin{bmatrix} 0 & 2 \\ 1 & 0 \end{bmatrix} \oplus \begin{bmatrix} 0 & 0 & 0 & -4 \\ 1 & 0 & 0 & 0 \\ 0 & 1 & 0 & 4 \\ 0 & 0 & 1 & 0 \end{bmatrix}$. Over $\mathbb{Q}$, A is an ED-RCF matrix and its elementary divisors are $x+1, x+1, (x+1)^3, x^2-2, (x^2-2)^2$. A is not an ED-RCF matrix over $\mathbb{C}$ because $x^2 - 2$ is not irreducible over $\mathbb{C}$.

   $$\mathrm{JCF}(A)=[-1] \oplus [-1] \oplus \begin{bmatrix} -1 & 1 & 0 \\ 0 & -1 & 1 \\ 0 & 0 & -1 \end{bmatrix} \oplus [\sqrt{2}] \oplus [-\sqrt{2}] \oplus \begin{bmatrix} \sqrt{2} & 1 \\ 0 & \sqrt{2} \end{bmatrix} \oplus \begin{bmatrix} -\sqrt{2} & 1 \\ 0 & -\sqrt{2} \end{bmatrix},$$

   where the order of the Jordan blocks has been chosen to emphasize the connection to $\mathrm{RCF}_{ED}(A) = A$.

2. Let $A = \begin{bmatrix} -2 & 2 & -2 & 1 & 1 \\ 6 & -2 & 2 & -2 & 0 \\ 0 & 0 & 0 & 0 & 1 \\ -12 & 7 & -8 & 5 & 4 \\ 0 & 0 & -1 & 0 & 2 \end{bmatrix} \in \mathbb{Q}^{5\times 5}$. We use Fact 5 to determine the elementary divisors rational canonical form of A. The following computations can be performed easily over $\mathbb{Q}$ in a computer algebra system such as Mathematica, Maple, or MATLAB® (see Chapters 71, 72, 73), or on a matrix-capable calculator. $p_A(x) = x^5 - 3x^4 + x^3 + 5x^2 - 6x + 2 = (x-1)^3(x^2-2)$. Table 6.2 gives the of ranks $h_i(A)^k$ where $h_i(x)$ is shown in the left column.

   $h(x) = x - 1$ The number of times $x - 1$ appears as an elementary divisor is $5 + 2 - 2 \cdot 3 = 1$.
   The number of $(x-1)^2$ appears as an elementary divisor is $3 + 2 - 2 \cdot 2 = 1$.

   $h(x) = x^2 - 2$ The number of times $x^2 - 2$ appears as an elementary divisor is $(5+3-2\cdot 3)/2 = 1$.

   Thus, $\mathrm{RCF}_{ED}(A) = C(x-1) \oplus C((x-1)^2) \oplus C(x^2-2) = [1] \oplus \begin{bmatrix} 0 & -1 \\ 1 & 2 \end{bmatrix} \oplus \begin{bmatrix} 0 & 2 \\ 1 & 0 \end{bmatrix}$.

TABLE 6.2 rank($h(A)^k$)

$k =$	1	2	3
$h_1(x) = x - 1$	3	2	2
$h_2(x) = x^2 - 2$	3	3	3

3. We find the projections $E_i, i = 1,2$ in Fact 12 for A in the previous example. From the elementary divisors of A, $q_A(x) = (x-1)^2(x^2-2)$. Let $h_1(x) = (x-1)^2, h_2(x) = x^2 - 2$. Then $f_1(x) = x^2-2$, $f_2(x) = (x-1)^2$. *Note:* normally the $f_i(x)$ will not be primary factors; this happens here because there are only two primary factors. If we choose $g_1(x) = -(2x-1), g_2(x) = 2x+3$, then $1 = f_1(x)g_1(x) + f_2(x)g_2(x)$ (g_1, g_2 can be found by the Euclidean algorithm). Then

$$E_1 = f_1(A)g_1(A) = \begin{bmatrix} -2 & 1 & -1 & 1 & 0 \\ 0 & 0 & 0 & 0 & 0 \\ 0 & 0 & 1 & 0 & 0 \\ -6 & 3 & -2 & 3 & 0 \\ 0 & 0 & 0 & 0 & 1 \end{bmatrix} \quad \text{and} \quad E_2 = f_2(A)g_2(A) = \begin{bmatrix} 3 & -1 & 1 & -1 & 0 \\ 0 & 1 & 0 & 0 & 0 \\ 0 & 0 & 0 & 0 & 0 \\ 6 & -3 & 2 & -2 & 0 \\ 0 & 0 & 0 & 0 & 0 \end{bmatrix},$$

and it is easy to verify that $E_1^2 = E_1, E_2^2 = E_2, E_1E_2 = E_2E_1 = 0$, and $E_1 + E_2 = I$.

6.5 Smith Normal Form on $F[x]^{n\times n}$

For a matrix $A \in F^{n\times n}$, the Smith normal form of $xI_n - A$ is an important tool for the computation of the invariant factors rational canonical form of A discussed in Section 6.6. In this section, Smith normal form is discussed only for matrices in $F[x]^{n\times n}$, and the emphasis is on finding the Smith normal form of $xI_n - A$, where $A \in F^{n\times n}$. Smith normal form is used more generally for matrices over principal ideal domains (see Section 23.2); it is not used extensively as a canonical form within $F^{n\times n}$, since the Smith normal form of a matrix $A \in F^{n\times n}$ of rank k is $I_k \oplus 0_{n-k}$.

Definitions:

Let F be a field. For $M \in F[x]^{n\times n}$, the following operations are the **elementary row and column operations** on M:

(a) Interchange rows i, j, denoted $R_i \leftrightarrow R_j$ (analogous column operation denoted $C_i \leftrightarrow C_j$).
(b) Add a $p(x)$ multiple of row j to row i, denoted $R_i + p(x)R_j \to R_i$ (analogous column operation denoted $C_i + p(x)C_j \to C_i$).
(c) Multiply row i by a nonzero element b of F, denoted $bR_i \to R_i$ (analogous column operation denoted $bC_i \to C_i$).

A **Smith normal matrix** in $F[x]^{n\times n}$ is a diagonal matrix $D = \text{diag}(1,\dots,1,a_1(x),\dots,a_s(x),0,\dots,0)$, where the $a_i(x)$ are monic nonconstant polynomials such that $a_i(x)$ divides $a_{i+1}(x)$ for $i = 1,\dots,s-1$.

The **Smith normal form** of $M \in F[x]^{n\times n}$ is the Smith normal matrix obtained from M by elementary row and column operations.

For $A \in F^{n\times n}$, the monic nonconstant polynomials of the Smith normal form of $xI_n - A$ are the **Smith invariant factors** of A.

Facts:

Facts requiring proof for which no specific reference is given can be found in [HK71, Chapter 7] or [DF04, Chapter 12].

1. Let $M \in F[x]^{n\times n}$. Then M has a unique Smith normal form.
2. Let $A \in F^{n\times n}$. There are no zeros on the diagonal of the Smith normal form of $xI_n - A$.
3. (*Division Property*) If $a(x), b(x) \in F[x]$ and $b(x) \neq 0$, then there exist polynomials $q(x), r(x)$ such that $a(x) = q(x)b(x) + r(x)$ and $r(x) = 0$ or $\deg r(x) < \deg b(x)$.
4. The Smith normal form of $M = xI - A$ and, thus, the Smith invariant factors of A can be computed as follows:

- For $k = 1, \ldots, n-1$
  - Use elementary row and column operations and the division property of $F[x]$ to place the greatest common divisor of the entries of $M[\{k, \ldots, n\}]$ in the kth diagonal position.
  - Use elementary row and column operations to create zeros in all nondiagonal positions in row k and column k.
- Make the nth diagonal entry monic by multiplying the last column by a nonzero element of F.

This process is illustrated in Example 1 below.

Examples:

1. Let $A = \begin{bmatrix} 1 & 1 & 1 & -1 \\ 0 & 3 & 2 & -2 \\ 2 & 0 & 4 & -2 \\ 4 & 0 & 6 & -3 \end{bmatrix}$. We use the method in Fact 4 above to find the Smith normal form of $M = xI - A$ and invariant factors of A.

   - $k = 1$: Use the row and column operations on M (in the order shown):
     $R_1 \leftrightarrow R_3, -\frac{1}{2}R_1 \to R_1, R_3 + (1-x)R_1 \to R_3, R_4 + 4R_1 \to R_4,$
     $C_3 + (-2 + \frac{x}{2})C_1 \to C_3, C_4 + C_1 \to C_4$
     to obtain $M_1 = \begin{bmatrix} 1 & 0 & 0 & 0 \\ 0 & x-3 & -2 & 2 \\ 0 & -1 & \frac{x^2}{2} - \frac{5x}{2} + 1 & x \\ 0 & 0 & 2-2x & x-1 \end{bmatrix}.$
   - $k = 2$: Use the row and column operations on M_1 (in the order shown):
     $R_3 \leftrightarrow R_2, -1R_2 \to R_2, R_3 + (3-x)R_2 \to R_3,$
     $C_3 + (1 - \frac{5x}{2} + \frac{x^2}{2})C_2 \to C_3, C_4 + xC_2 \to C_4$
     to obtain $M_2 = \begin{bmatrix} 1 & 0 & 0 & 0 \\ 0 & 1 & 0 & 0 \\ 0 & 0 & \frac{x^3}{2} - 4x^2 + \frac{17x}{2} - 5 & x^2 - 3x + 2 \\ 0 & 0 & 2-2x & x-1 \end{bmatrix}.$
   - $k = 3$ (and final step): Use the row and column operations on M_2 (in the order shown):
     $R_3 \leftrightarrow R_4, -\frac{1}{2}R_3 \to R_3, R_4 + \frac{-1}{2}(x-2)(x-5)R_3 \to R_4,$
     $C_4 + \frac{1}{2}C_3 \to C_4, 4C_4 \to C_4$
     to obtain the Smith normal form of M, $M_3 = \begin{bmatrix} 1 & 0 & 0 & 0 \\ 0 & 1 & 0 & 0 \\ 0 & 0 & x-1 & 0 \\ 0 & 0 & 0 & x^3 - 4x^2 + 5x - 2 \end{bmatrix}.$

   The Smith invariant factors of A are $x-1$, $x^3 - 4x^2 + 5x - 2$.

6.6 Rational Canonical Form: Invariant Factors

Like the elementary divisors version, the invariant factors rational canonical form does not require the field to be algebraically closed. It has two other advantages: This canonical form is unique (not just unique up to permutation), and (unlike elementary divisors rational canonical form) whether a matrix is in invariant factors rational canonical form is independent of the field (see Fact 2 below).

Definitions:

Let F be a field. An **invariant factors rational canonical form matrix (IF-RCF matrix)** is a block diagonal matrix of the form $C(a_1) \oplus \cdots \oplus C(a_s)$, where $a_i(x)$ divides $a_{i+1}(x)$ for $i = 1, \ldots, s-1$.

The **invariant factors** of the IF-RCF matrix $C(a_1) \oplus \cdots \oplus C(a_s)$ are the polynomials $a_i(x), i = 1, \ldots s$.

The **invariant factors rational canonical form** of matrix $A \in F^{n\times n}$, denoted $\mathrm{RCF}_{IF}(A)$, is the IF-RCF matrix that is similar to A.

The **invariant factors** of A are the invariant factors of $\mathrm{RCF}_{IF}(A)$.

Let V be a finite dimensional vector space over F and let T be a linear operator on V.

An **IF-RCF basis** for T is an ordered basis $\mathcal{B}$ of V, with respect to which the matrix ${}_{\mathcal{B}}[T]_{\mathcal{B}}$ of T is an IF-RCF matrix. In this case, ${}_{\mathcal{B}}[T]_{\mathcal{B}}$ is the **invariant factors rational canonical form of** T, denoted $\mathrm{RCF}_{IF}(T)$, and the **invariant factors** of T are the invariant factors of $\mathrm{RCF}_{IF}(T) = {}_{\mathcal{B}}[T]_{\mathcal{B}}$.

Facts:

Facts requiring proof for which no specific reference is given can be found in [HK71, Chapter 7] or [DF04, Chapter 12]. *Notation:* $A \in F^{n\times n}$.

1. Every square matrix A is similar to a unique IF-RCF matrix, $\mathrm{RCF}_{IF}(A)$.
2. $\mathrm{RCF}_{IF}(A)$ is independent of field. That is, if K is an extension field of F and A is considered as an element of $K^{n\times n}$, $\mathrm{RCF}_{IF}(A)$ is the same as when A is considered as an element of $F^{n\times n}$.
3. Let $B \in F^{n\times n}$. Then A, B are similar if and only if $\mathrm{RCF}_{IF}(A) = \mathrm{RCF}_{IF}(B)$.
4. The characteristic polynomial is the product of the invariant factors of A, i.e., $p_A(x) = a_1(x) \cdots a_s(x)$.
5. The minimal polynomial of A is the invariant factor of highest degree, i.e., $q_A(x) = a_s(x)$.
6. The elementary divisors of $A \in F^{n\times n}$ are the primary factors (over F) of the invariant factors of A.
7. The Smith invariant factors of A are the invariant factors of A.
8. [DF04, Chapter 12.2] $\mathrm{RCF}_{IF}(A)$ and a nonsingular matrix $S \in F^{n\times n}$ such that $S^{-1}AS = \mathrm{RCF}_{IF}(A)$ can be computed by Algorithm 2.

Algorithm 2: Rational Canonical Form (invariant factors)

1. Compute the Smith normal form D of $M = xI - A$ as in Fact 4 of section 6.5, keeping track of the elementary row operations, in the order performed (column operations need not be recorded).
2. The invariant factors are the nonconstant diagonal elements $a_1(x), \ldots, a_s(x)$ of D.
3. Let $d_1, \ldots, d_s$ denote the degrees of $a_1(x), \ldots, a_s(x)$.
4. Let $G = I$.
5. FOR $k = 1, \ldots,$ number of row operations performed in step 1
   (a) If the kth row operation is $R_i \leftrightarrow R_j$, then perform column operation $C_j \leftrightarrow C_i$ on G.
   (b) If the kth row operation is $R_i + p(x)R_j \to R_i$, then perform column operation $C_j - p(A)C_i \to C_j$ on G (*note index reversal*).
   (c) If the kth row operation is $bR_i \to R_i$, then perform column operation $\frac{1}{b}C_i \to C_i$ on G.
6. G will have 0s in the first $n - s$ columns; denote the remaining columns of G by $\mathbf{g}_1, \ldots, \mathbf{g}_s$.
7. Initially S has no columns.
8. FOR $k = 1, \ldots, s$
   (a) Insert $\mathbf{g}_k$ as the next column of S (working left to right).
   (b) FOR $i = 1, \ldots, d_k - 1$.
   Insert A times the last column inserted as the next column of S.
9. $\mathrm{RCF}_{IF}(A) = S^{-1}AS$.

9. Let V be an n-dimensional vector space over F, and let T be a linear operator on V. Facts 1, 2, 4 to 6 remain true when matrix A is replaced by linear operator T; in particular, $\text{RCF}_{IF}(T)$ exists and is unique (independent of the ordered basis of V used to compute it).

Examples:

1. We can use the elementary divisors already computed to find the invariant factors and IF-RCF of A in Example 2 of Section 6.4. The elementary divisors of A are $x-1, (x-1)^2, x^2-2$. We combine these, working down from the highest power of each irreducible polynomial.
$a_2(x) = (x-1)^2(x^2-2) = x^4 - 2x^3 - x^2 + 4x - 2, a_1(x) = x - 1$. Then
$$\text{RCF}_{IF}(A) = C(x-1) \oplus C(x^4 - 2x^3 - x^2 + 4x - 2) = \begin{bmatrix} 1 & 0 & 0 & 0 & 0 \\ 0 & 0 & 0 & 0 & 2 \\ 0 & 1 & 0 & 0 & -4 \\ 0 & 0 & 1 & 0 & 1 \\ 0 & 0 & 0 & 1 & 2 \end{bmatrix}.$$
2. By Fact 7, for the matrix A in Example 1 in Section 6.5, $\text{RCF}_{IF}(A) = C(x-1) \oplus C(x^3 - 4x^2 + 5x - 2)$.
3. We can use Algorithm 2 to find a matrix S such that $\text{RCF}_{IF}(A) = S^{-1}AS$ for the matrix A in Example 1.

- $k = 1$: Starting with $G = I_4$, perform the column operations (in the order shown):

  $C_1 \leftrightarrow C_3, -2C_1 \rightarrow C_1, C_1 - (I_4 - A)C_3 \rightarrow C_1, C_1 - 4C_4 \rightarrow C_1,$

  to obtain $G_1 = \begin{bmatrix} 0 & 0 & 1 & 0 \\ 0 & 1 & 0 & 0 \\ 0 & 0 & 0 & 0 \\ 0 & 0 & 0 & 1 \end{bmatrix}$.

- $k = 2$: Use column operations on G (in the order shown):

  $C_3 \leftrightarrow C_2, -1C_2 \rightarrow C_2, C_2 - (3I_4 - A)C_3 \rightarrow C_2,$

  to obtain $G = \begin{bmatrix} 0 & 0 & 0 & 0 \\ 0 & 0 & 1 & 0 \\ 0 & 0 & 0 & 0 \\ 0 & 0 & 0 & 1 \end{bmatrix}$.

- $k = 3$ (and final step of Fact 4 in Section 6.5):

  Use column operations on G (in the order shown):

  $C_3 \leftrightarrow C_4, -2C_3 \rightarrow C_3, C_3 + \frac{1}{2}(A - 2I_4)(A - 5I_4)C_4 \rightarrow C_3,$

  to obtain $G = [\mathbf{g}_1, \mathbf{g}_2, \mathbf{g}_3, \mathbf{g}_4] = \begin{bmatrix} 0 & 0 & -\frac{3}{2} & 0 \\ 0 & 0 & -1 & 1 \\ 0 & 0 & 1 & 0 \\ 0 & 0 & 0 & 0 \end{bmatrix}$.

Then $S = [\mathbf{g}_3, \mathbf{g}_4, A\mathbf{g}_4, A^2\mathbf{g}_4] = \begin{bmatrix} -\frac{3}{2} & 0 & 1 & 4 \\ -1 & 1 & 3 & 9 \\ 1 & 0 & 0 & 2 \\ 0 & 0 & 0 & 4 \end{bmatrix}$ and $\text{RCF}_{IF}(A) = S^{-1}AS = \begin{bmatrix} 1 & 0 & 0 & 0 \\ 0 & 0 & 0 & 2 \\ 0 & 1 & 0 & -5 \\ 0 & 0 & 1 & 4 \end{bmatrix}$.

Acknowledgment

The author thanks Jeff Stuart and Wolfgang Kliemann for helpful comments on an earlier version of this chapter.

References

[DF04] D. S. Dummit and R. M. Foote. *Abstract Algebra*, 3rd ed. John Wiley & Sons, New York, 2004.

[GV96] G. H. Golub and C. F. Van Loan. *Matrix Computations*, 3rd ed. Johns Hopkins University Press, Baltimore, 1996.

[HK71] K. H. Hoffman and R. Kunze. *Linear Algebra*, 2nd ed. Prentice Hall, Upper Saddle River, NJ, 1971.

[HJ85] R. Horn and C. R. Johnson. *Matrix Analysis*. Cambridge University Press, Cambridge, 1985.

[Mey00] C. D. Meyer. *Matrix Analysis and Applied Linear Algebra*. SIAM, Philadelphia, 2000.

7

Unitary Similarity, Normal Matrices, and Spectral Theory

Helene Shapiro
Swarthmore College

Unitary transformations preserve the inner product. Hence, they preserve the metric quantities that stem from the inner product, such as length, distance, and angle. While a general similarity preserves the algebraic features of a linear transformation, such as the characteristic and minimal polynomials, the rank, and the Jordan canonical form, unitary similarities also preserve metric features such as the norm, singular values, and the numerical range. Unitary similarities are desirable in computational linear algebra for stability reasons.

Normal transformations are those which have an orthogonal basis of eigenvectors and, thus, can be represented by diagonal matrices relative to an orthonormal basis. The class of normal transformations includes Hermitian, skew-Hermitian, and unitary transformations; studying normal matrices leads to a more unified understanding of all of these special types of transformations. Often, results that are discovered first for Hermitian matrices can be generalized to the class of normal matrices. Since normal matrices are unitarily similar to diagonal matrices, things that are obviously true for diagonal matrices often hold for normal matrices as well; for example, the singular values of a normal matrix are the absolute values of the eigenvalues. Normal matrices have two important properties — diagonalizability and an orthonormal basis of eigenvectors — that tend to make life easier in both theoretical and computational situations.

7.1 Unitary Similarity

In this subsection, all matrices are over the complex numbers and are square. All vector spaces are finite dimensional complex inner product spaces.

Definitions:

A matrix U is **unitary** if $U^*U = I$.

A matrix Q is **orthogonal** if $Q^T Q = I$.

Note: This extends the definition of orthogonal matrix given earlier in Section 5.2 for real matrices.

Matrices A and B are **unitarily similar** if $B = U^*AU$ for some unitary matrix U. The term **unitarily equivalent** is sometimes used in the literature.

The **numerical range** of A is $W(A) = \{\mathbf{v}^*A\mathbf{v} | \mathbf{v}^*\mathbf{v} = 1\}$.

The **Frobenius (Eulidean) norm** of the matrix A is $\|A\|_F = \left(\sum_{i,j=1}^n |a_{ij}|^2\right)^{1/2} = \left(tr(A^*A)\right)^{1/2}$. (See Chapter 37 for more information on norms.)

The **operator norm** of the matrix A induced by the vector 2-norm $\|\cdot\|_2$ is $\|A\|_2 = max\{\|A\mathbf{v}\| \,|\, \|\mathbf{v}\| = 1\}$; this norm is also called the **spectral norm.**

Facts:

Most of the material in this section can be found in one or more of the following: [HJ85, Chap. 2] [Hal87, Chap. 3] [Gan59, Chap. IX] [MM64, I.4, III.5]. Specific references are also given for some facts.

1. A real, orthogonal matrix is unitary.
2. The following are equivalent:
   - U is unitary.
   - U is invertible and $U^{-1} = U^*$.
   - The columns of U are orthonormal.
   - The rows of U are orthonormal.
   - For any vectors $\mathbf{x}$ and $\mathbf{y}$, we have $\langle U\mathbf{x}, U\mathbf{y}\rangle = \langle \mathbf{x}, \mathbf{y}\rangle$.
   - For any vector $\mathbf{x}$, we have $\|U\mathbf{x}\| = \|\mathbf{x}\|$.
3. If U is unitary, then U^*, U^T, and $\bar{U}$ are also unitary.
4. If U is unitary, then every eigenvalue of U has modulus 1 and $|\det(U)| = 1$. Also, $\|U\|_2 = 1$.
5. The product of two unitary matrices is unitary and the product of two orthogonal matrices is orthogonal.
6. The set of $n \times n$ unitary matrices, denoted $U(n)$, is a subgroup of $GL(n, \mathbb{C})$, called the unitary group. The subgroup of elements of $U(n)$ with determinant one is the special unitary group, denoted $SU(n)$. Similarly, the set of $n \times n$ real orthogonal matrices, denoted $O(n)$, is a subgroup of $GL(n, \mathbb{R})$, called the real, orthogonal group, and the subgroup of real, orthogonal matrices of determinant one is $SO(n)$, the special orthogonal group.
7. Let U be unitary. Then
   - $\|A\|_F = \|U^*AU\|_F$.
   - $\|A\|_2 = \|U^*AU\|_2$.
   - A and U^*AU have the same singular values, as well as the same eigenvalues.
   - $W(A) = W(U^*AU)$.
8. [Sch09] Any square, complex matrix A is unitarily similar to a triangular matrix. If $T = U^*AU$ is triangular, then the diagonal entries of T are the eigenvalues of A. The unitary matrix U can be chosen to get the eigenvalues in any desired order along the diagonal of T. Algorithm 1 below gives a method for finding U, assuming that one knows how to find an eigenvalue and eigenvector, e.g., by exact methods for small matrices (Section 4.3), and how to find an orthonormal basis containing the given vector, e.g., by the Gram-Schmidt process (Section 5.5). This algorithm is designed to illuminate the result, not for computation with large matrices in finite precision arithmetic; for such problems appropriate numerical methods should be used (cf. Section 43.2).

Algorithm 1: Unitary Triangularization
Input: $A \in \mathbb{C}^{n\times n}$.
Output: unitary U such that $U^*AU = T$ is triangular.
1. $A_1 = A$.
2. FOR $k = 1, \ldots, n-1$
   (a) Find an eigenvalue and normalized eigenvector $\mathbf{x}$ of the $(n+1-k)\times(n+1-k)$ matrix A_k.
   (b) Find an orthonormal basis $\mathbf{x}, \mathbf{y}_2, \ldots, \mathbf{y}_{n+1-k}$ for $\mathbb{C}^{n+1-k}$.
   (c) $U_k = [\mathbf{x}, \mathbf{y}_2, \ldots, \mathbf{y}_{n+1-k}]$.
   (d) $\tilde{U}_k = I_{k-1} \oplus U_k \qquad (\tilde{U}_1 = U_1)$.
   (e) $B_k = U_k^* A_k U_k$.
   (f) $A_{k+1} = B_k(1)$, the $(n-k)\times(n-k)$ matrix obtained from B_k by deleting the first row and column.
3. $U = \tilde{U}_1\tilde{U}_2, \ldots, \tilde{U}_{n-1}$.

9. (A strictly real version of the Schur unitary triangularization theorem) If A is a real matrix, then there is a real, orthogonal matrix Q such that Q^TAQ is block triangular, with the blocks of size 1×1 or 2×2. Each real eigenvalue of A appears as a 1×1 block of Q^TAQ and each nonreal pair of complex conjugate eigenvalues corresponds to a 2×2 diagonal block of Q^TAQ.
10. If $\mathcal{F}$ is a commuting family of matrices, then $\mathcal{F}$ is simultaneously unitarily triangularizable — i.e., there is a unitary matrix U such that U^*AU is triangular for every matrix A in $\mathcal{F}$. This fact has the analogous real form also.
11. [Lit53] [Mit53] [Sha91] Let $\lambda_1, \lambda_2, \cdots, \lambda_t$ be the distinct eigenvalues of A with multiplicities $m_1, m_2, \cdots, m_t$. Suppose U^*AU is block triangular with diagonal blocks $A_1, A_2, \ldots, A_t$, where A_i is size $m_i \times m_i$ and λ_i is the only eigenvalue of A_i for each i. Then the Jordan canonical form of A is the direct sum of the Jordan canonical forms of the blocks $A_1, A_2, \ldots, A_t$. *Note:* This conclusion also holds if the unitary similarity U is replaced by an ordinary similarity.
12. Let $\lambda_1, \lambda_2, \cdots, \lambda_n$ be the eigenvalues of the $n\times n$ matrix A and let $T = U^*AU$ be triangular. Then $\|A\|_F^2 = \sum_{i=1}^n |\lambda_i|^2 + \sum_{i<j} |t_{ij}|^2$. Hence, $\|A\|_F^2 \geq \sum_{i=1}^n |\lambda_i|^2$ and equality holds if and only if T is diagonal, or equivalently, if and only if A is normal (see Section 7.2).
13. A 2×2 matrix A with eigenvalues λ_1, λ_2 is unitarily similar to the triangular matrix $\begin{bmatrix} \lambda_1 & r \\ 0 & \lambda_2 \end{bmatrix}$, where $r = \sqrt{\|A\|_F^2 - (|\lambda_1|^2 + |\lambda_2|^2)}$. Note that r is real and nonnegative.
14. Two 2×2 matrices, A and B, are unitarily similar if and only if they have the same eigenvalues and $\|A\|_F = \|B\|_F$.
15. Any square matrix A is unitarily similar to a matrix in which all of the diagonal entries are equal to $\dfrac{tr(A)}{n}$.
16. [Spe40] Two $n\times n$ matrices, A and B, are unitarily equivalent if and only if tr $\omega(A, A^*) =$ tr $\omega(B, B^*)$ for every word $\omega(s,t)$ in two noncommuting variables.
17. [Pea62] Two $n\times n$ matrices, A and B, are unitarily equivalent if and only if tr $\omega(A, A^*) =$ tr $\omega(B, B^*)$ for every word $\omega(s,t)$ in two noncommuting variables of degree at most $2n^2$.

Examples:

1. The matrix $\frac{1}{\sqrt{2}}\begin{bmatrix} 1 & 1 \\ i & -i \end{bmatrix}$ is unitary but not orthogonal.

2. The matrix $\frac{1}{\sqrt{1+2i}}\begin{bmatrix} 1 & 1+i \\ 1+i & -1 \end{bmatrix}$ is orthogonal but not unitary.

3. Fact 13 shows that $A = \begin{bmatrix} 3 & 1 \\ 2 & 2 \end{bmatrix}$ is unitarily similar to $A = \begin{bmatrix} 4 & 1 \\ 0 & 1 \end{bmatrix}$.

4. For any nonzero r, the matrices $\begin{bmatrix} 3 & r \\ 0 & 2 \end{bmatrix}$ and $\begin{bmatrix} 3 & 0 \\ 0 & 2 \end{bmatrix}$ are similar, but not unitarily similar.

5. Let $A = \begin{bmatrix} -31 & 21 & 48 \\ -4 & 4 & 6 \\ -20 & 13 & 31 \end{bmatrix}$. Apply Algorithm 1 to A:

Step 1. $A_1 = A$.

Step 2. For

$k = 1$: (a) $p_{A_1}(x) = x^3 - 4x^2 + 5x - 2 = (x-2)(x-1)^2$, so the eigenvalues are 1, 1, 2. From the reduced row echelon form of $A - I_3$, we see that $[3, 0, 2]^T$ is an eigenvector for 1 and, thus, $\mathbf{x} = [\frac{3}{\sqrt{13}}, 0, \frac{2}{\sqrt{13}}]^T$ is a normalized eigenvector.

(b) One expects to apply the Gram–Schmidt process to a basis that includes $\mathbf{x}$ as the first vector to produce an orthonormal basis. In this example, it is obvious how to find an orthonormal basis for $\mathbb{C}^3$:

(c) $U_1 = \begin{bmatrix} \frac{3}{\sqrt{13}} & 0 & -\frac{2}{\sqrt{13}} \\ 0 & 1 & 0 \\ \frac{2}{\sqrt{13}} & 0 & \frac{3}{\sqrt{13}} \end{bmatrix}$.

(d) unnecessary.

(e) $B_1 = U_1^* A_1 U_1 = \begin{bmatrix} 1 & \frac{89}{\sqrt{13}} & 68 \\ 0 & 4 & 2\sqrt{13} \\ 0 & -\frac{3}{\sqrt{13}} & -1 \end{bmatrix}$.

(f) $A_2 = \begin{bmatrix} 4 & 2\sqrt{13} \\ -\frac{3}{\sqrt{13}} & -1 \end{bmatrix}$.

$k = 2$: (a) 1 is still an eigenvalue of A_2. From the reduced row echelon form of $A_2 - I_2$, we see that $[-2\sqrt{13}, 3]^T$ is an eigenvector for 1 and, thus, $\mathbf{x} = [-2\sqrt{\frac{13}{61}}, \frac{3}{\sqrt{61}}]^T$ is a normalized eigenvector.

(b) Again, the orthonormal basis is obvious:

(c) $U_2 = \begin{bmatrix} -2\sqrt{\frac{13}{61}} & \frac{3}{\sqrt{61}} \\ \frac{3}{\sqrt{61}} & 2\sqrt{\frac{13}{61}} \end{bmatrix}$.

(d) $\tilde{U}_2 = \begin{bmatrix} 1 & 0 & 0 \\ 0 & -2\sqrt{\frac{13}{61}} & \frac{3}{\sqrt{61}} \\ 0 & \frac{3}{\sqrt{61}} & 2\sqrt{\frac{13}{61}} \end{bmatrix}$.

(e) $B_2 = \begin{bmatrix} 1 & -\frac{29}{\sqrt{13}} \\ 0 & 2 \end{bmatrix}$.

(f) unnecessary.

Step 3. $U = \tilde{U}_1 \tilde{U}_2 = \begin{bmatrix} \frac{3}{\sqrt{13}} & -\frac{6}{\sqrt{793}} & -\frac{4}{\sqrt{61}} \\ 0 & -2\sqrt{\frac{13}{61}} & \frac{3}{\sqrt{61}} \\ \frac{2}{\sqrt{13}} & \frac{9}{\sqrt{793}} & \frac{6}{\sqrt{61}} \end{bmatrix}$. $T = U^* A U = \begin{bmatrix} 1 & \frac{26}{\sqrt{61}} & \frac{2035}{\sqrt{793}} \\ 0 & 1 & -\frac{29}{\sqrt{13}} \\ 0 & 0 & 2 \end{bmatrix}$.

6. [HJ85, p. 84] Schur's theorem tells us that every complex, square matrix is unitarily similar to a triangular matrix. However, it is not true that every complex, square matrix is similar to a triangular matrix via a complex, orthogonal similarity. For, suppose $A = QTQ^T$, where Q is complex orthogonal and T is triangular. Let $\mathbf{q}$ be the first column of Q. Then $\mathbf{q}$ is an eigenvector of A and $\mathbf{q}^T\mathbf{q} = 1$. However, the matrix $A = \begin{bmatrix} 1 & i \\ i & -1 \end{bmatrix}$ has no such eigenvector; A is nilpotent and any eigenvector of A is a scalar multiple of $\begin{bmatrix} 1 \\ i \end{bmatrix}$.

7.2 Normal Matrices and Spectral Theory

In this subsection, all matrices are over the complex numbers and are square. All vector spaces are finite dimensional complex inner product spaces.

Definitions:

The matrix A is **normal** if $AA^* = A^*A$.

The matrix A is **Hermitian** if $A^* = A$.

The matrix A is **skew-Hermitian** if $A^* = -A$.

The linear operator, T, on the complex inner product space V is **normal** if $TT^* = T^*T$.

Two orthogonal projections, P and Q, are **pairwise orthogonal** if $PQ = QP = 0$. (See Section 5.4 for information about orthogonal projection.)

The matrices A and B are said to have **Property L** if their eigenvalues $\alpha_k, \beta_k, (k = 1, \cdots, n)$ may be ordered in such a way that the eigenvalues of $xA + yB$ are given by $x\alpha_k + y\beta_k$ for all complex numbers x and y.

Facts:

Most of the material in this section can be found in one or more of the following: [HJ85, Chap. 2] [Hal87, Chap. 3] [Gan59, Chap. IX] [MM64, I.4, III.3.5, III.5] [GJSW87]. Specific references are also given for some facts.

1. Diagonal, Hermitian, skew-Hermitian, and unitary matrices are all normal. Note that real symmetric matrices are Hermitian, real skew-symmetric matrices are skew-Hermitian, and real, orthogonal matrices are unitary, so all of these matrices are normal.
2. If U is unitary, then A is normal if and only if U^*AU is normal.
3. Let T be a linear operator on the complex inner product space V. Let $\mathcal{B}$ be an ordered orthonormal basis of V and let $A = [T]_{\mathcal{B}}$. Then T is normal if and only if A is a normal matrix.
4. (Spectral Theorem) The following three versions are equivalent.
   - A matrix is normal if and only if it is unitarily similar to a diagonal matrix. (*Note:* This is sometimes taken as the definition of normal. See Fact 6 below for a strictly real version.)
   - The matrix A is normal if and only if there is an orthonormal basis of eigenvectors of A.
   - Let $\lambda_1, \lambda_2, \ldots, \lambda_t$ be the distinct eigenvalues of A with algebraic multiplicities $m_1, m_2, \ldots, m_t$. Then A is normal if and only if there exist t pairwise orthogonal, orthogonal projections $P_1, P_2, \ldots, P_t$ such that $\sum_{i=1}^{t} P_i = I$, $\text{rank}(P_i) = m_i$, and $A = \sum_{i=1}^{t} \lambda_i P_i$. (Note that the two orthogonal projections P and Q are pairwise orthogonal if and only if range(P) and range(Q) are orthogonal subspaces.)
5. (Principal Axes Theorem) A real matrix A is symmetric if and only if $A = QDQ^T$, where Q is a real, orthogonal matrix and D is a real, diagonal matrix. Equivalently, a real matrix A is symmetric

if and only if there is a real, orthonormal basis of eigenvectors of A. Note that the eigenvalues of A appear on the diagonal of D, and the columns of Q are eigenvectors of A. The Principal Axes Theorem follows from the Spectral Theorem, and the fact that all of the eigenvalues of a Hermitian matrix are real.

6. (A strictly real version of the Spectral Theorem) If A is a real, normal matrix, then there is a real, orthogonal matrix Q such that $Q^T AQ$ is block diagonal, with the blocks of size 1×1 or 2×2. Each real eigenvalue of A appears as a 1×1 block of $Q^T AQ$ and each nonreal pair of complex conjugate eigenvalues corresponds to a 2×2 diagonal block of $Q^T AQ$.
7. The following are equivalent. See also Facts 4 and 8. See [GJSW87] and [EI98] for more equivalent conditions.
   - A is normal.
   - A^* can be expressed as a polynomial in A.
   - For any B, $AB = BA$ implies $A^*B = BA^*$.
   - Any eigenvector of A is also an eigenvector of A^*.
   - Each invariant subspace of A is also an invariant subspace of A^*.
   - For each invariant subspace, $\mathcal{V}$, of A, the orthogonal complement, $\mathcal{V}^\perp$, is also an invariant subspace of A.
   - $\langle A\mathbf{x}, A\mathbf{y}\rangle = \langle A^*\mathbf{x}, A^*\mathbf{y}\rangle$ for all vectors $\mathbf{x}$ and $\mathbf{y}$.
   - $\langle A\mathbf{x}, A\mathbf{x}\rangle = \langle A^*\mathbf{x}, A^*\mathbf{x}\rangle$ for every vector $\mathbf{x}$.
   - $\|A\mathbf{x}\| = \|A^*\mathbf{x}\|$ for every vector $\mathbf{x}$.
   - $A^* = UA$ for some unitary matrix U.
   - $\|A\|_F^2 = \sum_{i=1}^n |\lambda_i|^2$, where $\lambda_1, \lambda_2, \cdots, \lambda_n$ are the eigenvalues of A.
   - The singular values of A are $|\lambda_1|, |\lambda_2|, \cdots, |\lambda_n|$, where $\lambda_1, \lambda_2, \cdots, \lambda_n$ are the eigenvalues of A.
   - If $A = UP$ is a polar decomposition of A, then $UP = PU$. (See Section 8.4.)
   - A commutes with a normal matrix with distinct eigenvalues.
   - A commutes with a Hermitian matrix with distinct eigenvalues.
   - The Hermitian matrix $AA^* - A^*A$ is semidefinite (i.e., it does not have both positive and negative eigenvalues).
8. Let $H = \dfrac{A + A^*}{2}$ and $K = \dfrac{A - A^*}{2i}$. Then H and K are Hermitian and $A = H + iK$. The matrix A is normal if and only if $HK = KH$.
9. If A is normal, then
   - A is Hermitian if and only if all of the eigenvalues of A are real.
   - A is skew-Hermitian if and only if all of the eigenvalues of A are pure imaginary.
   - A is unitary if and only if all of the eigenvalues of A have modulus 1.
10. The matrix U is unitary if and only if $U = \exp(iH)$ where H is Hermitian.
11. If Q is a real matrix with $\det(Q) = 1$, then Q is orthogonal if and only if $Q = \exp(K)$, where K is a real, skew-symmetric matrix.
12. (Cayley's Formulas/Cayley Transform) If U is unitary and does not have -1 as an eigenvalue, then $U = (I + iH)(I - iH)^{-1}$, where $H = i(I - U)(I + U)^{-1}$ is Hermitian.
13. (Cayley's Formulas/Cayley Transform, real version) If Q is a real, orthogonal matrix which does not have -1 as an eigenvalue, then $Q = (I - K)(I + K)^{-1}$, where $K = (I - Q)(I + Q)^{-1}$ is a real, skew-symmetric matrix.
14. A triangular matrix is normal if and only if it is diagonal. More generally, if the block triangular matrix, $\begin{bmatrix} B_{11} & B_{12} \\ 0 & B_{22} \end{bmatrix}$ (where the diagonal blocks, $B_{ii}, i = 1, 2$, are square), is normal, then $B_{12} = 0$.

15. Let A be a normal matrix. Then the diagonal entries of A are the eigenvalues of A if and only if A is diagonal.
16. If A and B are normal and commute, then AB is normal. However, the product of two noncommuting normal matrices need not be normal. (See Example 3 below.)
17. If A is normal, then $\rho(A) = \|A\|_2$. Consequently, if A is normal, then $\rho(A) \geq |a_{ij}|$ for all i and j. The converses of both of these facts are false (see Example 4 below).
18. [MM64, p. 168] [MM55] [ST80] If A is normal, then $W(A)$ is the convex hull of the eigenvalues of A. The converse of this statement holds when $n \leq 4$, but not for $n \geq 5$.
19. [WW49] [MM64, page 162] Let A be a normal matrix and suppose x is a vector such that $(Ax)_i = 0$ whenever $x_i = 0$. For each nonzero component, x_j, of x, define $\mu_j = \dfrac{(Ax)_j}{x_j}$. Note that μ_j is a complex number, which we regard as a point in the plane. Then any closed disk that contains all of the points μ_j must contain an eigenvalue of A.
20. [HW53] Let A and B be normal matrices with eigenvalues $\alpha_1, \cdots, \alpha_n$ and $\beta_1, \cdots, \beta_n$. Then

$$\min_{\sigma \in S_n} \sum_{i=1}^{n} |\alpha_i - \beta_{\sigma(i)}|^2 \leq \|A - B\|_F^2 \leq \max_{\sigma \in S_n} \sum_{i=1}^{n} |\alpha_i - \beta_{\sigma(i)}|^2,$$

where the minimum and maximum are over all permutations σ in the symmetric group S_n (i.e., the group of all permutations of $1, \ldots, n$).
21. [Sun82] [Bha82] Let A and B be $n \times n$ normal matrices with eigenvalues $\alpha_1, \cdots, \alpha_n$ and $\beta_1, \cdots, \beta_n$. Let Λ_A, Λ_B be the diagonal matrices with diagonal entries $\alpha_1, \cdots, \alpha_n$ and $\beta_1, \cdots, \beta_n$, respectively. Let $\|\cdot\|$ be any unitarily invariant norm. Then, if $A - B$ is normal, we have

$$\min_P \|\Lambda_A - P^{-1}\Lambda_B P\| \leq \|A - B\| \leq \max_P \|\Lambda_A - P^{-1}\Lambda_B P\|,$$

where the maximum and minimum are over all $n \times n$ permutation matrices P.

Observe that if A and B are Hermitian, then $A - B$ is also Hermitian and, hence, normal, so this inequality holds for all pairs of Hermitian matrices. However, Example 6 gives a pair of 2×2 normal matrices (with $A - B$ not normal) for which the inequality does not hold. Note that for the Frobenius norm, we get the Hoffman–Wielandt inequality (20), which does hold for all pairs of normal matrices.

For the operator norm, $\|\cdot\|_2$, this gives the inequality

$$\min_{\sigma \in S_n} \max_j |\alpha_j - \beta_{\sigma(j)}| \leq \|A - B\|_2 \leq \max_{\sigma \in S_n} \max_j |\alpha_j - \beta_{\sigma(j)}|$$

(assuming $A - B$ is normal), which, for the case of Hermitian A and B, is a classical result of Weyl [Wey12].
22. [OS90][BEK97][BDM83][BDK89][Hol92][AN86] Let A and B be normal matrices with eigenvalues $\alpha_1, \cdots, \alpha_n$ and $\beta_1, \cdots, \beta_n$, respectively. Using $\|A\|_2 \leq \|A\|_F \leq \sqrt{n}\|A\|_2$ together with the Hoffman–Wielandt inequality (20) yields

$$\frac{1}{\sqrt{n}} \min_{\sigma \in S_n} \max_j |\alpha_j - \beta_{\sigma(j)}| \leq \|A - B\|_2 \leq \sqrt{n} \max_{\sigma \in S_n} \max_j |\alpha_j - \beta_{\sigma(j)}|.$$

On the right-hand side, the factor $\sqrt{n}$ may be replaced by $\sqrt{2}$ and it is known that this constant is the best possible. On the left-hand side, the factor $\dfrac{1}{\sqrt{n}}$ may be replaced by the constant $\frac{1}{2.91}$, but the best possible value for this constant is still unknown. Thus, we have

$$\frac{1}{2.91} \min_{\sigma \in S_n} \max_j |\alpha_j - \beta_{\sigma(j)}| \leq \|A - B\|_2 \leq \sqrt{2} \max_{\sigma \in S_n} \max_j |\alpha_j - \beta_{\sigma(j)}|.$$

See also [Bha82], [Bha87], [BH85], [Sun82], [Sund82].

23. If A and B are normal matrices, then $AB = BA$ if and only if A and B have Property L. This was established for Hermitian matrices by Motzkin and Taussky [MT52] and then generalized to the normal case by Wiegmann [Wieg53]. For a stronger generalization see [Wiel53].
24. [Fri02] Let $a_{ij}, i = 1, \ldots, n, j = 1, \ldots, n$, be any set of $\frac{n(n+1)}{2}$ complex numbers. Then there exists an $n \times n$ normal matrix, N, such that $n_{ij} = a_{ij}$ for $i \leq j$. Thus, any upper triangular matrix A can be completed to a normal matrix.
25. [Bha87, p. 54] Let A be a normal $n \times n$ matrix and let B be an arbitrary $n \times n$ matrix such that $\|A - B\|_2 < \epsilon$. Then every eigenvalue of B is within distance ϵ of an eigenvalue of A. Example 7 below shows that this need not hold for an arbitrary pair of matrices.
26. There are various ways to measure the "nonnormality" of a matrix. For example, if A has eigenvalues $\lambda_1, \lambda_2, \ldots, \lambda_n$, the quantity $\sqrt{\|A\|_F^2 - \sum_{i=1}^n |\lambda_i|^2}$ is a natural measure of nonnormality, as is $\|A^*A - AA^*\|_2$. One could also consider $\|A^*A - AA^*\|$ for other choices of norm, or look at $\min\{\|A - N\| : N \text{ is normal}\}$. Fact 8 above suggests $\|HK - KH\|$ as a possible measure of nonnormality, while the polar decomposition (see Fact 7 above) $A = UP$ of A suggests $\|UP - PU\|$. See [EP87] for more measures of nonnormality and comparisons between them.
27. [Lin97] [FR96] For any $\epsilon > 0$ there is a $\delta > 0$ such that, for any $n \times n$ complex matrix A with $\|AA^* - A^*A\|_2 < \delta$, there is a normal matrix N with $\|N - A\|_2 < \epsilon$. Thus, a matrix which is approximately normal is close to a normal matrix.

Examples:

1. Let $A = \begin{bmatrix} 3 & 1 \\ 1 & 3 \end{bmatrix}$ and $U = \frac{1}{\sqrt{2}} \begin{bmatrix} 1 & 1 \\ 1 & -1 \end{bmatrix}$. Then $U^*AU = \begin{bmatrix} 4 & 0 \\ 0 & 2 \end{bmatrix}$ and $A = 4P_1 + 2P_2$, where the $P_i's$ are the pairwise orthogonal, orthogonal projection matrices
$$P_1 = U \begin{bmatrix} 1 & 0 \\ 0 & 0 \end{bmatrix} U^* = \frac{1}{2} \begin{bmatrix} 1 & 1 \\ 1 & 1 \end{bmatrix} \quad \text{and} \quad P_2 = U \begin{bmatrix} 0 & 0 \\ 0 & 1 \end{bmatrix} U^* = \frac{1}{2} \begin{bmatrix} 1 & -1 \\ -1 & 1 \end{bmatrix}.$$
2. $A = \begin{bmatrix} 1 & 4+2i & 6 \\ 0 & 8+2i & 0 \\ 2 & -2i & 4i \end{bmatrix} = H + iK$, where $H = \begin{bmatrix} 1 & 2+i & 4 \\ 2-i & 8 & i \\ 4 & -i & 0 \end{bmatrix}$ and $K = \begin{bmatrix} 0 & 1-2i & -2i \\ 1+2i & 2 & -1 \\ 2i & -1 & 4 \end{bmatrix}$ are Hermitian.
3. $A = \begin{bmatrix} 0 & 1 \\ 1 & 0 \end{bmatrix}$ and $B = \begin{bmatrix} 0 & 1 \\ 1 & 1 \end{bmatrix}$ are both normal matrices, but the product $AB = \begin{bmatrix} 1 & 1 \\ 0 & 1 \end{bmatrix}$ is not normal.
4. Let $A = \begin{bmatrix} 2 & 0 & 0 \\ 0 & 0 & 1 \\ 0 & 0 & 0 \end{bmatrix}$. Then $\rho(A) = 2 = \|A\|_2$, but A is not normal.
5. Let $Q = \begin{bmatrix} \cos\theta & \sin\theta \\ -\sin\theta & \cos\theta \end{bmatrix}$. Put $U = \frac{1}{\sqrt{2}} \begin{bmatrix} 1 & i \\ i & 1 \end{bmatrix}$ and $D = \begin{bmatrix} e^{i\theta} & 0 \\ 0 & e^{-i\theta} \end{bmatrix}$. Then $Q = UDU^* = U\left(\exp i \begin{bmatrix} \theta & 0 \\ 0 & -\theta \end{bmatrix}\right) U^* = \exp i \left(U \begin{bmatrix} \theta & 0 \\ 0 & -\theta \end{bmatrix} U^* \right)$. Put $H = U \begin{bmatrix} \theta & 0 \\ 0 & -\theta \end{bmatrix} U^* = \begin{bmatrix} 0 & -i\theta \\ i\theta & 0 \end{bmatrix}$.

   Then H is Hermitian and $Q = \exp(iH)$. Also, $K = iH = \begin{bmatrix} 0 & \theta \\ -\theta & 0 \end{bmatrix}$ is a real, skew-symmetric matrix and $Q = \exp(K)$.
6. Here is an example from [Sund82] showing that the condition that $A - B$ be normal cannot be dropped from 21. Let $A = \begin{bmatrix} 0 & 1 \\ 1 & 0 \end{bmatrix}$ and $B = \begin{bmatrix} 0 & -1 \\ 1 & 0 \end{bmatrix}$. Then A is Hermitian with eigenvalues ± 1

and B is skew-Hermitian with eigenvalues $\pm i$. So, we have $\|\Lambda_A - P^{-1}\Lambda_B P\|_2 = \sqrt{2}$, regardless of the permutation P. However, $A - B = \begin{bmatrix} 0 & 2 \\ 0 & 0 \end{bmatrix}$ and $\|A - B\|_2 = 2$.

7. This example shows that Fact 25 above does not hold for general pairs of matrices. Let $\alpha > \beta > 0$ and put $A = \begin{bmatrix} 0 & \alpha \\ \beta & 0 \end{bmatrix}$ and $B = \begin{bmatrix} 0 & \alpha - \beta \\ 0 & 0 \end{bmatrix}$. Then the eigenvalues of A are $\pm\sqrt{\alpha\beta}$ and both eigenvalues of B are zero. We have $A - B = \begin{bmatrix} 0 & \beta \\ \beta & 0 \end{bmatrix}$ and $\|A - B\|_2 = \beta$. But, since $\alpha > \beta$, we have $\sqrt{\alpha\beta} > \beta = \|A - B\|_2$.

References

[AN86] T. Ando and Y. Nakamura. "Bounds for the antidistance." Technical Report, Hokkaido University, Japan, 1986.

[BDK89] R. Bhatia, C. Davis, and P. Koosis. An extremal problem in Fourier analysis with applications to operator theory. *J. Funct. Anal.*, 82:138–150, 1989.

[BDM83] R. Bhatia, C. Davis, and A. McIntosh. Perturbation of spectral subspaces and solution of linear operator equations. *Linear Algebra Appl.*, 52/53:45–67, 1983.

[BEK97] R. Bhatia, L. Elsner, and G.M. Krause. Spectral variation bounds for diagonalisable matrices. *Aequationes Mathematicae*, 54:102–107, 1997.

[Bha82] R. Bhatia. Analysis of spectral variation and some inequalities. *Transactions of the American Mathematical Society*, 272:323–331, 1982.

[Bha87] R. Bhatia. *Perturbation Bounds for Matrix Eigenvalues.* Longman Scientific & Technical, Essex, U.K. (copublished in the United States with John Wiley & Sons, New York), 1987.

[BH85] R. Bhatia and J. A. R. Holbrook. Short normal paths and spectral variation. *Proc. Amer. Math. Soc.*, 94:377–382, 1985.

[EI98] L. Elsner and Kh.D. Ikramov. Normal matrices: an update. *Linear Algebra Appl.*, 285:291–303, 1998.

[EP87] L. Elsner and M.H.C Paardekooper. On measures of nonnormality of matrices. *Linear Algebra Appl.*, 92:107–124, 1987.

[Fri02] S. Friedland. Normal matrices and the completion problem. *SIAM J. Matrix Anal. Appl.*, 23:896–902, 2002.

[FR96] P. Friis and M. Rørdam. Almost commuting self-adjoint matrices — a short proof of Huaxin Lin's theorem. *J. Reine Angew. Math.*, 479:121–131, 1996.

[Gan59] F.R. Gantmacher. *Matrix Theory, Vol. I.* Chelsea Publishing, New York, 1959.

[GJSW87] R. Grone, C.R. Johnson, E.M. Sa, and H. Wolkowicz. Normal matrices. *Linear Algebra Appl.*, 87:213–225, 1987.

[Hal87] P.R. Halmos. *Finite-Dimensional Vector Spaces.* Springer-Verlag, New York, 1987.

[HJ85] R.A. Horn and C.R. Johnson. *Matrix Analysis.* Cambridge University Press, Cambridge, 1985.

[Hol92] J.A. Holbrook. Spectral variation of normal matrices. *Linear Algebra Appl.*, 174:131-–144, 1992.

[HOS96] J. Holbrook, M. Omladič, and P. Šemrl. Maximal spectral distance. *Linear Algebra Appl.*, 249:197–205, 1996.

[HW53] A.J. Hoffman and H.W. Wielandt. The variation of the spectrum of a normal matrix. *Duke Math. J.*, 20:37–39, 1953.

[Lin97] H. Lin. Almost commuting self-adjoint matrices and applications. *Operator algebras and their applications (Waterloo, ON, 1994/95), Fields Inst. Commun., 13, Amer. Math Soc., Providence, RI,* 193–233, 1997.

[Lit53] D.E. Littlewood. On unitary equivalence. *J. London Math. Soc.*, 28:314–322, 1953.

[Mir60] L. Mirsky. Symmetric guage functions and unitarily invariant norms. *Quart. J. Math. Oxford (2)*, 11:50–59, 1960.

[Mit53] B.E. Mitchell. Unitary transformations. *Can. J. Math*, 6:69–72, 1954.

[MM55] B.N. Moyls and M.D. Marcus. Field convexity of a square matrix. *Proc. Amer. Math. Soc.*, 6:981–983, 1955.

[MM64] M. Marcus and H. Minc. *A Survey of Matrix Theory and Matrix Inequalities.* Allyn and Bacon, Boston, 1964.

[MT52] T.S. Motzkin and O. Taussky Todd. Pairs of matrices with property L. *Trans. Amer. Math. Soc.*, 73:108–114, 1952.

[OS90] M. Omladič and P. Šemrl. On the distance between normal matrices. *Proc. Amer. Math. Soc.*, 110:591–596, 1990.

[Par48] W.V. Parker. Sets of numbers associated with a matrix. *Duke Math. J.*, 15:711–715, 1948.

[Pea62] C. Pearcy. A complete set of unitary invariants for operators generating finite W*-algebras of type I. *Pacific J. Math.*, 12:1405–1416, 1962.

[Sch09] I. Schur. Über die charakteristischen Wurzeln einer linearen Substitutionen mit einer Anwendung auf die Theorie der Intergralgleichungen. *Math. Ann.*, 66:488–510, 1909.

[Sha91] H. Shapiro. A survey of canonical forms and invariants for unitary similarity. *Linear Algebra Appl.*, 147:101–167, 1991.

[Spe40] W. Specht. Zur Theorie der Matrizen, II. *Jahresber. Deutsch. Math.-Verein.*, 50:19–23, 1940.

[ST80] H. Shapiro and O. Taussky. Alternative proofs of a theorem of Moyls and Marcus on the numerical range of a square matrix. *Linear Multilinear Algebra*, 8:337–340, 1980.

[Sun82] V.S. Sunder. On permutations, convex hulls, and normal operators. *Linear Algebra Appl.*, 48:403–411, 1982.

[Sund82] V.S. Sunder. Distance between normal operators. *Proc. Amer. Math. Soc.*, 84:483–484, 1982.

[Wey12] H. Weyl. Das assymptotische Verteilungsgesetz der Eigenwerte linearer partieller Diffferentialgleichungen. *Math. Ann.*, 71:441–479, 1912.

[Wieg53] N. Wiegmann. Pairs of normal matrices with property L. *Proc. Am. Math. Soc.*, 4: 35-36, 1953.

[Wiel53] H. Wielandt. Pairs of normal matrices with property L. *J. Res. Nat. Bur. Standards*, 51:89–90, 1953.

[WW49] A.G. Walker and J.D. Weston. Inclusion theorems for the eigenvalues of a normal matrix. *J. London Math. Soc.*, 24:28–31, 1949.

8

Hermitian and Positive Definite Matrices

Wayne Barrett
Brigham Young University

8.1 Hermitian Matrices

All matrices in this section are either real or complex, unless explicitly stated otherwise.

Definitions:

A matrix $A \in \mathbb{C}^{n\times n}$ is **Hermitian** or **self-adjoint** if $A^* = A$, or element-wise, $\bar{a}_{ij} = a_{ji}$, for $i, j = 1, \ldots, n$. The set of Hermitian matrices of order n is denoted by $\mathcal{H}_n$. Note that a matrix $A \in \mathbb{R}^{n\times n}$ is Hermitian if and only if $A^T = A$.

A matrix $A \in \mathbb{C}^{n\times n}$ is **symmetric** if $A^T = A$, or element-wise, $a_{ij} = a_{ji}$, for $i, j = 1, \ldots, n$. The set of real symmetric matrices of order n is denoted by $\mathcal{S}_n$. Since $\mathcal{S}_n$ is a subset of $\mathcal{H}_n$, all theorems for matrices in $\mathcal{H}_n$ apply to $\mathcal{S}_n$ as well.

Let V be a complex inner product space with inner product $\langle \mathbf{v}, \mathbf{w}\rangle$ and let $\mathbf{v}_1, \mathbf{v}_2, \ldots, \mathbf{v}_n \in V$. The matrix $G = [g_{ij}] \in \mathbb{C}^{n\times n}$ defined by $g_{ij} = \langle \mathbf{v}_i, \mathbf{v}_j\rangle$, $i, j \in \{1, 2, \ldots, n\}$ is called the **Gram matrix** of the vectors $\mathbf{v}_1, \mathbf{v}_2, \ldots, \mathbf{v}_n$.

The inner product $\langle \mathbf{x}, \mathbf{y}\rangle$ of two vectors $\mathbf{x}, \mathbf{y} \in \mathbb{C}^n$ will mean the standard inner product, i.e., $\langle \mathbf{x}, \mathbf{y}\rangle = \mathbf{y}^*\mathbf{x}$, unless stated otherwise. The term orthogonal will mean orthogonal with respect to this inner product, unless stated otherwise.

Facts:

For facts without a specific reference, see [HJ85, pp. 38, 101–104, 169–171, 175], [Lax96, pp. 80–83], and [GR01, pp. 169–171]. Many are an immediate consequence of the definition.

1. A real symmetric matrix is Hermitian, and a real Hermitian matrix is symmetric.
2. Let A, B be Hermitian.

(a) Then $A + B$ is Hermitian.

(b) If $AB = BA$, then AB is Hermitian.

(c) If $c \in \mathbb{R}$, then cA is Hermitian.

3. $A + A^*$, $A^* + A$, AA^*, and A^*A are Hermitian for all $A \in \mathbb{C}^{n\times n}$.
4. If $A \in \mathcal{H}_n$, then $\langle A\mathbf{x}, \mathbf{y}\rangle = \langle \mathbf{x}, A\mathbf{y}\rangle$ for all $\mathbf{x}, \mathbf{y} \in \mathbb{C}^n$.
5. If $A \in \mathcal{H}_n$, then $A^k \in \mathcal{H}_n$ for all $k \in \mathbb{N}$.
6. If $A \in \mathcal{H}_n$ is invertible, then $A^{-1} \in \mathcal{H}_n$.
7. The main diagonal entries of a Hermitian matrix are real.
8. All eigenvalues of a Hermitian matrix are real.
9. Eigenvectors corresponding to distinct eigenvalues of a Hermitian matrix are orthogonal.
10. *Spectral Theorem — Diagonalization version:* If $A \in \mathcal{H}_n$, there is a unitary matrix $U \in \mathbb{C}^{n\times n}$ such that $U^*AU = D$, where D is a real diagonal matrix whose diagonal entries are the eigenvalues of A. If $A \in \mathcal{S}_n$, the same conclusion holds with an orthogonal matrix $Q \in \mathbb{R}^{n\times n}$, i.e., $Q^TAQ = D$.
11. *Spectral Theorem — Orthonormal basis version:* If $A \in \mathcal{H}_n$, there is an orthonormal basis of $\mathbb{C}^n$ consisting of eigenvectors of A. If $A \in \mathcal{S}_n$, the same conclusion holds with $\mathbb{C}^n$ replaced by $\mathbb{R}^n$.
12. [Lay97, p. 447] *Spectral Theorem — Sum of rank one projections version:* Let $A \in \mathcal{H}_n$ with eigenvalues $\lambda_1, \lambda_2, \ldots, \lambda_n$, and corresponding orthonormal eigenvectors $\mathbf{u}_1, \mathbf{u}_2, \ldots, \mathbf{u}_n$. Then

$$A = \lambda_1 \mathbf{u}_1\mathbf{u}_1^* + \lambda_2 \mathbf{u}_2\mathbf{u}_2^* + \cdots + \lambda_n \mathbf{u}_n\mathbf{u}_n^*.$$

If $A \in \mathcal{S}_n$, then

$$A = \lambda_1 \mathbf{u}_1\mathbf{u}_1^T + \lambda_2 \mathbf{u}_2\mathbf{u}_2^T + \cdots + \lambda_n \mathbf{u}_n\mathbf{u}_n^T.$$

13. If $A \in \mathcal{H}_n$, then rank A equals the number of nonzero eigenvalues of A.
14. Each $A \in \mathbb{C}^{n\times n}$ can be written uniquely as $A = H + iK$, where $H, K \in \mathcal{H}_n$.
15. Given $A \in \mathbb{C}^{n\times n}$, then $A \in \mathcal{H}_n$ if and only if $\mathbf{x}^*A\mathbf{x}$ is real for all $\mathbf{x} \in \mathbb{C}^n$.
16. Any Gram matrix is Hermitian. Some examples of how Gram matrices arise are given in Chapter 66 and [Lax96, p. 124].
17. The properties given above for $\mathcal{H}_n$ and $\mathcal{S}_n$ are generally not true for symmetric matrices in $\mathbb{C}^{n\times n}$, but there is a substantial theory associated with them. (See [HJ85, sections 4.4 and 4.6].)

Examples:

1. The matrix $\begin{bmatrix} 3 & 2-i \\ 2+i & -5 \end{bmatrix} \in \mathcal{H}_2$ and $\begin{bmatrix} 6 & 0 & 2 \\ 0 & -1 & 5 \\ 2 & 5 & 3 \end{bmatrix} \in \mathcal{S}_3$.
2. Let D be an open set in $\mathbb{R}^n$ containing the point $\mathbf{x}_0$, and let $f : D \to \mathbb{R}$ be a twice continuously differentiable function on D. Define $H \in \mathbb{R}^{n\times n}$ by $h_{ij} = \dfrac{\partial^2 f}{\partial x_i \partial x_j}(\mathbf{x}_0.)$. Then H is a real symmetric matrix, and is called the **Hessian** of f.
3. Let $G = (V, E)$ be a simple undirected graph with vertex set $V = \{1, 2, 3, \ldots, n\}$. The $n \times n$ **adjacency matrix** $A(G) = [a_{ij}]$ (see Section 28.3) is defined by

$$a_{ij} = \begin{cases} 1 & \text{if } ij \in E \\ 0 & \text{otherwise.} \end{cases}$$

In particular, all diagonal entries of $A(G)$ are 0. Since ij is an edge of G if and only if ji is, the adjacency matrix is real symmetric. Observe that for each $i \in V$, $\sum_{j=1}^n a_{ij} = \delta(i)$, i.e., the sum of the ith row is the degree of vertex i.

8.2 Order Properties of Eigenvalues of Hermitian Matrices

Definitions:

Given $A \in \mathcal{H}_n$, the **Rayleigh quotient** $R_A : \mathbb{C}^n \backslash \{0\} \to \mathbb{R}$ is $R_A(x) = \dfrac{\mathbf{x}^* A\mathbf{x}}{\mathbf{x}^*\mathbf{x}} = \dfrac{\langle A\mathbf{x}, \mathbf{x}\rangle}{\langle \mathbf{x}, \mathbf{x}\rangle}$.

Facts:

For facts without a specific reference, see [HJ85, Sections 4.2, 4.3]; however, in that source the eigenvalues are labeled from smallest to greatest and the definition of majorizes (see Preliminaries) has a similar reversal of notation.

1. *Rayleigh–Ritz Theorem:* Let $A \in \mathcal{H}_n$, with eigenvalues $\lambda_1 \geq \lambda_2 \geq \cdots \geq \lambda_n$. Then

$$\lambda_n \leq \frac{\mathbf{x}^* A\mathbf{x}}{\mathbf{x}^*\mathbf{x}} \leq \lambda_1, \quad \text{for all nonzero } \mathbf{x} \in \mathbb{C}^n,$$

$$\lambda_1 = \max_{x \neq 0} \frac{\mathbf{x}^* A\mathbf{x}}{\mathbf{x}^*\mathbf{x}} = \max_{\|\mathbf{x}\|_2=1} \mathbf{x}^* A\mathbf{x},$$

and

$$\lambda_n = \min_{x \neq 0} \frac{\mathbf{x}^* A\mathbf{x}}{\mathbf{x}^*\mathbf{x}} = \min_{\|\mathbf{x}\|_2=1} \mathbf{x}^* A\mathbf{x}.$$

2. *Courant–Fischer Theorem:* Let $A \in M_n$ be a Hermitian matrix with eigenvalues $\lambda_1 \geq \lambda_2 \geq \ldots \geq \lambda_n$, and let k be a given integer with $1 \leq k \leq n$. Then

$$\max_{\mathbf{w}_1, \mathbf{w}_2, \ldots, \mathbf{w}_{n-k} \in \mathbb{C}^n} \quad \min_{\substack{\mathbf{x} \neq 0, \mathbf{x} \in \mathbb{C}^n \\ \mathbf{x} \perp \mathbf{w}_1, \mathbf{w}_2, \ldots, \mathbf{w}_{n-k}}} \frac{\mathbf{x}^* A\mathbf{x}}{\mathbf{x}^*\mathbf{x}} = \lambda_k$$

and

$$\min_{\mathbf{w}_1, \mathbf{w}_2, \ldots, \mathbf{w}_{k-1} \in \mathbb{C}^n} \quad \max_{\substack{\mathbf{x} \neq 0, \mathbf{x} \in \mathbb{C}^n \\ \mathbf{x} \perp \mathbf{w}_1, \mathbf{w}_2, \ldots, \mathbf{w}_{k-1}}} \frac{\mathbf{x}^* A\mathbf{x}}{\mathbf{x}^*\mathbf{x}} = \lambda_k.$$

3. (Also [Bha01, p. 291]) *Weyl Inequalities:* Let $A, B \in \mathcal{H}_n$ and assume that the eigenvalues of A, B and $A+B$ are arranged in decreasing order. Then for every pair of integers j, k such that $1 \leq j, k \leq n$ and $j + k \leq n + 1$,

$$\lambda_{j+k-1}(A + B) \leq \lambda_j(A) + \lambda_k(B)$$

and for every pair of integers j, k such that $1 \leq j, k \leq n$ and $j + k \geq n + 1$,

$$\lambda_{j+k-n}(A + B) \geq \lambda_j(A) + \lambda_k(B).$$

4. *Weyl Inequalities:* These inequalities are a prominent special case of Fact 3. Let $A, B \in \mathcal{H}_n$ and assume that the eigenvalues of A, B and $A + B$ are arranged in decreasing order. Then for each $j \in \{1, 2, \ldots, n\}$,

$$\lambda_j(A) + \lambda_n(B) \leq \lambda_j(A + B) \leq \lambda_j(A) + \lambda_1(B).$$

5. *Interlacing Inequalities:* Let $A \in \mathcal{H}_n$, let $\lambda_1 \geq \lambda_2 \geq \cdots \geq \lambda_n$ be the eigenvalues of A, and for any $i \in \{1, 2, \ldots, n\}$, let $\mu_1 \geq \mu_2 \geq \cdots \geq \mu_{n-1}$ be the eigenvalues of $A(i)$, where $A(i)$ is the principal

submatrix of A obtained by deleting its ith row and column. Then

$$\lambda_1 \geq \mu_1 \geq \lambda_2 \geq \mu_2 \geq \lambda_3 \geq \ldots \geq \lambda_{n-1} \geq \mu_{n-1} \geq \lambda_n.$$

6. Let $A \in \mathcal{H}_n$ and let B be any principal submatrix of A. If λ_k is the kth largest eigenvalue of A and μ_k is the kth largest eigenvalue of B, then $\lambda_k \geq \mu_k$.
7. Let $A \in \mathcal{H}_n$ with eigenvalues $\lambda_1 \geq \lambda_2 \geq \cdots \geq \lambda_n$. Let $\mathcal{S}$ be a k-dimensional subspace of $\mathbb{C}^n$ with $k \in \{1, 2, \ldots, n\}$. Then
   (a) If there is a constant c such that $\mathbf{x}^* A\mathbf{x} \geq c\mathbf{x}^*\mathbf{x}$ for all $\mathbf{x} \in \mathcal{S}$, then $\lambda_k \geq c$.
   (b) If there is a constant c such that $\mathbf{x}^* A\mathbf{x} \leq c\mathbf{x}^*\mathbf{x}$ for all $\mathbf{x} \in \mathcal{S}$, then $\lambda_{n-k+1} \leq c$.
8. Let $A \in \mathcal{H}_n$.
   (a) If $\mathbf{x}^* A\mathbf{x} \geq 0$ for all $\mathbf{x}$ in a k-dimensional subspace of $\mathbb{C}^n$, then A has at least k nonnegative eigenvalues.
   (b) If $\mathbf{x}^* A\mathbf{x} > 0$ for all nonzero $\mathbf{x}$ in a k-dimensional subspace of $\mathbb{C}^n$, then A has at least k positive eigenvalues.
9. Let $A \in \mathcal{H}_n$, let $\lambda = (\lambda_1, \lambda_2, \ldots, \lambda_n)$ be the vector of eigenvalues of A arranged in decreasing order, and let $\alpha = (a_1, a_2, \ldots, a_n)$ be the vector consisting of the diagonal entries of A arranged in decreasing order. Then $\lambda \succeq \alpha$. (See Preliminaries for the definition of $\succeq$.)
10. Let $\alpha = (a_1, a_2, \ldots, a_n)$, $\beta = (b_1, b_2, \ldots, b_n)$ be decreasing sequences of real numbers such that $\alpha \succeq \beta$. Then there exists an $A \in \mathcal{H}_n$ such that the eigenvalues of A are $a_1, a_2, \ldots, a_n$, and the diagonal entries of A are $b_1, b_2, \ldots, b_n$.
11. [Lax96, pp. 133–6] or [Bha01, p. 291] (See also Chapter 15.) Let $A, B \in \mathcal{H}_n$ with eigenvalues $\lambda_1(A) \geq \lambda_2(A) \geq \cdots \geq \lambda_n(A)$ and $\lambda_1(B) \geq \lambda_2(B) \geq \cdots \geq \lambda_n(B)$. Then
   (a) $|\lambda_i(A) - \lambda_i(B)| \leq ||A - B||_2,\ i = 1, \ldots, n.$
   (b) $\sum_{i=1}^{n} [\lambda_i(A) - \lambda_i(B)]^2 \leq ||A - B||_F^2.$

Examples:

1. Setting $\mathbf{x} = \mathbf{e}_i$ in the Rayleigh-Ritz theorem, we obtain $\lambda_n \leq a_{ii} \leq \lambda_1$. Thus, for any $A \in \mathcal{H}_n$, we have $\lambda_1 \geq \max\{a_{ii} \,|\, i \in \{1, 2, \ldots, n\}\}$ and $\lambda_n \leq \min\{a_{ii} \,|\, i \in \{1, 2, \ldots, n\}\}$.
2. Setting $\mathbf{x} = [1, 1, \ldots, 1]^T$ in the Rayleigh-Ritz theorem, we find that $\lambda_n \leq \frac{1}{n}\sum_{i,j=1}^{n} a_{ij} \leq \lambda_1$. If we take A to be the adjacency matrix of a graph, then this inequality implies that the largest eigenvalue of the graph is greater than or equal to its average degree.
3. The Weyl inequalities in Fact 3 above are a special case of the following general class of inequalities:

$$\sum_{k \in K} \lambda_k(A + B) \leq \sum_{i \in I} \lambda_i(A) + \sum_{j \in J} \lambda_j(B),$$

   where I, J, K are certain subsets of $\{1, 2, \ldots, n\}$. In 1962, A. Horn conjectured which inequalities of this form are valid for all Hermitian A, B, and this conjecture was proved correct in papers by A. Klyachko in 1998 and by A. Knutson and T. Tao in 1999. Two detailed accounts of the problem and its solution are given in [Bha01] and [Ful00].
4. Let $A = \begin{bmatrix} 1 & 1 & 1 & 0 \\ 1 & 1 & 1 & 0 \\ 1 & 1 & 1 & 1 \\ 0 & 0 & 1 & 1 \end{bmatrix}$ have eigenvalues $\lambda_1 \geq \lambda_2 \geq \lambda_3 \geq \lambda_4$. Since $A(4) = \begin{bmatrix} 1 & 1 & 1 \\ 1 & 1 & 1 \\ 1 & 1 & 1 \end{bmatrix}$ has eigenvalues 3, 0, 0, by the interlacing inequalities, $\lambda_1 \geq 3 \geq \lambda_2 \geq 0 \geq \lambda_3 \geq 0 \geq \lambda_4$. In particular, $\lambda_3 = 0$.

Applications:

1. To use the Rayleigh–Ritz theorem effectively to estimate the largest or smallest eigenvalue of a Hermitian matrix, one needs to take into account the relative magnitudes of the entries of the matrix. For example, let $A = \begin{bmatrix} 1 & 1 & 1 \\ 1 & 2 & 2 \\ 1 & 2 & 3 \end{bmatrix}$. In order to estimate λ_1, we should try to maximize the Rayleigh quotient. A vector $\mathbf{x} \in \mathbb{R}^3$ is needed for which no component is zero, but such that each component is weighted more than the last. In a few trials, one is led to $\mathbf{x} = [1, 2, 3]^T$, which gives a Rayleigh quotient of 5. So $\lambda_1 \geq 5$. This is close to the actual value of λ_1, which is $\frac{1}{4} \csc^2 \frac{\pi}{14} \approx 5.049$.
   This example is only meant to illustrate the method; its primary importance is as a tool for estimating the largest (smallest) eigenvalue of a large Hermitian matrix when it can neither be found exactly nor be computed numerically.
2. The interlacing inequalities can sometimes be used to efficiently find all the eigenvalues of a Hermitian matrix. The Laplacian matrix (from spectral graph theory, see Section 28.4) of a star is
$$L = \begin{bmatrix} n-1 & -1 & -1 & \cdots & -1 & -1 \\ -1 & 1 & 0 & \cdots & 0 & 0 \\ -1 & 0 & 1 & & 0 & 0 \\ \vdots & \vdots & & \ddots & & \vdots \\ -1 & 0 & 0 & & 1 & 0 \\ -1 & 0 & 0 & \cdots & 0 & 1 \end{bmatrix}.$$
   Since $L(1)$ is an identity matrix, the interlacing inequalities relative to $L(1)$ are: $\lambda_1 \geq 1 \geq \lambda_2 \geq 1 \geq \ldots \geq \lambda_{n-1} \geq 1 \geq \lambda_n$. Therefore, $n-2$ of the eigenvalues of L are equal to 1. Since the columns sum to 0, another eigenvalue is 0. Finally, since $\operatorname{tr} L = 2n - 2$, the remaining eigenvalue is n.
3. The sixth fact above is applied in spectral graph theory to establish the useful fact that the kth largest eigenvalue of a graph is greater than or equal to the kth largest eigenvalue of any induced subgraph.

8.3 Congruence

Definitions:

Two matrices $A, B \in \mathcal{H}_n$ are ***congruent** if there is an invertible matrix $C \in \mathbb{C}^{n\times n}$ such that $B = C^* AC$, denoted $A \overset{c}{\sim} B$. If C is real, then A and B are also called **congruent**.

Let $A \in \mathcal{H}_n$. The **inertia** of A is the ordered triple $\text{in}(A) = (\pi(A), \nu(A), \delta(A))$, where $\pi(A)$ is the number of positive eigenvalues of A, $\nu(A)$ is the number of negative eigenvalues of A, and $\delta(A)$ is the number of zero eigenvalues of A.

In the event that $A \in \mathbb{C}^{n\times n}$ has all real eigenvalues, we adopt the same definition for $\text{in}(A)$.

Facts:

The following can be found in [HJ85, pp. 221–223] and a variation of the last in [Lax96, pp. 77–78].

1. Unitary similarity is a special case of *congruence.
2. *Congruence is an equivalence relation.
3. For $A \in \mathcal{H}_n$, $\pi(A) + \nu(A) + \delta(A) = n$.
4. For $A \in \mathcal{H}_n$, $\text{rank}\, A = \pi(A) + \nu(A)$.
5. Let $A \in \mathcal{H}_n$ with inertia (r, s, t). Then A is *congruent to $I_r \oplus (-I_s) \oplus 0_t$. A matrix C that implements this *congruence is found as follows. Let U be a unitary matrix for which $U^* AU = D$

is a diagonal matrix with $d_{11}, \ldots, d_{rr}$ the positive eigenvalues, $d_{r+1,r+1}, \ldots, d_{r+s,r+s}$ the negative eigenvalues, and $d_{ii} = 0,\ k > r + s$. Let

$$s_i = \begin{cases} 1/\sqrt{d_{ii}}, & i = 1, \ldots, r \\ 1/\sqrt{-d_{ii}}, & i = r+1, \ldots, s \\ 1, & i > r + s \end{cases}$$

and let $S = \operatorname{diag}(s_1, s_2, \ldots, s_n)$. Then $C = US$.

6. *Sylvester's Law of Inertia:* Two matrices $A, B \in \mathcal{H}_n$ are *congruent if and only if they have the same inertia.

Examples:

1. Let $A = \begin{bmatrix} 0 & 0 & 3 \\ 0 & 0 & 4 \\ 3 & 4 & 0 \end{bmatrix}$. Since rank $A = 2$, $\pi(A) + \nu(A) = 2$, so $\delta(A) = 1$. Since tr $A = 0$, we have $\pi(A) = \nu(A) = 1$, and $\operatorname{in}(A) = (1, 1, 1)$. Letting

$$C = \begin{bmatrix} \frac{3}{5\sqrt{10}} & \frac{3}{5\sqrt{10}} & \frac{4}{5} \\ \frac{4}{5\sqrt{10}} & \frac{4}{5\sqrt{10}} & -\frac{3}{5} \\ \frac{1}{\sqrt{10}} & -\frac{1}{\sqrt{10}} & 0 \end{bmatrix}$$

we have

$$C^*AC = \begin{bmatrix} 1 & 0 & 0 \\ 0 & -1 & 0 \\ 0 & 0 & 0 \end{bmatrix}.$$

Now suppose

$$B = \begin{bmatrix} 0 & 1 & 0 \\ 1 & 0 & 1 \\ 0 & 1 & 0 \end{bmatrix}.$$

Clearly $\operatorname{in}(B) = (1, 1, 1)$ also. By Sylvester's law of inertia, B must be *congruent to A.

8.4 Positive Definite Matrices

Definitions:

A matrix $A \in \mathcal{H}_n$ is **positive definite** if $\mathbf{x}^* A\mathbf{x} > 0$ for all nonzero $\mathbf{x} \in \mathbb{C}^n$. It is **positive semidefinite** if $\mathbf{x}^* A\mathbf{x} \geq 0$ for all $\mathbf{x} \in \mathbb{C}^n$. It is **indefinite** if neither A nor $-A$ is positive semidefinite. The set of positive definite matrices of order n is denoted by PD_n, and the set of positive semidefinite matrices of order n by PSD_n. If the dependence on n is not significant, these can be abbreviated as PD and PSD. Finally, PD (PSD) are also used to abbreviate "positive definite" ("positive semidefinite").

Let k be a positive integer. If A, B are PSD and $B^k = A$, then B is called a **PSD k^{th} root** of A and is denoted $A^{1/k}$.

A **correlation matrix** is a PSD matrix in which every main diagonal entry is 1.

Facts:

For facts without a specific reference, see [HJ85, Sections 7.1 and 7.2] and [Fie86, pp. 51–57].

1. $A \in \mathcal{S}_n$ is PD if $\mathbf{x}^T A\mathbf{x} > 0$ for all nonzero $\mathbf{x} \in \mathbb{R}^n$, and is PSD if $\mathbf{x}^T A\mathbf{x} \geq 0$ for all $\mathbf{x} \in \mathbb{R}^n$.
2. Let $A, B \in \text{PSD}_n$.

   (a) Then $A + B \in \text{PSD}_n$.

   (b) If, in addition, $A \in \text{PD}_n$, then $A + B \in \text{PD}_n$.

   (c) If $c \geq 0$, then $cA \in \text{PSD}_n$.

   (d) If, in addition, $A \in \text{PD}_n$ and $c > 0$, then $cA \in \text{PD}_n$.

3. If $A_1, A_2, \ldots, A_k \in \text{PSD}_n$, then so is $A_1 + A_2 + \cdots + A_k$. If, in addition, there is an $i \in \{1, 2, \ldots, k\}$ such that $A_i \in \text{PD}_n$, then $A_1 + A_2 + \cdots + A_k \in \text{PD}_n$.
4. Let $A \in \mathcal{H}_n$. Then A is PD if and only if every eigenvalue of A is positive, and A is PSD if and only if every eigenvalue of A is nonnegative.
5. If A is PD, then tr $A > 0$ and det $A > 0$. If A is PSD, then tr $A \geq 0$ and det $A \geq 0$.
6. A PSD matrix is PD if and only if it is invertible.
7. *Inheritance Principle:* Any principal submatrix of a PD (PSD) matrix is PD (PSD).
8. All principal minors of a PD (PSD) matrix are positive (nonnegative).
9. Each diagonal entry of a PD (PSD) matrix is positive (nonnegative). If a diagonal entry of a PSD matrix is 0, then every entry in the row and column containing it is also 0.
10. Let $A \in \mathcal{H}_n$. Then A is PD if and only if every leading principal minor of A is positive. A is PSD if and only if every principal minor of A is nonnegative. (The matrix $\begin{bmatrix} 0 & 0 \\ 0 & -1 \end{bmatrix}$ shows that it is not sufficient that every leading principal minor be nonnegative in order for A to be PSD.)
11. Let A be PD (PSD). Then A^k is PD (PSD) for all $k \in \mathbb{N}$.
12. Let $A \in \text{PSD}_n$ and express A as $A = UDU^*$, where U is unitary and D is the diagonal matrix of eigenvalues. Given any positive integer k, there exists a unique **PSD** k^{th} **root** of A given by $A^{1/k} = UD^{1/k}U^*$. If A is real so is $A^{1/k}$. (See also Chapter 11.2.)
13. If A is PD, then A^{-1} is PD.
14. Let $A \in \text{PSD}_n$ and let $C \in \mathbb{C}^{n\times m}$. Then C^*AC is PSD.
15. Let $A \in \text{PD}_n$ and let $C \in \mathbb{C}^{n\times m}$, $n \geq m$. Then C^*AC is PD if and only if rank C $=$ m; i.e., if and only if C has linearly independent columns.
16. Let $A \in \text{PD}_n$ and $C \in \mathbb{C}^{n\times n}$. Then C^*AC is PD if and only if C is invertible.
17. Let $A \in \mathcal{H}_n$. Then A is PD if and only if there is an invertible $B \in \mathbb{C}^{n\times n}$ such that $A = B^*B$.
18. *Cholesky Factorization:* Let $A \in \mathcal{H}_n$. Then A is PD if and only if there is an invertible lower triangular matrix L with positive diagonal entries such that $A = LL^*$. (See Chapter 38 for information on the computation of the Cholesky factorization.)
19. Let $A \in \text{PSD}_n$ with rank A $=$ r $<$ n. Then A can be factored as $A = B^*B$ with $B \in \mathbb{C}^{r\times n}$. If A is a real matrix, then B can be taken to be real and $A = B^TB$. Equivalently, there exist vectors $\mathbf{v}_1, \mathbf{v}_2, \ldots, \mathbf{v}_n \in \mathbb{C}^r$ (or $\mathbb{R}^r$) such that $a_{ij} = \mathbf{v}_i^*\mathbf{v}_j$ (or $\mathbf{v}_i^T\mathbf{v}_j$). Note that A is the Gram matrix (see section 8.1) of the vectors $\mathbf{v}_1, \mathbf{v}_2, \ldots, \mathbf{v}_n$. In particular, any rank 1 PSD matrix has the form $\mathbf{x}\mathbf{x}^*$ for some nonzero vector $\mathbf{x} \in \mathbb{C}^n$.
20. [Lax96, p. 123]; see also [HJ85, p. 407] The Gram matrix G of a set of vectors $\mathbf{v}_1, \mathbf{v}_2, \ldots, \mathbf{v}_n$ is PSD. If $\mathbf{v}_1, \mathbf{v}_2, \ldots, \mathbf{v}_n$ are linearly independent, then G is PD.
21. [HJ85, p. 412] *Polar Form:* Let $A \in \mathbb{C}^{m\times n}, m \geq n$. Then A can be factored $A = UP$, where $P \in \text{PSD}_n$, rank $P =$ rank A, and $U \in \mathbb{C}^{m\times n}$ has orthonormal columns. Moreover, P is uniquely determined by A and equals $(A^*A)^{1/2}$. If A is real, then P and U are real. (See also Section 17.1.)
22. [HJ85, p. 400] Any matrix $A \in \text{PD}_n$ is diagonally congruent to a correlation matrix via the diagonal matrix $D = (1/\sqrt{a_{11}}, \ldots, 1/\sqrt{a_{nn}})$.

23. [BJT93] *Parameterization of Correlation Matrices in* $\mathcal{S}_3$: Let $0 \leq \alpha, \beta, \gamma \leq \pi$. Then the matrix

$$C = \begin{bmatrix} 1 & \cos\alpha & \cos\gamma \\ \cos\alpha & 1 & \cos\beta \\ \cos\gamma & \cos\beta & 1 \end{bmatrix}$$

is PSD if and only if $\alpha \leq \beta + \gamma, \quad \beta \leq \alpha + \gamma, \quad \gamma \leq \alpha + \beta, \quad \alpha + \beta + \gamma \leq 2\pi$. Furthermore, C is PD if and only if all of these inequalities are strict.

24. [HJ85, p. 472] and [Fie86, p. 55] Let $A = \begin{bmatrix} B & C \\ C^* & D \end{bmatrix} \in \mathcal{H}_n$, and assume that B is invertible. Then A is PD if and only if the matrices B and its Schur complement $S = D - C^* B^{-1} C$ are PD.

25. [Joh92] and [LB96, pp. 93–94] Let $A = \begin{bmatrix} B & C \\ C^* & D \end{bmatrix}$ be PSD. Then any column of C lies in the span of the columns of B.

26. [HJ85, p. 465] Let $A \in \text{PD}_n$ and $B \in \mathcal{H}_n$. Then

    (a) AB is diagonalizable.
    (b) All eigenvalues of AB are real.
    (c) $\text{in}(AB) = \text{in}(B)$.

27. Any diagonalizable matrix A with real eigenvalues can be factored as $A = BC$, where B is PSD and C is Hermitian.

28. If $A, B \in \text{PD}_n$, then every eigenvalue of AB is positive.

29. [Lax96, p. 120] Let $A, B \in \mathcal{H}_n$. If A is PD and $AB + BA$ is PD, then B is PD. It is not true that if A, B are both PD, then $AB + BA$ is PD as can be seen by the example $A = \begin{bmatrix} 1 & 2 \\ 2 & 5 \end{bmatrix}$, $B = \begin{bmatrix} 5 & 2 \\ 2 & 1 \end{bmatrix}$.

30. [HJ85, pp. 466–467] and [Lax96, pp. 125–126] The real valued function $f(X) = \log(\det X)$ is concave on the set PD_n; i.e., $f((1-t)X + tY) \geq (1-t)f(X) + tf(Y)$ for all $t \in [0,1]$ and all $X, Y \in PD_n$.

31. [Lax96, p. 129] If $A \in \text{PD}_n$ is real, $\displaystyle\int_{\mathbb{R}^n} e^{-\mathbf{x}^T A \mathbf{x}}\, d\mathbf{x} = \frac{\pi^{n/2}}{\sqrt{\det A}}$.

32. [Fie60] Let $A = [a_{ij}],\ B = [b_{ij}] \in \text{PD}_n$, with $A^{-1} = [\alpha_{ij}],\ B^{-1} = [\beta_{ij}]$. Then

$$\sum_{i,j=1}^{n} (a_{ij} - b_{ij})(\alpha_{ij} - \beta_{ij}) \leq 0,$$

with equality if and only if $A = B$.

33. [Ber73, p. 55] Consider PD_n to be a subset of $\mathbb{C}^{n^2}$ (or for real matrices of $\mathbb{R}^{n^2}$). Then the (topological) boundary of PD_n is PSD_n.

Examples:

1. If $A = [a]$ is 1×1, then A is PD if and only if $a > 0$, and is PSD if and only if $a \geq 0$; so PD and PSD matrices are a generalization of positive numbers and nonnegative numbers.
2. If one attempts to define PD (or PSD) for nonsymmetric real matrices according to the the usual definition, many of the facts above for (Hermitian) PD matrices no longer hold. For example, suppose $A = \begin{bmatrix} 0 & 1 \\ -1 & 0 \end{bmatrix}$. Then $\mathbf{x}^T A \mathbf{x} = 0$ for all $\mathbf{x} \in \mathbb{R}^2$. But $\sigma(A) = \{i, -i\}$, which does not agree with Fact 4 above.

3. The matrix $A = \begin{bmatrix} 17 & 8 \\ 8 & 17 \end{bmatrix}$ factors as $\frac{1}{\sqrt{2}}\begin{bmatrix} 1 & 1 \\ 1 & -1 \end{bmatrix}\begin{bmatrix} 25 & 0 \\ 0 & 9 \end{bmatrix}\frac{1}{\sqrt{2}}\begin{bmatrix} 1 & 1 \\ 1 & -1 \end{bmatrix}$, so $A^{1/2} = \frac{1}{\sqrt{2}}\begin{bmatrix} 1 & 1 \\ 1 & -1 \end{bmatrix}\begin{bmatrix} 5 & 0 \\ 0 & 3 \end{bmatrix}\frac{1}{\sqrt{2}}\begin{bmatrix} 1 & 1 \\ 1 & -1 \end{bmatrix} = \begin{bmatrix} 4 & 1 \\ 1 & 4 \end{bmatrix}$.
4. A self-adjoint linear operator on a complex inner product space V (see Section 5.3) is called **positive** if $\langle A\mathbf{x}, \mathbf{x}\rangle > 0$ for all nonzero $\mathbf{x} \in V$. For the usual inner product in $\mathbb{C}^n$ we have $\langle A\mathbf{x}, \mathbf{x}\rangle = \mathbf{x}^* A\mathbf{x}$, in which case the definition of positive operator and positive definite matrix coincide.
5. Let $X_1, X_2, \ldots, X_n$ be real-valued random variables on a probability space, each with mean zero and finite second moment. Define the matrix

$$a_{ij} = E(X_i X_j), \quad i, j \in \{1, 2, \ldots, n\}.$$

The real symmetric matrix A is called the **covariance matrix** of $X_1, X_2, \ldots, X_n$, and is necessarily PSD. If we let $X = (X_1, X_2, \ldots, X_n)^T$, then we may abbreviate the definition to $A = E(XX^T)$.

Applications:

1. [HFKLMO95, p. 181] or [MT88, p. 253] *Test for Maxima and Minima in Several Variables:* Let D be an open set in $\mathbb{R}^n$ containing the point $\mathbf{x}_0$, let $f : D \to \mathbb{R}$ be a twice continuously differentiable function on D, and assume that all first derivatives of f vanish at $\mathbf{x}_0$. Let H be the Hessian matrix of f (Example 2 of Section 8.1). Then

   (a) f has a relative minimum at $\mathbf{x}_0$ if $H(\mathbf{x}_0)$ is PD.
   (b) f has a relative maximum at $\mathbf{x}_0$ if $-H(\mathbf{x}_0)$ is PD.
   (c) f has a saddle point at $\mathbf{x}_0$ if $H(\mathbf{x}_0)$ is indefinite.

   Otherwise, the test is inconclusive.
2. Section 1.3 of the textbook [Str86] is an elementary introduction to real PD matrices emphasizing the significance of the Cholesky-like factorization LDL^T of a PD matrix. This representation is then used as a framework for many applications throughout the first three chapters of this text.
3. Let A be a real matrix in PD_n. A **multivariate normal distribution** is one whose probability density function in $\mathbb{R}^n$ is given by

$$f(\mathbf{x}) = \frac{1}{\sqrt{(2\pi)^n \det A}} e^{-\frac{1}{2}\mathbf{x}^T A^{-1}\mathbf{x}}.$$

It follows from Fact 31 above that $\int_{\mathbb{R}^n} f(\mathbf{x})\, d\mathbf{x} = 1$. A **Gaussian family** $X_1, X_2, \ldots X_n$, where each X_i has mean zero, is a set of random variables that have a multivariate normal distribution. The entries of the matrix A satisfy the identity $a_{ij} = E(X_i X_j)$, so the distribution is completely determined by its covariance matrix.

8.5 Further Topics in Positive Definite Matrices

Definitions:

Let $A, B \in F^{n \times n}$, where F is a field. The **Hadamard product** or **Schur product** of A and B, denoted $A \circ B$, is the matrix in $F^{n \times n}$ whose $(i, j)^{th}$ entry is $a_{ij} b_{ij}$.

A function $f : \mathbb{R} \to \mathbb{C}$ is called **positive semidefinite** if for each $n \in \mathbb{N}$ and all $x_1, x_2, \ldots, x_n \in \mathbb{R}$, the $n \times n$ matrix $[f(x_i - x_j)]$ is PSD.

Let $A, B \in \mathcal{H}_n$. We write $A \succ B$ if $A - B$ is PD, and $A \succeq B$ if $A - B$ is PSD. The partial ordering on $\mathcal{H}_n$ induced by $\succeq$ is called the **partial semidefinite ordering** or the **Loewner ordering.**

Let V be an n-dimensional inner product space over $\mathbb{C}$ or $\mathbb{R}$. A set $K \subseteq V$ is called a **cone** if

(a) For each $\mathbf{x}, \mathbf{y} \in K$, $\mathbf{x} + \mathbf{y} \in K$.
(b) If $\mathbf{x} \in K$ and $c \geq 0$, then $c\mathbf{x} \in K$.

A cone is frequently referred to as a convex cone. A cone K is **closed** if K is a closed subset of V, is **pointed** if $K \cap -K = \{0\}$, and is **full** if it has a nonempty interior. The set

$$K^* = \{\mathbf{y} \in V \mid \langle \mathbf{x}, \mathbf{y} \rangle \geq 0 \quad \forall\, \mathbf{x} \in K\}$$

is called the **dual space.**

Facts:

1. [HJ91, pp. 308–309]; also see [HJ85, p. 458] or [Lax96, pp. 124, 234] *Schur Product Theorem:* If $A, B \in \mathrm{PSD}_n$, then so is $A \circ B$. If $A \in \mathrm{PSD}_n$, $a_{ii} > 0$, $i = 1, \ldots, n$, and $B \in \mathrm{PD}_n$, then $A \circ B \in \mathrm{PD}_n$. In particular, if A and B are both PD, then so is $A \circ B$.
2. [HJ85, p. 459] *Fejer's Theorem:* Let $A = [a_{ij}] \in \mathcal{H}_n$. Then A is PSD if and only if
$$\sum_{i,j=1}^{n} a_{ij}\, b_{ij} \geq 0$$
for all matrices $B \in \mathrm{PSD}_n$.
3. [HJ91, pp. 245–246] If $A \in \mathrm{PD}_m$ and $B \in \mathrm{PD}_n$, then the Kronecker (tensor) product (see Section 10.4) $A \otimes B \in \mathrm{PD}_{mn}$. If $A \in \mathrm{PSD}_m$ and $B \in \mathrm{PSD}_n$, then $A \otimes B \in \mathrm{PSD}_{mn}$.
4. [HJ85, p. 477] or [Lax96, pp. 126–127, 131–132] *Hadamard's Determinantal Inequality:* If $A \in \mathrm{PD}_n$, then $\det A \leq \prod_{i=1}^{n} a_{ii}$. Equality holds if and only if A is a diagonal matrix.
5. [FJ00, pp. 199–200] or [HJ85, p. 478] *Fischer's Determinantal Inequality:* If $A \in \mathrm{PD}_n$ and α is any subset of $\{1, 2, \ldots, n\}$, then $\det A \leq \det A[\alpha] \det A[\alpha^c]$ (where $\det A[\emptyset] = 1$). Equality occurs if and only if $A[\alpha, \alpha^c]$ is a zero matrix. (See Chapter 1.2 for the definition of $A[\alpha]$ and $A[\alpha, \beta]$.)
6. [FJ00, pp. 199–200] or [HJ85, p. 485] *Koteljanskii's Determinantal Inequality:* Let $A \in \mathrm{PD}_n$ and let α, β be any subsets of $\{1, 2, \ldots, n\}$. Then $\det A[\alpha \cup \beta] \det A[\alpha \cap \beta] \leq \det A[\alpha] \det A[\beta]$. Note that if $\alpha \cap \beta = \emptyset$, Koteljanskii's inequality reduces to Fischer's inequality. Koteljanskii's inequality is also called the Hadamard–Fischer inequality.
For other determinantal inequalities for PD matrices, see [FJ00] and [HJ85, §7.8].
7. [Fel71, pp. 620–623] and [Rud62, pp. 19–21] *Bochner's Theorem:* A continuous function from $\mathbb{R}$ into $\mathbb{C}$ is positive semidefinite if and only if it is the Fourier transform of a finite positive measure.
8. [Lax96, p. 118] and [HJ85, p. 475, 470] Let $A, B, C, D \in \mathcal{H}_n$.
   (a) If $A \prec B$ and $C \prec D$, then $A + C \prec B + D$.
   (b) If $A \prec B$ and $B \prec C$, then $A \prec C$.
   (c) If $A \prec B$ and $S \in \mathbb{C}^{n \times n}$ is invertible, then $S^* A S \prec S^* B S$.

   The three statements obtained by replacing each occurrence of $\prec$ by $\preceq$ are also valid.
9. [Lax96, pp. 118–119, 121–122] and [HJ85, pp. 471–472] Let $A, B \in \mathrm{PD}_n$ with $A \prec B$. Then
   (a) $A^{-1} \succ B^{-1}$.
   (b) $A^{1/2} \prec B^{1/2}$.
   (c) $\det A < \det B$.
   (d) $\operatorname{tr} A < \operatorname{tr} B$.

   If $A \preceq B$, then statement (a) holds with $\succ$ replaced by $\succeq$, statement (b) holds with $\prec$ replaced by $\preceq$, and statements (c) and (d) hold with $<$ replaced by $\leq$.

10. [HJ85, pp. 182, 471–472] Let $A, B \in \mathcal{H}_n$ with eigenvalues $\lambda_1(A) \geq \lambda_2(A) \geq \cdots \geq \lambda_n(A)$ and $\lambda_1(B) \geq \lambda_2(B) \geq \cdots \geq \lambda_n(B)$. If $A \prec B$, then $\lambda_k(A) < \lambda_k(B)$, $k = 1, \ldots, n$. If $A \preceq B$, then $\lambda_k(A) \leq \lambda_k(B)$, $k = 1, \ldots, n$.
11. [HJ85, p. 474] Let A be PD and let $\alpha \subseteq \{1, 2, \ldots, n\}$. Then $A^{-1}[\alpha] \succeq (A[\alpha])^{-1}$.
12. [HJ85, p. 475] If A is PD, then $A^{-1} \circ A \succeq I \succeq (A^{-1} \circ A)^{-1}$.
13. [Hal83, p. 89] If K is a cone in an inner product space V, its dual space is a closed cone and is called the **dual cone** of K. If K is a closed cone, then $(K^*)^* = K$.
14. [Ber73, pp. 49–50, 55] and [HW87, p. 82] For each pair $A, B \in \mathcal{H}_n$, define $\langle A, B \rangle = \operatorname{tr}(AB)$.

    (a) $\mathcal{H}_n$ is an inner product space over the real numbers with respect to $\langle \cdot, \cdot \rangle$.
    (b) PSD_n is a closed, pointed, full cone in $\mathcal{H}_n$.
    (c) $(\mathrm{PSD}_n)^* = \mathrm{PSD}_n$.

Examples:

1. The matrix $C = [\cos|i - j|] \in \mathcal{S}_n$ is PSD, as can be verified with Fact 19 of section 8.4 and the addition formula for the cosine. But a quick way to see it is to consider the measure $\mu(x) = \frac{1}{2}[\delta(x+1) + \delta(x-1)]$; i.e., $\mu(E) = 0$ if $-1, 1 \notin E$, $\mu(E) = 1$ if $-1, 1 \in E$, and $\mu(E) = 1/2$ if exactly one of $-1, 1 \in E$. Since the Fourier transform of μ is $\cos t$, if we let $x_1, x_2, \ldots, x_n$ be $1, 2, \ldots, n$ in the definition of positive definite function, we see immediately by Bochner's Theorem that the matrix $[\cos(i - j)] = [\cos|i - j|] = C$ is PSD. By Hadamard's determinantal inequality $\det C \leq \prod_{i=1}^n c_{ii} = 1$.

2. Since $\begin{bmatrix} 1 & 1 \\ 1 & 2 \end{bmatrix} \prec \begin{bmatrix} 2 & 2 \\ 2 & 7 \end{bmatrix}$, taking inverses we have $\begin{bmatrix} .7 & -.2 \\ -.2 & .2 \end{bmatrix} \prec \begin{bmatrix} 2 & -1 \\ -1 & 1 \end{bmatrix}$.

3. The matrix $A = \begin{bmatrix} 1 & 1 & 1 \\ 1 & 2 & 2 \\ 1 & 2 & 3 \end{bmatrix}$ is *PD* with inverse $A^{-1} = \begin{bmatrix} 2 & -1 & 0 \\ -1 & 2 & -1 \\ 0 & -1 & 1 \end{bmatrix}$. Then $(A[\{1,3\}])^{-1} = \begin{bmatrix} 1.5 & -.5 \\ -.5 & .5 \end{bmatrix} \preceq \begin{bmatrix} 2 & 0 \\ 0 & 1 \end{bmatrix} = A^{-1}[\{1,3\}]$. Also, $A^{-1} \circ A = \begin{bmatrix} 2 & -1 & 0 \\ -1 & 4 & -2 \\ 0 & -2 & 3 \end{bmatrix} \succeq \begin{bmatrix} 1 & 0 & 0 \\ 0 & 1 & 0 \\ 0 & 0 & 1 \end{bmatrix} \succeq \frac{1}{13}\begin{bmatrix} 8 & 3 & 2 \\ 3 & 6 & 4 \\ 2 & 4 & 7 \end{bmatrix} = (A^{-1} \circ A)^{-1}$.

4. If $A \succeq B \succeq 0$, it does not follow that $A^2 \succeq B^2$. For example, if $A = \begin{bmatrix} 2 & 1 \\ 1 & 1 \end{bmatrix}$ and $B = \begin{bmatrix} 1 & 0 \\ 0 & 0 \end{bmatrix}$, then B and $A - B$ are PSD, but $A^2 - B^2$ is not.

Applications:

1. Hadamard's determinantal inequality can be used to obtain a sharp bound on the determinant of a matrix in $\mathbb{C}^{n \times n}$ if only the magnitudes of the entries are known. [HJ85, pp. 477–478] or [Lax96, p. 127].

   Hadamard's Determinantal Inequality for Matrices in $\mathbb{C}^{n \times n}$: Let $B \in \mathbb{C}^{n \times n}$. Then $|\det B| \leq \prod_{i=1}^n (\sum_{j=1}^n |b_{ij}|^2)^{1/2}$ with equality holding if and only if the rows of B are orthogonal.

   In the case that B is invertible, the inequality follows from Hadamard's determinantal inequality for positive definite matrices by using $A = BB^*$; if B is singular, the inequality is obvious.

   The inequality can be alternatively expressed as $|\det B| \leq \prod_{i=1}^n \|\mathbf{b}_i\|_2$, where $\mathbf{b}_i$ are the rows of B. If B is a real matrix, it has the geometric meaning that among all parallelepipeds with given side lengths $\|\mathbf{b}_i\|_2$, $i = 1, \ldots, n$, the one with the largest volume is rectangular.

   There is a corresponding inequality in which the right-hand side is the product of the lengths of the columns of B.

2. [Fel71, pp. 620–623] A special case of Bochner's theorem, important in probability theory, is: A continuous function ϕ is the characteristic function of a probability distribution if and only if it is positive semidefinite and $\phi(0) = 1$.
3. Understanding the cone PSD_n is important in semidefinite programming. (See Chapter 51.)

References

[BJT93] W. Barrett, C. Johnson, and P. Tarazaga. The real positive definite completion problem for a simple cycle. *Linear Algebra and Its Applications*, 192: 3–31 (1993).

[Ber73] A. Berman. *Cones, Matrices, and Mathematical Programming*. Springer-Verlag, Berlin, 1973.

[Bha97] R. Bhatia. *Matrix Analysis*. Springer-Verlag, New York, 1997.

[Bha01] R. Bhatia. Linear algebra to quantum cohomology: The story of Alfred Horn's inequalities. *The American Mathematical Monthly*, 108 (4): 289–318, 2001.

[FJ00] S. M. Fallat and C. R. Johnson. Determinantal inequalities: ancient history and recent advances. D. Huynh, S. Jain, and S. López-Permouth, Eds., *Algebra and Its Applications*, Contemporary Mathematics and Its Applications, American Mathematical Society, 259: 199–212, 2000.

[Fel71] W. Feller. *An Introduction to Probability Theory and Its Applications*, 2nd ed., Vol. II. John Wiley & Sons, New York, 1996.

[Fie60] M. Fiedler. A remark on positive definite matrices (Czech, English summary). *Casopis pro pest. mat.*, 85: 75–77, 1960.

[Fie86] M. Fiedler. *Special Matrices and Their Applications in Numerical Mathematics.* Martinus Nijhoff Publishers, Dordrecht, The Netherlands, 1986.

[Ful00] W. Fulton. Eigenvalues, invariant factors, highest weights, and Schubert calculus. *Bulletin of the American Mathematical Society*, 37: 209–249, 2000.

[GR01] C. Godsil and G. Royle. *Algebraic Graph Theory*. Springer-Verlag, New York, 2001.

[Hal83] M. Hall, Jr. *Combinatorial Theory*. John Wiley & Sons, New York, 1983.

[HFKLMO95] K. Heuvers, W. Francis, J. Kursti, D. Lockhart, D. Mak, and G. Ortner. *Linear Algebra for Calculus*. Brooks/Cole Publishing Company, Pacific Grove, CA, 1995.

[HJ85] R. A. Horn and C. R. Johnson. *Matrix Analysis*. Cambridge University Press, Cambridge, 1985.

[HJ91] R. A. Horn and C. R. Johnson. *Topics in Matrix Analysis*. Cambridge University Press, Cambridge, 1991.

[HW87] R. Hill and S. Waters. On the cone of positive semidefinite matrices. *Linear Algebra and Its Applications*, 90: 81–88, 1987.

[Joh92] C. R. Johnson. Personal communication.

[Lax96] P. D. Lax. *Linear Algebra*. John Wiley & Sons, New York, 1996.

[Lay97] D. Lay. *Linear Algebra and Its Applications*, 2nd ed., Addison-Wesley, Reading, MA., 1997.

[LB96] M. Lundquist and W. Barrett. Rank inequalities for positive semidefinite matrices. *Linear Algebra and Its Applications*, 248: 91–100, 1996.

[MT88] J. Marsden and A. Tromba. *Vector Calculus*, 3rd ed., W. H. Freeman and Company, New York, 1988.

[Rud62] W. Rudin. *Fourier Analysis on Groups*. Interscience Publishers, a division of John Wiley & Sons, New York, 1962.

[Str86] G. Strang. *Introduction to Applied Mathematics.* Wellesley-Cambridge Press, Wellesley, MA, 1986.

9

Nonnegative Matrices and Stochastic Matrices

Uriel G. Rothblum
Technion

Nonnegativity is a natural property of many measured quantities (physical and virtual). Consequently, nonnegative matrices arise in modelling transformations in numerous branches of science and engineering — these include probability theory (Markov chains), population models, iterative methods in numerical analysis, economics (input–output models), epidemiology, statistical mechanics, stability analysis, and physics. This section is concerned with properties of such matrices. The theory of the subject was originated in the pioneering work of Perron and Frobenius in [Per07a,Per07b,Fro08,Fro09, and Fro12]. There have been books, chapters in books, and hundreds of papers on the subject (e.g., [BNS89], [BP94], [Gan59, Chap. XIII], [Har02] [HJ85, Chap. 8], [LT85, Chap. 15], [Min88], [Sen81], [Var62, Chap. 1]). A brief outline of proofs of the classic result of Perron and a description of several applications of the theory can be found in the survey paper [Mac00]. Generalizations of many facts reported herein to cone-invariant matrices can be found in Chapter 26.

9.1 Notation, Terminology, and Preliminaries

Definitions:

For a positive integer n, $\langle n \rangle = \{1, \ldots, n\}$.

For a matrix $A \in \mathbb{C}^{m \times n}$:

A is **nonnegative (positive)**, written $A \geq 0$ ($A > 0$), if all of A's elements are nonnegative (positive).

A is **semipositive**, written $A \gneq 0$ if $A \geq 0$ and $A \neq 0$.

$|A|$ will denote the nonnegative matrix obtained by taking element-wise absolute values of A's coordinates.

For a square matrix $A = [a_{ij}] \in \mathbb{C}^{n\times n}$:

The k-**eigenspace** of A at a complex number λ, denoted $N_\lambda^k(A)$, is $\ker(A - \lambda I)^k$; a **generalized eigenvector** of P at λ is a vector in $\cup_{k=0}^{\infty} N_\lambda^k(A)$.

The **index** of A at λ, denoted $\nu_A(\lambda)$, is the smallest integer k with $N_\lambda^k(A) = N_\lambda^{k+1}(A)$.

The **ergodicity coefficient of** A, denoted $\tau(A)$, is $\max\{|\lambda| : \lambda \in \sigma(A) \text{ and } |\lambda| \neq \rho(A)\}$ (with the maximum over the empty set defined to be 0 and $\rho(A)$ being the spectral radius of A).

A **group inverse** of a square matrix A, denoted $A^{\#}$, is a matrix X satisfying $AXA = A$, $XAX = X$, and $AX = XA$ (whenever there exists such an X, it is unique).

The **digraph of** A, denoted $\Gamma(A)$, is the graph with vertex-set $V(A) = \langle n\rangle$ and arc-set $E(A) = \{(i, j) : i, j \in \langle n\rangle \text{ and } a_{ij} \neq 0\}$; in particular, $i = 1, \dots, n$ are called **vertices.**

Vertex $i \in \langle n\rangle$ **has access to** vertex $j \in \langle n\rangle$, written $i \mapsto j$, if either $i = j$ or $\Gamma(A)$ contains a simple walk (path) from i to j; we say that i and j **communicate**, written $i \sim j$, if each has access to the other.

A subset C of $\langle n\rangle$ is **final** if no vertex in C has access to a vertex not in C.

Vertex-communication is an equivalence relation. It partitions $\langle n\rangle$ into equivalence classes, called the **access equivalence classes** of A.

$\Gamma(A)$ is **strongly connected** if there is only one access equivalence class.

An access equivalence class C **has access to** an access equivalence class C', written $C \mapsto C'$ if some, or equivalently every, vertex in C has access to some, or equivalently every, vertex in C'; in this case we also write $i \mapsto C'$ and $C \mapsto i'$ when $i \in C$ and $i' \in C'$.

An access equivalence class C of A is **final** if its final as a subset of $\langle n\rangle$, that is, it does not have access to any access equivalence class but itself.

The **reduced digraph** of $\Gamma(A)$, denoted $R[\Gamma(A)]$, is the digraph whose vertex-set is the set of access equivalence classes of A and whose arcs are the pairs (C, C') with C and C' as distinct classes satisfying $C \mapsto C'$.

For a sequence $\{a_m\}_{m=0,1,\dots}$ of complex numbers and a complex number a:

a is a $(C, 0)$-**limit** of $\{a_m\}_{m=0,1,\dots}$, written $\lim_{m\to\infty} a_m = a\ (C, 0)$, if $\lim_{m\to\infty} a_m = a$ (in the sense of a regular limit).

a is the $(C, 1)$-**limit** of $\{a_m\}_{m=0,1,\dots}$, written $\lim_{m\to\infty} a_m = a\ (C, 1)$, if $\lim_{m\to\infty} m^{-1}\sum_{s=0}^{m-1} a_s = a$.

Inductively for $k = 2, 3, \dots$, a is a (C, k)-**limit** of $\{a_m\}_{m=0,1,\dots}$, written $\lim_{m\to\infty} a_m = a\ (C, k)$, if $\lim_{m\to\infty} m^{-1}\sum_{s=0}^{m-1} a_s = a\ (C, k-1)$.

For $0 \leq \beta < 1$, $\{a_m\}_{m=0,1,\dots}$ **converges geometrically to** a **with (geometric) rate** β if for each $\beta < \gamma < 1$, the set of real numbers $\{\frac{a_m - a}{\gamma^m} : m = 0, 1, \dots\}$ is bounded. (For simplicity, we avoid the reference of geometric convergence for (C, k)-limits.)

For a square nonnegative matrix P:

$\rho(P)$ (the spectral radius of P) is called the **Perron value** of P (see Facts 9.2–1(b) and 9.2–5(a) and 9.3–2(a)).

A **distinguished eigenvalue** of P is a (necessarily nonnegative) eigenvalue of P that is associated with a semipositive (right) eigenvector.

For more information about generalized eigenvectors, see Chapter 6.1. An example illustrating the digraph definitions is given in Figure 9.1; additional information about digraphs can be found in Chapter 29.

9.2 Irreducible Matrices

(See Chapter 27.3, Chapter 29.5, and Chapter 29.6 for additional information.)

Definitions:

A nonnegative square matrix P is **irreducible** if it is not permutation similar to any matrix having the (nontrivial) block-partition

$$\begin{bmatrix} A & B \\ 0 & C \end{bmatrix}$$

with A and C square.

The **period** of an irreducible nonnegative square matrix P (also known as the **index of imprimitivity** of P) is the greatest common divisor of lengths of the cycles of $\Gamma(P)$, the digraph of P.

An irreducible nonnegative square matrix P is **aperiodic** if its period is 1.

Note: We exclude from further consideration the (irreducible) trivial 0 matrix of dimension 1×1.

Facts:

Facts requiring proofs for which no specific reference is given can be found in [BP94, Chap. 2].

1. (*Positive Matrices — Perron's Theorem*) [Per07a, Per07b] Let P be a positive square matrix with spectral radius ρ and ergodicity coefficient τ.
   (a) P is irreducible and aperiodic.
   (b) ρ is positive and is a simple eigenvalue of P; in particular, the index of P at ρ is 1.
   (c) There exist positive right and left eigenvectors of P corresponding to ρ, in particular, ρ is a distinguished eigenvalue of both P and P^T.
   (d) ρ is the only distinguished eigenvalue of P.
   (e) ρ is the only eigenvalue λ of P with $|\lambda| = \rho$.
   (f) If $\mathbf{x} \in \mathbb{R}^n$ satisfies $\mathbf{x} \geq 0$ and either $(\rho I - P)\mathbf{x} \geq 0$ or $(\rho I - P)\mathbf{x} \leq 0$, then $(\rho I - P)\mathbf{x} = 0$.
   (g) If $\mathbf{v}$ and $\mathbf{w}$ are positive right and left eigenvectors of P corresponding to ρ (note that $\mathbf{w}$ is a row vector), then $\lim_{m\to\infty}(\frac{P}{\rho})^m = \frac{\mathbf{vw}}{\mathbf{wv}}$ and the convergence is geometric with rate $\frac{\tau}{\rho}$.
   (h) $Q \equiv \rho I - P$ has a group inverse; further, if $\mathbf{v}$ and $\mathbf{w}$ are positive right and left eigenvectors of P corresponding to ρ, then $Q + \frac{\mathbf{vw}}{\mathbf{wv}}$ is nonsingular, $Q^\# = (Q + \frac{\mathbf{vw}}{\mathbf{wv}})^{-1}(I - \frac{\mathbf{vw}}{\mathbf{wv}})$, and $\frac{\mathbf{vw}}{\mathbf{wv}} = I - QQ^\#$.
   (i) $\lim_{m\to\infty} \sum_{t=0}^{m-1}(\frac{P}{\rho})^t - m\frac{\mathbf{vw}}{\mathbf{wv}} = (\rho I - P)^\#$ and the convergence is geometric with rate $\frac{\tau}{\rho}$.
2. (*Characterizing Irreducibility*) Let P be a nonnegative $n \times n$ matrix with spectral radius ρ. The following are equivalent:
   (a) P is irreducible.
   (b) $\sum_{s=0}^{n-1} P^s > 0$.
   (c) $(I + P)^{n-1} > 0$.
   (d) The digraph of P is strongly connected, i.e., P has a single access equivalence class.
   (e) Every eigenvector of P corresponding to ρ is a scalar multiple of a positive vector.
   (f) For some $\mu > \rho$, $\mu I - P$ is nonsingular and $(\mu I - P)^{-1} > 0$.
   (g) For every $\mu > \rho$, $\mu I - P$ is nonsingular and $(\mu I - P)^{-1} > 0$.
3. (*Characterizing Aperiodicity*) Let P be an irreducible nonnegative $n \times n$ matrix. The following are equivalent:
   (a) P is aperiodic.
   (b) $P^m > 0$ for some m. (See Section 29.6.)
   (c) $P^m > 0$ for all $m \geq n$.
4. (*The Period*) Let P be an irreducible nonnegative $n \times n$ matrix with period q.
   (a) q is the greatest common divisor of $\{m : m \text{ is a positive integer and } (P^m)_{ii} > 0\}$ for any one, or equivalently all, $i \in \{1, \ldots, n\}$.

(b) There exists a partition $C_1, \ldots, C_q$ of $\{1, \ldots, n\}$ such that:

i. For $s, t = 1, \ldots, q$, $P[C_s, C_t] \neq 0$ if and only if $t = s + 1$ (with $q + 1$ identified with 1); in particular, P is permutation similar to a block rectangular matrix having a representation

$$\begin{bmatrix} 0 & P[C_1, C_2] & 0 & \ldots & 0 \\ 0 & 0 & P[C_2, C_3] & \ldots & 0 \\ \vdots & \vdots & \vdots & \ldots & \vdots \\ 0 & 0 & 0 & \ldots & P[C_{q-1}, C_q] \\ P[C_q, C_1] & 0 & 0 & \ldots & 0 \end{bmatrix}.$$

ii. $P^q[C_s]$ is irreducible for $s = 1, \ldots, q$ and $P^q[C_s, C_t] = 0$ for $s, t = 1, \ldots, n$ with $s \neq t$; in particular, P^q is permutation similar to a block diagonal matrix having irreducible blocks on the diagonal.

5. (*Spectral Properties — The Perron–Frobenius Theorem*) [Fro12] Let P be an irreducible nonnegative square matrix with spectral radius ρ and period q.

(a) ρ is positive and is a simple eigenvalue of P; in particular, the index of P at ρ is 1.

(b) There exist positive right and left eigenvectors of P corresponding to ρ; in particular, ρ is a distinguished eigenvalue of both P and P^T.

(c) ρ is the only distinguished eigenvalue of P and of P^T.

(d) If $\mathbf{x} \in \mathbb{R}^n$ satisfies $\mathbf{x} \geq 0$ and either $(\rho I - P)\mathbf{x} \geq 0$ or $(\rho I - P)\mathbf{x} \leq 0$, then $(\rho I - P)\mathbf{x} = 0$.

(e) The eigenvalues of P with modulus ρ are $\{\rho e^{(2\pi i)k/q} : k = 0, \ldots, q-1\}$ (here, i is the complex root of -1) and each of these eigenvalues is simple. In particular, if P is aperiodic ($q = 1$), then every eigenvalue $\lambda \neq \rho$ of P satisfies $|\lambda| < \rho$.

(f) $Q \equiv \rho I - P$ has a group inverse; further, if $\mathbf{v}$ and $\mathbf{w}$ are positive right and left eigenvectors of P corresponding to ρ, then $Q + \frac{\mathbf{vw}}{\mathbf{wv}}$ is nonsingular, $Q^\# = (Q + \frac{\mathbf{vw}}{\mathbf{wv}})^{-1}(I - \frac{\mathbf{vw}}{\mathbf{wv}})$, and $\frac{\mathbf{vw}}{\mathbf{wv}} = I - QQ^\#$.

6. (*Convergence Properties of Powers*) Let P be an irreducible nonnegative square matrix with spectral radius ρ, index ν, period q, and ergodicity coefficient τ. Also, let $\mathbf{v}$ and $\mathbf{w}$ be positive right and left eigenvectors of P corresponding to ρ and let $P^\star \equiv \frac{\mathbf{vw}}{\mathbf{wv}}$.

(a) $\lim_{m\to\infty} (\frac{P}{\rho})^m = P^\star$ (C,1).

(b) $\lim_{m\to\infty} \frac{1}{q} \sum_{t=m}^{m+q-1} (\frac{P}{\rho})^t = P^\star$ and the convergence is geometric with rate $\frac{\tau}{\rho} < 1$. In particular, if P is aperiodic ($q = 1$), then $\lim_{m\to\infty} (\frac{P}{\rho})^m = P^\star$ and the convergence is geometric with rate $\frac{\tau}{\rho} < 1$.

(c) For each $k = 0, \ldots, q-1$, $\lim_{m\to\infty} (\frac{P}{\rho})^{mq+k}$ exists and the convergence of these sequences to their limit is geometric with rate $(\frac{\tau}{\rho})^q < 1$.

(d) $\lim_{m\to\infty} \sum_{t=0}^{m-1} (\frac{P}{\rho})^t - mP^\star = (I - \rho^{-1}P)^\#$ (C,1); further, if P is aperiodic, this limit holds as a regular limit and the convergence is geometric with rate $\frac{\tau}{\rho} < 1$.

7. (*Bounds on the Perron Value*) Let P be an irreducible nonnegative $n \times n$ matrix with spectral radius ρ, let μ be a nonnegative scalar, and let $\diamond \in \{<, \lneq, \leq, =, \geq, \gneq, >\}$. The following are equivalent:

(a) $\rho \diamond \mu$.

(b) There exists a vector $\mathbf{u} \gneq 0$ in $\mathbb{R}^n$ with $P\mathbf{u} \diamond \mu\mathbf{u}$.

(c) There exists a vector $\mathbf{u} > 0$ in $\mathbb{R}^n$ with $P\mathbf{u} \diamond \mu\mathbf{u}$.

In particular,

$$\rho = \max_{\mathbf{x} \gneq 0} \min_{\{i : x_i > 0\}} \frac{(P\mathbf{x})_i}{x_i} = \min_{\mathbf{x} \gneq 0} \max_{\{i : x_i > 0\}} \frac{(P\mathbf{x})_i}{x_i}$$

$$= \max_{\mathbf{x} > 0} \min_i \frac{(P\mathbf{x})_i}{x_i} = \min_{\mathbf{x} > 0} \max_i \frac{(P\mathbf{x})_i}{x_i}.$$

Since $\rho(P^T) = \rho(P)$, the above properties (and characterizations) of ρ can be expressed by applying the above conditions to P^T.

Consider the sets $\Omega(P) \equiv \{\mu \geq 0 : \exists \mathbf{x} \gneq 0, P\mathbf{x} \geq \mu\mathbf{x}\}$, $\Omega_1(P) \equiv \{\mu \geq 0 : \exists \mathbf{x} > 0, P\mathbf{x} \geq \mu\mathbf{x}\}$, $\Sigma(P) \equiv \{\mu \geq 0 : \exists \mathbf{x} \gneq 0, P\mathbf{x} \leq \mu\mathbf{x}\}$, $\Sigma_1(P) \equiv \{\mu \geq 0 : \exists \mathbf{x} > 0, P\mathbf{x} \leq \mu\mathbf{x}\}$; these sets were named the Collatz–Wielandt sets in [BS75], giving credit to ideas used in [Col42], [Wie50]. The above properties (and characterizations) of ρ can be expressed through maximal/minimal elements of the Collatz–Wielandt sets of P and P^T. (For further details see Chapter 26.)

8. (*Bounds on the Spectral Radius*) Let A be a complex $n \times n$ matrix and let P be an irreducible nonnegative $n \times n$ matrix such that $|A| \leq P$.
   (a) $\rho(A) \leq \rho(P)$.
   (b) [Wie50], [Sch96] $\rho(A) = \rho(P)$ if and only if there exist a complex number μ with $|\mu| = 1$ and a complex diagonal matrix D with $|D| = I$ such that $A = \mu D^{-1} P D$; in particular, in this case $|A| = P$.
   (c) If A is real and μ and $\diamond \in \{<, \lneq\}$ satisfy the condition stated in 7b or 7c, then $\rho(A) < \mu$.
   (d) If A is real and μ and $\diamond \in \{\leq, =\}$ satisfy the condition stated in 7b or 7c, then $\rho(A) \leq \mu$.
9. (*Functional Inequalities*) Consider the function $\rho(.)$ mapping irreducible, nonnegative $n \times n$ matrices to their spectral radius.
   (a) $\rho(.)$ is strictly increasing in each element (of the domain matrices), i.e., if A and B are irreducible, nonnegative $n \times n$ matrices with $A \gneq B \geq 0$, then $\rho(A) > \rho(B)$.
   (b) [Coh78] $\rho(.)$ is (jointly) convex in the diagonal elements, i.e., if A and D are $n \times n$ matrices, with D diagonal, A and $A + D$ nonnegative and irreducible and if $0 < \alpha < 1$, then $\rho[\alpha A + (1-\alpha)(A + D)] \leq \alpha\rho(A) + (1-\alpha)\rho(A + D)$.

   For further functional inequalities that concern the spectral radius see Fact 8 of Section 9.3.
10. (*Taylor Expansion of the Perron Value*) [HRR92] The function $\rho(.)$ mapping irreducible nonnegative $n \times n$ matrices $X = [x_{ij}]$ to their spectral radius is differentiable of all orders and has a converging Taylor expansion. In particular, if P is an irreducible nonnegative $n \times n$ matrix with spectral radius ρ and corresponding positive right and left eigenvectors $\mathbf{v} = [v_i]$ and $\mathbf{w} = [w_j]$, normalized so that $\mathbf{wv} = 1$, and if F is an $n \times n$ matrix with $P + \epsilon F \geq 0$ for all sufficiently small positive ϵ, then $\rho(P + \epsilon F) = \sum_{k=0}^{\infty} \rho_k \epsilon^k$ with $\rho_0 = \rho, \rho_1 = \mathbf{w}F\mathbf{v}, \rho_2 = \mathbf{w}F(\rho I - P)^{\#}F\mathbf{v}, \rho_3 = \mathbf{w}F(\rho I - P)^{\#}(\mathbf{w}F\mathbf{v}I - F)(\rho I - P)^{\#}F\mathbf{v}$; in particular, $\frac{\partial\rho(X)}{\partial x_{ij}}|_{X=P} = w_i v_j$.

    An algorithm that iteratively generates all coefficients of the above Taylor expansion is available; see [HRR92].
11. (*Bounds on the Ergodicity Coefficient*) [RT85] Let P be an irreducible nonnegative $n \times n$ matrix with spectral radius ρ, corresponding positive right eigenvector $\mathbf{v}$, and ergodicity coefficient τ; let D be a diagonal $n \times n$ matrix with positive diagonal elements; and let $\|.\|$ be a norm on $\mathbb{R}^n$. Then

$$\tau \leq \max_{\mathbf{x} \in \mathbb{R}^n, \|\mathbf{x}\| \leq 1, \mathbf{x}^T D^{-1}\mathbf{v} = 0} \|\mathbf{x}^T D^{-1} P D\|.$$

Examples:

1. We illustrate Fact 1 using the matrix

$$P = \begin{bmatrix} \frac{1}{2} & \frac{1}{6} & \frac{1}{3} \\ \frac{1}{4} & \frac{1}{4} & \frac{1}{2} \\ \frac{3}{4} & \frac{1}{8} & \frac{1}{8} \end{bmatrix}.$$

The eigenvalues of P are $1, \frac{1}{48}\left(-3 - \sqrt{33}\right), \frac{1}{48}\left(-3 + \sqrt{33}\right)$, so $\rho(A) = 1$. Also, $\mathbf{v} = [1, 1, 1]^T$ and $\mathbf{w} = [57, 18, 32]$ are positive right and left eigenvectors, respectively, corresponding to eigenvalue

1 and

$$\frac{\mathbf{vw}}{\mathbf{wv}} = \begin{bmatrix} \frac{57}{107} & \frac{18}{107} & \frac{32}{107} \\ \frac{57}{107} & \frac{18}{107} & \frac{32}{107} \\ \frac{57}{107} & \frac{18}{107} & \frac{32}{107} \end{bmatrix}.$$

2. We illustrate parts of Facts 5 and 6 using the matrix

$$P \equiv \begin{bmatrix} 0 & 1 \\ 1 & 0 \end{bmatrix}.$$

The spectral radius of P is 1 with corresponding right and left eigenvectors $\mathbf{v} = (1,1)^T$ and $\mathbf{w} = (1,1)$, respectively, the period of P is 2, and $(I-P)^{\#} = \frac{I-P}{4}$. Evidently,

$$P^m = \begin{cases} I & \text{if } m \text{ is even} \\ P & \text{if } m \text{ is odd}. \end{cases}$$

In particular,

$$\lim_{m\to\infty} P^{2m} = I \quad \text{and} \quad \lim_{m\to\infty} P^{2m+1} = P$$

and

$$\frac{1}{2}[P^m + P^{m+1}] = \frac{I+P}{2} = \begin{bmatrix} .5 & .5 \\ .5 & .5 \end{bmatrix} = \mathbf{v}(\mathbf{wv})^{-1}\mathbf{w} \quad \text{for each } m = 0, 1, \ldots,$$

assuring that, trivially,

$$\lim_{m\to\infty} \frac{1}{2} \sum_{t=m}^{m+1} P^t = \begin{bmatrix} .5 & .5 \\ .5 & .5 \end{bmatrix} = \mathbf{v}(\mathbf{wv})^{-1}\mathbf{w}.$$

In this example, $\tau(P)$ is 0 (as the maximum over the empty set) and the convergence of the above sequences is geometric with rate 0. Finally,

$$\sum_{t=0}^{m-1} P^t = \begin{cases} \frac{m(I+P)}{2} & \text{if } m \text{ is even} \\ \frac{m(I+P)}{2} + \frac{I-P}{2} & \text{if } m \text{ is odd,} \end{cases}$$

implying that

$$\lim_{m\to\infty} P^m = \frac{(I+P)}{2} = \mathbf{v}(\mathbf{wv})^{-1}\mathbf{w} \text{ (C,1)}$$

and

$$\lim_{m\to\infty} \sum_{t=0}^{m-1} P^t - m\left(\frac{I+P}{2}\right) = \frac{I-P}{4} \text{ (C,1)}.$$

3. We illustrate parts of Fact 10 using the matrix P of Example 2 and $F \equiv \begin{bmatrix} 0 & 1 \\ 0 & 0 \end{bmatrix}$. Then $\rho(P+\epsilon F) = \sqrt{1+\epsilon} = 1 + \frac{1}{2}\epsilon + \sum_{k=2}^{\infty} \epsilon^k \frac{(-1)^{k+1}}{k2^{2k-2}} \binom{2k-3}{k-2}$.
4. [RT85, Theorem 4.1] and [Hof67] With $\|.\|$ as the 1-norm on $\mathbb{R}^n$ and $d_1, \ldots, d_n$ as the (positive) diagonal elements of D, the bound in Fact 11 on the coefficient of ergodicity $\tau(P)$ of P becomes

$$\max_{r,s=1,\ldots,n, r\neq s} \frac{1}{d_s v_r + d_r v_s} \left(\sum_{k=1}^{n} d_k |v_s P_{rk} - v_r P_{sk}| \right).$$

With $D = I$, a relaxation of this bound on $\tau(P)$ yields the expression

$$\leq \min\left\{\rho - \sum_{j=1}^{n} \min_i \left(\frac{P_{ij}v_j}{v_i}\right), \sum_{j=1}^{n} \max_i \left(\frac{P_{ij}v_j}{v_i}\right) - \rho\right\}.$$

5. [RT85, Theorem 4.3] For a positive vector $\mathbf{u} \in \mathbb{R}^n$, consider the function $M^{\mathbf{u}} : \mathbb{R}^n \to \mathbb{R}$ defined for $\mathbf{a} \in \mathbb{R}^n$ by

$$M^{\mathbf{u}}(\mathbf{a}) = \max\{\mathbf{x}^T\mathbf{a} : \mathbf{x} \in \mathbb{R}^n, \|\mathbf{x}\| \leq 1, \mathbf{x}^T\mathbf{u} = 0\}.$$

This function has a simple explicit representation obtained by sorting the ratios $\frac{a_j}{u_j}$, i.e., identifying a permutation $j(1), \ldots, j(n)$ of $1, \ldots, n$ such that

$$\frac{a_{j(1)}}{u_{j(1)}} \leq \frac{a_{j(2)}}{u_{j(2)}} \leq \cdots \leq \frac{a_{j(n)}}{u_{j(n)}}.$$

With $k^\star$ as the smallest integer in $\{1, \ldots, n\}$ such that $2\sum_{p=1}^{k^\star} u_{j(p)} > \sum_{t=1}^{n} u_t$ and

$$\mu \equiv 1 + \left(\sum_{t=1}^{n} u_t - 2\sum_{p=1}^{k^\star} u_{j(p)}\right),$$

we have that

$$M^{\mathbf{u}}(\mathbf{a}) = \sum_{p=1}^{k^\star - 1} a_{j(p)} + \mu a_{j(k^\star)} - \sum_{p=k^\star+1}^{n} a_{j(p)}.$$

With $\|.\|$ as the ∞-norm on $\mathbb{R}^n$ and $(D^{-1}PD)_1, \ldots, (D^{-1}PD)_n$ as the columns of $D^{-1}PD$, the bound in Fact 11 on the coefficient of ergodicity $\tau(P)$ of P becomes

$$\max_{r=1,\ldots,n} M^{D^{-1}\mathbf{w}}[(D^{-1}PD)_r].$$

9.3 Reducible Matrices

Definitions:

For a nonnegative $n \times n$ matrix P with spectral radius ρ:

A **basic class of** P is an access equivalence class B of P with $\rho(P[B]) = \rho$.

The **period** of an access equivalence class C of P (also known as the **index of imprimitivity** of C) is the period of the (irreducible) matrix $P[C]$.

The **period** of P (also known as the **index of imprimitivity** of P) is the least common multiple of the periods of its basic classes.

P is **aperiodic** if its period is 1.

The **index** of P, denoted ν_P, is $\nu_P(\rho)$.

The **co-index** of P, denoted $\bar{\nu}_P$, is $\max\{\nu_P(\lambda) : \lambda \in \sigma(P), |\lambda| = \rho \text{ and } \lambda \neq \rho\}$ (with the maximum over the empty set defined as 0).

The **basic reduced digraph** of P, denoted $R^*(P)$, is the digraph whose vertex-set is the set of basic classes of P and whose arcs are the pairs (B, B') of distinct basic classes of P for which there exists a simple walk in $R[\Gamma(P)]$ from B to B'.

The **height** of a basic class is the largest number of vertices on a simple walk in $R^*(P)$ which ends at B.

The **principal submatrix of** P at a distinguished eigenvalue λ, denoted $P[\lambda]$, is the principal submatrix of P corresponding to a set of vertices of $\Gamma(P)$ having no access to a vertex of an access equivalence class C that satisfies $\rho(P[C]) > \lambda$.

P is **convergent** or **transient** if $\lim_{m\to\infty} P^m = 0$.
P is **semiconvergent** if $\lim_{m\to\infty} P^m$ exists.
P is **weakly expanding** if $P\mathbf{u} \geq \mathbf{u}$ for some $\mathbf{u} > 0$.
P is **expanding** if for some $P\mathbf{u} > \mathbf{u}$ for some $\mathbf{u} > 0$.

An $n \times n$ **matrix polynomial of degree** d in the (integer) variable m is a polynomial in m with coefficients that are $n \times n$ matrices (expressible as $S(m) = \sum_{t=0}^{d} m^t B_t$ with $B_1, \ldots, B_d$ as $n \times n$ matrices and $B_d \neq 0$).

Facts:

Facts requiring proofs for which no specific reference is given can be found in [BP94, Chap. 2].

1. The set of basic classes of a nonnegative matrix is always nonempty.
2. (*Spectral Properties of the Perron Value*) Let P be a nonnegative $n \times n$ matrix with spectral radius ρ and index ν.

   (a) [Fro12] ρ is an eigenvalue of P.

   (b) [Fro12] There exist semipositive right and left eigenvectors of P corresponding to ρ, i.e., ρ is a distinguished eigenvalue of both P and P^T.

   (c) [Rot75] ν is the largest number of vertices on a simple walk in $R^*(P)$.

   (d) [Rot75] For each basic class B having height h, there exists a generalized eigenvector $\mathbf{v}^B$ in $N_\rho^h(P)$, with $(\mathbf{v}^B)_i > 0$ if $i \mapsto B$ and $(\mathbf{v}^B)_i = 0$ otherwise.

   (e) [Rot75] The dimension of $N_\rho^\nu(P)$ is the number of basic classes of P. Further, if $B_1, \ldots, B_p$ are the basic classes of P and $\mathbf{v}^{B_1}, \ldots, \mathbf{v}^{B_r}$ are generalized eigenvectors of P at ρ that satisfy the conclusions of Fact 2(d) with respect to $B_1, \ldots, B_r$, respectively, then $\mathbf{v}^{B_1}, \ldots, \mathbf{v}^{B_p}$ form a basis of $N_\rho^\nu(P)$.

   (f) [RiSc78, Sch86] If $B_1, \ldots, B_p$ is an enumeration of the basic classes of P with nondecreasing heights (in particular, $s < t$ assures that we do not have $B_t \mapsto B_s$), then there exist generalized eigenvectors $\mathbf{v}^{B_1}, \ldots, \mathbf{v}^{B_p}$ of P at ρ that satisfy the assumptions and conclusions of Fact 2(e) and a nonnegative $p \times p$ upper triangular matrix M with all diagonal elements equal to ρ, such that

   $$P[\mathbf{v}^{B_1}, \ldots, \mathbf{v}^{B_p}] = [\mathbf{v}^{B_1}, \ldots, \mathbf{v}^{B_p}]M$$

   (in particular, $\mathbf{v}^{B_1}, \ldots, \mathbf{v}^{B_p}$ is a basis of $N_\rho^\nu(P)$). Relationships between the matrix M and the Jordan Canonical Form of P are beyond the scope of the current review; see [Sch56], [Sch86], [HS89], [HS91a], [HS91b], [HRS89], and [NS94].

   (g) [Vic85], [Sch86], [Tam04] If $B_1, \ldots, B_r$ are the basic classes of P having height 1 and $\mathbf{v}^{B_1}, \ldots, \mathbf{v}^{B_r}$ are generalized eigenvectors of P at ρ that satisfy the conclusions of Fact 2(d) with respect to $B_1, \ldots, B_r$, respectively, then $\mathbf{v}^{B_1}, \ldots, \mathbf{v}^{B_r}$ are linearly independent, nonnegative eigenvectors of P at ρ that span the cone $(\mathbb{R}_0^+)^n \cap N_\rho^1(P)$; that is, each vector in the cone $(\mathbb{R}_0^+)^n \cap N_\rho^1(P)$ is a linear combination with nonnegative coefficients of $\mathbf{v}^{B_1}, \ldots, \mathbf{v}^{B_r}$ (in fact, the sets $\{\alpha \mathbf{v}^{B_s} : \alpha \geq 0\}$ for $s = 1, \ldots, r$ are the the extreme rays of the cone $(\mathbb{R}_0^+)^n \cap N_\rho^1(P)$).

3. (*Spectral Properties of Eigenvalues* $\lambda \neq \rho(P)$ *with* $|\lambda| = \rho(P)$) Let P be a nonnegative $n \times n$ matrix with spectral radius ρ, index ν, co-index $\bar{\nu}$, period q, and coefficient of ergodicity τ.

   (a) [Rot81a] The following are equivalent:

      i. $\{\lambda \in \sigma(P) \setminus \{\rho\} : |\lambda| = \rho\} = \emptyset$.

      ii. $\bar{\nu} = 0$.

      iii. P is aperiodic ($q = 1$).

(b) [Rot81a] If $\lambda \in \sigma(P) \setminus \{\rho\}$ and $|\lambda| = \rho$, then $(\frac{\lambda}{\rho})^h = 1$ for some $h \in \{2, \ldots, n\}$; further, $q = \min\{h = 2, \ldots, n : (\frac{\lambda}{\rho})^h = 1$ for each $\lambda \in \sigma(P) \setminus \{\rho\}$ with $|\lambda| = \rho\} \leq n$ (here the minimum over the empty set is taken to be 1).

(c) [Rot80] If $\lambda \in \sigma(P) \setminus \{\rho\}$ and $|\lambda| = \rho$, then $\nu_P(\lambda)$ is bounded by the largest number of vertices on a simple walk in $R^*(P)$ with each vertex corresponding to a (basic) access equivalence class C that has $\lambda \in \sigma(P[C])$; in particular, $\bar{\nu} \leq \nu$.

4. (*Distinguished Eigenvalues*) Let P be a nonnegative $n \times n$ matrix.

(a) [Vic85] λ is a distinguished eigenvalue of P if and only if there is a final set C with $\rho(P[C]) = \lambda$.

It is noted that the set of distinguished eigenvalues of P and P^T need not coincide (and the above characterization of distinguished eigenvalues is not invariant of the application of the transpose operator). (See Example 1 below.)

(b) [HS88b] If λ is a distinguished eigenvalue, $\nu_P(\lambda)$ is the largest number of vertices on a simple walk in $R^*(P[\lambda])$.

(c) [HS88b] If $\mu > 0$, then $\mu \leq \min\{\lambda : \lambda$ is a distinguished eigenvalue of $P\}$ if and only if there exists a vector $\mathbf{u} > 0$ with $P\mathbf{u} \geq \mu \mathbf{u}$.

(For additional characterizations of the minimal distinguished eigenvalue, see the concluding remarks of Facts 12(h) and 12(i).)

Additional properties of distinguished eigenvalues λ of P that depend on $P[\lambda]$ can be found in [HS88b] and [Tam04].

5. (*Convergence Properties of Powers*) Let P be a nonnegative $n \times n$ matrix with positive spectral radius ρ, index ν, co-index $\bar{\nu}$, period q, and coefficient of ergodicity τ (for the case where $\rho = 0$, see Fact 12(j) below).

(a) [Rot81a] There exists an $n \times n$ matrix polynomial $S(m)$ of degree $\nu - 1$ in the (integer) variable m such that $\lim_{m \to \infty}[(\frac{P}{\rho})^m - S(m)] = 0$ (C, p) for every $p \geq \bar{\nu}$; further, if P is aperiodic, this limit holds as a regular limit and the convergence is geometric with rate $\frac{\tau}{\rho} < 1$.

(b) [Rot81a] There exist matrix polynomials $S^0(m), \ldots, S^{q-1}(m)$ of degree $\nu - 1$ in the (integer) variable m, such that for each $k = 0, \ldots, q-1$, $\lim_{m \to \infty}[(\frac{P}{\rho})^{mq+k} - S^t(m)] = 0$ and the convergence of these sequences to their limit is geometric with rate $(\frac{\tau}{\rho})^q < 1$.

(c) [Rot81a] There exists a matrix polynomial $T(m)$ of degree ν in the (integer) variable m with $\lim_{m \to \infty}[\sum_{s=0}^{m-1}(\frac{P}{\rho})^s - T(m)] = 0$ (C, p) for every $p \geq \bar{\nu}$; further, if P is aperiodic, this limit holds as a regular limit and the convergence is geometric with rate $\frac{\tau}{\rho} < 1$.

(d) [FrSc80] The limit of $\frac{P^m}{\rho^m m^{\nu-1}}[I + \frac{P}{\rho} + \cdots + (\frac{P}{\rho})^{q-1}]$ exists and is semipositive.

(e) [Rot81b] Let $\mathbf{x} = [x_i]$ be a nonnegative vector in $\mathbb{R}^n$ and let $i \in \langle n \rangle$. With $K(i, \mathbf{x}) \equiv \{j \in \langle n \rangle : j \mapsto i\} \cap \{j \in \langle n \rangle : u \mapsto j$ for some $u \in \langle n \rangle$ with $x_u > 0\}$,

$$r(i|\mathbf{x}, P) \equiv \inf\{\alpha > 0 : \lim_{m \to \infty} \alpha^{-m}(P^m \mathbf{x})_i = 0\} = \rho(P[K(i, \mathbf{x})])$$

and if $r \equiv r(i|\mathbf{x}, P) > 0$,

$$k(i|\mathbf{x}, P) \equiv \inf\{k = 0, 1, \ldots : \lim_{m \to \infty} m^{-k} r^{-m}(P^m \mathbf{x})_i = 0\} = \nu_{P[K(i,\mathbf{x})]}(r).$$

Explicit expressions for the polynomials mentioned in Facts 5(a) to 5(d) in terms of characteristics of the underlying matrix P are available in Fact 12(a)ii for the case where $\nu = 1$ and in [Rot81a] for the general case. In fact, [Rot81a] provides (explicit) polynomial approximations of additional high-order partial sums of normalized powers of nonnegative matrices.

6. (*Bounds on the Perron Value*) Let P be a nonnegative $n \times n$ matrix with spectral radius ρ and let μ be a nonnegative scalar.

(a) For $\diamond \in \{<, \leq, =, \geq, >\}$,

$$[P\mathbf{u} \diamond \mu\mathbf{u} \text{ for some vector } \mathbf{u} > 0] \Rightarrow [\rho \diamond \mu] ;$$

further, the inverse implication holds for $\diamond$ as $<$, implying that

$$\rho = \max_{\mathbf{x} \gneq 0} \min_{\{i : x_i > 0\}} \frac{(A\mathbf{x})_i}{x_i}.$$

(b) For $\diamond \in \{\lneq, \leq, =, \geq, \gneq\}$,

$$[\rho \diamond \mu] \Rightarrow [P\mathbf{u} \diamond \mu\mathbf{u} \text{ for some vector } \mathbf{u} \gneq 0] ;$$

further, the inverse implication holds for $\diamond$ as $\geq$.

(c) $\rho < \mu$ if and only if $P\mathbf{u} < \rho\mathbf{u}$ for some vector $\mathbf{u} \geq 0$.

Since $\rho(P^T) = \rho(P)$, the above properties (and characterizations) of ρ can be expressed by applying the above conditions to P^T. (See Example 3 below.)
Some of the above results can be expressed in terms of the Collatz–Wielandt sets. (See Fact 7 of Section 9.2 and Chapter 26.)

7. (*Bounds on the Spectral Radius*) Let P be a nonnegative $n \times n$ matrix and let A be a complex $n \times n$ matrix such that $|A| \leq P$. Then $\rho(A) \leq \rho(P)$.
8. (*Functional Inequalities*) Consider the function $\rho(.)$ mapping nonnegative $n \times n$ matrices to their spectral radius.

   (a) $\rho(.)$ is nondecreasing in each element (of the domain matrices); that is, if A and B are nonnegative $n \times n$ matrices with $A \geq B \geq 0$, then $\rho(A) \geq \rho(B)$.

   (b) [Coh78] $\rho(.)$ is (jointly) convex in the diagonal elements; that is, if A and D are $n \times n$ matrices, with D diagonal, A and $A+D$ nonnegative, and if $0 < \alpha < 1$, then $\rho[\alpha A + (1-\alpha)(A+D)] \leq \alpha\rho(A) + (1-\alpha)\rho(A+D)$.

   (c) [EJD88] If $A = [a_{ij}]$ and $B = [b_{ij}]$ are nonnegative $n \times n$ matrices, $0 < \alpha < 1$ and $C = [c_{ij}]$ with $c_{ij} = a_{ij}^{\alpha} b_{ij}^{1-\alpha}$ for each $i, j = 1, \ldots, n$, then $\rho(C) \leq \rho(A)^{\alpha}\rho(B)^{1-\alpha}$.

   Further functional inequalities about $\rho(.)$ can be found in [EJD88] and [EHP90].
9. (*Resolvent Expansions*) Let P be a nonnegative square matrix with spectral radius ρ and let $\mu > \rho$. Then $\mu I - P$ is invertible and

$$(\mu I - P)^{-1} = \sum_{t=0}^{\infty} \frac{P^t}{\mu^{t+1}} \geq \frac{I}{\mu} + \frac{P}{\mu^2} \geq \frac{I}{\mu} \geq 0$$

   (the invertibility of $\mu I - P$ and the power series expansion of its inverse do not require nonnegativity of P).

   For explicit expansions of the resolvent about the spectral radius, that is, for explicit power series representations of $[(z+\rho)I - P]^{-1}$ with $|z|$ positive and sufficiently small, see [Rot81c], and [HNR90] (the latter uses such expansions to prove Perron–Frobenius-type spectral results for nonnegative matrices).
10. (*Puiseux Expansions of the Perron Value*) [ERS95] The function $\rho(.)$ mapping irreducible nonnegative $n \times n$ matrices $X = [x_{ij}]$ to their spectral radius has a converging Puiseux (fractional power series) expansion at each point; i.e., if P is a nonnegative $n \times n$ matrix and if F is an $n \times n$ matrix with $P + \epsilon F \geq 0$ for all sufficiently small positive ϵ, then $\rho(P + \epsilon F)$ has a representation $\sum_{k=0}^{\infty} \rho_k \epsilon^{k/q}$ with $\rho_0 = \rho(P)$ and q as a positive integer.
11. (*Bounds on the Ergodicity Coefficient*) [RT85, extension of Theorem 3.1] Let P be a nonnegative $n \times n$ matrix with spectral radius ρ, corresponding semipositive right eigenvector $\mathbf{v}$, and ergodicity

coefficient τ, let D be a diagonal $n \times n$ matrix with positive diagonal elements, and let $\|.\|$ be a norm on $\mathbb{R}^n$. Then

$$\tau \leq \max_{\mathbf{x}\in\mathbb{R}^n, \|\mathbf{x}\|\leq 1, \mathbf{x}^T D^{-1}\mathbf{v}=0} \|\mathbf{x}^T D^{-1} P D\|.$$

12. (*Special Cases*) Let P be a nonnegative $n \times n$ matrix with spectral radius ρ, index ν, and period q.
   (a) (Index 1) Suppose $\nu = 1$.
      i. $\rho I - P$ has a group inverse.
      ii. [Rot81a] With $P^\star \equiv I - (\rho I - P)(\rho I - P)^{\#}$, all of the convergence properties stated in Fact 6 of Section 9.2 apply.
      iii. If $\rho > 0$, then $\frac{P^m}{\rho^m}$ is bounded in m (element-wise).
      iv. $\rho = 0$ if and only if $P = 0$.
   (b) (*Positive eigenvector*) The following are equivalent:
      i. P has a positive right eigenvector corresponding to ρ.
      ii. The final classes of P are precisely its basic classes.
      iii. There is no vector $\mathbf{w}$ satisfying $\mathbf{w}^T P \lneq \rho \mathbf{w}^T$.

      Further, when the above conditions hold:
      i. $\nu = 1$ and the conclusions of Fact 12(a) hold.
      ii. If P satisfies the above conditions and $P \neq 0$, then $\rho > 0$ and there exists a diagonal matrix D having positive diagonal elements such that $S \equiv \frac{1}{\rho} D^{-1} P D$ is stochastic (that is, $S \geq 0$ and $S\mathbf{1} = \mathbf{1}$; see Chapter 4).
   (c) [Sch53] There exists a vector $\mathbf{x} > 0$ with $P\mathbf{x} \leq \rho x$ if and only if every basic class of P is final.
   (d) (*Positive generalized eigenvector*) [Rot75], [Sch86], [HS88a] The following are equivalent:
      i. P has a positive right generalized eigenvector at ρ.
      ii. Each final class of P is basic.
      iii. $P\mathbf{u} \geq \rho \mathbf{u}$ for some $\mathbf{u} > 0$.
      iv. Every vector $\mathbf{w} \geq 0$ with $\mathbf{w}^T P \leq \rho \mathbf{w}^T$ must satisfy $\mathbf{w}^T P = \rho \mathbf{w}^T$.
      v. ρ is the only distinguished eigenvalue of P.
   (e) (*Convergent/Transient*) The following are equivalent:
      i. P is convergent.
      ii. $\rho < 1$.
      iii. $I - P$ is invertible and $(I - P)^{-1} \geq 0$.
      iv. There exists a positive vector $\mathbf{u} \in R^n$ with $P\mathbf{u} < \mathbf{u}$.

      Further, when the above conditions hold, $(I - P)^{-1} = \sum_{t=0}^{\infty} P^t \geq I$.
   (f) (*Semiconvergent*) The following are equivalent:
      i. P is semiconvergent.
      ii. Either $\rho < 1$ or $\rho = \nu = 1$ and 1 is the only eigenvalue λ of P with $|\lambda| = 1$.
   (g) (*Bounded*) P^m is bounded in m (element-wise) if and only if either $\rho < 1$ or $\rho = 1$ and ν=1.
   (h) (*Weakly Expanding*) [HS88a], [TW89] [DR05] The following are equivalent:
      i. P is weakly expanding.
      ii. There is no vector $\mathbf{w} \in R^n$ with $\mathbf{w} \geq 0$ and $w^T P \lneq w^T$.
      iii. Every distinguished eigenvalue λ of P satisfies $\lambda \geq 1$.

iv. Every final class C of P has $\rho(P[C]) \geq 1$.

v. If C is a final set of P, then $\rho(P[C]) \geq 1$.

Given $\mu > 0$, the application of the above equivalence to $\frac{P}{\mu}$ yields characterizations of instances where each distinguished eigenvalue of P is bigger than or equal to μ.

(i) (*Expanding*) [HS88a], [TW89] [DR05] The following are equivalent:

i. P is expanding.

ii. There exists a vector $\mathbf{u} \in R^n$ with $\mathbf{u} \geq 0$ and $P\mathbf{u} > \mathbf{u}$.

iii. There is no vector $\mathbf{w} \in R^n$ with $\mathbf{w} \gneq 0$ and $w^T P \leq w^T$.

iv. Every distinguished eigenvalue λ of P satisfies $\lambda > 1$.

v. Every final class C of P has $\rho(P[C]) > 1$.

vi. If C is a final set of P, then $\rho(P[C]) > 1$.

Given $\mu > 0$, the application of the above equivalence to $\frac{P}{\mu}$ yields characterizations of instances where each distinguished eigenvalue of P is bigger than μ.

(j) (*Nilpotent*) The following are equivalent conditions:

i. P is nilpotent; that is, $P^m = 0$ for some positive integer m.

ii. P is permutation similar to an upper triangular matrix all of whose diagonal elements are 0.

iii. $\rho = 0$.

iv. $P^n = 0$.

v. $P^\nu = 0$.

(k) (*Symmetric*) Suppose P is symmetric.

i. $\rho = \max_{\mathbf{u} \gneq 0} \frac{\mathbf{u}^T P \mathbf{u}}{\mathbf{u}^T \mathbf{u}}$.

ii. $\rho = \frac{\mathbf{u}^T P \mathbf{u}}{\mathbf{u}^T \mathbf{u}}$ for $\mathbf{u} \gneq 0$ if and only if $\mathbf{u}$ is an eigenvector of P corresponding to ρ.

iii. [CHR97, Theorem 1] For $\mathbf{u}, \mathbf{w} \gneq 0$ with $w_i = \sqrt{u_i (Pu)_i}$ for $i = 1, \ldots, n$, $\frac{\mathbf{u}^T P \mathbf{u}}{\mathbf{u}^T \mathbf{u}} \leq \frac{\mathbf{w}^T P \mathbf{w}}{\mathbf{w}^T \mathbf{w}}$ and equality holds if and only if $\mathbf{u}[\mathcal{S}]$ is an eigenvector of $P[\mathcal{S}]$ corresponding to ρ, where $\mathcal{S} \equiv \{i : u_i > 0\}$.

Examples:

1. We illustrate parts of Fact 2 using the matrix

$$P = \begin{bmatrix} 2 & 2 & 2 & 0 & 0 & 0 \\ 0 & 2 & 0 & 0 & 0 & 0 \\ 0 & 0 & 1 & 2 & 0 & 0 \\ 0 & 0 & 0 & 1 & 1 & 0 \\ 0 & 0 & 0 & 1 & 1 & 1 \\ 0 & 0 & 0 & 0 & 0 & 1 \end{bmatrix}.$$

The eigenvalues of P are 2,1, and 0; so, $\rho(P) = 2 \in \sigma(P)$ as is implied by Fact 2(a). The vectors $\mathbf{v} = [1, 0, 0, 0, 0, 0]^T$ and $\mathbf{w} = [0, 0, 0, 1, 1, 1]$ are semipositive right and left eigenvectors corresponding to the eigenvalue 2; their existence is implied by Fact 2(b).
The basic classes are $B_1 = \{1\}$, $B_1 = \{2\}$ and $B_3 = \{4, 5\}$. The digraph corresponding to P, its reduced digraph, and the basic reduced digraph of P are illustrated in Figure 9.1. From Figure 9.1(c), the largest number of vertices in a simple walk in the basic reduced digraph of P is 2 (going from B_1 to either B_2 or B_3); hence, Fact 2(c) implies that $\nu_P(2) = 2$. The height of basic class B_1 is 1 and the height of basic classes B_2 and B_3 is 2. Semipositive generalized eigenvectors of P at (the eigenvalue)

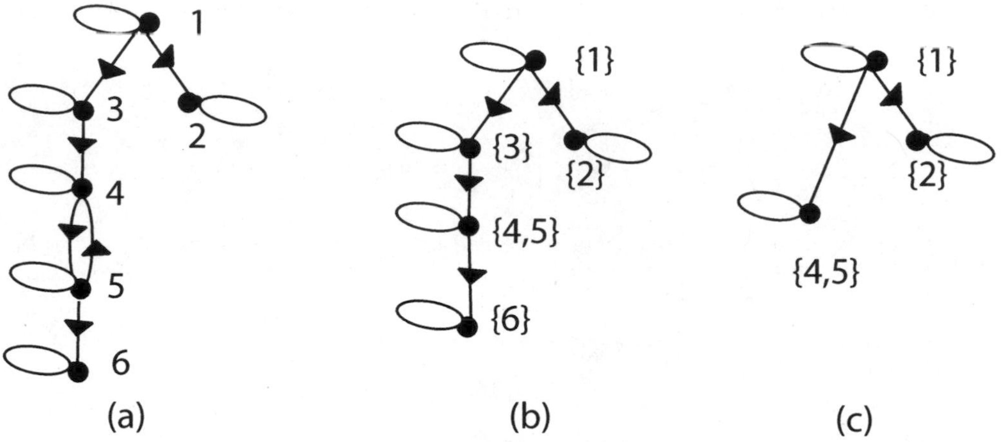

FIGURE 9.1 (a) The digraph $\Gamma(P)$, (b) reduced digraph $R[\Gamma(P)]$, and (c) basic reduced digraph $R^*(P)$.

2 that satisfy the assumptions of Fact 2(f) are $\mathbf{u}^{B_1} = [1,0,0,0,0,0]^T$, $\mathbf{u}^{B_2} = [1,1,0,0,0,0]^T$, and $\mathbf{u}^{B_3} = [1,0,2,1,1,0]^T$. The implied equality

$$P[\mathbf{u}^{B_1},\ldots,\mathbf{u}^{B_p}] = [\mathbf{u}^{B_1},\ldots,\mathbf{u}^{B_p}]M$$

of Fact 2(f) holds as

$$\begin{bmatrix} 2 & 2 & 2 & 0 & 0 & 0 \\ 0 & 2 & 0 & 0 & 0 & 0 \\ 0 & 0 & 1 & 2 & 0 & 0 \\ 0 & 0 & 0 & 1 & 1 & 0 \\ 0 & 0 & 0 & 1 & 1 & 1 \\ 0 & 0 & 0 & 0 & 0 & 1 \end{bmatrix} \begin{bmatrix} 1 & 1 & 1 \\ 0 & 1 & 0 \\ 0 & 0 & 2 \\ 0 & 0 & 1 \\ 0 & 0 & 1 \\ 0 & 0 & 0 \end{bmatrix} = \begin{bmatrix} 2 & 4 & 6 \\ 0 & 2 & 0 \\ 0 & 0 & 4 \\ 0 & 0 & 2 \\ 0 & 0 & 2 \\ 0 & 0 & 0 \end{bmatrix} = \begin{bmatrix} 1 & 1 & 1 \\ 0 & 1 & 0 \\ 0 & 0 & 2 \\ 0 & 0 & 1 \\ 0 & 0 & 1 \\ 0 & 0 & 0 \end{bmatrix} \begin{bmatrix} 2 & 2 & 4 \\ 0 & 2 & 0 \\ 0 & 0 & 2 \end{bmatrix}.$$

In particular, Fact 2(e) implies that $\mathbf{u}^{B_1}, \mathbf{u}^{B_2}, \mathbf{u}^{B_3}$ form a basis of $N_{\rho(P)}^{\nu(P)} = N_2^2$. We note that while there is only a single basic class of height 1, $\dim[N_\rho^1(P)] = 2$ and $\mathbf{u}^{B_1}, 2\mathbf{u}^{B_2} - \mathbf{u}^{B_3} = [-1,2,-2,-1,-1,0]^T$ form a basis of $N_\rho^1(P)$. Still, Fact 2(g) assures that $(\mathbb{R}_0^+)^n \cap N_\rho^1(P)$ is the cone $\{\alpha\mathbf{u}^{B_1} : \alpha \geq 0\}$ (consisting of its single ray).

Fact 4(a) and Figure 9.1 imply that the distinguished eigenvalues of P are 1 and 2, while 2 is the only distinguished eigenvalue of P^T.

2. Let $H = \begin{bmatrix} 0 & 1 \\ 1 & 0 \end{bmatrix}$; properties of H were demonstrated in Example 2 of section 9.2. We will demonstrate Facts 2(c), 5(b), and 5(a) on the matrix

$$P \equiv \begin{bmatrix} H & I \\ 0 & H \end{bmatrix}.$$

The spectral radius of P is 1 and its basic classes of P are $B_1 = \{1,2\}$ and $B_2 = \{3,4\}$ with B_1 having access to B_2. Thus, the index of 1 with respect to P, as the largest number of vertices on a walk of the marked reduced graph of P, is 2 (Fact 2(c)). Also, as the period of each of the two basic classes of P is 2, the period of P is 2. To verify the convergence properties of P, note that

$$P^m = \begin{cases} \begin{bmatrix} I & mH \\ 0 & I \end{bmatrix} & \text{if } m \text{ is even} \\ \begin{bmatrix} H & mI \\ 0 & H \end{bmatrix} & \text{if } m \text{ is odd,} \end{cases}$$

immediately providing matrix–polynomials $S^0(m)$ and $S^1(m)$ of degree 1 such that $\lim_{m\to\infty} P^{2m} - S^0(m) = 0$ and $\lim_{m\to\infty} P^{2m+1} - S^1(m) = 0$. In this example, $\tau(P)$ is 0 (as the maximum over the empty set) and the convergence of the above sequences is geometric with rate 0.
The above representation of P^m shows that

$$P^m = \begin{bmatrix} H^m & mH^{m+1} \\ 0 & H^m \end{bmatrix}$$

and Example 2 of section 9.2 shows that

$$\lim_{m\to\infty} H^m = \frac{I+H}{2} = \begin{bmatrix} .5 & .5 \\ .5 & .5 \end{bmatrix} \quad (\text{C},1).$$

We next consider the upper-right blocks of P^m. We observe that

$$\frac{1}{m}\sum_{t=0}^{m-1} P^t[B_1, B_2] = \begin{cases} \frac{mI}{4} + \frac{(m-2)H}{4} & \text{if } m \text{ is even} \\ \frac{(m-1)^2 I}{4m} + \frac{(m^2-1)H}{4m} & \text{if } m \text{ is odd,} \end{cases}$$
$$= \begin{cases} \frac{m(I+H)}{4} - \frac{H}{2} & \text{if } m \text{ is even} \\ \frac{m(I+H)}{4} - \frac{I}{2} + \frac{I-H}{4m} & \text{if } m \text{ is odd,} \end{cases}$$

implying that

$$\lim_{m\to\infty} \frac{1}{m}\sum_{t=0}^{m-1} P^t[B_1, B_2] - m\left(\frac{I+H}{4}\right) + \frac{I+H}{4} = 0 \ (\text{C},1).$$

As $m - 1 = \frac{1}{m}\sum_{t=0}^{m-1} t$ for each $m = 1, 2, \ldots$, the above shows that

$$\lim_{m\to\infty} \frac{1}{m}\sum_{t=0}^{m-1} \left\{ P^t[B_1, B_2] - t\left(\frac{I+H}{4}\right) \right\} = 0 \ (\text{C},1),$$

and, therefore (recalling that (C,1)-convergence implies (C,2)-convergence),

$$\lim_{m\to\infty} \left\{ P^m - \begin{bmatrix} .5 & .5 & -.25m & -.25m \\ .5 & .5 & -.25m & -.25m \\ 0 & 0 & .5 & .5 \\ 0 & 0 & .5 & .5 \end{bmatrix} \right\} = 0 \ (\text{C},2).$$

3. Fact 6 implies many equivalencies, in particular, as the spectral radius of a matrix equals that of its transpose. For example, for a nonnegative $n \times n$ matrix P with spectral radius ρ and nonnegative scalar μ, the following are equivalent:

   (a) $\rho < \mu$.

   (b) $P\mathbf{u} < \mu\mathbf{u}$ for some vector $\mathbf{u} > 0$.

   (c) $\mathbf{w}^T P < \mu\mathbf{w}^T$ for some vector $\mathbf{w} > 0$.

(d) $Pu < \rho u$ for some vector $u \geq 0$.

(e) $w^T P < \rho w^T$ for some vector $w \geq 0$.

(f) There is no vector $u \gneq 0$ satisfying $Pu \geq \mu u$.

(g) There is no vector $w \gneq 0$ satisfying $w^T P \geq \mu w^T$.

9.4 Stochastic and Substochastic Matrices

(For more information about stochastic matrices see Chapter 54 (including examples).)

Definitions:

A square $n \times n$ matrix $P = [p_{ij}]$ is **stochastic** if it is nonnegative and $P\mathbf{1} = \mathbf{1}$ where $\mathbf{1} = [1, \ldots, 1]^T \in R^n$. (Stochastic matrices are sometimes referred to as **row-stochastic**, while **column-stochastic** matrices are matrices whose transpose is (row-)stochastic.)

A square $n \times n$ matrix P is **doubly stochastic** if both P and its transpose are stochastic. The set of doubly stochastic matrices of order n is denoted Ω_n.

A square $n \times n$ matrix P is **substochastic** if it is nonnegative and $P\mathbf{1} \leq \mathbf{1}$.

A transient substochastic matrix is also called **stopping**.

An **ergodic class** of a stochastic matrix P is a basic class of P.

A **transient class** of a stochastic matrix P is an access equivalence class of P which is not ergodic.

A **state** of an $n \times n$ stochastic matrix P is an index $i \in \{1, \ldots, n\}$. Such a state is **ergodic** or **transient** depending on whether it belongs to an ergodic class or to a transient class.

A **stationary distribution** of a stochastic matrix P is a nonnegative vector π that satisfies $\pi^T \mathbf{1} = 1$ and $\pi^T P = \pi^T$.

Facts:

Facts requiring proofs for which no specific reference is given follow directly from facts in Sections 9.2 and 9.3 and/or can be found in [BP94, Chap. 8].

1. Let $P = [p_{ij}]$ be an $n \times n$ stochastic matrix.

(a) $\rho(P) = 1$, $\mathbf{1} \in R^n$ is a right eigenvector of P corresponding to 1 and the stationary distributions of P are nonnegative eigenvectors of P corresponding to 1.

(b) $\nu_P(1) = 1$.

(c) $I - P$ has a group inverse.

(d) The height of every ergodic class is 1.

(e) The final classes of P are precisely its ergodic classes.

(f)

i. For every ergodic class C, P has a unique stationary distribution π^C of P with $(\pi^C)_i > 0$ if $i \in C$ and $(\pi^C)_i = 0$ otherwise.

ii. If $C^1, \ldots, C^p$ are the ergodic classes of P, then the corresponding stationary distributions $\pi^{C^1}, \ldots, \pi^{C^p}$ (according to Fact 1(f)i above) form a basis of the set of left eigenvectors of P corresponding to the eigenvalue 1; further, every stationary distribution of P is a convex combination of these vectors.

(g)

i. Let T and R be the sets of transient and ergodic states of P, respectively. The matrix $I - P[T]$ is nonsingular and for each ergodic class C of P, the vector u^C given by

$$(\mathbf{u}^C)[K] = \begin{cases} e & \text{if } K = C \\ 0 & \text{if } K = R \setminus C \\ (I - P[T])^{-1} P[T, C] e & \text{if } K = T \end{cases}$$

is a right eigenvector of P corresponding to the eigenvalue 1; in particular, $(\mathbf{u}^C)_i > 0$ if i has access to C and $(\mathbf{u}^C)_i = 0$ if i does not have access to C.

ii. If $C^1, \ldots, C^p$ are the ergodic classes of P, then the corresponding vectors $\mathbf{u}^{C^1}, \ldots, \mathbf{u}^{C^p}$ (referred to in Fact 1(g)i above) form a basis of the set of right eigenvectors of P corresponding to the eigenvalue 1; further, $\sum_{t=1}^{p} \mathbf{u}^{C^t} = e$.

(h) Let $C^1, \ldots, C^p$ be the ergodic classes of P, $\boldsymbol{\pi}^{C^1}, \ldots, \boldsymbol{\pi}^{C^p}$ the corresponding stationary distributions (referred to in Fact 1(f)i above), and $\mathbf{u}^{C^1}, \ldots, \mathbf{u}^{C^p}$ the corresponding eigenvectors referred to in Fact 1(g)i above. Then the matrix

$$P^\star = \left[\mathbf{u}^{C^1}, \ldots, \mathbf{u}^{C^p}\right] \begin{bmatrix} \boldsymbol{\pi}^{C^1} \\ \vdots \\ \boldsymbol{\pi}^{C^p} \end{bmatrix}$$

is stochastic and satisfies $P^\star[\langle n \rangle, C] = 0$ if C is a transient class of P, $P^\star[i, C] \neq 0$ if C is an ergodic class and i has access to C, and $P^\star[i, C] = 0$ if C is an ergodic class and i does not have access to C.

(i) The matrix $P^\star$ from Fact 1(h) above has the representation $I - (I - P)^{\#}(I - P)$; further, $I - P + P^\star$ is nonsingular and $(I - P)^{\#} = (I - P + P^\star)^{-1}(I - P^\star)$.

(j) With $P^\star$ as the matrix from Fact 1(h) above, $\lim_{m\to\infty} P^m = P^\star$ (C,1); further, when P is aperiodic, this limit holds as a regular limit and the convergence is geometric with rate $\tau(P) < 1$.

(k) With $P^\star$ as the matrix from Fact 1(h) above, $\lim_{m\to\infty} \sum_{t=0}^{m-1} P^t - mP^\star = (I - P)^{\#}$ (C,1); further, when P is aperiodic, this limit holds as a regular limit and the convergence is geometric with rate $\tau(P) < 1$.

(l) With D a diagonal $n \times n$ matrix with positive diagonal elements and $\|.\|$ a norm on $\mathbb{R}^n$,

$$\tau(P) \leq \max_{\mathbf{x} \in \mathbb{R}^n, \|\mathbf{x}\| \leq 1, \mathbf{x}^T D^{-1} \mathbf{1} = 0} \|\mathbf{x}^T D^{-1} P D\|.$$

In particular, with $\|.\|$ as the 1-norm on $\mathbb{R}^n$ and $D = I$, the above bound specializes to

$$\tau(P) \leq \max_{r,s=1,\ldots,n, r\neq s} \sum_{k=1}^{n} \frac{|p_{rk} - p_{sk}|}{2} \leq \min \left\{ 1 - \sum_{k=1}^{n} \min_r p_{rk}, \sum_{k=1}^{n} \max_r p_{rk} - 1 \right\}$$

(cf. Fact 11 of section 9.3 and Example 4 of section 9.2).

(m) For every $0 < \alpha < 1$, $P_\alpha \equiv (1 - \alpha)I + \alpha P$ is an aperiodic stochastic matrix whose ergodic classes, transient classes, stationary distributions, and the vectors of Fact 1(g)i coincide with those of P. In particular, with $P^\star$ and $P_\alpha^\star$ as the matrices from Fact 1(h) corresponding to P and P_α, respectively, $\lim_{m\to\infty} (P_\alpha)^m = P_\alpha^\star = P^\star$.

2. Let P be an irreducible stochastic matrix with coefficient of ergodicity τ.
   (a) P has a unique stationary distribution, say $\boldsymbol{\pi}$. Also, up to scalar multiple, $\mathbf{1}$ is a unique right eigenvector or P corresponding to the eigenvalue 1.
   (b) With $\boldsymbol{\pi}$ as the unique stationary distribution of P, the matrix $P^\star$ from Fact 1(h) above equals $\mathbf{1}\boldsymbol{\pi}$.
3. A doubly stochastic matrix is a convex combination of permutation matrices (in fact, the $n \times n$ permutation matrices are the extreme points of the set Ω_n of $n \times n$ doubly stochastic matrices).
4. Let P be an $n \times n$ substochastic matrix.
   (a) $\rho(P) \leq 1$.
   (b) $\nu_P(1) \leq 1$.
   (c) $I - P$ has a group inverse.
   (d) The matrix $P^\star \equiv I - (I - P)^\#(I - P)$ is substochastic; further, $I - P + P^\star$ is nonsingular and $(I - P)^\# = (I - P + P^\star)^{-1}(I - P^\star)$.
   (e) With $P^\star$ as in Fact 4(d), $\lim_{m\to\infty} P^m = P^\star$ (C,1); further, when every access equivalence class C with $\rho(P[C]) = 1$ is aperiodic, this limit holds as a regular limit and the convergence is geometric with rate $\max\{|\lambda| : \lambda \in \sigma(P) \text{ and } |\lambda| \neq 1\} < 1$.
   (f) With $P^\star$ as the matrix from Fact 4(d) above, $\lim_{m\to\infty} \sum_{t=0}^{m-1} P^t - mP^\star = (I - P)^\#$ (C,1); further, when every access equivalence class C with $\rho(P[C]) = 1$ is aperiodic, this limit holds as a regular limit and the convergence is geometric with rate $\max\{|\lambda| : \lambda \in \sigma(P) \text{ and } |\lambda| \neq 1\} < 1$.
   (g) The following are equivalent:
      i. P is stopping.
      ii. $\rho(P) < 1$.
      iii. $I - P$ is invertible.
      iv. There exists a positive vector $\mathbf{u} \in R^n$ with $P\mathbf{u} < \mathbf{u}$.

   Further, when the above conditions hold, $(I - P)^{-1} = \sum_{t=0}^{\infty} P^t \geq 0$.

9.5 *M*-Matrices

Definitions:

An $n \times n$ real matrix $A = [a_{ij}]$ is a ***Z*-matrix** if its off-diagonal elements are nonpositive, i.e., if $a_{ij} \leq 0$ for all $i, j = 1, \ldots, n$ with $i \neq j$.

An ***M*$_0$-matrix** is a Z-matrix A that can be written as $A = sI - P$ with P as a nonnegative matrix and with s as a scalar satisfying $s \geq \rho(P)$.

An ***M*-matrix** A is a Z-matrix A that can be written as $A = sI - P$ with P as a nonnegative matrix and with s as a scalar satisfying $s > \rho(P)$.

A square real matrix A is an **inverse *M*-matrix** if it is nonsingular and its inverse is an M-matrix.

A square real matrix A is **inverse-nonnegative** if it is nonsingular and $A^{-1} \geq 0$ (the property is sometimes referred to as **inverse-positivity**).

A square real matrix A has a **convergent regular splitting** if A has a representation $A = M - N$ such that $N \geq 0$, M invertible with $M^{-1} \geq 0$ and $M^{-1}N$ is convergent.

A square complex matrix A is **positive stable** if the real part of each eigenvalue of A is positive; A is **nonnegative stable** if the real part of each eigenvalue of A is nonnegative.

An $n \times n$ complex matrix $A = [a_{ij}]$ is **strictly diagonally dominant (diagonally dominant)** if $|a_{ii}| > \sum_{j=1, j\neq i}^{n} |a_{ij}|$ ($|a_{ii}| \geq \sum_{j=1, j\neq i}^{n} |a_{ij}|$) for $i = 1, \ldots, n$.

An $n \times n$ M-matrix A **satisfies property C** if there exists a representation of A of the form $A = sI - P$ with $s > 0$, $P \geq 0$ and $\frac{P}{s}$ semiconvergent.

Facts:

Facts requiring proofs for which no specific reference is given follow directly from results about nonnegative matrices stated in Sections 9.2 and 9.3 and/or can be found in [BP94, Chap. 6].

1. Let A be an $n \times n$ real matrix with $n \geq 2$. The following are equivalent:
   (a) A is an M-matrix; that is, A is a Z-matrix that can be written as $sI - P$ with P nonnegative and $s > \rho(P)$.
   (b) A is a nonsingular M_0-matrix.
   (c) For each nonnegative diagonal matrix D, $A + D$ is inverse-nonnegative.
   (d) For each $\mu \geq 0$, $A + \mu I$ is inverse-nonnegative.
   (e) Each principal submatrix of A is inverse-nonnegative.
   (f) Each principal submatrix of A of orders $1, \ldots, n$ is inverse-nonnegative.
2. Let $A = [a_{ij}]$ be an $n \times n$ Z-matrix. The following are equivalent:*
   (a) A is an M-matrix.
   (b) Every real eigenvalue of A is positive.
   (c) $A + D$ is nonsingular for each nonnegative diagonal matrix D.
   (d) All of the principal minors of A are positive.
   (e) For each $k = 1, \ldots, n$, the sum of all the $k \times k$ principal minors of A is positive.
   (f) There exist lower and upper triangular matrices L and U, respectively, with positive diagonal elements such that $A = LU$.
   (g) A is permutation similar to a matrix satisfying condition 2(f).
   (h) A is positive stable.
   (i) There exists a diagonal matrix D with positive diagonal elements such that $AD + DA^T$ is positive definite.
   (j) There exists a vector $\mathbf{x} > 0$ with $A\mathbf{x} > 0$.
   (k) There exists a vector $\mathbf{x} > 0$ with $A\mathbf{x} \gneq 0$ and $\sum_{j=1}^{i} a_{ij}x_j > 0$ for $i = 1, \ldots, n$.
   (l) A is permutation similar to a matrix satisfying condition 2(k).
   (m) There exists a vector $\mathbf{x} > 0$ such that $A\mathbf{x} \gneq 0$ and the matrix $\hat{A} = [\hat{a}_{ij}]$ defined by

$$\hat{a}_{ij} = \begin{cases} 1 & \text{if either } a_{ij} \neq 0 \text{ or } (A\mathbf{x})_i \neq 0 \\ 0 & \text{otherwise} \end{cases}$$

   is irreducible.
   (n) All the diagonal elements of A are positive and there exists a diagonal matrix D such that AD is strictly diagonally dominant.
   (o) A is inverse-nonnegative.
   (p) Every representation of A of the form $A = M - N$ with $N \geq 0$ and M inverse-positive must have $M^{-1}N$ convergent (i.e., $\rho(M^{-1}N) < 1$).
   (q) For each vector $\mathbf{y} \geq 0$, the set $\{\mathbf{x} \geq 0 : A^T\mathbf{x} \leq \mathbf{y}\}$ is bounded and A is nonsingular.

*Each of the 17 conditions that are listed in Fact 2 is a representative of a set of conditions that are known to be equivalent for all matrices (not just Z-matrices); see [BP94, Theorem 6.2.3]. For additional characterizations of M-matrices, see [FiSc83].

3. Let A be an irreducible $n \times n$ Z-matrix with $n \geq 2$. The following are equivalent:
   (a) A is an M-matrix.
   (b) A is a nonsingular and $A^{-1} > 0$.
   (c) $A\mathbf{x} \gneq 0$ for some $\mathbf{x} > 0$.
4. Let $A = [a_{ij}]$ be an $n \times n$ M-matrix and let $B = [b_{ij}]$ be an $n \times n$ Z-matrix with $B \geq A$. Then:
   (a) B is an M-matrix.
   (b) $\det B \geq \det A$.
   (c) $A^{-1} \geq B^{-1}$.
   (d) $\det A \leq a_{11} \ldots a_{nn}$.
5. If P is an inverse M-matrix, then $P \geq 0$ and $\Gamma(P)$ is transitive; that is, if (v, u) and (u, w) are arcs of $\Gamma(P)$, then so is (v, w).
6. Let A be an $n \times n$ real matrix with $n \geq 2$. The following are equivalent:
   (a) A is a nonsingular M_0-matrix.
   (b) For each diagonal matrix D with positive diagonal elements, $A + D$ is inverse-nonnegative.
   (c) For each $\mu > 0$, $A + \mu I$ is inverse-nonnegative.
7. Let A be an $n \times n$ Z-matrix. The following are equivalent:*
   (a) A is an M_0-matrix.
   (b) Every real eigenvalue of A is nonnegative.
   (c) $A + D$ is nonsingular for each diagonal matrix D having positive diagonal elements.
   (d) For each $k = 1, \ldots, n$, the sum of all the $k \times k$ principal minors of A is nonnegative.
   (e) A is permutation similar to a matrix having a representation LU with L and U as lower and upper triangular matrices having positive diagonal elements.
   (f) A is nonnegative stable.
   (g) There exists a nonnegative matrix Y satisfying $YA^{k+1} = A^k$ for some $k \geq 1$.
   (h) A has a representation of the form $A = M - N$ with M inverse-nonnegative, $N \geq 0$ and $B \equiv M^{-1}N$ satisfying $\cap_{k=0}^{\infty}\text{range}(B^k) = \cap_{k=0}^{\infty}\text{range}(A^k)$ and $\rho(B) \leq 1$.
   (i) A has a representation of the form $A = M - N$ with M inverse-nonnegative, $M^{-1}N \geq 0$ and $B \equiv M^{-1}N$ satisfying $\cap_{k=0}^{\infty}\text{range}(B^k) = \cap_{k=0}^{\infty}\text{range}(A^k)$ and $\rho(B) \leq 1$.
8. Let A be an M_0-matrix.
   (a) A satisfies property C if and only if $\nu_A(0) \leq 1$.
   (b) A is permutation similar to a matrix having a representation LU with L as a lower triangular M-matrix and U as an upper triangular M_0 matrix.
9. [BP94, Theorem 8.4.2] If P is substochastic (see Section 9.4), then $I - P$ is an M_0-matrix satisfying property C.
10. Let A be an irreducible $n \times n$ singular M_0-matrix.
    (a) A has rank $n - 1$.
    (b) There exists a vector $\mathbf{x} > 0$ such that $A\mathbf{x} = 0$.

*Each of the 9 conditions that are listed in Fact 7 is a representative of a set of conditions that are known to be equivalent for all matrices (not just Z-matrices); see [BP94, Theorem 6.4.6]. For additional characterizations of M-matrices, see [FiSc83].

(c) A has property C.

(d) Each principal submatrix of A other than A itself is an M-matrix.

(e) $[A\mathbf{x} \geq 0] \Rightarrow [A\mathbf{x} = 0]$.

9.6 Scaling of Nonnegative Matrices

A scaling of a (usually nonnegative) matrix is the outcome of its pre- and post-multiplication by diagonal matrices having positive diagonal elements. Scaling problems concern the search for scalings of given matrices such that specified properties are satisfied. Such problems are characterized by:

(a) The class of matrices to be scaled.
(b) Restrictions on the pre- and post-multiplying diagonal matrices to be used.
(c) The target property.

Classes of matrices under (a) may refer to arbitrary rectangular matrices, square matrices, symmetric matrices, positive semidefinite matrices, etc. For possible properties of pre- and post-multiplying diagonal matrices under (b) see the following Definition subsection. Finally, examples for target properties under (c) include:

i. The specification of the row- and/or column-sums; for example, being stochastic or being doubly stochastic. See the following Facts subsection.
ii. The specification of the row- and/or column-maxima.
iii. (For a square matrix) being line-symmetric, that is, having each row-sum equal to the corresponding column-sum.
iv. Being optimal within a prescribed class of scalings under some objective function. One example of such optimality is to minimize the maximal element of a scalings of the form XAX^{-1}. Also, in numerical analysis, preconditioning a matrix may involve its replacement with a scaling that has a low ratio of largest to smallest element; so, a potential target property is to be a minimizer of this ratio among all scalings of the underlying matrix.

Typical questions that are considered when addressing scaling problems include:

(a) Characterizing existence of a scaling that satisfies the target property (precisely of approximately).
(b) Computing a scaling of a given matrix that satisfies the target property (precisely or approximately) or verifying that none exists.
(c) Determining complexity bounds for corresponding computation.

Early references that address scaling problems include [Kru37], which describes a heuristic for finding a doubly stochastic scaling of a positive square matrix, and Sinkhorn's [Sin64] pioneering paper, which provides a formal analysis of that problem. The subject has been intensively studied and an aspiration to provide a comprehensive survey of the rich literature is beyond the scope of the current review; consequently, we address only scaling problems where the target is to achieve, precisely or approximately, prescribed row- and column-sums.

Definitions:

Let $A = [a_{ij}]$ be an $m \times n$ matrix.

A **scaling** (sometimes referred to as an **equivalence-scaling** or a DAE**-scaling**) of A is any matrix of the form DAE where D and E are square diagonal matrices having positive diagonal elements; such a scaling is a **row-scaling** of A if $E = I$ and it is a **normalized-scaling** if $\det(D) = \det(E) = 1$.

If $m = n$, a scaling DAE of A is a **similarity-scaling** (sometimes referred to as a DAD^{-1} **scaling**) of A if $E = D^{-1}$, and DAE is a **symmetric-scaling** (sometimes referred to as a DAD **scaling**) of A if $E = D$.

The **support** (or **sparsity pattern**) of A, denoted Struct(A), is the set of indices ij with $a_{ij} \neq 0$; naturally, this definition applies to vectors.

Facts:

1. (*Prescribed-Line-Sum Scalings*) [RoSc89] Let $A = [a_{ij}] \in \mathbb{R}^{m\times n}$ be a nonnegative matrix, and let $\mathbf{r} = [r_i] \in \mathbb{R}^m$ and $\mathbf{c} = [c_j] \in \mathbb{R}^n$ be positive vectors.

   (a) The following are equivalent:

   i. There exists a scaling B of A with $B\mathbf{1} = \mathbf{r}$ and $\mathbf{1}^T B = \mathbf{c}^T$.

   ii. There exists nonnegative $m \times n$ matrix B having the same support as A with $B\mathbf{1} = \mathbf{r}$ and $\mathbf{1}^T B = \mathbf{c}^T$.

   iii. For every $I \subseteq \{1, \ldots, m\}$ and $J \subseteq \{1, \ldots, m\}$ for which $A[I^c, J] = 0$,

$$\sum_{i\in I} r_i \geq \sum_{j\in J} c_j$$

   and equality holds if and only if $A[I, J^c] = 0$.

   iv. $\mathbf{1}^T\mathbf{r} = \mathbf{1}^T\mathbf{r}$ and the following (geometric) optimization problem has an optimal solution:

$$\min \mathbf{x}^T A\mathbf{y}$$

$$\text{subject to}: \mathbf{x} = [x_i] \in \mathbb{R}^m, \mathbf{y} = [y_j] \in \mathbb{R}^n$$

$$\mathbf{x} \geq 0, \mathbf{y} \geq 0$$

$$\prod_{i=1}^{m} (x_i)^{r_i} = \prod_{j=1}^{n} (y_j)^{c_j} = 1.$$

   A standard algorithm for approximating a scaling of a matrix to one that has prescribed row- and column-sums (when one exists) is to iteratively scale rows and columns separately so as to achieve corresponding line-sums.

   (b) Suppose $\mathbf{1}^T\mathbf{r} = \mathbf{1}^T\mathbf{r}$ and $\bar{\mathbf{x}} = [\bar{x}_i]$ and $\bar{\mathbf{y}} = [\bar{\mathbf{y}}_j]$ form an optimal solution of the optimization problem of Fact 1(d). Let $\bar{\lambda} \equiv \frac{\bar{\mathbf{x}}^T A\bar{\mathbf{y}}}{\mathbf{1}^T\mathbf{r}}$ and let $\bar{X} \in \mathbb{R}^{m\times m}$ and $\bar{Y} \in \mathbb{R}^{n\times n}$ be the diagonal matrices having diagonal elements $\bar{X}_{ii} = \frac{\bar{x}_i}{\lambda}$ and $\bar{Y}_{jj} = \bar{y}_j$. Then $B \equiv \bar{X}A\bar{Y}$ is a scaling of A satisfying $B\mathbf{1} = \mathbf{r}$ and $\mathbf{1}^T B = \mathbf{c}^T$.

   (c) Suppose $\bar{X} \in \mathbb{R}^{m\times m}$ and $\bar{Y} \in \mathbb{R}^{n\times n}$ are diagonal matrices such that $B \equiv \bar{X}A\bar{Y}$ is a scaling of A satisfying $B\mathbf{1} = \mathbf{r}$ and $\mathbf{1}^T B = \mathbf{c}^T$. Then $\mathbf{1}^T\mathbf{r} = \mathbf{1}^T\mathbf{r}$ and with

$$\bar{\lambda} \equiv \prod_{i=1}^{m} (\bar{X}_{ii})^{-r_i/\mathbf{1}^T\mathbf{r}}$$

   and

$$\bar{\mu} \equiv \prod_{i=1}^{m} (\bar{Y}_{jj})^{-c_j/\mathbf{1}^T\mathbf{c}},$$

   the vectors $\bar{\mathbf{x}} = [\bar{x}_i] \in \mathbb{R}^m$ and $\bar{\mathbf{y}} = [\bar{y}_j] \in \mathbb{R}^n$ with $\bar{x}_i = \bar{\lambda}X_{ii}$ for $i = 1, \ldots, m$ and $\bar{y}_j = \bar{\mu}Y_{jj}$ for $j = 1, \ldots, n$ are optimal for the optimization problem of Fact 1(d).

2. (*Approximate Prescribed-Line-Sum Scalings*) [RoSc89] Let $A = [a_{ij}] \in \mathbb{R}^{m\times n}$ be a nonnegative matrix, and let $\mathbf{r} = [r_i] \in \mathbb{R}^m$ and $\mathbf{c} = [c_j] \in \mathbb{R}^n$ be positive vectors.

   (a) The following are equivalent:

   i. For every $\epsilon > 0$ there exists a scaling B of A with $\|B\mathbf{1} - \mathbf{r}\|_1 \leq \epsilon$ and $\|\mathbf{1}^T B - \mathbf{c}^T\|_1 \leq \epsilon$.

ii. There exists nonnegative $m \times n$ matrix $A' = [a'_{ij}]$ with Struct(A') $\subseteq$Struct(A) and $a'_{ij} = a_{ij}$ for each $ij \in$ Struct(A') such that A' has a scaling B satisfying $B\mathbf{1} = \mathbf{r}$ and $\mathbf{1}^T B = \mathbf{c}^T$.

iii. For every $\epsilon > 0$ there exists a matrix B having the same support as A and satisfying $\|B\mathbf{1} - \mathbf{r}\|_1 \leq \epsilon$ and $\|\mathbf{1}^T B - \mathbf{c}^T\|_1 \leq \epsilon$.

iv. There exists a matrix B satisfying Struct$(B) \subseteq$ Struct(A), $B\mathbf{1} = \mathbf{r}$ and $\mathbf{1}^T B = \mathbf{c}^T$.

v. For every $I \subseteq \{1, \ldots, m\}$ and $J \subseteq \{1, \ldots, m\}$ for which $A[I^c, J] = 0$,

$$\sum_{i \in I} r_i \geq \sum_{j \in j} c_j.$$

vi. $\mathbf{1}^T\mathbf{r} = \mathbf{1}^T\mathbf{r}$ and the objective of the optimization problem of Fact 2(a)iii is bounded away from zero.

See [NR99] for a reduction of the problem of finding a scaling of A that satisfies $\|B\mathbf{1} - \mathbf{r}\|_1 \leq \epsilon$ and $\|\mathbf{1}^T B - \mathbf{c}^T\|_1 \leq \epsilon$ for a given $\epsilon > 0$ to the approximate solution of geometric program that is similar to the one in Fact 1(a)iv and for the description of an (ellipsoid) algorithm that solves the latter with complexity bound of $O(1)(m+n)^4 \ln[2 + \frac{mn\sqrt{m^3+n^3}\ln(mn\beta)}{\epsilon^3}]$, where β is the ratio between the largest and smallest positive entries of A.

9.7 Miscellaneous Topics

In this subsection, we mention several important topics about nonnegative matrices that are not covered in detail in the current section due to size constraint; some relevant material appears in other sections.

9.7.1 Nonnegative Factorization and Completely Positive Matrices

A *nonnegative factorization* of a nonnegative matrix $A \in R^{m \times n}$ is a representation $A = LR$ of A with L and R as nonnegative matrices. The *nonnegative rank* of A is the smallest number of columns of L (rows of R) in such a factorization.

A square matrix A is *doubly nonnegative* if it is nonnegative and positive semidefinite. Such a matrix A is *completely positive* if it has a nonnegative factorization $A = BB^T$; the *CP-rank* of A is then the smallest number of columns of a matrix B in such a factorization.

Facts about nonnegative factorizations and completely positive matrices can be found in [CR93], [BSM03], and [CP05].

9.7.2 The Inverse Eigenvalue Problem

The *inverse eigenvalue problem* concerns the identification of necessary conditions and sufficient conditions for a finite set of complex numbers to be the spectrum of a nonnegative matrix.

Facts about the inverse eigenvalue problem can be found in [BP94, Sections 4.2 and 11.2] and Chapter 20.

9.7.3 Nonhomogenous Products of Matrices

A *nonhomogenous product* of nonnegative matrices is the finite matrix product of nonnegative matrices $P_1 P_2 \ldots P_m$, generalizing powers of matrices where the multiplicands are equal (i.e., $P_1 = P_2 = \cdots = P_m$); the study of such products focuses on the case where the multiplicands are taken from a prescribed set.

Facts about Perron–Frobenius type properties of nonhomogenous products of matrices can be found in [Sen81], and [Har02].

9.7.4 Operators Determined by Sets of Nonnegative Matrices in Product Form

A finite set of nonnegative $n \times n$ matrices $\{P_\delta : \delta \in \Delta\}$ is said to be in *product form* if there exists finite sets of row vectors $\Delta_1, \ldots, \Delta_n$ such that $\Delta = \prod_{i=1}^{n} \Delta_i$ and for each $\delta = (\delta_1, \ldots, \delta_n) \in \Delta$, P_δ is the matrix whose rows are, respectively, $\delta_1, \ldots, \delta_n$. Such a family determines the *operators* $P_\Delta^{\max}$ and $P_\Delta^{\min}$ on $\mathbb{R}^n$ with $P_\Delta^{\max} x = \max_{\delta\in\Delta} P_\delta x$ and $P_\Delta^{\min} x = \min_{\delta\in\Delta} P_\delta x$ for each $x \in \mathbb{R}^n$.

Facts about Perron–Frobenius-type properties of the operators corresponding to families of matrices in product form can be found in [Zij82], [Zij84], and [RW82].

9.7.5 Max Algebra over Nonnegative Matrices

Matrix operations under the max algebra are executed with the max operator replacing (real) addition and (real) addition replacing (real) multiplication.

Perron–Frobenius-type results and scaling results are available for nonnegative matrices when considered as operators under the max algebra; see [RSS94], [Bap98], [But03], [BS05], and Chapter 25.

Acknowledgment

The author wishes to thank H. Schneider for comments that were helpful in preparing this section.

References

[Bap98] R.B. Bapat, A max version of the Perron–Frobenius Theorem, *Lin. Alg. Appl.*, 3:18, 275–276, 1998.

[BS75] G.P. Barker and H. Schneider, Algebraic Perron–Frobenius Theory, *Lin. Alg. Appl.*, 11:219–233, 1975.

[BNS89] A. Berman, M. Neumann, and R.J. Stern, *Nonnegative Matrices in Dynamic Systems*, John Wiley & Sons, New York, 1989.

[BP94] A. Berman and R.J. Plemmons, *Nonnegative Matrices in the Mathematical Sciences*, Academic, 1979, 2nd ed., SIAM, 1994.

[BSM03] A. Berman and N. Shaked-Monderer, *Completely Positive Matrices*, World Scientific, Singapore, 2003.

[But03] P. Butkovic, Max-algebra: the linear algebra of combinatorics? *Lin. Alg. Appl.*, 367:313–335, 2003.

[BS05] P. Butkovic and H. Schneider, Applications of max algebra to diagonal scaling of matrices, *ELA*, 13:262–273, 2005.

[CP05] M. Chu and R. Plemmons, Nonnegative matrix factorization and applications, *Bull. Lin. Alg. Soc. — Image*, 34:2–7, 2005.

[Coh78] J.E. Cohen, Derivatives of the spectral radius as a function of non-negative matrix elements, *Mathematical Proceedings of the Cambridge Philosophical Society*, 83:183–190, 1978.

[CR93] J.E. Cohen and U.G. Rothblum, Nonnegative ranks, decompositions and factorizations of nonnegative matrices, *Lin. Alg. Appl.*, 190:149–168, 1993.

[Col42] L. Collatz, *Einschliessungssatz für die charakteristischen Zahlen von Matrizen, Math. Z.*, 48:221–226, 1942.

[CHR97] D. Coppersmith, A.J. Hoffman, and U.G. Rothblum, Inequalities of Rayleigh quotients and bounds on the spectral radius of nonnegative symmetric matrices, *Lin. Alg. Appl.*, 263:201–220, 1997.

[DR05] E.V. Denardo and U.G. Rothblum, Totally expanding multiplicative systems, *Lin. Alg. Appl.*, 406:142–158, 2005.

[ERS95] B.C. Eaves, U.G. Rothblum, and H. Schneider, Perron–Frobenius theory over real closed fields and fractional power series expansions, *Lin. Alg. Appl.*, 220:123–150, 1995.

[EHP90] L. Elsner, D. Hershkowitz, and A. Pinkus, Functional inequalities for spectral radii of nonnegative matrices, *Lin. Alg. Appl.*, 129:103–130, 1990.

[EJD88] L. Elsner, C. Johnson, and D. da Silva, The Perron root of a weighted geometric mean of nonnegative matrices, *Lin. Multilin. Alg.*, 24:1–13, 1988.

[FiSc83] M. Fiedler and H. Schneider, Analytic functions of M-matrices and generalizations, *Lin. Multilin. Alg.*, 13:185–201, 1983.

[FrSc80] S. Friedland and H. Schneider, The growth of powers of a nonnegative matrix, *SIAM J. Alg. Disc. Meth.*, 1:185–200, 1980.

[Fro08] G.F. Frobenius, *Über Matrizen aus positiven Elementen, S.-B. Preuss. Akad. Wiss. Berlin*, 471–476, 1908.

[Fro09] G.F. Frobenius, *Über Matrizen aus positiven Elementen*, II, *S.-B. Preuss. Akad. Wiss. Berlin*, 514–518, 1909.

[Fro12] G.F. Frobenius, *Über Matrizen aus nicht negativen Elementen, Sitzungsber. Kön. Preuss. Akad. Wiss. Berlin*, 456–457, 1912.

[Gan59] F.R. Gantmacher, *The Theory of Matrices*, Vol. II, Chelsea Publications, London, 1958.

[Har02] D.J. Hartfiel, *Nonhomogeneous Matrix Products*, World Scientific, River Edge, NJ, 2002.

[HJ85] R.A. Horn and C.R. Johnson, *Matrix Analysis*, Cambridge University Press, Cambridge, 1985.

[HNR90] R.E. Hartwig, M. Neumann, and N.J. Rose, An algebraic-analytic approach to nonnegative basis, *Lin. Alg. Appl.*, 133:77–88, 1990.

[HRR92] M. Haviv, Y. Ritov, and U.G. Rothblum, Taylor expansions of eigenvalues of perturbed matrices with applications to spectral radii of nonnegative matrices, *Lin. Alg. Appli.*, 168:159–188, 1992.

[HS88a] D. Hershkowitz and H. Schneider, Solutions of Z-matrix equations, *Lin. Alg. Appl.*, 106:25–38, 1988.

[HS88b] D. Hershkowitz and H. Schneider, On the generalized nullspace of M-matrices and Z-matrices, *Lin. Alg. Appl.*, 106:5–23, 1988.

[HRS89] D. Hershkowitz, U.G. Rothblum, and H. Schneider, The combinatorial structure of the generalized nullspace of a block triangular matrix, *Lin. Alg. Appl.*, 116:9–26, 1989.

[HS89] D. Hershkowitz and H. Schneider, Height bases, level bases, and the equality of the height and the level characteristic of an M-matrix, *Lin. Multilin. Alg.*, 25:149–171, 1989.

[HS91a] D. Hershkowitz and H. Schneider, Combinatorial bases, derived Jordan and the equality of the height and level characteristics of an M-matrix, *Lin. Multilin. Alg.*, 29:21–42, 1991.

[HS91b] D. Hershkowitz and H. Schneider, On the existence of matrices with prescribed height and level characteristics, *Israel Math J.*, 75:105–117, 1991.

[Hof67] A.J. Hoffman, Three observations on nonnegative matrices, *J. Res. Nat. Bur. Standards-B. Math and Math. Phys.*, 71B:39–41, 1967.

[Kru37] J. Kruithof, *Telefoonverkeersrekening, De Ingenieur*, 52(8):E15–E25, 1937.

[LT85] P. Lancaster and M. Tismenetsky, *The Theory of Matrices*, 2nd ed., Academic Press, New York, 1985.

[Mac00] C.R. MacCluer, The many proofs and applications of Perron's theorem, *SIAM Rev.*, 42:487–498, 2000.

[Min88] H. Minc, *Nonnegative Matrices*, John Wiley & Sons, New York, 1988.

[NR99] A. Nemirovski and U.G. Rothblum, On complexity of matrix scaling, *Lin. Alg. Appl.*, 302-303:435–460, 1999.

[NS94] M. Neumann and H. Schneider, Algorithms for computing bases for the Perron eigenspace with prescribed nonnegativity and combinatorial properties, *SIAM J. Matrix Anal. Appl.*, 15:578–591, 1994.

[Per07a] O. Perron, *Grundlagen für eine Theorie des Jacobischen Kettenbruchalogithmus, Math. Ann.*, 63:11–76, 1907.

[Per07b] O. Perron, *Zür Theorie der über Matrizen, Math. Ann.*, 64:248–263, 1907.

[RiSc78] D. Richman and H. Schneider, On the singular graph and the Weyr characteristic of an M-matrix, *Aequationes Math.*, 17:208–234, 1978.

[Rot75] U.G. Rothblum, Algebraic eigenspaces of nonnegative matrices, *Lin. Alg. Appl.*, 12:281–292, 1975.

[Rot80] U.G. Rothblum, Bounds on the indices of the spectral-circle eigenvalues of a nonnegative matrix, *Lin. Alg. Appl.*, 29:445–450, 1980.

[Rot81a] U.G. Rothblum, Expansions of sums of matrix powers, *SIAM Rev.*, 23:143–164, 1981.

[Rot81b] U.G. Rothblum, Sensitive growth analysis of multiplicative systems I: the stationary dynamic approach, *SIAM J. Alg. Disc. Meth.*, 2:25–34, 1981.

[Rot81c] U.G. Rothblum, Resolvant expansions of matrices and applications, *Lin. Alg. Appl.*, 38:33–49, 1981.

[RoSc89] U.G. Rothblum and H. Schneider, Scalings of matrices which have prespecified row-sums and column-sums via optimization, *Lin. Alg. Appl.*, 114/115:737–764, 1989.

[RSS94] U.G. Rothblum, H. Schneider, and M.H. Schneider, Scalings of matrices which have prespecified row-maxima and column-maxima, *SIAM J. Matrix Anal.*, 15:1–15, 1994.

[RT85] U.G. Rothblum and C.P. Tan, Upper bounds on the maximum modulus of subdominant eigenvalues of nonnegative matrices, *Lin. Alg. Appl.*, 66:45–86, 1985.

[RW82] U.G. Rothblum and P. Whittle, Growth optimality for branching Markov decision chains, *Math. Op. Res.*, 7:582–601, 1982.

[Sch53] H. Schneider, An inequality for latent roots applied to determinants with dominant principal diagonal, *J. London Math. Soc.*, 28:8–20, 1953.

[Sch56] H. Schneider, The elementary divisors associated with 0 of a singular M-matrix, *Proc. Edinburgh Math. Soc.*, 10:108–122, 1956.

[Sch86] H. Schneider, The influence of the marked reduced graph of a nonnegative matrix on the Jordan form and on related properties: a survey, *Lin. Alg. Appl.*, 84:161–189, 1986.

[Sch96] H. Schneider, Commentary on "*Unzerlegbare, nicht negative Matrizen*," in *Helmut Wielandt's "Mathematical Works," Vol. 2*, B. Huppert and H. Schneider, Eds., Walter de Gruyter Berlin, 1996.

[Sen81] E. Seneta, *Non-negative matrices and Markov chains*, Springer Verlag, New York, 1981.

[Sin64] R. Sinkhorn, A relationship between arbitrary positive and stochastic matrices, *Ann. Math. Stat.*, 35:876–879, 1964.

[Tam04] B.S. Tam, The Perron generalized eigenspace and the spectral cone of a cone-preserving map, *Lin. Alg. Appl.*, 393:375-429, 2004.

[TW89] B.S. Tam and S.F. Wu, On the Collatz-Wielandt sets associated with a cone-preserving map, *Lin. Alg. Appl.*, 125:77–95, 1989.

[Var62] R.S. Varga, *Matrix Iterative Analysis*, Prentice-Hall, Upper Saddle River, NJ, 1962, 2nd ed., Springer, New York, 2000.

[Vic85] H.D. Victory, Jr., On nonnegative solutions to matrix equations, *SIAM J. Alg. Dis. Meth.*, 6: 406–412, 1985.

[Wie50] H. Wielandt, *Unzerlegbare, nicht negative Matrizen, Mathematische Zeitschrift*, 52:642–648, 1950.

[Zij82] W.H.M. Zijm, *Nonnegative Matrices in Dynamic Programming*, Ph.D. dissertation, Mathematisch Centrum, Amsterdam, 1982.

[Zij84] W.H.M. Zijm, Generalized eigenvectors and sets of nonnegative matrices, *Lin. Alg. Appl.*, 59:91–113, 1984.

10

Partitioned Matrices

Robert Reams
Virginia Commonwealth University

10.1 Submatrices and Block Matrices

Definitions:

Let $A \in F^{m\times n}$. Then the row indices of A are $\{1,\dots,m\}$, and the column indices of A are $\{1,\dots,n\}$. Let α, β be nonempty sets of indices with $\alpha \subseteq \{1,\dots,m\}$ and $\beta \subseteq \{1,\dots,n\}$.

A **submatrix** $A[\alpha,\beta]$ is a matrix whose rows have indices α among the row indices of A, and whose columns have indices β among the column indices of A. $A(\alpha,\beta) = A[\alpha^c,\beta^c]$, where α^c is the complement of α.

A **principal submatrix** is a submatrix $A[\alpha,\alpha]$, denoted more compactly as $A[\alpha]$.

Let the set $\{1,\dots m\}$ be partitioned into the subsets $\alpha_1,\dots,\alpha_r$ in the usual sense of partitioning a set (so that $\alpha_i \cap \alpha_j = \emptyset$, for all $i \neq j$, $1 \leq i,j \leq r$, and $\alpha_1 \cup \cdots \cup \alpha_r = \{1,\dots,m\}$), and let $\{1,\dots,n\}$ be partitioned into the subsets $\beta_1,\dots,\beta_s$.

The matrix $A \in F^{m\times n}$ is said to be **partitioned** into the submatrices $A[\alpha_i,\beta_j]$, $1 \leq i \leq r$, $1 \leq j \leq s$.

A **block matrix** is a matrix that is partitioned into submatrices $A[\alpha_i,\beta_j]$ with the row indices $\{1,\dots,m\}$ and column indices $\{1,\dots,n\}$ partitioned into subsets sequentially, i.e., $\alpha_1 = \{1,\dots,i_1\}, \alpha_2 = \{i_1+1,\dots,i_2\}$, etc.

Each entry of a block matrix, which is a submatrix $A[\alpha_i,\beta_j]$, is called a **block**, and we will sometimes write $A = [A_{ij}]$ to label the blocks, where $A_{ij} = A[\alpha_i,\beta_j]$.

If the block matrix $A \in F^{m\times p}$ is partitioned with α_is and β_js, $1 \leq i \leq r$, $1 \leq j \leq s$, and the block matrix $B \in F^{p\times n}$ is partitioned with β_js and γ_ks, $1 \leq j \leq s$, $1 \leq k \leq t$, then the partitions of A and B are said to be **conformal** (or sometimes **conformable**).

Facts:

The following facts can be found in [HJ85]. This information is also available in many other standard references such as [LT85] or [Mey00].

1. Two block matrices $A = [A_{ij}]$ and $B = [B_{ij}]$ in $F^{m\times n}$, which are both partitioned with the same α_is and β_js, $1 \leq i \leq r$, $1 \leq j \leq s$, may be added block-wise, as with the usual matrix addition, so that the (i,j) block entry of $A+B$ is $(A+B)_{ij} = A_{ij} + B_{ij}$.

2. If the block matrix $A \in F^{m\times p}$ and the block matrix $B \in F^{p\times n}$ have conformal partitions, then we can think of A and B as having entries, which are blocks, so that we can then multiply A and B block-wise to form the $m \times n$ block matrix $C = AB$. Then $C_{ik} = \sum_{j=1}^{s} A_{ij}B_{jk}$, and the matrix C will be partitioned with the α_is and γ_ks, $1 \leq i \leq r, 1 \leq k \leq t$, where A is partitioned with α_is and β_js, $1 \leq i \leq r, 1 \leq j \leq s$, and B is partitioned with β_js and γ_ks, $1 \leq j \leq s, 1 \leq k \leq t$.
3. With addition and multiplication of block matrices described as in Facts 1 and 2 the usual properties of associativity of addition and multiplication of block matrices hold, as does distributivity, and commutativity of addition. The additive identity $\mathbf{0}$ and multiplicative identity I are the same under block addition and multiplication, as with the usual matrix addition and multiplication. The additive identity $\mathbf{0}$ has zero matrices as blocks; the multiplicative identity I has multiplicative identity submatrices as diagonal blocks and zero matrices as off-diagonal blocks.
4. If the partitions of A and B are conformal, the partitions of B and A are not necessarily conformal, even if BA is defined.
5. Let $A \in F^{n\times n}$ be a block matrix of the form $A = \begin{bmatrix} A_{11} & A_{12} \\ A_{21} & \mathbf{0} \end{bmatrix}$, where A_{12} is $k \times k$, and A_{21} is $(n-k) \times (n-k)$. Then $\det(A) = (-1)^{(n+1)k}\det(A_{12})\det(A_{21})$.

Examples:

1. Let the block matrix $A \in \mathbb{C}^{n\times n}$ given by $A = \begin{bmatrix} A_{11} & A_{12} \\ A_{21} & A_{22} \end{bmatrix}$ be Hermitian. Then A_{11} and A_{22} are Hermitian, and $A_{21} = A_{12}^*$.
2. If $A = [a_{ij}]$, then $A[\{i\},\{j\}]$ is the 1×1 matrix whose entry is a_{ij}. The submatrix $A(\{i\},\{j\})$ is the submatrix of A obtained by deleting row i and column j of A.

   Let $A = \begin{bmatrix} 1 & -2 & 5 & 3 & -1 \\ -3 & 0 & 1 & 6 & 1 \\ 2 & 7 & 4 & 5 & -7 \end{bmatrix}$.
3. Then $A[\{2\},\{3\}] = [a_{23}] = [1]$ and $A(\{2\},\{3\}) = \begin{bmatrix} 1 & -2 & 3 & -1 \\ 2 & 7 & 5 & -7 \end{bmatrix}$.
4. Let $\alpha = \{1,3\}$ and $\beta = \{1,2,4\}$. Then the submatrix $A[\alpha,\beta] = \begin{bmatrix} 1 & -2 & 3 \\ 2 & 7 & 5 \end{bmatrix}$, and the principal submatrix $A[\alpha] = \begin{bmatrix} 1 & 5 \\ 2 & 4 \end{bmatrix}$.
5. Let $\alpha_1 = \{1\}, \alpha_2 = \{2,3\}$ and let $\beta_1 = \{1,2\}, \beta_2 = \{3\}, \beta_3 = \{4,5\}$. Then the block matrix, with (i,j) block entry $A_{ij} = A[\alpha_i,\beta_j]$, $1 \leq i \leq 2, 1 \leq j \leq 3$, is

$$A = \begin{bmatrix} A_{11} & A_{12} & A_{13} \\ A_{21} & A_{22} & A_{23} \end{bmatrix} = \left[\begin{array}{cc|c|cc} 1 & -2 & 5 & 3 & -1 \\ \hline -3 & 0 & 1 & 6 & 1 \\ 2 & 7 & 4 & 5 & -7 \end{array}\right].$$

6. Let $B = \begin{bmatrix} B_{11} & B_{12} & B_{13} \\ B_{21} & B_{22} & B_{23} \end{bmatrix} = \left[\begin{array}{cc|c|cc} 2 & -1 & 0 & 6 & -2 \\ \hline -4 & 0 & 5 & 3 & 7 \\ 1 & 1 & -2 & 2 & 6 \end{array}\right]$. Then the matrices A (of this example) and B are partitioned with the same α_is and β_js, so they can be added block-wise

as $\begin{bmatrix} A_{11} & A_{12} & A_{13} \\ A_{21} & A_{22} & A_{23} \end{bmatrix} + \begin{bmatrix} B_{11} & B_{12} & B_{13} \\ B_{21} & B_{22} & B_{23} \end{bmatrix}$

$$= \begin{bmatrix} A_{11}+B_{11} & A_{12}+B_{12} & A_{13}+B_{13} \\ A_{21}+B_{21} & A_{22}+B_{22} & A_{23}+B_{23} \end{bmatrix}$$

$$= \left[\begin{array}{cc|c|cc} 3 & -3 & 5 & 9 & -3 \\ \hline -7 & 0 & 6 & 9 & 8 \\ 3 & 8 & 2 & 7 & -1 \end{array}\right].$$

7. The block matrices $A = \begin{bmatrix} A_{11} & A_{12} & A_{13} \\ A_{21} & A_{22} & A_{13} \end{bmatrix}$ and $B = \begin{bmatrix} B_{11} \\ B_{21} \\ B_{31} \end{bmatrix}$ have conformal partitions if the β_j index sets, which form the submatrices $A_{ij} = A[\alpha_i, \beta_j]$ of A, are the same as the β_j index sets, which form the submatrices $B_{jk} = B[\beta_j, \gamma_k]$ of B. For instance, the matrix

$$A = \begin{bmatrix} A_{11} & A_{12} & A_{13} \\ A_{21} & A_{22} & A_{23} \end{bmatrix} = \left[\begin{array}{cc|c|cc} 1 & -2 & 5 & 3 & -1 \\ \hline -3 & 0 & 1 & 6 & 1 \\ 2 & 7 & 4 & 5 & -7 \end{array}\right]$$

and the matrix $B = \begin{bmatrix} B_{11} \\ B_{21} \\ B_{31} \end{bmatrix} = \left[\begin{array}{cc} 4 & -1 \\ 3 & 9 \\ \hline 5 & 2 \\ \hline 2 & -8 \\ 7 & -1 \end{array}\right]$ have conformal partitions, so A and B can be multiplied block-wise to form the 3×2 matrix

$$AB = \begin{bmatrix} A_{11}B_{11} + A_{12}B_{21} + A_{13}B_{31} \\ A_{21}B_{11} + A_{22}B_{21} + A_{23}B_{31} \end{bmatrix}$$

$$= \begin{bmatrix} [1 \quad -2]\begin{bmatrix} 4 & -1 \\ 3 & 9 \end{bmatrix} + 5[5 \quad 2] + [3 \quad -1]\begin{bmatrix} 2 & -8 \\ 7 & -1 \end{bmatrix} \\ \begin{bmatrix} -3 & 0 \\ 2 & 7 \end{bmatrix}\begin{bmatrix} 4 & -1 \\ 3 & 9 \end{bmatrix} + \begin{bmatrix} 1 \\ 4 \end{bmatrix}[5 \quad 2] + \begin{bmatrix} 6 & 1 \\ 5 & -7 \end{bmatrix}\begin{bmatrix} 2 & -8 \\ 7 & -1 \end{bmatrix} \end{bmatrix}$$

$$= \begin{bmatrix} [-2 \quad -19] + [25 \quad 10] + [-1 \quad -23] \\ \begin{bmatrix} -12 & 3 \\ 29 & 61 \end{bmatrix} + \begin{bmatrix} 5 & 2 \\ 20 & 8 \end{bmatrix} + \begin{bmatrix} 19 & -49 \\ -39 & -33 \end{bmatrix} \end{bmatrix} = \begin{bmatrix} 22 & -32 \\ 12 & -44 \\ 10 & 36 \end{bmatrix}.$$

8. Let $A = \left[\begin{array}{cc|c} 1 & 2 & 3 \\ 4 & 5 & 6 \end{array}\right]$ and $B = \left[\begin{array}{c|c} 7 & 8 \\ 9 & 0 \\ \hline 1 & 2 \end{array}\right]$. Then A and B have conformal partitions. BA is defined, but B and A do not have conformal partitions.

10.2 Block Diagonal and Block Triangular Matrices

Definitions:

A matrix $A = [a_{ij}] \in F^{n\times n}$ is **diagonal** if $a_{ij} = 0$, for all $i \neq j$, $1 \leq i, j \leq n$.

A diagonal matrix $A = [a_{ij}] \in F^{n\times n}$ is said to be **scalar** if $a_{ii} = a$, for all i, $1 \leq i \leq n$, and some scalar $a \in F$, i.e., $A = aI_n$.

A matrix $A \in F^{n\times n}$ is **block diagonal** if A as a block matrix is partitioned into submatrices $A_{ij} \in F^{n_i\times n_j}$, so that $A = [A_{ij}]$, $\sum_{i=1}^{k} n_i = n$, and $A_{ij} = \mathbf{0}$, for all $i \neq j$, $1 \leq i, j \leq k$. Thus, $A = \begin{bmatrix} A_{11} & \mathbf{0} & \cdots & \mathbf{0} \\ \mathbf{0} & A_{22} & \cdots & \mathbf{0} \\ \vdots & & \ddots & \vdots \\ \mathbf{0} & \mathbf{0} & \cdots & A_{kk} \end{bmatrix}$. This block diagonal matrix A is denoted $A = \mathrm{diag}(A_{11}, \ldots, A_{kk})$, where A_{ii} is $n_i \times n_i$, (or sometimes denoted $A = A_{11} \oplus \cdots \oplus A_{kk}$, and called the **direct sum** of $A_{11}, \ldots, A_{kk}$).

A matrix $A = [a_{ij}] \in F^{n\times n}$ is **upper triangular** if $a_{ij} = 0$, for all $i > j 1 \leq i, j \leq n$.

An upper triangular matrix $A = [a_{ij}] \in F^{n\times n}$ is **strictly upper triangular** if $a_{ij} = 0$ for all $i \geq j$, $1 \leq i, j \leq n$.

A matrix $A \in F^{n\times n}$ is **lower triangular** if $a_{ij} = 0$ for all $i < j$, $1 \leq i, j \leq n$, i.e., if A^T is upper triangular.

A matrix $A \in F^{n\times n}$ is **strictly lower triangular** if A^T is strictly upper triangular.

A matrix is **triangular** it is upper or lower triangular.

A matrix $A \in F^{n\times n}$ is **block upper triangular**, if A as a block matrix is partitioned into the submatrices $A_{ij} \in F^{n_i\times n_j}$, so that $A = [A_{ij}]$, $\sum_{i=1}^{k} n_i = n$, and $A_{ij} = \mathbf{0}$, for all $i > j$, $1 \leq i, j \leq k$, i.e., considering the A_{ij} blocks as the entries of A, A is upper triangular. Thus, $A = \begin{bmatrix} A_{11} & A_{12} & \cdots & A_{1k} \\ \mathbf{0} & A_{22} & \cdots & A_{2k} \\ \vdots & & \ddots & \vdots \\ \mathbf{0} & \mathbf{0} & \cdots & A_{kk} \end{bmatrix}$, where each A_{ij} is $n_i \times n_j$, and $\sum_{i=1}^{k} n_i = n$. The matrix A is **strictly block upper triangular** if $A_{ij} = \mathbf{0}$, for all $i \geq j$, $1 \leq i, j \leq k$.

A matrix $A \in F^{n\times n}$ is **block lower triangular** if A^T is block upper triangular.

A matrix $A \in F^{n\times n}$ is **strictly block lower triangular** if A^T is strictly block upper triangular.

A matrix $A = [a_{ij}] \in F^{n\times n}$ is **upper Hessenberg** if $a_{ij} = 0$, for all $i - 2 \geq j$, $1 \leq i, j \leq n$, i.e., A has the form

$$A = \begin{bmatrix} a_{11} & a_{12} & a_{13} & \cdots & a_{1n-1} & a_{1n} \\ a_{21} & a_{22} & a_{23} & \cdots & a_{2n-1} & a_{2n} \\ 0 & a_{32} & a_{33} & \cdots & a_{3n-1} & a_{3n} \\ \vdots & \vdots & \ddots & \ddots & & \vdots \\ 0 & 0 & \cdots & a_{n-1n-2} & a_{n-1n-1} & a_{n-1n} \\ 0 & 0 & \cdots & 0 & a_{nn-1} & a_{nn} \end{bmatrix}.$$

A matrix $A = [a_{ij}] \in F^{n\times n}$ is **lower Hessenberg** if A^T is upper Hessenberg.

Facts:

The following facts can be found in [HJ85]. This information is also available in many other standard references such as [LT85] or [Mey00].

1. Let $D, D' \in F^{n\times n}$ be any diagonal matrices. Then $D + D'$ and DD' are diagonal, and $DD' = D'D$. If $D = \text{diag}(d_1, \ldots, d_n)$ is nonsingular, then $D^{-1} = \text{diag}(1/d_1, \ldots, 1/d_n)$.
2. Let $D \in F^{n\times n}$ be a matrix such that $DA = AD$ for all $A \in F^{n\times n}$. Then D is a scalar matrix.
3. If $A \in F^{n\times n}$ is a block diagonal matrix, so that $A = \text{diag}(A_{11}, \ldots, A_{kk})$, then $\text{tr}(A) = \sum_{i=1}^{k} \text{tr}(A_{ii})$, $\det(A) = \Pi_{i=1}^{k} \det(A_{ii})$, $\text{rank}(A) = \sum_{i=1}^{k} \text{rank}(A_{ii})$, and $\sigma(A) = \sigma(A_{11}) \cup \cdots \cup \sigma(A_{kk})$.
4. Let $A \in F^{n\times n}$ be a block diagonal matrix, so that $A = \text{diag}(A_{11}, A_{22} \ldots, A_{kk})$. Then A is nonsingular if and only if A_{ii} is nonsingular for each i, $1 \le i \le k$. Moreover, $A^{-1} = \text{diag}(A_{11}^{-1}, A_{22}^{-1} \ldots, A_{kk}^{-1})$.
5. See Chapter 4.3 for information on diagonalizability of matrices.
6. Let $A \in F^{n\times n}$ be a block diagonal matrix, so that $A = \text{diag}(A_{11}, \ldots, A_{kk})$. Then A is diagonalizable if and only if A_{ii} is diagonalizable for each i, $1 \le i \le k$.
7. If $A, B \in F^{n\times n}$ are upper (lower) triangular matrices, then $A + B$ and AB are upper (lower) triangular. If the upper (lower) triangular matrix $A = [a_{ij}]$ is nonsingular, then A^{-1} is upper (lower) triangular with diagonal entries $1/a_{11}, \ldots, 1/a_{nn}$.
8. Let $A \in F^{n\times n}$ be block upper triangular, so that $A = \begin{bmatrix} A_{11} & A_{12} & \cdots & A_{1k} \\ \mathbf{0} & A_{22} & \cdots & A_{2k} \\ \vdots & & \ddots & \vdots \\ \mathbf{0} & \mathbf{0} & \cdots & A_{kk} \end{bmatrix}$. Then $\text{tr}(A) = \sum_{i=1}^{k} \text{tr}(A_{ii})$, $\det(A) = \Pi_{i=1}^{k} \det(A_{ii})$, $\text{rank}(A) \ge \sum_{i=1}^{k} \text{rank}(A_{ii})$, and $\sigma(A) = \sigma(A_{11}) \cup \cdots \cup \sigma(A_{kk})$.
9. Let $A = (A_{ij}) \in F^{n\times n}$ be a block triangular matrix (either upper or lower triangular). Then A is nonsingular if and only if A_{ii} is nonsingular for each i, $1 \le i \le k$. Moreover, the $n_i \times n_i$ diagonal block entries of A^{-1} are A_{ii}^{-1}, for each i, $1 \le i \le k$.
10. *Schur's Triangularization Theorem:* Let $A \in \mathbb{C}^{n\times n}$. Then there is a unitary matrix $U \in \mathbb{C}^{n\times n}$ so that U^*AU is upper triangular. The diagonal entries of U^*AU are the eigenvalues of A.
11. Let $A \in \mathbb{R}^{n\times n}$. Then there is an orthogonal matrix $V \in \mathbb{R}^{n\times n}$ so that V^TAV is of upper Hessenberg form $\begin{bmatrix} A_{11} & A_{12} & \cdots & A_{1k} \\ \mathbf{0} & A_{22} & \cdots & A_{2k} \\ \vdots & & \ddots & \vdots \\ \mathbf{0} & \mathbf{0} & \cdots & A_{kk} \end{bmatrix}$, where each A_{ii}, $1 \le i \le k$, is 1×1 or 2×2. Moreover, when A_{ii} is 1×1, the entry of A_{ii} is an eigenvalue of A, whereas when A_{ii} is 2×2, then A_{ii} has two eigenvalues which are nonreal complex conjugates of each other, and are eigenvalues of A.
12. (For more information on unitary triangularization, see Chapter 7.2.)
13. Let $A = [A_{ij}] \in F^{n\times n}$ with $|\sigma(A)| = n$, where $\lambda_1, \ldots, \lambda_k \in \sigma(A)$ are the distinct eigenvalues of A. Then there is a nonsingular matrix $T \in F^{n\times n}$ so that $T^{-1}AT = \text{diag}(A_{11}, \ldots, A_{kk})$, where each $A_{ii} \in F^{n_i \times n_i}$ is upper triangular with all diagonal entries of A_{ii} equal to λ_i, for $1 \le i \le k$ (and $\sum_{i=1}^{k} n_i = n$).
14. Let $A \in F^{n\times n}$ be a block upper triangular matrix, of the form $A = \begin{bmatrix} A_{11} & A_{12} \\ \mathbf{0} & A_{22} \end{bmatrix}$, where A_{ij} is $n_i \times n_j$, $1 \le i, j \le 2$, and $\sum_{i=1}^{2} n_i = n$. (Note that any block upper triangular matrix can be said to have this form.) Let $\mathbf{x}$ be an eigenvector of A_{11}, with corresponding eigenvalue λ, so that $A_{11}\mathbf{x} = \lambda\mathbf{x}$, where $\mathbf{x}$ is a (column) vector with n_1 components. Then the (column) vector with n components $\begin{bmatrix} \mathbf{x} \\ \mathbf{0} \end{bmatrix}$ is an eigenvector of A with eigenvalue λ. Let $\mathbf{y}$ be a left eigenvector of A_{22}, with corresponding eigenvalue μ, so that $\mathbf{y}A_{22} = \mathbf{y}\mu$, where $\mathbf{y}$ is a row vector with n_2 components. Then the (row) vector with n components $[\mathbf{0} \quad \mathbf{y}]$ is a left eigenvector of A with eigenvalue μ.

Examples:

1. The matrix $A = \begin{bmatrix} 1 & 3 & 0 & 0 & 0 \\ 2 & 4 & 0 & 0 & 0 \\ 0 & 0 & 5 & 0 & 0 \\ 0 & 0 & 0 & 6 & 8 \\ 0 & 0 & 0 & 7 & 9 \end{bmatrix}$ is a block diagonal matrix, and the trace, determinant, and rank can be calculated block-wise, where $A_{11} = \begin{bmatrix} 1 & 3 \\ 2 & 4 \end{bmatrix}$, $A_{22} = 5$, and $A_{33} = \begin{bmatrix} 6 & 8 \\ 7 & 9 \end{bmatrix}$, as $\operatorname{tr}(A) = 25 = \sum_{i=1}^{3} \operatorname{tr}(A_{ii})$, $\det(A) = (-2)(5)(-2) = \Pi_{i=1}^{3}\det(A_{ii})$, and $\operatorname{rank}(A) = 5 = \sum_{i=1}^{3} \operatorname{rank}(A_{ii})$.

2. Let $A = \begin{bmatrix} a & b & c \\ 0 & d & e \\ 0 & 0 & f \end{bmatrix} \in F^{3\times 3}$, an upper triangular matrix. If a, d, f are nonzero, then $A^{-1} = \begin{bmatrix} \frac{1}{a} & -\frac{b}{ad} & \frac{be-cd}{adf} \\ 0 & \frac{1}{d} & -\frac{e}{df} \\ 0 & 0 & \frac{1}{f} \end{bmatrix}$.

3. The matrix $B = \begin{bmatrix} 1 & 3 & 0 & 0 & 0 \\ 2 & 4 & 5 & 0 & 0 \\ 0 & 0 & 0 & 6 & 0 \\ 0 & 0 & 0 & 0 & 7 \\ 0 & 0 & 0 & 0 & 8 \end{bmatrix}$ is not block diagonal. However, B is block upper triangular, with $B_{11} = \begin{bmatrix} 1 & 3 \\ 2 & 4 \end{bmatrix}$, $B_{22} = 0$, $B_{33} = \begin{bmatrix} 0 & 7 \\ 0 & 8 \end{bmatrix}$, $B_{12} = \begin{bmatrix} 0 \\ 5 \end{bmatrix}$, $B_{13} = \begin{bmatrix} 0 & 0 \\ 0 & 0 \end{bmatrix}$, and $B_{23} = [6 \quad 0]$.

   Notice that $4 = \operatorname{rank}(B) \geq \sum_{i=1}^{3} \operatorname{rank}(B_{ii}) = 2 + 0 + 1 = 3$.

4. The 4×4 matrix $\begin{bmatrix} 1 & 2 & 0 & 0 \\ 3 & 4 & 5 & 0 \\ 6 & 7 & 8 & 9 \\ 10 & 11 & 12 & 13 \end{bmatrix}$ is not lower triangular, but is lower Hessenberg.

10.3 Schur Complements

In this subsection, the square matrix $A = \begin{bmatrix} A_{11} & A_{12} \\ A_{21} & A_{22} \end{bmatrix}$ is partitioned as a block matrix, where A_{11} is nonsingular.

Definitions:

The **Schur complement** of A_{11} in A is the matrix $A_{22} - A_{21}A_{11}^{-1}A_{12}$, sometimes denoted A/A_{11}.

Facts:

1. [Zha99] $\begin{bmatrix} I & \mathbf{0} \\ -A_{21}A_{11}^{-1} & I \end{bmatrix} \begin{bmatrix} A_{11} & A_{12} \\ A_{21} & A_{22} \end{bmatrix} \begin{bmatrix} I & -A_{11}^{-1}A_{12} \\ \mathbf{0} & I \end{bmatrix} = \begin{bmatrix} A_{11} & \mathbf{0} \\ \mathbf{0} & A/A_{11} \end{bmatrix}$.

2. [Zha99] Let $A = \begin{bmatrix} A_{11} & A_{12} \\ A_{21} & A_{22} \end{bmatrix}$, where A_{11} is nonsingular. Then $\det(A) = \det(A_{11})\det(A/A_{11})$.

   Also, $\operatorname{rank}(A) = \operatorname{rank}(A_{11}) + \operatorname{rank}(A/A_{11})$.

3. [Zha99] Let $A = \begin{bmatrix} A_{11} & A_{12} \\ A_{21} & A_{22} \end{bmatrix}$. Then A is nonsingular if and only if both A_{11} and the Schur complement of A_{11} in A are nonsingular.

4. [HJ85] Let $A = \begin{bmatrix} A_{11} & A_{12} \\ A_{21} & A_{22} \end{bmatrix}$, where A_{11}, A_{22}, A/A_{11}, A/A_{22}, and A are nonsingular. Then $A^{-1} = \begin{bmatrix} (A/A_{22})^{-1} & -A_{11}^{-1}A_{12}(A/A_{11})^{-1} \\ -A_{22}^{-1}A_{21}(A/A_{22})^{-1} & (A/A_{11})^{-1} \end{bmatrix}$.

5. [Zha99] Let $A = \begin{bmatrix} A_{11} & A_{12} \\ A_{21} & A_{22} \end{bmatrix}$, where A_{11}, A_{22}, A/A_{11}, and A/A_{22} are nonsingular. An equation relating the Schur complements of A_{11} in A and A_{22} in A is $(A/A_{22})^{-1} = A_{11}^{-1} + A_{11}^{-1}A_{12}(A/A_{11})^{-1}A_{21}A_{11}^{-1}$.

6. [LT85] Let $A = \begin{bmatrix} A_{11} & A_{12} \\ A_{21} & A_{22} \end{bmatrix}$, where the $k \times k$ matrix A_{11} is nonsingular. Then $\operatorname{rank}(A) = k$ if and only if $A/A_{11} = \mathbf{0}$.

7. [Hay68] Let $A = \begin{bmatrix} A_{11} & A_{12} \\ A_{12}^* & A_{22} \end{bmatrix}$ be Hermitian, where A_{11} is nonsingular. Then the inertia of A is $\operatorname{in}(A) = \operatorname{in}(A_{11}) + \operatorname{in}(A/A_{11})$.

8. [Hay68] Let $A = \begin{bmatrix} A_{11} & A_{12} \\ A_{12}^* & A_{22} \end{bmatrix}$ be Hermitian, where A_{11} is nonsingular. Then A is positive (semi)definite if and only if both A_{11} and A/A_{11} are positive (semi)definite.

Examples:

1. Let $A = \begin{bmatrix} 1 & 2 & 3 \\ 4 & 5 & 6 \\ 7 & 8 & 10 \end{bmatrix}$. Then with $A_{11} = 1$, we have

$$\begin{bmatrix} 1 & \mathbf{0} \\ -\begin{bmatrix} 4 \\ 7 \end{bmatrix} & I_2 \end{bmatrix} \begin{bmatrix} 1 & 2 & 3 \\ 4 & 5 & 6 \\ 7 & 8 & 10 \end{bmatrix} \begin{bmatrix} 1 & -[2 \;\; 3] \\ \mathbf{0} & I_2 \end{bmatrix} = \begin{bmatrix} 1 & \mathbf{0} \\ \mathbf{0} & \begin{bmatrix} 5 & 6 \\ 8 & 10 \end{bmatrix} - \begin{bmatrix} 4 \\ 7 \end{bmatrix} 1^{-1} [2 \;\; 3] \end{bmatrix}$$
$$= \begin{bmatrix} 1 & \mathbf{0} \\ \mathbf{0} & \begin{bmatrix} -3 & -6 \\ -6 & -11 \end{bmatrix} \end{bmatrix}.$$

2. Let $A = \begin{bmatrix} 1 & 2 & 3 \\ 4 & 5 & 6 \\ 7 & 8 & 10 \end{bmatrix}$. With $A_{11} = 1$, and $A_{22} = \begin{bmatrix} 5 & 6 \\ 8 & 10 \end{bmatrix}$, then $A/A_{11} = \begin{bmatrix} -3 & -6 \\ -6 & -11 \end{bmatrix}$, $A/A_{22} = 1 - [2 \;\; 3] \begin{bmatrix} 5 & 6 \\ 8 & 10 \end{bmatrix}^{-1} \begin{bmatrix} 4 \\ 7 \end{bmatrix} = -\frac{3}{2}$, and

$$A^{-1} = \begin{bmatrix} (-\frac{3}{2})^{-1} & -[2 \;\; 3] \begin{bmatrix} -3 & -6 \\ -6 & -11 \end{bmatrix}^{-1} \\ -\begin{bmatrix} 5 & 6 \\ 8 & 10 \end{bmatrix}^{-1} \begin{bmatrix} 4 \\ 7 \end{bmatrix} (-\frac{3}{2})^{-1} & \begin{bmatrix} -3 & -6 \\ -6 & -11 \end{bmatrix}^{-1} \end{bmatrix}$$
$$= \begin{bmatrix} -\frac{2}{3} & \frac{1}{3}[-4 \;\; 3] \\ \begin{bmatrix} -\frac{2}{3} \\ 1 \end{bmatrix} & -\frac{1}{3} \begin{bmatrix} -11 & 6 \\ 6 & -3 \end{bmatrix} \end{bmatrix} = \begin{bmatrix} -\frac{2}{3} & -\frac{4}{3} & 1 \\ -\frac{2}{3} & \frac{11}{3} & -2 \\ 1 & -2 & 1 \end{bmatrix}.$$

3. Let $A = \begin{bmatrix} 1 & 2 & 3 \\ 4 & 5 & 6 \\ 7 & 8 & 10 \end{bmatrix}$.

Then, from Fact 5,

$$(A/A_{22})^{-1} = \left(-\frac{3}{2}\right)^{-1} = 1^{-1} + 1^{-1}[2 \quad 3] \begin{bmatrix} -3 & -6 \\ -6 & -11 \end{bmatrix}^{-1} \begin{bmatrix} 4 \\ 7 \end{bmatrix} 1^{-1},$$

$$= A_{11}^{-1} + A_{11}^{-1} A_{12} (A/A_{11})^{-1} A_{21} A_{11}^{-1}.$$

10.4 Kronecker Products

Definitions:

Let $A \in F^{m \times n}$ and $B \in F^{p \times q}$. Then the **Kronecker product** (sometimes called the **tensor product**) of A and B, denoted $A \otimes B$, is the $mp \times nq$ partitioned matrix $A \otimes B = \begin{bmatrix} a_{11}B & a_{12}B & \cdots & a_{1n}B \\ a_{21}B & a_{22}B & \cdots & a_{2n}B \\ \vdots & \vdots & \ddots & \vdots \\ a_{n1}B & a_{n2}B & \cdots & a_{nn}B \end{bmatrix}$.

Let $A \in F^{m \times n}$ and let the jth column of A, namely, $A[\{1, \ldots, m\}, \{j\}]$ be denoted a_j, for $1 \leq j \leq n$. The column vector with mn components, denoted vec(A), defined by $\text{vec}(A) = \begin{bmatrix} a_1 \\ a_2 \\ \vdots \\ a_n \end{bmatrix} \in F^{mn}$, is the **vec-function** of A, i.e., vec(A) is formed by stacking the columns of A on top of each other in their natural order.

Facts:

All of the following facts except those for which a specific reference is given can be found in [LT85].

1. Let $A \in F^{m \times n}$ and $B \in F^{p \times q}$. If $a \in F$, then $a(A \otimes B) = (aA) \otimes B = A \otimes (aB)$.
2. Let $A, B \in F^{m \times n}$ and $C \in F^{p \times q}$. Then $(A + B) \otimes C = A \otimes C + B \otimes C$.
3. Let $A \in F^{m \times n}$ and $B, C \in F^{p \times q}$. Then $A \otimes (B + C) = A \otimes B + A \otimes C$.
4. Let $A \in F^{m \times n}$, $B \in F^{p \times q}$, and $C \in F^{r \times s}$. Then $A \otimes (B \otimes C) = (A \otimes B) \otimes C$.
5. Let $A \in F^{m \times n}$ and $B \in F^{p \times q}$. Then $(A \otimes B)^T = A^T \otimes B^T$.
6. [MM64] Let $A \in \mathbb{C}^{m \times n}$ and $B \in \mathbb{C}^{p \times q}$. Then $\overline{(A \otimes B)} = \overline{A} \otimes \overline{B}$.
7. [MM64] Let $A \in \mathbb{C}^{m \times n}$ and $B \in \mathbb{C}^{p \times q}$. Then $(A \otimes B)^* = A^* \otimes B^*$.
8. Let $A \in F^{m \times n}$, $B \in F^{p \times q}$, $C \in F^{n \times r}$, and $D \in F^{q \times s}$. Then $(A \otimes B)(C \otimes D) = AC \otimes BD$.
9. Let $A \in F^{m \times n}$ and $B \in F^{p \times q}$. Then $A \otimes B = (A \otimes I_p)(I_n \otimes B) = (I_m \otimes B)(A \otimes I_q)$.
10. If $A \in F^{m \times m}$ and $B \in F^{n \times n}$ are nonsingular, then $A \otimes B$ is nonsingular and $(A \otimes B)^{-1} = A^{-1} \otimes B^{-1}$.
11. Let $A_1, A_2, \cdots, A_k \in F^{m \times m}$, and $B_1, B_2, \cdots, B_k \in F^{n \times n}$. Then $(A_1 \otimes B_1)(A_2 \otimes B_2) \cdots (A_k \otimes B_k) = (A_1 A_2 \cdots A_k) \otimes (B_1 B_2 \cdots B_k)$.
12. Let $A \in F^{m \times m}$ and $B \in F^{n \times n}$. Then $(A \otimes B)^k = A^k \otimes B^k$.

13. If $A \in F^{m\times m}$ and $B \in F^{n\times n}$, then there is an $mn \times mn$ permutation matrix P so that $P^T(A\otimes B)P = B \otimes A$.
14. Let $A, B \in F^{m\times n}$. Then $\text{vec}(aA + bB) = a\,\text{vec}(A) + b\,\text{vec}(B)$, for any $a, b \in F$.
15. If $A \in F^{m\times n}$, $B \in F^{p\times q}$, and $X \in F^{n\times p}$, then $\text{vec}(AXB) = (B^T \otimes A)\text{vec}(X)$.
16. If $A \in F^{m\times m}$ and $B \in F^{n\times n}$, then $\det(A\otimes B) = (\det(A))^n(\det(B))^m$, $\text{tr}(A\otimes B) = (\text{tr}(A))(\text{tr}(B))$, and $\text{rank}(A \otimes B) = (\text{rank}(A))(\text{rank}(B))$.
17. Let $A \in F^{m\times m}$ and $B \in F^{n\times n}$, with $\sigma(A) = \{\lambda_1,\ldots,\lambda_m\}$ and $\sigma(B) = \{\mu_1,\ldots,\mu_n\}$. Then $A\otimes B \in F^{mn\times mn}$ has eigenvalues $\{\lambda_s\mu_t | 1 \le s \le m, 1 \le t \le n\}$. Moreover, if the right eigenvectors of A are denoted $\mathbf{x}_i$, and the right eigenvectors of B are denoted $\mathbf{y}_j$, so that $A\mathbf{x}_i = \lambda_i\mathbf{x}_i$ and $B\mathbf{y}_j = \mu_j\mathbf{y}_j$, then $(A \otimes B)(\mathbf{x}_i \otimes \mathbf{y}_j) = \lambda_i\mu_j(\mathbf{x}_i \otimes \mathbf{y}_j)$.
18. Let $A \in F^{m\times m}$ and $B \in F^{n\times n}$, with $\sigma(A) = \{\lambda_1,\ldots,\lambda_m\}$ and $\sigma(B) = \{\mu_1,\ldots,\mu_n\}$. Then $(I_n \otimes A) + (B \otimes I_m)$ has eigenvalues $\{\lambda_s + \mu_t | 1 \le s \le m, 1 \le t \le n\}$.
19. Let $p(x, y) \in F[x, y]$ so that $p(x, y) = \sum_{i,j=1}^k a_{ij}x^iy^j$, where $a_{ij} \in F$, $1 \le i \le k$, $1 \le j \le k$. Let $A \in F^{m\times m}$ and $B \in F^{n\times n}$. Define $p(A; B)$ to be the $mn \times mn$ matrix $p(A; B) = \sum_{i,j=1}^k a_{ij}(A^i \otimes B^j)$. If $\sigma(A) = \{\lambda_1,\ldots,\lambda_m\}$ and $\sigma(B) = \{\mu_1,\ldots,\mu_n\}$, then $\sigma(p(A; B)) = \{p(\lambda_s, \mu_t) | 1 \le s \le m, 1 \le t \le n\}$.
20. Let $A_1, A_2 \in F^{m\times m}$, $B_1, B_2 \in F^{n\times n}$. If A_1 and A_2 are similar, and B_1 and B_2 are similar, then $A_1 \otimes B_1$ is similar to $A_2 \otimes B_2$.
21. If $A \in F^{m\times n}$, $B \in F^{p\times q}$, and $X \in F^{n\times p}$, then $\text{vec}(AX) = (I_p \otimes A)\text{vec}(X)$, $\text{vec}(XB) = (B^T \otimes I_n)\text{vec}(X)$, and $\text{vec}(AX + XB) = [(I_p \otimes A) + (B^T \otimes I_n)]\text{vec}(X)$.
22. If $A \in F^{m\times n}$, $B \in F^{p\times q}$, $C \in F^{m\times q}$, and $X \in F^{n\times p}$, then the equation $AXB = C$ can be written in the form $(B^T \otimes A)\text{vec}(X) = \text{vec}(C)$.
23. Let $A \in F^{m\times m}$, $B \in F^{n\times n}$, $C \in F^{m\times n}$, and $X \in F^{m\times n}$. Then the equation $AX + XB = C$ can be written in the form $[(I_n \otimes A) + (B^T \otimes I_m)]\text{vec}(X) = \text{vec}(C)$.
24. Let $A \in \mathbb{C}^{m\times m}$ and $B \in \mathbb{C}^{n\times n}$ be Hermitian. Then $A \otimes B$ is Hermitian.
25. Let $A \in \mathbb{C}^{m\times m}$ and $B \in \mathbb{C}^{n\times n}$ be positive definite. Then $A \otimes B$ is positive definite.

Examples:

1. Let $A = \begin{bmatrix} 1 & -1 \\ 0 & 2 \end{bmatrix}$ and $B = \begin{bmatrix} 1 & 2 & 3 \\ 4 & 5 & 6 \\ 7 & 8 & 9 \end{bmatrix}$. Then $A \otimes B = \begin{bmatrix} 1 & 2 & 3 & -1 & -2 & -3 \\ 4 & 5 & 6 & -4 & -5 & -6 \\ 7 & 8 & 9 & -7 & -8 & -9 \\ 0 & 0 & 0 & 2 & 4 & 6 \\ 0 & 0 & 0 & 8 & 10 & 12 \\ 0 & 0 & 0 & 14 & 16 & 18 \end{bmatrix}$.

2. Let $A = \begin{bmatrix} 1 & -1 \\ 0 & 2 \end{bmatrix}$. Then $\text{vec}(A) = \begin{bmatrix} 1 \\ 0 \\ -1 \\ 2 \end{bmatrix}$.

3. Let $A \in F^{m\times m}$ and $B \in F^{n\times n}$. If A is upper (lower) triangular, then $A \otimes B$ is block upper (lower) triangular. If A is diagonal then $A \otimes B$ is block diagonal. If both A and B are upper (lower) triangular, then $A \otimes B$ is (upper) triangular. If both A and B are diagonal, then $A \otimes B$ is diagonal.
4. Let $A \in F^{m\times n}$ and $B \in F^{p\times q}$. If $A \otimes B = \mathbf{0}$, then $A = \mathbf{0}$ or $B = \mathbf{0}$.
5. Let $A \in F^{m\times n}$. Then $A \otimes I_n = \begin{bmatrix} a_{11}I_n & a_{12}I_n & \cdots & a_{1n}I_n \\ a_{21}I_n & a_{22}I_n & \cdots & a_{2n}I_n \\ \vdots & \vdots & \ddots & \vdots \\ a_{m1}I_n & a_{n2}I_n & \cdots & a_{mn}I_n \end{bmatrix} \in F^{mn\times n^2}$. Let $B \in F^{p\times p}$. Then $I_n \otimes B = \text{diag}(B, B,\ldots, B) \in F^{np\times np}$, and $I_m \otimes I_n = I_{mn}$.

References

[Hay68] E. Haynsworth, Determination of the Inertia of a Partitioned Matrix, *Lin. Alg. Appl.*, 1:73–81 (1968).

[HJ85] R. A. Horn and C. R. Johnson, *Matrix Analysis*, Cambridge University Press, Combridge, 1985.

[LT85] P. Lancaster and M. Tismenetsky, *The Theory of Matrices, with Applications*, 2nd ed., Academic Press, San Diego, 1985.

[MM64] M. Marcus and H. Minc, *A Survey of Matrix Theory and Matrix Inequalities*, Prindle, Weber, & Schmidt, Boston, 1964.

[Mey00] C. Meyer, *Matrix Analysis and Applied Linear Algebra, SIAM*, 2000.

[Zha99] F. Zhang, *Matrix Theory*, Springer-Verlag, New York, 1999.

Advanced Linear Algebra

11

Functions of Matrices

Nicholas J. Higham
University of Manchester

Matrix functions are used in many areas of linear algebra and arise in numerous applications in science and engineering. The most common matrix function is the matrix inverse; it is not treated specifically in this chapter, but is covered in Section 1.1 and Section 38.2. This chapter is concerned with general matrix functions as well as specific cases such as matrix square roots, trigonometric functions, and the exponential and logarithmic functions.

The specific functions just mentioned can all be defined via power series or as the solution of nonlinear systems. For example, $\cos(A) = I - A^2/2! + A^4/4! - \cdots$. However, a general theory exists from which a number of properties possessed by all matrix functions can be deduced and which suggests computational methods. This chapter treats general theory, then specific functions, and finally outlines computational methods.

11.1 General Theory

Definitions:

A function of a matrix can be defined in several ways, of which the following three are the most generally useful.

- *Jordan canonical form definition.* Let $A \in \mathbb{C}^{n\times n}$ have the Jordan canonical form $Z^{-1}AZ = J_A = \operatorname{diag}\Big(J_1(\lambda_1), J_2(\lambda_2), \ldots, J_p(\lambda_p)\Big)$, where Z is nonsingular,

$$J_k(\lambda_k) = \begin{bmatrix} \lambda_k & 1 & & \\ & \lambda_k & \ddots & \\ & & \ddots & 1 \\ & & & \lambda_k \end{bmatrix} \in \mathbb{C}^{m_k\times m_k}, \qquad (11.1)$$

and $m_1 + m_2 + \cdots + m_p = n$. Then

$$f(A) := Zf(J_A)Z^{-1} = Z\,\mathrm{diag}(f(J_k(\lambda_k)))Z^{-1}, \tag{11.2}$$

where

$$f(J_k(\lambda_k)) := \begin{bmatrix} f(\lambda_k) & f'(\lambda_k) & \cdots & \dfrac{f^{(m_k-1)}(\lambda_k)}{(m_k-1)!} \\ & f(\lambda_k) & \ddots & \vdots \\ & & \ddots & f'(\lambda_k) \\ & & & f(\lambda_k) \end{bmatrix}. \tag{11.3}$$

- *Polynomial interpolation definition.* Denote by $\lambda_1, \ldots, \lambda_s$ the distinct eigenvalues of A and let n_i be the **index** of λ_i, that is, the order of the largest Jordan block in which λ_i appears. Then $f(A) := r(A)$, where r is the unique Hermite interpolating polynomial of degree less than $\sum_{i=1}^{s} n_i$ that satisfies the interpolation conditions

$$r^{(j)}(\lambda_i) = f^{(j)}(\lambda_i), \qquad j = 0\colon n_i - 1, \quad i = 1\colon s. \tag{11.4}$$

Note that in both these definitions the derivatives in (11.4) must exist in order for $f(A)$ to be defined. The function f is said to be **defined on the spectrum of** A if all the derivatives in (11.4) exist.

- *Cauchy integral definition.*

$$f(A) := \frac{1}{2\pi i}\int_\Gamma f(z)(zI - A)^{-1}\,dz, \tag{11.5}$$

where f is analytic inside a closed contour Γ that encloses $\sigma(A)$.

When the function f is multivalued and A has a repeated eigenvalue occurring in more than one Jordan block (i.e., A is derogatory), the Jordan canonical form definition has more than one interpretation. Usually, for each occurrence of an eigenvalue in different Jordan blocks the same branch is taken for f and its derivatives. This gives a **primary matrix function.** If different branches are taken for the same eigenvalue in two different Jordan blocks, then a **nonprimary matrix function** is obtained. A nonprimary matrix function is not expressible as a polynomial in the matrix, and if such a function is obtained from the Jordan canonical form definition (11.2) then it depends on the matrix Z. In most applications it is primary matrix functions that are of interest. For the rest of this section $f(A)$ is assumed to be a primary matrix function, unless otherwise stated.

Facts:

Proofs of the facts in this section can be found in one or more of [Hig], [HJ91], or [LT85], unless otherwise stated.

1. The Jordan canonical form and polynomial interpolation definitions are equivalent. Both definitions are equivalent to the Cauchy integral definition when f is analytic.
2. $f(A)$ is a polynomial in A and the coefficients of the polynomial depend on A.
3. $f(A)$ commutes with A.
4. $f(A^T) = f(A)^T$.
5. For any nonsingular X, $f(XAX^{-1}) = Xf(A)X^{-1}$.
6. If A is diagonalizable, with $Z^{-1}AZ = D = \mathrm{diag}(d_1, d_2, \ldots, d_n)$, then $f(A) = Zf(D)Z^{-1} = Z\,\mathrm{diag}(f(d_1), f(d_2), \ldots, f(d_n))Z^{-1}$.
7. $f\Big(\mathrm{diag}(A_1, A_2, \ldots, A_m)\Big) = \mathrm{diag}(f(A_1), f(A_2), \ldots, f(A_m))$.

8. Let f and g be functions defined on the spectrum of A.
   (a) If $h(t) = f(t) + g(t)$, then $h(A) = f(A) + g(A)$.
   (b) If $h(t) = f(t)g(t)$, then $h(A) = f(A)g(A)$.
9. Let $G(u_1, \ldots, u_t)$ be a polynomial in $u_1, \ldots, u_t$ and let $f_1, \ldots, f_t$ be functions defined on the spectrum of A. If $g(\lambda) = G(f_1(\lambda), \ldots, f_t(\lambda))$ takes zero values on the spectrum of A, then $g(A) = G(f_1(A), \ldots, f_t(A)) = 0$. For example, $\sin^2(A) + \cos^2(A) = I$, $(A^{1/p})^p = A$, and $e^{iA} = \cos A + i \sin A$.
10. Suppose f has a Taylor series expansion

$$f(z) = \sum_{k=0}^{\infty} a_k(z-\alpha)^k \qquad \left(a_k = \frac{f^{(k)}(\alpha)}{k!}\right)$$

with radius of convergence r. If $A \in \mathbb{C}^{n\times n}$, then $f(A)$ is defined and is given by

$$f(A) = \sum_{k=0}^{\infty} a_k(A-\alpha I)^k$$

if and only if each of the distinct eigenvalues $\lambda_1, \ldots, \lambda_s$ of A satisfies one of the conditions:
   (a) $|\lambda_i - \alpha| < r$.
   (b) $|\lambda_i - \alpha| = r$ and the series for $f^{n_i-1}(\lambda)$, where n_i is the index of λ_i, is convergent at the point $\lambda = \lambda_i$, $i = 1{:}s$.
11. [Dav73], [Des63], [GVL96, Theorem 11.1.3]. Let $T \in \mathbb{C}^{n\times n}$ be upper triangular and suppose that f is defined on the spectrum of T. Then $F = f(T)$ is upper triangular with $f_{ii} = f(t_{ii})$ and

$$f_{ij} = \sum_{(s_0,\ldots,s_k)\in S_{ij}} t_{s_0,s_1} t_{s_1,s_2} \cdots t_{s_{k-1},s_k} f[\lambda_{s_0}, \ldots, \lambda_{s_k}],$$

where $\lambda_i = t_{ii}$, S_{ij} is the set of all strictly increasing sequences of integers that start at i and end at j, and $f[\lambda_{s_0}, \ldots, \lambda_{s_k}]$ is the kth order divided difference of f at $\lambda_{s_0}, \ldots, \lambda_{s_k}$.

Examples:

1. For $\lambda_1 \neq \lambda_2$,

$$f\left(\begin{bmatrix} \lambda_1 & \alpha \\ 0 & \lambda_2 \end{bmatrix}\right) = \begin{bmatrix} f(\lambda_1) & \alpha \dfrac{f(\lambda_2) - f(\lambda_1)}{\lambda_2 - \lambda_1} \\ 0 & f(\lambda_2) \end{bmatrix}.$$

For $\lambda_1 = \lambda_2 = \lambda$,

$$f\left(\begin{bmatrix} \lambda & \alpha \\ 0 & \lambda \end{bmatrix}\right) = \begin{bmatrix} f(\lambda) & \alpha f'(\lambda) \\ 0 & f(\lambda) \end{bmatrix}.$$

2. Compute e^A for the matrix

$$A = \begin{bmatrix} -7 & -4 & -3 \\ 10 & 6 & 4 \\ 6 & 3 & 3 \end{bmatrix}.$$

We have $A = XJ_AX^{-1}$, where $J_A = [0] \oplus \begin{bmatrix} 1 & 1 \\ 0 & 1 \end{bmatrix}$ and

$$X = \begin{bmatrix} 1 & -1 & -1 \\ -1 & 2 & 0 \\ -1 & 0 & 3 \end{bmatrix}.$$

Hence, using the Jordan canonical form definition, we have

$$e^A = Xe_A^J X^{-1} = X\left([1] \oplus \left[\begin{smallmatrix} e & e \\ 0 & e \end{smallmatrix}\right]\right)X^{-1}$$

$$= \begin{bmatrix} 1 & -1 & -1 \\ -1 & 2 & 0 \\ -1 & 0 & 3 \end{bmatrix} \begin{bmatrix} 1 & 0 & 0 \\ 0 & e & e \\ 0 & 0 & e \end{bmatrix} \begin{bmatrix} 6 & 3 & 2 \\ 2 & 2 & 1 \\ 2 & 1 & 1 \end{bmatrix}$$

$$= \begin{bmatrix} 6-7e & 3-4e & 2-3e \\ -6+10e & -3+6e & -2+4e \\ -6+6e & -3+3e & -2+3e \end{bmatrix}.$$

3. Compute $\sqrt{A}$ for the matrix in Example 2. To obtain the square root, we use the polynomial interpolation definition. The eigenvalues of A are 0 and 1, with indices 1 and 2, respectively. The unique polynomial r of degree at most 2 satisfying the interpolation conditions $r(0) = f(0)$, $r(1) = f(1), r'(1) = f'(1)$ is

$$r(t) = f(0)(t-1)^2 + t(2-t)f(1) + t(t-1)f'(1).$$

With $f(t) = t^{1/2}$, taking the positive square root, we have $r(t) = t(2-t) + t(t-1)/2$ and, therefore,

$$A^{1/2} = A(2I - A) + A(A - I)/2 = \begin{bmatrix} -6 & -3.5 & -2.5 \\ 8 & 5 & 3 \\ 6 & 3 & 3 \end{bmatrix}.$$

4. Consider the $m_k \times m_k$ Jordan block $J_k(\lambda_k)$ in (11.1). The polynomial satisfying the interpolation conditions (11.4) is

$$r(t) = f(\lambda_k) + (t-\lambda_k)f'(\lambda_k) + \frac{(t-\lambda_k)^2}{2!}f''(\lambda_k) + \cdots + \frac{(t-\lambda_k)^{m_k-1}}{(m_k-1)!}f^{(m_k-1)}(\lambda_k),$$

which, of course, is the first m_k terms of the Taylor series of f about λ_k. Hence, from the polynomial interpolation definition,

$$\begin{aligned} f(J_k(\lambda_k)) &= r(J_k(\lambda_k)) \\ &= f(\lambda_k)I + (J_k(\lambda_k) - \lambda_k I)f'(\lambda_k) + \frac{(J_k(\lambda_k) - \lambda_k I)^2}{2!}f''(\lambda_k) + \cdots \\ &\quad + \frac{(J_k(\lambda_k) - \lambda_k I)^{m_k-1}}{(m_k-1)!}f^{(m_k-1)}(\lambda_k). \end{aligned}$$

The matrix $(J_k(\lambda_k) - \lambda_k I)^j$ is zero except for 1s on the jth superdiagonal. This expression for $f(J_k(\lambda_k))$ is, therefore, equal to that in (11.3), confirming the consistency of the first two definitions of $f(A)$.

11.2 Matrix Square Root

Definitions:

Let $A \in \mathbb{C}^{n\times n}$. Any X such that $X^2 = A$ is a **square root** of A.

Facts:

Proofs of the facts in this section can be found in one or more of [Hig], [HJ91], or [LT85], unless otherwise stated.

1. If $A \in \mathbb{C}^{n\times n}$ has no eigenvalues on $\mathbb{R}_0^-$ (the closed negative real axis) then there is a unique square root X of A each of whose eigenvalues is 0 or lies in the open right half-plane, and it is a primary matrix function of A. This is the **principal square root** of A and is written $X = A^{1/2}$. If A is real then $A^{1/2}$ is real. An integral representation is

$$A^{1/2} = \frac{2}{\pi} A \int_0^{\infty} (t^2 I + A)^{-1}\, dt.$$

2. A positive (semi)definite matrix $A \in \mathbb{C}^{n\times n}$ has a unique positive (semi)definite square root. (See also Section 8.3.)
3. [CL74] A singular matrix $A \in \mathbb{C}^{n\times n}$ may or may not have a square root. A necessary and sufficient condition for A to have a square root is that in the "ascent sequence" of integers $d_1, d_2, \ldots$ defined by

$$d_i = \dim(\ker(A^i)) - \dim(\ker(A^{i-1})),$$

   no two terms are the same odd integer.
4. $A \in \mathbb{R}^{n\times n}$ has a real square root if and only if A satisfies the condition in the previous fact and A has an even number of Jordan blocks of each size for every negative eigenvalue.
5. The $n \times n$ identity matrix I_n has 2^n diagonal square roots $\operatorname{diag}(\pm 1)$. Only two of these are primary matrix functions, namely I and $-I$. Nondiagonal but symmetric nonprimary square roots of I_n include any Householder matrix $I - 2\mathbf{v}\mathbf{v}^T/(\mathbf{v}^T\mathbf{v})$ $(\mathbf{v} \neq 0)$ and the identity matrix with its columns in reverse order. Nonsymmetric square roots of I_n are easily constructed in the form XDX^{-1}, where X is nonsingular but nonorthogonal and $D = \operatorname{diag}(\pm 1) \neq \pm I$.

Examples:

1. The Jordan block $\begin{bmatrix} 0 & 1 \\ 0 & 0 \end{bmatrix}$ has no square root. The matrix

$$\begin{bmatrix} 0 & 1 & 0 \\ 0 & 0 & 0 \\ 0 & 0 & 0 \end{bmatrix}$$

   has ascent sequence $2, 1, 0, \ldots$ and so does have a square root — for example, the matrix

$$\begin{bmatrix} 0 & 0 & 1 \\ 0 & 0 & 0 \\ 0 & 1 & 0 \end{bmatrix}.$$

11.3 Matrix Exponential

Definitions:

The exponential of $A \in \mathbb{C}^{n\times n}$, written e^A or $\exp(A)$, is defined by

$$e^A = I + A + \frac{A^2}{2!} + \cdots + \frac{A^k}{k!} + \cdots.$$

Facts:

Proofs of the facts in this section can be found in one or more of [Hig], [HJ91], or [LT85], unless otherwise stated.

1. $e^{(A+B)t} = e^{At}e^{Bt}$ holds for all t if and only if $AB = BA$.
2. The differential equation in $n \times n$ matrices
$$\frac{dY}{dt} = AY, \qquad Y(0) = C, \qquad A, Y \in \mathbb{C}^{n\times n},$$
has solution $Y(t) = e^{At}C$.
3. The differential equation in $n \times n$ matrices
$$\frac{dY}{dt} = AY + YB, \qquad Y(0) = C, \qquad A, B, Y \in \mathbb{C}^{n\times n},$$
has solution $Y(t) = e^{At}Ce^{Bt}$.
4. $A \in \mathbb{C}^{n\times n}$ is unitary if and only if it can be written $A = e^{iH}$, where H is Hermitian. In this representation H can be taken to be Hermitian positive definite.
5. $A \in \mathbb{R}^{n\times n}$ is orthogonal with $\det(A) = 1$ if and only if $A = e^S$ with $S \in \mathbb{R}^{n\times n}$ skew-symmetric.

Examples:

1. Fact 5 is illustrated by the matrix
$$A = \begin{bmatrix} 0 & \alpha \\ -\alpha & 0 \end{bmatrix},$$
for which
$$e^A = \begin{bmatrix} \cos\alpha & \sin\alpha \\ -\sin\alpha & \cos\alpha \end{bmatrix}.$$

11.4 Matrix Logarithm

Definitions:

Let $A \in \mathbb{C}^{n\times n}$. Any X such that $e^X = A$ is a **logarithm** of A.

Facts:

Proofs of the facts in this section can be found in one or more of [Hig], [HJ91], or [LT85], unless otherwise stated.

1. If A has no eigenvalues on $\mathbb{R}^-$, then there is a unique logarithm X of A all of whose eigenvalues lie in the strip $\{ z : -\pi < \text{Im}(z) < \pi \}$. This is the **principal logarithm** of A and is written $X = \log A$. If A is real, then $\log A$ is real.
2. If $\rho(A) < 1$,
$$\log(I + A) = A - \frac{A^2}{2} + \frac{A^3}{3} - \frac{A^4}{4} + \cdots.$$
3. $A \in \mathbb{R}^{n\times n}$ has a real logarithm if and only if A is nonsingular and A has an even number of Jordan blocks of each size for every negative eigenvalue.
4. $\exp(\log A) = A$ holds when log is defined on the spectrum of $A \in \mathbb{C}^{n\times n}$. But $\log(\exp(A)) = A$ does not generally hold unless the spectrum of A is restricted.
5. If $A \in \mathbb{C}^{n\times n}$ is nonsingular then $\det(A) = \exp(\text{tr}(\log A))$, where $\log A$ is any logarithm of A.

Examples:

For the matrix

$$A = \begin{bmatrix} 1 & 1 & 1 & 1 \\ 0 & 1 & 2 & 3 \\ 0 & 0 & 1 & 3 \\ 0 & 0 & 0 & 1 \end{bmatrix},$$

we have

$$\log(A) = \begin{bmatrix} 0 & 1 & 0 & 0 \\ 0 & 0 & 2 & 0 \\ 0 & 0 & 0 & 3 \\ 0 & 0 & 0 & 0 \end{bmatrix}.$$

11.5 Matrix Sine and Cosine

Definitions:

The sine and cosine of $A \in \mathbb{C}^{n \times n}$ are defined by

$$\cos(A) = I - \frac{A^2}{2!} + \cdots + \frac{(-1)^k}{(2k)!} A^{2k} + \cdots,$$

$$\sin(A) = A - \frac{A^3}{3!} + \cdots + \frac{(-1)^k}{(2k+1)!} A^{2k+1} + \cdots.$$

Facts:

Proofs of the facts in this subsection can be found in one or more of [Hig], [HJ91], or [LT85], unless otherwise stated.

1. $\cos(2A) = 2\cos^2(A) - I$.
2. $\sin(2A) = 2\sin(A)\cos(A)$.
3. $\cos^2(A) + \sin^2(A) = I$.
4. The differential equation

$$\frac{d^2y}{dt^2} + Ay = 0, \qquad y(0) = y_0, \quad y'(0) = y'_0$$

has solution

$$y(t) = \cos(\sqrt{A}t)y_0 + \left(\sqrt{A}\right)^{-1} \sin(\sqrt{A}t)y'_0,$$

where $\sqrt{A}$ denotes any square root of A.

Examples:

1. For

$$A = \begin{bmatrix} 0 & i\alpha \\ i\alpha & 0 \end{bmatrix},$$

we have

$$e^A = \begin{bmatrix} \cos\alpha & i\sin\alpha \\ i\sin\alpha & \cos\alpha \end{bmatrix}.$$

2. For

$$A = \begin{bmatrix} 1 & 1 & 1 & 1 \\ 0 & -1 & -2 & -3 \\ 0 & 0 & 1 & 3 \\ 0 & 0 & 0 & -1 \end{bmatrix},$$

we have

$$\cos(A) = \cos(1)I, \qquad \sin(A) = \begin{bmatrix} \sin(1) & \sin(1) & \sin(1) & \sin(1) \\ 0 & -\sin(1) & -2\sin(1) & -3\sin(1) \\ 0 & 0 & \sin(1) & 3\sin(1) \\ 0 & 0 & 0 & -\sin(1) \end{bmatrix},$$

and $\sin^2(A) = \sin(1)^2 I$, so $\cos(A)^2 + \sin(A)^2 = I$.

11.6 Matrix Sign Function

Definitions:

If $A = ZJ_AZ^{-1} \in \mathbb{C}^{n\times n}$ is a Jordan canonical form arranged so that

$$J_A = \begin{bmatrix} J_A^{(1)} & 0 \\ 0 & J_A^{(2)} \end{bmatrix},$$

where the eigenvalues of $J_A^{(1)} \in \mathbb{C}^{p\times p}$ lie in the open left half-plane and those of $J_A^{(2)} \in \mathbb{C}^{q\times q}$ lie in the open right half-plane, with $p + q = n$, then

$$\operatorname{sign}(A) = Z \begin{bmatrix} -I_p & 0 \\ 0 & I_q \end{bmatrix} Z^{-1}.$$

Alternative formulas are

$$\operatorname{sign}(A) = A(A^2)^{-1/2}, \tag{11.6}$$

$$\operatorname{sign}(A) = \frac{2}{\pi} A \int_0^\infty (t^2 I + A^2)^{-1}\, dt.$$

If A has any pure imaginary eigenvalues, then $\operatorname{sign}(A)$ is not defined.

Facts:

Proofs of the facts in this section can be found in [Hig].

Let $S = \operatorname{sign}(A)$ be defined. Then

1. $S^2 = I$ (S is involutory).
2. S is diagonalizable with eigenvalues ± 1.
3. $SA = AS$.
4. If A is real, then S is real.
5. If A is symmetric positive definite, then $\operatorname{sign}(A) = I$.

Examples:

1. For the matrix A in Example 2 of the previous subsection we have $\operatorname{sign}(A) = A$, which follows from (11.6) and the fact that A is involutory.

11.7 Computational Methods for General Functions

Many methods have been proposed for evaluating matrix functions. Three general approaches of wide applicability are outlined here. They have in common that they do not require knowledge of Jordan structure and are suitable for computer implementation. References for this subsection are [GVL96], [Hig].

1. *Polynomial and Rational Approximations*:

Polynomial approximations

$$p_m(X) = \sum_{k=0}^{m} b_k X^k, \qquad b_k \in \mathbb{C},\ X \in \mathbb{C}^{n \times n},$$

to matrix functions can be obtained by truncating or economizing a power series representation, or by constructing a best approximation (in some norm) of a given degree. How to most efficiently evaluate a polynomial at a matrix argument is a nontrivial question. Possibilities include Horner's method, explicit computation of the powers of the matrix, and a method of Paterson and Stockmeyer [GVL96, Sec. 11.2.4], [PS73], which is a combination of these two methods that requires fewer matrix multiplications.

Rational approximations $r_{mk}(X) = p_m(X)q_k(X)^{-1}$ are also widely used, particularly those arising from Padé approximation, which produces rationals matching as many terms of the Taylor series of the function at the origin as possible. The evaluation of rationals at matrix arguments needs careful consideration in order to find the best compromise between speed and accuracy. The main possibilities are

- Evaluating the numerator and denominator polynomials and then solving a multiple right-hand side linear system.
- Evaluating a continued fraction representation (in either top-down or bottom-up order).
- Evaluating a partial fraction representation.

Since polynomials and rationals are typically accurate over a limited range of matrices, practical methods involve a reduction stage prior to evaluating the polynomial or rational.

2. *Factorization Methods*:

Many methods are based on the property $f(XAX^{-1}) = Xf(A)X^{-1}$. If X can be found such that $B = XAX^{-1}$ has the property that $f(B)$ is easily evaluated, then an obvious method results. When A is diagonalizable, B can be taken to be diagonal, and evaluation of $f(B)$ is trivial. In finite precision arithmetic, though, this approach is reliable only if X is well conditioned, that is, if the condition number $\kappa(X) = \|X\|\|X^{-1}\|$ is not too large. Ideally, X will be unitary, so that in the 2-norm $\kappa_2(X) = 1$. For Hermitian A, or more generally normal A, the spectral decomposition $A = QDQ^*$ with Q unitary and D diagonal is always possible, and if this decomposition can be computed then the formula $f(A) = Qf(D)Q^*$ provides an excellent way of computing $f(A)$.

For general A, if X is restricted to be unitary, then the furthest that A can be reduced is to Schur form: $A = QTQ^*$, where Q is unitary and T upper triangular. This decomposition is computed by the QR algorithm. Computing a function of a triangular matrix is an interesting problem. While Fact 11 of section 11.1 gives an explicit formula for $F = f(T)$, the formula is not practically viable due to its exponential cost in n. Much more efficient is a recurrence of Parlett [Par76]. This is derived by starting with the observation that since F is representable as a polynomial in T, F is upper triangular, with diagonal elements $f(t_{ii})$. The elements in the strict upper triangle are determined by solving the equation $FT = TF$. Parlett's recurrence is:

Algorithm 1. Parlett's recurrence.

$f_{ii} = f(t_{ii}), i = 1{:}n$
for $j = 2{:}n$
  for $i = j-1{:}-1{:}1$

$$f_{ij} = t_{ij}\frac{f_{ii} - f_{jj}}{t_{ii} - t_{jj}} + \left(\sum_{k=i+1}^{j-1} f_{ik}t_{kj} - t_{ik}f_{kj}\right) \Big/ (t_{ii} - t_{jj})$$

  end
end

This recurrence can be evaluated in $2n^3/3$ operations. The recurrence breaks down when $t_{ii} = t_{jj}$ for some $i \neq j$. In this case, T can be regarded as a block matrix $T = (T_{ij})$, with square diagonal blocks, possibly of different sizes. T can be reordered so that no two diagonal blocks have an eigenvalue in common; reordering means applying a unitary similarity transformation to permute the diagonal elements whilst preserving triangularity. Then a block form of the recurrence can be employed. This requires the evaluation of the diagonal blocks $F_{ii} = f(T_{ii})$, where T_{ii} will typically be of small dimension. A general way to obtain F_{ii} is via a Taylor series. The use of the block Parlett recurrence in combination with a Schur decomposition represents the state of the art in evaluation of $f(A)$ for general functions [DH03].

3. *Iteration Methods*:

Several matrix functions f can be computed by iteration:

$$X_{k+1} = g(X_k), \qquad X_0 = A, \tag{11.7}$$

where, for reasons of computational cost, g is usually a polynomial or a rational function. Such an iteration might converge for all A for which f is defined, or just for a subset of such A. A standard means of deriving matrix iterations is to apply Newton's method to an algebraic equation satisfied by $f(A)$. The iterations most used in practice are quadratically convergent, but iterations with higher orders of convergence are known.

4. *Contour Integration*:

The Cauchy integral definition (11.5) provides a way to compute or approximate $f(A)$ via contour integration. While not suitable as a practical method for all functions or all matrices, this approach can be effective when numerical integration is done over a suitable contour using the repeated trapezium rule, whose high accuracy properties for periodic functions integrated over a whole period are beneficial [DH05], [TW05].

11.8 Computational Methods for Specific Functions

Some methods specialized to particular functions are now outlined. References for this section are [GVL96], [Hig].

1. *Matrix Exponential*:

A large number of methods have been proposed for the matrix exponential, many of them of pedagogic interest only or of dubious numerical stability. Some of the more computationally useful methods are surveyed in [MVL03]. Probably the best general-purpose method is the scaling and squaring method. In this method an integral power of 2, $\sigma = 2^s$ say, is chosen so that A/σ has norm not too far from 1. The exponential of the scaled matrix is approximated by an $[m/m]$ Padé approximant, $e^{A/2^s} \approx r_{mm}(A/2^s)$, and then s repeated squarings recover an approximation to e^A: $e^A \approx r_{mm}(A/2^s)^{2^s}$. Symmetries in the Padé approximant permit an efficient evaluation of $r_{mm}(A)$. The scaling and squaring method was originally developed in [MVL78] and [War77], and it is the method employed by MATLAB's expm function. How best to choose σ and m is described in [Hig05].

2. *Matrix Logarithm*:

The (principal) matrix logarithm can be computed using an inverse scaling and squaring method based on the identity $\log A = 2^k \log A^{1/2^k}$, where A is assumed to have no eigenvalues on $\mathbb{R}^-$. Square roots are taken to make $\| A^{1/2^k} - I \|$ small enough that an $[m/m]$ Padé approximant approximates $\log A^{1/2^k}$ sufficiently accurately, for some suitable m. Then $\log A$ is recovered by multiplying by 2^k. To reduce the cost of computing the square roots and evaluating the Padé approximant, a Schur decomposition can be computed initially so that the method works with a triangular matrix. For details, see [CHKL01], [Hig01], or [KL89, App. A].

3. *Matrix Cosine and Sine*:

A method analogous to the scaling and squaring method for the exponential is the standard method for computing the matrix cosine. The idea is again to scale A to have norm not too far from 1 and then compute a Padé approximant. The difference is that the scaling is undone by repeated use of the double-angle formula $\cos(2A) = 2\cos^2 A - I$, rather than by repeated squaring. The sine function can be obtained as $\sin(A) = \cos(A - \frac{\pi}{2} I)$. (See [SB80], [HS03], [HH05].)

4. *Matrix Square Root*:

The most numerically reliable way to compute matrix square roots is via the Schur decomposition, $A = QTQ^*$ [BH83]. Rather than use the Parlett recurrence, a square root U of the upper triangular factor T can be computed by directly solving the equation $U^2 = T$. The choices of sign in the diagonal of U, $u_{ii} = \sqrt{t_{ii}}$, determine which square root is obtained. When A is real, the real Schur decomposition can be used to compute real square roots entirely in real arithmetic [Hig87].

Various iterations exist for computing the principal square root when A has no eigenvalues on $\mathbb{R}^-$. The basic Newton iteration,

$$X_{k+1} = \frac{1}{2}(X_k + X_k^{-1}A), \qquad X_0 = A, \tag{11.8}$$

is quadratically convergent, but is numerically unstable unless A is extremely well conditioned and its use is not recommended [Hig86]. Stable alternatives include the Denman–Beavers iteration [DB76]

$$\begin{aligned} X_{k+1} &= \frac{1}{2}\left(X_k + Y_k^{-1}\right), & X_0 &= A,\\ Y_{k+1} &= \frac{1}{2}\left(Y_k + X_k^{-1}\right), & Y_0 &= I, \end{aligned}$$

for which $\lim_{k\to\infty} X_k = A^{1/2}$ and $\lim_{k\to\infty} Y_k = A^{-1/2}$, and the Meini iteration [Mei04]

$$\begin{aligned} Y_{k+1} &= -Y_k Z_k^{-1} Y_k, & Y_0 &= I - A,\\ Z_{k+1} &= Z_k + 2Y_{k+1}, & Z_0 &= 2(I + A), \end{aligned}$$

for which $Y_k \to 0$ and $Z_k \to 4A^{1/2}$. Both of these iterations are mathematically equivalent to (11.8) and, hence, are quadratically convergent.

An iteration not involving matrix inverses is the Schulz iteration

$$\begin{aligned} Y_{k+1} &= \tfrac{1}{2} Y_k(3I - Z_k Y_k), & Y_0 &= A,\\ Z_{k+1} &= \tfrac{1}{2}(3I - Z_k Y_k) Z_k, & Z_0 &= I, \end{aligned}$$

for which $Y_k \to A^{1/2}$ and $Z_k \to A^{-1/2}$ quadratically provided that $\| \mathrm{diag}(A - I, A - I) \| < 1$, where the norm is any consistent matrix norm [Hig97].

5. *Matrix Sign Function:*

The standard method for computing the matrix sign function is the Newton iteration

$$X_{k+1} = \frac{1}{2}(X_k + X_k^{-1}), \quad X_0 = A,$$

which converges quadratically to sign(A), provided A has no pure imaginary eigenvalues. In practice, a scaled iteration

$$X_{k+1} = \frac{1}{2}(\mu_k X_k + \mu_k^{-1} X_k^{-1}), \quad X_0 = A$$

is used, where the scale parameters μ_k are chosen to reduce the number of iterations needed to enter the regime where asymptotic quadratic convergence sets in. (See [Bye87], [KL92].)

The Newton–Schulz iteration

$$X_{k+1} = \frac{1}{2} X_k (3I - X_k^2), \quad X_0 = A,$$

involves no matrix inverses, but convergence is guaranteed only for $\|I - A^2\| < 1$.

A Padé family of iterations

$$X_{k+1} = X_k \, p_{\ell m}\left(1 - X_k^2\right) q_{\ell m}\left(1 - X_k^2\right)^{-1}, \quad X_0 = A$$

is obtained in [KL91], where $p_{\ell m}(\xi)/q_{\ell m}(\xi)$ is the $[\ell/m]$ Padé approximant to $(1-\xi)^{-1/2}$. The iteration is globally convergent to sign(A) for $\ell = m - 1$ and $\ell = m$, and for $\ell \geq m - 1$ is convergent when $\|I - A^2\| < 1$, with order of convergence $\ell + m + 1$ in all cases.

References

[BH83] Åke Björck and Sven Hammarling. A Schur method for the square root of a matrix. *Lin. Alg. Appl.*, 52/53:127–140, 1983.

[Bye87] Ralph Byers. Solving the algebraic Riccati equation with the matrix sign function. *Lin. Alg. Appl.*, 85:267–279, 1987.

[CHKL01] Sheung Hun Cheng, Nicholas J. Higham, Charles S. Kenney, and Alan J. Laub. Approximating the logarithm of a matrix to specified accuracy. *SIAM J. Matrix Anal. Appl.*, 22(4):1112–1125, 2001.

[CL74] G.W. Cross and P. Lancaster. Square roots of complex matrices. *Lin. Multilin. Alg.*, 1:289–293, 1974.

[Dav73] Chandler Davis. Explicit functional calculus. *Lin. Alg. Appl.*, 6:193–199, 1973.

[DB76] Eugene D. Denman and Alex N. Beavers, Jr. The matrix sign function and computations in systems. *Appl. Math. Comput.*, 2:63–94, 1976.

[Des63] Jean Descloux. Bounds for the spectral norm of functions of matrices. *Numer. Math.*, 15:185–190, 1963.

[DH03] Philip I. Davies and Nicholas J. Higham. A Schur–Parlett algorithm for computing matrix functions. *SIAM J. Matrix Anal. Appl.*, 25(2):464–485, 2003.

[DH05] Philip I. Davies and Nicholas J. Higham. Computing $f(A)b$ for matrix functions f. In Artan Boriçi, Andreas Frommer, Báalint Joó, Anthony Kennedy, and Brian Pendleton, Eds., *QCD and Numerical Analysis III*, Vol. 47 of *Lecture Notes in Computational Science and Engineering*, pp. 15–24. Springer-Verlag, Berlin, 2005.

[GVL96] Gene H. Golub and Charles F. Van Loan. *Matrix Computations, 3rd ed.*, Johns Hopkins University Press, Baltimore, MD, 1996.

[HH05] Gareth I. Hargreaves and Nicholas J. Higham. Efficient algorithms for the matrix cosine and sine. *Numerical Algorithms*, 40:383–400, 2005.

[Hig] Nicholas J. Higham. *Functions of a Matrix: Theory and Computation.* (Book in preparation.)

[Hig86] Nicholas J. Higham. Newton's method for the matrix square root. *Math. Comp.*, 46(174):537–549, 1986.

[Hig87] Nicholas J. Higham. Computing real square roots of a real matrix. *Lin. Alg. Appl.*, 88/89:405–430, 1987.

[Hig97] Nicholas J. Higham. Stable iterations for the matrix square root. *Num. Algor.*,15(2):227–242, 1997.

[Hig01] Nicholas J. Higham. Evaluating Padé approximants of the matrix logarithm. *SIAM J. Matrix Anal. Appl.*, 22(4):1126–1135, 2001.

[Hig05] Nicholas J. Higham. The scaling and squaring method for the matrix exponential revisited. *SIAM J. Matrix Anal. Appl.*, 26(4):1179–1193, 2005.

[HJ91] Roger A. Horn and Charles R. Johnson. *Topics in Matrix Analysis.* Cambridge University Press, Cambridge, 1991.

[HS03] Nicholas J. Higham and Matthew I. Smith. Computing the matrix cosine. *Num. Algor.*, 34:13–26, 2003.

[KL89] Charles S. Kenney and Alan J. Laub. Condition estimates for matrix functions. *SIAM J. Matrix Anal. Appl.*, 10(2):191–209, 1989.

[KL91] Charles S. Kenney and Alan J. Laub. Rational iterative methods for the matrix sign function. *SIAM J. Matrix Anal. Appl.*, 12(2):273–291, 1991.

[KL92] Charles S. Kenney and Alan J. Laub. On scaling Newton's method for polar decomposition and the matrix sign function. *SIAM J. Matrix Anal. Appl.*, 13(3):688–706, 1992.

[LT85] Peter Lancaster and Miron Tismenetsky. *The Theory of Matrices, 2nd ed.*, Academic Press, London, 1985.

[Mei04] Beatrice Meini. The matrix square root from a new functional perspective: theoretical results and computational issues. *SIAM J. Matrix Anal. Appl.*, 26(2):362–376, 2004.

[MVL78] Cleve B. Moler and Charles F. Van Loan. Nineteen dubious ways to compute the exponential of a matrix. *SIAM Rev.*, 20(4):801–836, 1978.

[MVL03] Cleve B. Moler and Charles F. Van Loan. Nineteen dubious ways to compute the exponential of a matrix, twenty-five years later. *SIAM Rev.*, 45(1):3–49, 2003.

[Par76] B. N. Parlett. A recurrence among the elements of functions of triangular matrices. *Lin. Alg. Appl.*, 14:117–121, 1976.

[PS73] Michael S. Paterson and Larry J. Stockmeyer. On the number of nonscalar multiplications necessary to evaluate polynomials. *SIAM J. Comput.*, 2(1):60–66, 1973.

[SB80] Steven M. Serbin and Sybil A. Blalock. An algorithm for computing the matrix cosine. *SIAM J. Sci. Statist. Comput.*, 1(2):198–204, 1980.

[TW05] L. N. Trefethen and J. A. C. Weideman. The fast trapezoid rule in scientific computing. Paper in preparation, 2005.

[War77] Robert C. Ward. Numerical computation of the matrix exponential with accuracy estimate. *SIAM J. Numer. Anal.*, 14(4):600–610, 1977.

12

Quadratic, Bilinear, and Sesquilinear Forms

Raphael Loewy
Technion

Bilinear forms are maps defined on $V \times V$, where V is a vector space, and are linear with respect to each of their variables. There are some similarities between bilinear forms and inner products that are discussed in Chapter 5. Basic properties of bilinear forms, symmetric bilinear forms, and alternating bilinear forms are discussed. The latter two types of forms satisfy additional symmetry conditions.

Quadratic forms are obtained from symmetric bilinear forms by equating the two variables. They are widely used in many areas. A canonical representation of a quadratic form is given when the underlying field is $\mathbb{R}$ or $\mathbb{C}$.

When the field is the complex numbers, it is standard to expect the form to be conjugate linear rather than linear in the second variable; such a form is called sesquilinear. The role of a symmetric bilinear form is played by a Hermitian sesquilinear form. The idea of a sesquilinear form can be generalized to an arbitrary automorphism, encompassing both bilinear and sesquilinear forms as φ-sesquilinear forms, where φ is an automorphism of the field.

Quadratic, bilinear, and φ-sesquilinear forms have applications to classical matrix groups. (See Chapter 67 for more information.)

12.1 Bilinear Forms

It is assumed throughout this section that V is a finite dimensional vector space over a field F.

Definitions:

A **bilinear form** on V is a map f from $V \times V$ into F which satisfies

$$f(a\mathbf{u}_1 + b\mathbf{u}_2, \mathbf{v}) = af(\mathbf{u}_1, \mathbf{v}) + bf(\mathbf{u}_2, \mathbf{v}), \quad \mathbf{u}_1, \mathbf{u}_2, \mathbf{v} \in V, \ a, b \in F,$$

and

$$f(\mathbf{u}, a\mathbf{v}_1 + b\mathbf{v}_2) = af(\mathbf{u}, \mathbf{v}_1) + bf(\mathbf{u}, \mathbf{v}_2), \quad \mathbf{u}, \mathbf{v}_1, \mathbf{v}_2 \in V, \ a, b \in F.$$

The space of all bilinear forms on V is denoted $B(V, V, F)$.

Let $\mathcal{B} = (\mathbf{w_1}, \mathbf{w_2}, \ldots, \mathbf{w_n})$ be an ordered basis of V and let $f \in B(V, V, F)$. **The matrix representing f relative to $\mathcal{B}$** is the matrix $A = [a_{ij}] \in F^{n\times n}$ such that $a_{ij} = f(\mathbf{w_i}, \mathbf{w_j})$.

The **rank** of $f \in B(V, V, F)$, $\text{rank}(f)$, is $\text{rank}(A)$, where A is a matrix representing f relative to an arbitrary ordered basis of V.

$f \in B(V, V, F)$ is **nondegenerate** if its rank is equal to dim V, and **degenerate** if it is not nondegenerate.

Let $A, B \in F^{n\times n}$. B is **congruent** to A if there exists an invertible $P \in F^{n\times n}$ such that $B = P^T AP$.

Let $f, g \in B(V, V, F)$. g is **equivalent** to f if there exists an ordered basis $\mathcal{B}$ of V such that the matrix of g relative to $\mathcal{B}$ is congruent to the matrix of f relative to $\mathcal{B}$.

Let T be a linear operator on V and let $f \in B(V, V, F)$. T **preserves** f if $f(T\mathbf{u}, T\mathbf{v}) = f(\mathbf{u}, \mathbf{v})$ for all $\mathbf{u}, \mathbf{v} \in V$.

Facts:

Let $f \in B(V, V, F)$. The following facts can be found in [HK71, Chap. 10].

1. f is a linear functional in each of its variables when the other variable is held fixed.
2. Let $\mathcal{B} = (\mathbf{w_1}, \mathbf{w_2}, \ldots, \mathbf{w_n})$ be an ordered basis of V and let

$$\mathbf{u} = \sum_{i=1}^{n} a_i \mathbf{w_i}, \qquad \mathbf{v} = \sum_{i=1}^{n} b_i \mathbf{w_i}.$$

Then,

$$f(\mathbf{u}, \mathbf{v}) = \sum_{i=1}^{n}\sum_{j=1}^{n} a_i b_j f(\mathbf{w_i}, \mathbf{w_j}).$$

3. Let A denote the matrix representing f relative to $\mathcal{B}$, and let $[\mathbf{u}]_{\mathcal{B}}$ and $[\mathbf{v}]_{\mathcal{B}}$ be the vectors in F^n that are the coordinate vectors of $\mathbf{u}$ and $\mathbf{v}$, respectively, with respect to $\mathcal{B}$. Then $f(\mathbf{u}, \mathbf{v}) = [\mathbf{u}]_{\mathcal{B}}^T A[\mathbf{v}]_{\mathcal{B}}$.
4. Let $\mathcal{B}$ and $\mathcal{B}'$ be ordered bases of V, and P be the matrix whose columns are the $\mathcal{B}$-coordinates of vectors in $\mathcal{B}'$. Let $f \in B(V, V, F)$. Let A and B denote the matrices representing f relative to $\mathcal{B}$ and $\mathcal{B}'$. Then

$$B = P^T AP.$$

5. The concept of rank of f, as given, is well defined.
6. The set $L = \{\mathbf{v} \in V : f(\mathbf{u}, \mathbf{v}) = 0 \text{ for all } \mathbf{u} \in V\}$ is a subspace of V and $\text{rank}(f) = \dim V - \dim L$. In particular, f is nondegenerate if and only if $L = \{0\}$.
7. Suppose that $\dim V = n$. The space $B(V, V, F)$ is a vector space over F under the obvious addition of two bilinear forms and multiplication of a bilinear form by a scalar. Moreover, $B(V, V, F)$ is isomorphic to $F^{n\times n}$.
8. Congruence is an equivalence relation on $F^{n\times n}$.
9. Let $f \in B(V, V, F)$ be nondegenerate. Then the set of all linear operators on V, which preserve f, is a group under the operation of composition.

Examples:

1. Let $A \in F^{n\times n}$. The map $f : F^n \times F^n \to F$ defined by

$$f(\mathbf{u}, \mathbf{v}) = \mathbf{u}^T A\mathbf{v} = \sum_{i=1}^{n}\sum_{j=1}^{n} a_{ij} u_i v_j, \quad \mathbf{u}, \mathbf{v} \in F^n,$$

is a bilinear form. Since $f(\mathbf{e_i}, \mathbf{e_j}) = a_{ij}$, $\quad i, j = 1, 2, \ldots, n$, f is represented in the standard basis of F^n by A. It follows that $\text{rank}(f) = \text{rank}(A)$, and f is nondegenerate if and only if A is invertible.

2. Let $C \in F^{m\times m}$ and rank$(C) = k$. The map $f : F^{m\times n} \times F^{m\times n} \to F$ defined by $f(A, B) = \text{tr}(A^T C B)$ is a bilinear form. This follows immediately from the basic properties of the trace function. To compute rank(f), let L be defined as in Fact 6, that is, $L = \{B \in F^{m\times n} : \text{tr}(A^T C B) = 0 \text{ for all } A \in F^{m\times n}\}$. It follows that $L = \{B \in F^{m\times n} : CB = 0\}$, which implies that $\dim L = n(m-k)$. Hence, rank$(f) = mn - n(m-k) = kn$. In particular, f is nondegenerate if and only if C is invertible.
3. Let $\mathbb{R}[x; n]$ denote the space of all real polynomials of the form $\sum_{i=0}^{n} a_i x^i$. Then $f(p(x), q(x)) = p(0)q(0) + p(1)q(1) + p(2)q(2)$ is a bilinear form on $\mathbb{R}[x; n]$. It is nondegenerate if $n = 2$ and degenerate if $n \geqslant 3$.

12.2 Symmetric Bilinear Forms

It is assumed throughout this section that V is a finite dimensional vector space over a field F.

Definitions:

Let $f \in B(V, V, F)$. Then f is **symmetric** if $f(\mathbf{u}, \mathbf{v}) = f(\mathbf{v}, \mathbf{u})$ for all $\mathbf{u}, \mathbf{v} \in V$.

Let f be a symmetric bilinear form on V, and let $\mathbf{u}, \mathbf{v} \in V$; $\mathbf{u}$ and $\mathbf{v}$ are **orthogonal with respect to** f if $f(\mathbf{u}, \mathbf{v}) = 0$.

Let f be a symmetric bilinear form on V. The **quadratic form corresponding to** f is the map $g : V \to F$ defined by $g(\mathbf{v}) = f(\mathbf{v}, \mathbf{v})$, $\mathbf{v} \in V$.

A symmetric bilinear form f on a real vector space V is **positive semidefinite** (**positive definite**) if $f(\mathbf{v}, \mathbf{v}) \geqslant 0$ for all $\mathbf{v} \in V$ ($f(\mathbf{v}, \mathbf{v}) > 0$ for all $0 \neq \mathbf{v} \in V$).

f is **negative semidefinite** (**negative definite**) if $-f$ is positive semidefinite (positive definite).

The **signature** of a real symmetric matrix A is the integer $\pi - \nu$, where (π, ν, δ) is the inertia of A. (See Section 8.3.)

The **signature** of a real symmetric bilinear form is the signature of a matrix representing the form relative to some basis.

Facts:

Additional facts about real symmetric matrices can be found in Chapter 8. Except where another reference is provided, the following facts can be found in [Coh89, Chap. 8], [HJ85, Chap. 4], or [HK71, Chap. 10].

1. A positive definite bilinear form is nondegenerate.
2. An inner product on a real vector space is a positive definite symmetric bilinear form. Conversely, a positive definite symmetric bilinear form on a real vector space is an inner product.
3. Let $\mathcal{B}$ be an ordered basis of V and let $f \in B(V, V, F)$. Let A be the matrix representing f relative to $\mathcal{B}$. Then f is symmetric if and only if A is a symmetric matrix, that is, $A = A^T$.
4. Let f be a symmetric bilinear form on V and let g be the quadratic form corresponding to f. Suppose that the characteristic of F is not 2. Then f can be recovered from g:
$$f(\mathbf{u}, \mathbf{v}) = \tfrac{1}{2}[g(\mathbf{u}+\mathbf{v}) - g(\mathbf{u}) - g(\mathbf{v})] \quad \text{for all } \mathbf{u}, \mathbf{v} \in V.$$
5. Let f be a symmetric bilinear form on V and suppose that the characteristic of F is not 2. Then there exists an ordered basis $\mathcal{B}$ of V such that the matrix representing f relative to it is diagonal; i.e., if $A \in F^{n\times n}$ is a symmetric matrix, then A is congruent to a diagonal matrix.
6. Suppose that V is a complex vector space and f is a symmetric bilinear form on V. Let $r = \text{rank}(f)$. Then there is an ordered basis $\mathcal{B}$ of V such that the matrix representing f relative to $\mathcal{B}$ is $I_r \oplus \mathbf{0}$. In matrix language, this fact states that if $A \in \mathbb{C}^{n\times n}$ is symmetric with rank$(A) = r$, then it is congruent to $I_r \oplus \mathbf{0}$.
7. The only invariant of $n \times n$ complex symmetric matrices under congruence is the rank.
8. Two complex $n \times n$ symmetric matrices are congruent if and only if they have the same rank.

9. (*Sylvester's law of inertia for symmetric bilinear forms*) Suppose that V is a real vector space and f is a symmetric bilinear form on V. Then there is an ordered basis $\mathcal{B}$ of V such that the matrix representing f relative to it has the form $I_\pi \oplus -I_\nu \oplus \mathbf{0}_\delta$. Moreover, π, ν, and δ do not depend on the choice of $\mathcal{B}$, but only on f.
10. (*Sylvester's law of inertia for matrices*) If $A \in \mathbb{R}^{n\times n}$ is symmetric, then A is congruent to the diagonal matrix $D = I_\pi \oplus -I_\nu \oplus \mathbf{0}_\delta$, where $(\pi, \nu, \delta) = \text{in}(A)$.
11. There are exactly two invariants of $n \times n$ real symmetric matrices under congruence, namely the rank and the signature.
12. Two real $n \times n$ symmetric matrices are congruent if and only if they have the same rank and the same signature.
13. The signature of a real symmetric bilinear form is well defined.
14. Two real symmetric bilinear forms are equivalent if and only if they have the same rank and the same signature.
15. [Hes68] Let $n \geqslant 3$ and let $A, B \in \mathbb{R}^{n\times n}$ be symmetric. Suppose that $\mathbf{x} \in \mathbb{R}^n$, $\mathbf{x}^T A\mathbf{x} = \mathbf{x}^T B\mathbf{x} = 0 \Rightarrow \mathbf{x} = 0$. Then $\exists$ $\mathbf{a}, \mathbf{b} \in \mathbb{R}$ such that $\mathbf{a}A + \mathbf{b}B$ is positive definite.
16. The group of linear operators preserving the form $f(\mathbf{u}, \mathbf{v}) = \sum_{i=1}^n u_i v_i$ on $\mathbb{R}^n$ is the real n-dimensional orthogonal group, while the group preserving the same form on $\mathbb{C}^n$ is the complex n-dimensional orthogonal group.

Examples:

1. Consider Example 1 in section 12.1. The map f is a symmetric bilinear form if and only if $A = A^T$. The quadratic form g corresponding to f is given by

$$g(\mathbf{u}) = \sum_{i=1}^{n}\sum_{j=1}^{n} a_{ij}u_i u_j, \quad \mathbf{u} \in F^n.$$

2. Consider Example 2 in section 12.1. The map f is a symmetric bilinear form if and only if $C = C^T$.
3. The symmetric bilinear form $f_\mathbf{a}$ on $\mathbb{R}^2$ given by

$$f_a(\mathbf{u}, \mathbf{v}) = u_1v_1 - 2u_1v_2 - 2u_2v_1 + au_2v_2, \quad \mathbf{u}, \mathbf{v} \in \mathbb{R}^2, \quad a \in \mathbb{R} \text{ is a parameter},$$

is an inner product on $\mathbb{R}^2$ if and only if $a > 4$.
4. Since we consider in this article only finite dimensional vector spaces, let V be any finite dimensional subspace of $C[0, 1]$, the space of all real valued, continuous functions on $[0, 1]$. Then the map $f : V \times V \to \mathbb{R}$ defined by

$$f(\mathbf{u}, \mathbf{v}) = \int_0^1 t^3 u(t)v(t)dt, \quad \mathbf{u}, \mathbf{v} \in V,$$

is a symmetric bilinear form on V.

Applications:

1. Conic sections: Consider the set of points (x_1, x_2) in $\mathbb{R}^2$, which satisfy the equation

$$ax_1^2 + bx_1x_2 + cx_2^2 + dx_1 + ex_2 + f = 0,$$

where $a, b, c, d, e, f \in \mathbb{R}$. The solution set is a conic section, namely an ellipse, hyperbola, parabola, or a degenerate form of those. The analysis of this equation depends heavily on the quadratic form $ax_1^2 + bx_1x_2 + cx_2^2$, which is represented in the standard basis of $\mathbb{R}^2$ by $A = \begin{bmatrix} a & b/2 \\ b/2 & c \end{bmatrix}$. If the solution of the quadratic equation above represents a nondegenerate conic section, then its type is determined by the sign of $4ac - b^2$. More precisely, the conic is an ellipse, hyperbola, or parabola if $4ac - b^2$ is positive, negative, or zero, respectively.

2. Theory of small oscillations: Suppose a mechanical system undergoes small oscillations about an equilibrium position. Let $x_1, x_2, \ldots, x_n$ denote the coordinates of the system, and let $\mathbf{x} = (x_1, x_2, \ldots, x_n)^T$. Then the kinetic energy of the system is given by a quadratic form (in the velocities $\dot{x}_1, \dot{x}_2, \ldots, \dot{x}_n$) $\frac{1}{2}\dot{\mathbf{x}}^T A\dot{\mathbf{x}}$, where A is a positive definite matrix. If $\mathbf{x} = 0$ is the equilibrium position, then the potential energy of the system is given by another quadratic form $\frac{1}{2}\mathbf{x}^T B\mathbf{x}$, where $B = B^T$. The equations of motion are $A\ddot{\mathbf{x}} + B\mathbf{x} = 0$. It is known that A and B can be simultaneously diagonalized, that is, there exists an invertible $P \in \mathbb{R}^{n\times n}$ such that $P^T AP$ and $P^T BP$ are diagonal matrices. This can be used to obtain the solution of the system.

12.3 Alternating Bilinear Forms

It is assumed throughout this section that V is a finite dimensional vector space over a field F.

Definitions:

Let $f \in B(V, V, F)$. Then f is **alternating** if $f(\mathbf{v}, \mathbf{v}) = 0$ for all $\mathbf{v} \in V$. f is **antisymmetric** if $f(\mathbf{u}, \mathbf{v}) = -f(\mathbf{v}, \mathbf{u})$ for all $\mathbf{u}, \mathbf{v} \in V$.

Let $A \in F^{n\times n}$. Then A is **alternating** if $a_{ii} = 0,\ i = 1, 2, \ldots, n$ and $a_{ji} = -a_{ij},\ 1 \leqslant i < j \leqslant n$.

Facts:

The following facts can be found in [Coh89, Chap. 8], [HK71, Chap. 10], or [Lan99, Chap. 15].

1. Let $f \in B(V, V, F)$ be alternating. Then f is antisymmetric because for all $\mathbf{u}, \mathbf{v} \in V$,
$$f(\mathbf{u}, \mathbf{v}) + f(\mathbf{v}, \mathbf{u}) = f(\mathbf{u}+\mathbf{v}, \mathbf{u}+\mathbf{v}) - f(\mathbf{u}, \mathbf{u}) - f(\mathbf{v}, \mathbf{v}) = 0.$$
The converse is true if the characteristic of F is not 2.
2. Let $A \in F^{n\times n}$ be an alternating matrix. Then $A^T = -A$. The converse is true if the characteristic of F is not 2.
3. Let $\mathcal{B}$ be an ordered basis of V and let $f \in B(V, V, F)$. Let A be the matrix representing f relative to $\mathcal{B}$. Then f is alternating if and only if A is an alternating matrix.
4. Let f be an alternating bilinear form on V and let $r = \text{rank}(f)$. Then r is even and there exists an ordered basis $\mathcal{B}$ of V such that the matrix representing f relative to it has the form
$$\underbrace{\begin{bmatrix} 0 & 1 \\ -1 & 0 \end{bmatrix} \oplus \begin{bmatrix} 0 & 1 \\ -1 & 0 \end{bmatrix} \oplus \cdots \oplus \begin{bmatrix} 0 & 1 \\ -1 & 0 \end{bmatrix}}_{r/2 - \text{times}} \oplus \mathbf{0}.$$
There is an ordered basis $\mathcal{B}_1$ where f is represented by the matrix $\begin{bmatrix} \mathbf{0} & I_{r/2} \\ -I_{r/2} & \mathbf{0} \end{bmatrix} \oplus \mathbf{0}$.
5. Let $f \in B(V, V, F)$ and suppose that the characteristic of F is not 2. Define:
$f_1 : V \times V \to F$ by $f_1(\mathbf{u}, \mathbf{v}) = \frac{1}{2}[f(\mathbf{u}, \mathbf{v}) + f(\mathbf{v}, \mathbf{u})]$, $\mathbf{u}, \mathbf{v} \in V$,
$f_2 : V \times V \to F$ by $f_2(\mathbf{u}, \mathbf{v}) = \frac{1}{2}[f(\mathbf{u}, \mathbf{v}) - f(\mathbf{v}, \mathbf{u})]$, $\mathbf{u}, \mathbf{v} \in V$.
Then f_1 (f_2) is a symmetric (alternating) bilinear form on V, and $f = f_1 + f_2$. Moreover, this representation of f as a sum of a symmetric and an alternating bilinear form is unique.
6. Let $A \in F^{n\times n}$ be an alternating matrix and suppose that A is invertible. Then n is even and A is congruent to the matrix $\begin{bmatrix} \mathbf{0} & I_{n/2} \\ -I_{n/2} & \mathbf{0} \end{bmatrix}$, so $\det(A)$ is a square in F. There exists a polynomial in $n(n-1)/2$ variables, called the **Pfaffian**, such that $\det(A) = a^2$, where $a \in F$ is obtained by substituting into the Pfaffian the entries of A above the main diagonal for the indeterminates.

7. Let f be an alternating nondegenerate bilinear form on V. Then $\dim V = 2m$ for some positive integer m. The group of all linear operators on V that preserve f is the symplectic group.

Examples:

1. Consider Example 1 in section 12.1. The map f is alternating if and only if the matrix A is an alternating matrix.
2. Consider Example 2 in section 12.1. The map f is alternating if and only if C is an alternating matrix.
3. Let $C \in F^{n \times n}$. Define $f : F^{n \times n} \to F^{n \times n}$ by $f(A, B) = \text{tr}(ACB - BCA)$. Then f is alternating.

12.4 φ-Sesquilinear Forms

This section generalizes Section 12.1, and is consequently very similar. This generalization is required by applications to matrix groups (see Chapter 67), but for most purposes such generality is not required, and the simpler discussion of bilinear forms in Section 12.1 is preferred. It is assumed throughout this section that V is a finite dimensional vector space over a field F and φ is an automorphism of F.

Definitions:

A **φ-sesquilinear form** on V is a map $f : V \times V \to F$, which is linear as a function in the first variable and φ-semilinear in the second, i.e.,

$$f(a\mathbf{u_1} + b\mathbf{u_2}, \mathbf{v}) = af(\mathbf{u_1}, \mathbf{v}) + bf(\mathbf{u_2}, \mathbf{v}), \quad \mathbf{u_1}, \mathbf{u_2}, \mathbf{v} \in V, \quad a, b \in F,$$

and

$$f(\mathbf{u}, a\mathbf{v_1} + b\mathbf{v_2}) = \varphi(a)f(\mathbf{u}, \mathbf{v_1}) + \varphi(b)f(\mathbf{u}, \mathbf{v_2}), \quad \mathbf{u}, \mathbf{v_1}, \mathbf{v_2} \in V, \quad a, b \in F.$$

In the case $F = \mathbb{C}$ and φ is complex conjugation, a φ-sesquilinear form is called a **sesquilinear form**. The space of all φ-sesquilinear forms on V is denoted $B(V, V, F, \varphi)$.

Let $\mathcal{B} = (\mathbf{w_1}, \mathbf{w_2}, \ldots, \mathbf{w_n})$ be an ordered basis of V and let $f \in B(V, V, F, \varphi)$. The **matrix representing f relative to $\mathcal{B}$** is the matrix $A = [a_{ij}] \in F^{n \times n}$ such that $a_{ij} = f(\mathbf{w_i}, \mathbf{w_j})$.

The **rank** of $f \in B(V, V, F, \varphi)$, rank$(f)$, is rank$(A)$, where A is a matrix representing f relative to an arbitrary ordered basis of V.

$f \in B(V, V, F, \varphi)$ is **nondegenerate** if its rank is equal to $\dim V$, and **degenerate** if it is not nondegenerate.

Let $A = [a_{ij}] \in F^{n \times n}$. $\varphi(A)$ is the $n \times n$ matrix whose i, j-entry is $\varphi(a_{ij})$.

Let $A, B \in F^{n \times n}$. B is **φ-congruent** to A if there exists an invertible $P \in F^{n \times n}$ such that $B = P^T A \varphi(P)$.

Let $f, g \in B(V, V, F, \varphi)$. g is **φ-equivalent** to f if there exists an ordered basis $\mathcal{B}$ of V such that the matrix of g relative to $\mathcal{B}$ is φ-congruent to the matrix of f relative to $\mathcal{B}$.

Let T be a linear operator on V and let $f \in B(V, V, F, \varphi)$. T **preserves** f if $f(T\mathbf{u}, T\mathbf{v}) = f(\mathbf{u}, \mathbf{v})$ for all $\mathbf{u}, \mathbf{v} \in V$.

Facts:

Let $f \in B(V, V, F, \varphi)$. The following facts can be obtained by obvious generalizations of the proofs of the corresponding facts in section 12.1; see that section for references.

1. A bilinear form is a φ-sesquilinear form with the automorphism being the identity map.
2. Let $\mathcal{B} = (\mathbf{w_1}, \mathbf{w_2}, \ldots, \mathbf{w_n})$ be an ordered basis of V and let

$$\mathbf{u} = \sum_{i=1}^{n} a_i \mathbf{w_i}, \quad \mathbf{v} = \sum_{i=1}^{n} b_i \mathbf{w_i}.$$

Then,

$$f(\mathbf{u},\mathbf{v}) = \sum_{i=1}^{n}\sum_{j=1}^{n} a_i\varphi(b_j) f(\mathbf{w_i},\mathbf{w_j}).$$

3. Let A denote the matrix representing the φ-sesquilinear f relative to $\mathcal{B}$, and let $[\mathbf{u}]_{\mathcal{B}}$ and $[\mathbf{v}]_{\mathcal{B}}$ be the vectors in F^n, which are the coordinate vectors of **u** and **v**, respectively, with respect to $\mathcal{B}$. Then $f(\mathbf{u},\mathbf{v}) = [\mathbf{u}]_{\mathcal{B}}^T A\varphi([\mathbf{v}]_{\mathcal{B}})$.
4. Let $\mathcal{B}$ and $\mathcal{B}'$ be ordered bases of V, and P be the matrix whose columns are the $\mathcal{B}$-coordinates of vectors in $\mathcal{B}'$. Let $f \in B(V,V,F,\varphi)$. Let A and B denote the matrices representing f relative to $\mathcal{B}$ and $\mathcal{B}'$. Then
$$B = P^T A\varphi(P).$$
5. The concept of rank of f, as given, is well defined.
6. The set $L = \{\mathbf{v} \in V : f(\mathbf{u},\mathbf{v}) = 0 \text{ for all } \mathbf{u} \in V\}$ is a subspace of V and $\text{rank}(f) = \dim V - \dim L$. In particular, f is nondegenerate if and only if $L = \{0\}$.
7. Suppose that $\dim V = n$. The space $B(V,V,F,\varphi)$ is a vector space over F under the obvious addition of two φ-sesquilinear forms and multiplication of a φ-sesquilinear form by a scalar. Moreover, $B(V,V,F,\varphi)$ is isomorphic to $F^{n\times n}$.
8. φ-Congruence is an equivalence relation on $F^{n\times n}$.
9. Let $f \in B(V,V,F,\varphi)$ be nondegenerate. Then the set of all linear operators on V which preserve f is a group under the operation of composition.

Examples:

1. Let $F = \mathbb{Q}(\sqrt{5}) = \{a + b\sqrt{5} : a,b \in \mathbb{Q}\}$ and $\varphi(a+b\sqrt{5}) = a - b\sqrt{5}$. Define the φ-sesquilinear form f on F^2 by $f(\mathbf{u},\mathbf{v}) = \mathbf{u}^T\varphi(\mathbf{v})$. $f([1+\sqrt{5},3]^T,[-2\sqrt{5},-1+\sqrt{5}]^T) = (1+\sqrt{5})(2\sqrt{5}) + 3(-1-\sqrt{5}) = 7-\sqrt{5}$.

   The matrix of f with respect to the standard basis is the identity matrix, rank $f = 2$, and f is nondegenerate.
2. Let $A \in F^{n\times n}$. The map $f : F^n \times F^n \to F$ defined by
$$f(\mathbf{u},\mathbf{v}) = \mathbf{u}^T A\varphi(\mathbf{v}) = \sum_{i=1}^{n}\sum_{j=1}^{n} a_{ij}u_i\varphi(v_j), \quad \mathbf{u},\mathbf{v} \in F^n,$$
is a φ-sesquilinear form. Since $f(\mathbf{e_i},\mathbf{e_j}) = a_{ij}, \quad i,j = 1,2,\ldots,n,$ f is represented in the standard basis of F^n by A. It follows that $\text{rank}(f) = \text{rank}(A)$, and f is nondegenerate if and only if A is invertible.

12.5 Hermitian Forms

This section closely resembles the results related to symmetric bilinear forms on real vector spaces. We assume here that V is a finite dimensional complex vector space.

Definitions:

A **Hermitian form** on V is a map $f : V \times V \to \mathbb{C}$, which satisfies

$$f(a\mathbf{u_1} + b\mathbf{u_2},\mathbf{v}) = af(\mathbf{u_1},\mathbf{v}) + bf(\mathbf{u_2},\mathbf{v}), \quad \mathbf{u},\mathbf{v} \in V, \quad a,b \in \mathbb{C},$$

and

$$f(\mathbf{v},\mathbf{u}) = \overline{f(\mathbf{u},\mathbf{v})}, \quad \mathbf{u},\mathbf{v} \in V.$$

A Hermitian form f on V is **positive semidefinite** (**positive definite**) if $f(\mathbf{v},\mathbf{v}) \geqslant 0$ for all $\mathbf{v} \in V$ ($f(\mathbf{v},\mathbf{v}) > 0$ for all $0 \neq \mathbf{v} \in V$).

f is **negative semidefinite** (**negative definite**) if $-f$ is positive semidefinite (positive definite).

The **signature** of a Hermitian matrix A is the integer $\pi - \nu$, where (π, ν, δ) is the inertia of A. (See Section 8.3.)

The **signature** of a Hermitian form is the signature of a matrix representing the form.

Let $A, B \in \mathbb{C}^{n\times n}$. B is ***congruent** to A if there exists an invertible $S \in \mathbb{C}^{n\times n}$ such that $B = S^*AS$ (where S^* denotes the Hermitian adjoint of S).

Let f, g be Hermitian forms on a finite dimensional complex vector space V. g is ***equivalent** to f if there exists an ordered basis $\mathcal{B}$ of V such that the matrix of g relative to $\mathcal{B}$ is *congruent to the matrix of f relative to $\mathcal{B}$.

Facts:

Except where another reference is provided, the following facts can be found in [Coh89, Chap. 8], [HJ85, Chap. 4], or [Lan99, Chap. 15]. Let f be a Hermitian form on V.

1. A Hermitian form is sesquilinear.
2. A positive definite Hermitian form is nondegenerate.
3. f is a linear functional in the first variable and conjugate linear in the second variable, that is,

$$f(\mathbf{u}, a\mathbf{v_1} + b\mathbf{v_2}) = \bar{a} f(\mathbf{u},\mathbf{v_1}) + \bar{b} f(\mathbf{u},\mathbf{v_2}).$$

4. $f(\mathbf{v},\mathbf{v}) \in \mathbb{R}$ for all $\mathbf{v} \in V$.
5. An inner product on a complex vector space is a positive definite Hermitian form. Conversely, a positive definite Hermitian form on a complex vector space is an inner product.
6. (*Polarization formula*)

$$\begin{aligned} 4f(\mathbf{u},\mathbf{v}) = {} & f(\mathbf{u}+\mathbf{v},\mathbf{u}+\mathbf{v}) - f(\mathbf{u}-\mathbf{v},\mathbf{u}-\mathbf{v}) + \\ & + i f(\mathbf{u}+i\mathbf{v},\mathbf{u}+i\mathbf{v}) - i f(\mathbf{u}-i\mathbf{v},\mathbf{u}-i\mathbf{v}). \end{aligned}$$

7. Let $\mathcal{B} = (\mathbf{w_1}, \mathbf{w_2}, \dots, \mathbf{w_n})$ be an ordered basis of V and let

$$\mathbf{u} = \sum_{i=1}^{n} a_i \mathbf{w_i}, \quad \mathbf{v} = \sum_{i=1}^{n} b_i \mathbf{w_i}.$$

Then

$$f(\mathbf{u},\mathbf{v}) = \sum_{i=1}^{n}\sum_{j=1}^{n} a_i \bar{b}_j f(\mathbf{w_i}, \mathbf{w_j}).$$

8. Let A denote the matrix representing f relative to the basis $\mathcal{B}$. Then

$$f(\mathbf{u},\mathbf{v}) = [\mathbf{u}]_{\mathcal{B}}^T A [\bar{\mathbf{v}}]_{\mathcal{B}}.$$

9. The matrix representing a Hermitian form f relative to any basis of V is a Hermitian matrix.
10. Let A, B be matrices that represent f relative to bases $\mathcal{B}$ and $\mathcal{B}'$ of V, respectively. Then B is *congruent to A.
11. (*Sylvester's law of inertia for Hermitian forms*, cf. 12.2) There exists an ordered basis $\mathcal{B}$ of V such that the matrix representing f relative to it has the form

$$I_\pi \oplus -I_\nu \oplus \mathbf{0}_\delta.$$

Moreover, π, ν, and δ depend only on f and not on the choice of $\mathcal{B}$.

12. (*Sylvester's law of inertia for Hermitian matrices*, cf. 12.2) If $A \in \mathbb{C}^{n\times n}$ is a Hermitian matrix, then A is *congruent to the diagonal matrix $D = I_\pi \oplus -I_\nu \oplus \mathbf{0}_\delta$, where $(\pi, \nu, \delta) = \text{in}(A)$.

13. There are exactly two invariants of $n \times n$ Hermitian matrices under *congruence, namely the rank and the signature.
14. Two Hermitian $n \times n$ matrices are *congruent if and only if they have the same rank and the same signature.
15. The signature of a Hermitian form is well-defined.
16. Two Hermitian forms are *equivalent if and only if they have the same rank and the same signature.
17. [HJ91, Theorem 1.3.5] Let $A, B \in \mathbb{C}^{n\times n}$ be Hermitian matrices. Suppose that $\mathbf{x} \in \mathbb{C}^n$, $\mathbf{x}^* A\mathbf{x} = \mathbf{x}^* B\mathbf{x} = 0 \Rightarrow \mathbf{x} = 0$. Then $\exists\ \mathbf{a}, \mathbf{b} \in \mathbb{R}$ such that $\mathbf{a}A + \mathbf{b}B$ is positive definite. This fact can be obtained from [HJ91], where it is stated in a slightly different form, using the decomposition of every square, complex matrix as a sum of a Hermitian matrix and a skew-Hermitian matrix.
18. The group of linear operators preserving the Hermitian form $f(u, v) = \sum_{i=1}^{n} u_i \bar{v}_i$ on $\mathbb{C}^n$ is the n-dimensional unitary group.

Examples:

1. Let $A \in \mathbb{C}^{n\times n}$ be a Hermitian matrix. The map $f : \mathbb{C}^n \times \mathbb{C}^n \to \mathbb{C}$ defined by $f(\mathbf{u}, \mathbf{v}) = \sum_{i=1}^{n} \sum_{j=1}^{n} a_{ij} u_i \bar{v}_j$ is a Hermitian form on $\mathbb{C}^n$.
2. Let $\psi_1, \psi_2, \ldots, \psi_k$ be linear functionals on V, and let $a_1, a_2, \ldots, a_k \in \mathbb{R}$. Then the map $f : V \times V \to \mathbb{C}$ defined by $f(\mathbf{u}, \mathbf{v}) = \sum_{i=1}^{k} a_i \psi_i(\mathbf{u}) \overline{\psi_i(\mathbf{v})}$ is a Hermitian form on V.
3. Let $H \in \mathbb{C}^{n\times n}$ be a Hermitian matrix.
   The map $f : \mathbb{C}^{n\times n} \times \mathbb{C}^{n\times n} \to \mathbb{C}$ defined by $f(A, B) = \mathrm{tr}(AHB^*)$ is a Hermitian form.

References

[Coh89] P. M. Cohn. *Algebra*, 2nd ed., Vol. 1, John Wiley & Sons, New York, 1989.
[Hes68] M. R. Hestenes. Pairs of quadratic forms. *Lin. Alg. Appl.*, 1:397–407, 1968.
[HJ85] R. A. Horn and C. R. Johnson. *Matrix Analysis*, Cambridge, University Press, Cambridge, 1985.
[HJ91] R. A. Horn and C. R. Johnson. *Topics in Matrix Analysis*, Cambridge University Press, Cambridge, New York 1991.
[HK71] K. H. Hoffman and R. Kunze. *Linear Algebra*, 2nd ed., Prentice-Hall, Upper Saddle River, NJ, 1971.
[Lan99] S. Lang. *Algebra*, 3rd ed., Addison-Wesley Publishing, Reading, MA, 1999.

13

Multilinear Algebra

José A. Dias da Silva
Universidade de Lisboa

Armando Machado
Universidade de Lisboa

13.1 Multilinear Maps

Unless otherwise stated, within this section V, U, and W as well as these letters with subscripts, superscripts, or accents, are finite dimensional vector spaces over a field F of characteristic zero.

Definitions:

A map φ from $V_1 \times \cdots \times V_m$ into U is a **multilinear map** (**m-linear map**) if it is linear on each coordinate, i.e., for every $\mathbf{v}_i, \mathbf{v}_i' \in V_i$, $i = 1, \ldots, m$ and for every $a \in F$ the following conditions hold:

(a) $\varphi(\mathbf{v}_1, \ldots, \mathbf{v}_i + \mathbf{v}_i', \ldots, \mathbf{v}_m) = \varphi(\mathbf{v}_1, \ldots, \mathbf{v}_i, \ldots, \mathbf{v}_m) + \varphi(\mathbf{v}_1, \ldots, \mathbf{v}_i', \ldots, \mathbf{v}_m)$;

(b) $\varphi(\mathbf{v}_1, \ldots, a\mathbf{v}_i, \ldots, \mathbf{v}_m) = a\varphi(\mathbf{v}_1, \ldots, \mathbf{v}_i, \ldots, \mathbf{v}_m)$.

The 2-linear maps and 3-linear maps are also called **bilinear** and **trilinear** maps, respectively.

If $U = F$ then a multilinear map into U is called a **multilinear form**.

The set of multilinear maps from $V_1 \times \cdots \times V_m$ into U, together with the operations defined as follows, is denoted $L(V_1, \ldots, V_m; U)$. For m-linear maps φ, ψ, and $a \in F$,

$$(\psi + \varphi)(v_1, \ldots, v_m) = \psi(v_1, \ldots, v_m) + \varphi(v_1, \ldots, v_m),$$
$$(a\varphi)(v_1, \ldots, v_m) = a\varphi(v_1, \ldots, v_m).$$

Let $(\mathbf{b}_{i1}, \ldots, \mathbf{b}_{in_i})$ be an ordered basis of V_i, $i = 1, \ldots, m$. The set of sequences $(j_1, \ldots, j_m)$, $1 \leq j_i \leq n_i$, $i = 1, \ldots, m$, will be identified with the set $\Gamma(n_1, \ldots, n_m)$ of maps α from $\{1, \ldots, m\}$ into $\mathbb{N}$ satisfying $1 \leq \alpha(i) \leq n_i$, $i = 1, \ldots, m$.

For $\alpha \in \Gamma(n_1, \ldots, n_m)$, the m-tuple of basis vectors $(\mathbf{b}_{1\alpha(1)}, \ldots, \mathbf{b}_{m,\alpha(m)})$ is denoted by $\mathbf{b}_\alpha$.

Unless otherwise stated $\Gamma(n_1, \ldots, n_m)$ is considered ordered by the lexicographic order. When there is no risk of confusion, Γ is used instead of $\Gamma(n_1, \ldots, n_m)$.

Let p, q be positive integers. If φ is an $(p+q)$-linear map from $W_1 \times \cdots \times W_p \times V_1 \times \cdots \times V_q$ into U, then for each choice of $\mathbf{w}_i$ in W_i, $i = 1, \ldots, p$, the map

$$(\mathbf{v}_1, \ldots, \mathbf{v}_q) \longmapsto \varphi(\mathbf{w}_1, \ldots, \mathbf{w}_p, \mathbf{v}_1, \ldots, \mathbf{v}_q),$$

from $V_1 \times \cdots \times V_q$ into U, is denoted $\varphi_{\mathbf{w}_1,\ldots,\mathbf{w}_p}$, i.e.

$$\varphi_{\mathbf{w}_1,\ldots,\mathbf{w}_p}(\mathbf{v}_1, \ldots, \mathbf{v}_q) = \varphi(\mathbf{w}_1, \ldots, \mathbf{w}_p, \mathbf{v}_1, \ldots, \mathbf{v}_q).$$

Let η be a linear map from U into U' and θ_i a linear map from V_i' into V_i, $i = 1, \ldots, m$. If $(\mathbf{v}_1, \ldots, \mathbf{v}_m) \mapsto \varphi(\mathbf{v}_1, \ldots, \mathbf{v}_m)$ is a multilinear map from $V_1 \times \cdots \times V_m$ into U, $L(\theta_1, \ldots, \theta_m; \eta)(\varphi)$ denotes the map from from $V_1' \times \cdots \times V_m'$ into U', defined by

$$(\mathbf{v}_1', \ldots, \mathbf{v}_m') \mapsto \eta(\varphi(\theta_1(\mathbf{v}_1'), \ldots, \theta_m(\mathbf{v}_m'))).$$

Facts:

The following facts can be found in [Mar73, Chap. 1] and in [Mer97, Chap. 5].

1. If φ is a multilinear map, then $\varphi(\mathbf{v}_1, \ldots, 0, \ldots, \mathbf{v}_m) = 0$.
2. The set $L(V_1, \ldots, V_m; U)$ is a vector space over F.
3. If φ is an $m-$linear map from $V_1 \times \cdots \times V_m$ into U, then for every integer p, $1 \leq p < m$, and $\mathbf{v}_i \in V_i$, $1 \leq i \leq p$, the map $\varphi_{\mathbf{v}_1,\ldots,\mathbf{v}_p}$ is an $(m-p)$-linear map.
4. Under the same assumptions than in (3.) the map $(\mathbf{v}_1, \ldots, \mathbf{v}_p) \mapsto \varphi_{\mathbf{v}_1,\ldots,\mathbf{v}_p}$ from $V_1 \times \cdots \times V_p$ into $L(V_{p+1}, \ldots, V_m; U)$, is p-linear. A linear isomorphism from $L(V_1, \ldots, V_p, V_{p+1}, \ldots, V_m; U)$ into $L(V_1, \ldots, V_p; L(V_{p+1}, \ldots, V_m; U))$ arises through this construction.
5. Let η be a linear map from U into U' and θ_i a linear map from V_i' into V_i, $i = 1, \ldots, m$. The map $L(\theta_1, \ldots, \theta_m; \eta)$ from $L(V_1, \ldots, V_m; U)$ into $L(V_1', \ldots, V_m'; U')$ is a linear map. When $m = 1$, and $U = U' = F$, then $L(\theta_1, I)$ is the dual or adjoint linear map θ_1^* from V_1^* into $V_1'^*$.
6. $|\Gamma(n_1, \ldots, n_m)| = \prod_{i=1}^m n_i$ where $|\ |$ denotes cardinality.
7. Let $(\mathbf{y}_\alpha)_{\alpha\in\Gamma}$ be a family of vectors of U. Then, there exists a unique m-linear map φ from $V_1 \times \cdots \times V_m$ into U satisfying $\varphi(\mathbf{b}_\alpha) = \mathbf{y}_\alpha$, for every $\alpha \in \Gamma$.
8. If $(\mathbf{u}_1, \ldots, \mathbf{u}_n)$ is a basis of U, then $(\varphi_{i,\alpha} : \alpha \in \Gamma, i = 1, \ldots, m)$ is a basis of $L(V_1, \ldots, V_m; U)$, where $\varphi_{i,\alpha}$ is characterized by the conditions $\varphi_{i,\alpha}(\mathbf{b}_\beta) = \delta_{\alpha,\beta}\mathbf{u}_i$. Moreover, if φ is an m-linear map from $V_1 \times \cdots \times V_m$ into U such that for each $\alpha \in \Gamma$,

   $$\varphi(\mathbf{b}_\alpha) = \sum_{i=1}^n a_{i,\alpha}\mathbf{u}_i,$$

   then

   $$\varphi = \sum_{\alpha,i} a_{i,\alpha}\varphi_{i,\alpha}.$$

Examples:

1. The map from F^m into F, $(a_1, \ldots, a_m) \rightarrow \prod_{i=1}^m a_i$, is an m-linear map.
2. Let V be a vector space over F. The map $(a, \mathbf{v}) \mapsto a\mathbf{v}$ from $F \times V$ into V is a bilinear map.
3. The map from $F^m \times F^m$ into F, $((a_1, \ldots, a_m), (b_1, \ldots, b_m)) \longmapsto \sum a_i b_i$, is bilinear.
4. Let U, V, and W be vector spaces over F. The map $(\theta, \eta) \mapsto \theta\eta$ from $L(V, W) \times L(U, V)$ into $L(U, W)$, given by composition, is bilinear.
5. The multiplication of matrices, $(A, B) \mapsto AB$, from $F^{m\times n} \times F^{n\times p}$ into $F^{m\times p}$, is bilinear. Observe that this example is the matrix counterpart of the previous one.

6. Let V and W be vector spaces over F. The evaluation map, from $L(V, W) \times V$ into W,

$$(\theta, \mathbf{v}) \longmapsto \theta(\mathbf{v}),$$

is bilinear.

7. The map

$$((a_{11}, a_{21}, \ldots, a_{m1}), \ldots, (a_{1m}, a_{2m}, \ldots, a_{mm})) \to \det([a_{ij}])$$

from the Cartesian product of m copies of F^m into F is m-linear.

13.2 Tensor Products

Definitions:

Let $V_1, \ldots, V_m, P$ be vector spaces over F. Let $\nu : V_1 \times \cdots \times V_m \longmapsto P$ be a multilinear map. The pair (ν, P) is called a **tensor product** of $V_1, \ldots, V_m$, or P is said to be **a tensor product of** $V_1, \ldots, V_m$ **with tensor multiplication** ν, if the following condition is satisfied:

Universal factorization property
If φ is a multilinear map from $V_1 \times \cdots \times V_m$ into the vector space U, then there exists a unique linear map, h, from P into U, that makes the following diagram commutative:

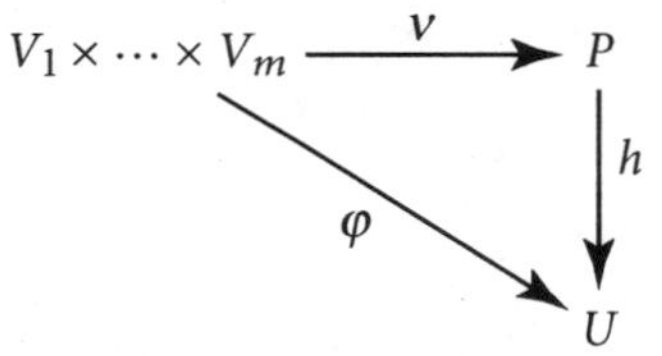

i.e., $h\nu = \varphi$.

If P is a tensor product of $V_1, \ldots, V_m$, with tensor multiplication ν, then P is denoted by $V_1 \otimes \cdots \otimes V_m$ and $\nu(\mathbf{v}_1, \ldots, \mathbf{v}_m)$ is denoted by $\mathbf{v}_1 \otimes \cdots \otimes \mathbf{v}_m$ and is called the **tensor product of the vectors** $\mathbf{v}_1, \ldots, \mathbf{v}_m$.

The elements of $V_1 \otimes \cdots \otimes V_m$ are called **tensors.** The tensors that are the tensor product of m vectors are called **decomposable tensors.**

When $V_1 = \cdots = V_m = V$, the vector space $V_1 \otimes \cdots \otimes V_m$ is called the mth **tensor power** of V and is denoted by $\bigotimes^m V$. It is convenient to define $\bigotimes^0 V = F$ and assume that 1 is the unique decomposable tensor of $\bigotimes^0 V$. When we consider simultaneously different models of tensor product, sometimes we use alternative forms to denote the tensor multiplication like $\otimes'$, $\tilde{\otimes}$, or $\hat{\otimes}$ to emphasize these different choices.

Within this section, $V_1, \ldots, V_m$ are finite dimensional vector spaces over F and $(\mathbf{b}_{i1}, \ldots, \mathbf{b}_{in_i})$ denotes a basis of V_i, $i = 1, \ldots, m$. When V is a vector space and $\mathbf{x}_1, \ldots, \mathbf{x}_k \in V$, Span$(\{\mathbf{x}_1, \ldots, \mathbf{x}_k\})$ denotes the subspace of V spanned by these vectors.

Facts:

The following facts can be found in [Mar73, Chap. 1] and in [Mer97, Chap. 5].

1. If $V_1 \otimes \cdots \otimes V_m$ and $V_1 \otimes' \cdots \otimes' V_m$ are two tensor products of $V_1, \ldots, V_m$, then the unique linear map h from $V_1 \otimes \cdots \otimes V_m$ into $V_1 \otimes' \cdots \otimes' V_m$ satisfying

$$h(\mathbf{v}_1 \otimes \cdots \otimes \mathbf{v}_m) = \mathbf{v}_1 \otimes' \cdots \otimes' \mathbf{v}_m$$

is an isomorphism.

2. If $(\nu(\mathbf{b}_\alpha))_{\alpha\in\Gamma(n_1,\dots,n_m)}$ is a basis of P, then the pair (ν, P) is a tensor product of $V_1,\dots,V_m$. This is often the most effective way to identify a model for the tensor product of vector spaces. It also implies the existence of a tensor product.
3. If P is the tensor product of $V_1,\dots,V_m$ with tensor multiplication ν, and $h : P \longmapsto Q$ is a linear isomorphism, then $(h\nu, Q)$ is a tensor product of $V_1,\dots,V_m$.
4. When $m = 1$, it makes sense to speak of a tensor product of one vector space V and V itself is used as a model for that tensor product with the identity as tensor multiplication, i.e., $\bigotimes^1 V = V$.
5. **Bilinear version of the universal property** — Given a multilinear map from $V_1 \times \cdots \times V_k \times U_1 \times \cdots \times U_m$ into W,

$$(\mathbf{v}_1,\dots,\mathbf{v}_k,\mathbf{u}_1,\dots,\mathbf{u}_m) \mapsto \varphi(\mathbf{v}_1,\dots,\mathbf{v}_k,\mathbf{u}_1,\dots,\mathbf{u}_m),$$

there exists a unique bilinear map χ from $(V_1 \otimes \cdots \otimes V_k) \times (U_1 \otimes \cdots \otimes U_m)$ into W satisfying

$$\chi(\mathbf{v}_1 \otimes \cdots \otimes \mathbf{v}_k, \mathbf{u}_1 \otimes \cdots \otimes \mathbf{u}_m) = \varphi(\mathbf{v}_1,\dots,\mathbf{v}_k,\mathbf{u}_1,\dots,\mathbf{u}_m),$$
$$\mathbf{v}_i \in V_i\ \mathbf{u}_j \in U_j,\ i = 1,\dots,k,\ j = 1,\dots,m.$$

6. Let $a \in F$ and $\mathbf{v}_i, \mathbf{v}_i' \in V_i$, $i = 1,\dots,m$. As the consequence of the multilinearity of $\otimes$, the following equalities hold:
   (a) $\mathbf{v}_1 \otimes \cdots \otimes (\mathbf{v}_i + \mathbf{v}_i') \otimes \cdots \otimes \mathbf{v}_m$
   $= \mathbf{v}_1 \otimes \cdots \otimes \mathbf{v}_i \otimes \cdots \otimes \mathbf{v}_m + \mathbf{v}_1 \otimes \cdots \otimes \mathbf{v}_i' \otimes \cdots \otimes \mathbf{v}_m,$
   (b) $a(\mathbf{v}_1 \otimes \cdots \otimes \mathbf{v}_m) = (a\mathbf{v}_1) \otimes \cdots \otimes \mathbf{v}_m = \cdots = \mathbf{v}_1 \otimes \cdots \otimes (a\mathbf{v}_m),$
   (c) $\mathbf{v}_1 \otimes \cdots \otimes 0 \otimes \cdots \otimes \mathbf{v}_m = 0.$
7. If one of the vector spaces V_i is zero, then $V_1 \otimes \cdots \otimes V_m = \{0\}$.
8. Write $\mathbf{b}_\alpha^\otimes$ to mean

$$\mathbf{b}_\alpha^\otimes := \mathbf{b}_{1\alpha(1)} \otimes \cdots \otimes \mathbf{b}_{m\alpha(m)}.$$

Then

$$(\mathbf{b}_\alpha^\otimes)_{\alpha\in\Gamma}$$

is a basis of $V_1 \otimes \cdots \otimes V_m$. This basis is said to be induced by the bases $(\mathbf{b}_{i1},\dots,\mathbf{b}_{in_i})$, $i = 1,\dots,m$.
9. The decomposable tensors span the tensor product $V_1 \otimes \cdots \otimes V_m$. Furthermore, if the set C_i spans V_i, $i = 1,\dots,m$, then the set $\{\mathbf{v}_1 \otimes \cdots \otimes \mathbf{v}_m : \mathbf{v}_i \in C_i,\ i = 1,\dots,m\}$ spans $V_1 \otimes \cdots \otimes V_m$.
10. $\dim(V_1 \otimes \cdots \otimes V_m) = \prod_{i=1}^m \dim(V_i)$.
11. The tensor product is commutative,

$$V_1 \otimes V_2 = V_2 \otimes V_1,$$

meaning that if $V_1 \otimes V_2$ is a tensor product of V_1 and V_2, then $V_1 \otimes V_2$ is also a tensor product of V_2 and V_1 with tensor multiplication $(\mathbf{v}_2,\mathbf{v}_1) \mapsto \mathbf{v}_1 \otimes \mathbf{v}_2$.

In general, with a similar meaning, for any $\sigma \in S_m$,

$$V_1 \otimes \cdots \otimes V_m = V_{\sigma(1)} \otimes \cdots \otimes V_{\sigma(m)}.$$

12. The tensor product is associative,

$$(V_1 \otimes V_2) \otimes V_3 = V_1 \otimes (V_2 \otimes V_3) = V_1 \otimes V_2 \otimes V_3,$$

meaning that:

(a) A tensor product $V_1 \otimes V_2 \otimes V_3$ is also a tensor product of $V_1 \otimes V_2$ and V_3 (respectively of V_1 and $V_2 \otimes V_3$) with tensor multiplication defined (uniquely by Fact 5 above) for $\mathbf{v}_i \in V_i, i = 1, 2, 3$, by $(\mathbf{v}_1 \otimes \mathbf{v}_2) \otimes \mathbf{v}_3 = \mathbf{v}_1 \otimes \mathbf{v}_2 \otimes \mathbf{v}_3$ (respectively by $\mathbf{v}_1 \otimes (\mathbf{v}_2 \otimes \mathbf{v}_3) = \mathbf{v}_1 \otimes \mathbf{v}_2 \otimes \mathbf{v}_3$).

(b) And, $V_1 \otimes V_2) \otimes V_3$ (respectively $V_1 \otimes (V_2 \otimes V_3)$ is a tensor product of V_1, V_2, V_3 with tensor multiplication defined by $\mathbf{v}_1 \otimes \mathbf{v}_2 \otimes \mathbf{v}_3 = (\mathbf{v}_1 \otimes \mathbf{v}_2) \otimes \mathbf{v}_3, \mathbf{v}_i \in V_i, i = 1, 2, 3$ (respectively $\mathbf{v}_1 \otimes \mathbf{v}_2 \otimes \mathbf{v}_3 = \mathbf{v}_1 \otimes (\mathbf{v}_2 \otimes \mathbf{v}_3), \mathbf{v}_i \in V_i, i = 1, 2, 3$).

In general, with an analogous meaning,

$$(V_1 \otimes \cdots \otimes V_k) \otimes (V_{k+1} \otimes \cdots \otimes V_m) = V_1 \otimes \cdots \otimes V_m,$$

for any k, $1 \leq k < m$.

13. Let W_i be a subspace of V_i, $i = 1, \ldots, m$. Then $W_1 \otimes \cdots \otimes W_m$ is a subspace of $V_1 \otimes \cdots \otimes V_m$, meaning that the subspace of $V_1 \otimes \cdots \otimes V_m$ spanned by the set of decomposable tensors of the form

$$\mathbf{w}_1 \otimes \cdots \otimes \mathbf{w}_m, \quad \mathbf{w}_i \in W_i,\ i = 1, \ldots, m$$

is a tensor product of $W_1, \ldots, W_m$ with tensor multiplication equal to the restriction of $\otimes$ to $W_1 \times \cdots \times W_m$.

From now on, the model for the tensor product described above is assumed when dealing with the tensor product of subspaces of V_i.

14. Let W_1, W_1' be subspaces of V_1 and W_2 and W_2' be subspaces of V_2. Then

(a) $(W_1 \otimes W_2) \cap (W_1' \otimes W_2') = (W_1 \cap W_1') \otimes (W_2 \cap W_2')$.

(b) $W_1 \otimes (W_2 + W_2') = (W_1 \otimes W_2) + (W_1 \otimes W_2')$,

$(W_1 + W_1') \otimes W_2 = (W_1 \otimes W_2) + (W_1' \otimes W_2)$.

(c) Assuming $W_1 \cap W_1' = \{0\}$,

$$(W_1 \oplus W_1') \otimes W_2 = (W_1 \otimes W_2) \oplus (W_1' \otimes W_2).$$

Assuming $W_2 \cap W_2' = \{0\}$,

$$W_1 \otimes (W_2 \oplus W_2') = (W_1 \otimes W_2) \oplus (W_1 \otimes W_2').$$

15. In a more general setting, if $W_{ij},\ j = 1, \ldots, p_i$ are subspaces of $V_i, i \in \{1, \ldots, m\}$, then

$$\left(\sum_{j=1}^{p_1} W_{1j}\right) \otimes \cdots \otimes \left(\sum_{j=1}^{p_m} W_{1j}\right) = \sum_{\gamma \in \Gamma(p_1 \cdots p_m)} W_{1\gamma(1)} \otimes \cdots \otimes W_{m\gamma(m)}.$$

If the sums of subspaces in the left-hand side are direct, then

$$\left(\bigoplus_{j=1}^{p_1} W_{1j}\right) \otimes \cdots \otimes \left(\bigoplus_{j=1}^{p_m} W_{1j}\right) = \bigoplus_{\gamma \in \Gamma(p_1, \ldots, p_m)} W_{1\gamma(1)} \otimes \cdots \otimes W_{m,\gamma(m)}.$$

Examples:

1. The vector space $F^{m\times n}$ of the $m \times n$ matrices over F is a tensor product of F^m and F^n with tensor multiplication (the usual tensor multiplication for $F^{m\times n}$) defined, for $(a_1,\ldots,a_m) \in F^m$ and $(b_1,\ldots,b_n) \in F^n$, by

$$(a_1,\ldots,a_m) \otimes (b_1,\ldots,b_n) = \begin{bmatrix} a_1 \\ \vdots \\ a_m \end{bmatrix} \begin{bmatrix} b_1 & \cdots & b_n \end{bmatrix}.$$

With this definition, $\mathbf{e}_i \otimes \mathbf{e}'_j = E_{ij}$ where $\mathbf{e}_i$, $\mathbf{e}'_j$, and E_{ij} are standard basis vectors of F^m, F^n, and $F^{m\times n}$.

2. The field F, viewed as a vector space over F, is an mth tensor power of F with tensor multiplication defined by

$$a_1 \otimes \cdots \otimes a_m = \prod_{i=1}^{m} a_i, \quad a_i \in F, \quad i = 1,\ldots,m.$$

3. The vector space V is a tensor product of F and V with tensor multiplication defined by

$$a \otimes \mathbf{v} = a\mathbf{v}, \quad a \in F, \quad \mathbf{v} \in V.$$

4. Let U and V be vector spaces over F. Then $L(V;U)$ is a tensor product $U \otimes V^*$ with tensor multiplication (the usual tensor multiplication for $L(V;U)$) defined by the equality $(\mathbf{u} \otimes f)(\mathbf{v}) = f(\mathbf{v})\mathbf{u}$, $\mathbf{u} \in U, \mathbf{v} \in V$.
5. Let $V_1,\ldots,V_m$ be vector spaces over F. The vector space $L(V_1,\ldots,V_m;U)$ is a tensor product $L(V_1,\ldots,V_m;F) \otimes U$ with tensor multiplication

$$(\varphi \otimes \mathbf{u})(\mathbf{v}_1,\ldots,\mathbf{v}_m) = \varphi(\mathbf{v}_1,\ldots,\mathbf{v}_m)\mathbf{u}.$$

6. Denote by $F^{n_1\times\cdots\times n_m}$ the set of all families with elements indexed in $\{1,\ldots,n_1\}\times\cdots\times\{1,\ldots,n_m\} = \Gamma(n_1,\ldots,n_m)$. The set $F^{n_1\times\cdots\times n_m}$ equipped with the sum and scalar product defined, for every $(j_1,\ldots,j_m) \in \Gamma(n_1,\ldots,n_m)$, by the equalities

$$(a_{j_1,\ldots,j_m}) + (b_{j_1,\ldots,j_m}) = (a_{j_1,\ldots,j_m} + b_{j_1,\ldots,j_m}),$$
$$\alpha(a_{j_1,\ldots,j_m}) = (\alpha a_{j_1,\ldots,j_m}), \quad \alpha \in F,$$

is a vector space over F. This vector space is a tensor product of $F^{n_1},\ldots,F^{n_m}$ with tensor multiplication defined by

$$(a_{11},\ldots,a_{1n_1}) \otimes \cdots \otimes (a_{m1},\ldots,a_{mn_m}) = \left(\prod_{i=1}^{m} a_{ij_i}\right)_{(j_1,\ldots,j_m)\in\Gamma}.$$

7. The vector space $L(V_1,\ldots,V_m;F)$ is a tensor product of $V_1^* = L(V_1;F),\ldots,V_m^* = L(V_m;F)$ with tensor multiplication defined by

$$g_1 \otimes \cdots \otimes g_m(\mathbf{v}_1,\ldots,\mathbf{v}_m) = \prod_{t=1}^{m} g_t(\mathbf{v}_t).$$

Very often, for example in the context of geometry, the factors of the tensor product are vector space duals. In those situations, this is the model of tensor product implicitly assumed.

8. The vector space

$$L(V_1,\ldots,V_m;F)^*$$

is a tensor product of $V_1, \ldots, V_m$ with tensor multiplication defined by

$$\mathbf{v}_1 \otimes \cdots \otimes \mathbf{v}_m(\psi) = \psi(\mathbf{v}_1, \ldots, \mathbf{v}_m).$$

9. The vector space $L(V_1, \ldots, V_m; F)$ is a tensor product $L(V_1, \ldots, V_k; F) \otimes L(V_{k+1}, \ldots, V_m; F)$ with tensor multiplication defined, for every $\mathbf{v}_i \in V_i$, $i = 1, \ldots, m$, by the equalities

$$(\varphi \otimes \psi)(\mathbf{v}_1, \ldots, \mathbf{v}_m) = \varphi(\mathbf{v}_1, \ldots, \mathbf{v}_k)\psi(\mathbf{v}_{k+1}, \ldots, \mathbf{v}_m).$$

13.3 Rank of a Tensor: Decomposable Tensors

Definitions:

Let $z \in V_1 \otimes \cdots \otimes V_m$. The tensor z has **rank** k if z is the sum of k decomposable tensors but it cannot be written as sum of l decomposable tensors, for any l less than k.

Facts:

The following facts can be found in [Bou89, Chap. II, §7.8]and [Mar73, Chap. 1].

1. The tensor $z = \mathbf{v}_1 \otimes \mathbf{w}_1 + \cdots + \mathbf{v}_t \otimes \mathbf{w}_t \in V \otimes W$ has rank t if and only if $(\mathbf{v}_1, \ldots, \mathbf{v}_t)$ and $(\mathbf{w}_1, \ldots, \mathbf{w}_t)$ are linearly independent.
2. If the model for the tensor product of F^m and F^n is the vector space of $m \times n$ matrices over F with the usual tensor multiplication, then the rank of a tensor is equal to the rank of the corresponding matrix.
3. If the model for the tensor product $U \otimes V^*$ is the vector space $L(V; U)$ with the usual tensor multiplication, then the rank of a tensor is equal to the rank of the corresponding linear map.
4. $\mathbf{x}_1 \otimes \cdots \otimes \mathbf{x}_m = 0$ if and only if $\mathbf{x}_i = 0$ for some $i \in \{1, \ldots, m\}$.
5. If $\mathbf{x}_i, \mathbf{y}_i$ are nonzero vectors of V_i, $i = 1, \ldots, m$, then

$$\text{Span}(\{\mathbf{x}_1 \otimes \cdots \otimes \mathbf{x}_m\}) = \text{Span}(\{\mathbf{y}_1 \otimes \cdots \otimes \mathbf{y}_m\})$$

   if and only if $\text{Span}(\{\mathbf{x}_i\}) = \text{Span}(\{\mathbf{y}_i\})$, $i = 1, \ldots, m$.

Examples:

1. Consider as a model of $F^m \otimes F^n$, the vector space of the $m \times n$ matrices over F with the usual tensor multiplication. Let A be a tensor of $F^m \otimes F^n$. If rank $A = k$ (using the matrix definition of rank), then

$$A = M \begin{bmatrix} I_k & 0 \\ 0 & 0 \end{bmatrix} N,$$

   where $M = [\mathbf{x}_1 \cdots \mathbf{x}_m]$ is an invertible matrix with columns $\mathbf{x}_1, \ldots, \mathbf{x}_m$ and

$$N = \begin{bmatrix} \mathbf{y}_1 \\ \mathbf{y}_2 \\ \vdots \\ \mathbf{y}_n \end{bmatrix}$$

   is an invertible matrix with rows $\mathbf{y}_1, \ldots, \mathbf{y}_n$. (See Chapter 2.) Then

$$A = \mathbf{x}_1 \otimes \mathbf{y}_1 + \cdots + \mathbf{x}_k \otimes \mathbf{y}_k$$

   has rank k as a tensor .

13.4 Tensor Product of Linear Maps

Definitions:

Let θ_i be a linear map from V_i into U_i, $i = 1, \ldots, m$. The unique linear map h from $V_1 \otimes \cdots \otimes V_m$ into $U_1 \otimes \cdots \otimes U_m$ satisfying, for all $\mathbf{v}_i \in V_i, i = 1, \ldots, m$,

$$h(\mathbf{v}_1 \otimes \cdots \otimes \mathbf{v}_m) = \theta_1(\mathbf{v}_1) \otimes \cdots \otimes \theta_m(\mathbf{v}_m)$$

is called the **tensor product** of $\theta_1, \ldots, \theta_m$ and is denoted by $\theta_1 \otimes \cdots \otimes \theta_m$.

Let $A_t = (a_{ij}^{(t)})$ be an $r_t \times s_t$ matrix over F, $t = 1, \ldots, m$. The **Kronecker product** of $A_1, \ldots, A_m$, denoted $A_1 \otimes \cdots \otimes A_m$, is the $(\prod_{t=1}^m r_t) \times (\prod_{t=1}^m s_t)$ matrix whose (α, β)-entry ($\alpha \in \Gamma(r_1, \ldots, r_m)$ and $\beta \in \Gamma(s_1, \ldots, s_m)$) is $\prod_{t=1}^m a_{\alpha(t)\beta(t)}^{(t)}$. (See also Section 10.4.)

Facts:

The following facts can be found in [Mar73, Chap. 2] and in [Mer97, Chap. 5].

Let θ_i be a linear map from V_i into U_i, $i = 1, \ldots, m$.

1. If η_i is a linear map from W_i into V_i, $i = 1, \ldots, m$,

$$(\theta_1 \otimes \cdots \otimes \theta_m)(\eta_1 \otimes \cdots \otimes \eta_m) = (\theta_1\eta_1) \otimes \cdots \otimes (\theta_m\eta_m).$$

2. $I_{V_1 \otimes \cdots \otimes V_m} = I_{V_1} \otimes \cdots \otimes I_{V_m}$.
3. $\mathrm{Ker}(\theta_1 \otimes \cdots \otimes \theta_m) = \mathrm{Ker}(\theta_1) \otimes V_2 \otimes \cdots \otimes V_m + V_1 \otimes \mathrm{Ker}(\theta_2) \otimes \cdots \otimes V_m + \cdots + V_1 \otimes \cdots \otimes V_{m-1} \otimes \mathrm{Ker}(\theta_m)$.
   In particular, $\theta_1 \otimes \cdots \otimes \theta_m$ is one to one if θ_i is one to one, $i = 1, \ldots, m$, [Bou89, Chap. II, §3.5].
4. $\theta_1 \otimes \cdots \otimes \theta_m(V_1 \otimes \cdots \otimes V_m) = \theta_1(V_1) \otimes \cdots \otimes \theta_m(V_m)$. In particular $\theta_1 \otimes \cdots \otimes \theta_m$ is onto if θ_i is onto, $i = 1, \ldots, m$.

   In the next three facts, assume that θ_i is a linear operator on the n_i-dimensional vector space V_i, $i = 1, \ldots, m$.
5. $\mathrm{tr}(\theta_1 \otimes \cdots \otimes \theta_m) = \prod_{i=1}^m \mathrm{tr}(\theta_i)$.
6. If $\sigma(\theta_i) = \{a_{i1}, \ldots, a_{in_i}\}$, $i = 1, \ldots, m$, then

$$\sigma(\theta_1 \otimes \cdots \otimes \theta_m) = \left\{\prod_{i=1}^m a_{i,\alpha(i)}\right\}_{\alpha \in \Gamma(n_1, \ldots, n_m)}.$$

7. $\det(\theta_1 \otimes \theta_2 \otimes \cdots \otimes \theta_m) = \det(\theta_1)^{n_2 \cdots n_m} \det(\theta_2)^{n_1 \cdot n_3 \cdots n_m} \cdots \det(\theta_m)^{n_1 \cdot n_2 \cdots n_{m-1}}$.
8. The map $\nu : (\theta_1, \ldots, \theta_m) \mapsto \theta_1 \otimes \cdots \otimes \theta_m$ is a multilinear map from $L(V_1; U_1) \times \cdots \times L(V_m; U_m)$ into $L(V_1 \otimes \cdots \otimes V_m; U_1 \otimes \cdots \otimes U_m)$.
9. The vector space $L(V_1 \otimes \cdots \otimes V_m; U_1 \otimes \cdots \otimes U_m))$ is a tensor product of the vector spaces $L(V_1; U_1), \ldots, L(V_m; U_m)$, with tensor multiplication $(\theta_1, \ldots, \theta_m) \mapsto \theta_1 \otimes \cdots \otimes \theta_m$:

$$L(V_1; U_1) \otimes \cdots \otimes L(V_m; U_m) = L(V_1 \otimes \cdots \otimes V_m; U_1 \otimes \cdots \otimes U_m).$$

10. As a consequence of (9.), choosing F as the model for $\bigotimes^m F$ with the product in F as tensor multiplication,

$$V_1^* \otimes \cdots \otimes V_m^* = (V_1 \otimes \cdots \otimes V_m)^*.$$

11. Let $(\mathbf{v}_{ij})_{j=1,\ldots,n_i}$ be an ordered basis of V_i and $(\mathbf{u}_{ij})_{j=1,\ldots,q_i}$ an ordered basis of U_i, $i = 1,\ldots,m$. Let A_i be the matrix of θ_i on the bases fixed in V_i and U_i. Then the matrix of $\theta_1 \otimes \cdots \otimes \theta_m$ on the bases $(\mathbf{v}_\alpha^\otimes)_{\alpha\in\Gamma(n_1,\ldots,n_m)}$ and $(\mathbf{u}_\alpha^\otimes)_{\alpha\in\Gamma(q_1,\ldots,q_r)}$ (induced by the bases $(\mathbf{v}_{ij})_{j=1,\ldots,n_i}$ and $(\mathbf{u}_{ij})_{j=1,\ldots,q_i}$, respectively) is the Kronecker product of $A_1,\ldots,A_m$,

$$A_1 \otimes \cdots \otimes A_m.$$

12. Let $n_1,\ldots,n_m, r_1,\ldots,r_m, t_1,\ldots,t_m$ be positive integers. Let A_i be an $n_i \times r_i$ matrix, and B_i be an $r_i \times t_i$ matrix, $i = 1,\ldots,m$. Then the following holds:

    (a) $(A_1 \otimes \cdots \otimes A_m)(B_1 \otimes \cdots \otimes B_m) = A_1B_1 \otimes \cdots \otimes A_mB_m$,

    (b) $(A_1 \otimes \cdots \otimes A_k) \otimes (A_{k+1} \otimes \cdots \otimes A_m) = A_1 \otimes \cdots \otimes A_m$.

Examples:

1. Consider as a model of $U \otimes V^*$, the vector space $L(V;U)$ with tensor multiplication defined by $(\mathbf{u} \otimes f)(\mathbf{v}) = f(\mathbf{v})\mathbf{u}$. Use a similar model for the tensor product of U' and V'^*. Let $\eta \in L(U;U')$ and $\theta \in L(V';V)$. Then, for all $\xi \in U \otimes V^* = L(V;U)$,

$$\eta \otimes \theta^*(\xi) = \eta\xi\theta.$$

2. Consider as a model of $F^m \otimes F^n$, the vector space of the $m \times n$ matrices over F with the usual tensor multiplication. Use a similar model for the tensor product of F^r and F^s. Identify the set of column matrices, $F^{m\times 1}$, with F^m and the set of row matrices, $F^{1\times n}$, with F^n. Let A be an $r \times m$ matrix over F. Let θ_A be the linear map from F^m into F^r defined by

$$\theta_A(a_1,\ldots,a_m) = A \begin{bmatrix} a_1 \\ a_2 \\ \vdots \\ a_m \end{bmatrix}.$$

   Let B be an $s \times n$ matrix. Then, for all $C \in F^{m\times n} = F^m \otimes F^n$, $\theta_A \otimes \theta_B(C) = ACB^T$.

3. For every $i = 1,\ldots,m$ consider the ordered basis $(\mathbf{b}_{i1},\ldots,\mathbf{b}_{in_i})$ fixed in V_i, and the basis $(\mathbf{b}'_{i1},\ldots,\mathbf{b}'_{is_i})$ fixed in U_i. Let θ_i be a linear map from V_i into U_i and let $A_i = (a^{(i)}_{jk})$ be the $s_i \times n_i$ matrix of θ_i with respect to the bases $(\mathbf{b}_{i1},\ldots,\mathbf{b}_{in_i})$, $(\mathbf{b}'_{i1},\ldots,\mathbf{b}'_{is_i})$. For every $z \in V_1 \otimes \cdots \otimes V_m$,

$$\begin{aligned} z &= \sum_{j_1=1}^{n_1}\sum_{j_2=1}^{n_2}\cdots\sum_{j_m=1}^{n_m} c_{j_1,\ldots,j_m}\mathbf{b}_{1j_1} \otimes \cdots \otimes \mathbf{b}_{m,j_m} \\ &= \sum_{\alpha\in\Gamma(n_1,\ldots,n_m)} c_\alpha \mathbf{b}_\alpha^\otimes. \end{aligned}$$

   Then, for $\beta = (i_1,\ldots,i_m) \in \Gamma(s_1,\ldots,s_m)$, the component $c'_{i_1,\ldots,i_m}$ of $\theta_1 \otimes \cdots \otimes \theta_m(z)$ on the basis element $\mathbf{b}'_{1i_1} \otimes \cdots \otimes \mathbf{b}'_{mi_m}$ of $U_1 \otimes \cdots \otimes U_m$ is

$$\begin{aligned} c'_\beta = c'_{i_1,\ldots,i_m} &= \sum_{j_1=1}^{n_1}\cdots\sum_{j_m=1}^{n_m} a^{(1)}_{i_1,j_1}\cdots a^{(m)}_{i_m,j_m} c_{j_1,\ldots,j_m} \\ &= \sum_{\gamma\in\Gamma(n_1,\ldots,n_m)} \left(\prod_{i=1}^m a^{(i)}_{\beta(i)\gamma(i)}\right) c_\gamma. \end{aligned}$$

4. If $A = [a_{ij}]$ is an $p \times q$ matrix over F and B is an $r \times s$ matrix over F, then the Kronecker product of A and B is the matrix whose partition in $r \times s$ blocks is

$$A \otimes B = \begin{bmatrix} a_{11}B & a_{12}B & \cdots & a_{1q}B \\ a_{21}B & a_{22}B & \cdots & a_{2q}B \\ \vdots & \vdots & \ddots & \vdots \\ a_{p1}B & a_{p2}B & \cdots & a_{pq}B \end{bmatrix}.$$

13.5 Symmetric and Antisymmetric Maps

Recall that we are assuming F to be of characteristic zero and that all vector spaces are finite dimensional over F. In particular, V and U denote finite dimensional vector spaces over F.

Definitions:

Let m be a positive integer. When $V_1 = V_2 = \cdots = V_m = V$ $L^m(V;U)$ denotes the vector space of the multilinear maps $L(V_1, \ldots, V_m; U)$. By convention $L^0(V;U) = U$.

An m-linear map $\psi \in L^m(V;U)$ is called **antisymmetric or alternating** if it satisfies

$$\psi(\mathbf{v}_{\sigma(1)}, \ldots, \mathbf{v}_{\sigma(m)}) = \operatorname{sgn}(\sigma)\psi(\mathbf{v}_1, \ldots, \mathbf{v}_m), \quad \sigma \in S_m,$$

where $\operatorname{sgn}(\sigma)$ denotes the sign of the permutation σ.

Similarly, an m-linear map $\varphi \in L^m(V;U)$ satisfying

$$\varphi(\mathbf{v}_{\sigma(1)}, \ldots, \mathbf{v}_{\sigma(m)}) = \varphi(\mathbf{v}_1, \ldots, \mathbf{v}_m)$$

for all permutations $\sigma \in S_m$ and for all $\mathbf{v}_1, \ldots, \mathbf{v}_m$ in V is called **symmetric.** Let $S^m(V;U)$ and $A^m(V;U)$ denote the subsets of $L^m(V;U)$ whose elements are respectively the symmetric and the antisymmetric m-linear maps. The elements of $A^m(V;F)$ are called **antisymmetric forms**. The elements of $S^m(V;F)$ are called **symmetric forms**.

Let $\Gamma_{m,n}$ be the set of all maps from $\{1, \ldots, m\}$ into $\{1, \ldots, n\}$, i.e,

$$\Gamma_{m,n} = \Gamma\underbrace{(n, \ldots, n)}_{m \text{ times}}.$$

The subset of $\Gamma_{m,n}$ of the strictly increasing maps α ($\alpha(1) < \cdots < \alpha(m)$) is denoted by $Q_{m,n}$. The subset of the increasing maps $\alpha \in \Gamma_{m,n}$ ($\alpha(1) \leq \cdots \leq \alpha(m)$) is denoted by $G_{m,n}$.

Let $A = [a_{ij}]$ be an $m \times n$ matrix over F. Let $\alpha \in \Gamma_{p,m}$ and $\beta \in \Gamma_{q,n}$. Then $A[\alpha|\beta]$ be the $p \times q$-matrix over F whose (i, j)-entry is $a_{\alpha(i),\beta(j)}$, i.e.,

$$A[\alpha|\beta] = [a_{\alpha(i),\beta(j)}].$$

The mth-tuple $(1, 2, \ldots, m)$ is denoted by ι_m. If there is no risk of confusion ι is used instead of ι_m.

Facts:

1. If $m > n$, we have $Q_{m,n} = \emptyset$. The cardinality of $\Gamma_{m,n}$ is n^m, the cardinality of $Q_{m,n}$ is $\binom{n}{m}$, and the cardinality of $G_{m,n}$ is $\binom{m+n-1}{m}$.
2. $A^m(V;U)$ and $S^m(V;U)$ are vector subspaces of $L^m(V;U)$.
3. Let $\psi \in L^m(V;U)$. The following conditions are equivalent:

   (a) ψ is an antisymmetric multilinear map.

   (b) For $1 \leq i < j \leq m$ and for all $\mathbf{v}_1, \ldots, \mathbf{v}_m \in V$,

$$\begin{aligned}&\psi(\mathbf{v}_1, \ldots, \mathbf{v}_{i-1}, \mathbf{v}_j, \mathbf{v}_{i+1}, \ldots, \mathbf{v}_{j-1}, \mathbf{v}_i, \mathbf{v}_{j+1}, \ldots, \mathbf{v}_m)\\ &\quad = -\psi(\mathbf{v}_1, \ldots, \mathbf{v}_{i-1}, \mathbf{v}_i, \mathbf{v}_{i+1}, \ldots, \mathbf{v}_{j-1}, \mathbf{v}_j, \mathbf{v}_{j+1}, \ldots, \mathbf{v}_m).\end{aligned}$$

   (c) For $1 \leq i < m$ and for all $\mathbf{v}_1, \ldots, \mathbf{v}_m \in V$,

$$\psi(\mathbf{v}_1, \ldots, \mathbf{v}_{i+1}, \mathbf{v}_i, \ldots, \mathbf{v}_m) = -\psi(\mathbf{v}_1, \ldots, \mathbf{v}_i, \mathbf{v}_{i+1}, \ldots, \mathbf{v}_m).$$

4. Let $\psi \in L^m(V;U)$. The following conditions are equivalent:

   (a) ψ is a symmetric multilinear map.

   (b) For $1 \leq i < j \leq m$ and for all $\mathbf{v}_1, \ldots, \mathbf{v}_m \in V$,

$$\begin{aligned}&\psi(\mathbf{v}_1, \ldots, \mathbf{v}_{i-1}, \mathbf{v}_j, \mathbf{v}_{i+1}, \ldots, \mathbf{v}_{j-1}, \mathbf{v}_i, \mathbf{v}_{j+1}, \ldots, \mathbf{v}_m)\\ &\quad = \psi(\mathbf{v}_1, \ldots, \mathbf{v}_{i-1}, \mathbf{v}_i, \mathbf{v}_{i+1}, \ldots, \mathbf{v}_{j-1}, \mathbf{v}_j, \mathbf{v}_{j+1}, \ldots, \mathbf{v}_m).\end{aligned}$$

   (c) For $1 \leq i < m$ and for all $\mathbf{v}_1, \ldots, \mathbf{v}_m \in V$,

$$\psi(\mathbf{v}_1, \ldots, \mathbf{v}_{i+1}, \mathbf{v}_i, \ldots, \mathbf{v}_m) = \psi(\mathbf{v}_1, \ldots, \mathbf{v}_i, \mathbf{v}_{i+1}, \ldots, \mathbf{v}_m).$$

5. When we consider $L^m(V;U)$ as the tensor product, $L^m(V;F) \otimes U$, with the tensor multiplication described in Example 5 in Section 13.2, we have

$$A^m(V;U) = A^m(V;F) \otimes U \quad \text{and} \quad S^m(V;U) = S^m(V;F) \otimes U.$$

6. **Polarization identity** [Dol04] If φ is a symmetric multilinear map, then for every m-tuple $(\mathbf{v}_1, \ldots, \mathbf{v}_m)$ of vectors of V, and for any vector $\mathbf{w} \in V$, the following identity holds:

$$\begin{aligned}&\varphi(\mathbf{v}_1, \ldots, \mathbf{v}_m) =\\ &= \frac{1}{2^m m!} \sum_{\varepsilon_1 \cdots \varepsilon_m} \varepsilon_1 \cdots \varepsilon_m \varphi(\mathbf{w} + \varepsilon_1 \mathbf{v}_1 + \cdots + \varepsilon_m \mathbf{v}_m, \ldots, \mathbf{w} + \varepsilon_1 \mathbf{v}_1 + \cdots + \varepsilon_m \mathbf{v}_m),\end{aligned}$$

   where $\varepsilon_i \in \{-1, +1\}$, $i = 1, \ldots, m$.

Examples:

1. The map

$$((a_{11}, a_{21}, \ldots, a_{m1}), \ldots, (a_{1m}, a_{2m}, \ldots, a_{mm})) \to \det([a_{ij}])$$

   from the Cartesian product of m copies of F^m into F is m-linear and antisymmetric.

2. The map

$$((a_{11}, a_{21}, \ldots, a_{m1}), \ldots, (a_{1m}, a_{2m}, \ldots, a_{mm})) \to \text{per}([a_{ij}])$$

from the Cartesian product of m copies of F^m into F is m-linear and symmetric.
3. The map $((a_1, \ldots, a_n), (b_1, \ldots, b_n)) \mapsto (a_i b_j - b_i a_j)$ from $F^n \times F^n$ into $F^{n \times n}$ is bilinear antisymmetric.
4. The map $((a_1, \ldots, a_n), (b_1, \ldots, b_n)) \mapsto (a_i b_j + b_i a_j)$ from $F^n \times F^n$ into $F^{n \times n}$ is bilinear symmetric.
5. The map χ from V^m into $A^m(V; F)^*$ defined by

$$\chi(\mathbf{v}_1, \ldots, \mathbf{v}_m)(\psi) = \psi(\mathbf{v}_1, \ldots, \mathbf{v}_m), \quad \mathbf{v}_1, \ldots, \mathbf{v}_m \in V,$$

is an antisymmetric multilinear map.
6. The map χ from V^m into $S^m(V; F)^*$ defined by

$$\chi(\mathbf{v}_1, \ldots, \mathbf{v}_m)(\psi) = \psi(\mathbf{v}_1, \ldots, \mathbf{v}_m), \quad \mathbf{v}_1, \ldots, \mathbf{v}_m \in V,$$

is a symmetric multilinear map.

13.6 Symmetric and Grassmann Tensors

Definitions:

Let $\sigma \in S_m$ be a permutation of $\{1, \ldots, m\}$. The unique linear map, from $\otimes^m V$ into $\otimes^m V$ satisfying

$$\mathbf{v}_1 \otimes \cdots \otimes \mathbf{v}_m \mapsto \mathbf{v}_{\sigma^{-1}(1)} \otimes \cdots \otimes \mathbf{v}_{\sigma^{-1}(m)}, \quad \mathbf{v}_1, \ldots, \mathbf{v}_m \in V,$$

is denoted $P(\sigma)$.

Let ψ be a multilinear form of $L^m(V; F)$ and σ an element of S_m. The multilinear form $(\mathbf{v}_1, \ldots, \mathbf{v}_m) \mapsto \psi(\mathbf{v}_{\sigma(1)}, \ldots, \mathbf{v}_{\sigma(m)})$ is denoted ψ_σ.

The linear operator Alt from $\otimes^m V$ into $\otimes^m V$ defined by

$$\text{Alt} := \frac{1}{m!} \sum_{\sigma \in S_m} \text{sgn}(\sigma) P(\sigma)$$

is called the **alternator**. In order to emphasize the degree of the domain of Alt, Alt_m is often used for the operator having $\bigotimes^m V$, as domain.

Similarly, the linear operator Sym is defined as the following linear combination of the maps $P(\sigma)$:

$$\text{Sym} = \frac{1}{m!} \sum_{\sigma \in S_m} P(\sigma).$$

As before, Sym_m is often written to mean the Sym operator having $\bigotimes^m V$, as domain.

The range of Alt is denoted by $\bigwedge^m V$, i.e., $\bigwedge^m V = \text{Alt}(\bigotimes^m V)$, and is called the **Grassmann space of degree** m associated with V or the m**th-exterior power** of V.

The range of Sym is denoted by $\bigvee^m V$, i.e., $\bigvee^m V = \text{Sym}(\bigotimes^m V)$, and is called the **symmetric space of degree** m associated with V or the m**th symmetric power** of V.

By convention

$$\bigotimes^0 V = \bigwedge^0 V = \bigvee^0 V = F.$$

Assume $m > 1$. The elements of $\bigwedge^m V$ that are the image under Alt of decomposable tensors of $\bigotimes^m V$ are called **decomposable elements of** $\bigwedge^m V$. If $\mathbf{x}_1, \ldots, \mathbf{x}_m \in V$, $\mathbf{x}_1 \wedge \cdots \wedge \mathbf{x}_m$ denotes the decomposable element of $\bigwedge^m V$,

$$\mathbf{x}_1 \wedge \cdots \wedge \mathbf{x}_m = m!\text{Alt}\,(\mathbf{x}_1 \otimes \cdots \otimes \mathbf{x}_m),$$

and $\mathbf{x}_1 \wedge \cdots \wedge \mathbf{x}_m$ is called the **exterior product** of $\mathbf{x}_1, \ldots, \mathbf{x}_m$. Similarly, the elements of $\bigvee^m V$ that are the image under Sym of decomposable tensors of $\bigotimes^m V$ are called **decomposable elements of** $\bigvee^m V$. If $\mathbf{x}_1, \ldots, \mathbf{x}_m \in V$, $\mathbf{x}_1 \vee \cdots \vee \mathbf{x}_m$ denotes the decomposable element of $\bigvee^m V$,

$$\mathbf{x}_1 \vee \cdots \vee \mathbf{x}_m = m!\text{Sym}\,(\mathbf{x}_1 \otimes \cdots \otimes \mathbf{x}_m),$$

and $\mathbf{x}_1 \vee \cdots \vee \mathbf{x}_m$ is called the **symmetric product** of $\mathbf{x}_1, \ldots, \mathbf{x}_m$.

Let $(\mathbf{b}_1, \ldots, \mathbf{b}_n)$ be a basis of V. If $\alpha \in \Gamma_{m,n}$, $\mathbf{b}_\alpha^\otimes$, $\mathbf{b}_\alpha^\wedge$, and $\mathbf{b}_\alpha^\vee$ denote respectively the tensors

$$\mathbf{b}_\alpha^\otimes = \mathbf{b}_{\alpha(1)} \otimes \cdots \otimes \mathbf{b}_{\alpha(m)},$$
$$\mathbf{b}_\alpha^\wedge = \mathbf{b}_{\alpha(1)} \wedge \cdots \wedge \mathbf{b}_{\alpha(m)},$$
$$\mathbf{b}_\alpha^\vee = \mathbf{b}_{\alpha(1)} \vee \cdots \vee \mathbf{b}_{\alpha(m)}.$$

Let n and m be positive integers. An n-**composition** of m is a sequence

$$\mu = (\mu_1, \ldots, \mu_n)$$

of nonnegative integers that sum to m. Let $C_{m,n}$ be the set of n-compositions of m.

Let $\lambda = (\lambda_1, \ldots, \lambda_n)$ be an n-composition of m. The integer $\lambda_1! \cdots \lambda_n!$ will be denoted by $\lambda!$.

Let $\alpha \in \Gamma_{m,n}$. The **multiplicity composition** of α is the n-tuple of the cardinalities of the fibers of α, $(|\alpha^{-1}(1)|, \ldots, |\alpha^{-1}(n)|)$, and is denoted by λ_α.

Facts:

The following facts can be found in [Mar73, Chap. 2], [Mer97, Chap. 5], and [Spi79, Chap. 7].

1. $\bigwedge^m V$ and $\bigvee^m V$ are vector subspaces of $\bigotimes^m V$.
2. The map $\sigma \mapsto P(\sigma)$ from the symmetric group of degree m into $L(\otimes^m V; \otimes^m V)$ is an F-representation of S_m, i.e., $P(\sigma\tau) = P(\sigma)P(\tau)$ for any $\sigma, \tau \in S_m$ and $P(I) = I_{\otimes^m V}$
3. Choosing $L^m(V; F)$, with the usual tensor multiplication, as the model for the tensor power, $\bigotimes^m V^*$, the linear operator $P(\sigma)$ acts on $L^m(V; F)$ by the following transformation

$$(P(\sigma)\psi) = \psi_\sigma.$$

4. The linear operators Alt and Sym are projections, i.e., $\text{Alt}^2 = \text{Alt}$ and $\text{Sym}^2 = \text{Sym}$.
5. If $m = 1$, we have

$$\text{Sym} = \text{Alt} = I_{\bigotimes^1 V} = I_V.$$

6. $\bigwedge^m V = \{z \in \bigotimes^m V \;:\; P(\sigma)(z) = \text{sgn}(\sigma)z,\ \forall\sigma \in S_m\}$.
7. $\bigvee^m V = \{z \in \bigotimes^m V \;:\; P(\sigma)(z) = z,\ \forall\sigma \in S_m\}$.
8. Choosing $L^m(V; F)$ as the model for the tensor power $\bigotimes^m V^*$ with the usual tensor multiplication,

$$\bigwedge^m V^* = A^m(V; F) \quad \text{and} \quad \bigvee^m V^* = S^m(V; F).$$

9.

$$\bigotimes^1 V = \bigwedge^1 V = \bigvee^1 V = V.$$

10. $\bigotimes^2 V = \bigwedge^2 V \oplus \bigvee^2 V$. Moreover for $z \in \bigotimes^2 V$,

$$z = \text{Alt}\,(z) + \text{Sym}\,(z).$$

The corresponding equality is no more true in $\bigotimes^m V$ if $m \neq 2$.
11. $\bigwedge^m V = \{0\}$ if $m > \dim(V)$.
12. If $m \geq 1$, any element of $\bigwedge^m V$ is a sum of decomposable elements of $\bigwedge^m V$.
13. If $m \geq 1$, any element of $\bigvee^m V$ is a sum of decomposable elements of $\bigvee^m V$.
14. $\text{Alt}\,(P(\sigma)z) = \text{sgn}(\sigma)\text{Alt}\,(z)$ and $\text{Sym}\,(P(\sigma)(z)) = \text{Sym}\,(z), z \in \bigotimes^m V$.
15. The map $\wedge$ from V^m into $\bigwedge^m V$ defined for $\mathbf{v}_1, \ldots, \mathbf{v}_m \in V$ by

$$\wedge(\mathbf{v}_1, \ldots, \mathbf{v}_m) = \mathbf{v}_1 \wedge \cdots \wedge \mathbf{v}_m$$

is an antisymmetric m-linear map.
16. The map $\vee$ from V^m into $\bigvee^m V$ defined for $\mathbf{v}_1, \ldots, \mathbf{v}_m \in V$ by

$$\vee(\mathbf{v}_1, \ldots, \mathbf{v}_m) = \mathbf{v}_1 \vee \cdots \vee \mathbf{v}_m$$

is a symmetric m-linear map.
17. **(Universal property for $\bigwedge^m V$)** Given an antisymmetric m-linear map ψ from V^m into U, there exists a unique linear map h from $\bigwedge^m V$ into U such that

$$\psi(\mathbf{v}_1, \ldots, \mathbf{v}_m) = h(\mathbf{v}_1 \wedge \cdots \wedge \mathbf{v}_m), \quad \mathbf{v}_1, \ldots, \mathbf{v}_m \in V,$$

i.e., there exists a unique linear map h that makes the following diagram commutative:

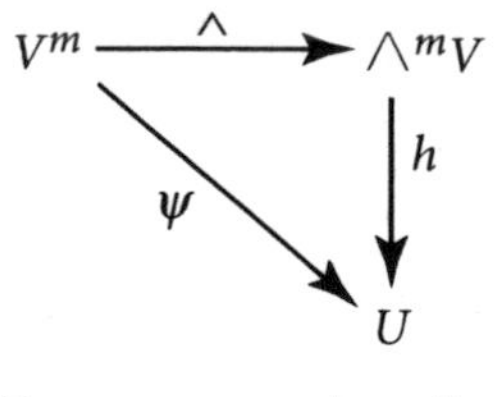

18. **(Universal property for $\bigvee^m V$)** Given a symmetric m-linear map φ from V^m into U, there exists a unique linear map h from $\bigvee^m V$ into U such that

$$\varphi(\mathbf{v}_1, \ldots, \mathbf{v}_m) = h(\mathbf{v}_1 \vee \cdots \vee \mathbf{v}_m), \quad \mathbf{v}_1, \ldots, \mathbf{v}_m \in V,$$

i.e., there exists a unique linear map h that makes the following diagram commutative:

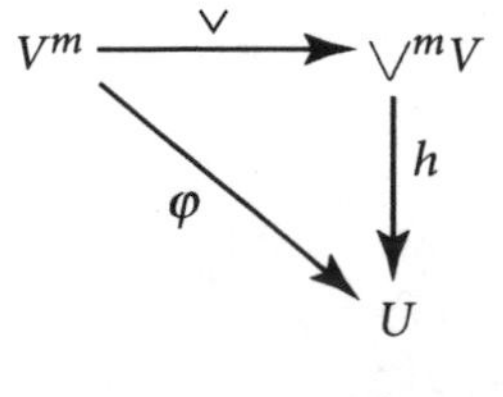

Let p and q be positive integers.

19. **(Universal property for $\bigotimes^m V$-bilinear version)** If ψ is a $(p+q)$-linear map from V^{p+q} into U, then there exists a unique bilinear map χ from $\bigotimes^p V \times \bigotimes^q V$ into U satisfying (recall Fact 5 in Section 13.2)

$$\chi(\mathbf{v}_1 \otimes \cdots \otimes \mathbf{v}_p, \mathbf{v}_{p+1} \otimes \cdots \otimes \mathbf{v}_{p+q}) = \psi(\mathbf{v}_1, \ldots, \mathbf{v}_{p+q}).$$

20. **(Universal property for $\bigwedge^m V$-bilinear version)** If ψ is a $(p+q)$-linear map from V^{p+q} into U antisymmetric in the first p variables and antisymmetric in the last q variables, then there exists a unique bilinear map χ from $\bigwedge^p V \times \bigwedge^q V$ into U satisfying

$$\chi(\mathbf{v}_1 \wedge \cdots \wedge \mathbf{v}_p, \mathbf{v}_{p+1} \wedge \cdots \wedge \mathbf{v}_{p+q}) = \psi(\mathbf{v}_1, \ldots, \mathbf{v}_{p+q}).$$

21. **(Universal property for $\bigvee^m V$-bilinear version)** If φ is a $(p+q)$-linear map from V^{p+q} into U symmetric in the first p variables and symmetric in the last q variables, then there exists a unique bilinear map χ from $\bigvee^p V \times \bigvee^q V$ into U satisfying

$$\chi(\mathbf{v}_1 \vee \cdots \vee \mathbf{v}_p, \mathbf{v}_{p+1} \vee \cdots \vee \mathbf{v}_{p+q}) = \varphi(\mathbf{v}_1, \ldots, \mathbf{v}_{p+q}).$$

22. If $(\mathbf{b}_1, \ldots, \mathbf{b}_n)$ is a basis of V, then $(\mathbf{b}_\alpha^\otimes)_{\alpha \in \Gamma_{m,n}}$ is a basis of $\otimes^m V$, $(\mathbf{b}_\alpha^\wedge)_{\alpha \in Q_{m,n}}$ is a basis of $\bigwedge^m V$, and $(\mathbf{b}_\alpha^\vee)_{\alpha \in G_{m,n}}$ is a basis of $\bigvee^m V$. These bases are said to be **induced** by the basis $(\mathbf{b}_1, \ldots, \mathbf{b}_n)$.
23. Assume $L^m(V; F)$ as the model for the tensor power of $\bigotimes^m V^*$, with the usual tensor multiplication. Let $(f_1, \ldots, f_n)$ be the dual basis of the basis $(\mathbf{b}_1, \ldots, \mathbf{b}_n)$. Then:

    (a) For every $\varphi \in L^m(V; F)$,

$$\varphi = \sum_{\alpha \in \Gamma_{m,n}} \varphi(\mathbf{b}_\alpha) f_\alpha^\otimes.$$

    (b) For every $\varphi \in A^m(V, F)$,

$$\varphi = \sum_{\alpha \in Q_{m,n}} \varphi(\mathbf{b}_\alpha) f_\alpha^\wedge.$$

    (c) For every $\varphi \in S^m(V, F)$,

$$\varphi = \sum_{\alpha \in G_{m,n}} \frac{1}{\lambda_\alpha!} \varphi(\mathbf{b}_\alpha) f_\alpha^\vee.$$

24. $\dim \bigotimes^m V = n^m$, $\dim \bigwedge^m V = \binom{n}{m}$, and $\dim \bigvee^m V = \binom{n+m-1}{m}$.
25. The family

$$((\mu_1 \mathbf{b}_1 + \cdots + \mu_n \mathbf{b}_n) \vee \cdots \vee (\mu_1 \mathbf{b}_1 + \cdots + \mu_n \mathbf{b}_n))_{\mu \in \mathcal{C}_{m,n}}$$

is a basis of $\bigvee^m V$ [Mar73, Chap. 3].

26. Let $\mathbf{x}_1, \ldots, \mathbf{x}_m$ be vectors of V and $g_1, \ldots, g_m$ forms of V^*. Let $a_{ij} = g_i(\mathbf{x}_j)$, $i, j = 1, \ldots, m$. Then, choosing $(\bigotimes^m V)^*$ as the model for $\bigotimes^m V^*$ with tensor multiplication as described in Fact 10 in Section 13.4,

$$g_1 \otimes \cdots \otimes g_m(\mathbf{x}_1 \wedge \cdots \wedge \mathbf{x}_m) = \det[a_{ij}].$$

27. Under the same conditions of the former fact,

$$g_1 \otimes \cdots \otimes g_m(\mathbf{x}_1 \vee \cdots \vee \mathbf{x}_m) = \mathrm{per}[a_{ij}].$$

28. Let $(f_1, \ldots, f_n)$ be the dual basis of the basis $(\mathbf{b}_1, \ldots, \mathbf{b}_n)$. Then, choosing $(\bigotimes^m V)^*$ as the model for $\bigotimes^m V^*$:

 (a)

$$(f_\alpha^{\otimes})_{\alpha \in \Gamma_{m,n}}$$

 is the dual basis of the basis $(\mathbf{b}_\alpha^{\otimes})_{\alpha \in \Gamma_{m,n}}$ of $\otimes^m V$.

 (b)

$$\left((f_\alpha^{\otimes})_{|\bigwedge^m V}\right)_{\alpha \in Q_{m,n}}$$

 is the dual basis of the basis $(\mathbf{b}_\alpha^{\wedge})_{\alpha \in Q_{m,n}}$ of $\bigwedge^m V$.

 (c)

$$\left(\frac{1}{\lambda_\alpha!}(f_\alpha^{\otimes})_{|\bigvee^m V}\right)_{\alpha \in G_{m,n}}$$

 is the dual basis of the basis $(\mathbf{b}_\alpha^{\vee})_{\alpha \in G_{m,n}}$ of $\bigvee^m V$.

Let $\mathbf{v}_1, \ldots, \mathbf{v}_m$ be vectors of V and $(\mathbf{b}_1, \ldots, \mathbf{b}_n)$ be a basis of V.

29. Let $A = [a_{ij}]$ be the $n \times m$ matrix over F such that $\mathbf{v}_j = \sum_{i=1}^n a_{ij}\mathbf{b}_i,\ j = 1, \ldots, m$. Then:

 (a)

$$\mathbf{v}_1 \otimes \cdots \otimes \mathbf{v}_m = \sum_{\alpha \in \Gamma_{m,n}} \left(\prod_{t=1}^m a_{\alpha(t),t}\right) \mathbf{b}_\alpha^{\otimes};$$

 (b)

$$\mathbf{v}_1 \wedge \cdots \wedge \mathbf{v}_m = \sum_{\alpha \in Q_{m,n}} \det A[\alpha|\iota]\mathbf{b}_\alpha^{\wedge};$$

 (c)

$$\mathbf{v}_1 \vee \cdots \vee \mathbf{v}_m = \sum_{\alpha \in G_{m,n}} \frac{1}{\lambda_\alpha!} \operatorname{per} A[\alpha|\iota]\mathbf{b}_\alpha^{\vee}.$$

30. $\mathbf{v}_1 \wedge \cdots \wedge \mathbf{v}_m = 0$ if and only if $(\mathbf{v}_1, \ldots, \mathbf{v}_m)$ is linearly dependent.
31. $\mathbf{v}_1 \vee \cdots \vee \mathbf{v}_m = 0$ if and only if one of the $\mathbf{v}_i$s is equal to 0.
32. Let $\mathbf{u}_1, \ldots, \mathbf{u}_m$ be vectors of V.

 (a) If $(\mathbf{v}_1, \ldots, \mathbf{v}_m)$ and $(\mathbf{u}_1, \ldots, \mathbf{u}_m)$ are linearly independent families, then

$$\text{Span}(\{\mathbf{u}_1 \wedge \cdots \wedge \mathbf{u}_m\}) = \text{Span}(\{\mathbf{v}_1 \wedge \cdots \wedge \mathbf{v}_m\})$$

 if and only if

$$\text{Span}(\{\mathbf{u}_1, \ldots, \mathbf{u}_m\}) = \text{Span}(\{\mathbf{v}_1, \ldots, \mathbf{v}_m\}).$$

 (b) If $(\mathbf{v}_1, \ldots, \mathbf{v}_m)$ and $(\mathbf{u}_1, \ldots, \mathbf{u}_m)$ are families of nonzero vectors of V, then

$$\text{Span}(\{\mathbf{v}_1 \vee \cdots \vee \mathbf{v}_m\}) = \text{Span}(\{\mathbf{u}_1 \vee \cdots \vee \mathbf{u}_m\})$$

if and only if there exists a permutation σ of S_m satisfying

$$\text{Span}(\{\mathbf{v}_i\}) = \text{Span}(\{\mathbf{u}_{\sigma(i)}\}), \quad i = 1, \ldots, m.$$

Examples:

1. If $m = 1$, we have

$$\text{Sym} = \text{Alt} = I_{\bigotimes^1 V} = I_V.$$

2. Consider as a model of $\bigotimes^2 F^n$, the vector space of the $n \times n$ matrices with the usual tensor multiplication. Then $\bigwedge^2 F^n$ is the subspace of the $n \times n$ antisymmetric matrices over F and $\bigvee^2 F^n$ is the subspace of the $n \times n$ symmetric matrices over F. Moreover, for $(a_1, \ldots, a_n), (b_1, \ldots, b_n) \in F^n$:

   (a) $(a_1, \ldots, a_n) \wedge (b_1, \ldots, b_n) = [a_i b_j - b_i a_j]_{i,j=1,\ldots,n}$.

   (b) $(a_1, \ldots, a_n) \vee (b_1, \ldots, b_n) = [a_i b_j + b_i a_j]_{i,j=1,\ldots,n}$.

   With these definitions, $\mathbf{e}_i \wedge \mathbf{e}_j = E_{ij} - E_{ji}$ and $\mathbf{e}_i \vee \mathbf{e}_j = E_{ij} + E_{ji}$, where $\mathbf{e}_i$, $\mathbf{e}_j$, and E_{ij} are standard basis vectors of F^m, F^n, and $F^{m \times n}$.
3. For $\mathbf{x} \in V$, $\mathbf{x} \vee \cdots \vee \mathbf{x} = m! \mathbf{x} \otimes \cdots \otimes \mathbf{x}$.

13.7 The Tensor Multiplication, the Alt Multiplication, and the Sym Multiplication

Next we will introduce "external multiplications" for tensor powers, Grassmann spaces, and symmetric spaces, Let p, q be positive integers.

Definitions:

The (p,q)-**tensor multiplication** is the unique bilinear map, $(z, z') \mapsto z \otimes z'$ from $(\bigotimes^p V) \times (\bigotimes^q V)$ into $\bigotimes^{p+q} V$, satisfying

$$(\mathbf{v}_1 \otimes \cdots \otimes \mathbf{v}_p) \otimes (\mathbf{v}_{p+1} \otimes \cdots \otimes \mathbf{v}_{p+q}) = \mathbf{v}_1 \otimes \cdots \otimes \mathbf{v}_{p+q}.$$

The (p,q)-**alt multiplication** (briefly alt multiplication) is the unique bilinear map (recall Fact 20 in section 13.6), $(z, z') \mapsto z \wedge z'$ from $(\bigwedge^p V) \times (\bigwedge^q V)$ into $\bigwedge^{p+q} V$, satisfying

$$(\mathbf{v}_1 \wedge \cdots \wedge \mathbf{v}_p) \wedge (\mathbf{v}_{p+1} \wedge \cdots \wedge \mathbf{v}_{p+q}) = \mathbf{v}_1 \wedge \cdots \wedge \mathbf{v}_{p+q}.$$

The (p,q)-**sym multiplication** (briefly sym multiplication) is the unique bilinear map (recall Fact 21 in section 13.6), $(z, z') \mapsto z \vee z'$ from $(\bigvee^p V) \times (\bigvee^q V)$ into $\bigvee^{p+q} V$, satisfying

$$(\mathbf{v}_1 \vee \cdots \vee \mathbf{v}_p) \vee (\mathbf{v}_{p+1} \vee \cdots \vee \mathbf{v}_{p+q}) = \mathbf{v}_1 \vee \cdots \vee \mathbf{v}_{p+q}.$$

These definitions can be extended to include the cases where either p or q is zero, taking as multiplication the scalar product.

Let m, n be positive integers satisfying $1 \leq m < n$. Let $\alpha \in Q_{m,n}$. We denote by α^c the element of $Q_{n-m,n}$ whose range is the complement in $\{1, \ldots, n\}$ of the range of α and by $\widetilde{\alpha}$ the permutation of S_n:

$$\widetilde{\alpha} = \begin{pmatrix} 1 & \cdots & m & m+1 & \cdots & n \\ \alpha(1) & \cdots & \alpha(m) & \alpha^c(1) & \cdots & \alpha^c(n) \end{pmatrix}.$$

Facts:

The following facts can be found in [Mar73, Chap. 2], [Mer97, Chap. 5], and in [Spi79, Chap. 7].

1. The value of the alt multiplication for arbitrary elements $z \in \bigwedge^p V$ and $z' \in \bigwedge^q V$ is given by
$$z \wedge z' = \frac{(p+q)!}{p!q!} \mathrm{Alt}_{\,p+q}(z \otimes z').$$
2. The product of $z \in \bigvee^p V$ and $z' \in \bigvee^q V$ by the sym multiplication is given by
$$z \vee z' = \frac{(p+q)!}{p!q!} \mathrm{Sym}_{\,p+q}(z \otimes z').$$
3. The alt-multiplication $z \wedge z'$ and the sym-multiplication $z \vee z'$ are not, in general, decomposable elements of any Grassmann or symmetric space of degree 2.
4. Let $0 \neq z \in \bigwedge^m V$. Then z is decomposable if and only if there exists a linearly independent family of vectors $\mathbf{v}_1, \ldots, \mathbf{v}_m$ satisfying $z \wedge \mathbf{v}_i = 0, i = 1, \ldots, m$.
5. If $\dim(V) = n$, all elements of $\bigwedge^{n-1} V$ are decomposable.
6. The multiplications defined in this subection are associative. Therefore,
$$z \otimes z' \otimes z'', z \in \bigotimes^p V, \quad z' \in \bigotimes^q V, \quad z'' \in \bigotimes^r V;$$
$$w \wedge w' \wedge w'', w \in \bigwedge^p V, \quad w' \in \bigwedge^q V, \quad w'' \in \bigwedge^r V;$$
$$y \vee y' \vee y'', y \in \bigvee^p V, \quad y' \in \bigvee^q V, \quad y'' \in \bigvee^r V$$
are meaningful as well as similar expressions with more than three factors.
7. If $w \in \bigwedge^p V, w' \in \bigwedge^q V$, then
$$w' \wedge w = (-1)^{pq} w \wedge w'.$$
8. If $y \in \bigvee^p V, y' \in \bigvee^q V$, then
$$y' \vee y = y \vee y'.$$

Examples:

1. When the vector space is the dual $V^* = L(V; F)$ of a vector space and we choose as the models of tensor powers of V^* the spaces of multilinear forms (with the usual tensor multiplication), then the image of the tensor multiplication $\varphi \otimes \psi$ ($\varphi \in L^p(V; F)$ and $\psi \in L^q(V; F)$) on $(\mathbf{v}_1, \ldots, \mathbf{v}_{p+q})$ is given by the equality
$$(\varphi \otimes \psi)(\mathbf{v}_1, \ldots, \mathbf{v}_{p+q}) = \varphi(\mathbf{v}_1, \ldots, \mathbf{v}_p)\psi(\mathbf{v}_{p+1}, \ldots, \mathbf{v}_{p+q}).$$
2. When the vector space is the dual $V^* = L(V; F)$ of a vector space and we choose as the models for the tensor powers of V^* the spaces of multilinear forms (with the usual tensor multiplication), the alt multiplication of $\varphi \in A^p(V; F)$ and $\psi \in A^q(V; F)$ takes the form
$$(\varphi \wedge \psi)(\mathbf{v}_1, \ldots, \mathbf{v}_{p+q}) = \frac{1}{p!q!} \sum_{\sigma \in S_{p+q}} \mathrm{sgn}(\sigma)\varphi(\mathbf{v}_{\sigma(1)}, \ldots, \mathbf{v}_{\sigma(p)})\psi(\mathbf{v}_{\sigma(p+1)}, \ldots, \mathbf{v}_{\sigma(p+q)}).$$
3. The equality in Example 2 has an alternative expression that can be seen as a "Laplace expansion" for antisymmetric forms

$$(\varphi \wedge \psi)(\mathbf{v}_1,\ldots,\mathbf{v}_{p+q}) = \sum_{\alpha\in Q_{p,p+q}} \operatorname{sgn}(\widetilde{\alpha})\varphi(\mathbf{v}_{\alpha(1)},\ldots,\mathbf{v}_{\alpha(p)})\psi(\mathbf{v}_{\alpha^c(1)},\ldots,\mathbf{v}_{\alpha^c(q)}).$$

4. In the case $p = 1$, the equality in Example 3 has the form

$$(\varphi \wedge \psi)(\mathbf{v}_1,\ldots,\mathbf{v}_{q+1}) = \sum_{j=1}^{q+1}(-1)^{j+1}\varphi(\mathbf{v}_j)\psi(\mathbf{v}_1,\ldots,\mathbf{v}_{j-1},\mathbf{v}_{j+1},\ldots,\mathbf{v}_{q+1}).$$

5. When the vector space is the dual $V^* = L(V; F)$ of a vector space and we choose as the models of tensor powers of V^* the spaces of multilinear forms (with the usual tensor multiplication), the value of sym multiplication of $\varphi \in S^p(V; F)$ and $\psi \in S^q(V; F)$ on $(\mathbf{v}_1,\ldots,\mathbf{v}_{p+q})$ is

$$(\varphi \vee \psi)(\mathbf{v}_1,\ldots,\mathbf{v}_{p+q}) = \frac{1}{p!q!}\sum_{\sigma\in S_{p+q}} \varphi(\mathbf{v}_{\sigma(1)},\ldots,\mathbf{v}_{\sigma(p)})\psi(\mathbf{v}_{\sigma(p+1)},\ldots,\mathbf{v}_{\sigma(p+q)}).$$

6. The equality in Example 5 has an alternative expression that can be seen as a "Laplace expansion" for symmetric forms

$$(\varphi \vee \psi)(\mathbf{v}_1,\ldots,\mathbf{v}_{p+q}) = \sum_{\alpha\in Q_{p,p+q}} \varphi(\mathbf{v}_{\alpha(1)},\ldots,\mathbf{v}_{\alpha(p)})\psi(\mathbf{v}_{\alpha^c(1)},\ldots,\mathbf{v}_{\alpha^c(q)}).$$

7. In the case $p = 1$, the equality in Example 6 has the form

$$(\varphi \vee \psi)(\mathbf{v}_1,\ldots,\mathbf{v}_{q+1}) = \sum_{j=1}^{q+1}\varphi(\mathbf{v}_j)\psi(\mathbf{v}_1,\ldots,\mathbf{v}_{j-1},\mathbf{v}_{j+1},\ldots,\mathbf{v}_{q+1}).$$

13.8 Associated Maps

Definitions:

Let $\theta \in L(V; U)$. The linear map $\theta \otimes \cdots \otimes \theta$ from $\bigotimes^m V$ into $\bigotimes^m U$ (the tensor product of m copies of θ) will be denoted by $\bigotimes^m \theta$. The subspaces $\bigwedge^m V$ and $\bigvee^m V$ are mapped by $\bigotimes^m \theta$ into $\bigwedge^m U$ and $\bigvee^m U$, respectively. The restriction of $\bigotimes^m \theta$ to $\bigwedge^m V$ and to $\bigvee^m V$ will be respectively denoted, $\bigwedge^m \theta$ e $\bigvee^m \theta$.

Facts:

The following facts can be found in [Mar73, Chap. 2].

1. Let $\mathbf{v}_1, \ldots, \mathbf{v}_m \in V$. The following properties hold:
   (a) $\bigwedge^m \theta(\mathbf{v}_1 \wedge \cdots \wedge \mathbf{v}_m) = \theta(\mathbf{v}_1) \wedge \cdots \wedge \theta(\mathbf{v}_m)$.
   (b) $\bigvee^m \theta(\mathbf{v}_1 \vee \cdots \vee \mathbf{v}_m) = \theta(\mathbf{v}_1) \vee \cdots \vee \theta(\mathbf{v}_m)$.
2. Let $\theta \in L(V;U)$ and $\eta \in L(W,V)$. The following equalities hold:
   (a) $\bigwedge^m(\theta\eta) = \bigwedge^m(\theta)\bigwedge^m(\eta)$.
   (b) $\bigvee^m(\theta\eta) = \bigvee^m(\theta)\bigvee^m(\eta)$.
3. $\bigwedge^m(I_V) = I_{\bigwedge^m V}$; $\bigvee^m(I_V) = I_{\bigvee^m V}$.
4. Let $\theta, \eta \in L(V;U)$ and assume that rank $(\theta) > m$. Then

$$\bigwedge^m \theta = \bigwedge^m \eta$$

   if and only if $\theta = a\eta$ and $a^m = 1$.
5. Let $\theta, \eta \in L(V;U)$. Then $\bigvee^m \theta = \bigvee^m \eta$ if and only if $\theta = a\eta$ and $a^m = 1$.
6. If θ is one-to-one (respectively onto), then $\bigwedge^m \theta$ and $\bigvee^m \theta$ are one-to-one (respectively onto).
   From now on θ is a linear operator on the n-dimensional vector space V.
7. Considering $\bigwedge^n \theta$ as an operator in the one-dimensional space $\bigwedge^n V$,

$$\left(\bigwedge^n \theta\right)(z) = \det(\theta)z, \text{ for all } z \in \bigwedge^n V.$$

8. If the characteristic polynomial of θ is

$$\mathbf{p}_\theta(x) = x^n + \sum_{i=1}^{n}(-1)^i a_i x^{n-i},$$

   then

$$a_i = \operatorname{tr}\left(\bigwedge^i \theta\right), \quad i = 1, \ldots, n.$$

9. If θ has spectrum $\sigma(\theta) = \{\lambda_1, \ldots, \lambda_n\}$, then

$$\sigma\left(\bigwedge^m \theta\right) = \left\{\prod_{i=1}^m \lambda_{\alpha(i)}\right\}_{\alpha \in Q_{m,n}}, \quad \sigma\left(\bigvee^m \theta\right) = \left\{\prod_{i=1}^m \lambda_{\alpha(i)}\right\}_{\alpha \in G_{m,n}}.$$

10.

$$\det\left(\bigwedge^m \theta\right) = \det(\theta)^{\binom{n-1}{m-1}}, \qquad \det\left(\bigvee^m \theta\right) = \det(\theta)^{\binom{m+n-1}{m-1}}.$$

Examples:

1. Let A be the matrix of the linear operator $\theta \in L(V;V)$ in the basis $(\mathbf{b}_1, \ldots, \mathbf{b}_n)$. The linear operator on $\bigwedge^m V$ whose matrix in the basis $(\mathbf{b}_\alpha^\wedge)_{\alpha \in Q_{m,n}}$ is the mth compound of A is $\bigwedge^m \theta$.

13.9 Tensor Algebras

Definitions:

Let A be an F-algebra and $(A_k)_{k\in\mathbb{N}}$ a family of vector subspaces of A. The algebra A is **graded** by $(A_k)_{k\in\mathbb{N}}$ if the following conditions are satisfied:

(a) $A = \bigoplus_{k\in\mathbb{N}} A_k$.
(b) $A_i A_j \subseteq A_{i+j}$ for every $i, j \in \mathbb{N}$.

The elements of A_k are known as **homogeneous of degree** k, and the elements of $\bigcup_{n\in\mathbb{N}} A_k$ are called **homogeneous.**

By condition (a), every element of A can be written uniquely as a sum of (a finite number of nonzero) homogeneous elements, i.e., given $u \in A$ there exist uniquely determined $\mathbf{u}_k \in A_k, k \in \mathbb{N}$ satisfying

$$\mathbf{u} = \sum_{k\in\mathbb{N}} \mathbf{u}_k.$$

These elements are called **homogeneous components of u**. The summand of degree k in the former equation is denoted by $[\mathbf{u}]_k$.

From now on V is a finite dimensional vector space over F of dimension n. As before $\bigotimes^k V$ denotes the kth-tensor power of V.

Denote by $\bigotimes V$ the external direct sum of the vector spaces $\bigotimes^k V$, $k \in \mathbb{N}$. If $z_i \in \bigotimes^i V$, z_i is identified with the sequence $z \in \bigotimes V$ whose ith coordinate is z_i and the remaining coordinates are 0. Therefore, after this identification,

$$\bigotimes V = \bigoplus_{k\in\mathbb{N}} \bigotimes^k V.$$

Consider in $\bigotimes V$ the multiplication $(x, y) \mapsto x \otimes y$ defined for $x, y \in \bigotimes V$ by

$$[x \otimes y]_k = \sum_{\substack{r,s\in\mathbb{N}\\ r+s=k}} [x]_r \otimes [y]_s, \quad k \in \mathbb{N},$$

where $[x]_r \otimes [y]_s$ is the (r, s)-tensor multiplication of $[x]_r$ and $[y]_s$ introduced in the definitions of Section 13.7. The vector space $\bigotimes V$ equipped with this multiplication is called the **tensor algebra** on V.

Denote by $\bigwedge V$ the external direct sum of the vector spaces $\bigwedge^k V$, $k \in \mathbb{N}$. If $z_i \in \bigwedge^i V$, z_i is identified with the sequence $z \in \bigwedge V$ whose ith coordinate is z_i and the remaining coordinates are 0. Therefore, after this identification,

$$\bigwedge V = \bigoplus_{k\in\mathbb{N}} \bigwedge^k V.$$

Recall that $\bigwedge^k V = \{0\}$ if $k > n$. Then

$$\bigwedge V = \bigoplus_{k=0}^{n} \bigwedge^k V$$

and the elements of $\bigwedge V$ can be uniquely written in the form

$$z_0 + z_1 + \cdots + z_n, \quad z_i \in \bigwedge^i V, \quad i = 0, \ldots, n.$$

Consider in $\bigwedge V$ the multiplication $(x, y) \mapsto x \wedge y$ defined, for $x, y \in \bigwedge V$, by

$$[x \wedge y]_k = \sum_{\substack{r,s\in\{0,\ldots,n\}\\ r+s=k}} [x]_r \wedge [y]_s, \quad k \in \{0, \ldots, n\},$$

where $[x]_r \wedge [y]_s$ is the (r, s)-alt multiplication of $[x]_r$ and $[y]_s$ referred in definitions of Section 13.7. The vector space $\bigwedge V$ equipped with this multiplication is called the **Grassmann algebra** on V.

Denote by $\bigvee V$ the external direct sum of the vector spaces $\bigvee^k V$, $k \in \mathbb{N}$.

If $z_i \in \bigvee^i V$, we identify z_i with the sequence $z \in \bigvee V$ whose ith coordinate is z_i and the remaining coordinates are 0. Therefore, after this identification

$$\bigvee V = \bigoplus_{k\in\mathbb{N}} \bigvee^k V.$$

Consider in $\bigvee V$ the multiplication $(x, y) \mapsto x \vee y$ defined for $x, y \in \bigvee V$ by

$$[x \vee y]_k = \sum_{\substack{r,s \in \mathbb{N} \\ r+s=k}} [x]_r \vee [y]_s, \quad k \in \mathbb{N},$$

where $[x]_r \vee [y]_s$ is the (r, s)-sym multiplication of $[x]_r$ and $[y]_s$ referred in definitions of Section 13.7.

The vector space $\bigvee V$ equipped with this multiplication is called the **symmetric algebra** on V.

Facts:

The following facts can be found in [Mar73, Chap. 3] and [Gre67, Chaps. II and III].

1. The vector space $\bigotimes V$ with the multiplication $(x, y) \mapsto x \otimes y$ is an algebra over F graded by $(\bigotimes^k V)_{k \in \mathbb{N}}$, whose identity is the identity of $F = \bigotimes^0 V$.
2. The vector space $\bigwedge V$ with the multiplication $(x, y) \mapsto x \wedge y$ is an algebra over F graded by $(\bigwedge^k V)_{k \in \mathbb{N}}$ whose identity is the identity of $F = \bigwedge^0 V$.
3. The vector space $\bigvee V$ with the multiplication $(x, y) \mapsto x \vee y$ is an algebra over F graded by $(\bigvee^k V)_{k \in \mathbb{N}}$ whose identity is the identity of $F = \bigvee^0 V$.
4. The F-algebra $\bigotimes V$ does not have zero divisors.
5. Let B be an F-algebra and θ a linear map from V into B satisfying $\theta(\mathbf{x})\theta(\mathbf{y}) = -\theta(\mathbf{y})\theta(\mathbf{x})$ for all $\mathbf{x}, \mathbf{y} \in V$. Then there exists a unique algebra homomorphism h from $\bigwedge V$ into B satisfying $h|V = \theta$.
6. Let B be an F-algebra and θ a linear map from V into B satisfying $\theta(\mathbf{x})\theta(\mathbf{y}) = \theta(\mathbf{y})\theta(\mathbf{x})$, for all $\mathbf{x}, \mathbf{y} \in V$. Then there exists a unique algebra homomorphism h from $\bigvee V$ into B satisfying $h|V = \theta$.
7. Let $(\mathbf{b}_1, \ldots, \mathbf{b}_n)$ be a basis of V. The symmetric algebra $\bigvee^m V$ is isomorphic to the algebra of polynomials in n indeterminates, $F[x_1, \ldots, x_n]$, by the algebra isomorphism whose restriction to V is the linear map that maps $\mathbf{b}_i$ into x_i, $i = 1, \ldots, n$.

Examples:

1. Let $x_1, \ldots, x_n$ be n distinct indeterminates. Let V be the vector space of the formal linear combinations with coefficients in F in the indeterminates $x_1, \ldots, x_n$. The tensor algebra on V is the algebra of the polynomials in the noncommuting indeterminates $x_1, \ldots, x_n$ ([Coh03], [Jac64]). This algebra is denoted by

$$F\langle x_1, \ldots, x_n \rangle.$$

The elements of this algebra are of the form

$$f(x_1, \ldots, x_n) = \sum_{m \in \mathbb{N}} \sum_{\alpha \in \Gamma_{m,n}} c_\alpha x_{\alpha(1)} \otimes \cdots \otimes x_{\alpha(m)},$$

with all but a finite number of the coefficients c_α equal to zero.

13.10 Tensor Product of Inner Product Spaces

Unless otherwise stated, within this section V, U, and W, as well as these letters subscripts, superscripts, or accents, are finite dimensional vector spaces over $\mathbb{R}$ or over $\mathbb{C}$, equipped with an inner product.

The inner product of V is denoted by $\langle\ ,\ \rangle_V$. When there is no risk of confusion $\langle\ ,\ \rangle$ is used instead. In this section F means either the field $\mathbb{R}$ or the field $\mathbb{C}$.

Definitions:

Let θ be a linear map from V into W. The notation θ^* will be used for the **adjoint** of θ (i.e., the linear map from W into V satisfying $\langle \theta(\mathbf{x}), \mathbf{y} \rangle = \langle \mathbf{x}, \theta^*(\mathbf{y}) \rangle$ for all $\mathbf{x} \in V$ and $\mathbf{y} \in W$).

The unique inner product $\langle\ ,\ \rangle$ on $V_1 \otimes \cdots \otimes V_m$ satisfying, for every $\mathbf{v}_i, \mathbf{u}_i \in V_i$, $i = 1, \ldots, m$,

$$\langle \mathbf{v}_1 \otimes \cdots \otimes \mathbf{v}_m, \mathbf{u}_1 \otimes \cdots \otimes \mathbf{u}_m \rangle = \prod_{i=1}^{m} \langle \mathbf{v}_i, \mathbf{u}_i \rangle_{V_i},$$

is called **induced inner product** associated with the inner products $\langle\ ,\ \rangle_{V_i}$, $i = 1, \ldots, m$.

For each $\mathbf{v} \in V$, let $f_{\mathbf{v}} \in V^*$ be defined by $f_{\mathbf{v}}(\mathbf{u}) = \langle \mathbf{u}, \mathbf{v} \rangle$. The inverse of the map $\mathbf{v} \rightarrow f_{\mathbf{v}}$ is denoted by ϱ_V (briefly ϱ). The inner product on V^*, defined by

$$\langle f, g \rangle = \langle \varrho(g), \varrho(f) \rangle_V,$$

is called the **dual** of $\langle\ ,\ \rangle_V$.

Let U, V be inner product spaces over F. We consider defined in $L(V; U)$ the **Hilbert–Schmidt** inner product, i.e., the inner product defined, for $\theta, \eta \in L(V; U)$, by $\langle \theta, \eta \rangle = \operatorname{tr}(\eta^* \theta)$.

From now on $V_1 \otimes \cdots \otimes V_m$ is assumed to be equipped with the inner product induced by the inner products $\langle\ ,\ \rangle_{V_i}$, $i = 1, \ldots, m$.

Facts:

The following facts can be found in [Mar73, Chap. 2].

1. The map $\mathbf{v} \rightarrow f_{\mathbf{v}}$ is bijective-linear if $F = \mathbb{R}$ and conjugate-linear (i.e., $c\mathbf{v} \mapsto \bar{c} f_{\mathbf{v}}$) if $F = \mathbb{C}$.
2. If $(\mathbf{b}_{i1}, \ldots, \mathbf{b}_{in_i})$ is an orthonormal basis of V_i, $i = 1, \ldots, m$, then $\{\mathbf{b}_\alpha^\otimes : \alpha \in \Gamma(n_1, \ldots, n_m)\}$ is an orthonormal basis of $V_1 \otimes \cdots \otimes V_m$.
3. Let $\theta_i \in L(V_i; W_i)$, $i = 1, \ldots, m$, with adjoint map $\theta_i^* \in L(W_i, V_i)$. Then,
$$(\theta_1 \otimes \cdots \otimes \theta_m)^* = \theta_1^* \otimes \cdots \otimes \theta_m^*.$$
4. If $\theta_i \in L(V_i; V_i)$ is Hermitian (normal, unitary), $i = 1, \ldots, m$, then $\theta_1 \otimes \cdots \otimes \theta_m$ is also Hermitian (normal, unitary).
5. Let $\theta \in L(V; V)$. If $\bigotimes^m \theta$ ($\bigvee^m \theta$) is normal, then θ is normal.
6. Let $\theta \in L(V; V)$. Assume that θ is a linear operator on V with rank greater than m. If $\bigwedge^m \theta$ is normal, then θ is normal.
7. If $\mathbf{u}_1, \ldots, \mathbf{u}_m, \mathbf{v}_1, \ldots, \mathbf{v}_m \in V$:
$$\langle \mathbf{u}_1 \wedge \cdots \wedge \mathbf{u}_m, \mathbf{v}_1 \wedge \cdots \wedge \mathbf{v}_m \rangle = m! \det \langle \mathbf{u}_i, \mathbf{v}_j \rangle,$$
$$\langle \mathbf{u}_1 \vee \cdots \vee \mathbf{u}_m, \mathbf{v}_1 \vee \cdots \vee \mathbf{v}_m \rangle = m! \operatorname{per} \langle \mathbf{u}_i, \mathbf{v}_j \rangle.$$
8. Let $(\mathbf{b}_1, \ldots, \mathbf{b}_n)$ be an orthonormal basis of V. Then the basis $(\mathbf{b}_\alpha^\otimes)_{\alpha \in \Gamma_{m,n}}$ is an orthonormal basis of $\bigotimes^m V$, $(\sqrt{\frac{1}{m!}} \mathbf{b}_\alpha^\wedge)_{\alpha \in Q_{m,n}}$ is an orthonormal basis of $\bigwedge^m V$, and $(\sqrt{\frac{1}{m! \lambda_\alpha!}} \mathbf{b}_\alpha^\vee)_{\alpha \in G_{m,n}}$ is an orthonormal basis of $\bigvee^m V$.

Examples:

The field F (recall that $F = \mathbb{R}$ or $F = \mathbb{C}$) has an inner product, $(a, b) \mapsto \langle a, b \rangle = a\bar{b}$. This inner product is called the **standard** inner product in F and it is the one assumed to equip F from now on.

1. When we choose F as the mth tensor power of F with the field multiplication as the tensor multiplication, then the canonical inner product is the inner product induced in $\bigotimes^m F$ by the canonical inner product.
2. When we assume V as the tensor product of F and V with the tensor multiplication $a \otimes \mathbf{v} = a\mathbf{v}$, the inner product induced by the canonical inner product of F and the inner product of V is the inner product of V.

3. Consider $L(V;U)$ as the tensor product of U and V^* by the tensor multiplication $(\mathbf{u} \otimes f)(\mathbf{v}) = f(\mathbf{v})\mathbf{u}$. Assume in V^* the inner product dual of the inner product of V. Then, if $(\mathbf{v}_1, \ldots, \mathbf{v}_n)$ is an orthonormal basis of V and $\theta, \eta \in L(V;U)$, we have

$$\langle \theta, \eta \rangle = \sum_{j=1}^{m} \langle \theta(\mathbf{v}_j), \eta(\mathbf{v}_j) \rangle = \operatorname{tr}(\eta^* \theta),$$

i.e., the associated inner product of $L(V;U)$ is the Hilbert–Schmidt one.

4. Consider $F^{m\times n}$ as the tensor product of F^m and F^n by the tensor multiplication described in Example 1 in section 13.2. Then if we consider in F^m and F^n the usual inner product we get in $F^{m\times n}$ as the induced inner product, the inner product

$$(A, B) \mapsto \operatorname{tr}(\overline{B}^T A) = \sum_{i,j} a_{ij} \overline{b_{i,j}}.$$

5. Assume that in V_i^* is defined the inner product dual of $\langle\ ,\ \rangle_{V_i}$, $i = 1, \ldots, m$. Then choosing $L(V_1, \ldots, V_m; F)$ as the tensor product of $V_1^*, \ldots, V_m^*$, with the usual tensor multiplication, the inner product of $L(V_1, \ldots, V_m; F)$ induced by the duals of inner products on V_i^*, $i = 1, \ldots, m$ is given by the equalities

$$\langle \varphi, \psi \rangle = \sum_{\alpha \in \Gamma} \varphi(\mathbf{b}_{1,\alpha(1)}, \ldots, \mathbf{b}_{m,\alpha(m)}) \overline{\psi(\mathbf{b}_{1,\alpha(1)}, \ldots, \mathbf{b}_{m,\alpha(m)})}.$$

13.11 Orientation and Hodge Star Operator

In this section, we assume that all vector spaces are real finite dimensional inner product spaces.

Definitions:

Let V be a one-dimensional vector space. The equivalence classes of the equivalence relation $\sim$, defined by the condition $\mathbf{v} \sim \mathbf{v}'$ if there exists a positive real number $a > 0$ such that $\mathbf{v}' = a\mathbf{v}$, partitions the set of nonzero vectors of V into two subsets.

Each one of these subsets is known as an open **half-line**.

An **orientation** of V is a choice of one of these subsets. The fixed open half-line is called the **positive half-line** and its vectors are known as **positive**. The other open half-line of V is called **negative half-line**, and its vectors are also called **negative**.

The field $\mathbb{R}$, regarded as one-dimensional vector space, has a "natural" orientation that corresponds to choose as positive half-line the set of positive numbers.

If V is an n-dimensional vector space, $\bigwedge^n V$ is a one-dimensional vector space (recall Fact 22 in section 13.6). An **orientation of** V is an orientation of $\bigwedge^n V$.

A basis $(\mathbf{b}_1, \ldots, \mathbf{b}_n)$ of V is said to be **positively oriented** if $\mathbf{b}_1 \wedge \cdots \wedge \mathbf{b}_n$ is positive and **negatively oriented** if $\mathbf{b}_1 \wedge \cdots \wedge \mathbf{b}_n$ is negative.

Throughout this section $\bigwedge^m V$ will be equipped with the inner product $\langle\ ,\ \rangle_\wedge$, a positive multiple of the induced the inner product, defined by

$$\langle z, w \rangle_\wedge = \frac{1}{m!} \langle z, w \rangle,$$

where the inner product on the right-hand side of the former equality is the inner product of $\bigotimes^m V$ induced by the inner product of V. This is also the inner product that is considered whenever the norm of antisymmetric tensors is referred.

The positive tensor of norm 1 of $\bigwedge^n V$, u_V, is called **fundamental tensor of** V or **element of volume** of V.

Let V be a real oriented inner product space . Let $0 \leq m \leq n$.

The **Hodge star operator** is the linear operator $\bigstar_m$ (denoted also by $\bigstar$) from $\bigwedge^m V$ into $\bigwedge^{n-m} V$ defined by the following condition:

$$\langle \bigstar_m(w), w' \rangle_\wedge u_V = w \wedge w', \text{ for all } w' \in \bigwedge^{n-m} V.$$

Let $n \geq 1$ and let V be an n-dimensional oriented inner product space over $\mathbb{R}$. The **external product on** V is the map

$$(\mathbf{v}_1, \dots, \mathbf{v}_{n-1}) \mapsto \mathbf{v}_1 \times \cdots \times \mathbf{v}_{n-1} = \bigstar_{n-1}(\mathbf{v}_1 \wedge \cdots \wedge \mathbf{v}_{n-1}),$$

from V^{n-1} into V.

Facts:

The following facts can be found in [Mar75, Chap. 1] and [Sch75, Chap. 1].

1. If $(\mathbf{b}_1, \dots, \mathbf{b}_n)$ is a positively oriented orthonormal basis of V, then $u_V = \mathbf{b}_1 \wedge \cdots \wedge \mathbf{b}_n$.
2. If $(\mathbf{b}_1, \dots, \mathbf{b}_n)$ is a positively oriented orthonormal basis of V, then

$$\bigstar_m \mathbf{b}^\wedge_\alpha = \operatorname{sgn}(\widetilde{\alpha}) \mathbf{b}^\wedge_{\alpha^c}, \quad \alpha \in Q_{m,n},$$

where $\widetilde{\alpha}$ and α^c are defined in Section 13.7.
3. Let $(\mathbf{v}_1, \dots, \mathbf{v}_n)$ and $(\mathbf{u}_1, \dots, \mathbf{u}_n)$ be two bases of V and $\mathbf{v}_j = \sum a_{ij} \mathbf{u}_i$, $j = 1, \dots, n$. Let $A = [a_{ij}]$. Since (recall Fact 29 in Section 13.6)

$$\mathbf{v}_1 \wedge \cdots \wedge \mathbf{v}_n = \det(A) \mathbf{u}_1 \wedge \cdots \wedge \mathbf{u}_n,$$

two bases have the same orientation if and only if their transition matrix has a positive determinant.
4. $\bigstar$ is an isometric isomorphism.
5. $\bigstar_0$ is the linear isomorphism that maps $1 \in \mathbb{R}$ onto the fundamental tensor.
6. $\bigstar_m \bigstar_{n-m} = (-1)^{m(n-m)} I_{\bigwedge^{n-m} V}$.

Let V be an n-dimensional oriented inner product space over $\mathbb{R}$.

7. If $m \neq 0$ and $m \neq n$, the Hodge star operator maps the set of decomposable elements of $\bigwedge^m V$ onto the set of decomposable elements of $\bigwedge^{n-m} V$.
8. Let $(\mathbf{x}_1, \dots, \mathbf{x}_m)$ be a linearly independent family of vectors of V. Then

$$\mathbf{y}_1 \wedge \cdots \wedge \mathbf{y}_{n-m} = \bigstar_m(\mathbf{x}_1 \wedge \cdots \wedge \mathbf{x}_m)$$

if and only if the following three conditions hold:

(a) $\mathbf{y}_1, \dots, \mathbf{y}_{n-m} \in \operatorname{Span}(\{\mathbf{x}_1, \dots, \mathbf{x}_m\})^\perp$;

(b) $\|\mathbf{y}_1 \wedge \cdots \wedge \mathbf{y}_{n-m}\| = \|\mathbf{x}_1 \wedge \cdots \wedge \mathbf{x}_m\|$;

(c) $(\mathbf{x}_1, \dots, \mathbf{x}_m, \mathbf{y}_1, \dots, \mathbf{y}_{n-m})$ is a positively oriented basis of V.

9. If $(\mathbf{v}_1, \ldots, \mathbf{v}_{n-1})$ is linearly independent, $\mathbf{v}_1 \times \cdots \times \mathbf{v}_{n-1}$ is completely characterized by the following three conditions:
   (a) $\mathbf{v}_1 \times \cdots \times \mathbf{v}_{n-1} \in \mathrm{Span}(\{\mathbf{v}_1, \ldots, \mathbf{v}_{n-1}\})^{\perp}$.
   (b) $\|\mathbf{v}_1 \times \cdots \times \mathbf{v}_{n-1}\| = \|\mathbf{v}_1 \wedge \cdots \wedge \mathbf{v}_{n-1}\|$.
   (c) $(\mathbf{v}_1, \ldots, \mathbf{v}_{n-1}, \mathbf{v}_1 \times \cdots \times \mathbf{v}_{n-1})$ is a positively oriented basis of V.
10. Assume $V^* = L(V; F)$, with $\dim(V) \geq 1$, is equipped with the dual inner product. Consider $L^m(V; F)$ as a model for the mth tensor power of V^* with the usual tensor multiplication. Then $\bigwedge^m V^* = A^m(V; F)$. If λ is an antisymmetric form in $A^m(V; F)$, then $\bigstar_m(\lambda)$ is the form whose value in $(\mathbf{v}_1, \ldots, \mathbf{v}_{n-m})$ is the component in the fundamental tensor of $\lambda \wedge \varrho^{-1}(\mathbf{v}_1) \wedge \cdots \wedge \varrho^{-1}(\mathbf{v}_{n-m})$, where ϱ is defined in the definition of section 13.10.

$$\bigstar_m(\lambda)(\mathbf{v}_1, \ldots, \mathbf{v}_{n-m}) u_{V^*} = \lambda \wedge \varrho^{-1}(\mathbf{v}_1) \wedge \cdots \wedge \varrho^{-1}(\mathbf{v}_{n-m}).$$

11. Assuming the above setting for the Hodge star operator, the external product of $\mathbf{v}_1, \ldots, \mathbf{v}_{n-1}$ is the image by ϱ of the form $(u_{V^*})_{\mathbf{v}_1, \ldots, \mathbf{v}_{n-1}}$ (recall that $(u_{V^*})_{\mathbf{v}_1, \ldots, \mathbf{v}_{n-1}}(\mathbf{v}_n) = u_{V^*}(\mathbf{v}_1, \ldots, \mathbf{v}_{n-1}, \mathbf{v}_n)$), i.e.,

$$\mathbf{v}_1 \times \cdots \times \mathbf{v}_{n-1} = \varrho((u_{V^*})_{\mathbf{v}_1, \ldots, \mathbf{v}_{n-1}}).$$

The preceeding formula can be unfolded by stating that for each $v \in V$, $\langle \mathbf{v}, \mathbf{v}_1 \times \cdots \times \mathbf{v}_{n-1} \rangle = u_{V^*}(\mathbf{v}_1, \ldots, \mathbf{v}_{n-1}, \mathbf{v})$.

Examples:

1. If V has dimension 0, the isomorphism $\bigstar_0$ from $\bigwedge^0 V = \mathbb{R}$ into $\bigwedge^0 V = \mathbb{R}$ is either the identity (in the case we choose the natural orientation of V) or $-I$ (in the case we fix the nonnatural orientation of V).
2. When V has dimension 2, the isomorphism $\bigstar_1$ is usually denoted by J. It has the property $J^2 = -I$ and corresponds to the positively oriented rotation of $\pi/2$.
3. Assume that V has dimension 2. Then the external product is the isomorphism J.
4. If $\dim(V) = 3$, the external product is the well-known cross product.

References

[Bou89] N. Bourbaki, *Algebra*, Springer-Verlag, Berlin (1989).

[Coh03] P. M. Cohn, *Basic Algebra–Groups Rings and Fields*, Springer-Verlag, London (2003).

[Dol04] Igor V. Dolgachev, *Lectures on Invarint Theory*. Online publication, 2004. Cambridge University Press, Cambridge-New York (1982).

[Gre67] W. H. Greub, *Multilinear Algebra*, Springer-Verlag, Berlin (1967).

[Jac64] Nathan Jacobson, *Structure of Rings*, American Mathematical Society Publications, Volume XXXVII, Providence, RI (1964).

[Mar73] Marvin Marcus, *Finite Dimensional Multilinear Algebra, Part I*, Marcel Dekker, New York (1973).

[Mar75] Marvin Marcus, *Finite Dimensional Multilinear Algebra, Part II*, Marcel Dekker, New York (1975).

[Mer97] Russell Merris, *Multilinear Algebra*, Gordon Breach, Amsterdam (1997).

[Sch75] Laurent Schwartz, *Les Tenseurs*, Hermann, Paris (1975).

[Spi79] Michael Spivak, *A Comprehensive Introduction to Differential Geometry*, Volume I, 2nd ed., Publish or Perish, Inc., Wilmington, DE (1979).

14

Matrix Equalities and Inequalities

Michael Tsatsomeros
Washington State University

In this chapter, we have collected classical equalities and inequalities regarding the eigenvalues, the singular values, the determinant, and the dimensions of the fundamental subspaces of a matrix. Also included is a section on identities for matrix inverses. The majority of these results can be found in comprehensive books on linear algebra and matrix theory, although some are of specialized nature. The reader is encouraged to consult, e.g., [HJ85], [HJ91], [MM92], or [Mey00] for details, proofs, and further bibliography.

14.1 Eigenvalue Equalities and Inequalities

The majority of the facts in this section concern general matrices; however, some classical and frequently used results on eigenvalues of Hermitian and positive definite matrices are also included. For the latter, see also Chapter 8 and [HJ85, Chap. 4]. Many of the definitions and some of the facts in this section are also given in Section 4.3.

Facts:

1. [HJ85, Chap. 1] Let $A \in F^{n\times n}$, where $F = \mathbb{C}$ or any algebraically closed field. Let $p_A(x) = \det(xI - A)$ be the characteristic polynomial of A, and $\lambda_1, \lambda_2, \ldots, \lambda_n$ be the eigenvalues of A. Denote by $S_k(\lambda_1, \ldots, \lambda_n)(k = 1, 2, \ldots, n)$ the kth elementary symmetric function of the eigenvalues (here abbreviated $S_k(\lambda)$), and by $S_k(A)$ the sum of all $k \times k$ principal minors of A. Then

   - The characteristic polynomial satisfies

$$\begin{aligned} p_A(x) &= (x - \lambda_1)(x - \lambda_2)\cdots(x - \lambda_n) \\ &= x^n - S_1(\lambda)x^{n-1} + S_2(\lambda)x^{n-2} + \cdots + (-1)^{n-1}S_{n-1}(\lambda)x + (-1)^n S_n(\lambda) \\ &= x^n - S_1(A)x^{n-1} + S_2(A)x^{n-2} + \cdots + (-1)^{n-1}S_{n-1}x + (-1)^n S_n(A). \end{aligned}$$

- $S_k(\lambda) = S_k(\lambda_1, \dots, \lambda_n) = S_k(A) (k = 1, 2, \dots, n)$.
- $\operatorname{tr} A = S_1(A) = \sum_{i=1}^n a_{ii} = \sum_{i=1}^n \lambda_i$ and $\det A = S_n(A) = \prod_{i=1}^n \lambda_i$.

2. [HJ85, (1.2.13)] Let $A(i)$ be obtained from $A \in \mathbb{C}^{n\times n}$ by deleting row and column i. Then
$$\frac{d}{dx} p_A(x) = \sum_{i=1}^n p_{A(i)}(x).$$

Facts 3 to 9 are collected, together with historical commentary, proofs, and further references, in [MM92, Chap. III].

3. (Hirsch and Bendixson) Let $A = [a_{ij}] \in \mathbb{C}^{n\times n}$ and λ be an eigenvalue of A. Denote $B = [b_{ij}] = (A + A^*)/2$ and $C = [c_{ij}] = (A - A^*)/(2i)$. Then the following inequalities hold:
$$|\lambda| \leq n \max_{i,j} |a_{ij}|,$$
$$|\mathrm{Re}\lambda| \leq n \max_{i,j} |b_{ij}|,$$
$$|\mathrm{Im}\lambda| \leq n \max_{i,j} |c_{ij}|.$$
Moreover, if $A + A^T \in \mathbb{R}^{n\times n}$, then
$$|\mathrm{Im}\lambda| \leq \max_{i,j} |c_{ij}| \sqrt{\frac{n(n-1)}{2}}.$$

4. (Pick's inequality) Let $A = [a_{ij}] \in \mathbb{R}^{n\times n}$ and λ be an eigenvalue of A. Denote $C = [c_{ij}] = (A - A^T)/2$. Then
$$|\mathrm{Im}\lambda| \leq \max_{i,j} |c_{ij}| \cot\left(\frac{\pi}{2n}\right).$$

5. Let $A = [a_{ij}] \in \mathbb{C}^{n\times n}$ and λ be an eigenvalue of A. Denote $B = [b_{ij}] = (A + A^*)/2$ and $C = [c_{ij}] = (A - A^*)/(2i)$. Then the following inequalities hold:
$$\min\{\mu : \mu \in \sigma(B)\} \leq \mathrm{Re}\lambda \leq \max\{\mu : \mu \in \sigma(B)\},$$
$$\min\{\nu : \nu \in \sigma(C)\} \leq \mathrm{Im}\lambda \leq \max\{\nu : \nu \in \sigma(C)\}.$$

6. (Schur's inequality) Let $A = [a_{ij}] \in \mathbb{C}^{n\times n}$ have eigenvalues λ_j $(j = 1, 2, \dots, n)$. Then
$$\sum_{j=1}^n |\lambda_j|^2 \leq \sum_{i,j=1}^n |a_{ij}|^2$$
with equality holding if and only if A is a normal matrix (i.e., $A^*A = AA^*$). (See Section 7.2 for more information on normal matrices.)

7. (Browne's Theorem) Let $A = [a_{ij}] \in \mathbb{C}^{n\times n}$ and $\lambda_j (j = 1, 2, \dots, n)$ be the eigenvalues of A ordered so that $|\lambda_1| \geq |\lambda_2| \geq \cdots \geq |\lambda_n|$. Let also $\sigma_1 \geq \sigma_2 \geq \cdots \geq \sigma_n$ be the singular values of A, which are real and nonnegative. (See Section 5.6 for the definition.) Then
$$\sigma_n \leq |\lambda_j| \leq \sigma_1 \quad (j = 1, 2, \dots, n).$$
In fact, the following more general statement holds:
$$\prod_{i=1}^k \sigma_{n-i+1} \leq \prod_{j=1}^k |\lambda_{t_j}| \leq \prod_{i=1}^k \sigma_i,$$
for every $k \in \{1, 2, \dots, n\}$ and every k-tuple $(t_1, t_2, \dots, t_k)$ of strictly increasing elements chosen from $\{1, 2, \dots, n\}$.

8. Let $A \in \mathbb{C}^{n\times n}$ and R_i, C_i $(i = 1, 2, \ldots, n)$ denote the sums of the absolute values of the entries of A in row i and column i, respectively. Also denote

$$R = \max_i \{R_i\} \quad \text{and} \quad C = \max_i \{C_i\}.$$

Let λ be an eigenvalue of A. Then the following inequalities hold:

$$|\lambda| \leq \max_i \frac{R_i + C_i}{2} \leq \frac{R + C}{2},$$
$$|\lambda| \leq \max_i \sqrt{R_i C_i} \leq \sqrt{RC},$$
$$|\lambda| \leq \min\{R, C\}.$$

9. (Schneider's Theorem) Let $A = [a_{ij}] \in \mathbb{C}^{n\times n}$ and $\lambda_j (j = 1, 2, \ldots, n)$ be the eigenvalues of A ordered so that $|\lambda_1| \geq |\lambda_2| \geq \cdots \geq |\lambda_n|$. Let $\mathbf{x} = [x_i]$ be any vector in $\mathbb{R}^n$ with positive entries and define the quantities

$$r_i = \sum_{j=1}^{n} \frac{|a_{ij}| x_j}{x_i} \qquad (i = 1, 2, \ldots, n).$$

Then

$$\prod_{j=1}^{k} |\lambda_j| \leq \prod_{j=1}^{k} r_{i_j} \qquad (k = 1, 2, \ldots, n)$$

for all n-tuples $(i_1, i_2, \ldots, i_n)$ of elements from $\{1, 2, \ldots, n\}$ such that

$$r_{i_1} \geq r_{i_2} \geq \cdots \geq r_{i_n}.$$

10. [HJ85, Theorem 8.1.18] For $A = [a_{ij}] \in \mathbb{C}^{n\times n}$, let its entrywise absolute value be denoted by $|A| = [|a_{ij}|]$. Let $B \in \mathbb{C}^{n\times n}$ and assume that $|A| \leq B$ (entrywise). Then

$$\rho(A) \leq \rho(|A|) \leq \rho(B).$$

11. [HJ85, Chap. 5, Sec. 6] Let $A \in \mathbb{C}^{n\times n}$ and $\|\cdot\|$ denote any matrix norm on $\mathbb{C}^{n\times n}$. (See Chapter 37). Then

$$\rho(A) \leq \|A\|$$

and

$$\lim_{k\longrightarrow\infty} \|A^k\|^{1/k} = \rho(A).$$

12. [HJ91, Corollary 1.5.5] Let $A = [a_{ij}] \in \mathbb{C}^{n\times n}$. The numerical range of $A \in \mathbb{C}^{n\times n}$ is $W(A) = \{v^* A v \in \mathbb{C} : v \in \mathbb{C}^n \text{ with } v^* v = 1\}$ and the numerical radius of $A \in \mathbb{C}^{n\times n}$ is $r(A) = \max\{|z| : z \in W(A)\}$. (See Chapter 18 for more information about the numerical range and numerical radius.) Then the following inequalities hold:

$$r(A^m) \leq [r(A)]^m \quad (m = 1, 2, \ldots),$$
$$\rho(A) \leq r(A) \leq \frac{\|A\|_1 + \|A\|_\infty}{2},$$
$$\frac{\|A\|_2}{2} \leq r(A) \leq \|A\|_2,$$
$$r(A) \leq r(|A|) = \frac{|A| + |A|^T}{2} \quad (\text{where } |A| = [|a_{ij}|]).$$

Moreover, the following statements are equivalent:

(a) $r(A) = \|A\|_2$.

(b) $\rho(A) = \|A\|_2$.

(c) $\|A^n\|_2 = \|A\|_2^n$.

(d) $\|A^k\|_2 = \|A\|_2^k \quad (k = 1, 2, \ldots)$.

Facts 13 to 15 below, along with proofs, can be found in [HJ85, Chap. 4].

13. (Rayleigh–Ritz) Let $A \in \mathbb{C}^{n\times n}$ be Hermitian (i.e., $A = A^*$) with eigenvalues $\lambda_1 \geq \lambda_2 \geq \cdots \geq \lambda_n$. Then

    (a) $\lambda_n x^*x \leq x^*Ax \leq \lambda_1 x^*x$ for all $x \in \mathbb{C}^n$.

    (b) $\lambda_1 = \max_{x\neq 0} \dfrac{x^*Ax}{x^*x} = \max_{x^*x=1} x^*Ax$.

    (c) $\lambda_n = \min_{x\neq 0} \dfrac{x^*Ax}{x^*x} = \min_{x^*x=1} x^*Ax$.

14. (Courant–Fischer) Let $A \in \mathbb{C}^{n\times n}$ be Hermitian with eigenvalues $\lambda_1 \geq \lambda_2 \geq \cdots \geq \lambda_n$. Let $k \in \{1, 2, \ldots, n\}$. Then

$$\lambda_k = \min_{w_1, w_2, \ldots, w_{k-1} \in \mathbb{C}^n} \quad \max_{\substack{x\neq 0, x\in \mathbb{C}^n \\ x \perp w_1, w_2, \ldots, w_{k-1}}} \frac{x^*Ax}{x^*x}$$
$$= \max_{w_1, w_2, \ldots, w_{n-k} \in \mathbb{C}^n} \quad \min_{\substack{x\neq 0, x\in \mathbb{C}^n \\ x \perp w_1, w_2, \ldots, w_{n-k}}} \frac{x^*Ax}{x^*x}.$$

15. (Weyl) Let $A, B \in \mathbb{C}^{n\times n}$ be Hermitian. Consider the eigenvalues of A, B, and $A + B$, denoted by $\lambda_i(A), \lambda_i(B), \lambda_i(A+B)$, respectively, arranged in decreasing order. Then the following hold:

    (a) For each $k \in \{1, 2, \ldots, n\}$,
$$\lambda_k(A) + \lambda_n(A) \leq \lambda_k(A+B) \leq \lambda_k(A) + \lambda_1(B).$$

    (b) For every pair $j, k \in \{1, 2, \ldots, n\}$ such that $j + k \geq n + 1$,
$$\lambda_{j+k-n}(A+B) \geq \lambda_j(A) + \lambda_k(B).$$

    (c) For every pair $j, k \in \{1, 2, \ldots, n\}$ such that $j + k \leq n + 1$,
$$\lambda_j(A) + \lambda_k(B) \geq \lambda_{j+k-1}(A+B).$$

Examples:

1. To illustrate several of the facts in this section, consider

$$A = \begin{bmatrix} 1 & -1 & 0 & 2 \\ 3 & 1 & -2 & 1 \\ 1 & 0 & 0 & -1 \\ -1 & 2 & 1 & 0 \end{bmatrix},$$

whose spectrum, $\sigma(A)$, consists of

$$\lambda_1 = -0.7112 + 2.6718i, \lambda_2 = -0.7112 - 2.6718i, \lambda_3 = 2.5506, \lambda_4 = 0.8719.$$

Note that the eigenvalues are ordered decreasingly with respect to their moduli (absolute values):

$$|\lambda_1| = |\lambda_2| = 2.7649 > |\lambda_3| = 2.5506 > |\lambda_4| = 0.8719.$$

The maximum and minimum eigenvalues of $(A + A^*)/2$ are 2.8484 and -1.495. Note that, as required by Fact 5, for every $\lambda \in \sigma(A)$,

$$-1.495 \leq |\lambda| \leq 2.8484.$$

To illustrate Fact 7, let $(t_1, t_2) = (1, 3)$ and compute the singular values of A:

$$\sigma_1 = 4.2418, \sigma_2 = 2.5334, \sigma_3 = 1.9890, \sigma_4 = 0.7954.$$

Then, indeed,

$$\sigma_4\sigma_3 = 1.5821 \leq |\lambda_1||\lambda_3| = 7.0522 \leq \sigma_1\sigma_2 = 10.7462.$$

Referring to the notation in Fact 8, we have $C = 6$ and $R = 7$. The spectral radius of A is $\rho(A) = 2.7649$ and, thus, the modulus of every eigenvalue of A is indeed bounded above by the quantities

$$\frac{R+C}{2} = \frac{13}{2} = 6.5, \quad \sqrt{RC} = 6.4807, \quad \min\{R, C\} = 6.$$

Letting B denote the entrywise absolute value of A, Facts 10 and 11 state that

$$\rho(A) = 2.7649 \leq \rho(B) = 4.4005 \quad \text{and} \quad \rho(A) = 2.7649 \leq \|A\|_2 = 4.2418.$$

Examples related to Fact 12 and the numerical range are found in Chapter 18. See also Example 2 that associates the numerical range with the location of the eigenvalues.

2. Consider the matrix

$$A = \begin{bmatrix} 1 & 0 & 0 \\ 0 & 0 & 1 \\ 0 & 0 & 0 \end{bmatrix}$$

and note that for every integer $m \geq 2$, A^m consists of zero entries, except for its $(1, 1)$ entry that is equal to 1. One may easily verify that

$$\rho(A) = 1, \quad \|A\|_\infty = \|A\|_1 = \|A\|_2 = 1.$$

By Fact 12, it follows that $r(A) = 1$ and all of the equivalent conditions (a) to (d) in that fact hold, despite A not being a normal matrix.

14.2 Spectrum Localization

This section presents results on classical inclusion regions for the eigenvalues of a matrix. The following facts, proofs, and details, as well as additional references, can be found in [MM92, Chap. III, Sec. 2], [HJ85, Chap. 6], and [Bru82].

Facts:

1. (Geršgorin) Let $A = [a_{ij}] \in \mathbb{C}^{n\times n}$ and define the quantities

$$R_i = \sum_{\substack{j=1 \\ j\neq i}}^{n} |a_{ij}| \quad (i = 1, 2, \ldots, n).$$

Consider the **Geršgorin** discs (centered at a_{ii} with radii R_i),

$$D_i = \{z \in \mathbb{C} : |z - a_{ii}| \leq R_i\} \quad (i = 1, 2, \ldots, n).$$

Then all the eigenvalues of A lie in the union of the Geršgorin discs; that is,

$$\sigma(A) \subset \bigcup_{i=1}^{n} D_i.$$

Moreover, if the union of k Geršgorin discs, G, forms a connected region disjoint from the remaining $n-k$ discs, then G contains exactly k eigenvalues of A (counting algebraic multiplicities).

2. (Lévy–Desplanques) Let $A = [a_{ij}] \in \mathbb{C}^{n\times n}$ be a strictly diagonally dominant matrix, namely,

$$|a_{ii}| > \sum_{\substack{j=1\\ j\neq i}}^{n} |a_{ij}| \quad (i = 1, 2, \ldots, n).$$

Then A is an invertible matrix.

3. (Brauer) Let $A = [a_{ij}] \in \mathbb{C}^{n\times n}$ and define the quantities

$$R_i = \sum_{\substack{j=1\\ j\neq i}}^{n} |a_{ij}| \quad (i = 1, 2, \ldots, n).$$

Consider the **ovals of Cassini**, which are defined by

$$V_{i,j} = \{z \in \mathbb{C} : |z - a_{ii}||z - a_{jj}| \leq R_i R_j\} \quad (i, j = 1, 2, \ldots, n, i \neq j).$$

Then all the eigenvalues of A lie in the union of the ovals of Cassini; that is,

$$\sigma(A) \subset \bigcup_{\substack{i,j=1\\ i\neq j}}^{n} V_{i,j}.$$

4. [VK99, Eq. 3.1] Denoting the union of the Geršgorin discs of $A \in \mathbb{C}^{n\times n}$ by $\Gamma(A)$ (see Fact 1) and the union of the ovals of Cassini of A by $K(A)$ (see Fact 2), we have that

$$\sigma(A) \subset K(A) \subseteq \Gamma(A).$$

That is, the ovals of Cassini provided at least as good a localization for the eigenvalues of A as do the Geršgorin discs.

5. Let $A = [a_{ij}] \in \mathbb{C}^{n\times n}$ such that

$$|a_{ii}||a_{kk}| > \sum_{\substack{j=1\\ j\neq i}}^{n} |a_{ij}| \sum_{\substack{j=1\\ j\neq k}}^{n} |a_{kj}| \quad (i, k = 1, 2, \ldots, n, i \neq k).$$

Then A is an invertible matrix.

6. Facts 1 to 5 can also be stated in terms of *column sums* instead of row sums.
7. (Ostrowski) Let $A = [a_{ij}] \in \mathbb{C}^{n\times n}$ and $\alpha \in [0, 1]$. Define the quantities

$$R_i = \sum_{\substack{j=1\\ j\neq i}}^{n} |a_{ij}|, \qquad C_i = \sum_{\substack{j=1\\ j\neq i}}^{n} |a_{ji}| \quad (i = 1, 2, \ldots, n).$$

Then all the eigenvalues of A lie in the union of the discs

$$D_i(\alpha) = \left\{z \in \mathbb{C} : |z - a_{ii}| \leq R_i^{\alpha} C_i^{1-\alpha}\right\} \quad (i = 1, 2, \ldots, n);$$

that is,

$$\sigma(A) \subset \bigcup_{i=1}^{n} D_i(\alpha).$$

8. Let $A \in \mathbb{C}^{n\times n}$ and consider the spectrum of A, $\sigma(A)$, as well as its numerical range, $W(A)$. Then

$$\sigma(A) \subset W(A).$$

In particular, if A is a normal matrix (i.e., $A^*A = AA^*$), then $W(A)$ is exactly equal to the convex hull of the eigenvalues of A.

Examples:

1. To illustrate Fact 1 (see also Facts 3 and 4) let

$$A = \begin{bmatrix} 3i & 1 & 0.5 & -1 & 0 \\ -1 & 2i & 1.5 & 0 & 0 \\ 1 & 2 & -7 & 0 & 1 \\ 0 & -1 & 0 & 10 & i \\ 1 & 0 & 1 & -1 & 1 \end{bmatrix}$$

and consider the Geršgorin discs of A displayed in Figure 14.1. Note that there are three connected regions of discs that are disjoint of each other. Each region contains as many eigenvalues (marked with +'s) as the number of discs it comprises. The ovals of Cassini are contained in the union of the Geršgorin discs. In general, although it is easy to verify whether a complex number belongs to an oval of Cassini or not, these ovals are generally difficult to draw. An interactive supplement to [VK99] (accessible at: `www.emis.math.ca/EMIS/journals/ETNA/vol.8.1999/pp15-20.dir/gershini.html`) allows one to draw and compare the Geršgorin discs and ovals of Cassini of 3×3 matrices.

2. To illustrate Fact 8, consider the matrices

$$A = \begin{bmatrix} 1 & -1 & 2 \\ 2 & -1 & 0 \\ -1 & 0 & 1 \end{bmatrix} \quad \text{and} \quad B = \begin{bmatrix} 2+2i & -2-i & -1-2i \\ 1+2i & -1-i & -1-2i \\ 2+i & -2-i & -1-i \end{bmatrix}.$$

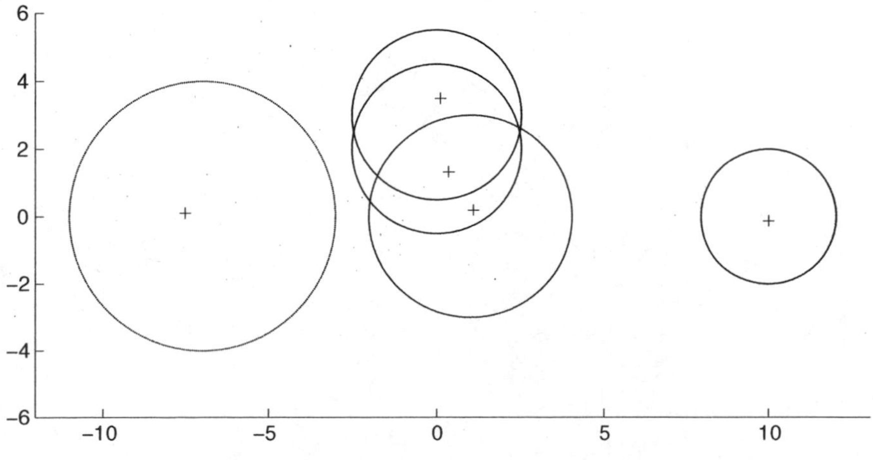

FIGURE 14.1 The Geršgorin disks of A.

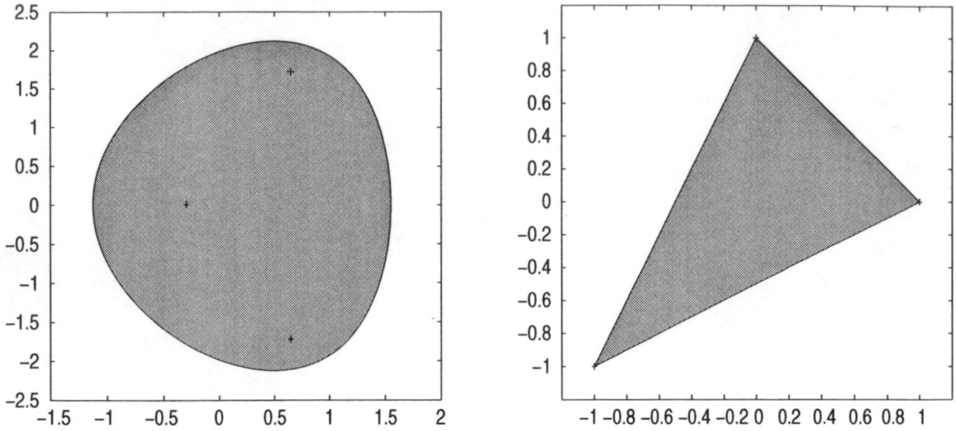

FIGURE 14.2 The numerical range of A and of the normal matrix B.

Note that B is a normal matrix with spectrum $\{1, i, -1-i\}$. As indicated in Figure 14.2, the numerical ranges of A and B contain the eigenvalues of A and B, respectively, marked with $+$'s. The numerical range of B is indeed the convex hull of the eigenvalues.

14.3 Inequalities for the Singular Values and the Eigenvalues

The material in this section is a selection of classical inequalities about the singular values. Extensive details and proofs, as well as a host of additional results on singular values, can be found in [HJ91, Chap. 3]. Definitions of many of the terms in this section are given in Section 5.6, Chapter 17, and Chapter 45; additional facts and examples are also given there.

Facts:

1. Let $A \in \mathbb{C}^{m\times n}$ and σ_1 be its largest singular value. Then
$$\sigma_1 = \|A\|_2.$$

2. Let $A \in \mathbb{C}^{m\times n}$, $q = \min\{m, n\}$. Denote the singular values of A by $\sigma_1 \geq \sigma_2 \geq \cdots \geq \sigma_q$ and let $k \in \{1, 2, \ldots, q\}$. Then
$$\begin{aligned}\sigma_k &= \min_{w_1,w_2,\ldots,w_{k-1}\in\mathbb{C}^n} \max_{\substack{\|x\|_2=1, x\in\mathbb{C}^n\\ x\perp w_1,w_2,\ldots,w_{k-1}}} \|Ax\|_2\\ &= \max_{w_1,w_2,\ldots,w_{n-k}\in\mathbb{C}^n} \min_{\substack{\|x\|_2=1, x\in\mathbb{C}^n\\ x\perp w_1,w_2,\ldots,w_{n-k}}} \|Ax\|_2\\ &= \min_{\substack{W\subseteq\mathbb{C}^n\\ \dim W=n-k+1}} \max_{\substack{x\in W\\ \|x\|_2=1}} \|Ax\|_2\\ &= \max_{\substack{W\subseteq\mathbb{C}^n\\ \dim W=k}} \min_{\substack{x\in W\\ \|x\|_2=1}} \|Ax\|_2,\end{aligned}$$
where the optimizations take place over all subspaces $W \subseteq \mathbb{C}^n$ of the indicated dimensions.

3. (Weyl) Let $A \in \mathbb{C}^{n\times n}$ have singular values $\sigma_1 \geq \sigma_2 \geq \cdots \geq \sigma_n$ and eigenvalues λ_j $(j = 1, 2, \ldots, n)$ be ordered so that $|\lambda_1| \geq |\lambda_2| \geq \cdots \geq |\lambda_n|$. Then
$$|\lambda_1\lambda_2\cdots\lambda_k| \leq \sigma_1\sigma_2\cdots\sigma_k \qquad (k = 1, 2, \ldots, n).$$
Equality holds in (3) when $k = n$.

4. (A. Horn) Let $A \in \mathbb{C}^{m\times p}$ and $B \in \mathbb{C}^{p\times n}$. Let also $r = \min\{m, p\}, s = \min\{p, n\}$, and $q = \min\{r, s\}$. Denote the singular values of A, B, and AB, respectively, by $\sigma_1 \geq \sigma_2 \geq \cdots \geq \sigma_r$, $\tau_1 \geq \tau_2 \geq \cdots \geq \tau_s$, and $\chi_1 \geq \chi_2 \geq \cdots \geq \chi_q$. Then

$$\prod_{i=1}^{k} \chi_i \leq \prod_{i=1}^{k} \sigma_i \tau_i \quad (k = 1, 2, \ldots, q).$$

Equality holds if $k = n = p = m$. Also for any $t > 0$,

$$\sum_{i=1}^{k} \chi_i^t \leq \sum_{i=1}^{k} (\sigma_i \tau_i)^t \quad (k = 1, 2, \ldots, q).$$

5. Let $A \in \mathbb{C}^{n\times n}$ have singular values $\sigma_1 \geq \sigma_2 \geq \cdots \geq \sigma_n$ and eigenvalues λ_j $(j = 1, 2, \ldots, n)$ ordered so that $|\lambda_1| \geq |\lambda_2| \geq \cdots \geq |\lambda_n|$. Then for any $t > 0$,

$$\sum_{i=1}^{k} |\lambda_i|^t \leq \sum_{i=1}^{k} \sigma_i^t \quad (k = 1, 2 \ldots, n).$$

In particular, for $t = 1$ and $k = n$ we obtain from the inequality above that

$$|\mathrm{tr} A| \leq \sum_{i=1}^{n} \sigma_i.$$

6. Let $A, B \in \mathbb{C}^{m\times n}$ and $q = \min\{m, n\}$. Denote the singular values of A, B, and $A + B$, respectively, by $\sigma_1 \geq \sigma_2 \geq \cdots \geq \sigma_q$, $\tau_1 \geq \tau_2 \geq \cdots \geq \tau_q$, and $\psi_1 \geq \psi_2 \geq \cdots \geq \psi_q$. Then the following inequalities hold:

   (a) $\psi_{i+j-1} \leq \sigma_i + \tau_j \quad (1 \leq i, j \leq q, \quad i + j \leq q + 1).$

   (b) $|\rho_i - \sigma_i| \leq \tau_1 \quad (i = 1, 2, \ldots, q).$

   (c) $\sum_{i=1}^{k} \psi_i \leq \sum_{i=1}^{k} \sigma_i + \sum_{i=1}^{k} \tau_i \quad (k = 1, 2, \ldots, q).$

7. Let $A \in \mathbb{C}^{n\times n}$ have eigenvalues λ_j $(j = 1, 2, \ldots, n)$ ordered so that $|\lambda_1| \geq |\lambda_2| \geq \cdots \geq |\lambda_n|$. Denote the singular values of A^k by $\sigma_1(A^k) \geq \sigma_2(A^k) \geq \cdots \geq \sigma_n(A^k)$. Then

$$\lim_{k\longrightarrow\infty} [\sigma_i(A^k)]^{1/k} = |\lambda_i| \qquad (i = 1, 2, \ldots, n).$$

Examples:

1. To illustrate Facts 1, 3, and 5, as well as gauge the bounds they provide, let

$$A = \begin{bmatrix} i & 2 & -1 & 0 \\ 2 & 1+i & 1 & 0 \\ 2i & 1 & 1 & 0 \\ 0 & 1-i & 1 & 0 \end{bmatrix},$$

whose eigenvalues and singular values ordered as required in Fact 3 are, respectively,

$$\lambda_1 = 2.6775 + 1.0227i, \ \lambda_2 = -2.0773 + 1.4685i, \ \lambda_3 = 1.3998 - 0.4912i, \ \lambda_4 = 0,$$

and

$$\sigma_1 = 3.5278, \sigma_2 = 2.5360, \sigma_3 = 1.7673, \sigma_4 = 0.$$

According to Fact 1, $\|A\|_2 = \sigma_1 = 3.5278$. The following inequalities hold according to Fact 3:

$$7.2914 = |\lambda_1\lambda_2| \leq \sigma_1\sigma_2 = 8.9465.$$
$$10.8167 = |\lambda_1\lambda_2\lambda_3| \leq \sigma_1\sigma_2\sigma_3 = 15.8114.$$

Finally, applying Fact 5 with $t = 3/2$ and $k = 2$, we obtain the inequality

$$8.9099 = |\lambda_1|^{3/2} + |\lambda_2|^{3/2} \leq \sigma_1^{3/2} + \sigma_2^{3/2} = 10.6646.$$

For $t = 1$ and $k = n$, we get

$$2.8284 = |2 + 2i| = |\mathrm{tr}(A)| \leq \sum_{j=1}^{4} \sigma_j = 7.8311.$$

14.4 Basic Determinantal Relations

The purpose of this section is to review some basic equalities and inequalities regarding the determinant of a matrix. For most of the facts mentioned here, see [Mey00, Chap. 6] and [HJ85, Chap. 0]. Definitions of many of the terms in this section are given in Sections 4.1 and 4.2; additional facts and examples are given there as well. Note that this section concludes with a couple of classical determinantal inequalities for positive semidefinite matrices; see Section 8.4 or [HJ85, Chap. 7] for more on this subject.

Following are some of the properties of determinants of $n \times n$ matrices, as well as classical formulas for the determinant of A and its submatrices.

Facts:

1. Let $A \in F^{n \times n}$. The following are basic facts about the determinant. (See also Chapter 4.1.)
   - $\det A = \det A^T$; if $F = \mathbb{C}$, then $\det A^* = \overline{\det A}$.
   - If B is obtained from A by multiplying one row (or column) by a scalar c, then $\det B = c \det A$.
   - $\det(cA) = c^n \det A$ for any scalar c.
   - $\det(AB) = \det A \det B$. If A is invertible, then $\det A^{-1} = (\det A)^{-1}$.
   - If B is obtained from A by adding nonzero multiples of one row (respectively, column) to other rows (respectively, columns), then $\det B = \det A$.
   - $\det A = \sum_{\sigma \in S_n} \mathrm{sgn}(\sigma) a_{1\sigma(1)} a_{2\sigma(2)} \cdots a_{n\sigma(n)}$, where the summation is taken over all permutations σ of n letters, and where $\mathrm{sgn}(\sigma)$ denotes the sign of the permutation σ.
   - Let A_{ij} denote the $(n-1) \times (n-1)$ matrix obtained from $A \in F^{n \times n}$ $(n \geq 2)$ by deleting row i and column j. The following formula is known as the **Laplace expansion of** $\det A$ **along column** j:

   $$\det A = \sum_{i=1}^{n} (-1)^{i+j} a_{ij} \det A_{ij} \quad (j = 1, 2, \ldots, n).$$

2. (Cauchy–Binet) Let $A \in F^{m,k}$, $B \in F^{k \times n}$ and consider the matrix $C = AB \in F^{m \times n}$. Let also $\alpha \subseteq \{1, 2, \ldots, m\}$ and $\beta \subseteq \{1, 2, \ldots, n\}$ have cardinality r, where $1 \leq r \leq \min\{m, k, n\}$. Then the submatrix of C whose rows are indexed by α and columns indexed by β satisfies

   $$\det C[\alpha, \beta] = \sum_{\substack{\gamma \subseteq \{1,2,\ldots,k\} \\ |\gamma| = r}} \det A[\alpha, \gamma] \det B[\gamma, \beta].$$

3. [Mey00, Sec. 6.1, p. 471] Let $A = [a_{ij}(x)]$ be an $n \times n$ matrix whose entries are complex differentiable functions of x. Let D_i $(i = 1, 2, \ldots, n)$ denote the $n \times n$ matrix obtained from A when the entries in its ith row are replaced by their derivatives with respect to x. Then

   $$\frac{d}{dx}(\det A) = \sum_{i=1}^{n} \det D_i.$$

4. Let $A = [a_{ij}]$ be an $n \times n$ matrix and consider its entries as independent variables. Then

$$\frac{\partial(\det A)}{\partial a_{ij}} = \det A(\{i\},\{j\}) \quad (i, j = 1, 2, \ldots, n),$$

where $A(\{i\},\{j\})$ denotes the submatrix of A obtained from A by deleting row i and column j.
5. [Mey00, Sec. 6.2] Let $A \in F^{n\times n}$ and $\alpha \subseteq \{1, 2, \ldots, n\}$. If the submatrix of A whose rows and columns are indexed by α, $A[\alpha]$, is invertible, then

$$\det A = \det A[\alpha] \det(A/A[\alpha]).$$

In particular, if A is partitioned in blocks as

$$A = \begin{bmatrix} A_{11} & A_{12} \\ A_{21} & A_{22} \end{bmatrix},$$

where A_{11} and A_{22} are square matrices, then

$$\det A = \begin{cases} \det A_{11} \det(A_{22} - A_{21}(A_{11})^{-1}A_{12}) & \text{if } A_{11} \text{ is invertible} \\ \det A_{22} \det(A_{11} - A_{12}(A_{22})^{-1}A_{21}) & \text{if } A_{22} \text{ is invertible.} \end{cases}$$

The following two facts for $F = \mathbb{C}$ can be found in [Mey00, Sec. 6.2, pp. 475, 483] and [Mey00, Exer. 6.2.15, p. 485], respectively. The proofs are valid for arbitrary fields.
6. Let $A \in F^{n\times n}$ be invertible and $c, d \in F^n$. Then

$$\det(A + cd^T) = \det(A)(1 + d^T A^{-1} c).$$

7. Let $A \in F^{n\times n}$ be invertible, $x, y \in F^n$. Then

$$\det \begin{bmatrix} A & x \\ y^T & -1 \end{bmatrix} = -\det(A + xy^T).$$

8. [HJ85, Theorem 7.8.1 and Corollary 7.8.2] (Hadamard's inequalities) Let $A = [a_{ij}] \in \mathbb{C}^{n\times n}$ be a positive semidefinite matrix. Then

$$\det A \le \prod_{i=1}^{n} a_{ii}.$$

If A is positive definite, equality holds if and only if A is a diagonal matrix.
For a general matrix $B = [b_{ij}] \in \mathbb{C}^{n\times n}$, applying the above inequality to B^*B and BB^*, respectively, one obtains

$$|\det B| \le \prod_{i=1}^{n} \left(\sum_{j=1}^{n} |b_{ij}|^2 \right)^{1/2} \quad \text{and} \quad |\det B| \le \prod_{j=1}^{n} \left(\sum_{i=1}^{n} |b_{ij}|^2 \right)^{1/2}.$$

If B is nonsingular, equalities hold, respectively, if and only if the rows or the columns of B are orthogonal.
9. [HJ85, Theorem 7.8.3] (Fischer's inequality) Consider a positive definite matrix

$$A = \begin{bmatrix} X & Y \\ Y^* & Z \end{bmatrix},$$

partitioned so that X, Z are square and nonvacuous. Then

$$\det A \le \det X \det Z.$$

Examples:

For examples relating to Facts 1, 2, and 5, see Chapter 4.

1. Let

$$A = \begin{bmatrix} 1 & 3 & -1 \\ 0 & 1 & 1 \\ -1 & 2 & 2 \end{bmatrix} \quad \text{and} \quad x = \begin{bmatrix} 1 \\ 1 \\ 1 \end{bmatrix}, y = \begin{bmatrix} 2 \\ 1 \\ -1 \end{bmatrix}.$$

Then, as noted by Fact 7,

$$\det \begin{bmatrix} A & x \\ y^T & -1 \end{bmatrix} = \begin{bmatrix} 1 & 3 & -1 & 1 \\ 0 & 1 & 1 & 1 \\ -1 & 2 & 2 & 1 \\ 2 & 1 & -1 & -1 \end{bmatrix} = -\det(A + xy^T) = \begin{bmatrix} 3 & 4 & -2 \\ 2 & 2 & 0 \\ 1 & 3 & 1 \end{bmatrix} = 10.$$

Next, letting $c = [121]^T$ and $d = [0 - 1 - 1]^T$, by Fact 6, we have

$$\det(A + cd^T) = \det(A)(1 + d^T A^{-1} c) = (-4) \cdot (-1) = 4.$$

2. To illustrate Facts 8 and 9, let

$$A = \left[\begin{array}{cc|c} 3 & 1 & 1 \\ 1 & 5 & 1 \\ \hline 1 & 1 & 3 \end{array}\right] = \begin{bmatrix} X & Y \\ Y^* & Z \end{bmatrix}.$$

Note that A is positive definite and so Hadamard's inequality says that

$$\det A \leq 3 \cdot 5 \cdot 3 = 45;$$

in fact, $\det A = 36$. Fischer's inequality gives a smaller upper bound for the determinant:

$$\det A \leq \det X \det Z = 13 \cdot 3 = 39.$$

3. Consider the matrix

$$B = \begin{bmatrix} 1 & 2 & -2 \\ 4 & -1 & 1 \\ 0 & 1 & 1 \end{bmatrix}.$$

The first inequality about general matrices in Fact 8 applied to B gives

$$|\det B| \leq \sqrt{9 \cdot 18 \cdot 2} = 18.$$

As the rows of B are mutually orthogonal, we have that $|\det B| = 18$; in fact, $\det B = -18$.

14.5 Rank and Nullity Equalities and Inequalities

Let A be a matrix over a field F. Here we present relations among the fundamental subspaces of A and their dimensions. As general references consult, e.g., [HJ85] and [Mey00, Sec. 4.2, 4.4, 4.5] (even though the matrices discussed there are complex, most of the proofs remain valid for any field). Additional material on rank and nullity can also be found in Section 2.4.

Facts:

1. Let $A \in F^{m \times n}$. Then $\text{rank}(A) = \dim \text{range} A = \dim \text{range} A^T$.
   If $F = \mathbb{C}$, then $\text{rank}(A) = \dim \text{range} A^* = \dim \text{range} \overline{A}$.

2. If $A \in \mathbb{C}^{m\times n}$, then range$A = (\ker A^*)^{\perp}$ and range$A^* = (\ker A)^{\perp}$.
3. If $A \in F^{m\times n}$ and rank$(A) = k$, then there exist $X \in F^{m\times k}$ and $Y \in F^{k\times n}$ such that $A = XY$.
4. Let $A, B \in F^{m\times n}$. Then rank$(A) =$ rank(B) if and only if there exist invertible matrices $X \in F^{m\times m}$ and $Y \in F^{n\times n}$ such that $B = XAY$.
5. (Dimension Theorem) Let $A \in F^{m\times n}$. Then

$$\text{rank}(A) + \text{null}(A) = n \qquad \text{and} \qquad \text{rank}(A) + \text{null}(A^T) = m.$$

   If $F = \mathbb{C}$, then rank$(A) +$ null$(A^*) = m$.
6. Let $A, B \in F^{m\times n}$. Then

$$\text{rank}(A) - \text{rank}(B) \le \text{rank}(A + B) \le \text{rank}(A) + \text{rank}(B).$$

7. Let $A \in F^{m\times n}$, $B \in F^{m\times k}$, and $C = [A|B] \in F^{m\times(n+k)}$. Then
   - rank$(C) =$ rank$(A) +$ rank$(B) - \dim(\text{range}A \cap \text{range}B)$.
   - null$(C) =$ null$(A) +$ null$(B) + \dim(\text{range}A \cap \text{range}B)$.
8. Let $A \in F^{m\times n}$ and $B \in F^{n\times k}$. Then
   - rank$(AB) =$ rank$(B) - \dim(\ker A \cap \text{range}B)$.
   - If $F = \mathbb{C}$, then rank$(AB) =$ rank$(A) - \dim(\ker B^* \cap \text{range}A^*)$.
   - Multiplication of a matrix from the left or right by an invertible matrix leaves the rank unchanged.
   - null$(AB) =$ null$(B) + \dim(\ker A \cap \text{range}B)$.
   - rank$(AB) \le \min\{\text{rank}(A), \text{rank}(B)\}$.
   - rank$(AB) \ge$ rank$(A) +$ rank$(B) - n$.
9. (Sylvester's law of nullity) Let $A, B \in \mathbb{C}^{n\times n}$. Then

$$\begin{aligned}\max\{\text{null}(A), \text{null}(B)\} &\le \text{null}(AB)\\ &\le \text{null}(A) + \text{null}(B).\end{aligned}$$

   The above fact is valid *only* for square matrices.
10. (Frobenius inequality) Let $A \in F^{m\times n}$, $B \in F^{n\times k}$, and $C \in F^{k\times p}$. Then

$$\text{rank}(AB) + \text{rank}(BC) \le \text{rank}(B) + \text{rank}(ABC).$$

11. Let $A \in \mathbb{C}^{m\times n}$. Then

$$\text{rank}(A^*A) = \text{rank}(A) = \text{rank}(AA^*).$$

   In fact,

$$\text{range}(A^*A) = \text{range}A^* \quad \text{and} \quad \text{range}A = \text{range}(AA^*),$$

   as well as

$$\ker(A^*A) = \ker A \quad \text{and} \quad \ker(AA^*) = \ker A^*.$$

12. Let $A \in F^{m\times n}$ and $B \in F^{k\times p}$. The rank of their direct sum is

$$\text{rank}(A \oplus B) = \text{rank}\begin{bmatrix} A & 0 \\ 0 & B \end{bmatrix} = \text{rank}(A) + \text{rank}(B).$$

13. Let $A = [a_{ij}] \in F^{m\times n}$ and $B \in F^{k\times p}$. The rank of the Kronecker product $A \otimes B = [a_{ij}B] \in F^{mk\times np}$ is

$$\text{rank}(A \otimes B) = \text{rank}(A)\text{rank}(B).$$

14. Let $A = [a_{ij}] \in F^{m\times n}$ and $B = [b_{ij}] \in F^{m\times n}$. The rank of the Hadamard product $A \circ B = [a_{ij}b_{ij}] \in F^{m\times n}$ satisfies

$$\text{rank}(A \circ B) \leq \text{rank}(A)\text{rank}(B).$$

Examples:

1. Consider the matrices

$$A = \begin{bmatrix} 1 & -1 & 1 \\ 2 & -1 & 0 \\ 3 & -2 & 1 \end{bmatrix}, \quad B = \begin{bmatrix} 2 & 3 & 4 \\ 0 & 0 & -1 \\ 2 & 1 & 2 \end{bmatrix}, \quad \text{and} \quad C = \begin{bmatrix} 1 & 2 & -1 & 1 \\ 1 & 2 & -1 & 1 \\ 2 & 4 & -2 & 2 \end{bmatrix}.$$

We have that

$$\text{rank}(A) = 2, \quad \text{rank}(B) = 3, \quad \text{rank}(C) = 1, \quad \text{rank}(A + B) = 3,$$
$$\text{rank}(AB) = 2, \quad \text{rank}(BC) = 1, \quad \text{rank}(ABC) = 1.$$

- As a consequence of Fact 5, we have

$$\text{null}(A) = 3 - 2 = 1, \quad \text{null}(B) = 3 - 3 = 0, \quad \text{null}(C) = 4 - 1 = 3,$$
$$\text{null}(A + B) = 3 - 3 = 0, \quad \text{null}(AB) = 3 - 2 = 1,$$
$$\text{null}(BC) = 3 - 1 = 2, \quad \text{null}(ABC) = 4 - 1 = 3.$$

- Fact 6 states that

$$-1 = 2 - 3 = \text{rank}(A) - \text{rank}(B) \leq \text{rank}(A + B) = 0 \leq \text{rank}(A) + \text{rank}(B) = 5.$$

- Since $\text{range}A \cap \text{range}B = \text{range}A$, Fact 7 states that

$$\text{rank}([A|B]) = \text{rank}(A) + \text{rank}(B) - \dim(\text{range}A \cap \text{range}B) = 2 + 3 - 2 = 3,$$
$$\text{null}([A|B]) = \text{null}(A) + \text{null}(B) + \dim(\text{range}A \cap \text{range}B) = 1 + 0 + 2 = 3.$$

- Since $\ker A \cap \text{range}B = \ker A$, Fact 8 states that

$$2 = \text{rank}(AB) = \text{rank}(B) - \dim(\ker A \cap \text{range}B) = 3 - 1 = 2.$$
$$2 = \text{rank}(AB) \leq \min\{\text{rank}(A), \text{rank}(B)\} = 2.$$
$$2 = \text{rank}(AB) \geq \text{rank}(A) + \text{rank}(B) - n = 2 + 3 - 3 = 2.$$

- Fact 9 states that

$$1 = \max\{\text{null}(A), \text{null}(B)\} \leq \text{null}(AB) = 1$$
$$\leq \text{Null}(A) + \text{null}(B) = 1.$$

Fact 9 can fail for nonsquare matrices. For example, if

$$D = [1 \quad 1],$$

then

$$1 = \max\{\text{null}(D), \text{null}(D^T)\} \not\leq \text{null}(DD^T) = 0.$$

- Fact 10 states that

$$3 = \text{rank}(AB) + \text{rank}(BC) \leq \text{rank}(B) + \text{rank}(ABC) = 4.$$

14.6 Useful Identities for the Inverse

This section presents facts and formulas related to inversion of matrices.

Facts:

1. [Oue81, (1.9)], [HJ85, p. 18] Recall that $A/A[\alpha]$ denotes the Schur complement of the principal submatrix $A[\alpha]$ in A. (See Section 4.2 and Section 10.3.) If $A \in F^{n\times n}$ is partitioned in blocks as

$$A = \begin{bmatrix} A_{11} & A_{12} \\ A_{21} & A_{22} \end{bmatrix},$$

where A_{11} and A_{22} are square matrices, then, provided that A, A_{11}, and A_{22} are invertible, we have that the Schur complements A/A_{11} and A/A_{22} are invertible and

$$A^{-1} = \begin{bmatrix} (A/A_{22})^{-1} & -A_{11}^{-1}A_{12}(A/A_{11})^{-1} \\ -(A/A_{11})^{-1}A_{21}A_{11}^{-1} & (A/A_{11})^{-1} \end{bmatrix}.$$

More generally, given an invertible $A \in F^{n\times n}$ and $\alpha \subseteq \{1, 2, \ldots, n\}$ such that $A[\alpha]$ and $A(\alpha)$ are invertible, A^{-1} is obtained from A by replacing

- $A[\alpha]$ by $(A/A(\alpha))^{-1}$,
- $A[\alpha, \alpha^c]$ by $-A[\alpha]^{-1}A[\alpha, \alpha^c](A/A[\alpha])^{-1}$,
- $A[\alpha^c, \alpha]$ by $-(A/A[\alpha])^{-1}A[\alpha^c, \alpha]A[\alpha]^{-1}$, and
- $A(\alpha)$ by $(A/A[\alpha])^{-1}$.

2. [HJ85, pp. 18–19] Let $A \in F^{n\times n}$, $X \in F^{n\times r}$, $R \in F^{r\times r}$, and $Y \in F^{r\times n}$. Let $B = A + XRY$. Suppose that A, B, and R are invertible. Then

$$B^{-1} = (A + XRY)^{-1} = A^{-1} - A^{-1}X(R^{-1} + YA^{-1}X)^{-1}YA^{-1}.$$

3. (Sherman–Morrison) Let $A \in F^{n\times n}$, $x, y \in F^n$. Let $B = A + xy^T$. Suppose that A and B are invertible. Then, if $y^T A^{-1}x \neq -1$,

$$B^{-1} = (A + xy^T)^{-1} = A^{-1} - \frac{1}{1 + y^T A^{-1}x} A^{-1}xy^T A^{-1}.$$

In particular, if $y^T x \neq -1$, then

$$(I + xy^T)^{-1} = I - \frac{1}{1 + y^T x} xy^T.$$

4. Let $A \in F^{n\times n}$. Then the adjugate of A (see Section 4.2) satisfies

$$(\text{adj}A)A = A(\text{adj}A) = (\det A)I.$$

If A is invertible, then

$$A^{-1} = \frac{1}{\det A}\text{adj}A.$$

5. Let $A \in F^{n\times n}$ be invertible and let its characteristic polynomial be $p_A(x) = x^n + a_{n-1}x^{n-1} + a_{n-2}x^{n-2} + \cdots + a_1 x + a_0$. Then,

$$A^{-1} = \frac{(-1)^{n+1}}{\det A}(A^{n+1} + a_1 A^n + a_2 A^{n-1} + \cdots + a_{n-1}A).$$

6. [Mey00, Sec. 7.10, p. 618] Let $A \in \mathbb{C}^{n\times n}$. The following statements are equivalent.

- The **Neumann series**, $I + A + A^2 + \ldots$, converges.
- $(I - A)^{-1}$ exists and $(I - A)^{-1} = \sum_{k=0}^{\infty} A^k$.
- $\rho(A) < 1$.
- $\lim_{k\to\infty} A^k = 0$.

Examples:

1. Consider the partitioned matrix

$$A = \begin{bmatrix} A_{11} & A_{12} \\ A_{21} & A_{22} \end{bmatrix} = \left[\begin{array}{cc|c} 1 & 3 & -1 \\ 0 & 2 & 1 \\ \hline -1 & -1 & 1 \end{array}\right].$$

Since

$$(A/A_{22})^{-1} = \begin{bmatrix} 0 & 2 \\ 1 & 3 \end{bmatrix}^{-1} = \begin{bmatrix} -1.5 & 1 \\ 0.5 & 0 \end{bmatrix} \quad \text{and} \quad (A/A_{11})^{-1} = (-1)^{-1} = -1,$$

by Fact 1, we have

$$A^{-1} = \begin{bmatrix} (A/A_{22})^{-1} & -A_{11}^{-1}A_{12}(A/A_{11})^{-1} \\ -(A/A_{11})^{-1}A_{21}A_{11}^{-1} & (A/A_{11})^{-1} \end{bmatrix} = \left[\begin{array}{cc|c} -1.5 & 1 & -2.5 \\ 0.5 & 0 & 0.5 \\ \hline -1 & 1 & -1 \end{array}\right].$$

2. To illustrate Fact 3, consider the invertible matrix

$$A = \begin{bmatrix} 1 & i & -1 \\ 1 & 0 & 1 \\ -2i & 1 & -2 \end{bmatrix}$$

and the vectors $x = y = [1\,1\,1]^T$. We have that

$$A^{-1} = \begin{bmatrix} 0.5i & 1+0.5i & 0.5 \\ -1-i & -1+i & i \\ -0.5i & -0.5i & -0.5 \end{bmatrix}.$$

Adding xy^T to A amounts to adding 1 to each entry of A; since

$$1 + y^T A^{-1} x = i \neq 0,$$

the resulting matrix is invertible and its inverse is given by

$$(A + xy^T)^{-1} = A^{-1} - \frac{1}{i} A^{-1} x y^T A^{-1}$$
$$= \begin{bmatrix} 2.5 & -0.5 - 0.5i & -1 - i \\ -2 + 2i & 1 & 2 \\ -1.5 - i & 0.5 + 0.5i & i \end{bmatrix}.$$

3. Consider the matrix

$$A = \begin{bmatrix} -1 & 1 & -1 \\ 1 & -1 & 3 \\ 1 & -1 & 2 \end{bmatrix}.$$

Since $A^3 = 0$, A is a nilpotent matrix and, thus, all its eigenvalues equal 0. That is, $\rho(A) = 0 < 1$. As a consequence of Fact 6, $I - A$ is invertible and

$$(I - A)^{-1} = I + A + A^2 = \begin{bmatrix} 1 & 0 & 1 \\ 2 & -1 & 5 \\ 1 & -1 & 3 \end{bmatrix}.$$

References

[Bru82] R.A. Brualdi. Matrices, eigenvalues, and directed graphs. *Lin. Multilin. Alg.*, 11:143–165, 1982.

[HJ85] R.A. Horn and C.R. Johnson. *Matrix Analysis.* Cambridge University Press, Cambridge, 1985.

[HJ91] R.A. Horn and C.R. Johnson. *Topics in Matrix Analysis.* Cambridge University Press, Cambridge, 1991.

[MM92] M. Marcus and H. Minc. *A Survey of Matrix Theory and Matrix Inequalities.* Dover Publications, New York, 1992.

[Mey00] C. D. Meyer. *Matrix Analysis and Applied Linear Algebra.* SIAM, Philadelphia, 2000.

[Oue81] D. Ouellette. Schur complements and statistics. *Lin. Alg. Appl.*, 36:187–295, 1981.

[VK99] R.S. Varga and A. Krautstengl. On Geršgorin-type problems and ovals of Cassini. *Electron. Trans. Numer. Anal.*, 8:15–20, 1999.

15

Matrix Perturbation Theory

Ren-Cang Li
University of Texas at Arlington

There is a vast amount of material in matrix (operator) perturbation theory. Related books that are worth mentioning are [SS90], [Par98], [Bha96], [Bau85], and [Kat70]. In this chapter, we attempt to include the most fundamental results up to date, except those for linear systems and least squares problems for which the reader is referred to Section 38.1 and Section 39.6.

Throughout this chapter, $\|\cdot\|_{\mathrm{UI}}$ denotes a general unitarily invariant norm. Two commonly used ones are the spectral norm $\|\cdot\|_2$ and the Frobenius norm $\|\cdot\|_{\mathrm{F}}$.

15.1 Eigenvalue Problems

The reader is referred to Sections 4.3, 14.1, and 14.2 for more information on eigenvalues and their locations.

Definitions:

Let $A \in \mathbb{C}^{n\times n}$. A scalar–vector pair $(\lambda, \mathbf{x}) \in \mathbb{C} \times \mathbb{C}^n$ is an **eigenpair** of A if $\mathbf{x} \neq 0$ and $A\mathbf{x} = \lambda\mathbf{x}$. A vector–scalar–vector triplet $(\mathbf{y}, \lambda, \mathbf{x}) \in \mathbb{C}^n \times \mathbb{C} \times \mathbb{C}^n$ is an **eigentriplet** if $\mathbf{x} \neq 0$, $\mathbf{y} \neq 0$, and $A\mathbf{x} = \lambda\mathbf{x}$, $\mathbf{y}^* A = \lambda \mathbf{y}^*$. The quantity

$$\mathrm{cond}(\lambda) = \frac{\|\mathbf{x}\|_2 \|\mathbf{y}\|_2}{|\mathbf{y}^*\mathbf{x}|}$$

is the **individual condition number** for λ, where $(\mathbf{y}, \lambda, \mathbf{x}) \in \mathbb{C}^n \times \mathbb{C} \times \mathbb{C}^n$ is an eigentriplet.

Let $\sigma(A) = \{\lambda_1, \lambda_2, \ldots, \lambda_n\}$, the multiset of A's eigenvalues, and set

$$\Lambda = \mathrm{diag}(\lambda_1, \lambda_2, \ldots, \lambda_n), \quad \Lambda_\tau = \mathrm{diag}(\lambda_{\tau(1)}, \lambda_{\tau(2)}, \ldots, \lambda_{\tau(n)}),$$

where τ is a **permutation** of $\{1, 2, \ldots, n\}$. For real Λ, i.e., all λ_j's are real,

$$\Lambda^{\uparrow} = \operatorname{diag}(\lambda_1^{\uparrow}, \lambda_2^{\uparrow}, \ldots, \lambda_n^{\uparrow}).$$

$\Lambda^{\uparrow}$ is in fact a Λ_τ for which the permutation τ makes $\lambda_{\tau(j)} = \lambda_j^{\uparrow}$ for all j.

Given two square matrices A_1 and A_2, the **separation** $\operatorname{sep}(A_1, A_2)$ between A_1 and A_2 is defined as [SS90, p. 231]

$$\operatorname{sep}(A_1, A_2) = \inf_{\|X\|_2=1} \|XA_1 - A_2X\|_2.$$

A is perturbed to $\widetilde{A} = A + \Delta A$. *The same notation is adopted for* $\widetilde{A}$, *except all symbols with tildes.*

Let $X, Y \in \mathbb{C}^{n\times k}$ with $\operatorname{rank}(X) = \operatorname{rank}(Y) = k$. The **canonical angles** between their column spaces are $\theta_i = \arccos \sigma_i$, where $\{\sigma_i\}_{i=1}^k$ are the singular values of $(Y^*Y)^{-1/2}Y^*X(X^*X)^{-1/2}$. Define the **canonical angle matrix** between X and Y as

$$\Theta(X, Y) = \operatorname{diag}(\theta_1, \theta_2, \ldots, \theta_k).$$

For $k = 1$, i.e., $\mathbf{x}, \mathbf{y} \in \mathbb{C}^n$ (both nonzero), we use $\angle(\mathbf{x}, \mathbf{y})$, instead, to denote the canonical angle between the two vectors.

Facts:

1. [SS90, p. 168] (**Elsner**) $\max_i \min_j |\widetilde{\lambda}_i - \lambda_j| \le (\|A\|_2 + \|\widetilde{A}\|_2)^{1-1/n}\|\Delta A\|_2^{1/n}$.
2. [SS90, p. 170] (**Elsner**) There exists a permutation τ of $\{1, 2, \ldots, n\}$ such that

$$\|\Lambda - \widetilde{\Lambda}_\tau\|_2 \le 2\left\lfloor \frac{n}{2} \right\rfloor (\|A\|_2 + \|\widetilde{A}\|_2)^{1-1/n}\|\Delta A\|_2^{1/n}.$$

3. [SS90, p. 183] Let $(\mathbf{y}, \mu, \mathbf{x})$ be an eigentriplet of A. ΔA changes μ to $\mu + \Delta\mu$ with

$$\Delta\mu = \frac{\mathbf{y}^*(\Delta A)\mathbf{x}}{\mathbf{y}^*\mathbf{x}} + O(\|\Delta A\|_2^2),$$

   and $|\Delta\mu| \le \operatorname{cond}(\mu)\|\Delta A\|_2 + O(\|\Delta A\|_2^2)$.
4. [SS90, p. 205] If A and $A + \Delta A$ are Hermitian, then

$$\|\Lambda^{\uparrow} - \widetilde{\Lambda}^{\uparrow}\|_{\text{UI}} \le \|\Delta A\|_{\text{UI}}.$$

5. [Bha96, p. 165] (**Hoffman–Wielandt**) If A and $A+\Delta A$ are normal, then there exists a permutation τ of $\{1, 2, \ldots, n\}$ such that $\|\Lambda - \widetilde{\Lambda}_\tau\|_F \le \|\Delta A\|_F$.
6. [Sun96] If A is normal, then there exists a permutation τ of $\{1, 2, \ldots, n\}$ such that $\|\Lambda - \widetilde{\Lambda}_\tau\|_F \le \sqrt{n}\|\Delta A\|_F$.
7. [SS90, p. 192] (**Bauer–Fike**) If A is diagonalizable and $A = X\Lambda X^{-1}$ is its eigendecomposition, then

$$\max_i \min_j |\widetilde{\lambda}_i - \lambda_j| \le \|X^{-1}(\Delta A)X\|_p \le \kappa_p(X)\|\Delta A\|_p.$$

8. [BKL97] Suppose both A and $\widetilde{A}$ are diagonalizable and have eigendecompositions $A = X\Lambda X^{-1}$ and $\widetilde{A} = \widetilde{X}\widetilde{\Lambda}\widetilde{X}^{-1}$.

   (a) There exists a permutation τ of $\{1, 2, \ldots, n\}$ such that

$$\|\Lambda - \widetilde{\Lambda}_\tau\|_F \le \sqrt{\kappa_2(X)\kappa_2(\widetilde{X})}\,\|\Delta A\|_F.$$

   (b) $\|\Lambda^{\uparrow} - \widetilde{\Lambda}^{\uparrow}\|_{\text{UI}} \le \sqrt{\kappa_2(X)\kappa_2(\widetilde{X})}\|\Delta A\|_{\text{UI}}$ for real Λ and $\widetilde{\Lambda}$.

9. [KPJ82] Let residuals $\mathbf{r} = A\widetilde{\mathbf{x}} - \widetilde{\mu}\widetilde{\mathbf{x}}$ and $\mathbf{s}^* = \widetilde{\mathbf{y}}^* A - \widetilde{\mu}\widetilde{\mathbf{y}}^*$, where $\|\widetilde{\mathbf{x}}\|_2 = \|\widetilde{\mathbf{y}}\|_2 = 1$, and let $\varepsilon = \max\{\|\mathbf{r}\|_2, \|\mathbf{s}\|_2\}$. The smallest error matrix ΔA in the 2-norm, for which $(\widetilde{\mathbf{y}}, \widetilde{\mu}, \widetilde{\mathbf{x}})$ is an exact eigentriplet of $\widetilde{A} = A + \Delta A$, satisfies $\|\Delta A\|_2 = \varepsilon$, and $|\widetilde{\mu} - \mu| \leq \text{cond}(\widetilde{\mu})\,\varepsilon + O(\varepsilon^2)$ for some $\mu \in \sigma(A)$.
10. [KPJ82], [DK70],[Par98, pp. 73, 244] Suppose A is Hermitian, and let residual $\mathbf{r} = A\widetilde{\mathbf{x}} - \widetilde{\mu}\widetilde{\mathbf{x}}$ and $\|\widetilde{\mathbf{x}}\|_2 = 1$.

    (a) The smallest Hermitian error matrix ΔA (in the 2-norm), for which $(\widetilde{\mu}, \widetilde{\mathbf{x}})$ is an exact eigenpair of $\widetilde{A} = A + \Delta A$, satisfies $\|\Delta A\|_2 = \|\mathbf{r}\|_2$.

    (b) $|\widetilde{\mu} - \mu| \leq \|\mathbf{r}\|_2$ for some eigenvalue μ of A.

    (c) Let μ be the closest eigenvalue in $\sigma(A)$ to $\widetilde{\mu}$ and $\mathbf{x}$ be its associated eigenvector with $\|\mathbf{x}\|_2 = 1$, and let $\eta = \min_{\mu \neq \lambda \in \sigma(A)} |\widetilde{\mu} - \lambda|$. If $\eta > 0$, then

    $$|\widetilde{\mu} - \mu| \leq \frac{\|\mathbf{r}\|_2^2}{\eta}, \quad \sin\angle(\widetilde{\mathbf{x}}, \mathbf{x}) \leq \frac{\|\mathbf{r}\|_2}{\eta}.$$

11. Let A be Hermitian, $X \in \mathbb{C}^{n\times k}$ have full column rank, and $M \in \mathbb{C}^{k\times k}$ be Hermitian having eigenvalues $\mu_1 \leq \mu_2 \leq \cdots \leq \mu_k$. Set

    $$R = AX - XM.$$

    There exist k eigenvalues $\lambda_{i_1} \leq \lambda_{i_2} \leq \cdots \leq \lambda_{i_k}$ of A such that the following inequalities hold. Note that subset $\{\lambda_{i_j}\}_{j=1}^k$ may be different at different occurrences.

    (a) [Par98, pp. 253–260], [SS90, Remark 4.16, p. 207] (**Kahan–Cao–Xie–Li**)

    $$\max_{1\leq j\leq k} |\mu_j - \lambda_{i_j}| \leq \frac{\|R\|_2}{\sigma_{\min}(X)},$$
    $$\sqrt{\sum_{j=1}^k (\mu_j - \lambda_{i_j})^2} \leq \frac{\|R\|_F}{\sigma_{\min}(X)}.$$

    (b) [SS90, pp. 254–257], [Sun91] If $X^*X = I$ and $M = X^*AX$, and if all but k of A's eigenvalues differ from every one of M's by at least $\eta > 0$ and $\varepsilon_F = \|R\|_F/\eta < 1$, then

    $$\sqrt{\sum_{j=1}^k (\mu_k - \lambda_{i_j})^2} \leq \frac{\|R\|_F^2}{\eta\sqrt{1 - \varepsilon_F^2}}.$$

    (c) [SS90, pp. 254–257], [Sun91] If $X^*X = I$ and $M = X^*AX$, and there is a number $\eta > 0$ such that either all but k of A's eigenvalues lie outside the open interval $(\mu_1 - \eta, \mu_k + \eta)$ or all but k of A's eigenvalues lie inside the closed interval $[\mu_\ell + \eta, \mu_{\ell+1} - \eta]$ for some $1 \leq \ell \leq k - 1$, and $\varepsilon = \|R\|_2/\eta < 1$, then

    $$\max_{1\leq j\leq k} |\mu_j - \lambda_{i_j}| \leq \frac{\|R\|_2^2}{\eta\sqrt{1 - \varepsilon^2}}.$$

12. [DK70] Let A be Hermitian and have decomposition

    $$\begin{bmatrix} X_1^* \\ X_2^* \end{bmatrix} A[X_1\ X_2] = \begin{bmatrix} A_1 & \\ & A_2 \end{bmatrix},$$

where $[X_1\ X_2]$ is unitary and $X_1 \in \mathbb{C}^{n\times k}$. Let $Q \in \mathbb{C}^{n\times k}$ have orthonormal columns and for a $k \times k$ Hermitian matrix M set

$$R = AQ - QM.$$

Let $\eta = \min|\mu - \nu|$ over all $\mu \in \sigma(M)$ and $\nu \in \sigma(A_2)$. If $\eta > 0$, then $\|\sin\Theta(X_1, Q)\|_F \leq \frac{\|R\|_F}{\eta}$.

13. [LL05] Let

$$A = \begin{bmatrix} M & E^* \\ E & H \end{bmatrix}, \quad \widetilde{A} = \begin{bmatrix} M & 0 \\ 0 & H \end{bmatrix}$$

be Hermitian, and set $\eta = \min|\mu - \nu|$ over all $\mu \in \sigma(M)$ and $\nu \in \sigma(H)$. Then

$$\max_{1\leq j\leq n} |\lambda_j^{\uparrow} - \widetilde{\lambda}_j^{\uparrow}| \leq \frac{2\|E\|_2^2}{\eta + \sqrt{\eta^2 + 4\|E\|_2^2}}.$$

14. [SS90, p. 230] Let $[X_1\ Y_2]$ be unitary and $X_1 \in \mathbb{C}^{n\times k}$, and let

$$\begin{bmatrix} X_1^* \\ Y_2^* \end{bmatrix} A[X_1\ Y_2] = \begin{bmatrix} A_1 & G \\ E & A_2 \end{bmatrix}.$$

Assume that $\sigma(A_1)\bigcap\sigma(A_2) = \emptyset$, and set $\eta = \operatorname{sep}(A_1, A_2)$. If $\|G\|_2\|E\|_2 < \eta^2/4$, then there is a unique $W \in \mathbb{C}^{(n-k)\times k}$, satisfying $\|W\|_2 \leq 2\|E\|_2/\eta$, such that $[\widetilde{X}_1\ \widetilde{Y}_2]$ is unitary and

$$\begin{bmatrix} \widetilde{X}_1^* \\ \widetilde{Y}_2^* \end{bmatrix} A[\widetilde{X}_1\ \widetilde{Y}_2] = \begin{bmatrix} \widetilde{A}_1 & \widetilde{G} \\ 0 & \widetilde{A}_2 \end{bmatrix},$$

where

$$\begin{aligned}
\widetilde{X}_1 &= (X_1 + Y_2 W)(I + W^* W)^{-1/2},\\
\widetilde{Y}_2 &= (Y_2 - X_1 W^*)(I + WW^*)^{-1/2},\\
\widetilde{A}_1 &= (I + W^* W)^{1/2}(A_1 + GW)(I + W^* W)^{-1/2},\\
\widetilde{A}_2 &= (I + WW^*)^{-1/2}(A_2 - WG)(I + WW^*)^{1/2}.
\end{aligned}$$

Thus, $\|\tan\Theta(X_1, \widetilde{X}_1)\|_2 < \frac{2\|E\|_2}{\eta}$.

Examples:

1. Bounds on $\|\Lambda - \widetilde{\Lambda}_\tau\|_{UI}$ are, in fact, bounds on $\lambda_j - \lambda_{\tau(j)}$ in disguise, only more convenient and concise. For example, for $\|\cdot\|_{UI} = \|\cdot\|_2$ (spectral norm), $\|\Lambda - \widetilde{\Lambda}_\tau\|_2 = \max_j |\lambda_j - \lambda_{\tau(j)}|$, and for $\|\cdot\|_{UI} = \|\cdot\|_F$ (Frobenius norm), $\|\Lambda - \widetilde{\Lambda}_\tau\|_F = \left[\sum_{j=1}^n |\lambda_j - \lambda_{\tau(j)}|^2\right]^{1/2}$.
2. Let $A, \widetilde{A} \in \mathbb{C}^{n\times n}$ as follows, where $\varepsilon > 0$.

$$A = \begin{bmatrix} \mu & 1 & & \\ & \mu & \ddots & \\ & & \ddots & 1 \\ & & & \mu \end{bmatrix}, \quad \widetilde{A} = \begin{bmatrix} \mu & 1 & & \\ & \mu & \ddots & \\ & & \ddots & 1 \\ \varepsilon & & & \mu \end{bmatrix}.$$

It can be seen that $\sigma(A) = \{\mu, \ldots, \mu\}$ (repeated n times) and the characteristic polynomial $\det(tI - \widetilde{A}) = (t - \mu)^n - \varepsilon$, which gives $\sigma(\widetilde{A}) = \{\mu + \varepsilon^{1/n} e^{2ij\pi/n}, 0 \leq j \leq n-1\}$. Thus,

$|\widetilde{\lambda} - \mu| = \varepsilon^{1/n} = \|\Delta A\|_2^{1/n}$. This shows that the fractional power $\|\Delta A\|_2^{1/n}$ in Facts 1 and 2 cannot be removed in general.

3. Consider

$$A = \begin{bmatrix} 1 & 2 & 3 \\ 0 & 4 & 5 \\ 0 & 0 & 4.001 \end{bmatrix} \quad \text{is perturbed by} \quad \Delta A = \begin{bmatrix} 0 & 0 & 0 \\ 0 & 0 & 0 \\ 0.001 & 0 & 0 \end{bmatrix}.$$

A's eigenvalues are easily read off, and

$$\lambda_1 = 1,\ \mathbf{x}_1 = [1, 0, 0]^T,\ \mathbf{y}_1 = [0.8285, -0.5523, 0.0920]^T,$$

$$\lambda_2 = 4,\ \mathbf{x}_2 = [0.5547, 0.8321, 0]^T,\ \mathbf{y}_2 = [0, 0.0002, -1.0000]^T,$$

$$\lambda_3 = 4.001,\ \mathbf{x}_3 = [0.5547, 0.8321, 0.0002]^T,\ \mathbf{y}_3 = [0, 0, 1]^T.$$

On the other hand, $\widetilde{A}$'s eigenvalues computed by MATLAB's `eig` are $\widetilde{\lambda}_1 = 1.0001$, $\widetilde{\lambda}_2 = 3.9427$, $\widetilde{\lambda}_3 \doteq 4.0582$. The following table gives $|\widetilde{\lambda}_j - \lambda_j|$ with upper bounds up to the 1st order by Fact 3.

j	$\text{cond}(\lambda_j)$	$\text{cond}(\lambda_j)\|\Delta A\|_2$	$\lvert\widetilde{\lambda}_j - \lambda_j\rvert$
1	1.2070	0.0012	0.0001
2	$6.0 \cdot 10^3$	6.0	0.057
3	$6.0 \cdot 10^3$	6.0	0.057

We see that $\text{cond}(\lambda_j)\|\Delta A\|_2$ gives a fairly good error bound for $j = 1$, but dramatically worse for $j = 2, 3$. There are two reasons for this: One is in the choice of ΔA and the other is that ΔA's order of magnitude is too big for the first order bound $\text{cond}(\lambda_j)\|\Delta A\|_2$ to be effective for $j = 2, 3$. Note that ΔA has the same order of magnitude as the difference between λ_2 and λ_3 and that is too big usually. For better understanding of this first order error bound, the reader may play with this example with $\Delta A = \varepsilon \frac{\mathbf{y}_j \mathbf{x}_j^*}{\|\mathbf{y}_j\|_2 \|\mathbf{x}_j^*\|_2}$ for various tiny parameters ε.

4. Let $\Sigma = \text{diag}(c_1, c_2, \ldots, c_k)$ and $\Gamma = \text{diag}(s_1, s_2, \ldots, s_k)$, where $c_j, s_j \geq 0$ and $c_j^2 + s_j^2 = 1$ for all j. The canonical angles between

$$X = Q \begin{bmatrix} I_k \\ 0 \\ 0 \end{bmatrix} V^*, \quad Y = Q \begin{bmatrix} \Sigma \\ \Gamma \\ 0 \end{bmatrix} U^*$$

are $\theta_j = \arccos c_j$, $j = 1, 2, \ldots, k$, where Q, U, V are unitary. On the other hand, every pair of $X, Y \in \mathbb{C}^{n \times k}$ with $2k \leq n$ and $X^*X = Y^*Y = I_k$, having canonical angles $\arccos c_j$, can be represented this way [SS90, p. 40].

5. Fact 13 is most useful when $\|E\|_2$ is tiny and the computation of A's eigenvalues is then decoupled into two smaller ones. In eigenvalue computations, we often seek unitary $[X_1\ X_2]$ such that

$$\begin{bmatrix} X_1^* \\ X_2^* \end{bmatrix} A [X_1\ X_2] = \begin{bmatrix} M & E^* \\ E & H \end{bmatrix}, \quad \begin{bmatrix} X_1^* \\ X_2^* \end{bmatrix} \widetilde{A} [X_1\ X_2] = \begin{bmatrix} M & 0 \\ 0 & H \end{bmatrix},$$

and $\|E\|_2$ is tiny. Since a unitarily similarity transformation does not alter eigenvalues, Fact 13 still applies.

6. [LL05] Consider the 2×2 Hermitian matrix

$$A = \begin{bmatrix} \alpha & \varepsilon \\ \varepsilon & \beta \end{bmatrix},$$

where $\alpha > \beta$ and $\varepsilon > 0$. It has two eigenvalues

$$\lambda_{\pm} = \frac{\alpha + \beta \pm \sqrt{(\alpha - \beta)^2 + 4\varepsilon^2}}{2},$$

and

$$0 < \begin{Bmatrix} \lambda_+ - \alpha \\ \beta - \lambda_- \end{Bmatrix} = \frac{2\varepsilon^2}{(\alpha - \beta) + \sqrt{(\alpha - \beta)^2 + 4\varepsilon^2}}.$$

The inequalities in Fact 13 become equalities for the example.

15.2 Singular Value Problems

The reader is referred to Section 5.6, Chapters 17 and 45 for more information on singular value decompositions.

Definitions:

$B \in \mathbb{C}^{m\times n}$ has a (first standard form) SVD $B = U\Sigma V^*$, where $U \in \mathbb{C}^{m\times m}$ and $V \in \mathbb{C}^{n\times n}$ are unitary, and $\Sigma = \text{diag}(\sigma_1, \sigma_2, \dots) \in \mathbb{R}^{m\times n}$ is leading diagonal (σ_j starts in the top-left corner) with all $\sigma_j \geq 0$.

Let $\text{SV}(B) = \{\sigma_1, \sigma_2, \dots, \sigma_{\min\{m,n\}}\}$, the set of B's singular values, and $\sigma_1 \geq \sigma_2 \geq \cdots \geq 0$, and let $\text{SV}_{\text{ext}}(B) = \text{SV}(B)$ unless $m > n$ for which $\text{SV}_{\text{ext}}(B) = \text{SV}(B)\bigcup\{0, \dots, 0\}$ (additional $m - n$ zeros).

A vector–scalar–vector triplet $(\mathbf{u}, \sigma, \mathbf{v}) \in \mathbb{C}^m \times \mathbb{R} \times \mathbb{C}^n$ is a **singular-triplet** if $\mathbf{u} \neq 0$, $\mathbf{v} \neq 0$, $\sigma \geq 0$, and $B\mathbf{v} = \sigma\mathbf{u}$, $B^*\mathbf{u} = \sigma\mathbf{v}$.

B is perturbed to $\widetilde{B} = B + \Delta B$. *The same notation is adopted for $\widetilde{B}$, except all symbols with tildes.*

Facts:

1. [SS90, p. 204] (**Mirsky**) $\|\Sigma - \widetilde{\Sigma}\|_{\text{UI}} \leq \|\Delta B\|_{\text{UI}}$.
2. Let residuals $\mathbf{r} = B\widetilde{\mathbf{v}} - \widetilde{\mu}\widetilde{\mathbf{u}}$ and $\mathbf{s} = B^*\widetilde{\mathbf{u}} - \widetilde{\mu}\widetilde{\mathbf{v}}$, and $\|\widetilde{\mathbf{v}}\|_2 = \|\widetilde{\mathbf{u}}\|_2 = 1$.

   (a) [Sun98] The smallest error matrix ΔB (in the 2-norm), for which $(\widetilde{\mathbf{u}}, \widetilde{\mu}, \widetilde{\mathbf{v}})$ is an exact singular-triplet of $\widetilde{B} = B + \Delta B$, satisfies $\|\Delta B\|_2 = \varepsilon$, where $\varepsilon = \max\{\|\mathbf{r}\|_2, \|\mathbf{s}\|_2\}$.

   (b) $|\widetilde{\mu} - \mu| \leq \varepsilon$ for some singular value μ of B.

   (c) Let μ be the closest singular value in $\text{SV}_{\text{ext}}(B)$ to $\widetilde{\mu}$ and $(\mathbf{u}, \sigma, \mathbf{v})$ be the associated singular-triplet with $\|\mathbf{u}\|_2 = \|\mathbf{v}\|_2 = 1$, and let $\eta = \min|\widetilde{\mu} - \sigma|$ over all $\sigma \in \text{SV}_{\text{ext}}(B)$ and $\sigma \neq \mu$. If $\eta > 0$, then $|\widetilde{\mu} - \mu| \leq \varepsilon^2/\eta$, and [SS90, p. 260]

   $$\sqrt{\sin^2 \angle(\widetilde{\mathbf{u}}, \mathbf{u}) + \sin^2 \angle(\widetilde{\mathbf{v}}, \mathbf{v})} \leq \frac{\sqrt{\|\mathbf{r}\|_2^2 + \|\mathbf{s}\|_2^2}}{\eta}.$$

3. [LL05] Let

   $$B = \begin{bmatrix} B_1 & F \\ E & B_2 \end{bmatrix} \in \mathbb{C}^{m\times n}, \quad \widetilde{B} = \begin{bmatrix} B_1 & 0 \\ 0 & B_2 \end{bmatrix},$$

   where $B_1 \in \mathbb{C}^{k\times k}$, and set $\eta = \min|\mu - \nu|$ over all $\mu \in \text{SV}(B_1)$ and $\nu \in \text{SV}_{\text{ext}}(B_2)$, and $\varepsilon = \max\{\|E\|_2, \|F\|_2\}$. Then

   $$\max_j |\sigma_j - \widetilde{\sigma}_j| \leq \frac{2\varepsilon^2}{\eta + \sqrt{\eta^2 + 4\varepsilon^2}}.$$

4. [SS90, p. 260] (**Wedin**) Let $B, \widetilde{B} \in \mathbb{C}^{m\times n}$ $(m \geq n)$ have decompositions

$$\begin{bmatrix} U_1^* \\ U_2^* \end{bmatrix} B[V_1 \;\; V_2] = \begin{bmatrix} B_1 & 0 \\ 0 & B_2 \end{bmatrix}, \quad \begin{bmatrix} \widetilde{U}_1^* \\ \widetilde{U}_2^* \end{bmatrix} \widetilde{B}[\widetilde{V}_1 \;\; \widetilde{V}_2] = \begin{bmatrix} \widetilde{B}_1 & 0 \\ 0 & \widetilde{B}_2 \end{bmatrix},$$

where $[U_1 \;\; U_2]$, $[V_1 \;\; V_2]$, $[\widetilde{U}_1 \;\; \widetilde{U}_2]$, and $[\widetilde{V}_1 \;\; \widetilde{V}_2]$ are unitary, and $U_1, \widetilde{U}_1 \in \mathbb{C}^{m\times k}$, $V_1, \widetilde{V}_1 \in \mathbb{C}^{n\times k}$. Set

$$R = B\widetilde{V}_1 - \widetilde{U}_1\widetilde{B}_1, \quad S = B^*\widetilde{U}_1 - \widetilde{V}_1\widetilde{B}_1.$$

If $\mathrm{sv}(\widetilde{B}_1) \bigcap \mathrm{sv}_{\mathrm{ext}}(B_2) = \emptyset$, then

$$\sqrt{\|\sin\Theta(U_1, \widetilde{U}_1)\|_F^2 + \|\sin\Theta(V_1, \widetilde{V}_1)\|_F^2} \leq \frac{\sqrt{\|R\|_F^2 + \|S\|_F^2}}{\eta},$$

where $\eta = \min |\widetilde{\mu} - \nu|$ over all $\widetilde{\mu} \in \mathrm{sv}(\widetilde{B}_1)$ and $\nu \in \mathrm{sv}_{\mathrm{ext}}(B_2)$.

Examples:

1. Let

$$B = \begin{bmatrix} 3\cdot 10^{-3} & 1 \\ 2 & 4\cdot 10^{-3} \end{bmatrix}, \quad \widetilde{B} = \begin{bmatrix} & 1 \\ 2 & \end{bmatrix} = [\mathbf{e}_2 \;\; \mathbf{e}_1] \begin{bmatrix} 2 & \\ & 1 \end{bmatrix} \begin{bmatrix} \mathbf{e}_1^T \\ \mathbf{e}_2^T \end{bmatrix}.$$

Then $\sigma_1 = 2.000012$, $\sigma_2 = 0.999988$, and $\widetilde{\sigma}_1 = 2$, $\widetilde{\sigma}_2 = 1$. Fact 1 gives

$$\max_{1\leq j\leq 2} |\sigma_j - \widetilde{\sigma}_j| \leq 4\cdot 10^{-3}, \quad \sqrt{\sum_{j=1}^{2} |\sigma_j - \widetilde{\sigma}_j|^2} \leq 5\cdot 10^{-3}.$$

2. Let B be as in the previous example, and let $\widetilde{\mathbf{v}} = \mathbf{e}_1$, $\widetilde{\mathbf{u}} = \mathbf{e}_2$, $\widetilde{\mu} = 2$. Then $\mathbf{r} = B\widetilde{\mathbf{v}} - \widetilde{\mu}\widetilde{\mathbf{u}} = 3\cdot 10^{-3}\mathbf{e}_1$ and $\mathbf{s} = B^*\widetilde{\mathbf{u}} - \widetilde{\mu}\widetilde{\mathbf{v}} = 4\cdot 10^{-3}\mathbf{e}_2$. Fact 2 applies. Note that, without calculating $\mathrm{sv}(B)$, one may bound η needed for Fact 2(c) from below as follows. Since B has two singular values that are near 1 and $\widetilde{\mu} = 2$, respectively, with errors no bigger than $4\cdot 10^{-3}$, then $\eta \geq 2 - (1 + 4\cdot 10^{-3}) = 1 - 4\cdot 10^{-3}$.
3. Let B and $\widetilde{B}$ be as in Example 1. Fact 3 gives $\max_{1\leq j\leq 2} |\sigma_j - \widetilde{\sigma}_j| \leq 1.6\cdot 10^{-5}$, a much better bound than by Fact 1.
4. Let B and $\widetilde{B}$ be as in Example 1. Note $\widetilde{B}$'s SVD there. Apply Fact 4 with $k = 1$ to give a similar bound as by Fact 2(c).
5. Since unitary transformations do not change singular values, Fact 3 applies to $B, \widetilde{B} \in \mathbb{C}^{m\times n}$ having decompositions

$$\begin{bmatrix} U_1^* \\ U_2^* \end{bmatrix} B[V_1 \;\; V_2] = \begin{bmatrix} B_1 & F \\ E & B_2 \end{bmatrix}, \quad \begin{bmatrix} U_1^* \\ U_2^* \end{bmatrix} \widetilde{B}[V_1 \;\; V_2] = \begin{bmatrix} B_1 & 0 \\ 0 & B_2 \end{bmatrix},$$

where $[U_1 \;\; U_2]$ and $[V_1 \;\; V_2]$ are unitary and $U_1 \in \mathbb{C}^{m\times k}$, $V_1 \in \mathbb{C}^{n\times k}$.

15.3 Polar Decomposition

The reader is referred to Chapter 17.1 for definition and for more information on polar decompositions.

Definitions:

$B \in \mathbb{F}^{m\times n}$ is perturbed to $\widetilde{B} = B + \Delta B$, and their polar decompositions are

$$B = QH, \quad \widetilde{B} = \widetilde{Q}\widetilde{H} = (Q + \Delta Q)(H + \Delta H),$$

where $\mathbb{F} = \mathbb{R}$ or $\mathbb{C}$. *ΔB is restricted to $\mathbb{F}$ for $B \in \mathbb{F}$.*

Denote the singular values of B and $\widetilde{B}$ as $\sigma_1 \geq \sigma_2 \geq \cdots$ and $\widetilde{\sigma}_1 \geq \widetilde{\sigma}_2 \geq \cdots$, respectively.

The **condition numbers for the polar factors** in the Frobenius norm are defined as

$$\text{cond}_{\mathbb{F}}(X) = \lim_{\delta \to 0} \sup_{\|\Delta B\|_{\text{F}} \leq \delta} \frac{\|\Delta X\|_{\text{F}}}{\delta}, \quad \text{for } X = H \text{ or } Q.$$

B is **multiplicatively perturbed** to $\widetilde{B}$ if $\widetilde{B} = D_{\text{L}}^* B D_{\text{R}}$ for some $D_{\text{L}} \in \mathbb{F}^{m \times m}$ and $D_{\text{R}} \in \mathbb{F}^{n \times n}$.

B is said to be **graded** if it can be scaled as $B = GS$ such that G is "well-behaved" (i.e., $\kappa_2(G)$ is of modest magnitude), where S is a scaling matrix, often diagonal but not required so for the facts below. Interesting cases are when $\kappa_2(G) \ll \kappa_2(B)$.

Facts:

1. [CG00] The condition numbers $\text{cond}_{\mathbb{F}}(Q)$ and $\text{cond}_{\mathbb{F}}(H)$ are tabulated as follows, where $\kappa_2(B) = \sigma_1/\sigma_n$.

		$\mathbb{R}$	$\mathbb{C}$
Factor Q	$m = n$	$2/(\sigma_{n-1} + \sigma_n)$	$1/\sigma_n$
	$m > n$	$1/\sigma_n$	$1/\sigma_n$
Factor H	$m \geq n$	$\dfrac{\sqrt{2(1+\kappa_2(B)^2)}}{1+\kappa_2(B)}$	

2. [Kit86] $\|\Delta H\|_{\text{F}} \leq \sqrt{2}\|\Delta B\|_{\text{F}}$.
3. [Li95] If $m = n$ and $\text{rank}(B) = n$, then

$$\|\Delta Q\|_{\text{UI}} \leq \frac{2}{\sigma_n + \widetilde{\sigma}_n} \|\Delta B\|_{\text{UI}}.$$

4. [Li95], [LS02] If $\text{rank}(B) = n$, then

$$\|\Delta Q\|_{\text{UI}} \leq \left(\frac{2}{\sigma_n + \widetilde{\sigma}_n} + \frac{1}{\max\{\sigma_n, \widetilde{\sigma}_n\}} \right) \|\Delta B\|_{\text{UI}},$$

$$\|\Delta Q\|_{\text{F}} \leq \frac{2}{\sigma_n + \widetilde{\sigma}_n} \|\Delta B\|_{\text{F}}.$$

5. [Mat93] If $B \in \mathbb{R}^{n \times n}$, $\text{rank}(B) = n$, and $\|\Delta B\|_2 < \sigma_n$, then

$$\|\Delta Q\|_{\text{UI}} \leq -\frac{2\|\Delta B\|_{\text{UI}}}{|||\Delta B|||_2} \ln \left(1 - \frac{|||\Delta B|||_2}{\sigma_n + \sigma_{n-1}} \right),$$

where $|||\cdot|||_2$ is the **Ky Fan 2-norm**, i.e., the sum of the first two largest singular values. (See Chapter 17.3.)

6. [LS02] If $B \in \mathbb{R}^{n \times n}$, $\text{rank}(B) = n$, and $\|\Delta B\|_2 < \sigma_n + \widetilde{\sigma}_n$, then

$$\|\Delta Q\|_{\text{F}} \leq \frac{4}{\sigma_{n-1} + \sigma_n + \widetilde{\sigma}_{n-1} + \widetilde{\sigma}_n} \|\Delta B\|_{\text{F}}.$$

7. [Li97] Let B and $\widetilde{B} = D_{\text{L}}^* B D_{\text{R}}$ having full column rank. Then

$$\|\Delta Q\|_{\text{F}} \leq \sqrt{\|I - D_{\text{L}}^{-1}\|_{\text{F}}^2 + \|D_{\text{L}} - I\|_{\text{F}}^2} + \sqrt{\|I - D_{\text{R}}^{-1}\|_{\text{F}}^2 + \|D_{\text{R}} - I\|_{\text{F}}^2}.$$

8. [Li97], [Li05] Let $B = GS$ and $\widetilde{B} = \widetilde{G}S$ and assume that G and B have full column rank. If $\|\Delta G\|_2 \|G^\dagger\|_2 < 1$, then

$$\|\Delta Q\|_F \le \gamma\, \|G^\dagger\|_2 \|\Delta G\|_F,$$
$$\|(\Delta H)S^{-1}\|_F \le \left(\gamma\, \|G^\dagger\|_2 \|G\|_2 + 1\right) \|\Delta G\|_F,$$

where $\gamma = \sqrt{1 + (1 - \|G^\dagger\|_2 \|\Delta G\|_2)^{-2}}$.

Examples:

1. Take both B and $\widetilde{B}$ to have orthonormal columns to see that some of the inequalities above on ΔQ are attainable.
2. Let

$$B = \frac{1}{\sqrt{2}} \begin{bmatrix} 2.01 & 502 \\ -1.99 & -498 \end{bmatrix} = \frac{1}{\sqrt{2}} \begin{bmatrix} 1 & 1 \\ 1 & -1 \end{bmatrix} \begin{bmatrix} 10^{-2} & 2 \\ 2 & 5 \cdot 10^2 \end{bmatrix}$$

and

$$\widetilde{B} = \begin{bmatrix} 1.4213 & 3.5497 \cdot 10^2 \\ -1.4071 & -3.5214 \cdot 10^2 \end{bmatrix}$$

obtained by rounding each entry of B to have five significant decimal digits. $B = QH$ can be read off above and $\widetilde{B} = \widetilde{Q}\widetilde{H}$ can be computed by $\widetilde{Q} = \widetilde{U}\widetilde{V}^*$ and $\widetilde{H} = \widetilde{V}\widetilde{\Sigma}\widetilde{V}^*$, where $\widetilde{B}$'s SVD is $\widetilde{U}\widetilde{\Sigma}\widetilde{V}^*$. One has

$$\mathrm{sv}(B) = \{5.00 \cdot 10^2, 2.00 \cdot 10^{-3}\},\ \mathrm{sv}(\widetilde{B}) = \{5.00 \cdot 10^2, 2.04 \cdot 10^{-3}\}$$

and

$\|\Delta B\|_2$	$\|\Delta B\|_F$	$\|\Delta Q\|_2$	$\|\Delta Q\|_F$	$\|\Delta H\|_2$	$\|\Delta H\|_F$
$3 \cdot 10^{-3}$	$3 \cdot 10^{-3}$	$2 \cdot 10^{-6}$	$3 \cdot 10^{-6}$	$2 \cdot 10^{-3}$	$2 \cdot 10^{-3}$

Fact 2 gives $\|\Delta H\|_F \le 3 \cdot 10^{-3}$ and Fact 6 gives $\|\Delta Q\|_F \le 10^{-5}$.

3. [Li97] and [Li05] have examples on the use of inequalities in Facts 7 and 8.

15.4 Generalized Eigenvalue Problems

The reader is referred to Section 43.1 for more information on generalized eigenvalue problems.

Definitions:

Let $A, B \in \mathbb{C}^{m\times n}$. A **matrix pencil** is a family of matrices $A - \lambda B$, parameterized by a (complex) number λ. The associated **generalized eigenvalue problem** is to find the nontrivial solutions of the equations

$$A\mathbf{x} = \lambda B\mathbf{x} \quad \text{and/or} \quad \mathbf{y}^* A = \lambda \mathbf{y}^* B,$$

where $\mathbf{x} \in \mathbb{C}^n$, $\mathbf{y} \in \mathbb{C}^m$, and $\lambda \in \mathbb{C}$.

$A - \lambda B$ is **regular** if $m = n$ and $\det(A - \lambda B) \ne 0$ for some $\lambda \in \mathbb{C}$.

In what follows, all pencils in question are assumed regular.

An eigenvalue λ is conveniently represented by a nonzero number pair, so-called a **generalized eigenvalue** $\langle \alpha, \beta \rangle$, interpreted as $\lambda = \alpha/\beta$. $\beta = 0$ corresponds to eigenvalue infinity.

A **generalized eigenpair** of $A - \lambda B$ refers to $(\langle\alpha, \beta\rangle, \mathbf{x})$ such that $\beta A\mathbf{x} = \alpha B\mathbf{x}$, where $\mathbf{x} \neq 0$ and $|\alpha|^2 + |\beta|^2 > 0$. A **generalized eigentriplet** of $A - \lambda B$ refers to $(\mathbf{y}, \langle\alpha, \beta\rangle, \mathbf{x})$ such that $\beta A\mathbf{x} = \alpha B\mathbf{x}$ and $\beta\mathbf{y}^* A = \alpha\mathbf{y}^* B$, where $\mathbf{x} \neq 0$, $\mathbf{y} \neq 0$, and $|\alpha|^2 + |\beta|^2 > 0$. The quantity

$$\text{cond}(\langle\alpha, \beta\rangle) = \frac{\|\mathbf{x}\|_2 \|\mathbf{y}\|_2}{\sqrt{|\mathbf{y}^* A\mathbf{x}|^2 + |\mathbf{y}^* B\mathbf{x}|^2}}$$

is the **individual condition number** for the generalized eigenvalue $\langle\alpha, \beta\rangle$, where $(\mathbf{y}, \langle\alpha, \beta\rangle, \mathbf{x})$ is a generalized eigentriplet of $A - \lambda B$.

$A - \lambda B$ is perturbed to $\widetilde{A} - \lambda\widetilde{B} = (A + \Delta A) - \lambda(B + \Delta B)$.

Let $\sigma(A, B) = \{\langle\alpha_1, \beta_1\rangle, \langle\alpha_2, \beta_2\rangle, \ldots, \langle\alpha_n, \beta_n\rangle\}$ be the set of the generalized eigenvalues of $A - \lambda B$, and set $Z = [A, B] \in \mathbb{C}^{2n\times n}$.

$A - \lambda B$ is **diagonalizable** if it is equivalent to a diagonal pencil, i.e., there are nonsingular $X, Y \in \mathbb{C}^{n\times n}$ such that $Y^* AX = \Lambda$, $Y^* BX = \Omega$, where $\Lambda = \text{diag}(\alpha_1, \alpha_2, \ldots, \alpha_n)$ and $\Omega = \text{diag}(\beta_1, \beta_2, \ldots, \beta_n)$.

$A - \lambda B$ is a **definite pencil** if both A and B are Hermitian and

$$\gamma(A, B) = \min_{\mathbf{x}\in\mathbb{C}^n, \|\mathbf{x}\|_2=1} |\mathbf{x}^* A\mathbf{x} + i\mathbf{x}^* B\mathbf{x}| > 0.$$

The same notation is adopted for $\widetilde{A} - \lambda\widetilde{B}$, except all symbols with tildes.

The **chordal distance** between two nonzero pairs $\langle\alpha, \beta\rangle$ and $\langle\widetilde{\alpha}, \widetilde{\beta}\rangle$ is

$$\chi\left(\langle\alpha, \beta\rangle, \langle\widetilde{\alpha}, \widetilde{\beta}\rangle\right) = \frac{|\widetilde{\beta}\alpha - \widetilde{\alpha}\beta|}{\sqrt{|\alpha|^2 + |\beta|^2}\sqrt{|\widetilde{\alpha}|^2 + |\widetilde{\beta}|^2}}.$$

Facts:

1. [SS90, p. 293] Let $(\mathbf{y}, \langle\alpha, \beta\rangle, \mathbf{x})$ be a generalized eigentriplet of $A - \lambda B$. $[\Delta A, \Delta B]$ changes $\langle\alpha, \beta\rangle = \langle\mathbf{y}^* A\mathbf{x}, \mathbf{y}^* B\mathbf{x}\rangle$ to

   $$\langle\widetilde{\alpha}, \widetilde{\beta}\rangle = \langle\alpha, \beta\rangle + \langle\mathbf{y}^*(\Delta A)\mathbf{x}, \mathbf{y}^*(\Delta B)\mathbf{x}\rangle + O(\varepsilon^2),$$

   where $\varepsilon = \|[\Delta A, \Delta B]\|_2$, and $\chi\left(\langle\alpha, \beta\rangle, \langle\widetilde{\alpha}, \widetilde{\beta}\rangle\right) \leq \text{cond}(\langle\alpha, \beta\rangle)\,\varepsilon + O(\varepsilon^2)$.
2. [SS90, p. 301], [Li88] If $A - \lambda B$ is diagonalizable, then

   $$\max_i \min_j \chi\left(\langle\alpha_i, \beta_i\rangle, \langle\widetilde{\alpha}_j, \widetilde{\beta}_j\rangle\right) \leq \kappa_2(X)\|\sin\Theta(Z^*, \widetilde{Z}^*)\|_2.$$

3. [Li94, Lemma 3.3] (**Sun**)

   $$\|\sin\Theta(Z^*, \widetilde{Z}^*)\|_{\text{UI}} \leq \frac{\|Z - \widetilde{Z}\|_{\text{UI}}}{\max\{\sigma_{\min}(Z), \sigma_{\min}(\widetilde{Z})\}},$$

   where $\sigma_{\min}(Z)$ is Z's smallest singular value.
4. The quantity $\gamma(A, B)$ is the minimum distance of the numerical range $W(A + iB)$ to the origin for definite pencil $A - \lambda B$.
5. [SS90, p. 316] Suppose $A - \lambda B$ is a definite pencil. If $\widetilde{A}$ and $\widetilde{B}$ are Hermitian and $\|[\Delta A, \Delta B]\|_2 < \gamma(A, B)$, then $\widetilde{A} - \lambda\widetilde{B}$ is also a definite pencil and there exists a permutation τ of $\{1, 2, \ldots, n\}$ such that

   $$\max_{1\leq j\leq n} \chi\left(\langle\alpha_j, \beta_j\rangle, \langle\widetilde{\alpha}_{\tau(j)}, \widetilde{\beta}_{\tau(j)}\rangle\right) \leq \frac{\|[\Delta A, \Delta B]\|_2}{\gamma(A, B)}.$$

6. [SS90, p. 318] Definite pencil $A - \lambda B$ is always diagonalizable: $X^* AX = \Lambda$ and $X^* BX = \Omega$, and with real spectra. Facts 7 and 10 apply.

7. [Li03] Suppose $A - \lambda B$ and $\widetilde{A} - \lambda \widetilde{B}$ are diagonalizable with real spectra, i.e.,

$$Y^* A X = \Lambda, \; Y^* B X = \Omega \quad \text{and} \quad \widetilde{Y}^* \widetilde{A} \widetilde{X} = \widetilde{\Lambda}, \; \widetilde{Y}^* \widetilde{B} \widetilde{X} = \widetilde{\Omega},$$

and all $\langle \alpha_j, \beta_j \rangle$ and all $\langle \widetilde{\alpha}_j, \widetilde{\beta}_j \rangle$ are real. Then the follow statements hold, where

$$\Xi = \operatorname{diag}(\chi(\langle \alpha_1, \beta_1 \rangle, \langle \widetilde{\alpha}_{\tau(1)}, \widetilde{\beta}_{\tau(1)} \rangle), \ldots, \chi(\langle \alpha_n, \beta_n \rangle, \langle \widetilde{\alpha}_{\tau(n)}, \widetilde{\beta}_{\tau(n)} \rangle))$$

for some permutation τ of $\{1, 2, \ldots, n\}$ (possibly depending on the norm being used). In all cases, the constant factor $\pi/2$ can be replaced by 1 for the 2-norm and the Frobenius norm.

(a) $\|\Xi\|_{\text{UI}} \leq \frac{\pi}{2} \sqrt{\kappa_2(X) \kappa_2(\widetilde{X})} \| \sin \Theta(Z^*, \widetilde{Z}^*) \|_{\text{UI}}$.

(b) If all $|\alpha_j|^2 + |\beta_j|^2 = |\widetilde{\alpha}_j|^2 + |\widetilde{\beta}_j|^2 = 1$ in their eigendecompositions, then $\|\Xi\|_{\text{UI}} \leq \frac{\pi}{2} \sqrt{\|X\|_2 \|Y^*\|_2 \|\widetilde{X}\|_2 \|\widetilde{Y}^*\|_2} \| [\Delta A, \; \Delta B] \|_{\text{UI}}$.

8. Let residuals $\mathbf{r} = \widetilde{\beta} A \widetilde{\mathbf{x}} - \widetilde{\alpha} B \widetilde{\mathbf{x}}$ and $\mathbf{s}^* = \widetilde{\beta} \widetilde{\mathbf{y}}^* A - \widetilde{\alpha} \widetilde{\mathbf{y}}^* B$, where $\|\widetilde{\mathbf{x}}\|_2 = \|\widetilde{\mathbf{y}}\|_2 = 1$. The smallest error matrix $[\Delta A, \; \Delta B]$ in the 2-norm, for which $(\widetilde{\mathbf{y}}, \langle \widetilde{\alpha}, \widetilde{\beta} \rangle, \widetilde{\mathbf{x}})$ is an exact generalized eigentriplet of $\widetilde{A} - \lambda \widetilde{B}$, satisfies $\| [\Delta A, \; \Delta B] \|_2 = \varepsilon$, where $\varepsilon = \max\{\|\mathbf{r}\|_2, \|\mathbf{s}\|_2\}$, and $\chi(\langle \alpha, \beta \rangle, \langle \widetilde{\alpha}, \widetilde{\beta} \rangle) \leq \operatorname{cond}(\langle \widetilde{\alpha}, \widetilde{\beta} \rangle) \, \varepsilon + O(\varepsilon^2)$ for some $\langle \alpha, \beta \rangle \in \sigma(A, B)$.
9. [BDD00, p. 128] Suppose A and B are Hermitian and B is positive definite, and let residual $\mathbf{r} = A\widetilde{\mathbf{x}} - \widetilde{\mu} B \widetilde{\mathbf{x}}$ and $\|\widetilde{\mathbf{x}}\|_2 = 1$.

(a) For some eigenvalue μ of $A - \lambda B$,

$$|\widetilde{\mu} - \mu| \leq \frac{\|\mathbf{r}\|_{B^{-1}}}{\|\widetilde{\mathbf{x}}\|_B} \leq \|B^{-1}\|_2 \|\mathbf{r}\|_2,$$

where $\|\mathbf{z}\|_M = \sqrt{\mathbf{z}^* M \mathbf{z}}$.

(b) Let μ be the closest eigenvalue to $\widetilde{\mu}$ among all eigenvalues of $A - \lambda B$ and $\mathbf{x}$ its associated eigenvector with $\|\mathbf{x}\|_2 = 1$, and let $\eta = \min |\widetilde{\mu} - \nu|$ over all other eigenvalues $\nu \neq \mu$ of $A - \lambda B$. If $\eta > 0$, then

$$|\widetilde{\mu} - \mu| \leq \frac{1}{\eta} \cdot \left(\frac{\|\mathbf{r}\|_{B^{-1}}}{\|\widetilde{\mathbf{x}}\|_B} \right)^2 \leq \|B^{-1}\|_2^2 \frac{\|\mathbf{r}\|_2^2}{\eta},$$

$$\sin \angle(\widetilde{\mathbf{x}}, \mathbf{x}) \leq \|B^{-1}\|_2 \sqrt{2\kappa_2(B)} \, \frac{\|\mathbf{r}\|_2}{\eta}.$$

10. [Li94] Suppose $A - \lambda B$ and $\widetilde{A} - \lambda \widetilde{B}$ are diagonalizable and have eigendecompositions

$$\begin{bmatrix} Y_1^* \\ Y_2^* \end{bmatrix} A [X_1, \; X_2] = \begin{bmatrix} \Lambda_1 & \\ & \Lambda_2 \end{bmatrix}, \; \begin{bmatrix} Y_1^* \\ Y_2^* \end{bmatrix} B [X_1, \; X_2] = \begin{bmatrix} \Omega_1 & \\ & \Omega_2 \end{bmatrix},$$

$$X^{-1} = [W_1, \; W_2]^*,$$

and the same for $\widetilde{A} - \lambda \widetilde{B}$ except all symbols with tildes, where $X_1, Y_1, W_1 \in \mathbb{C}^{n \times k}$, $\Lambda_1, \Omega_1 \in \mathbb{C}^{k \times k}$. Suppose $|\alpha_j|^2 + |\beta_j|^2 = |\widetilde{\alpha}_j|^2 + |\widetilde{\beta}_j|^2 = 1$ for $1 \leq j \leq n$ in the eigendecompositions, and set $\eta = \min \chi(\langle \alpha, \beta \rangle, \langle \widetilde{\alpha}, \widetilde{\beta} \rangle)$ taken over all $\langle \alpha, \beta \rangle \in \sigma(\Lambda_1, \Omega_1)$ and $\langle \widetilde{\alpha}, \widetilde{\beta} \rangle \in \sigma(\widetilde{\Lambda}_2, \widetilde{\Omega}_2)$. If $\eta > 0$, then

$$\left\| \sin \Theta(X_1, \widetilde{X}_1) \right\|_{\text{F}} \leq \frac{\|X_1^\dagger\|_2 \|\widetilde{W}_2^\dagger\|_2}{\eta} \left\| \widetilde{Y}_2^* (\widetilde{Z} - Z) \begin{bmatrix} X_1 & \\ & X_1 \end{bmatrix} \right\|_{\text{F}}.$$

15.5 Generalized Singular Value Problems

Definitions:

Let $A \in \mathbb{C}^{m\times n}$ and $B \in \mathbb{C}^{\ell\times n}$. A matrix pair $\{A, B\}$ is an (m,ℓ,n)-**Grassmann matrix pair** if $\operatorname{rank}\left(\begin{bmatrix} A \\ B \end{bmatrix}\right) = n$.

In what follows, all matrix pairs are (m,ℓ,n)-Grassmann matrix pairs.

A pair $\langle \alpha, \beta\rangle$ is a **generalized singular value** of $\{A, B\}$ if

$$\det(\beta^2 A^*A - \alpha^2 B^*B) = 0,\ \langle \alpha,\beta\rangle \neq \langle 0, 0\rangle,\ \alpha,\ \beta \geq 0,$$

i.e., $\langle \alpha,\beta\rangle = \langle \sqrt{\mu}, \sqrt{\nu}\rangle$ for some generalized eigenvalue $\langle \mu,\nu\rangle$ of matrix pencil $A^*A - \lambda B^*B$.

Generalized Singular Value Decomposition (GSVD) of $\{A, B\}$:

$$U^*AX = \Sigma_A, \quad V^*BX = \Sigma_B,$$

where $U \in \mathbb{C}^{m\times m}$, $V \in \mathbb{C}^{\ell\times\ell}$ are unitary, $X \in \mathbb{C}^{n\times n}$ is nonsingular, $\Sigma_A = \operatorname{diag}(\alpha_1, \alpha_2, \cdots)$ is **leading diagonal** (α_j starts in the top left corner), and $\Sigma_B = \operatorname{diag}(\cdots, \beta_{n-1}, \beta_n)$ is **trailing diagonal** (β_j ends in the bottom-right corner), α_j, $\beta_j \geq 0$ and $\alpha_j^2 + \beta_j^2 = 1$ for $1 \leq j \leq n$. (Set some $\alpha_j = 0$ and/or some $\beta_j = 0$, if necessary.)

$\{A, B\}$ is perturbed to $\{\widetilde{A}, \widetilde{B}\} = \{A + \Delta A, B + \Delta B\}$.

Let $\operatorname{SV}(A, B) = \{\langle\alpha_1,\beta_1\rangle, \langle\alpha_2,\beta_2\rangle, \ldots, \langle\alpha_n,\beta_n\rangle\}$ be the set of the generalized singular values of $\{A, B\}$, and set $Z = \begin{bmatrix} A \\ B \end{bmatrix} \in \mathbb{C}^{(m+\ell)\times n}$.

The same notation is adopted for $\{\widetilde{A}, \widetilde{B}\}$, except all symbols with tildes.

Facts:

1. If $\{A, B\}$ is an (m,ℓ,n)-Grassmann matrix pair, then $A^*A - \lambda B^*B$ is a definite matrix pencil.
2. [Van76] The GSVD of an (m,ℓ,n)-Grassmann matrix pair $\{A, B\}$ exists.
3. [Li93] There exist permutations τ and ω of $\{1, 2, \ldots, n\}$ such that

$$\max_{1\leq j\leq n} \chi\left(\langle\alpha_i,\beta_i\rangle, \langle\widetilde{\alpha}_{\tau(i)}, \widetilde{\beta}_{\tau(i)}\rangle\right) \leq \|\sin\Theta(Z,\widetilde{Z})\|_2,$$

$$\sqrt{\sum_{j=1}^{n}\left[\chi\left(\langle\alpha_i,\beta_i\rangle, \langle\widetilde{\alpha}_{\omega(i)}, \widetilde{\beta}_{\omega(i)}\rangle\right)\right]^2} \leq \|\sin\Theta(Z,\widetilde{Z})\|_F.$$

4. [Li94, Lemma 3.3] (**Sun**)

$$\|\sin\Theta(Z,\widetilde{Z})\|_{UI} \leq \frac{\|Z - \widetilde{Z}\|_{UI}}{\max\{\sigma_{\min}(Z), \sigma_{\min}(\widetilde{Z})\}},$$

where $\sigma_{\min}(Z)$ is Z's smallest singular value.

5. [Pai84] If $\alpha_i^2 + \beta_i^2 = \widetilde{\alpha}_i^2 + \widetilde{\beta}_i^2 = 1$ for $i = 1, 2, \ldots, n$, then there exists a permutation ϖ of $\{1, 2, \ldots, n\}$ such that

$$\sqrt{\sum_{j=1}^{n}\left[(\alpha_j - \widetilde{\alpha}_{\varpi(j)})^2 + (\beta_j - \widetilde{\beta}_{\varpi(j)})^2\right]} \leq \min_{Q \text{ unitary}} \|Z_0 - \widetilde{Z}_0 Q\|_F,$$

where $Z_0 = Z(Z^*Z)^{-1/2}$ and $\widetilde{Z}_0 = \widetilde{Z}(\widetilde{Z}^*\widetilde{Z})^{-1/2}$.

6. [Li93], [Sun83] Perturbation bounds on generalized singular subspaces (those spanned by one or a few columns of U, V, and X in GSVD) are also available, but it is quite complicated.

15.6 Relative Perturbation Theory for Eigenvalue Problems

Definitions:

Let scalar $\widetilde{\alpha}$ be an approximation to α, and $1 \le p \le \infty$. Define **relative distances** between α and $\widetilde{\alpha}$ as follows. For $|\alpha|^2 + |\widetilde{\alpha}|^2 \neq 0$,

$$d(\alpha,\widetilde{\alpha}) = \left|\frac{\widetilde{\alpha}}{\alpha} - 1\right| = \frac{|\widetilde{\alpha} - \alpha|}{|\alpha|}, \qquad \text{(classical measure)}$$

$$\varrho_p(\alpha,\widetilde{\alpha}) = \frac{|\widetilde{\alpha} - \alpha|}{\sqrt[p]{|\alpha|^p + |\widetilde{\alpha}|^p}}, \qquad \text{([Li98])}$$

$$\zeta(\alpha,\widetilde{\alpha}) = \frac{|\widetilde{\alpha} - \alpha|}{\sqrt{|\alpha\widetilde{\alpha}|}}, \qquad \text{([BD90], [DV92])}$$

$$\varsigma(\alpha,\widetilde{\alpha}) = |\ln(\widetilde{\alpha}/\alpha)|, \quad \text{for } \widetilde{\alpha}\alpha > 0, \qquad \text{([LM99a], [Li99b])}$$

and $d(0,0) = \varrho_p(0,0) = \zeta(0,0) = \varsigma(0,0) = 0$.

$A \in \mathbb{C}^{n\times n}$ is **multiplicatively perturbed** to $\widetilde{A}$ if $\widetilde{A} = D_{\mathrm{L}}^* A D_{\mathrm{R}}$ for some $D_{\mathrm{L}}, D_{\mathrm{R}} \in \mathbb{C}^{n\times n}$.

Denote $\sigma(A) = \{\lambda_1, \lambda_2, \ldots, \lambda_n\}$ and $\sigma(\widetilde{A}) = \{\widetilde{\lambda}_1, \widetilde{\lambda}_2, \ldots, \widetilde{\lambda}_n\}$.

$A \in \mathbb{C}^{n\times n}$ is said to be **graded** if it can be scaled as $A = S^* H S$ such that H is "well-behaved" (i.e., $\kappa_2(H)$ is of modest magnitude), where S is a scaling matrix, often diagonal but not required so for the facts below. Interesting cases are when $\kappa_2(H) \ll \kappa_2(A)$.

Facts:

1. [Bar00] $\varrho_p(\cdot,\cdot)$ is a metric on $\mathbb{C}$ for $1 \le p \le \infty$.
2. Let $A, \widetilde{A} = D^* A D \in \mathbb{C}^{n\times n}$ be Hermitian, where D is nonsingular.

   (a) [HJ85, p. 224] (**Ostrowski**) There exists t_j, satisfying

   $$\lambda_{\min}(D^* D) \le t_j \le \lambda_{\max}(D^* D),$$

   such that $\widetilde{\lambda}_j^{\uparrow} = t_j \lambda_j^{\uparrow}$ for $j = 1, 2, \ldots, n$ and, thus,

   $$\max_{1\le j\le n} d(\lambda_j^{\uparrow}, \widetilde{\lambda}_j^{\uparrow}) \le \|I - D^* D\|_2.$$

   (b) [LM99], [Li98]

   $$\left\|\operatorname{diag}\left(\varsigma(\lambda_1^{\uparrow}, \widetilde{\lambda}_1^{\uparrow}), \ldots, \varsigma(\lambda_n^{\uparrow}, \widetilde{\lambda}_n^{\uparrow})\right)\right\|_{\mathrm{UI}} \le \|\ln(D^* D)\|_{\mathrm{UI}},$$

   $$\left\|\operatorname{diag}\left(\zeta(\lambda_1^{\uparrow}, \widetilde{\lambda}_1^{\uparrow}), \ldots, \zeta(\lambda_n^{\uparrow}, \widetilde{\lambda}_n^{\uparrow})\right)\right\|_{\mathrm{UI}} \le \|D^* - D^{-1}\|_{\mathrm{UI}}.$$

3. [Li98], [LM99] Let $A = S^* H S$ be a positive semidefinite Hermitian matrix, perturbed to $\widetilde{A} = S^*(H + \Delta H)S$. Suppose H is positive definite and $\|H^{-1/2}(\Delta H)H^{-1/2}\|_2 < 1$, and set

   $$M = H^{1/2} S S^* H^{1/2}, \quad \widetilde{M} = DMD,$$

   where $D = \left[I + H^{-1/2}(\Delta H)H^{-1/2}\right]^{1/2} = D^*$. Then $\sigma(A) = \sigma(M)$ and $\sigma(\widetilde{A}) = \sigma(\widetilde{M})$, and the inequalities in Fact 2 above hold with D here. Note that

$$\|D - D^{-1}\|_{\text{UI}} \le \frac{\|H^{-1/2}(\Delta H)H^{-1/2}\|_{\text{UI}}}{\sqrt{1 - \|H^{-1/2}(\Delta H)H^{-1/2}\|_2}}$$
$$\le \frac{\|H^{-1}\|_2}{\sqrt{1 - \|H^{-1}\|_2 \|\Delta H\|_2}} \|\Delta H\|_{\text{UI}}.$$

4. [BD90], [VS93] Suppose A and $\widetilde{A}$ are Hermitian, and let $|A| = (A^2)^{1/2}$ be the positive semidefinite square root of A^2. If there exists $0 \le \delta < 1$ such that

$$|\mathbf{x}^*(\Delta A)\mathbf{x}| \le \delta \mathbf{x}^*|A|\mathbf{x} \quad \text{for all } \mathbf{x} \in \mathbb{C}^n,$$

then either $\widetilde{\lambda}_j^{\uparrow} = \lambda_j^{\uparrow} = 0$ or $1 - \delta \le \widetilde{\lambda}_j^{\uparrow}/\lambda_j^{\uparrow} \le 1 + \delta$.

5. [Li99a] Let Hermitian A, $\widetilde{A} = D^*AD$ have decompositions

$$\begin{bmatrix} X_1^* \\ X_2^* \end{bmatrix} A[X_1 \; X_2] = \begin{bmatrix} A_1 & \\ & A_2 \end{bmatrix}, \quad \begin{bmatrix} \widetilde{X}_1^* \\ \widetilde{X}_2^* \end{bmatrix} \widetilde{A}[\widetilde{X}_1 \; \widetilde{X}_2] = \begin{bmatrix} \widetilde{A}_1 & \\ & \widetilde{A}_2 \end{bmatrix},$$

where $[X_1 \; X_2]$ and $[\widetilde{X}_1 \; \widetilde{X}_2]$ are unitary and $X_1, \widetilde{X}_1 \in \mathbb{C}^{n\times k}$. If $\eta_2 = \min_{\mu\in\sigma(A_1),\widetilde{\mu}\in\sigma(\widetilde{A}_2)} \varrho_2(\mu, \widetilde{\mu}) > 0$, then

$$\|\sin\Theta(X_1, \widetilde{X}_1)\|_{\text{F}} \le \frac{\sqrt{\|(I - D^{-1})X_1\|_{\text{F}}^2 + \|(I - D^*)X_1\|_{\text{F}}^2}}{\eta_2}.$$

6. [Li99a] Let $A = S^*HS$ be a positive semidefinite Hermitian matrix, perturbed to $\widetilde{A} = S^*(H + \Delta H)S$, having decompositions, in notation, the same as in Fact 5. Let $D = \left[I + H^{-1/2}(\Delta H)H^{-1/2}\right]^{1/2}$. Assume H is positive definite and $\|H^{-1/2}(\Delta H)H^{-1/2}\|_2 < 1$. If $\eta_\zeta = \min_{\mu\in\sigma(A_1),\widetilde{\mu}\in\sigma(\widetilde{A}_2)} \zeta(\mu, \widetilde{\mu}) > 0$, then

$$\|\sin\Theta(X_1, \widetilde{X}_1)\|_{\text{F}} \le \frac{\|D - D^{-1}\|_{\text{F}}}{\eta_\zeta}.$$

Examples:

1. [DK90], [EI95] Let A be a real symmetric tridiagonal matrix with zero diagonal and off-diagonal entries $b_1, b_2, \ldots, b_{n-1}$. Suppose $\widetilde{A}$ is identical to A except for its off-diagonal entries which change to $\beta_1 b_1, \beta_2 b_2, \ldots, \beta_{n-1} b_{n-1}$, where all β_i are real and supposedly close to 1. Then $\widetilde{A} = DAD$, where $D = \text{diag}(d_1, d_2, \ldots, d_n)$ with

$$d_{2k} = \frac{\beta_1\beta_3 \cdots \beta_{2k-1}}{\beta_2\beta_4 \cdots \beta_{2k-2}}, \quad d_{2k+1} = \frac{\beta_2\beta_4 \cdots \beta_{2k}}{\beta_1\beta_3 \cdots \beta_{2k-1}}.$$

Let $\beta = \prod_{j=1}^{n-1} \max\{\beta_j, 1/\beta_j\}$. Then $\beta^{-1}I \le D^2 \le \beta I$, and Fact 2 and Fact 5 apply. Now if all $1 - \varepsilon \le \beta_j \le 1 + \varepsilon$, then $(1 - \varepsilon)^{n-1} \le \beta^{-1} \le \beta \le (1 + \varepsilon)^{n-1}$.

2. Let $A = SHS$ with $S = \text{diag}(1, 10, 10^2, 10^3)$, and

$$A = \begin{bmatrix} 1 & 1 & & \\ 1 & 10^2 & 10^2 & \\ & 10^2 & 10^4 & 10^4 \\ & & 10^4 & 10^6 \end{bmatrix}, \quad H = \begin{bmatrix} 1 & 10^{-1} & & \\ 10^{-1} & 1 & 10^{-1} & \\ & 10^{-1} & 1 & 10^{-1} \\ & & 10^{-1} & 1 \end{bmatrix}.$$

Suppose that each entry A_{ij} of A is perturbed to $A_{ij}(1+\delta_{ij})$ with $|\delta_{ij}| \le \varepsilon$. Then $|(\Delta H)_{ij}| \le \varepsilon |H_{ij}|$ and thus $\|\Delta H\|_2 \le 1.2\varepsilon$. Since $\|H^{-1}\|_2 \le 10/8$, Fact 3 implies

$$\zeta(\lambda_j^{\uparrow}, \widetilde{\lambda}_j^{\uparrow}) \le 1.5\varepsilon/\sqrt{1-1.5\varepsilon} \approx 1.5\varepsilon.$$

15.7 Relative Perturbation Theory for Singular Value Problems

Definitions:

$B \in \mathbb{C}^{m\times n}$ is **multiplicatively perturbed** to $\widetilde{B}$ if $\widetilde{B} = D_{\mathrm{L}}^* B D_{\mathrm{R}}$ for some $D_{\mathrm{L}} \in \mathbb{C}^{m\times m}$ and $D_{\mathrm{R}} \in \mathbb{C}^{n\times n}$.

Denote the singular values of B and $\widetilde{B}$ as

$$\mathrm{SV}(B) = \{\sigma_1, \sigma_2, \ldots, \sigma_{\min\{m,n\}}\}, \quad \mathrm{SV}(\widetilde{B}) = \{\widetilde{\sigma}_1, \widetilde{\sigma}_2, \ldots, \widetilde{\sigma}_{\min\{m,n\}}\}.$$

B is said to be (highly) **graded** if it can be scaled as $B = GS$ such that G is "well-behaved" (i.e., $\kappa_2(G)$ is of modest magnitude), where S is a scaling matrix, often diagonal but not required so for the facts below. Interesting cases are when $\kappa_2(G) \ll \kappa_2(B)$.

Facts:

1. Let B, $\widetilde{B} = D_{\mathrm{L}}^* B D_{\mathrm{R}} \in \mathbb{C}^{m\times n}$, where D_{L} and D_{R} are nonsingular.

   (a) [EI95] For $1 \le j \le n$, $\dfrac{\sigma_j}{\|D_{\mathrm{L}}^{-1}\|_2 \|D_{\mathrm{R}}^{-1}\|_2} \le \widetilde{\sigma}_j \le \sigma_j \|D_{\mathrm{L}}\|_2 \|D_{\mathrm{R}}\|_2$.

   (b) [Li98], [LM99]

$$\begin{aligned} &\|\mathrm{diag}\,(\zeta(\sigma_1, \widetilde{\sigma}_1), \ldots, \zeta(\sigma_n, \widetilde{\sigma}_n))\|_{\mathrm{UI}} \\ &\le \frac{1}{2}\|D_{\mathrm{L}}^* - D_{\mathrm{L}}^{-1}\|_{\mathrm{UI}} + \frac{1}{2}\|D_{\mathrm{R}}^* - D_{\mathrm{R}}^{-1}\|_{\mathrm{UI}}. \end{aligned}$$

2. [Li99a] Let B, $\widetilde{B} = D_{\mathrm{L}}^* B D_{\mathrm{R}} \in \mathbb{C}^{m\times n}$ $(m \ge n)$ have decompositions

$$\begin{bmatrix} U_1^* \\ U_2^* \end{bmatrix} B [V_1 \;\; V_2] = \begin{bmatrix} B_1 & 0 \\ 0 & B_2 \end{bmatrix}, \quad \begin{bmatrix} \widetilde{U}_1^* \\ \widetilde{U}_2^* \end{bmatrix} \widetilde{B} [\widetilde{V}_1 \;\; \widetilde{V}_2] = \begin{bmatrix} \widetilde{B}_1 & 0 \\ 0 & \widetilde{B}_2 \end{bmatrix},$$

   where $[U_1 \;\; U_2]$, $[V_1 \;\; V_2]$, $[\widetilde{U}_1 \;\; \widetilde{U}_2]$, and $[\widetilde{V}_1 \;\; \widetilde{V}_2]$ are unitary, and U_1, $\widetilde{U}_1 \in \mathbb{C}^{m\times k}$, V_1, $\widetilde{V}_1 \in \mathbb{C}^{n\times k}$. Set $\Theta_U = \Theta(U_1, \widetilde{U}_1)$, $\Theta_V = \Theta(V_1, \widetilde{V}_1)$. If $\mathrm{SV}(B_1) \bigcap \mathrm{SV}_{\mathrm{ext}}(\widetilde{B}_2) = \emptyset$, then

$$\begin{aligned} &\sqrt{\|\sin\Theta_U\|_{\mathrm{F}}^2 + \|\sin\Theta_V\|_{\mathrm{F}}^2} \\ &\le \frac{1}{\eta_2}\left[\|(I - D_{\mathrm{L}}^*)U_1\|_{\mathrm{F}}^2 + \|(I - D_{\mathrm{L}}^{-1})U_1\|_{\mathrm{F}}^2\right. \\ &\quad \left. + \|(I - D_{\mathrm{R}}^*)V_1\|_{\mathrm{F}}^2 + \|(I - D_{\mathrm{R}}^{-1})V_1\|_{\mathrm{F}}^2\right]^{1/2}, \end{aligned}$$

   where $\eta_2 = \min \varrho_2(\mu, \widetilde{\mu})$ over all $\mu \in \mathrm{SV}(B_1)$ and $\widetilde{\mu} \in \mathrm{SV}_{\mathrm{ext}}(\widetilde{B}_2)$.

3. [Li98], [Li99a], [LM99] Let $B = GS$ and $\widetilde{B} = \widetilde{G}S$ be two $m \times n$ matrices, where $\mathrm{rank}(G) = n$, and let $\Delta G = \widetilde{G} - G$. Then $\widetilde{B} = DB$, where $D = I + (\Delta G)G^{\dagger}$. Fact 1 and Fact 2 apply with $D_{\mathrm{L}} = D$ and $D_{\mathrm{R}} = I$. Note that

$$\|D^* - D^{-1}\|_{\mathrm{UI}} \le \left(1 + \frac{1}{1 - \|(\Delta G)G^{\dagger}\|_2}\right) \frac{\|(\Delta G)G^{\dagger}\|_{\mathrm{UI}}}{2}.$$

Examples:

1. [BD90], [DK90], [EI95] B is a real bidiagonal matrix with diagonal entries $a_1, a_2, \dots, a_n$ and off-diagonal (the one above the diagonal) entries are $b_1, b_2, \dots, b_{n-1}$. $\widetilde{B}$ is the same as B, except for its diagonal entries, which change to $\alpha_1 a_1, \alpha_2 a_2, \dots, \alpha_n a_n$, and its off-diagonal entries, which change to $\beta_1 b_1, \beta_2 b_2, \dots, \beta_{n-1} b_{n-1}$. Then $\widetilde{B} = D_{\mathrm{L}}^* B D_{\mathrm{R}}$ with

$$D_{\mathrm{L}} = \operatorname{diag}\left(\alpha_1, \frac{\alpha_1\alpha_2}{\beta_1}, \frac{\alpha_1\alpha_2\alpha_3}{\beta_1\beta_2}, \dots\right),$$
$$D_{\mathrm{R}} = \operatorname{diag}\left(1, \frac{\beta_1}{\alpha_1}, \frac{\beta_1\beta_2}{\alpha_1\alpha_2}, \dots\right).$$

   Let $\alpha = \prod_{j=1}^{n} \max\{\alpha_j, 1/\alpha_j\}$ and $\beta = \prod_{j=1}^{n-1} \max\{\beta_j, 1/\beta_j\}$. Then

$$(\alpha\beta)^{-1} \le \left(\|D_{\mathrm{L}}^{-1}\|_2 \|D_{\mathrm{R}}^{-1}\|_2\right)^{-1} \le \|D_{\mathrm{L}}\|_2 \|D_{\mathrm{R}}\|_2 \le \alpha\beta,$$

   and Fact 1 and Fact 2 apply. Now if all $1-\varepsilon \le \alpha_i, \beta_j \le 1+\varepsilon$, then $(1-\varepsilon)^{2n-1} \le (\alpha\beta)^{-1} \le (\alpha\beta) \le (1+\varepsilon)^{2n-1}$.

2. Consider block partitioned matrices

$$B = \begin{bmatrix} B_{11} & B_{12} \\ 0 & B_{22} \end{bmatrix},$$
$$\widetilde{B} = \begin{bmatrix} B_{11} & 0 \\ 0 & B_{22} \end{bmatrix} = B \begin{bmatrix} I & -B_{11}^{-1} B_{12} \\ 0 & I \end{bmatrix} = BD_{\mathrm{R}}.$$

   By Fact 2, $\zeta(\sigma_j, \widetilde{\sigma}_j) \le \frac{1}{2}\|B_{11}^{-1} B_{12}\|_2$. Interesting cases are when $\|B_{11}^{-1} B_{12}\|_2$ is tiny enough to be treated as zero and so $\mathrm{SV}(\widetilde{B})$ approximates $\mathrm{SV}(B)$ well. This situation occurs in computing the SVD of a bidiagonal matrix.

Author Note: Supported in part by the National Science Foundation under Grant No. DMS-0510664.

References

[BDD00] Z. Bai, J. Demmel, J. Dongarra, A. Ruhe, and H. van der Vorst (Eds). *Templates for the Solution of Algebraic Eigenvalue Problems: A Practical Guide.* SIAM, Philadelphia, 2000.

[BD90] J. Barlow and J. Demmel. Computing accurate eigensystems of scaled diagonally dominant matrices. *SIAM J. Numer. Anal.*, 27:762–791, 1990.

[Bar00] A. Barrlund. The p-relative distance is a metric. *SIAM J. Matrix Anal. Appl.*, 21(2):699–702, 2000.

[Bau85] H. Baumgärtel. *Analytical Perturbation Theory for Matrices and Operators.* Birkhäuser, Basel, 1985.

[Bha96] R. Bhatia. *Matrix Analysis.* Graduate Texts in Mathematics, Vol. 169. Springer, New York, 1996.

[BKL97] R. Bhatia, F. Kittaneh, and R.-C. Li. Some inequalities for commutators and an application to spectral variation. II. *Lin. Multilin. Alg.*, 43(1-3):207–220, 1997.

[CG00] F. Chatelin and S. Gratton. On the condition numbers associated with the polar factorization of a matrix. *Numer. Lin. Alg. Appl.*, 7:337–354, 2000.

[DK70] C. Davis and W. Kahan. The rotation of eigenvectors by a perturbation. III. *SIAM J. Numer. Anal.*, 7:1–46, 1970.

[DK90] J. Demmel and W. Kahan. Accurate singular values of bidiagonal matrices. *SIAM J. Sci. Statist. Comput.*, 11:873–912, 1990.

[DV92] J. Demmel and K. Veselić. Jacobi's method is more accurate than QR. *SIAM J. Matrix Anal. Appl.*, 13:1204–1245, 1992.

[EI95] S.C. Eisenstat and I.C.F. Ipsen. Relative perturbation techniques for singular value problems. *SIAM J. Numer. Anal.*, 32:1972–1988, 1995.

[HJ85] R.A. Horn and C.R. Johnson. *Matrix Analysis*. Cambridge University Press, Cambridge, 1985.

[KPJ82] W. Kahan, B.N. Parlett, and E. Jiang. Residual bounds on approximate eigensystems of nonnormal matrices. *SIAM J. Numer. Anal.*, 19:470–484, 1982.

[Kat70] T. Kato. *Perturbation Theory for Linear Operators*, 2nd ed., Springer-Verlag, Berlin, 1970.

[Kit86] F. Kittaneh. Inequalities for the schatten p-norm. III. *Commun. Math. Phys.*, 104:307–310, 1986.

[LL05] Chi-Kwong Li and Ren-Cang Li. A note on eigenvalues of perturbed Hermitian matrices. *Lin. Alg. Appl.*, 395:183–190, 2005.

[LM99] Chi-Kwong Li and R. Mathias. The Lidskii–Mirsky–Wielandt theorem — additive and multiplicative versions. *Numer. Math.*, 81:377–413, 1999.

[Li88] Ren-Cang Li. A converse to the Bauer-Fike type theorem. *Lin. Alg. Appl.*, 109:167–178, 1988.

[Li93] Ren-Cang Li. Bounds on perturbations of generalized singular values and of associated subspaces. *SIAM J. Matrix Anal. Appl.*, 14:195–234, 1993.

[Li94] Ren-Cang Li. On perturbations of matrix pencils with real spectra. *Math. Comp.*, 62:231–265, 1994.

[Li95] Ren-Cang Li. New perturbation bounds for the unitary polar factor. *SIAM J. Matrix Anal. Appl.*, 16:327–332, 1995.

[Li97] Ren-Cang Li. Relative perturbation bounds for the unitary polar factor. *BIT*, 37:67–75, 1997.

[Li98] Ren-Cang Li. Relative perturbation theory: I. Eigenvalue and singular value variations. *SIAM J. Matrix Anal. Appl.*, 19:956–982, 1998.

[Li99a] Ren-Cang Li. Relative perturbation theory: II. Eigenspace and singular subspace variations. *SIAM J. Matrix Anal. Appl.*, 20:471–492, 1999.

[Li99b] Ren-Cang Li. A bound on the solution to a structured Sylvester equation with an application to relative perturbation theory. *SIAM J. Matrix Anal. Appl.*, 21:440–445, 1999.

[Li03] Ren-Cang Li. On perturbations of matrix pencils with real spectra, a revisit. *Math. Comp.*, 72:715–728, 2003.

[Li05] Ren-Cang Li. Relative perturbation bounds for positive polar factors of graded matrices. *SIAM J. Matrix Anal. Appl.*, 27:424–433, 2005.

[LS02] W. Li and W. Sun. Perturbation bounds for unitary and subunitary polar factors. *SIAM J. Matrix Anal. Appl.*, 23:1183–1193, 2002.

[Mat93] R. Mathias. Perturbation bounds for the polar decomposition. *SIAM J. Matrix Anal. Appl.*, 14:588–597, 1993.

[Pai84] C.C. Paige. A note on a result of Sun Ji-Guang: sensitivity of the CS and GSV decompositions. *SIAM J. Numer. Anal.*, 21:186–191, 1984.

[Par98] B.N. Parlett. *The Symmetric Eigenvalue Problem*. SIAM, Philadelphia, 1998.

[SS90] G.W. Stewart and Ji-Guang Sun. *Matrix Perturbation Theory*. Academic Press, Boston, 1990.

[Sun83] Ji-Guang Sun. Perturbation analysis for the generalized singular value decomposition. *SIAM J. Numer. Anal.*, 20:611–625, 1983.

[Sun91] Ji-Guang Sun. Eigenvalues of Rayleigh quotient matrices. *Numer. Math.*, 59:603–614, 1991.

[Sun96] Ji-Guang Sun. On the variation of the spectrum of a normal matrix. *Lin. Alg. Appl.*, 246:215–223, 1996.

[Sun98] Ji-Guang Sun. Stability and accuracy, perturbation analysis of algebraic eigenproblems. Technical Report UMINF 98-07, Department of Computer Science, Umeå Univeristy, Sweden, 1998.

[Van76] C.F. Van Loan. Generalizing the singular value decomposition. *SIAM J. Numer. Anal.*, 13:76–83, 1976.

[VS93] Krešimir Veselić and Ivan Slapničar. Floating-point perturbations of Hermitian matrices. *Lin. Alg. Appl.*, 195:81–116, 1993.

16
Pseudospectra

Mark Embree
Rice University

Eigenvalues often provide great insight into the behavior of matrices, precisely explaining, for example, the *asymptotic* character of functions of matrices like A^k and e^{tA}. Yet many important applications produce matrices whose behavior cannot be explained by eigenvalues alone. In such circumstances further information can be gleaned from broader sets in the complex plane, such as the numerical range (see Chapter 18), the polynomial numerical hull [Nev93], [Gre02], and the subject of this section, pseudospectra.

The ε-pseudospectrum is a subset of the complex plane that always includes the spectrum, but can potentially contain points far from any eigenvalue. Unlike the spectrum, pseudospectra vary with choice of norm and, thus, for a given application one must take care to work in a physically appropriate norm. Unless otherwise noted, throughout this chapter we assume that $A \in \mathbb{C}^{n\times n}$ is a square matrix with complex entries, and that $\|\cdot\|$ denotes a vector space norm and the matrix norm it induces. When speaking of a norm associated with an inner product, we presume that adjoints and normal and unitary matrices are defined with respect to that inner product. All computational examples given here use the 2-norm.

For further details about theoretical aspects of this subject and the application of pseudospectra to a variety of problems see [TE05]; for applications in control theory, see [HP05]; and for applications in perturbation theory see [CCF96].

16.1 Fundamentals of Pseudospectra

Definitions:

The **ε-pseudospectrum** of a matrix $A \in \mathbb{C}^{n\times n}$, $\varepsilon > 0$, is the set

$$\sigma_\varepsilon(A) = \{z \in \mathbb{C} : z \in \sigma(A + E) \text{ for some } E \in \mathbb{C}^{n\times n} \text{ with } \|E\| < \varepsilon\}.$$

(This definition is sometimes written with a weak inequality, $\|E\| \le \varepsilon$; for matrices the difference has little significance, but the strict inequality proves to be convenient for infinite-dimensional operators.)

If $\|A\mathbf{v} - z\mathbf{v}\| < \varepsilon\|\mathbf{v}\|$ for some $\mathbf{v} \neq 0$, then z is an **ε-pseudoeigenvalue** of A with corresponding **ε-pseudoeigenvector $\mathbf{v}$**.

The **resolvent** of the matrix $A \in \mathbb{C}^{n\times n}$ at a point $z \notin \sigma(A)$ is the matrix $(zI - A)^{-1}$.

Facts: [TE05]

1. Equivalent definitions. The set $\sigma_\varepsilon(A)$ can be equivalently defined as:
   (a) The subset of the complex plane bounded within the $1/\varepsilon$ level set of the norm of the resolvent:
   $$\sigma_\varepsilon(A) = \{z \in \mathbb{C} : \|(zI - A)^{-1}\| > \varepsilon^{-1}\}, \tag{16.1}$$
   with the convention that $\|(zI - A)^{-1}\| = \infty$ when $zI - A$ is not invertible, i.e., when $z \in \sigma(A)$.
   (b) The set of all ε-pseudoeigenvalues of A:
   $$\sigma_\varepsilon(A) = \{z \in \mathbb{C} : \|A\mathbf{v} - z\mathbf{v}\| < \varepsilon \text{ for some unit vector } \mathbf{v} \in \mathbb{C}^n\}.$$
2. For finite $\varepsilon > 0$, $\sigma_\varepsilon(A)$ is a bounded open set in $\mathbb{C}$ containing no more than n connected components, and $\sigma(A) \subset \sigma_\varepsilon(A)$. Each connected component must contain at least one eigenvalue of A.
3. Pseudospectral mapping theorems.
   (a) For any $\alpha, \gamma \in \mathbb{C}$ with $\gamma \neq 0$, $\quad \sigma_\varepsilon(\alpha I + \gamma A) = \alpha + \sigma_{\varepsilon/\gamma}(A)$.
   (b) [Lui03] Suppose f is a function analytic on $\sigma_\varepsilon(A)$ for some $\varepsilon > 0$, and define $\gamma(\varepsilon) = \sup_{\|E\|\leq\varepsilon} \|f(A+E) - f(A)\|$. Then $f(\sigma_\varepsilon(A)) \subseteq \sigma_{\gamma(\varepsilon)}(f(A))$. See [Lui03] for several more inclusions of this type.
4. Stability of pseudospectra. For any $\varepsilon > 0$ and E such that $\|E\| < \varepsilon$,
   $$\sigma_{\varepsilon-\|E\|}(A) \subseteq \sigma_\varepsilon(A+E) \subseteq \sigma_{\varepsilon+\|E\|}(A).$$
5. Properties of pseudospectra as $\varepsilon \to 0$.
   (a) If λ is a eigenvalue of A with index k, then there exist constants d and C such that $\|(zI - A)^{-1}\| \leq C|z - \lambda|^{-k}$ for all z such that $|z - \lambda| < d$.
   (b) Any two matrices with the same ε-pseudospectra for all $\varepsilon > 0$ have the same minimal polynomial.
6. Suppose $\|\cdot\|$ is the natural norm in an inner product space.
   (a) The matrix A is normal (see Section 7.2) if and only if $\sigma_\varepsilon(A)$ equals the union of open ε-balls about each eigenvalue for all $\varepsilon > 0$.
   (b) For any $A \in \mathbb{C}^{n\times n}$, $\sigma_\varepsilon(A^*) = \overline{\sigma_\varepsilon(A)}$.
7. [BLO03] Suppose $\|\cdot\|$ is the natural norm in an inner product space. The point $z = x + \mathrm{i}y$, $x, y \in \mathbb{R}$, is on the boundary of $\sigma_\varepsilon(A)$ provided $\mathrm{i}y$ is an eigenvalue of the Hamiltonian matrix
   $$\begin{bmatrix} xI - A^* & \varepsilon I \\ -\varepsilon I & A - xI \end{bmatrix}.$$
   This fact implies that the boundary of $\sigma_\varepsilon(A)$ cannot contain a segment of any vertical line or, substituting $e^{\mathrm{i}\theta}A$ for A, a segment of any straight line.
8. The following results provide lower and upper bounds on the ε-pseudospectrum; Δ_δ denotes the open unit ball of radius δ in $\mathbb{C}$, and $\kappa(X) = \|X\|\|X^{-1}\|$.
   (a) For all $\varepsilon > 0$, $\sigma(A) + \Delta_\varepsilon \subseteq \sigma_\varepsilon(A)$.
   (b) For any nonsingular $S \in \mathbb{C}^{n\times n}$ and all $\varepsilon > 0$,
   $$\sigma_{\varepsilon/\kappa(S)}(SAS^{-1}) \subseteq \sigma_\varepsilon(A) \subseteq \sigma_{\varepsilon\kappa(S)}(SAS^{-1}).$$
   (c) (Bauer–Fike Theorems [BF60], [Dem97]) Let $\|\cdot\|$ denote a monotone norm. If A is diagonalizable, $A = V\Lambda V^{-1}$, then for all $\varepsilon > 0$,
   $$\sigma_\varepsilon(A) \subseteq \sigma(A) + \Delta_{\varepsilon\kappa(V)}.$$

If $A \in \mathbb{C}^{n\times n}$ has n distinct eigenvalues $\lambda_1, \ldots, \lambda_n$, then for all $\varepsilon > 0$,

$$\sigma_\varepsilon(A) \subseteq \cup_{j=1}^{N}(\lambda_j + \Delta_{\varepsilon n\kappa(\lambda_j)}),$$

where $\kappa(\lambda_j)$ here denotes the eigenvalue condition number of λ_j (i.e., $\kappa(\lambda_j) = 1/|\widehat{\mathbf{v}}_j^*\mathbf{v}_j|$, where $\widehat{\mathbf{v}}_j$ and $\mathbf{v}_j$ are unit-length left and right eigenvectors of A corresponding to the eigenvalue λ_j).

(d) If $\|\cdot\|$ is the natural norm in an inner product space, then for any $\varepsilon > 0, \sigma_\varepsilon(A) \subseteq W(A) + \Delta_\varepsilon$, where $W(\cdot)$ denotes the numerical range (Chapter 18).

(e) If $\|\cdot\|$ is the natural norm in an inner product space and U is a unitary matrix, then $\sigma_\varepsilon(U^*AU) = \sigma_\varepsilon(A)$ for all $\varepsilon > 0$.

(f) If $\|\cdot\|$ is unitarily invariant, then $\sigma_\varepsilon(A) \subseteq \sigma(A) + \Delta_{\varepsilon+\text{dep}(A)}$, where dep$(\cdot)$ denotes Henrici's departure from normality (i.e., the norm of the off-diagonal part of the triangular factor in a Schur decomposition, minimized over all such decompositions).

(g) (Geršgorin Theorem for pseudospectra [ET01]) Using the induced matrix 2-norm, for any $\varepsilon > 0$,

$$\sigma_\varepsilon(A) \subseteq \cup_{j=1}^{n}(a_{jj} + \Delta_{r_j+\varepsilon\sqrt{n}}),$$

where $r_j = \sum_{k=1,k\neq j}^{n} |a_{jk}|$.

9. The following results bound $\sigma_\varepsilon(A)$ by pseudospectra of smaller matrices. Here $\|\cdot\|$ is the natural norm in an inner product space.

(a) [GL02] If A has the block-triangular form

$$A = \begin{bmatrix} B & C \\ D & E \end{bmatrix},$$

then

$$\sigma_\varepsilon(A) \subseteq \sigma_\delta(B) \cup \sigma_\delta(E),$$

where $\delta = (\varepsilon + \|D\|)\sqrt{1 + \|C\|/(\varepsilon + \|D\|)}$. Provided $\varepsilon > \|D\|$,

$$\sigma_\gamma(B) \cup \sigma_\gamma(E) \subseteq \sigma_\varepsilon(A),$$

where $\gamma = \varepsilon - \|D\|$.

(b) If the columns of $V \in \mathbb{C}^{n\times m}$ form a basis for an invariant subspace of A, and $\widehat{V} \in \mathbb{C}^{n\times m}$ is such that $\widehat{V}^*V = I$, then $\sigma_\varepsilon(\widehat{V}^*AV) \subseteq \sigma_\varepsilon(A)$. In particular, if the columns of U form an orthonormal basis for an invariant subspace of A, then $\sigma_\varepsilon(U^*AU) \subseteq \sigma_\varepsilon(A)$.

(c) [ET01] If $U \in \mathbb{C}^{n\times m}$ has orthonormal columns and $AU = UH + R$, then $\sigma(H) \subseteq \sigma_\varepsilon(A)$ for $\varepsilon = \|R\|$.

(d) (Arnoldi factorization) If $AU = [U\ \mathbf{u}]H$, where $H \in \mathbb{C}^{(m+1)\times m}$ is an upper Hessenberg matrix ($h_{jk} = 0$ if $j > k+1$) and the columns of $[U\ \mathbf{u}] \in \mathbb{C}^{n\times(m+1)}$ are orthonormal, then $\sigma_\varepsilon(H) \subseteq \sigma_\varepsilon(A)$. (The ε-pseudospectrum of a rectangular matrix is defined in section 16.5 below.)

Examples:

The plots of pseudospectra that follow show the boundary of $\sigma_\varepsilon(A)$ for various values of ε, with the smallest values of ε corresponding to those boundaries closest to the eigenvalues. In all cases, $\|\cdot\|$ is the 2-norm.

1. The following three matrices all have the same eigenvalues, $\sigma(A) = \{1, \pm\text{i}\}$, yet their pseudospectra, shown in Figure 16.1, differ considerably:

$$\begin{bmatrix} 0 & -1 & 10 \\ 1 & 0 & 5 \\ 0 & 0 & 1 \end{bmatrix}, \quad \begin{bmatrix} 0 & -1 & 10 \\ 1 & 0 & 5\text{i} \\ 0 & 0 & 1 \end{bmatrix}, \quad \begin{bmatrix} 2 & -5 & 10 \\ 1 & -2 & 5\text{i} \\ 0 & 0 & 1 \end{bmatrix}.$$

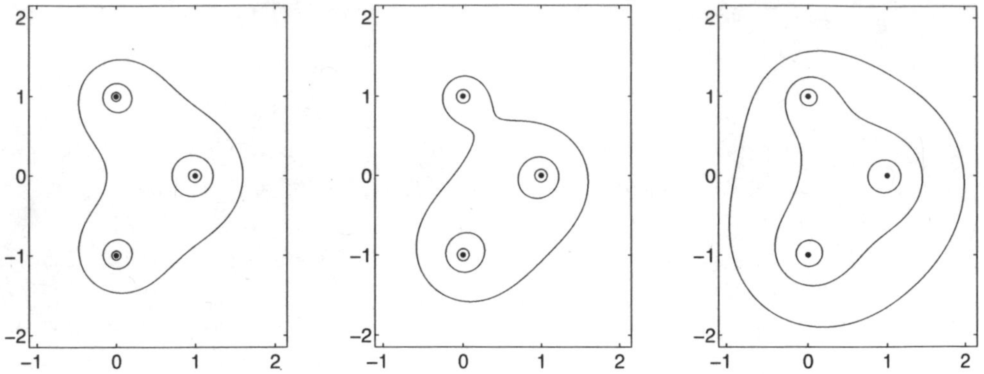

FIGURE 16.1 Spectra (solid dots) and ε-pseudospectra of the three matrices of Example 1, each with $\sigma(A) = \{1, \mathrm{i}, -\mathrm{i}\}$; $\varepsilon = 10^{-1}, 10^{-1.5}, 10^{-2}$.

2. [RT92] For any matrix that is zero everywhere except the first superdiagonal, $\sigma_\varepsilon(A)$ consists of an open disk centered at zero whose radius depends on ε for all $\varepsilon > 0$. Figure 16.2 shows pseudospectra for two such examples of dimension $n = 50$:

$$\begin{bmatrix} 0 & 1 & & & \\ & 0 & 1 & & \\ & & \ddots & \ddots & \\ & & & 0 & 1 \\ & & & & 0 \end{bmatrix}, \quad \begin{bmatrix} 0 & 3 & & & \\ & 0 & 3/2 & & \\ & & \ddots & \ddots & \\ & & & 0 & 3/(n-1) \\ & & & & 0 \end{bmatrix}.$$

Though these matrices have the same minimal polynomial, the pseudospectra differ considerably.

3. It is evident from Figure 16.1 that the components of $\sigma_\varepsilon(A)$ need not be convex. In fact, they need not be simply connected; that is, $\sigma_\varepsilon(A)$ can have "holes." This is illustrated in Figure 16.3 for the following examples, a circulant (hence, normal) matrix and a defective matrix constructed

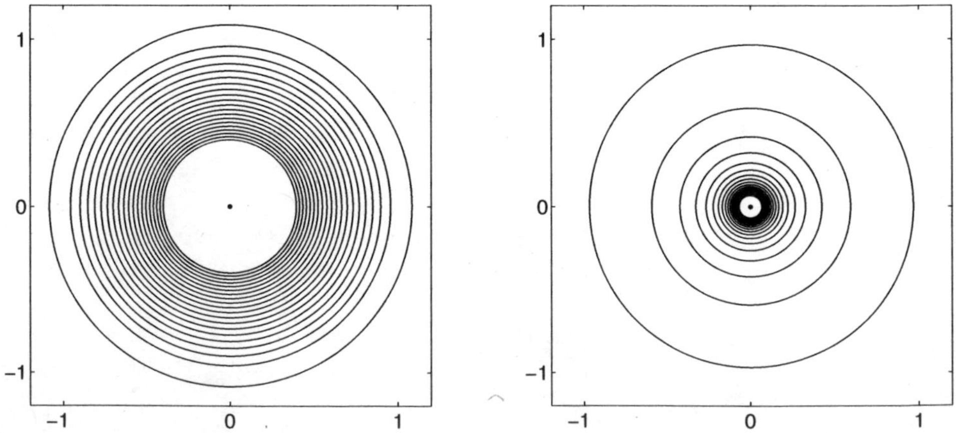

FIGURE 16.2 Spectra (solid dots) and ε-pseudospectra of the matrices in Example 2 for $\varepsilon = 10^{-1}, 10^{-2}, \ldots, 10^{-20}$.

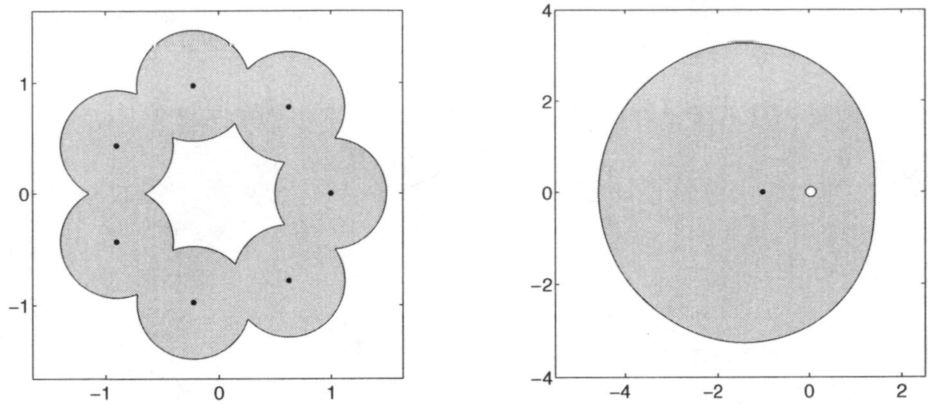

FIGURE 16.3 Spectra (solid dots) and ε-pseudospectra (gray regions) of the matrices in Example 3 for $\varepsilon = .5$ (left) and $\varepsilon = 10^{-3}$ (right). Both plotted pseudospectra are doubly connected.

by Demmel [Dem87]:

$$\begin{bmatrix} 0 & 1 & 0 & 0 & 0 & 0 & 0 \\ 0 & 0 & 1 & 0 & 0 & 0 & 0 \\ 0 & 0 & 0 & 1 & 0 & 0 & 0 \\ 0 & 0 & 0 & 0 & 1 & 0 & 0 \\ 0 & 0 & 0 & 0 & 0 & 1 & 0 \\ 0 & 0 & 0 & 0 & 0 & 0 & 1 \\ 1 & 0 & 0 & 0 & 0 & 0 & 0 \end{bmatrix}, \qquad \begin{bmatrix} -1 & -100 & -10000 \\ 0 & -1 & -100 \\ 0 & 0 & -1 \end{bmatrix}.$$

16.2 Toeplitz Matrices

Given the rich variety of important applications in which Toeplitz matrices arise, we are fortunate that so much is now understood about their spectral properties. Nonnormal Toeplitz matrices are prominent examples of matrices whose eigenvalues provide only limited insight into system behavior. The spectra of infinite-dimensional Toeplitz matrices are easily characterized, and one would hope to use these results to approximate the spectra of more recalcitrant large, finite-dimensional examples. For generic problems, the spectra of finite-dimensional Toeplitz matrices do not converge to the spectrum of the corresponding infinite-dimensional Toeplitz operator. However, the ε-pseudospectra do converge in the $n \to \infty$ limit for all $\varepsilon > 0$, and, moreover, for banded Toeplitz matrices this convergence is especially striking as the resolvent grows exponentially with n in certain regions. Comprehensive references addressing the pseudospectra of Toeplitz matrices include the books [BS99] and [BG05]. For a generalization of these results to "twisted Toeplitz matrices," where the entries on each diagonal are samples of a smoothly varying function, see [TC04].

Definitions:

A **Toeplitz operator** is a singly infinite matrix with constant entries on each diagonal:

$$T = \begin{bmatrix} a_0 & a_{-1} & a_{-2} & a_{-3} & \cdots \\ a_1 & a_0 & a_{-1} & a_{-2} & \ddots \\ a_2 & a_1 & a_0 & a_{-1} & \ddots \\ a_3 & a_2 & a_1 & a_0 & \ddots \\ \vdots & \ddots & \ddots & \ddots & \ddots \end{bmatrix}$$

for $a_0, a_{\pm 1}, a_{\pm 2}, \ldots \in \mathbb{C}$.

Provided it is well defined for all z on the unit circle $\mathbb{T}$ in the complex plane, the function $a(z) = \sum_{k=-\infty}^{\infty} a_k z^k$ is called the **symbol** of T.

The set $a(\mathbb{T}) \subset \mathbb{C}$ is called the **symbol curve**.

Given a symbol a, the corresponding n-dimensional **Toeplitz matrix** takes the form

$$T_n = \begin{bmatrix} a_0 & a_{-1} & a_{-2} & \cdots & a_{1-n} \\ a_1 & a_0 & a_{-1} & \ddots & \vdots \\ a_2 & a_1 & a_0 & \ddots & a_{-2} \\ \vdots & \ddots & \ddots & \ddots & a_{-1} \\ a_{n-1} & \cdots & a_2 & a_1 & a_0 \end{bmatrix} \in \mathbb{C}^{n\times n}.$$

For a symbol $a(z) = \sum_{k=-\infty}^{\infty} a_k z^k$, the set $\{T_n\}_{n\geq 1}$ is called a **family of Toeplitz matrices.**

A family of Toeplitz matrices with symbol a is **banded** if there exists some $m \geq 0$ such that $a_{\pm k} = 0$ for all $k \geq m$.

Facts:

1. [Böt94] (Convergence of pseudospectra) Let $\|\cdot\|$ denote any p norm. If the symbol a is a continuous function on $\mathbb{T}$, then

$$\lim_{n\to\infty} \sigma_\varepsilon(T_n) \to \sigma_\varepsilon(T)$$

as $n \to \infty$, where T is the infinite-dimensional Toeplitz operator with symbol a acting on the space ℓ_p, and its ε-pseudospectrum is a natural generalization of the first definition in section 16.1. The convergence of sets is understood in the Hausdorff sense [Hau91, p. 167], i.e., the distance between bounded sets $\Sigma_1, \Sigma_2 \subseteq \mathbb{C}$ is given by

$$d(\Sigma_1, \Sigma_2) = \max\left\{ \sup_{s_1\in\Sigma_1} \inf_{s_2\in\Sigma_2} |s_1 - s_2|,\ \sup_{s_2\in\Sigma_2} \inf_{s_1\in\Sigma_1} |s_2 - s_1| \right\}.$$

2. [BS99] Provided the symbol a is a continuous function on $\mathbb{T}$, the spectrum $\sigma(T)$ of the infinite dimensional Toeplitz operator T on ℓ_p comprises $a(\mathbb{T})$ together with all points $z \in \mathbb{C} \setminus a(\mathbb{T})$ that $a(\mathbb{T})$ encloses with winding number

$$\frac{1}{2\pi \mathrm{i}} \int_{a(\mathbb{T})} \frac{1}{\zeta - z}\, d\zeta$$

nonzero. From the previous fact, we deduce that $\|(zI - T_n)^{-1}\| \to \infty$ as $n \to \infty$ if $z \in \sigma(T)$ and that, for any fixed $\varepsilon > 0$, there exists some $N \geq 1$ such that $\sigma(T) \subseteq \sigma_\varepsilon(T_n)$ for all $n \geq N$.

3. [RT92] (Exponential growth of the resolvent) If the family of Toeplitz matrices T_n is banded, then for any fixed $z \in \mathbb{C}$ such that the winding number of $a(\mathbb{T})$ with respect to z is nonzero, there exists some $\gamma > 1$ and $N \geq 1$ such that $\|(zI - T_n)^{-1}\| \geq \gamma^n$ for all $n \geq N$.

Examples:

1. Consider the family of Toeplitz matrices with symbol

$$a(t) = t - \tfrac{1}{2} - \tfrac{1}{16}t^{-1}.$$

For any dimension n, the spectrum $\sigma(T_n(a))$ is contained in the line segment $[-\frac{1}{2} - \frac{1}{2}\mathrm{i}, -\frac{1}{2} + \frac{1}{2}\mathrm{i}]$ in the complex plane. This symbol was selected so that $\sigma(A)$ falls in both the left half-plane and the unit disk, while even for relatively small values of ε, $\sigma_\varepsilon(A)$ contains both points in the right half-plane and points outside the unit disk for all but the smallest values of n; see Figure 16.4 for $n = 50$.

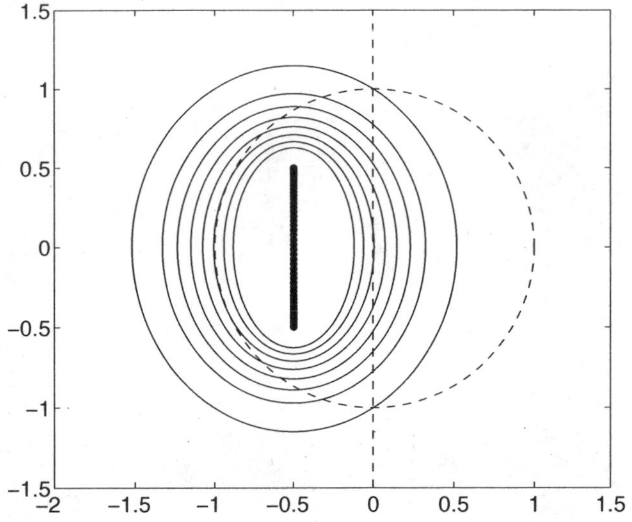

FIGURE 16.4 Spectrum (solid dots, so close they appear to be a thick line segment with real part $-1/2$) and ε-pseudospectra of the Toeplitz matrix T_{50} from Example 1; $\varepsilon = 10^{-1}, 10^{-3}, \ldots, 10^{-15}$. The dashed lines show the unit circle and the imaginary axis.

2. Pseudospectra of matrices with the symbols

$$a(t) = \mathrm{i}t^4 + t^2 + 2t + 5t^{-2} + \mathrm{i}t^{-5}$$

and

$$a(t) = 3\mathrm{i}t^4 + t + t^{-1} + 3\mathrm{i}t^{-4}$$

are shown in Figure 16.5.

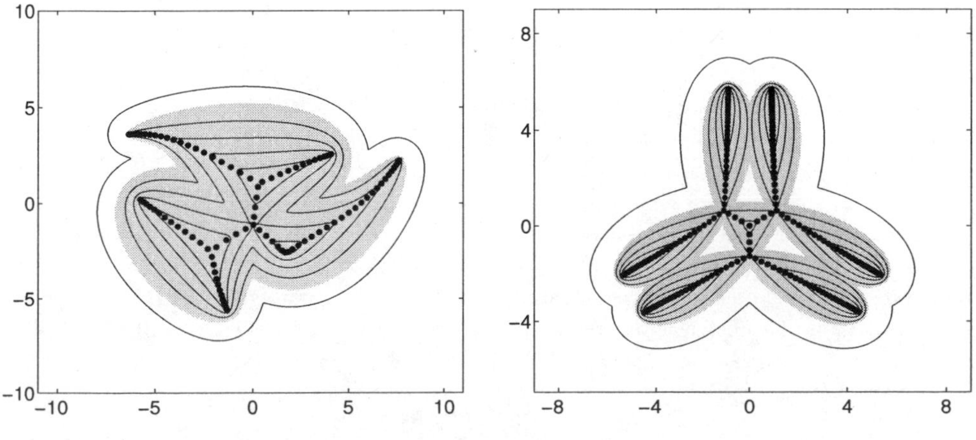

FIGURE 16.5 Spectra (solid dots) and ε-pseudospectra of Toeplitz matrices from Example 2 with the first symbol on the left ($n = 100$) and the second symbol on the right ($n = 200$), both with $\varepsilon = 10^0, 10^{-2}, \ldots, 10^{-8}$. In each plot, the gray region is the spectrum of the underlying infinite dimensional Toeplitz operator.

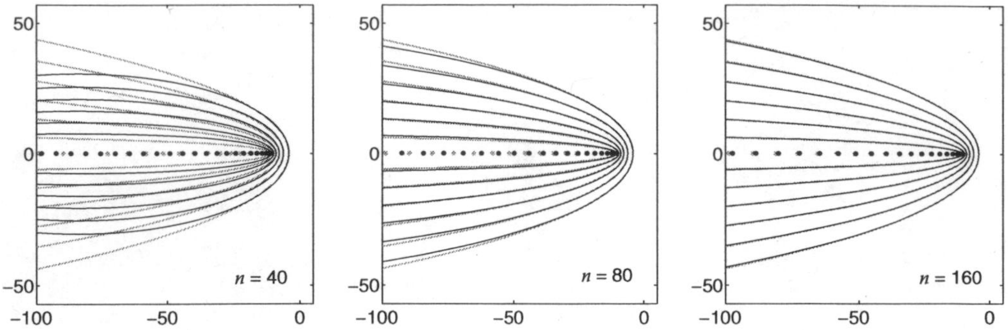

FIGURE 16.6 Spectra (solid dots) and ε-pseudospectra of Toeplitz matrices for the discretization of a convection–diffusion operator described in Application 1 with $\nu = 1/50$ and three values of n; $\varepsilon = 10^{-1}, 10^{-2}, \ldots, 10^{-6}$. The gray dots and lines in each plot show eigenvalues and pseudospectra of the differential operator to which the matrix spectra and pseudospectra converge.

Applications:

1. [RT94] Discretization of the one-dimensional convection-diffusion equation

$$\nu u''(x) + u'(x) = f(x), \quad u(0) = u(1) = 0$$

for $x \in [0, 1]$ with second-order centered finite differences on a uniform grid with spacing $h = 1/(n+1)$ between grid points results in an $n \times n$ Toeplitz matrix with symbol

$$a(t) = \left(\frac{\nu}{h^2} + \frac{1}{2h}\right)t - \left(\frac{2\nu}{h^2}\right) + \left(\frac{\nu}{h^2} - \frac{1}{2h}\right)t^{-1}.$$

On the right-most part of the spectrum, both the eigenvalues and pseudospectra of the discretization matrix converge to those of the underlying differential operator

$$\mathcal{L}u = \nu u'' + u'$$

whose domain is the space of functions that are square-integrable over $[0, 1]$ and satisfy the boundary conditions $u(0) = u(1) = 0$; see Figure 16.6.

16.3 Behavior of Functions of Matrices

In practice, pseudospectra are most often used to investigate the behavior of a function of a matrix. Does the solution $\mathbf{x}(t) = e^{tA}\mathbf{x}(0)$ or $\mathbf{x}_k = A^k\mathbf{x}_0$ of the linear dynamical system $\mathbf{x}'(t) = A\mathbf{x}(t)$ or $\mathbf{x}_{k+1} = A\mathbf{x}_k$ grow or decay as $t, k \to \infty$? Eigenvalues provide an answer: If $\sigma(A)$ lies in the open unit disk or left half-plane, the solution must eventually decay. However, the results described in this section show that if ε-pseudoeigenvalues of A extend well beyond the unit disk or left half-plane for small values of ε, then the system must exhibit transient growth for some initial states. While such growth is notable even for purely linear problems, it should spark special caution when observed for a dynamical system that arises from the linearization of a nonlinear system about a steady state based on the assumption that disturbances from that state are small in magnitude. This reasoning has been applied extensively in recent years in fluid dynamics; see, e.g., [TTRD93].

Definitions:

The **ε-pseudospectral abscissa** of A measures the rightmost extent of $\sigma_\varepsilon(A)$: $\alpha_\varepsilon(A) = \sup_{z\in\sigma_\varepsilon(A)} \operatorname{Re} z$.

The **ε-pseudospectral radius** of A measures the maximum magnitude in $\sigma_\varepsilon(A)$: $\rho_\varepsilon(A) = \sup_{z\in\sigma_\varepsilon(A)} |z|$.

Facts: [TE05, §§14–19]

1. For $\varepsilon > 0$, $\sup_{t\in\mathbb{R},t\geq 0} \|e^{tA}\| \geq \alpha_\varepsilon(A)/\varepsilon$.
2. For $\varepsilon > 0$, $\sup_{k\in\mathbb{N},k\geq 0} \|A^k\| \geq (\rho_\varepsilon(A) - 1)/\varepsilon$.
3. For any function f that is analytic on the spectrum of A,

$$\|f(A)\| \geq \max_{\lambda\in\sigma(A)} |f(\lambda)|.$$

   Equality holds when $\|\cdot\|$ is the natural norm in an inner product space in which A is normal.
4. In the special case of matrix exponentials e^{tA} and matrix powers A^k, this last fact implies that

$$\|e^{tA}\| \geq e^{t\alpha(A)}, \quad \|A^k\| \geq \rho(A)^k$$

   for all $t \geq 0$ and integers $k \geq 0$, where $\alpha(A) = \max_{\lambda\in\sigma(A)} \operatorname{Re}\lambda$ is the spectral abscissa and $\rho(A) = \max_{\lambda\in\sigma(A)} |\lambda|$ is the spectral radius.
5. Let Γ_ε be a finite union of Jordan curves containing $\sigma_\varepsilon(A)$ in their collective interior for some $\varepsilon > 0$, and suppose f is a function analytic on Γ_ε and its interior. Then

$$\|f(A)\| \leq \frac{L_\varepsilon}{2\pi\varepsilon} \max_{z\in\Gamma_\varepsilon} |f(z)|,$$

   where L_ε denotes the arc-length of Γ_ε.
6. In the special case of matrix exponentials e^{tA} and matrix powers A^k, this last fact implies that for all $t > 0$ and integers $k \geq 0$,

$$\|e^{tA}\| \leq \frac{L_\varepsilon}{2\pi\varepsilon} e^{t\alpha_\varepsilon(A)}, \quad \|A^k\| \leq \rho_\varepsilon(A)^{k+1}/\varepsilon.$$

   In typical cases, larger values of ε give superior bounds for small t and k, while smaller values of ε yield more descriptive bounds for larger t and k; see Figure 16.7.
7. Suppose $z \in \mathbb{C} \setminus \sigma(A)$ with $a \equiv \operatorname{Re} z$ and $\varepsilon \equiv 1/\|(zI - A)^{-1}\|$. Provided $a > \varepsilon$, then for any fixed $\tau > 0$,

$$\sup_{0<t\leq\tau} \|e^{tA}\| \geq \frac{ae^{\tau a}}{a + \varepsilon(e^{\tau a} - 1)}.$$

8. Suppose $z \in \mathbb{C} \setminus \sigma(A)$ with $a \equiv \operatorname{Re} z$ and $\varepsilon \equiv 1/\|(zI - A)^{-1}\|$, and that $\|e^{\tau A}\| \leq M$ for all $\tau > 0$ with $M \geq a/\varepsilon$. Then for any $t > 0$,

$$\|e^{tA}\| \geq e^{ta}(1 - \varepsilon M/a) + \varepsilon M/a.$$

9. Suppose $z \in \mathbb{C} \setminus \sigma(A)$ with $r \equiv |z|$ and $\varepsilon \equiv 1/\|(zI - A)^{-1}\|$. Provided $r > 1 + \varepsilon$, then for any fixed integer $\kappa \geq 1$,

$$\sup_{0<k\leq\kappa} \|A^k\| \geq \frac{r^\kappa(r - 1 - \varepsilon) + \varepsilon r^{\kappa-1}}{(r - 1 - \varepsilon) + \varepsilon r^{\kappa-1}}.$$

10. Suppose $z \in \mathbb{C} \setminus \sigma(A)$ with $r \equiv |z|$ and $\varepsilon \equiv 1/\|(zI - A)^{-1}\|$, and that $\|A^\kappa\| \leq M$ for all integers $\kappa \geq 0$ with $M \geq (r - 1)/\varepsilon$. Then for any integer $k \geq 0$,

$$\|A^k\| \geq r^k(r - 1 - \varepsilon M) + \varepsilon M.$$

11. For any $A \in \mathbb{C}^{n\times n}$,

$$\|A^k\| < e\,(k + 1) \sup_{\varepsilon>0} \frac{\rho_\varepsilon(A) - 1}{\varepsilon}.$$

12. (Kreiss Matrix Theorem) For any $A \in \mathbb{C}^{n\times n}$,

$$\sup_{\varepsilon>0} \frac{\rho_\varepsilon(A) - 1}{\varepsilon} \le \sup_{k\ge 0} \|A^k\| \le e n \sup_{\varepsilon>0} \frac{\rho_\varepsilon(A) - 1}{\varepsilon}.$$

13. [GT93] There exist matrices A and B such that, in the induced 2-norm, $\sigma_\varepsilon(A) = \sigma_\varepsilon(B)$ for all $\varepsilon > 0$, yet $\|f(A)\|_2 \ne \|f(B)\|_2$ for some polynomial f; see Example 2. That is, even if the 2-norm of the resolvents of A and B are identical for all $z \in \mathbb{C}$, the norms of other matrix functions in A and B need not agree. (Curiously, if the *Frobenius* norm of the resolvents of A and B agree for all $z \in \mathbb{C}$, then $\|f(A)\|_F = \|f(B)\|_F$ for all polynomials f.)

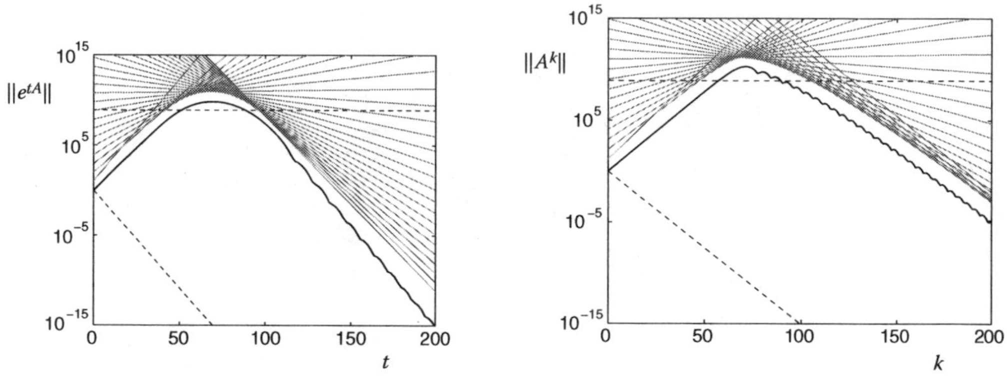

FIGURE 16.7 The functions $\|e^{tA}\|$ and $\|A^k\|$ exhibit transient growth before exponential decay for the Toeplitz matrix of dimension $n = 50$, whose pseudospectra were illustrated in Figure 16.4. The horizontal dashed lines show the lower bounds on maximal growth given in Facts 1 and 2, while the lower dashed lines show the lower bounds of Fact 4. The gray lines show the upper bounds in Fact 6 for $\varepsilon = 10^{-1}, 10^{-2}, \ldots, 10^{-28}$ (ordered by decreasing slope).

Examples:

1. Consider the tridiagonal Toeplitz matrix of dimension $n = 50$ from Example 1 of the last section, whose pseudospectra were illustrated in Figure 16.4. Since all the eigenvalues of this matrix are contained in both the left half-plane and the unit disk, $e^{tA} \to 0$ as $t \to \infty$ and $A^k \to 0$ as $k \to \infty$. However, $\sigma_\varepsilon(A)$ extends far into the right half-plane and beyond the unit disk even for ε as small as 10^{-7}. Consequently, the lower bounds in Facts 1 and 2 guarantee that $\|e^{tA}\|$ and $\|A^k\|$ exhibit transient growth before their eventual decay; results such as Fact 6 limit the extent of the transient growth. These bounds are illustrated in Figure 16.7. (For a similar example involving a different matrix, see the "Transient" demonstration in [Wri02b].)
2. [GT93] The matrices

$$A = \begin{bmatrix} 0 & 1 & 0 & 0 & 0 \\ 0 & 0 & 1 & 0 & 0 \\ 0 & 0 & 0 & 0 & 0 \\ 0 & 0 & 0 & 0 & \sqrt{2} \\ 0 & 0 & 0 & 0 & 0 \end{bmatrix}, \qquad B = \begin{bmatrix} 0 & 1 & 0 & 0 & 0 \\ 0 & 0 & 1 & 0 & 0 \\ 0 & 0 & 0 & 0 & 0 \\ 0 & 0 & 0 & 0 & 0 \\ 0 & 0 & 0 & 0 & 0 \end{bmatrix},$$

have the same 2-norm ε-pseudospectra for all $\varepsilon > 0$. However, $\|A\|_2 = \sqrt{2} > 1 = \|B\|_2$.

Applications:

1. Fact 5 leads to a convergence result for the GMRES algorithm (Chapter 41), which constructs estimates $\mathbf{x}_k$ to the solution $\mathbf{x}$ of the linear system $A\mathbf{x} = \mathbf{b}$. The kth residual $\mathbf{r}_k = \mathbf{b} - A\mathbf{x}_k$ is bounded by

$$\frac{\|\mathbf{r}_k\|_2}{\|\mathbf{r}_0\|_2} \leq \frac{L_\varepsilon}{2\pi\varepsilon} \min_{\substack{p\in\mathbb{C}[z;k] \\ p(0)=1}} \max_{z\in\Gamma_\varepsilon} |p(z)|,$$

where $\mathbb{C}[z;k]$ denotes the set of polynomials of degree k or less, Γ_ε is a finite union of Jordan curves containing $\sigma_\varepsilon(A)$ in their collective interior for some $\varepsilon > 0$, and L_ε is the arc-length of Γ_ε.
2. For further examples of the use of pseudospectra to analyze matrix iterations and the stability of discretizations of differential equations, see [TE05, §§24–34].

16.4 Computation of Pseudospectra

This section describes techniques for computing and approximating pseudospectra, focusing primarily on the induced matrix 2-norm, the case most studied in the literature and for which very satisfactory algorithms exist. For further details, see [Tre99], [TE05, §§39–44], or [Wri02a].

Facts: [TE05]

1. There are two general approaches to computing pseudospectra, both based on the expression for $\sigma_\varepsilon(A)$ in Fact 1(a) of Section 16.1. The most widely-used method computes the resolvent norm, $\|(zI - A)^{-1}\|_2$, on a grid of points in the complex plane and submits the results to a contour-plotting program; the second approach uses a curve-tracing algorithm to track the ε^{-1}-level curve of the resolvent norm ([Brü96]). Both approaches exploit the fact that the 2-norm of the resolvent is the reciprocal of the minimum singular value of $zI - A$. A third approach, based on the characterization of $\sigma_\varepsilon(A)$ as the set of all ε-pseudoeigenvalues, *approximates* $\sigma_\varepsilon(A)$ by the union of the eigenvalues of $A + E$ for randomly generated $E \in \mathbb{C}^{n\times n}$ with $\|E\| < \varepsilon$.
2. For dense matrices A, the computation of the minimum singular value of $zI - A$ requires $\mathcal{O}(n^3)$ floating point operations for each distinct value of z. Hence, the contour-plotting approach to computing pseudospectra based on a grid of $m \times m$ points in the complex plane, implemented via the most naive method, requires $\mathcal{O}(m^2n^3)$ operations.
3. [Lui97] Improved efficiency is obtained through the use of iterative methods for computing the minimum singular value of the resolvent. The most effective methods (inverse iteration or the inverse Lanczos method) require matrix-vector products of the form $(zI - A)^{-1}\mathbf{x}$ at each iteration. For dense A, this approach requires $\mathcal{O}(n^3)$ operations per grid point. One can decrease this labor to $\mathcal{O}(n^2)$ by first reducing A to Schur form, $A = UTU^*$, and then noting that $\|(zI - A)^{-1}\|_2 = \|(zI - T)^{-1}\|_2$. Vectors of the form $(zI - T)^{-1}\mathbf{x}$ can be computed in $\mathcal{O}(n^2)$ operations since T is triangular. As the inverse iteration and inverse Lanczos methods typically converge to the minimum singular value in a small number of iterations at each grid point, the total complexity of the contour-plotting approach is $\mathcal{O}(n^3 + m^2n^2)$.
4. For large-scale problems (say, $n > 1000$), the cost of preliminary triangularization can be prohibitive. Several alternatives are available: Use sparse direct or iterative methods to compute $(zI - A)^{-1}\mathbf{x}$ at each grid point, or reduce the dimension of the problem by replacing A with a smaller matrix, such as the $(m+1) \times m$ upper Hessenberg matrix in an Arnoldi decomposition, or U^*AU, where the columns of $U \in \mathbb{C}^{n\times m}$ form an orthonormal basis for an invariant subspace corresponding to physically relevant eigenvalues, with $m \ll n$. As per results stated in Fact 9 of Section 16.1, the pseudospectra of these smaller matrices provide a lower bounds on the pseudospectra of A.

5. [Wri02b] EigTool is a freely available MATLAB package based on a highly-efficient, robust implementation of the grid-based method with preliminary triangularization and inverse Lanczos iteration. For large-scale problems, EigTool uses ARPACK (Chapter 76), to compute a subspace that includes an invariant subspace associated with eigenvalues in a given region of the complex plane. The EigTool software, which was used to compute the pseudospectra shown throughout this section, can be downloaded from `http://www.comlab.ox.ac.uk/pseudospectra/eigtool`.
6. Curve-tracing algorithms can also benefit from iterative computation of the resolvent norm, though the standard implementation requires both left and right singular vectors associated with the minimal singular value ([Brü96]). Robust implementations require measures to ensure that all components of $\sigma_\varepsilon(A)$ have been located and to handle cusps in the boundary; see, e.g., [BG01].
7. Software for computing 2-norm pseudospectra can be used to compute pseudospectra in any norm induced by an inner product. Suppose the inner product of $\mathbf{x}$ and $\mathbf{y}$ is given by $(\mathbf{x}, \mathbf{y})_W = (W\mathbf{x}, \mathbf{y})$, where $(\cdot, \cdot)$ denotes the Euclidean inner product and $W = LL^*$, where L^* denotes the conjugate transpose of L. Then the W-norm pseudospectra of A are equal to the 2-norm pseudospectra of L^*AL^{-*}.
8. For norms not associated with inner products, all known grid-based algorithms require $\mathcal{O}(n^3)$ operations per grid point, typically involving the construction of the resolvent $(zI - A)^{-1}$. Higham and Tisseur ([HT00]) have proposed an efficient approach for approximating 1-norm pseudospectra using a norm estimator.
9. [BLO03],[MO05] There exist efficient algorithms, based on Fact 7 of section 16.1, for computing the 2-norm pseudospectral radius and abscissa without first determining the entire pseudospectrum.

16.5 Extensions

The previous sections address the standard formulation of the ε-pseudospectrum, the union of all eigenvalues of $A + E$ for a square matrix A and general complex perturbations E, with $\|E\| < \varepsilon$. Natural modifications restrict the structure of E or adapt the definition to more general eigenvalue problems. The former topic has attracted considerable attention in the control theory literature and is presented in detail in [HP05].

Definitions:

The **spectral value set**, or **structured ε-pseudospectrum**, of the matrix triplet (A, B, C), $A \in \mathbb{C}^{n\times n}$, $B \in \mathbb{C}^{n\times m}$, $C \in \mathbb{C}^{p\times n}$, for $\varepsilon > 0$ is the set

$$\sigma_\varepsilon(A; B, C) = \{z \in \mathbb{C} : z \in \sigma(A + BEC) \text{ for some } E \in C^{m\times p} \text{ with } \|E\| < \varepsilon\}.$$

The **real structured ε-pseudospectrum** of $A \in \mathbb{R}^{n\times n}$ is the set

$$\sigma_\varepsilon^{\mathrm{R}}(A) = \{z \in \sigma(A + E) : E \in \mathbb{R}^{n\times n}, \|E\| < \varepsilon\}.$$

The **ε-pseudospectrum of a rectangular matrix** $A \in \mathbb{C}^{n\times m}$ $(n \geq m)$ for $\varepsilon > 0$ is the set

$$\sigma_\varepsilon(A) = \{z \in \mathbb{C} : (A + E)\mathbf{x} = z\mathbf{x} \text{ for some } \mathbf{x} \neq 0 \text{ and } \|E\| < \varepsilon\}.$$

[Ruh95] For $A \in \mathbb{C}^{n\times n}$ and invertible $B \in \mathbb{C}^{n\times n}$, the **$\varepsilon$-pseudospectrum of the matrix pencil** $A - \lambda B$ (or generalized eigenvalue problem $A\mathbf{x} = \lambda B\mathbf{x}$) for $\varepsilon > 0$ is the set

$$\sigma_\varepsilon(A, B) = \sigma_\varepsilon(B^{-1}A).$$

[TH01] The **ε-pseudospectrum of the matrix polynomial** $P(\lambda)$ (or polynomial eigenvalue problem $P(\lambda)\mathbf{x} = \mathbf{0}$), where $P(\lambda) = \lambda^p A_p + \lambda^{p-1} A_{p-1} + \cdots + A_0$ and $\varepsilon > 0$, is the set

$$\sigma_\varepsilon(P) = \{z \in \mathbb{C} : z \in \sigma(P + E) \text{ for some } E(\lambda) = \lambda^p E_p + \cdots + E_0,\ \|E_j\| \le \varepsilon\alpha_j, \quad j = 0, \ldots, p\},$$

for values $\alpha_0, \ldots, \alpha_p$. For most applications, one would either take $\alpha_j = 1$ for all j, or $\alpha_j = \|A_j\|$. (This definition differs considerably from the one given for the pseudospectrum of a matrix pencil. In particular, when $p = 1$ the present definition *does not* reduce to the above definition for the pencil; see Fact 6 below.)

Facts:

1. [HP92, HK93] The above definition of the spectral value set $\sigma_\varepsilon(A; B, C)$ is equivalent to

$$\sigma_\varepsilon(A; B, C) = \{z \in \mathbb{C} : \|C(zI - A)^{-1}B\| > \varepsilon^{-1}\}.$$

2. [Kar03] The above definition of the real structured ε-pseudospectrum $\sigma_\varepsilon^{\mathrm{R}}(A)$ is equivalent to

$$\sigma_\varepsilon^{\mathrm{R}}(A) = \{z \in \mathbb{C} : r(A, z) < \varepsilon\},$$

where

$$r(A, z)^{-1} = \inf_{\gamma \in (0,1)} \sigma_2\left(\begin{bmatrix} \mathrm{Re}\,(zI - A)^{-1} & -\gamma\,\mathrm{Im}\,(zI - A)^{-1} \\ \gamma^{-1}\mathrm{Im}\,(zI - A)^{-1} & \mathrm{Re}\,(zI - A)^{-1} \end{bmatrix}\right)$$

and $\sigma_2(\cdot)$ denotes the second largest singular value. From this formulation, one can derive algorithms for computing $\sigma_\varepsilon^{\mathrm{R}}(A)$ akin to those used for computing $\sigma_\varepsilon(A)$.

3. The definition of $\sigma_\varepsilon^{\mathrm{R}}(A)$ suggests similar formulations that impose different restrictions upon E, such as a sparsity pattern, Toeplitz structure, nonnegativity or stochasticity of $A + E$, etc. Such structured pseudospectra are often difficult to compute or approximate.
4. [WT02] The above definition of the ε-pseudospectrum $\sigma_\varepsilon(A)$ of a rectangular matrix $A \in \mathbb{C}^{n\times m}$, $n \ge m$, is equivalent to

$$\sigma_\varepsilon(A) = \{z \in \mathbb{C} : \|(z\widetilde{I} - A)^\dagger\| > \varepsilon^{-1}\},$$

where $(\cdot)^\dagger$ denotes the Moore–Penrose pseudoinverse and $\widetilde{I}$ denotes the $n \times m$ matrix that has the $m \times m$ identity in the first m rows and is zero elsewhere.

5. The following facts apply to the ε-pseudospectrum of a rectangular matrix $A \in \mathbb{C}^{m\times n}$, $m \ge n$.

   (a) [WT02] It is possible that $\sigma_\varepsilon(A) = \emptyset$.

   (b) [BLO04] For $A \in \mathbb{C}^{m\times n}$, $m \ge n$, and any $\varepsilon > 0$, the set $\sigma_\varepsilon(A)$ contains no more than $2n^2 - n + 1$ connected components.

6. [TE05] Alternative definitions have been proposed for the pseudospectrum of the matrix pencil $A - \lambda B$. The definition presented above has the advantage that the pseudospectrum is invariant to premultiplication of the pencil by a nonsingular matrix, which is consistent with the fact that premultiplication of the differential equation $B\mathbf{x}' = A\mathbf{x}$ does not affect the solution $\mathbf{x}$. Here are two alternative definitions, *neither of which are equivalent to the previous definition.*

   (a) [Rie94] If B is Hermitian positive definite with Cholesky factorization $B = LL^*$, then the pseudospectrum of the pencil can be defined in terms of the standard pseudospectrum of a transformed problem:

$$\sigma_\varepsilon(A, B) = \sigma_\varepsilon(L^{-1}AL^{-*}).$$

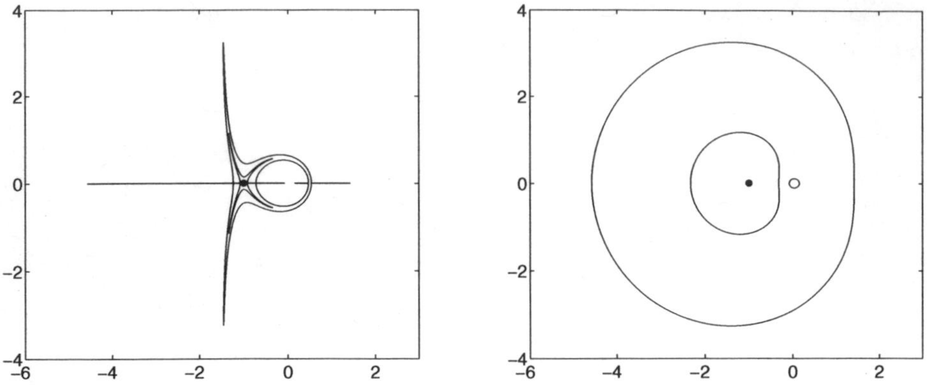

FIGURE 16.8 Spectrum (solid dot) and real structured ε-pseudospectra $\sigma_\varepsilon^{\mathrm{R}}(A)$ (left) and unstructured ε-pseudospectra $\sigma_\varepsilon(A)$ of the second matrix of Example 3 in section 16.1 for $\varepsilon = 10^{-3}, 10^{-4}$.

(b) [FGNT96, TH01] The following definition is more appropriate for the study of eigenvalue perturbations:

$$\sigma_\varepsilon(A, B) = \{z \in \mathbb{C} : (A + E_0)\mathbf{x} = z(B + E_1)\mathbf{x} \text{ for some } \mathbf{x} \neq \mathbf{0} \text{ and } E_0, E_1 \text{ with } \|E_0\| < \varepsilon\alpha_0, \|E_1\| < \varepsilon\alpha_1\},$$

where generally either $\alpha_j = 1$ for $j = 0, 1$, or $\alpha_0 = \|A\|$ and $\alpha_1 = \|B\|$. This is a special case of the definition given above for the pseudospectrum of a matrix polynomial.

7. [TH01] The above definition of the ε-pseudospectrum of a matrix polynomial, $\sigma_\varepsilon(P)$, is equivalent to

$$\sigma_\varepsilon(P) = \{z \in \mathbb{C} : \|P(z)^{-1}\| > 1/(\varepsilon\phi(|z|))\},$$

where $\phi(z) = \sum_{j=0}^{p} \alpha_k z^k$ for the same values of $\alpha_0, \ldots, \alpha_p$ used in the earlier definition.

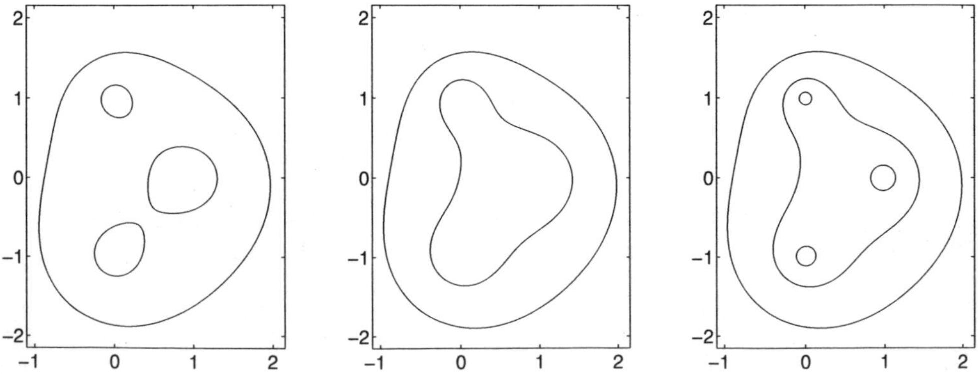

FIGURE 16.9 ε-pseudospectra of the rectangular matrix in Example 2 with $\delta = 0.02$ (left), $\delta = 0.01$ (middle), $\delta = 0.005$ (right), and $\varepsilon = 10^{-1}, 10^{-1.5}$, and 10^{-2}. Note that in the first two plots, $\sigma_\varepsilon(A) = \emptyset$ for $\varepsilon = 10^{-2}$.

Examples:

1. Figure 16.8 compares real structured ε-pseudospectra $\sigma_\varepsilon^{\mathrm{R}}(A)$ to the (unstructured) pseudospectra $\sigma_\varepsilon(A)$ for the second matrix in Example 3 of Section 16.1; cf. [TE05, Fig. 50.3].
2. Figure 16.9 shows pseudospectra of the rectangular matrix

$$A = \begin{bmatrix} 2 & -5 & 10 \\ 1 & -2 & 5i \\ 0 & 0 & 1 \\ \delta & \delta & \delta \end{bmatrix},$$

which is the third matrix in Example 1 of Section 16.1, but with an extra row appended.

References

[BF60] F.L. Bauer and C.T. Fike. Norms and exclusion theorems. *Numer. Math.*, 2:137–141, 1960.

[BG01] C. Bekas and E. Gallopoulos. Cobra: Parallel path following for computing the matrix pseudospectrum. *Parallel Comp.*, 27:1879–1896, 2001.

[BG05] A. Böttcher and S.M. Grudsky. *Spectral Properties of Banded Toeplitz Matrices.* SIAM, Philadelphia, 2005.

[BLO03] J.V. Burke, A.S. Lewis, and M.L. Overton. Robust stability and a criss-cross algorithm for pseudospectra. *IMA J. Numer. Anal.*, 23:359–375, 2003.

[BLO04] J.V. Burke, A.S. Lewis, and M.L. Overton. Pseudospectral components and the distance to uncontrollability. *SIAM J. Matrix Anal. Appl.*, 26:350–361, 2004.

[Bot94] Albrecht Böttcher. Pseudospectra and singular values of large convolution operators. *J. Int. Eqs. Appl.*, 6:267–301, 1994.

[Bru96] Martin Brühl. A curve tracing algorithm for computing the pseudospectrum. *BIT*, 36:441–454, 1996.

[BS99] Albrecht Böttcher and Bernd Silbermann. *Introduction to Large Truncated Toeplitz Matrices.* Springer-Verlag, New York, 1999.

[CCF96] Françoise Chaitin-Chatelin and Valérie Frayssé. *Lectures on Finite Precision Computations.* SIAM, Philadelphia, 1996.

[Dem87] James W. Demmel. A counterexample for two conjectures about stability. *IEEE Trans. Auto. Control*, AC-32:340–343, 1987.

[Dem97] James W. Demmel. *Applied Numerical Linear Algebra.* SIAM, Philadelphia, 1997.

[ET01] Mark Embree and Lloyd N. Trefethen. Generalizing eigenvalue theorems to pseudospectra theorems. *SIAM J. Sci. Comp.*, 23:583–590, 2001.

[FGNT96] Valérie Frayssé, Michel Gueury, Frank Nicoud, and Vincent Toumazou. Spectral portraits for matrix pencils. Technical Report TR/PA/96/19, CERFACS, Toulouse, August 1996.

[GL02] Laurence Grammont and Alain Largillier. On ε-spectra and stability radii. *J. Comp. Appl. Math.*, 147:453–469, 2002.

[Gre02] Anne Greenbaum. Generalizations of the field of values useful in the study of polynomial functions of a matrix. *Lin. Alg. Appl.*, 347:233–249, 2002.

[GT93] Anne Greenbaum and Lloyd N. Trefethen. Do the pseudospectra of a matrix determine its behavior? Technical Report TR 93-1371, Computer Science Department, Cornell University, Ithaca, NY, August 1993.

[Hau91] Felix Hausdorff. *Set Theory 4th ed.* Chelsea, New York, 1991.

[HK93] D. Hinrichsen and B. Kelb. Spectral value sets: A graphical tool for robustness analysis. *Sys. Control Lett.*, 21:127–136, 1993.

[HP92] D. Hinrichsen and A.J. Pritchard. On spectral variations under bounded real matrix perturbations. *Numer. Math.*, 60:509–524, 1992.

[HP05] Diederich Hinrichsen and Anthony J. Pritchard. *Mathematical Systems Theory I.* Springer-Verlag, Berlin, 2005.

[HT00] Nicholas J. Higham and Françoise Tisseur. A block algorithm for matrix 1-norm estimation, with an application to 1-norm pseudospectra. *SIAM J. Matrix Anal. Appl.*, 21:1185–1201, 2000.

[Kar03] Michael Karow. *Geometry of Spectral Value Sets.* Ph.D. thesis, Universität Bremen, Germany, 2003.

[Lui97] S.H. Lui. Computation of pseudospectra by continuation. *SIAM J. Sci. Comp.*, 18:565–573, 1997.

[Lui03] S.-H. Lui. A pseudospectral mapping theorem. *Math. Comp.*, 72:1841–1854, 2003.

[MO05] Emre Mengi and Michael L. Overton. Algorithms for the computation of the pseudospectral radius and the numerical radius of a matrix. *IMA J. Numer. Anal.*, 25:648–669, 2005.

[Nev93] Olavi Nevanlinna. *Convergence of Iterations for Linear Equations.* Birkhäuser, Basel, Germany, 1993.

[Rie94] Kurt S. Riedel. Generalized epsilon-pseudospectra. *SIAM J. Num. Anal.*, 31:1219–1225, 1994.

[RT92] Lothar Reichel and Lloyd N. Trefethen. Eigenvalues and pseudo-eigenvalues of Toeplitz matrices. *Lin. Alg. Appl.*, 162–164:153–185, 1992.

[RT94] Satish C. Reddy and Lloyd N. Trefethen. Pseudospectra of the convection-diffusion operator. *SIAM J. Appl. Math.*, 54:1634–1649, 1994.

[Ruh95] Axel Ruhe. The rational Krylov algorithm for large nonsymmetric eigenvalues — mapping the resolvent norms (pseudospectrum). Unpublished manuscript, March 1995.

[TC04] Lloyd N. Trefethen and S.J. Chapman. Wave packet pseudomodes of twisted Toeplitz matrices. *Comm. Pure Appl. Math.*, 57:1233–1264, 2004.

[TE05] Lloyd N. Trefethen and Mark Embree. *Spectra and Pseudospectra: The Behavior of Nonnormal Matrices and Operators.* Princeton University Press, Princeton, NJ, 2005.

[TH01] Françoise Tisseur and Nicholas J. Higham. Structured pseudospectra for polynomial eigenvalue problems, with applications. *SIAM J. Matrix Anal. Appl.*, 23:187–208, 2001.

[Tre99] Lloyd N. Trefethen. Computation of pseudospectra. *Acta Numerica*, 8:247–295, 1999.

[TTRD93] Lloyd N. Trefethen, Anne E. Trefethen, Satish C. Reddy, and Tobin A. Driscoll. Hydrodynamic stability without eigenvalues. *Science*, 261:578–584, 1993.

[Wri02a] Thomas G. Wright. *Algorithms and Software for Pseudospectra.* D.Phil. thesis, Oxford University, U.K., 2002.

[Wri02b] Thomas G. Wright. EigTool, 2002. Software available at: `http://www.comlab.ox.ac.uk/pseudospectra/eigtool`.

[WT02] Thomas G. Wright and Lloyd N. Trefethen. Pseudospectra of rectangular matrices. *IMA J. Num. Anal.*, 22:501–519, 2002.

17

Singular Values and Singular Value Inequalities

Roy Mathias
University of Birmingham

17.1 Definitions and Characterizations

Singular values and the singular value decomposition are defined in Chapter 5.6. Additional information on computation of the singular value decomposition can be found in Chapter 45. A brief history of the singular value decomposition and early references can be found in [HJ91, Chap. 3].

Throughout this chapter, $q = \min\{m, n\}$, and if $A \in \mathbb{C}^{n\times n}$ has real eigenvalues, then they are ordered $\lambda_1(A) \geq \cdots \geq \lambda_n(A)$.

Definitions:

For $A \in \mathbb{C}^{m\times n}$, define the **singular value vector** $\mathrm{sv}(A) = (\sigma_1(A), \ldots, \sigma_q(A))$.

For $A \in \mathbb{C}^{m\times n}$, define $r_1(A) \geq \cdots \geq r_m(A)$ and $c_1(A) \geq \cdots \geq c_n(A)$ to be the ordered Euclidean row and column lengths of A, that is, the square roots of the ordered diagonal entries of AA^* and A^*A.

For $A \in \mathbb{C}^{m\times n}$ define $|A|_{pd} = (A^*A)^{1/2}$. This is called the **spectral absolute value** of A. (This is also called the **absolute value**, but the latter term will not be used in this chapter due to potential confusion with the entry-wise absolute value of A, denoted $|A|$.)

A **polar decomposition** or **polar form** of the matrix $A \in \mathbb{C}^{m\times n}$ with $m \geq n$ is a factorization $A = UP$, where $P \in \mathbb{C}^{n\times n}$ is positive semidefinite and $U \in \mathbb{C}^{m\times n}$ satisfies $U^*U = I_n$.

Facts:

The following facts can be found in most books on matrix theory, for example [HJ91, Chap. 3] or [Bha97].

1. Take $A \in \mathbb{C}^{m\times n}$, and set
$$B = \begin{bmatrix} A & 0 \\ 0 & 0 \end{bmatrix}.$$
Then $\sigma_i(A) = \sigma_i(B)$ for $i = 1, \ldots, q$ and $\sigma_i(B) = 0$ for $i > q$. We may choose the zero blocks in B to ensure that B is square. In this way we can often generalize results on the singular values of square matrices to rectangular matrices. For simplicity of exposition, in this chapter we will sometimes state a result for square matrices rather than the more general result for rectangular matrices.
2. (Unitary invariance) Take $A \in \mathbb{C}^{m\times n}$. Then for any unitary $U \in \mathbb{C}^{m\times m}$ and $V \in \mathbb{C}^{n\times n}$,
$$\sigma_i(A) = \sigma_i(UAV), \quad i = 1, 2, \ldots, q.$$
3. Take $A, B \in \mathbb{C}^{m\times n}$. There are unitary matrices $U \in \mathbb{C}^{m\times m}$ and $V \in \mathbb{C}^{n\times n}$ such that $A = UBV$ if and only if $\sigma_i(A) = \sigma_i(B)$, $i = 1, 2, \ldots, q$.
4. Let $A \in \mathbb{C}^{m\times n}$. Then $\sigma_i^2(A) = \lambda_i(AA^*) = \lambda_i(A^*A)$ for $i = 1, 2, \ldots, q$.
5. Let $A \in \mathbb{C}^{m\times n}$. Let $\mathcal{S}_i$ denote the set of subspaces of $\mathbb{C}^n$ of dimension i. Then for $i = 1, 2, \ldots, q$,
$$\sigma_i(A) = \min_{\mathcal{X}\in\mathcal{S}_{n-i+1}} \max_{\mathbf{x}\in\mathcal{X}, \|\mathbf{x}\|_2=1} \|A\mathbf{x}\|_2 = \min_{\mathcal{Y}\in\mathcal{S}_{i-1}} \max_{\mathbf{x}\perp\mathcal{Y}, \|\mathbf{x}\|_2=1} \|A\mathbf{x}\|_2,$$
$$\sigma_i(A) = \max_{\mathcal{X}\in\mathcal{S}_{i}} \min_{\mathbf{x}\in\mathcal{X}, \|\mathbf{x}\|_2=1} \|A\mathbf{x}\|_2 = \max_{\mathcal{Y}\in\mathcal{S}_{n-i}} \min_{\mathbf{x}\perp\mathcal{Y}, \|\mathbf{x}\|_2=1} \|A\mathbf{x}\|_2.$$
6. Let $A \in \mathbb{C}^{m\times n}$ and define the Hermitian matrix
$$J = \begin{bmatrix} 0 & A \\ A^* & 0 \end{bmatrix} \in \mathbb{C}^{m+n, m+n}.$$
The eigenvalues of J are $\pm\sigma_1(A), \ldots, \pm\sigma_q(A)$ together with $|m - n|$ zeros. The matrix J is called the Jordan–Wielandt matrix. Its use allows one to deduce singular value results from results for eigenvalues of Hermitian matrices.
7. Take $m \geq n$ and $A \in \mathbb{C}^{m\times n}$. Let $A = UP$ be a polar decomposition of A. Then $\sigma_i(A) = \lambda_i(P)$, $i = 1, 2, \ldots, q$.
8. Let $A \in \mathbb{C}^{m\times n}$ and $1 \leq k \leq q$. Then
$$\sum_{i=1}^{k} \sigma_i(A) = \max\{\text{Re tr } U^*AV : U \in \mathbb{C}^{m\times k}, V \in \mathbb{C}^{n\times k}, U^*U = V^*V = I_k\},$$
$$\prod_{i=1}^{k} \sigma_i(A) = \max\{|\det U^*AV| : U \in \mathbb{C}^{m\times k}, V \in \mathbb{C}^{n\times k}, U^*U = V^*V = I_k\}.$$
If $m = n$, then
$$\sum_{i=1}^{n} \sigma_i(A) = \max\left\{\sum_{i=1}^{n} |(U^*AU)_{ii}| : U \in \mathbb{C}^{n\times n}, U^*U = I_n\right\}.$$
We cannot replace the n by a general $k \in \{1, \ldots, n\}$.

9. Let $A \in \mathbb{C}^{m\times n}$. A yields

   (a) $\sigma_i(A^T) = \sigma_i(A^*) = \sigma_i(\bar{A}) = \sigma_i(A)$, for $i = 1, 2, \ldots, q$.

   (b) Let $k = \text{rank}(A)$. Then $\sigma_i(A^\dagger) = \sigma_{k-i+1}^{-1}(A)$ for $i = 1, \ldots, k$, and $\sigma_i(A^\dagger) = 0$ for $i = k+1, \ldots, q$. In particular, if $m = n$ and A is invertible, then

   $$\sigma_i(A^{-1}) = \sigma_{n-i+1}^{-1}(A), \quad i = 1, \ldots, n.$$

   (c) For any $j \in \mathbb{N}$

   $$\sigma_i((A^*A)^j) = \sigma_i^{2j}(A), \quad i = 1, \ldots, q;$$
   $$\sigma_i((A^*A)^j A^*) = \sigma_i(A(A^*A)^j) = \sigma_i^{2j+1}(A) \quad i = 1, \ldots, q.$$

10. Let UP be a polar decomposition of $A \in \mathbb{C}^{m\times n}$ $(m \geq n)$. The positive semidefinite factor P is uniquely determined and is equal to $|A|_{pd}$. The factor U is uniquely determined if A has rank n. If A has singular value decomposition $A = U_1 \Sigma U_2^*$ $(U_1 \in \mathbb{C}^{m\times n}, U_2 \in \mathbb{C}^{n\times n})$, then $P = U_2 \Sigma U_2^*$, and U may be taken to be $U_1 U_2^*$.
11. Take $A, U \in \mathbb{C}^{n\times n}$ with U unitary. Then $A = U|A|_{pd}$ if and only if $A = |A^*|_{pd} U$.

Examples:

1. Take

$$A = \begin{bmatrix} 11 & -3 & -5 & 1 \\ 1 & -5 & -3 & 11 \\ -5 & 1 & 11 & -3 \\ -3 & 11 & 1 & -5 \end{bmatrix}.$$

The singular value decomposition of A is $A = U\Sigma V^*$, where $\Sigma = \text{diag}(20, 12, 8, 4)$, and

$$U = \frac{1}{2}\begin{bmatrix} -1 & 1 & -1 & 1 \\ -1 & -1 & 1 & 1 \\ 1 & -1 & -1 & 1 \\ 1 & 1 & 1 & 1 \end{bmatrix} \quad \text{and} \quad V = \frac{1}{2}\begin{bmatrix} -1 & 1 & -1 & 1 \\ 1 & 1 & 1 & 1 \\ 1 & -1 & -1 & 1 \\ -1 & -1 & 1 & 1 \end{bmatrix}.$$

The singular values of A are 20, 12, 8, 4. Let Q denote the permutation matrix that takes (x_1, x_2, x_3, x_4) to (x_1, x_4, x_3, x_2). Let $P = |A|_{pd} = QA$. The polar decomposition of A is $A = QP$. (To see this, note that a permutation matrix is unitary and that P is positive definite by Geršchgorin's theorem.) Note also that $|A|_{pd} \neq |A^*|_{pd} = AQ$.

17.2 Singular Values of Special Matrices

In this section, we present some matrices where the singular values (or some of the singular values) are known, and facts about the singular values of certain structured matrices.

Facts:

The following results can be obtained by straightforward computations if no specific reference is given.

1. Let $D = \text{diag}(\alpha_1, \ldots, \alpha_n)$, where the α_i are integers, and let H_1 and H_2 be Hadamard matrices. (See Chapter 32.2.) Then the matrix $H_1 D H_2$ has integer entries and has integer singular values $n|\alpha_1|, \ldots, n|\alpha_n|$.

2. (2×2 *matrix*) Take $A \in \mathbb{C}^{2\times 2}$. Set $D = |\det(A)|^2$, $N = \|A\|_F^2$. The singular values of A are

$$\sqrt{\frac{N \pm \sqrt{N^2 - 4D}}{2}}.$$

3. Let $X \in \mathbb{C}^{m\times n}$ have singular values $\sigma_1 \geq \cdots \geq \sigma_q$ ($q = \min\{m, n\}$). Set

$$A = \begin{bmatrix} I & 2X \\ 0 & I \end{bmatrix} \in \mathbb{C}^{m+n,m+n}.$$

The $m + n$ singular values of A are

$$\sigma_1 + \sqrt{\sigma_1^2 + 1}, \dots, \sigma_q + \sqrt{\sigma_q^2 + 1}, 1, \dots, 1, \sqrt{\sigma_q^2 + 1} - \sigma_q, \dots, \sqrt{\sigma_1^2 + 1} - \sigma_1.$$

4. [HJ91, Theorem 4.2.15] Let $A \in \mathbb{C}^{m_1\times n_1}$ and $B \in \mathbb{C}^{m_2\times n_2}$ have rank m and n. The nonzero singular values of $A \otimes B$ are $\sigma_i(A)\sigma_j(B)$, $i = 1, \dots, m$, $j = 1, \dots, n$.
5. Let $A \in \mathbb{C}^{n\times n}$ be normal with eigenvalues $\lambda_1, \dots, \lambda_n$, and let p be a polynomial. Then the singular values of $p(A)$ are $|p(\lambda_k)|$, $k = 1, \dots, n$. In particular, if A is a circulant with first row $a_0, \dots, a_{n-1}$, then A has singular values

$$\left| \sum_{j=0}^{n-1} a_j e^{-2\pi i jk/n} \right|, \quad k = 1, \dots, n.$$

6. Take $A \in \mathbb{C}^{n\times n}$ and nonzero $\mathbf{x} \in \mathbb{C}^n$. If $A\mathbf{x} = \lambda\mathbf{x}$ and $\mathbf{x}^* A = \lambda\mathbf{x}^*$, then $|\lambda|$ is a singular value of A. In particular, if A is doubly stochastic, then $\sigma_1(A) = 1$.
7. [Kit95] Let A be the companion matrix corresponding to the monic polynomial $p(t) = t^n + a_{n-1}t^{n-1} + \cdots + a_1 t + a_0$. Set $N = 1 + \sum_{i=0}^{n-1} |a_i|^2$. The n singular values of A are

$$\sqrt{\frac{N + \sqrt{N^2 - 4|a_0|^2}}{2}}, 1, \dots, 1, \sqrt{\frac{N - \sqrt{N^2 - 4|a_0|^2}}{2}}.$$

8. [Hig96, p. 167] Take $s, c \in \mathbb{R}$ such that $s^2 + c^2 = 1$. The matrix

$$A = \operatorname{diag}(1, s, \dots, s^{n-1}) \begin{bmatrix} 1 & -c & -c & \cdots & -c \\ & 1 & -c & \cdots & -c \\ & & \ddots & \ddots & \vdots \\ & & & \ddots & -c \\ & & & & 1 \end{bmatrix}$$

is called a Kahan matrix. If c and s are positive, then $\sigma_{n-1}(A) = s^{n-2}\sqrt{1 + c}$.
9. [GE95, Lemma 3.1] Take $0 = d_1 < d_2 < \cdots < d_n$ and $0 \neq z_i \in \mathbb{C}$. Let

$$A = \begin{bmatrix} z_1 & & & \\ z_2 & d_2 & & \\ \vdots & & \ddots & \\ z_n & & & d_n \end{bmatrix}.$$

The singular values of A satisfy the equation

$$f(t) = 1 + \sum_{i=1}^{n} \frac{|z_i|^2}{d_i^2 - t^2} = 0$$

and exactly one lies in each of the intervals $(d_1, d_2), \ldots, (d_{n-1}, d_n), (d_n, d_n + \|z\|_2)$. Let $\sigma_i = \sigma_i(A)$. The left and right ith singular vectors of A are $\mathbf{u}/\|\mathbf{u}\|_2$ and $\mathbf{v}/\|\mathbf{v}\|_2$ respectively, where

$$\mathbf{u} = \left[\frac{z_1}{d_1^2 - \sigma_i^2}, \cdots, \frac{z_n}{d_n^2 - \sigma_i^2}\right]^T \quad \text{and} \quad \mathbf{v} = \left[-1, \frac{d_2 z_2}{d_2^2 - \sigma_i^2}, \cdots, \frac{d_n z_n}{d_n^2 - \sigma_i^2}\right]^T.$$

10. (*Bidiagonal*) Take

$$B = \begin{bmatrix} \alpha_1 & \beta_1 & & \\ & \alpha_2 & \ddots & \\ & & \ddots & \beta_{n-1} \\ & & & \alpha_n \end{bmatrix} \in \mathbb{C}^{n \times n}.$$

If all the α_i and β_i are nonzero, then B is called an *unreduced bidiagonal matrix* and

(a) The singular values of B are distinct.

(b) The singular values of B depend only on the moduli of $\alpha_1, \ldots, \alpha_n, \beta_1, \ldots, \beta_{n-1}$.

(c) The largest singular value of B is a strictly increasing function of the modulus of each of the α_i and β_i.

(d) The smallest singular value of B is a strictly increasing function of the modulus of each of the α_i and a strictly decreasing function of the modulus of each of the β_i.

(e) (High relative accuracy) Take $\tau > 1$ and multiply one of the entries of B by τ to give $\hat{B}$. Then $\tau^{-1}\sigma_i(B) \leq \sigma_i(\hat{B}) \leq \tau\sigma_i(B)$.

11. [HJ85, Sec. 4.4, prob. 26] Let $A \in \mathbb{C}^{n \times n}$ be skew-symmetric (and possibly complex). The nonzero singular values of A occur in pairs.

17.3 Unitarily Invariant Norms

Throughout this section, $q = \min\{m, n\}$.

Definitions:

A vector norm $\|\cdot\|$ on $\mathbb{C}^{m \times n}$ is **unitarily invariant** (*u.i.*) if $\|A\| = \|UAV\|$ for any unitary $U \in \mathbb{C}^{m \times m}$ and $V \in \mathbb{C}^{n \times n}$ and any $A \in \mathbb{C}^{m \times n}$.

$\|\cdot\|_{UI}$ is used to denote a **general unitarily invariant norm.**

A function $g : \mathbb{R}^n \to \mathbb{R}_0^+$ is a **permutation invariant absolute norm** if it is a norm, and in addition $g(x_1, \ldots, x_n) = g(|x_1|, \ldots, |x_n|)$ and $g(\mathbf{x}) = g(P\mathbf{x})$ for all $\mathbf{x} \in \mathbb{R}^n$ and all permutation matrices $P \in \mathbb{R}^{n \times n}$. (Many authors call a permutation invariant absolute norm a symmetric gauge function.)

The **Ky Fan k norms** of $A \in \mathbb{C}^{m \times n}$ are

$$\|A\|_{K,k} = \sum_{i=1}^{k} \sigma_i(A), \quad k = 1, 2, \ldots, q.$$

The **Schatten-p norms** of $A \in \mathbb{C}^{m \times n}$ are

$$\|A\|_{S,p} = \left(\sum_{i=1}^{q} \sigma_i^p(A)\right)^{1/p} = (\operatorname{tr} |A|_{pd}^p)^{1/p} \quad 0 \leq p < \infty$$

$$\|A\|_{S,\infty} = \sigma_1(A).$$

The **trace norm** of $A \in \mathbb{C}^{m\times n}$ is

$$\|A\|_{\text{tr}} = \sum_{i=1}^{q} \sigma_i(A) = \|A\|_{K,q} = \|A\|_{S,1} = \text{tr}\,|A|_{pd}.$$

Other norms discussed in this section, such as the **spectral norm** $\|\cdot\|_2$ ($\|A\|_2 = \sigma_1(A) = \max_{\mathbf{x}\neq 0} \frac{\|A\mathbf{x}\|_2}{\|\mathbf{x}\|_2}$) and the **Frobenius norm** $\|\cdot\|_F$ ($\|A\|_F = (\sum_{i=1}^{q} \sigma_i^2(A))^{1/2} = (\sum_{i=1}^{m}\sum_{j=1}^{n} |a_{ij}|^2)^{1/2}$), are defined in Section 7.1. and discussed extensively in Chapter 37.

Warning: There is potential for considerable confusion. For example, $\|A\|_2 = \|A\|_{K,1} = \|A\|_{S,\infty}$, while $\|\cdot\|_\infty \neq \|\cdot\|_{S,\infty}$ (unless $m = 1$), and generally $\|A\|_2$, $\|A\|_{S,2}$ and $\|A\|_{K,2}$ are all different, as are $\|A\|_1$, $\|A\|_{S,1}$ and $\|A\|_{K,1}$. Nevertheless, many authors use $\|\cdot\|_k$ for $\|\cdot\|_{K,k}$ and $\|\cdot\|_p$ for $\|\cdot\|_{S,p}$.

Facts:

The following standard facts can be found in many texts, e.g., [HJ91, §3.5] and [Bha97, Chap. IV].

1. Let $\|\cdot\|$ be a norm on $\mathbb{C}^{m\times n}$. It is unitarily invariant if and only if there is a permutation invariant absolute norm g on $\mathbb{R}^q$ such that $\|A\| = g(\sigma_1(A), \ldots, \sigma_q(A))$ for all $A \in \mathbb{C}^{m\times n}$.
2. Let $\|\cdot\|$ be a unitarily invariant norm on $\mathbb{C}^{m\times n}$, and let g be the corresponding permutation invariant absolute norm g. Then the dual norms (see Chapter 37) satisfy $\|A\|^D = g^D(\sigma_1(A), \ldots, \sigma_q(A))$.
3. [HJ91, Prob. 3.5.18] The spectral norm and trace norm are duals, while the Frobenius norm is self dual. The dual of $\|\cdot\|_{S,p}$ is $\|\cdot\|_{S,\tilde{p}}$, where $1/p + 1/\tilde{p} = 1$ and

$$\|A\|_{K,k}^D = \max\left\{\|A\|_2, \frac{\|A\|_{tr}}{k}\right\}, \quad k = 1, \ldots, q.$$

4. For any $A \in \mathbb{C}^{m\times n}$, $q^{-1/2}\|A\|_F \leq \|A\|_2 \leq \|A\|_F$.
5. If $\|\cdot\|$ is a u.i. norm on $\mathbb{C}^{m\times n}$, then $N(A) = \|A^*A\|^{1/2}$ is a u.i. norm on $\mathbb{C}^{n\times n}$. A norm that arises in this way is called a *Q-norm.*
6. Let $A, B \in \mathbb{C}^{m\times n}$ be given. The following are equivalent

   (a) $\|A\|_{UI} \leq \|B\|_{UI}$ for all unitarily invariant norms $\|\cdot\|_{UI}$.

   (b) $\|A\|_{K,k} \leq \|B\|_{K,k}$ for $k = 1, 2, \ldots, q$.

   (c) $(\sigma_1(A), \ldots, \sigma_q(A)) \preceq_w (\sigma_1(B), \ldots, \sigma_q(B))$. ($\preceq_w$ is defined in Preliminaries)

   The equivalence of the first two conditions is Fan's Dominance Theorem.
7. The Ky–Fan-k norms can be represented in terms of an extremal problem involving the spectral norm and the trace norm. Take $A \in \mathbb{C}^{m\times n}$. Then

$$\|A\|_{K,k} = \min\{\|X\|_{\text{tr}} + k\|Y\|_2 : X + Y = A\} \quad k = 1, \ldots, q.$$

8. [HJ91, Theorem 3.3.14] Take $A, B \in \mathbb{C}^{m\times n}$. Then

$$|\text{tr}AB^*| \leq \sum_{i=1}^{q} \sigma_i(A)\sigma_i(B).$$

   This is an important result in developing the theory of unitarily invariant norms.

Examples:

1. The matrix A in Example 1 of Section 17.1 has singular values 20, 12, 8, and 4. So

$$\|A\|_2 = 20, \quad \|A\|_F = \sqrt{624}, \quad \|A\|_{tr} = 44;$$

$$\|A\|_{K,1} = 20, \quad \|A\|_{K,2} = 32, \quad \|A\|_{K,3} = 40, \quad \|A\|_{K,4} = 44;$$

$$\|A\|_{S,1} = 44, \quad \|A\|_{S,2} = \sqrt{624}, \quad \|A\|_{S,3} = \sqrt[3]{10304} = 21.7605, \quad \|A\|_{S,\infty} = 20.$$

17.4 Inequalities

Throughout this section, $q = \min\{m, n\}$ and if $A \in \mathbb{C}^{m\times n}$ has real eigenvalues, then they are ordered $\lambda_1(A) \geq \cdots \geq \lambda_n(A)$.

Definitions:

Pinching is defined recursively. If

$$A = \begin{bmatrix} A_{11} & A_{12} \\ A_{21} & A_{22} \end{bmatrix} \in \mathbb{C}^{m\times n}, \quad B = \begin{bmatrix} A_{11} & 0 \\ 0 & A_{22} \end{bmatrix} \in \mathbb{C}^{m\times n},$$

then B is a pinching of A. (Note that we do not require the A_{ii} to be square.) Furthermore, any pinching of B is a pinching of A.

For positive α, β, define the **measure of relative separation** $\chi(\alpha, \beta) = |\sqrt{\alpha/\beta} - \sqrt{\beta/\alpha}|$.

Facts:

The following facts can be found in standard references, for example [HJ91, Chap. 3], unless another reference is given.

1. (*Submatrices*) Take $A \in \mathbb{C}^{m\times n}$ and let B denote A with one of its rows or columns deleted. Then $\sigma_{i+1}(A) \leq \sigma_i(B) \leq \sigma_i(A), \quad i = 1, \ldots, q-1$.
2. Take $A \in \mathbb{C}^{m\times n}$ and let B be A with a row *and* a column deleted. Then

$$\sigma_{i+2}(A) \leq \sigma_i(B) \leq \sigma_i(A), \quad i = 1, \ldots, q-2.$$

   The $i+2$ cannot be replaced by $i+1$. (Example 2)
3. Take $A \in \mathbb{C}^{m\times n}$ and let B be an $(m-k) \times (n-l)$ submatrix of A. Then

$$\sigma_{i+k+l}(A) \leq \sigma_i(B) \leq \sigma_i(A), \quad i = 1, \ldots, q-(k+l).$$

4. Take $A \in \mathbb{C}^{m\times n}$ and let B be A with some of its rows and/or columns set to zero. Then $\sigma_i(B) \leq \sigma_i(A), \quad i = 1, \ldots, q$.
5. Let B be a pinching of A. Then $\mathrm{sv}(B) \preceq_w \mathrm{sv}(A)$. The inequalities $\prod_{i=1}^k \sigma_i(B) \leq \prod_{i=1}^k \sigma_i(A)$ and $\sigma_k(B) \leq \sigma_k(A)$ are not necessarily true for $k > 1$. (Example 1)
6. (*Singular values of* $A+B$) Let $A, B \in \mathbb{C}^{m\times n}$.

   (a) $\mathrm{sv}(A+B) \preceq_w \mathrm{sv}(A) + \mathrm{sv}(B)$, or equivalently

$$\sum_{i=1}^k \sigma_i(A+B) \leq \sum_{i=1}^k \sigma_i(A) + \sum_{i=1}^k \sigma_i(B), \quad i = 1, \ldots, q.$$

   (b) If $i + j - 1 \leq q$ and $i, j \in \mathbb{N}$, then $\sigma_{i+j-1}(A+B) \leq \sigma_i(A) + \sigma_j(B)$.

(c) We have the weak majorization $|\mathrm{sv}(A+B) - \mathrm{sv}(A)| \preceq_w \mathrm{sv}(B)$ or, equivalently, if $1 \le i_1 < \cdots < i_k \le q$, then

$$\sum_{j=1}^{k} |\sigma_{i_j}(A+B) - \sigma_{i_j}(A)| \le \sum_{j=1}^{k} \sigma_j(B),$$

$$\sum_{i=1}^{k} \sigma_{i_j}(A) - \sum_{j=1}^{k} \sigma_j(B) \le \sum_{j=1}^{k} \sigma_{i_j}(A+B) \le \sum_{i=1}^{k} \sigma_{i_j}(A) + \sum_{j=1}^{k} \sigma_j(B).$$

(d) [Tho75] (Thompson's Standard Additive Inequalities) If $1 \le i_1 < \cdots < i_k \le q$, $1 \le i_1 < \cdots < i_k \le q$ and $i_k + j_k \le q + k$, then

$$\sum_{s=1}^{k} \sigma_{i_s + j_s - s}(A+B) \le \sum_{s=1}^{k} \sigma_{i_s}(A) + \sum_{s=1}^{k} \sigma_{j_s}(B).$$

7. (*Singular values of* AB) Take $A, B \in \mathbb{C}^{n\times n}$.

(a) For all $k = 1, 2, \ldots, n$ and all $p > 0$, we have

$$\prod_{i=n}^{i=n-k+1} \sigma_i(A)\sigma_i(B) \le \prod_{i=n}^{i=n-k+1} \sigma_i(AB),$$

$$\prod_{i=1}^{k} \sigma_i(AB) \le \prod_{i=1}^{k} \sigma_i(A)\sigma_i(B),$$

$$\sum_{i=1}^{k} \sigma_i^p(AB) \le \sum_{i=1}^{k} \sigma_i^p(A)\sigma_i^p(B).$$

(b) If $i, j \in \mathbb{N}$ and $i + j - 1 \le n$, then $\sigma_{i+j-1}(AB) \le \sigma_i(A)\sigma_j(B)$.

(c) $\sigma_n(A)\sigma_i(B) \le \sigma_i(AB) \le \sigma_1(A)\sigma_i(B)$, $i = 1, 2, \ldots, n$.

(d) [LM99] Take $1 \le j_1 < \cdots < j_k \le n$. If A is invertible and $\sigma_{j_i}(B) > 0$, then $\sigma_{j_i}(AB) > 0$ and

$$\prod_{i=n-k+1}^{n} \sigma_i(A) \le \prod_{i=1}^{k} \max\left\{\frac{\sigma_{j_i}(AB)}{\sigma_{j_i}(B)}, \frac{\sigma_{j_i}(B)}{\sigma_{j_i}(AB)}\right\} \le \prod_{i=1}^{k} \sigma_i(A).$$

(e) [LM99] Take invertible $S, T \in \mathbb{C}^{n\times n}$. Set $\tilde{A} = SAT$. Let the singular values of A and $\tilde{A}$ be $\sigma_1 \ge \cdots \ge \sigma_n$ and $\tilde{\sigma}_1 \ge \cdots \ge \tilde{\sigma}_n$. Then

$$\|\mathrm{diag}(\chi(\sigma_1, \tilde{\sigma}_1), , \ldots, \chi(\sigma_n, \tilde{\sigma}_n))\|_{UI} \le \frac{1}{2}\left(\|S^* - S^{-1}\|_{UI} + \|T^* - T^{-1}\|_{UI}\right).$$

(f) [TT73] (Thompson's Standard Multiplicative Inequalities) Take $1 \le i_1 < \cdots < i_m \le n$ and $1 \le j_1 < \cdots < j_m \le n$. If $i_m + j_m \le m + n$, then

$$\prod_{s=1}^{m} \sigma_{i_s + j_s - s}(AB) \le \prod_{s=1}^{m} \sigma_{i_s}(A) \prod_{s=1}^{m} \sigma_{j_s}(B).$$

8. [Bha97, §IX.1] Take $A, B \in \mathbb{C}^{n\times n}$.

(a) If AB is normal, then

$$\prod_{i=1}^{k} \sigma_i(AB) \le \prod_{i=1}^{k} \sigma_i(BA), \quad k = 1, \ldots, q,$$

and, consequently, $\mathrm{sv}(AB) \preceq_w \mathrm{sv}(BA)$, and $\|AB\|_{UI} \le \|BA\|_{UI}$.

(b) If AB is Hermitian, then $\mathrm{sv}(AB) \preceq_w \mathrm{sv}(H(BA))$ and $\|AB\|_{UI} \leq \|H(BA)\|_{UI}$, where $H(X) = (X + X^*)/2$.

9. (*Term-wise singular value inequalities*) [Zha02, p. 28] Take $A, B \in \mathbb{C}^{m\times n}$. Then

$$2\sigma_i(AB^*) \leq \sigma_i(A^*A + B^*B), \quad i = 1, \dots, q$$

and, more generally, if $p, \tilde{p} > 0$ and $1/p + 1/\tilde{p} = 1$, then

$$\sigma_i(AB^*) \leq \sigma_i\left(\frac{(A^*A)^{p/2}}{p} + \frac{(B^*B)^{\tilde{p}/2}}{\tilde{p}}\right) = \sigma_i\left(\frac{|A|_{pd}^p}{p} + \frac{|B|_{pd}^{\tilde{p}}}{\tilde{p}}\right).$$

The inequalities $2\sigma_1(A^*B) \leq \sigma_1(A^*A + B^*B)$ and $\sigma_1(A+B) \leq \sigma_1(|A|_{pd} + |B|_{pd})$ are not true in general (Example 3), but we do have

$$\|A^*B\|_{UI}^2 \leq \|A^*A\|_{UI}\|B^*B\|_{UI}.$$

10. [Bha97, Prop. III.5.1] Take $A \in \mathbb{C}^{n\times n}$. Then $\lambda_i(A + A^*) \leq 2\sigma_i(A)$, $i = 1, 2, \dots, n$.

11. [LM02] (*Block triangular matrices*) Let $A = \begin{bmatrix} R & 0 \\ S & T \end{bmatrix} \in \mathbb{C}^{n\times n}$ ($R \in \mathbb{C}^{p\times p}$) have singular values $\alpha_1 \geq \cdots \geq \alpha_n$. Let $k = \min\{p, n-p\}$. Then

(a) If $\sigma_{\min}(R) \geq \sigma_{\max}(T)$, then

$$\begin{aligned} \sigma_i(R) &\leq \alpha_i, \quad i = 1, \dots, p \\ \alpha_i &\leq \sigma_{i-p}(T), \quad i = p+1, \dots, n. \end{aligned}$$

(b) $(\sigma_1(S), \dots, \sigma_k(S)) \preceq_w (\alpha_1 - \alpha_n, \cdots, \alpha_k - \alpha_{n-k+1})$.

(c) If A is invertible, then

$$(\sigma_1(T^{-1}SR^{-1}, \dots, \sigma_k(T^{-1}SR^{-1}) \preceq_w (\alpha_n^{-1} - \alpha_1^{-1}, \cdots, \alpha_{n-k+1}^{-1} - \alpha_k^{-1}),$$

$$(\sigma_1(T^{-1}S), \dots, \sigma_k(T^{-1}S)) \preceq_w \frac{1}{2}\left(\frac{\alpha_1}{\alpha_n} - \frac{\alpha_n}{\alpha_1}, \cdots, \frac{\alpha_k}{\alpha_{n-k+1}} - \frac{\alpha_{n-k+1}}{\alpha_k}\right).$$

12. [LM02] (*Block positive semidefinite matrices*) Let $A = \begin{bmatrix} A_{11} & A_{12} \\ A_{12}^* & A_{22} \end{bmatrix} \in \mathbb{C}^{n\times n}$ be positive definite with eigenvalues $\lambda_1 \geq \cdots \geq \lambda_n$. Assume $A_{11} \in \mathbb{C}^{p\times p}$. Set $k = \min\{p, n-p\}$. Then

$$\prod_{i=1}^{j} \sigma_i^2(A_{12}) \leq \prod_{i=1}^{j} \sigma_i(A_{11})\sigma_i(A_{22}), \quad j = 1, \dots, k,$$

$$\left(\sigma_1\left(A_{11}^{-1/2}A_{12}\right), \dots, \sigma_k\left(A_{11}^{-1/2}A_{12}\right)\right) \preceq_w \left(\sqrt{\lambda_1} - \sqrt{\lambda_n}, \dots, \sqrt{\lambda_k} - \sqrt{\lambda_{n-k+1}}\right),$$

$$\left(\sigma_1\left(A_{11}^{-1}A_{12}\right), \dots, \sigma_k\left(A_{11}^{-1}A_{12}\right)\right) \preceq_w \frac{1}{2}\left(\chi(\lambda_1, \lambda_n), \dots, \chi(\lambda_k, \lambda_{n-k+1})\right).$$

If $k = n/2$, then

$$\|A_{12}\|_{UI}^2 \leq \|A_{11}\|_{UI}\,\|A_{22}\|_{UI}.$$

13. (*Singular values and eigenvalues*) Let $A \in \mathbb{C}^{n\times n}$. Assume $|\lambda_1(A)| \geq \cdots \geq |\lambda_n(A)|$. Then

(a) $\prod_{i=1}^k |\lambda_i(A)| \leq \prod_{i=k}^k \sigma_i(A), \quad k = 1, \dots, n$, with equality for $k = n$.

(b) Fix $p > 0$. Then for $k = 1, 2, \ldots, n$,

$$\sum_{i=1}^{k} |\lambda_i^p(A)| \leq \sum_{i=1}^{k} \sigma_i^p(A).$$

Equality holds with $k = n$ if and only if equality holds for all $k = 1, 2, \ldots, n$, if and only if A is normal.

(c) [HJ91, p. 180] (Yamamoto's theorem) $\lim_{k\to\infty}(\sigma_i(A^k))^{1/k} = |\lambda_i(A)|, \quad i = 1, \ldots, n.$

14. [LM01] Let $\lambda_i \in \mathbb{C}$ and $\sigma_i \in \mathbb{R}_0^+$, $i = 1, \ldots, n$ be ordered in nonincreasing absolute value. There is a matrix A with eigenvalues $\lambda_1, \ldots, \lambda_n$ and singular values $\sigma_1, \ldots, \sigma_n$ if and only if

$$\prod_{i=1}^{k} |\lambda_i| \leq \prod_{i=1}^{k} \sigma_i, \quad k = 1, \ldots, n, \quad \text{with equality for } k = n.$$

In addition:

(a) The matrix A can be taken to be upper triangular with the eigenvalues on the diagonal in any order.

(b) If the complex entries in $\lambda_1, \ldots, \lambda_n$ occur in conjugate pairs, then A may be taken to be in real Schur form, with the 1×1 and 2×2 blocks on the diagonal in any order.

(c) There is a finite construction of the upper triangular matrix in cases (a) and (b).

(d) If $n > 2$, then A cannot always be taken to be bidiagonal. (Example 5)

15. [Zha02, Chap. 2] (*Singular values of* $A \circ B$) Take $A, B \in \mathbb{C}^{n\times n}$.

(a) $\sigma_i(A \circ B) \leq \min\{r_i(A), c_i(B)\} \cdot \sigma_1(B), \quad i = 1, 2, \ldots, n.$

(b) We have the following weak majorizations:

$$\sum_{i=1}^{k} \sigma_i(A \circ B) \leq \sum_{i=1}^{k} \min\{r_i(A), c_i(A)\}\sigma_i(B), \quad k = 1, \ldots, n,$$

$$\sum_{i=1}^{k} \sigma_i(A \circ B) \leq \sum_{i=1}^{k} \sigma_i(A)\sigma_i(B), \quad k = 1, \ldots, n,$$

$$\prod_{i=1}^{k} \sigma_i^2(A \circ B) \leq \prod_{i=1}^{k} \sigma_i((A^*A) \circ (B^*B)), \quad k = 1, \ldots, n.$$

(c) Take $X, Y \in \mathbb{C}^{n\times n}$. If $A = X^*Y$, then we have the weak majorization

$$\sum_{i=1}^{k} \sigma_i(A \circ B) \leq \sum_{i=1}^{k} c_i(X)c_i(Y)\sigma_i(B), \quad k = 1, \ldots, n.$$

(d) If B is positive semidefinite with diagonal entries $b_{11} \geq \cdots \geq b_{nn}$, then

$$\sum_{i=1}^{k} \sigma_i(A \circ B) \leq \sum_{i=1}^{k} b_{ii}\sigma_i(A), \quad k = 1, \ldots, n.$$

(e) If both A and B are positive definite, then so is $A \circ B$ (Schur product theorem). In this case the singular values of A, B and $A \circ B$ are their eigenvalues and BA has positive eigenvalues and we have the weak multiplicative majorizations

$$\prod_{i=k}^{n} \lambda_i(B)\lambda_i(A) \leq \prod_{i=k}^{n} b_{ii}\lambda_i(A) \leq \prod_{i=k}^{n} \lambda_i(BA) \leq \prod_{i=k}^{n} \lambda_i(A \circ B), \quad k = 1, 2, \ldots, n.$$

The inequalities are still valid if we replace $A \circ B$ by $A \circ B^T$. (Note B^T is not necessarily the same as $B^* = B$.)

16. Let $A \in \mathbb{C}^{m\times n}$. The following are equivalent:

    (a) $\sigma_1(A \circ B) \leq \sigma_1(B)$ for all $B \in \mathbb{C}^{m\times n}$.

    (b) $\sum_{i=1}^k \sigma_i(A \circ B) \leq \sum_{i=1}^k \sigma_i(B)$ for all $B \in \mathbb{C}^{m\times n}$ and all $k = 1, \ldots, q$.

    (c) There are positive semidefinite $P \in \mathbb{C}^{n\times n}$ and $Q \in \mathbb{C}^{m\times m}$ such that

$$\begin{bmatrix} P & A \\ A^* & Q \end{bmatrix}$$

    is positive semidefinite, and has diagonal entries at most 1.

17. (*Singular values and matrix entries*) Take $A \in \mathbb{C}^{m\times n}$. Then

$$(|a_{11}|^2, |a_{12}|^2, \ldots, |a_{mn}|^2) \preceq (\sigma_1^2(A), \ldots, \sigma_q^2(A), 0, \ldots, 0),$$

$$\sum_{i=1}^q \sigma_i^p(A) \leq \sum_{i=1}^m \sum_{j=1}^n |a_{ij}|^p, \quad 0 \leq p \leq 2,$$

$$\sum_{i=1}^m \sum_{j=1}^n |a_{ij}|^p \leq \sum_{i=1}^q \sigma_i^p(A), \quad 2 \leq p < \infty.$$

    If $\sigma_1(A) = |a_{ij}|$, then all the other entries in row i and column j of A are 0.

18. Take $\sigma_1 \geq \cdots \geq \sigma_n \geq 0$ and $\alpha_1 \geq \cdots \geq \alpha_n \geq 0$. Then

$$\exists A \in \mathbb{R}^{n\times n} \text{ s.t. } \sigma_i(A) = \sigma_i \quad \text{and} \quad c_i(A) = \alpha_i \quad \Leftrightarrow \quad (\alpha_1^2, \ldots, \alpha_n^2) \preceq (\sigma_1^2, \ldots, \sigma_n^2).$$

    This statement is still true if we replace $\mathbb{R}^{n\times n}$ by $\mathbb{C}^{n\times n}$ and/or $c_i(\cdot)$ by $r_i(\cdot)$.

19. Take $A \in \mathbb{C}^{n\times n}$. Then

$$\prod_{i=k}^n \sigma_i(A) \leq \prod_{i=k}^n c_i(A), \quad k = 1, 2, \ldots, n.$$

    The case $k = 1$ is Hadamard's Inequality: $|\det(A)| \leq \prod_{i=1}^n c_i(A)$.

20. [Tho77] Take $F = \mathbb{C}$ or $\mathbb{R}$ and $d_1, \ldots, d_n \in F$ such that $|d_1| \geq \cdots \geq |d_n|$, and $\sigma_1 \geq \cdots \geq \sigma_n \geq 0$. There is a matrix $A \in F^{n\times n}$ with diagonal entries $d_1, \ldots, d_n$ and singular values $\sigma_1, \ldots, \sigma_n$ if and only if

$$(|d_1|, \ldots, |d_n|) \preceq_w (\sigma_1(A), \ldots, \sigma_n(A)) \quad \text{and} \quad \sum_{j=1}^{n-1} |d_j| - |d_n| \leq \sum_{j=1}^{n-1} \sigma_j(A) - \sigma_n(A).$$

21. (*Nonnegative matrices*) Take $A = [a_{ij}] \in \mathbb{C}^{m\times n}$.

    (a) If $B = [|a_{ij}|]$, then $\sigma_1(A) \leq \sigma_1(B)$.

    (b) If A and B are real and $0 \leq a_{ij} \leq b_{ij} \ \forall \ i, j$, then $\sigma_1(A) \leq \sigma_1(B)$. The condition $0 \leq a_{ij}$ is essential. (Example 4)

    (c) The condition $0 \leq b_{ij} \leq 1 \ \forall \ i, j$ does not imply $\sigma_1(A \circ B) \leq \sigma_1(A)$. (Example 4)

22. (*Bound on* σ_1) Let $A \in \mathbb{C}^{m\times n}$. Then $\|A\|_2 = \sigma_1(A) \leq \sqrt{\|A\|_1 \|A\|_\infty}$.

23. [Zha99] (*Cartesian decomposition*) Let $C = A + iB \in \mathbb{C}^{n\times n}$, where A and B are Hermitian. Let A, B, C have singular values $\alpha_j, \beta_j, \gamma_i$, $j = 1, \ldots, n$. Then

$$(\gamma_1, \ldots, \gamma_n) \preceq_w \sqrt{2}(|\alpha_1 + i\beta_1|, \ldots, |\alpha_n + i\beta_n|) \preceq_w 2(\gamma_1, \ldots, \gamma_n).$$

Examples:

1. Take

$$A=\begin{bmatrix}1&1&1\\1&1&1\\1&1&1\end{bmatrix},\quad B=\begin{bmatrix}1&0&0\\0&1&1\\0&1&1\end{bmatrix},\quad C=\begin{bmatrix}1&0&0\\0&1&0\\0&0&1\end{bmatrix}.$$

   Then B is a pinching of A, and C is a pinching of both A and B. The matrices A, B, C have singular values $\alpha = (3,0,0)$, $\beta = (2,1,0)$, and $\gamma = (1,1,1)$. As stated in Fact 5, $\gamma \preceq_w \beta \preceq_w \alpha$. In fact, since the matrices are all positive semidefinite, we may replace $\preceq_w$ by $\preceq$. However, it is not true that $\gamma_i \le \alpha_i$ except for $i = 1$. Nor is it true that $|\det(C)| \le |\det(A)|$.
2. The matrices

$$A=\begin{bmatrix}11&-3&-5&1\\1&-5&-3&11\\-5&1&11&-3\\-3&11&1&-5\end{bmatrix},\quad B=\begin{bmatrix}11&-3&-5&1\\1&-5&-3&11\\-5&1&11&-3\end{bmatrix},\quad C=\begin{bmatrix}11&-3&-5\\1&-5&-3\\-5&1&11\end{bmatrix}$$

   have singular values $\alpha = (20,12,8,4)$, $\beta = (17.9,10.5,6.0)$, and $\gamma = (16.7,6.2,4.5)$ (to 1 decimal place). The singular values of B interlace those of A ($\alpha_4 \le \beta_3 \le \alpha_3 \le \beta_2 \le \alpha_2 \le \beta_1 \le \alpha_1$), but those of C do not. In particular, $\alpha_3 \not\le \gamma_2$. It is true that $\alpha_{i+2} \le \gamma_i \le \alpha_i$ $(i = 1,2)$.
3. Take

$$A=\begin{bmatrix}1&0\\1&0\end{bmatrix}\quad\text{and}\quad B=\begin{bmatrix}0&1\\0&1\end{bmatrix}.$$

   Then $\|A+B\|_2 = \sigma_1(A+B) = 2 \not\le \sqrt{2} = \sigma_1(|A|_{pd} + |B|_{pd}) = \|\, |A|_{pd} + |B|_{pd}\, \|_2$. Also, $2\sigma_1(A^*B) = 4 \not\le 2 = \sigma_1(A^*A + B^*B)$.
4. Setting entries of a matrix to zero can increase the largest singular value. Take

$$A=\begin{bmatrix}1&1\\-1&1\end{bmatrix},\quad\text{and}\quad B=\begin{bmatrix}1&1\\0&1\end{bmatrix}.$$

   Then $\sigma_1(A) = \sqrt{2} < (1+\sqrt{5})/2 = \sigma_1(B)$.
5. A bidiagonal matrix B cannot have eigenvalues 1, 1, 1 and singular values 1/2, 1/2, 4. If B is unreduced bidiagonal, then it cannot have repeated singular values. (See Fact 10, section 17.2.) However, if B were reduced, then it would have a singular value equal to 1.

17.5 Matrix Approximation

Recall that $\|\cdot\|_{UI}$ denotes a general unitarily invariant norm, and that $q = \min\{m,n\}$.

Facts:

The following facts can be found in standard references, for example, [HJ91, Chap. 3], unless another reference is given.

1. (*Best rank k approximation.*) Let $A \in \mathbb{C}^{m\times n}$ and $1 \le k \le q-1$. Let $A = U\Sigma V^*$ be a singular value decomposition of A. Let $\tilde{\Sigma}$ be equal to Σ except that $\tilde{\Sigma}_{ii} = 0$ for $i > k$, and let $\tilde{A} = U\tilde{\Sigma}V^*$. Then $\operatorname{rank}(\tilde{A}) \le k$, and

$$\|\Sigma - \tilde{\Sigma}\|_{UI} = \|A - \tilde{A}\|_{UI} = \min\{\|A - B\|_{UI} : \operatorname{rank}(B) \le k\}.$$

In particular, for the spectral norm and the Frobenius norm, we have

$$\sigma_{k+1}(A) = \min\{\|A - B\|_2 : \operatorname{rank}(B) \le k\},$$

$$\left(\sum_{i=k+1}^{q} \sigma_{k+1}^2(A)\right)^{1/2} = \min\{\|A - B\|_F : \operatorname{rank}(B) \le k\}.$$

2. [Bha97, p. 276] (*Best unitary approximation*) Take $A, W \in \mathbb{C}^{n\times n}$ with W unitary. Let $A = UP$ be a polar decomposition of A. Then

$$\|A - U\|_{UI} \le \|A - W\|_{UI} \le \|A + U\|_{UI}.$$

3. [GV96, §12.4.1] [HJ85, Ex. 7.4.8] (*Orthogonal Procrustes problem*) Let $A, B \in \mathbb{C}^{m\times n}$. Let B^*A have a polar decomposition $B^*A = UP$. Then

$$\|A - BU\|_F = \min\{\|A - BW\|_F : W \in \mathbb{C}^{n\times n}, W^*W = I\}.$$

This result is not true if $\|\cdot\|_F$ is replaced by $\|\cdot\|_{UI}$ ([Mat93, §4]).
4. [Hig89] (*Best PSD approximation*) Take $A \in \mathbb{C}^{n\times n}$. Set $A_H = (A + A^*)/2$, $B = (A_H + |A_H|)/2)$. Then B is positive semidefinite and is the unique solution to

$$\min\{\|A - X\|_F : X \in \mathbb{C}^{n\times n}, X \in \text{PSD}\}.$$

There is also a formula for the best PSD approximation in the spectral norm.
5. Let $A, B \in \mathbb{C}^{m\times n}$ have singular value decompositions $A = U_A\Sigma_A V_A^*$ and $B = U_B\Sigma_B V_B^*$. Let $U \in \mathbb{C}^{m\times m}$ and $V \in \mathbb{C}^{n\times n}$ be any unitary matrices. Then

$$\|\Sigma_A - \Sigma_B\|_{UI} \le \|A - UBV^*\|_{UI}.$$

17.6 Characterization of the Eigenvalues of Sums of Hermitian Matrices and Singular Values of Sums and Products of General Matrices

There are necessary and sufficient conditions for three sets of numbers to be the eigenvalues of Hermitian $A, B, C = A + B \in \mathbb{C}^{n\times n}$, or the singular values of $A, B, C = A + B \in \mathbb{C}^{m\times n}$, or the singular values of nonsingular A, B, $C = AB \in \mathbb{C}^{n\times n}$. The key results in this section were first proved by Klyachko ([Kly98]) and Knutson and Tao ([KT99]). The results presented here are from a survey by Fulton [Ful00]. Bhatia has written an expository paper on the subject ([Bha01]).

Definitions:

The inequalities are in terms of the sets T_r^n of triples (I, J, K) of subsets of $\{1, \ldots, n\}$ of the same cardinality r, defined by the following inductive procedure. Set

$$U_r^n = \left\{(I, J, K) \;\middle|\; \sum_{i\in I} i + \sum_{j\in J} j = \sum_{k\in K} k + r(r+1)/2\right\}.$$

When $r = 1$, set $T_1^n = U_1^n$. In general,

$$T_r^n = \Bigg\{(I, J, K) \in U_r^n \mid \text{for all } p < r \text{ and all}$$

$$(F, G, H) \text{ in } T_p^r, \sum_{f\in F} i_f + \sum_{g\in G} j_g \le \sum_{h\in H} k_h + p(p+1)/2\Bigg\}.$$

In this section, the vectors α, β, γ will have real entries ordered in nonincreasing order.

Facts:

The following facts are in [Ful00]:

1. A triple (α, β, γ) of real n-vectors occurs as eigenvalues of Hermitian $A, B, C = A + B \in \mathbb{C}^{n\times n}$ if and only if $\sum \gamma_i = \sum \alpha_i + \sum \beta_i$ and the inequalities
$$\sum_{k\in K} \gamma_k \leq \sum_{i\in I} \alpha_i + \sum_{j\in J} \beta_j$$
hold for every (I, J, K) in T_r^n, for all $r < n$. Furthermore, the statement is true if $\mathbb{C}^{n\times n}$ is replaced by $\mathbb{R}^{n\times n}$.
2. Take Hermitian $A, B \in \mathbb{C}^{n\times n}$ (not necessarily PSD). Let the vectors of eigenvalues of A, B, $C = A + B$ be α, β, and γ. Then we have the (nonlinear) inequality
$$\min_{\pi\in S_n} \prod_{i=1}^{n} (\alpha_i + \beta_{\pi(i)}) \leq \prod_{i=1}^{n} \gamma_i \leq \max_{\pi\in S_n} \prod_{i=1}^{n} (\alpha_i + \beta_{\pi(i)}).$$
3. Fix m, n and set $q = \min\{m, n\}$. For any subset X of $\{1, \ldots, m+n\}$, define $X_q = \{i : i \in X, i \leq q\}$ and $X_q' = \{i : i \leq q,\ m + n + 1 - i \in X\}$. A triple (α, β, γ) occurs as the singular values of $A, B, C = A + B \in \mathbb{C}^{m\times n}$, if and only if the inequalities
$$\sum_{k\in K_q} \gamma_k - \sum_{k\in K_q'} \gamma_k \leq \sum_{i\in I} \alpha_i - \sum_{i\in I_q'} \alpha_i + \sum_{j\in J_q} \beta_j - \sum_{j\in J_q'} \beta_j$$
are satisfied for all (I, J, K) in T_r^{m+n}, for all $r < m+n$. This statement is not true if $\mathbb{C}^{m\times n}$ is replaced by $\mathbb{R}^{m\times n}$. (See Example 1.)
4. A triple of positive real n-vectors (α, β, γ) occurs as the singular values of n by n matrices A,B, $C = AB \in \mathbb{C}^{n\times n}$ if and only if $\gamma_1 \cdots \gamma_n = \alpha_1 \cdots \alpha_n \beta_1 \cdots \beta_n$ and
$$\prod_{k\in K} \gamma_k \leq \prod_{i\in I} \alpha_i \cdot \prod_{j\in J} \beta_j$$
for all (I, J, K) in T_r^n, and all $r < n$. This statement is still true if $\mathbb{C}^{n\times n}$ is replaced by $\mathbb{R}^{n\times n}$.

Example:

1. There are $A, B, C = A + B \in \mathbb{C}^{2\times 2}$ with singular values $(1, 1)$, $(1, 0)$, and $(1, 1)$, but there are no $A, B, C = A + B \in \mathbb{R}^{2\times 2}$ with these singular values.

   In the complex case, take $A = \text{diag}(1, 1/2 + (\sqrt{3}/2)i)$, $B = \text{diag}(0, -1)$.

   Now suppose that A and B are real 2×2 matrices such that A and $C = A + B$ both have singular values $(1, 1)$. Then A and C are orthogonal. Consider $BC^T = AC^T - CC^T = AC^T - I$. Because AC^T is real, it has eigenvalues $\alpha, \bar{\alpha}$ and so BC^T has eigenvalues $\alpha - 1, \bar{\alpha} - 1$. Because AC^T is orthogonal, it is normal and, hence, so is BC^T, and so its singular values are $|\alpha - 1|$ and $|\bar{\alpha} - 1|$, which are equal and, in particular, cannot be $(1, 0)$.

17.7 Miscellaneous Results and Generalizations

Throughout this section F can be taken to be either $\mathbb{R}$ or $\mathbb{C}$.

Definitions:

Let $\mathcal{X}$, $\mathcal{Y}$ be subspaces of $\mathbb{C}^r$ of dimension m and n. The **principal angles** $0 \leq \theta_1 \leq \cdots \leq \theta_q \leq \pi/2$ between $\mathcal{X}$ and $\mathcal{Y}$ and **principal vectors** $\mathbf{u}_1, \ldots, \mathbf{u}_q$ and $\mathbf{v}_1, \ldots, \mathbf{v}_q$ are defined inductively:
$$\cos(\theta_1) = \max\{|\mathbf{x}^*\mathbf{y}| : \mathbf{x} \in \mathcal{X}, \max_{\mathbf{y}\in\mathcal{Y}}, \|\mathbf{x}\|_2 = \|\mathbf{y}\|_2 = 1\}.$$

Let $\mathbf{u}_1$ and $\mathbf{v}_1$ be a pair of maximizing vectors. For $k = 2, \ldots, q$,

$$\cos(\theta_k) = \max\{|\mathbf{x}^*\mathbf{y}| : \mathbf{x} \in \mathcal{X}, \mathbf{y} \in \mathcal{Y}, \|\mathbf{x}\|_2 = \|\mathbf{y}\|_2 = 1, \quad \mathbf{x}^*\mathbf{u}_i = \mathbf{y}^*\mathbf{v}_i = 0, \quad i = 1, \ldots, k-1\}.$$

Let $\mathbf{u}_k$ and $\mathbf{v}_k$ be a pair of maximizing vectors. (Principal angles are also called **canonical angles**, and the cosines of the principal angles are called **canonical correlations**.)

Facts:

1. (*Principal Angles*) Let $\mathcal{X}, \mathcal{Y}$ be subspaces of $\mathbb{C}^r$ of dimension m and n.

   (a) [BG73] The principal vectors obtained by the process above are not necessarily unique, but the principal angles are unique (and, hence, independent of the chosen principal vectors).

   (b) Let $m = n \leq r/2$ and X, Y be matrices whose columns form orthonormal bases for the subspaces $\mathcal{X}$ and $\mathcal{Y}$, respectively.

   i. The singular values of X^*Y are the cosines of the principal angles between the subspaces $\mathcal{X}$ and $\mathcal{Y}$.

   ii. There are unitary matrices $U \in \mathbb{C}^{r\times r}$ and V_X and $V_Y \in \mathbb{C}^{n\times n}$ such that

   $$UXV_X = \begin{bmatrix} I_n \\ 0_n \\ 0_{r-n,n} \end{bmatrix}, \quad UYV_Y = \begin{bmatrix} \Gamma \\ \Sigma \\ 0_{r-n,n} \end{bmatrix},$$

   where Γ and Σ are nonnegative diagonal matrices. Their diagonal entries are the cosines and sines respectively of the principal angles between $\mathcal{X}$ and $\mathcal{Y}$.

   (c) [QZL05] Take $m = n$. For any permutation invariant absolute norm g on $\mathbb{R}^m$,

   $$g(\sin(\theta_1), \ldots, \sin(\theta_m)), \quad g(2\sin(\theta_1/2), \ldots, 2\sin(\theta_m/2)), \quad \text{and} \quad g(\theta_1, \ldots, \theta_m)$$

   are metrics on the set of subspaces of dimension n of $\mathbb{C}^{r\times r}$.

2. [GV96, Theorem 2.6.2] (*CS decomposition*) Let $W \in F^{n\times n}$ be unitary. Take a positive integer l such that $2l \leq n$. Then there are unitary matrices $U_{11}, V_{11} \in F^{l\times l}$ and $U_{22}, V_{22} \in F^{(n-l)\times(n-l)}$ such that

   $$\begin{bmatrix} U_{11} & 0 \\ 0 & U_{22} \end{bmatrix} W \begin{bmatrix} V_{11} & 0 \\ 0 & V_{22} \end{bmatrix} = \begin{bmatrix} \Gamma & -\Sigma & 0 \\ \Sigma & \Gamma & 0 \\ 0 & 0 & I_{n-2l} \end{bmatrix},$$

   where $\Gamma = \text{diag}(\gamma_1, \ldots, \gamma_l)$ and $\Sigma = \text{diag}(\sigma_1, \ldots, \sigma_l)$ are nonnegative and $\Gamma^2 + \Sigma^2 = I$.

3. [GV96, Theorem 8.7.4] (*Generalized singular value decomposition*) Take $A \in F^{p\times n}$ and $B \in F^{m\times n}$ with $p \geq n$. Then there is an invertible $X \in F^{n\times n}$, unitary $U \in F^{p\times p}$ and $V \in F^{m\times m}$, and nonnegative diagonal matrices $\Sigma_A \in \mathbb{R}^{n\times n}$ and $\Sigma_B \in \mathbb{R}^{q\times q}$ ($q = \min\{m, n\}$) such that $A = U\Sigma_A X$ and $B = V\Sigma_B X$.

References

[And94] T. Ando. Majorization and inequalitites in matrix theory. *Lin. Alg. Appl.*, 199:17–67, 1994.

[Bha97] R. Bhatia. *Matrix Analysis*. Springer-Verlag, New York, 1997.

[Bha01] R. Bhatia. Linear algebra to quantum cohomology: the story of Alfred Horn's inequalities. *Amer. Math. Monthly*, 108(4):289–318, 2001.

[BG73] A. Björk and G. Golub. Numerical methods for computing angles between linear subspaces. *Math. Comp.*, 27:579–594, 1973.

[Ful00] W. Fulton. Eigenvalues, invariant factors, highest weights, and Schurbert calculus. *Bull. Am. Math. Soc.*, 37:255–269, 2000.

[GV96] G.H. Golub and C.F. Van Loan. *Matrix Computations*. The Johns Hopkins University Press, Baltimore, 3rd ed., 1996.

[GE95] Ming Gu and Stanley Eisenstat. A divide-and-conquer algorithm for the bidiagonal SVD. *SIAM J. Matrix Anal. Appl.*, 16:72–92, 1995.

[Hig96] N.J. Higham. *Accuracy and Stability of Numerical Algorithms*. SIAM, Philadelphia, 1996.

[Hig89] N.J. Higham. Matrix nearness problems and applications. In M.J.C. Gover and S. Barnett, Eds., *Applications of Matrix Theory*, pp. 1–27. Oxford University Press, U.K. 1989.

[HJ85] R.A. Horn and C.R. Johnson. *Matrix Analysis*. Cambridge University Press, Cambridge, 1985.

[HJ91] R.A. Horn and C.R. Johnson. *Topics in Matrix Analysis*. Cambridge University Press, Cambridge, 1991.

[Kit95] F. Kittaneh. Singular values of companion matrices and bounds on zeros of polynomials. *SIAM. J. Matrix Anal. Appl.*, 16(1):330–340, 1995.

[Kly98] A. A. Klyachko. Stable bundles, representation theory and Hermitian operators. *Selecta Math.*, 4(3):419–445, 1998.

[KT99] A. Knutson and T. Tao. The honeycomb model of $GL_n(C)$ tensor products i: proof of the saturation conjecture. *J. Am. Math. Soc.*, 12(4):1055–1090, 1999.

[LM99] C.-K. Li and R. Mathias. The Lidskii–Mirsky–Wielandt theorem — additive and multiplicative versions. *Numerische Math.*, 81:377–413, 1999.

[LM01] C.-K. Li and R. Mathias. Construction of matrices with prescribed singular values and eigenvalues. *BIT*, 41(1):115–126, 2001.

[LM02] C.-K. Li and R. Mathias. Inequalities on singular values of block triangular matrices. *SIAM J. Matrix Anal. Appl.*, 24:126–131, 2002.

[MO79] A.W. Marshall and I. Olkin. *Inequalities: Theory of Majorization and Its Applications*. Academic Press, London, 1979.

[Mat93] R. Mathias. Perturbation bounds for the polar decomposition. *SIAM J. Matrix Anal. Appl.*, 14(2):588–597, 1993.

[QZL05] Li Qiu, Yanxia Zhang, and Chi-Kwong Li. Unitarily invariant metrics on the Grassmann space. *SIAM J. Matrix Anal. Appl.*, 27(2):507–531, 2006.

[Tho75] R.C. Thompson. Singular value inequalities for matrix sums and minors. *Lin. Alg. Appl.*, 11(3):251–269, 1975.

[Tho77] R.C. Thompson. Singular values, diagonal elements, and convexity. *SIAM J. Appl. Math.*, 32(1):39–63, 1977.

[TT73] R.C. Thompson and S. Therianos. On the singular values of a matrix product-I, II, III. *Scripta Math.*, 29:99–123, 1973.

[Zha99] X. Zhan. Norm inequalities for Cartesian decompositions. *Lin. Alg. Appl.*, 286(1–3):297–301, 1999.

[Zha02] X. Zhan. *Matrix Inequalities*. Springer-Verlag, Berlin, Heidelberg, 2002. (Lecture Notes in Mathematics 1790.)

18

Numerical Range

Chi-Kwong Li
College of William and Mary

The numerical range $W(A)$ of an $n \times n$ complex matrix A is the collection of complex numbers of the form $\mathbf{x}^* A\mathbf{x}$, where $\mathbf{x} \in \mathbb{C}^n$ is a unit vector. It can be viewed as a "picture" of A containing useful information of A. Even if the matrix A is not known explicitly, the "picture" $W(A)$ would allow one to "see" many properties of the matrix. For example, the numerical range can be used to locate eigenvalues, deduce algebraic and analytic properties, obtain norm bounds, help find dilations with simple structure, etc. Related to the numerical range are the numerical radius of A defined by $w(A) = \max_{\mu \in W(A)} |\mu|$ and the distance of $W(A)$ to the origin denoted by $\widetilde{w}(A) = \min_{\mu \in W(A)} |\mu|$. The quantities $w(A)$ and $\widetilde{w}(A)$ are useful in studying perturbation, convergence, stability, and approximation problems.

Note that the spectrum $\sigma(A)$ can be viewed as another useful "picture" of the matrix $A \in M_n$. There are interesting relations between $\sigma(A)$ and $W(A)$.

18.1 Basic Properties and Examples

Definitions and Notation:

Let $A \in \mathbb{C}^{n\times n}$. The **numerical range** (also known as **the field of values**) of A is defined by

$$W(A) = \{\mathbf{x}^* A\mathbf{x} : \mathbf{x} \in \mathbb{C}^n, \mathbf{x}^*\mathbf{x} = 1\}.$$

The **numerical radius** of A and the **distance** of $W(A)$ to the origin are the quantities

$$w(A) = \max\{|\mu| : \mu \in W(A)\} \quad \text{and} \quad \widetilde{w}(A) = \min\{|\mu| : \mu \in W(A)\}.$$

Furthermore, let

$$\overline{W(A)} = \{\overline{a} : a \in W(A)\}.$$

Facts:

The following basic facts can be found in most references on numerical ranges such as [GR96], [Hal82], and [HJ91].

1. Let $A \in \mathbb{C}^{n\times n}$, $a, b \in \mathbb{C}$. Then $W(aA + bI) = aW(A) + b$.
2. Let $A \in \mathbb{C}^{n\times n}$. Then $W(U^*AU) = W(A)$ for any unitary $U \in \mathbb{C}^{n\times n}$.
3. Let $A \in \mathbb{C}^{n\times n}$. Suppose $k \in \{1, \dots, n-1\}$ and $X \in \mathbb{C}^{n\times k}$ satisfies $X^*X = I_k$. Then
$$W(X^*AX) \subseteq W(A).$$
In particular, for any $k \times k$ principal submatrix B of A, we have $W(B) \subseteq W(A)$.
4. Let $A \in \mathbb{C}^{n\times n}$. Then $W(A)$ is a compact convex set in $\mathbb{C}$.
5. If $A_1 \oplus A_2 \in M_n$, then $W(A) = \text{conv}\{W(A_1) \cup W(A_2)\}$.
6. Let $A \in M_n$. Then $W(A) = W(A^T)$ and $W(A^*) = \overline{W(A)}$.
7. If $A \in \mathbb{C}^{2\times 2}$ has eigenvalues λ_1, λ_2, then $W(A)$ is an elliptical disk with foci λ_1, λ_2, and minor axis with length $\{\text{tr}(A^*A) - |\lambda_1|^2 - |\lambda_2|^2\}^{1/2}$. Consequently, if $A = \begin{bmatrix} \lambda_1 & b \\ 0 & \lambda_2 \end{bmatrix}$, then the minor axis of the elliptical disk $W(A)$ has length $|b|$.
8. Let $A \in \mathbb{C}^{n\times n}$. Then $W(A)$ is a subset of a straight line if and only if there are $a, b \in \mathbb{C}$ with $a \neq 0$ such that $aA + bI$ is Hermitian. In particular, we have the following:
   (a) $A = aI$ if and only if $W(A) = \{a\}$.
   (b) $A = A^*$ if and only if $W(A) \subseteq \mathbb{R}$.
   (c) $A = A^*$ is positive definite if and only if $W(A) \subseteq (0, \infty)$.
   (d) $A = A^*$ is positive semidefinite if and only if $W(A) \subseteq [0, \infty)$.
9. If $A \in \mathbb{C}^{n\times n}$ is normal, then $W(A) = \text{conv}\,\sigma(A)$ is a convex polygon. The converse is true if $n \leq 4$.
10. Let $A \in \mathbb{C}^{n\times n}$. The following conditions are equivalent.
    (a) $W(A) = \text{conv}\,\sigma(A)$.
    (b) $W(A)$ is a convex polygon with vertices $\mu_1, \dots, \mu_k$.
    (c) A is unitarily similar to $\text{diag}(\mu_1, \dots, \mu_k) \oplus B$ such that $W(B) \subseteq \text{conv}\{\mu_1, \dots, \mu_k\}$.
11. Let $A \in \mathbb{C}^{n\times n}$. Then A is unitary if and only if all eigenvalues of A have modulus one and $W(A) = \text{conv}\,\sigma(A)$.
12. Suppose $A = (A_{ij})_{1\leq i,j\leq m} \in M_n$ is a block matrix such that $A_{11}, \dots, A_{mm}$ are square matrices and $A_{ij} = 0$ whenever $(i, j) \notin \{(1, 2), \dots, (m-1, m), (m, 1)\}$. Then $W(A) = cW(A)$ for any $c \in \mathbb{C}$ satisfying $c^m = 1$. If $A_{m,1}$ is also zero, then $W(A)$ is a circular disk centered at 0 with the radius equal to the largest eigenvalue of $(A + A^*)/2$.

Examples:

1. Let $A = \text{diag}(1, 0)$. Then $W(A) = [0, 1]$.
2. Let $A = \begin{bmatrix} 0 & 2 \\ 0 & 0 \end{bmatrix}$. Then $W(A)$ is the closed unit disk $\mathbf{D} = \{a \in \mathbb{C} : |a| \leq 1\}$.
3. Let $A = \begin{bmatrix} 2 & 2 \\ 0 & 1 \end{bmatrix}$. Then by Fact 7 above, $W(A)$ is the convex set whose boundary is the ellipse with foci 1 and 2 and minor axis 2, as shown in Figure 18.1.
4. Let $A = \text{diag}(1, i, -1, -i) \oplus \begin{bmatrix} 0 & 1 \\ 0 & 0 \end{bmatrix}$. By Facts 5 and 7, the boundary of $W(A)$ is the square with vertices $1, i, -1, -i$.

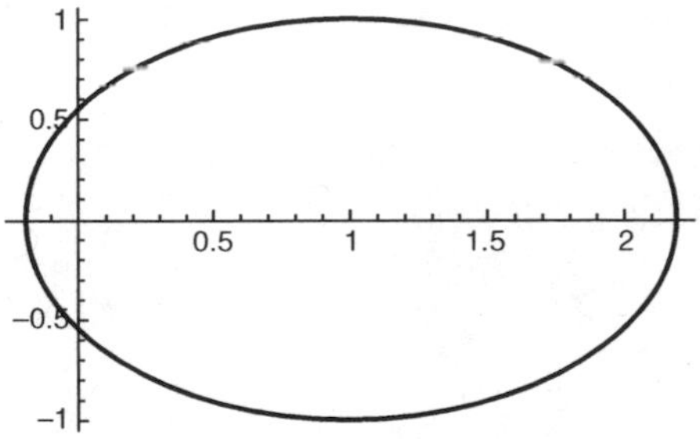

FIGURE 18.1 Numerical range of the matrix A in Example 3.

Applications:

1. By Fact 6, if A is real, then $W(A)$ is symmetric about the real axis, i.e., $W(A) = \overline{W(A)}$.
2. Suppose $A \in \mathbb{C}^{n\times n}$, and there are $a, b \in \mathbb{C}$ such that $(A - aI)(A - bI) = 0_n$. Then A is unitarily similar to a matrix of the form

$$aI_r \oplus bI_s \oplus \begin{bmatrix} a & d_1 \\ 0 & b \end{bmatrix} \oplus \cdots \oplus \begin{bmatrix} a & d_t \\ 0 & b \end{bmatrix}$$

with $d_1 \geq \cdots \geq d_t > 0$, where $r + s + 2t = n$. By Facts 1, 5, and 7, the set $W(A)$ is the elliptical disk with foci a, b and minor axis of length d, where

$$d = d_1 = \{(\|A\|_2^2 - |a|^2)(\|A\|_2^2 - |b|^2)\}^{1/2}/\|A\|_2$$

if $t \geq 1$, and $d = 0$ otherwise.
3. By Fact 12, if $A \in \mathbb{C}^{n\times n}$ is the basic circulant matrix $E_{12} + E_{23} + \cdots + E_{n-1,n} + E_{n1}$, then $W(A) = \text{conv}\{c \in \mathbb{C} : c^n = 1\}$; if $A \in M_n$ is the Jordan block of zero $J_n(0)$, then $W(A) = \{c \in \mathbb{C} : |c| \leq \cos(\pi/(n+1))\}$.
4. Suppose $A \in \mathbb{C}^{n\times n}$ is a primitive nonnegative matrix. Then A is permutationally similar to a block matrix (A_{ij}) as described in Fact 12 and, thus, $W(A) = cW(A)$ for any $c \in \mathbb{C}$ satisfying $c^m = 1$.

18.2 The Spectrum and Special Boundary Points

Definitions and Notation:

Let $\partial\mathcal{S}$ and int $(\mathcal{S})$ be the boundary and the interior of a convex compact subset $\mathcal{S}$ of $\mathbb{C}$.

A **support line** ℓ of $\mathcal{S}$ is a line that intersects $\partial\mathcal{S}$ such that $\mathcal{S}$ lies entirely within one of the closed half-planes determined by ℓ.

A boundary point μ of $\mathcal{S}$ is **nondifferentiable** if there is more than one support line of $\mathcal{S}$ passing through μ.

An eigenvalue λ of $A \in \mathbb{C}^{n\times n}$ is a **reducing eigenvalue** if A is unitarily similar to $[\lambda] \oplus A_2$.

Facts:

The following facts can be found in [GR96],[Hal82], and [HJ91].

1. Let $A \in \mathbb{C}^{n\times n}$. Then

$$\sigma(A) \subseteq W(A) \subseteq \{a \in \mathbb{C} : |a| \leq \|A\|_2\}.$$

2. Let $A, E \in \mathbb{C}^{n\times n}$. We have

$$\sigma(A+E) \subseteq W(A+E) \subseteq W(A)+W(E)$$
$$\subseteq \{a+b \in \mathbb{C} : a \in W(A), \quad b \in \mathbb{C} \quad \text{with} \quad |b| \leq \|E\|_2\}.$$

3. Let $A \in \mathbb{C}^{n\times n}$ and $a \in \mathbb{C}$. Then $a \in \sigma(A) \cap \partial W(A)$ if and only if A is unitarily similar to $aI_k \oplus B$ such that $a \notin \sigma(B) \cup \text{int}(W(B))$.
4. Let $A \in \mathbb{C}^{n\times n}$ and $a \in \mathbb{C}$. Then a is a nondifferentiable boundary point of $W(A)$ if and only if A is unitarily similar to $aI_k \oplus B$ such that $a \notin W(B)$. In particular, a is a reducing eigenvalue of A.
5. Let $A \in \mathbb{C}^{n\times n}$. If $W(A)$ has at least $n-1$ nondifferentiable boundary points or if at least $n-1$ eigenvalues of A (counting multiplicities) lie in $\partial W(A)$, then A is normal.

Examples:

1. Let $A = [1] \oplus \begin{bmatrix} 0 & 2 \\ 0 & 0 \end{bmatrix}$. Then $W(A)$ is the unit disk centered at the origin, and 1 is a reducing eigenvalue of A lying on the boundary of $W(A)$.
2. Let $A = [2] \oplus \begin{bmatrix} 0 & 2 \\ 0 & 0 \end{bmatrix}$. Then $W(A)$ is the convex hull of unit disk centered at the origin of the number 2, and 2 is a nondifferentiable boundary point of $W(A)$.

Applications:

1. By Fact 1, if $A \in \mathbb{C}^{n\times n}$ and $0 \notin W(A)$, then $0 \notin \sigma(A)$ and, thus, A is invertible.
2. By Fact 4, if $A \in \mathbb{C}^{n\times n}$, then $W(A)$ has at most n nondifferentiable boundary points.
3. While $W(A)$ does not give a very tight containment region for $\sigma(A)$ as shown in the examples in the last section. Fact 2 shows that the numerical range can be used to estimated the spectrum of the resulting matrix when A is under a perturbation E. In contrast, $\sigma(A)$ and $\sigma(E)$ usually do not carry much information about $\sigma(A+E)$ in general. For example, let $A = \begin{bmatrix} 0 & M \\ 0 & 0 \end{bmatrix}$ and $E = \begin{bmatrix} 0 & 0 \\ \varepsilon & 0 \end{bmatrix}$. Then $\sigma(A) = \sigma(E) = \{0\}$, $\sigma(A+E) = \{\pm\sqrt{M\varepsilon}\} \subseteq W(A+E)$, which is the elliptical disk with foci $\pm\sqrt{M\varepsilon}$ and length of minor axis equal to $||M| - |\varepsilon||$.

18.3 Location of the Numerical Range

Facts:

The following facts can be found in [HJ91].

1. Let $A \in \mathbb{C}^{n\times n}$ and $t \in [0, 2\pi)$. Suppose $\mathbf{x}_t \in \mathbb{C}^n$ is a unit eigenvector corresponding to the largest eigenvalue $\lambda_1(t)$ of $e^{it}A + e^{-it}A^*$, and

$$\mathcal{P}_t = \{a \in \mathbb{C} : e^{it}a + e^{-it}\bar{a} \leq \lambda_1(t)\}.$$

Then

$$e^{it}W(A) \subseteq \mathcal{P}_t, \quad \lambda_t = \mathbf{x}_t^* A \mathbf{x}_t \in \partial W(A) \cap \partial \mathcal{P}_t$$

and

$$W(A) = \cap_{r\in[0,2\pi)} e^{-ir}\mathcal{P}_r = \text{conv}\{\lambda_r : r \in [0, 2\pi)\}.$$

If $T = \{t_1, \ldots, t_k\}$ with $0 \leq t_1 < \cdots < t_k < 2\pi$ and $k > 2$ such that $t_k - t_1 > \pi$, then

$$P_T^O(A) = \cap_{r\in T} e^{-ir} P_r \quad \text{and} \quad P_T^I(A) = \text{conv}\{\lambda_r : r \in T\}$$

are two polygons in $\mathbb{C}$ such that

$$P_T^I(A) \subseteq W(A) \subseteq P_T^O(A).$$

Moreover, both the area $W(A)\setminus P_T^I(A)$ and the area of $P_T^O(A)\setminus W(A)$ converge to 0 as $\max\{t_j - t_{j-1} : 1 \leq j \leq k+1\}$ converges to 0, where $t_0 = 0, t_{k+1} = 2\pi$.

2. Let $A = (a_{ij}) \in \mathbb{C}^{n\times n}$. For each $j = 1, \ldots, n$, let

$$g_j = \sum_{i\neq j}(|a_{ij}| + |a_{ji}|)/2 \quad \text{and} \quad G_j(A) = \{a \in \mathbb{C} : |a - a_{jj}| \leq g_j\}.$$

Then

$$W(A) \subseteq \text{conv} \cup_{j=1}^n G_j(A).$$

Examples:

1. Let $A = \begin{bmatrix} 2 & 2 \\ 0 & 2 \end{bmatrix}$. Then $W(A)$ is the circular disk centered at 2 with radius 1 In Figure 18.2, $W(A)$ is approximated by $P_T^O(A)$ with $T = \{2k\pi/100 : 0 \leq k \leq 99\}$. If $T = \{0, \pi/2, \pi, 3\pi/2\}$, then the polygon $P_T^O(A)$ in Fact 1 is bounded by the four lines $\{3 + bi : b \in \mathbb{R}\}$, $\{a + i : a \in \mathbb{R}\}$, $\{1 + bi : b \in \mathbb{R}\}$, $\{a - i : a \in \mathbb{R}\}$, and the polygon $P_T^I(A)$ equals the convex hull of $\{2, 1+i, 0, 1-i\}$.
2. Let $A = \begin{bmatrix} 5i & 2 & 3 \\ 4 & -3i & -2 \\ 1 & 3 & 9 \end{bmatrix}$. In Figure 18.3, $W(A)$ is approximated by $P_T^O(A)$ with $T = \{2k\pi/100 : 0 \leq k \leq 99\}$. By Fact 2, $W(A)$ lies in the convex hull of the circles $G_1 = \{a \in \mathbb{C} : |a - 5i| \leq 5\}$, $G_2 = \{a \in \mathbb{C} : |a + 3i| \leq 5.5\}$, $G_3 = \{a \in \mathbb{C} : |a - 9| \leq 4.5\}$.

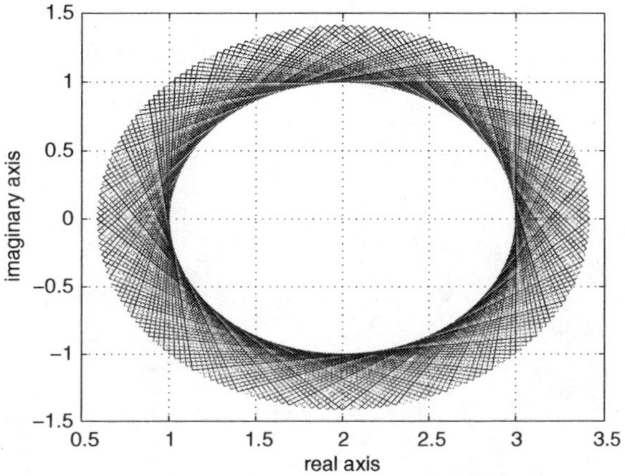

FIGURE 18.2 Numerical range of the matrix A in Example 1.

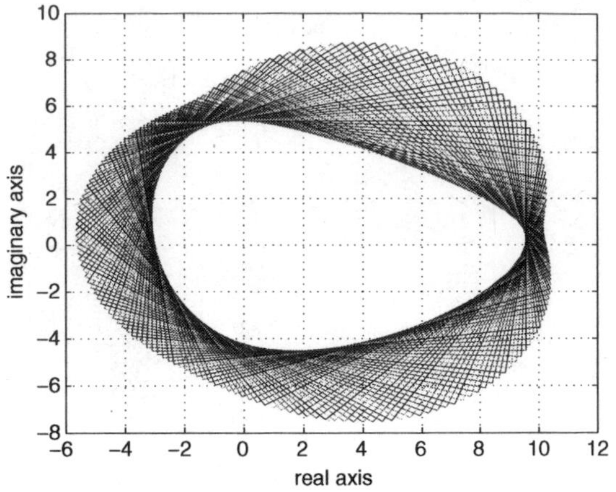

FIGURE 18.3 Numerical range of the matrix A in Example 2.

Applications:

1. Let $A = H + iG$, where $H, G \in \mathbb{C}^{n\times n}$ are Hermitian. Then

$$W(A) \subseteq W(H) + iW(G) = \{a + ib : a \in W(H),\ b \in W(G)\},$$

which is $P_T^O(A)$ for $T = \{0, \pi/2, \pi, 3\pi/2\}$.

2. Let $A = H + iG$, where $H, G \in \mathbb{C}^{n\times n}$ are Hermitian. Denote by $\lambda_1(X) \geq \cdots \geq \lambda_n(X)$ for a Hermitian matrix $X \in \mathbb{C}^{n\times n}$. By Fact 1,

$$w(A) = \max\{\lambda_1(\cos tH + \sin tG) : t \in [0, 2\pi)\}.$$

If $0 \notin W(A)$, then

$$\widetilde{w}(A) = \max\{\{\lambda_n(\cos tH + \sin tG) : t \in [0, 2\pi)\} \cup \{0\}\}.$$

3. By Fact 2, if $A = (a_{ij}) \in \mathbb{C}^{n\times n}$, then

$$w(A) \leq \max\{|a_{jj}| + g_j : 1 \leq j \leq n\}.$$

In particular, if A is nonnegative, then $w(A) = \lambda_1(A + A^T)/2$.

18.4 Numerical Radius

Definitions:

Let N be a vector norm on $\mathbb{C}^{n\times n}$. It is **submultiplicative** if

$$N(AB) \leq N(A)N(B) \quad \text{for all} \quad A, B \in \mathbb{C}^{n\times n}.$$

It is **unitarily invariant** if

$$N(UAV) = N(A) \quad \text{for all} \quad A \in \mathbb{C}^{n\times n} \text{ and unitary } U, V \in \mathbb{C}^{n\times n}.$$

It is **unitary similarity invariant** (also known as **weakly unitarily invariant**) if

$$N(U^*AU) = N(A) \quad \text{for all} \quad A \in \mathbb{C}^{n\times n} \text{ and unitary } U \in \mathbb{C}^{n\times n}.$$

Facts:

The following facts can be found in [GR96] and [HJ91].

1. The numerical radius $w(\cdot)$ is a unitary similarity invariant vector norm on $\mathbb{C}^{n\times n}$, and it is not unitarily invariant.
2. For any $A \in \mathbb{C}^{n\times n}$, we have

$$\rho(A) \leq w(A) \leq \|A\|_2 \leq 2w(A).$$

3. Suppose $A \in \mathbb{C}^{n\times n}$ is nonzero and the minimal polynomial of A has degree m. The following conditions are equivalent.
   (a) $\rho(A) = w(A)$.
   (b) There exists $k \geq 1$ such that A is unitarily similar to $\gamma U \oplus B$ for a unitary $U \in \mathbb{C}^{k\times k}$ and $B \in \mathbb{C}^{(n-k)\times(n-k)}$ with $w(B) \leq w(A) = \gamma$.
   (c) There exists $s \geq m$ such that $w(A^s) = w(A)^s$.
4. Suppose $A \in \mathbb{C}^{n\times n}$ is nonzero and the minimal polynomial of A has degree m. The following conditions are equivalent.
   (a) $\rho(A) = \|A\|_2$.
   (b) $w(A) = \|A\|_2$.
   (c) There exists $k \geq 1$ such that A is unitarily similar to $\gamma U \oplus B$ for a unitary $U \in \mathbb{C}^{k\times k}$ and a $B \in \mathbb{C}^{(n-k)\times(n-k)}$ with $\|B\|_2 \leq \|A\|_2 = \gamma$.
   (d) There exists $s \geq m$ such that $\|A^s\|_2 = \|A\|_2^s$.
5. Suppose $A \in \mathbb{C}^{n\times n}$ is nonzero. The following conditions are equivalent.
   (a) $\|A\|_2 = 2w(A)$.
   (b) $W(A)$ is a circular disk centered at origin with radius $\|A\|_2/2$.
   (c) $A/\|A\|_2$ is unitarily similar to $A_1 \oplus A_2$ such that $A_1 = \begin{bmatrix} 0 & 2 \\ 0 & 0 \end{bmatrix}$ and $w(A_2) \leq 1$.
6. The vector norm $4w$ on $\mathbb{C}^{n\times n}$ is submultiplicative, i.e.,

$$4w(AB) \leq (4w(A))(4w(B)) \quad \text{for all} \quad A, B \in \mathbb{C}^{n\times n}.$$

   The equality holds if

$$X = Y^T = \begin{bmatrix} 0 & 2 \\ 0 & 0 \end{bmatrix}.$$

7. Let $A \in \mathbb{C}^{n\times n}$ and k be a positive integer. Then

$$w(A^k) \leq w(A)^k.$$

8. Let N be a unitary similarity invariant vector norm on $\mathbb{C}^{n\times n}$ such that $N(A^k) \leq N(A)^k$ for any $A \in \mathbb{C}^{n\times n}$ and positive integer k. Then

$$w(A) \leq N(A) \quad \text{for all} \quad A \in \mathbb{C}^{n\times n}.$$

9. Suppose N is a unitarily invariant vector norm on $\mathbb{C}^{n\times n}$. Let

$$D = \begin{cases} 2I_k \oplus 0_k & \text{if } n = 2k, \\ 2I_k \oplus I_1 \oplus 0_k & \text{if } n = 2k+1. \end{cases}$$

Then $a = N(E_{11})$ and $b = N(D)$ are the best (largest and smallest) constants such that

$$aw(A) \leq N(A) \leq bw(A) \quad \text{for all} \quad A \in \mathbb{C}^{n\times n}.$$

10. Let $A \in \mathbb{C}^{n\times n}$. The following are equivalent:
    (a) $w(A) \leq 1$.
    (b) $\lambda_1(e^{it}A + e^{-it}A^*)/2 \leq 1$ for all $t \in [0, 2\pi)$.
    (c) There is $Z \in \mathbb{C}^{n\times n}$ such that $\begin{bmatrix} I_n + Z & A \\ A^* & I_n - Z \end{bmatrix}$ is positive semidefinite.
    (d) There exists $X \in \mathbb{C}^{2n\times n}$ satisfying $X^*X = I_n$ and

$$A = X^* \begin{bmatrix} 0_n & 2I_n \\ 0_n & 0_n \end{bmatrix} X.$$

18.5 Products of Matrices

Facts:

The following facts can be found in [GR96] and [HJ91].

1. Let $A, B \in \mathbb{C}^{n\times n}$ be such that $\widetilde{w}(A) > 0$. Then

$$\sigma(A^{-1}B) \subseteq \{b/a : a \in W(A), b \in W(B)\}.$$

2. Let $0 \leq t_1 < t_2 < t_1 + \pi$ and $S = \{re^{it} : r > 0,\ t \in [t_1, t_2]\}$. Then $\sigma(A) \subseteq S$ if and only if there is a positive definite $B \in \mathbb{C}^{n\times n}$ such that $W(AB) \subseteq S$.
3. Let $A, B \in \mathbb{C}^{n\times n}$.
    (a) If $AB = BA$, then $w(AB) \leq 2w(A)w(B)$.
    (b) If A or B is normal such that $AB = BA$, then $w(AB) \leq w(A)w(B)$.
    (c) If $A^2 = aI$ and $AB = BA$, then $w(AB) \leq \|A\|_2 w(B)$.
    (d) If $AB = BA$ and $AB^* = B^*A$, then $w(AB) \leq \min\{w(A)\|B\|_2, \|A\|_2 w(B)\}$.
4. Let A and B be square matrices such that A or B is normal. Then

$$W(A \circ B) \subseteq W(A \otimes B) = \text{conv}\,\{W(A)W(B)\}.$$

Consequently,

$$w(A \circ B) \leq w(A \otimes B) = w(A)w(B).$$

(See Chapter 8.5 and 10.4 for the definitions of t $A \circ B$ and $A \otimes B$.)
5. Let A and B be square matrices. Then

$$w(A \circ B) \leq w(A \otimes B) \leq \min\{w(A)\|B\|_2, \|A\|_2 w(B)\} \leq 2w(A)w(B).$$

6. Let $A \in \mathbb{C}^{n\times n}$. Then

$$w(A \circ X) \leq w(X) \quad \text{for all} \quad X \in \mathbb{C}^{n\times n}$$

if and only if $A = B^*WB$ such that W satisfies $\|W\| \leq 1$ and all diagonal entries of B^*B are bounded by 1.

Examples:

1. Let $A \in \mathbb{C}^{9\times 9}$ be the Jordan block of zero $J_9(0)$, and $B = A^3 + A^7$. Then $w(A) = w(B) = \cos(\pi/10) < 1$ and $w(AB) = 1 > \|A\|_2 w(B)$. So, even if $AB = BA$, we may not have $w(AB) \le \min\{w(A)\|B\|_2, \|A\|_2 w(B)\}$.

2. Let $A = \begin{bmatrix} 1 & 1 \\ 0 & 1 \end{bmatrix}$. Then $W(A) = \{a \in \mathbb{C} : |a-1| \le 1/2\}$ and

$$W(A^2) = \{a \in \mathbb{C} : |a-1| \le 1\},$$

whereas

$$\text{conv } W(A)^2 \subseteq \{se^{it} \in \mathbb{C} : s \in [0.25, 2.25],\ t \in [-\pi/3, \pi/3]\}.$$

So, $W(A^2) \not\subseteq \text{conv } W(A)^2$.

3. Let $A = \begin{bmatrix} 1 & 0 \\ 0 & -1 \end{bmatrix}$ and $B = \begin{bmatrix} 0 & 1 \\ 1 & 0 \end{bmatrix}$. Then $\sigma(AB) = \{i, -i\}$, $W(AB) = i[-1,1]$, and $W(A) = W(B) = W(A)W(B) = [-1,1]$. So, $\sigma(AB) \not\subseteq \text{conv } W(A)W(B)$.

Applications:

1. If $C \in \mathbb{C}^{n\times n}$ is positive definite, then $W(C^{-1}) = W(C)^{-1} = \{c^{-1} : c \in W(C)\}$. Applying Fact 1 with $A = C^{-1}$,

$$\sigma(CB) \subseteq W(C)W(B).$$

2. If $C \in \mathbb{C}^{n\times n}$ satisfies $\widetilde{w}(C) > 0$, then for every unit vector $\mathbf{x} \in \mathbb{C}^n$ $\mathbf{x}^* C^{-1}\mathbf{x} = \mathbf{y}^* C^* C^{-1} C\mathbf{y}$ with $\mathbf{y} = C^{-1}\mathbf{x}$ and, hence,

$$W(C^{-1}) \subseteq \{rb : r \ge 0, \quad b \in W(C^*)\} = \{r\overline{b} : r \ge 0, \quad b \in W(C)\}.$$

Applying this observation and Fact 1 with $A = C^{-1}$, we have

$$\sigma(AB) \subseteq \{rab : r \ge 0, \quad a \in W(A), \quad b \in W(B)\}.$$

18.6 Dilations and Norm Estimation

Definitions:

A matrix $A \in \mathbb{C}^{n\times n}$ has a **dilation** $B \in \mathbb{C}^{m\times m}$ if there is $X \in \mathbb{C}^{m\times n}$ such that $X^*X = I_n$ and $X^*BX = A$.

A matrix $A \in \mathbb{C}^{n\times n}$ is a **contraction** if $\|A\|_2 \le 1$.

Facts:

The following facts can be found in [CL00],[CL01] and their references.

1. A has a dilation B if and only if B is unitarily similar to a matrix of the form

$$\begin{bmatrix} A & * \\ * & * \end{bmatrix}.$$

2. Suppose $B \in \mathbb{C}^{3\times 3}$ has a reducing eigenvalue, or $B \in \mathbb{C}^{2\times 2}$. If $W(A) \subseteq W(B)$, then A has a dilation of the form $B \otimes I_m$.

3. Let $r \in [-1, 1]$. Suppose $A \in \mathbb{C}^{n\times n}$ is a contraction with

$$W(A) \subseteq \mathcal{S} = \{a \in \mathbb{C} : a + \bar{a} \leq 2r\}.$$

Then A has a unitary dilation $U \in \mathbb{C}^{2n\times 2n}$ such that $W(U) \subseteq \mathcal{S}$.

4. Let $A \in \mathbb{C}^{n\times n}$. Then

$$W(A) = \cap\{W(B) : B \in \mathbb{C}^{2n\times 2n} \text{ is a normal dilation of } A\}.$$

If A is a contraction, then

$$W(A) = \cap\{W(U) : U \in \mathbb{C}^{2n\times 2n} \text{ is a unitary dilation of } A\}.$$

5. Let $A \in \mathbb{C}^{n\times n}$.

   (a) If $W(A)$ lies in an triangle with vertices z_1, z_2, z_3, then

$$\|A\|_2 \leq \max\{|z_1|, |z_2|, |z_3|\}.$$

   (b) If $W(A)$ lies in an ellipse $\mathcal{E}$ with foci λ_1, λ_2, and minor axis of length b, then

$$\|A\|_2 \leq \{\sqrt{(|\lambda_1| + |\lambda_2|)^2 + b^2} + \sqrt{(|\lambda_1| - |\lambda_2|)^2 + b^2}\}/2.$$

   More generally, if $W(A)$ lies in the convex hull of the ellipse $\mathcal{E}$ and the point z_0, then

$$\|A\|_2 \leq \max\left\{|z_0|, \{\sqrt{(|\lambda_1| + |\lambda_2|)^2 + b^2} + \sqrt{(|\lambda_1| - |\lambda_2|)^2 + b^2}\}/2\right\}.$$

6. Let $A \in \mathbb{C}^{n\times n}$. Suppose there is $t \in [0, 2\pi)$ such that $e^{it}W(A)$ lies in a rectangle R centered at $z_0 \in \mathbb{C}$ with vertices $z_0 \pm \alpha \pm i\beta$ and $z_0 \pm \alpha \mp i\beta$, where $\alpha, \beta > 0$, so that $z_1 = z_0 + \alpha + i\beta$ has the largest magnitude. Then

$$\|A\|_2 \leq \begin{cases} |z_1| & \text{if } R \subseteq \text{conv}\,\{z_1, \bar{z}_1, -\bar{z}_1\}, \\ \alpha + \beta & \text{otherwise.} \end{cases}$$

   The bound in each case is attainable.

Examples:

1. Let $A = \begin{bmatrix} 0 & \sqrt{2} \\ 0 & 0 \end{bmatrix}$. Suppose

$$B = \begin{bmatrix} 0 & 1 & 0 \\ 0 & 0 & 1 \\ 0 & 0 & 0 \end{bmatrix} \quad \text{or} \quad B = \begin{bmatrix} 1 & 0 & 0 & 0 \\ 0 & i & 0 & 0 \\ 0 & 0 & -1 & 0 \\ 0 & 0 & 0 & -i \end{bmatrix}.$$

   Then $W(A) \subseteq W(B)$. However, A does not have a dilation of the form $B \otimes I_m$ for either of the matrices because

$$\|A\|_2 = \sqrt{2} > 1 = \|B\|_2 = \|B \otimes I_m\|_2.$$

   So, there is no hope to further extend Fact 1 in this section to arbitrary $B \in \mathbb{C}^{3\times 3}$ or normal matrix $B \in \mathbb{C}^{4\times 4}$.

18.7 Mappings on Matrices

Definitions:

Let $\phi : \mathbb{C}^{n\times n} \to \mathbb{C}^{m\times m}$ be a linear map. It is **unital** if $\phi(I_n) = I_m$; it is **positive** if $\phi(A)$ is positive semidefinite whenever A is positive semidefinite.

Facts:

The following facts can be found in [GR96] unless another reference is given.

1. [HJ91] Let $\mathcal{P}(\mathbb{C})$ be the set of subsets of $\mathbb{C}$. Suppose a function $F : \mathbb{C}^{n\times n} \to \mathcal{P}(\mathbb{C})$ satisfies the following three conditions.

   (a) $F(A)$ is compact and convex for every $A \in \mathbb{C}^{n\times n}$.

   (b) $F(aA + bI) = aF(A) + b$ for any $a, b \in \mathbb{C}$ and $A \in \mathbb{C}^{n\times n}$.

   (c) $F(A) \subseteq \{a \in \mathbb{C} : a + \bar{a} \geq 0\}$ if and only if $A + A^*$ is positive semidefinite.

   Then $F(A) = W(A)$ for all $A \in \mathbb{C}^{n\times n}$.
2. Use the usual topology on $\mathbb{C}^{n\times n}$ and the Hausdorff metric on two compact sets A, B of C defined by

$$d(\mathcal{A}, \mathcal{B}) = \max\left\{\max_{a\in\mathcal{A}} \min_{b\in\mathcal{B}} |a - b|, \max_{b\in\mathcal{B}} \min_{a\in\mathcal{A}} |a - b|\right\}$$

   The mapping $A \mapsto W(A)$ is continuous.
3. Suppose $f(x+iy) = (ax+by+c)+i(dx+ey+f)$ for some real numbers a, b, c, d, e, f. Define $f(H+iG) = (aH+bG+cI)+i(dH+eG+fI)$ for any two Hermitian matrices $H, G \in \mathbb{C}^{n\times n}$. We have

$$W(f(H+iG)) = f(W(A)) = \{f(x+iy) : x+iy \in W(A)\}.$$

4. Let $\mathbf{D} = \{a \in \mathbb{C} : |a| \leq 1\}$. Suppose $f : \mathbf{D} \to \mathbb{C}$ is analytic in the interior of $\mathbf{D}$ and continuous on the boundary of $\mathbf{D}$.

   (a) If $f(\mathbf{D}) \subseteq \mathbf{D}$ and $f(0) = 0$, then $W(f(A)) \subseteq \mathbf{D}$ whenever $W(A) \subseteq \mathbf{D}$.

   (b) If $f(\mathbf{D}) \subseteq \mathbb{C}_+ = \{a \in \mathbb{C} : a + \bar{a} \geq 0\}$, then $W(f(A)) \subseteq \mathbb{C}_+ \setminus \{(f(0) + \overline{f(0)})/2\}$ whenever $W(A) \subseteq \mathbf{D}$.
5. Suppose $\phi : \mathbb{C}^{n\times n} \to \mathbb{C}^{n\times n}$ is a unital positive linear map. Then $W(\phi(A)) \subseteq W(A)$ for all $A \in \mathbb{C}^{n\times n}$.
6. [Pel75] Let $\phi : \mathbb{C}^{n\times n} \to \mathbb{C}^{n\times n}$ be linear. Then

$$W(A) = W(\phi(A)) \quad \text{for all} \quad A \in \mathbb{C}^{n\times n}$$

   if and only if there is a unitary $U \in \mathbb{C}^{n\times n}$ such that ϕ has the form

$$X \mapsto U^*XU \quad \text{or} \quad X \mapsto U^*X^TU.$$

7. [Li87] Let $\phi : \mathbb{C}^{n\times n} \to \mathbb{C}^{n\times n}$ be linear. Then $w(A) = w(\phi(A))$ for all $A \in \mathbb{C}^{n\times n}$ if and only if there exist a unitary $U \in \mathbb{C}^{n\times n}$ and a complex unit μ such that ϕ has the form

$$X \mapsto \mu U^*XU \quad \text{or} \quad X \mapsto \mu U^*X^TU.$$

References

[CL00] M.D. Choi and C.K. Li, Numerical ranges and dilations, *Lin. Multilin. Alg.* 47 (2000), 35–48.

[CL01] M.D. Choi and C.K. Li, Constrained unitary dilations and numerical ranges, *J. Oper. Theory* 46 (2001), 435–447.

[GR96] K.E. Gustafson and D.K.M. Rao, *Numerical Range: the Field of Values of Linear Operators and Matrices*, Springer, New York, 1996.

[Hal82] P.R. Halmos, *A Hilbert Space Problem Book*, 2nd ed., Springer-Verlag, New York, 1982.

[HJ91]R.A. Horn and C.R. Johnson, *Topics in Matrix Analysis*, Cambridge University Press, New York, 1991.

[Li87]C.K. Li, Linear operators preserving the numerical radius of matrices, *Proc. Amer. Math. Soc.* 99 (1987), 105–118.

[Pel75] V. Pellegrini, Numerical range preserving operators on matrix algebras, *Studia Math.* 54 (1975), 143–147.

19

Matrix Stability and Inertia

Daniel Hershkowitz
Technion - Israel Institute of Technology

Much is known about spectral properties of square (complex) matrices. There are extensive studies of eigenvalues of matrices in certain classes. Some of the studies concentrate on the inertia of the matrices, that is, distribution of the eigenvalues in half-planes.

A special inertia case is of stable matrices, that is, matrices whose spectrum lies in the open left or right half-plane. These, and other related types of matrix stability, play an important role in various applications. For this reason, matrix stability has been intensively investigated in the past two centuries.

A. M. Lyapunov, called by F. R. Gantmacher "the founder of the modern theory of stability," studied the asymptotic stability of solutions of differential systems. In 1892, he proved a theorem that was restated (first, apparently, by Gantmacher in 1953) as a necessary and sufficient condition for stability of a matrix. In 1875, E. J. Routh introduced an algorithm that provides a criterion for stability. An independent solution was given by A. Hurwitz. This solution is known nowadays as the Routh–Hurwitz criterion for stability. Another criterion for stability, which has a computational advantage over the Routh–Hurwitz criterion, was proved in 1914 by Liénard and Chipart. The equivalent of the Routh–Hurwitz and Liénard–Chipart criteria was observed by M. Fujiwara. The related problem of requiring the eigenvalues to be within the unit circle was solved separately in the early 1900s by I. Schur and Cohn. The above-mentioned studies have motivated an intensive search for conditions for matrix stability.

An interesting question, related to stability, is the following one: Given a square matrix A, can we find a diagonal matrix D such that the matrix DA is stable? This question can be asked in full generality, as suggested above, or with some restrictions on the matrix D, such as positivity of the diagonal elements. A related problem is characterizing matrices A such that for every positive diagonal matrix D, the matrix DA is stable. Such matrices are called multiplicative D-stable matrices. This type of matrix stability, as well as two other related types, namely additive D-stability and Lyapunov diagonal (semi)stability, have important applications in many disciplines. Thus, they are very important to characterize. While regular stability is a spectral property (it is always possible to check whether a given matrix is stable or not by evaluating its eigenvalues), none of the other three types of matrix stability can be characterized by the spectrum of the matrix. This problem has been solved for certain classes of matrices. For example, for Z-matrices all the stability types are equivalent. Another case in which these characterization problems have been solved is the case of acyclic matrices.

Several surveys handle the above-mentioned types of matrix stability, e.g., the books [HJ91] and [KB00], and the articles [Her92], [Her98], and [BH85]. Finally, the mathematical literature has studies of other types of matrix stability, e.g., the above-mentioned Schur–Cohn stability (where all the eigenvalues lie within the unit circle), e.g., [Sch17] and [Zah92]; H-stability, e.g., [OS62], [Car68], and [HM98]; L_2-stability and strict H-stability, e.g., [Tad81]; and scalar stability, e.g., [HM98].

19.1 Inertia

Much is known about spectral properties of square matrices. In this chapter, we concentrate on the distribution of the eigenvalues in half-planes. In particular, we refer to results that involve the expression $AH + HA^*$, where A is a square complex matrix and H is a Hermitian matrix.

Definitions:

For a square complex matrix A, we denote by $\pi(A)$ the number of eigenvalues of A with positive real part, by $\delta(A)$ the number of eigenvalues of A on the imaginary axis, and by $\nu(A)$ the number of eigenvalues of A with negative real part. The **inertia** of A is defined as the triple $\text{in}(A) = (\pi(A), \nu(A), \delta(A))$.

Facts:

All the facts are proven in [OS62].

1. Let A be a complex square matrix. There exists a Hermitian matrix H such that the matrix $AH + HA^*$ is positive definite if and only if $\delta(A) = 0$. Furthermore, in such a case the inertias of A and H are the same.
2. Let $\{\lambda_1, \dots, \lambda_n\}$ be the eigenvalues of an $n \times n$ matrix A. If $\prod_{i,j=1}^{n}(\lambda_i + \overline{\lambda_j}) \neq 0$, then for any positive definite matrix P there exists a unique Hermitian matrix H such that $AH + HA^* = P$. Furthermore, the inertias of A and H are the same.
3. Let A be a complex square matrix. We have $\delta(A) = \pi(A) = 0$ if and only if there exists an $n \times n$ positive definite Hermitian matrix such that the matrix $-(AH + HA^*)$ is positive definite.

Examples:

1. It follows from Fact 1 above that a complex square matrix A has all of its eigenvalues in the right half-plane if and only if there exists a positive definite matrix H such that the matrix $AH + HA^*$ is positive definite. This fact, associating us with the discussion of the next section, is due to Lyapunov, originally proven in [L1892] for systems of differential equations. The matrix formulation is due to [Gan60].
2. In order to demonstrate that both the existence and uniqueness claims of Fact 2 may be false without the condition on the eigenvalues, consider the matrix

$$A = \begin{bmatrix} 1 & 0 \\ 0 & -1 \end{bmatrix},$$

for which the condition of Fact 2 is not satisfied. One can check that the only positive definite matrices P for which the equation $AH + HA^* = P$ has Hermitian solutions are matrices of the type $P = \begin{bmatrix} p_{11} & 0 \\ 0 & p_{22} \end{bmatrix}$, $p_{11}, p_{22} > 0$. Furthermore, for $P = \begin{bmatrix} 2 & 0 \\ 0 & 4 \end{bmatrix}$ it is easy to verify that the Hermitian solutions of $AH + HA^* = P$ are all matrices H of the type

$$\begin{bmatrix} 1 & c \\ \bar{c} & -2 \end{bmatrix}, \quad c \in \mathbb{C}.$$

If we now choose

$$A = \begin{bmatrix} 1 & 0 \\ 0 & -2 \end{bmatrix},$$

then here the condition of Fact 2 is satisfied. Indeed, for $H = \begin{bmatrix} a & c \\ \bar{c} & b \end{bmatrix}$ we have

$$AH + HA^* = \begin{bmatrix} 2a & -c \\ -\bar{c} & -4b \end{bmatrix},$$

which can clearly be solved uniquely for any Hermitian matrix P; specifically, for $P = \begin{bmatrix} 2 & 0 \\ 0 & 4 \end{bmatrix}$, the unique Hermitian solution H of $AH + HA^* = P$ is $\begin{bmatrix} 1 & 0 \\ 0 & -1 \end{bmatrix}$.

19.2 Stability

Definitions:

A complex polynomial is **negative stable [positive stable]** if its roots lie in the open left [right] half-plane. A complex square matrix A is **negative stable [positive stable]** if its characteristic polynomial is negative stable [positive stable].

We shall use the term **stable matrix** for positive stable matrix.

For an $n \times n$ matrix A and for an integer k, $1 \leq k \leq n$, we denote by $S_k(A)$ the sum of all principal minors of A of order k.

The **Routh–Hurwitz matrix** associated with A is defined to be the matrix

$$\begin{bmatrix}
S_1(A) & S_3(A) & S_5(A) & \cdot & \cdot & \cdot & \cdot & 0 & 0 \\
1 & S_2(A) & S_4(A) & & & & & \cdot & \cdot \\
0 & S_1(A) & S_3(A) & \cdot & & & & \cdot & \cdot \\
0 & 1 & S_2(A) & \cdot & \cdot & & & \cdot & \cdot \\
0 & 0 & S_1(A) & \cdot & \cdot & \cdot & & \cdot & \cdot \\
\cdot & \cdot & \cdot & & \cdot & \cdot & \cdot & 0 & 0 \\
\cdot & \cdot & \cdot & & & \cdot & \cdot & S_n(A) & 0 \\
\cdot & \cdot & \cdot & & & & \cdot & S_{n-1}(A) & 0 \\
0 & 0 & 0 & \cdot & \cdot & \cdot & \cdot & S_{n-2}(A) & S_n(A)
\end{bmatrix}.$$

A square complex matrix is a P**-matrix** if it has positive principal minors.

A square complex matrix is a P_0^+**-matrix** if it has nonnegative principal minors and at least one principal minor of each order is positive.

A principal minor of a square matrix is a **leading principal minor** if it is based on consecutive rows and columns, starting with the first row and column of the matrix.

An $n \times n$ real matrix A is **sign symmetric** if it satisfies

$$\det A[\alpha,\beta] \det A[\beta,\alpha] \geq 0, \qquad \forall \alpha,\beta \subseteq \{1,\ldots,n\},\ |\alpha| = |\beta|.$$

An $n \times n$ real matrix A is **weakly sign symmetric** if it satisfies

$$\det A[\alpha,\beta] \det A[\beta,\alpha] \geq 0, \qquad \forall \alpha,\beta \subseteq \{1,\ldots,n\},\ |\alpha| = |\beta| = |\alpha \cap \beta| + 1.$$

A square real matrix is a Z**-matrix** if it has nonpositive off-diagonal elements.

A Z-matrix with positive principal minors is an **M-matrix**. (See Section 24.5 for more information and an equivalent definition.)

Facts:

Lyapunov studied the asymptotic stability of solutions of differential systems. In 1892 he proved in his paper [L1892] a theorem which yields a necessary and sufficient condition for stability of a complex matrix. The matrix formulation of Lyapunov's Theorem is apparently due to Gantmacher [Gan60], and is given as Fact 1 below. The theorem in [Gan60] was proven for real matrices; however, as was also remarked in [Gan60], the generalization to the complex case is immediate.

1. **The Lyapunov Stability Criterion:** A complex square matrix A is stable if and only if there exists a positive definite Hermitian matrix H such that the matrix $AH + HA^*$ is positive definite.
2. [OS62] A complex square matrix A is stable if and only if for every positive definite matrix G there exists a positive definite matrix H such that the matrix $AH + HA^* = G$.
3. [R1877], [H1895] **The Routh–Hurwitz Stability Criterion:** An $n \times n$ complex matrix A with a real characteristic polynomial is stable if and only if the leading principal minors of the Routh–Hurwitz matrix associated with A are all positive.
4. [LC14] (see also [Fuj26]) **The Liénard–Chipart Stability Criterion:** Let A be an $n \times n$ complex matrix with a real characteristic polynomial. The following are equivalent:
   (a) A is stable.
   (b) $S_n(A), S_{n-2}(A), \ldots > 0$ and the odd order leading principal minors of the Routh–Hurwitz matrix associated with A are positive.
   (c) $S_n(A), S_{n-2}(A), \ldots > 0$ and the even order leading principal minors of the Routh–Hurwitz matrix associated with A are positive.
   (d) $S_n(A), S_{n-1}(A), S_{n-3}(A), \ldots > 0$ and the odd order leading principal minors of the Routh–Hurwitz matrix associated with A are positive.
   (e) $S_n(A), S_{n-1}(A), S_{n-3}(A), \ldots > 0$ and the even order leading principal minors of the Routh–Hurwitz matrix associated with A are positive.
5. [Car74] Sign symmetric P-matrices are stable.
6. [HK2003] Sign symmetric stable matrices are P-matrices.
7. [Hol99] Weakly sign symmetric P-matrices of order less than 6 are stable. Nevertheless, in general, weakly sign symetric P-matrices need not be stable.
8. (For example, [BVW78]) A Z-matrix is stable if and only if it is a P-matrix (that is, it is an M-matrix).
9. [FHR05] Let A be a stable real square matrix. Then either all the diagonal elements of A are positive or A has at least one positive diagonal element and one positive off-diagonal element.
10. [FHR05] Let ζ be an n-tuple of complex numbers, $n > 1$, consisting of real numbers and conjugate pairs. There exists a real stable $n \times n$ matrix A with exactly two positive entries such that ζ is the spectrum of A.

Examples:

1. Let

$$A = \begin{bmatrix} 2 & 2 & 3 \\ 2 & 5 & 4 \\ 3 & 4 & 5 \end{bmatrix}.$$

The Routh–Hurwitz matrix associated with A is

$$\begin{bmatrix} 12 & 1 & 0 \\ 1 & 16 & 0 \\ 0 & 12 & 1 \end{bmatrix}.$$

It is immediate to check that the latter matrix has positive leading principal minors. It, thus, follows that A is stable. Indeed, the eigenvalues of A are 1.4515, 0.0657, and 10.4828.

2. Stable matrices do not form a convex set, as is easily demonstrated by the stable matrices

$$\begin{bmatrix} 1 & 1 \\ 0 & 1 \end{bmatrix}, \quad \begin{bmatrix} 1 & 0 \\ 9 & 1 \end{bmatrix},$$

whose sum $\begin{bmatrix} 2 & 1 \\ 9 & 2 \end{bmatrix}$ has eigenvalues -1 and 5. Clearly, convex sets of stable matrices do exist. An example of such a set is the set of upper (or lower) triangular matrices with diagonal elements in the open right half-plane. Nevertheless, there is no obvious link between matrix stability and convexity or conic structure. Some interesting results on stable convex hulls can be found in [Bia85], [FB87], [FB88], [CL97], and [HS90]. See also the survey in [Her98].

3. In view of Facts 5 and 7 above, it would be natural to ask whether stability of a matrix implies that the matrix is a P-matrix or a weakly sign symmetric matrix. The answer to this question is negative as is demonstrated by the matrix

$$A = \begin{bmatrix} -1 & 1 \\ -5 & 3 \end{bmatrix}.$$

The eigenvalues of A are $1 \pm i$, and so A is stable. Nevertheless, A is neither a P-matrix nor a weakly sign symmetric matrix.

4. Sign symmetric P_0^+-matrices are not necessarily stable, as is demonstrated by the sign symmetric P_0^+-matrix

$$A = \begin{bmatrix} 1 & 0 & 0 & 0 & 0 \\ 0 & 1 & 0 & 0 & 0 \\ 0 & 0 & 0 & 1 & 0 \\ 0 & 0 & 0 & 0 & 1 \\ 0 & 0 & 1 & 0 & 0 \end{bmatrix}.$$

The matrix A is not stable, having the eigenvalues $\left\{e^{\pm\frac{2\pi i}{3}}, 1, 1, 1\right\}$.

5. A P-matrix is not necessarily stable as is demonstrated by the matrix

$$\begin{bmatrix} 1 & 0 & 3 \\ 3 & 1 & 0 \\ 0 & 3 & 1 \end{bmatrix}.$$

For extensive study of spectra of P-matrices look at [HB83], [Her83], [HJ86], [HS93], and [HK2003].

19.3 Multiplicative *D*-Stability

Multiplicative D-stability appears in various econometric models, for example, in the study of stability of multiple markets [Met45].

Definitions:

A real square matrix A is **multiplicative D-stable** if DA is stable for every positive diagonal matrix D.

In the literature, multiplicative D-stable matrices are usually referred to as just **D-stable** matrices.

A real square matrix A is **inertia preserving** if the inertia of AD is equal to the inertia of D for every nonsingular real diagonal matrix D.

The **graph** $G(A)$ of an $n \times n$ matrix A is the simple graph whose vertex set is $\{1, \ldots, n\}$, and where there is an edge between two vertices i and j $(i \neq j)$ if and only if $a_{ij} \neq 0$ or $a_{ji} \neq 0$. (See Chapter 28 more information on graphs.)

The matrix A is said to be **acyclic** if $G(A)$ is a forest.

Facts:

The problem of characterizing multiplicative D-stabity for certain classes and for matrices of order less than 5 is dealt with in several publications (e.g., [Cai76], [CDJ82], [Cro78], and [Joh74b]). However, in general, this problem is still open. Multiplicative D-stability is characterized in [BH84] for acyclic matrices. That result generalizes the handling of tridiagonal matrices in [CDJ82]. Characterization of multiplicative D-stability using cones is given in [HSh88]. See also the survey in [Her98].

1. Tridiagonal matrices are acyclic, since their graphs are paths or unions of disjoint paths.
2. [FF58] For a real square matrix A with positive leading principal minors there exists a positive diagonal matrix D such that DA is stable.
3. [Her92] For a complex square matrix A with positive leading principal minors there exists a positive diagonal matrix D such that DA is stable.
4. [Cro78] Multiplicative D-stable matrices are P_0^+-matrices.
5. [Cro78] A 2×2 real matrix is multiplicative D-stable if and only if it is a P_0^+-matrix.
6. [Cai76] A 3×3 real matrix A is multiplicative D-stable if and only if $A + D$ is multiplicative D-stable for every nonnegative diagonal matrix D.
7. [Joh75] A real square matrix A is multiplicative D-stable if and only if $A \pm iD$ is nonsingular for every positive diagonal matrix D.
8. (For example, [BVW78]) A Z-matrix is multiplicative D-stable if and only if it is a P-matrix (that is, it is an M-matrix).
9. [BS91] Inertia preserving matrices are multiplicative D-stable.
10. [BS91] An irreducible acyclic matrix is multiplicative D-stable if and only if it is inertia preserving.
11. [HK2003] Let A be a sign symmetric square matrix. The following are equivalent:

    (a) The matrix A is stable.

    (b) The matrix A has positive leading principal minors.

    (c) The matrix A is a P-matrix.

    (d) The matrix A is multiplicative D-stable.

    (e) There exists a positive diagonal matrix D such that the matrix DA is stable.

Examples:

1. In order to illustrate Fact 2, let

$$A = \begin{bmatrix} 1 & 1 & 1 \\ 0 & 1 & 1 \\ 4 & 1 & 2 \end{bmatrix}.$$

The matrix A is not stable, having the eigenvalues 4.0606 and $-0.0303 \pm 0.4953i$. Nevertheless, since A has positive leading minors, by Fact 2 there exists a positive diagonal matrix D such that the matrix DA is stable. Indeed, the eigenvalues of

$$\begin{bmatrix} 1 & 0 & 0 \\ 0 & 1 & 0 \\ 0 & 0 & 0.1 \end{bmatrix} \begin{bmatrix} 1 & 1 & 1 \\ 0 & 1 & 1 \\ 4 & 1 & 2 \end{bmatrix} = \begin{bmatrix} 1 & 1 & 1 \\ 0 & 1 & 1 \\ 0.4 & 0.1 & 0.2 \end{bmatrix}$$

are 1.7071, 0.2929, and 0.2.

2. In order to illustrate Fact 4, let

$$A = \begin{bmatrix} 1 & 1 & 0 \\ -1 & 0 & 1 \\ 0 & 1 & 2 \end{bmatrix}.$$

The matrix A is stable, having the eigenvalues $0.3376 \pm 0.5623i$ and 2.3247. Yet, we have det $A[\{2,3\}] < 0$, and so A is not a P_0^+-matrix. Indeed, observe that the matrix

$$\begin{bmatrix} 0.1 & 0 & 0 \\ 0 & 1 & 0 \\ 0 & 0 & 1 \end{bmatrix} \begin{bmatrix} 1 & 1 & 0 \\ -1 & 0 & 1 \\ 0 & 1 & 2 \end{bmatrix} = \begin{bmatrix} 0.1 & 0.1 & 0 \\ -1 & 0 & 1 \\ 0 & 1 & 2 \end{bmatrix}$$

is not stable, having the eigenvalues $-0.1540 \pm 0.1335i$ and 2.408.

3. While stability is a spectral property, and so it is always possible to check whether a given matrix is stable or not by evaluating its eigenvalues, multiplicative D-stability cannot be characterized by the spectrum of the matrix, as is demonstrated by the following two matrices

$$A = \begin{bmatrix} 1 & 0 \\ 0 & 2 \end{bmatrix}, \quad B = \begin{bmatrix} -1 & 2 \\ -3 & 4 \end{bmatrix}.$$

The matrices A and B have the same spectrum. Nevertheless, while A is multiplicative D-stable, B is not, since it is not a P_0^+-matrix. Indeed, the matrix

$$\begin{bmatrix} 5 & 0 \\ 0 & 1 \end{bmatrix} \begin{bmatrix} -1 & 2 \\ -3 & 4 \end{bmatrix} = \begin{bmatrix} -5 & 10 \\ -3 & 4 \end{bmatrix}$$

has eigenvalues $-0.5 \pm 3.1225i$.

4. It is shown in [BS91] that the converse of Fact 9 is not true, using the following example from [Har80]:

$$A = \begin{bmatrix} 1 & 0 & -50 \\ 1 & 1 & 0 \\ 1 & 1 & 1 \end{bmatrix}.$$

The matrix A is multiplicative D-stable (by the characterization of 3×3 multiplicative D-stable matrices, proven in [Cai76]). However, for $D = \text{diag}(-1, 3, -1)$ the matrix AD is stable and, hence, A is not inertia preserving. In fact, it is shown in [BS91] that even P-matrices that are both D-stable and Lyapunov diagonally semistable (see section 19.5) are not necessarily inertia preserving.

19.4 Additive D-Stability

Applications of additive D-stability may be found in linearized biological systems, e.g., [Had76].

Definitions:

A real square matrix A is said to be **additive D-stable** if $A + D$ is stable for every nonnegative diagonal matrix D.

In some references additive D-stable matrices are referred to as **strongly stable** matrices.

Facts:

The problem of characterizing additive D-stability for certain classes and for matrices of order less than 5 is dealt with in several publications (e.g., [Cai76], [CDJ82], [Cro78], and [Joh74b]). However, in general, this problem is still open. Additive D-stability is characterized in [Her86] for acyclic matrices. That result generalizes the handling of tridiagonal matrices in [Car84].

1. [Cro78] Additive D-stable matrices are P_0^+-matrices.
2. [Cro78] A 2×2 real matrix is additive D-stable if and only if it is a P_0^+-matrix.
3. [Cro78] A 3×3 real matrix A is additive D-stable if and only if it is a P_0^+-matrix and stable.
4. (For example, [BVW78]) A Z-matrix is additive D-stable if and only if it is a P-matrix (that is, it is an M-matrix).
5. An additive D-stable matrix need not be multiplicative D-stable (cf. Example 3).
6. [Tog80] A multiplicative D-stable matrix need not be additive D-stable.

Examples:

1. In order to illustrate Fact 1, let

$$A = \begin{bmatrix} 1 & 1 & 0 \\ -1 & 0 & 1 \\ 0 & 1 & 2 \end{bmatrix}.$$

The matrix A is stable, having the eigenvalues $0.3376 \pm 0.5623i$ and 2.3247. Yet, we have $\det A[2,3|2,3] < 0$, and so A is not a P_0^+-matrix. Indeed, observe that the matrix

$$\begin{bmatrix} 1 & 1 & 0 \\ -1 & 0 & 1 \\ 0 & 1 & 2 \end{bmatrix} + \begin{bmatrix} 2 & 0 & 0 \\ 0 & 0 & 0 \\ 0 & 0 & 0 \end{bmatrix} = \begin{bmatrix} 3 & 1 & 0 \\ -1 & 0 & 1 \\ 0 & 1 & 2 \end{bmatrix}$$

is not stable, having the eigenvalues $2.5739 \pm 0.3690i$ and -0.1479.

2. While stability is a spectral property, and so it is always possible to check whether a given matrix is stable or not by evaluating its eigenvalues, additive D-stability cannot be characterized by the spectrum of the matrix, as is demonstrated by the following two matrices:

$$A = \begin{bmatrix} 1 & 0 \\ 0 & 2 \end{bmatrix}, \quad B = \begin{bmatrix} -1 & 2 \\ -3 & 4 \end{bmatrix}.$$

The matrices A and B have the same spectrum. Nevertheless, while A is additive D-stable, B is not, since it is not a P_0^+-matrix. Indeed, the matrix

$$\begin{bmatrix} -1 & 2 \\ -3 & 4 \end{bmatrix} + \begin{bmatrix} 0 & 0 \\ 0 & 3 \end{bmatrix} = \begin{bmatrix} -1 & 2 \\ -3 & 7 \end{bmatrix}$$

has eigenvalues -0.1623 and 6.1623.

3. In order to demonstrate Fact 5, consider the matrix

$$A = \begin{bmatrix} 0.25 & 1 & 0 \\ -1 & 0.5 & 1 \\ 2.1 & 1 & 2 \end{bmatrix},$$

which is a P_0^+ matrix and is stable, having the eigenvalues $0.0205709 \pm 1.23009i$ and 2.70886. Thus, A is additively D-stable by Fact 3. Nevertheless, A is not multiplicative D-stable, as the eigenvalues of

$$\begin{bmatrix} 1 & 0 & 0 \\ 0 & 5 & 0 \\ 0 & 0 & 4 \end{bmatrix} \begin{bmatrix} 0.25 & 1 & 0 \\ -1 & 0.5 & 1 \\ 2.1 & 1 & 2 \end{bmatrix} = \begin{bmatrix} 0.25 & 1 & 0 \\ -5 & 2.5 & 5 \\ 8.4 & 4 & 8 \end{bmatrix}$$

are $-0.000126834 \pm 2.76183i$ and 10.7503.

19.5 Lyapunov Diagonal Stability

Lyapunov diagonally stable matrices play an important role in various applications, for example, predator–prey systems in ecology, e.g., [Goh76], [Goh77], and [RZ82]; dynamical systems, e.g., [Ara75]; and economic models, e.g., [Joh74a] and the references in [BBP78].

Definitions:

A real square matrix A is said to be **Lyapunov diagonally stable [semistable]** if there exists a positive diagonal matrix D such that $AD + DA^T$ is positive definite [semidefinite]. In this case, the matrix D is called a **Lyapunov scaling factor** of A.

In some references Lyapunov diagonally stable matrices are referred to as just **diagonally stable** matrices or as **Volterra–Lyapunov stable**.

An $n \times n$ matrix A is said to be an H**-matrix** if the **comparison matrix** $M(A)$ defined by

$$M(A)_{ij} = \begin{cases} |a_{ii}|, & i = j \\ -|a_{ij}|, & i \neq j \end{cases}$$

is an M-matrix.

A real square matrix A is said to be **strongly inertia preserving** if the inertia of AD is equal to the inertia of D for every (not necessarily nonsingular) real diagonal matrix D.

Facts:

The problem of characterizing Lyapunov diagonal stability is, in general, an open problem. It is solved in [BH83] for acyclic matrices. Lyapunov diagonal semistability of acyclic matrices is characterized in [Her88]. Characterization of Lyapunov diagonal stability and semistability using cones is given in [HSh88]; see also the survey in [Her98]. For a book combining theoretical results, applications, and examples, look at [KB00].

1. [BBP78], [Ple77] Lyapunov diagonally stable matrices are P-matrices.
2. [Goh76] A 2×2 real matrix is Lyapunov diagonally stable if and only if it is a P-matrix.
3. [BVW78] A real square matrix A is Lyapunov diagonally stable if and only if for every nonzero real symmetric positive semidefinite matrix H, the matrix HA has at least one positive diagonal element.
4. [QR65] Lyapunov diagonally stable matrices are multiplicative D-stable.
5. [Cro78] Lyapunov diagonally stable matrices are additive D-stable.
6. [AK72], [Tar71] A Z-matrix is Lyapunov diagonally stable if and only if it is a P-matrix (that is, it is an M-matrix).
7. [HS85a] An H-matrix A is Lyapunov diagonally stable if and only if A is nonsingular and the diagonal elements of A are nonnegative.
8. [BS91] Lyapunov diagonally stable matrices are strongly inertia preserving.
9. [BH83] Acyclic matrices are Lyapunov diagonally stable if and only if they are P-matrices.
10. [BS91] Acyclic matrices are Lyapunov diagonally stable if and only if they are strongly inertia preserving.

Examples:

1. Multiplicative D-stable and additive D-stable matrices are not necessarily diagonally stable, as is demonstrated by the matrix

$$\begin{bmatrix} 1 & -1 \\ 1 & 0 \end{bmatrix}.$$

2. Another example, given in [BH85] is the matrix

$$\begin{bmatrix} 0 & 1 & 0 & 0 \\ -1 & 1 & 1 & 0 \\ 0 & 1 & a & b \\ 0 & 0 & -b & 0 \end{bmatrix}, \quad a \geq 1, b \neq 0,$$

which is not Lyapunov diagonally stable, but is multiplicative D-stable if and only if $a > 1$, and is additive D-stable whenever $a = 1$ and $b \neq 1$.
3. Stability is a spectral property, and so it is always possible to check whether a given matrix is stable or not by evaluating its eigenvalues; Lyapunov diagonal stability cannot be characterized by the spectrum of the matrix, as is demonstrated by the following two matrices:

$$A = \begin{bmatrix} 1 & 0 \\ 0 & 2 \end{bmatrix}, \quad B = \begin{bmatrix} -1 & 2 \\ -3 & 4 \end{bmatrix}.$$

The matrices A and B have the same spectrum. Nevertheless, while A is Lyapunov diagonal stable, B is not, since it is not a P-matrix. Indeed, for every positive diagonal matrix D, the element of $AD + DA^T$ in the $(1, 1)$ position is negative and, hence, $AD + DA^T$ cannot be positive definite.
4. Let A be a Lyapunov diagonally stable matrix and let D be a Lyapunov scaling factor of A. Using continuity arguments, it follows that every positive diagonal matrix that is close enough to D is a Lyapunov scaling factor of A. Hence, a Lyapunov scaling factor of a Lyapunov diagonally stable matrix is not unique (up to a positive scalar multiplication). The Lyapunov scaling factor is not necessarily unique even in cases of Lyapunov diagonally semistable matrices, as is demonstrated by the zero matrix and the following more interesting example. Let

$$A = \begin{bmatrix} 2 & 2 & 3 \\ 2 & 2 & 3 \\ 1 & 1 & 2 \end{bmatrix}.$$

One can check that $D = \text{diag}(1, 1, d)$ is a scaling factor of A whenever $\frac{1}{9} \leq d \leq 1$. On the other hand, it is shown in [HS85b] that the identity matrix is the unique Lyapunov scaling factor of the matrix

$$\begin{bmatrix} 1 & 1 & 2 & 0 \\ 1 & 1 & 0 & 0 \\ 0 & 2 & 1 & 2 \\ 2 & 2 & 0 & 1 \end{bmatrix}.$$

Further study of Lyapunov scaling factors can be found in [HS85b], [HS85c], [SB87], [HS88], [SH88], [SB88], and [CHS92].

References

[Ara75] M. Araki. Applications of M-matrices to the stability problems of composite dynamical systems. *Journal of Mathematical Analysis and Applications* 52 (1975), 309–321.

[AK72] M. Araki and B. Kondo. Stability and transient behaviour of composite nonlinear systems. *IEEE Transactions on Automatic Control* AC-17 (1972), 537–541.

[BBP78] G.P. Barker, A. Berman, and R.J. Plemmons. Positive diagonal solutions to the Lyapunov equations. *Linear and Multilinear Algebra* 5 (1978), 249–256.

[BH83] A. Berman and D. Hershkowitz. Matrix diagonal stability and its implications. *SIAM Journal on Algebraic and Discrete Methods* 4 (1983), 377–382.

[BH84] A. Berman and D. Hershkowitz. Characterization of acyclic D-stable matrices. *Linear Algebra and Its Applications* 58 (1984), 17–31.

[BH85] A. Berman and D. Hershkowitz. Graph theoretical methods in studying stability. *Contemporary Mathematics* 47 (1985), 1–6.

[BS91] A. Berman and D. Shasha. Inertia preserving matrices. *SIAM Journal on Matrix Analysis and Applications* 12 (1991), 209–219.

[BVW78] A. Berman, R.S. Varga, and R.C. Ward. ALPS: Matrices with nonpositive off-diagonal entries. *Linear Algebra and Its Applications* 21 (1978), 233–244.

[Bia85] S. Bialas. A necessary and sufficient condition for the stability of convex combinations of stable polynomials or matrices. *Bulletin of the Polish Academy of Sciences. Technical Sciences* 33 (1985), 473–480.

[Cai76] B.E. Cain. Real 3×3 stable matrices. *Journal of Research of the National Bureau of Standards Section B* 8O (1976), 75–77.

[Car68] D. Carlson. A new criterion for H-stability of complex matrices. *Linear Algebra and Its Applications* 1 (1968), 59–64.

[Car74] D. Carlson. A class of positive stable matrices. *Journal of Research of the National Bureau of Standards Section B* 78 (1974), 1–2.

[Car84] D. Carlson. Controllability, inertia, and stability for tridiagonal matrices. *Linear Algebra and Its Applications* 56 (1984), 207–220.

[CDJ82] D. Carlson, B.N. Datta, and C.R. Johnson. A semidefinite Lyapunov theorem and the characterization of tridiagonal D-stable matrices. *SIAM Journal of Algebraic Discrete Methods* 3 (1982), 293–304.

[CHS92] D.H. Carlson, D. Hershkowitz, and D. Shasha. Block diagonal semistability factors and Lyapunov semistability of block triangular matrices. *Linear Algebra and Its Applications* 172 (1992), 1–25.

[CL97] N. Cohen and I. Lewkowicz. Convex invertible cones and the Lyapunov equation. *Linear Algebra and Its Applications* 250 (1997), 105–131.

[Cro78] G.W. Cross. Three types of matrix stability. *Linear Algebra and Its Applications* 20 (1978), 253–263.

[FF58] M.E. Fisher and A.T. Fuller. On the stabilization of matrices and the convergence of linear iterative processes. *Proceedings of the Cambridge Philosophical Society* 54 (1958), 417–425.

[FHR05] S. Friedland, D. Hershkowitz, and S.M. Rump. Positive entries of stable matrices. *Electronic Journal of Linear Algebra* 12 (2004/2005), 17–24.

[FB87] M. Fu and B.R. Barmish. A generalization of Kharitonov's polynomial framework to handle linearly independent uncertainty. Technical Report ECE-87-9, Department of Electrical and Computer Engineering, University of Wisconsin, Madison, 1987.

[FB88] M. Fu and B.R. Barmish. Maximal undirectional perturbation bounds for stability of polynomials and matrices. *Systems and Control Letters* 11 (1988), 173–178.

[Fuj26] M. Fujiwara. On algebraic equations whose roots lie in a circle or in a half-plane. *Mathematische Zeitschrift* 24 (1926), 161–169.

[Gan60] F.R. Gantmacher. *The Theory of Matrices*. Chelsea, New York, 1960.

[Goh76] B.S. Goh. Global stability in two species interactions. *Journal of Mathematical Biology* 3 (1976), 313–318.

[Goh77] B.S. Goh. Global stability in many species systems. *American Naturalist* 111 (1977), 135–143.

[Had76] K.P. Hadeler. Nonlinear diffusion equations in biology. In *Proceedings of the Conference on Differential Equations*, Dundee, 1976, Springer Lecture Notes.

[Har80] D.J. Hartfiel. Concerning the interior of the D-stable matrices. *Linear Algebra and Its Applications* 30 (1980), 201–207.

[Her83] D. Hershkowitz. On the spectra of matrices having nonnegative sums of principal minors. *Linear Algebra and Its Applications* 55 (1983), 81–86.

[Her86] D. Hershkowitz. Stability of acyclic matrices. *Linear Algebra and Its Applications* 73 (1986), 157–169.

[Her88] D. Hershkowitz. Lyapunov diagonal semistability of acyclic matrices. *Linear and Multilinear Algebra* 22 (1988), 267–283.

[Her92] D. Hershkowitz. Recent directions in matrix stability. *Linear Algebra and Its Applications* 171 (1992), 161–186.

[Her98] D. Hershkowitz. On cones and stability. *Linear Algebra and Its Applications* 275/276 (1998), 249–259.

[HB83] D. Hershkowitz and A. Berman. Localization of the spectra of P- and P_0-matrices. *Linear Algebra and Its Applications* 52/53 (1983), 383–397.

[HJ86] D. Hershkowitz and C.R. Johnson. Spectra of matrices with P-matrix powers. *Linear Algebra and Its Applications* 80 (1986), 159–171.

[HK2003] D. Hershkowitz and N. Keller. Positivity of principal minors, sign symmetry and stability. *Linear Algebra and Its Applications* 364 (2003) 105–124.

[HM98] D. Hershkowitz and N. Mashal. P^{α}-matrices and Lyapunov scalar stability. *Electronic Journal of Linear Algebra* 4 (1998), 39–47.

[HS85a] D. Hershkowitz and H. Schneider. Lyapunov diagonal semistability of real H-matrices. *Linear Algebra and Its Applications* 71 (1985), 119–149.

[HS85b] D. Hershkowitz and H. Schneider. Scalings of vector spaces and the uniqueness of Lyapunov scaling factors. *Linear and Multilinear Algebra* 17 (1985), 203–226.

[HS85c] D. Hershkowitz and H. Schneider. Semistability factors and semifactors. *Contemporary Mathematics* 47 (1985), 203–216.

[HS88] D. Hershkowitz and H. Schneider. On Lyapunov scaling factors for real symmetric matrices. *Linear and Multilinear Algebra* 22 (1988), 373–384.

[HS90] D. Hershkowitz and H. Schneider. On the inertia of intervals of matrices. *SIAM Journal on Matrix Analysis and Applications* 11 (1990), 565–574.

[HSh88] D. Hershkowitz and D. Shasha. Cones of real positive semidefinite matrices associated with matrix stability. *Linear and Multilinear Algebra* 23 (1988), 165–181.

[HS93] D. Hershkowitz and F. Shmidel. On a conjecture on the eigenvalues of P-matrices. *Linear and Multilinear Algebra* 36 (1993), 103–110.

[Hol99] O. Holtz. Not all GKK τ-matrices are stable, *Linear Algebra and Its Applications* 291 (1999), 235–244.

[HJ91] R.A. Horn and C.R.Johnson. *Topics in Matrix Analysis.* Cambridge University Press, Cambridge, 1991.

H1895 A. Hurwitz. Über die Bedingungen, unter welchen eine Gleichung nur Wurzeln mit negativen reellen Teilen besitzt. *Mathematische Annalen* 46 (1895), 273–284.

[Joh74a] C.R. Johnson. Sufficient conditions for D-stability. *Journal of Economic Theory* 9 (1974), 53–62.

[Joh74b] C.R. Johnson. Second, third and fourth order D-stability. *Journal of Research of the National Bureau of Standards Section B* 78 (1974), 11–13.

[Joh75] C.R. Johnson. A characterization of the nonlinearity of D-stability. *Journal of Mathematical Economics* 2 (1975), 87–91.

[KB00] Eugenius Kaszkurewicz and Amit Bhaya. *Matrix Diagonal Stability in Systems and Computation.* Birkhäuser, Boston, 2000.

[LC14] Liénard and Chipart. Sur la signe de la partie réelle des racines d'une equation algébrique. *Journal de Mathématiques Pures et Appliquées* (6) 10 (1914), 291–346.

[L1892] A.M. Lyapunov. Le Problème Général de la Stabilité du Mouvement. *Annals of Mathematics Studies 17*, Princeton University Press, NJ, 1949.

[Met45] L. Metzler. Stability of multiple markets: the Hick conditions. *Econometrica* 13 (1945), 277–292.

[OS62] A. Ostrowski and H. Schneider. Some theorems on the inertia of general matrices. *Journal of Mathematical Analysis and Applications* 4 (1962), 72–84.

[Ple77] R.J. Plemmons. M-matrix characterizations, I–non-singular M-matrices. *Linear Algebra and Its Applications* 18 (1977), 175–188.

[QR65] J. Quirk and R. Ruppert. Qualitative economics and the stability of equilibrium. *Review of Economic Studies* 32 (1965), 311–325.

[RZ82] R. Redheffer and Z. Zhiming. A class of matrices connected with Volterra prey–predator equations *SIAM Journal on Algebraic and Discrete Methods* 3 (1982), 122–134.

[R1877] E.J. Routh. *A Treatise on the Stability of a Given State of Motion*. Macmillan, London, 1877.

[Sch17] I. Schur. Über Potenzreihen, die im Innern des Einheitskreises beschrankt sind. *Journal für reine und angewandte Mathematik* 147 (1917), 205–232.

[SB87] D. Shasha and A. Berman. On the uniqueness of the Lyapunov scaling factors. *Linear Algebra and Its Applications* 91 (1987) 53–63.

[SB88] D. Shasha and A. Berman. More on the uniqueness of the Lyapunov scaling factors. *Linear Algebra and Its Applications* 107 (1988) 253–273.

[SH88] D. Shasha and D. Hershkowitz. Maximal Lyapunov scaling factors and their applications in the study of Lyapunov diagonal semistability of block triangular matrices. *Linear Algebra and Its Applications* 103 (1988), 21–39.

[Tad81] E. Tadmor. The equivalence of L_2-stability, the resolvent condition, and strict H-stability. *Linear Algebra and Its Applications* 41 (1981), 151–159.

[Tar71] L. Tartar. Une nouvelle characterization des matrices. *Revue Française d'Informatique et de Recherche Opérationnelle* 5 (1971), 127–128.

[Tog80] Y. Togawa. A geometric study of the D-stability problem. *Linear Algebra and Its Applications* 33(1980), 133–151.

[Zah92] Z. Zahreddine. Explicit relationships between Routh–Hurwitz and Schur–Cohn types of stability. *Irish Mathematical Society Bulletin* 29 (1992), 49–54.

Topics in Advanced Linear Algebra

20

Inverse Eigenvalue Problems

Alberto Borobia
UNED

In general, an **inverse eigenvalue problem** (IEP) consists of the construction of a matrix with prescribed structural and spectral constraints. This is a two-level problem: (1) on a theoretical level the target is to determine if the IEP is solvable, that is, to find necessary and sufficient conditions for the existence of at least one **solution matrix** (a matrix with the given constraints); and (2) on a practical level, the target is the effective construction of a solution matrix when the IEP is solvable. IEPs are classified into different types according to the specific constraints. We will consider three topics: IEPs with prescribed entries, nonnegative IEPs, and affine parameterized IEPs. Other important topics include pole assignment problems, Jacobi IEPs, inverse singular value problems, etc. For interested readers, we refer to the survey [CG02] where an account of IEPs with applications and extensive bibliography can be found.

20.1 IEPs with Prescribed Entries

The underlying question for an IEP with prescribed entries (PEIEPs) is to understand how the prescription of some entries of a matrix can have repercussions on its spectral properties. A classical result on this subject is the Schur–Horn Theorem allowing the construction of a real symmetric matrix with prescribed diagonal, prescribed eigenvalues, and subject to some restrictions (see Fact 1 below). Here we consider PEIEPs that require finding a matrix with some prescribed entries and with prescribed eigenvalues or characteristic polynomial; no structural constraints are imposed on the solution matrices.

Most of the facts of Sections 20.1 and 20.2 appear in [IC00], an excellent survey that describes finite step procedures for constructing solution matrices.

Definitions:

An **IEP with prescribed entries** (PEIEP) has the following standard formulation:

Given:

(a) A field F.

(b) n elements $\lambda_1, \ldots, \lambda_n$ of F (respectively, a monic polynomial $f \in F[x]$ of degree n).

(c) t elements $p_1, \ldots, p_t$ of F.

(d) A set $Q = \{(i_1, j_1), \ldots, (i_t, j_t)\}$ of t positions of an $n \times n$ matrix.

Find: A matrix $A = [a_{ij}] \in F^{n\times n}$ with $a_{i_k j_k} = p_k$ for $1 \leq k \leq t$ and such that $\sigma(A) = \{\lambda_1, \ldots, \lambda_n\}$ (respectively, such that $p_A(x) = f$).

Facts: [IC00]

1. (Schur–Horn Theorem) Given any real numbers $\lambda_1 \geq \cdots \geq \lambda_n$ and $d_1 \geq \cdots \geq d_n$ satisfying

$$\sum_{i=1}^{k} \lambda_i \geq \sum_{i=1}^{k} d_i \quad \text{for } k = 1, \ldots, n-1 \quad \text{and} \quad \sum_{i=1}^{n} \lambda_i = \sum_{i=1}^{n} d_i,$$

there exists a real symmetric $n \times n$ matrix with diagonal $(d_1, \ldots, d_n)$ and eigenvalues $\lambda_1, \ldots, \lambda_n$; and any Hermitian matrix satisfies these conditions on its eigenvalues and diagonal entries.

2. A finite step algorithm is provided in [CL83] for the construction of a solution matrix for the Schur–Horn Theorem.
3. Consider the following classes of PEIEPs:

(1.1)	F	$\lambda_1, \ldots, \lambda_n$	$p_1, \ldots, p_{n-1}$	$\lvert Q\rvert = n-1$
(1.2)	F	$f = x^n + c_1 x^{n-1} + \cdots + c_n$	$p_1, \ldots, p_{n-1}$	$\lvert Q\rvert = n-1$
(2.1)	F	$\lambda_1, \ldots, \lambda_n$	$p_1, \ldots, p_n$	$\lvert Q\rvert = n$
(2.2)	F	$f = x^n + c_1 x^{n-1} + \cdots + c_n$	$p_1, \ldots, p_n$	$\lvert Q\rvert = n$
(3.1)	F	$\lambda_1, \ldots, \lambda_n$	$p_1, \ldots, p_{2n-3}$	$\lvert Q\rvert = 2n-3$

- [dO73a] Each PEIEP of class (1.1) is solvable.
- [Dds74] Each PEIEP of class (1.2) is solvable except if all off-diagonal entries in one row or column are prescribed to be zero and f has no root on F.
- [dO73b] Each PEIEP of class (2.1) is solvable with the following exceptions: (1) all entries in the diagonal are prescribed and their sum is different from $\lambda_1 + \cdots + \lambda_n$; (2) all entries in one row or column are prescribed, with zero off-diagonal entries and diagonal entry different from $\lambda_1, \ldots, \lambda_n$; and (3) $n = 2$, $Q = \{(1,2),(2,1)\}$, and $x^2 - (\lambda_1 + \lambda_2)x + p_1 p_2 + \lambda_1\lambda_2 \in F[x]$ is irreducible over F.
- [Zab86] For $n > 4$, each PEIEP of class (2.2) is solvable with the following exceptions: (1) all entries in the diagonal are prescribed and their sum is different from-c_1; (2) all entries in a row or column are prescribed, with zero off-diagonal entries and diagonal entry which is not a root of f; and (3) all off-diagonal entries in one row or column are prescribed to be zero and f has no root on F. The case $n \leq 4$ is solved but there are more exceptions.
- [Her83] Each PEIEP of class (3.1) is solvable with the following exceptions: (1) all entries in the diagonal are prescribed and their sum is different from $\lambda_1 + \cdots + \lambda_n$; and (2) all entries in one row or column are prescribed, with zero off-diagonal entries and diagonal entry different from $\lambda_1, \ldots, \lambda_n$.
- [Her83] The result for PEIEPs of class (3.1) cannot be improved to $|Q| > 2n-3$ since a lot of specific nonsolvable situations appear, and, therefore, a closed result seems to be quite inaccessible.
- A gradient flow approach is proposed in [CDS04] to explore the existence of solution matrices when the set of prescribed entries has arbitrary cardinality.

4. The important case $\mathcal{Q} = \{(i, j) : i \neq j\}$ is discussed in section 20.9.
5. Let $\{p_{ij} : 1 \leq i \leq j < n\}$ be a set of $\frac{n^2+n}{2}$ elements of a field F. Define the set $\{r_1, \ldots, r_s\}$ of all those integers r such that $p_{ij} = 0$ whenever $1 \leq i \leq r < j \leq n$. Assume that $0 = r_0 < r_1 < \cdots < r_s < r_{s+1} = n$ and define $\beta_t = \sum_{r_{t-1} < k \leq r_t} p_{kk}$ for $t = 1, \ldots, s+1$. The following PEIEPs have been solved:

   - [BGRS90] Let $\lambda_1, \ldots, \lambda_n$ be n elements of F. Then there exists $A = [a_{ij}] \in F^{n \times n}$ with $a_{ij} = p_{ij}$ for $1 \leq i \leq j \leq n$ and $\sigma(A) = \{\lambda_1, \ldots, \lambda_n\}$ if and only if $\{1, \ldots, n\}$ has a partition $N_1 \cup \cdots \cup N_{s+1}$ such that $|N_t| = r_t - r_{t-1}$ and $\sum_{k \in N_t} \lambda_k = \beta_t$ for each $t = 1, \ldots, s+1$.
   - [Sil93] Let $f \in F[x]$ be a monic polynomial of degree n. Then there exists $A = [a_{ij}] \in F^{n \times n}$ with $a_{ij} = p_{ij}$ for $1 \leq i \leq j \leq n$ and $p_A(x) = f$ if and only if $f = f_1 \cdots f_{s+1}$, where $f_t = x^{r_t - r_{t-1}} - \beta_t x^{r_t - r_{t-1} - 1} + \cdots \in F[x]$ for $t = 1, \ldots, s+1$.

6. [Fil69] Let $d_1, \ldots, d_n$ be elements of a field F, and let $A \in F^{n \times n}$ with $A \neq \lambda I_n$ for all $\lambda \in F$ and $\operatorname{tr}(A) = \sum_{i=1}^n d_i$. Then A is similar to a matrix with diagonal $(d_1, \ldots, d_n)$.

Examples:

1. [dO73b] *Given*:

   (a) A field F.

   (b) $\lambda_1, \ldots, \lambda_n \in F$.

   (c) $p_1, \ldots, p_n \in F$.

   (d) $\mathcal{Q} = \{(1, 1), \ldots, (n, n)\}$.

   If $\sum_{i=1}^n \lambda_i = \sum_{i=1}^n p_i$, then $A = [a_{ij}] \in F^{n \times n}$ with

$$a_{ii} = p_i, \qquad a_{ij} = 0 \quad \text{if} \quad i \leq j - 2,$$
$$a_{i,i+1} = \sum_{k=1}^{i} \lambda_k - \sum_{k=1}^{i} p_k, \qquad a_{ij} = p_j - \lambda_{j+1} \quad \text{if} \quad i > j,$$

   has diagonal $(p_1, \ldots, p_n)$ and its spectrum is $\sigma(A) = \{\lambda_1, \ldots, \lambda_n\}$.

20.2 PEIEPs of 2 × 2 Block Type

In the 1970s, de Oliveira posed the problem of determining all possible spectra of a 2×2 block matrix A or all possible characteristic polynomials of A or all possible invariant polynomials of A when some of the blocks are prescribed and the rest vary (*invariant polynomial* is a synonym for *invariant factor*, cf. Section 6.6).

Definitions:

Let F be a field and let A be the 2×2 block matrix

$$A = \begin{bmatrix} A_{11} & A_{12} \\ A_{21} & A_{22} \end{bmatrix} \in F^{n \times n} \quad \text{with} \quad A_{11} \in F^{l \times l} \quad \text{and} \quad A_{22} \in F^{m \times m}.$$

Notation:

- $\deg(f)$: degree of $f \in F[x]$.
- $g | f$: polynomial g divides the polynomial f.
- $ip(B)$: invariant polynomials of the square matrix B.

Facts: [IC00]

1. [dO71] Let A_{11} and a monic polynomial $f \in F[x]$ of degree n be given. Let $ip(A_{11}) = g_1|\cdots|g_l$. Then $p_A(x) = f$ is possible except if $l > m$ and $g_1 \cdots g_{l-m}$ is not a divisor of f.
2. [dS79], [Tho79] Let A_{11} and n monic polynomials $f_1, \ldots, f_n \in F[x]$ with $f_1|\cdots|f_n$ and $\sum_{i=1}^{n} \deg(f_i) = n$ be given. Let $ip(A_{11}) = g_1|\cdots|g_l$. Then $ip(A) = f_1|\cdots|f_n$ is possible if and only if $f_i \mid g_i \mid f_{i+2m}$ for each $i = 1, \ldots, l$ where $f_k = 0$ for $k > n$.
3. [dO75] Let A_{12} and a monic polynomial $f \in F[x]$ of degree n be given. Then $p_A(x) = f$ is possible except if $A_{12} = 0$ and f has no divisor of degree l.
4. [Zab89], [Sil90] Let A_{12} and n monic polynomials $f_1, \ldots, f_n \in F[x]$ with $f_1|\cdots|f_n$ and $\sum_{i=1}^{n} \deg(f_i) = n$ be given. Let $r = \text{rank}(A_{12})$ and s the number of polynomials in $f_1, \ldots, f_n$ which are different from 1. Then $ip(A) = f_1|\cdots|f_n$ is possible if and only if $r \leq n - s$ with the following exceptions:
   (a) $r = 0$ and $\prod_{i=1}^{n} f_i$ has no divisor of degree l.
   (b) $r \geq 1$, $l - r$ odd and $f_{n-s+1} = \cdots = f_n$ with f_n irreducible of degree 2.
   (c) $r = 1$ and $f_{n-s+1} = \cdots = f_n$ with f_n irreducible of degree $k \geq 3$ and $k|l$.
5. [Wim74] Let A_{11}, A_{12}, and a monic polynomial $f \in F[x]$ of degree n be given. Let $h_1|\cdots|h_l$ be the invariant factors of $\left[xI_l - A_{11} \mid -A_{12} \right]$. Then $p_A(x) = f$ is possible if and only if $h_1 \cdots h_l | f$.
6. All possible invariant polynomials of A are characterized in [Zab87] when A_{11} and A_{12} are given. The statement of this result contains a majorization inequality involving the *controllability indices* of the pair (A_{11}, A_{12}).
7. [Sil87b] Let A_{11}, A_{22}, and n elements $\lambda_1, \ldots, \lambda_n$ of F be given. Assume that $l \geq m$ and let $ip(A_{11}) = g_1|\cdots|g_l$. Then $\sigma(A) = \{\lambda_1, \ldots, \lambda_n\}$ is possible if and only if all the following conditions are satisfied:
   (a) $\text{tr}(A_{11}) + \text{tr}(A_{22}) = \lambda_1 + \cdots + \lambda_n$.
   (b) If $l > m$, then $g_1 \cdots g_{l-m}|(x - \lambda_1) \cdots (x - \lambda_n)$.
   (c) If $A_{11} = aI_l$ and $A_{22} = dI_m$, then there exists a permutation τ of $\{1, \ldots, n\}$ such that $\lambda_{\tau(2i-1)} + \lambda_{\tau(2i)} = a + d$ for $1 \leq i \leq m$ and $\lambda_{\tau(j)} = a$ for $2m + 1 \leq j \leq n$.
8. [Sil87a] Let A_{12}, A_{21}, and n elements $\lambda_1, \ldots, \lambda_n$ of F be given. Then $\sigma(A) = \{\lambda_1, \ldots, \lambda_n\}$ is possible except if, simultaneously, $l = m = 1$, $A_{12} = [\,b\,]$, $A_{21} = [\,c\,]$ and the polynomial $x^2 - (\lambda_1 + \lambda_2)x + bc + \lambda_1\lambda_2 \in F[x]$ is irreducible over F.
9. Let A_{12}, A_{21}, and a monic polynomial $f \in F[x]$ of degree n be given:
   - [Fri77] If F is algebraically closed then $p_A(x) = f$ is always possible.
   - [MS00] If $F = \mathbb{R}$ and $n \geq 3$ then $p_A(x) = f$ is possible if and only if either $\min\{\text{rank}(A_{12}), \text{rank}(A_{21})\} > 0$ or f has a divisor of degree l.
   - If $F = \mathbb{R}$, $A_{12} = [\,b\,]$, $A_{21} = [\,c\,]$ and $f = x^2 + c_1x + c_2 \in \mathbb{R}[x]$ then $p_A(x) = f$ is possible if and only if $x^2 + c_1x + c_2 + bc$ has a root in $\mathbb{R}$.
10. [Sil91] Let A_{11}, A_{12}, A_{22}, and n elements $\lambda_1, \ldots, \lambda_n$ of F be given. Let $k_1|\cdots|k_l$ be the invariant factors of $\left[xI_l - A_{11} \mid -A_{12} \right]$, $h_1|\cdots|h_m$ the invariant factors of $\begin{bmatrix} -A_{12} \\ xI_m - A_{22} \end{bmatrix}$, and $g = k_1 \cdots k_l h_1 \cdots h_m$. Then $\sigma(A) = \{\lambda_1, \ldots, \lambda_n\}$ is possible if and only if all the following conditions hold:
    (a) $\text{tr}(A_{11}) + \text{tr}(A_{22}) = \lambda_1 + \cdots + \lambda_n$.
    (b) $g|(x - \lambda_1) \cdots (x - \lambda_n)$.
    (c) If $A_{11}A_{12} + A_{12}A_{22} = \eta A_{12}$ for some $\eta \in F$, then there exists a permutation τ of $\{1, \ldots, n\}$ such that $\lambda_{\tau(2i-1)} + \lambda_{\tau(2i)} = \eta$ for $1 \leq i \leq t$ where $t = \text{rank}(A_{12})$ and $\lambda_{\tau(2t+1)}, \ldots, \lambda_{\tau(n)}$ are the roots of g.
11. If a problem of block type is solved for prescribed characteristic polynomial then the solution for prescribed spectrum easily follows.

12. The book [GKvS95] deals with PEIEPs of block type from an operator point of view.
13. A description is given in [FS98] of all the possible characteristic polynomials of a square matrix with an arbitrary prescribed submatrix.

20.3 Nonnegative IEP (NIEP)

Nonnegative matrices appear naturally in many different mathematical areas, both pure and applied, such as numerical analysis, statistics, economics, social sciences, etc. One of the most intriguing problems in this field is the so-called **nonnegative IEP** (NIEP). Its origin goes back to A.N. Kolgomorov, who in 1938 posed the problem of determining which individual complex numbers belong to the spectrum of some $n \times n$ nonnegative matrix with its spectral radius normalized to be 1. Kolgomorov's problem was generalized in 1949 by H. R. Suleǐmanova, who posed the NIEP: To determine which n-tuples of complex numbers are spectra of $n \times n$ nonnegative matrices. For definitions and additional facts about nonnegative matrices, see Chapter 9.

Definitions:

Let Π_n denote the compact subset of $\mathbb{C}$ bounded by the regular n-sided polygon inscribed in the unit circle of $\mathbb{C}$ and with one vertex at $1 \in \mathbb{C}$.

Let Θ_n denote the subset of $\mathbb{C}$ composed of those complex numbers λ such that λ is an eigenvalue of some $n \times n$ row stochastic matrix.

A **circulant matrix** is a matrix in which every row is obtained by a single cyclic shift of the previous row.

Facts:

All the following facts appear in [Min88].

1. A complex nonzero number λ is an eigenvalue of a nonnegative $n \times n$ matrix with positive spectral radius ρ if and only if $\lambda/\rho \in \Theta_n$.
2. [DD45], [DD46] $\Theta_3 = \Pi_2 \cup \Pi_3$.
3. [Mir63] Each point in $\Pi_2 \cup \Pi_3 \cup \cdots \cup \Pi_n$ is an eigenvalue of a doubly stochastic $n \times n$ matrix.
4. [Kar51] The set Θ_n is symmetric relative to the real axis and is contained within the circle $|z| \leq 1$. It intersects $|z| = 1$ at the points $e^{\frac{2\pi i a}{b}}$ where a and b run over all integers satisfying $0 \leq a < b \leq n$. The boundary of Θ_n consists of the curvilinear arcs connecting these points in circular order. For $n \geq 4$, each arc is given by one of the following parametric equations:

$$z^q(z^p - t)^r = (1-t)^r,$$
$$(z^c - t)^d = (1-t)^d z^q,$$

where the real parameter t runs over the interval $0 \leq t \leq 1$, and c, d, p, q, r are natural numbers defined by certain rules (explicitly stated in [Min88]).

Examples:

1. [LL78] The circulant matrix

$$\frac{1}{3}\begin{bmatrix} 1+2r\cos\theta & 1-2r\cos(\frac{\pi}{3}+\theta) & 1-2r\cos(\frac{\pi}{3}-\theta) \\ 1-2r\cos(\frac{\pi}{3}-\theta) & 1+2r\cos\theta & 1-2r\cos(\frac{\pi}{3}+\theta) \\ 1-2r\cos(\frac{\pi}{3}+\theta) & 1-2r\cos(\frac{\pi}{3}-\theta) & 1+2r\cos\theta \end{bmatrix}$$

has spectrum $\{1, re^{i\theta}, re^{-i\theta}\}$, and it is doubly stochastic if and only if $re^{i\theta} \in \Pi_3$.

20.4 Spectra of Nonnegative Matrices

Definitions:

$\mathcal{N}_n \equiv \{\sigma = \{\lambda_1, \ldots, \lambda_n\} \subset \mathbb{C} : \exists A \geq 0$ with spectrum $\sigma\}$.

$\mathcal{R}_n \equiv \{\sigma = \{\lambda_1, \ldots, \lambda_n\} \subset \mathbb{R} : \exists A \geq 0$ with spectrum $\sigma\}$.

$\mathcal{S}_n \equiv \{\sigma = \{\lambda_1, \ldots, \lambda_n\} \subset \mathbb{R} : \exists A \geq 0$ symmetric with spectrum $\sigma\}$.

$\mathcal{R}_n^* \equiv \{(1, \lambda_2, \ldots, \lambda_n) \in \mathbb{R}^n : \{1, \lambda_2, \ldots, \lambda_n\} \in \mathcal{R}_n; 1 \geq \lambda_2 \geq \cdots \geq \lambda_n\}$.

$\mathcal{S}_n^* \equiv \{(1, \lambda_2, \ldots, \lambda_n) \in \mathbb{R}^n : \{1, \lambda_2, \ldots, \lambda_n\} \in \mathcal{S}_n; 1 \geq \lambda_2 \geq \cdots \geq \lambda_n\}$.

For any set $\sigma = \{\lambda_1, \ldots, \lambda_n\} \subset \mathbb{C}$, let

$$\rho(\sigma) = \max_{1 \leq i \leq n} |\lambda_i| \quad \text{and} \quad s_k = \sum_{i=1}^{n} \lambda_i^k \quad \text{for each } k \in \mathbb{N}.$$

A set $S \subset \mathbb{R}^n$ is **star-shaped** from $p \in S$ if every line segment drawn from p to another point in S lies entirely in S.

Facts:

Most of the following facts appear in [ELN04].

1. [Joh81] If $\sigma = \{\lambda_1, \ldots, \lambda_n\} \in \mathcal{N}_n$, then σ is the spectrum of a $n \times n$ nonnegative matrix with all row sums equal to $\rho(\sigma)$.
2. If $\sigma = \{\lambda_1, \ldots, \lambda_n\} \in \mathcal{N}_n$, then the following conditions hold:

   (a) $\rho(\sigma) \in \sigma$.

   (b) $\overline{\sigma} = \sigma$.

   (c) $s_i \geq 0$ for $i \geq 1$.

   (d) [LL78], [Joh81] $s_k^m \leq n^{m-1} s_{km}$ for $k, m \geq 1$.
3. $\mathcal{N}_n$ is known for $n \leq 3$, $\mathcal{R}_n$ and $\mathcal{S}_n$ are known for $n \leq 4$:
   - $\mathcal{N}_2 = \mathcal{R}_2 = \mathcal{S}_2 = \{\sigma = \{\lambda_1, \lambda_2\} \subset \mathbb{R} : s_1 \geq 0\}$.
   - $\mathcal{R}_3 = \mathcal{S}_3 = \{\sigma = \{\lambda_1, \lambda_2, \lambda_3\} \subset \mathbb{R} : \rho(\sigma) \in \sigma; s_1 \geq 0\}$.
   - [LL78] $\mathcal{N}_3 = \{\sigma = \{\lambda_1, \lambda_2, \lambda_3\} \subset \mathbb{C} : \overline{\sigma} = \sigma; \rho(\sigma) \in \sigma; s_1 \geq 0; s_1^2 \leq 3s_2\}$.
   - $\mathcal{R}_4 = \mathcal{S}_4 = \{\sigma = \{\lambda_1, \lambda_2, \lambda_3, \lambda_4\} \subset \mathbb{R} : \rho(\sigma) \in \sigma; s_1 \geq 0\}$.
4. (a) [JLL96] $\mathcal{R}_n$ and $\mathcal{S}_n$ are not always equal sets.

   (b) [ELN04] $\sigma = \{97, 71, -44, -54, -70\} \in \mathcal{R}_5$ but $\sigma \notin \mathcal{S}_5$.

   (c) [ELN04] provides symmetric matrices for all known elements of $\mathcal{S}_5$.
5. [Rea96] Let $\sigma = \{\lambda_1, \lambda_2, \lambda_3, \lambda_4\} \subset \mathbb{C}$ with $s_1 = 0$. Then $\sigma \in \mathcal{N}_4$ if and only if $s_2 \geq 0$, $s_3 \geq 0$ and $4s_4 \geq s_2^2$. Moreover, σ is the spectrum of

$$\begin{bmatrix} 0 & 1 & 0 & 0 \\ \frac{s_2}{4} & 0 & 1 & 0 \\ \frac{s_3}{4} & 0 & 0 & 1 \\ \frac{4s_4 - s_2^2}{16} & \frac{s_3}{12} & \frac{s_2}{4} & 0 \end{bmatrix}.$$

6. [LM99] Let $\sigma = \{\lambda_1, \lambda_2, \lambda_3, \lambda_4, \lambda_5\} \subset \mathbb{C}$ with $s_1 = 0$. Then $\sigma \in \mathcal{N}_5$ if and only if the following conditions are satisfied:

   (a) $s_i \geq 0$ for $i = 2, 3, 4, 5$.

   (b) $4s_4 \geq s_2^2$.

   (c) $12s_5 - 5s_2s_3 + 5s_3\sqrt{4s_4 - s_2^2} \geq 0$.

   The proof of the sufficient part is constructive.

7. (a) $\mathcal{R}_n^*$ and $\mathcal{S}_n^*$ are star-shaped from $(1, \ldots, 1)$.

   (b) [BM97] $\mathcal{R}_n^*$ is star-shaped from $(1, 0, \ldots, 0)$.

   (c) [KM01], [Mou03] $\mathcal{R}_n^*$ and $\mathcal{S}_n^*$ are not convex sets for $n \geq 5$.

Examples:

1. We show that $\sigma = \{5, 5, -3, -3, -3\} \notin \mathcal{N}_5$. Suppose A is a nonnegative matrix with spectrum σ. By the Perron–Frobenius Theorem, A is reducible and σ can be partitioned into two nonempty subsets, each one being the spectrum of a nonnegative matrix with Perron root equal to 5. This is not possible since one of the subsets must contain numbers with negative sum.
2. $\{6, 1, 1, -4, -4\} \in \mathcal{N}_5$ by Fact 6.

20.5 Nonzero Spectra of Nonnegative Matrices

For the definitions and additional facts about primitive matrices see Section 29.6 and Chapter 9.

Definitions:

The **Möbius function** $\mu : \mathbb{N} \mapsto \{-1, 0, 1\}$ is defined by $\mu(1) = 1$, $\mu(m) = (-1)^e$ if m is a product of e distinct primes, and $\mu(m) = 0$ otherwise.

The kth **net trace** of $\sigma = \{\lambda_1, \ldots, \lambda_n\} \subset \mathbb{C}$ is $\mathrm{tr}_k(\sigma) = \sum_{d|k} \mu(\frac{k}{d}) s_d$.

The set $\sigma = \{\lambda_1, \ldots, \lambda_n\} \subset \mathbb{C}$ with $0 \notin \sigma$ is the **nonzero spectrum** of a matrix if there exists a $t \times t$ matrix, $t \geq n$, whose spectrum is $\{\lambda_1, \ldots, \lambda_n, 0, \ldots, 0\}$ with $t - n$ zeros.

The set $\sigma = \{\lambda_1, \ldots, \lambda_n\} \subset \mathbb{C}$ has a **Perron value** if $\rho(\sigma) \in \sigma$ and there exists a unique index i with $\lambda_i = \rho(\sigma)$.

Facts:

1. [BH91] Spectral Conjecture: Let $\mathbb{S}$ be a unital subring of $\mathbb{R}$. The set $\sigma = \{\lambda_1, \ldots, \lambda_n\} \subset \mathbb{C}$ with $0 \notin \sigma$ is the nonzero spectrum of some primitive matrix over $\mathbb{S}$ if and only if the following conditions hold:

   (a) σ has a Perron value.

   (b) All the coefficients of the polynomial $\prod_{i=1}^{n}(x - \lambda_i)$ lie in $\mathbb{S}$.

   (c) If $\mathbb{S} = \mathbb{Z}$, then $\mathrm{tr}_k(\sigma) \geq 0$ for all positive integers k.

   (d) If $\mathbb{S} \neq \mathbb{Z}$, then $s_k \geq 0$ for all $k \in \mathbb{N}$ and $s_m > 0$ implies $s_{mp} > 0$ for all $m, p \in \mathbb{N}$.

2. [BH91] Subtuple Theorem: Let $\mathbb{S}$ be a unital subring of $\mathbb{R}$. Suppose that $\sigma = \{\lambda_1, \ldots, \lambda_n\} \subset \mathbb{C}$ with $0 \notin \sigma$ has $\rho(\sigma) = \lambda_1$ and satisfies conditions (a) to (d) of the spectral conjecture. If for some $j \leq n$ the set $\{\lambda_1, \ldots, \lambda_j\}$ is the nonzero spectrum of a nonnegative matrix over $\mathbb{S}$, then σ is the nonzero spectrum of a primitive matrix over $\mathbb{S}$.
3. The spectral conjecture is true for $\mathbb{S} = \mathbb{R}$ by the subtuple theorem.
4. [KOR00] The spectral conjecture is true for $\mathbb{S} = \mathbb{Z}$ and $\mathbb{S} = \mathbb{Q}$.
5. [BH91] The set $\sigma = \{\lambda_1, \ldots, \lambda_n\} \subset \mathbb{C}$ with $0 \notin \sigma$ is the nonzero spectrum of a positive matrix if and only if the following conditions hold:

   (a) σ has a Perron value.

   (b) All coefficients of $\prod_{i=1}^{n}(x - \lambda_i)$ are real.

   (c) $s_k > 0$ for all $k \in \mathbb{N}$.

Examples:

1. Let $\sigma_\epsilon = \{5, 4+\epsilon, -3, -3, -3\}$. Then:
   (a) σ_ϵ for $\epsilon < 0$ is not the nonzero spectrum of a nonnegative matrix since $s_1 < 0$.
   (b) σ_0 is the nonzero spectrum of a nonnegative matrix by Fact 2.
   (c) σ_1 is not the nonzero spectrum of a nonnegative matrix by arguing as in Example 1 of Section 20.4.
   (d) σ_ϵ for $\epsilon > 0, \epsilon \neq 1$, is the nonzero spectrum of a positive matrix by Fact 5.

20.6 Some Merging Results for Spectra of Nonnegative Matrices

Facts:

1. If $\{\lambda_1, \ldots, \lambda_n\} \in \mathcal{N}_n$ and $\{\mu_1, \ldots, \mu_m\} \in \mathcal{N}_m$, then $\{\lambda_1, \ldots, \lambda_n, \mu_1, \ldots, \mu_m\} \in \mathcal{N}_{n+m}$.
2. [Fie74] Let $\sigma = \{\lambda_1, \ldots, \lambda_n\} \in \mathcal{S}_n$ with $\rho(\sigma) = \lambda_1$ and $\tau = \{\mu_1, \ldots, \mu_m\} \in \mathcal{S}_m$ with $\rho(\tau) = \mu_1$. Then $\{\lambda_1 + \epsilon, \lambda_2, \ldots, \lambda_n, \mu_1 - \epsilon, \mu_2, \ldots, \mu_m\} \in \mathcal{S}_{n+m}$ for any $\epsilon \geq 0$ if $\lambda_1 \geq \mu_1$. The proof is constructive.
3. [Šmi04] Let A be a nonnegative matrix with spectrum $\{\lambda_1, \ldots, \lambda_n\}$ and maximal diagonal element d, and let $\tau = \{\mu_1, \ldots, \mu_m\} \in \mathcal{N}_m$ with $\rho(\tau) = \mu_1$. If $d \geq \mu_1$, then $\{\lambda_1, \ldots, \lambda_n, \mu_2, \ldots, \mu_m\} \in \mathcal{N}_{n+m-1}$. The proof is constructive.
4. Let $\sigma = \{\lambda_1, \ldots, \lambda_n\} \in \mathcal{N}_n$ with $\rho(\sigma) = \lambda_1$ and let $\epsilon \geq 0$. Then:
   (a) [Wuw97] $\{\lambda_1 + \epsilon, \lambda_2, \ldots, \lambda_n\} \in \mathcal{N}_n$.
   (b) If $\lambda_2 \in \mathbb{R}$, then not always $\{\lambda_1, \lambda_2 + \epsilon, \lambda_3, \ldots, \lambda_n\} \in \mathcal{N}_n$ (see the previous example).
   (c) [Wuw97] If $\lambda_2 \in \mathbb{R}$, then $\{\lambda_1 + \epsilon, \lambda_2 \pm \epsilon, \lambda_3, \ldots, \lambda_n\} \in \mathcal{N}_n$ (the proof is not constructive).

Examples:

1. Let $\sigma = \{\lambda_1, \ldots, \lambda_n\} \in \mathcal{N}_n$ with $\rho(\sigma) = \lambda_1$, and $\tau = \{\mu_1, \ldots, \mu_m\} \in \mathcal{N}_m$ with $\rho(\tau) = \mu_1$. By Fact 1 of section 20.4 there exists $A \geq 0$ with spectrum σ and row sums λ_1, and $B \geq 0$ with spectrum τ and row sums μ_1. [BMS04] If $\lambda_1 \geq \mu_1$ and $\epsilon \geq 0$, then the nonnegative matrix

$$\left[\begin{array}{c|c} A & \epsilon\, \mathbf{e}\mathbf{e}_1^T \\ \hline (\lambda_1 - \mu_1 + \epsilon)\, \mathbf{e}\mathbf{e}_1^T & B \end{array}\right] \geq 0$$

has row sums $\lambda_1 + \epsilon$ and spectrum $\{\lambda_1 + \epsilon, \lambda_2, \ldots, \lambda_n, \mu_1 - \epsilon, \mu_2, \ldots, \mu_m\}$.

20.7 Sufficient Conditions for Spectra of Nonnegative Matrices

Definitions:

The set $\{\lambda_1, \ldots, \lambda_{i-1}, \alpha, \beta, \lambda_{i+1}, \ldots, \lambda_n\}$ is a **negative subdivision** of $\{\lambda_1, \ldots, \lambda_n\}$ if $\alpha + \beta = \lambda_i$ with $\alpha, \beta, \lambda_i < 0$.

Facts:

Most of the following facts appear in [ELN04] and [SBM05].

1. [Sul49] Let $\sigma = \{\lambda_1, \ldots, \lambda_n\} \subset \mathbb{R}$ with $\lambda_1 \geq \cdots \geq \lambda_n$. Then $\sigma \in \mathcal{R}_n$ if

$$\textbf{(Su)} \begin{cases} \bullet\ \lambda_1 \geq 0 \geq \lambda_2 \geq \cdots \geq \lambda_n \\ \bullet\ \lambda_1 + \cdots + \lambda_n \geq 0 \end{cases}.$$

2. [BMS04] Complex version of (Su). Let $\sigma = \{\lambda_1, \ldots, \lambda_n\} \subset \mathbb{C}$ be a set that satisfies:

   (a) $\overline{\sigma} = \sigma$.

   (b) $\rho(\sigma) = \lambda_1$.

   (c) $\lambda_1 + \cdots + \lambda_n \geq 0$.

   (d) $\{\lambda_2, \ldots, \lambda_n\} \subset \{z \in \mathbb{C} : \mathrm{Re}z \leq 0, |\mathrm{Re}z| \geq |\mathrm{Im}z|\}$.

   Then $\sigma \in \mathcal{N}_n$ and the proof is constructive.

3. [Sou83] Let $\sigma = \{\lambda_1, \ldots, \lambda_n\} \subset \mathbb{R}$ with $\lambda_1 \geq \cdots \geq \lambda_n$. Then there exists a symmetric doubly stochastic matrix D such that $\lambda_1 D$ has spectrum σ if

$$\textbf{(Sou)} \quad \left\{ \frac{1}{n}\lambda_1 + \frac{n-m-1}{n(m+1)}\lambda_2 + \sum_{k=1}^{m} \frac{\lambda_{n-2k+2}}{(k+1)k} \geq 0 \quad \text{where} \quad m = \left[\tfrac{n-1}{2}\right] \right.$$

   and the proof is constructive.

4. [Kel71] Let $\sigma = \{\lambda_1, \ldots, \lambda_n\} \subset \mathbb{R}$ with $\lambda_1 \geq \cdots \geq \lambda_n$. Let r be the greatest index for which $\lambda_r \geq 0$ and let $\delta_i = \lambda_{n+2-i}$ for $2 \leq i \leq n-r+1$. Define $K = \{i : 2 \leq i \leq \min\{r, n-r+1\}$ and $\lambda_i + \delta_i < 0\}$. Then $\sigma \in \mathcal{R}_n$ if

$$\textbf{(Ke)} \quad \begin{cases} \bullet\ \lambda_1 + \sum_{i \in K, i<k} (\lambda_i + \delta_i) + \delta_k \geq 0 \text{ for all } k \in K \\ \bullet\ \lambda_1 + \sum_{i \in K} (\lambda_i + \delta_i) + \sum_{j=r+1}^{n-r+1} \delta_j \geq 0 \end{cases}.$$

5. [Bor95] Let $\sigma = \{\lambda_1, \ldots, \lambda_n\} \subset \mathbb{R}$ with $\lambda_1 \geq \cdots \geq \lambda_n$. Then $\sigma \in \mathcal{R}_n$ if

$$\textbf{(Bo)} \quad \begin{cases} \bullet\ \exists \tau = \{\beta_1, \ldots, \beta_d\} \subset \mathbb{R} \text{ with } d \leq n \text{ that satisfies conditions (Ke)} \\ \bullet\ \sigma \text{ is obtained from } \tau \text{ after } n-d \text{ negative subdivisions} \end{cases}.$$

6. [Sot03] Let $\sigma = \{\lambda_1, \ldots, \lambda_n\} \subset \mathbb{R}$. For any partition $\sigma = \sigma^{(1)} \cup \cdots \cup \sigma^{(t)}$ where $\sigma^{(k)} = \{\lambda_1^{(k)}, \ldots, \lambda_{n_k}^{(k)}\}$ with $\lambda_1^{(k)} \geq \cdots \geq \lambda_{n_k}^{(k)}$ define

$$R_j^{(k)} = \lambda_j^{(k)} + \lambda_{n_k-j+1}^{(k)} \text{ for } 2 \leq j \leq \left[\tfrac{n_k}{2}\right] \text{ and } R_{\frac{n_k+1}{2}}^{(k)} = \lambda_{\frac{n_k+1}{2}}^{(k)} \quad \text{if } n_k \text{ odd;}$$
$$T^{(k)} = \lambda_1^{(k)} + \lambda_{n_k}^{(k)} + \sum_{R_j^{(k)}<0} R_j^{(k)}.$$

   Then $\sigma \in \mathcal{R}_n$ if

$$\textbf{(Sot)} \quad \begin{cases} \text{there exists a partition } \sigma = \sigma^{(1)} \cup \cdots \cup \sigma^{(t)} \text{ such that} \\ \lambda_1^{(1)} + \displaystyle\sum_{T^{(k)}<0, k=2}^{t} T^{(k)} \geq \max\left\{\lambda_1^{(1)} - T^{(1)}, \max_{2 \leq k \leq t}\{\lambda_1^{(k)}\}\right\} \end{cases}$$

   and the proof is constructive.

7. [SBM05] (Su) $\Rightarrow$ (Ke) $\Rightarrow$ (Bo) $\Rightarrow$ (Sot) and no opposite implication is true.
8. [Rad96] (Sou) and (Ke) are not comparable (see Example 2 below).
9. [Rad96] If σ satisfies (Bo), then $\sigma \in \mathcal{S}_n$.
10. [RS03] Let $\sigma = \{\lambda_1, \ldots, \lambda_n\} \subset \mathbb{C}$ with $\overline{\sigma} = \sigma$ and $d = -\sum_{i=2}^{n} \lambda_i > 0$. Let $c_1, \ldots, c_n$ be defined by $(x-d)\prod_{i=2}^{n}(x-\lambda_i) = x^n + \sum_{k=1}^{n} c_k x^{n-k}$. If $\lambda_1 \geq d\left(1 + \sum_{c_k>0} \frac{c_k}{d^k}\right)$, then $\sigma \in \mathcal{N}_n$. The proof is constructive.

Examples:

1. If $\sigma = \{\lambda_1, \ldots, \lambda_n\} \subset \mathbb{R}$ satisfies (Su), then the companion matrix of the polynomial $\prod_{i=1}^{n}(x-\lambda_i)$ is nonnegative with spectrum σ.

2. $\{5, 3, -2, -2, -4\}$ satisfies (Ke), but not (Sou), and $\{5, 3, -2, -2, -2, -2\}$ satisfies (Sou), but not (Ke).
3. $\sigma = \{8, 4, -3, -3, -3, -3\}$ does not satisfies (Ke), but it satisfies (Bo) since σ is obtained from $\tau = \{8, 4, -6, -6\}$ after two negative subdivisions and τ satisfies (Ke).
4. $\sigma = \{9, 7, 2, 2, -5, -5, -5, -5\}$ does not satisfy (Bo), but it satisfies (Sot) with $\sigma^{(1)} = \{9, 2, -5, -5\}$ and $\sigma^{(2)} = \{7, 2, -5, -5\}$.
5. $\sigma = \{6, 1, 1, -4, -4\}$ does not satisfy (Sot), but $\sigma \in \mathcal{R}_5$ (Example 2 of section 20.4).

20.8 Affine Parameterized IEPs (PIEPs)

The set $F^{n\times n}$ of $n \times n$ matrices over the field F is naturally identified with the vector space F^{n^2}. An affine parameterized IEP requires finding within a given affine subspace of $F^{n\times n}$ a matrix with prescribed spectrum. Here we will consider that the given affine subspace is n-dimensional and $F = \mathbb{R}$ or $F = \mathbb{C}$. Especially interesting is the case where the affine subspace contains only real symmetric matrices. Some important motivating applications, including the Sturn–Liouville problem, inverse vibration problems, and nuclear spectroscopy are discussed in [FNO87].

Most of the facts of Sections 20.8, 20.9, and 20.10 appear in [Dai98].

Definitions:

An **affine parameterized IEP** (PIEP) has the following standard formulation:
Given: A field F; $n+1$ matrices $A, A_1, \ldots, A_n \in F^{n\times n}$; and n elements $\lambda_1, \ldots, \lambda_n \in F$.
Find: $\mathbf{c} = (c_1, \ldots, c_n)^T \in F^n$ such that $\{\lambda_1, \ldots, \lambda_n\}$ is the spectrum of the matrix

$$A(\mathbf{c}) = A + c_1 A_1 + \cdots + c_n A_n.$$

In particular, a PIEP($\mathbb{C}$) is a PIEP with $F = \mathbb{C}$, a PIEP($\mathbb{R}$) is a PIEP with $F = \mathbb{R}$, and a PIEP($\mathbb{R}$S) is a PIEP with $F = \mathbb{R}$ and with all given matrices symmetric.

Facts: [Dai98]

1. [Xu92] Almost all PIEP($\mathbb{C}$) are solvable.
2. [SY86] Almost all PIEP($\mathbb{R}$) and almost all PIEP($\mathbb{R}$S) are unsolvable in the presence of multiple eigenvalues.
3. All known sufficient conditions for the solvability of a PIEP($\mathbb{R}$) or a PIEP($\mathbb{R}$S) require that the eigenvalues should be sufficiently pairwise separated. An account of necessary and of sufficient conditions can be found in [Dai98].

20.9 Relevant PIEPs Which are Solvable Everywhere

Definitions:

Additive IEP (AIEP): Given $A \in \mathbb{C}^{n\times n}$ and $\lambda_1, \ldots, \lambda_n \in \mathbb{C}$, find a diagonal matrix $D \in \mathbb{C}^{n\times n}$ such that $\sigma(A + D) = \{\lambda_1, \ldots, \lambda_n\}$.

Multiplicative IEP (MIEP): Given $B \in \mathbb{C}^{n\times n}$ and $\lambda_1, \ldots, \lambda_n \in \mathbb{C}$, find a diagonal matrix $D \in \mathbb{C}^{n\times n}$ such that $\sigma(BD) = \{\lambda_1, \ldots, \lambda_n\}$.

Toeplitz IEP (ToIEP): Given $\lambda_1, \ldots, \lambda_n \in \mathbb{R}$, find $\mathbf{c} = [c_1, \ldots, c_n]^T \in \mathbb{R}^n$ such that $\left[t_{ij}\right]_{i,j=1}^n$ with $t_{ij} = c_{|i-j|+1}$ has spectrum $\{\lambda_1, \ldots, \lambda_n\}$.

Facts: [Dai98]

1. Each AIEP is a PIEP($\mathbb{C}$) with $A_k = \mathbf{e}_k \mathbf{e}_k^T$ for $k = 1, \ldots, n$. Each AIEP is also an IEP with prescribed (off-diagonal) entries.
2. [Fri77] For any $A \in \mathbb{C}^{n\times n}$ and any $\lambda_1, \ldots, \lambda_n \in \mathbb{C}$, the corresponding AIEP is solvable, with the number of solutions not exceeding $n!$. Moreover, for almost all $\lambda_1, \ldots, \lambda_n$ there are exactly $n!$ solutions.
3. Each MIEP is a PIEP($\mathbb{C}$) with $A = 0$ and $A_k = \mathbf{v}_k \mathbf{e}_k^T$ for $k = 1, \ldots, n$, where $\mathbf{v}_1, \ldots, \mathbf{v}_n \in \mathbb{C}^n$ and $B = [\, \mathbf{v}_1 \ \cdots \ \mathbf{v}_n \,] \in \mathbb{C}^{n\times n}$.
4. [Fri75] Assume that all principal minors of $B \in \mathbb{C}^{n\times n}$ are nonzero. For any $\lambda_1, \ldots, \lambda_n \in \mathbb{C}$, the corresponding MIEP is solvable, with the number of solutions not exceeding $n!$. Moreover, for almost all $\lambda_1, \ldots, \lambda_n$ there are exactly $n!$ solutions.
5. Each ToIEP is a PIEP($\mathbb{R}$S) with $A = 0$ and $A_k = [\, a_{ij}^{(k)} \,]_{i,j=1}^n$, where $a_{ij}^{(k)} = 1$ if $|i - j| + 1 = k$ and $a_{ij}^{(k)} = 0$ otherwise.
6. [Lan94] For any $\lambda_1, \ldots, \lambda_n \in \mathbb{R}$ the corresponding ToIEP is solvable.

20.10 Numerical Methods for PIEPs

Facts: [Dai98]

1. For a given PIEP($\mathbb{R}$S), it is possible to order both the eigenvalues $\lambda_1 \leq \cdots \leq \lambda_n$ and the eigenvalues $\lambda_1(\mathbf{c}) \leq \cdots \leq \lambda_n(\mathbf{c})$ of $A(\mathbf{c})$. Then solving the PIEP($\mathbb{R}$S) is equivalent to solving the nonlinear system

$$f(\mathbf{c}) = (\lambda_1(\mathbf{c}) - \lambda_1, \ldots, \lambda_n(\mathbf{c}) - \lambda_n)^T = 0.$$

   Assume that a solution $\mathbf{c}^*$ exists and that the given eigenvalues are distinct:

   - **Method 1a.** Newton's method provides a locally quadratically convergent (l.q.c.) algorithm, and it is usually l.q.c. even in the presence of multiple eigenvalues ([FNO87]). Each iteration in Newton's method involves the solution of an eigenvalue–eigenvector problem.
   - **Method 1b.** A Newton-like method is given in [FNO87] . Newton's method is modified by using the inverse power method to find approximate eigenvectors in each iteration. The new algorithm maintains l.q.c. (see [CXZ99]).
   - **Method 1c.** An inexact Newton-like method is given in [CCX03] . The last iterations of the inverse power method are truncated avoiding oversolving. The algorithm converges superlinearly, but the overall cost is reduced. In particular, for ToIEPs, this improved algorithm has better performance than specific known algorithms.

2. For a given PIEP($\mathbb{R}$) or PIEP($\mathbb{C}$), complex eigenvalues can appear. Assume that for the corresponding PIEP a solution $\mathbf{c}^*$ exists:

   - **Method 2.** In [BK81], Newton's method is applied to solve

$$f(\mathbf{c}) = (\det(A(\mathbf{c}) - \lambda_1 I_n), \ldots, \det(A(\mathbf{c}) - \lambda_n I_n))^T = 0.$$

     The algorithm is l.q.c. and is suitable for the case of distinct eigenvalues.
   - **Method 3.** Newton's method is applied in [Xu96] to solve

$$f(\mathbf{c}) = (\sigma_{min}(A(\mathbf{c}) - \lambda_1 I_n), \ldots, \sigma_{min}(A(\mathbf{c}) - \lambda_n I_n))^T = 0,$$

     where σ_{min} denotes the smallest singular value. The algorithm is l.q.c. under mild conditions even when multiple eigenvalues are present.

- **Method 4a.** An l.q.c. algorithm based on the QR decomposition theory is given in [Li92] that is suitable for the case of distinct eigenvalues.
- **Method 4b.** Method 4a is extended in [Dai99] to the case of multiple eigenvalues for any PIEP($\mathbb{R}$S). The new algorithm is l.q.c. and is based on QR-like decomposition theory and least square techniques. Methods 4a and 4b are given in a more general context.

3. All previous methods require starting from an initial point close to a solution in order to guarantee convergence.

   - **Method 5.** A homotopy approach has been considered for complex symmetric matrices (see [Chu90], [Xu93]), which in theory provides a globally convergent algorithm by which all solutions can be found.

References

[BGRS90] J.A. Ball, I. Gohberg, L. Rodman, and T. Shalom. On the eigenvalues of matrices with given upper triangular part. *Integ. Eq. Oper. Theory*, 13(4):488–497, 1990.

[BH91] M. Boyle and D. Handelman. The spectra of nonnegative matrices via symbolic dynamics. *Ann. of Math. (2)*, 133(2):249–316, 1991.

[BK81] F.W. Biegler-König. A Newton iteration process for inverse eigenvalue problems. *Numer. Math.*, 37(3):349–354, 1981.

[BM97] A. Borobia and J. Moro. On nonnegative matrices similar to positive matrices. *Lin. Alg. Appl.*, 266:365–379, 1997.

[BMS04] A. Borobia, J. Moro, and R. Soto. Negativity compensation in the nonnegative inverse eigenvalue problem. *Lin. Alg. Appl.*, 393:73–89, 2004.

[Bor95] A. Borobia. On the nonnegative eigenvalue problem. *Lin. Alg. Appl.*, 223/224:131–140, 1995.

[CCX03] R.H. Chan, H.L. Chung, and S.F. Xu. The inexact Newton-like method for inverse eigenvalue problem. *BIT*, 43(1):7–20, 2003.

[CDS04] M.T. Chu, F. Diele, and I. Sgura. Gradient flow methods for matrix completion with prescribed eigenvalues. *Lin. Alg. Appl.*, 379:85–112, 2004.

[CG02] M.T. Chu and G.H. Golub. Structured inverse eigenvalue problems. *Acta Numer.*, 11:1–71, 2002.

[Chu90] M.T. Chu. Solving additive inverse eigenvalue problems for symmetric matrices by the homotopy method. *IMA J. Numer. Anal.*, 10(3):331–342, 1990.

[CL83] N.N. Chan and K.H. Li. Diagonal elements and eigenvalues of a real symmetric matrix. *J. Math. Anal. Appl.*, 91(2):562–566, 1983.

[CXZ99] R.H. Chan, S.F. Xu, and H.M. Zhou. On the convergence of a quasi-Newton method for inverse eigenvalue problems. *SIAM J. Numer. Anal.*, 36(2):436–441 (electronic), 1999.

[Dai98] H. Dai. Some developments on parameterized inverse eigenvalue problems. *CERFACS Tech. Report TR/PA/98/34*,1998.

[Dai99] H. Dai. An algorithm for symmetric generalized inverse eigenvalue problems. *Lin. Alg. Appl.*, 296(1-3):79–98, 1999.

[DD45] N. Dmitriev and E. Dynkin. On the characteristic numbers of a stochastic matrix. *C.R. (Doklady) Acad. Sci. URSS*, 49:159–162, 1945.

[DD46] N. Dmitriev and E. Dynkin. On characteristic roots of stochastic matrix. *Izv. Akad. Nauk SSSR, Ser. Mat.*, 10:167–184, 1946.

[Dds74] J.A. Dias da Silva. Matrices with prescribed entries and characteristic polynomial. *Proc. Amer. Math. Soc.*, 45:31–37, 1974.

[dO71] G.N. de Oliveira. Matrices with prescribed characteristic polynomial and a prescribed submatrix. III. *Monatsh. Math.*, 75:441–446, 1971.

[dO73a] G.N. de Oliveira. Matrices with prescribed entries and eigenvalues. I. *Proc. Amer. Math. Soc.*, 37:380–386, 1973.

[dO73b] G.N. de Oliveira. Matrices with prescribed entries and eigenvalues. II. *SIAM J. Appl. Math.*, 21:414–417, 1973.

[dO75] G.N. de Oliveira. Matrices with prescribed characteristic polynomial and several prescribed submatrices. *Lin. Multilin. Alg.*, 2:357–364, 1975.

[dS79] E.M. de Sá. Imbedding conditions for λ-matrices. *Lin. Alg. Appl.*, 24:33–50, 1979.

[ELN04] P.D. Egleston, T.D. Lenker, and S.K. Narayan. The nonnegative inverse eigenvalue problem. *Lin. Alg. Appl.*, 379:475–490, 2004.

[Fie74] M. Fiedler. Eigenvalues of nonnegative symmetric matrices. *Lin. Alg. Appl.*, 9:119–142, 1974.

[Fil69] P.A. Fillmore. On similarity and the diagonal of a matrix. *Amer. Math. Monthly*, 76:167–169, 1969.

[FNO87] S. Friedland, J. Nocedal, and M.L. Overton. The formulation and analysis of numerical methods for inverse eigenvalue problems. *SIAM J. Numer. Anal.*, 24(3):634–667, 1987.

[Fri75] S. Friedland. On inverse multiplicative eigenvalue problems for matrices. *Lin. Alg. Appl.*, 12(2):127–137, 1975.

[Fri77] S. Friedland. Inverse eigenvalue problems. *Lin. Alg. Appl.*, 17(1):15–51, 1977.

[FS98] S. Furtado and F.C. Silva. On the characteristic polynomial of matrices with prescribed columns and the stabilization and observability of linear systems. *Electron. J. Lin. Alg.*, 4:19–31, 1998.

[GKvS95] I. Gohberg, M.A. Kaashoek, and F. van Schagen. *Partially Specified Matrices and Operators: Classification, Completion, Applications*, Vol. 79 of Operator Theory: Advances and Applications. Birkhäuser Verlag, Basel, Germany, 1995.

[Her83] D. Hershkowitz. Existence of matrices with prescribed eigenvalues and entries. *Lin. Multilin. Alg.*, 14(4):315–342, 1983.

[IC00] Kh.D. Ikramov and V.N. Chugunov. Inverse matrix eigenvalue problems. *J. Math. Sci. (New York)*, 98(1):51–136, 2000.

[JLL96] C.R. Johnson, T.J. Laffey, and R. Loewy. The real and the symmetric nonnegative inverse eigenvalue problems are different. *Proc. Amer. Math. Soc.*, 124(12):3647–3651, 1996.

[Joh81] C.R. Johnson. Row stochastic matrices similar to doubly stochastic matrices. *Lin. Multilin. Alg.*, 10(2):113–130, 1981.

[Kar51] F. Karpelevich. On the eigenvalues of a matrix with nonnegative elements (in russian). *Izv. Akad. Nauk SSR Ser. Mat.*, 14:361–383, 1951.

[Kel71] R.B. Kellogg. Matrices similar to a positive or essentially positive matrix. *Lin. Alg. Appl.*, 4:191–204, 1971.

[KOR00] K.H. Kim, N.S. Ormes, and F.W. Roush. The spectra of nonnegative integer matrices via formal power series. *J. Amer. Math. Soc.*, 13(4):773–806, 2000.

[KM01] J. Knudsen and J. McDonald. A note on the convexity of the realizable set of eigenvalues for nonnegative symmetryc matrices. *Electron. J. Lin. Alg.*, 8:110–114, 2001.

[Lan94] H.J. Landau. The inverse eigenvalue problem for real symmetric Toeplitz matrices. *J. Amer. Math. Soc.*, 7(3):749–767, 1994.

[Li92] R.C. Li. Algorithms for inverse eigenvalue problems. *J. Comput. Math.*, 10(2):97–111, 1992.

[LL78] R. Loewy and D. London. A note on an inverse problem for nonnegative matrices. *Lin. Multilin. Alg.*, 6(1):83–90, 1978.

[LM99] T.J. Laffey and E. Meehan. A characterization of trace zero nonnegative 5×5 matrices. *Lin. Alg. Appl.*, 302/303:295–302, 1999.

[Min88] H. Minc. *Nonnegative matrices.* Wiley-Interscience Series in Discrete Mathematics and Optimization. John Wiley & Sons, New York, 1988.

[Mir63] L. Mirsky. Results and problems in the theory of doubly-stochastic matrices. *Z. Wahrscheinlichkeitstheorie und Verw. Gebiete*, 1:319–334, 1962/1963.

[Mou03] B. Mourad. An inverse problem for symmetric doubly stochastic matrices. *Inverse Problems*, 19(4):821–831, 2003.

[MS00] I.T. Matos and F.C. Silva. A completion problem over the field of real numbers. *Lin. Alg. Appl.*, 320(1-3):63–77, 2000.

[Rad96] N. Radwan. An inverse eigenvalue problem for symmetric and normal matrices. *Lin. Alg. Appl.*, 248:101–109, 1996.

[Rea96] R. Reams. An inequality for nonnegative matrices and the inverse eigenvalue problem. *Lin. Multilin. Alg.*, 41(4):367–375, 1996.

[RS03] O. Rojo and R.L. Soto. Existence and construction of nonnegative matrices with complex spectrum. *Lin. Alg. Appl.*, 368:53–69, 2003.

[Sil87a] F.C. Silva. Matrices with prescribed characteristic polynomial and submatrices. *Portugal. Math.*, 44(3):261–264, 1987.

[Sil87b] F.C. Silva. Matrices with prescribed eigenvalues and principal submatrices. *Lin. Alg. Appl.*, 92:241–250, 1987.

[Sil90] F.C. Silva. Matrices with prescribed similarity class and a prescribed nonprincipal submatrix. *Portugal. Math.*, 47(1):103–113, 1990.

[Sil91] F.C. Silva. Matrices with prescribed eigenvalues and blocks. *Lin. Alg. Appl.*, 148:59–73, 1991.

[Sil93] F.C. Silva. Matrices with prescribed lower triangular part. *Lin. Alg. Appl.*, 182:27–34, 1993.

[Šmi04] H. Šmigoc. The inverse eigenvalue problem for nonnegative matrices. *Lin. Alg. Appl.*, 393:365–374, 2004.

[SBM05] R.L. Soto, A. Borobia, and J. Moro. On the comparison of some realizability criteria for the real nonnegative inverse eigenvalue problem. *Lin. Alg. Appl.*, 396:223–241, 2005.

[Sot03] R.L. Soto. Existence and construction of nonnegative matrices with prescribed spectrum. *Lin. Alg. Appl.*, 369:169–184, 2003.

[Sou83] G.W. Soules. Constructing symmetric nonnegative matrices. *Lin. Multilin. Alg.*, 13(3):241–251, 1983.

[Sul49] H.R. Suleĭmanova. Stochastic matrices with real characteristic numbers. *Doklady Akad. Nauk SSSR (N.S.)*, 66:343–345, 1949.

[SY86] J.G. Sun and Q. Ye. The unsolvability of inverse algebraic eigenvalue problems almost everywhere. *J. Comput. Math.*, 4(3):212–226, 1986.

[Tho79] R.C. Thompson. Interlacing inequalities for invariant factors. *Lin. Alg. Appl.*, 24:1–31, 1979.

[Wim74] H.K. Wimmer. Existenzsätze in der Theorie der Matrizen und lineare Kontrolltheorie. *Monatsh. Math.*, 78:256–263, 1974.

[Wuw97] G. Wuwen. Eigenvalues of nonnegative matrices. *Lin. Alg. Appl.*, 266:261–270, 1997.

[Xu92] S.F. Xu. The solvability of algebraic inverse eigenvalue problems almost everywhere. *J. Comput. Math.*, 10(Supplementary Issue):152–157, 1992.

[Xu93] S.F. Xu. A homotopy algorithm for solving the inverse eigenvalue problem for complex symmetric matrices. *J. Comput. Math.*, 11(1):7–19, 1993.

[Xu96] S.F. Xu. A smallest singular value method for solving inverse eigenvalue problems. *J. Comput. Math.*, 14(1):23–31, 1996.

[Zab86] I. Zaballa. Existence of matrices with prescribed entries. *Lin. Alg. Appl.*, 73:227–280, 1986.

[Zab87] I. Zaballa. Matrices with prescribed rows and invariant factors. *Lin. Alg. Appl.*, 87:113–146, 1987.

[Zab89] I. Zaballa. Matrices with prescribed invariant factors and off-diagonal submatrices. *Lin. Multilin. Alg.*, 25(1):39–54, 1989.

21

Totally Positive and Totally Nonnegative Matrices

Shaun M. Fallat
University of Regina

Total positivity has been a recurring theme in linear algebra and other aspects of mathematics for the past 80 years. Totally positive matrices, in fact, originated from studying small oscillations in mechanical systems [GK60], and from investigating relationships between the number of sign changes of the vectors $\mathbf{x}$ and $A\mathbf{x}$ for fixed A [Sch30]. Since then this class (and the related class of sign-regular matrices) has arisen in such a wide range of applications (see [GM96] for an incredible compilation of interesting articles dealing with a long list of relevant applications of totally positive matrices) that over the years many convenient points of view for total positivity have been offered and later defended by many prominent mathematicians.

After F.R. Gantmacher and M.G. Krein [GK60], totally positive matrices were seen and further developed in connection with spline functions, collocation matrices, and generating totally positive sequences (Polya frequency sequences). Then in 1968 came one of the most important and influential references in this area, namely the book *Total Positivity* by S. Karlin [Kar68]. Karlin approached total positivity by considering the analytic properties of totally positive functions. Along these lines, he studied totally positive kernels, sign-regular functions, and Polya frequency functions. Karlin also notes the importance of total positivity in the field of statistics.

The next significant view point can be seen in T. Ando's survey paper [And87]. Ando's contribution was to consider a multilinear approach to this subject, namely making use of skew-symmetric products and Schur complements as his underlying tools. More recently, it has become clear that factorizations of totally positive matrices are a fruitful avenue for research on this class. Coupled with matrix factorizations, totally positive matrices have taken on a new combinatorial form known as Planar networks. Karlin and G. McGregor [KM59] were some of the pioneers of this view point on total positivity (see also [Bre95][Lin73]), and there has since been a revolution of sorts with many new and exciting advances and additional applications (e.g., positive elements in reductive Lie groups, computer-aided geometric design, shape-preserving designs).

This area is not only a historically significant one in linear algebra, but it will continue to produce many important advances and spawn many more worthwhile applications.

21.1 Basic Properties

In this section, we present many basic, yet fundamental, properties associated with the important class of totally positive and totally nonnegative matrices.

Definitions:

An $m \times n$ real matrix A is **totally nonnegative** (TN) if the determinant of every square submatrix (i.e., minor) is nonnegative.

An $m \times n$ real matrix A is **totally positive** (TP) if every minor of A is positive.

An $n \times n$ matrix A is **oscillatory** if A is totally nonnegative and A^k is totally positive for some integer $k \geq 1$.

An $m \times n$ real matrix is **in double echelon form** if

(a) Each row of A has one of the following forms ($*$ indicates a nonzero entry):

1. $(*, *, \cdots, *)$,
2. $(*, \cdots, *, 0, \cdots, 0)$,
3. $(0, \cdots, 0, *, \cdots, *)$, or
4. $(0, \cdots, 0, *, \cdots, *, 0, \cdots, 0)$.

(b) The first and last nonzero entries in row $i+1$ are not to the left of the first and last nonzero entries in row i, respectively ($i = 1, 2, \ldots, n-1$).

Facts:

1. Every totally positive matrix is a totally nonnegative matrix.
2. Suppose A is a totally nonnegative (positive) rectangular matrix. Then
   (a) A^T, the transpose of A, is totally nonnegative (positive),
   (b) $A[\alpha, \beta]$ is totally nonnegative (positive) for any row index set α and column index set β.
3. (Section 4.2) Cauchy–Binet Identity: Since any $k \times k$ minor of the product AB is a sum of products of $k \times k$ minors of A and B, it follows that if all the $k \times k$ minors of two $n \times n$ matrices are positive, then all the $k \times k$ minors of their product are positive.
4. [GK02, p .74], [And87] The set of all totally nonnegative (positive) matrices is closed under multiplication.
5. Let A be TP (TN) and D_1 and D_2 be positive diagonal matrices. Then $D_1 A D_2$ is TP (TN).
6. [GK02, p .75] If A is a square invertible totally nonnegative (or is totally positive) matrix, then $SA^{-1}S$ is totally nonnegative (positive) for $S = \text{diag}(1, -1, \cdots, \pm 1)$. Hence, if A is a square invertible totally nonnegative matrix (or is totally positive), then the unsigned adjugate matrix of A (or the $(n-1)$st compound of A) is totally nonnegative (positive).
7. [And87], [Fal99] If A is a square totally nonnegative (positive) matrix, then, assuming $A[\alpha^c]$ is invertible, $A/A[\alpha^c] = A[\alpha] - A[\alpha, \alpha^c](A[\alpha^c])^{-1} A[\alpha^c, \alpha]$, the Schur complement of $A[\alpha^c]$ in A, is totally nonnegative (positive), for all index sets α based on consecutive indices. Recall that α^c denotes the complement of the set α.
8. [And87], [Fal01], [Whi52], [Kar68, p. 98] The closure of the totally positive matrices (in the usual topology on $\mathbb{R}^{m \times n}$) is the totally nonnegative matrices.
9. [Fal99], [JS00] Let $A = [\mathbf{a_1}, \mathbf{a_2}, \ldots, \mathbf{a_n}]$ be an $m \times n$ totally nonnegative (positive) matrix whose i^{th} column is $\mathbf{a_i}$ ($i = 1, 2, \ldots, n$). Suppose C denotes the set of all column vectors $\mathbf{b}$ for which

the $m \times (n+1)$ matrix $\hat{A} = [\mathbf{a_1}, \ldots, \mathbf{a_k}, \mathbf{b}, \mathbf{a_{k+1}}, \ldots, \mathbf{a_n}]$ is a totally nonnegative (positive) matrix (here k is fixed but arbitrary). Then C is a nonempty convex cone.

10. [Fal99] If A is an $m \times n$ totally nonnegative (positive) matrix, then increasing the (1,1) or the (m, n) entries of A results in a totally nonnegative (positive) matrix. In general these are the only two entries in a TN matrix with this property (see [FJS00]).
11. [And87] Let P denote the $n \times n$ permutation matrix induced by the permutation $i \to n - i + 1$, $(1 \le i \le n)$, and suppose A is an $n \times n$ totally nonnegative (positive) matrix. Then PAP is a totally nonnegative (positive) matrix.
12. Any irreducible tridiagonal matrix with nonzero main diagonal is in double echelon form.
13. [Fal99] Let A be an $m \times n$ totally nonnegative matrix with no zero rows or columns. Then A is in double echelon form.
14. [Rad68], [Fal99] An $n \times n$ totally nonnegative matrix $A = [a_{ij}]$ is irreducible if and only if $a_{i,i+1} > 0$ and $a_{i+1,i} > 0$, for $i = 1, 2, \ldots, n-1$.

Examples:

1. Consider the following 3×3 matrix: $A = \begin{bmatrix} 1 & 1 & 1 \\ 1 & 2 & 4 \\ 1 & 3 & 9 \end{bmatrix}$. It is not difficult to check that all minors of A are positive.
2. (Inverse tridiagonal matrix) From Fact 6 above, the inverse of a TN tridiagonal matrix is signature similar to a TN matrix. Such matrices are referred to as "single-pair" matrices in [GK02, pp. 78–80], are very much related to "Green's matrices" (see [Kar68, pp. 110–112]), and are similar to matrices of type D found in [Mar70a].
3. (Vandermonde matrix) Vandermonde matrices arise in the problem of determining a polynomial of degree at most $n-1$ that interpolates n data points. Suppose that n data points $(x_i, y_i)_{i=1}^n$ are given. The goal is to construct a polynomial $p(x) = a_0 + a_1 x + \cdots + a_{n-1}x^{n-1}$ that satisfies $p(x_i) = y_i$ for $i = 1, 2, \ldots, n$, which can be expressed as

$$\begin{bmatrix} 1 & x_1 & x_1^2 & \cdots & x_1^{n-1} \\ 1 & x_2 & x_2^2 & \cdots & x_2^{n-1} \\ \vdots & & \vdots & & \vdots \\ 1 & x_n & x_n^2 & \cdots & x_n^{n-1} \end{bmatrix} \begin{bmatrix} a_0 \\ a_1 \\ \vdots \\ a_{n-1} \end{bmatrix} = \begin{bmatrix} y_1 \\ y_2 \\ \vdots \\ y_n \end{bmatrix}. \tag{21.1}$$

The $n \times n$ coefficient matrix in (21.1) is called a *Vandermonde matrix*, and we denote it by $\mathcal{V}(x_1, \ldots, x_n)$. The determinant of the $n \times n$ Vandermonde matrix in (21.1) is given by the formula $\prod_{i>j}(x_i - x_j)$; see [MM64, pp. 15–16]. Thus, if $0 < x_1 < x_2 < \cdots < x_n$, then $\mathcal{V}(x_1, \ldots, x_n)$ has positive entries, positive leading principal minors, and positive determinant. More generally, it is known [GK02, p. 111] that if $0 < x_1 < x_2 < \cdots < x_n$, then $\mathcal{V}(x_1, \ldots, x_n)$ is TP. Example 1 above is a Vandermonde matrix.

4. Let $f(x) = \sum_{i=0}^n a_i x^i$ be an nth degree polynomial in x. The *Routh–Hurwitz matrix* is the $n \times n$ matrix given by

$$A = \begin{bmatrix} a_1 & a_3 & a_5 & a_7 & \cdots & 0 & 0 \\ a_0 & a_2 & a_4 & a_6 & \cdots & 0 & 0 \\ 0 & a_1 & a_3 & a_5 & \cdots & 0 & 0 \\ 0 & a_0 & a_2 & a_4 & \cdots & 0 & 0 \\ \vdots & \vdots & \vdots & \vdots & \cdots & \vdots & \vdots \\ 0 & 0 & 0 & 0 & \cdots & a_{n-1} & 0 \\ 0 & 0 & 0 & 0 & \cdots & a_{n-2} & a_n \end{bmatrix}.$$

A specific example of a Routh–Hurwitz matrix for an arbitrary polynomial of degree six, $f(x) = \sum_{i=0}^{6} a_i x^i$, is given by

$$A = \begin{bmatrix} a_1 & a_3 & a_5 & 0 & 0 & 0 \\ a_0 & a_2 & a_4 & a_6 & 0 & 0 \\ 0 & a_1 & a_3 & a_5 & 0 & 0 \\ 0 & a_0 & a_2 & a_4 & a_6 & 0 \\ 0 & 0 & a_1 & a_3 & a_5 & 0 \\ 0 & 0 & a_0 & a_2 & a_4 & a_6 \end{bmatrix}.$$

A polynomial $f(x)$ is *stable* if all the zeros of $f(x)$ have negative real parts. It is proved in [Asn70] that $f(x)$ is stable if and only if the Routh–Hurwitz matrix formed from f is totally nonnegative.

5. (Cauchy matrix) An $n \times n$ matrix $C = [c_{ij}]$ is called a *Cauchy matrix* if the entries of C are given by

$$c_{ij} = \frac{1}{x_i + y_j},$$

where $x_1, x_2, \ldots, x_n$ and $y_1, y_2, \ldots, y_n$ are two sequences of numbers (chosen so that c_{ij} is well-defined). A Cauchy matrix is totally positive if and only if $0 < x_1 < x_2 < \cdots < x_n$ and $0 < y_1 < y_2 < \cdots < y_n$ ([GK02, pp. 77–78]).

6. (Pascal matrix) Consider the 4×4 matrix $P_4 = \begin{bmatrix} 1 & 1 & 1 & 1 \\ 1 & 2 & 3 & 4 \\ 1 & 3 & 6 & 10 \\ 1 & 4 & 10 & 20 \end{bmatrix}$. The matrix P_4 is called the *symmetric* 4×4 *Pascal matrix* because of its connection with Pascal's triangle (see Example 4 of section 21.2 for a definition of P_n for general n). Then P_4 is TP, and the inverse of P_4 is given by

$$P_4^{-1} = \begin{bmatrix} 4 & -6 & 4 & -1 \\ -6 & 14 & -11 & 3 \\ 4 & -11 & 10 & -3 \\ -1 & 3 & -3 & 1 \end{bmatrix}.$$

Notice that the inverse of the 4×4 Pascal matrix is integral. Moreover, deleting the signs by forming $SP_4^{-1}S$, where $S = \text{diag}(1, -1, 1, -1)$, results in the TP matrix $\begin{bmatrix} 4 & 6 & 4 & 1 \\ 6 & 14 & 11 & 3 \\ 4 & 11 & 10 & 3 \\ 1 & 3 & 3 & 1 \end{bmatrix}$.

Applications:

1. (Tridiagonal matrices) When Gantmacher and Krein were studying the oscillatory properties of an elastic segmental continuum (no supports between the endpoints a and b) under small transverse oscillations, they were able to generate a system of linear equations that define the frequency of the oscillation (see [GK60]). The system of equations thus found can be represented in what is known as the influence-coefficient matrix, whose properties are analogous to those governing the segmental continuum. This process of obtaining the properties of the segmental continuum from the influence-coefficient matrix was only possible due to the inception of the theory of oscillatory matrices. A special case involves tridiagonal matrices (or Jacobi matrices as they were called in [GK02]). Tridiagonal matrices are not only interesting in their own right as a model example of oscillatory matrices, but they also naturally arise in studying small oscillations in certain mechanical systems, such as torsional oscillations of a system of disks fastened to a shaft. In [GK02, pp. 81–82] they prove that an irreducible tridiagonal matrix is totally nonnegative if and only if its entries are nonnegative and its leading principal minors are nonnegative.

21.2 Factorizations

Recently, there has been renewed interest in total positivity partly motivated by the so-called "bidiagonal factorization," namely, the fact that any totally positive matrix can be factored into entry-wise nonnegative bidiagonal matrices. This result has proven to be a very useful and tremendously powerful property for this class. (See Section 1.6 for basic information on LU factorizations.)

Definitions:

An **elementary bidiagonal matrix** is an $n \times n$ matrix whose main diagonal entries are all equal to one, and there is, at most, one nonzero off-diagonal entry and this entry must occur on the super- or subdiagonal. The lower elementary bidiagonal matrix whose elements are given by

$$c_{ij} = \begin{cases} 1, & \text{if } i = j, \\ \mu, & \text{if } i = k, \ j = k-1, \\ 0, & \text{otherwise} \end{cases}$$

is denoted by $E_k(\mu) = [c_{ij}]$ $(2 \le k \le n)$.

A triangular matrix is **ΔTP** if all of its nontrivial minors are positive. (Here a trivial minor is one which is zero only because of the zero pattern of a triangular matrix.)

Facts:

1. $(E_k(\mu))^{-1} = E_k(-\mu)$.
2. [Cry73] Let A be an $n \times n$ matrix. Then A is totally positive if and only if A has an LU factorization such that both L and U are $n \times n$ ΔTP matrices.
3. [And87], [Cry76] Let A be an $n \times n$ matrix. Then A is totally nonnegative if and only if A has an LU factorization such that both L and U are $n \times n$ totally nonnegative matrices.
4. [Whi52] Suppose $A = [a_{ij}]$ is an $n \times n$ matrix with $a_{j1}, a_{j+1,1} > 0$, and $a_{k1} = 0$ for $k > j+1$. Let B be the $n \times n$ matrix obtained from A by using row j to eliminate $a_{j+1,1}$. Then A is TN if and only if B is TN. Note that B is equal to $E_{j+1}(-a_{j+1,1}/a_{j1})A_j$ and, hence, $A = (E_{j+1}(-a_{j+1,1}/a_{j1}))^{-1}B = E_{j+1}(a_{j+1,1}/a_{j1})B$.
5. [Loe55], [GP96], [BFZ96], [FZ00], [Fal01] Let A be an $n \times n$ nonsingular totally nonnegative matrix. Then A can be written as

$$\begin{aligned} A = {} & (E_2(l_k))(E_3(l_{k-1})E_2(l_{k-2}))\cdots(E_n(l_{n-1})\cdots E_3(l_2)E_2(l_1))D \\ & (E_2^T(u_1)E_3^T(u_2)\cdots E_n^T(u_{n-1}))\cdots(E_2^T(u_{k-2})E_3^T(u_{k-1}))(E_2^T(u_k)), \end{aligned} \tag{21.2}$$

   where $k = \binom{n}{2}$; $l_i, u_j \ge 0$ for all $i, j \in \{1, 2, \ldots, k\}$; and D is a positive diagonal matrix.
6. [Cry76] Any $n \times n$ totally nonnegative matrix A can be written as

$$A = \prod_{i=1}^{M} L^{(i)} \prod_{j=1}^{N} U^{(j)}, \tag{21.3}$$

   where the matrices $L^{(i)}$ and $U^{(j)}$ are, respectively, lower and upper bidiagonal totally nonnegative matrices with at most one nonzero entry off the main diagonal.
7. [Cry76], [RH72] If A is an $n \times n$ totally nonnegative matrix, then there exists a totally nonnegative matrix S and a tridiagonal totally nonnegative matrix T such that

   (a) $TS = SA$.

   (b) The matrices A and T have the same eigenvalues.

   Moreover, if A is nonsingular, then S is nonsingular.

Examples:

1. Let P_4 be the matrix given in Example 6 of section 21.1. Then P_4 is TP, and a (unique up to a positive diagonal scaling) LU factorization of P_4 is given by

$$P_4 = LU = \begin{bmatrix} 1 & 0 & 0 & 0 \\ 1 & 1 & 0 & 0 \\ 1 & 2 & 1 & 0 \\ 1 & 3 & 3 & 1 \end{bmatrix} \begin{bmatrix} 1 & 1 & 1 & 1 \\ 0 & 1 & 2 & 3 \\ 0 & 0 & 1 & 3 \\ 0 & 0 & 0 & 1 \end{bmatrix}.$$

Observe that the rows of L, or the columns of U, come from the rows of Pascal's triangle (ignoring the zeros); hence, the name Pascal matrix (see Example 4 for a definition of P_n).

2. The 3×3 Vandermonde matrix A in Example 1 of Section 21.1 can be factored as

$$A = \begin{bmatrix} 1 & 0 & 0 \\ 0 & 1 & 0 \\ 0 & 1 & 1 \end{bmatrix} \begin{bmatrix} 1 & 0 & 0 \\ 1 & 1 & 0 \\ 0 & 0 & 1 \end{bmatrix} \begin{bmatrix} 1 & 0 & 0 \\ 0 & 1 & 0 \\ 0 & 1 & 1 \end{bmatrix} \begin{bmatrix} 1 & 0 & 0 \\ 0 & 1 & 0 \\ 0 & 0 & 2 \end{bmatrix} \begin{bmatrix} 1 & 0 & 0 \\ 0 & 1 & 2 \\ 0 & 0 & 1 \end{bmatrix} \begin{bmatrix} 1 & 1 & 0 \\ 0 & 1 & 0 \\ 0 & 0 & 1 \end{bmatrix} \begin{bmatrix} 1 & 0 & 0 \\ 0 & 1 & 1 \\ 0 & 0 & 1 \end{bmatrix}. \quad (21.4)$$

3. In fact, we can write $\mathcal{V}(x_1, x_2, x_3)$ as

$$\mathcal{V}(x_1, x_2, x_3) = \begin{bmatrix} 1 & 0 & 0 \\ 0 & 1 & 0 \\ 0 & 1 & 1 \end{bmatrix} \begin{bmatrix} 1 & 0 & 0 \\ 1 & 1 & 0 \\ 0 & 0 & 1 \end{bmatrix} \begin{bmatrix} 1 & 0 & 0 \\ 0 & 1 & 0 \\ 0 & \frac{(x_3 - x_2)}{(x_2 - x_1)} & 1 \end{bmatrix}$$
$$\begin{bmatrix} 1 & 0 & 0 \\ 0 & x_2 - x_1 & 0 \\ 0 & 0 & (x_3 - x_2)(x_3 - x_1) \end{bmatrix} \begin{bmatrix} 1 & 0 & 0 \\ 0 & 1 & x_2 \\ 0 & 0 & 1 \end{bmatrix} \begin{bmatrix} 1 & x_1 & 0 \\ 0 & 1 & 0 \\ 0 & 0 & 1 \end{bmatrix} \begin{bmatrix} 1 & 0 & 0 \\ 0 & 1 & x_1 \\ 0 & 0 & 1 \end{bmatrix}.$$

4. Consider the factorization (21.2) from Fact 5 of a 4×4 matrix in which all of the variables are equal to one. The resulting matrix is P_4, which is necessarily TP. On the other hand, consider the $n \times n$ matrix $P_n = [p_{ij}]$ whose first row and column entries are all ones, and for $2 \leq i, j \leq n$ let $p_{ij} = p_{i-1,j} + p_{i,j-1}$. In fact, the relation $p_{ij} = p_{i-1,j} + p_{i,j-1}$ implies [Fal01] that P_n can be written as

$$P_n = E_n(1) \cdots E_2(1) \begin{bmatrix} 1 & 0 \\ 0 & P_{n-1} \end{bmatrix} E_2^T(1) \cdots E_n^T(1).$$

Hence, by induction, P_n has the factorization (21.2) in which the variables involved are all equal to one. Consequently, the symmetric Pascal matrix P_n is TP for all $n \geq 1$. Furthermore, since in general $(E_k(\mu))^{-1} = E_k(-\mu)$ (Fact 1), it follows that P_n^{-1} is not only signature similar to a TP matrix, but it is also integral.

21.3 Recognition and Testing

In practice, how can one determine if a given $n \times n$ matrix is TN or TP? One could calculate *every* minor, but that would involve evaluating $\sum_{k=1}^{n} \binom{n}{k}^2 \sim 4^n/\sqrt{\pi n}$ determinants. Is there a smaller collection of minors whose nonnegativity or positivity implies the nonnegativity or positivity of all minors?

Definitions:

For $\alpha = \{i_1, i_2, \ldots, i_k\} \subseteq N = \{1, 2, \ldots, n\}$, with $i_1 < i_2 < \cdots < i_k$, the **dispersion of** α, denoted by $d(\alpha)$, is defined to be $\sum_{j=1}^{k-1}(i_{j+1} - i_j - 1) = i_k - i_1 - (k-1)$, with the convention that $d(\alpha) = 0$, when α is a singleton.

If α and β are two contiguous index sets with $|\alpha| = |\beta| = k$, then the minor $\det A[\alpha, \beta]$ is called **initial** if α or β is $\{1, 2, \ldots, k\}$. A minor is called a **leading principal minor** if it is an initial minor with both $\alpha = \beta = \{1, 2, \ldots, k\}$.

An upper right (lower left) **corner minor** of A is one of the form $\det A[\alpha, \beta]$ in which α consists of the first k (last k) and β consists of the last k (first k) indices, $k = 1, 2, \ldots, n$.

Facts:

1. The dispersion of a set α represents a measure of the "gaps" in the set α. In particular, observe that $d(\alpha) = 0$ if and only if α is a contiguous subset of N.
2. [Fek13] (Fekete's Criterion) An $m \times n$ matrix A is totally positive if and only if $\det A[\alpha, \beta] > 0$, for all $\alpha \subseteq \{1, 2, \ldots, m\}$ and $\beta \subseteq \{1, 2, \ldots, n\}$, with $|\alpha| = |\beta|$ and $d(\alpha) = d(\beta) = 0$. (Reduces the number of minors to be checked for total positivity to roughly n^3.)
3. [GP96], [FZ00] If all initial minors of A are positive, then A is TP. (Reduces the number of minors to be checked for total positivity to n^2.)
4. [SS95], [Fal04] Suppose that A is TN. Then A is TP if and only if all corner minors of A are positive.
5. [GP96] Let $A \in \mathbb{R}^{n \times n}$ be nonsingular.

   (a) A is TN if and only if for each $k = 1, 2, \ldots, n$,

      i. $\det A[\{1, 2, \ldots, k\}] > 0$.

      ii. $\det A[\alpha, \{1, 2, \ldots, k\}] \geq 0$, for every $\alpha \subseteq \{1, 2, \ldots, n\}$, $|\alpha| = k$.

      iii. $\det A[\{1, 2, \ldots, k\}, \beta] \geq 0$, for every $\beta \subseteq \{1, 2, \ldots, n\}$, $|\beta| = k$.

   (b) A is TP if and only if for each $k = 1, 2 \ldots, n$,

      i. $\det A[\alpha, \{1, 2, \ldots, k\}] > 0$, for every $\alpha \subseteq \{1, 2, \ldots, n\}$ with $|\alpha| = k$, $d(\alpha) = 0$.

      ii. $\det A[\{1, 2, \ldots, k\}, \beta] > 0$, for every $\beta \subseteq \{1, 2, \ldots, n\}$ with $|\beta| = k$, $d(\beta) = 0$.
6. [GK02, p. 100] An $n \times n$ totally nonnegative matrix $A = [a_{ij}]$ is oscillatory if and only if

   (a) A is nonsingular.

   (b) $a_{i,i+1} > 0$ and $a_{i+1,i} > 0$, for $i = 1, 2, \ldots, n-1$.
7. [Fal04] Suppose A is an $n \times n$ invertible totally nonnegative matrix. Then A is oscillatory if and only if a parameter from at least one of the bidiagonal factors E_k and E_k^T is positive, for each $k = 2, 3, \ldots, n$ in the elementary bidiagonal factorization of A given in Fact 5 of section 21.2.

Examples:

1. Unfortunately, Fekete's Criterion, Fact 2, does not hold in general if "totally positive" is replaced with "totally nonnegative" and "> 0" is replaced with "≥ 0." Consider the following simple example:

   $A = \begin{bmatrix} 1 & 0 & 2 \\ 1 & 0 & 1 \\ 2 & 0 & 1 \end{bmatrix}$. It is not difficult to verify that every minor of A based on contiguous row and column sets is nonnegative, but $\det A[\{1, 3\}] = -3$. For an invertible and irreducible example consider $A = \begin{bmatrix} 0 & 1 & 0 \\ 0 & 0 & 1 \\ 1 & 0 & 0 \end{bmatrix}$.

21.4 Spectral Properties

Approximately 60 years ago, Gantmacher and Krein [GK60], who were originally interested in oscillation dynamics, undertook a careful study into the theory of totally nonnegative matrices. Of the many topics they considered, one was the properties of the eigenvalues of totally nonnegative matrices.

Facts:

1. [GK02, pp. 86–91] Let A be an $n \times n$ oscillatory matrix. Then the eigenvalues of A are positive, real, and distinct. Moreover, an eigenvector $\mathbf{x_k}$ corresponding to the kth largest eigenvalue has exactly $k-1$ variations in sign for $k = 1, 2, \ldots, n$. Furthermore, assuming we choose the first entry of each eigenvector to be positive, the positions of the sign change in each successive eigenvector interlace. (See Preliminaries for the definition of interlace.)
2. [And87] Let A be an $n \times n$ totally nonnegative matrix. Then the eigenvalues of A are real and nonnegative.
3. [FGJ00] Let A be an $n \times n$ irreducible totally nonnegative matrix. Then the positive eigenvalues of A are distinct.
4. [GK02, pp. 107–108] If A is an $n \times n$ oscillatory matrix, then the eigenvalues of A are distinct and strictly interlace the eigenvalues of the two principal submatrices of order $n-1$ obtained from A by deleting the first row and column or the last row and column. If A is an $n \times n$ TN matrix, then nonstrict interlacing holds between the eigenvalues of A and the two principal submatrices of order $n-1$ obtained from A by deleting the first row and column or the last row and column.
5. [Pin98] If A is an $n \times n$ totally positive matrix with eigenvalues $\lambda_1 > \lambda_2 > \cdots > \lambda_n$ and $A(k)$ is the $(n-1) \times (n-1)$ principal submatrix obtained from A by deleting the kth row and column with eigenvalues $\mu_1 > \mu_2 > \cdots > \mu_{n-1}$, then for $j = 1, 2, \ldots, n-1$, $\lambda_{j-1} > \mu_j > \lambda_{j+1}$, where $\lambda_0 = \lambda_1$. In the usual Cauchy interlacing inequalities [MM64, p. 119] for positive semidefinite matrices, λ_{j-1} is replaced by λ_j. The nonstrict inequalities need not hold for TN matrices. The extreme cases $(j = 1, n-1)$ of this interlacing result were previously proved in [Fri85].
6. [Gar82] Let $n \geq 2$ and $A = [a_{ij}]$ be an oscillatory matrix. Then the main diagonal entries of A are majorized by the eigenvalues of A. (See Preliminaries for the definition of majorization.)

Examples:

1. Consider the 4×4 TP matrix $P_4 = \begin{bmatrix} 1 & 1 & 1 & 1 \\ 1 & 2 & 3 & 4 \\ 1 & 3 & 6 & 10 \\ 1 & 4 & 10 & 20 \end{bmatrix}$. Then the eigenvalues of P_4 are 26.305, 2.203, .454, and .038, with respective eigenvectors

$$\begin{bmatrix} .06 \\ .201 \\ .458 \\ .864 \end{bmatrix}, \begin{bmatrix} .53 \\ .64 \\ .392 \\ -.394 \end{bmatrix}, \begin{bmatrix} .787 \\ -.163 \\ -.532 \\ .265 \end{bmatrix}, \begin{bmatrix} .309 \\ -.723 \\ .595 \\ -.168 \end{bmatrix}.$$

2. The irreducible, singular TN (Hessenberg) matrix given by $H = \begin{bmatrix} 1 & 1 & 0 & 0 \\ 1 & 1 & 1 & 0 \\ 1 & 1 & 1 & 1 \\ 1 & 1 & 1 & 1 \end{bmatrix}$ has eigenvalues equal to 3, 1, 0, 0. Notice the positive eigenvalues of H are distinct.
3. Using the TP matrix P_4 in Example 1, the eigenvalues of $P_4(\{1\})$ are 26.213, 1.697, and .09. Observe that the usual Cauchy interlacing inequalities are satisfied in this case.

21.5 Deeper Properties

In this section, we explore more advanced topics that are not only interesting in their own right, but continue to demonstrate the delicate structure of these matrices.

Definitions:

For a given vector $\mathbf{c} = (c_1, c_2, \ldots, c_n)^T \in \mathbb{R}^n$ we define two quantities associated with the number of sign changes of the vector $\mathbf{c}$. These are:

$V^-(\mathbf{c})$ — the **number of sign changes in the sequence** $c_1, c_2, \ldots, c_n$ with the zero elements discarded; and

$V^+(\mathbf{c})$ — the **maximum number of sign changes in the sequence** $c_1, c_2, \ldots, c_n$, where the zero elements are arbitrarily assigned the values $+1$ and -1.

For example, $V^-((1,0,1,-1,0,1)^T) = 2$ and $V^+((1,0,1,-1,0,1)^T) = 4$.

We use $<, \leq$ to denote the usual entry-wise partial order on matrices; i.e., for $A = [a_{ij}]$, $B = [b_{ij}] \in \mathbb{R}^{m\times n}$, $A \leq (<)\ B$ means $a_{ij} \leq (<)\ b_{ij}$, for all i, j.

Let S be the signature matrix whose diagonal entries alternate in sign beginning with $+$. For $A, B \in \mathbb{R}^{m\times n}$, we write $A \overset{*}{\leq} B$ if and only if $SAS \leq SBS$ and $A \overset{*}{<} B$ if and only if $SAS < SBS$, and we call this the **"checkerboard" partial order** on real matrices.

Facts:

1. [Sch30] Let A be an $m \times n$ real matrix with $m \geq n$. If A is totally positive, then $V^+(A\mathbf{x}) \leq V^-(\mathbf{x})$, for all nonzero $\mathbf{x} \in \mathbb{R}^n$.
2. Let $A = [a_{ij}]$ be an $n \times n$ totally nonnegative matrix. Then:
   - Hadamard: [GK02, pp. 91–97], [Kot53]
     $\det A \leq \prod_{i=1}^n a_{ii}$,
   - Fischer: [GK02, pp. 91–97], [Kot53] Let $S \subseteq N = \{1, 2, \ldots, n\}$,
     $\det A \leq \det A[S] \cdot \det A[N \setminus S]$,
   - Koteljanskii: [Kot53] Let $S,\ T \subseteq N$,
     $\det A[S \cup T] \cdot \det A[S \cap T] \leq \det A[S] \cdot \det A[T]$.

   The above three determinantal inequalities also hold for positive semidefinite matrices. (See Chapter 8.5.)
3. [FGJ03] Let $\alpha_1, \alpha_2, \beta_1$, and β_2 be subsets of $\{1, 2, \ldots, n\}$, and let A be any TN matrix. Then

   $$\det A[\alpha_1] \det A[\alpha_2] \leq \det A[\beta_1] \det A[\beta_2]$$

   if and only if each index i has the same multiplicity in the multiset $\alpha_1 \cup \alpha_2$ and the multiset $\beta_1 \cup \beta_2$, and $\max(|\alpha_1 \cap L|, |\alpha_2 \cap L|) \geq \max(|\beta_1 \cap L|, |\beta_2 \cap L|)$, for every contiguous (i.e., $d(L) = 0$) $L \subseteq \{1, 2, \ldots, n\}$ (see [FGJ03] for other classes of principal-minor inequalities).
4. [FJS00] Let A be an $n \times n$ totally nonnegative matrix with $\det A(\{1\}) \neq 0$. Then $A - xE_{11}$ is totally nonnegative for all $x \in [0, \frac{\det A}{\det A(\{1\})}]$.
5. [FJ98] Let A be an $n \times n$ TN matrix partitioned as follows

   $$A = \begin{bmatrix} A_{11} & \mathbf{b} \\ \mathbf{c}^T & d \end{bmatrix},$$

   where A_{11} is $(n-1) \times (n-1)$. Suppose that $\operatorname{rank}(A_{11}) = p$. Then either $\operatorname{rank}([A_{11}, \mathbf{b}]) = p$ or $\operatorname{rank}[A_{11}^T, \mathbf{c}]^T = p$. See [FJ98] for other types of row and column inclusion results for TN matrices.
6. [CFJ01] Let T be an $n \times n$ totally nonnegative tridiagonal matrix. Then the Hadamard product of T with any other TN matrix is again TN.

7. [CFJ01][Mar70b] The Hadamard product of any two $n \times n$ tridiagonal totally nonnegative matrices is again totally nonnegative.
8. [CFJ01] Let $A = \begin{bmatrix} a & b & c \\ d & e & f \\ g & h & i \end{bmatrix}$ be a 3×3 totally nonnegative matrix. Then A has the property that $A \circ B$ is TN for all B TN if and only if

$$aei + gbf \geq afh + dbi,$$
$$aei + dch \geq afh + dbi.$$

9. [CFJ01] Let A be an $n \times n$ totally nonnegative matrix with the property that $A \circ B$ is TN for all B TN, and suppose B is any $n \times n$ totally nonnegative matrix. Then

$$\det(A \circ B) \geq \det B \prod_{i=1}^{n} a_{ii}.$$

10. [Ste91] Let $A = [a_{ij}] \in \mathbb{R}^{n \times n}$, and let S_n denote the symmetric group on n symbols. If χ is an irreducible character of S_n, then the corresponding matrix function

$$[a_{ij}] \mapsto \sum_{\omega \in S_n} \chi(\omega) \prod_{i=1}^{n} a_{i,\omega(i)}$$

is called an *immanant*. For example, if χ is the trivial character $\chi(w) = 1$, then the corresponding immanant is the permanent and, if χ is $\operatorname{sgn}(\omega)$, then the immanant is the determinant. Then every immanant of a totally nonnegative matrix is nonnegative.
11. [Gar96] If $A, B, C \in \mathbb{R}^{n \times n}$, $A \overset{*}{\leq} C \overset{*}{\leq} B$, and A and B are TP, then $\det C > 0$.
12. [Gar96] If $A, B, C \in \mathbb{R}^{n \times n}$, $A \overset{*}{\leq} C \overset{*}{\leq} B$, and A and B are TP, then C is TP.

Examples:

1. Let

$$P_4 = \begin{bmatrix} 1 & 1 & 1 & 1 \\ 1 & 2 & 3 & 4 \\ 1 & 3 & 6 & 10 \\ 1 & 4 & 10 & 20 \end{bmatrix} \quad \text{and} \quad \text{let } \mathbf{x} = \begin{bmatrix} 1 \\ -2 \\ -5 \\ 4 \end{bmatrix}.$$

Then $V^-(\mathbf{x}) = 2$ and $P_4\mathbf{x} = [-2, -2, 5, 23]^T$, so $V^+(P_4\mathbf{x}) = 1$. Hence, Schoenberg's variation diminishing property (Fact 1 of section 21.5) holds in this case.
2. Let

$$H = \begin{bmatrix} 1 & 1 & 0 & 0 \\ 1 & 1 & 1 & 0 \\ 1 & 1 & 1 & 1 \\ 1 & 1 & 1 & 1 \end{bmatrix} \quad \text{and} \quad \text{let } \mathbf{x} = \begin{bmatrix} 1 \\ -1 \\ -1 \\ 1 \end{bmatrix}.$$

Then $V^-(\mathbf{x}) = 2$ and $H\mathbf{x} = [0, -1, 0, 0]^T$, so $V^+(H\mathbf{x}) = 3$. Hence, Schoenberg's variation diminishing property (Fact 1 of section 21.5) does not hold in general for TN matrices.

3. If $0 < A < B$ and $A, B \in \mathbb{R}^{n\times n}$ are TP, then not all matrices in the interval between A and B need be TP. Let

$$A = \begin{bmatrix} 2 & 1 \\ 1 & 1 \end{bmatrix}, \quad B = \begin{bmatrix} 4 & 5 \\ 5 & 7 \end{bmatrix}.$$

Then A, B are TP and

$$C = \begin{bmatrix} 3 & 4 \\ 4 & 5 \end{bmatrix}$$

satisfies $A < C < B$, and C is not TP.

4. If TP is replaced by TN, then Fact 12 no longer holds. For example,

$$\begin{bmatrix} 2 & 0 & 1 \\ 0 & 0 & 0 \\ 1 & 0 & 1 \end{bmatrix} \overset{*}{\leq} \begin{bmatrix} 3 & 0 & 4 \\ 0 & 0 & 0 \\ 4 & 0 & 5 \end{bmatrix} \overset{*}{\leq} \begin{bmatrix} 4 & 0 & 5 \\ 0 & 0 & 0 \\ 5 & 0 & 7 \end{bmatrix};$$

however, both end matrices are TN while the middle matrix is not.

5. A polynomial $f(x)$ is *stable* if all the zeros of $f(x)$ have negative real parts, which is equivalent to the the Routh–Hurwitz matrix (Example 4 of Section 21.1) formed from f being totally nonnegative [Asn70]. Suppose $f(x) = \sum_{i=0}^{n} a_i x^i$ and $g(x) = \sum_{i=0}^{m} b_i x^i$ are two polynomials of degree n and m, respectively. Then the *Hadamard product of f and g* is the polynomial $(f \circ g)(x) = \sum_{i=0}^{k} a_i b_i x^i$, where $k = \min(m, n)$. In [GW96a], it is proved that the Hadamard product of stable polynomials is stable. Hence, the Hadamard product of two totally nonnegative Routh–Hurwitz matrices is in turn a totally nonnegative matrix ([GW96a]). See also [GW96b] for a list of other subclasses of TN matrices that are closed under Hadamard multiplication.

6. Let $A = \begin{bmatrix} 1 & 1 & 0 \\ 1 & 1 & 1 \\ 1 & 1 & 1 \end{bmatrix}$ and let $B = A^T$. Then A and B are TN, $A \circ B = \begin{bmatrix} 1 & 1 & 0 \\ 1 & 1 & 1 \\ 0 & 1 & 1 \end{bmatrix}$, and $\det(A \circ B) = -1 < 0$. Thus, $A \circ B$ is not TN.

7. (Polya matrix) Let $q \in (0, 1)$. The $n \times n$ Polya matrix Q has its $(i, j)^{th}$ entry equal to q^{-2ij}. Then Q is totally positive for all n (see [Whi52]). In fact, Q is diagonally equivalent to a TP Vandermonde matrix. Suppose Q represents the 3×3 Polya matrix. Then Q satisfies that $Q \circ A$ is TN for all A TN whenever $q \in (0, \sqrt{1/\mu})$, where $\mu = \frac{1+\sqrt{5}}{2}$ (the golden mean).

8. The bidiagonal factorization (21.2) from Fact 5 of section 21.2 for TN matrices was used in [Loe55] to show that for each nonsingular $n \times n$ TN matrix A there is a piece-wise continuous family of matrices $\Omega(t)$ of a special form such that the unique solution of the initial value problem

$$\frac{dA(t)}{dt} = \Omega(t)A(t), \; A(0) = I \tag{21.5}$$

has $A(1) = A$. Let $A(t) = [a_{ij}(t)]$ be a differentiable matrix-valued function of t such that $A(t)$ is nonsingular and TN for all $t \in [0, 1]$, and $A(0) = I$. Then

$$\Omega \equiv \left(\frac{dA(t)}{dt}\right)_{t=0} \tag{21.6}$$

is called an *infinitesimal element* of the semigroup of nonsingular TN matrices. By (21.2) every nonsingular TN matrix can be obtained from the solution of the initial value problem (21.5) in which all $\Omega(t)$ are infinitesimal elements.

9. If the main diagonal entries of a TP matrix are all ones, then it is not difficult to observe that as you move away from the diagonal there is a "drop off" effect in the entries of this matrix. Craven and

Csordas [CC98] have worked out an enticing sufficient "drop off" condition for a matrix to be TP. If $A = [a_{ij}]$ is an $n \times n$ matrix with positive entries and satisfies

$$a_{ij}a_{i+1,j+1} \geq c_0 a_{i,j+1}a_{i+1,j},$$

where $c_0 = 4.07959562349...$, then A is TP. This condition is particularly appealing for both Hankel and Toeplitz matrices. Recall that a Hankel matrix is an $(n+1) \times (n+1)$ matrix of the form

$$\begin{bmatrix} a_0 & a_1 & \cdots & a_n \\ a_1 & a_2 & \cdots & a_{n+1} \\ \vdots & \vdots & \cdots & \vdots \\ a_n & a_{n+1} & \cdots & a_{2n} \end{bmatrix}.$$

So, if the positive sequence $\{a_i\}$ satisfies $a_{k-1}a_{k+1} \geq c_0 a_k^2$, then the corresponding Hankel matrix is TP. An $(n+1) \times (n+1)$ Toeplitz matrix is of the form

$$\begin{bmatrix} a_0 & a_1 & a_2 & \cdots & a_n \\ a_{-1} & a_0 & a_1 & \cdots & a_{n-1} \\ a_{-2} & a_{-1} & a_0 & \cdots & a_{n-2} \\ \vdots & \vdots & \vdots & \cdots & \vdots \\ a_{-n} & a_{-(n-1)} & a_{-(n-2)} & \cdots & a_0 \end{bmatrix}.$$

Hence, if the positive sequence $\{a_i\}$ satisfies $a_k^2 \geq c_0 a_{k-1}a_{k+1}$, then the corresponding Toeplitz matrix is TP.

10. A sequence $a_0, a_1, \ldots$ of real numbers is called *totally positive* if the two-way infinite matrix given by

$$\begin{bmatrix} a_0 & 0 & 0 & \cdots \\ a_1 & a_0 & 0 & \cdots \\ a_2 & a_1 & a_0 & \cdots \\ \vdots & \vdots & \vdots & \end{bmatrix}$$

is TP. As usual, an infinite matrix is TP all of its minors are positive. Notice that the above matrix is a Toeplitz matrix. Studying the functions that generate totally positive sequences was a difficult and important step in the area of totally positive matrices; $f(x)$ generates the sequence $a_0, a_1, \ldots$ if $f(x) = a_0 + a_1 x + a_2 x^2 + \cdots$. In Aissen et al. [ASW52] (see also [Edr52]), it was shown that the above two-way infinite Toeplitz matrix is TP (i.e., the corresponding sequence is totally positive) if and only if the generating function $f(x)$ for the sequence $a_0, a_1, \ldots$ has the form

$$f(x) = e^{\gamma x} \frac{\prod_{\nu=1}^{\infty}(1+\alpha_\nu x)}{\prod_{\nu=1}^{\infty}(1-\beta_\nu x)},$$

where $\gamma, \alpha_\nu, \beta_\nu \geq 0$, and $\sum \alpha_\nu$ and $\sum \beta_\nu$ are convergent.

References

[ASW52] M. Aissen, I.J. Schoenberg, and A. Whitney. On generating functions of totally positive sequences, I. *J. Analyse Math.*, 2:93–103, 1952.

[And87] T. Ando. Totally positive matrices. *Lin. Alg. Appl.*, 90:165–219, 1987.

[Asn70] B.A. Asner. On the total nonnegativity of the Hurwitz matrix. *SIAM J. Appl. Math.*, 18:407–414, 1970.

[BFZ96] A. Berenstein, S. Fomin, and A. Zelevinsky. Parameterizations of canonical bases and totally positive matrices. *Adv. Math.*, 122:49–149, 1996.

[Bre95] F. Brenti. Combinatorics and total positivity. *J. Comb. Theory*, Series A, 71:175–218, 1995.

[CFJ01] A.S. Crans, S.M. Fallat, and C.R. Johnson. The Hadamard core of the totally nonnegative matrices. *Lin. Alg. Appl.*, 328:203–222, 2001.

[CC98] T. Craven and G. Csordas. A sufficient condition for strict total positivity of a matrix. *Lin. Multilin. Alg.*, 45:19–34, 1998.

[Cry73] C.W. Cryer. The LU-factorization of totally positive matrices. *Lin. Alg. Appl.*, 7:83–92, 1973.

[Cry76] C.W. Cryer. Some properties of totally positive matrices. *Lin. Alg. Appl.*, 15:1–25, 1976.

[Edr52] A. Edrei. On the generating functions of totally positive sequences, II. *J. Analyse Math.*, 2:104–109, 1952.

[Fal99] S.M. Fallat. *Totally Nonnegative Matrices*. Ph.D. dissertation, Department of Mathematics, College of William and Mary, Williamsburg, VA, 1999.

[Fal01] S.M. Fallat. Bidiagonal factorizations of totally nonnegative matrices. *Am. Math. Monthly*, 109:697–712, 2001.

[Fal04] S.M. Fallat. A remark on oscillatory matrices. *Lin. Alg. Appl.*, 393:139–147, 2004.

[FGJ00] S.M. Fallat, M.I. Gekhtman, and C.R. Johnson. Spectral structures of irreducible totally nonnegative matrices. *SIAM J. Matrix Anal. Appl.*, 22:627–645, 2000.

[FGJ03] S.M. Fallat, M.I. Gekhtman, and C.R. Johnson. Multiplicative principal-minor inequalities for totally nonnegative matrices. *Adv. App. Math.*, 30:442–470, 2003.

[FJ98] S.M. Fallat and C.R. Johnson. Olga, matrix theory and the Taussky unification problem. *Lin. Alg. Appl.*, 280:243–247, 1998.

[FJS00] S.M. Fallat, C.R. Johnson, and R.L. Smith. The general totally positive matrix completion problem with few unspecified entries. *Elect. J. Lin. Alg.*, 7:1–20, 2000.

[Fek13] M. Fekete. Über ein problem von Laguerre. *Rend. Circ. Mat. Palermo*, 34:89–100, 110–120, 1913.

[FZ00] S. Fomin and A. Zelevinsky. Total positivity: Tests and parameterizations. *Math. Intell.*, 22:23–33, 2000.

[Fri85] S. Friedland. Weak interlacing properties of totally positive matrices. *Lin. Alg. Appl.*, 71:247–266, 1985.

[GK60] F.R. Gantmacher and M.G. Krein. *Oszillationsmatrizen, Oszillationskerne und kleine Schwingungen Mechanischer Systeme.* Akademie-Verlag, Berlin, 1960.

[GK02] F.R. Gantmacher and M.G. Krein. *Oscillation Matrices and Kernels and Small Vibrations of Mechanical Systems.* AMS Chelsea Publishing, Providence, RI, revised edition, 2002. (Translation based on the 1941 Russian original, edited and with a preface by A. Eremenko.)

[Gar82] J. Garloff. Majorization between the diagonal elements and the eigenvalues of an oscillating matrix. *Lin. Alg. Appl.*, 47:181–184, 1982.

[Gar96] J. Garloff. Vertex implications for totally nonnegative matrices. *Total Positivity and Its Applications. Mathematics and Its Applications*, Vol. 359 (M. Gasca and C.A. Micchelli, Eds.), Kluwer Academic, Dordrecht, The Netherlands, 103–107, 1996.

[GW96a] J. Garloff and D.G. Wagner. Hadamard products of stable polynomials are stable. *J. Math. Anal. Appl.*, 202:797–809, 1996.

[GW96b] J. Garloff and D.G. Wagner. Preservation of total nonnegativity under Hadamard products and related topics. *Total Positivity and Its Applications. Mathematics and Its Applications*, Vol. 359 (M. Gasca and C.A. Micchelli, Eds.), Kluwer Academic, Dordrecht, The Netherlands, 97–102, 1996.

[GM96] M. Gasca and C.A. Micchelli. *Total Positivity and Its Applications. Mathematics and Its Applications*, Vol. 359, Kluwer Academic, Dordrecht, The Netherlands 1996.

[GP96] M. Gasca and J.M. Peña. On factorizations of totally positive matrices. *Total Positivity and Its Applications. Mathematics and Its Applications*, Vol. 359 (M. Gasca and C.A. Micchelli, Eds.), Kluwer Academic, Dordrecht, The Netherlands 109–130, 1996.

[JS00] C.R. Johnson and R.L. Smith. Line insertions in totally positive matrices. *J. Approx. Theory*, 105:305–312, 2000.

[Kar68] S. Karlin. *Total Positivity*. Vol. I, Stanford University Press, Stanford, CA, 1968.

[KM59] S. Karlin and G. McGregor. Coincidence probabilities. *Pacific J. Math.*, 9:1141–1164, 1959.

[Kot53] D.M. Koteljanskii. A property of sign-symmetric matrices (Russian). *Mat. Nauk (N.S.)*, 8:163–167, 1953; English transl.: *Translations of the AMS*, Series 2, 27:19–24, 1963.

[Lin73] B. Lindstrom. On the vector representations of induced matroids. *Bull. London Math. Soc.*, 5:85–90, 1973.

[Loe55] C. Loewner. On totally positive matrices. *Math. Z.*, 63:338–340, 1955.

[MM64] M. Marcus and H. Minc. *A Survey of Matrix Theory and Matrix Inequalities*. Dover Publications, New York, 1964.

[Mar70a] T.L. Markham. On oscillatory matrices. *Lin. Alg. Appl.*, 3:143–156, 1970.

[Mar70b] T.L. Markham. A semigroup of totally nonnegative matrices. *Lin. Alg. Appl.*, 3:157–164, 1970.

[Pin98] A. Pinkus. An interlacing property of eigenvalues of strictly totally positive matrices. *Lin. Alg. Appl.*, 279:201–206, 1998.

[Rad68] C.E. Radke. Classes of matrices with distinct, real characteristic values. *SIAM J. Appl. Math.*, 16:1192–1207, 1968.

[RH72] J.W. Rainey and G.J. Halbetler. Tridiagonalization of completely nonnegative matrices. *Math. Comp.*, 26:121–128, 1972.

[Sch30] I.J. Schoenberg. Uber variationsvermindernde lineare transformationen. *Math. Z.*, 32:321–328, 1930.

[SS95] B.Z. Shapiro and M.Z. Shapiro. On the boundary of totally positive upper triangular matrices. *Lin. Alg. Appl.*, 231:105–109, 1995.

[Ste91] J.R. Stembridge. Immanants of totally positive matrices are nonnegative. *Bull. London Math. Soc.*, 23:422–428, 1991.

[Whi52] A. Whitney. A reduction theorem for totally positive matrices. *J. Analyse Math.*, 2:88–92, 1952.

22

Linear Preserver Problems

Peter Šemrl
University of Ljubljana

Linear preservers are linear maps on linear spaces of matrices that leave certain subsets, properties, relations, functions, etc., invariant. Linear preserver problems ask what is the general form of such maps. Describing the structure of such maps often gives a deeper understanding of the matrix sets, functions, or relations under the consideration. Some of the linear preserver problems are motivated by applications (system theory, quantum mechanics, etc.).

22.1 Basic Concepts

Definitions:

Let $\mathcal{V}$ be a linear subspace of $F^{m\times n}$. Let f be a (scalar-valued, vector-valued, or set-valued) function on $\mathcal{V}$, $\mathcal{M}$ a subset of $\mathcal{V}$, and $\sim$ a relation defined on $\mathcal{V}$.

A linear map $\phi : \mathcal{V} \to \mathcal{V}$ is called a **linear preserver of function** f if $f(\phi(A)) = f(A)$ for every $A \in \mathcal{V}$.

A linear map $\phi : \mathcal{V} \to \mathcal{V}$ **preserves** $\mathcal{M}$ if $\phi(\mathcal{M}) \subseteq \mathcal{M}$.

The map ϕ **strongly preserves** $\mathcal{M}$ if $\phi(\mathcal{M}) = \mathcal{M}$.

A linear map $\phi : \mathcal{V} \to \mathcal{V}$ **preserves the relation** $\sim$ if $\phi(A) \sim \phi(B)$ whenever $A \sim B$, $A, B \in \mathcal{V}$.

If for every pair $A, B \in \mathcal{V}$ we have $\phi(A) \sim \phi(B)$ if and only if $A \sim B$, then ϕ **strongly preserves the relation** $\sim$.

Facts:

1. If linear maps $\phi : \mathcal{V} \to \mathcal{V}$ and $\psi : \mathcal{V} \to \mathcal{V}$ both preserve $\mathcal{M}$, then $\phi\psi : \mathcal{V} \to \mathcal{V}$ preserves $\mathcal{M}$. Consequently, the set of all linear transformations on $\mathcal{V}$ preserving $\mathcal{M}$ is a multiplicative semigroup.
2. The set of all bijective linear transformations strongly preserving $\mathcal{M}$ is a multiplicative group.
3. [BLL92, p. 41] Let $\mathcal{M} \subseteq \mathcal{V}$ be an algebraic subset and $\phi : \mathcal{V} \to \mathcal{V}$ a bijective linear map satisfying $\phi(\mathcal{M}) \subseteq \mathcal{M}$. Then ϕ strongly preserves $\mathcal{M}$.

Examples:

1. Let $R \in F^{m\times m}$ and $S \in F^{n\times n}$. The linear map $\phi: F^{m\times n} \to F^{m\times n}$ defined by $\phi(A) = RAS$, $A \in F^{m\times n}$, preserves the set of all matrices of rank at most one. In general, such a map does not preserve this set strongly.
2. If R and S in the previous example are invertible, then ϕ is a bijective linear map strongly preserving matrices of rank one. In fact, such a map ϕ strongly preserves matrices of rank k, $k = 1, \ldots, \min\{m, n\}$.
3. If $m = n$ and $R, S \in F^{n\times n}$, then the linear map $A \mapsto RA^T S$, $A \in F^{n\times n}$, preserves matrices of rank at most one. If both R and S are invertible, then ϕ strongly preserves the set of all matrices of rank k, $1 \leq k \leq n$.
4. Assume that $m = n$. Let $R \in F^{n\times n}$ be an invertible matrix, c a nonzero scalar, and $f : F^{n\times n} \to F$ a linear functional. Then both maps $A \mapsto cRAR^{-1} + f(A)I$, $A \in F^{n\times n}$, and $A \mapsto cRA^T R^{-1} + f(A)I$, $A \in F^{n\times n}$, strongly preserve commutativity; that is, if $\phi : F^{n\times n} \to F^{n\times n}$ is any of these two maps, then for every pair $A, B \in F^{n\times n}$ we have $\phi(A)\phi(B) = \phi(B)\phi(A)$ if and only if $AB = BA$.
5. Let $\|\cdot\|$ be any norm on $\mathbb{C}^{m\times n}$. A linear map $\phi : \mathbb{C}^{m\times n} \to \mathbb{C}^{m\times n}$ is called an isometry if $\|\phi(A)\| = \|A\|$, $A \in \mathbb{C}^{m\times n}$. Thus, isometries are linear preservers of norm functions.
6. Let $1 \leq k < \min\{m, n\}$. A matrix $A \in F^{m\times n}$ is of rank at most k if the determinant of every $(k+1) \times (k+1)$ submatrix of A is zero. Thus, the set of all $m \times n$ matrices of rank at most k is an algebraic subset of $F^{m\times n}$.
7. The set of all nilpotent $n \times n$ matrices is an algebraic subset of $F^{n\times n}$. More generally, given a polynomial $p \in F[X]$, the set of all matrices $A \in F^{n\times n}$ satisfying $p(A) = 0$ is an algebraic set. Hence, bijective linear maps on $F^{n\times n}$ preserving idempotent matrices, nilpotents, involutions, etc. preserve these sets strongly.

22.2 Standard Forms

Definitions:

Let $\mathcal{V}$ be a linear subspace of $F^{m\times n}$ and let P be a preserving property which makes sense for linear maps acting on $\mathcal{V}$ (P may be the property of preserving a certain subset of $\mathcal{V}$ or the property of preserving a certain relation on $\mathcal{V}$ or the property of preserving a certain function defined on $\mathcal{V}$). A **linear preserver problem** corresponding to the property P is the problem of characterizing all linear (bijective) maps on $\mathcal{V}$ satisfying this property. Very often, linear preservers have the standard forms (see the next section). Occasionally, there are interesting exceptional cases especially in low dimensions (see later examples).

Let $R \in F^{m\times m}$ and $S \in F^{n\times n}$ be invertible matrices. A map $\phi : F^{m\times n} \to F^{m\times n}$ is called an (R, S)-**standard map** if either $\phi(A) = RAS$, $A \in F^{m\times n}$, or $m = n$ and $\phi(A) = RA^T S$, $A \in F^{n\times n}$. In many cases we assume that R and S satisfy some additional assumptions.

Let $R \in F^{n\times n}$ be an invertible matrix, c a nonzero scalar, and $f : F^{n\times n} \to F$ a linear functional. A map $\phi : F^{n\times n} \to F^{n\times n}$ is called an (R, c, f)-**standard map** if either $\phi(A) = cRAR^{-1} + f(A)I$, $A \in F^{n\times n}$, or $\phi(A) = cRA^T R^{-1} + f(A)I$, $A \in F^{n\times n}$.

When we consider linear preservers on proper subspaces $\mathcal{V} \subset F^{m\times n}$, we usually have to modify the notion of standard maps. Let us consider the case when $\mathcal{V} = \mathcal{T}_n \subset F^{n\times n}$ is the subalgebra of all upper triangular matrices. The **flip map** $A \mapsto A^f$, $A \in \mathcal{T}_n$, is defined as the transposition over the antidiagonal; that is, $A^f = GA^T G$, where $G = E_{1n} + E_{2,n-1} + \ldots + E_{n1}$ (see Example 1). Standard maps on $\mathcal{T}_n$ are maps of the form

$$A \mapsto RAS, \quad A \in \mathcal{T}_n,$$

$$A \mapsto RA^f S, \quad A \in \mathcal{T}_n,$$

$$A \mapsto cRAR^{-1} + f(A)I, \quad A \in \mathcal{T}_n,$$

and

$$A \mapsto cRA^f R^{-1} + f(A)I, \quad A \in \mathcal{T}_n,$$

where R and S are invertible upper triangular matrices, c is a nonzero scalar, and f is a linear functional on $\mathcal{T}_n$.

Facts:

1. Every (R, S)-standard map is bijective. It strongly preserves matrices of rank k, $k = 1, \ldots, \min\{m, n\}$. Every (R, c, f)-standard map is either bijective, or its kernel is the one-dimensional subspace consisting of all scalar matrices.
2. Let $U \in \mathbb{C}^{m\times m}$ and $V \in \mathbb{C}^{n\times n}$ be unitary matrices. Then a (U, V)-standard map preserves singular values and, hence, all functions of singular values including unitarily invariant norms.
3. Let $R \in \mathbb{C}^{n\times n}$ be an invertible matrix. Then an (R, R^{-1})-standard map on $\mathbb{C}^{n\times n}$ preserves spectrum, idempotents, nilpotents, similarity, zero products, etc.
4. If $U \in \mathbb{C}^{n\times n}$ is a unitary matrix, then a (U, U^*)-standard map defined on $\mathbb{C}^{n\times n}$ strongly preserves the set of all orthogonal idempotents, the set of all normal matrices, numerical range, etc.
5. If $A, B \in \mathcal{T}_n$, then $(AB)^f = B^f A^f$.

Examples:

1.
$$\begin{bmatrix} a & b & c & d \\ 0 & e & f & g \\ 0 & 0 & h & i \\ 0 & 0 & 0 & j \end{bmatrix}^f = \begin{bmatrix} j & i & g & d \\ 0 & h & f & c \\ 0 & 0 & e & b \\ 0 & 0 & 0 & a \end{bmatrix}.$$

2. A map $\phi : \mathbb{C}^{2\times 2} \to \mathbb{C}^{2\times 2}$ given by
$$\phi\left(\begin{bmatrix} a & b \\ c & d \end{bmatrix}\right) = \begin{bmatrix} a & -b \\ c & d \end{bmatrix}$$
is a bijective linear map that strongly preserves commutativity but is not of a standard form. More generally, any bijective linear map $\phi : \mathbb{C}^{2\times 2} \to \mathbb{C}^{2\times 2}$ satisfying $\phi(I) = I$ strongly preserves commutativity [Kun99]. In higher dimensions there are no nonstandard bijective linear maps preserving commutativity [BLL92, p. 76].
3. Let $\mathcal{W} \subset \mathbb{C}^{n\times n}$ be any linear subspace of matrices with the property that $AB = BA$ for every pair $A, B \in \mathcal{W}$. Assume that $\phi : \mathbb{C}^{n\times n} \to \mathbb{C}^{n\times n}$ is a linear map whose range is contained in $\mathcal{W}$. Then ϕ is a nonstandard map preserving commutativity. If $n > 1$, then it does not preserve commutativity strongly. A map $\phi : \mathbb{C}^{2\times 2} \to \mathbb{C}^{2\times 2}$ defined by
$$\phi\left(\begin{bmatrix} a & b \\ c & d \end{bmatrix}\right) = \begin{bmatrix} a & b \\ -b & a \end{bmatrix}$$
is a concrete example of such map.
4. A map $\phi : \mathbb{R}^{4\times 4} \to \mathbb{R}^{4\times 4}$ given by
$$\phi\left(\begin{bmatrix} a & b & c & d \\ * & * & * & * \\ * & * & * & * \\ * & * & * & * \end{bmatrix}\right) = \begin{bmatrix} a & b & c & d \\ -b & a & -d & c \\ -c & d & a & -b \\ -d & -c & b & a \end{bmatrix}$$

preserves the real orthogonal group, that is, $\phi(O)$ is an orthogonal matrix for every orthogonal matrix $O \in \mathbb{R}^{4\times 4}$. Note that the right-hand side of the above equation is the standard matrix representation of the quaternion $a+bi+cj+dk$. Similar constructions with matrix representations of complex and Cayley numbers give nonstandard linear preservers of real orthogonal group on $\mathbb{R}^{2\times 2}$ and $\mathbb{R}^{8\times 8}$. If $\phi:\mathbb{R}^{n\times n} \to \mathbb{R}^{n\times n}$ is a linear preserver of orthogonal group and $n \notin \{2,4,8\}$, then ϕ is a (U, V)-standard map with U and V being orthogonal matrices [LP01, p. 601].

5. A linear map ϕ acting on 4×4 upper triangular matrices given by

$$\begin{bmatrix} a & b & c & d \\ 0 & e & f & g \\ 0 & 0 & h & i \\ 0 & 0 & 0 & j \end{bmatrix} \mapsto \begin{bmatrix} e & f & g & b \\ 0 & j & i & c \\ 0 & 0 & h & d \\ 0 & 0 & 0 & a \end{bmatrix}$$

is a nonstandard bijective linear map strongly preserving the set of invertible matrices. All that is important in this example is that A and $\phi(A)$ have the same diagonal entries up to a permutation.

22.3 Standard Linear Preserver Problems

Definitions:

Linear preserver problems ask what is the general form of (bijective) linear maps on matrix spaces having a certain preserving property. When the general form of linear preservers under the consideration is one of the standard forms, we speak of a **standard linear preserver problem**. The following list of some most important standard linear preserver results is far from being complete. Many more results can be found in the survey [BLL92].

Facts:

1. [BLL92, Theorem 2.2], [LT92, Prop. 3] Let m, n be positive integers and $\phi:\mathbb{C}^{m\times n} \to \mathbb{C}^{m\times n}$ a linear map. Assume that one of the following two conditions is satisfied:

   - Let k be a positive integer, $k \leq \min\{m, n\}$, and assume that $\operatorname{rank}\phi(A) = k$ whenever $\operatorname{rank} A = k$.
   - ϕ is invertible and $\operatorname{rank}\phi(A) = \operatorname{rank}\phi(B)$ whenever $\operatorname{rank} A = \operatorname{rank} B$.

   Then ϕ is an (R, S)-standard map for some invertible matrices $R \in \mathbb{C}^{m\times m}$ and $S \in \mathbb{C}^{n\times n}$.
2. [BLL92, p. 9] Let F be a field with more than three elements, k a positive integer, $k \leq \min\{m, n\}$, and $\phi: F^{m\times n} \to F^{m\times n}$ an invertible linear map such that $\operatorname{rank}\phi(A) = k$ whenever $\operatorname{rank} A = k$. Then ϕ is an (R, S)-standard map for some invertible matrices $R \in F^{m\times m}$ and $S \in F^{n\times n}$.
3. [BLL92, Theorem 2.6] Let $\phi: F^{m\times n} \to F^{p\times q}$ be a linear map such that $\operatorname{rank}\phi(A) \leq 1$ whenever $\operatorname{rank} A = 1$. Then either

   (a) $\phi(A) = RAS$ for some $R \in F^{p\times m}$ and some $S \in F^{n\times q}$ or
   (b) $\phi(A) = RA^T S$ for some $R \in F^{p\times n}$ and some $S \in F^{m\times q}$ or
   (c) The range of ϕ is contained in the set of all matrices of rank at most one.

4. [BLL92, p. 10] Let F be an infinite field with characteristic $\neq 2$, $\mathcal{S}_n$ the space of all $n \times n$ symmetric matrices over F, and let k be an integer, $1 \leq k \leq n$. If $\phi:\mathcal{S}_n \to \mathcal{S}_n$ is an invertible linear rank k preserver, then $\phi(A) = cRAR^T$ for every $A \in \mathcal{S}$. Here, c is a nonzero scalar and R an invertible $n \times n$ matrix.
5. [BLL92, Theorem 2.9] Let F be a field with characteristic $\neq 2$. Assume that F has more than three elements. Let $\phi:\mathcal{S}_n \to \mathcal{S}_m$ be a linear map such that $\operatorname{rank}\phi(A) \leq 1$ whenever $\operatorname{rank} A = 1$. Then either

(a) There exist an $m \times n$ matrix R and a scalar c such that $\phi(A) = cRAR^T$ or
(b) The range of ϕ is contained in a linear span of some rank one $m \times m$ symmetric matrix.

6. [BLL92, Theorem 2.7, Theorem 2.8, p. 25] Let ϕ be a real linear map acting on the real linear space $\mathcal{H}_n$ of all $n \times n$ Hermitian complex matrices. Let π, ν, δ be nonnegative integers with $\pi + \nu + \delta = n$ and denote by $G(\pi, \nu, \delta)$ the set of all $A \in \mathcal{H}_n$ with inertia $\text{in}(A) = (\pi, \nu, \delta)$. Assume that one of the following is satisfied:

   - $n \geq 3$, $k < n$, ϕ is invertible, and rank $\phi(A) \leq k$ whenever rank $A = k$.
   - $n \geq 2$, the range of ϕ is not one-dimensional, and rank $\phi(A) = 1$ whenever rank $A = 1$.
   - The triple (π, ν, δ) does not belong to the set $\{(n, 0, 0), (0, n, 0), (0, 0, n)\}$, ϕ is bijective, and $\phi(G(\pi, \nu, \delta)) \subseteq G(\pi, \nu, \delta)$.

   Then there exist an invertible $n \times n$ complex matrix R and a constant $c \in \{-1, 1\}$ such that either

   (a) $\phi(A) = cRAR^*$ for every $A \in \mathcal{H}_n$ or
   (b) $\phi(A) = cRA^T R^*$ for every $A \in \mathcal{H}_n$.

   Of course, if the third of the above conditions is satisfied, then the possibility $c = -1$ can occur only when $\pi = \nu$.
7. [BLL92, p. 76] Let $n \geq 3$ and let $\phi : F^{n\times n} \to F^{n\times n}$ be an invertible linear map such that $\phi(A)\phi(B) = \phi(B)\phi(A)$ whenever A and B commute. Then ϕ is an (R, c, f)-standard map for some invertible $R \in F^{n\times n}$, some nonzero scalar c, and some linear functional $f : F^{n\times n} \to F$.
8. [LP01, Theorem 2.3] Let $\phi : \mathbb{C}^{n\times n} \to \mathbb{C}^{n\times n}$ be a linear similarity preserving map. Then either

   (a) ϕ is an (R, c, f)-standard map or
   (b) $\phi(A) = (\text{tr } A)B$, $A \in \mathbb{C}^{n\times n}$.

   Here, $B, R \in \mathbb{C}^{n\times n}$ and R is invertible, c is a nonzero complex number, and $f : \mathbb{C}^{n\times n} \to \mathbb{C}$ is a functional of the form $f(A) = b\text{tr } A$, $A \in \mathbb{C}^{n\times n}$, for some $b \in \mathbb{C}$.
9. [BLL92, p. 77] Let ϕ be a real-linear unitary similarity preserving map on $\mathcal{H}_n$. Then either

   (a) $\phi(A) = (\text{tr } A)B$, $A \in \mathcal{H}_n$ or
   (b) $\phi(A) = cUAU^* + b(\text{tr } A)I$, $A \in \mathcal{H}_n$ or
   (c) $\phi(A) = cUA^T U^* + b(\text{tr } A)I$, $A \in \mathcal{H}_n$.

   Here, B is a Hermitian matrix, U is a unitary matrix, and b, c are real constants.
10. [BLL92, Theorem 4.7.6] Let $n > 2$ and let F be an algebraically closed field of characteristic zero. Let p be a polynomial of degree n with at least two distinct roots. Let us write p as $p(x) = x^k q(x)$ with $k \geq 0$ and $q(0) \neq 0$. Assume that $\phi : F^{n\times n} \to F^{n\times n}$ is an invertible linear map preserving the set of all matrices annihilated by p. Then either

    (a) $\phi(A) = cRAR^{-1}$, $A \in F^{n\times n}$ or
    (b) $\phi(A) = cRA^T R^{-1}$, $A \in F^{n\times n}$.

    Here, R is an invertible matrix and c is a constant permuting the roots of q; that is, $q(c\lambda) = 0$ for each $\lambda \in F$ satisfying $q(\lambda) = 0$.
11. [BLL92, p. 48] Let $sl_n \subset F^{n\times n}$ be the linear space of all trace zero matrices and $\phi : sl_n \to sl_n$ an invertible linear map preserving the set of all nilpotent matrices. Then there exist an invertible matrix $R \in F^{n\times n}$ and a nonzero scalar c such that either

    (a) $\phi(A) = cRAR^{-1}$, $A \in sl_n$ or
    (b) $\phi(A) = cRA^T R^{-1}$, $A \in sl_n$.

    When considering linear preservers of nilpotent matrices one should observe first that the linear span of all nilpotent matrices is sl_n and, therefore, it is natural to confine maps under consideration to this subspace of codimension one.

12. [GLS00, pp. 76, 78] Let F be an algebraically closed field of characteristic 0, m, n positive integers, and $\phi : F^{n\times n} \to F^{m\times m}$ a linear transformation. If ϕ is nonzero and maps idempotent matrices to idempotent matrices, then $m \geq n$ and there exist an invertible matrix $R \in F^{m\times m}$ and nonnegative integers k_1, k_2 such that $1 \leq k_1 + k_2$, $(k_1 + k_2)n \leq m$ and

$$\phi(A) = R(A \oplus \ldots \oplus A \oplus A^T \oplus \ldots \oplus A^T \oplus 0)R^{-1}$$

for every $A \in F^{n\times n}$. In the above block diagonal direct sum the matrix A appears k_1 times, A^T appears k_2 times, and 0 is the zero matrix of the appropriate size (possibly absent). If $p \in F[X]$ is a polynomial of degree > 1 with simple zeros (each zero has multiplicity one), ϕ is unital and maps every $A \in F^{n\times n}$ satisfying $p(A) = 0$ into some $m \times m$ matrix annihilated by p, then ϕ is of the above described form with $(k_1 + k_2)n = m$.
13. [BLL92, Theorem 4.6.2] Let $\phi : \mathbb{C}^{n\times n} \to \mathbb{C}^{n\times n}$ be a linear map preserving the unitary group. Then ϕ is a (U, V)-standard map for some unitary matrices $U, V \in \mathbb{C}^{n\times n}$.
14. [KH92] Let $\phi : \mathbb{C}^{n\times n} \to \mathbb{C}^{n\times n}$ be a linear map preserving normal matrices. Then either

    (a) $\phi(A) = cUAU^* + f(A)I$, $A \in \mathbb{C}^{n\times n}$ or

    (b) $\phi(A) = cUA^tU^* + f(A)I$, $A \in \mathbb{C}^{n\times n}$ or

    (c) the range of ϕ is contained in the set of normal matrices.

    Here, U is a unitary matrix, c is a nonzero scalar, and f is a linear functional on $\mathbb{C}^{n\times n}$.
15. [LP01, p. 595] Let $\|\cdot\|$ be a unitarily invariant norm on $\mathbb{C}^{m\times n}$ that is not a multiple of the Frobenius norm defined by $\|A\| = \sqrt{\operatorname{tr}(AA^*)}$. The group of linear preservers of $\|\cdot\|$ on $\mathbb{C}^{m\times n}$ is the group of all (U, V)-standard maps, where $U \in \mathbb{C}^{m\times m}$ and $V \in \mathbb{C}^{n\times n}$ are unitary matrices. Of course, if $\|\cdot\|$ is a mulitple of the Frobenious norm, then the group of linear preservers of $\|\cdot\|$ on $\mathbb{C}^{m\times n}$ is the group of all unitary operators, i.e., those linear operators $\phi : \mathbb{C}^{m\times n} \to \mathbb{C}^{m\times n}$ that preserve the usual inner product $\langle A, B\rangle = \operatorname{tr}(AB^*)$ on $\mathbb{C}^{m\times n}$.
16. [BLL92, p. 63–64] Let $\phi : \mathbb{C}^{n\times n} \to \mathbb{C}^{n\times n}$ be a linear map preserving the numerical radius. Then either

    (a) $\phi(A) = cUAU^*$, $A \in \mathbb{C}^{n\times n}$ or

    (b) $\phi(A) = cUA^TU^*$, $A \in \mathbb{C}^{n\times n}$.

    Here, U is a unitary matrix and c a complex constant with $|c| = 1$.
17. [BLL92, Theorem 4.3.1] Let $n > 2$ and let $\phi : F^{n\times n} \to F^{n\times n}$ be a linear map preserving the permanent. Then ϕ is an (R, S)-standard map, where R and S are each a product of a diagonal and a permutation matrix, and the product of the two diagonal matrices has determinant one.
18. [CL98] Let $\phi : \mathcal{T}_n \to \mathcal{T}_n$ be a linear rank one preserver. Then either

    (a) The range of ϕ is the space of all matrices of the form

$$\begin{bmatrix} * & * & \ldots & * \\ 0 & 0 & \ldots & 0 \\ \vdots & \vdots & \ddots & \vdots \\ 0 & 0 & \ldots & 0 \end{bmatrix}$$

    or

    (b) The range of ϕ is the space of all matrices of the form

$$\begin{bmatrix} 0 & 0 & \ldots & 0 & * \\ 0 & 0 & \ldots & 0 & * \\ \vdots & \vdots & \ddots & \vdots & \vdots \\ 0 & 0 & \ldots & 0 & * \end{bmatrix}$$

    or

(c) $\phi(A) = RAS$ for some invertible $R, S \in \mathcal{T}_n$ or

(d) $\phi(A) = RA^f S$ for some invertible $R, S \in \mathcal{T}_n$.

Examples:

1. Let $n \geq 2$. Then the linear map $\phi : \mathcal{T}_n \to \mathcal{T}_n$ defined by

$$\phi\left(\begin{bmatrix} a_{11} & a_{12} & \dots & a_{1n} \\ 0 & a_{22} & \dots & a_{2n} \\ \vdots & \vdots & \ddots & \vdots \\ 0 & 0 & \dots & a_{nn} \end{bmatrix}\right)$$

$$= \begin{bmatrix} a_{11} + a_{22} + \ldots + a_{nn} & a_{12} + a_{23} + \ldots + a_{n-1,n} & \dots & a_{1n} \\ 0 & 0 & \dots & 0 \\ \vdots & \vdots & \ddots & \vdots \\ 0 & 0 & \dots & 0 \end{bmatrix}$$

is an example of a singular preserver of rank one.

2. The most important example of a nonstandard linear preserver problem is the problem of characterizing linear maps on $n \times n$ real or complex matrices preserving the set of positive semidefinite matrices. Let $R_1, \ldots, R_r, S_1, \ldots, S_k$ be $n \times n$ matrices. Then the linear map ϕ given by $\phi(A) = R_1 A R_1^* + \cdots + R_r A R_r^* + S_1 A^T S_1^* + \cdots + S_k A^T S_k^*$ is a linear preserver of positive semidefinite matrices. Such a map is called decomposable. In general it cannot be reduced to a single congruence or a single congruence composed with the transposition. Moreover, there exist linear maps on the space of $n \times n$ matrices preserving positive semidefinite matrices that are not decomposable. There is no general structural result for such maps.

22.4 Additive, Multiplicative, and Nonlinear Preservers

Definitions:

A map $\phi : F^{m\times n} \to F^{m\times n}$ is **additive** if $\phi(A + B) = \phi(A) + \phi(B)$, $A, B \in F^{m\times n}$. An additive map $\phi : F^{m\times n} \to F^{m\times n}$ having a certain preserving property is called an **additive preserver**.

A map $\phi : F^{n\times n} \to F^{n\times n}$ is **multiplicative** if $\phi(AB) = \phi(A)\phi(B)$, $A, B \in F^{n\times n}$. A multiplicative map $\phi : F^{n\times n} \to F^{n\times n}$ having a certain preserving property is called a **multiplicative preserver**.

Two matrices $A, B \in F^{m\times n}$ are said to be **adjacent** if $\text{rank}\,(A - B) = 1$.

A map $\phi : F^{n\times n} \to F^{n\times n}$ is called a **local similarity** if for every $A \in F^{n\times n}$ there exists an invertible $R_A \in F^{n\times n}$ such that $\phi(A) = R_A A R_A^{-1}$.

Let $f : F \to F$ be an automorphism of the field F. A map $\phi : F^{n\times n} \to F^{n\times n}$ defined by $\phi([a_{ij}]) = [f(a_{ij})]$ is called a **ring automorphism of $F^{n\times n}$ induced by f**.

Facts:

1. [BS00] Let $n \geq 2$ and assume that $\phi : F^{n\times n} \to F^{n\times n}$ is a surjective additive map preserving rank one matrices. Then there exist a pair of invertible matrices $R, S \in F^{n\times n}$ and an automorphism f of the field F such that ϕ is a composition of an (R, S)-standard map and a ring automorphism of $F^{n\times n}$ induced by f.
2. [GLR03] Let $SL(n, \mathbb{C})$ denote the group of all $n \times n$ complex matrices A such that $\det A = 1$. A multiplicative map $\phi : SL(n, \mathbb{C}) \to \mathbb{C}^{n\times n}$ satisfies $\rho(\phi(A)) = \rho(A)$ for every $A \in SL(n, \mathbb{C})$ if and only if there exists $S \in SL(n, \mathbb{C})$ such that either

(a) $\phi(A) = SAS^{-1}$, $A \in SL(n, \mathbb{C})$ or

(b) $\phi(A) = S\overline{A}S^{-1}$, $A \in SL(n, \mathbb{C})$.

Here, $\overline{A}$ denotes the matrix obtained from A by applying the complex conjugation entrywise.

3. [PS98] Let $n \geq 3$ and let $\phi : \mathbb{C}^{n\times n} \to \mathbb{C}^{n\times n}$ be a continuous mapping. Then ϕ preserves spectrum, commutativity, and rank one matrices (no linearity, additivity, or multiplicativity is assumed) if and only if there exists an invertible matrix $R \in \mathbb{C}^{n\times n}$ such that ϕ is an (R, R^{-1})-standard map.
4. [BR00] Let $\phi : \mathbb{C}^{n\times n} \to \mathbb{C}^{n\times n}$ be a spectrum preserving C^1-diffeomorphism (again, we do not assume that ϕ is additive or multiplicative). Then ϕ is a local similarity.
5. [HHW04] Let $n \geq 2$. Then $\phi : \mathcal{H}_n \to \mathcal{H}_n$ is a bijective map such that $\phi(A)$ and $\phi(B)$ are adjacent for every adjacent pair $A, B \in \mathcal{H}_n$ if and only if there exist a nonzero real number c, an invertible $R \in \mathbb{C}^{n\times n}$, and $S \in \mathcal{H}_n$ such that either

   (a) $\phi(A) = cRAR^* + S$, $A \in \mathcal{H}_n$ or

   (b) $\phi(A) = cR\overline{A}R^* + S$, $A \in \mathcal{H}_n$.
6. [Mol01] Let $n \geq 2$ be an integer and $\phi : \mathcal{H}_n \to \mathcal{H}_n$ a bijective map such that $\phi(A) \leq \phi(B)$ if and only if $A \leq B$, $A, B \in \mathcal{H}_n$ (here, $A \leq B$ if and only if $B - A$ is a positive semidefinite matrix). Then there exist an invertible $R \in \mathbb{C}^{n\times n}$ and $S \in \mathcal{H}_n$ such that either

   (a) $\phi(A) = RAR^* + S$, $A \in \mathcal{H}_n$ or

   (b) $\phi(A) = R\overline{A}R^* + S$, $A \in \mathcal{H}_n$.

   This result has an infinite-dimensional analog important in quantum mechanics. In the language of quantum mechanics, the relation $A \leq B$ means that the expected value of the bounded observable A in any state is less than or equal to the expected value of B in the same state.

Examples:

1. We define a mapping $\phi : \mathbb{C}^{n\times n} \to \mathbb{C}^{n\times n}$ in the following way. For a diagonal matrix A with distinct diagonal entries, we define $\phi(A)$ to be the diagonal matrix obtained from A by interchanging the first two diagonal elements. Otherwise, let $\phi(A)$ be equal to A. Clearly, ϕ is a bijective mapping preserving spectrum, rank, and commutativity in both directions. This shows that the continuity assumption is indispensable in Fact 3 above.
2. Let $\phi : \mathbb{C}^{n\times n} \to \mathbb{C}^{n\times n}$ be a map defined by $\phi(0) = E_{12}$, $\phi(E_{12}) = 0$, and $\phi(A) = A$ for all $A \in \mathbb{C}^{n\times n} \setminus \{0, E_{12}\}$. Then ϕ is a bijective spectrum preserving map that is not a local similarity. More generally, we can decompose $\mathbb{C}^{n\times n}$ into the disjoint union of the classes of matrices having the same spectrum and then any bijection leaving each of this classes invariant preserves spectrum. Thus, the assumption on differentiability is essential in Fact 4 above.

References

[BR00] L. Baribeau and T. Ransford. Non-linear spectrum-preserving maps. *Bull. London Math. Soc.*, 32:8–14, 2000.

[BLL92] L. Beasley, C.-K. Li, M.H. Lim, R. Loewy, B. McDonald, S. Pierce (Ed.), and N.-K. Tsing. A survey of linear preserver problems. *Lin. Multlin. Alg.*, 33:1–129, 1992.

[BS00] J. Bell and A.R. Sourour. Additive rank-one preserving mappings on triangular matrix algebras. *Lin. Alg. Appl.*, 312:13–33, 2000.

[CL98] W.L. Chooi and M.H. Lim. Linear preservers on triangular matrices. *Lin. Alg. Appl.*, 269:241–255, 1998.

[GLR03] R.M. Guralnick, C.-K. Li, and L. Rodman. Multiplicative maps on invertible matrices that preserve matricial properties. *Electron. J. Lin. Alg.*, 10:291–319, 2003.

[GLS00] A. Guterman, C.-K. Li, and P. Šemrl. Some general techniques on linear preserver problems. *Lin. Alg. Appl.*, 315:61–81, 2000.

[HHW04] W.-L. Huang, R. Höfer, and Z.-X. Wan. Adjacency preserving mappings of symmetric and hermitian matrices. *Aequationes Math.*, 67:132–139, 2004.

[Kun99] C.M. Kunicki. Commutativity and normal preserving linear transformations on M_2. *Lin. Multilin. Alg.*, 45:341–347, 1999.

[KH92] C.M. Kunicki and R.D. Hill. Normal-preserving linear transformations. *Lin. Alg. Appl.*, 170:107–115, 1992.

[LP01] C.-K. Li and S. Pierce. Linear preserver problems. *Amer. Math. Monthly*, 108:591–605, 2001.

[LT92] C.-K. Li and N.-K. Tsing. Linear preserver problems: a brief introduction and some special techniques. *Lin. Alg. Appl.*, 162–164:217–235, 1992.

[Mol01] L. Molnár. Order-automorphisms of the set of bounded observables. *J. Math. Phys.*, 42:5904–5909, 2001.

[PS98] T. Petek and P. Šemrl. Characterization of Jordan homomorphisms on M_n using preserving properties. *Lin. Alg. Appl.*, 269:33–46, 1998.

23

Matrices over Integral Domains

Shmuel Friedland
University of Illinois at Chicago

In this chapter, we present some results on matrices over integral domains, which extend the well-known results for matrices over the fields discussed in Chapter 1 of this book. The general theory of linear algebra over commutative rings is extensively studied in the book [McD84]. It is mostly intended for readers with a thorough training in ring theory. The aim of this chapter is to give a brief survey of notions and facts about matrices over classical domains that come up in applications. Namely over the ring of integers, the ring of polynomials over the field, the ring of analytic functions in one variable on an open connected set, and germs of analytic functions in one variable at the origin. The last section of this chapter is devoted to the notion of strict equivalence of pencils.

Most of the results in this chapter are well known to the experts. A few new results are taken from the book in progress [Frixx], which are mostly contained in the preprint [Fri81].

23.1 Certain Integral Domains

Definitions:

A commutative ring without zero divisors and containing identity 1 is an **integral domain** and denoted by $\mathbb{D}$.

The **quotient field** F of a given integral domain $\mathbb{D}$ is formed by the set of equivalence classes of all quotients $\frac{a}{b}, b \neq 0$, where $\frac{a}{b} \equiv \frac{c}{d}$ if and only if $ad = bc$, such that

$$\frac{a}{b} + \frac{c}{d} = \frac{ad+bc}{bd}, \quad \frac{a}{b}\frac{c}{d} = \frac{ac}{bd}, \quad b, d \neq 0.$$

For $\mathbf{x} = [x_1, \ldots, x_n]^T \in \mathbb{D}^n, \alpha = [\alpha_1, \ldots, \alpha_n]^T \in \mathbb{Z}_+^n$ we define $\mathbf{x}^\alpha = x_1^{\alpha_1} \cdots x_n^{\alpha_n}$ and $|\alpha| = \sum_{i=1}^n |\alpha_i|$.

$\mathbb{D}[\mathbf{x}] = \mathbb{D}[x_1, \ldots, x_n]$ is the ring of all **polynomials** $p(\mathbf{x}) = p(x_1, \ldots, x_n)$ in n variables with coefficients in $\mathbb{D}$:

$$p(\mathbf{x}) = \sum_{|\alpha| \leq m} a_\alpha \mathbf{x}^\alpha.$$

The **total degree**, or simply the **degree** of $p(\mathbf{x}) \neq 0$, denoted by deg p, is the maximum $m \in \mathbb{Z}_+$ such that there exists $a_\alpha \neq 0$ such that $|\alpha| = m$. (deg $0 = -\infty$.)

A polynomial p is **homogeneous** if $a_\alpha = 0$ for all $|\alpha| <$ deg p.

A polynomial $p(x) = \sum_{i=0}^{n} a_i x^i \in \mathbb{D}[x]$ is **monic** if $a_n = 1$.

$F(\mathbf{x})$ denotes the quotient field of $F[\mathbf{x}]$, and is the **field of rational functions** over F in n variables.

Let $\Omega \subset \mathbb{C}^n$ be a nonempty path-connected set. Then $\mathrm{H}(\Omega)$ denotes the ring of analytic functions $f(\mathbf{z})$, such that for each $\zeta \in \Omega$ there exists an open neighborhood $O(\zeta, f)$ of ζ such that f is analytic on $O(f, \zeta)$. The addition and the product of functions are given by the standard identities: $(f+g)(\zeta) = f(\zeta)+g(\zeta), (fg)(\zeta) = f(\zeta)g(\zeta)$. If Ω is an open set, we assume that f is defined only on Ω. If Ω consists of one point ζ, then H_ζ stands for $\mathrm{H}(\{\zeta\})$.

Denote by $\mathcal{M}(\Omega), \mathcal{M}_\zeta$ the quotient fields of $\mathrm{H}(\Omega), \mathrm{H}_\zeta$ respectively.

For $a, d \in \mathbb{D}$, d **divides** a (or d is a **divisor** of a), denoted by $d|a$, if $a = db$ for some $b \in \mathbb{D}$.

$a \in \mathbb{D}$ is **unit** if $a|1$.

$a, b \in \mathbb{D}$ are **associates**, denoted by $a \equiv b$, if $a|b$ and $b|a$. Denote $\{\{a\}\} = \{b \in \mathbb{D} : b \equiv a\}$.

The associates of $a \in \mathbb{D}$ and the units are called **improper** divisors of a.

$a \in \mathbb{D}$ is **irreducible** if it is not a unit and every divisor of a is improper.

A nonzero, nonunit element $p \in \mathbb{D}$ is **prime** if for any $a, b \in \mathbb{D}$, $p|ab$ implies $p|a$ or $p|b$.

Let $a_1, \ldots, a_n \in \mathbb{D}$. Assume first that not all of $a_1, \ldots, a_n$ are equal to zero. An element $d \in \mathbb{D}$ is a **greatest common divisor** (g.c.d) of $a_1, \ldots, a_n$ if $d|a_i$ for $i = 1, \ldots, n$, and for any d' such that $d'|a_i, i = 1, \ldots, n$, $d'|d$. Denote by $(a_1, \ldots, a_n)$ any g.c.d. of $a_1, \ldots, a_n$. Then $\{\{(a_1, \ldots, a_n)\}\}$ is the equivalence class of all g.c.d. of $a_1, \ldots, a_n$. For $a_1 = \ldots = a_n = 0$, we define 0 to be the g.c.d. of $a_1, \ldots, a_n$, i.e. $(a_1, \ldots, a_n) = 0$.

$a_1, \ldots, a_n \in \mathbb{D}$ are **coprime** if $\{\{(a_1, \ldots, a_n)\}\} = \{\{1\}\}$.

$I \subseteq \mathbb{D}$ is an **ideal** if for any $a, b \in I$ and $p, q \in \mathbb{D}$ the element $pa + qb$ belongs to I.

An ideal $I \neq \mathbb{D}$ is **prime** if $ab \in I$ implies that either a or b is in I.

An ideal $I \neq \mathbb{D}$ is **maximal** if the only ideals that contain I are I and $\mathbb{D}$.

An ideal I is **finitely generated** if there exists k elements (**generators**) $p_1, \ldots, p_k \in I$ such that any $i \in I$ is of the form $i = a_1 p_1 + \cdots + a_k p_k$ for some $a_1, \ldots, a_k \in \mathbb{D}$.

An ideal is **principal** ideal if it is generated by one element p.

$\mathbb{D}$ is a **greatest common divisor** domain (GCDD), denoted by $\mathbb{D}_g$, if any two elements in $\mathbb{D}$ have a g.c.d.

$\mathbb{D}$ is a **unique factorization** domain (UFD), denoted by $\mathbb{D}_u$, if any nonzero, nonunit element a can be factored as a product of irreducible elements $a = p_1 \cdots p_r$, and this factorization is unique within order and unit factors.

$\mathbb{D}$ is a **principal ideal** domain (PID), denoted $\mathbb{D}_p$, if any ideal of $\mathbb{D}$ is principal.

$\mathbb{D}$ is a **Euclidean** domain (ED), denoted $\mathbb{D}_e$, if there exists a function $d : \mathbb{D}\backslash\{0\} \to \mathbb{Z}_+$ such that:

$$\text{for all } a, b \in \mathbb{D},\ ab \neq 0, \quad d(a) \leq d(ab);$$
$$\text{for any } a, b \in \mathbb{D},\ ab \neq 0, \text{ there exists } t, r \in \mathbb{D} \text{ such that}$$
$$a = tb + r, \text{ where either } r = 0 \text{ or } d(r) < d(b).$$

It is convenient to define $d(0) = \infty$. Let $a_1, a_2 \in \mathbb{D}_e$ and assume that $\infty > d(a_1) \geq d(a_2)$. **Euclid's algorithm** consists of a sequence $a_1, \ldots, a_{k+1}$, where $(a_1 \ldots a_k) \neq 0$, which is defined recursively as follows:

$$a_i = t_i a_{i+1} + a_{i+2}, \quad a_{i+2} = 0 \text{ or } d(a_{i+2}) < d(a_{i+1}) \quad \text{for} \quad i = 1, \ldots k-1.$$

[Hel43], [Kap49] A GCDD $\mathbb{D}$ is an **elementary divisor** domain (EDD), denoted by $\mathbb{D}_{ed}$, if for any three elements $a, b, c \in \mathbb{D}$ there exists $p, q, x, y \in \mathbb{D}$ such that $(a, b, c) = (px)a + (py)b + (qy)c$.

A GCDD $\mathbb{D}$ is a **Bezout** domain (BD), denoted by $\mathbb{D}_b$, if for any two elements $a, b \in \mathbb{D}$ $(a, b) = pa + qb$, for some $p, q \in \mathbb{D}$.

$p(x) = \sum_{i=0}^{m} a_i x^{m-i} \in \mathbb{Z}[x], a_0 \neq 0, m \geq 1$ is **primitive** if 1 is a g.c.d. of $a_0, \ldots, a_m$.

For $m \in \mathbb{N}$, the set of integers modulo m is denoted by $\mathbb{Z}_m$.

Facts:

Most of the facts about domains can be found in [ZS58] and [DF04]. More special results and references on the elementary divisor domains and rings are in [McD84]. The standard results on the domains of analytic functions can be found in [GR65]. More special results on analytic functions in one complex variable are in [Rud74].

1. Any integral domain satisfies *cancellation laws*: if $ab = ac$ or $ba = ca$ and $a \neq 0$, then $b = c$.
2. An integral domain such that any nonzero element is unit is a field F, and any field is an integral domain in which any nonzero element is unit.
3. A finite integral domain is a field.
4. $\mathbb{D}[\mathbf{x}]$ is an integral domain.
5. $\mathrm{H}(\Omega)$ is an integral domain.
6. Any prime element in $\mathbb{D}$ is irreducible.
7. In a UFD, any irreducible element is prime. This is not true in all integral domains.
8. Let $\mathbb{D}$ be a UFD. Then $\mathbb{D}[x]$ is a UFD. Hence, $\mathbb{D}[\mathbf{x}]$ is a UFD.
9. Let $a_1, a_2 \in \mathbb{D}_e$ and assume that $\infty > d(a_1) \geq d(a_2)$. Then Euclid's algorithm terminates in a finite number of steps, i.e., there exists $k \geq 3$ such that $a_1 \neq 0, \ldots, a_k \neq 0$ and $a_{k+1} = 0$. Hence, $a_k = (a_1, a_2)$.
10. An ED is a PID.
11. A PID is an EDD.
12. A PID is a UFD.
13. An EDD is a BD.
14. A BD is a GCDD.
15. A UFD is a GCDD.
16. The converses of Facts 10, 11, 12, 14, 15 are false (see Facts 28, 27, 21, 22).
17. [DF04, Chap. 8] An integral domain that is both a BD and a UFD is a PID.
18. An integral domain is a Bezout domain if and only if any finitely generated ideal is principal.
19. $\mathbb{Z}$ is an ED with $d(a) = |a|$.
20. Let $p, q \in \mathbb{Z}[x]$ be primitive polynomials. Then pq is primitive.
21. $F[x]$ is an ED with $d(p) =$ the degree of a nonzero polynomial. Hence, $F[x_1, \ldots, x_n]$ is a UFD. But $F[x_1, \ldots, x_n]$ is not a PID for $n \geq 2$.
22. $\mathbb{Z}[x_1, \ldots, x_m]$, $F[x_1, \ldots, x_n]$, and $\mathrm{H}(\Omega)$ (for a connected set $\Omega \subset \mathbb{C}^n$) are GCDDs, but for $m \geq 1$ and $n \geq 2$ these domains are not BDs.
23. [Frixx] (See Example 17 below.) Let $\Omega \subset \mathbb{C}$ be an open connected set. Then for $a, b \in \mathrm{H}(\Omega)$ there exists $p \in \mathrm{H}(\Omega)$ such that $(a, b) = pa + b$.
24. H_ζ, $\zeta \in \mathbb{C}$, is a UFD.
25. If $\Omega \subset \mathbb{C}^n$ is a connected open set, then $\mathrm{H}(\Omega)$ is not a UFD. (For $n = 1$ there is no prime factorization of an analytic function $f \in \mathrm{H}(\Omega)$ with an infinite countable number of zeros.)
26. Let $\Omega \subset \mathbb{C}$ be a compact connected set. Then $\mathrm{H}(\Omega)$ is an ED. Here, $d(a)$ is the number of zeros of a nonzero function $a \in \mathrm{H}(\Omega)$ counted with their multiplicities.
27. [Frixx] If $\Omega \subset \mathbb{C}$ is an open connected set, then $\mathrm{H}(\Omega)$ is an EDD. (See Example 17.) As $\mathrm{H}(\Omega)$ is not a UFD, it follows that $\mathrm{H}(\Omega)$ is not a PID. (Contrary to [McD84, Exc. II.E.10 (b), p. 144].)
28. [DF04, Chap. 8] $\mathbb{Z}[(1 + \sqrt{-19})/2]$ is a PID that is not an ED.

Examples:

1. $\{1, -1\}$ is the set of units in $\mathbb{Z}$. A g.c.d. of $a_1, \ldots, a_k \in \mathbb{Z}$ is **uniquely normalized** by the condition $(a_1, \ldots, a_k) \geq 0$.
2. A positive integer $p \in \mathbb{Z}$ is irreducible if and only if p is prime.
3. $\mathbb{Z}_m$ is an integral domain and, hence, a field with m elements if and only if p is a prime.
4. $\mathbb{Z} \supset I$ is a prime ideal if and only if all elements of I are divisible by some prime p.

5. $\{1, -1\}$ is the set of units in $\mathbb{Z}[x]$. A g.c.d. of $p_1, \ldots, p_k \in \mathbb{Z}[x]$, is **uniquely normalized** by the condition $(p_1, \ldots, p_k) = \sum_{i=0}^{m} a_i x^{m-i}$ and $a_0 > 0$.
6. Any prime element in $p(x) \in \mathbb{Z}[x], \deg p \geq 1$, is a primitive polynomial.
7. Let $p(x) = 2x+3, q(x) = 5x-3 \in \mathbb{Z}[x]$. Clearly p, q are primitive polynomials and $(p(x), q(x)) = 1$. However, 1 cannot be expressed as $1 = a(x)p(x) + b(x)q(x)$, where $a(x), b(x) \in \mathbb{Z}[x]$. Indeed, if this was possible, then $1 = a(0)p(0) + b(0)q(0) = 3(a(0) - b(0))$, which is impossible for $a(0), b(0) \in \mathbb{Z}$. Hence, $\mathbb{Z}[x]$ is not BD.
8. The field of quotients of $\mathbb{Z}$ is the field of rational numbers $\mathbb{Q}$.
9. Let $p(x), q(x) \in \mathbb{Z}[x]$ be two nonzero polynomials. Let $(p(x), q(x))$ be the g.c.d of p, q in $\mathbb{Z}[x]$. Use the fact that $p(x), q(x) \in \mathbb{Q}[x]$ to deduce that there exists a positive integer m and $a(x), b(x) \in \mathbb{Z}[x]$ such that $a(x)p(x) + b(x)q(x) = m(p(x), q(x))$. Furthermore, if $c(x)p(x) + d(x)q(x) = l(p(x), q(x))$ for some $c(x), d(x) \in \mathbb{Z}[x]$ and $0 \neq l \in \mathbb{Z}$, then $m|l$.
10. The set of real numbers $\mathbb{R}$ and the set of complex numbers $\mathbb{C}$ are fields.
11. A g.c.d. of $a_1, \ldots, a_k \in F[x]$ is **uniquely normalized** by the condition $(p_1, \ldots, p_k)$ is a monic polynomial.
12. A linear polynomial in $\mathbb{D}[\mathbf{x}]$ is irreducible.
13. Let $\Omega \subset \mathbb{C}$ be a connected set. Then each irreducible element of $\mathrm{H}(\Omega)$ is an associate of $z - \zeta$ for some $\zeta \in \Omega$.
14. For $\zeta \in \mathbb{C}$ H_ζ, every irreducible element is of the form $a(z - \zeta)$ for some $0 \neq a \in \mathbb{C}$. A g.c.d. of $a_1, \ldots, a_k \in \mathrm{H}_\zeta$ is **uniquely normalized** by the condition $(a_1, \ldots, a_k) = (z - \zeta)^m$ for some nonnegative integer m.
15. In $\mathrm{H}(\Omega)$, the set of functions which vanishes on a prescribed set $U \subseteq \Omega$, i.e.,
$$I(U) := \{f \in \mathrm{H}(\Omega) : \quad f(\zeta) = 0,\ \zeta \in U\},$$
is an ideal.
16. Let Ω be an open connected set in $\mathbb{C}$. [Rud74, Theorems 15.11, 15.13] implies the following:
    - $I(U) \neq \{0\}$ if and only if U is a countable set, with no accumulations points in Ω.
    - Let U be a countable subset of Ω with no accumulation points in Ω. Assume that for each $\zeta \in U$ one is given a nonnegative integer $m(\zeta)$ and $m(\zeta) + 1$ complex numbers $w_{0,\zeta}, \ldots, w_{m(\zeta),\zeta}$. Then there exists $f \in \mathrm{H}(\Omega)$ such that $f^{(n)}(\zeta) = n! w_{n,\zeta}, n = 0, \ldots, m(\zeta)$, for all $\zeta \in U$. Furthermore, if all $w_{n,\zeta} = 0$, then there exists $g \in \mathrm{H}(\Omega)$ such that all zeros of g are in U and g has a zero of order $m(\zeta) + 1$ at each $\zeta \in U$.
    - Let $a, b \in \mathrm{H}(\Omega), ab \neq 0$. Then there exists $f \in \mathrm{H}(\Omega)$ such that $a = cf, b = df$, where $c, d \in \mathrm{H}(\Omega)$ and c, d do not have a common zero in Ω.
    - Let $c, d \in \mathrm{H}(\Omega)$ and assume that c, d do not have a common zero in Ω. Let U be the zero set of c in Ω, and denote by $m(\zeta) \geq 1$ the multiplicity of the zero $\zeta \in U$ of c. Then there exists $g \in \mathrm{H}(\Omega)$ such that $(e^g)^{(n)}(\zeta) = d^{(n)}(\zeta)$ for $n = 0, \ldots, m(\zeta)$, for all $\zeta \in U$. Hence, $p = \frac{e^g - d}{c} \in \mathrm{H}(\Omega)$ and $e^g = pc + d$ is a unit in $\mathrm{H}(\Omega)$.
    - For $a, b \in \mathrm{H}(\Omega)$ there exists $p \in \mathrm{H}(\Omega)$ such $(a, b) = pa + b$.
    - For $a, b, c \in \mathrm{H}(\Omega)$ one has $(a, b, c) = p(a, b) + c = p(xa + b) + c$. Hence, $\mathrm{H}(\Omega)$ is EDD.
17. Let $I \subset \mathbb{C}[x, y]$ be the ideal given by given by the condition $p(0, 0) = 0$. Then I is generated by x and y, and $(x, y) = 1$. I is not principal and $\mathbb{C}[x, y]$ is not BD.
18. $\mathbb{D}[x, y]$ is not BD.
19. $\mathbb{Z}[\sqrt{-5}]$ is an integral domain that is not a GCDD, since $2 + 2\sqrt{-5}$ and 6 have no greatest common divisor. This can be seen by using the norm $N(a + b\sqrt{-5}) = a^2 + 5b^2$.

23.2 Equivalence of Matrices

In this section, we introduce matrices over an integral domain. Since any domain $\mathbb{D}$ can be viewed as a subset of its quotient field F, the notion of determinant, minor, rank, and adjugate in Chapters 1, 2, and 4 can be applied to these matrices. It is an interesting problem to determine whether one given matrix can be transformed to another by left multiplication, right multiplication, or multiplication on both sides, using only matrices invertible with the domain. These are equivalence relations and the problem is to characterize left (row) equivalence classes, right (columns) equivalence classes, and equivalence classes in $\mathbb{D}^{m\times n}$. For BD, the left equivalence classes are characterized by their Hermite normal form, which are attributed to Hermite. For EDD, the equivalence classes are characterized by their Smith normal form.

Definitions:

For a set S, denote by $S^{m\times n}$ the set of all $m \times n$ matrices $A = [a_{ij}]_{i=j=1}^{i=m,j=n}$, where each $a_{ij} \in S$.

For positive integers $p \leq q$, denote by $Q_{p,q}$ the set of all subsets $\{i_1, \ldots, i_p\} \subset \{1, 2, \ldots q\}$ of cardinality p, where we assume that $1 \leq i_1 < \ldots < i_p \leq q$.

$U \in \mathbb{D}^{n\times n}$ is **$\mathbb{D}$-invertible** (**unimodular**), if there exists $V \in \mathbb{D}^{n\times n}$ such that $UV = VU = I_n$.

GL$(n, \mathbb{D})$ denotes the group of $\mathbb{D}$-invertible matrices in $\mathbb{D}^{n\times n}$.

Let $A, B \in \mathbb{D}^{m\times n}$. Then A and B are **column equivalent**, **row equivalent**, and **equivalent** if the following conditions hold respectively:

$$\begin{aligned} &B = AP \quad \text{for some } P \in \mathbf{GL}(n, \mathbb{D}) \quad (A \sim_c B), \\ &B = QA \quad \text{for some } Q \in \mathbf{GL}(m, \mathbb{D}) \quad (A \sim_r B), \\ &B = QAP \quad \text{for some } P \in \mathbf{GL}(n, \mathbb{D}),\ Q \in \mathbf{GL}(m, \mathbb{D}) \quad (A \sim B). \end{aligned}$$

For $A \in \mathbb{D}_g^{m\times n}$, let

$$\begin{aligned} &\mu(\alpha, A) = \text{g.c.d. } \{\det A[\alpha, \theta],\ \theta \in Q_{k,n}\}, \quad \alpha \in Q_{k,m}, \\ &\nu(\beta, A) = \text{g.c.d. } \{\det A[\phi, \beta],\ \phi \in Q_{k,m}\}, \quad \beta \in Q_{k,n}, \\ &\delta_k(A) = \text{g.c.d. } \{\det A[\phi, \theta],\ \phi \in Q_{k,m},\ \theta \in Q_{k,n}\}. \end{aligned}$$

$\delta_k(A)$ is the k-th **determinant invariant** of A.

For $A \in \mathbb{D}_g^{m\times n}$,

$$\begin{aligned} &i_j(A) = \frac{\delta_j(A)}{\delta_{j-1}(A)}, \quad j = 1, \ldots, \text{rank } A, \quad (\delta_0(A) = 1), \\ &i_j(A) = 0 \quad \text{for rank } A < j \leq \min(m, n), \end{aligned}$$

are called the **invariant factors** of A. $i_j(A)$ is a **trivial** factor if $i_j(A)$ is unit in $\mathbb{D}_g$. We adopt the **normalization** $i_j(A) = 1$ for any trivial factor of A. For $\mathbb{D} = \mathbb{Z}, \mathbb{Z}[x], F[x]$, we adopt the normalizations given in the previous section in the Examples 1, 6, and 12, respectively.

Assume that $\mathbb{D}[x]$ is a GCDD. Then the invariant factors of $A \in \mathbb{D}[x]^{m\times n}$ are also called **invariant polynomials.**

$D = [d_{ij}] \in \mathbb{D}^{m\times n}$ is a **diagonal** matrix if $d_{ij} = 0$ for all $i \neq j$. The entries $d_{11}, \ldots, d_{\ell\ell}$, $\ell = \min(m, n)$, are called the **diagonal entries** of D. D is denoted as $D = \text{diag } (d_{11}, \ldots, d_{\ell\ell}) \in \mathbb{D}^{m\times n}$.

Denote by $\Pi_n \subset \mathbf{GL}(n, \mathbb{D})$ the group of $n \times n$ permutation matrices.

An $\mathbb{D}$-invertible matrix $U \in \mathbf{GL}(n, \mathbb{D})$ is **simple** if there exists $P, Q \in \Pi_n$ such that $U = P \begin{bmatrix} V & 0 \\ 0 & I_{n-2} \end{bmatrix} Q$, where $V = \begin{bmatrix} \alpha & \beta \\ \gamma & \delta \end{bmatrix} \in \mathbf{GL}(2, \mathbb{D})$, i.e., $\alpha\delta - \beta\gamma$ is $\mathbb{D}$-invertible.

U is **elementary** if U is of the above form and $V = \begin{bmatrix} \alpha & 0 \\ \gamma & \delta \end{bmatrix} \in \mathbf{GL}(2, \mathbb{D})$, i.e., α, δ are invertible.

For $A \in \mathbb{D}^{m\times n}$, the following row (column) operations are **elementary row operations**:

(a) Interchange any two rows (columns) of A.
(b) Multiply row (column) i by an invertible element a.
(c) Add to row (column) j b times row (column) i $(i \neq j)$.

For $A \in \mathbb{D}^{m\times n}$, the following row (column) operations are **simple row operations**:

(d) Replace row (column) i by a times row (column) i plus b times row (column) j, and row (column) j by c times row (column) i plus d times row (column) j, where $i \neq j$ and $ad - bc$ is invertible in $\mathbb{D}$.

$B = [b_{ij}] \in \mathbb{D}^{m\times n}$ is in **Hermite normal** form if the following conditions hold. Let $r = \text{rank } B$. First, the i-th row of B is a nonzero row if and only if $i \leq r$. Second, let b_{in_i} be the first nonzero entry in the i-th row for $i = 1,\dots,r$. Then $1 \leq n_1 < n_2 < \cdots < n_r \leq n$.

$B \in \mathbb{D}_g^{m\times n}$ is in **Smith normal** form if B is a diagonal matrix $B = \text{diag}(b_1,\dots,b_r,0,\dots,0)$, $b_i \neq 0$, for $i = 1,\dots,r$ and $b_{i-1}|b_i$ for $i = 2,\dots,r$.

Facts:

Most of the results of this section can be found in [McD84]. Some special results of this section are given in [Fri81] and [Frixx]. For information about equivalence over fields, see Chapter 1 and Chapter 2.

1. The cardinality of $Q_{p,q}$ is $\binom{q}{p}$.
2. If U is $\mathbb{D}$-invertible then det U is a unit in $\mathbb{D}$. Conversely, if det U is a unit then U is $\mathbb{D}$-invertible, and its inverse U^{-1} is given by $U^{-1} = (\det U)^{-1}\text{adj } U$.
3. For $A \in \mathbb{D}^{m\times n}$, the rank of A is the maximal size of the nonvanishing minor. (The rank of zero matrix is 0.)
4. Column equivalence, row equivalence, and equivalence of matrices are equivalence relations in $\mathbb{D}^{m\times n}$.
5. For any $A, B \in \mathbb{D}^{m\times n}$, one has $A \sim_r B \iff A^T \sim_c B^T$. Hence, it is enough to consider the row equivalence relation.
6. For $A, B \in \mathbb{D}_g^{m\times n}$, the Cauchy–Binet formula (Chapter 4) yields
$$\mu(\alpha, A) \equiv \mu(\alpha, B) \quad \text{for all } \alpha \in Q_{k,m} \quad \text{if } A \sim_c B,$$
$$\nu(\beta, A) \equiv \nu(\beta, B) \quad \text{for all } \beta \in Q_{k,n} \quad \text{if } A \sim_r B,$$
$$\delta_k(A) \equiv \delta_k(B) \quad \text{if } A \sim B,$$
for $k = 1,\dots,\min(m,n)$.
7. Any elementary matrix is a simple matrix, but not conversely.
8. The elementary row and column operations can be carried out by multiplications by A by suitable elementary matrices from the left and the right, respectively.
9. The simple row and column operations are carried out by multiplications by A by suitable simple matrices U from the left and right, respectively.
10. Let $\mathbb{D}_b$ be a Bezout domain, $A \in \mathbb{D}_b^{m\times n}$, rank $A = r$. Then A is row equivalent to $B = [b_{ij}] \in \mathbb{D}_b^{m\times n}$, in a Hermite normal form, which satisfies the following conditions.

    Let b_{in_i} be the first nonzero entry in the i-th row for $i = 1,\dots,r$. Then $1 \leq n_1 < n_2 < \cdots < n_r \leq n$ are uniquely determined and the elements b_{in_i}, $i = 1,\dots,r$ are uniquely determined, up to units, by the conditions
$$\nu((n_1,\dots,n_i), A) = b_{1n_1}\cdots b_{in_i}, \quad i = 1,\dots,r,$$
$$\nu(\alpha, A) = 0, \ \alpha \in Q_{i,n_i-1}, \quad i = 1,\dots,r.$$

    The elements b_{jn_i}, $j = 1,\dots,i-1$ are then successively uniquely determined up to the addition of arbitrary multiples of b_{in_i}. The remaining elements b_{ik} are now uniquely determined. The $\mathbb{D}$-invertible matrix Q, such that $B = QA$, can be given by a finite product of simple matrices.

    If b_{in_i} in the Hermite normal form is invertible, we assume the normalization conditions $b_{in_i} = 1$ and $b_{jn_i} = 0$ for $i < j$.
11. For Euclidean domains, we assume normalization conditions either $b_{jn_i} = 0$ or $d(b_{jn_i}) < d(b_{in_i})$ for $j < i$. Then for any $A \in \mathbb{D}_e^{m\times n}$, in a Hermite normal form $B = QA$, $Q \in \mathbf{GL}_m(\mathbb{D}_e)$ Q is a product of a finite elementary matrices.

12. $U \in \mathbf{GL}(n, \mathbb{D}_e)$ is a finite product of elementary $\mathbb{D}$-invertible matrices.
13. For $\mathbb{Z}$, we assume the normalization $b_{in_i} \geq 1$ and $0 \leq b_{jn_i} < b_{in_i}$ for $j < i$. For $F[x]$, we assume that b_{in_i} is a monic polynomial and $\deg b_{jn_i} < \deg b_{in_i}$ for $j < i$. Then for $\mathbb{D}_e = \mathbb{Z}, F[x]$, any $A \in \mathbb{D}_e^{m\times n}$ has a unique Hermite normal form.
14. $A, B \in \mathbb{D}_b$ are row equivalent if and only if A and B are row equivalent to the same Hermite normal form.
15. $A \in F^{m\times n}$ can be brought to its unique Hermite normal form, called the *reduced row echelon form* (RREF),

$$b_{in_i} = 1, \quad b_{jn_i} = 0, \quad j = 1, \ldots, i-1, \quad i = 1, \ldots, r = \text{rank } A,$$

by a finite number of elementary row operations. Hence, $A, B \in F^{m\times n}$ are row equivalent if and only if $r = \text{rank } A = \text{rank } B$ and they have the same RREF. (See Chapter 1.)
16. For $A \in \mathbb{D}_g^{m\times n}$ and $1 \leq p < q \leq \min(m,n)$, $\delta_p(A) | \delta_q(A)$.
17. For $A \in \mathbb{D}_g^{m\times n}$, $i_{j-1}(A) | i_j(A)$ for $j = 2, \ldots, \text{rank } A$.
18. Any $0 \neq A \in \mathbb{D}_{ed}^{m\times n}$ is equivalent to its Smith normal form $B = \text{diag}\,(i_1(A), \ldots, i_r(A), 0, \ldots, 0)$, where $r = \text{rank } A$ and $i_1(A), \ldots, i_r(A)$ are the invariants factors of A.
19. $A, B \in \mathbb{D}_{ed}^{m\times n}$ are equivalent if and only if A and B have the same rank and the same invariant factors.

Examples:

1. Let $A = \begin{bmatrix} 1 & a \\ 0 & 0 \end{bmatrix}, B = \begin{bmatrix} 1 & b \\ 0 & 0 \end{bmatrix} \in \mathbb{D}^{2\times 2}$ be two Hermite normal forms. It is straightforward to show that $A \sim_r B$ if and only if $a = b$. Assume that $\mathbb{D}$ is a BD and let $a \neq 0$. Then rank $A = 1, \nu((1), A) = 1, \{\{\nu((2), A)\}\} = \{\{a\}\}, \nu((1,2), A) = 0$. If $\mathbb{D}$ has other units than 1, it follows that $\nu(\beta, A)$ for all $\beta \in Q_{k,2}, k = 1, 2$ do not determine the row equivalence class of A.
2. Let $A = \begin{bmatrix} a & c \\ b & d \end{bmatrix} \in \mathbb{D}_b^{2\times 2}$. Then there exists $u, v \in \mathbb{D}_b$ such that $ua + vb = (a, b) = \nu((1), A)$. If $(a, b) \neq 0$, then $1 = (u, v)$. If $a = b = 0$, choose $u = 1, v = 0$. Hence, there exists $x, y \in \mathbb{D}_b$ such that $yu - xv = 1$. Thus $V = \begin{bmatrix} u & v \\ x & y \end{bmatrix} \in \mathbf{GL}(2, \mathbb{D}_b)$ and $VA = \begin{bmatrix} (a,b) & c' \\ b' & d' \end{bmatrix}$. Clearly $b' = xa + yb = (a, b)e$. Hence, $\begin{bmatrix} 1 & 0 \\ -e & 1 \end{bmatrix} VA = \begin{bmatrix} (a,b) & c' \\ 0 & f \end{bmatrix}$ is a Hermite normal form of A. This construction is easily extended to obtain a Hermite normal form for any $A \in \mathbb{D}_b^{m\times n}$, using simple row operations.
3. Let $A \in \mathbb{D}_e^{2\times 2}$ as in the previous example. Assume that $ab \neq 0$. Change the two rows of A if needed to assume that $d(a) \leq d(b)$. Let $a_1 = b, a_2 = a$ and do the first step of Euclid's algorithm: $a_1 = t_1 a_2 + a_3$, where $a_3 = 0$ or $d(a_3) < d(a_2)$. Let $V = \begin{bmatrix} 1 & 0 \\ -t_1 & 1 \end{bmatrix} \in \mathbf{GL}(2, \mathbb{D}_e)$ be an elementary matrix. Then $A_1 = VA = \begin{bmatrix} a_2 & * \\ a_3 & * \end{bmatrix}$. If $a_3 = 0$, then A_1 has a Hermite normal form. If $a_3 \neq 0$, continue as above. Since Euclid's algorithm terminates after a finite number of steps, it follows that A can be put into Hermite normal form by a finite number of elementary row operations. This statement holds similarly in the case $ab = 0$. This construction is easily extended to obtain a Hermite normal form for any $A \in \mathbb{D}_e^{m\times n}$ using elementary row operations.
4. Assume that $\mathbb{D}$ is a BD and let $A = \begin{bmatrix} a & b \\ 0 & c \end{bmatrix} \in \mathbb{D}^{2\times 2}$. Note that $\delta_1(A) = (a, b, c)$. If A is equivalent to a Smith normal form then there exists $V, U \in \mathbf{GL}(2, D)$ such that $VAU = \begin{bmatrix} (a,b,c) & * \\ * & * \end{bmatrix}$.

Assume that $V = \begin{bmatrix} p & q \\ \bar{q} & \bar{p} \end{bmatrix}, U = \begin{bmatrix} x & \bar{y} \\ y & \bar{x} \end{bmatrix}$. Then there exist $p, q, x, y \in \mathbb{D}$ such that $(px)a + (py)b + (qy)c = (a, b, c)$. Thus, if each $A \in \mathbb{D}^{2\times 2}$ is equivalent to Smith normal form in $\mathbb{D}^{2\times 2}$, then it follows that $\mathbb{D}$ is an EDD.

Conversely, suppose that $\mathbb{D}$ is an EDD. Then $\mathbb{D}$ is also a BD. Let $A \in \mathbb{D}^{2\times 2}$. First, bring A to an upper triangular Hermite normal form using simple row operations: $A_1 = WA = \begin{bmatrix} a & b \\ 0 & c \end{bmatrix}, W \in$ **GL**$(2, \mathbb{D})$. Note that $\delta_1(A) = \delta_1(A_1) = (a, b, c)$. Since $\mathbb{D}$ is an EDD, there exist $p, q, x, y \in \mathbb{D}$ such that $(px)a + (py)b + (qy)c = (a, b, c)$. If $(a, b, c) \neq (0, 0, 0)$, then $(p, q) = (x, y) = 1$. Otherwise $A = A_1 = 0$ and we are done. Hence, there exist $\bar{p}, \bar{q}, \bar{x}, \bar{y}$ such that $p\bar{p} - q\bar{q} = x\bar{x} - y\bar{y} = 1$. Let $V = \begin{bmatrix} p & q \\ \bar{q} & \bar{p} \end{bmatrix}, U = \begin{bmatrix} x & \bar{y} \\ y & \bar{x} \end{bmatrix}$. Thus, $G = VA_1U = \begin{bmatrix} \delta_1(A) & g_{12} \\ g_{21} & g_{22} \end{bmatrix}$. Since $\delta_1(G) = \delta_1(A)$, we deduce that $\delta_1(A)$ divides g_{12} and g_{21}. Apply appropriate elementary row and column operations to deduce that A is equivalent to a diagonal matrix $C = \operatorname{diag}(i_1(A), d_2)$. As $\delta_2(C) = i_1(A)d_2 = \delta_2(A)$, we see that C is has Smith normal form.

These arguments are easily extended to obtain a Smith normal form for any $A \in \mathbb{D}_{ed}^{m\times n}$, using simple row and column operations.

5. The converse of Fact 6 is false, as can be seen by considering

$A = \begin{bmatrix} 2 & 2 \\ x & x \end{bmatrix}, B = \begin{bmatrix} 1 & 1 \\ 0 & 0 \end{bmatrix} \in \mathbb{Z}[x]^{2\times 2}$. $\mu(\{1\}, A) = \mu(\{1\}, B) = 1$, $\mu(\{2\}, A) = \mu(\{2\}, B) = 1$, $\mu(\{1, 2\}, A) = \mu(\{1, 2\}, B) = 0$, but there does not exist a $\mathbb{D}$-invertible P such that $PA = B$.

6. Let $\mathbb{D}$ be an integral domain and assume that $p(x) = x^m + \sum_{i=1}^{m} a_i x^{m-i} \in \mathbb{D}[x]$ is a monic polynomial of degree $m \geq 2$. Let $C(p) \in \mathbb{D}^{m\times m}$ be the companion matrix (see Chapter 4). Then $\det(xI_m - C(p)) = p(x)$. Assume that $\mathbb{D}[x]$ is a GCDD. Let $C(p)(x) = xI_m - C(p) \in \mathbb{D}[x]^{m\times m}$. By deleting the last column and first row, we obtain a triangular $(m-1)\times(m-1)$ submatrix with -1s on the diagonal, so it follows that $\delta_i(C(p)(x)) = 1$ for $i = 1, \ldots, m-1$. Hence, the invariant factors of $C(p)(x)$ are $i_1(C(p)(x)) = \ldots = i_{m-1}(C(p)(x)) = 1$ and $i_m(C(p)(x)) = p(x)$. If $\mathbb{D}$ is a field, then $C(p)(x)$ is equivalent over $\mathbb{D}[x]$ to $\operatorname{diag}(1, \ldots, 1, p(x))$.

23.3 Linear Equations over Bezout Domains

Definitions:

M is a $\mathbb{D}$-**module** if **M** is an additive group with respect to the operation $+$ and **M** admits a multiplication by a **scalar** $a \in \mathbb{D}$, i.e., there exists a mapping $\mathbb{D} \times \mathbf{M} \to \mathbf{M}$ which satisfies the standard distribution properties and $1\mathbf{v} = \mathbf{v}$ for all $\mathbf{v} \in \mathbf{M}$ (the latter requirement sometimes results in the module being called unital). (For a field F, a module **M** is a vector space over F.)

M is **finitely generated** if there exist $\mathbf{v}_1, \ldots, \mathbf{v}_n \in \mathbf{M}$ so that every $\mathbf{v} \in \mathbf{M}$ is a **linear combination** of $\mathbf{v}_1, \ldots, \mathbf{v}_n$, over $\mathbb{D}$, i.e., $\mathbf{v} = a_1\mathbf{v}_1 + \ldots + a_n\mathbf{v}_n$ for some $a_1, \ldots, a_n \in \mathbb{D}$. $\mathbf{v}_1, \ldots, \mathbf{v}_n$ is a **basis** of **M** if every $\mathbf{v}$ can be written as a unique linear combination of $\mathbf{v}_1, \ldots, \mathbf{v}_n$. $\dim \mathbf{M} = n$ means that **M** has a basis of n elements.

$\mathbf{N} \subseteq \mathbf{M}$ is a $\mathbb{D}$-**submodule** of **M** if **N** is closed under the addition and multiplication by scalars.

$\mathbb{D}^n (= \mathbb{D}^{n\times 1})$ is a $\mathbb{D}$-module. It has a **standard** basis $\mathbf{e}_i$ for $i = 1, \ldots, n$, where $\mathbf{e}_i$ is the i-th column of the identity matrix I_n.

For any $A \in \mathbb{D}^{m\times n}$, the **range** of A, denoted by range(A), is the set of all linear combinations of the columns of A.

The **kernel** of A, denoted by $\ker(A)$, is the set of all solutions to the homogeneous equation $A\mathbf{x} = \mathbf{0}$.

Consider a system of m linear equations in n unknowns:

$\sum_{j=1}^{n} a_{ij}x_j = b_j$, $i = 1,\ldots,m$, where $a_{ij}, b_i \in \mathbb{D}$ for $i = 1,\ldots,m$, $j = 1,\ldots,n$. In matrix notation this system is $A\mathbf{x} = \mathbf{b}$, where $A = [a_{ij}] \in \mathbb{D}^{m\times n}$, $\mathbf{x} = [x_1,\ldots,x_n]^T \in \mathbb{D}^n$, $\mathbf{b} = [b_1,\ldots,b_m]^T \in \mathbb{D}^m$. A and $[A,\mathbf{b}] \in \mathbb{D}^{m\times(n+1)}$ are called the **coefficient** matrix and the **augmented** matrix, respectively.

Let $A \in \mathrm{H}_0^{m\times n}$. Then $A = A(z) = [a_{ij}(z)]_{i=j=1}^{m,n}$ and $A(z)$ has the McLaurin expansion $A(z) = \sum_{k=0}^{\infty} A_k z^k$, where $A_k \in \mathbb{C}^{m\times n}, k = 0,\ldots$. Here, each $a_{ij}(z)$ has convergent McLaurin series for $|z| < R(A)$ for some $R(A) > 0$.

The invariant factors of A are called the **local invariant polynomials** of A, which are normalized to be of the form $i_k(A) = z^{i_k}$ for $0 \le i_1 \le i_2 \le \ldots \le i_r$, where $r = \text{rank } A$.

The integer i_r is the **index** of A and is denoted by $\eta = \eta(A)$. For a nonnegative integer p, denote by $\mathcal{K}_p = \mathcal{K}_p(A)$ the number of local invariant polynomials of A whose degree is equal to p.

Facts:

For modules, see [ZS58], [DF04], and [McD84]. The solvability of linear systems over EDD can be traced to [Hel43] and [Kap49]. The results for BD can be found in [Fri81] and [Frixx]. The results for H_0 are given in [Fri80]. The general theory of solvability of the systems of equations over commutative rings is discussed in [McD84, Exc. I.G.7–I.G.8]. (See Chapter 1 for information about solvability over fields.)

1. The system $A\mathbf{x} = \mathbf{b}$ is solvable over a Bezout domain $\mathbb{D}_b$ if and only if $r = \text{rank } A = \text{rank } [A,\mathbf{b}]$ and $\delta_r(A) = \delta_r([A,\mathbf{b}])$, which is equivalent to the statement that A and $[A,\mathbf{b}]$ have the same set of invariant factors, up to invertible elements. For a field F, this result reduces to the equality $\text{rank } A = \text{rank } [A,\mathbf{b}]$.
2. For $A \in \mathbb{D}_b^{m\times n}$, range A and ker A are modules in $\mathbb{D}_b^m$ and $\mathbb{D}_b^n$ having finite bases with rank A and null A elements, respectively. Moreover, the basis of ker A can be completed to a basis of $\mathbb{D}_b^n$.
3. For $A, B \in \mathrm{H}_0^{m\times n}$, let $C(z) = A(z) + z^{k+1}B(z)$, where k is a nonnegative integer. Then A and C have the same local invariant polynomials up to degree k. Moreover, if k is equal to the index of A, and A and C have the same rank, then A is equivalent to C.
4. Consider a system of linear equations over H_0 $A(z)\mathbf{u}(z) = \mathbf{b}(z)$, where $A(z) \in \mathrm{H}_0^{m\times n}$ and $\mathbf{b}(z) = \sum_{k=0}^{\infty} \mathbf{b}_k z^k \in \mathrm{H}_0^m$, $\mathbf{b}_k \in \mathbb{C}^m$, $k = 0,\ldots$. Look for the power series solution $\mathbf{u}(z) = \sum_{k=0}^{\infty} \mathbf{u}_k z^k$, where $\mathbf{u}_k \in \mathbb{C}^n$, $k = 0,\ldots$. Then $\sum_{j=0}^{k} A_{k-j}\mathbf{u}_j = \mathbf{b}_k$, for $k = 0,\ldots$. This system is solvable for $k = 0,\ldots,q \in \mathbb{Z}_+$ if and only if $A(z)$ and $[A(z),\mathbf{b}(z)]$ have the same local invariant polynomials up to degree q.
5. Suppose that $A(z)\mathbf{u}(z) = \mathbf{b}(z)$ is solvable over H_0. Let $q = \eta(A)$ and suppose that $\mathbf{u}_0,\ldots,\mathbf{u}_q$ satisfies the system of equations, given in the previous fact, for $k = 0,\ldots,q$. Then there exists a solution $\mathbf{u}(z) \in \mathrm{H}_0^n$ satisfying $\mathbf{u}(0) = \mathbf{u}_0$.
6. Let $q \in \mathbb{Z}_+$ and $\mathbf{W}_q \subset \mathbb{C}^n$ be the subspace of all vectors $\mathbf{w}_0$ such that $\mathbf{w}_0,\ldots,\mathbf{w}_q$ is a solution to the homogenous system $\sum_{j=0}^{k} A_{k-j}\mathbf{w}_j = 0$, for $k = 0,\ldots,q$. Then $\dim \mathbf{W}_q = n - \sum_{j=0}^{q} \mathcal{K}_j(A)$. In particular, for $\eta = \eta(A)$ and any $\mathbf{w}_0 \in \mathbf{W}_\eta$ there exists $\mathbf{w}(z) \in \mathrm{H}_0^n$ such that $A(z)\mathbf{w}(z) = \mathbf{0}$, $\mathbf{w}(0) = \mathbf{w}_0$.

23.4 Strict Equivalence of Pencils

Definitions:

A matrix $A(x) \in \mathbb{D}[x]^{m\times n}$ is a **pencil** if $A(x) = A_0 + xA_1$, $A_0, A_1 \in \mathbb{D}^{m\times n}$.

A pencil $A(x)$ is **regular** if $m = n$ and det $A(x)$ is not the zero polynomial. Otherwise $A(x)$ is a **singular** pencil.

Associate with a pencil $A(x) = A_0 + xA_1 \in \mathbb{D}[x]^{m\times n}$ the **homogeneous** pencil $A(x_0, x_1) = x_0A_0 + x_1A_1 \in \mathbb{D}[x_0, x_1]^{m\times n}$.

Two pencils $A(x), B(x) \in \mathbb{D}[x]^{m\times n}$ are **strictly** equivalent, denoted by $A(x) \overset{s}{\sim} B(x)$, if $B(x) = QA(x)P$ for some $P \in \mathbf{GL}_n(\mathbb{D})$, $Q \in \mathbf{GL}_m(\mathbb{D})$. Similarly, two homogeneous pencils $A(x_0, x_1), B(x_0, x_1)$

$\in \mathbb{D}[x_0, x_1]^{m\times n}$ are **strictly** equivalent, denoted by $A(x_0, x_1)\overset{s}{\sim}B(x_0, x_1)$, if $B(x_0, x_1) = QA(x_0, x_1)P$ for some $P \in \mathbf{GL}_n(\mathbb{D})$, $Q \in \mathbf{GL}_m(\mathbb{D})$.

For a UFD $\mathbb{D}_u$ let $\delta_k(x_0, x_1)$, $i_k(x_0, x_1)$ be the invariant determinants and factors of $A(x_0, x_1)$, respectively, for $k = 1, \ldots, \text{rank } A(x_0, x_1)$. They are called **homogeneous** determinants and the invariant **homogeneous** polynomials (factors), respectively, of $A(x_0, x_1)$. (Sometimes, $\delta_k(x_0, x_1)$, $i_k(x_0, x_1)$, $k = 1, \ldots, \text{rank } A(x_0, x_1)$ are called the homogeneous determinants and the invariant homogeneous polynomials $A(x)$.)

Let $A(x) \in F[x]^{m\times n}$ and consider the module $\mathbf{M} \subset F[x]^n$ of all solutions of $A(x)\mathbf{w}(x) = 0$. The set of all solutions $\mathbf{w}(x)$ is an $F[x]$-module $\mathbf{M}$ with a finite basis $\mathbf{w}_1(x), \ldots, \mathbf{w}_s(x)$, where $s = n - \text{rank } A(x)$. Choose a basis $\mathbf{w}_1(x), \ldots, \mathbf{w}_s(x)$ in $\mathbf{M}$ such that $\mathbf{w}_k(x) \in \mathbf{M}$ has the lowest degree among all $\mathbf{w}(x) \in \mathbf{M}$, which are linearly independent over $F[x]$ of $\mathbf{w}_1, \ldots, \mathbf{w}_{k-1}(x)$ for $k = 1, \ldots, s$. Then the **column indices** $\alpha_1 \le \alpha_2 \le \ldots \le \alpha_s$ of $A(x)$ are given as $\alpha_k = \deg \mathbf{w}_k(x)$, $k = 1, \ldots, s$. The **row indices** $0 \le \beta_1 \le \beta_2 \le \ldots \le \beta_t$, $t = m - \text{rank } A(x)$, of $A(x)$, are the column indices of $A(x)^{\mathrm{T}}$.

Facts:

The notion of strict equivalence of $n \times n$ regular pencils over the fields goes back to K. Weierstrass [Wei67]. The notion of strict similarity of $m \times n$ matrices over the fields is due to L. Kronecker [Kro90]. Most of the details can be found in [Gan59]. Some special results are proven in [Frixx]. For information about matrix pencils over fields see Section 43.1.

1. Let $A_0, A_1, B_0, B_1 \in \mathbb{D}^{m\times n}$. Then $A_0 + xA_1\overset{s}{\sim}B_0 + xB_0 \iff x_0A_0 + x_1A_1\overset{s}{\sim}B_0x_0 + B_1x_1$.
2. Let $A_0, A_1 \in \mathbb{D}_u$. Then the invariant determinants and the invariant polynomials $\delta_k(x_0, x_1)$, $i_k(x_0, x_1)$, $k = 1, \ldots, \text{rank } x_0A_0 + x_1A_1$, of $x_0A_0 + x_1A_1$ are homogeneous polynomials. Moreover, if $\delta_k(x)$ and $i_k(x)$ are the invariant determinants and factors of the pencil $A_0 + xA_1$ for $k = 1, \ldots, \text{rank } A_0+xA_1$, then $\delta_k(x) = \delta_k(1, x)$, $i_k(x) = i_k(1, x)$, for $k = 1, \ldots, \text{rank } A_0+xA_1$.
3. [Wei67] Let $A_0 + xA_1 \in F[x]^{n\times n}$ be a regular pencil. Then a pencil $B_0 + xB_1 \in F[x]^{n\times n}$ is strictly equivalent to $A_0 + xA_1$ if and only if $A_0 + xA_1$ and $B_0 + xB_1$ have the same invariant polynomials over $F[x]$.
4. [Frixx] Let $A_0 + xA_1, B_0 + xB_1 \in \mathbb{D}[x]^{n\times n}$. Assume that $A_1, B_1 \in \mathbf{GL}_n(\mathbb{D})$. Then $A_0 + xA_1\overset{s}{\sim}B_0 + xB_1 \iff A_0 + xA_1 \sim B_0 + xB_1$.
5. [Gan59] The column (row) indices are independent of a particular allowed choice of a basis $\mathbf{w}_1(x), \ldots, \mathbf{w}_s(x)$.
6. For singular pencils the invariant homogeneous polynomials alone do not determine the class of strictly equivalent pencils.
7. [Kro90], [Gan59] The pencils $A(x), B(x) \in F[x]^{m\times n}$ are strictly equivalent if and only if they have the same invariant homogeneous polynomials and the same row and column indices.

References

[DF04] D.S. Dummit and R.M. Foote, *Abstract Algebra*, 3rd ed., John Wiley & Sons, New York, 2004.

[Fri80] S. Friedland, Analytic similarity of matrices, *Lectures in Applied Math.*, Amer. Math. Soc. 18 (1980), 43–85 (edited by C.I. Byrnes and C.F. Martin).

[Fri81] S. Friedland, *Spectral Theory of Matrices: I. General Matrices*, MRC Report, Madison, WI, 1981.

[Frixx] S. Friedland, *Matrices*, (a book in preparation).

[Gan59] F.R. Gantmacher, *The Theory of Matrices*, Vol. I and II, Chelsea Publications, New York, 1959. (Vol. I reprinted by AMS Chelsea Publishing, Providence 1998.)

[GR65] R. Gunning and H. Rossi, *Analytic Functions of Several Complex Variables*, Prentice-Hall, Upper Saddle Rever, NJ, 1965.

[Hel43] O. Helmer, The elementary divisor theorems for certain rings without chain conditions, *Bull. Amer. Math. Soc.* 49 (1943), 225–236.

[Kap49] I. Kaplansky, Elementary divisors and modules, *Trans. Amer. Math. Soc.* 66 (1949), 464–491.

[Kro90] L. Kronecker, Algebraische reduction der schaaren bilinear formen, *S-B Akad. Berlin*, 1890, 763–778.

[McD84] B.R. McDonald, *Linear Agebra over Commutative Rings*, Marcel Dekker, New York, 1984.

[Rud74] W. Rudin, *Real and Complex Analysis*, McGraw Hill, New York, 1974.

[Wei67] K. Weierstrass, Zur theorie der bilinearen un quadratischen formen, *Monatsch. Akad. Wiss. Berlin*, 310–338, 1867.

[ZS58] O. Zariski and P. Samuel, *Commutative Algebra* I, Van Nostrand, Princeton, 1958 (reprinted by Springer-Verlag, 1975).

24

Similarity of Families of Matrices

Shmuel Friedland
University of Illinois at Chicago

This chapter uses the notations, definitions, and facts given in Chapter 23. The aim of this chapter is to acquaint the reader with two difficult problems in matrix theory:

1. Similarity of matrices over integral domains, which are not fields.
2. Simultaneous similarity of tuples of matrices over $\mathbb{C}$.

Problem *1* is notoriously difficult. We show that for the local ring H_0 this problem reduces to a Problem *2* for certain kind of matrices. We then discuss certain special cases of Problem *2* as simultaneous similarity of tuples of matrices to upper triangular and diagonal matrices. The L-property of pairs of matrices, which is discussed next, is closely related to simultaneous similarity of pair of matrices to a diagonal pair. The rest of the chapter is devoted to a "solution" of the Problem *2*, by the author, in terms of basic notions of algebraic geometry.

24.1 Similarity of Matrices

The classical result of K. Weierstrass [Wei67] states that the similarity class of $A \in F^{n\times n}$ is determined by the invariant factors of $-A + xI_n$ over $F[x]$. (See Chapter 6 and Chapter 23.) For a given $A, B \in F^{n\times n}$, one can easily determine if A and B are similar, by considering *only* the ranks of three specific matrices associated with A, B [GB77]. It is well known that it is a difficult problem to determine if $A, B \in \mathbb{D}^{n\times n}$ are $\mathbb{D}$-similar for most integral domains that are not fields. The emphasis of this chapter is the similarity over the local field H_0. The subject of similarity of matrices over H_0 arises naturally in theory linear differential equations having singularity with respect to a parameter. It was studied by Wasow in [Was63], [Was77], and [Was78].

Definitions:

For $E \in \mathbb{D}^{m\times n}, G \in \mathbb{D}^{p\times q}$ extend the definition of the **tensor** or **Kronecker** product $E \otimes G \in \mathbb{D}^{mp\times nq}$ of E with G to the domain $\mathbb{D}$ in the obvious way. (See Section 10.4.)

$A, B \in \mathbb{D}^{m\times m}$ are called **similar**, denoted by $A \approx B$ if $B = QAQ^{-1}$, for some $Q \in \mathrm{GL}_m(\mathbb{D})$.

Let $A, B \in \mathrm{H}(\Omega)^{n\times n}$. Then A and B are called **analytically similar**, denoted as $A \overset{a}{\approx} B$, if A and B are similar over $\mathrm{H}(\Omega)$.

A and B are called **locally similar** if for any $\zeta \in \Omega$, the restrictions A_ζ, B_ζ of A, B to the local rings H_ζ, respectively, are similar over H_ζ.

A, B are called **point-wise** similar if $A(\zeta), B(\zeta)$ are similar matrices in $\mathbb{C}^{n\times n}$ for each $\zeta \in \Omega$.

A, B are called **rationally** similar, denoted as $A \overset{r}{\approx} B$, if A, B are similar over the quotient field $\mathcal{M}(\Omega)$ of $\mathrm{H}(\Omega)$.

Let $A, B \in \mathrm{H}_0^{n\times n}$:

$$A(x) = \sum_{k=0}^{\infty} A_k x^k, \quad |x| < R(A), \qquad B(x) = \sum_{k=0}^{\infty} B_k x^k, \quad |x| < R(B).$$

Then $\eta(A, B)$ and $\mathcal{K}_p(A, B)$ are the index and the number of local invariant polynomials of degree p of the matrix $I_n \otimes A(x) - B(x)^{\mathrm{T}} \otimes I_n$, respectively, for $p = 0, 1, \ldots$.

$\lambda(x)$ is called an **algebraic** function if there exists a monic polynomial $p(\lambda, x) = \lambda^n + \sum_{i=1}^{n} q_i(x)\lambda^{n-i} \in (\mathbb{C}[x])[\lambda]$ of λ-degree $n \geq 1$ such that $p(\lambda(x), x) = 0$ identically. Then $\lambda(x)$ is a multivalued function on $\mathbb{C}$, which has n **branches.** At each point $\zeta \in \mathbb{C}$ each branch $\lambda_j(x)$ of $\lambda(x)$ has **Puiseaux** expansion: $\lambda_j(x) = \sum_{i=0}^{\infty} b_{ji}(\zeta)(x-\zeta)^{\frac{i}{m}}$, which converges for $|x - \zeta| < R(\zeta)$, and some integer m depending on $p(x)$. $\sum_{i=0}^{m} b_{ji}(\zeta)(x-\zeta)^{\frac{i}{m}}$ is called the **linear part** of $\lambda_j(x)$ at ζ. Two distinct branches $\lambda_j(x)$ and $\lambda_k(x)$ are called **tangent** at $\zeta \in \mathbb{C}$ if the linear parts of $\lambda_j(x)$ and $\lambda_k(x)$ coincide at ζ. Each branch $\lambda_j(x)$ has Puiseaux expansion around ∞: $\lambda_j(x) = x^l \sum_{i=0}^{\infty} c_{ji} x^{-\frac{i}{m}}$, which converges for $|x| > R$. Here, l is the smallest nonnegative integer such that $c_{j0} \neq 0$ at least for some branch λ_j. $x^l \sum_{i=0}^{m} c_{ji} x^{-\frac{i}{m}}$ is called the **principal part** of $\lambda_j(x)$ at ∞. Two distinct branches $\lambda_j(x)$ and $\lambda_k(x)$ are called **tangent** at ∞ if the principal parts of $\lambda_j(x)$ and $\lambda_k(x)$ coincide at ∞.

Facts:

The standard results on the tensor products can be found in Chapter 10 or Chapter 13 or in [MM64]. Most of the results of this section related to the analytic similarity over H_0 are taken from [Fri80].

1. The similarity relation is an equivalence relation on $\mathbb{D}^{m\times m}$.
2. $A \approx B \iff A(x) = -A + xI \overset{s}{\sim} B(x) = -B + xI$.
3. Let $A, B \in F^{n\times n}$. Then A and B are similar if and only if the pencils $-A + xI$ and $-B + xI$ have the same invariant polynomials over $F[x]$.
4. If $E = [e_{ij}] \in \mathbb{D}^{m\times n}, G \in \mathbb{D}^{p\times q}$, then $E \otimes G$ can be viewed as the $m \times n$ block matrix $[e_{ij}G]_{i,j=1}^{m,n}$. Alternatively, $E \otimes G$ can be identified with the linear transformation
$$L(E, G) : \mathbb{D}^{q\times n} \to \mathbb{D}^{p\times m}, \quad X \to GXE^{\mathrm{T}}.$$
5. $(E \otimes G)(U \otimes V) = EU \otimes GV$ whenever the products EU and GV are defined. Also $(E \otimes G)^{\mathrm{T}} = E^{\mathrm{T}} \otimes G^{\mathrm{T}}$ (cf. §2.5.4)).
6. For $A, B \in \mathbb{D}^{n\times n}$, if A is similar to B, then
$$I_n \otimes A - A^{\mathrm{T}} \otimes I_n \sim I_n \otimes A - B^{\mathrm{T}} \otimes I_n \sim I_n \otimes B - B^{\mathrm{T}} \otimes I_n.$$
7. [Gur80] There are examples over Euclidean domains for which the reverse of the implication in Fact 6 does not hold.
8. [GB77] For $\mathbb{D} = F$, the reverse of the implication in Fact 6 holds.
9. [Fri80] Let $A \in F^{m\times m}$, $B \in F^{n\times n}$. Then
$$\text{null }(I_n \otimes A - B^{\mathrm{T}} \otimes I_m) \leq \frac{1}{2}(\text{null }(I_m \otimes A - A^{\mathrm{T}} \otimes I_m) + \text{null }(I_n \otimes B - B^{\mathrm{T}} \otimes I_n)).$$
Equality holds if and only if $m = n$ and A and B are similar.

10. Let $A \in F^{n\times n}$ and assume that $p_1(x), \ldots, p_k(x) \in F[x]$ are the nontrivial normalized invariant polynomials of $-A+xI$, where $p_1(x)|p_2(x)|\ldots|p_k(x)$ and $p_1(x)p_2(x)\ldots p_k(x) = \det(xI - A)$. Then $A \approx C(p_1) \oplus C(p_2) \oplus \ldots \oplus C(p_k)$ and $C(p_1) \oplus C(p_2) \oplus \ldots \oplus C(p_k)$ is called the *rational canonical form* of A (cf. Chapter 6.6).
11. For $A, B \in \mathrm{H}(\Omega)^{n\times n}$, analytic similarity implies local similarity, local similarity implies point-wise similarity, and point-wise similarity implies rational similarity.
12. For $n = 1$, all the four concepts in Fact 11 are equivalent. For $n \geq 2$, local similarity, point-wise similarity, and rational similarity, are distinct (see Example 2).
13. The equivalence of the three matrices in Fact 6 over $\mathrm{H}(\Omega)$ implies the point-wise similarity of A and B.
14. Let $A, B \in \mathrm{H}_0^{n\times n}$. Then A and B are analytically similar over H_0 if and only if A and B are rationally similar over H_0 and there exists $\eta(A, A) + 1$ matrices $T_0, \ldots, T_\eta \in \mathbb{C}^{n\times n}$ ($\eta = \eta(A, A)$), such that $\det T_0 \neq 0$ and

$$\sum_{i=0}^{k} A_i T_{k-i} - T_{k-i} B_i = 0, \quad k = 0, \ldots, \eta(A, A).$$

15. Suppose that the characteristic polynomial of $A(x)$ splits over H_0:

$$\det(\lambda I - A(x)) = \prod_{i=1}^{n} (\lambda - \lambda_i(x)), \quad \lambda_i(x) \in \mathrm{H}_0,\ i = 1, \ldots, n.$$

Then $A(x)$ is analytically similar to

$$\begin{aligned} &C(x) = \oplus_{i=1}^{\ell} C_i(x), \quad C_i(x) \in \mathrm{H}_0^{n_i\times n_i}, \\ &(\alpha_i I_{n_i} - C_i(0))^{n_i} = 0,\ \alpha_i = \lambda_{n_i}(0),\ \alpha_i \neq \alpha_j \quad \text{for} \quad i \neq j,\ i, j = 1, \ldots, \ell. \end{aligned}$$

16. Assume that the characteristic polynomial of $A(x) \in \mathrm{H}_0$ splits in H_0. Then $A(x)$ is analytically similar to a block diagonal matrix $C(x)$ of the form Fact 15 such that each $C_i(x)$ is an upper triangular matrix whose off-diagonal entries are polynomials in x. Moreover, the degree of each polynomial entry above the diagonal in the matrix $C_i(x)$ does not exceed $\eta(C_i, C_i)$ for $i = 1, \ldots, \ell$.
17. Let $P(x)$ and $Q(x)$ be matrices of the form

$$\begin{aligned} &P(x) = \oplus_{i=1}^{p} P_i(x),\ P_i(x) \in \mathrm{H}_0^{m_i\times m_i}, \\ &(\alpha_i I_{m_i} - P_i(0))^{m_i} = 0,\ \alpha_i \neq \alpha_j \quad \text{for} \quad i \neq j, i, j = 1, \ldots, p, \\ &Q(x) = \oplus_{j=1}^{q} Q_j(x),\ Q_j(x) \in \mathrm{H}_0^{n_j\times n_j}, \\ &(\beta_j I_{n_j} - Q_j(0))^{n_j} = 0,\ \beta_i \neq \beta_j \quad \text{for} \quad i \neq j,\ i, j = 1, \ldots, q. \end{aligned}$$

Assume furthermore that

$$\alpha_i = \beta_i,\ i = 1, \ldots, t,\ \alpha_j \neq \beta_j,\ i = t+1, \ldots, p,\ j = t+1, \ldots, q,\ 0 \leq t \leq \min(p, q).$$

Then the nonconstant local invariant polynomials of $I \otimes P(x) - Q(x)^{\mathrm{T}} \otimes I$ are the nonconstant local invariant polynomials of $I \otimes P_i(x) - Q_i(x)^{\mathrm{T}} \otimes I$ for $i = 1, \ldots, t$:

$$\mathcal{K}_p(P, Q) = \sum_{i=1}^{t} \mathcal{K}_p(P_i, Q_i), \quad p = 1, \ldots, .$$

In particular, if $C(x)$ is of the form in Fact 15, then

$$\eta(C, C) = \max_{1\leq i\leq \ell} \eta(C_i, C_i).$$

18. $A(x)\overset{a}{\approx}B(x) \iff A(y^m)\overset{a}{\approx}B(y^m)$ for any $2 \le m \in \mathbb{N}$.
19. [GR65] (*Weierstrass preparation theorem*) For any monic polynomial $p(\lambda, x) = \lambda^n + \sum_{i=1}^n a_i(x)\lambda^{n-i} \in \mathrm{H}_0[\lambda]$ there exists $m \in \mathbb{N}$ such that $p(\lambda, y^m)$ splits over H_0.
20. For a given rational canonical form $A(x) \in \mathrm{H}_0^{2\times 2}$ there are at most a countable number of analytic similarity classes. (See Example 3.)
21. For a given rational canonical form $A(x) \in \mathrm{H}_0^{n\times n}$, where $n \ge 3$, there may exist a family of distinct similarity classes corresponding to a finite dimensional variety. (See Example 4.)
22. Let $A(x) \in H_0^{n\times n}$ and assume that the characteristic polynomial of $A(x)$ splits in H_0 as in Fact 15. Let $B(x) = \operatorname{diag}(\lambda_1(x), \ldots, \lambda_n(x))$. Then $A(x)$ and $B(x)$ are not analytically similar if and only if there exists a nonnegative integer p such that

$$\mathcal{K}_p(A, A) + \mathcal{K}_p(B, B) < 2\mathcal{K}_p(A, B),$$
$$\mathcal{K}_j(A, A) + \mathcal{K}_j(B, B) = 2\mathcal{K}_j(A, B),\ j = 0, \ldots, p-1, \quad \text{if } p \ge 1.$$

In particular, $A(x)\overset{a}{\approx}B(x)$ if and only if the three matrices given in Fact 6 are equivalent over H_0.
23. [Fri78] Let $A(x) \in \mathbb{C}[x]^{n\times n}$. Then each eigenvalue $\lambda(x)$ of $A(x)$ is an algebraic function. Assume that $A(\zeta)$ is diagonalizable for some $\zeta \in \mathbb{C}$. Then the linear part of each branch of $\lambda_j(x)$ is linear at ζ, i.e., is of the form $\alpha + \beta x$ for some $\alpha, \beta \in \mathbb{C}$.
24. Let $A(x) \in \mathbb{C}[x]^{n\times n}$ be of the form $A(x) = \sum_{k=0}^{\ell} A_k x^k$, where $A_k \in \mathbb{C}^{n\times n}$ for $k = 0, \ldots, \ell$ and $\ell \ge 1$, $A_\ell \ne 0$. Then one of the following conditions imply that $A(x) = S(x)B(x)S^{-1}(x)$, where $S(x) \in \mathrm{GL}(n, \mathbb{C}[x])$ and $B(x) \in \mathbb{C}[x]^{n\times n}$ is a diagonal matrix of the form $\oplus_{i=1}^m \lambda_i(x) I_{k_i}$, where $k_1, \ldots, k_m \ge 1$. Furthermore, $\lambda_1(x), \ldots, \lambda_m(x)$ are m distinct polynomials satisfying the following conditions:

    (a) $\deg \lambda_1 = \ell \ge \deg \lambda_i(x), i = 2, \ldots, m-1$.

    (b) The polynomial $\lambda_i(x) - \lambda_j(x)$ has only simple roots in $\mathbb{C}$ for $i \ne j$. $(\lambda_i(\zeta) = \lambda_j(\zeta) \Rightarrow \lambda_i'(\zeta) \ne \lambda_j'(\zeta))$.

    i. The characteristic polynomial of $A(x)$ splits in $\mathbb{C}[x]$, i.e., all the eigenvalues of $A(x)$ are polynomials. $A(x)$ is point-wise diagonalizable in $\mathbb{C}$ and no two distinct eigenvalues are tangent at any $\zeta \in \mathbb{C}$.

    ii. $A(x)$ is point-wise diagonalizable in $\mathbb{C}$ and A_ℓ is diagonalizable. No two distinct eigenvalues are tangent at any point $\zeta \in \mathbb{C} \cup \{\infty\}$. Then $A(x)$ is strictly similar to $B(x)$, i.e., $S(x)$ can be chosen in $\mathrm{GL}(n, \mathbb{C})$. Furthermore, $\lambda_1(x), \ldots, \lambda_m(x)$ satisfy the additional condition:

    (c) For $i \ne j$, either $\frac{d^\ell \lambda_i}{d^\ell x}(0) \ne \frac{d^\ell \lambda_j}{d^\ell x}(0)$ or $\frac{d^\ell \lambda_i}{d^\ell x}(0) = \frac{d^\ell \lambda_j}{d^\ell x}(0)$ and $\frac{d^{\ell-1} \lambda_i}{d^{\ell=1} x}(0) \ne \frac{d^{\ell-1} \lambda_j}{d^{\ell-1} x}(0)$.

Examples:

1. Let

$$A = \begin{bmatrix} 1 & 0 \\ 0 & 5 \end{bmatrix}, \quad B = \begin{bmatrix} 1 & 1 \\ 0 & 5 \end{bmatrix} \in \mathbb{Z}^{2\times 2}.$$

Then $A(x)$ and $B(x)$ have the same invariant polynomials over $\mathbb{Z}[x]$ and A and B are not similar over $\mathbb{Z}$.

2. Let

$$A(z) = \begin{bmatrix} 0 & 1 \\ 0 & 0 \end{bmatrix}, \quad D(z) = \begin{bmatrix} z & 0 \\ 0 & 1 \end{bmatrix}.$$

Then $zA(z) = D(z)A(z)D(z)^{-1}$, i.e., $A(z), zA(z)$ are rationally similar. Clearly $A(z)$ and $zA(z)$ are not point-wise similar for any Ω containing 0. Now $zA(z), z^2A(z)$ are point-wise similar in $\mathbb{C}$, but they are not locally similar on H_0.

3. Let $A(x) \in \mathrm{H}_0^{2\times 2}$ and assume that $\det(\lambda I - A(x)) = (\lambda - \lambda_1(x))(\lambda - \lambda_2(x))$. Then $A(x)$ is analytically similar either to a diagonal matrix or to

$$B(x) = \begin{bmatrix} \lambda_1(x) & x^k \\ 0 & \lambda_2(x) \end{bmatrix}, \quad k = 0, \ldots, p \ (p \geq 0).$$

Furthermore, if $A(x) \stackrel{a}{\approx} B(x)$, then $\eta(A, A) = k$.

4. Let $A(x) \in \mathrm{H}_0^{3\times 3}$. Assume that

$$A(x) \stackrel{r}{\approx} C(p), \quad p(\lambda, x) = \lambda(\lambda - x^{2m})(\lambda - x^{4m}), \quad m \geq 1.$$

Then $A(x)$ is analytically similar to a matrix

$$B(x,a) = \begin{bmatrix} 0 & x^{k_1} & a(x) \\ 0 & x^{2m} & x^{k_2} \\ 0 & 0 & x^{4m} \end{bmatrix}, \quad 0 \leq k_1, k_2 \leq \infty \ (x^{\infty} = 0),$$

where $a(x)$ is a polynomial of degree $4m - 1$ at most. Furthermore, $B(x,a) \stackrel{a}{\approx} B(x,b)$ if and only if

(a) If $a(0) \neq 1$, then $b - a$ is divisible by x^m.

(b) If $a(0) = 1$ and $\frac{d^i a}{dx^i} = 0$, $i = 1, \ldots, k-1$, $\frac{d^k a}{dx^k} \neq 0$ for $1 \leq k < m$, then $b - a$ is divisible by x^{m+k}.

(c) If $a(0) = 1$ and $\frac{d^i a}{dx^i} = 0$, $i = 1, \ldots, m$, then $b - a$ is divisible by x^{2m}.

Then for $k_1 = k_2 = m$ and $a(0) \in \mathbb{C}\backslash\{1\}$, we can assume that $a(x)$ is a polynomial of degree less than m. Furthermore, the similarity classes of $A(x)$ are uniquely determined by such $a(x)$. These similarity classes are parameterized by $\mathbb{C}\backslash\{1\} \times \mathbb{C}^{m-1}$ (the Taylor coefficients of $a(x)$).

24.2 Simultaneous Similarity of Matrices

In this section, we introduce the notion of simultaneous similarity of matrices over a domain $\mathbb{D}$. The problem of simultaneous similarity of matrices over a field F, i.e., to describe the similarity class of a given m (≥ 2) tuple of matrices or to decide when a given two tuples of matrices are simultaneously similar, is in general a hard problem, which will be discussed in the next sections. There are some cases where this problem has a relatively simple solution. As shown below, the problem of analytic similarity of $A(x), B(x) \in \mathrm{H}_0^{n\times n}$ reduces to the problem of simultaneously similarity of certain 2-tuples of matrices.

Definitions:

For $A_0, \ldots, A_l \in \mathbb{D}^{n\times n}$ denote by $\mathcal{A}(A_0, \ldots, A_l) \subset \mathbb{D}^{n\times n}$ the minimal algebra in $\mathbb{D}^{n\times n}$ containing I_n and $A_0, \ldots, A_l$. Thus, every matrix $G \in \mathcal{A}(A_0, \ldots, A_l)$ is a noncommutative polynomial in $A_0, \ldots, A_l$.

For $l \geq 1$, $(A_0, A_1, \ldots, A_\ell), (B_0, \ldots, B_\ell) \in (\mathbb{D}^{n\times n})^{\ell+1}$ are called **simultaneously** similar, denoted by $(A_0, A_1, \ldots, A_\ell) \approx (B_0, \ldots, B_\ell)$, if there exists $P \in \mathrm{GL}(n, \mathbb{D})$ such that $B_i = P A_i P^{-1}, i = 0, \ldots, \ell$, i.e., $(B_0, B_1, \ldots, B_\ell) = P(A_0, A_1, \ldots, A_\ell)P^{-1}$.

Associate with $(A_0, A_1, \ldots, A_\ell), (B_0, \ldots, B_\ell) \in (\mathbb{D}^{n\times n})^{\ell+1}$ the matrix polynomials $A(x) = \sum_{i=0}^{\ell} A_i x^i$, $B(x) = \sum_{i=0}^{\ell} B_i x^i \in \mathbb{D}[x]^{n\times n}$. $A(x)$ and $B(x)$ are called **strictly** similar ($A \stackrel{s}{\approx} B$) if there exists $P \in \mathrm{GL}(n, \mathbb{D})$ such that $B(x) = P A(x) P^{-1}$.

Facts:

1. $A \stackrel{s}{\approx} B \iff (A_0, A_1, \ldots, A_\ell) \approx (B_0, \ldots, B_\ell)$.
2. $(A_0, \ldots, A_\ell) \in (\mathbb{C}^{n\times n})^{\ell+1}$ is simultaneously similar to a diagonal tuple $(B_0, \ldots, B_\ell) \in (\mathbb{C}^{n\times n})^{\ell+1}$, i.e., each B_i is a diagonal matrix if and only if $A_0, \ldots, A_\ell$ are $\ell + 1$ commuting diagonalizable matrices: $A_i A_j = A_j A_i$ for $i, j = 0, \ldots, \ell$.

3. If $A_0, \ldots, A_\ell \in \mathbb{C}^{n\times n}$ commute, then $(A_0, \ldots, A_\ell)$ is simultaneously similar to an upper triangular tuple $(B_0, \ldots, B_\ell)$.
4. Let $l \in \mathbb{N}$, $(A_0, \ldots, A_l), (B_0, \ldots, B_l) \in (\mathbb{C}^{n\times n})^{l+1}$, and $U = [U_{ij}]_{i,j=1}^{l+1}, V = [V_{ij}]_{i,j=1}^{l+1}, W = [W_{ij}]_{i,j=1}^{l+1} \in \mathbb{C}^{n(l+1)\times n(l+1)}$, $U_{ij}, V_{ij}, W_{ij} \in \mathbb{C}^{n\times n}, i, j = 1, \ldots, l+1$ be block upper triangular matrices with the following block entries:

$$U_{ij} = A_{j-i},\ V_{ij} = B_{j-i},\ W_{ij} = \delta_{(i+1)j} I_n, \quad i = 1, \ldots, l+1, j = i, \ldots, l+1.$$

Then the system in Fact 14 of section 24.1 is solvable with $T_0 \in \mathrm{GL}(n, \mathbb{C})$ if and only for $l = \kappa(A, A)$ the pairs (U, W) and (V, W) are simultaneously similar.
5. For $A_0, \ldots, A_\ell \in (\mathbb{C}^{n\times n})^{\ell+1}$ TFAE:
   - $(A_0, \ldots, A_\ell)$ is simultaneously similar to an upper triangular tuple $(B_0, \ldots, B_\ell) \in (\mathbb{C}^{n\times n})^{\ell+1}$.
   - For any $0 \le i < j \le \ell$ and $M \in \mathcal{A}(A_0, \ldots, A_\ell)$, the matrix $(A_i A_j - A_j A_i)M$ is nilpotent.
6. Let $\mathcal{X}_0 = \mathcal{A}(A_0, \ldots, A_\ell) \subseteq F^{n\times n}$ and define recursively

$$\mathcal{X}_k = \sum_{0\le i<j\le\ell} (A_i A_j - A_j A_i)\mathcal{X}_{k-1} \subseteq F^{n\times n}, \quad k = 1, \ldots.$$

Then $(A_0, \ldots, A_\ell)$ is simultaneously similar to an upper triangular tuple if and only if the following two conditions hold:

   - $A_i \mathcal{X}_k \subseteq \mathcal{X}_k, \quad i = 0, \ldots, \ell,\ k = 0, \ldots.$
   - There exists $q \ge 1$ such that $\mathcal{X}_q = \{0\}$ and $\mathcal{X}_k$ is a strict subspace of $\mathcal{X}_{k-1}$ for $k = 1, \ldots, q$.

Examples:

1. This example illustrates the construction of the matrices U and W in Fact 4. Let $A_0 = \begin{bmatrix} 1 & 2 \\ 3 & 4 \end{bmatrix}$, $A_1 = \begin{bmatrix} 5 & 6 \\ 7 & 8 \end{bmatrix}$, and $A_2 = \begin{bmatrix} 1 & -1 \\ -1 & 1 \end{bmatrix}$. Then

$$U = \begin{bmatrix} 1 & 2 & 5 & 6 & 1 & -1 \\ 3 & 4 & 7 & 8 & -1 & 1 \\ 0 & 0 & 1 & 2 & 5 & 6 \\ 0 & 0 & 3 & 4 & 7 & 8 \\ 0 & 0 & 0 & 0 & 1 & 2 \\ 0 & 0 & 0 & 0 & 3 & 4 \end{bmatrix} \quad \text{and} \quad W = \begin{bmatrix} 0 & 0 & 1 & 0 & 0 & 0 \\ 0 & 0 & 0 & 1 & 0 & 0 \\ 0 & 0 & 0 & 0 & 1 & 0 \\ 0 & 0 & 0 & 0 & 0 & 1 \\ 0 & 0 & 0 & 0 & 0 & 0 \\ 0 & 0 & 0 & 0 & 0 & 0 \end{bmatrix}.$$

24.3 Property *L*

Property L was introduced and studied in [MT52] and [MT55]. In this section, we consider only square pencils $A(x) = A_0 + A_1 x \in \mathbb{C}[x]^{n\times n}$, $A(x_0, x_1) \in \mathbb{C}[x_0, x_1]^{n\times n}$, where $A_1 \ne 0$.

Definitions:

A pencil $A(x) \in \mathbb{C}[x]^{n\times n}$ has **property** L if all the eigenvalues of $A(x_0, x_1)$ are linear functions. That is, $\lambda_i(x_0, x_1) = \alpha_i x_0 + \beta_i x_1$ is an eigenvalue of $A(x_0, x_1)$ of multiplicity n_i for $i = 1, \ldots, m$, where

$$n = \sum_{i=1}^{m} n_i, \quad (\alpha_i, \beta_i) \ne (\alpha_j, \beta_j), \quad \text{for} \quad 1 \le i < j \le m.$$

A pencil $A(x) = A_0 + A_1 x$ is **Hermitian** if A^0, A^1 are Hermitian.

Facts:

Most of the results of this section can be found in [MF80], [Fri81], and [Frixx].

1. For a pencil $A(x) = A_0 + xA_1 \in \mathbb{C}[x]^{n\times n}$ TFAE:
   - $A(x)$ has property L.
   - The eigenvalues of $A(x)$ are polynomials of degree 1 at most.
   - The characteristic polynomial of $A(x)$ splits into linear factors over $\mathbb{C}[x]$.
   - There is an ordering of the eigenvalues of A_0 and A_1, $\alpha_1, \ldots, \alpha_n$ and $\beta_1, \ldots, \beta_n$, respectively, such that the eigenvalues of $A_0x_0 + A_1x_1$ are $\alpha_1 x_0 + \beta_1 x_1, \ldots, \alpha_n x_0 + \beta_n x_1$.
2. A pencil in $A(x)$ has property L if one of the following conditions hold:
   - $A(x)$ is similar over $\mathbb{C}(x)$ to an upper triangular matrix $U(x) \in \mathbb{C}(x)^{n\times n}$.
   - $A(x)$ is strictly similar to an upper triangular pencil $U(x) = U_0 + U_1x$.
   - $A(x)$ is similar over $\mathbb{C}[x]$ to a diagonal matrix $B(x) \in \mathbb{C}[x]^{n\times n}$.
   - $A(x)$ is strictly similar to diagonal pencil.
3. If a pencil $A(x_0, x_1)$ has property L, then any two distinct eigenvalues are not tangent at any point of $\mathbb{C} \cup \infty$.
4. Assume that $A(x)$ is point-wise diagonalizable on $\mathbb{C}$. Then $A(x)$ has property L. Furthermore, $A(x)$ is similar over $\mathbb{C}[x]$ to a diagonal pencil $B(x) = B_0 + B_1x$. Suppose furthermore that A_1 is diagonalizable, i.e., $A(x_0, x_1)$ is point-wise diagonalizable on $\mathbb{C}^2$. Then $A(x)$ is strictly similar to a diagonal pencil $B(x)$, i.e., A_0 and A_1 are commuting diagonalizable matrices.
5. Let $A(x) = A_0 + A_1x \in \mathbb{C}[x]^{n\times n}$ such that A_1 and A_2 are diagonalizable and $A_0A_1 \neq A_1A_0$. Then exactly one of the following conditions hold:
   - $A(x)$ is not diagonalizable exactly at the points $\zeta_1, \ldots, \zeta_p$, where $1 \le p \le n(n-1)$.
   - For $n \ge 3$, $A(x) \in \mathbb{C}[x]^{n\times n}$ is diagonalizable exactly at the points $\zeta_1 = 0, \ldots, \zeta_q$ for some $q \ge 1$. (*We do not know if this condition is satisfied for some pencil.*)
6. Let $A(x) = A_0 + A_1x$ be a Hermitian pencil satisfying $A_0A_1 \neq A_1A_0$. Then there exists $2q$ distinct complex points $\zeta_1, \overline{\zeta}_1 \ldots, \zeta_q, \overline{\zeta}_q \in \mathbb{C}\backslash\mathbb{R}$, $1 \le q \le \frac{n(n-1)}{2}$ such that $A(x)$ is not diagonalizable if and only if $x \in \{\zeta_1, \overline{\zeta}_1, \ldots, \zeta_q, \overline{\zeta}_q\}$.

Examples:

1. This example illustrates the case $n = 2$ of Fact 5. Let

$$A_0 = \begin{bmatrix} 1 & 2 \\ 3 & 4 \end{bmatrix} \quad \text{and} \quad A_1 = \begin{bmatrix} 1 & 3 \\ -3 & 1 \end{bmatrix}, \quad \text{so} \quad A(x) = \begin{bmatrix} x+1 & 3x \\ -3x & x+2 \end{bmatrix}.$$

For $\zeta \in \mathbb{C}$, the only possible way $A(\zeta)$ can fail to be diagonalizable is if $A(\zeta)$ has repeated eigenvalues. The eigenvalues of $A(\zeta)$ are $\frac{1}{2}\left(2\zeta - \sqrt{1-36\zeta^2} + 3\right)$ and $\frac{1}{2}\left(2\zeta + \sqrt{1-36\zeta^2} + 3\right)$, so the only values of ζ at which it is possible that $A(\zeta)$ is not diagonalizeable are $\zeta = \pm\frac{1}{6}$, and in fact $A(\pm\frac{1}{6})$ is not diagonalizable.

24.4 Simultaneous Similarity Classification I

This section outlines the setting for the classification of conjugacy classes of $l+1$ tuples $(A_0, A_1, \ldots, A_l) \in (\mathbb{C}^{n\times n})^{l+1}$ under the simultaneous similarity. This classification depends on certain standard notions in algebraic geometry that are explained briefly in this section. A detailed solution to the classification of conjugacy classes of $l+1$ tuples is outlined in the next section.

Definitions:

$\mathcal{X} \subset \mathbb{C}^N$ is called an **affine algebraic variety** (called here a **variety**) if it is the zero set of a finite number of polynomial equations in $\mathbb{C}^N$.

$\mathcal{X}$ is irreducible if $\mathcal{X}$ does not decompose in a nontrivial way to a union of two varieties.

If $\mathcal{X}$ is a finite nontrivial union of irreducible varieties, these irreducible varieties are called the **irreducible** components of $\mathcal{X}$.

$x \in \mathcal{X}$ is called a **regular** (**smooth**) point of irreducible $\mathcal{X}$ if in the neighborhood of this point $\mathcal{X}$ is a complex manifold of a fixed dimension d, which is called the **dimension** of $\mathcal{X}$ and is denoted by dim $\mathcal{X}$. $\emptyset$ is an irreducible variety of dimension -1.

For a reducible variety $\mathcal{Y} \subset \mathbb{C}^N$, the **dimension** of $\mathcal{Y}$, denoted by dim $\mathcal{Y}$, is the maximum dimension of its irreducible components.

The set of singular (nonsmooth) points of $\mathcal{X}$ is denoted by $\mathcal{X}_s$.

A set $\mathcal{Z}$ is a **quasi-irreducible** variety if there exists a nonempty irreducible variety $\mathcal{X}$ and a strict subvariety $\mathcal{Y} \subset \mathcal{X}$ such that $\mathcal{Z} = \mathcal{X}\backslash\mathcal{Y}$. The dimension of $\mathcal{Z}$, denoted by dim $\mathcal{Z}$, is defined to be equal to the dimension of $\mathcal{X}$.

A quasi-irreducible variety $\mathcal{Z}$ is **regular** if $\mathcal{Z} \subset \mathcal{X}\backslash\mathcal{X}_s$.

A **stratification** of $\mathbb{C}^N$ is a decomposition of $\mathbb{C}^N$ to a finite disjoint union of $\mathcal{X}_1, \ldots, \mathcal{X}_p$ of regular quasi-irreducible varieties such that Cl $(\mathcal{X}_i)\backslash\mathcal{X}_i = \cup_{j\in\mathcal{A}_i}\mathcal{X}_j$ for some $\mathcal{A}_i \subset \{1, \ldots, p\}$ for $i = 1, \ldots, p$. (Cl $(\mathcal{X}_i) = \mathcal{X}_i \iff \mathcal{A}_i = \emptyset$.)

Denote by $\mathbb{C}[\mathbb{C}^N]$ the ring of polynomial in N variables with coefficients in $\mathbb{C}$.

Denote by $\mathcal{W}_{n,l+1,r+1}$ the finite dimensional vector space of multilinear polynomials in $(l+1)n^2$ variables of degree at most $r+1$. That is, the degree of each variable in any polynomial is at most 1. $N(n,l,r) :=$ dim $\mathcal{W}_{n,l+1,r+1}$. $\mathcal{W}_{n,l+1,r+1}$ has a standard basis $\mathbf{e}_1, \ldots, \mathbf{e}_{N(n,l,r)}$ in $\mathcal{W}_{n,l+1,r+1}$ consisting of monomials in $(l+1)n^2$ variables of degree $r+1$ at most, arranged in a lexicographical order.

Let $\mathcal{X} \subset \mathbb{C}^N$ be a quasi-irreducible variety. Denote by $\mathbb{C}[\mathcal{X}]$ the restriction of all polynomials $f(x) \in \mathbb{C}[\mathbb{C}^N]$ to $\mathcal{X}$, where $f, g \in \mathbb{C}[\mathbb{C}^N]$ are identified if $f - g$ vanishes on $\mathcal{X}$. Let $\mathbb{C}(\mathcal{X})$ denote the quotient field of $\mathbb{C}[\mathcal{X}]$.

A rational function $h \in \mathbb{C}(\mathcal{X})$ is **regular** if h is defined everywhere in $\mathcal{X}$. A regular rational function on $\mathcal{X}$ is an analytic function.

Denote by $\mathbf{A}$ the $l+1$ tuple $(A_0, \ldots, A_l) \in (\mathbb{C}^{n\times n})^{l+1}$. The group GL$(n,\mathbb{C})$ acts by conjugation on $(\mathbb{C}^{n\times n})^{l+1}$: $T\mathbf{A}T^{-1} = (TA_0T^{-1}, \ldots TA_lT^{-1})$ for any $\mathbf{A} \in (\mathbb{C}^{n\times n})^{l+1}$ and $T \in$ GL$(n,\mathbb{C})$.

Let orb $(\mathbf{A}) := \{T\mathbf{A}T^{-1} : T \in \text{GL}(n,\mathbb{C})\}$ be the **orbit** of $\mathbf{A}$ (under the action of GL$(n,\mathbb{C})$).

Let $\mathcal{X} \subset (\mathbb{C}^{n\times n})^{l+1}$ be a quasi-irreducible variety. $\mathcal{X}$ is called **invariant** (under the action of GL$(n,\mathbb{C})$) if $T\mathcal{X}T^{-1} = \mathcal{X}$ for all $T \in$ GL$(n,\mathbb{C})$.

Assume that $\mathcal{X}$ is an invariant quasi-irreducible variety. A rational function $h \in \mathbb{C}(\mathcal{X})$ is called **invariant** if h is the same value on any two points of a given orbit in $\mathcal{X}$, where h is defined. Denote by $\mathbb{C}[\mathcal{X}]^{\text{inv}} \subseteq \mathbb{C}[\mathcal{X}]$ and $\mathbb{C}(\mathcal{X})^{\text{inv}} \subseteq \mathbb{C}(\mathcal{X})$ the subdomain of invariant polynomials and subfield of invariant functions, respectively.

Facts:

For general background, consult for example [Sha77]. More specific details are given in [Fri83], [Fri85], and [Fri86].

1. An intersection of a finite or infinite number of varieties is a variety, which can be an empty set.
2. A finite union of varieties in $\mathbb{C}^N$ is a variety.
3. Every variety $\mathcal{X}$ is a finite nontrivial union of irreducible varieties.
4. Let $\mathcal{X} \subset \mathbb{C}^N$ be an irreducible variety. Then $\mathcal{X}$ is path-wise connected.
5. $\mathcal{X}_s$ is a proper subvariety of the variety $\mathcal{X}$ and dim $\mathcal{X}_s <$ dim $\mathcal{X}$.
6. dim $\mathbb{C}^N = N$ and $(\mathbb{C}^N)_s = \emptyset$. For any $\mathbf{z} \in \mathbb{C}^N$, the set $\{\mathbf{z}\}$ is an irreducible variety of dimension 0.
7. A quasi-irreducible variety $\mathcal{Z} = \mathcal{X}\backslash\mathcal{Y}$ is path-wise connected and its closure, denoted by Cl $(\mathcal{Z})$, is equal to $\mathcal{X}$. Cl $(\mathcal{Z})\backslash\mathcal{Z}$ is a variety of dimension strictly less than the dimension of $\mathcal{Z}$.

8. The set of all regular points of an irreducible variety $\mathcal{X}$, denoted by $\mathcal{X}_r := \mathcal{X}\backslash\mathcal{X}_s$, is a quasi-irreducible variety. Moreover, $\mathcal{X}_r$ is a path-wise connected complex manifold of complex dimension $\dim \mathcal{X}$.
9. $N(n,l,r) := \dim \mathcal{W}_{n,l+1,r+1} = \sum_{i=0}^{r+1} \binom{(l+1)n^2}{i}$.
10. For an irreducible $\mathcal{X}$, $\mathbb{C}[\mathcal{X}]$ is an integral domain.
11. For a quasi-irreducible $\mathcal{X}$, $\mathbb{C}[\mathcal{X}], \mathbb{C}(\mathcal{X})$ can be identified with $\mathbb{C}[\mathrm{Cl}\ (\mathcal{X})], \mathbb{C}(\mathrm{Cl}\ (\mathcal{X}))$, respectively.
12. For $\mathbf{A} \in (\mathbb{C}^{n\times n})^{l+1}$, orb $(\mathbf{A})$ is a quasi-irreducible variety in $(\mathbb{C}^{n\times n})^{l+1}$.
13. Let $\mathcal{X} \subset (\mathbb{C}^{n\times n})^{l+1}$ be a quasi-irreducible variety. $\mathcal{X}$ is invariant if $\mathbf{A} \in \mathcal{X} \iff \mathrm{orb}\ (\mathbf{A}) \subseteq \mathcal{X}$.
14. Let $\mathcal{X}$ be an invariant quasi-irreducible variety. The quotient field of $\mathbb{C}[\mathcal{X}]^{\mathrm{inv}}$ is a subfield of $\mathbb{C}(\mathcal{X})^{\mathrm{inv}}$, and in some interesting cases the quotient field of $\mathbb{C}[\mathcal{X}]^{\mathrm{inv}}$ is a strict subfield of $\mathbb{C}(\mathcal{X})^{\mathrm{inv}}$.
15. Assume that $\mathcal{X} \subset (\mathbb{C}^{n\times n})^{l+1}$ is an invariant quasi-irreducible variety. Then $\mathbb{C}[\mathcal{X}]^{\mathrm{inv}}$ and $\mathbb{C}(\mathcal{X})^{\mathrm{inv}}$ are finitely invariant generated. That is, there exists $f_1, \dots, f_i \in \mathbb{C}[\mathcal{X}]^{\mathrm{inv}}$ and $g_1, \dots, g_j \in \mathbb{C}(\mathcal{X})^{\mathrm{inv}}$ such that any polynomial in $\mathbb{C}[\mathcal{X}]^{\mathrm{inv}}$ is a polynomial in $f_1, \dots, f_i$, and any rational function in $\mathbb{C}(\mathcal{X})^{\mathrm{inv}}$ is a rational function in $g_1, \dots, g_j$.
16. (**Classification Theorem**) Let $n \geq 2$ and $l \geq 0$ be fixed integers. Then there exists a stratification $\cup_{i=1}^{p} \mathcal{X}_i$ of $(\mathbb{C}^{n\times n})^{l+1}$ with the following properties. For each $\mathcal{X}_i$ there exist m_i regular rational functions $g_{1,i}, \dots, g_{m_i,i} \in \mathbb{C}(\mathcal{X}_i)^{\mathrm{inv}}$ such that the values of $g_{j,i}$ for $j = 1, \dots, m_i$ on any orbit in $\mathcal{X}_i$ determines this orbit uniquely.
The rational functions $g_{1,i}, \dots, g_{m_i,i}$ are the generators of $\mathbb{C}(\mathcal{X}_i)^{\mathrm{inv}}$ for $i = 1, \dots, p$.

Examples:

1. Let $\mathcal{S}$ be an irreducible variety of scalar matrices $\mathcal{S} := \{A \in \mathbb{C}^{2\times 2} : A = \frac{\mathrm{tr}\ A}{2} I_2\}$ and $\mathcal{X} := \mathbb{C}^{2\times 2}\backslash\mathcal{S}$ be a quasi-irreducible variety. Then $\dim \mathcal{X} = 4$, $\dim \mathcal{S} = 1$, and $\mathbb{C}^{2\times 2} = \mathcal{X} \cup \mathcal{S}$ is a stratification of $\mathbb{C}^{2\times 2}$.
2. Let $\mathcal{U} \subset (\mathbb{C}^{2\times 2})^2$ be the set of all pairs $(A, B) \in (\mathbb{C}^{2\times 2})^2$, which are simultaneously similar to a pair of upper triangular matrices. Then $\mathcal{U}$ is a variety given by the zero of the following polynomial:

$$\mathcal{U} := \{(A,B) \in (\mathbb{C}^{2\times 2})^2 : (2\,\mathrm{tr}\ A^2 - (\mathrm{tr}\ A)^2)(2\,\mathrm{tr}\ B^2 - (\mathrm{tr}\ B)^2) - (2\,\mathrm{tr}\ AB - \mathrm{tr}\ A\,\mathrm{tr}\ B)^2 = 0\}.$$

Let $\mathcal{C} \subset \mathcal{U}$ be the variety of commuting matrices:

$$\mathcal{C} := \{(A,B) \in (\mathbb{C}^{2\times 2})^2 : AB - BA = 0\}.$$

Let $\mathcal{V}$ be the variety given by the zeros of the following three polynomials:

$$\mathcal{V} := \{(A,B) \in (\mathbb{C}^{2\times 2})^2 : 2\,\mathrm{tr}\ A^2 - (\mathrm{tr}\ A)^2 = 2\,\mathrm{tr}\ B^2 - (\mathrm{tr}\ B)^2 = 2\,\mathrm{tr}\ AB - \mathrm{tr}\ A\,\mathrm{tr}\ B = 0\}.$$

Then $\mathcal{V}$ is the variety of all pairs (A, B), which are simultaneously similar to a pair of the form $\left(\begin{bmatrix}\lambda & \alpha\\ 0 & \lambda\end{bmatrix}, \begin{bmatrix}\mu & \beta\\ 0 & \mu\end{bmatrix}\right)$. Hence, $\mathcal{V} \subset \mathcal{C}$. Let $\mathcal{W} := \{A \in (\mathbb{C}^{2\times 2}) : 2\,\mathrm{tr}\ A^2 - (\mathrm{tr}\ A)^2 = 0\}$ and $\mathcal{S} \subset \mathcal{W}$ be defined as in the previous example. Define the following quasi-irreducible varieties in $(\mathbb{C}^{2\times 2})^2$:

$$\mathcal{X}_1 := (\mathbb{C}^{2\times 2})^2\backslash\mathcal{U},\ \mathcal{X}_2 := \mathcal{U}\backslash\mathcal{C},\ \mathcal{X}_3 = \mathcal{C}\backslash\mathcal{V},\ \mathcal{X}_4 := \mathcal{V}\backslash(\mathcal{S}\times\mathcal{W}\cup\mathcal{W}\times\mathcal{S}),$$
$$\mathcal{X}_5 := \mathcal{S}\times(\mathcal{W}\backslash\mathcal{S}),\ \mathcal{X}_6 := (\mathcal{W}\backslash\mathcal{S})\times\mathcal{S},\ \mathcal{X}_7 = \mathcal{S}\times\mathcal{S}.$$

Then

$$\dim \mathcal{X}_1 = 8,\ \dim \mathcal{X}_2 = 7,\ \dim \mathcal{X}_3 = 6,\ \dim \mathcal{X}_4 = 5,$$
$$\dim \mathcal{X}_5 = \dim \mathcal{X}_6 = 4,\ \dim \mathcal{X}_7 = 2,$$

and $\cup_{i=1}^{7} \mathcal{X}_i$ is a stratification of $(\mathbb{C}^{2\times 2})^2$.
3. In the classical case of similarity classes in $\mathbb{C}^{n\times n}$, i.e., $l = 0$, it is possible to choose a fixed set of polynomial invariant functions as $g_j(A) = \mathrm{tr}\ (A^j)$ for $j = 1, \dots, n$. However, we still have to stratify $\mathbb{C}^{n\times n}$ to $\cup_{i=1}^{p} \mathcal{X}_i$, where each $A \in \mathcal{X}_i$ has some specific Jordan structures.

4. Consider the stratification $\mathbb{C}^{2\times 2} = \mathcal{X} \cup \mathcal{S}$ as in Example 1. Clearly $\mathcal{X}$ and $\mathcal{S}$ are invariant under the action of GL(2, $\mathbb{C}$). The invariant functions tr A, tr A^2 determine uniquely orb (A) on $\mathcal{X}$. The Jordan canonical for of any A in $\mathcal{X}$ is either consists of two distinct Jordan blocks of order 1 or one Jordan block of order 2. The invariant function tr A determines orb (A) for any $A \in \mathcal{S}$. It is possible to refine the stratification of $\mathbb{C}^{2\times 2}$ to three invariant components $\mathbb{C}^{2\times 2}\backslash\mathcal{W}, \mathcal{W}\backslash\mathcal{S}, \mathcal{S}$, where $\mathcal{W}$ is defined in Example 2. Each component contains only matrices with one kind of Jordan block. On the first component, tr A, tr A^2 determine the orbit, and on the second and third component, tr A determines the orbit.
5. To see the fundamental difference between similarity ($l = 0$) and simultaneous similarity $l \geq 1$, it is suffice to consider Example 2. Observe first that the stratification of $(\mathbb{C}^{2\times 2})^2 = \cup_{i=1}^{7}\mathcal{X}_i$ is invariant under the action of GL(2, $\mathbb{C}$). On $\mathcal{X}_1$ the five invariant polynomials tr A, tr A^2, tr B, tr B^2, tr AB, which are algebraically independent, determine uniquely any orbit in $\mathcal{X}_1$.

   Let $(A = [a_{ij}], B = [b_{ij}]) \in \mathcal{X}_2$. Then A and B have a unique one-dimensional common eigenspace corresponding to the eigenvalues λ_1, μ_1 of A, B, respectively. Assume that $a_{12}b_{21} - a_{21}b_{12} \neq 0$. Define

$$\lambda_1 = \alpha(A, B) := \frac{(b_{11} - b_{22})a_{12}a_{21} + a_{22}a_{12}b_{21} - a_{11}a_{21}b_{12}}{a_{12}b_{21} - a_{21}b_{12}},$$

$$\mu_1 = \alpha(B, A).$$

   Then tr A, tr B, $\alpha(A, B)$, $\alpha(B, A)$ are regular, algebraically independent, rational invariant functions on $\mathcal{X}_2$, whose values determine orb (A, B). Cl (orb (A, B)) contains an orbit generated by two diagonal matrices diag (λ_1, λ_2) and diag (μ_1, μ_2). Hence, $\mathbb{C}[\mathcal{X}_2]^{\text{inv}}$ is generated by the five invariant polynomials tr A, tr A^2, tr B, tr B^2, tr AB, which are algebraically dependent. Their values coincide exactly on two distinct orbits in $\mathcal{X}_2$. On $\mathcal{X}_3$ the above invariant polynomials separate the orbits.

   Any $(A = [a_{ij}], B = [b_{ij}]) \in \mathcal{X}_4$ is simultaneously similar a unique pair of the form $\left(\begin{bmatrix} \lambda & 1 \\ 0 & \lambda \end{bmatrix}, \begin{bmatrix} \mu & t \\ 0 & \mu \end{bmatrix}\right)$. Then $t = \gamma(A, B) := \frac{b_{12}}{a_{12}}$. Thus, tr A, tr B, $\gamma(A, B)$ are three algebraically independent regular rational invariant functions on $\mathcal{X}_4$, whose values determine a unique orbit in $\mathcal{X}_4$. Clearly $(\lambda I_2, \mu I_2) \in$ Cl (X_4). Then $\mathbb{C}[\mathcal{X}_4]^{\text{inv}}$ is generated by tr A, tr B. The values of tr $A = 2\lambda$, tr $B = 2\mu$ correspond to a complex line of orbits in $\mathcal{X}_4$. Hence, the classification problem of simultaneous similarity classes in $\mathcal{X}_4$ or $\mathcal{V}$ is a *wild* problem.

   On $\mathcal{X}_5$, $\mathcal{X}_6$, $\mathcal{X}_7$, the algebraically independent functions tr A, tr B determine the orbit in each of the stratum.

24.5 Simultaneous Similarity Classification II

In this section, we give an invariant stratification of $(\mathbb{C}^{n\times n})^{l+1}$, for $l \geq 1$, under the action of GL(n, $\mathbb{C}$) and describe a set of invariant regular rational functions on each stratum, which separate the orbits up to a finite number. We assume the nontrivial case $n > 1$. It is conjectured that the continuous invariants of the given orbit determine uniquely the orbit on each stratum given in the Classification Theorem.

Classification of simultaneous similarity classes of matrices is a known *wild* problem [GP69]. For another approach to classification of simultaneous similarity classes of matrices using *Belitskii* reduction see [Ser00]. See other applications of these techniques to classifications of linear systems [Fri85] and to canonical forms [Fri86].

Definitions:

For $\mathbf{A} = (A_0, \ldots, A_l), \mathbf{B} = (B_0, \ldots, B_l) \in (\mathbb{C}^{n\times n})^{l+1}$ let $L(\mathbf{B}, \mathbf{A}) : \mathbb{C}^{n\times n} \to (\mathbb{C}^{n\times n})^{l+1}$ be the linear operator given by $U \mapsto (B_0U - UA_0, \ldots, B_lU - UA_l)$. Then $L(\mathbf{B}, \mathbf{A})$ is represented by the $(l+1)n^2 \times n^2$ matrix $(I_n \otimes B_0^{\mathrm{T}} - A_0 \otimes I_n, \ldots, I_n \otimes B_l^{\mathrm{T}} - A_l \otimes I_n)^{\mathrm{T}}$, where $U \mapsto (I_n \otimes B_0^{\mathrm{T}} - A_0 \otimes I_n, \ldots, I_n \otimes B_l^{\mathrm{T}} - A_l \otimes I_n)^{\mathrm{T}}U$. Let $L(\mathbf{A}) := L(\mathbf{A}, \mathbf{A})$. The dimension of orb $(\mathbf{A})$ is denoted by dim orb $(\mathbf{A})$.

Let $\mathcal{S}_n := \{A \in \mathbb{C}^{n\times n} : A = \frac{\operatorname{tr} A}{n} I_n\}$ be the variety of scalar matrices. Let

$$\mathcal{M}_{n,l+1,r} := \{\mathbf{A} \in (\mathbb{C}^{n\times n})^{l+1} : \operatorname{rank} L(\mathbf{A}) = r\}, \quad r = 0, 1, \ldots, n^2 - 1.$$

Facts:

Most of the results in this section are given in [Fri83].

1. For $\mathbf{A} = (A_0, \ldots, A_l), \in (\mathbb{C}^{n\times n})^{l+1}$, dim orb $(\mathbf{A})$ is equal to the rank of $L(\mathbf{A})$.
2. Since any $U \in \mathcal{S}_n$ commutes with any $B \in \mathbb{C}^{n\times n}$ it follows that $\ker L(\mathbf{A}) \supset \mathcal{S}_n$. Hence, rank $L(\mathbf{A}) \leq n^2 - 1$.
3. $\mathcal{M}_{n,l+1,n^2-1}$ is a invariant quasi-irreducible variety of dimension $(l+1)n^2$, i.e., Cl $(\mathcal{M}_{n,l+1,n^2-1}) = (\mathbb{C}^{n\times n})^{l+1}$. The sets $\mathcal{M}_{n,l+1,r}, r = n^2 - 2, \ldots, 0$ have the decomposition to invariant quasi-irreducible varieties, each of dimension strictly less than $(l+1)n^2$.
4. Let $r \in [0, n^2 - 1]$, $\mathbf{A} \in \mathcal{M}_{n,l+1,r}$, and $\mathbf{B} = T\mathbf{A}T^{-1}$. Then $L(\mathbf{B}, \mathbf{A}) = \operatorname{diag}(I_n \otimes T, \ldots, I_n \otimes T)L(\mathbf{A})(I_n \otimes T^{-1})$, rank $L(\mathbf{B}, \mathbf{A}) = r$ and $\det L(\mathbf{B}, \mathbf{A})[\alpha, \beta] = 0$ for any $\alpha \in \mathrm{Q}_{r+1,(l+1)n^2}, \beta \in \mathrm{Q}_{r+1,n^2}$. (See Chapter 23.2)
5. Let $\mathbf{X} = (X_0, \ldots, X_l) \in (\mathbb{C}^{n\times n})^{l+1}$ with the indeterminate entries $X_k = [x_{k,ij}]$ for $k = 0, \ldots, l$. Each $\det L(\mathbf{X}, \mathbf{A})[\alpha, \beta], \alpha \in \mathrm{Q}_{r+1,(l+1)n^2}, \beta \in \mathrm{Q}_{r+1,n^2}$ is a vector in $\mathcal{W}_{n,l+1,r+1}$, i.e., it is a multilinear polynomial in $(l+1)n^2$ variables of degree $r+1$ at most. We identify $\det L(\mathbf{X}, \mathbf{A})[\alpha, \beta], \alpha \in \mathrm{Q}_{r+1,(l+1)n^2}, \beta \in \mathrm{Q}_{r+1,n^2}$ with the **row** vector $\mathbf{a}(\mathbf{A}, \alpha, \beta) \in \mathbb{C}^{N(n,l,r)}$ given by its coefficients in the basis $\mathbf{e}_1, \ldots, \mathbf{e}_{N(n,l,r)}$. The number of these vectors is $M(n,l,r) := \binom{(l+1)n^2}{r+1}\binom{n^2}{r+1}$. Let $R(\mathbf{A}) \in \mathbb{C}^{M(n,l,r)N(n,l,r)\times M(n,l,r)N(n,l,r)}$ be the matrix with the rows $\mathbf{a}(A, \alpha, \beta)$, where the pairs $(\alpha, \beta) \in \mathrm{Q}_{r+1,(l+1)n^2} \times \mathrm{Q}_{r+1,n^2}$ are listed in a lexicographical order.
6. All points on the orb $(\mathbf{A})$ satisfy the following polynomial equations in $\mathbb{C}[(\mathbb{C}^{n\times n})^{l+1}]$:

$$\begin{aligned} &\det L(\mathbf{X}, \mathbf{A})[\alpha, \beta] = 0, \\ &\text{for all } \alpha \in \mathrm{Q}_{r+1,(l+1)n^2}, \beta \in \mathrm{Q}_{r+1,n^2}. \end{aligned} \tag{24.1}$$

   Thus, the matrix $R(\mathbf{A})$ determines the above variety.
7. If $\mathbf{B} = T\mathbf{A}T^{-1}$, then $R(\mathbf{A})$ is row equivalent to $R(\mathbf{B})$. To each orb $(\mathbf{A})$ we can associate a unique reduced row echelon form $F(\mathbf{A}) \in \mathbb{C}^{M(n,l,r)N(n,l,r)\times M(n,l,r)N(n,l,r)}$ of $R(\mathbf{A})$. $\varrho(\mathbf{A}) :=$ rank $R(\mathbf{A})$ is the number of linearly independent polynomials given in (24.1). Let $\mathcal{I}(\mathbf{A}) = \{(1, j_1), \ldots, (\varrho(\mathbf{A}), j_{\varrho(\mathbf{A})})\} \subset \{1, \ldots, \varrho(\mathbf{A})\} \times \{1, \ldots, N(n,l,r)\}$ be the location of the pivots in the $M(n,l,r) \times N(n,l,r)$ matrix $F(\mathbf{A}) = [f_{ij}(\mathbf{A})]$. That is, $1 \leq j_1 < \ldots < j_{\varrho(\mathbf{A})} \leq N(n,l,r)$, $f_{ij_i}(\mathbf{A}) = 1$ for $i = 1, \ldots, \varrho(\mathbf{A})$ and $f_{ij} = 0$ unless $j \geq i$ and $i \in [1, \varrho(\mathbf{A})]$. The nontrivial entries $f_{ij}(\mathbf{A})$ for $j > i$ are rational functions in the entries of the $l+1$ tuple $\mathbf{A}$. Thus, $F(\mathbf{B}) = F(\mathbf{A})$ for $\mathbf{B} \in$ orb $(\mathbf{A})$. The numbers $r(\mathbf{A}) :=$ rank $L(\mathbf{A}), \varrho(\mathbf{A})$ and the set $\mathcal{I}(\mathbf{A})$ are called the **discrete** invariants of orb $(\mathbf{A})$. The rational functions $f_{ij}(\mathbf{A}), i = 1, \ldots, \varrho(\mathbf{A}), j = i+1, \ldots, N(n,l,r)$ are called the **continuous** invariants of orb $(\mathbf{A})$.
8. (*Classification Theorem for Simultaneous Similarity*) Let $l \geq 1, n \geq 2$ be integers. Fix an integer $r \in [0, n^2 - 1]$ and let $M(n,l,r), N(n,l,r)$ be the integers defined as above. Let $0 \leq \varrho \leq \min(M(n,l,r), N(n,l,r))$ and the set $\mathcal{I} = \{(1, j_1), \ldots, (\varrho, j_\varrho) \subset \{1, \ldots, \varrho\} \times \{1, \ldots, N(n,l,r)\}, 1 \leq j_1 < \ldots < j_\varrho \leq N(n,l,r)$ be given. Let $\mathcal{M}_{n,l+1.r}(\varrho, \mathcal{I})$ be the set of all $\mathbf{A} \in (\mathbb{C}^{n\times n})^{l+1}$ such that rank $L(\mathbf{A}) = r$, $\varrho(\mathbf{A}) = \varrho$, and $\mathcal{I}(\mathbf{A}) = \mathcal{I}$. Then $\mathcal{M}_{n,l+1.r}(\varrho, \mathcal{I})$ is invariant quasi-irreducible variety under the action of GL$(n, \mathbb{C})$. Suppose that $\mathcal{M}_{n,l+1.r}(\varrho, \mathcal{I}) \neq \emptyset$. Recall that for each $\mathbf{A} \in \mathcal{M}_{n,l+1.r}(\varrho, \mathcal{I})$ the continuous invariants of $\mathbf{A}$, which correspond to the entries $f_{ij}(\mathbf{A})$, $i = 1, \ldots, \varrho$, $j = i+1, \ldots, N(n,l,r)$ of the reduced row echelon form of $R(\mathbf{A})$, are regular rational invariant functions on $\mathcal{M}_{n,l+1.r}(\varrho, \mathcal{I})$. Then the values of the continuous invariants determine a finite number of orbits in $\mathcal{M}_{n,l+1.r}(\varrho, \mathcal{I})$.

   The quasi-irreducible variety $\mathcal{M}_{n,l+1.r}(\varrho, \mathcal{I})$ decomposes uniquely as a finite union of invariant regular quasi-irreducible varieties. The union of all these decompositions of $\mathcal{M}_{n,l+1.r}(\varrho, \mathcal{I})$ for all possible values r, ϱ, and the sets $\mathcal{I}$ gives rise to an invariant stratification of $(\mathbb{C}^{n\times n})^{l+1}$.

References

[Fri78] S. Friedland, Extremal eigenvalue problems, *Bull. Brazilian Math. Soc.* 9 (1978), 13–40.

[Fri80] S. Friedland, Analytic similarities of matrices, *Lectures in Applied Math.*, Amer. Math. Soc. 18 (1980), 43–85 (edited by C.I. Byrnes and C.F. Martin).

[Fri81] S. Friedland, A generalization of the Motzkin–Taussky theorem, *Lin. Alg. Appl.* 36 (1981), 103–109.

[Fri83] S. Friedland, Simultaneous similarity of matrices, *Adv. Math.*, 50 (1983), 189–265.

[Fri85] S. Friedland, Classification of linear systems, *Proc. of A.M.S. Conf. on Linear Algebra and Its Role in Systems Theory*, Contemp. Math. 47 (1985), 131–147.

[Fri86] S. Friedland, Canonical forms, *Frequency Domain and State Space Methods for Linear Systems*, 115–121, edited by C.I. Byrnes and A. Lindquist, North Holland, Amsterdam, 1986.

[Frixx] S. Friedland, *Matrices*, a book in preparation.

[GB77] M.A. Gauger and C.I. Byrnes, Characteristic free, improved decidability criteria for the similarity problem, *Lin. Multilin. Alg.* 5 (1977), 153–158.

[GP69] I.M. Gelfand and V.A. Ponomarev, Remarks on classification of a pair of commuting linear transformation in a finite dimensional vector space, *Func. Anal. Appl.* 3 (1969), 325–326.

[GR65] R. Gunning and H. Rossi, *Analytic Functions of Several Complex Variables*, Prentice-Hall, Upper Saddle River, NJ, 1965.

[Gur80] R.M. Guralnick, A note on the local-global principle for similarity of matrices, *Lin. Alg. Appl.* 30 (1980), 651–654.

[MM64] M. Marcus and H. Minc, *A Survey of Matrix Theory and Matrix Inequalities*, Prindle, Weber & Schmidt, Boston, 1964.

[MF80] N. Moiseyev and S. Friedland, The association of resonance states with incomplete spectrum of finite complex scaled Hamiltonian matrices, *Phys. Rev.* A 22 (1980), 619–624.

[MT52] T.S. Motzkin and O. Taussky, Pairs of matrices with property L, *Trans. Amer. Math. Soc.* 73 (1952), 108–114.

[MT55] T.S. Motzkin and O. Taussky, Pairs of matrices with property L, II, *Trans. Amer. Math. Soc.* 80 (1955), 387–401.

[Sha77] I.R. Shafarevich, *Basic Algebraic Geometry*, Springer-Verlag, Berlin-New York, 1977.

[Ser00] V.V. Sergeichuk, Canonical matrices for linear matrix problems, *Lin. Alg. Appl.* 317 (2000), 53–102.

[Was63] W. Wasow, On holomorphically similar matrices, *J. Math. Anal. Appl.* 4 (1963), 202–206.

[Was77] W. Wasow, Arnold's canonical matrices and asymptotic simplification of ordinary differential equations, *Lin. Alg. Appl.* 18 (1977), 163–170.

[Was78] W. Wasow, *Topics in Theory of Linear Differential Equations Having Singularities with Respect to a Parameter*, IRMA, Univ. L. Pasteur, Strasbourg, 1978.

[Wei67] K. Weierstrass, Zur theorie der bilinearen un quadratischen formen, *Monatsch. Akad. Wiss. Berlin*, 310–338, 1867.

25

Max-Plus Algebra

Marianne Akian
INRIA, France

Ravindra Bapat
Indian Statistical Institute

Stéphane Gaubert
INRIA, France

Max-plus algebra has been discovered more or less independently by several schools, in relation with various mathematical fields. This chapter is limited to finite dimensional linear algebra. For more information, the reader may consult the books [CG79], [Zim81], [CKR84], [BCOQ92], [KM97], [GM02], and [HOvdW06]. The collections of articles [MS92], [Gun98], and [LM05] give a good idea of current developments.

25.1 Preliminaries

Definitions:

The **max-plus semiring** $\mathbb{R}_{\max}$ is the set $\mathbb{R} \cup \{-\infty\}$, equipped with the **addition** $(a, b) \mapsto \max(a, b)$ and the **multiplication** $(a, b) \mapsto a + b$. The identity element for the addition, **zero**, is $-\infty$, and the identity element for the multiplication, **unit**, is 0. To illuminate the linear algebraic nature of the results, the generic notations $+\!\!\!+$, $\sum$, $\times$ (or concatenation), $\mathbb{0}$ and $\mathbb{1}$ are used for the addition, the sum, the multiplication, the zero, and the unit of $\mathbb{R}_{\max}$, respectively, so that when a, b belong to $\mathbb{R}_{\max}$, $a +\!\!\!+ b$ will mean $\max(a, b)$, $a \times b$ or ab will mean the usual sum $a + b$. We use blackboard (double struck) fonts to denote the max-plus operations (compare "$+\!\!\!+$" with "$+$").

The **min-plus semiring** $\mathbb{R}_{\min}$ is the set $\mathbb{R} \cup \{+\infty\}$ equipped with the addition $(a, b) \mapsto \min(a, b)$ and the multiplication $(a, b) \mapsto a + b$. The zero is $+\infty$, the unit 0. The name **tropical** is now also used essentially as a synonym of min-plus. Properly speaking, it refers to the **tropical semiring**, which is the subsemiring of $\mathbb{R}_{\min}$ consisting of the elements in $\mathbb{N} \cup \{+\infty\}$.

The **completed max-plus semiring** $\overline{\mathbb{R}}_{\max}$ is the set $\mathbb{R} \cup \{\pm\infty\}$ equipped with the addition $(a, b) \mapsto \max(a, b)$ and the multiplication $(a, b) \mapsto a+b$, with the convention that $-\infty+(+\infty) = +\infty+(-\infty) = -\infty$. The **completed min-plus semiring**, $\overline{\mathbb{R}}_{\min}$, is defined in a dual way.

Many classical algebraic definitions have max-plus analogues. For instance, $\mathbb{R}^n_{\max}$ is the set of n-dimensional **vectors** and $\mathbb{R}^{n\times p}_{\max}$ is the set of $n \times p$ **matrices** with entries in $\mathbb{R}_{\max}$. They are equipped with the vector and matrix operations, defined and denoted in the usual way. The $n \times p$ **zero matrix**, $\mathbf{0}_{np}$ or $\mathbf{0}$, has all its entries equal to $\mathbb{0}$. The $n \times n$ **identity matrix**, I_n or I, has diagonal entries equal to $\mathbb{1}$, and nondiagonal entries equal to $\mathbb{0}$. Given a matrix $A = (A_{ij}) \in \mathbb{R}_{\max}^{n\times p}$, we denote by $A_{i\cdot}$ and $A_{\cdot j}$ the i-th row and the j-th column of A. We also denote by A the linear map $\mathbb{R}_{\max}^{p} \to \mathbb{R}_{\max}^{n}$ sending a vector x to Ax. **Semimodules** and **subsemimodules** over the semiring $\mathbb{R}_{\max}$ are defined as the analogues of modules and submodules over rings. A subset F of a semimodule M over $\mathbb{R}_{\max}$ **spans** M, or is a **spanning family** of M, if every element $\mathbf{x}$ of M can be expressed as a finite linear combination of the elements of F, meaning that $\mathbf{x} = \sum_{\mathbf{f}\in F} \lambda_{\mathbf{f}}.\mathbf{f}$, where $(\lambda_{\mathbf{f}})_{\mathbf{f}\in F}$ is a family of elements of $\mathbb{R}_{\max}$ such that $\lambda_{\mathbf{f}} = \mathbb{0}$ for all but finitely many $\mathbf{f} \in F$. A semimodule is **finitely generated** if it has a finite spanning family.

The sets $\mathbb{R}_{\max}$ and $\overline{\mathbb{R}}_{\max}$ are ordered by the usual order of $\mathbb{R} \cup \{\pm\infty\}$. Vectors and matrices over $\overline{\mathbb{R}}_{\max}$ are ordered with the product ordering. The supremum and the infimum operations are denoted by $\vee$ and $\wedge$, respectively. Moreover, the sum of the elements of an arbitrary set X of scalars, vectors, or matrices with entries in $\overline{\mathbb{R}}_{\max}$ is by definition the supremum of X.

If $A \in \overline{\mathbb{R}}_{\max}^{n\times n}$, the **Kleene star** of A is the matrix $A^\star = I \oplus A \oplus A^2 \oplus \cdots$.

The **digraph** $\Gamma(A)$ associated to an $n \times n$ matrix A with entries in $\overline{\mathbb{R}}_{\max}$ consists of the vertices $1, \dots, n$, with an arc from vertex i to vertex j when $A_{ij} \neq \mathbb{0}$. The **weight** of a walk W given by $(i_1, i_2), \dots, (i_{k-1}, i_k)$ is $|W|_A := A_{i_1 i_2} \cdots A_{i_{k-1} i_k}$, and its **length** is $|W| := k - 1$. The matrix A is **irreducible** if $\Gamma(A)$ is strongly connected.

Facts:

1. When $A \in \overline{\mathbb{R}}_{\max}^{n\times n}$, the weight of a walk $W = ((i_1, i_2), \dots, (i_{k-1}, i_k))$ in $\Gamma(A)$ is given by the usual sum $|W|_A = A_{i_1 i_2} + \cdots + A_{i_{k-1} i_k}$, and $A^\star_{ij}$ gives the maximal weight $|W|_A$ of a walk from vertex i to vertex j. One can also define the matrix $A^\star$ when $A \in \overline{\mathbb{R}}_{\min}^{n\times n}$. Then, $A^\star_{ij}$ is the minimal weight of a walk from vertex i to vertex j. Computing $A^\star$ is the same as the all pairs' shortest path problem.
2. [CG79], [BCOQ92, Th. 3.20] If $A \in \overline{\mathbb{R}}_{\max}^{n\times n}$ and the weights of the cycles of $\Gamma(A)$ do not exceed $\mathbb{1}$, then $A^\star = I \oplus A \oplus \cdots \oplus A^{n-1}$.
3. [BCOQ92, Th. 4.75 and Rk. 80] If $A \in \overline{\mathbb{R}}_{\max}^{n\times n}$ and $\mathbf{b} \in \overline{\mathbb{R}}_{\max}^{n}$, then the smallest $\mathbf{x} \in \overline{\mathbb{R}}_{\max}^{n}$ such that $\mathbf{x} = A\mathbf{x} \oplus \mathbf{b}$ coincides with the smallest $\mathbf{x} \in \overline{\mathbb{R}}_{\max}^{n}$ such that $\mathbf{x} \geq A\mathbf{x} \oplus \mathbf{b}$, and it is given by $A^\star\mathbf{b}$.
4. [BCOQ92, Th. 3.17] When $A \in \mathbb{R}_{\max}^{n\times n}$, $\mathbf{b} \in \mathbb{R}_{\max}^{n}$, and when all the cycles of $\Gamma(A)$ have a weight strictly less than $\mathbb{1}$, then $A^\star\mathbf{b}$ is the unique solution $\mathbf{x} \in \mathbb{R}_{\max}^{n}$ of $\mathbf{x} = A\mathbf{x} \oplus \mathbf{b}$.
5. Let $A \in \mathbb{R}_{\max}^{n\times n}$ and $\mathbf{b} \in \mathbb{R}_{\max}^{n}$. Construct the sequence:

$$\mathbf{x}_0 = \mathbf{b},\ \mathbf{x}_1 = A\mathbf{x}_0 \oplus \mathbf{b},\ \mathbf{x}_2 = A\mathbf{x}_1 \oplus \mathbf{b}, \dots .$$

The sequence $\mathbf{x}_k$ is nondecreasing. If all the cycles of $\Gamma(A)$ have a weight less than or equal to $\mathbb{1}$, then, $\mathbf{x}_{n-1} = \mathbf{x}_n = \cdots = A^\star\mathbf{b}$. Otherwise, $\mathbf{x}_{n-1} \neq \mathbf{x}_n$. Computing the sequence $\mathbf{x}_k$ to determine $A^\star\mathbf{b}$ is a special instance of label correcting shortest path algorithm [GP88].
6. [BCOQ92, Lemma 4.101] For all $a \in \overline{\mathbb{R}}_{\max}^{n\times n}$, $b \in \overline{\mathbb{R}}_{\max}^{n\times p}$, $c \in \overline{\mathbb{R}}_{\max}^{p\times n}$, and $d \in \overline{\mathbb{R}}_{\max}^{p\times p}$, we have

$$\begin{bmatrix} a & b \\ c & d \end{bmatrix}^\star = \begin{bmatrix} a^\star \oplus a^\star b(ca^\star b \oplus d)^\star ca^\star & a^\star b(ca^\star b \oplus d)^\star \\ (ca^\star b \oplus d)^\star ca^\star & (ca^\star b \oplus d)^\star \end{bmatrix}.$$

This fact and the next one are special instances of well-known results of language theory [Eil74], concerning unambiguous rational identities. Both are valid in more general semirings.
7. [MY60] Let $A \in \overline{\mathbb{R}}_{\max}^{n\times n}$. Construct the sequence of matrices $A^{(0)}, \dots, A^{(n)}$ such that $A^{(0)} = A$ and

$$A^{(k)}_{ij} = A^{(k-1)}_{ij} \oplus A^{(k-1)}_{ik}(A^{(k-1)}_{kk})^\star A^{(k-1)}_{kj},$$

for $i, j = 1, \dots, n$ and $k = 1, \dots, n$. Then, $A^{(n)} = A \oplus A^2 \oplus \cdots$.

Example:

1. Consider the matrix

$$A = \begin{bmatrix} 4 & 3 \\ 7 & -\infty \end{bmatrix}.$$

The digraph $\Gamma(A)$ is

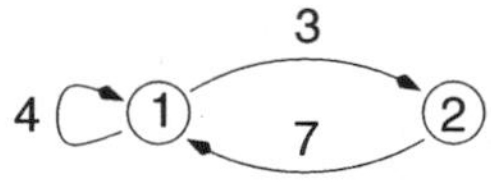

We have

$$A^2 = \begin{bmatrix} 10 & 7 \\ 11 & 10 \end{bmatrix}.$$

For instance, $A^2_{11} = A_{1\cdot}A_{\cdot 1} = [4\ 3][4\ 7]^T = \max(4+4, 3+7) = 10$. This gives the maximal weight of a walk of length 2 from vertex 1 to vertex 1, which is attained by the walk $(1,2),(2,1)$. Since there is one cycle with positive weight in $\Gamma(A)$ (for instance, the cycle $(1,1)$ has weight 4), and since A is irreducible, the matrix $A^\star$ has all its entries equal to $+\infty$. To get a Kleene star with finite entries, consider the matrix

$$C = (-5)A = \begin{bmatrix} -1 & -2 \\ 2 & -\infty \end{bmatrix}.$$

The only cycles in $\Gamma(A)$ are $(1,1)$ and $(1,2),(2,1)$ (up to a cyclic conjugacy). They have weights -1 and 0. Applying Fact 2, we get

$$C^\star = I \oplus C = \begin{bmatrix} 0 & -2 \\ 2 & 0 \end{bmatrix}.$$

Applications:

1. *Dynamic programming.* Consider a deterministic Markov decision process with a set of states $\{1,\dots,n\}$ in which one player can move from state i to state j, receiving a payoff of $A_{ij} \in \mathbb{R}\cup\{-\infty\}$. To every state i, associate an initial payoff $\mathbf{c}_i \in \mathbb{R}\cup\{-\infty\}$ and a terminal payoff $\mathbf{b}_i \in \mathbb{R}\cup\{-\infty\}$. The value in horizon k is by definition the maximum of the sums of the payoffs (including the initial and terminal payoffs) corresponding to all the trajectories consisting exactly of k moves. It is given by $\mathbf{c}A^k\mathbf{b}$, where the product and the power are understood in the max-plus sense. The special case where the initial state is equal to some given $m \in \{1,\dots,n\}$ (and where there is no initial payoff) can be modeled by taking $\mathbf{c} := \mathbf{e}_m$, the m-th max-plus basis vector (whose entries are all equal to $\mathbb{0}$, except the m-th entry, which is equal to $\mathbb{1}$). The case where the final state is fixed can be represented in a dual way. Deterministic Markov decision problems (which are the same as shortest path problems) are ubiquitous in operations research, mathematical economics, and optimal control.
2. [BCOQ92] *Discrete event systems.* Consider a system in which certain repetitive events, denoted by $1,\dots,n$, occur. To every event i is associated a dater function $x_i : \mathbb{Z} \to \mathbb{R}$, where $x_i(k)$ represents the date of the k-th occurrence of event i. Precedence constraints between the repetitive events are given by a set of arcs $E \subset \{1,\dots,n\}^2$, equipped with two valuations $\nu : E \to \mathbb{N}$ and $\tau : E \to \mathbb{R}$. If $(i,j) \in E$, the k-th execution of event i cannot occur earlier than τ_{ij} time units before the $(k-\nu_{ij})$-th execution of event j, so that $x_i(k) \geq \max_{j:\,(i,j)\in E} \tau_{ij} + x_j(k-\nu_{ij})$. This can be rewritten, using the max-plus notation, as

$$\mathbf{x}(k) \geq A_0\mathbf{x}(k) \oplus \cdots \oplus A_{\bar{\nu}}\mathbf{x}(k-\bar{\nu}),$$

where $\bar{\nu} := \max_{(i,j)\in E} \nu_{ij}$ and $\mathbf{x}(k) \in \mathbb{R}^n_{\max}$ is the vector with entries $x_i(k)$. Often, the dates $x_i(k)$ are only defined for positive k, then appropriate initial conditions must be incorporated in the model. One is particularly interested in the earliest dynamics, which, by Fact 3, is given by $\mathbf{x}(k) = A_0^\star A_1 \mathbf{x}(k-1) \oplus \cdots \oplus A_0^\star A_{\bar{\nu}} \mathbf{x}(k-\bar{\nu})$. The class of systems following dynamics of these forms is known in the Petri net literature as **timed event graphs**. It is used to model certain manufacturing systems [CDQV85], or transportation or communication networks [BCOQ92], [HOvdW06].

3. N. Bacaër [Bac03] observed that max-plus algebra appears in a familiar problem, crop rotation. Suppose n different crops can be cultivated every year. Assume for simplicity that the income of the year is a deterministic function, $(i, j) \mapsto A_{ij}$, depending only on the crop i of the preceding year, and of the crop j of the current year (a slightly more complex model in which the income of the year depends on the crops of the two preceding years is needed to explain the historical variations of crop rotations [Bac03]). The income of a sequence $i_1, \ldots, i_k$ of crops can be written as $\mathbf{c}_{i_1} A_{i_1 i_2} \cdots A_{i_{k-1} i_k}$, where $\mathbf{c}_{i_1}$ is the income of the first year. The maximal income in k years is given by $\mathbf{c}A^{k-1}\mathbf{b}$, where $\mathbf{b} = (\mathbb{1}, \ldots, \mathbb{1})$. We next show an example.

$$A = \begin{bmatrix} -\infty & 11 & 8 \\ 2 & 5 & 7 \\ 2 & 6 & 4 \end{bmatrix}$$

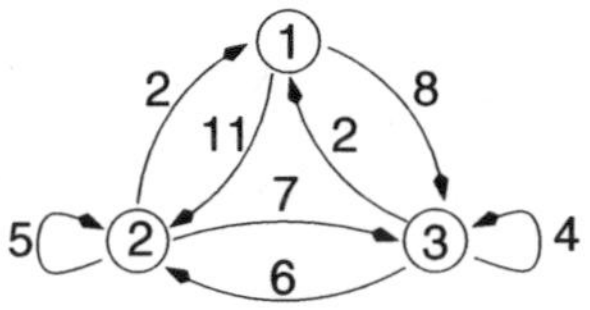

Here, vertices 1, 2, and 3 represent fallow (no crop), wheat, and oats, respectively. (We put no arc from 1 to 1, setting $A_{11} = -\infty$, to disallow two successive years of fallow.) The numerical values have no pretension to realism; however, the income of a year of wheat is 11 after a year of fallow, this is greater than after a year of cereal (5 or 6, depending on whether wheat or oats was cultivated). An initial vector coherent with these data may be $\mathbf{c} = [-\infty \ 11 \ 8]$, meaning that the income of the first year is the same as the income after a year of fallow. We have $\mathbf{c}A\mathbf{b} = 18$, meaning that the optimal income in 2 years is 18. This corresponds to the optimal walk $(2, 3)$, indicating that wheat and oats should be successively cultivated during these 2 years.

25.2 The Maximal Cycle Mean

Definitions:

1. The **maximal cycle mean**, $\rho_{\max}(A)$, of a matrix $A \in \mathbb{R}^{n\times n}_{\max}$, is the maximum of the weight-to-length ratio over all cycles c of $\Gamma(A)$, that is,

$$\rho_{\max}(A) = \max_{c \text{ cycle of } \Gamma(A)} \frac{|c|_A}{|c|} = \max_{k\geq 1} \max_{i_1,\ldots,i_k} \frac{A_{i_1 i_2} + \cdots + A_{i_k i_1}}{k}. \tag{25.1}$$

2. Denote by $\mathbb{R}^{n\times n}_+$ the set of real $n \times n$ matrices with nonnegative entries. For $A \in \mathbb{R}^{n\times n}_+$ and $p > 0$, $A^{(p)}$ is by definition the matrix such that $(A^{(p)})_{ij} = (A_{ij})^p$, and

$$\rho_p(A) := (\rho(A^{(p)}))^{1/p},$$

where ρ denotes the (usual) spectral radius. We also define $\rho_\infty(A) = \lim_{p\to+\infty} \rho_p(A)$.

Facts:

1. [CG79], [Gau92, Ch. IV], [BSvdD95] *Max-plus Collatz–Wielandt formula, I.* Let $A \in \mathbb{R}^{n\times n}_{\max}$ and $\lambda \in \mathbb{R}$. The following assertions are equivalent: (i) There exists $\mathbf{u} \in \mathbb{R}^n$ such that $A\mathbf{u} \leq \lambda\mathbf{u}$; (ii) $\rho_{\max}(A) \leq \lambda$. It follows that

$$\rho_{\max}(A) = \inf_{\mathbf{u}\in\mathbb{R}^n} \max_{1\leq i\leq n} (A\mathbf{u})_i \,/\, \mathbf{u}_i$$

(the product Au and the division by $\mathbf{u}_i$ should be understood in the max-plus sense). If $\rho_{\max}(A) > \mathbb{0}$, then this infimum is attained by some $\mathbf{u} \in \mathbb{R}^n$. If in addition A is irreducible, then Assertion (i) is equivalent to the following: (i') there exists $\mathbf{u} \in \mathbb{R}^n_{\max} \setminus \{\mathbf{0}\}$ such that $A\mathbf{u} \leq \lambda\mathbf{u}$.

2. [Gau92, Ch. IV], [BSvdD95] *Max-plus Collatz–Wielandt formula, II.* Let $\lambda \in \mathbb{R}_{\max}$. The following assertions are equivalent: (i) There exists $\mathbf{u} \in \mathbb{R}^n_{\max} \setminus \{\mathbf{0}\}$ such that $A\mathbf{u} \geq \lambda\mathbf{u}$; (ii) $\rho_{\max}(A) \geq \lambda$. It follows that

$$\rho_{\max}(A) = \max_{\mathbf{u} \in \mathbb{R}^n_{\max} \setminus \{\mathbf{0}\}} \min_{\substack{1 \leq i \leq n \\ \mathbf{u}_i \neq \mathbb{0}}} (A\mathbf{u})_i \,/\, \mathbf{u}_i .$$

3. [Fri86] For $A \in \mathbb{R}_+^{n \times n}$, we have $\rho_\infty(A) = \exp(\rho_{\max}(\log(A)))$, where log is interpreted entrywise.
4. [KO85] For all $A \in \mathbb{R}_+^{n \times n}$, and $1 \leq q \leq p \leq \infty$, we have $\rho_p(A) \leq \rho_q(A)$.
5. For all $A, B \in \mathbb{R}_+^{n \times n}$, we have

$$\rho(A \circ B) \leq \rho_p(A)\rho_q(B) \quad \text{for all} \quad p, q \in [1, \infty] \quad \text{such that} \quad \frac{1}{p} + \frac{1}{q} = 1.$$

This follows from the classical Kingman's inequality [Kin61], which states that the map $\log \circ \rho \circ \exp$ is convex (exp is interpreted entrywise). We have in particular $\rho(A \circ B) \leq \rho_\infty(A)\rho(B)$.

6. [Fri86] For all $A \in \mathbb{R}_+^{n \times n}$, we have

$$\rho_\infty(A) \leq \rho(A) \leq \rho_\infty(A)\rho(\hat{A}) \leq \rho_\infty(A)n,$$

where $\hat{A}$ is the pattern matrix of A, that is, $\hat{A}_{ij} = 1$ if $A_{ij} \neq 0$ and $\hat{A}_{ij} = 0$ if $A_{ij} = 0$.

7. [Bap98], [EvdD99] For all $A \in \mathbb{R}_+^{n \times n}$, we have $\lim_{k \to \infty}(\rho_\infty(A^k))^{1/k} = \rho(A)$.
8. [CG79] *Computing $\rho_{\max}(A)$ by linear programming.* For $A \in \mathbb{R}^{n \times n}_{\max}$, $\rho_{\max}(A)$ is the value of the linear program

$$\inf \lambda \text{ s.t. } \exists \mathbf{u} \in \mathbb{R}^n, \quad \forall (i, j) \in E, \quad A_{ij} + \mathbf{u}_j \leq \lambda + \mathbf{u}_i,$$

where $E = \{(i, j) \mid 1 \leq i, j \leq n,\ A_{ij} \neq \mathbb{0}\}$ is the set of arcs of $\Gamma(A)$.

9. *Dual linear program to compute $\rho_{\max}(A)$.* Let C denote the set of nonnegative vectors $x = (x_{ij})_{(i,j) \in E}$ such that

$$\forall 1 \leq i \leq n, \quad \sum_{1 \leq k \leq n,\, (k,i) \in E} x_{ki} = \sum_{1 \leq j \leq n, (i,j) \in E} x_{ij}, \quad \text{and} \quad \sum_{(i,j) \in E} x_{ij} = 1.$$

To every cycle c of $\Gamma(A)$ corresponds bijectively the extreme point of the polytope C that is given by $x_{ij} = 1/|c|$ if (i, j) belongs to c, and $x_{ij} = 0$ otherwise. Moreover, $\rho_{\max}(A) = \sup\{\sum_{(i,j) \in E} A_{ij} x_{ij} \mid x \in C\}$.

10. [Kar78] *Karp's formula.* If $A \in \mathbb{R}^{n \times n}_{\max}$ is irreducible, then, for all $1 \leq i \leq n$,

$$\rho_{\max}(A) = \max_{\substack{1 \leq j \leq n \\ A^n_{ij} \neq \mathbb{0}}} \min_{1 \leq k \leq n} \frac{(A^n)_{ij} - (A^{n-k})_{ij}}{k}. \tag{25.2}$$

To evaluate the right-hand side expression, compute the sequence $\mathbf{u}^0 = \mathbf{e}_i$, $\mathbf{u}^1 = \mathbf{u}^0 A$, $\mathbf{u}^n = \mathbf{u}^{n-1} A$, so that $\mathbf{u}^k = A^k_{i\cdot}$ for all $0 \leq k \leq n$. This takes a time $O(nm)$, where m is the number of arcs of $\Gamma(A)$. One can avoid storing the vectors $\mathbf{u}^0, \ldots, \mathbf{u}^n$, at the price of recomputing the sequence $\mathbf{u}^0, \ldots, \mathbf{u}^{n-1}$ once $\mathbf{u}^n$ is known. The time and space complexity of Karp's algorithm are $O(nm)$ and $O(n)$, respectively. The policy iteration algorithm of [CTCG+98] seems experimentally more efficient than Karp's algorithm. Other algorithms are given in particular in [CGL96], [BO93], and [EvdD99]. A comparison of maximal cycle mean algorithms appears in [DGI98]. When the entries of A take only two finite values, the maximal cycle mean of A can be computed in linear time [CGB95]. The Karp and policy iteration algorithms, as well as the general max-plus operations (full and sparse matrix products, matrix residuation, etc.) are implemented in the **Maxplus toolbox** of **Scilab**, freely available in the contributed section of the Web site www.scilab.org.

Example:

1. For the matrix A in Application 3 of section 25.1, we have $\rho_{\max}(A) = \max(5, 4, (2 + 11)/2, (2 + 8)/2, (7+6)/2, (11+7+2)/3, (8+6+2)/3) = 20/3$, which gives the maximal reward per year. This is attained by the cycle $(1, 2), (2, 3), (3, 1)$, corresponding to the rotation of crops: fallow, wheat, oats.

25.3 The Max-Plus Eigenproblem

The results of this section and of the next one constitute max-plus spectral theory. Early and fundamental contributions are due to Cuninghame–Green (see [CG79]), Vorobyev [Vor67], Romanovskiĭ [Rom67], Gondran and Minoux [GM77], and Cohen, Dubois, Quadrat, and Viot [CDQV83]. General presentations are included in [CG79], [BCOQ92], and [GM02]. The infinite dimensional max-plus spectral theory (which is not covered here) has been developed particularly after Maslov, in relation with Hamilton–Jacobi partial differential equations; see [MS92] and [KM97]. See also [MPN02], [AGW05], and [Fat06] for recent developments.

In this section and the next two, A denotes a matrix in $\mathbb{R}_{\max}^{n\times n}$.

Definitions:

An **eigenvector** of A is a vector $\mathbf{u} \in \mathbb{R}_{\max}^{n} \setminus \{\mathbb{0}\}$ such that $A\mathbf{u} = \lambda\mathbf{u}$, for some scalar $\lambda \in \mathbb{R}_{\max}$, which is called the **(geometric) eigenvalue** corresponding to $\mathbf{u}$. With the notation of classical algebra, the equation $A\mathbf{u} = \lambda\mathbf{u}$ can be rewritten as

$$\max_{1\le j\le n} A_{ij} + \mathbf{u}_j = \lambda + \mathbf{u}_i, \quad \forall 1 \le i \le n.$$

If λ is an eigenvalue of A, the set of vectors $\mathbf{u} \in \mathbb{R}_{\max}^{n}$ such that $A\mathbf{u} = \lambda\mathbf{u}$ is the **eigenspace** of A for the eigenvalue λ.

The **saturation digraph** with respect to $\mathbf{u} \in \mathbb{R}_{\max}^{n}$, $\text{Sat}(A, \mathbf{u})$, is the digraph with vertices $1, \dots, n$ and an arc from vertex i to vertex j when $A_{ij}\mathbf{u}_j = (A\mathbf{u})_i$.

A cycle $c = (i_1, i_2), \dots, (i_k, i_1)$ that attains the maximum in (25.1) is called **critical.** The **critical digraph** is the union of the critical cycles. The **critical vertices** are the vertices of the critical digraph.

The **normalized matrix** is $\tilde{A} = \rho_{\max}(A)^{-1}A$ (when $\rho_{\max}(A) \neq \mathbb{0}$).

For a digraph Γ, vertex i **has access** to a vertex j if there is a walk from i to j in Γ. The **(access equivalent) classes** of Γ are the equivalence classes of the set of its vertices for the relation "i has access to j and j has access to i." A class C **has access** to a class C' if some vertex of C has access to some vertex of C'. A class is **final** if it has access only to itself.

The **classes** of a matrix A are the classes of $\Gamma(A)$, and the **critical classes** of A are the classes of the critical digraph of A. A class C of A is **basic** if $\rho_{\max}(A[C, C]) = \rho_{\max}(A)$.

Facts:

The proof of most of the following facts can be found in particular in [CG79] or [BCOQ92, Sec. 3.7]; we give specific references when needed.

1. For any matrix A, $\rho_{\max}(A)$ is an eigenvalue of A, and any eigenvalue of A is less than or equal to $\rho_{\max}(A)$.
2. An eigenvalue of A associated with an eigenvector in $\mathbb{R}^n$ must be equal to $\rho_{\max}(A)$.
3. [ES75] *Max-plus diagonal scaling.* Assume that $\mathbf{u} \in \mathbb{R}^n$ is an eigenvector of A. Then the matrix B such that $B_{ij} = \mathbf{u}_i^{-1}A_{ij}\mathbf{u}_j$ has all its entries less than or equal to $\rho_{\max}(A)$, and the maximum of every of its rows is equal to $\rho_{\max}(A)$.
4. If A is irreducible, then $\rho_{\max}(A) > \mathbb{0}$ and it is the only eigenvalue of A. *From now on, we assume that $\Gamma(A)$ has at least one cycle, so that $\rho_{\max}(A) > \mathbb{0}$.*

5. For all critical vertices i of A, the column $\tilde{A}^\star_{\cdot i}$ is an eigenvector of A for the eigenvalue $\rho_{\max}(A)$. Moreover, if i and j belong to the same critical class of A, then $\tilde{A}^\star_{\cdot i} = \tilde{A}^\star_{\cdot j}\tilde{A}^\star_{ji}$.
6. *Eigenspace for the eigenvalue* $\rho_{\max}(A)$. Let $C_1, \ldots, C_s$ denote the critical classes of A, and let us choose arbitrarily one vertex $i_t \in C_t$, for every $t = 1, \ldots, s$. Then, the columns $\tilde{A}^\star_{\cdot, i_t}$, $t = 1, \ldots, s$ span the eigenspace of A for the eigenvalue $\rho_{\max}(A)$. Moreover, any spanning family of this eigenspace contains some scalar multiple of every column $\tilde{A}^\star_{\cdot, i_t}$, $t = 1, \ldots, s$.
7. Let C denote the set of critical vertices, and let $T = \{1, \ldots, n\} \setminus C$. The following facts are proved in a more general setting in [AG03, Th. 3.4], with the exception of (b), which follows from Fact 4 of Section 25.1.

   (a) The restriction $\mathbf{v} \mapsto \mathbf{v}[C]$ is an isomorphism from the eigenspace of A for the eigenvalue $\rho_{\max}(A)$ to the eigenspace of $A[C, C]$ for the same eigenvalue.

   (b) An eigenvector $\mathbf{u}$ for the eigenvalue $\rho_{\max}(A)$ is determined from its restriction $\mathbf{u}[C]$ by $\mathbf{u}[T] = (\tilde{A}[T, T])^\star \tilde{A}[T, C]\mathbf{u}[C]$.

   (c) Moreover, $\rho_{\max}(A)$ is the only eigenvalue of $A[C, C]$ and the eigenspace of $A[C, C]$ is stable by infimum and by convex combination in the usual sense.
8. *Complementary slackness.* If $\mathbf{u} \in \mathbb{R}^n_{\max}$ is such that $A\mathbf{u} \leq \rho_{\max}(A)\mathbf{u}$, then $(A\mathbf{u})_i = \rho_{\max}(A)\mathbf{u}_i$, for all critical vertices i.
9. *Critical digraph vs. saturation digraph.* Let $\mathbf{u} \in \mathbb{R}^n$ be such that $A\mathbf{u} \leq \rho_{\max}(A)\mathbf{u}$. Then, the union of the cycles of $\mathrm{Sat}(A, \mathbf{u})$ is equal to the critical digraph of A.
10. [CQD90], [Gau92, Ch. IV], [BSvdD95] *Spectrum of reducible matrices.* A scalar $\lambda \neq \mathbb{0}$ is an eigenvalue of A if and only if there is at least one class C of A such that $\rho_{\max}(A[C, C]) = \lambda$ and $\rho_{\max}(A[C, C]) \geq \rho_{\max}(A[C', C'])$ for all classes C' that have access to C.
11. [CQD90], [BSvdD95] The matrix A has an eigenvector in $\mathbb{R}^n$ if and only if all its final classes are basic.
12. [Gau92, Ch. IV] *Eigenspace for an eigenvalue* λ. Let $C^1, \ldots, C^m$ denote all the classes C of A such that $\rho_{\max}(A[C, C]) = \lambda$ and $\rho_{\max}(A[C', C']) \leq \lambda$ for all classes C' that have access to C. For every $1 \leq k \leq m$, let $C^k_1, \ldots, C^k_{s_k}$ denote the critical classes of the matrix $A[C^k, C^k]$. For all $1 \leq k \leq m$ and $1 \leq t \leq s_k$, let us choose arbitrarily an element $j_{k,t}$ in C^k_t. Then, the family of columns $(\lambda^{-1}A)^\star_{\cdot, j_{k,t}}$, indexed by all these k and t, spans the eigenspace of A for the eigenvalue λ, and any spanning family of this eigenspace contains a scalar multiple of every $(\lambda^{-1}A)^\star_{\cdot, j_{k,t}}$.
13. *Computing the eigenvectors.* Observe first that any vertex j that attains the maximum in Karp's formula (25.2) is critical. To compute one eigenvector for the eigenvalue $\rho_{\max}(A)$, it suffices to compute $\tilde{A}^\star_{\cdot j}$ for some critical vertex j. This is equivalent to a single source shortest path problem, which can be solved in $O(nm)$ time and $O(n)$ space. Alternatively, one may use the *policy iteration algorithm* of [CTCG+98] or the improvement in [EvdD99] of the *power algorithm* [BO93]. Once a particular eigenvector is known, the critical digraph can be computed from Fact 9 in $O(m)$ additional time.

Examples:

1. For the matrix A in Application 3 of section 25.1, the only critical cycle is $(1, 2), (2, 3), (3, 1)$ (up to a circular permutation of vertices). The critical digraph consists of the vertices and arcs of this cycle. By Fact 6, any eigenvector $\mathbf{u}$ of A is proportional to $\tilde{A}^\star_{\cdot 1} = [0\ -13/3\ -14/3]^T$ (or equivalently, to $\tilde{A}^\star_{\cdot 2}$ or $\tilde{A}^\star_{\cdot 3}$). Observe that an eigenvector yields a relative price information between the different states.
2. Consider the matrix and its associated digraph:

$$A = \left[\begin{array}{c|ccc|ccc|c} 0 & \cdot & 0 & \cdot & 7 & \cdot & \cdot & \cdot \\ \hline \cdot & \cdot & 3 & 0 & \cdot & \cdot & \cdot & \cdot \\ \cdot & 1 & \cdot & \cdot & \cdot & \cdot & \cdot & \cdot \\ \cdot & 2 & \cdot & \cdot & \cdot & \cdot & \cdot & 10 \\ \hline \cdot & \cdot & \cdot & \cdot & 1 & 0 & \cdot & \cdot \\ \cdot & \cdot & \cdot & \cdot & \cdot & \cdot & 0 & \cdot \\ \cdot & \cdot & \cdot & \cdot & -1 & 2 & \cdot & 23 \\ \hline \cdot & \cdot & \cdot & \cdot & \cdot & \cdot & \cdot & -3 \end{array}\right]$$

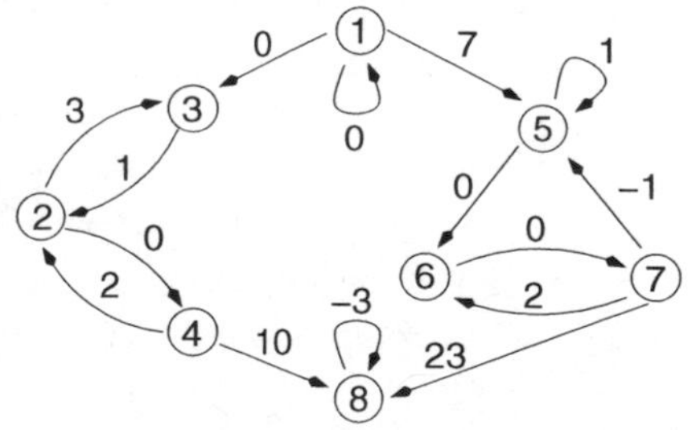

(We use $\cdot$ to represent the element $-\infty$.) The classes of A are $C^1 = \{1\}$, $C^2 = \{2,3,4\}$, $C^3 = \{5,6,7\}$, and $C^4 = \{8\}$. We have $\rho_{\max}(A) = \rho_{\max}(A[C^2,C^2]) = 2$, $\rho_{\max}(A[C^1,C^1]) = 0$, $\rho_{\max}(A[C^3,C^3]) = 1$, and $\rho_{\max}(A[C^4,C^4]) = -3$. The critical digraph is reduced to the critical cycle $(2,3)(3,2)$. By Fact 6, any eigenvector for the eigenvalue $\rho_{\max}(A)$ is proportional to $\tilde{A}^\star_{\cdot 2} = [-3\ 0\ -1\ 0\ -\infty\ -\infty\ -\infty\ -\infty]^T$. By Fact 10, the other eigenvalues of A are 0 and 1. By Fact 12, any eigenvector for the eigenvalue 0 is proportional to $A^\star_{\cdot 1} = \mathbf{e}_1$. Observe that the critical classes of $A[C^3,C^3]$ are $C^3_1 = \{5\}$ and $C^3_2 = \{6,7\}$. Therefore, by Fact 12, any eigenvector for the eigenvalue 1 is a max-plus linear combination of $(1^{-1}A)^\star_{\cdot 5} = [6\ -\infty\ -\infty\ -\infty\ 0\ -3\ -2\ -\infty]^T$ and $(1^{-1}A)^\star_{\cdot 6} = [5\ -\infty\ -\infty\ -\infty\ -1\ 0\ 1\ -\infty]^T$. The eigenvalues of A^T are 2, 1, and -3. So A and A^T have only two eigenvalues in common.

25.4 Asymptotics of Matrix Powers

Definitions:

A sequence $s_0, s_1, \ldots$ of elements of $\mathbb{R}_{\max}$ is **recognizable** if there exists a positive integer p, vectors $\mathbf{b} \in \mathbb{R}^{p\times 1}_{\max}$ and $\mathbf{c} \in \mathbb{R}^{1\times p}_{\max}$, and a matrix $M \in \mathbb{R}^{p\times p}_{\max}$ such that $s_k = \mathbf{c}M^k\mathbf{b}$, for all nonnegative integers k.

A sequence $s_0, s_1, \ldots$ of elements of $\mathbb{R}_{\max}$ is **ultimately geometric** with **rate** $\lambda \in \mathbb{R}_{\max}$ if $s_{k+1} = \lambda s_k$ for k large enough.

The **merge** of q sequences $s^1, \ldots, s^q$ is the sequence s such that $s_{kq+i-1} = s^i_k$, for all $k \geq 0$ and $1 \leq i \leq q$.

Facts:

1. [Gun94], [CTGG99] If every row of the matrix A has at least one entry different from $\mathbb{0}$, then, for all $1 \leq i \leq n$ and $\mathbf{u} \in \mathbb{R}^n$, the limit

$$\chi_i(A) = \lim_{k\to\infty} (A^k\mathbf{u})_i^{1/k}$$

exists and is independent of the choice of $\mathbf{u}$. The vector $\chi(A) = (\chi_i(A))_{1\leq i\leq n} \in \mathbb{R}^n$ is called the **cycle-time** of A. It is given by

$$\chi_i(A) = \max\{\rho_{\max}(A[C,C]) \mid C \text{ is a class of } A \text{ to which } i \text{ has access}\}.$$

In particular, if A is irreducible, then $\chi_i(A) = \rho_{\max}(A)$ for all $i = 1, \ldots, n$.

2. The following constitutes the cyclicity theorem, due to Cohen, Dubois, Quadrat, and Viot [CDQV83]. See [BCOQ92] and [AGW05] for more accessible accounts.
   (a) If A is irreducible, there exists a positive integer γ such that $A^{k+\gamma} = \rho_{\max}(A)^\gamma A^k$ for k large enough. The minimal value of γ is called the **cyclicity** of A.
   (b) Assume again that A is irreducible. Let $C_1, \ldots, C_s$ be the critical classes of A, and for $i = 1, \ldots, s$, let γ_i denote the g.c.d. (greatest common divisor) of the lengths of the critical cycles of A belonging to C_i. Then, the cyclicity γ of A is the l.c.m. (least common multiple) of $\gamma_1, \ldots, \gamma_s$.
   (c) Assume that $\rho_{\max}(A) \neq \mathbb{0}$. The **spectral projector** of A is the matrix $P := \lim_{k\to\infty} \tilde{A}^k \tilde{A}^\star = \lim_{k\to\infty} \tilde{A}^k \oplus \tilde{A}^{k+1} \oplus \cdots$. It is given by $P = \bigoplus_{i\in C} \tilde{A}^\star_{\cdot i}\tilde{A}^\star_{i\cdot}$, where C denotes the set of critical vertices of A. When A is irreducible, the limit is attained in finite time. If, in addition, A has cyclicity one, then $A^k = \rho_{\max}(A)^k P$ for k large enough.
3. Assume that A is irreducible, and let m denote the number of arcs of its critical digraph. Then, the cyclicity of A can be computed in $O(m)$ time from the critical digraph of A, using the algorithm of Denardo [Den77].

4. The smallest integer k such that $A^{k+\gamma} = \rho_{\max}(A)^{\gamma} A^k$ is called the **coupling time.** It is estimated in [HA99], [BG01], [AGW05] (assuming again that A is irreducible).
5. [AGW05, Th. 7.5] *Turnpike theorem.* Define a walk of $\Gamma(A)$ to be optimal if it has a maximal weight amongst all walks with the same ends and length. If A is irreducible, then the number of noncritical vertices of an optimal walk (counted with multiplicities) is bounded by a constant depending only on A.
6. [Mol88], [Gau94], [KB94], [DeS00] A sequence of elements of $\mathbb{R}_{\max}$ is recognizable if and only if it is a merge of ultimately geometric sequences. In particular, for all $1 \leq i, j \leq n$, the sequence $(A^k)_{ij}$ is a merge of ultimately geometric sequences.
7. [Sim78], [Has90], [Sim94], [Gau96] One can decide whether a finitely generated semigroup S of matrices with effective entries in $\mathbb{R}_{\max}$ is finite. One can also decide whether the set of entries in a given position of the matrices of S is finite (limitedness problem). However [Kro94], whether this set contains a given entry is undecidable (even when the entries of the matrices belong to $\mathbb{Z} \cup \{-\infty\}$).

Example:

1. For the matrix A in Application 3 of section 25.1, the cyclicity is 3, and the spectral projector is

$$P = \tilde{A}^{\star}_{\cdot 1}\tilde{A}^{\star}_{1\cdot} = \begin{bmatrix} 0 \\ -13/3 \\ -14/3 \end{bmatrix} \begin{bmatrix} 0 & 13/3 & 14/3 \end{bmatrix}^T = \begin{bmatrix} 0 & 13/3 & 14/3 \\ -13/3 & 0 & 1/3 \\ -14/3 & -1/3 & 0 \end{bmatrix}.$$

2. For the matrix A in Example 2 of Section 25.3, the cycle-time is $\chi(A) = [2\ 2\ 2\ 2\ 1\ 1\ 1\ -3]^T$. The cyclicity of $A[C^2, C^2]$ is 2 because there is only one critical cycle, which has length 2. Let $B := A[C^3, C^3]$. The critical digraph of B has two strongly connected components consisting, respectively, of the cycles $(5, 5)$ and $(6, 7), (7, 6)$. So B has cyclicity l.c.m. $(1, 2) = 2$. The sequence $s_k := (A^k)_{18}$ is such that $s_{k+2} = s_k + 4$, for $k \geq 24$, with $s_{24} = s_{25} = 51$. Hence, s_k is the merge of two ultimately geometric sequences, both with rate 4. To get an example where different rates appear, replace the entries A_{11} and A_{88} of A by $-\infty$. Then, the same sequence s_k is such that $s_{k+2} = s_k + 4$, for all even $k \geq 24$, and $s_{k+2} = s_k + 2$, for all odd $k \geq 5$, with $s_5 = 31$ and $s_{24} = 51$.

25.5 The Max-Plus Permanent

Definitions:

The (max-plus) **permanent** of A is per $A = \sum_{\sigma \in S_n} A_{1\sigma(1)} \cdots A_{n\sigma(n)}$, or with the usual notation of classical algebra, per $A = \max_{\sigma \in S_n} A_{1\sigma(1)} + \cdots + A_{n\sigma(n)}$, which is the value of the optimal assignment problem with weights A_{ij}.

A **max-plus polynomial function** P is a map $\mathbb{R}_{\max} \to \mathbb{R}_{\max}$ of the form $P(x) = \sum_{i=0}^{n} p_i x^i$ with $p_i \in \mathbb{R}_{\max}$, $i = 0, \ldots, n$. If $p_n \neq \mathbb{0}$, P is of **degree** n.

The **roots** of a nonzero max-plus polynomial function P are the points of nondifferentiability of P, together with the point $\mathbb{0}$ when the derivative of P near $-\infty$ is positive. The **multiplicity** of a root α of P is defined as the variation of the derivative of P at the point α, $P'(\alpha^+) - P'(\alpha^-)$, when $\alpha \neq \mathbb{0}$, and as its derivative near $-\infty$, $P'(\mathbb{0}^+)$, when $\alpha = \mathbb{0}$.

The (max-plus) **characteristic polynomial function** of A is the polynomial function P_A given by $P_A(x) = \text{per}(A \oplus xI)$ for $x \in \mathbb{R}_{\max}$. The **algebraic eigenvalues** of A are the roots of P_A.

Facts:

1. [CGM80] Any nonzero max-plus polynomial function P can be factored uniquely as $P(x) = a(x \oplus \alpha_1) \cdots (x \oplus \alpha_n)$, where $a \in \mathbb{R}$, n is the degree of P, and the α_i are the roots of P, counted with multiplicities.

2. [CG83], [ABG04, Th. 4.6 and 4.7]. The greatest algebraic eigenvalue of A is equal to $\rho_{\max}(A)$. Its multiplicity is less than or equal to the number of critical vertices of A, with equality if and only if the critical vertices can be covered by disjoint critical cycles.
3. Any geometric eigenvalue of A is an algebraic eigenvalue of A. (This can be deduced from Fact 2 of this section, and Fact 10 of section 25.3.)
4. [Yoe61] If $A \geq I$ and per $A = \mathbb{1}$, then $A^{\star}_{ij} = \text{per } A(j, i)$, for all $1 \leq i, j \leq n$.
5. [But00] Assume that all the entries of A are different from $\mathbb{0}$. The following are equivalent: (i) there is a vector $b \in \mathbb{R}^n$ that has a unique preimage by A; (ii) there is only one permutation σ such that $|\sigma|_A := A_{1\sigma(1)} \cdots A_{n\sigma(n)} = \text{per } A$. Further characterizations can be found in [But00] and [DSS05].
6. [Bap95] *Alexandroff inequality over* $\mathbb{R}_{\max}$. Construct the matrix B with columns $A_{\cdot 1}, A_{\cdot 1}, A_{\cdot 3}, \ldots, A_{\cdot n}$ and the matrix C with columns $A_{\cdot 2}, A_{\cdot 2}, A_{\cdot 3}, \ldots, A_{\cdot n}$. Then $(\text{per } A)^2 \geq (\text{per } B)(\text{per } C)$, or with the notation of classical algebra, $2 \times \text{per } A \geq \text{per } B + \text{per } C$.
7. [BB03] The max-plus characteristic polynomial function of A can be computed by solving $O(n)$ optimal assignment problems.

Example:

1. For the matrix A in Example 2 of section 25.3, the characteristic polynomial of A is the product of the characteristic polynomials of the matrices $A[C^i, C^i]$, for $i = 1, \ldots, 4$. Thus, $P_A(x) = (x \oplus 0)(x \oplus 2)^2 x (x \oplus 1)^3 (x \oplus (-3))$, and so, the algebraic eigenvalues of A are $-\infty, -3, 0, 1$, and 2, with respective multiplicities 1, 1, 1, 3, and 2.

25.6 Linear Inequalities and Projections

Definitions:

If $A \in \overline{\mathbb{R}}_{\max}^{n \times p}$, the **range** of A, denoted range A, is $\{A\mathbf{x} \mid \mathbf{x} \in \overline{\mathbb{R}}_{\max}^p\} \subset \overline{\mathbb{R}}_{\max}^n$. The **kernel** of A, denoted ker A, is the set of **equivalence classes modulo** A, which are the classes for the equivalence relation "$\mathbf{x} \sim \mathbf{y}$ if $A\mathbf{x} = A\mathbf{y}$."

The **support** of a vector $\mathbf{b} \in \overline{\mathbb{R}}_{\max}^n$ is $\text{supp}\,\mathbf{b} := \{i \in \{1, \ldots, n\} \mid \mathbf{b}_i \neq \mathbb{0}\}$.

The **orthogonal congruence** of a subset U of $\overline{\mathbb{R}}_{\max}^n$ is $U^{\perp} := \{(\mathbf{x}, \mathbf{y}) \in \overline{\mathbb{R}}_{\max}^n \times \overline{\mathbb{R}}_{\max}^n \mid \mathbf{u} \cdot \mathbf{x} = \mathbf{u} \cdot \mathbf{y}\ \forall \mathbf{u} \in U\}$, where "$\cdot$" denotes the max-plus scalar product. The **orthogonal space** of a subset C of $\overline{\mathbb{R}}_{\max}^n \times \overline{\mathbb{R}}_{\max}^n$ is $C^{\top} := \{\mathbf{u} \in \overline{\mathbb{R}}_{\max}^n \mid \mathbf{u} \cdot \mathbf{x} = \mathbf{u} \cdot \mathbf{y}\ \forall (\mathbf{x}, \mathbf{y}) \in C\}$.

Facts:

1. For all $a,\ b \in \overline{\mathbb{R}}_{\max}$, the maximal $c \in \overline{\mathbb{R}}_{\max}$ such that $ac \leq b$, denoted by $a \backslash b$ (or b / a), is given by $a \backslash b = b - a$ if $(a, b) \notin \{(-\infty, -\infty), (+\infty, +\infty)\}$, and $a \backslash b = +\infty$ otherwise.
2. [BCOQ92, Eq. 4.82] If $A \in \overline{\mathbb{R}}_{\max}^{n \times p}$ and $B \in \overline{\mathbb{R}}_{\max}^{n \times q}$, then the inequation $AX \leq B$ has a maximal solution $X \in \overline{\mathbb{R}}_{\max}^{p \times q}$ given by the matrix $A \backslash B$ defined by $(A \backslash B)_{ij} = \bigwedge_{1 \leq k \leq n} A_{ki} \backslash B_{kj}$. Similarly, for $A \in \overline{\mathbb{R}}_{\max}^{n \times p}$ and $C \in \overline{\mathbb{R}}_{\max}^{r \times p}$, the maximal solution $C / A \in \overline{\mathbb{R}}_{\max}^{r \times n}$ of $XA \leq C$ exists and is given by $(C / A)_{ij} = \bigwedge_{1 \leq k \leq p} C_{ik} / A_{jk}$.
3. The equation $AX = B$ has a solution if and only if $A(A \backslash B) = B$.
4. For $A \in \overline{\mathbb{R}}_{\max}^{n \times p}$, the map $A^{\sharp} :\ \mathbf{y} \in \overline{\mathbb{R}}_{\min}^n \to A \backslash \mathbf{y} \in \overline{\mathbb{R}}_{\min}^p$ is linear. It is represented by the matrix $-A^T$.
5. [BCOQ92, Table 4.1] For matrices A, B, C with entries in $\overline{\mathbb{R}}_{\max}$ and with appropriate dimensions, we have

$$\begin{aligned} A(A \backslash (AB)) &= AB, & A \backslash (A(A \backslash B)) &= A \backslash B, \\ (A \oplus B) \backslash C &= (A \backslash C) \wedge (B \backslash C), & A \backslash (B \wedge C) &= (A \backslash B) \wedge (A \backslash C), \\ (AB) \backslash C &= B \backslash (A \backslash C), & A \backslash (B / C) &= (A \backslash B) / C. \end{aligned}$$

The first five identities have dual versions, with $/$ instead of $\backslash$. Due to the last identity, we shall write $A \backslash B / C$ instead of $A \backslash (B / C)$.

6. [CGQ97] Let $A \in \overline{\mathbb{R}}_{\max}^{n\times p}$, $B \in \overline{\mathbb{R}}_{\max}^{n\times q}$ and $C \in \overline{\mathbb{R}}_{\max}^{r\times p}$. We have range $A \subset$ range $B \iff A = B(B \backslash A)$, and $\ker A \subset \ker C \iff C = (C / A)A$.
7. [CGQ96] Let $A \in \overline{\mathbb{R}}_{\max}^{n\times p}$. The map $\Pi_A := A \circ A^\sharp$ is a projector on the range of A, meaning that $(\Pi_A)^2 = \Pi_A$ and range Π_A = range A. Moreover, $\Pi_A(x)$ is the greatest element of the range of A, which is less than or equal to x. Similarly, the map $\Pi^A := A^\sharp \circ A$ is a projector on the range of $A^\sharp$, and $\Pi^A(x)$ is the smallest element of the range of $A^\sharp$ that is greater than or equal to x. Finally, every equivalence class modulo A meets the range of $A^\sharp$ at a unique point.
8. [CGQ04], [DS04] For any $A \in \overline{\mathbb{R}}_{\max}^{n\times p}$, the map $x \mapsto A(-x)$ is a bijection from range (A^T) to range (A), with inverse map $x \mapsto A^T(-x)$.
9. [CGQ96], [CGQ97] *Projection onto a range parallel to a kernel.* Let $B \in \overline{\mathbb{R}}_{\max}^{n\times p}$ and $C \in \overline{\mathbb{R}}_{\max}^{q\times n}$. For all $x \in \overline{\mathbb{R}}_{\max}^{n}$, there is a greatest ξ on the range of B such that $C\xi \leq Cx$. It is given by $\Pi_B^C(x)$, where $\Pi_B^C := \Pi_B \circ \Pi^C$. We have $(\Pi_B^C)^2 = \Pi_B^C$. Assume now that every equivalence class modulo C meets the range of B at a unique point. This is the case if and only if range (CB) = range C and $\ker(CB) = \ker B$. Then $\Pi_B^C(x)$ is the unique element of the range of B, which is equivalent to x modulo C, the map Π_B^C is a linear projector on the range of B, and it is represented by the matrix $(B / (CB))C$, which is equal to $B((CB) \backslash C)$.
10. [CGQ97] *Regular matrices.* Let $A \in \overline{\mathbb{R}}_{\max}^{n\times p}$. The following assertions are equivalent: (i) there is a linear projector from $\overline{\mathbb{R}}_{\max}^{n}$ to range A; (ii) $A = AXA$ for some $X \in \overline{\mathbb{R}}_{\max}^{p\times n}$; (iii) $A = A(A \backslash A / A)A$.
11. [Vor67], [Zim76, Ch. 3] (See also [But94], [AGK05].) *Vorobyev–Zimmermann covering theorem.* Assume that $A \in \mathbb{R}_{\max}^{n\times p}$ and $\mathbf{b} \in \overline{\mathbb{R}}_{\max}^{n}$. For $j \in \{1, \ldots, p\}$, let

$$S_j = \{i \in \{1, \ldots, n\} \mid A_{ij} \neq \mathbb{0} \text{ and } A_{ij} \backslash \mathbf{b}_i = (A \backslash \mathbf{b})_j\}.$$

The equation $A\mathbf{x} = \mathbf{b}$ has a solution if and only if $\cup_{1\leq j\leq p} S_j \supset \operatorname{supp} \mathbf{b}$ or equivalently $\cup_{j\in \operatorname{supp}(A \backslash \mathbf{b})} S_j \supset \operatorname{supp} \mathbf{b}$. It has a unique solution if and only if $\cup_{j\in \operatorname{supp}(A \backslash \mathbf{b})} S_j \supset \operatorname{supp} \mathbf{b}$ and $\cup_{j\in J} S_j \not\supset \operatorname{supp} \mathbf{b}$ for all strict subsets J of $\operatorname{supp}(A \backslash \mathbf{b})$.

12. [Zim77], [SS92], [CGQ04], [CGQS05], [DS04] *Separation theorem.* Let $A \in \overline{\mathbb{R}}_{\max}^{n\times p}$ and $\mathbf{b} \in \overline{\mathbb{R}}_{\max}^{n}$. If $\mathbf{b} \notin$ range A, then there exists $\mathbf{c}, \mathbf{d} \in \overline{\mathbb{R}}_{\max}^{n}$ such that the **halfspace** $H := \{\mathbf{x} \in \overline{\mathbb{R}}_{\max}^{n} \mid \mathbf{c} \cdot \mathbf{x} \geq \mathbf{d} \cdot \mathbf{x}\}$ contains range A but not $\mathbf{b}$. We can take $\mathbf{c} = -\mathbf{b}$ and $\mathbf{d} = -\Pi_A(\mathbf{b})$. Moreover, when A and $\mathbf{b}$ have entries in $\mathbb{R}_{\max}$, $\mathbf{c}, \mathbf{d}$ can be chosen with entries in $\mathbb{R}_{\max}$.
13. [GP97] For any $A \in \overline{\mathbb{R}}_{\max}^{n\times p}$, we have $((\text{range } A)^\perp)^\top$ = range A.
14. [LMS01], [CGQ04] A linear form defined on a finitely generated subsemimodule of $\overline{\mathbb{R}}_{\max}^{n}$ can be extended to $\overline{\mathbb{R}}_{\max}^{n}$. This is a special case of a max-plus analogue of the Riesz representation theorem.
15. [BH84], [GP97] Let $A, B \in \overline{\mathbb{R}}_{\max}^{n\times p}$. The set of solutions $\mathbf{x} \in \overline{\mathbb{R}}_{\max}^{p}$ of $A\mathbf{x} = B\mathbf{x}$ is a finitely generated subsemimodule of $\overline{\mathbb{R}}_{\max}^{p}$.
16. [GP97], [Gau98] Let X, Y be finitely generated subsemimodules of $\overline{\mathbb{R}}_{\max}^{n}$, $A \in \overline{\mathbb{R}}_{\max}^{n\times p}$ and $B \in \overline{\mathbb{R}}_{\max}^{r\times n}$. Then $X \cap Y$, $X \oplus Y := \{\mathbf{x} \oplus \mathbf{y} \mid \mathbf{x} \in X, \ \mathbf{y} \in Y\}$, and $X - Y := \{\mathbf{z} \in \overline{\mathbb{R}}_{\max}^{n} \mid \exists \mathbf{x} \in X, \mathbf{y} \in Y, \ \mathbf{x} = \mathbf{y} \oplus \mathbf{z}\}$ are finitely generated subsemimodules of $\overline{\mathbb{R}}_{\max}^{n}$. Also, $A^{-1}(X)$, $B(X)$, and $X^\perp$ are finitely generated subsemimodules of $\overline{\mathbb{R}}_{\max}^{p}$, $\overline{\mathbb{R}}_{\max}^{r}$, and $\overline{\mathbb{R}}_{\max}^{n} \times \overline{\mathbb{R}}_{\max}^{n}$, respectively. Similarly, if Z is a finitely generated subsemimodule of $\overline{\mathbb{R}}_{\max}^{n} \times \overline{\mathbb{R}}_{\max}^{n}$, then $Z^\top$ is a finitely generated subsemimodule of $\overline{\mathbb{R}}_{\max}^{n}$.
17. Facts 13 to 16 still hold if $\overline{\mathbb{R}}_{\max}$ is replaced by $\mathbb{R}_{\max}$.
18. When $A, B \in \mathbb{R}_{\max}^{n\times p}$, algorithms to find one solution of $A\mathbf{x} = B\mathbf{x}$ are given in [WB98] or [CGB03]. One can also use the general algorithm of [GG98] to compute a finite fixed point of a min-max function, together with the observation that $\mathbf{x}$ satisfies $A\mathbf{x} = B\mathbf{x}$ if and only if $\mathbf{x} = f(\mathbf{x})$, where $f(\mathbf{x}) = \mathbf{x} \wedge (A \backslash (B\mathbf{x})) \wedge (B \backslash (A\mathbf{x}))$.

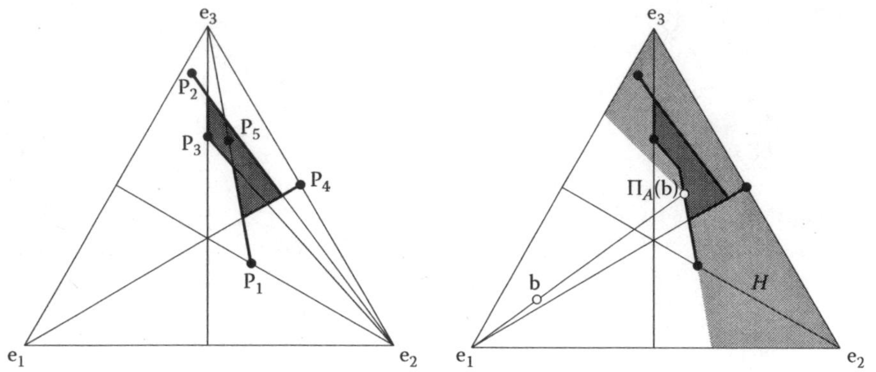

FIGURE 25.1 Projection of a point on a range.

Examples:

1. In order to illustrate Fact 11, consider

$$A = \begin{bmatrix} 0 & 0 & 0 & -\infty & 0.5 \\ 1 & -2 & 0 & 0 & 1.5 \\ 0 & 3 & 2 & 0 & 3 \end{bmatrix}, \quad \mathbf{b} = \begin{bmatrix} 3 \\ 0 \\ 0.5 \end{bmatrix}. \tag{25.3}$$

Let $\bar{\mathbf{x}} := A \setminus \mathbf{b}$. We have $\bar{\mathbf{x}}_1 = \min(-0+3, -1+0, -0+0.5) = -1$, and so, $S_1 = \{2\}$ because the minimum is attained only by the second term. Similarly, $\bar{\mathbf{x}}_2 = -2.5$, $S_2 = \{3\}$, $\bar{\mathbf{x}}_3 = -1.5$, $S_3 = \{3\}$, $\bar{\mathbf{x}}_4 = 0$, $S_4 = \{2\}$, $\bar{\mathbf{x}}_5 = -2.5$, $S_5 = \{3\}$. Since $\cup_{1 \le j \le 5} S_j = \{2, 3\} \not\supset \operatorname{supp} \mathbf{b} = \{1, 2, 3\}$, Fact 11 shows that the equation $Ax = \mathbf{b}$ has no solution. This also follows from the fact that $\Pi_A(\mathbf{b}) = A(A \setminus \mathbf{b}) = [-1\ 0\ 0.5]^T < \mathbf{b}$.

2. The range of the previous matrix A is represented in Figure 25.1 (left). A nonzero vector $\mathbf{x} \in \mathbb{R}^3_{\max}$ is represented by the point that is the barycenter with weights $(\exp(\beta \mathbf{x}_i))_{1 \le i \le 3}$ of the vertices of the simplex, where $\beta > 0$ is a fixed scaling parameter. Every vertex of the simplex represents one basis vector $\mathbf{e}_i$. Proportional vectors are represented by the same point. The i-th column of A, $A_{\cdot i}$, is represented by the point $\mathbf{p}_i$ on the figure. Observe that the broken segment from $\mathbf{p}_1$ to $\mathbf{p}_2$, which represents the semimodule generated by $A_{\cdot 1}$ and $A_{\cdot 2}$, contains $\mathbf{p}_5$. Indeed, $A_{\cdot 5} = 0.5 A_{\cdot 1} \oplus A_{\cdot 2}$. The range of A is represented by the closed region in dark grey and by the bold segments joining the points $\mathbf{p}_1, \mathbf{p}_2, \mathbf{p}_4$ to it.

   We next compute a half-space separating the point b defined in (25.3) from range A. Recall that $\Pi_A(\mathbf{b}) = [-1\ 0\ 0.5]^T$. So, by Fact 12, a half-space containing range A and not $\mathbf{b}$ is $H := \{\mathbf{x} \in \overline{\mathbb{R}}^3_{\max} (-3)\mathbf{x}_1 \oplus \mathbf{x}_2 \oplus (-0.5)\mathbf{x}_3 \ge 1\mathbf{x}_1 \oplus \mathbf{x}_2 \oplus (-0.5)\mathbf{x}_3\}$. We also have $H \cap \mathbb{R}^3_{\max} = \{\mathbf{x} \in \mathbb{R}^3_{\max} \mid \mathbf{x}_2 \oplus (-0.5)\mathbf{x}_3 \ge 1\mathbf{x}_1\}$. The set of nonzero points of $H \cap \mathbb{R}^3_{\max}$ is represented by the light gray region in Figure 25.1 (right).

25.7 Max-Plus Linear Independence and Rank

Definitions:

If M is a subsemimodule of $\mathbb{R}^n_{\max}$, $\mathbf{u} \in M$ is an **extremal generator** of M, or $\mathbb{R}_{\max}\mathbf{u} := \{\lambda.\mathbf{u} \mid \lambda \in \mathbb{R}_{\max}\}$ is an **extreme ray** of M, if $\mathbf{u} \ne \mathbf{0}$ and if $\mathbf{u} = \mathbf{v} \oplus \mathbf{w}$ with $\mathbf{v}, \mathbf{w} \in M$ imply that $\mathbf{u} = \mathbf{v}$ or $\mathbf{u} = \mathbf{w}$.

A family $\mathbf{u}_1, \ldots, \mathbf{u}_r$ of vectors of $\mathbb{R}^n_{\max}$ is **linearly independent in the Gondran–Minoux sense** if for all disjoints subsets I and J of $\{1, \ldots, r\}$, and all $\lambda_i \in \mathbb{R}_{\max}$, $i \in I \cup J$, we have $\sum_{i \in I} \lambda_i.\mathbf{u}_i \ne \sum_{j \in J} \lambda_j.\mathbf{u}_j$, unless $\lambda_i = \mathbb{0}$ for all $i \in I \cup J$.

For $A \in \mathbb{R}_{\max}^{n\times n}$, we define

$$\det{}^+ A := \sum_{\sigma\in S_n^+} A_{1\sigma(1)}\cdots A_{n\sigma(n)}, \qquad \det{}^- A := \sum_{\sigma\in S_n^-} A_{1\sigma(1)}\cdots A_{n\sigma(n)},$$

where S_n^+ and S_n^- are, respectively, the sets of even and odd permutations of $\{1,\ldots,n\}$. The **bideterminant** [GM84] of A is $(\det^+ A, \det^- A)$.

For $A \in \mathbb{R}_{\max}^{n\times p} \setminus \{\mathbb{0}\}$, we define

- The **row rank** (resp. the **column rank**) of A, denoted $\mathrm{rk_{row}}(A)$ (resp. $\mathrm{rk_{col}}(A)$), as the number of extreme rays of range A^T (resp. range A).
- The **Schein** rank of A as $\mathrm{rk_{Sch}}(A) := \min\{r \geq 1 \mid A = BC,\ \text{with } B \in \mathbb{R}_{\max}^{n\times r}, C \in \mathbb{R}_{\max}^{r\times p}\}$.
- The **strong rank** of A, denoted $\mathrm{rk_{st}}(A)$, as the maximal $r \geq 1$ such that there exists an $r \times r$ submatrix B of A for which there is only one permutation σ such that $|\sigma|_B = \mathrm{per}\, B$.
- The row (resp. column) **Gondran–Minoux rank** of A, denoted $\mathrm{rk_{GMr}}(A)$ (resp. $\mathrm{rk_{GMc}}$), as the maximal $r \geq 1$ such that A has r linearly independent rows (resp. columns) in the Gondran–Minoux sense.
- The **symmetrized rank** of A, denoted $\mathrm{rk_{sym}}(A)$, as the maximal $r \geq 1$ such that A has an $r \times r$ submatrix B such that $\det^+ B \neq \det^- B$.

(A new rank notion, **Kapranov rank**, which is not discussed here, has been recently studied [DSS05]. We also note that the Schein rank is called in this reference Barvinok rank.)

Facts:

1. [Hel88], [Mol88], [Wag91], [Gau98], [DS04] Let M be a finitely generated subsemimodule of $\mathbb{R}_{\max}^n$. A subset of vectors of M spans M if and only if it contains at least one nonzero element of every extreme ray of M.
2. [GM02] The columns of $A \in \mathbb{R}_{\max}^{n\times n}$ are linearly independent in the Gondran–Minoux sense if and only if $\det^+ A \neq \det^- A$.
3. [Plu90], [BCOQ92, Th. 3.78]. *Max-plus Cramer's formula.* Let $A \in \mathbb{R}_{\max}^{n\times n}$, and let $\mathbf{b}^-, \mathbf{b}^+ \in \mathbb{R}_{\max}^n$. Define the i-th positive Cramer's determinant by

$$D_i^+ := \det{}^+(A_{\cdot 1}\ldots A_{\cdot,i-1}\mathbf{b}^+ A_{\cdot,i+1}\ldots A_{\cdot n}) +\!\!\!\!+ \det{}^-(A_{\cdot 1}\ldots A_{\cdot,i-1}\mathbf{b}^- A_{\cdot,i+1}\ldots A_{\cdot n}),$$

   and the i-th negative Cramer's determinant, D_i^-, by exchanging $\mathbf{b}^+$ and $\mathbf{b}^-$ in the definition of D_i^+. Assume that $\mathbf{x}^+, \mathbf{x}^- \in \mathbb{R}_{\max}^n$ have disjoint supports. Then $A\mathbf{x}^+ +\!\!\!\!+ \mathbf{b}^- = A\mathbf{x}^- +\!\!\!\!+ \mathbf{b}^+$ implies that

$$(\det{}^+ A)\mathbf{x}_i^+ +\!\!\!\!+ (\det{}^- A)\mathbf{x}_i^- +\!\!\!\!+ D_i^- = (\det{}^- A)\mathbf{x}_i^+ +\!\!\!\!+ (\det{}^+ A)\mathbf{x}_i^- +\!\!\!\!+ D_i^+ \quad \forall 1 \leq i \leq n. \tag{25.4}$$

   The converse implication holds, and the vectors $\mathbf{x}^+$ and $\mathbf{x}^-$ are uniquely determined by (25.4), if $\det^+ A \neq \det^- A$, and if $D_i^+ \neq D_i^-$ or $D_i^+ = D_i^- = \mathbb{0}$, for all $1 \leq i \leq n$. This result is formulated in a simpler way in [Plu90], [BCOQ92] using the **symmetrization** of the max-plus semiring, which leads to more general results. We note that the converse implication relies on the following semiring analogue of the classical adjugate identity: $A\,\mathrm{adj}^+ A +\!\!\!\!+ \det^- A\, I = A\,\mathrm{adj}^- A +\!\!\!\!+ \det^+ A\, I$, where $\mathrm{adj}^\pm A := (\det^\pm A(j,i))_{1\leq i,j\leq n}$. This identity, as well as analogues of many other determinantal identities, can be obtained using the general method of [RS84]. See, for instance, [GBCG98], where the derivation of the Binet–Cauchy identity is detailed.
4. For $A \in \mathbb{R}_{\max}^{n\times p}$, we have

$$\mathrm{rk_{st}}(A) \leq \mathrm{rk_{sym}}(A) \leq \begin{Bmatrix} \mathrm{rk_{GMr}}(A) \\ \mathrm{rk_{GMc}}(A) \end{Bmatrix} \leq \mathrm{rk_{Sch}}(A) \leq \begin{Bmatrix} \mathrm{rk_{row}}(A) \\ \mathrm{rk_{col}}(A) \end{Bmatrix}.$$

The second inequality follows from Fact 2, the third one from Facts 2 and 3. The other inequalities are immediate. Moreover, all these inequalities become equalities if A is regular [CGQ06].

Examples:

1. The matrix A in Example 1 of section 25.6 has column rank 4: The extremal rays of range A are generated by the first four columns of A. All the other ranks of A are equal to 3.

References

[ABG04] M. Akian, R. Bapat, and S. Gaubert. Min-plus methods in eigenvalue perturbation theory and generalised Lidskiĭ-Višik-Ljusternik theorem. *arXiv:math.SP/0402090*, 2004.

[AG03] M. Akian and S. Gaubert. Spectral theorem for convex monotone homogeneous maps, and ergodic control. *Nonlinear Anal.*, 52(2):637–679, 2003.

[AGK05] M. Akian, S. Gaubert, and V. Kolokoltsov. Set coverings and invertibility of functional Galois connections. In *Idempotent Mathematics and Mathematical Physics*, Contemp. Math., pp. 19–51. Amer. Math. Soc., 2005.

[AGW05] M. Akian, S. Gaubert, and C. Walsh. Discrete max-plus spectral theory. In *Idempotent Mathematics and Mathematical Physics*, Contemp. Math., pp. 19–51. Amer. Math. Soc., 2005.

[Bac03] N. Bacaër. Modèles mathématiques pour l'optimisation des rotations. *Comptes Rendus de l'Académie d'Agriculture de France*, 89(3):52, 2003. Electronic version available on www.academie-agriculture.fr.

[Bap95] R.B. Bapat. Permanents, max algebra and optimal assignment. *Lin. Alg. Appl.*, 226/228:73–86, 1995.

[Bap98] R.B. Bapat. A max version of the Perron-Frobenius theorem. In *Proceedings of the Sixth Conference of the International Linear Algebra Society (Chemnitz, 1996)*, vol. 275/276, pp. 3–18, 1998.

[BB03] R.E. Burkard and P. Butkovič. Finding all essential terms of a characteristic maxpolynomial. *Discrete Appl. Math.*, 130(3):367–380, 2003.

[BCOQ92] F. Baccelli, G. Cohen, G.-J. Olsder, and J.-P. Quadrat. *Synchronization and Linearity.* John Wiley & Sons, Chichester, 1992.

[BG01] A. Bouillard and B. Gaujal. Coupling time of a (max,plus) matrix. In *Proceedings of the Workshop on Max-Plus Algebras, a satellite event of the first IFAC Symposium on System, Structure and Control (Praha, 2001).* Elsevier, 2001.

[BH84] P. Butkovič and G. Hegedüs. An elimination method for finding all solutions of the system of linear equations over an extremal algebra. *Ekonom.-Mat. Obzor*, 20(2):203–215, 1984.

[BO93] J.G. Braker and G.J. Olsder. The power algorithm in max algebra. *Lin. Alg. Appl.*, 182:67–89, 1993.

[BSvdD95] R.B. Bapat, D. Stanford, and P. van den Driessche. Pattern properties and spectral inequalities in max algebra. *SIAM J. of Matrix Ana. Appl.*, 16(3):964–976, 1995.

[But94] P. Butkovič. Strong regularity of matrices—a survey of results. *Discrete Appl. Math.*, 48(1):45–68, 1994.

[But00] P. Butkovič. Simple image set of (max, +) linear mappings. *Discrete Appl. Math.*, 105(1-3):73–86, 2000.

[CDQV83] G. Cohen, D. Dubois, J.-P. Quadrat, and M. Viot. Analyse du comportement périodique des systèmes de production par la théorie des dioïdes. Rapport de recherche 191, INRIA, Le Chesnay, France, 1983.

[CDQV85] G. Cohen, D. Dubois, J.-P. Quadrat, and M. Viot. A linear system theoretic view of discrete event processes and its use for performance evaluation in manufacturing. *IEEE Trans. on Automatic Control*, AC–30:210–220, 1985.

[CG79] R.A. Cuninghame-Green. *Minimax Algebra*, vol. 166 of *Lect. Notes in Econom. and Math. Systems.* Springer-Verlag, Berlin, 1979.

[CG83] R.A. Cuninghame-Green. The characteristic maxpolynomial of a matrix. *J. Math. Ana. Appl.*, 95:110–116, 1983.

[CGB95] R.A. Cuninghame-Green and P. Butkovič. Extremal eigenproblem for bivalent matrices. *Lin. Alg. Appl.*, 222:77–89, 1995.

[CGB03] R.A. Cuninghame-Green and P. Butkovič. The equation $A \otimes x = B \otimes y$ over (max, +). *Theoret. Comp. Sci.*, 293(1):3–12, 2003.

[CGL96] R.A. Cuninghame-Green and Y. Lin. Maximum cycle-means of weighted digraphs. *Appl. Math. JCU*, 11B:225–234, 1996.

[CGM80] R.A. Cuninghame-Green and P.F.J. Meijer. An algebra for piecewise-linear minimax problems. *Discrete Appl. Math*, 2:267–294, 1980.

[CGQ96] G. Cohen, S. Gaubert, and J.-P. Quadrat. Kernels, images and projections in dioids. In *Proceedings of WODES'96*, pp. 151–158, Edinburgh, August 1996. IEE.

[CGQ97] G. Cohen, S. Gaubert, and J.-P. Quadrat. Linear projectors in the max-plus algebra. In *Proceedings of the IEEE Mediterranean Conference*, Cyprus, 1997. IEEE.

[CGQ04] G. Cohen, S. Gaubert, and J.-P. Quadrat. Duality and separation theorems in idempotent semimodules. *Lin. Alg. Appl.*, 379:395–422, 2004.

[CGQ06] G. Cohen, S. Gaubert, and J.-P. Quadrat. Regular matrices in max-plus algebra. Preprint, 2006.

[CGQS05] G. Cohen, S. Gaubert, J.-P. Quadrat, and I. Singer. Max-plus convex sets and functions. In *Idempotent Mathematics and Mathematical Physics*, Contemp. Math., pp. 105–129. Amer. Math. Soc., 2005.

[CKR84] Z.Q. Cao, K.H. Kim, and F.W. Roush. *Incline Algebra and Applications.* Ellis Horwood, New York, 1984.

[CQD90] W. Chen, X. Qi, and S. Deng. The eigen-problem and period analysis of the discrete event systems. *Sys. Sci. Math. Sci.*, 3(3), August 1990.

[CTCG+98] J. Cochet-Terrasson, G. Cohen, S. Gaubert, M. McGettrick, and J.-P. Quadrat. Numerical computation of spectral elements in max-plus algebra. In *Proc. of the IFAC Conference on System Structure and Control*, Nantes, France, July 1998.

[CTGG99] J. Cochet-Terrasson, S. Gaubert, and J. Gunawardena. A constructive fixed point theorem for min-max functions. *Dyn. Stabil. Sys.*, 14(4):407–433, 1999.

[Den77] E.V. Denardo. Periods of connected networks and powers of nonnegative matrices. *Math. Oper. Res.*, 2(1):20–24, 1977.

[DGI98] A. Dasdan, R.K. Gupta, and S. Irani. An experimental study of minimum mean cycle algorithms. Technical Report 32, UCI-ICS, 1998.

[DeS00] B. De Schutter. On the ultimate behavior of the sequence of consecutive powers of a matrix in the max-plus algebra. *Lin. Alg. Appl.*, 307(1-3):103–117, 2000.

[DS04] M. Develin and B. Sturmfels. Tropical convexity. *Doc. Math.*, 9:1–27, 2004. (Erratum pp. 205–206)

[DSS05] M. Develin, F. Santos, and B. Sturmfels. On the rank of a tropical matrix. In *Combinatorial and Computational Geometry*, vol. 52 of *Math. Sci. Res. Inst. Publ.*, pp. 213–242. Cambridge Univ. Press, Cambridge, 2005.

[Eil74] S. Eilenberg. *Automata, Languages, and Machines, Vol. A.* Academic Press, New York, 1974. *Pure and Applied Mathematics*, Vol. 58.

[ES75] G.M. Engel and H. Schneider. Diagonal similarity and equivalence for matrices over groups with 0. *Czechoslovak Math. J.*, 25(100)(3):389–403, 1975.

[EvdD99] L. Elsner and P. van den Driessche. On the power method in max algebra. *Lin. Alg. Appl.*, 302/303:17–32, 1999.

[Fat06] A. Fathi. Weak KAM theorem in Lagrangian dynamics. Lecture notes, 2006, to be published by Cambridge University Press.

[Fri86] S. Friedland. Limit eigenvalues of nonnegative matrices. *Lin. Alg. Appl.*, 74:173–178, 1986.

[Gau92] S. Gaubert. *Théorie des systèmes linéaires dans les dioïdes.* Thèse, École des Mines de Paris, July 1992.

[Gau94] S. Gaubert. Rational series over dioids and discrete event systems. In *Proc. of the 11th Conf. on Anal. and Opt. of Systems: Discrete Event Systems*, vol. 199 of *Lect. Notes in Control and Inf. Sci*, Sophia Antipolis, Springer, London, 1994.

[Gau96] S. Gaubert. On the Burnside problem for semigroups of matrices in the (max,+) algebra. *Semigroup Forum*, 52:271–292, 1996.

[Gau98] S. Gaubert. Exotic semirings: examples and general results. Support de cours de la 26ième École de Printemps d'Informatique Théorique, Noirmoutier, 1998.

[GBCG98] S. Gaubert, P. Butkovič, and R. Cuninghame-Green. Minimal (max,+) realization of convex sequences. *SIAM J. Cont. Optimi.*, 36(1):137–147, January 1998.

[GG98] S. Gaubert and J. Gunawardena. The duality theorem for min-max functions. *C. R. Acad. Sci. Paris.*, 326, Série I:43–48, 1998.

[GM77] M. Gondran and M. Minoux. Valeurs propres et vecteurs propres dans les dioïdes et leur interprétation en théorie des graphes. *E.D.F., Bulletin de la Direction des Études et Recherches, Série C, Mathématiques Informatique*, 2:25–41, 1977.

[GM84] M. Gondran and M. Minoux. Linear algebra in dioids: a survey of recent results. *Ann. Disc. Math.*, 19:147–164, 1984.

[GM02] M. Gondran and M. Minoux. *Graphes, dioïdes et semi-anneaux*. Éditions TEC & DOC, Paris, 2002.

[GP88] G. Gallo and S. Pallotino. Shortest path algorithms. *Ann. Op. Res.*, 13:3–79, 1988.

[GP97] S. Gaubert and M. Plus. Methods and applications of (max,+) linear algebra. In *STACS'97*, vol. 1200 of *Lect. Notes Comput. Sci.*, pp. 261–282, Lübeck, March 1997. Springer.

[Gun94] J. Gunawardena. Cycle times and fixed points of min-max functions. In *Proceedings of the 11th International Conference on Analysis and Optimization of Systems*, vol. 199 of *Lect. Notes in Control and Inf. Sci*, pp. 266–272. Springer, London, 1994.

[Gun98] J. Gunawardena, Ed. *Idempotency*, vol. 11 of *Publications of the Newton Institute*. Cambridge University Press, Cambridge, UK, 1998.

[HA99] M. Hartmann and C. Arguelles. Transience bounds for long walks. *Math. Oper. Res.*, 24(2):414–439, 1999.

[Has90] K. Hashiguchi. Improved limitedness theorems on finite automata with distance functions. *Theoret. Comput. Sci.*, 72:27–38, 1990.

[Hel88] S. Helbig. On Carathéodory's and Kreĭn-Milman's theorems in fully ordered groups. *Comment. Math. Univ. Carolin.*, 29(1):157–167, 1988.

[HOvdW06] B. Heidergott, G.-J. Olsder, and J. van der Woude, *Max Plus at work*, Princeton University Press, 2000.

[Kar78] R.M. Karp. A characterization of the minimum mean-cycle in a digraph. *Discrete Math.*, 23:309–311, 1978.

[KB94] D. Krob and A. Bonnier Rigny. A complete system of identities for one letter rational expressions with multiplicities in the tropical semiring. *J. Pure Appl. Alg.*, 134:27–50, 1994.

[Kin61] J.F.C. Kingman. A convexity property of positive matrices. *Quart. J. Math. Oxford Ser. (2)*, 12:283–284, 1961.

[KM97] V.N. Kolokoltsov and V.P. Maslov. *Idempotent analysis and its applications*, vol. 401 of *Mathematics and Its Applications*. Kluwer Academic Publishers Group, Dordrecht, 1997.

[KO85] S. Karlin and F. Ost. Some monotonicity properties of Schur powers of matrices and related inequalities. *Lin. Alg. Appl.*, 68:47–65, 1985.

[Kro94] D. Krob. The equality problem for rational series with multiplicities in the tropical semiring is undecidable. *Int. J. Alg. Comp.*, 4(3):405–425, 1994.

[LM05] G.L. Litvinov and V.P. Maslov, Eds. *Idempotent Mathematics and Mathematical Physics*. Number 377 in Contemp. Math. Amer. Math. Soc., 2005.

[LMS01] G.L. Litvinov, V.P. Maslov, and G.B. Shpiz. Idempotent functional analysis: an algebraic approach. *Math. Notes*, 69(5):696–729, 2001.

[Mol88] P. Moller. *Théorie algébrique des Systèmes à Événements Discrets*. Thèse, École des Mines de Paris, 1988.

[MPN02] J. Mallet-Paret and R. Nussbaum. Eigenvalues for a class of homogeneous cone maps arising from max-plus operators. *Disc. Cont. Dynam. Sys.*, 8(3):519–562, July 2002.

[MS92] V.P. Maslov and S.N. Samborskii, Eds. *Idempotent analysis*, vol. 13 of *Advances in Soviet Mathematics.* Amer. Math. Soc., Providence, RI, 1992.

[MY60] R. McNaughton and H. Yamada. Regular expressions and state graphs for automata. *IRE trans on Elec. Comp.*, 9:39–47, 1960.

[Plu90] M. Plus. Linear systems in (max, +)-algebra. In *Proceedings of the 29th Conference on Decision and Control*, Honolulu, Dec. 1990.

[Rom67] I.V. Romanovskiĭ. Optimization of stationary control of discrete deterministic process in dynamic programming. *Kibernetika*, 3(2):66–78, 1967.

[RS84] C. Reutenauer and H. Straubing. Inversion of matrices over a commutative semiring. *J. Alg.*, 88(2):350–360, June 1984.

[Sim78] I. Simon. Limited subsets of the free monoid. In *Proc. of the 19th Annual Symposium on Foundations of Computer Science*, pp. 143–150. IEEE, 1978.

[Sim94] I. Simon. On semigroups of matrices over the tropical semiring. *Theor. Infor. and Appl.*, 28(3-4):277–294, 1994.

[SS92] S.N. Samborskiĭ and G.B. Shpiz. Convex sets in the semimodule of bounded functions. In *Idempotent Analysis*, pp. 135–137. Amer. Math. Soc., Providence, RI, 1992.

[Vor67] N.N. Vorob′ev. Extremal algebra of positive matrices. *Elektron. Informationsverarbeit. Kybernetik*, 3:39–71, 1967. (In Russian)

[Wag91] E. Wagneur. Moduloïds and pseudomodules. I. Dimension theory. *Disc. Math.*, 98(1):57–73, 1991.

[WB98] E.A. Walkup and G. Borriello. A general linear max-plus solution technique. In *Idempotency*, vol. 11 of *Publ. Newton Inst.*, pp. 406–415. Cambridge Univ. Press, Cambridge, 1998.

[Yoe61] M. Yoeli. A note on a generalization of boolean matrix theory. *Amer. Math. Monthly*, 68:552–557, 1961.

[Zim76] K. Zimmermann. *Extremální Algebra*. Ekonomický ùstav ČSAV, Praha, 1976. (in Czech).

[Zim77] K. Zimmermann. A general separation theorem in extremal algebras. *Ekonom.-Mat. Obzor*, 13(2):179–201, 1977.

[Zim81] U. Zimmermann. Linear and combinatorial optimization in ordered algebraic structures. *Ann. Discrete Math.*, 10:viii, 380, 1981.

26

Matrices Leaving a Cone Invariant

Bit-Shun Tam
Tamkang University

Hans Schneider
University of Wisconsin

Generalizations of the Perron–Frobenius theory of nonnegative matrices to linear operators leaving a cone invariant were first developed for operators on a Banach space by Krein and Rutman [KR48], Karlin [Kar59], and Schaefer [Sfr66], although there are early examples in finite dimensions, e.g., [Sch65] and [Bir67]. In this chapter, we describe a generalization that is sometimes called the geometric spectral theory of nonnegative linear operators in finite dimensions, which emerged in the late 1980s. Motivated by a search for geometric analogs of results in the previously developed combinatorial spectral theory of (reducible) nonnegative matrices (for reviews see [Sch86] and [Her99]), this area is a study of the Perron–Frobenius theory of a nonnegative matrix and its generalizations from the cone-theoretic viewpoint. The treatment is linear-algebraic and cone-theoretic (geometric) with the facial and duality concepts and occasionally certain elementary analytic tools playing the dominant role. The theory is particularly rich when the underlying cone is polyhedral (finitely generated) and it reduces to the nonnegative matrix case when the cone is simplicial.

26.1 Perron—Frobenius Theorem for Cones

We work with cones in a real vector space, as "cone" is a real concept. To deal with cones in $\mathbb{C}^n$, we can identify the latter space with $\mathbb{R}^{2n}$. For a discussion on the connection between the real and complex case of the spectral theory, see [TS94, Sect. 8].

Definitions:

A **proper cone** K in a finite-dimensional real vector space V is a closed, pointed, full convex cone, *viz.*

- $K + K \subseteq K$, viz. $\mathbf{x}, \mathbf{y} \in K \Longrightarrow \mathbf{x} + \mathbf{y} \in K$.
- $\mathbb{R}^+ K \subseteq K$, viz. $\mathbf{x} \in K, \alpha \in \mathbb{R}^+ \Longrightarrow \alpha\mathbf{x} \in K$.
- K is closed in the usual topology of V.

- $K \cap (-K) = \{\mathbf{0}\}$, viz. $\mathbf{x}, -\mathbf{x} \in K \Longrightarrow \mathbf{x} = \mathbf{0}$.
- $\text{int}K \neq \emptyset$, where $\text{int}K$ is the interior of K.

Usually, the unqualified term **cone** is defined by the first two items in the above definition. However, in this chapter we call a proper cone simply a **cone.** We denote by K a cone in $\mathbb{R}^n$, $n \geq 2$.

The vector $\mathbf{x} \in \mathbb{R}^n$ is **K-nonnegative,** written $\mathbf{x} \geq^K \mathbf{0}$, if $\mathbf{x} \in K$.

The vector $\mathbf{x}$ is **K-semipositive,** written $\mathbf{x} \gneq^K \mathbf{0}$, if $\mathbf{x} \geq^K \mathbf{0}$ and $\mathbf{x} \neq \mathbf{0}$.

The vector $\mathbf{x}$ is **K-positive,** written $\mathbf{x} >^K \mathbf{0}$, if $\mathbf{x} \in \text{int } K$.

For $\mathbf{x}, \mathbf{y} \in \mathbb{R}^n$, we write $\mathbf{x} \geq^K \mathbf{y}$ ($\mathbf{x} \gneq^K \mathbf{y}$, $\mathbf{x} >^K \mathbf{y}$) if $\mathbf{x} - \mathbf{y}$ is K-nonnegative (K-semipositive, K-positive).

The matrix $A \in \mathbb{R}^{n\times n}$ is **K-nonnegative,** written $A \geq^K \mathbf{0}$, if $AK \subseteq K$.

The matrix A is **K-semipositive,** written $A \gneq^K \mathbf{0}$, if $A \geq^K \mathbf{0}$ and $A \neq \mathbf{0}$.

The matrix A is **K-positive,** written $A >^K \mathbf{0}$, if $A(K \setminus \{\mathbf{0}\}) \subseteq \text{int } K$.

For $A, B \in \mathbb{R}^{n\times n}$, $A \geq^K B$ ($A \gneq^K B$, $A >^K B$) means $A - B \geq^K \mathbf{0}$ ($A - B \gneq^K \mathbf{0}$, $A - B >^K \mathbf{0}$).

A **face** F of a cone $K \subseteq \mathbb{R}^n$ is a subset of K, which is a cone in the linear span of F such that $\mathbf{x} \in F$, $\mathbf{x} \geq^K \mathbf{y} \geq^K \mathbf{0} \Longrightarrow \mathbf{y} \in F$.

(In this chapter, F will always denote a face rather than a field, since the only fields involved are $\mathbb{R}$ and $\mathbb{C}$.) Thus, F satisfies all definitions of a cone except that its interior may be empty.

A face F of K is a **trivial face** if $F = \{\mathbf{0}\}$ or $F = K$.

For a subset S of a cone K, the intersection of all faces of K including S is called the **face of K generated by** S and is denoted by $\Phi(S)$. If $S = \{\mathbf{x}\}$, then $\Phi(S)$ is written simply as $\Phi(\mathbf{x})$.

For faces F, G of K, their **meet** and **join** are given respectively by $F \wedge G = F \cap G$ and $F \vee G = \Phi(F \cup G)$.

A vector $\mathbf{x} \in K$ is an **extreme vector** if either $\mathbf{x}$ is the zero vector or $\mathbf{x}$ is nonzero and $\Phi(\mathbf{x}) = \{\lambda\mathbf{x} : \lambda \geq 0\}$; in the latter case, the face $\Phi(\mathbf{x})$ is called an **extreme ray.**

If P is K-nonnegative, then a face F of K is a **P-invariant face** if $PF \subseteq F$.

If P is K-nonnegative, then P is **K-irreducible** if the only P-invariant faces are the trivial faces.

If K is a cone in $\mathbb{R}^n$, then a cone, called the **dual cone** of K, is denoted and given by

$$K^* = \{\mathbf{y} \in \mathbb{R}^n : \mathbf{y}^T\mathbf{x} \geq 0 \text{ for all } \mathbf{x} \in K\}.$$

If A is an $n \times n$ complex matrix and $\mathbf{x}$ is a vector in $\mathbb{C}^n$, then the **local spectral radius of A at $\mathbf{x}$** is denoted and given by $\rho_{\mathbf{x}}(A) = \limsup_{m\to\infty} \|A^m\mathbf{x}\|^{1/m}$, where $\|\cdot\|$ is any norm of $\mathbb{C}^n$. For $A \in \mathbb{C}^{n\times n}$, its spectral radius is denoted by $\rho(A)$ (or ρ) (cf. Section 4.3).

Facts:

Let K be a cone in $\mathbb{R}^n$.

1. The condition $\text{int}K \neq \emptyset$ in the definition of a cone is equivalent to $K - K = V$, *viz.*, for all $\mathbf{z} \in V$ there exist $\mathbf{x}, \mathbf{y} \in K$ such that $\mathbf{z} = \mathbf{x} - \mathbf{y}$.
2. A K-positive matrix is K-irreducible.
3. [Van68], [SV70] Let P be a K-nonnegative matrix. The following are equivalent:
   (a) P is K-irreducible.
   (b) $\sum_{i=0}^{n-1} P^i >^K \mathbf{0}$.
   (c) $(I + P)^{n-1} >^K \mathbf{0}$.
   (d) No eigenvector of P (for any eigenvalue) lies on the boundary of K.
4. (Generalization of Perron–Frobenius Theorem) [KR48], [BS75] Let P be a K-irreducible matrix with spectral radius ρ. Then
   (a) ρ is positive and is a simple eigenvalue of P.
   (b) There exists a (up to a scalar multiple) unique K-positive (right) eigenvector $\mathbf{u}$ of P corresponding to ρ.
   (c) $\mathbf{u}$ is the only K-semipositive eigenvector for P (for any eigenvalue).
   (d) $K \cap (\rho I - P)\mathbb{R}^n = \{\mathbf{0}\}$.

5. (Generalization of Perron–Frobenius Theorem) Let P be a K-nonnegative matrix with spectral radius ρ. Then
   (a) ρ is an eigenvalue of P.
   (b) There is a K-semipositive eigenvector of P corresponding to ρ.
6. If P, Q are K-nonnegative and $Q \stackrel{K}{\leq} P$, then $\rho(Q) \leq \rho(P)$. Further, if P is K-irreducible and $Q \stackrel{K}{\lneq} P$, then $\rho(Q) < \rho(P)$.
7. P is K-nonnegative (K-irreducible) if and only if P^T is K^*-nonnegative (K^*-irreducible).
8. If A is an $n \times n$ complex matrix and $\mathbf{x}$ is a vector in $\mathbf{C}^n$, then the local spectral radius $\rho_{\mathbf{x}}(A)$ of A at $\mathbf{x}$ is equal to the spectral radius of the restriction of A to the A-cyclic subspace generated by $\mathbf{x}$, i.e., span$\{A^i\mathbf{x} : i = 0, 1, \ldots\}$. If $\mathbf{x}$ is nonzero and $\mathbf{x} = \mathbf{x}_1 + \cdots + \mathbf{x}_k$ is the representation of $\mathbf{x}$ as a sum of generalized eigenvectors of A corresponding, respectively, to distinct eigenvalues $\lambda_1, \ldots, \lambda_k$, then $\rho_{\mathbf{x}}(A)$ is also equal to $\max_{1 \leq i \leq k} |\lambda_i|$.
9. Barker and Schneider [BS75] developed Perron–Frobenius theory in the setting of a (possibly infinite-dimensional) vector space over a fully ordered field without topology. They introduced the concepts of irreducibility and strong irreducibility, and show that these two concepts are equivalent if the underlying cone has ascending chain condition on faces. See [ERS95] for the role of real closed-ordered fields in this theory.

Examples:

1. The nonnegative orthant $(\mathbb{R}_0^+)^n$ in $\mathbb{R}^n$ is a cone. Then $\mathbf{x} \geq^K \mathbf{0}$ if and only if $\mathbf{x} \geq \mathbf{0}$, *viz.* the entries of $\mathbf{x}$ are nonnegative, and F is a face of $(\mathbb{R}_0^+)^n$ if and only if F is of the form F_J for some $J \subseteq \{1, \ldots, n\}$, where

$$F_J = \{\mathbf{x} \in (\mathbb{R}_0^+)^n \ : \ x_i = 0, \ i \notin J\}.$$

   Further, $P \geq^K \mathbf{0}$ ($P \gneq^K \mathbf{0}$, $P >^K \mathbf{0}$, P is K-irreducible) if and only if $P \geq \mathbf{0}$ ($P \gneq \mathbf{0}$, $P > \mathbf{0}$, P is irreducible) in the sense used for nonnegative matrices, cf. Chapter 9.
2. The nontrivial faces of the Lorentz (ice cream) cone K_n in $\mathbb{R}^n$, *viz.*

$$K_n = \{\mathbf{x} \in \mathbb{R}^n : (x_1^2 + \cdots + x_{n-1}^2)^{1/2} \leq x_n\},$$

   are precisely its extreme rays, each generated by a nonzero boundary vector, that is, one for which the equality holds above. The matrix

$$P = \begin{bmatrix} -1 & 0 & 0 \\ 0 & 0 & 0 \\ 0 & 0 & 1 \end{bmatrix}$$

   is K_3-irreducible [BP79, p. 22].

26.2 Collatz–Wielandt Sets and Distinguished Eigenvalues

Collatz–Wielandt sets were apparently first defined in [BS75]. However, they are so-called because they are closely related to Wielandt's proof of the Perron–Frobenius theorem for irreducible nonnegative matrices, [Wie50], which employs an inequality found in Collatz [Col42]. See also [Sch96] for further remarks on Collatz–Wielandt sets and related max-min and min-max characterizations of the spectral radius of nonnegative matrices and their generalizations.

Definitions:

Let P be a K-nonnegative matrix.

The **Collatz–Wielandt sets** associated with P ([BS75], [TW89], [TS01], [TS03], and [Tam01]) are defined by

$$\begin{aligned}\Omega(P) &= \{\omega \geq 0 :\ \exists \mathbf{x} \in K\backslash\{\mathbf{0}\},\ P\mathbf{x} \geq^K \omega\mathbf{x}\}.\\ \Omega_1(P) &= \{\omega \geq 0 :\ \exists \mathbf{x} \in \text{int } K,\ P\mathbf{x} \geq^K \omega\mathbf{x}\}.\\ \Sigma(P) &= \{\sigma \geq 0 :\ \exists \mathbf{x} \in K\backslash\{\mathbf{0}\},\ P\mathbf{x} \,{}^K\!\!\leq \sigma\mathbf{x}\}.\\ \Sigma_1(P) &= \{\sigma \geq 0 :\ \exists \mathbf{x} \in \text{int } K,\ P\mathbf{x} \,{}^K\!\!\leq \sigma\mathbf{x}\}.\end{aligned}$$

For a K-nonnegative vector $\mathbf{x}$, the **lower and upper Collatz–Wielandt numbers** of $\mathbf{x}$ with respect to P are defined by

$$\begin{aligned}r_P(x) &= \sup\ \{\omega \geq 0 :\ P\mathbf{x} \geq^K \omega\mathbf{x}\},\\ R_P(x) &= \inf\ \{\sigma \geq 0 :\ P\mathbf{x} \,{}^K\!\!\leq \sigma\mathbf{x}\},\end{aligned}$$

where we write $R_P(\mathbf{x}) = \infty$ if no σ exists such that $P\mathbf{x} \,{}^K\!\!\leq \sigma\mathbf{x}$.

A (nonnegative) eigenvalue of P is a **distinguished eigenvalue** for K if it has an associated K-semipositive eigenvector.

The **Perron space** $N_\rho^\nu(P)$ (or N_ρ^ν) is the subspace consisting of all $\mathbf{u} \in \mathbb{R}^n$ such that $(P - \rho I)^k\mathbf{u} = \mathbf{0}$ for some positive integer k. (See Chapter 6.1 for a more general definition of $N_\lambda^\nu(A)$.)

If F is a P-invariant face of K, then the restriction of P to span$F$ is written as $P|F$. The spectral radius of $P|F$ is written as $\rho[F]$, and if λ is an eigenvalue of $P|F$, its index is written as $\nu_\lambda[F]$.

A cone K in $\mathbb{R}^n$ is **polyhedral** if it is the set of linear combinations with nonnegative coefficients of vectors taken from a finite subset of $\mathbb{R}^n$, and is **simplicial** if the finite subset is linearly independent.

Facts:

Let P be a K-nonnegative matrix.

1. [TW89] A real number λ is a distinguished eigenvalue of P for K if and only if $\lambda = \rho_{\mathbf{b}}(P)$ for some K-semipositive vector $\mathbf{b}$.
2. [Tam90] Consider the following conditions:

   (a) ρ is the only distinguished eigenvalue of P for K.

   (b) $\mathbf{x} \geq^K \mathbf{0}$ and $P\mathbf{x} \,{}^K\!\!\leq \rho\mathbf{x}$ imply that $P\mathbf{x} = \rho\mathbf{x}$.

   (c) The Perron space of P^T contains a K^*-positive vector.

   (d) $\rho \in \Omega_1(P^T)$.

   Conditions (a), (b), and (c) are always equivalent and are implied by condition (d). When K is polyhedral, condition (d) is also an equivalent condition.
3. [Tam90] The following conditions are equivalent:

   (a) $\rho(P)$ is the only distinguished eigenvalue of P for K and the index of $\rho(P)$ is one.

   (b) For any vector $\mathbf{x} \in \mathbb{R}^n$, $P\mathbf{x} \,{}^K\!\!\leq \rho(P)\mathbf{x}$ implies that $P\mathbf{x} = \rho(P)\mathbf{x}$.

   (c) $K \cap (\rho I - P)\mathbb{R}^n = \{\mathbf{0}\}$.

   (d) P^T has a K^*-positive eigenvector (corresponding to $\rho(P)$).
4. [TW89] The following statements all hold:

   (a) [BS75] If P is K-irreducible, then
   $\sup\ \Omega(P) = \sup\ \Omega_1(P) = \inf\ \Sigma(P) = \inf\ \Sigma_1(P) = \rho(P)$.

   (b) $\sup\ \Omega(P) = \inf\ \Sigma_1(P) = \rho(P)$.

   (c) $\inf\ \Sigma(P)$ is equal to the least distinguished eigenvalue of P for K.

(d) sup $\Omega_1(P) = \inf \Sigma(P^T)$ and, hence, is equal to the least distinguished eigenvalue of P^T for K^*.

(e) sup $\Omega(P) \in \Omega(P)$ and inf $\Sigma(P) \in \Sigma(P)$.

(f) When K is polyhedral, we have sup $\Omega_1(P) \in \Omega_1(P)$. For general cones, we may have sup $\Omega_1(P) \notin \Omega_1(P)$.

(g) [Tam90] When K is polyhedral, $\rho(P) \in \Omega_1(P)$ if and only if $\rho(A)$ is the only distinguished eigenvalue of P^T for K^*.

(h) [TS03] $\rho(P) \in \Sigma_1(P)$ if and only if $\Phi((N^1_\rho(P) \cap K) \cup C) = K$, where C is the set $\{\mathbf{x} \in K : \rho_{\mathbf{x}}(P) < \rho(P)\}$ and $N^1_\rho(P)$ is the Perron eigenspace of P.

5. In the irreducible nonnegative matrix case, statement (b) of the preceding fact reduces to the well-known max-min and min-max characterizations of $\rho(P)$ due to Wielandt. Schaefer [Sfr84] generalized the result to irreducible compact operators in L^p-spaces and more recently Friedland [Fri90], [Fri91] also extended the characterizations in the settings of a Banach space or a C^*-algebra.
6. [TW89, Theorem 2.4(i)] For any $\mathbf{x} \geq^K \mathbf{0}, r_P(\mathbf{x}) \leq \rho_{\mathbf{x}}(P) \leq R_P(\mathbf{x})$. (This fact extends the well-known inequality $r_P(\mathbf{x}) \leq \rho(P) \leq R_P(\mathbf{x})$ in the nonnegative matrix case, due to Collatz [Col42] under the assumption that $\mathbf{x}$ is a positive vector and due to Wielandt [Wie50] under the assumption that P is irreducible and $\mathbf{x}$ is semipositive. For similar results concerning a nonnegative linear continuous operator in a Banach space, see [FN89].)
7. A discussion on estimating $\rho(P)$ or $\rho_x(P)$ by a convergent sequence of (lower or upper) Collatz–Wielandt numbers can be found in [TW89, Sect. 5] and [Tam01, Subsect. 3.1.4].
8. [GKT95, Corollary 3.2] If K is strictly convex (i.e., each boundary vector is extreme), then P has at most two distinguished eigenvalues. This fact supports the statement that the spectral theory of nonnegative linear operators depends on the geometry of the underlying cone.

26.3 The Peripheral Spectrum, the Core, and the Perron–Schaefer Condition

In addition to using Collatz–Wielandt sets to study Perron–Frobenius theory, we may also approach this theory by considering the core (whose definition will be given below). This geometric approach started with the work of Pullman [Pul71], who succeeded in rederiving the Frobenius theorem for irreducible nonnegative matrices. Naturally, this approach was also taken up in geometric spectral theory. It was found that there are close connections between the core, the peripheral spectrum, the Perron–Schaefer condition, and the distinguished faces of a K-nonnegative linear operator. This led to a revival of interest in the Perron–Schaefer condition and associated conditions for the existence of a cone K such that a preassigned matrix is K-nonnegative. (See [Bir67], [Sfr66], [Van68], [Sch81].) The study has also led to the identification of necessary and equivalent conditions for a collection of Jordan blocks to correspond to the peripheral eigenvalues of a nonnegative matrix. (See [TS94] and [McD03].) The local Perron–Schaefer condition was identified in [TS01] and has played a role in the subsequent work. In the course of this investigation, methods were found for producing invariant cones for a matrix with the Perron–Schaefer condition, see [TS94], [Tam06]. These constructions may also be useful in the study of allied fields, such as linear dynamical systems. There invariant cones for matrices are often encountered. (See, for instance, [BNS89].)

Definitions:

If P is K-nonnegative, then a nonzero P-invariant face F of K is a **distinguished face** (associated with λ) if for every P-invariant face G, with $G \subset F$, we have $\rho[G] < \rho[F]$ (and $\rho[F] = \lambda$).

If λ is an eigenvalue of $A \in \mathbf{C}^{n\times n}$, then $\ker(A - \lambda I)^k$ is denoted by $N^k_\lambda(A)$ for $k = 1, 2, \ldots$, the **index** of λ is denoted by $\nu_A(\lambda)$ (or ν_λ when A is clear), and the **generalized eigenspace** at λ is denoted by $N^{\nu}_\lambda(A)$. See Chapter 6.1 for more information.

Let $A \in \mathbb{C}^{n\times n}$.

The **order** of a generalized eigenvector $\mathbf{x}$ for λ is the smallest positive integer k such that $(A - \lambda I)^k\mathbf{x} = \mathbf{0}$. The maximal order of all K-semipositive generalized eigenvectors in $N_\lambda^\nu(A)$ is denoted by ord_λ.

The matrix A satisfies the **Perron–Schaefer condition** ([Sfr66], [Sch81]) if

- $\rho = \rho(A)$ is an eigenvalue of A.
- If λ is an eigenvalue of A and $|\lambda| = \rho$, then $\nu_A(\lambda) \leq \nu_A(\rho)$.

If K is a cone and P is K-nonnegative, then the set $\bigcap_{i=0}^{\infty} P^i K$, denoted by $\mathrm{core}_K(P)$, is called the **core** of P relative to K.

An eigenvalue λ of A is called **a peripheral eigenvalue** if $|\lambda| = \rho(A)$. The peripheral eigenvalues of A constitute the **peripheral spectrum** of A.

Let $\mathbf{x} \in \mathbb{C}^n$. Then A satisfies the **local Perron–Schaefer condition at x** if there is a generalized eigenvector $\mathbf{y}$ of A corresponding to $\rho_{\mathbf{x}}(A)$ that appears as a term in the representation of $\mathbf{x}$ as a sum of generalized eigenvectors of A. Furthermore, the order of $\mathbf{y}$ is equal to the maximum of the orders of the generalized eigenvectors that appear in the representation and correspond to eigenvalues with modulus $\rho_{\mathbf{x}}(A)$.

Facts:

1. [Sfr66, Chap. V] Let K be a cone in $\mathbb{R}^n$ and let P be a K-nonnegative matrix. Then P satisfies the Perron–Schaefer condition.
2. [Sch81] Let K be a cone in $\mathbb{R}^n$ and let P be a K-nonnegative matrix with spectral radius ρ. Then P has at least m linearly independent K-semipositive eigenvectors corresponding to ρ, where m is the number of Jordan blocks in the Jordan form of P of maximal size that correspond to ρ.
3. [Van68] Let $A \in \mathbb{R}^{n\times n}$. Then there exists a cone K in $\mathbb{R}^n$ such that A is K-nonnegative if and only if A satisfies the Perron–Schaefer condition.
4. [TS94] Let $A \in \mathbb{R}^{n\times n}$ that satisfies the Perron–Schaefer condition. Let m be the number of Jordan blocks in the Jordan form of A of maximal size that correspond to $\rho(A)$. Then for each positive integer k, $m \leq k \leq \dim N_\rho^1(A)$, there exists a cone K in $\mathbb{R}^n$ such that A is K-nonnegative and dim span$(N_\rho^1(A) \cap K) = k$.
5. Let $A \in \mathbb{R}^{n\times n}$. Let k be a nonnegative integer and let $\omega_k(A)$ consist of all linear combinations with nonnegative coefficients of $A^k, A^{k+1}, \ldots$. The closure of $\omega_k(A)$ is a cone in its linear span if and only if A satisfies the Perron–Schaefer condition. (For this fact in the setting of complex matrices see [Sch81].)
6. Necessary and sufficient conditions involving $\omega_k(A)$ so that $A \in \mathbb{C}^{n\times n}$ has a positive (nonnegative) eigenvalue appear in [Sch81]. For the corresponding real versions, see [Tam06].
7. [Pul71], [TS94] If K is a cone and P is K-nonnegative, then $\mathrm{core}_K(P)$ is a cone in its linear span and $P(\mathrm{core}_K(P)) = \mathrm{core}_K(P)$. Furthermore, $\mathrm{core}_K(P)$ is polyhedral (or simplicial) whenever K is. So when $\mathrm{core}_K(P)$ is polyhedral, P permutes the extreme rays of $\mathrm{core}_K(P)$.
8. For a K-nonnegative matrix P, a characterization of K-irreducibility (as well as K-primitivity) of P in terms of $\mathrm{core}_K(P)$, which extends the corresponding result of Pullman for a nonnegative matrix, can be found in [TS94].
9. [Pul71] If P is an irreducible nonnegative matrix, then the permutation induced by P on the extreme rays of $\mathrm{core}_{(\mathbb{R}_0^+)^n}(P)$ is a single cycle of length equal to the number of distinct peripheral eigenvalues of P. (This fact can be regarded as a geometric characterization of the said quantity (cf. the known combinatorial characterization, see Fact 5(c) of Chapter 9.2), whereas part (b) of the next fact is its extension.)
10. [TS94, Theorem 3.14] For a K-nonnegative matrix P, if $\mathrm{core}_K(P)$ is a nonzero simplicial cone, then:

    (a) There is a one-to-one correspondence between the set of distinguished faces associated with nonzero eigenvalues and the set of cycles of the permutation τ_P induced by P on the extreme rays of $\mathrm{core}_K(P)$.

(b) If σ is a cycle of the induced permutation ι_P, then the peripheral eigenvalues of the restriction of P to the linear span of the distinguished P-invariant face F corresponding to σ are simple and are exactly $\rho[F]$ times all the d_σth roots of unity, where d_σ is the length of the cycle σ.

11. [TS94] If P is K-nonnegative and $\text{core}_K(P)$ is nonzero polyhedral, then:

    (a) $\text{core}_K(P)$ consists of all linear combinations with nonnegative coefficients of the distinguished eigenvectors of positive powers of P corresponding to nonzero distinguished eigenvalues.

    (b) $\text{core}_K(P)$ does not contain a generalized eigenvector of any positive powers of P other than eigenvectors.

    This fact indicates that we cannot expect that the index of the spectral radius of a nonnegative linear operator can be determined from a knowledge of its core.
12. A complete description of the core of a nonnegative matrix (relative to the nonnegative orthant) can be found in [TS94, Theorem 4.2].
13. For $A \in \mathbb{R}^{n\times n}$, in order that there exists a cone K in $\mathbb{R}^n$ such that $AK = K$ and A has a K-positive eigenvector, it is necessary and sufficient that A is nonzero, diagonalizable, all eigenvalues of A are of the same modulus, and $\rho(A)$ is an eigenvalue of A. For further equivalent conditions, see [TS94, Theorem 5.9].
14. For $A \in \mathbb{R}^{n\times n}$, an equivalent condition given in terms of the peripheral eigenvalues of A so that there exists a cone K in $\mathbb{R}^n$ such that A is K-nonnegative and (a) K is polyhedral, or (b) $\text{core}_K(A)$ is polyhedral (simplicial or a single ray) can be found in [TS94, Theorems 7.9, 7.8, 7.12, 7.10].
15. [TS94, Theorem 7.12] Let $A \in \mathbb{R}^{n\times n}$ with $\rho(A) > 0$ that satisfies the Perron–Schaefer condition. Let S denote the multiset of peripheral eigenvalues of A with maximal index (i.e., $\nu_A(\rho)$), the multiplicity of each element being equal to the number of corresponding blocks in the Jordan form of A of order $\nu_A(\rho)$. Let T be the multiset of peripheral eigenvalues of A for which there are corresponding blocks in the Jordan form of A of order less than $\nu_A(\rho)$, the multiplicity of each element being equal to the number of such corresponding blocks. The following conditions are equivalent:

    (a) There exists a cone K in $\mathbb{R}^n$ such that A is K-nonnegative and $\text{core}_K(A)$ is simplicial.

    (b) There exists a multisubset $\widetilde{T}$ of T such that $S \cup \widetilde{T}$ is the multiset union of certain complete sets of roots of unity multiplied by $\rho(A)$.

16. McDonald [McD03] refers to the condition (b) that appears in the preceding result as the **Tam–Schneider condition**. She also provides another condition, called the **extended Tam–Schneider condition**, which is necessary and sufficient for a collection of Jordan blocks to correspond to the peripheral spectrum of a nonnegative matrix.
17. [TS01] If P is K-nonnegative and $\mathbf{x}$ is K-semipositive, then P satisfies the local Perron–Schaefer condition at $\mathbf{x}$.
18. [Tam06] Let A be an $n \times n$ real matrix, and let $\mathbf{x}$ be a given nonzero vector of $\mathbb{R}^n$. The following conditions are equivalent:

    (a) A satisfies the local Perron–Schaefer condition at $\mathbf{x}$.

    (b) The restriction of A to $\text{span}\{A^i\mathbf{x} : i = 0, 1, \ldots\}$ satisfies the Perron–Schaefer condition.

    (c) For every (or, for some) nonnegative integer k, the closure of $\omega_k(A, \mathbf{x})$, where $\omega_k(A, \mathbf{x})$ consists of all linear combinations with nonnegative coefficients of $A^k\mathbf{x}$, $A^{k+1}\mathbf{x}, \ldots$, is a cone in its linear span.

    (d) There is a cone C in a subspace of $\mathbb{R}^n$ containing $\mathbf{x}$ such that $AC \subseteq C$.
19. The local Perron–Schaefer condition has played a role in the work of [TS01], [TS03], and [Tam04]. Further work involving this condition and the cones $\omega_k(A, \mathbf{x})$ (defined in the preceding fact) will appear in [Tam06].
20. One may apply results on the core of a nonnegative matrix to rederive simply many known results on the limiting behavior of Markov chains. An illustration can be found in [Tam01, Sec. 4.6].

26.4 Spectral Theory of K-Reducible Matrices

In this section, we touch upon the geometric version of the extensive combinatorial spectral theory of reducible nonnegative matrices first found in [Fro12, Sect. 11] and continued in [Sch56]. Many subsequent developments are reviewed in [Sch86] and [Her99]. Results on the geometric spectral theory of reducible K-nonnegative matrices may be largely found in a series of papers by B.S. Tam, some jointly with Wu and H. Schneider ([TW89], [Tam90], [TS94], [TS01], [TS03], [Tam04]). For a review containing considerably more information than this section, see [Tam01].

In some studies, the underlying cone is lattice-ordered (for a definition and much information, see [Sfr74]) and, in some studies, the Frobenius form of a reducible nonnegative matrix is generalized; see the work by Jang and Victory [JV93] on positive eventually compact linear operators on Banach lattices. However in the geometric spectral theory the Frobenius normal form of a nonnegative reducible matrix is not generalized as the underlying cone need not be lattice-ordered. Invariant faces are considered instead of the classes that play an important role in combinatorial spectral theory of nonnegative matrices; in particular, distinguished faces and semidistinguished faces are used in place of distinguished classes and semidistinguished classes, respectively. (For definitions of the preceding terms, see [TS01].)

It turns out that the various results on a reducible nonnegative matrix are extended to a K-nonnegative matrix in different degrees of generality. In particular, the Frobenius–Victory theorem ([Fro12], [Vic85]) is extended to a K-nonnegative matrix on a general cone. The following are extended to a polyhedral cone: The Rothblum index theorem ([Rot75]), a characterization (in terms of the accessibility relation between basic classes) for the spectral radius to have geometric multiplicity 1, for the spectral radius to have index 1 ([Sch56]), and a majorization relation between the (spectral) height characteristic and the (combinatorial) level characteristic of a nonnegative matrix ([HS91b]). Various conditions are used to generalize the theorem on equivalent conditions for equality of the two characteristics ([RiS78], [HS89], [HS91a]). Even for polyhedral cones there is no complete generalization for the nonnegative-basis theorem, not to mention the preferred-basis theorem ([Rot75], [RiS78], [Sch86], [HS88]). There is a natural conjecture for the latter case ([Tam04]). The attempts to carry out the extensions have also led to the identification of important new concepts or tools. For instance, the useful concepts of semidistinguished faces and of spectral pairs of faces associated with a K-nonnegative matrix are introduced in [TS01] in proving the cone version of some of the combinatorial theorems referred to above. To achieve these ends certain elementary analytic tools are also brought in.

Definitions:

Let P be a K-nonnegative matrix.

A nonzero P-invariant face F is a **semidistinguished face** if F contains in its relative interior a generalized eigenvector of P and if F is not the join of two P-invariant faces that are properly included in F.

A K-**semipositive Jordan chain** for P of length m (corresponding to $\rho(P)$) is a sequence of m K-semipositive vectors $\mathbf{x}, (P - \rho(P)I)\mathbf{x}, \ldots, (P - \rho(P)I)^{m-1}\mathbf{x}$ such that $(P - \rho(P)I)^m\mathbf{x} = \mathbf{0}$.

A basis for $N_\rho^\nu(P)$ is called a K-**semipositive basis** if it consists of K-semipositive vectors.

A basis for $N_\rho^\nu(P)$ is called a K-**semipositive Jordan basis** for P if it is composed of K-semipositive Jordan chains for P.

The set $C(P, K) = \{\mathbf{x} \in K : (P - \rho(P)I)^i\mathbf{x} \in K$ for all positive integers $i\}$ is called the **spectral cone** of P (for K corresponding to $\rho(P)$).

Denote ν_ρ by ν.

The **height characteristic of** P is the ν-tuple $\eta(P) = (\eta_1, \ldots, \eta_\nu)$ given by:

$$\eta_k = \dim(N_\rho^k(P)) - \dim(N_\rho^{k-1}(P)).$$

The **level characteristic of** P is the ν-tuple $\lambda(P) = (\lambda_1, \ldots, \lambda_\nu)$ given by:

$$\lambda_k = \dim \operatorname{span}(N_\rho^k(P) \cap K) - \dim \operatorname{span}(N_\rho^{k-1}(P) \cap K).$$

The **peak characteristic of** P is the ν-tuple $\xi(P) = (\xi_1, ..., \xi_\nu)$ given by:

$$\xi_k = \dim(P - \rho(P)I)^{k-1}(N_\rho^k \cap K).$$

If $A \in \mathbb{C}^{n\times n}$ and $\mathbf{x}$ is a nonzero vector of $\mathbb{C}^n$, then the **order of x relative to** A, denoted by $\text{ord}_A(\mathbf{x})$, is defined to be the maximum of the orders of the generalized eigenvectors, each corresponding to an eigenvalue of modulus $\rho_{\mathbf{x}}(A)$ that appear in the representation of $\mathbf{x}$ as a sum of generalized eigenvectors of A.

The ordered pair $(\rho_{\mathbf{x}}(A), \text{ord}_A(\mathbf{x}))$ is called the **spectral pair of x relative to** A and is denoted by $\text{sp}_A(\mathbf{x})$. We also set $\text{sp}_A(\mathbf{0}) = (0, 0)$ to take care of the zero vector $\mathbf{0}$.

We use $\preceq$ to denote the lexicographic ordering between ordered pairs of real numbers, i.e., $(a, b) \preceq (c, d)$ if either $a < c$, or $a = c$ and $b \leq d$. In case $(a, b) \preceq (c, d)$ but $(a, b) \neq (c, d)$, we write $(a, b) \prec (c, d)$.

Facts:

1. If $A \in \mathbb{C}^{n\times n}$ and $\mathbf{x}$ is a vector of $\mathbb{C}^n$, then $\text{ord}_A(\mathbf{x})$ is equal to the size of the largest Jordan block in the Jordan form of the restriction of A to the A-cyclic subspace generated by $\mathbf{x}$ for a peripheral eigenvalue.
   Let P be a K-nonnegative matrix.
2. In the nonnegative matrix case, the present definition of the level characteristic of P is equivalent to the usual graph-theoretic definition; see [NS94, (3.2)] or [Tam04, Remark 2.2].
3. [TS01] For any $\mathbf{x} \in K$, the smallest P-invariant face containing $\mathbf{x}$ is equal to $\Phi(\hat{\mathbf{x}})$, where $\hat{\mathbf{x}} = (I + P)^{n-1}\mathbf{x}$. Furthermore, $\text{sp}_P(\mathbf{x}) = \text{sp}_P(\hat{\mathbf{x}})$. In the nonnegative matrix case, the said face is also equal to F_J, where F_J is as defined in Example 1 of Section 26.1 and J is the union of all classes of P having access to $\text{supp}(\mathbf{x}) = \{i : \mathbf{x}_i > 0\}$. (For definitions of classes and the accessibility relation, see Chapter 9.)
4. [TS01] For any face F of K, P-invariant or not, the value of the spectral pair $\text{sp}_P(\mathbf{x})$ is independent of the choice of $\mathbf{x}$ from the relative interior of F. This common value, denoted by $\text{sp}_A(F)$, is referred to as the **spectral pair of** F **relative to** A.
5. [TS01] For any faces F, G of K, we have
   (a) $\text{sp}_P(F) = \text{sp}_P(\hat{F})$, where $\hat{F}$ is the smallest P-invariant face of K, including F.
   (b) If $F \subseteq G$, then $\text{sp}_P(F) \preceq \text{sp}_P(G)$. If F, G are P-invariant faces and $F \subset G$, then $\text{sp}_P(F) \preceq \text{sp}_P(G)$; *viz.* either $\rho[F] < \rho[G]$ or $\rho[F] = \rho[G]$ and $\nu_{\rho[F]}[F] \leq \nu_{\rho[G]}[G]$.
6. [TS01] If K is a cone with the property that the dual cone of each of its faces is a facially exposed cone, for instance, when K is a polyhedral cone, a perfect cone, or equals $P(n)$ (see [TS01] for definitions), then for any nonzero P-invariant face G, G is semidistinguished if and only if $\text{sp}_P(F) \prec \text{sp}_P(G)$ for all P-invariant faces F properly included in G.
7. [Tam04] (Cone version of the Frobenius–Victory theorem, [Fro12], [Vic85], [Sch86])
   (a) For any real number λ, λ is a distinguished eigenvalue of P if and only if $\lambda = \rho[F]$ for some distinguished face F of K.
   (b) If F is a distinguished face, then there is (up to multiples) a unique eigenvector $\mathbf{x}$ of P corresponding to $\rho[F]$ that lies in F. Furthermore, $\mathbf{x}$ belongs to the relative interior of F.
   (c) For each distinguished eigenvalue λ of P, the extreme vectors of the cone $N_\lambda^1(P) \cap K$ are precisely all the distinguished eigenvectors of P that lie in the relative interior of certain distinguished faces of K associated with λ.
8. Let P be a nonnegative matrix. The Jordan form of P contains only one Jordan block corresponding to $\rho(P)$ if and only if any two basic classes of P are comparable (with respect to the accessibility relation); all Jordan blocks corresponding to $\rho(P)$ are of size 1 if and only if no two basic classes are comparable ([Schn56]). An extension of these results to a K-nonnegative matrix on a class of cones that contains all polyhedral cones can be found in [TS01, Theorems 7.2 and 7.1].

9. [Tam90, Theorem 7.5] If K is polyhedral, then:

   (a) There is a K-semipositive Jordan chain for P of length ν_ρ; thus, there is a K-semipositive vector in $N_\rho^\nu(P)$ of order ν_ρ, *viz.* $\text{ord}_\rho = \nu_\rho$.

   (b) The Perron space $N_\rho^\nu(P)$ has a basis consisting of K-semipositive vectors.

   However, when K is nonpolyhedral, there need not exist a K-semipositive vector in $N_\rho^\nu(P)$ of order ν_ρ, *viz.* $\text{ord}_\rho < \nu_\rho$. For a general distinguished eigenvalue λ, we always have $\text{ord}_\lambda \leq \nu_\lambda$, no matter whether K is polyhedral or not.
10. Part (b) of the preceding fact is not yet a complete cone version of the nonnegative-basis theorem, as the latter theorem guarantees the existence of a basis for the Perron space that consists of semipositive vectors that satisfy certain combinatorial properties. For a conjecture on a cone version of the nonnegative-basis theorem, see [Tam04, Conj. 9.1].
11. [TS01, Theorem 5.1] (Cone version of the (combinatorial) generalization of the Rothblum index theorem, [Rot75], [HS88]).

    Let K be a polyhedral cone. Let λ be a distinguished eigenvalue of P for K. Then there is a chain $F_1 \subset F_2 \subset \ldots \subset F_k$ of $k = \text{ord}_\lambda$ distinct semidistinguished faces of K associated with λ, but there is no such chain with more than ord_λ members. When K is a general cone, the maximum cardinality of a chain of semidistinguished faces associated with a distinguished eigenvalue λ may be less than, equal to, or greater than ord_λ; see [TS01, Ex. 5.3, 5.4, 5.5].
12. For $K = (\mathbb{R}_0^+)^n$, *viz.* P is a nonnegative matrix, characterizations of different types of P-invariant faces (in particular, the distinguished and semidistinguished faces) are given in [TS01] (in terms of the concept of an initial subset for P; see [HS88] or [TS01] for definition of an initial subset).
13. [Tam04] The spectral cone $C(P, K)$ is always invariant under $P - \rho(P)I$ and satisfies:

$$N_\rho^1(P) \cap K \subseteq C(A, K) \subseteq N_\rho^\nu(P) \cap K.$$

    If K is polyhedral, then $C(A, K)$ is a polyhedral cone in $N_\rho^\nu(P)$.
14. (Generalization of corresponding results on nonnegative matrices, [NS94]) We always have $\xi_k(P) \leq \eta_k(P)$ and $\xi_k(P) \leq \lambda_k(P)$ for $k = 1, \ldots, \nu_\rho$.
15. [Tam04, Theorem 5.9] Consider the following conditions:

   (a) $\eta(P) = \lambda(P)$.

   (b) $\eta(P) = \xi(P)$.

   (c) For each k, $k = 1, \ldots, \nu_\rho$, $N_\rho^k(P)$ contains a K-semipositive basis.

   (d) There exists a K-semipositive Jordan basis for P.

   (e) For each k, $k = 1, \ldots, \nu_\rho$, $N_\rho^k(P)$ has a basis consisting of vectors taken from $N_\rho^k(P) \cap C(P, K)$.

   (f) For each k, $k = 1, \ldots, \nu_\rho$, we have

$$\eta_k(P) = \dim(P - \rho(P)I)^{k-1}[N_\rho^k(P) \cap C(P, K)].$$

    Conditions (a) to (c) are equivalent and so are conditions (d) to (f). Moreover, we always have (a)$\Longrightarrow$(d), and when K is polyhedral, conditions (a) to (f) are all equivalent.
16. As shown in [Tam04], the level of a nonzero vector $\mathbf{x} \in N_\rho^\nu(P)$ can be defined to be the smallest positive integer k such that $\mathbf{x} \in \text{span}(N_\rho^k(P) \cap K)$; when there is no such k the level is taken to be ∞. Then the concepts of K-semipositive level basis, height-level basis, peak vector, etc., can be introduced and further conditions can be added to the list given in the preceding result.
17. [Tam04, Theorem 7.2] If K is polyhedral, then $\lambda(P) \preceq \eta(P)$.
18. Cone-theoretic proofs for the preferred-basis theorem for a nonnegative matrix and for a result about the nonnegativity structure of the principal components of a nonnegative matrix can be found in [Tam04].

26.5 Linear Equations over Cones

Given a K-nonnegative matrix P and a vector $\mathbf{b} \in K$, in this section we consider the solvability of following two linear equations over cones and some consequences:

$$(\lambda I - P)\mathbf{x} = \mathbf{b}, \quad \mathbf{x} \in K \tag{26.1}$$

and

$$(P - \lambda I)\mathbf{x} = \mathbf{b}, \quad \mathbf{x} \in K. \tag{26.2}$$

Equation (26.1) has been treated by several authors in finite-dimensional as well as infinite-dimensional settings, and several equivalent conditions for its solvability have been found. (See [TS03] for a detailed historical account.) The study of Equation (26.2) is relatively new. A treatment of the equation by graph-theoretic arguments for the special case when $\lambda = \rho(P)$ and $K = (\mathbb{R}_0^+)^n$ can be found in [TW89]. The general case is considered in [TS03]. It turns out that the solvability of Equation (26.2) is a more delicate problem. It depends on whether λ is greater than, equal to, or less than $\rho_{\mathbf{b}}(P)$.

Facts:

Let P be a K-nonnegative matrix, let $\mathbf{0} \neq \mathbf{b} \in \mathbf{K}$, and let λ be a given positive real number.

1. [TS03, Theorem 3.1] The following conditions are equivalent:

   (a) Equation (26.1) is solvable.

   (b) $\rho_{\mathbf{b}}(P) < \lambda$.

   (c) $\lim_{m\to\infty} \sum_{j=0}^{m} \lambda^{-j} P^j \mathbf{b}$ exists.

   (d) $\lim_{m\to\infty} (\lambda^{-1} P)^m \mathbf{b} = \mathbf{0}$.

   (e) $\langle \mathbf{z}, \mathbf{b} \rangle = 0$ for each generalized eigenvector $\mathbf{z}$ of P^T corresponding to an eigenvalue with modulus greater than or equal to λ.

   (f) $\langle \mathbf{z}, \mathbf{b} \rangle = 0$ for each generalized eigenvector $\mathbf{z}$ of P^T corresponding to a distinguished eigenvalue of P for K that is greater than or equal to λ.

2. For a fixed λ, the set $(\lambda I - P)K \cap K$, which consists of precisely all vectors $\mathbf{b} \in K$ for which Equation (26.1) has a solution, is equal to $\{\mathbf{b} \in K : \rho_{\mathbf{b}}(P) < \lambda\}$ and is a face of K.
3. For a fixed λ, the set $(P - \lambda I)K \cap K$, which consists of precisely all vectors $\mathbf{b} \in K$ for which Equation (26.2) has a solution, is, in general, not a face of K.
4. [TS03, Theorem 4.1] When $\lambda > \rho_{\mathbf{b}}(P)$, Equation (26.2) is solvable if and only if λ is a distinguished eigenvalue of P for K and $\mathbf{b} \in \Phi(N_\lambda^1(P) \cap K)$.
5. [TS03, Theorem 4.5] When $\lambda = \rho_{\mathbf{b}}(P)$, if Equation (26.2) is solvable, then
   $\mathbf{b} \in (P - \rho_{\mathbf{b}}(P)I)\Phi$
   $(N^{\nu}_{\rho_{\mathbf{b}}(P)}(P) \cap K)$.
6. [TS03, Theorem 4.19] Let r denote the largest real eigenvalue of P less than $\rho(P)$. (If no such eigenvalues exist, take $r = -\infty$.) Then for any λ, $r < \lambda < \rho(P)$, we have

   $$\Phi((P - \lambda I)K \cap K) = \Phi(N_\rho^\nu(P) \cap K).$$

   Thus, a necessary condition for Equation (26.2) to have a solution is that $\mathbf{b} \overset{K}{\leq} \mathbf{u}$ for some $\mathbf{u} \in N_\rho^\nu(P) \cap K$.
7. [TS03, Theorem 5.11] Consider the following conditions:

   (a) $\rho(P) \in \Sigma_1(P^T)$.

   (b) $N_\rho^\nu(P) \cap K = N_\rho^1(P) \cap K$, and P has no eigenvectors in $\Phi(N_\rho^1(P) \cap K)$ corresponding to an eigenvalue other than $\rho(P)$.

(c) $K \cap (P - \rho(P)I)K = \{\mathbf{0}\}$ (equivalently, $\mathbf{x} \geq^K \mathbf{0}$, $P\mathbf{x} \geq^K \rho(P)\mathbf{x}$ imply that $P\mathbf{x} = \rho(P)\mathbf{x}$).
We always have (a)$\Longrightarrow$ (b)$\Longrightarrow$(c). When K is polyhedral, conditions (a), (b), and (c) are equivalent. When K is nonpolyhedral, the missing implications do not hold.

26.6 Elementary Analytic Results

In geometric spectral theory, besides the linear-algebraic method and the cone-theoretic method, certain elementary analytic methods have also been called into play; for example, the use of Jordan form or the components of a matrix. This approach may have begun with the work of Birkhoff [Bir67] and it was followed by Vandergraft [Van68] and Schneider [Sch81]. Friedland and Schneider [FS80] and Rothblum [Rot81] have also studied the asymptotic behavior of the powers of a nonnegative matrix, or their variants, by elementary analytic methods. The papers [TS94] and [TS01] in the series also need a certain kind of analytic argument in their proofs; more specifically, they each make use of the K-nonnegativity of a certain matrix, either itself a component or a matrix defined in terms of the components of a given K-nonnegative matrix (see Facts 3 and 4 in this section). In [HNR90], Hartwig, Neumann, and Rose offer a (linear) algebraic-analytic approach to the Perron–Frobenius theory of a nonnegative matrix, one which utilizes the resolvent expansion, but does not involve the Frobenius normal form. Their approach is further developed by Neumann and Schneider ([NS92], [NS93], [NS94]). By employing the concept of spectral cone and combining the cone-theoretic methods developed in the earlier papers of the series with this algebraic-analytic method, Tam [Tam04] offers a unified treatment to reprove or extend (or partly extend) several well-known results in the combinatorial spectral theory of nonnegative matrices. The proofs given in [Tam04] rely on the fact that if K is a cone in $\mathbb{R}^n$, then the set $\pi(K)$ that consists of all K-nonnegative matrices is a cone in the matrix space $\mathbb{R}^{n\times n}$ and if, in addition, K is polyhedral, then so is $\pi(K)$ ([Fen53, p. 22], [SV70], [Tam77]). See [Tam01, Sec. 6.5] and [Tam04, Sec. 9] for further remarks on the use of the cone $\pi(K)$ in the study of the spectral properties of K-nonnegative matrices.

In this section, we collect a few elementary analytic results (whose proofs rely on the Jordan form), which have proved to be useful in the study of the geometric spectral theory. In particular, Facts 3, 4, and 5 identify members of $\pi(K)$. As such, they can be regarded as nice results, which are difficult to come by for the following reason: If K is nonsimplicial, then $\pi(K)$ must contain matrices that are not nonnegative linear combinations of its rank-one members ([Tam77]). However, not much is known about such matrices ([Tam92]).

Definitions:

Let P be a K-nonnegative matrix. Denote ν_ρ by ν.

The **principal eigenprojection** of P, denoted by $Z_P^{(0)}$, is the projection of $\mathbb{C}^n$ onto the Perron space N_ρ^ν along the direct sum of other generalized eigenspaces of P.

For $k = 0, \ldots, \nu$, the **kth principal component of** P is given by

$$Z_P^{(k)} = (P - \rho(P))^k Z_P^{(0)}.$$

The kth component of P corresponding to an eigenvalue λ is defined in a similar way.

For $k = 0, \ldots, \rho$, the **kth transform principal component of** P is given by:

$$J_P^{(k)}(\varepsilon) = Z_P^{(k)} + Z_P^{(k+1)}/\varepsilon + \cdots + Z_P^{(\nu-1)}/\varepsilon^{\nu-k-1} \quad \text{for all } \varepsilon \in \mathbb{C}\backslash\{0\}.$$

Facts:

Let P be a K-nonnegative matrix. Denote ν_ρ by ν.

1. [Kar59], [Sch81] $Z_P^{(\nu-1)}$ is K-nonnegative.
2. [TS94, Theorem 4.19(i)] The sum of the νth components of P corresponding to its peripheral eigenvalues is K-nonnegative; it is the limit of a convergent subsequence of $((\nu-1)!P^k/[\rho^{k-\nu+1}k^{\nu-1}])$.

3. [Tam04, Theorem 3.6(i)] If K is a polyhedral cone, then for $k = 0, \ldots, \nu - 1$, $J_P^{(k)}(\varepsilon)$ is K-nonnegative for all sufficiently small positive ε.

26.7 Splitting Theorems and Stability

Splitting theorems for matrices have played a large role in the study of convergence of iterations in numerical linear algebra; see [Var62]. Here we present a cone version of a splitting theorem which is proven in [Sch65] and applied to stability (inertia) theorems for matrices. A closely related result is generalized to operators on a partially ordered Banach space in [DH03] and [Dam04]. There it is used to describe stability properties of (stochastic) control systems and to derive non-local convergence results for Newton's method applied to nonlinear operator equation of Riccati type. We also discuss several kinds of positivity for operators involving a cone that are relevant to the applications mentioned.

For recent splitting theorems involving cones, see [SSA05]. For applications of theorems of the alternative for cones to the stability of matrices, see [CHS97]. Cones occur in many parts of stability theory; see, for instance, [Her98].

Definitions:

Let K be a cone in $\mathbb{R}^n$ and let $A \in \mathbb{R}^{n\times n}$.

A is **positive stable** if $\text{spec}(A) \subseteq \mathbf{C}^+$ *viz.* the spectrum of A is contained in the open right half-plane.

A is K-**inverse nonnegative** if A is nonsingular and A^{-1} is K-nonnegative.

A is K-**resolvent nonnegative** if there exists an $\alpha_0 \in \mathbb{R}$ such that, for all $\alpha > \alpha_0$, $\alpha I - A$ is K-inverse nonnegative.

A is **cross-positive** on K if for all $\mathbf{x} \in K, \mathbf{y} \in K^*, \mathbf{y}^T\mathbf{x} = 0$ implies $\mathbf{y}^T A\mathbf{x} \geq 0$.

A is a Z-**matrix** if all of its off-diagonal entries are nonpositive.

Facts:

Let K be a cone in $\mathbb{R}^n$.

1. A is K-resolvent nonnegative if and only if A is cross-positive on K. Other equivalent conditions and also Perron-Frobenius type theorems for the class of cross-positive matrices can be found in [Els74], [SV70] or [BNS89].
2. When K is $(\mathbb{R}_0^+)^n$, A is cross-positive on K if and only if $-A$ is a Z-matrix.
3. [Sch65], [Sch97]. Let $T = R - P$ where $R, P \in \mathbb{R}^{n\times n}$ and suppose that P is K-nonnegative. If R satisfies $R(\text{int}K) \supseteq \text{int}K$ or $R(\text{int}K) \cap \text{int}K = \emptyset$, then the following are equivalent:

   (a) T is K-inverse nonnegative.

   (b) For all $\mathbf{y} >^K \mathbf{0}$ there exists (unique) $\mathbf{x} >^K \mathbf{0}$ such that $\mathbf{y} = T\mathbf{x}$.

   (c) There exists $\mathbf{x} >^K \mathbf{0}$ such that $T\mathbf{x} >^K \mathbf{0}$.

   (d) There exists $\mathbf{x} \geq^K \mathbf{0}$ such that $T\mathbf{x} >^K \mathbf{0}$.

   (e) R is K-inverse nonnegative and $\rho(R^{-1}P) < 1$.
4. Let $T \in \mathbb{R}^{n\times n}$. If $-T$ is K-resolvent nonnegative, then T satisfies $T(\text{int}K) \supseteq \text{int}K$ or $T(\text{int}K) \cap \text{int}K = \emptyset$. But the converse is false, see Example 1 below.
5. [DH03, Theorem 2.11], [Dam04, Theorem 3.2.10]. Let T, R, P be as given in Fact 3. If $-R$ is K-resolvent nonnegative, then conditions (a)–(e) of Fact 3 are equivalent. Moreover, the following are additional equivalent conditions:

   (f) T is positive stable.

   (g) R is positive stable and $\rho(R^{-1}P) < 1$.
6. If K is $(\mathbb{R}_0^+)^n$, $R = \alpha I$ and P is a nonnegative matrix, then $T = R - P$ is a Z-matrix. It satisfies the equivalent conditions (a)–(g) of Facts 3 and 5 if and only if it is an M-matrix [BP79, Chapter 6].

7. (Special case of Fact 5 with $P = 0$). Let $T \in \mathbb{R}^{n\times n}$. If $-T$ is K-resolvent nonnegative, then conditions (a)–(d) of Fact 3 and conditions (f) of Fact 5 are equivalent.
8. In [GT06] a matrix T is called a Z-tranformation on K if $-T$ is cross-positive on K. Many properties on Z-matrices, such as being a P-matrix, a Q-matrix (which has connection with the linear complementarity problem), an inverse-nonnegative matrix, a positive stable matrix, a diagonally stable matrix, etc., are extended to Z-transformations. For a Z-transformation, the equivalence of these properties is examined for various kinds of cones, particularly for symmetric cones in Euclidean Jordan algebras.
9. [Schn65], [Schn97]. (Special case of Fact 3 with K equal to the cone of positive semi-definite matrices in the real space of $n \times n$ Hermitian matrices, and $R(H) = AHA^*$, $P(H) = \Sigma_{k=1}^{s} C_k H C_K^*$). Let $A, C_k, k = 1, \ldots, s$ be complex $n \times n$ matrices which can be simultaneously upper triangularized by similarity. Then there exists a natural correspondence $\alpha_i, \gamma_i^{(k)}$ of the eigenvalues of $A, C_k, k = 1, \ldots s$. For Hermitian H, let $T(H) = AHA^* - \Sigma_{k=1}^{s} C_k H C_k^*$. Then the following are equivalent:
   (a) $|\alpha_i|^2 - \Sigma_{k=1}^{s} |\gamma_i^{(k)}|^2 > 0, i = 1, \ldots, n$.
   (b) For all positive definite G there exists a (unique) positive definite H such that $T(H) = G$.
   (c) There exists a positive definite H such that $T(H)$ is positive definite.
10. Gantmacher-Lyapunov [Gan59, Chapter XV] (Special case of Fact 9 with A replaced by $A + I, s = 2, C_1 = A, C_2 = I$, and special case of Fact 7 with K equal to the cone of positive semi-definite matrices in the real space of $n \times n$ Hermitian matrices and $T(H) = AH + HA^*$).
   Let $A \in \mathbb{C}^{n\times n}$. The following are equivalent:
   (a) For all positive definite G there exists a (unique) positive definite H such that $AH + HA^* = G$.
   (b) There exists a positive definite H such that $AH + HA^*$ is positive definite.
   (c) A is positive stable.
11. Stein [Ste52](Special case of Fact 9 with $A = I, s = 1, C_1 = C$, and special case of Fact 7 with $T(H) = H - CHC^*$).
   Let $C \in \mathbb{C}^{n\times n}$. The following are equivalent:
   (a) There exists a positive definite H such that $H - CHC^*$ is positive definite.
   (b) The spectrum of C is contained in the open unit disk.

Examples:

Let $K = (\mathbb{R}_0^+)^2$ and take $T = \begin{bmatrix} 0 & 1 \\ 1 & 0 \end{bmatrix}$. Then $TK = K$ and so $T(\text{int}K) \supseteq \text{int}K$. Note that $\langle T\mathbf{e}_1, \mathbf{e}_2\rangle = 1 > 0$ whereas $\langle \mathbf{e}_1, \mathbf{e}_2\rangle = 0$; so $-T$ is not cross-positive on K and hence not K-resolvent nonnegative. Since the eigenvalues of T are -1 and 1, T is not positive stable. This example tells us that the converse of Fact 4 is false. It also shows that to the list of equivalent conditions of Fact 3 we cannot add condition (f) of Fact 5.

References

[Bir67] G. Birkhoff, Linear transformations with invariant cones, *Amer. Math. Month.* 74 (1967), 274–276.

[BNS89] A. Berman, M. Neumann, and R.J. Stern, *Nonnegative Matrices in Dynamic Systems*, John Wiley & Sons, New York, 1989.

[BP79] A. Berman and R.J. Plemmons, *Nonnegative Matrices in the Mathematical Sciences*, Academic Press, 1979, 2nd ed., SIAM, 1994.

[BS75] G.P. Barker and H. Schneider, Algebraic Perron-Frobenius Theory, *Lin. Alg. Appl.* 11 (1975), 219–233.

[CHS97] B. Cain, D. Hershkowitz, and H. Schneider, Theorems of the alternative for cones and Lyapunov regularity, *Czech. Math. J.* 47(122) (1997), 467–499.

[Col42] L. Collatz, Einschliessungssatz für die charakteristischen Zahlen von Matrizen, *Math. Z.* 48 (1942), 221–226.

[Dam04] T. Damm, *Rational Matrix Equations in Stochastic Control*, Lecture Notes in Control and Information Sciences, 297, Springer, 2004.

[DH03] T. Damm and D. Hinrichsen, Newton's method for concave operators with resolvent positive derivatives in ordered Banach spaces, *Lin. Alg. Appl.* 363 (2003), 43–64.

[Els74] L. Elsner, Quasimonotonie und Ungleichungen in halbgeordneten Räumen, *Lin. Alg. Appl.* 8 (1974), 249–261.

[ERS95] B.C Eaves, U.G. Rothblum, and H. Schneider, Perron-Frobenius theory over real closed ordered fields and fractional power series expansions, *Lin. Alg. Appl.* 220 (1995), 123–150.

[Fen53] W. Fenchel, *Convex Cones, Sets and Functions*, Princeton University Notes, Princeton, NJ, 1953.

[FN89] K.-H. Förster and B. Nagy, On the Collatz–Wielandt numbers and the local spectral radius of a nonnegative operator, *Lin. Alg. Appl.* 120 (1989), 193–205.

[Fri90] S. Friedland, Characterizations of spectral radius of positive operators, *Lin. Alg. Appl.* 134 (1990), 93–105.

[Fri91] S. Friedland, Characterizations of spectral radius of positive elements on C^* algebras, *J. Funct. Anal.* 97 (1991), 64–70.

[Fro12] G.F. Frobenius, Über Matrizen aus nicht negativen Elementen, *Sitzungsber. Kön. Preuss. Akad. Wiss. Berlin* (1912), 456–477, and *Ges. Abh.*, Vol. 3, 546–567, Springer-Verlag, Berlin, 1968.

[FS80] S. Friedland and H. Schneider, The growth of powers of a nonnegative matrix, *SIAM J. Alg. Dis. Meth.* 1 (1980), 185–200.

[Gan59] F.R. Gantmacher, *The Theory of Matrices*, Chelsea, London, 1959.

[GKT95] P. Gritzmann, V. Klee, and B.S. Tam, Cross-positive matrices revisited, *Lin. Alg. Appl.* 223-224 (1995), 285–305.

[GT06] M. Seetharama Gowda and Jiyuan Tao, Z-transformations on proper and symmetric cones, Research Report (#TRGOW06–01, Department of Mathematics and Statistics, University of Maryland, Baltimore County, January 2006).

[HNR90] R.E. Hartwig, M. Neumann, and N.J. Rose, An algebraic-analytic approach to nonnegative basis, *Lin. Alg. Appl.* 133 (1990), 77–88.

[Her98] D. Hershkowitz, On cones with stability, *Lin. Alg. Appl.* 275-276 (1998), 249–259.

[Her99] D. Hershkowitz, The combinatorial structure of generalized eigenspaces — from nonnegative matrices to general matrices, *Lin. Alg. Appl.* 302-303 (1999), 173–191.

[HS88] D. Hershkowitz and H. Schneider, On the generalized nullspace of M-matrices and Z-matrices, *Lin. Alg. Appl.*, 106 (1988), 5–23.

[HS89] D. Hershkowitz and H. Schneider, Height bases, level bases, and the equality of the height and the level characteristic of an M-matrix, *Lin. Multilin. Alg.* 25 (1988), 149–171.

[HS91a] D. Hershkowitz and H. Schneider, Combinatorial bases, derived Jordan sets, and the equality of the height and the level characteristics of an M-matrix, *Lin. Multilin. Alg.* 29 (1991), 21–42.

[HS91b] D. Hershkowitz and H. Schneider, On the existence of matrices with prescribed height and level characteristics, *Israel J. Math.* 75 (1991), 105–117.

[JV93] R.J. Jang and H.D. Victory, Jr., On the ideal structure of positive, eventually compact linear operators on Banach lattices, *Pacific J. Math.* 157 (1993), 57–85.

[Kar59] S. Karlin, Positive operators, *J. Math. and Mech.* 8 (1959), 907–937.

[KR48] M.G. Krein and M.A. Rutman, Linear operators leaving invariant a cone in a Banach space, *Amer. Math. Soc. Transl.* 26 (1950), and Ser. 1, Vol. 10 (1962), 199–325 [originally *Uspekhi Mat. Nauk* 3 (1948), 3–95].

[McD03] J. J. McDonald, The peripheral spectrum of a nonnegative matrix, *Lin. Alg. Appl.* 363 (2003), 217–235.

[NS92] M. Neumann and H. Schneider, Principal components of minus M-matrices, *Lin. Multilin. Alg.* 32 (1992), 131–148.

[NS93] M. Neumann and H. Schneider, Corrections and additions to: "Principal components of minus M-matrices," *Lin. Multilin. Alg.* 36 (1993), 147–149.

[NS94] M. Neumann and H. Schneider, Algorithms for computing bases for the Perron eigenspace with prescribed nonnegativity and combinatorial properties, *SIAM J. Matrix Anal. Appl.* 15 (1994), 578–591.

[Pul71] N.J. Pullman, A geometric approach to the theory of nonnegative matrices, *Lin. Alg. Appl.* 4 (1971), 297–312.

[Rot75] U.G. Rothblum, Algebraic eigenspaces of nonnegative matrices, *Lin. Alg. Appl.* 12 (1975), 281–292.

[Rot81] U.G. Rothblum, Expansions of sums of matrix powers, *SIAM Review* 23 (1981), 143–164.

[RiS78] D.J. Richman and H. Schneider, On the singular graph and the Weyr characteristic of an M-matrix, *Aequationes Math.* 17 (1978), 208–234.

[Sfr66] H.H. Schaefer, *Topological Vector Spaces*, Macmillan, New York, 1966, 2nd ed., Springer, New York, 1999.

[Sfr74] H.H. Schaefer, *Banach Lattices and Positive Operators*, Springer, New York, 1974.

[Sfr84] H.H. Schaefer, A minimax theorem for irreducible compact operators in L^P-spaces, *Israel J. Math.* 48 (1984), 196–204.

[Sch56] H. Schneider, The elementary divisors associated with 0 of a singular M-matrix, *Proc. Edinburgh Math. Soc.*(2) 10 (1956), 108–122.

[Sch65] H. Schneider, Positive operators and an inertia theorem, *Numer. Math.* 7 (1965), 11–17.

[Sch81] H. Schneider, Geometric conditions for the existence of positive eigenvalues of matrices, *Lin. Alg. Appl.* 38 (1981), 253–271.

[Sch86] H. Schneider, The influence of the marked reduced graph of a nonnegative matrix on the Jordan form and on related properties: a survey, *Lin. Alg. Appl.* 84 (1986), 161–189.

[Sch96] H. Schneider, Commentary on "Unzerlegbare, nicht negative Matrizen," in: *Helmut Wielandt's "Mathematical Works"*, Vol. 2, B. Huppert and H. Schneider (Eds.), de Gruyter, Berlin, 1996.

[Sch97] H. Schneider, Lyapunov revisited: variations on a matrix theme, in *Operators, Systems and Linear Algebra*, U. Helmke, Pratzel-Wolters, E. Zerz (Eds.), B.G. Teubner, Stuttgart, 1997, pp. 175–181.

[SSA05] T.I. Seidman, H. Schneider, and M. Arav, Comparison theorems using general cones for norms of iteration matrices, *Lin. Alg. Appl.* 399 (2005), 169–186.

[Ste52] P. Stein, Some general theorems on iterants, *J. Res. Nat. Bur. Stand.* 48 (1952), 82–83.

[SV70] H. Schneider and M. Vidyasagar, Cross-positive matrices, *SIAM J. Num. Anal.* 7 (1970), 508–519.

[Tam77] B.S. Tam, Some results of polyhedral cones and simplicial cones, *Lin. Multilin. Alg.* 4 (1977), 281–284.

[Tam90] B.S. Tam, On the distinguished eigenvalues of a cone-preserving map, *Lin. Alg. Appl.* 131 (1990), 17–37.

[Tam92] B.S. Tam, On the structure of the cone of positive operators, *Lin. Alg. Appl.* 167 (1992), 65–85.

[Tam01] B.S. Tam, A cone-theoretic approach to the spectral theory of positive linear operators: the finite dimensional case, *Taiwanese J. Math.* 5 (2001), 207–277.

[Tam04] B.S. Tam, The Perron generalized eigenspace and the spectral cone of a cone-preserving map, *Lin. Alg. Appl.* 393 (2004), 375–429.

[Tam06] B.S. Tam, On local Perron-Frobenius theory, in preparation.

[TS94] B.S. Tam and H. Schneider, On the core of a cone-preserving map, *Trans. Am. Math. Soc.* 343 (1994), 479–524.

[TS01] B.S. Tam and H. Schneider, On the invariant faces associated with a cone-preserving map, *Trans. Am. Math. Soc.* 353 (2001), 209–245.

[TS03] B.S. Tam and H. Schneider, Linear equations over cones, Collatz–Wielandt numbers and alternating sequences, *Lin. Alg. Appl.* 363 (2003), 295–332.

[TW89] B.S. Tam and S.F. Wu, On the Collatz–Wielandt sets associated with a cone-preserving map, *Lin. Alg. Appl.* 125 (1989), 77–95.

[Van68] J.S. Vandergraft, Spectral properties of matrices which have invariant cones, *SIAM J. Appl. Math.* 16 (1968), 1208–1222.

[Var62] R.S. Varga, *Matrix Iterative Analysis*, Prentice-Hall, Upper Saddle River, NJ, 1962, 2nd ed., Springer New York, 2000.

[Vic85] H.D. Victory, Jr., On nonnegative solutions to matrix equations, *SIAM J. Alg. Dis. Meth.* 6 (1985), 406–412.

[Wie50] H. Wielandt, Unzerlegbare, nicht negative Matrizen, *Math. Z.* 52 (1950), 642–648.

II

Combinatorial Matrix Theory and Graphs

Matrices and Graphs

Topics in Combinatorial Matrix Theory

Matrices and Graphs

27

Combinatorial Matrix Theory

Richard A. Brualdi
University of Wisconsin

27.1 Combinatorial Structure and Invariants

The **combinatorial structure of a matrix** generally refers to the locations of the nonzero entries of a matrix, or it might be used to refer to the locations of the zero entries. To study and take advantage of the combinatorial structure of a matrix, graphs are used as models. Associated with a matrix are several graphs that represent the combinatorial structure of a matrix in various ways. The type of graph (undirected graph, bipartite graph, digraph) used depends on the kind of matrices (symmetric, rectangular, square) being studied ([BR91], [Bru92], [BS04]). Conversely, associated with a graph, bipartite graph, or digraph are matrices that allow one to consider it as an algebraic object. These matrices — their algebraic properties — can often be used to obtain combinatorial information about a graph that is not otherwise obtainable. These are two of three general aspects of **combinatorial matrix theory**. A third aspect concerns intrinsic combinatorial properties of matrices viewed simply as an array of numbers.

Definitions:

Let $A = [a_{ij}]$ be an $m \times n$ matrix.

A **strong combinatorial invariant** of A is a quantity or property that does not change when the rows and columns of A are permuted, that is, which is shared by all matrices of the form PAQ, where P is a permutation matrix of order m and Q is a permutation matrix of order n.

A less restrictive definition can be considered when A is a square matrix of order n.

A **weak combinatorial invariant** is a quantity or property that does not change when the rows and columns are simultaneously permuted, that is, which is shared by all matrices of the form PAP^T where P is a permutation matrix of order n.

The $(0, 1)$-matrix obtained from A by replacing each nonzero entry with a 1 is the **pattern** of A. (In those situations where the actual value of the nonzero entries is unimportant, one may replace a matrix with its pattern, that is, one may assume that A itself is a $(0, 1)$-matrix.)

A **line** of a matrix is a row or column.

A **zero line** is a line of all zeros.

The **term rank** of a (0, 1)-matrix A is the largest size $\varrho(A)$ of a collection of 1s of A with no two 1s in the same line.

A **cover** of A is a collection of lines that contain all the 1s of A.

A **minimum cover** is a cover with the smallest number of lines. The number of lines in a minimum line cover of A is denoted by $c(A)$.

A **co-cover** of A is a collection of 1s of A such that each line of A contains at least one of the 1s.

A **minimum co-cover** is a co-cover with the smallest number of 1s. The number of 1s in a minimum co-cover is denoted by $c^*(A)$.

The quantity $\varrho^*(A)$ is the **largest size of a zero submatrix** of A, that is, the maximum of $r + s$ taken over all integers r and s with $0 \leq r \leq m$ and $0 \leq s \leq n$ such that A has an $r \times s$ zero (possibly vacuous) submatrix.

Facts:

The following facts are either elementary or can be found in Chapters 1 and 4 of [BR91].

1. These are strong combinatorial invariants:
   (a) The number of rows (respectively, columns) of a matrix.
   (b) The quantity $\max\{r, s\}$ taken over all $r \times s$ zero submatrices $(0 \leq r, s)$.
   (c) The maximum value of $r + s$ taken over all $r \times s$ zero submatrices $(0 \leq r, s)$.
   (d) The number of zeros (respectively, nonzeros) in a matrix.
   (e) The number of zero rows (respectively, zero columns) of a matrix.
   (f) The multiset of row sums (respectively, column sums) of a matrix.
   (g) The rank of a matrix.
   (h) The permanent (see Chapter 31) of a matrix.
   (i) The singular values of a matrix.
2. These are weak combinatorial invariants:
   (a) The largest order of a principal submatrix that is a zero matrix.
   (b) The number of A zeros on the main diagonal of a matrix.
   (c) The maximum value of $p + q$ taken over all $p \times q$ zero submatrices that do not meet the main diagonal.
   (d) Whether or not for some integer r with $1 \leq r \leq n$, the matrix A of order n has an $r \times n - r$ zero submatrix that does not meet the main diagonal of A.
   (e) Whether or not A is a symmetric matrix.
   (f) The trace $\text{tr}(A)$ of a matrix A.
   (g) The determinant $\det A$ of a matrix A.
   (h) The eigenvalues of a matrix.
   (i) The multiset of elements on the main diagonal of a matrix.
3. $\varrho(A), c(A), \varrho^*(A)$, and $c^*(A)$ are all strong combinatorial invariants.
4. $\rho(A) = c(A)$.
5. A matrix A has a co-cover if and only if it does not have any zero lines. If A does not have any zero lines, then $\varrho^*(A) = c^*(A)$.
6. If A is an $m \times n$ matrix without zero lines, then $\varrho(A) + \varrho^*(A) = c(A) + c^*(A) = m + n$.

7. $\text{rank}(A) \leq \varrho(A)$.
8. Let A be an $m \times n$ (0,1)-matrix. Then there are permutation matrices P and Q such that

$$PAQ = \begin{bmatrix} A_1 & X & Y & Z \\ O & A_2 & O & O \\ O & S & A_3 & O \\ O & T & O & O \end{bmatrix},$$

where A_1, A_2, and A_3 are square, possibly vacuous, matrices with only 1s on their main diagonals, and $\rho(A)$ is the sum of the orders of A_1, A_2, and A_3. The rows, respectively columns, of A that are in every minimum cover of A are the rows, respectively columns, that meet A_1, respectively A_2. These rows and columns together with either the rows that meet A_3 or the columns that meet A_3 form minimum covers of A.

Examples:

1. Let

$$A = \left[\begin{array}{cc|cc|c|cc} \mathbf{1} & 1 & 1 & 1 & 1 & 1 & 1 \\ 0 & \mathbf{1} & 0 & 1 & 0 & 0 & 1 \\ \hline 0 & 0 & \mathbf{1} & 0 & 0 & 0 & 0 \\ 0 & 0 & 1 & \mathbf{1} & 0 & 0 & 0 \\ \hline 0 & 0 & 1 & 1 & \mathbf{1} & 0 & 0 \\ \hline 0 & 0 & 1 & 0 & 0 & 0 & 0 \\ 0 & 0 & 1 & 1 & 0 & 0 & 0 \end{array}\right].$$

Then $\varrho(A) = c(A) = 5$ with the five **1**s in different lines, and rows 1, 2, and 5 and columns 3 and 4 forming a cover. The matrix is partitioned in the form given in Fact 8.

27.2 Square Matrices and Strong Combinatorial Invariants

In this section, we consider the strong combinatorial structure of square matrices.

Definitions:

Let A be a $(0, 1)$-matrix of order n.

A collection of n nonzero entries in A no two on the same line is a **diagonal** of A (this term is also applied to nonnegative matrices).

The next definitions are concerned with the existence of certain zero submatrices in A.

A is **partly decomposable** provided there exist positive integers p and q with $p + q = n$ such that A has a $p \times q$ zero submatrix. Equivalently, there are permutation matrices P and Q and an integer k with $1 \leq k \leq n - 1$ such that

$$PAQ = \begin{bmatrix} B & C \\ O_{k,n-k} & D \end{bmatrix}.$$

A is a **Hall matrix** provided there does not exist positive integers p and q with $p + q > n$ such that A has a $p \times q$ zero submatrix.

A has **total support** provided $A \neq O$ and each 1 of A is on a diagonal of A.

A is **fully indecomposable** provided it is not partly decomposable.

A is **nearly decomposable** provided it is fully indecomposable and each matrix obtained from A by replacing a 1 with a 0 is partly decomposable.

Facts:

Unless otherwise noted, the following facts can be found in Chapter 4 of [BR91].

1. [BS94] Each of the following properties is equivalent to the matrix A of order n being a Hall matrix:
   (a) $\rho(A) = n$, that is, A has a diagonal (*Frobenius–König theorem*).
   (b) For all nonempty subsets L of $\{1, 2, \ldots, n\}$, $A[\{1, 2, \ldots, n\}, L]$ has at least $|L|$ nonzero rows.
   (c) For all nonempty subsets K of $\{1, 2, \ldots, n\}$, $A[K, \{1, 2, \ldots, n\}]$ has at least $|K|$ nonzero columns.
2. Each of the following properties is equivalent to the matrix A of order n being a fully indecomposable matrix:
   (a) $\rho(A) = n$ and the only minimum line covers are the set of all rows and the set of all columns.
   (b) For all nonempty subsets L of $\{1, 2, \ldots, n\}$, $A[\{1, 2, \ldots, n\}, L]$ has at least $|L| + 1$ nonzero rows.
   (c) For all nonempty subsets K of $\{1, 2, \ldots, n\}$, $A[K, \{1, 2, \ldots, n\}]$ has at least $|K| + 1$ nonzero columns.
   (d) The term rank $\rho(A(i, j))$ of the matrix $A(i, j)$ obtained from A by deleting row i and column j equals $n - 1$ for all $i, j = 1, 2, \ldots, n$.
   (e) A^{n-1} is a positive matrix.
   (f) The determinant $\det A \circ X$ of the Hadamard product of A with a matrix $X = [x_{ij}]$ of distinct indeterminates over a field F is irreducible in the ring $F[\{x_{ij} : 1 \leq i, j \leq n\}]$.
3. Each of the following properties is equivalent to the matrix A of order n having total support:
   (a) $A \neq O$ and the term rank $\rho(A(i, j))$ equals $n - 1$ for all $i, j = 1, 2, \ldots, n$ with $a_{ij} \neq 0$.
   (b) There are permutation matrices P and Q such that PAQ is a direct sum of fully indecomposable matrices.
4. (*Dulmage–Mendelsohn Decomposition theorem*) If the matrix A of order n has term rank equal to n, then there exist permutation matrices P and Q and an integer $t \geq 1$ such that

$$PAQ = \begin{bmatrix} A_1 & A_{12} & \cdots & A_{1t} \\ O & A_2 & \cdots & A_{2t} \\ \vdots & \vdots & \ddots & \vdots \\ O & O & \cdots & A_t \end{bmatrix},$$

   where $A_1, A_2, \ldots, A_t$ are square fully indecomposable matrices. The matrices $A_1, A_2, \ldots, A_t$ are called the **fully indecomposable components** of A and they are uniquely determined up to permutations of their rows and columns. The matrix A has total support if and only if $A_{ij} = O$ for all i and j with $i < j$; A is fully indecomposable if and only if $t = 1$.
5. (*Inductive structure of fully indecomposable matrices*) If A is a fully indecomposable matrix of order n, then there exist permutation matrices P and Q and an integer $k \geq 2$ such that

$$PAQ = \begin{bmatrix} B_1 & O & \cdots & O & E_1 \\ E_2 & B_2 & \cdots & O & O \\ \vdots & \vdots & \ddots & \vdots & \vdots \\ O & O & \cdots & B_{k-1} & O \\ O & O & \cdots & E_k & B_k \end{bmatrix},$$

   where $B_1, B_2, \ldots, B_k$ are fully indecomposable and $E_1, E_2, \ldots, E_k$ each contain at least one nonzero entry. Conversely, a matrix of such a form is fully indecomposable.

6. (*Inductive structure of nearly decomposable matrices*) If A is a nearly decomposable $(0,1)$-matrix, then there exist permutation matrices P and Q and an integer p with $1 \leq p \leq n-1$ such that

$$PAQ = \left[\begin{array}{cccccc|c} 1 & 0 & 0 & \cdots & 0 & 0 & \\ 1 & 1 & 0 & \cdots & 0 & 0 & \\ 0 & 1 & 1 & \cdots & 0 & 0 & \\ \vdots & \vdots & \vdots & \ddots & \vdots & \vdots & F_1 \\ 0 & 0 & 0 & \cdots & 1 & 0 & \\ 0 & 0 & 0 & \cdots & 1 & 1 & \\ \hline & & & F_2 & & & A' \end{array}\right],$$

where A' is a nearly decomposable matrix of order $n-p$, the matrix F_1 has exactly one 1 and this 1 occur in its first row, and the matrix F_2 has exactly one 1 and this 1 occurs in its last column. If $n-p \geq 2$, and the 1 in F_2 is in its column j and the 1 in F_2 is in its row i, then the (i,j) entry of A' is 0.
7. The number of nonzero entries in a nearly decomposable matrix A of order $n \geq 3$ is between $2n$ and $3(n-1)$.

Examples:

1. Let

$$A_1 = \begin{bmatrix} 1 & 0 & 0 \\ 1 & 0 & 0 \\ 1 & 1 & 1 \end{bmatrix}, \; A_2 = \begin{bmatrix} 1 & 1 & 0 \\ 1 & 1 & 0 \\ 1 & 1 & 1 \end{bmatrix}, \; A_3 = \begin{bmatrix} 1 & 1 & 0 \\ 1 & 1 & 0 \\ 0 & 0 & 1 \end{bmatrix}, \; A_4 = \begin{bmatrix} 1 & 1 & 0 \\ 0 & 1 & 1 \\ 1 & 0 & 1 \end{bmatrix}.$$

Then A_1 is partly decomposable and not a Hall matrix. The matrix A_2 is a Hall matrix and is partly decomposable, but does not have total support. The matrix A_3 has total support. The matrix A_4 is nearly decomposable.

27.3 Square Matrices and Weak Combinatorial Invariants

In this section, we restrict our attention to the weak combinatorial structure of square matrices.

Definitions:

Let A be a matrix of order n.

B is **permutation similar** to A if there exists a permutation matrix P such that $B = P^T AP \; (= P^{-1}AP)$.

A is **reducible** provided $n \geq 2$ and for some integer r with $1 \leq r \leq n-1$, there exists an $r \times (n-r)$ zero submatrix which does not meet the main diagonal of A, that is, provided there is a permutation matrix P and an integer r with $1 \leq r \leq n-1$ such that

$$PAP^T = \begin{bmatrix} B & C \\ O_{r,n-r} & D \end{bmatrix}.$$

A is **irreducible** provided that A is not reducible.

A is **completely reducible** provided there exists an integer $k \geq 2$ and a permutation matrix P such that $PAP^T = A_1 \oplus A_2 \oplus \cdots \oplus A_k$ where $A_1, A_2, \ldots, A_k$ are irreducible.

A is **nearly reducible** provided A is irreducible and each matrix obtained from A by replacing a nonzero entry with a zero is reducible.

A **Frobenius normal form** of A is a block upper triangular matrix with irreducible diagonal blocks that is permutation similar to A; the diagonal blocks are called the **irreducible components** of A. (cf. Fact 27.3.)

The following facts can be found in Chapter 3 of [BR91].

Facts:

1. (*Frobenius normal form*) There is a permutation matrix P and an integer $r \geq 1$ such that

$$PAP^T = \begin{bmatrix} A_1 & A_{12} & \cdots & A_{1r} \\ O & A_2 & \cdots & A_{2r} \\ \vdots & \vdots & \ddots & \vdots \\ O & O & \cdots & A_r \end{bmatrix},$$

where $A_1, A_2, \ldots, A_t$ are square irreducible matrices. The matrices $A_1, A_2, \ldots, A_r$ are the **irreducible components** of A and they are uniquely determined up to simultaneous permutations of their rows and columns.

2. There exists a permutation matrix Q such that AQ is irreducible if and only if A has at least one nonzero element in each line.
3. If A does not have any zeros on its main diagonal, then A is irreducible if and only if A is fully indecomposable. The matrix A is fully indecomposable if and only if there is a permutation matrix Q such that AQ has no zeros on its main diagonal and AQ is irreducible.
4. (*Inductive structure of irreducible matrices*) Let A be an irreducible matrix of order $n \geq 2$. Then there exists a permutation matrix P and an integer $m \geq 2$ such that

$$PAP^T = \begin{bmatrix} A_1 & O & \cdots & O & E_1 \\ E_2 & A_2 & \cdots & O & O \\ \vdots & \vdots & \ddots & \vdots & \vdots \\ O & O & \cdots & A_{m-1} & O \\ O & O & \cdots & E_m & A_m \end{bmatrix},$$

where $A_1, A_2, \ldots, A_m$ are irreducible and $E_1, E_2, \ldots, E_m$ each have at least one nonzero entry.

5. (*Inductive structure of nearly reducible matrices*) If A is a nearly reducible (0, 1)-matrix, then there exist permutation matrix P and an integer m with $1 \leq m \leq n-1$ such that

$$PAP^T = \left[\begin{array}{cccccc|c} 0 & 0 & 0 & \cdots & 0 & 0 & \\ 1 & 0 & 0 & \cdots & 0 & 0 & \\ 0 & 1 & 0 & \cdots & 0 & 0 & F_1 \\ \vdots & \vdots & \vdots & \ddots & \vdots & \vdots & \\ 0 & 0 & 0 & \cdots & 1 & 0 & \\ \hline & & & F_2 & & & A' \end{array}\right],$$

where A' is a nearly reducible matrix of order m, the matrix F_1 has exactly one 1 and it occurs in the first row and column j of F_1 with $1 \leq j \leq m$, and the matrix F_2 has exactly one 1 and it occurs in the last column and row i of F_2 where $1 \leq i \leq m$. The element in position (i, j) of A' is 0.

6. The number of nonzero entries in a nearly reducible matrix of order $n \geq 2$ is between n and $2(n-1)$

Examples:

1. Let

$$A_1 = \begin{bmatrix} 1 & 0 & 0 \\ 1 & 1 & 1 \\ 1 & 1 & 1 \end{bmatrix}, \quad A_2 = \begin{bmatrix} 1 & 1 & 1 \\ 1 & 0 & 1 \\ 1 & 0 & 1 \end{bmatrix}, \quad A_3 = \begin{bmatrix} 1 & 0 & 0 \\ 0 & 1 & 1 \\ 0 & 1 & 1 \end{bmatrix}, \quad A_4 = \begin{bmatrix} 0 & 1 & 0 \\ 0 & 0 & 1 \\ 1 & 0 & 0 \end{bmatrix}.$$

Then A_1 is reducible but not completely reducible, and A_2 is irreducible. (Both A_1 and A_2 are partly decomposable.) The matrix A_3 is completely reducible. The matrix A_4 is nearly reducible.

27.4 The Class $\mathcal{A}(R,S)$ of (0,1)-Matrices

In the next definition, we introduce one of the most important and widely studied classes of (0, 1)-matrices (see Chapter 6 of [Rys63] and [Bru80]).

Definitions:

Let $A = [a_{ij}]$ be an $m \times n$ matrix.

The **row sum vector** of A is $R = (r_1, r_2, \ldots, r_m)$, where $r_i = \sum_{j=1}^{n} a_{ij}$, $(i = 1, 2, \ldots, n)$.

The **column sum vector** of A is $S = (s_1, s_2, \ldots, s_n)$, where $s_j = \sum_{i=1}^{m} a_{ij}$, $(j = 1, 2, \ldots, n)$.

A real vector $(c_1, c_2, \ldots, c_n)$ is **monotone** provided $c_1 \geq c_2 \geq \cdots \geq c_n$.

The **class of all** $m \times n$ **(0, 1)-matrices with row sum vector** R **and column sum vector** S is denoted by $\mathcal{A}(R, S)$.

The class $\mathcal{A}(R, S)$ is a **monotone class** provided R and S are both monotone vectors.

An **interchange** is a transformation on a (0, 1)-matrix that replaces a submatrix equal to the identity matrix I_2 by the submatrix

$$L_2 = \begin{bmatrix} 0 & 1 \\ 1 & 0 \end{bmatrix}$$

or vice versa.

If $\theta(A)$ is any real numerical quantity associated with a matrix A, then the **extreme values** of θ are $\bar{\theta}(R, S)$ and $\tilde{\theta}(R, S)$, defined by

$$\bar{\theta}(R, S) = \max\{\theta(A) : A \in \mathcal{A}(R, S)\} \text{ and } \tilde{\theta}(R, S) = \min\{\theta(A) : A \in \mathcal{A}(R, S)\}.$$

Let $T = [t_{kl}]$ be the $(m + 1) \times (n + 1)$ matrix defined by

$$t_{kl} = kl - \sum_{j=1}^{l} s_j + \sum_{i=k+1}^{m} r_i, \quad (k = 0, 1, \ldots, m; l = 0, 1, \ldots, n).$$

The matrix T is the **structure matrix** of $\mathcal{A}(R, S)$.

Facts:

The following facts can be found in Chapter 6 of [Rys63], [Bru80], Chapter 6 of [BR91], and Chapters 3 and 4 of [Bru06].

1. A class $\mathcal{A}(R, S)$ can be transformed into a monotone class by row and column permutations.
2. Let $U = (u_1, u_2, \ldots, u_n)$ and $V = (v_1, v_2, \ldots, v_n)$ be monotone, nonnegative integral vectors. $U \preceq V$ if and only if $V^* \preceq U^*$, and $U^{**} = U$ or U extended with 0s.
3. (*Gale–Ryser theorem*) $\mathcal{A}(R, S)$ is nonempty if and only if $S \preceq R^*$.
4. Let the monotone class $\mathcal{A}(R, S)$ be nonempty, and let A be a matrix in $\mathcal{A}(R, S)$. Let $K = \{1, 2, \ldots, k\}$ and $L = \{1, 2, \ldots, l\}$. Then t_{kl} equals the number of 0s in the submatrix $A[K, L]$ plus the number of 1s in the submatrix $A(K, L)$; in particular, we have $t_{kl} \geq 0$.
5. (*Ford–Fulkerson theorem*) The monotone class $\mathcal{A}(R, S)$ is nonempty if and only if its structure matrix T is a nonnegative matrix.
6. If A is in $\mathcal{A}(R, S)$ and B results from A by an interchange, then B is in $\mathcal{A}(R, S)$. Each matrix in $\mathcal{A}(R, S)$ can be transformed to every other matrix in $\mathcal{A}(R, S)$ by a sequence of interchanges.
7. The maximum and minimum term rank of a nonempty monotone class $\mathcal{A}(R, S)$ satisfy:

$$\bar{\rho}(R, S) = \min\{t_{kl} + k + l; k = 0, 1, \ldots, m, l = 0, 1, \ldots, n\},$$
$$\tilde{\rho}(R, S) = \min\{k + l : \phi_{kl} \geq t_{kl}, k = 0, 1, \ldots, m, l = 0, 1, \ldots, n\},$$

where

$$\phi_{kl} = \min\{t_{i_1, l+j_2} + t_{k+i_2, j_1} + (k - i_1)(l - j_1)\},$$

the minimum being taken over all integers i_1, i_2, j_1, j_2 such that $0 \leq i_1 \leq k \leq k + i_2 \leq m$ and $0 \leq j_1 \leq l \leq l + j_2 \leq n$.

8. Let $\text{tr}(A)$ denote the trace of a matrix A. The maximum and minimum trace of a nonempty monotone class $\mathcal{A}(R, S)$ satisfy:

$$\overline{\text{tr}}(R, S) = \min\{t_{kl} + \max\{k, l\} : 0 \leq k \leq m, 0 \leq l \leq n\},$$
$$\tilde{\text{tr}}(R, S) = \max\{\min\{k, l\} - t_{kl} : 0 \leq k \leq m, 0 \leq l \leq n\}.$$

9. Let k and n be integers with $0 \leq k \leq n$, and let $\mathcal{A}(n, k)$ denote the class $\mathcal{A}(R, S)$, where $R = S = (k, k, \ldots, k)$ (n k's). Let $\tilde{\nu}(n, k)$ and $\bar{\nu}(n, k)$ denote the minimum and maximum rank, respectively, of matrices in $\mathcal{A}(n, k)$.

   (a) $\bar{\nu}(n, k) = \begin{cases} 0, \text{ if } k = 0, \\ 1, \text{ if } k = n, \\ 3, \text{ if } k = 2 \text{ and } n = 4, \\ n, \text{ otherwise.} \end{cases}$

   (b) $\tilde{\nu}(n, k) = \tilde{\nu}(n, n - k)$ if $1 \leq k \leq n - 1$.

   (c) $\tilde{\nu}(n, k) \geq \lceil n/k \rceil$, $(1 \leq k \leq n - 1)$, with equality if and only if k divides n.

   (d) $\tilde{\nu}(n, k) \leq \lfloor n/k \rfloor + k$, $(1 \leq k \leq n)$.

   (e) $\tilde{\nu}(n, 2) = n/2$ if n is even, and $(n + 3)/2$ if n is odd.

   (f) $\tilde{\nu}(n, 3) = n/3$ if 3 divides n and $\lfloor n/3 \rfloor + 3$ otherwise.

Additional properties of $\mathcal{A}(R, S)$ can be found in [Bru80] and in Chapters 3 and 4 of [Bru06].

Examples:

1. Let $R = (7, 3, 2, 2, 1, 1)$ and $S = (5, 5, 3, 1, 1, 1)$. Then $R^* = (6, 4, 2, 1, 1, 1, 1)$. Since $5 + 5 + 3 > 6 + 4 + 2$, $S \not\preceq R^*$ and, by Fact 3, $\mathcal{A}(R, S) = \emptyset$.
2. Let $R = S = (2, 2, 2, 2, 2)$. Then the matrices

$$A = \begin{bmatrix} 1 & 1 & 0 & 0 & 0 \\ 0 & 1 & 1 & 0 & 0 \\ 0 & 0 & 1 & 1 & 0 \\ 0 & 0 & 0 & 1 & 1 \\ 1 & 0 & 0 & 0 & 1 \end{bmatrix} \text{ and } B = \begin{bmatrix} 1 & 0 & 0 & 0 & 1 \\ 0 & 1 & 1 & 0 & 0 \\ 0 & 1 & 0 & 1 & 0 \\ 0 & 0 & 1 & 1 & 0 \\ 1 & 0 & 0 & 0 & 1 \end{bmatrix}$$

   are in $\mathcal{A}(R, S)$. Then A can be transformed to B by two interchanges:

$$\begin{bmatrix} 1 & \mathbf{1} & 0 & 0 & \mathbf{0} \\ 0 & 1 & 1 & 0 & 0 \\ 0 & 0 & 1 & 1 & 0 \\ 0 & \mathbf{0} & 0 & 1 & \mathbf{1} \\ 1 & 0 & 0 & 0 & 1 \end{bmatrix} \to \begin{bmatrix} 1 & 0 & 0 & 0 & 1 \\ 0 & 1 & 1 & 0 & 0 \\ 0 & \mathbf{0} & \mathbf{1} & 1 & 0 \\ 0 & \mathbf{1} & \mathbf{0} & 1 & 0 \\ 1 & 0 & 0 & 0 & 1 \end{bmatrix} \to \begin{bmatrix} 1 & 0 & 0 & 0 & 1 \\ 0 & 1 & 1 & 0 & 0 \\ 0 & 1 & 0 & 1 & 0 \\ 0 & 0 & 1 & 1 & 0 \\ 1 & 0 & 0 & 0 & 1 \end{bmatrix}.$$

27.5 The Class $\mathcal{T}(R)$ of Tournament Matrices

In the next definition, we introduce another important class of $(0, 1)$-matrices.

Definitions:

A $(0, 1)$-matrix $A = [a_{ij}]$ of order n is a **tournament matrix** provided $a_{ii} = 0$, $(1 \leq i \leq n)$ and $a_{ij} + a_{ji} = 1$, $(1 \leq i < j \leq n)$, that is, provided $A + A^T = J_n - I_n$.

The digraph of a tournament matrix is called a **tournament**.

Thinking of n teams $p_1, p_2, \ldots, p_n$ playing in a **round-robin tournament**, we have that $a_{ij} = 1$ signifies that team p_i beats team p_j.

The row sum vector R is also called the **score vector** of the tournament (matrix).

A **transitive tournament matrix** is one for which $a_{ij} = a_{jk} = 1$ implies $a_{ik} = 1$.

The **class of all tournament matrices with score vector** R is denoted by $\mathcal{T}(R)$.

A δ**-interchange** is a transformation on a tournament matrix that replaces a principal submatrix of order 3 equal to

$$\begin{bmatrix} 0 & 0 & 1 \\ 1 & 0 & 0 \\ 0 & 1 & 0 \end{bmatrix} \text{ with } \begin{bmatrix} 0 & 1 & 0 \\ 0 & 0 & 1 \\ 1 & 0 & 0 \end{bmatrix}$$

or vice versa.

The following facts can be found in Chapters 2 and 5 of [Bru06].

Facts:

1. The row sum vector $R = (r_1, r_2, \ldots, r_n)$ and column sum vector $S = (s_1, s_2, \ldots, s_n)$ of a tournament matrix of order n satisfy $r_i + s_i = n - 1$, $(1 \leq i \leq n)$; in particular, the column sum vector is determined by the row sum vector.
2. If A is a tournament matrix and P is a permutation matrix, then PAP^T is a tournament matrix. Thus, one may assume without loss of generality that R is nondecreasing, that is, $r_1 \leq r_2 \leq \cdots \leq r_n$, so that the teams are ordered from worst to best.
3. (*Landau's theorem*) If $R = (r_1, r_2, \ldots, r_n)$ is a nondecreasing, nonnegative integral vector, then $\mathcal{T}(R)$ is nonempty if and only if

$$\sum_{i=1}^{k} r_i \geq \binom{k}{2}, \quad (1 \leq k \leq n)$$

   with equality when $k = n$. (A binomial coefficient $\binom{k}{s}$ is 0 if $k < s$.)
4. Let $R = (r_1, r_2, \ldots, r_n)$ be a nondecreasing nonnegative integral vector. The following are equivalent:

   (a) There exists an irreducible matrix in $\mathcal{T}(R)$.

   (b) $\mathcal{T}(R)$ is nonempty and every matrix in $\mathcal{T}(R)$ is irreducible.

   (c) $\sum_{i=1}^{k} r_i \geq \binom{k}{2}$, $(1 \leq k \leq n)$ with equality if and only if $k = n$.
5. If A is in $\mathcal{T}(R)$ and B results from A by a δ-interchange, then B is in $\mathcal{T}(R)$. Each matrix in $\mathcal{T}(R)$ can be transformed to every other matrix in $\mathcal{T}(R)$ by a sequence of δ-interchanges.
6. The rank, and so term rank, of a tournament matrix of order n is at least $n - 1$.
7. (*Strengthened Landau inequalities*) Let $R = (r_1, r_2, \ldots, r_n)$ be a nondecreasing nonnegative integral vector. Then $\mathcal{T}(R)$ is nonempty if and only if

$$\sum_{i \in I} r_i \geq \frac{1}{2} \sum_{i \in I} (i - 1) + \frac{1}{2} \binom{|I|}{2}, \quad (I \subseteq \{1, 2, \ldots, n\}),$$

   with equality if $I = \{1, 2, \ldots, n\}$.
8. Let $R = (r_1, r_2, \ldots, r_n)$ be a nondecreasing, nonnegative integral vector such that $\mathcal{T}(R)$ is nonempty. Then there exists a tournament matrix A such that the principal submatrices made out of the even-indexed and odd-indexed, respectively, rows and columns are transitive tournament matrices, that is, a matrix A in $\mathcal{T}(R)$ such that $A[\{1, 3, 5, \ldots\}]$ and $A[\{2, 4, 6, \ldots\}]$ are transitive tournament matrices.

Examples:

1. The following are tournament matrices:

$$A_1 = \begin{bmatrix} 0 & 1 & 0 \\ 0 & 0 & 1 \\ 1 & 0 & 0 \end{bmatrix} \text{ and } A_2 = \begin{bmatrix} 0 & 0 & 0 & 0 \\ 1 & 0 & 0 & 0 \\ 1 & 1 & 0 & 0 \\ 1 & 1 & 1 & 0 \end{bmatrix}.$$

   The matrix A_2 is a transitive tournament matrix.

2. Let $R = (2, 2, 2, 2, 3, 4)$. A tournament matrix in $\mathcal{T}(R)$ satisfying Fact 7 is

$$\begin{bmatrix} 0 & 0 & 1 & 0 & 0 & 0 \\ 1 & 0 & 1 & 0 & 0 & 0 \\ 0 & 0 & 0 & 1 & 0 & 1 \\ 1 & 1 & 0 & 0 & 1 & 0 \\ 1 & 1 & 1 & 0 & 0 & 0 \\ 1 & 1 & 0 & 1 & 1 & 0 \end{bmatrix},$$

   since the two submatrices

$$A[|[1,3,5\}] = \begin{bmatrix} 0 & 0 & 1 \\ 1 & 0 & 1 \\ 0 & 0 & 0 \end{bmatrix} \text{ and } A[\{2,4,6\}] = \begin{bmatrix} 0 & 1 & 0 \\ 0 & 0 & 0 \\ 1 & 1 & 0 \end{bmatrix}$$

   are transitive tournament matrices.

27.6 Convex Polytopes of Doubly Stochastic Matrices

Doubly stochastic matrices (see Chapter 9.4) are widely studied because of their connection with probability theory, as every doubly stochastic matrix is the transition matrix of a Markov chain. The reader is referred to Chapter 28 for graph terminology.

Definitions:

Since a convex combination $cA + (1 - c)B$ of two doubly stochastic matrices A and B, where $0 \leq c \leq 1$, is doubly stochastic, the set Ω_n of doubly stochastic matrices of order n is a convex polytope in $\mathbb{R}^{n^2}$. Ω_n is also called the **assignment polytope** because of its appearance in the classical assignment problem.

If A is a $(0, 1)$-matrix of order n, then $\mathcal{F}(A)$ is the convex polytope of all doubly stochastic matrices whose patterns P satisfy $P \leq A$ (entrywise), that is, that have 0s at least wherever A has 0s.

The **dimension** of a convex polytope $\mathcal{P}$ is the smallest dimension of an affine space containing it and is denoted by $\dim \mathcal{P}$.

The **graph** $G(\mathcal{P})$ of a convex polytope $\mathcal{P}$ has the extreme points of $\mathcal{P}$ as its vertices and the pairs of extreme points of one-dimensional faces of $\mathcal{P}$ as its edges.

The **chromatic index** of a graph is the smallest integer t such that the edges can be partitioned into sets $E_1, E_2, \ldots, E_t$ such that no two edges in the same E_i meet.

A **scaling** of a matrix A is a matrix of the form $D_1 A D_2$, where D_1 and D_2 are diagonal matrices with positive diagonal entries.

If $D_1 = D_2$, then the scaling $D_1 A D_2$ is a **symmetric scaling**.

Let $A = [a_{ij}]$ have order n. A **diagonal product** of A is the product of the entries on a diagonal of A, that is, $a_{1j_1} a_{2j_2} \cdots a_{nj_n}$ where $j_1, j_2, \ldots, j_n$ is a permutation of $\{1, 2, \ldots, n\}$.

The following facts can be found in Chapter 9 of [Bru06].

Facts:

1. (*Birkhoff's theorem*) The extreme points of Ω_n are the permutation matrices of order n. Thus, each doubly stochastic matrix is a convex combination of permutation matrices.
2. The patterns of matrices in Ω_n are precisely the $(0, 1)$-matrices of order n with total support.
3. The faces of Ω_n are the sets $\mathcal{F}(A)$, where A is a $(0, 1)$-matrix of order n with total support. Ω_n is a face of itself with $\Omega_n = \mathcal{F}(J_n)$. The dimension of $\mathcal{F}(A)$ satisfies

$$\dim \mathcal{F}(A) = t - 2n + k,$$

where t is the number of 1s of A and k is the number of fully indecomposable components of A. The number of extreme points of $\mathcal{F}(A)$ (this number is the permanent $\operatorname{per}(A)$ of A), is at least $t - 2n + k + 1$. If A is fully indecomposable and $\dim \mathcal{F}(A) = d$, then $\mathcal{F}(A)$ has at most $2^{d-1} + 1$ extreme points. In general, $\mathcal{F}(A)$ has at most 2^d extreme points.
4. The graph $G(\Omega_n)$ has the following properties:
   (a) The number of vertices of $G(\Omega_n)$ is $n!$.
   (b) The degree of each vertex of $G(\Omega_n)$ is

$$d_n = \sum_{k=2}^{n} \binom{n}{k} (k-1)!.$$

   (c) $G(\Omega_n)$ is connected and its diameter equals 1 if $n = 1, 2$, and 3, and equals 2 if $n \geq 4$.
   (d) $G(\Omega_n)$ has a Hamilton cycle.
   (e) The chromatic index of $G(\Omega_n)$ equals d_n.
5. (*Hardy, Littlewood, Pólya theorem*) Let $U = (u_1, u_2, \dots, u_n)$ and $V = (v_1, v_2, \dots, v_n)$ be monotone, nonnegative integral vectors. Then $U \preceq V$ if and only if there is a doubly stochastic matrix A such that $U = VA$.
6. The set Υ_n of symmetric doubly stochastic matrices of order n is a subpolytope of Ω_n whose extreme points are those matrices A such that there is a permutation matrix P for which PAP^T is a direct sum of matrices each of which is either the identity matrix I_1 of order 1, the matrix $\begin{bmatrix} 0 & 1 \\ 1 & 0 \end{bmatrix}$, or an odd order matrix of the type:

$$\begin{bmatrix} 0 & 1/2 & 0 & 0 & \cdots & 0 & 1/2 \\ 1/2 & 0 & 1/2 & 0 & \cdots & 0 & 0 \\ 0 & 1/2 & 0 & 1/2 & \cdots & 0 & 0 \\ 0 & 0 & 1/2 & 0 & \cdots & 0 & 0 \\ \vdots & \vdots & \vdots & \vdots & \ddots & \vdots & \vdots \\ 0 & 0 & 0 & 0 & \cdots & 0 & 1/2 \\ 1/2 & 0 & 0 & 0 & \cdots & 1/2 & 0 \end{bmatrix}.$$

7. Let A be a nonnegative matrix. Then there is a scaling $B = D_1 A D_2$ of A that is doubly stochastic if and only if A has total support. If A is fully indecomposable, then the doubly stochastic matrix B is unique and the diagonal matrices D_1 and D_2 are unique up to reciprocal scalar factors.
8. Let A be a nonnegative symmetric matrix with no zero lines. Then there is a symmetric scaling $B = DAD$ such that B is doubly stochastic if and only if A has total support. If A is fully indecomposable, then the doubly stochastic matrix B and the diagonal matrix D are unique.
9. Distinct doubly stochastic matrices of order n do not have proportional diagonal products; that is, if $A = [a_{ij}]$ and $B = [b_{ij}]$ are doubly stochastic matrices of order n with $A \neq B$, there does not exist a constant c such that $a_{1j_1} a_{2j_2} \cdots a_{nj_n} = c b_{1j_1} b_{2j_2} \cdots b_{nj_n}$ for all permutations $j_1, j_2, \dots, j_n$ of $\{1, 2, \dots, n\}$.

The subpolytopes of Ω_n consisting of (i) the convex combinations of the $n! - 1$ nonidentity permutation matrices of order n and (ii) the permutation matrices corresponding to the even permutations of order n have been studied (see [Bru06]).

Polytopes of matrices more general than Ω_n have also been studied, for instance, the nonnegative generalizations of $\mathcal{A}(R, S)$ consisting of all nonnegative matrices with a given row sum vector R and a given column sum vector S (see Chapter 8 of [Bru06]).

Examples:

1. Ω_2 consists of all matrices of the form

$$\begin{bmatrix} a & 1-a \\ 1-a & a \end{bmatrix}, \quad (0 \leq a \leq 1).$$

All permutation matrices are doubly stochastic.

2. The matrix

$$\begin{bmatrix} 1/2 & 1/4 & 1/4 \\ 1/6 & 1/3 & 1/2 \\ 1/3 & 5/12 & 1/4 \end{bmatrix}$$

is a doubly stochastic matrix of order 3.

3. If

$$A = \begin{bmatrix} 1 & 1 & 0 \\ 0 & 1 & 1 \\ 1 & 1 & 1 \end{bmatrix},$$

then

$$\begin{bmatrix} 1/2 & 1/2 & 0 \\ 0 & 1/2 & 1/2 \\ 1/2 & 0 & 1/2 \end{bmatrix}$$

is in $\mathcal{F}(A)$.

References

[BR97] R.B. Bapat and T.E.S. Raghavan, *Nonnegative Matrices and Applications*, Encyclopedia of Mathematical Sciences, No. 64, Cambridge University Press, Cambridge, 1997.

[Bru80] R.A. Brualdi, Matrices of zeros and ones with fixed row and column sum vectors, *Lin. Alg. Appl.*, 33: 159–231, 1980.

[Bru92] R.A. Brualdi, The symbiotic relationship of combinatorics and matrix theory, *Lin. Alg. Appl.*, 162–164: 65–105, 1992.

[Bru06] R.A. Brualdi, *Combinatorial Matrix Classes*, Encyclopedia of Mathematics and Its Applications, Vol. 108, Cambridge Universty Press, Cambridge, 2006.

[BR91] R.A. Brualdi and H.J. Ryser, *Combinatorial Matrix Theory*, Encyclopedia of Mathematics and its Applications, Vol. 39, Cambridge University Press, Cambridge, 1991.

[BS94] R.A. Brualdi and B.L. Shader, Strong Hall matrices, *SIAM J. Matrix Anal. Appl.*, 15: 359–365, 1994.

[BS04] R.A. Brualdi and B.L. Shader, Graphs and matrices, in *Topics in Algebraic Graph Theory*, L. Beineke and R. Wilson, Eds., Cambridge University press, Cambridge, 2004, 56–87.

[Rys63] H.J. Ryser, *Combinatorial Mathematics*, Carus Mathematical Monograph No. 14, Mathematical Association of America, Washington, D.C., 1963.

28

Matrices and Graphs

Willem H. Haemers
Tilburg University

The first two sections of this chapter "Matrices and Graphs" give a short introduction to graph theory. Unfortunately much graph theoretic terminology is not standard, so we had to choose. We allow, for example, graphs to have multiple edges and loops, and call a graph simple if it has none of these. On the other hand, we assume that graphs are finite.

For all nontrivial facts, references are given, sometimes to the original source, but often to text books or survey papers. A recent global reference for this chapter is [BW04]. (This book was not available to the author when this chapter was written, so it is not referred to in the text below.)

28.1 Graphs: Basic Notions

Definitions:

A **graph** $G = (V, E)$ consists of a finite set $V = \{v_1, \ldots, v_n\}$ of **vertices** and a finite multiset E of **edges**, where each edge is a pair $\{v_i, v_j\}$ of vertices (not necessarily distinct). If $v_i = v_j$, the edge is called a **loop**. A vertex v_i of an edge is called an **endpoint** of the edge.

The **order** of graph G is the number of vertices of G.

A **simple graph** is a graph with no loops where each edge has multiplicity at most one.

Two graphs (V, E) and (V', E') are **isomorphic** whenever there exist bijections $\phi : V \to V'$ and $\psi : E \to E'$, such that $v \in V$ is an endpoint of $e \in E$ if and only if $\phi(v)$ is an endpoint of $\psi(e)$.

A **walk** of **length** ℓ in a graph is an alternating sequence $(v_{i_0}, e_{i_1}, v_{i_1}, e_{i_2}, \ldots, e_{i_\ell}, v_{i_\ell})$ of vertices and edges (not necessarily distinct), such that $v_{i_{j-1}}$ and v_{i_j} are endpoints of e_{i_j} for $j = 1, \ldots, \ell$.

A **path** of **length** ℓ in a graph is a walk of length ℓ with all vertices distinct.

A **cycle** of **length** ℓ in a graph is a walk $(v_{i_0}, e_{i_1}, v_{i_1}, e_{i_2}, \ldots, e_{i_\ell}, v_{i_\ell})$ with $v_{i_0} = v_{i_\ell}$, $\ell \neq 0$, and $v_{i_1}, \ldots, v_{i_\ell}$ all distinct.

A **Hamilton cycle** in a graph is a cycle that includes all vertices.

A graph (V, E) is **connected** if $V \neq \emptyset$ and there exists a walk between any two distinct vertices of V.

The **distance** between two vertices v_i and v_j of a graph G (denoted by $d_G(v_i, v_j)$ or $d(v_i, v_j)$) is the length of a shortest path between v_i and v_j. ($d(v_i, v_j) = 0$ if $i = j$, and $d(v_i, v_j)$ is infinite if there is no path between v_i and v_j.)

The **diameter** of a connected graph G is the largest distance that occurs between two vertices of G.

A **tree** is a connected graph with no cycles.

A **forest** is a graph with no cycles.

A graph (V', E') is a **subgraph** of a graph (V, E) if $V' \subseteq V$ and $E' \subseteq E$. If E' contains all edges from E with endpoints in V', (V', E') is an **induced** subgraph of (V, E).

A **spanning subgraph** of a connected graph (V, E) is a subgraph (V', E') with $V' = V$, which is connected.

A **spanning tree** of a connected graph (V, E) is a spanning subgraph, which is a tree.

A **connected component** of a graph (V, E) is an induced subgraph (V', E'), which is connected and such that there exists no edge in E with one endpoint in V' and one outside V'. A connected component with one vertex and no edge is called an **isolated vertex**.

Two graphs (V, E) and (V', E') are **disjoint** if V and V' are disjoint sets.

Two vertices u and v are **adjacent** if there exists an edge with endpoints u and v. A vertex adjacent to v is called a **neighbor** of v.

The **degree** or **valency** of a vertex v of a graph G (denoted by $\delta_G(v)$ or $\delta(v)$) is the number of times that v occurs as an endpoint of an edge (that is, the number of edges containing v, where loops count as 2).

A graph (V, E) is **bipartite** if the vertex set V admits a partition into two parts, such that no edge of E has both endpoints in one part (thus, there are no loops). More information on bipartite graphs is given in Chapter 30.

A simple graph (V, E) is **complete** if E consists of all unordered pairs from V. The (isomorphism class of the) complete graph on n vertices is denoted by K_n.

A graph (V, E) is **empty** if $E = \emptyset$. If also $V = \emptyset$, it is called the **null graph**.

A bipartite simple graph (V, E) with nonempty parts V_1 and V_2 is **complete bipartite** if E consists of all unordered pairs from V with one vertex in V_1 and one in V_2. The (isomorphism class of the) complete bipartite graph is denoted by K_{n_1,n_2}, where $n_1 = |V_1|$ and $n_2 = |V_2|$.

The (isomorphism class of the) simple graph that consists only of vertices and edges of a path of length ℓ is called **the path of length** ℓ, and denoted by $P_{\ell+1}$.

The (isomorphism class of the) simple graph that consists only of vertices and edges of a cycle of length ℓ is called **the cycle of length** ℓ, and denoted by C_ℓ.

The **complement** of a simple graph $G = (V, E)$ is the simple graph $\overline{G} = (V, \overline{E})$, where $\overline{E}$ consists of all unordered pairs from V that are not in E.

The **union** $G \cup G'$ of two graphs $G = (V, E)$ and $G' = (V', E')$ is the graph with vertex set $V \cup V'$, and edge (multi)set $E \cup E'$.

The **intersection** $G \cap G'$ of two graphs $G = (V, E)$ and $G' = (V', E')$ is the graph with vertex set $V \cap V'$, and edge (multi)set $E \cap E'$.

The **join** $G + G'$ of two disjoint graphs $G = (V, E)$ and $G' = (V', E')$ is the union of $G \cup G'$ and the complete bipartite graph with vertex set $V \cup V'$ and partition $\{V, V'\}$.

The **(strong) product** $G \cdot G'$ of two simple graphs $G = (V, E)$ and $G' = (V', E')$ is the simple graph with vertex set $V \times V'$, where two distinct vertices are adjacent whenever in both coordinate places the vertices are adjacent or equal in the corresponding graph. The strong product of ℓ copies of a graph G is denoted by G^ℓ.

Facts:

The facts below are elementary results that can be found in almost every introduction to graph theory, such as [Har69] or [Wes01].

1. For any graph, the sum of its degrees equals twice the number of edges; therefore, the number of vertices with odd degree is even.
2. For any simple graph, at least two vertices have the same degree.
3. A graph G is bipartite if and only if G has no cycles of odd length.
4. A tree with n vertices has $n - 1$ edges.

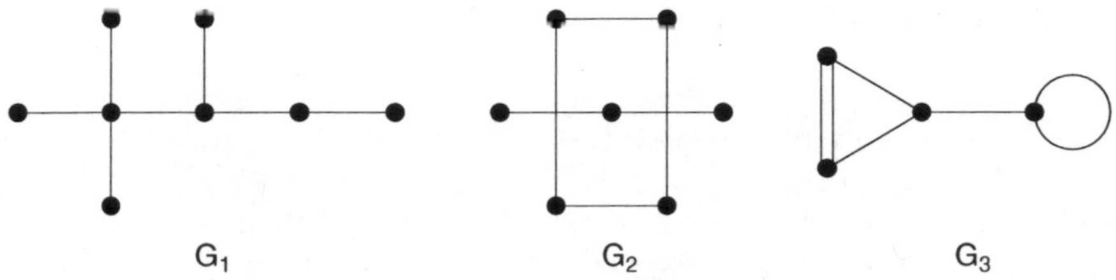

FIGURE 28.1 Three graphs. (Vertices are represented by points and an edge is represented by a line segment between the endpoints, or a loop.)

5. A graph is a tree if and only if there is a unique path between any two vertices.
6. A graph G is connected if and only if G cannot be expressed as the union of two or more mutually disjoint connected graphs.

Examples:

1. Consider the complete bipartite graph $K_{3,3} = (V_1 \cup V_2, E)$ with parts $V_1 = \{v_1, v_2, v_3\}$ and $V_2 = \{v_4, v_5, v_6\}$. Then

$$v_1, \{v_1, v_5\}, v_5, \{v_5, v_2\}, v_2, \{v_2, v_6\}, v_6, \{v_6, v_3\}, v_3$$

   is a path of length 4 between v_1 and v_3,

$$v_1, \{v_1, v_5\}, v_5, \{v_5, v_2\}, v_2, \{v_2, v_5\}, v_5, \{v_5, v_3\}, v_3$$

   is a walk of length 4, which is not a path, and

$$v_1, \{v_1, v_5\}, v_5, \{v_5, v_2\}, v_2, \{v_2, v_6\}, v_6, \{v_6, v_1\}, v_1$$

   is a cycle of length 4.
2. Graphs G_1 and G_2 in Figure 28.1 are simple, but G_3 is not.
3. Graphs G_1 and G_2 are bipartite, but G_3 is not.
4. Graph G_1 is a tree. Its diameter equals 4.
5. Graph G_2 is not connected; it is the union of two disjoint graphs, a path P_3 and a cycle C_4. The complement $\overline{G_2}$ is connected and can be expressed as the join of $\overline{P_3}$ and $\overline{C_4}$.
6. Graph G_3 contains three kinds of cycles, cycles of length 3 (corresponding to the two triangles), cycles of length 2 (corresponding to the pair of multiple edges), and one of length 1 (corresponding to the loop).

28.2 Special Graphs

A graph G is **regular** (or k-**regular**) if every vertex of G has the same degree (equal to k).

A graph G is **walk-regular** if for every vertex v the number of walks from v to v of length ℓ depends only on ℓ (not on v).

A simple graph G is **strongly regular** with parameters (n, k, λ, μ) whenever G has n vertices and

- G is k-regular with $1 \leq k \leq n-2$.
- Every two adjacent vertices of G have exactly λ common neighbors.
- Every two distinct nonadjacent vertices of G have exactly μ common neighbors.

An **embedding** of a graph in $\mathbb{R}^n$ consists of a representation of the vertices by distinct points in $\mathbb{R}^n$, and a representation of the edges by curve segments between the endpoints, such that a curve segment intersects another segment or itself only in an endpoint. (A curve segment between $\mathbf{x}$ and $\mathbf{y}$ is the range of a continuous map ϕ from $[0, 1]$ to $\mathbb{R}^n$ with $\phi(0) = \mathbf{x}$ and $\phi(1) = \mathbf{y}$.)

A graph is **planar** if it admits an embedding in $\mathbb{R}^2$.

A graph is **outerplanar** if it admits an embedding in $\mathbb{R}^2$, such that the vertices are represented by points on the unit circle, and the representations of the edges are contained in the unit disc.

A graph G is **linklessly embeddable** if it admits an embedding in $\mathbb{R}^3$, such that no two disjoint cycles of G are linked. (Two disjoint Jordan curves in $\mathbb{R}^3$ are linked if there is no topological 2-sphere in $\mathbb{R}^3$ separating them.)

Deletion of an edge e from a graph $G = (V, E)$ is the operation that deletes e from E and results in the subgraph $G - e = (V, E \setminus \{e\})$ of G.

Deletion of a vertex v from a graph $G = (V, E)$ is the operation that deletes v from V and all edges with endpoint v from E. The resulting subgraph of G is denoted by $G - v$.

Contraction of an edge e of a graph (V, E) is the operation that merges the endpoints of e in V, and deletes e from E.

A **minor** of a graph G is any graph that can be obtained from G by a sequence of edge deletions, vertex deletions, and contractions.

Let G be a simple graph. The **line graph** $L(G)$ of G has the edges of G as vertices, and vertices of $L(G)$ are adjacent if the corresponding edges of G have an endpoint in common.

The **cocktail party graph** $CP(a)$ is the graph obtained by deleting a disjoint edges from the complete graph K_{2a}. (Note that $CP(0)$ is the null graph.)

Let G be a simple graph with vertex set $\{v_1, \ldots, v_n\}$, and let $a_1, \ldots, a_n$ be nonnegative integers. The **generalized line graph** $L(G; a_1, \ldots, a_n)$ consists of the disjoint union of the line graph $L(G)$ and the cocktail party graphs $CP(a_1), \ldots, CP(a_n)$, together with all edges joining a vertex $\{v_i, v_j\}$ of $L(G)$ with each vertex of $CP(a_i)$ and $CP(a_j)$.

Facts:

If no reference is given, the fact is trivial or a classical result that can be found in almost every introduction to graph theory, such as [Har69] or [Wes01].

1. [God93, p. 81] A strongly regular graph is walk-regular.
2. A walk-regular graph is regular.
3. The complement of a strongly regular graph with parameters (n, k, λ, μ) is strongly regular with parameters $(n, n-k-1, n-2k+\mu-2, n-2k+\lambda)$.
4. Every graph can be embedded in $\mathbb{R}^3$.
5. [RS04] (Robertson, Seymour) For every graph property $\mathcal{P}$ that is closed under taking minors, there exists a finite list of graphs such that a graph G has property $\mathcal{P}$ if and only if no graph from the list is a minor of G.
6. The graph properties: planar, outerplanar, and linklessly embeddable are closed undertaking minors.
7. (Kuratowski, Wagner) A graph G is planar if and only if no minor of G is isomorphic to K_5 or $K_{3,3}$.
8. [CRS04, p. 8] A regular generalized line graph is a line graph or a cocktail party graph.
9. (Whitney) The line graphs of two connected nonisomorphic graphs G and G' are nonisomorphic, unless $\{G, G'\} = \{K_3, K_{1,3}\}$.

Examples:

1. Graph G_3 of Figure 28.1 is regular of degree 3.
2. The complete graph K_n is walk-regular and regular of degree $n-1$.
3. The complete bipartite graph $K_{k,k}$ is regular of degree k, walk-regular and strongly regular with parameters $(2k, k, 0, k)$.
4. Examples of outerplanar graphs are all trees, C_n, and $\overline{P_5}$.
5. Examples of graphs that are planar, but not outerplanar are: K_4, $CP(3)$, $\overline{C_6}$, and $K_{2,n-2}$ for $n \geq 5$.

6. Examples of graphs that are not planar, but linklessly embeddable are: K_5, and $K_{3,n-3}$ for $n \geq 6$.
7. The Petersen graph (Figure 28.2) and K_n for $n \geq 6$ are not linklessly embeddable.
8. The complete graph K_5 can be obtained from the Petersen graph by contraction with respect to five mutually disjoint edges. Therefore, K_5 is a minor of the Petersen graph.
9. The cycle C_9 is a subgraph of the Petersen graph and, therefore, the Petersen graph has every cycle C_ℓ with $\ell \leq 9$ as a minor.
10. Figure 28.3 gives a simple graph G, the line graph $L(G)$, and the generalized line graph $L(G; 2, 1, 0, 0, 0)$ (the vertices of G are ordered from left to right).
11. For $n \geq 4$ and $k \geq 2$ the line graphs $L(K_n)$ and $L(K_{k,k})$ and their complements are strongly regular. The complement of $L(K_5)$ is the Petersen graph.

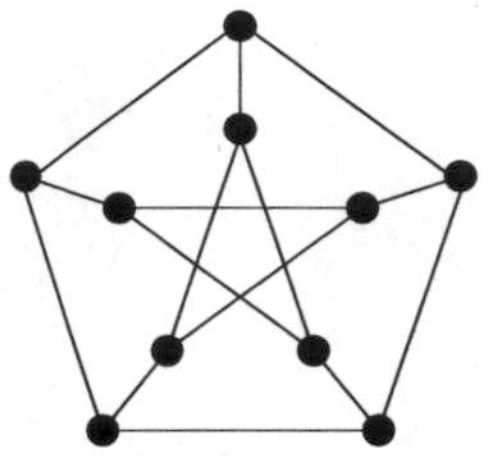

FIGURE 28.2 The Petersen graph.

28.3 The Adjacency Matrix and Its Eigenvalues

Definitions:

The **adjacency matrix** $\mathcal{A}_G$ of a graph G with vertex set $\{v_1, \dots, v_n\}$ is the symmetric $n \times n$ matrix, whose (i, j)th entry is equal to the number of edges between v_i and v_j.

The **eigenvalues** of a graph G are the eigenvalues of its adjacency matrix.

The **spectrum** $\sigma(G)$ of a graph G is the multiset of eigenvalues (that is, the eigenvalues with their multiplicities).

Two graphs are **cospectral** whenever they have the same spectrum.

A graph G is **determined by its spectrum** if every graph cospectral with G is isomorphic to G.

The **characteristic polynomial** $p_G(x)$ of a graph G is the characteristic polynomial of its adjacency matrix $\mathcal{A}_G$, that is, $p_G(x) = \det(xI - \mathcal{A}_G)$.

A **Hoffman polynomial** of a graph G is a polynomial $h(x)$ of minimum degree such that $h(\mathcal{A}_G) = J$.

The **main angles** of a graph G are the cosines of the angles between the eigenspaces of $\mathcal{A}_G$ and the all-ones vector **1**.

Facts:

If no reference is given, the fact is trivial or a standard result in algebraic graph theory that can be found in the classical books [Big74] and [CDS80].

1. If $\mathcal{A}_G$ is the adjacency matrix of a simple graph G, then $J - I - \mathcal{A}_G$ is the adjacency matrix of the complement of G.
2. If $\mathcal{A}_G$ and $\mathcal{A}_{G'}$ are adjacency matrices of simple graphs G and G', respectively, then $((\mathcal{A}_G + I) \otimes (\mathcal{A}_{G'} + I)) - I$ is the adjacency matrix of the strong product $G \cdot G'$.
3. Isomorphic graphs are cospectral.

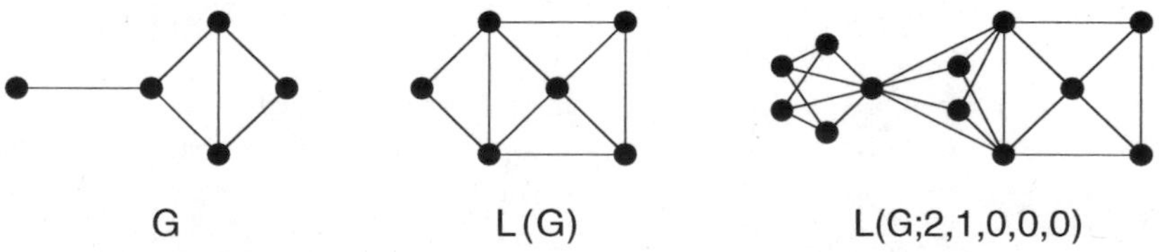

FIGURE 28.3 A graph with its line graph and a generalized line graph.

4. Let G be a graph with vertex set $\{v_1, \dots, v_n\}$ and adjacency matrix $\mathcal{A}_G$. The number of walks of length ℓ from v_i to v_j equals $(\mathcal{A}_G^{\ell})_{ij}$, i.e., the i, j-entry of $\mathcal{A}_G^{\ell}$.
5. The eigenvalues of a graph are real numbers.
6. The adjacency matrix of a graph is diagonalizable.
7. If $\lambda_1 \geq \dots \geq \lambda_n$ are the eigenvalues of a graph G, then $|\lambda_i| \leq \lambda_1$. If $\lambda_1 = \lambda_2$, then G is disconnected. If $\lambda_1 = -\lambda_n$ and G is not empty, then at least one connected component of G is nonempty and bipartite.
8. [CDS80, p. 87] If $\lambda_1 \geq \dots \geq \lambda_n$ are the eigenvalues of a graph G, then G is bipartite if and only if $\lambda_i = -\lambda_{n+1-i}$ for $i = 1, \dots, n$. For more information on bipartite graphs see Chapter 30.
9. If G is a simple k-regular graph, then the largest eigenvalue of G equals k, and the multiplicity of k equals the number of connected components of G.
10. [CDS80, p. 94] If $\lambda_1 \geq \dots \geq \lambda_n$ are the eigenvalues of a simple graph G with n vertices and m edges, then $\sum_i \lambda_i^2 = 2m \leq n\lambda_1$. Equality holds if and only if G is regular.
11. [CDS80, p. 95] A simple graph G has a Hoffman polynomial if and only if G is regular and connected.
12. [CRS97, p. 99] Suppose G is a simple graph with n vertices, r distinct eigenvalues $\nu_1, \dots, \nu_r$, and main angles $\beta_1, \dots, \beta_r$. Then the complement $\overline{G}$ of G has characteristic polynomial

$$p_{\overline{G}}(x) = (-1)^n p_G(-x-1)\left(1 - n\sum_{i=1}^{r} \beta_i^2/(x+1+\nu_i)\right).$$

13. [CDS80, p. 103], [God93, p. 179] A connected simple regular graph is strongly regular if and only if it has exactly three distinct eigenvalues. The eigenvalues $(\nu_1 > \nu_2 > \nu_3)$ and parameters (n, k, λ, μ) are related by $\nu_1 = k$ and

$$\nu_2, \nu_3 = \frac{1}{2}\left(\lambda - \mu \pm \sqrt{(\mu-\lambda)^2 + 4(k-\mu)}\right).$$

14. [BR91, p. 150], [God93, p. 180] The multiplicities of the eigenvalues ν_1, ν_2, and ν_3 of a connected strongly regular graph with parameters (n, k, λ, μ) are 1 and

$$\frac{1}{2}\left(n - 1 \pm \frac{(n-1)(\mu-\lambda) - 2k}{\sqrt{(\mu-\lambda)^2 + 4(k-\mu)}}\right) \quad \text{(respectively)}.$$

15. [GR01, p. 190] A regular simple graph with at most four distinct eigenvalues is walk-regular.
16. [CRS97, p. 79] Cospectral walk-regular simple graphs have the same main angles.
17. [Sch73] Almost all trees are cospectral with another tree.
18. [DH03] The number of nonisomorphic simple graphs on n vertices, not determined by the spectrum, is asymptotically bounded from below by $n^3 g_{n-1}(\frac{1}{24} - o(1))$, where g_{n-1} denotes the number of nonisomorphic simple graphs on $n-1$ vertices.
19. [DH03] The complete graph, the cycle, the path, the regular complete bipartite graph, and their complements are determined by their spectrum.
20. [DH03] Suppose G is a regular connected simple graph on n vertices, which is determined by its spectrum. Then also the complement $\overline{G}$ of G is determined by its spectrum, and if $n+1$ is not a square, also the line graph $L(G)$ of G is determined by its spectrum.
21. [CRS04, p. 7] A simple graph G is a generalized line graph if and only if the adjacency matrix $\mathcal{A}_G$ can be expressed as $\mathcal{A}_G = C^T C - 2I$, where C is an integral matrix with exactly two nonzero entries in each column. (It follows that the nonzero entries are ± 1.)
22. [CRS04, p. 7] A generalized line graph has smallest eigenvalue at least -2.
23. [CRS04, p. 85] A connected simple graph with more than 36 vertices and smallest eigenvalue at least -2 is a generalized line graph.
24. [CRS04, p. 90] There are precisely 187 connected regular simple graphs with smallest eigenvalue at least -2 that are not a line graph or a cocktail party graph. Each of these graphs has smallest eigenvalue equal to -2, at most 28 vertices, and degree at most 16.

$$\begin{bmatrix} 0 & 1 & 0 & 1 & 0 \\ 1 & 0 & 0 & 0 & 1 \\ 0 & 0 & 0 & 0 & 0 \\ 1 & 0 & 0 & 0 & 1 \\ 0 & 1 & 0 & 1 & 0 \end{bmatrix} \quad \begin{bmatrix} 0 & 0 & 1 & 0 & 0 \\ 0 & 0 & 1 & 0 & 0 \\ 1 & 1 & 0 & 1 & 1 \\ 0 & 0 & 1 & 0 & 0 \\ 0 & 0 & 1 & 0 & 0 \end{bmatrix}$$

FIGURE 28.4 Two cospectral graphs with their adjacency matrices.

Examples:

1. Figure 28.4 gives a pair of nonisomorphic bipartite graphs with their adjacency matrices. Both matrices have spectrum $\{2,\ 0^3,\ -2\}$ (exponents indicate multiplicities), so the graphs are cospectral.
2. The main angles of the two graphs of Figure 28.4 (with the given ordering of the eigenvalues) are $2/\sqrt{5},\ 1/\sqrt{5},\ 0$ and $3/\sqrt{10},\ 0, 1/\sqrt{10}$, respectively.
3. The spectrum of K_{n_1,n_2} is $\{\sqrt{n_1 n_2},\ 0^{n-2},\ -\sqrt{n_1 n_2}\}$.
4. By Fact 14, the multiplicities of the eigenvalues of any strongly regular graph with parameters $(n, k, 1, 1)$ would be nonintegral, so no such graph can exist (this result is known as the Friendship theorem).
5. The Petersen graph has spectrum $\{3,\ 1^5,\ -2^4\}$ and Hoffman polynomial $(x-1)(x+2)$. It is one of the 187 connected regular graphs with least eigenvalue -2, which is neither a line graph nor a cocktail party graph.
6. The eigenvalues of the path P_n are $2\cos\frac{i\pi}{n+1}$ $(i = 1, \ldots, n)$.
7. The eigenvalues of the cycle C_n are $2\cos\frac{2i\pi}{n}$ $(i = 1, \ldots, n)$.

28.4 Other Matrix Representations

Definitions:

Let G be a simple graph with adjacency matrix $\mathcal{A}_G$. Suppose D is the diagonal matrix with the degrees of G on the diagonal (with the same vertex ordering as in $\mathcal{A}_G$). Then $L_G = D - \mathcal{A}_G$ is the **Laplacian matrix** of G (often abbreviated to the **Laplacian**, and also known as **admittance matrix**), and the matrix $|L_G| = D + \mathcal{A}_G$ is (sometimes) called the **signless Laplacian matrix.**

The **Laplacian eigenvalues** of a simple graph G are the eigenvalues of the Laplacian matrix L_G.

If $\mu_1 \leq \mu_2 \leq \ldots \leq \mu_n$ are the Laplacian eigenvalues of G, then μ_2 is called the **algebraic connectivity** of G. (See section 28.6 below.)

Let G be simple graph with vertex set $\{v_1, \ldots, v_n\}$. A symmetric real matrix $M = [m_{ij}]$ is called a **generalized Laplacian** of G, whenever $m_{ij} < 0$ if v_i and v_j are adjacent, and $m_{ij} = 0$ if v_i and v_j are nonadjacent and distinct (nothing is required for the diagonal entries of M).

Let G be a graph without loops with vertex set $\{v_1, \ldots, v_n\}$ and edge set $\{e_1, \ldots, e_m\}$. The **(vertex-edge) incidence matrix** of G is the $n \times m$ matrix N_G defined by $(N_G)_{ij} = 1$ if vertex v_i is an endpoint of edge e_j and $(N_G)_{ij} = 0$ otherwise.

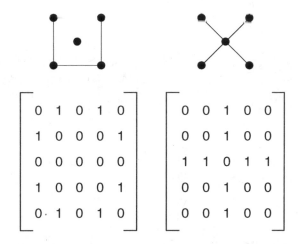

An **oriented (vertex-edge) incidence matrix** of G is a matrix N'_G obtained from N_G by replacing a 1 in each column by a -1, and thereby orienting each edge of G.

If $\mathcal{A}_G$ is the adjacency matrix of a simple graph G, then $S_G = J - I - 2\mathcal{A}_G$ is the **Seidel matrix** of G.

Let G be a simple graph with Seidel matrix S_G, and let I' be a diagonal matrix with ± 1 on the diagonal. Then the simple graph G' with Seidel matrix $S_{G'} = I'S_G I'$ is **switching equivalent** to G. The graph operation that changes G into G' is called **Seidel switching.**

Facts:

In all facts below, G is a simple graph. If no reference is given, the fact is trivial or a classical result that can be found in [BR91].

1. Let G be a simple graph. The Laplacian matrix L_G and the signless Laplacian $|L_G|$ are positive semidefinite.
2. The nullity of L_G is equal to the number of connected components of G.
3. The nullity of $|L_G|$ is equal to the number of connected components of G that are bipartite.
4. [DH03] The Laplacian and the signless Laplacian of a graph G have the same spectrum if and only if G is bipartite.
5. (Matrix-tree theorem) Let G be a graph with Laplacian matrix L_G, and let c_G denote the number of spanning trees of G. Then $\text{adj}(L_G) = c_G J$.
6. Suppose N_G is the incidence matrix of G. Then $N_G N_G^T = |L_G|$ and $N_G^T N_G - 2I = \mathcal{A}_{L(G)}$.
7. Suppose N'_G is an oriented incidence matrix of G. Then $N'_G N_G'^T = L_G$.
8. If $\mu_1 \leq \ldots \leq \mu_n$ are the Laplacian eigenvalues of G, and $\overline{\mu}_1 \leq \ldots \leq \overline{\mu}_n$ are the Laplacian eigenvalues of $\overline{G}$, then $\mu_1 = \overline{\mu}_1 = 0$ and $\overline{\mu}_i = n - \mu_{n+2-i}$ for $i = 2, \ldots, n$.
9. [DH03] If $\mu_1 \leq \ldots \leq \mu_n$ are the Laplacian eigenvalues of a graph G with n vertices and m edges, then $\sum_i \mu_i = 2m \leq \sqrt{n \sum_i \mu_i(\mu_i - 1)}$ with equality if and only if G is regular.
10. [DH98] A connected graph G has at most three distinct Laplacian eigenvalues if and only if there exist integers μ and $\overline{\mu}$, such that any two distinct nonadjacent vertices have exactly μ common neighbors, and any two adjacent vertices have exactly $\overline{\mu}$ common nonneighbors.
11. If G is k-regular and $\mathbf{v} \notin \text{span}\{\mathbf{1}\}$, then the following are equivalent:
    - λ is an eigenvalue of $\mathcal{A}_G$ with eigenvector $\mathbf{v}$.
    - $k - \lambda$ is an eigenvalue of L_G with eigenvector $\mathbf{v}$.
    - $k + \lambda$ is an eigenvalue of $|L_G|$ with eigenvector $\mathbf{v}$.
    - $-1 - 2\lambda$ is an eigenvalue of S_G with eigenvector $\mathbf{v}$.
12. [DH03] Consider a simple graph G with n vertices and m edges. Let $\nu_1 \leq \ldots \leq \nu_n$ be the eigenvalues of $|L_G|$, the signless Laplacian of G. Let $\lambda_1 \geq \ldots \geq \lambda_m$ be the eigenvalues of $L(G)$, the line graph of G. Then $\lambda_i = \nu_{n-i+1} - 2$ if $1 \leq i \leq \min\{m, n\}$, and $\lambda_i = -2$ if $\min\{m, n\} < i \leq m$.
13. [GR01, p. 298] Let G be a connected graph, let M be a generalized Laplacian of G, and let $\mathbf{v}$ be an eigenvector for M corresponding to the second smallest eigenvalue of M. Then the subgraph of G induced by the vertices corresponding to the positive entries of $\mathbf{v}$ is connected.
14. The Seidel matrices of switching equivalent graphs have the same spectrum.

FIGURE 28.5 Graphs with cospectral Laplacian matrices.

FIGURE 28.6 Graphs with cospectral signless Laplacian matrices.

Examples:

1. The Laplacian eigenvalues of the Petersen graph are $\{0,\ 2^5,\ 5^4\}$.
2. The two graphs of Figure 28.5 are nonisomorphic, but the Laplacian matrices have the same spectrum. Both Laplacian matrices have $12J$ as adjugate, so both have 12 spanning trees. They are not cospectral with respect to the adjacency matrix because one is bipartite and the other one is not.
3. Figure 28.6 gives two graphs with cospectral signless Laplacian matrices. They are not cospectral with respect to the adjacency matrix because one is bipartite and the other one is not. They also do not have cospectral Laplacian matrices because the numbers of components differ.
4. The eigenvalues of the Laplacian and the signless Laplacian matrix of the path P_n are $2 + 2\cos\frac{i\pi}{n}$ $(i = 1, \ldots, n)$.
5. The complete bipartite graph K_{n_1,n_2} is Seidel switching equivalent to the empty graph on $n = n_1 + n_2$ vertices. The Seidel matrices have the same spectrum, being $\{n-1, -1^{n-1}\}$.

28.5 Graph Parameters

Definitions:

A subgraph G' on n' vertices of a simple graph G is a **clique** if G' is isomorphic to the complete graph $K_{n'}$. The largest value of n' for which a clique with n' vertices exists is called the **clique number** of G and is denoted by $\omega(G)$.

An induced subgraph G' on n' vertices of a graph G is a **coclique** or **independent set of vertices** if G' has no edges. The largest value of n' for which a coclique with n' vertices exists is called the **vertex independence number** of G and is denoted by $\iota(G)$. Note that the standard notation for the vertex independence number of G is $\alpha(G)$, but $\iota(G)$ is used here due to conflict with the use of $\alpha(G)$ to denote the algebraic connectivity of G in Chapter 36.

The **Shannon capacity** $\Theta(G)$ of a simple graph G is defined by $\Theta(G) = \sup_\ell \sqrt[\ell]{\iota(G^\ell)}$

A **vertex coloring** of a graph is a partition of the vertex set into cocliques. A coclique in such a partition is called a **color class.**

The **chromatic number** $\chi(G)$ of a graph G is the smallest number of color classes of any vertex coloring of G. (The chromatic number is not defined if G has loops.)

For a simple graph $G = (V, E)$, the **conductance** or **isoperimetric number** $\Phi(G)$ is defined to be the minimum value of $\partial(V')/|V'|$ over any subset $V' \subset V$ with $|V'| \leq |V|/2$, where $\partial(V')$ equals the number of edges in E with one endpoint in V' and one endpoint outside V'.

An infinite family of graphs with constant degree and isoperimetric number bounded from below is called a family of **expanders.**

A symmetric real matrix M is said to satisfy the **Strong Arnold Hypothesis** provided there does not exist a symmetric nonzero matrix X with zero diagonal, such that $MX = \mathbf{0}$, $M \circ X = \mathbf{0}$.

The **Colin de Verdière parameter** $\mu(G)$ of a simple graph G is the largest nullity of any generalized Laplacian M of G satisfying the following:

- M has exactly one negative eigenvalue of multiplicity 1.
- The Strong Arnold Hypothesis.

Consider a simple graph G with vertex set $\{v_1, \ldots, v_n\}$. The **Lovász parameter** $\vartheta(G)$ is the minimum value of the largest eigenvalue $\lambda_1(M)$ of any real symmetric $n \times n$ matrix $M = [m_{ij}]$, which satisfies $m_{ij} = 1$ if v_i and v_j are nonadjacent (including the diagonal).

Consider a simple graph G with vertex set $\{v_1, \ldots, v_n\}$. The integer $\eta(G)$ is defined to be the smallest rank of any $n \times n$ matrix M (over any field), which satisfies $m_{ii} \neq 0$ for $i = 1, \ldots, n$ and $m_{ij} = 0$, if v_i and v_j are distinct nonadjacent vertices.

Facts:

In the facts below, all graphs are simple.

1. [Big74, p. 13] A connected graph with r distinct eigenvalues (for the adjacency, the Laplacian or the signless Laplacian matrix) has diameter at most $r - 1$.
2. [CDS80, pp. 90–91], [God93, p. 83] The chromatic number $\chi(G)$ of a graph G with adjacency eigenvalues $\lambda_1 \geq \ldots \geq \lambda_n$ satisfies: $1 - \lambda_1/\lambda_n \leq \chi(G) \leq 1 + \lambda_1$.
3. [CDS80, p. 88] For a graph G, let m_+ and m_- denote the number of nonnegative and nonpositive adjacency eigenvalues, respectively. Then $\iota(G) \leq \min\{m_+, m_-\}$.
4. [GR01, p. 204] If G is a k-regular graph with adjacency eigenvalues $\lambda_1 \geq \ldots \geq \lambda_n$, then $\omega(G) \leq \frac{n(\lambda_2+1)}{n-k+\lambda_2}$ and $\iota(G) \leq \frac{-n\lambda_n}{k-\lambda_n}$.
5. [Moh97] Suppose G is a graph with maximum degree Δ and algebraic connectivity μ_2. Then the isoperimetric number $\Phi(G)$ satisfies $\mu_2/2 \leq \Phi(G) \leq \sqrt{\mu_2(2\Delta - \mu_2)}$.
6. [HLS99] The Colin de Verdière parameter $\mu(G)$ is minor monotonic, that is, if H is a minor of G, then $\mu(H) \leq \mu(G)$.
7. [HLS99] If G has at least one edge, then $\mu(G) = \max\{\mu(H) \mid H \text{ is a component of } G\}$.
8. [HLS99] The Colin de Verdière parameter $\mu(G)$ satisfies the following:
   - $\mu(G) \leq 1$ if and only if G is the disjoint union of paths.
   - $\mu(G) \leq 2$ if and only if G is outerplanar.
   - $\mu(G) \leq 3$ if and only if G is planar.
   - $\mu(G) \leq 4$ if and only if G is linklessly embeddable.
9. (Sandwich theorems)[Lov79], [Hae81] The parameters $\vartheta(G)$ and $\eta(G)$ satisfy: $\iota(G) \leq \vartheta(G) \leq \chi(\overline{G})$ and $\iota(G) \leq \eta(G) \leq \chi(\overline{G})$.
10. [Lov79], [Hae81] The parameters $\vartheta(G)$ and $\eta(G)$ satisfy: $\vartheta(G \cdot H) = \vartheta(G)\vartheta(H)$ and $\eta(G \cdot H) \leq \eta(G)\eta(H)$.
11. [Lov79], [Hae81] The Shannon capacity $\Theta(G)$ of a graph G satisfies: $\iota(G) \leq \Theta(G), \Theta(G) \leq \vartheta(G)$, and $\Theta(G) \leq \eta(G)$.
12. [Lov79], [Hae81] If G is a k-regular graph with eigenvalues $k = \lambda_1 \geq \ldots \geq \lambda_n$, then $\vartheta(G) \leq -n\lambda_n/(k - \lambda_n)$. Equality holds if G is strongly regular.
13. [Lov79] The Lovász parameter $\vartheta(G)$ can also be defined as the maximum value of $\text{tr}(MJ_n)$, where M is any positive semidefinite $n \times n$ matrix, satisfying $\text{tr}(M) = 1$ and $m_{ij} = 0$ if v_i and v_j are adjacent vertices in G.

Examples:

1. Suppose G is the Petersen graph. Then $\iota(G) = 4$, $\vartheta(G) = 4$ (by Facts 9 and 12). Thus, $\Theta(G) = 4$ (by Fact 11). Moreover, $\chi(G) = 3$, $\chi(\overline{G}) = 5$, $\mu(G) = 5$ (take $M = L_G - 2I$), and $\eta(G) = 4$ (take $M = \mathcal{A}_G + I$ over the field with two elements).
2. The isoperimetric number $\Phi(G)$ of the Petersen graph equals 1. Indeed, $\Phi(G) \geq 1$, by Fact 5, and any pentagon gives $\Phi(G) \leq 1$.
3. $\mu(K_n) = n - 1$ (take $M = -J$).
4. If G is the empty graph with at least two vertices, then $\mu(G) = 1$. (M must be a diagonal matrix with exactly one negative entry, and the Strong Arnold Hypothesis forbids two or more diagonal entries to be 0.)

5. By Fact 12, $\vartheta(C_5) = \sqrt{5}$. If $(v_1, \ldots, v_5)$ are the vertices of C_5, cyclically ordered, then (v_1, v_1), (v_2, v_3), (v_3, v_5), (v_4, v_2), (v_5, v_4) is a coclique of size 5 in $C_5 \cdot C_5$. Thus, $\iota(C_5 \cdot C_5) \geq 5$ and, therefore, $\Theta(C_5) = \sqrt{5}$.

28.6 Association Schemes

Definitions:

A set of graphs $G_0, \ldots, G_d$ on a common vertex set $V = \{v_1, \ldots, v_n\}$ is an **association scheme** if the adjacency matrices $A_0, \ldots, A_d$ satisfy:

- $A_0 = I$.
- $\sum_{i=0}^{d} A_i = J$.
- $\text{span}\{A_0, \ldots, A_d\}$ is closed under matrix multiplication.

The numbers $p_{i,j}^k$ defined by $A_i A_j = \sum_{i=0}^{d} p_{i,j}^k A_k$ are called the **intersection numbers** of the association scheme.

The (associative) algebra spanned by $A_0, \ldots, A_d$ is the **Bose–Mesner algebra** of the association scheme.

Consider a connected graph $G_1 = (V, E_1)$ with diameter d. Define $G_i = (V, E_i)$ to be the graph wherein two vertices are adjacent if their distance in G_1 equals i. If $G_0, \ldots, G_d$ is an association scheme, then G_1 is a **distance-regular graph**.

Let V' be a subset of the vertex set V of an association scheme. The **inner distribution** $\mathbf{a} = [a_0, \ldots, a_d]^T$ of V' is defined by $a_i|V'| = \mathbf{c}^T A_i \mathbf{c}$, where $\mathbf{c}$ is the characteristic vector of V' (that is, $c_i = 1$ if $v_i \in V$ and $c_i = 0$ otherwise).

Facts:

Facts 1 to 7 below are standard results on association schemes that can be found in any of the following references: [BI84], [BCN89], [God93].

1. Suppose $G_0, \ldots, G_d$ is an association scheme. For any three integers $i, j, k \in \{0, \ldots, d\}$ and for any two vertices x and y adjacent in G_k, the number of vertices z adjacent to x in G_i and to y in G_j equals the intersection number p_{ij}^k. In particular, G_i is regular of degree $k_i = p_{ii}^0$ $(i \neq 0)$.
2. The matrices of a Bose–Mesner algebra $\mathcal{A}$ can be diagonalized simultaneously. In other words, there exists a nonsingular matrix S such that SAS^{-1} is a diagonal matrix for every $A \in \mathcal{A}$.
3. A Bose–Mesner algebra has a basis $\{E_0 = \frac{1}{n}J, E_1, \ldots, E_d\}$ of idempotents, that is, $E_i E_j = \delta_{i,j} E_i$ ($\delta_{i,j}$ is the Kronecker symbol).
4. The change-of-coordinates matrix $P = [p_{ij}]$ defined by $A_j = \sum_i p_{ij} E_i$ satisfies:
   - p_{ij} is an eigenvalue of A_j with eigenspace range(E_i).
   - $p_{i0} = 1$, $p_{0i} = k_i$ (the degree of G_i $(i \neq 0)$).
   - $nk_j(P^{-1})_{ji} = m_i p_{ij}$, where $m_i = \text{rank}(E_i)$ (the multiplicity of eigenvalue p_{ij}).
5. (Krein condition) The Bose–Mesner algebra of an association scheme is closed under Hadamard multiplication. The numbers $q_{i,j}^k$, defined by $E_i \circ E_j = \sum_k q_{i,j}^k E_k$, are nonnegative.
6. (Absolute bound) The multiplicities $m_0 = 1, m_1, \ldots, m_d$ of an association scheme satisfy

$$\sum_{k:q_{i,j}^k > 0} m_k \leq m_i m_j \quad \text{and} \quad \sum_{k:q_{i,i}^k > 0} m_k \leq m_i(m_i + 1)/2.$$

7. A connected strongly regular graph is distance-regular with diameter two.
8. [BCN89, p. 55] Let V' be a subset of the vertex set V of an association scheme with change-of-coordinates matrix P. The inner distribution $\mathbf{a}$ of V' satisfies $\mathbf{a}^T P^{-1} \geq \mathbf{0}$.

Examples:

1. The change-of-coordinates matrix P of a strongly regular graph with eigenvalues k, ν_2, and ν_3 is equal to

$$\begin{bmatrix} 1 & k & n-k-1 \\ 1 & \nu_2 & -\nu_2-1 \\ 1 & \nu_3 & -\nu_3-1 \end{bmatrix}.$$

2. A strongly regular graph with parameters $(28, 9, 0, 4)$ cannot exist, because it violates Facts 5 and 6.
3. The Hamming association scheme $H(d, q)$ has vertex set $V = Q^d$, the set of all vectors with d entries from a finite set Q of size q. Two such vectors are adjacent in G_i if they differ in exactly i coordinate places. The graph G_1 is distance-regular. The matrix P of a Hamming association scheme can be expressed in terms of Kravčuk polynomials, which gives

$$p_{ij} = \sum_{k=0}^{j} (-1)^k (q-1)^{j-k} \binom{i}{k} \binom{d-i}{j-k}.$$

4. An error correcting code with minimum distance δ is a subset V' of the vertex set V of a Hamming association scheme, such that V' induces a coclique in $G_1, \ldots, G_{\delta-1}$. If $\mathbf{a}$ is the inner distribution of V', then $a_0 = 1$, $a_1 = \cdots = a_{\delta-1} = 0$, and $|V'| = \sum_i a_i$. Therefore, by Fact 8, the linear programming problem "Maximize $\sum_{i \geq \delta} a_i$, subject to $\mathbf{a}^T P^{-1} \geq 0$ (with $a_0 = 1$ and $a_1 = \cdots = a_{\delta-1} = 0$)" leads to an upper bound for the size of an error correcting code with given minimum distance. This bound is known as Delsarte's Linear Programming Bound.
5. The Johnson association scheme $J(d, \ell)$ has as vertex set V all subsets of size d of a set of size ℓ ($\ell \geq 2d$). Two vertices are adjacent in G_i if the intersection of the corresponding subsets has size $d - i$. The graph G_1 is distance-regular. The matrix P of a Johnson association scheme can be expressed in terms of Eberlein polynomials, which gives:

$$p_{ij} = \sum_{k=0}^{j} (-1)^k \binom{i}{k} \binom{d-i}{j-k} \binom{\ell-d-i}{j-k}.$$

References

[BW04] L.W. Beineke and R.J. Wilson (Eds.). *Topics in Algebraic Graph Theory.* Cambridge University Press, Cambridge, 2005.

[BI84] Eiichi Bannai and Tatsuro Ito. *Algebraic Combinatorics I: Association Schemes.* The Benjamin/Cummings Publishing Company, London, 1984.

[Big74] N.L. Biggs, *Algebraic Graph Theory.* Cambridge University Press, Cambridge, 1974. (2nd ed., 1993.)

[BCN89] A.E. Brouwer, A.M. Cohen, and A. Neumaier. *Distance-Regular Graphs.* Springer, Heidelberg, 1989.

[BR91] Richard A. Brualdi and Herbert J. Ryser. *Combinatorial Matrix Theory.* Cambridge University Press, Cambridge, 1991.

[CDS80] Dragǒs M. Cvetković, Michael Doob, and Horst Sachs. *Spectra of Graphs: Theory and Applications.* Deutscher Verlag der Wissenschaften, Berlin, 1980; Academic Press, New York, 1980. (3rd ed., Johann Abrosius Barth Verlag, Heidelberg-Leipzig, 1995.)

[CRS97] Dragǒs Cvetković, Peter Rowlinson, and Slobodan Simić. *Eigenspaces of Graphs.* Cambridge University Press, Cambridge, 1997.

[CRS04] Dragǒs Cvetković, Peter Rowlinson, and Slobodan Simić. *Spectral Generalizations of Line Graphs: On graphs with Least Eigenvalue −2.* Cambridge University Press, Cambridge, 2004.

[DH98] Edwin R. van Dam and Willem H. Haemers. Graphs with constant μ and $\overline{\mu}$. *Discrete Math.* 182: 293–307, 1998.

[DH03] Edwin R. van Dam and Willem H. Haemers. Which graphs are determined by their spectrum? *Lin. Alg. Appl.*, 373: 241–272, 2003.

[God93] C.D. Godsil. *Algebraic Combinatorics.* Chapman and Hall, New York, 1993.

[GR01] Chris Godsil and Gordon Royle. *Algebraic Graph Theory.* Springer-Verlag, New York, 2001.

[Hae81] Willem H. Haemers. An upper bound for the Shannon capacity of a graph. *Colloqua Mathematica Societatis János Bolyai 25* (proceedings "Algebraic Methods in Graph Theory," Szeged, 1978). North-Holland, Amsterdam, 1981, pp. 267–272.

[Har69] Frank Harary. *Graph Theory.* Addison-Wesley, Reading, MA, 1969.

[HLS99] Hein van der Holst, László Lovász, and Alexander Schrijver. The Colin de Verdière graph parameter. *Graph Theory and Combinatorial Biology* (L. Lovász, A. Gyárfás, G. Katona, A. Recski, L. Székely, Eds.). János Bolyai Mathematical Society, Budapest, 1999, pp. 29–85.

[Lov79] László Lovász. On the Shannon Capacity of a graph. *IEEE Trans. Inform. Theory,* 25: 1–7, 1979.

[Moh97] Bojan Mohar. Some applications of Laplace eigenvalues of Graphs. *Graph Symmetry: Algebraic Methods and Applications* (G. Hahn, G. Sabidussi, Eds.). Kluwer Academic Publishers, Dordrecht, 1997, pp. 225–275.

[RS04] Neil Robertson and P.D. Seymour. Graph Minors XX: Wagner's conjecture. *J. Combinatorial Theory, Ser. B.* 92: 325–357, 2004.

[Sch73] A.J. Schwenk. Almost all trees are cospectral, in *New directions in the theory of graphs* (F. Harary, Ed.). Academic Press, New York 1973, pp. 275–307.

[Wes01] Douglas West. *Introduction to Graph Theory.* 2nd ed., Prentice Hall, Upper Saddle River, NJ, 2001.

29

Digraphs and Matrices

Jeffrey L. Stuart
Pacific Lutheran University

Directed graphs, often called digraphs, have much in common with graphs, which were the subject of the previous chapter. While digraphs are of interest in their own right, and have been the subject of much research, this chapter focuses on those aspects of digraphs that are most useful to matrix theory. In particular, it will be seen that digraphs can be used to understand how the zero–nonzero structure of square matrices affects matrix products, determinants, inverses, and eigenstructure. Basic material on digraphs and their adjacency matrices can be found in many texts on graph theory, nonnegative matrix theory, or combinatorial matrix theory. For all aspects of digraphs, except their spectra, see [BG00]. Perhaps the most comprehensive single source for results, proofs, and references to original papers on the interplay between digraphs and matrices is [BR91, Chapters 3 and 9]. Readers preferring a matrix analytic rather than combinatorial approach to irreducibility, primitivity, and their consequences, should consult [BP94, Chapter 2].

29.1 Digraphs

Definitions:

A **directed graph** $\Gamma = \Gamma(V, E)$ consists of a finite, nonempty set V of **vertices** (sometimes called **nodes**), together with a multiset E of elements of $V \times V$, whose elements are called **arcs** (sometimes called **edges**, **directed edges**, or **directed arcs**).

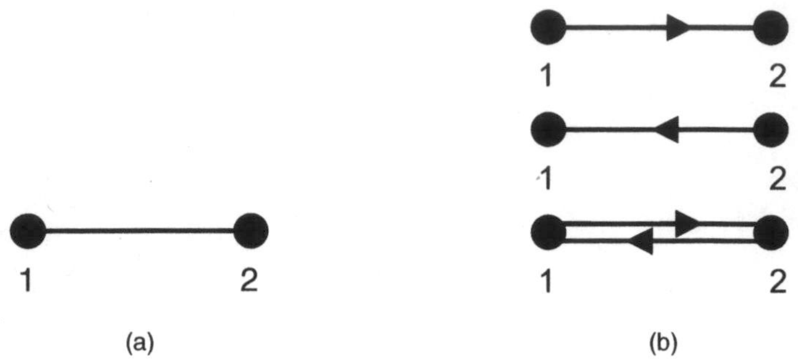

FIGURE 29.1

A **loop** is an arc of the form (v, v) for some vertex v.

If there is more than one arc (u, v) for some u and v in V, then Γ is called a **directed multigraph**.

If there is at most one arc (u, v) for each u and v in V, then Γ is called a **digraph**.

If the digraph Γ contains no loops, then Γ is called a **simple** digraph.

A **weighted digraph** is a digraph Γ with a **weight function** $w : E \to \mathbf{F}$, where the set **F** is often the real or complex numbers.

A **subdigraph** Γ' of a digraph Γ is a digraph $\Gamma' = \Gamma'(V', E')$ such that $V' \subseteq V$ and $E' \subseteq E$.

A **proper** subdigraph Γ' of a digraph Γ is a subdigraph of Γ such that $V' \subset V$ or $E' \subset E$.

If V' is a nonempty subset of V, then the **induced subdigraph of** Γ **induced by** V' is the digraph with vertex set V' whose arcs are those arcs in E that lie in $V' \times V'$.

A **walk** is a sequence of arcs $(v_0, v_1), (v_1, v_2), \ldots, (v_{k-1}, v_k)$, where one or more vertices may be repeated.

The **length** of a walk is the number of arcs in the walk. (Note that some authors define the length to be the number of vertices rather than the number of arcs.)

A **simple walk** is a walk in which all vertices, except possibly v_0 and v_k, are distinct. (Note that some authors use **path** to mean what we call a simple walk.)

A **cycle** is a simple walk for which $v_0 = v_k$. A cycle of length k is called a k**-cycle.**

A **generalized cycle** is either a cycle passing through all vertices in V or else a union of cycles such that every vertex in V lies on exactly one cycle.

Let $\Gamma = \Gamma(V, E)$ be a digraph. The **undirected graph** G **associated with the digraph** Γ is the undirected graph with vertex set V, whose edge set is determined as follows: There is an edge between vertices u and v in G if and only if at least one of the arcs (u, v) and (v, u) is present in Γ.

The digraph Γ is **connected** if the associated undirected graph G is connected.

The digraph Γ is a **tree** if the associated undirected graph G is a tree.

The digraph Γ is a **doubly directed tree** if the associated undirected graph is a tree and if whenever (i, j) is an arc in Γ, (j, i) is also an arc in Γ.

Examples:

1. The key distinction between a graph on a vertex set V and a digraph on the same set is that for a graph we refer to the *edge between vertices* u *and* v, whereas for a digraph, we have two arcs, the *arc from* u *to* v and the *arc from* v *to* u. Thus, there is one connected, simple graph on two vertices, K_2 (see Figure 29.1a), but there are three possible connected, simple digraphs on two vertices (see Figure 29.1b). Note that all graphs in Figure 29.1 are trees, and that the graph in Figure 29.1a is the undirected graph associated with each of the digraphs in Figure 29.1b. The third graph in Figure 29.1b is a doubly directed tree.

2. Let Γ be the digraph in Figure 29.2. Then (1,1), (1,1), (1,3), (3,1), (1,2) is a walk of length 5 from vertex 1 to vertex 2. $(1, 2), (2, 3)$ is a simple walk of length 2. $(1, 1)$ is a 1-cycle; $(1, 3), (3, 1)$ is a 2-cycle; and $(1, 2), (2, 3), (3, 1)$ is a 3-cycle. $(1, 1), (2, 3), (3, 2)$ and $(1, 2), (2, 3), (3, 1)$ are two generalized cycles. Not all digraphs contain generalized cycles; consider the digraph obtained by deleting the arc $(2, 3)$ from Γ, for example. Unless we are emphasizing a particular vertex on a cycle, such as all cycles starting (and ending) at vertex v, we view cyclic permutations of a cycle as equivalent. That is, in Figure 29.2, we would speak of *the* 3-cycle, although technically $(1, 2), (2, 3), (3, 1)$; $(2, 3), (3, 1), (1, 2)$; and $(3, 1), (1, 2), (2, 3)$ are distinct cycles.

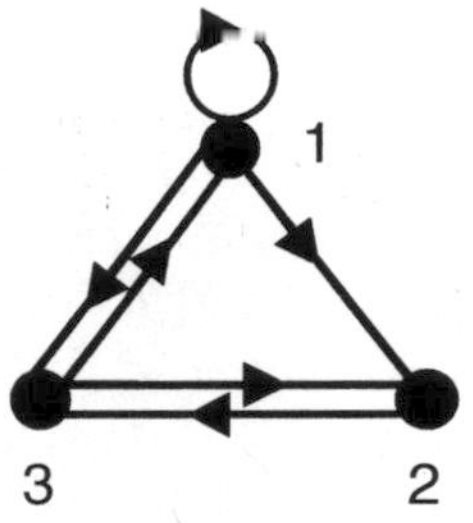

FIGURE 29.2

29.2 The Adjacency Matrix of a Directed Graph and the Digraph of a Matrix

If Γ is a directed graph on n vertices, then there is a natural way that we can record the arc information for Γ in an $n \times n$ matrix. Conversely, if A is an $n \times n$ matrix, we can naturally associate a digraph Γ on n vertices with A.

Definitions:

Let Γ be a digraph with vertex set V. Label the vertices in V as $v_1, v_2, \ldots, v_n$. Once the vertices have been ordered, the **adjacency matrix** for Γ, denoted $\mathcal{A}_\Gamma$, is the 0, 1-matrix whose entries a_{ij} satisfy: $a_{ij} = 1$ if (v_i, v_j) is an arc in Γ, and $a_{ij} = 0$ otherwise. When the set of vertex labels is $\{1, 2, \ldots, n\}$, the default labeling of the vertices is $v_i = i$ for $1 \leq i \leq n$.

Let A be an $n \times n$ matrix. Let V be the set $\{1, 2, \ldots, n\}$. Construct a digraph denoted $\Gamma(A)$ on V as follows. For each i and j in V, let (i, j) be an arc in Γ exactly when $a_{ij} \neq 0$. $\Gamma(A)$ is called the **digraph of the matrix** A. Commonly Γ is viewed as a weighted digraph with weight function $w((i, j)) = a_{ij}$ for all (i, j) with $a_{ij} \neq 0$.

Facts: [BR91, Chap. 3]

1. If a digraph H is obtained from a digraph Γ by adding or removing an arc, then $\mathcal{A}_H$ is obtained by changing the corresponding entry of $\mathcal{A}_\Gamma$ to a 1 or a 0, respectively. If a digraph H is obtained from a digraph Γ by deleting the ith vertex in the ordered set V and by deleting all arcs in E containing the ith vertex, then $\mathcal{A}_H = \mathcal{A}_\Gamma(i)$. That is, $\mathcal{A}_H$ is obtained by deleting row i and column i of $\mathcal{A}_\Gamma$.
2. Given one ordering of the vertices in V, any other ordering of those vertices is simply a permutation of the original ordering. Since the rows and columns of $A = \mathcal{A}_\Gamma$ are labeled by the ordered vertices, reordering the vertices in Γ corresponds to simultaneously permuting the rows and columns of A. That is, if P is the permutation matrix corresponding to a permutation of the vertices of V, then the new adjacency matrix is PAP^T. Since $P^T = P^{-1}$ for a permutation matrix, all algebraic properties preserved by similarity transformations are invariant under changes of the ordering of the vertices.
3. Let Γ be a digraph with vertex set V. The Jordan canonical form of an adjacency matrix for Γ is independent of the ordering applied to the vertices in V. Consequently, all adjacency matrices for Γ have the same rank, trace, determinant, minimum polynomial, characteristic polynomial, and spectrum.
4. If A is an $n \times n$ matrix and if $v_i = i$ for $1 \leq i \leq n$, then A and $\mathcal{A}_{\Gamma(A)}$ have the same zero–nonzero pattern and, hence, $\Gamma(A) = \Gamma(\mathcal{A}_{\Gamma(A)})$.

Examples:

1. For the digraph Γ given in Figure 29.3, if we order the vertices as $v_i = i$ for $i = 1, 2, \ldots, 7$, then

$$A = \mathcal{A}_\Gamma = \begin{bmatrix} 0 & 1 & 1 & 1 & 0 & 0 & 0 \\ 1 & 0 & 0 & 0 & 1 & 0 & 0 \\ 0 & 0 & 0 & 0 & 0 & 1 & 0 \\ 0 & 0 & 1 & 0 & 0 & 0 & 0 \\ 0 & 1 & 0 & 0 & 0 & 0 & 0 \\ 0 & 0 & 1 & 0 & 0 & 1 & 1 \\ 0 & 0 & 0 & 0 & 0 & 0 & 1 \end{bmatrix}.$$

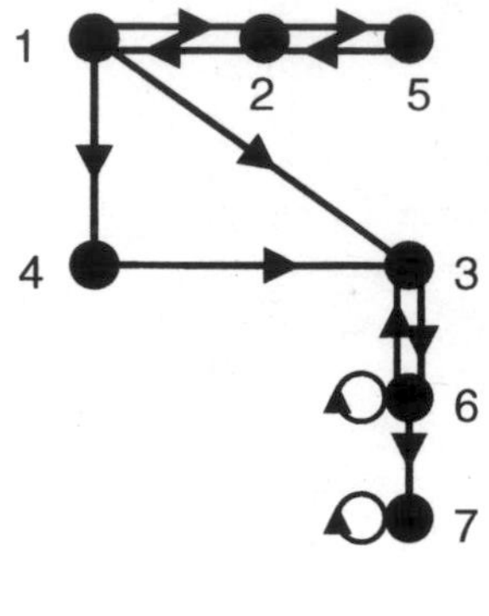

FIGURE 29.3

If we reorder the vertices in the digraph Γ given in Figure 29.3 so that $v_1, v_2, \cdots, v_7$ is the sequence $1, 2, 5, 4, 3, 6, 7$, then the new adjacency matrix is

$$B = \mathcal{A}_\Gamma = \left[\begin{array}{ccc|c|cc|c} 0 & 1 & 0 & 1 & 1 & 0 & 0 \\ 1 & 0 & 1 & 0 & 0 & 0 & 0 \\ 0 & 1 & 0 & 0 & 0 & 0 & 0 \\ \hline 0 & 0 & 0 & 0 & 1 & 0 & 0 \\ \hline 0 & 0 & 0 & 0 & 0 & 1 & 0 \\ 0 & 0 & 0 & 0 & 1 & 1 & 1 \\ \hline 0 & 0 & 0 & 0 & 0 & 0 & 1 \end{array}\right].$$

2. If A is the 3×3 matrix

$$A = \begin{bmatrix} 3 & -2 & 5 \\ 0 & 0 & -11 \\ 9 & -6 & 0 \end{bmatrix},$$

then $\Gamma(A)$ is the digraph given in Figure 29.2. Up to permutation similarity,

$$\mathcal{A}_\Gamma = \begin{bmatrix} 1 & 1 & 1 \\ 0 & 0 & 1 \\ 1 & 1 & 0 \end{bmatrix}.$$

29.3 Walk Products and Cycle Products

For a square matrix A, $a_{12}a_{23}$ is nonzero exactly when both a_{12} and a_{23} are nonzero. That is, exactly when both $(1, 2)$ and $(2, 3)$ are arcs in $\Gamma(A)$. Note also that $a_{12}a_{23}$ is one summand in $(A^2)_{13}$. Consequently, there is a close connection between powers of a matrix A and walks in its digraph $\Gamma(A)$. In fact, the signs (complex arguments) of walk products play a fundamental role in the study of the matrix sign patterns for real matrices (matrix ray patterns for complex matrices). See Chapter 33 [LHE94] or [Stu03].

Definitions:

Let A be an $n \times n$ matrix. Let W given by $(v_0, v_1), (v_1, v_2), \ldots, (v_{k-1}, v_k)$ be a walk in $\Gamma(A)$. The **walk product** for the walk W is

$$\prod_{j=1}^{k} a_{v_{j-1}, v_j},$$

and is often denoted by $\prod_W a_{ij}$. This product is a generic summand of the (v_0, v_k)-entry of A^k. If $s_1, s_2 \ldots, s_n$ are scalars, $\prod_W s_i$ denotes the ordinary product of the s_i over the index set $v_0, v_1, v_2, \ldots, v_{k-1}, v_k$.

If W is a cycle in the directed graph $\Gamma(A)$, then the walk product for W is called a **cycle product.**

Let A be an $n \times n$ real or complex matrix. Define $|A|$ to be the matrix obtained from A by replacing a_{ij} with $|a_{ij}|$ for all i and j.

Facts:

1. Let A be a square matrix. The walk W given by $(v_0, v_1), (v_1, v_2), \ldots, (v_{k-1}, v_k)$ occurs in $\Gamma(A)$ exactly when $a_{v_0 v_1} a_{v_1 v_2} \cdots a_{v_{k-1} v_k}$ is nonzero.
2. [BR91, Sec. 3.4] [LHE94] Let A be a square matrix. For each positive integer k, and for all i and j, the (i, j)-entry of A^k is the sum of the walk products for all length k walks in $\Gamma(A)$ from i to j. Further, there is a walk of length k from i to j in $\Gamma(A)$ exactly when the (i, j)-entry of $|A|^k$ is nonzero.
3. [BR91, Sec. 3.4] [LHE94] Let A be a square, real or complex matrix. For all positive integers k, $\Gamma(A^k)$ is a subdigraph of $\Gamma(|A|^k)$. Further, $\Gamma(A^k)$ is a proper subgraph of $\Gamma(|A|^k)$ exactly when additive cancellation occurs in summing products for length k walks from some i to some j.
4. [LHE94] Let A be a square, real matrix. The sign pattern of the kth power of A is determined solely by the sign pattern of A when the signs of the entries in A are assigned so that for each ordered pair of vertices, all products of length k walks from the first vertex to the second have the same sign.
5. [FP69] Let A and B be irreducible, real matrices with $\Gamma(A) = \Gamma(B)$. There exists a nonsingular, real diagonal matrix D such that $B = DAD^{-1}$ if and only if the cycle product for every cycle in $\Gamma(A)$ equals the cycle product for the corresponding cycle in $\Gamma(B)$.
6. Let A be an irreducible, real matrix. There exists a nonsingular, real diagonal matrix D such that DAD^{-1} is nonnegative if and only if the cycle product for every cycle in $\Gamma(A)$ is positive.

Examples:

1. If $A = \begin{bmatrix} 1 & 1 \\ 1 & -1 \end{bmatrix}$, then $(A^2)_{12} = a_{11}a_{12} + a_{12}a_{22} = (1)(1) + (1)(-1) = 0$, whereas $(|A|^2)_{12} = 2$.
2. If A is the matrix in Example 2 of the previous section, then $\Gamma(A)$ contains four cycles: the loop $(1, 1)$; the two 2-cycles $(1, 3), (3, 1)$ and $(2, 3), (3, 2)$; and the 3-cycle $(1, 2), (2, 3), (3, 1)$. Each of these cycles has a positive cycle product and using $D = diag(1, -1, 1)$, DAD^{-1} is nonnegative.

29.4 Generalized Cycle Products

If the matrix A is 2×2, then $\det(A) = a_{11}a_{22} - a_{12}a_{22}$. Assuming that the entries of A are nonzero, the two summands $a_{11}a_{22}$ and $a_{12}a_{21}$ are exactly the walk products for the two generalized cycles of $\Gamma(A)$. From Chapter 4.1, the determinant of an $n \times n$ matrix A is the sum of all terms of the form $(-1)^{sign(\sigma)} a_{1j_1} a_{2j_2} \cdots a_{nj_n}$, where $\sigma = (j_1, j_2, \ldots, j_n)$ is a permutation of the ordered set $\{1, 2, \ldots, n\}$. Such a summand is nonzero precisely when $(1, j_1), (2, j_2), \cdots, (n, j_n)$ are all arcs in $\Gamma(A)$. For this set of arcs, there is exactly one arc originating at each of the n vertices and exactly one arc terminating at each of the n vertices. Hence, the arcs correspond to a generalized cycle in $\Gamma(A)$. See [BR91, Sect. 9.1]. Since the eigenvalues of A are the roots of $\det(\lambda I - A)$, it follows that results connecting the cycle structure of a matrix to its determinant should play a key role in determining the spectrum of the matrix. For further results connecting determinants and generalized cycles, see [MOD89] or [BJ86]. Generalized cycles play a crucial role in the study of the nonsingularity of sign patterns. (See Chapter 33 or [Stu91]). In general, there are fewer results for the spectra of digraphs than for the spectra of graphs. There are generalizations of Geršgorin's Theorem for spectral inclusion regions for complex matrices that depend on directed cycles. (See Chapter 14.2 [Bru82] or [BR91, Sect. 3.6], or especially, [Var04]).

Definitions:

Let A be an $n \times n$ matrix. Let $\sigma = \begin{pmatrix} 1 & 2 & \cdots & n \\ j_1 & j_2 & \cdots & j_n \end{pmatrix}$ be a permutation of the ordered set $\{1, 2, \ldots, n\}$. When (i, j_i) is an arc in $\Gamma(A)$ for each i, the cycle or vertex disjoint union of cycles with arcs $(1, j_1)$, $(2, j_2), \ldots, (n, j_n)$ is called the **generalized cycle induced by** σ.

The product of the cycle products for the cycle(s) comprising the generalized cycle induced by σ is called the **generalized cycle product** corresponding to σ.

Facts:

1. The entries in A that correspond to a generalized cycle are a diagonal of A and vice versa.
2. If σ is a permutation of the ordered set $\{1, 2, \ldots, n\}$, then the nonzero entries of the $n \times n$ permutation matrix P corresponding to σ are precisely the diagonal of P corresponding to the generalized cycle induced by σ.
3. [BR91, Sec. 9.1] Let A be an $n \times n$ real or complex matrix. Then $\det(A)$ is the sum over all permutations of the ordered set $\{1, 2, \ldots, n\}$ of all of the signed generalized cycle products for $\Gamma(A)$ where the sign of a generalized cycle is determined as $(-1)^{n-k}$, where k is the number of disjoint cycles in the generalized cycle. If $\Gamma(A)$ contains no generalized cycle, then A is singular. If $\Gamma(A)$ contains at least one generalized cycle, then A is nonsingular unless additive cancellation occurs in the sum of the signed generalized cycle products.
4. [Cve75] [Har62] Let A be an $n \times n$ real or complex matrix. The coefficient of x^{n-k} in $\det(xI - A)$ is the sum over all induced subdigraphs H of $\Gamma(A)$ on k vertices of the signed generalized cycle products for H.
5. [BP94, Chap. 3], [BR97, Sec. 1.8] Let A be an $n \times n$ real or complex matrix. Let g be the greatest common divisor of the lengths of all cycles in $\Gamma(A)$. Then $\det(xI - A) = x^k p(x^g)$ for some nonnegative integer k and some polynomial $p(z)$ with $p(0) \neq 0$. Consequently, the spectrum of A is invariant under $\frac{2\pi}{g}$ rotations of the complex plane. Further, if A is nonsingular, then g divides n, and $\det(xI - A) = p(x^g)$ for some polynomial $p(z)$ with $p(0) = (-1)^n \det(A)$.

Examples:

1. If A is the matrix in Example 2 of the previous section, then $\Gamma(A)$ contains two generalized cycles — the loop $(1, 1)$ together with the 2-cycle $(2, 3), (3, 2)$; and the 3-cycle $(1, 2), (2, 3), (3, 1)$. The corresponding generalized cycle products are $(3)(-11)(-6) = 198$ and $(-2)(-11)(9) = 198$, with corresponding signs -1 and 1, respectively. Thus, $\det(A) = 0$ is a consequence of additive cancellation.
2. If $A = \begin{bmatrix} 1 & 0 \\ 0 & -1 \end{bmatrix}$, then the only cycles in $\Gamma(A)$ are loops, so $g = 1$. The spectrum of A is clearly invariant under $\frac{2\pi}{g}$ rotations, but it is also invariant under rotations through the smaller angle of π.

29.5 Strongly Connected Digraphs and Irreducible Matrices

Irreducibility of a matrix, which can be defined in terms of permutation similarity (see Section 27.3), and which Frobenius defined as an algebraic property in his extension of Perron's work on the spectra of positive matrices, is equivalent to the digraph property of being strongly connected, defined in this section. Today, most discussions of the celebrated Perron–Frobenius Theorem (see Chapter 9) use digraph theoretic terminology.

Definitions:

Vertex u **has access to** vertex v in a digraph Γ if there exists a walk in Γ from u to v. By convention, every vertex has access to itself even if there is no walk from that vertex to itself.

If u and v are vertices in a digraph Γ such that u has access to v, and such that v has access to u, then u and v are **access equivalent** (or u and v **communicate**).

Access equivalence is an equivalence relation on the vertex set V of Γ that partitions V into **access equivalence classes.**

If V_1 and V_2 are nonempty, disjoint subsets of V, then V_1 **has access to** V_2 if some vertex in v_1 in V_1 has access in Γ to some vertex v_2 in V_2.

For a digraph Γ, the subdigraphs induced by each of the access equivalence classes of V are the **strongly connected components** *of* Γ.

When all of the vertices of Γ lie in a single access equivalence class, Γ is **strongly connected.**

Let $V_1, V_2, \ldots, V_k$ be the access equivalence classes for some digraph Γ. Define a new digraph, $R(\Gamma)$, called the **reduced digraph** (also called the **condensation digraph**) for Γ as follows. Let $W = \{1, 2, \ldots, k\}$ be the vertex set for $R(\Gamma)$. If $i, j \in W$ with $i \neq j$, then (i, j) is an arc in $R(\Gamma)$ precisely when V_i has access to V_j.

Facts: [BR91, Chap. 3]

1. [BR91, Sec. 3.1] A digraph Γ is strongly connected if and only if there is a walk from each vertex in Γ to every other vertex in Γ.
2. The square matrix A is irreducible if and only if $\Gamma(A)$ is strongly connected. A reducible matrix A is completely reducible if $\Gamma(A)$ is a disjoint union of two or more strongly connected digraphs.
3. [BR91, Sec. 3.1] Suppose that V_1 and V_2 are distinct access equivalence classes for some digraph Γ. If any vertex in V_1 has access to any vertex in V_2, then every vertex in V_1 has access to every vertex in V_2. Further, exactly one of the following holds: V_1 has access to V_2, V_2 has access to V_1, or neither has access to the other. Consequently, access induces a partial order on the access equivalence classes of vertices.
4. [BR91, Lemma 3.2.3] The access equivalence classes for a digraph Γ can be labelled as $V_1, V_2, \ldots, V_k$ so that whenever there is an arc from a vertex in V_i to a vertex in V_j, $i \leq j$.
5. [Sch86] If Γ is a digraph, then $R(\Gamma)$ is a simple digraph that contains no cycles. Further, the vertices in $R(\Gamma)$ can always be labelled so that if (i, j) is an arc in $R(\Gamma)$, then $i < j$.
6. [BR91, Theorem 3.2.4] Suppose that Γ is not strongly connected. Then there exists at least one ordering of the vertices in V so that $\mathcal{A}_\Gamma$ is block upper triangular, where the diagonal blocks of $\mathcal{A}_\Gamma$ are the adjacency matrices of the strongly connected components of Γ.
7. [BR91, Theorem 3.2.4] Let A be a square matrix. Then A has a Frobenius normal form. (See Chapter 27.3.)
8. The Frobenius normal form of a square matrix A is not necessarily unique. The set of Frobenius normal forms for A is preserved by permutation similarities that correspond to permutations that reorder the vertices within the access equivalence classes of $\Gamma(A)$. If B is a Frobenius normal form for A, then all the arcs in $R(\Gamma(B))$ satisfy (i, j) is an edge implies $i < j$. Let σ be any permutation of the vertices of $R(\Gamma(B))$ such that $(\sigma(i), \sigma(j))$ is an edge in $R(\Gamma(B))$ implies $\sigma(i) < \sigma(j)$. Let B be block partitioned by the access equivalence classes of $\Gamma(B)$. Applying the permutation similarity corresponding to σ to the blocks of B produces a Frobenius normal form for A. All Frobenius normal forms for A are produced using combinations of the above two types of permutations.
9. [BR91, Sect. 3.1] [BP94, Chap. 3] Let A be an $n \times n$ matrix for $n \geq 2$. The following are equivalent:

   (a) A is irreducible.

   (b) $\Gamma(A)$ is strongly connected.

   (c) For each i and j, there is a positive integer k such that $(|A|^k)_{ij} > 0$.

   (d) There does not exist a permutation matrix P such that

$$PAP^T = \left[\begin{array}{c|c} A_{11} & A_{12} \\ \hline \mathbf{0} & A_{22} \end{array}\right],$$

   where A_{11} and A_{22} are square matrices.

10. Let A be a square matrix. A is completely reducible if and only if there exists a permutation matrix P such that PAP^T is a direct sum of at least two irreducible matrices.
11. All combinations of the following transformations preserve irreducibility, reducibility, and complete reducibility: Scalar multiplication by a nonzero scalar; transposition; permutation similarity; left or right multiplication by a nonsingular, diagonal matrix.
12. Complex conjugation preserves irreducibility, reducibility, and complete reducibility for square, complex matrices.

Examples:

1. In the digraph Γ in Figure 29.3, vertex 4 has access to itself and to vertices 3, 6, and 7, but not to any of vertices 1, 2, or 5. For Γ, the access equivalence classes are: $V_1 = \{1, 2, 5\}$, $V_2 = \{3, 6\}$, $V_3 = \{4\}$, and $V_4 = \{7\}$. The strongly connected components of Γ are given in Figure 29.4, and $R(\Gamma)$ is given in Figure 29.5. If the access equivalence classes for Γ are relabeled so that $V_2 = \{4\}$ and $V_3 = \{3, 6\}$, then the labels on vertices 2 and 3 switch in the reduced digraph given in Figure 29.5. With this labeling, if (i, j) is an arc in the reduced digraph, then $i \leq j$.

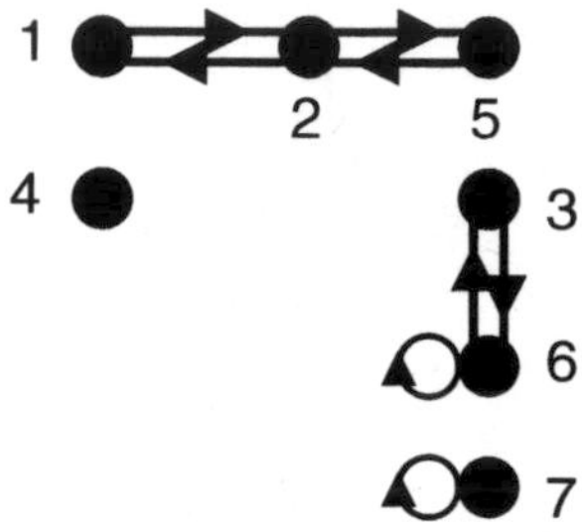

FIGURE 29.4

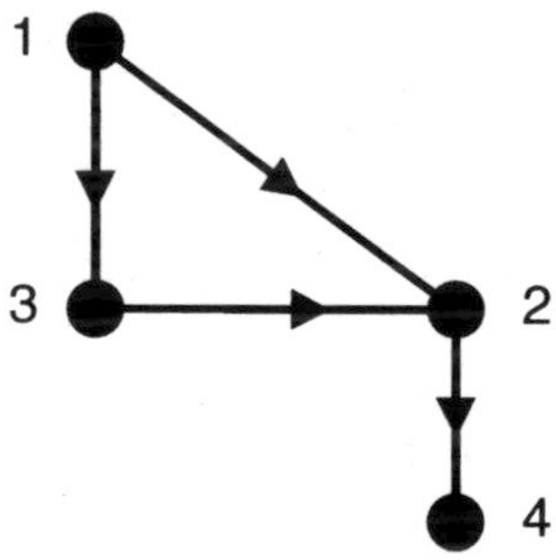

FIGURE 29.5

2. If A and B are the adjacency matrices of Example 1 of Section 29.2, then $A(3, 4, 6, 7) = A[1, 2, 5]$ is the adjacency matrix for the largest strongly connected component of the digraph Γ in Figure 29.3 using the first ordering of V; and using the second ordering of V, B is block-triangular and the irreducible, diagonal blocks of B are the adjacency matrices for each of the four strongly connected components of Γ. B is a Frobenius normal form for A.
3. The matrix A in Example 2 of Section 29.2 is irreducible, hence, it is its own Frobenius normal form.

29.6 Primitive Digraphs and Primitive Matrices

Primitive matrices, defined in this section, are necessarily irreducible. Unlike irreducibility, primitivity depends not just on the matrix A, but also its powers, and, hence, on the signs of the entries of the original matrix. Consequently, most authors restrict discussions of primitivity to nonnegative matrices. Much work has been done on bounding the exponent of a primitive matrix; see [BR91, Sec. 3.5] or [BP94, Sec. 2.4]. One consequence of the fifth fact stated below is that powers of sparse matrices and inverses of sparse matrices can experience substantial fill-in.

Definitions:

A digraph Γ with at least two vertices is **primitive** if there is a positive integer k such that for every pair of vertices u and v (not necessarily distinct), there exists at least one walk of length k from u to v. A digraph on a single vertex is **primitive** if there is a loop on that vertex.

The **exponent** of a digraph Γ (sometimes called **the index of primitivity**) is the smallest value of k that works in the definition of primitivity.

A digraph Γ is **imprimitive** if it is not primitive. This includes the simple digraph on one vertex.

If A is a square, nonnegative matrix such that $\Gamma(A)$ is primitive with exponent k, then A is called a **primitive matrix** *with* **exponent** k.

Facts:

1. A primitive digraph must be strongly connected, but not conversely.
2. A strongly connected digraph with at least one loop is primitive.
3. [BP94, Chap. 3] [BR91, Sections 3.2 and 3.4] Let Γ be a strongly connected digraph with at least two vertices. The following are equivalent:

   (a) Γ is primitive.

   (b) The greatest common divisor of the cycle lengths for Γ is 1 (i.e., Γ is aperiodic, cf. Chapter 9.2).

   (c) There is a smallest positive integer k such that for each $t \geq k$ and each pair of vertices u and v in Γ, there is a walk of length t from u to v.

4. [BR91, Sect. 3.5] Let Γ be a primitive digraph with $n \geq 2$ vertices and exponent k. Then

   (a) $k \leq (n-1)^2 + 1$.

   (b) If s is the length of the shortest cycle in Γ, then $k \leq n + s(n-2)$.

   (c) If Γ has $p \geq 1$ loops, then $k \leq 2n - p - 1$.

5. [BR91, Theorem 3.4.4] Let A be an $n \times n$ nonnegative matrix with $n \geq 2$. The matrix A is primitive if and only if there exists a positive integer k such that A^k is positive. When such a positive integer k exists, the smallest such k is the exponent of A. Further, if A^k is positive, then A^h is positive for all integers $h \geq k$. A nonnegative matrix A with the property that some power of A is positive is also called **regular**. See Chapter 9 and Chapter 54 for more information about primitive matrices and their uses.
6. [BP94, Chap. 3] If A is an irreducible, nonnegative matrix with positive trace, then A is primitive.
7. [BP94, Chap. 6] Let A be a nonnegative, tridiagonal matrix with all entries on the first superdiagonal and on the first subdiagonal positive, and at least one entry on the main diagonal positive. Then A is primitive and, hence, some power of A is positive. Further, if $s > \rho(A)$ where $\rho(A)$ is the spectral radius of A, then the tridiagonal matrix $sI - A$ is a nonsingular M-matrix with a positive inverse.

Examples:

1. The digraph Γ in Figure 29.2 is primitive with exponent 3; the strongly connected digraph in Figure 29.1*b* is not primitive.
2. Let $A_1 = \begin{bmatrix} 1 & 1 \\ 1 & 1 \end{bmatrix}$ and let $A_2 = \begin{bmatrix} 1 & -1 \\ 1 & -1 \end{bmatrix}$. Note that $A_1 = |A_2|$. Clearly, $\Gamma(A_1) = \Gamma(A_2)$ is an irreducible, primitive digraph with exponent 1. For all positive integers k, A_1^k is positive, so it makes sense to call A_1 a primitive matrix. In contrast, $A_2^k = \mathbf{0}$ for all integers $k \geq 2$.

29.7 Irreducible, Imprimitive Matrices and Cyclic Normal Form

While most authors restrict discussions of matrices with primitive digraphs to nonnegative matrices, many authors have exploited results for imprimitive digraphs to understand the structure of real and complex matrices with imprimitive digraphs.

Definitions:

Let A be an irreducible $n \times n$ matrix with $n \geq 2$ such that $\Gamma(A)$ is imprimitive. The greatest common divisor $g > 1$ of the lengths of all cycles in $\Gamma(A)$ is called the **index of imprimitivity of** A (or **period** of A).

If there is a permutation matrix P such that

$$PAP^T = \begin{bmatrix} \mathbf{0} & A_1 & \mathbf{0} & \cdots & \mathbf{0} \\ \mathbf{0} & \mathbf{0} & A_2 & \cdots & \mathbf{0} \\ \vdots & \vdots & \vdots & \ddots & \vdots \\ \mathbf{0} & \mathbf{0} & \mathbf{0} & \cdots & A_{g-1} \\ A_g & \mathbf{0} & \mathbf{0} & \cdots & \mathbf{0} \end{bmatrix},$$

where each of the diagonal blocks is a square zero matrix, then the matrix PAP^T is called a **cyclic normal form** for A.

By convention, when A is primitive, A is said to be its own cyclic normal form.

Facts:

1. [BP94, Sec. 2.2] [BR91, Sections 3.4] [Min88, Sec. 3.3–3.4] Let A be an irreducible matrix with index of imprimitivity $g > 1$. Then there exists a permutation matrix P such that PAP^T is a cyclic normal form for A. Further, the cyclic normal form is unique up to cyclic permutation of the blocks A_j and permutations within the partition sets of V of $\Gamma(A)$ induced by the partitioning of PAP^T. Finally, if A is real or complex, then $|A_1|\,|A_2| \cdots |A_g|$ is irreducible and nonzero.
2. [BP94, Sec. 3.3] [BR91, Sec. 3.4] If A is an irreducible matrix with index of imprimitivity $g > 1$, and if there exists a permutation matrix P and a positive integer k such that

$$PAP^T = \begin{bmatrix} \mathbf{0} & A_1 & \mathbf{0} & \cdots & \mathbf{0} \\ \mathbf{0} & \mathbf{0} & A_2 & \cdots & \mathbf{0} \\ \vdots & \vdots & \vdots & \ddots & \vdots \\ \mathbf{0} & \mathbf{0} & \mathbf{0} & \cdots & A_{k-1} \\ A_k & \mathbf{0} & \mathbf{0} & \cdots & \mathbf{0} \end{bmatrix},$$

where each diagonal block is square zero matrix, then k divides g. Conversely, if A is real or complex, if PAP^T has the specified form for some positive integer k, if PAP^T has no zero rows and no zero columns, and if $|A_1|\,|A_2| \cdots |A_k|$ is irreducible, then A is irreducible, and k divides g.
3. [BR91, Sec. 3.4] Let A be an irreducible, nonnegative matrix with index of imprimitivity $g > 1$. Let m be a positive integer. Then A^m is irreducible if and only if m and g are relatively prime. If A^m is reducible, then it is completely reducible, and it is permutation similar to a direct sum of r irreducible matrices for some positive integer r. Further, either each of these summands is primitive (when $g/r = 1$), or each of these summands has index of imprimitivity $g/r > 1$.
4. [Min88, Sec. 3.4] Let A be an irreducible, nonnegative matrix in cyclic normal form with index of imprimitivity $g > 1$. Suppose that for $1 \leq i \leq k-1$, A_i is $n_i \times n_{i+1}$ and that A_g is $n_g \times n_1$. Let $k = \min\left(n_1, n_2, \ldots, n_g\right)$. Then 0 is an eigenvalue for A with multiplicity at least $n - gk$; and if A is nonsingular, then each $n_i = n/g$.
5. [Min88, Sec. 3.4] If A is an irreducible, nonnegative matrix in cyclic normal form with index of imprimitivity g, then for $j = 1, 2, \ldots, g$, reading the indices modulo g, $B_j = \prod_{i=j}^{j+g-1} A_i$ is irreducible. Further, all of the matrices B_j have the same nonzero eigenvalues. If the nonzero eigenvalues (not necessarily distinct) of B_1 are $\omega_1, \omega_2, \ldots, \omega_m$ for some positive integer m, then the spectrum of A consists of 0 with multiplicity $n - gm$ together with the complete set of gth roots of each of the ω_i.

6. Let A be a square matrix. Then A has a Frobenius normal form for which each irreducible, diagonal block is in cyclic normal form.
7. Let A be a square matrix. If A is reducible, then the spectrum of A (which is a multiset) is the union of the spectra of the irreducible, diagonal blocks of any Frobenius normal form for A.
8. Explicit, efficient algorithms for computing the index of imprimitivity and the cyclic normal form for an imprimitive matrix and for computing the Frobenius normal form for a matrix can be found in [BR91, Sec. 3.7].
9. All results stated here for block upper triangular forms have analogs for block lower triangular forms.

Examples:

1. If A is the 4×4 matrix $A = \begin{bmatrix} \mathbf{0} & A_1 \\ A_2 & \mathbf{0} \end{bmatrix}$, where $A_1 = I_2$ and $A_2 = \begin{bmatrix} 0 & 1 \\ 1 & 0 \end{bmatrix}$, then A and $A_1 A_2 = A_2$ are irreducible but $g \neq 2$. In fact, $g = 4$ since A is actually a permutation matrix corresponding to the permutation (1324). Also note that when $A_1 = A_2 = I_2$, A is completely reducible since $\Gamma(A)$ consists of two disjoint cycles.
2. If M is the irreducible matrix $M = \begin{bmatrix} 0 & -1 & 0 & -1 \\ 6 & 0 & 3 & 0 \\ 0 & 2 & 0 & 2 \\ 6 & 0 & 3 & 0 \end{bmatrix}$, then $g = 2$, and using the permutation matrix $Q = \begin{bmatrix} 0 & 0 & 0 & 1 \\ 0 & 1 & 0 & 0 \\ 0 & 0 & 1 & 0 \\ 1 & 0 & 0 & 0 \end{bmatrix}$, $N = QMQ^T = \left[\begin{array}{cc|cc} 0 & 0 & 3 & 6 \\ 0 & 0 & 3 & 6 \\ \hline 2 & 2 & 0 & 0 \\ -1 & -1 & 0 & 0 \end{array}\right]$ is a cyclic normal form for M. Note that $|N_1|\,|N_2| = \begin{bmatrix} 3 & 6 \\ 3 & 6 \end{bmatrix} \begin{bmatrix} 2 & 2 \\ 1 & 1 \end{bmatrix} = \begin{bmatrix} 12 & 12 \\ 12 & 12 \end{bmatrix}$ is irreducible even though $N_1 N_2 = \begin{bmatrix} 0 & 0 \\ 0 & 0 \end{bmatrix}$ is not irreducible.
3. The matrix B in Example 1 of Section 29.2 is a Frobenius normal form for the matrix A in that example. Observe that B_{11} is imprimitive with $g = 2$, but B_{11} is not in cyclic normal form. The remaining diagonal blocks, B_{22} and B_{33}, have primitive digraphs, hence, they are in cyclic normal form. Let $Q = \begin{bmatrix} 0 & 1 & 0 \\ 1 & 0 & 0 \\ 0 & 0 & 1 \end{bmatrix}$. Then $QB_{11}Q^T = \left[\begin{array}{c|cc} 0 & 3 & 6 \\ \hline 2 & 0 & 0 \\ -1 & 0 & 0 \end{array}\right]$ is a cyclic normal form for B_{11}. Using the permutation matrix $P = Q \bigoplus I_1 \bigoplus I_2 \bigoplus I_1$, PBP^T is a Frobenius normal form for A with each irreducible, diagonal block in cyclic normal form.
4. Let $A = \begin{bmatrix} 2 & 1 & 0 \\ 1 & 2 & 0 \\ 0 & 0 & 3 \end{bmatrix}$ and let $B = \begin{bmatrix} 2 & 1 & 1 \\ 1 & 2 & 0 \\ 0 & 0 & 3 \end{bmatrix}$. Observe that A and B are each in Frobenius normal form, each with two irreducible, diagonal blocks, and that $A_{11} = B_{11}$ and $A_{22} = B_{22}$. Consequently, $\sigma(A) = \sigma(B) = \sigma(A_{11}) \cup \sigma(A_{22}) = \{1, 3, 3\}$. However, B has only one independent eigenvector for eigenvalue 3, whereas A has two independent eigenvectors for eigenvalue 3. The underlying cause for this difference is the difference in the access relations as captured in the reduced digraphs $R(\Gamma(A))$ (two isolated vertices) and $R(\Gamma(B))$ (two vertices joined by a single arc). The role that the reduced digraph of a matrix plays in connections between the eigenspaces for each of the irreducible, diagonal blocks of a matrix and those of the entire matrix is discussed in [Sch86] and in [BP94, Theorem 2.3.20]. For connections between $R(\Gamma(A))$ and the structure of the generalized eigenspaces for A, see also [Rot75].

29.8 Minimally Connected Digraphs and Nearly Reducible Matrices

Replacing zero entries in an irreducible matrix with nonzero entries preserves irreducibility, and equivalently, adding arcs to a strongly connected digraph preserves strong connectedness. Consequently, it is of interest to understand how few nonzero entries are needed in a matrix and in what locations to guarantee irreducibility; or equivalently, how few arcs are needed in a digraph and between which vertices to guarantee strong connectedness. Except as noted, all of the results in this section can be found in both [BR91, Sec. 3.3] and [Min88, Sec. 4.5]. Further results on nearly irreducible matrices and on their connections to nearly decomposable matrices can be found in [BH79].

Definitions:

A digraph is **minimally connected** if it is strongly connected and if the deletion of any arc in the digraph produces a subdigraph that is not strongly connected.

Facts:

1. A digraph Γ is minimally connected if and only if $\mathcal{A}_\Gamma$ is nearly reducible.
2. A matrix A is nearly reducible if and only if $\Gamma(A)$ is minimally connected.
3. The only minimally connected digraph on one vertex is the simple digraph on one vertex.
4. The only nearly reducible 1×1 matrix is the zero matrix.
5. [BR91, Theorem 3.3.5] Let Γ be a minimally connected digraph on n vertices with $n \geq 2$. Then Γ has no loops, at least n arcs, and at most $2(n-1)$ arcs. When Γ has exactly n arcs, Γ is an n-cycle. When Γ has exactly $2(n-1)$ arcs, Γ is a doubly directed tree.
6. Let A be an $n \times n$ nearly reducible matrix with $n \geq 2$. Then A has no nonzeros on its main diagonal, A has at least n nonzero entries, and A has at most $2(n-1)$ nonzero entries. When A has exactly n nonzero entries, $\Gamma(A)$ is an n-cycle. When A has exactly $2(n-1)$ nonzero entries, $\Gamma(A)$ is a doubly directed tree.
7. [Min88, Theorem 4.5.1] Let A be an $n \times n$ nearly reducible matrix with $n \geq 2$. Then there exists a positive integer m and a permutation matrix P such that

$$PAP^T = \left[\begin{array}{cccccc|c} 0 & 1 & 0 & \cdots & & 0 & \mathbf{0}^T \\ 0 & 0 & 1 & & \cdots & 0 & \mathbf{0}^T \\ \vdots & \vdots & \ddots & & \ddots & \vdots & \vdots \\ 0 & 0 & \cdots & & 0 & 1 & \mathbf{0}^T \\ 0 & 0 & \cdots & & 0 & 0 & \mathbf{v}^T \\ \hline \mathbf{u} & \mathbf{0} & \cdots & & \mathbf{0} & \mathbf{0} & B \end{array}\right],$$

where the upper left matrix is $m \times m$, B is nearly reducible, $\mathbf{0}$ is a $(n-m) \times 1$ vector, both of the vectors $\mathbf{u}$ and $\mathbf{v}$ are $(n-m) \times 1$, and each of $\mathbf{u}$ and $\mathbf{v}$ contains a single nonzero entry.

Examples:

1. Let Γ be the third digraph in Figure 29.1b. Then Γ is minimally connected. The subdigraph Γ' obtained by deleting arc $(1,2)$ from Γ is no longer strongly connected, however, it is still connected since its associated undirected graph is the graph in Figure 29.1a.

2. For $n \geq 1$, let $A^{(n)}$ be the $(n+1) \times (n+1)$ matrix be given by

$$A^{(n)} = \left[\begin{array}{c|cccc} 0 & 1 & 1 & \cdots & 1 \\ \hline 1 & & & & \\ 1 & & & & \\ \vdots & & & \mathbf{0}_{n\times n} & \\ 1 & & & & \end{array}\right].$$

Then $A^{(n)}$ is nearly reducible. The digraph $\Gamma(A^{(n)})$ is called a **rosette**, and has the most arcs possible for a minimally connected digraph on $n+1$ vertices. Suppose that $n \geq 2$, and let $P = \begin{bmatrix} 0 & 1 \\ 1 & 0 \end{bmatrix} \bigoplus I_{n-2}$. Then

$$PA^{(n)}P^T = \left[\begin{array}{c|cccc} 0 & 1 & 0 & \cdots & 0 \\ \hline 1 & & & & \\ 0 & & & & \\ \vdots & & & A^{(n-1)} & \\ 0 & & & & \end{array}\right]$$

is the decomposition for $A^{(n)}$ given in Fact 7 with $m = 1$.

References

[BG00] J. Bang-Jensen and G. Gutin. *Digraphs: Theory, Algorithms and Applications.* Springer-Verlag, London, 2000.

[BH79] R.A. Brualdi and M.B. Hendrick. A unified treatment of nearly reducible and nearly decomposable matrices. *Lin. Alg. Appl.*, 24 (1979) 51–73.

[BJ86] W.W. Barrett and C.R. Johnson. Determinantal Formulae for Matrices with Sparse Inverses, II: Asymmetric Zero Patterns. *Lin. Alg. Appl.*, 81 (1986) 237–261.

[BP94] A. Berman and R.J. Plemmons. *Nonnegative Matrices in the Mathematical Sciences.* SIAM, Philadelphia, 1994.

[Bru82] R.A. Brualdi. Matrices, eigenvalues and directed graphs. *Lin. Multilin. Alg. Applics.*, 8 (1982) 143–165.

[BR91] R.A. Brualdi and H.J. Ryser. *Combinatorial Matrix Theory.* Cambridge University Press, Cambridge, 1991.

[BR97] R.B. Bapat and T.E.S. Raghavan. *Nonnegative Matrices and Applications.* Cambridge University Press, Cambridge, 1997.

[Cve75] D.M. Cvetković. The determinant concept defined by means of graph theory. *Mat. Vesnik*, 12 (1975) 333–336.

[FP69] M. Fiedler and V. Ptak. Cyclic products and an inequality for determinants. *Czech Math J.*, 19 (1969) 428–450.

[Har62] F. Harary. The determinant of the adjacency matrix of a graph. *SIAM Rev.*, 4 (1962) 202–210.

[LHE94] Z. Li, F. Hall, and C. Eschenbach. On the period and base of a sign pattern matrix. *Lin. Alg. Appl.*, 212/213 (1994) 101–120.

[Min88] H. Minc. *Nonnegative Matrices.* John Wiley & Sons, New York, 1988.

[MOD89] J. Maybee, D. Olesky, and P. van den Driessche. Matrices, digraphs and determinants. *SIAM J. Matrix Anal. Appl.*, 4, (1989) 500–519.

[Rot75] U.G. Rothblum. Algebraic eigenspaces of nonnegative matrices. *Lin. Alg. Appl.*, 12 (1975) 281–292.

[Sch86] H. Schneider. The influence of the marked reduced graph of a nonnegative matrix on the Jordan form and related properties: a survey. *Lin. Alg. Appl.*, 84 (1986) 161–189.

[Stu91] J.L. Stuart. The determinant of a Hessenberg L-matrix, *SIAM J. Matrix Anal.*, 12 (1991) 7–15.

[Stu03] J.L. Stuart. Powers of ray pattern matrices. *Conference proceedings of the SIAM Conference on Applied Linear Algebra*, July 2003, at http://www.siam.org/meetings/la03/proceedings/stuartjl.pdf .

[Var04] R.S. Varga. *Gershgorin and His Circles*. Springer, New York, 2004.

30

Bipartite Graphs and Matrices

Bryan L. Shader
University of Wyoming

An $m \times n$ matrix is naturally associated with a bipartite graph, and the structure of the matrix is reflected by the combinatorial properties of the associated bipartite graph. This section discusses the fundamental structural theorems for matrices that arise from this association, and describes their implications for linear algebra.

30.1 Basics of Bipartite Graphs

This section introduces the various properties and families of bipartite graphs that have special significance for linear algebra.

Definitions:

A graph G is **bipartite** provided its vertices can be partitioned into disjoint subsets U and V such that each edge of G has the form $\{u, v\}$, where $u \in U$ and $v \in V$. The set $\{U, V\}$ is a **bipartition** of G.

A **complete bipartite graph** is a simple bipartite graph with bipartition $\{U, V\}$ such that each $\{u, v\}$ $(u \in U, v \in V)$ is an edge. The complete bipartite graph with $|U| = m$ and $|V| = n$ is denoted by $K_{m,n}$.

A **chordal graph** is one in which every cycle of length 4 or more has a chord, that is, an edge joining two nonconsecutive vertices on the cycle.

A **chordal bipartite graph** is a bipartite graph in which every cycle of length 6 or more has a chord.

A bipartite graph is **quadrangular** provided it is simple and each pair of vertices with a common neighbor lies on a cycle of length 4.

A **weighted bipartite graph** consists of a simple bipartite graph G and a function $w : E \to X$, where E is the edge set of G and X is a set (usually $\mathbb{Z}, \mathbb{R}, \mathbb{C}, \{-1, 1\}$, or a set of indeterminates). A **signed bipartite graph** is a weighted bipartite graph with $X = \{-1, 1\}$. In a signed bipartite graph, the **sign** of a set α of edges, denoted $\operatorname{sgn}(\alpha)$, is the product of the weights of the edges in α. The set α is **positive** or **negative** depending on whether $\operatorname{sgn}(\alpha)$ is $+1$ or -1.

Let G be a bipartite graph with bipartition $\{U, V\}$ and let $u_1, u_2, \ldots, u_m$ and $v_1, v_2, \ldots, v_n$ be orderings of the distinct elements of U and V, respectively. The **biadjacency matrix** of G is the $m \times n$ matrix $\mathcal{B}_G = [b_{ij}]$, where b_{ij} is the multiplicity of the edge $\{u_i, v_j\}$. Note that if U, respectively, V, is empty, then $\mathcal{B}_G$ is a matrix with no rows, respectively, no columns. For a weighted bipartite graph, b_{ij} is defined to be the weight of the edge $\{u_i, v_j\}$ if present, and 0 otherwise. For a signed bipartite graph, b_{ij} is the sign of the edge $\{u_i, v_j\}$ if present, and 0 otherwise.

Let N'_G be an oriented incidence matrix of simple graph G. The **cut space** of G is the column space of ${N'_G}^T$, and the **cut lattice** of G is the set of integer vectors in the cut space of G. The **flow space** of G is $\{x \in \mathbb{R}^m : N'_G x = 0\}$, and the **flow lattice** of G is $\{x \in \mathbb{Z}^m : N'_G x = 0\}$.

A **matching** of G is a set M of mutually disjoint edges. If M has k edges, then M is a k-**matching**, and if each vertex of G is in some (and hence exactly one) edge of M, then M is a **perfect matching**.

Facts:

Unless otherwise noted, the following can be found in [BR91, Chap. 3] or [Big93]. In the references, the results are stated and proven for simple graphs, but still hold true for graphs.

1. A bipartite graph has no loops. It has more than one bipartition if and only if the graph is disconnected. Each forest (and, hence, each tree and each path) is bipartite. The cycle C_n is bipartite if and only if n is even.
2. The following statements are equivalent for a graph G:

   (a) G is bipartite.

   (b) The vertices of G can be labeled with the colors red and blue so that each edge of G has a red vertex and a blue vertex.

   (c) G has no cycles of odd length.

   (d) There exists a permutation matrix P such that $P^T \mathcal{A}_G P$ has the form
   $$\begin{bmatrix} O & B \\ B^T & O \end{bmatrix}.$$

   (e) G is loopless and every minor of the vertex-edge incidence matrix N_G of G is 0, 1, or -1.

   (f) The characteristic polynomial $p_{\mathcal{A}_G}(x) = \sum_{i=0}^{n} c_i x^{n-i}$ of $\mathcal{A}_G$ satisfies $c_k = 0$ for each odd integer k.

   (g) $\sigma(G) = -\sigma(G)$ (as multisets), where $\sigma(G)$ is the spectrum of $\mathcal{A}_G$.
3. The connected graph G is bipartite if and only if $-\rho(\mathcal{A}_G)$ is an eigenvalue of $\mathcal{A}_G$.
4. The bipartite graph G disconnected if and only if there exist permutation matrices P and Q such that $P\mathcal{B}_G Q$ has the form
   $$\begin{bmatrix} B_1 & O \\ O & B_2 \end{bmatrix},$$
   where both B_1 and B_2 have at least one column or at least one row. More generally, if G is bipartite and has k connected components, then there exist permutation matrices P and Q such that
   $$P\mathcal{B}_G Q = \begin{bmatrix} B_1 & O & \cdots & O \\ O & B_2 & \cdots & O \\ \vdots & \vdots & \ddots & \vdots \\ O & O & \cdots & B_k \end{bmatrix},$$
   where the B_i are the biadjacency matrices of the connected components of G.
5. [GR01] If G is a simple graph with n vertices, m edges, and c components, then its cut space has dimension $n - c$, and its flow space has dimension $m - n + c$. If G is a plane graph, then the edges of G can be oriented and ordered so that the flow space of G equals the cut space of its dual graph. The norm, $\mathbf{x}^T\mathbf{x}$, is even for each vector $\mathbf{x}$ in the cut lattice of G if and only if each vertex has even degree. The norm of each vector in the flow lattice of G is even if and only if G is bipartite.

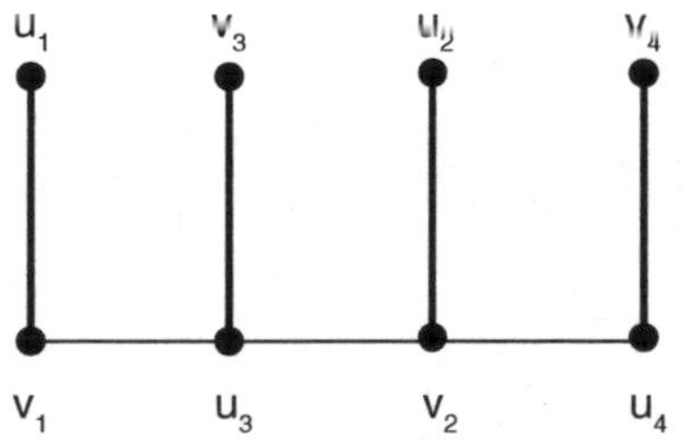

FIGURE 30.1

6. [Bru66], [God85], [Sim89] Let G be a bipartite graph with a unique perfect matching. Then there exist permutation matrices P and Q such that $P\mathcal{B}_G Q$ is a square, lower triangular matrix with all 1s on its main diagonal. If G is a tree, then the inverse of $P\mathcal{B}_G Q$ is a $(0, 1, -1)$-matrix. Let n be the order of $\mathcal{B}_G$, and let H be the simple graph with vertices $1, 2, \ldots, n$ and $\{i, j\}$ an edge if and only if $i \neq j$ and either the (i, j)- or (j, i)-entry of $P\mathcal{B}_G Q$ is nonzero. If H is bipartite, then $(P\mathcal{B}_G Q)^{-1}$ is diagonally similar to a nonnegative matrix, which equals $P\mathcal{B}_G Q$ if and only if G can be obtained by appending a pendant edge to each vertex of a bipartite graph.

Examples:

1. Up to matrix transposition and permutations of rows and columns, the biadjacency matrix of the path P_{2n}, the path P_{2n+1}, and the cycle C_{2n} are

$$\begin{bmatrix} 1 & 1 & 0 & \cdots & 0 \\ 0 & 1 & 1 & \cdots & 0 \\ \vdots & & \ddots & \ddots & \vdots \\ 0 & 0 & \cdots & 1 & 1 \\ 0 & 0 & \cdots & 0 & 1 \end{bmatrix}_{n\times n}, \quad \begin{bmatrix} 1 & 1 & 0 & \cdots & 0 & 0 \\ 0 & 1 & 1 & \cdots & 0 & 0 \\ \vdots & & \ddots & \ddots & \vdots & 0 \\ 0 & 0 & \cdots & 1 & 1 & 0 \\ 0 & 0 & \cdots & 0 & 1 & 1 \end{bmatrix}_{n\times(n+1)}, \quad \begin{bmatrix} 1 & 1 & 0 & \cdots & 0 \\ 0 & 1 & 1 & \cdots & 0 \\ \vdots & \vdots & \ddots & \ddots & \vdots \\ 0 & 0 & \cdots & 1 & 1 \\ 1 & 0 & \cdots & 0 & 1 \end{bmatrix}_{n\times n}.$$

2. The biadjacency matrix of the complete bipartite graph $K_{m,n}$ is $J_{m,n}$, the $m \times n$ matrix of all ones.
3. Up to row and column permutations, the biadjacency matrix of the graph obtained from $K_{n,n}$ by removing the edges of a perfect matching is $J_n - I_n$.
4. Let G be the bipartite graph (Figure 30.1).
   Then

$$\mathcal{B}_G = \begin{bmatrix} 1 & 0 & 0 & 0 \\ 0 & 1 & 0 & 0 \\ 1 & 1 & 1 & 0 \\ 0 & 1 & 0 & 1 \end{bmatrix}$$

   G has a unique perfect matching, and the graph H defined in Fact 6 is the path 1–3–2–4. Hence, $\mathcal{B}_G$ is diagonally similar to a nonnegative matrix. Also, since G is obtained from the bipartite graph v_1—u_3—v_2—u_4 by appending pendant vertices to each vertex, $\mathcal{B}_G^{-1}$ is diagonally similar to $\mathcal{B}_G$. Indeed,

$$S\mathcal{B}_G^{-1}S = S \begin{bmatrix} 1 & 0 & 0 & 0 \\ 0 & 1 & 0 & 0 \\ -1 & -1 & 1 & 0 \\ 0 & -1 & 0 & 1 \end{bmatrix} S^{-1} = \mathcal{B}_G,$$

   where S is the diagonal matrix with main diagonal $(1, 1, -1, -1)$.

30.2 Bipartite Graphs Associated with Matrices

This section presents some of the ways that matrices have been associated to bipartite graphs and surveys resulting consequences.

Definitions:

The **bigraph** of the $m \times n$ matrix $A = [a_{ij}]$ is the simple graph with vertex set $U \cup V$, where $U = \{1, 2, \ldots, m\}$ and $V = \{1', 2', \ldots, n'\}$, and edge set $\{\{i, j'\} : a_{ij} \neq 0\}$. If A is a nonnegative integer matrix, then the **multi-bigraph** of A has vertex set $U \cup V$ and edge $\{i, j'\}$ of multiplicity a_{ij}. If A is a general matrix, then the **weighted bigraph** of A has vertex set $U \cup V$ and edge $\{i, j'\}$ of weight a_{ij}. If A is a real matrix, then the **signed bigraph** of A is obtained by weighting the edge $\{i, j'\}$ of the bigraph by $+1$ if $a_{ij} > 0$, and by -1 if $a_{ij} < 0$.

The **(zero) pattern** of the $m \times n$ matrix $A = [a_{ij}]$ is the $m \times n$ $(0, 1)$-matrix whose (i, j)-entry is 1 if and only if $a_{ij} \neq 0$.

The **sign pattern** of the real $m \times n$ matrix $A = [a_{ij}]$ is the $m \times n$ matrix whose (i, j)-entry is $+$, 0, or $-$, depending on whether a_{ij} is positive, zero, or negative. (See Chapter 33 for more information on sign patterns.)

A $(0, 1)$-matrix is a **Petrie matrix** provided the 1s in each of its columns occur in consecutive rows. A $(0, 1)$-matrix A has the **consecutive ones property** if there exists a permutation P such that PA is a Petrie matrix.

The **directed bigraph** of the real $m \times n$ matrix $A = [a_{ij}]$ is the directed graph with vertices $1, 2, \ldots, m, 1', 2', \ldots, n'$, the arc (i, j') if and only if $a_{ij} > 0$, and the arc (j', i) if and only if $a_{ij} < 0$.

An $m \times n$ matrix A is a **generic matrix** with respect to the field F provided its nonzero elements are independent indeterminates over the field F. The matrix A can be viewed as a matrix whose elements are in the ring of polynomials in these indeterminates with coefficients in F.

Let A be an $n \times n$ matrix with each diagonal entry nonzero. The **bipartite fill-graph** of A, denoted $G^+(A)$, is the simple bipartite graph with vertex set $\{1, 2, \ldots, n\} \cup \{1', 2', \ldots, n'\}$ with an edge joining i and j' if and only if there exists a path from i to j in the digraph, $\Gamma(A)$, of A each of whose intermediate vertices has label less than $\min\{i, j\}$. If A is symmetric, then (by identifying vertices i and i' for $i = 1, 2, \ldots, n$ and deleting loops), $G^+(A)$ can be viewed as a simple graph, and is called the **fill-graph** of A.

The square matrix B has a **perfect elimination ordering** provided there exist permutation matrices P and Q such that the bipartite fill-graph, $G^+(PBQ)$, and the bigraph of PBQ are the same.

Associated with the $n \times n$ matrix $A = [a_{ij}]$ is the sequence $H_0, H_1, \ldots, H_{n-1}$ of bipartite graphs as defined by:

1. H_0 consists of vertices $1, 2, \ldots, n$, and $1', 2', \ldots, n'$, and edges of the form $\{i, j'\}$, where $a_{ij} \neq 0$.
2. For $k = 1, \ldots, n-1$, H_k is the graph obtained from H_{k-1} by deleting vertices k and k' and inserting each edge of the form $\{r, c'\}$, where $r > k$, $c > k$, and both $\{r, k'\}$ and $\{k, c'\}$ are edges of H_{k-1}.

The **4-cockades** are the bipartite graphs recursively defined by: A 4-cycle is a 4-cockade, and if G is a 4-cockade and e is an edge of G, then the graph obtained from G by identifying e with an edge of a 4-cycle disjoint from G is a 4-cockade. A **signed 4-cockade** is a 4-cockade whose edges are weighted by ± 1 in such a way that every 4-cycle is negative.

Facts:

General references for bipartite graphs associated with matrices are [BR91, Chap. 3] and [BS04].

1. [Rys69] (See also [BR91, p. 18].) If A is an $m \times n$ $(0, 1)$-matrix such that each entry of AA^T is positive, then either A has a column with no zeros or the bigraph of A has a chordless cycle of length 6. The converse is not true.
2. [RT76], [GZ98] If $G = (V, E)$ is a connected quadrangular graph, then $|E| \leq 2|V| - 4$. The connected quadrangular graphs with $|E| = 2|V| - 4$ are characterized in the first reference.

3. [RT76], If A is an $m \times n$ $(0, 1)$-matrix such that no entry of AA^T or $A^T A$ is 1, and the bigraph of A is connected, then A has at most $2(m + n) - 4$ nonzero entries.
4. [Tuc70] The $(0, 1)$-matrix A has the consecutive ones property if and only if it does not have a submatrix whose rows and columns can be permuted to have one of the following forms for $k \geq 1$.

$$\begin{bmatrix} 1 & 1 & 0 & \cdots & 0 \\ 0 & 1 & 1 & & 0 \\ \vdots & & \ddots & \ddots & \vdots \\ 0 & 0 & & 1 & 1 \\ 1 & 0 & \cdots & 0 & 1 \end{bmatrix}_{(k+2)\times(k+2)}, \qquad \begin{bmatrix} 1 & 0 & \cdots & 0 & 0 \\ 1 & 1 & & 0 & 1 \\ 0 & 1 & \ddots & \vdots & \vdots \\ \vdots & & \ddots & 1 & 1 \\ 0 & 0 & \cdots & 1 & 0 \\ 0 & 0 & \cdots & 0 & 1 \end{bmatrix}_{(k+3)\times(k+2)},$$

$$\begin{bmatrix} 1 & 0 & \cdots & 0 & 0 & 1 \\ 1 & 1 & & 0 & 1 & 1 \\ 0 & 1 & \ddots & \vdots & \vdots & \vdots \\ \vdots & & \ddots & 1 & 1 & 1 \\ 0 & 0 & \cdots & 1 & 1 & 0 \\ 0 & 0 & \cdots & 0 & 1 & 1 \end{bmatrix}_{(k+3)\times(k+3)}, \qquad \begin{bmatrix} 1 & 1 & 0 & 0 & 0 & 0 \\ 0 & 0 & 1 & 1 & 0 & 0 \\ 0 & 0 & 0 & 0 & 1 & 1 \\ 0 & 1 & 0 & 1 & 0 & 1 \end{bmatrix}_{4\times 6}, \qquad \begin{bmatrix} 1 & 1 & 0 & 0 & 0 \\ 1 & 1 & 1 & 1 & 0 \\ 0 & 0 & 1 & 1 & 0 \\ 1 & 0 & 0 & 1 & 1 \end{bmatrix}_{4\times 5}.$$

5. [ABH99] Let A be a $(0, 1)$-matrix and let $L = D - AA^T$, where D is the diagonal matrix whose ith diagonal entry is the ith row sum of AA^T. Then L is a symmetric, singular matrix each of whose eigenvalues is nonnegative. Let $\mathbf{v}$ be a eigenvector of L corresponding to the second smallest eigenvalue of L. If A has the consecutive ones property and the entries of $\mathbf{v}$ are distinct, then PA is a Petrie matrix, where P is the permutation matrix such that the entries of $P\mathbf{v}$ are in increasing order. In addition, the reference gives a recursive method for finding a P such that PA is a Petrie matrix when the elements of $\mathbf{v}$ are not distinct.
6. The directed bigraph of the real matrix A contains at most one of the arcs (i, j') or (j', i).
7. [FG81] The directed bigraph of the real matrix A is strongly connected if and only if there do not exist subsets α and β such that $A[\alpha, \beta] \geq 0$ and $A(\alpha, \beta) \leq 0$. Here, either α or β may be the empty set, and a vacuous matrix M satisfies both $M \geq 0$ and $M \leq 0$.
8. [FG81] If $A = [a_{ij}]$ is a fully indecomposable, $n \times n$ sign pattern, then the following are equivalent:

   (a) There is a matrix $\widehat{A}$ with sign pattern A such that $\widehat{A}$ is invertible and its inverse is a positive matrix.

   (b) There do not exist subsets α and β such that $A[\alpha, \beta] \geq O$ and $A(\alpha, \beta) \leq O$.

   (c) The bipartite directed graph of A is strongly connected.

   (d) There exists a matrix with sign pattern A each of whose line sums is 0.

   (e) There exists a rank $n - 1$ matrix with sign pattern A each of whose line sums is 0.

9. [Gol80] Up to relabeling of vertices, G is the fill-graph of some $n \times n$ symmetric matrix if and only if G is chordal.
10. [GN93] Let A be an $n \times n$ $(0, 1)$-matrix with each diagonal entry equal to 1. Suppose that B is a matrix with zero pattern A, and that B can be factored as $B = LU$, where $L = [\ell_{ij}]$ is a lower triangular matrix and $U = [u_{ij}]$ is an upper triangular matrix. If $i \neq j$ and either $\ell_{ij} \neq 0$ or $u_{ij} \neq 0$, then $\{i, j'\}$ is an edge of $G^+(A)$. Moreover, if B is a generic matrix with zero pattern A, then such a factorization $B = LU$ exists, and for each edge $\{i, j'\}$ of $G^+(A)$ either $\ell_{ij} \neq 0$ or $u_{ij} \neq 0$.

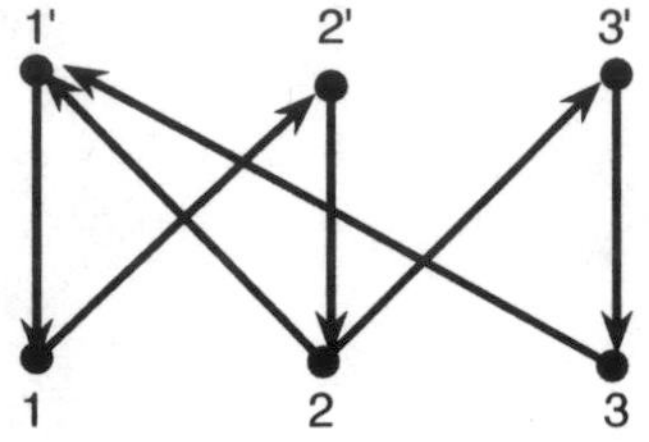

FIGURE 30.2

11. [GG78] If the bigraph of the generic, square matrix A is chordal bipartite, then A has a perfect elimination ordering and, hence, there exist permutation matrices P and Q such that performing Gaussian elimination on PAQ has no fill-in. The converse is not true; see Example 3.
12. [GN93] If A is a generic $n \times n$ matrix with each diagonal entry nonzero, and $\alpha = \{1, 2, \ldots, r\}$, then the bigraph of the Schur complement of $A[\alpha]$ in A is the bigraph H_r defined above.
13. [DG93] For each matrix with a given pattern, small relative perturbations in the nonzero entries cause only small relative perturbations in the singular values (independent of the values of the matrix entries) if and only if the bigraph of the pattern is a forest. The singular values of such a matrix can be computed to high relative accuracy.
14. [DES99] If the signed bipartite graph of the real matrix is a signed 4-cockade, then small relative perturbations in the nonzero entries cause only small relative perturbations in the singular values (independent of the values of the matrix entries). The singular values of such a matrix can be computed to high relative accuracy.

Examples:

1. Let

$$A = \begin{bmatrix} - & + & 0 \\ + & - & + \\ + & 0 & - \end{bmatrix}.$$

The directed bigraph of A is (Figure 30.2).

Since this is strongly connected, Fact 7 implies that there do not exist subsets α and β such that $A[\alpha, \beta] \geq O$ and $A(\alpha, \beta) \leq O$. Also, there is a matrix with sign pattern A whose inverse is positive. One such matrix is

$$\begin{bmatrix} -3/2 & 2 & 0 \\ 1 & -2 & 1 \\ 1 & 0 & -1 \end{bmatrix}.$$

2. A signed 4-cockade on 8 vertices (unlabeled edges have sign $+1$) and its biadjacency matrix are (Figure 30.3)

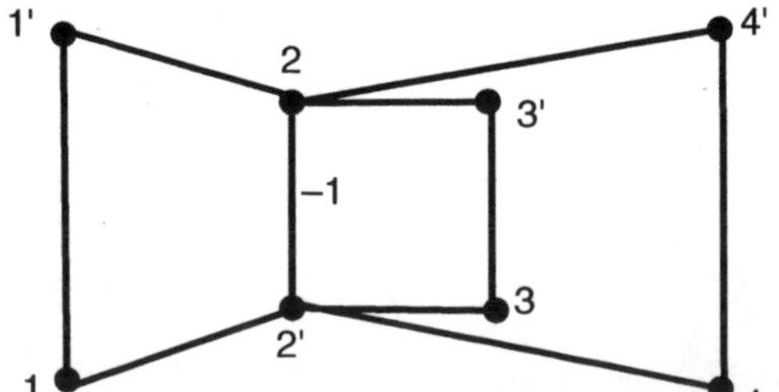

$$\begin{bmatrix} 1 & 1 & 0 & 0 \\ 1 & -1 & 1 & 1 \\ 0 & 1 & 1 & 0 \\ 0 & 1 & 0 & 1 \end{bmatrix}.$$

FIGURE 30.3

3. Both the bipartite fill-graph and the bigraph of the matrix (Figure 30.4) below

$$\begin{bmatrix} 1 & 0 & 0 & 0 & 0 & 0 \\ 0 & 1 & 0 & 0 & 0 & 0 \\ 0 & 0 & 1 & 0 & 0 & 0 \\ 1 & 1 & 0 & 1 & 0 & 0 \\ 0 & 1 & 1 & 0 & 1 & 0 \\ 1 & 0 & 1 & 0 & 0 & 1 \end{bmatrix}$$

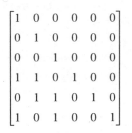

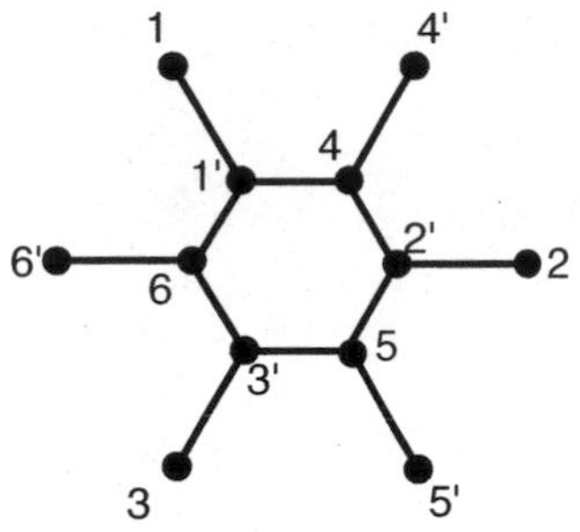

FIGURE 30.4

are the graph illustrated. Since its bigraph has a chordless 6-cycle, this example shows that the converse to Fact 11 is false.

4. Let

$$A = \begin{bmatrix} x_1 & x_2 & x_3 & x_4 \\ x_5 & x_6 & 0 & 0 \\ x_7 & 0 & x_8 & 0 \\ x_9 & 0 & 0 & x_{10} \end{bmatrix},$$

where $x_1, \ldots, x_{10}$ are independent indeterminates. The bigraph of A is chordal bipartite. The biadjacency matrix of H_1 is J_3. Thus, by Fact 12, the pattern of the Schur complement of $A[\{1\}]$ in A is J_3. The bipartite fill-graph of A has biadjacency matrix J_4. If

$$P = \begin{bmatrix} 0 & 0 & 0 & 1 \\ 0 & 0 & 1 & 0 \\ 0 & 1 & 0 & 0 \\ 1 & 0 & 0 & 0 \end{bmatrix},$$

then the bipartite fill-graph of PAP^T and the bigraph of PAP^T are the same. Hence, it is possible to perform Gaussian elimination (without pivoting) on PAP^T without any fill-in.

Applications:

1. [Ken69], [ABH99] Petrie matrices are named after the archaeologist Flinders Petrie and were first introduced in the study of seriation, that is, the chronological ordering of archaeological sites. If the rows of the matrix A represent archaeological sites ordered by their historical time period, the columns of A represent artifacts, and $a_{ij} = 1$ if and only if artifact j is present at site i, then one would expect A to be a Petrie matrix. More recently, matrices with the consecutive ones property have arisen in genome sequencing (see [ABH99]).
2. [BBS91], [Sha97] If U is a unitary matrix and A is the pattern of U, then the bigraph of A is quadrangular. If U is fully indecomposable, then U has at most $4n-4$ nonzero entries. The matrices achieving equality are characterized in the first reference. (See Fact 2 for more on quadrangular graphs.)

30.3 Factorizations and Bipartite Graphs

This section discusses the combinatorial interpretations and applications of certain matrix factorizations.

Definitions:

A **biclique** of a graph G is a subgraph that is a complete bipartite graph. For disjoint subsets X and Y of vertices, $B(X, Y)$ denotes the biclique consisting of all edges of the form $\{x, y\}$ such that $x \in X$ and $y \in Y$ (each of multiplicity 1). If G is bipartite with bipartition $\{U, V\}$, then it is customary to take $X \subseteq U$ and $Y \subseteq V$.

A **biclique partition** of $G = (V, E)$ is a collection $B(X_1, Y_1), \ldots, B(X_k, Y_k)$ of bicliques of G whose edges partition E.

A **biclique cover** of $G = (V, E)$ is a collection of bicliques such that each edge of E is in at least one biclique.

The **biclique partition number of** G, denoted $\text{bp}(G)$, is the smallest k such that there is a partition of G into k bicliques. The **biclique cover number of** G, denoted $\text{bc}(G)$, is the smallest k such that there is a cover of G by k bicliques. If G does not have a biclique partition, respectively, cover, then $\text{bp}(G)$, respectively, $\text{bc}(G)$, is defined to be infinite.

If G is a graph, then $n_+(G)$, respectively, $n_-(G)$, denotes the number of positive, respectively, negative, eigenvalues of $\mathcal{A}_G$ (including multiplicity).

If $X \subseteq \{1, 2, \ldots, n\}$, then the **characteristic vector** of X is the $n \times 1$ vector $\vec{\mathbf{X}} = [\mathbf{x_i}]$, where $x_i = 1$ if $i \in X$, and $x_i = 0$ otherwise.

The **nonnegative integer rank** of the nonnegative integer matrix A is the minimum k such that there exist an $m \times k$ nonnegative integer matrix B and a $k \times n$ nonnegative integer matrix C with $A = BC$.

The **(0,1)-Boolean algebra** consists of the elements 0 and 1, endowed with the operations defined by $0+0=0, 0+1=1=1+0, 1+1=1, 0*1=0=1*0, 0*0=0$, and $1*1=1$. A **Boolean matrix** is a matrix whose entries belong to the (0,1)-Boolean algebra. Addition and multiplication of Boolean matrices is defined as usual, except Boolean arithmetic is used.

The **Boolean rank** of the $m \times n$ Boolean matrix A is the minimum k such that there exists an $m \times k$ Boolean matrix B and a $k \times n$ Boolean matrix C such that $A = BC$.

Let G be a bipartite graph with bipartition $\{\{1, 2, \ldots, m\}, \{1', 2', \ldots, n'\}\}$. Then $\mathcal{M}(G)$ denotes the set of all $m \times n$ matrices $A = [a_{ij}]$ such that if $a_{ij} \neq 0$, then $\{i, j'\}$ is an edge of G, that is, the bigraph of A is a subgraph of G. The graph G **supports rank decompositions** provided each matrix $A \in \mathcal{M}(G)$ is the sum of $\text{rank}(A)$ elements of $\mathcal{M}(G)$ each having rank 1.

If G is a signed bipartite graph, then $\mathcal{M}(G)$ denotes the set of all matrices $A = [a_{ij}]$ such that if $a_{ij} > 0$, then $\{i, j'\}$ is a positive edge of G, and if $a_{ij} < 0$, then $\{i, j'\}$ is a negative edge of G. The signed bigraph G **supports rank decompositions** provided each matrix $A \in \mathcal{M}(G)$ is the sum of $\text{rank}(A)$ elements of $\mathcal{M}(G)$ each having rank 1.

Facts:

1. [GP71]

   - A graph has a biclique partition (and, hence, cover) if and only if it has no loops.
   - For every graph G, $\text{bc}(G) \leq \text{bp}(G)$.
   - Every simple graph G with n vertices has a biclique partition with at most $n-1$ bicliques, namely, $B(\{i\}, \{j : \{i, j\} \text{ is an edge of } G \text{ and } j > i\})$ $(i = 1, 2, \ldots, n-1)$.

2. [CG87] Let G be a bipartite graph with bipartition (U, V), where $|U| = m$ and $|V| = n$. Let $B(X_1, Y_1), B(X_2, Y_2), \ldots, B(X_k, Y_k)$ be bicliques with $X_i \subseteq U$ and $Y_i \subseteq V$ for all i. The following are equivalent:

   (a) $B(X_1, Y_1), B(X_2, Y_2), \ldots, B(X_k, Y_k)$ is a biclique partition of G.

(b) $\sum_{i=1}^{k} \vec{\mathbf{X_i}}\vec{\mathbf{Y_i}}^T = \mathcal{B}_G$.

(c) $XY^T = \mathcal{B}_G$, where X is the $n \times k$ matrix whose ith column is $\vec{\mathbf{X_i}}$, and Y is the $n \times k$ matrix whose ith column is $\vec{\mathbf{Y_i}}$.

3. [CG87] For a simple bipartite graph G, $\text{bp}(G)$ equals the nonnegative integer rank of $\mathcal{B}_G$.
4. [CG87]
   - Let G be the bipartite graph obtained from $K_{n,n}$ by removing a perfect matching. Then $\text{bp}(G) = n$. Furthermore, if $B(X_i, Y_i)$ $(i = 1, 2, \ldots, n)$ is a biclique partition of G, then there exist positive integers r and s such that $rs = n - 1$, $|X_i| = r$ and $|Y_i| = s$ $(i = 1, 2, \ldots, n)$, k is in exactly r of the X_i's and exactly s of the Y_i's $(k = 1, 2, \ldots, n)$, and $X_i \cap Y_j = 1$ for $i \neq j$.
   - In matrix terminology, if X and Y are $n \times n$ (0, 1)-matrices such that $XY^T = J_n - I_n$, then there exist integers r and s such that $rs = n - 1$, X has constant line sums r, Y has constant line sums s, and $Y^T X = J_n - I_n$.
   - In particular, if $n - 1$ is prime, then either X is a permutation matrix and $Y = (J_n - I_n)X$, or Y is a permutation matrix and $X = (J_n - I_n)Y$.
5. [BS04, see p. 67] Let G be a graph on n vertices with adjacency matrix $\mathcal{A}_G$, and let $B(X_1, Y_1)$, $B(X_2, Y_2), \ldots, B(X_k, Y_k)$ be bicliques of G. Then the following are equivalent:

   (a) $B(X_1, Y_1), B(X_2, Y_2), \ldots, B(X_k, Y_k)$ is a biclique partition of G.

   (b) $\sum_{i=1}^{k} \vec{\mathbf{X_i}}\vec{\mathbf{Y_i}}^T + \sum_{i=1}^{k} \vec{\mathbf{Y_i}}\vec{\mathbf{X_i}}^{\mathbf{T}} = \mathcal{A}_G$.

   (c) $XY^T + YX^T = \mathcal{A}_G$, where X is the $n \times k$ matrix whose ith column is $\vec{\mathbf{X_i}}$, and Y is the $n \times k$ matrix whose ith column is $\vec{\mathbf{Y_i}}$.

   (d) $\mathcal{A}_G = M \begin{bmatrix} O & I_m \\ I_m & O \end{bmatrix} M^T$, where M is the $n \times 2k$ matrix $\begin{bmatrix} X & Y \end{bmatrix}$ formed from the matrices X and Y defined in (c).
6. [CH89] The bicliques $B(X_1, Y_1), B(X_2, Y_2), \ldots, B(X_k, Y_k)$ partition K_n if and only if XY^T is an $n \times n$ tournament matrix, where X is the $n \times k$ matrix whose ith column is $\vec{\mathbf{X_i}}$, and Y is the $n \times k$ matrix whose ith column is $\vec{\mathbf{Y_i}}$. Thus, $\text{bp}(K_n)$ is the minimum nonnegative integer rank among all the $n \times n$ tournament matrices.
7. [CH89] The rank of an $n \times n$ tournament matrix is at least $n - 1$.
8. (Attributed to Witsenhausen in [GP71])

$$\text{bp}(K_n) = n - 1,$$

   that is, it is impossible to partition the complete graph into $n - 2$ or fewer bicliques.
9. [GP71] *The Graham–Pollak Theorem*: If G is a loopless graph, then

$$\text{bp}(G) \geq \max\{n_+(G), n_-(G)\}. \tag{30.1}$$

   The graph G is **eigensharp** if equality holds in (30.1). It is conjectured in [CGP86] that for all λ, and n sufficiently large, the complete graph λK_n with each edge of multiplicity k is eigensharp.
10. [ABS91] If $B(X_1, Y_1)$, $B(X_2, Y_2), \ldots, B(X_k, Y_k)$ is a biclique partition of G, then there exists an acyclic subgraph of G with $\max\{(n_+(G), n_-(G)\}$ edges no two in the same $B(X_i, Y_i)$.
   In particular, for each biclique partition of K_n there exists a spanning tree no two of whose edges belong to the same biclique of the partition.
11. [CH89] For all positive integers r and s with $2 \leq r < s$, the edges of the complete graph K_{2rs} cannot be partitioned into copies of the complete bipartite graph $K_{r,s}$.

12. [Hof01] If m and n are positive integers with $2m \leq n$, and $G_{m,n}$ is the graph obtained from the complete graph K_n by duplicating the edges of an m-matching, then $n_+(G) = n - m - 1$, $n_-(G) = m + 1$, and $\text{bp}(G) \geq n - m + \lfloor\sqrt{2m}\rfloor - 1$.
13. [CGP86] Let A be an $m \times n$ $(0,1)$-matrix with bigraph G and let $B(X_1, Y_1)$, $B(X_2, Y_2)$, ... , $B(X_k, Y_k)$ be bicliques. The following are equivalent:

    (a) $B(X_1, Y_1), B(X_2, Y_2), \ldots, B(X_k, Y_k)$ is a biclique cover of G.

    (b) $\sum_{i=1}^{k} \vec{\mathbf{X}}_{\mathbf{i}}\vec{\mathbf{Y}}_{\mathbf{i}}^{\mathbf{T}} = A$ (using Boolean arithmetic).

    (c) $XY^T = B$ (using Boolean arithmetic), where X is the $m \times k$ matrix whose jth column is $\vec{\mathbf{X}}_{\mathbf{j}}$, and Y is the $m \times k$ matrix whose j column is $\vec{\mathbf{Y}}_{\mathbf{j}}$.
14. [CGP86] The Boolean rank of a $(0,1)$-matrix A equals the biclique cover number of its bigraph.
15. [CSS87] Let k be a positive integer and let $t(k)$ be the largest integer n such that there exists an $n \times n$ tournament matrix with Boolean rank k. Then for $k \geq 2$, $t(k) < k^{\log_2(2k)}$, and $n(n^2+n+1)+2 \leq t(n^2+n+1)$.

    It is still an open problem to determine the minimum Boolean rank among $n \times n$ tournament matrices.
16. [DHM95, JM97] The bipartite graph G supports rank decompositions if and only if G is chordal bipartite.
17. [GMS96] The signed bipartite graph G support rank decompositions if and only if

$$\text{sgn}(\gamma) = (-1)^{(\ell(\gamma)/2)-1} \tag{30.2}$$

    for every cycle γ of G of length $\ell(\gamma) \geq 6$. Additionally, every matrix in $\mathcal{M}(G)$ has its rank equal to its term rank if and only if (30.2) holds for every cycle of G.

Examples:

1. Below, the edges of different textures form the bicliques (Figure 30.5) in a biclique partition of the graph $G_{2,4}$ obtained from K_4 by duplicating two disjoint edges.
2. Let n be an integer and r and s positive integers with $n - 1 = rs$. Then $XY^T = J_n - I_n$, where $X = I + C^s + C^{2s} + \cdots + C^{s(r-1)}$, $Y = C + C^2 + C^3 + \cdots + C^s$, and C is the $n \times n$ permutation matrix with 1s in positions $(1,2), (2,3), \ldots, (n-1,n)$, and $(n,1)$.

    This shows that for each pair of positive integers r and s with $rs = n-1$, there is a biclique partition of $J_n - I_n$ with X_i and Y_i satisfying the conditions in Fact 4.
3. For n odd, $B(\{i\}, \{i+1, i+2, \ldots, i+\frac{n-1}{2}\})$ $(i = 1, 2, \ldots, n)$ is a partition of K_n into bicliques each isomorphic to $K_{1,\frac{n-1}{2}}$, where the indices are read mod n (see Fact 11).

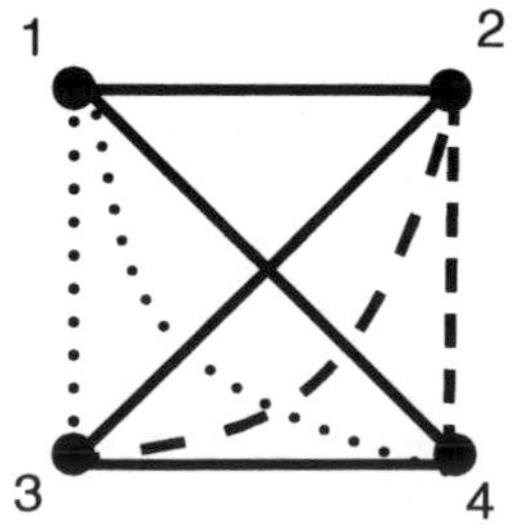

FIGURE 30.5

References

[ABS91] N. Alon, R.A. Brualdi, and B.L. Shader. Multicolored forests in bipartite decompositions of graphs. *J. Combin. Theory Ser. B*, 53:143–148, 1991.

[ABH99] J. Atkins, E. Boman, and B. Hendrickson. A spectral algorithm for seriation and the consecutive ones problem. *SIAM J. Comput.*, 28:297–310, 1999.

[BBS91] L.B. Beasley, R.A. Brualdi, and B.L. Shader. Combinatorial orthogonality. *Combinatorial and Graph-Theoretical Problems in Linear Algebra*, The IMA Volumes in Mathematics and Its Applications, vol. 50, Springer-Verlag, New York, 207–218, 1991.

[Big93] N. Biggs. *Algebraic Graph Theory*. Cambridge University Press, Cambridge, 2nd ed., 1993.

[BR91] R.A. Brualdi and H.J. Ryser. *Combinatorial Matrix Theory*. Cambridge University Press, Cambridge, 1991.

[Bru66] R.A. Brualdi. Permanent of the direct product of matrices. *Pac. J. Math.*, 16:471–482, 1966.

[BS04] R.A. Brualdi and B.L. Shader. Graphs and matrices. *Topics in Algebraic Graph Theory* (L. Beineke and R. J. Wilson, Eds.), Cambridge University Press, Cambridge, 56–87, 2004.

[CG87] D. de Caen and D.A. Gregory. On the decomposition of a directed graph into complete bipartite subgraphs. *Ars Combin.*, 23:139–146, 1987.

[CGP86] D. de Caen, D.A. Gregory, and N.J. Pullman. The Boolean rank of zero-one matrices. *Proceedings of the Third Caribbean Conference on Combinatorics and Computing*, 169–173, 1981.

[CH89] D. de Caen and D.G. Hoffman. Impossibility of decomposing the complete graph on n points into $n-1$ isomorphic complete bipartite graphs. *SIAM J. Discrete Math.*, 2:48–50, 1989.

[CSS87] K.L. Collins, P.W. Shor, and J.R. Stembridge. A lower bound for $(0, 1, *)$ tournament codes. *Discrete Math.*, 63:15–19, 1987.

[DES99] J. Demmel, M. Gu, S. Eisenstat, I. Slapnicar, K. Veselic, and Z. Drmac. Computing the singular value decompostion with high relative accuracy. *Lin. Alg. Appls.*, 119:21–80, 1999.

[DG93] J. Demmel and W. Gragg. On computing accurate singular values and eigenvalues of matrices with acyclic graphs. *Lin. Alg. Appls.*, 185:203–217, 1993.

[DHM95] K.R. Davidson, K.J. Harrison, and U.A. Mueller. Rank decomposability in incident spaces. *Lin. Alg. Appls.*, 230:3–19, 1995.

[FG81] M. Fiedler and R. Grone. Characterizations of sign patterns of inverse-positive matrices. *Lin. Alg. Appls.*, 40:237–45, 1983.

[Gol80] M.C. Golumbic. *Algorithmic Graph Theory and Perfect Graphs*. Academic Press, New York, 1980.

[GG78] M.C. Golumbic and C.F. Goss. Perfect elimination and chordal bipartite graphs. *J. Graph Theory*, 2:155–263, 1978.

[GMS96] D.A. Gregory, K.N. vander Meulen, and B.L. Shader. Rank decompositions and signed bigraphs. *Lin. Multilin. Alg.*, 40:283–301, 1996.

[GN93] J.R. Gilbert and E.G. Ng. Predicting structure in nonsymmetric sparse matrix factorizations. *Graph Theory and Sparse Matrix Computations*, The IMA Volumes in Mathematics and Its Applications, v. 56, 1993, 107–140.

[God85] C.D. Godsil. Inverses of trees. *Combinatorica*, 5:33–39, 1985.

[GP71] R.L. Graham and H.O. Pollak. On the addressing problem for loop switching. *Bell Sys. Tech. J.*, 50:2495–2519, 1971.

[GR01] C.D. Godsil and G. Royle. *Algebraic Graph Theory*, Graduate Texts in Mathematics, 207, Springer-Verlag, Heidelberg 2001.

[GZ98] P.M. Gibson and G.-H. Zhang. Combinatorially orthogonal matrices and related graphs. *Lin. Alg. Appls.*, 282:83–95, 1998.

[Hof01] A.J. Hoffman. On a problem of Zaks. *J. Combin. Theory Ser. A*, 93:371–377, 2001.

[JM97] C.R. Johnson and J. Miller. Rank decomposition under combinatorial constraints. *Lin. Alg. Appls.*, 251:97–104, 1997.

[Ken69] D.G. Kendall. Incidence matrices, interval graphs and seriation in archaeology. *Pac. J. Math.*, 28:565–570, 1969.

[RT76] K.B. Reid and C. Thomassen. Edge sets contained in circuits. *Israel J. Math.*, 24:305–319, 1976.

[Rys69] H.J. Ryser. Combinatorial configurations. *SIAM J. Appl. Math.*, 17:593-602, 1969.

[Sha97] B.L. Shader. A simple proof of Fiedler's conjecture concerning orthogonal matrices. *Rocky Mountain J. Math.*, 27:1239–1243, 1997.

[Sim89] R. Simion. Solution to a problem of C.D. Godsil regarding bipartite graphs with unique perfect matching. *Combinatorica*, 9:85–89, 1989.

[Tuc70] A. Tucker. Characterizing the consecutive 1's property. *Proc. Second Chapel Hill Conf. on Combinatorial Mathematics and Its Applications* (Univ. North Carolina, Chapel Hill), 472–477, 1970.

Topics in Combinatorial Matrix Theory

31

Permanents

Ian M. Wanless

Monash University

The permanent is a matrix function introduced (independently) by Cauchy and Binet in 1812. At first sight it seems to be a simplified version of the determinant, but this impression is misleading. In some important respects the permanent is much less tractable than the determinant. Nonetheless, permanents have found a wide range of applications from pure combinatorics (e.g., counting problems involving permutations) right through to applied science (e.g., modeling subatomic particles). For further reading see [Min78], [Min83], [Min87], [CW05], and the references therein.

31.1 Basic Concepts

Definitions:

Let $A = [a_{ij}]$ be an $m \times n$ matrix over a commutative ring, $m \leq n$. Let S be the set of all injective functions from $\{1, 2, \ldots, m\}$ to $\{1, 2, \ldots, n\}$ (in particular, if $m = n$, then S is the symmetric group on $\{1, 2, \ldots, n\}$). The **permanent** of A is defined by

$$\operatorname{per}(A) = \sum_{\sigma \in S} \prod_{i=1}^{m} a_{i\sigma(i)}.$$

Two matrices A and B are **permutation equivalent** if there exist permutation matrices P and Q such that $B = PAQ$.

Facts:

For facts for which no specific reference is given and for background reading on the material in this subsection, see [Min78].

1. If A is any square matrix, then $\text{per}(A) = \text{per}(A^T)$.
2. Our definition implies that $\text{per}(A) = 0$ for all $m \times n$ matrices A, where $m > n$. Some authors prefer to define $\text{per}(A) = \text{per}(A^T)$ in this case.
3. If A is any $m \times n$ matrix and P and Q are permutation matrices of respective orders m and n, then $\text{per}(PAQ) = \text{per}(A)$. That is, the permanent is invariant under permutation equivalence.
4. If A is any $m \times n$ matrix and c is any scalar, then $\text{per}(cA) = c^m \,\text{per}(A)$.
5. The permanent is a multilinear function of the rows. If $m = n$, it is also a multilinear function of the columns.
6. It is *not* in general true that $\text{per}(AB) = \text{per}(A)\,\text{per}(B)$.
7. If M has block decomposition $M = \begin{bmatrix} A & \mathbf{0} \\ B & C \end{bmatrix}$, where either A or C is square, then $\text{per}(M) = \text{per}(A)\,\text{per}(C)$.
8. Let A be an $m \times n$ matrix with $m \leq n$. Then $\text{per}(A) = 0$ if A contains an $s \times (n - s + 1)$ submatrix of zeroes, for some $s \in \{1, 2, \ldots, m\}$.
9. (Laplace expansion) If A is an $m \times n$ matrix, $2 \leq m \leq n$, and $\alpha \in Q_{r,m}$, where $1 \leq r < m \leq n$ then

$$\text{per}(A) = \sum_{\beta \in Q_{r,n}} \text{per}\left(A[\alpha, \beta]\right) \text{per}\left(A(\alpha, \beta)\right).$$

   In particular, for any $i \in \{1, 2, \ldots, m\}$,

$$\text{per}(A) = \sum_{j=1}^{n} a_{ij} \,\text{per}\left(A(i, j)\right).$$

10. If A and B are $n \times n$ matrices and s and t are arbitrary scalars, then

$$\text{per}(sA + tB) = \sum_{r=0}^{n} s^r t^{n-r} \sum_{\alpha, \beta \in Q_{r,n}} \text{per}\left(A[\alpha, \beta]\right) \text{per}\left(B(\alpha, \beta)\right)$$

   (where we interpret the permanent of a 0×0 matrix to be 1).
11. (Binet–Cauchy) If A and B are $m \times n$ and $n \times m$ matrices, respectively, where $m \leq n$, then

$$\text{per}(AB) = \sum_{\alpha} \frac{1}{\mu(\alpha)} \text{per}\left(A[\{1, 2, \ldots, m\}, \alpha]\right) \text{per}\left(B[\alpha, \{1, 2, \ldots, m\}]\right).$$

   The sum is over all nondecreasing sequences α of m integers chosen from the set $\{1, 2, \ldots, n\}$ and $\mu(\alpha) = \alpha_1! \alpha_2! \cdots \alpha_n!$, where α_i denotes the number of occurrences of the integer i in α.
12. [MM62], [Bot67] Let F be a field and $m \geq 3$ an integer. Let T be a linear transformation for which $\text{per}\left(T(X)\right) = \text{per}(X)$ for all $X \in F^{m \times m}$. Then there exist permutation matrices P and Q and diagonal matrices D_1 and D_2 such that $\text{per}(D_1 D_2) = 1$ and either $T(X) = D_1 P X Q D_2$ for all $X \in F^{m \times m}$ or $T(X) = D_1 P X^T Q D_2$ for all $X \in F^{m \times m}$.
13. (Alexandrov's inequality) Let $A = [a_{ij}] \in \mathbb{R}^{n \times n}$ and $1 \leq r < s \leq n$. If $a_{ij} \geq 0$ whenever $j \neq s$, then

$$(\text{per}(A))^2 \geq \sum_{i=1}^{n} a_{ir} \,\text{per}(A(i, s)) \sum_{i=1}^{n} a_{is} \,\text{per}(A(i, r)). \tag{31.1}$$

   Moreover, if $a_{ij} > 0$ whenever $j \neq s$, then equality holds in (31.1) iff there exists $c \in \mathbb{R}$ such that $a_{ir} = c a_{is}$ for all i.

14. If G is a balanced bipartite graph (meaning the two parts have equal size), then $\text{per}(\mathcal{B}_G)$ counts perfect matchings (also known as 1-factors) in G.
15. If D is a directed graph, then $\text{per}(\mathcal{A}_D)$ counts the cycle covers of D. A cycle cover is a set of disjoint cycles which include every vertex exactly once.

Examples:

1.

$$\text{per}\begin{bmatrix}1 & 2 & 3\\ 4 & 5 & 6\\ 7 & 8 & 9\end{bmatrix} = 1\ \text{per}\begin{bmatrix}5 & 6\\ 8 & 9\end{bmatrix} + 2\ \text{per}\begin{bmatrix}4 & 6\\ 7 & 9\end{bmatrix} + 3\ \text{per}\begin{bmatrix}4 & 5\\ 7 & 8\end{bmatrix}$$
$$= 1\cdot5\cdot9+1\cdot6\cdot8+2\cdot4\cdot9+2\cdot6\cdot7+3\cdot4\cdot8+3\cdot5\cdot7 = 450.$$

2. $\text{per}\begin{bmatrix}0 & 2 & 3 & 4\\ 5 & 6 & 0 & 8\\ 9 & 0 & 0 & 12\end{bmatrix} = 2\cdot5\cdot12+2\cdot8\cdot9+3\cdot5\cdot12+3\cdot6\cdot9+3\cdot6\cdot12+3\cdot8\cdot9+4\cdot6\cdot9 = 1254.$
3. If $A = \begin{bmatrix}2 & 3\\ 2 & -2\end{bmatrix}$ and $B = \begin{bmatrix}-4 & -2\\ 1 & -1\end{bmatrix}$, then $AB = \begin{bmatrix}-5 & -7\\ -10 & -2\end{bmatrix}$. Hence, $80 = \text{per}(AB) \neq \text{per}(A)\,\text{per}(B) = 2 \times 2 = 4$.
4. Below is a bipartite graph G (Figure 31.1) and its biadjacency matrix $\mathcal{B}_G$.

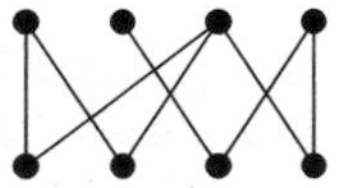

FIGURE 31.1

$$\begin{bmatrix}1 & 1 & 0 & 0\\ 0 & 0 & 1 & 0\\ 1 & 1 & 0 & 1\\ 0 & 0 & 1 & 1\end{bmatrix}$$

Now, $\text{per}(\mathcal{B}_G) = 2$, which means that G has two perfect matchings (Figure 31.2 and Figure 31.3).

and

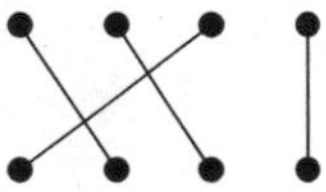

FIGURE 31.2 FIGURE 31.3

5. The matrix in the previous example can also be interpretted as $\mathcal{A}_D$ for the directed graph D (Figure 31.4). It has two cycle covers (Figure 31.5 and Figure 31.6).

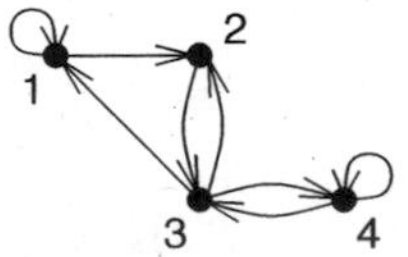

FIGURE 31.4

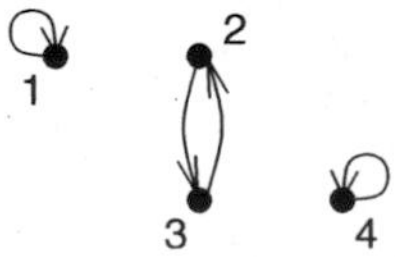

FIGURE 31.5

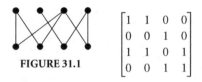

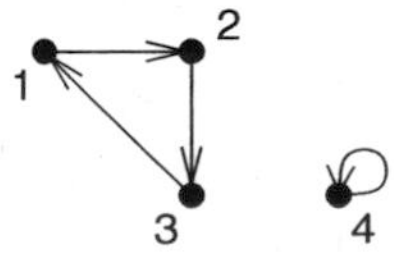

FIGURE 31.6

31.2 Doubly Stochastic Matrices

Facts:

For facts for which no specific reference is given and for background reading on the material in this subsection, see [Min78].

1. If $A \in \Omega_n$, then $\text{per}(A) \leq 1$ with equality iff A is a permutation matrix.
2. [Ego81], [Fal81] If $A \in \Omega_n$ and $A \neq \frac{1}{n}J_n$, then $\text{per}(A) > \text{per}(\frac{1}{n}J_n) = n!/n^n$.

3. [Fri82] If $A \in \Omega_n$ and A has a $p \times q$ submatrix of zeros (where $p + q \leq n$), then

$$\operatorname{per}(A) \geq \frac{(n-p)!}{(n-p)^{n-p}} \frac{(n-q)!}{(n-q)^{n-q}} \frac{(n-p-q)^{n-p-q}}{(n-p-q)!}.$$

Equality is achieved by any matrix permutation equivalent to the matrix $A = [a_{ij}]$ defined by

$$a_{ij} = \begin{cases} 0 & \text{if } i \leq p \text{ and } j \leq q, \\ \frac{1}{n-q} & \text{if } i \leq p \text{ and } j > q, \\ \frac{1}{n-p} & \text{if } i > p \text{ and } j \leq q, \\ \frac{n-p-q}{(n-p)(n-q)} & \text{if } i > p \text{ and } j > q. \end{cases}$$

If $p + q \neq n - 1$, then no other matrix achieves equality.

4. [BN66] If A is any $n \times n$ row substochastic matrix and $1 \leq r \leq n$, then $\operatorname{per}(A) \leq m_r$, where m_r is the maximum permanent over all $r \times r$ submatrices of A.
5. [Min78, p. 41] If $A = [a_{ij}]$ is a fully indecomposable matrix, then the matrix $S = [s_{ij}]$ defined by $s_{ij} = a_{ij} \operatorname{per}\left(A(i, j)\right) / \operatorname{per}(A)$ is doubly stochastic and has the same zero pattern as A.

Examples:

1. The minimum value of the permanent in Ω_5 is 24/625, which is (uniquely) achieved by

$$\tfrac{1}{5}J_5 = \begin{bmatrix} 1/5 & 1/5 & 1/5 & 1/5 & 1/5 \\ 1/5 & 1/5 & 1/5 & 1/5 & 1/5 \\ 1/5 & 1/5 & 1/5 & 1/5 & 1/5 \\ 1/5 & 1/5 & 1/5 & 1/5 & 1/5 \\ 1/5 & 1/5 & 1/5 & 1/5 & 1/5 \end{bmatrix}.$$

2. In the previous example, if we require that two specified entries in the same row must be zero, then the minimum value that the permanent can take is 1/24, which is (uniquely, up to permutation equivalence) achieved by

$$\begin{bmatrix} 0 & 0 & 1/3 & 1/3 & 1/3 \\ 1/4 & 1/4 & 1/6 & 1/6 & 1/6 \\ 1/4 & 1/4 & 1/6 & 1/6 & 1/6 \\ 1/4 & 1/4 & 1/6 & 1/6 & 1/6 \\ 1/4 & 1/4 & 1/6 & 1/6 & 1/6 \end{bmatrix}.$$

3. A nonnegative matrix of order $n \geq 3$ can have zero permanent even if we insist that each row and column sum is at least one. For example,

$$\operatorname{per} \begin{bmatrix} 1 & 1 & 0 \\ 0 & 0 & 1 \\ 0 & 0 & 1 \end{bmatrix} = 0.$$

4. Suppose n identical balls are placed, one ball per bucket, in n labeled buckets on the back of a truck. When the truck goes over a bump the balls are flung into the air, but then fall back into the buckets. Suppose that the probability that the ball from bucket i lands in bucket j is p_{ij}. Then the matrix $P = [p_{ij}]$ is row stochastic and $\operatorname{per}(P)$ is the probability that we end up with one ball in each bucket. That is, $\operatorname{per}(P)$ is the *permanence* of the initial state.

31.3 Binary Matrices

Definitions:

A **binary matrix** is a matrix in which each entry is either 0 or 1.

Λ_n^k is the set of $n \times n$ binary matrices in which each row and column sum is k.

For an $m \times n$ binary matrix M the **complement** M^c is defined by $M^c = J_{mn} - M$.

A system of distinct representatives (**SDR**) for the finite sets $S_1, S_2, \ldots, S_n$ is a choice of $x_1, x_2, \ldots, x_n$ with the properties that $x_i \in S_i$ for each i and $x_i \neq x_j$ whenever $i \neq j$.

The **incidence matrix** for subsets $S_1, S_2, \ldots, S_m$ of a finite set $\{x_1, x_2, \ldots, x_n\}$ is the $m \times n$ binary matrix $M = [m_{ij}]$ in which $m_{ij} = 1$ iff $x_j \in S_i$.

$P_{(12\cdots n)}$ is the permutation matrix for the full cycle permutation $(12\cdots n)$.

Facts:

For facts for which no specific reference is given and for background reading on the material in this section, see [Min78].

1. The number of SDRs for a set of sets with incidence matrix M is $\text{per}(M)$.
2. If $M \in \Lambda_n^k$, then $\frac{1}{k}M \in \Omega_n$, so the results of the previous subsection apply.
3. [Min78, p. 52] Let A be an $m \times n$ binary matrix where $m \leq n$. Suppose A has at least t positive entries in each row. If $t < m$ and $\text{per}(A) > 0$, then $\text{per}(A) \geq t!$. If $t \geq m$, then $\text{per}(A) \geq t!/(t-m)!$.
4. If $A \in \Lambda_n^k$, then there exist permutation matrices $P_1, P_2, \ldots, P_k$ such that
$$A = \sum_{i=1}^{k} P_i.$$
5. [Brè73], [Sch78] Let A be any $n \times n$ binary matrix with row sums $r_1, r_2, \ldots, r_n$. Then
$$\text{per}(A) \leq \prod_{i=1}^{n} (r_i!)^{1/r_i} \qquad (31.2)$$
with equality iff A is permutation equivalent to a direct sum of square matrices each of which contains only 1s.
6. [MW98] If $m \geq 5$, then $(J_k \oplus J_k \oplus \cdots \oplus J_k)^c$ (where there are m copies of J_k) maximizes the permanent in Λ_{mk}^{mk-k}. The result is not true for $m = 3$.
7. [Wan99b] For each $k \geq 1$ there exists N such that for all $n \geq N$ a matrix M maximizes the permanent in Λ_n^k iff $M \oplus J_k$ maximizes the permanent in Λ_{n+k}^k.
8. [Wan03] If $n = tk + r$ with $0 \leq r < k \leq n$, then $k!^t r! \leq \max_{A \in \Lambda_n^k} \text{per}(A) \leq (k!)^{n/k}$.
9. [Wan03] If $k = o(n)$ as $n \to \infty$, then $\left(\max_{A \in \Lambda_n^k} \text{per}(A)\right)^{1/n} \sim (k!)^{1/k}$.
10. [GM90] Suppose $0 \leq k = O(n^{1-\delta})$ for a constant $\delta > 0$ as $n \to \infty$. Then
$$\text{per}(A) = n!\left(\frac{n-k}{n}\right)^n \exp\left(\frac{k}{2n} + \frac{3k^2 - k}{6n^2} + \frac{2k^3 - k}{4n^3}\right.$$
$$\left. + \frac{15k^4 + 70k^3 - 105k^2 + 32k}{60n^4} + \frac{z}{n^4} + \frac{2z(2k-1)}{n^5} + O\left(\frac{k^5}{n^5}\right)\right)$$
for all $A \in \Lambda_n^{n-k}$, where z denotes the number of 2×2 submatrices of A that contain only zeros. In particular, if $0 \leq k = O(n^{1-\delta})$ for a constant $\delta > 0$ as $n \to \infty$, then $\text{per}(A)$ is asymptotically equal to $n!(1 - k/n)^n$ for all $A \in \Lambda_n^{n-k}$.
11. [Wan06] If $2 \leq k \leq n$ as $n \to \infty$, then
$$\left(\min_{A \in \Lambda_n^k} \text{per}(A)\right)^{1/n} \sim \frac{(k-1)^{k-1}}{k^{k-2}}.$$

12. [BN65] For any integers $k \geq 1$ and $n \geq \log_2 k + 1$ there exists a binary matrix A of order n such that $\text{per}(A) = k$.
13. The permanent of a square binary matrix counts **permutations with restricted positions**. This means that for each point being permuted there is some set of allowable images, while other images are forbidden.

Examples:

1. If $D_n = I_n^c$, then

$$\text{per}(D_n) = n!\left(1 - \frac{1}{1!} + \frac{1}{2!} - \cdots + (-1)^n \frac{1}{n!}\right)$$

is the number of **derangements** of n things, that is, the number of permutations of n points that leave no point in its original place. The 4th derangement number is

$$\text{per}(D_4) = \text{per}\begin{bmatrix} 0 & 1 & 1 & 1 \\ 1 & 0 & 1 & 1 \\ 1 & 1 & 0 & 1 \\ 1 & 1 & 1 & 0 \end{bmatrix} = 9.$$

2. The number of ways that n married couples can sit around a circular table with men and women alternating and so that nobody sits next to their spouse is $2\,n!\,M_n$, where M_n is known as the n-th **menagé number** and is given by

$$M_n = \text{per}\left((I_n + P_{(12\cdots n)})^c\right) = \sum_{r=0}^{n}(-1)^r \frac{2n}{2n-r}\binom{2n-r}{r}(n-r)!.$$

The 5th menagé number is $M_5 = \text{per}\begin{bmatrix} 0 & 0 & 1 & 1 & 1 \\ 1 & 0 & 0 & 1 & 1 \\ 1 & 1 & 0 & 0 & 1 \\ 1 & 1 & 1 & 0 & 0 \\ 0 & 1 & 1 & 1 & 0 \end{bmatrix} = 13.$

3. SDRs are important for many combinatorial problems. For example, a $k \times n$ **Latin rectangle** (where $k \leq n$) is a $k \times n$ matrix in which n symbols occur in such a way that each symbol occurs exactly once in each row and at most once in each column. The number of extensions of a given $k \times n$ Latin rectangle R to a $(k+1) \times n$ Latin rectangle is the number of SDRs of the sets $S_1, S_2, \ldots, S_n$ defined so that S_i consists of the symbols not yet used in column i of R.
4. [CW05] For $n \leq 11$ the minimum values of $\text{per}(A)$ for $A \in \Lambda_n^k$ are as follows:

k	$n=2$	3	4	5	6	7	8	9	10	11
1	1	1	1	1	1	1	1	1	1	1
2	2	2	2	2	2	2	2	2	2	2
3	—	6	9	12	17	24	33	42	60	83
4	—	—	24	44	80	144	248	440	764	1316
5	—	—	—	120	265	578	1249	2681	5713	12105
6	—	—	—	—	720	1854	4738	12000	30240	75510
7	—	—	—	—	—	5040	14833	43386	126117	364503
8	—	—	—	—	—	—	40320	133496	439792	1441788
9	—	—	—	—	—	—	—	362880	1334961	4890740
10	—	—	—	—	—	—	—	—	3628800	14684570
11	—	—	—	—	—	—	—	—	—	39916800

5. [MW98] For $n \leq 11$ the maximum values of per(A) for $A \in \Lambda_n^k$ are as follows:

k	$n=2$	3	4	5	6	7	8	9	10	11
1	1	1	1	1	1	1	1	1	1	1
2	2	2	4	4	8	8	16	16	32	32
3	—	6	9	13	36	54	81	216	324	486
4	—	—	24	44	82	148	576	1056	1968	3608
5	—	—	—	120	265	580	1313	2916	14400	31800
6	—	—	—	—	720	1854	4752	12108	32826	86400
7	—	—	—	—	—	5040	14833	43424	127044	373208
8	—	—	—	—	—	—	40320	133496	440192	1448640
9	—	—	—	—	—	—	—	362880	1334961	4893072
10	—	—	—	—	—	—	—	—	3628800	14684570
11	—	—	—	—	—	—	—	—	—	39916800

31.4 Nonnegative Matrices

Definitions:

Δ_n^k is the set of $n \times n$ matrices of nonnegative integers in which each row and column sum is k.

Facts:

1. [Min78, p. 33] Let A be an $m \times n$ nonnegative matrix with $m \leq n$. Then per$(A) = 0$ iff A contains an $s \times (n - s + 1)$ submatrix of zeros, for some $s \in \{1, 2, \ldots, m\}$.
2. [Min78, p. 38] Let A be a nonnegative matrix of order $n \geq 2$. Then A is fully indecomposable iff per $(A(i, j)) > 0$ for all i, j.
3. [Sch98] $\left(\frac{(k-1)^{k-1}}{k^{k-2}}\right)^n \leq \min_{A \in \Delta_n^k} \text{per}(A) \leq \frac{k^{2n}}{\binom{kn}{n}}$.
4. [Min83] It is conjectured that $\min_{A \in \Delta_n^k} \text{per}(A) = \min_{A \in \Lambda_n^k} \text{per}(A)$.
5. [Sou03] Let Γ denote the gamma function and let A be a nonnegative matrix of order n. In row i of A, let m_i and r_i denote, respectively, the largest entry and the total of the entries. Then,

$$\text{per}(A) \leq \prod_{i=1}^{n} m_i \left(\Gamma\left(\frac{r_i}{m_i} + 1\right)\right)^{m_i/r_i}. \tag{31.3}$$

6. [Min78, p. 62] Let A be a nonnegative matrix of order n. Define s_i to be the sum of the i smallest entries in row i of A. Similarly, define S_i to be the sum of the i largest entries in row i of A. Then

$$\prod_{i=1}^{n} s_i \leq \text{per}(A) \leq \prod_{i=1}^{n} S_i.$$

7. [Gib72] If A is nonnegative and π is a root of per$(zI - A)$, then $|\pi| \leq \rho(A)$.

Examples:

1. [BB67] If A is row substochastic, the roots of per$(zI - A) = 0$ satisfy $|z| \leq 1$.
2. If Soules' bound (31.3) is applied to matrices in Λ_n^k it reduces to Brègman's bound (31.2).

31.5 (± 1)-Matrices

Facts:

1. [KS83] If A is a (± 1)-matrix of order n, then $\text{per}(A)$ is divisible by $2^{n-\lfloor \log_2(n+1) \rfloor}$.
2. [Wan74] If H is an $n \times n$ Hadamard matrix, then $|\text{per}(H)| \leq |\det(H)| = n^{n/2}$.
3. [KS83], [Wan05] There is no solution to $|\text{per}(A)| = |\det(A)|$ among the nonsingular (± 1)-matrices of order n when $n \in \{2, 3, 4\}$ or $n = 2^k - 1$ for $k \geq 2$, but there are solutions when $n \in \{5, \ldots, 20\} \setminus \{7, 15\}$.
4. [Wan05] There exists a (± 1)-matrix A of order n satisfying $\text{per}(A) = 0$ iff $n + 1$ is not a power of 2.

Examples:

1. The 11×11 matrix $A = [a_{ij}]$ defined by

$$a_{ij} = \begin{cases} -1 & \text{if } j - i \in \{1, 2, 3, 5, 7, 10\}, \\ +1 & \text{otherwise,} \end{cases}$$

   satisfies $\text{per}(A) = 0$. No smaller (± 1)-matrix of order $n \equiv 3 \bmod 4$ has this property.
2. The following matrix has per = det = 16, and is the smallest example (excluding the trivial case of order 1) for which per = det among ± 1-matrices.

$$\begin{bmatrix} +1 & +1 & +1 & +1 & +1 \\ +1 & -1 & -1 & +1 & +1 \\ +1 & -1 & +1 & +1 & +1 \\ +1 & +1 & +1 & -1 & -1 \\ +1 & +1 & +1 & -1 & +1 \end{bmatrix}.$$

31.6 Matrices over $\mathbb{C}$

Facts:

1. If $A = [a_{ij}] \in \mathbb{C}^{m \times n}$ and $B = [b_{ij}] \in \mathbb{R}^{m \times n}$ satisfy $b_{ij} = |a_{ij}|$, then $|\text{per}(A)| \leq \text{per}(B)$.
2. If $A \in \mathbb{C}^{n \times n}$, then $\text{per}(\overline{A}) = \overline{\text{per}(A)} = \text{per}(A^*)$.
3. [Min78, p. 113] If $A \in \mathbb{C}^{n \times n}$ is normal with eigenvalues $\lambda_1, \lambda_2, \ldots, \lambda_n$, then

$$|\text{per}(A)| \leq \frac{1}{n} \sum_{i=1}^{n} |\lambda_i|^n.$$

4. [Min78, p. 115] Let $A \in \mathbb{C}^{n \times n}$ and let $\lambda_1, \ldots, \lambda_n$ be the eigenvalues of AA^*. Then

$$|\text{per}(A)|^2 \leq \frac{1}{n} \sum_{i=1}^{n} \lambda_i^n.$$

5. [Lie66] If $A = \begin{bmatrix} B & C \\ C^* & D \end{bmatrix} \in \text{PD}_n$, then $\text{per}(A) \geq \text{per}(B)\,\text{per}(D)$.
6. [JP87] Suppose $\alpha \subseteq \beta \subseteq \{1, 2, \ldots, n\}$. Then for any $A \in \text{PD}_n$,

$$\det(A[\beta, \beta])\,\text{per}(A(\beta, \beta)) \leq \det(A[\alpha, \alpha])\,\text{per}(A(\alpha, \alpha)).$$

   (We interpret det or per of a 0×0 matrix to be 1.)

7. [MN62] If $A \in \mathbb{C}^{m\times n}$ and $B \in \mathbb{C}^{n\times m}$, then

$$|\operatorname{per}(AB)|^2 \le \operatorname{per}(AA^*)\operatorname{per}(B^*B).$$

8. [Bre59] If $A = [a_{ij}] \in \mathbb{C}^{n\times n}$ satisfies $|a_{ii}| > \sum_{j\neq i} |a_{ij}|$ for each i, then $\operatorname{per}(A) \neq 0$.

Examples:

1. If U is a unitary matrix, then $|\operatorname{per}(U)| \le 1$.

31.7 Subpermanents

Definitions:

The k-th **subpermanent sum**, $\operatorname{per}_k(A)$ of an $m \times n$ matrix A, is defined to be the sum of the permanents of all order k submatrices of A. That is,

$$\operatorname{per}_k(A) = \sum_{\substack{\alpha\in Q_{k,m}\\ \beta\in Q_{k,n}}} \operatorname{per}(A[\alpha,\beta]).$$

By convention, we define $\operatorname{per}_0(A) = 1$.

Facts:

For facts for which no specific reference is given and for background reading on the material in this subsection see [Min78].

1. For each k, per_k is invariant under permutation equivalence and transposition.
2. If A is any $m \times n$ matrix and c is any scalar, then $\operatorname{per}_k(cA) = c^k \operatorname{per}_k(A)$.
3. [BN66] If A is any $n \times n$ row substochastic matrix and $1 \le r \le n$, then $\operatorname{per}_r(A) \le \binom{n}{r}$.
4. [Fri82] $\operatorname{per}_k(A) \ge \operatorname{per}_k(\frac{1}{n}J_n)$ for every $A \in \Omega_n$ and integer k.
5. [Wan03] If $A \in \Lambda_n^k$ and $i \le k$, then $\operatorname{per}_i(A) \ge \binom{n}{i}\dfrac{k!}{(k-i)!}$.
6. [Wan03] For $1 \le i,k \le n$ and $A \in \Lambda_n^k$,

$$\binom{kn}{i} - kn(k-1)\binom{kn-2}{i-2} \le \operatorname{per}_i(A) \le \binom{kn}{i}.$$

7. [Wan03] For $A \in \Lambda_n^k$, let $\xi_i = \left(\operatorname{per}_i(A)/\binom{n}{i}\right)^{1/i}$. Then

$$\frac{(k-1)^{k-1}}{k^{k-2}} \le \xi_n \le \xi_{n-1} \le \cdots \le \xi_1 = k.$$

8. [Nij76], [HLP52, p. 104] Let A be a nonnegative $m\times n$ matrix with $\operatorname{per}(A) \neq 0$. For $1 \le i \le m-1$,

$$\frac{\operatorname{per}_i(A)}{\operatorname{per}_{i-1}(A)} \ge \frac{(i+1)(m-i+1)}{i(m-i)}\frac{\operatorname{per}_{i+1}(A)}{\operatorname{per}_i(A)} > \frac{\operatorname{per}_{i+1}(A)}{\operatorname{per}_i(A)}.$$

9. [Wan99b] For $A \in \Lambda_n^k$,

$$\frac{\operatorname{per}_i(A)}{\operatorname{per}_{i+1}(A)} \ge \frac{i+1}{(n-i)^2}.$$

10. [Wan99a] Let $k \geq 0$ be an integer. There exists no polynomial $p_k(n)$ such that for all n and $A \in \Lambda_n^3$,

$$\frac{\mathrm{per}_{n-k-1}(A)}{\mathrm{per}_{n-k}(A)} \leq p_k(n).$$

11. If G is a bipartite graph, then $\mathrm{per}_k(\mathcal{B}_G)$ counts the k-matchings in G. A k-matching in G is a set of k edges in G such that no two edges share a vertex.

Examples:

1. For any matrix A the sum of the entries in A is $\mathrm{per}_1(A)$.
2. Any $m \times n$ matrix A has $\mathrm{per}_m(A) = \mathrm{per}(A)$.
3. $\mathrm{per}_k(J_n) = k!\binom{n}{k}^2$.
4. [Wan99a] Let $A \in \Lambda_n^k$ have s submatrices which are copies of J_2. Then
   - $\mathrm{per}_1(A) = kn$.
   - $\mathrm{per}_2(A) = \frac{1}{2}kn(kn - 2k + 1)$.
   - $\mathrm{per}_3(A) = \frac{1}{6}kn(k^2n^2 - 6k^2n + 3kn + 10k^2 - 12k + 4)$.
   - $\mathrm{per}_4(A) = \frac{1}{24}kn\big(k^3n^3 - 12k^3n^2 + 6k^2n^2 + 52k^3n - 60k^2n + 19kn - 84k^3 + 168k^2 - 120k + 30\big) + s$.
   - $\mathrm{per}_5(A) = \frac{1}{120}kn\big[k^4n^4 - 20k^4n^3 + 10k^3n^3 + 160k^4n^2 - 180k^3n^2 + 55k^2n^2 - 620k^4n + 1180k^3n - 800k^2n + 190kn + 1008k^4 - 2880k^3 + 3240k^2 - 1680k + 336\big] + (nk - 8k + 8)s$.
5. The subpermanent sums are also known as **rook numbers** since they count the number of placements of rooks in mutually nonattacking positions on a chessboard. Let A be a binary matrix in which each 1 denotes a permitted position on the board and each 0 denotes a forbidden position for a rook. Then $\mathrm{per}_i(A)$ is the number of placements of i rooks on permitted squares so that no two rooks occupy the same row or column. For example, the number of ways of putting 4 nonattacking rooks on the white squares of a standard chessboard is

$$\mathrm{per}_4\begin{bmatrix} 1&0&1&0&1&0&1&0\\ 0&1&0&1&0&1&0&1\\ 1&0&1&0&1&0&1&0\\ 0&1&0&1&0&1&0&1\\ 1&0&1&0&1&0&1&0\\ 0&1&0&1&0&1&0&1\\ 1&0&1&0&1&0&1&0\\ 0&1&0&1&0&1&0&1 \end{bmatrix} = 8304.$$

31.8 Rook Polynomials

Definitions:

Let A be an $m \times n$ binary matrix. The polynomials $\rho_1(A,x) = \sum_{i=0}^m \mathrm{per}_i(A)x^i$ and $\rho_2(A,x) = \sum_{i=0}^m (-1)^i \mathrm{per}_i(A)x^{m-i}$ are both called **rook polynomials** because they are generating functions for the rook numbers.

Let $\ell_k(x)$ be the kth Laguerre polynomial, normalized to be monic. That is,

$$\ell_k(x) = (-1)^k k! \sum_{i=0}^{k} \binom{k}{i} \frac{(-x)^i}{i!}.$$

Facts:

For facts for which no specific reference is given and for background reading on the material in this section, see [Rio58].

1. When A is an $m \times n$ binary matrix, $\rho_1(A,x) = (-x)^m \rho_2(A, -\frac{1}{x})$.
2. If $A = B \oplus C$, then $\rho_1(A,x) = \rho_1(B,x)\rho_1(C,x)$ and $\rho_2(A,x) = \rho_2(B,x)\rho_2(C,x)$.
3. [HL72], [Nij76] For any nonnegative matrix A, all the roots of $\rho_1(A,x)$ and $\rho_2(A,x)$ are real.
4. [HL72] For $2 \leq k \leq n$ and $A \in \Lambda_n^k$, the roots of $\rho_1(A,x)$ are less than $-1/(4k-4)$, while the roots of $\rho_2(A,x)$ lie in the interval $(0, 4k-4)$.
5. [JR80], [God81] For a square binary matrix A, with complement A^c,
$$\mathrm{per}(A) = \int_0^\infty e^{-x} \rho_2(A^c, x)\,\mathrm{d}x.$$
6. [God81] For an $n \times n$ binary matrix M,
$$\rho_2(M,x) = \sum_{i=0}^{n} \mathrm{per}_i(M^c)\ell_{n-i}(x).$$
7. [Sze75, p. 100] For any j, k let $\delta_{j,k}$ denote the Kronecker delta. Then
$$\int_0^\infty e^{-x}\ell_j(x)\ell_k(x)\,\mathrm{d}x = \delta_{j,k}k!^2.$$

Examples:

1. $\rho_1(P,x) = (x+1)^n$ and $\rho_2(P,x) = (x-1)^n$ for a permutation matrix $P \in \Lambda_n^1$.
2. $\rho_2(J_k, x) = \ell_k(x)$.
3. Let $C_n = I_n + P_{(12\cdots n)}$. Then
$$\rho_1(C_n, x) = \sum_{i=0}^{n} \frac{2n}{2n-i}\binom{2n-i}{i}x^i$$
and $\rho_2(C_n, 4x^2) = 2T_{2n}(x)$, where $T_n(x)$ is the nth Chebyshev polynomial of the first kind.
4. $\rho_2(I_n^c, x) = \sum_{i=0}^{n} \binom{n}{i}\ell_i(x)$ for any integer $n \geq 1$.
5. The ideas of this chapter allow a quick computation of the permanent of many matrices that are built from blocks of ones by recursive use of direct sums and complementation. For example,
$$\begin{aligned}\mathrm{per}\left((J_k \oplus I_{k+1}^c)^c\right) &= \int_0^\infty e^{-x}\rho_2(J_k \oplus I_{k+1}^c)\,\mathrm{d}x = \int_0^\infty e^{-x}\rho_2(J_k)\rho_2(I_{k+1}^c)\,\mathrm{d}x \\ &= \int_0^\infty e^{-x}\ell_k(x)\sum_{i=0}^{k+1}\binom{k+1}{i}\ell_i(x)\,\mathrm{d}x = \binom{k+1}{k}k!^2 = (k+1)!\,k!.\end{aligned}$$

31.9 Evaluating Permanents

Facts:

For facts for which no specific reference is given and for background reading on the material in this subsection, see [Min78].

1. Since the permanent is *not* invariant under elementary row operations, it cannot be calculated by Gaussian elimination.
2. (Ryser's formula) If $A = [a_{ij}]$ is any $n \times n$ matrix,
$$\mathrm{per}(A) = \sum_{r=1}^{n}(-1)^r \sum_{\alpha \in Q_{r,n}} \prod_{i=1}^{n} \sum_{j \in \alpha} a_{ij}.$$

3. [NW78] A straightforward implementation of Ryser's formula has time complexity $\Theta(n^2 2^n)$. By enumerating the α in Gray Code order (i.e., by choosing an ordering in which any two consecutive α's differ by a single entry), Nijenhuis/Wilf improved this to $\Theta(n2^n)$. They cut the execution time by a further factor of two by exploiting the relationship between the term corresponding to α and that corresponding to $\{1, 2, \ldots, n\} \setminus \alpha$. This second savings is not always desirable when calculating permanents of integer matrices, since it introduces fractions.

Ryser/Nijenhuis/Wilf (RNW) Algorithm for calculating per(A) for $A = [a_{ij}]$ of order n.

```
p := −1;
for i from 1 to n do
      x_i := a_in − (1/2) Σ_{j=1}^{n} a_ij;
      p := p * x_i;
      g_i := 0;
s := −1;
for k from 2 to 2^{n−1} do
      if k is even then
            j := 1;
      else
            j := 2;
            while g_{j−1} = 0 do
                  j := j + 1;
      z := 1 − 2 * g_j;
      g_j := 1 − g_j;
      s := −s;
      t := s;
      for i from 1 to n do
            x_i := x_i + z * a_ij;
            t := t * x_i;
      p := p + t;
return(2(−1)^n p)
```

4. For sufficiently sparse matrices, a simple enumeration of nonzero diagonals by backtracking, or a recursive Laplace expansion, will be faster than RNW.
5. A hybrid approach is to use Laplace expansion to expand any rows or columns that have very few nonzero entries, then employ RNW.
6. [DL92] The calculation of the permanent is a $\#P$-complete problem. This is still true if attention is restricted to matrices in Λ_n^3. So, it is extremely unlikely that a polynomial time algorithm for calculating permanents exists.
7. [Lub90] As a result of the above, much work has been done on approximation algorithms for permanents.

31.10 Connections between Determinants and Permanents

Definitions:

For any partition λ of n let χ_λ denote the irreducible character of the symmetric group S_n associated with λ by the standard bijection (see Section 68.6 or [Mac95, p. 114]) between partitions and irreducible characters.

Let ε_n be the identity in S_n.

The matrix function f_λ defined by

$$f_\lambda(M) = \frac{1}{\chi_\lambda(\varepsilon_n)} \sum_{\sigma \in S_n} \chi_\lambda(\sigma) \prod_{i=1}^{n} m_{i\sigma(i)},$$

for each $M = [m_{ij}] \in \mathbb{C}^{n\times n}$, is called a **normalized immanant** (without the factor of $1/\chi_\lambda(\varepsilon_n)$ it is an **immanant**).

The partial order $\lhd$ is defined on the set of partitions of an integer n by stating that $\lambda \lhd \mu$ means that $f_\lambda(H) \leq f_\mu(H)$ for all $H \in \text{PD}_n$.

Facts:

For facts for which no specific reference is given and for background reading on the material in this subsection, see [Mer97].

1. If $\lambda = (n)$, then χ_λ is the principal/trivial character and f_λ is the permanent.
2. If $\lambda = (1^n)$, then χ_λ is the alternating character and f_λ is the determinant.
3. [Sch18] $f_\lambda(H)$ is a nonnegative real number for all $H \in \text{PD}_n$ and all λ.
4. [MM61] For $n \geq 3$ there is no linear transformation T such that $\text{per}(A) = \det\big(T(A)\big)$ for every $A \in \mathbb{R}^{n\times n}$. In particular, there is no way of affixing minus signs to some entries that will convert the permanent into a determinant.
5. [Lev73] For all sufficiently large n there exists a fully indecomposable matrix $A \in \Lambda_n^3$ such that $\text{per}(A) = \det(A)$.
6. [Sch18], [Mar64] If $A = [a_{ij}] \in \text{PD}_n$, then $\det(A) \leq \prod_{i=1}^{n} a_{ii} \leq \text{per}(A)$. Equality holds iff A is diagonal or A has a zero row/column.
7. [Sch18] For arbitrary λ, if $A \in \text{PD}_n$, then $f_\lambda(A) \geq \det(A)$. In other words $(1^n) \lhd \lambda$ for all partitions λ of n.
8. [Hey88] The hook immanants are linearly ordered between det and per. That is, $(1^n) \lhd (2, 1^{n-2}) \lhd (3, 1^{n-3}) \lhd \cdots \lhd (n-1, 1) \lhd (n)$.
9. A special case of the **permanental dominance conjecture** asserts that $\lambda \lhd (n)$ for all partitions λ of n. This has been proven only for special cases, which include (a) $n \leq 13$, (b) partitions with no more than three parts which exceed 2, and (c) the hook immanants mentioned above.
10. For two partitions λ, μ of n to satisfy $\lambda \lhd \mu$, it is necessary but not sufficient that μ majorizes λ.

Examples:

1. Although $\mu = (3, 1)$ majorizes $\lambda = (2, 2)$, neither $\lambda \lhd \mu$ nor $\mu \lhd \lambda$. This is demonstrated by taking $A = J_2 \oplus J_2$ and $B = J_1 \oplus J_3$ and noting that $f_\lambda(A) = 2 > \frac{4}{3} = f_\mu(A)$ but $f_\lambda(B) = 0 < 2 = f_\mu(B)$.

References

[Bot67] P. Botta, Linear transformations that preserve the permanent, *Proc. Amer. Math. Soc.* 18:566–569, 1967.

[Brè73] L.M. Brègman, Some properties of nonnegative matrices and their permanents, *Soviet Math. Dokl.* 14:945–949, 1973.

[Bre59] J.L. Brenner, Relations among the minors of a matrix with dominant principal diagonal, *Duke Math. J.* 26:563–567, 1959.

[BB67] J.L. Brenner and R.A. Brualdi, Eigenschaften der Permanentefunktion, *Arch. Math.* 18:585–586, 1967.

[BN65] R.A. Brualdi and M. Newman, Some theorems on the permanent, *J. Res. Nat. Bur. Standards Sect. B* 69B:159–163, 1965.

[BN66] R.A. Brualdi and M. Newman, Inequalities for the permanental minors of non-negative matrices, *Canad. J. Math.* 18:608–615, 1966.

[BR91] R.A. Brualdi and H.J. Ryser, *Combinatorial matrix theory*, Encyclopedia Math. Appl. 39, Cambridge University Press, Cambridge, 1991.

[CW05] G.-S. Cheon and I.M. Wanless, An update on Minc's survey of open problems involving permanents, *Lin. Alg. Appl.* 403:314–342, 2005.

[DL92] P. Dagum and M. Luby, Approximating the permanent of graphs with large factors, *Theoret. Comput. Sci.* 102:283–305, 1992.

[Ego81] G.P. Egorychev, Solution of the van der Waerden problem for permanents, *Soviet Math. Dokl.* 23:619–622, 1981.

[Fal81] D.I. Falikman, Proof of the van der Waerden conjecture regarding the permanent of a doubly stochastic matrix, *Math. Notes* 29:475–479, 1981.

[Fri82] S. Friedland, A proof of the generalized van der Waerden conjecture on permanents, *Lin. Multilin. Alg.* 11:107–120, 1982.

[Gib72] P.M. Gibson, Localization of the zeros of the permanent of a characteristic matrix, *Proc. Amer. Math. Soc.* 31:18–20, 1972.

[God81] C.D. Godsil, Hermite polynomials and a duality relation for the matchings polynomial, *Combinatorica* 1:257–262, 1981.

[GM90] C.D. Godsil and B.D. McKay, Asymptotic enumeration of Latin rectangles, *J. Combin. Theory Ser. B* 48:19–44, 1990.

[HLP52] G. Hardy, J.E. Littlewood, and G. Pólya, *Inequalities* (2nd ed.), Cambridge University Press, Cambridge, 1952.

[Hey88] P. Heyfron, Immanant dominance orderings for hook partitions, *Lin. Multilin. Alg.* 24:65–78, 1988.

[HL72] O.J. Heilmann and E.H. Lieb, Theory of monomer-dimer systems, *Comm. Math. Physics* 25:190–232, 1972.

[HKM98] S.-G. Hwang, A.R. Kräuter, and T.S. Michael, An upper bound for the permanent of a nonnegative matrix, *Lin. Alg. Appl.* 281:259–263, 1998.

[JP87] C.R. Johnson and S. Pierce, Permanental dominance of the normalized single-hook immanants on the positive semi-definite matrices, *Lin. Multilin. Alg.* 21:215–229, 1987.

[JR80] S.A. Joni and G.-C. Rota, A vector space analog of permutations with restricted position, *J. Combin. Theory Ser. A* 29:59–73, 1980.

[KS83] A.R. Kräuter and N. Seifter, On some questions concerning permanents of $(1, -1)$-matrices, *Israel J. Math.* 45:53–62, 1983.

[Lev73] R.B. Levow, Counterexamples to conjectures of Ryser and de Oliveira, *Pacific J. Math.* 44:603–606, 1973.

[Lie66] E.H. Lieb, Proofs of some conjectures on permanents, *J. Math. Mech.* 16:127–134, 1966.

[Lub90] M. Luby, A survey of approximation algorithms for the permanent, *Sequences* (Naples/Positano, 1988), 75–91, Springer, New York, 1990.

[Mac95] I. G. Macdonald, *Symmetric Functions and Hall Polynomials* (2nd ed.), Oxford University Press, Oxford, 1995.

[Mar64] M. Marcus, The Hadamard theorem for permanents, *Proc. Amer. Math. Soc.* 65:967–973, 1964.

[MM62] M. Marcus and F. May, The permanent function, *Can. J. Math.* 14:177–189, 1962.

[MM61] M. Marcus and H. Minc, On the relation between the determinant and the permanent, *Ill. J. Math.* 5:376–381, 1961.

[MN62] M. Marcus and M. Newman, Inequalities for the permanent function, *Ann. Math.* 675:47–62, 1962.

[MW98] B.D. McKay and I.M. Wanless, Maximising the permanent of (0, 1)-matrices and the number of extensions of Latin rectangles, *Electron. J. Combin.* 5: R11, 1998.

[Mer97] R. Merris, *Multilinear Algebra*, Gordon and Breach, Amsterdam, 1997.

[Min78] H. Minc, *Permanents*, Encyclopedia Math. Appl. 6, Addison-Wesley, Reading, MA, 1978.

[Min83] H. Minc, Theory of permanents 1978–1981, *Lin. Multilin. Alg.* 12:227–263, 1983.

[Min87] H. Minc, Theory of permanents 1982–1985, *Lin. Multilin. Alg.* 21:109–148, 1987.

[Nij76] A. Nijenhuis, On permanents and the zeros of rook polynomials, *J. Combin. Theory Ser. A* 21:240–244, 1976.

[NW78] A. Nijenhuis and H.S. Wilf, *Combinatorial Algorithms for Computers and Calculators* (2nd ed.), Academic Press, New York–London, 1978.

[Rio58] J. Riordan, *An Introduction to Combinatorial Analysis*, John Wiley & Sons, New York, 1958.

[Sch78] A. Schrijver, A short proof of Minc's conjecture, *J. Combin. Theory Ser. A* 25:80–83, 1978.

[Sch98] A. Schrijver, Counting 1-factors in regular bipartite graphs, *J. Combin. Theory Ser. B* 72:122–135, 1998.

[Sch18] I. Schur, Über endliche Gruppen und Hermitesche Formen, *Math. Z.* 1:184–207, 1918.

[Sou03] G.W. Soules, New permanental upper bounds for nonnegative matrices, *Lin. Multilin. Alg.* 51:319–337, 2003.

[Sze75] G. Szegö, *Orthogonal Polynomials* (4th ed.), American Mathematical Society, Providence, RI, 1975.

[Wan74] E.T.H. Wang, On permanents of $(1, -1)$-matrices, *Israel J. Math.* 18:353–361, 1974.

[Wan99a] I.M. Wanless, The Holens-Đoković conjecture on permanents fails, *Lin. Alg. Appl.* 286:273–285, 1999.

[Wan99b] I.M. Wanless, Maximising the permanent and complementary permanent of (0,1)-matrices with constant line sum, *Discrete Math.* 205:191–205, 1999.

[Wan03] I.M. Wanless, A lower bound on the maximum permanent in Λ_n^k, *Lin. Alg. Appl.* 373:153–167, 2003.

[Wan05] I.M. Wanless, Permanents of matrices of signed ones, *Lin. Multilin. Alg.* 53:427–433, 2005.

[Wan06] I.M. Wanless, Addendum to Schrijver's work on minimum permanents, *Combinatorica*, (to appear).

32

D-Optimal Matrices

Michael G. Neubauer
California State University/Northridge

William Watkins
California State University/Northridge

An $m \times n$ matrix W whose entries are all 1 or -1 is called a (± 1)-design matrix; if the entries of W are 0 or 1, then W is a $(0, 1)$-design matrix. Each design matrix corresponds to a weighing design. That is, a scheme for estimating the weights of n objects in m weighings. Since the weights of n objects cannot be estimated in fewer than n weighings, we consider only those pairs (m, n) with $m \geq n$. The rows of W encode a two-pan or one-pan weighing design with n objects $x_1, \ldots, x_n$ being weighed in m weighings. If $W \in \{\pm 1\}^{m\times n}$, an entry of 1 in the (i, j)-th position of W indicates that object x_j is put in the right pan in the i-th weighing while an entry of -1 means that x_j is placed in the left pan. If $W \in \{0, 1\}^{m\times n}$, an entry of 1 in the (i, j)-th position indicates that object x_j is included in the i-th weighing while an entry of 0 means that the object is not included. In the presence of errors for the scale, we can expect only to find estimators $\hat{w}_1, \ldots, \hat{w}_n$ for the actual weights $w_1, \ldots, w_n$ of the objects. We want to choose a weighing design that is optimal with respect to some condition, an idea going back to Hotelling [Hot44] and Mood [Moo46]. See also [HS79] and [Slo79]. Under certain assumptions on the error of the scale, we can express optimality conditions in terms of $W^T W$ (see [Puk93]). The value of $\det W^T W$ is inversely proportional to the volume of the confidence region of the estimators of the weights of the objects. Thus, matrices for which $\det W^T W$ is large correspond to weighing designs that are desirable.

32.1 Introduction

This section includes basic definitions; facts and examples can be found in the following sections.

Definitions:

A matrix $W \in \{\pm 1\}^{m\times n}$ is a (± 1)-**design matrix**; $W \in \{0, 1\}^{m\times n}$ is a $(0, 1)$-**design matrix**.

A matrix $W \in \{\pm 1\}^{m\times n}$ (respectively, $W \in \{0, 1\}^{m\times n}$) is called **D-optimal** if $\det W^T W$ is maximal over all matrices in $\{\pm 1\}^{m\times n}$ (respectively, $\{0, 1\}^{m\times n}$).

$\alpha(m, n) = \max\{\det W^T W | W \in \{\pm 1\}^{m\times n}\}$.

$\beta(m, n) = \max\{\det W^T W | W \in \{0, 1\}^{m\times n}\}$.

We write $\alpha(n)$ and $\beta(n)$ for $\alpha(n, n)$ and $\beta(n, n)$.

32.2 The (± 1) and $(0, 1)$ Square Case

Definitions:

For a matrix $V \in \{0, 1\}^{(n-1)\times(n-1)}$, define

$$W_V = \left[\begin{array}{c|ccc} 1 & 1 & \cdots & 1 \\ \hline -1 & & & \\ \vdots & & 2V - J_{n-1} & \\ -1 & & & \end{array}\right] \in \{\pm 1\}^{n\times n}.$$

For a matrix $W \in \{\pm 1\}^{n\times n}$, define $V_W \in \{0, 1\}^{(n-1)\times(n-1)}$ by

$$V_W = \frac{1}{2}\left(W(1) + J_{n-1}\right),$$

where $W(1)$ is obtained from W by deleting the first row and column.

A **signature matrix** is a ± 1 diagonal matrix.

A **Hadamard matrix** of order n is a matrix $H_n \in \{\pm 1\}^{n\times n}$ with $H_n H_n^T = nI_n$.

A **2-design** with parameters (v, k, λ) (also called a (v, k, λ)-**design**) is a collection of k-subsets B_i, called **blocks**, of a finite set X with cardinality v, such that each 2-subset of X is contained in exactly λ blocks.

The (± 1)-**incidence matrix** $W = (w_{ij})$ of a 2-design is a matrix whose rows are indexed by the elements x_i of X and whose columns are indexed by the blocks B_j. The entry $w_{ij} = -1$ if $x_i \in B_j$ and $w_{ij} = +1$ otherwise.

Facts:

1. For any $W \in \{\pm 1\}^{n\times n}$, there exist signature matrices S_1, S_2 such that for $W' = (S_1 W S_2)$, $W'_{1j} = 1$ for $j = 1, \ldots, n$ and $W'_{i1} = -1$ for $i = 2, \ldots, n$. W is D-optimal if and only if W' is D-optimal.
2. $\det W_V = 2^{n-1} \det V$.
3. [Wil46] The (± 1) square case in dimension n is related to the $(0, 1)$ square case in dimension $n - 1$ by the previous two facts, and $\alpha(n) = 4^{n-1}\beta(n - 1)$. Facts are stated here for $\alpha(n)$ only, since the facts for $\beta(n)$ can be easily derived from these.
4. [Had93], [CRC96], [BJL93], [WPS72] Hadamard matrices
   - A necessary condition for the existence of a Hadamard matrix of order n is $n = 1, n = 2$, or $n \equiv 0 \pmod 4$.
   - Let H_m and H_n be two Hadamard matrices of orders m and n, respectively. Then $H_m \otimes H_n$ is a Hadamard matrix of order mn.
   - There exist infinitely many values n for which a Hadamard matrix H_n exists.
   - It is conjectured that for all $n = 4k$ there exists a Hadamard matrix H_n.
   - The smallest n for which the existence of a Hadamard matrix is in question (at the time of the writing of this chapter) is $n = 668$.
5. [Had93], [CRC96], [BJL93], [WPS72] D-optimal (± 1)-matrices: the case $n = 4k$
   - $\alpha(4k) \leq (4k)^{4k}$.
   - A necessary and sufficient condition for equality to occur in this case is the existence of a Hadamard matrix of order n.
6. [Bar33], [Woj64], [Ehl64a], [Coh67], [Neu97] D-optimal (± 1)-matrices: the case $n = 4k + 1$
   - $\alpha(4k + 1) \leq (8k + 1)\,(4k)^{4k}$.
   - For equality to occur in this case, it is necessary and sufficient that $8k + 1$ is the square of an integer and that there exists a matrix $W \in \{\pm 1\}^{m\times n}$ with $W^T W = (n - 1)I_n + J_n$.

- Equality occurs for infinitely many values of $n = 4k + 1$. A.E. Brouwer ([Bro83]) constructed an infinite family of 2-designs with parameters $(n = 2q^2 + 2q + 1, q^2, (q^2 - q)/2)$. The (± 1)-incidence matrix W_n of such a design satisfies $W_n^T W_n = (n-1)I_n + J_n$.
- The results in [MK82] and [CMK87] provide upper bounds, which are stronger than $(8k + 1)(4k)^{4k}$ in case $8k + 1$ is not the square of an integer.

7. [Ehl64a], [Woj64] D-optimal (± 1)-matrices: the case $n = 4k + 2$
   - $\alpha(4k+2) \leq (8k+2)^2\,(4k)^{4k}$.
   - For equality to occur in this case, it is necessary that $n - 1$ is the sum of two squares and that there exists a matrix $W_n \in \{\pm 1\}^{n\times n}$ such that

$$W_n^T W_n = \begin{bmatrix} (n-2)I_{\frac{n}{2}} + 2J_{\frac{n}{2}} & 0_n \\ 0_n & (n-2)I_{\frac{n}{2}} + 2J_{\frac{n}{2}} \end{bmatrix}. \qquad (32.1)$$

   - It is conjectured that the bound is attained whenever this is the case.
   - The bound is attained infinitely often.
   - If $n - 1$ is a square and there exists a matrix $W_{\frac{n}{2}} \in \{\pm 1\}^{\frac{n}{2}\times\frac{n}{2}}$ such that $W_{\frac{n}{2}}^T W_{\frac{n}{2}} = \frac{n-2}{2} I_{\frac{n}{2}} + J_{\frac{n}{2}}$, then construct the matrix $W_n = W_{\frac{n}{2}} \otimes H_2$ where

   $H_2 = \begin{bmatrix} 1 & 1 \\ 1 & -1 \end{bmatrix}$. Then $W_n \in \{\pm 1\}^{n\times n}$ satisfies Equation (32.1) and attains the bound. Such a matrix $W_{\frac{n}{2}}$ exists if $\frac{n}{2} = 2q^2 + 2q + 1$.

8. [Ehl64b] D-optimal (± 1)-matrices: The case $n = 4k + 3$

$$\alpha(4k+3) \leq (4k)^{4k+3-s}(4k+4r)^u(4k+4+4r)^v\left(1 - \frac{ur}{4k+4r} - \frac{v(r+1)}{4k+4+4r}\right),$$

   where $s = 5$ for $k = 1$, $s = 5$ or 6 for $k = 2$, $s = 6$ for $3 \leq k \leq 14$, and $s = 7$ for $k \geq 15$ and where $r = \lfloor (4k+3)/s \rfloor$, $4k + 3 = rs + v$, and $u = s - v$.
   - This case is the least well understood of the four.
   - For equality to occur for $n \geq 63$, it is necessary that $n = 112j^2 \pm 28j + 7$ and that there exists a matrix $W_n \in \{\pm 1\}^{n\times n}$ with

$$W_n^T W_n = I_7 \otimes \left[(n-3)I_{\frac{n}{7}} + 4J_{\frac{n}{7}}\right] - J_n.$$

   - However, it is not known if this bound is attainable for any $n \geq 63$.
   - The best lower bound seems to be the one in [NR97]. In [NR97], an infinite family of matrices is constructed whose determinants attain about 37% of the bound above.

9. See [OS] for (± 1)-matrices with largest known determinant for $n \leq 103$.

Examples:

1. The following matrices are Hadamard matrices in $\{\pm 1\}^{4\times 4}$ and $\{\pm 1\}^{12\times 12}$:

$$H_4 = \begin{bmatrix} 1 & 1 & 1 & 1 \\ 1 & -1 & 1 & -1 \\ 1 & 1 & -1 & -1 \\ 1 & -1 & -1 & 1 \end{bmatrix}$$

$$H_{12} = \begin{bmatrix} 1 & 1 & 1 & 1 & -1 & -1 & -1 & 1 & 1 & 1 & -1 & -1 \\ 1 & 1 & 1 & -1 & 1 & -1 & 1 & -1 & 1 & -1 & 1 & -1 \\ 1 & 1 & 1 & -1 & -1 & 1 & 1 & 1 & -1 & -1 & -1 & 1 \\ 1 & -1 & -1 & 1 & -1 & -1 & 1 & -1 & -1 & -1 & -1 & -1 \\ -1 & 1 & -1 & -1 & 1 & -1 & -1 & 1 & -1 & -1 & -1 & -1 \\ -1 & -1 & 1 & -1 & -1 & 1 & -1 & -1 & 1 & -1 & -1 & -1 \\ -1 & 1 & 1 & 1 & -1 & -1 & -1 & -1 & -1 & -1 & 1 & 1 \\ 1 & -1 & 1 & -1 & 1 & -1 & -1 & -1 & -1 & 1 & -1 & 1 \\ 1 & 1 & -1 & -1 & -1 & 1 & -1 & -1 & -1 & 1 & 1 & -1 \\ 1 & -1 & -1 & -1 & -1 & -1 & -1 & 1 & 1 & -1 & 1 & 1 \\ -1 & 1 & -1 & -1 & -1 & -1 & 1 & -1 & 1 & 1 & -1 & 1 \\ -1 & -1 & 1 & -1 & -1 & -1 & 1 & 1 & -1 & 1 & 1 & -1 \end{bmatrix}$$

$H_4H_4^T = 4I_4$ and $H_{12}H_{12}^T = 12I_{12}$.

2. Let $A = J_5 - 2I_5$. Then $A^TA = 4I_5 + J_5$ and $\det(A^TA)$ achieves the upper bound in Fact 6.
3. Let

$$W = \begin{bmatrix} A & A \\ A & -A \end{bmatrix} = A \otimes H_2 \in \{\pm 1\}^{10\times 10},$$

where $A = J_5 - 2I_5$. Then

$$W^TW = \begin{bmatrix} 8I_5 + 2J_5 & 0 \\ 0 & 8I_5 + 2J_5 \end{bmatrix},$$

and, hence, $\det(W^TW)$ achieves the upper bound in Fact 7.

4. To obtain the upper bound in Fact 6 for $n = 13$, let $V \in \{0,1\}^{13\times 13}$ be the $(0,1)$ line-point incidence matrix for a projective plane of order 3. Then $V^TV = 3I_{13} + J_{13}$ and the matrix $W = J_{13} - 2V \in \{\pm 1\}^{13\times 13}$ satisfies $W^TW = 12I_{13} + J_{13}$ and its determinant attains the upper bound.
5. For $n \geq 11$ and $n \equiv 3 \pmod 4$, no (± 1)-matrix is known to have a determinant that equals the upper bound in Fact 8. However, the following matrix $W \in \{\pm 1\}^{15\times 15}$, which is listed in [OS], satisfies

$$\det(W^TW) = 174755568785817600,$$

which is about 94% of the upper bound 185454889323724800 in Fact 8 with $k = 3$, $s = 6$.

$$W = \begin{bmatrix} -1 & -1 & -1 & -1 & 1 & 1 & 1 & 1 & 1 & -1 & 1 & 1 & 1 & 1 & -1 \\ -1 & -1 & -1 & 1 & -1 & 1 & 1 & -1 & 1 & 1 & 1 & 1 & -1 & 1 & 1 \\ -1 & -1 & -1 & 1 & 1 & -1 & 1 & 1 & -1 & 1 & 1 & -1 & 1 & 1 & 1 \\ -1 & 1 & 1 & -1 & -1 & 1 & 1 & 1 & -1 & 1 & 1 & -1 & -1 & -1 & -1 \\ 1 & 1 & -1 & 1 & -1 & -1 & 1 & 1 & 1 & -1 & 1 & -1 & -1 & -1 & -1 \\ 1 & -1 & 1 & -1 & 1 & -1 & 1 & -1 & 1 & 1 & 1 & -1 & -1 & -1 & -1 \\ 1 & 1 & 1 & 1 & 1 & 1 & -1 & 1 & 1 & 1 & 1 & -1 & -1 & 1 & -1 \\ -1 & 1 & 1 & -1 & -1 & -1 & -1 & -1 & 1 & -1 & 1 & -1 & 1 & 1 & 1 \\ 1 & 1 & 1 & -1 & -1 & -1 & 1 & 1 & 1 & 1 & -1 & 1 & 1 & 1 & 1 \\ 1 & -1 & 1 & -1 & -1 & -1 & -1 & 1 & -1 & -1 & 1 & 1 & -1 & 1 & 1 \\ 1 & 1 & -1 & -1 & -1 & -1 & -1 & -1 & -1 & 1 & 1 & 1 & 1 & 1 & -1 \\ 1 & 1 & -1 & -1 & 1 & 1 & 1 & -1 & -1 & -1 & -1 & -1 & -1 & 1 & 1 \\ 1 & -1 & 1 & 1 & -1 & 1 & 1 & -1 & -1 & -1 & -1 & -1 & 1 & 1 & -1 \\ -1 & 1 & 1 & 1 & 1 & -1 & 1 & -1 & -1 & -1 & -1 & 1 & -1 & 1 & -1 \\ 1 & 1 & 1 & 1 & 1 & 1 & 1 & -1 & -1 & -1 & 1 & 1 & 1 & -1 & 1 \end{bmatrix}.$$

32.3 The (± 1) Nonsquare Case

Facts:

1. [Had93] $\alpha(4k, n) \leq (4k)^n$.
2. [Pay74] $\alpha(4k+1, n) \leq (4k+n)(4k)^{n-1}$.
3. [Pay74]

$$\alpha(4k+2, n) \leq \begin{cases} (4k+n)^2(4k)^{n-2}, & \text{if } n \text{ is even,} \\ (4k+n+1)(4k+n-1)(4k)^{n-2}, & \text{if } n \text{ is odd.} \end{cases} \tag{32.2}$$

4. [GK80b] If $m = 4k-1 \geq 2n-5$, then $\alpha(m, n) \leq (4k-n)(4k)^{n-1}$.

Examples:

1. If $m = 4k$, equality can be achieved by taking W_0 to be the matrix consisting of any n columns of a Hadamard matrix of order $4k$. Then $W_0^T W_0 = 4kI_n$ and, hence, $\det W_0^T W_0 = \alpha(4k, n)$.
2. If $m = 4k+1$, adjoin a row of all 1s to W_0 and call the new matrix W_1. We have $W_1 \in \{\pm 1\}^{(4k+1)\times n}$ and $W_1^T W_1 = mI_n + J$. Hence, $\det W_1^T W_1 = \alpha(4k+1, n)$.
3. If $m = 4k+2$, let $r_1 = (1, 1, \ldots, 1)$ and $r_2 = (1, \ldots, 1, -1, \ldots, -1)$ be n-tuples with $r_1 \cdot r_2 = 0$, if n is even, and $r_1 \cdot r_2 = 1$, if n is odd. Adjoin rows r_1 and r_2 to W_0. Call the resulting matrix W_2. Then $W_2 \in \{\pm 1\}^{(4k+2)\times n}$ and

$$W_2^T W_2 = \begin{bmatrix} 4kI_l + 2J_l & 0_l \\ 0_l & 4kI_l + 2J_l \end{bmatrix} \quad \text{if } n = 2l \text{ is even}$$

and

$$W_2^T W_2 = \begin{bmatrix} 4kI_{l+1} + 2J_{l+1} & 0_{l+1,l} \\ 0_{l,l+1} & 4kI_l + 2J_l \end{bmatrix} \quad \text{if } n = 2l+1 \text{ is odd.}$$

Thus, $\det W_2^T W_2 = \alpha(4k+2, n)$.
4. If $m = 4k-1$, we may assume without loss of generality that the first row of W_0 is an all 1s row. Remove this first row of W_0 and call that matrix W_{-1}. Note that $W_{-1} \in \{\pm 1\}^{(4k-1)\times n}$ and that $W_{-1}^T W_{-1} = 4kI_n - J_n$. Hence, $\det W_{-1}^T W_{-1} = \alpha(4k-1, n)$.
5. It is not necessary to have a Hadamard matrix H_m of order m. All we require is the existence of a matrix $W \in \{\pm 1\}^{4k\times n}$ with $W^T W = mI_n$. See [GK80a] for details.
6. Upper bounds on $\alpha(4k-1, n)$ when $m = 4k-1 \leq 2n-5$ are given in [GK80b], [KF84], [SS91].

32.4 The (0, 1) Nonsquare Case: Regular D-Optimal Matrices

Definitions:

Let W be a (0, 1)-design matrix in $\{0, 1\}^{m\times n}$. For n odd, W is **balanced** if every row of W has exactly $(n+1)/2$ ones; for n even, W is **balanced** if every row of W has exactly $n/2$ ones or exactly $(n+2)/2$ ones.

A design matrix $W \in \{0, 1\}^{m\times n}$ is **regular** if it is balanced and $W^T W = t(I + J)$ for some integer t.

Facts:

1. [HKL96] If n is odd, then $\beta(m,n) \le (n+1)\left(\frac{(n+1)m}{4n}\right)^n$, with equality if and only if W is regular.
2. [NWZ97] If n is even, then $\beta(m,n) \le (n+1)\left(\frac{(n+2)m}{4(n+1)}\right)^n$, with equality if and only if W is regular.
3. [NWZ98a] A regular matrix exists in $\{0,1\}^{m\times n}$ only if

$$\begin{aligned} 2(n+1) \text{ divides } m \text{ for } n &\equiv 0 \pmod 4,\\ 2n \text{ divides } m \text{ for } n &\equiv 1 \pmod 4,\\ n+1 \text{ divides } m \text{ for } n &\equiv 2 \pmod 4,\\ n \text{ divides } m \text{ for } n &\equiv 3 \pmod 4. \end{aligned}$$

4. [NWZ98a] If $n = 4t-1$ and $H \in \{0,1\}^{m\times n}$ is the incidence matrix for a $(4t-1, 2t-1, t-1)$-design, then $W = J - H$ is a regular D-optimal matrix and $[\overbrace{W^T, \cdots, W^T}^{k}]^T$ is a regular D-optimal matrix in $\{0,1\}^{kn\times n}$. Let W_1 be the matrix obtained by deleting any column from W. Then $[\overbrace{W_1^T, \cdots, W_1^T}^{k}]^T$ is a regular D-optimal matrix in $\{0,1\}^{kn\times(n-1)}$.
5. [NWZ98a] If $n = 4t+1$ is a power of a prime integer, then a D-optimal regular matrix $W_2 \in \{0,1\}^{2n\times n}$ exists. Let $W_3 \in \{0,1\}^{2n\times(n-1)}$ be the matrix obtained by deleting any column from W_2. Then $[\overbrace{W_2^T, \cdots, W_2^T}^{k}]^T$ and $[\overbrace{W_3^T, \cdots, W_3^T}^{k}]^T$ are regular D-optimal matrices.

Examples:

1. Let $n = 4$ and $m = 10$. The following matrix is balanced and regular:

$$W^T = \begin{bmatrix} 1 & 1 & 1 & 0 & 1 & 1 & 1 & 0 & 0 & 0\\ 1 & 1 & 0 & 1 & 1 & 0 & 0 & 1 & 1 & 0\\ 1 & 0 & 1 & 1 & 0 & 1 & 0 & 1 & 0 & 1\\ 0 & 1 & 1 & 1 & 0 & 0 & 1 & 0 & 1 & 1 \end{bmatrix}, \qquad W^TW = \begin{bmatrix} 6 & 3 & 3 & 3\\ 3 & 6 & 3 & 3\\ 3 & 3 & 6 & 3\\ 3 & 3 & 3 & 6 \end{bmatrix}.$$

The inequality in Fact 2 is attained at W:

$$\beta(10,4) = 5\left(\frac{6\cdot 10}{4\cdot 5}\right)^4 = 405 = \det(W^TW).$$

Thus W is D-optimal.

2. A regular matrix exists for the case $n = 9, m = 18$. (Fact 3, where $n \equiv 1 \pmod 4$ and $2n = 18$ divides $m = 18$.) The regular matrix is constructed from the Galois field $GF(9)$ with nine elements. Choosing $GF(9) = \mathbb{Z}_3/(x^2+1)$, the element $\theta = x+2$ of order 8, generates the nonzero elements in $GF(9)$. The nonzero quadratic residues of $GF(9)$ are $Q = \{1, \theta^2, \theta^4, \theta^6\}$. Define $K_1 \in \{0,1\}^{9\times 9}$ by

$$(K_1)_{\rho,\tau} = \begin{cases} 1, & \text{if } \tau \in \rho + Q,\\ 0, & \text{if otherwise,} \end{cases}$$

where the rows and columns ρ, τ are indexed by $\{0, 1, \theta, \theta^2, \ldots, \theta^7\}$. Then

$$K_1 = \begin{bmatrix} 0 & 1 & 0 & 1 & 0 & 1 & 0 & 1 & 0 \\ 1 & 0 & 0 & 0 & 0 & 1 & 1 & 0 & 1 \\ 0 & 0 & 0 & 1 & 1 & 1 & 0 & 0 & 1 \\ 1 & 0 & 1 & 0 & 0 & 0 & 0 & 1 & 1 \\ 0 & 0 & 1 & 0 & 0 & 1 & 1 & 1 & 0 \\ 1 & 1 & 1 & 0 & 1 & 0 & 0 & 0 & 0 \\ 0 & 1 & 0 & 0 & 1 & 0 & 0 & 1 & 1 \\ 1 & 0 & 0 & 1 & 1 & 0 & 1 & 0 & 0 \\ 0 & 1 & 1 & 1 & 0 & 0 & 1 & 0 & 0 \end{bmatrix}.$$

Define K_2 in the same way but with the nonzero quadratic nonresidues $R = \{\theta, \theta^3, \theta^5.\theta^7\}$ in place of Q. Then the matrix $[K_1, K_2] \in \{0, 1\}^{9\times 18}$ satisfies

$$K_1 K_1^T + K_2 K_2^T = 5I_9 + 3J_9.$$

Let $W = [J_9 - K_1, J_9 - K_2]$. Then W is a D-optimal regular design matrix: $WW^T = 5(I_9 + J_9)$.

3. Let $t = 2$ and $n = 7$. The following matrix H is the incidence matrix for a $(7, 3, 1)$-design:

$$H = \begin{bmatrix} 0 & 1 & 0 & 1 & 0 & 1 & 0 \\ 1 & 0 & 0 & 1 & 1 & 0 & 0 \\ 0 & 0 & 1 & 1 & 0 & 0 & 1 \\ 1 & 1 & 1 & 0 & 0 & 0 & 0 \\ 0 & 1 & 0 & 0 & 1 & 0 & 1 \\ 1 & 0 & 0 & 0 & 0 & 1 & 1 \\ 0 & 0 & 1 & 0 & 1 & 1 & 0 \end{bmatrix}.$$

Then (Fact 4) $W = J_7 - H$ is a regular D-optimal matrix in $\{0, 1\}^{7\times 7}$ and $[W^T, \ldots, W^T]^T$ is a regular D-optimal matrix in $\{0, 1\}^{7k\times 7}$.

32.5 The (0, 1) Nonsquare Case: Nonregular D-Optimal Matrices

It is clear from Fact 3 in section 32.4 that for most pairs (m, n), no regular D-optimal matrix exists. For example, if $n = 7$, then the only values of m for which a regular D-optimal matrix exists are $m = 7t$. Thus, for $m = 7t + r$, with $0 \le r \le 6$, a D-optimal matrix cannot be regular unless $r = 0$. The only values of n for which $\beta(m, n)$ is known for all values of m are $n = 2, 3, 4, 5, 6$.

Facts:

1. [HKL96] $n = 2$, $m = 3t + r$ with $r = 0, 1, 2$:

$$\beta(m, 2) = \begin{cases} 3t^2 \quad , & \text{for } r = 0 \\ 3t^2 + 2t, & \text{for } r = 1 \end{cases}.$$

2. [HKL96] $n = 3$, $m = 3t + r$ with $r = 0, 1, 2$:

$$\beta(m, 3) = 4t^{3-r}(t + 1)^r.$$

3. [NWZ98b] $n = 4$, $m = 10t + r$ with $0 \leq r \leq 9$:

$$\beta(m,4) = \begin{cases} 405\,t^4 & , \quad \text{for } r{=}0 \\ 81\,t^3\,(2+5\,t) & , \quad \text{for } r{=}1 \\ 3\,t\,(1+3\,t)^2\,(2+15\,t) & , \quad \text{for } r{=}2 \\ 3\,t\,(1+3\,t)^2\,(8+15\,t) & , \quad \text{for } r{=}3 \\ 3\,(1+3\,t)^3\,(3+5\,t) & , \quad \text{for } r{=}4 \\ (1+3\,t)^2\,\left(19+60\,t+45\,t^2\right) & , \quad \text{for } r{=}5 \\ 3\,(2+3\,t)^3\,(2+5\,t) & , \quad \text{for } r{=}6 \\ 3\,(1+t)\,(2+3\,t)^2\,(7+15\,t) & , \quad \text{for } r{=}7 \\ 3\,(1+t)\,(2+3\,t)^2\,(13+15\,t), & \quad \text{for } r{=}8 \\ 81\,(1+t)^3\,(3+5\,t) & , \quad \text{for } r{=}9 \end{cases}.$$

4. [NWZ98b] In the case $n = 5$, all D-optimal matrices are balanced except when $m = 5, 6, 7, 8, 15, 16, 17, 27$. For $m = 10t + r$ with $0 \leq r \leq 9$ and m not equal to any of the exceptional values, we have

$$\beta(m,5) = \begin{cases} 1458\,t^5 & , \quad \text{for } r{=}0 \\ 729\,t^4\,(1+2\,t) & , \quad \text{for } r{=}1 \\ 162\,t^3\,(1+3\,t)\,(2+3\,t) & , \quad \text{for } r{=}2 \\ 27\,t^2\,(1+3\,t)^2\,(5+6\,t) & , \quad \text{for } r{=}3 \\ 54\,t\,(1+t)\,(1+3\,t)^3 & , \quad \text{for } r{=}4 \\ 9\,(1+t)\,(1+3\,t)^2\,\left(1+15\,t+18\,t^2\right), & \quad \text{for } r{=}5 \\ 54\,(1+t)^2\,(1+3\,t)^3 & , \quad \text{for } r{=}6 \\ 27\,(1+t)^2\,(1+3\,t)^2\,(5+6\,t) & , \quad \text{for } r{=}7 \\ 162\,(1+t)^3\,(1+3\,t)\,(2+3\,t) & , \quad \text{for } r{=}8 \\ 729\,(1+t)^4\,(1+2\,t) & , \quad \text{for } r{=}8 \end{cases}.$$

The values of $\beta(m, 5)$ for the eight exceptional values of m are

$$\begin{array}{llll} \beta(5,5) = 25 & \beta(6,5) = 64 & \beta(7,5) = 192 & \beta(8,5) = 384 \\ \beta(15,5) = 9880 & \beta(16,5) = 13975 & \beta(17,5) = 19500 & \beta(27,5) = 202752. \end{array}$$

Each of these is greater than the value of the corresponding polynomial above. For example, if $m = 16$ so that $t = 1$ and $r = 6$, then $54\,(1+t)^2\,(1+3\,t)^3 = 13824$, which is less than 13975.

5. [NWZ00] For $n = 6$, all D-optimal matrices are balanced except when $m = 6, 8, 9, 13$. For $m = 7t + r$ with $0 \leq r \leq 6$ and $m \neq 6, 8, 9, 13$:

$$\beta(m,6) = \begin{cases} 448\,t^6 & , \quad \text{for } r{=}0 \\ 16\,t^4\,(1+2\,t)\,(5+14\,t) & , \quad \text{for } r{=}1 \\ 4\,t^2\,(1+2\,t)^3\,(3+14\,t) & , \quad \text{for } r{=}2 \\ (1+2\,t)^5\,(1+14\,t) & , \quad \text{for } r{=}3 \\ (1+2\,t)^5\,(13+14\,t) & , \quad \text{for } r{=}4 \\ 4\,(1+t)^2\,(1+2\,t)^3\,(11+14\,t), & \quad \text{for } r{=}5 \\ 16\,(1+t)^4\,(1+2\,t)\,(9+14\,t)\,, & \quad \text{for } r{=}6 \end{cases}$$

The values of $\beta(6, m)$ for the four exceptional values of m are

$$\beta(6,6) = 81 \qquad \beta(8,6) = 832 \qquad \beta(9,6) = 1620 \qquad \beta(13,6) = 16512.$$

As in the case for $n = 5$, the values of $\beta(m, 6)$ exceed the value of the corresponding polynomial.

Examples:

1. Design matrices are exhibited for the above cases in the sources listed. For example, if $n = 4$, let

$$W_0 = \begin{bmatrix} 1 & 1 & 1 & 0 & 1 & 1 & 1 & 0 & 0 & 0 \\ 1 & 1 & 0 & 1 & 1 & 0 & 0 & 1 & 1 & 0 \\ 1 & 0 & 1 & 1 & 0 & 1 & 0 & 1 & 0 & 1 \\ 0 & 1 & 1 & 1 & 0 & 0 & 1 & 0 & 1 & 1 \end{bmatrix}^T,$$

v_i be the ith row of W_0. If $r = 3$, then the matrix $[\overbrace{W_0^T, \cdots, W_0^T}^{t}, v_8^T, v_9^T, v_{10}^T]^T$ is a D-optimal matrix in $\{0, 1\}^{(4t+3)\times 4}$.

32.6 The (0, 1) Nonsquare Case: Large m

Facts:

1. [NWZ98a] For each value of n, all D-optimal matrices in $\{0, 1\}^{m\times n}$ are balanced for sufficiently large values of m.
2. [NW02], [AFN03] In addition to the values $n = 2, 3, 4, 5, 6$, for which $\beta(m, n)$ is known for all m, the only other values of n for which $\beta(m, n)$ is known for all sufficiently large values of m are $n = 7, 11, 15, 19, 23, 27$.

Examples:

1. [NW02]

$$\beta(7t + r, 7) = 2^{10} t^{7-r} (t + 1)^r,$$

for sufficiently large values of $m = 7t + r$.

32.7 The (0, 1) Nonsquare Case: $n \equiv -1 \pmod 4$

The theory for D-optimal (0, 1)-designs is most developed for the cases where $n \equiv -1 \pmod 4$.

Definitions:

For an $n \times n$ matrix A, the **trace-sequence** A is $(\text{trace}(A), \text{trace}(A^2), \cdots, \text{trace}(A^n))$.

$\mathcal{G}(v, \delta)$ is the set of all δ-regular graphs on v vertices.

Let graph G be a graph in $\mathcal{G}(v, \delta)$ and let $\mathcal{A}_G$ be the adjacency matrix of G. The graph G is **trace-minimal** if the **trace-sequence** of its adjacency matrix $(\text{trace}(\mathcal{A}_G), \text{trace}(\mathcal{A}_G^2), \cdots, \text{trace}(\mathcal{A}_G^n))$ is least in lexicographic order among all graphs in $\mathcal{G}(v, \delta)$.

Facts:

1. [AFN03] If $n \equiv -1 \pmod 4$, then for each $0 \leq r < n$ and all sufficiently large values of t, $\beta(nt + r, n)$ is a polynomial in t of degree n. These polynomials are related to the adjacency matrices $\mathcal{A}_G$ of certain regular graphs G.

2. [AFN03], [AFN06] The polynomial $\beta(nt + r, n)$ depends on a trace-minimal graph in $\mathcal{G}(v, \delta)$. Once a trace-minimal graph G is found in the appropriate graph class $\mathcal{G}(v, \delta)$, the polynomial $\beta(nt + r, n)$ can be computed. There are four theorems [AFN03] governing this situation; one for each congruence class of $r \pmod 4$.
3. [AFN03] Trace-minimal graphs are known for all of the graph classes necessary to obtain formulas for $\beta(nt + r, n)$ for $n = 3, 7, 11, 15, 19, 23$, and 27 and t sufficiently large.
4. [AFN06] Let G be a connected strongly regular graph with no three cycles. Then G is trace-minimal.
5. [AFN06] The following graphs are trace-minimal in their graph class:

Graph Class	G
$\mathcal{G}(v, 0)$	Graph with v vertices and no edges
$\mathcal{G}(2v, 1)$	vK_2, a matching of $2v$ vertices
$\mathcal{G}(v, 2)$	C_v, the cycle graph on v vertices
$\mathcal{G}(2v, v)$	$K_{v,v}$, the complete bipartite graph with v vertices in each set of the bipartition
$\mathcal{G}(2v, 2v-2)$	$K_{2v} - vK_2$, the complement of a matching
$\mathcal{G}(v, v-1)$	K_v, the complete graph on v vertices

6. [AFN06] Let G be a connected regular graph with girth g such that $\mathcal{A}_G$ has $k+1$ distinct eigenvalues. If g is even, then $g \leq 2k$ with equality only if G is trace-minimal. If g is odd, then $g \leq 2k + 1$ with equality only if G is trace-minimal.

Examples:

1. Let $n = 4p - 1 \equiv -1 \pmod 4$ and $r = 4d + 2 \equiv 2 \pmod 4$. Let G be a trace-minimal graph in $\mathcal{G}(2p, p + d)$. Then

$$\beta(nt + r, n) = \frac{4t[p_{\mathcal{A}_G}(pt + d)]^2}{(t-1)^2},$$

for sufficiently large values of t. Taking $n = 15, r = 10$, we have $p = 4, d = 2$. The appropriate graph class is $\mathcal{G}(8, 6)$. There is only one graph G in this class, namely the complement of the matching $4K_2$. Thus, it is trace-minimal. Since $p_{\mathcal{A}_G}(x) = (x - 6)x^4(x + 2)^3$,

$$\beta(15t + 10, 15) = \frac{4t[p_{\mathcal{A}_G}(4t + 2)]^2}{(t-1)^2} = 16(4t)(4t + 2)^8(4t + 4)^6,$$

for sufficiently large t.

2. Let $n = 4p - 1 \equiv -1 \pmod 4$ and $r = 4d + 1 \equiv 1 \pmod 4$. Let G be a trace-minimal graph in $\mathcal{G}(2p, d)$. Then

$$\beta(nt + r, n) = \frac{4(t+1)[p_{\mathcal{A}_G}(pt + d)]^2}{t^2},$$

for sufficiently large t. Taking $n = 15, r = 9$ we have $p = 4, d = 2$. The appropriate graph class is $\mathcal{G}(8, 2)$. There are three (nonisomorphic) graphs in this class: C_8, $C_5 \cup C_3$, and $C_4 \cup C_4$, where C_k stands for a k-cycle graph. The trace sequences for these three graphs are

$$(\text{trace}(\mathcal{A}_G), \text{trace}(\mathcal{A}_G^2), \cdots, \text{trace}(\mathcal{A}_G^8)) = \begin{cases} (0, 16, 0, 48, 0, 160, 0, 576), & \text{for } G = C_8 \\ (0, 16, 6, 48, 40, 166, 196, 608), & \text{for } G = C_5 \cup C_3 \\ (0, 16, 0, 64, 0, 256, 0, 1024), & \text{for } G = C_4 \cup C_4. \end{cases}$$

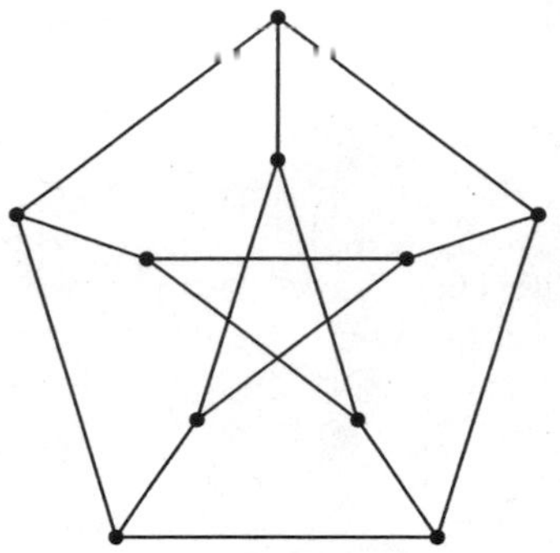

FIGURE 32.1

Thus, C_8 is the only trace-minimal graph in the graph class $\mathcal{G}(8, 2)$. The characteristic polynomial for $\mathcal{A}_{C_8}$ is

$$(x-2)x^2(x+2)(x^2-2)^2.$$

Thus,

$$\beta(15t+9, 15) = \frac{4(t+1)[p_{\mathcal{A}_G}(4t+2)]^2}{t^2} = 16(4t+2)^4(4t+4)^3(16t^2+16t+2)^4.$$

3. The Petersen graph (Figure 32.1) is an example of a strongly regular graph. (See Fact 4.) It is trace-minimal in $\mathcal{G}(10, 3)$:
4. Let P be the projective geometry with seven points, 1, 2, 3, 4, 5, 6, 7 and seven lines, 123, 147, 156, 257, 246, 367, 345 (Figure 32.2): The line-point incidence matrix for P is:

$$N = \begin{bmatrix} 1 & 1 & 1 & 0 & 0 & 0 & 0 \\ 1 & 0 & 0 & 1 & 0 & 0 & 1 \\ 1 & 0 & 0 & 0 & 1 & 1 & 0 \\ 0 & 1 & 0 & 0 & 1 & 0 & 1 \\ 0 & 1 & 0 & 1 & 0 & 1 & 0 \\ 0 & 0 & 1 & 0 & 0 & 1 & 1 \\ 0 & 0 & 1 & 1 & 1 & 0 & 0 \end{bmatrix}.$$

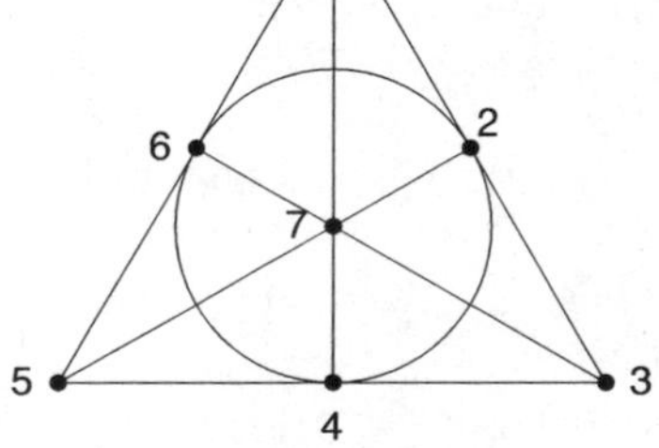

FIGURE 32.2

Let G be the incidence graph of P having 14 vertices and adjacency matrix given by

$$\mathcal{A}_G = \begin{bmatrix} 0 & N \\ N^T & 0 \end{bmatrix}.$$

Then G is trace-minimal by Fact 6: G is a regular graph of degree 3. The girth of G is $g = 6$. The characteristic polynomial of $\mathcal{A}_G$ is $(x-3)(x+3)(x^2-2)^6$ and so $\mathcal{A}_G$ has $k+1 = 4$ distinct eigenvalues. Since $2k = g$, it follows that G is trace-minimal in $\mathcal{G}(14, 3)$.

32.8 Balanced (0, 1)-Matrices and (±1)-Matrices

Let $n = 2k - 1$ be odd. There is a connection between balanced $(0, 1)$-design matrices and (± 1)-design matrices.

Facts:

1. [NWZ98a] Let W be a balanced design matrix in $\{0, 1\}^{m\times n}$, so that each row of W contains exactly k ones and $k-1$ zeros. Let q be a positive integer and

$$L(W) = \begin{bmatrix} J_{n,1} & J_{n,m} - 2W^T \\ J_{q,1} & J_{q,m} \end{bmatrix}.$$

Then $\det L(W)^T L(W) = q4^m \det W^T W$. It follows that for sufficiently large m, if W is a balanced $(0, 1)$-design matrix and $L(W)$ is a D-optimal design matrix, then W is also D-optimal.

References

[AFN03] B.M. Ábrego, S. Fernández-Merchant, M.G. Neubauer, and W. Watkins. D-optimal weighing designs for $n \equiv -1 \pmod 4$ objects and a large number of weighings. *Lin. Alg. Appl.*, 374:175–218, 2003.

[AFN06] B.M. Ábrego, S. Fernández-Merchant, M.G. Neubauer, and W. Watkins. Trace-minimal graphs and D-optimal weighing designs. *Lin. Alg. Appl.*, 412/2-3:161–221, 2006.

[Bar33] G. Barba. Intorno al teorema di Hadamard sui determinanti a valore massimo. *Giorn. Mat. Battaglia*, 71:70–86, 1933.

[BJL93] T. Beth, D. Jungnickel, and H. Lenz. *Design Theory*, Cambridge University Press, Cambridge, 1993.

[Bro83] A.E. Brouwer. An Infinite Series of Symmetric Designs. Report ZW202/83, Math. Zentrum Amsterdam, 1983.

[CMK87] T. Chadjipantelis, S. Kounias, and C. Moyssiadis. The maximum determinant of 21×21 $(+1, -1)$-matrices and D-optimal designs. *J. Statist. Plann. Inference*, 16:121–128, 1987.

[Coh67] J.H.E. Cohn. On determinants with elements ± 1. *J. London Math. Soc.*, 42:436–442, 1967.

[CRC96] *The CRC Handbook of Combinatorial Designs*, edited by C.J. Colburn and J.H. Dinitz. CRC Press, Inc., Boca Raton, FL, 1996.

[Ehl64a] H. Ehlich. Determinantenabschätzungen für binäre Matrizen. *Math. Zeitschrift*, 83:123–132, 1964.

[Ehl64b] H. Ehlich. Determinantenabschätzungen für binäre Matrizen mit $n \equiv 3 \bmod 4$. *Math. Zeitschrift*, 84:438–447, 1964.

[GK80a] Z. Galil and J. Kiefer. D-optimum weighing designs. *Ann. Stat.*, 8:1293–1306, 1980.

[GK80b] Z. Galil and J. Kiefer. Optimum weighing designs. *Recent Developments in Statistical Inference and Data Analysis* (K. Matsuita, Ed.), North Holland, Amsterdam, 1980.

[GK82] Z. Galil and J. Kiefer. Construction methods D-optimum weighing designs when $n \equiv 3$ mod 4. *Ann. Stat.*, 10:502–510, 1982.

[Had93] J. Hadamard. Résolution d'une question relative aux déterminants. *Bull. Sci. Math.*, 2:240–246, 1893.

[Hot44] H. Hotelling. Some improvements in weighing and other experimental techniques. *Ann. Math. Stat.*, 15:297–306, 1944.

[HKL96] M. Hudelson, V. Klee, and D. Larman. Largest j-simplices in d-cubes: some relatives of the Hadamard determinant problem. *Lin. Alg. Appl.*, 241:519–598, 1996.

[HS79] M. Harwit and N.J.A. Sloane. *Hadamard Transform Optics*, Academic Press, New York, 1979.

[KF84] S. Kounias and N. Farmakis. A construction of D-optimal weighing designs when $n \equiv 3$ mod 4. *J. Statist. Plann. Inference*, 10:177–187, 1984.

[Moo46] A.M. Mood. On Hotelling's weighing problem. *Ann. Math. Stat.*, 17:432–446, 1946.

[MK82] C. Moyssiadis and S. Kounias. The exact D-optimal first order saturated design with 17 observations. *J. Statist. Plann. Inference*, 7:13–27, 1982.

[Neu97] M. Neubauer. An inequality for positive definite matrices with applications to combinatorial matrices. *Lin. Alg. Appl.*, 267:163–174, 1997.

[NR97] M. Neubauer and A.J. Radcliffe. The maximum determinant of (± 1)-matrices. *Lin. Alg. Appl.*, 257:289–306, 1997.

[NW02] M. Neubauer and W. Watkins. D-optimal designs for seven objects and a large number of weighings. *Lin. Multilin. Alg.*, 50:61–74, 2002.

[NWZ97] M. Neubauer, W. Watkins, and J. Zeitlin. Maximal j-simplices in the real d-dimensional unit cube. *J. Comb. Th. A*, 80:1–12, 1997.

[NWZ98a] M. Neubauer, W. Watkins, and J. Zeitlin. Notes on D-optimal designs. *Lin. Alg. Appl.*, 280: 109–127, 1998.

[NWZ98b] M. Neubauer, W. Watkins, and J. Zeitlin. Maximal D-optimal weighing designs for 4 and 5 objects. *Elec. J. Lin. Alg.*, 4:48–72, 1998.

[NWZ00] M. Neubauer, W. Watkins, and J. Zeitlin. D-optimal weighing designs for 6 objects. *Metrika*, 52:185–211, 2000.

[OS] W. Orrick and B. Solomon. The Hadamard maximal determinant problem. http://www.indiana.edu/~maxdet/.

[Pay74] S.E. Payne. On maximizing det(A^TA). *Discrete Math.*, 10:145–158, 1974.

[Puk93] F. Pukelsheim. *Optimal Design of Experiments*, John Wiley & Sons, New York, 1993.

[SS91] Y.S. Sathe and R.G. Shenoy. Further results on construction methods for some A- and D-optimal weighing designs when $N \equiv 3 \pmod 4$. *J. Statist. Plann. Inference*, 28:339–352, 1991.

[Slo79] N.J.A. Sloane. Multiplexing methods in spetroscopy. *Math. Mag.*, 52:71–80, 1979.

[Wil46] J. Williamson. Determinants whose elements are 0 and 1. *Amer. Math. Monthly*, 53:427–434, 1946.

[Woj64] M. Wojtas. On Hadamard's inequality for the determinants of order non-divisible by 4. *Colloq. Math.*, 12:73–83, 1964.

[WPS72] W.D. Wallis, A. Penfold Street, and J. Seberry Wallis. *Combinatorics: Room Squares, Sum-Free Sets, Hadamard Matrices*, Lecture Notes in Mathematics 292, Springer-Verlag, Berlin, 1972.

33

Sign Pattern Matrices

Frank J. Hall
Georgia State University
Zhongshan Li
Georgia State University

The origins of sign pattern matrices are in the book [Sam47] by the Nobel Economics Prize winner P. Samuelson, who pointed to the need to solve certain problems in economics and other areas based only on the signs of the entries of the matrices. The study of sign pattern matrices has become somewhat synonymous with qualitative matrix analysis. The dissertation of C. Eschenbach [Esc87], directed by C.R. Johnson, studied sign pattern matrices that "require" or "allow" certain properties and summarized the work on sign patterns up to that point. In 1995, Richard Brualdi and Bryan Shader produced a thorough treatment [BS95] on sign pattern matrices from the sign-solvability vantage point. There is such a wealth of information contained in [BS95] that it is not possible to represent all of it here. Since 1995 there has been a considerable number of papers on sign patterns and some generalized notions such as ray patterns. We remark that in this chapter we mostly use $\{+, -, 0\}$ notation for sign patterns, whereas in the literature $\{1, -1, 0\}$ notation is also commonly used, such as in [BS95]. We further note that because of the interplay between sign pattern matrices and graph theory, the study of sign patterns is regarded as a part of combinatorial matrix theory.

33.1 Basic Concepts

Definitions:

A **sign pattern matrix** (or **sign pattern**) is a matrix whose entries come from the set $\{+, -, 0\}$. For a real matrix B, $\text{sgn}(B)$ is the sign pattern whose entries are the signs of the corresponding entries in B.

If A is an $m \times n$ sign pattern matrix, the **sign pattern class** (or **qualitative class**) of A, denoted $Q(A)$, is the set of all $m \times n$ real matrices B with $\text{sgn}(B) = A$. If C is a real matrix, its qualitative class is given by $Q(C) = Q(\text{sgn}(C))$.

A **generalized sign pattern** $\tilde{A}$ is a matrix whose entries are from the set $\{+, -, 0, \#\}$, where # indicates an ambiguous sum (the result of adding + with −). The qualitative class of $\tilde{A}$ is defined by allowing the # entries to be completely free. Two generalized sign patterns are **compatible** if there is a common real matrix in their qualitative classes.

A **subpattern** $\hat{A}$ of a sign pattern A is a sign pattern obtained by replacing some (possibly none) of the nonzero entries in A with 0; this fact is denoted by $\hat{A} \preceq A$.

A **diagonal pattern** is a square sign pattern all of whose off-diagonal entries are zero. Similarly, standard matrix terms such as "tridiagonal" and "upper triangular" can be applied to sign patterns having the required pattern of zero entries.

A **permutation pattern** is a square sign pattern matrix with entries 0 and +, where the entry + occurs precisely once in each row and in each column.

A **permutational similarity** of the (square) sign pattern A is a product of the form $P^T AP$, where P is a permutation pattern.

A **permutational equivalence** of the sign pattern A is a product of the form $P_1 A P_2$, where P_1 and P_2 are permutation patterns.

The **identity** pattern of order n, denoted I_n, is the $n \times n$ diagonal pattern with + diagonal entries.

A **signature pattern** is a diagonal sign pattern matrix, each of whose diagonal entries is + or −.

A **signature similarity** of the (square) sign pattern A is a product of the form SAS, where S is a signature pattern.

If P is a property referring to a real matrix, then a sign pattern A **requires** P if every real matrix in $Q(A)$ has property P, or **allows** P if some real matrix in $Q(A)$ has property P.

The **digraph** of an $n \times n$ sign pattern $A = [a_{ij}]$, denoted $\Gamma(A)$, is the digraph with vertex set $\{1, 2, \ldots, n\}$, where (i, j) is an arc iff $a_{ij} \neq 0$. (See Chapter 29 for more information on digraphs.)

The **signed digraph** of an $n \times n$ sign pattern $A = [a_{ij}]$, denoted $D(A)$, is the digraph with vertex set $\{1, 2, \ldots, n\}$, where (i, j) is an arc (bearing a_{ij} as its sign) iff $a_{ij} \neq 0$.

If $A = [a_{ij}]$ is an $n \times n$ sign pattern matrix, then a **(simple) cycle** of **length** k (or a k-**cycle**) in A is a formal product of the form $\gamma = a_{i_1 i_2} a_{i_2 i_3} \ldots a_{i_k i_1}$, where each of the elements is nonzero and the index set $\{i_1, i_2, \ldots, i_k\}$ consists of distinct indices. The **sign** (positive or negative) of a simple cycle in a sign pattern A is the actual product of the entries in the cycle, following the obvious rules that multiplication is commutative and associative, and $(+)(+) = +$, $(+)(-) = -$.

A **composite cycle** γ in A is a product of simple cycles, say $\gamma = \gamma_1 \gamma_2 \ldots \gamma_m$, where the index sets of the γ_i's are mutually disjoint. If the length of γ_i is l_i, then the length of γ is $\sum_{i=1}^m l_i$, and the **signature** of γ is $(-)^{\sum_{i=1}^m (l_i - 1)}$. A cycle γ is **odd** (**even**) when the length of the simple or composite cycle γ is odd (even).

If $A = [a_{ij}]$ is an $n \times n$ sign pattern matrix, then a **path** of length k in A is a formal product of the form $a_{i_1 i_2} a_{i_2 i_3} \ldots a_{i_k i_{k+1}}$, where each of the elements is nonzero and the indices $i_1, i_2, \ldots, i_{k+1}$ are distinct.

Facts:

1. Simple cycles and paths in an $n \times n$ sign pattern matrix A correspond to simple cycles and paths in the digraph of A. In particular, the path $a_{i_1 i_2} a_{i_2 i_3} \ldots a_{i_k i_{k+1}}$ in A corresponds to the path $i_1 \rightarrow i_2 \rightarrow \ldots \rightarrow i_{k+1}$ in the digraph of A.
2. If A is an $n \times n$ sign pattern, then each nonzero term in $\det(A)$ is the product of the signature of a composite cycle γ of length n in A with the actual product of the entries in γ.
3. Two generalized sign patterns are compatible if and only if the signs of each position whose sign is specified in both are equal.

Examples:

1. The matrix $\begin{bmatrix} 0 & 5 & -4 \\ -2 & -1 & 7 \end{bmatrix}$ is in $Q(A)$, where $A = \begin{bmatrix} 0 & + & - \\ - & - & + \end{bmatrix}$.

2. If $A = [a_{ij}] = \begin{bmatrix} + & + & 0 & - & + \\ 0 & - & - & - & + \\ + & - & - & + & + \\ - & - & + & - & + \\ + & - & 0 & - & - \end{bmatrix}$,

then the composite cycle $\gamma = (a_{12}a_{23}a_{31})(a_{45}a_{54})$ has length 5 and negative signature, and yields the term $-a_{12}a_{23}a_{31}a_{45}a_{54} = -$ in $\det(A)$.

3. If $A = \begin{bmatrix} + & + \\ + & - \end{bmatrix}$, then $A^2 = \begin{bmatrix} + & \# \\ \# & + \end{bmatrix}$, which is compatible with $\begin{bmatrix} + & - \\ 0 & \# \end{bmatrix}$.

33.2 Sign Nonsingularity

Definitions:

A square sign pattern A is **sign nonsingular** (SNS) if every matrix $B \in Q(A)$ is nonsingular.

A **strong sign nonsingular** sign pattern, abbreviated an S^2NS-pattern, is an SNS-pattern A such that the matrix B^{-1} is in the same sign pattern class for all $B \in Q(A)$.

A **self-inverse sign pattern** is an S^2NS-pattern A such that $B^{-1} \in Q(A)$ for every matrix $B \in Q(A)$.

A **maximal** sign nonsingular sign pattern matrix is an SNS-pattern A where no zero entry of A can be set nonzero so that the resulting pattern is SNS.

A **nearly sign nonsingular** (NSNS) sign pattern matrix is a square pattern A having at least two nonzero terms in the expansion of its determinant, with precisely one nonzero term having opposite sign to the others.

A square sign pattern A is **sign singular** if every matrix $B \in Q(A)$ is singular.

The **zero pattern** of a sign pattern A, denoted $|A|$, is the $(0,+)$-pattern obtained by replacing each $-$ entry in A by a $+$.

Since a sign pattern may be represented by any real matrix in its qualitative class, many concepts defined on sign patterns (such as SNS and S^2NS) may be applied to real matrices.

Facts:

Most of the following facts can be found in [BS95, Chaps. 1–4 and 6–8].

1. The $n \times n$ sign pattern A is sign nonsingular if and only if $\det(A) = +$ or $\det(A) = -$, that is, in the standard expansion of $\det(A)$ into $n!$ terms, there is at least one nonzero term, and all the nonzero terms have the same sign.
2. An $n \times n$ pattern A is an SNS-pattern iff for any $n \times n$ signature pattern D and any $n \times n$ permutation patterns P_1, P_2, DP_1AP_2 is an SNS-pattern.
3. [BMQ68] For any SNS-pattern A, there exist a signature pattern D and a permutation pattern P such that DPA has negative diagonal entries.
4. [BMQ68] An $n \times n$ sign pattern A with negative main diagonal is SNS iff the actual product of entries of every simple cycle in A is negative.
5. [Gib71] If an $n \times n$, $n \geq 3$, sign pattern A is SNS, then A has at least $\binom{n-1}{2}$ zero entries, with exactly this number iff there exist permutation patterns P_1 and P_2 such that P_1AP_2 has the same zero/nonzero pattern as the Hessenberg pattern given in Example 1 below.
6. The fully indecomposable maximal SNS-patterns of order ≤ 9 are given in [LMV96]. [GOD96] An $n \times n$ sign pattern A is a fully indecomposable maximal SNS-pattern with $\binom{n-1}{2}$ zero entries iff A is equivalent (namely, one can be be transformed into the other by any combination of transposition, multiplication by permutation patterns, and multiplication by signature patterns) to the pattern given in Example 1 below. For $n \geq 5$, there is precisely one equivalence class of fully indecomposable

maximal SNS-patterns with $\binom{n-1}{2}+1$ zero entries, and there are precisely two such equivalence classes having $\binom{n-1}{2}+2$ zero entries.

7. [BS95, Corollary 1.2.8] If A is an $n \times n$ sign pattern, then A is an S^{2}NS-pattern iff
   (a) A is an SNS-pattern, and
   (b) For each i and j with $a_{ij}=0$, the submatrix $A(i,j)$ of A of order $n-1$ obtained by deleting row i and column j is either an SNS-pattern or a sign singular pattern.
8. [BMQ68] If A is an $n \times n$ sign pattern with negative main diagonal, then A is an S^{2}NS-pattern iff
   (a) The actual product of entries of every simple cycle in A is negative, and
   (b) The actual product of entries of every simple path in A is the same, for any paths with the same initial row index and the same terminal column index.
9. [LLM95] An irreducible sign pattern A is NSNS iff there exists a permutation pattern P and a signature pattern S such that $B = APS$ satisfies:
   (a) $b_{ii} < 0$ for $i = 1, 2, \ldots, n$.
   (b) The actual product of entries of every cycle of length at least 2 of $D(B)$ is positive.
   (c) $D(B)$ is intercyclic (namely, any two cycles of lengths at least two have a common vertex).
10. [Bru88] An $n \times n$ sign pattern A is SNS iff $\text{per}(|B|) = |\det(B)|$, where B is the $(1,-1,0)$-matrix in $Q(A)$ and $|B|$ is obtained from B by replacing every -1 entry with 1.
11. [Tho86] The problem of determining whether a given sign pattern A is an SNS-pattern is equivalent to the problem of determining whether a certain digraph related to $D(A)$ has an even cycle.

For further reading, see [Kas63], [Bru88], [BS91], [BC92], [EHJ93], [BCS94a], [LMO96], [SS01], and [SS02].

Examples:

1. For $n \geq 2$, the following Hessenberg pattern is SNS:

$$H_n = \begin{bmatrix} - & + & 0 & \cdots & 0 & 0 \\ - & - & + & \cdots & 0 & 0 \\ - & - & - & \cdots & 0 & 0 \\ \vdots & \vdots & \vdots & \ddots & \vdots & \vdots \\ - & - & - & \cdots & - & + \\ - & - & - & \cdots & - & - \end{bmatrix}.$$

2. [BS95, p. 11] For $n \geq 2$, the following Hessenberg pattern is S^{2}NS:

$$G_n = \begin{bmatrix} + & - & 0 & \cdots & 0 & 0 \\ 0 & + & - & \cdots & 0 & 0 \\ 0 & 0 & + & \cdots & 0 & 0 \\ \vdots & \vdots & \vdots & \ddots & \vdots & \vdots \\ 0 & 0 & 0 & \cdots & + & - \\ + & 0 & 0 & \cdots & 0 & + \end{bmatrix}.$$

3.

$$A = \begin{bmatrix} - & + & 0 & 0 \\ 0 & - & + & 0 \\ 0 & 0 & - & + \\ - & 0 & 0 & - \end{bmatrix}$$

is S^2NS with inverse pattern

$$\begin{bmatrix} - & - & - & - \\ + & - & - & - \\ + & + & - & - \\ + & + & + & - \end{bmatrix}.$$

4. [BS95, p. 114] The following patterns are maximal SNS-patterns:

$$\begin{bmatrix} - & + & 0 \\ - & - & + \\ - & - & - \end{bmatrix}, \quad \begin{bmatrix} - & + & 0 & + \\ - & - & + & 0 \\ 0 & - & - & + \\ - & 0 & - & - \end{bmatrix}.$$

33.3 Sign-Solvability, L-Matrices, and S^*-Matrices

Definitions:

A system of linear equations $Ax = b$ (where A and b are both sign patterns or both real matrices) is **sign-solvable** if for each $\tilde{A} \in Q(A)$ and for each $\tilde{b} \in Q(b)$, the system $\tilde{A}x = \tilde{b}$ is consistent and

$$\{\tilde{x} : \text{there exist } \tilde{A} \in Q(A) \text{ and } \tilde{b} \in Q(b) \text{ with } \tilde{A}\tilde{x} = \tilde{b}\}$$

is entirely contained in one qualitative class.

A sign pattern or real matrix A is an L**-matrix** if for every $B \in Q(A)$, the rows of B are linearly independent.

A **barely** L**-matrix** is an L-matrix that is not an L-matrix if any column is deleted.

An $n \times (n+1)$ matrix B is an S^***-matrix** provided that each of the $n+1$ matrices obtained by deleting a column of B is an SNS matrix.

An $n \times (n+1)$ matrix B is an S**-matrix** if it is an S^*-matrix and the kernel of every matrix in $Q(B)$ contains a vector all of whose coordinates are positive.

A **signing** of order k is a nonzero $(0, 1, -1)$- or $(0, +, -)$-diagonal matrix of order k.

A signing $D' = \text{diag}(d_1', d_2', \ldots, d_k')$ is an **extension** of the signing D if $D' \neq D$ and $d_i' = d_i$ whenever $d_i \neq 0$.

A **strict signing** is a signing where all the diagonal entries are nonzero.

Let $D = \text{diag}(d_1, d_2, \ldots, d_k)$ be a signing of order k and let A be an $m \times n$ (real or sign pattern) matrix. If $k = m$, then DA is a **row signing** of the matrix A, and if D is strict, then DA is a **strict row signing** of the matrix A. If $k = n$, then AD is a **column signing** of the matrix A, and if D is strict, then AD is a **strict column signing** of the matrix A.

A real or sign pattern vector is **balanced** provided either it is a zero vector or it has both a positive entry and a negative entry. A vector v is **unsigned** if it is not balanced. A **balanced row signing** of the matrix A is a row signing of A in which all the columns are balanced. A **balanced column signing** of the matrix A is a column signing of A in which all the rows are balanced.

Facts:

Most of the following facts can be found in [BS95, Chaps. 1–3].

1. A square sign pattern A is an L-matrix iff A is an SNS matrix.
2. The linear system $Ax = 0$ is sign-solvable iff A^T is an L-matrix
3. If $Ax = b$ is sign-solvable, then A^T is an L-matrix.
4. $Ax = b$ is sign-solvable for all b if A is a square matrix and there exists a permutation matrix P such that PA is an invertible diagonal matrix.
5. Sign-solvability has been studied using signed digraphs; see [Man82], [Han83], [BS95, Chap. 3], and [Sha00].
6. [BS95] Let A be a matrix of order n and let b be an $n \times 1$ vector. Then $Ax = b$ is sign-solvable iff A is an SNS-matrix and for each i, $1 \leq i \leq n$, the matrix $A(i \leftarrow b)$, obtained from A by replacing the i-th column by b, is either an SNS matrix or has an identically zero determinant.
7. [BS95] If $AX = B$ is a sign-solvable linear system where A and B are square matrices of order n and B does not have an identically zero determinant, then A is an S^2NS matrix.
8. [BS95] Let $Ax = b$ be a linear system such that A has no zero rows. Then the linear system $Ax = b$ is sign-solvable and the vectors in its qualitative solution class have no zero coordinates iff the matrix $[A \mid -b]$ is an S^*-matrix.
9. An $n \times (n+1)$ matrix B is an S^*-matrix iff there exists a vector w with no zero coordinates such that the kernels of the matrices $\tilde{B} \in Q(B)$ are contained in $\{0\} \cup Q(w) \cup Q(-w)$.
10. [Man82] and [KLM84] Let $A = [a_{ij}]$ be an $m \times n$ matrix and let b be an $m \times 1$ vector. Assume that $z = (z_1, z_2, \ldots, z_n)^T$ is a solution of the linear system $Ax = b$. Let

$$\beta = \{j \,:\, z_j \neq 0\} \text{ and } \alpha = \{i \,:\, a_{ij} \neq 0 \text{ for some } j \in \beta\}.$$

Then $Ax = b$ is sign-solvable iff the matrix $[A[\alpha, \beta] \mid -b[\alpha]]$ is an S^*-matrix and the matrix $A(\alpha, \beta)^T$ is an L-matrix.

11. [KLM84] A matrix A is an L-matrix iff every row signing of A contains a unisigned column.
12. [BCS94a] An $m \times n$ matrix A is a barely L-matrix iff
    (a) A is an L-matrix.
    (b) For each $i = 1, 2, \ldots, n$, there is a row signing of A such that column i is the only unisigned column.
13. [BCS94a] An $m \times n$ matrix A is an S^*-matrix iff $n = m + 1$ and every row signing of A contains at least two unisigned columns.
14. [BCS94a] An $m \times n$ matrix A is an S^*-matrix iff $n = m + 1$ and there exists a strict signing D such that AD and $A(-D)$ are the only balanced column signings of A.
15. [KLM84] A matrix A is an S-matrix iff A is an S^*-matrix and every row of A is balanced.
16. Let A be an $m \times n$ sign pattern that does not have a $p \times q$ zero submatrix for any positive integers p and q with $p + q \geq m$. Then A is an L-matrix iff every strict row signing of A has a unisigned column.

For further reading, see [Sha95], [Sha99], [KS00], and [SR04].

Examples:

1. $\begin{bmatrix} + & - & + & + \\ + & + & - & + \\ + & + & + & - \end{bmatrix}$ is an L-matrix by Fact 12, and is a barely L-matrix by Fact 5 of Section 33.2.

2. [BS95, p. 65] The $m \times (m+1)$ matrix

$$H'_{m+1} = \begin{bmatrix} - & + & 0 & \cdots & 0 & 0 & 0 \\ - & - & + & \cdots & 0 & 0 & 0 \\ - & - & - & \cdots & 0 & 0 & 0 \\ \vdots & \vdots & \vdots & \ddots & \vdots & \vdots & \vdots \\ - & - & - & \cdots & - & + & 0 \\ - & - & - & \cdots & - & - & + \end{bmatrix}$$

is an S-matrix. Every strict column signing $H'_{m+1}D$ of H'_{m+1} is an S^*-matrix, with only two such strict column signings yielding S-matrices (namely, when $D = \pm I$).

Applications:

[BS95, Sec. 1.1] In supply and demand analysis in economics, linear systems, where the coefficients as well as the constants have prescribed signs, arise naturally. For instance, the sign-solvable linear system

$$\begin{bmatrix} + & - \\ - & - \end{bmatrix} \begin{bmatrix} x_1 \\ x_2 \end{bmatrix} = \begin{bmatrix} 0 \\ - \end{bmatrix}$$

arises from the study of a market with one product, where the price and quantity are determined by the intersection of its supply and demand curves.

33.4 Stability

Definitions:

A **negative stable** (respectively, **negative semistable**) real matrix is a square matrix B where each of the eigenvalues of B has negative (respectively, nonpositive) real part. In this section the term *(semi)stable* will mean *negative (semi)stable*. More information on matrix stability can be found in Section 9.5 and Chapter 19.

A **sign stable** (respectively, **sign semistable**) sign pattern matrix is a square sign pattern A where every matrix $B \in Q(A)$ is stable (respectively, semistable).

A **potentially stable** sign pattern matrix is a square sign pattern A where some matrix $B \in Q(A)$ is stable.

An $n \times n$ sign pattern matrix A allows a **properly signed nest** if there exists $B \in Q(A)$ and a permutation matrix P such that $\text{sgn}(\det(P^T B P[\{1, \ldots, k\}])) = (-1)^k$ for $k = 1, \ldots, n$.

A **minimally potentially stable** sign pattern matrix is a potentially stable, irreducible pattern such that replacing any nonzero entry by zero results in a pattern that is not potentially stable.

Facts:

Many of the following facts can be found in [BS95, Chap. 10].

1. A square sign pattern A is sign stable (respectively, sign semistable) iff each of the irreducible components of A is sign stable (respectively, sign semistable).
2. [QR65] If A is an $n \times n$ irreducible sign pattern, then A is sign semistable iff
   (a) A has nonpositive main diagonal entries.
   (b) If $i \neq j$, then $a_{ij}a_{ji} \leq 0$.
   (c) The digraph of A is a doubly directed tree.

3. If A is an $n \times n$ irreducible sign pattern, then A is sign stable iff
   (a) A has nonpositive main diagonal entries.
   (b) If $i \neq j$, then $a_{ij}a_{ji} \leq 0$.
   (c) The digraph of A is a doubly directed tree.
   (d) A does not have an identically zero determinant.
   (e) There does not exist a nonempty subset β of $[1, 2, \ldots, n]$ such that each diagonal element of $A[\beta]$ is zero, each row of $A[\beta]$ contains at least one nonzero entry, and no row of $A[\bar{\beta}, \beta]$ contains exactly one nonzero entry.

   The original version of this result was in terms of matchings and colorings in a graph ([JKD77, Theorem 2]); the restatement given here comes from [BS95, Theorem 10.2.2].
4. An efficient algorithm for determining whether a pattern is sign stable is given in [KD77], and the sign stable patterns have been characterized in finitely computable terms in [JKD87].
5. The characterization of the potentially stable patterns is a very difficult open question.
6. The potentially stable tree sign patterns (see Section 33.5) for dimensions less than five are given in [JS89].
7. If an $n \times n$ sign pattern A allows a properly signed nest, then A is potentially stable.
8. In [JMO97], sufficient conditions are determined for an $n \times n$ zero–nonzero pattern to allow a nested sequence of nonzero principal minors, and a method is given to sign a pattern that meets these conditions so that it allows a properly signed nest. It is also shown that if A is a tree sign pattern that has exactly one nonzero diagonal entry, then A is potentially stable iff A allows a properly signed nest.
9. In [LOD02], a measure of the relative distance to the unstable matrices for a stable matrix is defined and extended to a potentially stable sign pattern, and the minimally potentially stable patterns are studied.

Examples:

1. [BS95] The pattern $\begin{bmatrix} - & + \\ - & - \end{bmatrix}$ is sign stable, while the pattern $\begin{bmatrix} 0 & + \\ - & 0 \end{bmatrix}$ is sign semistable, but not sign stable.
2. [JMO97] The matrix $\begin{bmatrix} -1 & 1 & 0 \\ -1 & -1 & 1 \\ 0 & -3 & 1 \end{bmatrix}$ has a (leading) properly signed nest, so that the pattern $\begin{bmatrix} - & + & 0 \\ - & - & + \\ 0 & - & + \end{bmatrix}$ is potentially stable.
3. [JMO97] The matrix $B = \begin{bmatrix} -3 & 1 & 0 \\ 0 & 0 & 1 \\ 8 & -3 & 0 \end{bmatrix}$ has -1 as a triple eigenvalue, and so is stable. Thus, the sign pattern $A = \text{sgn}(B)$ is potentially stable, but it does not have a properly signed nest.
4. [LOD02] The $n \times n$ tridiagonal sign pattern A with $a_{11} = -, a_{i,i+1} = +, a_{i+1,i} = -$ for $i = 1, \ldots, n-1$, and all other entries 0, is minimally potentially stable.

Applications:

[BS95, sec. 10.1] The theory of sign stability is very important in population biology ([Log92]). For instance, a general ecosystem consisting of n different populations can be modeled by a linear system of differential equations. The entries of the coefficient matrix of this linear system reflect the effects on the ecosystem due to a small perturbation. The signs of the entries of the coefficient matrix can often be determined from general principles of ecology, while the actual magnitudes are difficult to determine and can only be approximated. The sign stability of the coefficient matrix determines the stability of an equilibrium state of the system.

33.5 Other Eigenvalue Characterizations or Allowing Properties

Definitions:

A **bipartite sign pattern matrix** is a sign pattern matrix whose digraph is bipartite.

A **combinatorially symmetric sign pattern matrix** is a square sign pattern A, where $a_{ij} \neq 0$ iff $a_{ji} \neq 0$.

The **graph of a combinatorially symmetric $n \times n$ sign pattern matrix** $A = [a_{ij}]$ is the graph with vertex set $\{1, 2, \ldots, n\}$, where $\{i, j\}$ is an edge iff $a_{ij} \neq 0$.

A **tree sign pattern** (t.s.p.) matrix is a combinatorially symmetric sign pattern matrix whose graph is a tree (possibly with loops).

An **n-cycle pattern** is a sign pattern A where the digraph of A is an n-cycle.

A **k-consistent sign pattern matrix** is a sign pattern A where every matrix $B \in Q(A)$ has exactly k real eigenvalues.

Facts:

1. [EJ91] An $n \times n$ sign pattern A requires all real eigenvalues iff each irreducible component of A is a symmetric t.s.p. matrix.
2. [EJ91] An $n \times n$ sign pattern A requires all nonreal eigenvalues iff each irreducible component of A
   (a) Is bipartite.
   (b) Has all negative simple cycles.
   (c) Is SNS.
3. [EJ93] For an $n \times n$ sign pattern A, the following statements are equivalent for each positive integer $k \geq 2$:
   (a) The minimum algebraic multiplicity of the eigenvalue 0 occurring among matrices in $Q(A)$ is k.
   (b) A requires an eigenvalue with algebraic multiplicity k, with k maximal.
   (c) The maximum composite cycle length in A is $n - k$.
4. If an $n \times n$ sign pattern A does not require an eigenvalue with algebraic multiplicity 2 (namely, if A allows n distinct eigenvalues), then A allows diagonalizability.
5. Sign patterns that require all distinct eigenvalues have many nice properties, such as requiring diagonalizability. In [LH02], a number of sufficient and/or necessary conditions for a sign pattern to require all distinct eigenvalues are established. Characterization of patterns that require diagonalizability is still open.
6. [EHL94a] If a sign pattern A requires all distinct eigenvalues, then A is k-consistent for some k.
7. [EHL94a] Let A be an $n \times n$ sign pattern and let A_I denote the sign pattern obtained from A by replacing all the diagonal entries by $+$. Then A_I requires n distinct real eigenvalues iff A is permutation similar to a symmetric irreducible tridiagonal sign pattern.
8. [LHZ97] A 3×3 nonnegative irreducible nonsymmetric sign pattern A allows normality iff $AA^T = A^T A$ and A is not permutation similar to $\begin{bmatrix} + & + & 0 \\ + & 0 & + \\ + & + & + \end{bmatrix}$.

9. [LHZ97] If $\begin{bmatrix} A_1 & A_2 \\ A_3 & A_4 \end{bmatrix}$ allows normality, where A_1 is square, then the square pattern

$$\begin{bmatrix} A_1 & A_2 & A_2 & \dots & A_2 \\ A_3 & A_4 & A_4 & \dots & A_4 \\ A_3 & A_4 & A_4 & \dots & A_4 \\ \vdots & \vdots & \vdots & \ddots & \vdots \\ A_3 & A_4 & A_4 & \dots & A_4 \end{bmatrix}$$

also allows normality. Parallel results hold for allowing idempotence and for allowing nilpotence of index 2. (See [HL99] and [EL99].)
10. [EL99] Suppose an $n \times n$ sign pattern A allows nilpotence. Then A^n is compatible with 0, and for $1 \leq k \leq n-1$, $\text{tr}(A^k)$ is compatible with 0. Further, for each m, $1 \leq m \leq n$, $E_m(A)$ (the sum of all principal minors of order m) is compatible with 0.
11. [HL99] Let A be a 5×5 irreducible symmetric sign pattern such that A^2 is compatible with A. Then A allows a symmetric idempotent matrix unless A can be obtained from the following by using permutation similarity and signature similarity:

$$\begin{bmatrix} + & + & + & + & 0 \\ + & + & - & 0 & - \\ + & - & + & + & + \\ + & 0 & + & + & - \\ 0 & - & + & - & + \end{bmatrix}.$$

12. [SG03] Let A be an $n \times n$ sign pattern. If the maximum composite cycle length in A is equal to the maximum rank (see section 33.6) of A, then A allows diagonalizability.
13. [SG03] Every combinatorially symmetric sign pattern allows diagonalizability.
14. A nonzero $n \times n$ ($n \geq 2$) sign pattern that requires nilpotence does not allow diagonalizability.
15. Complete characterization of sign patterns that allow diagonalizability is still open.

For further reading, see [Esc93a], [Yeh96], and [KMT96].

Examples:

1. By Fact 1, the pattern

$$\begin{bmatrix} * & + & 0 & 0 \\ + & * & - & - \\ 0 & - & * & 0 \\ 0 & - & 0 & * \end{bmatrix},$$

where each $*$ entry could be 0, $+$, or $-$, requires all real eigenvalues.
2. [LH02] Up to equivalence (negation, transposition, permutational similarity, and signature similarity), the 3×3 irreducible sign patterns that require 3 distinct eigenvalues are the irreducible tridiagonal symmetric sign patterns, irreducible tridiagonal skew-symmetric sign patterns, and 3-cycle sign patterns, together with the following:

$$\begin{bmatrix} + & + & 0 \\ 0 & 0 & + \\ + & 0 & 0 \end{bmatrix}, \begin{bmatrix} 0 & + & 0 \\ - & 0 & + \\ + & - & 0 \end{bmatrix}, \begin{bmatrix} 0 & + & 0 \\ - & 0 & + \\ + & 0 & 0 \end{bmatrix}, \begin{bmatrix} 0 & + & - \\ - & 0 & + \\ + & - & 0 \end{bmatrix}, \begin{bmatrix} + & + & 0 \\ 0 & 0 & + \\ + & - & 0 \end{bmatrix}.$$

3. [EL99] The sign pattern $A = \begin{bmatrix} + & + & 0 \\ - & - & - \\ - & + & + \end{bmatrix}$ does not allow nilpotence, though it satisfies many "obvious" necessary conditions.
4. [LHZ97] The $n \times n$ sign pattern $\begin{bmatrix} + & + & + & \cdots & + \\ + & + & 0 & \cdots & 0 \\ \vdots & \vdots & \vdots & \ddots & \vdots \\ + & + & 0 & \cdots & 0 \end{bmatrix}$ allows normality.

33.6 Inertia, Minimum Rank

Definitions:

Let A be a sign pattern matrix.

The **minimal rank** of A, denoted mr(A), is defined by $\text{mr}(A) = \min\{ \text{rank } B : B \in Q(A) \}$.

The **maximal rank** of A, MR(A), is given by $\text{MR}(A) = \max\{ \text{rank } B : B \in Q(A) \}$.

The **term rank** of A is the maximum number of nonzero entries of A no two of which are in the same row or same column.

For a symmetric sign pattern A, smr(A), the **symmetric minimal rank** of A is $\text{smr}(A) = \min\{ \text{rank } B : B = B^T, B \in Q(A) \}$.

The **symmetric maximal rank** of A, SMR(A), is $\text{SMR}(A) = \max\{ \text{rank } B : B = B^T, B \in Q(A) \}$.

For a symmetric sign pattern A, the **(symmetric) inertia set of** A is $\text{in}(A) = \{ \text{in}(B) : B = B^T \in Q(A) \}$.

A **requires unique inertia** if $\text{in}(B_1) = \text{in}(B_2)$ for all symmetric matrices $B_1, B_2 \in Q(A)$.

A sign pattern A of order n is an **inertially arbitrary pattern** (IAP) if every possible ordered triple (p, q, z) of nonnegative integers $p, q,$ and z with $p + q + z = n$ can be achieved as the inertia of some $B \in Q(A)$.

A **spectrally arbitrary pattern** (SAP) is a sign pattern A of order n such that every monic real polynomial of degree n can be achieved as the characteristic polynomial of some matrix $B \in Q(A)$.

Facts:

1. MR(A) is equal to the term rank of A.
2. Starting with a real matrix whose rank is mr(A) and changing one entry at a time to eventually reach a real matrix whose rank is MR(A), all ranks between mr(A) and MR(A) are achieved by real matrices.
3. [HLW01] For every symmetric sign pattern A, MR(A)=SMR(A).
4. [HS93] A sign pattern A requires a fixed rank r iff A is permutationally equivalent to a sign pattern of the form $\begin{bmatrix} X & Y \\ Z & 0 \end{bmatrix}$, where X is $k \times (r - k), 0 \leq k \leq r$, and Y and Z^T are L-matrices.
5. [HLW01] A symmetric sign pattern A requires a unique inertia iff smr(A)=SMR(A).
6. [HLW01] For the symmetric sign pattern $A = \begin{bmatrix} 0 & A_1 \\ A_1^T & 0 \end{bmatrix}$ of order n, we have
$$\text{in}(A) = \{(k, k, n - 2k) : \text{mr}(A_1) \leq k \leq \text{MR}(A_1)\}.$$
In particular, $2\,\text{mr}(A_1) = \text{smr}(A)$.
7. [DJO00], [EOD03] Let T_n be the $n \times n$ tridiagonal sign pattern with each superdiagonal entry positive, each subdiagonal entry negative, the $(1, 1)$ entry negative, the (n, n) entry positive, and every other entry zero. It is conjectured that T_n is an SAP for all $n \geq 2$. It is shown that for $3 \leq n \leq 16$, T_n is an SAP.

8. [GS01] Let S_n be the $n \times n$ $(n \geq 2)$ sign pattern with each strictly upper (resp., lower) triangular entry positive (resp., negative), the $(1, 1)$ entry negative, the (n, n) entry positive, and all other diagonal entries zero. Then S_n is inertially arbitrary.
   [ML02] Further, if the $(1, n)$ and $(n, 1)$ entries of S_n are replaced by zero, the resulting sign pattern is also an IAP.
9. [CV05] Not every inertially arbitrary sign pattern is spectrally arbitrary.
10. [MOT03] Suppose $1 \leq p \leq n - 1$. Then every $n \times n$ sign pattern with p positive columns and $n - p$ negative columns is spectrally arbitrary.

For further reading see [CHL03], [SSG04], [Hog05], and references [Nyl96], [JD99], [BFH04], [BF04], and [BHL04] in Chapter 34.

Examples:

1. [HLW01] Let

$$A = \begin{bmatrix} + & 0 & + & + \\ 0 & + & + & + \\ + & + & - & 0 \\ + & + & 0 & - \end{bmatrix}.$$

   Then $\text{in}(A) = (2, 2, 0)$, $\text{smr}(A) = SMR(A) = 4$, but $3 = \text{mr}(A) < \text{smr}(A) = 4$.
2. [HL01] Let J_n be the $n \times n$ sign pattern with all entries equal to $+$. Then

$$\text{in}(J_n) = \{(s, t, n - s - t) : s \geq 1,\ t \geq 0,\ s + t \leq n\}.$$

3. [Gao01], [GS03], [HL01] Let A be the $n \times n$ $(n \geq 3)$ sign pattern all of whose diagonal entries are zero and all of whose off-diagonal entries are $+$. Then

$$\text{in}(A) = \{(s, t, n - s - t) : s \geq 1,\ t \geq 2,\ s + t \leq n\}.$$

33.7 Patterns That Allow Certain Types of Inverses

Definitions:

A sign pattern matrix A is **nonnegative (positive)**, denoted $A \geq 0 (A > 0)$, if all of its entries are nonnegative (positive).

An **inverse nonnegative (inverse positive)** sign pattern matrix is a square sign pattern A that allows an entrywise nonnegative (positive) inverse.

Let both B and X be real matrices or nonnegative sign pattern matrices. Consider the following conditions.

(1) $BXB = B$.
(2) $XBX = X$.
(3) BX is symmetric.
(4) XB is symmetric.

For a real matrix B, there is a unique matrix X satisfying all four conditions above and it is called the **Moore–Penrose inverse** of B, and denoted by $B^{\dagger}$. More generally, let $B\{i, j, \ldots, l\}$ denote the set of matrices X satisfying conditions $(i), (j), \ldots, (l)$ from among conditions (1)–(4). A matrix $X \in B\{i, j, \ldots, l\}$ is called an **$(i, j, \ldots, l)$-inverse** of B. For example, if (1) holds, X is called a (1)-inverse of B; if (1) and (2) hold, X is called a (1, 2)-inverse of B, and so forth. See Section 5.7.

For a nonnegative sign pattern matrix B, if there is a nonnegative sign pattern X satisfying (1) to (4), then X is unique and it is called the Moore–Penrose inverse of B. An $(i, j, \ldots, l)$-inverse of B is defined similarly as in the preceding paragraph.

Facts:

The first three facts below are contained in [BS95, Chap. 9].

1. [JLR79] Let A be an $n \times n$ $(+, -)$-pattern, where $n \geq 2$. Then A is inverse positive iff A is not permutationally equivalent to a pattern of the form $\begin{bmatrix} A_{11} & A_{12} \\ A_{21} & A_{22} \end{bmatrix}$, where $A_{12} < 0, A_{21} > 0$, the blocks A_{11}, A_{22} are square or rectangular, and one (but not both) of A_{11}, A_{22} may be empty.
2. The above result in [JLR79] is generalized in [FG81] to $(+, -, 0)$-patterns, and additional equivalent conditions are established. Let e denote the column vector of all ones, J the $n \times n$ matrix all of whose entries equal 1, and A' the matrix obtained from A by replacing all negative entries with zeros. For an $n \times n$ fully indecomposable sign pattern A, the following are equivalent:
   (a) A is inverse positive.
   (b) A is not permutationally equivalent to a pattern of the form $\begin{bmatrix} A_{11} & A_{12} \\ A_{21} & A_{22} \end{bmatrix}$, where $A_{12} \leq 0, A_{21} \geq 0$, the blocks A_{11}, A_{22} are square or rectangular, and one (but not both) of A_{11}, A_{22} may be empty.
   (c) The pattern $\begin{bmatrix} 0 & A \\ -A^T & 0 \end{bmatrix}'$ is irreducible.
   (d) There exists $B \in Q(A)$ such that $Be = B^T e = 0$.
   (e) There exists a doubly stochastic matrix D such that $(D - \frac{1}{n}J) \in Q(A)$.
3. [Joh83] For an $n \times n$ fully indecomposable sign pattern A, the following are equivalent: A is inverse nonnegative; A is inverse positive; $-A$ is inverse nonnegative; $-A$ is inverse positive.
4. [EHL97] If an $n \times n$ sign pattern A allows B and B^{-1} to be in $Q(A)$, then
   (a) $\text{MR}(A) = n$.
   (b) A^2 is compatible with I.
   (c) adj A is compatible with A and $\det(A)$ is compatible with $+$, or, adj A is compatible with $-A$ and $\det(A)$ is compatible with $-$, where adj A is the adjoint of A.
5. In [EHL94b], the class $\mathcal{G}$ of all square patterns A that allow $B, C \in Q(A)$ where $BCB = B$ is investigated; it is shown for nonnegative patterns that $\mathcal{G}$ coincides with the class of square patterns that allow $B \in Q(A)$ where $B^3 = B$.
6. [HLR04] An $m \times n$ nonnegative sign pattern A has a nonnegative $(1, 3)$-inverse (Moore–Penrose inverse) iff A allows a nonnegative $(1, 3)$-inverse (Moore–Penrose inverse).

For further reading, see [BF87], [BS95, Theorem 9.2.6], [SS01], and [SS02].

Examples:

1. By Facts 2 and 3, the sign pattern

$$A = \begin{bmatrix} + & 0 & + & + \\ 0 & + & + & + \\ - & - & - & 0 \\ - & - & 0 & - \end{bmatrix}$$

is not inverse nonnegative.

2. [EHL97] The sign pattern

$$A = \begin{bmatrix} + & + & + & + \\ 0 & + & + & + \\ 0 & + & - & - \\ 0 & - & + & + \end{bmatrix}$$

satisfies all the necessary conditions in Fact 4, but it does not allow an inverse pair B and B^{-1} in $Q(A)$.

33.8 Complex Sign Patterns and Ray Patterns

Definitions:

A **complex sign pattern matrix** is a matrix of the form $A = A_1 + iA_2$ for some $m \times n$ sign patterns A_1 and A_2, and the **sign pattern class** or **qualitative class** of A is

$$Q(A) = \{B_1 + iB_2 \ : \ B_1 \in Q(A_1) \text{ and } B_2 \in Q(A_2)\}.$$

Many definitions for sign patterns, such as SNS, extend in the obvious way to complex sign patterns.

The **determinantal region** of a complex sign pattern A is the set

$$S_A = \{\det(B) \ : \ B \in Q(A)\}.$$

A **ray pattern** is a matrix each of whose entries is either 0 or a ray in the complex plane of the form $\{re^{i\theta} \ : \ r > 0\}$ (which is represented by $e^{i\theta}$). The **ray pattern class** of an $m \times n$ ray pattern A is

$$Q(A) = \{B = [b_{pq}] \in M_{m\times n}(\mathbb{C}) : b_{pq} = 0 \text{ iff } a_{pq} = 0, \text{ and otherwise } \arg b_{pq} = \arg a_{pq}\}.$$

For $\alpha < \beta$, the **open sector** from the ray $e^{i\alpha}$ to the ray $e^{i\beta}$ is the set of rays $\{re^{i\theta} \ : \ r > 0, \alpha < \theta < \beta\}$.

The **determinantal region** of a ray pattern A is the set

$$R_A = \{\det(B) \ : \ B \in Q(A)\}.$$

An $n \times n$ ray pattern A is **ray nonsingular** if the Hadamard product $X \circ A$ is nonsingular for every entrywise positive $n \times n$ matrix X.

A **cyclically real** ray pattern is a square ray pattern A where the actual products of every cycle in A is real.

Facts:

1. [EHL98], [SS05] For a complex sign pattern A, the boundaries of S_A are always on the axes on the complex plane.
2. [EHL98], [SS05] For a sign nonsingular complex sign pattern A, S_A is either entirely contained in an axis of the complex plane or is an open sector in the complex plane with boundary rays on the axes.
3. [SS05] For a complex sign pattern or ray pattern A, the region $S_A\backslash\{0\}$ (or $R_A\backslash\{0\}$) is an open set (in fact, a disjoint union of open sectors) in the complex plane, except in the cases that S_A (R_A) is entirely contained in a line through the origin.
4. The results of [MOT97], [LMS00], and [LMS04] show that there is an entrywise nonzero ray nonsingular ray pattern of order n if and only if $1 \leq n \leq 4$.
5. [EHL00] An irreducible ray pattern A is cyclically real iff A is diagonally similar to a real sign pattern. More generally, a ray pattern A is diagonally similar to a real sign pattern iff A and $A + A^*$ are both cyclically real.

Examples:

1. [EHL98] If $A = \begin{bmatrix} + & - \\ + & + \end{bmatrix} + i \begin{bmatrix} + & 0 \\ 0 & - \end{bmatrix}$, then A is sign nonsingular and S_A is the open sector from the ray $e^{-i\pi/2}$ to the ray $e^{i\pi/2}$.

2. [MOT97] The ray pattern $\begin{bmatrix} e^{i\pi/2} & + & + & + \\ + & e^{i\pi/2} & + & + \\ + & + & e^{i\pi/2} & + \\ + & + & + & e^{i\pi/2} \end{bmatrix}$ is ray nonsingular.

3. [EHL00] Let

$$A = \begin{bmatrix} 0 & e^{-i\theta_1} & 0 & -e^{-i\theta_1} \\ 0 & + & -e^{-i\theta_2} & 0 \\ e^{i(\theta_1+\theta_2)} & -e^{i\theta_2} & 0 & -e^{i\theta_2} \\ e^{i\theta_1} & - & 0 & - \end{bmatrix},$$

where θ_1 and θ_2 are arbitrary. Then A is cyclically real, and A is diagonally similar (via the diagonal ray pattern $S = \text{diag}(+, e^{i\theta_1}, e^{i(\theta_1+\theta_2)}, -e^{i\theta_1})$) to

$$\begin{bmatrix} 0 & + & 0 & + \\ 0 & + & - & 0 \\ + & - & 0 & + \\ - & + & 0 & - \end{bmatrix}.$$

33.9 Powers of Sign Patterns and Ray Patterns

Definitions:

Let J_n (or simply J) denote the all + sign (ray) pattern of order n.

A square sign pattern or ray pattern A is **powerful** if all the powers $A^1, A^2, A^3, \ldots,$ are unambiguously defined, that is, no entry in any A^k is a sum involving two or more distinct rays. For a powerful pattern A, the smallest positive integers $l = l(A)$ and $p = p(A)$ such that $A^l = A^{l+p}$ are called the **base** and **period** of A, respectively.

A square sign pattern or ray pattern A is k-**potent** if k is the smallest positive integer such that $A = A^{k+1}$.

Facts:

1. [LHE94] An irreducible sign pattern A with index of imprimitivity h (see Section 29.7) is powerful iff all cycles of A with lengths odd multiples of h have the same sign and all cycles (if any) of A with lengths even multiples of h are positive (see [SS04]). A sign pattern A is powerful iff for every positive integer d and for every pair of matrices $B, C \in Q(A)$, $\text{sgn}(B^d) = \text{sgn}(C^d)$.
2. [LHE94] Let A be an irreducible powerful sign pattern, with index of imprimitivity h. Then the base and the period of A are given by $l(A) = l(|A|)$, $p(A) = h$ if A does not have any negative cycles, and $p(A) = 2h$ if A has a negative cycle.
3. [Esc93b] The only irreducible idempotent sign pattern of order $n \geq 2$ is the all + sign pattern.
4. [LHE94], [SEK99], [SBS02] Every k-potent irreducible sign or ray pattern matrix is powerful.
5. [EL97] The maximum number of − entries in the square of a sign pattern of order n is $n^2 - 2$; the maximum number of − entries in the square of a $(+, -)$ sign pattern of order n is $\lfloor n^2/2 \rfloor$.
6. [HL01b] Let A be an $n \times n$ $(n \geq 3)$ sign pattern. If A^2 has only one entry that is not nonpositive, then A^2 has at most $n^2 - n$ negative entries.
7. [LHS02] Let A be an irreducible ray pattern. Then A is powerful iff A is diagonally similar to a subpattern of $e^{i\alpha} J$ for some $\alpha \in \mathbb{R}$, where J is the all + ray pattern.

8. [LHS05] Suppose that $A = \begin{bmatrix} A_{11} & A_{12} \\ 0 & A_{22} \end{bmatrix}$ is a powerful ray pattern, where A_{11} (resp., A_{22}) is irreducible with index of imprimitivity h_1 (resp., h_2) and $0 \neq A_{11} \preceq c_1 J_{n_1}, 0 \neq A_{22} \preceq c_2 J_{n_2}$. If $A_{12} \neq 0$, then $\left(\frac{c_2}{c_1}\right)^{\text{lcm}(h_1,h_2)} = 1$.

For further reading, the structures of k-potent sign patterns or ray patterns are studied in [SEK99], [Stu99], [SBS02], [LG01], and [Stu03].

Examples:

1. [LHE94] The reducible sign pattern $A = \begin{bmatrix} 0 & + & + & + \\ 0 & + & 0 & + \\ 0 & 0 & - & - \\ 0 & 0 & 0 & 0 \end{bmatrix}$ satisfies $A^2 = \begin{bmatrix} 0 & + & - & \# \\ 0 & + & 0 & + \\ 0 & 0 & + & + \\ 0 & 0 & 0 & 0 \end{bmatrix}$ and $A^3 = A$. Thus, A is 2-potent and yet A is not powerful.
2. [SEK99] Let P_n be the $n \times n$ circulant permutation sign pattern with (1, 2) entry equal to +. Let Q_n be the sign pattern obtained from P_n by replacing the + in the (n, 1) position with a −. Then P_n is n-potent and Q_n is $2n$-potent.
3. [SBS02] Suppose that $3|k$. Let $A = \omega \begin{bmatrix} 0 & J_{p\times q} & 0 \\ 0 & 0 & J_{q\times r} \\ J_{r\times p} & 0 & 0 \end{bmatrix}$, where $J_{m\times n}$ denotes the all ones $m \times n$ matrix and ω^3 is a primitive $k/3$-th root of unity. Then A is a k-potent ray pattern.

33.10 Orthogonality

Definitions:

A square sign pattern A is **potentially orthogonal (PO)** if A allows an orthogonal matrix.

A square sign pattern A that does not have a zero row or zero column is **sign potentially orthogonal (SPO)** if every pair of rows and every pair of columns allows orthogonality.

Two vectors $\mathbf{x} = [x_1, \ldots, x_n]$ and $\mathbf{y} = [y_1, \ldots, y_n]$ are **combinatorially orthogonal** if $|\{i : x_i y_i \neq 0\}| \neq 1$.

Facts:

1. Every PO sign pattern is SPO.
2. [BS94], [Wat96] For $n \leq 4$, every $n \times n$ SPO sign pattern is PO.
3. [Wat96] There is a 5×5 fully indecomposable SPO sign pattern that is not PO.
4. [JW98] There is a 6×6 $(+,-)$ sign pattern that is SPO but not PO.
5. [BBS93] Let A be an $n \times n$ fully indecomposable sign pattern whose rows are combinatorially orthogonal and whose columns are combinatorially orthogonal. Then A has at least $4(n-1)$ nonzero entries. This implies that a conjecture of Fiedler [Fie64], which says a fully indecomposable orthogonal matrix of order n has at least $4(n-1)$ nonzero entries, is true.
6. [CJL99] For $n \geq 2$, there is an $n \times n$ fully indecomposable orthogonal matrix with k zero entries iff $0 \leq k \leq (n-2)^2$.
7. [EHH99] Let S be any skew symmetric sign pattern of order n all of whose off-diagonal entries are nonzero. Then $I + S$ is PO.
8. [EHH99] It is an open question as to whether every sign pattern A that allows an inverse in $Q(A^T)$ is PO.

For further reading see [Lim93], [Sha98], [CS99], and [CHR03].

Examples:

1. [BS94] Every $3 \times 3 \pm$ SPO sign pattern can be obtained from the following sign pattern by using permutation equivalence and multiplication by signature patterns:

$$\begin{bmatrix} + & + & + \\ + & + & - \\ + & - & + \end{bmatrix}.$$

2. [Wat96] The sign pattern

$$\begin{bmatrix} - & + & 0 & + & - \\ + & + & - & 0 & - \\ 0 & + & + & + & + \\ + & 0 & - & + & + \\ - & - & - & + & + \end{bmatrix}$$

is SPO but not PO.

33.11 Sign-Central Patterns

Definitions:

A real matrix B is **central** if the zero vector is in the convex hull of the columns of B. A real or sign pattern matrix A is **sign-central** if A requires centrality.

A **minimal sign-central matrix** A is a sign-central matrix that is not sign-central if any column of A is deleted.

A **tight sign-central** matrix is a sign-central matrix A for which the Hadamard (entrywise) product of any two columns of A contains a negative component.

A **nearly sign-central matrix** is a matrix that is not sign-central but can be augmented to a sign-central matrix by adjoining a column.

Facts:

1. [AB94] An $m \times n$ matrix A is sign-central iff the matrix DA has a nonnegative column vector for every strict signing D of order m.
2. [HKK03] Every tight sign-central matrix is a minimal sign-central matrix.
3. [LC00] If A is nearly sign-central and $[A \mid \alpha]$ is sign-central, then $[A \mid \alpha']$ is also sign-central for every $\alpha' \neq 0$ obtained from α by zeroing out some of its entries.

For further reading, see [BS95, Sect. 5.4], [DD90], [LLS97], and [BJS98].

Examples:

1. [BS95, p. 100], [HKK03] For each positive integer m, the $m \times 2^m \pm$ sign pattern E_m such that each m-tuple of $+$'s and $-$'s is a column of E_m, is a tight sign-central sign pattern.

References

[AB94] T. Ando and R.A. Brualdi, Sign-central matrices, *Lin. Alg. Appl.* 208/209:283–295, 1994.

[BJS98]M. Bakonyi, C.R. Johnson, and D.P. Stanford, Sign pattern matrices that require domain-range pairs with given sign patterns, *Lin. Multilin. Alg.* 44:165–178, 1998.

[BMQ68] L. Bassett, J.S. Maybee, and J. Quirk, Qualitative economics and the scope of the correspondence principle, *Econometrica* 36:544–563, 1968.

[BBS93] L.B. Beasley, R.A. Brualdi, and B.L. Shader, Combinatorial orthogonality, *Combinatorial and Graph Theoretic Problems in Linear Algebra*, IMA Vol. Math. Appl. 50, Springer-Verlag, New York, 1993:207–218.

[BS94] L.B. Beasley and D. Scully, Linear operators which preserve combinatorial orthogonality, *Lin. Alg. Appl.* 201:171–180, 1994.

[BF87] M.A. Berger and A. Felzenbaum, Sign patterns of matrices and their inverse, *Lin. Alg. Appl.* 86:161–177, 1987.

[Bru88] R.A. Brualdi, Counting permutations with restricted positions: permanents of (0, 1)-matrices. A tale in four parts, *Lin. Alg. Appl.* 104:173–183, 1988.

[BC92] R.A. Brualdi and K.L. Chavey, Sign-nonsingular matrix pairs, *SIAM J. Matrix Anal. Appl.* 13:36–40, 1992.

[BCS93] R.A. Brualdi, K.L. Chavey, and B.L. Shader, Conditional sign-solvability, *Math. Comp. Model.* 17:141–148, 1993.

[BCS94a] R.A. Brualdi, K.L. Chavey, and B.L. Shader, Bipartite graphs and inverse sign patterns of strong sign-nonsingular matrices, *J. Combin. Theory, Ser. B* 62:133–152, 1994.

[BCS94b] R.A. Brualdi, K.L. Chavey, and B.L. Shader, Rectangular L-matrices, *Lin. Alg. Appl.* 196:37–61, 1994.

[BS91] R.A. Brualdi and B.L. Shader, On sign-nonsingular matrices and the conversion of the permanent into the determinant, in *Applied Geometry and Discrete Mathematics* (P. Gritzmann and B. Sturmfels, Eds.), Amer. Math. Soc., Providence, RI, 117–134, 1991.

[BS95] R.A. Brualdi and B.L. Shader, *Matrices of Sign-Solvable Linear Systems*, Cambridge University Press, Cambridge, 1995.

[CV05] M.S. Cavers and K.N. Vander Meulen, Spectrally and inertially arbitrary sign patterns, *Lin. Alg. Appl.* 394:53–72, 2005.

[CHL03] G. Chen, F.J. Hall, Z. Li, and B. Wei, On ranks of matrices associated with trees, *Graphs Combin.* 19(3):323–334, 2003.

[CJL99] G.-S. Cheon, C.R. Johnson, S.-G. Lee, and E.J. Pribble, The possible numbers of zeros in an orthogonal matrix, *Elect. J. Lin. Alg.* 5:19–23, 1999.

[CS99] G.-S. Cheon and B.L. Shader, How sparse can a matrix with orthogonal rows be? *J. of Comb. Theory, Ser. A* 85:29–40, 1999.

[CHR03] G.-S. Cheon, S.-G. Hwang, S. Rim, B.L. Shader, and S. Song, Sparse orthogonal matrices, *Lin. Alg. Appl.* 373:211–222, 2003.

[DD90] G.V. Davydov and I.M. Davydova, Solubility of the system $Ax = 0, x \geq 0$ with indefinite coefficients, *Soviet Math. (Iz. VUZ)* 43(9):108–112, 1990.

[DJO00] J.H. Drew, C.R. Johnson, D.D. Olesky, and P. van den Driessche, Spectrally arbitrary patterns, *Lin. Alg. Appl.* 308:121–137, 2000.

[EOD03] L. Elsner, D.D. Olesky, and P. van den Driessche, Low rank perturbations and the spectrum of a tridiagonal sign pattern, *Lin. Alg. Appl.* 374:219–230, 2003.

[Esc87] C.A. Eschenbach, *Eigenvalue Classification in Qualitative Matrix Analysis*, doctoral dissertation directed by C.R. Johnson, Clemson University, 1987.

[Esc93a] C.A. Eschenbach, Sign patterns that require exactly one real eigenvalue and patterns that require $n - 1$ nonreal eigenvalues, *Lin. and Multilin. Alg.* 35:213–223, 1993.

[Esc93b] C.A. Eschenbach, Idempotence for sign pattern matrices, *Lin. Alg. Appl.* 180:153–165, 1993.

[EHJ93] C.A. Eschenbach, F.J. Hall, and C.R. Johnson, Self-inverse sign patterns, in *IMA Vol. Math. Appl. 50*, Springer-Verlag, New York, 245–256, 1993.

[EHH99] C.A. Eschenbach, F.J. Hall, D.L. Harrell, and Z. Li, When does the inverse have the same sign pattern as the inverse? *Czech. Math. J.* 49:255–275, 1999.

[EHL94a] C.A. Eschenbach, F.J. Hall, and Z. Li, Eigenvalue frequency and consistent sign pattern matrices, *Czech. Math. J.* 44:461–479, 1994.

[EHL94b] C.A. Eschenbach, F.J. Hall, and Z. Li, Sign pattern matrices and generalized inverses, *Lin. Alg. Appl.* 211:53–66, 1994.

[EHL97] C.A. Eschenbach, F.J. Hall, and Z. Li, Some sign patterns that allow a real inverse pair B and B^{-1}, *Lin. Alg. Appl.* 252:299–321, 1997.

[EHL98] C.A. Eschenbach, F.J. Hall, and Z. Li, From real to complex sign pattern matrices, *Bull. Aust. Math. Soc.* 57:159–172, 1998.

[EHL00] C.A. Eschenbach, F.J. Hall, and Z. Li, Eigenvalue distribution of certain ray patterns, *Czech. Math. J.* 50(125):749–762, 2000.

[EJ91] C.A. Eschenbach and C.R. Johnson, Sign patterns that require real, nonreal or pure imaginary eigenvalues, *Lin. Multilin. Alg.* 29:299–311, 1991.

[EJ93] C.A. Eschenbach and C.R. Johnson, Sign patterns that require repeated eigenvalues, *Lin. Alg. Appl.* 190:169–179, 1993.

[EL97] C.A. Eschenbach and Z. Li, How many negative entries can A^2 have? *Lin. Alg. Appl.* 254:99–117, 1997.

[EL99] C.A. Eschenbach and Z. Li, Potentially nilpotent sign pattern matrices, *Lin. Alg. Appl.* 299:81–99, 1999.

[Fie64] M. Fiedler (Ed.), *Proceedings: Theory of Graphs and Its Applications*, Publishing House of the Czech. Acad. of Sc., Prague, 1964.

[FG81] M. Fiedler and R. Grone, Characterizations of sign patterns of inverse positive matrices, *Lin. Alg. Appl.* 40:237–245, 1981.

[Gao01] Y. Gao, *Sign Pattern Matrices*, Ph.D. dissertation, University of Science and Technology of China, 2001.

[GS01] Y. Gao and Y. Shao, Inertially arbitrary patterns, *Lin. Multilin. Alg.* 49(2):161–168, 2001.

[GS03] Y. Gao and Y. Shao, The inertia set of nonnegative symmetric sign pattern with zero diagonal, *Czech. Math. J.* 53(128):925–934, 2003.

[Gib71] P.M. Gibson, Conversion of the permanent into the determinant, *Proc. Amer. Math. Soc.* 27:471–476, 1971.

[GOD96] B. C. J. Green, D.D. Olesky, and P. van den Driessche, Classes of sign nonsingular matrices with a specified number of zero entries, *Lin. Alg. Appl.* 248:253–275, 1996.

[HL99] F.J. Hall and Z. Li, Sign patterns of idempotent matrices, *J. Korean Math. Soc.* 36:469–487, 1999.

[HL01] F.J. Hall and Z. Li, Inertia sets of symmetric sign pattern matrices, *Num. Math J. Chin. Univ.* 10:226–240, 2001.

[HLW01] F.J. Hall, Z. Li, and D. Wang, Symmetric sign pattern matrices that require unique inertia, *Lin. Alg. Appl.* 338:153–169, 2001.

[HLR04] F.J. Hall, Z. Li, and B. Rao, From Boolean to sign pattern matrices, *Lin. Alg. Appl.*, 393: 233–251, 2004.

[Han83] P. Hansen, Recognizing sign-solvable graphs, *Discrete Appl. Math.* 6:237–241, 1983.

[HS93] D. Hershkowitz and H. Schneider, Ranks of zero patterns and sign patterns, *Lin. Multilin. Alg.* 34:3–19, 1993.

[Hog05] L. Hogben, Spectral graph theory and the inverse eigenvalue problem of a graph, *Elect. Lin. Alg.* 14:12–31, 2005.

[HL01b] Y. Hou and J. Li, Square nearly nonpositive sign pattern matrices, *Lin. Alg. Appl.* 327:41–51, 2001.

[HKK03] S.-G. Hwang, I.-P. Kim, S.-J. Kim, and X. Zhang, Tight sign-central matrices, *Lin. Alg. Appl.* 371:225–240, 2003.

[JKD77] C. Jeffries, V. Klee, and P. van den Driessche, When is a matrix sign stable? *Can. J. Math.* 29:315–326, 1977.

[JKD87] C. Jeffries, V. Klee, and P. van den Driessche, Qualitative stability of linear systems, *Lin. Alg. Appl.* 87:1–48, 1987.

[Joh83] C.R. Johnson, Sign patterns of inverse nonnegative matrices, *Lin. Alg. Appl.* 55:69–80, 1983.

[JLR79] C.R. Johnson, F.T. Leighton, and H.A. Robinson, Sign patterns of inverse-positive matrices, *Lin. Alg. Appl.* 24:75–83, 1979.

[JMO97] C.R. Johnson, J.S. Maybee, D.D. Olesky, and P. van den Driessche, Nested sequences of principal minors and potential stability, *Lin. Alg. Appl.* 262:243–257, 1997.

[JS89] C.R. Johnson and T.A. Summers, The potentially stable tree sign patterns for dimensions less than five, *Lin. Alg. Appl.* 126:1–13, 1989.

[JW98] C.R. Johnson and C. Waters, Sign patterns occuring in orthogonal matrices, *Lin. Multilin. Alg.* 44:287–299, 1998.

[Kas63] P.W. Kasteleyn, Dimer statistics and phase transitions, *J. Math. Phys.* 4:287–293, 1963.

[KS00] S. Kim and B.L. Shader, Linear systems with signed solutions, *Lin. Alg. Appl.* 313:21–40, 2000.

[KMT96] S.J. Kirkland, J.J. McDonald, and M.J. Tsatsomeros, Sign patterns which require a positive eigenvalue, *Lin. Multilin. Alg.* 41:199-210, 1996.

[KLM84] V. Klee, R. Ladner, and R. Manber, Sign-solvability revisited, *Lin. Alg. Appl.* 59:131–147, 1984.

[KD77] V. Klee and P. van den Driessche, Linear algorithms for testing the sign stability of a matrix and for finding Z-maximum matchings in acyclic graphs, *Numer. Math.* 28:273–285, 1977.

[LLM95] G. Lady, T. Lundy, and J. Maybee, Nearly sign-nonsingular matrices, *Lin. Alg. Appl.* 220:229–248, 1995.

[LC00] G.-Y. Lee and G.-S. Cheon, A characterization of nearly sign-central matrices, *Bull. Korean Math. Soc.* 37:771–778, 2000.

[LLS97] G.-Y. Lee, S.-G. Lee, and S.-Z. Song, Linear operators that strongly preserve the sign-central matrices, *Bull. Korean Math. Soc.* 34:51–61, 1997.

[LMS00] G.Y. Lee, J.J. McDonald, B.L. Shader, and M.J. Tstsomeros, Extremal properties of ray-nonsingular matrices, *Discrete Math.* 216:221–233, 2000.

[LMS04] C.-K. Li, T. Milligan, and B.L. Shader, Non-existence of 5×5 full ray-nonsingular matrices, *Elec. J. Lin. Alg.* 11:212–240, 2004.

[LG01] J. Li and Y. Gao, The structure of tripotent sign pattern matrices, *Appl. Math. J. of Chin. Univ. Ser. B* 16(1):1–7, 2001.

[LH02] Z. Li and L. Harris, Sign patterns that require all distinct eigenvalues, *JP J. Alg. Num. Theory Appl.* 2:161–179, 2002.

[LEH96] Z. Li, C.A. Eschenbach, and F.J. Hall, The structure of nonnegative cyclic matrices, *Lin. Multilin. Alg.* 41:23–33, 1996

[LHE94] Z. Li, F.J. Hall, and C.A. Eschenbach, On the period and base of a sign pattern matrix, *Lin. Alg. Appl.* 212/213:101–120, 1994.

[LHS02] Z. Li, F.J. Hall, and J.L. Stuart, Irreducible powerful ray pattern matrices, *Lin. Alg. Appl.* 342:47–58, 2002.

[LHS05] Z. Li, F.J. Hall, and J.L. Stuart, Reducible powerful ray pattern matrices, *Lin. Alg. Appl.* 399:125–140, 2005.

[LHZ97] Z. Li, F.J. Hall, and F. Zhang, Sign patterns of nonnegative normal matrices, *Lin. Alg. Appl.* 254:335–354, 1997.

[Lim93] C.C. Lim, Nonsingular sign patterns and the orthogonal group, *Lin. Alg. Appl.* 184:1–12, 1993.

[LOD02] Q. Lin, D.D. Olesky, and P. van den Driessche, The distance of potentially stable sign patterns to the unstable matrices, *SIAM J. Matrix Anal. Appl.* 24:356–367, 2002.

[Log92] D. Logofet, *Matrices and Graphs: Stability Problems in Mathematical Ecology*, CRC Press, Boca Raton, FL, 1992.

[LMO96] T. Lundy, J.S. Maybee, D.D. Olesky, and P. van den Driessche, Spectra and inverse sign patterns of nearly sign-nonsingular matrices, *Lin. Alg. Appl.* 249:325–339, 1996.

[LMV96] T.J. Lundy, J. Maybee, and J. Van Buskirk, On maximal sign-nonsingular matrices, *Lin. Alg. Appl.* 247:55–81, 1996.

[Man82] R. Manber, Graph-theoretic approach to qualitative solvability of linear systems, *Lin. Alg. Appl.* 48:457–470, 1982.

[MOT97] J.J. McDonald, D.D. Olesky, M. Tsatsomeros, and P. van den Driessche, Ray patterns of matrices and nonsingularity, *Lin. Alg. Appl.* 267:359–373, 1997.

[MOT03] J.J. McDonald, D.D. Olesky, M. Tsatsomeros, and P. van den Driessche, On the spectra of striped sign patterns, *Lin. Multilin. Alg.* 51:39–48, 2003.

[ML02] Z. Miao and J. Li, Inertially arbitrary $(2r - 1)$-diagonal sign patterns, *Lin. Alg. Appl.* 357:133–141, 2002.

[QR65] J. Quirk and R. Ruppert, Qualitative economics and the stability of equilibrium, *Rev. Econ. Stud.* 32:311–326, 1965.

[Sam47] P.A. Samuelson, *Foundations of Economic Analysis*, Harvard University Press, Cambridge, MA, 1947, Atheneum, New York, 1971.

[Sha95] B.L. Shader, Least squares sign-solvability, *SIAM J. Matrix Anal. Appl.* 16 (4);1056–1073, 1995.

[Sha98] B.L. Shader, Sign-nonsingular matrices and orthogonal sign-patterns, *Ars Combin.* 48:289–296, 1998.

[SS04] H. Shan and J. Shao, Matrices with totally signed powers, *Lin. Alg. Appl.* 376:215–224, 2004.

[Sha99] J. Shao, On sign inconsistent linear systems, *Lin. Alg. Appl.* 296:245–257, 1999.

[Sha00] J. Shao, On the digraphs of sign solvable linear systems, *Lin. Alg. Appl.* 331:115–126, 2000.

[SR04] J. Shao and L. Ren, Some properties of matrices with signed null spaces, *Discrete Math.* 279:423–435, 2004.

[SS01] J. Shao and H. Shan, Matrices with signed generalized inverses, *Lin. Alg. Appl.* 322:105–127, 2001.

[SS02] J. Shao and H. Shan, The solution of a problem on matrices having signed generalized inverses, *Lin. Alg. Appl.* 345:43–70, 2002.

[SS05] J. Shao and H. Shan, The determinantal regions of complex sign pattern matrices and ray pattern matrices, *Lin. Alg. Appl.* 395:211–228, 2005.

[SG03] Y. Shao and Y. Gao, Sign patterns that allow diagonalizability, *Lin. Alg. Appl.* 359:113–119, 2003.

[SSG04] Y. Shao, L. Sun, and Y. Gao, Inertia sets of two classes of symmetric sign patterns, *Lin. Alg. Appl.* 384:85–95, 2004.

[SEK99] J. Stuart, C. Eschenbach, and S. Kirkland, Irreducible sign k-potent sign pattern matrices, *Lin. Alg. Appl.* 294:85–92, 1999.

[Stu99] J. Stuart, Reducible sign k-potent sign pattern matrices, *Lin. Alg. Appl.* 294:197–211, 1999.

[Stu03] J. Stuart, Reducible pattern k-potent ray pattern matrices, *Lin. Alg. Appl.* 362:87–99, 2003.

[SBS02] J. Stuart, L. Beasley, and B. Shader, Irreducible pattern k-potent ray pattern matrices, *Lin. Alg. Appl.* 346:261–271, 2002.

[Tho86] C. Thomassen, Sign-nonsingular matrices and even cycles in directed graphs, *Lin. Alg. Appl.* 75:27–41, 1986.

[Wat96] C. Waters, Sign pattern matrices that allow orthogonality, *Lin. Alg. Appl.* 235:1–13, 1996.

[Yeh96] L. Yeh, Sign patterns that allow a nilpotent matrix, *Bull. Aust. Math. Soc.* 53:189–196, 1996.

34

Multiplicity Lists for the Eigenvalues of Symmetric Matrices with a Given Graph

Charles R. Johnson
College of William and Mary

António Leal Duarte
Universidade de Coimbra

Carlos M. Saiago
Universidade Nova de Lisboa

This chapter assumes basic terminology from graph theory in Chapter 28; a good general graph theory reference is [CL96]. For standard terms or concepts from matrix analysis, see Part 1: Basic Linear Algebra, particularly Chapter 4.3, and Chapter 8; a good general matrix reference is [HJ85]. As we will be interested in properties of A that are permutation similarity invariant, primarily eigenvalues and their multiplicities, we will generally view a graph as unlabeled, except when referencing by labels is convenient.

For a given simple graph G on n vertices, let $\mathcal{S}(G)$ (respectively, $\mathcal{H}(G)$) denote the set of all $n \times n$ real symmetric (respectively, complex Hermitian) $n \times n$ matrices $A = [a_{ij}]$ such that for $i \neq j$, $a_{ij} \neq 0$ if and only if there is an edge between i and j.

Our primary interest lies in the following very general question. Given G, what are all the possible lists of multiplicities for the eigenvalues that occur among matrices in $\mathcal{S}(G)$ (respectively, $\mathcal{H}(G)$)? Much of our focus here is on the case in which $G = T$ is a tree.

It is important to distinguish two possible interpretations of "multiplicity list." Since the eigenvalues of a real symmetric or complex Hermitian matrix are real numbers, they may be placed in numerical order. If the multiplicities are placed in an order corresponding to the numerical order of the underlying eigenvalues, then we refer to such a way of listing the multiplicities as *ordered multiplicities*. If, alternatively, the multiplicities are simply listed in nonincreasing order of the values of the multiplicities themselves, we refer to such a list as *unordered multiplicities*. For example, if A has eigenvalues $-3, 0, 0, 1, 2, 2, 2, 5, 7$, the list of ordered multiplicities is $(1, 2, 1, 3, 1, 1)$, while the list of unordered multiplicities is $(3, 2, 1, 1, 1, 1)$. In either case, such a list means that there are exactly 6 different eigenvalues, of which 4 have multiplicity 1.

If a graph G is not connected, then the multiplicity lists for G may be deduced from those of its components via superposition. Also, graphs with many edges admit particularly rich collections of multiplicity lists. For example, the complete graph admits all multiplicity lists with the given number of eigenvalues, except the list in which all eigenvalues are the same. For these reasons, a natural beginning for the study of multiplicity lists for $\mathcal{S}(G)$ or $\mathcal{H}(G)$ is the case in which $G = T$, a tree. In addition, trees present several attractive features for this problem, so much of the research in this area, and this chapter, focuses on trees.

34.1 Multiplicities and Parter Vertices

Definitions:

Let G be a simple graph.

For an $n \times n$ real symmetric or complex Hermitian matrix $A = [a_{ij}]$, the **graph** of A, denoted by $G(A)$, is the simple graph on n vertices, labeled $1, 2, \ldots, n$, with an edge between i and j if and only if $a_{ij} \neq 0$.

Let $\mathcal{S}(G)$ (respectively, $\mathcal{H}(G)$), denote the set of all $n \times n$ real symmetric (respectively, complex Hermitian) matrices A such that $G(A) = G$ (where G has n vertices). No restriction is placed upon the diagonal entries of A by G, except that they are real.

If G' is the subgraph of G induced by β, $A(G')$ can be used to denote $A(\beta)$ and $A[G']$ to denote $A[\beta]$.

Given a tree T, the components of $T \setminus \{j\}$ are called **branches** of T at j.

If $A \in \mathcal{H}(G)$, $\lambda \in \sigma(A)$, and j is an index such that $\alpha_{A(j)}(\lambda) = \alpha_A(\lambda) + 1$, then j is called a **Parter index** or **Parter vertex** (for λ, A and G) (where $\alpha_A(\lambda)$ denotes the multiplicity of λ). Some authors refer to such a vertex as a **Parter–Wiener vertex** or a **Wiener vertex**.

If j is a Parter vertex for an eigenvalue λ of an $A \in \mathcal{H}(G)$ such that λ occurs as an eigenvalue of at least three direct summands of $A(j)$, j is called a **strong Parter vertex**.

A **downer vertex** i in a graph G (for $\lambda \in \sigma(A)$ and $A \in \mathcal{H}(G)$) is a vertex i such that $\alpha_{A(i)}(\lambda) = \alpha_A(\lambda) - 1$.

A **downer branch** of a tree T at j is a branch T_i at j, determined by a neighbor i of j such that i is a downer vertex in T_i (for λ and $A[T_i]$).

If a branch of a tree at a vertex j is a path and the neighbor of j in this branch is a pendant vertex of this path, the branch is a **pendant path** at j.

Facts:

1. If $A \in \mathcal{H}(G)$, then trivially $A(i) \in \mathcal{H}(G \setminus \{i\})$.
2. [HJ85] (Interlacing Inequalities) If an $n \times n$ Hermitian matrix A has eigenvalues $\lambda_1 \leq \lambda_2 \leq \cdots \leq \lambda_n$ and $A(i)$ has eigenvalues $\beta_{i,1} \leq \beta_{i,2} \leq \cdots \leq \beta_{i,n-1}$, then $\lambda_1 \leq \beta_{i,1} \leq \lambda_2 \leq \beta_{i,2} \leq \cdots \leq \beta_{i,n-1} \leq \lambda_n$, $i = 1, \ldots, n$.
3. If λ is an eigenvalue of an Hermitian matrix A, then $\alpha_A(\lambda) - 1 \leq \alpha_{A(i)}(\lambda) \leq \alpha_A(\lambda) + 1$ for $i = 1, \ldots, n$.
4. If T is a tree, then any matrix of $\mathcal{H}(T)$ is diagonally unitarily similar to one in $\mathcal{S}(T)$.
5. [JLS03a] (Parter–Wiener Theorem: Generalization) Let T be a tree and $A \in \mathcal{S}(T)$. Suppose that there exists an index i and a real number λ such that $\lambda \in \sigma(A) \cap \sigma(A(i))$. Then
   - There is in T a Parter vertex j for λ.
   - If $\alpha_A(\lambda) \geq 2$, then j may be chosen so that $\delta_T(j) \geq 3$ and so that there are at least three components T_1, T_2, and T_3 of $T \setminus \{j\}$ such that $\alpha_{A[T_k]}(\lambda) \geq 1$, $k = 1, 2, 3$.
   - If $\alpha_A(\lambda) = 1$, then j may be chosen so that there are two components T_1 and T_2 of $T \setminus \{j\}$ such that $\alpha_{A[T_k]}(\lambda) = 1$, $k = 1, 2$.
6. [JLS03a] For $A \in \mathcal{S}(T)$, T a tree, j is a Parter vertex for λ if and only if there is a downer branch at j for λ.

7. [JL] Suppose that G is a simple graph on n vertices that is not a tree. Then
   - There is a matrix $A \in \mathcal{S}(G)$ with an eigenvalue λ such that there is an index j so that $\alpha_A(\lambda) = \alpha_{A(j)}(\lambda) = 1$ and $\alpha_{A(i)}(\lambda) \leq 1$ for every $i = 1, \ldots, n$.
   - There is a matrix $B \in \mathcal{S}(G)$ with an eigenvalue λ such that $\alpha_B(\lambda) \geq 2$ and $\alpha_{B(i)}(\lambda) = \alpha_B(\lambda) - 1$, for every $i = 1, \ldots, n$.
8. [JLSSW03] Let T be a tree and $\lambda_1 < \lambda_2$ be eigenvalues of $A \in \mathcal{S}(T)$ that share a Parter vertex in T. Then there is at least one $\lambda \in \sigma(A)$ such that $\lambda_1 < \lambda < \lambda_2$.
9. Let T be a tree and $A \in \mathcal{S}(T)$. If T' is a pendant path of a vertex j of T, then T' is a downer branch for each eigenvalue of the direct summand $A[T']$.
10. Let T be a path and $A \in \mathcal{S}(T)$. Then each eigenvalue of A has multiplicity 1.
11. Let A be an $n \times n$ irreducible real symmetric tridiagonal matrix. Then
    - A has distinct eigenvalues.
    - In $A(i)$, there are at most $\min\{i-1, n-i\}$ interlacing equalities and this number may occur.
    - For each interlacing equality that does occur, the relevant eigenvalues must be an eigenvalue (of multiplicity 1) of both irreducible principal submatrices of $A(i)$.

    See Example 3 below.

Examples:

1. In general, one might expect that in passing from A to $A(i)$, multiplicities typically decline. However, Fact 5 is counter to this intuition in the case for trees. A rather complete statement has evolved through a series of papers ([Par60], [Wie84], [JLS03a]). In particular, Fact 5 says that when T is a tree and $\alpha_A(\lambda) \geq 2$, there must be a strong Parter vertex because, by interlacing, the hypothesis $\lambda \in \sigma(A) \cap \sigma(A(i))$ must be satisfied for any i. However, i itself need not be a Parter vertex. Even when $\alpha_A(\lambda) \geq 2$, it can happen that $\alpha_{A(j)}(\lambda) = \alpha_A(\lambda) + 1$ with $\delta_T(j) = 1$ or $\delta_T(j) = 2$ or λ appears in only one or two components of $T \setminus \{j\}$, even if $\delta_T(j) \geq 3$. There may, as well, be several Parter vertices and even several strong Parter vertices. Much information about Parter vertices may be found in [JLSSW03] and [JLS03a]. Let $\lambda, \mu \in \mathbb{R}$, $\lambda \neq \mu$, and consider real symmetric matrices whose graphs are the following trees, assuming that every diagonal entry corresponds to the label of the corresponding vertex.
   - The vertex v is a Parter vertex for λ in real symmetric matrices for which the graph is each of the trees in Figure 34.1. We also note that, depending on the tree T, several different vertices of T could be Parter for an eigenvalue of the same matrix in $\mathcal{S}(T)$. The matrices $A[T_1]$ and $A[T_2]$ each have u and v as Parter vertices for λ.

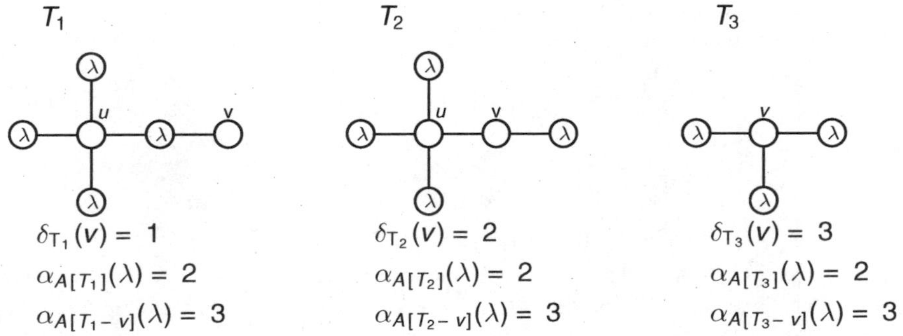

FIGURE 34.1 Examples of Parter vertices.

- Also, depending on the tree T, the same vertex could be a Parter vertex for different eigenvalues of a matrix in $\mathcal{S}(T)$. The vertex v is a Parter vertex for λ and μ in a real symmetric matrix A for which the graph is the tree in Figure 34.2. Such a matrix A has λ and μ as eigenvalues with $\alpha_A(\lambda) = 2 = \alpha_A(\mu)$. Since it is clear that we have $\alpha_{A(v)}(\lambda) = 3 = \alpha_{A(v)}(\mu)$, it means that v is Parter for λ and μ.

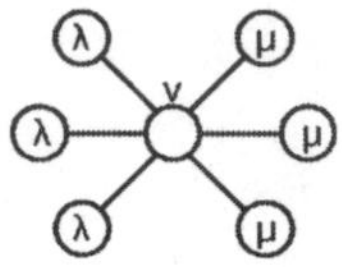

FIGURE 34.2 Vertex v is a Parter vertex for λ and μ.

2. Though a notion of "Parter vertex" can be defined for nontrees, Fact 7 is the converse to Fact 5 that shows that its remarkable conclusions are generally valid only for trees.
   Consider the matrix J_3 whose graph is the cycle C_3 (the possible multiplicities for the eigenvalues of a matrix whose graph is a cycle was studied in [Fer80]), which is not a tree. The matrix J_3 has eigenvalues 0, 0, 3. Since the removal of any vertex from C_3 leaves a path, we conclude that there is no Parter vertex for the multiple eigenvalue 0.
3. If a graph G is a path on n vertices, then G is a tree and if the vertices are labeled consecutively, any matrix in $\mathcal{S}(G)$ is an irreducible tridiagonal matrix. Conversely, the graph of an irreducible real symmetric tridiagonal matrix is a path. The very special spectral structure of such matrices has been of interest for some time for a variety of reasons. Two well-known classical facts are that all eigenvalues are distinct (i.e., all 1s is the only multiplicity list) and, if a pendant vertex is deleted, the interlacing inequalities are strict. Both statements follow from Fact 5, but more can be gotten from Fact 5 as well. If A is $n \times n$ real symmetric and $1 \leq i \leq n$, then as many as $n - 1$ of the eigenvalues of $A(i)$ might coincide with some eigenvalue of A. We refer to such an occurrence as an "interlacing equality." If a pendant vertex is removed from a path, no interlacing equalities can occur, but if an interior vertex is removed, interlacing equalities can occur. The complete picture in this regard may be also be deduced from Fact 5.

34.2 Maximum Multiplicity and Minimum Rank

Definitions:

Let G be a simple graph.

The **maximum multiplicity** of G, $M(G)$, is the maximum multiplicity for a single eigenvalue among matrices in $\mathcal{S}(G)$.

The **minimum rank** of G is $\text{mr}(G) = \min_{A \in \mathcal{S}(G)} \text{rank}(A)$.

The **path cover number** of G, $P(G)$, is the minimum number of induced paths of G that do not intersect, but do cover all vertices of G.

$\Delta(T) = \max[p - q]$ over all ways in which q vertices may be deleted from G, so as to leave p paths. Isolated vertices count as (degenerate) paths.

The **maximum rank deficiency** for G is $m(G) = n - mr(G)$, where the order of G is n.

Facts:

Let G be a simple connected graph of order n.

1. If G is a path, $P(G) = 1$; otherwise $P(G) > 1$.
2. There may be many minimum path covers. See Example 2.
3. Maximizing sets of removed vertices (used in the computation of $\Delta(G)$) are not unique; even the number of removed vertices is not unique. See Example 3.

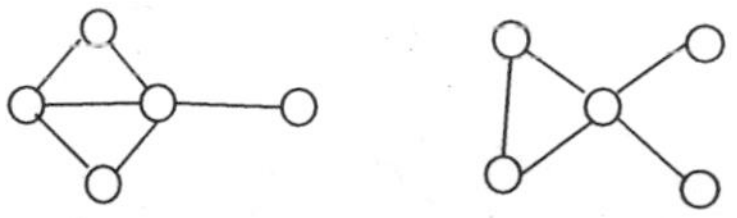

FIGURE 34.3 Two of the forbidden graphs for $\mathrm{mr}(G) = 2$.

4. [Fie69] $M(G) = 1$ if and only if G is a path.
5. $M(G) = n - 1$ if and only if G is the complete graph K_n.
6. $M(G) = m(G)$.
7. [JL99] For a tree T, $M(T) = \Delta(T) = P(T) = m(T)$.
8. [JS02] For any tree T, let H_T denote the subgraph of T induced by the vertices of degree at least 3. For a tree T, $\Delta(T)$ can be computed by the following algorithm. See Example 4.

Algorithm 1: Computation of $\Delta(T)$
Given a tree T.

1. Set $Q = \emptyset$ and $T' = T$.
2. While $H_{T'} \neq \emptyset$:
   Remove from T' all vertices v of $H_{T'}$ such that $\delta_{T'}(v) - \delta_{H_{T'}}(v) \geq 2$
   and add these vertices to Q.
3. $\Delta(T) = p - |Q|$ where p is the number of components (all of which are paths) in $T \backslash Q$.

9. [JL99], [BFH04] $\Delta(G) \leq M(G)$ and $\Delta(G) \leq P(G)$.
10. [BFH04], [BFH05] If $n \geq 2$, $P(K_n) < M(K_n)$, where K_n is the complete graph on n vertices. If G is unicyclic (i.e., has a unique cycle), then $P(G) \geq M(G)$ and strict inequality is possible.
11. [BFH04] Minimum rank (and, thus, maximum multiplicity) of a graph with a cut vertex can be computed from the minimum ranks of induced subgraphs.
12. [BHL04] If H is an induced subgraph of G, then $\mathrm{mr}(H) \leq \mathrm{mr}(G)$. Furthermore, $\mathrm{mr}(G) =2$ (i.e., $M(G) = n - 2$) if and only if G does not contain as an induced subgraph one of the following four forbidden graphs: the path on 4 vertices P_4, the complete tripartite graph $K_{3,3,3}$, the two graphs shown in Figure 34.3. Other characterizations are also given.

Examples:

1. Considering the tree T in Figure 34.4, we have $P(T) = 2$ (e.g., 1-3-2 and 5-4-6 constitute a minimal path cover of the vertices) and, of course, $\Delta(T) = 2$, as removal of vertex 4 leaves the 3 paths 1-3-2, 5, and 6 (and neither can be improved upon). Note that if submatrices $A[\{1,2,3\}]$, $A[\{5\}]$, and $A[\{6\}]$ of $A \in \mathcal{S}(T)$ are constructed so that λ is an eigenvalue of each (this is always possible and no higher multiplicity in any of them is possible), then $\alpha_A(\lambda) \geq 3 - 1 = 2$, which is the maximum possible.

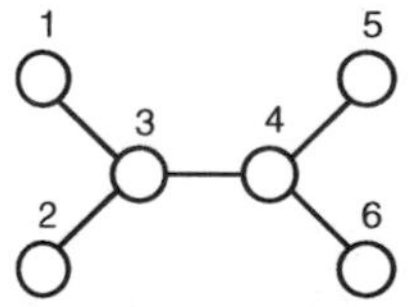

FIGURE 34.4 A tree with path cover number 2.

2. Consider the tree T on 12 vertices in Figure 34.5. It is not difficult to see that the path cover number of T, $P(T)$, is 4. However, it can be achieved by different collections of paths. For example, $P(T)$ can be achieved from the collection of 4 paths of T, 1-2-3, 4-5-6, 7-8 and 12-9-10-11. Similarly,

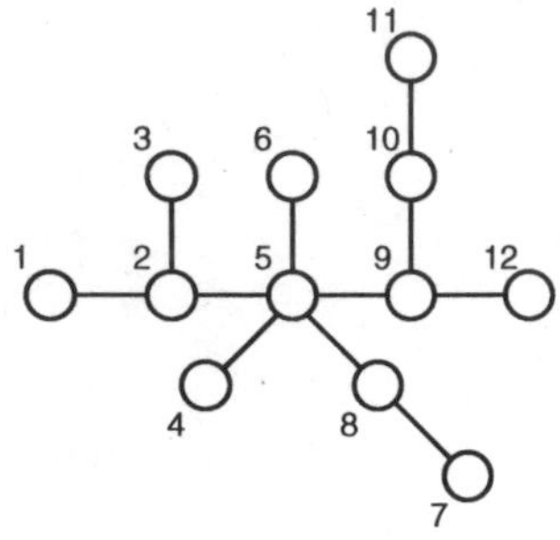

FIGURE 34.5 A tree with path cover number 4.

the paths of T, 1-2-3, 4-5-8-7, 6 and 12-9-10-11, form a collection of vertex disjoint paths (each one is an induced subgraph of T) that cover all the vertices of T.

3. Consider the tree T on 12 vertices in Figure 34.5. As we can see in Table 34.1, $\Delta(T)$ can be achieved for $q = 1, 2, 3$. When $q = 2$, there are 3 different sets of vertices whose removal from T leaves 6 components (paths), i.e., $p - q = 4$.
4. The algorithm in Fact 8 applied to the tree T in Figure 34.6 gives, in one step, a subset of vertices of T, $Q = \{v_2, v_3, v_4, v_5\}$, with cardinality 4 and such that $T \backslash Q$ has 13 components, each of which is a path. Therefore, $\Delta(T) = 13 - 4 = 9$.

   Note that, in any stage of the process to determine $\Delta(T)$, we may not choose just any vertex with degree greater than or equal to 3.

TABLE 34.1 $\Delta(T)$ for the tree T in Figure 34.5

Removed Vertices from T	q	p	$p - q$ $(= \Delta(T))$
5	1	5	4
5, 2	2	6	4
5, 9	2	6	4
5, 10	2	6	4
2, 5, 9	3	7	4
2, 5, 10	3	7	4

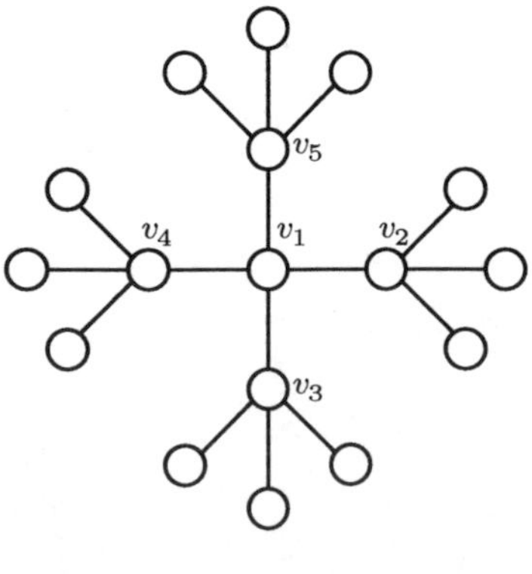

FIGURE 34.6 A tree T with $\Delta(T) = 9$.

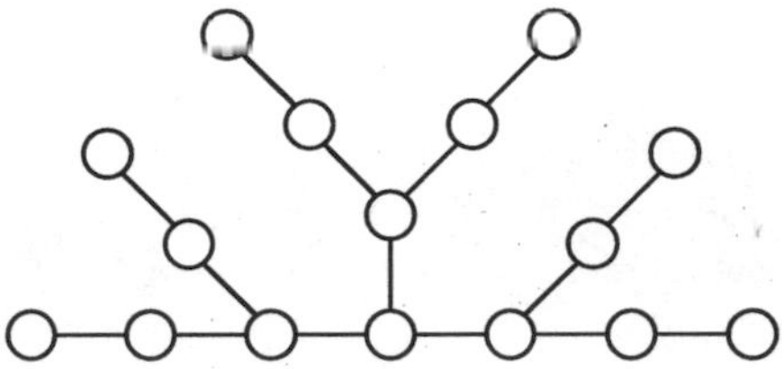

FIGURE 34.7 A tree of diameter 6 for which the minimum number of distinct eigenvalues is $8 > 6 + 1$.

34.3 The Minimum Number of Distinct Eigenvalues

Definitions:

Let T be a tree.

The **diameter** of T, $d(T)$, is the maximum number of edges in a path occurring as an induced subgraph of T.

The **minimum number of distinct eigenvalues** of T, $\mathcal{N}(T)$, is the minimum, among $A \in \mathcal{S}(T)$, number of distinct eigenvalues of A.

Facts:

1. [JL02a] Let T be a tree. Then $\mathcal{N}(T) \geq d(T) + 1$.
2. [JSa2] If T is a tree such that $d(T) < 5$, then there exist matrices in $\mathcal{S}(T)$ attaining as few distinct eigenvalues as $d(T) + 1$.

Examples:

1. Since each entry in a multiplicity list represents a distinct eigenvalue, the "length" of a list represents the number of different eigenvalues. This number can be as large as n (the number of vertices), of course, but it cannot be too small. Restrictions upon length limit the possible multiplicity lists. Just as a path has many distinct eigenvalues, a long (chordless) path occurring as an induced subgraph of a tree forces a large number of distinct eigenvalues.
2. For many trees T, there exist matrices in $\mathcal{S}(T)$ attaining as few distinct eigenvalues as $d(T) + 1$. However, for the tree T in Figure 34.7, $d(T) = 6$ and, in [BF04], the authors have shown that $\mathcal{N}(T) = 8 > d(T)+1$. It is not known how to deduce the minimum number of distinct eigenvalues from the structure of the tree, in general.

34.4 The Number of Eigenvalues Having Multiplicity 1

Definitions:

Given a tree T, let $\mathcal{U}(T)$ be the minimum number of 1s among multiplicity lists occurring for T.

Facts:

1. [JLS03a] For any tree T, the largest and smallest eigenvalues of any $A \in \mathcal{S}(T)$ necessarily have multiplicity 1.
2. [JLS03a] For any tree T on $n \geq 2$ vertices, $\mathcal{U}(T) \geq 2$ and, for each n, there exist trees T for which $\mathcal{U}(T) = 2$.
3. Let T be a tree on n vertices. $\mathcal{U}(T) \geq 2\mathcal{N}(T) - n$. In particular, $\mathcal{U}(T) \geq 2(d(T) + 1) - n$.

Examples:

1. As with the length of lists, it is relatively easy to have many 1s in a multiplicity list. The more interesting issue is how few of 1s may occur among lists for a given tree T. It certainly depends upon the tree, as the star (see Figure 34.8) may have just two 1s, while a path (see Figure 34.9) always has as many as the number of vertices.

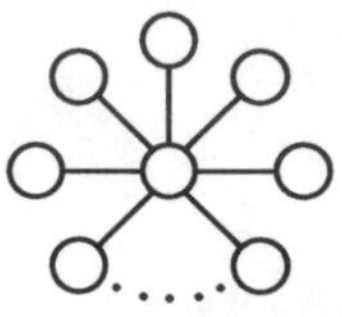

FIGURE 34.8 A star.

2. If T has a diameter that is large relative to its number of vertices (a path is an extremal example), then it may have to have a minimum number of distinct eigenvalues, which forces $\mathcal{U}(T)$ to be much greater than 2. However, $\mathcal{U}(T)$ may be greater than 2 for other reasons. For example, for the tree T in Figure 34.10, $d(T) = 4$, $n = 8$, but $\mathcal{U}(T) = 3$. It is not known how $\mathcal{U}(T)$ is determined by T, and it appears to be quite subtle.

34.5 Existence/Construction of Trees with Given Multiplicities

Definitions:

For a given graph G, the collection of all **multiplicity lists** is denoted by $\mathcal{L}(G)$. If it is not clear from the context, we will distinguish the unordered lists as $\mathcal{L}_u(G)$ from the ordered lists $\mathcal{L}_o(G)$.

General Inverse Eigenvalue Problem (GIEP) for $\mathcal{S}(T)$: Given a vertex v of a tree T, what are all the sequences of real numbers that may occur as eigenvalues of A and $A(v)$, as A runs over $\mathcal{S}(T)$?

Inverse Eigenvalue Problem (IEP) for $\mathcal{S}(T)$: What are all possible spectra that occur among matrices in $\mathcal{S}(T)$, T being a tree?

A tree T has **equivalence of the ordered multiplicity lists and the IEP** if a spectrum occurs for some matrix in $\mathcal{S}(T)$ whenever it is consistent with some list of ordered multiplicities of $\mathcal{L}_o(T)$.

Facts:

1. [Lea89] Let T be a tree on n vertices and v be a vertex of T. Let $\lambda_1, \lambda_2, \ldots, \lambda_n$ and $\mu_1, \mu_2, \ldots, \mu_{n-1}$ be real numbers. If

$$\lambda_1 < \mu_1 < \lambda_2 < \cdots < \mu_{n-1} < \lambda_n,$$

then there exists a matrix A in $\mathcal{S}(T)$ with eigenvalues $\lambda_1, \lambda_2, \ldots, \lambda_n$, and such that, $A(v)$ has eigenvalues $\mu_1, \mu_2, \ldots, \mu_{n-1}$.

2. For any tree T on n vertices and any given sequence of n distinct real numbers, there exists a matrix

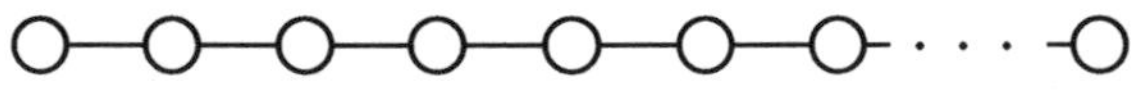

FIGURE 34.9 A path.

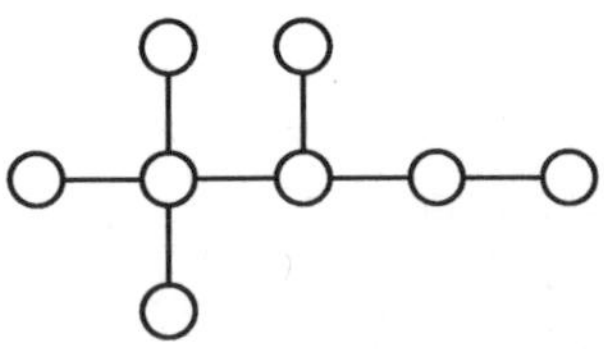

FIGURE 34.10 A tree T on 8 vertices with $d(T) = 4$ and $\mathcal{U}(T) = 3$.

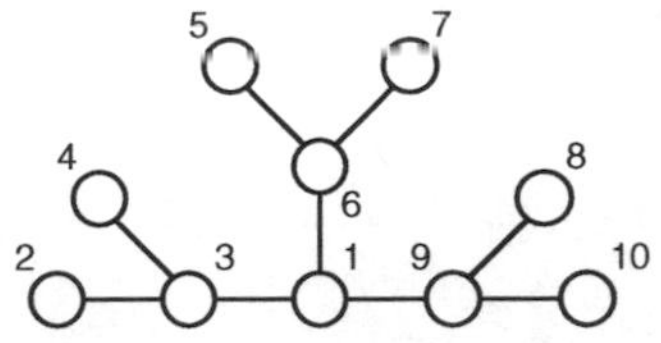

FIGURE 34.11 A tree for which there is not equivalence between ordered multiplicity lists and the IEP.

in $\mathcal{S}(T)$ having these numbers as eigenvalues.

3. Any path has equivalence of the ordered multiplicity lists and the IEP.
4. [BF04] There exists a tree for which the equivalence of the ordered multiplicity lists and the IEP is not verified. (See Example 1.)

Examples:

1. For remarkably many small trees and many of the families of trees to be discussed in the next section, the ordered multiplicity lists are equivalent to the IEP. This is not always the case. Extremal multiplicity lists for large numbers of vertices can force numerical relations upon the eigenvalues, as in the tree T shown in Figure 34.11.
   Let $\mathcal{A}$ be the adjacency matrix of T (i.e., the 0,1-matrix in $\mathcal{S}(T)$ with all diagonal entries 0). The Parter–Wiener Theorem (Fact 5 Section 34.1) guarantees that $\alpha_{\mathcal{A}}(0) = 4$, and likewise guarantees two eigenvalues of multiplicity 2 (the nonzero eigenvalues of $\mathcal{A}[\{2,3,4\}] = \mathcal{A}[\{5,6,7\}] = \mathcal{A}[\{8,9,10\}]$, so the ordered multiplicity list of $\mathcal{A}$ is $(1,2,4,2,1)$. In fact, direct computation shows that $\sigma(\mathcal{A}) = (-\sqrt{5}, -\sqrt{2}, -\sqrt{2}, 0, 0, 0, 0, \sqrt{2}, \sqrt{2}, \sqrt{5})$.
   However, it is not possible to prescribe arbitrary real numbers as the eigenvalues with this ordered multiplicity list. If $A = [a_{ij}] \in \mathcal{S}(T)$ has eigenvalues $\lambda_1 < \lambda_2 < \lambda_3 < \lambda_4 < \lambda_5$ with multiplicities $1, 2, 4, 2, 1$, respectively, then $\lambda_1 + \lambda_5 = \lambda_2 + \lambda_4$. The method used to establish this restriction comes from [BF04]. By the Parter–Wiener Theorem and an examination of subsets of vertices, $a_{11} = \lambda_3$. The restriction then follows from comparison of the traces of A and $A(1)$.
   It is not known for which trees the determination of all possible ordered multiplicity lists is equivalent to the solution of the IEP. Even when the two are not equivalent for some ordered list, they may be for all other ordered lists.
2. In the construction of multiplicity lists for a tree, it is often useful (and, perhaps necessary) to know the solution of the GIEP (or some weak form of it) for some of the subtrees of the tree.
   It is often more difficult (than giving necessary restrictions) to construct matrices $A \in \mathcal{S}(T)$ with a given, especially extremal, multiplicity list, even when that list does occur. There are three basic approaches besides ad hoc methods and computer assisted solution of equations. They are

   (a) Manipulation of polynomials, viewing the nonzero entries as variables and targeting a desired characteristic polynomial (see [Lea89] for an initial reference; this method, based on some nice formulas for the characteristic polynomial in the case of a tree (see, e.g., [Par60], [MOD89]), can be quite tedious for larger, more complicated trees).

   (b) Careful use of the implicit function theorem (initiated in [JSW]).

   (c) Division of the tree into understood parts and using the interlacing inequalities to give lower bounds that are forced to be attained by known constraints (this is along the lines of the brief discussion in Example 1 Section 34.2 involving $\Delta(T)$, but for larger trees can lead to complicated simultaneity conditions).

   As an example of method (c) and its subtleties (see also Example 2 Section 34.7), consider again the tree T in Figure 34.4. Since $P(T) = 2$, the maximum multiplicity is 2, and because $d(T) = 3$, there must be at least four distinct eigenvalues, two of which have multiplicity 1. This leaves the question of whether the list $(2, 2, 1, 1)$ (which would have to be the ordered list $(1, 2, 2, 1)$) can occur. It can, but this is the nontrivial example with smallest number of vertices. Suppose that the two multiple eigenvalues are λ and μ. We want $A \in \mathcal{S}(T)$ with $\alpha_A(\lambda) = 2$ and $\alpha_A(\mu) = 2$. Each must have a Parter vertex, which must be either vertex 3 or 4. One must be for λ (and not μ) and the other for μ (and not λ), as two consecutive eigenvalues cannot share a Parter vertex (Fact 8 section 34.1). So assume that 3 is Parter for λ and 4 for μ. Then, we must have $A[\{1\}] = \lambda = A[\{2\}]$ and $\lambda \in \sigma(A[\{4, 5, 6\}])$; and $A[\{5\}] = \mu = A[\{6\}]$ and $\mu \in \sigma(A[\{1, 2, 3\}])$. A calculation (or other methods) shows this can be achieved simultaneously.

34.6 Generalized Stars

Definitions:

A tree T in which there is at most one vertex of degree greater than two is a **generalized star**.

In a generalized star, a vertex v is a **central vertex** if its neighbors are pendant vertices of their branches, and each branch is a path. (Note that, under this definition, a path is a (degenerate) generalized star, in which any vertex is a central vertex. When referring to a path as a generalized star, one vertex has been fixed as the central vertex.)

For a central vertex v of a generalized star T, each branch of T at v is called an **arm** of T; the **lengths** of an arm are the number of vertices in the arm.

Supposing that v is a central vertex of a generalized star T, with $\delta_T(v) = k$. Denote by $T_1, \ldots, T_k$ its arms and by $l_1, \ldots, l_k$ the lengths of $T_1, \ldots, T_k$, respectively.

A **star** on n vertices is a tree in which there is a vertex of degree $n - 1$.

Let $u = (u_1, \ldots, u_b)$, $u_1 \geq \cdots \geq u_b$, and $v = (v_1, \ldots, v_c)$, $v_1 \geq \cdots \geq v_c$, be two nonincreasing partitions of integers M and N, respectively. If $M < N$, denote by u_e the partition of N obtained from u appending 1s to the partition u. Note that if $M = N$, then $u_e = u$.

Facts:

1. A star is trivially a generalized star.
2. If u and v are two nonincreasing partitions of integers M and N, respectively, $M \leq N$, such that $u_1 + \cdots + u_s \leq v_1 + \cdots + v_s$ for all s (interpreting u_s or v_s as 0 when s exceeds b or c, respectively), then trivially v majorizes u_e, denoted $u_e \preceq v$. See Preliminaries.
3. [JLS03b] Let T be a generalized star on n vertices with central vertex v of degree k, $l_1, \ldots, l_k$ be the lengths of the arms $T_1, \ldots, T_k$, and $f(x), g_1(x), \ldots, g_k(x)$ be monic polynomials with all their roots real in which $\deg f = n$, $\deg g_1 = l_1, \ldots, \deg g_k = l_k$. There exists $A \in \mathcal{S}(T)$ such that A has characteristic polynomial $f(x)$ and $A[T_i]$ has characteristic polynomial $g_i(x)$ if and only if
   - Each $g_i(x)$ has only simple roots.
   - If λ is a root of $g_1(x) \cdots g_k(x)$ of multiplicity $m \geq 1$, then λ is a root of $f(x)$ of multiplicity $m - 1$.
   - The roots of $f(x)$ that are not roots of $g_1(x) \cdots g_k(x)$ are simple and strictly interlace the set of roots of $g_1(x) \cdots g_k(x)$ (multiple roots counting only once).
4. [JL02b] Let T be a generalized star on n vertices with central vertex of degree s and arm lengths $l_1 \geq \cdots \geq l_s$. Then $(p_1, \ldots, p_r) \in \mathcal{L}_u(T)$ if and only if
   - $\sum_{i=1}^{r} p_i = n$.
   - $r \geq l_1 + l_2 + 1$.
   - $p_h = p_{h+1} = \cdots = p_r = 1$, in which $h = \lceil \frac{r+1}{2} \rceil$.
   - $(p_1, p_2, \ldots, p_{r-l_1-1}) \preceq (l_1^* - 1, \ldots, l_{l_1}^* - 1)$.

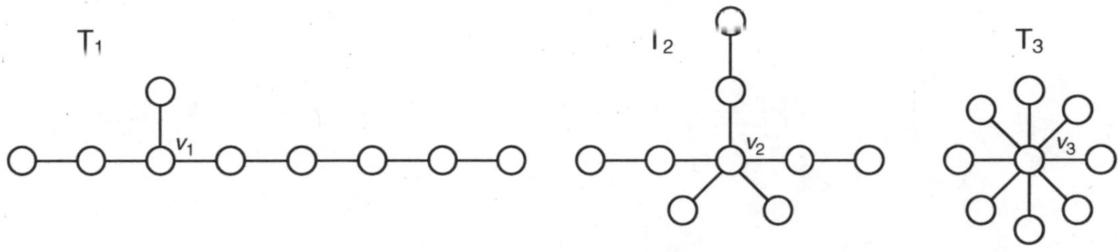

FIGURE 34.12 T_1, T_2, and T_3 are generalized stars on 9 vertices with central vertices v_1, v_2, and v_3, respectively.

5. [JLS03b] Let T be a generalized star on n vertices with central vertex of degree s and arm lengths $l_1 \geq \cdots \geq l_s$. Let $\lambda_1 < \cdots < \lambda_r$ be any sequence of real numbers. Then there exists a matrix $A \in \mathcal{S}(T)$ with distinct eigenvalues $\lambda_1 < \cdots < \lambda_r$ and list of ordered multiplicities $q = (q_1, \ldots, q_r)$ if and only if q satisfies the following conditions:
   - $\sum_{i=1}^{r} q_i = n$.
   - If $q_i > 1$, then $1 < i < r$ and $q_{i-1} = 1 = q_{i+1}$.
   - $(q_{i_1} + 1, \ldots, q_{i_h} + 1)_e \preceq (l_1, \ldots, l_s)^*$, in which $q_{i_1} \geq \cdots \geq q_{i_h}$ are the entries of the r-tuple $(q_1, \ldots, q_r)$ greater than 1.

   (That is, when T is a generalized star, there is equivalence of the ordered multiplicity lists and the IEP.)

Examples:

1. Let T_1, T_2, and T_3 be the generalized stars in Figure 34.12. We have

$$\begin{aligned}
\mathcal{L}_u(T_1) &= \{(1,1,1,1,1,1,1,1,1), (2,1,1,1,1,1,1,1)\},\\
\mathcal{L}_u(T_2) &= \{(1,1,1,1,1,1,1,1,1), (2,1,1,1,1,1,1,1), (2,2,1,1,1,1,1),\\
&\quad (3,1,1,1,1,1,1), (3,2,1,1,1,1), (3,3,1,1,1), (4,1,1,1,1,1),\\
&\quad (4,2,1,1,1)\},
\end{aligned}$$

and

$$\begin{aligned}
\mathcal{L}_u(T_3) &= \{(1,1,1,1,1,1,1,1,1), (2,1,1,1,1,1,1,1), (2,2,1,1,1,1,1),\\
&\quad (3,1,1,1,1,1,1), (3,2,1,1,1,1), (3,3,1,1,1), (4,1,1,1,1,1),\\
&\quad (4,2,1,1,1), (5,1,1,1,1), (6,1,1,1), (7,1,1)\}.
\end{aligned}$$

34.7 Double Generalized Stars

Definitions:

A **double generalized star** is a tree resulting from joining the central vertices of two generalized stars T_1 and T_2 by an edge. Such a tree will be denoted by $D(T_1, T_2)$.

A **double star** is a double generalized star $D(T_1, T_2)$ in which T_1 and T_2 are stars.

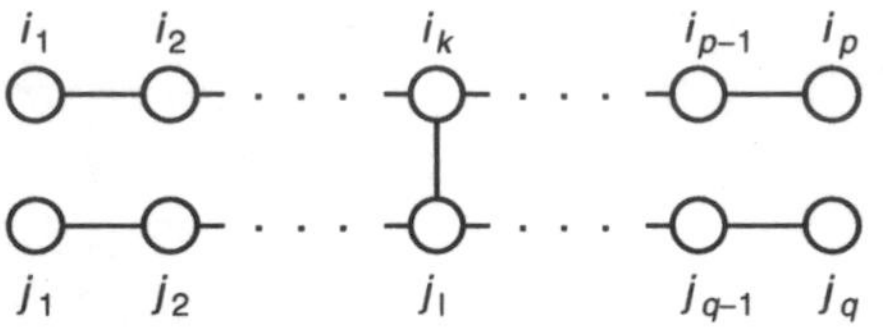

FIGURE 34.13 A double path.

A **double path** is a double generalized star $D(T_1, T_2)$ in which T_1 and T_2 are paths. When we refer to a double path T on $n = p + q$ vertices we suppose T is represented as in Figure 34.13, in which the only constraint on the connecting edge $\{i_k, j_l\}$ is that not both $k \in \{1, p\}$ and $l \in \{1, q\}$. The upper (i) path has $k - 1$ vertices to the left of the connecting vertex and another $p - k$ vertices to the right; set $s_1 = \min\{k - 1, p - k\}$, $s_2 = \min\{l - 1, q - l\}$, and $s = \min\{q, p, s_1 + s_2\}$.

Let G be a tree. Let v be a vertex of G of degree k and $G_1, \ldots, G_k$ be the components of $G \setminus \{v\}$ having order $l_1, \ldots, l_k$, respectively. To the tree G is **associated the generalized star**, $S_v(G)$, with central vertex v of degree k, and with arms $T_1, \ldots, T_k$ of lengths $l_1, \ldots, l_k$, respectively.

Let u_1 and u_2 be adjacent vertices of a tree G. Denote by G_{u_1} the connected component of $G \backslash \{u_2\}$ that contains u_1 and by G_{u_2} the connected component of $G \backslash \{u_1\}$ that contains u_2. Put $S_1 = S_{u_1}(G_{u_1})$ and $S_2 = S_{u_2}(G_{u_2})$. Now, to the tree G is **associated the double generalized star** $D(S_1, S_2)$, which is denoted by $D_{u_1,u_2}(G)$.

Given a vertex v of a tree T and an eigenvalue λ of a matrix $A \in \mathcal{S}(T)$, λ is an **upward eigenvalue** of A at v if $\alpha_{A(v)}(\lambda) = \alpha_A(\lambda) + 1$, and $\alpha_A(\lambda)$ is an **upward multiplicity** of A at v.

If $q = q(A) = (q_1, \ldots, q_r)$ is the list of ordered multiplicities of A, define the **list of upward multiplicities** of A at v, denoted by $\hat{q}$, as the list with the same entries as q but in which any upward multiplicity q_i of A at v is marked as $\hat{q}_i$ in $\hat{q}$.

Given a generalized star T with central vertex v, we denote by $\hat{\mathcal{L}}_o(T)$ the **set of all lists of upward multiplicities** at v occurring among matrices in $\mathcal{S}(T)$.

Facts:

1. A double path is a tree whose path cover number is 2.
2. [JLS03b] Let T be a generalized star on n vertices with central vertex v of degree k and arm lengths $l_1 \geq \cdots \geq l_k$. Let $\lambda_1 < \cdots < \lambda_r$ be any sequence of real numbers. Then there exists a matrix A in $\mathcal{S}(T)$ with distinct eigenvalues $\lambda_1 < \cdots < \lambda_r$ and a list of upward multiplicities $\hat{q} = (\widehat{q_1, \ldots, q_r})$ if and only if $\hat{q}$ satisfies the following conditions:

   - $\sum_{i=1}^{r} q_i = n$.
   - If q_i is an upward multiplicity in $\hat{q}$, then $1 < i < r$ and neither q_{i-1} nor q_{i-1} is an upward multiplicity in $\hat{q}$.
   - $(q_{i_1} + 1, \ldots, q_{i_h} + 1)_e \preceq (l_1, \ldots, l_k)^*$, in which $q_{i_1} \geq \cdots \geq q_{i_h}$ are the upward multiplicities of $\hat{q}$.

3. [JLS03b] (Superposition Principle) Let $D(T_1, T_2)$ be a double generalized star, $\hat{b} = (\widehat{b_1, \ldots, b_{s_1}}) \in \hat{\mathcal{L}}_o(T_1)$, and $\hat{c} = (\widehat{c_1, \ldots, c_{s_2}}) \in \hat{\mathcal{L}}_o(T_2)$. Construct any $b^+ = (b_1^+, \ldots, b_{s_1+t_1}^+)$ and $c^+ = (c_1^+, \ldots, c_{s_2+t_2}^+)$ subject to the following conditions:

   - $t_1, t_2 \in \mathbb{N}_0$ and $s_1 + t_1 = s_2 + t_2$.
   - b^+ (respectively, c^+) is obtained from $\hat{b}$ (respectively, $\hat{c}$) by inserting t_1 (respectively, t_2) 0s.
   - b_i^+ and c_i^+ cannot both be 0.
   - If $b_i^+ > 0$ and $c_i^+ > 0$, then at least one of the b_i^+ or c_i^+ must be an upward multiplicity of $\hat{b}$ or $\hat{c}$.

Then $b^+ + c^+ \in \mathcal{L}_o(D(T_1, T_2))$. Moreover, $a \in \mathcal{L}_o(D(T_1, T_2))$ if and only if there are $\hat{b} \in \hat{\mathcal{L}}_o(T_1)$, $\hat{c} \in \hat{\mathcal{L}}_o(T_2)$ such that $a = b^+ + c^+$.

4. Let T be a tree, v be a vertex of T, and v_1, v_2 be adjacent vertices of T. Then
   - $\mathcal{L}_u(S_v(T)) \subseteq \mathcal{L}_u(T)$, $\mathcal{L}_o(S_v(T)) \subseteq \mathcal{L}_o(T)$.
   - $\mathcal{L}_u(D_{v_1,v_2}(T)) \subseteq \mathcal{L}_u(T)$, $\mathcal{L}_o(D_{v_1,v_2}(T)) \subseteq \mathcal{L}_o(T)$.
5. [JL02b] Let T be a double path on $n = p + q$ vertices and suppose that $A \in \mathcal{S}(T)$. Then
   - The maximum multiplicity of an eigenvalue of A is 2.
   - The diameter of G is $\max\{p, q, p+q-(s_1+s_2)\} - 1$, so that A has at least $\max\{p, q, p+q-(s_1+s_2)\}$ distinct eigenvalues.
   - A has at most s multiplicity 2 eigenvalues.
   - The possible list of unordered multiplicities for T, $\mathcal{L}_u(T)$, consists of all partitions of $p+q$ into parts each one not greater than two and with at most s equal to 2.
   - Any list in $\mathcal{L}_u(T)$ has at least $n - 2s$ 1s.

Examples:

1. Let T_1 and T_2 be the stars in Figure 34.14 with central vertices v_1 and v_2, respectively, and G be the double star $D(T_1, T_2)$. By Fact 2, we have that

$$\hat{\mathcal{L}}_o(T_1) = \{(1, \hat{2}, 1), (1, \hat{1}, 1, 1), (1, 1, \hat{1}, 1), (1, 1, 1, 1)\}$$

and

$$\hat{\mathcal{L}}_o(T_2) = \{(1, \hat{1}, 1), (1, 1, 1)\}.$$

Applying the Superposition Principle (Fact 3) to the lists of upward multiplicities of T_1 and T_2, it follows that

$$\begin{aligned}\mathcal{L}_o(G) = \{&(1,3,2,1), (1,2,3,1), (1,3,1,1,1), (1,1,3,1,1), (1,1,1,3,1),\\ &(1,2,2,1,1), (1,2,1,2,1), (1,1,2,2,1), (1,2,1,1,1,1),\\ &(1,1,2,1,1,1), (1,1,1,2,1,1), (1,1,1,1,2,1), (1,1,1,1,1,1,1)\}.\end{aligned}$$

For example, $(1,3,2,1) \in \mathcal{L}_o(G)$ because $\hat{b} = (1, \hat{2}, 1) \in \hat{\mathcal{L}}_o(T_1)$, $\hat{c} = (1, \hat{1}, 1) \in \hat{\mathcal{L}}_o(T_2)$, and

$$(1,3,2,1) = b^+ + c^+ = (1, \hat{2}, 1, 0) + (0, 1, \hat{1}, 1).$$

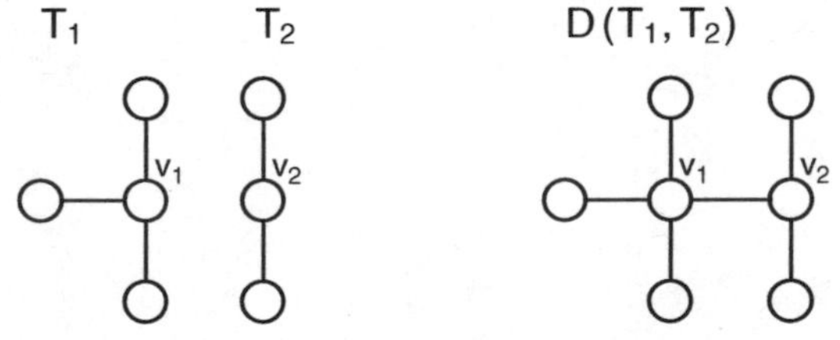

FIGURE 34.14 Stars and a double star.

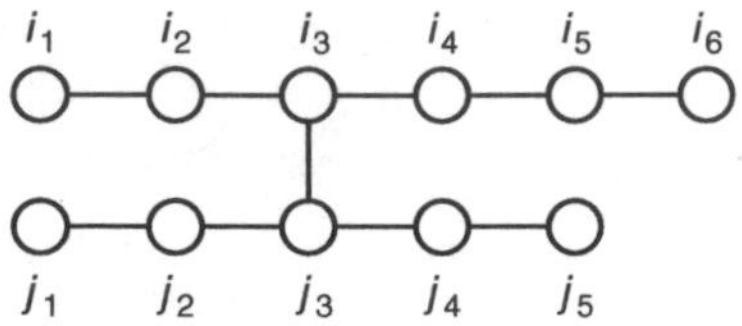

FIGURE 34.15 A double path on 11 vertices.

2. Consider the double path T on 11 vertices in Figure 34.15. Let T_1 be the path with vertices $i_1, \ldots, i_6$ and T_2 be the path with vertices $j_1, \ldots, j_5$ (subgraphs of T induced by the mentioned vertices). Because T_1 and T_2 are generalized stars with central vertices i_3 and j_3, respectively, from Fact 2 we conclude that $\hat{b} = (1, \hat{1}, 1, \hat{1}, 1, 1) \in \hat{\mathcal{L}}_o(T_1)$ and $\hat{c} = (1, \hat{1}, 1, \hat{1}, 1) \in \hat{\mathcal{L}}_o(T_2)$. Since, for example,

$$(1, 2, 2, 2, 2, 1, 1) = b^+ + c^+ = (0, 1, \hat{1}, 1, \hat{1}, 1, 1) + (1, \hat{1}, 1, \hat{1}, 1, 0, 0),$$

by the Superposition Principle, we conclude that $(1, 2, 2, 2, 2, 1, 1) \in \mathcal{L}_o(T)$ and, therefore, $(2, 2, 2, 2, 1, 1, 1) \in \mathcal{L}_u(T)$. We may construct a matrix $A \in \mathcal{S}(T)$ with list of multiplicities $(2, 2, 2, 2, 1, 1, 1)$ in the following way.

Pick real numbers $\lambda_1 > \mu_1 > \lambda_2 > \mu_2 > \lambda_3 > \mu_3 > \lambda_4$. Construct A_1 with graph T_1 such that A_1 has eigenvalues $\lambda_1, \mu_1, \lambda_2, \mu_2, \lambda_3, \lambda_4$ and such that the eigenvalues of $A_1[\{i_1, i_2\}]$ and $A_1[\{i_4, i_5, i_6\}]$ are μ_1, μ_2 and μ_1, μ_2, μ_3, respectively; construct A_2 with graph T_2 such that A_2 has eigenvalues $\mu_1, \lambda_2, \mu_2, \lambda_3, \mu_3$ and such that the eigenvalues of both $A_2[\{j_1, j_2\}]$ and $A_2[\{j_4, j_5\}]$ are λ_2, λ_3. According to Fact 3 section 34.6, these constructions are possible. Now construct A with graph T and such that $A[T_1] = A_1$ and $A[T_2] = A_2$. Then i_3 is a strong Parter vertex for μ_1, μ_2, while j_3 is a strong Parter for λ_2, λ_3 and so $(2, 2, 2, 2, 1, 1, 1)$ is the list of unordered multiplicities of A.

3. Regarding Fact 4, the results for generalized stars or double generalized stars may be extended to a general tree T by associating with T either a generalized star or a double generalized star according to the given definitions. (See also [JLS03b, Theorem 10] for a corresponding result for the GIEP.)

Note that, under our definition, there are many different possibilities to associate either a generalized star or a double generalized star to a given tree T, and so Fact 3 provides many possible lists for $\mathcal{L}(T)$. The natural question is to ask whether all the elements of $\mathcal{L}(T)$ can be obtained in this manner. The answer is no. It suffices to note that the path cover number of T will be, in general, strictly greater than that of either of $S_v(T)$ or $D_{v_1,v_2}(T)$ for any possible choice of v, v_1, and v_2. So any lists for which the maximum multiplicity occurs cannot generally be obtained from the inclusions in Fact 4. For example, the tree T in Figure 34.16 has path cover number 3, which is strictly greater than the maximum path cover number 2, of any generalized star or double generalized star associated with T.

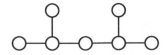

FIGURE 34.16 A tree with path cover number 3.

34.8 Vines

Definitions:

A **binary tree** is a tree in which no vertex has degree greater than 3.

A **vine** is a binary tree in which every degree 3 vertex is adjacent to at least one vertex of degree 1 and no two vertices of degree 3 are adjacent.

Facts:

1. [JSW] Let T be a vine on n vertices. The set $\mathcal{L}_u(T)$ consists of all sequences that are majorized by the sequence $s = (P(T), 1, \ldots, 1)$ (s being a partition of n).
   (The description of $\mathcal{L}_u(T)$ was given by using the implicit function theorem technique referred to in section 34.5.)

References

[BF04] F. Barioli and S.M. Fallat. On two conjectures regarding an inverse eigenvalue problem for acyclic symmetric matrices. *Elec. J. Lin. Alg.* 11:41–50 (2004).

[BFH04] F. Barioli, S. Fallat, and L. Hogben. Computation of minimal rank and path cover number for graphs. *Lin. Alg. Appl.* 392: 289–303 (2004).

[BFH05] F. Barioli, S. Fallat, and L. Hogben. On the difference between the maximum multiplicity and path cover number for tree-like graphs. *Lin. Alg. Appl.* 409: 13–31 (2005).

[BHL04] W.W. Barrett, H. van der Holst, and R. Loewy. Graphs whose minimal rank is two. *Elec. J. Lin. Alg.* 11:258–280 (2004).

[BL05] A. Bento and A. Leal-Duarte. On Fiedler's characterization of tridiagonal matrices over arbitrary fields. *Lin. Alg. Appl.* 401:467–481 (2005).

[BG87] D. Boley and G.H. Golub. A survey of inverse eigenvalue problems. *Inv. Prob.* 3:595–622 (1987).

[CL96] C. Chartrand and L. Lesniak. *Graphs & Digraphs.* Chapman & Hall, London, 1996.

[Chu98] M.T. Chu. Inverse eigenvalue problems. *SIAM Rev.* 40:1–39 (1998).

[CDS95] D. Cvetković, M. Doob, and H. Sachs. *Spectra of Graphs.* Johann Ambrosius Barth Verlag, Neidelberg, 1995.

[FP57] K. Fan and G. Pall. Imbedding conditions for Hermitian and normal matrices. *Can. J. Math.* 9:298–304 (1957).

[Fer80] W. Ferguson. The construction of Jacobi and periodic Jacobi matrices with prescribed spectra. *Math. Comp.* 35:1203–1220 (1980).

[Fie69] M. Fiedler. A characterization of tridiagonal matrices. *Lin. Alg. Appl.* 2:191–197 (1969).

[GM74] J. Genin and J. Maybee. Mechanical vibration trees. *J. Math. Anal. Appl.* 45:746–763 (1974).

[HJ85] R. Horn and C.R. Johnson. *Matrix Analysis.* Cambridge University Press, New York, 1985.

[JL99] C.R. Johnson and A. Leal-Duarte. The maximum multiplicity of an eigenvalue in a matrix whose graph is a tree. *Lin. Multilin. Alg.* 46:139–144 (1999).

[JL02a] C.R. Johnson and A. Leal-Duarte. On the minimum number of distinct eigenvalues for a symmetric matrix whose graph is a given tree. *Math. Inequal. Appl.* 5(2):175–180 (2002)

[JL02b] C.R. Johnson and A. Leal-Duarte. On the possible multiplicities of the eigenvalues of an Hermitian matrix whose graph is a given tree. *Lin. Alg. Appl.* 348:7–21 (2002).

[JL] C.R. Johnson and A. Leal-Duarte. Converse to the Parter–Wiener theorem: the case of non-trees. *Discrete Math.* (to appear).

[JLSSW03] C.R. Johnson, A. Leal-Duarte, C.M. Saiago, B.D. Sutton, and A.J. Witt. On the relative position of multiple eigenvalues in the spectrum of an Hermitian matrix with a given graph. *Lin. Alg. Appl.* 363:147–159 (2003).

[JLS03a] C.R. Johnson, A. Leal-Duarte, and C.M. Saiago. The Parter–Wiener theorem: refinement and generalization. *SIAM J. Matrix Anal. Appl.* 25(2):352–361 (2003).

[JLS03b] C.R. Johnson, A. Leal-Duarte, and C.M. Saiago. Inverse eigenvalue problems and lists of multiplicities of eigenvalues for matrices whose graph is a tree: the case of generalized stars and double generalized stars. *Lin. Alg. Appl.* 373:311–330 (2003).

[JLS] C.R. Johnson, R. Loewy, and P. Smith. The graphs for which the maximum multiplicity of an eigenvalue is two, (manuscript).

[JS02] C.R. Johnson and C.M. Saiago. Estimation of the maximum multiplicity of an eigenvalue in terms of the vertex degrees of the graph of a matrix. *Elect. J. Lin. Alg.* 9:27–31 (2002).

[JSa] C.R. Johnson and C.M. Saiago. The trees for which maximum multiplicity implies the simplicity of other eigenvalues. *Discrete Mathematics*, (to appear).

[JSb] C.R. Johnson and C.M. Saiago. Branch duplication for the construction of multiple eigenvalues in an Hermitian matrix whose graph is a tree. *Lin. Multilin. Alg.* (to appear).

[JS04] C.R. Johnson and B.D. Sutton. Hermitian matrices, eigenvalue multiplicities, and eigenvector components. *SIAM J. Matrix Anal. Appl.* 26(2):390–399 (2004).

[JSW] C.R. Johnson, B.D. Sutton, and A. Witt. Implicit construction of multiple eigenvalues for trees, (preprint).

[Lea89] A. Leal-Duarte. Construction of acyclic matrices from spectral data. *Lin. Alg. Appl.* 113:173–182 (1989).

[Lea92] A. Leal-Duarte. *Desigualdades Espectrais e Problemas de Existência em Teoria de Matrizes.* Dissertação de Doutoramento, Coimbra, 1992.

[MOD89] J.S. Maybee, D.D. Olesky, P. Van Den Driessche, and G. Wiener. Matrices, digraphs, and determinants. *SIAM J. Matrix Anal. Appl.* 10(4):500–519 (1989).

[Nyl96] P. Nylen. Minimum-rank matrices with prescribed graph. *Lin. Alg. Appl.* 248:303–316 (1996).

[Par60] S. Parter. On the eigenvalues and eigenvectors of a class of matrices. *J. Soc. Ind. Appl. Math.* 8:376–388 (1960).

[Sai03] C.M. Saiago. *The Possible Multiplicities of the Eigenvalues of an Hermitian Matrix Whose Graph Is a Tree.* Dissertação de Doutoramento, Universidade Nova de Lisboa, 2003.

[She04] D. Sher. Observations on the multiplicities of the eigenvalues of an Hermitian matrix with a tree graph. University of William and Mary, Research Experiences for Undergraduates program, summer 2004. (Advisor: C.R. Johnson)

[Wie84] G. Wiener. Spectral multiplicity and splitting results for a class of qualitative matrices. *Lin. Alg. Appl.* 61:15–29 (1984).

35

Matrix Completion Problems

Leslie Hogben
Iowa State University

Amy Wangsness
Fitchburg State College

A partial matrix is a rectangular array of numbers in which some entries are specified while others are free to be chosen. A completion of a partial matrix is a specific choice of values for the unspecified entries. A matrix completion problem asks whether a partial matrix (or family of partial matrices with a given pattern of specified entries) has a completion of a specific type, such as a positive definite matrix. In some cases, a "best" completion is sought.

Matrix completion problems arise in applications whenever a full set of data is not available, but it is known that the full matrix of data must have certain properties. Such applications include molecular biology and chemistry (see Chapter 60), seismic reconstruction problems, mathematical programming, and data transmission, coding, and image enhancement problems in electrical and computer engineering.

A matrix completion problem for a family of partial matrices with a given pattern of specified entries is usually studied by means of graphs or digraphs. If a pattern of specified entries does not always allow completion to the desired type of matrix, conditions on the entries that will allow such completion are sought. A question of finding the "best completion" often involves optimization techniques. Matrix completion results are usually constructive, with the result established by giving a specific construction for a completion.

In this chapter, we focus on completion problems involving classes of matrices that generalize the positive definite matrices, and emphasize graph theoretic techniques that allow completion of families of matrices. This chapter is organized by class of matrices, with the symmetric classes first. The authors also maintain a Web page containing updated information [HW].

Organizing the information by classes of matrices provides easy access to results about a particular class, but obscures techniques that apply to the matrix completion problems of many classes and relationships between completion problems for different classes. For information on these subjects, see for example [FJT00], [Hog01], and [Hog03a].

35.1 Introduction

All matrices and partial matrices discussed here are square. Graphs allow loops but do not allow multiple edges. The definitions and terminology about graphs and digraphs given in Chapter 28 and Chapter 29 is used here; however, the association between matrices and digraphs is different from the association in those chapters, where an arc is associated with a nonzero entry. Here an arc is associated with a specified entry.

Definitions:

A **partial matrix** is a square array in which some entries are specified and others are not. An unspecified entry is denoted by ? or by x_{ij}. An ordinary matrix is considered a partial matrix, as is a matrix with no specified entries.

A **completion** of a partial matrix is a choice of values for the unspecified entries.

A partial matrix B is **combinatorially symmetric** (also called **positionally symmetric**) if b_{ij} specified implies b_{ji} specified.

Let B be an $n \times n$ partial matrix. The **digraph** of B, $\mathcal{D}(B) = (V, E)$, has vertices $V = \{1, \ldots, n\}$, and for each i and j in V, the arc $(i, j) \in E$ exactly when b_{ij} is specified.

Let B be an $n \times n$ combinatorially symmetric partial matrix. The **graph** of B, $\mathcal{G}(B) = (V, E)$, has vertices $V = \{1, \ldots, n\}$, and for each i and j in V, the edge $\{i, j\} \in E$ exactly when b_{ij} is specified.

A connected graph or digraph is **nonseparable** if it does not have a cut-vertex.

A **block** of a graph or digraph is a maximal nonseparable sub(di)graph. This use of "block" is for graphs and digraphs and differs from "block" in a block matrix.

A graph (respectively, digraph) is a **clique** if every vertex has a loop and for any two distinct vertices u, v, the edge $\{u, v\}$ is present (respectively, both arcs $(u, v), (v, u)$ are present).

A graph or digraph is **block-clique** (also called **1-chordal**) if every block is a clique.

A digraph $G = (V, E)$ is **symmetric** if $(i, j) \in E$ implies $(j, i) \in E$ for all $i, j \in V$.

A digraph $G = (V, E)$ is **asymmetric** if $(i, j) \in E$ implies $(j, i) \notin E$ for all distinct $i, j \in V$.

A **simple cycle** in a digraph is an induced subdigraph that is a cycle.

A digraph (respectively, graph) G has the **X-completion property** (where X is a type of matrix) if every partial X-matrix B such that $\mathcal{D}(B) = G$ (respectively, $\mathcal{G}(B) = G$) can be completed to an X-matrix. In the literature, the phrase "has X-completion" is sometimes used for "has the X-completion property."

A class X is **closed under permutation similarity** if whenever A is an X-matrix and P is a permutation matrix, then $P^T AP$ is an X-matrix.

A class X is **hereditary** (or **closed under taking principal submatrices**) if whenever A is an X-matrix and $\alpha \subseteq \{1, \ldots, n\}$, then $A[\alpha]$ is an X-matrix.

A class X is **closed under matrix direct sums** if whenever $A_1, A_2, \ldots, A_k$ are X-matrices, then $A_1 \oplus A_2 \oplus \cdots \oplus A_k$ is an X-matrix.

A class X has the **triangular property** if whenever A is a block triangular matrix and every diagonal block is an X-matrix, then A is an X matrix.

A partial matrix B is in **pattern block triangular form** if the adjacency matrix of $\mathcal{D}(B)$ is in block triangular form.

Note: Many matrix terms, such as size, entry, submatrix, etc., are applied in the obvious way to partial matrices.

Facts:

Let X be one of the following classes of matrices: positive (semi)definite matrices, Euclidean distance matrices, (symmetric) M-matrices, (symmetric) M_0-matrices, (symmetric) inverse M-matrices, completely positive matrices, doubly nonnegative matrices, (strictly) copositive matrices, P-matrices, P_0-matrices, $P_{0,1}$-matrices, nonnegative P-matrices, nonnegative P_0-matrices, positive P-matrices, entry (weakly) sign symmetric P-matrices, entry (weakly) sign symmetric P_0-matrices, entry (weakly) sign symmetric $P_{0,1}$-matrices. (The definitions of these classes can be found in the relevant sections.)

Proofs of the facts below can be found in [Hog01] for most of the classes discussed, and the proofs given there apply to all the classes X listed above. Most of these facts are also in the original papers discussing the completion problem for a specific class; those references are listed in the section devoted to the class. In the literature, when it is assumed that a partial matrix B has every diagonal entry specified (equivalently, every vertex of the graph $\mathcal{G}(B)$ or digraph $\mathcal{D}(B)$ has a loop), it is customary to suppress all the loops and treat $\mathcal{G}(B)$ or $\mathcal{D}(B)$ as a simple graph or digraph. That is not done in this chapter because of the danger of confusion. Also, in some references, such as [Hog01], a mark is used to indicate a specified vertex instead of a loop. This has no effect on the results (but requires translation of the notation). If X is a symmetric class of matrices, then there is no loss of generality in assuming every partial matrix is combinatorially symmetric and this assumption is standard practice.

1. If B is a combinatorially symmetric partial matrix, then $\mathcal{D}(B)$ is a symmetric digraph and $\mathcal{G}(B)$ is the graph associated with $\mathcal{D}(B)$. Combinatorially symmetric partial matrices are usually studied by means of graphs rather than digraphs, and it is understood that the graph represents the associated symmetric digraph.
2. Each of the classes X listed at the beginning of the facts is closed under permutation similarity. This fact is not true for the classes of totally nonnegative and totally positive matrices. (See [FJS00] for information about matrix completion problems for these matrices.)
3. Applying a permutation similarity to a partial matrix B corresponds to renumbering the vertices of the digraph $\mathcal{D}(B)$ (or graph $\mathcal{G}(B)$ if B is combinatorially symmetric).
4. Renumbering the vertices of a graph or digraph does not affect whether it has the X-completion property. It is customary to use unlabeled (di)graph diagrams. This fact is not true for the classes of totally nonnegative and totally positive matrices.
5. Each of the classes X listed at the beginning of the facts is hereditary.
6. Let B be a partial matrix and $\alpha \subseteq \{1, \ldots, n\}$. The digraph of the principal submatrix $B[\alpha]$ is isomorphic to the subdigraph of $\mathcal{D}(B)$ induced by α (and is customarily identified with it). The same is true for the graph if B is combinatorially symmetric.
7. If a graph or digraph G has the X-completion property, then every induced subgraph or induced subdigraph of G has the X-completion property.
8. Each of the classes X listed at the beginning of the facts is closed under matrix direct sums.
9. Let B be a partial matrix such that all specified entries are contained in diagonal blocks $B_1, B_2, \ldots, B_k$. The connected components of $\mathcal{D}(B)$ are isomorphic to the $\mathcal{D}(B_i)$, $i = 1, \ldots, k$. The same is true for $\mathcal{G}(B)$ if B is combinatorially symmetric.
10. A graph or digraph G has the X-completion property if and only if every connected component of G has the X-completion property.
11. If X has the triangular property, B is a partial matrix in pattern block triangular form, and each pattern diagonal block can be completed to an X-matrix, then B can be completed to an X-matrix.
12. If X has the triangular property and is closed under permutation similarity, then a graph or digraph G has the X-completion property if and only if every strongly connected component of G has the X-completion property.
13. A block-clique graph is chordal.
14. A block-clique digraph is symmetric.

Examples:

1. Graphs (a) through (n) will be used in the examples in the following sections.

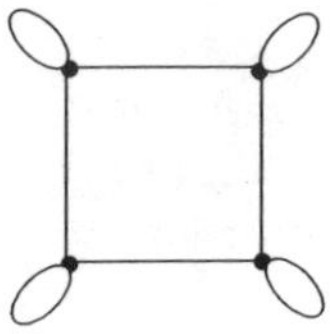

(a)

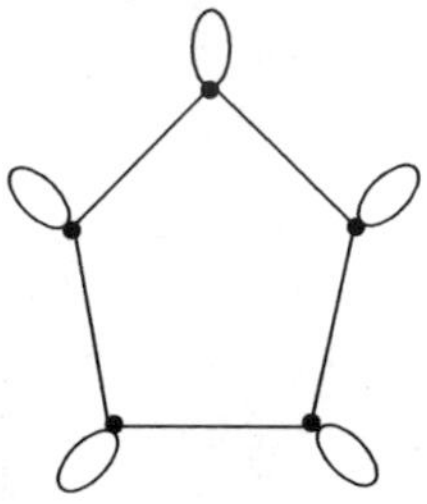

(b)

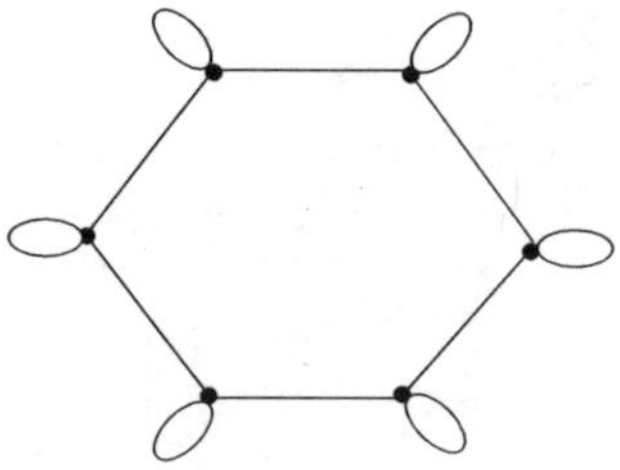

(c)

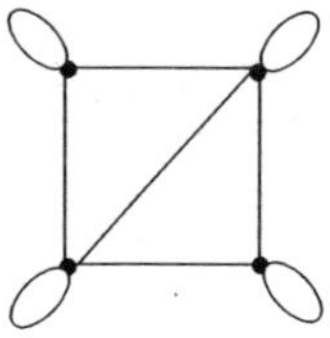

(d)

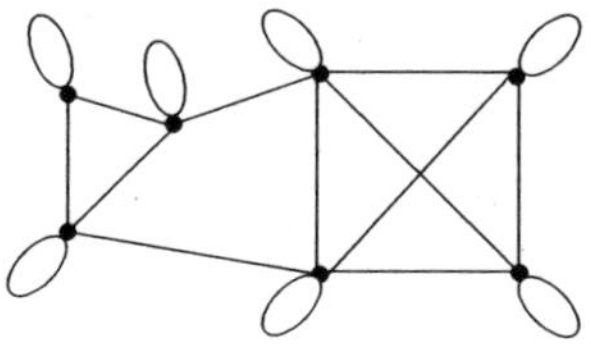

(e)

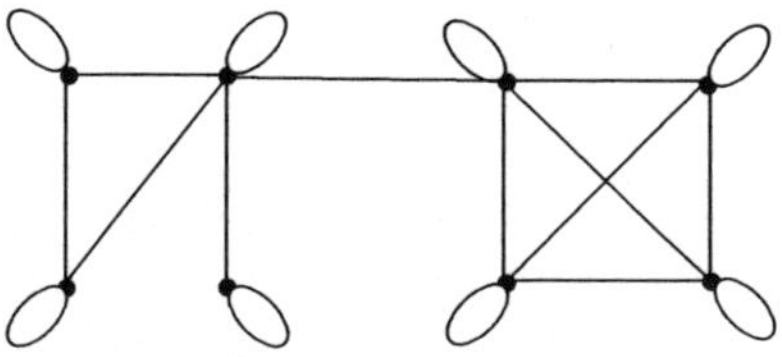

(f)

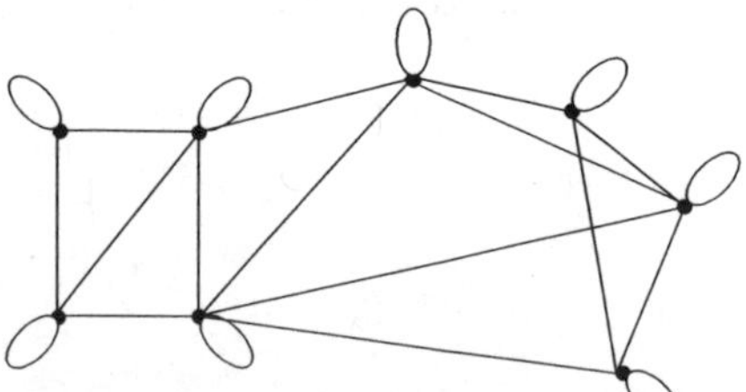

(g)

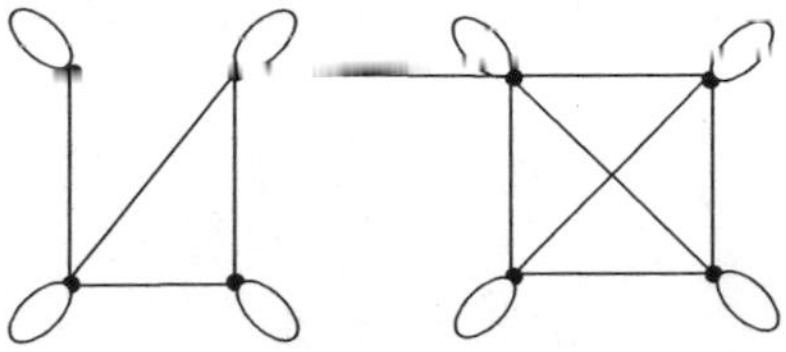

(h)

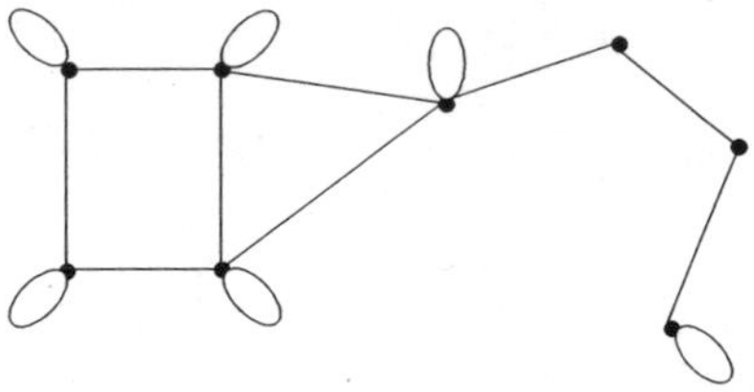

(i)

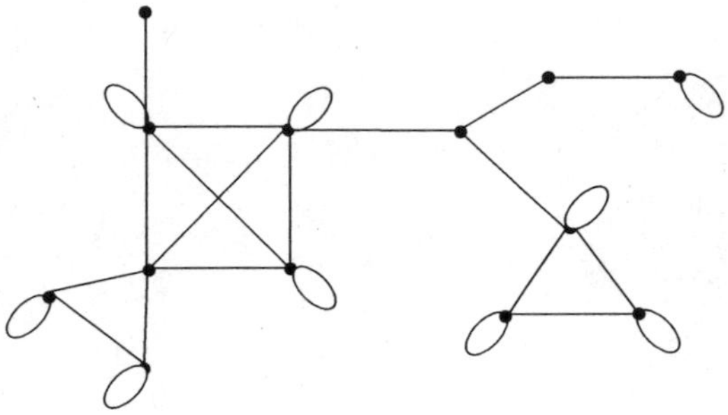

(j)

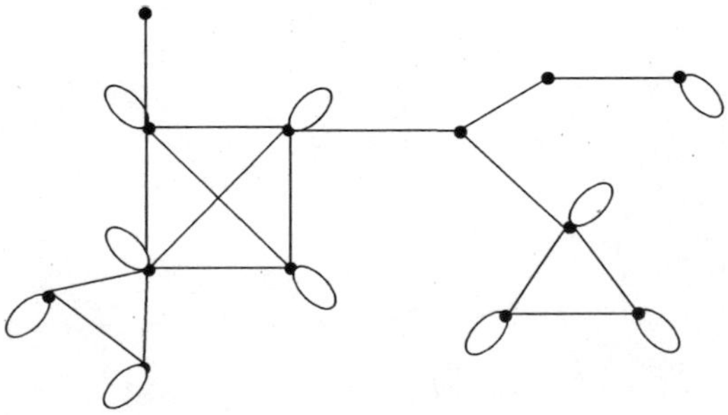

(k)

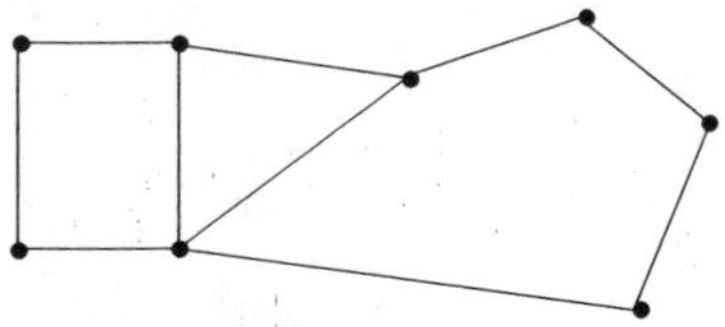

(l)

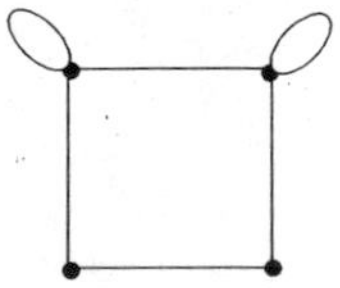

(m)

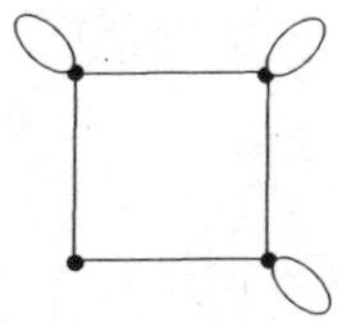

(n)

2. The matrix $\begin{bmatrix} 1 & 3 & ? & 0 \\ -1 & 1 & -7 & ? \\ ? & -7 & 1 & 2 \\ 8 & ? & 0 & 1 \end{bmatrix}$ is a partial matrix specifying the graph 1a with vertices numbered 1, 2, 3, 4 clockwise from upper left (or any other numbering around the cycle in order).

3. The graph 1f is block-clique, and this is the only block-clique graph in Example 1.
4. The following digraphs ((a) through (r)) will be used in the examples in the following sections. Note that when both arcs (i, j) and (j, i) are present, the arrows are omitted.

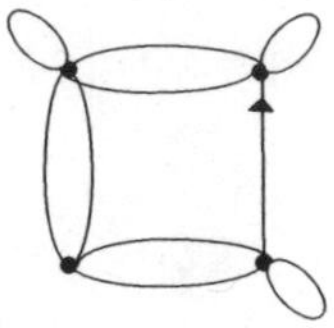

(a)

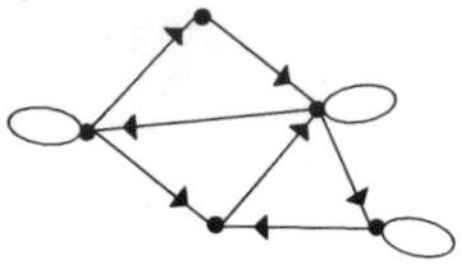

(b)

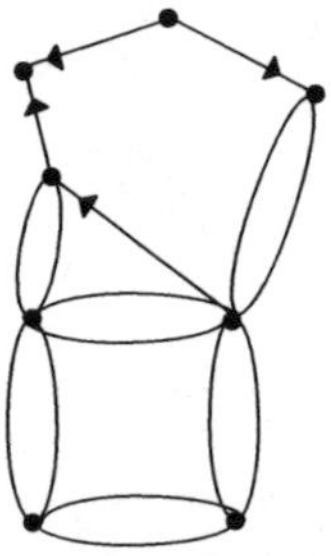

(c)

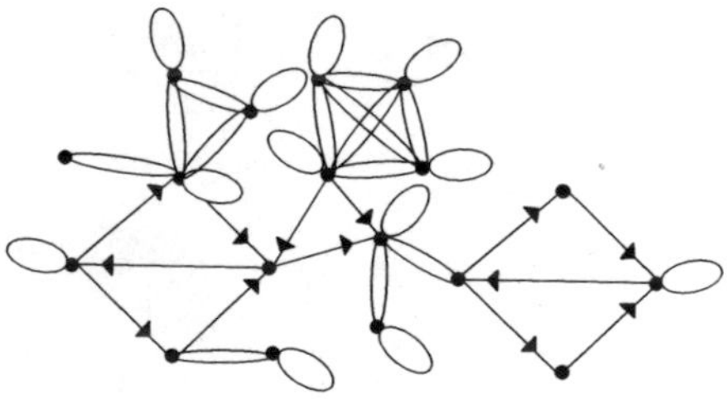

(d)

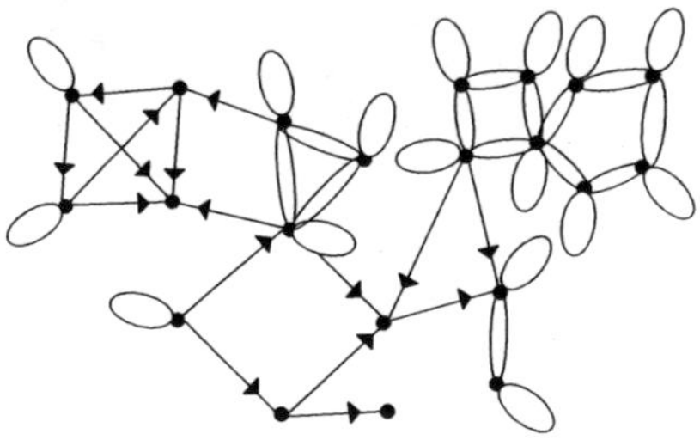

(e)

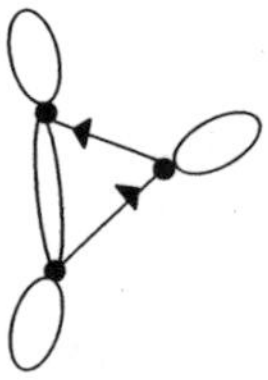

(f)

(g)

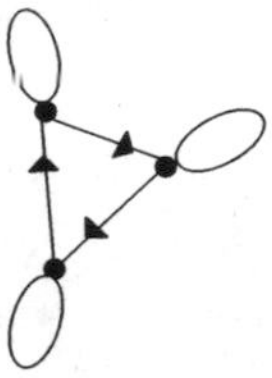

(h)

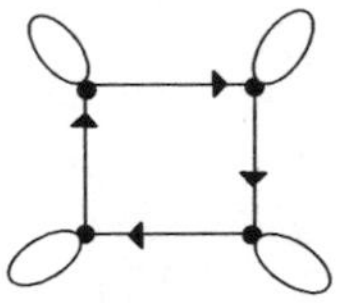

(i)

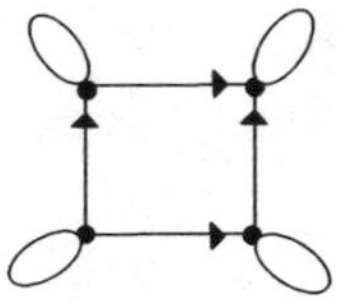

(j)

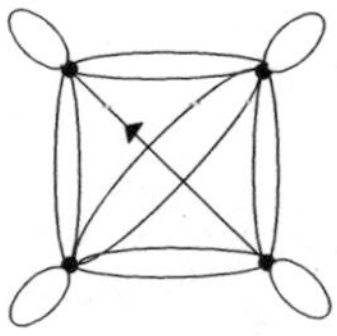

(k)

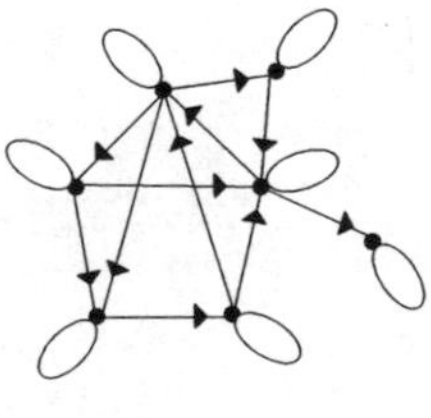

(l)

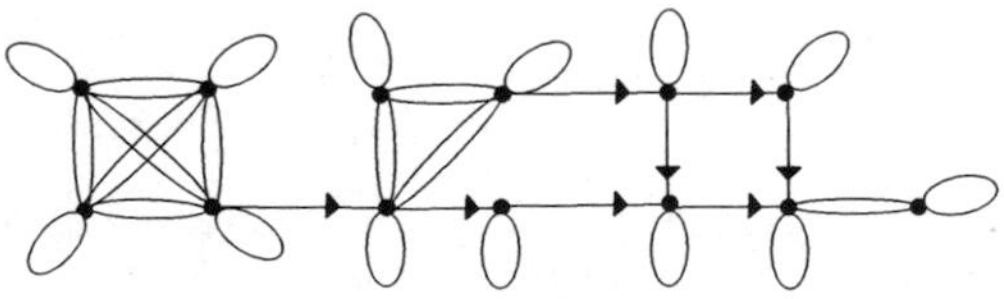

(m)

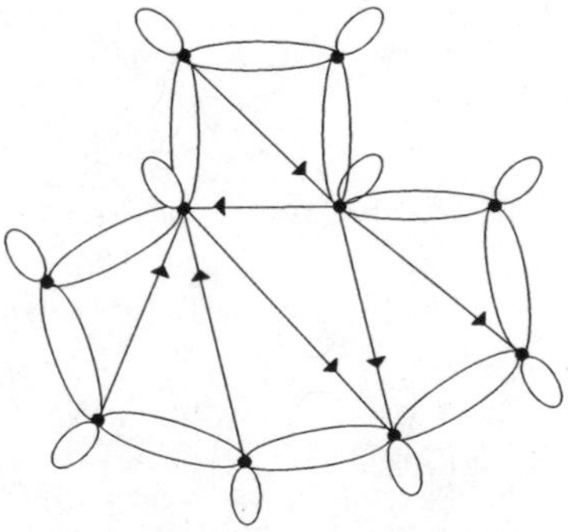

(n)

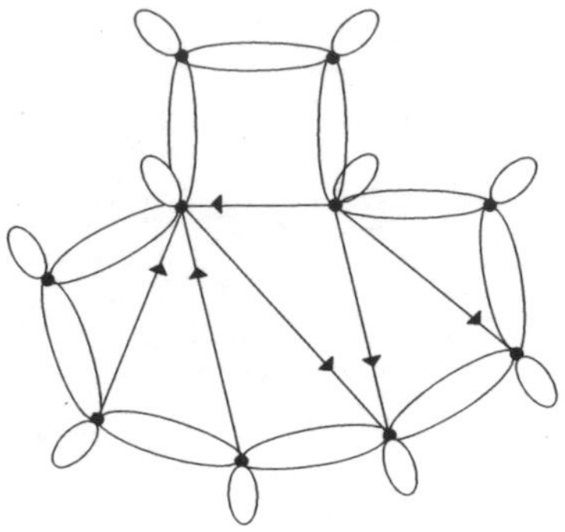

(o)

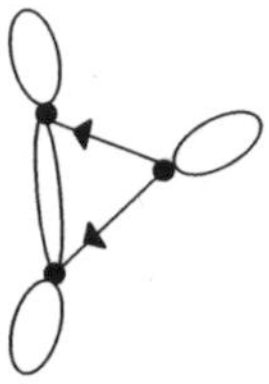

(p)

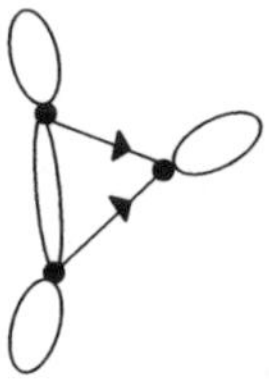

(q)

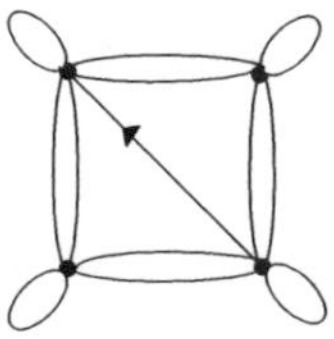

(r)

5. None of the digraphs in Example 4 are symmetric. (We diagram a symmetric digraph by its associated graph.) Digraphs 4b, 4h, 4i, 4j, and 4l are asymmetric.

35.2 Positive Definite and Positive Semidefinite Matrices

In this section, all matrices are real or complex.

Definitions:

The matrix A is **positive definite** (respectively, **positive semidefinite**) if A is Hermitian and for all $\mathbf{x} \neq 0$, $\mathbf{x}^* A\mathbf{x} > 0$ (respectively, $\mathbf{x}^* A\mathbf{x} \geq 0$).

The partial matrix B is a **partial positive definite matrix** (respectively, **partial positive semidefinite matrix**) if every fully specified principal submatrix of B is a positive definite matrix (respectively, positive semidefinite matrix), and whenever b_{ij} is specified then so is b_{ji} and $b_{ji} = \overline{b_{ij}}$.

Facts:

1. A Hermitian matrix A is positive definite (respectively, positive semidefinite) if and only if it is positive stable (respectively, positive semistable) if and only if all principal minors are positive (respectively, nonnegative). There are many additional characterizations. (See Chapter 8.4 for more information.)
2. [GJS84] A graph that has a loop at every vertex has the positive definite (positive semidefinite) completion property if and only if it is chordal. (For information on how to construct such a completion, see [GJS84] and [DG81].)

3. [GJS84] A graph has the positive definite completion property if and only if the subgraph induced by the vertices with loops has the positive definite completion property.
4. [Hog01] A graph G has the positive semidefinite completion property if and only if for each connected component H of G, either H has a loop at every vertex and is chordal, or H has no loops.
5. [GJS84] If B is a partial positive definite matrix with all diagonal entries specified such that $\mathcal{G}(B)$ is chordal, then there is a unique positive definite completion A of B that maximizes the determinant, and this completion has the property that whenever the i, j-entry of B is unspecified, the i, j-entry of A^{-1} is zero.
6. [Fie66] Let C be a partial positive semidefinite matrix such that every diagonal entry is specified and $\mathcal{G}(B)$ with loops suppressed is a cycle. If any diagonal entry is 0, then B can be completed to a positive semidefinite matrix. If every diagonal entry is nonzero, there is a positive diagonal matrix D such that every diagonal entry of $C = DBD$ is equal to 1. Let the specified off-diagonal entries of C be denoted $c_1, \ldots, c_n$. Then C (and, hence, B) can be completed to a positive semidefinite matrix if and only if the following *cycle conditions* are satisfied:

$$2 \max_{1 \leq k \leq n} \arccos|c_k| \leq \Sigma_{k=1}^{n} \arccos|c_k| \qquad \text{for } c_1 \ldots c_n > 0,$$
$$\Sigma_{k=1}^{n} \arccos|c_k| \geq \pi \qquad \text{for } c_1 \ldots c_n \leq 0.$$

(See [BJL96] for additional information.)

Examples:

The graphs for the examples can be found in Example 1 of Section 35.1.

1. The graphs 1d, 1f, and 1h have both the positive definite and positive semidefinite completion properties by Fact 2.
2. The graphs 1a, 1b, 1c, 1e, and 1g have neither the positive definite nor the positive semidefinite completion property by Fact 2.
3. The graphs 1j, 1k, 1l, 1m, and 1n have the positive definite completion property by Facts 3 and 2.
4. The graph 1i does not have the positive definite completion property by Facts 3 and 2.
5. The graph 1l has the positive semidefinite completion property by Fact 4.
6. The graphs 1i, 1j, 1k, 1m, and 1n do not have the positive semidefinite completion property by Fact 4.
7. The partial matrix $B = \begin{bmatrix} 1 & .3 & ? & ? & -.1 \\ .3 & 1 & 1 & ? & ? \\ ? & 1 & 1 & .2 & ? \\ ? & ? & .2 & 1 & 1 \\ -.1 & ? & ? & 1 & 1 \end{bmatrix}$ can be completed to a positive semidefinite matrix by Fact 6 because

$$\Sigma_{k=1}^{n} \arccos|c_k| = 4.10617 \geq \pi.$$

35.3 Euclidean Distance Matrices

In this section, all matrices are real.

Definitions:

The matrix $A = [a_{ij}]$ is a **Euclidean distance matrix** if there exist vectors $\mathbf{x}_1, \ldots, \mathbf{x}_n \in \mathbb{R}^d$ (for some $d \geq 1$) such that $a_{ij} = \|\mathbf{x}_i - \mathbf{x}_j\|_2$ for all $i, j = 1, \ldots, n$.

The partial matrix B is a **partial Euclidean distance matrix** if every diagonal entry is specified and equal to 0, every fully specified principal submatrix of B is a Euclidean distance matrix, and whenever b_{ij} is specified then so is b_{ji} and $b_{ji} = b_{ij}$.

Facts:

1. Every Euclidean distance matrix has all diagonal elements equal to 0. There is no loss of generality by considering only a graph that has loops at every vertex, and requiring all diagonal entries of partial Euclidean distance matrices to be 0.
2. [Lau98] A graph with a loop at every vertex has the Euclidean distance completion property if and only if it is chordal.
3. [Lau98] A graph with a loop at every vertex has the Euclidean distance completion property if and only if it has the positive semidefinite completion property. There is a method for transforming the Euclidean distance completion problem into the positive semidefinite completion problem via the Schoenberg transform that provides additional information about conditions on entries that are sufficient to guarantee completion.

Examples:

The graphs for the examples can be found in Example 1 of Section 35.1.

1. The graphs 1d, 1f, and 1h have the Euclidean distance completion property by Fact 2.
2. The graphs 1a, 1b, 1c, 1e, and 1g do not have the Euclidean distance completion property by Fact 2.

35.4 Completely Positive and Doubly Nonnegative Matrices

In this section, all matrices are real.

Definitions:

The matrix A is a **completely positive matrix** if $A = CC^T$ for some nonnegative $n \times m$ matrix C.

A matrix is a **doubly nonnegative matrix** if it is positive semidefinite and every entry is nonnegative.

The partial matrix B is a **partial completely positive matrix** (respectively, **partial doubly nonnegative matrix**) if every fully specified principal submatrix of B is a completely positive matrix (respectively, doubly nonnegative matrix), and whenever b_{ij} is specified then so is b_{ji} and $b_{ji} = b_{ij}$, and all specified off-diagonal entries are nonnegative.

Facts:

1. A completely positive matrix is doubly nonnegative.
2. [DJ98] A graph that has a loop at every vertex has the completely positive completion property (respectively, doubly nonnegative completion property) if and only if it is block-clique.
3. [Hog02] A graph G has the completely positive completion property (respectively, doubly nonnegative completion property) if and only if for every connected component H of G, H is block-clique, or H has no loops.
4. A graph has the completely positive completion property if and only if it has the doubly nonnegative completion property.
5. [DJK00] A partial matrix that satisfies the conditions of Fact 6 of Section 35.2 can be completed to a CP- (respectively, DN-) matrix.

Examples:

The graphs for the examples can be found in Example 1 of Section 35.1.

1. The graph 1f has both the completely positive completion property and the doubly nonnegative completion property by Fact 2.

2. The graphs 1a, 1b, 1c, 1d, 1e, 1g, and 1h have neither the completely positive completion property nor the doubly nonnegative completion property by Fact 2.
3. The graph 1l has both the completely positive completion property and the doubly nonnegative completion property by Fact 3.
4. The graphs 1i, 1j, 1k, 1m, and 1n have neither the completely positive completion property nor the doubly nonnegative completion property by Fact 3.

35.5 Copositive and Strictly Copositive Matrices

In this section, all matrices are real.

Definitions:

The symmetric matrix A is **strictly copositive** if $\mathbf{x}^T A\mathbf{x} > 0$ for all $\mathbf{x} \geq 0$ and $\mathbf{x} \neq 0$; A is **copositive** if $\mathbf{x}^T A\mathbf{x} \geq 0$ for all $\mathbf{x} \geq 0$.

The partial matrix B is a **partial strictly copositive matrix** (respectively, **partial copositive matrix**) if every fully specified principal submatrix of B is a strictly copositive matrix (respectively, copositive matrix) and whenever b_{ij} is specified then so is b_{ji} and $b_{ji} = b_{ij}$.

Facts:

1. If A is (strictly) copositive, then so is $A + M$ for any symmetric nonnegative matrix M.
2. [HJR05] [Hog] Every partial strictly copositive matrix can be completed to a strictly copositive matrix using the method described in Facts 4 and 5 below.
3. [HJR05] Every partial copositive matrix that has every diagonal entry specified can be completed to a copositive matrix using the completion described in Fact 4 below. There exists a partial copositive matrix with an unspecified diagonal entry that cannot be completed to a copositive matrix (see Example 2 below).
4. [HJR05] Let B be a partial copositive matrix with every diagonal entry specified. For each pair of unspecified off-diagonal entries, set $x_{ij} = x_{ji} = \sqrt{b_{ii}b_{jj}}$. The resulting matrix is copositive, and is strictly copositive if B is a partial strictly copositive matrix.
5. [Hog] Any completion of a partial strictly copositive matrix omitting only one diagonal entry found by Algorithm 1 is a strictly copositive matrix. If B is a partial strictly copositive matrix that omits some diagonal entries, values for these entries can be chosen one at a time using Algorithm 1, using the largest value obtained by considering all principal submatrices that are completed by the choice of that diagonal entry, to obtain a partial strictly copositive matrix with specified diagonal that agrees with B on every specified entry of B.

Algorithm 1: Completing one unspecified diagonal entry

Let $B = \begin{bmatrix} x_{11} & \mathbf{b}^T \\ \mathbf{b} & B_1 \end{bmatrix}$ be a partial strictly copositive $n \times n$ matrix having all entries except the 1,1-entry specified. Let $\|\cdot\|$ be a vector norm.
Complete B by choosing a value for x_{11} as follows:

1. $\beta = \min_{\mathbf{y}\in\mathbb{R}^{n-1},\mathbf{y}\geq 0,\|\mathbf{y}\|=1} \mathbf{b}^T\mathbf{y}$.
2. $\gamma = \min_{\mathbf{y}\in\mathbb{R}^{n-1},\mathbf{y}\geq 0,\|\mathbf{y}\|=1} \mathbf{y}^T B_1\mathbf{y}$.
3. $x_{11} > \frac{\beta^2}{\gamma}$.

6. [HJR05], [Hog] Every graph has the strictly copositive completion property.
7. [HJR05], [Hog] A graph has the copositive completion property if and only if for each connected component H of G, either H has a loop at every vertex, or H has no loops.

Examples:

1. The partial matrix $B = \begin{bmatrix} x_{11} & -5 & 1 & x_{14} & x_{15} & x_{16} \\ -5 & 1 & -2 & x_{24} & x_{25} & 1 \\ 1 & -2 & 5 & 1 & -1 & -1 \\ x_{14} & x_{24} & 1 & 1 & x_{45} & 1 \\ x_{15} & x_{25} & -1 & x_{45} & x_{55} & x_{56} \\ x_{16} & 1 & -1 & 1 & x_{56} & 3 \end{bmatrix}$ is a partial strictly copositive matrix.

   We use the method in Facts 4 and 5 to complete B to a strictly copositive matrix:
   Select index 5. The only principal submatrix completed by a choice of b_{55} is $B[\{3,5\}]$. Any value that makes $x_{55}b_{33} > b_{35}^2$ will work; we choose $x_{55} = 1$.
   Select index 1. The only principal submatrices completed by a choice of b_{11} are principal submatrices of $B[\{1,2,3\}] = \begin{bmatrix} x_{11} & \mathbf{b}^T \\ \mathbf{b} & B[\{2,3\}] \end{bmatrix}$. Apply Algorithm 1 (using $\|\cdot\|_1$):

   1. $\beta = \min_{\|\mathbf{y}\|_1=1} \mathbf{b}^T\mathbf{y} = -5$.
   2. $\gamma = \min_{\|\mathbf{y}\|_1=1} \mathbf{y}^T B[\{2,3\}]\mathbf{y} = \frac{1}{10}$.
   3. Choose $x_{11} > \frac{\beta^2}{\gamma}$; we choose $b_{11} = 256$.

   Then by Fact 4, $B = \begin{bmatrix} 256 & -5 & 1 & 16 & 16 & 16\sqrt{3} \\ -5 & 1 & -2 & 1 & 1 & 1 \\ 1 & -2 & 5 & 1 & -1 & -1 \\ 16 & 1 & 1 & 1 & 1 & 1 \\ 16 & 1 & -1 & 1 & 1 & \sqrt{3} \\ 16\sqrt{3} & 1 & -1 & 1 & \sqrt{3} & 3 \end{bmatrix}$ is a strictly copositive matrix.

2. $B = \begin{bmatrix} x_{11} & -1 \\ -1 & 0 \end{bmatrix}$ is a partial copositive matrix that cannot be completed to a copositive matrix because once x_{11} is chosen (clearly $x_{11} > 0$), then $\mathbf{v} = [\frac{1}{x_{11}}, 1]^T$ results in $\mathbf{v}^T B\mathbf{v} = \frac{-1}{x_{11}} < 0$.
3. The graphs 1a, 1b, 1c, 1d, 1e, 1f, 1g, 1h, and 1l have copositive completion by Fact 7, and these are the only graphs in Example 1 of section 35.1 that have the copositive completion property.

35.6 *M*- and *M*$_0$-Matrices

In this section, all matrices are real.

Definitions:

The matrix A is an M-**matrix** (respectively, M_0-**matrix**) if there exist a nonnegative matrix P and a real number $s > \rho(P)$ (respectively, $s \geq \rho(P)$) such that $A = sI - P$.

The partial matrix B is a **partial** M-**matrix** (respectively, **partial** M_0-**matrix**) if every fully specified principal submatrix of B is an M-matrix (respectively, M_0-matrix) and every specified off-diagonal entry of B is nonpositive.

If B is a partial matrix that includes all diagonal entries, the **zero completion** of B, denoted B_0, is obtained by setting all unspecified (off-diagonal) entries to 0.

Facts:

1. For a Z-matrix A (i.e., every off-diagonal entry of A is nonpositive), the following are equivalent:
   (a) A is an M-matrix.
   (b) Every principal minor of A is positive.
   (c) A is positive stable.
   (d) A^{-1} is nonnegative.

   The analogs of the first three conditions are equivalent to A being an M_0-matrix. See Section 9.5 for more information about M- and M_0-matrices.
2. A principal submatrix of an M- (M_0-) matrix is an M- (M_0-) matrix (cf. Fact 5 in Section 35.1).
3. [JS96], [Hog01] A partial M- (M_0-) matrix B that includes all diagonal entries can be completed to an M- (M_0-) matrix if and only if the zero completion B_0 is an M- (M_0-) matrix.
4. [Hog98b], [Hog01] A digraph G with a loop at every vertex has the M- (M_0-) completion property if and only if every strongly connected induced subdigraph of G is a clique.
5. [Hog98b] A digraph G has the M-completion property if and only if the subdigraph induced by the vertices of G that have loops has M-completion.
6. [Hog01] A digraph G has the M_0-completion property if and only if for every strongly connected induced subdigraph H of G, either H is a clique or H has no loops.
7. [Hog02] Symmetric M- and M_0-matrices and partial matrices are defined in the obvious way. A graph has the symmetric M-completion property if and only if every connected component of the subgraph induced by the vertices with loops is a clique. A graph has the symmetric M_0-completion property if and only if every connected component is either a clique or has no loops.

Examples:

The graphs and digraphs for the examples can be found in Examples 1 and 4 of Section 35.1.

1. Even if the digraph of a partial M-matrix does not have the M-matrix completion property, the matrix may still have an M-matrix completion. By Fact 3 it is easy to determine whether there is an M-completion. For example, complete the partial M-matrix $B = \begin{bmatrix} 1 & ? & ? & -2 \\ -0.5 & 2 & ? & ? \\ ? & -1 & 1 & ? \\ ? & ? & -1 & 1 \end{bmatrix}$ to B_0 by setting every unspecified entry to 0. To determine whether B_0 is an M-matrix, compute the eigenvalues: $\sigma(B_0) = \{2.38028, 1.21945 \pm 0.914474i, 0.180827\}$. If we change the values of entries we can obtain a partial M-matrix that does not have an M-matrix completion, e.g., $C = \begin{bmatrix} 1 & ? & ? & -2 \\ -2.5 & 2 & ? & ? \\ ? & -1 & 1 & ? \\ ? & ? & -1 & 1 \end{bmatrix}$. Then $\sigma(C_0) = \{2.82397, 1.23609 \pm 1.43499i, -0.296153\}$, so by Fact 3, C cannot be completed to an M-matrix. Note $\mathcal{D}(B) = \mathcal{D}(C)$ is the digraph 4i, which does not have the M-matrix completion property by Fact 4.
2. The digraphs 4j, 4m, 4p, and 4q have the M- and M_0-completion properties by Fact 4.
3. The graphs 1a, 1b, 1c, 1d, 1e, 1f, 1g, and 1h and the digraphs 4f, 4g, 4h, 4i, 4k, 4l, 4n, 4o, and 4r have neither the M-completion property nor the M_0-completion property by Fact 4.
4. The graphs 1j, 1l, and 1m and the digraphs 4a, 4b, 4c, and 4d have the M-completion property by Facts 5 and 4. Of these (di)graphs, only 1l and 4c have the M_0 completion property, by Fact 6.
5. The graphs 1i, 1k, and 1n and the digraph 4e do not have the M-completion property (respectively, the M_0-completion property) by Facts 5 and 4 (respectively, Fact 6).

35.7 Inverse M-Matrices

In this section, all matrices are real.

Definitions:

The matrix A is an **inverse M-matrix** if A is the inverse of an M-matrix.

The partial matrix B is a **partial inverse M-matrix** if every fully specified principal submatrix of B is an inverse M-matrix and every specified entry of B is nonnegative.

A digraph is **cycle-clique** if the induced subdigraph of every cycle is a clique.

An **alternate path to a single arc** in a digraph G is a path of length greater than 1 between vertices i and j such that the arc (i, j) is in G.

A digraph G is **path-clique** if the induced subdigraph of every alternate path to a single arc is a clique.

A digraph is **homogeneous** if it is either symmetric or asymmetric.

Facts:

1. A matrix A is an inverse M-matrix if and only if all entries of A are nonnegative and all off-diagonal entries of A^{-1} are nonpositive. There are many equivalent characterizations of inverse M-matrices; see Section 9.5 for more information.
2. [JS96] Let $B = \begin{bmatrix} B_{11} & \mathbf{b}_{12} & ? \\ \mathbf{b}_{21}^T & b_{22} & \mathbf{b}_{23}^T \\ ? & \mathbf{b}_{32} & B_{33} \end{bmatrix}$ be an $n \times n$ partial inverse M matrix, where B_{11} and B_{33} are square matrices of size k and $n-k-1$, and all entries of the submatrices shown are specified. Then $A = \begin{bmatrix} B_{11} & \mathbf{b}_{12} & \mathbf{b}_{12} b_{22}^{-1} \mathbf{b}_{21}^T \\ \mathbf{b}_{21}^T & b_{22} & \mathbf{b}_{23}^T \\ \mathbf{b}_{32} b_{22}^{-1} \mathbf{b}_{21}^T & \mathbf{b}_{32} & B_{33} \end{bmatrix}$ is the unique inverse M-completion of B such that A^{-1} has zeros in all the positions where B has unspecified entries. This method can be used to complete a partial inverse M-matrix whose digraph is block-clique.
3. [JS96], [JS99], [Hog98a], [Hog00], [Hog02] A symmetric digraph with a loop at every vertex has the inverse M-completion property if and only if it is block-clique. A digraph obtained from a block-clique digraph by deleting loops from vertices not contained in any block of order greater than 2 also has the inverse M-completion property, and any symmetric digraph that has the inverse M-completion property has that form. The same is true for the symmetric inverse M-completion property (with the obvious definition).
4. [Hog98a] A digraph with a loop at every vertex has the inverse M-completion property if and only if G is path-clique and cycle-clique.
5. [Hog00], [Hog01] A digraph G has the inverse M-completion property if and only if it is path-clique and every strongly connected nonseparable induced subdigraph has the inverse M-completion property. A strongly connected nonseparable digraph is homogeneous. A simple cycle with at least one vertex that does not have a loop has the inverse M-completion property.

Examples:

The graphs and digraphs for the examples can be found in Examples 1 and 4 of Section 35.1.

1. Let $B = \begin{bmatrix} 3 & 1 & ? & ? \\ 4 & 2 & 4 & 2 \\ ? & 1 & 5 & 1 \\ ? & 1 & 2 & 2 \end{bmatrix}$. The completion given by Fact 2 is $A = \begin{bmatrix} 3 & 1 & 2 & 1 \\ 4 & 2 & 4 & 2 \\ 2 & 1 & 5 & 1 \\ 2 & 1 & 2 & 2 \end{bmatrix}$, and $A^{-1} = \begin{bmatrix} 1 & -\frac{1}{2} & 0 & 0 \\ -2 & \frac{7}{3} & -\frac{2}{3} & -1 \\ 0 & -\frac{1}{6} & \frac{1}{3} & 0 \\ 0 & -\frac{1}{2} & 0 & 1 \end{bmatrix}$.

2. By Fact 3, the graphs 1f and 1k have the inverse M-completion property, and these are the only graphs in Example 1 that do.
3. The digraphs 4j and 4m have the inverse M-completion property by Fact 4.
4. The digraphs 4f, 4g, 4h, 4i, 4k, 4l, 4n, 4o, 4p, 4q, and 4r do not have the inverse M-completion property by Fact 4.
5. The digraphs 4a, 4b, 4c, 4d, and 4e do not have the inverse M-completion property by Fact 5.

35.8 P-, $P_{0,1}$-, and P_0-Matrices

In this section, all matrices are real.

Definitions:

The matrix A is a P-**matrix** (respectively, P_0-**matrix**, $P_{0,1}$-**matrix**) if every principal minor of A is positive (respectively, nonnegative, nonnegative and every diagonal element is positive).

The partial matrix B is a **partial** P-**matrix** (respectively, **partial** P_0-**matrix**, **partial** $P_{0,1}$-**matrix**) if every fully specified principal submatrix of B is a P-matrix (respectively, P_0-matrix, $P_{0,1}$-matrix).

Facts:

1. A positive definite matrix, M-matrix, or inverse M-matrix is a P-matrix. A positive semidefinite matrix or M_0-matrix is a P_0-matrix. See [HJ91] for more information on P-, $P_{0,1}$-, and P_0-matrices. A principal submatrix of a P- ($P_{0,1}$-, P_0-) matrix is a P- ($P_{0,1}$-, P_0-) matrix (cf. Fact 5 in section 35.1).
2. [Hog03a] If a digraph has the P_0-completion property, then it has the $P_{0,1}$-completion property. If a digraph has the $P_{0,1}$-completion property, then it has the P-completion property.
3. [JK96], [Hog01] A digraph has the P-completion property if and only if the subdigraph induced by the vertices that have loops has the P-completion property.
4. [JK96] Every symmetric digraph has the P-completion property. The P-completion of a combinatorially symmetric partial P-matrix can be accomplished by selecting one pair of unspecified entries at a time and choosing the entries of opposite sign and large enough magnitude to make the determinants of all principal matrices completed positive.
5. [JK96] Every order 3 digraph has the P-completion property, but there is an order 4 digraph (see the digraph 4k in Example 4 of Section 35.1) that does not have the P-completion property. [DH00] extended a revised version of this example, digraph 4r, to a family of digraphs, called *minimally chordal symmetric Hamiltonian*, that do not have the P-completion property. The digraph 4n is another example of a digraph in this family (so both 4r and 4n do not have the P-completion property).
6. [DH00] A digraph that can be made symmetric by adding arcs one at a time so that at each stage at most one order 3 induced subdigraph (and no larger) becomes a clique, has the P-completion property.
7. [CDH02] A partial P- ($P_{0,1}$-, P_0-) matrix whose digraph is asymmetric can be completed to a P- ($P_{0,1}$-, P_0-) matrix as follows: If $i \neq j$ and b_{ji} is specified, then set $x_{ij} = -b_{ij}$. Otherwise, set $x_{ij} = 0$. Every asymmetric digraph has the P- ($P_{0,1}$-, P_0-) completion property.
8. [FJT00] Let $B = \begin{bmatrix} B_{11} & \mathbf{b}_{12} & ? \\ \mathbf{b}_{21}^T & b_{22} & \mathbf{b}_{23}^T \\ ? & \mathbf{b}_{32} & B_{33} \end{bmatrix}$ be an $n \times n$ partial P- ($P_{0,1}$-) matrix, where B_{11} and B_{33} are square matrices of size k and $n-k-1$, and all entries of the submatrices shown are specified. Then $A = \begin{bmatrix} B_{11} & \mathbf{b}_{12} & \mathbf{b}_{12} b_{22}^{-1} \mathbf{b}_{21}^T \\ \mathbf{b}_{21}^T & b_{22} & \mathbf{b}_{23}^T \\ 0 & \mathbf{b}_{32} & B_{33} \end{bmatrix}$ is a P- ($P_{0,1}$-) completion of B. This method can be used to complete any partial P- ($P_{0,1}$-) matrix whose digraph is block-clique. (See [Hog01] for the details of the analogous completion for P_0-matrices.)

9. [FJT00] Every block-clique digraph has the P- ($P_{0,1}$-, P_0-) completion property.
10. [Hog01] A digraph G has the P- (respectively, $P_{0,1}$-, P_0-) completion property if and only if every strongly connected and nonseparable induced subdigraph of G has the P- (respectively, $P_{0,1}$-, P_0-) completion property. A method to obtain such a completion is given in Algorithm 2.

Algorithm 2:
Let B be a partial P- ($P_{0,1}$-, P_0-) matrix such that every strongly connected and nonseparable induced subdigraph K of $\mathcal{D}(B)$ has the P- ($P_{0,1}$-, P_0-) property.

1. For each such K, complete the principal submatrix of B corresponding to K to obtain a partial P- ($P_{0,1}$-, P_0-) matrix B_1 such that each strongly connected induced subdigraph S of $\mathcal{D}(B_1)$ is block-clique.
2. For each such S, complete the principal submatrix of B_1 corresponding to S to obtain a partial P- ($P_{0,1}$-, P_0-) matrix B_2.
3. Set any remaining unspecified entries to 0.

11. [Hog01] A digraph that omits all loops has the P- ($P_{0,1}$-, P_0-) completion property. Each connected component of a symmetric digraph that has the P_0-completion property must have a loop at every vertex or omit all loops.
12. [Hog01], [CDH02], [JK96] A symmetric n-cycle with a loop at every vertex has the P_0-completion property if and only if $n \neq 4$. A symmetric n-cycle with a loop at every vertex has the P- and $P_{0,1}$-completion properties for all n.
13. [CDH02] All order 2, order 3, and order 4 digraphs with a loop at every vertex have been classified as having or not having the P_0-completion property. There are some order 4 digraphs that (in 2005) have not been classified as to the P- and $P_{0,1}$-completion properties.

Examples:

The graphs and digraphs for the examples can be found in Examples 1 and 4 of Section 35.1.

1. It is easy to verify that $B = \begin{bmatrix} 1 & 2 & x_{13} & -1 \\ 1 & 3 & -2 & x_{24} \\ x_{31} & 2 & 1 & 1 \\ -1 & x_{42} & 1 & 2 \end{bmatrix}$ is a partial P-matrix. The graph 1d, called the *double triangle*, is interpreted as the digraph of B. Let $x_{24} = y$ and $x_{42} = -y$. A choice of y completes three principal minors, $\det B[\{2,4\}] = 6+y^2$, $\det B[\{1,2,4\}] = -1-y+y^2$, and $\det B[\{2,3,4\}] = 11 + 4y + y^2$. The choice $y = 2$ makes all three minors positive. Let $x_{24} = z$ and $x_{42} = -z$. With $y = 2$, a choice of z completes four principal minors, $\det B[\{1,3\}] = 1 + z^2$, $\det B[\{1,2,3\}] = 5 + 6z + 3z^2$, $\det B[\{1,3,4\}] = 2z^2$, and $\det B = 10 + 18z + 10z^2$, so setting $z = 0$ completes B to the P-matrix $\begin{bmatrix} 1 & 2 & 0 & -1 \\ 1 & 3 & -2 & 2 \\ 0 & 2 & 1 & 1 \\ -1 & -2 & 1 & 2 \end{bmatrix}$. Any partial P-matrix specifying the double triangle can be completed in a similar manner, so the double triangle has the P-completion property.

2. The double triangle 1d does not have the P_0-completion property because $B = \begin{bmatrix} 1 & 2 & 1 & ? \\ -1 & 0 & 0 & -2 \\ -1 & 0 & 0 & -1 \\ ? & 1 & 1 & 1 \end{bmatrix}$ cannot be completed to a P_0-matrix ([JK96]). This implies the graphs 1g and 1h do not have the

P_0-completion property by Fact 7 in Section 35.1. The double triangle does have the $P_{0,1}$-completion property [Wan05].

3. All graphs in Example 1 have the P-completion property by Fact 4.
4. The digraphs 4c and 4d have the P-completion property by Fact 3. Fact 3 can also be applied in conjunction with other facts to several other digraphs in Example 4.
5. The digraphs 4f, 4g, 4h, 4p, and 4q have the P-completion property by Fact 5.
6. The digraphs 4a, 4b, 4c, 4d, 4i, 4j, 4l, 4m, and 4o have the P-completion property by Fact 6.
7. The digraphs 4b, 4h, 4i, 4j, and 4l have the P-, $P_{0,1}$-, and P_0-completion properties by Fact 7.
8. The completion of $\begin{bmatrix} 1 & -1 & ? & ? \\ 3 & 2 & -4 & 2 \\ ? & 1 & 5 & -1 \\ ? & 1 & 2 & 2 \end{bmatrix}$ given by Fact 8 is $\begin{bmatrix} 1 & -1 & 2 & -1 \\ 3 & 2 & -4 & 2 \\ 0 & 1 & 5 & -1 \\ 0 & 1 & 2 & 2 \end{bmatrix}$.
9. The graph 1f has the P_0- and $P_{0,1}$-completion properties by Fact 9. (It also has the P-completion property but one would not normally cite Fact 9 for that.)
10. The digraphs 4m, 4p, and 4q have the P-, $P_{0,1}$-, and P_0-completion properties by Fact 10. Fact 10 can also be applied in conjunction with other facts to several other digraphs in Example 4.
11. The graph 1l and the digraph 4c have the P-, P_0-, and $P_{0,1}$-completion properties by Fact 11.
12. The graphs 1i, 1j, 1k, 1m, and 1n do not have the P_0-completion property by Fact 11.
13. The graphs 1b and 1c have the P_0- ($P_{0,1}$)-completion property by Fact 12.
14. The graph 1a does not have the P_0-completion property, but does have the $P_{0,1}$-completion property, by Fact 12.
15. The graphs 1e and 1g do not have the P_0-completion property by Fact 12 and by Fact 7 in Section 35.1.

35.9 Positive P-, Nonnegative P-, Nonnegative $P_{0,1}$-, and Nonnegative P_0-Matrices

In this section, all matrices are real.

Definitions:

The matrix A is a **positive** (respectively, **nonnegative**) P**-matrix** if A is a P-matrix and every entry of A is positive (respectively, nonnegative). The matrix A is a **nonnegative** P_0**-matrix** (respectively, **nonnegative** $P_{0,1}$**-matrix**) if A is a P_0-matrix (respectively, $P_{0,1}$-matrix) and every entry of A is nonnegative.

The partial matrix B is a **partial positive** P**-matrix** (respectively, **partial nonnegative** P**-matrix, partial nonnegative** P_0**-matrix, partial nonnegative** $P_{0,1}$**-matrix**) if and only if every fully specified principal submatrix of B is a positive P-matrix (respectively, nonnegative P-matrix, nonnegative P_0-matrix, partial nonnegative $P_{0,1}$-matrix) and all specified entries are positive (respectively, nonnegative, nonnegative, nonnegative).

Facts:

1. [Hog03a], [Hog03b] If a digraph has the nonnegative P_0-completion property, then it has the nonnegative $P_{0,1}$-completion property. If a digraph has the nonnegative $P_{0,1}$-completion property, then it has the nonnegative P-completion property. If a digraph has the nonnegative P-completion property, then it has the positive P-completion property.
2. [Hog01] A digraph has the positive (respectively, nonnegative) P-completion property if and only if the subdigraph induced by vertices that have loops has the positive (respectively, nonnegative) P-completion property.

3. [FJT00], [Hog01], [CDH03] All order 2 and order 3 digraphs that have a loop at every vertex have the positive P- (nonnegative P-, nonnegative P_0-, nonnegative $P_{0,1}$-) completion property.
4. [BEH06] Suppose G is a digraph such that by adding arcs one at a time so that at each stage at most one order 3 induced subdigraph (and no larger) becomes a clique, it is possible to obtain a digraph G' that has the positive P- (respectively, nonnegative P-, nonnegative P_0-, nonnegative $P_{0,1}$-) completion property. Then G has the positive P- (respectively, nonnegative P-, nonnegative P_0-, nonnegative $P_{0,1}$-) completion property.
5. [FJT00] A block-clique digraph has the positive P- (nonnegative P-, nonnegative P_0-, nonnegative $P_{0,1}$-) completion property. See Fact 8 of section 35.8 for information on the construction.
6. [Hog01] A digraph G has the positive P- (respectively, nonnegative P-, nonnegative P_0-, nonnegative $P_{0,1}$-) completion property if and only if every strongly connected and nonseparable induced subdigraph of G has the positive P- (respectively, nonnegative P-, nonnegative P_0-, nonnegative $P_{0,1}$-) completion property. See Algorithm 2 of section 35.8 for information on the construction.
7. [Hog01] A digraph that omits all loops has the positive P- (nonnegative P-, nonnegative $P_{0,1}$-, nonnegative P_0-) completion property. Each connected component of a symmetric digraph that has the nonnegative P_0-completion property must have a loop at every vertex or omit all loops.
8. [CDH03] An order 4 digraph with a loop at every vertex has the nonnegative P_0-completion property if and only if it does not contain a 4-cycle or is the clique on 4-vertices. This characterization does not extend to higher order digraphs.
9. [BEH06], [JTU03] All order 4 digraphs that have a loop at every vertex have been classified as to the positive P- (nonnegative P-) completion property.
10. [CDH03], [FJT00] A symmetric n-cycle that has a loop at every vertex has the nonnegative P_0-completion property if and only if $n \neq 4$. A symmetric n-cycle that has a loop at every vertex has the positive P- (nonnegative P-, nonnegative $P_{0,1}$-) completion property.
11. [BEH06] A minimally chordal symmetric Hamiltonian digraph (cf. Fact 5 in Section 35.8) has neither the positive nor the nonnegative P-completion property.

Examples:

The graphs and digraphs for the examples can be found in Examples 1 and 4 of Section 35.1.

1. The digraphs 4f, 4g, 4h, 4p, and 4q have the positive P- (nonnegative P-, nonnegative P_0-, nonnegative $P_{0,1}$-) completion property by Fact 3.
2. The graph 1f has the positive P- (nonnegative P-, nonnegative P_0-, nonnegative $P_{0,1}$-) completion property by Fact 5.
3. The graphs 1j, 1k, 1l, 1m, and 1n and the digraphs 4c and 4d have the positive P- (nonnegative P-) completion property by Facts 2 and 5.
4. The digraphs 4j, 4m, 4p, and 4q have the positive P- (nonnegative P-, nonnegative P_0-, nonnegative $P_{0,1}$-) completion property by Fact 6.
5. The graph 1l and the digraph 4c have the positive P-, nonnegative P-, nonnegative P_0-, and nonnegative $P_{0,1}$-completion properties by Fact 7.
6. The graphs 1i, 1j, 1k, 1m, and 1n do not have the nonnegative P_0-completion property by Fact 7.
7. The graphs 1a and 1d and the digraphs 4i, 4k, and 4r do not have the nonnegative P_0-completion property by Fact 8. The graphs 1e, 1g, and 1h and the digraphs 4l, 4n, and 4o do not have the nonnegative P_0-completion property by Fact 8, using Fact 7 of Section 35.1.
8. By Fact 10, the graph 1a does not have and the graphs 1b and 1c do have the nonnegative P_0-completion property.
9. The graphs 1a, 1b, and 1c have the positive P- (nonnegative P-, nonnegative $P_{0,1}$-) completion property by Fact 10.

35.10 Entry Sign Symmetric P-, Entry Sign Symmetric P_0-, Entry Sign Symmetric $P_{0,1}$-, Entry Weakly Sign Symmetric P-, and Entry Weakly Sign Symmetric P_0-Matrices

In this section, all matrices are real. In the literature, *entry sign symmetric* is often called *sign symmetric*; as defined in Chapter 19.2, the latter term is used for a different condition.

Definitions:

The matrix A is an **entry sign symmetric** P- (respectively, $P_{0,1}$-, P_0-) **matrix** if and only if A is a P-matrix (respectively, $P_{0,1}$-matrix, P_0-matrix) and for all i, j, either $a_{ij}a_{ji} > 0$ or $a_{ij} = a_{ji} = 0$.

The matrix A is an **entry weakly sign symmetric** P- (respectively, $P_{0,1}$-, P_0-) **matrix** if and only if A is a P-matrix (respectively, $P_{0,1}$-matrix, P_0-matrix) and for all i, j, $a_{ij}a_{ji} \geq 0$.

The partial matrix B is a **partial entry sign symmetric** P- (respectively, $P_{0,1}$-, P_0-) **matrix** if and only if every fully specified principal submatrix of B is an entry sign symmetric P- (respectively, $P_{0,1}$-, P_0-) matrix and if both b_{ij} and b_{ji} are specified then $b_{ij}b_{ji} > 0$ or $b_{ij} = b_{ji} = 0$.

The partial matrix B is a **partial entry weakly sign symmetric** P- (respectively, $P_{0,1}$-, P_0-) **matrix** if and only if every fully specified principal submatrix of B is an entry weakly sign symmetric P- (respectively, $P_{0,1}$-, P_0-) matrix and if both b_{ij} and b_{ji} are specified then $b_{ij}b_{ji} \geq 0$.

Facts:

1. [Hog03a] Any pattern that has the entry sign symmetric P_0- (respectively, entry weakly sign symmetric P_0-) completion property also has the entry sign symmetric $P_{0,1}$- (respectively, entry weakly sign symmetric $P_{0,1}$-) completion property. Any pattern that has the entry sign symmetric $P_{0,1}$- (respectively, entry weakly sign symmetric $P_{0,1}$-) completion property also has the entry sign symmetric P- (respectively, entry weakly sign symmetric P-) completion property.
2. [Hog01] A digraph G has the (weakly) entry sign symmetric P-completion property if and only if the subdigraph of G induced by vertices that have loops has the entry (weakly) sign symmetric P-completion property.
3. [FJT00], [Hog01] A digraph G has the entry sign symmetric P_0-completion property if and only if for every connected component H of G, either H omits all loops or H has a loop at every vertex and is block-clique.
4. [FJT00] A symmetric digraph with a loop at every vertex has the entry sign symmetric $P_{0,1}$-completion property if and only if every connected component is block-clique.
5. [FJT00] A block-clique digraph has the entry sign symmetric P- (entry sign symmetric $P_{0,1}$-, entry sign symmetric P_0-, entry weakly sign symmetric P-, entry weakly sign symmetric $P_{0,1}$-, entry weakly sign symmetric P_0-) completion property. (See Fact 8 of Section 35.8 for information on the construction.)
6. [Hog01] A digraph G has the entry sign symmetric P- (entry weakly sign symmetric P-, entry weakly sign symmetric $P_{0,1}$-, entry weakly sign symmetric P_0-) completion property if and only if every strongly connected and nonseparable induced subdigraph of G has the entry sign symmetric P- (entry weakly sign symmetric P-, entry weakly sign symmetric $P_{0,1}$-, entry weakly sign symmetric P_0-) completion property. (See Algorithm 2 of Section 35.8 for information on the construction.)
7. Fact 6 is not true for the entry sign symmetric P_0-matrices (cf. Fact 3) or for the entry sign symmetric $P_{0,1}$-matrices [Wan05]. In particular, the digraphs 4p and 4q in Example 4 of Section 35.1 have neither the entry sign symmetric P_0-completion property nor the entry sign symmetric $P_{0,1}$-completion property.
8. [DHH03] A symmetric n-cycle with a loop at every vertex has the entry sign symmetric P- (entry weakly sign symmetric P-, entry weakly sign symmetric $P_{0,1}$-, entry weakly sign symmetric P_0-) completion property if and only if $n \neq 4$ and $n \neq 5$.

9. [DHH03] An order 3 digraph G with a loop at every vertex has the entry sign symmetric P- (entry weakly sign symmetric P-, entry weakly sign symmetric $P_{0,1}$-, entry weakly sign symmetric P_0-) completion property if and only if its digraph does not contain a 3-cycle or is a clique.
10. [DHH03], [Wan05] All order 4 digraphs that have a loop at every vertex have been classified as to the entry sign symmetric P- (entry sign symmetric $P_{0,1}$-, entry sign symmetric P_0-, entry weakly sign symmetric P-, entry weakly sign symmetric $P_{0,1}$-, entry weakly sign symmetric P_0-) completion property.

Examples:

The graphs and digraphs for the examples can be found in Examples 1 and 4 of section 35.1.

1. By Fact 3, the graphs 1f and 1l and the digraph 4c have the entry sign symmetric P_0-completion property and none of the other graphs or digraphs pictured do.
2. The graphs 1a, 1b, 1c, 1d, 1e, 1g, and 1h do not have the entry sign symmetric $P_{0,1}$-completion proper by Fact 4.
3. The graph 1f has the entry sign symmetric P- (entry sign symmetric $P_{0,1}$-, entry sign symmetric P_0-, entry weakly sign symmetric P-, entry sign symmetric $P_{0,1}$-, entry weakly sign symmetric P_0-) completion proper by Fact 5.
4. The graphs 1j, 1k, 1l, 1m, and 1n and the digraphs 4c and 4d and have the entry (weakly) sign symmetric P-completion property by Facts 2 and 5.
5. The digraphs 4j, 4m, 4p, and 4q have the entry sign symmetric P- (entry weakly sign symmetric P-, entry weakly sign symmetric $P_{0,1}$-, entry weakly sign symmetric P_0-) completion property by Fact 6.
6. The graphs 1a and 1b do not have the entry sign symmetric P- (entry weakly sign symmetric P-, entry weakly sign symmetric $P_{0,1}$-, entry weakly sign symmetric P_0-) completion property by Fact 8.
7. The graph 1c has the entry sign symmetric P- (entry weakly sign symmetric P-, entry weakly sign symmetric $P_{0,1}$-, entry weakly sign symmetric P_0-) completion property by Fact 8.
8. The digraphs 4f, 4g, 4h, 4k, 4l, 4n, 4o, and 4r do not have the entry sign symmetric P- (entry weakly sign symmetric P-, entry weakly sign symmetric $P_{0,1}$-, entry weakly sign symmetric P_0-) completion property by Fact 9 and Fact 7 in section 35.1.

References

[BJL96] W.W. Barrett, C.R. Johnson, and R. Loewy. The real positive definite completion problem: cycle completabilitly. *Memoirs AMS*, 584: 1–69, 1996.

[BEH06] J. Bowers, J. Evers, L. Hogben, S. Shaner, K. Snider, and A. Wangsness. On completion problems for various classes of P-matrices. *Lin. Alg. Appl.*, 413: 342–354, 2006.

[CDH03] J.Y. Choi, L.M. DeAlba, L. Hogben, B. Kivunge, S. Nordstrom, and M. Shedenhelm. The non-negative P_0-matrix completion problem. *Elec. J. Lin. Alg.*, 10:46–59, 2003.

[CDH02] J.Y. Choi, L.M. DeAlba, L. Hogben, M. Maxwell, and A. Wangsness. The P_0-matrix completion problem. *Elec. J. Lin. Alg.*, 9:1–20, 2002.

[DH00] L. DeAlba and L. Hogben. Completions of P-matrix Patterns. *Lin. Alg. Appl.*, 319:83–102, 2000.

[DHH03] L.M. DeAlba, T.L. Hardy, L. Hogben, and A. Wangsness. The (weakly) sign symmetric P-matrix completion problems. *Elec. J. Lin. Alg.*, 10: 257–271, 2003.

[DJ98] J.H. Drew and C.R. Johnson. The completely positive and doubly nonnegative completion problems. *Lin. Multilin. Alg.*, 44:85–92, 1998.

[DJK00] J.H. Drew, C.R. Johnson, S.J. Kilner, and A.M. McKay. The cycle completable graphs for the completely positive and doubly nonnegative completion problems. *Lin. Alg. Appl.*, 313:141–154, 2000.

[DG81] H. Dym and I. Gohberg. Extensions of band matrices with band inverses. *Lin. Alg. Appl.*, 36: 1–24, 1981.

[FJS00] S.M. Fallat, C.R. Johnson, and R.L. Smith. The general totally positive matrix completion problem with few unspecified entries. *Elec. J. Lin. Alg.*, 7: 1–20, 2000.

[FJT00] S.M. Fallat, C.R. Johnson, J.R. Torregrosa, and A.M. Urbano. P-matrix completions under weak symmetry assumptions. *Lin. Alg. Appl.*, 312:73–91, 2000.

[Fie66] M. Fiedler. Matrix inequalities. *Numer. Math.*, 9:109–119, 1966.

[GJS84] R. Grone, C.R. Johnson, E.M. Sá, and H. Wolkowicz. Positive definite completions of partial Hermitian matrices. *Lin. Alg. Appl.*, 58:109–124, 1984.

[Hog98a] L. Hogben. Completions of inverse M-matrix patterns. *Lin. Alg. Appl.*, 282:145–160, 1998.

[Hog98b] L. Hogben. Completions of M-matrix patterns. *Lin. Alg. Appl.*, 285: 143–152, 1998.

[Hog00] L. Hogben. Inverse M-matrix completions of patterns omitting some diagonal positions. *Lin. Alg. Appl.*, 313:173–192, 2000.

[Hog01] L. Hogben. Graph theoretic methods for matrix completion problems. *Lin. Alg. Appl.*, 328:161–202, 2001.

[Hog02] L. Hogben. The symmetric M-matrix and symmetric inverse M-matrix completion problems. *Lin. Alg. Appl.*, 353:159–168, 2002.

[Hog03a] L. Hogben. Matrix completion problems for pairs of related classes of matrices. *Lin. Alg. Appl.*, 373:13–29, 2003.

[Hog03b] L. Hogben. Relationships between the completion problems for various classes of matrices. *Proceedings of SIAM International Conference of Applied Linear Algebra*, 2003, available electronically at: http://www.siam.org/meetings/la03/proceedings/.

[Hog] L. Hogben. Completions of partial strictly copositive matrices omitting some diagonal entries, to appear in *Lin. Alg. Appl.*

[HJR05] L. Hogben, C.R. Johnson, and R. Reams. The copositive matrix completion problem. *Lin. Alg. Appl.*, 408:207–211, 2005.

[HW] L. Hogben and A. Wangsness: Matrix Completions Webpage: http://orion.math.iastate.edu/lhogben/MC/homepage.html.

[HJ91] R. Horn and C.R. Johnson. *Topics in Matrix Analysis.* Cambridge University Press, Cambridge, 1991.

[JK96] C. Johnson and B. Kroschel. The combinatorially symmetric P-matrix completion problem. *Elec. J. Lin. Alg.*, 1:59–63, 1996.

[JS96] C.R. Johnson and R.L. Smith. The completion problem for M-matrices and inverse M-matrices. *Lin. Alg. Appl.*, 290:241–243, 1996.

[JS99] C.R. Johnson and R.L. Smith. The symmetric inverse M-matrix completion problem. *Lin. Alg. Appl.*, 290:193–212, 1999.

[JS00] C. Johnson and R.L. Smith. The positive definite completion problem relative to a subspace. *Lin. Alg. Appl.*, 307:1–14, 2000.

[JTU03] C. Jordán, J.R. Torregrosa, and A.M. Urbano. Completions of partial P-matrices with acyclic or non-acyclic associated graph. *Lin. Alg. Appl.*, 368:25–51, 2003.

[Lau98] M. Laurent. A connection between positive semidefinite and Euclidean distance matrix completion problems. *Lin. Alg. Appl.*, 273:9–22, 1998.

[Wan05] A. Wangsness. The matrix completion prioblem regarding various classes of $P_{0,1}$-matrices. Ph.D. thesis, Iowa State University, 2005.

36 Algebraic Connectivity

Steve Kirkland
University of Regina

36.1 Algebraic Connectivity for Simple Graphs: Basic Theory

Let G be a simple graph on $n \geq 2$ vertices with Laplacian matrix L_G, and label the eigenvalues of L_G as $0 = \mu_1 \leq \mu_2 \leq \ldots \leq \mu_n$. Throughout this chapter, we consider only Laplacian matrices for graphs on at least two vertices. Henceforth, we use the term *graph* to refer to a simple graph.

Definitions:

The **algebraic connectivity of** G, denoted $\alpha(G)$, is given by $\alpha(G) = \mu_2$.

A **Fiedler vector** is an eigenvector of L_G corresponding to $\alpha(G)$.

Given graphs $G_1 = (V_1, E_1)$ and $G_2 = (V_2, E_2)$, their **product**, $G_1 \times G_2$, is the graph with vertex set $V_1 \times V_2$, with vertices (u_1, u_2) and (w_1, w_2) adjacent if and only if either u_1 is adjacent to w_1 in G_1 and $u_2 = w_2$, or u_2 is adjacent to w_2 in G_2 and $u_1 = w_1$.

Facts:

1. [Fie89] Let G be a graph of order n with Laplacian matrix L_G. Then $\alpha(G) = \min\{\sum_{i<j,\{i,j\}\in E} (x_i - x_j)^2 | \sum_{1\leq i\leq n} x_i{}^2 = 1, \sum_{1\leq i\leq n} x_i = 0\} = \min\{x^T L_G x | x^T x = 1, x^T \mathbf{1} = 0\}$.
2. [Fie73] The algebraic connectivity of a graph is nonnegative, and is equal to 0 if and only if the graph is disconnected.
3. [Fie73] Let G be a connected graph on n vertices with vertex connectivity $\kappa_v(G)$, and suppose that $G \neq K_n$. Then $\alpha(G) \leq \kappa_v(G)$. (See Fact 13 below for a discussion of the equality case.)
4. [Fie73] Suppose that G is a graph on n vertices, and let $\overline{G}$ denote the complement of G. Then $\alpha(\overline{G}) = n - \mu_n$, where μ_n denotes the largest Laplacian eigenvalue for G.

5. ([GR01], p. 280) If G is a graph on $n \geq 2$ vertices that is regular of degree k, then denoting the eigenvalues of $\mathcal{A}_G$ by $\lambda_1 \leq \ldots \leq \lambda_n$, we have $\alpha(G) = k - \lambda_{n-1}$.
6. [Mer94] Suppose that G_1 and G_2 are two graphs on n_1 and n_2 vertices, respectively, with $n_1, n_2 \geq 2$. Then $\alpha(G_1 + G_2) = \min\{\alpha(G_1) + n_2, \alpha(G_2) + n_1\}$. Similarly, if H is a graph on $k \geq 2$ vertices, then $\alpha(H + K_1) = \alpha(H) + 1$.
7. [Fie73] Suppose that G_1 and G_2 are graphs, each of which has at least two vertices. Then $\alpha(G_1 \times G_2) = \min\{\alpha(G_1), \alpha(G_2)\}$.
8. [Fie73] Suppose that the graph $\hat{G}$ is formed from the graph G by adding an edge not already present in G. Then $\alpha(G) \leq \alpha(\hat{G})$.
9. [FK98] Let G and H be graphs, and suppose that the graph $\hat{G}$ is formed from $G \cup H$ as follows: Fix a vertex v of G and a subset S of the vertex set for H, and for each $w \in S$, add in the edge between v and w. Then $\alpha(\hat{G}) \leq \alpha(G)$.
10. [Fie73] Let G be a graph, and suppose that the graph $\hat{G}$ is formed from G by deleting a collection of k vertices and all edges incident with them. Then $\alpha(\hat{G}) \geq \alpha(G) - k$.
11. [GMS90] Let G be a graph on $n \geq 3$ vertices and suppose that the edge e of G is not on any 3-cycles. Form $\hat{G}$ from G by deleting e and identifying the two vertices incident with it. Then $\alpha(\hat{G}) \geq \alpha(G)$.
12. [Fie73] If G is a graph on n vertices, then $\alpha(G) \leq n$. Equality holds if and only if $G = K_n$.
13. [KMNS02] Let G be a connected graph on n vertices with vertex connectivity $\kappa_v(G)$, and suppose that $G \neq K_n$. Then $\alpha(G) = \kappa_v(G)$ if and only if G can be written as $G = G_1 + G_2$, where G_1 is disconnected, G_2 has $\kappa_v(G)$ vertices, and $\alpha(G_2) \geq 2\kappa_v(G) - n$.
14. [Fie89] If G is a connected graph on n vertices with edge connectivity $\kappa_e(G)$, then $2(1 - cos(\frac{\pi}{n}))\kappa_e(G) \leq \alpha(G)$. Equality holds if and only if $G = P_n$.

Examples:

1. ([GK69], p. 138) For $n \geq 2$, the algebraic connectivity of the path P_n is $2(1 - cos(\frac{\pi}{n}))$, and it is a simple eigenvalue of the corresponding Laplacian matrix.
2. The following can be deduced from basic results on circulant matrices. If $n \geq 3$, the algebraic connectivity of the cycle C_n is $2(1 - cos(\frac{2\pi}{n}))$, and it is an eigenvalue of multiplicity 2 of the corresponding Laplacian matrix.
3. The algebraic connectivity of K_n is n, and it is an eigenvalue of multiplicity $n - 1$ of the corresponding Laplacian matrix.
4. If $m \leq n$ and $2 \leq n$, then $\alpha(K_{m,n}) = m$. If $1 \leq m < n$, then m is an eigenvalue of multiplicity $n - 1$ of the corresponding Laplacian matrix, while if $2 \leq m = n$, then m is an eigenvalue of multiplicity $2m - 2$ of the corresponding Laplacian matrix.
5. The algebraic connectivity of the Petersen graph is 2, and it is an eigenvalue of multiplicity 5 of the corresponding Laplacian matrix.
6. The algebraic connectivity of the ladder on 6 vertices, shown in Figure 36.1, is 1.
7. Graphs with large algebraic connectivity arise in the study of the class of so-called expander graphs. (See Section 28.5 or [Alo86] for definitions and discussion.)
8. A Fiedler vector for a graph provides a heuristic for partitioning its vertex set so that the number of edges between the two parts is small (see [Moh92] for a discussion), and that heuristic has applications to sparse matrix computations (see [PSL90]).

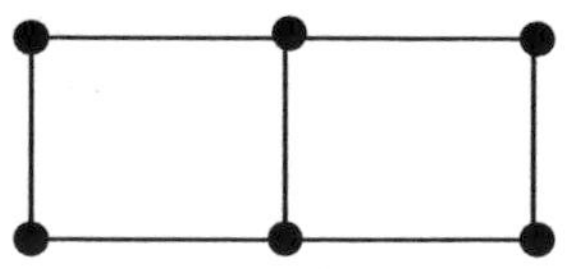

FIGURE 36.1

36.2 Algebraic Connectivity for Simple Graphs: Further Results

The term *graph* means *simple graph* in this section.

Definitions:

Let $G = (V, E)$ be a graph, and let $X, V\backslash X$ be a nontrivial partitioning of V. The corresponding **edge cut** is the set E_X of edges of G that have one end point in X and the other end point in $V\backslash X$.

Suppose that G_1 and G_2 are graphs. A graph H is formed by **appending** G_2 **at vertex** v **of** G_1 if H is constructed from $G_1 \cup G_2$ by adding an edge between v and a vertex of G_2.

Suppose that $g, n \in \mathbb{N}$ with $n > g \geq 3$. The graph $C_{n,g}$ is formed by appending the cycle C_g at a pendent vertex of the path P_{n-g}.

Suppose that $g, n \in \mathbb{N}$ with $n > g \geq 3$. Let $D_{g,n-g}$ denote the graph formed from the cycle C_g by appending $n - g$ isolated vertices at a single vertex of that cycle.

For a connected graph G, a vertex v is a **cut-vertex** if $G - v$, the graph formed from G by deleting the vertex v and all edges incident with it, is disconnected.

A graph is **unicyclic** if it contains precisely one cycle.

Facts:

1. [Moh91] If G is a connected graph on n vertices with diameter d, then $\alpha(G) \geq \frac{4}{dn}$.
2. [AM85] If G is a connected graph on n vertices with diameter d and maximum degree Δ, then $d \leq 2\lceil\sqrt{\frac{2\Delta}{\alpha(G)}}log_2 n\rceil$.
3. [Moh91] If G is a connected graph on n vertices with diameter d and maximum degree Δ, then $d \leq 2\lceil\frac{\Delta+\alpha(G)}{4\alpha(G)}ln(n-1)\rceil$.
4. [Moh92] Let $G = (V, E)$ be a graph, let $X, V\backslash X$ be a nontrivial partitioning of its vertex set, and let E_X denote the corresponding edge cut. Then $\alpha(G) \leq \frac{|V||E_X|}{|X||V\backslash X|}$ and ([FKP03a]) if $\alpha(G) = \frac{|V||E_X|}{|X||V\backslash X|}$, then necessarily there are integers d_1, d_2 such that:
   - Each vertex in X is adjacent to precisely d_1 vertices in $V\backslash X$.
   - Each vertex in $V\backslash X$ is adjacent to precisely d_2 vertices in X.
   - $|X|d_1 = |V\backslash X|d_2$.
   - $\alpha(G) = d_1 + d_2$.
5. [Moh89] Let $G = (V, E)$ be a graph on at least four vertices, with maximum degree Δ. Then $\frac{\alpha(G)}{2} \leq \Phi(G) \leq \sqrt{\alpha(G)(2\Delta - \alpha(G))}$, where $\Phi(G)$ is the isoperimetric number of G. (See Section 32.5.)
6. [FK98] Among all connected graphs on n vertices with girth 3, the algebraic connectivity is uniquely minimized by $C_{n,3}$.
7. [FKP02] If $g \geq 4$ and $n \geq 3g - 1$, then among all connected graphs on n vertices with girth g, the algebraic connectivity is uniquely minimized by $C_{n,g}$.
8. [Kir00] Let G be a graph on n vertices, and suppose that G has k cut-vertices, where $2 \leq k \leq \frac{n}{2}$. Then $\alpha(G) \leq \frac{2(n-k)}{n-k+2+\sqrt{(n-k)^2+4}}$. For each such k and n, there is a graph on n vertices with k cut-vertices such that equality is attained in the upper bound on α.
9. [Kir01] Suppose that $n \geq 5$ and $\frac{n}{2} < k$, and let G be a graph on n vertices with k cut-vertices. Let $q = \lfloor\frac{k}{n-k}\rfloor$ and $l = k - (n-k)q$.
   - If $l = 1$, then $\alpha(G) \leq 2(1 - cos(\frac{\pi}{2q+3}))$.
   - If $l = 0$, let θ_0 be the unique element of $\left[\frac{\pi}{2q+3}, \frac{\pi}{2q+1}\right]$ such that $(n-k-1)cos((2q+1)\theta_0/2) + cos((2q+3)\theta_0/2) = 0$. Then $\alpha(G) \leq 2(1 - cos(\theta_0))$.

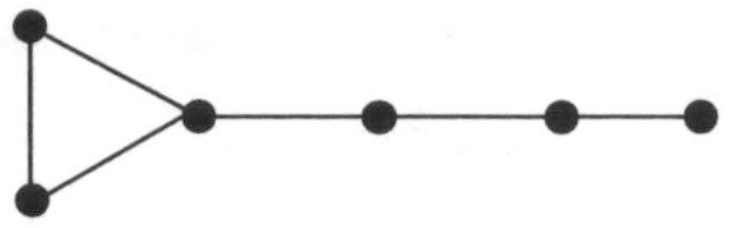

FIGURE 36.2 The graph $C_{6,3}$.

- If $2 \leq l$, let θ_0 be the unique element of $\left[\frac{\pi}{2q+5}, \frac{\pi}{2q+3}\right]$ such that $(n-k-1)cos((2q+3)\theta_0/2) + cos((2q+5)\theta_0/2) = 0$. Then $\alpha(G) \leq 2(1 - cos(\theta_0))$.

For each k and n with $\frac{n}{2} < k$, there is a graph on n vertices with k cut-vertices such that equality is attained in the corresponding upper bound on α.

10. [FK98] Over the class of unicyclic graphs on n vertices having girth 3, the algebraic connectivity is uniquely maximized by $D_{3,n-3}$.
11. [FKP03b] Over the class of unicyclic graphs on n vertices having girth 4, the algebraic connectivity is uniquely maximized by $D_{4,n-4}$.
12. [FKP03b] Fix $g \geq 5$. There is an N such that if $n > N$, then over the class of unicyclic graphs on n vertices having girth g, the algebraic connectivity is uniquely maximized by $D_{g,n-g}$.
13. [Kir03] For each real number $r \geq 0$, there is a sequence of graphs G_k with distinct algebraic connectivities, such that $\alpha(G_k)$ converges monotonically to r as $k \to \infty$.

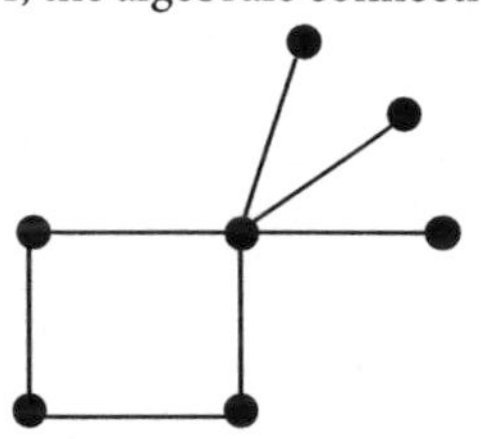

FIGURE 36.3 The graph $D_{4,3}$.

Examples:

1. The graph $C_{6,3}$ is shown in Figure 36.2; its algebraic connectivity is approximately 0.3249.
2. The graph $D_{4,3}$ is shown in Figure 36.3.
3. [FK98] The algebraic connectivity of $D_{3,n-3}$ is 1, and it is an eigenvalue of multiplicity $n-3$ of the corresponding Laplacian matrix.
4. Consider the graph G pictured in Figure 36.4. Its algebraic connectivity is 2, so that from Fact 5 above, we deduce that $1 \leq \Phi(G) \leq 2\sqrt{2}$. It turns out that the value of $\Phi(G)$ is $\frac{3}{2}$.

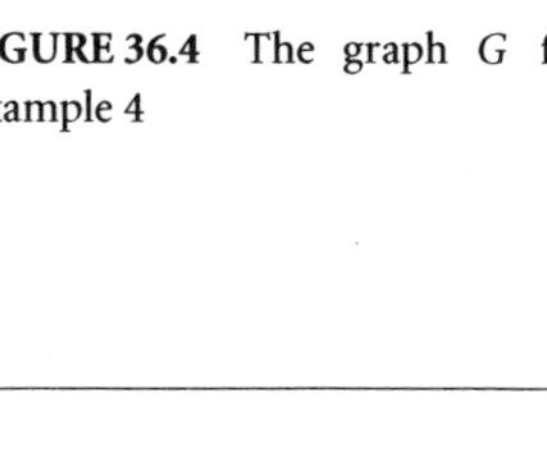

FIGURE 36.4 The graph G for Example 4

36.3 Algebraic Connectivity for Trees

The term *graph* means *simple graph* in this section.

Definitions:

Let T be a tree, and let $\mathbf{y}$ be a Fiedler vector for T. A **characteristic vertex** of T is a vertex i satisfying one of the following conditions:

I $y_i = 0$, and vertex i is adjacent to a vertex j such that $y_j \neq 0$.
II There is a vertex j adjacent to vertex i such that $y_i y_j < 0$.

A tree is called **type I** if it has a single characteristic vertex, and **type II** if it has two characteristic vertices.

Let T be a tree and suppose that v is a vertex of T. The **bottleneck matrix** of a branch of T at v is the inverse of the principal submatrix of the Laplacian matrix corresponding to the vertices of that branch.

A branch at vertex v is called a **Perron branch at** v if the Perron value of the corresponding bottleneck matrix is maximal among all branches at v.

Suppose that T is a type I tree with characteristic vertex i. A branch at i is called an **active branch** if, for some Fiedler vector **y**, the entries in **y** corresponding to the vertices in the branch are nonzero.

Suppose that $n \geq d+1$. Denote by $P(n, d)$ the tree constructed as follows: Begin with the path P_{d+1}, on vertices $1, \ldots, d+1$, labeled so that vertices 1 and $d+1$ are pendent, while for each $i = 2, \ldots, d$, vertex i is adjacent to vertices $i-1$ and $i+1$, then append $n-d-1$ isolated vertices at vertex $\lfloor \frac{d}{2} \rfloor + 1$ of that path.

Let $T(k, l, d)$ be the tree on $n = d+k+l$ vertices constructed from the path P_d by appending k isolated vertices at one end vertex of P_d, and appending l isolated vertices at the other end vertex of P_d.

Facts:

1. The bottleneck matrix for a branch is the inverse of an irreducible M-matrix, and so is entrywise positive (see Chapter 9.5 and/or [KNS96]).
2. [Fie89] If T is a tree on n vertices, then $2(1 - cos(\frac{\pi}{n})) \leq \alpha(T)$; equality holds if and only if $G = P_n$.
3. [Mer87] If T is a tree on $n \geq 3$ vertices, then $\alpha(G) \leq 1$; equality holds if and only if $G = K_{1,n-1}$.
4. [GMS90] If T is a tree with diameter d, then $\alpha(T) \leq 2(1 - cos(\frac{\pi}{d+1}))$.
5. [Fie75b] Let T be a tree on vertices labeled $1, \ldots, n$, and suppose that **y** is a Fiedler vector of T. Then exactly one of two cases can occur.

   - There are no zero entries in **y**. Then T contains a unique edge $\{i, j\}$ such that $y_j < 0 < y_i$. As we move along any path in T that starts at i and does not contain j, the corresponding entries in **y** are positive and increasing. As we move along any path in T that starts at j and does not contain i, the corresponding entries in **y** are negative and decreasing. In this case T is type II.
   - The vector **y** has at least one zero entry. Then there is a unique vertex i of T such that $y_i = 0$ and i is adjacent to a vertex j such that $y_j \neq 0$. As we move along any path in T that starts at vertex i, the corresponding entries in **y** are either increasing, decreasing, or identically zero. In this case T is type I.

6. [KNS96] Let T be a tree and **y** be a Fiedler vector for T. Let P be a path in T that starts at a characteristic vertex i of T, and does not contain any other characteristic vertices of T. If, as we move along P away from i the entries of **y** are increasing, then they are also concave down. If, as we move along P away from i the entries of **y** are decreasing, then they are also concave up.
7. [Mer87] Let T be a tree. Then each Fiedler vector for T identifies the same vertex (or vertices) as the characteristic vertex (or vertices). Consequently, the type of the tree is independent of the particular choice of the Fiedler vector.
8. [Fie75a] If T is a type II tree, then $\alpha(T)$ is a simple eigenvalue.
9. [KNS96] A tree T is type I if and only if at some vertex i, there are two or more Perron branches. In that case, i is the characteristic vertex of T, $\alpha(T)$ is the reciprocal of the Perron value of the bottleneck matrix of a Perron branch at i, and ([KF98]) the multiplicity of $\alpha(T)$ is one less than the number of Perron branches at i.
10. [KNS96] A tree T is type II if and only if there is a unique Perron branch at every vertex. In this case, the characteristic vertices of T are the unique adjacent vertices i, j such that the Perron branch at vertex i is the branch containing vertex j, and the Perron branch at vertex j is the branch containing vertex i. Letting B_i and B_j denote the bottleneck matrices for the Perron branches at vertices i and j, respectively, $\exists! \gamma \in (0, 1)$ such that $\rho(B_i - \gamma J) = \rho(B_j - (1-\gamma)J)$, and the common Perron value for these matrices is $1/\alpha(T)$.
11. The following is a consequence of results in [GM87] and [KNS96]. Suppose that T is a type I tree with characteristic vertex i; then a branch at i is an active branch if and only if it is a Perron branch at i.
12. [FK98] Among all trees on n vertices with diameter d, the algebraic connectivity is maximized by $P(n, d)$.
13. [FK98] Let T be a tree on n vertices with diameter d. Then $\alpha(T) \geq \alpha(T(\lceil \frac{n-d+1}{2} \rceil, \lceil \frac{n-d+1}{2} \rceil, d-1))$, with equality holding if and only if $T = T(\lceil \frac{n-d+1}{2} \rceil, \lceil \frac{n-d+1}{2} \rceil, d-1)$.

Examples:

1. [GM90] The algebraic connectivity of $T(k,l,2)$ is the smallest root of the polynomial $x^3 - (k + l + 4)x^2 + (2k + 2l + kl + 5)x - (k + l + 2)$.
2. Let T be the tree constructed as follows: At a single vertex v, append $k \geq 2$ copies of the path P_2 and $l \geq 0$ pendent vertices. For each branch at v consisting of a path on two vertices, the bottleneck matrix can be written as $\begin{bmatrix} 2 & 1 \\ 1 & 1 \end{bmatrix}$, which has Perron value $\frac{3+\sqrt{5}}{2}$, while each branch at v consisting of a single vertex has bottleneck matrix equal to $[1]$. Then $\alpha(T) = \frac{3-\sqrt{5}}{2}$, and is an eigenvalue of multiplicity $k-1$ of the corresponding Laplacian matrix. In particular, T is a type I tree with characteristic vertex v.
3. [FK98] If $d \geq 4$ is even, $\alpha(P(n,d)) = 2(1 - \cos(\frac{\pi}{d+1}))$, and is a simple eigenvalue of the corresponding Laplacian matrix.
4. Consider the tree $T(2,2,2)$ shown in Figure 36.5. At both of its nonpendent vertices, the bottleneck matrix for the corresponding Perron component can be written as $\begin{bmatrix} 2 & 1 & 1 \\ 1 & 2 & 1 \\ 1 & 1 & 1 \end{bmatrix}$.

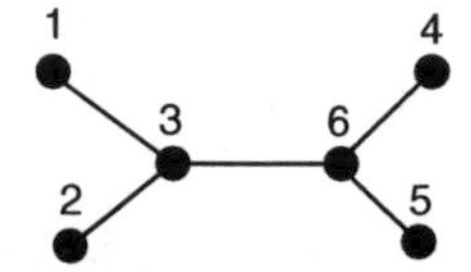

FIGURE 36.5 The tree $T(2,2,2)$.

It follows that $1/\alpha(T) = \rho\left(\begin{bmatrix} 2 & 1 & 1 \\ 1 & 2 & 1 \\ 1 & 1 & 1 \end{bmatrix} - 1/2J\right) = (5 + \sqrt{17})/4$. Hence, the algebraic connectivity for $T(2,2,2)$ is $(5 - \sqrt{17})/2$, and it is a type II tree, with the two nonpendent vertices as the characteristic vertices. $[\frac{3+\sqrt{17}}{4}, \frac{3+\sqrt{17}}{4}, 1, -\frac{3+\sqrt{17}}{4}, -\frac{3+\sqrt{17}}{4}, -1]^T$ is a Fiedler vector.
5. [GM87] Consider the tree T shown in Figure 36.6; this is a type I tree with characteristic vertex v. The numbers above the vertices are the vertex numbers and the numbers below are (to four decimal places) the entries in a Fiedler vector for T. The bottleneck matrix for the branch at v on 5 vertices can be written as $\begin{bmatrix} 3 & 1 & 2 & 1 & 1 \\ 1 & 3 & 1 & 2 & 1 \\ 2 & 1 & 2 & 1 & 1 \\ 1 & 2 & 1 & 2 & 1 \\ 1 & 1 & 1 & 1 & 1 \end{bmatrix}$, while that for the branch at v on 4 vertices can be written $\begin{bmatrix} 3 & 2 & 2 & 1 \\ 2 & 3 & 2 & 1 \\ 2 & 2 & 2 & 1 \\ 1 & 1 & 1 & 1 \end{bmatrix}$. The algebraic connectivity is the smallest root of the polynomial $x^3 - 6x^2 + 8x - 1$, and is approximately 0.1392.

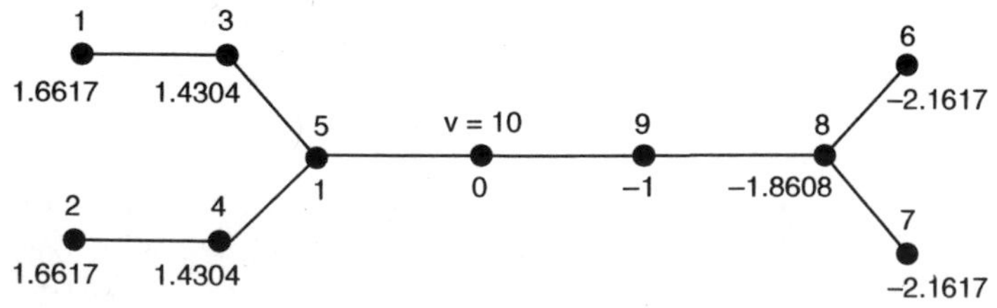

FIGURE 36.6 The tree T in Example 5.

36.4 Fiedler Vectors and Algebraic Connectivity for Weighted Graphs

The term *graph* means *simple graph* in this section.

Definitions:

Let $G = (V, E)$ be a graph on $n \geq 2$ vertices, and suppose that for each edge $e \in E$ we have an associated positive number $w(e)$.

- Then $w(e)$ is the **weight** of e.
- The function $w : E \to \mathbb{R}^+$ is a **weight function** on G.
- The graph G, together with the function $w : E \to \mathbb{R}^+$, is a **weighted graph**, and is denoted G_w.

The **Laplacian matrix** $L(G_w)$ for the weighted graph G_w is the $n \times n$ matrix $L(G_w) = [l_{i,j}]$ such that $l_{i,j} = 0$ if vertices i and j are distinct and not adjacent in G, $l_{i,j} = -w(e)$ if $e = \{i, j\} \in E$, and $l_{i,i} = \sum w(e)$, where the sum is taken over all edges e incident with vertex i.

Throughout this chapter, we only consider Laplacian matrices for weighted graphs on at least two vertices. The notation $L(G_w)$ for the Laplacian matrix of the weighted graph G_w should not be confused with the notation for the line graph introduced in Section 28.2.

Let G_w be a weighted graph on n vertices, and denote the eigenvalues of $L(G_w)$ by $0 = \mu_1 \leq \mu_2 \leq \ldots \leq \mu_n$. The **algebraic connectivity of** G_w is μ_2 and is denoted $\alpha(G_w)$.

Let G_w be a weighted graph. A **Fiedler vector for** G_w is an eigenvector of $L(G_w)$ corresponding to the eigenvalue $\alpha(G_w)$.

Let G_w be a weighted graph and $\mathbf{y}$ be a Fiedler vector for G_w. For each vertex i of G_w, the **valuation of** i is equal to the corresponding entry in $\mathbf{y}$. A vertex is **valuated positively, negatively,** or **zero** accordingly, as the corresponding entry in $\mathbf{y}$ is positive, negative, or zero.

Facts:

1. [Fie75b] Let G_w be a weighted graph and let $\mathbf{y}$ be a Fiedler vector for G_w. Then exactly one of the following holds.

   - Case A—There is a single block B_0 in G_w containing both positively valuated vertices and negatively valuated vertices. For every other block of G_w, the vertices are either all valuated positively or all valuated negatively or all valuated zero. Along any path in G that starts at a vertex $k \in B_0$ and contains at most two cut-vertices from each block, the valuations of the cut-vertices form either an increasing sequence, a decreasing sequence, or an all zero sequence, according as $y_k > 0$, $y_k < 0$, or $y_k = 0$, respectively. In the case where $y_k = 0$, then each vertex on that path is valuated zero.
   - Case B — No block in G_w has both positively and negatively valuated vertices. Then there is a unique vertex i of G_w having zero valuation that is adjacent to a vertex j with nonzero valuation. The vertex i is a cut-vertex. Each block of G_w contains (with the exception of i) only vertices of positive valuation, or only vertices of negative valuation, or only vertices of zero valuation. Along any path that starts at vertex i and contains at most two cut-vertices from each block, the valuations of the cut-vertices form either an increasing sequence, a decreasing sequence, or an all zero sequence. Any path in G_w that contains both positively and negatively valuated vertices must pass through vertex i.

2. [KF98] Let G_w be a weighted graph. If case A holds for some Fiedler vector, then case A holds for every Fiedler vector and, moreover, every Fiedler vector identifies the same block of G_w having vertices of both positive and negative valuations. Similarly, if case B holds for some Fiedler vector, then case B holds for every Fiedler vector and, moreover, every vector identifies the same vertex i that is valuated zero and is adjacent to a vertex with nonzero valuation.

3. [KF98] Suppose that for a weighted graph G_w, case A holds, and let B_0 be the unique block of G_w containing both positively and negatively valuated vertices. Fix an edge $e \in B_0$, and let $\hat{w}$ be the weighting of G formed from w by replacing $w(e)$ by t, where $0 < t < w(e)$. Then for the weighted graph $G_{\hat{w}}$, case A also holds, and B_0 is still the block of G containing both positively and negatively valuated vertices. Similarly, if $B_0 - e$ is a block of $G - e$, then case A still holds for the weighting $\hat{w}$ of $G - e$ arising from setting $\hat{w}(f) = w(f)$ for each $f \in E\backslash\{e\}$, and $B_0 - e$ is still the block containing both positively and negatively valuated vertices.
4. [KF98] Suppose that G is a graph, that w is weight function on G, and consider the weighted graph G_w. Suppose that B is a block of G_w that is valuated all nonnegatively or all nonpositively by some Fiedler vector. Let $\hat{G}_w$ be the weighted graph formed from G_w by either adding a weighted edge to B, or raising the weight of an existing edge in B, and denote the modified block by $\hat{B}$. Then every Fiedler vector for $\hat{G}_w$ valuates the vertices of $\hat{B}$ all nonnegatively, or all nonpositively.
5. [FN02] Suppose that G_w is a weighted graph on n vertices with Laplacian matrix $L(G_w)$. For each nonempty subset U of the vertex set V, let $L(G_w)[U]$ denote the principal submatrix of $L(G_w)$ on the rows and columns corresponding to the vertices in U, and let $\tau(L(G_w)[U])$ denote its smallest eigenvalue. Then $\alpha(G_w) \geq \min\{\max\{\tau(L(G_w)[U]), \tau(L(G_w)[V\setminus U])\} | U \subset V, 0 < |U| \leq [\frac{n}{2}]\}$.
6. [KN04] Suppose that $G = (V, E)$ is a graph, that w is weight function on G, and consider the weighted graph G_w. Fix an edge $e = \{i, j\} \in E$. For each $t \geq w(e)$, let w_t be the weight function on G constructed from w by setting the weight of the edge e to t.

   - If some Fiedler vector $\mathbf{y}$ for G_w has the property that $y_i = y_j$, then for any $t \geq w(e), \alpha(G_{w_t}) = \alpha(G_w)$. In particular, if $\alpha(G_w)$ is an eigenvalue of multiplicity at least two, then $\alpha(G_{w_t}) = \alpha(G_w)$ for all $t \geq w(e)$.
   - If $\alpha(G_w)$ is a simple eigenvalue, then $\exists t_0 > 0$ such that for each $t \in [0, t_0), \alpha(G_{w_t})$ is a twice differentiable function of t, with $0 \leq \frac{d\alpha(G_{w_t})}{dt} \leq 2$ and $\frac{d^2\alpha(G_{w_t})}{dt^2} \leq 0$. Further, let $\mathbf{y}(t)$ denote a Fiedler vector for G_{w_t} of norm 1 that is analytic for $t \in [0, t_0)$; then $|(\mathbf{y}(t))_i - (\mathbf{y}(t))_j|$ is nonincreasing for $t \in [0, t_0)$.

Examples:

1. Consider the weighting of K_n in which each edge has weight 1. Then any vector that is orthogonal to $\mathbf{1}_n$ is a Fiedler vector.
2. Consider the weighting of $K_{1,n-1}$ in which edge has weight 1. If i and j are pendent vertices of $K_{1,n-1}$, then $\mathbf{e}_i - \mathbf{e}_j$ is a Fiedler vector.
3. Consider the weighting of the path P_n in which each edge has weight 1, and label the vertices of P_n so that vertices 1 and n are pendent, while for each $i = 2, \ldots, n-1$, vertex i is adjacent to vertices $i-1$ and $i+1$. Then the vector $\mathbf{y}$ with $y_i = cos((2i-1)\pi/(2n))$ is a Fiedler vector.
4. Suppose that $a, b > 0$, and let T_w be the weighting of the path P_4 arising by assigning the pendent edges weight a, and the nonpendent edge weight b. The corresponding Laplacian matrix can be written as $\begin{bmatrix} a & 0 & -a & 0 \\ 0 & a & 0 & -a \\ -a & 0 & a+b & -b \\ 0 & -a & -b & a+b \end{bmatrix}$, which has eigenvalues $0, a+b \pm \sqrt{a^2+b^2}$, and $2a$. In particular, $\alpha(T_w) = a + b - \sqrt{a^2+b^2}$, with corresponding Fiedler vector $\begin{bmatrix} a \\ -a \\ \sqrt{a^2+b^2} - b \\ b - \sqrt{a^2+b^2} \end{bmatrix}$.

5. Figure 36.7 shows a weighted graph G_w, where the parameters a, b, c are the weights of the edges, and where the vertices are labeled $1, \ldots, 4$, as indicated in that figure. The corresponding Laplacian matrix is $L(G_w) = \begin{bmatrix} b & 0 & 0 & -b \\ 0 & a+c & -c & -a \\ 0 & -c & a+c & -a \\ -b & -a & -a & 2a+b \end{bmatrix}$. The eigenvalues of $L(G_w)$ are 0, $a+2c$, and $\frac{3a+2b\pm\sqrt{9a^2-4ab+4b^2}}{2}$, so that $\alpha(G_w) = \min\{a+2c, \frac{3a+2b-\sqrt{9a^2-4ab+4b^2}}{2}\}$. If $\alpha(G_w) = a+2c$, then $\begin{bmatrix} 0 \\ 1 \\ -1 \\ 0 \end{bmatrix}$ is a Fiedler vector for G_w, while if $\alpha(G_w) = \frac{3a+2b-\sqrt{9a^2-4ab+4b^2}}{2}$, then $\begin{bmatrix} \frac{3a+\sqrt{9a^2-4ab+4b^2}}{2} \\ -\frac{a+2b+\sqrt{9a^2-4ab+4b^2}}{4} \\ -\frac{a+2b+\sqrt{9a^2-4ab+4b^2}}{4} \\ b-a \end{bmatrix}$ is a Fiedler vector for G_w.

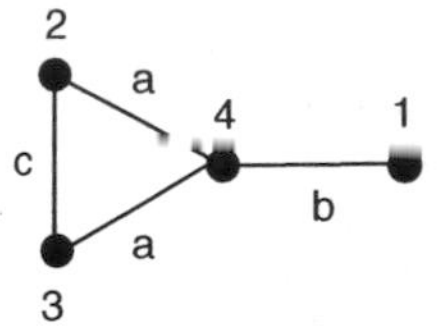

FIGURE 36.7 The weighted graph G_w in Example 5.

36.5 Absolute Algebraic Connectivity for Simple Graphs

The term *graph* means *simple graph* in this section.

Definitions:

Let $G = (V, E)$ be a graph on $n \geq 2$ vertices. Let W denote the set of all nonnegative weightings of G with the property that for each weighting w, we have $\sum_{e \in E} w_e = |E|$ (observe that here we relax the restriction that each edge weight be positive, and allow zero weights). The **absolute algebraic connectivity of** G is denoted by $\hat{\alpha}(G)$, and is given by $\hat{\alpha}(G) = \max\{\alpha(G_w) | w \in W\}$.

Let $T = (V, E)$ be a tree. For each vertex u of T, define $S(u) = \sum_{k \in V}(d(u, k))^2$.

Facts:

1. [Fie90] For a graph $G, \hat{\alpha}(G) = 0$ if and only if G is disconnected.
2. [Fie90] If G is a graph and Ω is its automorphism group, then $\hat{\alpha}(G) = \max\{\alpha(G_w) | w \in W_0\}$, where W_0 is the subclass of W consisting of weightings for which equivalent edges in Ω have the same weight.
3. [KP02] If G is a graph with m edges, and H is the graph formed from G by adding an edge not already present in G, then $\frac{m+1}{m}\hat{\alpha}(G) \leq \hat{\alpha}(H)$.
4. [Fie90] Let T be a tree on n vertices. Then one of the following two cases occurs:

   - There is a vertex v of T such that for any vertex $k \neq v$, $S(v) \leq S(k) - n$. In this case, $\hat{\alpha}(T) = \frac{n-1}{S(v)}$.
   - There is an edge $\{u, v\}$ of T such that $|S(u) - S(v)| < n$. In this case, $\hat{\alpha}(T) = \frac{4n(n-1)}{4nS(u)-(n+S(u)-S(v))^2}$.

   Weightings of trees that yield the maximum value $\hat{\alpha}$ are also discussed in [Fie90].
5. [Fie93] If T is a tree on n vertices, then $\frac{12}{n(n+1)} \leq \hat{\alpha}(T) \leq 1$. Equality holds in the lower bound if $T = P_n$, and equality holds in the upper bound if $T = K_{1,n-1}$.
6. [KP02] Suppose that G is a graph on $n \geq 7$ vertices that has a cut-vertex. Then $\hat{\alpha}(G) \leq \left(\frac{n^2-3n+4}{n-3}\right)\left(1 - \frac{4}{2(n-1)-(n-3)\sqrt{2(n-2)/(n-1)}}\right)$. Further, the upper bound in attained by the graph formed by appending a pendent vertex at a vertex of K_{n-1}; a weighting of that graph that yields the maximum value $\hat{\alpha}$ is also given in [KP02].

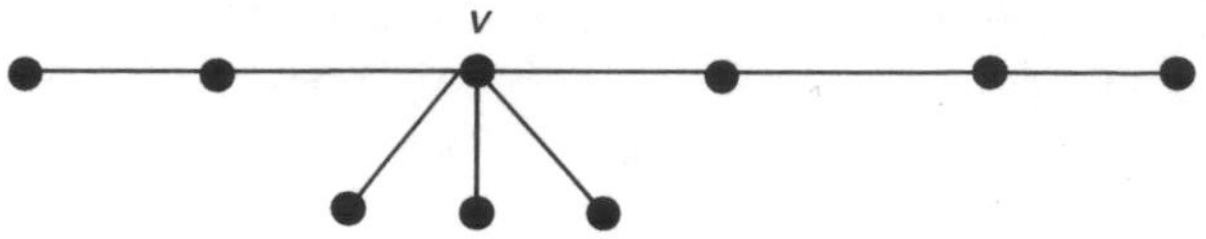

FIGURE 36.8 The tree $P(9,5)$.

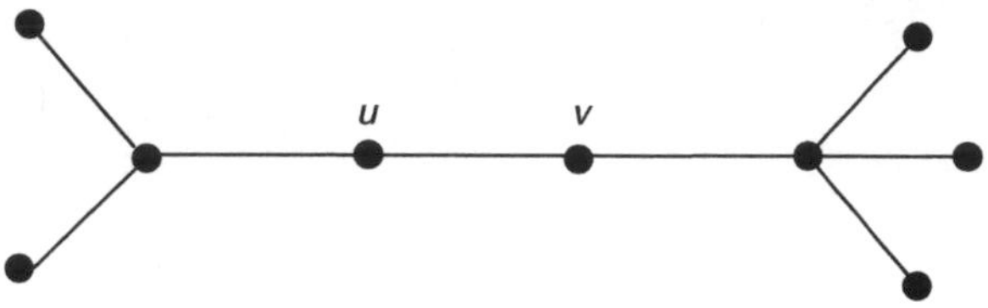

FIGURE 36.9 The tree $T(2,3,4)$.

Examples:

1. The absolute algebraic connectivity of K_n is n.
2. If $m \leq n$ and $2 \leq n$, then $\hat{\alpha}(K_{m,n}) = m$.
3. [KP02] Let e be an edge of K_n. Then $\hat{\alpha}(K_n - e) = \frac{(n-2)(n^2-n-2)}{n^2-3n+4}$.
4. Consider the tree $P(9,5)$ shown in Figure 36.8; we have $S(v) = 22$, and $S(v) \leq S(k) - 9$ for any vertex $k \neq v$, so that the first case of Fact 4 above holds. Hence, $\hat{\alpha}(P(9,5)) = \frac{4}{11}$.
5. Consider the tree $T(2,3,4)$ pictured in Figure 36.9. We have $S(u) = 41$ and $S(v) = 36$, so that the second case of Fact 4 above holds. Hence, $\hat{\alpha}(T(2,3,4)) = \frac{9}{40}$.

36.6 Generalized Laplacians and Multiplicity

The term *graph* means *simple graph* in this section.

Definitions:

Let $G = (V, E)$ be a graph on $n \geq 2$ vertices. A **generalized Laplacian** of G is a symmetric $n \times n$ matrix M with the property that for each $i \neq j, m_{i,j} < 0$ if $\{i, j\} \in E$ and $m_{i,j} = 0$ if $\{i, j\} \notin E$.

We only consider generalized Laplacian matrices for graphs on at least two vertices.

For any symmetric matrix B, let $\tau(B)$ denote its smallest eigenvalue, let $\mu(B)$ denote its second smallest eigenvalue, and let $\lambda(B)$ denote its largest eigenvalue.

Let G be a graph, and let v be a vertex of G. Suppose that M is a generalized Laplacian matrix for G. Let $C_1, \ldots, C_k$ denote the components of $G - v$; a component C_j is called a **Perron component at** v if $\tau(M[C_j])$ is minimal among all $\tau(M[C_i]), i = 1, \ldots, k$.

Facts:

1. [BKP01] Let G be a connected graph and let v be a cut-vertex of G. Suppose that M is a generalized Laplacian matrix for G. If there are $p \geq 2$ Perron components at v, say $C_1, \ldots, C_p$, then $\mu(M) = \tau(M[C_1])$. Further, $\mu(M)$ is an eigenvalue of M of multiplicity $p - 1$.
2. [BKP01] Let G be a connected graph, and let v be a cut-vertex of G. Let M be a generalized Laplacian matrix for G. Suppose that there is a unique Perron component A_0 at v; denote the other connected components at v by $B_1, \ldots, B_k$, and set $A_1 = G \setminus A_0$. Let $\mathbf{z}$ be an eigenvector of M corresponding to $\tau(M)$, and let $\mathbf{z_0}, \mathbf{z_1}$ be the subvectors of $\mathbf{z}$ corresponding to the vertices in A_0, A_1, respectively. For each $i = 1, \ldots, k$, denote the principal submatrix of M corresponding to

the vertices of B_i by $M[B_i]$. Let $S = (M[B_1])^{-1} \oplus \ldots \oplus (M[B_k])^{-1} \oplus [0]$, and let M_0 denote the principal submatrix of M corresponding to the vertices of A_0. There is a unique $\gamma > 0$ such that $\lambda(M_0^{-1} - \gamma \mathbf{z_0}\mathbf{z_0}^T) = \lambda(S + \gamma \mathbf{z_1}\mathbf{z_1}^T) = 1/\mu(M)$. Further, the multiplicity of $\mu(M)$ as an eigenvalue of M coincides with the multiplicity of $1/\mu(M)$ as an eigenvalue of $M_0^{-1} - \gamma \mathbf{z_0}\mathbf{z_0}^T$.

Examples:

1. Consider the generalized Laplacian matrix $M = \begin{bmatrix} 0 & -1 & -1 & -1 & -2 \\ -1 & 3 & -1 & 0 & 0 \\ -1 & -1 & 3 & 0 & 0 \\ -1 & 0 & 0 & 2 & 0 \\ -2 & 0 & 0 & 0 & 2 \end{bmatrix}$. There are three Perron components at vertex 1, induced by the vertex sets $\{2, 3\}$, $\{4\}$, and $\{5\}$. We have $\mu(M) = 2$, which is an eigenvalue of multiplicity two.
2. Consider the generalized Laplacian matrix $M = \begin{bmatrix} -3 & -1 & -1 & -1 & -2 \\ -1 & 2 & -1 & 0 & 0 \\ -1 & -1 & 2 & 0 & 0 \\ -1 & 0 & 0 & 2 & 0 \\ -2 & 0 & 0 & 0 & 2 \end{bmatrix}$. The unique Perron component at vertex 1 is induced by the vertex set $\{2, 3\}$. We have $\mu(M) = \frac{-3+\sqrt{29}}{2}$, which is a simple eigenvalue.

References

[Alo86] N. Alon. Eigenvalues and expanders. *Combinatorica*, 6: 83–96, 1986.

[AM85] N. Alon and V.D. Milman. λ_1, isoperimetric inequalities for graphs, and superconcentrators. *J. Combin. Theory B*, 38: 73–88, 1985.

[BKP01] R. Bapat, S. Kirkland, and S. Pati. The perturbed Laplacian matrix of a graph. *Lin. Multilin. Alg.*, 49: 219–242, 2001.

[FK98] S. Fallat and S. Kirkland. Extremizing algebraic connectivity subject to graph theoretic constraints. *Elec. J. Lin. Alg.*, 3: 48–74, 1998.

[FKP02] S. Fallat, S. Kirkland, and S. Pati. Minimizing algebraic connectivity over connected graphs with fixed girth. *Dis. Math.*, 254: 115–142, 2002.

[FKP03a] S. Fallat, S. Kirkland, and S. Pati. On graphs with algebraic connectivity equal to minimum edge density. *Lin. Alg. Appl.*, 373: 31–50, 2003.

[FKP03b] S. Fallat, S. Kirkland, and S. Pati. Maximizing algebraic connectivity over unicyclic graphs. *Lin. Multilin. Alg.*, 51: 221–241, 2003.

[Fie73] M. Fiedler. Algebraic connectivity of graphs. *Czech. Math. J.*, 23(98): 298–305, 1973.

[Fie75a] M. Fiedler. Eigenvectors of acyclic matrices. *Czech. Math. J.*, 25(100): 607–618, 1975.

[Fie75b] M. Fiedler. A property of eigenvectors of nonnegative symmetric matrices and its application to graph theory. *Czech. Math. J.*, 25(100): 619–633, 1975.

[Fie89] M. Fiedler. Laplacians of graphs and algebraic connectivity. In *Combinatorics and Graph Theory*, Eds. Z. Skuplén and M. Borowiecki Banach Center Publication 25: 57–70. PWN, Warsaw, 1989.

[Fie90] M. Fiedler. Absolute algebraic connectivity of trees. *Lin. Multilin. Alg.*, 26: 85–106, 1990.

[Fie93] M. Fiedler. Some minimax problems for graphs. *Dis. Math.*, 121: 65–74, 1993.

[FN02] S. Friedland and R. Nabben. On Cheeger-type inequalities for weighted graphs. *J. Graph Theory*, 41: 1–17, 2002.

[GR01] C. Godsil and G. Royle. *Algebraic Graph Theory.* Springer, New York, 2001.

[GK69] R. Gregory and D. Karney. *A Collection of Matrices for Testing Computational Algorithms.* Wiley-Interscience, New York, 1969.

[GM87] R. Grone and R. Merris. Algebraic connectivity of trees. *Czech. Math. J.*, 37(112): 660–670, 1987.

[GM90] R. Grone and R. Merris. Ordering trees by algebraic connectivity. *Graphs Comb.*, 6: 229–237, 1990.

[GMS90] R. Grone, R. Merris, and V. Sunder. The Laplacian spectrum of a graph. *SIAM J. Matrix Anal. Appl.*, 11: 218–238, 1990.

[Kir00] S. Kirkland. A bound on the algebraic connectivity of a graph in terms of the number of cutpoints. *Lin. Multilin. Alg.*, 47: 93–103, 2000.

[Kir01] S. Kirkland. An upper bound on the algebraic connectivity of a graph with many cutpoints. *Elec. J. Lin. Alg.*, 8: 94–109, 2001.

[Kir03] S. Kirkland. A note on limit points for algebraic connectivity. *Lin. Alg. Appl.*, 373: 5–11, 2003.

[KF98] S. Kirkland and S. Fallat. Perron components and algebraic connectivity for weighted graphs. *Lin. Multilin. Alg.*, 44: 131–148, 1998.

[KMNS02] S. Kirkland, J. Molitierno, M. Neumann, and B. Shader. On graphs with equal algebraic and vertex connectivity. *Lin. Alg. Appl.*, 341: 45–56, 2002.

[KN04] S. Kirkland and M. Neumann. On algebraic connectivity as a function of an edge weight. *Lin. Multilin. Alg.*, 52: 17–33, 2004.

[KNS96] S. Kirkland, M. Neumann, and B. Shader. Characteristic vertices of weighted trees via Perron values. *Lin. Multilin. Alg.*, 40: 311–325, 1996.

[KP02] S. Kirkland and S. Pati. On vertex connectivity and absolute algebraic connectivity for graphs. *Lin. Multilin. Alg.*, 50: 253–284, 2002.

[Mer87] R. Merris. Characteristic vertices of trees. *Lin. Multilin. Alg.*, 22: 115–131, 1987.

[Mer94] R. Merris. Laplacian matrices of graphs: a survey. *Lin. Alg. Appl.*, 197,198: 143–176, 1994.

[Moh89] B. Mohar. Isoperimetric numbers of graphs. *J. Comb. Theory B*, 47: 274–291, 1989.

[Moh91] B. Mohar. Eigenvalues, diameter, and mean distance in graphs. *Graphs Comb.*, 7: 53–64, 1991.

[Moh92] B. Mohar. Laplace eigenvalues of graphs — a survey. *Dis. Math.*, 109: 171–183, 1992.

[PSL90] A. Pothen, H. Simon, and K-P. Liou. Partitioning sparse matrices with eigenvectors of graphs. *SIAM J. Matrix Anal. Appl.*, 11: 430–452, 1990.

III

Numerical Methods

Numerical Methods for Linear Systems

Numerical Methods for Eigenvalues

Computational Linear Algebra

Numerical Methods for Linear Systems

37

Vector and Matrix Norms, Error Analysis, Efficiency, and Stability

Ralph Byers
University of Kansas

Biswa Nath Datta
Northern Illinois University

Calculations are subject to errors. There may be modeling errors, measurement errors, manufacturing errors, noise, equipment is subject to wear and damage, etc. In preparation for computation, data must often be perturbed by rounding it to fit a particular finite precision, floating-point format. Further errors may be introduced during a computation by using finite precision arithmetic and by truncating an infinite process down to a finite number of steps.

This chapter outlines aspects of how such errors affect the results of a mathematical computation with an emphasis on matrix computations. Topics include:

- Vector and matrix norms and seminorms that are often used to analyze errors and express error bounds.
- Floating point arithmetic.
- Condition numbers that measure how much data errors may affect computational results.
- Numerical stability, an assessment of whether excessive amounts of data may be lost to rounding errors during a finite precision computation.

For elaborative discussions of vector and matrix norms, consult [HJ85]. See [Ove01] for a textbook introduction to IEEE standard floating point arithmetic. Complete details of the standard are available in [IEEE754] and [IEEE854]. Basic concepts of numerical stability and conditioning appear in numerical linear algebra books, e.g., [GV96], [Ste98], [TB97], and [Dat95]. The two particularly authoritative books on these topics are the classical book by Wilkinson [Wil65] and the more recent one by Higham [Hig96]. For perturbation analysis, see the classic monograph [Kat66] and the more modern treatments in [SS90] and [Bha96].

The set $\mathbb{C}^n = \mathbb{C}^{n\times 1}$ is the complex vector space of n-row, 1-column matrices, and $\mathbb{R}^n = \mathbb{R}^{n\times 1}$ is the real vector space of n-row, 1-column matrices. Unless otherwise specified, $\mathbb{F}^n$ is either $\mathbb{R}^n$ or $\mathbb{C}^n$, $\mathbf{x}$ and $\mathbf{y}$ are members of $\mathbb{F}^n$, and $\alpha \in \mathbb{F}$ is a scalar $\alpha \in \mathbb{R}$ or $\alpha \in \mathbb{C}$, respectively. For $\mathbf{x} \in \mathbb{R}^n$, $\mathbf{x}^T$ is the one row n-column transpose of $\mathbf{x}$. For $\mathbf{x} \in \mathbb{C}^n$, $\mathbf{x}^*$ is the one row n-column complex-conjugate-transpose of $\mathbf{x}$. A and B are members of $\mathbb{F}^{m\times n}$. For $A \in \mathbb{R}^{m\times n}$, $A^* \in \mathbb{R}^{n\times m}$ is the transpose of A. For $A \in \mathbb{C}^{m\times n}$, $A^* \in \mathbb{C}^{n\times m}$ is the complex-conjugate-transpose of A.

37.1 Vector Norms

Most uses of vector norms involve $\mathbb{R}^n$ or $\mathbb{C}^n$, so the focus of this section is on those vector spaces. However, the definitions given here can be extended in the obvious way to any finite dimensional real or complex vector space.

Let $\mathbf{x}, \mathbf{y} \in \mathbb{F}^n$ and $\alpha \in \mathbb{F}$, where $\mathbb{F}$ is either $\mathbb{R}$ or $\mathbb{C}$.

Definitions:

A vector **norm** is a real-valued function on $\mathbb{F}^n$ denoted $\|\mathbf{x}\|$ with the following properties for all $\mathbf{x}, \mathbf{y} \in \mathbb{F}^n$ and all scalars $\alpha \in \mathbb{F}$.

- **Positive definiteness:** $\|\mathbf{x}\| \geq 0$ and $\|\mathbf{x}\| = 0$ if and only if $\mathbf{x}$ is the zero vector.
- **Homogeneity:** $\|\alpha\mathbf{x}\| = |\alpha|\,\|\mathbf{x}\|$.
- **Triangle inequality:** $\|\mathbf{x} + \mathbf{y}\| \leq \|\mathbf{x}\| + \|\mathbf{y}\|$.

For $\mathbf{x} = [x_1, x_2, x_3, \ldots, x_n]^* \in \mathbb{F}^n$, the following are commonly encountered vector norms.

- **Sum-norm or 1-norm:** $\|\mathbf{x}\|_1 = |x_1| + |x_2| + \cdots + |x_n|$.
- **Euclidean norm or 2-norm:** $\|\mathbf{x}\|_2 = \sqrt{|x_1|^2 + |x_2|^2 + \cdots + |x_n|^2}$.
- **Sup-norm or ∞-norm:** $\|\mathbf{x}\|_\infty = \max\limits_{1\leq i\leq n} |x_i|$.
- **Hölder norm or p-norm:** For $p \geq 1$, $\|\mathbf{x}\|_p = (|x_1|^p + \cdots + |x_n|^p)^{\frac{1}{p}}$.

If $\|\cdot\|$ is a vector norm on $\mathbb{F}^n$ and $M \in F^{n\times n}$ is a nonsingular matrix, then $\|y\|_M \equiv \|My\|$ is an **M-norm** or **energy norm**. (Note that this notation is ambiguous, since $\|\cdot\|$ is not specified; it either doesn't matter or must be stated explicitly when used.)

A vector norm $\|\cdot\|$ is **absolute** if for all $\mathbf{x} \in \mathbb{F}^n$, $\|\,|\mathbf{x}|\,\| = \|\mathbf{x}\|$, where $|[x_1, \ldots, x_n]^*| = [|x_1|, \ldots, |x_n|]^*$.

A vector norm $\|\cdot\|$ is **monotone** if for all $\mathbf{x}, \mathbf{y} \in \mathbb{F}^n$, $|\mathbf{x}| \leq |\mathbf{y}|$ implies $\|\mathbf{x}\| \leq \|\mathbf{y}\|$.

A vector norm $\|\cdot\|$ is **permutation invariant** if $\|P\mathbf{x}\| = \|\mathbf{x}\|$ for all $\mathbf{x} \in \mathbb{R}^n$ and all permutation matrices $P \in \mathbb{R}^{n\times n}$.

Let $\|\cdot\|$ be a vector norm. The **dual norm** is defined by $\|\mathbf{y}\|^D = \max_{\mathbf{x}\neq 0} \frac{|\mathbf{y}^*\mathbf{x}|}{\|\mathbf{x}\|}$.

The unit disk corresponding to a vector norm $\|\cdot\|$ is the set $\{\mathbf{x} \in \mathbb{F}^n \mid \|\mathbf{x}\| \leq 1\}$.

The unit sphere corresponding to a vector norm $\|\cdot\|$ is the set $\{\mathbf{x} \in \mathbb{F}^n \mid \|\mathbf{x}\| = 1\}$.

Facts:

For proofs and additional background, see, for example, [HJ85, Chap. 5].

Let $\mathbf{x}, \mathbf{y} \in \mathbb{F}^n$ and $\alpha \in \mathbb{F}$, where $\mathbb{F}$ is either $\mathbb{R}$ or $\mathbb{C}$.

1. The commonly encountered norms, $\|\cdot\|_1, \|\cdot\|_2, \|\cdot\|_\infty, \|\cdot\|_p$, are permutation invariant, absolute, monotone vector norms.
2. If $M \in F^{n\times n}$ is a nonsingular matrix and $\|\cdot\|$ is a vector norm, then the M-norm $\|\cdot\|_M$ is a vector norm.
3. If $\|\cdot\|$ is a vector norm, then $\big|\, \|\mathbf{x}\| - \|\mathbf{y}\| \,\big| \leq \|\mathbf{x} - \mathbf{y}\|$.
4. A sum of vector norms is a vector norm.
5. $\lim\limits_{k\to\infty} \mathbf{x}_k = \mathbf{x}_*$ if and only if in any norm $\lim\limits_{k\to\infty} \|\mathbf{x}_k - x_*\| = 0$.

6. **Cauchy–Schwartz inequality:**

   (a) $|\mathbf{x}^*\mathbf{y}| \leq \|\mathbf{x}\|_2\|\mathbf{y}\|_2$

   (b) $|\mathbf{x}^*\mathbf{y}| = \|\mathbf{x}\|_2\|\mathbf{y}\|_2$ if and only if there exist scalars α and β, not both zero, for which $\alpha x = \beta y$.
7. **Hölder inequality:** If $p \geq 1$ and $q \geq 1$ satisfy $\frac{1}{p} + \frac{1}{q} = 1$, then $|\mathbf{x}^*\mathbf{y}| \leq \|\mathbf{x}\|_p \, \|\mathbf{y}\|_q$.
8. If $\|\cdot\|$ is a vector norm on $\mathbb{F}^n$, then its dual $\|\cdot\|^D$ is also a vector norm on $\mathbb{F}^n$, and $\|\cdot\|^{DD} = \|\cdot\|$.
9. If $p > 0$ and $q > 0$ satisfy $\frac{1}{p} + \frac{1}{q} = 1$, then $\|\cdot\|_p^D = \|\cdot\|_q$. In particular, $\|\cdot\|_2^D = \|\cdot\|_2$. Also, $\|\cdot\|_1^D = \|\cdot\|_\infty$.
10. If $\|\cdot\|$ is a vector norm on $\mathbb{F}^n$, then for any $\mathbf{x} \in \mathbb{F}^n$, $|\mathbf{x}^*\mathbf{y}| \leq \|\mathbf{x}\| \, \|\mathbf{y}\|^D$.
11. A vector norm is absolute if and only if it is monotone.
12. **Equivalence of norms:** All vector norms on $\mathbb{F}^n$ are equivalent in the sense that for any two vector norms $\|\cdot\|_\mu$ and $\|\cdot\|_\nu$ there constants $\alpha > 0$ and $\beta > 0$ such that for all $\mathbf{x} \in \mathbb{F}^n$, $\alpha\|\mathbf{x}\|_\mu \leq \|\mathbf{x}\|_\nu \leq \beta\|\mathbf{x}\|_\mu$. The constants α and β are independent of $\mathbf{x}$ but typically depend on the dimension n.
    In particular,

    (a) $\|\mathbf{x}\|_2 \leq \|\mathbf{x}\|_1 \leq \sqrt{n}\|\mathbf{x}\|_2$.

    (b) $\|\mathbf{x}\|_\infty \leq \|\mathbf{x}\|_2 \leq \sqrt{n}\|\mathbf{x}\|_\infty$.

    (c) $\|\mathbf{x}\|_\infty \leq \|\mathbf{x}\|_1 \leq n\|\mathbf{x}\|_\infty$.
13. A set $D \subset \mathbb{F}^n$ is the unit disk of a vector norm if and only if it has the following properties.

    (a) **Point-wise bounded:** For every vector $\mathbf{x} \in \mathbb{F}^n$ there is a number $\delta > 0$ for which $\delta\mathbf{x} \notin D$.

    (b) **Absorbing:** For every vector $\mathbf{x} \in \mathbb{F}^n$ there is a number $\tau > 0$ for which $|\alpha| \leq \tau$ implies $\alpha\mathbf{x} \in D$.

    (c) **Convex:** For every pair of vectors $\mathbf{x}$, $\mathbf{y} \in D$ and every number t, $0 \leq t \leq 1$, $t\mathbf{x} + (1-t)\mathbf{y} \in D$.

Examples:

1. Let $\mathbf{x} = [1, 1, -2]^*$. Then $\|\mathbf{x}\|_1 = 4$, $\|\mathbf{x}\|_2 = \sqrt{1^2 + 1^2 + (-2)^2} = \sqrt{6}$, and $\|\mathbf{x}\|_\infty = 2$.
2. Let $M = \begin{bmatrix} 1 & 2 \\ 3 & 4 \end{bmatrix}$. Using the 1-norm, $\left\|\begin{bmatrix} 0 \\ 1 \end{bmatrix}\right\|_M = \left\|\begin{bmatrix} 2 \\ 4 \end{bmatrix}\right\|_1 = 6$.

37.2 Vector Seminorms

Definitions:

A **vector seminorm** is a real-valued function on $\mathbb{F}^n$, denoted $\nu(\mathbf{x})$, with the following properties for all $\mathbf{x}, \mathbf{y} \in \mathbb{F}^n$ and all scalars $\alpha \in \mathbb{F}$.

1. **Positiveness:** $\nu(\mathbf{x}) \geq 0$.
2. **Homogeneity:** $\|\alpha\mathbf{x}\| = |\alpha|\|\mathbf{x}\|$.
3. **Triangle inequality:** $\|\mathbf{x} + \mathbf{y}\| \leq \|\mathbf{x}\| + \|\mathbf{y}\|$.

Vector norms are *a fortiori* also vector seminorms.

The unit disk corresponding to a vector seminorm $\|\cdot\|$ is the set $\{\mathbf{x} \in \mathbb{F}^n \mid \nu(\mathbf{x}) \leq 1\}$.

The unit sphere corresponding to a vector seminorm $\|\cdot\|$ is the set $\{\mathbf{x} \in \mathbb{F}^n \mid \nu(\mathbf{x}) = 1\}$.

Facts:

For proofs and additional background, see, for example, [HJ85, Chap. 5].

Let $\mathbf{x}, \mathbf{y} \in \mathbb{F}^n$ and $\alpha \in \mathbb{F}$, where $\mathbb{F}$ is either $\mathbb{R}$ or $\mathbb{C}$.

1. $\nu(0) = 0$.
2. $\nu(\mathbf{x} - \mathbf{y}) \geq |\nu(\mathbf{x}) - \nu(\mathbf{y})|$.

3. A sum of vector seminorms is a vector seminorm. If one of the summands is a vector norm, then the sum is a vector norm.
4. A set $D \subset \mathbb{F}^n$ is the unit disk of a seminorm if and only if it has the following properties.

   (a) **Absorbing:** For every vector $\mathbf{x} \in \mathbb{F}^n$ there is a number $\tau > 0$ for which $|\alpha| \leq \tau$ implies $\alpha x \in D$.

   (b) **Convex:** For every pair of vectors $\mathbf{x}$, $\mathbf{y} \in D$ and every number $t, 0 \leq t \leq 1, t\mathbf{x}+(1-t)\mathbf{y} \in D$.

Examples:

1. For $\mathbf{x} = [x_1, x_2, x_3, \ldots, x_n]^T \in \mathbb{F}^n$, the function $\nu(x) = |x_1|$ is a vector seminorm that is not a vector norm. For $n \geq 2$, this seminorm is not equivalent to any vector norm $\|\cdot\|$, since $\|\mathbf{e}_2\| > 0$ but $\nu(\mathbf{e}_2) = 0$, for $\mathbf{e}_2 = [0, 1, 0, \ldots, 0]^T$.

37.3 Matrix Norms

Definitions:

A **matrix norm** is a family of real-valued functions on $\mathbb{F}^{m\times n}$ for all positive integers m and n, denoted uniformly by $\|A\|$ with the following properties for all matrices A and B and all scalars $\alpha \in \mathbb{F}$.

- **Positive definiteness:** $\|A\| \geq 0$; $\|A\| = 0$ only if $A = 0$.
- **Homogeneity:** $\|\alpha A\| = |\alpha|\|A\|$.
- **Triangle inequality:** $\|A + B\| \leq \|A\| + \|B\|$, where A and B are compatible for matrix addition.
- **Consistency:** $\|AB\| \leq \|A\| \, \|B\|$, where A and B are compatible for matrix multiplication.

If $\|\cdot\|$ is a family of vector norms on $\mathbb{F}^n$ for $n = 1, 2, 3, \ldots$, then the matrix norm on $\mathbb{F}^{m\times n}$ **induced** by (or **subordinate** to) $\|\cdot\|$ is $\|A\| = \max_{\mathbf{x}\neq 0} \dfrac{\|A\mathbf{x}\|}{\|\mathbf{x}\|}$. Induced matrix norms are also called **operator norms** or **natural norms**. The matrix norm $\|A\|_p$ denotes the norm induced by the Hölder vector norm $\|\mathbf{x}\|_p$.

The following are commonly encountered matrix norms.

- **Maximum absolute column sum norm:** $\|A\|_1 = \max_{1\leq j\leq n} \sum_{i=1}^{m} |a_{ij}|$.
- **Spectral norm:** $\|A\|_2 = \sqrt{\rho(A^*A)}$, where $\rho(A^*A)$ is the largest eigenvalue of A^*A.
- **Maximum absolute row sum norm:** $\|A\|_\infty = \max_{1\leq i\leq m} \sum_{j=1}^{n} |a_{ij}|$.
- **Euclidean norm** or **Frobenius norm:** $\|A\|_F = \sqrt{\sum_{i,j=1}^{n} |a_{ij}|^2}$.

Let $\mathcal{M} = \{M_n \in F^{n\times n} : n \geq 1\}$ be a family of nonsingular matrices and let $\|\cdot\|$ be a family of vector norms. Define a family of vector norms by $\|\mathbf{x}\|_{\mathcal{M}}$ for $\mathbf{x} \in \mathbb{F}^n$ by $\|\mathbf{x}\|_{\mathcal{M}} = \|M_n\mathbf{x}\|$. This family of vector norms is also called the **$\mathcal{M}$-norm** and denoted by $\|\cdot\|_{\mathcal{M}}$. (Note that this notation is ambiguous, since $\|\cdot\|$ is not specified; it either does not matter or must be stated explicitly when used.)

A matrix norm $\|\cdot\|$ is **minimal** if for any matrix norm $\|\cdot\|_\nu$, $\|A\|_\nu \leq \|A\|$ for all $A \in F^{n\times n}$ implies $\|\cdot\|_\nu = \|\cdot\|$.

A matrix norm is **absolute** if as a vector norm, each member of the family is absolute.

Facts:

For proofs and additional background, see, for example, [HJ85, Chap. 5]. Let $\mathbf{x}, \mathbf{y} \in \mathbb{F}^n$, $A, B \in \mathbb{F}^{m\times n}$, and $\alpha \in \mathbb{F}$, where $\mathbb{F}$ is either $\mathbb{R}$ or $\mathbb{C}$.

1. A matrix norm is a family of vector norms, but not every family of vector norms is a matrix norm (see Example 2).
2. The commonly encountered norms, $\|\cdot\|_1, \|\cdot\|_2, \|\cdot\|_\infty, \|\cdot\|_F$, and norms induced by vector norms are matrix norms. Furthermore,

   (a) $\|A\|_1$ is the matrix norm induced by the vector norm $\|\cdot\|_1$.

   (b) $\|A\|_2$ is the matrix norm induced by the vector norm $\|\cdot\|_2$.

   (c) $\|A\|_\infty$ is the matrix norm induced by the vector norm $\|\cdot\|_\infty$.

   (d) $\|A\|_F$ is not induced by any vector norm.

   (e) If $\mathcal{M} = \{M_n\}$ is a family of nonsingular matrices and $\|\cdot\|$ is an induced matrix norm, then for $A \in \mathbb{F}^{m\times n}$, $\|A\|_{\mathcal{M}} = \|M_m A M_n^{-1}\|$.

3. If $\|\cdot\|$ is the matrix norm induced by a family of vector norms $\|\cdot\|$, then $\|I_n\| = 1$ for all positive integers n (where I_n is the $n \times n$ identity matrix).
4. If $\|\cdot\|$ is the matrix norm induced by a family of vector norms $\|\cdot\|$, then for all $A \in \mathbb{F}^{m\times n}$ and all $\mathbf{x} \in \mathbb{F}^n$, $\|A\mathbf{x}\| \le \|A\|\,\|\mathbf{x}\|$.
5. For all $A \in \mathbb{F}^{m\times n}$ and all $\mathbf{x} \in \mathbb{F}^n$, $\|A\mathbf{x}\|_F \le \|A\|_F \|x\|_2$.
6. $\|\cdot\|_1, \|\cdot\|_\infty, \|\cdot\|_F$ are absolute norms. However, for some matrices A, $\|\,|A|\,\|_2 \ne \|A\|_2$ (see Example 3).
7. A matrix norm is minimal if and only if it is an induced norm.
8. All matrix norms are equivalent in the sense that for any two matrix norms $\|\cdot\|_\mu$ and $\|\cdot\|_\nu$, there exist constants $\alpha > 0$ and $\beta > 0$ such that for all $A \in \mathbb{F}^{m\times n}$, $\alpha\|A\|_\mu \le \|A\|_\nu \le \beta\|A\|_\mu$. The constants α and β are independent of A but typically depend on n and m. In particular,

   (a) $\frac{1}{\sqrt{n}}\|A\|_\infty \le \|A\|_2 \le \sqrt{m}\|A\|_\infty$.

   (b) $\|A\|_2 \le \|A\|_F \le \sqrt{n}\|A\|_2$.

   (c) $\frac{1}{\sqrt{m}}\|A\|_1 \le \|A\|_2 \le \sqrt{n}\|A\|_1$.

9. $\|A\|_2 \le \sqrt{\|A\|_1\|A\|_\infty}$.
10. $\|AB\|_F \le \|A\|_F\|B\|_2$ and $\|AB\|_F \le \|A\|_2\|B\|_F$ whenever A and B are compatible for matrix multiplication.
11. $\|A\|_2 \le \|A\|_F$ and $\|A\|_2 = \|A\|_F$ if and only if A has rank less than or equal to 1.
12. If $A = \mathbf{x}\mathbf{y}^*$ for some $\mathbf{x} \in \mathbb{F}^n$ and $\mathbf{y} \in \mathbb{F}^m$, then $\|A\|_2 = \|A\|_F = \|\mathbf{x}\|_2\|\mathbf{y}\|_2$.
13. $\|A\|_2 = \|A^*\|_2$ and $\|A\|_F = \|A^*\|_F$.
14. If $U \in F^{n\times n}$ is a unitary matrix, i.e., if $U^* = U^{-1}$, then the following hold.

   (a) $\|U\|_2 = 1$ and $\|U\|_F = \sqrt{n}$.

   (b) If $A \in \mathbb{F}^{m\times n}$, then $\|AU\|_2 = \|A\|_2$ and $\|AU\|_F = \|A\|_F$.

   (c) If $A \in \mathbb{F}^{n\times m}$, then $\|UA\|_2 = \|A\|_2$ and $\|UA\|_F = \|A\|_F$.

15. For any matrix norm $\|\cdot\|$ and any $A \in F^{n\times n}$, $\rho(A) \le \|A\|$, where $\rho(A)$ is the spectral radius of A. This need not be true for a vector norm on matrices (see Example 2).
16. For any $A \in F^{n\times n}$ and $\varepsilon > 0$, there exists a matrix norm $\|\cdot\|$ such that $\|A\| < \rho(A)+\varepsilon$. A method for finding such a norm is given in Example 5.
17. For any matrix norm $\|\cdot\|$ and $A \in F^{n\times n}$, $\lim_{k\to\infty}\|A^k\|^{1/k} = \rho(A)$.
18. For $A \in F^{n\times n}$, $\lim_{k\to\infty} A^k = 0$ if and only if $\rho(A) < 1$.

Examples:

1. If $A = \begin{bmatrix} 1 & -2 \\ 3 & -4 \end{bmatrix}$, then $\|A\|_1 = 6$, $\|A\|_\infty = 7$, $\|A\|_2 = \sqrt{15 + \sqrt{221}}$, and $\|A\|_F = \sqrt{30}$.
2. The family of matrix functions defined for $A \in \mathbb{F}^{m\times n}$ by
$$\nu(A) = \max_{\substack{1 \le i \le m \\ 1 \le j \le n}} |a_{ij}|$$
is *not* a matrix norm because consistency fails. For example, if $J = \begin{bmatrix} 1 & 1 \\ 1 & 1 \end{bmatrix}$, then $\nu(J^2) = 2 > 1 = \nu(J)\nu(J)$. Note that ν is a family of vector norms on matrices (it is the ∞ norm on the n^2-tuple of entries), and $\nu(J) = 1 < 2 = \rho(J)$.
3. If $A = \begin{bmatrix} 3 & 4 \\ -4 & 3 \end{bmatrix}$, then $\|A\|_2 = 5$ but $\|\,|A|\,\|_2 = 7$.
4. If A is perturbed by an error matrix E and U is unitary (i.e., $U^* = U^{-1}$), then $U(A+E) = UA+UE$ and $\|UE\|_2 = \|E\|_2$. Numerical analysts often use unitary matrices in numerical algorithms because multiplication by unitary matrices does not magnify errors.
5. Given $A \in F^{n\times n}$ and $\varepsilon > 0$, we show how an $\mathcal{M}$-norm can be constructed such that $\|A\|_{\mathcal{M}} < \rho+\varepsilon$, where ρ is the spectral radius of A. The procedure below determines M_n where $A \in F^{n\times n}$. The procedure is illustrated with the matrix $A = \begin{bmatrix} -38 & 13 & 52 \\ 3 & 0 & -4 \\ -30 & 10 & 41 \end{bmatrix}$ and with $\varepsilon = 0.1$. The norm used to construct the $\mathcal{M}$-norm will be the 1-norm; note the 1-norm of $A = 97$.

   (a) Determine ρ: The characteristic polynomial of A is $p_A(x) = \det(A-xI) = x^3-3x^2+3x-1 = (x-1)^3$, so $\rho = 1$.

   (b) Find a unitary matrix U such that $T = UAU^*$ is triangular. Using the method in Example 5 of Chapter 7.1, we find
$$U = \begin{bmatrix} \frac{1}{\sqrt{10}} & \frac{6}{\sqrt{65}} & -\frac{3}{\sqrt{26}} \\ \frac{3}{\sqrt{10}} & -\frac{2}{\sqrt{65}} & \frac{1}{\sqrt{26}} \\ 0 & \sqrt{\frac{5}{13}} & 2\sqrt{\frac{2}{13}} \end{bmatrix} \approx \begin{bmatrix} 0.316228 & 0.744208 & -0.588348 \\ 0.948683 & -0.248069 & 0.196116 \\ 0. & 0.620174 & 0.784465 \end{bmatrix} \text{ and}$$
$$T = U^*AU = \begin{bmatrix} 1 & 0 & 2\sqrt{65} \\ 0 & 1 & 26\sqrt{10} \\ 0 & 0 & 1 \end{bmatrix} \approx \begin{bmatrix} 1 & 0 & 16.1245 \\ 0 & 1 & 82.2192 \\ 0 & 0 & 1 \end{bmatrix}.$$

   (c) Find a diagonal matrix $\operatorname{diag}(1, \alpha, \alpha^2, \ldots, \alpha^{n-1})$ such that $\|DTD^{-1}\|_1 < \rho + \varepsilon$ (this is always possible, since $\lim_{\alpha\to\infty} \|DTD^{-1}\|_1 = \rho$).

   In the example, for $\alpha = 1000$, $DTD^{-1} \approx \begin{bmatrix} 1 & 0 & 0.0000161245 \\ 0 & 1 & 0.0822192 \\ 0 & 0 & 1 \end{bmatrix}$ and $\|DTD^{-1}\|_1 \approx 1.08224 < 1.1$.

   (d) Then $\|DU^*AUD^{-1}\|_1 < \rho + \varepsilon$. That is, $\|A\|_{\mathcal{M}} < 2.1$, where $M_3 = DU^*$
$$\approx \begin{bmatrix} 0.316228 & 0.948683 & 0. \\ 744.208 & -248.069 & 620.174 \\ -588348. & 196116. & 784465. \end{bmatrix}.$$

37.4 Conditioning and Condition Numbers

Data have limited precision. Measurements are inexact, equipment wears, manufactured components meet specifications only to some error tolerance, floating point arithmetic introduces errors. Consequently, the results of nontrivial calculations using data of limited precision also have limited precision. This section summarizes the topic of **conditioning:** How much errors in data can affect the results of a calculation.

(See [Ric66] for an authoritative treatment of conditioning.)

Definitions:

Consider a computational problem to be the task of evaluating a function $P : \mathbb{R}^n \to \mathbb{R}^m$ at a nominal data point $\mathbf{z} \in \mathbb{R}^n$, which, because data errors are ubiquitous, is known only to within a small relative-to-$\|\mathbf{z}\|$ error ε.

If $\hat{\mathbf{z}} \in \mathbb{F}^n$ is an approximation to $\mathbf{z} \in \mathbb{F}^n$, the **absolute error** in $\hat{\mathbf{z}}$ is $\|\mathbf{z} - \hat{\mathbf{z}}\|$ and the **relative error** in $\hat{\mathbf{z}}$ is $\|\mathbf{z} - \hat{\mathbf{z}}\|/\|\mathbf{z}\|$. If $\mathbf{z} = 0$, then the relative error is undefined.

The data $\mathbf{z}$ are **well-conditioned** if small relative perturbations of $\mathbf{z}$ cause small relative perturbations of $P(\mathbf{z})$. The data are **ill-conditioned** or **badly conditioned** if some small relative perturbation of $\mathbf{z}$ causes a large relative perturbation of $P(\mathbf{z})$. Precise meanings of "small" and "large" are dependent on the precision required in the context of the computational task.

Note that it is the data $\mathbf{z}$—not the solution $P(\mathbf{z})$—that is ill-conditioned or well-conditioned.

If $\mathbf{z} \neq 0$ and $P(\mathbf{z}) \neq 0$, then the **relative condition number**, or simply **condition number cond**$(\mathbf{z}) =$ **cond**$_P(\mathbf{z})$ of the data $\mathbf{z} \in \mathbb{F}^n$ with respect to the computational task of evaluating $P(\mathbf{z})$ may be defined as

$$\mathbf{cond}_P(\mathbf{z}) = \lim_{\varepsilon \to 0} \sup \left\{ \left(\frac{\|P(\mathbf{z} + \delta\mathbf{z}) - P(\mathbf{z})\|}{\|P(\mathbf{z})\|} \right) \left(\frac{\|\mathbf{z}\|}{\|\delta\mathbf{z}\|} \right) \;\middle|\; \|\delta\mathbf{z}\| \leq \varepsilon \right\}. \tag{37.1}$$

Sometimes it is useful to extend the definition to $\mathbf{z} = 0$ or to an isolated root of $P(\mathbf{z})$ by $\mathbf{cond}_P(\mathbf{z}) = \limsup_{\mathbf{x} \to \mathbf{z}} \mathbf{cond}_P(\mathbf{x})$.

Note that although the condition number depends on P and on the choice of norm, **cond**$(\mathbf{z}) =$ **cond**$_P(\mathbf{z})$ is the condition number of the data $\mathbf{z}$—not the condition number of the solution $P(\mathbf{z})$ and not the condition number of an algorithm that may be used to evaluate $P(\mathbf{z})$.

Facts:

For proofs and additional background, see, for example, [Dat95], [GV96], [Ste98], or [Wil65].

1. Because rounding errors are ubiquitous, a finite precision computational procedure can at best produce $P(\mathbf{z} + \delta\mathbf{z})$ where, in a suitably chosen norm, $\|\delta\mathbf{z}\| \leq \varepsilon\|\mathbf{z}\|$ and ε is a modest multiple of the unit round of the floating point system. (See section 37.6.)
2. The relative condition number determines the tight, asymptotic relative error bound
$$\frac{\|P(\mathbf{z} + \delta\mathbf{z}) - P(\mathbf{z})\|}{\|P(\mathbf{z})\|} \leq \mathbf{cond}_P(\mathbf{z})\frac{\|\delta\mathbf{z}\|}{\|\mathbf{z}\|} + o\left(\frac{\|\delta\mathbf{z}\|}{\|\mathbf{z}\|}\right)$$
as $\delta\mathbf{z}$ tends to zero. Very roughly speaking, if the larger components of the data $\mathbf{z}$ have p correct significant digits and the condition number is $\mathbf{cond}_P(\mathbf{z}) \approx 10^s$, then the larger components of the result $P(\mathbf{z})$ have $p - s$ correct significant digits.
3. [Hig96, p. 9] If $P(\mathbf{x})$ has a Frechet derivative $D(\mathbf{z})$ at $\mathbf{z} \in \mathbb{F}^n$, then the relative condition number is
$$\mathbf{cond}_P(\mathbf{z}) = \frac{\|D(\mathbf{z})\| \, \|\mathbf{z}\|}{\|P(\mathbf{z})\|}.$$
In particular, if $f(x)$ is a smooth real function of a real variable x, then $\mathbf{cond}_f(z) = |zf'(z)/f(z)|$.

Examples:

1. If $P(x) = \sin(x)$ and the nominal data point $z = 22/7$ may be in error by as much as $\pi - 22/7 \approx .00126$, then $P(z) = \sin(z)$ may be in error by as much as 100%. With such an uncertainty in $z = 22/7$, $\sin(z)$ may be off by 100%, i.e., $\sin(z)$ may have relative error equal to one. In most circumstances, $z = 22/7$ is considered to be ill-conditioned.

   The condition number of $z \in \mathbb{R}$ with respect to $\sin(z)$ is $\mathbf{cond}_{\sin}(z) = |z\cot(z)|$, and, in particular, $\mathbf{cond}(22/7) \approx 2485.47$. If $z = 22/7$ is perturbed to $z + \delta z = \pi$, then the asymptotic relative error bound in Fact 2 becomes

$$\begin{aligned}\left|\frac{\sin(z+\delta z)-\sin(z)}{\sin(z)}\right| &\leq \mathbf{cond}(z)\left|\frac{\delta z}{z}\right| + o(|(\delta z)/z|)\\ &= 0.9999995\ldots + o(|(\delta z)/z|).\end{aligned}$$

   The actual relative error in $\sin(z)$ is $\left|\frac{\sin(z+\delta z)-\sin(z)}{\sin(z)}\right| = 1$.
2. *Subtractive Cancellation:* For $\mathbf{x} \in \mathbb{R}^2$, define $P(\mathbf{x})$ by $P(\mathbf{x}) = [1, -1]\mathbf{x}$. The gradient of $P(x)$ is $\nabla P(x) = [1, -1]$ independent of $\mathbf{x}$, so, using the ∞-norm, Fact 3 gives

$$\mathbf{cond}_P(\mathbf{x}) = \frac{\|\nabla f\|_\infty \|\mathbf{x}\|_\infty}{\|f(\mathbf{x})\|_\infty} = \frac{2\max\{|x_1|,|x_2|\}}{|x_1 - x_2|}.$$

   Reflecting the trouble associated with subtractive cancellation, $\mathbf{cond}_P(\mathbf{x})$ shows that $\mathbf{x}$ is ill-conditioned when $x_1 \approx x_2$.
3. *Conditioning of Matrix–Vector Multiplication:* More generally, for a fixed matrix $A \in \mathbb{F}^{m\times n}$ that is not subject to perturbation, define $P(\mathbf{x}) : \mathbb{F}^n \to \mathbb{F}^n$ by $P(\mathbf{x}) = A\mathbf{x}$. The relative condition number of $\mathbf{x} \in \mathbb{F}^n$ is

$$\mathbf{cond}(\mathbf{x}) = \|A\| \frac{\|\mathbf{x}\|}{\|A\mathbf{x}\|}, \tag{37.2}$$

   where the matrix norm is the operator norm induced by the chosen vector norm. If A is square and nonsingular, then $\mathbf{cond}(\mathbf{x}) \leq \|A\|\,\|A^{-1}\|$.
4. *Conditioning of the Polynomial Zeros:* Let $q(x) = x^2 - 2x + 1$ and consider the computational task of determining the roots of $q(x)$ from the power basis coefficients $[1, -2, 1]$. Formally, the computational problem is to evaluate the function $P : \mathbb{R}^3 \to \mathbb{C}$ that maps the power basis coefficients of quadratic polynomials to their roots. If $q(x)$ is perturbed to $q(x)+\varepsilon$, then the roots change from a double root at $x = 1$ to $x = 1 \pm \sqrt{\varepsilon}$. A relative error of ε in the data $[1, -2, 1]$ induces a relative error of $\sqrt{|\varepsilon|}$ in the roots. In particular, the roots suffer an infinite rate of change at $\varepsilon = 0$. The condition number of the coefficients $[1, -2, 1]$ is infinite (with respect to root finding).

   The example illustrates the fact that the problem of calculating the roots of a polynomial q from its coefficients is highly ill-conditioned when q has multiple or near multiple roots. Although it is common to say that "multiple roots are ill-conditioned," strictly speaking, this is incorrect. It is the coefficients that are ill-conditioned because they are the initial data for the calculation.
5. [Dat95, p. 81], [Wil64, Wil65] *Wilkinson Polynomial:* Let $w(x)$ be the degree 20 polynomial

$$w(x) = (x-1)(x-2)\ldots(x-20) = x^{20} - 210x^{19} + 20615x^{18}\cdots + 2432902008176640000.$$

   The roots of $w(x)$ are the integes 1, 2, 3, ..., 20. Although distinct, the roots are highly ill-conditioned functions of the power basis coefficients. For simplicity, consider only perturbations to the coefficient of x^{19}. Perturbing the coefficient of x^{19} from -210 to $-210 - 2^{-23} \approx 210 - 1.12 \times 10^{-7}$ drastically changes some of the roots. For example, the roots 16 and 17 become a complex conjugate pair approximately equal to $16.73 \pm 2.81i$.

Let $P_{16}(z)$ be the root of $\hat{w}(x) = w(x) + (z - 210)x^{19}$ nearest 16 and let $P_{17}(z)$ be the root nearest 17. So, for $z = 210$, $P_{16}(z) = 16$ and $P_{17}(z) = 17$. The condition numbers of $z = 210$ with respect to P_{16} and P_{17} are $\mathbf{cond}_{16}(210) = 210(16^{19}/(16w'(16))) \approx 3 \times 10^{10}$ and $\mathbf{cond}_{17}(210) = 210(17^{19}/(17w'(17))) \approx 2 \times 10^{10}$, respectively. The condition numbers are so large that even perturbations as small as 2^{-23} are outside the asymptotic region in which $o\left(\frac{\|\delta \mathbf{z}\|}{\|\mathbf{z}\|}\right)$ is negligible in Fact 2.

37.5 Conditioning of Linear Systems

This section applies conditioning concepts to the computational task of finding a solution to the system of linear equations $A\mathbf{x} = \mathbf{b}$ for a given matrix $A \in \mathbb{R}^{n \times n}$ and right-hand side vector $\mathbf{b} \in \mathbb{R}^n$.

Throughout this section, $A \in \mathbb{R}^{n \times n}$ is nonsingular. Let the matrix norm $\| \cdot \|$ be an operator matrix norm induced by the vector norm $\| \cdot \|$. Use $\|A\| + \|\mathbf{b}\|$ to measure the magnitude of the data A and b. If $E \in \mathbb{R}^{n \times n}$ is a perturbation of A and $\mathbf{r} \in \mathbb{R}^n$ is a perturbation of $\mathbf{b}$, then $\|E\| + \|r\|$ is the magnitude of the perturbation to the linear system $A\mathbf{x} = \mathbf{b}$.

Definitions:

The norm-wise **condition number** of a nonsingular matrix A (for solving a linear system) is $\kappa(A) = \|A^{-1}\| \, \|A\|$. If A is singular, then by convention, $\kappa(A) = \infty$. For a specific norm $\| \cdot \|_\mu$, the condition number of A is denoted $\kappa_\mu(A)$.

Facts:

For proofs and additional background, see, for example, [Dat95], [GV96], [Ste98], or [Wil65].

1. *Properties of the Condition Number:*

   (a) $\kappa(A) \geq 1$.

   (b) $\kappa(AB) \leq \kappa(A)\kappa(B)$.

   (c) $\kappa(\alpha A) = \kappa(A)$, for all scalars $\alpha \neq 0$.

   (d) $\kappa_2(A) = 1$ if and only if A is a nonzero scalar multiple of an orthogonal matrix, i.e., $A^T A = \alpha I$ for some scalar α.

   (e) $\kappa_2(A) = \kappa_2(A^T)$.

   (f) $\kappa_2(A^T A) = (\kappa_2(A))^2$.

   (g) $\kappa_2(A) = \|A\|_2 \|A^{-1}\|_2 = \sigma_{\max}/\sigma_{\min}$, where $\sigma_{\max}$ and $\sigma_{\min}$ are the largest and smallest singular values of A.

2. For the p-norms (including $\| \cdot \|_1$, $\| \cdot \|_2$, and $\| \cdot \|_\infty$),

$$\frac{1}{\kappa(A)} = \min \left\{ \frac{\|\delta A\|}{\|A\|} \;\middle|\; A + \delta A \text{ is singular} \right\}.$$

   So, $\kappa(A)$ is one over the relative-to-$\|A\|$ distance from A to the nearest singular matrix, and, in particular, $\kappa(A)$ is large if and only if a small-relative-to-$\|A\|$ perturbation of A is singular.

3. Regarding A as fixed and not subject to errors, it follows from Equation 37.2 that the condition number of $\mathbf{b}$ with respect to solving $A\mathbf{x} = \mathbf{b}$ as defined in Equation 37.1 is

$$\mathbf{cond}(\mathbf{b}) = \frac{\|A^{-1}\| \, \|\mathbf{b}\|}{\|A^{-1}\mathbf{b}\|} \leq \kappa(A).$$

   If the matrix norm is $\|A^{-1}\|$ is induced by the vector norm $\|b\|$, then equality is possible.

4. Regarding $\mathbf{b}$ as fixed and not subject to errors, the condition number of A with respect to solving $A\mathbf{x} = \mathbf{b}$ as defined in Equation 37.1 is $\mathbf{cond}(A) = \|A^{-1}\| \, \|A\| = \kappa(A)$.
5. $\kappa(A) \leq \mathbf{cond}([A, b]) \leq \left(\frac{(\|A\|+\|\mathbf{b}\|)^2}{\|A\| \, \|\mathbf{b}\|}\right) \kappa(A)$, where $\mathbf{cond}([A, \mathbf{b}])$ is the condition number of the data $[A, \mathbf{b}]$ with respect to solving $A\mathbf{x} = \mathbf{b}$ as defined in Equation 37.1. Hence, the data $[A, \mathbf{b}]$ are norm-wise ill-conditioned for the problem of solving $A\mathbf{x} = \mathbf{b}$ if and only if $\kappa(A)$ is large.
6. If $\mathbf{r} = b - A(\mathbf{x} + \delta\mathbf{x})$, then the 2-norm and Frobenius norm smallest perturbation $\delta A \in \mathbb{R}^{n \times n}$ satisfying $(A + \delta A)(x + \delta x) = \mathbf{b}$ is $\delta A = \frac{\mathbf{r}\mathbf{x}^T}{\mathbf{x}^T\mathbf{x}}$ and $\|\delta A\|_2 = \|\delta A\|_F = \frac{\|\mathbf{r}\|_2}{\|\mathbf{x}\|_2}$.
7. Let δA and $\delta\mathbf{b}$ be perturbations of the data A and $\mathbf{b}$, respectively. If $\|A^{-1}\delta A\| < 1$, then $A + \delta A$ is nonsingular, there is a unique solution $\mathbf{x} + \delta\mathbf{x}$ to $(A + \delta A)(\mathbf{x} + \delta\mathbf{x}) = (\mathbf{b} - \delta\mathbf{b})$, and

$$\frac{\|\delta\mathbf{x}\|}{\|\mathbf{x}\|} \leq \frac{\|A\| \, \|A^{-1}\|}{(1 - \|A^{-1}\delta A\|)} \left(\frac{\|\delta A\|}{\|A\|} + \frac{\|\delta\mathbf{b}\|}{\|\mathbf{b}\|}\right).$$

Examples:

1. *An Ill-Conditioned Linear System:* For $\varepsilon \in \mathbb{R}$, let $A = \begin{bmatrix} 1 & 1 \\ 1 & 1+\varepsilon \end{bmatrix}$ and $\mathbf{b} = \begin{bmatrix} 1 \\ 1 \end{bmatrix}$. For $\varepsilon \neq 0$, A is nonsingular and $\mathbf{x} = \begin{bmatrix} 1 \\ 0 \end{bmatrix}$ satisfies $A\mathbf{x} = \mathbf{b}$. The system of equations is ill-conditioned when ε is small because some small changes in the data cause a large change in the solution. For example, perturbing $\mathbf{b}$ to $\mathbf{b}+\delta\mathbf{b}$, where $\delta\mathbf{b} = \begin{bmatrix} 0 \\ \varepsilon \end{bmatrix} \in \mathbb{R}^2$, changes the solution $\mathbf{x}$ to $\mathbf{x}+\delta\mathbf{x} = \begin{bmatrix} 0 \\ 1 \end{bmatrix}$ independent of the choice of ε no matter how small.

   Using the 1-norm, $\kappa_1(A) = \|A^{-1}\|_1 \|A\|_1 = (2 + \varepsilon)^2\varepsilon^{-1}$. As ε tends to zero, the perturbation $\delta\mathbf{b}$ tends to zero, but the condition number $\kappa_1(A)$ explodes to infinity.

   Geometrically, $\mathbf{x}$ is gives the coordinates of the intersection of the two lines $x + y = 1$ and $x + (1 + \varepsilon)y = 1$. If ε is small, then these lines are nearly parallel, so a small change in them may move the intersection a long distance.

   Also notice that the singular matrix $\begin{bmatrix} 1 & 1 \\ 1 & 1 \end{bmatrix}$ is a ε perturbation of A.

2. *A Well-Conditioned Linear System Problem:* Let $A = \begin{bmatrix} 1 & 1 \\ 1 & -1 \end{bmatrix}$. For $\mathbf{b} \in \mathbb{R}^2$, the solution to $A\mathbf{x} = \mathbf{b}$ is $\mathbf{x} = \frac{1}{2}\begin{bmatrix} b_1 + b_2 \\ b_1 - b_2 \end{bmatrix}$. In particular, perturbing $\mathbf{b}$ to $\mathbf{b} + \delta\mathbf{b}$ changes $\mathbf{x}$ to $\mathbf{x} + \delta\mathbf{x}$ with $\|\delta\mathbf{x}\|_1 \leq \|\mathbf{b}\|_1$ and $\|\delta\mathbf{x}\|_2 = \|\delta\mathbf{b}\|_2$, i.e., $\mathbf{x}$ is perturbed by no more than $\mathbf{b}$ is perturbed. This is a well-conditioned system of equations.

   The 1-norm condition number of A is $\kappa_1(A) = 2$, and the 2-norm condition number is $\kappa_2(A) = 1$, which is as small as possible.

   Geometrically, $\mathbf{x}$ gives the coordinates of the intersection of the perpendicular lines $x + y = 1$ and $x - y = 1$. Slighly perturbing the lines only slightly perturbs their intersection.

   Also notice that for both the 1-norm and 2-norm $\min_{\|x\|=1} \|A\mathbf{x}\| = 1$, so no small-relative-to-$\|A\|$ perturbation of A is singular. If $A + \delta A$ is singular, then $\|\delta A\| \geq 1$.
3. *Some Well-known Ill-conditioned Matrices:*

   (a) The upper triangular matrices $B_n \in \mathbb{R}^n$ of the form

$$B_3 = \begin{bmatrix} 1 & -1 & -1 \\ 0 & 1 & -1 \\ 0 & 0 & 1 \end{bmatrix}$$

have ∞-norm condition number $\kappa_\infty = n2^{n-1}$. Replacing the $(n, 1)$ entry by -2^{2-n} makes B_n singular. Note that the determinant $\det(B_n) = 1$ gives no indication of how nearly singular the matrices B_n are.

(b) *The Hilbert matrix:* The order n Hilbert matrix $H_n \in \mathbb{R}^{n\times n}$ is defined by $h_{ij} = 1/(i + j - 1)$. The Hilbert matrix arises naturally in calculating best L_2 polynomial approximations. The following table lists the 2-norm condition numbers to the nearest power of 10 of selected Hilbert matrices.

n:	1	2	3	4	5	6	7	8	9	10
$\kappa_2(H_n)$:	1	10	10^3	10^4	10^5	10^7	10^8	10^{10}	10^{11}	10^{13}

(c) *Vandermonde matrix:* The Vandermonde matrix corresponding to $\mathbf{x} \in \mathbb{R}^n$ is $V_{\mathbf{x}} \in \mathbb{R}^{n\times n}$ given by $v_{ij} = x_i^{n-j}$. Vandermonde matrices arise naturally in polynomial interpolation computations. The following table lists the 2-norm condition numbers to the nearest power of 10 of selected Vandermonde matrices.

n:	1	2	3	4	5	6	7	8	9	10
$\kappa_2\left(V_{[1,2,3,\ldots,n]}\right)$:	1	10	10	10^3	10^4	10^5	10^7	10^9	10^{10}	10^{12}

37.6 Floating Point Numbers

Most scientific and engineering computations rely on **floating point arithmetic**. At this writing, the IEEE 754 standard of binary floating point arithmetic [IEEE754] and the IEEE 854 standard of radix-independent floating point arithmetic [IEEE854] are the most widely accepted standards for floating point arithmetic. The still incomplete revised floating point arithmetic standard [IEEE754r] is planned to incorporate both [IEEE754] and [IEEE854] along with extensions, revisions, and clarifications. See [Ove01] for a textbook introduction to IEEE standard floating point arithmetic.

Even 20 years after publication of [IEEE754], implementations of floating point arithmetic vary in so many different ways that few axiomatic statements hold for all of them. Reflecting this unfortunate state of affairs, the summary of floating point arithmetic here is based upon IEEE 754r draft standard [IEEE754r] (necessarily omitting most of it), with frequent digressions to nonstandard floating point arithmetic.

In this section, the phrase *standard-conforming* refers to the October 20, 2005 IEEE 754r draft standard.

Definitions:

A p-digit, radix b **floating point number** with exponent bounds $e_{\max}$ and $e_{\min}$ is a real number of the form $x = \pm\left(\frac{m}{b^{p-1}}\right) b^e$, where e is an integer **exponent**, $e_{\min} \le e \le e_{\max}$, and m is a p−digit, base b integer **significand**. The related quantity m/b^p is called the **mantissa**. Virtually all floating point systems allow $m = 0$ and $b^{p-1} \le m < b^p$. Standard-conforming, floating point systems allow all significands $0 \le m < b^p$. If two or more different choices of significand m and exponent e yield the same floating point number, then the largest possible significand m with smallest possible exponent e is preferred.

In addition to finite floating point numbers, standard-conforming, floating point systems include elements that are not numbers, including ∞, $-\infty$, and not-a-number elements collectively called NaNs. Invalid or indeterminate arithmetic operations like $0/0$ or $\infty - \infty$ as well as arithmetic operations involving NaNs result in NaNs.

The representation $\pm(m/b^{p-1})b^e$ of a floating point number is said to be **normalized** or **normal**, if $b^{p-1} \le m < b^p$.

Floating point numbers of magnitude less than $b^{e_{\min}}$ are said to be **subnormal**, because they are too small to be normalized. The term **gradual underflow** refers to the use of subnormal floating point numbers. Standard-conforming, floating point arithmetic allows gradual underflow.

For $x \in \mathbb{R}$, a **rounding mode** maps x to a floating point number $\mathbf{fl}(x)$. Except in cases of overflow discussed below, $\mathrm{fl}(x)$ is either the smallest floating point number greater than or equal to x or the largest floating point number less than or equal to x. Standard-conforming, floating point arithmetic allows program control over which choice is used. The default rounding mode in standard conforming arithmetic is **round-to-nearest, ties-to-even** in which, except for overflow (described below), $\mathrm{fl}(x)$ is the nearest floating point number to x. In case there are two floating point numbers equally distant from x, $\mathrm{fl}(x)$ is the one with even significand.

Underflow occurs in $\mathrm{fl}(x) = 0$ when $0 < |x| \leq b^{e_{\min}}$. Often, underflows are set quietly to zero. **Gradual underflow** occurs when $\mathrm{fl}(x)$ is a subnormal floating point number. **Overflow** occurs when $|x|$ equals or exceeds a threshold at or near the largest floating point number $(b - b^{1-p})b^{e_{\max}}$. Standard-conforming arithmetic allows some, very limited program control over the overflow and underflow threshold, whether to set overflows to $\pm\infty$ and whether to trap program execution on overflow or underflow in order to take corrective action or to issue error messages. In the default round-to-nearest, ties-to-even rounding mode, overflow occurs if $|x| \geq (b - \frac{1}{2}b^{1-p})b^{e_{\max}}$, and in that case, $\mathrm{fl}(x) = \pm\infty$ with the sign chosen to agree with the sign of x. By default, program execution continues without traps or interruption.

A variety of terms describe the precision with which a floating point system models real numbers.

- The **precision** is the number p of base-b digits in the significand.
- **Big M** is the largest integer M with the property that all integers $1, 2, 3, \ldots, M$ are floating point numbers, but $M + 1$ is not a floating point number. If the exponent upper bound $e_{\max}$ is greater than the precision p, then $M = b^p$.
- The **machine epsilon,** $\epsilon = b^{1-p}$, is the distance between the number one and the next larger floating point number.
- The **unit round** $u = \inf\{\delta > 0 \mid \mathrm{fl}(1+\delta) > 1\}$. Depending on the rounding mode, u may be as large as the machine epsilon ϵ. In round-to-nearest, ties-to-even rounding mode, $u = \frac{1}{2}\epsilon$.

In standard-conforming, floating point arithmetic, if α and β are floating point numbers, then **floating point addition** $\oplus$, **floating point subtraction** $\ominus$, **floating point multiplication** $\otimes$, and **floating point division** $\oslash$ are defined by

$$\alpha \oplus \beta = \mathrm{fl}(\alpha + \beta), \tag{37.3}$$

$$\alpha \ominus \beta = \mathrm{fl}(\alpha - \beta), \tag{37.4}$$

$$\alpha \otimes \beta = \mathrm{fl}(\alpha \times \beta), \tag{37.5}$$

$$\alpha \oslash \beta = \mathrm{fl}(\alpha \div \beta), \tag{37.6}$$

The IEEE 754r [IEEE754r] standard also includes a fused addition-multiply operation that evaluates $\alpha\beta + \gamma$ with only one rounding error.

In particular, if the exact, infinitely precise value of $\alpha + \beta$, $\alpha - \beta$, $\alpha \times \beta$, or $\alpha \div \beta$ is also a floating point number, then the corresponding floating point arithmetic operation occurs without rounding error. Floating point sums, products, and differences of small integers have zero rounding error.

Nonstandard-conforming, floating point arithmetics do not always conform to this definition, but often they do. Even when they deviate, it is nearly always the case that if $\bullet$ is one of the arithmetic operations $+$, $-$, $\times$, or $\div$ and $\odot$ is the corresponding nonstandard floating point operation, then $\alpha \odot \beta$ is a floating point number satisfying $\alpha \odot \beta = \alpha(1 + \delta\alpha) \bullet \beta(1 + \delta\beta)$ with $|\delta\alpha| \leq b^{2-p}$ and $|\delta\beta| \leq b^{2-p}$.

If $\bullet$ is one of the arithmetic operations $+$, $-$, $\times$, or $\div$ and $\odot$ is the corresponding floating point operation, then the **rounding error** in $\alpha \odot \beta$ is $(\alpha \bullet \beta) - (\alpha \cdot \beta)$, i.e., rounding error is the difference between the exact, infinitely precise arithmetic operation and the floating point arithmetic operation. In more extensive calculations, **rounding error** refers to the cumulative effect of the rounding errors in the individual floating point operations.

In machine computation, **truncation error** refers to the error made by replacing an infinite process by a finite process, e.g., truncating an infinite series of numbers to a finite partial sum.

Many computers implement floating point numbers of two or more different precisions. Typical **single precision** floating point numbers have machine epsilon roughly 10^{-7} and precision roughly 7 decimal digits or 24 binary digits. Typical **double precision** floating point numbers have machine epsilon roughly 10^{-16} and precision roughly 16 decimal digits or 53 binary digits. Specification of IEEE standard arithmetic [IEEE754r] includes these three precisions. See the table in Example 1. In addition, it is not unusual to also implement **extended precision** floating point numbers with even greater precision.

If $\hat{x} \in \mathbb{F}$ is an approximation to $x \in \mathbb{F}$, the **absolute error** in $\hat{x}$ is $|x - \hat{x}|$ and the **relative error** in $\hat{x}$ is $|(x - \hat{x})/x|$. If $x = 0$, then the relative error is undefined.

Subtractive cancellation of significant digits occurs in floating point sums when the relative error in the rounding-error-corrupted approximate sum is substantially greater than the relative error in the summands. In cases of subtractive cancellation, the sum has magnitude substantially smaller than the magnitude of the individual summands.

Facts:

For proofs and additional background, see, for example, [Ove01].

In this section, we make the following assumptions.

- The numbers α, β, and γ are p-digit radix b floating point numbers with exponent bounds $e_{\max}$ and $e_{\min}$.
- The floating point arithmetic operations satisfy Equation 37.3 to Equation 37.6.
- In the absence of overflows, the rounding mode $\text{fl}(x)$ maps $x \in \mathbb{R}$ to the closest floating point number or to one of the two closest in case of a tie. In particular, the unit round is $\frac{1}{2}b^{1-p}$.

Standard-conforming arithmetic in round-to-nearest, ties-to-even rounding mode satisfies these assumptions.

For vectors $\mathbf{x} \in \mathbb{R}^n$ and matrices $M \in \mathbb{R}^{m \times n}$, the notation $|\mathbf{x}|$ and $|M|$ indicates the vector and matrix whose entries are the absolute values of the corresponding entries of $\mathbf{x}$ and M, respectively. For $\mathbf{x}$, $\mathbf{y} \in \mathbb{R}^n$ the inequality $\mathbf{x} \leq \mathbf{y}$ represents the n scalar inequalities $x_i \leq y_i$, $i = 1, 2, 3, \ldots, n$. Similarly, for $A, B \in \mathbb{R}^{m \times n}$, the inequality $A \leq B$ represents the mn inequalities $a_{ij} \leq b_{ij}$, $i = 1, 2, 3, \ldots, m$ and $j = 1, 2, 3, \ldots, n$.

If e is an arithmetic expression involving only floating point numbers, the notation $\text{fl}(e)$ represents the expression obtained by replacing each arithmetic operation by the corresponding floating point arithmetic operation 37.3, 37.4, 37.5, or 37.6. Note that $\text{fl}(\cdot)$ is not a function, because its value may depend on the order of operations in e—not value of e.

1. At this writing, the only radixes in common use on computers and calculators are $b = 2$ and $b = 10$.
2. The commutative properties of addition and multiplication hold for floating point addition 37.3 and floating point multiplication 37.5, i.e., $\alpha \oplus \beta = \beta \oplus \alpha$ and $\alpha \otimes \beta = \beta \otimes \alpha$. (Examples below show that the associative and distributive properties of addition and multiplication *do not* in general hold for floating point arithmetic.)
3. In the absence of overflow or any kind of underflow, $\text{fl}(x) = x(1 + \delta)$ with $|\delta| < \frac{1}{2}b^{1-p}$.
4. *Rounding in Arithmetic Operations:* If $\bullet$ is an arithmetic operation and $\odot$ is the corresponding floating point operation, then

$$\alpha \odot \beta = (\alpha \bullet \beta)(1 + \delta)$$

   with $|\delta| \leq \frac{1}{2}b^{1-p}$. The error δ may depend on α and β as well as the arithmetic operation.
5. *Differences of Nearby Floating Point Numbers:* If $\alpha > 0$ and $\beta > 0$ are normalized floating point numbers and $\frac{1}{2} < \alpha/\beta < 2$, then $\text{fl}(\alpha - \beta) = \alpha \ominus \beta = \alpha - \beta$, i.e., there is zero rounding error in floating point subtraction of nearby numbers.

6. *Product of Floating Point Numbers:* If α_i, $i = 1, 2, 3, \ldots$ n are n floating point numbers, then

$$\mathrm{fl}\left(\prod_{i=1}^{n} \alpha_i\right) = \left(\prod_{i=1}^{n} \alpha_i\right)(1+\delta)^n$$

with $|\delta| < \frac{1}{2}b^{1-p}$. Consequently, rounding errors in floating point products create only minute relative errors.

7. [Ste98], [Wil65] *Dot (or Inner) Product of Floating Point Numbers:* Let $\mathbf{x} \in \mathbb{R}^n$ and $\mathbf{y} \in \mathbb{R}^n$ be vectors of floating point numbers and let s be the finite precision dot product $s = \mathrm{fl}(\mathbf{x}^T\mathbf{y}) = x_1 \otimes y_1 \oplus x_2 \otimes y_2 \oplus x_3 \otimes y_3 \oplus \cdots \oplus x_n \otimes y_n$ evaluated in the order shown, multiplications first followed by additions from left to right. If $n < 0.1/(\frac{1}{2}b^{1-p})$ and neither overflow nor any kind of underflow occurs, then the following hold.

   (a) $s = \mathrm{fl}(\mathbf{x}^T\mathbf{y}) = x_1y_1(1+\delta_1) + x_2y_2(1+\delta_2) + x_3y_3(1+\delta_3) + \cdots + x_ny_n(1+\delta_3)$ with $|\delta_1| < 1.06n(\frac{1}{2}b^{1-p})$ and $|\delta_j| < 1.06(n-j+2)(\frac{1}{2}b^{1-p}) \leq 1.06n(\frac{1}{2}b^{1-p})$, for $j = 2, 3, 4, \ldots, n$.

   (b) $s = \mathrm{fl}(\mathbf{x}^T\mathbf{y}) = \hat{\mathbf{x}}^T\hat{\mathbf{y}}$ for some vectors $\hat{\mathbf{x}}, \hat{\mathbf{y}} \in \mathbb{R}^n$ satisfying $|\mathbf{x} - \hat{\mathbf{x}}| \leq |\mathbf{x}|(1 + 1.06n(\frac{1}{2}b^{1-p}))$ and $|\mathbf{y} - \hat{\mathbf{y}}| \leq |\mathbf{y}|(1 + 1.06n(\frac{1}{2}b^{1-p}))$. So, s is the mathematically correct product of the vectors $\hat{\mathbf{x}}$ and $\hat{\mathbf{y}}$ each of whose entries differ from the corresponding entries of $\mathbf{x}$ or $\mathbf{y}$ by minute relative errors.

   There are infinitely many choices of $\hat{\mathbf{x}}$ and $\hat{\mathbf{y}}$. In the notation of Fact 7(a), two are $\hat{x}_j = x_j(1+\delta_j)$, $\hat{y}_j = y_j$ and $\hat{x}_j = x_j(1+\delta_j)^{1/2}$, $\hat{y}_j = y_j(1+\delta_j)^{1/2}$, $j = 1, 2, 3, \ldots, n$.

   (c) If $x^Ty \neq 0$, then the relative error in s is bounded as

$$\left|\frac{s - x^Ty}{x^Ty}\right| \leq 1.06n(\tfrac{1}{2}b^{1-p})\frac{|x|^T|y|}{|x^Ty|}.$$

   The bound shows that if there is little cancellation in the sum, then s has small relative error. The bound allows the possibility that s has large relative error when there is substantial cancellation in the sum. Indeed this is often the case.

8. *Rounding Error Bounds for Floating Point Matrix Operations:* In the following, A and B are matrices each of whose entries is a floating point number, $\mathbf{x}$ is a vector of floating point numbers, and c is a floating point number. The matrices A and B are compatible for matrix addition or matrix multiplication as necessary and E is an error matrix (usually different in each case) whose entries may or may not be floating point numbers. The integer dimension n is assumed to satisfy $n < 0.1/(\frac{1}{2}b^{1-p})$.

   If neither overflows nor any kind of underflow occurs, then

   (a) $\mathrm{fl}(cA) = cA + E$ with $|E| \leq \frac{1}{2}b^{1-p}|cA|$.

   (b) $\mathrm{fl}(A+B) = (A+B) + E$ with $|E| \leq \frac{1}{2}b^{1-p}|A+B|$.

   (c) If $\mathbf{x} \in \mathbb{R}^n$ and $A \in \mathbb{R}^{m\times n}$, then $\mathrm{fl}(A\mathbf{x}) = (A+E)\mathbf{x}$ with $|E| \leq 1.06n(\frac{1}{2}b^{1-p})$.

   (d) *Matrix multiplication:* If $A \in \mathbb{R}^{m\times n}$ and $B \in \mathbb{R}^{n\times q}$, $\mathrm{fl}(AB) = AB + E$ with $|E| \leq 1.06n(\frac{1}{2}b^{1-p})$ $|A|\,|B|$.

   Note that if $|AB| \approx |A|\,|B|$, then each entry in $\mathrm{fl}(AB)$ is correct to within a minute relative error. Otherwise, subtractive cancellation is possible and some entries of $\mathrm{fl}(AB)$ may have large relative errors.

   (e) Let $\|\cdot\|$ be a matrix norm satisfying $\|E\| \leq \|\,|E|\,\|$. (All of $\|\cdot\|_1$, $\|\cdot\|_2$, $\|\cdot\|_p$, $\|\cdot\|_\infty$ and $\|\cdot\|_F$ satisfy this requirement.) If $A \in \mathbb{R}^{m\times n}$ and $B \in \mathbb{R}^{n\times q}$, then $\mathrm{fl}(AB) = AB + E$ with $\|E\| \leq 1.06n(\frac{1}{2}b^{1-p})\|\,|A|\,\|\,\|\,|B|\,\|$.

(f) *Matrix multiplication by an orthogonal matrix:* If $Q \in \mathbb{R}^{n\times n}$ is an orthogonal matrix, i.e., if $Q^T = Q^{-1}$, and if $A \in \mathbb{R}^{m\times n}$, then

(i) $\text{fl}(QA) = QA+E$ with $\|E\|_2 \leq 1.06n^2\left(\frac{1}{2}b^{1-p}\right)\|A\|_2$ and $\|E\|_F \leq 1.06n^{3/2}\left(\frac{1}{2}b^{1-p}\right)\|A\|_F$.

(ii) $\text{fl}(QA) = Q(A+\hat{E})$ with $\|\hat{E}\|_2 \leq 1.06n^2\left(\frac{1}{2}b^{1-p}\right)\|A\|_2$ and $\|\hat{E}\|_F \leq 1.06n^{3/2}\left(\frac{1}{2}b^{1-p}\right)\|A\|_F$.

Note that this bound shows that if one of the factors is an orthogonal matrix, then subtractive cancellation in floating point matrix multiplication is limited to those entries (if any) that have magnitude substantially smaller than the other, possibly nonorthogonal factor. Many particularly successful numerical methods derive their robustness in the presence of rounding errors from this observation. (See, for example, [Dat95], [GV96], [Ste98], or [Wil65].)

Examples:

1. The following table lists typical floating point systems in common use at the time of this writing.

	radix b	precision p	$e_{\min}$	$e_{\max}$
Some calculators	10	14	-99	99
Some calculators	10	14	-999	999
IEEE 754r decimal 64 [IEEE754r]	10	16	-383	384
IEEE 754r decimal 128 [IEEE754r]	10	34	-6143	6144
IEEE 754r binary 32 (Single) [IEEE754r]	2	24	-126	127
IEEE 754r binary 64 (Double) [IEEE754r]	2	53	-1022	1023
IEEE 754r binary 128 (Double Extended) [IEEE754r]	2	113	-16382	16383

2. Consider $p = 5$ digit, radix $b = 10$ floating-point arithmetic with round-to-nearest, ties-to-even rounding mode.

(a) *Floating point addition is not associative:* If $\alpha = 1$, $\beta = 10^5$, and $\gamma = -10^5$, then $(\alpha \oplus \beta) \oplus \gamma = 0$ but $\alpha \oplus (\beta \oplus \gamma) = 1$. (This is also an example of subtractive cancellation.)

(b) *Floating point multiplication/division is not associative:* If $\alpha = 3$, $\beta = 1$, and $\gamma = 3$, then $\alpha \otimes (\beta \oslash \gamma) = .99999$ but $(\alpha \otimes \beta) \oslash \gamma = 1$. If $\alpha = 44444$, $\beta = 55555$ and $\gamma = 66666$, then $\alpha \otimes (\beta \otimes \gamma) = 1.6460 \times 10^{14}$ but $(\alpha \otimes \beta) \otimes \gamma = 1.6461 \times 10^{14}$. Although different, both expressions have minute relative error. It is generally the case that floating point products have small relative errors. (See Fact 6.)

(c) *Floating point multiplication does not distribute across floating point addition:* If $\alpha = 9$, $\beta = 1$, and $\gamma = -.99999$, then $\alpha \otimes (\beta \oplus \gamma) = 9.0000 \times 10^{-5}$ but $(\alpha \otimes \beta) \oplus (\alpha \otimes \gamma) = 1.0000 \times 10^{-4}$.

3. *Subtractive Cancellation:* In $p = 5$, radix $b = 10$ arithmetic with round-to-nearest, ties-to-even rounding, the expression $(\sqrt{1+10^{-4}} - 1)/10^{-4}$ evaluates as follows. (Here we assume that the floating point evaluation of the square root gives the same result as rounding the exact square root to $p = 5$, radix $b = 10$ digits.)

$$\begin{aligned}(\text{fl}(\sqrt{1.0000 \oplus 10^{-4}}) \ominus 1) \oslash 10^{-4} &= (\text{fl}(\sqrt{1.0001}) \ominus 1.0000) \oslash 10^{-4}\\ &= (\text{fl}(1.000049999\ldots) \ominus 1.0000) \oslash 10^{-4}\\ &= (1.0000 \ominus 1.0000) \oslash 10^{-4}\\ &= 0.0000.\end{aligned}$$

In exact, infinite precision arithmetic, $(\sqrt{1+10^{-4}} - 1)/10^{-4} = .4999875\ldots$, so the relative error is 1. Note that zero rounding error occurs in the subtraction. The only nonzero rounding error is the square root, which has minute relative error roughly 5×10^{-5}, but this is enough to ruin the final result. Subtractive cancellation did not cause the large relative error. It only exposed the unfortunate fact that the result was ruined by earlier rounding errors.

4. *Relative vs. Absolute Errors:* Consider $\hat{x}_1 = 31416$ as an approximation to $x_1 = 10000\pi$ and $\hat{x}_2 = 3.07$ as an approximation to $x_2 = \pi$, The absolute errors are nearly the same: $|\hat{x}_1 - x_1| \approx -0.0735$ and $|\hat{x}_2 - x_2| \approx -0.0716$. On the other hand, the relative error in the first case, $\frac{|\hat{x}_1 - x_1|}{|x_1|} \approx 2 \times 10^{-6}$, is much smaller than the relative error in the second case, $\frac{|\hat{x}_2 - x_2|}{|x_2|} \approx 2 \times 10^{-2}$. The smaller relative error shows that $\hat{x}_1 = 31416$ is a better approximation to 10000π than $\hat{x}_2 = 3.07$ is to π. The absolute errors gave no indication of this.

37.7 Algorithms and Efficiency

In this section, we introduce efficiency of algorithms.

Definitions:

An **algorithm** is a precise set of instructions to perform a task. The algorithms discussed here perform mathematical tasks that transform an initial data set called the **input** into a desired result final data set called the **output** using an ordered list of arithmetic operations, comparisons, and decisions. For example, the Gaussian elimination algorithm (see Chapter 39.3) solves the linear system of equations $A\mathbf{x} = \mathbf{b}$ by transforming the given, nonsingular matrix $A \in \mathbb{R}^{n\times n}$ and right-hand-side $\mathbf{b} \in \mathbb{R}^n$ into a vector $\mathbf{x}$ satisfying $A\mathbf{x} = \mathbf{b}$.

One algorithm is more **efficient** than another if it accomplishes the same task with a lower cost of computation. Cost of computation is usually dominated by the direct and indirection cost of execution time. So, in general, in a given computational environment, the more efficient algorithm finishes its task sooner. (The very real economic cost of algorithm development and implementation is not a part of the efficiency of an algorithm.) However, the amount of primary and secondary memory required by an algorithm or the expense of and availability of the necessary equipment to execute the algorithm may also be significant part of the cost. In this sense, an algorithm that can accomplish its task on an inexpensive, programmable calculator is more efficient than one that needs a supercomputer.

For this discussion, a floating point operation or **flop** consists of a floating point addition, subtraction, multiplication, division, or square root along with any necessary subscripting and loop index overhead. In Fortran `A(I,J) = A(I,J) + C*A(K,J)` performs two flops. (Note that this is a slightly different definition of flop than is used in computer engineering.)

Formal algorithms are often specified in terms of an informal computer program called **pseudo-code**.

On early digital computers, computation time was heavily dominated by evaluating floating point operations. So, traditionally, numerical analysts compare the efficiency of two algorithms by counting the number of floating point operations each of them executes. If n measures the input data set size, e.g., an input matrix $A \in \mathbb{R}^{n\times n}$, then an $O(n^p)$ algorithm is one that, for some positive constant c, performs cn^p plus a sum of lower powers of n floating point operations.

Facts:

For proofs and additional background, see, for example, [Dat95], [GV96], [Ste98], or [Wil65].

In choosing a numerical method, efficiency must be balanced against considerations like robustness against rounding error and likelyhood of failure.

Despite tradition, execution time has never been more than roughly proportional to the amount of floating point arithmetic. On modern computers with fast floating point arithmetic, multiple levels of cache memory, overlapped instruction execution, and parallel processing, execution time is correlated more closely with the number of cache misses (i.e., references to main RAM memory) than it is to the number of floating point operations. In addition, the relative execution time of algorithms depends strongly on the environment in which they are executed.

Nevertheless, for algorithms highly dominated by floating point arithmetic, flop counts are still useful. Despite the complexities of modern computers, flop counts typically expose the rate at which execution time increases as the size of the problem increases. Solving linear equations $A\mathbf{x} = \mathbf{b}$ for given $A \in \mathbb{R}^{n\times n}$ and $\mathbf{b} \in \mathbb{R}^n$ by Gaussian elimination with partial pivoting is an $O(n^3)$ algorithm. For larger values of n, solving a $2n$-by-$2n$ system of equations takes roughly eight times longer than an n-by-n system.

1. Triangular back substitution is an $O(n^2)$ algorithm to solve a triangular system of linear equations $T\mathbf{x} = b$, $\mathbf{b} \in \mathbb{R}^n$ and $T \in \mathbb{R}^{n\times n}$, with $t_{ij} = 0$ whenever $i > j$ [GV96]. (See the pseudo-code algorithm below.)
2. [GV96] Gaussian elimination with partial pivoting (cf. Algorithm 1, Section 38.3) is an $O(n^3)$ algorithm to solve a system of equations $A\mathbf{x} = \mathbf{b}$ with $A \in \mathbb{R}^{n\times n}$ and $\mathbf{b} \in \mathbb{R}^n$.
3. Because of the need to repeatedly search an entire submatrix, Gaussian elimination with complete pivoting is an $O(n^4)$ algorithm to solve a system of equations $A\mathbf{x} = \mathbf{b}$ with $A \in \mathbb{R}^{n\times n}$ and $\mathbf{b} \in \mathbb{R}^n$ [GV96]. Hence, complete pivoting is not competitive with $O(n^3)$ methods like Gaussian elimination with partial pivoting.
4. [GV96] The QR-factorization by Householder's method is an $O(n^3)$ algorithm to solve a system of equations $A\mathbf{x} = \mathbf{b}$ with $A \in \mathbb{R}^{n\times n}$ and $\mathbf{b} \in \mathbb{R}^n$.
5. [GV96] The QR factorization by Householder's method to solve the least squares problem $\min \|A\mathbf{x} - \mathbf{b}\|_2$ for given $A \in \mathbb{R}^{m\times n}$ and $\mathbf{b} \in \mathbb{R}^m$ is an $O(n^2(m - n/3))$ algorithm.
6. [GV96] The singular value decomposition using the Golub–Kahan–Reinsch algorithm to solve the least squares problem $\min \|A\mathbf{x} - \mathbf{b}\|_2$ for given $A \in \mathbb{R}^{m\times n}$ and $\mathbf{b} \in \mathbb{R}^m$ is an $O(m^2n + mn^2 + n^3)$ algorithm.
7. [GV96] The implicit, double-shift QR iteration algorithm to find all eigenvalues of a given matrix $A \in \mathbb{R}^{n\times n}$ is an $O(n^3)$ algorithm.
8. Cramer's rule for solving the system of equations $A\mathbf{x} = \mathbf{b}$ for given $A \in \mathbb{R}^{n\times n}$ and $\mathbf{b} \in \mathbb{R}^n$ in which determinants are evaluated using minors and cofactors is an $O((n + 1)n!)$ algorithm and is impractical for all but small values of n.
9. Cramer's rule for solving the system of equations $A\mathbf{x} = \mathbf{b}$ for given $A \in \mathbb{R}^{n\times n}$ and $\mathbf{b} \in \mathbb{R}^n$ in which determinants are evaluated using Gaussian elimination is an $O(n^4)$ algorithm and is not competitive with $O(n^3)$ methods like Gaussian elimination with partial pivoting.

Examples:

1. It takes roughly 4.6 seconds on a 2GHz Pentium workstation, using Gaussian elimination with partial pivoting, to solve the $n = 2000$ linear equations in 2000 unknowns $A\mathbf{x} = \mathbf{b}$ in which the entries of $\mathbf{A}$ and $\mathbf{b}$ are normally distributed pseudo-random numbers with mean zero and variance one. It takes roughly 34 seconds to solve a similar $n = 4000$ system of equations. This is consistent with the estimate that Gaussian elimination with partial pivoting is an $O(n^3)$ algorithm.
2. This is an example of a formal algorithm specified in pseudo-code. Consider the problem of solving for $\mathbf{y}$ in the upper triangular system of equations $T\mathbf{y} = \mathbf{b}$, where $\mathbf{b} \in \mathbb{R}^n$ is a given right-hand-side vector and $T \in \mathbb{R}^{n\times n}$ is a given nonsingular upper triangular matrix; i.e., $t_{ij} = 0$ for $i > j$ and $t_{ii} \neq 0$ for $i = 1, 2, 3, \ldots, n$.

   Input: A nonsingular, upper triangular matrix $T \in \mathbb{R}^{n\times n}$ and a vector $\mathbf{b} \in \mathbb{R}^n$.

   Output: The vector $\mathbf{y} \in \mathbb{R}^n$ satisfying $T\mathbf{y} = \mathbf{b}$.

   Step 1. $y_n \leftarrow b_n/t_{nn}$

   Step 2. For $i = n - 1, n - 2, \ldots, 2, 1$ do

   2.1 $s_i \leftarrow b_i - \sum_{j=i+1}^{n} t_{ij}y_j$

   2.2 $y_i = s_i/t_{ii}$

37.8 Numerical Stability and Instability

Numerically stable algorithms, despite rounding and truncation errors, produce results that are roughly as accurate as the errors in the input data allow. *Numerically unstable* algorithms allow rounding and truncation errors to produce results that are substantially less accurate than the errors in the input data allow. This section concerns numerical stability and instability and is loosely based on [Bun87].

Definitions:

A computational problem is the task of evaluating a function $f : \mathbb{R}^n \to \mathbb{R}^m$ at a particular data point $\mathbf{x} \in \mathbb{R}^n$. A numerical algorithm that is subject to rounding and truncation errors evaluates a perturbed function $\hat{f}(\mathbf{x})$. Throughout this section ε represents a modest multiple of the unit round.

The **forward error** is $f(\mathbf{x}) - \hat{f}(\mathbf{x})$, the difference between the mathematically exact function evaluation and the perturbed function evaluation.

The **backward error** is a vector $\mathbf{e} \in \mathbb{R}^n$ of smallest norm for which $f(\mathbf{x}+\mathbf{e}) = \hat{f}(\mathbf{x})$. If no such $\mathbf{e}$ exists, then the backward error is undefined. This definition is illustrated in Figure 37.1 [Ste98, p. 123].

An algorithm is **forward stable** if, despite rounding and truncation errors,

$$\frac{\|f(\mathbf{x}) - \hat{f}(\mathbf{x})\|}{\|f(\mathbf{x})\|} \leq \varepsilon$$

for all valid input data $\mathbf{x}$. In a forward stable algorithm, the forward relative error is small for all valid input data $\mathbf{x}$ despite rounding and truncation errors in the algorithm.

An algorithm is **backward stable** or **strongly stable** if the backward relative error $\mathbf{e}$ exists and satisfies the relative error bound $\|\mathbf{e}\| \leq \varepsilon\|\mathbf{x}\|$ for all valid input data $\mathbf{x}$ despite rounding and truncation errors.

In this context, "small" means a modest multiple of the size of the errors in the data $\mathbf{x}$. If rounding errors are the only relevant errors, then "small" means a modest multiple of the unit round.

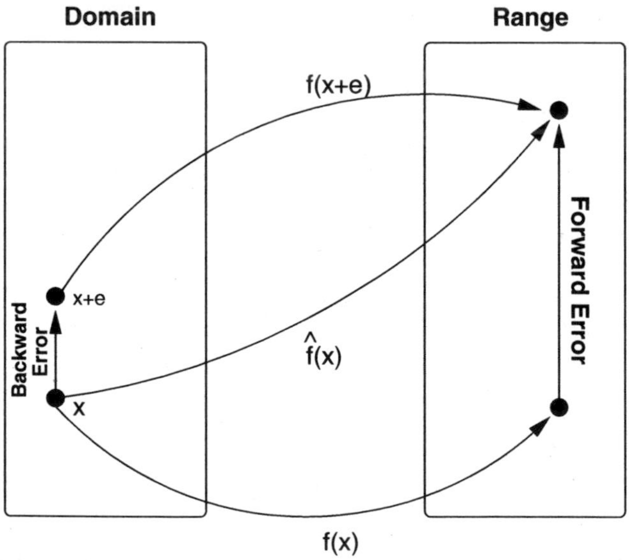

FIGURE 37.1 A computational problem is the task of evaluating a function f at a particular data point $\mathbf{x}$. A numerical algorithm that is subject to rounding and truncation errors evaluates a perturbed function $\hat{f}(\mathbf{x})$. The **forward error** is $f(\mathbf{x}) - \hat{f}(\mathbf{x})$, the difference between the mathematically exact function evaluation and the perturbed function evaluation. The **backward error** is a vector $\mathbf{e} \in \mathbb{R}^n$ of smallest norm for which $f(\mathbf{x}+\mathbf{e}) = \hat{f}(\mathbf{x})$ [Ste98, p. 123]. If no such vector e exists, then there is no backward error.

A finite precision numerical algorithm is **weakly numerically stable** if rounding and truncation errors cause it to evaluate a perturbed function $\hat{f}(\mathbf{x})$ satisfying the relative error bound

$$\frac{\|f(\mathbf{x}) - \hat{f}(\mathbf{x})\|}{\|f(\mathbf{x})\|} \leq \varepsilon \mathbf{cond}_f(\mathbf{x})$$

for all valid input data $\mathbf{x}$. In a weakly stable algorithm, the *magnitude* of the forward error is no greater than magnitude of an error that could be induced by perturbing the data by small multiple of the unit round. Note that this does *not* imply that there is a small backward error or even that a backward error exists. (An even weaker kind of "weak stability" requires the relative error bound only when $\mathbf{x}$ is well-conditioned [Bun87].)

An algorithm is **numerically stable** if rounding errors and truncation errors cause it to evaluate a perturbed function $\hat{f}(\mathbf{x})$ satisfying

$$\frac{\|f(\mathbf{x}+\mathbf{e}) - \hat{f}(\mathbf{x})\|}{\|f(\mathbf{x})\|} \leq \varepsilon$$

for some small relative-to-$\|\mathbf{x}\|$ backward error $\mathbf{e}$, $\|\mathbf{e}\| \leq \varepsilon\|\mathbf{x}\|$. In a numerically stable algorithm, $\hat{f}(\mathbf{x})$ lies near a function with small backward error.

Figure 37.2 illustrates the definitions of stability. The black dot on the left represents a nominal data point $\mathbf{x}$. The black dot on the right is the exact, unperturbed value of $f(\mathbf{x})$. The shaded region on the left represents the small relative-to-$\|\mathbf{x}\|$ perturbations of $\mathbf{x}$. The shaded region on the right is its exact image under $f(\mathbf{x})$.

In a weakly stable numerical method, the computed function value $\hat{f}(\mathbf{x})$ lies inside the large circle with radius equal to the longest distance from the black dot $f(\mathbf{x})$ to the furthest point in the shaded region containing it. The error in a weakly stable algorithm is no larger than would have been obtained from a backward stable algorithm. However, the actual result may or may not correspond to a small perturbation of the data.

In a numerically stable algorithm, $\hat{f}(\mathbf{x})$ lies either near or inside the shaded region on the right.

In a backward stable algorithm, $\hat{f}(\mathbf{x})$ lies in the shaded region, but, if the data are ill-conditioned as in the illustration, $\hat{f}(\mathbf{x})$ may have a large relative error. (To avoid clutter, there is no arrow illustrating a backward stable algorithm.)

In a forward stable algorithm, $\hat{f}(\mathbf{x})$ has a small relative error, but $\hat{f}(\mathbf{x})$ may or may not correspond to a small perturbation of the data.

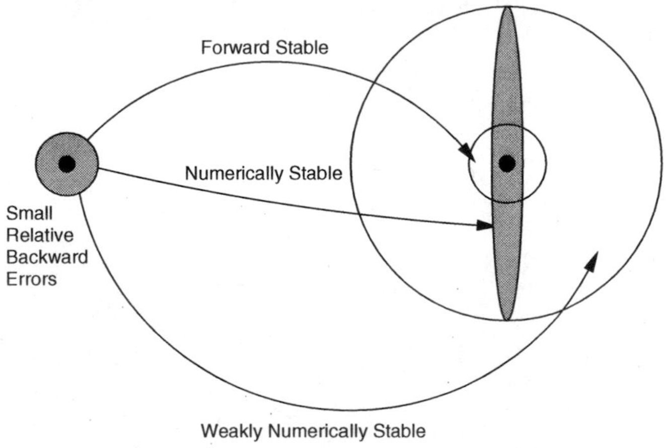

FIGURE 37.2 The black dot on the left represents a nominal data point $\mathbf{x}$. The black dot on the right is the exact, unperturbed value of $f(\mathbf{x})$. The shaded region on the left represents the small relative-to-$\|\mathbf{x}\|$ perturbations of $\mathbf{x}$. The shaded region on the right is its exact image under $f(\mathbf{x})$. The diagram illustrates the error behavior of a weakly numerically stable algorithm, a numerically stable algorithm, and a forward stable algorithm.

Facts:

1. A numerically stable algorithm applied to an ill-conditioned problem may produce inaccurate results. For ill-conditioned problems even small errors in the input data may lead to large errors in the computed solution. Just as no numerical algorithm can reliably correct errors in the data or create information not originally implicit in the data, nor can a numerical algorithm reliably calculate accurate solutions to ill-conditioned problems.
2. Rounding and truncation errors in a backward stable algorithm are equivalent to a further perturbation of the data. The computed results of a backward stable algorithm are realistic in the sense that they are what would have been obtained in exact arithmetic from an extra rounding-error-small relative perturbation of the data. Typically this extra error is negligible compared to other errors already present in the data.
3. The forward error that occurs in a backward stable algorithm obeys the asymptotic condition number bound in Fact 2 of Section 37.4. **Backward error analysis** is based on this observation.
4. [Dat95], [GV96] *Some Well-Known Backward Stable Algorithms:*

   (a) Fact 4 in Section 37.6 implies that a single floating point operation is both forward and backward stable.

   (b) Fact 7b in section 37.6 shows that the given naive dot product algorithm is backward stable. The algorithm is not, in general, forward stable because there may be cancellation of significant digits in the summation.

   (c) *Gaussian elimination:* Gaussian elimination with complete pivoting is backward stable. Gaussian elimination with partial pivoting is not, strictly speaking, backward stable. However, linear equations for which the algorithm exhibits instability are so extraordinarily rare that the algorithm is said to be "backward stable in practice" [Dat95, GV96].

   (d) *Triangular back substitution:* The back-substitution algorithm in Example 2 in Section 37.7 is backward stable. It can be shown that the rounding error corrupted computed solution $\hat{x}$ satisfies $(T + E)\hat{\mathbf{x}} = \mathbf{b}$, where $| e_{ij} | \leq \epsilon | t_{ij} |$, $i, j = 1, 2, 3, \ldots, n$. Thus, the computed solution $\hat{x}$ solves a nearby system. The back-substitution process is, therefore, backward stable.

   (e) *QR factorization:* The Householder and Givens methods for factorization of $A = QR$, where Q is orthogonal and R is upper triangular, are backward stable.

   (f) *SVD computation:* The Golub–Kahan–Reinsch algorithm is a backward stable algorithm for finding the singular value decomposition $A = U\Sigma V^T$, where Σ is diagonal and U and V are orthogonal.

   (g) *Least-square problem:* The Householder QR factorization, the Givens QR factorization, and Singular Value Decomposition (SVD) methods for solving linear least squares problems are backward stable.

   (h) *Eigenvalue computations:* The implicit double-shift QR iteration is backward stable.

Examples:

1. An example of an algorithm that is forward stable but not backward stable is the natural computation of the outer product $A = \mathbf{x}\mathbf{y}^T$ from vectors $\mathbf{x}, \mathbf{y} \in \mathbb{R}^n$: for $i, j = 1, 2, 3, \ldots n$, set $a_{ij} \leftarrow x_i \otimes y_i$. This algorithm produces the correctly rounded value of the exact outer product, so it is forward stable. However, in general, rounding errors perturb the rank 1 matrix $\mathbf{x}\mathbf{y}^T$ into a matrix of higher rank. So, the rounding error perturbed outer product is not equal to the outer product of any pair of vectors; i.e., there is no backward error, so the algorithm is not backward stable.
2. *Backward vs. Forward Errors:* Consider the problem of evaluating $f(x) = e^x$ at $x = 1$. One numerical method is to sum several terms of the Taylor series for e^x. If $f(x)$ is approximated by the truncated Taylor series $\hat{f}(x) = 1 + x + \frac{x^2}{2} + \frac{x^3}{3!}$, then $f(1) = e \approx 2.7183$ and $\hat{f}(1) \approx 2.6667$.

The forward error is $f(1) - \hat{f}(1) \approx 2.7183 - 2.6667 = 0.0516$. The backward error is $1 - y$, where $f(y) = \hat{f}(1)$, i.e., the backward error is $1 - \ln(\hat{f}(1)) \approx .0192$.

3. *A Numerically Unstable Algorithm:* Consider the computational problem of evaluating the function $f(x) = \ln(1 + x)$ for x near zero. The naive approach is to use $\text{fl}(\ln(1 \oplus x))$, i.e., add one to x in finite precision arithmetic and evaluate the natural logarithm of the result also in finite precision arithmetic. (For this discussion, assume that if z is a floating point number, then $\ln(z)$ returns the correctly rounded exact value of $\ln(z)$.) Applying this very simple algorithm to $x = 10^{-16}$ in $p = 16$, radix $b = 10$ arithmetic, gives $\text{fl}(\ln(1 \oplus 10^{-16})) = \text{fl}(\ln(1)) = 0$. The exact value of $\ln(1 + 10^{-16}) \approx 10^{-16}$, so the rounding error corrupted result has relative error 1. It is 100% incorrect.

   However, the function $f(x) = \ln(1 + x)$ is well-conditioned when x is near zero. Moreover, $\lim_{x\to 0} \textbf{cond}_f(x) = 1$. The large relative error is not due to ill-conditioning. It demonstrates that this simple algorithm is numerically unstable for x near zero.

4. An alternative algorithm to evaluate $f(x) = \ln(1+x)$ that does not suffer the above gross numerical instability for x near zero is to sum several terms of the Taylor series

$$\ln(1 + x) = x - \frac{x^2}{2} + \frac{x^3}{3} - \cdots .$$

   Although it is adequate for many purposes, this method can be improved. Note also that the series does not converge for $|x| > 1$ and converges slowly if $|x| \approx 1$, so some other method (perhaps $\text{fl}(\ln(1 \oplus x))$) is needed when x is not near zero.

5. *Gaussian Elimination without Pivoting:* Gaussian elimination without pivoting is not numerically stable. For example, consider solving a system of two equations in two unknowns

$$\begin{aligned} 10^{-10}x_1 + x_2 &= 1 \\ x_1 + 2x_2 &= 3 \end{aligned}$$

   using $p = 9$ digit, radix $b = 10$ arithmetic. Eliminating x_2 from the second equation, we obtain

$$\begin{aligned} 10^{-10}x_1 + \qquad\qquad x_2 &= \quad 1 \\ (2 \ominus 10^{10})x_2 &= 3 \ominus 10^{10}, \end{aligned}$$

   which becomes

$$\begin{aligned} 10^{-10}x_1 + \qquad x_2 &= \quad 1 \\ -10^{10}x_2 &= -10^{10}, \end{aligned}$$

   giving $x_2 = 1, x_1 = 0$. The exact solution is $x_1 = (1 - 2 \times 10^{-10}) \approx 1, x_2 = (1 - 3 \times 10^{-10})/1 - 2 \times 10^{-10}) \approx 1$.

   The ∞-norm condition number of the coefficient matrix $A = \begin{bmatrix} 10^{-10} & 1 \\ 1 & 2 \end{bmatrix}$ is $\kappa(A) \approx 9$, so the large error in the rounding error corrupted solution is not due to ill-conditioning. Hence, Gaussian elimination without pivoting is numerically unstable.

6. *An Unstable Algorithm for Eigenvalue Computations:* Finding the eigenvalues of a matrix by finding the roots of its characteristic polynomial is a numerically unstable process because the roots of the characteristic polynomial may be ill-conditioned when the eigenvalues of the corresponding matrix are well-conditioned. Transforming a matrix to companion form often requires an ill-conditioned similarity transformation, so even calculating the coefficients of the characteristic polynomial may be an unstable process. A well-known example is the diagonal matrix $A = \text{diag}(1, 2, 3, \ldots, 20)$. The Wielandt-Hoffman theorem [GV96] shows that perturbing A to a nearby matrix $A + E$ perturbs the eigenvalues by no more than $\|E\|_F$. However, the characteristic polynomial is the infamous Wilkinson polynomial discussed in Example 5 of Section 37.4, which has highly ill-conditioned roots.

Author Note

The contribution of Ralph Byers' material is partially supported by the National Sciences Foundation Award 0098150.

References

[Bha96] R. Bhatia, *Matrix Analysis*, Springer, New York, 1996.

[Bun87] J. Bunch, The weak and strong stability of algorithms in numerical linear algebra, *Linear Algebra and Its Applications*, 88, 49–66, 1987.

[Dat95] B.N. Datta, *Numerical Linear Algebra and Applications*, Brooks/Cole Publishing Company, Pacific Grove, CA, 1995. (Section edition to be published in 2006.)

[GV96] G.H. Golub and C.F. Van Loan, *Matrix Computations*, 3rd ed., Johns Hopkins University Press, Baltimore, MD, 1996.

[Hig96] N.J. Higham, *Accuracy and Stability of Numerical Algorithms*, SIAM, Philadelphia, 1996.

[HJ85] R.A. Horn and C.R. Johnson, *Matrix Analysis*, Cambridge University Press, New York, 1985.

[IEEE754] IEEE 754-1985. "IEEE Standard for Binary Floating-Point Arithmetic" (ANSI/IEEE Std 754-1985), The Institute of Electrical and Electronics Engineers, Inc., New York, 1985. Reprinted in *SIGPLAN Notices*, 22(2): 9–25, 1987.

[IEEE754r] (Draft) "Standard for Floating Point Arithmetic P754/D0.15.3—2005," October 20. The Institute of Electrical and Electronics Engineers, Inc., New York, 2005.

[IEEE854] IEEE 854-1987. "IEEE Standard for Radix-Independent Floating-Point Arithmetic." The Institute of Electrical and Electronics Engineers, Inc., New York, 1987.

[Kat66] T. Kato, *Perturbation Theory for Linear Operators*, Springer-Verlag, New York, 1966. (A corrected second edition appears in 1980, which was republished in the *Classics in Mathematics* series in 1995.)

[Ove01] M. Overton, *Numerical Computing with IEEE Floating Point Arithmetic*, Society for Industrial and Applied Mathematics, Philadelphia, PA, 2001.

[Ric66] J. Rice, A theory of condition, SIAM J. Num. Anal., 3, 287–310, 1966.

[Ste98] G.W. Stewart, *Matrix Algorithms, Vol. 1,* Basic Decompositions, SIAM, Philadelphia, 1998.

[SS90] G.W. Stewart and J.G. Sun, *Matrix Perturbation Theory*, Academic Press, New York, 1990.

[TB97] L.N. Trefethan and D. Bau, *Numerical Linear Algebra*, SIAM, Philadelphia, 1997.

[Wil65] J.H. Wilkinson, *The Algebraic Eigenvalue Problem,* Clarendon Press, Oxford, U.K., 1965.

[Wil64] J.H. Wilkinson, *Rounding Errors in Algebraic Processes*, Prentice-Hall, Inc., Upper Saddle River, NJ, 1963. Reprinted by Dover Publications, Mineola, NY, 1994.

38

Matrix Factorizations and Direct Solution of Linear Systems

Christopher Beattie
Virginia Polytechnic Institute and State University

The need to solve systems of linear equations arises often within diverse disciplines of science, engineering, and finance. The expression "direct solution of linear systems" refers generally to computational strategies that can produce solutions to linear systems after a predetermined number of arithmetic operations that depends only on the structure and dimension of the coefficient matrix. The evolution of computers has and continues to influence the development of these strategies as well as fostering particular styles of perturbation analysis suited to illuminating their behavior. Some general themes have become dominant, as a result; others have been pushed aside. For example, Cramer's Rule may be properly thought of as a direct solution strategy for solving linear systems; however it requires a much larger number of arithmetic operations than Gauss elimination and is generally much more susceptable to the deleterious effects of rounding. Most current approaches for the direct solution of a linear system, $A\mathbf{x} = \mathbf{b}$, are patterned after Gauss elimination and favor systematically decoupling the system of equations. Zeros are introduced systematically into the coefficient matrix, transforming it into triangular form; the resulting triangular system is easily solved. The entire process can be viewed in this way:

1. Find invertible matrices $\{S_i\}_{i=1}^{\rho}$ such that $S_\rho \dots S_2 S_1 A = U$ is triangular; then
2. Calculate a modified right-hand side $\mathbf{y} = S_\rho \dots S_2 S_1 \mathbf{b}$; and then
3. Determine the solution set to the triangular system $U\mathbf{x} = \mathbf{y}$.

The matrices $S_1, S_2, \dots S_\rho$ are typically either row permutations of lower triangular matrices (Gauss transformations) or unitary matrices and so have readily available inverses. Evidently A can be written as $A = NU$, where $N = (S_\rho \dots S_2 S_1)^{-1}$, A solution framework may built around the availability of decompositions such as this:

1. Find a decompostion $A = NU$ such that U is triangular and $N\mathbf{y} = \mathbf{b}$ is easily solved;
2. Solve $N\mathbf{y} = \mathbf{b}$; then
3. Determine the solution set to triangular system $U\mathbf{x} = \mathbf{y}$.

38.1 Perturbations of Linear Systems

In the computational environment afforded by current computers, the finite representation of real numbers creates a small but persistant source of errors that may on occasion severely degrade the overall accuracy of a calculation. This effect is of fundamental concern in assessing strategies for solving linear systems.

Rounding errors can be introduced into the solution process for linear systems often before any calculations are performed — as soon as data are stored within the computer and represented within the internal floating point number system of the computer. Further errors that may be introduced in the course of computation often may be viewed in aggregate as an effective additional contribution to the initial representation error. Inevitably then the linear system for which a solution is computed deviates slightly from the "true" linear system and it becomes of critical interest to determine whether such deviations will have a significant effect on the accuracy of the final computed result.

Definitions:

Let $A \in \mathbb{C}^{n\times n}$ be a nonsingular matrix, $\mathbf{b} \in \mathbb{C}^n$, and then denote by $\hat{\mathbf{x}} = A^{-1}\mathbf{b}$ the unique solution of the linear system $A\mathbf{x} = \mathbf{b}$.

Given **data perturbations** $\delta A \in \mathbb{C}^{n\times n}$ and $\delta\mathbf{b} \in \mathbb{C}^n$ to A and $\mathbf{b}$, respectively, the **solution perturbation,** $\delta\mathbf{x} \in \mathbb{C}^n$ satisfies the associated **perturbed linear system** $(A + \delta A)(\hat{\mathbf{x}} + \delta\mathbf{x}) = \mathbf{b} + \delta\mathbf{b}$ (presuming then that the perturbed system is consistent).

For any $\tilde{\mathbf{x}} \in \mathbb{C}^n$, the **residual vector** associated with the linear system $A\mathbf{x} = \mathbf{b}$ is defined as $\mathbf{r}(\tilde{\mathbf{x}}) = \mathbf{b} - A\tilde{\mathbf{x}}$.

For any $\tilde{\mathbf{x}} \in \mathbb{C}^n$, the associated (norm-wise) **relative backward error of the linear system** $A\mathbf{x} = \mathbf{b}$ (with respect to the the p-norm) is

$$\eta_p(A, \mathbf{b}; \tilde{\mathbf{x}}) = \min\left\{ \varepsilon \;\middle|\; \begin{array}{l} \text{there exist } \delta A, \delta\mathbf{b} \text{ such that} \\ \quad (A + \delta A)\tilde{\mathbf{x}} = \mathbf{b} + \delta\mathbf{b} \text{ with} \end{array} \begin{array}{l} \|\delta A\|_p \le \varepsilon \|A\|_p \\ \|\delta b\|_p \le \varepsilon \|\mathbf{b}\|_p \end{array} \right\}$$

for $1 \le p \le \infty$.

For any $\tilde{\mathbf{x}} \in \mathbb{C}^n$, the associated **component-wise relative backward error of the linear system** $A\mathbf{x} = \mathbf{b}$ is

$$\omega(A, \mathbf{b}; \tilde{\mathbf{x}}) = \min\left\{ \varepsilon \;\middle|\; \begin{array}{l} \text{there exist } \delta A, \delta\mathbf{b} \text{ such that} \\ \quad (A + \delta A)\tilde{\mathbf{x}} = \mathbf{b} + \delta\mathbf{b} \text{ with} \end{array} \begin{array}{l} |\delta A| \le \varepsilon |A| \\ |\delta\mathbf{b}| \le \varepsilon |\mathbf{b}| \end{array} \right\},$$

where the absolute values and inequalities applied to vectors and matrices are interpretted component-wise: $|B| \le |A|$ means $|b_{ij}| \le |a_{ij}|$ for all index pairs i, j.

The (norm-wise) **condition number of the linear system** $A\mathbf{x} = \mathbf{b}$ (relative to the the p-norm) is

$$\kappa_p(A, \hat{\mathbf{x}}) = \|A^{-1}\|_p \frac{\|\mathbf{b}\|_p}{\|\hat{\mathbf{x}}\|_p}$$

for $1 \le p \le \infty$.

The **matrix condition number** relative to the the p-norm of A is

$$\kappa_p(A) = \|A\|_p \|A^{-1}\|_p$$

for $1 \le p \le \infty$.

The **Skeel condition number of the linear system** $A\mathbf{x} = \mathbf{b}$ is

$$\operatorname{cond}(A, \hat{\mathbf{x}}) = \frac{\|\,|A^{-1}|\,|A|\,|\hat{\mathbf{x}}|\,\|_\infty}{\|\hat{\mathbf{x}}\|_\infty}.$$

The **Skeel matrix condition number** is $\operatorname{cond}(A) = \|\,|A^{-1}|\,|A|\,\|_\infty$.

Facts: [Hig96],[StS90]

1. For any $\tilde{\mathbf{x}} \in \mathbb{C}^n$, $\tilde{\mathbf{x}}$ is the exact solution to any one of the family of perturbed linear systems

$$(A + \delta A_\theta)\tilde{\mathbf{x}} = \mathbf{b} + \delta \mathbf{b}_\theta,$$

where $\theta \in \mathbb{C}$, $\delta \mathbf{b}_\theta = (\theta - 1)\,\mathbf{r}(\tilde{\mathbf{x}})$, $\delta A_\theta = \theta\, \mathbf{r}(\tilde{\mathbf{x}})\tilde{\mathbf{y}}^*$, and $\tilde{\mathbf{y}} \in \mathbb{C}^n$ is any vector such that $\tilde{\mathbf{y}}^*\tilde{\mathbf{x}} = 1$. In particular, for $\theta = 0$, $\delta A = 0$ and $\delta \mathbf{b} = -\mathbf{r}(\tilde{\mathbf{x}})$; for $\theta = 1$, $\delta A = \mathbf{r}(\tilde{\mathbf{x}})\tilde{\mathbf{y}}^*$ and $\delta \mathbf{b} = 0$.

2. (Rigal–Gaches Theorem) For any $\tilde{\mathbf{x}} \in \mathbb{C}^n$,

$$\eta_p(A, \mathbf{b}; \tilde{\mathbf{x}}) = \frac{\|\mathbf{r}(\tilde{\mathbf{x}})\|_p}{\|A\|_p \|\tilde{\mathbf{x}}\|_p + \|\mathbf{b}\|_p}.$$

If $\tilde{\mathbf{y}}$ is the dual vector to $\tilde{\mathbf{x}}$ with respect to the p-norm ($\tilde{\mathbf{y}}^*\tilde{\mathbf{x}} = \|\tilde{\mathbf{y}}\|_q \, \|\tilde{\mathbf{x}}\|_p = 1$ with $\frac{1}{p} + \frac{1}{q} = 1$), then $\tilde{\mathbf{x}}$ is an exact solution to the perturbed linear system $(A + \delta A_{\tilde{\theta}})\tilde{\mathbf{x}} = \mathbf{b} + \delta \mathbf{b}_{\tilde{\theta}}$ with data perturbations as in (1) and $\tilde{\theta} = \frac{\|A\|_p \|\tilde{\mathbf{x}}\|_p}{\|A\|_p \|\tilde{\mathbf{x}}\|_p + \|\mathbf{b}\|_p}$, and as a result

$$\frac{\|\delta A_{\tilde{\theta}}\|_p}{\|A\|_p} = \frac{\|\delta \mathbf{b}_{\tilde{\theta}}\|_p}{\|\mathbf{b}\|_p} = \eta_p(A, \mathbf{b}; \tilde{\mathbf{x}}).$$

3. (Oettli–Prager Theorem) For any $\tilde{\mathbf{x}} \in \mathbb{C}^n$,

$$\omega(A, \mathbf{b}; \tilde{\mathbf{x}}) = \max_i \frac{|r_i|}{(|A|\,|\tilde{\mathbf{x}}| \; + \; |\mathbf{b}|)_i}.$$

If $D_1 = \operatorname{diag}\left(\frac{r_i}{(|A|\,|\tilde{\mathbf{x}}| \; + \; |\mathbf{b}|)_i}\right)$ and $D_2 = \operatorname{diag}(sign(\tilde{\mathbf{x}})_i)$, then $\tilde{\mathbf{x}}$ is an exact solution to the perturbed linear system $(A + \delta A)\tilde{\mathbf{x}} = \mathbf{b} + \delta \mathbf{b}$ with $\delta A = D_1\,|A|\,D_2$ and $\delta \mathbf{b} = -D_1\,|\mathbf{b}|$

$$|\delta A| \le \omega(A, \mathbf{b}; \tilde{\mathbf{x}})\,|A| \quad \text{and} \quad |\delta \mathbf{b}| \le \omega(A, \mathbf{b}; \tilde{\mathbf{x}})\,|A|$$

and no smaller constant can be used in place of $\omega(A, \mathbf{b}; \tilde{\mathbf{x}})$.

4. The reciprocal of $\kappa_p(A)$ is the smallest norm-wise relative distance of A to a singular matrix, i.e.,

$$\frac{1}{\kappa_p(A)} = \min \left\{ \frac{\|\delta A\|_p}{\|A\|_p} \;\middle|\; A + \delta A \text{ is singular} \right\}.$$

In particular, the perturbed coefficient matrix $A + \delta A$ is nonsingular if

$$\frac{\|\delta A\|_p}{\|A\|_p} < \frac{1}{\kappa_p(A)}.$$

5. $1 \le \kappa_p(A, \hat{\mathbf{x}}) \le \kappa_p(A)$ and $1 \le \operatorname{cond}(A, \hat{\mathbf{x}}) \le \operatorname{cond}(A) \le \kappa_\infty(A)$.
6. $\operatorname{cond}(A) = \min \left\{ \kappa_\infty(DA) \,\middle|\, D \text{ diagonal} \right\}$.
7. If $\delta A = 0$, then

$$\frac{\|\delta \mathbf{x}\|_p}{\|\hat{\mathbf{x}}\|_p} \le \kappa_p(A, \hat{\mathbf{x}}) \frac{\|\delta \mathbf{b}\|_p}{\|\mathbf{b}\|_p}.$$

8. If $\delta \mathbf{b} = 0$ and $A + \delta A$ is nonsingular, then

$$\frac{\|\delta \mathbf{x}\|_p}{\|\hat{\mathbf{x}} + \delta \mathbf{x}\|_p} \le \kappa_p(A) \frac{\|\delta A\|_p}{\|A\|_p}.$$

9. If $\|\delta A\|_p \le \epsilon \|A\|_p$, $\|\delta \mathbf{b}\|_p \le \epsilon \|\mathbf{b}\|_p$, and $\epsilon < \dfrac{1}{\kappa_p(A)}$, then

$$\frac{\|\delta \mathbf{x}\|_p}{\|\hat{\mathbf{x}}\|_p} \le \frac{2\epsilon\, \kappa_p(A)}{1 - \epsilon\, \kappa_p(A)}.$$

10. If $|\delta A| \leq \epsilon |A|$, $|\delta \mathbf{b}| \leq \epsilon\, |\mathbf{b}|$, and $\epsilon < \dfrac{1}{\text{cond}(A)}$, then
$$\frac{\|\delta \mathbf{x}\|_\infty}{\|\hat{\mathbf{x}}\|_\infty} \leq \frac{2\,\epsilon\, \text{cond}(A, \hat{\mathbf{x}})}{1 - \epsilon\, \text{cond}(A)}.$$

Examples:

1. Let $A = \begin{bmatrix} 1000 & 999 \\ 999 & 998 \end{bmatrix}$ so $A^{-1} = \begin{bmatrix} -998 & 999 \\ 999 & -1000 \end{bmatrix}$. Then $\|A\|_1 = \|A^{-1}\|_1 = 1999$ so that $\kappa_1(A) \approx 3.996 \times 10^6$. Consider
$$\mathbf{b} = \begin{bmatrix} 1999 \\ 1997 \end{bmatrix} \text{ associated with a solution } \hat{\mathbf{x}} = \begin{bmatrix} 1 \\ 1 \end{bmatrix}.$$
A perturbation of right-hand side $\delta \mathbf{b} = \begin{bmatrix} -0.01 \\ 0.01 \end{bmatrix}$ constitutes a relative change in the right-hand side of $\frac{\|\delta \mathbf{b}\|_1}{\|\mathbf{b}\|_1} \approx 5.005 \times 10^{-6}$ yet it produces a perturbed solution $\hat{\mathbf{x}} + \delta \mathbf{x} = \begin{bmatrix} 20.97 \\ -18.99 \end{bmatrix}$ constituting a relative change $\frac{\|\delta \mathbf{x}\|_1}{\|\hat{\mathbf{x}}\|_1} = 19.98 \leq 20 = \kappa_1(A) \frac{\|\delta \mathbf{b}\|_1}{\|\mathbf{b}\|_1}$. The bound determined by the condition number is very nearly achieved. Note that the same perturbed solution $\hat{\mathbf{x}} + \delta \mathbf{x}$ could be produced by a change in the coefficient matrix
$$\delta A = \tilde{\mathbf{r}} \tilde{\mathbf{y}}^* = -\begin{bmatrix} -0.01 \\ 0.01 \end{bmatrix} \begin{bmatrix} \frac{1}{39.96} & -\frac{1}{39.96} \end{bmatrix} = (1/3996) \begin{bmatrix} 1 & -1 \\ -1 & 1 \end{bmatrix}$$
constituting a relative change $\frac{\|\delta A\|_1}{\|A\|_1} \approx 2.5 \times 10^{-7}$. Then $(A + \delta A)(\hat{\mathbf{x}} + \delta \mathbf{x}) = \mathbf{b}$.
2. Let $n = 100$ and A be tridiagonal with diagonal entries equal to -2 and all superdiagonal and subdiagonal entries equal to 1 (associated with a centered difference approximation to the second derivative). Let $\mathbf{b}$ be a vector with a quadratic variation in entries
$$b_k = (k-1)(100-k)/10{,}000.$$
Then
$$\kappa_2(A, \hat{\mathbf{x}}) \approx 1, \quad \text{but} \quad \kappa_2(A) \approx 4.1336 \times 10^3.$$
Since the elements of $\mathbf{b}$ do not have an exact binary representation, the linear system that is presented to any computational algorithm will be $A\mathbf{x} = \mathbf{b} + \delta \mathbf{b}$ with $\|\delta \mathbf{b}\|_2 \leq \epsilon \|\mathbf{b}\|_2$, where ϵ is the unit roundoff error. For example, if the linear system data is stored in IEEE single precision format, $\epsilon \approx 6 \times 10^{-8}$. The matrix condition number, $\kappa_2(A)$, would yield a bound of $(6 \times 10^{-8})(4.1336 \times 10^3) \approx 2.5 \times 10^{-4}$ anticipating the loss of more than 4 significant digits in solution components even if all computations were done on the stored data with no further error. However, the condition number of the linear system, $\kappa_2(A, \hat{\mathbf{x}})$, is substantially smaller and the predicted error for the system is roughly the same as the initial representation error $\approx 6 \times 10^{-8}$, indicating that the solution will be fairly insensitive to the consequences of rounding of the right-hand side data—assuming no further errors occur. But, in fact, this conclusion remains true even if further errors occur, if whatever computational algorithm that is used produces small backward error, as might be asserted if, say, a final residual satisfies $\|\mathbf{r}\|_2 \leq \mathcal{O}(\epsilon)\, \|\mathbf{b}\|_2$. This situation changes substantially if the right-hand side is changed to
$$b_k = (-1)^k (k-1)(100-k)/10{,}000,$$

which only introduces a sign variation in **b**. In this case, $\kappa_2(A, \hat{\mathbf{x}}) \approx \kappa_2(A)$, and the components of the computed solution can be expected to lose about 4 significant digits purely on the basis of errors that are made in the initial representation. Additional errors made in the course of the computation can hardly be expected to improve this situation.

38.2 Triangular Linear Systems

Systems of linear equations for which the unknowns may be solved for one at a time in sequence may be reordered to produce linear systems with triangular coefficient matrices. Such systems can be solved both with remarkable accuracy and remarkable efficiency. Triangular systems are the archetype for easily solvable systems of linear equations; as such, they often constitute an intermediate goal for strategies of solving linear systems.

Definitions:

A linear system of equations $T\mathbf{x} = \mathbf{b}$ with $T \in \mathbb{C}^{n\times n}$ (representing n equations in n unknowns) is a **triangular system** if $T = [t_{ij}]$ is either an **upper triangular matrix** ($t_{ij} = 0$ for $i > j$) or a **lower triangular matrix** ($t_{ij} = 0$ for $i < j$).

Facts: [Hig96], [GV96]

1. [GV96, pp. 88–90]

> **Algorithm 1**: Row-oriented forward-substitution for solving lower triangular system
> Input: $L \in \mathbb{R}^{n\times n}$ with $\ell_{ij} = 0$ for $i < j$; $\mathbf{b} \in \mathbb{R}^n$
> Output: solution vector $\mathbf{x} \in \mathbb{R}^n$ that satisfies $L\mathbf{x} = \mathbf{b}$
>
> $x_1 \leftarrow b_1/\ell_{1,1}$
> For $k = 2 : n$
>
> $x_k \leftarrow (b_k - L_{k,1:k-1} \cdot x_{1:k-1})/\ell_{k,k}$
>
> end

2. [GV96, pp. 88–90]

> **Algorithm 2**: Column-oriented back-substitution for solving upper triangular system
> Input: $U \in \mathbb{R}^{n\times n}$ with $u_{ij} = 0$ for $i > j$; $\mathbf{b} \in \mathbb{R}^n$
> Output: solution vector $\mathbf{x} \in \mathbb{R}^n$ that satisfies $U\mathbf{x} = \mathbf{b}$
>
> For $k = n$ down to 2 by steps of -1
>
> $x_k \leftarrow b_k/u_{k,k}$
> $b_{1:k-1} \leftarrow b_{1:k-1} - x_k U_{1:k-1,k}$
>
> end
>
> $x_1 \leftarrow b_1/u_{1,1}$

3. Algorithm 1 involves as a core calculation, dot products of portions of coefficient matrix rows with corresponding portions of the emerging solution vector. This can incur a performance penalty for large n from accumulation of dot products using a scalar recurrence. A "column-oriented" reformulation may have better performance for large n. Algorithm 2 is a "column-oriented" formulation for solving upper triangular systems.

4. The solution of triangular systems using either Algorithm 1 or 2 is *componentwise backward stable.* In particular the computed result, $\tilde{\mathbf{x}}$, produced either by Algorithm 1 or 2 in solving a triangular system, $T\mathbf{x} = \mathbf{b}$, will be the exact result of a perturbed system $(T + \delta T)\tilde{\mathbf{x}} = b$, where $|\delta T| \leq \frac{n\epsilon}{1 - n\epsilon}|T|$ and ϵ is the unit roundoff error.
5. The error in the solution of a triangular system, $T\mathbf{x} = \mathbf{b}$, using either Algorithm 1 or 2 satisfies

$$\frac{\|\tilde{\mathbf{x}} - \hat{\mathbf{x}}\|_\infty}{\|\hat{\mathbf{x}}\|_\infty} \leq \frac{n\,\epsilon\,\mathrm{cond}(T, \hat{\mathbf{x}})}{1 - n\,\epsilon\,(\mathrm{cond}(T) + 1)}.$$

6. If $T = [t_{ij}]$ is an lower triangular matrix satisfying $|t_{ii}| \geq |t_{ij}|$ for $j \leq i$, the computed solution to the linear system $T\mathbf{x} = \mathbf{b}$ produced by either Algorithm 1 or the variant of Algorithm 2 for lower triangular systems satisfies

$$|\hat{x}_i - \tilde{x}_i| \leq \frac{2^i\, n\epsilon}{1 - n\epsilon} \max_{j \leq i} |\tilde{x}_j|,$$

where $\tilde{x}_i$ are the components of the computed solution, $\tilde{\mathbf{x}}$, and $\hat{x}_i$ are the components of the exact solution, $\hat{\mathbf{x}}$. Although this bound degrades exponentially with i, it shows that early solution components will be computed to high accuracy relative to those components already computed.

Examples:

1. Use Algorithm 2 to solve the triangular system

$$\begin{bmatrix} 1 & 2 & -3 \\ 0 & 2 & -6 \\ 0 & 0 & 3 \end{bmatrix} \begin{bmatrix} x_1 \\ x_2 \\ x_3 \end{bmatrix} = \begin{bmatrix} 1 \\ 1 \\ 1 \end{bmatrix}.$$

$k = 3$ step: Solve for $x_3 = 1/3$. Update right-hand side:

$$\begin{bmatrix} 1 & 2 \\ 0 & 2 \\ 0 & 0 \end{bmatrix} \begin{bmatrix} x_1 \\ x_2 \end{bmatrix} = \begin{bmatrix} 1 \\ 1 \\ 1 \end{bmatrix} - (1/3) \begin{bmatrix} -3 \\ -6 \\ 3 \end{bmatrix} = \begin{bmatrix} 2 \\ 3 \\ 0 \end{bmatrix}.$$

$k = 2$ step: Solve for $x_2 = 3/2$. Update right-hand side:

$$\begin{bmatrix} 1 \\ 0 \\ 0 \end{bmatrix} \begin{bmatrix} x_1 \end{bmatrix} = \begin{bmatrix} 2 \\ 3 \\ 0 \end{bmatrix} - (3/2) \begin{bmatrix} 2 \\ 2 \\ 0 \end{bmatrix} = \begin{bmatrix} -1 \\ 0 \\ 0 \end{bmatrix}.$$

$k = 1$ step: Solve for $x_1 = -1$.

2. ([Hig96, p. 156]) For $\epsilon > 0$, consider $T = \begin{bmatrix} 1 & 0 & 0 \\ \varepsilon & \varepsilon & 0 \\ 0 & 1 & 1 \end{bmatrix}$. Then $T^{-1} = \begin{bmatrix} 1 & 0 & 0 \\ -1 & \frac{1}{\varepsilon} & 0 \\ 1 & -\frac{1}{\varepsilon} & 1 \end{bmatrix}$, and so $\mathrm{cond}(T) = 5$, even though

$$\kappa_\infty(T) = 2(2 + \frac{1}{\varepsilon}) \approx \frac{2}{\varepsilon} + \mathcal{O}(1).$$

Thus, linear systems having T as a coefficient matrix will be solved to high relative accuracy, independent of both right-hand side and size of ϵ, despite the poor conditioning of T (as measured by κ_∞) as ϵ becomes small. However, note that

$$\mathrm{cond}(T^T) = 1 + \frac{2}{\varepsilon} \quad \text{and} \quad \kappa_\infty(T^T) = (1 + \varepsilon)\frac{2}{\varepsilon} \approx \frac{2}{\varepsilon} + \mathcal{O}(1).$$

So, linear systems having T^T as a coefficient matrix may have solutions that are sensitive to perturbations and indeed, $\text{cond}(T^T, \hat{\mathbf{x}}) \approx \text{cond}(T^T)$ for any right-hand side $\mathbf{b}$ with $b_3 \neq 0$ yielding solutions that are sensitive to perturbations for small ϵ.

38.3 Gauss Elimination and LU Decomposition

Gauss elimination is an elementary approach to solving systems of linear equations, yet it still constitutes the core of the most sophisticated of solution strategies. In the kth step, a transformation matrix, M_k, (a "Gauss transformation") is designed so as to introduce zeros into A — typically into a portion of the kth column — without harming zeros that have been introduced in earlier steps. Typically, successive applications of Gauss transformations are interleaved with row interchanges. Remarkably, this reduction process can be viewed as producing a decomposition of the coefficient matrix $A = NU$, where U is a triangular matrix and N is a row permutation of a lower triangular matrix.

Definitions:

For each index k, a **Gauss vector** is a vector in $\mathbb{C}^n$ with the leading k entries equal to zero: $\boldsymbol{\ell}_k = [\underbrace{0, \ldots, 0}_{k}, \ell_{k+1}, \ldots, \ell_n]^T$. The entries $\ell_{k+1}, \ldots, \ell_n$ are **Gauss multipliers** and the associated matrix

$$M_k = I - \boldsymbol{\ell}_k \mathbf{e}_k^T$$

is called a **Gauss transformation.**

For the pair of indices (i, j), with $i \leq j$ the associated **permutation matrix,** $\Pi_{i,j}$ is an $n \times n$ identity matrix with the ith row and j^{th} row interchanged. Note that $\Pi_{i,i}$ is the identity matrix.

A matrix $U \in \mathbb{C}^{m\times n}$ is in **row-echelon form** if (1) the first nonzero entry of each row has a strictly smaller column index than all nonzero entries having a strictly larger row index and (2) zero rows occur at the bottom. The first nonzero entry in each row of U is called a **pivot.** Thus, the determining feature of row echelon form is that pivots occur to the left of all nonzero entries in lower rows.

A matrix $A \in \mathbb{C}^{m\times n}$ has an ***LU* decomposition** if there exists a unit lower triangular matrix $L \in \mathbb{C}^{m\times m}$ ($L_{i,j} = 0$ for $i < j$ and $L_{i,i} = 1$ for all i) and an upper triangular matrix $U \in \mathbb{C}^{m\times n}$ ($U_{i,j} = 0$ for $i > j$) such that $A = LU$.

Facts: [GV96]

1. Let $\mathbf{a} \in \mathbb{C}^n$ be a vector with a nonzero component in the rth entry, $a_r \neq 0$. Define the Gauss vector, $\boldsymbol{\ell}_r = [\underbrace{0, \ldots, 0}_{r}, \frac{a_{r+1}}{a_r}, \ldots, \frac{a_n}{a_r}]^T$. The associated Gauss transformation $M_r = I - \boldsymbol{\ell}_r \mathbf{e}_r^T$ introduces zeros into the last $n - r$ entries of $\mathbf{a}$:
$$M_r \mathbf{a} = [a_1, \ldots, a_r, 0, \ldots, 0]^T.$$
2. If $A \in \mathbb{C}^{m\times n}$ with $\text{rank}(A) = \rho \geq 1$ has ρ leading principal submatrices nonsingular, $A_{1:r,1:r}$, $r = 1, \ldots, \rho$, then there exist Gauss transformations $M_1, M_2, \ldots, M_\rho$ so that
$$M_\rho M_{\rho-1} \cdots M_1 A = U$$
with U upper triangular. Each Gauss transformation M_r introduces zeros into the rth column.
3. Gauss transformations are unit lower triangular matrices. They are invertible, and for the Gauss transformation, $M_r = I - \boldsymbol{\ell}_r \mathbf{e}_r^T$,
$$M_r^{-1} = I + \boldsymbol{\ell}_r \mathbf{e}_r^T.$$

4. If Gauss vectors $\boldsymbol{\ell}_1, \boldsymbol{\ell}_2, \ldots, \boldsymbol{\ell}_{n-1}$ are given with

$$\boldsymbol{\ell}_1 = \begin{Bmatrix} 0 \\ \ell_{21} \\ \ell_{31} \\ \vdots \\ \ell_{n1} \end{Bmatrix}, \quad \boldsymbol{\ell}_2 = \begin{Bmatrix} 0 \\ 0 \\ \ell_{32} \\ \vdots \\ \ell_{n2} \end{Bmatrix}, \ldots, \quad \boldsymbol{\ell}_{n-1} = \begin{Bmatrix} 0 \\ 0 \\ \vdots \\ 0 \\ \ell_{n,n-1} \end{Bmatrix},$$

then the product of Gauss transformations $M_{n-1}M_{n-2}\cdots M_2M_1$ is invertible and has an explicit inverse

$$(M_{n-1}M_{n-2}\ldots M_2M_1)^{-1} = I + \sum_{k=1}^{n-1} \boldsymbol{\ell}_k \mathbf{e}_k^T = \begin{bmatrix} 1 & 0 & \ldots & 0 & 0 \\ \ell_{21} & 1 & & & 0 \\ \ell_{31} & \ell_{32} & \ddots & & 0 \\ \vdots & & & 1 & 0 \\ \ell_{n1} & \ell_{n2} & \ldots & \ell_{n,n-1} & 1 \end{bmatrix}.$$

5. If $A \in \mathbb{C}^{m\times n}$ with $\text{rank}(A) = \rho$ has ρ leading principal submatrices nonsingular, $A_{1:r,1:r}$, $r = 1, \ldots, \rho$, then A has an LU decomposition: $A = LU$, with L unit lower triangular and U upper triangular. The (i, j) entry of L, $L_{i,j}$ with $i > j$ is the Gauss multiplier used to introduce a zero into the corresponding (i, j) entry of A. If, additionally, $\rho = m$, then the LU decomposition is unique.
6. Let $\mathbf{a}$ be an arbitrary vector in $\mathbb{C}^n$. For any index r, there is an index $\mu \geq r$, a permutation matrix $\Pi_{r,\mu}$, and a Gauss transformation M_r so that

$$M_r\Pi_{r,\mu}\mathbf{a} = [a_1, \ldots, a_{r-1}, a_\mu, \underbrace{0, \ldots, 0}_{n-r}]^T.$$

The index μ is chosen so that $a_\mu \neq 0$ out of the set $\{a_r, a_{r+1}, \ldots, a_n\}$. If $a_r \neq 0$, then $\mu = k$ and $\Pi_{r,\mu} = I$ is a possible choice; if each element is zero, $a_r = a_{r+1} = \cdots = a_n = 0$, then $\mu = k$, $\Pi_{r,\mu} = I$, and $M_r = I$ is a possible choice.
7. For every matrix $A \in \mathbb{C}^{m\times n}$ with $\text{rank}(A) = \rho$, there exists a sequence of ρ indices $\mu_1, \mu_2, \ldots, \mu_\rho$ with $i \leq \mu_i \leq m$ for $i = 1, \ldots, \rho$ and Gauss transformations $M_1, M_2, \ldots, M_\rho$ so that

$$M_\rho\Pi_{\rho,\mu_\rho}M_{\rho-1}\Pi_{\rho-1,\mu_{\rho-1}}\cdots M_1\Pi_{1,\mu_1}A = U$$

with U upper triangular and in row echelon form. Each pair of transformations $M_r\Pi_{r,\mu_r}$ introduces zeros below the rth pivot.
8. For $r < i < j$, $\Pi_{i,j}M_r = \widetilde{M}_r\Pi_{i,j}$, where $\widetilde{M}_r = I - \tilde{\boldsymbol{\ell}}_r\mathbf{e}_r^T$ and $\tilde{\boldsymbol{\ell}}_r = \Pi_{i,j}\boldsymbol{\ell}_r$ (i.e., the i and j entries of $\boldsymbol{\ell}_r$ are interchanged to form $\tilde{\boldsymbol{\ell}}_r$).
9. For every matrix $A \in \mathbb{C}^{m\times n}$ with $\text{rank}(A) = \rho$, there is a row permutation of A that has an LU decomposition: $PA = LU$, with a permutation matrix P, unit lower triangular matrix L, and an upper triangular matrix U that is in row echelon form. P can be chosen as $P = \Pi_{\rho,\mu_\rho}\Pi_{\rho-1,\mu_{\rho-1}}\ldots\Pi_{1,\mu_1}$ from (7), though in general there can be many other possibilities as well.
10. Reduction of A with Gauss transformations (or equivalently, calculation of an LU factorization) must generally incorporate row interchanges. As a practical matter, these row interchanges commonly are chosen so as to bring the largest magnitude entry within the column being reduced up into the pivot location. This strategy is called "partial pivoting." In particular, if zeros are to be introduced into the kth column below the rth row (with $r \leq k$), then one seeks an index μ_r such that $r \leq \mu_r \leq m$ and $|A_{\mu_r,k}| = \max_{r\leq i\leq m}|A_{i,k}|$. When $\mu_1, \mu_2, \ldots, \mu_\rho$ in (7) are chosen in this way, the reduction process is called "Gaussian Elimination with Partial Pivoting" (GEPP) or, within the context of factorization, the permuted LU factorization (PLU).

11. [GV96, p. 115]

Algorithm 1: GEPP/PLU decomposition of a rectangular matrix (outer product)

Input: $A \in \mathbb{R}^{m\times n}$
Output: $L \in \mathbb{R}^{m\times m}$ (unit lower triangular matrix)
$U \in \mathbb{R}^{m\times n}$ (upper triangular matrix - row echelon form)
$P \in \mathbb{R}^{m\times m}$ (permutation matrix) so that $PA = LU$
(P is represented with an index vector $\mathbf{p}$ such that $\mathbf{y} = P\mathbf{z} \Leftrightarrow y_j = z_{p_j}$)
$L \leftarrow I_m$; $U \leftarrow 0 \in \mathbb{R}^{m\times n}$;
$\mathbf{p} = [1, 2, 3, \ldots, m]$
$r \leftarrow 1$;
For $k = 1$ to n
  Find μ such that $r \leq \mu \leq m$ and $|A_{\mu,k}| = \max_{r\leq i\leq m} |A_{i,k}|$
  If $A_{\mu,k} \neq 0$, then
    Exchange $A_{\mu,k:n} \leftrightarrow A_{r,k:n}$, $L_{\mu,1:r-1} \leftrightarrow L_{r,1:r-1}$, and $p_\mu \leftrightarrow p_r$
    $L_{r+1:m,r} \leftarrow A_{r+1:m,k}/A_{r,k}$
    $U_{r,k:n} \leftarrow A_{r,k:n}$
    For $i = r + 1$ to m
      For $j = k + 1$ to n
        $A_{i,j} \leftarrow A_{i,j} - L_{i,r}U_{r,j}$
    $r \leftarrow r + 1$

12. [GV96, p. 115]

Algorithm 2: GEPP/PLU decomposition of a rectangular matrix (gaxpy)

Input: $A \in \mathbb{R}^{m\times n}$
Output: $L \in \mathbb{R}^{m\times m}$ (unit lower triangular matrix),
$U \in \mathbb{R}^{m\times n}$ (upper triangular matrix - row echelon form), and
$P \in \mathbb{R}^{m\times m}$ (permutation matrix) so that $PA = LU$
(P is represented with an index vector $\boldsymbol{\pi}$ that records row interchanges
$\pi_r = \mu$ means row r and row $\mu > r$ were interchanged)
$L \leftarrow I_m \in \mathbb{R}^{m\times m}$; $U \leftarrow 0 \in \mathbb{R}^{m\times n}$; and $r \leftarrow 1$;
For $j = 1$ to n
  $\mathbf{v} \leftarrow A_{1:m,j}$
  If $r > 1$, then
    for $i = 1$ to $r - 1$,
      Exchange $v_i \leftrightarrow v_{\pi_i}$
    Solve the triangular system, $L_{1:r-1,1:r-1} \cdot \mathbf{z} = \mathbf{v}_{1:r-1}$;
    $U_{1:r-1,j} \leftarrow \mathbf{z}$;
    Update $\mathbf{v}_{r:m} \leftarrow \mathbf{v}_{r:m} - L_{r:m,1:r-1} \cdot \mathbf{z}$;
  Find μ such that $|v_\mu| = \max_{r\leq i\leq m} |v_i|$
  If $v_\mu \neq 0$, then
    $\pi_r \leftarrow \mu$
    Exchange $v_\mu \leftrightarrow v_r$
    For $i = 1$ to $r - 1$,
      Exchange $L_{\mu,i} \leftrightarrow L_{r,i}$
    $L_{r+1:m,r} \leftarrow \mathbf{v}_{r+1:m}/v_r$
    $U_{r,j} \leftarrow v_r$
    $r \leftarrow r + 1$

13. The condition for skipping reduction steps ($A_{\mu,k} \neq 0$ in Algorithm 1 and $v_\mu \neq 0$ in Algorithm 2) indicates deficiency of column rank and the potential for an infinite number of solutions. These

conditions are sensitive to rounding errors that may occur in the calculation of those columns and as such, GEPP/PLU is applied for the most part in full column rank settings (rank$(A) = n$), guaranteeing that no zero pivots are encountered and that no reduction steps are skipped.

14. Both Algorithms 1 and 2 require approximately $\frac{2}{3}\rho^3 + \rho m(n-\rho) + \rho n(m-\rho)$ arithmetic operations. Algorithm 1 involves as a core calculation the updating of a submatrix having ever diminishing size. For large matrix dimension, the contents of this submatrix, $A_{r+1:m,k+1:n}$, may be widely scattered through computer memory and a performance penalty can occur in gathering the data for computation (which can be costly relative to the number of arithmetic operations that must be performed). Algorithm 2 is a reorganization that avoids excess data motion by delaying updates to columns until the step within which they have zeros introduced. This forces modifications to the matrix entries to be made just one column at a time and the necessary data motion can be more efficient.
15. Other strategies for avoiding the adverse effects of small pivots exist. Some are more aggressive than partial pivoting in producing the largest possible pivot, others are more restrained.

    "Complete pivoting" uses both row and column permutations to bring in the largest possible pivot: If zeros are to be introduced into the kth column in row entries $r+1$ to m, then one seeks indices μ and ν such that $r \leq \mu \leq m$ and $k < \nu \leq n$ such that $|A_{\mu,\nu}| = \max_{\substack{r \leq i \leq m \\ k < j \leq n}} |A_{i,j}|$. Gauss elimination with complete pivoting produces a unit lower triangular matrix $L \in \mathbb{R}^{m \times m}$, an upper triangular matrix $U \in \mathbb{R}^{m \times n}$, and two permutation matrices, P and Q, so that $PAQ = LU$.

    "Threshold pivoting" identifies pivot candidates in each step that achieve a significant (predetermined) fraction of the magnitude of the pivot that would have been used in that step for partial pivoting: Consider all $\hat{\mu}$ such that $r \leq \hat{\mu} \leq m$ and $|A_{\hat{\mu},k}| \geq \tau \cdot \max_{r \leq i \leq m} |A_{i,k}|$, where $\tau \in (0, 1)$ is a given threshold. This allows pivots to be chosen on the basis of other criteria such as influence on sparsity while still providing some protection from instability. τ can often be chosen quite small ($\tau = 0.1$ or $\tau = 0.025$ are typical values).
16. If $\hat{P} \in \mathbb{R}^{m \times m}$, $\hat{L} \in \mathbb{R}^{m \times m}$, and $\hat{U} \in \mathbb{R}^{m \times n}$ are the computed permutation matrix and LU factors from either Algorithm 1 or 2 on $A \in \mathbb{R}^{m \times n}$, then

$$\hat{L}\hat{U} = \hat{P}(A + \delta A) \quad \text{with} \quad |\delta A| \leq \frac{2n\epsilon}{1 - n\epsilon} |\hat{L}||\hat{U}|$$

    and for the particular case that $m = n$ and A is nonsingular, if an approximate solution, $\hat{\mathbf{x}}$, to $A\mathbf{x} = \mathbf{b}$ is computed by solving the two triangular linear systems, $\hat{L}\mathbf{y} = \hat{P}\mathbf{b}$ and $\hat{U}\hat{\mathbf{x}} = \mathbf{y}$, then $\hat{\mathbf{x}}$ is the exact solution to a perturbed linear system:

$$(A + \delta A)\hat{\mathbf{x}} = \mathbf{b} \quad \text{with} \quad |\delta A| \leq \frac{2n\epsilon}{1 - n\epsilon} \hat{P}^T |\hat{L}||\hat{U}|.$$

    Furthermore, $|L_{i,j}| \leq 1$ and $|U_{i,j}| \leq 2^{i-1} \max_{k \leq i} |A_{k,j}|$, so

$$\|\delta A\|_\infty \leq \frac{2^n n^2 \epsilon}{1 - n\epsilon} \|A\|_\infty.$$

Examples:

1. Using Algorithm 1, find a permuted LU factorization of

$$A = \begin{bmatrix} 1 & 1 & 2 & 2 \\ 2 & 2 & 4 & 6 \\ -1 & -1 & -1 & 1 \\ 1 & 1 & 3 & 1 \end{bmatrix}.$$

Setup:	$\mathbf{p} = [1\ 2\ 3\ 4], r \leftarrow 1$
$k = 1$ *step*:	$\mu \leftarrow 2\ \mathbf{p} = [2\ 1\ 3\ 4]$
	Permuted A: $\begin{bmatrix} 2 & 2 & 4 & 6 \\ 1 & 1 & 2 & 3 \\ -1 & -1 & -2 & 1 \\ 1 & 1 & 3 & 1 \end{bmatrix}$
LU snapshot:	$L = \begin{bmatrix} 1 & 0 & 0 & 0 \\ \frac{1}{2} & 1 & 0 & 0 \\ -\frac{1}{2} & 0 & 1 & 0 \\ \frac{1}{2} & 0 & 0 & 1 \end{bmatrix}$ and $U = \begin{bmatrix} 2 & 2 & 4 & 6 \\ 0 & 0 & 0 & 0 \\ 0 & 0 & 0 & 0 \\ 0 & 0 & 0 & 0 \end{bmatrix}$.
	Updated $A_{2:4,2:4}$: $\begin{bmatrix} 0 & 0 & -1 \\ 0 & 1 & 4 \\ 0 & 1 & -2 \end{bmatrix}$
	$r \leftarrow 2$
$k = 2$ *step*:	$\mu \leftarrow 2, \lvert A_{2,2} \rvert = \max_{2 \le i \le 4} \lvert A_{i,2} \rvert = 0$
$k = 3$ *step*:	$\mu \leftarrow 3, \mathbf{p} = [2\ 3\ 1\ 4], \lvert A_{3,3} \rvert = \max_{2 \le i \le 4} \lvert A_{i,3} \rvert = 1$
	Permuted $A_{2:4,3:4}$: $\begin{bmatrix} 1 & 4 \\ 0 & -1 \\ 1 & -2 \end{bmatrix}$
LU snapshot:	$L = \begin{bmatrix} 1 & 0 & 0 & 0 \\ -\frac{1}{2} & 1 & 0 & 0 \\ \frac{1}{2} & 0 & 1 & 0 \\ \frac{1}{2} & 1 & 0 & 1 \end{bmatrix}$ and $U = \begin{bmatrix} 2 & 2 & 4 & 6 \\ 0 & 0 & 1 & 4 \\ 0 & 0 & 0 & 0 \\ 0 & 0 & 0 & 0 \end{bmatrix}$.
	Updated $A_{3:4,4}$: $\begin{bmatrix} -1 \\ -6 \end{bmatrix}$
	$r \leftarrow 3$
$k = 4$ *step*:	$\mu \leftarrow 4, \mathbf{p} = [2\ 3\ 4\ 1], \lvert A_{4,4} \rvert = \max_{3 \le i \le 4} \lvert A_{i,4} \rvert = 6$
	Permuted $A_{2:4,3:4}$: $\begin{bmatrix} -6 \\ -1 \end{bmatrix}$
LU snapshot:	$L = \begin{bmatrix} 1 & 0 & 0 & 0 \\ -\frac{1}{2} & 1 & 0 & 0 \\ \frac{1}{2} & 1 & 1 & 0 \\ \frac{1}{2} & 0 & \frac{1}{6} & 1 \end{bmatrix}$ and $U = \begin{bmatrix} 2 & 2 & 4 & 6 \\ 0 & 0 & 1 & 4 \\ 0 & 0 & 0 & -6 \\ 0 & 0 & 0 & 0 \end{bmatrix}$.

The permutation matrix associated with $\mathbf{p} = [2\ 3\ 4\ 1]$ is

$$P = \begin{bmatrix} 0 & 1 & 0 & 0 \\ 0 & 0 & 1 & 0 \\ 0 & 0 & 0 & 1 \\ 1 & 0 & 0 & 0 \end{bmatrix}$$

and

$$PA = \begin{bmatrix} 2 & 2 & 4 & 6 \\ -1 & -1 & - & 1 \\ 1 & 1 & 3 & 1 \\ 1 & 1 & 2 & 2 \end{bmatrix} = \begin{bmatrix} 1 & 0 & 0 & 0 \\ -\frac{1}{2} & 1 & 0 & 0 \\ \frac{1}{2} & 1 & 1 & 0 \\ \frac{1}{2} & 0 & \frac{1}{6} & 1 \end{bmatrix} \begin{bmatrix} 2 & 2 & 4 & 6 \\ 0 & 0 & 1 & 4 \\ 0 & 0 & 0 & -6 \\ 0 & 0 & 0 & 0 \end{bmatrix} = L \cdot U.$$

2. Using Algorithm 2, solve the system of linear equations

$$\begin{bmatrix} 1 & 3 & 1 \\ 2 & 2 & -1 \\ 2 & -1 & 0 \end{bmatrix} \begin{bmatrix} x_1 \\ x_2 \\ x_3 \end{bmatrix} = \begin{bmatrix} 1 \\ -3 \\ 3 \end{bmatrix}.$$

Phase 1: Find permuted LU decomposition.

$r \leftarrow 1$

---

j = 1 *step:* $\mathbf{v} \leftarrow \begin{bmatrix} 1 \\ 2 \\ 2 \end{bmatrix}$. $\pi_1 \leftarrow \mu = 2$. Permuted **v**: $\begin{bmatrix} 2 \\ 1 \\ 2 \end{bmatrix}$

LU snapshot: $\boldsymbol{\pi} = [2]$ $L = \begin{bmatrix} 1 & 0 & 0 \\ \frac{1}{2} & 1 & 0 \\ 1 & 0 & 1 \end{bmatrix}$ and $U = \begin{bmatrix} 2 & 0 & 0 \\ 0 & 0 & 0 \\ 0 & 0 & 0 \end{bmatrix}$.

$r \leftarrow 2$

---

j = 2 *step:* $\mathbf{v} \leftarrow \begin{bmatrix} 3 \\ 2 \\ -1 \end{bmatrix}$. Permuted **v**: $\begin{bmatrix} 2 \\ 3 \\ -1 \end{bmatrix}$.

Solve $1 \cdot \mathbf{z} = 2$. $U_{1,2} \leftarrow \mathbf{z} = [2]$.

$$\begin{bmatrix} \mathbf{v}_2 \\ \mathbf{v}_3 \end{bmatrix} \leftarrow \begin{bmatrix} 2 \\ -3 \end{bmatrix} = \begin{bmatrix} 3 \\ -1 \end{bmatrix} - \begin{bmatrix} \frac{1}{2} \\ 1 \end{bmatrix} [2]$$

$\pi_2 \leftarrow \mu = 3$. $L_{3,2} \leftarrow -\frac{2}{3}$ and $U_{2,2} \leftarrow -3$

LU snapshot: $\boldsymbol{\pi} = [2, 3]$ $L = \begin{bmatrix} 1 & 0 & 0 \\ 1 & 1 & 0 \\ \frac{1}{2} & -\frac{2}{3} & 1 \end{bmatrix}$ and $U = \begin{bmatrix} 2 & 2 & 0 \\ 0 & -3 & 0 \\ 0 & 0 & 0 \end{bmatrix}$.

$r \leftarrow 3$

---

j = 3 *step:* $\mathbf{v} \leftarrow \begin{bmatrix} 1 \\ -1 \\ 0 \end{bmatrix}$ Permuted **v**: $\begin{bmatrix} -1 \\ 0 \\ 1 \end{bmatrix}$. Solve $\begin{bmatrix} 1 & 0 \\ 1 & 1 \end{bmatrix} \cdot \mathbf{z} = \begin{bmatrix} -1 \\ 0 \end{bmatrix}$,

$\begin{bmatrix} U_{1,3} \\ U_{2,3} \end{bmatrix} \leftarrow \mathbf{z} = \begin{bmatrix} -1 \\ 1 \end{bmatrix}$. $v_3 \leftarrow 2\frac{1}{6} = 1 - [\frac{1}{2}, -\frac{2}{3}] \cdot \begin{bmatrix} -1 \\ 1 \end{bmatrix}$

$\pi_3 \rightarrow 3$ and $U_{3,3} \leftarrow 2\frac{1}{6}$

LU snapshot: $\boldsymbol{\pi} = [2, 3, 3]$ $L = \begin{bmatrix} 1 & 0 & 0 \\ 1 & 1 & 0 \\ \frac{1}{2} & -\frac{2}{3} & 1 \end{bmatrix}$ and $U = \begin{bmatrix} 2 & 2 & -1 \\ 0 & -3 & 1 \\ 0 & 0 & 2\frac{1}{6} \end{bmatrix}$.

The permutation matrix associated with $\boldsymbol{\pi}$ is $P = \begin{bmatrix} 0 & 1 & 0 \\ 0 & 0 & 1 \\ 1 & 0 & 0 \end{bmatrix}$ and

$$PA = \begin{bmatrix} 2 & 2 & -1 \\ 2 & -1 & 0 \\ 1 & 3 & 1 \end{bmatrix} = \begin{bmatrix} 1 & 0 & 0 \\ 1 & 1 & 0 \\ \frac{1}{2} & -\frac{2}{3} & 1 \end{bmatrix} \begin{bmatrix} 2 & 2 & -1 \\ 0 & -3 & 1 \\ 0 & 0 & 2\frac{1}{6} \end{bmatrix} = L \cdot U.$$

Phase 2: Solve the lower triangular system $L\mathbf{y} = P\mathbf{b}$.

$$\begin{bmatrix} 1 & 0 & 0 \\ 1 & 1 & 0 \\ \frac{1}{2} & -\frac{2}{3} & 1 \end{bmatrix} \begin{bmatrix} y_1 \\ y_2 \\ y_3 \end{bmatrix} = \begin{bmatrix} -3 \\ 3 \\ 1 \end{bmatrix} \quad \Rightarrow \quad y_1 = -3,\ y_2 = 6,\ y_3 = 6\frac{1}{2}.$$

Phase 3: Solve the upper triangular system $U\mathbf{x} = \mathbf{y}$.

$$\begin{bmatrix} 2 & 2 & -1 \\ 0 & -3 & 1 \\ 0 & 0 & 2\frac{1}{6} \end{bmatrix} \begin{bmatrix} x_1 \\ x_2 \\ x_3 \end{bmatrix} = \begin{bmatrix} -3 \\ 6 \\ 6\frac{1}{2} \end{bmatrix} \quad \Rightarrow \quad x_1 = 1,\ x_2 = -1,\ x_3 = 3.$$

38.4 Orthogonalization and QR Decomposition

The process of transforming an arbitrary linear system into a triangular system may also be approached by systematically introducing zeros into the coefficient matrix with unitary transformations: Given a system $A\mathbf{x} = \mathbf{b}$, find unitary matrices $V_1, V_2, \cdots V_\ell$ such that $V_\ell \ldots V_2 V_1 A = T$ is triangular; calculate $\mathbf{y} = V_\ell \cdots V_2 V_1 \mathbf{b}$; solve the triangular system $T\mathbf{x} = \mathbf{y}$.

There are two different types of rudimentary unitary transformations that are described here: Householder transformations and Givens transformations.

Definitions:

Let $\mathbf{v} \in \mathbb{C}^n$ be a nonzero vector. The matrix $H = I - \frac{2}{\|\mathbf{v}\|_2^2}\mathbf{v}\mathbf{v}^*$ is called a **Householder transformation** (or **Householder reflector**). In this context, $\mathbf{v}$, is called a **Householder vector**.

For $\theta, \vartheta \in [0, 2\pi)$, let $G(i, j, \theta, \vartheta)$ be an $n \times n$ identity matrix modified so that the (i,i) and (j,j) entries are replaced by $c = \cos(\theta)$, the (i,j) entry is replaced by $s = e^{\imath\vartheta}\sin(\theta)$, and the (j,i) entry is replaced by $-\bar{s} = -e^{-\imath\vartheta}\sin(\theta)$:

$$G(i,j,\theta,\vartheta) = \begin{bmatrix} 1 & \cdots & 0 & \cdots & 0 & \cdots & 0 \\ \vdots & \ddots & \vdots & & \vdots & & \vdots \\ 0 & \cdots & c & \cdots & s & \cdots & 0 \\ \vdots & & \vdots & \ddots & \vdots & & \vdots \\ 0 & \cdots & -\bar{s} & \cdots & c & \cdots & 0 \\ \vdots & & \vdots & & \vdots & \ddots & \vdots \\ 0 & \cdots & 0 & \cdots & 0 & \cdots & 1 \end{bmatrix}.$$

$G(i, j, \theta, \vartheta)$ is called a **Givens transformation** (or **Givens rotation**).

Facts: [GV96]

1. Householder transformations are unitary matrices.
2. Let $\mathbf{a} \in \mathbb{C}^n$ be a nonzero vector. Define $\mathbf{v} = \text{sign}(a_1)\|\mathbf{a}\|\mathbf{e}_1 + \mathbf{a}$ with $\mathbf{e}_1 = [1, 0, \ldots, 0]^T \in \mathbb{C}^n$. Then the Householder transformation $H = I - \dfrac{2}{\|\mathbf{v}\|_2^2}\mathbf{v}\mathbf{v}^*$ satisfies
$$H\mathbf{a} = \alpha\mathbf{e}_1 \quad \text{with} \quad \alpha = -\text{sign}(a_1)\|\mathbf{a}\|.$$
3. [GV96,pp. 210–213]

Algorithm 3: Householder QR Factorization:
Input: matrix $A \in \mathbb{C}^{m\times n}$ with $m \geq n$
Output: the QR factorization $A = QR$, where the upper triangular part of R is stored in the upper triangular part of A
$Q = I_m$
For $k = 1 : n$
  $\mathbf{x} = A_{k:m,k}$
  $\mathbf{v}_k = \text{sign}(x_1)\|\mathbf{x}\|\mathbf{e}_1 + \mathbf{x}$, where $\mathbf{e}_1 \in \mathbb{C}^{m-k+1}$
  $\mathbf{v}_k = \mathbf{v}_k / \|\mathbf{v}_k\|$
  $A_{k:m,k:n} = (I_{m-k+1} - 2\mathbf{v}_k\mathbf{v}_k^*)A_{k:m,k:n}$
  $Q_{1:k-1,k:m} = Q_{1:k-1,k:m}(I_{m-k+1} - 2\mathbf{v}_k\mathbf{v}_k^*)$
  $Q_{k:m,k:m} = Q_{k:m,k:m}(I_{m-k+1} - 2\mathbf{v}_k\mathbf{v}_k^*)$

4. [GV96, p. 215] A Givens rotation is a unitary matrix.
5. [GV96, pp. 216–221] For any scalars $x, y \in \mathbb{C}$, there exists a Givens rotation $G \in \mathbb{C}^{2\times 2}$ such that

$$G\begin{bmatrix} x \\ y \end{bmatrix} = \begin{bmatrix} c & s \\ -\bar{s} & c \end{bmatrix}\begin{bmatrix} x \\ y \end{bmatrix} = \begin{bmatrix} r \\ 0 \end{bmatrix},$$

where c, s, and r can be computed via

- If $y = 0$ (includes the case $x = y = 0$), then $c = 1,\ s = 0,\ r = x$.
- If $x = 0$ (y must be nonzero), then $c = 0,\ s = \text{sign}(\bar{y}),\ r = |y|$.
- If both x and y are nonzero, then $c = |x|/\sqrt{|x|^2 + |y|^2}$, $s = \text{sign}(x)\bar{y}/\sqrt{|x|^2 + |y|^2}$, $r = \text{sign}(x)\sqrt{|x|^2 + |y|^2}$.

6. [GV96, pp. 226–227]

Algorithm 4: Givens QR Factorization

Input: matrix $A \in \mathbb{C}^{m\times n}$ with $m \geq n$

Output: the QR factorization $A = QR$, where the upper triangular part of R is stored in the upper triangular part of A

$Q = I_m$

For $k = 1 : n$

  For $i = k + 1 : m$

    $[x, y] = [A_{kk}, A_{ik}]$

    Compute $G = \begin{bmatrix} c & s \\ -\bar{s} & c \end{bmatrix}$ via Fact 5.

    $\begin{bmatrix} A_{k,k:n} \\ A_{i,k:n} \end{bmatrix} = G \begin{bmatrix} A_{k,k:n} \\ A_{i,k:n} \end{bmatrix}$

    $[Q_{1:m,k}, Q_{1:m,i}] = [Q_{1:m,k}, Q_{1:m,i}]G^*$

7. [GV96, p. 212] In many applications, it is not necessary to compute Q explicitly in Algorithm 3 and Algorithm 4. See also [TB97, p. 74] for details.
8. [GV96, pp. 225–227] If $A \in \mathbb{R}^{m\times n}$ with $m \geq n$, then the cost of Algorithm 3 without explicitly computing Q is $2n^2(m - n/3)$ flops and the cost of Algorithm 4 without explicitly computing Q is $3n^2(m - n/3)$ flops.
9. [Mey00, p. 349] Algorithm 3 and Algorithm 4 are numerically stable for computing the QR factorization.

Examples:

1. We shall use Givens rotations to transform $A = \begin{bmatrix} 1 & 1 \\ 1 & 2 \\ 1 & 3 \end{bmatrix}$ to upper triangular form, as in Algorithm 4.

First, to annihilate the element in position (2,1), we use Fact 5 with $(x, y) = (1, 1)$ and obtain $c = s = 1/\sqrt{2}$; hence:

$$A^{(1)} = G_1 A = \begin{bmatrix} 0.7071 & 0.7071 & 0 \\ -0.7071 & 0.7071 & 0 \\ 0 & 0 & 1 \end{bmatrix}\begin{bmatrix} 1 & 1 \\ 1 & 2 \\ 1 & 3 \end{bmatrix} = \begin{bmatrix} 1.4142 & 2.1213 \\ 0 & 0.7071 \\ 1 & 3 \end{bmatrix}.$$

Next, to annihilate the element in position (3,1), we use $(x, y) = (1.4142, 1)$ in Fact 5 and get

$$A^{(2)} = G_2 A^{(1)} = \begin{bmatrix} 0.8165 & 0 & 0.5774 \\ 0 & 1 & 0 \\ -0.5774 & 0 & 0.8165 \end{bmatrix} A^{(1)} = \begin{bmatrix} 1.7321 & 3.4641 \\ 0 & 0.7071 \\ 0 & 1.2247 \end{bmatrix}.$$

Finally, we annihilate the element in position (3,2) using $(x, y) = (.7071, 1.2247)$:

$$A^{(3)} = G_3 A^{(2)} = \begin{bmatrix} 1 & 0 & 0 \\ 0 & 0.5000 & 0.8660 \\ 0 & -0.8660 & 0.5000 \end{bmatrix} A^{(2)} = \begin{bmatrix} 1.7321 & 3.4641 \\ 0 & 1.4142 \\ 0 & 0 \end{bmatrix}.$$

As a result, $R = A^{(3)}$ and $\widehat{R}$ consists of the first two rows of $A^{(3)}$. The matrix Q can be computed as the product $G_1^T G_2^T G_3^T$.

2. We shall use Householder reflections to transform A from Example 1 to upper triangular form as in Algorithm 3. First, let $\mathbf{a} = A_{:,1} = \begin{bmatrix} 1 & 1 & 1 \end{bmatrix}^T$, $\gamma_1 = -\sqrt{3}$, $\tilde{\mathbf{a}} = \begin{bmatrix} -\sqrt{3} & 0 & 0 \end{bmatrix}^T$, and $\mathbf{u}_1 = \begin{bmatrix} 0.8881 & 0.3251 & 0.3251 \end{bmatrix}^T$; then

$$A^{(1)} = \left(I - 2\mathbf{u}_1\mathbf{u}_1^T\right) A = A - \mathbf{u}_1 \overbrace{\begin{bmatrix} 3.0764 & 5.0267 \end{bmatrix}}^{2\mathbf{u}_1^T A} = \begin{bmatrix} -1.7321 & -3.4641 \\ 0 & 0.3660 \\ 0 & 1.3660 \end{bmatrix}.$$

Next, $\gamma_2 = -\|A^{(1)}_{2:3,2}\|_2$, $\mathbf{u}_2 = \begin{bmatrix} 0 & 0.7934 & 0.6088 \end{bmatrix}^T$, and

$$A^{(2)} = \left(I - 2\mathbf{u}_2\mathbf{u}_2^T\right) A^{(1)} = A^{(1)} - \mathbf{u}_2 \overbrace{\begin{bmatrix} 0 & 2.2439 \end{bmatrix}}^{2\mathbf{u}_2^T A^{(1)}} = \begin{bmatrix} -1.7321 & -3.4641 \\ 0 & -1.4142 \\ 0 & 0 \end{bmatrix}.$$

Note that $R = A^{(2)}$ has changed sign as compared with Example 1. The matrix Q can be computed as $\left(I - 2\mathbf{u}_1\mathbf{u}_1^T\right)\left(I - 2\mathbf{u}_2\mathbf{u}_2^T\right)$. Therefore, we have full information about the transformation if we store the vectors $\mathbf{u}_1$ and $\mathbf{u}_2$.

38.5 Symmetric Factorizations

Real symmetric matrices ($A = A^T$) and their complex analogs, Hermitian matrices (Chapter 8), are specified by roughly half the number of parameters than general $n \times n$ matrices, so one could anticipate benefits that take advantage of this structure.

Definitions:

An $n \times n$ matrix, A, is **Hermitian** if $A = A^* = \bar{A}^T$.

$A \in \mathbb{C}^{n\times n}$ is **positive-definite** if $\mathbf{x}^* A\mathbf{x} > 0$ for all $\mathbf{x} \in \mathbb{C}^n$ with $\mathbf{x} \neq 0$.

The **Cholesky decomposition** (or **Cholesky factorization**) of a positive-definite matrix A is $A = G\,G^*$ with $G \in \mathbb{C}^{n\times n}$ lower triangular and having positive diagonal entries.

Facts: [Hig96], [GV96]

1. A positive-definite matrix is Hermitian. Note that the similar but weaker assertion for a matrix $A \in \mathbb{R}^{n\times n}$ that "$\mathbf{x}^T A\mathbf{x} > 0$ for all $\mathbf{x} \in \mathbb{R}^n$ with $\mathbf{x} \neq 0$" does not imply that $A = A^T$.

2. If $A \in \mathbb{C}^{n\times n}$ is positive-definite, then A has an LU decomposition, $A = LU$, and the diagonal of U, $\{u_{11}, u_{22}, \ldots, u_{nn}\}$, has strictly positive entries.
3. If $A \in \mathbb{C}^{n\times n}$ is positive-definite, then the LU decomposition of A satisfies $A = LU$ with $U = DL^*$ and $D = \text{diag}(U)$. Thus, A can be written as $A = LDL^*$ with L unit lower triangular and D diagonal with positive diagonal entries. Furthermore, A has a Cholesky decomposition $A = GG^*$ with $G \in \mathbb{C}^{n\times n}$ lower triangular. Indeed, if
$$\widehat{D} = diag(\{\sqrt{u_{11}}, \sqrt{u_{22}}, \ldots, \sqrt{u_{nn}}\})$$
then $\widehat{D}\widehat{D} = D$ and $G = L\,\widehat{D}$.
4. [GV96, p. 144] The Cholesky decomposition of a positive-definite matrix A can be computed directly:

> **Algorithm 1**: Cholesky decomposition of a positive-definite matrix
> Input: $A \in \mathbb{C}^{n\times n}$ positive definite
> Output: $G \in \mathbb{C}^{n\times n}$ (lower triangular matrix so that $A = GG^*$)
> $G \leftarrow 0 \in \mathbb{C}^{n\times n}$;
> For $j = 1$ to n
> $\quad \mathbf{v} \leftarrow A_{j:n,j}$
> $\quad$ for $k = 1$ to $j-1$,
> $\quad\quad \mathbf{v} \leftarrow \mathbf{v} - \overline{G_{j,k}}G_{j:n,k}$
> $\quad G_{j:n,j} \leftarrow \frac{1}{\sqrt{v_1}}\mathbf{v}$

5. Algorithm 1 requires approximately $n^3/3$ floating point arithmetic operations and n floating point square roots to complete (roughly half of what is required for an LU decomposition).
6. If $A \in \mathbb{R}^{n\times n}$ is symmetric and positive-definite and Algorithm 1 runs to completion producing a computed Cholesky factor $\hat{G} \in \mathbb{R}^{n\times n}$, then
$$\hat{G}\hat{G}^T = A + \delta A \quad \text{with} \quad |\delta A| \leq \frac{(n+1)\epsilon}{1-(n+1)\epsilon}\,|\hat{G}|\,|\hat{G}^T|.$$
Furthermore, if an approximate solution, $\hat{\mathbf{x}}$, to $A\mathbf{x} = \mathbf{b}$ is computed by solving the two triangular linear systems $\hat{G}\mathbf{y} = \mathbf{b}$ and $\hat{G}^T\hat{\mathbf{x}} = \mathbf{y}$, and a scaling matrix is defined as $\Delta = \text{diag}(\sqrt{a_{ii}})$, then the scaled error $\Delta(\mathbf{x} - \hat{\mathbf{x}})$ satisfies
$$\frac{\|\Delta(\mathbf{x}-\hat{\mathbf{x}})\|_2}{\|\Delta\mathbf{x}\|_2} \leq \frac{\kappa_2(H)\epsilon}{1-\kappa_2(H)\epsilon},$$
where $A = \Delta H \Delta$. If $\kappa_2(H) \ll \kappa_2(A)$, then it is quite likely that the entries of $\Delta\hat{\mathbf{x}}$ will have mostly the same magnitude and so the error bound suggests that all entries of the solution will be computed to high relative accuracy.
7. If $A \in \mathbb{C}^{n\times n}$ is Hermitian and has all leading principal submatrices nonsingular, then A has an LU decomposition that can be written as $A = LU = LDL^*$ with L unit lower triangular and D diagonal with real diagonal entries. Furthermore, the number of positive and negative entries of D is equal to the number of positive and negative eigenvalues of A, respectively (the *Sylvester law of inertia*).
8. Note that it may not be prudent to compute the LU (or LDL^T) decomposition of a Hermitian indefinite matrix A without pivoting, yet the use of pivoting will likely eliminate the advantages symmetry might offer. An alternative is a block LDL^T decomposition that incorporates a diagonal pivoting strategy (see [GV96] for details).

Examples:

1. Calculate the Cholesky decomposition of the 3×3 Hilbert matrix,

$$A = \begin{bmatrix} 1 & \frac{1}{2} & \frac{1}{3} \\ \frac{1}{2} & \frac{1}{3} & \frac{1}{4} \\ \frac{1}{3} & \frac{1}{4} & \frac{1}{5} \end{bmatrix}.$$

Setup:	$G \leftarrow \begin{bmatrix} 0 & 0 & 0 \\ 0 & 0 & 0 \\ 0 & 0 & 0 \end{bmatrix}.$
$j = 1$ *step*:	$\mathbf{v} \leftarrow [1, \frac{1}{2}, \frac{1}{3}]^T$
G snapshot:	$G = \begin{bmatrix} 1 & 0 & 0 \\ \frac{1}{2} & 0 & 0 \\ \frac{1}{3} & 0 & 0 \end{bmatrix}$
$j = 2$ *step*:	$\mathbf{v} \leftarrow \begin{bmatrix} \frac{1}{3} \\ \frac{1}{4} \end{bmatrix} - \frac{1}{2}\begin{bmatrix} \frac{1}{2} \\ \frac{1}{3} \end{bmatrix} = \begin{bmatrix} \frac{1}{12} \\ \frac{1}{12} \end{bmatrix}$
G snapshot:	$G = \begin{bmatrix} 1 & 0 & 0 \\ \frac{1}{2} & \frac{1}{2\sqrt{3}} & 0 \\ \frac{1}{3} & \frac{1}{2\sqrt{3}} & 0 \end{bmatrix}$
$j = 3$ *step*:	$\mathbf{v} \leftarrow \frac{1}{5} - \left(\frac{1}{3}\right)^2 - \left(\frac{1}{2\sqrt{3}}\right)^2 = \frac{1}{180} = \left(\frac{1}{6\sqrt{5}}\right)^2$
G snapshot:	$G = \begin{bmatrix} 1 & 0 & 0 \\ \frac{1}{2} & \frac{1}{2\sqrt{3}} & 0 \\ \frac{1}{3} & \frac{1}{2\sqrt{3}} & \frac{1}{6\sqrt{5}} \end{bmatrix}$

References

[Dem97] J. Demmel. *Applied Numerical Linear Algebra*. SIAM, Philadelphia, 1997.

[GV96] G.H. Golub and C.F. Van Loan. *Matrix Computations*. 3rd ed., Johns Hopkins University Press, Baltimore, MD, 1996.

[Hig96] N. J. Higham. *Accuracy and Stability of Numerical Algorithms*. SIAM, Philadelphia, 1996.

[Mey00] C. Meyer. *Matrix Analysis and Applied Linear Algebra*. SIAM, Philadelphia, 2000.

[StS90] G.W. Stewart and J.-G. Sun. *Matrix Perturbation Theory*. Academic Press, San Diego, CA, 1990.

[TB97] L.N. Trefethen and D. Bau. *Numerical Linear Algebra*. SIAM, Philadelphia, 1997.

39

Least Squares Solution of Linear Systems

Per Christian Hansen
Technical University of Denmark

Hans Bruun Nielsen
Technical University of Denmark

39.1 Basic Concepts

(See Chapter 5 for additional information.)

Definitions:

Given a vector $\mathbf{b} \in \mathbb{R}^m$ and a matrix $A \in \mathbb{R}^{m\times n}$ with $m > n$, the **least squares problem** is to find a vector $\mathbf{x}_0 \in \mathbb{R}^n$ that minimizes the Euclidean length of the difference between $A\mathbf{x}$ and $\mathbf{b}$:

$$\textbf{Problem LS: } \text{Find } \mathbf{x}_0 \text{ satisfying } \|\mathbf{b} - A\mathbf{x}_0\|_2 = \min \|\mathbf{b} - A\mathbf{x}\|_2.$$

Such an $\mathbf{x}_0$ is called a **least squares solution.**

For any vector $\mathbf{x}$ the vector $\mathbf{r} = \mathbf{r}(\mathbf{x}) = \mathbf{b} - A\mathbf{x}$ is the **residual vector**. The residual of a least squares solution is denoted by $\mathbf{r}_0$. The least squares problem is **consistent** if $\mathbf{b} \in \text{range}(A)$.

A **basic solution,** $\mathbf{x}_{0B}$, is a least squares solution with at least $n - \text{rank}(A)$ zero components. The **minimum-norm least squares solution,** $\mathbf{x}_{0M}$, is the least squares solution of minimum Euclidean norm.

In a **weighted least squares problem**, we are also given weights $w_i \geq 0$ for $i = 1, \ldots, m$, and the objective is to minimize $\|W(\mathbf{b} - A\mathbf{x})\|_2$, where $W = \text{diag}(w_1, \ldots, w_m)$. This is an important special case of the **generalized least squares problem**:

$$\begin{aligned}\textbf{Problem GLS: } & \text{Let } A\mathbf{x} + B\mathbf{v} = \mathbf{b}, \text{ where } B \in \mathbb{R}^{m\times p} \text{ with } p \leq m.\\ & \text{Find } \mathbf{x}_G, \mathbf{v}_G \text{ such that } \|\mathbf{v}\|_2 \text{ is minimized.}\end{aligned}$$

Note that $B\mathbf{v}$ plays the role of the residual vector.

In the **total least squares problem**, we allow for errors both in the vector $\mathbf{b}$ and in the matrix A:

Problem TLS: Let $(A + E)\mathbf{x} + \mathbf{r} = \mathbf{b}$, where $E \in \mathbb{R}^{m\times n}$ with $n \leq m$.
Find $\mathbf{x}_T$ such that $\|(E, \mathbf{r})\|_F$ is minimized.

$\|\cdot\|_F$ denotes the Frobenius norm.

In this chapter, $A_{i,:}$ and $A_{:,j}$ denote the vectors given by the elements in the ith row and jth column of matrix A, respectively. Similarly, $A_{k:l,:}$ (or $A_{:,k:l}$) is the submatrix consisting of rows (or columns) k through l of A.

In the examples in this chapter, the computation is done with about 16 digits accuracy, but the displayed results are rounded to fewer digits.

Facts:

(See, e.g., Chapters 1 and 2 in [Bjo96].)

1. If $m > n = \text{rank}(A)$, then the least squares solution $\mathbf{x}_0$ is analytically equivalent to the solution to the *normal equations* $A^T A\,\mathbf{x} = A^T\mathbf{b}$.
2. If the least squares problem is consistent, then $\mathbf{r}_0 = \mathbf{0}$.
3. The least squares solution is unique if $m \geq n$ and A has full rank.
4. If the system is underdetermined $(m < n)$ or if A is rank deficient, then the solution to Problem LS is not unique. Also, a basic solution is not unique.
5. The minimum-norm least squares solution $\mathbf{x}_{0M}$ is always unique.
6. If $m > n = \text{rank}(A)$, then the least squares solution can be written as $\mathbf{x}_0 = A^\dagger b$, where the matrix $A^\dagger$ is the Moore–Penrose generalized inverse or pseudoinverse of A. (See Section 5.7.) In general, $A^\dagger$ produces the minimum-norm solution: $\mathbf{x}_{0M} = A^\dagger\mathbf{b}$.
7. If B is nonsingular, then $\mathbf{x}_G$ minimizes $\|B^{-1}(\mathbf{b} - A\mathbf{x})\|_2$.
8. If the covariance matrix for $\mathbf{b}$ has the Cholesky factorization $\text{Cov}(\mathbf{b}) = C^T C$, then $\mathbf{x}_G$ is the *best linear unbiased estimate* (BLUE) in the general linear model with $B = C^T$.
9. If $\text{Cov}(\mathbf{b})$ has full rank, then $\mathbf{x}_G$ is the solution to the least squares problem $\min \|C^{-T}(\mathbf{b} - A\mathbf{x})\|_2$.
10. In particular, if $\text{Cov}(\mathbf{b}) = \sigma^2 I$, then $\mathbf{x}_G = \mathbf{x}_0$ and $\text{Cov}(\mathbf{x}_0) = \sigma^2(A^T A)^{-1}$.
11. An English translation of the original work on least squares problems by C. F. Gauss is available in [Gau95].

Examples:

1. Consider problem LS with $m = 3$ and $n = 2$:

$$\min \left\| \begin{bmatrix} 1 & 1 \\ 1 & 2 \\ 1 & 3 \end{bmatrix} \begin{bmatrix} x_1 \\ x_2 \end{bmatrix} - \begin{bmatrix} 0.75 \\ 1.13 \\ 1.39 \end{bmatrix} \right\|_2 .$$

The associated normal equations and the least squares solution are

$$\begin{bmatrix} 3 & 6 \\ 6 & 14 \end{bmatrix} \begin{bmatrix} x_1 \\ x_2 \end{bmatrix} = \begin{bmatrix} 3.27 \\ 7.18 \end{bmatrix}, \qquad \mathbf{x}_0 = \begin{bmatrix} 0.45 \\ 0.32 \end{bmatrix}.$$

The residual vector corresponding to $\mathbf{x}_0$ is

$$\mathbf{r}_0 = \mathbf{r}(\mathbf{x}_0) = \mathbf{b} - A\mathbf{x}_0 = \begin{bmatrix} -0.02 \\ 0.04 \\ -0.02 \end{bmatrix},$$

and $A^T\mathbf{r}_0 = \mathbf{0}$.

2. If we use the weights $w_1 = 10$ and $w_2 = w_3 = 1$, the problem is changed to

$$\min \left\| \begin{bmatrix} 10 & 10 \\ 1 & 2 \\ 1 & 3 \end{bmatrix} \begin{bmatrix} x_1 \\ x_2 \end{bmatrix} - \begin{bmatrix} 7.5 \\ 1.13 \\ 1.39 \end{bmatrix} \right\|_2$$

whose least squares solution and corresponding residual are

$$\mathbf{x}_0 = \begin{bmatrix} 0.41838 \\ 0.33186 \end{bmatrix}, \qquad \mathbf{r}_0 = \begin{bmatrix} -0.00024 \\ 0.04790 \\ -0.02395 \end{bmatrix}.$$

Note that the first component of $\mathbf{r}_0$ is reduced when w_1 is increased from 1 to 10.

39.2 Least Squares Data Fitting

Definitions:

Given m data points (t_i, y_i), $i = 1, \ldots, m$, and n linearly independent functions f_j, $j = 1, \ldots, n$ (with $m > n$), find the linear combination

$$F(\mathbf{x}, t) = \sum_{j=1}^{n} x_j f_j(t)$$

that minimizes the sum of squared residuals $y_i - F(\mathbf{x}, t_i)$ at the data points:

$$\min_{\mathbf{x}} \sum_{i=1}^{m} (y_i - F(\mathbf{x}, t_i))^2.$$

The coefficients x_i are the components of the least squares solution to $\min \|\mathbf{b} - A\mathbf{x}\|_2$, where the columns $A_{:,j}$ of A are samples of f_j at t_i and the elements of $\mathbf{b}$ are the values y_i:

$$A_{ij} = f_j(t_i), \qquad b_i = y_i, \qquad i = 1, \ldots, m \qquad j = 1, \ldots, n.$$

The solution $F(\mathbf{x}_0, t)$ is said to fit the data in the least squares sense.

Facts:

(See, e.g., Chapter 4 in [Bjo96].)

1. The fit can be made more robust to outliers by solving a weighted least squares problem $\min \|W(\mathbf{b} - A\mathbf{x})\|_2$ with $W = \text{diag}(w_1, \ldots, w_m)$, $w_i = \psi(r_i) = \psi(b_i - A_{i,:}\mathbf{x})$; ψ being a convex function. This problem is usually solved by an iteratively reweighted least squares algorithm.
2. In orthogonal distance fitting, instead of minimizing the residuals one minimizes the orthogonal distances between the fitting function F and the data points. Important examples are fitting of a circle, an arc, or an ellipse to data points.

Examples:

1. Given $f_1(t) = 1$ and $f_2(t) = t$, find the least squares fit to the data points (1, 0.75), (2, 1.13), and (3, 1.39). We get the A, $\mathbf{b}$, and $\mathbf{x}_0$ from Example 1 in section 39.1, and

$$F(\mathbf{x}_0, t) = 0.45 + 0.32t.$$

If the third data point is changed to (3, 13.9), then the least squares solution changes to $\mathbf{x}_0 = (-7.890,\ 6.575)^T$, and the least squares fit becomes $F(\mathbf{x}_0, t) = -7.890 + 6.575t$; this illustrates the sensitivity to outliers.

39.3 Geometric and Algebraic Aspects

Definitions:

The columns of $A \in \mathbb{R}^{m\times n}$ span the **range of** A, while the **nullspace** or **kernel** of A is the set of solutions to the homogeneous system $A\mathbf{x} = \mathbf{0}$:

$$\text{range}(A) = \{\mathbf{z} = A\mathbf{x} \mid \mathbf{x} \in \mathbb{R}^n\}, \qquad \ker(A) = \{\mathbf{x} \in \mathbb{R}^n \mid A\mathbf{x} = \mathbf{0}\}.$$

The four **fundamental subspaces** associated with Problem LS are range(A), ker(A^T), ker(A), and range(A^T). (See Section 2.4 for more information.)

Facts:

The first three facts can be found in [Str88, Sec. 2.4]; the remaining facts are discussed in [Bjo96, Chap. 1]. p denotes the rank of A: $p = \text{rank}(A)$.

1. If $p = n < m$, then the vector $\mathbf{0}$ of all zeros is the only element in ker(A).
2. The spaces range(A) and ker(A^T) are subspaces of $\mathbb{R}^m$ with dimensions p and $m-p$, respectively. The two spaces are orthogonal complements, i.e., $\mathbf{y}^T\mathbf{z} = 0$ for any pair ($\mathbf{y} \in$ range(A), $\mathbf{z} \in$ ker(A^T)), and range$(A) \oplus$ ker$(A^T) = \mathbb{R}^m$.
3. The spaces ker(A) and range(A^T) are subspaces of $\mathbb{R}^n$ with dimensions $n-p$ and p, respectively. The two spaces are orthogonal complements.
4. The least squares residual vector $\mathbf{r}_0 = \mathbf{b} - A\mathbf{x}_0$ is an element in ker(A^T). Combining this with the definition of $\mathbf{r}$, we get the so-called **augmented system** associated with Problem LS:

$$\begin{bmatrix} I & A \\ A^T & 0 \end{bmatrix} \begin{bmatrix} \mathbf{r} \\ \mathbf{x} \end{bmatrix} = \begin{bmatrix} \mathbf{b} \\ \mathbf{0} \end{bmatrix}.$$

   If $p = n$, then the augmented system is nonsingular and the solution components are $\mathbf{r}_0$ and $\mathbf{x}_0$.
5. The vector $A\mathbf{x}_0$ is the orthogonal projection of $\mathbf{b}$ onto range(A).
6. The vector $\mathbf{r}_0$ is the orthogonal projection of $\mathbf{b}$ onto ker(A^T).
7. If $p < n$, then the columns in A can be reordered such that $A\,\Pi = \begin{bmatrix} \widetilde{A} & \widehat{A} \end{bmatrix}$, where Π is a permutation matrix and the submatrix $\widetilde{A}$ has p columns, range(A) = range$(\widetilde{A})$. The permutation is not unique.
8. The orthogonal projectors onto range(A) and ker(A) are given by $AA^\dagger$ and $I - A^\dagger A$, respectively.

Examples:

1. The Figure 39.1 illustrates Facts 5 and 6 in the case $m = 3$, $n = p = 2$.

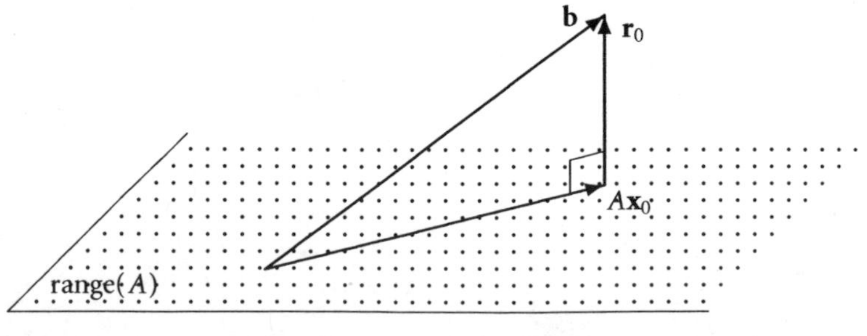

FIGURE 39.1

2. For the problem in Example 1 in Section 39.1, both range(A) and ker(A^T) are subspaces of $\mathbb{R}^3$ given by, respectively,

$$\text{range}(A) = \alpha \begin{bmatrix} 1 \\ 1 \\ 1 \end{bmatrix} + \beta \begin{bmatrix} 1 \\ 2 \\ 3 \end{bmatrix}, \qquad \ker(A^T) = \gamma \begin{bmatrix} 1 \\ -2 \\ 1 \end{bmatrix},$$

with $\alpha, \beta, \gamma \in \mathbb{R}$.

3. Fact 4 can be used to derive the normal equations:

$$\mathbf{r} = \mathbf{b} - A\mathbf{x} \in \ker(A^T) \quad \Rightarrow \quad A^T(\mathbf{b} - A\mathbf{x}) = \mathbf{0}\,.$$

39.4 Orthogonal Factorizations

(See Section 5.5 and Section 38.4 for additional information on orthogonal factorizations.)

Definitions:

The real matrix Q is **orthogonal** if it is square and satisfies $Q^T Q = I$.

A **QR factorization** of a matrix $A \in \mathbb{R}^{m\times n}$ with $m \geq n$ has the form

$$A = QR = Q \begin{bmatrix} \widehat{R} \\ 0 \end{bmatrix} = \widehat{Q}\,\widehat{R},$$

where $Q \in \mathbb{R}^{m\times m}$ is orthogonal, $R \in \mathbb{R}^{m\times n}$, $\widehat{R} \in \mathbb{R}^{n\times n}$ is upper triangular, and $\widehat{Q} = Q_{:,1:n}$. The form $A = \widehat{Q}\widehat{R}$ is the so-called "reduced" (or "skinny") QR factorization.

The **singular value decomposition** (SVD) of $A \in \mathbb{R}^{m\times n}$ has the form

$$A = U\,\Sigma\,V^T,$$

where $U \in \mathbb{R}^{m\times m}$ and $V \in \mathbb{R}^{n\times n}$ are orthogonal matrices and $\Sigma = \text{diag}(\sigma_1, \ldots, \sigma_r) \in \mathbb{R}^{m\times n}$ has diagonal elements

$$\sigma_1 \geq \sigma_2 \geq \cdots \geq \sigma_r \geq 0, \quad r = \min\{m, n\}\,.$$

Letting $\mathbf{u}_j$ and $\mathbf{v}_j$ denote the jth column in U and V, respectively, we can write

$$A = \sum_{j=1}^{r} \sigma_j \mathbf{u}_j \mathbf{v}_j^T\,.$$

For $k < r$ the matrix $\sum_{j=1}^{k} \sigma_j \mathbf{u}_j \mathbf{v}_j^T$ is called the **truncated SVD** approximation to A. (See Sections 5.6, 17, and 45 for more information about the singular value decomposition.)

Facts:

Except for Facts 1 and 8 see [Bjo96, Chap. 1]. Also see Section 39.5.

1. [Str88, Chap. 3]. A QR factorization preserves rank: rank(R) = rank($\widehat{R}$) = rank(A).
2. A QR factorization is not unique, but two factorizations Q_1R_1 and Q_2R_2 always satisfy $\widehat{R}_1 = D\,\widehat{R}_2$, where D is diagonal with $D_{ii} = \pm 1$.
3. The triangular factor $\widehat{R}$ and the upper triangular Cholesky factor C for the normal equations matrix A^TA always satisfy $\widehat{R} = \widetilde{D}C$, where $\widetilde{D}$ is diagonal with $\widetilde{D}_{ii} = \pm 1$.
4. If A has full rank and has the QR factorization $A = QR$, then $\mathbf{x}_0$ can be found by back substitution in the upper triangular system $\widehat{R}\,\mathbf{x} = Q_{:,1:n}^T\mathbf{b}$.

5. If Q has not been saved, then we can use forward and back substitution to solve the seminormal equations: $\widehat{R}^T \widehat{R}\,\mathbf{x} = A^T\mathbf{b}$. For reasons of numerical stability, this must be followed by one step of iterative refinement; the complete process is called the *corrected seminormal equations* method.
6. Let $\text{rank}(A) = p \leq n \leq m$ and $A = QR$. The columns in $Q_{:,1:p}$ and $Q_{:,p+1:m}$ are orthonormal bases of range(A) and $\ker(A^T)$, respectively.
7. Let $A = U\,\Sigma\,V^T$. Then $p = \text{rank}(A)$ is equal to the number of strictly positive singular values: $\sigma_1 \geq \cdots \geq \sigma_p > 0$, $\sigma_{p+1} = \cdots = \sigma_{\min\{m,n\}} = 0$. The columns in $U_{:,1:p}$ and $U_{:,p+1:m}$ are orthonormal bases of range(A) and $\ker(A^T)$, respectively, and the columns in $V_{:,1:p}$ and $V_{:,p+1:n}$ are orthonormal bases of range(A^T) and $\ker(A)$, respectively.
8. [GV96, Chap. 12]. The TLS solution can be computed as follows: compute the SVD of the coefficient matrix A augmented with the right-hand side $\mathbf{b}$, i.e., $[\,A\,,\,\mathbf{b}\,] = \widetilde{U}\,\widetilde{\Sigma}\,\widetilde{V}^T$. If the smallest singular value $\widetilde{\sigma}_{n+1}$ is simple, and if $\beta = \widetilde{V}_{n+1,n+1} \neq 0$, then

$$\mathbf{x}_{\mathrm{T}} = -\beta^{-1}\widetilde{V}_{1:n,n+1}, \quad E_{\mathrm{T}} = -\widetilde{\sigma}_{n+1}\,\widetilde{U}_{:,n+1}\,\widetilde{V}^T_{1:n,n+1}, \quad \mathbf{r}_{\mathrm{T}} = -\widetilde{\sigma}_{n+1}\,\beta\,\widetilde{U}_{:,n+1}\,.$$

Examples:

1. For the problem from Example 1 in Section 39.1, we find $[\,A\,,\,\mathbf{b}\,] = \widetilde{U}\,\widetilde{\Sigma}\,\widetilde{V}^T$ with

$$\widetilde{\Sigma} = \text{diag}\begin{bmatrix} 4.515 \\ 0.6198 \\ 0.0429 \end{bmatrix}, \quad \widetilde{U}_{:,3} = \begin{bmatrix} 0.4248 \\ -0.8107 \\ 0.4029 \end{bmatrix}, \quad \widetilde{V}_{:,3} = \begin{bmatrix} 0.3950 \\ 0.2796 \\ -0.8751 \end{bmatrix}.$$

Thus,

$$\mathbf{x}_{\mathrm{T}} = \begin{bmatrix} 0.4513 \\ 0.3195 \end{bmatrix}, \quad E_{\mathrm{T}} = \begin{bmatrix} -0.0072 & -0.0051 \\ 0.0137 & 0.0097 \\ -0.0068 & -0.0048 \end{bmatrix}, \quad \mathbf{r}_{\mathrm{T}} = \begin{bmatrix} -0.015939 \\ 0.030421 \\ -0.015118 \end{bmatrix}.$$

In Example 1 in section 39.1 we found $\mathbf{x}_0 = \begin{bmatrix} 0.4500 & 0.3200 \end{bmatrix}^T$. The difference between $\mathbf{x}_0$ and $\mathbf{x}_{\mathrm{T}}$ is small because the problem is almost consistent and A is well conditioned; see Section 39.6 The elements in $\mathbf{r}_{\mathrm{T}}$ are about 80% of the elements in $\mathbf{r}_0$ given in Example 1 in Section 39.1.

39.5 Least Squares Algorithms

Definitions:

By least squares algorithms we mean algorithms for computing the least squares solution efficiently and stably on a computer. The algorithms should take into account the size of the matrix and, if applicable, also its structure.

Facts:

For real systems the following facts can be found in, e.g., [Bjo96], [Bjo04], and [LH95].

1. The algorithm which is least sensitive to the influence of rounding errors is based on the QR factorization of A:

   (a) Compute the reduced QR factorization $A = \widehat{Q}\widehat{R}$.

   (b) Compute the vector $\boldsymbol{\beta} = \widehat{Q}^T\mathbf{b}$ (can be computed during the QR factorization algorithm without forming $\widehat{Q}$ explicitly).

   (c) Compute $\mathbf{x} = \widehat{R}^{-1}\boldsymbol{\beta}$ via back substitution.

   The use of this algorithm was first suggested in [Gol65].

2. If A is well conditioned, the normal equations can be used instead:
   (a) Compute the normal equation system $M = A^T A$ and $\mathbf{d} = A^T \mathbf{b}$.
   (b) Compute the Cholesky factorization $M = C^T C$ and $\mathbf{y} = C^{-T}\mathbf{d}$ (the vector $\mathbf{y}$ can be computed during the Cholesky algorithm).
   (c) Compute $\mathbf{x} = C^{-1}\mathbf{y}$ via back substitution.
3. If A or $A^T A$ is a Toeplitz matrix, use an algorithm that utilizes this structure to obtain the computational complexity $O(mn)$.
4. If A is large and sparse, use a sparse QR factorization algorithm that avoids storing the matrix Q. If solving a system with the same A but a different right-hand side, use the corrected seminormal equations.
5. Alternatively, if A is large and sparse, it may be preferable to use the augmented system approach because it may lead to less fill-in. Then a symmetric indefinite solver, such as the LDL^T factorization, must be used, cf. [GV96, Sec. 4.4].
6. If A is large and the matrix-vector multiplications with A and A^T can be computed easily, then use the conjugate gradient algorithm on the normal equations. Several implementations are available; CGLS is the classical formulation; LSQR is more accurate for ill-conditioned matrices.

39.6 Sensitivity

Definitions:

For $A \in \mathbb{R}^{m\times n}$ with $p = \mathrm{rank}(A) \le \min(m, n)$ the **condition number** for the least squares problem is given by $\kappa(A) = \sigma_1/\sigma_p$, where $\sigma_1 \ge \sigma_2 \ge \cdots \ge \sigma_p > 0$ are the nonzero singular values of A.

Let $\mathbf{x}_{0M}$ and $\widetilde{\mathbf{x}}_{0M}$ denote the minimum-norm least squares solutions to the problems $\min \|\mathbf{b} - A\mathbf{x}\|_2$ and $\min \|\widetilde{\mathbf{b}} - \widetilde{A}\mathbf{x}\|_2$, the latter being a perturbation of the former. Define the quantities

$$\delta_A = \|\widetilde{A} - A\|_2, \qquad \delta_b = \|\widetilde{\mathbf{b}} - \mathbf{b}\|_2, \qquad \eta = \frac{\delta_A}{\sigma_p} = \kappa(A)\,\frac{\delta_A}{\|A\|_2}.$$

Facts:

(See, e.g., [Bjo96, Sec. 1.4] or [LH95, Chap. 9].)

1. If $\eta < 1$, then the rank is not changed by the perturbation $\mathrm{rank}(\widetilde{A}) = \mathrm{rank}(A)$.
2. If $\eta < 1$, then the relative perturbation in $\widetilde{\mathbf{x}}_{0M}$ is bounded as

$$\frac{\|\widetilde{\mathbf{x}}_{0M} - \mathbf{x}_{0M}\|_2}{\|\mathbf{x}_{0M}\|_2} \le \frac{\kappa(A)}{1-\eta}\left(\frac{\delta_A}{\|A\|_2} + \frac{\delta_b + \eta\|\mathbf{r}_0\|_2}{\|A\|_2\|\mathbf{x}_{0M}\|_2}\right) + \eta, \qquad \mathbf{r}_0 = \mathbf{b} - A\mathbf{x}_{0M}\,.$$

   If $\mathrm{rank}(A) = n$, then the last term η is omitted. If the problem is consistent, i.e., $\mathbf{r}_0 = \mathbf{0}$, then the relative error can be expected to grow linearly with κ. For $\mathbf{r}_0 \ne \mathbf{0}$ the contribution $\eta\|\mathbf{r}_0\|_2$ and the definition of η show that the relative error may grow as $\kappa(A)^2$.
3. The condition number for the normal equations matrix is $\kappa(A^T A) = \kappa(A)^2$. Due to the finite computer precision, information may be lost when the normal equations are formed; see, e.g., [Bjo96, Sec. 2.2] and Example 2 below.
4. *Component-wise perturbation theory* applies when component-wise perturbation bounds are available for the errors in A and $\mathbf{b}$; if

$$|\widetilde{A} - A| \le \epsilon|A| \qquad \text{and} \qquad |\widetilde{\mathbf{b}} - \mathbf{b}| \le \epsilon|\mathbf{b}|\,,$$

   where the absolute values and inequalities are interpreted componentwise, then

$$|\widetilde{\mathbf{x}} - \mathbf{x}| = \epsilon\,|A^\dagger|\,(|\mathbf{b}| + |A|\,|\mathbf{x}|) + \epsilon\,|(A^T A)^{-1}|\,|E^T|\,|\mathbf{r}|$$

and

$$\|\widetilde{\mathbf{x}} - \mathbf{x}\|_\infty \leq \epsilon \left\| |A^\dagger| \left(|A|\,|\mathbf{x}| + |\mathbf{b}|\right)\right\|_\infty + \epsilon \left\| |(A^T A)^{-1}|\,|A^T|\,|\mathbf{r}|\right\|_\infty .$$

See [Bjo96, Secs. 1.4.5–6] for further references about component-wise perturbation analysis and *a posteriori* error estimation.

Examples:

1. We consider the problem from Example 1 in Section 39.1 and two perturbed versions of it,

$$A = \begin{bmatrix} 1 & 1 \\ 1 & 2 \\ 1 & 3 \end{bmatrix}, \quad \mathbf{b} = \begin{bmatrix} 0.75 \\ 1.13 \\ 1.39 \end{bmatrix}, \quad \widetilde{A} = \begin{bmatrix} 0.8 & 1.1 \\ 0.95 & 2 \\ 1.1 & 2.95 \end{bmatrix}, \quad \widetilde{\mathbf{b}} = \begin{bmatrix} 0.79 \\ 1.23 \\ 1.30 \end{bmatrix}.$$

The matrix has full rank, so the least squares solution is unique (and equal to the minimum-norm least squares solution). The vectors

$$\mathbf{x}_0 = \begin{bmatrix} 0.4500 \\ 0.3200 \end{bmatrix}, \quad \widetilde{\mathbf{x}}_1 = \begin{bmatrix} 0.5967 \\ 0.2550 \end{bmatrix}, \quad \widetilde{\mathbf{x}}_2 = \begin{bmatrix} 0.8935 \\ 0.1280 \end{bmatrix}$$

are the minimizers of $\|\mathbf{b} - A\mathbf{x}\|_2$, $\|\widetilde{\mathbf{b}} - A\mathbf{x}\|_2$, and $\|\widetilde{\mathbf{b}} - \widetilde{A}\mathbf{x}\|_2$, respectively.

The matrix has condition number $\kappa(A) = 6.793$, and $\eta = \kappa(A) * \|\widetilde{A} - A\|_2 / \|A\|_2 = 0.4230 < 1$. The relative errors and their upper bounds are

$$\frac{\|\widetilde{\mathbf{x}}_1 - \mathbf{x}_0\|_2}{\|\mathbf{x}_0\|_2} = 0.291 \leq \kappa(A) \frac{\|\widetilde{\mathbf{b}} - \mathbf{b}\|_2}{\|A\|_2 \|\mathbf{x}_0\|_2} = 0.423$$

$$\frac{\|\widetilde{\mathbf{x}}_2 - \mathbf{x}_0\|_2}{\|\mathbf{x}_0\|_2} = 0.875 \leq \frac{1}{1-\eta}\left(\eta + \kappa(A) \frac{\|\widetilde{\mathbf{b}} - \mathbf{b}\|_2 + \eta \|\mathbf{b} - A\mathbf{x}_0\|_2}{\|A\|_2 \|\mathbf{x}_0\|_2}\right) = 1.575 .$$

2. Consider the following matrix A and the corresponding normal equation matrix:

$$A = \begin{bmatrix} 1 & 1 \\ \delta & 0 \\ 0 & \delta \end{bmatrix}, \quad A^T A = \begin{bmatrix} 1+\delta^2 & 1 \\ 1 & 1+\delta^2 \end{bmatrix}.$$

If $|\delta| \leq \sqrt{\epsilon}$ (where ϵ is the machine precision), then the quantity $1 + \delta^2$ is represented by 1 on the computer and, therefore, the computed $A^T A$ is singular. If we use Householder transformations to compute the QR factorization, we get $R = \begin{bmatrix} -1 & -1 \\ 0 & \delta\sqrt{2} \end{bmatrix}$, so information about δ is preserved.

39.7 Up- and Downdating of QR Factorization

Definitions:

Given $A \in \mathbb{R}^{m \times n}$ with $m > n$ and its QR factorization, as well as a row vector $\mathbf{a}^T$ with $\mathbf{a} \in \mathbb{R}^n$, **updating** of the factorization means computing the QR factorization of the augmented matrix $\begin{bmatrix} A \\ \mathbf{a}^T \end{bmatrix} = \widetilde{A} = \widetilde{Q}\widetilde{R}$ from the QR factors of A. Similarly, **downdating** means computing the QR factors of A from those of $\widetilde{A}$. Up- and downdating algorithms require only $O(mn)$ flops (compared to the $O(mn^2)$ flops of recomputing the QR factors).

Facts:

The following facts can be found in Chapter 3 of [Bjo96].

1. The matrix R can be updated to $\widetilde{R}$ without knowledge of Q by means a sequence of n Givens transformations $G_1, \ldots, G_n$.
2. The updating of Q to $\widetilde{Q}$ then takes the form

$$\widetilde{Q} = \begin{bmatrix} Q & 0 \\ 0 & 1 \end{bmatrix} G_1^T \cdots G_n^T .$$

3. Downdating of $\widetilde{R}$ to R requires the first row $\widetilde{Q}_{1,:}$ of $\widetilde{Q}$. Let $\widetilde{G}_1, \ldots, \widetilde{G}_{m-n}$ be Givens rotations such that $\widehat{R} = G_{m-n} \cdots G_1 (\,\widetilde{R}\ \ \widetilde{Q}_{1,:}^T\,)$ is upper triangular. Then $R = \widehat{R}_{2:n+1,2:n}$.
4. If $\widetilde{Q}$ is not available, then its first row $\widetilde{Q}_{1,:}$ can be computed by the LINPACK/ Saunders algorithm or via hyperbolic rotations. If A is available, the corrected seminormal equations provide a more accurate algorithm.
5. Up- and downdating algorithms are also available for the cases where a column is appended to or deleted from A.
6. Up- and downdating of the Cholesky factor under a rank-one modification $\widetilde{M} = M \pm \mathbf{a}\mathbf{a}^T$ is analytically equivalent to updating R from the QR factorization. In the downdating case the matrix $\widetilde{M}$ must be positive (semi)definite.

Examples:

1. Let

$$A = \begin{bmatrix} 1 & 1 \\ 1 & 2 \\ 1 & 3 \end{bmatrix} = QR = \begin{bmatrix} 0.5774 & -0.7071 & -0.4082 \\ 0.5774 & 0 & 0.8165 \\ 0.5774 & 0.7071 & -0.4082 \end{bmatrix} \begin{bmatrix} 1.732 & 3.464 \\ 0 & 1.414 \\ 0 & 0 \end{bmatrix}.$$

If $\mathbf{a}^T = (\,1\ \ 4\,)$, then

$$\widetilde{A} = \begin{bmatrix} 1 & 1 \\ 1 & 2 \\ 1 & 3 \\ 1 & 4 \end{bmatrix} = \widetilde{Q}\,\widetilde{R} \qquad \text{with} \qquad \widetilde{R} = \begin{bmatrix} 2 & 5 \\ 0 & 2.236 \\ 0 & 0 \\ 0 & 0 \end{bmatrix}.$$

The updated factor $\widetilde{R}$ is computed by augmenting R with $\mathbf{a}^T$ and applying two left Givens rotations G_1 and G_2 to row pairs (1,4) and (2,4), respectively:

$$\begin{bmatrix} R \\ a^T \end{bmatrix} = \begin{bmatrix} 1.732 & 3.464 \\ 0 & 1.414 \\ 0 & 0 \\ 1 & 4 \end{bmatrix} \xrightarrow{G_1} \begin{bmatrix} 2 & 5 \\ 0 & 1.414 \\ 0 & 0 \\ 0 & 1.732 \end{bmatrix} \xrightarrow{G_2} \begin{bmatrix} 2 & 5 \\ 0 & 2.236 \\ 0 & 0 \\ 0 & 0 \end{bmatrix}.$$

39.8 Damped Least Squares

Definitions:

The **damped least squares** solution is the solution to the problem $(A^TA + \alpha I)\mathbf{x} = A^T\mathbf{b}$, where $\alpha > 0$ and A and $\mathbf{b}$ are real. Damped least squares is also known as **ridge regression** and **Tikhonov** (or **Phillips**) **regularization**.

Facts:

(See, e.g., [Han98] for further details.)

1. The two formulations

$$\min\{\|\mathbf{b} - A\mathbf{x}\|_2^2 + \alpha\|\mathbf{x}\|_2^2\} \qquad \text{and} \qquad \min\left\|\begin{bmatrix}\mathbf{b}\\ \mathbf{0}\end{bmatrix} - \begin{bmatrix}A\\ \sqrt{\alpha}\, I\end{bmatrix}\mathbf{x}\right\|_2$$

are analytically equivalent.
2. The damping (controlled by the parameter α) reduces the variance of the solution, at the cost of introducing bias.
3. If $\text{Cov}(\mathbf{b}) = I$, then the covariance matrix for the damped least squares solution $\mathbf{x}_\alpha$ is

$$\text{Cov}(\mathbf{x}_\alpha) = \left(A^T A + \alpha I\right)^{-1} A^T A \left(A^T A + \alpha I\right)^{-1} = V\,\Sigma^2\,(\Sigma^2 + \alpha I)^{-2}\,V^T\,,$$

where Σ and V are from the SVD of A. Hence,

$$\|\text{Cov}(\mathbf{x}_\alpha)\|_2 = \max_i\{\sigma_i^2/(\sigma_i^2 + \alpha)^2\} \leq (4\alpha)^{-1}\,,$$

while $\|\text{Cov}(\mathbf{x}_0)\|_2 = \|(A^T A)^{-1}\|_2 = \sigma_n^{-2}$, which can be much larger.
4. The expected value of $\mathbf{x}_\alpha$ is

$$\mathcal{E}(\mathbf{x}_\alpha) = \left(A^T A + \alpha I\right)^{-1} A^T A\,\mathbf{x}_0\,,$$

which introduces a bias because $\mathcal{E}(\mathbf{x}_\alpha) \neq \mathcal{E}(\mathbf{x}_0)$ when $\alpha > 0$.
5. The damped least squares problem can take the more general form

$$\min\{\|\mathbf{b} - A\mathbf{x}\|_2^2 + \alpha\|B\mathbf{x}\|_2^2\} \qquad \Leftrightarrow \qquad (A^T A + \alpha B^T B)\,\mathbf{x} = A^T\mathbf{b}\,,$$

where $\|B\,\cdot\,\|_2$ defines a (semi)norm. The solution to this problem is unique when the nullspaces of A and B intersect trivially.
6. [Han98, Chap. 7]. Some algorithms for computing α are the discrepancy principle, generalized cross validation, and the L-curve criterion.

Examples:

1. Let $\delta = 10^{-5}$ and consider the matrix and vectors

$$A = \begin{bmatrix}1 & 1\\ 1 & 1+\delta\\ 1 & 1+2\delta\end{bmatrix}, \qquad \mathbf{x}_0 = \begin{bmatrix}1\\ 1\end{bmatrix}, \qquad \mathbf{b} = A\,\mathbf{x}_0, \qquad \widetilde{\mathbf{b}} = \mathbf{b} + \begin{bmatrix}0\\ 0\\ \delta\end{bmatrix}.$$

Obviously $\mathbf{x}_0$ is the least squares solution to $A\mathbf{x} + \mathbf{r} = \mathbf{b}$ with $\mathbf{r}_0 = \mathbf{0}$. The minimizer of $\|\widetilde{\mathbf{b}} - A\mathbf{x}\|_2$ is $\widetilde{\mathbf{x}}_0 = (\,0.5\;\;1.5\,)^T$, showing that for this problem the least squares solution is very sensitive to perturbations (the condition number is $\kappa(A) = 2.4{\cdot}10^5$). Using $\alpha = 10^{-8}$, we obtain the damped least squares solutions

$$\mathbf{x}_\alpha = \begin{bmatrix}0.999995\\ 1.000005\end{bmatrix} \qquad \text{and} \qquad \widetilde{\mathbf{x}}_\alpha = \begin{bmatrix}0.995\\ 1.005\end{bmatrix}.$$

Comparing the damped and the undamped least squares solutions $\widetilde{\mathbf{x}}_\alpha$ and $\widetilde{\mathbf{x}}_0$ to the perturbed least squares problem, we see that $\widetilde{\mathbf{x}}_\alpha$ is a better approximation to the unperturbed solution $\mathbf{x}_0$ than $\widetilde{\mathbf{x}}_0$.

39.9 Rank Revealing Decompositions

Definitions:

A **rank revealing decomposition** is a two-sided orthogonal decomposition of the form

$$A = U\,\widehat{R}\,V^T = U\begin{bmatrix} R \\ 0 \end{bmatrix} V^T,$$

where U and V are orthogonal, and R is upper triangular and reveals the (numerical) rank of A in the size of its diagonal elements.

The **numerical rank** k_τ of A, with respect to the threshold τ, is defined as

$$k_\tau = \min \operatorname{rank}(A + E) \quad \text{subject to} \quad \|E\|_2 \le \tau.$$

Facts:

(See, e.g., [Bjo96, Sec. 1.7.3–6], [Han98, Sec. 2.2], and [Ste98, Chap. 5].)

1. [GV96, p. 73] The numerical rank k_τ is equal to the number of singular values greater than τ, i.e., $\sigma_{k_\tau} > \tau \ge \sigma_{k_\tau+1}$.
2. The singular value decomposition is rank revealing with the middle matrix $R = \Sigma$. The SVD is difficult to update.
3. If A is exactly rank deficient with $\operatorname{rank}(A) = p$, then there always exists a **pivoted QR factorization** $A\,\Pi = Q\widehat{R}$ with $\widehat{R}$ of the form

$$\widehat{R} = \begin{bmatrix} R_{11} & R_{12} \\ 0 & 0 \end{bmatrix}, \qquad R_{11} \in \mathbb{R}^{p\times p}, \qquad \operatorname{rank}(\widehat{R}) = \operatorname{rank}(R_{11}) = p\,,$$

   and a **complete orthogonal decomposition** of the form $A = U\,\widehat{R}\,V^T$, where

$$\widehat{R} = \begin{bmatrix} R_{11} & 0 \\ 0 & 0 \end{bmatrix}, \qquad R_{11} \in \mathbb{R}^{p\times p}, \qquad \operatorname{rank}(\widehat{R}) = \operatorname{rank}(R_{11}) = p\,.$$

   The pseudoinverse of A is $A^\dagger = V_{:,1:p} R_{11}^{-1} U_{:,1:p}^T$.
4. A basic solution can be computed from the pivoted QR factorization as $\mathbf{x}_{0B} = \Pi\, R_{11}^{-1}\, Q_{:,1:p}^T \mathbf{b}$. The minimum-norm least squares solution is given in terms of the complete orthogonal decomposition as $\mathbf{x}_{0M} = V_{:,1:p} R_{11}^{-1} U_{:,1:p}^T \mathbf{b}$.
5. The **rank revealing QR** (RRQR) **decomposition** is a pivoted QR factorization $A\,\Pi = Q\,\widehat{R}$ such that

$$\sigma_i / c_i \le \sigma_i(R_{1:i,1:i}) \le \sigma_i \le \|R_{1:i,1:i}\|_2 \le c_i \sigma_i, \qquad i = 1, \ldots, n,$$

   where σ_i is the ith singular value of A,

$$c_i = \sqrt{i(n-i) + \min(i, n-i)}\,,$$

   and $\sigma_i(R_{1:i,1:i})$ denotes the smallest singular value of $R_{1:i,1:i}$. The RRQR factorization can be used to estimate the numerical rank k_τ. The RRQR factorization is not unique.
6. The **URV decomposition** is a two-sided orthogonal decomposition $A = U\widehat{R}\,V^T$, such that, for $k = 1, \ldots, n$,

$$\sigma_i \check{c}_k \le \sigma_i(R_{1:i,1:i}) \le \sigma_i, \quad i = 1, \ldots, k$$

and

$$\sigma_i \le \sigma_{i-k}(R_{k+1:n,k+1:n}) \le \sigma_i/\check{c}_k, \quad i = k+1, \ldots, n,$$

where

$$\check{c}_k = \left(1 - \frac{\| R_{k+1:n,k+1:n} \|_2^2}{\sigma_k(R_{1:i,1:i})^2 - \| R_{k+1:n,k+1:n} \|_2^2}\right)^{1/2}.$$

There is also a ULV decomposition with a lower triangular middle matrix; both can be used to estimate the numerical rank of A.

7. The RRQR, URV, and ULV decompositions can be updated, at slightly more cost than the QR factorization.

Examples:

1. The rank of A is revealed by the zero element in the (3,3) position of R:

$$A = \begin{bmatrix} 1 & 2 & 3 \\ 2 & 3 & 4 \\ 3 & 4 & 5 \\ 4 & 5 & 6 \end{bmatrix} = QR \quad \text{with} \quad R = \begin{bmatrix} 5.477 & 7.303 & 9.129 \\ 0 & 0.816 & 1.633 \\ 0 & 0 & 0 \end{bmatrix}.$$

Here the QR factorization is rank revealing ($U = Q$ and $V = I$).

2. Pivoting must be used to ensure that a QR factorization is rank revealing. The "standard column pivoting" often works well in connection with Householder transformations; here the pivot column in each stage is chosen to maximize the norm of the leading column of the submatrix $A^{(k)}_{k:m,k:n}$ to be reduced. Example:

$$A = \begin{bmatrix} 4 & 2 & 2 \\ 2 & 1 & 2 \\ 0 & 0 & 1 \end{bmatrix},$$

$$A^{(1)} = H_1 A = \begin{bmatrix} -4.4721 & -2.2361 & -2.6833 \\ 0 & 0 & 0.8944 \\ 0 & 0 & 1 \end{bmatrix},$$

$$A^{(1)}\Pi = \begin{bmatrix} -4.4721 & -2.6833 & -2.2361 \\ 0 & 0.8944 & 0 \\ 0 & 1 & 0 \end{bmatrix}, \quad \Pi = \begin{bmatrix} 1 & 0 & 0 \\ 0 & 0 & 1 \\ 0 & 1 & 0 \end{bmatrix},$$

$$\widehat{R} = A^{(2)} = H_2 A^{(1)}\Pi = \begin{bmatrix} -4.4721 & -2.6833 & -2.2361 \\ 0 & -1.3416 & 0 \\ 0 & 0 & 0 \end{bmatrix}.$$

3. The standard column pivoting strategy is not guaranteed to reveal the numerical rank; hence, the development of the RRQR and URV decompositions.

References

[Bjo96] Å. Björck. *Numerical Methods for Least Squares Problems*. SIAM, Philadelphia, 1996.

[Bjo04] Å. Björck. The calculation of linear least squares problems. *Acta Numerica* (2004):1–53, Cambridge University Press, Cambridge, 2004.

[Gau95] C.F. Gauss. *Theory of the Combination of Observations Least Subject to Errors.* (Translated by G. W. Stewart.) SIAM, Philadelphia, 1995.

[Gol65] G.H. Golub. Numerical methods for solving least squares problems. *Numer. Math.*, 7:206–216, 1965.

[GV96] G.H. Golub and C.F. Van Loan. *Matrix Computations*, 3rd ed., Johns Hopkins University Press, Baltimore, MD, 1996.

[Han98] P.C. Hansen. *Rank-Deficient and Discrete Ill-Posed Problems: Numerical Aspects of Linear Inversion.* SIAM, Philadelphia, 1998.

[LH95] C.L. Lawson and R.J. Hanson. *Solving Least Squares Problems.* Classics in Applied Mathematics, SIAM, Philadelphia, 1995.

[Ste98] G.W. Stewart. *Matrix Algorithms Volume I: Basic Decompositions.* SIAM, Philadelphia, 1998.

[Str88] G. Strang. *Linear Algebra and Its Applications*, 3rd ed., Saunders College Publishing, Fort Worth, TX, 1988.

40

Sparse Matrix Methods

Esmond G. Ng
Lawrence Berkeley National Laboratory

40.1 Introduction

Let A be an n by n nonsingular matrix and $\mathbf{b}$ be an n-vector. As discussed in Chapter 38, Matrix Factorizations and Direct Solution of Linear Systems, the solution of the system of linear equations $A\mathbf{x} = \mathbf{b}$ using Gaussian elimination requires $O(n^3)$ operations, which typically include additions, subtractions, multiplications, and divisions. The solution also requires $O(n^2)$ words of storage. The computational complexity is based on the assumption that every element of the matrix has to be stored and operated on. However, linear systems that arise in many scientific and engineering applications can be large; that is, n can be large. It is not uncommon for n to be over hundreds of thousands or even millions. Fortunately, for these linear systems, it is often the case that most of the elements in the matrix A will be zero.

Following is a simple example that illustrates where the zero elements come from. Consider the Laplace equation defined on a unit square:

$$\frac{\partial^2 u}{\partial x^2} + \frac{\partial^2 u}{\partial y^2} = 0.$$

Assume that u is known along the boundary. Suppose the square domain is discretized into a $(k+2)$ by $(k+2)$ mesh with evenly spaced mesh points, as shown in Figure 40.1. Also suppose that the mesh points are labeled from 0 to $k+1$ in the x and y directions. For $0 \le i \le k+1$, let the variables in the x direction be denoted by x_i. Similarly, for $0 \le j \le k+1$, let the variables in the y direction be denoted by y_j. The solution at (x_i, y_j) will be denoted by $u_{i,j} = u(x_i, y_j)$. To solve the Laplace equation numerically, the partial derivatives at (x_i, y_j) will be approximated, for example, using second-order centered difference approximations:

$$\left.\frac{\partial^2 u}{\partial x^2}\right|_{(x_i,y_j)} \approx \frac{u_{i-1,j} - 2u_{i,j} + u_{i+1,j}}{h^2},$$

$$\left.\frac{\partial^2 u}{\partial y^2}\right|_{(x_i,y_j)} \approx \frac{u_{i,j-1} - 2u_{i,j} + u_{i,j+1}}{h^2},$$

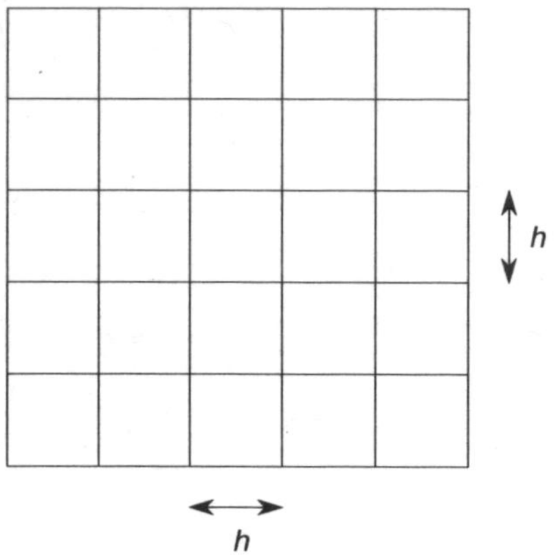

FIGURE 40.1 Discretization of a unit square.

where $h = \frac{1}{k+1}$ is the spacing between two mesh points in each direction. Here, it is assumed that $u_{0,j}, u_{k+1,j}, u_{i,0}, u_{i,k+1}$ are given by the boundary condition, for $1 \leq i, j \leq k$. Using the difference approximations, the Laplace equation at each mesh point (x_i, y_j), $1 \leq i, j \leq k$, is approximated by the following linear equation:

$$u_{i-1,j} + u_{i,j-1} - 4u_{i,j} + u_{i+1,j} + u_{i,j+1} = 0, \qquad \text{for } 1 \leq i, j \leq k.$$

This leads to a system of k^2 by k^2 linear equations in k^2 unknowns. The solution to the linear system provides the approximate solution $u_{i,j}$, $1 \leq i, j \leq k$, at the mesh points. Note that each equation has at most five unknowns. Thus, the coefficient matrix of the linear system, which is k^2 by k^2 and has k^4 elements, has at most $5k^2$ nonzero elements. It is therefore crucial, for the purpose of efficiency (both in terms of operations and storage), to take advantage of the zero elements as much as possible when solving the linear system. The goal is to compute the solution without storing and operating on most of the zero elements of the matrix. This chapter will discuss some of techniques for exploiting the zero elements in Gaussian elimination.

Throughout this chapter, the matrices are assumed to be real. However, most of the discussions are also applicable to complex matrices, with the exception of those on real symmetric positive definite matrices. The discussions related to real symmetric positive definite matrices are applicable to Hermitian positive definite matrices (which are complex but not symmetric).

40.2 Sparse Matrices

Definitions:

A matrix A is **sparse** if substantial savings in either operations or storage can be achieved when the zero elements of A are exploited during the application of Gaussian elimination to A.

The number of nonzero elments in a matrix A is denoted by $\boldsymbol{nnz}(A)$.

Facts: [DER89], [GL81]

1. Let A be an n by n sparse matrix. It often takes much less than n^2 words to store the nonzero elements in A.
2. Let T be an n by n sparse triangular matrix. The number of operations required to solve the triangular system of linear equations $T\mathbf{x} = \mathbf{b}$ is $O(nnz(T))$.

Examples:

1. A tridiagonal matrix $T = [t_{i,j}]$ has the form

$$T = \begin{bmatrix} t_{1,1} & t_{1,2} & & & & & \\ t_{2,1} & t_{2,2} & t_{2,3} & & & & \\ & t_{3,2} & t_{3,3} & t_{3,4} & & & \\ & \ddots & \ddots & \ddots & & & \\ & & t_{n-2,n-3} & t_{n-2,n-2} & t_{n-2,n-1} & \\ & & & t_{n-1,n-2} & t_{n-1,n-1} & t_{n-1,n} \\ & & & & t_{n,n-1} & t_{n,n} \end{bmatrix},$$

where $t_{i,i} \neq 0$ $(1 \leq i \leq n)$, and $t_{i,i+1} \neq 0$, and $t_{i+1,i} \neq 0$, $(1 \leq i \leq n-1)$. The matrix T is an example of a sparse matrix. If Gaussian elimination is applied to T with partial pivoting for numerical stability, $t_{i,i+2}$, $1 \leq i \leq n-2$, may become nonzero. Thus, in the worst case, there will be at most $5n$ nonzero elements in the triangular factorization (counting the 1s on the diagonal of the lower triangular factor). The number of operations required in Gaussian elimination is at most $5n$.

2. Typically only the nonzero elements of a sparse matrix have to be stored. One of the common ways to store a sparse matrix A is the *compressed column storage* (CCS) scheme. The nonzero elements are stored in an array (e.g., VAL) column by column, along with an integer array (e.g., IND) that stores the corresponding row subscripts. Another integer array (e.g., COLPTR) will be used to provide the index k where VAL(k) contains the first nonzero element in column k of A. Suppose that A is given below:

$$A = \begin{bmatrix} 11 & 0 & 16 & 19 & 22 \\ 0 & 13 & 0 & 0 & 23 \\ 12 & 14 & 17 & 0 & 0 \\ 0 & 0 & 18 & 20 & 24 \\ 0 & 15 & 0 & 21 & 25 \end{bmatrix}.$$

The CCS scheme for storing the nonzero elements is depicted in Figure 40.2. For example, the nonzero elements in column 4 and the corresponding row subscripts can be found in VAL(s) and IND(s), where $s = \text{COLPTR}(4), \text{COLPTR}(4)+1, \text{COLPTR}(4)+2, \cdots, \text{COLPTR}(5)-1$. Note that, for this example, COLPTR$(6) = 16$, which is one more than the number of nonzero elements in A. This is used to indicate the end of the set of nonzero elements.

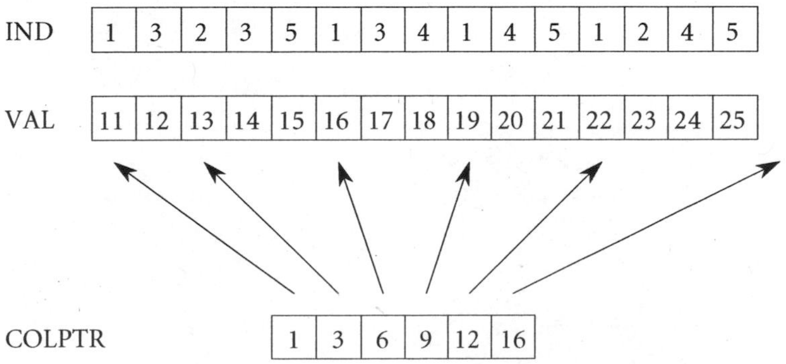

FIGURE 40.2 An example of a compressed column storage scheme.

3. The *compressed row storage* (CRS) scheme is another possibility of storing the nonzero elements of a sparse matrix. It is very similar to the CCS scheme, except that the nonzero elements are stored by rows.
4. Let $A = [a_{i,j}]$ be an n by n sparse matrix. Let $\mathbf{b} = [b_i]$ and $\mathbf{c} = [c_i]$ be n-vectors. The following algorithm computes the product $\mathbf{c} = A\mathbf{b}$, with the assumption that the nonzero elements of A are stored using the CRS scheme.

   for $i = 1, 2, \cdots, n$ **do**

   $c_i \leftarrow 0$

   for $s = \text{ROWPTR}(i), \text{ROWPTR}(i) + 1, \cdots, \text{ROWPTR}(s + 1) - 1$ **do**

   $c_i \leftarrow c_i + \text{VAL}(s) b_{\text{IND}(s)}$.

5. Let $A = [a_{i,j}]$ be an n by n sparse matrix. Let $\mathbf{b} = [b_i]$ and $\mathbf{c} = [c_i]$ be n-vectors. The following algorithm computes the product $\mathbf{c} = A\mathbf{b}$, with the assumption that the nonzero elements of A are stored using the CCS scheme.

   for $i = 1, 2, \cdots, n$ **do**

   $c_i \leftarrow 0$

   for $i = 1, 2, \cdots, n$ **do**

   for $s = \text{COLPTR}(i), \text{COLPTR}(i) + 1, \cdots, \text{COLPTR}(s + 1) - 1$ **do**

   $c_{\text{IND}(s)} \leftarrow c_{\text{IND}(s)} + \text{VAL}(s) b_i$.

40.3 Sparse Matrix Factorizations

Mathematically, computing a triangular factorization of a sparse matrix using Gaussian elimination is really no different from that of a dense matrix. However, the two are very different algorithmically. Sparse triangular factorizations can be quite complicated because of the need to preserve the zero elements as much as possible.

Definitions:

Let A be a sparse matrix. An element of a sparse matrix A is a **fill element** if it is zero in A but becomes nonzero during Gaussian elimination.

The **sparsity structure** of a matrix $A = [a_{i,j}]$ refers to the set $\text{Struct}(A) = \{(i, j) : a_{i,j} \neq 0\}$.

Consider applying Gaussian elimination to a matrix A with row and column pivoting:

$$A = P_1 M_1 P_2 M_2 \cdots P_{n-1} M_{n-1} U Q_{n-1} \cdots Q_2 Q_1,$$

where P_i and Q_i $(1 \leq i \leq n - 1)$, are, respectively, the row and column permutations due to pivoting, M_i $(1 \leq i \leq n - 1)$ is a Gauss transformation (see Chapter 38), and U is the upper triangular factor. Let $L = M_1 + M_2 + \cdots + M_{n-1} - (n - 2)I$, where I is the identity matrix. Note that L is a lower triangular matrix. The matrix $F = L + U - I$ is referred to as the **fill matrix**.

The matrix A is said to have a **zero-free diagonal** if all the diagonal elements of A are nonzero. The zero-free diagonal is also known as a **maximum transversal**.

No **exact numerical cancellation** between two numbers u and v means that $u + v$ (or $u - v$) is nonzero regardless of the values of u and v.

Facts: [DER89], [GL81]

In the following discussion, A is a sparse nonsingular matrix and F is its fill matrix.

1. [Duf81], [DW88] There exists a (row) permutation matrix P_r such that $P_r A$ has a zero-free diagonal. Similarly, there exists a (column) permutation matrix P_c such that $A P_c$ has a zero-free diagonal.

2. It is often true that there are more nonzero elements in F than in A.
3. The sparsity structure of F depends on the sparsity structure of A, as well as the pivot sequence needed to maintain numerical stability in Gaussian elimination. This means that $\text{Struct}(F)$ is known only during numerical factorization. As a result, the storage scheme cannot be created in advance to accommodate the fill elements that occur during Gaussian elimination.
4. If A is symmetric and positive definite, then pivoting for numerical stability is not needed during Gaussian elimination. Assume that exact numerical cancellations do not occur during Gaussian elimination. Then $\text{Struct}(A) \subseteq \text{Struct}(F)$.
5. Let A be symmetric and positive definite. Assume that exact numerical cancellations do not occur during Gaussian elimination. Then $\text{Struct}(F)$ is determined solely by $\text{Struct}(A)$. This implies that $\text{Struct}(F)$ can be computed before any numerical factorization proceeds. Knowing $\text{Struct}(F)$ in advance allows a storage scheme to be set up prior to numerical factorization.
6. [GN85] Suppose that A is a nonsymmetric matrix. Consider applying Gaussian elimination to a matrix A with partial pivoting:

$$A = P_1 M_1 P_2 M_2 \cdots P_{n-1} M_{n-1} U,$$

where P_i $(1 \leq i \leq n-1)$, is a row permutation due to pivoting, M_i $(1 \leq i \leq n-1)$ is a Gauss transformation, and U is the upper triangular factor. Let F denote the corresponding fill matrix; that is, $F = M_1 + M_2 + \cdots + M_{n-1} + U - (n-1)I$. The matrix product $A^T A$ is symmetric and positive definite and, hence, has a Cholesky factorization $A^T A = L_C L_C^T$. Assume that A has a zero-free diagonal. Then $\text{Struct}(F) \subseteq \text{Struct}(L_C + L_C^T)$. This result holds for *every* legitimate sequence of pivots $\{P_1, P_2, \cdots, P_{n-1}\}$. Thus, $\text{Struct}(L_C)$ and $\text{Struct}(L_C^T)$ can serve as *upper bounds* on $\text{Struct}(L)$ and $\text{Struct}(U)$, respectively. When A is irreducible, then $\text{Struct}(L_C^T)$ is a *tight* bound on $\text{Struct}(U)$: for a given $(i, j) \in \text{Struct}(L_C^T)$, there is an assignment of numerical values to the nonzero elements of A so that $U_{i,j}$ is nonzero; this is referred to as an *one-at-time* result [GN93]. However, $\text{Struct}(L_C)$ is *not* a tight bound on $\text{Struct}(L)$. A tight bound on $\text{Struct}(L)$ can be found in [GN87] and [GN93].

Examples:

Some of the properties of sparse matrices, such as zero-free diagonals and reducible/irreducible matrices, depend only on the sparsity structures of the matrices; the actual values of the nonzero elements are irrelevant. As a result, for the examples illustrating such properties, only the sparsity structures will be taken into consideration. The numerical values will not be shown. Nonzero elements will be indicated by $\times$.

1. Following is an example of a reducible matrix:

$$A = \begin{bmatrix}
\times & \times & \times & \times & \times & & \times & & & \\
 & & & & \times & & \times & \times & \times & \\
 & \times & \times & & \times & & & & & \\
 & \times & & \times & & \times & & & \times & \times \\
 & \times & & & \times & & & & \times & \\
 & \times & \times & & & & & & & \times \\
\times & \times & \times & \times & & & \times & & \times & \times \\
 & \times & & & \times & & & & \times & \\
\times & & & & \times & & & & \times & \times \\
 & & & & & & & & & \times
\end{bmatrix}.$$

The matrix A can be put into a block upper triangular form using the following permutations:

$$\pi_r = [2, 4, 7, 1, 9, 6, 8, 5, 3, 10],$$
$$\pi_c = [8, 6, 4, 7, 1, 3, 9, 5, 2, 10].$$

Here, $\pi_r(i) = j$ means that row i of the permuted matrix comes from row j of the original matrix. Similarly, $\pi_c(i) = j$ means that column i of the permuted matrix comes from column j of the original matix. When π_r and π_c are applied to the identity matrix (separately), the corresponding permutation matrices are obtained:

$$P_r = \begin{bmatrix} & \times & & & & & & & & \\ & & & \times & & & & & & \\ & & & & & & \times & & & \\ \times & & & & & & & & & \\ & & & & & & & & \times & \\ & & & & & \times & & & & \\ & & & & & & & \times & & \\ & & & & \times & & & & & \\ & & \times & & & & & & & \\ & & & & & & & & & \times \end{bmatrix}, \quad P_c = \begin{bmatrix} & & & & \times & & & & & \\ & & & & & & & & \times & \\ & & & & & \times & & & & \\ & & \times & & & & & & & \\ & & & & & & & \times & & \\ & \times & & & & & & & & \\ & & & \times & & & & & & \\ \times & & & & & & & & & \\ & & & & & & \times & & & \\ & & & & & & & & & \times \end{bmatrix}.$$

The permuted matrix $P_r A P_c$ is shown below:

$$P_r A P_c = \left[\begin{array}{c|c|cc|c|cccc|c} \times & & & \times & & & \times & \times & & \\ \hline & \times & \times & & & & \times & & \times & \times \\ \hline & & \times & \times & \times & \times & \times & & \times & \times \\ & & \times & \times & \times & \times & & \times & \times & \\ \hline & & & & \times & & \times & \times & & \times \\ \hline & & & & & \times & & & \times & \times \\ & & & & & & \times & \times & \times & \\ & & & & & & \times & \times & \times & \\ & & & & & \times & & \times & \times & \\ \hline & & & & & & & & & \times \end{array}\right].$$

The block triangular form has four 1×1 blocks, one 2×2 block, and one 4×4 block on the diagonal.

2. The advantage of the block triangular form is that only the diagonal blocks have to be factored in the solution of a linear system. As an example, suppose that A is lower block triangular:

$$A = \begin{bmatrix} A_{1,1} & & & \\ A_{2,1} & A_{2,2} & & \\ \vdots & \vdots & \ddots & \\ A_{m,1} & A_{m,2} & \cdots & A_{m,m} \end{bmatrix}.$$

Here, m is the number of blocks on the diagonal of A. Consider the solution of the linear system $A\mathbf{x} = \mathbf{b}$. Suppose that $\mathbf{b}$ and $\mathbf{x}$ are partitioned according to the block structure of A:

$$\mathbf{b} = \begin{bmatrix} \mathbf{b}_1 \\ \mathbf{b}_2 \\ \vdots \\ \mathbf{b}_m \end{bmatrix} \quad \text{and} \quad \mathbf{x} = \begin{bmatrix} \mathbf{x}_1 \\ \mathbf{x}_2 \\ \vdots \\ \mathbf{x}_m \end{bmatrix}.$$

Then the solution can be obtained using a block substitution scheme:

$$A_{k,k}\mathbf{x}_k = \mathbf{b}_k - \sum_{i=1}^{k-1} A_{k,i}\mathbf{x}_i, \quad \text{for } 1 \le i \le m.$$

Note that only the diagonal blocks $A_{k,k}$, $1 \le k \le m$ have to be factored; Gaussian elimination does not have to be applied to the entire matrix A. This can result in substantial savings in storage and operations.

3. In this example, the matrix A has zero elements on the diagonal:

$$A = \begin{bmatrix}
 & & & & \times & & \times & \times & & \times \\
 & & & \times & & & & & & \times \\
 & & & \times & \times & & & \times & & \times \\
 & & \times & & \times & & \times & & \times & \\
 & \times & & & & \times & \times & \times & \times & \times \\
 & \times & & & & & \times & & \times & \times \\
 & & \times & \times & & & & \times & \times & \times \\
 & & \times & \times & \times & \times & \times & \times & & \\
 & & & & & \times & & & \times & \\
\times & \times & & & & \times & & & & \times
\end{bmatrix}.$$

The following permutation matrix P, when applied to the columns of A, will produce a matrix with a zero-free diagonal:

$$P = \begin{bmatrix}
 & & & & & & & & & \times \\
 & & & & & \times & & & & \\
 & & & & & & & \times & & \\
 & \times & & & & & & & & \\
 & & & \times & & & & & & \\
 & & & & & & & & \times & \\
\times & & & & & & & & & \\
 & & \times & & & & & & & \\
 & & & & \times & & & & & \\
 & & & & & & \times & & &
\end{bmatrix}.$$

The sparsity structure of the permuted matrix is shown below:

$$AP = \begin{bmatrix} \times & & \times & \times & & & \times & & & \\ & \times & & & & & \times & & & \\ & \times & \times & \times & & & \times & & & \\ \times & & & \times & \times & & & \times & & \\ \times & & \times & & \times & \times & \times & & \times & \\ \times & & & & \times & \times & \times & & & \\ & \times & \times & & \times & & \times & \times & & \\ \times & \times & \times & \times & & & & \times & \times & \\ & & & & \times & & & & \times & \\ & & & & & \times & \times & & \times & \times \end{bmatrix}.$$

The permutation matrix P is obtained by applying the following permutation π to the columns of the identity matrix:

$$\pi = [7, 4, 8, 5, 9, 2, 10, 3, 6, 1].$$

Again, $\pi(i) = j$ means that column i of the permuted matrix comes from column j of the original matrix.

4. Consider the following matrix:

$$A = \begin{bmatrix} 10 & 0 & 1 & 0 & 1 & 0 & 0 \\ 1 & 10 & 0 & 0 & 1 & 0 & 0 \\ 0 & 0 & 10 & 0 & 0 & 1 & 0 \\ 0 & 0 & 1 & 10 & 0 & 0 & 1 \\ 0 & 1 & 0 & 0 & 10 & 0 & 1 \\ 1 & 0 & 0 & 0 & 0 & 10 & 0 \\ 0 & 0 & 0 & 0 & 1 & 0 & 10 \end{bmatrix},$$

which is diagonal dominant. The matrix can be factored without pivoting for stability. The triangular factors are given below:

$$L = \begin{bmatrix} 1 & & & & & & \\ \frac{1}{10} & 1 & & & & & \\ 0 & 0 & 1 & & & & \\ 0 & 0 & \frac{1}{10} & 1 & & & \\ 0 & \frac{1}{10} & \frac{1}{1000} & 0 & 1 & & \\ \frac{1}{10} & 0 & -\frac{1}{100} & 0 & -\frac{10}{991} & 1 & \\ 0 & 0 & 0 & 0 & \frac{100}{991} & \frac{1}{99199} & 1 \end{bmatrix}, U = \begin{bmatrix} 10 & 0 & 1 & 0 & 1 & 0 & 0 \\ & 10 & -\frac{1}{10} & 0 & \frac{9}{10} & 0 & 0 \\ & & 10 & 0 & 0 & 1 & 0 \\ & & & 10 & 0 & -\frac{1}{10} & 1 \\ & & & & \frac{991}{100} & -\frac{1}{1000} & 1 \\ & & & & & \frac{99199}{9910} & \frac{10}{991} \\ & & & & & & \frac{981980}{99199} \end{bmatrix}.$$

Note that A has 18 nonzero elements, whereas $L + U$ has 26 nonzero elements, showing that there are more nonzero elements in the fill matrix than in A.

5. Consider the matrix A in the previous example. Suppose $\hat{A} = [\hat{a}_{i,j}]$ is obtained from A by swapping the $(1, 1)$ and $(2, 1)$ elements:

$$\hat{A} = \begin{bmatrix} 1 & 0 & 1 & 0 & 1 & 0 & 0 \\ 10 & 10 & 0 & 0 & 1 & 0 & 0 \\ 0 & 0 & 10 & 0 & 0 & 1 & 0 \\ 0 & 0 & 1 & 10 & 0 & 0 & 1 \\ 0 & 1 & 0 & 0 & 10 & 0 & 1 \\ 1 & 0 & 0 & 0 & 0 & 10 & 0 \\ 0 & 0 & 0 & 0 & 1 & 0 & 10 \end{bmatrix}.$$

When Gaussian elimination with partial pivoting is applied to $\hat{A}$, rows 1 and 2 will be interchanged at step 1 of the elimination since $|\hat{a}_{2,1}| > |\hat{a}_{1,1}|$. It can be shown that no more interchanges are needed in the subsequent steps. The triangular factors are given by

$$\hat{L} = \begin{bmatrix} 1 & 0 & 0 & 0 & 0 & 0 & 0 \\ \frac{1}{10} & 1 & 0 & 0 & 0 & 0 & 0 \\ 0 & 0 & 1 & 0 & 0 & 0 & 0 \\ 0 & 0 & \frac{1}{10} & 1 & 0 & 0 & 0 \\ 0 & -1 & \frac{1}{10} & 0 & 1 & 0 & 0 \\ \frac{1}{10} & 1 & -\frac{1}{10} & 0 & -\frac{10}{109} & 1 & 0 \\ 0 & 0 & 0 & 0 & \frac{10}{109} & \frac{10}{10999} & 1 \end{bmatrix}, \hat{U} = \begin{bmatrix} 10 & 10 & 0 & 0 & 1 & 0 & 0 \\ 0 & -1 & 1 & 0 & \frac{9}{10} & 0 & 0 \\ 0 & 0 & 10 & 0 & 0 & 1 & 0 \\ 0 & 0 & 0 & 10 & 0 & -\frac{1}{10} & 1 \\ 0 & 0 & 0 & 0 & \frac{109}{10} & -\frac{1}{10} & 1 \\ 0 & 0 & 0 & 0 & 0 & \frac{10999}{1090} & \frac{10}{109} \\ 0 & 0 & 0 & 0 & 0 & 0 & \frac{108980}{10999} \end{bmatrix}.$$

Even though A (in the previous example) and $\hat{A}$ have the same sparsity structures, their fill matrices have different numbers of nonzero elements. For $\hat{A}$, there are 27 nonzero elements in the fill matrix of $\hat{A}$. While this example is small, it illustrates that the occurrence of fill elements in Gaussian elimination generally depends on both the sparsity structure and the values of the nonzero elements of the matrix.

6. Consider the example in the last two examples again. Note that A has a zero-free diagonal. Both A and $\hat{A}$ have the same sparsity structure. The sparsity structure of $A^T A$ is the same as that of $\hat{A}^T \hat{A}$:

$$A^T A = \begin{bmatrix} \times & \times & \times & & \times & \times & \\ \times & \times & & & \times & & \times \\ \times & & \times & \times & \times & \times & \times \\ & & \times & \times & & & \times \\ \times & \times & \times & & \times & & \times \\ \times & & \times & & & \times & \\ & \times & \times & \times & \times & & \times \end{bmatrix}.$$

(Again, $\times$ denotes a nonzero element.) The Cholesky factor L_C of the symmetric positive definite matrix $A^T A$ has the form:

$$L_C = \begin{bmatrix} \times & & & & & & \\ \times & \times & & & & & \\ \times & \times & \times & & & & \\ & & \times & \times & & & \\ \times & \times & \times & \times & \times & & \\ \times & \times & \times & \times & \times & \times & \\ & \times & \times & \times & \times & \times & \times \end{bmatrix}.$$

Note that $\text{Struct}(L) \neq \text{Struct}(\hat{L})$, but $\text{Struct}(L) \subset \text{Struct}(L_C)$ and $\text{Struct}(\hat{L}) \subset \text{Struct}(L_C)$. Similarly, $\text{Struct}(U) \neq \text{Struct}(\hat{U})$, but $\text{Struct}(U) \subset \text{Struct}(L_C^T)$ and $\text{Struct}(\hat{U}) \subset \text{Struct}(L_C^T)$. This example illustrates that when A has a zero-free diagonal the sparsity structure of the Cholesky factor of $A^T A$ indeed contains the sparsity structure of L (or U^T), irrespective of the choice of the pivot sequence.

7. It is shown in an earlier example that fill elements do occur in Gaussian elimination of sparse matrices. In order to allow for these fill elements during Gaussian elimination, the simple storage schemes (CCS or CRS) may not be sufficient. More sophisticated storage schemes are often needed. The choice of a storage scheme depends on the choice of the factorization algorithm. There are many implementations of sparse Gaussian elimination, such as *profile methods*, *left-looking methods*, *right-looking methods*, and *frontal/multifrontal methods* [DER89], [GL81]. Describing these implementations is beyond the scope of this chapter. Following is a list of pointers to some of the implementations.
   (a) Factorization of sparse symmetric positive definite matrices: [NP93], [RG91].
   (b) Factorization of sparse symmetric indefinite matrices: [AGL98], [DR83], [DR95], [Duf04], [Liu87a], [Liu87b].
   (c) Factorization of sparse nonsymmetric matrices: [DEG+99], [Duf77], [DR96], [GP88].

40.4 Modeling and Analyzing Fill

A key component of sparse Gaussian elimination is the exploitation of the sparsity structure of the given matrix and its triangular factors. Graphs are useful in understanding and analyzing how fill elements are introduced in sparse Gaussian elimination. Some basic graph-theoretical tools and results are described in this section. Others can be found in [EL92], [GL81], [GL93], [GNP94], [Liu86], [Liu90], and [Sch82].

In this and the next sections, all graphs (bipartite graphs, directed graphs, and undirected graphs) are simple. That is, loops and multiple edges are not allowed. More information on graphs can be found in Chapter 28 (graphs), Chapter 29 (digraphs), and Chapter 30 (bipartite graphs).

Definitions:

Consider a sparse nonsymmetric matrix $A = [a_{i,j}]$. Let $R = \{r_1, r_2, \cdots, r_n\}$ be a set of "row" vertices associated with the rows of A. Similarly, let $C = \{c_1, c_2, \cdots, c_n\}$ be a set of "column" vertices associated with the columns of A. The **bipartite graph** or **bigraph** of A, denoted by $H(A) = (R, C, E)$, can be associated with the sparsity structure of A. There is an edge $\{r_i, c_j\} \in E$ if and only $a_{i,j} \neq 0$.

Let $A = [a_{i,j}]$ be a sparse nonsymmetric matrix. Suppose that A has a zero-free diagonal, and assume that the pivots are always chosen from the diagonal. Then the sparsity structure of A can be represented by a **directed graph** or **digraph**, denoted by $\Gamma(A) = (X, E)$. Here, $X = \{x_1, x_2, \cdots, x_n\}$, with x_i, $1 \leq i \leq n$, representing column i and row i of A. There is a directed edge or arc $(x_i, x_j) \in E$ if and only if $a_{i,j} \neq 0$, for $i \neq j$. (The nonzero elements on the diagonal of A are not represented.)

Suppose $A = [a_{i,j}]$ is symmetric and positive definite, and assume that the pivots are always chosen from the diagonal. Then an **undirected graph** or **graph** (when the context is clear) $G(A) = (X, E)$ can be used to represent the sparsity structure of A. Let $X = \{x_1, x_2, \cdots, x_n\}$, with x_i representing row i and column i of A. There is an (undirected) edge $\{x_i, x_j\} \in E$ if and only if $a_{i,j} \neq 0$ (and, hence, $a_{j,i} \neq 0$), for $i \neq j$. (The nonzero elements on the diagonal of A are not represented.)

A **path** in a graph (which can be a bigraph, digraph, or undirected graph) is a sequence of *distinct* vertices $(x_{s_1}, x_{s_2}, \cdots, x_{s_t})$ such that there is an edge between every pair of consecutive vertices. For bigraphs and undirected graphs, $\{x_{s_p}, x_{s_{p+1}}\}$, $1 \leq p \leq t-1$, is an (undirected) edge. For digraphs, the path is a directed path, and $(x_{s_p}, x_{s_{p+1}})$, $1 \leq p \leq t-1$, is an arc.

Let A be an n by n matrix. After k $(1 \leq k \leq n)$ steps of Gaussian elimination, the matrix remaining to be factored is the trailing submatrix that consists of the elements in the last $(n-k)$ rows and the last

$(n-k)$ columns of A. The graph (bipartite, digraph, or undirected graph) associated with this $(n-k)$ by $(n-k)$ trailing matrix is the k th **elimination graph**.

Facts:

1. [PM83] Let $H^{(0)}, H^{(1)}, H^{(2)}, \cdots, H^{(n)}$ be the sequence of elimination bigraphs associated with the Gaussian elimination of a sparse nonsymmetric matrix $A = [a_{i,j}]$. The initial bigraph $H^{(0)}$ is the bigraph of A. Each elimination bigraph can be obtained from the previous one through a simple transformation. Suppose that the nonzero element $a_{s,t}$ is chosen as the pivot at step k, $1 \leq k \leq n$. Then the edge corresponding to $a_{s,t}$, $\{r_s, c_t\}$, is removed from $H^{(k-1)}$, together with all the edges incident to r_s and c_t. To obtain the next elimination bigraph $H^{(k)}$ from the modified $H^{(k-1)}$, an edge $\{r, c\}$ is added if there is a path (r, c_t, r_s, c) in the original $H^{(k-1)}$ and if $\{r, c\}$ is not already in $H^{(k-1)}$. The new edges added to create $H^{(k)}$ correspond to the fill elements introduced when $a_{s,t}$ is used to eliminate row s and column t. The bigraph $H^{(k)}$ represents the sparsity structure of the matrix remaining to be factored after row s and column t are eliminated.
2. [RT78] Let A be a sparse nonsymmetric matrix. Suppose that A has a zero-free diagonal and assume that pivots are restricted to the diagonal. Then Gaussian elimination can be modeled using elimination digraphs. Let $\Gamma^{(0)}, \Gamma^{(1)}, \Gamma^{(2)}, \cdots, \Gamma^{(n)}$ be the sequence of elimination digraphs. The initial digraph $\Gamma^{(0)}$ is the digraph of A. Each elimination digraph can be obtained from the previous one through a simple transformation. Without loss of generality, assume that Gaussian elimination proceeds from row/column 1 to row/column n. Consider step k, $1 \leq k \leq n$. Vertex x_k is removed from $\Gamma^{(k-1)}$, together with all the arcs of the form (x_k, u) and (v, x_k), where u and v are vertices in $\Gamma^{(k-1)}$. To obtain the next elimination digraph $\Gamma^{(k)}$ from the modified $\Gamma^{(k-1)}$, an arc (r, c) is added if there is a directed path (r, x_k, c) in the original $\Gamma^{(k-1)}$ and if (r, c) is not already in $\Gamma^{(k-1)}$. The new arcs added to create $\Gamma^{(k)}$ correspond to the fill elements introduced by Gaussian elimination at step k. The digraph $\Gamma^{(k)}$ represents the sparsity structure of the matrix remaining to be factored after k steps of Gaussian elimination.
3. [Ros72] For a symmetric positive definite matrix A, the elimination graphs associated with Gaussian elimination can be represented by (undirected) graphs. Let the elimination graphs be denoted by $G^{(0)}, G^{(1)}, G^{(2)}, \cdots$, and $G^{(n)}$. The initial graph $G^{(0)}$ is the graph of A. Each elimination graph can be obtained from the previous one through a simple transformation. Without loss of generality, assume that Gaussian elimination proceeds from row/column 1 to row/column n. Consider step k, $1 \leq k \leq n$. Vertex x_k is removed from $G^{(k-1)}$, together with all its incident edges. To obtain the next elimination graph $G^{(k)}$ from the modified $G^{(k-1)}$, an (undirected) edge $\{r, c\}$ is added if there is a path (r, x_k, c) in the original $G^{(k-1)}$ and if $\{r, c\}$ is not already in $G^{(k-1)}$. The new edges added to create $G^{(k)}$ correspond to the fill elements introduced by Gaussian elimination at step k. The graph $G^{(k)}$ represents the sparsity structure of the matrix remaining to be factored after k steps of Gaussian elimination.
4. [GL80b], [RTL76] Consider an n by n symmetric positive definite matrix A. Let $G = (X, E)$ be the (undirected) graph of A and let $X = \{x_1, x_2, \cdots, x_n\}$. Denote the Cholesky factor of A by L. For $i > j$, the (i, j) element of L is nonzero if and only there is a path $(x_i, x_{s_1}, x_{s_2}, \cdots, x_{s_t}, x_j)$ in G such that $s_p < j\ (< i)$, for $1 \leq p \leq t$. Such a path is sometimes referred to as a *fill path*. Note that t can be zero, which corresponds to $\{x_i, x_j\} \in E$.

Examples:

1. As one may observe, the operations involved in the generation of the elimination bigraphs, digraphs, and graphs are very similar. Thus, only one example will be illustrated here. Consider the nonsymmetric matrix $A = [a_{i,j}]$ in Figure 40.3. The bigraph H of A is depicted in Figure 40.4. Suppose that $a_{1,1}$ is used to elimination row 1 and column 1. After the first step of Gaussian elimination, the remaining matrix is shown in Figure 40.5. Following the recipe given above, the edges $\{r_1, c_1\}$, $\{r_1, c_3\}$, $\{r_1, c_6\}$, $\{r_4, c_1\}$, and $\{r_7, c_1\}$ are to be removed from the bigraph H in Figure 40.4. Then r_1

$$A = \begin{bmatrix} \times & & \times & & & \times & \\ & & & \times & & & \\ & & \times & & & & \\ \times & & & \times & \times & & \\ & \times & & & & & \times \\ & & \times & & & \times & \times \\ \times & & & & \times & & \end{bmatrix}.$$

FIGURE 40.3 A sparse nonsymmetric matrix A.

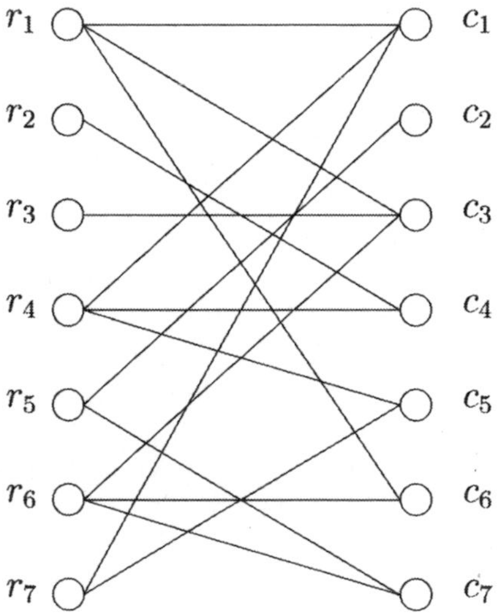

FIGURE 40.4 The bigraph of the matrix A in Figure 40.3.

$$A' = \begin{bmatrix} & & \times & & & \\ & \times & & & & \\ & + & \times & \times & + & \\ \times & & & & & \times \\ & \times & & & \times & \times \\ & + & & \times & + & \end{bmatrix}.$$

FIGURE 40.5 The remaining matrix after the first step of Gaussian elimination on the matrix A in Figure 40.3.

and c_1 are also removed from the bigraph H. The new bigraph is obtained by adding to H the following edges:

(a) $\{r_4, c_3\}$ (because of the path (r_4, c_1, r_1, c_3) in the original H).
(b) $\{r_4, c_6\}$ (because of the path (r_4, c_1, r_1, c_6) in the original H).
(c) $\{r_7, c_3\}$ (because of the path (r_7, c_1, r_1, c_3) in the original H).
(d) $\{r_7, c_6\}$ (because of the path (r_7, c_1, r_1, c_6) in the original H).

The new bigraph is shown in Figure 40.6, in which the new edges are shown as dashed lines. Note that the new bigraph is exactly the bigraph of A'.

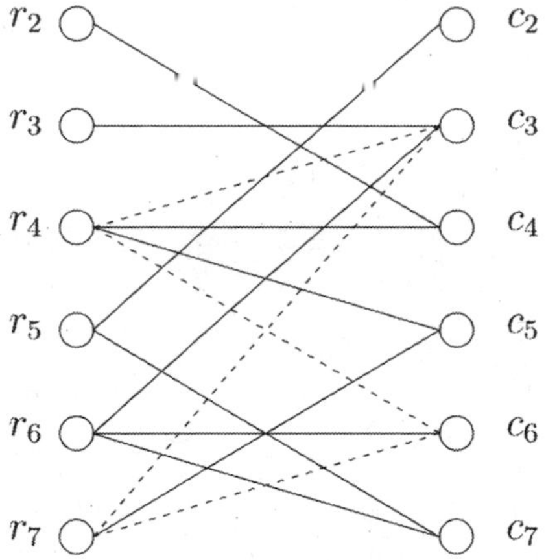

FIGURE 40.6 The bigraph of the matrix A' in Figure 40.5.

2. Let $A = [a_{i,j}]$ be a symmetric and positive definite matrix:

$$A = \begin{bmatrix} \times & & \times & & & & & \times \\ & \times & \times & & \times & & & \\ \times & \times & \times & & & & & \\ & & & \times & \times & & \times & \\ & \times & & \times & \times & \times & & \\ & & & & \times & \times & & \times \\ & & & \times & & & \times & \\ \times & & & & & \times & & \times \end{bmatrix}.$$

Assume that exact numerical cancellations do not occur. Then the sparsity structure of the Cholesky factor $L = [\ell_{i,j}]$ of A is given below:

$$L = \begin{bmatrix} \times & & & & & & & \\ & \times & & & & & & \\ \times & \times & \times & & & & & \\ & & & \times & & & & \\ & \times & + & \times & \times & & & \\ & & & & \times & \times & & \\ & & & \times & + & + & \times & \\ \times & & + & & + & \times & + & \times \end{bmatrix}.$$

The symbol $+$ represents a fill element. The (undirected) graph G of A is shown in Figure 40.7. There are several fill paths in G. The fill path $(x_5, x_2, x_3, x_1, x_8)$ corresponds to the fill element $\ell_{8,5}$ in L. Another example is the fill path (x_7, x_4, x_5, x_6), which corresponds to the fill element $\ell_{7,6}$ in L.

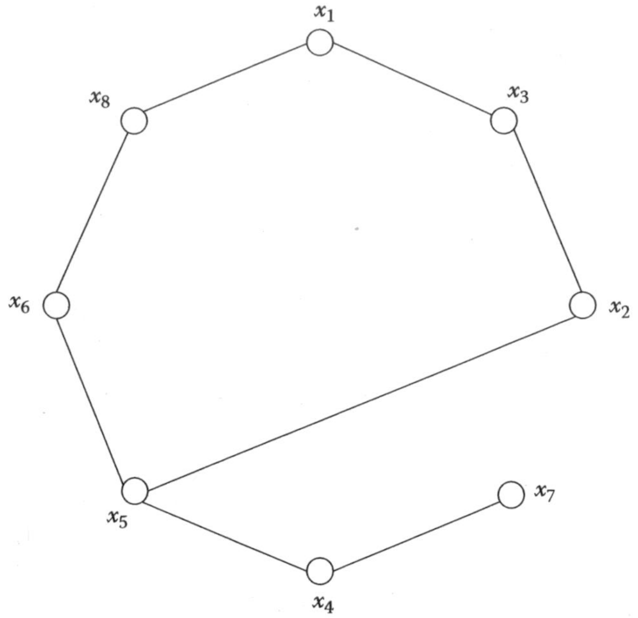

FIGURE 40.7 An example illustrating fill paths.

40.5 Effect of Reorderings

Let A be a sparse nonsingular matrix. As noted above, the occurrence of fill elements in Gaussian elimination generally depends on the values of the nonzero elements in A (which affect the choice of pivots if numerical stability is a concern) and the sparsity structure of A. This section will consider some techniques that will help preserve the sparsity structure in the factorization of A.

Definitions:

Let A be a sparse matrix. Suppose that G is a graph associated with A as described in the previous section; the graph can be a bipartite graph, directed graph, or undirected graph, depending on whether A is nonsymmetric or symmetric and whether the pivots are chosen to be on the diagonal. The **fill graph** G^F of A is G, together with all the additional edges corresponding to the fill elements that occur during Gaussian elimination.

An **elimination ordering** (or **elimination sequence**) for the rows (or columns) of a matrix is a bijection $\alpha : \{1, 2, \cdots, n\} \to \{1, 2, \cdots, n\}$. It specifies the order in which the rows (or columns) of the matrix are eliminated during Gaussian elimination.

A **perfect elimination ordering** is an elimination ordering that does not produce any fill elements during Gaussian elimination.

Consider an n by n sparse matrix A. Let f_i and ℓ_i be the column indices of the *first* and *last* nonzero elements in row i of A, respectively. The **envelope** of A is the set

$$\{(i, j) : f_i \leq j \leq \ell_i, \quad \text{for } 1 \leq i \leq n\}.$$

That is, all the elements between the first and last nonzero elements of every row are in the envelope. The **profile** of a matrix is the number of elements in the envelope.

Facts:

The problem of reordering a sparse matrix is combinatorial in nature. The facts stated below are some of fundamental ones. Others can be found, for example, in [GL81] and [Gol04].

1. [GL81] An elimination ordering for the rows/columns of the matrix corresponds to a permutation of the rows/columns.
2. [DER89], [GL81] The choice of an elimination ordering will affect the number of fill elements in the triangular factors.
3. [GL81], [Ros72] When A is sparse symmetric positive definite, an elimination ordering α can be determined by analyzing the sparsity structure of A. Equivalently, α can be obtained by analyzing the sequence of elimination graphs. The elimination ordering provides a symmetric permutation of A. Let P denote the permutation matrix corresponding to α. Let G^F be the fill graph of the matrix PAP^T. There exists a perfect elimination ordering for G^F and G^F is a chordal graph.
4. [Yan81] For sparse symmetric positive definite matrices, finding the optimal elimination ordering (i.e., an elimination ordering that minimizes the number of fill elements in the Cholesky factor) is NP-complete. This implies that almost all reordering techniques are heuristic in nature.
5. [DER89] When A is a sparse nonsymmetric matrix, the elimination orderings for the rows and columns of A have to be chosen to preserve the sparsity structure and to maintain numerical stability.
6. [GN85] Suppose that A is a sparse nonsymmetric matrix. If partial pivoting (by rows) is used to maintain numerical stability, then an elimination ordering for the columns of A can be chosen to preserve the sparsity structure.

Examples:

1. Consider the following two diagonally dominant matrices:

$$A = \begin{bmatrix} 7 & 1 & -1 & 1 & -1 \\ -1 & 7 & 0 & 0 & 0 \\ 1 & 0 & 7 & 0 & 0 \\ -1 & 0 & 0 & 7 & 0 \\ 1 & 0 & 0 & 0 & 7 \end{bmatrix} \quad \text{and} \quad B = \begin{bmatrix} 7 & 0 & 0 & 0 & 1 \\ 0 & 7 & 0 & 0 & -1 \\ 0 & 0 & 7 & 0 & 1 \\ 0 & 0 & 0 & 7 & -1 \\ -1 & 1 & -1 & 1 & 7 \end{bmatrix}.$$

Applying Gaussian elimination to A produces the following triangular factors:

$$A = \begin{bmatrix} 1 & 0 & 0 & 0 & 0 \\ -\frac{1}{7} & 1 & 0 & 0 & 0 \\ \frac{1}{7} & -\frac{1}{50} & 1 & 0 & 0 \\ -\frac{1}{7} & \frac{1}{50} & -\frac{1}{51} & 1 & 0 \\ \frac{1}{7} & -\frac{1}{50} & \frac{1}{51} & -\frac{1}{52} & 0 \end{bmatrix} \begin{bmatrix} 7 & 1 & -1 & 1 & -1 \\ 0 & \frac{50}{7} & -\frac{1}{7} & \frac{1}{7} & -\frac{1}{7} \\ 0 & 0 & \frac{357}{50} & -\frac{7}{50} & \frac{7}{50} \\ 0 & 0 & 0 & \frac{364}{51} & -\frac{7}{51} \\ 0 & 0 & 0 & 0 & \frac{371}{52} \end{bmatrix}.$$

Applying Gaussian elimination to B, on the other hand, produces the following triangular factors:

$$B = \begin{bmatrix} 1 & 0 & 0 & 0 & 0 \\ 0 & 1 & 0 & 0 & 0 \\ 0 & 0 & 1 & 0 & 0 \\ 0 & 0 & 0 & 1 & 0 \\ -\frac{1}{7} & \frac{1}{7} & -\frac{1}{7} & \frac{1}{7} & 1 \end{bmatrix} \begin{bmatrix} 7 & 0 & 0 & 0 & 1 \\ 0 & 7 & 0 & 0 & -1 \\ 0 & 0 & 7 & 0 & 1 \\ 0 & 0 & 0 & 7 & -1 \\ 0 & 0 & 0 & 0 & \frac{53}{7} \end{bmatrix}.$$

The two matrices A and B have the *same* numbers of nonzero elements, but their respective triangular factors have *very different* numbers of fill elements. In fact, the triangular factors of B

have *no* fill elements. Note that B can be obtained by permuting the rows and columns of A. Let P be the following permutation matrix:

$$P = \begin{bmatrix} 0 & 0 & 0 & 0 & 1 \\ 0 & 0 & 0 & 1 & 0 \\ 0 & 0 & 1 & 0 & 0 \\ 0 & 1 & 0 & 0 & 0 \\ 1 & 0 & 0 & 0 & 0 \end{bmatrix}.$$

Then $B = PAP^T$; that is, B is obtained by *reversing* the order of the rows and columns. This example illustrates that permuting the rows and columns of a sparse matrix may have a drastic effect on the sparsity structures of the triangular factors in Gaussian elimination.

2. A popular way to preserve the sparsity structure of a sparse matrix during Gaussian elimination is to find elimination orderings so that the nonzero elements in the permuted matrix are near the diagonal. This can be accomplished, for example, by permuting the matrix so that it has a small envelope or profile. The Cuthill–McKee algorithm [CM69] and the reverse Cuthill–McKee algorithms [Geo71], [LS76] are well-known heuristics for producing reorderings that reduce the profile of a sparse symmetric matrix. The permuted matrix can be factored using the *envelope* or *profile* method [GL81], which is similar to the factorization methods for band matrices. The storage scheme for the profile method is very simple. The permuted matrix is stored by rows. If the symmetric matrix is positive definite, then for every row of the permuted matrix, all the elements between the first nonzero element and the diagonal elements are stored. It is easy to show that the fill elements in the triangular factor can only occur inside the envelope of the lower triangular part of the permuted matrix.
3. Although the profile method is easy to implement, it is not designed to reduce the number of fill elements in Gaussian elimination. The nested dissection algorithm [Geo73], which is based on a divide-and-conquer idea, is a well-known heuristic for preserving the sparsity structure. Let A be a symmetric positive definite matrix, and let $G = (X, E)$ be the (undirected) graph of A. Without loss of generality, assume that G is connected. Let $S \subseteq X$. Suppose that S is removed from G. Also assume that all edges incident to every vertex in S are removed from G. Denote the remaining graph by $G(X - S)$. If S is chosen so that $G(X - S)$ contains one or more disconnected components, then the set S is referred to as a *separator*. Consider the following reordering strategy: renumber the vertices of the disconnected components of $G(X - S)$ first, and renumber of the vertices of S last. Now pick vertex x in one component and vertex y in another component. The renumbering scheme ensures that there is *no* fill path between x and y in G. This is a heuristic way to limit the creation of fill elements. The renumbering corresponds to a symmetric permutation of the rows and columns of the matrix. Consider the example in Figure 40.8. The removal of the vertices in S (together with the incident edges) divides the mesh into two joint meshes (labeled G_1 and G_2). Suppose that the mesh points in G_1 are renumbered before those in G_2. Then the matrix on the right in Figure 40.8 shows the sparsity structure of the permuted matrix. Note the block structure of the permuted matrix. The blocks A_1 and A_2 correspond to mesh points in G_1 and G_2, respectively. The block A_S corresponds to the mesh points in S. The nonzero elements in the off-diagonal blocks C_1/C_2 correspond to the edges between G_1/G_2 and S. The unlabeled blocks are entirely zero. This is referred to as the "dissection" strategy. When the strategy is applied recursively to the disconnected components, more zero (but smaller) blocks will be created in A_1 and A_2. The resulting reordering is called a "nested dissection" ordering. It can be shown that, when the nested dissection algorithm is applied to a k by k mesh (like the example in the Introduction), the number of nonzero elements in the Cholesky factor will be $O(k^2 \log k)$ and the number of operations required to compute the Cholesky factor will be $O(k^3)$. Incidentally, for the k by k mesh, it has been proved that the number of nonzero elements in the Cholesky factor and the number of operations required to compute the triangular factorization are at least $O(k^2 \log k)$ and $O(k^3)$, respectively [HMR73], [Geo73]. Thus, nested

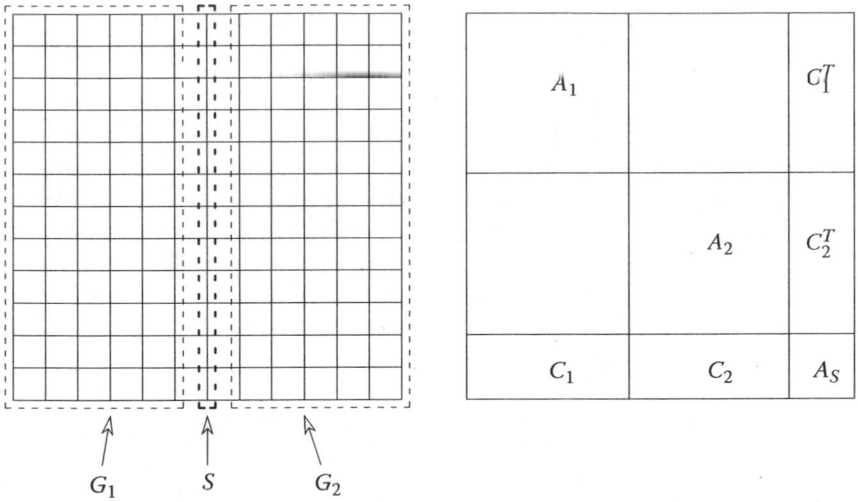

FIGURE 40.8 An example of the dissection strategy.

dissection orderings can be optimal asymptotically. In recent years, higher quality nested dissection orderings have been obtained for general sparse symmetric matrices by using more sophisticated graph partitioning techniques to generate the separators [HR98], [PSL90], [Sch01].

4. The nested dissection algorithm is a "top-down" algorithm since it identifies the vertices (i.e., rows/columns) to be reordered last. The minimum degree algorithm [TW67] is a "bottom-up" algorithm. It is a heuristic that is best described using the elimination graphs. Let $G^{(0)}$ be the (undirected) graph of a sparse symmetric positive definite matrix. The minimum degree algorithm picks the vertex x_m with the *smallest* degree (i.e., the smallest number of incident edges) to be eliminated; that is, x_m is to be reordered as the first vertex. Then x_m, together with all the edges incident to x_m, are eliminated from $G^{(0)}$ to generate the next elimination graph $G^{(1)}$. This process is repeated until all the vertices are eliminated. The order in which the vertices are eliminated is a minimum degree ordering. Note that if several vertices have the minimum degree in the current elimination graph, then ties have to be broken. It is well known that the quality of a minimum degree ordering can be influenced by the choice of the tie-breaking strategy [BS90]. Several efficient implementations of the minimum degree algorithm are available [ADD96], [GL80b], [GL80a], [Liu85]. An excellent survey of the minimum degree algorithm can be found in [GL89].
5. The minimum deficiency algorithm [TW67] is another bottom-up strategy. It is similar to the minimum degree algorithm, except that the vertex whose elimination would introduce the *fewest* fill elements will be eliminated at each step. In general, the minimum deficiency algorithm is much more expensive to implement than the minimum degree algorithm. This is because the former one needs the *look-ahead* to predict the number of fill elements that would be introduced. However, inexpensive approximations to the minimum deficiency algorithms have been proposed recently [NR99], [RE98].
6. For a sparse nonsymmetric matrix $A = [a_{i,j}]$, there are analogs of the minimum degree and minimum deficiency algorithms. The Markowitz scheme [Mar57] is the nonsymmetric version of the minimum degree algorithm for sparse nonsymmetric matrices. Recall that $nnz(A_{i,1:n})$ is the number of nonzero elements in row i of A and $nnz(A_{1:n,j})$ is the number of nonzero elements in column j of A. For each nonzero $a_{i,j}$ in A, define its "Markowitz" cost to be the product $[nnz(A_{i,1:n}) - 1][nna(A_{1:n,j}) - 1]$, which would be the number of nonzero elements in the rank-1 update if $a_{i,j}$ were chosen as pivot. At each step of Gaussian elimination, the nonzero element that has the *smallest* Markowitz cost will be chosen as the pivot. After the elimination, the Markowitz costs of all the nonzero elements, including the fill elements, are updated to reflect the change in

the sparsity structure before proceeding to the next step. If A is symmetric and positive definite, and if pivots are chosen from the diagonal, then the Markowitz scheme is the same as the minimum degree algorithm.

7. The Markowitz scheme for sparse nonsymmetric matrices attempts to preserve the sparsity structure by minimizing the number of nonzero elements introduced into the triangular factors at each step of Gaussian elimination. The resulting pivots may not lead to a numerically stable factorization. For example, the magnitude of a pivot may be too small compared to the magnitudes of the other nonzero elements in the matrix. To enhance numerical stability, a modified Markowitz scheme is often used. Denote the matrix to be factored by $A = [a_{i,j}]$. Let $s = \max\{|a_{i,j}| : a_{i,j} \neq 0\}$. Let τ be a given tolerance; e.g., τ can be 0.01 or 0.001. Without loss of generality, consider the first step of Gaussian elimination of A. Instead of considering all nonzero elements in A, let $C = \{a_{i,j} : |a_{i,j}| \geq \tau\, s\}$. Thus, C is the subset of nonzero elements whose magnitudes are larger than or equal to $\tau\, s$; it is the set of candidate pivots. Then the pivot search is limited to applying the Markowitz scheme to the nonzero elements in C. This is a compromise between preserving sparsity and maintaining numerical stability. The two parameters τ and s are usually fixed throughout the entire Gaussian elimination process. The modified scheme is often referred to as the *threshold pivoting* scheme [Duf77].
8. As noted earlier, if Gaussian elimination with partial pivoting is used to factor a sparse nonsymmetric matrix A, then the columns may be permuted to preserve sparsity. Suppose that A has a zero-free diagonal, and let L_C be the Cholesky factor of the symmetric positive definite matrix $A^T A$. Since $\text{Struct}(L) \subseteq \text{Struct}(L_C)$ and $\text{Struct}(U) \subseteq \text{Struct}(L_C^T)$, a possibility is to make sure that L_C is sparse. In other words, one can choose a permutation P_{L_C} for $A^T A$ to reduce the number of nonzero elements in the Cholesky factor of $P_{L_C}^T(A^T A)P_{L_C}$. Note that $P_{L_C}^T(A^T A)P_{L_C} = (AP_{L_C})^T(AP_{L_C})$. Thus, P_{L_C} can be applied to the columns of A [GN85], [GN87].

Author Note

This work was supported by the Director, Office of Science, U.S. Department of Energy under contract no. DE-AC03-76SF00098.

References

[ADD96] Patrick R. Amestoy, Timothy A. Davis, and Iain S. Duff. An approximate minimum degree ordering algorithm. *SIAM J. Matrix Anal. Appl.*, 17(4):886–905, 1996.

[AGL98] Cleve Ashcraft, Roger G. Grimes, and John G. Lewis. Accurate symmetric indefinite linear equation solvers. *SIAM J. Matrix Anal. Appl.*, 20(2):513–561, 1998.

[BS90] Piotr Berman and Georg Schnitger. On the performance of the minimum degree ordering for Gaussian elimination. *SIAM J. Matrix Anal. Appl.*, 11(1):83–88, 1990.

[CM69] E. Cuthill and J. McKee. Reducing the bandwidth of sparse symmetric matrices. In *Proceedings of the 24th ACM National Conference*, pp. 157–172. ACM, Aug. 1969.

[DEG+99] James W. Demmel, Stanley C. Eisenstat, John R. Gilbert, Xiaoye S. Li, and Joseph W.H. Liu. A supernodal approach to sparse partial pivoting. *SIAM J. Matrix Anal. Appl.*, 20(3):720–755, 1999.

[DER89] I.S. Duff, A.M. Erisman, and J.K. Reid. *Direct Methods for Sparse Matrices.* Oxford University Press, Oxford, 1989.

[DR83] I.S. Duff and J.K. Reid. The multifrontal solution of indefinite sparse symmetric linear equations. *ACM Trans. Math. Software*, 9(3):302–325, 1983.

[DR95] I.S. Duff and J.K. Reid. MA47, a Fortran code for direct solution of indefinite sparse symmetric linear systems. Technical Report RAL 95-001, Rutherford Appleton Laboratory, Oxfordshire, U.K., 1995.

[DR96] I.S. Duff and J.K. Reid. The design of MA48: A code for the direct solution of sparse unsymmetric linear systems of equations. *ACM Trans. Math. Software*, 22(2):187–226, 1996.

[Duf77] I.S. Duff. MA28 — a set of Fortran subroutines for sparse unsymmetric linear equations. Technical Report AERE R-8730, Harwell, Oxfordshire, U.K., 1977.

[Duf81] I.S. Duff. On algorithms for obtaining a maximum transversal. *ACM Trans. Math. Software*, 7(3):315–330, 1981.

[Duf04] I.S. Duff. MA57 — a code for the solution of sparse symmetric definite and indefinite systems. *ACM Trans. Math. Software*, 30(2):118–144, 2004.

[DW88] I. S. Duff and Torbjörn Wiberg. Remarks on implementation of $O(n^{\frac{1}{2}}\tau)$ assignment algorithms. *ACM Trans. Math. Software*, 14(3):267–287, 1988.

[EL92] Stanley C. Eisenstat and Joseph W.H. Liu. Exploiting structural symmetry in unsymmetric sparse symbolic factorization. *SIAM J. Matrix Anal. Appl.*, 13(1):202–211, 1992.

[Geo71] John Alan George. *Computer Implementation of the Finite Element Method.* Ph.D. thesis, Dept. of Computer Science, Stanford University, CA, 1971.

[Geo73] Alan George. Nested dissection of a regular finite element mesh. *SIAM J. Numer. Anal.*, 10(2):345–363, 1973.

[GL80a] Alan George and Joseph W.H. Liu. A fast implementation of the minimum degree algorithm using quotient graphs. *ACM Trans. Math. Software*, 6(3):337–358, 1980.

[GL80b] Alan George and Joseph W.H. Liu. A minimal storage implementation of the minimum degree algorithm. *SIAM J. Numer. Anal.*, 17(2):282–299, 1980.

[GL81] Alan George and Joseph W-H. Liu. *Computer Solution of Large Sparse Positive Definite Systems.* Prentice-Hall, Upper Saddle River, NJ, 1981.

[GL89] Alan George and Joseph W.H. Liu. The evolution of the minimum degree ordering algorithm. *SIAM Review*, 31(1):1–19, 1989.

[GL93] John R. Gilbert and Joseph W.H. Liu. Elimination structures for unsymmetric sparse lu factors. *SIAM J. Matrix Anal. Appl.*, 14(2):334–352, 1993.

[GN85] Alan George and Esmond Ng. An implementation of Gaussian elimination with partial pivoting for sparse systems. *SIAM J. Sci. Stat. Comput.*, 6(2):390–409, 1985.

[GN87] Alan George and Esmond Ng. Symbolic factorization for sparse Gaussian elimination with partial pivoting. *SIAM J. Sci. Stat. Comput.*, 8(6):877–898, 1987.

[GN93] John R. Gilbert and Esmond G. Ng. Predicting structure in nonsymmetric sparse matrix factorizations. In Alan George, John R. Gilbert, and Joseph W.H. Liu, Eds., *Graph Theory and Sparse Matrix Computation*, vol. IMA #56, pp. 107–140. Springer-Verlag, Heidelberg, 1993.

[GNP94] John R. Gilbert, Esmond G. Ng, and Barry W. Peyton. An efficient algorithm to compute row and column counts for sparse Cholesky factorization. *SIAM J. Matrix Anal. Appl.*, 15(4):1075–1091, 1994.

[Gol04] M.C. Golumbic. *Algorithmic Graph Theory and Perfect Graphs*, vol. 57 *Annuals of Discrete Mathematics.* North Holland, NY, 2nd ed., 2004.

[GP88] John R. Gilbert and Tim Peierls. Sparse partial pivoting in time proportional to arithmetic operations. *SIAM J. Sci. Stat. Comput.*, 9(5):862–874, 1988.

[HMR73] Alan J. Hoffman, Michael S. Martin, and Donald J. Rose. Complexity bounds for regular finite difference and finite element grids. *SIAM J. Numer. Anal.*, 10(2):364–369, 1973.

[HR98] Bruce Hendrickson and Edward Rothberg. Improving the run time and quality of nested dissection ordering. *SIAM J. Sci. Comput.*, 20(2):468–489, 1998.

[Liu85] Joseph W.H. Liu. Modification of the minimum-degree algorithm by multiple elimination. *ACM Trans. Math. Software*, 11(2):141–153, 1985.

[Liu86] Joseph W. Liu. A compact row storage scheme for Cholesky factors using elimination trees. *ACM Trans. Math. Software*, 12(2):127–148, 1986.

[Liu87a] Joseph W.H. Liu. On threshold pivoting in the multifrontal method for sparse indefinite systems. *ACM Trans. Math. Software*, 13(3):250–261, 1987.

[Liu87b] Joseph W.H. Liu. A partial pivoting strategy for sparse symmetric matrix decomposition. *ACM Trans. Math. Software,* 13(2):173–182, 1987.

[Liu90] Joseph W.H. Liu. The role of elimination trees in sparse factorization. *SIAM J. Matrix Anal. Appl.*, 11(1):134–172, 1990.

[LS76] Wai-Hung Liu and Andrew H. Sherman. Comparative analysis of the Cuthill–McKee and reverse Cuthill–McKee ordering algorithms for sparse matrices. *SIAM J. Numer. Anal.*, 13(2):198–213, 1976.

[Mar57] H.M. Markowitz. The elimination form of the inverse and its application to linear programming. *Management Sci.*, 3(3):255–269, 1957.

[NP93] Esmond G. Ng and Barry W. Peyton. Block sparse Cholesky algorithms on advanced uniprocessor computers. *SIAM J. Sci. Comput.*, 14(5):1034–1056, 1993.

[NR99] Esmond G. Ng and Padma Raghavan. Performance of greedy ordering heuristics for sparse Cholesky factorization. *SIAM J. Matrix Anal. Appl.*, 20(4):902–914, 1999.

[PM83] G. Pagallo and C. Maulino. A bipartite quotient graph model for unsymmetric matrices. In V. Pereyra and A. Reinoza, Eds., *Numerical Methods*, vol. 1005, *Lecture Notes in Mathematics*, pp. 227–239. Springer-Verlag, Heidelberg, 1983.

[PSL90] Alex Pothen, Horst D. Simon, and Kang-Pu Liou. Partitioning sparse matrices with eigenvectors of graphs. *SIAM J. Matrix Anal. Appl.*, 11(3):430–452, 1990.

[RE98] Edward Rothberg and Stanley C. Eisenstat. Node selection strategies for bottom-up sparse matrix ordering. *SIAM J. Matrix Anal. Appl.*, 19(3):682–695, 1998.

[RG91] Edward Rothberg and Anoop Gupta. Efficient sparse matrix factorization on high-performance workstations — exploiting the memory hierarchy. *ACM Trans. Math. Software,* 17:313–334, 1991.

[Ros72] D.J. Rose. A graph-theoretic study of the numerical solution of sparse positive definite systems of linear equations. In R.C. Read, Ed., *Graph Theory and Computing*, pp. 183–217. Academic Press, New York, 1972.

[RT78] Donald J. Rose and Robert Endre Tarjan. Algorithmic aspects of vertex elimination of directed graphs. *SIAM J. Appl. Math.*, 34(1):176–197, 1978.

[RTL76] Donald J. Rose, R. Endre Tarjan, and George S. Lueker. Algorithmic aspects of vertex elimination on graphs. *SIAM J. Comput.*, 5(2):266–283, 1976.

[Sch82] Robert Schreiber. A new implementation of sparse Gaussian elimination. *ACM Trans. Math. Software,* 8(3):256–276, 1982.

[Sch01] Jürgen Schulze. Towards a tighter coupling of bottom-up and top-down sparse matrix ordering methods. *BIT Num. Math.*, 41(4):800–841, 2001.

[TW67] W.F. Tinney and J.W. Walker. Direct solutions of sparse network equations by optimally ordered triangular factorization. *Proc. IEEE,* 55:1801–1809, 1967.

[Yan81] Mihalis Yannakakis. Computing the minimum fill-in is NP-complete. *SIAM J. Alg. Disc. Meth.*, 2(1):77–79, 1981.

41

Iterative Solution Methods for Linear Systems

Anne Greenbaum
University of Washington

Given an n by n nonsingular matrix A and an n-vector $\mathbf{b}$, the linear system $A\mathbf{x} = \mathbf{b}$ can always be solved for $\mathbf{x}$ by Gaussian elimination. The work required is approximately $2n^3/3$ operations (additions, subtractions, multiplications, and divisions), and, in general, n^2 words of storage are required. This is often acceptable if n is of moderate size, say $n \leq 1000$, but for much larger values of n, say, $n \approx 10^6$, both the work and storage for Gaussian elimination may become prohibitive.

Where do such large linear systems arise? They may occur in many different areas, but one important source is the numerical solution of partial differential equations (PDEs). Solutions to PDEs can be approximated by replacing derivatives by finite difference quotients. For example, to solve the equation

$$\frac{\partial}{\partial x}\left(a(x,y,z)\frac{\partial u}{\partial x}\right) + \frac{\partial}{\partial y}\left(a(x,y,z)\frac{\partial u}{\partial y}\right) + \frac{\partial}{\partial z}\left(a(x,y,z)\frac{\partial u}{\partial z}\right) = f(x,y,z) \quad \text{in } \Omega$$

$$u = g \quad \text{on boundary of } \Omega,$$

on a three-dimensional region Ω, where a, f, and g are given functions with a bounded away from 0, one might first divide the region Ω into small subregions of width h in each direction, and then replace each partial derivative by a centered difference approximation; e.g.,

$$\frac{\partial}{\partial x}\left(a\frac{\partial u}{\partial x}\right)(x,y,z) \approx \frac{1}{h^2}[a(x+h/2,y,z)(u(x+h,y,z)-u(x,y,z)) \\ -a(x-h/2,y,z)(u(x,y,z)-u(x-h,y,z))],$$

with similar approximations for $\partial/\partial y(a\partial u/\partial y)$ and $\partial/\partial z(a\partial u/\partial z)$. If the resulting finite difference approximation to the differential operator is set equal to the right-hand side value $f(x_i, y_j, z_k)$ at each of the interior mesh points (x_i, y_j, z_k), $i = 1, \ldots, n_1$, $j = 1, \ldots, n_2$, $k = 1, \ldots, n_3$, then this gives a system of $n = n_1n_2n_3$ linear equations for the n unknown values of u at these mesh points. If u_{ijk} denotes the approximation to $u(x_i, y_j, z_k)$, then the equations are

$$\begin{aligned}
&\frac{1}{h^2}[a(x_i + h/2, y_j, z_k)(u_{i+1,j,k} - u_{ijk}) - a(x_i - h/2, y_j, z_k)(u_{ijk} - u_{i-1,j,k}) \\
&\quad + a(x_i, y_j + h/2, z_k)(u_{i,j+1,k} - u_{ijk}) - a(x_i, y_j - h/2, z_k)(u_{ijk} - u_{i,j-1,k}) \\
&\quad + a(x_i, y_j, z_k + h/2)(u_{i,j,k+1} - u_{ijk}) - a(x_i, y_j, z_k - h/2)(u_{ijk} - u_{i,j,k-1})] \\
&\quad = f(x_i, y_j, z_k).
\end{aligned}$$

The formula must be modified near the boundary of the region, where known boundary values are added to the right-hand side. Still the result is a system of linear equations for the unknown interior values of u. If $n_1 = n_2 = n_3 = 100$, then the number of equations and unknowns is $100^3 = 10^6$.

Notice, however, that the system of linear equations is *sparse*; each equation involves only a few (in this case seven) of the unknowns. The actual form of the system matrix A depends on the numbering of equations and unknowns. Using the *natural ordering*, equations and unknowns are ordered first by i, then j, then k. The result is a *banded* matrix, whose bandwidth is approximately n_1n_2, since unknowns in any z plane couple only to those in the same and adjacent z planes. This results in some savings for Gaussian elimination. Only entries inside the band need be stored because these are the only ones that *fill in* (become nonzero, even if originally they were zero) during the process. The resulting work is about $2(n_1n_2)^2n$ operations, and the storage required is about n_1n_2n words. Still, this is too much when $n_1 = n_2 = n_3 = 100$. Different orderings can be used to further reduce fill in, but another option is to use iterative methods.

Because the matrix is so sparse, matrix-vector multiplication is very cheap. In the above example, the product of the matrix with a given vector can be accomplished with just $7n$ multiplications and $6n$ additions. The nonzeros of the matrix occupy only $7n$ words and, in this case, they are so simple that they hardly need be stored at all. If the linear system $Ax = b$ could be solved iteratively, using only matrix-vector multiplication and, perhaps, solution of some much simpler linear systems such as diagonal or sparse triangular systems, then a tremendous savings might be achieved in both work and storage.

This section describes how to solve such systems iteratively. While iterative methods are appropriate for sparse systems like the one above, they also may be useful for *structured systems*. If matrix-vector multiplication can be performed rapidly, and if the structure of the matrix is such that it is not necessary to store the entire matrix but only certain parts or values in order to carry out the matrix-vector multiplication, then iterative methods may be faster and require less storage than Gaussian elimination or other methods for solving $A\mathbf{x} = \mathbf{b}$.

41.1 Krylov Subspaces and Preconditioners

Definitions:

An **iterative method** for solving a linear system $A\mathbf{x} = \mathbf{b}$ is an algorithm that starts with an initial guess $\mathbf{x_0}$ for the solution and successively modifies that guess in an attempt to obtain improved approximate solutions $\mathbf{x_1}, \mathbf{x_2}, \ldots$.

The **residual** at step k of an iterative method for solving $A\mathbf{x} = \mathbf{b}$ is the vector $\mathbf{r_k} \equiv \mathbf{b} - A\mathbf{x_k}$, where $\mathbf{x_k}$ is the approximate solution generated at step k. The initial residual is $\mathbf{r_0} \equiv \mathbf{b} - A\mathbf{x_0}$, where $\mathbf{x_0}$ is the initial guess for the solution.

The **error** at step k is the difference between the true solution $A^{-1}\mathbf{b}$ and the approximate solution $\mathbf{x_k}$: $\mathbf{e_k} \equiv A^{-1}\mathbf{b} - \mathbf{x_k}$.

A **Krylov space** is a space of the form span$\{\mathbf{q}, A\mathbf{q}, A^2\mathbf{q}, \ldots, A^{k-1}\mathbf{q}\}$, where A is an n by n matrix and $\mathbf{q}$ is an n-vector. This space will be denoted as $K_k(A, \mathbf{q})$.

A **preconditioner** is a matrix M designed to improve the performance of an iterative method for solving the linear system $A\mathbf{x} = \mathbf{b}$. Linear systems with coefficient matrix M should be easier to solve than the original linear system, since such systems will be solved at each iteration.

The matrix $M^{-1}A$ (for **left preconditioning**) or AM^{-1} (for **right preconditioning**) or $L^{-1}AL^{-*}$ (for **Hermitian preconditioning**, when $M = LL^*$) is sometimes referred to as the **preconditioned iteration matrix**.

Another name for a preconditioner is a **splitting**; that is, if A is written in the form $A = M - N$, then this is referred to as a splitting of A, and iterative methods based on this splitting are equivalent to methods using M as a preconditioner.

A **regular splitting** is one for which M is nonsingular with $M^{-1} \geq 0$ (elementwise) and $M \geq A$ (elementwise).

Facts:

The following facts and general information on Krylov spaces and precondtioners can be found, for example, in [Axe95], [Gre97], [Hac94], [Saa03], and [Vor03].

1. An iterative method may obtain the exact solution at some stage (in which case it might be considered a direct method), but it may still be thought of as an iterative method because the user is interested in obtaining a good approximate solution before the exact solution is reached.
2. Each iteration of an iterative method usually requires one or more matrix-vector multiplications, using the matrix A and possibly its Hermitian transpose A^*. An iteration may also require the solution of a preconditioning system $M\mathbf{z} = \mathbf{r}$.
3. The residual and error vector at step k of an iterative method are related by $\mathbf{r_k} = A\mathbf{e_k}$.
4. All of the iterative methods to be described in this chapter generate approximate solutions $\mathbf{x_k}$, $k = 1, 2, \ldots$, such that $\mathbf{x_k} - \mathbf{x_0}$ lies in the Krylov space $\text{span}\{\mathbf{z_0}, C\mathbf{z_0}, \ldots, C^{k-1}\mathbf{z_0}\}$, where $\mathbf{z_0}$ is the initial residual, possibly multiplied by a preconditioner, and C is the preconditioned iteration matrix.
5. The *Jacobi*, *Gauss-Seidel*, and *SOR (successive overrelaxation)* methods use the simple iteration

$$\mathbf{x_k} = \mathbf{x_{k-1}} + M^{-1}(\mathbf{b} - A\mathbf{x_{k-1}}), \quad k = 1, 2, \ldots,$$

with different preconditioners M. For the Jacobi method, M is taken to be the diagonal of A, while for the Gauss-Seidel method, M is the lower triangle of A. For the SOR method, M is of the form $\omega^{-1}D - L$, where D is the diagonal of A, $-L$ is the strict lower triangle of A, and ω is a relaxation parameter. Subtracting each side of this equation from the true solution $A^{-1}\mathbf{b}$, we find that the error at step k is

$$\mathbf{e_k} = (I - M^{-1}A)\mathbf{e_{k-1}} = \ldots = (I - M^{-1}A)^k\mathbf{e_0}.$$

Subtracting each side of this equation from $\mathbf{e_0}$, we find that $\mathbf{x_k}$ satisfies

$$\begin{aligned}\mathbf{e_0} - \mathbf{e_k} = \mathbf{x_k} - \mathbf{x_0} &= [I - (I - M^{-1}A)^k]\mathbf{e_0} \\ &= \left[\sum_{j=1}^{k} \binom{k}{j} (-1)^{j-1}(M^{-1}A)^{j-1}\right] \mathbf{z_0},\end{aligned}$$

where $\mathbf{z_0} = M^{-1}A\mathbf{e_0} = M^{-1}\mathbf{r_0}$. Thus, $\mathbf{x_k} - \mathbf{x_0}$ lies in the Krylov space

$$\text{span}\{\mathbf{z_0}, (M^{-1}A)\mathbf{z_0}, \ldots, (M^{-1}A)^{k-1}\mathbf{z_0}\}.$$

6. Standard *multigrid methods* for solving linear systems arising from partial differential equations are also of the form $\mathbf{x_k} = \mathbf{x_{k-1}} + M^{-1}\mathbf{r_{k-1}}$. For these methods, computing $M^{-1}\mathbf{r_{k-1}}$ involves restricting the residual to a coarser grid or grids, solving (or iterating) with the linear system on those grids, and then prolonging the solution back to the finest grid.

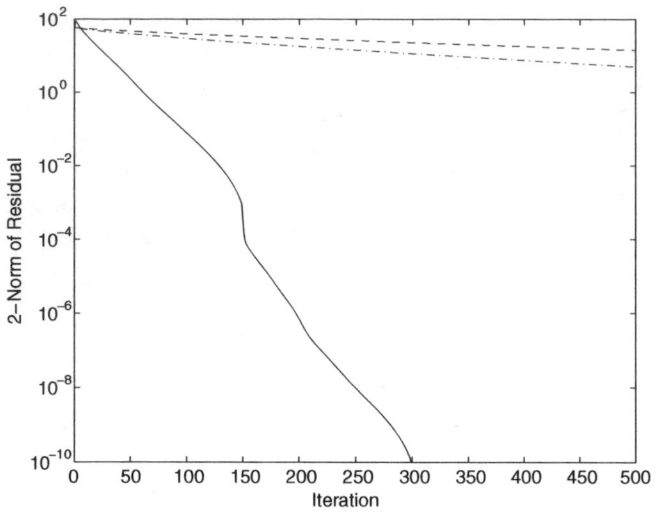

FIGURE 41.1 Convergence of iterative methods for the problem given in the introduction with $a(x, y, z) = 1 + x + 3yz$, $h = 1/50$. Jacobi (dashed), Gauss–Seidel (dash-dot), and SOR with $\omega = 1.9$ (solid).

Applications:

1. Figure 41.1 shows the convergence of the Jacobi, Gauss–Seidel, and SOR (with $\omega = 1.9$) iterative methods for the problem described at the beginning of this chapter, using a mildly varying coefficient $a(x, y, z) = 1 + x + 3yz$ on the unit cube $\Omega = [0, 1] \times [0, 1] \times [0, 1]$ with homogeneous Dirichlet boundary conditions, $u = 0$ on $\partial\Omega$. The right-hand side function f was chosen so that the solution to the differential equation would be $u(x, y, z) = x(1 - x)y^2(1 - y)z(1 - z)^2$. The region was discretized using a $50 \times 50 \times 50$ mesh, and the natural ordering of nodes was used, along with a zero initial guess.

41.2 Optimal Krylov Space Methods for Hermitian Problems

Throughout this section, we let A and b denote the already preconditioned matrix and right-hand side vector, and we assume that A is Hermitian. Note that if the original coefficient matrix is Hermitian, then this requires Hermitian positive definite preconditioning (preconditioner of the form $M = LL^*$ and preconditioned matrix of the form $L^{-1}AL^{-*}$) in order to maintain this property.

Definitions:

The **Minimal Residual (MINRES)** algorithm generates, at each step k, the approximation $\mathbf{x_k}$ with $\mathbf{x_k} - \mathbf{x_0} \in K_k(A, \mathbf{r_0})$ for which the 2-norm of the residual, $\|\mathbf{r_k}\| \equiv \langle \mathbf{r_k}, \mathbf{r_k} \rangle^{1/2}$, is minimal.

The **Conjugate Gradient (CG)** algorithm for Hermitian positive definite matrices generates, at each step k, the approximation $\mathbf{x_k}$ with $\mathbf{x_k} - \mathbf{x_0} \in K_k(A, \mathbf{r_0})$ for which the A-norm of the error, $\|\mathbf{e_k}\|_A \equiv \langle \mathbf{e_k}, A\mathbf{e_k} \rangle^{1/2}$, is minimal. (Note that this is sometimes referred to as the $A^{1/2}$-norm of the error, e.g., in Chapter 37 of this book.)

The **Lanczos** algorithm for Hermitian matrices is a short recurrence for constructing an orthonormal basis for a Krylov space.

Facts:

The following facts can be found in any of the general references [Axe95], [Gre97], [Hac94], [Saa03], and [Vor03].

1. The Lanczos algorithm [Lan50] is implemented as follows:

> **Lanczos Algorithm.** (For Hermitian matrices A)
>
> Given $\mathbf{q_1}$ with $\|\mathbf{q_1}\| = 1$, set $\beta_0 = 0$. For $j = 1, 2, \ldots,$
>
> $\tilde{\mathbf{q}}_{\mathbf{j+1}} = A\mathbf{q_j} - \beta_{j-1}\mathbf{q_{j-1}}$. Set $\alpha_j = \langle \tilde{\mathbf{q}}_{\mathbf{j+1}}, \mathbf{q_j} \rangle$, $\tilde{\mathbf{q}}_{\mathbf{j+1}} \longleftarrow \tilde{\mathbf{q}}_{\mathbf{j+1}} - \alpha_j \mathbf{q_j}$.
>
> $\beta_j = \|\tilde{\mathbf{q}}_{\mathbf{j+1}}\|$, $\mathbf{q_{j+1}} = \tilde{\mathbf{q}}_{\mathbf{j+1}}/\beta_j$.

2. It can be shown by induction that the Lanczos vectors $\mathbf{q_1}, \mathbf{q_2}, \ldots$ produced by the above algorithm are orthogonal. Gathering the first k vectors together as the columns of an n by k matrix Q_k, this recurrence can be written succinctly in the form

$$AQ_k = Q_k T_k + \beta_k \mathbf{q_{k+1}} \boldsymbol{\xi}_{\mathbf{k}}^T,$$

where $\boldsymbol{\xi}_{\mathbf{k}} \equiv (0, \ldots, 0, 1)^T$ is the kth unit vector and T_k is the tridiagonal matrix of recurrence coefficients:

$$T_k \equiv \begin{pmatrix} \alpha_1 & \beta_1 & & \\ \beta_1 & \ddots & \ddots & \\ & \ddots & \ddots & \beta_{k-1} \\ & & \beta_{k-1} & \alpha_k \end{pmatrix}.$$

The above equation is sometimes written in the form

$$AQ_k = Q_{k+1} \underline{T}_k,$$

where $\underline{T}_k$ is the $k+1$ by k matrix whose top k by k block is T_k and whose bottom row is zero except for the last entry which is β_k.

3. If the initial vector $\mathbf{q_1}$ in the Lanczos algorithm is taken to be $\mathbf{q_1} = \mathbf{r_0}/\|\mathbf{r_0}\|$, then the columns of Q_k span the Krylov space $K_k(A, \mathbf{r_0})$. Both the MINRES and CG algorithms take the approximation $\mathbf{x_k}$ to be of the form $\mathbf{x_0} + Q_k \mathbf{y_k}$ for a certain vector $\mathbf{y_k}$. For the MINRES algorithm, $\mathbf{y_k}$ is the solution of the $k+1$ by k least squares problem

$$\min_{\mathbf{y}} \|\beta \xi_1 - \underline{T}_k \mathbf{y}\|,$$

where $\beta \equiv \|\mathbf{r_0}\|$ and $\xi_1 \equiv (1, 0, \ldots, 0)^T$ is the first unit vector. For the CG algorithm, $\mathbf{y_k}$ is the solution of the k by k tridiagonal system

$$T_k \mathbf{y} = \beta \xi_1.$$

4. The following algorithms are standard implementations of the CG and MINRES methods.

Conjugate Gradient Method (CG).
(For Hermitian Positive Definite Problems)

Given an initial guess $\mathbf{x_0}$, compute $\mathbf{r_0} = \mathbf{b} - A\mathbf{x_0}$ and set $\mathbf{p_0} = \mathbf{r_0}$. For $k = 1, 2, \ldots,$

Compute $A\mathbf{p_{k-1}}$.

Set $\mathbf{x_k} = \mathbf{x_{k-1}} + a_{k-1}\mathbf{p_{k-1}}$, where $a_{k-1} = \frac{\langle \mathbf{r_{k-1}}, \mathbf{r_{k-1}} \rangle}{\langle \mathbf{p_{k-1}}, A\mathbf{p_{k-1}} \rangle}$.

Compute $\mathbf{r_k} = \mathbf{r_{k-1}} - a_{k-1} A\mathbf{p_{k-1}}$.

Set $\mathbf{p_k} = \mathbf{r_k} + b_{k-1}\mathbf{p_{k-1}}$, where $b_{k-1} = \frac{\langle \mathbf{r_k}, \mathbf{r_k} \rangle}{\langle \mathbf{r_{k-1}}, \mathbf{r_{k-1}} \rangle}$.

Minimal Residual Algorithm (MINRES). (For Hermitian Problems)

Given $\mathbf{x_0}$, compute $\mathbf{r_0} = \mathbf{b} - A\mathbf{x_0}$ and set $\mathbf{q_1} = \mathbf{r_0}/\|\mathbf{r_0}\|$.
Initialize $\xi = (1, 0, \ldots, 0)^T$, $\beta = \|\mathbf{r_0}\|$. For $k = 1, 2, \ldots,$

Compute $\mathbf{q_{k+1}}$, $\alpha_k \equiv T(k,k)$, and $\beta_k \equiv T(k+1,k) \equiv T(k,k+1)$ using the Lanczos algorithm.

Apply rotations F_{k-2} and F_{k-1} to the last column of T; that is,

$$\begin{pmatrix} T(k-2,k) \\ T(k-1,k) \end{pmatrix} \leftarrow \begin{pmatrix} c_{k-2} & s_{k-2} \\ -\bar{s}_{k-2} & c_{k-2} \end{pmatrix} \begin{pmatrix} 0 \\ T(k-1,k) \end{pmatrix}, \quad \text{if } k > 2,$$

$$\begin{pmatrix} T(k-1,k) \\ T(k,k) \end{pmatrix} \leftarrow \begin{pmatrix} c_{k-1} & s_{k-1} \\ -\bar{s}_{k-1} & c_{k-1} \end{pmatrix} \begin{pmatrix} T(k-1,k) \\ T(k,k) \end{pmatrix}, \quad \text{if } k > 1.$$

Compute the kth rotation, c_k and s_k, to annihilate the $(k+1,k)$ entry of T:
$c_k = |T(k,k)|/\sqrt{|T(k,k)|^2 + |T(k+1,k)|^2}$, $\bar{s}_k = c_k T(k+1,k)/T(k,k)$.

Apply k^{th} rotation to ξ and to last column of T:

$$\begin{pmatrix} \xi(k) \\ \xi(k+1) \end{pmatrix} \leftarrow \begin{pmatrix} c_k & s_k \\ -\bar{s}_k & c_k \end{pmatrix} \begin{pmatrix} \xi(k) \\ 0 \end{pmatrix}.$$

$$T(k,k) \leftarrow c_k T(k,k) + s_k T(k+1,k), \quad T(k+1,k) \leftarrow 0.$$

Compute $\mathbf{p_{k-1}} = [\mathbf{q_k} - T(k-1,k)\mathbf{p_{k-2}} - T(k-2,k)\mathbf{p_{k-3}}]/T(k,k)$, where undefined terms are zero for $k \leq 2$.

Set $\mathbf{x_k} = \mathbf{x_{k-1}} + a_{k-1}\mathbf{p_{k-1}}$, where $a_{k-1} = \beta\xi(k)$.

5. In exact arithmetic, both the CG and the MINRES algorithms obtain the exact solution in at most n steps, since the affine space $\mathbf{x_0} + K_n(A, \mathbf{r_0})$ contains the true solution.

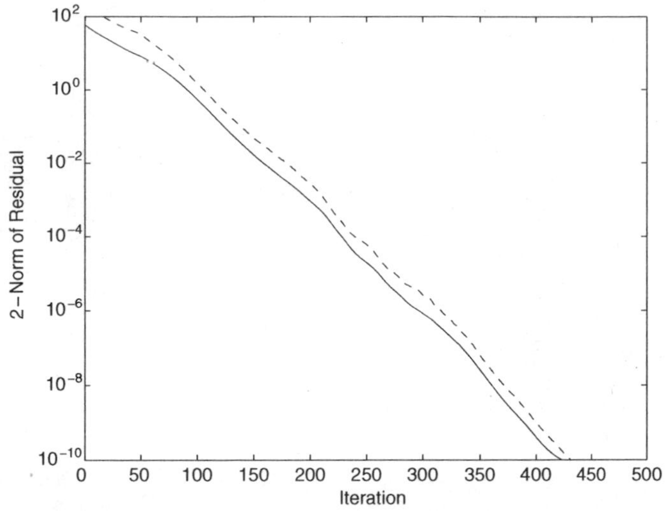

FIGURE 41.2 Convergence of MINRES (solid) and CG (dashed) for the problem given in the introduction with $a(x, y, z) = 1 + x + 3yz$, $h = 1/50$.

Applications:

1. Figure 41.2 shows the convergence (in terms of the 2-norm of the residual) of the (unpreconditioned) CG and MINRES algorithms for the same problem used in the previous section.

   Note that the 2-norm of the residual decreases monotonically in the MINRES algorithm, but not in the CG algorithm. Had we instead plotted the A-norm of the error, then the CG convergence curve would have been below that for MINRES.

41.3 Optimal and Nonoptimal Krylov Space Methods for Non-Hermitian Problems

In this section, we again let A and $\mathbf{b}$ denote the already preconditioned matrix and right-hand side vector. The matrix A is assumed to be a general nonsingular n by n matrix.

Definitions:

The **Generalized Minimal Residual (GMRES)** algorithm generates, at each step k, the approximation $\mathbf{x_k}$ with $\mathbf{x_k} - \mathbf{x_0} \in K_k(A, \mathbf{r_0})$ for which the 2-norm of the residual is minimal.

The **Full Orthogonalization Method (FOM)** generates, at each step k, the approximation $\mathbf{x_k}$ with $\mathbf{x_k} - \mathbf{x_0} \in K_k(A, \mathbf{r_0})$ for which the residual is orthogonal to the Krylov space $K_k(A, \mathbf{r_0})$.

The **Arnoldi algorithm** is a method for constructing an orthonormal basis for a Krylov space that requires saving all of the basis vectors and orthogonalizing against them at each step.

The **restarted GMRES algorithm**, GMRES(j), is defined by simply restarting GMRES every j steps, using the latest iterate as the initial guess for the next GMRES cycle. Sometimes partial information from the previous GMRES cycle is retained and used after the restart.

The **non-Hermitian (or two-sided) Lanczos algorithm** uses a pair of three-term recurrences involving A and A^* to construct **biorthogonal** bases for the Krylov spaces $K_k(A, \mathbf{r_0})$ and $K_k(A^*, \hat{\mathbf{r}}_0)$, where $\hat{\mathbf{r}}_0$ is a given vector with $\langle \mathbf{r_0}, \hat{\mathbf{r}}_0 \rangle \neq 0$. If the vectors $\mathbf{v_1}, \ldots, \mathbf{v_k}$ are the basis vectors for $K_k(A, \mathbf{r_0})$, and $\mathbf{w_1}, \ldots, \mathbf{w_k}$ are the basis vectors for $K_k(A^*, \hat{\mathbf{r}}_0)$, then $\langle \mathbf{v_i}, \mathbf{w_j} \rangle = 0$ for $i \neq j$.

In the **BiCG** (biconjugate gradient) method, the approximate solution $\mathbf{x_k}$ is chosen so that the residual $\mathbf{r_k}$ is orthogonal to $K_k(A^*, \hat{\mathbf{r}}_0)$.

In the **QMR** (quasi-minimal residual) algorithm, the approximate solution $\mathbf{x_k}$ is chosen to minimize a quantity that is related to (but not necessarily equal to) the residual norm.

The **CGS** (conjugate gradient squared) algorithm constructs an approximate solution $\mathbf{x_k}$ for which $\mathbf{r_k} = \varphi_k^2(A)\mathbf{r_0}$, where $\varphi_k(A)$ is the kth degree polynomial constructed in the BiCG algorithm; that is, the BiCG residual at step k is $\varphi_k(A)\mathbf{r_0}$.

The **BiCGSTAB** algorithm combines CGS with a one or more step residual norm minimizing method to smooth out the convergence.

Facts:

1. The Arnoldi algorithm [Arn51] is implemented as follows:

Arnoldi Algorithm.

Given $\mathbf{q_1}$ with $\|\mathbf{q_1}\| = 1$. For $j = 1, 2, \ldots$,

$\tilde{\mathbf{q}}_{\mathbf{j+1}} = A\mathbf{q_j}$. For $i = 1, \ldots, j$, $h_{ij} = \langle \tilde{\mathbf{q}}_{\mathbf{j+1}}, \mathbf{q_i} \rangle$, $\tilde{\mathbf{q}}_{\mathbf{j+1}} \longleftarrow \tilde{\mathbf{q}}_{\mathbf{j+1}} - h_{ij}\mathbf{q_i}$.

$h_{j+1,j} = \|\tilde{\mathbf{q}}_{\mathbf{j+1}}\|$, $\mathbf{q_{j+1}} = \tilde{\mathbf{q}}_{\mathbf{j+1}}/h_{j+1,j}$.

2. Unlike the Hermitian case, if A is non-Hermitian then there is no known algorithm for finding the optimal approximations from successive Krylov spaces, while performing only $O(n)$ operations per iteration. In fact, a theorem due to Faber and Manteuffel [FM84] shows that for most non-Hermitian matrices A there is no short recurrence that generates these optimal approximations for successive values $k = 1, 2, \ldots$. Hence, the current options for non-Hermitian problems are either to perform extra work ($O(nk)$ operations at step k) and use extra storage ($O(nk)$ words to perform k iterations) to find optimal approximations from the successive Krylov subspaces or to settle for nonoptimal approximations. The (full) GMRES (generalized minimal residual) algorithm [SS86] finds the approximation for which the 2-norm of the residual is minimal, at the cost of this extra work and storage, while other non-Hermitian iterative methods (e.g., BiCG [Fle75], CGS [Son89], QMR [FN91], BiCGSTAB [Vor92], and restarted GMRES [SS86], [Mor95], [DeS99]) generate nonoptimal approximations.
3. Similar to the MINRES algorithm, the GMRES algorithm uses the Arnoldi iteration defined above to construct an orthonormal basis for the Krylov space $K_k(A, \mathbf{r_0})$.

   If Q_k is the n by k matrix with the orthonormal basis vectors $\mathbf{q_1}, \ldots, \mathbf{q_k}$ as columns, then the Arnoldi iteration can be written simply as

$$AQ_k = Q_k H_k + h_{k+1,k}\mathbf{q_{k+1}}\boldsymbol{\xi_k}^T = Q_{k+1}\underline{H}_k.$$

   Here H_k is the k by k upper Hessenberg matrix with (i, j) entry equal to h_{ij}, and $\underline{H}_k$ is the $k+1$ by k matrix whose upper k by k block is H_k and whose bottom row is zero except for the last entry, which is $h_{k+1,k}$.

   If $\mathbf{q_1} = \mathbf{r_0}/\|\mathbf{r_0}\|$, then the columns of Q_k span the Krylov space $K_k(A, \mathbf{r_0})$, and the GMRES approximation is taken to be of the form $\mathbf{x_k} = \mathbf{x_0} + Q_k\mathbf{y_k}$ for some vector $\mathbf{y_k}$. To minimize the 2-norm of the residual, the vector $\mathbf{y_k}$ is chosen to solve the least squares problem

$$\min_{\mathbf{y}} \|\beta\xi_1 - \underline{H}_k\mathbf{y}\|, \quad \beta \equiv \|\mathbf{r_0}\|.$$

The GMRES algorithm [SS86] can be implemented as follows:

Generalized Minimal Residual Algorithm (GMRES).
Given $\mathbf{x_0}$, compute $\mathbf{r_0} = \mathbf{b} - A\mathbf{x_0}$ and set $\mathbf{q_1} = \mathbf{r_0}/\|\mathbf{r_0}\|$.

Initialize $\xi = (1, 0, \ldots, 0)^T$, $\beta = \|\mathbf{r_0}\|$. For $k = 1, 2, \ldots$,
Compute $\mathbf{q_{k+1}}$ and $h_{i,k} \equiv H(i,k)$, $i = 1, \ldots, k+1$, using the Arnoldi algorithm.

Apply rotations $F_1, \ldots, F_{k-1}$ to the last column of H; that is,

For $i = 1, \ldots, k-1$,

$$\begin{pmatrix} H(i,k) \\ H(i+1,k) \end{pmatrix} \leftarrow \begin{pmatrix} c_i & s_i \\ -\bar{s}_i & c_i \end{pmatrix} \begin{pmatrix} H(i,k) \\ H(i+1,k) \end{pmatrix}.$$

Compute the kth rotation, c_k and s_k, to annihilate the $(k+1, k)$ entry of H:
$c_k = |H(k,k)|/\sqrt{|H(k,k)|^2 + |H(k+1,k)|^2}$, $\bar{s}_k = c_k H(k+1,k)/H(k,k)$.
Apply k^{th} rotation to ξ and to last column of H:

$$\begin{pmatrix} \xi(k) \\ \xi(k+1) \end{pmatrix} \leftarrow \begin{pmatrix} c_k & s_k \\ -\bar{s}_k & c_k \end{pmatrix} \begin{pmatrix} \xi(k) \\ 0 \end{pmatrix}$$

$$H(k,k) \leftarrow c_k H(k,k) + s_k H(k+1,k), \quad H(k+1,k) \leftarrow 0.$$

If residual norm estimate $\beta|\xi(k+1)|$ is sufficiently small, then
Solve upper triangular system $H_{k\times k}\, \mathbf{y_k} = \beta\, \xi_{k\times 1}$.
Compute $\mathbf{x_k} = \mathbf{x_0} + Q_k \mathbf{y_k}$.

4. The (full) GMRES algorithm described above may be impractical because of increasing storage and work requirements, if the number of iterations needed to solve the linear system is large. In this case, the restarted GMRES algorithm or one of the algorithms based on the non-Hermitian Lanczos process may provide a reasonable alternative. The BiCGSTAB algorithm [Vor92] is often among the most effective iteration methods for solving non-Hermitian linear systems. The algorithm can be written as follows:

BiCGSTAB.

Given $\mathbf{x_0}$, compute $\mathbf{r_0} = \mathbf{b} - A\mathbf{x_0}$ and set $\mathbf{p_0} = \mathbf{r_0}$. Choose $\hat{\mathbf{r}}_0$ such that $\langle \mathbf{r_0}, \hat{\mathbf{r}}_0 \rangle \neq 0$.
For $k = 1, 2, \ldots$,

Compute $A\mathbf{p_{k-1}}$.

Set $\mathbf{x_{k-1/2}} = \mathbf{x_{k-1}} + a_{k-1}\mathbf{p_{k-1}}$, where $a_{k-1} = \frac{\langle \mathbf{r_{k-1}}, \hat{\mathbf{r}}_0 \rangle}{\langle A\mathbf{p_{k-1}}, \hat{\mathbf{r}}_0 \rangle}$.

Compute $\mathbf{r_{k-1/2}} = \mathbf{r_{k-1}} - a_{k-1} A\mathbf{p_{k-1}}$.

Compute $A\mathbf{r_{k-1/2}}$.

Set $\mathbf{x_k} = \mathbf{x_{k-1/2}} + \omega_k \mathbf{r_{k-1/2}}$, where $\omega_k = \frac{\langle \mathbf{r_{k-1/2}}, A\mathbf{r_{k-1/2}} \rangle}{\langle A\mathbf{r_{k-1/2}}, A\mathbf{r_{k-1/2}} \rangle}$.

Compute $\mathbf{r_k} = \mathbf{r_{k-1/2}} - \omega_k A\mathbf{r_{k-1/2}}$.

Compute $\mathbf{p_k} = \mathbf{r_k} + b_k(\mathbf{p_{k-1}} - \omega_k A\mathbf{p_{k-1}})$, where $b_k = \frac{a_{k-1}}{\omega_k} \frac{\langle \mathbf{r_k}, \hat{\mathbf{r}}_0 \rangle}{\langle \mathbf{r_{k-1}}, \hat{\mathbf{r}}_0 \rangle}$.

5. The non-Hermitian Lanczos algorithm can break down if $\langle \mathbf{v_i}, \mathbf{w_i} \rangle = 0$, but neither $\mathbf{v_i}$ nor $\mathbf{w_i}$ is zero. In this case **look-ahead** strategies have been devised to skip steps at which the Lanczos vectors are undefined. See, for instance, [PTL85], [Nac91], and [FN91]. These look-ahead procedures are used in the QMR algorithm.
6. When A is Hermitian and $\hat{\mathbf{r}}_0 = \mathbf{r}_0$, the BiCG method reduces to the CG algorithm, while the QMR method reduces to the MINRES algorithm.
7. The question of which iterative method to use is, of course, an important one. Unfortunately, there is no straightforward answer. It is problem dependent and may depend also on the type of machine being used. If matrix-vector multiplication is very expensive (e.g., if A is dense and has no special properties to enable fast matrix-vector multiplication), then full GMRES is probably the method of choice because it requires the fewest matrix-vector multiplications to reduce the residual norm to a desired level. If matrix-vector multiplication is not so expensive or if storage becomes a problem for full GMRES, then a restarted GMRES algorithm, some variant of the QMR method, or some variant of BiCGSTAB may be a reasonable alternative. With a sufficiently good preconditioner, each of these iterative methods can be expected to find a good approximate solution quickly. In fact, with a sufficiently good preconditioner M, an even simpler iteration method such as $\mathbf{x_k} = \mathbf{x_{k-1}} + M^{-1}(\mathbf{b} - A\mathbf{x_{k-1}})$ may converge in just a few iterations, and this avoids the cost of inner products and other things in the more sophisticated Krylov space methods.

Applications:

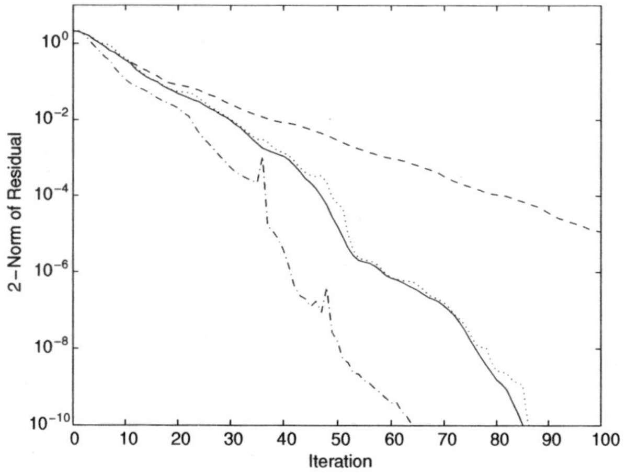

FIGURE 41.3 Convergence of full GMRES (solid), restarted GMRES (restarted every 10 steps) (dashed), QMR (dotted), and BiCGSTAB (dash-dot) for a problem from neutron transport. For GMRES (full or restarted), the number of matrix-vector multiplications is the same as the number of iterations, while for QMR and BiCGSTAB, the number of matrix-vector multiplications is *twice* the number of iterations.

1. To illustrate the behavior of iterative methods for solving non-Hermitian linear systems, we have taken a simple problem involving the Boltzmann transport equation in one dimension:

$$\mu \frac{\partial \psi}{\partial x} + \sigma_T \psi - \sigma_s \phi = f, \quad x \in [a, b], \quad \mu \in [-1, 1],$$

where

$$\phi(x) = \frac{1}{2} \int_{-1}^{1} \psi(x, \mu')\, d\mu',$$

with boundary conditions

$$\psi(b, \mu) = \psi_b(\mu), \quad -1 \le \mu < 0,$$
$$\psi(a, \mu) = \psi_a(\mu), \quad 0 < \mu \le 1.$$

The difference method used is described in [Gre97], and a test problem from [ML82] was solved. Figure 41.3 shows the convergence of full GMRES, restarted GMRES (restarted every 10 steps), QMR, and BiCGSTAB. One should keep in mind that each iteration of the QMR algorithm requires *two* matrix-vector multiplications, one with A and one with A^*. Still, the QMR approximation at iteration k lies in the k-dimensional affine space $\mathbf{x}_0 + \text{span}\{\mathbf{r}_0, A\mathbf{r}_0, \ldots, A^{k-1}\mathbf{r}_0\}$. Each iteration of the BiCGSTAB algorithm requires *two* matrix-vector multiplications with A, and the approximate solution generated at step k lies in the $2k$-dimensional affine space $\mathbf{x}_0 + \text{span}\{\mathbf{r}_0, A\mathbf{r}_0, \ldots, A^{2k-1}\mathbf{r}_0\}$. The full GMRES algorithm finds the optimal approximation from this space at step $2k$. Thus, the GMRES residual norm at step $2k$ is guaranteed to be less than or equal to the BiCGSTAB residual norm at step k, and each requires the *same* number of matrix-vector multiplications to compute.

41.4 Preconditioners

Definitions:

An **incomplete Cholesky decomposition** is a preconditioner for a Hermitian positive definite matrix A of the form $M = LL^*$, where L is a sparse lower triangular matrix. The entries of L are chosen so that certain entries of LL^* match those of A. If L is taken to have the same sparsity pattern as the lower triangle of A, then its entries are chosen so that LL^* matches A in the positions where A has nonzeros.

A **modified incomplete Cholesky decomposition** is a preconditioner of the same form $M = LL^*$ as the incomplete Cholesky preconditioner, but the entries of L are modified so that instead of having M match as many entries of A as possible, the preconditioner M has certain other properties, such as the same row sums as A.

An **incomplete LU decomposition** is a preconditioner for a general matrix A of the form $M = LU$, where L and U are sparse lower and upper triangular matrices, respectively. The entries of L and U are chosen so that certain entries of LU match the corresponding entries of A.

A **sparse approximate inverse** is a sparse matrix M^{-1} constructed to approximate A^{-1}.

A **multigrid** preconditioner is a preconditioner designed for problems arising from partial differential equations discretized on grids. Solving the preconditioning system $Mz = r$ entails *restricting* the residual to coarser grids, performing relaxation steps for the linear system corresponding to the same differential operator on the coarser grids, and *prolonging* solutions back to finer grids.

An **algebraic multigrid** preconditioner is a preconditioner that uses principles similar to those used for PDE problems on grids, when the "grid" for the problem is unknown or nonexistent and only the matrix is available.

Facts:

1. If A is an M-matrix, then for every subset S of off-diagonal indices there exists a lower triangular matrix $L = [l_{ij}]$ with unit diagonal and an upper triangular matrix $U = [u_{ij}]$ such that $A = LU - R$, where

$$l_{ij} = 0 \text{ if } (i, j) \in S, \quad u_{ij} = 0 \text{ if } (i, j) \in S, \quad \text{and} \quad r_{ij} = 0 \text{ if } (i, j) \notin S.$$

   The factors L and U are unique and the splitting $A = LU - R$ is a regular splitting [Var60, MV77]. The idea of generating such approximate factorizations was considered by a number of people, one of the first of whom was Varga [Var60]. The idea became popular when it was used by Meijerink and van der Vorst to generate preconditioners for the conjugate gradient method and related iterations [MV77]. It has proved a successful technique in a range of applications and is now widely used with many variations. For example, instead of specifying the sparsity pattern of L, one might begin to compute the entries of the exact L-factor and set entries to 0 if they fall below some threshold (see, e.g., [Mun80]).
2. For a real symmetric positive definite matrix A arising from a standard finite difference or finite element approximation for a second order self-adjoint elliptic partial differential equation on a grid

with spacing h, the condition number of A is $O(h^{-2})$. When A is preconditioned using the incomplete Cholesky decomposition LL^T, where L has the same sparsity pattern as the lower triangle of A, the condition number of the preconditioned matrix $L^{-1}AL^{-T}$ is still $O(h^{-2})$, but the constant multiplying h^{-2} is smaller. When A is preconditioned using the modified incomplete Cholesky decomposition, the condition number of the preconditioned matrix is $O(h^{-1})$ [DKR68, Gus78].

3. For a general matrix A, the incomplete LU decomposition can be used as a preconditioner in a non-Hermitian matrix iteration such as GMRES, QMR, or BiCGSTAB. At each step of the preconditioned algorithm one must solve a linear system $M\mathbf{z} = \mathbf{r}$. This is accomplished by first solving the lower triangular system $L\mathbf{y} = \mathbf{r}$ and then solving the upper triangular system $U\mathbf{z} = \mathbf{y}$.
4. One difficulty with incomplete Cholesky and incomplete LU decompositions is that the solution of the triangular systems may not parallelize well. In order to make better use of parallelism, sparse approximate inverses have been proposed as preconditioners. Here, a sparse matrix M^{-1} is constructed directly to approximate A^{-1}, and each step of the iteration method requires computation of a matrix-vector product $\mathbf{z} = M^{-1}\mathbf{r}$. For an excellent recent survey of all of these preconditioning methods see [Ben02].
5. Multigrid methods have the very desirable property that for many problems arising from elliptic PDEs the number of cycles required to reduce the error to a desired fixed level is independent of the grid size. This is in contrast to methods such as ICCG and MICCG (incomplete and modified incomplete Cholesky decomposition used as preconditioners in the CG algorithm). Early developers of multigrid methods include Fedorenko [Fed61] and later Brandt [Bra77]. A very readable and up-to-date introduction to the subject can be found in [BHM00].
6. Algebraic multigrid methods represent an attempt to use principles similar to those used for PDE problems on grids, when the origin of the problem is not necessarily known and only the matrix is available. An example is the AMG code by Ruge and Stüben [RS87]. The AMG method attempts to achieve mesh-independent convergence rates, just like standard multigrid methods, without making use of the underlying grid. A related class of preconditioners are *domain decomposition* methods. (See [QV99] and [SBG96] for recent surveys.)

41.5 Preconditioned Algorithms

Facts:

1. It is easy to modify the algorithms of the previous sections to use left preconditioning: Simply replace A by $M^{-1}A$ and $\mathbf{b}$ by $M^{-1}\mathbf{b}$ wherever they appear. Since one need not actually compute M^{-1}, this is equivalent to solving linear systems with coefficient matrix M for the preconditioned quantities. For example, letting $\mathbf{z_k}$ denote the preconditioned residual $M^{-1}(\mathbf{b} - A\mathbf{x_k})$, the left-preconditioned BiCGSTAB algorithm is as follows:

Left-Preconditioned BiCGSTAB.
Given $\mathbf{x_0}$, compute $\mathbf{r_0} = \mathbf{b} - A\mathbf{x_0}$, solve $M\mathbf{z_0} = \mathbf{r_0}$, and set $\mathbf{p_0} = \mathbf{z_0}$.
Choose $\hat{\mathbf{z}}_0$ such that $\langle \mathbf{z_0}, \hat{\mathbf{z}}_0 \rangle \neq 0$. For $k = 1, 2, \ldots,$
Compute $A\mathbf{p_{k-1}}$ and solve $M\mathbf{q_{k-1}} = A\mathbf{p_{k-1}}$.
Set $\mathbf{x_{k-1/2}} = \mathbf{x_{k-1}} + a_{k-1}\mathbf{p_{k-1}}$, where $a_{k-1} = \frac{\langle \mathbf{z_{k-1}}, \hat{\mathbf{z}}_0 \rangle}{\langle \mathbf{q_{k-1}}, \hat{\mathbf{z}}_0 \rangle}$.
Compute $\mathbf{r_{k-1/2}} = \mathbf{r_{k-1}} - a_{k-1}A\mathbf{p_{k-1}}$ and $\mathbf{z_{k-1/2}} = \mathbf{z_{k-1}} - a_{k-1}\mathbf{q_{k-1}}$.
Compute $A\mathbf{z_{k-1/2}}$ and solve $M\mathbf{s_{k-1/2}} = A\mathbf{z_{k-1/2}}$.
Set $\mathbf{x_k} = \mathbf{x_{k-1/2}} + \omega_k \mathbf{z_{k-1/2}}$, where $\omega_k = \frac{\langle \mathbf{z_{k-1/2}}, \mathbf{s_{k-1/2}} \rangle}{\langle \mathbf{s_{k-1/2}}, \mathbf{s_{k-1/2}} \rangle}$.
Compute $\mathbf{r_k} = \mathbf{r_{k-1/2}} - \omega_k A\mathbf{z_{k-1/2}}$ and $\mathbf{z_k} = \mathbf{z_{k-1/2}} - \omega_k \mathbf{s_{k-1/2}}$.
Compute $\mathbf{p_k} = \mathbf{z_k} + b_k(\mathbf{p_{k-1}} - \omega_k \mathbf{q_{k-1}})$, where $b_k = \frac{a_{k-1}}{\omega_k} \frac{\langle \mathbf{z_k}, \hat{\mathbf{z}}_0 \rangle}{\langle \mathbf{z_{k-1}}, \hat{\mathbf{z}}_0 \rangle}$.

2. Right or Hermitian preconditioning requires a little more thought since we want to generate approximations $\mathbf{x_k}$ to the solution of the original linear system, not the modified one $AM^{-1}\mathbf{y} = \mathbf{b}$ or $L^{-1}AL^{-*}\mathbf{y} = L^{-1}\mathbf{b}$.
If the CG algorithm is applied directly to the problem $L^{-1}AL^{-*}\mathbf{y} = L^{-1}\mathbf{b}$, then the iterates satisfy

$$\mathbf{y_k} = \mathbf{y_{k-1}} + a_{k-1}\hat{\mathbf{p}}_{\mathbf{k-1}}, \quad a_{k-1} = \frac{\langle \hat{\mathbf{r}}_{\mathbf{k-1}}, \hat{\mathbf{r}}_{\mathbf{k-1}} \rangle}{\langle \hat{\mathbf{p}}_{\mathbf{k-1}}, L^{-1}AL^{-*}\hat{\mathbf{p}}_{\mathbf{k-1}} \rangle},$$

$$\hat{\mathbf{r}}_{\mathbf{k}} = \hat{\mathbf{r}}_{\mathbf{k-1}} - a_{k-1}L^{-1}AL^{-*}\hat{\mathbf{p}}_{\mathbf{k-1}},$$

$$\hat{\mathbf{p}}_{\mathbf{k}} = \hat{\mathbf{r}}_{\mathbf{k}} + b_{k-1}\hat{\mathbf{p}}_{\mathbf{k-1}}, \quad b_{k-1} = \frac{\langle \hat{\mathbf{r}}_{\mathbf{k}}, \hat{\mathbf{r}}_{\mathbf{k}} \rangle}{\langle \hat{\mathbf{r}}_{\mathbf{k-1}}, \hat{\mathbf{r}}_{\mathbf{k-1}} \rangle}.$$

Defining

$$\mathbf{x_k} \equiv L^{-*}\mathbf{y_k}, \quad \mathbf{r_k} \equiv L\hat{\mathbf{r}}_{\mathbf{k}}, \quad \mathbf{p_k} \equiv L^{-*}\hat{\mathbf{p}}_{\mathbf{k}},$$

we obtain the following preconditioned CG algorithm for $A\mathbf{x} = \mathbf{b}$:

Preconditioned Conjugate Gradient Method (PCG).
(For Hermitian Positive Definite Problems, with Hermitian Positive Definite Preconditioners)
Given an initial guess $\mathbf{x_0}$, compute $\mathbf{r_0} = \mathbf{b} - A\mathbf{x_0}$ and solve
$M\mathbf{z_0} = \mathbf{r_0}$. Set $\mathbf{p_0} = \mathbf{z_0}$. For $k = 1, 2, \ldots,$
Compute $A\mathbf{p_{k-1}}$.
Set $\mathbf{x_k} = \mathbf{x_{k-1}} + a_{k-1}\mathbf{p_{k-1}}$, where $a_{k-1} = \frac{\langle \mathbf{r_{k-1}}, \mathbf{z_{k-1}} \rangle}{\langle \mathbf{p_{k-1}}, A\mathbf{p_{k-1}} \rangle}$.
Compute $\mathbf{r_k} = \mathbf{r_{k-1}} - a_{k-1}A\mathbf{p_{k-1}}$.
Solve $M\mathbf{z_k} = \mathbf{r_k}$.
Set $\mathbf{p_k} = \mathbf{z_k} + b_{k-1}\mathbf{p_{k-1}}$, where $b_{k-1} = \frac{\langle \mathbf{r_k}, \mathbf{z_k} \rangle}{\langle \mathbf{r_{k-1}}, \mathbf{z_{k-1}} \rangle}$.

Applications:

1. Figure 41.4 shows the convergence (in terms of the 2-norm of the residual) of the ICCG algorithm (CG with incomplete Cholesky decomposition as a preconditioner, where the sparsity pattern of

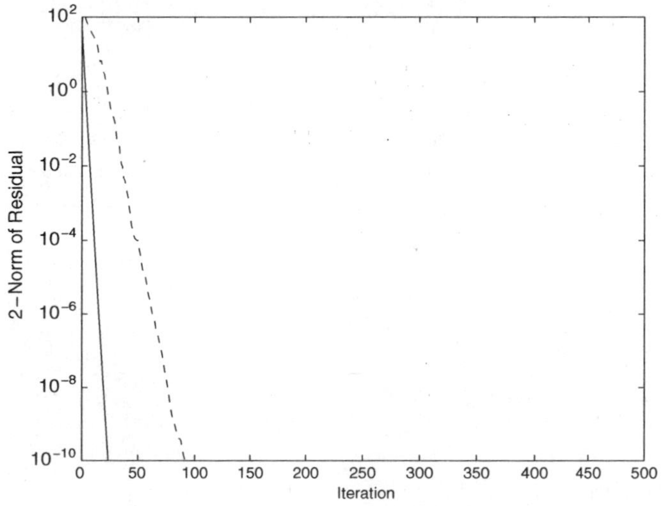

FIGURE 41.4 Convergence of ICCG (dashed) and multigrid (solid) for the problem given in the introduction with $a(x, y, z) = 1 + x + 3yz$, $h = 1/50$ (for ICCG), and $h = 1/64$ (for multigrid).

L was taken to match that of the lower triangle of A) and a multigrid method (using standard restriction and prolongation operators and the Gauss-Seidel algorithm for relaxation) for the same problem used in sections 2 and 3. The horizontal axis represents iterations for ICCG or cycles for the multigrid method. Each multigrid V-cycle costs about twice as much as an ICCG iteration, since it performs two triangular solves and two matrix vector multiplications on each coarse grid, while doing one of each on the fine grid. There is also a cost for the restriction and prolongation operations. The grid size was taken to be 64^3 for the multigrid method since this enabled easy formation of coarser grids by doubling h. The coarsest grid, on which the problem was solved directly, was taken to be of size $4 \times 4 \times 4$. It should be noted that the number of multigrid cycles can be reduced by using the multigrid procedure as a preconditioner for the CG algorithm (although this would require a different relaxation method in order to maintain symmetry). The only added expense is the cost of inner products in the CG algorithm.

41.6 Convergence Rates of CG and MINRES

In this section, we again let A and $\mathbf{b}$ denote the already preconditioned matrix and right-hand side vector, and we assume that A is Hermitian (and also positive definite for CG).

Facts:

1. The CG error vector and the MINRES residual vector at step k can be written in the form

$$\mathbf{e_k} = P_k^C(A)\mathbf{e_0}, \quad \mathbf{r_k} = P_k^M(A)\mathbf{r_0},$$

where P_k^C and P_k^M are the kth degree polynomials with value 1 at the origin that minimize the A-norm of the error in the CG algorithm and the 2-norm of the residual in the MINRES algorithm, respectively. In other words, the error $\mathbf{e_k}$ in the CG approximation satisfies

$$\|\mathbf{e_k}\|_A = \min_{p_k} \|p_k(A)\mathbf{e_0}\|_A$$

and the residual $\mathbf{r_k}$ in the MINRES algorithm satisfies

$$\|\mathbf{r_k}\| = \min_{p_k} \|p_k(A)\mathbf{r_0}\|,$$

where the minimum is taken over all polynomials p_k of degree k or less with $p_k(0) = 1$.

2. Let an eigendecomposition of A be written as $A = U\Lambda U^*$, where U is a unitary matrix and $\Lambda = \operatorname{diag}(\lambda_1, \ldots, \lambda_n)$ is a diagonal matrix of eigenvalues. If A is positive definite, define $A^{1/2}$ to be $U\Lambda^{1/2}U^*$. Then the A-norm of a vector $\mathbf{v}$ is just the 2-norm of the vector $A^{1/2}\mathbf{v}$. The equalities in Fact 1 imply

$$\begin{aligned}\|\mathbf{e_k}\|_A &= \min_{p_k} \|A^{1/2}p_k(A)\mathbf{e_0}\| = \min_{p_k} \|p_k(A)A^{1/2}\mathbf{e_0}\| \\ &= \min_{p_k} \|Up_k(\Lambda)U^*A^{1/2}\mathbf{e_0}\| \leq \min_{p_k} \|p_k(\Lambda)\| \; \|\mathbf{e_0}\|_A,\end{aligned}$$

$$\|\mathbf{r_k}\| = \min_{p_k} \|Up_k(\Lambda)U^*\mathbf{r_0}\| \leq \min_{p_k} \|p_k(\Lambda)\| \; \|\mathbf{r_0}\|.$$

These bounds are *sharp*; that is, for each step k there is an initial vector for which equality holds [Gre79, GG94, Jou96], but the initial vector that gives equality at step k may be different from the one that results in equality at some other step j.

The problem of describing the convergence of these algorithms therefore reduces to one in approximation theory — How well can one approximate zero on the set of eigenvalues of A using a kth degree polynomial with value 1 at the origin? While there is no simple expression for the maximum value of the minimax polynomial on a discrete set of points, this minimax polynomial can be calculated if the eigenvalues of A are known and, more importantly, this sharp upper bound provides intuition as to what constitute "good" and "bad" eigenvalue distributions. Eigenvalues tightly clustered around a single point (away from the origin) are good, for instance, because the polynomial $(1 - z/c)^k$ is small in absolute value at all points near c. Widely spread eigenvalues, especially if they lie on both sides of the origin, are bad, because a low degree polynomial with value 1 at the origin cannot be small at a large number of such points.

3. Since one usually has only limited information about the eigenvalues of A, it is useful to have error bounds that involve only a few properties of the eigenvalues. For example, in the CG algorithm for Hermitian positive definite problems, knowing only the largest and smallest eigenvalues of A, one can obtain an error bound by considering the minimax polynomial on the *interval* from λ_{min} to λ_{max}; i.e., the Chebyshev polynomial shifted to the interval and scaled to have value 1 at the origin. The result is

$$\frac{\|\mathbf{e_k}\|_A}{\|\mathbf{e_0}\|_A} \le 2\left[\left(\frac{\sqrt{\kappa}-1}{\sqrt{\kappa}+1}\right)^k + \left(\frac{\sqrt{\kappa}+1}{\sqrt{\kappa}-1}\right)^k\right]^{-1} \le 2\left(\frac{\sqrt{\kappa}-1}{\sqrt{\kappa}+1}\right)^k,$$

where $\kappa = \lambda_{max}/\lambda_{min}$ is the ratio of largest to smallest eigenvalue of A.

If additional information is available about the interior eigenvalues of A, one often can improve on this estimate while maintaining a simpler expression than the sharp bound in Fact 2. Suppose, for example, that A has just a few eigenvalues that are much larger than the others, say, $\lambda_1 \le \cdots \le \lambda_{n-\ell} << \lambda_{n-\ell+1} \le \cdots \le \lambda_n$. Consider a polynomial p_k that is the product of a factor of degree ℓ that is zero at $\lambda_{n-\ell+1}, \ldots \lambda_n$ and the $(k-\ell)$th degree scaled and shifted Chebyshev polynomial on the interval $[\lambda_1, \lambda_{n-\ell}]$:

$$p_k(x) = \left[T_{k-\ell}\left(\frac{2x - \lambda_{n-\ell} - \lambda_1}{\lambda_{n-\ell} - \lambda_1}\right) \Big/ T_{k-\ell}\left(\frac{-\lambda_{n-\ell} - \lambda_1}{\lambda_{n-\ell} - \lambda_1}\right)\right] \cdot \left(\prod_{i=n-\ell+1}^{n} \frac{\lambda_i - x}{\lambda_i}\right).$$

Since the second factor is zero at $\lambda_{n-\ell+1}, \ldots, \lambda_n$ and less than one in absolute value at each of the other eigenvalues, the maximum absolute value of this polynomial on $\{\lambda_1, \ldots, \lambda_n\}$ is less than the maximum absolute value of the first factor on $\{\lambda_1, \ldots, \lambda_{n-\ell}\}$. It follows that

$$\frac{\|\mathbf{e_k}\|_A}{\|\mathbf{e_0}\|_A} \le 2\left(\frac{\sqrt{\kappa_{n-\ell}}-1}{\sqrt{\kappa_{n-\ell}}+1}\right)^{k-\ell}, \quad \kappa_{n-\ell} \equiv \frac{\lambda_{n-\ell}}{\lambda_1}.$$

Analogous results hold for the 2-norm of the residual in the MINRES algorithm applied to a Hermitian positive definite linear system. For estimates of the convergence rate of the MINRES algorithm applied to indefinite linear systems see, for example, [Fis96].

41.7 Convergence Rate of GMRES

Facts:

1. Like MINRES for Hermitian problems, the GMRES algorithm for general linear systems produces a residual at step k whose 2-norm satisfies $\|\mathbf{r_k}\| = \min_{p_k} \|p_k(A)\mathbf{r_0}\|$, where the minimum is over all kth degree polynomials p_k with $p_k(0) = 1$. To derive a bound on this expression that is

independent of the direction of $\mathbf{r_0}$, we could proceed as in the previous section by employing an eigendecomposition of A. To this end, assume that A is diagonalizable and let $A = V\Lambda V^{-1}$ be an eigendecomposition, where $\Lambda = \text{diag}(\lambda_1, \ldots, \lambda_n)$ is a diagonal matrix of eigenvalues and the columns of V are right eigenvectors of A. Then it follows that

$$\|\mathbf{r_k}\| = \min_{p_k} \|Vp_k(\Lambda)V^{-1}\mathbf{r_0}\| \leq \kappa(V) \min_{p_k} \|p_k(\Lambda)\| \cdot \|\mathbf{r_0}\|,$$

where $\kappa(V) = \|V\| \cdot \|V^{-1}\|$ is the condition number of the eigenvector matrix V. We can assume that the columns of V have been scaled to make this condition number as small as possible. As in the Hermitian case, the polynomial that minimizes $\|Vp_k(\Lambda)V^{-1}\mathbf{r_0}\|$ is not necessarily the one that minimizes $\|p_k(\Lambda)\|$, and it is not clear whether this bound is sharp. It turns out that if A is a normal matrix, then $\kappa(V) = 1$ and the bound is sharp [GG94, Jou96]. In this case, as in the Hermitian case, the problem of describing the convergence of GMRES reduces to a problem in approximation theory — How well can one approximate zero on the set of *complex* eigenvalues using a kth degree polynomial with value 1 at the origin?

If the matrix A is nonnormal but has a fairly well-conditioned eigenvector matrix V, then the above bound, while not necessarily sharp, gives a reasonable estimate of the actual size of the residual. In this case again, it is A's eigenvalue distribution that essentially determines the behavior of GMRES.

2. In general, however, the behavior of GMRES cannot be determined from eigenvalues alone. In fact, it is shown in [GS94b] and [GPS96] that *any* nonincreasing sequence of residual norms can be obtained with the GMRES method applied to a problem whose coefficient matrix has any desired eigenvalues. Thus, for example, eigenvalues tightly clustered around 1 are not necessarily good for nonnormal matrices; one can construct a nonnormal matrix whose eigenvalues are equal to 1 or as tightly clustered around 1 as one might like, for which the GMRES algorithm makes no progress until step n (when it must find the exact solution).

   The convergence behavior of the GMRES algorithm for nonnormal matrices is a topic of current research. The analysis must involve quantities other than the eigenvalues. Some partial results have been obtained in terms of the *field of values* [EES83, Eie93, BGT05], in terms of the ϵ*-pseudospectrum* [TE05], and in terms of the *polynomial numerical hull of degree k* [Gre02, Gre04].

41.8 Inexact Preconditioners and Finite Precision Arithmetic, Error Estimation and Stopping Criteria, Text and Reference Books

There are a number of topics of current or recent research that will not be covered in this article. Here we list a few of these with sample references.

Facts:

1. The effects of finite precision arithmetic on both the convergence rate and the ultimately attainable accuracy of iterative methods have been studied. Example papers include [DGR95, Gre89, Gre97b, GRS97, GS92, GS00, Pai76, Vor90], and [Woz80].
2. A related topic is inexact preconditioners. Suppose the preconditioning system $M\mathbf{z} = \mathbf{r}$ is solved inexactly, perhaps by using an iterative method inside the outer iteration for $A\mathbf{x} = \mathbf{b}$. How accurately should the preconditioning system be solved in order to obtain the best overall performance of the process? The answer is surprising. See [BF05, ES04], and [SS03] for recent discussions of this question.
3. Another related idea is the use of different preconditioners at different steps of the iteration, sometimes called *flexible* iterative methods. See, for example, [Saa93].

4. Finally, there is the important question of when an iterative method should be stopped. Sometimes one wishes to stop when the 2-norm of the error or the A-norm of the error in the CG method reaches a certain threshold. But one cannot compute these quantities directly. For discussions of estimating the A-norm of the error in the CG algorithm, as well as its connections to Gauss quadrature, see, for instance, [GM97, GS94, HS52], and [ST02]
5. A number of text and reference books on iterative methods are available. These include [Axe95, Gre97, Hac94, Saa03], and [Vor03]

References

[Arn51] W.E. Arnoldi. The principal of minimized iterations in the solution of the matrix eigenvalue problem. *Q. Appl. Math.*, 9:17–29, 1951.

[Axe95] O. Axelsson. *Iterative Solution Methods.* Cambridge University Press, Cambridge, 1995.

[BGT05] B. Beckermann, S.A. Goreinov, and E.E. Tyrtyshnikov. Some remarks on the Elman estimate for GMRES. to appear.

[Ben02] M. Benzi. Preconditioning techniques for large linear systems: A survey. *J. Comput. Phys.*, 182:418-477, 2002.

[BF05] A. Bouras and V. Frayssé. Inexact matrix-vector products in Krylov methods for solving linear systems: A relaxation strategy. *SIAM J. Matrix Anal. Appl.*, 26:660-678, 2005.

[Bra77] A. Brandt. Multi-level adaptive solutions to boundary-value problems. *Math. Comput.*, 31:333-390, 1977.

[BHM00] W.L. Briggs, V.E. Henson, and S.F. McCormick. *A Multigrid Tutorial.* SIAM, Philadelphia, 2000.

[DeS99] E. De Sturler. Truncation strategies for optimal Krylov subspace methods. *SIAM J. Numer. Anal.*, 36:864-889, 1999.

[DGR95] J. Drkošová, A. Greenbaum, M. Rozložník, and Z. Strakoš. Numerical stability of the GMRES method. *BIT*, 3:309-330, 1995.

[DKR68] T. Dupont, R.P. Kendall, and H.H. Rachford, Jr. An approximate factorization procedure for solving self-adjoint elliptic difference equations. *SIAM J. Numer. Anal.*, 5:559-573, 1968.

[Eie93] M. Eiermann. Fields of values and iterative methods. *Lin. Alg. Appl.*, 180:167-197, 1993.

[EES83] S. Eisenstat, H. Elman, and M. Schultz. Variational iterative methods for nonsymmetric systems of linear equations. *SIAM J. Numer. Anal.*, 20:345-357, 1983.

[ES04] J. van den Eshof and G.L.G. Sleijpen. Inexact Krylov subspace methods for linear systems. *SIAM J. Matrix Anal. Appl.*, 26:125–153, 2004.

[FM84] V. Faber and T. Manteuffel. Necessary and sufficient conditions for the existence of a conjugate gradient method. *SIAM J. Numer. Anal.*, 21:352-362, 1984.

[Fed61] R. P. Fedorenko. A relaxation method for solving elliptic difference equations. *USSR Comput. Math. Math. Phys.*, 1:1092-1096, 1961.

[Fis96] B. Fischer. *Polynomial Based Iteration Methods for Symmetric Linear Systems.* Wiley-Teubner, Leipzig, 1996.

[Fle75] R. Fletcher. Conjugate gradient methods for indefinite systems, in *Proc. of the Dundee Biennial Conference on Numerical Analysis*, G.A. Watson, Ed. Springer-Verlag, Berlin/New York, 1975.

[FN91] R.W. Freund and N.M. Nachtigal. QMR: A quasi-minimal residual method for non-Hermitian linear systems. *Numer. Math.*, 60:315-339, 1991.

[GM97] G.H. Golub and G. Meurant. Matrices, moments, and quadrature II: How to compute the norm of the error in iterative methods. *BIT*, 37:687-705, 1997.

[GS94] G.H. Golub and Z. Strakoš. Estimates in quadratic formulas. *Numer. Algorithms*, 8:241-268, 1994.

[Gre79] A. Greenbaum. Comparison of splittings used with the conjugate gradient algorithm. *Numer. Math.*, 33:181-194, 1979.

[Gre89] A. Greenbaum. Behavior of slightly perturbed Lanczos and conjugate gradient recurrences. *Lin. Alg. Appl.*, 113:7-63, 1989.

[Gre97] A. Greenbaum. *Iterative Methods for Solving Linear Systems.* SIAM, Philadelphia, 1997.

[Gre97b] A. Greenbaum. Estimating the attainable accuracy of recursively computed residual methods. *SIAM J. Matrix Anal. Appl.*, 18:535-551, 1997.

[Gre02] A. Greenbaum. Generalizations of the field of values useful in the study of polynomial functions of a matrix. *Lin. Alg. Appl.*, 347:233-249, 2002.

[Gre04] A. Greenbaum. Some theoretical results derived from polynomial numerical hulls of Jordan blocks. *Electron. Trans. Numer. Anal.*, 18:81-90, 2004.

[GG94] A. Greenbaum and L. Gurvits. Max-min properties of matrix factor norms *SIAM J. Sci. Comput.*, 15:348-358, 1994.

[GPS96] A. Greenbaum, V. Pták, and Z. Strakoš. Any non-increasing convergence curve is possible for GMRES *SIAM J. Matrix Anal. Appl.*, 17:465-469, 1996.

[GRS97] A. Greenbaum, M. Rozložník, and Z. Strakoš. Numerical behavior of the modified Gram-Schmidt GMRES implementation. *BIT*, 37:706-719, 1997.

[GS92] A. Greenbaum and Z. Strakoš. Behavior of finite precision Lanczos and conjugate gradient computations. *SIAM J. Matrix Anal. Appl.*, 13:121-137, 1992.

[GS94b] A. Greenbaum and Z. Strakoš. Matrices that generate the same Krylov residual spaces. in *Recent Advances in Iterative Methods*, pp. 95-118, G. Golub, A. Greenbaum, and M. Luskin, Eds. Springer-Verlag, Berlin, New York, 1994.

[Gus78] I. Gustafsson. A class of 1st order factorization methods. *BIT*, 18:142-156, 1978.

[GS00] M. Gutknecht and Z. Strakoš. Accuracy of two three-term and three two-term recurrences for Krylov space solvers. *SIAM J. Matrix Anal. Appl.*, 22:213-229, 2000.

[Hac94] W. Hackbusch. *Iterative Solution of Large Sparse Systems of Equations*. Springer-Verlag, Berlin, New York, 1994.

[HS52] M.R. Hestenes and E. Stiefel. Methods of conjugate gradients for solving linear systems. *J. Res. Natl. Bur. Standards*, 49:409-435, 1952.

[Jou96] W. Joubert. A robust GMRES-based adaptive polynomial preconditioning algorithm for nonsymmetric linear systems. *SIAM J. Sci. Comput.*, 15:427-439, 1994.

[Lan50] C. Lanczos. An iteration method for the solution of the eigenvalue problem of linear differential and integral operators. *J. Res. Natl. Bur. Standards*, 45:255-282, 1950.

[ML82] D.R. McCoy and E.W. Larsen. Unconditionally stable diffusion-synthetic acceleration methods for the slab geometry discrete ordinates equations, part II: Numerical results *Nuclear Sci. Engrg.*, 82:64-70, 1982.

[MV77] J.A. Meijerink and H.A. van der Vorst. An iterative solution method for linear systems of which the coefficient matrix is a symmetric M-matrix. *Math. Comp.*, 31:148-162, 1977.

[Mor95] R.B. Morgan. A restarted GMRES method augmented with eigenvectors. *SIAM J. Matrix Anal. Appl.*, 16:1154-1171, 1995.

[Mun80] N. Munksgaard. Solving sparse symmetric sets of linear equations by preconditioned conjugate gradients. *ACM Trans. Math. Software*, 6:206-219, 1980.

[Nac91] N. Nachtigal. A look-ahead variant of the Lanczos algorithm and its application to the quasi-minimal residual method for non-Hermitian linear systems. PhD dissertation, Massachusetts Institute of Technology, 1991.

[Pai76] C.C. Paige. Error analysis of the Lanczos algorithm for tridiagonalizing a symmetric matrix. *J. Inst. Math. Appl.*, 18:341-349, 1976.

[PTL85] B.N. Parlett, D.R. Taylor, and Z.A. Liu. A look-ahead Lanczos algorithm for unsymmetric matrices. *Math. Comp.*, 44:105-124, 1985.

[QV99] A. Quarteroni and A. Valli. *Domain Decomposition Methods for Partial Differential Equations*. Clarendon, Oxford, 1999.

[RS87] J.W. Ruge and K. Stüben. Algebraic multigrid, in *Multigrid Methods*, S.F. McCormick, Ed., p. 73. SIAM, Philadelphia, 1987.

[Saa93] Y. Saad. A flexible inner-outer preconditioned GMRES algorithm. *SIAM J. Sci. Comput.*, 14:461–469, 1993.

[Saa03] Y. Saad. *Iterative Methods for Sparse Linear Systems*, 2nd Edition. SIAM, Philadelphia, 2003.

[SS86] Y. Saad and M.H. Schultz. GMRES: A generalized minimal residual algorithm for solving nonsymmetric linear systems. *SIAM J. Sci. Stat. Comput.*, 7:856-869, 1986.

[SS03] V. Simoncini and D.B. Szyld. Theory of inexact Krylov subspace methods and applications to scientific computing. *SIAM J. Sci. Comp.*, 28:454–477, 2003.

[SBG96] B.F. Smith, P.E. Bjorstad, and W.D. Gropp. *Parallel Multilevel Methods for Elliptic Partial Differential Equations.* Cambridge Univ. Press, Cambridge/New York/Melbourne, 1996.

[Son89] P. Sonneveld. CGS, a fast Lanczos-type solver for nonsymmetric linear systems. *SIAM J. Sci. Stat. Comput.*, 10:36-52, 1989.

[ST02] Z. Strakoš and P. Tichý. On error estimation in the conjugate gradient method and why it works in finite precision computations. *Electron. Trans. Numer. Anal.*, 13:56-80, 2002.

[TE05] L.N. Trefethen and M. Embree. *Spectra and Pseudospectra*, Princeton Univ. Press, Princeton, 2005.

[Var60] R.S. Varga. Factorization and normalized iterative methods, in *Boundary Problems in Differential Equations*, R.E. Langer, Ed., pp. 121-142. 1960.

[Vor90] H.A. van der Vorst. The convergence behavior of preconditioned CG and CG-S in the presence of rounding errors, in *Preconditioned Conjugate Gradient Methods*, O. Axelsson and L. Kolotilina, Eds., Lecture Notes in Mathematics 1457. Springer-Verlag, Berlin, 1990.

[Vor92] H.A. van der Vorst. Bi-CGSTAB: A fast and smoothly converging variant of Bi-CG for the solution of nonsymmetric linear systems. *SIAM J. Sci. Comput.*, 13:631–644, 1992.

[Vor03] H.A. van der Vorst. *Iterative Krylov Methods for Large Linear Systems.* Cambridge University Press, Cambridge, 2003.

[Woz80] H. Wozniakowski. Roundoff error analysis of a new class of conjugate gradient algorithms. *Lin. Alg. Appl.*, 29:507–529, 1980.

Numerical Methods for Eigenvalues

42

Symmetric Matrix Eigenvalue Techniques

Ivan Slapničar
University of Split, Croatia

The eigenvalue decomposition (EVD) is an infinite iterative procedure — finding eigenvalues is equivalent to finding zeros of the characteristic polynomial, and, by the results of Abel and Galois, there is no algebraic formula for roots of the polynomial of degree greater than four. However, the number of arithmetic operations required to compute EVD to some prescribed accuracy is also finite — EVD of a general symmetric matrix requires $O(n^3)$ operations, while for matrices with special structure this number can be smaller. For example, the EVD of a tridiagonal matrix can be computed in $O(n^2)$ operations (see Sections 42.5 and 42.6).

Basic methods for the symmetric eigenvalue computations are the power method, the inverse iteration method, and the QR iteration method (see Section 42.1). Since direct application of those methods to a general symmetric matrix requires $O(n^4)$ operations, the most commonly used algorithms consist of two steps: the given matrix is first reduced to tridiagonal form, followed by the computation of the EVD of the tridiagonal matrix by QR iteration, the divide and conquer method, bisection and inverse iteration, or the method of multiple relatively robust representations. Two other methods are the Jacobi method, which does not require tridiagonalization, and the Lanczos method, which computes only a part of the tridiagonal matrix.

Design of an efficient algorithm must take into account the target computer, the desired speed and accuracy, the specific goal (whether all or some eigenvalues and eigenvectors are desired), and the matrix size and structure (small or large, dense or sparse, tridiagonal, etc.). For example, if only some eigenvalues and eigenvectors are required, one can use the methods of Sections 42.5, 42.6, and 42.8. If high relative accuracy is desired and the matrix is positive definite, the Jacobi method is the method of choice. If the matrix is sparse, the Lanczos method should be used. We shall cover the most commonly used algorithms, like those which are implemented in LAPACK (see Chapter 75) and MATLAB®(see Chapter 71). The algorithms provided in this chapter are intended to assist the reader in understanding the methods. Since the actual software is very complex, the reader is advised to use professional software in practice.

Efficient algorithms should be designed to use BLAS, and especially BLAS 3, as much as possible (see Chapter 74). The reasons are twofold: First, calling predefined standardized routines makes programs shorter and more easily readable, and second, processor vendors can optimize sets of standardized routines for their processor beyond the level given by compiler optimization. Examples of such optimized libraries are the *Intel Math Kernel Library* and *AMD Core Math Library*. Both libraries contain processor optimized BLAS, LAPACK, and FFT routines.

This chapter deals only with the computation of EVD of real symmetric matrices. The need to compute EVD of a complex Hermitian matrix (see Chapter 8) does not arise often in applications, and it is theoretically and numerically similar to the real symmetric case addressed here. All algorithms described in this chapter have their Hermitian counterparts (see e.g., [ABB99], [LSY98], and Chapters 71, 75, and 76).

The chapter is organized as follows: In Section 42.1, we describe basic methods for EVD computations. These methods are necessary to understand algorithms of Sections 42.3 to 42.6. In Section 42.2, we describe tridiagonalization by Householder reflections and Givens rotations. In Sections 42.3 to 42.6, we describe methods for computing the EVD of a tridiagonal matrix—QR iteration, the divide and conquer method, bisection and inverse iteration, and the method of multiple relatively robust representations, respectively. The Jacobi method is described in Section 42.7 and the Lanczos method is described in Section 42.8. For each method, we also describe the existing LAPACK or Matlab implementations. The respective timings of the methods are given in Section 42.9.

42.1 Basic Methods

Definitions:

The **eigenvalue decomposition** (EVD) of a real symmetric matrix $A = [a_{ij}]$ is given by $A = U\Lambda U^T$, where U is a $n \times n$ real orthonormal matrix, $U^TU = UU^T = I_n$, and $\Lambda = \text{diag}(\lambda_1, \ldots, \lambda_n)$ is a real diagonal matrix.

The numbers λ_i are the **eigenvalues** of A, the columns $\mathbf{u}_i$, $i = 1, \ldots, n$, of U are the **eigenvectors** of A, and $A\mathbf{u}_i = \lambda_i \mathbf{u}_i$, $i = 1, \ldots, n$.

If $|\lambda_1| > |\lambda_2| \geq \cdots \geq |\lambda_n|$, we say that λ_1 is the **dominant eigenvalue**.

Deflation is a process of reducing the size of the matrix whose EVD is to be determined, given that one eigenvector is known (see Fact 4 below for details).

The **shifted matrix** of the matrix A is the matrix $A - \mu I$, where μ is the **shift**.

The simplest method for computing the EVD (also in the unsymmetric case) is the **power method**: given starting vector $\mathbf{x}_0$, the method computes the sequences

$$\nu_k = \mathbf{x}_k^T A\mathbf{x}_k, \qquad \mathbf{x}_{k+1} = A\mathbf{x}_k / \|A\mathbf{x}_k\|, \qquad k = 0, 1, 2, \ldots, \tag{42.1}$$

until convergence. Normalization of $\mathbf{x}_k$ can be performed in any norm and serves the numerical stability of the algorithm (avoiding overflow or underflow).

Inverse iteration is the power method applied to the inverse of a shifted matrix, starting from $\mathbf{x}_0$:

$$\nu_k = \mathbf{x}_k^T A\mathbf{x}_k, \quad \mathbf{v}_{k+1} = (A - \mu I)^{-1}\mathbf{x}_k, \quad \mathbf{x}_{k+1} = \mathbf{v}_{k+1}/\|\mathbf{v}_{k+1}\|, \quad k = 0, 1, 2, \ldots. \tag{42.2}$$

Given starting $n \times p$ matrix X_0 with orthonormal columns, the **orthogonal iteration** (also **subspace iteration**) forms the sequence of matrices

$$Y_{k+1} = AX_k, \qquad Y_{k+1} = X_{k+1}R_{k+1} \quad \text{(QR factorization)}, \qquad k = 0, 1, 2, \ldots, \tag{42.3}$$

where $X_{k+1}R_{k+1}$ is the reduced QR factorization of Y_{k+1} (X_{k+1} is an $n \times p$ matrix with orthonormal columns and R_{k+1} is a upper triangular $p \times p$ matrix).

Starting from the matrix $A_0 = A$, the **QR iteration** forms the sequence of matrices

$$A_k = Q_k R_k \quad \text{(QR factorization)}, \qquad A_{k+1} = R_k Q_k, \qquad k = 0, 1, 2, \ldots \tag{42.4}$$

Given the shift μ, the **shifted QR iteration** forms the sequence of matrices

$$A_k - \mu I = Q_k R_k \quad \text{(QR factorization)}, \quad A_{k+1} = R_k Q_k + \mu I, \quad k = 0, 1, 2, \ldots \tag{42.5}$$

Facts:

The Facts 1 to 14 can be found in [GV96, §8.2], [Par80, §4, 5], [Ste01, §2.1, 2.2.1, 2.2.2], and [Dem97, §4].

1. If λ_1 is the dominant eigenvalue and if $\mathbf{x}_0$ is not orthogonal to $\mathbf{u}_1$, then in Equation 42.1 $\nu_k \to \lambda_1$ and $\mathbf{x}_k \to \mathbf{u}_1$. In other words, the power method converges to the dominant eigenvalue and its eigenvector.
2. The convergence of the power method is linear in the sense that

$$|\lambda_1 - \nu_k| = O\left(\left|\frac{\lambda_2}{\lambda_1}\right|^k\right), \qquad \|\mathbf{u}_1 - \mathbf{x}_k\|_2 = O\left(\left|\frac{\lambda_2}{\lambda_1}\right|^k\right).$$

   More precisely,

$$|\lambda_1 - \nu_k| \approx \left|\frac{c_2}{c_1}\right| \left|\frac{\lambda_2}{\lambda_1}\right|^k,$$

   where c_i is the coefficient of the i-th eigenvector in the linear combination expressing the starting vector $\mathbf{x}_0$.
3. Since λ_1 is not readily available, the convergence is in practice determined using residuals. If $\|A\mathbf{x}_k - \nu_k \mathbf{x}_k\|_2 \leq tol$, where tol is a user prescribed stopping criterion, then $|\lambda_1 - \nu_k| \leq tol$.
4. After computing the dominant eigenpair, we can perform deflation to reduce the given EVD to the one of size $n-1$. Let $Y = [\mathbf{u}_1 \quad X]$ be an orthogonal matrix. Then

$$\begin{bmatrix} \mathbf{u}_1 & X \end{bmatrix}^T A \begin{bmatrix} \mathbf{u}_1 & X \end{bmatrix} = \begin{bmatrix} \lambda_1 & 0 \\ 0 & A_1 \end{bmatrix},$$

   where $A_1 = X^T A X$.
5. The EVD of the shifted matrix $A - \mu I$ is given by $U(\Lambda - \mu I)U^T$. Sometimes we can choose shift μ such that the shifted matrix $A - \mu I$ has better ratio between the dominant eigenvalue and the absolutely closest one, than the original matrix. In this case, applying the power method to the shifted matrix will speed up the convergence.
6. Inverse iteration requires solving the system of linear equations $(A - \mu I)\mathbf{v}_{k+1} = \mathbf{x}_k$ for $\mathbf{v}_{k+1}$ in each step. At the beginning, we must compute the LU factorization of $A - \mu I$, which requires $2n^3/3$ operations and in each subsequent step we must solve two triangular systems, which requires $2n^2$ operations.
7. If μ is very close to some eigenvalue of A, then the eigenvalues of the shifted matrix satisfy $|\lambda_1| \gg |\lambda_2| \geq \cdots \geq |\lambda_n|$, so the convergence of the inverse iteration method is very fast.
8. If μ is very close to some eigenvalue of A, then the matrix $A - \mu I$ is nearly singular, so the solutions of linear systems may have large errors. However, these errors are almost entirely in the direction of the dominant eigenvector so the inverse iteration method is both fast and accurate.
9. We can further increase the speed of convergence of inverse iterations by substituting the shift μ with the Rayleigh quotient ν_k in each step, at the cost of computing new LU factorization each time. See Chapter 8.2 for more information about the Rayleigh quotient.
10. If

$$|\lambda_1| \geq \cdots \geq |\lambda_p| > |\lambda_{p+1}| \geq \cdots \geq |\lambda_n|,$$

then the subspace iteration given in Equation 42.3 converges such that

$$X_k \to [\mathbf{u}_1, \ldots, \mathbf{u}_p], \qquad X_k^T A X_k \to \text{diag}(\lambda_1, \ldots, \lambda_p),$$

at a speed which is proportional to $|\lambda_{p+1}/\lambda_p|^k$.

11. If $|\lambda_1| > |\lambda_2| > \cdots > |\lambda_n|$, then the sequence of matrices A_k generated by the QR iteration given in Equation 42.4 converges to diagonal matrix Λ. However, this result is not of practical use, since the convergence may be very slow and each iteration requires $O(n^3)$ operations. Careful implementation, like the one described in section 42.3, is needed to construct an useful algorithm.
12. The QR iteration is equivalent to orthogonal iteration starting with the matrix $X_0 = I$. More precisely, the matrices X_k from Equation 42.3 and A_k from Equation 42.4 satisfy $X_k^T A X_k = A_k$.
13. Matrices A_k and A_{k+1} from Equations 42.4 and Equation 42.5 are orthogonally similar. In both cases

$$A_{k+1} = Q_k^T A_k Q_k.$$

14. The QR iteration method is essentially equivalent to the power method and the shifted QR iteration method is essentially equivalent to the inverse power method on the shifted matrix.
15. [Wil65, §3, 5, 6, 7] [TB97, §V] Let $U\Lambda U^T$ and $\tilde{U}\tilde{\Lambda}\tilde{U}^T$ be the exact and the computed EVDs of A, respectively, such that the diagonals of Λ and $\tilde{\Lambda}$ are in the same order. Numerical methods generally compute the EVD with the errors bounded by

$$|\lambda_i - \tilde{\lambda}_i| \leq \phi\epsilon \|A\|_2, \qquad \|\mathbf{u}_i - \tilde{\mathbf{u}}_i\|_2 \leq \psi\epsilon \frac{\|A\|_2}{\min_{j\neq i} |\lambda_i - \tilde{\lambda}_j|},$$

where ϵ is machine precision and ϕ and ψ are slowly growing polynomial functions of n which depend upon the algorithm used (typically $O(n)$ or $O(n^2)$).

Examples:

1. The eigenvalue decomposition of the matrix

$$A = \begin{bmatrix} 4.5013 & 0.6122 & 2.1412 & 2.0390 \\ 0.6122 & 2.6210 & -0.4941 & -1.2164 \\ 2.1412 & -0.4941 & 1.1543 & -0.1590 \\ 2.0390 & -1.2164 & -0.1590 & -0.9429 \end{bmatrix}$$

computed by the MATLAB command `[U,Lambda]=eig(A)` is $A = U\Lambda U^T$ with (properly rounded to four decimal places)

$$U \begin{bmatrix} -0.3697 & 0.2496 & 0.1003 & -0.8894 \\ 0.2810 & -0.0238 & 0.9593 & -0.0153 \\ 0.3059 & -0.8638 & -0.1172 & -0.3828 \\ 0.8311 & 0.4370 & -0.2366 & -0.2495 \end{bmatrix}, \quad \Lambda = \begin{bmatrix} -2.3197 & 0 & 0 & 0 \\ 0 & 0.6024 & 0 & 0 \\ 0 & 0 & 3.0454 & 0 \\ 0 & 0 & 0 & 6.0056 \end{bmatrix}.$$

2. Let A, U, and Λ be as in the Example 1, and set $\mathbf{x}_0 = [1 \quad 1 \quad 1 \quad 1]^T$. The power method in Equation 42.1 gives $\mathbf{x}_6 = [0.8893 \quad 0.0234 \quad 0.3826 \quad 0.2496]^T$. By setting $\mathbf{u}_1 = -U_{:,4}$ we have $\|\mathbf{u}_1 - \mathbf{x}_6\|_2 = 0.0081$. Here (Fact 2), $c_2 = 0.7058$, $c_1 = -1.5370$, and

$$\left|\frac{c_2}{c_1}\right| \left|\frac{\lambda_2}{\lambda_1}\right|^6 = 0.0078.$$

Similarly, $\|\mathbf{u}_1 - \mathbf{x}_{50}\|_2 = 1.3857 \cdot 10^{-15}$. However, for a different (bad) choice of the starting vector, $\mathbf{x}_0 = [0 \quad 1 \quad 0 \quad 0]^T$, where $c_2 = 0.9593$ and $c_1 = -0.0153$, we have $\|\mathbf{u}_1 - \mathbf{x}_6\|_2 = 0.7956$.

3. The deflation matrix Y and the deflated matrix A_1 (Fact 4) for the above example are equal to (correctly rounded):

$$Y = \begin{bmatrix} -0.8894 & -0.0153 & -0.3828 & -0.2495 \\ -0.0153 & 0.9999 & -0.0031 & -0.0020 \\ -0.3828 & -0.0031 & 0.9224 & -0.0506 \\ -0.2495 & -0.0020 & -0.0506 & 0.9670 \end{bmatrix},$$

$$A_1 = \begin{bmatrix} 6.0056 & 0 & 0 & 0 \\ 0 & 2.6110 & -0.6379 & -1.3154 \\ 0 & -0.6379 & 0.2249 & -0.8952 \\ 0 & -1.3154 & -0.8952 & -1.5078 \end{bmatrix}.$$

4. Let A and $\mathbf{x}_0$ be as in Example 2. For the shift $\mu = 6$, the inverse iteration method in Equation 42.2 gives $\|\mathbf{u}_1 - \mathbf{x}_6\|_2 = 6.5187 \cdot 10^{-16}$, so the convergence is much faster than in Example 2 (Fact 7).
5. Let A be as in Example 1. Applying six steps of the QR iteration in Equation 42.4 gives

$$A_6 = \begin{bmatrix} 6.0055 & -0.0050 & -0.0118 & -0.0000 \\ -0.0050 & 3.0270 & 0.3134 & 0.0002 \\ -0.0118 & 0.3134 & -2.3013 & -0.0017 \\ -0.0000 & 0.0002 & -0.0017 & 0.6024 \end{bmatrix}.$$

and applying six steps of the shifted QR iteration in Equation 42.5 with $\mu = 6$ gives

$$A_6 = \begin{bmatrix} -2.3123 & 0.1452 & -0.0215 & -0.0000 \\ 0.1452 & 0.6623 & 0.4005 & 0.0000 \\ -0.0215 & 0.4005 & 2.9781 & -0.0000 \\ 0.0000 & 0.0000 & 0.0000 & 6.0056 \end{bmatrix}.$$

In this case both methods converge. The convergence towards the matrix where the eigenvalue nearest to the shift can be deflated is faster for the shifted iterations.

42.2 Tridiagonalization

The QR iteration in Equation 42.4 in Section 42.1 and the shifted QR iteration in Equation 42.5 in Section 42.1 require $O(n^3)$ operations (one QR factorization) for each step, which makes these algorithms highly unpractical. However, if the starting matrix is tridiagonal, one step of these iterations requires only $O(n)$ operations. As a consequence, the practical algorithm consists of three steps:

1. Reduce A to tridiagonal form T by orthogonal similarities, $X^T AX = T$.
2. Compute the EVD of T, $T = Q\Lambda Q^T$.
3. Multiply $U = XQ$.

The EVD of A is then $A = U\Lambda U^T$. Reduction to tridiagonal form can be performed by using Householder reflectors or Givens rotations and it is a finite process requiring $O(n^3)$ operations. Reduction to tridiagonal form is a considerable compression of data since an EVD of T can be computed very quickly. The EVD of T can be efficiently computed by various methods such as QR iteration, the divide and conquer method (DC), bisection and inverse iteration, or the method of multiple relatively robust representations (MRRR). These methods are described in subsequent sections.

Facts:

All the following facts, except Fact 6, can be found in [Par80, §7], [TB97, pp. 196–201], [GV96, §8.3.1], [Ste01, pp. 158–162], and [Wil65, pp. 345–367].

1. Tridiagonal form is not unique (see Examples 1 and 2).
2. The reduction of A to tridiagonal matrix by Householder reflections is performed as follows. Let us partition A as
$$A = \left[\begin{array}{c|c} a_{11} & \mathbf{a}^T \\ \hline \mathbf{a} & B \end{array}\right].$$
Let H be the appropriate Householder reflection (see Chapter 38.4), that is,
$$\mathbf{v} = \mathbf{a} + \operatorname{sign}(a_{21})\|\mathbf{a}\|_2\mathbf{e}_1, \qquad H = I - 2\frac{\mathbf{v}\mathbf{v}^T}{\mathbf{v}^T\mathbf{v}},$$
and let
$$H_1 = \left[\begin{array}{c|c} 1 & \mathbf{0}^T \\ \hline \mathbf{0} & H \end{array}\right].$$
Then
$$H_1AH_1 = \left[\begin{array}{c|c} a_{11} & \mathbf{a}^T H \\ \hline H\mathbf{a} & HBH \end{array}\right] = \left[\begin{array}{c|c} a_{11} & \nu\mathbf{e}_1^T \\ \hline \nu\mathbf{e}_1 & A_1 \end{array}\right], \qquad \nu = -\operatorname{sign}(a_{21})\|\mathbf{a}\|_2.$$
This step annihilates all elements in the first column below the first subdiagonal and all elements in the first row to the right of the first subdiagonal. Applying this procedure recursively yields the triangular matrix $T = X^T AX$, $X = H_1H_2\cdots H_{n-2}$.
3. H does not depend on the normalization of $\mathbf{v}$. The normalization $v_1 = 1$ is useful since $\mathbf{a}_{2:n}$ can be overwritten by $\mathbf{v}_{2:n}$ and v_1 does not need to be stored.
4. Forming H explicitly and then computing $A_1 = HBH$ requires $O(n^3)$ operations, which would ultimately yield an $O(n^4)$ algorithm. However, we do not need to form the matrix H explicitly — given $\mathbf{v}$, we can overwrite B with HBH in just $O(n^2)$ operations by using one matrix-vector multiplication and two rank-one updates.
5. The entire tridiagonalization algorithm is as follows:

Algorithm 1: Tridiagonalization by Householder reflections
Input: real symmetric $n \times n$ matrix A
Output: the main diagonal and sub- and superdiagonal of A are overwritten by T, the Householder vectors are stored in the lower triangular part of A below the first subdiagonal

for $j = 1 : n-2$
  $\mu = \operatorname{sign}(a_{j+1,j})\|A_{j+1:n,j}\|_2$
  if $\mu \neq 0$, **then**
    $\beta = a_{j+1,j} + \mu$
    $\mathbf{v}_{j+2:n} = A_{j+2:n,j}/\beta$
  endif
  $a_{j+1,j} = -\mu$
  $a_{j,j+1} = -\mu$
  $v_{j+1} = 1$
  $\gamma = -2/\mathbf{v}_{j+1:n}^T\mathbf{v}_{j+1:n}$
  $\mathbf{w} = \gamma A_{j+1:n,j+1:n}\mathbf{v}_{j+1:n}$
  $\mathbf{q} = \mathbf{w} + \frac{1}{2}\gamma\mathbf{v}_{j+1:n}(\mathbf{v}_{j+1:n}^T\mathbf{w})$
  $A_{j+1:n,j+1:n} = A_{j+1:n,j+1:n} + \mathbf{v}_{j+1:n}\mathbf{q}^T + \mathbf{q}\mathbf{v}_{j+1:n}^T$
  $A_{j+2:n,j} = \mathbf{v}_{j+2:n}$
endfor

6. [DHS89] When symmetry is exploited in performing rank-2 update, Algorithm 1 requires $4n^3/3$ operations. Another important enhancement is the derivation of the block-version of the algorithm. Instead of performing rank-2 update on B, thus obtaining A_1, we can accumulate p transformations and perform rank-$2p$ update. In the first p steps, the algorithm is modified to update only columns and rows $1, \ldots, p$, which are needed to compute the first p Householder vectors. Then the matrix A is updated by $A - UV^T - VU^T$, where U and V are $n \times p$ matrices. This algorithm is rich in matrix–matrix multiplications (roughly one half of the operations is performed using BLAS 3 routines), but it requires extra workspace for U and V.
7. If the matrix X is needed explicitly, it can be computed from the stored Householder vectors by Algorithm 2. In order to minimize the operation count, the computation starts from the smallest matrix and the size is gradually increased, that is, the algorithm computes the sequence of matrices

$$H_{n-2}, \quad H_{n-3}H_{n-2}, \ldots, \quad X = H_1 \cdots H_{n-2}.$$

A column-oriented version is possible as well, and the operation count in both cases is $4n^3/3$. If the Householder matrices H_i are accumulated in the order in which they are generated, the operation count is $2n^3$.

Algorithm 2: Computation of the tridiagonalizing matrix X
Input: output from Algorithm 1
Output: matrix X such that $X^T AX = T$, where A is the input of Algorithm 1
and T is tridiagonal.

$X = I_n$
for $j = n - 2 : -1 : 1$
  $v_{j+1} = 1$
  $\mathbf{v}_{j+2:n} = A_{j+2:n,j}$
  $\gamma = -2/\mathbf{v}_{j+1:n}^T \mathbf{v}_{j+1:n}$
  $\mathbf{w} = \gamma X_{j+1:n,j+1:n}^T \mathbf{v}_{j+1:n}$
  $X_{j+1:n,j+1:n} = X_{j+1:n,j+1:n} + \mathbf{v}_{j+1:n}\mathbf{w}^T$
endfor

8. The error bounds for Algorithms 1 and 2 are as follows: The matrix $\tilde{T}$ computed by Algorithm 1 is equal to the matrix, which would be obtained by exact tridiagonalization of some perturbed matrix $A + E$ (backward error), where $\|E\|_2 \leq \psi\epsilon\|A\|_2$ and ψ is a slowly increasing function of n. The matrix $\tilde{X}$ computed by Algorithm 2 satisfies $\tilde{X} = X + F$, where $\|F\|_2 \leq \phi\epsilon$ and ϕ is a slowly increasing function of n.
9. Givens rotation parameters c and s are computed as in Fact 5 of Section 38.4. Tridiagonalization by Givens rotations is performed as follows:

Algorithm 3: Tridiagonalization by Givens rotations
Input: real symmetric $n \times n$ matrix A
Output: the matrix X such that $X^T AX = T$ is tridiagonal, main diagonal
and sub- and superdiagonal of A are overwritten by T

$X = I_n$
for $j = 1 : n - 2$
  for $i = j + 2 : n$
    set $x = a_{j+1,j}$ and $y = a_{i,j}$
    compute $G = \begin{bmatrix} c & s \\ -s & c \end{bmatrix}$ via Fact 5 of Section 38.4

Algorithm 3: Tridiagonalization by Givens rotations (Continued)

$$\begin{bmatrix} A_{j+1,j:n} \\ A_{i,j:n} \end{bmatrix} = G \begin{bmatrix} A_{j+1,j:n} \\ A_{i,j:n} \end{bmatrix}$$

$$\begin{bmatrix} A_{j:n,j+1} & A_{j:n,i} \end{bmatrix} = \begin{bmatrix} A_{j:n,j+1} & A_{j:n,i} \end{bmatrix} G^T$$

$$\begin{bmatrix} X_{1:n,j+1} & X_{1:n,i} \end{bmatrix} = \begin{bmatrix} X_{1:n,j+1} & X_{1:n,i} \end{bmatrix} G^T$$

endfor

endfor

10. Algorithm 3 requires $(n-1)(n-2)/2$ plane rotations, which amounts to $4n^3$ operations if symmetry is properly exploited. The operation count is reduced to $8n^3/3$ if fast rotations are used. Fast rotations are obtained by factoring out absolutely larger of c and s from G.
11. The Givens rotations in Algorithm 3 can be performed in different orderings. For example, the elements in the first column and row can be annihilated by rotations in the planes $(n-1, n)$, $(n-2, n-1), \ldots$ $(2, 3)$. Since Givens rotations act more selectively than Householder reflectors, they can be useful if A has some special structure. For example, Givens rotations are used to efficiently tridiagonalize symmetric band matrices (see Example 4).
12. Error bounds for Algorithm 3 are the same as the ones for Algorithms 1 and 2 (Fact 8), but with slightly different functions ψ and ϕ.

Examples:

1. Algorithms 1 and 2 applied to the matrix A from Example 1 in Section 42.1 give

$$T = \begin{bmatrix} 4.5013 & -3.0194 & 0 & 0 \\ -3.0194 & -0.3692 & 1.2804 & 0 \\ 0 & 1.2804 & 0.5243 & -0.9303 \\ 0 & 0 & -0.9303 & 2.6774 \end{bmatrix},$$

$$X = \begin{bmatrix} 1 & 0 & 0 & 0 \\ 0 & -0.2028 & 0.4417 & -0.8740 \\ 0 & -0.7091 & -0.6817 & -0.1800 \\ 0 & -0.6753 & 0.5833 & 0.4514 \end{bmatrix}.$$

2. Tridiagonalization is implemented in the MATLAB function `T = hess(A)` (`[X,T] = hess(A)` if X is to be computed, as well). In fact, the function `hess` is more general and it computes the Hessenberg form of a general square matrix. For the same matrix A as above, the matrices T and X computed by `hess` are:

$$T = \begin{bmatrix} 2.6562 & 1.3287 & 0 & 0 \\ 1.3287 & 2.4407 & 2.4716 & 0 \\ 0 & 2.4716 & 3.1798 & 2.3796 \\ 0 & 0 & 2.3796 & -0.9429 \end{bmatrix}, \quad X = \begin{bmatrix} 0.4369 & 0.2737 & 0.8569 & 0 \\ 0.7889 & 0.3412 & -0.5112 & 0 \\ -0.4322 & 0.8993 & -0.0668 & 0 \\ 0 & 0 & 0 & 1.0000 \end{bmatrix}.$$

3. The block version of tridiagonal reduction is implemented in the LAPACK subroutine DSYTRD (file `dsytrd.f`). The computation of X is implemented in the subroutine DORGTR. The size of the required extra workspace (in elements) is $lwork = nb * n$, where nb is the optimal block size (here, $nb = 64$), and it is determined automatically by the subroutines. The timings are given in Section 42.9.

4. Computation of Givens rotation in Algorithm 3 is implemented in the MATLAB functions `planerot` and `givens`, BLAS 1 subroutine DROTG, and LAPACK subroutine DLARTG. These implementations avoid unnecessary overflow or underflow by appropriately scaling x and y. Plane rotations (multiplications with G) are implemented in the BLAS 1 subroutine DROT. LAPACK subroutines DLAR2V, DLARGV, and DLARTV generate and apply multiple plane rotations. LAPACK subroutine DSBTRD tridiagonalizes a symmetric band matrix by using Givens rotations.

42.3 Implicitly Shifted QR Method

This method is named after the fact that, for a tridiagonal matrix, each step of the shifted QR iterations given by Equation 42.5 in Section 42.1 can be elegantly implemented without explicitly computing the shifted matrix $A_k - \mu I$.

Definitions:

Wilkinson's shift μ is the eigenvalue of the bottom right 2×2 submatrix of T, which is closer to $t_{n,n}$.

Facts:

The following facts can be found in [GV96, pp. 417–422], [Ste01, pp. 163–171], [TB97, pp. 211–224], [Par80, §8], [Dem97, §5.3.1], and [Wil65, §8.50, 8.54].

$T = [t_{ij}]$ is a real symmetric tridiagonal matrix of order n and $T = Q\Lambda Q^T$ is its EVD.

1. The stable formula for the Wilkinson's shift is

$$\mu = t_{n,n} - \frac{t_{n,n-1}^2}{\tau + \operatorname{sign}(\tau)\sqrt{\tau^2 + t_{n,n-1}^2}}, \qquad \tau = \frac{t_{n-1,n-1} - t_{n,n}}{2}.$$

2. The following recursive function implements the implicitly shifted QR method given by Equation 42.5:

Algorithm 4: Implicitly shifted QR method for tridiagonal matrices
Input: real symmetric tridiagonal $n \times n$ matrix T
Output: the diagonal of T is overwritten by its eigenvalues
function $T = QR\_iteration(T)$
  repeat % one sweep
    compute a suitable shift μ
    set $x = t_{11} - \mu$ and $y = t_{21}$
    compute $G = \begin{bmatrix} c & s \\ -s & c \end{bmatrix}$ via Fact 5 of Chapter 38.4
    $\begin{bmatrix} T_{1,1:3} \\ T_{2,1:3} \end{bmatrix} = G \begin{bmatrix} T_{1,1:3} \\ T_{2,1:3} \end{bmatrix}$
    $\begin{bmatrix} T_{1:3,1} & T_{1:3,2} \end{bmatrix} = \begin{bmatrix} T_{1:3,1} & T_{1:3,2} \end{bmatrix} G^T$
    for $i = 2 : n - 1$
      set $x = t_{i,i-1}$ and $y = t_{i+1,i-1}$
      compute $G = \begin{bmatrix} c & s \\ -s & c \end{bmatrix}$ via Fact 5 of Section 38.4

Algorithm 4: Implicitly shifted QR method for tridiagonal matrices (Continued)

$$\begin{bmatrix} T_{i,i-1:i+2} \\ T_{i+1,i-1:i+2} \end{bmatrix} = G \begin{bmatrix} T_{i,i-1:i+2} \\ T_{i+1,i-1:i+2} \end{bmatrix}$$

$$\begin{bmatrix} T_{i-1:i+2,i} & T_{i-1:i+2,i+1} \end{bmatrix} = \begin{bmatrix} T_{i-1:i+2,i} & T_{i-1:i+2,i+1} \end{bmatrix} G^T$$

endfor

until $|t_{i,i+1}| \leq \epsilon\sqrt{|t_{i,i} \cdot t_{i+1,i+1}|}$ for some i % deflation

set $t_{i+1,i} = 0$ and $t_{i,i+1} = 0$

$T_{1:i,1:i} = QR\_iteration(T_{1:i,1:i})$

$T_{i+1:n,i+1:n} = QR\_iteration(T_{i+1:n,i+1:n})$

3. Wilkinson's shift (Fact 1) is the most commonly used shift. With Wilkinson's shift, the algorithm always converges in the sense that $t_{n-1,n} \to 0$. The convergence is quadratic, that is, $|[T_{k+1}]_{n-1,n}| \leq c|[T_k]_{n-1,n}|^2$ for some constant c, where T_k is the matrix after k-th sweep. Even more, the convergence is usually cubic. However, it can also happen that some $t_{i,i+i}$, $i \neq n-1$, becomes sufficiently small before $t_{n-1,n}$, so the practical program has to check for deflation at each step.
4. The plane rotation parameters at the start of the sweep are computed as if the shifted matrix $T - \mu I$ has been formed. Since the rotation is applied to the original T and not to $T - \mu I$, this creates new nonzero elements at the positions $(3, 1)$ and $(1, 3)$, the so-called **bulge**. The subsequent rotations simply chase the bulge out of the lower right corner of the matrix. The rotation in the $(2, 3)$ plane sets the elements $(3, 1)$ and $(1, 3)$ back to zero, but it generates two new nonzero elements at positions $(4, 2)$ and $(2, 4)$; the rotation in the $(3, 4)$ plane sets the elements $(4, 2)$ and $(2, 4)$ back to zero, but it generates two new nonzero elements at positions $(5, 3)$ and $(3, 5)$, etc. The procedure is illustrated in Figure 42.1: "x" denotes the elements that are transformed by the current plane rotation, "$*$" denotes the newly generated nonzero elements (the bulge), and 0 denotes the zeros that are reintroduced by the current plane rotation.

   The effect of this procedure is the following. At the end of the first sweep, the resulting matrix T_1 is equal to the the matrix that would have been obtained by factorizing $T - \mu I = QR$ and computing $T_1 = RQ + \mu I$ as in Equation 42.5.
5. Since the convergence of Algorithm 4 is quadratic (or even cubic), an eigenvalue is isolated after just a few steps, which requires $O(n)$ operations. This means that $O(n^2)$ operations are needed to compute all eigenvalues.
6. If the eigenvector matrix Q is desired, the plane rotations need to be accumulated similarly to the accumulation of X in Algorithm 3. This accumulation requires $O(n^3)$ operations (see Example 2 below and Fact 5 in Section 42.9). Another, usually faster, algorithm to compute Q is given in Fact 9 in Section 42.9.

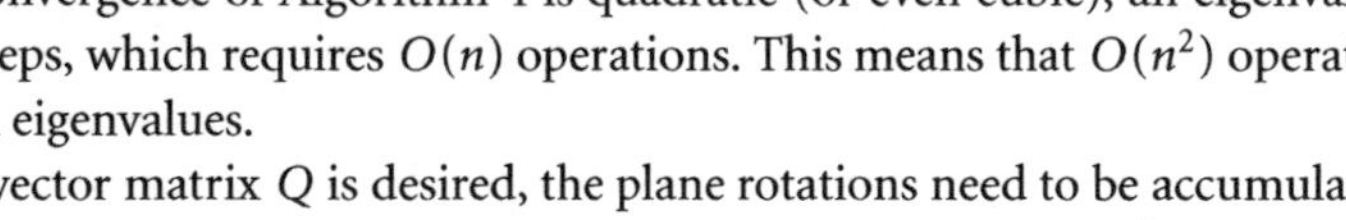

FIGURE 42.1 Chasing the bulge in one sweep of the implicit QR iteration for $n = 6$.

7. The computed eigenvalue decomposition $T = Q\Lambda Q^T$ satisfies the error bounds from Fact 15 in section 42.1 with A replaced by T and U replaced by Q. The deflation criterion implies $|t_{i,i+1}| \leq \epsilon \|T\|_F$, which is within these bounds.
8. Combining Algorithms 1, 2, and 4 we get the the following algorithm:

```
Algorithm 5: Real symmetric eigenvalue decomposition
Input: real symmetric n × n matrix A
Output: eigenvalue matrix Λ and, optionally, eigenvector matrix U of A
if only eigenvalues are required, then
    Compute T by Algorithm 1
    T = QR_iteration(T)       % Algorithm 4
    Λ = diag(T)
else
    Compute T by Algorithm 1
    Compute X by Algorithm 2
    T = QR_iteration(T)       % with rotations accumulated in Q
    Λ = diag(T)
    U = XQ
endif
```

9. The EVD computed by Algorithm 5 satisfies the error bounds given in Fact 15 in section 42.1. However, the algorithm tends to perform better on matrices, which are graded downwards, that is, on matrices that exhibit systematic decrease in the size of the matrix elements as we move along the diagonal. For such matrices the tiny eigenvalues can usually be computed with higher relative accuracy (although counterexamples can be easily constructed). If the tiny eigenvalues are of interest, it should be checked whether there exists a symmetric permutation that moves larger elements to the upper left corner, thus converting the given matrix to the one that is graded downwards.

Examples:

1. For the matrix T from Example 1 in section 42.2, after one sweep of Algorithm 4, we have

$$T = \begin{bmatrix} 2.9561 & 3.9469 & 0 & 0 \\ 3.9469 & 0.8069 & -0.7032 & 0 \\ 0 & -0.7032 & 0.5253 & 0.0091 \\ 0 & 0 & 0.0091 & 3.0454 \end{bmatrix}.$$

2. Algorithm 4 is implemented in the LAPACK subroutine DSTEQR. This routine can compute just the eigenvalues, or both eigenvalues and eigenvectors. To avoid double indices, the diagonal and subdiagonal entries of T are stored in one dimensional vectors, $d_i = T_{ii}$ and $e_i = T_{i+1,i}$, respectively. The timings are given in Section 42.9.
3. Algorithm 5 is implemented in the Matlab routine `eig`. The command `Lambda = eig(A)` returns only the eigenvalues, `[U,Lambda]=eig(A)` returns the eigenvalues and the eigenvectors (see Example 1 in Section 42.1).
4. The LAPACK implementation of Algorithm 5 is given in the subroutine DSYEV. To compute only eigenvalues, DSYEV calls DSYTRD and DSTEQR without eigenvector option. To compute both eigenvalues and eigenvectors, DSYEV calls DSYTRD, DORGTR, and DSTEQR with the eigenvector option. The timings are given in Section 42.9.

42.4 Divide and Conquer Method

This is currently the fastest method for computing the EVD of a real symmetric tridiagonal matrix T. It is based on splitting the given tridiagonal matrix into two matrices, then computing the EVDs of the smaller matrices and computing the final EVD from the two EVDs. The method was first introduced in [Cup81], but numerically stable and efficient implementation was first derived in [GE95].

Facts:

The following facts can be found in [Dem97, pp. 216–228], [Ste01, pp. 171–185], and [GE95].
$T = [t_{ij}]$ is a real symmetric tridiagonal matrix of order n and $T = U\Lambda U^T$ is its EVD.

1. Let T be partitioned as

$$T = \left[\begin{array}{cccc|cccc} d_1 & e_1 & & & & & & \\ e_1 & d_2 & e_2 & & & & & \\ & \ddots & \ddots & \ddots & & & & \\ & & e_{k-1} & d_k & e_k & & & \\ \hline & & & e_k & d_{k+1} & e_{k+1} & & \\ & & & & \ddots & \ddots & \ddots & \\ & & & & & e_{n-2} & d_{n-1} & e_{n-1} \\ & & & & & & e_{n-1} & d_n \end{array}\right] \equiv \begin{bmatrix} T_1 & e_k\mathbf{e}_k\mathbf{e}_1^T \\ e_k\mathbf{e}_1\mathbf{e}_k^T & T_2 \end{bmatrix}.$$

We assume that T is unreduced, that is, $e_i \neq 0$ for all i. Further, we assume that $e_i > 0$ for all i, which can be easily be attained by diagonal similarity with a diagonal matrix of signs (see Example 1 below). Let

$$\hat{T}_1 = T_1 - e_k\mathbf{e}_k\mathbf{e}_k^T, \qquad \hat{T}_2 = T_2 - e_k\mathbf{e}_1\mathbf{e}_1^T. \tag{42.6}$$

In other words, $\hat{T}_1$ is equal to T_1 except that d_k is replaced by $d_k - e_k$, and $\hat{T}_2$ is equal to T_2 except that d_{k+1} is replaced by $d_{k+1} - e_k$.

Let $\hat{T}_i = \hat{U}_i\hat{\Lambda}_i\hat{U}_i^T$, $i = 1, 2$, be the respective EVDs and let $\mathbf{v} = \begin{bmatrix} \hat{U}_1^T\mathbf{e}_k \\ \hat{U}_2^T\mathbf{e}_1 \end{bmatrix}$ ($\mathbf{v}$ consists of the last column of $\hat{U}_1^T$ and the first column of $\hat{U}_2^T$). Set $\hat{U} = \hat{U}_1 \oplus \hat{U}_2$ and $\hat{\Lambda} = \hat{\Lambda}_1 \oplus \hat{\Lambda}_2$. Then

$$T = \begin{bmatrix} \hat{U}_1 & \\ & \hat{U}_2 \end{bmatrix} \left[\begin{bmatrix} \hat{\Lambda}_1 & \\ & \hat{\Lambda}_2 \end{bmatrix} + e_k\mathbf{v}\mathbf{v}^T \right] \begin{bmatrix} \hat{U}_1^T & \\ & \hat{U}_2^T \end{bmatrix} = \hat{U}(\hat{\Lambda} + e_k\mathbf{v}\mathbf{v}^T)\hat{U}^T. \tag{42.7}$$

If

$$\hat{\Lambda} + e_k\mathbf{v}\mathbf{v}^T = X\Lambda X^T$$

is the EVD of the rank-one modification of the diagonal matrix $\hat{\Lambda}$, then $T = U\Lambda U^T$, where $U = \hat{U}X$ is the EVD of T. Thus, the original tridiagonal eigenvalue problem is reduced to two smaller tridiagonal eigenvalue problems and one eigenvalue problem for the rank-one update of a diagonal matrix.

2. If the matrix $\hat{\Lambda} + e_k\mathbf{v}\mathbf{v}^T$ is permuted such that $\hat{\lambda}_1 \geq \cdots \geq \hat{\lambda}_n$, then λ_i and $\hat{\lambda}_i$ are interlaced, that is,

$$\lambda_1 \geq \hat{\lambda}_1 \geq \lambda_2 \geq \hat{\lambda}_2 \geq \cdots \geq \lambda_{n-1} \geq \hat{\lambda}_{n-1} \geq \lambda_n \geq \hat{\lambda}_n.$$

Moreover, if $\hat{\lambda}_{i-1} = \hat{\lambda}_i$ for some i, then one eigenvalue is obviously known exactly, that is, $\lambda_i = \hat{\lambda}_i$. In this case, λ_i can be deflated by applying to $\hat{\Lambda} + e_k\mathbf{v}\mathbf{v}^T$ a plane rotation in the $(i-1, i)$ plane, where the Givens rotation parameters c and s are computed from v_{i-1} and v_i as in Fact 5 of Section 38.4.

3. If all $\hat{\lambda}_i$ are different, then the eigenvalues λ_i of $\hat{\Lambda} + e_k \mathbf{v}\mathbf{v}^T$ are solutions of the so-called secular equation,

$$1 + e_k \sum_{i=1}^{n} \frac{v_i^2}{\hat{\lambda}_i - \lambda} = 0.$$

   The eigenvalues can be computed by bisection, or by some faster zero finder of the Newton type, and they need to be computed as accurately as possible.
4. Once the eigenvalues λ_i of $\hat{\Lambda} + e_k \mathbf{v}\mathbf{v}^T$ are known, the corresponding eigenvectors are

$$\mathbf{x}_i = (\hat{\Lambda} - \lambda_i I)^{-1}\mathbf{v}.$$

5. Each λ_i and $\mathbf{x}_i$ in Facts 3 and 4 is computed in $O(n)$ operations, respectively, so the overall computational cost for computing the EVD of $\hat{\Lambda} + e_k \mathbf{v}\mathbf{v}^T$ is $O(n^2)$.
6. The accuracy of the computed EVD is given by Fact 15 in section 42.1. However, if some eigenvalues are too close, they may not be computed with sufficient relative accuracy. As a consequence, the eigenvectors computed by using Fact 4 may not be sufficiently orthogonal. One remedy to this problem is to solve the secular equation from Fact 3 in double of the working precision. A better remedy is based on the solution of the following inverse eigenvalue problem. If $\hat{\lambda}_1 > \cdots > \hat{\lambda}_n$ and $\lambda_1 > \hat{\lambda}_1 > \lambda_2 > \hat{\lambda}_2 > \cdots > \lambda_{n-1} > \hat{\lambda}_{n-1} > \lambda_n > \hat{\lambda}_n$, then λ_i are the exact eigenvalues of the matrix $\hat{\Lambda} + e_k \hat{\mathbf{v}}\hat{\mathbf{v}}^T$, where

$$\hat{v}_i = \operatorname{sign} v_i \sqrt{\frac{\prod_{j=1}^{n}(\lambda_j - \hat{\lambda}_i)}{\prod_{j=1, j\neq i}^{n}(\hat{\lambda}_j - \hat{\lambda}_i)}}.$$

   Instead of computing $\mathbf{x}_i$ according to Fact 4, we compute $\hat{\mathbf{x}}_i = (\hat{\Lambda} - \lambda_i I)^{-1}\hat{\mathbf{v}}$. The eigenvector matrix of T is now computed as $U = \hat{U}\hat{X}$, where $\hat{X} = \begin{bmatrix} \mathbf{x}_1 & \cdots & \mathbf{x}_n \end{bmatrix}$, instead of $U = \hat{U}X$ as in Fact 1. See also Fact 8.
7. The algorithm for the divide and conquer method is the following:

> **Algorithm 6:** Divide and conquer method
> Input: real symmetric tridiagonal $n \times n$ matrix T with $t_{i-1,i} > 0$ for all i
> Output: eigenvalue matrix Λ and eigenvector matrix U of T
> **function** $(\Lambda, U) = \mathit{Divide\_and\_Conquer}(T)$
>   **if** $n = 1$, **then**
> $U = 1$
> $\Lambda = T$
>   **else**
> $k = floor(n/2)$
>   form $\hat{T}_1$ and $\hat{T}_2$ = as in Equation 42.6 in Fact 1
> $(\hat{\Lambda}_1, \hat{U}_1) = \mathit{Divide\_and\_Conquer}(\hat{T}_1)$
> $(\hat{\Lambda}_2, \hat{U}_2) = \mathit{Divide\_and\_Conquer}(\hat{T}_2)$
>   form $\hat{\Lambda} + e_k \mathbf{v}\mathbf{v}^T$ as in Equation 42.7 in Fact 1
>   compute the eigenvalues λ_i via Fact 3
>   compute $\hat{\mathbf{v}}$ via Fact 6
> $\hat{\mathbf{x}}_i = (\hat{\Lambda} - \lambda_i I)^{-1}\hat{\mathbf{v}}$
> $U = \begin{bmatrix} \hat{U}_1 & \\ & \hat{U}_2 \end{bmatrix} \hat{X}$
>   **endif**

8. The rationale for the approach of Fact 6 and Algorithm 6 is the following: The computations of $\hat{\mathbf{v}}$ and $\hat{\mathbf{x}}_i$ involve only subtractions of exact quantities, so there is no cancellation. Thus, all entries of each $\hat{\mathbf{x}}_i$ are computed with high relative accuracy so $\hat{\mathbf{x}}_i$ are mutually orthonormal to working

precision. Also, the transition from the matrix $\hat{\Lambda} + e_k \mathbf{v}\mathbf{v}^T$ to the matrix $\hat{\Lambda} + e_k \hat{\mathbf{v}}\hat{\mathbf{v}}^T$ induces only perturbations that are bounded by $\epsilon \| T \|$. Thus, the EVD computed by Algorithm 6 satisfies the error bounds given in Fact 15 in section 42.1, producing at the same time numerically orthogonal eigenvectors. For details see [Dem97, pp. 224–226] and [GE95].

9. Although Algorithm 6 requires $O(n^3)$ operations (this is due to the computation of U in the last line), it is in practice usually faster than Algorithm 4 from Fact 2 in section 42.3. This is due to deflations which are performed when solving the secular equation from Fact 3, resulting in matrix $\hat{X}$ having many zeros.
10. The operation count of Algorithm 6 can be reduced to $O(n^2 \log n)$ if the Fast Multipole Method, originally used in particle simulation, is used for solving secular equation from Fact 3 and for multiplying $\hat{U}\hat{X}$ in the last line of Algorithm 6. For details see [Dem97, pp. 227–228] and [GE95].

Examples:

1. Let T be the matrix from Example 1 in section 42.2 pre- and postmultiplied by the matrix $D = \text{diag}(1, -1, -1, 1)$:

$$T = \begin{bmatrix} 4.5013 & 3.0194 & 0 & 0 \\ 3.0194 & -0.3692 & 1.2804 & 0 \\ 0 & 1.2804 & 0.5243 & 0.9303 \\ 0 & 0 & 0.9303 & 2.6774 \end{bmatrix}.$$

The EVDs of the matrices $\hat{T}_1$ and $\hat{T}_2$ from Equation 42.6 in Fact 1 are

$$\hat{T}_1 = \begin{bmatrix} 4.5013 & 3.0194 \\ 3.0194 & -1.6496 \end{bmatrix}, \quad \hat{U}_1 = \begin{bmatrix} 0.3784 & -0.9256 \\ -0.9256 & -0.3784 \end{bmatrix}, \quad \hat{\Lambda}_1 = \begin{bmatrix} -2.8841 & 0 \\ 0 & 5.7358 \end{bmatrix},$$

$$\hat{T}_2 = \begin{bmatrix} -0.7561 & 0.9303 \\ 0.9303 & 2.6774 \end{bmatrix}, \quad \hat{U}_2 = \begin{bmatrix} -0.9693 & -0.2458 \\ 0.2458 & -0.9693 \end{bmatrix}, \quad \hat{\Lambda}_2 = \begin{bmatrix} -0.9920 & 0 \\ 0 & 2.9132 \end{bmatrix},$$

so, in Equation 42.7 in Fact 1, we have

$$\hat{\Lambda} = \text{diag}(-2.8841, 5.7358, -0.9920, 2.9132),$$
$$\mathbf{v} = [-0.9256 \quad -0.3784 \quad -0.9693 \quad -0.2458]^T.$$

2. Algorithm 6 is implemented in the LAPACK subroutine DSTEDC. This routine can compute just the eigenvalues or both, eigenvalues and eigenvectors. The routine requires workspace of approximately n^2 elements. The timings are given in Section 42.9.

42.5 Bisection and Inverse Iteration

The bisection method is convenient if only part of the spectrum is needed. If the eigenvectors are needed, as well, they can be efficiently computed by the inverse iteration method (see Facts 7 and 8 in Section 42.1).

Facts:

The following facts can be found in [Dem97, pp. 228–213] and [Par80, pp. 65–75].

A is a real symmetric $n \times n$ matrix and T is a real symmetric tridiagonal $n \times n$ matrix.

1. (Sylvester's theorem) For a real nonsingular matrix X, the matrices A and $X^T AX$ have the same inertia. (See also Section 8.3.)

2. Let $\alpha, \beta \in \mathbb{R}$ with $\alpha < \beta$. The number of eigenvalues of A in the interval $[\alpha, \beta)$ is equal to $\nu(A - \beta I) - \nu(A - \alpha I)$. By systematically choosing the intervals $[\alpha, \beta)$, the bisection method pinpoints each eigenvalue of A to any desired accuracy.
3. In the factorization $T - \mu I = LDL^T$, where $D = \text{diag}(d_1, \ldots, d_n)$ and L is the unit lower bidiagonal matrix, the elements of D are computed by the recursion

$$d_1 = t_{11} - \mu, \quad d_i = (t_{ii} - \mu) - t_{i,i-1}^2/d_{i-1}, \quad i = 2, \ldots n,$$

and the subdiagonal elements of L are given by $l_{i+1,i} = t_{i+1,i}/d_i$. By Fact 1 the matrices T and D have the same inertia, thus the above recursion enables an efficient implementation of the bisection method for T.
4. The factorization from Fact 3 is essentially Gaussian elimination without pivoting. Nevertheless, if $d_i \neq 0$ for all i, the above recursion is very stable (see [Dem97, Lemma 5.4] for details).
5. Even when $d_{i-1} = 0$ for some i, if the IEEE arithmetic is used, the computation will continue and the inertia will be computed correctly. Namely, in that case, we would have $d_i = -\infty$, $l_{i+1,i} = 0$, and $d_{i+1} = t_{i+1,i+1} - \mu$. For details see [Dem97, pp. 230–231] and the references therein.
6. Computing one eigenvalue of T by using the recursion from Fact 3 and bisection requires $O(n)$ operations. For a computed eigenvalue the corresponding eigenvector is computed by inverse iteration given by Equation 42.2. The convergence is very fast (Fact 7 in Section 42.1), so the cost of computing each eigenvector is also $O(n)$ operations. Therefore, the overall cost for computing all eigenvalues and eigenvectors is $O(n^2)$ operations.
7. Both, bisection and inverse iteration are highly parallel since each eigenvalue and eigenvector can be computed independently.
8. If some of the eigenvalues are too close, the corresponding eigenvectors computed by inverse iteration may not be sufficiently orthogonal. In this case, it is necessary to orthogonalize these eigenvectors (for example, by the modified Gram–Schmidt procedure). If the number of close eigenvalues is too large, the overall operation count can increase to $O(n^3)$.
9. The EVD computed by bisection and inverse iteration satisfies the error bounds from Fact 15 in Section 42.1.

Examples:

1. The bisection method for tridiagonal matrices is implemented in the LAPACK subroutine DSTEBZ. This routine can compute all eigenvalues in a given interval or the eigenvalues from λ_l to λ_k, where $l < k$, and the eigenvalues are ordered from smallest to largest. Inverse iteration (with reorthogonalization) is implemented in the LAPACK subroutine DSTEIN. The timings for computing half of the largest eigenvalues and the corresponding eigenvectors are given in Section 42.9.

42.6 Multiple Relatively Robust Representations

The computation of the tridiagonal EVD which satisfies the error bounds of Fact 15 in section 42.1 such that the eigenvectors are orthogonal to working precision, all in $O(n^2)$ operations, has been the "holy grail" of numerical linear algebra for a long time. The method of Multiple Relatively Robust Representations (MRRR) does the job, except in some exceptional cases. The key idea is to implement inverse iteration more carefully. The practical algorithm is quite elaborate and only main ideas are described here.

Facts:

The following facts can be found in [Dhi97], [DP04], and [DPV04].

$T = [t_{ij}]$ denotes a real symmetric tridiagonal matrix of order n. D, D_+, and D_- are diagonal matrices with the i-th diagonal entry denoted by d_i, $D_+(i)$, and $D_-(i)$, respectively. L and L_+ are unit lower bidiagonal matrices and U_- is a unit upper bidiagonal matrix, where we denote $(L)_{i+1,i}$ by l_i, $(L_+)_{i+1,i}$ by $L_+(i)$, and $(U_-)_{i,i+1}$ by $U_-(i)$.

1. Instead of working with the given T, the MRRR method works with the factorization $T = LDL^T$ (computed, for example, as in Fact 3 in Section 42.5 with $\mu = 0$). If T is positive definite, then all eigenvalues of LDL^T are determined to high relative accuracy in the sense that small relative changes in the elements of L and D cause only small relative changes in the eigenvalues. If T is indefinite, then the tiny eigenvalues of LDL^T are determined to high relative accuracy in the same sense. The bisection method based on Algorithms 7a and 7b computes the well determined eigenvalues of LDL^T to high relative accuracy, that is, the computed eigenvalue $\hat{\lambda}$ satisfies $|\lambda - \hat{\lambda}| = O(n\epsilon|\hat{\lambda}|)$.
2. The MRRR method is based on the following three algorithms:

Algorithm 7a: Differential stationary qd transform
Input: factors L and D of T and the computed eigenvalue $\hat{\lambda}$
Output: matrices D_+ and L_+ such that $LDL^T - \hat{\lambda}I = L_+D_+L_+^T$ and vector $\mathbf{s}$
$s_1 = -\hat{\lambda}$
for $i = 1 : n-1$
    $D_+(i) = s_i + d_i$
    $L_+(i) = (d_i l_i)/D_+(i)$
    $s_{i+1} = L_+(i)l_i s_i - \hat{\lambda}$
endfor
$D_+(n) = s_n + d_n$

Algorithm 7b: Differential progressive qd transform
Input: factors L and D of T and the computed eigenvalue $\hat{\lambda}$
Output: matrices D_- and U_- such that $LDL^T - \hat{\lambda}I = U_-D_-U_-^T$ and vector $\mathbf{p}$
$p_n = d_n - \hat{\lambda}$
for $i = n-1 : -1 : 1$
    $D_-(i+1) = d_i l_i^2 + p_{i+1}$
    $t = d_i/D_-(i+1)$
    $U_-(i) = l_i t$
    $p_i = p_{i+1}t - \hat{\lambda}$
endfor
$D_-(1) = p_1$

Algorithm 7c: Eigenvector computation
Input: output of Algorithms 7a and 7b and the computed eigenvalue $\hat{\lambda}$
Output: index r and the eigenvector $\mathbf{u}$ such that $LDL^T\mathbf{u} = \hat{\lambda}\mathbf{u}$.
for $i = 1 : n-1$
    $\gamma_i = s_i + \frac{d_i}{D_-(i+1)} p_{i+1}$
endfor
$\gamma_n = s_n + p_n + \hat{\lambda}$
find r such that $|\gamma_r| = \min_i |\gamma_i|$
$u_r = 1$
for $i = r-1 : -1 : 1$
    $u_i = -L_+(i)u_{i+1}$
endfor
for $i = r : n-1$
    $u_{i+1} = -U_-(i)u_i$
endfor
$\mathbf{u} = \mathbf{u}/\|\mathbf{u}\|_2$

3. Algorithm 7a is accurate in the sense that small relative perturbations (of the order of few ϵ) in the elements l_i, d_i, and the computed elements $L_+(i)$ and $D_+(i)$ make $LDL^T - \hat{\lambda}I = L_+D_+L_+^T$ an exact equality. Similarly, Algorithm 7b is accurate in the sense that small relative perturbations in the elements l_i, d_i, and the computed elements $U_-(i)$ and $D_-(i)$ make $LDL^T - \hat{\lambda}I = U_-D_-U_-^T$ an exact equality.
4. The idea behind the Algorithm 7c is the following: Index r is the index of the column of the matrix $(LDL^T - \hat{\lambda}I)^{-1}$ with the largest norm. Since the matrix $LDL^T - \hat{\lambda}I$ is nearly singular, the eigenvector is computed in just one step of inverse iteration given by Equation 42.2 starting from the vector $\gamma_r \mathbf{e}_r$. Further, $LDL^T - \hat{\lambda}I = N\Delta N^T$, where $N\Delta N^T$ is the the so-called twisted factorization obtained from L_+, D_+, U_-, and D_-:

$$\Delta = \text{diag}(D_+(1), \ldots, D_+(r-1), \gamma_r, D_-(r+1), \ldots, D_-(n)),$$

$$N_{ii} = 1, \quad N_{i+1,i} = L_+(i), \quad i = 1, \ldots, r-1, \quad N_{i,i+1} = U_-(i), \quad i = r, \ldots, n-1.$$

Since $\Delta \mathbf{e}_r = \gamma_r \mathbf{e}_r$ and $N\mathbf{e}_r = \mathbf{e}_r$, solving $N\Delta N^T \mathbf{u} = \gamma_r \mathbf{e}_r$ is equivalent to solving $N^T\mathbf{u} = \mathbf{e}_r$, which is exactly what is done by Algorithm 7c.
5. If an eigenvalue λ is well separated from other eigenvalues in the relative sense (the quantity $\min_{\mu \in \sigma(A), \mu \neq \lambda} |\lambda - \mu|/|\lambda|$ is large, say greater than 10^{-3}), then the computed vector $\hat{\mathbf{u}}$ satisfies $\|\sin\Theta(\mathbf{u}, \hat{\mathbf{u}})\|_2 = O(n\epsilon)$. If all eigenvalues are well separated from each other, then the computed EVD satisfies error bounds of Fact 15 in Section 42.1 and the computed eigenvectors are numerically orthogonal, that is, $|\hat{\mathbf{u}}_i^T \hat{\mathbf{u}}_j| = O(n\epsilon)$ for $i \neq j$.
6. If there is a cluster of poorly separated eigenvalues which is itself well separated from the rest of $\sigma(A)$, the MRRR method chooses a shift μ which is near one end of the cluster and computes a new factorization $LDL^T - \mu I = L_+D_+L_+^T$. The eigenvalues within the cluster are then recomputed by bisection as in Fact 1 and their corresponding eigenvectors are computed by Algorithms 7a, 7b, and 7c. When properly implemented, this procedure results in the computed EVD, which satisfies the error bounds of Fact 15 in Section 42.1 and the computed eigenvectors are numerically orthogonal.

Examples:

1. The MRRR method is implemented in the LAPACK subroutine DSTEGR. This routine can compute just the eigenvalues, or both eigenvalues and eigenvectors. The timings are given in Section 42.9.

42.7 Jacobi Method

The Jacobi method is the oldest method for EVD computations [Jac846]. The method does not require tridiagonalization. Instead, the method computes a sequence of orthogonally similar matrices which converge to Λ. In each step a simple plane rotation which sets one off-diagonal element to zero is performed.

Definitions:

A is a real symmetric matrix of order x and $A = U\Lambda U^T$ is its EVD.

The **Jacobi method** forms a sequence of matrices,

$$A_0 = A, \qquad A_{k+1} = G(i_k, j_k, c, s) A_k G(i_k, j_k, c, s)^T, \qquad k = 1, 2, \ldots,$$

where $G(i_k, j_k, c, s)$ is the plane rotation matrix defined in Chapter 38.4. The parameters c and s are chosen such that $[A_{k+1}]_{i_k j_k} = [A_{k+1}]_{j_k i_k} = 0$ and are computed as described in Fact 1.

The plane rotation with c and s as above is also called the **Jacobi rotation.**

The **off-norm** of A is defined as $\text{off}(A) = (\sum_i \sum_{j \neq i} a_{ij}^2)^{1/2}$, that is, off-norm is the Frobenius norm of the matrix consisting of all off-diagonal elements of A.

The choice of **pivot elements** $[A_k]_{i_k j_k}$ is called the **pivoting strategy**.

The **optimal pivoting** strategy, originally used by Jacobi, chooses pivoting elements such that $|[A_k]_{i_k j_k}| = \max_{i<j} |[A_k]_{ij}|$.

The **row cyclic** pivoting strategy chooses pivot elements in the systematic row-wise order,

$$(1, 2), (1, 3), \ldots, (1, n), (2, 3), (2, 4), \ldots, (2, n), (3, 4), \ldots, (n-1, n).$$

Similarly, the column-cyclic strategy chooses pivot elements column-wise.
One pass through all matrix elements is called **cycle** or **sweep**.

Facts:

The Facts 1 to 8 can be found in [Wil65, pp. 265–282], [Par80, §9], [GV96, §8.4], and [Dem97, §5.3.5].

1. The Jacobi rotations parameters c and s are computed as follows: If $[A_k]_{i_k j_k} = 0$, then $c = 1$ and $s = 0$, otherwise

$$\tau = \frac{[A_k]_{i_k i_k} - [A_k]_{j_k j_k}}{2[A_k]_{i_k j_k}}, \qquad t = \frac{\operatorname{sign}(\tau)}{|\tau| + \sqrt{1+\tau^2}}, \qquad c = \frac{1}{\sqrt{1+t^2}}, \qquad s = c \cdot t.$$

2. After each rotation, the off-norm decreases, that is,

$$\operatorname{off}^2(A_{k+1}) = \operatorname{off}^2(A_k) - 2[A_k]^2_{i_k j_k}.$$

With the appropriate pivoting strategy, the method converges in the sense that

$$\operatorname{off}(A_k) \to 0, \qquad A_k \to \Lambda, \qquad \prod_{k=1}^{\infty} R^T_{(i_k, j_k)} \to U.$$

3. For the optimal pivoting strategy the square of the pivot element is greater than the average squared element, $[A_k]^2_{i_k j_k} \geq \operatorname{off}^2(A) \frac{1}{n(n-1)}$. Thus,

$$\operatorname{off}^2(A_{k+1}) \leq \left(1 - \frac{2}{n(n-1)}\right) \operatorname{off}^2(A_k)$$

and the method converges.
4. For the row cyclic and the column cyclic pivoting strategies, the method converges. The convergence is ultimately quadratic in the sense that $\operatorname{off}(A_{k+n(n-1)/2}) \leq \gamma \operatorname{off}^2(A_k)$ for some constant γ, provided $\operatorname{off}(A_k)$ is sufficiently small.
5. We have the following algorithm:

Algorithm 8: Jacobi method with row-cyclic pivoting strategy
Input: real symmetric $n \times n$ matrix A
Output: the eigenvalue matrix Λ and the eigenvector matrix U

```
U = I_n
repeat      % one cycle
   for i = 1 : n − 1
      for j = i + 1 : n
         compute c and s according to Fact 1
         [A_{i,1:n}; A_{j,1:n}] = G(i, j, c, s) [A_{i,1:n}; A_{j,1:n}]
         [A_{1:n,i}  A_{1:n,j}] = [A_{1:n,i}  A_{1:n,j}] G(i, j, c, s)^T
         [U_{1:n,i}  U_{1:n,j}] = [U_{1:n,i}  U_{1:n,j}] G(i, j, c, s)^T
      endfor
   endfor
until off(A) ≤ tol   for some user defined stopping criterion tol
Λ = diag(A)
```

6. Detailed implementation of the Jacobi method can be found in [Rut66] and [WR71].
7. The EVD computed by the Jacobi method satisfies the error bounds from Fact 15 in Section 42.1.
8. The Jacobi method is suitable for parallel computation. There exist convergent parallel strategies which enable simultaneous execution of several rotations.
9. [GV96, p. 429] The Jacobi method is simple, but it is slower than the methods based on tridiagonalization. It is conjectured that standard implementations require $O(n^3 \log n)$ operations. More precisely, each cycle clearly requires $O(n^3)$ operations and it is conjectured that $\log n$ cycles are needed until convergence.
10. [DV92], [DV05] If A is positive definite, the method can be modified such that it reaches the speed of the methods based on tridiagonalization and at the same time computes the eigenvalues with high relative accuracy. (See Chapter 46 for details.)

Examples:

1. Let A be the matrix from Example 1 in section 42.1. After executing two cycles of Algorithm 8, we have

$$A = \begin{bmatrix} 6.0054 & -0.0192 & 0.0031 & 0.0003 \\ -0.0192 & 3.0455 & -0.0005 & -0.0000 \\ 0.0031 & -0.0005 & 0.6024 & -0.0000 \\ 0.0003 & -0.0000 & 0.0000 & -2.3197 \end{bmatrix}.$$

42.8 Lanczos Method

If the matrix A is large and sparse and if only some eigenvalues and their eigenvectors are desired, sparse matrix methods are the methods of choice. For example, the power method can be useful to compute the eigenvalue with the largest modulus. The basic operation in the power method is matrix-vector multiplication, and this can be performed very fast if A is sparse. Moreover, A need not be stored in the computer — the input for the algorithm can be just a program which, given some vector $\mathbf{x}$, computes the product $A\mathbf{x}$. An "improved" version of the power method, which efficiently computes several eigenvalues (either largest in modulus or near some target value μ) and the corresponding eigenvectors, is the Lanczos method.

Definitions:

A is a real symmetric matrix of order n.

Given a nonzero vector $\mathbf{x}$ and an index $k < n$, the **Krylov matrix** is defined as $K_k = [\mathbf{x} \quad A\mathbf{x} \quad A^2\mathbf{x} \quad \cdots \quad A^{k-1}\mathbf{x}]$.

Facts:

The following facts can be found in [Par80, §13], [GV96, §9], [Dem97, §7], and [Ste01, §5.3].

1. The Lanczos method is based on the following observation. If $K_k = XR$ is the QR factorization of the matrix K_k (see Sections 5.5 and 38.4), then the $k \times k$ matrix $T = X^T AX$ is tridiagonal. The matrices X and T can be computed by using only matrix-vector products in just $O(kn)$ operations. Let $T = Q\Lambda Q^T$ be the EVD of T (computed by any of the methods from Sections 42.3 to 42.6). Then λ_i approximate well some of the largest and smallest eigenvalues of A. The columns of the matrix $U = XQ$ approximate the corresponding eigenvectors of A. We have the following algorithm:

```
Algorithm 9: Lanczos method
Input: real symmetric n × n matrix A, unit vector x and index k < n
Output: matrices Λ and U
X(:,1) = x
for i = 1 : k
    z = A X(:,i)
    t(i,i) = X(:,i)^T z
    if i = 1, then
        z = z − t(i,i) X(:,i)
    else
        z = z − t(i,i) X(:,i) − t(i,i−1) X(:,i−1)
    endif
    μ = ||z||_2
    if μ = 0, then
        stop
    else
        t(i+1,i) = μ
        t(i,i+1) = μ
        X(:,i+1) = z/μ
    endif
endfor
compute the EVD of the tridiagonal matrix, T(1 : k, 1 : k) = Q Λ Q^T
U = X Q
```

2. As j increases, the largest (smallest) eigenvalues of the matrix $T_{1:j,1:j}$ converge towards some of the largest (smallest) eigenvalues of A (due to the Cauchy interlace property). The algorithm can be redesigned to compute only largest or smallest eigenvalues. Also, by using shift and invert strategy, the method can be used to compute eigenvalues near some specified value. In order to obtain better approximations, k should be greater than the number of required eigenvalues. On the other side, in order to obtain better accuracy and efficacy, k should be as small as possible (see Facts 3 and 4 below).
3. The eigenvalues of A are approximated from the matrix $T_{1:k,1:k}$, thus, the last element $\nu = t_{k+1,k}$ is not needed. However, this element provides key information about accuracy at no extra computational cost. The exact values of residuals are as follows: $\|AU - U\Lambda\|_2 = \nu$ and, in particular, $\|AU_{:,i} - \lambda_i U_{:,i}\|_2 = \nu|q_{ki}|$, $i = 1, \dots, k$. Further, there are k eigenvalues $\tilde{\lambda}_1, \dots, \tilde{\lambda}_k$ of A such that $|\lambda_i - \tilde{\lambda}_i| \leq \nu$. For the corresponding eigenvectors, we have $\sin 2\Theta(\mathbf{u}_i, \tilde{\mathbf{u}}_i) \leq 2\nu / \min_{j\neq i} |\lambda_i - \tilde{\lambda}_j|$. In practical implementations of Algorithm 9, ν is usually used to determine the index k.
4. Although theoretically very elegant, the Lanczos method has inherent numerical instability in the floating point arithmetic, and so it must be implemented carefully (see, e.g., [LSY98]). Since the Krylov vectors are, in fact, generated by the power method, they converge towards an eigenvector of A. Thus, as k increases, the Krylov vectors become more and more parallel. As a consequence, the recursion in Algorithm 9, which computes the orthogonal bases X for the subspace range K_k, becomes numerically unstable and the computed columns of X cease to be sufficiently orthogonal. This affects both the convergence and the accuracy of the algorithm. For example, it can happen that T has several eigenvalues which converge towards some simple eigenvalue of A (these are the so called ghost eigenvalues).

   The loss of orthogonality is dealt with by using the full reorthogonalization procedure. In each step, the new $\mathbf{z}$ is orthogonalized against all previous columns of X. In Algorithm 9, the formula $\mathbf{z} = \mathbf{z} - t_{ii} X_{:,i} - t_{i,i-1} X_{:,i-1}$ is replaced by $\mathbf{z} = \mathbf{z} - \sum_{j=1}^{i-1} (\mathbf{z}^T X(:, j)) X(:, j)$. To obtain better orthogonality, the latter formula is usually executed twice.

The full reorthogonalization raises the operation count to $O(k^2n)$. The selective reorthogonalization is the procedure in which the current $\mathbf{z}$ is orthogonalized against some selected columns of X. This is the way to attain sufficient numerical stability and not increase the operation count too much. The details of selective reorthogonalization procedures are very subtle and can be found in the references. (See also Chapter 44.)

5. The Lanczos method is usually used for sparse matrices. Sparse matrix A is stored in the sparse format in which only values and indices of nonzero elements are stored. The number of operations required to multiply some vector by A is also proportional to the number of nonzero elements. (See also Chapter 43.)

Examples:

1. Let A be the matrix from Example 1 in section 42.1 and let $x = [1/2 \quad 1/2 \quad 1/2 \quad 1/2]^T$. For $k = 2$, the output of Algorithm 9 is

$$\Lambda = \begin{bmatrix} -2.0062 & \\ & 5.7626 \end{bmatrix}, \qquad U = \begin{bmatrix} -0.4032 & -0.8804 \\ 0.4842 & -0.2749 \\ 0.3563 & -0.3622 \\ 0.6899 & -0.1345 \end{bmatrix},$$

with $\nu = 1.4965$ (c.f. Fact 3). For $k = 3$, the output is

$$\Lambda = \begin{bmatrix} -2.3107 & 0 & 0 \\ 0 & 2.8641 & 0 \\ 0 & 0 & 5.9988 \end{bmatrix}, \qquad U = \begin{bmatrix} 0.3829 & -0.0244 & 0.8982 \\ -0.2739 & -0.9274 & 0.0312 \\ -0.3535 & -0.1176 & 0.3524 \\ -0.8084 & 0.3541 & 0.2607 \end{bmatrix},$$

with $\nu = 0.6878$.

2. The Lanczos method is implemented in the ARPACK routine `DSDRV*`, where * denotes the computation mode [LSY98, App. A]. The routines from ARPACK are implemented in the MATLAB command `eigs`. Generation of a sparse symmetric $10,000 \times 10,000$ matrix with 10% nonzero elements with the MATLAB command `A=sprandsym(10000,0.1)` takes 15 seconds on a processor described in Fact 1 in secton 42.9. The computation of 100 largest eigenvalues and the corresponding eigenvectors with `[U,Lambda]=eigs(A,100,'LM',opts)` takes 140 seconds. Here, index $k = 200$ is automatically chosen by the algorithm. (See also Chapter 76.)

42.9 Comparison of Methods

In this section, we give timings for the LAPACK implementations of the methods described in Sections 42.2 to 42.6. The timing for the Lanczos method is given in Example 2 in Section 42.8.

Definitions:

A measure of processor's **efficacy** or **speed** is the number of floating-point operations per second (flops).

Facts:

A is an $n \times n$ real symmetric matrix and $A = U\Lambda U^T$ is its EVD. T is a tridiagonal $n \times n$ real symmetric matrix and $T = Q\Lambda Q^T$ is its EVD. $T = X^T AX$ is the reduction of A to a tridiagonal from Section 42.2.

1. Our tests were performed on the Intel Xeon processor running at 2.8 MHz with 2 Mbytes of cache memory. This processor performs up to 5 Gflops (5 billion operations per second). The peak performance is attained for the matrix multiplication with the BLAS 3 subroutine DGEMM.

TABLE 42.1 Execution times(s) for LAPACK routines for various matrix dimensions n.

Routine	Input	Output	Example	$n = 500$	$n = 1000$	$n = 2000$
DSYTRD	A	T	2.3	0.10	0.78	5.5
DSYTRD/DORGTR		T, X		0.17	1.09	8.6
DSTEQR	T	Λ	3.2	0.03	0.11	0.44
		Λ, Q		0.32	2.23	15.41
DSYEV	A	Λ	3.4	0.12	0.85	5.63
		Λ, U		0.46	3.13	22.30
DSTEDC	T	Λ	4.2	0.02	0.08	0.28
		Λ, Q		0.05	0.12	0.36
DSTEBZ	T	Λ	5.1	0.21	0.81	3.15
DSTEIN		Q		0.04	0.17	0.72
DSTEGR	T	Λ	6.1	0.07	0.25	0.87
		Λ, Q		0.09	0.35	1.29

2. Our test programs were compiled with the Intel `ifort` FORTRAN compiler (version 9.0) and linked with the Intel Math Kernel Library (version 8.0.2).
3. Timings for the methods are given in Table 42.1. The execution times for DSTEBZ (bisection) and DSTEIN (inverse iteration) are for computing one half of the eigenvalues (the largest ones) and the corresponding eigenvectors, respectively.
4. The performance attained for practical algorithms is lower than the peak performance from Fact 1. For example, by combining Facts 6 and 7 in Section 42.2 with Table 42.1, we see that the tridiagonalization routines DSYTRD and DORGTR attain the speed of 2 Gflops.
5. The computation times for the implicitly shifted QR routine, DSTEQR, grow with n^2 when only eigenvalues are computed, and with n^3 when eigenvalues and eigenvectors are computed, as predicted in Facts 5 and 6 in Section 42.3.
6. The execution times for DSYEV are approximately equal to the sums of the timings for DSYTRD (tridiagonalization), DORGTR (computing X), and DSTEQR with the eigenvector option (computing the EVD of T).
7. The divide and conquer method, implemented in DSTEDC, is the fastest method for computing the EVD of a tridiagonal matrix.
8. DSTEBZ and DSTEIN (bisection and inverse iteration) are faster, especially for larger dimensions, than DSTEQR (tridiagonal QR iteration), but slower than DSTEDC (divide and conquer) and DSTEGR (multiple relatively robust representations).
9. Another algorithm to compute the EVD of T is to use DSTEQR to compute only the eigenvalues and then use DSTEIN (inverse iteration) to compute the eigenvectors. This is usually considerably faster than computing both, eigenvalues and eigenvectors, by DSTEQR.
10. The executions times for DSTEGR are truly proportional to $O(n^2)$.
11. The new LAPACK release, in which some of the above mentioned routines are improved with respect to speed and/or accuracy, is announced for the second half of 2006.

References

[ABB99] E. Anderson, Z. Bai, and C. Bischof, *LAPACK Users' Guide*, 3rd ed., SIAM, Philadelphia, 1999.

[Cup81] J.J.M. Cuppen, A divide and conquer method for the symmetric tridiagonal eigenproblem, *Numer. Math.*, 36:177–195, 1981.

[Dem97] J.W. Demmel, *Applied Numerical Linear Algebra*, SIAM, Philadelphia, 1997.

[DV92] J.W. Demmel and K. Veselić, Jacobi's method is more accurate than QR, *SIAM J. Matrix Anal. Appl.*, 13:1204–1245, 1992.

[Dhi97] I.S. Dhillon, *A New $O(N^2)$ Algorithm for the Symmetric Tridiagonal Eigenvalue/Eigenvector Problem*, Ph.D. thesis, University of California, Berkeley, 1997.

[DP04] I.S. Dhillon and B.N. Parlett, Orthogonal eigenvectors and relative gaps, *SIAM J. Matrix Anal. Appl.*, 25:858–899, 2004.

[DPV04] I.S. Dhillon, B.N. Parlett, and C. Vömel, "The Design and Implementation of the MRRR Algorithm," Tech. Report UCB/CSD-04-1346, University of California, Berkeley, 2004.

[DHS89] J.J. Dongarra, S.J. Hammarling, and D.C. Sorensen, Block reduction of matrices to condensed forms for eigenvalue computations, *J. Comp. Appl. Math.*, 27:215–227, 1989.

[DV05] Z. Drmač and K. Veselić, New fast and accurate Jacobi SVD algorithm: I, Technical report, University of Zagreb, 2005, also LAPACK Working Note #169.

[GV96] G.H. Golub and C.F. Van Loan, *Matrix Computations*, 3rd ed., The John Hopkins University Press, Baltimore, MD, 1996.

[GE95] M. Gu and S.C. Eisenstat, A divide-and-conquer algorithm for the symmetric tridiagonal eigenproblem, *SIAM J. Matrix Anal. Appl.*, 16:79–92, 1995.

[Jac846] C.G.J. Jacobi, Über ein leichtes Verfahren die in der Theorie der Säcularstörungen vorkommenden Gleichungen numerisch aufzulösen, *Crelles Journal für Reine und Angew.* Math., 30:51–95, 1846.

[LSY98] R.B. Lehoucq, D.C. Sorensen, and C. Yang, *ARPACK Users' Guide: Solution of Large-Scale Eigenvalue Problems with Implicitly Restarted Arnoldi Methods*, SIAM, Philadelphia, 1998.

[Par80] B.N. Parlett, *The Symmetric Eigenvalue Problem*, Prentice-Hall, Upper Saddle River, NJ, 1980.

[Rut66] H. Rutishauser, The Jacobi method for real symmetric matrices, *Numerische Mathematik*, 9:1–10,1966.

[Ste01] G.W. Stewart, *Matrix Algorithms, Vol. II: Eigensystems*, SIAM, Philadelphia, 2001.

[TB97] L.N. Trefethen and D. Bau, III, *Numerical Linear Algebra*, SIAM, Philadelphia, 1997.

[Wil65] J.H. Wilkinson, *The Algebraic Eigenvalue Problem*, Clarendon Press, Oxford, U.K., 1965.

[WR71] J.H. Wilkinson and C. Reinsch, *Handbook for Automatic Computation, Vol. II, Linear Algebra*, Springer, New York, 1971.

43

Unsymmetric Matrix Eigenvalue Techniques

David S. Watkins
Washington State University

The definitions and basic properties of eigenvalues and eigenvectors are given in Section 4.3. A natural generalization is presented here in Section 43.1. Algorithms for computation of eigenvalues, eigenvectors, and their generalizations will be discussed in Sections 43.2 and 43.3. Although the characteristic equation is important in theory, it plays no role in practical eigenvalue computations.

If a large fraction of a matrix's entries are zeros, the matrix is called **sparse**. A matrix that is not sparse is called **dense**. Dense matrix techniques are methods that store the matrix in the conventional way, as an array, and operate on the array elements. Any matrix that is not too big to fit into a computer's main memory can be handled by dense matrix techniques, regardless of whether the matrix is dense or not. However, since the time to compute the eigenvalues of an $n \times n$ matrix by dense matrix techniques is proportional to n^3, the user may have to wait awhile for the results if n is very large. Dense matrix techniques do not exploit the zeros in a matrix and tend to destroy them. With modern computers, dense matrix techniques can be applied to matrices of dimension up to 1000 or more. If a matrix is very large and sparse, and only a portion of the spectrum is needed, sparse matrix techniques (Section 43.3) are preferred.

The usual approach is to preprocess the matrix into Hessenberg form and then to effect a similarity transformation to triangular form: $T = S^{-1}AS$ by an iterative method. This yields the eigenvalues of A as the main-diagonal entries of T. For $k = 1, \ldots, n-1$, the first k columns of S span an invariant subspace. The eigenvectors of an upper-triangular matrix are easily computed by back substitution, and the eigenvectors of A can be deduced from the eigenvectors of T [GV96, § 7.6], [Wat02, § 5.8]. If a matrix A is very large and sparse, only a partial similarity transformation is possible because a complete similarity transformation would require too much memory and take too long to compute.

43.1 The Generalized Eigenvalue Problem

Many matrix eigenvalue problems are most naturally viewed as generalized eigenvalue problems.

Definitions:

Given $A \in \mathbb{C}^{n\times n}$ and $B \in \mathbb{C}^{n\times n}$, the nonzero vector $\mathbf{v} \in \mathbb{C}^n$ is called an **eigenvector** of the pair (A, B) if there are scalars $\mu, \nu \in \mathbb{C}$, not both zero, such that

$$\nu A\mathbf{v} = \mu B\mathbf{v}.$$

Then, the scalar $\lambda = \mu/\nu$ is called the **eigenvalue** of (A, B) associated with the eigenvector $\mathbf{v}$. If $\nu = 0$, then the eigenvalue is ∞ by convention.

The expression $A - xB$, with indeterminate x, is called a **matrix pencil**. Whether we refer to the pencil $A - xB$ or the pair (A, B), we are speaking of the same object. The pencil (or the pair (A, B)) is called **singular** if $A - \lambda B$ is singular for all $\lambda \in \mathbb{C}$. The pencil is **regular** if there exists a $\lambda \in \mathbb{C}$ such that $A - \lambda B$ is nonsingular. We will restrict our attention to regular pencils.

The **characteristic polynomial** of the pencil $A - xB$ is $\det(xB - A)$, and the **characteristic equation** is $\det(xB - A) = 0$.

Two pairs (A, B) and (C, D) are **strictly equivalent** if there exist nonsingular matrices S_1 and S_2 such that $C - \lambda D = S_1(A - \lambda B)S_2$ for all $\lambda \in \mathbb{C}$. If S_1 and S_2 can be taken to be unitary, then the pairs are **strictly unitarily equivalent**.

A pair (A, B) is called **upper triangular** if both A and B are upper triangular.

Facts:

The following facts are discussed in [GV96, § 7.7] and [Wat02, § 6.7].

1. When $B = I$, the generalized eigenvalue problem for the pair (A, B) reduces to the standard eigenvalue problem for the matrix A.
2. λ is an eigenvalue of (A, B) if and only if $A - \lambda B$ is singular.
3. λ is an eigenvalue of (A, B) if and only if $\ker(\lambda B - A) \neq \{0\}$.
4. The eigenvalues of (A, B) are exactly the solutions of the characteristic equation $\det(xB - A) = 0$.
5. The characteristic polynomial $\det(xB - A)$ is a polynomial in x of degree $\leq n$.
6. The pair (A, B) (or the pencil $A - xB$) is singular if and only if $\det(\lambda B - A) = 0$ for all λ, if and only if the characteristic polynomial $\det(xB - A)$ is equal to zero.
7. If the pair (A, B) is regular, then $\det(xB - A)$ is a nonzero polynomial of degree $k \leq n$. (A, B) has k finite eigenvalues.
8. The degree of $\det(xB - A)$ is exactly n if and only if B is nonsingular.
9. If B is nonsingular, then the eigenvalues of (A, B) are exactly the eigenvalues of the matrices AB^{-1} and $B^{-1}A$.
10. If $\lambda \neq 0$, then λ is an eigenvalue of (A, B) if and only if λ^{-1} is an eigenvalue of (B, A).
11. Zero is an eigenvalue of (A, B) if and only if A is a singular matrix.
12. Infinity is an eigenvalue of (A, B) if and only if B is a singular matrix.
13. Two pairs that are strictly equivalent have the same eigenvalues.
14. If $C - \lambda D = S_1(A - \lambda B)S_2$, then $\mathbf{v}$ is an eigenvector of (A, B) if and only if $S_2^{-1}\mathbf{v}$ is an eigenvector of (C, D).
15. (Schur's Theorem) Every $A \in \mathbb{C}^{n \times n}$ is unitarily similar to an upper triangular matrix S.
16. (Generalized Schur Theorem) Every pair (A, B) is strictly unitarily equivalent to an upper triangular pair (S, T).
17. The characteristic polynomial of an upper triangular pair (S, T) is $\prod_{k=1}^{n}(\lambda t_{kk} - s_{kk})$. The eigenvalues of (S, T) are $\lambda_k = s_{kk}/t_{kk}$, $k = 1, \ldots, n$. If $t_{kk} = 0$ and $s_{kk} \neq 0$, then $\lambda_k = \infty$. If $t_{kk} = 0$ and $s_{kk} = 0$ for some k, the pair (S, T) is singular.

Examples:

1. Let $A = \begin{bmatrix} 1 & 2 \\ 3 & 4 \end{bmatrix}$ and $B = \begin{bmatrix} 1 & 2 \\ 0 & 1 \end{bmatrix}$. Then the characteristic polynomial of the pair (A, B) is $x^2 + x - 2 = 0$, and the eigenvalues are 1 and -2.

2. Since the pencil

$$\begin{bmatrix} 2 & 5 \\ 0 & 7 \end{bmatrix} - x \begin{bmatrix} 5 & 1 \\ 0 & 3 \end{bmatrix}$$

is upper triangular, its characteristic polynomial is $(5x - 2)(3x - 7)$, and its eigenvalues are $2/5$ and $7/3$.
3. The pencil

$$\begin{bmatrix} 0 & 0 \\ 0 & 1 \end{bmatrix} - x \begin{bmatrix} 1 & 0 \\ 0 & 0 \end{bmatrix}$$

has characteristic equation $x = 0$. It is a regular pencil with eigenvalues 0 and ∞.

43.2 Dense Matrix Techniques

The steps that are usually followed for solving the unsymmetric eigenvalue problem are preprocessing, eigenvalue computation with the QR Algorithm, and eigenvector computation.

The most widely used public domain software for this problem is from LAPACK [ABB99], and Chapter 75. Versions in FORTRAN and C are available. The most popular proprietary software is MATLAB, which uses computational routines from LAPACK. Several of LAPACK's computational routines will be mentioned in this section. LAPACK also has a number of driver routines that call the computational routines to perform the most common tasks, thereby making the user's job easier. A very easy way to use LAPACK routines is to use MATLAB.

This section presents algorithms for the reader's edification. However, the reader is strongly advised to use well-tested software written by experts whenever possible, rather than writing his or her own code. The actual software is very complex and addresses details that cannot be discussed here.

Definitions:

A matrix $A \in \mathbb{C}^{n \times n}$ is called **upper Hessenberg** if $a_{ij} = 0$ whenever $i > j + 1$. This means that every entry below the first subdiagonal of A is zero. An upper Hessenberg matrix is called **unreduced upper Hessenberg** if $a_{j+1,j} \neq 0$ for $j = 1, \ldots, n - 1$.

Facts:

The following facts are proved in [Dem97], [GV96], [Kre05], or [Wat02].

1. Preprocessing is a two step process involving balancing the matrix and transforming by unitary similarity to upper Hessenberg form.
2. The first step, which is optional, is to balance the matrix. The balancing operation begins by performing a permutation similarity transformation that exposes any obvious eigenvalues. The remaining submatrix is irreducible. It then performs a diagonal similarity transformation $D^{-1}AD$ that attempts to make the norms of the ith row and ith column as nearly equal as possible, $i = 1, \ldots, n$. This has the effect of reducing the overall norm of the matrix and in diminishing the effects of roundoff errors [Osb60]. The scaling factors in D are taken to be powers of the base of floating point arithmetic (usually 2). No roundoff errors are caused by this transformation.
3. All modern balancing routines, including the code GEBAL in LAPACK, are derived from the code in Parlett and Reinsch [PR69]. See also [Kre05].

Algorithm 1: Balancing an Irreducible Matrix. An irreducible matrix $A \in \mathbb{C}^{n\times n}$ is input. On output, A has been overwritten by $D^{-1}AD$, where D is diagonal.

$b \leftarrow$ base of floating point arithmetic (usually 2)
$D \leftarrow I_n$
done $\leftarrow 0$
while done $= 0$
  done $\leftarrow 1$
  for $j = 1 : n$
    $c \leftarrow \sum_{i\neq j} |a_{ij}|, \quad r \leftarrow \sum_{k\neq j} |a_{jk}|$
    $s \leftarrow c + r, \quad f \leftarrow 1$
    while $b\,c < r$
      $c \leftarrow b\,c, \quad r \leftarrow r/b, \quad f \leftarrow b\,f$
    while $, b\,r < c$
      $c \leftarrow c/b, \quad r \leftarrow b\,r, \quad f \leftarrow f/b$
    if $c + r < 0.95\,s$
      done $\leftarrow 0, \quad d_{jj} \leftarrow f\,d_{jj}$
      $A_{1:n,j} \leftarrow f\,A_{1:n,j}, \quad A_{j,1:n} \leftarrow (1/f)\,A_{j,1:n}$
end

4. In most cases balancing will have little effect on the outcome of the computation, but sometimes it results in greatly improved accuracy [BDD00, § 7.2].
5. The second preprocessing step is to transform the matrix to upper Hessenberg form. This is accomplished by a sequence of $n-2$ steps. On the jth step, zeros are introduced into the jth column.
6. For every $\mathbf{x} \in \mathbb{C}^n$ there is a unitary matrix U such that $U\mathbf{x} = \alpha\mathbf{e}_1$, for some scalar $\alpha \in \mathbb{C}$, where $\mathbf{e}_1$ is the vector having a 1 in the first position and zeros elsewhere. U can be chosen to be a rank-one modification of the identity matrix: $U = I + \mathbf{u}\mathbf{v}^*$. (See Section 38.4 for a discussion of Householder and Givens matrices.)
7. **Algorithm 2. Unitary Similarity Transformation to Upper Hessenberg Form.** A general matrix $A \in \mathbb{C}^{n\times n}$ is input. On output, A has been overwritten by an upper Hessenberg matrix Q^*AQ. The unitary transforming matrix Q has also been generated.

$Q \leftarrow I_n$
for $j = 1 : n-2$
  Let $\mathbf{x} = A_{j+1:n,j} \in \mathbb{C}^{n-j}$.
  Build unitary $U \in \mathbb{C}^{n-j\times n-j}$ such that $U^*\mathbf{x} = \gamma\mathbf{e}_1$.
  $A_{j+1:n,j:n} \leftarrow U^* A_{j+1:n,j:n}$
  $A_{1:n,j+1:n} \leftarrow A_{1:n,j+1:n}U$
  $Q_{1:n,j+1:n} \leftarrow Q_{1:n,j+1:n}U$
end

8. The cost of the reduction to Hessenberg form is proportional to n^3 for large n; that is, it is $O(n^3)$.
9. For large matrices, efficient cache use can be achieved by processing several columns at a time. This allows the processor(s) to run at much closer to maximum speed. See [GV96, p. 225], [Wat02, p. 210], and the LAPACK code GEHRD [ABB99].

10. Once the matrix is in upper Hessenberg form, if any of the subdiagonal entries $a_{j+1,j}$ is zero, the matrix is block upper triangular with a $j \times j$ block and an $n - j \times n - j$ block, and the eigenvalue problem decouples to two independent problems of smaller size. Thus, we always work with unreduced upper Hessenberg matrices.
11. In practice we set an entry $a_{j+1,j}$ to zero whenever
$$|a_{j+1,j}| < \epsilon(|a_{jj}| + |a_{j+1,j+1}|),$$
where ϵ is the computer's unit roundoff.
12. If $T \in \mathbb{C}^{n\times n}$ is upper triangular and nonsingular, then T^{-1} is upper triangular. If $H \in \mathbb{C}^{n\times n}$ is upper Hessenberg, then TH, HT, and THT^{-1} are upper Hessenberg.
13. The standard method for computing the eigenvalues of a Hessenberg matrix is the QR algorithm, an iterative method that produces a sequence of unitarily similar matrices that converges to upper triangular form.
14. The most basic version of the QR algorithm starts with $A_0 = A$, an unreduced upper Hessenberg matrix, and generates a sequence (A_m) as follows: Given A_{m-1}, a decomposition $A_{m-1} = Q_m R_m$, where Q_m is unitary and R_m is upper triangular, is computed. Then the factors are multiplied back together in reverse order to yield $A_m = R_m Q_m$. Equivalently $A_m = Q_m^* A_{m-1} Q_m$.
15. Upper Hessenberg form is preserved by iterations of the QR algorithm.
16. The QR algorithm can also be applied to non-Hessenberg matrices, but the operations are much more economical in the Hessenberg case.
17. The basic QR algorithm converges slowly, so shifts of origin are used to accelerate convergence:
$$A_{m-1} - \mu_m I = Q_m R_m, \quad R_m Q_m + \mu_m I = Q_m^* A_{m-1} Q_m = A_m,$$
where $\mu_m \in \mathbb{C}$ is a shift chosen to approximate an eigenvalue.
18. Often it is convenient to take several steps at once:

Algorithm 3. Explicit QR iteration of degree k.

Choose k shifts $\mu_1, \ldots \mu_k$.
Let $p(A) = (A - \mu_1 I)(A - \mu_2 I) \cdots (A - \mu_k I)$.
Compute a QR decomposition $p(A) = QR$.
$A \leftarrow Q^* A Q$

19. A QR iteration of degree k is equivalent to k iterations of degree 1 with shifts $\mu_1, \ldots, \mu_k$ applied in succession in any order [Wat02]. Upper Hessenberg form is preserved. In practice k is never taken very big; typical values are 1, 2, 4, and 6.
20. One important application of multiple steps is to complex shifts applied to real matrices. Complex arithmetic is avoided by taking $k = 2$ and shifts related by $\mu_2 = \overline{\mu}_1$.
21. The usual choice of k shifts is the set of eigenvalues of the lower right-hand $k \times k$ submatrix of the current iterate. With this choice of shifts at each iteration, the entry $a_{n-k+1,n-k}$ typically converges to zero quadratically [WE91], isolating a $k \times k$ submatrix after only a few iterations. However, convergence is not guaranteed, and failures do occasionally occur. No shifting strategy that guarantees convergence in all cases is known. For discussions of shifting strategies and convergence see [Wat02] or [WE91].
22. After each iteration, all of the subdiagonal entries should be checked to see if any of them can be set to zero. The objective is to break the big problem into many small problems in as few iterations as possible. Once a submatrix of size 1×1 has been isolated, an eigenvalue has been found. The eigenvalues of a 2×2 submatrix can be found by careful use of the quadratic formula. Complex conjugate eigenvalues of real matrices are extracted in pairs.

23. The explicit QR iteration shown above is expensive and never used in practice. Instead the iteration is performed implicitly.

> **Algorithm 4: Implicit QR iteration of degree k (chasing the bulge).**
> Choose k shifts $\mu_1, \ldots \mu_k$.
> $\mathbf{x} \leftarrow \mathbf{e}_1$ % first column of identity matrix
> for $j = 1 : k$
> $\left[\mathbf{x} \leftarrow (A - \mu_k I)\mathbf{x}\right.$
> end % $\mathbf{x}$ is the first column of $p(A)$.
> $\hat{\mathbf{x}} \leftarrow \mathbf{x}_{1:k+1}$ % $\mathbf{x}_{k+2:n} = 0$
> Let $U \in \mathbb{C}^{k+1 \times k+1}$ be unitary with $U^*\mathbf{x} = \alpha \mathbf{e}_1$
> $A_{1:k+1,1:n} \leftarrow U^* A_{1:k+1,1:n}$
> $A_{1:n,1:k+1} \leftarrow A_{1:n,1:k+1} U$
> Return A to upper Hessenberg form as in Algorithm 2 (Fact 7).

24. The initial transformation in the implicit QR iteration disturbs the upper Hessenberg form of A, making a bulge in the upper left-hand corner. The size of the bulge is equal to k. In the case $k = 2$, the pattern of nonzeros is

$$\begin{bmatrix} * & * & * & * & * & * \\ * & * & * & * & * & * \\ * & * & * & * & * & * \\ * & * & * & * & * & * \\ & & & * & * & * \\ & & & & * & * \end{bmatrix}.$$

The subsequent reduction to Hessenberg form chases the bulge down through the matrix and off the bottom. The equivalence of the explicit and implicit QR iterations is demonstrated in [GV96, § 7.5] and [Wat02, § 5.7]. For this result it is crucial that the matrix is unreduced upper Hessenberg.

25. For a fixed small value of k, the implicit QR iteration requires only $O(n^2)$ work. Typically only a small number of iterations, independent of n, are needed per eigenvalue found; the total number of iterations is $O(n)$. Thus, the implicit QR algorithm is considered to be an $O(n^3)$ process.

26. The main unsymmetric QR routine in LAPACK [ABB99] is HSEQR, a multishift implicit QR algorithm with $k = 6$. For processing small submatrices (50×50 and under), HSEQR calls LAHQR, a multishift QR code with $k = 2$. Future versions of LAPACK will include improved QR routines that save work by doing aggressive early deflation [BBM02b] and make better use of cache by chasing bulges in bunches and aggregating the transforming matrices [BBM02a].

27. If eigenvectors are wanted, the aggregate similarity transformation matrix S, the product of all transformations from start to finish, must be accumulated. $T = S^{-1}AS$, where A is the original matrix and T is the final upper triangular matrix. In the real case, T will not quite be upper triangular. It is **quasi-triangular** with a 2×2 block along the main diagonal for each complex conjugate pair of eigenvalues. This causes complications in the descriptions of the algorithms, but does not cause any practical problems

28. The eigenvectors of T are computed by back substitution [Wat02, § 5.8]. For each eigenvector $\mathbf{x}$ of T, $S\mathbf{x}$ is an eigenvector of A. The total additional cost of the eigenvector computation is $O(n^3)$. In LAPACK these tasks are performed by the routines HSEQR and TREVC.

29. Invariant subspaces can also be computed. The eigenvalues of A are $\lambda_1 = t_{11}, \ldots, \lambda_n = t_{nn}$. If $\lambda_1, \ldots, \lambda_k$ are disjoint from $\lambda_{k+1}, \ldots, \lambda_n$, then, because T is upper triangular, the first k columns of S span the invariant subspace associated with $\{\lambda_1, \ldots, \lambda_k\}$.

30. If an invariant subspace associated with k eigenvalues that are not at the top of T is wanted, then those k eigenvalues must be moved to the top by a sequence of swapping operations. Each operation is a unitary similarity transformation that reverses the positions of two adjacent main-diagonal entries of T. The transformations are applied to S as well. Once the desired eigenvalues have been moved to the top, the first k columns of the transformed S span the desired invariant subspace. For details see [BD93] and [GV96, § 7.6]. In LAPACK these tasks are performed by the routines TREXC and TRSEN.
31. An important difference between the symmetric and unsymmetric eigenvalue problems is that in the unsymmetric case, the eigenvalues can be ill conditioned. That is, a small perturbation in the entries of A can cause a large change in the eigenvalues. Suppose λ is an eigenvalue of A of algebraic multiplicity 1, and let E be a perturbation that is small in the sense that $\|E\|_2 \ll \|A\|_2$. Then $A + E$ has an eigenvalue $\lambda + \delta$ near λ. A condition number for λ is the smallest number κ such that

$$|\delta| \leq \kappa \|E\|_2$$

for all small perturbations E. If $\mathbf{x}$ and $\mathbf{y}$ are eigenvectors of A and A^T, respectively, associated with λ, then [Wat02, § 6.5]

$$\kappa \approx \frac{\|\mathbf{x}\|_2 \ \|\mathbf{y}\|_2}{|\mathbf{y}^T\mathbf{x}|}.$$

If $\kappa \gg 1$, λ is ill conditioned. If κ is not much bigger than 1, λ is well conditioned.
32. Condition numbers can also be defined for eigenvectors and invariant subspaces [GV96, § 7.2], [Wat02, § 6.5]. Eigenvectors associated with a tight cluster of eigenvalues are always ill conditioned. A more meaningful object is the invariant subspace associated with all of the eigenvalues in the cluster. This space will usually be well conditioned, even though the eigenvectors are ill conditioned. The LAPACK routines TRSNA and TRSEN compute condition numbers for eigenvalues, eigenvectors, and invariant subspaces.
33. The invariant subspace associated with $\{\lambda_1, \ldots, \lambda_k\}$ will certainly be ill conditioned if any of the eigenvalues $\lambda_{k+1}, \ldots, \lambda_n$ are close to any of $\lambda_1, \ldots, \lambda_k$. A necessary (but not sufficient) condition for well conditioning is that $\lambda_1, \ldots, \lambda_k$ be well separated from $\lambda_{k+1}, \ldots, \lambda_n$. A related practical fact is that if two eigenvalues are very close together, it may not be possible to swap them stably by LAPACK's TREXC.
34. (Performance) A 3.0 GHz Pentium 4 machine with 1 GB main memory and 1 MB cache computed the complete eigensystem of a random 1000×1000 real matrix using MATLAB in 56 seconds. This included balancing, reduction to upper Hessenberg form, triangularization by the QR algorithm, and back solving for the eigenvectors. All computed eigenpairs $(\lambda, \mathbf{v})$ satisfied $\|A\mathbf{v} - \lambda\mathbf{v}\|_1 < 10^{-15}\|A\|_1\|\mathbf{v}\|_1$.
35. (Generalized eigenvalue problem) The steps for solving the dense, unsymmetric, generalized eigenvalue problem $A\mathbf{v} = \lambda B\mathbf{v}$ are analogous to those for solving the standard problem. First (optionally) the pair (A, B) is balanced (by routine GGBAL in LAPACK). Then it is transformed by a strictly unitary equivalence to a condensed form in which A is upper Hessenberg and B is upper triangular. Then the QZ algorithm completes the reduction to triangular form. Details are given in [GV96, § 7.7] and [Wat02, § 6.7]. In LAPACK, the codes GGHRD and HGEQZ reduce the pair to Hessenberg-triangular form and perform the QZ iterations, respectively.
36. Once A has been reduced to triangular form, the eigenvalues are $\lambda_j = a_{jj}/b_{jj}$, $j = 1. \ldots, n$. The eigenvectors can be obtained by routines analogous to those used for the standard problem (LAPACK codes TGEVC and GGBAK), and condition numbers can be computed (LAPACK codes TGSNA and TGSEN).

Examples:

1. The matrix

$$A = \begin{bmatrix} -5.5849 \times 10^{-01} & -2.4075 \times 10^{+07} & -6.1644 \times 10^{+14} & 6.6275 \times 10^{+00} \\ -7.1724 \times 10^{-09} & -2.1248 \times 10^{+00} & -3.6183 \times 10^{+06} & 2.6435 \times 10^{-06} \\ -4.1508 \times 10^{-16} & -2.1647 \times 10^{-07} & 1.6229 \times 10^{-01} & -7.6315 \times 10^{-14} \\ 4.3648 \times 10^{-03} & 1.2614 \times 10^{+06} & -1.1986 \times 10^{+13} & -6.2002 \times 10^{-01} \end{bmatrix}$$

was balanced by Algorithm 1 (Fact 3) to produce

$$B = \begin{bmatrix} -0.5585 & -0.3587 & -1.0950 & 0.1036 \\ -0.4813 & -2.1248 & -0.4313 & 2.7719 \\ -0.2337 & -1.8158 & 0.1623 & -0.6713 \\ 0.2793 & 1.2029 & -1.3627 & -0.6200 \end{bmatrix}.$$

2. The matrix B of Example 1 was reduced to upper Hessenberg form by Algorithm 2 (Fact 7) to yield

$$H = \begin{bmatrix} -0.5585 & 0.7579 & 0.0908 & -0.8694 \\ 0.6036 & -3.2560 & -0.0825 & -1.8020 \\ 0 & 0.9777 & 1.2826 & -0.8298 \\ 0 & 0 & -1.5266 & -0.6091 \end{bmatrix}.$$

3. Algorithm 4 (Fact 23) was applied to the matrix H of Example 2, with $k = 1$ and shift $\mu_1 = h_{44} = -0.6091$, to produce

$$\begin{bmatrix} -3.1238 & -0.5257 & 1.0335 & 1.6798 \\ -1.3769 & 0.3051 & -1.5283 & 0.1296 \\ 0 & -1.4041 & 0.3261 & -1.0462 \\ 0 & 0 & -0.0473 & -0.6484 \end{bmatrix}.$$

The process was repeated twice again (with $\mu_1 = h_{44}$) to yield

$$\begin{bmatrix} -3.1219 & 0.7193 & 1.2718 & -1.4630 \\ 0.8637 & 1.8018 & 0.0868 & -0.3916 \\ 0 & 0.6770 & -1.2385 & 1.1642 \\ 0 & 0 & -0.0036 & -0.5824 \end{bmatrix}$$

and

$$\begin{bmatrix} -3.0939 & -0.6040 & 1.3771 & 1.2656 \\ -0.8305 & 1.8532 & -0.3517 & 0.5050 \\ 0 & 0.2000 & -1.3114 & -1.3478 \\ 0 & 0 & 0.00003 & -0.5888 \end{bmatrix}.$$

The (4,4) entry is an eigenvalue of A correct to four decimal places.

This matrix happens to have a real eigenvalue. If it had not, Algorithm 4 could have been used with $k = 2$ to extract the complex eigenvalues in pairs.

4. For an example of an ill-conditioned eigenvalue (Fact 31) consider a matrix

$$A = \begin{bmatrix} 1 & t \\ 0 & 1+\epsilon \end{bmatrix},$$

where t is large or ϵ is small or both. Since A is upper triangular, its eigenvalues are 1 and $1 + \epsilon$.

Eigenvectors of A and A^T associated with the eigenvalue 1 are

$$\mathbf{x} = \begin{bmatrix} 1 \\ 0 \end{bmatrix} \quad \text{and} \quad \mathbf{y} = \begin{bmatrix} 1 \\ -t/\epsilon \end{bmatrix},$$

respectively. Since $\|\mathbf{x}\|_2 = 1$, $\|\mathbf{y}\|_2 = \sqrt{1 + t^2/\epsilon^2}$, and $|\mathbf{y}^T\mathbf{x}| = 1$, the condition number of eigenvalue $\lambda = 1$ is $\kappa = \sqrt{1 + t^2/\epsilon^2} \approx t/\epsilon$. Thus if, for example, $t = 10^7$ and $\epsilon = 10^{-7}$, we have $\kappa \approx 10^{14}$.

5. This example illustrates Fact 32 on the ill conditioning of eigenvectors associated with a tight cluster of eigenvalues. Given a positive number ϵ that is as small as you please, the matrices

$$A_1 = \begin{bmatrix} 2+\epsilon & 0 & 0 \\ 0 & 2-\epsilon & 0 \\ 0 & 0 & 1 \end{bmatrix}$$

and

$$A_2 = \begin{bmatrix} 2 & \epsilon & 0 \\ \epsilon & 2 & 0 \\ 0 & 0 & 1 \end{bmatrix}$$

both have eigenvalues $1, 2+\epsilon$, and $2-\epsilon$, and they are very close together: $\|A_1 - A_2\|_2 = \sqrt{2}\epsilon$. However, unit eigenvectors associated with clustered eigenvalues $2+\epsilon$ and $2-\epsilon$ for A_1 are

$$\mathbf{e}_1 = \begin{bmatrix} 1 \\ 0 \\ 0 \end{bmatrix} \quad \text{and} \quad \mathbf{e}_2 = \begin{bmatrix} 0 \\ 1 \\ 0 \end{bmatrix},$$

while unit eigenvectors for A_2 are

$$\frac{1}{\sqrt{2}}\begin{bmatrix} 1 \\ 1 \\ 0 \end{bmatrix} \quad \text{and} \quad \frac{1}{\sqrt{2}}\begin{bmatrix} 1 \\ -1 \\ 0 \end{bmatrix}.$$

Thus, the tiny perturbation of order ϵ from A_1 to A_2 changes the eigenvectors completely; the eigenvectors are ill conditioned. In contrast the two-dimensional invariant subspace associated with the cluster $2+\epsilon, 2-\epsilon$ is $Span(\mathbf{e}_1, \mathbf{e}_2)$ for both A_1 and A_2, and it is well conditioned.

43.3 Sparse Matrix Techniques

If the matrix A is large and sparse and just a few eigenvalues are needed, sparse matrix techniques are appropriate. Some examples of common tasks are: (1) find the few eigenvalues of largest modulus, (2) find the few eigenvalues with largest real part, and (3) find the few eigenvalues nearest some target value τ. The corresponding eigenvectors might also be wanted. These tasks are normally accomplished by computing the low-dimensional invariant subspace associated with the desired eigenvalues. Then the information about the eigenvalues and eigenvectors is extracted from the invariant subspace.

The most widely used method for the sparse unsymmetric eigenvalue problem is the implicitly restarted Arnoldi method, as implemented in ARPACK [LSY98], which is discussed in Chapter 76. A promising variant is the Krylov–Schur algorithm of Stewart [Ste01]. MATLAB's sparse eigenvalue command "eigs" calls ARPACK.

Definitions:

Given a subspace $\mathcal{S}$ of $\mathbb{C}^n$, a vector $\mathbf{v} \in \mathcal{S}$ is called a **Ritz vector** of A from $\mathcal{S}$ if there is a $\theta \in \mathbb{C}$ such that $A\mathbf{v} - \theta\mathbf{v} \perp \mathcal{S}$. The scalar θ is the **Ritz value** associated with $\mathcal{S}$. The pair $(\theta, \mathbf{v})$ is a **Ritz pair**.

Facts:

1. [Wat02, § 6.1] Let $\mathbf{v}_1, \ldots, \mathbf{v}_m$ be a basis for a subspace $\mathcal{S}$ of $\mathbb{C}^n$, and let $V = [\mathbf{v}_1 \cdots \mathbf{v}_m]$. Then $\mathcal{S}$ is invariant under A if and only if there is a $B \in \mathbb{C}^{m\times m}$ such that $AV = VB$.
2. [Wat02, § 6.1] If $AV = VB$, then the eigenvalues of B are eigenvalues of A. If $\mathbf{x}$ is an eigenvector of B associated with eigenvalue μ, then $V\mathbf{x}$ is an eigenvector of A associated with μ.
3. [Wat02, §6.4] Let $\mathbf{v}_1, \ldots, \mathbf{v}_m$ be an orthonormal basis of $\mathcal{S}$, $V = [\mathbf{v}_1 \cdots \mathbf{v}_m]$, and $B = V^*AV$. Then the Ritz values of A associated with $\mathcal{S}$ are exactly the eigenvalues of B. If $(\theta, \mathbf{x})$ is an eigenpair of B, then $(\theta, V\mathbf{x})$ is a Ritz pair of A, and conversely.
4. If A is very large and sparse, it is essential to store A in a sparse data structure, in which only the nonzero entries of A are stored. One simple structure stores two integers n and nnz, which represent the dimension of the matrix and the number of nonzeros in the matrix, respectively. The matrix entries are stored in an array ent of length nnz, and the row and column indices are stored in two integer arrays of length nnz called row and col, respectively. For example, if the nonzero entry a_{ij} is stored in $ent(m)$, then this is indicated by setting $row(m) = i$ and $col(m) = j$. The space needed to store a matrix in this data structure is proportional to nnz.
5. Many operations that are routinely applied to dense matrices are impossible if the matrix is stored sparsely. Similarity transformations are out of the question because they quickly turn the zeros to nonzeros, transforming the sparse matrix to a full matrix.
6. One operation that is always possible is to multiply the matrix by a vector. This requires one pass through the data structure, and the work is proportional to nnz.

Algorithm 5. Sparse Matrix-Vector Multiply. Multiply A by $\mathbf{x}$ and store the result in $\mathbf{y}$.

$\mathbf{y} \leftarrow 0$
for $m = 1 : nnz$
  $[\mathbf{y}(row(m)) \leftarrow \mathbf{y}(row(m)) + ent(m) * \mathbf{x}(col(m))]$
end

7. Because the matrix-vector multiply is so easy, many sparse matrix methods access the matrix A in only this way. At each step, A is multiplied by one or several vectors, and this is the only way A is used.
8. The following standard methodology is widely used. A starting vector $\mathbf{v}_1$ is chosen, and the algorithm adds one vector per step, so that after $j - 1$ steps it has produced j orthonormal vectors $\mathbf{v}_1, \ldots, \mathbf{v}_j$. Let $V_j = [\mathbf{v}_1, \ldots, \mathbf{v}_j] \in \mathbb{C}^{n\times j}$ and let $\mathcal{S}_j = Span(V_j) = Span(\mathbf{v}_1, \ldots, \mathbf{v}_j)$. The jth step uses information from $\mathcal{S}_j$ to produce $\mathbf{v}_{j+1}$. The Ritz values of A associated with $\mathcal{S}_j$ are the eigenvalues of the $j \times j$ matrix $B_j = V_j^* A V_j$. The Ritz pair $(\theta, \mathbf{w})$ for which θ has the largest modulus is an estimate of the largest eigenvalue of A, and $\mathbf{x} = V_j\mathbf{w}$ is an estimate of the associated eigenvector. The residual $\mathbf{r}_j = A\mathbf{x} - \mathbf{x}\theta$ gives an indication of the quality of the approximate eigenpair.
9. Several methods use the residual $\mathbf{r}_j$ to help decide on the next basis vector $\mathbf{v}_{j+1}$. These methods typically use $\mathbf{r}_j$ to determine another vector $\mathbf{s}_j$, which is then orthonormalized against $\mathbf{v}_1, \ldots, \mathbf{v}_j$ to produce $\mathbf{v}_{j+1}$. The choice $\mathbf{s}_j = \mathbf{r}_j$ leads to a method that is equivalent to the Arnoldi process. However, Arnoldi's process should not be implemented this way in practice. (See Chapter 44.) The choice $\mathbf{s}_j = (D - \theta I)^{-1}\mathbf{r}_j$, where D is the diagonal part of A, gives Davidson's method. The Jacobi–Davidson methods have more elaborate ways of choosing $\mathbf{s}_j$. (See [BDD00, § 7.12] for details.)
10. Periodic purging is employed to keep the dimension of the active subspace from becoming too large. Given m vectors, the purging process keeps the most promising k-dimensional subspace of

$\mathcal{S}_m = Span(V_m)$ and discards the rest. Again, let $B_m = V_m^* A V_m$, and let $B_m = U_m T_m U_m^*$ be a unitary similarity transformation to upper triangular form. The Ritz values lie on the main diagonal of T_m and can be placed in any order. Place the k most promising Ritz values at the top. Let $\tilde{V}_m = V_m U_m$, and let $\tilde{V}_k$ denote the $n \times k$ submatrix of $\tilde{V}_m$ consisting of the first k columns. The columns of $\tilde{V}_k$ are the vectors that are kept.

11. After each purge, the algorithm can be continued from step k. Once the basis has been expanded back to m vectors, another purge can be carried out. After a number of cycles of expansion and purging, the invariant subspace associated with the desired eigenvalues will have been found.
12. When purging is carried out in connection with the Arnoldi process, it is called an implicit restart, and there are some extra details. (See Chapter 44 and [Ste01]).
13. The implicitly restarted Arnoldi process is well suited for computing the eigenvalues on the periphery of the spectrum of A. Thus, it is good for computing the eigenvalues of maximum modulus or those of maximum or minimum real part.
14. For computing interior eigenvalues, the shift-and-invert strategy is often helpful. Suppose the eigenvalues nearest some target value τ are sought. The matrix $(A - \tau I)^{-1}$ has the same eigenvectors as A, but the eigenvalues are different. If $\lambda_1, \ldots, \lambda_n$ are the eigenvalues of A, then $(\lambda_1 - \tau)^{-1} \ldots, (\lambda_n - \tau)^{-1}$ are the eigenvalues of $(A - \tau I)^{-1}$. The eigenvalues of $(A - \tau I)^{-1}$ of largest modulus correspond to the eigenvalues of A closest to τ. These can be computed by applying the implicitly restarted Arnoldi process to $(A - \tau I)^{-1}$. This is feasible whenever operations of the type $\mathbf{w} \leftarrow (A - \tau I)^{-1}\mathbf{x}$ can be performed efficiently. If a sparse decomposition $A - \tau I = PLU$ can be computed, as described in Chapter 41, then that decomposition can be used to perform the operation $\mathbf{w} \leftarrow (A - \tau I)^{-1}\mathbf{x}$ by back solves. If the LU factors take up too much space to fit into memory, this method cannot be used.
15. Another option for solving $(A - \tau I)\mathbf{w} = \mathbf{x}$ is to use an iterative method, as described in Chapter 41. However, this is very computationally intensive, as the systems must be solved to high accuracy if the eigenvalues are to be computed accurately.
16. The shift-and-invert strategy can also be applied to the generalized eigenvalue problem $A\mathbf{v} = \lambda B\mathbf{v}$. The implicitly restarted Arnoldi process is applied to the operator $(A - \tau B)^{-1}B$ to find eigenvalues near τ.
17. If the matrix is too large for the shift-and-invert strategy, Jacobi–Davidson methods can be considered. These also require the iterative solution of linear systems. In this family of methods, inaccurate solution of the linear systems may slow convergence of the algorithm, but it will not cause the eigenvalues to be computed inaccurately.
18. Arnoldi-based and Jacobi–Davidson algorithms are described in [BDD00]. A brief overview is given in [Wat02, § 6.4]. Balancing of sparse matrices is discussed in [BDD00, § 7.2].

References

[ABB99] E. Anderson, Z. Bai, C. Bischof, S. Blackford, J. Demmel, J. Dongarra, J. Du Croz, A. Greenbaum, S. Hammarling, A. McKenney, and D. Sorensen. *LAPACK Users' Guide,* 3rd ed., SIAM, Philadelphia, 1999, www.netlib.org/lapack/lug/.

[BD93] Z. Bai and J. W. Demmel. On swapping diagonal blocks in real Schur form. *Lin. Alg. Appl.* 186:73–95, 1993.

[BDD00] Z. Bai, J. Demmel, J. Dongarra, A. Ruhe, and H. van der Vorst. *Templates for the Solution of Algebraic Eigenvalue Problems, a Practical Guide.* SIAM, Philadelphia, 2000.

[BBM02a] K. Braman, R. Byers, and R. Mathias. The multi-shift QR algorithm part I: maintaining well-focused shifts and level 3 performance. *SIAM J. Matrix Anal. Appl.* 23:929–947, 2002.

[BBM02b] K. Braman, R. Byers, and R. Mathias. The multi-shift QR algorithm part II: aggressive early deflation. *SIAM J. Matrix Anal. Appl.* 23:948–973, 2002.

[Dem97] J.W. Demmel. *Applied Numerical Linear Algebra.* SIAM, Philadelphia, 1997.

[GV96] G.H. Golub and C.F. Van Loan. *Matrix Computations,* 3rd ed., The Johns Hopkins University Press, Baltimore, MD, 1996.

[Kre05] D. Kressner. *Numerical Methods for General and Structured Eigenproblems.* Springer, New York, 2005.

[LSY98] R.B. Lehoucq, D.C. Sorensen, and C.Yang. *ARPACK Users' Guide.* SIAM, Philadelphia, 1998.

[Osb60] E.E. Osborne. On pre-conditioning of matrices. *J. Assoc. Comput. Mach.* 7:338–345 1960.

[PR69] B.N. Parlett and C. Reinsch. Balancing a matrix for calculation of eigenvalues and eigenvectors. *Numer. Math.* 13:293–304, 1969. Also published as contribution II/11 in [WR71].

[Ste01] G.W. Stewart. A Krylov–Schur algorithm for large eigenproblems. *SIAM J. Matrix Anal. Appl.*, 23:601–614, 2001.

[Wat02] D.S. Watkins. *Fundamentals of Matrix Computations,* 2nd ed., John Wiley & Sons, New York, 2002.

[WE91] D.S. Watkins and L. Elsner. Convergence of algorithms of decomposition type for the eigenvalue problem. *Lin. Alg. Appl.* 143:19–47, 1991.

[WR71] J.H. Wilkinson and C. Reinsch. *Handbook for Automatic Computation, Vol. II, Linear Algebra,* Springer-Verlag, New York, 1971.

44

The Implicitly Restarted Arnoldi Method

D. C. Sorensen
Rice University

The implicitly restarted Arnoldi method (IRAM) [Sor92] is a variant of Arnoldi's method for computing a selected subset of eigenvalues and corresponding eigenvectors for large matrices. Implicit restarting is a synthesis of the implicitly shifted QR iteration and the Arnoldi process that effectively limits the dimension of the Krylov subspace required to obtain good approximations to desired eigenvalues. The space is repeatedly expanded and contracted with each new Krylov subspace generated by an updated starting vector obtained by implicit application of a matrix polynomial to the old starting vector. This process is designed to filter out undesirable components in the starting vector in a way that enables convergence to the desired invariant subspace. This method has been implemented and is freely available as ARPACK. The MATLAB® function `eigs` is based upon ARPACK. Use of this software is described in Chapter 76.

In this article, all matrices, vectors, and scalars are complex and the algorithms are phrased in terms of complex arithmetic. However, when the matrix (or matrix pair) happens to be real then the computations may be organized so that only real arithmetic is required. Multiplication of a vector $\mathbf{x}$ by a scalar λ is denoted by $\mathbf{x}\lambda$ so that the eigenvector–eigenvalue relation is $A\mathbf{x} = \mathbf{x}\lambda$. This convention provides for direct generalizations to the more general invariant subspace relations $AX = XH$, where X is an $n \times k$ matrix and H is a $k \times k$ matrix with $k < n$. More detailed discussion of all facts and definitions may be found in the overview article [Sor02].

44.1 Krylov Subspace Projection

The classic power method is the simplest way to compute the dominant eigenvalue and corresponding eigenvector of a large matrix. Krylov subspace projection provides a way to extract additional eigen-information from the power method iteration by considering all possible linear combinations of the sequence of vectors produced by the power method.

Definitions:

The best approximate eigenvectors and corresponding eigenvalues are extracted from the **Krylov subspace**

$$\mathcal{K}_k(A, \mathbf{v}) := \text{span}\{\mathbf{v}, A\mathbf{v}, A^2\mathbf{v}, \ldots, A^{k-1}\mathbf{v}\}.$$

The approximate eigenpairs are constructed through a Galerkin condition. An approximate eigenvector $\mathbf{x} \in \mathcal{S}$ is called a **Ritz vector** with corresponding **Ritz value** θ if the **Galerkin condition**

$$\mathbf{w}^*(A\mathbf{x} - \mathbf{x}\theta) = 0, \quad \text{for all } \mathbf{w} \in \mathcal{K}_k(A, \mathbf{v})$$

is satisfied.

Facts: [Sor92], [Sor02]

1. Every $\mathbf{w} \in \mathcal{K}_k$ is of the form $\mathbf{w} = \phi(A)\mathbf{v}_1$ for some polynomial ϕ of degree less than k and $\mathcal{K}_{j-1} \subset \mathcal{K}_j$ for $j = 2, 3, \ldots, k$.
2. If a sequence of orthogonal bases $V_k = [\mathbf{v}_1, \mathbf{v}_2, \ldots, \mathbf{v}_k]$ has been constructed with $\mathcal{K}_k = \text{range}(V_k)$ and $V_k^* V_k = I_k$, then a new basis vector $\mathbf{v}_{k+1}$ is obtained by the **projection formulas**

$$\mathbf{h}_k = V_k^* A\mathbf{v}_k,$$
$$\mathbf{f}_k = A\mathbf{v}_k - V_k\mathbf{h}_k,$$
$$\mathbf{v}_{k+1} = \mathbf{f}_k / \|\mathbf{f}_k\|_2.$$

   The vector $\mathbf{h}_k$ is constructed to achieve $V_k^*\mathbf{f}_k = 0$ so that $\mathbf{v}_{k+1}$ is a vector of unit length that is orthogonal to the columns of V_k.
3. The columns of $V_{k+1} = [V_k, \mathbf{v}_{k+1}]$ provide an orthonormal basis for $\mathcal{K}_{k+1}(A, \mathbf{v}_1)$.
4. The basis vectors are of the form $\mathbf{v}_j = \phi_{j-1}(A)\mathbf{v}_1$, where ϕ_{j-1} is a polynomial of degree $j-1$ for each $j = 1, 2, \ldots, k+1$.
5. This construction fails when $\mathbf{f}_k = 0$, but then

$$AV_k = V_k H_k,$$

   where $H_k = V_k^* A V_k = [\mathbf{h}_1, \mathbf{h}_2, \ldots, \mathbf{h}_k]$ (with a slight abuse of notation). This "good breakdown" happens precisely when $\mathcal{K}_k$ is an invariant subspace of A. Hence, $\sigma(H_k) \subset \sigma(A)$.

44.2 The Arnoldi Factorization

The projection formulas given above result in the fundamental Arnoldi method for constructing an orthonormal basis for $\mathcal{K}_k$.

Definitions:

The relations between the matrix A, the basis matrix V_k and the residual vector $\mathbf{f}_k$ may be concisely expressed as

$$AV_k = V_k H_k + \mathbf{f}_k \mathbf{e}_k^*,$$

where $V_k \in \mathbb{C}^{n \times k}$ has orthonormal columns, $V_k^* \mathbf{f}_k = 0$, and $H_k = V_k^* A V_k$ is a $k \times k$ upper Hessenberg matrix with nonnegative subdiagonal elements.

The above expression shall be called a **k-step Arnoldi factorization** of A. When A is Hermitian, H_k will be real, symmetric, and tridiagonal and then the relation is called a **k-step Lanczos factorization** of A. The columns of V_k are referred to as **Arnoldi vectors** or **Lanczos vectors**, respectively. The Hessenberg matrix H_k is called **unreduced** if all subdiagonal elements are nonzero.

Facts: [Sor92], [Sor02]

1. The explicit steps needed to form a k-step Arnoldi factorization are shown in Algorithm 1.

Algorithm 1: k**-step Arnoldi factorization.** A square matrix A, a nonzero vector $\mathbf{v}$ and a positive integer $k \le n$ are input.

Output is an $n \times k$ ortho-normal matrix V_k, an upper Hessenberg matrix H_k and a vector $\mathbf{f}_k$ such that $AV_k = V_k H_k + \mathbf{f}_k \mathbf{e}_k^T$.

$\mathbf{v}_1 = \mathbf{v}/\|\mathbf{v}\|_2$;
$\mathbf{w} = A\mathbf{v}_1$; $\alpha_1 = \mathbf{v}_1^* \mathbf{w}$;
$\mathbf{f}_1 \leftarrow \mathbf{w} - \mathbf{v}_1 \alpha_1$;
$V_1 \leftarrow [\mathbf{v}_1]$; $H_1 \leftarrow [\alpha_1]$;
for $j = 1, 2, 3, \ldots k-1$,
  $\beta_j = \|\mathbf{f}_j\|_2$; $\mathbf{v}_{j+1} \leftarrow \mathbf{f}_j/\beta_j$;
  $V_{j+1} \leftarrow [V_j, \mathbf{v}_{j+1}]$;
  $\hat{H}_j \leftarrow \begin{bmatrix} H_j \\ \beta_j \mathbf{e}_j^* \end{bmatrix}$;
  $\mathbf{w} \leftarrow A\mathbf{v}_{j+1}$;
  $\mathbf{h} \leftarrow V_{j+1}^* \mathbf{w}$;
  $\mathbf{f}_{j+1} \leftarrow \mathbf{w} - V_{j+1}\mathbf{h}$;
  $H_{j+1} \leftarrow [\hat{H}_j, \mathbf{h}]$;
end

2. Ritz pairs satisfying the Galerkin condition (see Section 44.1) are derived from the eigenpairs of the small projected matrix H_k. If $H_k \mathbf{y} = \mathbf{y}\theta$ with $\|\mathbf{y}\|_2 = 1$, then the vector $\mathbf{x} = V_k \mathbf{y}$ is a vector of unit norm that satisfies

$$\|A\mathbf{x} - \mathbf{x}\theta\|_2 = \|(AV_k - V_k H_k)\mathbf{y}\|_2 = |\beta_k \mathbf{e}_k^* \mathbf{y}|,$$

where $\beta_k = \|\mathbf{f}_k\|_2$.
3. If $(\mathbf{x}, \theta)$ is a Ritz pair constructed as shown in Fact 2, then

$$\theta = \mathbf{y}^* H_k \mathbf{y} = (V_k \mathbf{y})^* A (V_k \mathbf{y}) = \mathbf{x}^* A\mathbf{x}$$

is always a Rayleigh quotient (assuming $\|\mathbf{y}\|_2 = 1$).
4. The Rayleigh quotient residual $\mathbf{r}(\mathbf{x}) := A\mathbf{x} - \mathbf{x}\theta$ satisfies $\|\mathbf{r}(\mathbf{x})\|_2 = |\beta_k \mathbf{e}_k^* \mathbf{y}|$. When A is Hermitian, this relation provides computable rigorous bounds on the accuracy of the approximate eigenvalues [Par80]. When A is non-Hermitian, one needs additional sensitivity information. Nonnormality effects may corrupt the accuracy. In exact arithmetic, these Ritz pairs are eigenpairs of A whenever $\mathbf{f}_k = 0$. However, even with a very small residual these may be far from actual eigenvalues when A is highly nonnormal.
5. The orthogonalization process is based upon the *classical Gram–Schmidt* (CGS) scheme. This process is notoriously unstable and will fail miserably in this application without modification.

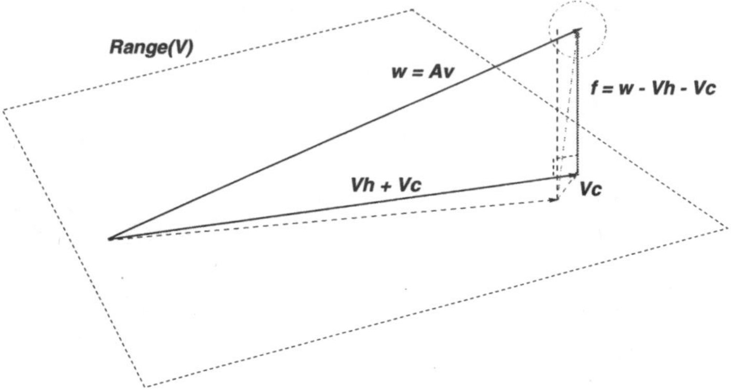

FIGURE 44.1 DGKS Correction.

The iterative refinement technique proposed by Daniel, Gragg, Kaufman, and Stewart (DGKS) [DGK76] provides an excellent way to construct a vector $\mathbf{f}_{j+1}$ that is numerically orthogonal to V_{j+1}. It amounts to computing a correction

$$\mathbf{c} = V_{j+1}^*\mathbf{f}_{j+1}; \;\; \mathbf{f}_{j+1} \leftarrow \mathbf{f}_{j+1} - V_{j+1}\mathbf{c}; \;\; \mathbf{h} \leftarrow \mathbf{h} + \mathbf{c};$$

just after computing $\mathbf{f}_{j+1}$ if necessary, i.e., when $\mathbf{f}_{j+1}$ is not sufficiently orthogonal to the columns of V_{j+1}. This formulation is crucial to both accuracy and performance. It provides numerically orthogonal basis vectors and it may be implemented using the Level 2 BLAS operation _GEMV [DDH88]. This provides a significant performance advantage on virtually every platform from workstation to supercomputer.

6. The *modified Gram–Schmidt* (MGS) process will generally fail to produce orthogonal vectors and cannot be implemented with Level 2 BLAS in this setting. ARPACK relies on a restarting scheme wherein the goal is to reach a state of dependence in order to obtain $\mathbf{f}_k = 0$. MGS is completely inappropriate for this situation, but the CGS with DGKS correction performs beautifully.
7. Failure to maintain orthogonality leads to numerical difficulties in the Lanczos/Arnoldi process. Loss of orthogonality typically results in the presence of spurious copies of the approximate eigenvalue.

Examples:

1. Figure 44.1 illustrates how the DGKS mechanism works. When the vector $\mathbf{w} = A\mathbf{v}$ is nearly in the range(V) then the projection $V\mathbf{h}$ is possibly inaccurate, but vector $= \mathbf{w} - \mathbf{Vh}$ is not close to range(V) and can be safely orthogonalized to compute the correction $\mathbf{c}$ accurately. The corrected vector $\mathbf{f} \leftarrow \mathbf{f} - V\mathbf{c}$ will be numerically orthogonal to the columns of V in almost all cases. Additional corrections might be necessary in very unusual cases.

44.3 Restarting the Arnoldi Process

The number of Arnoldi steps required to calculate eigenvalues of interest to a specified accuracy cannot be pre-determined. Usually, eigen-information of interest does not appear until k gets very large. In Figure 44.2 the distribution in the complex plane of the Ritz values (shown in grey dots) is compared with the spectrum (shown as +s). The original matrix is a normally distributed random matrix of order 200 and the Ritz values are from a ($k = 50$)-step Arnoldi factorization. Eigenvalues at the extremes of the spectrum of A are clearly better approximated than the interior eigenvalues.

For large problems, it is intractable to compute and store a numerically orthogonal basis set V_k for large k. Storage requirements are $\mathcal{O}(n \cdot k)$ and arithmetic costs are $\mathcal{O}(n \cdot k^2)$ flops to compute the basis vectors

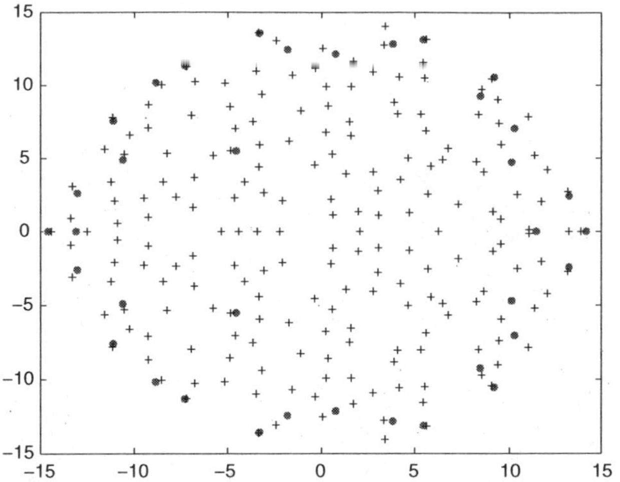

FIGURE 44.2 Typical distribution of Ritz values.

plus $\mathcal{O}(k^3)$ flops to compute the eigensystem of H_k. Thus, restarting schemes have been developed that iteratively replace the starting vector $\mathbf{v}_1$ with an "improved" starting vector $\mathbf{v}_1^+$ and then compute a new Arnoldi factorization of fixed length k to limit the costs. Beyond this, there is an interest in forcing $\mathbf{f}_k = 0$ and, thus, producing an invariant subspace. However, this is useful only if the spectrum $\sigma(H_k)$ has the desired properties.

The structure of $\mathbf{f}_k$ suggests the restarting strategy. The goal will be to iteratively force $\mathbf{v}_1$ to be a linear combination of eigenvectors of interest.

Facts: [Sor92], [Sor02]

1. If $\mathbf{v} = \sum_{j=1}^{k} \mathbf{q}_j \gamma_j$ where $A\mathbf{q}_j = \mathbf{q}_j \lambda_j$ and

$$AV = VH + \mathbf{f}\mathbf{e}_k^T$$

   is a k-step Arnoldi factorization with unreduced H, then $\mathbf{f} = 0$ and $\sigma(H) = \{\lambda_1, \lambda_2, \ldots, \lambda_k\}$.
2. Since $\mathbf{v}_1$ determines the subspace $\mathcal{K}_k$, this vector must be constructed to select the eigenvalues of interest. The starting vector must be forced to become a linear combination of eigenvectors that span the desired invariant subspace. There is a necessary and sufficient condition for $\mathbf{f}$ to vanish that involves Schur vectors and does not require diagonalizability.

44.4 Polynomial Restarting

Polynomial restarting strategies replace $\mathbf{v}_1$ by

$$\mathbf{v}_1 \leftarrow \psi(A)\mathbf{v}_1,$$

where ψ is a polynomial constructed to damp unwanted components from the starting vector. If $\mathbf{v}_1 = \sum_{j=1}^{n} \mathbf{q}_j \gamma_j$ where $A\mathbf{q}_j = \mathbf{q}_j \lambda_j$, then

$$\mathbf{v}_1^+ = \psi(A)\mathbf{v}_1 = \sum_{j=1}^{n} \mathbf{q}_j \gamma_j \psi(\lambda_j),$$

where the polynomial ψ has also been normalized to give $\|\mathbf{v}_1\|_2 = 1$. Motivated by the structure of $\mathbf{f}_k$, the idea is to force the starting vector to be closer and closer to an invariant subspace by constructing ψ so that $|\psi(\lambda)|$ is as small as possible on a region containing the unwanted eigenvalues.

An iteration is defined by repeatedly restarting until the updated Arnoldi factorization eventually contains the desired eigenspace. An explicit scheme for restarting was proposed by Saad in [Saa92]. One of the more successful choices is to use Chebyshev polynomials in order to damp unwanted eigenvector components.

Definitions:

The polynomial ψ is sometimes called a **filter polynomial**, which may also be specified by its roots. The roots of the filter polynomial may also be referred to as **shifts**. This terminology refers to their usage in an implicitly shifted QR-iteration. One straightforward choice of shifts is to find the eigenvalues θ_j of the projected matrix H and sort these into two sets according to a given criterion: the wanted set $\mathcal{W} = \{\theta_j : j = 1, 2, \ldots, k\}$ and the unwanted set $\mathcal{U} = \{\theta_j : j = k+1, k+2, \ldots, k+p\}$. Then one specifies the polynomial ψ as the polynomial with these unwanted Ritz values as it roots. This choice of roots, called **exact shifts**, was suggested in [Sor92].

Facts: [Sor92], [Sor02]

1. Morgan [Mor96] found a remarkable property of this strategy. If exact shifts are used to define $\psi(\tau) = \prod_{j=k+1}^{k+p}(\tau - \theta_j)$ and if $\hat{\mathbf{q}}_j$ denotes a Ritz vector of unit length corresponding to θ_j, then the Krylov space generated by $\mathbf{v}_1^+ = \psi(A)\mathbf{v}_1$ satisfies

$$\mathcal{K}_m(A, \mathbf{v}_1^+) = Span\{\hat{\mathbf{q}}_1, \hat{\mathbf{q}}_2, \ldots, \hat{\mathbf{q}}_k, A\hat{\mathbf{q}}_j, A^2\hat{\mathbf{q}}_j, \ldots, A^p\hat{\mathbf{q}}_j\},$$

   for any $j = 1, 2, \ldots, k$. Thus, polynomial restarting with exact shifts will generate a new subspace that contains all of the possible choices of updated staring vector consisting of linear combinations of the wanted Ritz vectors.
2. Exact shifts tend to perform remarkably well in practice and have been adopted as the shift selection of choice in ARPACK when no other information is available. However, there are many other possibilities such as the use of Leja points for certain containment regions or intervals [BCR96].

44.5 Implicit Restarting

There are a number of schemes used to implement polynomial restarting. We shall focus on an implicit restarting scheme.

Definitions:

A straightforward way to implement polynomial restarting is to explicitly construct the starting vector $\mathbf{v}_1^+ = \psi(A)\mathbf{v}_1$ by applying $\psi(A)$ through a sequence of matrix-vector products. This is called **explicit restarting.** A more efficient and numerically stable alternative is **implicit restarting**. This technique applies a sequence of implicitly shifted QR steps to an m-step Arnoldi or Lanczos factorization to obtain a truncated form of the implicitly shifted QR-iteration.

On convergence, the IRAM iteration (see Algorithm 2) gives an orthonormal matrix V_k and an upper Hessenberg matrix H_k such that $AV_k \approx V_k H_k$. If $H_k Q_k = Q_k R_k$ is a Shur decompositon of H_k, then we call $\hat{V}_k \equiv V_k Q_k$ a **Schur basis** for the Krylov subspace $\mathcal{K}_k(A, \mathbf{v}_1)$. Note that if $AV_k = V_k H_k$ exactly, then $\hat{V}_k$ would form the leading k columns of a unitary matrix $\hat{V}$ and R_k would form the leading $k \times k$ block of an upper triangular matrix R, where $A\hat{V} = \hat{V}R$ is a complete Schur decomposition. We refer to this as a **partial Schur decomposition** of A.

Algorithm 2: IRAM iteration
Input is an $n \times k$ ortho-normal matrix V_k, an upper Hessenberg matrix H_k and a vector $\mathbf{f}_k$
such that $AV_k = V_k H_k + \mathbf{f}_k \mathbf{e}_k^T$.
Output is an $n \times k$ ortho-normal matrix V_k, an upper triangular
matrix H_k such that $AV_k = V_k H_k$.

repeat until convergence,
    Beginning with the k-step factorization,
    apply p additional steps of the Arnoldi process
    to compute an $m = k + p$ step Arnoldi factorization
        $AV_m = V_m H_m + \mathbf{f}_m \mathbf{e}_m^*$.
    Compute $\sigma(H_m)$ and select p shifts $\mu_1, \mu_2, \ldots \mu_p$;
    $Q = I_m$;
    for $j = 1, 2, \ldots, p$,
        Factor $[Q_j, R_j] = \text{qr}(H_m - \mu_j I)$;
        $H_m \leftarrow Q_j^* H_m Q_j$;
        $Q \leftarrow QQ_j$;
    end
    $\hat{\beta}_k = H_m(k+1, k)$; $\sigma_k = Q(m, k)$;
    $\mathbf{f}_k \leftarrow \mathbf{v}_{k+1}\hat{\beta}_k + \mathbf{f}_m \sigma_k$;
    $V_k \leftarrow V_m Q(:, 1:k)$; $H_k \leftarrow H_m(1:k, 1:k)$;
end

Facts: [Sor92], [Sor02]

1. Implicit restarting avoids numerical difficulties and storage problems normally associated with Arnoldi and Lanczos processes. The algorithm is capable of computing a few (k) eigenvalues with user specified features such as largest real part or largest magnitude using $2nk + \mathcal{O}(k^2)$ storage. The computed Schur basis vectors for the desired k-dimensional eigenspace are numerically orthogonal to working precision.
2. Desired eigen-information from a high-dimensional Krylov space is continually compressed into a fixed size k-dimensional subspace through an implicitly shifted QR mechanism. An Arnoldi factorization of length $m = k + p$,
$$AV_m = V_m H_m + \mathbf{f}_m \mathbf{e}_m^*,$$
is compressed to a factorization of length k that retains the eigen-information of interest. Then the factorization is expanded once more to m-steps and the compression process is repeated.
3. QR steps are used to apply p linear polynomial factors $A - \mu_j I$ implicitly to the starting vector $\mathbf{v}_1$. The first stage of this shift process results in
$$AV_m^+ = V_m^+ H_m^+ + \mathbf{f}_m \mathbf{e}_m^* Q,$$
where $V_m^+ = V_m Q$, $H_m^+ = Q^* H_m Q$, and $Q = Q_1 Q_2 \cdots Q_p$. Each Q_j is the orthogonal matrix associated with implicit application of the shift $\mu_j = \theta_{k+j}$. Since each of the matrices Q_j is Hessenberg, it turns out that the first $k-1$ entries of the vector $\mathbf{e}_m^* Q$ are zero (i.e., $\mathbf{e}_m^* Q = [\sigma \mathbf{e}_k^T, \hat{\mathbf{q}}^*]$). Hence, the leading k columns remain in an Arnoldi relation and provide an updated k-step Arnoldi factorization
$$AV_k^+ = V_k^+ H_k^+ + \mathbf{f}_k^+ \mathbf{e}_k^*,$$
with an updated residual of the form $\mathbf{f}_k^+ = V_m^+ \mathbf{e}_{k+1} \hat{\beta}_k + \mathbf{f}_m \sigma$. Using this as a starting point, it is possible to apply p additional steps of the Arnoldi process to return to the original m-step form.

4. Virtually any explicit polynomial restarting scheme can be applied with implicit restarting, but considerable success has been obtained with exact shifts. Exact shifts result in H_k^+ having the k wanted Ritz values as its spectrum. As convergence takes place, the subdiagonals of H_k tend to zero and the most desired eigenvalue approximations appear as eigenvalues of the leading $k \times k$ block of R as a partial Schur decomposition of A. The basis vectors V_k tend to numerically orthogonal Schur vectors.
5. The basic IRAM iteration is shown in Algorithm 2.

Examples:

1. The expansion and contraction process of the IRAM iteration is visualized in Figure 44.3.

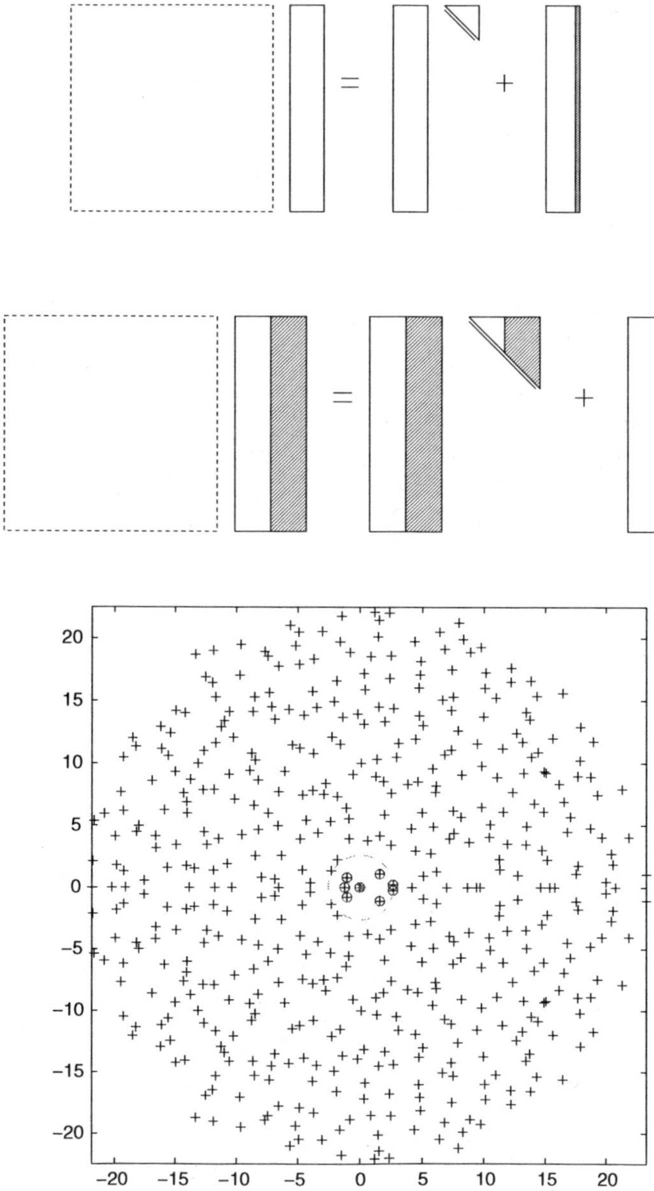

FIGURE 44.3 Visualization of IRAM.

44.6 Convergence of IRAM

IRAM converges linearly. An intuitive explanation follows. If $\mathbf{v}_1$ is expressed as a linear combination of eigenvectors $\{\mathbf{q}_j\}$ of A, then

$$\mathbf{v}_1 = \sum_{j=1}^{n} \mathbf{q}_j \gamma_j \Rightarrow \psi(A)\mathbf{v}_1 = \sum_{j=1}^{n} \mathbf{q}_j \psi(\lambda_j)\gamma_j.$$

Applying the same polynomial (i.e., using the same shifts) repeatedly for ℓ iterations will result in the j-th original expansion coefficient being attenuated by a factor

$$\left(\frac{\psi(\lambda_j)}{\psi(\lambda_1)}\right)^{\ell},$$

where the eigenvalues have been ordered according to decreasing values of $|\psi(\lambda_j)|$. The leading k eigenvalues become dominant in this expansion and the remaining eigenvalues become less and less significant as the iteration proceeds. Hence, the starting vector $\mathbf{v}_1$ is forced into an invariant subspace as desired. The adaptive choice of ψ provided with the exact shift mechanism further enhances the isolation of the wanted components in this expansion. Hence, the wanted eigenvalues are approximated ever better as the iteration proceeds. Making this heuristic argument precise has turned out to be quite difficult. Some fairly sophisticated analysis is required to understand convergence of these methods.

44.7 Convergence in Gap: Distance to a Subspace

To fully discuss convergence we need some notion of nearness of subspaces. When nonnormality is present or when eigenvalues are clustered, the distance between the computed subspace and the desired subspace is a better measure of success than distance between eigenvalues. The subspaces carry uniquely defined Ritz values with them, but these can be very sensitive to perturbations in the nonnormal setting.

Definitions:

A notion of distance that is useful in our setting is the **containment gap** between the subspaces $\mathcal{W}$ and $\mathcal{V}$:

$$\delta(\mathcal{W}, \mathcal{V}) := \max_{\mathbf{w} \in \mathcal{W}} \min_{\mathbf{v} \in \mathcal{V}} \frac{\|\mathbf{w} - \mathbf{v}\|_2}{\|\mathbf{w}\|_2}.$$

Note: $\delta(\mathcal{W}, \mathcal{V})$ is the sine of the largest canonical angle between $\mathcal{W}$ and the closest subspace of $\mathcal{V}$ with the same dimension as $\mathcal{W}$.

In keeping with the terminology developed in [BER04] and [BES05], $\mathcal{X}_g$ shall be the invariant subspace of A associated with the so called "good" eigenvalues (the desired eigenvalues) and $\mathcal{X}_b$ is the complementary subspace. $\mathbf{P}_g$ and $\mathbf{P}_b$ are the spectral projectors with respect to these spaces.

It is desirable to have **convergence in Gap** for the Krylov method, meaning

$$\delta(\mathcal{K}_m(A, \mathbf{v}_1^{(\ell)}), \mathcal{X}_g) \to 0.$$

Fundamental quantities required to study convergence.

1. **Minimal Polynomial for $\mathcal{X}_g$:**

$$a_g := \textit{minimal polynomial of } A \textit{ with respect to } \mathbf{P}_g \mathbf{v}_1,$$

which is the monic polynomial of least degree s.t. $a_g(A)\mathbf{P}_g\mathbf{v}_1 = \mathbf{0}$.

2. **Nonnormality constant $\kappa(\Omega)$:** The smallest positive number s.t.

$$\|f(A)\,\Pi_{\mathcal{U}}\|_2 \leq \kappa(\Omega) \max_{z\in\Omega} |f(z)|$$

uniformly for all functions f analytic on Ω. This constant and its historical origins are discussed in detail in [BER04].

3. **ε-pseudospectrum of A:**

$$\Lambda_\varepsilon(A) := \{z \in \mathbf{C} : \|(zI - A)^{-1}\|_2 \geq \varepsilon^{-1}\}.$$

Facts: [BER04], [BES05]

1. Two fundamental convergence questions:
   - What is the gap $\delta(\mathcal{U}_g, \mathcal{K}_k(A, \mathbf{v}_1))$ as k increases?
   - How does $\delta(\mathcal{U}_g, \mathcal{K}_m(A, \widehat{\mathbf{v}}_1))$ depend on $\widehat{\mathbf{v}}_1 = \Phi(A)\mathbf{v}_1$, and how can we optimize the asymptotic behavior?

   Key ingredients to convergence behavior are the nonnormality of A and the distribution of $\mathbf{v}_1$ w. r. t. $\mathcal{U}_g$. The goal of restarting is to attain the unrestarted iteration performance, but within **restricted subspace dimensions.**

2. **Convergence with no restarts:** In [BES05], it is shown that

$$\delta(\mathcal{U}_g, \mathcal{K}_\ell(A, \mathbf{v}_1)) \leq C_o C_b \min_{p\in\mathcal{P}_{\ell-2m}} \max_{z\in\Omega_b} \left|1 - a_g(z)p(z)\right|,$$

   where the compact set $\Omega_g \subseteq \mathbf{C} \setminus \Omega_b$ contains all the good eigenvalues.

$$C_o := \max_{\psi\in\mathcal{P}_{m-1}} \frac{\|\psi(A)\mathbf{P}_b\mathbf{v}_1\|_2}{\|\psi(A)\mathbf{P}_g\mathbf{v}_1\|_2}, \qquad C_b := \kappa(\Omega_b).$$

3. Rate of convergence estimates are obtained from complex approximation theory. Construct conformal map $\mathcal{G}$ taking the exterior of Ω_b to the exterior of the unit disk with $\mathcal{G}(\infty) = \infty$ and $\mathcal{G}'(\infty) > 0$. Define $\rho := \left(\min_{j=1,\ldots,L} |\mathcal{G}(\lambda_j)|\right)^{-1}$. Then (Gaier, Walsh)

$$\limsup_{k\to\infty} \min_{p\in\mathcal{P}_k} \max_{z\in\Omega_b} \left|\frac{1}{a_g(z)} - p(z)\right|^{1/k} = \rho.$$

   The image of $\{|z| = \rho^{-1}\}$ is a curve $\mathcal{C} := \mathcal{G}^{-1}(\{|z| = \rho^{-1}\})$ around Ω_b. This critical curve passes through a good eigenvalue "closest to" Ω_b The curve contains at least one good eigenvalue, with all bad and no good eigenvalues in its interior.

4. Convergence with the exact shift strategy has not yet been fully analyzed. However, convergence rates have been established for restarts with asymptotically optimal points. These are the Fejér, Fekete, or Leja points for Ω_b. In [BES05], computational experiments are shown that indicate that exact shifts behave very much like optimal points for certain regions bounded by pseudo-spectral level curves or lemniscates.

5. Let Ψ_M interpolate $1/a_g(z)$ at the M restart shifts:

$$\delta(\mathcal{U}_g, \mathcal{K}_\ell(A, \widehat{\mathbf{v}}_1)) \leq C_o C_g \max_{z\in\Omega_b} \left|1 - \Psi_M(z)a_g(z)\right| \leq C_o\, C_g\, C_r\, r^M$$

   for any $r > \rho$ (see [Gai87], [FR89]). Here, $\widehat{\mathbf{v}}_1 = \Phi(A)\mathbf{v}_1$, where Φ is the aggregate restart polynomial (its roots are all the implicit restart shifts that have been applied). The subspace dimension is $\ell = 2m$, the restart degree is m, and the aggregate degree is $M = \nu m$.

44.8 The Generalized Eigenproblem

In many applications, the generalized eigenproblem $A\mathbf{x} = M\mathbf{x}\lambda$ arises naturally. A typical setting is a finite element discretization of a continuous problem where the matrix M arises from inner products of basis functions. In this case, M is symmetric and positive (semi) definite, and for some algorithms this property is a necessary condition. Generally, algorithms are based upon transforming the generalized problem to a standard problem.

44.9 Krylov Methods with Spectral Transformations

Definitions:

A very successful scheme for converting the generalized problem to a standard problem that is amenable to a Krylov or a subspace iteration method is to use the **spectral transformation** suggested by Ericsson and Ruhe [ER80],

$$(A - \sigma M)^{-1} M x = \mathbf{x}\nu.$$

Facts: [Sor92], [Sor02]

1. An eigenvector $\mathbf{x}$ of the spectral transformation is also an eigenvector of the original problem $A\mathbf{x} = M\mathbf{x}\lambda$, with the corresponding eigenvalue given by $\lambda = \sigma + \frac{1}{\nu}$.
2. There is generally rapid convergence to eigenvalues near the shift σ because they are transformed to extremal well-separated eigenvalues. Perhaps an even more influential aspect of this transformation is that eigenvalues far from σ are damped (mapped near zero).
3. One strategy is to choose σ to be a point in the complex plane that is near eigenvalues of interest and then compute the eigenvalues ν of largest magnitude of the spectral trasformation matrix. It is not necessary to have σ extremely close to an eigenvalue. This transformation together with the implicit restarting technique is usually adequate for computing a significant number of eigenvalues near σ.
4. Even when $M = I$, one generally must use the shift-invert spectral transformation to find interior eigenvalues. The extreme eigenvalues of the transformed operator A_σ are generally large and well separated from the rest of the spectrum. The eigenvalues ν of largest magnitude will transform back to eigenvalues λ of the original A that are in a disk about the point σ. This is illustrated in Figure 44.4, where the + symbols are the eigenvalues of A and the circled ones are the computed eigenvalues in the disk (dashed circle) centered at the point σ.
5. With shift-invert, the Arnoldi process is applied to the matrix $A_\sigma := (A - \sigma M)^{-1} M$. Whenever a matrix-vector product $\mathbf{w} \leftarrow A_\sigma \mathbf{v}$ is required, the following steps are performed:
   - $\mathbf{z} = M\mathbf{v}$,
   - Solve $(A - \sigma M)\mathbf{w} = \mathbf{z}$ for $\mathbf{w}$.

   The matrix $A - \sigma M$ is factored initially with a sparse direct **LU**-decomposition or in a symmetric indefinite factorization and this single factorization is used repeatedly to apply the matrix operator A_σ as required.
6. The scheme is modified to preserve symmetry when A and M are both symmetric and M is positive (semi)definite. One can utilize a weighted M (semi) inner product in the Lanczos/Arnoldi process [ER80], [GLS94], [MS97]. This amounts to replacing the computation of $\mathbf{h} \leftarrow V_{j+1}^* \mathbf{w}$ and $\beta_j = \|\mathbf{f}_j\|_2$ with $\mathbf{h} \leftarrow V_{j+1}^* M\mathbf{w}$ and $\beta_j = \sqrt{\mathbf{f}_j^* M\mathbf{f}_j}$, respectively, in the Arnoldi process described in Algorithm 1.
7. The matrix operator A_σ is self-adjoint with respect to this (semi)inner product, i.e., $\langle A_\sigma \mathbf{x}, \mathbf{y}\rangle = \langle \mathbf{x}, A_\sigma \mathbf{y}\rangle$ for all vectors $\mathbf{x}, \mathbf{y}$, where $\langle \mathbf{w}, \mathbf{v}\rangle := \sqrt{\mathbf{w}^* M\mathbf{v}}$. This implies that the projected Hessenberg

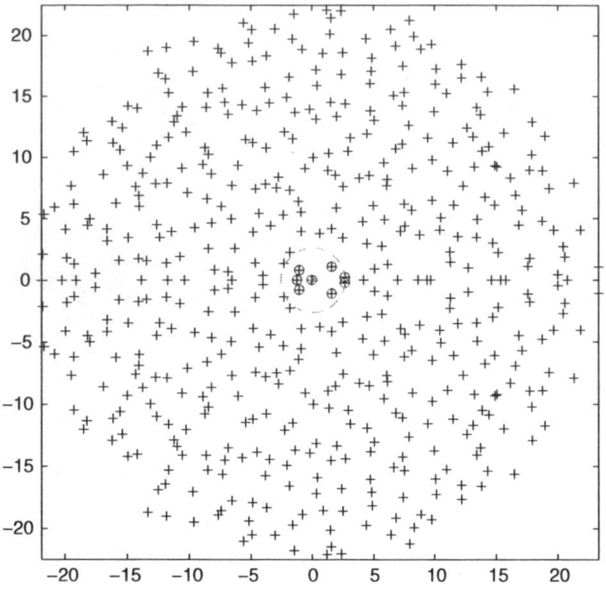

FIGURE 44.4 Eigenvalues from shift-invert.

matrix H is actually symmetric and tridiagonal and the standard three-term Lanczos recurrence is recovered with this inner product.

8. There is a subtle aspect to this approach when M is singular. The most pathological case, when $\text{null}(A) \cap \text{null}(M) \neq \{0\}$, is not treated here. However, when M is singular there may be infinite eigenvalues of the pair (A, M) and the presence of these can introduce large perturbations to the computed Ritz values and vectors. To avoid these difficulties, a purging operation has been suggested by Ericsson and Ruhe [ER80]. If $\mathbf{x} = V\mathbf{y}$ with $H\mathbf{y} = \mathbf{y}\theta$, then

$$A_\sigma \mathbf{x} = VH\mathbf{y} + \mathbf{f}\mathbf{e}_k^T\mathbf{y} = \mathbf{x}\theta + \mathbf{f}\mathbf{e}_k^T\mathbf{y}.$$

Replacing the $\mathbf{x}$ with the improved eigenvector approximation $\mathbf{x} \leftarrow (\mathbf{x} + \frac{1}{\theta}\mathbf{f}\mathbf{e}_k^T\mathbf{y})$ and renormalizing has the effect of purging undesirable components without requiring any additional matrix vector products with A_σ.

9. The residual error of the purged vector $\mathbf{x}$ with respect to the original problem is

$$\|A\mathbf{x} - M\mathbf{x}\lambda\|_2 = \|M\mathbf{f}\|_2 \frac{|\mathbf{e}_k^T\mathbf{y}|}{|\theta|^2},$$

where $\lambda = \sigma + 1/\theta$. Since $|\theta|$ is usually quite large under the spectral transformation, this new residual is generally considerably smaller than the original.

References

[BCR96] J. Baglama, D. Calvetti, and L. Reichel, Iterative methods for the computation of a few eigenvalues of a large symmetric matrix, *BIT*, 36 (3), 400–440 (1996).

[BER04] C.A. Beattie, M. Embree, and J. Rossi, Convergence of restarted Krylov subspaces to invariant subspaces, *SIAM J. Matrix Anal. Appl.*, 25, 1074–1109 (2004).

[BES05] C.A. Beattie, M. Embree, and D.C. Sorensen, Convergence of polynomial restart Krylov methods for eigenvalue computation, *SIAM Review*, 47 (3), 492–515 (2005).

[DGK76] J. Daniel, W.B. Gragg, L. Kaufman, and G.W. Stewart, Reorthogonalization and stable algorithms for updating the Gram–Schmidt QR factorization, *Math. Comp.*, 30, 772–795 (1976).

[DDH88] J.J. Dongarra, J. DuCroz, S. Hammarling, and R. Hanson, An extended set of Fortran basic linear algebra subprograms, *ACM Trans. Math. Softw.*, **14**, 1–17 (1988).

[ER80] T. Ericsson and A. Ruhe, The spectral transformation Lanczos method for the numerical solution of large sparse generalized symmetric eigenvalue problems, *Math. Comp.*, 35, (152), 1251–1268 (1980).

[FR89] B. Fischer and L. Reichel, Newton interpolation in Fejér and Chebyshev points, *Math. Comp.*, 53, 265–278 (1989).

[Gai87] D. Gaier, *Lectures on Complex Approximation*, Birkhäuser, Boston (1987).

[GLS94] R.G. Grimes, J.G. Lewis, and H.D. Simon, A shifted block Lanczos algorithm for solving sparse symmetric generalized eigenproblems, *SIAM J. Matrix Anal. Appl.*, 15 (1), 228–272 (1994).

[LSY98] R. Lehoucq, D.C. Sorensen, and C. Yang, *ARPACK Users Guide: Solution of Large Scale Eigenvalue Problems with Implicitly Restarted Arnoldi Methods*, SIAM Publications, Philadelphia (1998). (Software available at: http://www.caam.rice.edu/software/ARPACK.)

[MS97] K. Meerbergen and A. Spence, Implicitly restarted Arnoldi with purification for the shift–invert transformation, *Math. Comp.*, 218, 667–689 (1997).

[Mor96] R.B. Morgan, On restarting the Arnoldi method for large nonsymmetric eigenvalue problems, *Math. Comp.*, 65, 1213–1230 (1996).

[Par80] B.N. Parlett, *The Symmetric Eigenvalue Problem*, Prentice-Hall, Upper Saddle River, NJ (1980).

[Saa92] Y. Saad, *Numerical Methods for Large Eigenvalue Problems*, Manchester University Press, Manchester, U.K. (1992).

[Sor92] D.C. Sorensen, Implicit application of polynomial filters in a k-step Arnoldi method, *SIAM J. Matrix Anal. Applic.*, 13: 357–385 (1992).

[Sor02] D.C. Sorensen, Numerical methods for large eigenvalue problems, *Acta Numerica*, **11**, 519–584 (2002)

45

Computation of the Singular Value Decomposition

Alan Kaylor Cline
The University of Texas at Austin

Inderjit S. Dhillon
The University of Texas at Austin

45.1 Singular Value Decomposition

Definitions:

Given a complex matrix A having m rows and n columns, if σ is a nonnegative scalar and $\mathbf{u}$ and $\mathbf{v}$ are nonzero m- and n-vectors, respectively, such that

$$A\mathbf{v} = \sigma\mathbf{u} \quad \text{and} \quad A^*\mathbf{u} = \sigma\mathbf{v},$$

then σ is a **singular value** of A and $\mathbf{u}$ and $\mathbf{v}$ are corresponding **left** and **right singular vectors,** respectively. (For generality it is assumed that the matrices here are complex, although given these results, the analogs for real matrices are obvious.)

If, for a given positive singular value, there are exactly t linearly independent corresponding right singular vectors and t linearly independent corresponding left singular vectors, the singular value has **multiplicity** t and the space spanned by the right (left) singular vectors is the corresponding **right (left) singular space.**

Given a complex matrix A having m rows and n columns, the matrix product $U\Sigma V^*$ is a **singular value decomposition** for a given matrix A if

- $U$and V, respectively, have orthonormal columns.
- Σ has nonnegative elements on its principal diagonal and zeros elsewhere.
- $A = U\Sigma V^*$.

Let p and q be the number of rows and columns of Σ. U is $m \times p$, $p \leq m$, and V is $n \times q$ with $q \leq n$.

There are three **standard forms** of the SVD. All have the ith diagonal value of Σ denoted σ_i and ordered as follows: $\sigma_1 \geq \sigma_2 \geq \cdots \geq \sigma_k$, and r is the index such that $\sigma_r > 0$ and either $k = r$ or $\sigma_{r+1} = 0$.

1. $p = m$ and $q = n$. The matrix Σ is $m \times n$ and has the same dimensions as A (see Figures 45.1 and 45.2).
2. $p = q = min\{m, n\}$.The matrix Σ is square (see Figures 45.3 and 45.4).
3. If $p = q = r$, the matrix Σ is square. This form is called a **reduced SVD** and denoted is by $\hat{U}\hat{\Sigma}\hat{V}^*$ (see Figures 45.5 and 45.6).

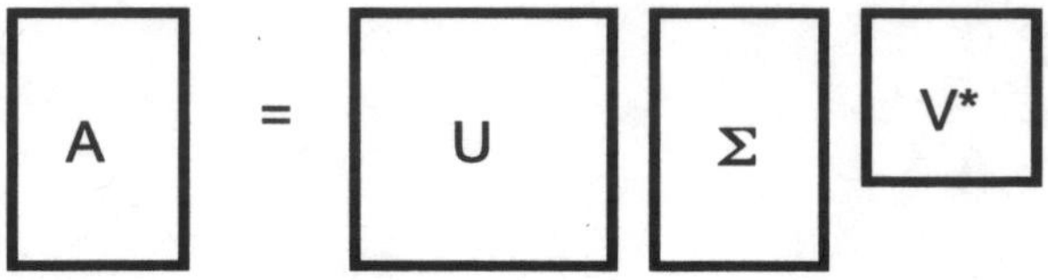

FIGURE 45.1 The first form of the singular value decomposition where $m \geq n$.

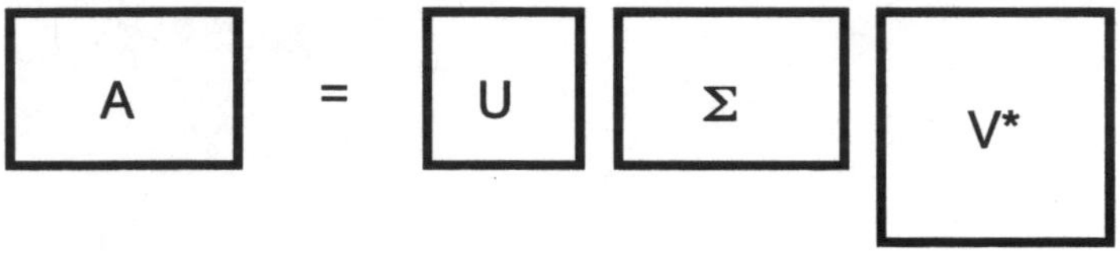

FIGURE 45.2 The first form of the singular value decomposition where $m < n$.

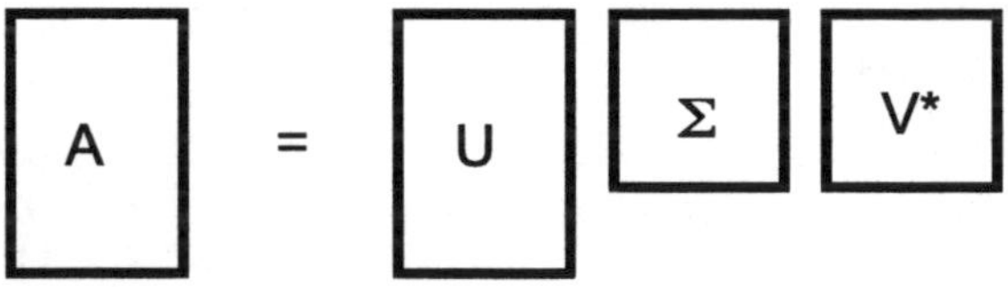

FIGURE 45.3 The second form of the singular value decomposition where $m \geq n$.

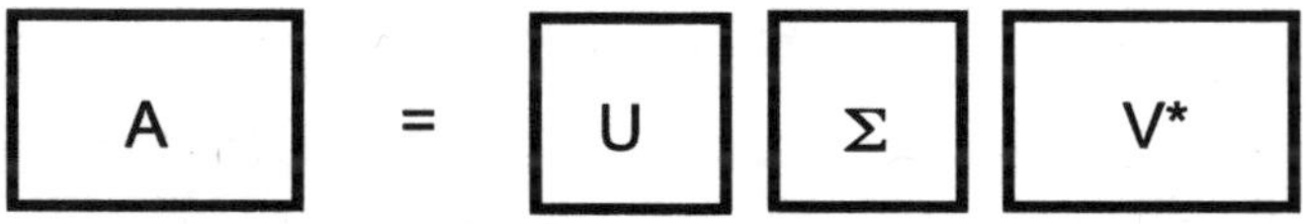

FIGURE 45.4 The second form of the singular value decomposition where $m < n$.

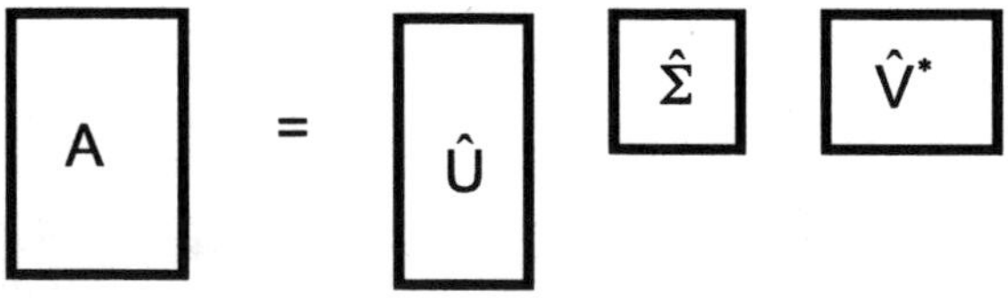

FIGURE 45.5 The third form of the singular value decomposition where $r \leq n \leq m$.

FIGURE 45.6 The third form of the singular value decomposition where $r \leq m < n$.

Facts:

The results can be found in [GV96, pp. 70–79]. Additionally, see Chapter 5.6 for introductory material and examples of SVDs, Chapter 17 for additional information on singular value decomposition, Chapter 15 for information on perturbations of singular values and vectors, and Section 39.9 for information about numerical rank.

1. If $U\Sigma V^*$ is a singular value decomposition for a given matrix A, then the diagonal elements $\{\sigma_i\}$ of Σ are singular values of A. The columns $\{\mathbf{u}_i\}_{i=1}^{p}$ of U and $\{\mathbf{v}_i\}_{i=1}^{q}$ of V are left and right singular vectors of A, respectively.
2. If $m \geq n$, the first standard form of the SVD can be found as follows:

   (a) Let $A^*A = V\Lambda V^*$ be an eigenvalue decomposition for the Hermitian, positive semidefinite $n \times n$ matrix A^*A such that Λ is diagonal (with the diagonal entries in nonincreasing order) and V is unitary.

   (b) Let the $m \times n$ matrix Σ have zero off-diagonal elements and for $i = 1, \ldots, n$ let σ_i, the ith diagonal element of Σ, equal $+\sqrt{\lambda_i}$, the positive square root of the ith diagonal element of Λ.

   (c) For $i = 1, \ldots, n$, let the $m \times m$ matrix U have ith column, $\mathbf{u}_i$, equal to $1/\sigma_i A\mathbf{v}_i$ if $\sigma_i \neq 0$. If $\sigma_i = 0$, let $\mathbf{u}_i$ be of unit length and orthogonal to all $\mathbf{u}_j$ for $j \neq i$, then $U\Sigma V^*$ is a singular decomposition of A.

3. If $m < n$ the matrix A^* has a singular value decomposition $U\Sigma V^*$ and $V\Sigma^T U^*$ is a singular value decomposition for A. The diagonal elements of Σ are the square roots of the eigenvalues of AA^*. The eigenvalues of A^*A are those of AA^* plus $n - m$ zeros. The notation Σ^T rather than Σ^* is used because in this case the two are identical and the transpose is more suggestive. All elements of Σ are real so that taking complex conjugates has no effect.
4. The value of r, the number of nonzero singular values, is the rank of A .
5. If A is real, then U and V (in addition to Σ) can be chosen real in any of the forms of the SVD.
6. The range of A is exactly the subspace of $\mathbb{C}^m$ spanned by the r columns of U that correspond to the positive singular values.
7. In the first form, the null space of A is that subspace of $\mathbb{C}^n$ spanned by the $n - r$ columns of V that correspond to zero singular values.
8. In reducing from the first form to the third (reduced) form, a basis for the null space of A has been discarded if columns of Vhave been deleted. A basis for the space orthogonal to the range of A (i.e., the null space of A^*) has been discarded if columns of U have been deleted.
9. In the first standard form of the SVD, U and V are unitary.
10. The second form can be obtained from the first form simply by deleting columns $n+1, \ldots, m$ of U and the corresponding rows of S, if $m > n$, or by deleting columns $m+1, \ldots, n$ of V and the corresponding columns of S, if $m < n$. If $m \neq n$, then only one of U and V is square and either $UU^* = I_m$ or $VV^* = I_n$ fails to hold. Both $U^*U = I_p$ and $V^*V = I_p$.
11. The reduced (third) form can be obtained from the second form by taking only the $r \times r$ principle submatrix of Σ, and only the first r columns of U and V . If A is rank deficient (i.e., $r < \min\{m, n\}$), then neither U nor V is square and neither U^*U nor V^*V is an identity matrix.
12. If $p < m$, let $\tilde{U}$ be an $m \times (m - p)$matrix of columns that are mutually orthonormal to one another as well as to the columns of $U$and define the $m \times m$ unitary matrix

$$\widehat{U} = \begin{bmatrix} U & \tilde{U} \end{bmatrix}.$$

If $q < n$, let $\tilde{V}$ be an $n \times (n - q)$matrix of columns that are mutually orthonormal to one another as well as to the columns of $V$and define the $n \times n$ unitary matrix

$$\widehat{V} = \begin{bmatrix} V & \tilde{V} \end{bmatrix}.$$

Let $\widehat{\Sigma}$ be the $m \times n$ matrix

$$\widehat{\Sigma} = \begin{bmatrix} \Sigma & 0 \\ 0 & 0 \end{bmatrix}.$$

Then

$$A = \widehat{U}\widehat{\Sigma}\widehat{V}^*, \; A\widehat{V} = \widehat{U}\widehat{\Sigma}^*, \; A^* = \widehat{V}\widehat{\Sigma}^T\widehat{U}^*, \text{ and } A^*\widehat{U} = \widehat{V}\widehat{\Sigma}^T.$$

13. Let $U\Sigma V^*$be a singular value decomposition for A, an $m \times n$ matrix of rank r. Then:
    (a) There are exactly r positive elements of Σ and they are the square roots of the r positive eigenvalues of A^*A (and also AA^*) with the corresponding multiplicities.
    (b) The columns of V are eigenvectors of A^*A; more precisely, $\mathbf{v}_j$ is a normalized eigenvector of A^*A corresponding to the eigenvalue σ_j^2, and $\mathbf{u}_j$ satisfies $\sigma_j \mathbf{u}_j = A\mathbf{v}_j$.
    (c) Alternatively, the columns of U are eigenvectors of AA^*; more precisely, $\mathbf{u}_j$ is a normalized eigenvector of AA^* corresponding to the eigenvalue σ_j^2, and $\mathbf{v}_j$ satisfies $\sigma_j v_j = A^*\mathbf{u}_j$.
14. The singular value decomposition $U\Sigma V^*$ is not unique. If $U\Sigma V^*$ is a singular value decomposition, so is $(-U)\Sigma(-V^*)$. The singular values may be arranged in any order if the columns of singular vectors in U and V are reordered correspondingly.
15. If the singular values are in nonincreasing order then the only option for the construction of Σ is the choice for its dimensions p and q and these must satisfy $r \leq p \leq m$ and $r \leq q \leq n$.
16. If A is square and if the singular values are ordered in a nonincreasing fashion, the matrix Σ is unique.
17. Corresponding to a simple (i.e., nonrepeated) singular value σ_j, the left and right singular vectors, $\mathbf{u}_j$ and $\mathbf{v}_j$, are unique up to scalar multiples of modulus one. That is, if $\mathbf{u}_j$ and $\mathbf{v}_j$ are singular vectors, then for any real value of θ so are $e^{i\theta}\mathbf{u}_j$ and $e^{i\theta}\mathbf{v}_j$, but no other vectors are singular vectors corresponding to σ_j.
18. Corresponding to a repeated singular value, the associated left singular vectors $\mathbf{u}_j$ and right singular vectors $\mathbf{v}_j$ may be selected in any fashion such that they span the proper subspace. Thus, if $\mathbf{u}_{j_1}, \ldots, \mathbf{u}_{j_r}$ and $\mathbf{v}_{j_1}, \ldots, \mathbf{v}_{j_r}$ are the left and right singular vectors corresponding to a singular value σ_j of multiplicity s, then so are $\mathbf{u}'_{j_1}, \ldots, \mathbf{u}'_{j_r}$ and $\mathbf{v}'_{j_1}, \ldots, \mathbf{v}'_{j_r}$ if and only if there exists an $s \times s$ unitary matrix Q such that $[\mathbf{u}'_{j_1}, \ldots, \mathbf{u}'_{j_r}] = [\mathbf{u}_{j_1}, \ldots, \mathbf{u}_{j_r}]\, Q$ and $[\mathbf{v}'_{j_1}, \ldots, \mathbf{v}'_{j_r}] = [\mathbf{v}_{j_1}, \ldots, \mathbf{v}_{j_r}]\, Q$.

Examples:

For examples illustrating SVD see Section 5.6.

45.2 Algorithms for the Singular Value Decomposition

Generally, algorithms for computing singular values are analogs of algorithms for computing eigenvalues of symmetric matrices. See Chapter 42 and Chapter 46 for additional information. The idea is always to find square roots of eigenvalues of $A^T A$ without actually computing $A^T A$. As before, we assume the matrix A whose singular values or singular vectors we seek is $m \times n$. All algorithms assume $m \geq n$; if $m < n$, the algorithms may be applied to A^T. To avoid undue complication, all algorithms will be presented as if the matrix is real. Nevertheless, each algorithm has an extension for complex matrices. Algorithm 1 is presented in three parts. It is analogous to the QR algorithm for symmetric matrices. The developments for it can be found in [GK65], [GK68], [BG69], and [GR70]. Algorithm 1a is a Householder reduction of a matrix to bidiagonal form. Algorithm 1c is a step to be used iteratively in Algorithm 1b. Algorithm 2 computes the singular values and singular vectors of a bidiagonal matrix to high relative accuracy [DK90], [Dem97].

Algorithm 3 gives a "Squareroot-free" method to compute the singular values of a bidiagonal matrix to high relative accuracy — it is the method of choice when only singular values are desired [Rut54], [Rut90], [FP94], [PM00]. Algorithm 4 computes the singular values of an $n \times n$ bidiagonal matrix by the bisection method, which allows k singular values to be computed in $O(kn)$ time. By specifying the input tolerance *tol* appropriately, Algorithm 4 can also compute the singular values to high relative accuracy. Algorithm 5 computes the SVD of a bidiagonal by the divide and conquer method [GE95]. The most recent method, based on the method of multiple relatively robust representations (not presented here), is the fastest and allows computation of k singular values as well as the corresponding singular vectors of a bidiagonal matrix in $O(kn)$ time [DP04a], [DP04b], [GL03], [WLV05]. All of the above mentioned methods first reduce the matrix to bidiagonal form. The following algorithms iterate directly on the input matrix. Algorithms 6 and 7 are analogous to the Jacobi method for symmetric matrices. Algorithm 6 — also known as the "one-sided Jacobi method for SVD" — can be found in [Hes58] and Algorithm 7 can be found in [Kog55] and [FH60]. Algorithm 7 begins with an orthogonal reduction of the $m \times n$ input matrix so that all the nonzeros lie in the upper $n \times n$ portion. (Although this algorithm was named biorthogonalization in [FH60], it is not the biorthogonalization found in certain iterative methods for solving linear equations.) Many of the algorithms require a tolerance ε to control termination. It is suggested that ε be set to a small multiple of the unit round off precision ε_o.

Algorithm 1a: Householder reduction to bidiagonal form:
Input: m, n, A where A is $m \times n$.
Output: B, U, V so that B is upper bidiagonal, U and V are products of Householder matrices, and $A = UBV^T$.

1. $B \leftarrow A$. (This step can be omitted if A is to be overwritten with B.)
2. $U = I_{m\times n}$.
3. $V = I_{n\times n}$.
4. For $k = 1, \ldots, n$
   a. Determine Householder matrix Q_k with the property that:
      - Left multiplication by Q_k leaves components $1, \ldots, k-1$ unaltered, and
      - $$Q_k \begin{bmatrix} 0 \\ \vdots \\ 0 \\ b_{k-1,k} \\ b_{k,k} \\ b_{k+1,k} \\ \vdots \\ b_{m,k} \end{bmatrix} = \begin{bmatrix} 0 \\ \vdots \\ 0 \\ b_{k-1,k} \\ s \\ 0 \\ \vdots \\ 0 \end{bmatrix}, \text{ where } s = \pm\sqrt{\sum_{i=k}^{m} b_{i,k}^2}.$$
   b. $B \leftarrow Q_k B$.
   c. $U \leftarrow U Q_k$.
   d. If $k \leq n-2$, determine Householder matrix P_{k+1} with the property that:
      - Right multiplication by P_{k+1} leaves components $1, \ldots, k$ unaltered, and
      - $[0 \cdots 0 \;\; b_{k,k} \;\; b_{k,k+1} \;\; b_{k,k+2} \cdots b_{k,n}]\, P_{k+1} = [0 \cdots 0 \;\; b_{k,k} \;\; s \;\; 0 \cdots 0]$, where $s = \pm\sqrt{\sum_{j=k+1}^{n} b_{k,j}^2}$.
   e. $B \leftarrow B P_{k+1}$.
   f. $V \leftarrow P_{k+1} V$.

Algorithm 1b: Golub–Reinsch SVD:
Input: m, n, A where A is $m \times n$.
Output: Σ, U, V so that Σ is diagonal, U and V have orthonormal columns, U is $m \times n$, V is $n \times n$, and $A = U \Sigma V^T$.

1. Apply Algorithm 1a to obtain B, U, V so that B is upper bidiagonal, U and V are products of Householder matrices, and $A = U B V^T$.
2. Repeat:
   a. If for any $i = 1, \dots, n-1$, $|b_{i,i+1}| \leq \varepsilon \left(|b_{i,i}| + |b_{i+1,i+1}|\right)$, set $b_{i,i+1} = 0$.
   b. Determine the smallest p and the largest q so that B can be blocked as

$$B = \begin{bmatrix} B_{1,1} & 0 & 0 \\ 0 & B_{2,2} & 0 \\ 0 & 0 & B_{3,3} \end{bmatrix} \quad \begin{matrix} p \\ n-p-q \\ q \end{matrix}$$

   where $B_{3,3}$ is diagonal and $B_{2,2}$ has no zero superdiagonal entry.
   c. If $q = n$, set $\Sigma =$ the diagonal portion of B **STOP**.
   d. If for $i = p+1, \dots, n-q-1$, $b_{i,i} = 0$, then
      Apply Givens rotations so that $b_{i,i+1} = 0$ and $B_{2,2}$ is still upper bidiagonal. (For details, see [GL96, p. 454].)
   else
      Apply Algorithm 1c to n, B, U, V, p, q.

Algorithm 1c: Golub–Kahan SVD step:
Input: n, B, Q, P, p, q where B is $n \times n$ and upper bidiagonal, Q and P have orthogonal columns, and $A = Q B P^T$.
Output: B, Q, P so that B is upper bidiagonal, $A = Q B P^T$, Q and P have orthogonal columns, and the output B has smaller off-diagonal elements than the input B. In storage, B, Q, and P are overwritten.

1. Let $B_{2,2}$ be the diagonal block of B with row and column indices $p+1, \dots, n-q$.
2. Set $C =$ lower, right 2×2 submatrix of $B_{2,2}^T B_{2,2}$.
3. Obtain eigenvalues λ_1, λ_2 of C. Set $\mu =$ whichever of λ_1, λ_2 that is closer to $c_{2,2}$.
4. $k = p+1, \alpha = b_{k,k}^2 - \mu, \beta = b_{k,k} b_{k,k+1}$.
5. For $k = p+1, \dots, n-q-1$
   a. Determine $c = \cos(\theta)$ and $s = \sin(\theta)$ with the property that:

$$[\alpha \quad \beta] \begin{bmatrix} c & s \\ -s & c \end{bmatrix} = \begin{bmatrix} \sqrt{\alpha^2 + \beta^2} & 0 \end{bmatrix}.$$

   b. $B \leftarrow B R_{k,k+1}(c, s)$ where $R_{k,k+1}(c, s)$ is the Givens rotation matrix that acts on columns k and $k+1$ during right multiplication.
   c. $P \leftarrow P R_{k,k+1}(c, s)$.
   d. $\alpha = b_{k,k}, \beta = b_{k+1,k}$.
   e. Determine $c = \cos(\theta)$ and $s = \sin(\theta)$ with the property that:

$$\begin{bmatrix} c & -s \\ s & c \end{bmatrix} \begin{bmatrix} \alpha \\ \beta \end{bmatrix} = \begin{bmatrix} \sqrt{\alpha^2 + \beta^2} \\ 0 \end{bmatrix}.$$

   f. $B \leftarrow R_{k,k+1}(c, -s) B$, where $R_{k,k+1}(c, -s)$ is the Givens rotation matrix that acts on rows k and $k+1$ during left multiplication.
   g. $Q \leftarrow Q R_{k,k+1}(c, s)$.
   h. if $k \leq n - q - 1 \alpha = b_{k,k+1}, \beta = b_{k,k+2}$.

Algorithm 2a: High Relative Accuracy Bidiagonal SVD:
Input: n, B where B is an $n \times n$ upper bidiagonal matrix.
Output: Σ is an $n \times n$ diagonal matrix, U and V are orthogonal $n \times n$ matrices, and $B = U\Sigma V^T$.

1. Compute $\underline{\sigma}$ to be a reliable underestimate of $\sigma_{\min}(B)$ (for details, see [DK90]).
2. Compute $\overline{\sigma} = \max_i(b_{i,i}, b_{i,i+1})$.
3. Repeat:

   a. For all $i = 1, \ldots, n-1$, set $b_{i,i+1} = 0$ if a relative convergence criterion is met (see [DK90] for details).

   b. Determine the smallest p and largest q so that B can be blocked as

   $$B = \begin{bmatrix} B_{1,1} & 0 & 0 \\ 0 & B_{2,2} & 0 \\ 0 & 0 & B_{3,3} \end{bmatrix} \begin{matrix} p \\ n-p-q \\ q \end{matrix}$$

   where $B_{3,3}$ is diagonal and $B_{2,2}$ has no zero superdiagonal entry.

   c. If $q = n$, set $\Sigma =$ the diagonal portion of B. **STOP**.

   d. If for $i = p+1, \ldots, n-q-1$, $b_{i,i} = 0$, then
      Apply Givens rotations so that $b_{i,i+1} = 0$ and $B_{2,2}$ is still upper bidiagonal. (For details, see [GV96, p. 454].)
   else
      Apply Algorithm 2b with $n, B, U, V, p, q, \overline{\sigma}, \underline{\sigma}$ as inputs.

Algorithm 2b: Demmel–Kahan SVD step:
Input: $n, B, Q, P, p, q, \overline{\sigma}, \underline{\sigma}$ where B is $n \times n$ and upper bidiagonal, Q and P have orthogonal columns such that $A = QBP^T$, $\overline{\sigma} \approx ||B||$ and $\underline{\sigma}$ is an underestimate of $\sigma_{\min}(B)$.
Output: B, Q, P so that B is upper bidiagonal, $A = QBP^T$, Q and P have orthogonal columns, and the output B has smaller off-diagonal elements than the input B. In storage, B, Q, and P are overwritten.

1. Let $B_{2,2}$ be the diagonal block of B with row and column indices $p+1, \ldots, n-q$.
2. If $tol^*\underline{\sigma} \le \varepsilon_0\overline{\sigma}$, then

   a. $c' = c = 1$.

   b. For $k = p+1, n-q-1$

   - $\alpha = c b_{k,k}$; $\beta = b_{k,k+1}$.
   - Determine c and s with the property that:

   $$[\alpha \quad \beta]\begin{bmatrix} c & s \\ -s & c \end{bmatrix} = [r \quad 0], \text{ where } r = \sqrt{\alpha^2 + \beta^2}.$$

   - If $k \ne p+1$, $b_{k-1,k} = s'r$.
   - $P \leftarrow P R_{k,k+1}(c, s)$, where $R_{k,k+1}(c, s)$ is the Givens rotation matrix that acts on columns k and $k+1$ during right multiplication.
   - $\alpha = c'r$, $\beta = s b_{k+1,k+1}$.

(continued)

Algorithm 2b: Demmel–Kahan SVD step: (Continued)

- Determine c' and s' with the property that:

$$\begin{bmatrix} c' & -s' \\ s' & c' \end{bmatrix} \begin{bmatrix} \alpha \\ \beta \end{bmatrix} = \begin{bmatrix} \sqrt{\alpha^2+\beta^2} \\ 0 \end{bmatrix}.$$

- $Q \leftarrow QR_{k,k+1}(c,-s)$, where $R_{k,k+1}(c,-s)$ is the Givens rotation matrix that acts on rows k and $k+1$ during left multiplication.
- $b_{k,k} = \sqrt{\alpha^2+\beta^2}$.

c. $b_{n-q-1,n-q} = (b_{n-q,n-q}c)s'$; $b_{n-q,n-q} = (b_{n-q,n-q}c)c'$.

Else

d. Apply Algorithm 1c to n, B, Q, P, p, q.

Algorithm 3a: High Relative Accuracy Bidiagonal Singular Values:
Input: n, B where B is an $n \times n$ upper bidiagonal matrix.
Output: Σ is an $n \times n$ diagonal matrix containing the singular values of B.

1. Square the diagonal and off-diagonal elements of B to form the arrays **s** and **e**, respectively, i.e., for $i = 1, \ldots, n-1$, $s_i = b_{i,i}^2$, $e_i = b_{i,i+1}^2$, end for $s_n = b_{n,n}^2$.
2. Repeat:

a. For all $i = 1, \ldots, n-1$, set $e_i = 0$ if a relative convergence criterion is met (see [PM00] for details).

b. Determine the smallest p and largest q so that B can be blocked as

$$B = \begin{bmatrix} B_{1,1} & 0 & 0 \\ 0 & B_{2,2} & 0 \\ 0 & 0 & B_{3,3} \end{bmatrix} \begin{matrix} p \\ n-p-q \\ q \end{matrix}$$

where $B_{3,3}$ is diagonal and $B_{2,2}$ has no zero superdiagonal entry.

c. If $q = n$, set $\Sigma = \sqrt{\text{diag}(\mathbf{s})}$. **STOP.**

d. If for $i = p+1, \ldots, n-q-1$, $s_i = 0$ then
  Apply Givens rotations so that $e_i = 0$ and $B_{2,2}$ is still upper bidiagonal. (For details, see [GV96, p. 454].)
else
  Apply Algorithm 3b with inputs n, **s**, **e**.

Algorithm 3b: Differential quotient-difference (dqds) step:
Input: n, **s**, **e** where **s** and **e** are the squares of the diagonal and superdiagonal entries, respectively, of an $n \times n$ upper bidiagonal matrix.
Output: **s** and **e** are overwritten on output.

1. Choose μ by using a suitable shift strategy. The shift μ should be smaller than $\sigma_{\min}(B)^2$. See [FP94,PM00] for details.
2. $d = s_1 - \mu$.

Algorithm 3b: Differential quotient-difference (dqds) step: (Continued)

1. For $k = 1, \ldots, n-1$
   a. $s_k = d + e_k$.
   b. $t = s_{k+1}/s_k$.
   c. $e_k = e_k t$.
   d. $d = dt - \mu$.
   e. If $d < 0$, go to step 1.
2. $s_n = d$.

Algorithm 4a: Bidiagonal Singular Values by Bisection:
Input: n, B, α, β, tol where $n \times n$ is a bidiagonal matrix, $[\alpha, \beta)$ is the input interval, and *tol* is the tolerance for the desired accuracy of the singular values.
Output: **w** is the output array containing the singular values of B that lie in $[\alpha, \beta)$.

1. $n_\alpha = Negcount(n, B, \alpha)$.
2. $n_\beta = Negcount(n, B, \beta)$.
3. If $n_\alpha = n_\beta$, there are no singular values in $[\alpha, \beta)$. **STOP**.
4. Put $[\alpha, n_\alpha, \beta, n_\beta]$ onto *Worklist*.
5. While *Worklist* is not empty do
   a. Remove $[low, n_{low}, up, n_{up}]$ from *Worklist*.
   b. $mid = (low + up)/2$.
   c. If $(up - low < tol)$, then
      - For $i = n_{low} + 1, n_{up}, w(i - n_a) = mid$;

      Else
      - $n_{mid} = Negcount(n, B, mid)$.
      - If $n_{mid} > n_{low}$ then
        Put $[low, n_{low}, mid, n_{mid}]$ onto *Worklist*.
      - If $n_{up} > n_{mid}$ then
        Put $[mid, n_{mid}, up, n_{up}]$ onto *Worklist*.

Algorithm 4b: *Negcount* (n, B, μ):
Input: The $n \times n$ bidiagonal matrix B and a number μ.
Output: *Negcount*, i.e., the number of singular values smaller than μ is returned.

1. $t = -\mu$.
2. For $k = 1, \ldots, n-1$
   $d = b_{k,k}^2 + t$.
   If $(d < 0)$ then $Negcount = Negcount + 1$
   $t = t * (b_{k,k+1}^2/d) - \mu$.

   End for
3. $d = b_{n,n}^2 + t$.
4. If $(d < 0)$, then $Negcount = Negcount + 1$.

Algorithm 5: DC_SVD(n, B, Σ, U, V): Divide and Conquer Bidiagonal SVD:
Input: n, B where B is an $(n+1) \times n$ lower bidiagonal matrix.
Output: Σ is an $n \times n$ diagonal matrix, U is an $(n+1) \times (n+1)$ orthogonal matrix, V is an orthogonal $n \times n$ matrix, so that $B = U\Sigma V^T$.

1. If $n < n_0$, then apply Algorithm 1b with inputs $n+1, n, B$ to get outputs Σ, U, V.
Else

$$\text{Let } B = \begin{pmatrix} B_1 & \alpha_k \mathbf{e}_k & 0 \\ 0 & \beta_k \mathbf{e}_1 & B_2 \end{pmatrix}, \quad \text{where } k = n/2.$$

a. Call DC_SVD($k-1, B_1, \Sigma_1, U_1, W_1$).

b. Call DC_SVD($n-k, B_2, \Sigma_2, U_2, W_2$).

c. Partition $U_i = (Q_i \quad \mathbf{q}_i)$, for $i = 1, 2$, where $\mathbf{q}_i$ is a column vector.

d. Extract $l_1 = Q_1^T \mathbf{e}_k, \lambda_1 = \mathbf{q}_1^T \mathbf{e}_k, l_2 = Q_2^T \mathbf{e}_1, \lambda_2 = \mathbf{q}_2^T \mathbf{e}_1$.

e. Partition B as

$$B = \begin{pmatrix} c_0 \mathbf{q}_1 & Q_1 & 0 & -s_0 \mathbf{q}_1 \\ s_0 \mathbf{q}_2 & 0 & Q_2 & c_0 \mathbf{q}_2 \end{pmatrix} \begin{pmatrix} r_0 & 0 & 0 \\ \alpha_k l_1 & \Sigma_1 & 0 \\ \beta_k l_2 & 0 & \Sigma_2 \\ 0 & 0 & 0 \end{pmatrix} \begin{pmatrix} 0 & W_1 & 0 \\ 1 & 0 & 0 \\ 0 & 0 & W_2 \end{pmatrix}^T = (Q \quad q) \begin{pmatrix} M \\ 0 \end{pmatrix} W^T$$

where $r_0 = \sqrt{(\alpha_k \lambda_1)^2 + (\beta_k \lambda_2)^2}, c_0 = \alpha_k \lambda_1 / r_0, s_0 = \beta_k \lambda_2 / r_0$.

f. Compute the singular values of M by solving the secular equation

$$f(w) = 1 + \sum_{k=1}^{n} \frac{z_k^2}{d_k^2 - w^2} = 0,$$

and denote the computed singular values by $\hat{w}_1, \hat{w}_2, \ldots, \hat{w}_n$.

g. For $i = 1, \ldots, n$, compute

$$\hat{z}_i = \sqrt{(\hat{w}_n^2 - d_i^2) \prod_{k=1}^{i-1} \frac{(\hat{w}_k^2 - d_i^2)}{(d_k^2 - d_i^2)} \prod_{k=1}^{n-1} \frac{(\hat{w}_k^2 - d_i^2)}{(d_{k+1}^2 - d_i^2)}}.$$

h. For $i = 1, \ldots, n$, compute the singular vectors

$$\mathbf{u}_i = \left(\frac{\hat{z}_1}{d_1^2 - \hat{w}_i^2}, \cdots, \frac{\hat{z}_n}{d_n^2 - \hat{w}_i^2} \right) \Bigg/ \sqrt{\sum_{k=1}^{n} \frac{\hat{z}_k^2}{(d_k^2 - \hat{w}_i^2)^2}},$$

$$\mathbf{v}_i = \left(-1, \frac{d_2 \hat{z}_2}{d_2^2 - \hat{w}_i^2}, \cdots, \frac{d_n \hat{z}_n}{d_n^2 - \hat{w}_i^2} \right) \Bigg/ \sqrt{1 + \sum_{k=2}^{n} \frac{(d_k \hat{z}_k)^2}{(d_k^2 - \hat{w}_i^2)^2}}$$

and let $U = [\mathbf{u}_1, \ldots, \mathbf{u}_n], V = [\mathbf{v}_1, \ldots, \mathbf{v}_n]$.

i. Return $\Sigma = \begin{pmatrix} \text{diag}(\hat{w}_1, \hat{w}_2, \ldots, \hat{w}_n) \\ 0 \end{pmatrix}, U \leftarrow (QU \quad q), V \leftarrow WV.$

Algorithm 6: Biorthogonalization SVD:
Input: m, n, A where A is $m \times n$.
Output: Σ, U, V so that Σ is diagonal, U and V have orthonormal columns, U is $m \times n$, V is $n \times n$, and $A = U\Sigma V^T$.

1. $U \leftarrow A$. (This step can be omitted if A is to be overwritten with U.)
2. $V = I_{n\times n}$.
3. Set $N^2 = \left(\sum_{i=1}^{n}\sum_{j=1}^{n} u_{i,j}^2\right)$, $s = 0$, and *first* = **true**.
4. Repeat until $s^{1/2} \leq \varepsilon^2 N^2$ and *first* = **false**.
   a. Set $s = 0$ and *first* = **false**.
   b. For $i = 1, \dots, n-1$.
      i. For $j = i+1, \dots, n$
         - $s \leftarrow s + \left(\sum_{k=1}^{m} u_{k,i}u_{k,j}\right)^2$.
         - Determine d_1, d_2, $c = \cos(\theta)$, and $s = \sin(\varphi)$ such that:

$$\begin{bmatrix} c & -s \\ s & c \end{bmatrix} \begin{bmatrix} \sum_{k=1}^{m} u_{k,i}^2 & \sum_{k=1}^{m} u_{k,i}u_{k,i} \\ \sum_{k=1}^{m} u_{k,i}u_{k,i} & \sum_{k=1}^{m} u_{k,j}^2 \end{bmatrix} \begin{bmatrix} c & s \\ -s & c \end{bmatrix} = \begin{bmatrix} d_1 & 0 \\ 0 & d_2 \end{bmatrix}.$$

         - $U \leftarrow UR_{i,j}(c,s)$ where $R_{i,j}(c,s)$ is the Givens rotation matrix that acts on columns i and j during right multiplication.
         - $V \leftarrow VR_{i,j}(c,s)$.
5. For $i = 1, \dots, n$:
   a. $\sigma_i = \sqrt{\sum_{k=1}^{m} u_{k,i}^2}$.
   b. $U \leftarrow U\Sigma^{-1}$.

Algorithm 7: Jacobi Rotation SVD:
Input: m, n, A where A is $m \times n$.
Output: Σ, U, V so that Σ is diagonal, U and V have orthonormal columns, U is $m \times n$, V is $n \times n$, and $A = U\Sigma V^T$.

1. $B \leftarrow A$. (This step can be omitted if A is to be overwritten with B.)
2. $U = I_{m\times n}$.
3. $V = I_{n\times n}$.
4. If $m > n$, compute the QR factorization of B using Householder matrices so that $B \leftarrow QA$, where B is upper triangular, and let $U \leftarrow UQ$. (See A6 for details.)
5. Set $N^2 = \sum_{i=1}^{n}\sum_{j=1}^{n} b_{i,j}^2$, $s = 0$, and *first* = **true**.
6. Repeat until $s \leq \varepsilon^2 N^2$ and *first* = **false**.
   a. Set $s = 0$ and *first* = **false**.
   b. For $i = 1, \dots, n-1$

(continued)

Algorithm 7: Jacobi Rotation SVD: (Continued)

i. For $j = i + 1, \ldots, n$:
   - $s = s + b_{i,j}^2 + b_{j,i}^2$.
   - Determine $d_1, d_2, c = \cos(\theta)$ and $s = \sin(\varphi)$ with the property that d_1 and d_2 are positive and

$$\begin{bmatrix} c & -s \\ s & c \end{bmatrix} \begin{bmatrix} b_{i,i} & b_{i,j} \\ b_{j,i} & b_{j,j} \end{bmatrix} \begin{bmatrix} \hat{c} & \hat{s} \\ -\hat{s} & \hat{c} \end{bmatrix} = \begin{bmatrix} d_1 & 0 \\ 0 & d_2 \end{bmatrix}.$$

   - $B \leftarrow R_{i,j}(c, s) B R_{i,j}(\hat{c}, -\hat{s})$ where $R_{i,j}(c, s)$ is the Givens rotation matrix that acts on rows i and j during left multiplication and $R_{i,j}(\hat{c}, -\hat{s})$ is the Givens rotation matrix that acts on columns i and j during right multiplication.
   - $U \leftarrow U R_{i,j}(c, s)$.
   - $V \leftarrow V R_{i,j}(\hat{c}, \hat{s})$.

7. Set Σ to the diagonal portion of B.

References

[BG69] P.A. Businger and G.H. Golub. Algorithm 358: singular value decomposition of a complex matrix, *Comm. Assoc. Comp. Mach.*, 12:564–565, 1969.

[Dem97] J.W. Demmel. *Applied Numerical Linear Algebra*, SIAM, Philadephia, 1997.

[DK90] J.W. Demmel and W. Kahan. Accurate singular values of bidiagonal matrices, *SIAM J. Stat. Comp.*, 11:873–912, 1990.

[DP04a] I.S. Dhillon and B.N. Parlett., Orthogonal eigenvectors and relative gaps, *SIAM J. Mat. Anal. Appl.*, 25:858–899, 2004.

[DP04b] I.S. Dhillon and B.N. Parlett. Multiple representations to compute orthogonal eigenvectors of symmetric tridiagonal matrices, *Lin. Alg. Appl.*, 387:1–28, 2004.

[FH60] G.E. Forsythe and P. Henrici. The cyclic Jacobi method for computing the principle values of a complex matrix, *Proc. Amer. Math. Soc.*, 94:1–23, 1960.

[FP94] V. Fernando and B.N. Parlett. Accurate singular values and differential qd algorithms, *Numer. Math.*, 67:191–229, 1994.

[GE95] M. Gu and S.C. Eisenstat. A divide-and-conquer algorithm for the bidiagonal SVD, *SIAM J. Mat. Anal. Appl.*, 16:79–92, 1995.

[GK65] G.H. Golub and W. Kahan. Calculating the Singular Values and Pseudoinverse of a Matrix, *SIAM J. Numer. Anal., Ser. B*, 2:205–224, 1965.

[GK68] G.H. Golub and W. Kahan. Least squares, singular values, and matrix approximations, *Aplikace Matematiky*, 13:44–51 1968.

[GL03] B. Grosser and B. Lang. An $O(n^2)$ algorithm for the bidiagonal SVD, *Lin. Alg. Appl.*, 358:45–70, 2003.

[GR70] G.H. Golub and C. Reinsch. Singular value decomposition and least squares solutions, *Numer. Math.*, 14:403–420, 1970.

[GV96] G.H. Golub and C.F. Van Loan. *Matrix Computations*, 3rd ed., The Johns Hopkins University Press, Baltimore, MD, 1996.

[Hes58] M.R. Hestenes. Inversion of matrices by biorthogonalization and related results, *J. SIAM*, 6:51–90, 1958.

[Kog55] E.G. Kogbetliantz. Solutions of linear equations by diagonalization of coefficient matrix, *Quart. Appl. Math.*, 13:123–132, 1955.

[PM00] B.N. Parlett and O.A. Marques. An implementation of the dqds algorithm (positive case), *Lin. Alg. Appl.*, 309:217–259, 2000.

[Rut54] H. Rutishauser. Der Quotienten-Differenzen-Algorithmus, *Z. Angnew Math. Phys.*, 5:233–251, 1954.

[Rut90] H. Rutishauser. *Lectures on Numerical Mathematics*, Birkhauser, Boston, 1990, (English translation of *Vorlesungen uber numerische mathematic*, Birkhauser, Basel, 1996).

[WLV05] P.R. Willems, B. Lang, and C. Voemel. Computing the bidiagonal SVD using multiple relatively robust representations, LAPACK Working Note 166, TR# UCB/CSD-05-1376, University of California, Berkeley, 2005.

[Wil65] J.H. Wilkinson. *The Algebraic Eigenvalue Problem*, Clarendon Press, Oxford, U.K., 1965.

46

Computing Eigenvalues and Singular Values to High Relative Accuracy

Zlatko Drmač
University of Zagreb

To compute the eigenvalues and singular values to high relative accuracy means to have a guaranteed number of accurate digits in all computed approximate values. If $\tilde{\lambda}_i$ is the computed approximation of λ_i, then the desirable high relative accuracy means that $|\lambda_i - \tilde{\lambda}_i| \leq \varepsilon\, |\lambda_i|$, where $0 \leq \varepsilon \ll 1$ independent of the ratio $|\lambda_i|/\max_j |\lambda_j|$. This is not always possible. The proper course of action is to first determine classes of matrices and classes of perturbations under which the eigenvalues (singular values) undergo only small relative changes. This means that the development of highly accurate algorithms is determined by the framework established by the perturbation theory.

Input to the perturbation theory is a perturbation matrix whose size is usually measured in a matrix norm. This may not be always adequate in numerical solutions of real world problems. If numerical information stored in the matrix represents a physical system, then choosing different physical units can give differently scaled rows and columns of the matrix, but representing the same physical system. Different scalings may be present because matrix entries represent quantities of different physical nature. It is possible that the smallest matrix entries are determined as accurately as the biggest ones. Choosing the most appropriate scaling from the application's point of view can be difficult task. It is then desirable that simple change of reference units do not cause instabilities in numerical algorithms because changing the description (scaling the data matrix) does not change the underlying physical system. This issue is often overlooked or ignored in numerical computations and it can be the cause of incorrectly computed and misinterpreted results with serious consequences.

This chapter describes algorithms for computation of the singular values and the eigenvalues of symmetric matrices to high relative accuracy. Accurate computation of eigenvectors and singular vectors is not considered. Singular values are discussed in sections 46.1 to 46.3; algorithms for computation of singular values are also given in Chapter 45.2.

Algorithms for computing the eigenvalues of symmetric matrices are given in Chapter 42. Sections 45.4 to 45.5 discuss numerical issues concerning the accuracy of the computed approximate eigenvalues, divided into the cases positive definite and symmetric indefinite (a negative definite matrix A is handled by applying methods for positive definite matrices to $-A$).

For symmetric H (which may contain initial uncertainty), perturbation theory determines whether or not it is numerically feasible to compute even the smallest eigenvalues with high relative accuracy. State of the art perturbation theory, which is a necessary prerequisite for algorithmic development, is given in Chapter 15. The insights from the perturbation theory are then capitalized in the development of numerical algorithms capable of achieving optimal theoretical accuracy. Unlike in the case of standard accuracy, positive definite and indefinite matrices are analyzed separately.

All matrices are assumed to be over the real field $\mathbb{R}$. Additional relevant preparatory results can be found in Chapter 8 and Chapter 17.

46.1 Accurate SVD and One-Sided Jacobi SVD Algorithm

Numerical computation of the SVD inevitably means computation with errors. How many digits in the computed singular values are provably accurate is answered by perturbation theory adapted to a particular algorithm. For dense matrices with no additional structure, the Jacobi SVD algorithm is proven to be more accurate than any other method that first bidiagonalizes the matrix. The simplest form is the one-sided Jacobi SVD introduced by Hestenes [Hes58]. It represents an implicit form of the classical Jacobi algorithm [Jac46] for symmetric matrices, with properly adjusted stopping criterion. Basic properties of classical Jacobi algorithm are listed in Section 42.7. Detailed analysis and implementation details can be found in [DV92], [Drm94], [Mat96], and [Drm97].

Definitions:

Two vectors $\mathbf{x}, \mathbf{y} \in \mathbb{R}^m$ are **numerically orthogonal** if $|\mathbf{x}^T\mathbf{y}| \leq \varepsilon \|\mathbf{x}\|_2 \|\mathbf{y}\|_2$, where $\varepsilon \geq 0$ is at most round-off unit ϵ times a moderate function of m.

The square matrix $\tilde{U}$ is a **numerically orthogonal matrix** if each pair of its columns is numerically orthogonal.

A numerical algorithm that computes approximations $\tilde{U} \approx U$, $\tilde{V} \approx V$ and $\tilde{\Sigma} \approx \Sigma$ of the SVD $A = U\Sigma V^T$ of A is **backward stable** if $\tilde{U}$, $\tilde{V}$ are numerically orthogonal and $\tilde{U}\tilde{\Sigma}\tilde{V}^T = A + \delta A$, with **backward error** δA small compared to A.

Facts:

1. There is no loss of generality in considering real tall matrices, i.e., $m \geq n$. In case $m < n$, consider A^T with the SVD $A^T = V\Sigma^T U^T$.
2. Let $\tilde{U} \approx U$, $\tilde{V} \approx V$, $\tilde{\Sigma} \approx \Sigma$ be the approximations of the SVD of $A = U\Sigma V^T$, computed by a backward stable algorithm as $A + \delta A = \tilde{U}\tilde{\Sigma}\tilde{V}^T$. Since the orthogonality of $\tilde{U}$ and $\tilde{V}$ cannot be guaranteed, the product $\tilde{U}\tilde{\Sigma}\tilde{V}^T$ in general does not represent an SVD. However, there exist orthogonal $\hat{U}$ close to $\tilde{U}$ and an orthogonal $\hat{V}$, close to $\tilde{V}$, such that $(I + E_1)(A + \delta A)(I + E_2) = \hat{U}\tilde{\Sigma}\hat{V}^T$, where E_1, E_2, which represent departure from orthogonality of $\tilde{U}$, $\tilde{V}$, are small in norm.
3. Different algorithms produce differently structured δA. Consider the singular values $\sigma_1 \geq \cdots \geq \sigma_n$ and $\tilde{\sigma}_1 \geq \cdots \geq \tilde{\sigma}_n$ of A and $A + \delta A$, respectively.
   - If $\|\delta A\|_2 \leq \varepsilon \|A\|_2$ and without any additional structure of δA (δA small in norm), then the best error bound in the singular values is $\max_{1 \leq i \leq n} |\sigma_i - \tilde{\sigma}_i| \leq \|\delta A\|_2$, i.e., $\max_{1 \leq i \leq n} \frac{|\sigma_i - \tilde{\sigma}_i|}{\sigma_i} \leq \kappa_2(A)\, \varepsilon$.

- [DV92] Let, for all i, the ith column $\delta \mathbf{a}_i$ of δA satisfy $\|\delta \mathbf{a}_i\|_2 \leq \varepsilon \|\mathbf{a}_i\|_2$, where $\mathbf{a}_i$ is the ith column of A. (δA is column-wise small perturbation.) Then $A + \delta A = (B + \delta B)D$, where $A = BD$, $D = \text{diag}(\|\mathbf{a}_1\|_2, \ldots, \|\mathbf{a}_n\|_2)$, and $\|\delta B\|_F \leq \sqrt{n}\varepsilon$. Let A have full column rank and let p be the rank of δA. Then $\max_{1 \leq i \leq n} \frac{|\sigma_i - \tilde{\sigma}_i|}{\sigma_i} \leq \|\delta B B^\dagger\|_2 \leq \sqrt{p}\varepsilon \|B^\dagger\|_2$.
- [vdS69] $\|B^\dagger\|_2 \leq \kappa_2(B) \leq \sqrt{n} \min_{S=\text{diag}} \kappa_2(AS) \leq \sqrt{n}\kappa_2(A)$. This implies that a numerical algorithm with column-wise small backward error δA computes more accurate singular values than an algorithm with backward error that is only small in norm.

4. In the process of computation of the SVD of A, an algorithm can produce intermediate matrix with singular values highly sensitive even to smallest entry-wise rounding errors.
5. The matrix A may have initial uncertainty on input and, in fact, $A = A_0 + \delta A_0$, where A_0 is the true unknown data matrix and δA_0 the initial error already present in A. If δA generated by the algorithm is comparable with δA_0, the computed SVD is as accurate as warranted by the data.
6. [Hes58] If $H = A^T A$, then the classical Jacobi algorithm can be applied to H implicitly. The one-sided (or implicit) Jacobi SVD method starts with general $m \times n$ matrix $A^{(0)} = A$ and it generates the sequence $A^{(k+1)} = A^{(k)} V^{(k)}$, where the matrix $V^{(k)}$ is the plane rotation as in the classical symmetric Jacobi method for the matrix $H^{(k)} = (A^{(k)})^T A^{(k)}$. Only $H^{(k)}[\{i_k, j_k\}]$ is needed to determine $V^{(k)}$, where (i_k, j_k) is the pivot pair determined by pivot strategy.
7. In the one-sided Jacobi SVD algorithm applied to $A \in \mathbb{R}^{m \times n}$, $m \geq n$, $A^{(k)}$ tends to $U\Sigma$, with diagonal $\Sigma = \text{diag}(\sigma_1, \ldots, \sigma_n)$ carrying the singular values. If A has full column rank, then the columns of U are the n corresponding left singular vectors. If $\text{rank}(A) < n$, then for each $\sigma_i = 0$ the ith column of U is zero. The m-rank(A) left singular vectors from the orthogonal complement of range(A) cannot be computed using one-sided Jacobi SVD. The accumulated product $V^{(0)} V^{(1)} \cdots$ converges to orthogonal matrix V of right singular vectors.
8. Simple implementation of Jacobi rotation in the one-sided Jacobi SVD algorithm (one-sided Jacobi rotation) is given in Algorithm 1. At any moment in the algorithm, $d_1, \ldots, d_n$ contain the squared Euclidean norms of the columns of current $A^{(k)}$, and ξ stores the Euclidean inner product of the pivot columns in the kth step.

Algorithm 1: One-sided Jacobi rotation

ROTATE $(A_{1:m,i}, A_{1:m,j}, d_i, d_j, \xi, [V_{1:m,i}, V_{1:m,j}])$

1: $\vartheta = \dfrac{d_j - d_i}{2 \cdot \xi}$; $t = \dfrac{\text{sign}(\vartheta)}{|\vartheta| + \sqrt{1 + \vartheta^2}}$; $c = \dfrac{1}{\sqrt{1 + t^2}}$; $s = t \cdot c$;

2: $[A_{1:m,i} \ \ A_{1:m,j}] = [A_{1:m,i} \ \ A_{1:m,j}] \begin{bmatrix} c & s \\ -s & c \end{bmatrix}$;

3: $d_i = d_i - t \cdot \xi$; $d_j = d_j + t \cdot \xi$;

4: **if** V is wanted, **then**

5: $[V_{1:n,i} \ \ V_{1:n,j}] = [V_{1:n,i} \ \ V_{1:n,j}] \begin{bmatrix} c & s \\ -s & c \end{bmatrix}$

6: **end if**

In case of cancellation in computation of the smaller of d_i, d_j, the value is refreshed by explicit computation of the squared norm of the corresponding column.

9. [dR89], [DV92] Numerical convergence of the one-sided Jacobi SVD algorithm (Algorithm 2) is declared at step k if $\max_{i \neq j} \frac{|(A^{(k)}_{1:m,i})^T A^{(k)}_{1:m,j}|}{\|A^{(k)}_{1:m,i}\|_2 \|A^{(k)}_{1:m,j}\|_2} \leq \zeta \approx m\epsilon$. This stopping criterion guarantees that computed approximation $\tilde{A}^{(k)}$ of $A^{(k)}$ can be written as $\tilde{U}\tilde{\Sigma}$, where the columns of $\tilde{U}$ are numerically orthogonal, and $\tilde{\Sigma}$ is diagonal.

Algorithm 2: One-sided Jacobi SVD (de Rijk's row-cyclic pivoting)
$(U, \Sigma, [V]) = \mathrm{SVD0}(A)$
$\zeta = m\epsilon$; $\hat{p} = n(n-1)/2$; $s = 0$; $convergence =$ **false** ;
if V is wanted, **then initialize** $V = I_n$ **end if**
for $i = 1$ **to** n **do** $d_i = A_{1:m,i}^T A_{1:m,i}$ **end for**;
repeat
$s = s + 1$; $p = 0$;
for $i = 1$ **to** $n - 1$ **do**
  find index i_0 such that $d_{i_0} = \max_{i \le \ell \le n} d_\ell$; **swap**$(d_i, d_{i_0})$;
  swap$(A_{1:m,i}, A_{1:m,i_0})$; **swap**$(V_{1:m,i}, V_{1:m,i_0})$;
  for $j = i + 1$ **to** n **do**
    $\xi = A_{1:m,i}^T A_{1:m,j}$;
    if $|\xi| > \zeta \sqrt{d_i d_j}$ **then** *% apply Jacobi rotation*
      call ROTATE$(A_{1:m,i}, A_{1:m,j}, d_i, d_j, \xi, [V_{1:m,i}, V_{1:m,j}])$;
    else $p = p + 1$ **end if**
  end for
end for
if $p = \hat{p}$, **then** convergence=**true**; **go to** ▶ **end if**
until $s > 30$
▶ **if** *convergence*, **then** *% numerical orthogonality reached*
  for $i = 1$ **to** n **do** $\Sigma_{ii} = \sqrt{d_i}$; $U_{1:m,i} = A_{1:m,i} \Sigma_{ii}^{-1}$ **end for**
else
  Error: *Numerical convergence did not occur after* 30 *sweeps.*
end if

10. [DV92], [Mat96], [Drm97] Let $\tilde{A}^{(k)}$, $k = 0, 1, 2, \ldots$ be the matrices computed by Algorithm 2 in floating point arithmetic. Write each $\tilde{A}^{(k)}$ as $\tilde{A}^{(k)} = B^{(k)} D^{(k)}$, with diagonals $D^{(k)}$ and $B^{(k)}$ with columns of unit Euclidean norms. Then

- $\tilde{A}^{(k+1)} = (\tilde{A}^{(k)} + \delta\tilde{A}^{(k)})\check{V}^{(k)}$, where $\check{V}^{(k)}$ is the exact plane rotation transforming pivot columns, and $\delta\tilde{A}^{(k)}$ is zero except in the i_kth and the j_kth columns.
- $\tilde{A}^{(k)} + \delta\tilde{A}^{(k)} = (B^{(k)} + \delta B^{(k)})D^{(k)}$, with $\|\delta B^{(k)}\|_F < c_k\epsilon$, where c_k is a small factor (e.g., $c_k < 20$) that depends on the implementation of the rotation.
- The above holds as long as the Euclidean norms of the i_kth and the j_kth columns of $\tilde{A}^{(k)}$ do not underflow or overflow. It holds even if the computed angle is so small that the computed value of $\tan\phi_k$ underflows (gradually or flushed to zero). This, however, requires special implementation of the Jacobi rotation.
- If all matrices $\tilde{A}^{(j)}$, $j = 0, \ldots, k-1$ are nonsingular, then for all $i = 1, \ldots, n$,
$$\max\left\{0, \prod_{j=0}^{k-1}(1 - \eta_j)\right\} \le \frac{\sigma_i(\tilde{A}^{(k)})}{\sigma_i(A)} \le \prod_{j=0}^{k-1}(1 + \eta_j),$$
where $\eta_j = \|\delta\tilde{A}^{(j)}(\tilde{A}^{(j)})^\dagger)\|_2 \le c_j\epsilon\|(B^{(j)})^\dagger\|_2$, and $\sigma_i(\cdot)$ stands for the ith largest singular value of a matrix. The accuracy of Algorithm 2 is determined by $\beta \equiv \max_{k\ge 0}\|(B^{(k)})^\dagger\|_2$. It is observed in practice that β never exceeds $\|B^\dagger\|_2$ too much.
- If A initially contains column-wise small uncertainty and if $\beta/\|B^\dagger\|_2$ is moderate, then Algorithm 2 computes as accurate approximations of the singular values as warranted by the data.

Examples:

1. Using orthogonal factorizations in finite precision arithmetic does not guarantee high relative accuracy in SVD computation. Bidiagonalization, as a preprocessing step in state-of-the-art SVD

algorithms, is an example. We use MATLAB with roundoff $\epsilon = \texttt{eps} \approx 2.22 \cdot 10^{-16}$. Let $\xi = 10/\epsilon$. Then floating point bidiagonalization of

$$A = \begin{pmatrix} 1 & 1 & \boxed{1} \\ 0 & 1 & \xi \\ 0 & -1 & \xi \end{pmatrix} \text{ yields } \tilde{A}^{(1)} = \begin{pmatrix} 1 & \alpha & 0 \\ 0 & \beta & \beta \\ 0 & \boxed{\beta} & \beta \end{pmatrix}, \quad \tilde{A}^{(2)} = \begin{pmatrix} 1 & \alpha & 0 \\ 0 & \gamma & \gamma \\ 0 & 0 & 0 \end{pmatrix},$$

$\alpha \approx 1.4142135\text{e}{+}0$, $\beta \approx 3.1845258\text{e}{+}16$, $\gamma \approx 4.5035996\text{e}{+}16$. The matrices $\tilde{A}^{(1)}$ and $\tilde{A}^{(2)}$ are computed using Givens plane rotations at the positions indicated by $\boxed{\cdot}$. The computed matrix $\tilde{A}^{(1)}$ entry-wise approximates with high relative accuracy the matrix $A^{(1)}$ from the exact computation (elimination of A_{13}). However, A and $A^{(1)}$ are nonsingular and $\tilde{A}^{(1)}$ is singular. The smallest singular value of A is lost — $\tilde{A}^{(1)}$ carries no information about σ_3. Transformation from $\tilde{A}^{(1)}$ to $\tilde{A}^{(2)}$ is again perfectly entry-wise accurate. But, even exact SVD of the bidiagonal $\tilde{A}^{(2)}$ cannot restore the information lost in the process of bidiagonalization. (See Fact 4 above.)

On the other hand, let $\hat{A}^{(1)}$ denote the matrix obtained from A by perturbing A_{13} to zero. This changes the third column $\mathbf{a}_3$ of A by $\|\delta\mathbf{a}_3\|_2 \leq (\epsilon/14)\|\mathbf{a}_3\|_2$ and causes, at most, $\epsilon/5$ relative perturbation in the singular values. (See Fact 3 above. Here, $A = BD$ with $\kappa(B) < 2$.) If we apply Givens rotation from the left to annihilate the $(3, 2)$ entry in $\hat{A}^{(1)}$, the introduced perturbation is column-wise small. Also, the largest singular value decouples; the computed matrix is $\hat{A}^{(2)} = \begin{bmatrix} 1 & 1 \\ 0 & \alpha \end{bmatrix} \bigoplus \tilde{\sigma}_1$.

46.2 Preconditioned Jacobi SVD Algorithm

The accuracy properties and the run-time efficiency of the one-sided Jacobi SVD algorithm can be enhanced using appropriate preprocessing and preconditioning. More details can be found in [Drm94], [Drm99], [DV05a], and [DV05b].

Definitions:

QR factorization with column pivoting of $A \in \mathbb{R}^{m\times n}$ is the factorization $A\Pi = Q\begin{bmatrix} R \\ 0 \end{bmatrix}$, where Π is a permutation matrix, Q is orthogonal, and R is $n \times n$ upper triangular.

The QR factorization with **Businger–Golub pivoting** chooses Π to guarantee that $r_{kk}^2 \geq \sum_{i=k}^{j} r_{ij}^2$ for all $j = k, \ldots, n$ and $k = 1, \ldots, n$.

The QR factorization with **Powell–Reid's complete (row and column) pivoting** computes the QR factorization using row and column permutations, $\Pi_r A \Pi_c = Q\begin{bmatrix} R \\ 0 \end{bmatrix}$, where Π_c is as in Businger–Golub pivoting and Π_r enhances numerical stability in case of A with differently scaled rows.

QR factorization with pivoting of A is **rank revealing** if small singular values of A are revealed by correspondingly small diagonal entries of R.

Facts:

1. [Drm94], [Hig96]. Let the QR factorization $A = Q\begin{bmatrix} R \\ 0 \end{bmatrix}$ of $A \in \mathbb{R}^{m\times n}$, $m \geq n$, be computed using the Givens or the Householder algorithm in the IEEE floating point arithmetic with rounding relative error $\epsilon < 10^{-7}$. Let the computed approximations of Q and R be $\tilde{Q}$ and $\tilde{R}$, respectively. Then there exist an orthogonal matrix $\hat{Q}$ and a backward perturbation δA such that

$$A + \delta A = \hat{Q}\begin{bmatrix} \tilde{R} \\ 0 \end{bmatrix}, \quad \|\tilde{Q}_{1:m,i} - \hat{Q}_{1:m,i}\|_2 \leq \varepsilon_{qr}, \quad \|\delta A_{1:m,i}\|_2 \leq \varepsilon_{qr}\|A_{1:m,i}\|_2$$

holds for all $i = 1, \ldots, n$, with $\varepsilon_{qr} \leq O(mn)\epsilon$. This remains true if the factorization is computed with pivoting. It suffices to assume that the matrix A is already prepermuted on input, and that the factorization itself is performed without pivoting.

2. [Drm94], [Mat96], [Drm99] Let $A = BD$, where D is diagonal and B has unit columns in Euclidean norm.

   - If $\|B^\dagger\|_2$ is moderate, then the computed $\tilde{R}$ allows approximation of the singular values of A to high relative accuracy. If $\tilde{R}$ is computed with Businger–Golub pivoting, then Algorithm 2 applied to $\tilde{R}^T$ converges swiftly and computes the singular values of A with relative accuracy determined by $\|B^\dagger\|_2$.
   - Let A have full column rank and let its QR factorization be computed by Businger–Golub column pivoting. Write $R = ST$, where S is the diagonal matrix and T has unit rows in Euclidean norm. Then $\|T^{-1}\|_2 \leq n\|B^\dagger\|_2$.

3. [DV05a] The following algorithm carefully combines the properties of the one–sided Jacobi SVD and the properties of the QR factorization (Facts 1 and 2):

Algorithm 3: Preconditioned Jacobi SVD

$(U, \Sigma, [V]) = \text{SVD1}(A)$

Input: $A \in \mathbb{R}^{m\times n}$, $m \geq n$.

1: $(P_r A)P_c = Q_1 \begin{bmatrix} R_1 \\ 0 \end{bmatrix}$; % $R_1 \in \mathbb{R}^{\rho\times n}$, *rank revealing QRF.*

2: $R_1^T = Q_2 \begin{bmatrix} R_2 \\ 0 \end{bmatrix}$; % $R_2 \in \mathbb{R}^{\rho\times\rho}$.

3: $(U_2, \Sigma_2, V_2) = \text{SVD0}(R_2^T)$;% *one-sided Jacobi SVD on* R_2^T.

Output: $U = P_r^T Q_1 \begin{bmatrix} U_2 & 0 \\ 0 & I_{m-\rho} \end{bmatrix}$; $\Sigma = \begin{bmatrix} \Sigma_2 & 0 \\ 0 & 0 \end{bmatrix}$; $V = P_c Q_2 \begin{bmatrix} V_2 & 0 \\ 0 & I_{n-\rho} \end{bmatrix}$.

4. [Drm99], [DV05a], [DV05b]

   - Let A and the computed matrix $\tilde{R}_1 \approx R_1$ in Algorithm 3 be of full column rank. Let $\tilde{R}_2$ be computed approximation of R_2 in Step 2. Then, there exist perturbations δA, $\delta \tilde{R}_1$, and there exist orthogonal matrices $\hat{Q}_1$, $\hat{Q}_2$ such that

$$(I + \delta A A^\dagger)A(I + \tilde{R}_1^{-1}\delta \tilde{R}_1) = \hat{Q}_1 \begin{bmatrix} \tilde{R}_2^T \\ 0 \end{bmatrix} \hat{Q}_2^T.$$

   Thus, the first two steps of Algorithm 3 preserve the singular values if $\|B^\dagger\|_2$ is moderate.

   - Let the one-sided Jacobi SVD algorithm with row cyclic pivot strategy be applied to $\tilde{R}_2^T$. Let the stopping criterion be satisfied at the matrix $(\tilde{R}_2^T)^{(k)}$ during the sth sweep. Write $(\tilde{R}_2^T)^{(k)} = \tilde{U}_2\tilde{\Sigma}$, where $\tilde{\Sigma}$ is diagonal and $\tilde{U}_2$ is numerically orthogonal. Then there exist an orthogonal matrix $\hat{V}_2$ and a backward error $\delta \tilde{R}_2$ such that $\tilde{U}_2\tilde{\Sigma} = (\tilde{R}_2 + \delta\tilde{R}_2)^T \hat{V}_2$, $\|\delta(\tilde{R}_2)_{1:\rho,i}\|_2 \leq \varepsilon_J \|(\tilde{R}_2)_{1:\rho,i}\|_2$, $i = 1, \ldots, \rho$, with $\varepsilon_J \leq (1 + 6\epsilon)^{s(2n-3)} - 1$. If $\hat{U}_2$ is the closest orthogonal matrix to $\tilde{U}_2$, then $\tilde{R}_2^T = \hat{V}_2\tilde{\Sigma}\hat{U}_2^T(I + E)$, where the dominant part of $\|E\|_2$ is $\|\delta\tilde{R}_2\tilde{R}_2^{-1}\|_2$. Similar holds for any serial or parallel convergent pivot strategy.
   - Assembling the above yields

$$(I + \delta A A^\dagger)A(I + \tilde{R}_1^{-1}\delta \tilde{R}_1) = \hat{Q}_1 \begin{bmatrix} \hat{V}_2 & 0 \\ 0 & I \end{bmatrix} \begin{bmatrix} \tilde{\Sigma} \\ 0 \end{bmatrix} \hat{U}_2^T \hat{Q}_2^T (I + \hat{Q}_2 E \hat{Q}_2^T).$$

The upper bound on the maximal relative error $\|\Sigma^{-1}(\Sigma - \tilde{\Sigma})\|_2$ is dominated by $\|\delta A A^{\dagger}\|_2 + \|\tilde{R}_1^{-1}\delta\tilde{R}_1\|_2 + \|\delta\tilde{R}_2\tilde{R}_2^{-1}\|_2$.

- Let A have the structure $A = BD$, where full-column rank B has equilibrated columns and D is arbitrary diagonal scaling. The only relevant condition number for relative accuracy of Algorithm 3 is $\|B^{\dagger}\|_2$.

5. [DV05a], [DV05b] Algorithm 3 outperforms Algorithm 2 both in speed and accuracy.
6. [Drm99], [Drm00b], [Hig00], [DV05a] It is possible that $\kappa_2(B)$ is large, but A is structured as $A = D_1 C D_2$ with diagonal scalings D_1, D_2 (diagonal entries in nonincreasing order) and well conditioned C. In that case, desirable backward error for the first QR factorization (Step 1) is $P_r D_1(C + \delta C)D_2 P_c$ with satisfactory bound on $\|\delta C\|_2$. This is nearly achieved if the QR factorization is computed with Powell–Reid's complete pivoting or its simplification, which replaces row pivoting with initial sorting. Although theoretical understanding of this fact is not complete, initial row sorting (descending in $\|\cdot\|_\infty$ norm) is highly recommended.

Examples:

1. The key step of the preconditioning is in transposing the computed triangular factors. If A is a 100×100 Hilbert matrix, written as $A = BD$ (cf. Fact 2,), then $\kappa_2(B) > 10^{19}$. If in Step 2 of Algorithm 3, we write R_1^T as $R_1^T = B_1 D_1$, where D_1 is diagonal and B_1 has unit columns in Euclidean norm, then $\kappa_2(B_1) < 50$.
2. Take $\gamma = 10^{-20}$, $\delta = 10^{-40}$ and consider the matrix

$$A = \begin{bmatrix} 1 & \gamma & \gamma \\ -\gamma & \gamma & \gamma^2 \\ 0 & \delta & 0 \end{bmatrix} = \begin{bmatrix} 1 & 0 & 0 \\ 0 & \gamma & 0 \\ 0 & 0 & \delta \end{bmatrix} \begin{bmatrix} 1 & \gamma & 1 \\ -1 & 1 & 1 \\ 0 & 1 & 0 \end{bmatrix} \begin{bmatrix} 1 & 0 & 0 \\ 0 & 1 & 0 \\ 0 & 0 & \gamma \end{bmatrix},$$

with singular values nearly $\sigma_1 \approx 1$, $\sigma_2 \approx \gamma$, $\sigma_3 \approx 2\gamma\delta$, cf. [DGE99]. A cannot be written as $A = BD$ with diagonal D and well-conditioned B. Algorithm 2 computes no accurate digit of σ_3, while Algorithm 3 approximates all singular values to nearly full accuracy. (See Fact 5.) The computed R_1^T in Step 2 is

$$\tilde{R}_1^T = \begin{bmatrix} -1.000000000000000e+0 & 0 & 0 \\ -1.000000000000000e-20 & -1.000000000000000e-20 & 0 \\ -1.000000000000000e-20 & -1.000000000000000e-40 & -2.000000000000000e-60 \end{bmatrix}.$$

Note that the neither the columns of A nor the columns of $\tilde{R}_1$ reveal the singular value of order 10^{-60}. On the other hand, the Euclidean norms of the column of $\tilde{R}_1^T$ nicely approximate the singular values of A. (See Facts 2, 3.) If the order of the rows of A is changed, $\tilde{R}_1$ is computed with zero third row.

46.3 Accurate SVD from a Rank Revealing Decomposition: Structured Matrices

In some cases, the matrix A is given implicitly as the product of two or three matrices or it can be factored into such a product using more general nonorthogonal transformations. For instance, some specially structured matrices allow Gaussian eliminations with complete pivoting $P_1 A P_2 = LDU$ to compute entry-wise accurate $\tilde{L} \approx L$, $\tilde{D} \approx D$, $\tilde{U} \approx U$. Moreover, it is possible that the triple $\tilde{L}, \tilde{D}, \tilde{U}$ implicitly defines the SVD of A to high relative accuracy, and that direct application of any SVD algorithm directly to A does not return accurate SVD. For more information on matrices and bipartite graphs, see Chapter 30. For more information on sign pattern matrices, see Chapter 33. For more information on totally nonnegative (TN) matrices, see Chapter 21.

Definitions:

The singular values of A are said to be **perfectly well determined to high relative accuracy** if the following holds: No matter what the entries of the matrix are, changing an arbitrary nonzero entry $a_{k\ell}$ to $\theta a_{k\ell}$, with arbitrary $\theta \neq 0$, will cause perturbation $\sigma_i \rightsquigarrow \tilde{\sigma}_i \in [\theta_l \sigma_i, \theta_u \sigma_i]$, $\theta_l = \min\{|\theta|, 1/|\theta|\}$, $\theta_u = \max\{|\theta|, 1/|\theta|\}$, $i = 1, \ldots, n$.

The **sparsity pattern** Struct(A) of A is defined as the set of indices (k, ℓ) of the entries of A permitted to be nonzero.

The **bipartite graph** $\mathcal{G}(\mathcal{S})$ of the sparsity pattern $\mathcal{S}$ is the graph with vertices partitioned into row vertices $r_1, \ldots, r_m$ and column vertices $c_1, \ldots, c_n$, where r_k and r_l are connected if and only if $(k, l) \in \mathcal{S}$.

If $\mathcal{G}(\mathcal{S})$ is acyclic, matrices with sparsity pattern $\mathcal{S}$ are **biacyclic.**

The **sign** pattern sgn(A) prescribes locations and signs of nonzero entries of A.

A sign pattern S is **total signed compound** (TSC) if every square submatrix of every matrix A such that sgn$(A) = S$ is either **sign nonsingular** (nonsingular and determinant expansion is the sum of monomials of like sign) or **sign singular** (determinant expansion degenerates to sum of monomials, which are all zero); cf. Chapter 33.2.

$A \in \mathbb{R}^{m\times n}$ is **diagonally scaled totally unimodular (DSTU)** if there exist diagonal D_1, D_2 and **totally unimodular** Z (all minors of Z are from $\{-1, 0, 1\}$) such that $A = D_1 Z D_2$.

A decomposition $A = XDY^T$ with diagonal matrix D is called an **RRD** if X and Y are full-column rank, well-conditioned matrices. (RRD is an abbreviation for "rank revealing decomposition," but that term is defined slightly differently in Chapter 39.9, where X, Y are required to be orthogonal and D is replaced by an upper triangular matrix.)

Facts:

1. [DG93] The singular values of A are perfectly well determined to high relative accuracy if and only if the bipartite graph $\mathcal{G}(S)$ is acyclic (forest of trees). Sparsity pattern $\mathcal{S}$ with acyclic $\mathcal{G}(S)$ allows at most $m + n - 1$ nonzero entries. A bisection algorithm computes all singular values of biacyclic matrices to high relative accuracy.
2. [DK90], [FP94] Bidiagonal matrices are a special case of acyclic sparsity pattern. Let $n \times n$ bidiagonal matrix B be perturbed by $b_{ii} \rightsquigarrow b_{ii}\varepsilon_{2i-1}$, $b_{i,i+1} \rightsquigarrow b_{i,i+1}\varepsilon_{2i}$ for all admissible is, and let $\varepsilon = \prod_{i=1}^{2n-1} \max(|\varepsilon_i|, 1/|\varepsilon_i|)$, where all $\varepsilon_i \neq 0$. If $\tilde{\sigma}_1 \geq \cdots \geq \tilde{\sigma}_n$ are the singular values of the perturbed matrix, then for all i, $\sigma_i/\varepsilon \leq \tilde{\sigma}_i \leq \varepsilon\sigma_i$. The singular values of bidiagonal matrices are efficiently computed to high relative accuracy by the zero shift QR algorithm and the differential qd algorithm.
3. [DGE99] Let $\mathcal{S}_\pm$ be a sparsity-and-sign pattern. Let each matrix A with pattern $\mathcal{S}_\pm$ have the property that small relative changes of its (nonzero) entries cause only small relative perturbations of its singular values. This is equivalent with $\mathcal{S}_\pm$ being total signed compound (TSC).
4. [Drm98a], [DGE99] Let A be given by an RRD $A = XDY^T$. Without loss of generality assume that X and Y have equilibrated (e.g., unit in Euclidean norm) columns. Algorithm 4 computes the SVD of A to high relative accuracy.

Algorithm 4: RRD Jacobi SVD

$(U, \Sigma, V) = \text{SVD2}(X, D, Y)$

Input: $X \in \mathbb{R}^{m\times p}$, $Y \in \mathbb{R}^{n\times p}$ full column rank, $D \in \mathbb{R}^{p\times p}$ diagonal.

1: $\Upsilon = YD$; $\Upsilon P = Q\begin{bmatrix} R \\ 0 \end{bmatrix}$ *% rank revealing QR factorization.*

2: $Z = (XP)R^T$; *% explicit standard matrix multiplication.*

3: $(U_z, \Sigma_z, V_z) = \text{SVD1}(Z)$ *% one-sided Jacobi SVD on Z.*

Output: $U = U_z$; $\Sigma = \begin{bmatrix} \Sigma_z & 0 \\ 0 & 0 \end{bmatrix}$; $V = Q\begin{bmatrix} V_z & 0 \\ 0 & I_{n-p} \end{bmatrix}$. $XDY^T = U\Sigma V^T$.

5. [Drm98a], [DGE99], [DM04]

   - In Step 2 of Algorithm 4, the matrix R^T has the structure $R^T = LS$, where S is diagonal scaling, L has unit columns, and $\|L^{-1}\|_2$ is bounded by a function of p independent of input data. The upper bound on $\kappa_2(L)$ depends on pivoting P in Step 1 and can be $O(n^{1+(1/4)\log_2 n})$. For the Businger–Golub, pivoting the upper bound is $O(2^n)$, but in practice one can expect an $O(n)$ bound.
   - The product $Z = (XP)R^T$ must not be computed by fast (e.g., Strassen) algorithm if all singular values are wanted to high relative accuracy. The reason is that fast matrix multiplication algorithms can produce larger component-wise perturbations than the standard $O(n^3)$ algorithm.
   - The computed $\tilde{U}$, $\tilde{\Sigma}$, $\tilde{V}$ satisfy: There exist orthogonal $\hat{U} \approx \tilde{U}$, orthogonal $\hat{V} \approx \tilde{V}$, and E_1, E_2 such that $\hat{U}\tilde{\Sigma}\hat{V}^T = (I + E_1)A(I + E_2)$, $\|E_1\|_2 \leq O(\epsilon)\kappa_2(X)$, $\|E_2\|_2 \leq O(\epsilon)\kappa_2(L)\kappa_2(Y)$. The relative errors in the computed singular values are bounded by $O(\epsilon)\kappa_2(L)\max\{\kappa_2(X), \kappa_2(Y)\}$. Here, $O(\epsilon)$ denotes machine precision ϵ multiplied by a moderate polynomial in m, n, p.

6. [DGE99] Classes of matrices with accurate LDU factorization and, thus, accurate SVD computation by Algorithm 4, or other algorithms tailored for special classes of matrices (e.g., QR and qd algorithms for bidiagonal matrices), include:

   - *Acyclic matrices:* Accurate LDU factorization with pivoting uses the correspondence between the monomials in determinant expansion and perfect matchings in $\mathcal{G}(S)$.
   - *Total signed compound (TSC) matrices:* LDU factorization with complete pivoting $P_r A P_c = LDU$ of an TSC matrix A can be computed in a forward stable way. Cancellations in Gaussian eliminations can be avoided by computing some elements as quotients of minors. All entries of L, D, U are computed to high relative accuracy. The behavior of $\kappa(L)$ and $\kappa(U)$ is not fully analyzed, but they behave well in practice.
   - *Diagonally scaled totally unimodular (DSTU) matrices:* This class includes acyclic and certain finite element (e.g., mass-spring systems) matrices. Gaussian elimination with complete pivoting is modified by replacing dangerous cancellations by exactly predicted exact zeros. All entries of L, D, U are computed to high relative accuracy, and $\kappa(L)$, $\kappa(U)$ are at most $O(mn)$ and $O(n^2)$, respectively.

7. [Dem99] In some cases, new classes of matrices and accurate SVD computation are derived from relations with previously solved problems or by suitable matrix representations. The following cases are analyzed and solved with $O(n^3)$ complexity:

   - *Cauchy matrices:* $C = C(x, y)$, $c_{ij} = 1/(x_i + y_j)$, and the parameters x_is, y_is are the initial data. The algorithm extends to Cauchy-like matrices $C' = D_1 C D_2$, where C is a Cauchy matrix and D_1, D_2 are diagonal scalings. The required LDU factorization is computed with $\frac{4}{3}n^3$ operations and given as input to Algorithm 4. It should be noted that this does not imply that Cauchy matrices determine their singular values perfectly well. The statement is: For a Cauchy matrix C, given by set of parameters x_i's, y_i's stored as floating point numbers in computer memory, all singular values of C can be computed in a forward stable way. Thus, for example, singular values of a notoriously ill-conditioned Hilbert matrix can be computed to high relative accuracy.
   - *Vandermonde matrices:* $V = [v_{ij}]$, $v_{ij} = x_i^{j-1}$. An accurate algorithm exploits the fact that postmultiplication of V by the discrete Fourier transform matrix gives a Cauchy-like matrix. The finite precision arithmetic requires a guard digit and extra precision to tabulate certain constants.
   - *Unit displacement rank matrix X:* Solution of matrix equation $AX + XB = d_1 d_2^T$, where A, B are diagonal, or normal matrices with known accurate spectral decompositions. The class of unit displacement rank matrices generalizes the Cauchy-like matrices (A and B diagonal) and Vandermonde matrices (A diagonal and B circular shift).

8. [DK04] Weakly diagonally dominant M-matrices given with $a_{ij} \leq 0$ for $i \neq j$ and with the row-sums $s_i = \sum_{j=1}^{n} a_{ij} \geq 0$, $i = 1, \ldots, n$ known with small relative errors, allow accurate LDU factorization with complete pivoting.
9. [Koe05] Totally nonnegative (TN) matrix A can be expressed as the product of nonnegative bidiagonal matrices. If A is given implicitly by this bidiagonal representation, then all its singular values (and eigenvalues, too) can be computed from the bidiagonals to high relative accuracy. Accurate bidiagonal representation for given TN matrix A is possible for totally positive cases (all minors positive), provided certain pivotal minors can be computed accurately.
10. [DK05a] The singular values of the matrix V with entries $v_{ij} = P_i(x_j)$, where the P_is are orthonormal polynomials and the x_js are the nodes, can be computed to high relative accuracy.
11. [Drm00a] Accurate SVD of the RRD XDY^T extends to the triple product XSY^T, where S is not necessarily diagonal, but it can be accurately factored by Gaussian eliminations with pivoting.

Examples:

1. An illustration of the power of the algorithms described in this section is the example of a 100×100 Hilbert matrix described in [Dem99]. Its singular values range over 150 orders of magnitude and are computed using the package Mathematica with 200-decimal digit software floating point arithmetic. The computed singular values are rounded to 16 digits and used as reference values. The singular values computed in IEEE double precision floating point ($\epsilon \approx 10^{-16}$) by the algorithms described in this section agree with the reference values with relative error less than $34 \cdot \epsilon$.
2. [DGE99] Examples of sign patterns of TSC matrices:

$$\begin{bmatrix} + & + & 0 & 0 & 0 \\ + & - & + & 0 & 0 \\ 0 & + & + & + & 0 \\ 0 & 0 & + & - & + \\ 0 & 0 & 0 & + & + \end{bmatrix}, \quad \begin{bmatrix} + & + & + & + & + \\ + & - & 0 & 0 & 0 \\ + & 0 & - & 0 & 0 \\ + & 0 & 0 & - & 0 \\ + & 0 & 0 & 0 & - \end{bmatrix}.$$

TSC matrices must be sparse because an $m \times n$ TSC matrix can have at most $\frac{3}{2}(m+n) - 2$ nonzero entries.

46.4 Positive Definite Matrices

Definitions:

The **Cholesky factorization with pivoting** of symmetric positive definite $n \times n$ matrix H computes the Cholesky factorization $P^T H P = LL^T$, where the permutation matrix P is such that $\ell_{ii}^2 \geq \sum_{k=i}^{j} \ell_{jk}^2$, $1 \leq i \leq j \leq n$.

The **component-wise relative distance** between H and its component-wise relative perturbation $\tilde{H}$ is $\text{reldist}(H, \tilde{H}) = \max_{i,j} \dfrac{|h_{ij} - \tilde{h}_{ij}|}{|h_{ij}|}$, where $0/0 = 0$.

A **diagonally scaled representation** of symmetric matrix H with positive diagonal entries is a factored representation $H = DAD$ with $D = \text{diag}(\sqrt{h_{11}}, \ldots, \sqrt{h_{nn}})$, $a_{ij} = \dfrac{h_{ij}}{\sqrt{h_{ii}h_{jj}}}$.

Facts:

1. [Dem89], [Dem92] Let $H = DAD$ be a diagonally scaled representation of positive definite H, and let $\lambda_{\min}(A)$ denote the minimal eigenvalue of A.

   - If δH is a symmetric perturbation such that $H + \delta H$ is not positive definite, then
$$\max_{1 \leq i,j \leq n} \frac{|\delta h_{ij}|}{\sqrt{h_{ii}h_{jj}}} \geq \frac{\lambda_{\min}(A)}{n} = \frac{1}{n\|A^{-1}\|_2}.$$

- If $\delta H = -\lambda_{\min}(A)D^2$, then $\max_{i,j} \frac{|\delta h_{ij}|}{\sqrt{h_{ii}h_{jj}}} = \lambda_{\min}(A)$, $\text{reldist}(H, H+\delta H) = \lambda_{\min}(A)$, and $H + \delta H$ is singular.

2. [Dem89] Let $H = DAD$ be an $n \times n$ symmetric matrix with positive diagonal entries, stored in the machine memory. Let H be the input matrix in the Cholesky factorization algorithm. Then the following holds:
   - If the Cholesky algorithm successfully completes all operations and computes lower triangular matrix $\tilde{L}$, then there exists symmetric backward perturbation δH such that $\tilde{L}\tilde{L}^T = H + \delta H$ and $|\delta h_{ij}| \leq \eta_C \sqrt{h_{ii}h_{jj}}$, $\eta_C \leq O(n\epsilon)$.
   - If $\lambda_{\min}(A) > n\eta_C$, then the Cholesky algorithm will succeed and compute $\tilde{L}$.
   - If $\lambda_{\min}(A) < \epsilon$, then there exists simulation of rounding errors in which the Cholesky algorithm fails to complete all operations.
   - If $\lambda_{\min}(A) \leq -n\eta_C$, then it is certain that the Cholesky algorithm will fail.
3. [DV92] If $H = DAD \in \text{PD}_n$ is perturbed to a positive definite $\tilde{H} = H + \delta H$, then
   - $\max_{1 \leq i \leq n} \frac{|\lambda_i - \tilde{\lambda}_i|}{\lambda_i} \leq \|L^{-1}\delta H L^{-T}\|_2 \leq \frac{\|D^{-1}\delta H D^{-1}\|_2}{\lambda_{\min}(A)} = \|A^{-1}\|_2 \|D^{-1}\delta H D^{-1}\|_2.$
   - Let $\delta H = \eta D^2$, with any $\eta \in (0, \lambda_{\min}(A))$. Then for some index i it holds that $\frac{|\lambda_i - \tilde{\lambda}_i|}{\lambda_i} \geq \sqrt[n]{1 + \frac{\eta}{\lambda_{\min}(A)}}.$
4. [DV92], [VS93] Let $H = DAD$ be positive definite and $c > 0$ constant such that for all $\varepsilon \in (0, 1/c)$ and for all symmetric component-wise relative perturbations δH with $|\delta h_{ij}| \leq \varepsilon |h_{ij}|$, $1 \leq i, j \leq n$, the ordered eigenvalues λ_i and $\tilde{\lambda}_i$ of H and $H + \delta H$ satisfy $\max_{1 \leq i \leq n} \frac{|\lambda_i - \tilde{\lambda}_i|}{\lambda_i} \leq c\varepsilon$. Then $\|A^{-1}\|_2 < (1+c)/2$. The same holds for more general perturbations, e.g., δH with $|\delta h_{ij}| \leq \varepsilon\sqrt{h_{ii}h_{jj}}$.
5. [vdS69] It holds that $\|A^{-1}\|_2 \leq \kappa_2(A) \leq n \min_{D=\text{diag}} \kappa_2(DHD) \leq n\kappa_2(H)$.
6. For a general dense positive definite $H = DAD$ stored in the machine memory, eigenvalue computation with high relative accuracy is numerically feasible if and only if $\lambda_{\min}(A)$ is not smaller than the machine round-off unit. It is possible that matrix H is theoretically positive definite and that errors in computing its entries as functions of some parameters cause the stored matrix to be indefinite. Failure of the Cholesky algorithm is a warning that the matrix is entry-wise close to a symmetric matrix that is not positive definite.
7. [VH89] If $P^T HP = LL^T$ is the Cholesky factorization with pivoting of positive definite H, then the SVD $L = U\Sigma V^T$ of L is computed very efficiently by the one-sided Jacobi SVD algorithm, and H is diagonalized as $H = (PU)\Sigma^2(PU)^T$.

Algorithm 5: Positive definite Jacobi EVD

$(\lambda, U) = \text{EIG}_+(H)$

Input: $H \in \text{PD}_n$;

$P^T HP = LL^T$; % *Cholesky factorization with pivoting.*

if L computed successfully, **then**

$(U, \Sigma) = \text{SVD0}(L)$ % *One-sided Jacobi SVD on* L. V *is not computed.*

$\lambda_i = \Sigma_{ii}^2$, $U_{1:n,i} = PU_{1:n,i}$, $i = 1, \ldots, n$;

Output: $\lambda = (\lambda_1, \ldots, \lambda_n)$; U

else

Error: H is not numerically positive definite

end if

8. [DV92], [Mat96], [Drm98b]. Let $\tilde{\lambda}_1 \geq \cdots \geq \tilde{\lambda}_n$ be the approximations of the eigenvalues $\lambda_1 \geq \cdots \geq \lambda_n$ of $H = DAD$, computed by Algorithm 5. Then
   - The computed approximate eigenvalues of H are the exact eigenvalues of a nearby symmetric positive definite matrix $H + \delta H$, where $\max_{1\leq i,j\leq n} \frac{|\delta h_{ij}|}{\sqrt{h_{ii}h_{jj}}} \leq \varepsilon$, and ε is bounded by $O(n)$ times the round-off unit ϵ.
   - $\max_{1\leq i\leq n} \frac{|\lambda_i - \tilde{\lambda}_i|}{\lambda_i} \leq n\varepsilon \|A^{-1}\|_2$. The dominant part in the forward relative error is committed during the Cholesky factorization. The one-sided Jacobi SVD contributes to this error with, at most, $O(n)\epsilon\sqrt{\|A^{-1}\|_2} + O(n^2)\epsilon$.
9. Numerical properties of Algorithm 5, given in Fact 8, are better appreciated if compared with algorithms that first reduce H to tridiagonal matrix T and then diagonalize T. For such triadiagonalization based procedures the following hold:
   - The computed approximate eigenvalues of H are the exact eigenvalues of a nearby symmetric matrix $H + \delta H$, where $\|\delta H\|_2 \leq \varepsilon \|H\|_2$ and ε is bounded by a low degree polynomial in n times the round-off unit ϵ.
   - The computed eigenvalue approximations $\tilde{\lambda}_i \approx \lambda_i$ satisfy the *absolute error* bound $|\lambda_i - \tilde{\lambda}_i| \leq \varepsilon \|H\|_2$, that is, $\max_{1\leq i\leq n} \frac{|\lambda_i - \tilde{\lambda}_i|}{\lambda_i} \leq \varepsilon \kappa_2(H)$.
10. In some applications it might be possible to work with a positive definite matrix implicitly as $H = C^T C$, where only a full column rank C is explicitly formed. Then the spectral computation with H is replaced with the SVD of C. The Cholesky factor L of H is computed implicitly from the QR factorization of C. This implicit formulation has many numerical advantages and it should be the preferred way of computation with positive definite matrices. An example is natural factor formulation of stiffness matrices in finite element computations.
11. [Drm98a], [Drm98b] Generalized eigenvalues of $HM - \lambda I$ and $H - \lambda M$ can be computed to high relative accuracy if $H = D_H A_H D_H$, $M = D_M A_M D_M$ with diagonal D_H, D_M and moderate $\|A_H^{-1}\|_2$, $\|A_M^{-1}\|_2$.

Examples:

1. In this numerical experiment we use MATLAB 6.5, Release 13 (on a Pentium 4 machine under MS Windows[R] 2000), and the function `eig(·)` for eigenvalue computation. Let

$$H = \begin{bmatrix} 10^{40} & 10^{29} & 10^{19} \\ 10^{29} & 10^{20} & 10^{9} \\ 10^{19} & 10^{9} & 1 \end{bmatrix}.$$

   The sensitivity of the eigenvalues of H and the accuracy of numerical algorithms can be illustrated by applications of the algorithms to various functions and perturbations of H. Let $P^T HP$ be obtained from H by permutation similarity with permutation matrix P. Let $H + \Delta H$ be obtained from H by changing H_{22} into $-H_{22} = -10^{20}$, and let $H + \delta H$ be obtained from H by multiplying H_{13} and H_{31} by $1 + \epsilon$, where $\epsilon \approx 2.22 \cdot 10^{-16}$ is the round-off unit in MATLAB. For the sake of experiment, the eigenvalues of numerically computed $(H^{-1})^{-1}$ are also examined. The values returned by `eig()` are shown in Table 46.1. All six approximations of the spectrum of H are with small absolute error, $\max_{1\leq i\leq 3} \frac{|\lambda_i - \tilde{\lambda}_i|}{\|H\|_2} \leq O(\epsilon)$. Some results might be different if a different version of MATLAB or operating system is used.
2. Let H be the matrix from Example 1. H is positive definite, $\kappa_2(H) > 10^{40}$, and its $H = DAD$ representation with $D = \text{diag}(10^{20}, 10^{10}, 1)$ gives $\kappa_2(A) < 1.4$, $\|A^{-1}\|_2 < 1.2$. The Cholesky factor

TABLE 46.1

	eig(H)	eig(P^THP), $P \simeq (2,1,3)$	eig(inv(inv(H)))
$\tilde{\lambda}_1$	1.000000000000000e + 040	1.000000000000000e + 040	1.000000000000000e + 040
$\tilde{\lambda}_2$	−8.100009764062724e + 019	9.900000000000000e + 019	9.900000000000000e + 019
$\tilde{\lambda}_3$	−3.966787845610502e + 023	9.818181818181818e − 001	9.818181818181817e − 001
	eig($H + \Delta H$)	eig(P^THP), $P \simeq (3,2,1)$	eig($H + \delta H$)
$\tilde{\lambda}_1$	1.000000000000000e + 040	1.000000000000000e + 040	1.000000000000000e + 040
$\tilde{\lambda}_2$	−8.100009764062724e + 019	9.900000000000000e + 019	1.208844819952007e + 024
$\tilde{\lambda}_3$	−3.966787845610502e + 023	9.818181818181819e − 001	9.899993299416013e − 001

L of H is successfully computed in MATLAB by the chol(·) function. The matrix L^TL, which is implicitly diagonalized in Algorithm 5, reads

$$\hat{H} = L^TL = \begin{bmatrix} 1.000000000e+40 & 9.949874371e+18 & 9.908673886e-02 \\ 9.949874371e+18 & 9.900000000e+19 & 8.962732759e-02 \\ 9.908673886e-02 & 8.962732759e-02 & 9.818181818e-01 \end{bmatrix}.$$

The diagonal of $\hat{H}$ approximates the eigenvalues of H with all shown digits correct. To see that, write $\hat{H}$ as $\hat{H} = \hat{D}\hat{A}\hat{D}$, where $\hat{D} = \text{diag}(\sqrt{\hat{H}_{11}}, \sqrt{\hat{H}_{22}}, \sqrt{\hat{H}_{33}})$ and

$$\hat{A} = \begin{bmatrix} 1.000000000e+00 & 1.000000000e-11 & 1.000000000e-21 \\ 1.000000000e-11 & 1.000000000e+00 & 9.090909091e-12 \\ 1.000000000e-21 & 9.090909091e-12 & 1.000000000e+00 \end{bmatrix}$$

with $\|\hat{A} - I_3\| < 1.4 \cdot 10^{-11}$, $\|\hat{A}^{-1}\|_2 \approx 1$. Algorithm 5 computes the eigenvalues of H, as $\lambda_1 \approx 1.0e+40, \lambda_2 \approx 9.900000000000002e+19, \lambda_3 \approx 9.818181818181817e-01$.

3. Smallest eigenvalues can be irreparably damaged simply by computing and storing matrix entries. This rather convincing example is discussed in [DGE99]. The stiffness matrix of a mass spring system with 3 masses,

$$K = \begin{bmatrix} k_1 + k_2 & -k_2 & 0 \\ -k_2 & k_2 + k_3 & -k_3 \\ 0 & -k_3 & k_3 \end{bmatrix}, \quad k_1, k_2, k_3 \text{ spring constants,}$$

is computed with $k_1 = k_3 = 1$ and $k_2 = \epsilon/2$, where ϵ is the round-off unit. Then the true and the computed assembled matrix are, respectively,

$$K = \begin{bmatrix} 1+\epsilon/2 & -\epsilon/2 & 0 \\ -\epsilon/2 & 1+\epsilon/2 & -1 \\ 0 & -1 & 1 \end{bmatrix}, \quad \tilde{K} = \begin{bmatrix} 1 & -\epsilon/2 & 0 \\ -\epsilon/2 & 1 & -1 \\ 0 & -1 & 1 \end{bmatrix}.$$

$\tilde{K}$ is the component-wise relative perturbation of K with reldist$(K, \tilde{K}) = \epsilon/(2+\epsilon) < \epsilon/2$. K is positive definite with minimal eigenvalue near $\epsilon/4$, $\tilde{K}$ is indefinite with minimal eigenvalue near $-\epsilon^2/8$. MATLAB'S function chol(·) fails to compute the Cholesky factorization of $\tilde{K}$ and reports that the matrix is not positive definite.

On the other hand, writing $K = A^TA$ with

$$A = \begin{bmatrix} \sqrt{k_1} & 0 & 0 \\ -\sqrt{k_2} & \sqrt{k_2} & 0 \\ 0 & -\sqrt{k_3} & \sqrt{k_3} \end{bmatrix} = \begin{bmatrix} \sqrt{k_1} & 0 & 0 \\ 0 & \sqrt{k_2} & 0 \\ 0 & 0 & \sqrt{k_3} \end{bmatrix} \begin{bmatrix} 1 & 0 & 0 \\ -1 & 1 & 0 \\ 0 & -1 & 1 \end{bmatrix}$$

clearly separates physical parameters and the geometry of the connections. Since A is bidiagonal, for any choice of k_1, k_2, k_3, the eigenvalues of K can be computed as squared singular values of A to nearly the same number of accurate digits to which the spring constants are given.

46.5 Accurate Eigenvalues of Symmetric Indefinite Matrices

Relevant relative perturbation theory for floating point computation with symmetric indefinite matrices is presented in [BD90], [DV92], [VS93], [DMM00], and [Ves00]. Full review of perturbation theory is given in Chapter 15.

Definitions:

Bunch–Parlett factorization of symmetric H is the factorization $P^T H P = L B L^T$, where P is a permutation matrix, B is a block-diagonal matrix with diagonal blocks of size 1×1 or 2×2, and L is a full column rank unit lower triangular matrix, where the diagonal blocks in L that correspond to 2×2 blocks in B are 2×2 identity matrices.

A **symmetric rank revealing decomposition (SRRD)** of H is a decomposition $H = XDX^T$, where D is diagonal and X is a full column rank well-conditioned matrix.

Let $\mathcal{J}$ denote a nonsingular symmetric matrix. Matrix B is **$\mathcal{J}$-orthogonal** if $F^T \mathcal{J} F = \mathcal{J}$. (Warning: this is a nonstandard usage of F, since in this book F usually denotes a field.)

The **hyperbolic SVD** decomposition of the matrix pair $(G, \mathcal{J})$ is a decomposition of G, $G = W \Sigma F^{-1}$, where W is orthogonal, Σ is diagonal, and F is $\mathcal{J}$-orthogonal.

Facts:

1. If $H = U \Sigma V^T$ is the SVD of an $n \times n$ symmetric H, where $\Sigma = \oplus_{i=1}^m \sigma_{j_i} I_{n_i}$, then the matrix $V^T U$ is block-diagonal with m symmetric and orthogonal blocks of sizes $n_i \times n_i$, $i = 1, \ldots, m$ along its diagonal. The n_i eigenvalues of the ith block are from $\{-1, 1\}$ and they give the signs of n_i eigenvalues of H with absolute value σ_{j_i}.
2. If $H = G \mathcal{J} G^T$ is a factorization with full column rank $G \in \mathbb{R}^{n \times r}$ and $\mathcal{J} = \operatorname{diag}(\pm 1)$, then the eigenvalue problems $Hx = \lambda x$ and $(G^T G) y = \lambda \mathcal{J} y$ are equivalent. If F is the $\mathcal{J}$orthogonal eigenvector matrix of the pencil $G^T G - \lambda \mathcal{J}$ ($F^T \mathcal{J} F = \mathcal{J}$, $F^T (G^T G) F = \Sigma^2 = \operatorname{diag}(\sigma_1^2, \ldots, \sigma_r^2)$), then the matrix $\Sigma^2 \mathcal{J} = \operatorname{diag}(\lambda_1, \ldots, \lambda_r)$ contains the nonzero eigenvalues of H with the columns of $(GF)\Sigma^{-1}$ as corresponding eigenvectors.
3. [Ves00] Let H have factorization $H = G \mathcal{J} G^T$ as in Fact 2. Suppose that H is perturbed implicitly by changing $G \rightsquigarrow G + \delta G$, thus $\tilde{H} = (G + \delta G) \mathcal{J} (G + \delta G)^T$. Write $G = BD$, where D is diagonal and B has unit columns in the Euclidean norm, and let $\delta B = \delta G D^{-1}$. Let $\theta \equiv \|\delta B B^\dagger\|_2 < 1$. Then $\tilde{H}$ has the same number of zero eigenvalues as H and $\max_{\lambda_i \neq 0} \dfrac{|\delta \lambda_i|}{|\lambda_i|} \leq 2\theta + \theta^2$.
4. [Sla98], [Sla02] The factorization $H = G \mathcal{J} G^T$ in Fact 2 is computed by a modification of Bunch–Parlett factorization. Let $\tilde{G}$ be the computed factor and let $\tilde{\mathcal{J}} = \operatorname{diag}(\pm 1)$ be the computed signature matrix. Then
   - A backward stability relation $H + \delta H = \tilde{G} \tilde{\mathcal{J}} \tilde{G}^T$ holds with the entry-wise bound $|\delta H| \leq O(n)\epsilon(|H| + |\tilde{G}||\tilde{G}|^T)$.
   - Let $\tilde{G}$ have rank n. If $\tilde{\mathcal{J}} = \mathcal{J}$ and if $\hat{\lambda}_1 \geq \cdots \geq \hat{\lambda}_n$ are the exact eigenvalues of the pencil $\tilde{G}^T \tilde{G} - \lambda \mathcal{J}$, then, for all i, $|\lambda_i - \hat{\lambda}_i| \leq \dfrac{O(n^2)\epsilon}{\sigma_{\min}^2 (D^{-1} \tilde{G} \tilde{F})} |\lambda_i|$, where D denotes a diagonal matrix such that $D^{-1} \tilde{G}$ has unit rows in Euclidean norm, $\tilde{F}$ is the eigenvector matrix of $\tilde{G}^T \tilde{G} - \lambda \mathcal{J}$, and $\sigma_{\min}(\cdot)$ denotes the minimal singular value of a matrix.

5. [Ves93] The one-sided (implicit) $\mathcal{J}$-symmetric Jacobi algorithm essentially computes the hyperbolic SVD of $(G, \mathcal{J})$, $(W, \Sigma, F) = \text{HSVD}(G, \mathcal{J})$. It follows the structure of the one-sided Jacobi SVD (Algorithm 2) with the following modifications:

   - On input, A is replaced with the pair $(G, \mathcal{J})$.
   - In step k, $G^{(k+1)} = G^{(k)}V^{(k)}$ is computed from $G^{(k)}$ using Jacobi plane rotations exactly as in Algorithm 2 if $\mathcal{J}_{ii}$ and $\mathcal{J}_{jj}$ are of the same sign. (Here, $i = i_k$, $j = j_k$ are the pivot indices in $G^{(k)}$.)
   - If $\mathcal{J}_{ii}$ and $\mathcal{J}_{jj}$ have opposite signs, then the Jacobi rotation is replaced with hyperbolic rotation

$$\begin{bmatrix} \cosh\zeta_k & \sinh\zeta_k \\ \sinh\zeta_k & \cosh\zeta_k \end{bmatrix}, \quad \tanh 2\zeta_k = -\frac{2\xi}{d_i + d_j},$$

   $\xi = (G^{(k)})^T_{1:n,i}(G^{(k)})_{1:n,j}$, $d_\ell = (G^{(k)})^T_{1:n,\ell}(G^{(k)})_{1:n,\ell}$, $\ell = i, j$. The tangent is determined as $\tanh\zeta_k = \dfrac{\text{sign}(\tanh 2\zeta_k)}{|\tanh 2\zeta_k| + \sqrt{\tanh^2 2\zeta_k - 1}}$.

   - The limit of $G^{(k)}$ is $W\Sigma$, and the accumulated product $V^{(0)}V^{(1)}\cdots V^{(k)}\cdots$ converges to $\mathcal{J}$-orthogonal F, and $G = W\Sigma F^{-1}$ is the hyperbolic SVD of G.
   - The iterations are stopped at index k if $\max\limits_{i \neq j} \dfrac{|(G^{(k)})^T_{1:n,i}(G^{(k)})_{1:n,j}|}{\|(G^{(k)})_{1:n,i}\|_2 \|(G^{(k)})_{1:n,j}\|_2} \leq \tau$. The tolerance τ is usually set to $n\epsilon$.

6. [Ves93] The eigenvalue problem of a symmetric indefinite matrix can be implicitly solved as a hyperbolic SVD problem. Algorithm 6 uses the factorization $H = G\mathcal{J}G^T$ (Fact 2) and hyperbolic SVD $\text{HSVD}(G, \mathcal{J})$ (Fact 5) to compute the eigenvalues and eigenvectors of H.

Algorithm 6: Hyperbolic Jacobi

$(\lambda, U) = \text{EIG0}(H)$

Input: $H \in \mathcal{S}_n$

1: $H = G\mathcal{J}G^T$, $\mathcal{J} = I_p \oplus (I_{n-p})$; % *Bunch–Parlett factorization (modified).*

2: $(W, \Sigma, F) = \text{HSVD}(G, \mathcal{J})$ % *One-sided $\mathcal{J}$-symmetric Jacobi algorithm.*

3: $\lambda_i = \mathcal{J}_{ii} \cdot \Sigma^2_{ii}$; $U_{1:n,i} = W_{1:n,i}$, $i = 1, \ldots, n$;

Output: $\lambda = (\lambda_1, \ldots, \lambda_n)$; U

7. [Sla02] Let $\tilde{G}^{(0)} = \tilde{G}$, $\tilde{G}^{(k)}$, $k = 1, 2, \ldots$ be the sequence of matrices computed by the one-sided $\mathcal{J}$-symmetric Jacobi algorithm in floating point arithmetic. Write each $\tilde{G}^{(k)}$ as $\tilde{G}^{(k)} = B^{(k)}D^{(k)}$, where $D^{(k)}$ is the diagonal matrix with Euclidean column norms of $\tilde{G}^{(k)}$ along its diagonal.

   - $\tilde{G}^{(k+1)}$ is the result of an exactly $\mathcal{J}$-orthogonal plane transformation of $\tilde{G}^{(k)} + \delta\tilde{G}^{(k)} = (B^{(k)} + \delta B^{(k)})D^{(k)}$, where $\|\delta B^{(k)}\|_F \leq c_k\epsilon$ with moderate factor c_k.
   - Let $\tilde{\lambda}_1 \geq \cdots \geq \tilde{\lambda}_n$ be computed as $\mathcal{J}_{ii}(\tilde{G}^{(k)})^T_{1:n,i}(\tilde{G}^{(k)})_{1:n,i}$, $i = 1, \ldots, n$, where $\tilde{G}^{(k)}$ is the first matrix which satisfies stopping criterion. If k' is the index of last applied rotation, then $\max\limits_{1 \leq i \leq n} \dfrac{|\hat{\lambda}_i - \tilde{\lambda}_i|}{|\hat{\lambda}_i|} \leq 2\eta + \eta^2$, with $\hat{\lambda}_i$ as in Fact 4, and $\eta = O(\epsilon)\sum\limits_{j=0}^{k'} \|(B^{(j)})^\dagger\|_2 + O(n^2)\epsilon + O(\epsilon^2)$.

8. [DMM03] Accurate diagonalization of indefinite matrices can be derived from their accurate SVD decomposition. (See Fact 1.)

Algorithm 7: Signed SVD

$(\lambda, Q) = \text{EIG1}(H)$

Input: $H \in \mathcal{S}_n$

1: $H = XDY^T$; % *rank revealing decomposition.*

2: $(U, \Sigma, V) = \text{SVD2}(X, D, Y)$; % *accurate SVD.*

3: Recover the signs of the eigenvalues, $\lambda_i = \pm\sigma_i$, using the structure of $V^T U$;

4. Recover the eigenvector matrix Q using the structure of $V^T U$.

Output: $\lambda = (\lambda_1, \ldots, \lambda_n)$; Q

Let $H \approx \tilde{U}\tilde{\Sigma}\tilde{V}^T$, $\tilde{\Sigma} = \text{diag}(\tilde{\sigma}_1, \ldots, \tilde{\sigma}_n)$, be the SVD computed by Algorithm 7.

- If all computed singular values $\tilde{\sigma}_i$, $i = 1, \ldots, n$, are well separated, Algorithm 7 chooses $\tilde{\lambda}_i = \tilde{\sigma}_i \text{sign}(\tilde{V}_{1:n,i}^T \tilde{U}_{1:n,i})$, with the eigenvector $\tilde{Q}_{1:n,i} = \tilde{V}_{1:n,i}$.
- If singular values are clustered, then clusters are determined by perturbation theory and the signs are determined inside each individual cluster.

9. [DMM03] If the rank revealing factorization in Step 1 of Algorithm 7 is computed as $H \approx \tilde{X}\tilde{D}\tilde{Y}^T$ with $\|\tilde{X} - X\| \leq \xi$, $\|\tilde{Y} - Y\| \leq \xi$, $\max_{i=1:n} |d_{ii} - \tilde{d}_{ii}|/|d_{ii}| \leq \xi$, then the computed eigenvalues $\tilde{\lambda}_i$ satisfy $|\lambda_i - \tilde{\lambda}_i| \leq \xi' \kappa(L) \max\{\kappa(X), \kappa(Y)\} |\lambda_i|$, $i = 1, \ldots, n$. Here, ξ and ξ' are bounded by the round-off ϵ times moderate functions of the dimensions, and it is assumed that the traces of the diagonal blocks of $V^T U$ (Cf. Fact 1) are computed correctly.
10. [DK05b] An accurate symmetric rank revealing decomposition $H = XDX^T$ can be given as input to Algorithm 7 or it can replace Step 1 in Algorithm 6 by defining $G = X\sqrt{|D|}$, $\mathcal{J} = \text{diag}(\text{sign}(d_{11}), \ldots, \text{sign}(d_{nn}))$. Once an accurate SRRD is available, the eigenvalues are computed to high relative accuracy. For the following structured matrices, specially tailored algorithms compute accurate SRRDs:
    - Symmetric scaled Cauchy matrices, $H = DCD$, D diagonal, C Cauchy matrix.
    - Symmetric Vandermonde matrices, $V = (\nu^{(i-1)(j-1)})_{i,j=1}^n$, where $\nu \in \mathbb{R}$.
    - Symmetric totally nonnegative matrices.

Examples:

1. [Ves96] Initial factorization or rank revealing decomposition is the key for success or failure of the algorithms presented in this section. Let $\varepsilon = 10^{-7}$ and

$$H = \begin{bmatrix} 1 & 1 & 1 \\ 1 & 0 & 0 \\ 1 & 0 & \varepsilon \end{bmatrix} = \begin{bmatrix} 1 & 0 & 0 \\ 0 & 1 & 0 \\ 0 & 1 & \sqrt{\varepsilon} \end{bmatrix} \begin{bmatrix} 0 & 1 & 0 \\ 1 & 0 & 0 \\ 0 & 0 & 1 \end{bmatrix} \begin{bmatrix} 1 & 0 & 0 \\ 0 & 1 & 1 \\ 0 & 0 & \sqrt{\varepsilon} \end{bmatrix}.$$

This factorization implies that all eigenvalues of H are determined to high relative accuracy. Whether or not they will be determined to that accuracy by Algorithm 6 (or any other algorithm depending on initial factorization) depends on the factorization. In Algorithm 6, the factorization in Step 1 chooses to start with 1×1 pivot, which after the first step gives

$$\begin{bmatrix} 1 & 0 & 0 \\ 1 & 1 & 0 \\ 1 & 0 & 1 \end{bmatrix} \begin{bmatrix} 1 & 0 & 0 \\ 0 & -1 & -1 \\ 0 & -1 & -1+\varepsilon \end{bmatrix} \begin{bmatrix} 1 & 1 & 1 \\ 0 & 1 & 0 \\ 0 & 0 & 1 \end{bmatrix}.$$

The 2×2 Schur complement is ill-conditioned (entry-wise ε close to singularity, the condition number behaving as $1/\varepsilon$) and the smallest eigenvalue is lost.

Author Note: This work is supported by the Croatian Ministry of Science, Education and Sports (Grant 0037120) and by a Volkswagen–Stiftung grant.

References

[BD90] J. Barlow and J. Demmel. Computing accurate eigensystems of scaled diagonally dominant matrices. *SIAM J. Num. Anal.*, 27(3):762–791, 1990.

[Dem89] J. Demmel. On floating point errors in Cholesky. LAPACK Working Note 14, Computer Science Department, University of Tennessee, October 1989.

[Dem92] J. Demmel. The component-wise distance to the nearest singular matrix. *SIAM J. Matrix Anal. Appl.*, 13(1):10–19, 1992.

[Dem99] J. Demmel. Accurate singular value decompositions of structured matrices. *SIAM J. Matrix Anal. Appl.*, 21(2):562–580, 1999.

[DG93] J. Demmel and W. Gragg. On computing accurate singular values and eigenvalues of acyclic matrices. *Lin. Alg. Appl.*, 185:203–218, 1993.

[DGE99] J. Demmel, M. Gu, S. Eisenstat, I. Slapničar, K. Veselić, and Z. Drmač. Computing the singular value decomposition with high relative accuracy. *Lin. Alg. Appl.*, 299:21–80, 1999.

[DK90] J. Demmel and W. Kahan. Accurate singular values of bidiagonal matrices. *SIAM J. Sci. Stat. Comp.*, 11(5):873–912, 1990.

[DK04] J. Demmel and P. Koev. Accurate SVDs of weakly diagonally dominant M-matrices. *Num. Math.*, 98:99–104, 2004.

[DK05a] J. Demmel and P. Koev. Accurate SVDs of polynomial Vandermonde matrices involving orthonormal polynomials. *Lin. Alg. Appl.*, (to appear).

[DK05b] F.M. Dopico and P. Koev. Accurate eigendecomposition of symmetric structured matrices. *SIAM J. Matrix Anal. Appl.*, (to appear).

[DM04] F.M. Dopico and J. Moro. A note on multiplicative backward errors of accurate SVD algorithms. *SIAM J. Matrix Anal. Appl.*, 25(4):1021–1031, 2004.

[DMM00] F.M. Dopico, J. Moro, and J.M. Molera. Weyl-type relative perturbation bounds for eigensystems of Hermitian matrices. *Lin. Alg. Appl.*, 309:3–18, 2000.

[DMM03] F.M. Dopico, J.M. Molera, and J. Moro. An orthogonal high relative accuracy algorithm for the symmetric eigenproblem. *SIAM J. Matrix Anal. Appl.*, 25(2):301–351, 2003.

[dR89] P.P.M. de Rijk. A one-sided Jacobi algorithm for computing the singular value decomposition on a vector computer. *SIAM J. Sci. Stat. Comp.*, 10(2):359–371, 1989.

[Drm94] Z. Drmač. *Computing the Singular and the Generalized Singular Values.* Ph.D. thesis, Lehrgebiet Mathematische Physik, Fernuniversität Hagen, Germany, 1994.

[Drm97] Z. Drmač. Implementation of Jacobi rotations for accurate singular value computation in floating point arithmetic. *SIAM J. Sci. Comp.*, 18:1200–1222, 1997.

[Drm98a] Z. Drmač. Accurate computation of the product induced singular value decomposition with applications. *SIAM J. Numer. Anal.*, 35(5):1969–1994, 1998.

[Drm98b] Z. Drmač. A tangent algorithm for computing the generalized singular value decomposition. *SIAM J. Num. Anal.*, 35(5):1804–1832, 1998.

[Drm99] Z. Drmač. *A posteriori* computation of the singular vectors in a preconditioned Jacobi SVD algorithm. *IMA J. Num. Anal.*, 19:191–213, 1999.

[Drm00a] Z. Drmač. New accurate algorithms for singular value decomposition of matrix triplets. *SIAM J. Matrix Anal. Appl.*, 21(3):1026–1050, 2000.

[Drm00b] Z. Drmač. On principal angles between subspaces of Euclidean space. *SIAM J. Matrix Anal. Appl.*, 22:173–194, 2000.

[DV92] J. Demmel and K. Veselić. Jacobi's method is more accurate than QR. *SIAM J. Matrix Anal. Appl.*, 13(4):1204–1245, 1992.

[DV05a] Z. Drmač and K. Veselić. New fast and accurate Jacobi SVD algorithm: I. Technical report, Department of Mathematics, University of Zagreb, Croatia, June 2005. LAPACK Working Note 169.

[DV05b] Z. Drmač and K. Veselić. New fast and accurate Jacobi SVD algorithm: II. Technical report, Department of Mathematics, University of Zagreb, Croatia, June 2005. LAPACK Working Note 170.

[FP94] K.V. Fernando and B.N. Parlett. Accurate singular values and differential qd algorithms. *Num. Math.*, 67:191–229, 1994.

[Hes58] M.R. Hestenes. Inversion of matrices by biorthogonalization and related results. *J. SIAM*, 6(1):51–90, 1958.

[Hig96] N.J. Higham. *Accuracy and Stability of Numerical Algorithms.* SIAM, Philadelphia, 1996.

[Hig00] N.J. Higham. QR factorization with complete pivoting and accurate computation of the SVD. *Lin. Alg. Appl.*, 309:153–174, 2000.

[Jac46] C.G.J. Jacobi. Über ein leichtes Verfahren die in der Theorie der Säcularstörungen vorkommenden Gleichungen numerisch aufzulösen. *Crelle's Journal für Reine und Angew. Math.*, 30:51–95, 1846.

[Koe05] P. Koev. Accurate eigenvalues and SVDs of totally nonnegative matrices. *SIAM J. Matrix Anal. Appl.*, 27(1):1–23, 2005.

[Mat96] R. Mathias. Accurate eigensystem computations by Jacobi methods. *SIAM J. Matrix Anal. Appl.*, 16(3):977–1003, 1996.

[Sla98] I. Slapničar. Component-wise analysis of direct factorization of real symmetric and Hermitian matrices. *Lin. Alg. Appl.*, 272:227–275, 1998.

[Sla02] I. Slapničar. Highly accurate symmetric eigenvalue decomposition and hyperbolic SVD. *Lin. Alg. Appl.*, 358:387–424, 2002.

[vdS69] A. van der Sluis. Condition numbers and equilibration of matrices. *Num. Math.*, 14:14–23, 1969.

[Ves93] K. Veselić. A Jacobi eigenreduction algorithm for definite matrix pairs. *Num. Math.*, 64:241–269, 1993.

[Ves96] K. Veselić. Note on the accuracy of symmetric eigenreduction algorithms. *ETNA*, 4:37–45, 1996.

[Ves00] K. Veselić. Perturbation theory for the eigenvalues of factorised symmetric matrices. *Lin. Alg. Appl.*, 309:85–102, 2000.

[VH89] K. Veselić and V. Hari. A note on a one-sided Jacobi algorithm. *Num. Math.*, 56:627–633, 1989.

[VS93] K. Veselić and I. Slapničar. Floating point perturbations of Hermitian matrices. *Lin. Alg. Appl.*, 195:81–116, 1993.

Computational Linear Algebra

47

Fast Matrix Multiplication

Dario A. Bini
Università di Pisa

Multiplying matrices is an important problem from both the theoretical and the practical point of view. Determining the arithmetic complexity of this problem, that is, the minimum number of arithmetic operations sufficient for computing an $n \times n$ matrix product, is still an open issue. Other important computational problems like computing the inverse of a nonsingular matrix, solving a linear system, computing the determinant, or more generally the coefficients of the characteristic polynomial of a matrix have a complexity related to that of matrix multiplication. Certain combinatorial problems, like the all pair shortest distance problem of a digraph, are strictly related to matrix multiplication. This chapter deals with fast algorithms for multiplication of unstructured matrices. Fast algorithms for structured matrix computations are presented in Chapter 48.

47.1 Basic Concepts

Let $A = [a_{i,j}]$, $B = [b_{i,j}]$, and $C = [c_{i,j}]$ be $n \times n$ matrices over the field F such that $C = AB$, that is, C is the matrix product of A and B.

Facts:

1. The elements of the matrices A, B, and C are related by the following equations:

$$c_{i,j} = \sum_{k=1}^{n} a_{i,k} b_{k,j}, \quad i, j = 1, \ldots, n.$$

2. Each element $c_{i,j}$ is the scalar product of the ith row of A and the jth column of B and can be computed by means of n multiplications and $n - 1$ additions. This computation is described in Algorithm 1.
3. The overall cost of computing the n^2 elements of C is n^3 multiplications and $n^3 - n^2$ additions, that is $2n^3 - n^2$ arithmetic operations.

Algorithm 1. Conventional matrix multiplication
Input: the elements $a_{i,j}$ and $b_{i,j}$ of two $n \times n$ matrices A and B;
Output: the elements $c_{i,j}$ of the matrix product $C = AB$;
for $i = 1$ **to** n **do**
  for $j = 1$ **to** n **do**
    $c_{i,j} = 0$
    for $k = 1$ **to** n **do**
      $c_{i,j} = c_{i,j} + a_{i,k} b_{k,j}$

4. [Hig02, p. 71] The computation of $c_{i,j}$ by means of Algorithm 1 is element-wise forward numerically stable. More precisely, if $\widetilde{C}$ denotes the matrix actually computed by performing Algorithm 1 in floating point arithmetic with machine precision ϵ, then $|C - \widetilde{C}| \leq n\epsilon |A||B| + O(\epsilon^2)$, where $|A|$ denotes the matrix with elements $|a_{i,j}|$ and the inequality holds element-wise.

Examples:

1. For 2×2 matrices Algorithm 1 requires 8 multiplications and 4 additions. For 5×5 matrices Algorithm 1 requires 125 multiplications and 100 additions.

47.2 Fast Algorithms

Definitions:

Define $\omega \in \mathbb{R}$ as the infimum of the real numbers w such that there exists an algorithm for multiplying $n \times n$ matrices with $O(n^w)$ arithmetic operations. ω is called the **exponent of matrix multiplication complexity**.

Algorithms that do not use commutativity of multiplication, are called **noncommutative algorithms.**

Algorithms for multiplying the $p \times p$ matrices A and B of the form

$$m_k = \left(\sum_{i,j=1}^{p} \alpha_{i,j,k} a_{i,j} \right) \left(\sum_{i,j=1}^{p} \beta_{i,j,k} b_{i,j} \right), \quad k = 1, \ldots t,$$

$$c_{i,j} = \sum_{k=1}^{t} \gamma_{i,j,k} m_k,$$

where $\alpha_{i,j,k}, \beta_{i,j,k}, \gamma_{i,j,k}$ are given scalar constants, are called **bilinear noncommutative algorithms** with t **nonscalar multiplications.**

Facts:

1. [Win69] *Winograd's commutative algorithm.* For moderately large values of n, it is possible to compute the product of $n \times n$ matrices with less than $2n^3 - n^2$ arithmetic operations by means of the following simple identities where $n = 2m$ is even:

$$u_i = \sum_{k=1}^{m} a_{i,2k-1} a_{i,2k}, \quad v_i = \sum_{k=1}^{m} b_{2k-1,j} b_{2k,j}, \quad i = 1, n,$$

$$w_{i,j} = \sum_{k=1}^{m} (a_{i,2k-1} + b_{2k,j})(a_{i,2k} + b_{2k-1,j}), \quad i, j = 1, n,$$

$$c_{i,j} = w_{i,j} - u_i - b_j, \quad i, j = 1, n.$$

2. The number of arithmetic operations required is $n^3 + 4n^2 - 2n$, which is less than $2n^3 - n^2$ already for $n \geq 8$. This formula, which for large values of n is faster by a factor of about 2 with respect to the

conventional algorithm, relies on the commutative property of multiplication. It can be extended to the case where n is odd.

3. [Str69] *Strassen's formula.* It is possible to multiply 2×2 matrices with only 7 multiplications instead of 8, but with a higher number of additions by means of the following identities:
$$m_1 = (a_{1,2} - a_{2,2})(b_{2,1} + b_{2,2}), \qquad m_5 = a_{1,1}(b_{1,2} - b_{2,2}),$$
$$m_2 = (a_{1,1} + a_{2,2})(b_{1,1} + b_{2,2}), \qquad m_6 = a_{2,2}(b_{2,1} - b_{1,1}),$$
$$m_3 = (a_{1,1} - a_{2,1})(b_{1,1} + b_{1,2}), \qquad m_7 = (a_{2,1} + a_{2,2})b_{1,1},$$
$$m_4 = (a_{1,1} + a_{1,2})b_{2,2},$$
$$c_{1,1} = m_1 + m_2 - m_4 + m_6, \quad c_{1,2} = m_4 + m_5,$$
$$c_{2,1} = m_6 + m_7, \qquad c_{2,2} = m_2 - m_3 + m_5 - m_7.$$
4. The overall number of arithmetic operations required by the Strassen formula is higher than the number of arithmetic operations required by the conventional matrix multiplication described in Algorithm 1 of section 47.1. However, the decrease from 8 to 7 of the number of multiplications provides important consequences.
5. The identities of Fact 3 do not exploit the commutative property of multiplication like the identities of Fact 1, therefore they are still valid if the scalar factors are replaced by matrices.
6. Strassen's formula provides a bilinear noncommutative algorithm where the constants $\alpha_{i,j,k}$, $\beta_{i,j,k}$, $\gamma_{i,j,k}$ are in the set $\{0, 1, -1\}$.
7. If n is even, say $n = 2m$, then A, B, and C can be partitioned into four $m \times m$ blocks, that is,
$$A = \begin{bmatrix} A_{1,1} & A_{1,2} \\ A_{2,1} & A_{2,2} \end{bmatrix}, \quad B = \begin{bmatrix} B_{1,1} & B_{1,2} \\ B_{2,1} & B_{2,2} \end{bmatrix}, \quad A = \begin{bmatrix} C_{1,1} & C_{1,2} \\ C_{2,1} & C_{2,2} \end{bmatrix},$$
where $A_{i,j}, B_{i,j}, C_{i,j} \in F^{m \times m}$, $i, j = 1, 2$, so that an $n \times n$ matrix product can be viewed as a 2×2 block matrix product. More specifically it holds that
$$C_{1,1} = A_{1,1}B_{1,1} + A_{1,2}B_{2,1}, \quad C_{1,2} = A_{1,1}B_{1,2} + A_{1,2}B_{2,2},$$
$$C_{2,1} = A_{2,1}B_{1,1} + A_{2,2}B_{2,1}, \quad C_{2,2} = A_{2,1}B_{1,2} + A_{2,2}B_{2,2}.$$
8. If $n = 2m$, the four blocks $C_{i,j}$ of Fact 7 can be computed by means of Strassen's formula with $7\ m \times m$ matrix multiplications and $18\ m \times m$ matrix additions, i.e., with $7(2m^3 - m^2) + 18m^2$ arithmetic operations. The arithmetic cost of matrix multiplication is reduced roughly by a factor of 7/8.
9. [Str69] *Strassen's algorithm.* Furthermore, if m is even then the seven $m \times m$ matrix products of Fact 8 can be computed once again by means of Strassen's formula. If $n = 2^k$, k a positive integer, Strassen's formula can be repeated recursively until the size of the blocks is 1. Algorithm 2 synthesizes this computation.
10. The number $M(k)$ of arithmetic operations required by Algorithm 2 to multiply $2^k \times 2^k$ matrices is such that
$$M(k) = 7M(k-1) + 18\left(2^{k-1}\right)^2,$$
$$M(0) = 1,$$
which provides $M(k) = 7 \cdot 7^k - 6 \cdot 4^k = 7n^{\log_2 7} - 6n^2$, where $n = 2^k$ and $\log_2 7 = 2.8073\ldots < 3$. This yields the bound $\omega \leq \log_2 7$ on the exponent ω of matrix multiplication complexity.
11. In practice it is not convenient to carry out Strassen's algorithm up to matrices of size 1. In fact, for 2×2 matrices Strassen's formula requires much more operations than the conventional multiplication formula (see Example 1). Therefore, in the actual implementation the recursive iteration of Strassen's algorithm is stopped at size $p = 2^r$, where p is the largest value such that the conventional method applied to $p \times p$ matrices is faster than Strassen's method.

12. Strassen's algorithm can be carried out even though n is not an integer power of 2. Assume that $2^{k-1} < n < 2^k$, set $p = 2^k - n$, and embed the matrices A, B, C into matrices $\widehat{A}, \widehat{B}, \widehat{C}$ of size 2^k in the following way:

$$\widehat{A} = \begin{bmatrix} A & \mathbf{0}_{np} \\ \mathbf{0}_{pn} & \mathbf{0}_{pp} \end{bmatrix}, \widehat{B} = \begin{bmatrix} B & \mathbf{0}_{np} \\ \mathbf{0}_{pn} & \mathbf{0}_{pp} \end{bmatrix}, \widehat{C} = \begin{bmatrix} C & \mathbf{0}_{np} \\ \mathbf{0}_{pn} & \mathbf{0}_{pp} \end{bmatrix}.$$

Then one has $\widehat{C} = \widehat{A}\widehat{B}$ so that Strassen's algorithm can be applied to $\widehat{A}$ and $\widehat{B}$ in order to compute C. Even in this case the cost of Strassen's algorithm is still $O(n^{\log_2 7})$.

Algorithm 2. Strassen's algorithm

Procedure Strassen(k,A,B)
Input: the elements $a_{i,j}$ and $b_{i,j}$ of the $2^k \times 2^k$ matrices A and B;
Output: the elements $c_{i,j}$ of the $2^k \times 2^k$ matrix $C = AB$;
If $k = 0$ **then**
  output $c_{1,1} = a_{1,1}b_{1,1}$;
else
  partition A, B, and C into $2^{k-1} \times 2^{k-1}$ blocks $A_{i,j}, B_{i,j}, C_{i,j}$, respectively, where $i, j = 1, 2$;
  compute

$$\begin{array}{llll} P_1 = A_{1,2} - A_{2,2}, & Q_1 = B_{2,1} + B_{2,2}, & P_5 = A_{1,1}, & Q_5 = B_{1,2} - B_{2,2}, \\ P_2 = A_{1,1} + A_{2,2}, & Q_2 = B_{1,1} + B_{2,2}, & P_6 = A_{2,2}, & Q_6 = B_{2,1} - B_{1,1}, \\ P_3 = A_{1,1} - A_{2,1}, & Q_3 = B_{1,1} + B_{1,2}, & P_7 = A_{2,1} + A_{2,2}, & Q_7 = B_{1,1}, \\ P_4 = A_{1,1} + A_{1,2}, & Q_4 = B_{2,2}, & & \end{array}$$

  for $i = 1$ **to** 7 **do**
    $M_i =$ Strassen $(k-1, P_i, Q_i)$;
  compute

$$\begin{array}{ll} C_{1,1} = M_1 + M_2 - M_4 + M_6, & C_{1,2} = M_4 + M_5, \\ C_{2,1} = M_6 + M_7, & C_{2,2} = M_2 - M_3 + M_5 - M_7. \end{array}$$

13. [Hig02, p. 440] *Numerical stability of Strassen's algorithm.* Let $\widetilde{C}$ be the $n \times n$ matrix obtained by performing Strassen's algorithm in floating point arithmetic with machine precision ϵ where $n = 2^k$. Then the following bound holds: $\max_{i,j} |\widetilde{c}_{i,j} - c_{i,j}| \le \gamma_n \epsilon \max_{i,j} |a_{i,j}| \max_{i,j} |b_{i,j}| + O(\epsilon^2)$, where $\gamma_n = 6n^{\log_2 12}$ and $\log_2 12 \approx 3.585$. Thus, Strassen's algorithm has slightly less favorable stability properties than the conventional algorithm: the error bound does not hold component-wise but only norm-wise, and the multiplicative factor $6n^{\log_2 12}$ is larger than the factor n^2 of the bound given in Fact 4 of section 47.1.

Examples:

1. For $n = 16$, applying the basic Strassen algorithm to the 2×2 block matrices with blocks of size 8 and computing the seven products with the conventional algorithm requires $7 * (2 * 8^3 - 8^2) + 18 * 8^2 = 7872$ arithmetic operations. Using the conventional algorithm requires $2 * 16^3 - 16^2 = 7936$ arithmetic operations. Thus, it is convenient to stop the recursion of Algorithm 2 when $n = 16$. A similar analysis can be performed if the Winograd commutative formula of Fact 1 is used.

47.3 Other Algorithms

Facts:

1. [Win71] *Winograd's formula.* The following identities enable one to compute the product of 2×2 matrices with 7 multiplications and with 15 additions; this is the minimum number of additions among bilinear noncommutative algorithms for multiplying 2×2 matrices with 7 multiplications:
$$s_1 = a_{2,1} + a_{2,2}, \quad s_2 = s_1 - a_{1,1}, \quad s_3 = a_{1,1} - a_{2,1}, \quad s_4 = a_{1,2} - s_2,$$
$$t_1 = b_{1,2} - b_{1,1}, \quad t_2 = b_{2,2} - t_1, \quad t_3 = b_{2,2} - b_{1,2}, \quad t_4 = t_2 - b_{2,1},$$
$$m_1 = a_{1,1}b_{1,1}, \quad m_2 = a_{1,2}b_{2,1}, \quad m_3 = s_4 b_{2,2}, \quad m_4 = a_{2,2}t_4,$$
$$m_5 = s_1 t_1, \quad m_6 = s_2 t_2, \quad m_7 = s_3 t_3,$$
$$u_1 = m_1 + m_6, \quad u_2 = u_1 + m_7, \quad u_3 = u_1 + m_5,$$
$$c_{1,1} = m_1 + m_2, \quad c_{1,2} = u_3 + m_3,$$
$$c_{2,1} = u_2 - m_4, \quad c_{2,2} = u_2 + m_5.$$
The numerical stability of the recursive version of Winograd's formula is slightly inferior since the error bound of Fact 13 of Section 47.2 holds with $\gamma_n = 12n^{\log_2 18}$, $\log_2 18 \approx 4.17$.
2. [Win71], [BD78], [AS81] No algorithm exists for multiplying 2×2 matrices with less than 7 multiplications. The number of nonscalar multiplications needed for multiplying $n \times n$ matrices is at least $2n^2 - 1$.
3. If $n = 3^k$, the matrices A, B, and C can be partitioned into 9 blocks of size $n/3$ so that $n \times n$ matrix multiplication is reduced to computing the product of 3×3 block matrices. Formulas for 3×3 matrix multiplication that do not use the commutative property can be recursively used for general $n \times n$ matrix multiplication.
4. In general, if there exists a bilinear noncommutative formula for computing the product of $q \times q$ matrices with t nonscalar multiplications, then matrices of size $n = q^k$ can be multiplied with the cost of $O(q^t) = O(n^{\log_q t})$ arithmetic operations.
5. There exist algorithms for multiplying 3×3 matrices that require 25 multiplications [Gas71], 24 multiplications [Fid72], 23 multiplications [Lad76]. None of these algorithms beats Strassen's algorithm since $\log_3 21 < \log_2 7 < \log_3 22$. No algorithm is known for multiplying 3×3 matrices with less than 23 nonscalar multiplications.
6. [HM73] *Rectangular matrix multiplication and the duality property.* If there exists a bilinear noncommutative algorithm for multiplying two (rectangular) matrices of size $n_1 \times n_2$ and $n_2 \times n_3$, respectively, with t nonscalar multiplications, then there exist bilinear noncommutative algorithms for multiplying matrices of size $n_{\sigma_1} \times n_{\sigma_2}$ and $n_{\sigma_2} \times n_{\sigma_3}$ with t nonscalar multiplications for any permutation $\sigma = (\sigma_1, \sigma_2, \sigma_3)$.
7. If there exists a bilinear noncommutative algorithm for multiplying $n_1 \times n_2$ and $n_2 \times n_3$ matrices with t nonscalar multiplications, then square matrices of size $q = n_1 n_2 n_3$ can be multiplied with t^3 multiplications.
8. From Fact 7 and Fact 4 it follows that if there exists a bilinear noncommutative algorithm for multiplying $n_1 \times n_2$ and $n_2 \times n_3$ matrices with t nonscalar multiplications, then $n \times n$ matrices can be multiplied with $O(n^w)$ arithmetic operations, where $w = \log_{n_1 n_2 n_3} t^3$.
9. [HK71] There exist bilinear noncommutative algorithms for multiplying matrices of size 2×2 and 2×3 with 11 multiplications; there exist algorithms for multiplying matrices of size 2×3 and 3×3 with 15 multiplications.
10. There are several implementations of fast algorithms for matrix multiplication based on Strassen's formula and on Winograd's formula. In 1970, R. Brent implemented Strassen's algorithm on an IBM 360/67 (see [Hig02, p. 436]). This implementation was faster than the conventional algorithm already for $n \geq 110$. In 1988, D. Bailey provided a Fortran implementation for the Cray-2. Fortran codes, based on the Winograd variant have been provided since the late 1980s. For detailed comments and for more bibliography in this regard, we refer the reader to [Hig02, Sect. 23.3].

47.4 Approximation Algorithms

Matrices can be multiplied faster if we allow that the matrix product can be affected by some arbitrarily small nonzero error. Throughout this section, the underlying field F is $\mathbb{R}$ or $\mathbb{C}$ and we introduce a parameter $\lambda \in F$ that represents a nonzero number with small modulus. Multiplication by λ and λ^{-1} is negligible in the complexity estimate for two reasons: firstly, by choosing λ equal to a power of 2, multiplication by λ can be accomplished by shifting the exponent in the base-two representation of floating point numbers. This operation has a cost lower than the cost of multiplication. Secondly, in the block application of matrix multiplication algorithms, multiplication by λ corresponds to multiplying an $m \times m$ matrix by the scalar λ. This operation costs only m^2 arithmetic operations like matrix addition.

Definitions:

Algorithms for multiplying the $p \times p$ matrices A and B of the form

$$m_k = \left(\sum_{i,j=1}^{p} \alpha_{i,j,k} a_{i,j}\right)\left(\sum_{i,j=1}^{p} \beta_{i,j,k} a_{i,j}\right), \quad k = 1, \dots t,$$

$$c_{i,j} = \sum_{k=1}^{t} \gamma_{i,j,k} m_k + \lambda p_{i,j}(\lambda),$$

where $\alpha_{i,j,k}, \beta_{i,j,k}, \gamma_{i,j,k}$ are given rational functions of λ and $p_{i,j}(\lambda)$ are polynomials, are called **Arbitrary Precision Approximating (APA) algorithms** with t **nonscalar multiplications** [Bin80], [BCL79].

Facts:

1. [BLR80] The matrix-vector product

$$\begin{bmatrix} f_1 \\ f_2 \end{bmatrix} = \begin{bmatrix} a & b \\ 0 & a \end{bmatrix} \begin{bmatrix} x \\ y \end{bmatrix} = \begin{bmatrix} ax + by \\ ay \end{bmatrix}$$

cannot be computed with less than three multiplications. However, the following APA algorithm approximates f_1 and f_2 with two nonscalar multiplications:

$$\begin{aligned} &m_1 = (a + \lambda b)(x + \lambda^{-1} y), \quad m_2 = ay, \\ &f_1 \approx f_1 + \lambda bx = m_1 - \lambda^{-1} m_2, \\ &f_2 = m_2. \end{aligned}$$

The algorithm is not defined for $\lambda = 0$, but for $\lambda \to 0$ the output of the algorithm converges to the exact solution if performed in exact arithmetic.

2. [BCL79] Consider the 2×2 matrix product $C = AB$ where $a_{1,2} = 0$, i.e.,

$$\begin{bmatrix} c_{1,1} & c_{1,2} \\ c_{2,1} & c_{2,2} \end{bmatrix} = \begin{bmatrix} a_{1,1} & 0 \\ a_{2,1} & a_{2,2} \end{bmatrix} \begin{bmatrix} b_{1,1} & b_{1,2} \\ b_{2,1} & b_{2,2} \end{bmatrix}.$$

The elements $c_{i,j}$ can be approximated with 5 nonscalar multiplications by means of the following identities:

$$\begin{aligned} &m_1 = (a_{1,1} + a_{2,2})(b_{1,2} + \lambda b_{2,1}), \quad m_2 = a_{1,1}(b_{1,1} + \lambda b_{2,1}), \quad m_3 = a_{2,2}(b_{1,2} + \lambda b_{2,2}), \\ &m_4 = (a_{1,1} + \lambda a_{2,1})(b_{1,1} - b_{1,2}), \quad m_5 = (\lambda a_{2,1} - a_{2,2}) b_{1,2}, \\ &c_{1,1} = m_2 - \lambda a_{1,1} b_{2,1}, \qquad c_{1,2} = m_1 - m_3 - \lambda(a_{1,1} b_{2,1} + a_{2,2} b_{2,1} - a_{2,2} b_{2,2}), \\ &c_{2,1} = \lambda^{-1}(m_1 - m_2 + m_4 + m_5), \quad c_{2,2} = \lambda^{-1}(m_3 + m_5). \end{aligned}$$

3. Formulas of Fact 2 can be suitably adjusted to the case where only the element $a_{2,1}$ of the matrix A is zero.

4. The product of a 3×2 matrix and a 2×2 matrix can be approximated with 10 nonscalar multiplications by simply combining the formulas of Fact 2 and of Fact 3 in the following way:

$$\begin{bmatrix} \cdot & \cdot \\ \cdot & \cdot \\ \cdot & \cdot \end{bmatrix} = \left(\begin{bmatrix} \# & \# \\ 0 & \# \\ 0 & 0 \end{bmatrix} + \begin{bmatrix} 0 & 0 \\ \star & 0 \\ \star & \star \end{bmatrix} \right) \begin{bmatrix} \cdot & \cdot \\ \cdot & \cdot \end{bmatrix}.$$

5. Facts 4, 7, and 8 of section 47.3 are still valid for APA algorithms. By Fact 7 of Section 47.3 it follows that 1000 nonscalar multiplications are sufficient to approximate the product of 12×12 matrices.
6. By Fact 8 of Section 47.3 it follows that $O(n^{\log_{12} 1000})$ arithmetic operations are sufficient to approximate the product of $n \times n$ matrices, where $\log_{12} 1000 = 2.7798 \ldots < \log_2 7$.
7. [BLR80] A rounding error analysis of an APA algorithm shows that the relative error in the output is bounded by $\alpha\lambda^h + \epsilon\beta\lambda^{-k}$, where α and β are positive constant depending on the input values, h and k are positive constants depending on the algorithm and ϵ is the machine precision. This bound grows to infinity as λ converges to zero. Asymptotically, in ϵ the minimum bound is $\gamma\epsilon^{\frac{h}{h+k}}$, for a constant γ.
8. [Bin80] *From approximate to exact computations.* Given an APA algorithm for approximating a $k \times k$ matrix product with t nonscalar multiplications, there exists an algorithm for $n \times n$ *exact* matrix multiplication requiring $O(n^w \log n)$ arithmetic operations where $w = \log_k t$. In particular, by Fact 5 an $O(n^w \log n)$ algorithm exists for exact matrix multiplication with $w = \log_{12} 1000$. This provides the bound $\omega \leq \log_{12} 1000$ for the exponent ω of matrix multiplication defined in Section 47.2.
9. The exact algorithm is obtained from the approximate algorithm by applying the approximate algorithm with $O(\log n)$ different values (not necessarily small) of λ and then taking a suitable linear combination of the $O(\log n)$ different values obtained in this way in order to get the exact product. More details on this approach, which is valid for any APA algorithm that does not use the commutative property of multiplication, can be found in [Bin80].

Examples:

1. The algorithm of Fact 1 computes $f_1(\lambda) = f_1 + \lambda bx$ that is close to f_1 for a small lambda. On the other hand, one has $f_1 = (f_1(1) + f_1(-1))/2$; i.e., a linear combination of the values computed by the APA algorithm with $\lambda = 1$ and with $\lambda = -1$ provides exactly f_1.

47.5 Advanced Techniques

More advanced techniques have been introduced for designing fast algorithms for $n \times n$ matrix multiplication. The asymptotically fastest algorithm currently known requires $O(n^{2.38})$ arithmetic operations, but it is faster than the conventional algorithm only for huge values of n. Finding the infimum ω of the numbers w for which there exist $O(n^w)$ complexity algorithms is still an open problem. In this section, we provide a list of the main techniques used for designing asymptotically fast algorithms for $n \times n$ matrix multiplication.

Facts:

1. [Pan78] *Trilinear aggregating technique.* Different schemes for (approximate) matrix multiplication are based on the technique of trilinear aggregating by V. Pan. This technique is very versatile: Different algorithms based on this technique have been designed for fast matrix multiplication of several sizes [Pan84]; in particular, an algorithm for multiplying 70×70 matrices with 143,640 multiplications which leads to the bound $\omega < 2.795$.

2. [Sch81] *Partial matrix multiplication.* In the expression

$$c_{i,j} = \sum_{r=1}^{m} a_{i,j} b_{i,j}, \quad i, j = 1, \ldots, m,$$

there are m^3 terms which are summed up. A partial matrix multiplication is encountered if some $a_{i,j}$ or some $b_{i,j}$ are zero, or if not all the elements $c_{i,j}$ are computed so that in the above expression there are less than m^3 terms, say, $k < m^3$. A. Schönhage [Sch81] has proved that if there exists a noncommutative bilinear (APA) algorithm for computing (approximating) partial $m \times m$ matrix multiplication with k terms that uses t nonscalar multiplications, then $\omega \leq 3 \log_k t$. This result applied to the formula of Fact 2 of Section 47.4 provides the bound $\omega \leq 3 \log_6 5 < 2.695$.
3. [Sch81] *Disjoint matrix multiplication* A. Schönhage has proven that it is possible to approximate two disjoint matrix products with less multiplications than the number of multiplications needed for computing separately these products. In particular, he has provided an APA algorithm for simultaneously multiplying a 4×1 matrix by a 1×4 and a 1×9 matrix by a 9×1 with only 17 nonscalar multiplications. Observe that 16 multiplications are *needed* for the former product and 9 multiplications are *needed* for the latter.
4. [Sch81] *The τ-theorem.* A. Schönhage has proven the τ-theorem. Namely, if the set of disjoint matrix products of size $m_i \times n_i$ times $n_i \times p_i$, for $i = 1, \ldots, k$ can be approximated with t nonscalar multiplications by a bilinear noncommutative APA algorithm, then $\omega \leq 3\tau$, where τ solves the equation $\sum_{i=1}^{k} (m_i n_i p_i)^\tau = t$. From the disjoint matrix multiplication of Fact 3 it follows the equation $16^\tau + 9^\tau = 17$, which yields $\omega \leq 3\tau = 2.5479\ldots$.
5. [Str87], [Str88] *Asymptotic spectrum: the laser method.* In 1988, V. Strassen introduced a powerful and sophisticated method, which, by taking tensor powers of set of bilinear forms that apparently are not completely related to matrix multiplication, provides some scheme for fast matrix multiplication. The name *laser method* was motivated from the fact that by tensor powering a set of "incoherent" bilinear forms it is possible to obtain a "coherent" set of bilinear forms.
6. *Lower bounds.* At least $2n^2 - 1$ multiplications are needed for multiplying $n \times n$ matrices by means of noncommutative algorithms [BD78]. If $n \geq 3$, then $2n^2 + n - 2$ multiplications are needed [Bla03]. The lower bound $\frac{5}{2}n^2 - 3n$ has been proved in [Bla01]. The nonlinear asymptotic lower bound $n^2 \log n$ has been proved in [Raz03]. At least $n^2 + 2n - 2$ multiplications are needed for approximating the product of $n \times n$ matrices by means of a noncommutative APA algorithm [Bin84]. The lower bound turns to $n^2 + \frac{3}{2}n - 2$ multiplications if commutativity is allowed. The product of 2×2 matrices can be approximated with 6 multiplications by means of a commutative algorithm [Bin84], 5 multiplications are needed. Seven multiplications are needed for approximating 2×2 matrix product by means of noncommutative algorithms [Lan05].
7. *History of matrix multiplication complexity.* After the 1969 paper by V. Strassen [Str69], where it was shown that $O(n^{\log_2 7})$ operations were sufficient for $n \times n$ matrix multiplication and inversion, the exponent ω of matrix multiplication complexity remained stuck at $2.807\ldots$ for almost 10 years until when V. Pan, relying on the technique of trilinear aggregating, provided a bilinear noncommutative algorithm for 70×70 matrix multiplication using 143,640 products. This led to the upper bound $\omega \leq \log_{70} 143640 \approx 2.795$. A few months later, Bini, Capovani, Lotti, and Romani [BCL79] introduced the concept of APA algorithms, and presented a scheme for approximating a 12×12 matrix product with 1000 products. In [Bin80] Bini showed that from any APA algorithm for matrix multiplication it is possible to obtain an exact algorithm with almost the same asymptotic complexity. This led to the bound $\omega \leq \log_{12} 1000 \approx 2.7798$. The technique of partial matrix multiplication was introduced by Schönhage [Sch81] in 1981 together with the τ-theorem, yielding the bound $\omega \leq 2.55$, a great improvement with respect to the previous estimates. This bound relies on the tools of trilinear aggregating and of approximate algorithms. Based on the techniques so far developed, V. Pan obtained the bound $\omega < 2.53$ in [Pan80] and one year later, F. Romani obtained the bound $\omega < 2.52$ in [Rom82]. The landmark bound $\omega < 2.5$ was obtained

TABLE 47.1 Main steps in the history of fast matrix multiplication

$\omega <$			
2.81	1969	[Str69]	Bilinear algorithms
2.79	1979	[Pan78]	Trilinear aggregating
2.78	1979	[BCL79],[Bin80]	Approximate algorithms
2.55	1981	[Sch81]	τ-theorem
2.53	1981	[Pan80]	
2.52	1982	[Rom82]	
2.50	1982	[CW72]	Refinement of the τ-theorem
2.48	1987,1988	[Str87],[Str88]	Laser method
2.38	1990	[CW82]	

by Coppersmith and Winograd [CW72] in 1982 by means of a refinement of the τ-theorem. In [Str87] Strassen introduced the powerful *laser method* and proved the bound $\omega < 2.48$. The laser method has been perfected by Coppersmith and Winograd [CW82], who proved the best estimate known so far, i.e., $\omega < 2.38$. Table 47.1 synthesizes this picture together with the main concepts used.

47.6 Applications

Some of the main applications of matrix multiplication are outlined in this section. For more details the reader is referred to [Pan84].

Definitions:

A square matrix A is **strongly nonsingular** if all its principal submatrices are nonsingular.

Facts:

1. *Classic matrix inversion.* Given a nonsingular $n \times n$ matrix A, the elements of A^{-1} can be computed by means of Gaussian elimination in $O(n^3)$ arithmetic operations.
2. *Inversion formula.* Let $n = 2m$ and partition the $n \times n$ nonsingular matrix A into four square blocks of size m as

$$A = \begin{bmatrix} A_{1,1} & A_{1,2} \\ A_{2,1} & A_{2,2} \end{bmatrix}.$$

Assume that $A_{1,1}$ is nonsingular. Denote $S = A_{2,2} - A_{2,1}A_{1,1}^{-1}A_{1,2}$ the Schur complement of $A_{1,1}$ in A and $R = A_{1,1}^{-1}$. Then S is nonsingular and the inverse of A can be written as

$$A^{-1} = \begin{bmatrix} R + RA_{1,2}S^{-1}A_{2,1}R & -RA_{1,2}S^{-1} \\ -S^{-1}A_{2,1}R & S^{-1} \end{bmatrix}.$$

Moreover, $\det A = \det S \det A_{1,1}$.
3. *Fast matrix inversion.* Let $n = 2^k$, with k positive integer and assume that A is strongly nonsingular. Then also $A_{1,1}$ and the Schur complement S are strongly nonsingular and the inversion formula of Fact 2 can be applied again to S by partitioning S into four square blocks of size $n/4$, recursively repeating this procedure until the size of the blocks is 1. The algorithm obtained in this way is described in Algorithm 3. Denoting by $\mathcal{I}(n)$ the complexity of this algorithm for inverting a strongly nonsingular $n \times n$ matrix and denoting by $\mathcal{M}(n)$ the complexity of the algorithm used

for $n \times n$ matrix multiplication, we obtain the expression

$$\mathcal{I}(n) = 2\mathcal{I}(n/2) + 6\mathcal{M}(n/2) + n^2/2,$$
$$\mathcal{I}(1) = 1.$$

If $\mathcal{M}(n) = O(n^w)$ with $w \geq 2$, then one deduces that $\mathcal{I}(n) = O(n^w)$. That is, the complexity of matrix inversion is not asymptotically larger than the complexity of matrix multiplication.

4. The complexity of matrix multiplication is not asymptotically greater than the complexity of matrix inversion. This property follows from the simple identity

$$\begin{bmatrix} I & A & 0 \\ 0 & I & B \\ 0 & 0 & I \end{bmatrix}^{-1} = \begin{bmatrix} I & -A & AB \\ 0 & I & -B \\ 0 & 0 & I \end{bmatrix}.$$

5. Combining Facts 3 and 4, we deduce that matrix multiplication and matrix inversion have the same asymptotical complexity.

Algorithm 3: Fast matrix inversion
Procedure Fast_Inversion(k, A)
Input: the elements $a_{i,j}$ of the $2^k \times 2^k$ strongly nonsingular matrix A;
Output: the elements $b_{i,j}$ of $B = A^{-1}$.
If $k = 0$, **then**
  output $b_{1,1} = a_{1,1}^{-1}$;
else
  partition A, into $2^{k-1} \times 2^{k-1}$ blocks $A_{i,j}, i, j = 1, 2$;
  set $R =$ Fast_Inversion$(k - 1, A_{1,1})$;
  compute $S = A_{2,2} - A_{2,1} R A_{1,2}$ and $V =$ Fast_Inversion$(k - 1, S)$
  output

$$\begin{bmatrix} R + RA_{1,2} V A_{2,1} R & -RA_{1,2} V \\ -V A_{2,1} R & V \end{bmatrix}$$

6. The property of strong singularity for A is not a great restriction if $F = \mathbb{R}$ or $F = \mathbb{C}$. In fact, if A is nonsingular then $A^{-1} = (A^* A)^{-1} A^*$ and the matrix $A^* A$ is strongly nonsingular.
7. *Computing the determinant.* Let A be strongly nonsingular. From Fact 2, $\det A = \det S \det A_{1,1}$. Therefore, an algorithm for computing $\det A$ can be recursively designed by computing S and then by applying recursively this algorithm to S and to $A_{1,1}$ until the size of the blocks is 1. Denoting by $\mathcal{D}(n)$ the complexity of this algorithm one has

$$\mathcal{D}(n) = 2\mathcal{D}(n/2) + \mathcal{I}(n/2) + 2\mathcal{M}(n/2) + n^2/4,$$

hence, if $\mathcal{M}(n) = O(n^w)$ with $w \geq 2$, then $\mathcal{D}(n) = O(n^w)$. In [BS83] it is shown that $\mathcal{M}(n) = O(\mathcal{D}(n))$

8. *Computing the characteristic polynomial.* The coefficients of the characteristic polynomial $p(x) = \det(A - xI)$ of the matrix A can be computed with the same asymptotic complexity of matrix multiplication.
9. *Combinatorial problems.* The complexity of some combinatorial problems is related to matrix multiplication, in particular, the complexity of the *all pair shortest distance problem* of finding the shortest distances $d(i, k)$ from i to k for all pairs (i, k) of vertices of a given digraph. We refer the reader to section 18 of [Pan84] for more details. The problem of Boolean matrix multiplication can be reduced to that of general matrix multiplication.

References

[AS81] A. Alder and V. Strassen. On the algorithmic complexity of associative algebras. *Theoret. Comput. Sci.*, 15(2):201–211, 1981.

[BS83] W. Baur and V. Strassen. The complexity of partial derivatives. *Theoret. Comput. Sci.*, 22(3):317–330, 1983.

[Bin84] D. Bini. On commutativity and approximation. *Theoret. Comput. Sci.*, 28(1-2):135–150, 1984.

[Bin80] D. Bini. Relations between exact and approximate bilinear algorithms. Applications. *Calcolo*, 17(1):87–97, 1980.

[BCL79] D. Bini, M. Capovani, G. Lotti, and F. Romani. $O(n^{2.7799})$ complexity for $n \times n$ approximate matrix multiplication. *Inform. Process. Lett.*, 8(5):234–235, 1979.

[BLR80] D. Bini, G. Lotti, and F. Romani. Approximate solutions for the bilinear forms computational problem. *SIAM J. Comp.*, 9:692–697, 1980.

[Bla03] M. Bläser. On the complexity of the multiplication of matrices of small formats. *J. Complexity*, 19(1):43–60, 2003.

[Bla01] M. Bläser. A $\frac{5}{2}n^2$-lower bound for the multiplicative complexity of $n \times n$-matrix multiplication. Lecture Notes in Comput. Sci., 2010, 99–109, Springer, Berlin, 2001.

[BD78] R.W. Brockett and D. Dobkin. On the optimal evaluation of a set of bilinear forms. *Lin. Alg. Appl.*, 19(3):207–235, 1978.

[CW82] D. Coppersmith and S. Winograd. On the asymptotic complexity of matrix multiplication. *SIAM J. Comp.*, 11(3):472–492, 1982.

[CW72] D. Coppersmith and S. Winograd. Matrix multiplication via arithmetic progressions. *J. Symb. Comp.*, 9(3):251–280, 1972.

[Fid72] C.M. Fiduccia. On obtaining upper bounds on the complexity of matrix multiplication. In R.E. Miller and J.W. Thatcher, Eds., *Complexity of Computer Computations.* Plenum Press, New York, 1972.

[Gas71] N. Gastinel. Sur le calcul des produits de matrices. *Numer. Math.*, 17:222–229, 1971.

[Hig02] N.J. Higham. *Accuracy and Stability of Numerical Algorithms.* Society for Industrial and Applied Mathematics (SIAM), Philadelphia, PA, 2nd ed., 2002.

[HM73] J. Hopcroft and J. Musinski. Duality applied to the complexity of matrix multiplication and other bilinear forms. *SIAM J. Comp.*, 2:159–173, 1973.

[HK71] J.E. Hopcroft and L.R. Kerr. On minimizing the number of multiplications necessary for matrix multiplication. *SIAM J. Appl. Math.*, 20:30–36, 1971.

[Kel85] W. Keller-Gehrig. Fast algorithms for the characteristic polynomial. *Theor. Comp. Sci.*, 36:309–317, 1985.

[Lad76] J.D. Laderman. A noncommutative algorithm for multiplying 3×3 matrices using 23 muliplications. *Bull. Amer. Math. Soc.*, 82(1):126–128, 1976.

[Lan05] J.M. Landsberg. The border rank of the multiplication of 2×2 matrices is seven. *J. Am. Math. Soc.* (Electronic, September 26, 2005).

[Pan84] V. Pan. How to multiply matrices faster, Vol. 179 of Lecture Notes in Computer Science. Springer-Verlag, Berlin, 1984.

[Pan78] V. Ya. Pan. Strassen's algorithm is not optimal. Trilinear technique of aggregating, uniting and canceling for constructing fast algorithms for matrix operations. In *19th Annual Symposium on Foundations of Computer Science (Ann Arbor, MI., 1978)*, pp. 166–176. IEEE, Long Beach, CA, 1978.

[Pan80] V. Ya. Pan. New fast algorithms for matrix operations. *SIAM J. Comp.*, 9(2):321–342, 1980.

[Raz03] R. Raz. On the complexity of matrix product. *SIAM J. Comp.*, 32(5):1356–1369, 2003.

[Rom82] F. Romani. Some properties of disjoint sums of tensors related to matrix multiplication. *SIAM J. Comp.*, 11(2):263–267, 1982.

[Sch81] A. Schönhage. Partial and total matrix multiplication. *SIAM J. Comp.*, 10(3):434–455, 1981.

[Str69] V. Strassen. Gaussian elimination is not optimal. *Numer. Math.*, 13:354–356, 1969.

[Str87] V. Strassen. Relative bilinear complexity and matrix multiplication. *J. Reine Angew. Math.*, 375/376:406–443, 1987.

[Str88] V. Strassen. The asymptotic spectrum of tensors. *J. Reine Angew. Math.*, 384:102–152, 1988.

[Win69] S. Winograd. The number of multiplications involved in computing certain functions. In *Information Processing 68 (Proc. IFIP Congress, Edinburgh, 1968), Vol. 1: Mathematics, Software*, pages 276–279. North-Holland, Amsterdam, 1969.

[Win71] S. Winograd. On multiplication of 2 × 2 matrices. *Lin. Alg. Appl.*, 4:381–388, 1971.

48

Structured Matrix Computations

Michael Ng
Hong Kong Baptist University

48.1 Structured Matrices

In various application fields, the matrices encountered have special structures that can be exploited to facilitate the solution process. Sparsity is one of these features. However, the matrices we consider in this chapter are mostly dense matrices with a special structure. Structured matrices have been around for a long time and are encountered in various fields of application. (See [GS84, KS98, BTY01].) Some interesting families are listed below. For simplicity, we give the definitions for real square matrices of size n.

Definitions:

Toeplitz matrices: Matrices with constant diagonals, i.e., $[T]_{i,j} = t_{i-j}$ for all $1 \leq i, j \leq n$:

$$T = \begin{bmatrix} t_0 & t_{-1} & \cdots & t_{2-n} & t_{1-n} \\ t_1 & t_0 & t_{-1} & & t_{2-n} \\ \vdots & t_1 & t_0 & \ddots & \vdots \\ t_{n-2} & & \ddots & \ddots & t_{-1} \\ t_{n-1} & t_{n-2} & \cdots & t_1 & t_0 \end{bmatrix}.$$

(See Chapter 16.2 for additional information on families of Toeplitz matrices.)

Lower shift matrix: The matrix with ones on the first subdiagonal and zeros elsewhere:

$$Z_n = \begin{bmatrix} 0 & 0 & \cdots & 0 & 0 \\ 1 & 0 & 0 & & 0 \\ \vdots & 1 & 0 & \ddots & \vdots \\ 0 & & \ddots & \ddots & 0 \\ 0 & 0 & \cdots & 1 & 0 \end{bmatrix}.$$

Circulant matrices: Toeplitz matrices where each column is a circular shift of its preceding column:

$$C = \begin{bmatrix} c_0 & c_{n-1} & \cdots & & c_2 & c_1 \\ c_1 & c_0 & c_{n-1} & & & c_2 \\ c_2 & c_1 & c_0 & & & \vdots \\ \vdots & \ddots & \ddots & \ddots & \ddots & \\ c_{n-2} & & & \ddots & \ddots & c_{n-1} \\ c_{n-1} & c_{n-2} & \cdots & c_2 & c_1 & c_0 \end{bmatrix}.$$

***n*-Cycle matrix:** The $n \times n$ matrix with ones along the subdiagonal and in the 1, n-entry, and zeros elsewhere, i.e.,

$$C_n = \begin{bmatrix} 0 & 0 & \cdots & 0 & 1 \\ 1 & 0 & 0 & & 0 \\ \vdots & 1 & 0 & \ddots & \vdots \\ 0 & & \ddots & \ddots & 0 \\ 0 & 0 & \cdots & 1 & 0 \end{bmatrix}.$$

Hankel matrices: Matrices with constant elements along their antidiagonals, i.e.,

$$H = \begin{bmatrix} h_0 & h_1 & \cdots & h_{n-2} & h_{n-1} \\ h_1 & h_2 & \iddots & \iddots & h_n \\ \vdots & \iddots & \iddots & \iddots & \vdots \\ h_{n-2} & \iddots & \iddots & \iddots & h_{2n-2} \\ h_{n-1} & h_n & \cdots & h_{2n-2} & h_{2n-1} \end{bmatrix}.$$

Anti-identity matrix: The matrix with ones along the antidiagonal and zeros elsewhere, i.e.,

$$P_n = \begin{bmatrix} 0 & 0 & \cdots & 0 & 1 \\ 0 & 0 & \iddots & 1 & 0 \\ \vdots & \iddots & \iddots & \iddots & \vdots \\ 0 & 1 & \iddots & \iddots & 0 \\ 1 & 0 & \cdots & 0 & 0 \end{bmatrix}.$$

Cauchy matrices: Given vectors $\mathbf{x} = [x_1, \ldots, x_n]^T$ and $\mathbf{y} = [y_1, \ldots, y_n]^T$, the Cauchy matrix $C(\mathbf{x}, \mathbf{y})$ has i, j-entry equal to $\frac{1}{x_i + y_j}$.

Vandermonde matrices: A matrix having each row equal to successive powers of a number, i.e.,

$$V = \begin{bmatrix} 1 & v_1^1 & \cdots & v_1^{n-2} & v_1^{n-1} \\ 1 & v_2^1 & \cdots & v_2^{n-2} & v_2^{n-1} \\ \vdots & \vdots & \ddots & \vdots & \vdots \\ 1 & v_n^1 & \cdots & v_n^{n-2} & v_n^{n-1} \end{bmatrix}.$$

Block matrices: An $m \times m$ block matrix with $n \times n$ blocks is a matrix of the form

$$\begin{bmatrix} A^{(1,1)} & A^{(1,2)} & \cdots & A^{(1,m)} \\ A^{(2,1)} & A^{(2,2)} & \cdots & A^{(2,m)} \\ \vdots & \vdots & \ddots & \vdots \\ A^{(m-1,1)} & A^{(m-1,2)} & \cdots & A^{(m-1,m)} \\ A^{(m,1)} & A^{(m,2)} & \cdots & A^{(m,m)} \end{bmatrix},$$

where each block $A^{(i,j)}$ is an $n \times n$ matrix.

Toeplitz-block matrices: $mn \times mn$ block matrices where each block $\{A^{(i,j)}\}_{i,j=1}^{m}$ is an $n \times n$ Toeplitz matrix.

Block-Toeplitz matrices: $mn \times mn$ block matrices of the form

$$T = \begin{bmatrix} A^{(0)} & A^{(-1)} & \cdots & A^{(1-m)} \\ A^{(1)} & A^{(0)} & \cdots & A^{(2-m)} \\ \vdots & \vdots & \ddots & \vdots \\ A^{(m-1)} & A^{(m-2)} & \cdots & A^{(0)} \end{bmatrix},$$

where $\{A^{(i)}\}_{i=1-m}^{m-1}$ are arbitrary $n \times n$ matrices.

Block-Toeplitz–Toeplitz-block (BTTB) matrices: The blocks $A^{(i)}$ are themselves Toeplitz matrices.

Block matrices for other structured matrices such as the block-circulant matrices or the circulant-block matrices can be defined similarly.

Facts:

1. The transpose of a Toeplitz matrix is a Toeplitz matrix.
2. Any linear combination of Toeplitz matrices is a Toeplitz matrix.
3. The lower shift shift matrices Z_n are Toeplitz matrices. The $n \times n$ Toeplitz matrix $T = [t_{ij}]$ with $t_{ij} = t_{i-j}$ satisfies $T = t_0 I_n + \Sigma_{k=1}^{n-1} t_k Z_n^k + \Sigma_{k=1}^{n-1} t_{-k} (Z_n^T)^k$.
4. Every circulant matrix is a Toeplitz matrix, but not conversely.
5. The transpose of a circulant matrix is a circulant matrix.
6. Any linear combination of circulant matrices is a circulant matrix.
7. The n-cycle matrix C_n is a circulant matrix. The $n \times n$ circulant matrix $C = [c_{ij}]$ with $c_{i1} = c_{i-1}$ satisfies $C = \Sigma_{k=0}^{n-1} c_k C_n^k$
8. An important property of circulant matrices is that they can diagonalized by discrete Fourier transform matrices (see Section 47.3 and Chapter 58.3). Thus circulant matrices are normal.
9. A Hankel matrix is symmetric.
10. Any linear combination of Hankel matrices is a Hankel matrix.
11. Multiplication of a Toeplitz matrix and the anti-identity matrix P_n is a Hankel matrix, and multiplication of a Hankel matrix and P_n is a Toeplitz matrix. For H as in the definition,

$$P_n H = \begin{bmatrix} h_{n-1} & h_n & \cdots & h_{2n-2} & h_{2n-1} \\ h_{n-2} & h_{n-1} & h_n & & h_{2n-2} \\ \vdots & h_{n-2} & h_{n-1} & \ddots & \vdots \\ h_1 & & \ddots & \ddots & h_n \\ h_0 & h_1 & \cdots & h_{n-2} & h_{n-1} \end{bmatrix},$$

$$HP_n = \begin{bmatrix} h_{n-1} & h_{n-2} & \cdots & h_1 & h_0 \\ h_n & h_{n-1} & h_{n-2} & & h_1 \\ \vdots & h_n & h_{n-1} & \ddots & \vdots \\ h_{2n-2} & & \ddots & \ddots & h_{-1} \\ h_{2n-1} & h_{2n-2} & \cdots & h_n & h_{n-1} \end{bmatrix}.$$

12. The Kronecker product $A \otimes T$ of any matrix A and a Toeplitz matrix T is a Toeplitz-block matrix and Kronecker product $T \otimes A$ is a block-Toeplitz matrix.
13. Most of the applications, such as partial differential equations and image processing, are concerned [Jin02], [Ng04] with two-dimensional problems where the matrices will have block structures.

Examples:

1. $\begin{bmatrix} 1 & 2 & 1 & 1 \\ 3 & 1 & 1 & 1 \\ -5 & 2 & 1 & 2 \\ 0 & -5 & 3 & 1 \end{bmatrix}$ is a BTTB matrix with $A^{(0)} = \begin{bmatrix} 1 & 2 \\ 3 & 1 \end{bmatrix}$, $A^{(-1)} = \begin{bmatrix} 1 & 1 \\ 1 & 1 \end{bmatrix}$, $A^{(1)} = \begin{bmatrix} -5 & 2 \\ 0 & 5 \end{bmatrix}$.
2. For $\mathbf{x} = \begin{bmatrix} 1 \\ 1 \\ 5 \end{bmatrix}$ and $\mathbf{y} = \begin{bmatrix} 1 \\ 2 \\ 3 \end{bmatrix}$, $C(\mathbf{x}, \mathbf{y}) = \begin{bmatrix} \frac{1}{2} & \frac{1}{3} & \frac{1}{4} \\ \frac{1}{2} & \frac{1}{3} & \frac{1}{4} \\ \frac{1}{6} & \frac{1}{7} & \frac{1}{8} \end{bmatrix}$.

48.2 Direct Toeplitz Solvers

Most of the early work on Toeplitz solvers was focused on direct methods. These systems arise in a variety of applications in mathematics and engineering. In fact, Toeplitz structure was one of the first structures analyzed in signal processing.

Definitions:

Toeplitz systems: A system of linear equations with a Toeplitz coefficient matrix.

Facts:

1. Given an $n \times n$ Toeplitz system $Tx = b$, a straightforward application of the Gaussian elimination method will result in an algorithm of $O(n^3)$ complexity.
2. However, since the matrix is determined by only $(2n - 1)$ entries rather than n^2 entries, it is to be expected that a solution can be obtained in less than $O(n^3)$ operations. There are a number of fast Toeplitz solvers that can reduce the complexity to $O(n^2)$ operations. The original references for these algorithms are Schur [Sch17], Levinson [Lev46], Durbin [Dur60], and Trench [Tre64].
3. In the 1980s, superfast algorithms of complexity $(n \log^2 n)$ operations for Toeplitz systems were proposed by a different group of researchers: Bitmead and Anderson [BA80], Brent et al. [BGY80], Morf [Mor80], de Hoog [Hoo87], and Ammar and Gragg [AG88]. The key to these direct methods is to solve the system recursively. In this section, we will give a brief summary of the development of these methods. We refer the reader to the works cited for more details.
4. If the Toeplitz matrix T has a singular or ill-conditioned principal submatrix, then a breakdown or near-breakdown can occur in the direct Toeplitz solvers. Such breakdowns will cause numerical instability in subsequent steps of the algorithms and result in inaccurately computed solutions. The question of how to avoid breakdowns or near-breakdowns by skipping over singular submatrices

or ill-conditioned submatrices has been studied extensively, and various such algorithms have been proposed. (See Chan and Hansen [CH92].)

5. The fast direct Toeplitz solvers are in general numerically unstable for indefinite systems. Look-ahead methods are numerically stable and, although it may retain the $O(n^2)$ complexity, it requires $O(n^3)$ operations in the worst case.
6. The stability properties of direct methods for symmetric positive definite Toeplitz systems were discussed in Sweet [Swe84], Bunch [Bun85], Cybenko [Cyb87], and Bojanczyk et al. [BBH95].
7. Gohberg et al. [GKO95] have shown how to perform Gaussian elimination in a fast way for matrices having special displacement structures. Such matrices include Toeplitz, Vandermonde, Hankel, and Cauchy matrices. They have shown how to incorporate partial pivoting into Cauchy solvers. They pointed out that although pivoting cannot be incorporated directly for Toeplitz matrices, Toeplitz problems can be transformed by simple orthogonal operations to Cauchy problems. The solutions to the original problems can be recovered from those of the transformed systems by the reverse orthogonal operations. Thus, fast Gaussian elimination with partial pivoting can be carried out for Toeplitz systems.

   Brent and Sweet [BS95] gave a rounding error analysis on the Cauchy and Toeplitz variants of the recent method of Gohberg et al. [GK095]. It has been shown that the error growth depends on the growth in certain auxiliary vectors, the generators, which are computed by the Gohberg algorithm. In certain circumstances, growth in the generators can be large, so the error growth is much larger than would be encountered when using normal Gaussian elimination with partial pivoting.

48.3 Iterative Toeplitz Solvers

A circulant matrix is a special form of Toeplitz matrix where each row of the matrix is a circular shift of its preceding row. Because of the periodicity, circulant systems can be solved quickly via a deconvolution by discrete Fast Fourier Transforms (FFTs) [Ng04]. Circulant approximations to Toeplitz matrices have been used for some time in signal and image processing [Ng04]. However, in these applications, the circulant approximation so obtained is used to replace the given Toeplitz matrix in subsequent computations. In effect, the matrix equation is changed and, hence, so is the solution.

Development of solely circulant-based iterative methods for Toeplitz systems started in the 1970s. Rino [Rin70] developed a method for generating a series expansion solution to Toeplitz systems by writing a Toeplitz matrix as the sum of a circulant matrix and another Toeplitz matrix and presented a method for choosing the circulant matrix. Silverman and Pearson [SP73] applied similar methods to deconvolution.

In 1986, Strang [Str86] and Olkin [Olk86] independently proposed to precondition Toeplitz matrices by circulant matrices in conjugate gradient iterations. Their motivation was to exploit the fast inversion of circulant matrices. Numerical results in [SE87] and [Olk86] show that the method converges very fast for a wide range of Toeplitz matrices. This has later been proved theoretically in [CS89] and in other papers for other circulant preconditioners [Ng04]. Circulant approximations are used here only as preconditioners for Toeplitz systems and the solutions to the Toeplitz systems are unchanged.

One of the main important results of this methodology is that the complexity of solving a large class of $n \times n$ Toeplitz systems can be reduced to $O(n \log n)$ operations, provided that a suitable preconditioner is used. Besides the reduction of the arithmetic complexity, there are important types of Toeplitz matrix where the fast direct Toeplitz solvers are notoriously unstable, e.g., indefinite and certain non-Hermitian Toeplitz matrices. Therefore, iterative methods provide alternatives for solving these Toeplitz systems.

48.4 Linear Systems with Matrix Structure

This section provides some examples of the latest developments on iterative methods for the iterative solution of linear systems of equations with structured coefficient matrices such as Toeplitz-like, Toeplitz-plus-Hankel, and Toeplitz-plus-band matrices. We would like to make use of their structure to construct some good preconditioners for such matrices.

Facts:

1. Toeplitz-like systems: Let A be an $n \times n$ structured matrix with respect to Z_n (the lower shift matrix):

$$\nabla A_n = A_n - Z_n A_n Z_n^* = GSG^*$$

for some $n \times r$ generator matrix G and $r \times r$ signature matrix $S = (I_p \oplus -I_q)$. If we partition the columns of G into two sets $\{x_i\}_{i=0}^{p-1}$ and $\{y_i\}_{i=0}^{q-1}$,

$$G = [x_0\, x_1 \ldots x_{p-1}\, y_0\, y_1 \ldots\, y_{q-1}] \quad \text{with } p + q = r,$$

then we know from the representation that we can express A as a linear combination of lower triangular Toeplitz matrices,

$$A = \sum_{i=0}^{p-1} L(x_i)L^*(x_i) - \sum_{i=0}^{q-1} L(y_i)L^*(y_i).$$

For example, if $T_{m,n}$ is an $m \times n$ Toeplitz matrix with $m \geq n$, then $T_{m,n}^* T_{m,n}$ is in general not a Toeplitz matrix. However, $T_{m,n}^* T_{m,n}$ does have a small displacement rank, $r \leq 4$, and the displacement representation of $T_{m,n}^* T_{m,n}$ is

$$T_{m,n}^* T_{m,n} = L_n(x_0)L_n(x_0)^* - L_n(y_0)L_n(y_0)^* + L_n(x_1)L_n(x_1)^* - L_n(y_1)L_n(y_1)^*,$$

where

$$x_0 = T_{m,n}^* T_{m,n} e_1 / ||T_{m,n} e_1||, \quad y_0 = Z_n Z_n^* x_0,$$

$$x_1 = [0, t_{-1}, t_{-2}, \cdots, t_{1-n}]^*, \quad \text{and} \quad y_1 = [0, t_{m-1}, t_{m-2}, \cdots, t_{m-n+1}]^*.$$

2. For structured matrices with displacement representations, it was suggested in [CNP94] to define the displacement preconditioner to be the circulant approximation of the factors in the displacement representation of A, i.e., the circulant approximation C of A is

$$\sum_{i=0}^{p-1} C(L(x_i))C^*(L_n(x_i)) - \sum_{i=0}^{q-1} C(L_n(y_i))C^*(L_n(y_i)).$$

Here, $C(X)$ denotes some circulant approximations to X.
3. The displacement preconditioner approach is applied to Toeplitz least squares and Toeplitz-plus-Hankel least squares problems [Ng04].
4. The systems of linear equations with Toeplitz-plus-Hankel coefficient matrices arise in many signal processing applications. For example, the inverse scattering problem can be formulated as Toeplitz-plus-Hankel systems of equations (see [Ng04].)

   The product of P_n and H and the product of H and P_n both give Toeplitz matrices.

   Premultiplying P_n to a vector v corresponds to reversing the order of the elements in v. Since

$$Hv = HP_nP_nv$$

and HP_n is a Toeplitz matrix, the Hankel matrix-vector products Hv can be done efficiently using FFTs. A Toeplitz-plus-Hankel matrix can be expressed as $T + H = T + P_nP_nH$.
5. Given circulant preconditioners $C^{(1)}$ and $C^{(2)}$ for Toeplitz matrices T and P_nH, respectively, it was proposed in [KK93] to use

$$M = C^{(1)} + P_nC^{(2)}$$

as a preconditioner for the Toeplitz-plus-Hankel matrix $T + H$. With the equality $P_n^2 = I$, we have

$$Mz = C^{(1)}z + P_n C^{(2)} P_n P_n z = v,$$

which is equivalent to

$$P_n M z = P_n C^{(1)} P_n P_n z + C^{(2)} z = P_n v.$$

By using these two equations, the solution of $z = M^{-1}v$ can be determined.

6. We consider the solution of systems of the form $(T + B)x = b$, where T is a Toeplitz matrix and B is a banded matrix with bandwidth $2b + 1$ independent of the size of the matrix. These systems appear in solving Fredholm integro-differential equations of the form

$$L\{x(\theta)\} + \int_\alpha^\beta K(\phi - \theta)x(\phi)d\phi = b(\theta).$$

Here, $x(\theta)$ is the unknown function to be found, $K(\theta)$ is a convolution kernel, L is a differential operator and $b(\theta)$ is a given function. After discretization, K will lead to a Toeplitz matrix, L to a banded matrix, and $b(\theta)$ to the right-hand side vector [Ng04]. Toeplitz-plus-band matrices also appear in signal processing literature and have been referred to as periheral innovation matrices.

Unlike Toeplitz systems, there exist no fast direct solvers for solving Toeplitz-plus-band systems. It is mainly because the displacement rank of the matrix $T + B$ can take any value between 0 and n. Hence, fast Toeplitz solvers that are based on small displacement rank of the matrices cannot be applied. Conjugate gradient methods with circulant preconditioners do not work for Toeplitz-plus-band systems either. The main reason is that when the eigenvalues of B are not clustered, the matrix $C(T) + B$ cannot be inverted easily.

In [CN93], it was proposed to use the matrix $E + B$ to precondition $T + B$, where E is the band-Toeplitz preconditioner such that E is spectrally equivalent to T. Note that E is a banded matrix, and the banded system $(E + B)y = z$ can be solved by using any band matrix solver.

7. Banded preconditioners are successfully applied to precondition Sinc–Galerkin systems (Toeplitz-plus-band systems) arising from the Sinc–Galerkin method to partial differential equations (see [Ng04].)
8. In most of applications we simply use the circulant or other transform-based preconditioners [Ng04]. We can extend the results for point circulant or point transform-based preconditioners to block circulant preconditioners or block transform-based preconditioners [Jin02] for block-Toeplitz, Toeplitz-block, and Toeplitz-block–block-Toeplitz matrices.
9. Consider the system $(A \otimes B)x = b$, where A is an m-by-m Hermitian positive definite matrix and B is an n-by-n Hermitian positive definite Toeplitz matrix. By using a circulant approximation $C(B)$ to B, the preconditioned system becomes

$$(A \otimes C(B))^{-1}(A \otimes B)x = (A \otimes C(B))^{-1}b,$$

or

$$(I \otimes C(B)^{-1}B)\mathbf{x} = (A^{-1} \otimes C(B)^{-1})b.$$

When B is a Hermitian positive definite Toeplitz matrix, $C(B)$ can be obtained in $O(n)$ operation [Jin02], [Ng04]. The initialization cost is about $O(m^3 + m^2 n + mn \log n)$ operations. Moreover, since the cost of multiplying By becomes $O(n \log n)$, we see that the cost per iteration is equal to $O(mn \log n)$ when iterative methods are employed.

10. When Toeplitz matrices have full rank, Toeplitz least squares problems

$$\min \|Tx - b\|_2^2$$

are equivalent to solving the normal equation matrices

$$T^* T x = T^* b.$$

Circulant preconditioners can be applied effectively and efficiently to solving Toeplitz least squares problems if Toeplitz matrices have full rank. When Toeplitz matrices do not have full rank, it is still an open research problem to find efficient algorithms for solving rank-deficient Toeplitz least squares problem. One possibility is to consider the generalized inverses of Toeplitz matrices. In the literature, computing the inverses and the generalized inverses of structured matrices are important practical computational problems. (See, for instance, Pan and Rami [PR01] and Bini et al. [BCV03].)

11. Instead of Toeplitz least squares problems $\min \|Tx - b\|_2^2$, we are interested in the 1-norm problem, i.e., $\min \|Tx - b\|_1$. The advantage of using the 1-norm is that the solution is more robust than using the 2-norm in statistical estimation problems. In particular, a small number of outliers have less influence on the solution. It is interesting to develop efficient algorithms for solving 1-norm Toeplitz least squares problems. Fu et al. [FNN06] have considered the least absolute deviation (LAD) solution of image restoration problems.
12. It is interesting to find good preconditioners for Toeplitz-related systems with large displacement rank. Good examples are Toeplitz-plus-band systems studied. Direct Toeplitz-like solvers cannot be employed because of the large displacement rank. However, iterative methods are attractive since coefficient matrix–vector products can be computed efficiently at each iteration. For instance, for the Toeplitz-plus-band matrix, its matrix-vector product can be computed in $O(n \log n)$ operations. The main concern is how to design good preconditioners for such Toeplitz-related systems with large displacement rank. Recently, Lin et al. [LNC05] proposed and developed factorized banded inverse preconditioners for matrices with Toeplitz structure. Also, Lin et al. [LCN04] studied incomplete factorization-based preconditioners for Toeplitz-like systems with large displacement ranks in image processing.

48.5 Total Least Squares Problems

1. The least squares problem $Tf \approx g$ is

$$\min_f \|Tf - g\|_2.$$

If the matrix T is known exactly, but the vector g is corrupted by random errors that are uncorrelated with zero mean and equal variance, then the least squares solution provides the best unbiased estimate of f. However, if T is also corrupted by errors, then the total least squares (TLS) method may be more appropriate. The TLS problem minimizes

$$\min_{\hat{T},\hat{g}} \|[T\ g] - [\hat{T}\ \hat{g}]\|_F^2$$

with the constraint $\hat{T} f = \hat{g}$. If the smallest singular value of T is larger than the smallest singular value σ^2 of $[T\ g]$, then there exists a unique TLS solution f_{TLS}, which can be represented as the solution to the normal equations:

$$(T^T T - \sigma^2 I) f = T^T g,$$

or as the solution to the eigenvalue problem:

$$\begin{bmatrix} T^T T & T^T g \\ g^T T & g^T g \end{bmatrix} \begin{bmatrix} f \\ -1 \end{bmatrix} = \sigma^2 \begin{bmatrix} f \\ -1 \end{bmatrix}.$$

Kamm and Nagy [KN98] proposed using Newton and Rayleigh quotient iterations for large TLS Toeplitz problems. Their method is a modification of a method suggested by Cybenko and Van Loan [CV86] for computing the minimum eigenvalue of a symmetric positive definite Toeplitz matrix. Specifically, first note that the TLS solution f_{TLS} solves the eigenvalue problem. Moreover, this eigenvalue problem is equivalent to

$$T^T T f - T^T g = \sigma^2 f$$

and

$$g^T T f - g^T g = -\sigma^2,$$

which can be combined to obtain the following secular equation for σ^2:

$$g^T g - g^T T(T^T T - \sigma^2 I)^{-1} T^T g - \sigma^2 = 0.$$

Therefore, σ^2 is the smallest root of the rational equation

$$h(\sigma^2) = g^T g - g^T T(T^T T - \sigma^2 I)^{-1} T^T g - \sigma^2$$

and can be found using Newton's method. Note that if σ^2 is less than the smallest singular value $\hat{\sigma}^2$ of T, then the matrix $T^T T - \sigma^2 I$ is positive definite. Assume for now that the initial estimate is within the interval $[\sigma^2, \hat{\sigma}^2)$. An analysis given in [CV86] shows that subsequent Newton iterates will remain within this interval and will converge from the right to σ^2.

2. In the above computation, $\hat{T}$ is not necessary to have Toeplitz structure. In another development, Ng [NPP00] presented an iterative, regularized, and constrained total least squares algorithm by requiring $\hat{T}$ to be Toeplitz. Preliminary numerical tests are reported on some simulated optical imaging problems. The numerical results showed that the regularized constrained TLS method is better than the regularized least squares method.
3. Other interesting areas are to design efficient algorithms based on preconditioning techniques for finding eigenvalues and singular values of Toeplitz-like matrices. Ng [Ng00] has employed preconditioned Lanczos methods for the minimum eigenvalue of a symmetric positive definite Toeplitz matrix.

References

[AG88] G. Ammar and W. Gragg, Superfast solution of real positive definite Toeplitz systems, *SIAM J. Matrix Anal. Appl.*, 9:61–67, 1988.

[BTY01] D. Bini, E. Tyrtyshnikov, and P. Yalamov, *Structured Matrices: Recent Developments in Theory and Computation*, Nova Science Publishers, Inc., New York, 2001.

[BCV03] D. Bini, G. Codevico, and M. Van Barel, Solving Toeplitz least squares problems by means of Newton's iterations, *Num. Algor.*, 33:63–103, 2003.

[BA80] R. Bitmead and B. Anderson, Asymptotically fast solution of Toeplitz and related systems of linear equations, *Lin. Alg. Appl.*, 34:103–116, 1980.

[BBH95] A. Bojanczyk, R. Brent, F. de Hoog, and D. Sweet, On the stability of the Bareiss and related Toeplitz factorization algorithms, *SIAM J. Matrix Anal. Appl.*, 16:40–57, 1995.

[BGY80] R. Brent, F. Gustavson, and D. Yun, Fast solution of Toeplitz systems of equations and computation of Padé approximants, *J. Algor.*, 1:259–295, 1980.

[BS95] R. Brent and D. Sweet, Error analysis of a partial pivoting method for structured matrices, Technical report, The Australian National University, Canberra, 1995.

[Bun85] J. Bunch, Stability of methods for solving Toeplitz systems of equations, *SIAM J. Sci. Stat. Comp.*, 6:349–364, 1985.

[CNP94] R. Chan, J. Nagy, and R. Plemmons, Displacement preconditioner for Toeplitz least squares iterations, *Elec. Trans. Numer. Anal.*, 2:44–56, 1994.

[CN93] R. Chan and M. Ng, Fast iterative solvers for Toeplitz-plus-band systems, *SIAM J. Sci. Comput.*, 14:1013–1019, 1993.

[CS89] R. Chan and G. Strang, Toeplitz equations by conjugate gradients with circulant preconditioner, *SIAM J. Sci. Stat. Comput.*, 10:104–119, 1989.

[CH92] T. Chan and P. Hansen, A look-ahead Levinson algorithm for general Toeplitz systems, *IEEE Trans. Signal Process.*, 40:1079–1090, 1992.

[Cyb87] G. Cybenko, Fast Toeplitz orthogonalization using inner products, *SIAM J. Sci. Stat. Comput.*, 8:734–740, 1987.

[CV86] G. Cybenko and C. Van Loan, Computing the minimum eigenvalue of a symmetric positive definite Toeplitz matrix, *SIAM J. Sci. Stat. Comput.*, 7:123–131, 1986.

[Dur60] J. Durbin, The fitting of time series models, *Rev. Int. Stat.*, 28:233–244, 1960.

[FNN06] H. Fu, M. Ng, M. Nikolova, and J. Barlow, Efficient minimization methods of mixed l2-l1 and l1-l1 norms for image restoration, *SIAM J. Sci. Comput.*, 27:1881–1902, 2006.

[GKO95] I. Gohberg, T. Kailath, and V. Olshevsky, Fast Gaussian elimination with partial pivoting for matrices with displacement structure, *Math. Comp.*, 64:65–72, 1995.

[GS84] U. Grenander and G. Szegö, *Toeplitz Forms and Their Applications*, Chelsea Publishing, 2nd ed., New York, 1984.

[Hoo87] F. de Hoog, A new algorithm for solving Toeplitz systems of equations, *Lin. Alg. Appl.*, 88/89:123–138, 1987.

[Jin02] X. Jin, *Developments and Applications of Block Toeplitz Iterative Solvers*, Kluwer Academic Publishers, New York, 2002.

[KS98] T. Kailath and A. Sayed, *Fast Reliable Algorithms for Matrices with Structured*, SIAM, Philadelphia, 1998.

[KN98] J. Kamm and J. Nagy, A total least squares method for Toeplitz systems of equations, *BIT*, 38:560–582, 1998.

[KK93] T. Ku and C. Kuo, Preconditioned iterative methods for solving Toeplitz-plus-Hankel systems, *SIAM J. Numer. Anal.*, 30:824–845, 1993.

[Lev46] N. Levinson, The Wiener rms (root mean square) error criterion in filter design and prediction, *J. Math. Phys.*, 25:261–278, 1946.

[LCN04] F. Lin, W. Ching, and M. Ng, Preconditioning regularized least squares problems arising from high-resolution image reconstruction from low-resolution frames, *Lin. Alg. Appl.*, 301:149–168, 2004.

[LNC05] F. Lin, M. Ng, and W. Ching, Factorized banded inverse preconditioners for matrices with Toeplitz structure, *SIAM J. Sci. Comput.*, 26:1852–1870, 2005.

[Mor80] M. Morf, Doubling algorithms for Toeplitz and related equations, in *Proc. IEEE Internat. Conf. on Acoustics, Speech and Signal Processing*, pp. 954–959, Denver, CO, 1980.

[Ng00] M. Ng, Preconditioned Lanczos methods for the minimum eigenvalue of a symmetric positive definite Toeplitz matrix, *SIAM J. Sci. Comput.*, 21:1973–1986, 2000.

[Ng04] M. Ng, *Iterative Methods for Toeplitz Systems*, Oxford University Press, UK, 2004.

[NPP00] M. Ng, R. Plemmons, and F. Pimentel, A new approach to constrained total least squares image restoration, *Lin. Alg. Appl.*, 316: 237–258, 2000.

[Olk86] J. Olkin, Linear and nonlinear deconvolution problems, Technical report, Rice University, Houston, TX, 1986.

[PR01] V. Pan and Y. Rami, Newton's iteration for the inversion of structured matrices, in D. Bini, E. Tyrtyshnikov, and P. Yalmov, Eds., *Structured Matrices: Recent Developments in Theory and Computation*, pp. 79–90, Nova Science Pub., Hauppauge, NY, 2001.

[Rin70] C. Rino, The inversion of covariance matrices by finite Fourier transforms, *IEEE Trans. Inform. Theory*, 16:230–232, 1970.

[Sch17] I. Schur, Über potenzreihen, die im innern des einheitskreises besschrankt sind, *J. Reine Angew Math.*, 147:205–232, 1917.

[SP73] H. Silverman and A. Pearson, On deconvolution using the discrete Fourier transform, *IEEE Trans. Audio Electroacous.*, 21:112–118, 1973.

[Str86] G. Strang, A proposal for Toeplitz matrix calculations, *Stud. Appl. Math.*, 74:171–176, 1986.

[SE87] G. Strang and A. Edelman, The Toeplitz-circulant eigenvalue problem $ax = \lambda cx$, in L. Bragg and J. Dettman, Eds., *Oakland Conf. on PDE and Appl. Math.*, New York, Longman Sci. Tech., 1987.

[Swe84] D. Sweet, Fast Toeplitz orthogonalization, *Numer. Math.*, 43:1–21, 1984.

[Tre64] W. Trench, An algorithm for the inversion of finite Toeplitz matrices, *SIAM J. Appl. Math.*, 12:515–522, 1964.

49

Large-Scale Matrix Computations

Roland W. Freund
University of California, Davis

Computational problems, especially in science and engineering, often involve large matrices. Examples of such problems include large sparse systems of linear equations [FGN92],[Saa03],[vdV03], e.g., arising from discretizations of partial differential equations, eigenvalue problems for large matrices [BDD00], [LM05], linear time-invariant dynamical systems with large state-space dimensions [FF94],[FF95],[Fre03], and large-scale linear and nonlinear optimization problems [KR91],[Wri97],[NW99],[GMS05]. The large matrices in these problems exhibit special structures, such as sparsity, that can be exploited in computational procedures for their solution. Roughly speaking, computational problems involving matrices are called "large-scale" if they can be solved only by methods that exploit these special matrix structures.

In this section, as in Chapter 44, multiplication of a vector $\mathbf{v}$ by a scalar λ is denoted by $\mathbf{v}\lambda$ rather than $\lambda\mathbf{v}$.

49.1 Basic Concepts

Many of the most efficient algorithms for large-scale matrix computations are based on approximations of the given large matrix by small matrices obtained via Petrov–Galerkin projections onto suitably chosen small-dimensional subspaces. In this section, we present some basic concepts of such projections.

Definitions:

Let $C \in \mathbb{C}^{n\times n}$ and let $V_j = [\mathbf{v}_1 \quad \mathbf{v}_2 \quad \cdots \quad \mathbf{v}_j] \in \mathbb{C}^{n\times j}$ be a matrix with orthonormal columns, i.e.,

$$\mathbf{v}_i^*\mathbf{v}_k = \begin{cases} 0 & \text{if } i \neq k, \\ 1 & \text{if } i = k, \end{cases} \quad \text{for all} \quad i, k = 1, 2, \ldots, j.$$

The matrix

$$C_j := V_j^* C V_j \in \mathbb{C}^{j\times j}$$

is called the **orthogonal Petrov–Galerkin projection** of C onto the subspace

$$S = \text{span}\{\mathbf{v}_1, \mathbf{v}_2, \ldots, \mathbf{v}_j\}$$

of $\mathbb{C}^n$ spanned by the columns of V_j.

Let $C \in \mathbb{C}^{n\times n}$, and let $V_j = [\mathbf{v}_1 \quad \mathbf{v}_2 \quad \cdots \quad \mathbf{v}_j] \in \mathbb{C}^{n\times j}$ and $W_j = [\mathbf{w}_1 \quad \mathbf{w}_2 \quad \cdots \quad \mathbf{w}_j] \in \mathbb{C}^{n\times j}$ be two matrices such that $W_j^T V_j$ is nonsingular. The matrix

$$C_j := (W_j^T V_j)^{-1} W_j^T C V_j \in \mathbb{C}^{j\times j}$$

is called the **oblique Petrov–Galerkin projection** of C onto the subspace

$$S = \text{span}\{\mathbf{v}_1, \mathbf{v}_2, \ldots, \mathbf{v}_j\}$$

of $\mathbb{C}^n$ spanned by the columns of V_j and orthogonally to the subspace

$$T = \text{span}\{\mathbf{w}_1, \mathbf{w}_2, \ldots, \mathbf{w}_j\}$$

of $\mathbb{C}^n$ spanned by the columns of W_j.

A **flop** is the work associated with carrying out any one of the elementary operations $a + b$, $a - b$, ab, or a/b, where $a, b \in \mathbb{C}$, in floating-point arithmetic.

Let $A = [a_{ik}] \in \mathbb{C}^{m\times n}$ be a given matrix. Matrix-vector multiplications with A are said to be **fast** if for any $\mathbf{x} \in \mathbb{C}^n$, the computation of $\mathbf{y} = A\mathbf{x}$ requires significantly fewer than $2mn$ flops.

A matrix $A = [a_{ik}] \in \mathbb{C}^{m\times n}$ is said to be **sparse** if only a small fraction of its entries a_{ik} are nonzero.

For a sparse matrix $A = [a_{ik}] \in \mathbb{C}^{m\times n}$, $nnz(A)$ denotes the number of nonzero entries of A.

A matrix $A = [a_{ik}] \in \mathbb{C}^{m\times n}$ is said to be **dense** if most of its entries a_{ik} are nonzero.

Facts:

The following facts on sparse matrices can be found in [Saa03, Chap. 3] and the facts on computing Petrov–Galerkin projections of matrices in [Saa03, Chap. 6].

1. For a sparse matrix $A = [a_{ik}] \in \mathbb{C}^{m\times n}$, only its nonzero or potentially nonzero entries a_{ik}, together with their row and column indices i and k, need to be stored.
2. Matrix-vector multiplications with a sparse matrix $A = [a_{ik}] \in \mathbb{C}^{m\times n}$ are fast. More precisely, for any $\mathbf{x} \in \mathbb{C}^n$, $\mathbf{y} = A\mathbf{x}$ can be computed with at most $2nnz(A)$ flops.
3. If $C \in \mathbb{C}^{n\times n}$ and $j \ll n$, the computational cost for computing the orthogonal Petrov–Galerkin projection of C onto the j-dimensional subspace $S = \text{span}\{\mathbf{v}_1, \mathbf{v}_2, \ldots, \mathbf{v}_j\}$ of $\mathbb{C}^n$ is dominated by the j matrix-vector products $\mathbf{y}_i = C\mathbf{v}_i$, $i = 1, 2, \ldots, j$.
4. If $C \in \mathbb{C}^{n\times n}$ and $j \ll n$, the computational cost for computing the oblique Petrov–Galerkin projection of C onto the j-dimensional subspace $S = \text{span}\{\mathbf{v}_1, \mathbf{v}_2, \ldots, \mathbf{v}_j\}$ of $\mathbb{C}^n$ and orthogonally to the j-dimensional subspace $T = \text{span}\{\mathbf{w}_1, \mathbf{w}_2, \ldots, \mathbf{w}_j\}$ of $\mathbb{C}^n$ is dominated by the j matrix-vector products $\mathbf{y}_i = C\mathbf{v}_i$, $i = 1, 2, \ldots, j$.
5. If matrix-vector products with a large matrix $C \in \mathbb{C}^{n\times n}$ are fast, then orthogonal and oblique Petrov–Galerkin projections C_j of C can be generated with low computational cost.

49.2 Sparse Matrix Factorizations

In this section, we present some basic concepts of sparse matrix factorizations. A more detailed description can be found in [DER89].

Definitions:

Let $A \in \mathbb{C}^{n\times n}$ be a sparse nonsingular matrix. A **sparse LU factorization** of A is a factorization of the form

$$A = PLUQ,$$

where $P, Q \in \mathbb{R}^{n\times n}$ are permutation matrices, $L \in \mathbb{C}^{n\times n}$ is a sparse unit lower triangular matrix, and $U \in \mathbb{C}^{n\times n}$ is a sparse nonsingular upper triangular matrix.

Fill-in of a sparse LU factorization $A = PLUQ$ is the set of nonzero entries of L and U that appear in positions (i, k) where $a_{ik} = 0$.

Let $A = A^* \in \mathbb{C}^{n\times n}$, $A \succ 0$, be a sparse Hermitian positive definite matrix. A **sparse Cholesky factorization** of A is a factorization of the form

$$A = PLL^* P^T,$$

where $P \in \mathbb{R}^{n\times n}$ is a permutation matrix and $L \in \mathbb{C}^{n\times n}$ is a sparse lower triangular matrix.

Fill-in of a sparse Cholesky factorization $A = PLL^* P^T$ is the set of nonzero entries of L that appear in positions (i, k) where $a_{ik} = 0$.

Let $T \in \mathbb{C}^{n\times n}$ be a sparse nonsingular (upper or lower) triangular matrix, and let $\mathbf{b} \in \mathbb{C}^n$. A **sparse triangular solve** is the solution of a linear system

$$T\mathbf{x} = \mathbf{b}$$

with a sparse triangular coefficient matrix T.

Facts:

The following facts can be found in [DER89].

1. The permutation matrices P and Q in a sparse LU factorization of A allow for reorderings of the rows and columns of A. These reorderings serve two purposes. First, they allow for pivoting for numerical stability in order to avoid division by the number 0 or by numbers close to 0, which would result in breakdowns or numerical instabilities in the procedure used for the computation of the factorization. Second, the reorderings allow for pivoting for sparsity, the goal of which is to minimize the amount of fill-in.
2. For Cholesky factorizations of matrices $A = A^* \succ 0$, the positive definiteness of A implies that pivoting for numerical stability is not needed. Therefore, the permutation matrix P in a sparse Cholesky factorization serves the single purpose of pivoting for sparsity.
3. For both sparse LU and sparse Cholesky factorizations, the problem of "optimal" pivoting for sparsity, i.e., finding reorderings that minimize the amount of fill-in, is NP-complete. This means that for practical purposes, minimizing the amount of fill-in of factorizations of large sparse matrices is impossible in general. However, there are a large number of pivoting strategies that — while not minimizing fill-in — efficiently limit the amount of fill-in for many important classes of large sparse matrices. (See, e.g., [DER89].)
4. A sparse triangular solve with the matrix T requires at most $2nnz(T)$ flops.
5. Not every large sparse matrix A has a sparse LU factorization with limited amounts of fill-in. For example, LU or Cholesky factorizations of sparse matrices A arising from discretization of partial differential equations for three-dimensional problems are often prohibitive due to the large amount of fill-in.

Examples:

1. Given a sparse LU factorization $A = PLUQ$ of a sparse nonsingular matrix $A \in \mathbb{C}^{n\times n}$, the solution $\mathbf{x}$ of the linear system $A\mathbf{x} = \mathbf{b}$ with any right-hand side $\mathbf{b} \in \mathbb{C}^n$ can be computed as follows:

$$\begin{aligned} &\text{Set} \quad \mathbf{c} = P^T\mathbf{b},\\ &\text{Solve} \quad L\mathbf{z} = \mathbf{c} \quad \text{for} \quad \mathbf{z},\\ &\text{Solve} \quad U\mathbf{y} = \mathbf{z} \quad \text{for} \quad \mathbf{y},\\ &\text{Set} \quad \mathbf{x} = Q^T\mathbf{y}. \end{aligned}$$

Since P and Q are permutation matrices, the first and the last steps are just reorderings of the entries of the vectors $\mathbf{b}$ and $\mathbf{y}$, respectively. Therefore, the main computational cost is the two triangular solves with L and U, which requires at most $2(nnz(L) + nnz(U))$ flops.

2. Given a sparse Cholesky factorization $A = PLL^*P^T$ of a sparse Hermitian positive definite matrix $A \in \mathbb{C}^{n\times n}$, the solution $\mathbf{x}$ of the linear system $A\mathbf{x} = \mathbf{b}$ with any right-hand side $\mathbf{b} \in \mathbb{C}^n$ can be computed as follows:

$$\begin{aligned} &\text{Set} \quad \mathbf{c} = P^T\mathbf{b},\\ &\text{Solve} \quad L\mathbf{z} = \mathbf{c} \quad \text{for} \quad \mathbf{z},\\ &\text{Solve} \quad L^*\mathbf{y} = \mathbf{z} \quad \text{for} \quad \mathbf{y},\\ &\text{Set} \quad \mathbf{x} = P^T\mathbf{y}. \end{aligned}$$

Since P is a permutation matrix, the first and the last steps are just reorderings of the entries of the vectors $\mathbf{b}$ and $\mathbf{y}$, respectively. Therefore, the main computational cost is the two triangular solves with L and L^*, which requires at most $4nnz(L)$ flops.

3. In large-scale matrix computations, sparse factorizations are often not applied to a given sparse matrix $A \in \mathbb{C}^{n\times n}$, but to a suitable "approximation" $A_0 \in \mathbb{C}^{n\times n}$ of A. For example, if sparse factorizations of A itself are prohibitive due to excessive fill-in, such approximations A_0 can often be obtained by computing an "incomplete" factorization of A that simply discards unwanted fill-in entries. Given a sparse LU factorization

$$A_0 = PLUQ$$

of a sparse nonsingular matrix $A_0 \in \mathbb{C}^{n\times n}$, which in some sense approximates the original matrix $A \in \mathbb{C}^{n\times n}$, one then uses iterative procedures that only involve matrix-vector products with the matrix

$$C := A_0^{-1}A = Q^TU^{-1}L^{-1}P^TA,$$

or possibly its transpose C^T. In the context of solving linear systems $A\mathbf{x} = \mathbf{b}$, the matrix A_0 is called a **preconditioner**, and the matrix C is called the **preconditioned** coefficient matrix.

In general, the matrix $C = A_0^{-1}A$ is full. However, if C is only used in the form of matrix-vector products, then there is no need to explicitly form C. Instead, for any $\mathbf{v} \in \mathbb{C}^n$, the result of the matrix-vector product $\mathbf{y} = C\mathbf{v}$ can be computed as follows:

$$\begin{aligned} &\text{Set} \quad \mathbf{c} = A\mathbf{v},\\ &\text{Set} \quad \mathbf{d} = P^T\mathbf{c},\\ &\text{Solve} \quad L\mathbf{f} = \mathbf{d} \quad \text{for} \quad \mathbf{f},\\ &\text{Solve} \quad U\mathbf{z} = \mathbf{f} \quad \text{for} \quad \mathbf{z},\\ &\text{Set} \quad \mathbf{y} = Q^T\mathbf{z}. \end{aligned}$$

Since P and Q are permutation matrices, the second and the last steps are just reorderings of the entries of the vectors $\mathbf{c}$ and $\mathbf{z}$, respectively. Therefore, the main computational cost is the matrix-vector product with the sparse matrix A in the first step, the triangular solve with L in the third step, and the triangular solve with U in the fourth step, which requires a total of at most $2(nnz(A) + nnz(L) + nnz(U))$ flops. Similarly, each matrix product with C^T can be computed with at most $2(nnz(A) + nnz(L) + nnz(U))$ flops. In particular, matrix-vector products with both C and C^T are fast.

4. For sparse Hermitian matrices $A = A^* \in \mathbb{C}^{n\times n}$, preconditioning is often applied in a symmetric manner. Suppose

$$A_0 = PLL^* P^T$$

is a sparse Cholesky factorization of a sparse matrix $A_0 = A_0^* \in \mathbb{C}^{n\times n}$, $A_0 \succ 0$, which in some sense approximates the original matrix A. Then the symmetrically preconditioned matrix C is defined as

$$C := (PL)^{-1} A(L^* P^T)^{-1} = L^{-1} P^T A P(L^*)^{-1}.$$

Note that $C = C^*$ is a Hermitian matrix. For any $\mathbf{v} \in \mathbb{C}^n$, the result of the matrix-vector product $\mathbf{y} = C\mathbf{v}$ can be computed as follows:

$$\begin{aligned} \text{Solve} \quad & L^*\mathbf{c} = \mathbf{v} \quad \text{for} \quad \mathbf{c}, \\ \text{Set} \quad & \mathbf{d} = P\mathbf{c}, \\ \text{Set} \quad & \mathbf{f} = A\mathbf{d}, \\ \text{Set} \quad & \mathbf{z} = P^T\mathbf{f}, \\ \text{Solve} \quad & L\mathbf{y} = \mathbf{z} \quad \text{for} \quad \mathbf{y}. \end{aligned}$$

The main computational cost is the triangular solve with L^* in the first step, the matrix-vector product with the sparse matrix A in the third step, and the triangular solve with L in the last step, which requires a total of at most $2(nnz(A) + 2nnz(L))$ flops. In particular, matrix-vector products with C are fast.

49.3 Krylov Subspaces

Petrov–Galerkin projections are often used in conjunction with Krylov subspaces. In this section, we present the basic concepts of Krylov subspaces. In the following, it is assumed that $C \in \mathbb{C}^{n\times n}$ and $\mathbf{r} \in \mathbb{C}^n$, $\mathbf{r} \neq \mathbf{0}$.

Definitions:

The sequence

$$\mathbf{r}, C\mathbf{r}, C^2\mathbf{r}, \ldots, C^{j-1}\mathbf{r}, \ldots$$

is called the **Krylov sequence** induced by C and $\mathbf{r}$.

Let $j \geq 1$. The subspace

$$K_j(C, \mathbf{r}) := \operatorname{span}\{\, \mathbf{r}, C\mathbf{r}, C^2\mathbf{r}, \ldots, C^{j-1}\mathbf{r} \,\}$$

of $\mathbb{C}^n$ spanned by the first j vectors of the Krylov sequence is called the **jth Krylov subspace** induced by C and $\mathbf{r}$.

A sequence of linearly independent vectors

$$\mathbf{v}_1, \mathbf{v}_2, \ldots, \mathbf{v}_j \in \mathbb{C}^n$$

is said to be a **nested basis** for the jth Krylov subspace $K_j(C, \mathbf{r})$ if

$$\operatorname{span}\{\, \mathbf{v}_1, \mathbf{v}_2, \ldots, \mathbf{v}_i \,\} = K_i(C, \mathbf{r}) \quad \text{for all} \quad i = 1, 2, \ldots, j.$$

Let $p(\lambda) = c_0 + c_1\lambda + c_2\lambda^2 + \cdots + c_{d-1}\lambda^{d-1} + \lambda^d$ be a monic polynomial of degree d with coefficients in $\mathbb{C}$. The **minimal polynomial of** C **with respect to r** is the unique monic polynomial of smallest possible degree for which $p(C)\mathbf{r} = \mathbf{0}$.

The **grade of** C **with respect to r**, $d(C, \mathbf{r})$, is the degree of the minimal polynomial of C and $\mathbf{r}$.

Facts:

The following facts can be found in [Hou75, Sect. 1.5], [SB02, Sect. 6.3], or [Saa03, Sect. 6.2].

1. The vectors
$$\mathbf{r}, C\mathbf{r}, C^2\mathbf{r}, \ldots, C^{j-1}\mathbf{r}$$
are linearly independent if and only if $j \leq d(C, \mathbf{r})$.
2. Let $d = d(C, \mathbf{r})$. The vectors
$$\mathbf{r}, C\mathbf{r}, C^2\mathbf{r}, \ldots, C^{d-2}\mathbf{r}, C^{d-1}\mathbf{r}, C^j\mathbf{r}$$
are linearly dependent for all $j > d$.
3. The dimension of the jth Krylov subspace $K_j(C, \mathbf{r})$ is given by
$$\dim K_j(C, \mathbf{r}) = \begin{cases} j & \text{if } j \leq d(C, \mathbf{r}), \\ d(C, \mathbf{r}) & \text{if } j > d(C, \mathbf{r}). \end{cases}$$
4. $d(C, \mathbf{r}) = \operatorname{rank}\,[\mathbf{r} \quad C\mathbf{r} \quad C^2\mathbf{r} \quad \cdots \quad C^{n-1}\mathbf{r}]$.

49.4 The Symmetric Lanczos Process

In this section, we assume that $C = C^* \in \mathbb{C}^{n\times n}$ is a Hermitian matrix and that $\mathbf{r} \in \mathbb{C}^n$, $\mathbf{r} \neq \mathbf{0}$ is a nonzero starting vector. We discuss the **symmetric Lanczos process** [Lan50] for constructing a nested basis for the Krylov subspace $K_j(C, \mathbf{r})$ induced by C and $\mathbf{r}$.

Algorithm (Symmetric Lanczos process)
Compute $\beta_1 = \|\mathbf{r}\|_2$, and set $\mathbf{v}_1 = \mathbf{r}/\beta_1$, and $\mathbf{v}_0 = \mathbf{0}$.
For $j = 1, 2, \ldots$, do:

1) Compute $\mathbf{v} = C\mathbf{v}_j$, and set $\mathbf{v} = \mathbf{v} - \mathbf{v}_{j-1}\beta_j$.
2) Compute $\alpha_j = \mathbf{v}_j^*\mathbf{v}$, and set $\mathbf{v} = \mathbf{v} - \mathbf{v}_j\alpha_j$.
3) Compute $\beta_{j+1} = \|\mathbf{v}\|_2$.
   If $\beta_{j+1} = 0$, stop.
   Otherwise, set $\mathbf{v}_{j+1} = \mathbf{v}/\beta_{j+1}$.

end for

Facts:

The following facts can be found in [CW85], [SB02, Sect. 6.5.3], or [Saa03, Sect. 6.6].

1. In exact arithmetic, the algorithm stops after a finite number of iterations. More precisely, it stops when $j = d(C, \mathbf{r})$ is reached.

2. The **Lanczos vectors**

$$\mathbf{v}_1, \mathbf{v}_2, \ldots, \mathbf{v}_j$$

generated during the first j iterations of the algorithm form a nested basis for the jth Krylov subspace $K_j(C, \mathbf{r})$.

3. The Lanczos vectors satisfy the three-term recurrence relations

$$\mathbf{v}_{i+1}\beta_{i+1} = C\mathbf{v}_i - \mathbf{v}_i\alpha_i - \mathbf{v}_{i-1}\beta_i, \quad i = 1, 2, \ldots, j.$$

4. These three-term recurrence relations can be written in compact matrix form as follows:

$$CV_j = V_j T_j + \beta_{j+1}\mathbf{v}_{j+1}\mathbf{e}_j^T = V_{j+1}T_j^{(e)}.$$

Here, we set

$$V_j = [\mathbf{v}_1 \quad \mathbf{v}_2 \quad \cdots \quad \mathbf{v}_j], \quad \mathbf{e}_j^T = [0 \quad 0 \quad \cdots \quad 0 \quad 1] \in \mathbb{R}^{1\times j},$$

$$T_j = \begin{bmatrix} \alpha_1 & \beta_2 & 0 & \cdots & 0 \\ \beta_2 & \alpha_2 & \beta_3 & \ddots & \vdots \\ 0 & \beta_3 & \ddots & \ddots & 0 \\ \vdots & \ddots & \ddots & \ddots & \beta_j \\ 0 & \cdots & 0 & \beta_j & \alpha_j \end{bmatrix}, \quad T_j^{(e)} = \begin{bmatrix} T_j \\ \beta_{j+1}\mathbf{e}_j^T \end{bmatrix}, \quad \text{and} \quad V_{j+1} = [V_j \quad \mathbf{v}_{j+1}].$$

Note that $T_j \in \mathbb{C}^{j\times j}$ and $T_j^{(e)} \in \mathbb{C}^{(j+1)\times j}$ are tridiagonal matrices.

5. In exact arithmetic, the Lanczos vectors are orthonormal. Since the Lanczos vectors are the columns of V_j, this orthonormality can be stated compactly as follows:

$$V_j^* V_j = I_j \quad \text{and} \quad V_j^* \mathbf{v}_{j+1} = \mathbf{0}.$$

6. These orthogonality relations, together with the above compact form of the three-term recurrence relations, imply that

$$T_j = V_j^* C V_j.$$

Thus, the jth **Lanczos matrix** T_j is the orthogonal Petrov–Galerkin projection of C onto the jth Krylov subspace $K_j(C, \mathbf{r})$.

7. The computational cost of each jth iteration of the symmetric Lanczos process is fixed, and it is dominated by the matrix-vector product $\mathbf{v} = C\mathbf{v}_j$. In particular, the computational cost for generating the orthogonal Petrov–Galerkin projection T_j of C is dominated by the j matrix-vector products with C.
8. If C is a sparse matrix or a preconditioned matrix with a sparse preconditioner, then the matrix–vector products with C are fast. In this case, the symmetric Lanczos process is a very efficient procedure for computing orthogonal Petrov–Galerkin projections T_j of C onto Krylov subspaces $K_j(C, \mathbf{r})$.
9. The three-term recurrence relations used to generate the Lanczos vectors explicitly enforce orthogonality only among each set of three consecutive vectors, $\mathbf{v}_{j-1}$, $\mathbf{v}_j$, and $\mathbf{v}_{j+1}$. As a consequence, in finite-precision arithmetic, round-off error will usually cause loss of orthogonality among all Lanczos vectors $\mathbf{v}_1, \mathbf{v}_2, \ldots, \mathbf{v}_{j+1}$.
10. For applications of the Lanczos process in large-scale matrix computations, this loss of orthogonality is often benign, and only delays convergence. More precisely, in such applications, the Lanczos matrix $T_j \in \mathbb{C}^{j\times j}$ for some $j \ll n$ is used to obtain an approximate solution of a matrix problem involving the large matrix $C \in \mathbb{C}^{n\times n}$. Due to round-off error and the resulting loss of orthogonality, the number j of iterations that is needed to obtain a satisfactory approximate solution is larger than the number of iterations that would be needed in exact arithmetic.

49.5 The Nonsymmetric Lanczos Process

In this section, we assume that $C \in \mathbb{C}^{n\times n}$ is a general square matrix, and that $\mathbf{r} \in \mathbb{C}^n$, $\mathbf{r} \neq \mathbf{0}$, and $\mathbf{l} \in \mathbb{C}^n$, $\mathbf{l} \neq \mathbf{0}$, is a pair of right and left nonzero starting vectors. The **nonsymmetric Lanczos process** [Lan50] is an extension of the symmetric Lanczos process that simultaneously constructs a nested basis for the Krylov subspace $K_j(C,\mathbf{r})$ induced by C and $\mathbf{r}$, and a nested basis for the Krylov subspace $K_j(C^T,\mathbf{l})$ induced by C^T and $\mathbf{l}$. In the context of the nonsymmetric Lanczos process, $K_j(C,\mathbf{r})$ is called the jth **right** Krylov subspace, and $K_j(C^T,\mathbf{l})$ is called the jth **left** Krylov subspace.

Algorithm (Nonsymmetric Lanczos process)
Compute $\rho_1 = \|\mathbf{r}\|_2$, $\eta_1 = \|\mathbf{l}\|_2$, and set $\mathbf{v}_1 = \mathbf{r}/\beta_1$, $\mathbf{w}_1 = \mathbf{l}/\eta_1$, $\mathbf{v}_0 = \mathbf{w}_0 = \mathbf{0}$, and $\delta_0 = 1$.
For $j = 1,2,\ldots,$ do:

1) Compute $\delta_j = \mathbf{w}_j^T\mathbf{v}_j$.
   If $\delta_j = 0$, stop.
2) Compute $\mathbf{v} = C\mathbf{v}_j$, and set $\beta_j = \eta_j\delta_j/\delta_{j-1}$ and $\mathbf{v} = \mathbf{v} - \mathbf{v}_{j-1}\beta_j$.
3) Compute $\alpha_j = \mathbf{w}_j^T\mathbf{v}$, and set $\mathbf{v} = \mathbf{v} - \mathbf{v}_j\alpha_j$.
4) Compute $\mathbf{w} = C^T\mathbf{w}_j$, and set $\gamma_j = \rho_j\delta_j/\delta_{j-1}$ and $\mathbf{w} = \mathbf{w} - \mathbf{w}_j\alpha_j - \mathbf{w}_{j-1}\gamma_j$.
5) Compute $\rho_{j+1} = \|\mathbf{v}\|_2$ and $\eta_{j+1} = \|\mathbf{w}\|_2$.
   If $\rho_{j+1} = 0$ or $\eta_{j+1} = 0$, stop.
   Otherwise, set $\mathbf{v}_{j+1} = \mathbf{v}/\rho_{j+1}$ and $\mathbf{w}_{j+1} = \mathbf{w}/\eta_{j+1}$.

end for

Facts:

The following facts can be found in [SB02, Sect. 8.7.3] or [Saa03, Sect. 7.1].

1. The occurrence of $\delta_j = 0$ in Step 1 of the nonsymmetric Lanczos process is called an **exact breakdown**. In finite-precision arithmetic, one also needs to check for $\delta_j \approx 0$, which is called a **near-breakdown**. It is possible to continue the nonsymmetric Lanczos process even if an exact breakdown or a near-breakdown has occurred, by using so-called "look-ahead" techniques. (See, e.g., [FGN93] and the references given there.) However, in practice, exact breakdowns and even near-breakdowns are fairly rare and, therefore, here we consider only the basic form of the nonsymmetric Lanczos process without look-ahead.
2. In exact arithmetic and if no exact breakdowns occur, the algorithm stops after a finite number of iterations. More precisely, it stops when $j = \min\{\, d(C,\mathbf{r}), d(C^T,\mathbf{l})\,\}$ is reached.
3. The **right Lanczos vectors** and the **left Lanczos vectors**

$$\mathbf{v}_1, \mathbf{v}_2, \ldots, \mathbf{v}_j \quad \text{and} \quad \mathbf{w}_1, \mathbf{w}_2, \ldots, \mathbf{w}_j$$

   generated during the first j iterations of the algorithm form a nested basis for the jth right Krylov subspace $K_j(C,\mathbf{r})$ and the jth left Krylov subspace $K_j(C^T,\mathbf{l})$, respectively.
4. The right and left Lanczos vectors satisfy the three-term recurrence relations

$$\mathbf{v}_{i+1}\rho_{i+1} = C\mathbf{v}_i - \mathbf{v}_i\alpha_i - \mathbf{v}_{i-1}\beta_i, \quad i = 1,2,\ldots,j,$$

   and

$$\mathbf{w}_{i+1}\eta_{i+1} = C^T\mathbf{w}_i - \mathbf{w}_i\alpha_i - \mathbf{w}_{i-1}\gamma_i, \quad i = 1,2,\ldots,j,$$

   respectively.

5. These three-term recurrence relations can be written in compact matrix form as follows:

$$CV_j = V_j T_j + \rho_{j+1}\mathbf{v}_{j+1}\mathbf{e}_j^T = V_{j+1}T_j^{(e)},$$
$$C^T W_j = W_j \tilde{T}_j + \eta_{j+1}\mathbf{w}_{j+1}\mathbf{e}_j^T.$$

Here, we set

$$V_j = [\mathbf{v}_1 \quad \mathbf{v}_2 \quad \cdots \quad \mathbf{v}_j], \quad W_j = [\mathbf{w}_1 \quad \mathbf{w}_2 \quad \cdots \quad \mathbf{w}_j],$$

$$T_j = \begin{bmatrix} \alpha_1 & \beta_2 & 0 & \cdots & 0 \\ \rho_2 & \alpha_2 & \beta_3 & \ddots & \vdots \\ 0 & \rho_3 & \ddots & \ddots & 0 \\ \vdots & \ddots & \ddots & \ddots & \beta_j \\ 0 & \cdots & 0 & \rho_j & \alpha_j \end{bmatrix}, \qquad \tilde{T}_j = \begin{bmatrix} \alpha_1 & \gamma_2 & 0 & \cdots & 0 \\ \eta_2 & \alpha_2 & \gamma_3 & \ddots & \vdots \\ 0 & \eta_3 & \ddots & \ddots & 0 \\ \vdots & \ddots & \ddots & \ddots & \gamma_j \\ 0 & \cdots & 0 & \eta_j & \alpha_j \end{bmatrix},$$

$$\mathbf{e}_j^T = [0 \quad 0 \quad \cdots \quad 0 \quad 1] \in \mathbb{R}^{1\times j}, \quad \text{and} \quad T_j^{(e)} = \begin{bmatrix} T_j \\ \rho_{j+1}\mathbf{e}_j^T \end{bmatrix}.$$

Note that T_j, $\tilde{T}_j \in \mathbb{C}^{j\times j}$, and $T_j^{(e)} \in \mathbb{C}^{(j+1)\times j}$ are tridiagonal matrices.

6. The matrix $T_j^{(e)}$ has full rank, i.e., rank $T_j^{(e)} = j$.
7. In exact arithmetic, the right and left Lanczos vectors are **biorthogonal** to each other, i.e.,

$$\mathbf{w}_i^T \mathbf{v}_k = \begin{cases} 0 & \text{if } i \neq k, \\ \delta_i & \text{if } i = k, \end{cases} \qquad \text{for all} \quad i, k = 1, 2, \ldots, j.$$

Since the right and left Lanczos vectors are the columns of V_j and W_j, respectively, the biorthogonality can be stated compactly as follows:

$$W_j^T V_j = D_j, \quad W_j^T \mathbf{v}_{j+1} = \mathbf{0}, \quad \text{and} \quad V_j^T \mathbf{w}_{j+1} = \mathbf{0}.$$

Here, D_j is the diagonal matrix

$$D_j = \operatorname{diag}(\delta_1, \delta_2, \ldots, \delta_j).$$

Note that D_j is nonsingular, as long as no exact breakdowns occur.

8. These biorthogonality relations, together with the above compact form of the three-term recurrence relations, imply that

$$T_j = D_j^{-1} V_j^* C V_j = \left(W_j^T V_j\right)^{-1} W_j^T C V_j.$$

Thus, the jth **Lanczos matrix** T_j is the oblique Petrov–Galerkin projection of C onto the jth right Krylov subspace $K_j(C, \mathbf{r})$, and orthogonally to the jth left Krylov subspace $K_j(C^T, \mathbf{l})$.

9. The matrices T_j and $\tilde{T}_j^T$ are diagonally similar:

$$\tilde{T}_j^T = D_j T_j D_j^{-1}.$$

10. The computational cost of each jth iteration of the nonsymmetric Lanczos process is fixed, and it is dominated by the matrix-vector product $\mathbf{v} = C\mathbf{v}_j$ with C and by the matrix-vector product $\mathbf{w} = C^T\mathbf{w}_j$ with C^T. In particular, the computational cost for generating the oblique Petrov–Galerkin projection T_j of C is dominated by the j matrix-vector products with C and the j matrix-vector products with C^T.

11. If C is a sparse matrix or a preconditioned matrix with a sparse preconditioner, then the matrix-vector products with C and C^T are fast. In this case, the nonsymmetric Lanczos process is a very efficient procedure for computing oblique Petrov–Galerkin projections T_j of C onto right Krylov subspaces $\mathcal{K}_j(C, \mathbf{r})$, and orthogonally to left Krylov subspaces $\mathcal{K}_j(C^T, \mathbf{l})$.
12. The three-term recurrence relations, which are used to generate the right and left Lanczos vectors, explicitly enforce biorthogonality only between three consecutive right vectors, $\mathbf{v}_{j-1}, \mathbf{v}_j, \mathbf{v}_{j+1}$, and three consecutive left vectors, $\mathbf{w}_{j-1}$, $\mathbf{w}_j$, $\mathbf{w}_{j+1}$. As a consequence, in finite-precision arithmetic, round-off error will usually cause loss of biorthogonality between all right vectors $\mathbf{v}_1, \mathbf{v}_2, \ldots, \mathbf{v}_{j+1}$ and all left vectors $\mathbf{w}_1, \mathbf{w}_2, \ldots, \mathbf{w}_{j+1}$.
13. For applications of the Lanczos process in large-scale matrix computations, this loss of orthogonality is often benign, and only delays convergence. More precisely, in such applications, the Lanczos matrix $T_j \in \mathbb{C}^{j\times j}$ for some $j \ll n$ is used to obtain an approximate solution of a matrix problem involving the large matrix $C \in \mathbb{C}^{n\times n}$. Due to round-off error and the resulting loss of biorthogonality, the number j of iterations that is needed to obtain a satisfactory approximate solution is larger than the number of iterations that would be needed in exact arithmetic. (See, e.g., [CW86].)
14. If $C = C^*$ is a Hermitian matrix and $\mathbf{l} = \overline{\mathbf{r}}$, i.e., the left starting vector $\mathbf{l}$ is the complex conjugate of the right starting vector $\mathbf{r}$, then the right and left Lanczos vectors satisfy

$$\mathbf{w}_i = \overline{\mathbf{v}_i} \quad \text{for all} \quad i = 1, 2, \ldots, j+1,$$

and the nonsymmetric Lanczos process reduces to the symmetric Lanczos process.

49.6 The Arnoldi Process

The **Arnoldi process** [Arn51] is another extension of the symmetric Lanczos process for Hermitian matrices to general square matrices. Unlike the nonsymmetric Lanczos process, which produces bases for both right and left Krylov subspaces, the Arnoldi process generates basis vectors only for the right Krylov subspaces. However, these basis vectors are constructed to be orthonormal, resulting in a numerical procedure that is much more robust than the nonsymmetric Lanczos process.

In this section, we assume that $C \in \mathbb{C}^{n\times n}$ is a general square matrix, and that $\mathbf{r} \in \mathbb{C}^n$, $\mathbf{r} \neq \mathbf{0}$, is a nonzero starting vector.

Algorithm (Arnoldi process)
Compute $\rho_1 = \|\mathbf{r}\|_2$, and set $\mathbf{v}_1 = \mathbf{r}/\rho_1$.
For $j = 1, 2, \ldots,$ do:

1) Compute $\mathbf{v} = C\mathbf{v}_j$.
2) For $i = 1, 2, \ldots, j$, do:
   Compute $h_{ij} = \mathbf{v}^*\mathbf{v}_i$, and set $\mathbf{v} = \mathbf{v} - \mathbf{v}_j h_{ij}$.
   end for
3) Compute $h_{j+1,j} = \|\mathbf{v}\|_2$.
   If $h_{j+1,j} = 0$, stop.
   Otherwise, set $\mathbf{v}_{j+1} = \mathbf{v}/h_{j+1,j}$.

end for

Facts:

The following facts can be found in [Saa03, Sect. 6.3].

1. In exact arithmetic, the algorithm stops after a finite number of iterations. More precisely, it stops when $j = d(C, \mathbf{r})$ is reached.

2. The **Arnoldi vectors**

$$\mathbf{v}_1, \mathbf{v}_2, \ldots, \mathbf{v}_j$$

generated during the first j iterations of the algorithm form a nested basis for the jth Krylov subspace $K_j(C, \mathbf{r})$.

3. The Arnoldi vectors satisfy the $(i+1)$-term recurrence relations

$$\mathbf{v}_{i+1} h_{i+1,i} = C\mathbf{v}_i - \mathbf{v}_i h_{ii} - \mathbf{v}_{i-1} h_{i-1,i} - \cdots - \mathbf{v}_2 h_{2i} - \mathbf{v}_1 h_{1i}, \quad i = 1, 2, \ldots, j.$$

These $(i+1)$-term recurrence relations can be written in compact matrix form as follows:

$$C V_j = V_j H_j + h_{j+1,j} \mathbf{v}_{j+1} \mathbf{e}_j^T = V_{j+1} H_j^{(e)}.$$

Here, we set

$$V_j = [\mathbf{v}_1 \quad \mathbf{v}_2 \quad \cdots \quad \mathbf{v}_j], \quad \mathbf{e}_j^T = [0 \quad 0 \quad \cdots \quad 0 \quad 1] \in \mathbb{R}^{1 \times j},$$

$$H_j = \begin{bmatrix} h_{11} & h_{12} & h_{13} & \cdots & h_{1j} \\ h_{21} & h_{22} & h_{23} & \ddots & \vdots \\ 0 & h_{32} & \ddots & \ddots & h_{j-2,j} \\ \vdots & \ddots & \ddots & \ddots & h_{j-1,j} \\ 0 & \cdots & 0 & h_{j,j-1} & h_{jj} \end{bmatrix},$$

$$H_j^{(e)} = \begin{bmatrix} H_j \\ h_{j+1,j} \mathbf{e}_j^T \end{bmatrix}, \quad \text{and} \quad V_{j+1} = [V_j \quad \mathbf{v}_{j+1}].$$

Note that $H_j \in \mathbb{C}^{j \times j}$ and $H_j^{(e)} \in \mathbb{C}^{(j+1) \times j}$ are upper Hessenberg matrices.

4. The matrix $H_j^{(e)}$ has full rank, i.e., rank $H_j^{(e)} = j$.
5. Since the Arnoldi vectors are the columns of V_j, this orthonormality can be stated compactly as follows:

$$V_j^* V_j = I_j \quad \text{and} \quad V_j^* \mathbf{v}_{j+1} = \mathbf{0}.$$

6. These orthogonality relations, together with the above compact form of the recurrence relations, imply that

$$H_j = V_j^* C V_j.$$

Thus, the jth **Arnoldi matrix** H_j is the orthogonal Petrov–Galerkin projection of C onto the jth Krylov subspace $K_j(C, \mathbf{r})$.

7. As in the case of the symmetric Lanczos process, each jth iteration of the Arnoldi process requires only a single matrix-vector product $\mathbf{v} = C\mathbf{v}_j$. If C is a sparse matrix or a preconditioned matrix with a sparse preconditioner, then the matrix-vector products with C are fast.
8. However, unlike the Lanczos process, the additional computations in each jth iteration do increase with j. In particular, each jth iteration requires the computation of j inner products of vectors of length n, and the computation of j SAXPY-type updates of the form $\mathbf{v} = \mathbf{v} - \mathbf{v}_j h_{ij}$ with vectors of length n.
9. For most large-scale matrix computations, the increasing work per iteration limits the number of iterations that the Arnoldi process can be run. Therefore, in practice, the Arnoldi process is usually combined with restarting; i.e., after a number of iterations (with the matrix C and starting vector $\mathbf{r}$), the algorithm is started again with the same matrix C, but a different starting vector, say $\mathbf{r}_1$.
10. On the other hand, the $(i+1)$-term recurrence relations used to generate the Arnoldi vectors explicitly enforce orthogonality among the first $i+1$ vectors, $\mathbf{v}_1, \mathbf{v}_2, \ldots, \mathbf{v}_{i+1}$. As a result, the Arnoldi process is much less susceptible to round-off error in finite-precision arithmetic than the Lanczos process.

49.7 Eigenvalue Computations

In this section, we consider the problem of computing a few eigenvalues, and possibly eigenvectors, of a large matrix $C \in \mathbb{C}^{n\times n}$. We assume that matrix-vector products with C are fast. In this case, orthogonal and, in the non-Hermitian case, oblique Petrov–Galerkin projections of C onto Krylov subspaces $K_j(C, \mathbf{r})$ can be computed efficiently, as long as $j \ll n$.

Facts:

The following facts can be found in [CW85], [CW86], and [BDD00].

1. Assume that $C = C^* \in \mathbb{C}^{n\times n}$ is a Hermitian matrix. We choose any nonzero starting vector $\mathbf{r} \in \mathbb{C}^n$, $\mathbf{r} \neq \mathbf{0}$, e.g., a vector with random entries, and run the symmetric Lanczos process. After j iterations of the algorithm, we have computed the jth Lanczos matrix T_j, which — in exact arithmetic — is the orthogonal Petrov–Galerkin projection of C onto the jth Krylov subspace $K_j(C, \mathbf{r})$. Neglecting the last term in the compact form of the three-term recurrence relations used in the first j iterations of the symmetric Lanczos process, we obtain the approximation
$$CV_j \approx V_j T_j.$$
This approximation suggests to use the j eigenvalues $\lambda_i^{(j)}$, $i = 1, 2, \ldots, j$, of the jth Lanczos matrix $T_j \in \mathbb{C}^{j\times j}$ as approximate eigenvalues of the original matrix C. Furthermore, if one is also interested in approximate eigenvectors, then the above approximation suggests to use
$$\mathbf{x}_i^{(j)} = V_j \mathbf{z}_i^{(j)} \in \mathbb{C}^n, \quad \text{where} \quad T_j \mathbf{z}_i^{(j)} = \mathbf{z}_i^{(j)} \lambda_i^{(j)}, \quad \mathbf{z}_i^{(j)} \neq \mathbf{0},$$
as an approximate eigenvector of C corresponding to the approximate eigenvalue $\lambda_i^{(j)}$ of C.
2. Assume that $C \in \mathbb{C}^{n\times n}$ is a general square matrix. Here one can use either the nonsymmetric Lanczos process or the Arnoldi process to obtain approximate eigenvalues.
3. In the case of the nonsymmetric Lanczos process, one chooses any nonzero starting vectors $\mathbf{r} \in \mathbb{C}^n$, $\mathbf{r} \neq \mathbf{0}$, and $\mathbf{l} \in \mathbb{C}^n$, $\mathbf{l} \neq \mathbf{0}$, $\mathbf{r} \in \mathbb{C}^n$, $\mathbf{r} \neq \mathbf{0}$. In analogy to the symmetric case, the eigenvalues of Lanczos matrix $T_j \in \mathbb{C}^{j\times j}$ computed by j iterations of the nonsymmetric Lanczos process are used as approximate eigenvalues of the original matrix C. Corresponding approximate right eigenvectors are given by the same formula as above. Furthermore, one can also obtain approximate left eigenvectors from the left eigenvectors of T_j and the first j left Lanczos vectors. A discussion of many practical aspects of using the nonsymmetric Lanczos process for eigenvalue computations can be found in [CW86].
4. In the case of the Arnoldi process, one only needs to choose a single nonzero starting vector $\mathbf{r} \in \mathbb{C}^n$, $\mathbf{r} \neq \mathbf{0}$. Here, one has the approximation
$$CV_j \approx V_j H_j,$$
where V_j is the matrix containing the first j Arnoldi vectors as columns and H_j is the jth Arnoldi matrix. The eigenvalues $\lambda_i^{(j)}$, $i = 1, 2, \ldots, j$, of $H_j \in \mathbb{C}^{j\times j}$ are used as approximate eigenvalues of C. Furthermore, for each i,
$$\mathbf{x}_i^{(j)} = V_j \mathbf{z}_i^{(j)} \in \mathbb{C}^n, \quad \text{where} \quad H_j \mathbf{z}_i^{(j)} = \mathbf{z}_i^{(j)} \lambda_i^{(j)}, \quad \mathbf{z}_i^{(j)} \neq \mathbf{0},$$
is an approximate eigenvector of C corresponding to the approximate eigenvalue $\lambda_i^{(j)}$ of C.

49.8 Linear Systems of Equations

In this section, we consider the problem of solving large systems of linear equations,
$$C\mathbf{x} = \mathbf{b},$$
where $C \in \mathbb{C}^{n\times n}$ is a nonsingular matrix and $\mathbf{b} \in \mathbb{C}^n$. We assume that any possible preconditioning was already applied and so, in general, C is a preconditioned version of the original coefficient matrix.

In particular, the matrix C may actually be dense. However, we assume that matrix-vector products with C and possibly C^T are fast. This is the case when C is a preconditioned version of a sparse matrix A and a preconditioner A_0 that allows a sparse LU or Cholesky factorization.

Facts:

The following facts can be found in [FGN92] or [Saa03].

1. Let $\mathbf{x}_0 \in \mathbb{C}^n$ be an arbitrary initial guess for the solution of the linear system, and denote by

$$\mathbf{r}_0 = \mathbf{b} - C\mathbf{x}_0$$

the corresponding residual vector. A Krylov subspace-based iterative method for the solution of the above linear system constructs a sequence of approximate solutions of the form

$$\mathbf{x}_j \in \mathbf{x}_0 + K_j(C, \mathbf{r}_0), \quad j = 1, 2, \ldots,$$

i.e., the jth iterate is an additive correction of the initial guess, where the correction is chosen from the jth Krylov subspace $K_j(C, \mathbf{r}_0)$ induced by the coefficient matrix C and the initial residual $\mathbf{r}_0$. Now let $V_j \in \mathbb{C}^{n\times j}$ be a matrix the columns of which form a nested basis for $K_j(C, \mathbf{r}_0)$. Then, any possible jth iterate can be parametrized in the form

$$\mathbf{x}_j = \mathbf{x}_0 + V_j\mathbf{z}_j, \quad \text{where} \quad \mathbf{z}_j \in \mathbb{C}^j.$$

Moreover, the corresponding residual vector is given by

$$\mathbf{r}_j = \mathbf{b} - C\mathbf{x}_j = \mathbf{r}_0 - CV_j\mathbf{z}_j.$$

Different Krylov subspace-based iterative methods are then obtained by specifying the choice of the basis matrix V_j and the choice of the parameter vector $\mathbf{z}_j$.

2. The **biconjugate gradient algorithm** (BCG) [Lan52] employs the nonsymmetric Lanczos process to generate nested bases for the right Krylov subspaces $K_j(C, \mathbf{r}_0)$ and the left Krylov subspaces $K_j(C^T, \mathbf{l})$. Here, $\mathbf{l} \in \mathbb{C}^n, \mathbf{l} \neq \mathbf{0}$, is an arbitrary nonzero starting vector. The biorthogonality of the right and left Lanczos vectors is exploited to construct the jth iterate $\mathbf{x}_j$ such that the corresponding residual vector $\mathbf{r}_j$ is orthogonal to the left Lanczos vectors, i.e., $W_j^T\mathbf{r}_j = \mathbf{0}$. Using the recurrence relations of the Lanczos process and the above relation for $\mathbf{r}_j$, one can show that the defining condition $W_j^T\mathbf{r}_j = \mathbf{0}$ is equivalent to $\mathbf{z}_j$ being the solution of the linear system

$$T_j\mathbf{z}_j = \mathbf{e}_1^{(j)}\rho_1,$$

where $\mathbf{e}_1^{(j)}$ denotes the first unit vector of length j. Moreover, the corresponding iterates $\mathbf{x}_j$ can be obtained via a simple update from the previous iterate $\mathbf{x}_{j-1}$, resulting in an elegant overall computational procedure. Unfortunately, in general, it cannot be guaranteed that all Lanczos matrices T_j are nonsingular. As a result, BCG iterates $\mathbf{x}_j$ may not exist for every j. More precisely, BCG breaks down if T_j is singular, and it exhibits erratic convergence behavior when T_j is nearly singular.

3. The possible breakdowns and the erratic convergence behavior can be avoided by replacing the $j \times j$ linear system $T_j\mathbf{z}_j = \mathbf{e}_1^{(j)}\rho_1$ by the $(j+1) \times j$ least-squares problem

$$\min_{\mathbf{z}\in\mathbb{C}^j} \left\| \mathbf{e}_1^{(j+1)}\rho_1 - T_j^{(e)}\mathbf{z} \right\|_2.$$

Since $T_j^{(e)} \in \mathbb{C}^{(j+1)\times j}$ always has full rank j, the above least-squares problem has a unique solution $\mathbf{z}_j$. The resulting iterative procedure is the **quasi-minimal residual method** (QMR) [FN91].

4. The **generalized minimal residual algorithm** (GMRES) [SS86] uses the Arnoldi process to generate orthonormal basis vectors for the Krylov subspaces $K_j(C, \mathbf{r}_0)$. The orthonormality of the columns of the Arnoldi basis matrix V_j allows one to choose $\mathbf{z}_j$ such that the residual vector $\mathbf{r}_j$ has the smallest possible norm, i.e.,

$$\|\mathbf{r}_j\|_2 = \|\mathbf{r}_0 - CV_j\mathbf{z}_j\|_2 = \min_{\mathbf{z}\in\mathbb{C}^j} \|\mathbf{r}_0 - CV_j\mathbf{z}\|_2.$$

Using the compact form of the recurrence relations used to generate the Arnoldi vectors, one readily verifies that the above minimal residual property is equivalent to $\mathbf{z}_j$ being the solution of the least-squares problem

$$\min_{\mathbf{z}\in\mathbb{C}^j} \left\| \mathbf{e}_1^{(j+1)} \rho_1 - H_j^{(e)} \mathbf{z} \right\|_2,$$

where $H_j^{(e)} \in \mathbb{C}^{(j+1)\times j}$ is an upper Hessenberg matrix.

5. The idea of quasi-minimization of the residual vector can also be applied to Lanczos-type iterations that, in each jth step, perform two matrix–vector products with C, instead of one product with C and one product with C^T. The resulting algorithm is called the **transpose-free quasi-minimal residual method** (TFQMR) [Fre93]. We stress that QMR and TFQMR produce different sequences of iterates and, thus, QMR and TFQMR are not mathematically equivalent algorithms.

49.9 Dimension Reduction of Linear Dynamical Systems

In this section, we discuss the application of the nonsymmetric Lanczos process to a large-scale matrix problem that arises in dimension reduction of time-invariant linear dynamical systems. A more detailed description can be found in [Fre03].

Definitions:

Let $A, E \in \mathbb{C}^{n\times n}$. The matrix pencil $A - sE$, $s \in \mathbb{C}$, is said to be **regular** if the matrix $A - sE$ is singular only for finitely many values $s \in \mathbb{C}$.

A **single-input, single-output, time-invariant linear dynamical system** is a system of differential-algebraic equations (DAEs) of the form

$$E\frac{d}{dt}\mathbf{x} = A\mathbf{x} + \mathbf{b}u(t),$$

$$y(t) = \mathbf{l}^T\mathbf{x}(t),$$

together with suitable initial conditions. Here, $A, E \in \mathbb{C}^{n\times n}$ are given matrices such that $A - sE$ is a regular matrix pencil, $\mathbf{b} \in \mathbb{C}^n$, $\mathbf{b} \neq \mathbf{0}$, and $\mathbf{l} \in \mathbb{C}^n$, $\mathbf{l} \neq \mathbf{0}$, are given nonzero vectors, $\mathbf{x}(t) \in \mathbb{C}^n$ is the vector of **state variables**, $u(t) \in \mathbb{C}$ is the given input function, $y(t) \in \mathbb{C}$ is the output function, and n is the **state-space dimension.**

The rational function

$$H : \mathbb{C} \mapsto \mathbb{C} \cup \infty, \quad H(s) := \mathbf{l}^T (sE - A)^{-1} \mathbf{b},$$

is called the **transfer function** of the above time-invariant linear dynamical system.

A **reduced-order model** of state-space dimension j $(< n)$ of the above system is a single-input, single-output, time-invariant linear dynamical system of the form

$$E_j\frac{d}{dt}\mathbf{z} = A_j\mathbf{z} + \mathbf{b}_j u(t),$$

$$y(t) = \mathbf{l}_j^T\mathbf{z}(t),$$

where $A_j, E_j \in \mathbb{C}^{j\times j}$ and $\mathbf{b}_j, \mathbf{l}_j \in \mathbb{C}^j$, together with suitable initial conditions.

Let $s_0 \in \mathbb{C}$ be such that the matrix $A - s_0E$ is nonsingular. A reduced-order model of state-space dimension j of the above system is said to be a **Padé model** about the expansion point s_0 if the matrices A_j, E_j and the vectors $\mathbf{b}_j, \mathbf{l}_j$ are chosen such that the Taylor expansions about s_0 of the transfer function H of the original system and of the reduced-order transfer function

$$H_j : \mathbb{C} \mapsto \mathbb{C} \cup \infty, \quad H_j(s) := \mathbf{l}_j^T(sE_j - A_j)^{-1}\mathbf{b}_j,$$

agree in as many leading Taylor coefficients as possible, i.e.,

$$H_j(s) = H(s) + \mathcal{O}\left((s - s_0)^{q(j)}\right),$$

where $q(j)$ is as large as possible.

Facts:

The following facts can be found in [FF94], [FF95], or [Fre03].

1. In the "generic" case, $q(j) = 2j$.
2. In the general case, $q(j) \geq 2j$; the case $q(j) > 2j$ occurs only in certain degenerate situations.
3. The transfer function H can be rewritten in terms of a single square matrix $C \in \mathbb{C}^{n\times n}$ as follows:

$$H(s) = \mathbf{l}^T\left(I_n + (s - s_0)C\right)^{-1}\mathbf{r}, \quad \text{where} \quad C := \left(s_0 E - A\right)^{-1} E, \quad \mathbf{r} := \left(s_0 E - A\right)^{-1}\mathbf{b}.$$

   Note that the matrix C can be viewed as a preconditioned version of the matrix E using the "shift-and-invert" preconditioner $s_0 E - A$.
4. In many cases, the state-space dimension n of the original time-invariant linear dynamical system is very large, but the large square matrices $A, E \in \mathbb{C}^{n\times n}$ are sparse. Furthermore, these matrices are usually such that sparse LU factorizations of the shift-and-invert preconditioner $s_0 E - A$ can be computed with limited amounts of fill-in. In this case, matrix-vector products with the preconditioned matrix C and its transpose C^T are fast.
5. The above definition of Padé models suggests the computation of these reduced-order models by first explicitly generating the leading $q(j)$ Taylor coefficients of H about the expansion point s_0, and then constructing the Padé model from these. However, this process is extremely ill-conditioned and numerically unstable. (See the discussion in [FF94] or [FF95].)
6. A much more stable way to compute Padé models without explicitly generating the Taylor coefficients is based on the nonsymmetric Lanczos process. The procedure is simply as follows: One uses the vectors $\mathbf{r}$ and $\mathbf{l}$ from the above representation of the transfer function H as right and left starting vectors, and applies the nonsymmetric Lanczos process to the preconditioned matrix C. After j iterations, the algorithm has produced the $j \times j$ tridiagonal Lanczos matrix T_j. The reduced-order model defined by

$$A_j := s_0 T_j - I_j, \quad E_j := T_j, \quad \mathbf{b}_j := (\mathbf{l}^T\mathbf{r})\,\mathbf{e}_1^{(j)}, \quad \mathbf{l}_j := \mathbf{e}_1^{(j)}$$

   is a Padé model of state-space dimension j about the expansion point s_0. Here, $\mathbf{e}_1^{(j)}$ denotes the first unit vector of length j.
7. In the large-scale case, Padé models of state-space dimension $j \ll n$ often provide very accurate approximations of the original system of state-space dimension n. In particular, this is the case for applications in VLSI circuit simulation. (See [FF95],[Fre03], and the references given there.)
8. Multiple-input multiple-output time-invariant linear dynamical systems are extensions of the above single-input single-output case with the vectors $\mathbf{b}$ and $\mathbf{l}$ replaced by matrices $B \in \mathbb{C}^{n\times m}$ and $L \in \mathbb{C}^{n\times p}$, respectively, where m is the number of inputs and p is the number of outputs. The approach outlined in this section can be extended to the general multiple-input multiple-output case. A suitable extension of the nonsymmetric Lanczos process that can handle multiple right and left starting vectors is needed in this case. For a discussion of such a Lanczos-type algorithm and its application in dimension reduction of general multiple-input multiple-output time-invariant linear dynamical systems, we refer the reader to [Fre03] and the references given there.

References

[Arn51] W.E. Arnoldi. The principle of minimized iterations in the solution of the matrix eigenvalue problem. *Quart. Appl. Math.*, 9:17–29, 1951.

[BDD00] Z. Bai, J. Demmel, J. Dongarra, A. Ruhe, and H. van der Vorst, Eds., *Templates for the Solution of Algebraic Eigenvalue Problems: A Practical Guide.* SIAM Publications, Philadelphia, PA, 2000.

[CW85] J.K. Cullum and R.A. Willoughby. *Lanczos Algorithms for Large Symmetric Eigenvalue Computations, Vol. 1, Theory.* Birkhäuser, Basel, Switzerland, 1985.

[CW86] J.K. Cullum and R.A. Willoughby. A practical procedure for computing eigenvalues of large sparse nonsymmetric matrices. In J.K. Cullum and R.A. Willoughby, Eds., *Large Scale Eigenvalue Problems*, pp. 193–240. North-Holland, Amsterdam, The Netherlands, 1986.

[DER89] I.S. Duff, A.M. Erisman, and J.K. Reid. *Direct Methods for Sparse Matrices.* Oxford University Press, Oxford, U.K., 1989.

[FF94] P. Feldmann and R.W. Freund. Efficient linear circuit analysis by Padé approximation via the Lanczos process. In *Proceedings of EURO-DAC '94 with EURO-VHDL '94*, pp. 170–175, Los Alamitos, CA, 1994. IEEE Computer Society Press.

[FF95] P. Feldmann and R.W. Freund. Efficient linear circuit analysis by Padé approximation via the Lanczos process. *IEEE Trans. Comp.-Aid. Des.*, 14:639–649, 1995.

[Fre93] R.W. Freund. A transpose-free quasi-minimal residual algorithm for non-Hermitian linear systems. *SIAM J. Sci. Comp.*, 14:470–482, 1993.

[Fre03] R.W. Freund. Model reduction methods based on Krylov subspaces. *Acta Numerica*, 12:267–319, 2003.

[FGN92] R.W. Freund, G.H. Golub, and N.M. Nachtigal. Iterative solution of linear systems. *Acta Numerica*, 1:57–100, 1992.

[FGN93] R.W. Freund, M.H. Gutknecht, and N.M. Nachtigal. An implementation of the look-ahead Lanczos algorithm for non-Hermitian matrices. *SIAM J. Sci. Comp.*, 14:137–158, 1993.

[FN91] R.W. Freund and N.M. Nachtigal. QMR: a quasi-minimal residual method for non-Hermitian linear systems. *Num. Math.*, 60:315–339, 1991.

[GMS05] P.E. Gill, W. Murray, and M.A. Saunders. SNOPT: An SQP algorithm for large-scale constrained optimization. *SIAM Rev.*, 47:99–131, 2005.

[Hou75] A.S. Householder. *The Theory of Matrices in Numerical Analysis.* Dover Publications, New York, 1975.

[KR91] N.K. Karmarkar and K.G. Ramakrishnan. Computational results of an interior point algorithm for large scale linear programming. *Math. Prog.*, 52:555–586, 1991.

[Lan50] C. Lanczos. An iteration method for the solution of the eigenvalue problem of linear differential and integral operators. *J. Res. Nat. Bur. Stand.*, 45:255–282, 1950.

[Lan52] C. Lanczos. Solution of systems of linear equations by minimized iterations. *J. Res. Nat. Bur. Stand.*, 49:33–53, 1952.

[LM05] A.N. Langville and C.D. Meyer. A survey of eigenvector methods for web information retrieval. *SIAM Rev.*, 47(1):135–161, 2005.

[NW99] J. Nocedal and S.J. Wright. *Numerical Optimization.* Springer-Verlag, New York, 1999.

[Saa03] Y. Saad. *Iterative Methods for Sparse Linear Systems.* SIAM Publications, Philadelphia, PA, 2nd ed., 2003.

[SS86] Y. Saad and M.H. Schultz. GMRES: A generalized minimal residual algorithm for solving nonsymmetric linear systems. *SIAM J. Sci. Statist. Comp.*, 7(3):856–869, 1986.

[SB02] J. Stoer and R. Bulirsch. *Introduction to Numerical Analysis.* Springer-Verlag, New York, 3rd ed., 2002.

[vdV03] H.A. van der Vorst. *Iterative Krylov Methods for Large Linear Systems.* Cambridge University Press, Cambridge, 2003.

[Wri97] S.J. Wright. *Primal-dual interior-point methods.* SIAM Publications, Philadelphia, PA, 1997.

IV

Applications

Applications to Optimization

Applications to Probability and Statistics

Applications to Analysis

Applications to Physical and Biological Sciences

Applications to Computer Science

Applications to Geometry

Applications to Algebra

Applications to Optimization

50

Linear Programming

Leonid N. Vaserstein
Penn State

We freely use the textbook [Vas03]. Additional references, including references to Web sites with software, can be found in [Vas03], [Ros00, Sec. 15.1], and INFORMS Resources (http://www.informs.org/Resources/).

50.1 What Is Linear Programming?

Definitions:

Optimization is maximization or minimization of a real-valued function, called the **objective function,** on a set, called the **feasible region** or the set of **feasible solutions**.

The values of function on the set are called **feasible values.**

An **optimal solution** (optimizer) is a feasible solution where the objective function reaches an **optimal value** (optimum), i.e., maximal or minimal value, respectively.

An optimization problem is **infeasible** if there are no feasible solutions, i.e., the feasible region is empty.

It is **unbounded** if the feasible values are arbitrary large, in the case of a maximization problem, or arbitrary small, in the case of a minimization problem.

A **mathematical program** is an optimization problem where the feasible region is a subset of $\mathbb{R}^n$, a finite dimensional real space; i.e., the objective function is a function of one or several real variables.

A **linear form** in variables $x_1, \ldots, x_n$ is $c_1x_1 + \cdots + c_nx_n$, where c_i are given numbers.

An **affine function** is a linear form plus a given number.

The term *linear function,* which is not used here, means a linear form in some textbooks and an affine function in others.

A **linear constraint** is one of the following three constraints: $f \le g$, $f = g$, $f \ge g$, where f and g are affine functions. In standard form, f is a linear form and g is a given number.

A **linear program** is a mathematical program where the objective function is an affine function and the feasible region is given by a finite system of linear constraints, all of them to be satisfied.

Facts:

For background reading on the material in this subsection see [Vas03].

1. Linear constraints with equality signs are known as linear equations. The main tool of simplex method, pivot steps, allows us to solve any system of linear equations.
2. In some textbooks, the objective function in a linear program is required to be a linear form. Dropping a constant in the objective function does not change optimal solutions, but the optimal value changes in the obvious way.
3. Solving an optimization problem usually means finding the optimal value and an optimal solution or showing that they do not exist. By comparison, solving a system of linear equations usually means finding all solutions. In both cases, in real life we only find approximate solutions.
4. Linear programs with one and two variables can be solved graphically. In the case of one variable x, the feasible region has one of the following forms: Empty, a point $x = a$, a finite interval $a \leq x \leq b$, with $a < b$, a ray $x \geq a$ or $x \leq a$, the whole line. The objective function $f = cx + d$ is represented by a straight line. Depending on the sign of c and whether we want maximize or minimize f, we move in the feasible region to the right or to the left as far as we can in search for an optimal solution.

   In the case of two variables, the feasible region is a closed convex set with finitely many vertices (corners). Together with the feasible region, we can draw in plane levels of the objective function. Unless the objective function is constant, every level (where the function takes a certain value) is a straight line. This picture allows us to see whether the program is feasible and bounded. If it is, the picture allows us to find a vertex which is optimal.
5. Linear programming is about formulating, collecting data, and solving linear programs, and also about analyzing and implementing solutions in real life.
6. Linear programming is an important part of mathematical programming. In its turn, mathematical programming is a part of *operations research*. *Systems engineering* and *management science* are engineering and business versions of operations research.
7. Every linear program is either infeasible, or unbounded, or has an optimal solution.

Examples:

1. Here are 3 linear forms in $x, y, z : 2x - 3y + 5z, x + z, y$.
2. Here are 3 affine functions of $x, y, z : 2x - 3y + 5z - 1, x + z + 3, y$.
3. Here are 3 functions of x, y, z that are not affine: $xy, x^2 + z^3, \sin z$.
4. The constraint $|x| \leq 1$ is not linear, but it is equivalent to a system of two linear constraints, $x \leq 1, x \geq -1$.
5. (An infeasible linear program) Here is a linear program:
   Maximize $x + y$ subject to $x \leq -1, x \geq 0$.

   It has two variables and two linear constraints. The objective function is a linear form. The program is infeasible.
6. (An unbounded linear program) Here is a linear program:
   Maximize $x + y$.

   This program has two variables and no constraints. The objective function is a linear form. The program is unbounded.
7. Here is a linear program:
   Minimize $x + y$ subject to $x \geq 1, y \geq 2$.

   This program has two variables and two linear constraints. The objective function is a linear form. The optimal value (the maximum) is 3. An optimal solution is $x = 1, y = 2$. It is unique.

50.2 Setting Up (Formulating) Linear Programs

Examples:

1. Finding the maximum of n given numbers $c_1, \ldots, c_n$ does not look like a linear program. However, it is equivalent to the following linear program with n variables x_i and $n+1$ linear constraints: $c_1x_1 + \cdots + c_nx_n \to \max$, all $x_i \geq 0$, $x_1 + \cdots + x_n = 1$.

   An equivalent linear program with one variable y and n linear constraints is $y \to \min$, $y \geq c_i$ for all i.

2. (Diet problem [Vas03, Ex. 2.1]). The general idea is to select a mix of different foods in such a way that basic nutritional requirements are satisfied at minimum cost. Our example is drastically simplified.

   According to the recommendations of a nutritionist, a person's daily requirements for protein, vitamin A, and calcium are as follows: 50 grams of protein, 4000 IUs (international units) of vitamin A, 1000 milligrams of calcium. For illustrative purposes, let us consider a diet consisting only of apples (raw, with skin), bananas (raw), carrots (raw), dates (domestic, natural, pitted, chopped), and eggs (whole, raw, fresh) and let us, if we can, determine the amount of each food to be consumed in order to meet the recommended dietary allowances (RDA) at minimal cost.

Food	Unit	Protein (g)	Vit. A (IU)	Calcium (mg)
Apple	1 medium (138 g)	0.3	73	9.6
Banana	1 medium (118 g)	1.2	96	7
Carrot	1 medium (72 g)	0.7	20253	19
Dates	1 cup (178 g)	3.5	890	57
Egg	1 medium (44 g)	5.5	279	22

   Since our goal is to meet the RDA with minimal cost, we also need to compile the costs of these foods:

Food		Cost (in cents)
1	apple	10
1	banana	15
1	carrot	5
1 cup	dates	60
1	egg	8

   Using these data, we can now set up a linear program. Let a, b, c, d, e be variables representing the quantities of the five foods we are going to use in the diet. The objective function to be minimized is the total cost function (in cents),

$$C = 10a + 15b + 5c + 60d + 8e,$$

   where the coefficients represent cost per unit of the five items under consideration.

   What are the constraints? Obviously,

$$a, b, c, d, e \geq 0. \qquad (i)$$

   These constraints are called *nonnegativity constraints.*

   Then, to ensure that the minimum daily requirements of protein, vitamin A, and calcium are satisfied, it is necessary that

$$\begin{cases} 0.3a + 1.2b + 0.7c + 3.5d + 5.5e \geq 50 \\ 73a + 96b + 20253c + 890d + 279e \geq 4000 \\ 9.6a + 7b + 19c + 57d + 22e \geq 1000, \end{cases} \qquad (ii)$$

where, for example, in the first constraint, the term $0.3a$ expresses the number of grams of protein in each apple multiplied by the quantity of apples needed in the diet, the second term $1.2b$ expresses the number of grams of protein in each banana multiplied by the quantity of bananas needed in the diet, and so forth.

Thus, we have a linear program with 5 variables and 8 linear constraints.

3. (Blending problem [Vas03, Ex. 2.2]). Many coins in different countries are made from cupronickel (75% copper, 25% nickel). Suppose that the four available alloys (scrap metals) A, B, C, D to be utilized to produce the coin contain the percentages of copper and nickel shown in the following table:

Alloy	A	B	C	D
% copper	90	80	70	60
% nickel	10	20	30	40
$/lb	1.2	1.4	1.7	1.9

The cost in dollars per pound of each alloy is given as the last row in the same table.

Notice that none of the four alloys contains the desired percentages of copper and nickel. Our goal is to combine these alloys into a new blend containing the desired percentages of copper and nickel for cupronickel while minimizing the cost. This lends itself to a linear program.

Let $a,\ b,\ c,\ d$ be the amounts of alloys A, B, C, D in pounds to make a pound of the new blend. Thus,

$$a,\ b,\ c,\ d \geq 0. \qquad (i)$$

Since the new blend will be composed exclusively from the four alloys, we have

$$a + b + c + d = 1. \qquad (ii)$$

The conditions on the composition of the new blend give

$$\begin{cases} .9a & + & .8b & + & .7c & + & .6d & = & .75 \\ .1a & + & .2b & + & .3c & + & .4d & = & .25. \end{cases} \qquad (iii)$$

For example, the first equality states that 90% of the amount of alloy A, plus 80% of the amount of alloy B, plus 70% of the amount of alloy C, plus 60% of the amount of alloy D will give the desired 75% of copper in a pound of the new blend. Likewise, the second equality gives the desired amount of nickel in the new blend.

Taking the preceding constraints into account, we minimize the cost function

$$C = 1.2a + 1.4b + 1.7c + 1.9d.$$

In this problem, all the constraints, except (i), are *equalities.* In fact, there are three linear equations and four unknowns. However, the three equations are not independent. For example, the sum of the equations in (iii) gives (ii). Thus, (ii) is redundant.

In general, a constraint is said to be *redundant* if it follows from the other constraints of our system. Since it contributes no new information regarding the solutions of the linear program, it can be dropped from consideration without changing the feasible set.

4. (Manufacturing problem [Vas03, Ex. 2.3]). We are now going to state a program in which the objective function, a profit function, is to be maximized. A factory produces three products: P1, P2, and P3. The unit of measure for each product is the standard-sized boxes into which the product is placed. The profit per box of P1, P2, and P3 is $2, $3, and $7, respectively. Denote by $x_1,\ x_2,\ x_3$ the number of boxes of P1, P2, and P3, respectively. So the profit function we want to maximize is

$$P = 2x_1 + 3x_2 + 7x_3.$$

The five resources used are raw materials R1 and R2, labor, working area, and time on a machine. There are 1200 lbs of R1 available, 300 lbs of R2, 40 employee-hours of labor, 8000 m^2 of working area, and 8 machine-hours on the machine.

The amount of each resource needed for a box of each of the products is given in the following table (which also includes the aforementioned data):

Resource	Unit	P1	P2	P3	‖ Available
R1	lb	40	20	60	‖ 1200
R2	lb	4	1	6	‖ 300
Labor	hour	.2	.7	2	‖ 40
Area	m^2	100	100	800	‖ 8000
Machine	hour	.1	.3	.6	‖ 8
Profit	$	2	3	7	‖ → max

As we see from this table, to produce a box of P1 we need 40 pounds of R1, 4 pounds of R2, 0.2 hours of labor, 100 m^2 of working area, and 0.1 hours on the machine. Also, the amount of resources needed to produce a box of P2 and P3 can be deduced from the table. The constraints are

$$x_1, x_2, x_3 \geq 0, \qquad (i)$$

and

$$\left\{\begin{array}{rcrcrcrl} 40x_1 & + & 20x_2 & + & 60x_3 & \leq & 1200 & \text{(pounds of R1)} \\ 4x_1 & + & x_2 & + & 6x_3 & \leq & 300 & \text{(pounds of R2)} \\ .2x_1 & + & .7x_2 & + & 2x_3 & \leq & 40 & \text{(hours of labor)} \\ 100x_1 & + & 100x_2 & + & 800x_3 & \leq & 8000 & \text{(area in m}^2\text{)} \\ .1x_1 & + & .3x_2 & + & .8x_3 & \leq & 8 & \text{(machine).} \end{array}\right. \qquad (ii)$$

5. (Transportation problem [Vas03, Ex. 2.4]). Another concern that manufacturers face daily is *transportation costs* for their products. Let us look at the following hypothetical situation and try to set it up as a linear program. A manufacturer of widgets has warehouses in Atlanta, Baltimore, and Chicago. The warehouse in Atlanta has 50 widgets in stock, the warehouse in Baltimore has 30 widgets in stock, and the warehouse in Chicago has 50 widgets in stock. There are retail stores in Detroit, Eugene, Fairview, Grove City, and Houston. The retail stores in Detroit, Eugene, Fairview, Grove City, and Houston need at least 25, 10, 20, 30, 15 widgets, respectively. Obviously, the manufacturer needs to ship widgets to all five stores from the three warehouses and he wants to do this in the cheapest possible way. This presents a perfect backdrop for a linear program to minimize shipping cost. To start, we need to know the cost of shipping one widget from each warehouse to each retail store. This is given by a shipping cost table.

	1.D	2.E	3.F	4.G	5.H
1. Atlanta	55	30	40	50	40
2. Baltimore	35	30	100	45	60
3. Chicago	40	60	95	35	30

Thus, it costs $30 to ship one unit of the product from Baltimore to Eugene (E), $95 from Chicago to Fairview (F), and so on.

In order to set this up as a linear program, we introduce variables that represent the number of units of product shipped from each warehouse to each store. We have numbered the warehouses according to their alphabetical order and we have enumerated the stores similarly. Let x_{ij}, for all $1 \leq i \leq 3$, $1 \leq j \leq 5$, represent the number of widgets shipped from warehouse #i

to store $\#j$. This gives us 15 unknowns. The objective function (the quantity to be minimized) is the shipping cost given by

$$\begin{aligned} C = {} & 55x_{11} + 30x_{12} + 40x_{13} + 50x_{14} + 40x_{15} \\ & +35x_{21} + 30x_{22} + 100x_{23} + 45x_{24} + 60x_{25} \\ & +40x_{31} + 60x_{32} + 95x_{33} + 35x_{34} + 30x_{35}, \end{aligned}$$

where $55x_{11}$ represents the cost of shipping one widget from the warehouse in Atlanta to the retail store in Detroit (D) multiplied by the number of widgets that will be shipped, and so forth.

What are the constraints? First, our 15 variables satisfy the condition that

$$x_{ij} \geq 0, \qquad \text{for all } 1 \leq i \leq 3,\ 1 \leq j \leq 5, \tag{i}$$

since shipping a negative amount of widgets makes no sense. Second, since the warehouse $\#i$ cannot ship more widgets than it has in stock, we get

$$\begin{cases} x_{11} + x_{12} + x_{13} + x_{14} + x_{15} \leq 50 \\ x_{21} + x_{22} + x_{23} + x_{24} + x_{25} \leq 30 \\ x_{31} + x_{32} + x_{33} + x_{34} + x_{35} \leq 50. \end{cases} \tag{ii}$$

Next, working with the amount of widgets that each retail store needs, we obtain the following five constraints:

$$\begin{cases} x_{11} + x_{21} + x_{31} \geq 25 \\ x_{12} + x_{22} + x_{32} \geq 10 \\ x_{13} + x_{23} + x_{33} \geq 20 \\ x_{14} + x_{24} + x_{34} \geq 30 \\ x_{15} + x_{25} + x_{35} \geq 15. \end{cases} \tag{iii}$$

The problem is now set up. It is a linear program with 15 variables and 23 linear constraints.

6. (Job assignment problem [Vas03, Ex. 2.5]). Suppose that a production manager must assign n workers to do n jobs. If every worker could perform each job at the same level of skill and efficiency, the job assignments could be issued arbitrarily. However, as we know, this is seldom the case. Thus, each of the n workers is evaluated according to the time he or she takes to perform each job. The time, given in hours, is expressed as a number greater than or equal to zero. Obviously, the goal is to assign workers to jobs in such a way that the total time is as small as possible. In order to set up the notation, we let c_{ij} be the time it takes for worker $\#i$ to perform job $\#j$. Then the times could naturally be written in a table. For example, take $n = 3$ and let the times be given as in the following table:

	a	b	c
A	10	70	40
B	20	60	10
C	10	20	90

We can examine all six assignments and find that the minimum value of the total time is 40. So, we conclude that the production manager would be wise to assign worker **A** to job **a**, worker **B** to job **c**, and worker **C** to job **b**.

In general, this method of selection is not good. The total number of possible ways of assigning jobs is $n! = n \times (n-1) \times (n-2) \times \cdots \times 2 \times 1$. This is an enormous number even for moderate n. For $n = 70$,

$$\begin{aligned} n! = {} & 11978571669969891796072783721689098736458938142 5 \\ & 46425857555362864628009582789845319680000000000000000. \end{aligned}$$

It has been estimated that if a Sun Workstation computer had started solving this problem at the time of the Big Bang, by looking at all possible job assignments, then by now it would not yet have finished its task.

Although is not obvious, the job assignment problem (with any number n of workers and jobs) can be expressed as a linear program. Namely, we set

$$x_{ij} = 0 \quad \text{or} \quad 1 \qquad (i)$$

depending on whether the worker i is assigned to do the job j. The total time is then

$$\sum_i \sum_j c_{ij} x_{ij} \text{(to be minimized)}. \qquad (ii)$$

The condition that every worker i is assigned to exactly one job is

$$\sum_j x_{ij} = 1 \quad \text{for all} \quad i. \qquad (iii)$$

The condition that exactly one worker is assigned to every job j is

$$\sum_i x_{ij} = 1 \quad \text{for all} \quad j. \qquad (iv)$$

The constraints (i) are not linear. If we replace them by linear constraints $x_{i,j} \geq 0$, then we obtain a linear program with n^2 variables and $n^2 + 2n$ linear constraints. Mathematically, it is a transportation problem with every demand and supply equal 1. As such, it can be solved by the simplex method (see below). (When $n = 70$ it takes seconds.)

The simplex method for transportation problem does not involve any divisions. Therefore, an optimal solution it gives integral values for all components $x_{i,j}$. Thus, the conditions (i) hold, and the simplex method solves the job assignment problem. (Another way to express this is that all vertices of the feasible region (iii) to (iv) satisfy (i).)

50.3 Standard and Canonical Forms for Linear Programs

Definitions:

LP in **canonical form** [Vas03] is $\mathbf{c}^T\mathbf{x} + d \to \min, \mathbf{x} \geq 0, A\mathbf{x} \leq \mathbf{b}$, where $\mathbf{x}$ is a column of distinct (decision) variables, $\mathbf{c}^T$ is a given row, d is a given number, A is a given matrix, and $\mathbf{b}$ is a given column.

LP in **standard form** [Vas03] is $\mathbf{c}^T\mathbf{x} + d \to \min, \mathbf{x} \geq 0, A\mathbf{x} = \mathbf{b}$, where $\mathbf{x}, \mathbf{c}, d, A$, and $\mathbf{b}$ are as in the previous paragraph.

The **slack variable** for a constraint $f \leq c$ is $s = c - f \geq 0$. The **surplus variable** for a constraint $f \geq c$ is $s = f - c \geq 0$.

Facts:

For background reading on the material in this section, see [Vas03].

1. Every LP can be written in normal as well as standard form using the following five little tricks (*normal* is a term used by some texts):
   (a) Maximization and minimization problems can be converted to each other. Namely, the problems $f(\mathbf{x}) \to \min, \mathbf{x} \in S$ and $-f(\mathbf{x}) \to \max, \mathbf{x} \in S$ are equivalent in the sense that they have the same optimal solutions and $\max = -\min$.
   (b) The equation $f = g$ is equivalent to the system of two inequalities, $f \geq g, g \leq f$.
   (c) The inequality $f \leq g$ is equivalent to the inequality $-f \geq -g$.

(d) The inequality $f \leq g$ is equivalent to $f + x = g, x \geq 0$, where $x = g - f$ is a new variable called a *slack variable.*

(e) A variable x unconstrained in sign can be replaced in our program by two new nonnegative variables, $x = x' - x''$, where $x', x'' \geq 0$.

2. The same tricks are sufficient for rewriting any linear program in different standard, canonical, and normal forms used in different textbooks and software packages. In most cases, all decision variables in these forms are assumed to be nonnegative.

Examples:

1. The canonical form $\mathbf{c}^T\mathbf{x}+d \to \min, \mathbf{x} \geq 0, A\mathbf{x} \leq \mathbf{b}$ can be rewritten in the following standard form:

$$\mathbf{c}^T\mathbf{x} + d \to \min,\ \mathbf{x} \geq 0,\ \mathbf{y} \geq 0,\ A\mathbf{x} + \mathbf{y} = \mathbf{b}.$$

2. The standard form $\mathbf{c}^T\mathbf{x}+d \to \min, \mathbf{x} \geq 0, A\mathbf{x} = \mathbf{b}$ can be rewritten in the following canonical form:

$$\mathbf{c}^T\mathbf{x} + d \to \min,\ \mathbf{x} \geq 0,\ A\mathbf{x} \leq \mathbf{b},\ -A\mathbf{x} \leq -\mathbf{b}.$$

3. The diet problem (Example 2 in Section 50.2) can be put in canonical form by replacing (ii) with

$$\begin{cases} -0.3a - 1.2b - 0.7c - 3.5d - 5.5e \geq -50 \\ -73a - 96b - 20253c - 890d - 279e \geq -4000 \\ -9.6a + 7b - 19c - 57d - 22e \geq -1000. \end{cases} \qquad (ii)$$

4. The blending problem (Example 3 in Section 50.2) is in standard form.

50.4 Standard Row Tableaux

Definitions:

A **standard row tableau** (of [Vas03]) is

$$\begin{matrix} & \mathbf{x}^T & 1 & \\ & \left[\begin{matrix} A \\ \mathbf{c}^T \end{matrix}\right. & \left.\begin{matrix} \mathbf{b} \\ d \end{matrix}\right] & \begin{matrix} = \mathbf{u} \\ \to \min,\ \mathbf{x} \geq 0,\ \mathbf{u} \geq 0 \end{matrix} \end{matrix} \qquad \text{(SRT)}$$

with given matrix A, columns $\mathbf{b}$ and $\mathbf{c}$, and number d, where all decision variables in $\mathbf{x}, \mathbf{u}$ are distinct.

This tableau means the following linear program:

$$A\mathbf{x} + \mathbf{b} = \mathbf{u} \geq 0,\ \mathbf{x} \geq 0,\ \mathbf{c}^T\mathbf{x} + d \to \min.$$

The **basic solution** for the standard tableau (SRT) is $\mathbf{x} = 0, \mathbf{u} = \mathbf{b}$. The corresponding value for the objective function is d.

A standard tableau (SRT) is **row feasible** if $\mathbf{b} \geq 0$, i.e., the basic solution is feasible. Graphically, a feasible tableau looks like

$$\begin{matrix} & \oplus & 1 & \\ & \left[\begin{matrix} * \\ * \end{matrix}\right. & \left.\begin{matrix} \oplus \\ * \end{matrix}\right] & \begin{matrix} = \oplus \\ \to \min, \end{matrix} \end{matrix}$$

where $\oplus$ stands for nonnegative entries or variables.

A standard tableau (SRT) is **optimal** if $\mathbf{b} \geq \mathbf{0}$ and $\mathbf{c} \geq \mathbf{0}$. Graphically, an optimal tableau looks like

$$\begin{array}{cc} \oplus & 1 \end{array}$$
$$\begin{bmatrix} * & \oplus \\ \oplus & * \end{bmatrix} \begin{array}{l} = \oplus \\ \to \min. \end{array}$$

A **bad row** in a standard tableau is a row of the form $[\ominus -] = \oplus$, where $\ominus$ stands for nonpositive entries, $-$ stands for a negative number, and $\oplus$ stands for a nonnegative variable.

A **bad column** in a standard tableau is a column of the form $\begin{array}{c} \oplus \\ \begin{bmatrix} \oplus \\ - \end{bmatrix} \end{array}$, where $\oplus$ stands for a nonnegative variable and nonnegative numbers and $-$ stands for a negative number.

Facts:

For background reading on the material in this section see [Vas03].

1. Standard row tableaux are used in simplex method; see Section 50.6 below.
2. The canonical form above can be written in a standard tableau as follows:

$$\begin{array}{cc} \mathbf{x}^T & 1 \end{array}$$
$$\begin{bmatrix} -A & \mathbf{b} \\ \mathbf{c}^T & d \end{bmatrix} \begin{array}{l} = \mathbf{u} \\ \to \min,\ \mathbf{x} \geq 0,\ \mathbf{u} \geq 0 \end{array}$$

   where $\mathbf{u} = \mathbf{b} - A\mathbf{x} \geq 0$.
3. The standard form above can be transformed into canonical form $\mathbf{c}^T\mathbf{x} + d \to \min, \mathbf{x} \geq 0, -A\mathbf{x} \leq -\mathbf{b}$, $A\mathbf{x} \leq \mathbf{b}$, which gives the following standard tableau:

$$\begin{array}{cc} \mathbf{x}^T & 1 \end{array}$$
$$\begin{bmatrix} -A & \mathbf{b} \\ A & -\mathbf{b} \\ \mathbf{c}^T & d \end{bmatrix} \begin{array}{l} = \mathbf{u} \\ = \mathbf{v} \\ \to \min,\ \mathbf{x} \geq 0;\ \mathbf{u}, \mathbf{v} \geq 0. \end{array}$$

   To get a smaller standard tableau, we can solve the system of linear equations $A\mathbf{x} = \mathbf{b}$. If there are no solutions, the linear program is infeasible. Otherwise, we can write the answer in the form $\mathbf{y} = B\mathbf{z} + \mathbf{b}'$, where the column $\mathbf{y}$ contains some variables in $\mathbf{x}$ and the column $\mathbf{z}$ consists of the rest of variables. This gives the standard tableau

$$\begin{array}{cc} \mathbf{z}^T & 1 \end{array}$$
$$\begin{bmatrix} B & \mathbf{b}' \\ c'^T & d' \end{bmatrix} \begin{array}{l} = \mathbf{y} \\ \to \min,\ \mathbf{y} \geq 0,\ \mathbf{z} \geq 0, \end{array}$$

   where $\mathbf{c}'^T\mathbf{z} + d'$ is the objective function $\mathbf{c}^T\mathbf{x} + \mathbf{b}$ expressed in the terms of $\mathbf{z}$.
4. The basic solution of an optimal tableau is optimal.
5. An optimal tableau allows us to describe all optimal solutions as follows. The variables on top with nonzero last entries in the corresponding columns must be zeros. Crossing out these columns and the last row, we obtain a system of linear constraints on the remaining variables describing the set of all optimal solutions. In particular, the basic solution is the only optimal solution if all entries in the $\mathbf{c}$-part are positive.
6. A bad row shows that the linear program is infeasible.
7. A bad column in a feasible tableau shows that the linear program is is unbounded.

Examples:

1. The linear programs in Example 1 of Section 50.2 and written in an SRT and in an SCT in Example 1 of Section 50.8 below.
2. Consider the blending problem, Example 3 in Section 50.2. We solve the system of linear equations for a, b and the objective function f and, hence, obtain the standard tableau

$$\begin{array}{ccc} c & d & 1 \end{array}$$
$$\begin{bmatrix} 1 & 2 & -0.5 \\ -2 & -3 & 1.5 \\ 0.1 & 0.1 & 1.5 \end{bmatrix} \begin{array}{l} = a \\ = b \\ = f \to \min. \end{array}$$

The tableau is not optimal and has no bad rows or columns. The associated LP can be solved graphically, working in the (c, d)-plane. In Example 1 of Section 50.6, we will solve this LP by the simplex method.

50.5 Pivoting

Definitions:

Given a system of linear equations $\mathbf{y} = A\mathbf{z} + \mathbf{b}$ solved for m variables, a **pivot step** solves it for a subset of m variables which differs from $\mathbf{y}$ in one variable.

Variables in $\mathbf{y}$ are called **basic variables.** The variables in $\mathbf{z}$ are called **nonbasic variables.**

Facts:

For background reading on the material in this section see [Vas03].

1. Thus, a pivot step switches a basic and a nonbasic variable. In other words, one variable leaves the basis and another variable enters the basis.
2. Here is the pivot rule:

$$\begin{array}{cc} x & y \end{array} \qquad\qquad \begin{array}{cc} u & y \end{array}$$
$$\begin{bmatrix} \alpha^* & \beta \\ \gamma & \delta \end{bmatrix} \begin{array}{l} = u \\ = v \end{array} \mapsto \begin{bmatrix} 1/\alpha & -\beta/\alpha \\ \gamma/\alpha & \delta - \beta\gamma/\alpha \end{bmatrix} \begin{array}{l} = x \\ = v. \end{array}$$

   We switch the two variables x and u. The pivot entry α marked by * must be nonzero, β represents any entry in the pivot row that is not the pivot entry, γ represents any entry in the pivot column that is not the pivot entry, and δ represents any entry outside the pivot row and column.
3. A way to solve any of linear equations $A\mathbf{x} = \mathbf{b}$ is to write it in the row tableau

$$\begin{array}{c} \mathbf{x}^T \\ [A] = \mathbf{b} \end{array}$$

   and move as many constants from the right margin to the top margin by pivot steps. After this we drop the rows that read $c = c$ with a constant c. If one of the rows reads $c_1 = c_2$ with distinct constants c_1, c_2, the system is infeasible. The terminal tableau has no constants remaining at the right margin. If no rows are left, the answer is $0 = 0$ (every $\mathbf{x}$ is a solution). If no variables are left at the right margin, we obtain the unique solution $\mathbf{x} = \mathbf{b}'$. Otherwise, we obtain the answer in the form $\mathbf{y} = B\mathbf{z} + \mathbf{b}'$ with nonempy disjoint sets of variables $\mathbf{y}$, $\mathbf{z}$.

   This method requires somewhat more computations than the Gauss elimination, but it solves the system with parametric $\mathbf{b}$.

Examples:

1. Here is a way to solve the system $\begin{cases} x + 2y = 3 \\ 4x + 7y = 5 \end{cases}$

    by two pivot steps:

$$
\begin{array}{cc} x & y \end{array} \\
\begin{bmatrix} 1^* & 2 \\ 4 & 7 \end{bmatrix} \begin{matrix} = 3 \\ = 5 \end{matrix}
\mapsto
\begin{array}{cc} 3 & y \end{array} \\
\begin{bmatrix} 1 & -2 \\ 4 & -1^* \end{bmatrix} \begin{matrix} = x \\ = 5 \end{matrix}
\mapsto
\begin{array}{cc} 3 & 5 \end{array} \\
\begin{bmatrix} -7 & 2 \\ 4 & -1 \end{bmatrix} \begin{matrix} = x \\ = y \end{matrix},
$$

i.e., $x = -11, y = 7$.

50.6   Simplex Method

The simplex method is the most common method for solving linear programs. It was suggested by Fourier for linear programs arising from linear approximation (see below). Important early contributions to linear programming were made by L. Walras and L. Kantorovich. The fortuitous synchronization of the advent of the computer and George B. Dantzig's reinvention of the simplex method in 1947 contributed to the explosive development of linear programming with applications to economics, business, industrial engineering, actuarial sciences, operations research, and game theory. For their work in linear programming, P. Samuelson (b. 1915) was awarded the Nobel Prize in Economics in 1970, and L. Kantorovich (1912–1986) and T. C. Koopmans (1910–1985) received the Nobel Prize in Economics in 1975.

Now we give the simplex method in terms of standard tableaux. The method consists of finitely many pivot steps, separated in two phases (stages). In Phase 1, we obtain either a feasible tableau or a bad row. In Phase 2, we work with feasible tableaux. We obtain either a bad column or an optimal tableau. Following is the scheme of the simplex method, where LP stands for the linear program:

$$
\begin{bmatrix}
& & & (bad\ row), & & & \\
& \text{Phase 1} & \nearrow & \text{LP is infeasible} & & & \\
\begin{pmatrix} initial \\ tableau \end{pmatrix} & - - - - - & & & & (bad\ column), & \\
& \text{pivot steps} & \searrow & \text{Phase 2} & \nearrow & \text{LP is unbounded} & \\
& & \begin{pmatrix} feasible \\ tableau \end{pmatrix} & - - - - - & & & \\
& & & \text{pivot steps} & \searrow & & \\
& & & & \begin{pmatrix} optimal \\ tableau \end{pmatrix}. & &
\end{bmatrix}
$$

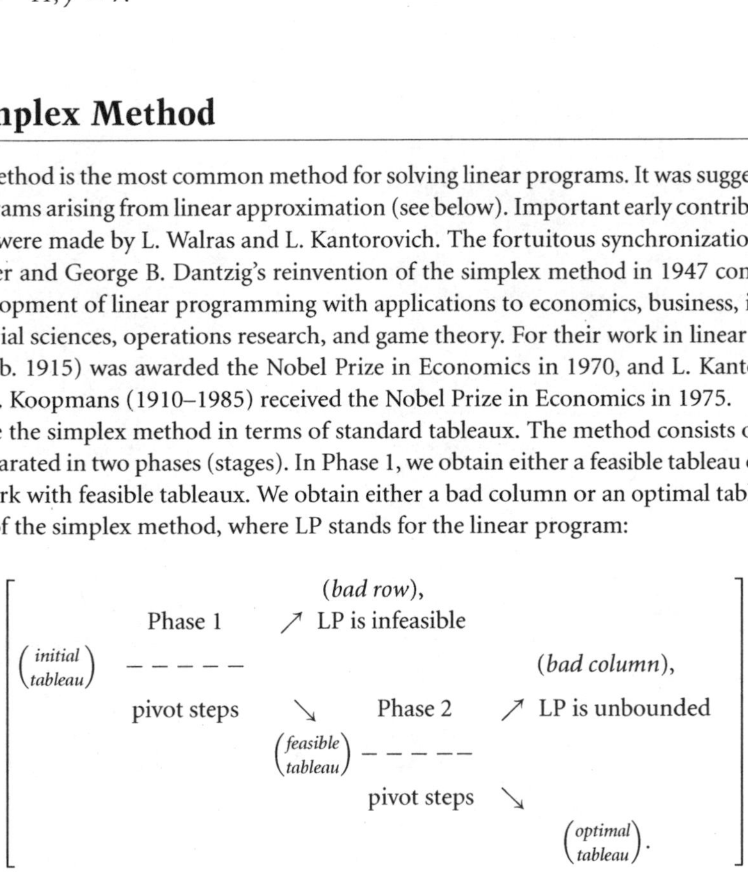

Both phases are similar and can be reduced to each other. We start Definitions with Phase 2 and Phase 1 in detail, dividing a programming loop (including a pivot step) for each phase into four substeps.

Definitions:

Phase 2 of simplex method. We start with a feasible tableau (SRT).

1. Is the tableau optimal? If yes, we write the answer: min $= d$ at $\mathbf{x} = 0, \mathbf{u} = \mathbf{b}$.
2. Are there any bad columns? If yes, the linear program is unbounded.
3. (Choosing a pivot entry) We choose a negative entry in the **c**-part, say, c_j in column j. Then we consider all negative entries $a_{i,j}$ above to choose our pivot entry. For every $a_{i,j} < 0$ we compute $b_i/a_{i,j}$ where b_i is the last entry in the row i. Then we maximize $b_i/a_{i,j}$ to obtain our pivot entry $a_{i,j}$.
4. Pivot and go to Substep 1.

Phase 1 of simplex method. We start with a standard tableau (SRT).

1. Is the tableau feasible? If yes, go to Phase 2.
2. Are there bad rows? If yes, the LP is infeasible.
3. (Choosing a pivot entry) Let the first negative number in the **b**-part be in the row i. Consider the subtableau consisting of the first i rows with the ith row multiplied by -1, and choose the pivot entry as in Substep 3 of Phase 2.
4. Pivot and go to Substep 1.

A pivot step (in Phase 1 or 2) is **degenerate** if the last entry in the pivot row is 0, i.e., the (basic) feasible solution stays the same.

A **cycle** in the simplex method (Phase 1 or 2) is a finite sequence of pivot steps which starts and ends with the same tableau.

Facts:

For background reading on the material in this section see [Vas03].

1. In Phase 2, the current value of the objective function (the last entry in the last row) either improves (decreases) or stays the same (if and only if the pivot step is degenerate).
2. In Phase 1, the first negative entry in the last column increases or stays the same (if and only if the pivot step is degenerate).
3. Following this method, we either terminate in finitely many pivot steps or, after a while, all our steps are degenerate, i.e., the basic solution does not change and we have a cycle.
4. The basic solutions in a cycle are all the same.
5. To prevent cycling, we can make a perturbation (small change in the **b**-part) such that none of the entries in this part of the tableau is ever 0 (see [Vas03] for details).
6. Another approach was suggested by Bland. We can make an ordered list of our variables, and then whenever there is a choice we choose the variable highest on the list (see [Vas03] for a reference and the proof that this rule prevents cycling). Bland's rule also turns the simplex method into a deterministic algorithm, i.e., it eliminates freedom of choices allowed by simplex method. Given an initial tableau and an ordered list of variables, Bland's rule dictates a unique finite sequence of pivot steps resulting in a terminal tableau.
7. With Bland's rule, the simplex method (both phases) terminates in at most $\binom{m+n}{n} - 1$ pivot steps where m is the number of basis variables and n is the number of nonbasic variables (so the tableau has $m+1$ rows and $n+1$ columns). The number $\binom{m+n}{n}$ here is an upper bound for the number of all standard tableaux that can be obtained from the initial tableau by pivot steps, up to permutation of the variables on the top (between themselves) and permutation of the variables at the right margin (between themselves).
8. If all data for an LP are rational numbers, then all entries of all standard tableaux are rational numbers. In particular, all basic solutions are rational. If the LP is feasible and bounded, then there is an optimal solution in rational numbers. However, like for a system of linear equations, the numerators and denominators could be so large that finding them could be impractical.

Examples:

1. Consider the blending problem, Example 3 in Section 50.2.
   We start with the standard tableau in Example 2 of Section 50.3. The simplex method allows two choices for the first pivot entry, namely, 1 or 2 in the first row. We pick 1 as the pivot entry:

$$
\begin{array}{ccc}
c & d & 1 \\
\end{array}
\begin{bmatrix}
1^* & 2 & -0.5 \\
-2 & -3 & 1.5 \\
0.1 & 0.1 & 1.5
\end{bmatrix}
\begin{array}{l}
= a \\
= b \\
= f \to \min
\end{array}
\qquad \mapsto \qquad
\begin{array}{ccc}
a & d & 1 \\
\end{array}
\begin{bmatrix}
1 & -2 & 0.5 \\
-2 & 1 & 0.5 \\
0.1 & -0.1 & 1.55
\end{bmatrix}
\begin{array}{l}
= c \\
= b \\
= f \to \min .
\end{array}
$$

Now the tableau is feasible, and we can proceed with Phase 2:

$$\begin{array}{c} \begin{matrix} a & d & 1 \end{matrix} \\ \left[\begin{matrix} 1 & -2^* & 0.5 \\ -2 & 1 & 0.5 \\ 0.1 & -0.1 & 1.55 \end{matrix}\right] \end{array} \begin{matrix} = c \\ = b \\ = f \to \min \end{matrix} \quad \mapsto \quad \begin{array}{c} \begin{matrix} a & c & 1 \end{matrix} \\ \left[\begin{matrix} 0.5 & -0.5 & 0.25 \\ -1.5 & -0.5 & 0.75 \\ 0.15 & 0.05 & 1.525 \end{matrix}\right] \end{array} \begin{matrix} = d \\ = b \\ = f \to \min \end{matrix} .$$

The tableau is optimal, so min $= 1.525$ at $a = 0, b = 0.75, c = 0$, and $d = 0.25$.

50.7 Geometric Interpretation of Phase 2

Definitions:

A subset S of $\mathbf{R}^N$ is **closed** if S contains the limit of any convergent sequence in S.

A **convex linear combination** or **mixture** of two points $\mathbf{x}, \mathbf{y}$, is $\alpha\mathbf{x} + (1 - \alpha)\mathbf{y}$, where $0 \leq \alpha \leq 1$. A particular case of a mixture is the **halfsum** $\mathbf{x}/2 + \mathbf{y}/2$.

The **line segment** connecting distinct points $\mathbf{x}, \mathbf{y}$ consists of all mixtures of $\mathbf{x}, \mathbf{y}$.

A set S is **convex** if it contains all mixtures of its points.

An **extreme point** in a convex set is a point that is not the halfsum of any two distinct points in the set, i.e., the set stays convex after removing the point.

In the case of a closed convex set with finitely many extreme points, the extreme points are called **vertices**.

Two extreme points in a convex set are **adjacent** if the set stays convex after deleting the line segment connecting the points.

Facts:

For background reading on the material in this section see [Vas03].

1. The feasible region for any linear program is a closed convex set with finitely many extreme points (vertices).
2. Consider a linear program and the associated feasible tableaux. The vertices of the feasible region are the basic solutions of the feasible tableaux. Permutation of rows or columns does not change the basic solution.
3. A degenerate pivot step does not change the basic solution. Any nondegenerate pivot step takes us from a vertex to an adjacent vertex with a better value for the objective function.
4. The set of optimal solutions for any linear program is a closed convex set with finitely many extreme points.
5. The number of pivot steps in Phase 2 without cycles (say, with Bland's rule) is less than the number of vertices in the feasible region.
6. For (SRT) with $m + 1$ rows and $n + 1$ columns, the number of vertices in the feasible region is at most $\binom{m+n}{m}$. When $n = 1$, this upper bound can be improved to 2. When $n = 2$, this upper bound can be improved to $m + 2$. It is unknown (in 2005) whether, for arbitrary m, n, a bound exists that is a polynomial in $m + n$.
7. The total number of pivot steps in both phases (with Bland's rule) is less than $\binom{m+n}{m}$.

50.8 Duality

Definitions:

A **standard column tableau** has the form

$$\begin{matrix} -y \\ 1 \end{matrix} \begin{array}{c} \left[\begin{matrix} A & b \\ \mathbf{c}^T & d \end{matrix}\right] \\ \begin{matrix} \| & \downarrow \\ \mathbf{v}^T & \max . \end{matrix} \end{array} \quad \mathbf{y} \geq 0,\ \mathbf{v} \geq 0 \qquad \text{(SCT)}$$

The **associated LP** is $-\mathbf{y}^T\mathbf{b} + d \to \max$, $-\mathbf{y}^T A + \mathbf{c}^T = \mathbf{v}^T \geq 0$, $\mathbf{y} \geq 0$.

The **basic solution** associated with (SCT) is $\mathbf{y} = 0$, $\mathbf{v} = \mathbf{c}$.

The tableau (SCT) is **column feasible** if $\mathbf{c} \geq 0$, i.e., the basic solution is feasible. (So, a tableau is optimal if and only if it is both row and column feasible.)

The linear program in (SCT) is **dual** to that in (SRT). We can write both as the row and the column problem with the same matrix:

$$\begin{array}{r} \\ -\mathbf{y} \\ 1 \\ \\ \end{array}\begin{array}{c} \begin{array}{cc} \mathbf{x}^T & 1 \end{array} \\ \left[\begin{array}{cc} A & \mathbf{b} \\ \mathbf{c}^T & d \end{array}\right] \\ \begin{array}{cc} = \mathbf{v}^T & = w \end{array} \end{array}\begin{array}{l} \\ = \mathbf{u} \\ = z \to \min \\ \to \max. \end{array} \qquad \begin{array}{l} \mathbf{x} \geq 0,\ \mathbf{u} \geq 0 \\ \mathbf{y} \geq 0,\ \mathbf{v} \geq 0 \end{array} \qquad \text{(ST)}$$

Facts:

For background reading on the material in this section, see [Vas03].

1. The linear program in the standard row tableau (SRT) can be written in the following standard column tableau:

$$\begin{array}{r} -\mathbf{x} \\ 1 \end{array}\left[\begin{array}{cc} -A^T & \mathbf{c} \\ \mathbf{b}^T & -d \end{array}\right] \quad \mathbf{x} \geq 0, \mathbf{u} \geq 0$$
$$\begin{array}{cc} \| & \downarrow \\ \mathbf{u}^T & \max. \end{array}$$

Then the dual problem becomes the row problem with the same matrix. This shows that the dual of the dual is the primal program.

2. The pivot rule for column and row tableaux is the same inside tableaux:

$$\begin{array}{r} -x \\ -u \end{array}\left[\begin{array}{cc} \alpha^* & \beta \\ \gamma & \delta \end{array}\right] \to \begin{array}{r} -u \\ -u \end{array}\left[\begin{array}{cc} 1/\alpha & -\beta/\alpha \\ \gamma/\alpha & \delta - \beta\gamma/\alpha \end{array}\right].$$
$$\begin{array}{cc} \| & \| \\ u & v \end{array} \qquad\qquad \begin{array}{cc} \| & \| \\ x & v \end{array}$$

3. If a linear program has an optimal solution, then the simplex method produces an optimal tableau. This tableau is also optimal for the dual program, hence both programs have the same optimal values (the duality theorem).
4. The duality theorem is a deep fact with several interpretations and applications. Geometrically, the duality theorem means that certain convex sets can be separated from points outside them by hyperplanes. We will see another interpretation of the duality theorem in the theory of matrix games (see below). For problems in economics, the dual problems and the duality theorem also have important economic interpretations (see examples below).
5. If a linear program is unbounded, then (after writing it in a standard row tableau) the simplex method produces a row feasible tableau with a bad column. The bad column shows that the dual problem is infeasible.
6. There is a standard tableau that has both a bad row and a bad column, hence there is an infeasible linear program such that the dual program is also infeasible.
7. Here is another way to express the duality theorem: Given a feasible solution for a linear program and a feasible solution for the dual problem, they are both optimal if and only if the feasible values are the same.
8. Given a feasible solution for a linear program and a feasible solution for the dual problem, they are both optimal if and only if for every pair of dual variables at least one value is 0, i.e., the *complementary slackness* condition $\mathbf{v}^T\mathbf{x} + \mathbf{y}^T\mathbf{u} = 0$ in terms of (ST) holds.
9. More precisely, given a feasible solution $(\mathbf{x}, \mathbf{u})$ for the row program in (ST) and a feasible solution $(\mathbf{y}, \mathbf{v})$ for the column program, the difference $(\mathbf{c}^T\mathbf{x} + d) - (-\mathbf{y}^T\mathbf{u} + d)$ between the corresponding feasible values is $\mathbf{v}^T\mathbf{x} + \mathbf{y}^T\mathbf{u}$.

10. All pairs $(\mathbf{x}, \mathbf{u}), (\mathbf{y}, \mathbf{v})$ of optimal solutions for the row and column programs in (ST) are described by the following system of linear constraints: $A\mathbf{x} + \mathbf{b} = \mathbf{u} \geq 0, \mathbf{x} \geq 0, -\mathbf{y}^T A + \mathbf{c} = \mathbf{v} \geq 0, \mathbf{y} \geq 0, \mathbf{v}^T \mathbf{x} + \mathbf{y}^T \mathbf{u} = 0$. This is a way to show that solving a linear program can be reduced to finding a feasible solution for a finite system of linear constraints.

Examples:

(Generalizations of Examples 1 to 4 in section 50.2 above and their duals):

1. The linear programs $c_1x_1 + \cdots + c_nx_n \to \max$, all $x_i \geq 0, x_1 + \cdots + x_n = 1$, and $y \to \min$, $y \geq c_i$ for all i in Example 1 are dual to each other. To see this, we use the standard tricks: $y = y' - y''$ with $y', y'' \geq 0$; $x_1 + \cdots + x_n = 1$ is equivalent to $x_1 + \cdots + x_n \leq 1, -x_1 - \cdots - x_n \leq -1$.
   We obtain the following standard tableau:

$$\begin{array}{c} \\ -\mathbf{x} \\ 1 \\ \end{array} \begin{array}{c} \begin{matrix} y' & y'' & 1 \end{matrix} \\ \left[\begin{matrix} \mathbf{J} & -\mathbf{J} & -\mathbf{c} \\ 1 & -1 & 0 \end{matrix}\right] \\ \begin{matrix} = \oplus & = \oplus & \to \end{matrix} \end{array} \begin{array}{l} \\ = \oplus \\ = y \to \min \\ \max, \end{array}$$

   where $\mathbf{J}$ is the column of n ones, $\mathbf{c} = [c_1, \ldots, c_n]^T$, and $\mathbf{x} = [x_1, \ldots, x_n]^T$.
2. Consider the general diet problem (a generalization of Example 2):

$$A\mathbf{x} \geq \mathbf{b},\ \mathbf{x} \geq 0,\ C = \mathbf{c}^T\mathbf{x} \to \min,$$

   where m variables in $\mathbf{x}$ represent different foods and n constraints in the system $A\mathbf{x} \geq \mathbf{b}$ represent ingredients. We want to satisfy given requirements $\mathbf{b}$ in ingredients using given foods at minimal cost C.
   On the other hand, we consider a warehouse that sells the ingredients at prices $y_1, \ldots, y_n \geq 0$. Its objective is to maximize the profit $P = \mathbf{y}^T\mathbf{b}$, matching the price for each food: $\mathbf{y}^T A \leq \mathbf{c}^T$.
   We can write both problems in a standard tableau using slack variables $\mathbf{u} = A\mathbf{x} - \mathbf{b} \geq 0$ and $\mathbf{v}^T = \mathbf{c}^T - \mathbf{y}^t A \geq 0$:

$$\begin{array}{c} \\ -\mathbf{y} \\ 1 \\ \end{array} \begin{array}{c} \begin{matrix} \mathbf{x}^T & 1 \end{matrix} \\ \left[\begin{matrix} A & -\mathbf{b} \\ \mathbf{c}^T & 0 \end{matrix}\right] \\ \begin{matrix} = \mathbf{v}^T & = P & \to \end{matrix} \end{array} \begin{array}{l} \\ = \mathbf{u} \\ = C \to \min \\ \max. \end{array} \qquad \begin{array}{l} \\ \mathbf{x} \geq 0, \mathbf{u} \geq 0 \\ \mathbf{y} \geq 0, \mathbf{v} \geq 0 \\ \\ \end{array}$$

   So, these two problems are dual to each other. In particular, the simplex method solves both problems and if both problems are feasible, then $\min(C) = \max(P)$.
   The optimal prices for the ingredients in the dual problem are also called *dual* or *shadow* prices. These prices tell us how the optimal value reacts to small changes in $\mathbf{b}$; see Section 50.9 below.
3. Consider the general mixing problem (a generalization of Example 3):

$$A\mathbf{x} = \mathbf{b}, \mathbf{x} \geq 0,\ C = \mathbf{c}^T\mathbf{x} \to \min,$$

   where m variables in $\mathbf{x}$ represent different alloys and n constraints in $A\mathbf{x} = \mathbf{b}$ represent elements. We want to satisfy given requirements b in elements using given alloys at minimal cost C.
   On the other hand, consider a dealer who buys and sells the elements at prices $y_1, \ldots, y_n$. A positive price means that the dealer sells, and negative price means that the dealer buys. Dealer's objective is to maximize the profit $P = \mathbf{y}^T\mathbf{b}$ matching the price for each alloy: $\mathbf{y}^T A \leq c$.
   To write the problems in standard tableaux, we use the standard tricks and artificial variables:

$$\mathbf{u}' = A\mathbf{x} - \mathbf{b} \geq 0,\ \mathbf{u}'' = -A\mathbf{x} + \mathbf{b} \geq 0;$$
$$\mathbf{v}^T = \mathbf{c}^T - \mathbf{y}^T A \geq 0;\ \mathbf{y} = \mathbf{y}' - \mathbf{y}'',\ \mathbf{y}' \geq 0,\ \mathbf{y}'' \geq 0.$$

Now we manage to write both problems in the same standard tableau:

$$\begin{array}{c} \\ -\mathbf{y}' \\ -\mathbf{y}'' \\ 1 \end{array} \begin{array}{c} \begin{array}{cc} \mathbf{x}^T & 1 \end{array} \\ \left[\begin{array}{cc} A & -\mathbf{b} \\ -A & \mathbf{b} \\ \mathbf{c}^T & 0 \end{array}\right] \\ = \mathbf{v}^T \;\; = P \end{array} \begin{array}{l} \\ = \mathbf{u}' \\ = \mathbf{u}'' \\ = C \to \min \\ \to \max. \end{array} \quad \begin{array}{l} \mathbf{x} \geq 0;\ \mathbf{u}', \mathbf{u}'' \geq 0 \\ \mathbf{y}', \mathbf{y}'' \geq 0;\ \mathbf{v} \geq 0 \end{array}$$

4. Consider a generalization of the manufacturing problem in Example 4:

$$P = \mathbf{c}^T\mathbf{x} \to \max,\ A\mathbf{x} \leq \mathbf{b},\ \mathbf{x} \geq 0,$$

where the variables in $\mathbf{x}$ are the amounts of products, P is the profit (or revenue) you want to maximize, constraints $A\mathbf{x} \leq \mathbf{b}$ correspond to resources (e.g., labor of different types, clean water you use, pollutants you emit, scarce raw materials), and the given column $\mathbf{b}$ consists of amounts of resources you have. Then the dual problem

$$\mathbf{y}^T\mathbf{b} \to \min,\ \mathbf{y}^T A \geq \mathbf{c}^T,\ \mathbf{y} \geq 0$$

admits the following interpretation. Your competitor, Bob, offers to buy you out at the following terms: You go out of business and he buys all resources you have at price $\mathbf{y}^T \geq 0$, matching your profit for every product you may want to produce, and he wants to minimize his cost.

Again Bob's optimal prices are your resource shadow prices by the duality theorem. The shadow price for a resource shows the increase in your profit per unit increase in the quantity b_0 of the resource available or decrease in the profit when the limit b_o decreases by one unit. While changing b_0, we do not change the limits for the other resources and any other data for our program. There are only finitely many values of b_0 for which the downward and upward shadow prices are different. One of these values could be the borderline between the values of b_0 for which the corresponding constraint is binding or nonbinding (in the sense that dropping of this constraint does not change the optimal value).

The shadow price of a resource cannot increase when supply b_0 of this resource increases (the law of diminishing returns, see the next section).

5. (General transportation problem and its dual). We have m warehouses and n retail stores. The warehouse #i has a_i widgets available and the store #j needs b_j widgets.

It is assumed that the following *balance condition* holds:

$$\sum_{i=1}^{m} a_i = \sum_{j=1}^{n} b_j.$$

If total supply is greater than total demand, the problem can be reduced to the one with the balance condition by introducing a fictitious store where the surplus can be moved at zero cost. If total supply is less than total demand, the program is infeasible.

The cost of shipping a widget from warehouse #i to store #j is denoted by c_{ij} and the number of widgets shipped from warehouse #i to store #j is denoted by x_{ij}. The linear program can be stated as follows:

$$\left\{ \begin{array}{ll} \text{minimize} & C(x_{11}, \ldots, x_{mn}) = \sum_{i=1}^{m}\sum_{j=1}^{n} c_{ij}\, x_{ij} \\ \text{subject to} & \sum_{i=1}^{n} x_{ij} \geq b_j, \quad j = 1, \ldots, m \\ & \sum_{j=1}^{m} x_{ij} \leq a_i, \quad i = 1, \ldots, n \\ & x_{ij} \geq 0,\ i = 1, \ldots, n,\ j = 1, \ldots, m. \end{array} \right.$$

The dual program is

$$-\sum_{i=1}^{m} a_i u_i + \sum_{j=1}^{n} b_j v_j \to \max,\ w_{ij} = c_{ij} + u_i - v_j \geq 0 \quad \text{for all} \quad i, j; u \geq 0, v \geq 0.$$

So, what is a possible meaning of the dual problem? The control variables u_i, v_j of the dual problem are called *potentials* or "zones." While the potentials correspond to the constraints on each retail store and each warehouse (or to the corresponding slack variables), there are other variables w_{ij} in the dual problem that correspond to the decision variable x_{ij} of the primal problem.

Imagine that you want to be a mover and suggest a simplified system of tariffs. Instead of mn numbers $c_{i,j}$, we use $m + n$ "zones." Namely, you assign a "zone" $u_i \geq 0$ $i = 1, 2$ to each of the warehouses and a "zone" $v_j \geq 0$ $j = 1, 2$ to each of the retail stores. The price you charge is $v_j - u_i$ instead of $c_{i,j}$. To beat competition, you want $v_j - u_i \leq c_{i,j}$ for all i, j. Your profit is $-\sum a_i u_i + \sum b_j v_j$ and you want to maximize it.

The simplex method for any transportation problem can be implemented using m by n tables rather than $mn + 1$ by $m + n + 1$ tableaux. Since all pivot entries are ± 1, no division is used. In particular, if all a_i, b_j are integers, we obtain an optimal solution with integral $x_{i,j}$.

Phase 1 is especially simple. If $m, n \geq 2$, we choose any position and write down the maximal possible number (namely, the minimum of supply and demand). Then we cross out the row or column and adjust the demand or supply respectively. If $m = 1 < n$, we cross out the column. If $n = 1 < m$, we cross out the row. If $m = n = 1$, then we cross out both the row and column. Thus, we find a feasible solution in $m + n - 1$ steps, which correspond to pivot steps, and the $m + n - 1$ selected positions correspond to the basic variables.

Every transportation problem with the balance condition has an optimal solution. (See [Vas03] for details.)

50.9 Sensitivity Analysis and Parametric Programming

Sensitivity analysis is concerned with how small changes in data affect the optimal value and optimal solutions, while large changes are studied in parametric programming.

Consider the linear program given by (SRT) with the last column being an affine function of a parameter t, i.e., we replace b, d by affine functions $b + b_1 t, d + d_1 t$ of a parameter t. Then the optimal value becomes a function $f(t)$ of t.

Facts:

For background reading on the material in this section, see [Vas03].

1. The function $f(t)$ is defined on a closed convex set S on the line, i.e., S is one of the following "intervals": empty set, a point, an interval $a \leq t \leq b$ with $a < b$, a ray $t \geq a$, a ray $t \leq a$, the whole line.
2. The function $f(t)$ is piece-wise linear, i.e., the interval S is a finite union of subintervals S_k with $f(t)$ being an affine function on each subinterval.
3. The function $f(t)$ is *convex* in the sense that the set of points in the plane below the plot is a convex set. In particular, the function $f(t)$ is continuous.
4. In parametric programming, there are methods for computing $f(t)$. The set S is covered by a finite set of tableaux optimal for various values of t. The number of these tableaux is at most $\binom{m+n}{m}$.
5. Suppose that the LP with $t = 0$ (i.e., the LP given by (SRT)) has an optimal tableau T_0. Let $\mathbf{x} = \mathbf{x}^{(0)}, \mathbf{u} = \mathbf{u}^{(0)}$ be the corresponding optimal solution (the basic solution), and let $\mathbf{y} = \mathbf{y}^{(0)}, \mathbf{v} = \mathbf{v}^{(0)}$ be the corresponding optimal solution for the dual problem (see (ST) for notation). Assume that the $\mathbf{b}$-part of T_0 has no zero entries. Then the function $f(t)$ is affine in an interval containing 0. Its slope, i.e., derivative $f'(0)$ at $t = 0$, is $f'(0) = d_1 + \mathbf{b_1}^T \mathbf{v}^{(0)}$. Thus, we can easily compute $f'(0)$

from T_0. In other words, the optimal tableaus give the partial derivatives of the optimal value with respect to the components of given **b** and d.
6. If we pivot the tableau with a parameter, the last column stays an affine function of the parameter and the rest of the tableau stays independent of parameter.
7. Similar facts are true if we introduce a parameter into the last row rather than into the last column.
8. If both the last row and the last column are affine functions of parameter t, then after pivot steps, the A-part stays independent of t, the **b**-part and **c**-part stay affine functions of t, but the d-part becomes a quadratic polynomial in t. So the optimal value is a piece-wise quadratic function of t.
9. If we want to maximize, say, profit (rather than minimize, say, cost), then the optimal value is concave (convex upward), i.e., the set of points below the graph is convex. The fact that the slope is nonincreasing is referred to as the *law of diminishing returns.*

50.10 Matrix Games

Matrix games are very closely related with linear programming.

Definitions:

A matrix game is given by a matrix A, the **payoff matrix,** of real numbers.

There are two players. The players could be humans, teams, computers, or animals. We call them *He* and *She.* He chooses a row, and She chooses a column. The corresponding entry in the matrix represents what she pays to him (the payoff of the row player). Games like chess, football, and blackjack can be thought of as (very large) matrix games. Every row represents a strategy, i.e., his decision what to do in every possible situation. Similarly, every column corresponds to a strategy for her. The rows are his **(pure) strategies** and columns are her (pure) strategies.

A **mixed strategy** is a mixture of pure strategies. In other words, a mixed strategy for him is a probability distribution on the rows and a mixed strategy for her is a probability distribution on the columns.

We write his mixed strategy as columns $\mathbf{p} = (p_i)$ with $\mathbf{p} \geq 0, \sum p_i = 1$. We write her mixed strategy as rows $\mathbf{q}^T = (q_j)$ with $\mathbf{q} \geq 0, \sum q_j = 1$.

The corresponding payoff is $\mathbf{p}^T A\mathbf{q}$, the mathematical expectation.

A pair of strategies (his strategy, her strategy) is an **equilibrium** or a **saddle point** if neither player can gain by changing his or her strategy. In other words, no player can do better by a unilateral change. (The last sentence can be used to define the term "equilibrium" in any game, while "saddle point" is used only for zero-sum two-player games; sometimes it is restricted to the pure joint strategies.)

In other words, at an equilibrium, the payoff $\mathbf{p}^T A\mathbf{q}$ has maximum as function of $\mathbf{p}$ and minimum as function of q.

His mixed strategy p is **optimal** if his worst case payoff $\min(\mathbf{p}^T A)$ is maximal. The maximum is the **value of the game** (for him). Her mixed strategy $\mathbf{q}^T$ is **optimal** if her worst case payoff $\min(-A\mathbf{q})$ is maximal. The maximum is the value of the game for her.

If the payoff matrix is skew-symmetric, the matrix game is called **symmetric.**

Facts:

For background reading on the material in this section, see [Vas03].

1. A pair (i, j) of pure strategies is a a saddle point if and only if the entry $a_{i,j}$ of the payoff matrix $A = a_{i,j}$ is both largest in its column and the smallest in its row.
2. Every matrix game has an equilibrium.
3. A pair (his strategy, her strategy) is an equilibrium if and only if both strategies are optimal.
4. His payoff $\mathbf{p}^T A\mathbf{q}$ at any equilibrium $(\mathbf{p}, \mathbf{q})$ equals his value of the game and equals the negative of her value of the game (the *minimax theorem*).
5. To solve a matrix game means to find an equilibrium and the value of the game.

6. Solving a matrix game can be reduced to solving the following dual pair of linear programs written in a standard tableau:

$$\begin{array}{r} \\ -\mathbf{P} \\ -\lambda' \\ -\lambda'' \\ 1 \end{array} \begin{array}{c} \begin{array}{cccc} \mathbf{q}^T & \mu' & \mu'' & 1 \end{array} \\ \left[\begin{array}{cccc} -A & \mathbf{1}_m & -\mathbf{1}_m & 0 \\ \mathbf{1}_n^T & 0 & 0 & -1 \\ -\mathbf{1}_n^T & 0 & 0 & 1 \\ 0 & 1 & -1 & 0 \end{array}\right] \\ \begin{array}{cccc} \| & \| & \| & \| \\ * & * & * & \lambda \rightarrow \end{array} \end{array} \begin{array}{l} \\ = * \geq 0 \\ = * \geq 0 \\ = * \geq 0 \\ = \mu \rightarrow \min \\ \\ \max. \end{array}$$

Here, A is the m by n payoff matrix, $\mathbf{1}_n^T$ is the row of n ones, $\mathbf{1}_m$ is the column of m ones, $\mathbf{p}$ is his mixed strategy, $\mathbf{q}^T$ is her mixed strategy. Note that her problem is the row problem and his problem is the column problem. Their problems are dual to each other. Since both problems have feasible solutions (take, for example, $\mathbf{p} = \mathbf{1}_m/m, \mathbf{q} = \mathbf{1}_n/n$), the duality theorem says that $\min(f) = \max(g)$. That is,

$$\text{his value of the game} = \max(\lambda) = \min(\mu) = -\text{her value of the game.}$$

Thus, the minimax theorem follows from the duality theorem.

7. We can save two rows and two columns and get a row feasible tableau as follows:

$$\begin{array}{r} \\ -\mathbf{p}/(\lambda + c) \\ 1 \end{array} \begin{array}{c} \begin{array}{cc} \mathbf{q}/(\mu + c) & 1 \end{array} \\ \left[\begin{array}{cc} -A - c\mathbf{1}_m\mathbf{1}_n^T & \mathbf{1}_m \\ -\mathbf{1}_n^T & 0 \end{array}\right] \\ \begin{array}{cc} \| & \| \\ * & -1/(\lambda + c) \end{array} \end{array} \begin{array}{l} \\ = * \geq 0 \\ = -1/(\mu + c) \rightarrow \min \\ \\ \rightarrow \max. \end{array}$$

Here we made sure that the value of game is positive by adding a number c to all entries of A, i.e., replacing A by $A + c\mathbf{1}_m\mathbf{1}_n^T$. E.g., $c = 1 - \min(A)$. Again his and her problems are dual to each other, since they share the same standard tableau. Since the tableau is feasible, we bypass Phase 1.

8. An arbitrary dual pair (ST) of linear programs can reduced to a symmetric matrix game, with the following payoff matrix

$$M = \begin{bmatrix} 0 & -A & -\mathbf{b} \\ A^T & 0 & -\mathbf{c} \\ \mathbf{b}^T & \mathbf{c}^T & 0 \end{bmatrix}.$$

Its size is $(m + n + 1) \times (m + n + 1)$, where $m \times n$ is the size of the matrix A.

9. The definition of equilibria makes sense for any game (not only for two-player zero-sum games). However, finding equilibria is not the same as solving the game when there are equilibria with different payoffs or when cooperation between players is possible and makes sense.

10. For any symmetric game, the value is 0, and the optimal strategies for the row and column players are the transposes of each other.

Examples:

1. [Vas03, Ex. 19.3] Solve the matrix game

$$A = \begin{bmatrix} 5 & 0 & 6 & 1 & -2 \\ 2 & 1 & 2 & 1 & 2 \\ -9 & 0 & 5 & 2 & -9 \\ -9 & -8 & 0 & 4 & 2 \end{bmatrix}.$$

We mark the maximal entries in every column by $*$. Then we mark the minimal entries in each row by $'$. The positions marked by both $*$ and $'$ are exactly the saddle points:

$$A = \begin{bmatrix} 5^* & 0 & 6^* & 1 & -2' \\ 2 & 1^{*\prime} & 2 & 1' & 2^* \\ -9' & 0 & 5 & 2 & -9' \\ -9' & -8 & 0 & 4^* & 2^* \end{bmatrix}.$$

In this example, the position $(i, j) = (2, 2)$ is the only saddle point. The corresponding payoff (the value of the game) is 1.

2. This small matrix game is known as Heads and Tails or Matching Pennies. We will call the players He and She. He chooses: heads (H) or tails (T). Independently, she chooses: H or T. If they choose the same, he pays her a penny. Otherwise, she pays him a penny. Here is his payoff in cents:

$$\begin{array}{cc} & \quad She \\ & \begin{array}{cc} H & \quad T \end{array} \\ He \begin{array}{c} H \\ T \end{array} & \begin{bmatrix} -1 & 1 \\ 1 & -1 \end{bmatrix}. \end{array}$$

There is no equilibrium in pure strategies. The only equilibrium in mixed strategies is $([1/2, 1/2]^T, [1/2, 1/2])$. The value of game is 0. The game is not symmetric in the usual sense. However the game is symmetric in the following sense: if we switch the players and also switch H and T for a player, then we get the same game.

3. Another game is Rock, Scissors, Paper. In this game two players simultaneously choose *Rock*, *Scissors*, or *Paper*, usually by a show of hand signals on the count of three, and the payoff function is defined by the rules *Rock breaks Scissors*, *Scissors cuts Paper*, *Paper covers Rock*, and every strategy ties against itself. Valuing a win at 1, a tie at 0, and a loss at -1, we can represent the game with the following matrix, where, for both players, strategy 1 is *Rock*, strategy 2 is *Scissors*, and strategy 3 is *Paper*:

$$A = \begin{bmatrix} 0 & 1 & -1 \\ -1 & 0 & 1 \\ 1 & -1 & 0 \end{bmatrix}.$$

The only optimal strategy for the column player is $\mathbf{q}^T = [1/3, 1/3, 1/3]$. Since the game is symmetric, the value is 0, and $\mathbf{q}$ is the only optimal strategy for the row player.

50.11 Linear Approximation

Definitions:

An l^p**-best linear approximation** (fit) of a given column $\mathbf{w}$ with n entries by the columns of a given m by n matrix A is $A\mathbf{x}$, where $\mathbf{x}$ is an optimal solution for $\|\mathbf{w} - A\mathbf{x}\|_p \to \min$. (See Chapter 37 for information about $\| \cdot \|_p$.)

In other words, we want the vector $\mathbf{w} - A\mathbf{x}$ of **residuals** (offsets, errors) to be smallest in a certain sense.

Facts:

For background reading on the material in this section, see [Vas03].

1. Most common values for p are $2, 1, \infty$. In statistics, usually $p = 2$ and the first column of the matrix A is the column of ones. In simple regression analysis, $n = 2$. In multiple regression, $n \geq 3$.

In time series analysis, the second column of A is an arithmetic progression representing time; typically, this column is $[1, 2, \ldots, m]^T$. (See Chapter 52 for more information.)

2. If $n = 1$ and the matrix A is a column of ones, we want to approximate given numbers w_i by one number Y. When $p = 2$, the best fit is the arithmetic mean $(\sum w_i)/m$. When $p = 1$, the best fits are the medians. When $p = \infty$, the best fit is the midrange $(\min(w_i) + \max(w_i))/2$.
3. When $p = 2$, the l^2-norm is the usual Euclidean norm, the most common way to measure the size of a vector, and this norm is used in Euclidean geometry. The best fit is known as the least squares fit. To find it, we drop a perpendicular from w onto the column space of A. In other words, we want the vector $\mathbf{w} - A\mathbf{x}$ to be orthogonal to all columns of A; that is, $A^T(\mathbf{w} - A\mathbf{x}) = 0$. This gives a system of n linear equations $A^T A\mathbf{x} = A^T\mathbf{w}$ for n unknowns in the column $\mathbf{x}$. The system always has a solution. Moreover, the best fit $A\mathbf{x}$ is the same for all solutions $\mathbf{x}$. In the case when w belongs to the column space, the best fit is w (this is true for all p). Otherwise, $\mathbf{x}$ is unique. (See Section 5.8 or Chapter 39 for more information about least squares methods.)
4. The best l^∞-fit is also known as the least-absolute-deviation fit and the Chebyshev approximation.
5. When $p = 1$, finding the best fit can be reduced to a linear program. Namely, we reduce the optimization problem with the objective function $\|\mathbf{e}\|_1 \to \min$, where $\mathbf{e} = (e_i) = \mathbf{w} - A\mathbf{x}$, to a linear program using m additional variables u_i such that $|e_i| \leq u_i$ for all i. We obtain the following linear program with $m + n$ variables a_j, u_i and $2m$ linear constraints:

$$\sum u_i \to min, \; -u_i \leq w_i - A_i\mathbf{x} \leq u_i \quad \text{for} \quad i = 1, \ldots, m,$$

where A_i is the ith row of the given matrix A.

6. When $p = \infty$, finding the best fit can also be reduced to a linear program. Namely reduce the optimization problem with the objective function $\|e\|_\infty \to \min$, where $\mathbf{e} = (e_i) = \mathbf{w} - A\mathbf{x}$, to a linear program using an additional variable u such that $e_i| \leq u$ for all i. A similar trick was used when we reduced solving matrix games to linear programming. We obtain the following linear program with $n + 1$ variables a_j, u and $2m$ linear constraints:

$$t \to \min, \; -u \leq w_i - A_i\mathbf{x} \leq u \quad \text{for} \quad i = 1, \ldots, m,$$

where A_i is the ith row of the given matrix A.

7. Any linear program can be reduced to finding a best l^∞-fit.

Examples:

1. [Vas03, Prob. 22.7] Find the best l^p-fit $w = ch^2$ for $p = 1, 2, \infty$ given the following data:

i	1	2	3
Height h in m	1.6	1.5	1.7
Weight w in kg	65	60	70

Compare the optimal values for c with those for the best fits of the form $w/h^2 = c$ with the same p (the number w/h^2 in kg/m^2 is known as BMI, the body mass index). Compare the minimums with those for the best fits of the form $w = b$ with the same p.

Solution

Case $p = 1$. We could convert this problem to a linear program with four variables and then solve it by the simplex method (see Fact 6 above). But we can just solve graphically the nonlinear program with the objective function

$$f(c) = |65 - 1.6^2c| + |60 - 1.5^2c| + |70 - 1.7^2c|$$

to be minimized and no constraints. The function $f(c)$ is piece-wise affine and convex. The optimal solution is $c = x_1 = 65/1.6^2 \approx 25.39$. This c equals the median of the three observed BMIs $65/1.6^2, 60/1.5^2$ and $70/1.7^2$. The optimal value is

$$\min = |65 - c_1 1.6^2| + |60 - c_1 1.5^2| + |70 - c_1 1.7^2| = 6.25.$$

To compare this with the best l^1-fit for the model $w = b$, we compute the median $b = x_1 = 65$ and the corresponding optimal values:

$$\min = |65 - x_1| + |60 - x_1| + |70 - x_1| = 10.$$

So the model $w = ch^2$ is better than $w = b$ for our data with the l^1-approach.

Case $p = 2$. Our optimization problem can be reduced to solving a linear equation for c (see Fact 3 above). Also we can solve the problem using calculus, taking advantage of the fact that our objective function

$$f(c) = (65 - 1.6^2 c)^2 + (60 - 1.5^2 c)^2 + (70 - 1.7^2 c)^2$$

is differentiable. The optimal solution is $c = x_2 = 2518500/99841 \approx 25.225$. This x_2 is not the mean of the observed BMIs, which is about 25.426. The optimal value is $\min \approx 18$.

The mean of w_i is 65, and the corresponding minimal value is $5^2 + 0^2 + 5^2 = 50$. So, again the model $w = ch^2$ is better than $w = b$.

Case $p = \infty$. We could reduce this problem to a linear program with two variables and then solve it by graphical method or simplex method (see Fact 7 above). But we can do a graphical method with one variable. The objective function to minimize now is

$$f(c) = \max(|65 - 1.6^2 c|, |60 - 1.5^2 c|, |70 - 1.7^2 c|).$$

This objective function $f(c)$ is piecewise affine and convex. The optimal solution is

$$c_\infty = 6500/257 \approx 25.29.$$

It differs from the midrange of the BMIs, which is about 25.44. The optimal value is ≈ 3.

On the other hand, the midrange of the weights w_i is 65, which gives $\min = 5$ for the model $w = b$ with the best l^∞-fit. So, again the model $w = ch^2$ is better than $w = b$.

2. [Vas03, Ex. 2, p. 255] A student is interested in the number w of integer points $[x, y]$ in the disc $x^2 + y^2 \leq r^2$ of radius r. He computed w for some r:

r	1	2	3	4	5	6	7	8	9
w	5	13	29	45	81	113	149	197	253

The student wants to approximate w by a simple formula $w = ar + b$ with constants a, b. But you feel that the area of the disc, πr^2 would be a better approximation and, hence, the best l^2-fit of the form $w = ar^2$ should work even better for the numbers above.

Compute the best l^2-fit (the least squares fit) for both models, $w = ar + b$ and $w = ar^2$, and find which is better. Also compare both optimal values with

$$\sum_{i=1}^{9} (w_i - \pi i^2)^2.$$

Solution

For our data, the equation $w = ar + b$ is the system of linear equations $A\mathbf{x} = \mathbf{w}$, where

$$\mathbf{w} = [5, 13, 29, 45, 81, 113, 149, 197, 253]^T,$$

$$A = \begin{bmatrix} 1 & 2 & 3 & 4 & 5 & 6 & 7 & 8 & 9 \\ 1 & 1 & 1 & 1 & 1 & 1 & 1 & 1 & 1 \end{bmatrix}^T, \text{ and } \mathbf{x} = \begin{bmatrix} a \\ b \end{bmatrix}.$$

The least-squares solution is the solution to $A^T A\mathbf{x} = A^T w$, i.e.,

$$\begin{bmatrix} 285 & 45 \\ 45 & 9 \end{bmatrix} \begin{bmatrix} a \\ b \end{bmatrix} = \begin{bmatrix} 6277 \\ 885 \end{bmatrix}.$$

So, $a = 463/15, b = -56$. The error

$$\sum_{i=1}^{9} (w_i - 463i/15 + 56)^2 = 48284/15 \approx 3218.93.$$

For $w = ar^2$, the system is $Ba = w$, where w is as above and $B = [1^2, 3^2, 3^2, 4^2, 5^2, 6^2, 7^2, 8^2, 9^2]^T$. The equation $B^T Ba = B^T w$ is $415398a = 47598$, hence $a = 23799/7699 \approx 3.09118$. The error

$$\sum_{i=1}^{9} (w_i - 23799i^2/7699)^2 = 2117089/7699 \approx 274.982.$$

So, the model $w = ar^2$ gives a much better least-squares fit than the model $w = ar + b$ although it has one parameter instead of two.

The model $w = \pi r^2$ without parameters gives the error

$$\sum_{i=1}^{9} (w_i - \pi i)^2 = 147409 - 95196\pi + 15398\pi^2 \approx 314.114.$$

It is known that $w_i / h_i^2 \to \pi$ as $i \to \infty$. Moreover, $|w_i - \pi r_i^2|/r_i$ is bounded as $i \to \infty$. It follows that $a \to \pi$ for the the best L_p-fit $w = ar^2$ for any $p \geq 1$ as we use data for $r = 1, 2, \ldots n$ with $n \to \infty$.

50.12 Interior Point Methods

Definitions:

A point $\mathbf{x}$ in a convex subset $X \subset \mathbf{R}^n$ is **interior** if

$$\{\mathbf{x} + (\mathbf{x} - \mathbf{y})\delta/\|\mathbf{y} - \mathbf{x}\|_2 : \mathbf{y} \in X, \mathbf{y} \neq \mathbf{x}\} \subset X$$

for some $\delta > 0$.

The **boundary** of X is the points in X, which are not interior.

An **interior solution** for a linear program is an interior point of the feasible region.

An interior point method for solving bounded linear programs starts with given interior solutions and produces a sequence of interior solutions which converges to an optimal solution.

An exterior point method for linear programs starts with a point outside the feasible region and produces a sequence of points which converges to feasible solution (in the case when the LP is feasible).

Facts:

For background reading on the material in this section see [Vas03].

1. In linear programming, the problem of finding a feasible solution (i.e., Phase 1) and the problem of finding an optimal solution starting from a feasible solution (i.e., Phase 2) can be reduced to each other. So, the difference between interior point methods and exterior point methods is not so sharp.
2. The simplex method, Phase 1, can be considered as an exterior point method which (theoretically) converges in finitely many steps. The ellipsoid method [Ha79] is truly an exterior point method. Hačijan proved an exponential convergence for the method.

3. In the simplex method, Phase 2, we travel from a vertex to an adjacent vertex and reach an optimal vertex (if the LP is bounded) in finitely many steps. The vertices are not interior solutions unless the feasible region consists of single point. If an optimal solution for a linear program is interior, then all feasible solutions are optimal.
4. The first interior point method was given by Brown in the context of matrix games. The convergence was proved by J. Robinson. J. von Neumann suggested a similar method with better convergence.
5. Karmarkar suggested an interior point method with exponential convergence. After him, many similar methods were suggested. They can be interpreted as follows. We modify our objective function by adding a penalty for approaching the boundary. Then we use Newton's method. The recent progress is related to understanding of properties of convex functions which make minimization by Newton's method efficient. Recent interior methods beat simplex methods for very large problems.
6. [Vas03, Sec. A8] On the other hand, it is known that for LPs with small numbers of variables (or, by duality, with a small number of constraints), there are faster methods than the simplex method.

   For example, consider an (SRT) with only two variables on top and m variables at the right margin (basis variables). The feasible region S has at most $m + 2$ vertices. Phase 2, starting with any vertex, terminates in at most $m + 1$ pivot steps.

   At each pivot step, it takes at most two comparisons to check whether the tableau is optimal or to find a pivot column. Then in m sign checks, at most m divisions, and at most $m - 1$ comparisons we find a pivot entry or a bad column.

   Next, we pivot to compute the new $3m + 3$ entries of the tableau. In one division, we find the new entry in the pivot row that is not the last entry (the last entry was computed before) and in $2m$ multiplications and $2m$ additions, we find the new entries outside the pivot row and column. Finally, we find the new entries in the pivot column in $m + 1$ divisions. So a pivot step, including finding a pivot entry and pivoting, can be done in $8m + 3$ *operations*—arithmetic operations and comparisons.

   Thus, Phase 2 can be done in $(m + 1)(8m + 3)$ operations. While small savings in this number are possible (e.g., at the $(m + 1)$-th pivot step we need to compute only the last column of the tableau), it is unlikely that any substantial reduction of this number for any modification of the simplex method can be achieved (in the worst case). Concerning the number of pivot steps, for any $m \geq 1$ it is clearly possible for S to be a bounded convex $(m + 2)$-gon, in which case for any vertex there is a linear objective function such that the simplex method requires exactly $\lfloor 1 + n/2 \rfloor$ pivot steps with only one choice of pivot entry at each step. It is also possible to construct an $(m+2)$-gon, an objective point, and an initial vertex such that m pivot steps, with unique choice of pivot entry at each step, are required. There is an example with two choices at the first step such that the first choice leads to the optimal solution while the second choice leads to m additional pivot steps with unique choice.

   Using a fast median search instead of the simplex method, it is possible to have Phase 2 done in $\leq 100m + 100$ operations.

References

[Ha79] L.G. Hačijan, A polynomial algorithm in linear programming. (Russian) *Dokl. Akad. Nauk SSSR* 244 (1979), 5, 1093–1096.

[Ros00] K.H. Rosen, Ed., *Handbook of Discrete and Combinatorial Mathematics,* CRC Press, Boca Raton, FL, 2000.

[Vas03] L.N. Vaserstein (in collaboration with C.C. Byrne), *Introduction to Linear Programming,* Prentice-Hall, Upper Saddle River, NJ, 2003.

51

Semidefinite Programming

Henry Wolkowicz
University of Waterloo

51.1 Introduction

Semidefinite programming, SDP, refers to optimization problems where variables X in the objective function and/or constraints can be symmetric matrices restricted to the cone of positive semidefinite matrices. (We restrict ourselves to real symmetric matrices, $\mathcal{S}_n$, since the vast majority of applications are for the real case. The complex case requires using the complex inner-product space.) An example of a *simple linear SDP* is

$$\text{(SDP)} \qquad \begin{array}{rll} p^* = & \min & \operatorname{tr} CX \\ & \text{subject to} & TX = \mathbf{b} \\ & & X \succeq 0, \end{array}$$

where $T : \mathcal{S}_n \to \mathbb{R}^m$. The details are given in the definitions in section 51.2; the SDP relaxation of the Max-Cut problem is given in Example 1 in this section. The linear SDP is a generalization of the standard *linear programming problem* (see Chapter 50), (LP) $\min \mathbf{c}^{\mathrm{T}}\mathbf{x}$, s.t. $A\mathbf{x} = b, \mathbf{x} \geq \mathbf{0}$, where $A \in \mathbb{R}^{m \times n}$, and the element-wise nonnegativity constraint $\mathbf{x} \geq \mathbf{0}$ is replaced by the positive semidefinite constraint, $X \succeq 0$. These cone constraints are the *hard constraints* for these linear problems, i.e., they introduce a combinatorial element into the problem. In the LP case, this refers to finding the *active constraints*, i.e., finding the *active set* at the optimum $\{j : x_j^* = 0\}$. For SDP, this translates to finding the set of zero eigenvalues of the optimum X^*. However, for SDP this is further complicated by the unknown eigenvectors.

But there are surprisingly many parallels between semidefinite and linear programming as well as interesting differences. The parallels include: elegant and strong duality theory, many important applications, and the successful interior-point approaches for handling the hard constraints. The differences include possible existence of duality gaps, as well as possible failure of strict complementarity. Since LP is such a well-known area, we emphasize the similarities and differences between LP and SDP as we progress through this chapter.

The study of SDP, or linear matrix inequalities (LMI), dates back to the work of Lyapunov on the stability of trajectories from differential equations in 1890, [Lya47]. More recently, applications in control theory appear in the work of [Yak62]. More details are given in [BEF94].

Cone Programming also called *generalized linear programming*, is a generalization of SDP [BF63]. Other books dealing with problems over cones that date back to the 1960s are [Lue69], [Hol75], and [Jah86]. More recently, SDP falls into the class of *symmetric or homogeneous cone programming* [Fay97], [SA01], which includes optimization over combinations of: the nonnegative orthant; the cone of positive semidefinite matrices, and the Lorentz (or second-order) cone.

Positive definite and Euclidean matrix completion problems are feasibility questions in SDP. Research dates back to the early 1980s [DG81], [GJM84]. This continues to be an active area of research [Lau98], [Joh90]. (More recently, positive definite completion theory is being used to solve large scale instances of SDP [BMZ02], [FKM01].)

Combinatorial optimization applications fueled the popularity of SDP in the 1980s, [Lov79] and the strong approximation results in [GW95] and Example 1 in this section. A related survey paper is [Ren99]. SDP continues to attract new applications in, e.g., molecular conformation [AKW99], sensor localization [Jin05], [SY06], and optimization over polynomials [HL03].

The fact that SDP is a convex program that can be solved to any desired accuracy in polynomial time follows from the seminal work in [NN94].

Examples:

1. Suppose that we are given the weighted undirected graph $G = (V, E)$ with vertex set $V = \{1, \dots, n\}$ and nonnegative edge weights w_{ij}, $ij \in E$ (with $w_{ij} = 0$, $ij \notin E$). The Max-Cut problem finds the partition of the vertices into two parts so as to maximize the sum of the weights on the edges that are cut (vertices in different parts). This problem arises in, e.g., physics and VLSI design. We can model this problem as a ± 1 program with quadratic functions.

$$\text{(MC)} \qquad \begin{array}{rll} \mu^* = & \max & \frac{1}{4}\sum_{ij=1}^{n} w_{ij}(1 - x_i x_j) = \frac{1}{4}\mathbf{x}^T L\mathbf{x} \\ & \text{subject to} & x_j^2 = 1, \quad j = 1, \dots, n, \end{array}$$

where L is the Laplacian matrix of the graph, $L_{ij} = \begin{cases} \sum_{k \neq i} w_{ik} & \text{if } i = j; \\ -w_{ij} & \text{if } i \neq j. \end{cases}$ We consider the simple example with vertex set $V = \{1, 2, 3\}$ and nonnegative edge weights $w_{12} = 1, w_{13} = 2, w_{23} = 2$. Here, the Laplacian matrix $L = \begin{bmatrix} 3 & -1 & -2 \\ -1 & 3 & -2 \\ -2 & -2 & 4 \end{bmatrix}$. The optimal solution places vertices 1 and 2 in one part and vertex 3 in the other; thus it cuts the edges 13 and 23. An optimal solution is $\mathbf{x}^* = \begin{bmatrix} 1 \\ 1 \\ -1 \end{bmatrix}$. with optimal value $\mu^* = 4$.

The well known semidefinite relaxation for the Max-Cut problem is obtained using the commutativity of the trace $\mathbf{x}^T L\mathbf{x} = \text{tr}\, \text{L}\mathbf{x}\mathbf{x}^{\text{T}} = \text{tr}\, \text{LX}$ and discarding the (hard) rank one constraint on X.

$$\text{(MCSDP)} \qquad \begin{array}{rll} \mu^* \leq \nu^* = & \max & \frac{1}{4}\text{tr}\, \text{LX} \\ & \text{subject to} & \text{diag}\,(\text{X}) = \text{e} \\ & & X \succeq 0, \end{array}$$

where the linear transformation $\text{diag}\,(\text{X}) : \mathcal{S}_\text{n} \rightarrow \mathbb{R}^\text{n}$ denotes the diagonal of X, and e is the vector of ones. The optimal solution to MCSDP is $X^* = \begin{bmatrix} 1 & 1 & -1 \\ 1 & 1 & -1 \\ -1 & -1 & 1 \end{bmatrix}$. (This is verified using duality; see Example 3 in section 51.4 below.) In fact, here X^* is rank one, and column one of X^* provides the optimal solution $\mathbf{x}^*$ of MC and the optimal value $\mu^* = \nu^* = 4$.

For this simple example, X^* was rank one and the SDP relaxation obtained the exact solution of the NP hard MC problem. This is not true in general. However, the MCSDP relaxation is a surprisingly strong relaxation for MC. A randomized approximation algorithm based on MCSDP provides a ± 1 vector $\mathbf{x}$ with objective value at least .87856 times the optimal value μ^* [GW94]. Empirical tests obtain even stronger results.

51.2 Specific Notation and Preliminary Results

Note that in this section all matrices are real.

Definitions:

For general $m \times n$ matrices $M, N \in \mathbb{R}^{m\times n}$, we use the matrix inner product $\langle M, N\rangle = \mathrm{tr}\, N^T M$. The corresponding norm is $\|M\|_F = \sqrt{\mathrm{tr}\, M^T M}$, the **Frobenius norm**.

$\mathcal{S}_n$ denotes the space of **symmetric** $n \times n$ **matrices.**

The set of **positive semidefinite (respectively definite) matrices** is denoted by PSD (respectively, PD); $X \succeq 0$ means $X \in$ PSD.

The **linear SDP** is

$$\text{(SDP)} \qquad p^* = \begin{array}{ll} \min & \mathrm{tr}\, CX \\ \text{subject to} & TX = \mathbf{b} \\ & X \succeq 0, \end{array}$$

where the variable $X \in \mathcal{S}_n$, the linear transformation $T : \mathcal{S}_n \to \mathbb{R}^m$, the vector $\mathbf{b} \in \mathbb{R}^m$, and the matrix $C \in \mathcal{S}_n$ are (given) problem parameters. The constraint $TX = \mathbf{b}$ can be written concretely as $\mathrm{tr}\, A_k X = b_k$, for some given $A_k \in \mathcal{S}_n$, $k = 1, \ldots, m$. Thus, SDP can be expressed explicitly as

$$\text{(SDP)} \qquad p^* = \begin{array}{ll} \min & \sum_{ij} C_{ij} X_{ij} \\ \text{subject to} & \sum_{ij} A_{ijk} X_{ij} = b_k, \quad k = 1, \ldots, m \\ & X \succeq 0, \end{array}$$

where A_{ijk} denotes the ij element of the symmetric matrix A_k.

For $M \in \mathbb{R}^{m\times n}$, $\mathrm{vec}\,(M) \in \mathbb{R}^{mn}$ is the vector formed from the columns of M.

For $S \in \mathcal{S}_n$, $\mathrm{svec}\,(S) \in \mathbb{R}^{\frac{n(n+1)}{2}}$ is the vector formed from the upper triangular part of S, taken columnwise, with the strict upper triangular part multiplied by $\sqrt{2}$. The inverse of svec is denoted sMat.

For $U \in \mathcal{S}_n$, the **symmetric Kronecker product** is defined by

$$(M \overset{s}{\otimes} N)\mathrm{svec}\,(U) = \mathrm{svec}\left(\frac{1}{2}(NUM^T + MUN^T)\right).$$

For a linear transformation between two vector spaces $T : V \to W$, the **adjoint linear transformation** is denoted $T^{\mathrm{adj}} : W \to V$, i.e., it is the unique linear transformation that satisfies the adjoint equation for the inner products in the respective vector spaces

$$\langle T\mathbf{v}, \mathbf{w}\rangle_W = \langle \mathbf{v}, T^{\mathrm{adj}}\mathbf{w}\rangle_V, \quad \forall\, \mathbf{v} \in V,\ \mathbf{w} \in W.$$

If $V = W$, then this reduces to the definition of the **adjoint of a linear operator** given in Chapter 5.3.

We follow the standard notation used in SDP: for $\mathbf{y} \in \mathbb{R}^n$, $\mathrm{Diag}\,(\mathbf{y})$ denotes the diagonal matrix formed using the vector $\mathbf{y}$, while the adjoint linear transformation is $\mathrm{diag}\,(X) = \mathrm{Diag}^{\mathrm{adj}}(X) \in \mathbb{R}^n$.

For an inner product space V, the **polar cone** of $S \subseteq V$ is

$$S^+ = \{\phi : \langle \phi, \mathbf{s}\rangle \geq 0,\ \forall \mathbf{s} \in S\}.$$

Given a set $K \subseteq \mathbb{R}^n$, we let cone (K) denote the convex cone generated by K. (See Chapter 9.1 for more information on cones.) A set K is a convex cone if it is closed under nonnegative scalar multiplication and addition, i.e.,

$$\alpha K \subseteq K, \forall \alpha \geq 0, \quad K + K \subseteq K.$$

For a convex cone K, the convex cone $\mathcal{F} \subseteq K$ is a **face** of K (denoted $\mathcal{F} \lhd K$) if

$$\mathbf{x}, \mathbf{y} \in K, \mathbf{z} = \alpha \mathbf{x} + (1-\alpha)\mathbf{y} \in \mathcal{F}, 0 < \alpha < 1 \quad \text{implies} \quad \mathbf{x}, \mathbf{y} \in \mathcal{F}.$$

Facts:

1. In the space of real, symmetric matrices, $\mathcal{S}_n$, the matrix inner product reduces to $\langle X, Y \rangle = \mathrm{tr}\, XY$.
2. For general compatible matrices M, N, the trace is commutative, i.e., $\mathrm{tr}\, MN^T = \mathrm{tr}\, N^T M = \sum_{i=1}^n \sum_{j=1}^n M_{ij} N_{ij}$.
3. svec (and sMat) is an isometry between $\mathcal{S}_n$ (equipped with the Frobenius norm) and $\mathbb{R}^{\frac{n(n+1)}{2}}$ (equipped with the Euclidean norm). The svec transformation is used in algorithms when solving symmetric matrix equations. Using the isometry (rather than the ordinary vec) provides additional robustness.
4. PSD is a closed convex cone and the interior of PSD consists of the positive definite matrices. The following are equivalent: (a) $A \succeq 0 \quad (A \succ 0)$, (b) the vector of eigenvalues $\lambda(A) \geq 0 \quad (\lambda(A) > 0)$, (c) all principal minors ≥ 0 (all leading principal minors > 0).
5. The Kronecker product $K \otimes L$ is easily formed using the blocks $K_{ij} L$. It is useful in changing matrix equations into vector equations (see [HJ94]), (here K is the unknown matrix) $\mathrm{vec}\,(NKM^T) = (M \otimes N)\mathrm{vec}\,(K)$. Similarly, the symmetric Kronecker product $M \overset{s}{\otimes} N$ changes matrix equations with symmetric matrix variables $U \in \mathcal{S}_n$ to vector equations. (See the definition in section 51.2.) By extending the definition from $\mathcal{S}_n$ to $\mathbb{R}^{n \times n}$, the symmetric Kronecker product can be expressed explicitly using the standard Kronecker product see ([TW05]),

$$8M \overset{s}{\otimes} N = (I + A)(N \otimes M^T + M \otimes N^T) + (N \otimes M^T + M \otimes N^T)(I + A),$$

where A is the matrix representation of the transpose transformation. Matrix equations with symmetric matrix variables arise when finding the search direction in interior-point methods for SDP; see section 51.6. Surprisingly (see [TW05]), for $A, B \in \mathcal{S}_n$,

$$A \overset{s}{\otimes} B \succeq 0 \iff A \otimes B \succeq 0;$$

$$A \overset{s}{\otimes} B \succ 0 \iff A \otimes B \succ 0.$$

6. If $A \in \mathbb{R}^{m \times n}$ and the linear transformation $T\mathbf{v} = A\mathbf{v}$, then $T^{\mathrm{adj}} \cong A^T$.
7. The polar cone is a closed convex cone and the second polar $(K^+)^+ = K$ if and only if K is a closed convex cone.

Examples:

1.

$$A = \begin{bmatrix} 1 & 2 & 3 \\ 2 & 4 & 5 \\ 3 & 5 & 6 \end{bmatrix}, \mathbf{v} = \begin{bmatrix} 1 \\ 2\sqrt{2} \\ 4 \\ 3\sqrt{2} \\ 5\sqrt{2} \\ 6 \end{bmatrix}, \quad \begin{array}{c} \mathrm{svec}\,(A) = \mathbf{v} \\ \|A\|_F = \|\mathbf{v}\|_2 = \sqrt{129}. \end{array}$$

2. The polars: (a) $(\mathbb{R}^n)^+ = \{\mathbf{0}\}$; $\{\mathbf{0}\}^+ = \mathbb{R}^n$, (b) for the unit vector $\mathbf{e}_i \in \mathbb{R}^n$, $\{\mathbf{e_i}\}^+ = \{\mathbf{v} \in \mathbb{R}^n : v_i \geq 0\}$, (c) $(\text{cone}\{(1 \quad 1)^T, (1 \quad 0)^T\})^+ = \text{cone}\{[0 \quad 1]^T, [1 \quad -1]^T\}$; (d) let $\mathcal{I} = \{\mathbf{v} \in \mathbb{R}^3 : v_3^2 \leq v_1 v_2, v_1 \geq 0, v_2 \geq 0\}$ be the so-called ice-cream cone [Ber73], i.e., the cone of vectors that makes an angle at most 45 degrees with the vector $(1 \quad 1 \quad 1)^T$. Then $\mathcal{I} = \mathcal{I}^+$.

51.3 Geometry

Extending LP to SDP can be thought of as replacing a polyhedral cone (the nonnegative orthant $(\mathbb{R}^+)^n$) by the nonpolyhedral cone PSD. We now see that many of the geometric properties follow through from $K = (\mathbb{R}^+)^n$ to $K = \text{PSD}$.

Definitions:

A cone K is called **self-polar** if it equals its polar $K = K^+$

A face $\mathcal{F} \lhd K$ is called **facially exposed** if $\mathcal{F} = K \cap \phi^\perp$, for some $\phi \in K^+$.

A face $\mathcal{F} \lhd K$ is called **projectionally exposed** if $\mathcal{F}$ is the image of K under an orthogonal projection, $\mathcal{F} = P(K)$.

Facts:

1. Both $(\mathbb{R}^+)^n$ and PSD are closed convex cones with nonempty interior. And both are self-polar cones, i.e., $((\mathbb{R}^+)^n)^+ = (\mathbb{R}^+)^n$, $(\text{PSD})^+ = \text{PSD}$. (See [Ber73].)
2. Suppose that $\hat{X} \in \text{relint}\,\mathcal{F} \subseteq \text{PSD}$ and $\text{rank}\,\hat{X} = \text{r}$. Then $\hat{X} = PD_r P^T$, where P is the $n \times r$ matrix with orthonormal columns of eigenvectors corresponding to nonzero (positive) eigenvalues. We get that $\mathcal{F} \lhd \text{PSD}$ if and only if $\mathcal{F} = P\mathcal{S}_+^r P^T$. Equivalently, faces $\mathcal{F} \lhd \text{PSD}$ are characterized by $\{Y \in \text{PSD} : \mathcal{R}(Y) \subseteq \mathcal{R}(\hat{X})\}$, where $\hat{X}$ is any matrix in the relative interior of $\mathcal{F}$ and $\mathcal{R}$ denotes the range. Equivalently, we could use $\mathcal{N}(Y) \supset \mathcal{N}(\hat{X})$, where $\mathcal{N}$ denotes nullspace. (See [Boh48], [BC75].)
3. Both $(\mathbb{R}^+)^n$ and PSD are facially and projectionally exposed, i.e., all faces are both facially and projectionally exposed [BW81], [BLP87], [ST90].
4. The following relates to closure conditions that are needed for strong duality results. Suppose that $\mathcal{F} \lhd \text{PSD}$. Then (see [RTW97]),

$$\text{PSD} + \mathcal{F}^\perp \text{ is closed; } \text{PSD} + \text{Span}\,\mathcal{F} \text{ is not closed.}$$

Examples:

1. Consider the face $\mathcal{F} = \{X \in \text{PSD} : X\mathbf{e} = 0\}$, i.e., the face of PSD of centered matrices, positive semidefinite matrices with row and column sums equal to 0. Let $P = (\mathbf{e} \quad V)$ be an orthogonal matrix, i.e., $V^T e = 0$, $V^T V = I$. Then we get $\mathcal{F} = V\{X \in \mathcal{S}_{n-1} : X \succeq 0\}V^T$, i.e., a one-to-one mapping between the face $\mathcal{F}$ and the semidefinite cone in $\mathcal{S}_{n-1}$. This relationship is used in [AKW99] to map Euclidean distance matrices to positive semidefinite matrices of full rank.

51.4 Duality and Optimality Conditions

Duality lies behind efficient algorithms in optimization. Unlike LP, strong duality can fail for SDP if a *constraint qualification* does not hold. We consider duality the linear *primal* SDP introduced above a game-theoretic approach.

Definitions:

The corresponding **Lagrangian function** (or payoff function from primal player X, who plays matrix X, to dual player Y, who plays vector y) is $L(X, \mathbf{y}) = \text{tr}\,\text{CX} + \mathbf{y}^\text{T}(\mathbf{b} - \text{TX})$.

The **dual problem** is

$$\text{DSDP} \qquad \begin{array}{lll} d^* = & \text{maximize} & \mathbf{b}^T\mathbf{y} \\ & \text{subject to} & T^{\text{adj}}\mathbf{y} \preceq C \\ & (\text{equivalently} & T^{\text{adj}}\mathbf{y} + Z = C, \quad Z \succeq 0). \end{array}$$

Weak duality $p^* \geq d^*$ holds. If $p^* = d^*$ and d^* is attained, then **Strong duality** holds.
Complementary slackness holds if $\operatorname{tr} ZX = 0$; **strict complementarity** holds if, in addition, $Z + X \succ 0$.

Facts:

1. The optimal strategy of player X over all possible strategies by the dual player Y has optimal value greater or equal to that of the optimal strategy for player Y over all strategies for player X, i.e.,

$$p^* = \min_{X \succeq 0} \max_{\mathbf{y}} L(X, \mathbf{y}) \geq d^* = \max_{\mathbf{y}} \min_{X \succeq 0} b^T\mathbf{y} + \operatorname{tr} X(C - T^{\text{adj}}\mathbf{y}).$$

   The first equality in the min–max problem can be seen from the *hidden constraint* for the inner maximization problem. The inequality follows from interchanging min and max. The hidden constraint in the inner minimization of the max–min problem yields our dual problem and weak duality. However, *strong duality* can fail. (See [RTW97].)
2. For $X, Z \succeq 0$, complementary slackness $\operatorname{tr} ZX = 0$ is equivalent to $ZX = 0$.
3. *Characterization of Optimality Theorem*
   The primal-dual pair $X, (\mathbf{y}, Z)$, with $X \succeq 0, Z \succeq 0$, is primal-dual optimal for SDP and DSDP if and only if

$$(\text{OPT}) \qquad \begin{array}{rl} T^{\text{adj}}\mathbf{y} + Z - C = 0 & \text{(dual feasibility)} \\ TX - \mathbf{b} = 0 & \text{(primal feasibility)} \\ ZX = 0 & \text{(complementary slackness)}. \end{array}$$

4. The theorem in the previous fact is used to derive efficient primal-dual interior-point methods.
5. Again we note the similarity between the above optimality conditions and those for LP, where Z, X are diagonal matrices. However, in contrast to LP, the Goldman–Tucker theorem [GT56] can fail, i.e., a strict complementary solution may not exist. This means that there may be no optimal solution with $X+Z \succ 0$ positive definite; see [Sha00], [WW05]. This can result in loss of superlinear convergence for numerical algorithms for SDP. Moreover, in LP, Z, X are diagonal matrices and so is ZX. This means that the optimality conditions OPT are a square system. However, for SDP, the product of two symmetric matrices ZX is not necessarily symmetric. Therefore, the optimality conditions are an overdetermined nonlinear system of equations. This gives rise to many subtle algorithmic difficulties.

Examples:

1. The lack of closure noted in Fact 4 in section 51.3 is illustrated by the sequence

$$\begin{bmatrix} \frac{1}{i} & 1 \\ 1 & i \end{bmatrix} + \begin{bmatrix} 0 & 0 \\ 0 & -i \end{bmatrix} \to \begin{bmatrix} 0 & 1 \\ 1 & 0 \end{bmatrix} \notin \text{PSD} + \text{Span}\left\{ \begin{bmatrix} 0 & 0 \\ 0 & 1 \end{bmatrix} \right\}.$$

   This gives rise to an SDP with a duality gap. Let $C = \begin{bmatrix} 0 & 1 \\ 1 & 0 \end{bmatrix}$ and $A_1 = \begin{bmatrix} 0 & 0 \\ 0 & 1 \end{bmatrix}$. Then a primal SDP is $0 = \min \operatorname{tr} CX$ s.t. $\operatorname{tr} A_1 X = 0, X \succeq 0$. But the dual program has optimal value $-\infty = \max 0\mathbf{y}$ s.t. $\mathbf{y}A_1 \preceq C$, i.e., it is infeasible.

2. For fixed $\alpha > 0$, we use parameters $\mathbf{b} = \begin{bmatrix} 0 \\ 1 \end{bmatrix}$ and $C = \begin{bmatrix} \alpha & 0 & 0 \\ 0 & 0 & 0 \\ 0 & 0 & 0 \end{bmatrix}$, $A_1 = \begin{bmatrix} 0 & 0 & 0 \\ 0 & 1 & 0 \\ 0 & 0 & 0 \end{bmatrix}$, $A_2 = \begin{bmatrix} 1 & 0 & 0 \\ 0 & 0 & 1 \\ 0 & 1 & 0 \end{bmatrix}$. Then we get the primal and dual SDP pair with a finite duality gap: $\alpha = \min \alpha U_{11}$ s.t. $U_{22} = 0,\ U_{11} + 2U_{23} = 1,\ U \succeq 0; 0 = \max y_2$ s.t. $\begin{bmatrix} y_2 & 0 & 0 \\ 0 & y_1 & y_2 \\ 0 & y_2 & 0 \end{bmatrix} \preceq \begin{bmatrix} \alpha & 0 & 0 \\ 0 & 0 & 0 \\ 0 & 0 & 0 \end{bmatrix}$.
3. We now verify that $X^* = \begin{bmatrix} 1 & 1 & -1 \\ 1 & 1 & -1 \\ -1 & -1 & 1 \end{bmatrix}$ is optimal for MCSDP in Example 1 in Section 51.1.

   The dual of MCSDP is $\min e^T\mathbf{y}$ s.t. $\text{Diag}(\mathbf{y}) \succeq \frac{1}{4}\text{L}$. Since $\mathbf{y} = [1 \quad 1 \quad 2]^T$ is feasible for this dual, optimality follows from checking strong duality $\text{tr}\,\text{LX}^* = \text{e}^\text{T}\text{y} = 4$. Equivalently, one can check complementary slackness $(\text{Diag}(\mathbf{y}) - \text{L})\text{X}^* = 0$.

51.5 Strong Duality without a Constraint Qualification

Definitions:

A **constraint qualification**, CQ, is a condition on the constraints that guarantees that strong duality holds at an optimum point.

Slater's Constraint Qualification (strict feasibility) is defined as: $\exists X \succ 0,\ TX = \mathbf{b}$.

Facts:

1. Suppose that Slater's CQ holds. Then strong duality holds for SDP and DSDP, i.e., $p^* = d^*$ and d^* is attained.
2. [BW81] *Duality Theorem*: Suppose that p^* is finite. Let $\mathcal{F}_e$ denote the minimal face of PSD that contains the feasible set of SDP. Consider the extended dual program

$$\text{DSDP}_e \qquad \begin{array}{lll} d_e^* = & \text{maximize} & \mathbf{b}^T\mathbf{y} \\ & \text{subject to} & T^{\text{adj}}\mathbf{y} + Z = C \\ & & Z \in \mathcal{F}_e^+. \end{array}$$

   Then $p^* = d_e^*$ and d_e^* is attained, i.e., strong duality holds between SDP and DSDP_e.
3. An algorithm for finding the minimal face $\mathcal{F}_e$ defined above is given in [BW81].

Examples:

We can apply the strong duality results to the two examples given in Section 51.4.

1. For the first example, replace PSD in the primal with the minimal face $\mathcal{F} = \text{PSD} \cap \begin{bmatrix} 1 & 1 \\ 1 & 0 \end{bmatrix}^\perp = \text{cone}\{A_1\}$; so $\mathcal{F}^+ = \{A_1\}^+$ and the dual is now feasible.
2. For the second example, replace PSD in the primal with the minimal face $\mathcal{F} = \text{PSD} \cap \begin{bmatrix} 0 & 0 & 0 \\ 0 & 0 & 0 \\ 0 & 0 & 1 \end{bmatrix}^\perp$; so U_{23} is no longer restricted to be 0 in the dual.
3. The strong duality results are needed in many SDP applications to combinatorial problems where Slater's constraint qualification (strict feasibility) fails, e.g., [WZ99], [ZKR98].

51.6 A Primal-Dual Interior-Point Algorithm

As mentioned above, it was the development of efficient polynomial-time algorithms for SDP at the end of the 1980s that spurred the increased interest in the field. We look at the specific method derived in [HRV96]. (See also [KSH97], [Mon97].)

We assume that Slater's constraint qualification (strict feasibility) holds for both primal and dual problems.

Definitions:

For each $\mu > 0$, the log-barrier approach applied to DSDP requires the solution of the **log-barrier problem**

$$\text{DSDP}_\mu \qquad \begin{array}{rll} d_\mu^* = & \text{maximize} & \mathbf{b}^T\mathbf{y} + \mu \log\det Z \\ & \text{subject to} & T^{\text{adj}}\mathbf{y} + Z = C, \quad Z \succ 0. \end{array}$$

The **Lagrangian** is $L(X, \mathbf{y}, Z) = \mathbf{b}^T\mathbf{y} + \mu \log\det Z + \operatorname{tr} X(C - T^{\text{adj}}\mathbf{y} - Z)$, where we use X for the Lagrange multiplier vector.

Facts:

1. The derivative, with respect to $(X, \mathbf{y}, Z)$ of the Lagrangian, yields the optimality conditions for the barrier problem:

$$\text{OPT}_\mu \qquad \begin{array}{rl} T^{\text{adj}}\mathbf{y} + Z - C = 0 & \text{(dual feasibility)} \\ TX - b = 0 & \text{(primal feasibility)} \\ \mu Z^{-1} - X = 0 & \text{(perturbed complementary slackness).} \end{array}$$

   We let $X_\mu, \mathbf{y}_\mu, Z_\mu$ denote the unique solution of OPT_μ, i.e., the unique optimum of the log-barrier problem. The set $\{X_\mu, \mathbf{y}_\mu, Z_\mu : \mu \downarrow 0\}$ is called the **central path**. Successful primal-dual interior-point methods are actually path-following methods, i.e., they follow the central path to the optimum at $\mu = 0$.

2. The last equation in the optimality conditions OPT_μ is very nonlinear and leads to ill-conditioning as μ gets close to zero. Therefore, it is replaced by the more familiar

$$\text{OPT}_\mu \qquad ZX - \mu I = 0 \quad \text{(perturbed complementary slackness).}$$

   We denote the resulting optimality conditions using

$$F_\mu(X, \mathbf{y}, Z) = 0.$$

   This system has the same appearance as the one used in LP. However, it is an overdetermined nonlinear system, i.e., $F_\mu : \mathcal{S}_n \times \mathbb{R}^m \times \mathcal{S}_n \to \mathcal{S}_n \times \mathbb{R}^m \times \mathbb{R}^{n \times n}$, since the product ZX is not necessarily symmetric. Therefore, the successful application of Newton's method done in LP must be modified. A Gauss–Newton method is used in [KMR01]. Another approach is to symmetrize the last equation of OPT_μ. A general symmetrization scheme is presented in [MZ98].

3. We derive and present (arguably) the most popular algorithm for SDP, the **HKM method** [HRV96], [KSH97], [Mon97].

   We consider a current estimate X, y, Z of the optimum of the log-barrier problem DSDP_μ, where both $X \succ 0$, $Z \succ 0$. Since we want $ZX = \mu I$, we set $\mu = \frac{1}{n}\operatorname{tr} ZX$. We use a centering parameter $0 < \sigma$ that is adaptive in that it decreases (respectively increases) according to the decrease (respectively increase) in the duality gap estimate at each iteration. Small σ signifies an aggressive step in decreasing μ to 0. We replace $\mu \leftarrow \sigma\mu$ in the optimality conditions. We denote the optimality conditions using $F_\mu = F_\mu(X, \mathbf{y}, Z) = \begin{bmatrix} R_d \\ \mathbf{r_p} \\ R_c \end{bmatrix}$, i.e., using three residuals. To derive the HKM method, we solve for the Newton direction $d_N = \begin{bmatrix} \Delta X \\ \Delta \mathbf{y} \\ \Delta Z \end{bmatrix}$ in the Newton equation

$$F'_\mu(X, \mathbf{y}, Z) = \begin{bmatrix} 0 & T^{\mathrm{adj}}\cdot & I\cdot \\ T\cdot & 0 & 0 \\ Z\cdot & 0 & \cdot X \end{bmatrix} d_N = -\begin{bmatrix} R_d \\ \mathbf{r_p} \\ R_c \end{bmatrix} = -F_\mu.$$

To solve the above linearized system efficiently, we use block elimination. We use the first equation to solve for

$$\Delta Z = -R_d - T^{\mathrm{adj}}\,\Delta\mathbf{y}.$$

We then substitute into the third equation and solve for

$$\Delta X = Z^{-1}\left((R_d + T^{\mathrm{adj}}\,\Delta\mathbf{y})X - R_c\right).$$

We then substitute this into the second equation and obtain the so-called *Schur complement* system, which is similar to the normal equation in LP:

$$\begin{aligned} TZ^{-1}T^{\mathrm{adj}}\,\Delta\mathbf{y}X &= -TZ^{-1}(R_dX - R_c) - \mathbf{r_p} \\ &= -TZ^{-1}(R_dX - ZX + \mu I) - TX + \mathbf{b} \\ &= \mathbf{b} - TZ^{-1}(R_dX + \mu I). \end{aligned}$$

This is a positive definite linear system. Once $\Delta\mathbf{y}$ is found, we can backsolve for $\Delta X, \Delta Z$.

We now observe another major difference with LP and a major difficulty for SDP. Though ΔZ is now symmetric, this is not true, in general, for ΔX, since the product of symmetric matrices is not necessarily symmetric. Therefore, we symmetrize $\Delta X \leftarrow \frac{1}{2}(\Delta X + \Delta X^T)$. We have now obtained the search direction $(\Delta X, \Delta\mathbf{y}, \Delta Z)$.

Next, a line search is performed that maintains X, Z positive definite, i.e., we find the next iterate (X, y, Z) using the primal and dual steplengths $X \leftarrow X + \alpha_p \Delta X \succ 0, Z \leftarrow Z + \alpha_d \Delta Z \succ 0$ and set $\mathbf{y} \leftarrow \mathbf{y} + \alpha_d \Delta\mathbf{y}$. We then update σ, μ and repeat the iteration, i.e., we go back to Fact 3 above.

4. See [HRV96] for more details on the HKM algorithm, including a predictor-corrector approach. Details on different symmetrization schemes that lead to various search directions and convergence proofs can be found in [MT00]. Algorithms based on bundle methods that handle large-scale SDP can be found in [HO00].

51.7 Applications of SDP

There are a surprising number of important applications for SDP, several of which have already been mentioned. Early applications in engineering involved solutions of linear matrix inequalities, LMIs, e.g., Lyapunov and Riccatti equations; see [BEF94]. Semidefinite programming plays an important role in combinatorial optimization where it is used to solve nonlinear convex relaxations of NP-hard problems [Ali95], [WZ99], [ZKR98]. This includes strong theoretical results that characterize the quality of the bounds; see [GR00]. Matrix completion problems have been mentioned above. This includes Euclidean Distance Matrix completion problems with applications to, e.g., molecular conformation; see [AKW99]. Further applications to control theory, finance, nonlinear programming, etc. can be found in [BEF94] and [WSV00].

Facts:

1. SDP arises naturally when finding relaxations for hard combinatorial problems. We saw one simple case, the Max-Cut problem, MC, in the introduction in Example 1, i.e., MC was modeled as a quadratic maximization problem and the binary ± 1 variables were modeled using the quadratic constraints $x_j^2 = 1$.
2. In addition to ± 1 binary variables in Fact 1, we can model $0, 1$ variables using the quadratic constraints $x_j^2 - x_j = 0$. Linear constraints $A\mathbf{x} - \mathbf{b} = 0$ can be modeled using the quadratic constraint $\|A\mathbf{x} - \mathbf{b}\|^2 = 0$. Thus, in general, many hard combinatorial optimization problems are equivalent to a quadratically constrained quadratic program, QQP. Denote the quadratic function $q_i(\mathbf{y}) = \frac{1}{2}\mathbf{y}^T Q_i \mathbf{y} + \mathbf{y}^T b_i + c_i, \mathbf{y} \in \mathbb{R}^n, i = 0, 1, \ldots, m$. Then we have

$$\text{(QQP)} \quad q^* = \begin{array}{ll} \min & q_0(\mathbf{y}) \\ \text{subject to} & q_i(\mathbf{y}) = 0, \quad i = 1, \ldots m. \end{array}$$

(For simplicity, we restrict to equality constraints.)

3. SDP arises naturally from the Lagrangian relaxation of QQP in Fact 2 above. We write the Lagrangian $L(\mathbf{y}, \mathbf{x})$, with Lagrange multiplier vector $\mathbf{x}$.

$$L(\mathbf{y},\mathbf{x}) = \underbrace{\tfrac{1}{2} y^T\left(Q_0 - \textstyle\sum_{i=1}^m x_i Q_i\right) y}_{\text{quadratic in } \mathbf{y}} + \underbrace{y^T\left(b_0 - \textstyle\sum_{i=1}^m x_i b_i\right)}_{\text{linear in } \mathbf{y}} + \underbrace{\left(c_0 - \textstyle\sum_{i=1}^m x_i c_i\right)}_{\text{constant in } y}.$$

Weak duality follows from the primal-dual pair relationship

$$q^* = \min_y \max_x L(y, x) \geq d^* = \max_x \min_y L(y, x).$$

We now homogenize the Lagrangian by adding y_0 to the linear term: $y_0 y^T(b_0 - \sum_{i=1}^m x_i b_i), y_0^2 = 1$. Then, after moving the constraint on y_0 into the Lagrangian, the dual program becomes

$$\begin{aligned} d^* &= \max_x \min_y \quad L(y, x) \\ &= \max_{x,t} \min_y \quad \tfrac{1}{2} y^T \left(Q_0 - \textstyle\sum_{i=1}^m x_i Q_i\right) y (+t y_0^2) \\ &\qquad\qquad + y_0 y^T \left(b_0 - \textstyle\sum_{i=1}^m x_i b_i\right) \\ &\qquad\qquad + \left(c_0 - \textstyle\sum_{i=1}^m x_i c_i\right)(-t). \end{aligned}$$

We now exploit the *hidden constraint* from the inner minimization of a homogeneous quadratic function, i.e., the Hessian of the Lagrangian must be positive semidefinite. This gives rise to the SDP

$$\textbf{(D)} \quad \begin{array}{lll} d^* = & \sup & -t - \sum_{i=1}^m x_i c_i \\ & \text{s.t.} & T \begin{bmatrix} t \\ \boldsymbol{x} \end{bmatrix} \preceq B \\ & & \boldsymbol{x} \in \mathbb{R}^m, \quad t \in \mathbb{R}, \end{array}$$

where the linear transformation $T : \mathbb{R}^{m+1} \to \mathcal{S}_{n+1}$ and the matrix B are defined by

$$B = \begin{bmatrix} 0 & b_0^T \\ b_0 & Q_0 \end{bmatrix}, \quad T \begin{bmatrix} t \\ \boldsymbol{x} \end{bmatrix} = \begin{bmatrix} -t & \sum_{i=1}^m x_i b_i^T \\ \sum_{i=1}^m x_i b_i & \sum_{i=1}^m x_i Q_i \end{bmatrix}.$$

The dual, DD, of the program D gives rise to the SDP relaxation of QQP .

$$
\textbf{(DD)} \qquad \begin{array}{rll} p^* = & \inf & \operatorname{tr} BU \\ & \text{s.t.} & T^{\text{adj}} U = \begin{bmatrix} -1 \\ -c \end{bmatrix} \\ & & U \succeq 0. \end{array}
$$

References

[AKW99] A. Alfakih, A. Khandani, and H. Wolkowicz. Solving Euclidean distance matrix completion problems via semidefinite programming. *Comput. Optim. Appl.*, 12(1–3):13–30, 1999. (Computational optimization—a tribute to Olvi Mangasarian, Part I.)

[Ali95] F. Alizadeh. Interior point methods in semidefinite programming with applications to combinatorial optimization. *SIAM J. Optim.*, 5:13–51, 1995.

[BC75] G.P. Barker and D. Carlson. Cones of diagonally dominant matrices. *Pac. J. Math.*, 57:15–32, 1975.

[BEF94] S. Boyd, L. El Ghaoui, E. Feron, and V. Balakrishnan. *Linear Matrix Inequalities in System and Control Theory*, Vol. 15 of *Studies in Applied Mathematics.* SIAM, Philadelphia, PA, June 1994.

[Ber73] A. Berman. *Cones, Matrices and Mathematical Programming.* Springer-Verlag, Berlin, New York, 1973.

[BF63] R. Bellman and K. Fan. On systems of linear inequalities in Hermitian matrix variables. In Convexity, V. L. Klee, editor, *Proceedings of Symposia in Pure Mathematics,* Vol. 7, *AMS*, Providence, RI, 1963.

[BLP87] G.P. Barker, M. Laidacker, and G. Poole. Projectionally exposed cones. *SIAM J. Alg. Disc. Meth.*, 8(1):100–105, 1987.

[BMZ02] S. Burer, R.D.C. Monteiro, and Y. Zhang. Solving a class of semidefinite programs via nonlinear programming. *Math. Prog.*, 93(1, Ser. A):97–122, 2002.

[Boh48] F. Bohnenblust. Joint positiveness of matrices. Technical report, UCLA, 1948. (Manuscript.)

[BW81] J.M. Borwein and H. Wolkowicz. Regularizing the abstract convex program. *J. Math. Anal. Appl.*, 83(2):495–530, 1981.

[BW81] J.M. Borwein and H. Wolkowicz. Characterization of optimality for the abstract convex program with finite-dimensional range. *J. Aust. Math. Soc. Ser. A*, 30(4):390–411, 1980/81.

[DG81] H. Dym and I. Gohberg. Extensions of band matrices with band inverses. *Lin. Alg. Appl.*, 36:1–24, 1981.

[Fay97] L. Faybusovich. Euclidean Jordan algebras and interior-point algorithms. *Positivity*, 1(4):331–357, 1997.

[FKM01] K. Fukuda, M. Kojima, K. Murota, and K. Nakata. Exploiting sparsity in semidefinite programming via matrix completion. I. General framework. *SIAM J. Optim.*, 11(3):647–674 (electronic), 2000/01.

[GJM84] B. Grone, C.R. Johnson, E. Marques de Sa, and H. Wolkowicz. Positive definite completions of partial Hermitian matrices. *Lin. Alg. Appl.*, 58:109–124, 1984.

[GR00] M.X. Goemans and F. Rendl. Combinatorial optimization. In H. Wolkowicz, R. Saigal, and L. Vandenberghe, Eds., *Handbook of Semidefinite Programming: Theory, Algorithms, and Applications.* Kluwer Academic Publishers, Boston, MA, 2000.

[GT56] A.J. Goldman and A.W. Tucker. Theory of linear programming. In *Linear Inequalities and Related Systems*, pp. 53–97. Princeton University Press, Princeton, N.J., 1956. Annals of Mathematics Studies, No. 38.

[GW94] M.X. Goemans and D.P. Williamson. 878-approximation algorithms for MAX CUT and MAX 2SAT. In *ACM Symposium on Theory of Computing (STOC)*, 1994.

[GW95] M.X. Goemans and D.P. Williamson. Improved approximation algorithms for maximum cut and satisfiability problems using semidefinite programming. *J. Assoc. Comput. Mach.*, 42(6):1115–1145, 1995.

[HJ94] R.A. Horn and C.R. Johnson. *Topics in Matrix Analysis*. Cambridge University Press, Cambridge, 1994. Corrected reprint of the 1991 original.

[HL03] D. Henrion and J.B. Lasserre. GloptiPoly: global optimization over polynomials with MATLAB and SeDuMi. *ACM Trans. Math. Software*, 29(2):165–194, 2003.

[HO00] C. Helmberg and F. Oustry. Bundle methods to minimize the maximum eigenvalue function. In H. Wolkowicz, R. Saigal, and L. Vandenberghe, Eds., *Handbook of Semidefinite Programming: Theory, Algorithms, and Applications*. Kluwer Academic Publishers, Boston, MA, 2000.

[Hol75] R.B. Holmes. *Geometric Functional Analysis and Its Applications*. Springer-Verlag, Berlin, 1975.

[HRV96] C. Helmberg, F. Rendl, R.J. Vanderbei, and H. Wolkowicz. An interior-point method for semidefinite programming. *SIAM J. Optim.*, 6(2):342–361, 1996.

[Jah86] J. Jahn. *Mathematical Vector Optimization in Partially Ordered Linear Spaces*. Peter Lang, Frankfurt am Main, 1986.

[Jin05] H. Jin. *Scalable Sensor Localization Algorithms for Wireless Sensor Networks*. Ph.D. thesis, Toronto University, Ontario, Canada, 2005.

[Joh90] C.R. Johnson. Matrix completion problems: a survey. In *Matrix Theory and Applications* (Phoenix, AZ, 1989), pp. 171–198. American Mathematical Society, Providence, RI, 1990.

[KMR01] S. Kruk, M. Muramatsu, F. Rendl, R.J. Vanderbei, and H. Wolkowicz. The Gauss–Newton direction in linear and semidefinite programming. *Optimiz. Meth. Softw.*, 15(1):1–27, 2001.

[KSH97] M. Kojima, S. Shindoh, and S. Hara. Interior-point methods for the monotone semidefinite linear complementarity problem in symmetric matrices. *SIAM J. Optim.*, 7(1):86–125, 1997.

[Lau98] M. Laurent. A tour d'horizon on positive semidefinite and Euclidean distance matrix completion problems. In *Topics in Semidefinite and Interior-Point Methods*, Vol. 18 of *The Fields Institute for Research in Mathematical Sciences, Communications Series*, pp. 51–76, Providence, RI, 1998. American Mathematical Society.

[Lov79] L. Lovász. On the Shannon capacity of a graph. *IEEE Trans. Inform. Theory*, 25:1–7, 1979.

[Lue69] D.G. Luenberger. *Optimization by Vector Space Methods*. John Wiley & Sons, New York, 1969.

[Lya47] A.M. Lyapunov. *Problème général de la stabilité du mouvement*, volume 17 of *Annals of Mathematics Studies*. Princeton University Press, Princeton, NJ, 1947.

[Mon97] R.D.C. Monteiro. Primal-dual path-following algorithms for semidefinite programming. *SIAM J. Optim.*, 7(3):663–678, 1997.

[MT00] R.D.C. Monteiro and M.J. Todd. Path-following methods. In *Handbook of Semidefinite Programming*, pp. 267–306. Kluwer Academic Publications, Boston, MA, 2000.

[MZ98] R.D.C. Monteiro and Y. Zhang. A unified analysis for a class of long-step primal-dual path-following interior-point algorithms for semidefinite programming. *Math. Prog.*, 81(3, Ser. A):281–299, 1998.

[NN94] Y.E. Nesterov and A.S. Nemirovski. *Interior Point Polynomial Algorithms in Convex Programming*. SIAM Publications. SIAM, Philadelphia, 1994.

[Ren99] F. Rendl. Semidefinite programming and combinatorial optimization. *Appl. Numer. Math.*, 29:255–281, 1999.

[RTW97] M.V. Ramana, L. Tunçel, and H. Wolkowicz. Strong duality for semidefinite programming. *SIAM J. Optim.*, 7(3):641–662, 1997.

[SA01] S.H. Schmieta and F. Alizadeh. Associative and Jordan algebras, and polynomial time interior point algorithms for symmetric cones. *INFORMS*, 26(3):543–564, 2001. Available at URL: `ftp://rutcor.rutgers.edu/pub/rrr/reports99/12.ps`.

[Sha00] A. Shapiro. Duality and optimality conditions. In *Handbook of Semidefinite Programming: Theory, Algorithms, and Applications*. Kluwer Academic Publishers, Boston, MA, 2000.

[ST90] C.H. Sung and B.-S. Tam. A study of projectionally exposed cones. *Lin. Alg. Appl.*, 139:225–252, 1990.

[SY06] A.M. So and Y. Ye. Theory of semidefinite programming for sensor network localization. *Math. Prog.*, to appear.

[TW05] L. Tunçel and H. Wolkowicz. Strengthened existence and uniqueness conditions for search directions in semidefinite programming. *Lin. Alg. Appl.*, 400:31–60, 2005.

[WSV00] H. Wolkowicz, R. Saigal, and L. Vandenberghe, Eds. *Handbook of Semidefinite Programming: Theory, Algorithms, and Applications*. Kluwer Academic Publishers, Boston, MA, 2000.

[WW05] H. Wei and H. Wolkowicz. Generating and solving hard instances in semidefinite programming. Technical Report CORR 2006-01, University of Waterloo, Waterloo, Ontario, 2006.

[WZ99] H. Wolkowicz and Q. Zhao. Semidefinite programming relaxations for the graph partitioning problem. *Discrete Appl. Math.*, 96/97:461–479, 1999. (Selected for the special Editors' Choice, Edition 1999.)

[Yak62] V.A. Yakubovich. The solution of certain matrix inequalities in automatic control theory. *Sov. Math. Dokl.*, 3:620–623, 1962. In Russian, 1961.

[ZKR98] Q. Zhao, S.E. Karisch, F. Rendl, and H. Wolkowicz. Semidefinite programming relaxations for the quadratic assignment problem. *J. Comb. Optim.*, 2(1):71–109, 1998.

Applications to Probability and Statistics

52

Random Vectors and Linear Statistical Models

Simo Puntanen
University of Tampere

George P. H. Styan
McGill University

52.1 Introduction and Mise-En-Scène

Linear algebra is used extensively in statistical science, in particular in linear statistical models, see, e.g., Graybill [Gra01], Ravishanker and Dey [RD02], Rencher [Ren00], Searle [Sea97], Seber and Lee [SL03], Sengupta and Jammalamadaka [SJ03], and Stapleton [Sta95]; as well as in applied economics, see, e.g., Searle and Willett [SW01]; econometrics, see, e.g., Davidson and MacKinnon [DM04], Magnus and Neudecker [MN99], and Rao and Rao [RR98]; Markov chain theory, see, e.g., Chapter 54 or Kemeny and Snell [KS83]; multivariate statistical analysis, see, e.g., Anderson [And03], Kollo and von Rosen [KvR05], and Seber [Seb04]; psychometrics, see, e.g., Takane [Tak04] and Takeuchi, Yanai, and Mukherjee [TYM82]; and random matrix theory, see, e.g., Bleher and Its [BI01] and Mehta [Meh04]. Moreover, there are several books on linear algebra and matrix theory written by (and mainly for) statisticians, see, e.g., Bapat [Bap00], Graybill [Gra83], Hadi [Had96], Harville [Har97], [Har01], Healy [Hea00], Rao and Rao [RR98], Schott [Sch05], Searle [Sea82], and Seber [Seb06].

As Miller and Miller [MM04, p. 1] and Wackerly, Mendenhall, and Scheaffer [WMS02, p. 2] point out: "Statistics is a theory of information, with inference making as its objective" and may be viewed as encompassing "the science of basing inferences on observed data and the entire problem of making decisions in the face of uncertainty."

In this chapter we present an introduction (Section 52.2) to the use of linear algebra in studying properties of random vectors (Section 52.3) and in linear statistical models (Section 52.4); for further uses, see, e.g., [PSS05] and [PS05], and for an introduction to the use of linear algebra in multivariate statistical analysis see Chapter 53 or [PSS05].

We begin (in Section 52.2) with a brief introduction to the basic concepts of statistics and random variables [MM04], [WMS02], [WS00], with special emphasis (in Section 52.3) on random vectors — vectors where each entry is a random variable.

All matrices (with at least two rows and two columns) considered in this chapter are nonrandom.

52.2 Introduction to Statistics and Random Variables

Notation:

In this chapter most uppercase, light-face italic letters, in particular X, Y, and Z, denote scalar random variables, but the notation $P(A)$ is reserved for the probability of the set A. Throughout this chapter, the uppercase, light-face roman letter E denotes expectation.

Definitions:

The focus in statistics is on making inferences concerning a large body of data called a **population** based on a subset collected from it called a **sample**. An **experiment** associated with this sample is repeatable and its outcome is not predetermined. A **simple event** is one associated with an experiment, which cannot be decomposed, and a simple event corresponds to one and only one **sample point**. The **sample space** associated with an experiment is the set of all possible sample points.

Suppose that S is a sample space associated with an experiment. To every subset A of S we assign a real number $P(A)$, called the **probability** of A, so that the following axioms hold: (1) $P(A) \geq 0$, (2) $P(S) = 1$, (3) If $A_1, A_2, A_3, \ldots$ form a sequence of pairwise mutually exclusive subsets in S, i.e., $A_i \cap A_j = \emptyset$ if $i \neq j$, then $P(A_1 \cup A_2 \cup A_3 \cup \cdots) = \sum_{i=1}^{\infty} P(A_i)$.

A **random variable** is a real-valued function for which the domain is a sample space.

A random variable which can assume a finite or countably infinite number of values is **discrete**.

The **probability** $P(Y = y)$ that the random variable Y takes on the value y is defined as the sum of the probabilities of all sample points in S that have the value y, and the **probability function** of Y is the set of all the probabilities $P(Y = y)$.

If the random variable Y has the probability function $P(Y = 0) = p$ and $P(Y = 1) = 1 - p$ for some real number p in the interval $[0, 1]$, then Y is a **Bernoulli random variable.**

The **cumulative distribution function** (cdf) is $F(y) = P(Y \leq y)$ of the random variable Y for $-\infty < y < \infty$. A random variable Y is **continuous** when its cdf $F(y)$ is continuous for $-\infty < y < \infty$, and then its **probability density function** (pdf) $f(y) = dF(y)/dy$.

Suppose that the continuous random variable Y has pdf $f(y) = 1$ for $0 \leq y \leq 1$ and $f(y) = 0$ otherwise. Then Y follows a **uniform distribution** on the interval $[0, 1]$.

The **expectation** (or **expected value** or **mean** or **mean value**) $\mathrm{E}(Y)$ of the random variable Y is $\mathrm{E}(Y) = \sum_y yP(Y = y)$ when Y is discrete with probability function $P(Y = y)$ and is $\mathrm{E}(Y) = \int_{-\infty}^{+\infty} yf(y)dy$ when Y is continuous with pdf $f(y)$.

The **variance** σ^2 of the random variable Y is

$$\sigma^2 = \mathrm{var}(Y) = \mathrm{E}\big((Y - \mu)^2\big),$$

where $\mu = \mathrm{E}(Y)$ and the **standard deviation** $\sigma = \sqrt{\sigma^2}$.

Facts:

1. The variance $\sigma^2 = \mathrm{var}(Y) = \mathrm{E}(Y^2) - \mathrm{E}^2(Y)$.
2. For any random variable Y, the expectation of its square $\mathrm{E}(Y^2) \geq \mathrm{E}^2(Y)$ with equality if and only if the random variable $Y = \mathrm{E}(Y)$ with probability 1.
3. If Y is a Bernoulli random variable with probability function $P(Y = 0) = p$ and $P(Y = 1) = 1-p$, then the expectation $\mathrm{E}(Y) = 1 - p$ and the variance $\mathrm{var}(Y) = p(1 - p)$.
4. If the random variable Z follows a uniform distribution on the interval $[0, 1]$, then the expectation $\mathrm{E}(Z) = 1/2$ and the variance $\mathrm{var}(Z) = 1/12$.

Examples:

1. Every person's blood type is A, B, AB, or O. In addition, each individual either has the Rhesus (Rh) factor (+) or does not (–). A medical technician records a person's blood type and Rh factor. The sample space for this experiment is {A+, B+, AB+, O+, A−, B−, AB−, O−} with eight sample points.
2. Consider the experiment of tossing a single fair coin and define the random variable $Y = 0$ if the outcome is "heads," and $Y = 1$ if the outcome is "tails." Then Y is a Bernoulli random variable, and $P(Y = 0) = \frac{1}{2} = P(Y = 1)$, $\mathrm{E}(Y) = \frac{1}{2}$, $\mathrm{var}(Y) = \frac{1}{4}$.
3. Suppose that a bus always arrives at a particular stop in the interval between 12 noon and 1 p.m. and that the probability that the bus will arrive in any given subinterval of time is proportional only to the length of the subinterval. Let Y denote the length of time that a person arriving at the stop at 12 noon must wait for the bus to arrive, and let us code 12 noon as 0 and measure the time in hours. Then the random variable Y follows a uniform distribution on the interval $[0, 1]$.

52.3 Random Vectors: Basic Definitions and Facts

Linear algebra is extensively used in the study of random vectors, where we consider the simultaneous behavior of two or more random variables assembled as a vector. In this section all vectors and matrices are real.

Notation:

In this section uppercase, light-face italic letters, such as X, Y, and Z, denote scalar random variables and lowercase bold roman letters, such as **x**, **y**, and **z**, denote random vectors. Uppercase, light-face italic letters such as A and B denote nonrandom matrices.

Definitions:

Let $A \in \mathbb{R}^{n\times k}$ and $B \in \mathbb{R}^{n\times q}$. Then the **partitioned matrix** $[A \mid B]$ is the $n \times (k+q)$ matrix formed by placing A next to B.

A $k \times 1$ **random vector y** is a vector $\mathbf{y} = [Y_1, \ldots, Y_k]^T$ of k random variables $Y_1, \ldots, Y_k$. The **expectation** (or **expected value** or **mean vector**) of **y** is the $k \times 1$ vector $\mathrm{E}(\mathbf{y}) = [\mathrm{E}(Y_1), \ldots, \mathrm{E}(Y_k)]^T$. Sometimes, for clarity, a vector of constants (belonging to $\mathbb{R}^n$) is called a **nonrandom vector** and a matrix of constants a **nonrandom matrix**. (Random matrices are not considered in this chapter.)

The **covariance** $\mathrm{cov}(Y, Z)$ between the two random variables Y and Z is $\mathrm{cov}(Y, Z) = \mathrm{E}\big((Y-\mu)(Z-\nu)\big)$, where $\mu = \mathrm{E}(Y)$ and $\nu = \mathrm{E}(Z)$.

The **correlation** (or **correlation coefficient** or **product-moment correlation**) $\mathrm{cor}(Y, Z)$ between the two random variables Y and Z is $\mathrm{cor}(Y, Z) = \mathrm{cov}(Y, Z)/\sqrt{\mathrm{var}(Y)\mathrm{var}(Z)}$.

The **covariance matrix** (or **variance-covariance matrix** or **dispersion matrix**) of the $k \times 1$ random vector $\mathbf{y} = [Y_1, \ldots, Y_k]^T$ is the $k \times k$ matrix $\mathrm{var}(\mathbf{y}) = \Sigma$ of variances and covariances of all the entries of **y**:

$$\begin{aligned}\mathrm{var}(\mathbf{y}) = \Sigma = [\sigma_{ij}] = [\mathrm{cov}(Y_i, Y_j)] &= [\mathrm{E}(Y_i - \mu_i)(Y_j - \mu_j)] \\ &= \mathrm{E}\big((\mathbf{y} - \boldsymbol{\mu})(\mathbf{y} - \boldsymbol{\mu})^T\big),\end{aligned}$$

where $\boldsymbol{\mu} = \mathrm{E}(\mathbf{y})$. The determinant $\det \Sigma$ is the **generalized variance** of the random vector **y**. The variances σ_{ii} are often denoted as σ_i^2 and, in this, chapter, we will assume that they are all positive. If $\sigma_i = 0$, then the random variable $Y_i = \mathrm{E}(Y_i)$ with probability 1, and then we interpret Y_i as a constant. In statistics it is quite common to denote **standard deviations** as σ_i. (The reader should note that in all the other chapters of this book except in the two statistics chapters, σ_i denotes the ith largest singular value.)

The **cross-covariance matrix** $\text{cov}(\mathbf{y}, \mathbf{z})$ between the $k \times 1$ random vector $\mathbf{y} = [Y_1, \ldots, Y_k]^T$ and the $q \times 1$ random vector $\mathbf{z} = [Z_1, \ldots, Z_q]^T$ is the $k \times q$ matrix of all the covariances $\text{cov}(Y_i, Z_j); i = 1, \ldots, k$ and $j = 1, \ldots, q$:

$$\text{cov}(\mathbf{y}, \mathbf{z}) = [\text{cov}(Y_i, Z_j)] = [\text{E}(Y_i - \mu_i)(Z_j - \nu_j)] = \text{E}\big((\mathbf{y} - \boldsymbol{\mu})(\mathbf{z} - \boldsymbol{\nu})^T\big),$$

where $\boldsymbol{\mu} = [\mu_i] = \text{E}(\mathbf{y})$ and $\boldsymbol{\nu} = [\nu_j] = \text{E}(\mathbf{z})$. The random vectors $\mathbf{y}$ and $\mathbf{z}$ are **uncorrelated** whenever the cross-covariance matrix $\text{cov}(\mathbf{y}, \mathbf{z}) = \mathbf{0}$.

The **correlation matrix** $\text{cor}(\mathbf{y}) = R$, say, of the $k \times 1$ random vector $\mathbf{y} = [Y_1, \ldots, Y_k]^T$, is the $k \times k$ matrix of correlations of all the entries in $\mathbf{y}$: $\text{cor}(\mathbf{y}) = R = [\rho_{ij}] = [\text{cor}(Y_i, Y_j)] = [\frac{\sigma_{ij}}{\sigma_i \sigma_j}]$, where $\sigma_i = \sqrt{\sigma_{ii}} =$ standard deviation of Y_i; $\sigma_i, \sigma_j > 0$.

Let $\mathbf{1}_k$ denote the $k \times 1$ column vector with every entry equal to 1. Then $J_k = \mathbf{1}_k \mathbf{1}_k^T$ is the $k \times k$ **all-ones matrix** (with all k^2 entries equal to 1) and $C_k = I_k - \frac{1}{k} J_k$ is the $k \times k$ **centering matrix**.

Suppose that the real positive numbers $p_1, p_2, \ldots, p_k$ are such that $\sum_{i=1}^k p_i = 1$. Then the $k \times 1$ random vector $\mathbf{y} = [Y_1, \ldots, Y_k]^T$ follows a **multinomial distribution** with parameters n and $p_1, p_2, \ldots, p_k$ if the joint probability function of $Y_1, Y_2, \ldots, Y_k$ is given by

$$P(Y_1 = y_1, Y_2 = y_2, \ldots, Y_k = y_k) = \frac{n!}{y_1! y_2! \cdots y_k!} p_1^{y_1} p_2^{y_2} \cdots p_k^{y_k},$$

where for each i, $y_i = 0, 1, 2, \ldots, n$ and $\sum_{i=1}^k y_i = n$. When $k = 2$ the distribution is **binomial**, and when $k = 3$ the distribution is **trinomial.**

Let the symmetric matrices $A \in \mathbb{R}^{k \times k}$ and $B \in \mathbb{R}^{k \times k}$. Then $A \succeq B$ means $A - B$ is positive semidefinite and $A \succ B$ means $A - B$ is positive definite. The partial ordering induced by $\succeq$ is called the **partial semidefinite ordering** (or **Loewner partial ordering** or **Löwner partial ordering**). (See Section 8.5 for more information.)

Let the $(k + q) \times 1$ random vector $\mathbf{x}$ have covariance matrix Σ. Consider the following partitioning:

$$\mathbf{x} = \begin{bmatrix} \mathbf{y} \\ \mathbf{z} \end{bmatrix}, \qquad \text{E}(\mathbf{x}) = \begin{bmatrix} \boldsymbol{\mu} \\ \boldsymbol{\nu} \end{bmatrix}, \qquad \text{var}(\mathbf{x}) = \Sigma = \begin{bmatrix} \Sigma_{yy} & \Sigma_{yz} \\ \Sigma_{zy} & \Sigma_{zz} \end{bmatrix},$$

where $\mathbf{y}$ and $\mathbf{z}$ have k and q elements, respectively. Then the **partial covariance matrix** $\Sigma_{zz \cdot y}$ of the $q \times 1$ random vector $\mathbf{z}$ after **adjusting for** (or **controlling for** or **removing the effect of** or **allowing for**) the $k \times 1$ random vector $\mathbf{y}$ is the (uniquely defined) **generalized Schur complement**

$$\Sigma / \Sigma_{yy} = [\sigma_{ij \cdot \mathbf{y}}] = \Sigma_{zz} - \Sigma_{zy} \Sigma_{yy}^{-} \Sigma_{yz}$$

of Σ_{yy} in Σ; any **generalized inverse** Σ_{yy}^{-} satisfying $\Sigma_{yy} = \Sigma_{yy} \Sigma_{yy}^{-} \Sigma_{yy}$ may be chosen.

When Σ_{yy} is positive definite, we use the inverse Σ_{yy}^{-1} instead of a generalized inverse Σ_{yy}^{-} and refer to $\Sigma / \Sigma_{yy} = \Sigma_{zz} - \Sigma_{zy} \Sigma_{yy}^{-1} \Sigma_{yz}$ as the **Schur complement** of Σ_{yy} in Σ.

The $q \times 1$ random vector $\mathbf{e}_{\mathbf{z} \cdot \mathbf{y}} = \mathbf{z} - \boldsymbol{\nu} - \Sigma_{zy} \Sigma_{yy}^{-} (\mathbf{y} - \boldsymbol{\mu})$ is the (uniquely defined) vector of **residuals** of the $q \times 1$ random vector $\mathbf{z}$ from its **regression** on the $k \times 1$ random vector $\mathbf{y}$.

The ijth entry of the **partial correlation matrix** of $q \times 1$ random vector $\mathbf{z} = [Z_1, \ldots, Z_q]^T$ after adjusting for the $k \times 1$ random vector $\mathbf{y}$ is the **partial correlation coefficient** between z_i and z_j after adjusting for $\mathbf{y}$:

$$\rho_{ij \cdot \mathbf{y}} = \frac{\sigma_{ij \cdot \mathbf{y}}}{\sqrt{\sigma_{ii \cdot \mathbf{y}} \sigma_{jj \cdot \mathbf{y}}}}; \; i, j = 1, \ldots, q,$$

which is well defined provided the diagonal entries of the associated partial covariance matrix are all positive.

Facts:

1. [WMS02, Th. 5.13, p. 265] Let the $k \times 1$ random vector $\mathbf{y} = [Y_1, \ldots, Y_k]^T$ follow a multinomial distribution with parameters n and $p_1, \ldots, p_k$ and let the $k \times 1$ vector $\mathbf{p} = [p_1, \ldots, p_k]^T$. Then:

   - The random variable Y_i can be represented as the sum of n independently and identically distributed Bernoulli random variables with parameter p_i; $i = 1, \ldots, k$.
   - The expectation $\mathrm{E}(\mathbf{y}) = n\mathbf{p}$ and the covariance matrix

   $$\mathrm{var}(\mathbf{y}) = n\big(\mathrm{diag}(\mathbf{p}) - \mathbf{p}\mathbf{p}^T\big) = \Psi_k,$$

   say, where $\mathrm{diag}(\mathbf{p})$ is the $k \times k$ diagonal matrix formed from the $k \times 1$ nonrandom vector $\mathbf{p}$.
   - The covariance matrix Ψ_k is singular since all its row (and column) totals are 0, and the $\mathrm{rank}(\Psi_k) = k - 1$.

2. When the $k \times 1$ multinomial probability vector $\mathbf{p} = \mathbf{1}_k/k$, then the multinomial covariance matrix $\Psi_k = \frac{n}{k}C_k$, where C_k is the $k \times k$ centering matrix.
3. The $k \times k$ covariance matrix $\mathrm{var}(\mathbf{y}) = \Sigma = \mathrm{E}(\mathbf{y}\mathbf{y}^T) - \boldsymbol{\mu}\boldsymbol{\mu}^T$, where $\boldsymbol{\mu} = \mathrm{E}(\mathbf{y})$.
4. The $k \times k$ correlation matrix $\mathrm{cor}(\mathbf{y}) = [\mathrm{diag}(\Sigma)]^{-1/2}\Sigma[\mathrm{diag}(\Sigma)]^{-1/2}$, where $\Sigma = \mathrm{var}(\mathbf{y})$.
5. The $k \times q$ cross-covariance matrix

   $$\mathrm{cov}(\mathbf{y}, \mathbf{z}) = \Sigma_{yz} = \Sigma_{zy}^T = \big(\mathrm{cov}(\mathbf{z}, \mathbf{y})\big)^T = \mathrm{E}(\mathbf{y}\mathbf{z}^T) - \boldsymbol{\mu}\boldsymbol{\nu}^T,$$

   where $\boldsymbol{\mu} = \mathrm{E}(\mathbf{y})$ and $\boldsymbol{\nu} = \mathrm{E}(\mathbf{z})$.
6. [RM71, Lemma 2.2.4, p. 21]: The product AB^-C (for $A \neq \mathbf{0}$, $C \neq \mathbf{0}$) is invariant with respect to the choice of $B^- \iff \mathrm{range}(C) \subseteq \mathrm{range}(B)$ and $\mathrm{range}(A^T) \subseteq \mathrm{range}(B^T)$.
7. Consider the $(k + q) \times (k + q)$ covariance matrix

   $$\Sigma = \begin{bmatrix} \Sigma_{yy} & \Sigma_{yz} \\ \Sigma_{zy} & \Sigma_{zz} \end{bmatrix}.$$

   Then the range (or column space) $\mathrm{range}(\Sigma_{yz}) \subseteq \mathrm{range}(\Sigma_{yy})$ and $\mathrm{range}(\Sigma_{zy}) \subseteq \mathrm{range}(\Sigma_{zz})$, and, hence, the matrix $\Sigma_{zy}\Sigma_{yy}^-\Sigma_{yz}$ and the generalized Schur complement $\Sigma/\Sigma_{yy} = \Sigma_{zz} - \Sigma_{zy}\Sigma_{yy}^-\Sigma_{yz}$ are invariant (unique) with respect to the choice of generalized inverse Σ_{yy}^-.
8. Let the $k \times 1$ random vector $\mathbf{x} = [X_1, \ldots, X_k]^T$. Then the $k \times 1$ centered vector $C_k\mathbf{x} = [X_1 - \bar{X}, \ldots, X_k - \bar{X}]^T$, where C_k is the $k \times k$ centering matrix and the arithmetic mean (or average) $\bar{X} = \sum_{i=1}^k X_i/k$.
9. A nonsingular positive semidefinite matrix is positive definite.
10. A covariance matrix is always symmetric and positive semidefinite.
11. A cross-covariance matrix is usually rectangular.
12. A correlation matrix is always symmetric and positive semidefinite.
13. The diagonal entries of a correlation matrix are all equal to 1 and the off-diagonal entries are all at most equal to 1 in absolute value.
14. [PS05, p. 168] The (generalized) Schur complement of the leading principal submatrix of a positive semidefinite matrix is positive semidefinite.
15. Let $\mathbf{y}$ be a $k \times 1$ random vector with covariance matrix Σ. Then the variance $\mathrm{var}(\mathbf{a}^T\mathbf{y}) = \mathbf{a}^T\Sigma\mathbf{a}$ for all nonrandom $\mathbf{a} \in \mathbb{R}^k$. (Since variance must be nonnegative this fact shows that a covariance matrix must be positive semidefinite.)
16. Let $\mathbf{y}$ be a $k \times 1$ random vector with expectation $\boldsymbol{\mu} = \mathrm{E}(\mathbf{y})$, covariance matrix $\Sigma = \mathrm{var}(\mathbf{y})$, and let the matrix $A \in \mathbb{R}^{n\times k}$ and the nonrandom vector $\mathbf{b} \in \mathbb{R}^n$. Then the expectation $\mathrm{E}(A\mathbf{y} + \mathbf{b}) = A\mathrm{E}(\mathbf{y}) + \mathbf{b} = A\boldsymbol{\mu} + \mathbf{b}$ and the covariance matrix $\mathrm{var}(A\mathbf{y} + \mathbf{b}) = A\mathrm{var}(\mathbf{y})A^T = A\Sigma A^T$.
17. Let $\mathbf{y}$ be a $k \times 1$ random vector with expectation $\boldsymbol{\mu} = \mathrm{E}(\mathbf{y})$, covariance matrix $\Sigma = \mathrm{var}(\mathbf{y})$, and let the matrix $A \in \mathbb{R}^{k\times k}$, not necessarily symmetric. Then $\mathrm{E}(\mathbf{y}^T A\mathbf{y}) = \boldsymbol{\mu}^T A\boldsymbol{\mu} + \mathrm{tr}(A\Sigma)$.

18. [Rao73a, p. 522] Let $\mathbf{y}$ be a $k \times 1$ random vector with expectation $\boldsymbol{\mu} = \mathrm{E}(\mathbf{y})$ and covariance matrix $\Sigma = \mathrm{var}(\mathbf{y})$, and let $[\boldsymbol{\mu} \mid \Sigma]$ denote the $k \times (k+1)$ partitioned matrix with $\boldsymbol{\mu}$ as its first column. Then $\mathbf{y} - \boldsymbol{\mu} \in \mathrm{range}(\Sigma)$ and $\mathbf{y} \in \mathrm{range}([\boldsymbol{\mu} \mid \Sigma])$, both with probability 1.
19. Let the $(k+q) \times 1$ random vector $\mathbf{x}$ have covariance matrix Σ. Consider the following partitioning:

$$\mathbf{x} = \begin{bmatrix} \mathbf{y} \\ \mathbf{z} \end{bmatrix}, \qquad \mathrm{E}(\mathbf{x}) = \begin{bmatrix} \boldsymbol{\mu} \\ \boldsymbol{\nu} \end{bmatrix}, \qquad \Sigma = \begin{bmatrix} \Sigma_{yy} & \Sigma_{yz} \\ \Sigma_{zy} & \Sigma_{zz} \end{bmatrix},$$

where $\mathbf{y}$ and $\mathbf{z}$ have k and q components, respectively. Then

- The variance $\mathrm{var}(\mathbf{a}^T\mathbf{y} + \mathbf{b}^T\mathbf{z}) = \mathbf{a}^T\Sigma_{yy}\mathbf{a} + 2\mathbf{a}^T\Sigma_{yz}\mathbf{b} + \mathbf{b}^T\Sigma_{zz}\mathbf{b}$ for all nonrandom $\mathbf{a} \in \mathbb{R}^k$ and for all nonrandom $\mathbf{b} \in \mathbb{R}^q$.
- The covariance matrix $\mathrm{var}(A\mathbf{y} + B\mathbf{z}) = A\Sigma_{yy}A^T + A\Sigma_{yz}B^T + B\Sigma_{zy}A^T + B\Sigma_{zz}B^T$ for all $A \in \mathbb{R}^{n\times k}$ and all $B \in \mathbb{R}^{n\times q}$.
- [PS05b, pp. 187–188] For any $A \in \mathbb{R}^{q\times k}$ the covariance matrix

$$\mathrm{var}(\mathbf{z} - A\mathbf{y}) \succeq \mathrm{var}(\mathbf{z} - \Sigma_{zy}\Sigma_{yy}^{-}\mathbf{y})$$

with respect to the partial semidefinite ordering, and the partial covariance matrix

$$\mathrm{var}(\mathbf{z} - \Sigma_{zy}\Sigma_{yy}^{-}\mathbf{y}) = \Sigma_{zz} - \Sigma_{zy}\Sigma_{yy}^{-}\Sigma_{yz} = \Sigma_{zz\cdot y} = \Sigma/\Sigma_{yy},$$

the generalized Schur complement of Σ_{yy} in Σ.
- Let $q = k$. Then the covariance matrix $\mathrm{var}(\mathbf{y} + \mathbf{z}) = \mathrm{var}(\mathbf{y}) + \mathrm{var}(\mathbf{z})$ if and only if $\mathrm{cov}(\mathbf{y}, \mathbf{z}) = -\mathrm{cov}(\mathbf{z}, \mathbf{y})$, i.e., the cross-covariance matrix $\mathrm{cov}(\mathbf{y}, \mathbf{z})$ is skew-symmetric; the condition that $\mathrm{cov}(\mathbf{y}, \mathbf{z}) = \mathbf{0}$ is sufficient, but not necessary (unless $k = 1$).
- The vector $\Sigma_{zy}\Sigma_{yy}^{-}\mathbf{y}$ is not necessarily invariant with respect to the choice of generalized inverse Σ_{yy}^{-}, but its covariance matrix $\mathrm{var}(\Sigma_{zy}\Sigma_{yy}^{-}\mathbf{y}) = \Sigma_{zy}\Sigma_{yy}^{-}\Sigma_{yz}$ is invariant (and, hence, unique).

Examples:

1. Let the 4×1 random vector $\mathbf{x}$ have covariance matrix Σ. Consider the following partitioning:

$$\mathbf{x} = \begin{bmatrix} \mathbf{y} \\ \mathbf{z} \end{bmatrix}, \qquad \Sigma = \begin{bmatrix} 1 & 0 & a & b \\ 0 & 1 & c & d \\ a & c & 1 & 0 \\ b & d & 0 & 1 \end{bmatrix},$$

where $\mathbf{y}$ and $\mathbf{z}$ each have 2 components. Then $\mathrm{var}(\mathbf{y}+\mathbf{z}) = \mathrm{var}(\mathbf{y}) + \mathrm{var}(\mathbf{z})$ if and only if $a = d = 0$ and $c = -b$, with $b^2 \leq 1$.
2. Let the $k \times 1$ random vector $\mathbf{y} = [Y_1, \ldots, Y_k]^T$ follow a multinomial distribution with parameters n and $\mathbf{p} = [p_1, \ldots, p_k]^T$, with $p_1 + \cdots + p_k = 1$ and $p_1 > 0, \ldots, p_k > 0$, and let the $k \times k$ matrix $A = [a_{ij}]$. Then the expectation $\mathrm{E}(\mathbf{y}^T A\mathbf{y}) = n(n-1)\mathbf{p}^T A\mathbf{p} + n\sum_{i=1}^{k} a_{ii}\,p_i$.
3. Let the 3×1 random vector $\mathbf{y} = [Y_1, Y_2, Y_3]^T$ follow a trinomial distribution with parameters n and p_1, p_2, p_3, with $p_1 + p_2 + p_3 = 1$ and $p_1 > 0, p_2 > 0, p_3 > 0$, and let the 3×1 vector $\mathbf{p} = [p_1, p_2, p_3]^T$. Then:

- The expectation $\mathrm{E}(\mathbf{y}) = n[p_1, p_2, p_3]^T$ and the covariance matrix

$$\Psi_3 = \mathrm{var}(\mathbf{y}) = n \begin{bmatrix} p_1(1-p_1) & -p_1p_2 & -p_1p_3 \\ -p_1p_2 & p_2(1-p_2) & -p_2p_3 \\ -p_1p_3 & -p_2p_3 & p_3(1-p_3) \end{bmatrix},$$

which has rank equal to 2 since Ψ_3 is singular and the determinant of the top left-hand corner of Ψ_3 equals $n^2 p_1 p_2 p_3 > 0$.

- The partial covariance matrix of Y_1 and Y_2 adjusting for Y_3 is the Schur complement $\Psi_3/\big(p_3(1-p_3)\big) = nS$, say, where

$$S = \begin{bmatrix} p_1(1-p_1) & -p_1 p_2 \\ -p_1 p_2 & p_2(1-p_2) \end{bmatrix}$$

$$- \begin{bmatrix} -p_1 p_3 \\ -p_2 p_3 \end{bmatrix} \frac{1}{p_3(1-p_3)} \begin{bmatrix} -p_1 p_3 & -p_2 p_3 \end{bmatrix} = \frac{p_1 p_2}{p_1 + p_2} \begin{bmatrix} 1 & -1 \\ -1 & 1 \end{bmatrix},$$

which has rank equal to 1, and so $\text{rank}(\Psi_3) = 2$.

- When $p_1 = p_2 = p_3 = 1/3$, then the covariance matrix $\Psi_3 = (n/3)C_3$ and the partial covariance matrix of Y_1 and Y_2 adjusting for Y_3 is $(n/3)C_2$; here, C_h is the $h \times h$ centering matrix, $h = 2, 3$.

4. [PS05, p. 183] If the 3×3 symmetric matrix

$$R_3 = \begin{bmatrix} 1 & r_{12} & r_{13} \\ r_{12} & 1 & r_{23} \\ r_{13} & r_{23} & 1 \end{bmatrix} = \begin{bmatrix} R_2 & \mathbf{r}_2 \\ \mathbf{r}_2^T & 1 \end{bmatrix}$$

is a correlation matrix, then $r_{ij}^2 \leq 1$ for all $1 \leq i < j \leq 3$. But not all symmetric matrices with diagonal elements all equal to 1 and all off-diagonal elements r_{ij} such that $r_{ij}^2 \leq 1$ are correlation matrices. For example, consider R_3 with $r_{13}^2 \leq 1$ and $r_{23}^2 \leq 1$. Then R_3 is a correlation matrix if and only if

$$r_{13}r_{23} - \sqrt{(1-r_{13}^2)(1-r_{23}^2)} \leq r_{12} \leq r_{13}r_{23} + \sqrt{(1-r_{13}^2)(1-r_{23}^2)}.$$

When $r_{13} = 0$ and $r_{12} = r_{23} = r$, say, then this condition becomes $r^2 \leq 1/2$ and so the matrix

$$\begin{bmatrix} 1 & 0.8 & 0 \\ 0.8 & 1 & 0.8 \\ 0 & 0.8 & 1 \end{bmatrix}$$

is not a correlation matrix.

When $r_{12}^2 \leq 1$, then the matrix R_3 is a correlation matrix if and only if any one of the following conditions holds:

- $\det(R_3) = 1 - r_{12}^2 - r_{13}^2 - r_{23}^2 + 2r_{12}r_{13}r_{23} \geq 0$.
- (i) $\mathbf{r}_2 \in \text{range}(R_2)$ and (ii) $1 \geq \mathbf{r}_2^T R_2^- \mathbf{r}_2$ for some and, hence, for every generalized inverse R_2^-.

5. Let the random vector $\mathbf{x}$ be 2×1 and write

$$\mathbf{x} = \begin{bmatrix} Y \\ Z \end{bmatrix}, \qquad \text{E}(\mathbf{x}) = \begin{bmatrix} \mu \\ \nu \end{bmatrix}, \qquad \text{var}(\mathbf{x}) = \Sigma = \begin{bmatrix} \sigma_y^2 & \sigma_{yz} \\ \sigma_{yz} & \sigma_z^2 \end{bmatrix},$$

with $\sigma_y^2 > 0$. Then the residual vector $\mathbf{e}_{\mathbf{z}\cdot\mathbf{y}} = \mathbf{z} - \boldsymbol{\nu} - \Sigma_{zy}\Sigma_{yy}^-(\mathbf{y} - \boldsymbol{\mu})$ of the random vector $\mathbf{z}$ from its regression on $\mathbf{y}$ becomes the scalar residual

$$e_{Z\cdot Y} = Z - \nu - \frac{\sigma_{yz}}{\sigma_y^2}(Y - \mu)$$

of the random variable Z from its regression on Y. The matrix of partial covariances of the random vector $\mathbf{z}$ after adjusting for $\mathbf{y}$ becomes the single partial variance

$$\sigma_{z\cdot y}^2 = \sigma_z^2 - \frac{\sigma_{yz}^2}{\sigma_y^2} = \sigma_z^2(1 - \rho_{yz}^2)$$

of the random variable Z after adjusting for the random variable Y; here, the correlation coefficient $\rho_{yz} = \sigma_{yz}/(\sigma_y \sigma_z)$.

52.4 Linear Statistical Models: Basic Definitions and Facts

Notation:

In this section, the uppercase, light-face italic letter X is reserved for the nonrandom $n \times p$ model matrix and V is reserved for an $n \times n$ covariance matrix. The uppercase, light-face italic letter H is reserved for the (symmetric idempotent) $n \times n$ hat matrix $X(X^T X)^- X^T$ and $M = I - H$ is reserved for the (symmetric idempotent) $n \times n$ residual matrix. The lowercase, bold-face roman letter $\mathbf{y}$ is reserved for an observable $n \times 1$ random vector and $\mathbf{x}$ is reserved for a column of the $n \times p$ model matrix X.

Definitions:

The **general linear model** (or **Gauss–Markov model** or **Gauß–Markov model**) is the model

$$\mathscr{M} = \{\mathbf{y},\ X\boldsymbol{\beta},\ \sigma^2 V\}$$

defined by the equation $\mathbf{y} = X\boldsymbol{\beta} + \boldsymbol{\varepsilon}$, where $\mathrm{E}(\mathbf{y}) = X\boldsymbol{\beta}$, $\mathrm{E}(\boldsymbol{\varepsilon}) = \mathbf{0}$, $\mathrm{var}(\mathbf{y}) = \mathrm{var}(\boldsymbol{\varepsilon}) = \sigma^2 V$. The vector $\mathbf{y}$ is an $n \times 1$ observable random vector, $\boldsymbol{\varepsilon}$ is an $n \times 1$ unobservable random error vector, X is a known $n \times p$ **model matrix** (or **design matrix**, particularly when its entries are $-1, 0$, or $+1$), $\boldsymbol{\beta}$ is a $p \times 1$ vector of unknown parameters, V is a known $n \times n$ positive semidefinite matrix, and σ^2 is an unknown positive constant. The realization of the $n \times 1$ observable random vector $\mathbf{y}$ will also be denoted by $\mathbf{y}$.

The classical theory of linear statistical models covers the **full-rank model**, where X has full column rank and V is positive definite. In the full-rank model, the **ordinary least squares estimator**

$$\mathrm{OLSE}(\boldsymbol{\beta}) = \hat{\boldsymbol{\beta}} = (X^T X)^{-1} X^T \mathbf{y} = X^\dagger \mathbf{y}$$

and the **generalized least squares estimator** (or **Aitken estimator**)

$$\mathrm{GLSE}(\boldsymbol{\beta}) = \tilde{\boldsymbol{\beta}} = (X^T V^{-1} X)^{-1} X^T V^{-1} \mathbf{y},$$

where $X^\dagger$ denotes the Moore–Penrose inverse of X.

When either X or V is (or both X and V are) rank deficient, then it is usually assumed that $\mathrm{rank}(X) < \mathrm{rank}(V)$. The model $\mathscr{M} = \{\mathbf{y},\ X\boldsymbol{\beta},\ \sigma^2 V\}$ is called a **weakly singular model** (or **Zyskind–Martin model**) whenever $\mathrm{range}(X) \subseteq \mathrm{range}(V)$, and then $\mathrm{rank}(X) < \mathrm{rank}(V)$, and is **consistent** if the realization $\mathbf{y}$ satisfies $\mathbf{y} \in \mathrm{range}([X \mid V])$.

Let $\hat{\boldsymbol{\beta}}$ be any vector minimizing $\|\mathbf{y} - X\boldsymbol{\beta}\|^2 = (\mathbf{y} - X\boldsymbol{\beta})^T(\mathbf{y} - X\boldsymbol{\beta})$. Then $\hat{\mathbf{y}} = X\hat{\boldsymbol{\beta}} = \mathrm{OLSE}(X\boldsymbol{\beta}) =$ the **ordinary least squares estimator** (OLSE) of $X\boldsymbol{\beta}$. When $\mathrm{rank}(X) < p$, then $\hat{\boldsymbol{\beta}}$ is an **ordinary least squares solution** to $\min_{\boldsymbol{\beta}}(\mathbf{y} - X\boldsymbol{\beta})^T(\mathbf{y} - X\boldsymbol{\beta})$. Moreover, $\hat{\boldsymbol{\beta}}$ is any solution to the **normal equations** $X^T X\hat{\boldsymbol{\beta}} = X^T \mathbf{y}$. The vector of OLS **residuals** is $\mathbf{e} = \mathbf{y} - \hat{\mathbf{y}} = \mathbf{y} - X\hat{\boldsymbol{\beta}}$ and the **residual sum of squares** $SSE = \mathbf{e}^T\mathbf{e} = (\mathbf{y} - \hat{\mathbf{y}})^T(\mathbf{y} - \hat{\mathbf{y}})$.

The **coefficient of determination** (or **coefficient of multiple determination** or **squared multiple correlation**) $R^2 = 1 - (SSE/\mathbf{y}^T C_n \mathbf{y})$ identifies the proportion of variance explained in a **multiple linear regression** where the model matrix $X = [\mathbf{1}_n \mid \mathbf{x}_{[1]} \mid \cdots \mid \mathbf{x}_{[p-1]}]$ with $p - 1$ **regressor vectors** (or **regressors**) $\mathbf{x}_{[1]}, \ldots, \mathbf{x}_{[p-1]}$ each $n \times 1$. In **simple linear regression** $p = 2$ and the model matrix $X = [\mathbf{1}_n \mid \mathbf{x}]$

with the single regressor vector $\mathbf{x}$. The **sample correlation coefficient** $r = \mathbf{x}^T C_n \mathbf{y}/\sqrt{\mathbf{x}^T C_n \mathbf{x} \cdot \mathbf{y}^T C_n \mathbf{y}}$, where it is usually assumed that $\mathbf{x}$ is an $n \times 1$ nonrandom vector (such as a regressor vector) and $\mathbf{y}$ is a realization of the $n \times 1$ random vector $\mathbf{y}$.

Let the matrix $A \in \mathbb{R}^{k\times n}$ and let the matrix $K \in \mathbb{R}^{k\times p}$. Then the linear estimator $A\mathbf{y}$ is a **linear unbiased estimator** (LUE) of $K\boldsymbol{\beta}$ if $\mathrm{E}(A\mathbf{y}) = K\boldsymbol{\beta}$ for all $\boldsymbol{\beta} \in \mathbb{R}^p$. Let the matrix $B \in \mathbb{R}^{k\times n}$. Then the LUE $B\mathbf{y}$ of $K\boldsymbol{\beta}$ is the **best linear unbiased estimator** (BLUE) of $K\boldsymbol{\beta}$ if it has the smallest covariance matrix (in the positive semidefinite ordering) in that $\mathrm{var}(A\mathbf{y}) \succeq \mathrm{var}(B\mathbf{y})$ for all LUEs $A\mathbf{y}$ of $K\boldsymbol{\beta}$.

The **hat matrix** $H = X(X^TX)^-X^T$ associated with the model matrix X is so named since $\hat{\mathbf{y}} = H\mathbf{y}$. The **residual matrix** $M = I - H$ and vector of OLS residuals is $\mathbf{e} = \mathbf{y} - \hat{\mathbf{y}} = \mathbf{y} - H\mathbf{y} = M\mathbf{y}$. Let the nonrandom vector $\mathbf{a} \in \mathbb{R}^n$. Then the linear estimator $\mathbf{a}^T\mathbf{y}$, which is unbiased for 0, i.e., $\mathrm{E}(\mathbf{a}^T\mathbf{y}) = 0$, is a **linear zero function**.

The **Watson efficiency** ϕ under the full-rank model $\mathscr{M} = \{\mathbf{y}, X\boldsymbol{\beta}, \sigma^2 V\}$, with the $n \times p$ model matrix X having full column rank equal to $p < n$ and with the $n \times n$ covariance matrix V positive definite, measures the relative efficiency of the $\mathrm{OLSE}(\boldsymbol{\beta}) = \hat{\boldsymbol{\beta}}$ vs. the $\mathrm{BLUE}(\boldsymbol{\beta}) = \tilde{\boldsymbol{\beta}}$ and is defined by the ratio of the corresponding generalized variances:

$$\phi = \frac{\det[\mathrm{var}(\tilde{\boldsymbol{\beta}})]}{\det[\mathrm{var}(\hat{\boldsymbol{\beta}})]} = \frac{\det^2(X^TX)}{\det(X^TVX)\cdot\det(X^TV^{-1}X)}.$$

The **Bloomfield–Watson efficiency** ψ under the general linear model $\mathscr{M} = \{\mathbf{y}, X\boldsymbol{\beta}, \sigma^2 V\}$ with no rank assumptions measures the relative efficiency of the $\mathrm{OLSE}(X\boldsymbol{\beta}) = X\hat{\boldsymbol{\beta}}$ vs. the $\mathrm{BLUE}(\boldsymbol{\beta}) = \tilde{\boldsymbol{\beta}}$ and is defined by: $\psi = \frac{1}{2}\|HV - VH\|^2 = \|HVM\|^2$, where the norm $\|A\| = \mathrm{tr}^{1/2}(A^TA)$ is defined for any $k \times q$ matrix A.

The $n \times n$ covariance matrix $(1-\rho)I_n + \rho\mathbf{1}_n\mathbf{1}_n^T = (1-\rho)I_n + \rho J_n$ has **intraclass correlation structure** (or **equicorrelation structure**) and is the **intraclass correlation matrix** (or the **equicorrelation matrix**). The parameter ρ is the **intraclass correlation** (or **intraclass correlation coefficient**).

Facts:

The following facts, except for those with a specific reference, can be found in [Gro04], [PS89], or [SJ03, §4.1–4.3]. Throughout this set of facts, X denotes the $n \times p$ nonrandom model matrix.

1. The hat matrix $H = X(X^TX)^-X^T$ associated with the model matrix X is invariant (unique) with respect to choice of generalized inverse $(X^TX)^-$ and is a symmetric idempotent matrix: $H = H^T = H^2$, and $\mathrm{rank}(H) = \mathrm{tr}(H) = \mathrm{rank}(X)$. Moreover, the hat matrix H is the orthogonal projector onto $\mathrm{range}(X)$.
2. If the $p \times p$ matrix Q is nonsingular, then the hat matrix associated with the model matrix XQ equals the hat matrix associated with the model matrix X.
3. The residual sum of squares $SSE = \mathbf{y}^TM\mathbf{y} = (\mathbf{y} - \hat{\mathbf{y}})^T(\mathbf{y} - \hat{\mathbf{y}}) = \mathbf{y}^T\mathbf{y} - \mathbf{y}^TX\hat{\boldsymbol{\beta}}$, where M is the residual matrix and $\hat{\boldsymbol{\beta}} = \mathrm{OLSE}(\boldsymbol{\beta})$.
4. In simple linear regression the coefficient of determination $R^2 = r^2$, the square of the sample correlation coefficient. In multiple linear regression with model matrix $X = [\mathbf{1}_n \mid X_0] = [\mathbf{1}_n \mid \mathbf{x}_{[1]} \mid \cdots \mid \mathbf{x}_{[p-1]}]$ and $(p-1) \times 1$ nonrandom vector $\mathbf{a} \in \mathbb{R}^p$,

$$R^2 = \max_{\mathbf{a}} r_{\mathbf{a}}^2 = \max_{\mathbf{a}} \frac{(\mathbf{a}^TX_0^TC_n\mathbf{y})^2}{\mathbf{a}^TX_0^TC_nX_0\mathbf{a}\cdot\mathbf{y}^TC_n\mathbf{y}},$$

the square of the sample correlation coefficient $r_{\mathbf{a}}$ between the variables whose observed values are in vectors $\mathbf{y}$ and $X_0\mathbf{a}$.

5. The vector $X\hat{\boldsymbol{\beta}}$ is invariant (unique) with respect to the choice of least squares solution $\hat{\boldsymbol{\beta}}$, but $\hat{\boldsymbol{\beta}}$ is unique if and only if X has full column rank equal to $p \leq n$, and then $\hat{\boldsymbol{\beta}} = \mathrm{OLSE}(\boldsymbol{\beta}) = (X^TX)^{-1}X^T\mathbf{y} = X^\dagger\mathbf{y}$, where $X^\dagger$ is the Moore–Penrose inverse of X. The covariance matrix $\mathrm{var}(\hat{\boldsymbol{\beta}}) = \sigma^2(X^TX)^{-1}X^TVX(X^TX)^{-1}$.

6. The Watson efficiency ϕ is always positive, and $\phi \leq 1$ with equality if and only if OLSE($\boldsymbol{\beta}$) = BLUE($\boldsymbol{\beta}$).
7. [DLL02, p. 477], [Gus97, p. 67] *Bloomfield–Watson–Knott Inequality*. The Watson efficiency

$$\phi = \frac{\det^2(X^T X)}{\det(X^T V X) \cdot \det(X^T V^{-1} X)} \geq \prod_{i=1}^{m} \frac{4\lambda_i \lambda_{n-i+1}}{(\lambda_i + \lambda_{n-i+1})^2},$$

for all $n \times p$ model matrices X with full column rank p. Here $m = \min(p, n-p)$ and $\lambda_1 \geq \cdots \geq \lambda_n$ denote the necessarily positive eigenvalues of the $n \times n$ positive definite covariance matrix V. The ratios $4\lambda_i \lambda_{n-i+1}/(\lambda_i + \lambda_{n-i+1})^2$ in the lower bound for the Watson efficiency are the squared *antieigenvalues* of the covariance matrix V.
8. [DLL02, p. 454] Let $p = 1$ and set the $n \times 1$ model matrix $X = \mathbf{x}$. Then the Bloomfield–Watson–Knott Inequality is the *Kantorovich Inequality* (or *Frucht–Kantorovich Inequality*):

$$\frac{(\mathbf{x}^T \mathbf{x})^2}{\mathbf{x}^T V \mathbf{x} \cdot \mathbf{x}^T V^{-1} \mathbf{x}} \geq \frac{4\lambda_1 \lambda_n}{(\lambda_1 + \lambda_n)^2},$$

where λ_1 and λ_n are, respectively, the largest and smallest eigenvalues of the $n \times n$ positive definite covariance matrix V.
9. The Bloomfield–Watson efficiency

$$\begin{aligned} \psi &= \frac{1}{2}\|HV - VH\|^2 = \|HVM\|^2 = \operatorname{tr}(HVMVH) = \operatorname{tr}(HVMV) \\ &= \operatorname{tr}(HV^2 - HVHV) = \operatorname{tr}(HV^2) - \operatorname{tr}\big((HV)^2\big) \geq 0, \end{aligned}$$

with equality if and only if OLSE($\boldsymbol{\beta}$) = BLUE($\boldsymbol{\beta}$) if and only if the Watson efficiency $\phi = 1$.
10. [DLL02, p. 473] The *Bloomfield–Watson Trace Inequality*. Let A be a nonrandom symmetric $n \times n$ matrix, not necessarily positive semidefinite. Then for all the nonrandom matrices $U \in \mathbb{R}^{n \times p}$ that satisfy $U^T U = I_p$:

$$\operatorname{tr}(U^T A^2 U) - \operatorname{tr}\big((U^T A U)^2\big) \leq \frac{1}{4} \sum_{i=1}^{\min(p,n-p)} (\alpha_i - \alpha_{n-i+1})^2,$$

where $\alpha_1 \geq \cdots \geq \alpha_n$ denote the eigenvalues of the $n \times n$ matrix A.
11. The Bloomfield–Watson efficiency

$$\psi = \operatorname{tr}(HV^2) - \operatorname{tr}\big((HV)^2\big) \leq \frac{1}{4} \sum_{i=1}^{\min(p,n-p)} (\lambda_i - \lambda_{n-i+1})^2,$$

for all $n \times n$ hat matrices H with rank p (and so for all $n \times p$ model matrices X with full column rank p). Here, $\lambda_1 \geq \cdots \geq \lambda_n$ denote the necessarily positive eigenvalues of the $n \times n$ positive definite covariance matrix V.
12. The $n \times n$ intraclass correlation matrix $R_{\text{ic}} = (1-\rho)I_n - \rho \mathbf{1}_n \mathbf{1}_n^T$ has eigenvalues $1 - \rho$ with multiplicity $n - 1$ and $1 + \rho(n-1)$ with multiplicity 1, and so R_{ic} is singular if and only if $\rho = -1/(n-1)$ or $\rho = 1$.
13. The intraclass correlation coefficient ρ is such that $-1/(n-1) \leq \rho \leq 1$ and the $n \times n$ intraclass correlation matrix is positive definite if and only if $-1/(n-1) < \rho < 1$.
14. The inverse of the $n \times n$ positive definite intraclass correlation matrix

$$\big((1-\rho)I_n - \rho \mathbf{1}_n \mathbf{1}_n^T\big)^{-1} = \frac{1}{1-\rho}\left(I_n - \frac{\rho}{1+\rho(n-1)} \mathbf{1}_n \mathbf{1}_n^T\right).$$

15. *Gauss–Markov Theorem* (or *Gauß–Markov Theorem*). In the full-rank model $\{\mathbf{y},\, X\boldsymbol{\beta},\, \sigma^2 V\}$, the generalized least squares estimator $\tilde{\boldsymbol{\beta}} = \mathrm{GLSE}(\boldsymbol{\beta}) = (X^T V^{-1} X)^{-1} X^T V^{-1}\mathbf{y} = \mathrm{BLUE}(\boldsymbol{\beta})$. In the full-rank model $\{\mathbf{y},\, X\boldsymbol{\beta},\, \sigma^2 I\}$, the ordinary least-squares estimator $\mathrm{OLSE}(\boldsymbol{\beta}) = \hat{\boldsymbol{\beta}} = (X^T X)^{-1} X^T \mathbf{y} = X^{\dagger}\mathbf{y} = \mathrm{BLUE}(\boldsymbol{\beta})$.
16. In the model $\{\mathbf{y},\, X\boldsymbol{\beta},\, \sigma^2 V\}$, where V is positive definite, but with X possibly with less than full column rank, the

$$\mathrm{BLUE}(X\boldsymbol{\beta}) = X(X^T V^{-1} X)^{-} X^T V^{-1}\mathbf{y}.$$

17. [Sea97, §5.4] Let the matrix $K \in \mathbb{R}^{k\times p}$. Then $K\boldsymbol{\beta}$ is estimable $\iff \exists$ matrix $A \in \mathbb{R}^{n\times k}$: $K^T = X^T A \iff \mathrm{range}(K^T) \subseteq \mathrm{range}(X^T) \iff K\hat{\boldsymbol{\beta}}$ is invariant for any choice of $\hat{\boldsymbol{\beta}} = (X^T X)^{-} X^T \mathbf{y}$.
18. [Rao73b, p. 282] Consider the general linear model $\{\mathbf{y},\, X\boldsymbol{\beta},\, \sigma^2 V\}$, where X and V need not be of full rank. Let the matrix $G \in \mathbb{R}^{n\times n}$. Then $G\mathbf{y} = \mathrm{BLUE}(X\boldsymbol{\beta}) \iff G[X \mid VM] = [X \mid \mathbf{0}]$, where the residual matrix $M = I - H$. Let the matrix $A \in \mathbb{R}^{k\times n}$ and the matrix $K \in \mathbb{R}^{k\times p}$. Then the corresponding condition for $A\mathbf{y}$ to be the BLUE of an estimable parametric function $K\boldsymbol{\beta}$ is $A[X \mid VM] = [K \mid \mathbf{0}]$.
19. Let G_1 and G_2 both be $n\times n$. If $G_1\mathbf{y}$ and $G_2\mathbf{y}$ are two BLUEs of $X\boldsymbol{\beta}$ under the model $\{\mathbf{y},\, X\boldsymbol{\beta},\, \sigma^2 V\}$, then $G_1\mathbf{y} = G_2\mathbf{y}$ for all $\mathbf{y} \in \mathrm{range}([X \mid V])$. The matrix G yielding the BLUE is unique if and only if $\mathrm{range}([X \mid V]) = \mathbb{R}^n$.
20. Every linear zero function can be written as $\mathbf{b}^T M\mathbf{y}$ for some nonrandom $\mathbf{b} \in \mathbb{R}^n$. Let the matrix $G \in \mathbb{R}^{n\times n}$. Then an unbiased estimator $G\mathbf{y} = \mathrm{BLUE}(X\boldsymbol{\beta})$ if and only if $G\mathbf{y}$ is uncorrelated with every linear zero function.
21. [Rao71] Let the matrix $A \in \mathbb{R}^{n\times n}$. Then the linear estimator $A\mathbf{y} = \mathrm{BLUE}(X\boldsymbol{\beta})$ under the model $\{\mathbf{y}, X\boldsymbol{\beta}, \sigma^2 V\}$ if and only if there exists a matrix Λ so that A is a solution to *Pandora's box*

$$\begin{bmatrix} V & X \\ X^T & \mathbf{0} \end{bmatrix}\begin{bmatrix} A^T \\ \Lambda \end{bmatrix} = \begin{bmatrix} \mathbf{0} \\ X^T \end{bmatrix}.$$

22. [Rao71] Let the $(n+p)\times(n+p)$ matrix B be defined as any generalized inverse:

$$B = \begin{bmatrix} V & X \\ X^T & \mathbf{0} \end{bmatrix}^{-} = \begin{bmatrix} B_1 & B_2 \\ B_3 & -B_4 \end{bmatrix}.$$

Let $\mathbf{k}^T\boldsymbol{\beta}$ be estimable; then the $\mathrm{BLUE}(\mathbf{k}^T\boldsymbol{\beta}) = \mathbf{k}^T\tilde{\boldsymbol{\beta}} = \mathbf{k}^T B_2^T\mathbf{y} = \mathbf{k}^T B_3\mathbf{y}$, the variance $\mathrm{var}(\mathbf{k}^T\tilde{\boldsymbol{\beta}}) = \sigma^2\mathbf{k}^T B_4\mathbf{k}$, and the quadratic form $\mathbf{y}^T B_1\mathbf{y}/f$ is an unbiased estimator of σ^2 with $f = \mathrm{rank}([V \mid X]) - \mathrm{rank}(X)$.
23. [PS89] In the model $\{\mathbf{y}, X\boldsymbol{\beta}, \sigma^2 V\}$ with no rank assumptions, the $\mathrm{OLSE}(X\boldsymbol{\beta}) = \mathrm{BLUE}(X\boldsymbol{\beta})$ if and only if any one of the following equivalent conditions holds:
    - $HV = VH$.
    - $HV = HVH$.
    - $HVM = \mathbf{0}$.
    - $X^T VL = \mathbf{0}$, where the $n\times l$ matrix L has $\mathrm{range}(L) = \mathrm{range}(M)$.
    - $\mathrm{range}(VX) \subseteq \mathrm{range}(X)$.
    - $\mathrm{range}(VX) = \mathrm{range}(X) \cap \mathrm{range}(V)$.
    - $HVH \preceq V$, i.e., $V - HVH$ is positive semidefinite.
    - $\mathrm{rank}(V - HVH) = \mathrm{rank}(V) - \mathrm{rank}(HVH)$.
    - $\mathrm{rank}(V - HVH) = \mathrm{rank}(V) - \mathrm{rank}(VX)$.
    - $\mathrm{range}(X)$ has a basis consisting of r eigenvectors of V, where $r = \mathrm{rank}(X)$.
    - V can be expressed as $V = \alpha I + XAX^T + LBL^T$, where $\alpha \in \mathbb{R}$, $\mathrm{range}(L) = \mathrm{range}(M)$, and the $p\times p$ matrices A and B are symmetric, and such that V is positive semidefinite.

More conditions can be obtained by replacing V with its Moore–Penrose inverse $V^{\dagger}$ and the hat matrix H with the residual matrix $M = I - H$.

24. Suppose that the positive definite covariance matrix V has h distinct eigenvalues: $\lambda_{\{1\}} > \lambda_{\{2\}} > \cdots > \lambda_{\{h\}} > 0$ with multiplicities $m_1, \ldots, m_h$, $\sum_{i=1}^{h} m_i = n$, and with associated orthonormalized sets of eigenvectors $U_{\{1\}}, \ldots, U_{\{h\}}$, respectively, $n \times m_1, \ldots, n \times m_h$. Then $\text{OLSE}(X\boldsymbol{\beta}) = \text{BLUE}(X\boldsymbol{\beta})$ if and only if any one of the following equivalent conditions holds:

    - $\text{rank}(U_{\{1\}}^T X) + \cdots + \text{rank}(U_{\{h\}}^T X) = \text{rank}(X)$.
    - $U_{\{i\}}^T H U_{\{i\}} = (U_{\{i\}}^T H U_{\{i\}})^2$ for all $i = 1, \ldots, h$.
    - $U_{\{i\}}^T H U_{\{j\}} = \mathbf{0}$ for all $i \neq j$; $i, j = 1, \ldots, h$.

25. [Rao73b] Let the $p \times p$ matrix U be such that the $n \times n$ matrix $W = V + XUX^T$ has $\text{range}(W) = \text{range}([X \mid V])$. Then the $\text{BLUE}(X\boldsymbol{\beta}) = X(X^T W^- X)^- X^T W^- \mathbf{y}$.
26. When V is nonsingular, the $n \times n$ matrix G such that $G\mathbf{y}$ is the BLUE of $X\boldsymbol{\beta}$ is unique, but when V is singular this may not be so. However, the numerical value of $\text{BLUE}(X\boldsymbol{\beta})$ is unique with probability 1.
27. [SJ03, §7.4] The residual vector associated with the $\text{BLUE}(X\boldsymbol{\beta})$ is

$$\tilde{\mathbf{e}} = \mathbf{y} - X\tilde{\boldsymbol{\beta}} = VM(MVM)^- M\mathbf{y} = M\mathbf{y} + HVM(MVM)^- M\mathbf{y},$$

which is invariant (unique) with respect to choice of generalized inverse $(MVM)^-$. The weighted sum of squares of BLUE residuals, which is needed when estimating σ^2, can be written as $(\mathbf{y} - X\tilde{\boldsymbol{\beta}})^T V^- (\mathbf{y} - X\tilde{\boldsymbol{\beta}}) = \tilde{\mathbf{e}}^T V^- \tilde{\mathbf{e}} = \mathbf{y}^T M(MVM)^- M\mathbf{y}$.

Examples:

1. Let $n = 3$ and $p = 2$ with the model matrix $X = \begin{bmatrix} 1 & 1 \\ 1 & 0 \\ 1 & -1 \end{bmatrix}$. Then X has full column rank equal to 2, the matrix $X^T X$ is nonsingular, and the hat matrix is

$$H = X(X^T X)^- X^T = X(X^T X)^{-1} X^T = \frac{1}{6} \begin{bmatrix} 5 & 2 & -1 \\ 2 & 2 & 2 \\ -1 & 2 & 5 \end{bmatrix}$$

with $\text{rank}(H) = \text{tr}(H) = 2$. The $\text{OLSE}(\boldsymbol{\beta})$ is

$$\hat{\boldsymbol{\beta}} = (X^T X)^{-1} X^T \mathbf{y} = \begin{bmatrix} \frac{1}{3}(y_1 + y_2 + y_3) \\ \frac{1}{2}(y_1 - y_3) \end{bmatrix},$$

where $\mathbf{y} = [y_1, y_2, y_3]^T$. The vector of OLS residuals is

$$M\mathbf{y} = \frac{1}{6} \begin{bmatrix} 1 & -2 & 1 \\ -2 & 4 & -2 \\ 1 & -2 & 1 \end{bmatrix} \begin{bmatrix} y_1 \\ y_2 \\ y_3 \end{bmatrix} = \frac{1}{6}(y_1 - 2y_2 + y_3) \begin{bmatrix} 1 \\ -2 \\ 1 \end{bmatrix}$$

with residual sum of squares $SSE = (y_1 - 2y_2 + y_3)^2/6$.

Now let the variance $\sigma^2 = 1$ and let the covariance matrix

$$V = \begin{bmatrix} 1 & 0 & 0 \\ 0 & \delta & 0 \\ 0 & 0 & 1 \end{bmatrix}$$

with $\delta > 0$. Then V is positive definite and

$$\text{BLUE}(\boldsymbol{\beta}) = \text{GLSE}(\boldsymbol{\beta}) = \tilde{\boldsymbol{\beta}} = \begin{bmatrix} \frac{1}{2\delta+1}(\delta y_1 + y_2 + \delta y_3) \\ \frac{1}{2}(y_1 - y_3) \end{bmatrix},$$

while the covariance matrices are

$$\text{var}(\tilde{\boldsymbol{\beta}}) = \begin{bmatrix} \frac{\delta}{2\delta+1} & 0 \\ 0 & \frac{1}{2} \end{bmatrix}, \quad \text{var}(\hat{\boldsymbol{\beta}}) = \begin{bmatrix} \frac{\delta+2}{9} & 0 \\ 0 & \frac{1}{2} \end{bmatrix}.$$

Hence, the Watson efficiency

$$\phi = \frac{9\delta}{(2\delta+1)(\delta+2)} \leq 1$$

with equality if and only if $\delta = 1$. As $\delta \to 0$ or $\delta \to \infty$, the Watson efficiency $\phi \to 0$.

Since the eigenvalues of V here are 1 (multiplicity 2) and δ (multiplicity 1), we find that the lower bound for ϕ in the Bloomfield–Watson–Knott Inequality [here, $m = \min(p, n-p) = \min(2,1) = 1$] is equal to $4\delta/(1+\delta)^2$, and it is easy to show that this is less than $9\delta/\big((2\delta+1)(\delta+2)\big)$ (unless $\delta = 1$, and then both these ratios are equal to 1).

The Bloomfield–Watson efficiency

$$\psi = \frac{2}{9}(1-\delta)^2 \geq 0$$

with equality if and only if $\delta = 1$. As $\delta \to 0$ the Bloomfield–Watson efficiency $\psi \to 2/9$, and as $\delta \to \infty$ the Bloomfield–Watson efficiency $\psi \to \infty$. We note that when $\delta = 0$, then Σ is singular, but ψ is well-defined and equal to 2/9. We find that the upper bound for ψ in the Bloomfield–Watson Trace Inequality is $(1-\delta)^2/4$ and certainly $(1-\delta)^2/4 \geq 2(1-\delta)^2/9$ with equality if and only if $\delta = 1$.

2. Let $n = 3$ and $p = 1$ with the model matrix $X = \mathbf{x} = \begin{bmatrix} 1 \\ 2 \\ 3 \end{bmatrix}$ and with $\boldsymbol{\beta} = \beta$, a scalar. The hat matrix

$$H = X(X^T X)^{-1} X^T = \frac{1}{\mathbf{x}^T\mathbf{x}}\mathbf{x}\mathbf{x}^T = \frac{1}{14}\begin{bmatrix} 1 & 2 & 3 \\ 2 & 4 & 6 \\ 3 & 6 & 9 \end{bmatrix}$$

and so

$$\hat{\beta} = \text{OLSE}(\beta) = \frac{\mathbf{x}^T\mathbf{y}}{\mathbf{x}^T\mathbf{x}} = \frac{y_1 + 2y_2 + 3y_3}{14},$$

where the realization $\mathbf{y} = [y_1, y_2, y_3]^T$.

Let the covariance matrix have intraclass correlation structure:

$$V = \begin{bmatrix} 1 & \rho & \rho \\ \rho & 1 & \rho \\ \rho & \rho & 1 \end{bmatrix}$$

with $-1/2 < \rho < 1$. Then V is positive definite and its inverse

$$V^{-1} = \frac{1}{(1-\rho)(1+2\rho)}\begin{bmatrix} 1+\rho & -\rho & -\rho \\ -\rho & 1+\rho & -\rho \\ -\rho & -\rho & 1+\rho \end{bmatrix}.$$

And so

$$\tilde{\beta} = \text{BLUE}(\beta) = \text{GLSE}(\beta) = \frac{\mathbf{x}^T V^{-1}\mathbf{y}}{\mathbf{x}^T V^{-1}\mathbf{x}} = \frac{(1-4\rho)y_1 + 2(1-\rho)y_2 + 3y_3}{6(1-\rho)}.$$

The variances are (with $\sigma^2 = 1$):

$$\text{var}(\tilde{\beta}) = \frac{1}{\mathbf{x}^T V^{-1}\mathbf{x}} = \frac{(1-\rho)(1+2\rho)}{2(7-4\rho)}; \quad \text{var}(\hat{\beta}) = \frac{\mathbf{x}^T V\mathbf{x}}{(\mathbf{x}^T\mathbf{x})^2} = \frac{7+11\rho}{98},$$

and so the Watson efficiency

$$\phi = \frac{\text{var}(\tilde{\beta})}{\text{var}(\hat{\beta})} = \frac{49(1-\rho)(1+2\rho)}{(7-4\rho)(7+11\rho)} \to 0$$

as $\rho \to -1/2$ or as $\rho \to 1$. As $\rho \to 0$ the Watson efficiency $\phi \to 1$.

Since the eigenvalues of V here are $1-\rho$ (multiplicity 2) and $1+2\rho$ (multiplicity 1), we find that the lower bound for ϕ in the Bloomfield–Watson–Knott Inequality (which is now the Kantorovich Inequality) is $4(1-\rho)(1+2\rho)/(2+\rho)^2$ and it is easy to show that this is less than $\phi = 49(1-\rho)(1+2\rho)/\big((7-4\rho)(7+11\rho)\big)$ (unless $\rho = 0$ and then both these ratios are equal to 1).

The Bloomfield–Watson efficiency $\psi = 54\rho^2/49$, which is well defined for all ρ; when $\rho = -1/2$, then $\psi = 27/98$ and when $\rho = 1$, then $\psi = 54/49$. We find that the upper bound for ψ in the Bloomfield–Watson Trace Inequality is $9\rho^2/4$ and certainly $9\rho^2/4 \geq \psi = 54\rho^2/49$, with equality if and only if $\rho = 0$.

3. [DLL02, p. 475] Let A be a nonrandom symmetric $n \times n$ matrix, not necessarily positive semidefinite, and let the nonrandom vector $\mathbf{u} = [u_1, \ldots, u_n]^T$ be such that $\mathbf{u}^T\mathbf{u} = 1$. Then from the Bloomfield–Watson Trace Inequality with $U = \mathbf{u}$, we obtain the special case

$$\mathbf{u}^T A^2\mathbf{u} - \left(\mathbf{u}^T A\mathbf{u}\right)^2 \leq \frac{1}{4}(\alpha_1 - \alpha_n)^2,$$

where α_1 and α_n are, respectively, the largest and smallest eigenvalues of A. Now let $a_1, \ldots, a_n$ denote n nonrandom scalars, not necessarily all positive, and let $\bar{a} = \sum_{i=1}^n a_i/n$. Then the Popoviciu–Nair Inequality

$$\frac{1}{n}\sum_{i=1}^{n}(a_i - \bar{a})^2 \leq \frac{1}{4}(a_{\max} - a_{\min})^2$$

follows directly from the special case above of the Bloomfield–Watson Trace Inequality with $\mathbf{u} = \mathbf{1}/\sqrt{n}$ and $A = \text{diag}\{a_i\}$.

4. When the covariance matrix V has intraclass correlation structure, then the OLSE$(X\boldsymbol{\beta})$ = BLUE $(X\boldsymbol{\beta})$ if and only if $X^T\mathbf{1}_n = \mathbf{0}$ or $X\mathbf{f} = \mathbf{1}_n$ for some nonrandom $p \times 1$ vector $\mathbf{f}$, and so OLSE $(X\boldsymbol{\beta})$ = BLUE$(X\boldsymbol{\beta})$ when the columns of X are centered or when $\mathbf{1}_n \in \text{range}(X)$ as in Example 1 above, where $X = \begin{bmatrix} 1 & 1 \\ 1 & 0 \\ 1 & -1 \end{bmatrix}$ with first column equal to $\mathbf{1}_3$.

5. Let the $n \times p$ model matrix $X = [\mathbf{1} \mid X_0]$, where $\mathbf{1}$ is the $n \times 1$ column vector with every entry equal to 1 and X_0 is $n \times (p-1)$. Then the hat matrix associated with X,

$$H = \frac{1}{n}\mathbf{1}_n\mathbf{1}_n^T + C_n X_0 (X_0^T C_n X_0)^- X_0^T C_n$$

coincides with the hat matrix associated with $X_c = [\mathbf{1}_n \mid C_n X_0]$, since $X_c = XQ$, with the $p \times p$ matrix $Q = \begin{bmatrix} 1 & -\mathbf{q}^T \\ \mathbf{0} & I_{p-1} \end{bmatrix}$ for some $\mathbf{q}$.

When $p = 2$, we may write $X_0 = \mathbf{x}$, an $n \times 1$ vector, and if X now has rank equal to 2, then $\mathbf{x}^T C_n \mathbf{x} > 0$ and the hat matrix

$$H = \frac{1}{n}\mathbf{1}_n\mathbf{1}_n^T + \frac{1}{\mathbf{x}^T C_n \mathbf{x}} C_n \mathbf{x}\mathbf{x}^T C_n = \frac{1}{n} J_n + \frac{1}{\mathbf{x}^T C_n \mathbf{x}} C_n \mathbf{x}\mathbf{x}^T C_n.$$

And so the quadratic forms

$$\mathbf{y}^T H \mathbf{y} = n\bar{y}^2 + \frac{(\mathbf{x}^T C_n \mathbf{y})^2}{\mathbf{x}^T C_n \mathbf{x}}, \quad \mathbf{y}^T M \mathbf{y} = \mathbf{y}^T C_n \mathbf{y} + \frac{(\mathbf{x}^T C_n \mathbf{y})^2}{\mathbf{x}^T C_n \mathbf{x}},$$

where $\bar{y} = \sum_{i=1}^n y_i/n = \mathbf{1}_n^T \mathbf{y}/n$.

Acknowledgments

We are very grateful to Ka Lok Chu, Anne Greenbaum, Leslie Hogben, Jarkko Isotalo, Augustyn Markiewicz, George A. F. Seber, Evelyn Matheson Styan, Götz Trenkler, and Kimmo Vehkalahti for their help. This research was supported in part by the Natural Sciences and Engineering Research Council of Canada.

References

[And03] T.W. Anderson. *An Introduction to Multivariate Statistical Analysis,* 3rd ed. John Wiley & Sons, New York, 2003. (2nd ed. 1984, 1st ed. 1958.)

[Bap00] R.B. Bapat. *Linear Algebra and Linear Models*, 2nd ed. Springer, Heidelberg, 2000. (Original ed. Hindustan Book Agency, Delhi, 1993.)

[BI01] Pavel Bleher and Alexander Its, Eds., *Random Matrix Models and Their Applications.* Cambridge University Press, Cambridge, 2001.

[DM04] Russell Davidson and James G. MacKinnon. *Econometric Theory and Methods.* Oxford University Press, Oxford, U.K., 2004.

[DLL02] S.W. Drury, Shuangzhe Liu, Chang-Yu Lu, Simo Puntanen, and George P.H. Styan. Some comments on several matrix inequalities with applications to canonical correlations: historical background and recent developments. *Sankhyā: Ind. J. Stat., Ser. A*, 64:453–507, 2002.

[Gra83] Franklin A. Graybill. *Matrices with Applications in Statistics,* 2nd ed. Wadsworth, Belmont, CA, 1983. (Original ed. = *Introduction to Matrices with Applications in Statistics*, 1969.)

[Gra01] Franklin A. Graybill. *Theory and Application of the Linear Model.* Duxbury Classics Library Reprint Edition, 2001. (Paperback reprint of 1976 edition.)

[Gro04] Jürgen Groß. The general Gauss–Markov model with possibly singular covariance matrix. *Stat. Pap.*, 45:311–336, 2004.

[Gus97] Karl E. Gustafson and Duggirala K.M. Rao. *Numerical Range: The Field of Values of Linear Operators and Matrices.* Springer, New York, 1997.

[Had96] Ali S. Hadi. *Matrix Algebra as a Tool.* Duxbury, 1996.

[Har97] David A. Harville. *Matrix Algebra from a Statistician's Perspective.* Springer, New York, 1997.

[Har01] David A. Harville. *Matrix Algebra: Exercises and Solutions.* Springer, New York, 2001.

[Hea00] M.J.R. Healy. *Matrices for Statistics,* 2nd ed. Oxford University Press, Oxford, U.K. 2000. (Original ed. 1986.)

[KS83] John G. Kemeny and J. Laurie Snell. *Finite Markov Chains: With a New Appendix "Generalization of a Fundamental Matrix."* Springer, Heidelberg, 1983. (Original ed.: *Finite Markov Chains*, Van Nostrand, Princeton, NJ, 1960; reprint edition: Springer, Heidelberg, 1976.)

[KvR05] Tõnu Kollo and Dietrich von Rosen. *Advanced Multivariate Statistics with Matrices.* Springer, Heidleberg, 2005.

[MN99] Jan R. Magnus and Heinz Neudecker. *Matrix Differential Calculus with Applications in Statistics and Econometrics*, Revised ed. John Wiley & Sons, New York, 1999. (Original ed. 1988.)

[Meh04] Madan Lal Mehta. *Random Matrices,* 3rd ed. Elsevier, Amsterdam/New York 2004. (Original ed.: *Random Matrices and the Statistical Theory of Energy Levels*, Academic Press, New York 1967; Revised and enlarged 2nd ed.: *Random Matrices,* Academic Press, New York 1991.)

[MM04] Irwin Miller and Marylees Miller. *John E. Freund's Mathematical Statistics with Applications,* 7th ed. Pearson Prentice Hall, Upper Saddle River, NJ, 2004. (Original: *Mathematical Statistics* by John E. Freund, Prentice-Hall, Englewood-Cliffs, NJ, 1962; 2nd ed., 1971; 3rd ed.: *Mathematical Statistics* by John E. Freund & Ronald E. Walpole; 4th ed., 1987; 5th ed.: *Mathematical Statistics* by John E. Freund, 1992; 6th ed.: *John E. Freund's Mathematical Statistics* by Irwin Miller and Marylees Miller, 1999.)

[PSS05] Simo Puntanen, George A.F. Seber, and George P.H. Styan. *Definitions and Facts for Linear Statistical Models and Multivariate Statistical Analysis Using Linear Algebra.* Report A 357, Dept. of Mathematics, Statistics & Philosophy, University of Tampere, Tampere, Finland, 2005.

[PS89] Simo Puntanen and George P.H. Styan. The equality of the ordinary least squares estimator and the best linear unbiased estimator (with comments by Oscar Kempthorne and Shayle R. Searle, and with reply by the authors). *Am. Stat.*, 43:153–164, 1989.

[PS05] Simo Puntanen and George P. H. Styan. Schur complements in statistics and probability. Chapter 6 and Bibliography in *The Schur Complement and Its Applications* (Fuzhen Zhang, Ed.), Springer, New York, pp. 163–226, 259–288, 2005.

[Rao71] C. Radhakrishna Rao. Unified theory of linear estimation. *Sankhyā: Ind. J. Stat., Series A*, 33:371–394, 1971. (Corrigendum: 34:194 and 477, 1972.)

[Rao73a] C. Radhakrishna Rao. *Linear Statistical Inference and Its Applications,* 2nd ed. John Wiley & Sons, New York, 1973. (Original ed. 1965.)

[Rao73b] C. Radhakrishna Rao. Representations of best linear unbiased estimators in the Gauss–Markoff model with a singular dispersion matrix. *J. Multivar. Anal.*, 3:276–292, 1973.

[RM71] C. Radhakrishna Rao and Sujit Kumar Mitra. *Generalized Inverse of Matrices and Its Applications.* John Wiley & Sons, New York, 1971.

[RR98] C. Radhakrishna Rao and M. Bhaskara Rao. *Matrix Algebra and Its Applications to Statistics and Econometrics.* World Scientific, Singapore 1998.

[RD02] Nalini Ravishanker and Dipak K. Dey. *A First Course in Linear Model Theory.* Chapman & Hall/CRC, Boca Raton, FL 2002.

[Ren00] Alvin C. Rencher. *Linear Models in Statistics.* John Wiley & Sons, New York, 2000.

[Sch05] James R. Schott. *Matrix Analysis for Statistics*, 2nd ed., John Wiley & Sons, New York, 2005. (Original ed. 1997.)

[Sea82] Shayle R. Searle. *Matrix Algebra Useful for Statistics.* John Wiley & Sons, New York, 1982.

[Sea97] Shayle R. Searle. *Linear Models*, Wiley, Classics Library Reprint Edition. John Wiley & Sons, New York, 1997. (Paperback reprint of 1971 edition.)

[SW01] Shayle R. Searle and Lois Schertz Willett. *Matrix Algebra for Applied Economics.* John Wiley & Sons, New York, 2001.

[Seb04] George A.F. Seber. *Multivariate Observations,* Reprint ed., John Wiley & Sons, New York, 2004. (Original ed. 1984.)

[Seb06] George A.F. Seber. *Handbook of Linear Algebra for Statisticians.* John Wiley & Sons, New York, 2006, to appear.

[SL03] George A.F. Seber and Alan J. Lee. *Linear Regression Analysis,* 2nd ed., John Wiley & Sons, New York, 2003. (Original ed. by George A.F. Seber, 1977.)

[SJ03] Debasis Sengupta and Sreenivasa Rao Jammalamadaka. *Linear Models: An Integrated Approach.* World Scientific, Singapore, 2003.

[Sta95] James H. Stapleton. *Linear Statistical Models.* John Wiley & Sons, New York, 1995.

[Tak04] Yoshio Takane. Matrices with special reference to applications in psychometrics. *Lin. Alg. Appl.*, 388:341–361, 2004.

[TYM82] Kei Takeuchi, Haruo Yanai, and Bishwa Nath Mukherjee. *The Foundations of Multivariate Analysis: A Unified Approach by Means of Projection onto Linear Subspaces.* John Wiley & Sons, New York, Eastern, 1982.

[WMS02] Dennis D. Wackerly, William Mendenhall, III, and Richard L. Scheaffer. *Mathematical Statistics with Applications*, 6th edition. Duxbury Thomson Learning, Pacific Grove, CA, 2002. (Original ed. by William Mendenhall and Richard L. Scheaffer, Duxbury, North Scituate, MA, 1973; 2nd ed. by William Mendenhall, Richard L. Scheaffer, and Dennis D. Wackerly, Duxbury, Boston, MA, 1981; 3rd ed. by William Mendenhall, Dennis D. Wackerly, and Richard L. Scheaffer, Duxbury, Boston, MA, 1986; 4th ed., PWS-Kent, Boston, MA, 1990; 5th ed., Duxbury, Belmont, CA, 1996.)

[WS00] Christopher J. Wild and George A. F. Seber. *Chance Encounters: A First Course in Data Analysis and Inference.* John Wiley & Sons, New York, 2000.

53

Multivariate Statistical Analysis

Simo Puntanen
University of Tampere

George A. F. Seber
University of Auckland

George P. H. Styan
McGill University

Vectors and matrices arise naturally in the analysis of statistical data and we have seen this in Chapter 52. For example, suppose we take a random sample of n females and measure their heights x_i $(i = 1, 2, \ldots, n)$, giving us the vector $\mathbf{x} = [x_1, x_2, \ldots, x_n]^T$. We can treat $\mathbf{x}$ as simply a random vector, which can give rise to different values depending on the sample chosen, or as a vector of observed values (the data) taken on by the random vector from which we calculate a statistic like a sample mean (average). We use the former approach when we want to make inferences about the female population. In this case, we find that the value of a given x_i will depend on how it varies across the population, and this is described by its statistical "distribution," which is defined by a univariate function of one variable called a probability density function (pdf). For example, x_i may follow the well-known normal distribution, which seems to apply to many naturally occurring measurements. This distribution has a probability density function totally characterized by two (unknown) parameters, the population mean μ and the population variance σ^2, where σ is the standard deviation; we write $x_i \sim \mathrm{N}_1(\mu, \sigma^2)$, where "$\sim$" means "distributed as." (Throughout this chapter σ will always refer to a standard deviation and not to a singular value.) When the sample is random, the choice of any female does not affect the choice of any other, so that technically we say that the x_i are statistically independent and they all have the same distribution.

Three important aspects of statistical inference are: (1) estimating μ from the data and deriving the distributional properties of the estimate from the assumed underlying pdf; (2) finding a so-called confidence interval for μ, which is an interval containing the true value of μ with a given probability, e.g., 0.95 or, when typically expressed as a percentage, 95%; and (3) testing a hypothesis (called the null hypothesis H_0) about whether $\boldsymbol{\mu}$ is equal to some predetermined value μ_0, say. Symbolically, we write $H_0 : \mu = \mu_0$. To test H_0, we need to have a test statistic and be able to derive its distribution when H_0 is true and when it is false, called the noncentral distribution in the latter case. Noncentral distributions are usually quite complex, but they have a role in determining how good a test is in detecting departures from the null hypothesis.

In multivariate analysis we are interested in measuring not just a single characteristic or variable on each female, but we may also want to record weight, income, blood pressure, and so on, so that each x_i gets replaced by a d-dimensional vector $\mathbf{x}_i$ of d variables. Then $\mathbf{x}$ from the sample of recorded heights gets replaced by a matrix X with rows $\mathbf{x}_i'$. In this case the probability density function for $\mathbf{x}_i$ is said to be multivariate and the population mean is now a vector $\boldsymbol{\mu}$, say. We can look at X as either a matrix of numbers, or a random matrix, which we use for carrying out statistical inferences. In doing this, the univariate normal distribution extends naturally to the multivariate normal distribution, which has a pdf totally characterized by its mean $\boldsymbol{\mu}$ and its covariance matrix $\text{var}(\mathbf{x}) = \Sigma$. One can then derive other distributions based on various functions of X, which can then be used for inference.

The vectors $\mathbf{x}_i$ can be regarded as a cluster of points in $\mathbb{R}^d$. Such a set of points can also come from sampling a mixed population of men and women so that there would be two overlapping clusters. Alternatively, the population of females could be a mixture of races, each with its own characteristics, leading to several overlapping clusters. In order to study such clusters, we need techniques to somehow reduce the dimension of the data (hopefully, to two dimensions) in some optimal fashion, but still retain as much of the original variation as possible. We will introduce just three dimension-reducing techniques below. The first method, called principal component analysis, is used for a single cluster. The second method, called discriminant coordinates, is used for several clusters and the reduction in dimension is carried out to maximize group separation. The third method, called canonical variates, is also used for a single cluster, but utilizes the internal variation within the vectors. Looking at how one variable within a vector depends on the other variables lends itself to a range of other techniques, such as multivariate linear regression where we might compare the internal relationships in one cluster with those in others. For example, comparing how blood pressure depends on other variables for both males and females.

We finally introduce just one other topic called metric multidimensional scaling. Instead of having observations on n people or objects, we simply have measures of similarity or dissimilarity between each pair of objects. The challenge then is to try and represent these objects as a cluster in a low-dimensional space so that the inter-point Euclidean distances will closely reproduce the dissimilarity measures. Once we have the cluster we can then examine it to try and uncover any underlying structure or clustering of the objects.

It is hoped that these few applications will demonstrate the richness of the subject and its interplay between statistics and matrices. In this chapter we assume some of the basic statistical ideas defined in Chapter 52. However, there has to be a change in notation as we now need to continually distinguish between random and nonrandom vectors and matrices.

53.1 Data Matrix

Notation:

The latter part of the alphabet from u to z, upper or lower case, together with Q we reserve for random quantities, and the remainder of the alphabet for nonrandom quantities. All quantities in this section are real, though in practice some of the theory can be extended to complex quantities, as, for example, in the theory of time series.

Definitions:

If $\mathbf{w} = [\mathbf{y}^T, \mathbf{z}^T]^T$ is a random vector, then $\mathbf{y}$ and $\mathbf{z}$ are said to be **uncorrelated** if and only if their cross-covariance $\text{cov}(\mathbf{y}, \mathbf{z}) = \mathbf{0}$. We say that $\mathbf{y}$ and $\mathbf{z}$ are **statistically independent** if and only if the probability density function (pdf) of $\mathbf{w}$ is the product of the pdfs of $\mathbf{y}$ and $\mathbf{z}$.

Let $X = [x_{ij}] = \begin{bmatrix} \mathbf{x}_1^T \\ \mathbf{x}_2^T \\ \cdots \\ \mathbf{x}_n^T \end{bmatrix} = [\mathbf{x}^{(1)}, \mathbf{x}^{(2)}, \ldots, \mathbf{x}^{(d)}]$ be an $n \times d$ matrix of random variables with rows $\mathbf{x}_i^T$ such that all $\mathbf{x}_i$ have the same covariance matrix Σ and are uncorrelated, i.e., $\text{cov}(\mathbf{x}_r, \mathbf{x}_s) = \delta_{rs}\Sigma$, where $\delta_{rs} = 1$ for $r = s$ and $\delta_{rs} = 0$ for $r \neq s$. We call a matrix with the above properties a **data matrix**. As mentioned in the introduction, the $\mathbf{x}_i$ often constitute a random sample, which we now formally define. A **random sample** of vectors $\mathbf{x}_i$ $(i = 1, 2, \ldots, n)$ of size n consists of n random vectors that are **independently and identically distributed (i.i.d.)**, that is, they are statistically independent of each other and have the same pdf. When this is the case, we see from the data matrix above that the n elements of $\mathbf{x}^{(j)}$ form a random sample from the jth characteristic or variable, being just the jth elements in each $\mathbf{x}_i$. However, $\mathbf{x}^{(j)}$ and $\mathbf{x}^{(k)}$ $(j \neq k)$ can be correlated.

The **moment generating function** or m.g.f. of any random vector $\mathbf{x}$ is defined to be $M_{\mathbf{x}}(\mathbf{t}) = \text{E}(e^{\mathbf{t}^T\mathbf{x}})$. It can be used for finding such things as the mean and covariance matrix, or the distribution of related random variables.

Facts:

1. For a random variable x, $M_x(t) = \text{E}(e^{xt}) = \sum_{r=0}^{\infty} \text{E}(x^r)\frac{t^r}{r!}$. The coefficient of $\frac{t^r}{r!}$ in the power series expansion of $M_x(t)$ gives us $\text{E}(x^r)$ from which we can get $\mu = \text{E}(x)$ and $\text{var}(x) = \text{E}(x^2) - \mu^2$.
2. Statistical independence implies zero covariance but not generally vice versa except for one notable exception, the multivariate normal distribution (see next section).

53.2 Multivariate Normal Distribution

Definitions:

Let $\mathbf{x} = [x_i] = [x_1, x_2, \ldots, x_d]^T$ be a $d \times 1$ random vector with mean $\boldsymbol{\mu}$ and positive definite covariance matrix Σ. Then $\mathbf{x}$ is said to follow a **(nonsingular) multivariate normal distribution** when its pdf is

$$f(\mathbf{x} : \boldsymbol{\mu}, \Sigma) = (2\pi)^{-d/2}(\det \Sigma)^{-1/2} \exp\{-\tfrac{1}{2}(\mathbf{x} - \boldsymbol{\mu})^T \Sigma^{-1/2}(\mathbf{x} - \boldsymbol{\mu})\},$$

where $-\infty < x_j < \infty$, $j = 1, 2, \ldots, d$. We write $\mathbf{x} \sim \text{N}_d(\boldsymbol{\mu}, \Sigma)$. When $d = 1$ we have the univariate normal distribution. When Σ is singular, the pdf of $\mathbf{x}$ does not exist but it can be defined via a transformation $\mathbf{x} = A\mathbf{y}$ where $\mathbf{y}$ does have a nonsingular distribution. This is mentioned further below.

The **noncentral χ^2-distribution** with p degrees of freedom and noncentrality parameter δ, denoted by $\chi^2_{p,\delta}$, is the distribution of $\mathbf{x}^T\mathbf{x}$ when $\mathbf{x} \sim \text{N}_p(\boldsymbol{\mu}, I_p)$ and $\delta = \boldsymbol{\mu}^T\boldsymbol{\mu}$. The distribution is said to be **central** whenever $\delta = 0$, and we denote it by χ^2_p.

The **noncentral F-distribution** with m, n degrees of freedom and noncentrality parameter δ, denoted by $F(m, n; \delta)$, is the distribution of the so-called **F-ratio** $F = \frac{u/m}{v/n}$, where $u \sim \chi^2_{m,\delta}$, $v \sim \chi^2_n$, and u and v are independent. The distribution is **central** whenever $\delta = 0$ and we then write $F \sim F(m, n)$.

Facts:

All the following facts except those with a specific reference can be found in [And03, Ch. 2], [Rao73, Sec. 8a], or [SL03, Ch. 1].

We can extend our definition to include the singular multivariate normal distribution using one of the first two facts below as a definition, each of which includes both the nonsingular and singular cases. When we write $\mathbf{x} \sim \text{N}_d(\boldsymbol{\mu}, \Sigma)$, we include both possibilities unless otherwise stated.

1. Assuming for the trivial case that $y \sim N_1(b, 0)$ means $y = b$ with probability 1, then the random $d \times 1$ vector $\mathbf{x}$ follows a multivariate normal distribution if and only if $y = \mathbf{a}^T\mathbf{x}$ is univariate normal for all $d \times 1$ vectors $\mathbf{a}$.
2. A random $d \times 1$ vector $\mathbf{x}$ with mean $\boldsymbol{\mu}$ and covariance matrix Σ follows a multivariate normal distribution of rank $m \leq d$ when it has the same distribution as $A\mathbf{z} + \boldsymbol{\mu}$, where A is any $d \times m$ matrix satisfying $\Sigma = AA^T$ and $\mathbf{z} \sim N_m(\mathbf{0}, I_m)$.
3. Given $\mathbf{x} \sim N_d(\boldsymbol{\mu}, \Sigma)$, then the m.g.f. of $\mathbf{x}$ is $M_{\mathbf{x}}(\mathbf{t}) = e^{\mathbf{t}^T\boldsymbol{\mu} + \frac{1}{2}\mathbf{t}^T\Sigma\mathbf{t}}$. When Σ is positive definite, this m.g.f. uniquely determines the nonsingular multivariate normal distribution.
4. Given $\mathbf{x} \sim N_d(\boldsymbol{\mu}, \Sigma)$ and A $m \times d$, then $A\mathbf{x} \sim N_m(A\boldsymbol{\mu}, A\Sigma A^T)$. The distribution is nonsingular if and only if $A\Sigma$ has full row rank m $(m \leq d)$, which is assured when Σ is positive definite and A has rank m.
5. Any subset of the components of a multivariate normal random variable is multivariate normal.
6. If the cross-covariance matrix of any two vectors which contain disjoint subsets of $\mathbf{x}$ is zero, then the two vectors are statistically independent.
7. If the cross-covariance matrix $\text{cov}(A\mathbf{x}, B\mathbf{x}) = \mathbf{0}$, then $A\mathbf{x}$ and $B\mathbf{x}$ are statistically independent.
8. If Σ is positive definite, then $(\mathbf{x} - \boldsymbol{\mu})^T\Sigma^{-1}(\mathbf{x} - \boldsymbol{\mu}) \sim \chi^2_d$, the central χ^2-distribution with d degrees of freedom.
9. Let $\mathbf{x}$ be an $n \times 1$ random vector with $E(\mathbf{x}) = \boldsymbol{\mu}$ and $\text{var}(\mathbf{x}) = \Sigma$, and having any distribution. If A is any symmetric $n \times n$ matrix, then $E[(\mathbf{x} - \mathbf{a})^T A(\mathbf{x} - \mathbf{a})] = \text{tr}(A\Sigma) + (\boldsymbol{\mu} - \mathbf{a})^T A(\boldsymbol{\mu} - \mathbf{a})$.
10. Let $\mathbf{x} = [x_i]$ be an $n \times 1$ random vector with $E(x_i) = \theta_i$ and $\mu_r = E(x_i - \theta_i)^r$; $r = 2, 3, 4$. If A is any $n \times n$ symmetric matrix and $\mathbf{a}$ is the column vector of the diagonal elements of A, then

$$\text{var}(\mathbf{x}^T A\mathbf{x}) = (\mu_4 - 3\mu_2^2)\mathbf{a}^T\mathbf{a} + 2\mu_2^2\text{tr}A^2 + 4\mu_2\boldsymbol{\theta}^T A^2\boldsymbol{\theta} + 4\mu_3\boldsymbol{\theta}^T A\mathbf{a}.$$

Examples:

1. If $x \sim N_1(0, \sigma^2)$, then $M_x(t) = e^{\frac{1}{2}\sigma^2 t^2}$ from Fact 3 above. Since $\mu = 0$, $\mu_r = E(x^r)$ can be found by expanding $M_x(t)$ and finding the coefficient of $t^r/r!$. For example, $\mu_2 = \sigma^2, \mu_3 = 0$, and $\mu_4 = 3\mu_2^2$.
2. If the x_i $(i = 1, 2, \ldots, n)$ are i.i.d. as $N_1(0, \sigma^2)$, then by either multiplying pdfs together for independent random variables or using m.g.f.s we find that $\mathbf{x} \sim N_n(\mathbf{0}, \sigma^2 I_n)$. Substituting the results from the previous example into Fact 10 we get $\text{var}(\mathbf{x}^T A\mathbf{x}) = 2\sigma^4\text{tr}A^2$.

53.3 Inference for the Multivariate Normal

Definitions:

Given a random sample $\{\mathbf{x}_i\}$ that are i.i.d., from the nonsingular multivariate normal distribution, the **likelihood function** is defined to be the joint pdf of the sample expressed as a function of the unknown parameters, namely,

$$L(\mathbf{u}, \Sigma) = \prod_{i=1}^{n} f(\mathbf{x}_i; \boldsymbol{\mu}, \Sigma).$$

The parameter estimates that maximize this function are called the **maximum likelihood estimates**. The **sample mean** of the sample is defined to be $\bar{\mathbf{x}} = \sum_{i=1}^{n} \mathbf{x}_i/n$, and is not to be confused with the complex conjugate.

Facts:

1. The maximum likelihood estimates of $\boldsymbol{\mu}$ and Σ are, respectively, $\widehat{\boldsymbol{\mu}} = \bar{\mathbf{x}}$ and $\widehat{\Sigma} = Q/n = \sum_{i=1}^{n}(\mathbf{x}_i - \bar{\mathbf{x}})(\mathbf{x}_i - \bar{\mathbf{x}})^T/n$. Here, "$\widehat{\ }$" denotes "estimate of" in statistics.
2. $Q = X^T C X$, where X is the data matrix previously defined and the centering matrix $C = I_n - \frac{1}{n}\mathbf{1}\mathbf{1}^T$ is symmetric and idempotent. Also, $CX = [\mathbf{x}_1 - \bar{\mathbf{x}}, \ldots, \mathbf{x}_n - \bar{\mathbf{x}}]^T = \tilde{X}$, say, and $Q = \tilde{X}^T\tilde{X}$.

53.4 Principal Component Analysis

Definitions:

Let $\mathbf{x}$ be a random d-dimensional vector with mean $\boldsymbol{\mu}$ and positive definite covariance matrix Σ. Let $T = [\mathbf{t}_1, \mathbf{t}_2, \ldots, \mathbf{t}_d]$ be an orthogonal matrix such that $T^T\Sigma T = \Lambda = \text{diag}(\lambda_1, \lambda_2, \ldots, \lambda_d)$, where $\lambda_1 \geq \lambda_2 \geq \cdots \geq \lambda_d > 0$ are the ordered eigenvalues of Σ. The sum $\text{tr}\Sigma$ is sometimes called the **total variance**. If $\mathbf{y} = [y_j] = T^T(\mathbf{x} - \boldsymbol{\mu})$, then $y_j = \mathbf{t}_j^T(\mathbf{x} - \boldsymbol{\mu})$ $(j = 1, 2, \ldots, d)$ is called the jth **population principal component** of $\mathbf{x}$, and $z_j = \lambda_j^{-1/2} y_j$ is called the jth **standardized population principal component.**

In practice, $\boldsymbol{\mu}$ and Σ are unknown and have to be estimated from a sample $\mathbf{x}_1, \mathbf{x}_2, \ldots, \mathbf{x}_n$. Assuming that the underlying distribution is multivariate normal we can use the estimates of the previous section, $\bar{\mathbf{x}}$ and $\widehat{\Sigma} = \tilde{X}^T\tilde{X}/n$. Carrying out a similar factorization on $\widehat{\Sigma}$ as we did for Σ, we obtain the eigenvalues $\hat{\lambda}_1 \geq \hat{\lambda}_2 \geq \cdots \geq \hat{\lambda}_d > 0$ and an orthogonal matrix $\widehat{T} = [\hat{\mathbf{t}}_1, \hat{\mathbf{t}}_2, \ldots, \hat{\mathbf{t}}_d]$ of corresponding eigenvectors. For each observation $\mathbf{x}_i$, we can define a vector $\widehat{\mathbf{y}}_i = \widehat{T}^T(\mathbf{x}_i - \bar{\mathbf{x}})$ of **sample principal components** or **estimated principal components** yielding $\widehat{Y}^T = [\widehat{\mathbf{y}}_1, \widehat{\mathbf{y}}_2, \ldots, \widehat{\mathbf{y}}_n] = \widehat{T}^T\tilde{X}^T$.

Although the above method is mainly used descriptively, asymptotic inference for large n can be carried out on the assumption that the underlying distribution is normal.

Facts:

The following facts can be found in [Chr01, Sec. 3.1–3.3], [Rao73, Sec. 8g2], or in [Seb04, Sec. 5.2].

1. As $\text{var}(\mathbf{y}) = \Lambda$, which is diagonal, the y_j are uncorrelated and $\text{var}(y_j) = \lambda_j$.
2. $\sum_{j=1}^{d}\text{var}(y_j) = \sum_{j=1}^{d}\text{var}(x_j) = \text{tr}\Sigma$. We can use $\lambda_j/\text{tr}\Sigma$ to measure the relative magnitude of λ_j. If the λ_i $(i = k+1, \ldots, d)$ are relatively small so that the corresponding y_i are "small" (with zero means and small variances), then $\mathbf{y}_{(k)} = [y_1, y_2, \ldots, y_k]^T$ can be regarded as a k-dimensional approximation for $\mathbf{y}$. Thus, $\mathbf{y}_{(k)}$ can be used as a "proxy" for $\mathbf{x}$ in terms of explaining a major part of the total variance.
3. Let $T_{(k)} = [\mathbf{t}_1, \ldots, \mathbf{t}_k]$. Then:
   - $\max\limits_{\mathbf{a}^T\mathbf{a}=1} \text{var}(\mathbf{a}^T\mathbf{x}) = \text{var}(\mathbf{t}_1^T\mathbf{x}) = \text{var}\left(\mathbf{t}_1^T(\mathbf{x} - \boldsymbol{\mu})\right) = \text{var}(y_1) = \lambda_1$, so that y_1 is the normalized linear combination of the elements of $\mathbf{x} - \boldsymbol{\mu}$ with maximum variance λ_1.
   - $\max\limits_{\mathbf{a}^T\mathbf{a}=1, T_{(k-1)}^T\mathbf{a}=\mathbf{0}} \text{var}(\mathbf{a}^T\mathbf{x}) = \text{var}(\mathbf{t}_k^T\mathbf{x}) = \text{var}(y_k) = \lambda_k$, so that $\mathbf{t}_k^T(\mathbf{x} - \boldsymbol{\mu})$ is the normalized linear combination of the elements $\mathbf{x} - \boldsymbol{\mu}$ uncorrelated with $y_1, y_2, \ldots, y_{k-1}$, with maximum variance λ_k.
4. $\hat{y}_{ij} = \hat{\mathbf{t}}_j^T(\mathbf{x}_i - \bar{\mathbf{x}})$, the score of the ith individual on the jth sample principal component, is related to the orthogonal projection of $\mathbf{x}_i - \bar{\mathbf{x}}$ onto range$(\hat{\mathbf{t}}_j)$, namely $P_{\hat{\mathbf{t}}_j}(\mathbf{x}_i - \bar{\mathbf{x}}) = \hat{\mathbf{t}}_j\hat{\mathbf{t}}_j^T(\mathbf{x}_i - \bar{\mathbf{x}}) = \hat{y}_{ij}\hat{\mathbf{t}}_j$.

Examples:

1. Suppose we assume that each $\mathbf{x}_i$ is just an observed vector, i.e., a constant. Let $\mathbf{v}$ be a discrete random vector taking the values $\mathbf{x}_i$ $(i = 1, 2, \ldots, n)$ with probability $\frac{1}{n}$. Then $\text{E}(\mathbf{v}) = \sum_{i=1}^{n}\mathbf{x}_i\, P(\mathbf{v} = \mathbf{x}_i) =$

$\sum_{i=1}^{n} \mathbf{x}_i/n = \bar{\mathbf{x}}$ and, similarly, $\text{var}(\mathbf{v}) = \widehat{\Sigma}$. This means that the sample principal components for $\mathbf{x}$ are the population components for $\mathbf{v}$ so that all the optimal properties of population principal components hold correspondingly for the sample principal components.

2. Entomologists are interested in the number of distinct taxa present in a population of winged aphids as they are difficult to identify. Forty aphids were trapped and 19 variables on each aphid were measured. A principal component analysis was carried out and the first two components accounted for 85% of the estimate of the total variance giving an effective reduction from 19 to 2 dimensions. The 2-dimensional $\widehat{\mathbf{y}}_i$ were plotted showing the presence of four groups. Also, the $\widehat{\mathbf{t}}_r$ $(r = 1, 2)$ suggested which two linear combinations of the 19 measurements gave the best discrimination. As the aphids came from slightly different populations, the original assumption that the sample is from a homogeneous population is not quite true, but it does provide a starting place for further study, e.g., use discriminant coordinates described in the next section with four groups.

53.5 Discriminant Coordinates

Definitions:

Suppose we have n d-dimensional observations of which n_i belong to group i from the ith underlying population $(i = 1, 2, \ldots, g;\ n = \sum_{i=1}^{g} n_i)$. Let $\mathbf{x}_{ij}$ be the jth observation in group i, and define

$$\bar{\mathbf{x}}_{i\cdot} = \frac{1}{n_i}\sum_{j=1}^{n_i} \mathbf{x}_{ij} \quad \text{and} \quad \bar{\mathbf{x}}_{\cdot\cdot} = \frac{1}{n}\sum_{i=1}^{g}\sum_{j=1}^{n_i} \mathbf{x}_{ij}\,.$$

Let $W_g = \sum_{i=1}^{g}\sum_{j=1}^{n_i}(\mathbf{x}_{ij} - \bar{\mathbf{x}}_{i\cdot})(\mathbf{x}_{ij} - \bar{\mathbf{x}}_{i\cdot})^T$, the **within-groups matrix**, and let $W_b = \sum_{i=1}^{g} n_i(\bar{\mathbf{x}}_{i\cdot} - \bar{\mathbf{x}}_{\cdot\cdot})(\bar{\mathbf{x}}_{i\cdot} - \bar{\mathbf{x}}_{\cdot\cdot})^T$, the **between-groups matrix**. Since W_g and W_b are positive definite with probability 1, the eigenvalues of $W_g^{-1}W_b$, which are the same as those of $W_g^{-1/2}W_b W_g^{-1/2}$, are positive and distinct with probability 1, say $\lambda_1 > \lambda_2 > \cdots > \lambda_d > 0$.

Let $W_g^{-1}W_b\mathbf{c}_r = \lambda_r\mathbf{c}_r$ be suitably scaled eigenvectors and define $C^T = [\mathbf{c}_1, \mathbf{c}_2, \ldots, \mathbf{c}_k]$ $(k \leq d)$. If we define $\mathbf{y}_{ij} = C\mathbf{x}_{ij}$, then the k elements of $\mathbf{z}_{ij}$ are called the first k **discriminant coordinates** (or **canonical variates**). These coordinates are determined so as to emphasize group separation, but with decreasing effectiveness so that k has to be determined. The coordinates can be computed using an appropriate transformation combined with a principal component analysis. Typically the $\mathbf{c}_i$ are scaled so that $CSC^T = I_r$, where $S = W_g/(n - g)$.

Examples:

1. Trivariate measurements $\mathbf{x}_{ij}$ were taken on the skulls of six collections (four subspecies with three collections apparently from the same species) of anteaters. Using six groups of data from six underlying populations in the above theory, we found that $[\lambda_1, \lambda_2, \lambda_3] = [2.400, 0.905, 0.052]$. Since λ_3 was small, we chose $k = 2$. We could, therefore, transform the 3-dimensional observations $\mathbf{x}_{ij}$ into the 2-dimensional $\mathbf{y}_{ij} = C\mathbf{x}_{ij}$ and still account for most of the variation in the data. The 2-dimensional reductions were then plotted to help us look for any patterns. As an aid to our search, it was possible to draw a circle with center the reduced mean of each group so that the circle contained any reduced random observation from its population with probability 0.95. The center of the circle locates the center of gravity of the group and the boundary of the circle shows where the bulk of the observations are expected to lie. The closeness of these circles indicated how "close" the groups and subspecies were. We found that three of the circles almost overlapped completely, confirming a common underlying species, and the remaining circles had little overlap, suggesting the presence of four species.

53.6 Canonical Correlations and Variates

Definitions:

Let $\mathbf{z} = \begin{bmatrix} \mathbf{x} \\ \mathbf{y} \end{bmatrix}$ denote a d-dimensional random vector with mean $\boldsymbol{\mu}$ and positive definite covariance matrix Σ. Let $\mathbf{x}$ and $\mathbf{y}$ have dimensions d_1 and $d_2 = d - d_1$, respectively, and consider the partition

$$\Sigma = \begin{bmatrix} \Sigma_{11} & \Sigma_{12} \\ \Sigma_{21} & \Sigma_{22} \end{bmatrix},$$

where Σ_{ii} is $d_i \times d_i$ and $\Sigma_{21} = \Sigma_{12}^T$ has rank r. Let ρ_1^2 be the maximum value of the squared correlation between arbitrary linear combinations $\boldsymbol{\alpha}^T\mathbf{x}$ and $\boldsymbol{\beta}^T\mathbf{y}$, and let $\boldsymbol{\alpha} = \mathbf{a}_1$ and $\boldsymbol{\beta} = \mathbf{b}_1$ be the corresponding maximizing values of $\boldsymbol{\alpha}$ and $\boldsymbol{\beta}$. Then the positive square root $\sqrt{\rho_1^2}$ is called the **first (population) canonical correlation** between $\mathbf{x}$ and $\mathbf{y}$, and $u_1 = \mathbf{a}_1^T\mathbf{x}$ and $v_1 = \mathbf{b}_1^T\mathbf{y}$ are called the **first (population) canonical variables.**

Let ρ_2^2 be the maximum value of the squared correlation between $\boldsymbol{\alpha}^T\mathbf{x}$ and $\boldsymbol{\beta}^T\mathbf{y}$, where $\boldsymbol{\alpha}^T\mathbf{x}$ is uncorrelated with $\mathbf{a}_1^T\mathbf{x}$ and $\boldsymbol{\beta}^T\mathbf{y}$ is uncorrelated with $\mathbf{b}_1^T\mathbf{y}$, and let $u_2 = \mathbf{a}_2^T\mathbf{x}$ and $v_2 = \mathbf{b}_2^T\mathbf{y}$ be the maximizing values. Then the positive square root $\sqrt{\rho_2^2}$ is called the **second canonical correlation**, and u_2 and v_2 are called the **second canonical variables.** Continuing in this manner, we obtain r pairs of canonical variables $\mathbf{u} = [u_1, u_2, \ldots, u_r]^T$ and $\mathbf{v} = [v_1, v_2, \ldots, v_r]^T$. We can then regard $\mathbf{u}$ and $\mathbf{v}$ as lower-dimensional "representations" of $\mathbf{x}$ and $\mathbf{y}$.

Facts:

1. [Seb04, Sec. 5.7] If $m = \text{rank}\Sigma_{12} \geq 1$ and Σ is positive definite, then $\Sigma_{11}^{-1}\Sigma_{12}\Sigma_{22}^{-1}\Sigma_{21}$ has m positive eigenvalues $\rho_1^2 \geq \rho_2^2 \geq \cdots \geq \rho_m^2$, say, and $\rho_1^2 < 1$. Moreover, $\Sigma_{11}^{-1}\Sigma_{12}\Sigma_{22}^{-1}\Sigma_{21}$ and $\Sigma_{22}^{-1}\Sigma_{21}\Sigma_{11}^{-1}\Sigma_{12}$ have the same (nonzero) eigenvalues. Let $\mathbf{a}_1, \mathbf{a}_2, \ldots, \mathbf{a}_m$ and $\mathbf{b}_1, \mathbf{b}_2, \ldots, \mathbf{b}_m$ be the corresponding eigenvectors of $\Sigma_{11}^{-1}\Sigma_{12}\Sigma_{22}^{-1}\Sigma_{21}$ and $\Sigma_{22}^{-1}\Sigma_{21}\Sigma_{11}^{-1}\Sigma_{12}$, respectively. Suppose that $\boldsymbol{\alpha}$ and $\boldsymbol{\beta}$ are arbitrary vectors such that for $r \leq m$, $\boldsymbol{\alpha}^T\mathbf{x}$ is uncorrelated with each $\mathbf{a}_j^T\mathbf{x}$ $(j = 1, 2, \ldots, r-1)$, and $\boldsymbol{\beta}^T\mathbf{y}$ is uncorrelated with each $\mathbf{b}_j^T\mathbf{y}$ $(j = 1, 2, \ldots, r-1)$. Let $u_j = \mathbf{a}_j^T\mathbf{x}$ and $v_j = \mathbf{b}_j^T\mathbf{y}$, for $j = 1, 2, \ldots, r$. Then:
   - The maximum squared correlation between $\boldsymbol{\alpha}^T\mathbf{x}$ and $\boldsymbol{\beta}^T\mathbf{y}$ is given by ρ_r^2 and it occurs when $\boldsymbol{\alpha} = \mathbf{a}_r$ and $\boldsymbol{\beta} = \mathbf{b}_r$.
   - $\text{cov}(u_j, u_k) = 0$ for $j \neq k$, and $\text{cov}(v_j, v_k) = 0$ for $j \neq k$.
   - The squared correlation between u_j and v_j is ρ_j^2.
   - Since ρ_j^2 is independent of scale, we can scale $\mathbf{a}_j$ and $\mathbf{b}_j$ such that $\mathbf{a}_j^T\Sigma_{11}\mathbf{a}_j = 1$ and $\mathbf{b}_j^T\Sigma_{22}\mathbf{b}_j = 1$. The u_j and v_j then have unit variances.
   - If the $d_1 \times d_2$ matrix Σ_{12} has row full rank, and $d_1 < d_2$, then $m = d_1$. All the eigenvalues of $\Sigma_{11}^{-1}\Sigma_{12}\Sigma_{22}^{-1}\Sigma_{21}$ are then positive, while $\Sigma_{22}^{-1}\Sigma_{21}\Sigma_{11}^{-1}\Sigma_{12}$ has d_1 positive eigenvalues and $d_2 - d_1$ zero eigenvalues. The rank of Σ_{12} can vary as there may be constraints on Σ_{12}, such as $\Sigma_{12} = \mathbf{0}$ (rank 0) or $\Sigma_{12} = a\mathbf{1}_{d_1}\mathbf{1}_{d_1}^T$ (rank 1).
2. [Bri01, p. 370]. Suppose $\mathbf{x}$ and $\mathbf{y}$ have means $\boldsymbol{\mu}_{\mathbf{x}}$ and $\boldsymbol{\mu}_{\mathbf{y}}$, respectively. Let $\mathbf{u} = A(\mathbf{x} - \boldsymbol{\mu}_{\mathbf{x}})$ and $\mathbf{v} = B(\mathbf{y} - \boldsymbol{\mu}_{\mathbf{y}})$, where A and B are any matrices, each with r rows that are linearly independent, satisfying $A\Sigma_{11}A^T = I_r$ and $B\Sigma_{22}B^T = I_r$. Then $\text{E}[(\mathbf{u} - \mathbf{v})^T(\mathbf{u} - \mathbf{v})]$ is minimized when $\mathbf{u}$ and $\mathbf{v}$ are vectors of the canonical variables.
3. [BS93], [SS80] When Σ is singular, then the (population) canonical correlations are the positive square roots of the eigenvalues of $\Sigma_{11}^{-}\Sigma_{12}\Sigma_{22}^{-}\Sigma_{21}$; any generalized inverses Σ_{11}^{-} and Σ_{22}^{-} may be chosen. There will be u canonical correlations equal to 1, where $u = \text{rank}(\Sigma_{11}) + \text{rank}(\Sigma_{22}) - \text{rank}(\Sigma)$.

53.7 Estimation of Correlations and Variates

Definitions:

Let $\mathbf{z}_1, \mathbf{z}_2, \ldots, \mathbf{z}_n$ be a random sample and let $\bar{\mathbf{z}} = \frac{1}{n}\sum_{i=1}^{n} \mathbf{z}_i = \begin{bmatrix} \bar{\mathbf{x}} \\ \bar{\mathbf{y}} \end{bmatrix}$ be the sample mean and $\widehat{\Sigma} = \frac{1}{n}\sum_{i=1}^{n}(\mathbf{z}_i - \bar{\mathbf{z}})(\mathbf{z}_i - \bar{\mathbf{z}})^T$, where $\widehat{\Sigma}$ is partitioned in the same way as Σ, namely

$$n\widehat{\Sigma} = \begin{bmatrix} \tilde{X}^T\tilde{X}, & \tilde{X}^T\tilde{Y} \\ \tilde{Y}^T\tilde{X}, & \tilde{Y}^T\tilde{Y} \end{bmatrix} = \begin{bmatrix} Q_{11} & Q_{12} \\ Q_{21} & Q_{22} \end{bmatrix},$$

say. Assuming that $d_1 \leq d_2$, let $r_1^2 > r_2^2 > \cdots > r_{d_1}^2 > 0$ be the eigenvalues of $Q_{11}^{-1}Q_{12}Q_{22}^{-1}Q_{21}$, with corresponding eigenvectors $\hat{\mathbf{a}}_1, \hat{\mathbf{a}}_2, \ldots, \hat{\mathbf{a}}_{d_1}$. We define $u_{ij} = \hat{\mathbf{a}}_j^T(\mathbf{x}_i - \bar{\mathbf{x}})$, the ith element of $\mathbf{u}_j = \tilde{X}\hat{\mathbf{a}}_j$, where $\hat{\mathbf{a}}_j$ is scaled so that $1 = n^{-1}\hat{\mathbf{a}}_j^T\tilde{X}^T\tilde{X}\hat{\mathbf{a}}_j = \sum_{i=1}^{n} u_{ij}^2/n = \mathbf{u}_j^T\mathbf{u}_j/n$. Then $\sqrt{r_j^2}$ is the jth **sample canonical correlation.** Moreover, u_{ij} is the jth **sample canonical variable** of $\mathbf{x}_i$. In a similar fashion, we define $v_{ij} = \hat{\mathbf{b}}_j^T(\mathbf{y}_i - \bar{\mathbf{y}})$, the ith element of $\mathbf{v}_j = \tilde{Y}\hat{\mathbf{b}}_j$, to be the jth sample canonical variable of $\mathbf{y}_i$. Here, $\hat{\mathbf{b}}_1, \hat{\mathbf{b}}_2, \ldots, \hat{\mathbf{b}}_{d_1}$ are the corresponding eigenvectors of $Q_{22}^{-1}Q_{21}Q_{11}^{-1}Q_{12}$, scaled so that $\mathbf{v}_j^T\mathbf{v}_j/n = 1$. The u_{ij} and v_{ij} are the **scores** of the ith observation on the jth canonical variables.

Facts:

1. When Σ is positive definite and $n - 1 \geq d$, then, with probability 1, $n\widehat{\Sigma}$ is positive definite and $\operatorname{rank} Q_{12} = d_1$.
2. The Canonical correlations $\sqrt{r_j^2}$ are all distinct with probability 1.
3. r_j^2 is the square of the sample correlation between the canonical variables whose values are in the vectors $\mathbf{u}_j$ and $\mathbf{v}_j$.

Examples:

1. Length and breadth head measurements were carried out on the first and second sons of 25 families to see what relationships existed. Here, $\mathbf{x}$ refers to the two measurements on the first son and $\mathbf{y}$ to the second. It was found that $r_1 = 0.7885$ and $r_2 = 0.0537$, indicating that just one pair of sample canonical variables (u_1, v_1) would give a reasonable reduction in dimension from 2 to 1. Plotting u_1 against v_1 using the 21 pairs (u_{i1}, v_{i1}) gave a reasonably linear plot. The first linear combinations (u_1, v_1) of the two measurements suggested that they could be interpreted as a measure of "girth" while the second (u_2, v_2) could be interpreted as "shape" measurements. We can conclude that there is a strong correlation between the head sizes of first and second brothers, but not between the head shapes.

53.8 Matrix Quadratic Forms

To carry out inference for multivariate normal distributions we now require further theory.

Definitions:

If X is an $n \times d$ data matrix and $A = [a_{ij}]$ is a symmetric $n \times n$ matrix, then the expression $X^TAX = \sum_{i=1}^{n}\sum_{j=1}^{n} a_{ij}\mathbf{x}_i\mathbf{x}_j^T$ is said to be a **matrix quadratic form**. We have already had one example, namely, $Q = X^TCX$ in section 53.3. When the $\{\mathbf{x}_i\}$ are a random sample from a distribution with mean $\boldsymbol{\mu}$ and covariance matrix Σ we define $S = Q/(n-1)$ to be the **sample covariance matrix**. (Some authors define it as $\widehat{\Sigma}$ in Section 53.3)

If the $\mathbf{x}_i$ are i.i.d. as $\mathrm{N}_d(\mathbf{0}, \Sigma)$, with $d \leq n$ and $\Sigma = [\sigma_{ij}]$ positive semidefinite, then $W = X^TX$ is said to a have a **Wishart distribution** with n degrees of freedom and scale matrix Σ, and we write $W \sim \mathrm{W}_d(n, \Sigma)$. If Σ is positive definite then the distribution is said to be **nonsingular**.

Suppose $\mathbf{y} \sim N_d(\mathbf{0}, \Sigma)$ independently of $W \sim W_d(m, \Sigma)$ and that both distributions are nonsingular. Then **Hotelling's T^2 distribution** is defined to be the distribution of $T^2 = m\mathbf{y}^T W^{-1}\mathbf{y}$ and we denote this distribution by $T^2 \sim T^2_{d,m}$.

Finally, in inference we are usually interested in **contrasts $\mathbf{a}^T\boldsymbol{\mu}$**, in the μ_i, where $\mathbf{a}$ is a vector whose entries sum to 0, i.e., $\mathbf{a}^T\mathbf{1}_d = 0$. Examples are $\mu_1 - \mu_2 = [1, -1, 0, \ldots, 0]\boldsymbol{\mu}$ and $\mu_1 - \frac{1}{2}(\mu_2 + \mu_3)$.

Facts:

All facts, unless otherwise indicated, appear in [Seb04, Secs. 2.3–2.4, 3.3–3.4, 3.6.2].

1. $E(S) = \Sigma$, irrespective of the distribution of the $\mathbf{x}_i$.
2. [HS79] Let X be an $n \times d$ data matrix and let A and B be matrices of appropriate sizes. If vec is the usual "stacking" operator, then:
   - $\operatorname{cov}[\operatorname{vec}(AXB), \operatorname{vec}(CXD)] = (B^T\Sigma D) \otimes (AC^T)$.
   - If $U = AXB$ and $V = CXD$, then U and V are pairwise uncorrelated, that is, $\operatorname{cov}(u_{ij}, v_{rs}) = 0$ for all i, j, r, and s, if $AC^T = 0$ and/or $B^T\Sigma D = \mathbf{0}$.
3. Suppose $W = [w_{ij}]$ has a Wishart distribution $W_d(m, \Sigma)$, with Σ possibly singular. Then the following hold:
   - $E(W) = m\Sigma$.
   - Let A be a $q \times d$ matrix with rank $q \leq d$. Then $AWA^T \sim W_q(m, A\Sigma A^T)$. When Σ is positive definite, then this distribution is nonsingular.
   - When $\sigma_{jj} > 0$, then $w_{jj}/\sigma_{jj} \sim \chi^2_m$, $(j = 1, 2, \ldots, d)$. However, the w_{jj} $(j = 1, \ldots, d)$ are not statistically independent.
   - When Σ is positive definite, then $\det \Sigma > 0$ and $\det W/\det \Sigma$ is distributed as the product of d independent chi-square variables with respective degrees of freedom $m, m-1, \ldots, m-d+1$.
4. (See [SK79, p. 97], [Sty89].) Let $W \sim W_d(m, \Sigma)$ with Σ possibly singular, and let A be an $n \times n$ symmetric matrix. Then:
   - $E(WAW) = m[\Sigma A^T \Sigma + \operatorname{tr}(A\Sigma)\Sigma] + m^2\Sigma A\Sigma$.
   - If $m > d+1$ and Σ is nonsingular, then

$$E(WAW^{-1}) = \frac{1}{m-d-1}[m\Sigma A\Sigma^{-1} - A^T - (\operatorname{tr}A)I],$$

$$E(W^{-1}AW) = \frac{1}{m-d-1}[m\Sigma^{-1}A\Sigma - A^T - (\operatorname{tr}A)I].$$

   - If $m > d+3$ and Σ is nonsingular, then

$$E(W^{-1}AW^{-1}) = \frac{(m-d-2)\Sigma^{-1}A\Sigma^{-1} + \Sigma^{-1}A^T\Sigma^{-1} - (\operatorname{tr}A\Sigma^{-1})\Sigma^{-1}}{(m-d)(m-d-1)(m-d-3)}.$$

5. Suppose that the rows $\mathbf{x}_i^T$ of X are uncorrelated with $\operatorname{var}(\mathbf{x}_i) = \Sigma_i$ for $i = 1, 2, \ldots, n$, and A is an $n \times n$ symmetric matrix. Then:
   - $E(X^TAX) = \sum_{i=1}^{n} a_{ii}\Sigma_i + E(X^T)AE(X)$.
   - If $\Sigma_i = \Sigma$ for all i, then $E(X^TAX) = (\operatorname{tr}A)\Sigma + E(X^T)AE(X)$.
6. [Das71, Th. 5], [EP73, Th. 2.3] Suppose that the rows of X are independent and A is an $n \times n$ positive semidefinite matrix of rank $r \geq d$. If for each $\mathbf{x}_i$ and all $\mathbf{b}$ and c with $\mathbf{b} \neq \mathbf{0}$ the probability $\Pr(\mathbf{b}^T\mathbf{x}_i = c) = 0$, then X^TAX is positive definite with probability 1.

7. Suppose that rows of X are i.i.d. as $\mathrm{N}_d(\mathbf{0}, \Sigma)$, with Σ possibly singular, and let A and B be $n \times n$ symmetric matrices.
   - If A has rank $r \leq d$, then $X^T AX \sim \mathrm{W}_d(r, \Sigma)$ if and only if $A = A^2$.
   - Suppose $X^T AX$ and $X^T BX$ have Wishart distributions. They are statistically independent if and only if $AB = \mathbf{0}$.
8. Referring to the definition of Hotelling's T^2, $\frac{m-d+1}{md} T^2_{d,m} \sim F(d, m-d+1)$, the F-distribution with d and $m-d+1$ degrees of freedom. When $\mathbf{y} \sim \mathrm{N}_d(\boldsymbol{\theta}, \Sigma)$, then $F \sim F(d, m-d+1; \delta)$, the corresponding noncentral F-distribution with noncentrality parameter $\delta = \boldsymbol{\theta}^T \Sigma \boldsymbol{\theta}$.
9. Suppose that the rows of X are i.i.d. as nonsingular $\mathrm{N}_d(\boldsymbol{\mu}, \Sigma)$. Then:
   - $\bar{\mathbf{x}} \sim \mathrm{N}_d(\boldsymbol{\mu}, \Sigma/n)$.
   - $Q = (n-1)S \sim \mathrm{W}_d(n-1, \Sigma)$.
   - $\bar{\mathbf{x}}$ and S are statistically independent.
   - $T^2 = n(\bar{\mathbf{x}} - \boldsymbol{\mu})^T S^{-1}(\bar{\mathbf{x}} - \boldsymbol{\mu}) \sim T^2_{d,n-1}$. This statistic can be used for testing the null hypothesis $H_0 : \boldsymbol{\mu} = \boldsymbol{\mu}_0$.
10. Given $H_0 : \boldsymbol{\mu} \in \mathcal{V}$, where $\mathcal{V}$ is a p-dimensional vector subspace of $\mathbb{R}^d$, then:
    - $T^2_{\min} = \min_{\boldsymbol{\mu} \in \mathcal{V}} T^2 \sim T^2_{d-p,n-1}$.
    - If $H_0 : \boldsymbol{\mu} = K\boldsymbol{\beta}$, where K is a known $d \times p$ matrix of rank p and $\boldsymbol{\beta}$ is a vector of p unknown parameters, then $\mathcal{V} = \mathrm{range}(K)$, and $T^2_{\min} = n(\bar{\mathbf{x}}^T S^{-1}\bar{\mathbf{x}} - \bar{\mathbf{x}}^T S^{-1} K\boldsymbol{\beta}^*)$, where $\boldsymbol{\beta}^* = (K^T S^{-1} K)^{-1} K^T S^{-1} \bar{\mathbf{x}}$.
11. If $H_0 : A\boldsymbol{\mu} = \mathbf{0}$, where A is $(d-p) \times d$ of rank $d-p$, then $\mathcal{V} = \ker(A)$ (also called the null space) and $T^2_{\min} = n(A\bar{\mathbf{x}})^T (ASA^T)^{-1} A\bar{\mathbf{x}}$.
    - Let A be a $q \times d$ matrix of rank $q \leq d$. Then the quadratic $n(A\bar{\mathbf{x}} - A\boldsymbol{\mu})^T (ASA^T)^{-1}(A\bar{\mathbf{x}} - A\boldsymbol{\mu}) \sim T^2_{q,n-1}$. This can be used for testing $H_0 : A\boldsymbol{\mu} = \mathbf{c}$.
    - If A is a matrix with rows which are contrasts so that $A\mathbf{1}_d = \mathbf{0}$, then

$$n\bar{\mathbf{x}}^T A^T (ASA^T)^{-1} A\bar{\mathbf{x}} = n\bar{\mathbf{x}}^T S^{-1}\bar{\mathbf{x}} - \frac{n(\bar{\mathbf{x}}^T S^{-1}\mathbf{1}_d)^2}{\mathbf{1}_d^T S^{-1}\mathbf{1}_d}.$$

12. Let $\mathbf{v}_1, \mathbf{v}_2, \ldots, \mathbf{v}_{n_1}$ be a random sample from $\mathrm{N}_d(\boldsymbol{\mu}_1, \Sigma)$ and let $\mathbf{w}_1, \mathbf{w}_2, \ldots, \mathbf{w}_{n_2}$ be an independent random sample from $\mathrm{N}_d(\boldsymbol{\mu}_2, \Sigma)$, both distributions being nonsingular. Let $Q_1 = \sum_{i=1}^{n_1} (\mathbf{v}_i - \bar{\mathbf{v}})(\mathbf{v}_i - \bar{\mathbf{v}})^T$ and $Q_2 = \sum_{i=1}^{n_2} (\mathbf{w}_i - \bar{\mathbf{w}})(\mathbf{w}_i - \bar{\mathbf{w}})^T$. If $\boldsymbol{\theta} = \boldsymbol{\mu}_1 - \boldsymbol{\mu}_2$ and A is a $q \times d$ matrix of rank $q \leq d$, then:
    - $\bar{\mathbf{z}} = \bar{\mathbf{v}} - \bar{\mathbf{w}} \sim \mathrm{N}_d\left(\boldsymbol{\theta}, (\frac{1}{n_1} + \frac{1}{n_2})\Sigma\right)$.
    - $Q = Q_1 + Q_2 \sim \mathrm{W}_d(n_1 + n_2 - 2, \Sigma)$.
    - The statistic

$$\frac{n_1 n_2}{n_1 + n_2}(A\bar{\mathbf{z}} - A\boldsymbol{\theta})^T (AS_p A^T)^{-1}(A\bar{\mathbf{z}} - A\boldsymbol{\theta}) \sim T^2_{q,n_1+n_2-2},$$

    where $S_p = Q/(n_1 + n_2 - 2)$. This can be used to test $H_0 : A\boldsymbol{\theta} = \mathbf{0}$. When A is a certain $(d-1) \times d$ contrast matrix, then the methodology relating to H_0 is called *profile analysis*.

Examples:

1. Suppose the rows of X are i.i.d. nonsingular $\mathrm{N}_d(\mathbf{0}, \Sigma)$. Then $\bar{\mathbf{x}} = X^T \mathbf{1}_n / n$ and $W_1 = n\bar{\mathbf{x}}\bar{\mathbf{x}}^T = X^T AX$, where $A = \mathbf{1}_n \mathbf{1}_n^T / n$ is symmetric and idempotent of rank 1. Also, from Section 53.3, $W_2 = (\mathbf{x}_i - \bar{\mathbf{x}})(\mathbf{x}_i - \bar{\mathbf{x}})^T = X^T CX$, where $C = I_n - A$ is symmetric and idempotent of rank $n-1$. Since $AC = \mathbf{0}$, we have from Fact 7 above that $W_1 \sim \mathrm{W}_d(1, \Sigma)$, $W_2 \sim \mathrm{W}_d(n-1, \Sigma)$, and W_1 and W_2 are statistically independent.

2. (cf. Fact 10) Suppose we have a group of 30 animals all subject to the same conditions, and the length of each animal is observed at d points in time $t_1, t_2, \ldots, t_d$. The d lengths for the ith animal will give us a d-dimensional vector, $\mathbf{x}_i$ say, and we assume that the $\mathbf{x}_i$ are all i.i.d., as nonsingular $\mathrm{N}_d(\boldsymbol{\mu}, \Sigma)$. We are interested in testing the hypothesis H_0 that the "growth curve" is a third degree polynomial so that $H_0 : \mu_j = \beta_0 + \beta_1 t_j + \beta_2 t_j^2 + \beta_3 t_j^3$ $(j = 1, 2, \ldots, d)$ or $H_0 : \boldsymbol{\mu} = K\boldsymbol{\beta}$, where the jth row of K is $[1, t_j, t_j^2, t_j^3]$ and $\boldsymbol{\beta} = [\beta_0, \beta_1, \beta_2, \beta_3]^T$ is a vector of unknown parameters. We then test H_0 by calculating $T^2_{\min}$ given in Fact 10, and this statistic, which has a Hotellings T^2 distribution when H is true, can then be converted to an F-statistic using Fact 8. If this F-value is significantly large we reject H_0.
3. (cf. Fact 11) The lengths of the femur and humerus on the left- and right-hand sides are measured for $n = 100$ males aged over 40 years to see if men are symmetrical with respect to the lengths of these bones. For the ith male, we therefore have a four-dimensional observation vector $\mathbf{x}_i$. Assuming that the population of measurements is multivariate normal with mean $\boldsymbol{\mu} = [\mu_1, \mu_2, \mu_3, \mu_4]^T$, where μ_1 and μ_2 refer to the left side, then we are interested in testing the contrasts $\mu_1 - \mu_3 = 0$ and $\mu_2 - \mu_4 = 0$. Thus, A has rows $[1, 0, -1, 0]$ and $[0, 1, 0, -1]$, giving us $H_0 : A\mathbf{1}_4 = \mathbf{0}$. We can then test H_0 using $T^2_{\min}$ defined in Fact 11, and this T^2 statistic is then converted to an F-statistic, as in the previous example.

53.9 Multivariate Linear Model: Least Squares Estimation

Definitions:

Let Y be an $n \times d$ data matrix which comes from an experimental design giving rise to n observations, the rows of Y. Then we can use observations on the jth characteristic (variable) $\mathbf{y}^{(j)}$, the jth column of Y, to construct a linear regression model $\mathbf{y}^{(j)} = \boldsymbol{\theta}^{(j)} + \mathbf{u}^{(j)} = K\boldsymbol{\beta}^{(j)} + \mathbf{u}^{(j)}$ as in Section 52.2 (with K replaced by X there). Because the design is the same for each variable, the design matrix K will be independent of j $(j = 1, 2, \ldots, d)$, though the models will not be independent as the $\mathbf{y}^{(j)}$ are not independent. Putting all d regression models together, we get $Y = \Theta + \mathbf{U}$, where $\Theta = KB$, B is $p \times d$ matrix of unknown parameters, K is $n \times p$ of rank r $(r \leq p)$, and $U = [\mathbf{u}^{(1)}, \ldots, \mathbf{u}^{(d)}] = [\mathbf{u}_1, \ldots, \mathbf{u}_n]^T$. We shall assume that the $\mathbf{u}_i$ are a random sample from a distribution with mean $\mathbf{0}$ and covariance matrix $\Sigma = [\sigma_{ij}]$. Then $Y = KB + U$ is called a **multivariate linear model**. When $d = 1$, this reduces to the **univariate linear model** or **Gauss–Markov model**; see Chapter 52 and [Seb04, ch. 8].

Now, for the jth model, $\widehat{\boldsymbol{\theta}}^{(j)} = P\mathbf{y}^{(j)}$ is the (ordinary) least squares estimate of $\boldsymbol{\theta}^{(j)}$, where $P = K(K^TK)^-K^T$, and $\widehat{\Theta} = PY$ is defined to be the **least squares estimate of** Θ. When $r = p$, then setting $\widehat{\Theta} = K\widehat{B}$ we have $\widehat{B} = (K^TK)^{-1}K^T\widehat{\Theta} = (K^TK)^{-1}K^TY$, called the **the least squares estimate of** B. If $r < p$, then $\widehat{B}$ is not unique and is given by $\widehat{B} = (K^TK)^-K^TY$, where $(K^TK)^-$ is any generalized inverse of K^TK.

If K has less than full rank, then each of the d associated univariate models also has less than full rank. Moreover, $\mathbf{a}_i^T\boldsymbol{\beta}^{(j)}$ is estimable for each $i = 1, 2, \ldots, q$ and each model $j = 1, 2, \ldots, d$ if $\mathbf{a}_i \in \mathrm{range}(K^T)$. Let $A = [\mathbf{a}_1, \mathbf{a}_2, \ldots, \mathbf{a}_q]^T$. Combining these linear combinations we say that AB is **estimable** when $A = LK$ for some $q \times n$ matrix L.

Facts:

All the following facts can be found in [Seb04, Ch. 8].

1. If $K^T = [\mathbf{k}_1, \ldots, \mathbf{k}_n]$ then, transposing the model, $\mathbf{y}_i = B^T\mathbf{k}_i + \mathbf{u}_i$.
2. The cross-covariance $\mathrm{cov}(\mathbf{y}_r, \mathbf{y}_s) = \mathrm{cov}(\mathbf{u}_r, \mathbf{u}_s) = \delta_{rs}\Sigma$, where $\delta_{rs} = 1$ when $r = s$ and 0, otherwise.
3. The cross-covariance $\mathrm{cov}(\mathbf{y}^{(j)}, \mathbf{y}^{(k)}) = \mathrm{cov}(\mathbf{u}^{(j)}, \mathbf{u}^{(k)}) = \sigma_{jk}I_d$ for all $j, k = 1, \ldots, d$.
4. If K has full rank p, then $\hat{\boldsymbol{\beta}}^{(j)} = (K^TK)^{-1}K^T\mathbf{y}^{(j)}$ and $\mathrm{cov}(\hat{\boldsymbol{\beta}}^{(j)}, \hat{\boldsymbol{\beta}}^{(k)}) = \sigma_{jk}(K^TK)^{-1}$ (all $j, k = 1, \ldots, d$).

5. Let $F(\Theta) = (Y - \Theta)^T(Y - \Theta)$. Then:
   - $F(\widehat{\Theta}) = Y^T(I_n - P)Y = U^T(I_n - P)U$.
   - $\mathrm{E}[U^T(I_n - P)U] = (n - r)\Sigma$.
6. $F(\Theta) - F(\widehat{\Theta})$ is positive semidefinite for all $\Theta = XB$, and equal to 0 if and only if $\Theta = \widehat{\Theta}$. We say that $\widehat{\Theta}$ is the minimum of $F(\Theta)$. Then:
   - $\mathrm{tr}F(\Theta) \geq \mathrm{tr}F(\widehat{\Theta})$.
   - $\det F(\Theta) \geq \det F(\widehat{\Theta})$.
   - $||F(\Theta)|| \geq ||F(\widehat{\Theta})||$, where $||A|| = \{\mathrm{tr}(AA^T)\}^{1/2}$.

   Any of these three results could be used as a definition of $\widehat{\Theta}$.
7. *Multivariate Gauss–Markov Theorem.* If $\phi = \sum_{j=1}^{d} \mathbf{h}_j^T \boldsymbol{\theta}^{(j)}$, a linear combination of all the elements of Θ, then $\hat{\phi} = \sum_{j=1}^{d} \mathbf{h}_j^T \hat{\boldsymbol{\theta}}^{(j)}$ is the linear unbiased estimate with minimum variance or BLUE (best linear unbiased estimate) of ϕ. (See also Chapter 52.)

Examples:

1. [Seb04, Sec. 8.6.4] By setting $V^T = [\mathbf{v}_1, \mathbf{v}_2, \ldots, \mathbf{v}_{n_1}]$, $W^T = [\mathbf{w}_1, \mathbf{w}_2, \ldots, \mathbf{w}_{n_2}]$, and $Y = \begin{bmatrix} V \\ W \end{bmatrix}$, we see that the *two-sample problem* mentioned in Fact 12 of Section 53.8 is a special case of the multivariate model with

$$XB = \begin{bmatrix} \mathbf{1}_{n_1} & \mathbf{0} \\ \mathbf{0} & \mathbf{1}_{n_2} \end{bmatrix} \begin{bmatrix} \boldsymbol{\mu}_1^T \\ \boldsymbol{\mu}_2^T \end{bmatrix}.$$

53.10 Multivariate Linear Model: Statistical Inference

Definitions:

Let $Y = \Theta + U$, where $\Theta = KB$, be a multivariate linear model. We now assume that the underlying distribution (of the $\mathbf{u}_i$) is a nonsingular multivariate normal distribution $\mathrm{N}_d(\mathbf{0}, \Sigma)$.

Facts:

The following facts appear in [Seb04, Sec. 8.6].

1. The likelihood function for Y, i.e., the pdf of vec Y, can be expressed in the form

$$(2\pi)^{-nd/2}(\det \Sigma)^{-n/2} \exp\{\mathrm{tr}[-\frac{1}{2}(Y - \Theta)^T \Sigma^{-1}(Y - \Theta)]\}.$$

2. The maximum likelihood estimates of Σ and Θ that maximize the likelihood function are $\widehat{\Theta}$ (the least squares estimate) and $\widehat{\Sigma} = E/n$, where $E = (Y - \widehat{\Theta})^T(Y - \widehat{\Theta})$; E is usually called the residual matrix or error matrix. If the $n \times p$ matrix K has full rank (i.e., rank p), then the maximum likelihood estimate of B is $\widehat{B}$.
3. Let E be the residual matrix and assume $n - r \geq d$. Then
   - E is positive definite with probability 1.
   - $E \sim \mathrm{W}_d(n - r, \Sigma)$.
   - E is statistically independent of $\widehat{\Theta}$, and of $\widehat{B}$ if K has full rank.
   - The maximum value of the likelihood function is $(2\pi)^{-nd/2}(\det \widehat{\Sigma})^{-n/2}e^{-nd/2}$.
   - If K has full rank, then $\widehat{\boldsymbol{\beta}}^{(j)} \sim \mathrm{N}_n(\boldsymbol{\beta}^{(j)}, \sigma_{jj}(K^TK)^{-1})$.

4. Suppose that K has rank $r < p$. Let A be a known $q \times p$ matrix of rank q and let AB be estimable. We are interested in testing $H_0 : AB = C$, where C is known.
   - The minimum, E_H say, of $(Y - KB)^T(Y - KB)$ subject to $AB = C$ occurs when B equals
   $$\widehat{B}_H = \widehat{B} - (K^TK)^- A^T[A(K^TK)^- A^T]^{-1}(A\widehat{B} - C).$$
   Although $\widehat{B}_H$ is not unique, $\widehat{\Theta}_H = K\widehat{B}_H$ is unique. Moreover, $E_H = (Y - \widehat{\Theta}_H)^T(Y - \widehat{\Theta}_H)$ is positive definite with probability 1.
   - $H = E_H - E = (A\widehat{B} - C)^T[A(K^TK)^- A^T]^{-1}(A\widehat{B} - C)$ is positive definite with probability 1, and H and E are statistically independent (both irrespective of whether H_0 is true or not).
   - $\mathrm{E}(H) = q\Sigma + (AB - C)^T[A(X^TX)^- A^T]^{-1}(AB - C) = q\Sigma + D$, where D is positive definite; D is zero when H_0 is true. This means that $\mathrm{E}(H)$, and, therefore, H itself, tends to be "inflated" when H_0 is false so that a test statistic for H_0 can be based on a suitable function of H.
   - When H_0 is true, $H \sim \mathrm{W}_d(q, \Sigma)$.
   - Let $E_H^{1/2}$ be the positive definite square root of E_H. Then when H_0 is true, $V = E_H^{-1/2} H E_H^{-1/2}$, a scaled function of H, has a d-dimensional multivariate Beta distribution with parameters q and $n - r$.

5. [Seb04, Sec. 8.6.2] Four different criteria are usually computed for testing H_0, and these are essentially based on the eigenvalues of V given above, which measure the scaled magnitude of H in some sense. They detect different kinds of departures from H_0 and are all significantly large when H_0 is false.
   - *Roy's maximum root statistic* ϕ_{max}, the maximum eigenvalue of HE^{-1}.
   - *Wilks' Lambda* or *likelihood ratio statistic* $\Lambda = (\det E / \det E_H)^{n/2}$.
   - *Lawley–Hotelling trace statistic* $(n - r)\mathrm{tr}(HE^{-1})$.
   - *Pillai's trace statistic* $\mathrm{tr}(HE_H^{-1})$.

Example:

1. [Seb04, Sec. 8.7.2] To test the general linear hypothesis $H_0 : ABD = \mathbf{0}$, where A is $q \times p$ of rank $q \leq p$ and D is $d \times v$ of rank $v \leq d$, we let $Y_D = YD$ so that the linear model $Y = KB + U$ is transformed to
$$Y_D = KBD + UD = K\Lambda + U_0,$$
say, where the rows of U_0 are i.i.d. $\mathrm{N}_v(\mathbf{0}, D^T\Sigma D)$. Then H_0 becomes $A\Lambda = \mathbf{0}$ and
   - H becomes $H_D = D^THD = (A\widehat{B}D)^T[A(K^TK)^{-1}A^T]^{-1}A\widehat{B}D$ and E becomes $E_D = D^TED \sim \mathrm{W}_v(n - r, D^T\Sigma D)$.
   - When $ABD = \mathbf{0}$ is true, then $H_D \sim \mathrm{W}_v(q, D^T\Sigma D)$.

   We can now use Facts 4 and 5 above to test H_0 using H_D and E_D instead of H and E. This hypothesis arises in carrying out a so-called profile analysis of more than two populations.

53.11 Metric Multidimensional Scaling

Definitions:

Given a set of n objects, a **proximity measure** d_{rs} is a measure of the "closeness" of objects r and s; here, "closeness" does not necessarily refer to physical distance. We shall consider only one such measure as there are several. A proximity d_{rs} is called a **(symmetric) dissimilarity** if $d_{rr} = 0$, $d_{rs} \geq 0$, and $d_{rs} = d_{sr}$, for all $r, s = 1, 2, \ldots, n$; the matrix $D = [d_{rs}]$ is called a **dissimilarity matrix**. We say that D is **Euclidean** if there exists a p-dimensional configuration of points $\mathbf{y}_1, \mathbf{y}_2, \ldots, \mathbf{y}_n$ for some p such that all interpoint Euclidean distances satisfy $||\mathbf{y}_r - \mathbf{y}_s|| = d_{rs}$.

Facts:

1. Let $A = [a_{ij}]$ be a symmetric $n \times n$ matrix, where $a_{rs} = -\frac{1}{2}d_{rs}^2$. Define $b_{rs} = a_{rs} - \bar{a}_{r\cdot} - \bar{a}_{\cdot s} + \bar{a}_{\cdot\cdot}$, where $\bar{a}_{r\cdot}$ is the average of the elements of A in the rth row, $\bar{a}_{\cdot s}$ is the average for the sth column, and $\bar{a}_{\cdot\cdot}$ is the average of all the elements. Hence, $B = [b_{rs}] = CAC$, where $C = I - \frac{1}{n}\mathbf{1}\mathbf{1}^T$ is the usual centering matrix. Then $D = [d_{rs}]$ is Euclidean if and only if B is positive semidefinite; see [Seb04, p. 236].
2. If D is not Euclidean, then some eigenvalues of B will be negative. However, if the first k eigenvalues are comparatively large and positive, and the remaining positive or negative eigenvalues are near zero, then the rows of $Y_k = [\mathbf{y}^{(1)}, \mathbf{y}^{(2)}, \ldots, \mathbf{y}^{(k)}]$ will give a reasonable configuration. If the original objects are d-dimensional points $\mathbf{x}_i$ $(i = 1, 2, \ldots, n)$ to begin with so that $||\mathbf{x}_r - \mathbf{x}_s||^2 = d_{rs}^2$, then D is Euclidean and the n rows of Y_k will give a k-dimensional reduction of a d-dimensional system of points.

Example:

1. The migration pattern that occurred with the colonization of islands in the Pacific Ocean can be investigated linguistically and values of d_{rs} can be constructed based on linguistic information. For example, the proportion p_{rs} of the words for, say, 50 items that islands r and s have in common can be used as a measure of similarity (with $p_{rr} = 1$), which we can convert into a dissimilarity $d_{rs} = (2 - 2p_{rs})^{1/2}$. Then $a_{rs} = -\frac{1}{2}d_{rs}^2 = p_{rs} - 1$ and, when computing b_{rs}, the -1 drops out so that we can leave it out and set $a_{rs} = p_{rs}$. If A is positive semidefinite then so is B, which implies that D is Euclidean. Using B, we can then find a Y_k from which we can construct a lower dimensional map to see which islands are closest together linguistically. The same method can also be applied to blood groups using a different d_{rs}.

Acknowledgment

We are grateful to Ka Lok Chu, Jarkko Isotalo, Evelyn Mathason Styan, Kimmo Vahkalahti Andrei Volodin, and Douglas P. Wiens for their help. The research was supported in part by the National Sciences and Engineering Council of Canada.

References

[And03] T.W. Anderson. *An Introduction to Multivariate Statistical Analysis,* 3rd edition. John Wiley & Sons, New York 2003. (2nd ed. 1984, original ed. 1958)

[BS93] Jerzy K. Baksalary and George P.H. Styan. Around a formula for the rank of a matrix product with some statistical applications. In *Graphs, Matrices, and Designs: Festschrift in Honor of Norman J. Pullman* (Rolf S. Rees, Ed.), Lecture Notes in Pure and Applied Mathematics 139, Marcel Dekker, New York, pp. 1–18, 1993.

[Bri01] David R. Brillinger. *Time Series: Data Analysis and Theory.* Classics in Applied Mathematics 36, SIAM, Philadelphia, 2001. [Unabridged republication of the Expanded Edition, Holden-Day, 1981; original ed. Holt, Rinehart and Winston, 1975.]

[Chr01] Ronald Christensen. *Advanced Linear Modeling: Multivariate, Time Series, and Spatial Data; Nonparametric Regression and Response Surface Maximization,* 2nd ed. Springer, New York, 2001. (Original ed.: *Linear Models for Multivariate, Time Series, and Spatial Data,* 1991.)

[Das71] Somesh Das Gupta. Nonsingularity of the sample covariance matrix. *Sankhyā Ser. A* 33:475–478, 1971.

[EP73] Morris L. Eaton and Michael D. Perlman. The non-singularity of generalized sample covariance matrices. *Ann. Statist.* 1:710–717, 1973.

[HS79] Harold V. Henderson and S.R. Searle. Vec and vech operators for matrices, with some uses in Jacobians and multivariate statistics. *Can. J. Statist.* 7:65–81, 1979.

[Rao73] C. Radhakrishna Rao. *Linear Statistical Inference and Its Applications,* 2nd ed. John Wiley & Sons, New York, 1973. (1st ed. 1965)

[Seb04] George A.F. Seber. *Multivariate Observations,* Reprint edition. John Wiley & Sons, New York, 2004. (Original ed. 1984.)

[SL03] George A.F. Seber and Alan J. Lee. *Linear Regression Analysis,* 2nd ed. John Wiley & Sons, New York, 2003. (Original ed. by George A. F. Seber, 1977.)

[SS80] V. Seshadri and G.P.H. Styan. Canonical correlations, rank additivity and characterizations of multivariate normality. In *Analytic Function Methods in Probability Theory: Proceedings of the Colloquium on the Methods of Complex Analysis in the Theory of Probability and Statistics held at the Kossuth L. University, Debrecen, Hungary, August 29–September 2, 1977* (B. Gyires, Ed.), Colloquia Mathematica Societatis János Bolyai 21, North-Holland, Amsterdam pp. 331–344, 1980.

[SK79] M.S. Srivastava and C.G. Khatri. *An Introduction to Multivariate Statistics.* North-Holland, Amsterdam 1979.

[Sty89] George P.H. Styan. Three useful expressions for expectations involving a Wishart matrix and its inverse. In *Statistical Data Analysis and Inference* (Yadolah Dodge, Ed.), North-Holland, Amsterdam pp. 283–296, 1989.

54

Markov Chains

Beatrice Meini
Università di Pisa

Markov chains are encountered in several applications arising in different contexts, and model many real problems which evolve in time. Throughout, we denote by $\mathrm{P}[X = j]$ the probability that the random variable X takes the value j, and by $\mathrm{P}[X = j|Y = i]$ the conditional probability that X takes the value j, given that the random variable Y takes the value i. Moreover, we denote by $\mathrm{E}[X]$ the expected value of the random variable X and by $\mathrm{E}[X|A]$ the conditional expectation of X, given the event A.

54.1 Basic Concepts

Definitions:

Given a denumerable set E, a **discrete stochastic process** on E is a family $\{X_t : t \in T\}$ of random variables X_t indexed by some denumerable set T and with values in E, i.e., $X_t \in E$ for all $t \in T$. Here, E is the **state space**, and T is the **time space**.

The discrete stochastic process $\{X_n : n \in \mathbb{N}\}$ is a **Markov chain** if

$$\mathrm{P}[X_{n+1} = j_{n+1}|X_0 = j_0, X_1 = j_1, \ldots, X_n = j_n] = \mathrm{P}[X_{n+1} = j_{n+1}|X_n = j_n], \qquad (54.1)$$

for any $(n+2)$-tuple of states $\{j_0, \ldots, j_{n+1}\} \in E$, and for all time $n \in \mathbb{N}$.

A Markov chain $\{X_n : n \in \mathbb{N}\}$ is **homogeneous** if

$$\mathrm{P}[X_{n+1} = j|X_n = i] = \mathrm{P}[X_1 = j|X_0 = i], \qquad (54.2)$$

for all states $i, j \in E$ and for all time $n \in \mathbb{N}$.

Given a homogeneous Markov chain $\{X_n : n \in \mathbb{N}\}$ we define the **transition matrix** of the Markov chain to be the matrix $P = [p_{i,j}]_{i,j \in E}$ such that

$$p_{i,j} = \mathrm{P}[X_1 = j|X_0 = i], \qquad \text{for all } i, j \text{ in } E.$$

Throughout, unless differently specified, for the sake of notational simplicity we will indicate with the term **Markov chain** a homogeneous Markov chain.

A **finite Markov chain** is a Markov chain with finite space state E; an **infinite Markov chain** is a Markov chain with infinite space state E.

A row vector $\boldsymbol{\pi} = (\pi_i)_{i\in E}$ such that

$$\boldsymbol{\pi} P = \boldsymbol{\pi} \tag{54.3}$$

is an **invariant vector.**

If the invariant vector $\boldsymbol{\pi}$ is such that

$$\pi_i \geq 0 \text{ for all } i, \quad \text{and} \quad \sum_{i\in E} \pi_i = 1, \tag{54.4}$$

then $\boldsymbol{\pi}$ is an **invariant probability vector**, or **stationary distribution.** The definition of stationary distribution extends the definition given in Section 9.4 to the case of an infinite matrix P.

Facts:

1. The Markov property in Equation (54.1) means that the state X_n of the system at time n is sufficient to determine which state might be occupied at time $n+1$, and the past history $X_0, X_1, \ldots, X_{n-1}$ does not influence the future state at time $n+1$.
2. In a homogeneous Markov chain, the property in Equation (54.2) means that the laws which govern the evolution of the system are independent of the time n; therefore, the evolution of the Markov chain is ruled by the transition matrix P, whose (i, j)th entry represents the probability to change from state i to state j in one time unit.
3. The number of rows and columns of the transition matrix P is equal to the cardinality of E. In particular, if the set E is finite, P is a finite matrix (see Examples 1 and 5); if the set E is infinite, P is an infinite matrix, i.e., a matrix with an infinite number of rows and columns (see Examples 2–4).
4. The matrix P is row stochastic, i.e., it is a matrix with nonnegative entries such that $\sum_{j\in E} p_{i,j} = 1$ for any $i \in E$, i.e., the sum of the entries on each row is equal to 1.
5. If $|E| < \infty$, the matrix P has spectral radius 1. Moreover, the vector $\mathbf{1}_{|E|}$ is a right eigenvector of P corresponding to the eigenvalue 1; any nonzero invariant vector is a left eigenvector of P corresponding to the eigenvalue 1; the invariant probability vector $\boldsymbol{\pi}$ is a nonnegative left eigenvector of P corresponding to the eigenvalue 1, normalized so that $\boldsymbol{\pi}\mathbf{1}_{|E|} = 1$.
6. In the analysis of Markov chains, we may encounter matrix products $A = BC$, where $B = [b_{i,j}]_{i,j\in E}$ and $C = [c_{i,j}]_{i,j\in E}$ are nonnegative matrices such that $\sum_{j\in E} b_{i,j} \leq 1$ and $\sum_{j\in E} c_{i,j} \leq 1$ for any $i \in E$ (see, for instance, Fact 7 below). The (i, j)th entry of A, given by $a_{i,j} = \sum_{h\in E} b_{i,h}c_{h,j}$, is well defined also if the set E is infinite, since $0 \leq \sum_{h\in E} b_{i,h}c_{h,j} \leq \sum_{h\in E} b_{i,h} \leq 1$. Moreover, A is a nonnegative matrix such that $\sum_{j\in E} a_{i,j} \leq 1$ for any $i \in E$. Indeed, $\sum_{j\in E} a_{i,j} = \sum_{j\in E}\sum_{h\in E} b_{i,h}c_{h,j} = \sum_{h\in E} b_{i,h} \sum_{j\in E} c_{h,j} \leq \sum_{h\in E} b_{i,h} \leq 1$. Similarly, the product $\boldsymbol{v} = \boldsymbol{u}B$, where $\boldsymbol{u} = (u_i)_{i\in E}$ is a nonnegative row vector such that $\sum_{i\in E} u_i \leq 1$, is well defined also in the case where E is infinite; moreover, $\boldsymbol{v} = (v_i)_{i\in E}$ is a nonnegative vector such that $\sum_{i\in E} v_i \leq 1$.
7. [Nor99, Theorem 1.1.3] The dynamic behavior of a homogeneous Markov chain is completely characterized by the transition matrix P:

$$\mathrm{P}[X_{n+k} = j | X_n = i] = (P^k)_{i,j},$$

for all times $n \geq 0$, all intervals of time $k \geq 0$, and all pairs of states i and j in E.
8. In addition to the system dynamics, one must choose the starting point X_0. Let $\boldsymbol{\pi}^{(0)} = (\pi_i^{(0)})_{i\in E}$ be a probability distribution on E, i.e., a nonnegative row vector such that the sum of its components is equal to one. Assume that $\pi_i^{(0)} = \mathrm{P}[X_0 = i]$, and define the row vector $\boldsymbol{\pi}^{(n)} = (\pi_i^{(n)})_{i\in E}$ to be the probability vector of the Markov chain at time $n \geq 1$, that is, $\pi_i^{(n)} = \mathrm{P}[X_n = i | X_0]$.

Then

$$\pi^{(n+1)} = \pi^{(n)} P, \qquad n \geq 0, \tag{54.5}$$

$$\pi^{(n)} = \pi^{(0)} P^n, \qquad n \geq 0. \tag{54.6}$$

9. If the initial distribution $\pi^{(0)}$ coincides with the invariant probability vector π, then $\pi^{(n)} = \pi$ for any $n \geq 0$.
10. In certain applications (see for instance, Example 5) we are interested in the asymptotic behavior of the Markov chain. In particular, we would like to compute, if it exists, the vector $\lim_{n\to\infty} \pi^{(n)}$. From Equation (54.5) of Fact 8 one deduces that, if such limit exists, it coincides with the invariant probability vector, i.e., $\lim_{n\to\infty} \pi^{(n)} = \pi$.

Examples:

1. *Random walk on* $\{0, 1, \ldots, k\}$: Consider a particle which moves on the interval $[0, k]$ in unit steps at integer instants of time; let $X_n \in \{0, 1, \ldots, k\}$, $n \geq 0$, be the position of the particle at time n and let $0 < p < 1$. Assume that, if the particle is in the open interval $(0, k)$, at the next unit time it will move to the right with probability p and to the left with probability $q = 1 - p$; if the particle is in position 0, it will move to the right with probability 1; if the particle is in position k, it will move to the left with probability 1. Clearly, the discrete stochastic process $\{X_n : n \in \mathbb{N}\}$ is a homogeneous Markov chain with space state the set $E = \{0, 1, \ldots, k\}$. The transition matrix is the $(k+1) \times (k+1)$ matrix $P = [p_{i,j}]_{i,j=0,\ldots,k}$, given by

$$P = \begin{bmatrix} 0 & 1 & 0 & \ldots & 0 \\ q & 0 & p & \ddots & \vdots \\ 0 & \ddots & \ddots & \ddots & 0 \\ \vdots & \ddots & q & 0 & p \\ 0 & \ldots & 0 & 1 & 0 \end{bmatrix}.$$

From Equation (54.6) of Fact 8 one has that, if the particle is in position 0 at time 0, the probability vector at time $n = 10$ is the vector

$$[\pi_0^{(10)}, \pi_1^{(10)}, \ldots, \pi_k^{(10)}] = [1, 0, \ldots, 0] P^{10}.$$

2. *Random walk on* $\mathbb{N}$: If we allow the particle to move on $\mathbb{N}$, we have a homogeneous Markov chain with state space the set $E = \mathbb{N}$. The transition matrix is semi-infinite and is given by

$$P = \begin{bmatrix} 0 & 1 & 0 & 0 & \ldots \\ q & 0 & p & 0 & \ddots \\ 0 & q & 0 & p & \ddots \\ 0 & 0 & q & 0 & \ddots \\ \vdots & \ddots & \ddots & \ddots & \ddots \end{bmatrix}.$$

3. *Random walk on* $\mathbb{Z}$: If we allow the particle to move on $\mathbb{Z}$, we still have a homogeneous Markov chain, with state space the set $E = \mathbb{Z}$. The transition matrix is bi-infinite and is given by

$$P = \begin{bmatrix} \ddots & \ddots & \ddots & \ddots & \ddots \\ \ddots & 0 & p & 0 & \ddots \\ \ddots & q & 0 & p & \ddots \\ \ddots & 0 & q & 0 & \ddots \\ \ddots & \ddots & \ddots & \ddots & \ddots \end{bmatrix}.$$

4. *A simple queueing system* [BLM05, Example 1.3]: Simple queues consist of one server which attends to one customer at a time, in order of their arrivals. We assume that time is discretized into intervals of length one, that a random number of customers join the system during each interval, that customers do not leave the queue, and that the server removes one customer from the queue at the end of each interval, if there is any. Defining α_n as the number of new arrivals during the interval $[n-1, n)$ and X_n as the number of customers in the system at time n, we have

$$X_{n+1} = \begin{cases} X_n + \alpha_{n+1} - 1 & \text{if } X_n + \alpha_{n+1} \geq 1 \\ 0 & \text{if } X_n + \alpha_{n+1} = 0. \end{cases}$$

If $\{\alpha_n\}$ is a collection of independent random variables, then X_{n+1} is conditionally independent of $X_0, \ldots, X_{n-1}$ if X_n is known. If, in addition, the α_n's are identically distributed, then $\{X_n\}$ is homogeneous. The state space is $\mathbb{N}$ and the transition matrix is

$$P = \begin{bmatrix} q_0 + q_1 & q_2 & q_3 & q_4 & \cdots \\ q_0 & q_1 & q_2 & q_3 & \ddots \\ & q_0 & q_1 & q_2 & \ddots \\ 0 & & \ddots & \ddots & \ddots \end{bmatrix},$$

where q_i is the probability $\mathrm{P}[\alpha = i]$ that i new customers join the queue during a unit time interval, α denoting any of the identically distributed random variables α_n. Markov chains having transition matrix of the form

$$P = \begin{bmatrix} B_1 & B_2 & B_3 & B_4 & \cdots \\ A_0 & A_1 & A_2 & A_3 & \ddots \\ & A_0 & A_1 & A_2 & \ddots \\ 0 & & \ddots & \ddots & \ddots \end{bmatrix},$$

where A_i, B_{i+1}, $i \geq 0$, are nonnegative $k \times k$ matrices, are called M/G/1-type Markov chains, and model a large variety of queuing problems. (See [Neu89], [LR99], and [BLM05].)

5. *Search engines* (see Section 63.5, [PBM99], [ANT02] and [Mol04, Section 2.11]): PageRank is used by Google to sort, in order of relevance, the pages on the Web that match the query of the user. From the Web page *Google Technology* at `http://www.google.com/technology/`: "The heart of our software is PageRank^{TM}, a system for ranking Web pages developed by our founders Larry Page and Sergey Brin at Stanford University. And while we have dozens of engineers working to improve every aspect of Google on a daily basis, PageRank continues to provide the basis for all

of our Web search tools." Surfing on the Web is seen as a random walk, where either one starts from a Web page and goes from one page to the next page by randomly following a link (if any), or one simply chooses a random page from the Web. Let E be the set of Web pages that can be reached by following a sequence of hyperlinks starting from a given Web page. If k is the number of Web pages, i.e., $k = |E|$, the connectivity matrix is defined as the $k \times k$ matrix $G = [g_{i,j}]$ such that $g_{i,j} = 1$ if there is a hyperlink from page i to page j, and zero otherwise. Let $r_i = \sum_j g_{i,j}$ and $c_i = \sum_j g_{j,i}$ be the row and column sums of G; the quantities r_i and c_i are called the out-degree and the in-degree, respectively, of the page i. Let $0 < q < 1$ and let $P = [p_{i,j}]$ be the $k \times k$ stochastic matrix such that $p_{i,j} = qg_{i,j}/r_i + (1-q)/k, i, j = 1, \ldots, k$. The value q is the probability that the random walk on E follows a link and, therefore, $1 - q$ is the probability that an arbitrary page is chosen. The matrix P is the transition matrix of the Markov chain that models the random walk on E. The importance of a Web page is related to the probability to reach such page during this randow walk as the time tends to infinity. Therefore, Google's PageRank is determined by the invariant probability vector $\boldsymbol{\pi}$ of P: The larger the value of an entry π_i of $\boldsymbol{\pi}$, the higher the relevance of the ith Web page in the set E.

54.2 Irreducible Classes

Some of the Definitions and Facts given in this section extend the corresponding Definitions and Facts of Chapter 9 and Chapter 29. Indeed, in these chapters, it is assumed that the matrices have a finite size, while in the framework of Markov chains we may encounter infinite matrices.

Definitions:

The **transition graph** of a Markov chain with transition matrix P is the digraph of P, $\Gamma(P)$. That is, the digraph defined as follows: To each state in E there corresponds a vertex of the digraph and one defines a directed arc from vertex i to vertex j for each pair of states such that $p_{i,j} > 0$. More information about digraphs can be found in Chapter 9 and Chapter 29.

A **closed walk** in a digraph is a walk in which the first vertex equals the last vertex.

State i **leads** to j (or i **has access** to j) if there is a walk from i to j in the transition graph. States i and j **communicate** if i leads to j and j leads to i.

A Markov chain is called **irreducible** if the transition graph is strongly connected, i.e., if all the states communicate. A Markov chain is called **reducible** if it is not irreducible.

The strongly connected components of the transition graph are the **communicating classes of states**, or of the Markov chain. Communicating classes are also called **access equivalence classes** or **irreducible classes**.

A communicating class C is a **final class** if for every state i in C, there is no state j outside of C such that i leads to j. If, on the contrary, there is a state in C that leads to some state outside of C, the class is a **passage class**. A single state that forms a final class by itself is **absorbing**. In Section 9.4, passage classes are called transient classes, and final classes are called ergodic classes; in fact, transient and ergodic states of Markov chains will be introduced in Section 54.3, and for finite Markov chains the states in a passage class are transient, and the states in a final class are ergodic, cf. Section 54.4.

A state i is **periodic** with period $\delta \geq 2$ if all closed walks through i in the transition graph have a length that is a multiple of δ. A state i is **aperiodic** if it is not periodic.

A Markov chain is **periodic** with period δ if all states are periodic and have the same period δ.

Facts:

1. A Markov chain is irreducible if and only if the transition matrix P is irreducible, i.e., if P is not permutation similar to a block triangular matrix, cf. Section 9.2, Section 27.1.
2. If we adopt the convention that each state communicates with itself, then the relation *communicates* is an equivalence relation and the communicating classes are the equivalence classes of this relation, cf. Section 9.1.

3. A Markov chain is irreducible if and only if the states form one single communicating class, cf. Section 9.1, Section 9.2.
4. If a Markov chain with transition matrix P has $K \geq 2$ communicating classes, denoted by $C_1, C_2, \ldots, C_K$, then the states may be permuted so that the transition matrix $P' = \Pi P \Pi^T$ associated with the permuted states is block triangular:

$$P' = \begin{bmatrix} P_{1,1} & P_{1,2} & \ldots & P_{1,K} \\ & P_{2,2} & \ddots & \vdots \\ & & \ddots & P_{K-1,K} \\ 0 & & & P_{K,K} \end{bmatrix}$$

where $P_{i,j}$ is the submatrix of transition probabilities from the states of C_i to C_j, the diagonal blocks are irreducible square matrices, and Π is the permutation matrix associated with the rearrangement, cf. Section 9.2.
5. [Çin75, Theorem (3.16), Chap. 5] Periodicity is a class property and all states in a communicating class have the same period. Thus, for irreducible Markov chains, either all states are aperiodic, or all have the same period δ, which we may call the period of the Markov chain itself.

Examples:

1. Figure 54.1 is the transition graph associated with the Markov chain on $E = \{1, 2, 3\}$ with transition matrix

$$P = \begin{bmatrix} 1 & 0 & 0 \\ \frac{1}{3} & \frac{1}{3} & \frac{1}{3} \\ \frac{1}{4} & \frac{1}{2} & \frac{1}{4} \end{bmatrix}.$$

The two sets $C_1 = \{1\}$ and $C_2 = \{2, 3\}$ are two communicating classes. C_2 is a passage class, while C_1 is a final class, therefore state 1 is absorbing.
2. Figure 54.2 is the transition graph associated with the Markov chain on $E = \{k \in \mathbb{N} : k \geq 1\}$ with transition matrix having following structure:

$$P = \begin{bmatrix} 0 & * & & & & & 0 \\ 0 & 0 & * & & & & \\ * & 0 & 0 & * & & & \\ 0 & 0 & 0 & 0 & * & & \\ 0 & 0 & * & 0 & 0 & * & \\ 0 & 0 & 0 & 0 & 0 & 0 & \ddots \\ 0 & 0 & 0 & 0 & * & 0 & \ddots \\ \vdots & \ddots & \ddots & \ddots & \ddots & \ddots & \ddots \end{bmatrix},$$

FIGURE 54.1 Transition graph of the Markov chain of Example 1.

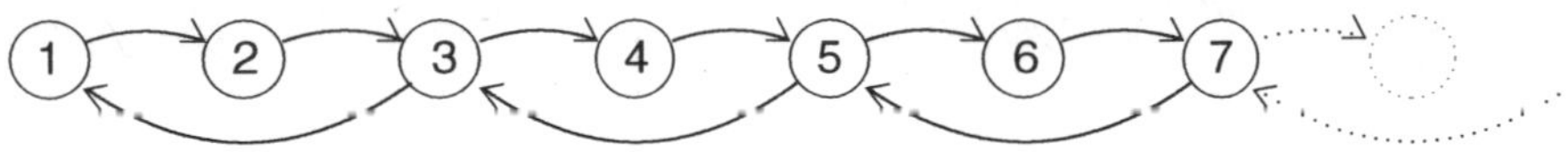

FIGURE 54.2 Graph of an infinite irreducible periodic Markov chain of period 3.

where "$*$" denotes a nonzero element. This is an example of a periodic irreducible Markov chain of period 3.

54.3 Classification of the States

Definitions:

Let T_j be the time of the first visit to j, without taking into account the state at time 0, i.e.,

$$T_j = \min\{n \geq 1 : X_n = j\}.$$

Define

$$f_j = \mathrm{P}[T_j < \infty | X_0 = j],$$

that is, f_j is the probability that, starting from j, the Markov chain returns to j in a finite time.

A state $j \in E$ is **transient** if $f_j < 1$.

A state $j \in E$ is **recurrent** if $f_j = 1$. A recurrent state $j \in E$ is **positive recurrent** if the expected return time $\mathrm{E}[T_j | X_0 = j]$ is finite; it is **null recurrent** if the expected return time $\mathrm{E}[T_j | X_0 = j]$ is infinite. Positive recurrent states are also called **ergodic**.

A Markov chain is **positive/null recurrent** or **transient** if all its states are positive/null recurrent or transient, respectively.

A **regular Markov chain** is a positive recurrent Markov chain that is aperiodic.

The matrix $R = \sum_{n=0}^{\infty} P^n$ is the **potential matrix** of the Markov chain.

Facts:

1. Transient states may be visited only a finite number of times by the Markov chain. On the contrary, once the Markov chain has visited one recurrent state, it will return to it over and over again; if j is null recurrent the expected time between two successive visits to j is infinite.
2. [Çin75, Cor. (2.13), Chap. 5] For the potential matrix $R = [r_{i,j}]$ one has $r_{j,j} = 1/(1 - f_j)$, where we set $1/0 = \infty$.
3. [Çin75, Sec. 3, Chap. 5] The (j, j)th entry of R is the expected number of returns of the Markov chain to state j. A state j is recurrent if and only if $r_{j,j} = \infty$. A state j is transient if and only if $r_{j,j} < \infty$.
4. [Nor99, Theorems 1.5.4, 1.5.5, 1.7.7] The nature of a state is a class property. More specifically, the states in a passage class are transient; in a final class the states are either all positive recurrent, or all null recurrent, or all transient.
5. [Nor99, Theorem 1.5.6] If a final class contains a finite number of states only, then all its states are positive recurrent.
6. From Facts 4 and 5 one has that, for a communicating class with a finite number of states, either all the states are transient or all the states are positive recurrent.
7. From Fact 4 above, if a Markov chain is irreducible, the states are either all positive recurrent, or all null recurrent, or all transient. Therefore, an irreducible Markov chain is either positive recurrent or null recurrent or transient.

8. [Nor99, Secs. 1.7 and 1.8] Assume that the Markov chain is irreducible. The Markov chain is positive recurrent if and only if there exists a strictly positive invariant probability vector, that is, a row vector $\pi = (\pi_i)_{i \in E}$ such that $\pi_i > 0$ that satisfies Equations (54.3), (54.4) of Section 54.1. The invariant vector π is unique among nonnegative vectors, up to a multiplicative constant.
9. [Nor99, Secs. 1.7 and 1.8], [BLM05, Sec. 1.5] If the Markov chain is irreducible and null recurrent, there exists a strictly positive invariant vector, unique up to a multiplicative constant, such that the sum of its elements is not finite. Thus, there always exists an invariant vector for the transition matrix of a recurrent Markov chain. Some transient Markov chains also have an invariant vector (with infinite sum of the entries, like in the null recurrent case) but some do not.
10. [Çin75, Theorem (3.2), Chap. 5], [Nor99, Secs. 1.7 and 1.8] If j is a transient or null recurrent state, then for any $i \in E$, $\lim_{n\to\infty}(P^n)_{i,j} = 0$. If j is a positive recurrent and aperiodic state, then $\lim_{n\to\infty}(P^n)_{j,j} > 0$. If j is periodic with period δ, then $\lim_{n\to\infty}(P^{n\delta})_{j,j} > 0$.
11. [Nor99, Secs. 1.7 and 1.8] Assume that the Markov chain is irreducible, aperiodic, and positive recurrent. Then $\lim_{n\to\infty}(P^n)_{i,j} = \pi_j > 0$ for all j, independently of i, where $\pi = (\pi_i)_{i \in E}$ is the stationary distribution.
12. Facts 3, 4, 6, and 10 provide a criterium to classify the states. First identify the communicating classes. If a communicating class contains a finite number of states, the states are all positive recurrent if the communicating class is final; the states are all transient if the communicating class is a passage class. If a communicating class has infinitely many states, we apply Fact 3 to determine if they are recurrent or transient; for recurrent states we use Fact 10 to determine if they are null or positive recurrent.

Examples:

1. Let P be the transition matrix of Example 1 of Section 54.2. Observe that with positive probability the Markov chain moves from state 2 or 3 to state 1, and when the Markov chain will be in state 1, it will remain there forever. Indeed, according to Facts 4 and 5, states 2 and 3 are transient since they belong to a passage class, and state 1 is positive recurrent since it is absorbing. Moreover, if we partition the matrix P into the 2×2 block matrix

$$P = \begin{bmatrix} 1 & 0 \\ A & B \end{bmatrix},$$

where

$$A = \begin{bmatrix} \frac{1}{3} \\ \frac{1}{4} \end{bmatrix}, \quad B = \begin{bmatrix} \frac{1}{3} & \frac{1}{3} \\ \frac{1}{2} & \frac{1}{4} \end{bmatrix},$$

we may easily observe that

$$P^n = \begin{bmatrix} 1 & 0 \\ \sum_{i=0}^{n-1} B^i A & B^n \end{bmatrix}.$$

Since $\|B\|_1 = \frac{5}{6}$, then $\rho(B) < 1$, and therefore $\lim_{n\to\infty} B^n = 0$ and $\sum_{n=0}^{\infty} B^n = (I - B)^{-1}$. A simple computation leads to

$$\lim_{n\to\infty} P^n = \begin{bmatrix} 1 & 0 & 0 \\ 1 & 0 & 0 \\ 1 & 0 & 0 \end{bmatrix}, \quad R = \sum_{n=0}^{\infty} P^n = \begin{bmatrix} \infty & 0 & 0 \\ \infty & \frac{9}{4} & 1 \\ \infty & \frac{3}{2} & 2 \end{bmatrix},$$

in accordance with Facts 3 and 10.

2. [BLM05, Ex. 1.19] The transition matrix

$$P = \begin{bmatrix} 0 & 1 & & & 0 \\ \frac{1}{2} & 0 & \frac{1}{2} & & \\ & \frac{1}{2} & 0 & \frac{1}{2} & \\ & & & & \\ 0 & & \ddots & \ddots & \ddots \end{bmatrix}$$

is irreducible, and $\pi = [\frac{1}{2}, 1, 1, \ldots]$ is an invariant vector. The vector π has "infinite" mass, that is, the sum of its components is infinite. In fact, the Markov chain is actually null recurrent (see [BLM05, Sec. 1.5]).

3. [BLM05, Ex. 1.20] For the transition matrix

$$P = \begin{bmatrix} 0 & 1 & & & 0 \\ \frac{1}{4} & 0 & \frac{3}{4} & & \\ & \frac{1}{4} & 0 & \frac{3}{4} & \\ & & & & \\ 0 & & \ddots & \ddots & \ddots \end{bmatrix}$$

one has $\pi P = \pi$ with $\pi = [1, 4, 12, 36, 108, \ldots]$. The vector π has unbounded elements. In this case the Markov chain is transient.

4. [BLM05, Ex. 1.21] For the transition matrix

$$P = \begin{bmatrix} 0 & 1 & & & 0 \\ \frac{3}{4} & 0 & \frac{1}{4} & & \\ & \frac{3}{4} & 0 & \frac{1}{4} & \\ & & & & \\ 0 & & \ddots & \ddots & \ddots \end{bmatrix}$$

one has $\pi P = \pi$ with $\pi = \frac{4}{3}[\frac{1}{4}, \frac{1}{3}, \frac{1}{9}, \frac{1}{27}, \ldots]$ and $\sum_i \pi_i = 1$. In this case the Markov chain is positive recurrent by Fact 8.

54.4 Finite Markov Chains

The transition matrix of a finite Markov chain is a finite dimensional stochastic matrix; the reader is advised to consult Section 9.4 for properties of finite stochastic matrices.

Definitions:

A $k \times k$ matrix A is **weakly cyclic of index** δ if there exists a permutation matrix Π such that $A' = \Pi A \Pi^T$ has the block form

$$A' = \begin{bmatrix} 0 & 0 & \ldots & 0 & A_{1,\delta} \\ A_{2,1} & 0 & \ddots & & 0 \\ 0 & A_{3,2} & \ddots & \ddots & \vdots \\ \vdots & \ddots & \ddots & 0 & 0 \\ 0 & \ldots & 0 & A_{\delta,\delta-1} & 0 \end{bmatrix},$$

where the zero diagonal blocks are square.

Facts:

1. If P is the transition matrix of a finite Markov chain, there exists an invariant probability vector π. If P is irreducible, the vector π is strictly positive and unique, cf. Section 9.4.
2. For a finite Markov chain no state is null recurrent, and not all states are transient. The states belonging to a final class are positive recurrent, and the states belonging to a passage class are transient, cf. Section 9.4.
3. [BP94, Theorem (3.9), Chap. 8] Let P be the transition matrix of a finite Markov chain. The Markov chain is
   (a) Positive recurrent if and only if P is irreducible.
   (b) Regular if and only if P is primitive.
   (c) Periodic if and only of P is irreducible and periodic.
4. [BP94, Theorem (3.16), Chap. 8] Let P be the transition matrix of a finite irreducible Markov chain. Then the Markov chain is periodic if and only if P is a weakly cyclic matrix.
5. If P is the transition matrix of a regular Markov chain, then there exists $\lim_{n\to\infty} P^n = \mathbf{1}\pi$, where π is the probability invariant vector. If the Markov chain is periodic, the sequence $\{P^n\}_{n\geq 0}$ is bounded, but not convergent.
6. [Mey89] Assume that $E = \{1, 2, \ldots, k\}$ and that P is irreducible. Let $1 \leq m \leq k-1$ and $\alpha = \{1, \ldots, m\}$, $\beta = \{m+1, \ldots, k\}$. Then
   (a) The matrix $I - P[\alpha]$ is an M-matrix.
   (b) The matrix $P' = P[\alpha] + P[\alpha,\beta](I - P[\beta])^{-1}P[\beta,\alpha]$, such that $I - P'$ is the Schur complement of $I - P[\beta]$ in the matrix

$$\begin{bmatrix} I - P[\alpha] & -P[\alpha,\beta] \\ -P[\beta,\alpha] & I - P[\beta] \end{bmatrix},$$

   is a stochastic irreducible matrix.
   (c) If we partition the invariant probability vector π as $\pi = [\pi_\alpha, \pi_\beta]$, with $\pi_\alpha = (\pi_i)_{i\in\alpha}$ and $\pi_\beta = (\pi_i)_{i\in\beta}$, we have $\pi_\alpha P' = \pi_\alpha$, $\pi_\beta = \pi_\alpha P[\alpha,\beta](I - P[\beta])^{-1}$.

Examples:

1. The matrix

$$P = \begin{bmatrix} 0 & 1 \\ 1 & 0 \end{bmatrix}$$

   is the transition matrix of a periodic Markov chain of period 2. In fact, P is an irreducible matrix of period 2.
2. The Markov chain with transition matrix

$$P = \begin{bmatrix} \frac{1}{5} & \frac{1}{5} \\ 1 & 0 \end{bmatrix}$$

   is regular. In fact, $P^2 > 0$, i.e., P is primitive.
3. Let

$$P = \begin{bmatrix} \frac{1}{2} & \frac{1}{2} & 0 & 0 \\ \frac{1}{4} & \frac{1}{2} & \frac{1}{4} & 0 \\ \frac{1}{3} & 0 & \frac{1}{3} & \frac{1}{3} \\ 0 & 0 & \frac{1}{2} & \frac{1}{2} \end{bmatrix}$$

be the transition matrix of a Markov chain. The matrix P is irreducible and aperiodic, therefore the Markov chain is regular. The vector $\pi = \frac{1}{13}[4, 4, 3, 2]$ is the stationary distribution. According to Fact 5, one has

$$\lim_{n\to\infty} P^n = \frac{1}{13}\begin{bmatrix}1\\1\\1\\1\end{bmatrix}\begin{bmatrix}4 & 4 & 3 & 2\end{bmatrix}.$$

54.5 Censoring

Definitions:

Partition the state space E into two disjoint subsets, α and β, and denote by $\{t_0, t_1, t_2, \ldots\}$ the time when the Markov chain visits the set α:

$$t_0 = \min\{n \geq 0 : X_n \in \alpha\}, \qquad t_{k+1} = \min\{n \geq t_k + 1 : X_n \in \alpha\},$$

for $k \geq 0$. The **censored process** restricted to the subset α is the sequence $\{Y_n\}_{n\geq 0}$, where $Y_n = X_{t_n}$, of successive states visited by the Markov chain in α.

Facts:

1. [BLM05, Sec. 1.6] The censored process $\{Y_n\}$ is a Markov chain.
2. [BLM05, Sec. 1.6] Arrange the states so that the transition matrix can be partitioned as

$$P = \begin{bmatrix} P[\alpha] & P[\alpha,\beta] \\ P[\beta,\alpha] & P[\beta] \end{bmatrix}.$$

   Then the transition matrix of $\{Y_n\}$ is

$$P' = P[\alpha] + \sum_{n=0}^{+\infty}(P[\alpha,\beta]P[\beta]^n P[\beta,\alpha])$$

   provided that $\sum_{n=0}^{+\infty}(P[\alpha,\beta]P[\beta]^n P[\beta,\alpha])$ is convergent. If the series $S' = \sum_{n=0}^{+\infty} P[\beta]^n$ is convergent, then we may rewrite P' as $P' = P[\alpha] + P[\alpha,\beta]S'P[\beta,\alpha]$.
3. [BLM05, Theorem 1.23] Assume that the Markov chain is irreducible and positive recurrent. Partition the stationary distribution π as $\pi = [\pi_\alpha, \pi_\beta]$, with $\pi_\alpha = (\pi_i)_{i\in\alpha}$ and $\pi_\beta = (\pi_i)_{i\in\beta}$. Then one has $\pi_\alpha P' = \pi_\alpha$, $\pi_\beta = \pi_\alpha P[\alpha,\beta]S'$, where P' and S' are defined in Fact 2.
4. [BLM05, Secs. 1.6, 3.5, 4.5] In the case of finite Markov chains, censoring is equivalent to Schur complementation (see Fact 6 of Section 54.4). Censoring is at the basis of several numerical methods for computing the invariant probability vector π; indeed, from Fact 3, if the invariant probability vector π_α associated with the censored process is available, the vector π_β can be easily computed from π_α. A smart choice of the sets α and β can lead to an efficient computation of the stationary distribution. As an example of this fact, Ramaswami's recursive formula for computing the vector π associated with an M/G/1-type Markov chain is based on successive censorings, where the set A is finite, and $\beta = \mathbb{N} - \alpha$.

54.6 Numerical Methods

Definitions:

Let A be a $k \times k$ matrix. A splitting $A = M - N$, where $\det M \neq 0$, is a **regular splitting** if $M^{-1} \geq 0$ and $N \geq 0$. A regular splitting is said to be **semiconvergent** if the matrix $M^{-1}N$ is semiconvergent.

Facts:

1. The main computational problem in Markov chains is the computation of the invariant probability vector $\boldsymbol{\pi}$. If the space state E is finite and the Markov chain is irreducible, classical techniques for solving linear systems can be adapted and specialized to this purpose. We refer the reader to the book [Ste94] for a comprehensive treatment on these numerical methods; in facts below we analyze the methods based on LU factorization and on regular splittings of M_0-matrices. If the space state E is infinite, general numerical methods for computing $\boldsymbol{\pi}$ can be hardly designed. Usually, when the state space is infinite, the matrix P has some structures which are specific of the real problem modeled by the Markov chain, and numerical methods for computing $\boldsymbol{\pi}$ which exploit the structure of P can be designed. Quite common structures arising in queueing problems are the (block) tridiagonal structure, the (block) Hessenberg structure, and the (block) Toeplitz structure (see, for instance, Example 4 Section 54.1). We refer the reader to the book [BLM05] for a treatment on the numerical solution of Markov chains, where the matrix P is infinite and structured.
2. [BP94, Cor. (4.17), Chap. 6], [Ste94, Sec. 2.3] If P is a $k \times k$ irreducible stochastic matrix, then the matrix $I - P$ has a unique LU factorization $I - P = LU$, where L is a lower triangular matrix with unit diagonal entries and U is upper triangular. Moreover, L and U are M_0-matrices, U is singular, $U\mathbf{1}_k = 0$, and $(U)_{k,k} = 0$.
3. [Ste94, Sec. 2.5] The computation of the LU factorization of an M_0-matrix by means of Gaussian elimination involves additions of nonnegative numbers in the computation of the off-diagonal entries of L and U, and subtractions of nonnegative numbers in the computation of the diagonal entries of U. Therefore, numerical cancellation cannot occur in the computation of the off-diagonal entries of L and U. In order to avoid possible cancellation errors in computing the diagonal entries of U, Grassmann, Taksar, and Heyman have introduced in [GTH85] a simple trick, which fully exploits the property that U is an M_0-matrix such that $U\mathbf{1}_k = 0$. At the general step of the elimination procedure, the diagonal entries are not updated by means of the classical elimination formulas, but are computed as minus the sum of the off-diagonal entries. Details are given in Algorithm 1.
4. If $I - P = LU$ is the LU factorization of $I - P$, the invariant probability vector $\boldsymbol{\pi}$ is computed by solving the two triangular systems $\boldsymbol{y}L = \boldsymbol{x}$ and $\boldsymbol{x}U = 0$, where $\boldsymbol{x}$ and $\boldsymbol{y}$ are row vectors, and then by normalizing the solution $\boldsymbol{y}$. From Fact 2, the nontrivial solution of the latter system is $\boldsymbol{x} = \alpha \boldsymbol{e}_k^T$, for any $\alpha \neq 0$. Therefore, if we choose $\alpha = 1$, the vector $\boldsymbol{\pi}$ can be simply computed by solving the system $\boldsymbol{y}L = \boldsymbol{e}_k^T$ by means of back substitution, and then by setting $\boldsymbol{\pi} = (1/\boldsymbol{y}\mathbf{1})\boldsymbol{y}$, as described in Algorithm 2.
5. [BP94, Cor. (4.17), Chap. 6] If P is a $k \times k$ irreducible stochastic matrix, the matrix $I - P^T$ has a unique LU factorization $I - P^T = LU$, where L is a lower triangular matrix with unit diagonal entries, and U is upper triangular. Moreover, both L and U are M_0-matrices, U is singular, and $(U)_{k,k} = 0$. Since L is nonsingular, the invariant probability vector can be computed by calculating a nonnegative solution of the system $U\boldsymbol{y} = 0$ by means of back substitution, and by setting $\boldsymbol{\pi} = (1/\|\boldsymbol{y}\|_1)\boldsymbol{y}^T$.
6. If P is a $k \times k$ irreducible stochastic matrix, and if $\hat{P}$ is the $(k-1) \times (k-1)$ matrix obtained by removing the ith row and the ith column of P, where $i \in \{1, \ldots, k\}$, then the matrices $I - \hat{P}^T$ and $I - \hat{P}$ have a unique LU factorization, where both the factors L and U are M-matrices. Therefore, if $i = k$ and if the matrix $I - P^T$ is partitioned as
$$I - P^T = \begin{bmatrix} I - \hat{P}^T & \boldsymbol{a} \\ \boldsymbol{b}^T & c \end{bmatrix},$$

one finds that the vector $\hat{\pi} = [\pi_1, \ldots, \pi_{k-1}]^T$ solves the nonsingular system $(I - \hat{P}^T)\hat{\pi} = -\pi_k \boldsymbol{a}$. Therefore, the vector $\boldsymbol{\pi}$ can be computed by solving the system $(I - \hat{P}^T)\boldsymbol{y} = -\boldsymbol{a}$, say by means of LU factorization, and then by setting $\boldsymbol{\pi} = [\boldsymbol{y}^T, 1]/(1 + \|\boldsymbol{y}\|_1)$.

7. [Ste94, Sec. 3.6] Let P be a finite irreducible stochastic matrix, and let $I - P^T = M - N$ be a regular splitting of $I - P^T$. Then the matrix $H = M^{-1}N$ has spectral radius 1 and 1 is a simple eigenvalue of H.
8. [Ste94, Theorem 3.3], [BP94], [Var00] Fact 7 above is not sufficient to guarantee that any regular splitting of $I - P^T$, where P is a finite stochastic irreducible matrix, is semiconvergent (see the Example below). In order to be semiconvergent, the matrix H must not have eigenvalues of modulus 1 different from 1, and the Jordan blocks associated with 1 must have dimension 1.
9. An iterative method for computing $\boldsymbol{\pi}$ based on a regular splitting $I - P^T = M - N$ consists in generating the sequence $\boldsymbol{x}_{n+1} = M^{-1}N\boldsymbol{x}_n$, for $n \geq 0$, starting from an initial column vector $\boldsymbol{x}_0$. If the regular splitting is semiconvergent, by choosing $\boldsymbol{x}_0 \geq 0$ such that $\|\boldsymbol{x}_0\|_1 = 1$, the sequence $\{\boldsymbol{x}_n\}_n$ converges to $\boldsymbol{\pi}^T$. The convergence is linear and the asymptotic rate of convergence is the second largest modulus eigenvalue θ of $M^{-1}N$.
10. [Ste94, Theorem 3.6] If P is a finite stochastic matrix, the methods of Gauss–Seidel, Jacobi, and SOR, for $0 < \omega \leq 1$, applied to $I - P^T$, are based on regular splittings.
11. If E is finite, the invariant probability vector is, up to a multiplicative constant, the nonnegative left eigenvector associated with the dominating eigenvalue of P. The power method applied to the matrix P^T is the iterative method defined by the trivial regular splitting $I - P^T = M - N$, where $M = I$, $N = P^T$. If P is primitive, the power method is convergent.
12. [Ste94, Theorems 3.6, 3.7] If P is a $k \times k$ irreducible stochastic matrix, and if $\hat{P}$ is the $(k-1)\times(k-1)$ matrix obtained by removing the ith row and the ith column of P, where $i \in \{1, \ldots, k\}$, then any regular splitting of the matrix $I - \hat{P}^T = M - N$ is convergent. In particular, the methods of Gauss–Seidel, Jacobi, and SOR, for $0 < \omega \leq 1$, applied to $I - \hat{P}^T$, being based on regular splittings, are convergent.
13. [Ste94, Theorem 3.17] Let P be a $k \times k$ irreducible stochastic matrix and let $\epsilon > 0$. Then the splitting $I - P^T = M_\epsilon - N_\epsilon$, where $M_\epsilon = D - L + \epsilon I$ and $N_\epsilon = U + \epsilon I$, is a semiconvergent regular splitting for every $\epsilon > 0$.
14. [Ste94, Theorem 3.18] Let P be a $k \times k$ irreducible stochastic matrix, let $I - P^T = M - N$ be a regular splitting of $I - P^T$, and let $H = M^{-1}N$. Then the matrix $H_\alpha = (1 - \alpha)I + \alpha H$ is semiconvergent for any $0 < \alpha < 1$.

Algorithms:

1. The following algorithm computes the LU factorization of $I - P$ by using the Grassmann, Taksar, Heyman (GTH) trick of Fact 3:

Computation of the LU factorization of $I - P$ with GTH trick
Input: the $k \times k$ irreducible stochastic matrix P.
Output: the matrix $A = [a_{i,j}]$ such that $a_{i,j} = l_{i,j}$ for $i > j$,
and $a_{i,j} = u_{i,j}$ for $i \leq j$, where $L = [l_{i,j}]$, $U = [u_{i,j}]$ are the factors
of the LU factorization of $I - P$.
Computation:
1– set $A = I - P$;
2– for $m = 1, \ldots, k - 1$:
(a) for $i = m + 1, \ldots, k$, set $a_{i,m} = a_{i,m}/a_{m,m}$;
(b) for $i, j = m + 1, \ldots, k$, set $a_{i,j} = a_{i,j} - a_{i,m}a_{m,j}$ if $i \neq j$;
(c) for $i = m + 1, \ldots, k$ set $a_{i,i} = -\sum_{l=m+1, l\neq i}^{k} a_{l,i}$.

2. The algorithm derived by Fact 4 above is the following:

> **Computation of π through LU factorization**
> Input: the factor $L = [l_{i,j}]$ of the LU factorization of the matrix $I - P$, where P is a $k \times k$ irreducible stochastic matrix.
> Output: the invariant probability vector $\boldsymbol{\pi} = (\pi_i)_{i=1,k}$.
> Computation:
> 1– set $\pi_k = 1$;
> 2– for $j = k-1, \ldots, 1$ set $\pi_j = -\sum_{i=j+1}^{k} \pi_i l_{i,j}$;
> 3– set $\boldsymbol{\pi} = (\sum_{i=1}^{k} \pi_i)^{-1} \boldsymbol{\pi}$.

Examples:

The matrix

$$P = \begin{bmatrix} 0.5 & 0 & 0 & 0.5 \\ 0.2 & 0.8 & 0 & 0 \\ 0 & 0.6 & 0.4 & 0 \\ 0 & 0 & 0.1 & 0.9 \end{bmatrix}$$

is the transition matrix of an irreducible and aperiodic Markov chain. The splitting $I - P^T = M - N$ associated with the Jacobi method is a regular splitting. One may easily verify that the iteration matrix is

$$M^{-1}N = D \begin{bmatrix} 0 & 1 & 0 & 0 \\ 0 & 0 & 1 & 0 \\ 0 & 0 & 0 & 1 \\ 1 & 0 & 0 & 0 \end{bmatrix} D^{-1},$$

where $D = \mathrm{diag}(1, \frac{5}{2}, \frac{5}{6}, 5)$. Therefore, the regular splitting is not semiconvergent since the iteration matrix has period 4 (see Fact 8 above).

References

[ANT02] A. Arasu, J. Novak, A. Tomkins, and J. Tomlin. PageRank computation and the structure of the Web. `http://www2002.org/CDROM/poster/173.pdf`, 2002.

[BLM05] D.A. Bini, G. Latouche, and B. Meini. *Numerical Methods for Structured Markov Chains.* Series on Numerical Mathematics and Scientific Computation. Oxford University Press Inc., New York, 2005.

[BP94] A. Berman and R.J. Plemmons. *Nonnegative Matrices in the Mathematical Sciences*, Vol. 9 of *Classics in Applied Mathematics.* Society for Industrial and Applied Mathematics (SIAM), Philadelphia, 1994. (Revised reprint of the 1979 original.)

[Çin75] E. Çinlar. *Introduction to Stochastic Processes.* Prentice-Hall, Upper Saddle River, N.J., 1975.

[GTH85] W.K. Grassmann, M.I. Taksar, and D.P. Heyman. Regenerative analysis and steady state distributions for Markov chains. *Oper. Res.*, 33(5):1107–1116, 1985.

[LR99] G. Latouche and V. Ramaswami. *Introduction to Matrix Analytic Methods in Stochastic Modeling.* ASA-SIAM Series on Statistics and Applied Probability. Society for Industrial and Applied Mathematics (SIAM), Philadelphia, 1999.

[Mey89] C.D. Meyer. Stochastic complementation, uncoupling Markov chains, and theory of nearly reducible systems. *SIAM Rev.*, 31(2):240–272, 1989.

[Mol04] C. Moler. *Numerical computing with MATLAB.* `http://www.mathworks.com/moler`, 2004. (Electronic edition.)

[Neu89] M.F. Neuts. *Structured Stochastic Matrices of M/G/1 Type and Their Applications*, Vol. 5 of *Probability: Pure and Applied.* Marcel Dekker, New York, 1989.

[Nor99] J.R. Norris. *Markov Chains.* Cambridge Series on Statistical and Probabilistic Mathematics. Cambridge University Press, Cambridge, U.K., 3rd ed., 1999.

[PBM99] L. Page, S. Brin, R. Motwani, and T. Winograd. The pagerank citation ranking: Bringing order to the Web, 1999. `http://dbpubs.stanford.edu:8090/pub/1999-66`.

[Ste94] W.J. Stewart. *Introduction to the Numerical Solution of Markov Chains.* Princeton University Press, Princeton, NJ, 1994.

[Var00] R.S. Varga. *Matrix Iterative Analysis*, Vol. 27 of *Springer Series in Computational Mathematics.* Springer-Verlag, Berlin, expanded edition, 2000.

Applications to Analysis

55

Differential Equations and Stability

Volker Mehrmann
Technische Universität, Berlin

Tatjana Stykel
Technische Universität, Berlin

Differential equations and differential-algebraic equations arise in numerous branches of science and engineering that include biology, chemistry, medicine, structural mechanics, and electrical engineering. This chapter is concerned with linear differential(-algebraic) equations with constant coefficients that can be analyzed completely via techniques from linear algebra. We discuss the existence and uniqueness of solutions of such equations as well as the stability theory.

55.1 Linear Differential Equations with Constant Coefficients: Basic Concepts

Definitions:

A **linear differential equation** in an unknown function $x : \mathbb{R} \to \mathbb{C}, t \mapsto x(t)$, has the form

$$\dot{x} = ax + f,$$

where $a \in \mathbb{C}$ and the inhomogeneity $f : \mathbb{R} \to \mathbb{C}$ is a given function. Here $\dot{x}$ denotes the derivative of $x(t)$ with respect to t.

A **linear differential equation of order** k for an unknown function $x : \mathbb{R} \to \mathbb{C}$ has the form

$$a_k x^{(k)} + \cdots + a_0 x = f,$$

where $a_0, \ldots, a_k \in \mathbb{C}$, $a_k \neq 0$ and $f : \mathbb{R} \to \mathbb{C}$. Here $x^{(k)}$ denotes the k-th derivative of $x(t)$ with respect to t.

A **system of linear differential(-algebraic) equations with constant coefficients** has the form

$$E\dot{\mathbf{x}} = A\mathbf{x} + \mathbf{f},$$

where $E, A \in \mathbb{C}^{m\times n}$ are **coefficient matrices,** $\mathbf{x} : \mathbb{R} \to \mathbb{C}^n$ is a vector-valued function of unknowns, and the inhomogeneity $\mathbf{f} : \mathbb{R} \to \mathbb{C}^m$ is a given vector-valued function.

If $E = I_n$ and $A \in \mathbb{C}^{n\times n}$, then $E\dot{\mathbf{x}} = A\mathbf{x} + \mathbf{f}$ is a system of **ordinary differential equations**; otherwise it is a system of **differential-algebraic equations.**

A **homogeneous** system has $\mathbf{f}(t) \equiv 0$; otherwise the system is **inhomogeneous**.

A **system of linear differential-algebraic equations of order** k for an unknown function $\mathbf{x} : \mathbb{R} \to \mathbb{C}^n$ has the form $A_k\mathbf{x}^{(k)} + \cdots + A_0\mathbf{x} = \mathbf{f}$, where $A_0, \ldots, A_k \in \mathbb{C}^{m\times n}$, $A_k \neq 0$ and $\mathbf{f} : \mathbb{R} \to \mathbb{C}^m$.

A continuously differentiable function $\mathbf{x} : \mathbb{R} \to \mathbb{C}^n$ is a **solution** of $E\dot{\mathbf{x}} = A\mathbf{x} + \mathbf{f}$ with a sufficiently often differentiable function $\mathbf{f}$ if it satisfies the equation pointwise.

A solution of $E\dot{\mathbf{x}} = A\mathbf{x} + \mathbf{f}$ that also satisfies an initial condition $\mathbf{x}(t_0) = \mathbf{x}_0$ with $t_0 \in \mathbb{R}$ and $\mathbf{x}_0 \in \mathbb{C}^n$ is a solution of the **initial value problem.**

If the initial value problem $E\dot{\mathbf{x}} = A\mathbf{x} + \mathbf{f}$, $\mathbf{x}(t_0) = \mathbf{x}_0$ has a solution, then the initial condition is called **consistent.**

Facts:

1. [Cam80, p. 33] Let $E, A \in \mathbb{C}^{n\times n}$. If E is nonsingular, then the system $E\dot{\mathbf{x}} = A\mathbf{x} + \mathbf{f}$ is equivalent to the system of ordinary differential equations $\dot{\mathbf{x}} = E^{-1}A\mathbf{x} + E^{-1}\mathbf{f}$.
2. [Arn92, p. 105] The k-th order system of differential-algebraic equations

$$A_k\mathbf{y}^{(k)} + \cdots + A_0\mathbf{y} = \mathbf{g}$$

can be rewritten as a first-order system $E\dot{\mathbf{x}} = A\mathbf{x} + \mathbf{f}$, where $\mathbf{f} = [\,0, \ldots, 0, \mathbf{g}^T\,]^T$ and

$$E = \begin{bmatrix} I & & & \\ & \ddots & & \\ & & I & \\ & & & A_k \end{bmatrix}, \quad A = \begin{bmatrix} 0 & I & & 0 \\ \vdots & \ddots & \ddots & \\ 0 & \cdots & 0 & I \\ -A_0 & \cdots & -A_{k-2} & -A_{k-1} \end{bmatrix}, \quad \mathbf{x} = \begin{bmatrix} \mathbf{y} \\ \vdots \\ \mathbf{y}^{(k-2)} \\ \mathbf{y}^{(k-1)} \end{bmatrix}.$$

Applications:

1. Consider a mass–spring–damper model as shown in Figure 55.1. This model is described by the equation

$$m\ddot{x} + d\dot{x} + kx = 0,$$

where m is a mass, k is a spring constant, and d is a damping parameter. Since $m \neq 0$, we obtain the following first-order system of ordinary differential equations

$$\begin{bmatrix} \dot{x} \\ \dot{v} \end{bmatrix} = \begin{bmatrix} 0 & 1 \\ -k/m & -d/m \end{bmatrix} \begin{bmatrix} x \\ v \end{bmatrix},$$

where the velocity is denoted by v.

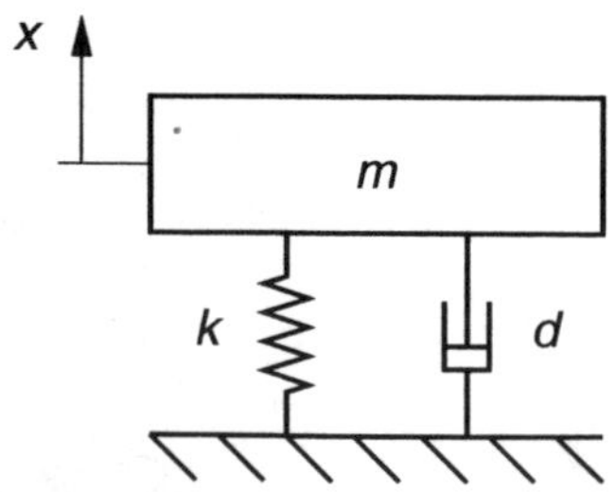

FIGURE 55.1 A mass–spring–damper model.

2. Consider a one-dimensional heat equation

$$\frac{\partial}{\partial t}T(t,\xi) = c\frac{\partial^2}{\partial \xi^2}T(t,\xi), \quad (t,\xi) \in (0,t_e) \times (0,l),$$

together with an initial condition $T(0,\xi) = g(\xi)$ and Cauchy boundary conditions

$$\alpha_1 T(t,0) + \alpha_2 \frac{\partial}{\partial n}T(t,0) = u(t),$$

$$\beta_1 T(t,l) + \beta_2 \frac{\partial}{\partial n}T(t,l) = v(t).$$

Here $T(t,\xi)$ is the temperature field in a thin beam of length l, $c > 0$ is the heat conductivity of the material, $g(\xi)$, $u(t)$, and $v(t)$ are given functions, and $\frac{\partial}{\partial n}$ denotes the derivative in the direction of the outward normal. A spatial discretization by a finite difference method with $n+1$ equidistant grid points leads to the initial value problem $\dot{\mathbf{x}} = A_{a,b}\mathbf{x} + \mathbf{f}$, $\mathbf{x}(0) = \mathbf{x}_0$, where

$$\mathbf{x}(t) = [\,T(t,h),\ T(t,2h),\ \ldots,\ T(t,nh)\,]^T, \quad \mathbf{x}_0 = [\,g(h),\ g(2h),\ \ldots,\ g(nh)\,]^T,$$

$$\mathbf{f}(t) = [cu(t)/(h^2\alpha_1 - h\alpha_2),\ 0,\ \ldots,\ 0,\ cv(t)/(h^2\beta_1 + h\beta_2)\,]^T,$$

and

$$A_{a,b} = \frac{c}{h^2}\begin{bmatrix} -a & 1 & & & \\ 1 & -2 & 1 & & \\ & \ddots & \ddots & \ddots & \\ & & 1 & -2 & 1 \\ & & & 1 & -b \end{bmatrix} \in \mathbb{R}^{n\times n}$$

with $h = l/(n+1)$, $a = (2h\alpha_1 - \alpha_2)/(h\alpha_1 - \alpha_2)$, and $b = (2h\beta_1 + \beta_2)/(h\beta_1 + \beta_2)$.

3. A simple pendulum as shown in Figure 55.2 describes the movement of a mass point with mass m and Cartesian coordinates (x, y) under the influence of gravity in a distance l around the origin.

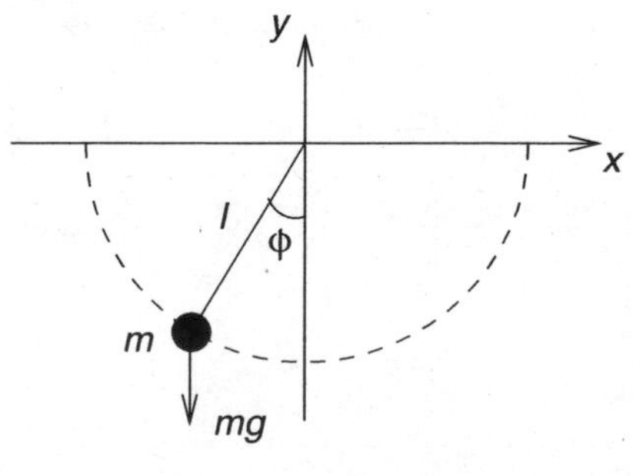

FIGURE 55.2 A simple pendulum.

The equations of motion have the form

$$\begin{aligned} m\ddot{x} + 2x\lambda &= 0, \\ m\ddot{y} + 2y\lambda + mg &= 0, \\ x^2 + y^2 - l^2 &= 0, \end{aligned}$$

where λ is a Lagrange multiplier. Transformation of this system into the first-order form by introducing new variables $v = \dot{x}$ and $w = \dot{y}$ and linearization at the equilibrium $x_e = 0$, $y_e = -l$, $v_e = 0$, $w_e = 0$, and $\lambda_e = mg/(2l)$ yields the homogeneous first-order linear differential-algebraic system

$$\begin{bmatrix} 1 & 0 & 0 & 0 & 0 \\ 0 & 1 & 0 & 0 & 0 \\ 0 & 0 & m & 0 & 0 \\ 0 & 0 & 0 & m & 0 \\ 0 & 0 & 0 & 0 & 0 \end{bmatrix} \begin{bmatrix} \dot{\widetilde{x}} \\ \dot{\widetilde{y}} \\ \dot{\widetilde{v}} \\ \dot{\widetilde{w}} \\ \dot{\widetilde{\lambda}} \end{bmatrix} = \begin{bmatrix} 0 & 0 & 1 & 0 & 0 \\ 0 & 0 & 0 & 1 & 0 \\ -2\lambda_e & 0 & 0 & 0 & 0 \\ 0 & -2\lambda_e & 0 & 0 & 2l \\ 0 & -2l & 0 & 0 & 0 \end{bmatrix} \begin{bmatrix} \widetilde{x} \\ \widetilde{y} \\ \widetilde{v} \\ \widetilde{w} \\ \widetilde{\lambda} \end{bmatrix},$$

where $\widetilde{x} = x - x_e$, $\widetilde{y} = y - y_e$, $\widetilde{v} = v - v_e$, $\widetilde{w} = w - w_e$, and $\widetilde{\lambda} = \lambda - \lambda_e$.

The motion of the pendulum can also be described by the ordinary differential equation

$$\ddot{\varphi} = -\omega^2 \sin(\varphi),$$

where φ is an angle between the vertical axis and the pendulum and $\omega = \sqrt{g/l}$ is an angular frequency of the motion. By introducing a new variable $\psi = \dot{\varphi}$ and linearization at the equilibrium $\varphi_e = 0$ and $\psi_e = 0$, we obtain the first-order homogeneous system

$$\begin{bmatrix} \dot{\widetilde{\varphi}} \\ \dot{\widetilde{\psi}} \end{bmatrix} = \begin{bmatrix} 0 & 1 \\ -\omega^2 & 0 \end{bmatrix} \begin{bmatrix} \widetilde{\varphi} \\ \widetilde{\psi} \end{bmatrix}.$$

4. Consider a simple RLC electrical circuit as shown in Figure 55.3. Using Kirchoff's and Ohm's laws, the circuit can be described by the system $E\dot{\mathbf{x}} = A\mathbf{x} + \mathbf{f}$ with

$$E = \begin{bmatrix} L & 0 & 0 & 0 \\ 0 & 0 & 1 & 0 \\ 0 & 0 & 0 & 0 \\ 0 & 0 & 0 & 0 \end{bmatrix}, \quad A = \begin{bmatrix} 0 & 1 & 0 & 0 \\ 1/C & 0 & 0 & 0 \\ -R & 0 & 0 & 1 \\ 0 & 1 & 1 & 1 \end{bmatrix}, \quad \mathbf{x} = \begin{bmatrix} i \\ v_L \\ v_C \\ v_R \end{bmatrix}, \quad \mathbf{f} = \begin{bmatrix} 0 \\ 0 \\ 0 \\ -v \end{bmatrix}.$$

Here R, L, and C are the resistance, inductance, and capacitance, respectively; v_R, v_L, and v_C are the corresponding voltage drops, i is the current, and v is the voltage source. From the last two equations

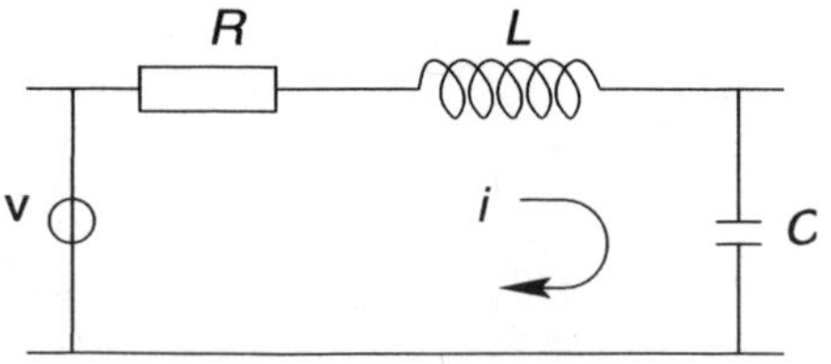

FIGURE 55.3 A simple RLC circuit.

in the system, we find $v_R = Ri$ and $v_L = v - v_C - Ri$. Substituting v_L in the first equation and introducing a new variable $w_C = i/C$, we obtain the system of ordinary differential equations

$$\begin{bmatrix} \dot{v}_C \\ \dot{w}_C \end{bmatrix} = \begin{bmatrix} 0 & 1 \\ -1/(LC) & -R/L \end{bmatrix} \begin{bmatrix} v_C \\ w_C \end{bmatrix} + \begin{bmatrix} 0 \\ v/(LC) \end{bmatrix}.$$

This shows the relationship to the mass–spring–damper model as in Application 1.

55.2 Linear Ordinary Differential Equations

Facts:

The following facts can be found in [Gan59a, pp. 116–124, 153–154].

1. Let $J_A = T^{-1}AT$ be in Jordan canonical form. Then $e^{At} = Te^{J_A t}T^{-1}$. (See Chapter 6 and Chapter 11 for more information on the Jordan canonical form and the matrix exponential.)
2. Every solution of the homogeneous system $\dot{\mathbf{x}} = A\mathbf{x}$ has the form $\mathbf{x}(t) = e^{At}\mathbf{v}$ with $\mathbf{v} \in \mathbb{C}^n$.
3. The initial value problem $\dot{\mathbf{x}} = A\mathbf{x}$, $\mathbf{x}(t_0) = \mathbf{x}_0$ has the unique solution $\mathbf{x}(t) = e^{A(t-t_0)}\mathbf{x}_0$.
4. The initial value problem $\dot{\mathbf{x}} = A\mathbf{x} + \mathbf{f}$, $\mathbf{x}(t_0) = \mathbf{x}_0$ has a unique solution for every initial vector $\mathbf{x}_0$ and every continuous inhomogeneity $\mathbf{f}$. This solution is given by

$$\mathbf{x}(t) = e^{A(t-t_0)}\mathbf{x}_0 + \int_{t_0}^{t} e^{A(t-\tau)}\mathbf{f}(\tau)\,d\tau.$$

Examples:

1. Let

$$A = \begin{bmatrix} 3 & 3 & 1 \\ 0 & 0 & 0 \\ -1 & -1 & 1 \end{bmatrix}, \qquad \mathbf{x}_0 = \begin{bmatrix} 1 \\ 2 \\ 3 \end{bmatrix}, \qquad \mathbf{f}(t) = \begin{bmatrix} -3t^2 - 3t \\ 2t \\ t^2 + t - 1 \end{bmatrix}.$$

For

$$T = \begin{bmatrix} 1 & 0 & -1 \\ 0 & 0 & 1 \\ -1 & 1 & 0 \end{bmatrix} \quad \text{and} \quad T^{-1} = \begin{bmatrix} 1 & 1 & 0 \\ 1 & 1 & 1 \\ 0 & 1 & 0 \end{bmatrix},$$

we have

$$J_A = T^{-1}AT = \left[\begin{array}{cc|c} 2 & 1 & 0 \\ 0 & 2 & 0 \\ \hline 0 & 0 & 0 \end{array}\right].$$

Then

$$\begin{aligned} e^{At} &= \begin{bmatrix} 1 & 0 & -1 \\ 0 & 0 & 1 \\ -1 & 1 & 0 \end{bmatrix} \left[\begin{array}{cc|c} e^{2t} & te^{2t} & 0 \\ 0 & e^{2t} & 0 \\ \hline 0 & 0 & 1 \end{array}\right] \begin{bmatrix} 1 & 1 & 0 \\ 1 & 1 & 1 \\ 0 & 1 & 0 \end{bmatrix} \\ &= \begin{bmatrix} (1+t)e^{2t} & (1+t)e^{2t} - 1 & te^{2t} \\ 0 & 1 & 0 \\ -te^{2t} & -te^{2t} & (1-t)e^{2t} \end{bmatrix}. \end{aligned}$$

Every solution of the homogeneous system $\dot{\mathbf{x}} = A\mathbf{x}$ has the form

$$\mathbf{x}(t) = e^{At}\mathbf{v} = \begin{bmatrix} ((1+t)v_1 + (1+t)v_2 + tv_3)e^{2t} - v_2 \\ v_2 \\ (-tv_1 - tv_2 + (1-t)v_3)e^{2t} \end{bmatrix}$$

with $\mathbf{v} = [v_1, v_2, v_3]^T$. The solution of the initial value problem $\dot{\mathbf{x}} = A\mathbf{x}$, $\mathbf{x}(0) = \mathbf{x}_0$ has the form

$$\mathbf{x}(t) = e^{At}\mathbf{x}_0 = \begin{bmatrix} (3+6t)e^{2t} - 2 \\ 2 \\ (3-6t)e^{2t} \end{bmatrix}.$$

The initial value problem $\dot{\mathbf{x}} = A\mathbf{x} + \mathbf{f}$, $\mathbf{x}(0) = \mathbf{x}_0$ has the solution

$$\mathbf{x}(t) = \begin{bmatrix} (3+6t)e^{2t} + t - 2 \\ t^2 + 2 \\ (3-6t)e^{2t} + 1 \end{bmatrix}.$$

Applications:

1. Consider the matrix A from the mass–spring–damper example

$$A = \begin{bmatrix} 0 & 1 \\ -k/m & -d/m \end{bmatrix}.$$

The Jordan canonical form of A is given by $J_A = T^{-1}AT = \mathrm{diag}(\lambda_1, \lambda_2)$, where

$$T = \begin{bmatrix} 1 & 1 \\ \lambda_1 & \lambda_2 \end{bmatrix}, \qquad T^{-1} = \frac{1}{\lambda_2 - \lambda_1}\begin{bmatrix} \lambda_2 & -1 \\ -\lambda_1 & 1 \end{bmatrix}$$

and

$$\lambda_1 = \frac{-d - \sqrt{d^2 - 4km}}{2m}, \qquad \lambda_2 = \frac{-d + \sqrt{d^2 - 4km}}{2m}$$

are the eigenvalues of A. We have

$$e^{At} = T\mathrm{diag}(e^{\lambda_1 t}, e^{\lambda_2 t})T^{-1} = \frac{1}{\lambda_2 - \lambda_1}\begin{bmatrix} \lambda_2 e^{\lambda_1 t} - \lambda_1 e^{\lambda_2 t} & e^{\lambda_2 t} - e^{\lambda_1 t} \\ \lambda_1\lambda_2(e^{\lambda_1 t} - e^{\lambda_2 t}) & \lambda_2 e^{\lambda_2 t} - \lambda_1 e^{\lambda_1 t} \end{bmatrix}.$$

The solution of the mass–spring–damper model with the initial conditions $x(0) = x_0$ and $v(0) = 0$ is given by

$$x(t) = \frac{x_0}{\lambda_2 - \lambda_1}\left(\lambda_2 e^{\lambda_1 t} - \lambda_1 e^{\lambda_2 t}\right), \qquad v(t) = \frac{\lambda_1\lambda_2 x_0}{\lambda_2 - \lambda_1}\left(e^{\lambda_1 t} - e^{\lambda_2 t}\right).$$

2. Since the matrix $A_{a,b}$ in the semidiscretized heat equation is symmetric, there exists an orthogonal matrix U such that $U^T A_{a,b} U = \mathrm{diag}(\lambda_1, \ldots, \lambda_n)$, where $\lambda_1, \ldots, \lambda_n \in \mathbb{R}$ are the eigenvalues of $A_{a,b}$. (See Chapter 7.2 and Chapter 45.) In this case $e^{A_{a,b}t} = U\mathrm{diag}(e^{\lambda_1 t}, \ldots, e^{\lambda_n t})U^T$.

55.3 Linear Differential-Algebraic Equations

Definitions:

A **Drazin inverse** A^D of a matrix $A \in \mathbb{C}^{n\times n}$ is defined as the unique solution of the system of matrix equations

$$A^D A A^D = A^D, \qquad A A^D = A^D A, \qquad A^{k+1} A^D = A^k,$$

where k is a smallest nonnegative integer such that $\operatorname{rank}(A^{k+1}) = \operatorname{rank}(A^k)$.

Let $E, A \in \mathbb{C}^{m\times n}$. A pencil of the form

$$\lambda E - A = \operatorname{diag}\left(\mathcal{L}_{n_1}, \ldots, \mathcal{L}_{n_p}, \mathcal{M}_{m_1}, \ldots, \mathcal{M}_{m_q}, \mathcal{J}_k, \mathcal{N}_s\right)$$

is called pencil in **Kronecker canonical form** if the block entries have the following properties: every entry $\mathcal{L}_{n_j} = \lambda L_{n_j} - R_{n_j}$ is a bidiagonal block of size $n_j \times (n_j + 1)$, $n_j \in \mathbb{N}$, where

$$L_{n_j} = \begin{bmatrix} 1 & 0 & & \\ & \ddots & \ddots & \\ & & 1 & 0 \end{bmatrix}, \qquad R_{n_j} = \begin{bmatrix} 0 & 1 & & \\ & \ddots & \ddots & \\ & & 0 & 1 \end{bmatrix};$$

every entry $\mathcal{M}_{m_j} = \lambda L_{m_j}^T - R_{m_j}^T$ is a bidiagonal block of size $(m_j+1)\times m_j$, $m_j \in \mathbb{N}$; the entry $\mathcal{J}_k = \lambda I_k - A_k$ is a block of size $k \times k$, $k \in \mathbb{N}$, where A_k is in Jordan canonical form; the entry $\mathcal{N}_s = \lambda N_s - I_s$ is a block of size $s \times s$, $s \in \mathbb{N}$, where $N_s = \operatorname{diag}(N_{s_1}, \ldots, N_{s_r})$; and

$$N_{s_j} = \begin{bmatrix} 0 & 1 & & \\ & \ddots & \ddots & \\ & & \ddots & 1 \\ & & & 0 \end{bmatrix}$$

is a nilpotent Jordan block with index of nilpotency s_j.

The numbers $n_1, \ldots, n_p$ are called the **right Kronecker indices** of the pencil $\lambda E - A$.

The numbers $m_1, \ldots, m_q$ are called the **left Kronecker indices** of the pencil $\lambda E - A$.

The number $\nu = \max_{1\leq j\leq r} s_j$ is called the **index** of the pencil $\lambda E - A$.

A matrix pencil $\lambda E - A$ with $E, A \in \mathbb{C}^{m\times n}$ is called **regular**, if $m = n$ and $\det(\lambda E - A) \neq 0$ for some $\lambda \in \mathbb{C}$. Otherwise, the pencil is called **singular**.

Let $E, A \in \mathbb{C}^{m,n}$. Subspaces $W_l \subset \mathbb{C}^m$ and $W_r \subset \mathbb{C}^n$ are called **left** and **right reducing subspaces** of the pencil $\lambda E - A$ if $W_l = E W_r + A W_r$ and $\dim(W_l) = \dim(W_r) - p$, where p is the number of $\mathcal{L}_{n_j}$ blocks in the Kronecker canonical form.

Let $\lambda E - A$ be a regular pencil. Subspaces $W_l, W_r \subset \mathbb{C}^n$ are called **left** and **right deflating subspaces** of $\lambda E - A$ if $W_l = E W_r + A W_r$ and $\dim(W_l) = \dim(W_r)$.

Let $W_1, W_2 \subset \mathbb{C}^n$ be subspaces such that $W_1 \cap W_2 = \{\mathbf{0}\}$ and $W_1 + W_2 = \mathbb{C}^n$. A matrix $P \in \mathbb{C}^{n,n}$ is called a **projection** onto W_1 along W_2 if $P^2 = P$, $\operatorname{range}(P) = W_1$, and $\ker(P) = W_2$.

Let $\lambda E - A$ be a regular pencil. If $T_l, T_r \in \mathbb{C}^{n\times n}$ are nonsingular matrices such that $T_l^{-1}(\lambda E - A)T_r$ is in Kronecker canonical form, then

$$P_l = T_l \begin{bmatrix} I_k & \mathbf{0} \\ \mathbf{0} & \mathbf{0} \end{bmatrix} T_l^{-1}, \qquad P_r = T_r \begin{bmatrix} I_k & \mathbf{0} \\ \mathbf{0} & \mathbf{0} \end{bmatrix} T_r^{-1}$$

are the **spectral projections** onto the left and right deflating subspaces of $\lambda E - A$ corresponding to the finite eigenvalues along the left and right deflating subspaces corresponding to the eigenvalue at infinity.

Facts:

1. [Cam80, p. 8] If $A \in \mathbb{C}^{n\times n}$ is nonsingular, then $A^D = A^{-1}$.
2. [Cam80, p. 8] Let

$$J_A = T^{-1}AT = \begin{bmatrix} A_1 & \mathbf{0} \\ \mathbf{0} & A_0 \end{bmatrix}$$

be in Jordan canonical form, where A_1 contains all the Jordan blocks associated with the nonzero eigenvalues, and A_0 contains all the Jordan blocks associated with the eigenvalue 0. Then

$$A^D = T\begin{bmatrix} A_1^{-1} & \mathbf{0} \\ \mathbf{0} & \mathbf{0} \end{bmatrix}T^{-1}.$$

3. [Gan59b, pp. 29–37] For every matrix pencil $\lambda E - A$ with $E, A \in \mathbb{C}^{m\times n}$ there exist nonsingular matrices $T_l \in \mathbb{C}^{m\times m}$ and $T_r \in \mathbb{C}^{n\times n}$ such that $T_l^{-1}(\lambda E - A)T_r$ is in Kronecker canonical form. The Kronecker canonical form is unique up to permutation of the diagonal blocks, i.e., the kind, size, and number of the blocks are characteristic for the pencil $\lambda E - A$. (For more information on matrix pencils, see Section 43.1.)
4. [Gan59b, p. 47] If $\mathbf{f}(t) = [f_1(t), \ldots, f_{n_j}(t)]^T$ is an n_j-times continuously differentiable vector-valued function and $g(t)$ is an arbitrary (n_j+1)-times continuously differentiable function, then the system $L_{n_j}\dot{\mathbf{x}} = R_{n_j}\mathbf{x} + \mathbf{f}$ has a continuously differentiable solution of the form

$$\mathbf{x}(t) = \begin{bmatrix} x_1(t) \\ x_2(t) \\ \vdots \\ x_{n_j+1}(t) \end{bmatrix} = \begin{bmatrix} g(t) \\ g^{(1)}(t) - f_1(t) \\ \vdots \\ g^{(n_j)}(t) - \sum_{i=1}^{n_j} f_i^{(n_j-i)}(t) \end{bmatrix}.$$

A consistent initial condition has to satisfy this defining equation at t_0.
5. [Gan59b, p. 47] A system of differential-algebraic equations $L_{m_j}^T\dot{\mathbf{x}} = R_{m_j}^T\mathbf{x} + \mathbf{f}$ with a vector-valued function $\mathbf{f}(t) = [f_1(t), \ldots, f_{m_j+1}(t)]^T$ has a unique solution if and only if $\mathbf{f}$ is m_j-times continuously differentiable and $\sum_{i=1}^{m_j+1} f_i^{(i-1)}(t) \equiv 0$. If this holds, then the solution is given by

$$\mathbf{x}(t) = \begin{bmatrix} x_1(t) \\ \vdots \\ x_{m_j}(t) \end{bmatrix} = \begin{bmatrix} -\sum_{i=1}^{m_j} f_{i+1}^{(i-1)}(t) \\ \vdots \\ -f_{m_j+1}(t) \end{bmatrix}.$$

A consistent initial condition has to satisfy this defining equation at t_0.
6. [Gan59b, p. 48] A system $N_{s_j}\dot{\mathbf{x}} = \mathbf{x} + \mathbf{f}$ has a unique continuously differentiable solution $\mathbf{x}$ if $\mathbf{f}$ is s_j-times continuously differentiable. This solution is given by

$$\mathbf{x}(t) = -\sum_{i=0}^{s_j-1} N_{s_j}^i \mathbf{f}^{(i)}(t).$$

A consistent initial condition has to satisfy this defining equation at t_0.

7. [Cam80, pp. 37–39] If the pencil $\lambda E - A$ is regular of index ν, then for every ν-times differentiable inhomogeneity $\mathbf{f}$ there exists a solution of the differential-algebraic system $E\dot{\mathbf{x}} = A\mathbf{x} + \mathbf{f}$. Every solution of this system has the form

$$\mathbf{x}(t) = e^{\hat{E}^D\hat{A}(t-t_0)}\hat{E}^D\hat{E}\mathbf{v} + \int_{t_0}^{t} e^{\hat{E}^D\hat{A}(t-\tau)}\hat{E}^D\hat{\mathbf{f}}(\tau)\,d\tau - (I - \hat{E}^D\hat{E})\sum_{j=0}^{\nu-1}(\hat{E}\hat{A}^D)^j\hat{A}^D\hat{\mathbf{f}}^{(j)}(t),$$

where $\mathbf{v} \in \mathbb{C}^n$, $\hat{E} = (\lambda_0 E - A)^{-1}E$, $\hat{A} = (\lambda_0 E - A)^{-1}A$, and $\hat{\mathbf{f}} = (\lambda_0 E - A)^{-1}\mathbf{f}$ for some $\lambda_0 \in \mathbb{C}$ such that $\lambda_0 E - A$ is nonsingular.
8. [Cam80, pp. 37–39] If the pencil $\lambda E - A$ is regular of index ν and if $\mathbf{f}$ is ν-times differentiable, then the initial value problem $E\dot{\mathbf{x}} = A\mathbf{x} + \mathbf{f}$, $\mathbf{x}(t_0) = \mathbf{x}_0$ possesses a solution if and only if there exists $\mathbf{v} \in \mathbb{C}^n$ that satisfies

$$\mathbf{x}_0 = \hat{E}^D\hat{E}\mathbf{v} - (I - \hat{E}^D\hat{E})\sum_{j=0}^{\nu-1}(\hat{E}\hat{A}^D)^j\hat{A}^D\hat{\mathbf{f}}^{(j)}(t_0).$$

If such a $\mathbf{v}$ exists, then the solution is unique.
9. [KM06, p. 21] The existence of a unique solution of $E\dot{\mathbf{x}} = A\mathbf{x} + \mathbf{f}$, $\mathbf{x}(t_0) = \mathbf{x}_0$ does not imply that the pencil $\lambda E - A$ is regular.
10. [Cam80, pp. 41–44] If the pencil $\lambda E - A$ is singular, then the initial value problem $E\dot{\mathbf{x}} = A\mathbf{x} + \mathbf{f}$, $\mathbf{x}(t_0) = \mathbf{x}_0$ may have no solutions or the solution, if it exists, may not be unique.
11. [Sty02, pp. 23–26] Let the pencil $\lambda E - A$ be regular of index ν. If

$$T_l^{-1}(\lambda E - A)T_r = \begin{bmatrix} \lambda I - A_k & \mathbf{0} \\ \mathbf{0} & \lambda N_s - I \end{bmatrix}$$

is in Kronecker canonical form, then the solution of the initial value problem $E\dot{\mathbf{x}} = A\mathbf{x} + \mathbf{f}$, $\mathbf{x}(t_0) = \mathbf{x}_0$ can be represented as

$$\mathbf{x}(t) = \mathcal{F}(t - t_0)E\mathbf{x}_0 + \int_{t_0}^{t}\mathcal{F}(t-\tau)\mathbf{f}(\tau)\,d\tau + \sum_{j=0}^{\nu-1}F_{-j-1}\mathbf{f}^{(j)}(t),$$

where

$$\mathcal{F}(t) = T_r\begin{bmatrix} e^{A_k t} & \mathbf{0} \\ \mathbf{0} & \mathbf{0} \end{bmatrix}T_l^{-1}, \qquad F_{-j} = T_r\begin{bmatrix} \mathbf{0} & \mathbf{0} \\ \mathbf{0} & -N_s^{j-1} \end{bmatrix}T_l^{-1}.$$

Examples:

1. The system

$$\begin{bmatrix} 1 & 0 \\ 0 & 0 \end{bmatrix}\dot{\mathbf{x}} = \begin{bmatrix} 1 & 0 \\ 0 & 0 \end{bmatrix}\mathbf{x} + \begin{bmatrix} 0 \\ g(t) \end{bmatrix}, \qquad \mathbf{x}(0) = \begin{bmatrix} 1 \\ 0 \end{bmatrix}$$

has no solution if $g(t) \not\equiv 0$. For $g(t) \equiv 0$, this system has the solution $\mathbf{x}(t) = [\, e^t, \ \phi(t)\,]^T$, where $\phi(t)$ is a differentiable function such that $\phi(0) = 0$.
2. The system

$$\begin{bmatrix} 1 & 0 \\ 0 & 1 \\ 0 & 0 \end{bmatrix}\dot{\mathbf{x}} = \begin{bmatrix} 0 & 0 \\ 1 & 0 \\ 0 & 1 \end{bmatrix}\mathbf{x} + \begin{bmatrix} -\sin(t) \\ -\cos(t) \\ 0 \end{bmatrix}, \qquad \mathbf{x}(0) = \begin{bmatrix} 1 \\ 0 \end{bmatrix}$$

has a unique solution $\mathbf{x}(t) = [\, \cos(t), \ 0\,]^T$, but the pencil $\lambda E - A$ is singular.

Applications:

1. The pencil in the linearized pendulum example has the Kronecker canonical form

$$\text{diag}(\mathcal{J}_2, \mathcal{N}_3) = \lambda \left[\begin{array}{cc|ccc} 1 & 0 & 0 & 0 & 0 \\ 0 & 1 & 0 & 0 & 0 \\ \hline 0 & 0 & 0 & 1 & 0 \\ 0 & 0 & 0 & 0 & 1 \\ 0 & 0 & 0 & 0 & 0 \end{array}\right] - \left[\begin{array}{cc|ccc} -i\sqrt{g/l} & 0 & 0 & 0 & 0 \\ 0 & i\sqrt{g/l} & 0 & 0 & 0 \\ \hline 0 & 0 & 1 & 0 & 0 \\ 0 & 0 & 0 & 1 & 0 \\ 0 & 0 & 0 & 0 & 1 \end{array}\right].$$

This pencil is regular of index 3. Since the linearized pendulum system is homogeneous, it has a unique solution for every consistent initial condition.

2. The pencil of the circuit equation has the Kronecker canonical form

$$\text{diag}(\mathcal{J}_2, \mathcal{N}_2) = \lambda \left[\begin{array}{cc|cc} 1 & 0 & 0 & 0 \\ 0 & 1 & 0 & 0 \\ \hline 0 & 0 & 0 & 0 \\ 0 & 0 & 0 & 0 \end{array}\right] - \left[\begin{array}{cc|cc} \lambda_1 & 0 & 0 & 0 \\ 0 & \lambda_2 & 0 & 0 \\ \hline 0 & 0 & 1 & 0 \\ 0 & 0 & 0 & 1 \end{array}\right],$$

with

$$\lambda_1 = -\frac{R}{2L} - \sqrt{\frac{R^2}{4L^2} - \frac{1}{LC}}, \qquad \lambda_2 = -\frac{R}{2L} + \sqrt{\frac{R^2}{4L^2} - \frac{1}{LC}}.$$

This pencil is regular of index 1. Hence, there exists a unique continuous solution for every continuous voltage source $v(t)$ and for every consistent initial condition.

55.4 Stability of Linear Ordinary Differential Equations

The notion of stability is used to study the behavior of dynamical systems under initial perturbations around equilibrium points. In this section, we consider the stability of linear homogeneous ordinary differential equations with constant coefficients only. For extensions of this concept to general nonlinear systems, see, e.g., [Ces63] and [Hah67].

Definitions:

The equilibrium $\mathbf{x}_e(t) \equiv 0$ of the system $\dot{\mathbf{x}} = A\mathbf{x}$ is called **stable in the sense of Lyapunov**, or simply **stable**, if for every $\varepsilon > 0$ there exists a $\delta = \delta(\varepsilon) > 0$ such that any solution $\mathbf{x}$ of $\dot{\mathbf{x}} = A\mathbf{x}$, $\mathbf{x}(t_0) = \mathbf{x}_0$ with $\|\mathbf{x}_0\|_2 < \delta$ satisfies $\|\mathbf{x}(t)\|_2 < \varepsilon$ for all $t \geq t_0$.

The equilibrium $\mathbf{x}_e(t) \equiv 0$ of the system $\dot{\mathbf{x}} = A\mathbf{x}$ is called **asymptotically stable** if it is stable and $\lim_{t\to\infty} \mathbf{x}(t) = 0$ for any solution $\mathbf{x}$ of $\dot{\mathbf{x}} = A\mathbf{x}$.

The equilibrium $\mathbf{x}_e(t) \equiv 0$ of the system $\dot{\mathbf{x}} = A\mathbf{x}$ is called **unstable** if it is not stable.

The equilibrium $\mathbf{x}_e(t) \equiv 0$ of the system $\dot{\mathbf{x}} = A\mathbf{x}$ is called **exponentially stable** if there exist $\alpha > 0$ and $\beta > 0$ such that the solution $\mathbf{x}$ of $\dot{\mathbf{x}} = A\mathbf{x}$, $\mathbf{x}(t_0) = \mathbf{x}_0$ satisfies $\|\mathbf{x}(t)\|_2 \leq \alpha\, e^{-\beta(t-t_0)}\|\mathbf{x}_0\|_2$ for all $t \geq t_0$.

Facts:

1. [Gan59a, pp. 125–129] The equilibrium $\mathbf{x}_e(t) \equiv 0$ of the system $\dot{\mathbf{x}} = A\mathbf{x}$ is stable if and only if all the eigenvalues of A have nonpositive real part and those with zero real part have the same algebraic and geometric multiplicities. If at least one of these conditions is violated, then the equilibrium $\mathbf{x}_e(t) \equiv 0$ of $\dot{\mathbf{x}} = A\mathbf{x}$ is unstable.
2. [Gan59a, pp. 125–129] The equilibrium $\mathbf{x}_e(t) \equiv 0$ of the system $\dot{\mathbf{x}} = A\mathbf{x}$ is asymptotically stable if and only if all the eigenvalues of A have negative real part.

3. [Ces63, p. 22] Let $p_A(\lambda) = \det(\lambda I - A) = \lambda^n + a_1\lambda^{n-1} + \cdots + a_n$ be the characteristic polynomial of $A \in \mathbb{R}^{n,n}$. If the equilibrium $\mathbf{x}_e(t) \equiv 0$ of the system $\dot{\mathbf{x}} = A\mathbf{x}$ is asymptotically stable, then $a_j > 0$ for $j = 1, \ldots, n$.
4. [Gan59b, pp. 185–189] The equilibrium $\mathbf{x}_e(t) \equiv 0$ of the system $\dot{\mathbf{x}} = A\mathbf{x}$ is asymptotically stable if and only if the Lyapunov equation $A^*X + XA = -Q$ has a unique Hermitian, positive definite solution X for every Hermitian, positive definite matrix Q.
5. [God97] Let H be a Hermitian, positive definite solution of the Lyapunov equation

$$A^*H + HA = -I$$

and let $\mathbf{x}$ be a solution of the initial value problem $\dot{\mathbf{x}} = A\mathbf{x}$, $\mathbf{x}(0) = \mathbf{x}_0$. Then in terms of the original data,

$$\|\mathbf{x}(t)\|_2 \leq \sqrt{\kappa(A)}\, e^{-t\|A\|_2/\kappa(A)} \|\mathbf{x}_0\|_2,$$

where $\kappa(A) = 2\|A\|_2\|H\|_2$.
6. [Hah67, pp. 113–117] The equilibrium $\mathbf{x}_e(t) \equiv 0$ of the system $\dot{\mathbf{x}} = A\mathbf{x}$ is exponentially stable if and only if it is asymptotically stable.
7. [Hah67, p. 16] If the equilibrium $\mathbf{x}_e(t) \equiv 0$ of the homogeneous system $\dot{\mathbf{x}} = A\mathbf{x}$ is asymptotically stable, then all the solutions of the inhomogeneous system $\dot{\mathbf{x}} = A\mathbf{x} + \mathbf{f}$ with a bounded inhomogeneity $\mathbf{f}$ are bounded.

Examples:

1. Consider the linear system $\dot{\mathbf{x}} = A\mathbf{x}$ with

$$A = \begin{bmatrix} -1 & 0 \\ 0 & 2 \end{bmatrix}.$$

For the initial condition $\mathbf{x}(0) = [\,1,\ 0\,]^T$, this system has the solution $\mathbf{x}(t) = [\,e^{-t},\ 0\,]^T$ that is bounded for all $t \geq 0$. However, this does not mean that the equilibrium $\mathbf{x}_e(t) \equiv 0$ is stable. For linear systems with constant coefficients, stability means that the solution $\mathbf{x}(t)$ remains bounded for all time and for all initial conditions, but not just for some specific initial condition. If we can find at least one initial condition that causes one of the states to approach infinity with time, then the equilibrium is unstable. For the above system, we can choose, for example, $\mathbf{x}(0) = [\,1,\ 1\,]^T$. In this case $\mathbf{x}(t) = [\,e^{-t},\ e^{2t}\,]^T$ is unbounded, which proves that the equilibrium $\mathbf{x}_e(t) \equiv 0$ is unstable.
2. Consider the linear system $\dot{\mathbf{x}} = A\mathbf{x}$ with

$$A = \begin{bmatrix} -0.1 & -1 \\ 1 & -0.1 \end{bmatrix}.$$

The eigenvalues of A are $-0.1 \pm i$, and, hence, the equilibrium $\mathbf{x}_e(t) \equiv 0$ is asymptotically stable. Indeed, the solution of this system is given by $\mathbf{x}(t) = e^{At}\mathbf{x}(0)$, which can be written in the real form as

$$x_1(t) = e^{-0.1t}(x_1(0)\cos(t) - x_2(0)\sin(t)),$$

$$x_2(t) = e^{-0.1t}(x_1(0)\sin(t) + x_2(0)\cos(t)).$$

(See also Chapter 56.) Thus, for all initial conditions $x_1(0)$ and $x_2(0)$, the solution tends to zero as $t \to \infty$. The phase portrait for $x_1(0) = 1$ and $x_2(0) = 0$ is presented in Figure 55.4.
3. Consider the linear system $\dot{\mathbf{x}} = A\mathbf{x}$ with

$$A = \begin{bmatrix} 0 & -1 \\ 1 & 0 \end{bmatrix}.$$

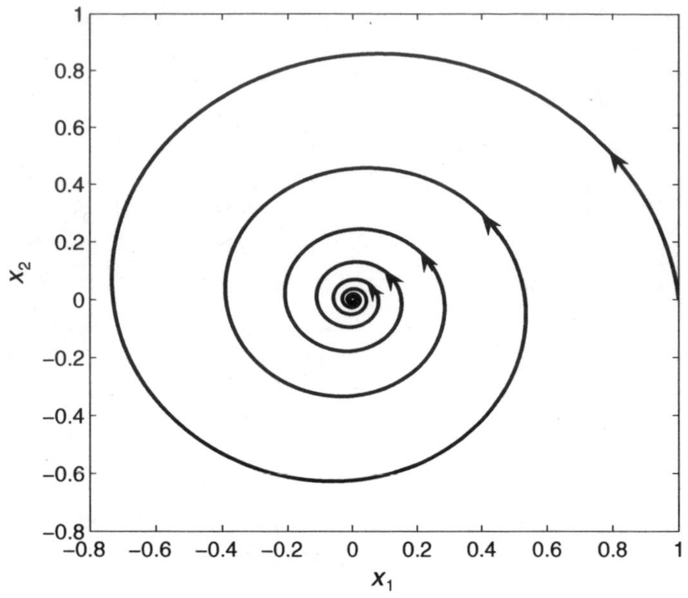

FIGURE 55.4 Asymptotic stability.

The matrix A has the eigenvalues $\pm i$. The solution of this system $\mathbf{x}(t) = e^{At}\mathbf{x}(0)$ can be written in the real form as

$$x_1(t) = x_1(0)\cos(t) - x_2(0)\sin(t),$$
$$x_2(t) = x_1(0)\sin(t) + x_2(0)\cos(t).$$

It remains bounded for all initial values $x_1(0)$ and $x_2(0)$, and, hence, the equilibrium $\mathbf{x}_e(t) \equiv 0$ is stable. The phase portrait for $x_1(0) = 1$ and $x_2(0) = 0$ is given in Figure 55.5.

4. Consider the linear system $\dot{\mathbf{x}} = A\mathbf{x}$ with

$$A = \begin{bmatrix} 0.1 & -1 \\ 1 & 0.1 \end{bmatrix}.$$

The eigenvalues of A are $0.1 \pm i$. The solution of this system in the real form is given by

$$x_1(t) = e^{0.1t}(x_1(0)\cos(t) - x_2(0)\sin(t)),$$
$$x_2(t) = e^{0.1t}(x_1(0)\sin(t) + x_2(0)\cos(t)).$$

It is unbounded for all nontrivial initial conditions. Thus, the equilibrium $\mathbf{x}_e(t) \equiv 0$ is unstable. The phase portrait of the solution with $x_1(0) = 1$ and $x_2(0) = 0$ is shown in Figure 55.6.

5. Consider the linear system $\dot{\mathbf{x}} = A\mathbf{x}$ with

$$A = \begin{bmatrix} a & b \\ c & d \end{bmatrix} \in \mathbb{R}^{2\times 2}.$$

The characteristic polynomial of the matrix A is given by $p_A(\lambda) = \lambda^2 - (a+d)\lambda + (ad - bc)$ and the eigenvalues of A have the form

$$\lambda_1 = \frac{a+d}{2} + \frac{\sqrt{(a+d)^2 - 4(ad-bc)}}{2}, \qquad \lambda_2 = \frac{a+d}{2} - \frac{\sqrt{(a+d)^2 - 4(ad-bc)}}{2}.$$

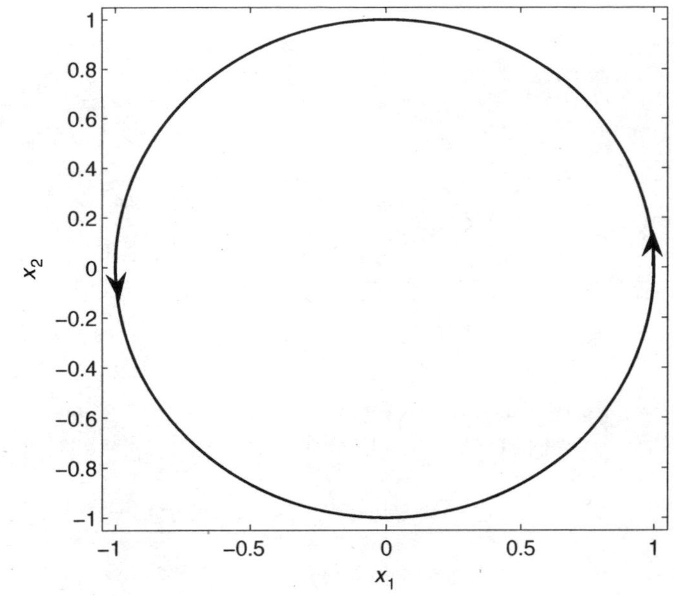

FIGURE 55.5 Stability.

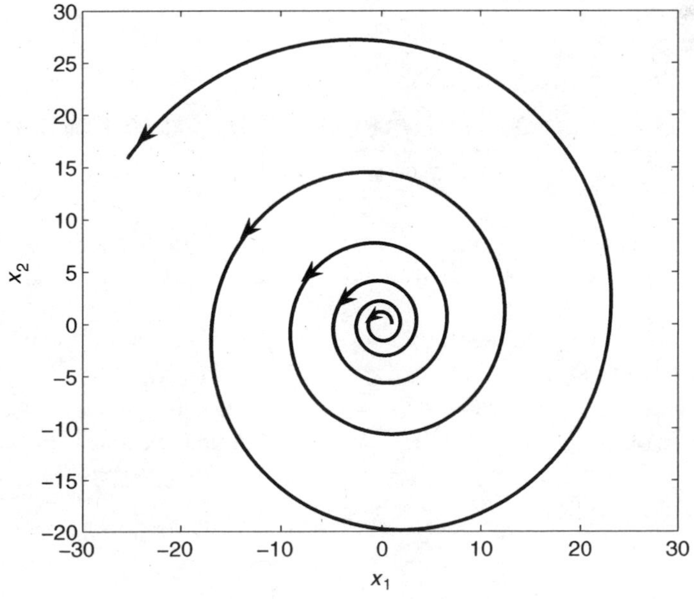

FIGURE 55.6 Instability.

We have the following cases:

$a + d < 0,$	$ad - bc > 0$	$\text{Re}(\lambda_1) < 0, \text{Re}(\lambda_2) < 0,$	Asymptotically stable
$a + d < 0,$	$ad - bc = 0$	$\lambda_1 = 0, \lambda_2 < 0$	Stable
$a + d < 0,$	$ad - bc < 0$	$\lambda_1 > 0, \lambda_2 < 0$	Unstable
$a + d = 0,$	$ad - bc > 0$	$\lambda_1 = i\alpha, \lambda_2 = -i\alpha, \alpha$ - real	Stable
$a + d = 0,$	$ad - bc = 0$	$\lambda_1 = 0, \lambda_2 = 0$	Unstable
$a^2 + b^2 + c^2 + d^2 \neq 0$			
$a = 0,\ b = 0,\ c = 0,\ d = 0$		$\lambda_1 = 0, \lambda_2 = 0$	Stable
$a + d = 0,$	$ad - bc < 0$	$\lambda_1 > 0, \lambda_2 < 0$	Unstable
$a + d > 0,$	$ad - bc \leq 0$	$\lambda_1 > 0, \lambda_2 \leq 0$	Unstable
$a + d > 0,$	$ad - bc > 0$	$\text{Re}(\lambda_1) > 0, \text{Re}(\lambda_2) > 0$	Unstable

Applications:

1. Consider the semidiscretized heat equation; see Application 2 in Section 55.1. Let $\alpha_1 = \beta_1 = 1$ and $\alpha_2 = \beta_2 = 0$. Then $a = b = 2$ and the matrix $A_{2,2}$ has the eigenvalues

$$\lambda_j(A_{2,2}) = -\frac{4c}{h^2}\sin^2\frac{j\pi}{2(n+1)}.$$

In this case, the equilibrium $\mathbf{x}_e(t) \equiv 0$ of the system $\dot{\mathbf{x}} = A_{2,2}\mathbf{x}$ is asymptotically stable. However, for $\alpha_1 = \beta_1 = 0$ and $\alpha_2 = \beta_2 = 1$, we have $a = b = 1$. Then the matrix $A_{1,1}$ has a simple zero eigenvalue and, hence, the equilibrium $\mathbf{x}_e(t) \equiv 0$ of $\dot{\mathbf{x}} = A_{1,1}\mathbf{x}$ is only stable.

2. Consider the mass–spring–damper model with $m > 0$, $d \geq 0$, and $k \geq 0$. The coefficient matrix of this model has eigenvalues

$$\lambda_1 = \frac{-d - \sqrt{d^2 - 4km}}{2m}, \qquad \lambda_2 = \frac{-d + \sqrt{d^2 - 4km}}{2m}.$$

For $d = 0$, the equilibrium $\mathbf{x}_e(t) \equiv 0$ is unstable if $k = 0$, and it is stable if $k > 0$. For $d > 0$, the equilibrium $\mathbf{x}_e(t) \equiv 0$ is stable if $k = 0$, and it is asymptotically stable if $k > 0$.

55.5 Stability of Linear Differential-Algebraic Equations

Definitions:

The equilibrium $\mathbf{x}_e(t) \equiv 0$ of the system $E\dot{\mathbf{x}} = A\mathbf{x}$ is called **stable in the sense of Lyapunov**, or simply **stable**, if for every $\varepsilon > 0$ there exists a $\delta = \delta(\varepsilon) > 0$ such that any solution $\mathbf{x}$ of $E\dot{\mathbf{x}} = A\mathbf{x}$, $\mathbf{x}(t_0) = P_r\mathbf{x}_0$ with $\|P_r\mathbf{x}_0\|_2 < \delta$ satisfies $\|\mathbf{x}(t)\|_2 < \varepsilon$ for all $t \geq t_0$.

The equilibrium $\mathbf{x}_e(t) \equiv 0$ of the system $E\dot{\mathbf{x}} = A\mathbf{x}$ is called **asymptotically stable** if it is stable and $\lim_{t\to\infty} \mathbf{x}(t) = 0$ for every solution $\mathbf{x}$ of $E\dot{\mathbf{x}} = A\mathbf{x}$.

The equilibrium $\mathbf{x}_e(t) \equiv 0$ of the system $E\dot{\mathbf{x}} = A\mathbf{x}$ is called **unstable** if it is not stable.

The equilibrium $\mathbf{x}_e(t) \equiv 0$ of the system $E\dot{\mathbf{x}} = A\mathbf{x}$ is called **exponentially stable** if there exist constants $\alpha > 0$ and $\beta > 0$ such that the solution $\mathbf{x}$ of $E\dot{\mathbf{x}} = A\mathbf{x}$, $\mathbf{x}(t_0) = P_r\mathbf{x}_0$ satisfies $\|\mathbf{x}(t)\|_2 \leq \alpha\, e^{-\beta(t-t_0)}\|P_r\mathbf{x}_0\|_2$ for all $t \geq t_0$.

Facts:

1. [Dai89, pp. 68–69] If the pencil $\lambda E - A$ is regular, then the equilibrium $\mathbf{x}_e(t) \equiv 0$ of the system $E\dot{\mathbf{x}} = A\mathbf{x}$ is stable if and only if all finite eigenvalues of the pencil $\lambda E - A$ have nonpositive real part and those with zero real part have the same algebraic and geometric multiplicities.

2. [Dai89, pp. 68–69] If the pencil $\lambda E - A$ is regular, then the equilibrium $\mathbf{x}_e(t) \equiv 0$ of the system $E\dot{\mathbf{x}} \doteq A\mathbf{x}$ is asymptotically stable if and only if all finite eigenvalues of $\lambda E - A$ have negative real part.
3. [Sty02, p. 48] Let Q be a Hermitian matrix such that $\mathbf{v}^* Q\mathbf{v} > 0$ for all nonzero vectors $\mathbf{v} \in \text{range}(P_r)$. The equilibrium $\mathbf{x}_e(t) \equiv 0$ of $E\dot{\mathbf{x}} = A\mathbf{x}$ is asymptotically stable if the generalized Lyapunov equation $E^*XA + A^*XE = -Q$ has a Hermitian, positive semidefinite solution X.
4. [Sty02, pp. 49–52] The equilibrium $\mathbf{x}_e(t) \equiv 0$ of the system $E\dot{\mathbf{x}} = A\mathbf{x}$ is asymptotically stable and the pencil $\lambda E - A$ is of index at most one if and only if the generalized Lyapunov equation $E^*XA + A^*XE = -E^*QE$, with Hermitian, positive definite Q has a Hermitian, positive semidefinite solution X.
5. [TMK95] The equilibrium $\mathbf{x}_e(t) \equiv 0$ of the system $E\dot{\mathbf{x}} = A\mathbf{x}$ is asymptotically stable and the pencil $\lambda E - A$ is of index at most one if and only if the generalized Lyapunov equation

$$A^*X + Y^*A = -Q, \qquad Y^*E = E^*X,$$

with Hermitian, positive definite Q has a solution (X, Y) such that E^*X is Hermitian, positive semidefinite.
6. [Sty02, pp. 52–54] The equilibrium $\mathbf{x}_e(t) \equiv 0$ of the system $E\dot{\mathbf{x}} = A\mathbf{x}$ is asymptotically stable if and only if the projected generalized Lyapunov equation

$$E^*XA + A^*XE = -P_r^*QP_r, \qquad X = P_l^*XP_l$$

has a unique Hermitian, positive semidefinite solution X for every Hermitian, positive definite matrix Q.
7. [Sty02] The equilibrium $\mathbf{x}_e(t) \equiv 0$ of the system $E\dot{\mathbf{x}} = A\mathbf{x}$ is asymptotically stable if and only if the projected generalized Lyapunov equation

$$EYA^* + AYE^* = -P_lQP_l^*, \qquad Y = P_rYP_r^*$$

has a unique Hermitian, positive semidefinite solution Y for every Hermitian, positive definite matrix Q.
8. [Sty02, pp. 28–31] Let H be a symmetric, positive semidefinite solution of the projected generalized Lyapunov equation $E^*HA + A^*HE = -P_r^*P_r$, $H = P_l^*HP_l$ and let $\mathbf{x}$ be a solution of the initial value problem $E\dot{\mathbf{x}} = A\mathbf{x}$, $\mathbf{x}(0) = P_r\mathbf{x}_0$. Then in terms of the original data,

$$\|\mathbf{x}(t)\|_2 \leq \sqrt{\kappa(E, A)\|E\|_2\|(EP_r + A(I - P_r))^{-1}\|_2}\, e^{-t\|A\|_2/(\kappa(E,A)\|E\|_2)}\|P_r\mathbf{x}_0\|_2,$$

where $\kappa(E, A) = 2\|E\|_2\|A\|_2\|H\|_2$.
9. From the previous fact it follows that the equilibrium $\mathbf{x}_e(t) \equiv 0$ of the system $E\dot{\mathbf{x}} = A\mathbf{x}$ is exponentially stable if and only if it is asymptotically stable.

Applications:

1. The finite eigenvalues of the pencil $\lambda E - A$ in the RLC electrical circuit example are given by

$$\lambda_1 = -\frac{R}{2L} - \sqrt{\frac{R^2}{4L^2} - \frac{1}{LC}}, \qquad \lambda_2 = -\frac{R}{2L} + \sqrt{\frac{R^2}{4L^2} - \frac{1}{LC}}.$$

Hence, the equilibrium $\mathbf{x}_e(t) \equiv 0$ of the system $E\dot{\mathbf{x}} = A\mathbf{x}$ is asymptotically stable.
2. The pencil $\lambda E - A$ in the linearized pendulum example has the finite eigenvalues $\lambda_1 = -i\sqrt{g/l}$ and $\lambda_2 = i\sqrt{g/l}$. In this case the equilibrium $\mathbf{x}_e(t) \equiv 0$ of the system $E\dot{\mathbf{x}} = A\mathbf{x}$ is stable but not asymptotically stable.

Examples:

1. The generalized Lyapunov equation $E^*XA + A^*XE = -Q$ with

$$E = \begin{bmatrix} 1 & 0 \\ 0 & 0 \end{bmatrix}, \qquad A = \begin{bmatrix} -1 & 0 \\ 0 & 1 \end{bmatrix}, \qquad Q = \begin{bmatrix} 1 & 0 \\ 0 & 1 \end{bmatrix}$$

has no solution, although the finite eigenvalue of $\lambda E - A$ is negative and $\lambda E - A$ has index one.

2. The generalized Lyapunov equation $E^*XA + A^*XE = -E^*QE$ with

$$E = \begin{bmatrix} 1 & 0 & 0 \\ 0 & 0 & 1 \\ 0 & 0 & 0 \end{bmatrix}, \qquad A = \begin{bmatrix} -2 & 0 & 0 \\ 0 & 1 & 0 \\ 0 & 0 & 1 \end{bmatrix}, \qquad Q = \begin{bmatrix} 1 & 0 & 0 \\ 0 & 2 & 0 \\ 0 & 0 & 2 \end{bmatrix}$$

has no Hermitian, positive semidefinite solution, although the finite eigenvalue of $\lambda E - A$ is negative.

3. The generalized Lyapunov equation $A^*X + Y^*A = -Q$, $Y^*E = E^*X$ with

$$E = \begin{bmatrix} 1 & 0 & 0 \\ 0 & 0 & 1 \\ 0 & 0 & 0 \end{bmatrix}, \qquad A = \begin{bmatrix} -1 & 0 & 0 \\ 0 & 1 & 0 \\ 0 & 0 & 1 \end{bmatrix}, \qquad Q = \begin{bmatrix} 1 & 0 & 0 \\ 0 & 1 & 0 \\ 0 & 0 & 1 \end{bmatrix}$$

has no solution, although the finite eigenvalue of $\lambda E - A$ is negative.

References

[Arn92] V.I. Arnold. *Ordinary Differential Equations.* Springer-Verlag, Berlin, 1992.

[Cam80] S.L. Campbell. *Singular Systems of Differential Equations.* Pitman, San Francisco, 1980.

[Ces63] L. Cesari. *Asymptotic Behavior and Stability Problems in Ordinary Differential Equations.* Springer-Verlag, Berlin, 1963.

[Dai89] L. Dai. *Singular Control Systems.* Lecture Notes in Control and Information Sciences, 118, Springer-Verlag, Berlin, 1989.

[Gan59a] F.R. Gantmacher. *The Theory of Matrices.* Vol. 1. Chelsea Publishing Co., New York, 1959.

[Gan59b] F.R. Gantmacher. *The Theory of Matrices.* Vol. 2. Chelsea Publishing Co., New York, 1959.

[God97] S.K. Godunov. *Ordinary Differential Equations with Constant Coefficients.* Translations of Mathematical Monographs 169, AMS, Providence, RI, 1997.

[Hah67] W. Hahn. *Stability of Motion.* Springer-Verlag, Berlin, 1967.

[KM06] P. Kunkel and V. Mehrmann. *Differential-Algebraic Equations. Analysis and Numerical Solution.* EMS Publishing House, Zürich, Switzerland, 2006.

[Sty02] T. Stykel. *Analysis and Numerical Solution of Generalized Lyapunov Equations.* Ph.D. thesis, Institut für Mathematik, Technische Universität Berlin, 2002.

[TMK95] K. Takaba, N. Morihira, and T. Katayama. A generalized Lyapunov theory for descriptor systems. *Syst. Cont. Lett.*, 24:49–51, 1995.

56

Dynamical Systems and Linear Algebra

Fritz Colonius
Universität Augsburg

Wolfgang Kliemann
Iowa State University

Linear algebra plays a key role in the theory of dynamical systems, and concepts from dynamical systems allow the study, characterization, and generalization of many objects in linear algebra, such as similarity of matrices, eigenvalues, and (generalized) eigenspaces. The most basic form of this interplay can be seen as a matrix A gives rise to a continuous time dynamical system via the linear ordinary differential equation $\dot{\mathbf{x}} = A\mathbf{x}$, or a discrete time dynamical system via iteration $\mathbf{x}_{n+1} = A\mathbf{x}_n$. The properties of the solutions are intimately related to the properties of the matrix A. Matrices also define nonlinear systems on smooth manifolds, such as the sphere $\mathbb{S}^{d-1}$ in $\mathbb{R}^d$, the Grassmann manifolds, or on classical (matrix) Lie groups. Again, the behavior of such systems is closely related to matrices and their properties. And the behavior of nonlinear systems, e.g., of differential equations $\dot{\mathbf{y}} = f(\mathbf{y})$ in $\mathbb{R}^d$ with a fixed point $\mathbf{y}_0 \in \mathbb{R}^d$, can be described locally around $\mathbf{y}_0$ via the linear differential equation $\dot{\mathbf{x}} = D_{\mathbf{y}} f(\mathbf{y}_0)\mathbf{x}$.

Since A. M. Lyapunov's thesis in 1892, it has been an intriguing problem how to construct an appropriate linear algebra for time varying systems. Note that, e.g., for stability of the solutions of $\dot{\mathbf{x}} = A(t)\mathbf{x}$, it is not sufficient that for all $t \in \mathbb{R}$ the matrices $A(t)$ have only eigenvalues with negative real part (see [Hah67], Chapter 62). Of course, Floquet theory (see [Flo83]) gives an elegant solution for the periodic case, but it is not immediately clear how to build a linear algebra around Lyapunov's "order numbers" (now called Lyapunov exponents). The multiplicative ergodic theorem of Oseledets [Ose68] resolves the issue for measurable linear systems with stationary time dependencies, and the Morse spectrum together with Selgrade's theorem [Sel75] clarifies the situation for continuous linear systems with chain transitive time dependencies.

This chapter provides a first introduction to the interplay between linear algebra and analysis/topology in continuous time. Section 56.1 recalls facts about d-dimensional linear differential equations $\dot{\mathbf{x}} = A\mathbf{x}$, emphasizing eigenvalues and (generalized) eigenspaces. Section 56.2 studies solutions in Euclidian space $\mathbb{R}^d$ from the point of view of topological equivalence and conjugacy with related characterizations of the matrix A. Section 56.3 presents, in a fairly general set-up, the concepts of chain recurrence and Morse decompositions for dynamical systems. These ideas are then applied in section 56.4 to nonlinear systems on Grassmannian and flag manifolds induced by a single matrix A, with emphasis on characterizations of the matrix A from this point of view. Section 56.5 introduces linear skew product flows as a way to model time varying linear systems $\dot{\mathbf{x}} = A(t)\mathbf{x}$ with, e.g., periodic, measurable ergodic, and continuous chain transitive time dependencies. The following sections 56.6, 56.7, and 56.8 develop generalizations of (real parts of) eigenvalues and eigenspaces as a starting point for a linear algebra for classes of time varying linear systems, namely periodic, random, and robust systems. (For the corresponding generalization of the imaginary parts of eigenvalues see, e.g., [Arn98] for the measurable ergodic case and [CFJ06] for the continuous, chain transitive case.) Section 56.9 introduces some basic ideas to study genuinely nonlinear systems via linearization, emphasizing invariant manifolds and Grobman–Hartman-type results that compare nonlinear behavior locally to the behavior of associated linear systems.

Notation:

In this chapter, the set of $d \times d$ real matrices is denoted by $gl(d, \mathbb{R})$ rather than $\mathbb{R}^{d\times d}$.

56.1 Linear Differential Equations

Linear differential equations can be solved explicitly if one knows the eigenvalues and a basis of eigenvectors (and generalized eigenvectors, if necessary). The key idea is that of the Jordan form of a matrix. The real parts of the eigenvectors determine the exponential behavior of the solutions, described by the Lyapunov exponents and the corresponding Lyapunov subspaces.

For information on matrix functions, including the matrix exponential, see Chapter 11. For information on the Jordan canonical form see Chapter 6. Systems of first order linear differential equations are also discussed in Chapter 55.

Definitions:

For a matrix $A \in gl(d, \mathbb{R})$, the **exponential** $e^A \in \mathrm{GL}(d, \mathbb{R})$ is defined by $e^A = I + \sum_{n=1}^{\infty} \frac{1}{n!} A^n \in GL(d, \mathbb{R})$, where $I \in gl(d, \mathbb{R})$ is the identity matrix.

A **linear differential equation** (with constant coefficients) is given by a matrix $A \in gl(d, \mathbb{R})$ via $\dot{\mathbf{x}}(t) = A\mathbf{x}(t)$, where $\dot{\mathbf{x}}$ denotes differentiation with respect to t. Any function $\mathbf{x} : \mathbb{R} \longrightarrow \mathbb{R}^d$ such that $\dot{\mathbf{x}}(t) = A\mathbf{x}(t)$ for all $t \in \mathbb{R}$ is called a solution of $\dot{\mathbf{x}} = A\mathbf{x}$.

The **initial value problem** for a linear differential equation $\dot{\mathbf{x}} = A\mathbf{x}$ consists in finding, for a given **initial value** $\mathbf{x}_0 \in \mathbb{R}^d$, a solution $\mathbf{x}(\cdot, \mathbf{x}_0)$ that satisfies $\mathbf{x}(0, \mathbf{x}_0) = \mathbf{x}_0$.

The distinct (complex) eigenvalues of $A \in gl(d, \mathbb{R})$ will be denoted $\mu_1, \ldots, \mu_r$. (For definitions and more information about eigenvalues, eigenvectors, and eigenspaces, see Section 4.3. For information about generalized eigenspaces, see Chapter 6.) The real version of the generalized eigenspace is denoted by $E(A, \mu_k) \subset \mathbb{R}^d$ or simply E_k for $k = 1, \ldots, r \leq d$.

The **real Jordan form** of a matrix $A \in gl(d, \mathbb{R})$ is denoted by $J_A^{\mathbb{R}}$. Note that for any matrix A there is a matrix $T \in GL(d, \mathbb{R})$ such that $A = T^{-1} J_A^{\mathbb{R}} T$.

Let $\mathbf{x}(\cdot, \mathbf{x}_0)$ be a solution of the linear differential equation $\dot{\mathbf{x}} = A\mathbf{x}$. Its **Lyapunov exponent** for $\mathbf{x}_0 \neq 0$ is defined as $\lambda(\mathbf{x}_0) = \limsup_{t\to\infty} \frac{1}{t} \log \|\mathbf{x}(t, \mathbf{x}_0)\|$, where log denotes the natural logarithm and $\|\cdot\|$ is any norm in $\mathbb{R}^d$.

Let $\mu_k = \lambda_k + i\nu_k$, $k = 1, \ldots, r$, be the distinct eigenvalues of $A \in gl(d, \mathbb{R})$. We order the distinct real parts of the eigenvalues as $\lambda_1 < \ldots < \lambda_l$, $1 \leq l \leq r \leq d$, and define the **Lyapunov space** of λ_j as $L(\lambda_j) = \bigoplus E_k$, where the direct sum is taken over all generalized real eigenspaces associated to eigenvalues with real part equal to λ_j. Note that $\bigoplus_{j=1}^{l} L(\lambda_j) = \mathbb{R}^d$.

The **stable, center, and unstable subspaces** associated with the matrix $A \in gl(d,\mathbb{R})$ are defined as $L^- = \bigoplus\{L(\lambda_j), \lambda_j < 0\}$, $L^0 = \bigoplus\{L(\lambda_j), \lambda_j = 0\}$, and $L^+ = \bigoplus\{L(\lambda_j), \lambda_j > 0\}$, respectively.

The zero solution $\mathbf{x}(t,0) \equiv 0$ is called **exponentially stable** if there exists a neighborhood $U(0)$ and positive constants $a, b > 0$ such that $\mathbf{x}(t,\mathbf{x}_0) \le a\|\mathbf{x}_0\|e^{-bt}$ for all $t \in \mathbb{R}$ and $\mathbf{x}_0 \in U(0)$.

Facts:

Literature: [Ama90], [HSD04].

1. For each $A \in gl(d,\mathbb{R})$ the solutions of $\dot{\mathbf{x}} = A\mathbf{x}$ form a d-dimensional vector space $sol(A) \subset C^\infty(\mathbb{R},\mathbb{R}^d)$ over $\mathbb{R}$, where $C^\infty(\mathbb{R},\mathbb{R}^d) = \{f : \mathbb{R} \longrightarrow \mathbb{R}^d,\ f$ is infinitely often differentiable$\}$. Note that the solutions of $\dot{\mathbf{x}} = A\mathbf{x}$ are even real analytic.
2. For each initial value problem given by $A \in gl(d,\mathbb{R})$ and $\mathbf{x}_0 \in \mathbb{R}^d$, the solution $\mathbf{x}(\cdot,\mathbf{x}_0)$ is unique and given by $\mathbf{x}(t,\mathbf{x}_0) = e^{At}\mathbf{x}_0$.
3. Let $\mathbf{v}_1, \ldots, \mathbf{v}_d \in \mathbb{R}^d$ be a basis of $\mathbb{R}^d$. Then the functions $\mathbf{x}(\cdot,\mathbf{v}_1), \ldots, \mathbf{x}(\cdot,\mathbf{v}_d)$ form a basis of the solution space $sol(A)$. The matrix function $X(\cdot) := [\mathbf{x}(\cdot,\mathbf{v}_1), \ldots, \mathbf{x}(\cdot,\mathbf{v}_d)]$ is called a fundamental matrix of $\dot{\mathbf{x}} = A\mathbf{x}$, and it satisfies $\dot{X}(t) = AX(t)$.
4. Let $A \in gl(d,\mathbb{R})$ with distinct eigenvalues $\mu_1, \ldots, \mu_r \in \mathbb{C}$ and corresponding multiplicities $n_k = \alpha(\mu_k)$, $k = 1, \ldots, r$. If E_k are the corresponding generalized real eigenspaces, then $\dim E_k = n_k$ and $\bigoplus_{k=1}^r E_k = \mathbb{R}^d$, i.e., every matrix has a set of generalized real eigenvectors that form a basis of $\mathbb{R}^d$.
5. If $A = T^{-1}J_A^{\mathbb{R}}T$, then $e^{At} = T^{-1}e^{J_A^{\mathbb{R}}t}T$, i.e., for the computation of exponentials of matrices it is sufficient to know the exponentials of Jordan form matrices.
6. Let $\mathbf{v}_1, \ldots, \mathbf{v}_d$ be a basis of generalized real eigenvectors of A. If $\mathbf{x}_0 = \sum_{i=1}^d \alpha_i\, \mathbf{v}_i$, then $\mathbf{x}(t,\mathbf{x}_0) = \sum_{i=1}^d \alpha_i \mathbf{x}(t,\mathbf{v}_i)$ for all $t \in \mathbb{R}$. This reduces the computation of solutions to $\dot{\mathbf{x}} = A\mathbf{x}$ to the computation of solutions for Jordan blocks; see the examples below or [HSD04, Chap. 5] for a discussion of this topic.
7. Each generalized real eigenspace E_k is invariant for the linear differential equation $\dot{\mathbf{x}} = A\mathbf{x}$, i.e., for $\mathbf{x}_0 \in E_k$ it holds that $\mathbf{x}(t,\mathbf{x}_0) \in E_k$ for all $t \in \mathbb{R}$.
8. The Lyapunov exponent $\lambda(\mathbf{x}_0)$ of a solution $\mathbf{x}(\cdot,\mathbf{x}_0)$ (with $\mathbf{x}_0 \neq 0$) satisfies $\lambda(\mathbf{x}_0) = \lim_{t\to\pm\infty} \frac{1}{t}\log\|\mathbf{x}(t,\mathbf{x}_0)\| = \lambda_j$ if and only if $\mathbf{x}_0 \in L(\lambda_j)$. Hence, associated to a matrix $A \in gl(d,\mathbb{R})$ are exactly l Lyapunov exponents, the distinct real parts of the eigenvalues of A.
9. The following are equivalent:

   (a) The zero solution $\mathbf{x}(t,0) \equiv 0$ of the differential equation $\dot{\mathbf{x}} = A\mathbf{x}$ is asymptotically stable.

   (b) The zero solution is exponentially stable

   (c) All Lyapunov exponents are negative.

   (d) $L^- = \mathbb{R}^d$.

Examples:

1. Let $A = \operatorname{diag}(a_1, \ldots, a_d)$ be a diagonal matrix. Then the solution of the linear differential equation $\dot{\mathbf{x}} = A\mathbf{x}$ with initial value $\mathbf{x}_0 \in \mathbb{R}^d$ is given by $$\mathbf{x}(t,\mathbf{x}_0) = e^{At}\mathbf{x}_0 = \begin{bmatrix} e^{a_1 t} & & & \\ & \cdot & & \\ & & \cdot & \\ & & & \cdot \\ & & & & e^{a_d t} \end{bmatrix} \mathbf{x}_0.$$
2. Let $\mathbf{e}_1 = (1, 0, \ldots, 0)^T, \ldots, \mathbf{e}_d = (0, 0, \ldots, 1)^T$ be the standard basis of $\mathbb{R}^d$. Then $\{\mathbf{x}(\cdot,\mathbf{e}_1), \ldots, \mathbf{x}(\cdot,\mathbf{e}_d)\}$ is a basis of the solution space $sol(A)$.
3. Let $A = \operatorname{diag}(a_1, \ldots, a_d)$ be a diagonal matrix. Then the standard basis $\{\mathbf{e}_1, \ldots, \mathbf{e}_d\}$ of $\mathbb{R}^d$ consists of eigenvectors of A.

4. Let $A \in gl(d,\mathbb{R})$ be diagonalizable, i.e., there exists a transformation matrix $T \in GL(d,\mathbb{R})$ and a diagonal matrix $D \in gl(d,\mathbb{R})$ with $A = T^{-1}DT$. Then the solution of the linear differential equation $\dot{\mathbf{x}} = A\mathbf{x}$ with initial value $x_0 \in \mathbb{R}^d$ is given by $\mathbf{x}(t,\mathbf{x}_0) = T^{-1}e^{Dt}T\mathbf{x}_0$, where e^{Dt} is given in Example 1.
5. Let $B = \begin{bmatrix} \lambda & -\nu \\ \nu & \lambda \end{bmatrix}$ be the real Jordan block associated with a complex eigenvalue $\mu = \lambda + i\nu$ of the matrix $A \in gl(d,\mathbb{R})$. Let $\mathbf{y}_0 \in E(A,\mu)$, the real eigenspace of μ. Then the solution $\mathbf{y}(t,\mathbf{y}_0)$ of $\dot{\mathbf{y}} = B\mathbf{y}$ is given by $\mathbf{y}(t,\mathbf{y}_0) = e^{\lambda t}\begin{bmatrix} \cos \nu t & -\sin \nu t \\ \sin \nu t & \cos \nu t \end{bmatrix}\mathbf{y}_0$. According to Fact 6 this is also the $E(A,\mu)$-component of the solutions of $\dot{\mathbf{x}} = J_A^{\mathbb{R}}\mathbf{x}$.
6. Let B be a Jordan block of dimension n associated with the real eigenvalue μ of a matrix $A \in gl(d,\mathbb{R})$. Then for

$$B = \begin{bmatrix} \mu & 1 & & & \\ & \cdot & \cdot & & \\ & & \cdot & \cdot & \\ & & & \cdot & 1 \\ & & & & \mu \end{bmatrix} \quad \text{one has } e^{Bt} = e^{\mu t}\begin{bmatrix} 1 & t & \frac{t^2}{2!} & \cdot & \cdot & \frac{t^{n-1}}{(n-1)!} \\ & \cdot & \cdot & \cdot & & \cdot \\ & & \cdot & \cdot & \cdot & \cdot \\ & & & \cdot & \cdot & \frac{t^2}{2!} \\ & & & & \cdot & t \\ & & & & & 1 \end{bmatrix}.$$

In other words, for $\mathbf{y}_0 = [y_1,\ldots,y_n]^T \in E(A,\mu)$, the jth component of the solution of $\dot{\mathbf{y}} = B\mathbf{y}$ reads $y_j(t,\mathbf{y}_0) = e^{\mu t}\sum_{k=j}^{n}\frac{t^{k-j}}{(k-j)!}y_k$. According to Fact 6 this is also the $E(A,\mu)$-component of $e^{J_A^{\mathbb{R}}t}$.
7. Let B be a real Jordan block of dimension $n = 2m$ associated with the complex eigenvalue $\mu = \lambda + i\nu$ of a matrix $A \in gl(d,\mathbb{R})$. Then with $D = \begin{bmatrix} \lambda & -\nu \\ \nu & \lambda \end{bmatrix}$ and $I = \begin{bmatrix} 1 & 0 \\ 0 & 1 \end{bmatrix}$, for

$$B = \begin{bmatrix} D & I & & & \\ & \cdot & \cdot & & \\ & & \cdot & \cdot & \\ & & & \cdot & I \\ & & & & D \end{bmatrix} \quad \text{one has } e^{Bt} = e^{\lambda t}\begin{bmatrix} \widehat{D} & t\widehat{D} & \frac{t^2}{2!}\widehat{D} & \cdot & \cdot & \frac{t^{n-1}}{(n-1)!}\widehat{D} \\ & \cdot & \cdot & \cdot & & \cdot \\ & & \cdot & \cdot & \cdot & \cdot \\ & & & \cdot & \cdot & \frac{t^2}{2!}\widehat{D} \\ & & & & \cdot & t\widehat{D} \\ & & & & & \widehat{D} \end{bmatrix},$$

where $\widehat{D} = \begin{bmatrix} \cos \nu t & -\sin \nu t \\ \sin \nu t & \cos \nu t \end{bmatrix}$. In other words, for $\mathbf{y}_0 = [y_1, z_1, \ldots, y_m, z_m]^T \in E(A,\mu)$, the jth components, $j = 1,\ldots,m$, of the solution of $\dot{\mathbf{y}} = B\mathbf{y}$ read

$$y_j(t,\mathbf{y}_0) = e^{\mu t}\sum_{k=j}^{m}\frac{t^{k-j}}{(k-j)!}(y_k \cos \nu t - z_k \sin \nu t),$$

$$z_j(t,\mathbf{y}_0) = e^{\mu t}\sum_{k=j}^{m}\frac{t^{k-j}}{(k-j)!}(z_k \cos \nu t + y_k \sin \nu t).$$

According to Fact 6, this is also the $E(A,\mu)$-component of $e^{J_A^{\mathbb{R}}t}$.
8. Using these examples and Facts 5 and 6, it is possible to compute explicitly the solutions to any linear differential equation in $\mathbb{R}^d$.
9. Recall that for any matrix A there is a matrix $T \in GL(d,\mathbb{R})$ such that $A = T^{-1}J_A^{\mathbb{R}}T$, where $J_A^{\mathbb{R}}$ is the real Jordan canonical form of A. The exponential behavior of the solutions of $\dot{\mathbf{x}} = A\mathbf{x}$ can be read off from the diagonal elements of $J_A^{\mathbb{R}}$.

56.2 Linear Dynamical Systems in $\mathbb{R}^d$

The solutions of a linear differential equation $\dot{\mathbf{x}} = A\mathbf{x}$, where $A \in gl(d, \mathbb{R})$, define a (continuous time) dynamical system, or linear flow in $\mathbb{R}^d$. The standard concepts for comparison of dynamical systems are equivalences and conjugacies that map trajectories into trajectories. For linear flows in $\mathbb{R}^d$ these concepts lead to two different classifications of matrices, depending on the smoothness of the conjugacy or equivalence.

Definitions:

The real square matrix A is **hyperbolic** if it has no eigenvalues on the imaginary axis.

A **continuous dynamical system** over the "time set" $\mathbb{R}$ with state space M, a complete metric space, is defined as a map $\Phi : \mathbb{R} \times M \longrightarrow M$ with the properties

(i) $\Phi(0, x) = x$ for all $x \in M$,
(ii) $\Phi(s + t, x) = \Phi(s, \Phi(t, x))$ for all $s, t \in \mathbb{R}$ and all $x \in M$,
(iii) Φ is continuous (in both variables).

The map Φ is also called a (continuous) flow.

For each $x \in M$ the set $\{\Phi(t, x), t \in \mathbb{R}\}$ is called the **orbit (or trajectory)** of the system through x.

For each $t \in \mathbb{R}$ the **time-t map** is defined as $\varphi_t = \Phi(t, \cdot) : M \longrightarrow M$. Using time-$t$ maps, the properties (i) and (ii) above can be restated as (i)$'$ $\varphi_0 = id$, the identity map on M, (ii)$'$ $\varphi_{s+t} = \varphi_s \circ \varphi_t$ for all $s, t \in \mathbb{R}$.

A **fixed point (or equilibrium)** of a dynamical system Φ is a point $x \in M$ with the property $\Phi(t, x) = x$ for all $t \in \mathbb{R}$.

An orbit $\{\Phi(t, x), t \in \mathbb{R}\}$ of a dynamical system Φ is called **periodic** if there exists $\widehat{t} \in \mathbb{R}, \widehat{t} > 0$ such that $\Phi(\widehat{t} + s, x) = \Phi(s, x)$ for all $s \in \mathbb{R}$. The infimum of the positive $\widehat{t} \in \mathbb{R}$ with this property is called the **period** of the orbit. Note that an orbit of period 0 is a fixed point.

Denote by $C^k(X, Y)$ $(k \geq 0)$ the set of k-times **differentiable functions** between C^k-manifolds X and Y, with C^0 denoting continuous.

Let $\Phi, \Psi : \mathbb{R} \times M \longrightarrow M$ be two continuous dynamical systems of class C^k $(k \geq 0)$, i.e., for $k \geq 1$ the state space M is at least a C^k-manifold and Φ, Ψ are C^k-maps. The flows Φ and Ψ are:

(i) C^k**–equivalent** $(k \geq 1)$ if there exists a (local) C^k-diffeomorphism $h : M \to M$ such that h takes orbits of Φ onto orbits of Ψ, **preserving the orientation** (but not necessarily parametrization by time), i.e.,

   (a) For each $x \in M$ there is a strictly increasing and continuous parametrization map $\tau_x : \mathbb{R} \to \mathbb{R}$ such that $h(\Phi(t, x)) = \Psi(\tau_x(t), h(x))$ or, equivalently,

   (b) For all $x \in M$ and $\delta > 0$ there exists $\varepsilon > 0$ such that for all $t \in (0, \delta)$, $h(\Phi(t, x)) = \Psi(t', h(x))$ for some $t' \in (0, \varepsilon)$.

(ii) C^k**-conjugate** $(k \geq 1)$ if there exists a (local) C^k-diffeomorphism $h : M \to M$ such that $h(\Phi(t, x)) = \Psi(t, h(x))$ for all $x \in M$ and $t \in \mathbb{R}$.

Similarly, the flows Φ and Ψ are C^0**-equivalent** if there exists a (local) homeomorphism $h : M \to M$ satisfying the properties of (i) above, and they are C^0**-conjugate** if there exist a (local) homeomorphism $h : M \to M$ satisfying the properties of (ii) above. Often, C^0-equivalence is called **topological equivalence,** and C^0-conjugacy is called **topological conjugacy** or simply **conjugacy**.

Warning: While this terminology is standard in dynamical systems, the terms *conjugate* and *equivalent* are used differently in linear algebra. *Conjugacy* as used here is related to matrix similarity (cf. Fact 6), *not* to matrix conjugacy, and *equivalence* as used here is *not* related to matrix equivalence.

Facts:

Literature: [HSD04], [Rob98].

1. If the flows Φ and Ψ are C^k-conjugate, then they are C^k-equivalent.
2. Each time-t map φ_t has an inverse $(\varphi_t)^{-1} = \varphi_{-t}$, and $\varphi_t : M \longrightarrow M$ is a homeomorphism, i.e., a continuous bijective map with continuous inverse.
3. Denote the set of time-t maps again by $\Phi = \{\varphi_t, t \in \mathbb{R}\}$. A dynamical system is a group in the sense that $(\Phi, \circ)$, with $\circ$ denoting composition of maps, satisfies the group axioms, and $\varphi : (\mathbb{R}, +) \longrightarrow (\Phi, \circ)$, defined by $\varphi(t) = \varphi_t$, is a group homomorphism.
4. Let M be a C^∞-differentiable manifold and X a C^∞-vector field on M such that the differential equation $\dot{x} = X(x)$ has unique solutions $x(t, x_0)$ for all $x_0 \in M$ and all $t \in \mathbb{R}$, with $x(0, x_0) = x_0$. Then $\Phi(t, x_0) = x(t, x_0)$ defines a dynamical system $\Phi : \mathbb{R} \times M \longrightarrow M$.
5. A point $x_0 \in M$ is a fixed point of the dynamical system Φ associated with a differential equation $\dot{x} = X(x)$ as above if and only if $X(x_0) = 0$.
6. For two linear flows Φ (associated with $\dot{\mathbf{x}} = A\mathbf{x}$) and Ψ (associated with $\dot{\mathbf{x}} = B\mathbf{x}$) in $\mathbb{R}^d$, the following are equivalent:
   - Φ and Ψ are C^k-conjugate for $k \geq 1$.
   - Φ and Ψ are linearly conjugate, i.e., the conjugacy map h is a linear operator in GL($\mathbb{R}^d$).
   - A and B are similar, i.e., $A = TBT^{-1}$ for some $T \in GL(d, \mathbb{R})$.
7. Each of the statements in Fact 6 implies that A and B have the same eigenvalue structure and (up to a linear transformation) the same generalized real eigenspace structure. In particular, the C^k-conjugacy classes are exactly the real Jordan canonical form equivalence classes in $gl(d, \mathbb{R})$.
8. For two linear flows Φ (associated with $\dot{\mathbf{x}} = A\mathbf{x}$) and Ψ (associated with $\dot{\mathbf{x}} = B\mathbf{x}$) in $\mathbb{R}^d$, the following are equivalent:
   - Φ and Ψ are C^k-equivalent for $k \geq 1$.
   - Φ and Ψ are linearly equivalent, i.e., the equivalence map h is a linear map in GL($\mathbb{R}^d$).
   - $A = \alpha TBT^{-1}$ for some positive real number α and $T \in GL(d, \mathbb{R})$.
9. Each of the statements in Fact 8 implies that A and B have the same real Jordan structure and their eigenvalues differ by a positive constant. Hence, the C^k-equivalence classes are real Jordan canonical form equivalence classes modulo a positive constant.
10. The set of hyperbolic matrices is open and dense in $gl(d, \mathbb{R})$. A matrix A is hyperbolic if and only if it is structurally stable in $gl(d, \mathbb{R})$, i.e., there exists a neighborhood $U \subset gl(d, \mathbb{R})$ of A such that all $B \in U$ are topologically equivalent to A.
11. If A and B are hyperbolic, then the associated linear flows Φ and Ψ in $\mathbb{R}^d$ are C^0-equivalent (and C^0-conjugate) if and only if the dimensions of the stable subspaces (and, hence, the dimensions of the unstable subspaces) of A and B agree.

Examples:

1. Linear differential equations: For $A \in gl(d, \mathbb{R})$ the solutions of $\dot{\mathbf{x}} = A\mathbf{x}$ form a continuous dynamical system with time set $\mathbb{R}$ and state space $M = \mathbb{R}^d$: Here $\Phi : \mathbb{R} \times \mathbb{R}^d \longrightarrow \mathbb{R}^d$ is defined by $\Phi(t, \mathbf{x}_0) = \mathbf{x}(t, \mathbf{x}_0) = e^{At}\mathbf{x}_0$.
2. Fixed points of linear differential equations: A point $\mathbf{x}_0 \in \mathbb{R}^d$ is a fixed point of the dynamical system Φ associated with the linear differential equation $\dot{\mathbf{x}} = A\mathbf{x}$ if and only if $\mathbf{x}_0 \in \ker A$, the kernel of A.
3. Periodic orbits of linear differential equations: The orbit $\Phi(t, \mathbf{x}_0) := \mathbf{x}(t, \mathbf{x}_0), t \in \mathbb{R}$ is periodic with period $\widehat{t} > 0$ if and only if $\mathbf{x}_0$ is in the eigenspace of a nonzero complex eigenvalue with zero real part.

4. For each matrix $A \in gl(d,\mathbb{R})$ its associated linear flow in $\mathbb{R}^d$ is C^k-conjugate (and, hence, C^k-equivalent) for all $k \geq 0$ to the dynamical system associated with the Jordan form $J_A^{\mathbb{R}}$.

56.3 Chain Recurrence and Morse Decompositions of Dynamical Systems

A matrix $A \in gl(d,\mathbb{R})$ and, hence, a linear differential equation $\dot{\mathbf{x}} = A\mathbf{x}$ maps subspaces of $\mathbb{R}^d$ into subspaces of $\mathbb{R}^d$. Therefore, the matrix A also defines dynamical systems on spaces of subspaces, such as the Grassmann and the flag manifolds. These are nonlinear systems, but they can be studied via linear algebra, and vice versa; the behavior of these systems allows for the investigation of certain properties of the matrix A. The key topological concepts for the analysis of systems on compact spaces like the Grassmann and flag manifolds are chain recurrence, Morse decompositions, and attractor–repeller decompositions. This section concentrates on the first two approaches, the connection to attractor–repeller decompositions can be found, e.g., in [CK00, App. B2].

Definitions:

Given a dynamical system $\Phi : \mathbb{R} \times M \longrightarrow M$, for a subset $N \subset M$ the **α-limit set** is defined as $\alpha(N) = \{y \in M$, there exist sequences x_n in N and $t_n \to -\infty$ in $\mathbb{R}$ with $\lim_{n\to\infty} \Phi(t_n, x_n) = y\}$, and similarly the **$\omega$-limit set** of N is defined as $\omega(N) = \{y \in M$, there exist sequences x_n in N and $t_n \to \infty$ in $\mathbb{R}$ with $\lim_{n\to\infty} \Phi(t_n, x_n) = y\}$.

For a flow Φ on a complete metric space M and $\varepsilon, T > 0$, an (ε, T)-**chain** from $x \in M$ to $y \in M$ is given by

$$n \in \mathbb{N},\ x_0 = x, \ldots, x_n = y,\ T_0, \ldots, T_{n-1} > T$$

with

$$\mathrm{d}(\Phi(T_i, x_i), x_{i+1}) < \varepsilon \text{ for all } i,$$

where d is the metric on M.

A set $K \subset M$ is **chain transitive** if for all $x, y \in K$ and all $\varepsilon, T > 0$ there is an (ε, T)-chain from x to y.

The **chain recurrent set** $\mathcal{CR}$ is the set of all points that are chain reachable from themselves, i.e., $CR = \{x \in M$, for all $\varepsilon, T > 0$ there is an (ε, T)-chain from x to $x\}$.

A set $\mathcal{M} \subset M$ is a **chain recurrent component**, if it is a maximal (with respect to set inclusion) chain transitive set. In this case $\mathcal{M}$ is a connected component of the chain recurrent set $\mathcal{CR}$.

For a flow Φ on a complete metric space M, a compact subset $K \subset M$ is called **isolated invariant**, if it is invariant and there exists a neighborhood N of K, i.e., a set N with $K \subset \operatorname{int} N$, such that $\Phi(t, x) \in N$ for all $t \in \mathbb{R}$ implies $x \in K$.

A **Morse decomposition** of a flow Φ on a complete metric space M is a finite collection $\{\mathcal{M}_i,\ i = 1, \ldots, l\}$ of nonvoid, pairwise disjoint, and isolated compact invariant sets such that

(i) For all $x \in M$, $\omega(x), \alpha(x) \subset \bigcup_{i=1}^{l} \mathcal{M}_i$; and

(ii) Suppose there are $\mathcal{M}_{j_0}, \mathcal{M}_{j_1}, \ldots, \mathcal{M}_{j_n}$ and $x_1, \ldots, x_n \in M \setminus \bigcup_{i=1}^{l} \mathcal{M}_i$ with $\alpha(x_i) \subset \mathcal{M}_{j_{i-1}}$ and $\omega(x_i) \subset \mathcal{M}_{j_i}$ for $i = 1, \ldots, n$; then $\mathcal{M}_{j_0} \neq \mathcal{M}_{j_n}$.

The elements of a Morse decomposition are called **Morse sets**.

A Morse decomposition $\{\mathcal{M}_i,\ i = 1, \ldots, l\}$ is **finer** than another decomposition $\{\mathcal{N}_j,\ j = 1, \ldots, n\}$, if for all $\mathcal{M}_i$ there exists an index $j \in \{1, \ldots, n\}$ such that $\mathcal{M}_i \subset \mathcal{N}_j$.

Facts:

Literature: [Rob98], [CK00], [ACK05].

1. For a Morse decomposition $\{\mathcal{M}_i, i = 1,\ldots,l\}$ the relation $\mathcal{M}_i \prec \mathcal{M}_j$, given by $\alpha(x) \subset \mathcal{M}_i$ and $\omega(x) \subset \mathcal{M}_j$ for some $x \in M \backslash \cup_{i=1}^{l} \mathcal{M}_i$, induces an order.
2. Let $\Phi, \Psi : \mathbb{R} \times M \longrightarrow M$ be two dynamical systems on a state space M and let $h : M \to M$ be a topological equivalence for Φ and Ψ. Then

   (i) The point $p \in M$ is a fixed point of Φ if and only if $h(p)$ is a fixed point of Ψ;

   (ii) The orbit $\Phi(\cdot, p)$ is closed if and only if $\Psi(\cdot, h(p))$ is closed;

   (iii) If $K \subset M$ is an α-(or ω-) limit set of Φ from $p \in M$, then $h[K]$ is an α-(or ω-) limit set of Ψ from $h(p) \in M$.

   (iv) Given, in addition, two dynamical systems $\Theta_{1,2} : \mathbb{R} \times N \longrightarrow N$, if $h : M \to M$ is a topological conjugacy for the flows Φ and Ψ on M, and $g : N \to N$ is a topological conjugacy for Θ_1 and Θ_2 on N, then the product flows $\Phi \times \Theta_1$ and $\Psi \times \Theta_2$ on $M \times N$ are topologically conjugate via $h \times g : M \times N \longrightarrow M \times N$. This result is, in general, not true for topological equivalence.
3. Topological equivalences (and conjugacies) on a compact metric space M map chain transitive sets onto chain transitive sets.
4. Topological equivalences map invariant sets onto invariant sets, and minimal closed invariant sets onto minimal closed invariant sets.
5. Topological equivalences map Morse decompositions onto Morse decompositions.

Examples:

1. Dynamical systems in $\mathbb{R}^1$: Any limit set $\alpha(x)$ and $\omega(x)$ from a single point x of a dynamical system in $\mathbb{R}^1$ consists of a single fixed point. The chain recurrent components (and the finest Morse decomposition) consist of single fixed points or intervals of fixed points. Any Morse set consists of fixed points and intervals between them.
2. Dynamical systems in $\mathbb{R}^2$: A nonempty, compact limit set of a dynamical system in $\mathbb{R}^2$, which contains no fixed points, is a closed, i.e., a periodic orbit (Poincaré–Bendixson). Any nonempty, compact limit set of a dynamical system in $\mathbb{R}^2$ consists of fixed points, connecting orbits (such as homoclinic or heteroclinic orbits), and periodic orbits.
3. Consider the following dynamical system Φ in $\mathbb{R}^2 \backslash \{0\}$, given by a differential equation in polar form for $r > 0, \theta \in [0, 2\pi)$, and $a \neq 0$:

$$\dot{r} = 1 - r, \ \dot{\theta} = a.$$

   For each $\mathbf{x} \in \mathbb{R}^2 \backslash \{0\}$ the ω-limit set is the circle $\omega(\mathbf{x}) = \mathbb{S}^1 = \{(r,\theta), r = 1, \theta \in [0, 2\pi)\}$. The state space $\mathbb{R}^2 \backslash \{0\}$ is not compact, and α-limit sets exist only for $\mathbf{y} \in \mathbb{S}^1$, for which $\alpha(\mathbf{y}) = \mathbb{S}^1$.
4. Consider the flow Φ from the previous example and a second system Ψ, given by

$$\dot{r} = 1 - r, \ \ \dot{\theta} = b$$

   with $b \neq 0$. Then the flows Φ and Ψ are topologically equivalent, but not conjugate if $b \neq a$.
5. An example of a flow for which the limit sets from points are strictly contained in the chain recurrent components can be obtained as follows: Let $M = [0, 1] \times [0, 1]$. Let the flow Φ on M be defined such that all points on the boundary are fixed points, and the orbits for points $(x, y) \in (0, 1) \times (0, 1)$ are straight lines $\Phi(\cdot, (x, y)) = \{(z_1, z_2), z_1 = x, z_2 \in (0, 1)\}$ with $\lim_{t\to\pm\infty} \Phi(t, (x, y)) = (x, \pm 1)$. For this system, each point on the boundary is its own α- and ω-limit set. The α-limit sets for points in the interior $(x, y) \in (0, 1) \times (0, 1)$ are of the form $\{(x, -1)\}$, and the ω-limit sets are $\{(x, +1)\}$.

The only chain recurrent component for this system is $M = [0,1] \times [0,1]$, which is also the only Morse set.

56.4 Linear Systems on Grassmannian and Flag Manifolds

Definitions:

The ***k*th Grassmannian** $\mathbb{G}_k$ of $\mathbb{R}^d$ can be defined via the following construction: Let $F(k,d)$ be the set of k-frames in $\mathbb{R}^d$, where a ***k*-frame** is an ordered set of k linearly independent vectors in $\mathbb{R}^d$. Two k-frames $X = [\mathbf{x}_1, \ldots, \mathbf{x}_k]$ and $Y = [\mathbf{y}_1, \ldots, \mathbf{y}_k]$ are said to be equivalent, $X \sim Y$, if there exists $T \in GL(k,\mathbb{R})$ with $X^T = TY^T$, where X and Y are interpreted as $d \times k$ matrices. The quotient space $\mathbb{G}_k = F(k,d)/\sim$ is a compact, $k(d-k)$-dimensional differentiable manifold. For $k = 1$, we obtain the projective space $\mathbb{P}^{d-1} = \mathbb{G}_1$ in $\mathbb{R}^d$.

The ***k*th flag** of $\mathbb{R}^d$ is given by the following k–sequences of subspace inclusions,

$$\mathbb{F}_k = \{F_k = (V_1, \ldots, V_k),\ V_i \subset V_{i+1} \text{ and } \dim V_i = i \text{ for all } i\}.$$

For $k = d$, this is the **complete flag** $\mathbb{F} = \mathbb{F}_d$.

Each matrix $A \in gl(d,\mathbb{R})$ defines a map on the subspaces of $\mathbb{R}^d$ as follows: Let $V = \text{Span}(\{\mathbf{x}_1, \ldots, \mathbf{x}_k\})$. Then $AV = \text{Span}(\{A\mathbf{x}_1, \ldots, A\mathbf{x}_k\})$.

Denote by $\mathbb{G}_k\Phi$ and $\mathbb{F}_k\Phi$ the induced flows on the Grassmannians and the flags, respectively.

Facts:

Literature: [Rob98], [CK00], [ACK05].

1. Let $\mathbb{P}\Phi$ be the projection onto $\mathbb{P}^{d-1}$ of a linear flow $\Phi(t,x) = e^{At}x$. Then $\mathbb{P}\Phi$ has l chain recurrent components $\{\mathcal{M}_1, \ldots, \mathcal{M}_l\}$, where l is the number of different Lyapunov exponents (i.e., of different real parts of eigenvalues) of A. For each Lyapunov exponent λ_i, $\mathcal{M}_i = \mathbb{P}L_i$, the projection of the ith Lyapunov space onto $\mathbb{P}^{d-1}$. Furthermore $\{\mathcal{M}_1, \ldots, \mathcal{M}_l\}$ defines the finest Morse decomposition of $\mathbb{P}\Phi$ and $\mathcal{M}_i \prec \mathcal{M}_j$ if and only if $\lambda_i < \lambda_j$.
2. For A, $B \in gl(d,\mathbb{R})$, let $\mathbb{P}\Phi$ and $\mathbb{P}\Psi$ be the associated flows on $\mathbb{P}^{d-1}$ and suppose that there is a topological equivalence h of $\mathbb{P}\Phi$ and $\mathbb{P}\Psi$. Then the chain recurrent components $\mathcal{N}_1, \ldots, \mathcal{N}_n$ of $\mathbb{P}\Psi$ are of the form $\mathcal{N}_i = h[\mathcal{M}_i]$, where $\mathcal{M}_i$ is a chain recurrent component of $\mathbb{P}\Phi$. In particular, the number of chain recurrent components of $\mathbb{P}\Phi$ and $\mathbb{P}\Psi$ agree, and h maps the order on $\{\mathcal{M}_1, \ldots, \mathcal{M}_l\}$ onto the order on $\{\mathcal{N}_1, \ldots, \mathcal{N}_l\}$.
3. For A, $B \in gl(d,\mathbb{R})$ let $\mathbb{P}\Phi$ and $\mathbb{P}\Psi$ be the associated flows on $\mathbb{P}^{d-1}$ and suppose that there is a topological equivalence h of $\mathbb{P}\Phi$ and $\mathbb{P}\Psi$. Then the projected subspaces corresponding to real Jordan blocks of A are mapped onto projected subspaces corresponding to real Jordan blocks of B preserving the dimensions. Furthermore, h maps projected eigenspaces corresponding to real eigenvalues and to pairs of complex eigenvalues onto projected eigenspaces of the same type. This result shows that while C^0-equivalence of projected linear flows on $\mathbb{P}^{d-1}$ determines the number l of distinct Lyapunov exponents, it also characterizes the Jordan structure within each Lyapunov space (but, obviously, not the size of the Lyapunov exponents nor their sign). It imposes very restrictive conditions on the eigenvalues and the Jordan structure. Therefore, C^0-equivalences are not a useful tool to characterize l. The requirement of mapping orbits into orbits is too strong. A weakening leads to the following characterization.
4. Two matrices A and B in $gl(d,\mathbb{R})$ have the same vector of the dimensions d_i of the Lyapunov spaces (in the natural order of their Lyapunov exponents) if and only if there exist a homeomorphism h : $\mathbb{P}^{d-1} \to \mathbb{P}^{d-1}$ that maps the finest Morse decomposition of $\mathbb{P}\Phi$ onto the finest Morse decomposition of $\mathbb{P}\Psi$, i.e., h maps Morse sets onto Morse sets and preserves their orders.

5. Let $A \in gl(d, \mathbb{R})$ with associated flows Φ on $\mathbb{R}^d$ and $\mathbb{F}_k\Phi$ on the k-flag.
   (i) For every $k \in \{1, \ldots, d\}$ there exists a unique finest Morse decomposition $\{{}_k\mathcal{M}_{i_j}\}$ of $\mathbb{F}_k\Phi$, where $i_j \in \{1, \ldots, d\}^k$ is a multi-index, and the number of chain transitive components in $\mathbb{F}_k$ is bounded by $\frac{d!}{(d-k)!}$.
   (ii) Let $\mathcal{M}_i$ with $i \in \{1, \ldots, d\}^k$ be a chain recurrent component in $\mathbb{F}_{k-1}$. Consider the $(d-k+1)$-dimensional vector bundle $\pi : \mathcal{W}(\mathcal{M}_i) \to \mathcal{M}_i$ with fibers
   $$\mathcal{W}(\mathcal{M}_i)_{F_{k-1}} = \mathbb{R}^d / V_{k-1} \text{ for } F_k = (V_1, \ldots, V_{k-1}) \in \mathcal{M}_i \subset \mathbb{F}_{k-1}.$$
   Then every chain recurrent component ${}_{\mathbb{P}}\mathcal{M}_{i_j}$, $j = 1, \ldots, k_i \leq d - k + 1$, of the projective bundle $\mathbb{P}\mathcal{W}(\mathcal{M}_i)$ determines a chain recurrent component ${}_k\mathcal{M}_{i_j}$ on $\mathbb{F}_k$ via
   $${}_k\mathcal{M}_{i_j} = \left\{F_k = (F_{k-1}, V_k) \in \mathbb{F}_k : F_{k-1} \in \mathcal{M}_i \text{ and } \mathbb{P}(V_k / V_{k-1}) \in {}_{\mathbb{P}}\mathcal{M}_{i_j}\right\}.$$
   Every chain recurrent component in $\mathbb{F}_k$ is of this form; this determines the multiindex i_j inductively for $k = 2, \ldots, d$.
6. On every Grassmannian $\mathbb{G}_i$ there exists a finest Morse decomposition of the dynamical system $\mathbb{G}_i\Phi$. Its Morse sets are given by the projection of the chain recurrent components from the complete flag $\mathbb{F}$.
7. Let $A \in gl(d, \mathbb{R})$ be a matrix with flow Φ on $\mathbb{R}^d$. Let L_i, $i = 1, \ldots, l$, be the Lyapunov spaces of A, i.e., their projections $\mathbb{P}L_i = \mathcal{M}_i$ are the finest Morse decomposition of $\mathbb{P}\Phi$ on the projective space. For $k = 1, \ldots, d$ define the index set
   $$I(k) = \{(k_1, \ldots, k_m) : k_1 + \ldots + k_m = k \text{ and } 0 \leq k_i \leq d_i = \dim L_i\}.$$
   Then the finest Morse decomposition on the Grassmannian $\mathbb{G}_k$ is given by the sets
   $$\mathcal{N}^k_{k_1,\ldots,k_m} = \mathbb{G}_{k_1} L_1 \oplus \ldots.. \oplus \mathbb{G}_{k_m} L_m, \ (k_1, \ldots, k_m) \in I(k).$$
8. For two matrices $A, B \in gl(d, \mathbb{R})$ the vector of the dimensions d_i of the Lyapunov spaces (in the natural order of their Lyapunov exponents) are identical if and only if certain graphs defined on the Grassmannians are isomorphic; see [ACK05].

Examples:

1. For $A \in gl(d, \mathbb{R})$ let Φ be its linear flow in $\mathbb{R}^d$. The flow Φ projects onto a flow $\mathbb{P}\Phi$ on $\mathbb{P}^{d-1}$, given by the differential equation
   $$\dot{s} = h(s, A) = (A - s^T A s\, I)\, s, \text{ with } s \in \mathbb{P}^{d-1}.$$
   Consider the matrices
   $$A = \text{diag}(-1, -1, 1) \text{ and } B = \text{diag}(-1, 1, 1).$$
   We obtain the following structure for the finest Morse decompositions on the Grassmannians for A:
   $\mathbb{G}_1$: $\mathcal{M}_1 = \{\text{Span}(\mathbf{e}_1, \mathbf{e}_2)\}$ and $\mathcal{M}_3 = \{\text{Span}(\mathbf{e}_3)\}$
   $\mathbb{G}_2$: $\mathcal{M}_{1,2} = \{\text{Span}(\mathbf{e}_1, \mathbf{e}_2)\}$ and $\mathcal{M}_{1,3} = \{\{\text{Span}(\mathbf{x}, \mathbf{e}_3)\} : \mathbf{x} \in \text{Span}(\mathbf{e}_1, \mathbf{e}_2)\}$
   $\mathbb{G}_3$: $\mathcal{M}_{1,2,3} = \{\text{Span}(\mathbf{e}_1, \mathbf{e}_2, \mathbf{e}_3)\}$
   and for B we have
   $\mathbb{G}_1$: $\mathcal{N}_1 = \{\text{Span}(\mathbf{e}_1)\}$ and $\mathcal{N}_2 = \{\text{Span}(\mathbf{e}_2, \mathbf{e}_3)\}$
   $\mathbb{G}_2$: $\mathcal{N}_{1,2} = \{\text{Span}(\mathbf{e}_1, \mathbf{x}) : \mathbf{x} \in \text{Span}(\mathbf{e}_2, \mathbf{e}_3)\}$ and $\mathcal{N}_{2,3} = \{\text{Span}(\mathbf{e}_2, \mathbf{e}_3)\}$
   $\mathbb{G}_3$: $\mathcal{N}_{1,2,3} = \{\text{Span}(\mathbf{e}_1, \mathbf{e}_2, \mathbf{e}_3)\}$.

On the other hand, the Morse sets in the full flag are given for A and B by

$$\begin{bmatrix} \mathcal{M}_{1,2,3} \\ \mathcal{M}_{1,2} \\ \mathcal{M}_1 \end{bmatrix} \preceq \begin{bmatrix} \mathcal{M}_{1,2,3} \\ \mathcal{M}_{1,3} \\ \mathcal{M}_1 \end{bmatrix} \preceq \begin{bmatrix} \mathcal{M}_{1,2,3} \\ \mathcal{M}_{1,3} \\ \mathcal{M}_3 \end{bmatrix} \text{ and } \begin{bmatrix} \mathcal{N}_{1,2,3} \\ \mathcal{N}_{1,2} \\ \mathcal{N}_1 \end{bmatrix} \preceq \begin{bmatrix} \mathcal{N}_{1,2,3} \\ \mathcal{N}_{1,2} \\ \mathcal{N}_2 \end{bmatrix} \preceq \begin{bmatrix} \mathcal{N}_{1,2,3} \\ \mathcal{N}_{2,3} \\ \mathcal{N}_2 \end{bmatrix},$$

respectively. Thus, in the full flag, the numbers and the orders of the Morse sets coincide, while on the Grassmannians (together with the projection relations between different Grassmannians) one can distinguish also the dimensions of the corresponding Lyapunov spaces. (See [ACK05] for a precise statement.)

56.5 Linear Skew Product Flows

Developing a linear algebra for time varying systems $\dot{\mathbf{x}} = A(t)\mathbf{x}$ means defining appropriate concepts to generalize eigenvalues, linear eigenspaces and their dimensions, and certain normal forms that characterize the behavior of the solutions of a time varying system and that reduce to the constant matrix case if $A(t) \equiv A \in gl(d,\mathbb{R})$. The eigenvalues and eigenspaces of the family $\{A(t), t \in \mathbb{R}\}$ do not provide an appropriate generalization; see, e.g., [Hah67], Chapter 62. For certain classes of time varying systems it turns out that the Lyapunov exponents and Lyapunov spaces introduced in section 56.1 capture the key properties of (real parts of) eigenvalues and of the associated subspace decomposition of $\mathbb{R}^d$. These systems are linear skew product flows for which the base is a (nonlinear) system θ_t that enters into the linear dynamics of a differential equation in the form $\dot{\mathbf{x}} = A(\theta_t)\mathbf{x}$. Examples for this type of systems include periodic and almost periodic differential equations, random differential equations, systems over ergodic or chain recurrent bases, linear robust systems, and bilinear control systems. This section concentrates on periodic linear differential equations, random linear dynamical systems, and robust linear systems. It is written to emphasize the correspondences between the linear algebra in Section 56.1, Floquet theory, the multiplicative ergodic theorem, and the Morse spectrum and Selgrade's theorem.
Literature: [Arn98], [BK94], [CK00], [Con97], [Rob98].

Definitions:

A (continuous time) **linear skew-product flow** is a dynamical system with state space $M = \Omega \times \mathbb{R}^d$ and flow $\Phi : \mathbb{R}\times\Omega\times\mathbb{R}^d \longrightarrow \Omega\times\mathbb{R}^d$, where $\Phi = (\theta,\varphi)$ is defined as follows: $\theta : \mathbb{R}\times\Omega \longrightarrow \Omega$ is a dynamical system, and $\varphi : \mathbb{R}\times\Omega\times\mathbb{R}^d \longrightarrow \mathbb{R}^d$ is linear in its $\mathbb{R}^d$-component, i.e., for each $(t,\omega) \in \mathbb{R}\times\Omega$ the map $\varphi(t,\omega,\cdot) : \mathbb{R}^d \longrightarrow \mathbb{R}^d$ is linear. Skew-product flows are called measurable (continuous, differentiable) if $\Omega = (\theta,\varphi)$ is a measurable space (topological space, differentiable manifold) and Φ is measurable (continuous, differentiable). For the time-t maps, the notation $\theta_t = \theta(t,\cdot) : \Omega \longrightarrow \Omega$ is used again.

Note that the **base component** $\theta : \mathbb{R}\times\Omega \longrightarrow \Omega$ is a dynamical system itself, while the **skew-component** φ is not a dynamical system. The skew-component φ is often called a **co-cycle** over θ.

Let $\Phi : \mathbb{R}\times\Omega\times\mathbb{R}^d \longrightarrow \Omega\times\mathbb{R}^d$ be a linear skew-product flow. For $\mathbf{x}_0 \in \mathbb{R}^d$, $\mathbf{x}_0 \neq 0$, the **Lyapunov exponent** is defined as $\lambda(\mathbf{x}_0,\omega) = \limsup_{t\to\infty} \frac{1}{t} \log \|\varphi(t,\omega,\mathbf{x}_0)\|$, where log denotes the natural logarithm and $\|\cdot\|$ is any norm in $\mathbb{R}^d$.

Examples:

1. Time varying linear differential equations: Let $A : \mathbb{R} \longrightarrow gl(d,\mathbb{R})$ be a uniformly continuous function and consider the linear differential equation $\dot{\mathbf{x}}(t) = A(t)\mathbf{x}(t)$. The solutions of this differential equation define a dynamical system via $\Phi : \mathbb{R}\times\mathbb{R}\times\mathbb{R}^d \longrightarrow \mathbb{R}\times\mathbb{R}^d$, where $\theta : \mathbb{R}\times\mathbb{R} \longrightarrow \mathbb{R}$ is given by $\theta(t,\tau) = t+\tau$, and $\varphi : \mathbb{R}\times\mathbb{R}\times\mathbb{R}^d \longrightarrow \mathbb{R}^d$ is defined as $\varphi(t,\tau,\mathbf{x}_0) = X(t+\tau,\tau)\mathbf{x}_0$. Here $X(t,\tau)$ is a fundamental matrix of the differential equation $\dot{X}(t) = A(t)X(t)$ in $gl(d,\mathbb{R})$. Note that for $\varphi(t,\tau,\cdot) : \mathbb{R}^d \longrightarrow \mathbb{R}^d$, $t \in \mathbb{R}$, we have $\varphi(t+s,\tau) = \varphi(t,\theta(s,\tau)) \circ \varphi(s,\tau)$ and, hence, the

solutions of $\dot{\mathbf{x}}(t) = A(t)\mathbf{x}(t)$ themselves do not define a flow. The additional component θ "keeps track of time."

2. Metric dynamical systems: Let $(\Omega, \mathcal{F}, P)$ be a probability space, i.e., a set Ω with σ-algebra $\mathcal{F}$ and probability measure P. Let $\theta : \mathbb{R} \times \Omega \longrightarrow \Omega$ be a measurable flow such that the probability measure P is invariant under θ, i.e., $\theta_t P = P$ for all $t \in \mathbb{R}$, where for all measurable sets $X \in \mathcal{F}$ we define $\theta_t P(X) = P\{\theta_t^{-1}(X)\} = P(X)$. Flows of this form are often called metric dynamical systems.
3. Random linear dynamical systems: A random linear dynamical system is a skew-product flow $\Phi : \mathbb{R} \times \Omega \times \mathbb{R}^d \longrightarrow \Omega \times \mathbb{R}^d$, where $(\Omega, \mathcal{F}, P, \theta)$ is a metric dynamical system and each $\varphi : \mathbb{R} \times \Omega \times \mathbb{R}^d \longrightarrow \mathbb{R}^d$ is linear in its $\mathbb{R}^d$-component. Examples for random linear dynamical systems are given, e.g., by linear stochastic differential equations or linear differential equations with stationary background noise; see [Arn98].
4. Robust linear systems: Consider a linear system with time varying perturbations of the form $\dot{\mathbf{x}} = A(u(t))\mathbf{x} := A_0\mathbf{x} + \sum_{i=1}^{m} u_i(t)A_i\mathbf{x}$, where $A_0, \ldots, A_m \in gl(d, \mathbb{R})$, $u \in \mathcal{U} = \{u : \mathbb{R} \longrightarrow U$, integrable on every bounded interval$\}$, and $U \subset \mathbb{R}^m$ is compact, convex with $0 \in \text{int } U$. A robust linear system defines a linear skew-product flow via the following construction: We endow $\mathcal{U}$ with the weak*-topology of $L^\infty(\mathbb{R}, U)^*$ to make it a compact, metrizable space. The base component is defined as the shift $\theta : \mathbb{R} \times \mathcal{U} \longrightarrow \mathcal{U}$, $\theta(t, u(\cdot)) = u(\cdot + t)$, and the skew-component consists of the solutions $\varphi(t, u(\cdot), \mathbf{x})$, $t \in \mathbb{R}$ of the perturbed differential equation. Then $\Phi : \mathbb{R} \times \mathcal{U} \times \mathbb{R}^d \longrightarrow \mathcal{U} \times \mathbb{R}^d$, $\Phi(t, u, \mathbf{x}) = (\theta(t, u), \varphi(t, u, \mathbf{x}))$ defines a continuous linear skew-product flow. The functions u can also be considered as (open loop) controls.

56.6 Periodic Linear Differential Equations: Floquet Theory

Definitions:

A **periodic linear differential equation** $\dot{\mathbf{x}} = A(\theta_t)\mathbf{x}$ is given by a matrix function $A : \mathbb{R} \longrightarrow gl(d, \mathbb{R})$ that is continuous and periodic (of period $\widehat{t} > 0$). As above, the solutions define a dynamical system via $\Phi : \mathbb{R} \times \mathbb{S}^1 \times \mathbb{R}^d \longrightarrow \mathbb{S}^1 \times \mathbb{R}^d$, if we identify $\mathbb{R}$ mod$\widehat{t}$ with the circle $\mathbb{S}^1$.

Facts:

Literature: [Ama90], [GH83], [Hah67], [Sto92], [Wig96].

1. Consider the periodic linear differential equation $\dot{\mathbf{x}} = A(\theta_t)\mathbf{x}$ with period $\widehat{t} > 0$. A fundamental matrix $X(t)$ of the system is of the form $X(t) = P(t)e^{Rt}$ for $t \in \mathbb{R}$, where $P(\cdot)$ is a nonsingular, differentiable, and $\widehat{t}$-periodic matrix function and $R \in gl(d, \mathbb{C})$.
2. Let $X(\cdot)$ be a fundamental solution with $X(0) = I \in GL(d, \mathbb{R})$. The matrix $X(\widehat{t}) = e^{R\widehat{t}}$ is called the monodromy matrix of the system. Note that R is, in general, not uniquely determined by X, and does not necessarily have real entries. The eigenvalues α_j, $j = 1, \ldots, d$ of $X(\widehat{t})$ are called the characteristic multipliers of the system, and the eigenvalues $\mu_j = \lambda_j + i\nu_j$ of R are the characteristic exponents. It holds that $\mu_j = \frac{1}{\widehat{t}} \log \alpha_j + \frac{2m\pi i}{\widehat{t}}$, $j = 1, \ldots, d$ and $m \in \mathbb{Z}$. This determines uniquely the real parts of the characteristic exponents $\lambda_j = \text{Re } \mu_j = \log |\alpha_j|$, $j = 1, \ldots, d$. The λ_j are called the Floquet exponents of the system.
3. Let $\Phi = (\theta, \varphi) : \mathbb{R} \times \mathbb{S}^1 \times \mathbb{R}^d \longrightarrow \mathbb{S}^1 \times \mathbb{R}^d$ be the flow associated with a periodic linear differential equation $\dot{\mathbf{x}} = A(t)\mathbf{x}$. The system has a finite number of Lyapunov exponents λ_j, $j = 1, \ldots, l \leq d$. For each exponent λ_j and each $\tau \in \mathbb{S}^1$ there exists a splitting $\mathbb{R}^d = \bigoplus_{j=1}^{l} L(\lambda_j, \tau)$ of $\mathbb{R}^d$ into linear subspaces with the following properties:

   (a) The subspaces $L(\lambda_j, \tau)$ have the same dimension independent of τ, i.e., for each $j = 1, \ldots, l$ it holds that $\dim L(\lambda_j, \sigma) = \dim L(\lambda_j, \tau) =: d_i$ for all $\sigma, \tau \in \mathbb{S}^1$.

(b) The subspaces $L(\lambda_j, \tau)$ are invariant under the flow Φ, i.e., for each $j = 1, \ldots, l$ it holds that $\varphi(t,\tau)L(\lambda_j,\tau) = L(\lambda_j, \theta(t,\tau)) = L(\lambda_j, t+\tau)$ for all $t \in \mathbb{R}$ and $\tau \in \mathbb{S}^1$.

(c) $\lambda(\mathbf{x},\tau) = \lim_{t\to\pm\infty} \frac{1}{t} \log \|\varphi(t,\tau,\mathbf{x})\| = \lambda_j$ if and only if $\mathbf{x} \in L(\lambda_j,\tau)\backslash\{0\}$.

4. The Lyapunov exponents of the system are exactly the Floquet exponents. The linear subspaces $L(\lambda_j, \cdot)$ are called the Lyapunov spaces (or sometimes the Floquet spaces) of the periodic matrix function $A(t)$.
5. For each $j = 1, \ldots, l \leq d$ the map $L_j : \mathbb{S}^1 \longrightarrow \mathbb{G}_{d_j}$ defined by $\tau \longmapsto L(\lambda_j, \tau)$ is continuous.
6. These facts show that for periodic matrix functions $A : \mathbb{R} \longrightarrow gl(d,\mathbb{R})$ the Floquet exponents and Floquet spaces replace the real parts of eigenvalues and the Lyapunov spaces, concepts that are so useful in the linear algebra of (constant) matrices $A \in gl(d,\mathbb{R})$. The number of Lyapunov exponents and the dimensions of the Lyapunov spaces are constant for $\tau \in \mathbb{S}^1$, while the Lyapunov spaces themselves depend on the time parameter τ of the periodic matrix function $A(t)$, and they form periodic orbits in the Grassmannians $\mathbb{G}_{d_j}$ and in the corresponding flag.
7. As an application of these results, consider the problem of stability of the zero solution of $\dot{\mathbf{x}}(t) = A(t)\mathbf{x}(t)$ with period $\widehat{t} > 0$: The stable, center, and unstable subspaces associated with the periodic matrix function $A : \mathbb{R} \longrightarrow gl(d,\mathbb{R})$ are defined as $L^-(\tau) = \bigoplus\{L(\lambda_j,\tau), \lambda_j < 0\}$, $L^0(\tau) = \bigoplus\{L(\lambda_j,\tau), \lambda_j = 0\}$, and $L^+(\tau) = \bigoplus\{L(\lambda_j,\tau), \lambda_j > 0\}$, respectively, for $\tau \in \mathbb{S}^1$. The zero solution $\mathbf{x}(t,0) \equiv 0$ of the periodic linear differential equation $\dot{\mathbf{x}} = A(t)\mathbf{x}$ is asymptotically stable if and only if it is exponentially stable if and only if all Lyapunov exponents are negative if and only if $L^-(\tau) = \mathbb{R}^d$ for some (and hence for all) $\tau \in \mathbb{S}^1$.
8. Another approach to the study of time-dependent linear differential equations is via transforming an equation with bounded coefficients into an equation of known type, such as equations with constant coefficients. Such transformations are known as Lyapunov transformations; see [Hah67, Secs. 61–63].

Examples:

1. Consider the $\widehat{t}$-periodic differential equation $\dot{\mathbf{x}} = A(t)\mathbf{x}$. This equation has a nontrivial $\widehat{t}$-periodic solution iff the system has a characteristic multiplier equal to 1; see Example 2.3 for the case with constant coefficients ([Ama90, Prop. 20.12]).
2. Let H be a continuous quadratic form in $2d$ variables $x_1, \ldots, x_d, y_1, \ldots, y_d$ and consider the Hamiltonian system

$$\dot{x}_i = \frac{\partial H}{\partial y_i}, \; \dot{y}_i = -\frac{\partial H}{\partial x_i}, \; i = 1, \ldots, d.$$

Using $\mathbf{z}^T = [\mathbf{x}^T, \mathbf{y}^T]$, we can set $H(\mathbf{x},\mathbf{y},t) = \mathbf{z}^T A(t)\mathbf{z}$, where $A = \begin{bmatrix} A_{11} & A_{12} \\ A_{12}^T & A_{22} \end{bmatrix}$ with A_{11} and A_{22} symmetric, and, hence, the equation takes the form

$$\dot{\mathbf{z}} = \begin{bmatrix} A_{12}^T(t) & A_{22}(t) \\ -A_{11}(t) & -A_{12}(t) \end{bmatrix} \mathbf{z} =: P(t)\mathbf{z}.$$

Note that $-P^T(t) = QP(t)Q^{-1}$ with $Q = \begin{bmatrix} 0 & -I \\ I & 0 \end{bmatrix}$, where I is the $d \times d$ identity matrix. Assume that H is $\widehat{t}$-periodic, then the equation for z and its adjoint have the same Floquet exponents and for each exponent λ its negative $-\lambda$ is also a Floquet exponent. Hence, the fixed point $0 \in \mathbb{R}^{2d}$ cannot be exponentially stable ([Hah67, Sec. 60]).
3. Consider the periodic linear oscillator

$$\ddot{y} + q_1(t)\dot{y} + q_2(t)y = 0.$$

Using the substitution $y = z\exp(-\frac{1}{2}\int q_1(u)du)$ one obtains Hill's differential equation

$$\ddot{z} + p(t)z = 0,\ p(t) := q_2(t) - \frac{1}{4}q_1(t)^2 - \frac{1}{2}\dot{q}_1(t).$$

Its characteristic equation is $\lambda^2 - 2a\lambda + 1 = 0$, with a still to be determined. The multipliers satisfy the relations $\alpha_1\alpha_2 = 1$ and $\alpha_1 + \alpha_2 = 2a$. The exponential stability of the system can be analyzed using the parameter a: If $a^2 > 1$, then one of the multipliers has absolute value > 1 and, hence, the system has an unbounded solution. If $a^2 = 1$, then the system has a nontrivial periodic solution according to Example 1. If $a^2 < 1$, then the system is stable. The parameter a can often be expressed in form of a power series; see [Hah67, Sec. 62] for more details. A special case of Hill's equation is the Mathieu equation

$$\ddot{z} + (\beta_1 + \beta_2 \cos 2t)z = 0,$$

with β_1, β_2 real parameters. For this equation numerically computed stability diagrams are available; see [Sto92, Secs. VI. 3 and 4].

56.7 Random Linear Dynamical Systems

Definitions:

Let $\theta : \mathbb{R} \times \Omega \longrightarrow \Omega$ be a metric dynamical system on the probability space $(\Omega, \mathcal{F}, P)$. A set $\Delta \in \mathcal{F}$ is called P-**invariant under** θ if $P[(\theta^{-1}(t, \Delta) \setminus \Delta) \cup (\Delta \setminus \theta^{-1}(t, \Delta))] = 0$ for all $t \in \mathbb{R}$. The flow θ is called **ergodic**, if each invariant set $\Delta \in \mathcal{F}$ has P-measure 0 or 1.

Facts:

Literature: [Arn98], [Con97].

1. (Oseledets Theorem, Multiplicative Ergodic Theorem) Consider a random linear dynamical system $\Phi = (\theta, \varphi) : \mathbb{R} \times \Omega \times \mathbb{R}^d \longrightarrow \Omega \times \mathbb{R}^d$ and assume

$$\sup_{0 \le t \le 1} \log^+ \|\varphi(t, \omega)\| \in \mathcal{L}^1(\Omega, \mathcal{F}, \mathbb{P}) \text{ and } \sup_{0 \le t \le 1} \log^+ \left\|\varphi(t, \omega)^{-1}\right\| \in \mathcal{L}^1(\Omega, \mathcal{F}, \mathbb{P}),$$

where $\|\cdot\|$ is any norm on $GL(d, \mathbb{R})$, $\mathcal{L}^1$ is the space of integrable functions, and $\log^+$ denotes the positive part of log, i.e.,

$$\log^+(x) = \begin{cases} \log(x) & \text{for } \log(x) > 0 \\ 0 & \text{for } \log(x) \le 0. \end{cases}$$

Then there exists a set $\widehat{\Omega} \subset \Omega$ of full P-measure, invariant under the flow $\theta : \mathbb{R} \times \Omega \longrightarrow \Omega$, such that for each $\omega \in \widehat{\Omega}$ there is a splitting $\mathbb{R}^d = \bigoplus_{j=1}^{l(\omega)} L_j(\omega)$ of $\mathbb{R}^d$ into linear subspaces with the following properties:

(a) The number of subspaces is θ-invariant, i.e., $l(\theta(t, \omega)) = l(\omega)$ for all $t \in \mathbb{R}$, and the dimensions of the subspaces are θ-invariant, i.e., $\dim L_j(\theta(t, \omega)) = \dim L_j(\omega) =: d_j(\omega)$ for all $t \in \mathbb{R}$.

(b) The subspaces are invariant under the flow Φ, i.e., $\varphi(t, \omega)L_j(\omega) \subset L_j(\theta(t, \omega))$ for all $j = 1, \ldots, l(\omega)$.

(c) There exist finitely many numbers $\lambda_1(\omega) < \ldots < \lambda_{l(\omega)}(\omega)$ in $\mathbb{R}$ (with possibly $\lambda_1(\omega) = -\infty$), such that for each $\mathbf{x} \in \mathbb{R}^d \backslash \{0\}$ the Lyapunov exponent $\lambda(\mathbf{x}, \omega)$ exists as a limit and

$\lambda(\mathbf{x}, \omega) = \lim_{t \to \pm\infty} \frac{1}{t} \log \|\varphi(t, \tau, \mathbf{x})\| = \lambda_j(\omega)$ if and only if $\mathbf{x} \in L_j(\omega) \backslash \{0\}$. The subspaces $L_j(\omega)$ are called the Lyapunov (or sometimes the Oseledets) spaces of the system Φ.

2. The following maps are measurable: $l : \Omega \longrightarrow \{1, \ldots, d\}$ with the discrete σ-algebra, and for each $j = 1, \ldots, l(\omega)$ the maps $L_j : \Omega \longrightarrow \mathbb{G}_{d_j}$ with the Borel σ-algebra, $d_j : \Omega \longrightarrow \{1, \ldots, d\}$ with the discrete σ-algebra, and $\lambda_j : \Omega \longrightarrow \mathbb{R} \cup \{-\infty\}$ with the (extended) Borel σ-algebra.
3. If the base flow $\theta : \mathbb{R} \times \Omega \longrightarrow \Omega$ is ergodic, then the maps l, d_j, and λ_j are constant on $\widehat{\Omega}$, but the Lyapunov spaces $L_j(\omega)$ still depend (in a measurable way) on $\omega \in \widehat{\Omega}$.
4. As an application of these results, we consider random linear differential equations: Let $(\Gamma, \mathcal{E}, Q)$ be a probability space and $\xi : \mathbb{R} \times \Gamma \longrightarrow \mathbb{R}^m$ a stochastic process with continuous trajectories, i.e., the functions $\xi(\cdot, \gamma) : \mathbb{R} \longrightarrow \mathbb{R}^m$ are continuous for all $\gamma \in \Gamma$. The process ξ can be written as a measurable dynamical system in the following way: Define $\Omega = C(\mathbb{R}, \mathbb{R}^m)$, the space of continuous functions from $\mathbb{R}$ to $\mathbb{R}^m$. We denote by $\mathcal{F}$ the σ-algebra on Ω generated by the cylinder sets, i.e., by sets of the form $Z = \{\omega \in \Omega,\ \omega(t_1) \in F_1, \ldots, \omega(t_n) \in F_n,\ n \in \mathbb{N},\ F_i$ Borel sets in $\mathbb{R}^m\}$. The process ξ induces a probability measure P on $(\Omega, \mathcal{F})$ via $P(Z) = Q\{\gamma \in \Gamma, \xi(t_i, \gamma) \in F_i$ for $i = 1, \ldots, n\}$. Define the shift $\theta : \mathbb{R} \times \Omega \longrightarrow \mathbb{R} \times \Omega$ as $\theta(t, \omega(\cdot)) = \omega(t + \cdot)$. Then $(\Omega, \mathcal{F}, P, \theta)$ is a measurable dynamical system. If ξ is stationary, i.e., if for all $n \in \mathbb{N}$, and $t, t_1, \ldots, t_n \in \mathbb{R}$, and all Borel sets $F_1, \ldots, F_n$ in $\mathbb{R}^m$, it holds that $Q\{\gamma \in \Gamma, \xi(t_i, \gamma) \in F_i$ for $i = 1, \ldots, n\} = Q\{\gamma \in \Gamma,$ $\xi(t_i + t, \gamma) \in F_i$ for $i = 1, \ldots, n\}$, then the shift θ on Ω is P-invariant, and $(\Omega, \mathcal{F}, P, \theta)$ is a metric dynamical system.
5. Let $A : \Omega \longrightarrow gl(d, \mathbb{R})$ be measurable with $A \in \mathcal{L}^1$. Consider the random linear differential equation $\dot{\mathbf{x}}(t) = A(\theta(t, \omega))\mathbf{x}(t)$, where $(\Omega, \mathcal{F}, P, \theta)$ is a metric dynamical system as described before. We understand the solutions of this equation to be ω-wise. Then the solutions define a random linear dynamical system. Since we assume that $A \in \mathcal{L}^1$, this system satisfies the integrability conditions of the Multiplicative Ergodic Theorem.
6. Hence, for random linear differential equations $\dot{\mathbf{x}}(t) = A(\theta(t, \omega))\mathbf{x}(t)$ the Lyapunov exponents and the associated Oseledets spaces replace the real parts of eigenvalues and the Lyapunov spaces of constant matrices $A \in gl(d, \mathbb{R})$. If the "background" process ξ is ergodic, then all the quantities in the Multiplicative Ergodic Theorem are constant, except for the Lyapunov spaces that do, in general, depend on chance.
7. The problem of stability of the zero solution of $\dot{\mathbf{x}}(t) = A(\theta(t, \omega))\mathbf{x}(t)$ can now be analyzed in analogy to the case of a constant matrix or a periodic matrix function: The stable, center, and unstable subspaces associated with the random matrix process $A(\theta(t, \omega))$ are defined as $L^-(\omega) = \bigoplus\{L_j(\omega),$ $\lambda_j(\omega) < 0\}$, $L^0(\omega) = \bigoplus\{L_j(\omega), \lambda_j(\omega) = 0\}$, and $L^+(\omega) = \bigoplus\{L_j(\omega), \lambda_j(\omega) > 0\}$, respectively for $\omega \in \widehat{\Omega}$. We obtain the following characterization of stability: The zero solution $\mathbf{x}(t, \omega, 0) \equiv 0$ of the random linear differential equation $\dot{\mathbf{x}}(t) = A(\theta(t, \omega))\mathbf{x}(t)$ is P-almost surely exponentially stable if and only if P-almost surely all Lyapunov exponents are negative if and only if $P\{\omega \in \Omega,$ $L^-(\omega) = \mathbb{R}^d\} = 1$.

Examples:

1. The case of constant matrices: Let $A \in gl(d, \mathbb{R})$ and consider the dynamical system $\varphi : \mathbb{R} \times \mathbb{R}^d \longrightarrow \mathbb{R}^d$ generated by the solutions of the linear differential equation $\dot{\mathbf{x}} = A\mathbf{x}$. The flow φ can be considered as the skew-component of a random linear dynamical system over the base flow given by $\Omega = \{0\}$, $\mathcal{F}$ the trivial σ-algebra, P the Dirac measure at $\{0\}$, and $\theta : \mathbb{R} \times \Omega \longrightarrow \Omega$ defined as the constant map $\theta(t, \omega) = \omega$ for all $t \in \mathbb{R}$. Since the flow is ergodic and satisfies the integrability condition, we can recover all the results on Lyapunov exponents and Lyapunov spaces for φ from the Multiplicative Ergodic Theorem.
2. Weak Floquet theory: Let $A : \mathbb{R} \longrightarrow gl(d, \mathbb{R})$ be a continuous, periodic matrix function. Define the base flow as follows: $\Omega = \mathbb{S}^1$, $\mathcal{B}$ is the Borel σ-algebra on $\mathbb{S}^1$, P is the uniform distribution on $\mathbb{S}^1$, and θ is the shift $\theta(t, \tau) = t + \tau$. Then $(\Omega, \mathcal{F}, P, \theta)$ is an ergodic metric dynamical system. The solutions $\varphi(\cdot, \tau, \mathbf{x})$ of $\dot{\mathbf{x}} = A(t)\mathbf{x}$ define a random linear dynamical system $\Phi : \mathbb{R} \times \Omega \times \mathbb{R}^d \longrightarrow \Omega \times \mathbb{R}^d$ via

$\Phi(t, \omega, \mathbf{x}) = (\theta(t, \omega), \varphi(t, \omega, \mathbf{x}))$. With this set-up, the Multiplicative Ergodic Theorem recovers the results of Floquet Theory with P-probability 1.

3. Average Lyapunov exponent: In general, Lyapunov exponents for random linear systems are difficult to compute explicitly—numerical methods are usually the way to go. In the ergodic case, the average Lyapunov exponent $\overline{\lambda} := \frac{1}{d}\sum d_j\lambda_j$ is given by $\overline{\lambda} = \frac{1}{d} tr\, E(A \mid \mathcal{I})$, where $A : \Omega \longrightarrow gl(d, \mathbb{R})$ is the random matrix of the system, and $E(\cdot, \mathcal{I})$ is the conditional expectation of the probability measure P given the σ-algebra $\mathcal{I}$ of invariant sets on Ω. As an example, consider the linear oscillator with random restoring force

$$\ddot{y}(t) + 2\beta\dot{y}(t) + (1 + \sigma f(\theta(t, \omega)))y(t) = 0,$$

where $\beta, \sigma \in \mathbb{R}$ are positive constants and $f : \Omega \to \mathbb{R}$ is in $\mathcal{L}^1$. We assume that the background process is ergodic. Using the notation $x_1 = y$ and $x_2 = \dot{y}$ we can write the equation as

$$\dot{\mathbf{x}}(t) = A(\theta(t, \omega)\mathbf{x}(t) = \begin{bmatrix} 0 & 1 \\ -1 - \sigma f(\theta(t, \omega)) & -2\beta \end{bmatrix} \mathbf{x}(t).$$

For this system we obtain $\overline{\lambda} = -\beta$ ([Arn98, Remark 3.3.12]).

56.8 Robust Linear Systems

Definitions:

Let $\Phi : \mathbb{R}\times\mathcal{U}\times\mathbb{R}^d \longrightarrow \mathcal{U}\times\mathbb{R}^d$ be a linear skew-product flow with continuous base flow $\theta : \mathbb{R}\times\mathcal{U} \longrightarrow \mathcal{U}$. Throughout this section, $\mathcal{U}$ is compact and θ is chain recurrent on $\mathcal{U}$. Denote by $\mathcal{U} \times \mathbb{P}^{d-1}$ the projective bundle and recall that Φ induces a dynamical system $\mathbb{P}\Phi : \mathbb{R} \times \mathcal{U} \times \mathbb{P}^{d-1} \longrightarrow \mathcal{U} \times \mathbb{P}^{d-1}$. For ε, $T > 0$ an (ε, T)-chain ζ of $\mathbb{P}\Phi$ is given by $n \in \mathbb{N}$, $T_0, \ldots, T_n \geq T$, and $(u_0, p_0), \ldots, (u_n, p_n) \in \mathcal{U} \times \mathbb{P}^{d-1}$ with $d(\mathbb{P}\Phi(T_i, u_i, p_i), (u_{i+1}, p_{i+1})) < \varepsilon$ for $i = 0, \ldots, n-1$.

Define the **finite time exponential growth rate** of such a chain ζ (or **chain exponent**) by

$$\lambda(\zeta) = \left(\sum_{i=0}^{n-1} T_i\right)^{-1} \sum_{i=0}^{n-1} (\log \|\varphi(T_i, x_i, u_i)\| - \log \|x_i\|),$$

where $x_i \in \mathbb{P}^{-1}(p_i)$.

Let $\mathcal{M} \subset \mathcal{U} \times \mathbb{P}^{d-1}$ be a chain recurrent component of the flow $\mathbb{P}\Phi$. Define the **Morse spectrum over** $\mathcal{M}$ as

$$\Sigma_{Mo}(\mathcal{M}) = \left\{ \begin{array}{c} \lambda \in \mathbb{R}, \text{ there exist sequences } \varepsilon_n \to 0,\ T_n \to \infty \text{ and} \\ (\varepsilon_n, T_n)\text{-chains } \zeta_n \text{ in } \mathcal{M} \text{ such that } \lim \lambda(\zeta_n) = \lambda \end{array} \right\}$$

and the **Morse spectrum of the flow** as

$$\Sigma_{Mo}(\Phi) = \left\{ \begin{array}{c} \lambda \in \mathbb{R}, \text{ there exist sequences } \varepsilon_n \to 0,\ T_n \to \infty \text{ and } (\varepsilon_n, T_n)\text{-} \\ \text{chains } \zeta_n \text{ in the chain recurrent set of } \mathbb{P}\Phi \text{ such that } \lim \lambda(\zeta_n) = \lambda \end{array} \right\}.$$

Define the **Lyapunov spectrum over** $\mathcal{M}$ as

$$\Sigma_{Ly}(\mathcal{M}) = \{\lambda(u, x),\ (u, x) \in \mathcal{M},\ x \neq 0\}$$

and the **Lyapunov spectrum of the flow** Φ as

$$\Sigma_{Ly}(\Phi) = \{\lambda(u, \mathbf{x}),\ (u, \mathbf{x}) \in \mathcal{U} \times \mathbb{R}^d, \mathbf{x} \neq 0\}.$$

Facts:

Literature: [CK00], [Gru96], [HP05].

1. The projected flow $\mathbb{P}\Phi$ has a finite number of chain-recurrent components $\mathcal{M}_1, \ldots, \mathcal{M}_l, l \leq d$. These components form the finest Morse decomposition for $\mathbb{P}\Phi$, and they are linearly ordered $\mathcal{M}_1 \prec \ldots \prec \mathcal{M}_l$. Their lifts $\mathbb{P}^{-1}\mathcal{M}_i \subset \mathcal{U} \times \mathbb{R}^d$ form a continuous subbundle decomposition of $\mathcal{U} \times \mathbb{R}^d = \bigoplus_{i=1}^{l} \mathbb{P}^{-1}\mathcal{M}_i$.
2. The Lyapunov spectrum and the Morse spectrum are defined on the Morse sets, i.e., $\Sigma_{Ly}(\Phi) = \bigcup_{i=1}^{l} \Sigma_{Ly}(\mathcal{M}_i)$ and $\Sigma_{Mo}(\Phi) = \bigcup_{i=1}^{l} \Sigma_{Mo}(\mathcal{M}_i)$.
3. For each Morse set $\mathcal{M}_i$ the Lyapunov spectrum is contained in the Morse spectrum, i.e., $\Sigma_{Ly}(\mathcal{M}_i) \subset \Sigma_{Mo}(\mathcal{M}_i)$ for $i = 1, \ldots, l$.
4. For each Morse set, its Morse spectrum is a closed, bounded interval $\Sigma_{Mo}(\mathcal{M}_i) = [\kappa_i^*, \kappa_i]$, and $\kappa_i^*, \kappa_i \in \Sigma_{Ly}(\mathcal{M})$ for $i = 1, \ldots, l$.
5. The intervals of the Morse spectrum are ordered according to the order of the Morse sets, i.e., $\mathcal{M}_i \prec \mathcal{M}_j$ is equivalent to $\kappa_i^* < \kappa_j^*$ and $\kappa_i < \kappa_j$.
6. As an application of these results, consider robust linear systems of the form $\Phi : \mathbb{R} \times \mathcal{U} \times \mathbb{R}^d \longrightarrow \mathcal{U} \times \mathbb{R}^d$, given by a perturbed linear differential equation $\dot{\mathbf{x}} = A(u(t))\mathbf{x} := A_0\mathbf{x} + \sum_{i=1}^{m} u_i(t) A_i \mathbf{x}$, with $A_0, \ldots, A_m \in gl(d, \mathbb{R})$, $u \in \mathcal{U} = \{u : \mathbb{R} \longrightarrow U$, integrable on every bounded interval$\}$ and $U \subset \mathbb{R}^m$ is compact, convex with $0 \in intU$. Explicit equations for the induced perturbed system on the projective space $\mathbb{P}^{d-1}$ can be obtained as follows: Let $\mathbb{S}^{d-1} \subset \mathbb{R}^d$ be the unit sphere embedded into $\mathbb{R}^d$. The projected system on $\mathbb{S}^{d-1}$ is given by

$$\dot{s}(t) = h(u(t), s(t)), \ u \in \mathcal{U}, \ s \in \mathbb{S}^{d-1}$$

where

$$h(u, s) = h_0(s) + \sum_{i=1}^{m} u_i h_i(s) \text{ with } h_i(s) = \left(A_i - s^T A_i s \cdot I\right) s, \ i = 0, 1, \ldots, m.$$

Define an equivalence relation on $\mathbb{S}^{d-1}$ via $s_1 \sim s_2$ if $s_1 = -s_2$, identifying opposite points. Then the projective space can be identified as $\mathbb{P}^{d-1} = \mathbb{S}^{d-1}/\sim$. Since $h(u, s) = -h(u, -s)$, the differential equation also describes the projected system on $\mathbb{P}^{d-1}$. For the Lyapunov exponents one obtains in the same way

$$\lambda(u, \mathbf{x}) = \limsup_{t \to \infty} \frac{1}{t} \log \|\mathbf{x}(t)\| = \limsup_{t \to \infty} \frac{1}{t} \int_0^t q(u(\tau), s(\tau))\, d\tau$$

with

$$q(u, s) = q_0(s) + \sum_{i=1}^{m} u_i q_i(s) \text{ with } q_i(s) = \left(A_i - s^T A_i s \cdot I\right) s, \ i = 0, 1, \ldots, m.$$

For a constant perturbation $u(t) \equiv u \in \mathbb{R}$ for all $t \in \mathbb{R}$ the corresponding Lyapunov exponents $\lambda(u, \mathbf{x})$ of the flow Φ are the real parts of the eigenvalues of the matrix $A(u)$ and the corresponding Lyapunov spaces are contained in the subbundles $\mathbb{P}^{-1}\mathcal{M}_i$. Similarly, if a perturbation $u \in \mathcal{U}$ is periodic, the Floquet exponents of $\dot{\mathbf{x}} = A(u(\cdot))\mathbf{x}$ are part of the Lyapunov (and, hence, of the Morse) spectrum of the flow Φ, and the Floquet spaces are contained in $\mathbb{P}^{-1}\mathcal{M}_i$. The systems treated in this example can also be considered as "bilinear control systems" and studied relative to their control behavior and (exponential) stabilizability — this is the point of view taken in [CK00].
7. For robust linear systems "generically" the Lyapunov spectrum and the Morse spectrum agree see [CK00] for a precise definition of "generic" in this context.
8. Of particular interest is the upper spectral interval $\Sigma_{Mo}(\mathcal{M}_l) = [\kappa_l^*, \kappa_l]$, as it determines the robust stability of $\dot{\mathbf{x}} = A(u(t))\mathbf{x}$ (and stabilizability of the system if the set $\mathcal{U}$ is interpreted as a

set of admissible control functions; see [Gru96]). The stable, center, and unstable subbundles of $\mathcal{U} \times \mathbb{R}^d$ associated with the perturbed linear system $\dot{\mathbf{x}} = A(u(t))\mathbf{x}$ are defined as $L^- = \bigoplus\{\mathbb{P}^{-1}\mathcal{M}_j, \kappa_j < 0\}$, $L^0 = \bigoplus\{\mathbb{P}^{-1}\mathcal{M}_j, 0 \in [\kappa_j^*, \kappa_j]\}$, and $L^+ = \bigoplus\{\mathbb{P}^{-1}\mathcal{M}_j, \kappa_j^* > 0\}$, respectively. The zero solution of $\dot{\mathbf{x}} = A(u(t))\mathbf{x}$ is exponentially stable for all perturbations $u \in \mathcal{U}$ if and only if $\kappa_l < 0$ if and only if $L^- = \mathcal{U} \times \mathbb{R}^d$.

Examples:

1. In general, it is not possible to compute the Morse spectrum and the associated subbundle decompositions explicitly, even for relatively simple systems, and one has to revert to numerical algorithms; compare [CK00, App. D]. Let us consider, e.g., the linear oscillator with uncertain restoring force

$$\begin{bmatrix} \dot{x}_1 \\ \dot{x}_2 \end{bmatrix} = \begin{bmatrix} 0 & 1 \\ -1 & -2b \end{bmatrix} \begin{bmatrix} x_1 \\ x_2 \end{bmatrix} + u(t) \begin{bmatrix} 0 & 0 \\ -1 & 0 \end{bmatrix} \begin{bmatrix} x_1 \\ x_2 \end{bmatrix}$$

with $u(t) \in [-\rho, \rho]$ and $b > 0$. Figure 56.1 shows the spectral intervals for this system depending on $\rho \geq 0$.

2. We consider robust linear systems as described in Fact 6, with varying perturbation range by introducing the family $U^\rho = \rho U$ for $\rho \geq 0$. The resulting family of systems is

$$\dot{\mathbf{x}}^\rho = A(u^\rho(t))\mathbf{x}^\rho := A_0\mathbf{x}^\rho + \sum_{i=1}^{m} u_i^\rho(t) A_i \mathbf{x}^\rho,$$

with $u^\rho \in \mathcal{U}^\rho = \{u : \mathbb{R} \longrightarrow U^\rho$, integrable on every bounded interval$\}$. The corresponding maximal spectral value $\kappa_l(\rho)$ is continuous in ρ and we define the (asymptotic) stability radius of this family as $r = \inf\{\rho \geq 0$, there exists $u_0 \in \mathcal{U}^\rho$ such that $\dot{x}^\rho = A(u_0(t))x^\rho$ is not exponentially stable$\}$. This stability radius is based on asymptotic stability under all time varying perturbations. Similarly one can introduce stability radii based on time invariant perturbations (with values in $\mathbb{R}^m$ or $\mathbb{C}^m$) or on quadratic Lyapunov functions ([CK00], Chapter 11 and [HP05]).

3. Linear oscillator with uncertain damping: Consider the oscillator

$$\ddot{y} + 2(b + u(t))\dot{y} + (1 + c)y = 0$$

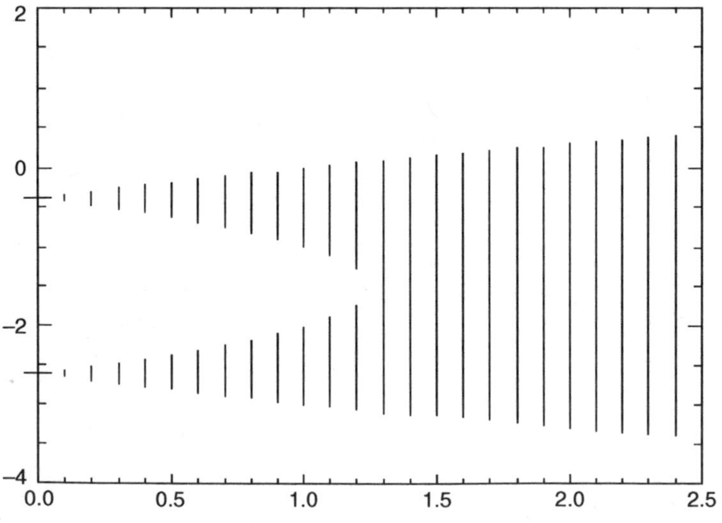

FIGURE 56.1 Spectral intervals depending on $\rho \geq 0$ for the system in Example 1.

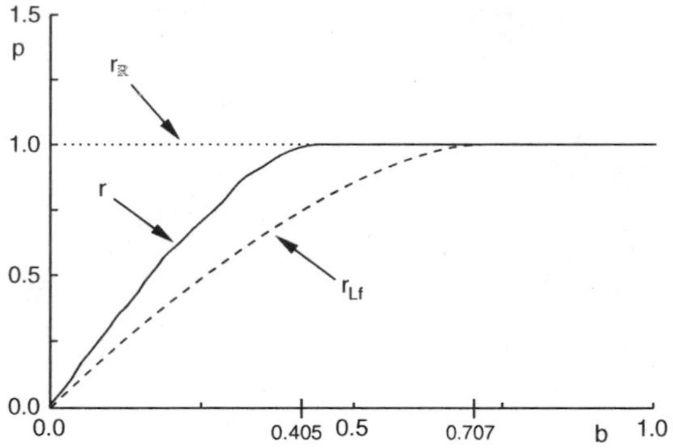

FIGURE 56.2 Stability radii for the system in Example 4.

with $u(t) \in [-\rho, \rho]$ and $c \in \mathbb{R}$. In equivalent first-order form the system reads

$$\begin{bmatrix} \dot{x}_1 \\ \dot{x}_2 \end{bmatrix} = \begin{bmatrix} 0 & 1 \\ -1-c & -2b \end{bmatrix} \begin{bmatrix} x_1 \\ x_2 \end{bmatrix} + u(t) \begin{bmatrix} 0 & 0 \\ 0 & -2 \end{bmatrix} \begin{bmatrix} x_1 \\ x_2 \end{bmatrix}.$$

Clearly, the system is not exponentially stable for $c \leq -1$ with $\rho = 0$, and for $c > -1$ with $\rho \geq b$. It turns out that the stability radius for this system is

$$r(c) = \begin{cases} 0 & \text{for} \quad c \leq -1 \\ b & \text{for} \quad c > -1. \end{cases}$$

4. Linear oscillator with uncertain restoring force: Here we look again at a system of the form

$$\begin{bmatrix} \dot{x}_1 \\ \dot{x}_2 \end{bmatrix} = \begin{bmatrix} 0 & 1 \\ -1 & -2b \end{bmatrix} \begin{bmatrix} x_1 \\ x_2 \end{bmatrix} + u(t) \begin{bmatrix} 0 & 0 \\ -1 & 0 \end{bmatrix} \begin{bmatrix} x_1 \\ x_2 \end{bmatrix}$$

with $u(t) \in [-\rho, \rho]$ and $b > 0$. (For $b \leq 0$ the system is unstable even for constant perturbations.) A closed form expression of the stability radius for this system is not available and one has to use numerical methods for the computation of (maximal) Lyapunov exponents (or maxima of the Morse spectrum); compare [CK00, App. D]. Figure 56.2 shows the (asymptotic) stability radius r, the stability radius under constant real perturbations $r_{\mathbb{R}}$, and the stability radius based on quadratic Lyapunov functions r_{Lf}, all in dependence on $b > 0$; see [CK00, Ex. 11.1.12].

56.9 Linearization

The local behavior of the dynamical system induced by a nonlinear differential equation can be studied via the linearization of the flow. At a fixed point of the nonlinear system the linearization is just a linear differential equation as studied in Sections 56.1 to 56.4. If the linearized system is hyperbolic, then the theorem of Hartman and Grobman states that the nonlinear flow is topologically conjugate to the linear flow. The invariant manifold theorem deals with those solutions of the nonlinear equation that are asymptotically attracted to (or repelled from) a fixed point. Basically these solutions live on manifolds that are described by nonlinear changes of coordinates of the linear stable (and unstable) subspaces.

Fact 4 below describes the simplest form of the invariant manifold theorem at a fixed point. It can be extended to include a "center manifold" (corresponding to the Lyapunov space with exponent 0). Furthermore, (local) invariant manifolds can be defined not just for the stable and unstable subspace, but for all Lyapunov spaces; see [BK94], [CK00], and [Rob98] for the necessary techniques and precise statements.

Both the Grobman–Hartman theorem as well as the invariant manifold theorem can be extended to time varying systems, i.e., to linear skew product flows as described in Sections 56.5 to 56.8. The general situation is discussed in [BK94], the case of linearization at a periodic solution is covered in [Rob98], random dynamical systems are treated in [Arn98], and robust systems (control systems) are the topic of [CK00].

Definitions:

A (nonlinear) **differential equation** in $\mathbb{R}^d$ is of the form $\dot{\mathbf{y}} = f(\mathbf{y})$, where f is a vector field on $\mathbb{R}^d$. Assume that f is at least of class C^1 and that for all $\mathbf{y}_0 \in \mathbb{R}^d$ the solutions $\mathbf{y}(t, \mathbf{y}_0)$ of the initial value problem $\mathbf{y}(0, \mathbf{y}_0) = \mathbf{y}_0$ exist for all $t \in \mathbb{R}$.

A point $\mathbf{p} \in \mathbb{R}^d$ is a **fixed point** of the differential equation $\dot{\mathbf{y}} = f(\mathbf{y})$ if $\mathbf{y}(t, \mathbf{p}) = \mathbf{p}$ for all $t \in \mathbb{R}$.

The **linearization** of the equation $\dot{\mathbf{y}} = f(\mathbf{y})$ at a fixed point $\mathbf{p} \in \mathbb{R}^d$ is given by $\dot{\mathbf{x}} = D_{\mathbf{y}} f(\mathbf{p})\mathbf{x}$, where $D_{\mathbf{y}} f(\mathbf{p})$ is the Jacobian (matrix of partial derivatives) of f at the point $\mathbf{p}$.

A fixed point $\mathbf{p} \in \mathbb{R}^d$ of the differential equation $\dot{\mathbf{y}} = f(\mathbf{y})$ is called **hyperbolic** if $D_{\mathbf{y}} f(\mathbf{p})$ has no eigenvalues on the imaginary axis, i.e., if the matrix $D_{\mathbf{y}} f(\mathbf{p})$ is hyperbolic.

Consider a differential equation $\dot{\mathbf{y}} = f(\mathbf{y})$ in $\mathbb{R}^d$ with flow $\Phi : \mathbb{R} \times \mathbb{R}^d \longrightarrow \mathbb{R}^d$, hyperbolic fixed point $\mathbf{p}$ and neighborhood $U(\mathbf{p})$. In this situation the **local stable manifold** and the **local unstable manifold** are defined as

$$W^s_{loc}(\mathbf{p}) = \{\mathbf{q} \in U\colon \lim_{t\to\infty} \Phi(t, \mathbf{q}) = \mathbf{p}\} \text{ and } W^u_{loc}(\mathbf{p}) = \{\mathbf{q} \in U\colon \lim_{t\to-\infty} \Phi(t, \mathbf{q}) = \mathbf{p}\},$$

respectively.

The local stable (and unstable) manifolds can be extended to **global invariant manifolds** by following the trajectories, i.e.,

$$W^s(\mathbf{p}) = \bigcup_{t\geq 0} \Phi(-t, W^s_{loc}(\mathbf{p})) \text{ and } W^u(\mathbf{p}) = \bigcup_{t\geq 0} \Phi(t, W^u_{loc}(\mathbf{p})).$$

Facts:

Literature: [Arn98], [AP90], [BK94], [CK00], [Rob98].

See Facts 3 and 4 in Section 56.2 for dynamical systems induced by differential equations and their fixed points.

1. (Hartman–Gobman) Consider a differential equation $\dot{\mathbf{y}} = f(\mathbf{y})$ in $\mathbb{R}^d$ with flow $\Phi : \mathbb{R} \times \mathbb{R}^d \longrightarrow \mathbb{R}^d$. Assume that the equation has a hyperbolic fixed point $\mathbf{p}$ and denote the flow of the linearized equation $\dot{\mathbf{x}} = D_{\mathbf{y}} f(\mathbf{p})\mathbf{x}$ by $\Psi : \mathbb{R} \times \mathbb{R}^d \longrightarrow \mathbb{R}^d$. Then there exist neighborhoods $U(\mathbf{p})$ of $\mathbf{p}$ and $V(\mathbf{0})$ of the origin in $\mathbb{R}^d$, and a homeomorphism $h : U(\mathbf{p}) \longrightarrow V(\mathbf{0})$ such that the flows $\Phi\ |_{U(\mathbf{p})}$ and $\Psi\ |_{V(\mathbf{0})}$ are (locally) C^0-conjugate, i.e., $h(\Phi(t, \mathbf{y})) = \Psi(t, h(\mathbf{y}))$ for all $\mathbf{y} \in U(\mathbf{p})$ and $t \in \mathbb{R}$ as long as the solutions stay within the respective neighborhoods.
2. Consider two differential equations $\dot{\mathbf{y}} = f_i(\mathbf{y})$ in $\mathbb{R}^d$ with flows $\Phi_i : \mathbb{R} \times \mathbb{R}^d \longrightarrow \mathbb{R}^d$ for $i = 1, 2$. Assume that Φ_i has a hyperbolic fixed point $\mathbf{p}_i$ and the flows are C^k-conjugate for some $k \geq 1$ in neighborhoods of the $\mathbf{p}_i$. Then $\sigma(D_{\mathbf{y}} f_1(\mathbf{p}_1)) = \sigma(D_{\mathbf{y}} f_2(\mathbf{p}_2))$, i.e., the eigenvalues of the linearizations agree; compare Facts 5 and 6 in Section 56.2 for the linear situation.
3. Consider two differential equations $\dot{\mathbf{y}} = f_i(\mathbf{y})$ in $\mathbb{R}^d$ with flows $\Phi_i : \mathbb{R} \times \mathbb{R}^d \longrightarrow \mathbb{R}^d$ for $i = 1, 2$. Assume that Φ_i has a hyperbolic fixed point $\mathbf{p}_i$ and the number of negative (or positive) Lyapunov exponents of $D_{\mathbf{y}} f_i(\mathbf{p}_i)$ agrees. Then the flows Φ_i are locally C^0-conjugate around the fixed points.
4. (Invariant Manifold Theorem) Consider a differential equation $\dot{\mathbf{y}} = f(\mathbf{y})$ in $\mathbb{R}^d$ with flow $\Phi : \mathbb{R} \times \mathbb{R}^d \longrightarrow \mathbb{R}^d$. Assume that the equation has a hyperbolic fixed point p and denote the linearized equation by $\dot{\mathbf{x}} = D_{\mathbf{y}} f(p)\mathbf{x}$.

   (i) There exists a neighborhood $U(\mathbf{p})$ in which the flow Φ has a local stable manifold $W^s_{loc}(\mathbf{p})$ and a local unstable manifold $W^u_{loc}(\mathbf{p})$.

(ii) Denote by L^- (and L^+) the stable (and unstable, respectively) subspace of $D_{\mathbf{y}} f(\mathbf{p})$; compare the definitions in Section 56.1. The dimensions of L^- (as a linear subspace of $\mathbb{R}^d$) and of $W^s_{loc}(\mathbf{p})$ (as a topological manifold) agree, similarly for L^+ and $W^u_{loc}(\mathbf{p})$.

(iii) The stable manifold $W^s_{loc}(\mathbf{p})$ is tangent to the stable subspace L^- at the fixed point $\mathbf{p}$, similarly for $W^u_{loc}(\mathbf{p})$ and L^+.

5. Consider a differential equation $\dot{\mathbf{y}} = f(\mathbf{y})$ in $\mathbb{R}^d$ with flow $\Phi : \mathbb{R} \times \mathbb{R}^d \longrightarrow \mathbb{R}^d$. Assume that the equation has a hyperbolic fixed point $\mathbf{p}$. Then there exists a neighborhood $U(\mathbf{p})$ on which Φ is C^0-equivalent to the flow of a linear differential equation of the type

$$\dot{\mathbf{x}}_s = -\mathbf{x}_s, \mathbf{x}_s \in \mathbb{R}^{d_s},$$

$$\dot{\mathbf{x}}_u = \mathbf{x}_u, \mathbf{x}_u \in \mathbb{R}^{d_u},$$

where d_s and d_u are the dimensions of the stable and the unstable subspace of $D_{\mathbf{y}} f(\mathbf{p})$, respectively, with $d_s + d_u = d$.

Examples:

1. Consider the nonlinear differential equation in $\mathbb{R}$ given by $\ddot{z} + z - z^3 = 0$, or in first-order form in $\mathbb{R}^2$

$$\begin{bmatrix} \dot{y}_1 \\ \dot{y}_2 \end{bmatrix} = \begin{bmatrix} y_2 \\ -y_1 + y_1^3 \end{bmatrix} = f(\mathbf{y}).$$

The fixed points of this system are $\mathbf{p}_1 = [0,0]^T$, $\mathbf{p}_2 = [1,0]^T$, $\mathbf{p}_3 = [-1,0]^T$. Computation of the linearization yields

$$D_{\mathbf{y}} f = \begin{bmatrix} 0 & 1 \\ -1 + 3y_1^2 & 0 \end{bmatrix}.$$

Hence, the fixed point $\mathbf{p}_1$ is not hyperbolic, while $\mathbf{p}_2$ and $\mathbf{p}_3$ have this property.

2. Consider the nonlinear differential equation in $\mathbb{R}$ given by $\ddot{z} + \sin(z) + \dot{z} = 0$, or in first-order form in $\mathbb{R}^2$

$$\begin{bmatrix} \dot{y}_1 \\ \dot{y}_2 \end{bmatrix} = \begin{bmatrix} y_2 \\ -\sin(y_1) - y_2 \end{bmatrix} = f(\mathbf{y}).$$

The fixed points of the system are $\mathbf{p}_n = [n\pi, 0]^T$ for $n \in \mathbb{Z}$. Computation of the linearization yields

$$D_{\mathbf{y}} f = \begin{bmatrix} 0 & 1 \\ -\cos(y_1) & -1 \end{bmatrix}.$$

Hence, for the fixed points $\mathbf{p}_n$ with n even the eigenvalues are $\mu_1, \mu_2 = -\frac{1}{2} \pm i\sqrt{\frac{3}{4}}$ with negative real part (or Lyapunov exponent), while at the fixed points $\mathbf{p}_n$ with n odd one obtains as eigenvalues $\nu_1, \nu_2 = -\frac{1}{2} \pm \sqrt{\frac{5}{4}}$, resulting in one positive and one negative eigenvalue. Hence, the flow of the differential equation is locally C^0-conjugate around all fixed points with even n, and around all fixed points with odd n, while the flows around, e.g., $\mathbf{p}_0$ and $\mathbf{p}_1$ are not conjugate.

References

[Ama90] H. Amann, *Ordinary Differential Equations*, Walter de Gruyter, Berlin, 1990.

[Arn98] L. Arnold, *Random Dynamical Systems*, Springer-Verlag, Heidelberg, 1998.

[AP90] D.K. Arrowsmith and C.M. Place, *An Introduction to Dynamical Systems*, Cambridge University Press, Cambridge, 1990.

[ACK05] V. Ayala, F. Colonius, and W. Kliemann, Dynamical characterization of the Lyapunov form of matrices, *Lin. Alg. Appl.* 420 (2005), 272–290.

[BK94] I.U. Bronstein and A.Ya Kopanskii, *Smooth Invariant Manifolds and Normal Forms*, World Scientific, Singapore, 1994.

[CFJ06] F. Colonius, R. Fabbri, and R. Johnson, Chain recurrence, growth rates and ergodic limits, to appear in Ergodic Theory and Dynamical Systems (2006).

[CK00] F. Colonius and W. Kliemann, *The Dynamics of Control*, Birkhäuser, Boston, 2000.

[Con97] N. D. Cong, *Topological Dynamics of Random Dynamical Systems*, Oxford Mathematical Monographs, Clarendon Press, Oxford, U.K., 1997.

[Flo83] G. Floquet, Sur les équations différentielles linéaires à coefficients périodiques, *Ann. École Norm. Sup.* 12 (1883), 47–88.

[Gru96] L. Grüne, Numerical stabilization of bilinear control systems, *SIAM J. Cont. Optimiz.* 34 (1996), 2024–2050.

[GH83] J. Guckenheimer and P. Holmes, *Nonlinear Oscillations, Dynamical Systems, and Bifurcation of Vector Fields*, Springer-Verlag, Heidelberg, 1983.

[Hah67] W. Hahn, *Stability of Motion*, Springer-Verlag, Heidelberg, 1967.

[HP05] D. Hinrichsen and A.J. Pritchard, *Mathematical Systems Theory*, Springer-Verlag, Heidelberg, 2005.

[HSD04] M.W. Hirsch, S. Smale, and R.L. Devaney, *Differential Equations, Dynamical Systems and an Introduction to Chaos*, Elsevier, Amsterdom, 2004.

[Lya92] A.M. Lyapunov, *The General Problem of the Stability of Motion*, Comm. Soc. Math. Kharkov (in Russian), 1892. Problème Géneral de la Stabilité de Mouvement, Ann. Fac. Sci. Univ. Toulouse 9 (1907), 203–474, reprinted in *Ann. Math. Studies* 17, Princeton (1949), in English, Taylor & Francis 1992.

[Ose68] V.I. Oseledets, A multiplicative ergodic theorem. Lyapunov characteristic numbers for dynamical systems, *Trans. Moscow Math. Soc.* 19 (1968), 197–231.

[Rob98] C. Robinson, *Dynamical Systems*, 2nd ed., CRC Press, Boca Paton, FL, 1998.

[Sel75] J. Selgrade, Isolated invariant sets for flows on vector bundles, *Trans. Amer. Math. Soc.* 203 (1975), 259–390.

[Sto92] J.J. Stoker, *Nonlinear Vibrations in Mechanical and Electrical Systems*, John Wiley & Sons, New York, 1950 (reprint Wiley Classics Library, 1992).

[Wig96] S. Wiggins, *Introduction to Applied Nonlinear Dynamical Systems and Applications*, Springer-Verlag, Heidelberg, 1996.

57

Control Theory

Peter Benner
Technische Universität Chemnitz

Given a dynamical system described by the ordinary differential equation (ODE)

$$\dot{\mathbf{x}}(t) = \mathbf{f}(t, \mathbf{x}(t), \mathbf{u}(t)), \ \ \mathbf{x}(t_0) = \mathbf{x}^0,$$

where $\mathbf{x}$ is the state of the system and $\mathbf{u}$ serves as input, the major problem in control theory is to steer the state from $\mathbf{x}^0$ to some desired state; i.e., for a given initial value $\mathbf{x}(t_0) = \mathbf{x}^0$ and target $\mathbf{x}^1$, can we find a piecewise continuous or L_2 (i.e., square-integrable, Lebesgue measurable) control function $\hat{\mathbf{u}}$ such that there exists $t_1 \geq t_0$ with $\mathbf{x}(t_1; \hat{\mathbf{u}}) = \mathbf{x}^1$, where $\mathbf{x}(t; \hat{\mathbf{u}})$ is the solution trajectory of the ODE given above for $\mathbf{u} \equiv \hat{\mathbf{u}}$? Often, the target is $\mathbf{x}^1 = 0$, in particular if $\mathbf{x}$ describes the deviation from a nominal path. A weaker demand is to asymptotically stabilize the system, i.e., to find an admissible control function $\hat{\mathbf{u}}$ (i.e., a piecewise continuous or L_2 function $\hat{\mathbf{u}} : [t_0, t_1] \mapsto \mathcal{U}$) such that $\lim_{t\to\infty} \mathbf{x}(t; \hat{\mathbf{u}}) = 0$).

Another major problem in control theory arises from the fact that often, not all states are available for measurements or observations. Thus, we are faced with the question: Given partial information about the states, is it possible to reconstruct the solution trajectory from the measurements/observations? If this is the case, the states can be estimated by state observers. The classical approach leads to the Luenberger observer, but nowadays most frequently the famous Kalman–Bucy filter [KB61] is used as it can be considered as an optimal state observer in a least-squares sense and allows for stochastic uncertainties in the system.

Analyzing the above questions concerning controllability, observability, etc. for general control systems is beyond the scope of linear algebra. Therefore, we will mostly focus on linear time-invariant (LTI) systems that can be analyzed with tools relying on linear algebra techniques. (For further reading, see, e.g., [Lev96], [Mut99], and [Son98].)

Once the above questions are settled, it is interesting to ask how the desired control objectives can be achieved in an optimal way. The linear-quadratic regulator (LQR) problem is equivalent to a dynamic optimization problem for linear differential equations. Its significance for control theory was fully discovered first by Kalman in 1960 [Kal60]. One of its main applications is to steer the solution of the underlying linear differential equation to a desired reference trajectory with minimal cost given full information on the states. If full information is not available, then the states can be estimated from the measurements or observations using a Kalman–Bucy filter. This leads to the linear-quadratic Gaussian (LQG) control problem. The latter problem and its solution were first described in the classical papers [Kal60] and [KB61] and are nowadays contained in any textbook on control theory.

In the past decades, the interest has shifted from optimal control to robust control. The question raised is whether a given control law is still able to achieve a desired performance in the presence of uncertain disturbances. In this sense, the LQR control law has some robustness, while the LQG design cannot be considered to be robust [Doy78]. The H_∞ control problem aims at minimizing the worst-case error that can occur if the system is perturbed by exogenous perturbations. It is, thus, one example of a *robust control* problem. We will only introduce the standard H_∞ control problem, though there exist many other robust control problems and several variations of the H_∞ control problem; see [GL95], [PUA00], [ZDG96].

Many of the above questions lead to methods that involve the solution of linear and nonlinear matrix equations, in particular Lyapunov, Sylvester, and Riccati equations. For instance, stability, controllability, and observability of LTI systems can be related to solutions of Lyapunov equations (see, e.g., [LT85, Sec. 13] and [HJ91]), while the LQR, LQG, and H_∞ control problems lead to the solution of algebraic Riccati equations, (see, e.g., [AKFIJ03], [Dat04], [LR95], [Meh91], and [Sim96]). Therefore, we will provide the most relevant properties of these matrix equations.

The concepts and solution techniques contained in this chapter and many other control-related algorithms are implemented in the MATLAB® Control System Toolbox, the Subroutine Library in Control SLICOT [BMS+99], and many other computer-aided control systems design tools.

Finally, we note that all concepts described in this section are related to *continuous-time* systems. Analogous concepts hold for *discrete-time systems* whose dynamics are described by difference equations; see, e.g., [Kuc91].

57.1 Basic Concepts

Definitions:

Given vector spaces $\mathcal{X}$ (the **state space**), $\mathcal{U}$ (the **input space**), and $\mathcal{Y}$ (the **output space**) and measurable functions $\mathbf{f}, \mathbf{g} : [t_0, t_f] \times \mathcal{X} \times \mathcal{U} \mapsto \mathbb{R}^n$, a **control system** is defined by

$$\dot{\mathbf{x}}(t) = \mathbf{f}(t, \mathbf{x}(t), \mathbf{u}(t)),$$
$$\mathbf{y}(t) = \mathbf{g}(t, \mathbf{x}(t), \mathbf{u}(t)),$$

where the differential equation is called the **state equation**, the second equation is called the **observer equation**, and $t \in [t_0, t_f]$ ($t_f \in [\,0, \infty\,]$).

Here,

$$\mathbf{x} : [t_0, t_f] \mapsto \mathcal{X} \text{ is the } \textbf{state (vector)},$$
$$\mathbf{u} : [t_0, t_f] \mapsto \mathcal{U} \text{ is the } \textbf{control (vector)},$$
$$\mathbf{y} : [t_0, t_f] \mapsto \mathcal{Y} \text{ is the } \textbf{output (vector)}.$$

A control system is called **autonomous (time-invariant)** if

$$\mathbf{f}(t, \mathbf{x}, \mathbf{u}) \equiv \mathbf{f}(\mathbf{x}, \mathbf{u}) \text{ and } \mathbf{g}(t, \mathbf{x}, \mathbf{u}) \equiv \mathbf{g}(\mathbf{x}, \mathbf{u}).$$

The number of state-space variables n is called the **order** or **degree** of the system.

Let $\mathbf{x}^1 \in \mathbb{R}^n$. A control system with initial value $\mathbf{x}(t_0) = \mathbf{x}^0$ is **controllable to $\mathbf{x}^1$ in time** $t_1 > t_0$ if there exists an **admissible** control function $\mathbf{u}$ (i.e., a piecewise continuous or L_2 function $\mathbf{u} : [t_0, t_1] \mapsto \mathcal{U}$) such that $\mathbf{x}(t_1; \mathbf{u}) = \mathbf{x}^1$. (Equivalently, $(t_1, \mathbf{x}^1)$ is **reachable from** $(t_0, \mathbf{x}^0)$.)

A control system with initial value $\mathbf{x}(t_0) = \mathbf{x}^0$ is **controllable to $\mathbf{x}^1$** if there exists $t_1 > t_0$ such that $(t_1, \mathbf{x}^1)$ is reachable from $(t_0, \mathbf{x}^0)$.

If the control system is controllable to all $\mathbf{x}^1 \in \mathcal{X}$ for all $(t_0, \mathbf{x}^0)$ with $\mathbf{x}^0 \in \mathcal{X}$, it is **(completely) controllable.**

A control system is **linear** if $\mathcal{X} = \mathbb{R}^n, \mathcal{U} = \mathbb{R}^m, \mathcal{Y} = \mathbb{R}^p$, and

$$\mathbf{f}(t, \mathbf{x}, \mathbf{u}) = A(t)\mathbf{x}(t) + B(t)\mathbf{u}(t),$$
$$\mathbf{g}(t, \mathbf{x}, \mathbf{u}) = C(t)\mathbf{x}(t) + D(t)\mathbf{u}(t),$$

where $A : [t_0, t_f] \mapsto \mathbb{R}^{n\times n}, B : [t_0, t_f] \mapsto \mathbb{R}^{n\times m}, C : [t_0, t_f] \mapsto \mathbb{R}^{p\times n}, D : [t_0, t_f] \mapsto \mathbb{R}^{p\times m}$ are smooth functions.

A linear **time-invariant system** (**LTI system**) has the form

$$\begin{aligned}\dot{\mathbf{x}}(t) &= A\mathbf{x}(t) + B\mathbf{u}(t),\\ \mathbf{y}(t) &= C\mathbf{x}(t) + D\mathbf{u}(t),\end{aligned}$$

with $A \in \mathbb{R}^{n\times n}$, $B \in \mathbb{R}^{n\times m}$, $C \in \mathbb{R}^{p\times n}$, and $D \in \mathbb{R}^{p\times m}$.

An LTI system is **(asymptotically) stable** if the corresponding linear homogeneous ODE $\dot{\mathbf{x}} = A\mathbf{x}$ is (asymptotically) stable. (For a definition of (asymptotic) stability confer Chapter 55 and Chapter 56.)

An LTI system is **stabilizable (by state feedback)** if there exists an admissible control in the form of a **state feedback**

$$\mathbf{u}(t) = F\mathbf{x}(t), \quad F \in \mathbb{R}^{m\times n},$$

such that the unique solution of the corresponding **closed-loop** ODE

$$\dot{\mathbf{x}}(t) = (A + BF)\mathbf{x}(t) \tag{57.1}$$

is asymptotically stable.

An LTI system is **observable (reconstructible)** if for two solution trajectories $\mathbf{x}(t)$ and $\tilde{\mathbf{x}}(t)$ of its state equation, it holds that

$$C\mathbf{x}(t) = C\tilde{\mathbf{x}}(t) \quad \forall t \leq t_0 \ (\forall t \geq t_0)$$

implies $\mathbf{x}(t) = \tilde{\mathbf{x}}(t) \quad \forall t \leq t_0 \ (\forall t \geq t_0)$.

An LTI system is **detectable** if for any solution $\mathbf{x}(t)$ of $\dot{\mathbf{x}} = A\mathbf{x}$ with $C\mathbf{x}(t) \equiv 0$ we have $\lim_{t\to\infty} \mathbf{x}(t) = 0$.

Facts:

1. For LTI systems, all controllability and reachability concepts are equivalent. Therefore, we only speak of controllability of LTI systems.
2. Observability implies that one can obtain all necessary information about the LTI system from the output equation.
3. Detectability weakens observability in the same sense as stabilizability weakens controllability: Not all of $\mathbf{x}$ can be observed, but the unobservable part is asymptotically stable.
4. Observability (detectability) and controllability (stabilizability) are dual concepts in the following sense: an LTI system is observable (detectable) if and only if the dual system

$$\dot{\mathbf{z}}(t) = A^T\mathbf{z}(t) + C^T\mathbf{v}(t)$$

is controllable (stabilizable). This fact is sometimes called the **duality principle of control theory.**

Examples:

1. A fundamental problem in robotics is to control the position of a single-link rotational joint using a motor placed at the "pivot." A simple mathematical model for this is the pendulum [Son98]. Applying a torque $\mathbf{u}$ as external force, this can serve as a means to control the motion of the pendulum (Figure 57.1).

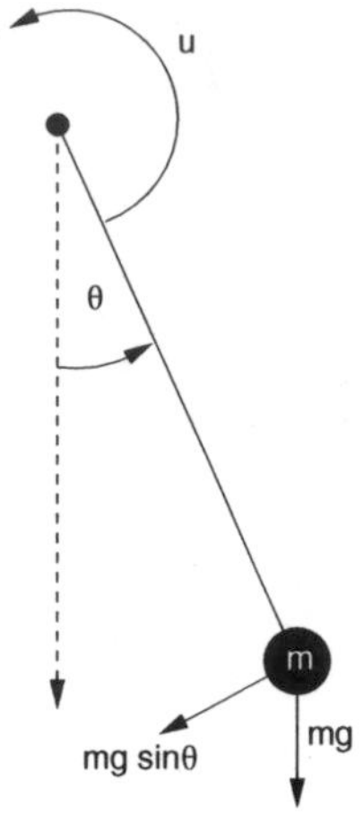

FIGURE 57.1 Pendulum as mathematical model of a single-link rotational joint .

If we neglect friction and assume that the mass is concentrated at the tip of the pendulum, Newton's law for rotating objects

$$m\ddot{\Theta}(t) + mg\sin\Theta(t) = \mathbf{u}(t)$$

describes the counter clockwise movement of the angle between the vertical axis and the pendulum subject to the control $\mathbf{u}(t)$. This is a first example of a (nonlinear) control system if we set

$$\mathbf{x}(t) = \begin{bmatrix} x_1(t) \\ x_2(t) \end{bmatrix} = \begin{bmatrix} \Theta(t) \\ \dot{\Theta}(t) \end{bmatrix},$$

$$\mathbf{f}(t, \mathbf{x}, \mathbf{u}) = \begin{bmatrix} x_2 \\ -mg\sin(x_1) \end{bmatrix}, \quad \mathbf{g}(t, \mathbf{x}, \mathbf{u}) = x_1,$$

where we assume that only $\Theta(t)$ can be measured, but not the angular velocity $\dot{\Theta}(t)$.

For $\mathbf{u}(t) \equiv 0$, the stationary position $\Theta = \pi, \dot{\Theta} = 0$ is an unstable equilibrium, i.e., small perturbations will lead to unstable motion. The objective now is to apply a torque (control $\mathbf{u}$) to correct for deviations from this unstable equilibrium, i.e., to keep the pendulum in the upright position, (Figure 57.2).

2. Scaling the variables such that $m = 1 = g$ and assuming a small perturbation $\Theta - \pi$ in the inverted pendulum problem described above, we have

$$\sin\Theta = -(\Theta - \pi) + o((\Theta - \pi)^2).$$

(Here, $\mathbf{g}(x) = o(x)$ if $\lim\limits_{x \to \infty} \frac{\mathbf{g}(x)}{x} = 0$.) This allows us to linearize the control system in order to obtain a linear control system for $\varphi(t) := \Theta(t) - \pi$:

$$\ddot{\varphi}(t) - \varphi(t) = \mathbf{u}(t).$$

This can be written as an LTI system, assuming only positions can be observed, with

$$\mathbf{x} = \begin{bmatrix} \varphi \\ \dot{\varphi} \end{bmatrix}, \quad A = \begin{bmatrix} 0 & 1 \\ 1 & 0 \end{bmatrix}, \quad B = \begin{bmatrix} 0 \\ 1 \end{bmatrix}, \quad C = \begin{bmatrix} 1 & 0 \end{bmatrix}, \quad D = 0.$$

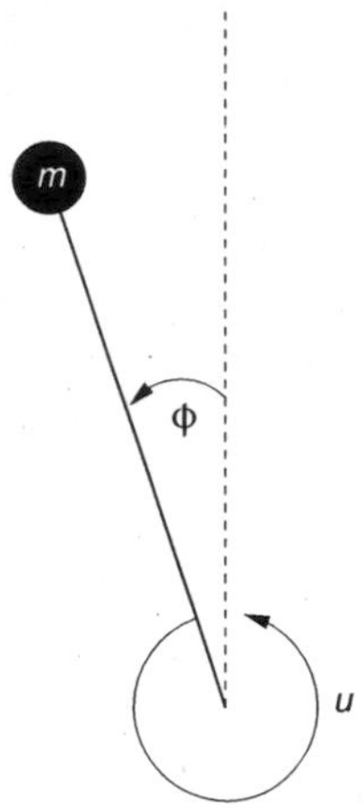

FIGURE 57.2 Inverted pendulum; apply control to move to upright position.

Now the objective translates to: Given initial values $x_1(0) = \varphi(0)$, $x_2(0) = \dot{\varphi}(0)$, find $\mathbf{u}(t)$ to bring $\mathbf{x}(t)$ to zero "as fast as possible." It is usually an additional goal to avoid overshoot and oscillating behavior as much as possible.

57.2 Frequency-Domain Analysis

So far, LTI systems are treated in state-space. In systems and control theory, it is often beneficial to use the **frequency domain** formalism obtained from applying the Laplace transformation to its state and observer equations.

Definitions:

The rational matrix function

$$G(s) = C(sI - A)^{-1}B + D \in \mathbb{R}^{p\times m}[s]$$

is called the **transfer function** of the LTI system defined in section 57.1.

In a **frequency domain analysis**, $G(s)$ is evaluated for $s = i\omega$, where $\omega \in [0, \infty]$ has the physical interpretation of a frequency and the input is considered as a signal with frequency ω.

The L_∞**-norm** of a transfer function is the operator norm induced by the frequency domain analogue of the L_2-norm that applies to Laplace transformed input functions $\mathbf{u} \in L_2(-\infty, \infty; \mathbb{R}^m)$, where $L_2(a, b; \mathbb{R}^m)$ is the Lebesgue space of square-integrable, measurable functions on the interval $(a, b) \subset \mathbb{R}$ with values in $\mathbb{R}^m$.

The $p \times m$-matrix-valued functions G for which $\|G\|_{L_\infty}$ is bounded form the space L_∞.

The subset of L_∞ containing all $p \times m$-matrix-valued functions that are analytical and bounded in the open right-half complex plane form the **Hardy space** H_∞.

The H_∞**-norm** of $G \in H_\infty$ is defined as

$$\|G\|_{H_\infty} = \operatorname*{ess\,sup}_{\omega\in\mathbb{R}} \sigma_{\max}(G(i\omega)), \tag{57.2}$$

where $\sigma_{\max}(M)$ is the maximum singular value of the matrix M and $\operatorname{ess\,sup}_{t\in M} h(t)$ is the essential supremum of a function h evaluated on the set M, which is the function's supremum on $M \setminus L$ where L is a set of Lebesgue measure zero.

For $T \in \mathbb{R}^{n\times n}$ nonsingular, the mapping implied by

$$(A, B, C, D) \mapsto (TAT^{-1}, TB, CT^{-1}, D)$$

is called a **state-space transformation.**

(A, B, C, D) is called a **realization** of an LTI system if its transfer function can be expressed as $G(s) = C(sI_n - A)^{-1}B + D$.

The minimum number $\hat{n}$ so that there exists no realization of a given LTI system with $n < \hat{n}$ is called the **McMillan degree** of the system.

A realization with $n = \hat{n}$ is a **minimal realization.**

Facts:

1. If $\mathbf{X}, \mathbf{Y}, \mathbf{U}$ are the Laplace transforms of $\mathbf{x}, \mathbf{y}, \mathbf{u}$, respectively, s is the Laplace variable, and $\mathbf{x}(0) = 0$, the state and observer equation of an LTI system transform to

$$\begin{aligned} s\mathbf{X}(s) &= A\mathbf{X}(s) + B\mathbf{U}(s), \\ \mathbf{Y}(s) &= C\mathbf{X}(s) + D\mathbf{U}(s). \end{aligned}$$

   Thus, the resulting input–output relation

$$\mathbf{Y}(s) = \left(C(sI - A)^{-1}B + D\right)\mathbf{U}(s) = G(s)\mathbf{U}(s) \tag{57.3}$$

   is completely determined by the transfer function of the LTI system.
2. As a consequence of the maximum modulus theorem, H_∞ functions must be bounded on the imaginary axis so that the essential supremum in the definition of the H_∞-norm simplifies to a supremum for rational functions G.
3. The transfer function of an LTI system is invariant w.r.t. state-space transformations:

$$D + (CT^{-1})(sI - TAT^{-1})^{-1}(TB) = C(sI_n - A)^{-1}B + D = G(s).$$

   Consequently, there exist infinitely many realizations of an LTI system.
4. Adding zero inputs/outputs does not change the transfer function, thus the order n of the system can be increased arbitrarily.

Examples:

1. The LTI system corresponding to the inverted pendulum has the transfer function

$$G(s) = \begin{bmatrix} 1 & 0 \end{bmatrix} \begin{bmatrix} s & -1 \\ -1 & s \end{bmatrix}^{-1} \begin{bmatrix} 0 \\ 1 \end{bmatrix} + [0] = \frac{1}{s^2 - 1}.$$

2. The L_∞-norm of the transfer function corresponding to the inverted pendulum is

$$\|G\|_{L_\infty} = 1.$$

3. The transfer function corresponding to the inverted pendulum is not in H_∞ as $G(s)$ has a pole at $s = 1$ and, thus, is not bounded in the right-half plane.

57.3 Analysis of LTI Systems

In this section, we provide characterizations of the properties of LTI systems defined in the introduction. Controllability and the related concepts can be checked using several algebraic criteria.

Definitions:

A matrix $A \in \mathbb{R}^{n\times n}$ is **Hurwitz** or **(asymptotically) stable** if all its eigenvalues have strictly negative real part.

The **controllability matrix** corresponding to an LTI system is

$$\mathcal{C}(A,B) = [B, AB, A^2B, \ldots, A^{n-1}B] \in \mathbb{R}^{n\times n\cdot m}.$$

The **observability matrix** corresponding to an LTI system is

$$\mathcal{O}(A,C) = \begin{bmatrix} C \\ CA \\ CA^2 \\ \vdots \\ CA^{n-1} \end{bmatrix} \in \mathbb{R}^{np\times n}.$$

The following transformations are **state-space transformations:**

- **Change of basis:**

$$\begin{array}{llll} \mathbf{x} \mapsto P\mathbf{x} & \text{for} & P \in \mathbb{R}^{n\times n} & \text{nonsingular,} \\ \mathbf{u} \mapsto Q\mathbf{u} & \text{for} & Q \in \mathbb{R}^{m\times m} & \text{nonsingular,} \\ \mathbf{y} \mapsto R\mathbf{y} & \text{for} & R \in \mathbb{R}^{p\times p} & \text{nonsingular.} \end{array}$$

- **Linear state feedback:** $\mathbf{u} \mapsto F\mathbf{x} + \mathbf{v}$, $F \in \mathbb{R}^{m\times n}$, $\mathbf{v} : [t_0, t_f] \mapsto \mathbb{R}^m$.
- **Linear output feedback:** $\mathbf{u} \mapsto G\mathbf{y} + \mathbf{v}$, $G \in \mathbb{R}^{m+p}$, $\mathbf{v} : [t_0, t_f] \mapsto \mathbb{R}^m$.

The **Kalman decomposition** of (A, B) is

$$V^T AV = \begin{bmatrix} A_1 & A_2 \\ 0 & A_3 \end{bmatrix}, \quad V^T B = \begin{bmatrix} B_1 \\ 0 \end{bmatrix}, \quad V \in \mathbb{R}^{n\times n} \text{ orthogonal,}$$

where (A_1, B_1) is controllable.

The **observability Kalman decomposition** of (A, C) is

$$W^T AW = \begin{bmatrix} A_1 & 0 \\ A_2 & A_3 \end{bmatrix}, \quad CW = [C_1 \ 0], \quad W \in \mathbb{R}^{n\times n} \text{ orthogonal,}$$

where (A_1, C_1) is observable.

Facts:

1. An LTI system is asymptotically stable if and only if A is Hurwitz.
2. For a given LTI system, the following are equivalent.

   (a) The LTI system is controllable.

   (b) The controllability matrix corresponding to the LTI system has full (row) rank, i.e., rank $\mathcal{C}(A, B) = n$.

(c) **(Hautus–Popov test)** If $\mathbf{p} \neq 0$ and $\mathbf{p}^* A = \lambda \mathbf{p}^*$, then $\mathbf{p}^* B \neq 0$.

(d) $\text{rank}([\lambda I - A, B]) = n \; \forall \lambda \in \mathbb{C}$.

The essential part of the proof of the above characterizations (which is "d)⇒b)") is an application of the Cayley–Hamilton theorem (Section 4.3).

3. For a given LTI system, the following are equivalent:

(a) The LTI system is stabilizable, i.e., $\exists F \in \mathbb{R}^{m \times n}$ such that $A + BF$ is Hurwitz.

(b) **(Hautus–Popov test)** If $\mathbf{p} \neq 0$, $\mathbf{p}^* A = \lambda \mathbf{p}^*$, and $\text{Re}(\lambda) \geq 0$, then $\mathbf{p}^* B \neq 0$.

(c) $\text{rank}([A - \lambda I, B]) = n, \quad \forall \lambda \in \mathbb{C}$ with $\text{Re}(\lambda) \geq 0$.

(d) In the Kalman decomposition of (A, B), A_3 is Hurwitz.

4. Using the change of basis $\tilde{\mathbf{x}} = V^T \mathbf{x}$ implied by the Kalman decomposition, we obtain

$$\dot{\tilde{\mathbf{x}}}_1 = A_1 \tilde{\mathbf{x}}_1 + A_2 \tilde{\mathbf{x}}_2 + B_1 \mathbf{u},$$
$$\dot{\tilde{\mathbf{x}}}_2 = A_3 \tilde{\mathbf{x}}_2.$$

Thus, $\tilde{\mathbf{x}}_2$ is not controllable. The eigenvalues of A_3 are, therefore, called **uncontrollable modes.**

5. For a given LTI system, the following are equivalent:

(a) The LTI system is observable.

(b) The observability matrix corresponding to the LTI system has full (column) rank, i.e., $\text{rank}\, \mathcal{O}(A, C) = n$.

(c) **(Hautus–Popov test)**, $\mathbf{p} \neq 0$, $A\mathbf{p} = \lambda \mathbf{p} \Longrightarrow C^T \mathbf{p} \neq 0$.

(d) $\text{rank} \begin{bmatrix} \lambda I - A \\ C \end{bmatrix} = n, \quad \forall \lambda \in \mathbb{C}$.

6. For a given LTI system, the following are equivalent:

(a) The LTI system is detectable.

(b) The dual system $\dot{\mathbf{z}} = A^T \mathbf{z} + C^T \mathbf{v}$ is stabilizable.

(c) **(Hautus–Popov test)** $\mathbf{p} \neq 0$, $A\mathbf{p} = \lambda \mathbf{p}$, $\text{Re}(\lambda) \geq 0 \Longrightarrow C^T \mathbf{p} \neq 0$.

(d) $\text{rank} \begin{bmatrix} \lambda I - A \\ C \end{bmatrix} = n, \quad \forall \lambda \in \mathbb{C}$ with $\text{Re}(\lambda) \geq 0$.

(e) In the observability Kalman decomposition of (A, C), A_3 is Hurwitz.

7. Using the change of basis $\tilde{\mathbf{x}} = W^T \mathbf{x}$ implied by the observability Kalman decomposition we obtain

$$\begin{aligned} \dot{\tilde{\mathbf{x}}}_1 &= A_1 \tilde{\mathbf{x}}_1 + B_1 \mathbf{u}, \\ \tilde{\mathbf{x}}_2 &= A_2 \tilde{\mathbf{x}}_1 + A_3 \tilde{\mathbf{x}}_2 + B_2 \mathbf{u}, \\ \mathbf{y} &= C_1 \tilde{\mathbf{x}}_1. \end{aligned}$$

Thus, $\tilde{\mathbf{x}}_2$ is not observable. The eigenvalues of A_3 are, therefore, called **unobservable modes.**

8. The characterizations of observability and detectability are proved using the duality principle and the characterizations of controllability and stabilizability.
9. If an LTI system is controllable (observable, stabilizable, detectable), then the corresponding LTI system resulting from a state-space transformation is controllable (observable, stabilizable, detectable).

10. For $A \in \mathbb{R}^{n \times n}$, $B \in \mathbb{R}^{n \times m}$ there exist $P \in \mathbb{R}^{n \times n}$, $Q \in \mathbb{R}^{m \times m}$ orthogonal such that

$$PAP^T = \left[\begin{array}{cccccc|c} A_{11} & & & & & A_{1,s-1} & A_{1,s} \\ A_{21} & \ddots & & & & \vdots & \vdots \\ 0 & \ddots & \ddots & & & \vdots & \vdots \\ \vdots & \ddots & & & & & \\ 0 & \cdots & 0 & A_{s-1,s-2} & & A_{s-1,s-1} & A_{s-1,s} \\ 0 & \cdots & 0 & 0 & & 0 & A_{ss} \end{array}\right] \begin{array}{c} n_1 \\ n_2 \\ \\ \\ n_{s-1} \\ n_s \end{array},$$
$$\qquad\qquad n_1 \qquad\qquad\qquad n_{s-2} \qquad n_{s-1} \qquad n_s$$

$$PBQ = \left[\begin{array}{cc} B_1 & 0 \\ 0 & 0 \\ \vdots & \vdots \\ 0 & 0 \end{array}\right] \begin{array}{c} n_1 \\ n_2 \\ \vdots \\ n_s \end{array},$$
$$\qquad n_1 \quad m - n_1$$

where $n_1 \geq n_2 \geq \ldots \geq n_{s-1} \geq n_s \geq 0$, $n_{s-1} > 0$, $A_{i,i-1} = [\Sigma_{i,i-1} \; 0] \in \mathbb{R}^{n_i \times n_{i-1}}$, $\Sigma_{i,i-1} \in \mathbb{R}^{n_i \times n_i}$ nonsingular for $i = 1, \ldots, s-1$, $\Sigma_{s-1,s-2}$ is diagonal, and B_1 is nonsingular.

Moreover, this transformation to **staircase form** can be computed by a finite sequence of singular value decompositions.

11. An LTI system is controllable if in the staircase form of (A, B), $n_s = 0$.
12. An LTI system is observable if $n_s = 0$ in the staircase form of (A^T, C^T).
13. An LTI system is stabilizable if in the staircase form of (A, B), A_{ss} is Hurwitz.
14. An LTI system is detectable if in the staircase form of (A^T, C^T), A_{ss} is Hurwitz.
15. In case $m = 1$, the staircase form of (A, B) is given by

$$PAP^T = \begin{bmatrix} a_{11} & \cdots & \cdots & a_{1,n} \\ a_{21} & & & \vdots \\ & \ddots & & \vdots \\ & & a_{n,n-1} & a_{n,n} \end{bmatrix}, \quad PB = \begin{bmatrix} b_1 \\ 0 \\ \vdots \\ 0 \end{bmatrix}$$

and is called the **controllability Hessenberg form**. The corresponding staircase from of (A^T, C^T) in case $p = 1$ is called the **observability Hessenberg form**.

Examples:

1. The LTI system corresponding to the inverted pendulum problem is not asymptotically stable as A is not Hurwitz: $\sigma(A) = \{\pm 1\}$.
2. The LTI system corresponding to the inverted pendulum problem is controllable as the controllability matrix

$$\mathcal{C}(A, B) = \begin{bmatrix} 0 & 1 \\ 1 & 0 \end{bmatrix}$$

has full rank. Thus, it is also stabilizable.

3. The LTI system corresponding to the inverted pendulum problem is observable as the observability matrix

$$\mathcal{O}(A, C) = \begin{bmatrix} 1 & 0 \\ 0 & 1 \end{bmatrix}$$

has full rank. Thus, it is also detectable.

57.4 Matrix Equations

A fundamental role in many tasks in control theory is played by matrix equations. We, therefore, review their most important properties. More details can be found in [AKFIJ03], [HJ91], [LR95], and [LT85].

Definitions:

A linear matrix equation of the form

$$AX + XB = W, \qquad A \in \mathbb{R}^{n\times n}, B \in \mathbb{R}^{m\times m}, W \in \mathbb{R}^{n\times m},$$

is called **Sylvester equation.**

A linear matrix equation of the form

$$AX + XA^T = W, \qquad A \in \mathbb{R}^{n\times n}, W = W^T \in \mathbb{R}^{n\times n},$$

is called **Lyapunov equation.**

A quadratic matrix equation of the form

$$0 = Q + A^T X + XA - XGX, \qquad A \in \mathbb{R}^{n\times n}, G = G^T, Q = Q^T \in \mathbb{R}^{n\times n},$$

is called **algebraic Riccati equation (ARE).**

Facts:

1. The Sylvester equation is equivalent to the linear system of equations

$$\left[(I_m \otimes A) + (B^T \otimes I_n)\right] \operatorname{vec}(X) = \operatorname{vec}(W),$$

where $\otimes$ and vec denote the Kronecker product and the vec-operator defined in Section 10.4. Thus, the Sylvester equation has a unique solution if and only if $\sigma(A) \cap \sigma(-B) = \emptyset$.

2. The Lyapunov equation is equivalent to the linear system of equations

$$[(I_m \otimes A) + (A \otimes I_n)] \operatorname{vec}(X) = \operatorname{vec}(W).$$

Thus, it has a unique solution if and only if $\sigma(A) \cap \sigma(-A^T) = \emptyset$. In particular, this holds if A is Hurwitz.

3. If G and Q are positive semidefinite with (A, G) stabilizable and (A, Q) detectable, then the ARE has a unique positive semidefinite solution X_* with the property that $\sigma(A - GX_*)$ is Hurwitz.
4. If the assumptions given above are not satisfied, there may or may not exist a stabilizing solution with the given properties. Besides, there may exist a continuum of solutions, a finite number of solutions, or no solution at all. The solution theory for AREs is a vast topic by itself; see the monographs [AKFIJ03], [LR95] and [Ben99], [Dat04], [Meh91], and [Sim96] for numerical algorithms to solve these equations.

Examples:

1. For

$$A = \begin{bmatrix} 1 & 2 \\ 0 & 1 \end{bmatrix}, \quad B = \begin{bmatrix} 2 & -1 \\ 1 & 0 \end{bmatrix}, \quad W = \begin{bmatrix} -1 & 0 \\ 0 & -1 \end{bmatrix},$$

a solution of the Sylvester equation is

$$X = \frac{1}{4} \begin{bmatrix} -3 & 3 \\ 1 & -3 \end{bmatrix}.$$

Note that $\sigma(A) = \sigma(B) = \{1, 1\}$ so that $\sigma(A) \cap \sigma(-B) = \emptyset$. Thus, this Sylvester equation has the unique solution X given above.

2. For

$$A = \begin{bmatrix} 0 & 1 \\ 0 & 0 \end{bmatrix}, \quad G = \begin{bmatrix} 0 & 0 \\ 0 & 1 \end{bmatrix}, \quad Q = \begin{bmatrix} 1 & 0 \\ 0 & 2 \end{bmatrix},$$

the stabilizing solution of the associated ARE is

$$X_* = \begin{bmatrix} 2 & 1 \\ 1 & 2 \end{bmatrix}$$

and the spectrum of the closed-loop matrix

$$A - GX_* = \begin{bmatrix} 0 & 1 \\ -1 & -2 \end{bmatrix}$$

is $\{-1, -1\}$.

3. Consider the ARE

$$0 = C^T C + A^T X + XA - XBB^T X$$

corresponding to an LTI system with

$$A = \begin{bmatrix} -1 & 0 \\ 0 & 0 \end{bmatrix}, \quad B = \begin{bmatrix} 1 \\ 0 \end{bmatrix}, \quad C = \begin{bmatrix} \sqrt{2} & 0 \end{bmatrix}, \quad D = 0.$$

For this ARE, $X = \begin{bmatrix} -1 + \sqrt{3} & 0 \\ 0 & \xi \end{bmatrix}$ is a solution for all $\xi \in \mathbb{R}$. It is positive semidefinite for all $\xi \geq 0$, but this ARE does not have a stabilizing solution as the LTI system is neither stabilizable nor detectable.

57.5 State Estimation

In this section, we present the two most famous approaches to state observation, that is, finding a function $\hat{\mathbf{x}}(t)$ that approximates the state $\mathbf{x}(t)$ of a given LTI system if only its inputs $\mathbf{u}(t)$ and outputs $\mathbf{y}(t)$ are known. While the first approach (the Luenberger observer) assumes a deterministic system behavior, the Kalman–Bucy filter allows for uncertainty in the system, modeled by white-noise, zero-mean stochastic processes.

Definitions:

Given an LTI system with $D = 0$, a **state observer** is a function

$$\hat{\mathbf{x}} : [0, \infty) \mapsto \mathbb{R}^n$$

such that for some nonsingular matrix $Z \in \mathbb{R}^{n \times n}$ and $\mathbf{e}(t) = \hat{\mathbf{x}}(t) - Z\mathbf{x}(t)$, we have

$$\lim_{t \to \infty} \mathbf{e}(t) = 0.$$

Given an LTI system with stochastic disturbances

$$\begin{aligned} \dot{\mathbf{x}}(t) &= A\mathbf{x}(t) + B\mathbf{u}(t) + \tilde{B}\mathbf{w}(t), \\ \mathbf{y}(t) &= C\mathbf{x}(t) + \mathbf{v}(t), \end{aligned}$$

where A, B, C are as before, $\tilde{B} \in \mathbb{R}^{n \times \tilde{m}}$, and $\mathbf{w}(t), \mathbf{v}(t)$ are white-noise, zero-mean stochastic processes with corresponding covariance matrices $W = W^T \in \mathbb{R}^{\tilde{m} \times \tilde{m}}$ (positive semidefinite), $V = V^T \in \mathbb{R}^{p \times p}$ (positive definite), the problem to minimize the mean square error

$$E\left[\|\mathbf{x}(t) - \hat{\mathbf{x}}(t)\|_2^2\right]$$

over all state observers is called the **optimal estimation problem.** (Here, $E[r]$ is the **expected value** of r.)

Facts:

1. A state observer, called the **Luenberger observer**, is obtained as the solution of the dynamical system

$$\dot{\hat{\mathbf{x}}}(t) = H\hat{\mathbf{x}}(t) + F\mathbf{y}(t) + G\mathbf{u}(t),$$

where $H \in \mathbb{R}^{n \times n}$ and $F \in \mathbb{R}^{n \times p}$ are chosen so that H is Hurwitz and the **Sylvester observer equation**

$$HX - XA + FC = 0$$

has a nonsingular solution X. Then $G = XB$ and the matrix Z in the definition of the state observer equals the solution X of the Sylvester observer equation.

2. Assuming that

   - $\mathbf{w}$ and $\mathbf{v}$ are uncorrelated stochastic processes,
   - the initial state $\mathbf{x}^0$ is a Gaussian zero-mean random variable, uncorrelated with $\mathbf{w}$ and $\mathbf{v}$,
   - (A, B) is controllable and (A, C) is observable,

   the solution to the optimal estimation problem is given by the **Kalman–Bucy** filter, defined as the solution of the linear differential equation

$$\dot{\hat{\mathbf{x}}}(t) = (A - Y_* C^T V^{-1} C)\hat{\mathbf{x}}(t) + B\mathbf{u}(t) + Y_* C^T V^{-1}\mathbf{y}(t),$$

where Y_* is the unique stabilizing solution of the **filter ARE:**

$$0 = \tilde{B} W \tilde{B}^T + AY + YA^T - YC^T V^{-1} CY.$$

3. Under the same assumptions as above, the stabilizing solution of the filter ARE can be shown to be symmetric positive definite.

Examples:

1. A Luenberger observer for the LTI system corresponding to the inverted pendulum problem can be constructed as follows: Choose $H = \text{diag}(-2, -\frac{1}{2})$ and $F = \begin{bmatrix} 2 & 1 \end{bmatrix}^T$. Then the Sylvester observer equation has the unique solution

$$X = \frac{1}{3}\begin{bmatrix} 4 & -2 \\ -2 & 4 \end{bmatrix}.$$

Note that X is nonsingular. Thus, we get $G = XB = \frac{1}{3}\begin{bmatrix} -2 & 4 \end{bmatrix}$.

2. Consider the inverted pendulum with disturbances v, w and $\tilde{B} = \begin{bmatrix} 1 & 1 \end{bmatrix}^T$. Assume that $V = W = 1$. The Kalman–Bucy filter is determined via the filter ARE, yielding

$$Y_* = (1 + \sqrt{2})\begin{bmatrix} 1 & 1 \\ 1 & 1 \end{bmatrix}.$$

Thus, the state estimation obtained from the Kalman filter is given by the solution of

$$\dot{\hat{\mathbf{x}}}(t) = \begin{bmatrix} -1 - \sqrt{2} & 1 \\ -\sqrt{2} & 0 \end{bmatrix}\hat{\mathbf{x}}(t) + \begin{bmatrix} 0 \\ 1 \end{bmatrix}\mathbf{u}(t) + (1 + \sqrt{2})\begin{bmatrix} 1 \\ 1 \end{bmatrix}\mathbf{y}(t).$$

57.6 Control Design for LTI Systems

This section provides the background for some of the most important control design methods.

Definitions:

A **(feedback) controller** for an LTI system is given by another LTI system

$$\begin{aligned} \dot{\mathbf{r}}(t) &= E\mathbf{r}(t) + F\mathbf{y}(t), \\ \mathbf{u}(t) &= H\mathbf{r}(t) + K\mathbf{y}(t), \end{aligned}$$

where $E \in \mathbb{R}^{N \times N}$, $F \in \mathbb{R}^{N \times p}$, $H \in \mathbb{R}^{m \times N}$, $K \in \mathbb{R}^{m \times p}$, and the "output" $\mathbf{u}(t)$ of the controller serves as the input for the original LTI system.

If E, F, H are zero matrices, a controller is called **static feedback**, otherwise it is called a **dynamic compensator**.

A static feedback control law is a **state feedback** if in the controller equations, the output function $\mathbf{y}(t)$ is replaced by the state $\mathbf{x}(t)$, otherwise it is called **output feedback**.

The **closed-loop system** resulting from inserting the control law $\mathbf{u}(t)$ obtained from a dynamic compensator into the LTI system is illustrated by the block diagram in Figure 57.3, where $\mathbf{w}$ is as in the definition of LTI systems with stochastic disturbances in Section 57.5 and $\mathbf{z}$ will only be needed later when defining the H_∞ control problem.

The linear-quadratic optimization (optimal control) problem

$$\min_{\mathbf{u} \in L_2(0,\infty;\mathcal{U})} \mathcal{J}(\mathbf{u}), \quad \text{where } \mathcal{J}(\mathbf{u}) = \frac{1}{2}\int_0^\infty \left(\mathbf{y}(t)^T Q\mathbf{y}(t) + \mathbf{u}(t)^T R\mathbf{u}(t)\right)\, dt$$

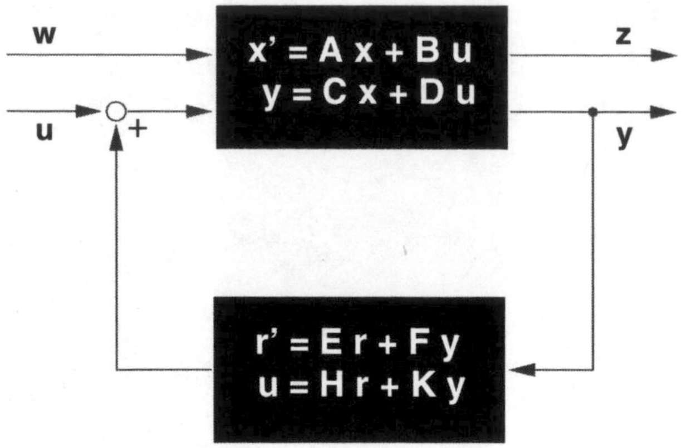

FIGURE 57.3 Closed-loop diagram of an LTI system and a dynamic compensator.

subject to the dynamical constraint given by an LTI system is called the **linear-quadratic regulator (LQR) problem.**

The linear-quadratic optimization (optimal control) problem

$$\min_{\mathbf{u}\in L_2(0,\infty;\mathcal{U})} \mathcal{J}(\mathbf{u}), \quad \text{where } \mathcal{J}(\mathbf{u}) = \lim_{t_f\to\infty} \frac{1}{2t_f} E\left[\int_{-t_f}^{t_f} \left(\mathbf{y}(t)^T Q\mathbf{y}(t) + \mathbf{u}(t)^T R\mathbf{u}(t)\right) dt\right]$$

subject to the dynamical constraint given by an LTI system with stochastic disturbances is called the **linear-quadratic Gaussian (LQG) problem.**

Consider an LTI system where inputs and outputs are split into two parts, so that instead of $B\mathbf{u}(t)$ we have

$$B_1\mathbf{w}(t) + B_2\mathbf{u}(t),$$

and instead of $\mathbf{y}(t) = C\mathbf{x}(t) + D\mathbf{u}(t)$, we write

$$\begin{aligned} \mathbf{z}(t) &= C_1\mathbf{x}(t) + D_{11}\mathbf{w}(t) + D_{12}\mathbf{u}(t), \\ \mathbf{y}(t) &= C_2\mathbf{x}(t) + D_{21}\mathbf{w}(t) + D_{22}\mathbf{u}(t), \end{aligned}$$

where $\mathbf{u}(t) \in \mathbb{R}^{m_2}$ denotes the control input, $\mathbf{w}(t) \in \mathbb{R}^{m_1}$ is an **exogenous input** that may include noise, linearization errors, and unmodeled dynamics, $\mathbf{y}(t) \in \mathbb{R}^{p_2}$ contains **measured outputs**, while $\mathbf{z}(t) \in \mathbb{R}^{p_1}$ is the **regulated output** or an **estimation error.** Let $G = \begin{bmatrix} G_{11} & G_{12} \\ G_{21} & G_{22} \end{bmatrix}$ denote the corresponding transfer function such that

$$\begin{bmatrix} \mathbf{Z} \\ \mathbf{Y} \end{bmatrix} = \begin{bmatrix} G_{11} & G_{12} \\ G_{21} & G_{22} \end{bmatrix} \begin{bmatrix} \mathbf{W} \\ \mathbf{U} \end{bmatrix},$$

where $\mathbf{Y}, \mathbf{Z}, \mathbf{U}, \mathbf{W}$ denote the Laplace transforms of $\mathbf{y}, \mathbf{z}, \mathbf{u}, \mathbf{w}$.

The **optimal H_∞ control problem** is then to determine a dynamic compensator

$$\begin{aligned} \dot{\mathbf{r}}(t) &= E\mathbf{r}(t) + F\mathbf{y}(t), \\ \mathbf{u}(t) &= H\mathbf{r}(t) + K\mathbf{y}(t), \end{aligned}$$

with $E \in \mathbb{R}^{N\times N}$, $F \in \mathbb{R}^{N\times p_2}$, $H \in \mathbb{R}^{m_2\times N}$, $K \in \mathbb{R}^{m_2\times p_2}$ and transfer function $M(s) = H(sI - E)^{-1}F + K$ such that the resulting closed-loop system

$$\begin{aligned}\dot{\mathbf{x}}(t) &= (A + B_2KZ_1C_2)\mathbf{x}(t) + (B_2Z_2H)\mathbf{r}(t) + (B_1 + B_2KZ_1D_{21})\mathbf{w}(t),\\ \dot{\mathbf{r}}(t) &= FZ_1C_2\mathbf{x}(t) + (E + FZ_1D_{22}H)\mathbf{r}(t) + FZ_1D_{21}\mathbf{w}(t),\\ \mathbf{z}(t) &= (C_1 + D_{12}Z_2KC_2)\mathbf{x}(t) + D_{12}Z_2H\mathbf{r}(t) + (D_{11} + D_{12}KZ_1D_{21})\mathbf{w}(t),\end{aligned}$$

with $Z_1 = (I - D_{22}K)^{-1}$ and $Z_2 = (I - KD_{22})^{-1}$,

- is **internally stable**, i.e., the solution of the system with $\mathbf{w}(t) \equiv 0$ is asymptotically stable, and
- the closed-loop transfer function $T_{zw}(s) = G_{22}(s) + G_{21}(s)M(s)(I - G_{11}(s)M(s))^{-1}G_{12}(s)$ from $\mathbf{w}$ to $\mathbf{z}$ is minimized in the H_∞-norm.

The **suboptimal H_∞ control problem** is to find an internally stabilizing controller so that

$$\|T_{zw}\|_{H_\infty} < \gamma,$$

where $\gamma > 0$ is a robustness threshold.

Facts:

1. If $D = 0$ and the LTI system is both stabilizable and detectable, the weighting matrix Q is positive semidefinite, and R is positive definite, then the solution of the LQR problem is given by the state feedback controller

$$\mathbf{u}_*(t) = -R^{-1}B^TX_*\mathbf{x}(t), \qquad t \geq 0,$$

where X_* is the unique stabilizing solution of the LQR ARE,

$$0 = C^TQC + A^TX + XA - XBR^{-1}B^TX.$$

2. The LQR problem does not require an observer equation — inserting $\mathbf{y}(t) = C\mathbf{x}(t)$ into the cost functional, we obtain a problem formulation depending only on states and inputs:

$$\begin{aligned}\mathcal{J}(\mathbf{u}) &= \frac{1}{2}\int_0^\infty \left(\mathbf{y}(t)^TQ\mathbf{y}(t) + \mathbf{u}(t)^TR\mathbf{u}(t)\right)dt\\ &= \frac{1}{2}\int_0^\infty \left(\mathbf{x}(t)^TC^TQC\mathbf{x}(t) + \mathbf{u}(t)^TR\mathbf{u}(t)\right)dt.\end{aligned}$$

3. Under the given assumptions, it can also be shown that X_* is symmetric and the unique positive semidefinite matrix among all solutions of the LQR ARE.
4. The assumptions for the feedback solution of the LQR problem can be weakened in several aspects; see, e.g., [Gee89] and [SSC95].
5. Assuming that
   - $\mathbf{w}$ and $\mathbf{v}$ are uncorrelated stochastic processes,
   - the initial state $\mathbf{x}^0$ is a Gaussian zero-mean random variable, uncorrelated with $\mathbf{w}$ and $\mathbf{v}$,
   - (A, B) is controllable and (A, C) is observable,

   the solution to the LQG problem is given by the feedback controller

$$\mathbf{u}(t) = -R^{-1}B^TX_*\hat{\mathbf{x}}(t),$$

where X_* is the solution of the LQR ARE and $\hat{\mathbf{x}}$ is the Kalman–Bucy filter

$$\dot{\hat{\mathbf{x}}}(t) = (A - BR^{-1}B^T X_* - Y_* C^T V^{-1} C)\hat{\mathbf{x}}(t) + Y_* C^T V^{-1}\mathbf{y}(t),$$

corresponding to the closed-loop system resulting from the LQR solution with Y_* being the stabilizing solution of the corresponding filter ARE.

6. In principle, there is no restriction on the degree N of the H_∞ controller, although smaller dimensions N are preferred for practical implementation and computation.
7. The state-space solution to the H_∞ suboptimal control problem [DGKF89] relates H_∞ control to AREs: under the assumptions that

   - (A, B_k) is stabilizable and (A, C_k) is detectable for $k = 1, 2$,
   - $D_{11} = 0$, $D_{22} = 0$, and

   $$D_{12}^T \begin{bmatrix} C_1 & D_{12} \end{bmatrix} = \begin{bmatrix} 0 & I \end{bmatrix}, \quad \begin{bmatrix} B_1 \\ D_{21} \end{bmatrix} D_{21}^T = \begin{bmatrix} 0 \\ I \end{bmatrix},$$

   a suboptimal H_∞ controller exists if and only if the AREs

   $$0 = C_1^T C_1 + AX + XA^T + X\left(\frac{1}{\gamma^2} B_1 B_1^T - B_2 B_2^T\right) X,$$

   $$0 = B_1^T B_1 + A^T Y + YA + Y\left(\frac{1}{\gamma^2} C_1 C_1^T - C_2 C_2^T\right) Y$$

   both have positive semidefinite stabilizing solutions X_∞ and Y_∞, respectively, satisfying the spectral radius condition

   $$\rho(XY) < \gamma^2.$$

8. The solution of the optimal H_∞ control problem can be obtained by a bisection method (or any other root-finding method) minimizing γ based on the characterization of an H_∞ suboptimal controller given in Fact 7, starting from γ_0 for which no suboptimal H_∞ controller exists and γ_1 for which the above conditions are satisfied.
9. The assumptions made for the state-space solution of the H_∞ control problem can mostly be relaxed.
10. The robust numerical solution of the H_∞ control problem is a topic of ongoing research — the solution via AREs may suffer from several difficulties in the presence of roundoff errors and should be avoided if possible. One way out is a reformulation of the problem using structured generalized eigenvalue problems; see [BBMX99b], [CS92] and [GL97].
11. Once a (sub-)optimal γ is found, it remains to determine a realization of the H_∞ controller. One possibility is the **central (minimum entropy) controller** [ZDG96]:

    $$E = A + \frac{1}{\gamma^2} B_1 B_1^T - B_2 B_2^T X_\infty - Z_\infty Y_\infty C_2^T C_2,$$

    $$F = Z_\infty Y_\infty C_2^T, \quad K = -B_2^T X_\infty, \quad H = 0,$$

    where

    $$Z_\infty = \left(I - \frac{1}{\gamma^2} Y_\infty X_\infty\right)^{-1}.$$

Examples:

1. The cost functional in the LQR and LQG problems values the energy needed to reach the desired state by the weighting matrix R on the inputs. Thus, usually

$$R = \mathrm{diag}(\rho_1, \ldots, \rho_m).$$

The weighting on the states or outputs in the LQR or LQG problems is usually used to penalize deviations from the desired state of the system and is often also given in diagonal form. Common examples of weighting matrices are $R = \rho I_m$, $Q = \gamma I_p$ for $\rho, \gamma > 0$.

2. The solution to the LQR problem for the inverted pendulum with $Q = R = 1$ is given via the stabilizing solution of the LQR ARE, which is

$$X_* = \begin{bmatrix} 2\sqrt{1+\sqrt{2}} & 1+\sqrt{2} \\ 1+\sqrt{2} & \sqrt{2}\sqrt{1+\sqrt{2}} \end{bmatrix},$$

resulting in the state feedback law

$$\mathbf{u}(t) = -\begin{bmatrix} 1+\sqrt{2} & \sqrt{2}\sqrt{1+\sqrt{2}} \end{bmatrix} \mathbf{x}(t).$$

The eigenvalues of the closed-loop system are (up to four digits) $\sigma(A - BR^{-1}B^T X_*) = \{-1.0987 \pm 0.4551i\}$.

3. The solution to the LQG problem for the inverted pendulum with Q, R as above and uncertainties $\mathbf{v}, \mathbf{w}$ with $\tilde{B} = \begin{bmatrix} 1 & 1 \end{bmatrix}^T$ is obtained by combining the LQR solution derived above with the Kalman–Bucy filter obtained as in the examples part of the previous section.
Thus, we get the LQG control law

$$\mathbf{u}(t) = -\begin{bmatrix} 1+\sqrt{2} & \sqrt{2}\sqrt{1+\sqrt{2}} \end{bmatrix} \hat{\mathbf{x}}(t),$$

where $\hat{\mathbf{x}}$ is the solution of

$$\dot{\hat{\mathbf{x}}}(t) = -\begin{bmatrix} 1+\sqrt{2} & -1 \\ 1+2\sqrt{2} & \sqrt{2}\sqrt{1+\sqrt{2}} \end{bmatrix} \mathbf{x}(t) + (1+\sqrt{2}) \begin{bmatrix} 1 \\ 1 \end{bmatrix} \mathbf{y}(t).$$

References

[AKFIJ03] H. Abou-Kandil, G. Freiling, V. Ionescu, and G. Jank. *Matrix Riccati Equations in Control and Systems Theory*. Birkhäuser, Basel, Switzerland, 2003.

[Ben99] P. Benner. Computational methods for linear-quadratic optimization. *Supplemento ai Rendiconti del Circolo Matematico di Palermo, Serie II*, No. 58:21–56, 1999.

[BBMX99] P. Benner, R. Byers, V. Mehrmann, and H. Xu. Numerical methods for linear-quadratic and H_∞ control problems. In G. Picci and D.S. Gilliam, Eds., *Dynamical Systems, Control, Coding, Computer Vision: New Trends, Interfaces, and Interplay*, Vol. 25 of *Progress in Systems and Control Theory*, pp. 203–222. Birkhäuser, Basel, 1999.

[BMS+99] P. Benner, V. Mehrmann, V. Sima, S. Van Huffel, and A. Varga. SLICOT — a subroutine library in systems and control theory. In B.N. Datta, Ed., *Applied and Computational Control, Signals, and Circuits*, Vol. 1, pp. 499–539. Birkhäuser, Boston, MA, 1999.

[CS92] B.R. Copeland and M.G. Safonov. A generalized eigenproblem solution for singular H^2 and H^∞ problems. In *Robust Control System Techniques and Applications, Part 1*, Vol. 50 of *Control Dynam. Systems Adv. Theory Appl.*, pp. 331–394. Academic Press, San Diego, CA, 1992.

[Dat04] B.N. Datta. *Numerical Methods for Linear Control Systems*. Elsevier Academic Press, Amsterdom, 2004.

[Doy78] J. Doyle. Guaranteed margins for LQG regulators. *IEEE Trans. Automat. Control*, 23:756–757, 1978.

[DGKF89] J. Doyle, K. Glover, P.P. Khargonekar, and B.A. Francis. State-space solutions to standard H_2 and H_∞ control problems. *IEEE Trans. Automat. Cont.*, 34:831–847, 1989.

[GL97] P. Gahinet and A.J. Laub. Numerically reliable computation of optimal performance in singular H_∞ control. *SIAM J. Cont. Optim.*, 35:1690–1710, 1997.

[Gee89] T. Geerts. All optimal controls for the singular linear–quadratic problem without stability; a new interpretation of the optimal cost. *Lin. Alg. Appl.*, 116:135–181, 1989.

[GL95] M. Green and D.J.N Limebeer. *Linear Robust Control*. Prentice-Hall, Upper Saddle River, NJ, 1995.

[HJ91] R.A. Horn and C.R. Johnson. *Topics in Matrix Analysis*. Cambridge University Press, Cambridge, 1991.

[Kal60] R.E. Kalman. Contributions to the theory of optimal control. *Boletin Sociedad Matematica Mexicana*, 5:102–119, 1960.

[KB61] R.E. Kalman and R.S. Bucy. New results in linear filtering and prediction theory. *Trans. ASME, Series D*, 83:95–108, 1961.

[Kuc91] V. Kučera. *Analysis and Design of Discrete Linear Control Systems*. Academia, Prague, Czech Republic, 1991.

[Lev96] W.S. Levine, Ed. *The Control Handbook*. CRC Press, Boca Raton, FL, 1996.

[LR95] P. Lancaster and L. Rodman. *The Algebraic Riccati Equation*. Oxford University Press, Oxford, U.K., 1995.

[LT85] P. Lancaster and M. Tismenetsky. *The Theory of Matrices*. Academic Press, Orlando, FL, 2nd ed., 1985.

[Meh91] V. Mehrmann. *The Autonomous Linear Quadratic Control Problem, Theory and Numerical Solution*. Number 163 in Lecture Notes in Control and Information Sciences. Springer-Verlag, Heidelberg, July 1991.

[Mut99] A.G.O. Mutambara. *Design and Analysis of Control Systems*. CRC Press, Boca Raton, FL, 1999.

[PUA00] I.R. Petersen, V.A. Ugrinovskii, and A.V.Savkin. *Robust Control Design Using H^∞ Methods*. Springer-Verlag, London, 2000.

[SSC95] A. Saberi, P. Sannuti, and B.M. Chen. *H_2 Optimal Control*. Prentice-Hall, Hertfordshire, U.K., 1995.

[Sim96] V. Sima. *Algorithms for Linear-Quadratic Optimization*, Vol. 200 of *Pure and Applied Mathematics*. Marcel Dekker, Inc., New York, 1996.

[Son98] E.D. Sontag. *Mathematical Control Theory*. Springer-Verlag, New York, 2nd ed., 1998.

[ZDG96] K. Zhou, J.C. Doyle, and K. Glover. *Robust and Optimal Control*. Prentice-Hall, Upper Saddle River, NJ, 1996.

58

Fourier Analysis

Kenneth Howell
University of Alabama in Huntsville

58.1 Introduction

Fourier analysis has been employed with great success in a wide range of applications. The underlying theory is based on a small set of linear transforms on particular linear spaces. It may, in fact, be best to refer to two parallel theories of Fourier analysis—the function/functional theory and the discrete theory—according to whether these linear spaces are infinite or finite dimensional. The function/functional theory involves infinite dimensional spaces of functions on $\mathbb{R}^n$ and can be further divided into two "subtheories"—one concerned with functions that are periodic on $\mathbb{R}^n$ (or can be treated as periodic extensions of functions on finite subregions) and one concerned with more general functions and functionals on $\mathbb{R}^n$. This theory played the major role in applications up to half a century or so ago. However, the tools from the discrete theory (involving vectors in $\mathbb{C}^N$ instead of functions on $\mathbb{R}^n$) are much more easily implemented on digital computers. This became of interest both because much of the function/functional theory analysis can be closely approximated within the discrete theory and because the discrete theory is a natural setting for functions known only by samplings. In addition, "fast" algorithms for computing the discrete transforms were developed, allowing discrete analysis to be done very quickly even with very large data sets. For these reasons the discrete theory has become extremely important in modern applications, and its utility, in turn, has greatly extended the applicability of Fourier analysis.

In the first part of this chapter, elements of the function/functional theory are presented and illustrated. For expediency, attention will be restricted to functions and functionals over subsets of $\mathbb{R}^1$. A corresponding review of the analogous elements of the discrete theory then follows, with a discussion of the relations between the two theories following that. Finally, one of the fast algorithms is described and the extent to which this algorithm improves speed and applicability is briefly discussed.

The development here is necessarily abbreviated and covers only a small fraction of the theory and applications of Fourier analysis. The reader interested in more complete treatments of the subject is encouraged to consult the references given throughout this chapter.

58.2 The Function/Functional Theory

Definitions:

The following function spaces often arise in Fourier analysis. In each case $\mathcal{I}$ is a subinterval of $\mathbb{R}$, and all functions on $\mathcal{I}$ are assumed to be complex valued.

- $\mathcal{C}^0(\mathcal{I})$, the normed linear space of bounded and continuous functions on $\mathcal{I}$ with the norm $\|f\| = \sup\{|f(x)| : x \in \mathcal{I}\}$.
- $\mathcal{L}^1(\mathcal{I})$, the normed linear space of absolutely integrable functions on $\mathcal{I}$ with the norm $\|f\| = \int_{\mathcal{I}} |f(x)|\, dx$.
- $\mathcal{L}^2(\mathcal{I})$, the inner product space of square integrable functions on $\mathcal{I}$ with the inner product $\langle f, g\rangle = \int_{\mathcal{I}} f(x)\overline{g(x)}\, dx$.

If $\mathcal{I}$ is not specified, then $\mathcal{I} = \mathbb{R}$.

Let $\phi \in \mathcal{L}^1(\mathbb{R})$. The **Fourier transform** $\mathcal{F}[\phi]$ and the **inverse Fourier transform** $\mathcal{F}^{-1}[\phi]$ of ϕ are the functions

$$\mathcal{F}[\phi]|_x = \int_{-\infty}^{\infty} \phi(y)e^{-i2\pi xy} dy \qquad \text{and} \qquad \mathcal{F}^{-1}[\phi]|_x = \int_{-\infty}^{\infty} \phi(y)e^{i2\pi xy} dy\,.$$

Additionally, the terms "Fourier transform" and "inverse Fourier transform" can refer to the processes for computing these functions from ϕ.

A function ϕ on $\mathbb{R}$ is said to be **periodic** if there is a positive value p, called a **period** for ϕ, such that

$$\phi(x + p) = \phi(x) \quad \forall x \in \mathbb{R}\,.$$

The smallest period, if it exists, is called the **fundamental period.**

The **Fourier series** for a suitably integrable, periodic function ϕ with period $p > 0$ is the infinite series

$$\sum_{k=-\infty}^{\infty} c_k\, e^{i2\pi\omega_k x}$$

where

$$\omega_k = \frac{k}{p} \qquad \text{and} \qquad c_k = \frac{1}{p}\int_0^p \phi(y)e^{-i2\pi\omega_k y} dy.$$

The theory for Fourier transforms and series can be generalized so that the requirement of ϕ being "suitably integrable" can be greatly relaxed (see [How01, Chap. 20, 30–34] or [Str94, Chap. 1–4]). Within this generalized theory the **delta function** at $a \in \mathbb{R}$, δ_a, is the functional limit

$$\delta_a(x) = \lim_{\epsilon\to 0^+} \frac{1}{2\epsilon}\text{pulse}_\epsilon(x - a) \qquad \text{where} \qquad \text{pulse}_\epsilon(s) = \begin{cases} 1 & \text{if } |s| \le \epsilon \\ 0 & \text{if } |s| > \epsilon \end{cases}.$$

That is, δ_a is the (generalized) function such that

$$\int_{-\infty}^{\infty} \psi(x)\delta_a(x)\, dx = \lim_{\epsilon\to 0^+} \int_{-\infty}^{\infty} \psi(x)\frac{1}{2\epsilon}\text{pulse}_\epsilon(x - a)\, dx$$

whenever ψ is a function continuous at a.

An **array of delta functions** is any expression of the form

$$\sum_{k=-\infty}^{\infty} c_k \delta_{k\Delta x}$$

where $\Delta x > 0$ is fixed (the **spacing** of the array) and $c_k \in \mathbb{C}$ for each $k \in \mathbb{Z}$. If the array is also periodic with period p, then the corresponding **index period** is the positive integer N such that

$$p = N\Delta x \qquad \text{and} \qquad \phi_{k+N} = \phi_k \quad \forall\, k \in \mathbb{Z}.$$

The **convolution** $\phi * \psi$ of a suitably integrable pair of functions ϕ and ψ on $\mathbb{R}$ is the function given by

$$\phi * \psi(x) = \int_{-\infty}^{\infty} \phi(x-y)\psi(y)\,dy\,.$$

Facts:

All the following facts except those with specific reference can be found in [How01] or [Str94].

1. Warning: Slight variations of the above integral formulas, e.g.,

$$\frac{1}{\sqrt{2\pi}}\int_{-\infty}^{\infty}\phi(y)e^{-ixy}dy \qquad \text{and} \qquad \frac{1}{\sqrt{2\pi}}\int_{-\infty}^{\infty}\phi(y)e^{ixy}dy,$$

are also often used to define, respectively, $\mathcal{F}[\phi]$ and $\mathcal{F}^{-1}[\phi]$ (or even $\mathcal{F}^{-1}[\phi]$ and $\mathcal{F}[\phi]$). There is little difference in the resulting "Fourier theories," but the formulas resulting from using different versions of the Fourier transforms do differ in details. Thus, when computing Fourier transforms using different tables or software, it is important to take into account the specific integral formulas on which the table or software is based.
2. Both $\mathcal{F}$ and $\mathcal{F}^{-1}$ are continuous, linear mappings from $\mathcal{L}^1(\mathbb{R})$ into $\mathcal{C}^0(\mathbb{R})$. Moreover, if both ϕ and $\mathcal{F}[\phi]$ are absolutely integrable, then $\mathcal{F}^{-1}[\mathcal{F}[\phi]] = \phi$.
3. In the generalized theory (which will be assumed hereafter), $\mathcal{F}$ and $\mathcal{F}^{-1}$ are defined on a linear space of (generalized) functions that contains all elements of $\mathcal{L}^1(\mathbb{R})$ and $\mathcal{L}^2(\mathbb{R})$, all Fourier transforms of elements of $\mathcal{L}^1(\mathbb{R})$ and $\mathcal{L}^2(\mathbb{R})$, all piecewise continuous periodic functions, the delta functions, and all periodic arrays of delta functions. $\mathcal{F}$ and $\mathcal{F}^{-1}$ are one-to-one linear mappings from this space onto this space, and are inverses of each other.
4. Every nonconstant, piecewise continuous periodic function has a fundamental period, and every period of such a function is an integral multiple of that fundamental period. Moreover, any two Fourier series for a single periodic function computed using two different periods will be identical after simplification (e.g., after eliminating zero-valued terms).
5. The Fourier series $\sum_{k\in\mathbb{Z}} c_k e^{i2\pi\omega_k x}$ for a periodic function ϕ with period p can be written in *trigonometric form*

$$c_0 + \sum_{k=1}^{\infty}\left[a_k\cos(2\pi\omega_k x) + b_k\sin(2\pi\omega_k x)\right]$$

where

$$a_k = c_k + c_{-k} = \frac{2}{p}\int_0^p \phi(x)\cos(2\pi\omega_k x)\,dx$$

and

$$b_k = ic_k - ic_{-k} = \frac{2}{p}\int_0^p \phi(x)\sin(2\pi\omega_k x)\,dx\,.$$

6. On $\mathcal{I} = (0, p)$ (or any other interval $\mathcal{I}$ of length p), the exponentials

$$\left\{e^{i2\pi\omega_k x} : \omega_k = k/p,\ k \in \mathbb{Z}\right\}$$

form an orthogonal set in $\mathcal{L}^2(\mathcal{I})$, and the Fourier series $\sum_{k\in\mathbb{Z}} c_k\, e^{i2\pi\omega_k x}$ for a periodic function ϕ with period p is simply the expansion of ϕ with respect to this orthogonal set. That is,

$$c_k = \frac{\langle \phi(x), e^{i2\pi\omega_k x}\rangle}{\|e^{i2\pi\omega_k x}\|^2} \qquad \forall k \in \mathbb{Z}.$$

Similar comments apply regarding the trigonometric form for the Fourier series and the set

$$\{1,\ \cos(2\pi\omega_k x),\ \sin(2\pi\omega_k x)\ :\ \omega_k = k/p,\ k \in \mathbb{N}\}.$$

7. A periodic function ϕ can be identified with its Fourier series $\sum_{k\in\mathbb{Z}} c_k e^{i2\pi\omega_k x}$ under a wide range of criteria. In particular:
   - If ϕ is smooth, then its Fourier series converges uniformly to ϕ.
   - If ϕ is piecewise smooth, then its Fourier series converges pointwise to ϕ everywhere ϕ is continuous.
   - If ϕ is square-integrable on $(0, p)$, then

$$\lim_{(M,N)\to(-\infty,\infty)} \int_0^p \left|\phi(x) - \sum_{k=M}^{N} c_k\, e^{i2\pi\omega_k x}\right|^2 dx = 0.$$

   - Within the generalized theory,

$$\lim_{(M,N)\to(-\infty,\infty)} \int_{-\infty}^{\infty} \psi(x)\left[\phi(x) - \sum_{k=M}^{N} c_k\, e^{i2\pi\omega_k x}\right] dx = 0$$

   whenever ψ is a sufficiently smooth function vanishing sufficiently rapidly at infinity (e.g., any Gaussian function).

8. [BH95, p. 186] Suppose ϕ is a periodic function with Fourier series $\sum_{k\in\mathbb{Z}} c_k e^{i2\pi\omega_k x}$. If ϕ is m-times differentiable on $\mathbb{R}$ and $\phi^{(m)}$ is piecewise continuous for some $m \in \mathbb{N}$, then there is a finite constant β such that

$$|c_k| \le \frac{\beta}{|k|^m} \qquad \forall k \in \mathbb{Z}.$$

9. For each $a \in \mathbb{R}$ and function ϕ continuous at a:

$$\int_{-\infty}^{\infty} \phi(x)\delta_a(x)\, dx = \phi(a) \qquad \text{and} \qquad \phi\delta_a = \phi(a)\delta_a.$$

10. For each $a \in \mathbb{R}$:
    - $\mathcal{F}[\delta_a]|_x = e^{-i2\pi ax}$ and $\mathcal{F}^{-1}[e^{-i2\pi ax}] = \delta_a$.
    - $\mathcal{F}[e^{i2\pi ax}]|_x = \delta_a$ and $\mathcal{F}^{-1}[\delta_a]|_x = e^{i2\pi ax}$.

11. A function ϕ is periodic with period p and Fourier series $\sum_{k\in\mathbb{Z}} c_k e^{i2\pi\omega_k x}$ if and only if its transforms are the arrays

$$\mathcal{F}[\phi] = \sum_{k=-\infty}^{\infty} c_k \delta_{k\Delta\omega} \qquad \text{and} \qquad \mathcal{F}^{-1}[\phi] = \sum_{k=-\infty}^{\infty} c_{-k}\delta_{k\Delta\omega}$$

    with spacing $\Delta\omega = 1/p$.

12. If

$$\phi = \sum_{k=-\infty}^{\infty} \phi_k \delta_{k\Delta x}$$

    is a periodic array with spacing Δx, period p, and corresponding index period N, then:

- The Fourier series for ϕ is given by

$$\sum_{n=-\infty}^{\infty} \Phi_n e^{i2\pi\omega_n x}$$

where $\omega_n = n/p = n/(N\Delta x)$ and

$$\Phi_n = \frac{1}{N\Delta x}\sum_{k=0}^{N-1} \phi_k\, e^{-i2\pi kn/N}.$$

- The Fourier transforms of ϕ are also periodic arrays of delta functions with index period N. Both have spacing $\Delta\omega = 1/p$ and period $P = 1/\Delta x$. These transforms are given by

$$\mathcal{F}[\phi] = \sum_{n=-\infty}^{\infty} \Phi_n \delta_{n\Delta\omega} \qquad \text{and} \qquad \mathcal{F}^{-1}[\phi] = \sum_{n=-\infty}^{\infty} \Phi_{-n}\delta_{n\Delta\omega}.$$

13. (*Convolution identities*) Let f, F, g, and G be functions with $F = \mathcal{F}[f]$ and $G = \mathcal{F}[g]$. Then, provided the convolutions exist,

$$\mathcal{F}[fg] = F * G \qquad \text{and} \qquad \mathcal{F}[f * g] = FG.$$

Examples:

1. For $\alpha > 0$,

$$\begin{aligned}\mathcal{F}[\text{pulse}_\alpha]|_x &= \int_{-\infty}^{\infty} \text{pulse}_\alpha(y)e^{-i2\pi xy}\,dy \\ &= \int_{-\alpha}^{\alpha} e^{-i2\pi xy}\,dy \\ &= \frac{e^{i2\pi\alpha x} - e^{-i2\pi\alpha x}}{i2\pi x} = \frac{\sin(2\pi\alpha x)}{\pi x}.\end{aligned}$$

2. Let f be the periodic function,

$$f(x) = \begin{cases} |x| & \text{for } -1 \le x < 1 \\ f(x-2) & \text{for all } x \in \mathbb{R} \end{cases}.$$

The period of this function is $p = 2$, and its Fourier series $\sum_{k\in\mathbb{Z}} c_k e^{i2\pi\omega_k x}$ is given by

$$\omega_k = \frac{k}{2} \qquad (\text{so } e^{i2\pi\omega_k x} = e^{i\pi kx})$$

and

$$\begin{aligned} c_k &= \frac{1}{2}\int_0^2 \left\{\begin{matrix} |x| & \text{if } x < 1 \\ |x-2| & \text{if } 1 < x \end{matrix}\right\} e^{-i\pi kx}\,dx \\ &= \begin{cases} 1/2 & \text{if } k = 0 \\ (k\pi)^{-2}\left[(-1)^k - 1\right] & \text{if } k \neq 0 \end{cases}.\end{aligned}$$

Since f is continuous and piecewise smooth, f equals its Fourier series,

$$f(x) = \frac{1}{2} + \sum_{k\in\mathbb{Z}\setminus\{0\}} \frac{(-1)^k - 1}{k^2\pi^2} e^{i\pi kx} \qquad \forall x \in \mathbb{R}.$$

Moreover,

$$\mathcal{F}[f] = \frac{1}{2}\delta_0 + \sum_{k\in\mathbb{Z}\setminus\{0\}} \frac{(-1)^k - 1}{k^2\pi^2} \delta_{k/2}.$$

3. Let

$$\phi = \sum_{k=-\infty}^{\infty} \phi_k \delta_{\Delta x}$$

be the periodic array with spacing $\Delta x = 1/3$, index period $N = 4$, and coefficients

$$\phi_0 = 0, \quad \phi_1 = 1, \quad \phi_2 = 2, \quad \text{and} \quad \phi_3 = 1.$$

The period p of ϕ is then

$$p = N\Delta x = 4 \cdot \frac{1}{3} = \frac{4}{3}.$$

Its Fourier transform $\Phi = \mathcal{F}[\phi]$ is also a periodic array,

$$\Phi = \sum_{n=-\infty}^{\infty} \Phi_n \delta_{n\Delta\omega},$$

with index period $N = 4$. The spacing $\Delta\omega$ and period P of Φ are determined, respectively, from the period p and spacing Δx of ϕ by

$$\Delta\omega = \frac{1}{p} = \frac{1}{4/3} = \frac{3}{4} \qquad \text{and} \qquad P = \frac{1}{\Delta x} = \frac{1}{1/3} = 3.$$

The coefficients are given by

$$\begin{aligned}
\Phi_n &= \frac{1}{N\Delta x} \sum_{k=0}^{N-1} \phi_k\, e^{-i2\pi kn/N} \\
&= \frac{1}{N\Delta x} \left[\phi_0 e^{-i2\pi 0n/4} + \phi_1 e^{-i2\pi 1n/4} + \phi_2 e^{-i2\pi 2n/4} + \phi_3 e^{-i2\pi 3n/4}\right] \\
&= \frac{1}{4(1/3)} \left[0e^0 + 1e^{-in\pi/2} + 2e^{-in\pi} + 1e^{-i3n\pi/2}\right] \\
&= \frac{3}{4}\left[0 + (-i)^n + 2 + i^n\right].
\end{aligned}$$

Thus,

$$\Phi_0 = 3, \quad \Phi_1 = \frac{3}{2}, \quad \Phi_2 = 0 \quad \text{and} \quad \Phi_3 = \frac{3}{2}.$$

Applications:

1. [BC01, p. 155] or [Col88, pp. 159–160] (*Partial differential equations*) Using polar coordinates, the steady-state temperature u at position (r, θ) on a uniform, insulated disk of radius 1 satisfies

$$r^2 \frac{\partial^2 u}{\partial r^2} + r\frac{\partial u}{\partial r} + \frac{\partial^2 u}{\partial \theta^2} = 0 \qquad \text{for } 0 < r < 1\,.$$

As a function of θ, u is periodic with period 2π and has (equivalent) Fourier series

$$\sum_{k=-\infty}^{\infty} c_k e^{ikx} \qquad \text{and} \qquad c_0 + \sum_{k=1}^{\infty} [a_k \cos(kx) + b_k \sin(kx)]$$

where the coefficients are functions of r. If u satisfies the boundary condition

$$u(1,\theta) = f(\theta) \qquad \text{for } 0 \le \theta < 2\pi$$

for some function f on $[0, 2\pi)$, then the coefficients in the series can be determined, yielding

$$u(r,\theta) = \sum_{k=-\infty}^{\infty} r^k \gamma_k e^{ik\theta} = \gamma_0 + \sum_{k=1}^{\infty} \left[r^k \alpha_k \cos(k\theta) + r^k \beta_k \sin(k\theta)\right]$$

where

$$\gamma_k = \frac{1}{2\pi}\int_0^{2\pi} f(\theta) e^{-ik\theta}\, d\theta\,,$$

$$\alpha_k = \frac{1}{\pi}\int_0^{2\pi} f(\theta)\cos(k\theta)\, d\theta, \qquad \text{and} \qquad \beta_k = \frac{1}{\pi}\int_0^{2\pi} f(\theta)\sin(k\theta)\, d\theta\,.$$

2. (*Systems analysis*) In systems analysis, a "system" S transforms any "input" function f_I to a corresponding "output" function $f_O = S[f_I]$. Often, the output of S can be described by the convolution formula

$$f_O = h * f_I$$

where h is some fixed function called the impulse response of the system. The corresponding Fourier transform $H = \mathcal{F}[h]$ is the system's transfer function. By the convolution identity, the output is also given by

$$f_O = \mathcal{F}^{-1}[HF_I] \qquad \text{where } F_I = \mathcal{F}[f_I]\,.$$

Two such systems are

- A delayed output system, for which

$$h(x) = \delta_T(x) \qquad \text{and} \qquad H(\omega) = e^{-i2\pi T\omega}$$

  for some $T > 0$. Then

$$f_O(x) = f_I * \delta_T(x) = \int_{-\infty}^{\infty} f_I(x-y)\delta_T(y)dy = f_I(x-T)\,.$$

- [ZTF98, p. 178] or [How01, p. 477] An ideal low pass filter, for which

$$h(x) = \frac{\sin(2\pi\Omega x)}{\pi x} \qquad \text{and} \qquad H(y) = \text{pulse}_\Omega(y)$$

for some $\Omega > 0$. If an input function f_I is periodic with Fourier series $\sum_{k\in\mathbb{Z}} c_k e^{i2\pi\omega_k x}$, then

$$\begin{aligned} HF_I &= \text{pulse}_\Omega \cdot \sum_{k=-\infty}^{\infty} c_k \delta_{\omega_k} \\ &= \sum_{k-\infty}^{\infty} c_k \, \text{pulse}_\Omega(\omega_k)\delta_{\omega_k} \\ &= \sum_{k-\infty}^{\infty} c_k \begin{Bmatrix} 1 & \text{if } |\omega_k| \le \Omega \\ 0 & \text{if } |\omega_k| > \Omega \end{Bmatrix} \delta_{\omega_k} = \sum_{|\omega_k|\le\Omega} c_k \delta_{\omega_k} \,. \end{aligned}$$

Thus,

$$f_O(x) = \mathcal{F}^{-1}[HF_I]|_x = \mathcal{F}^{-1}\left[\sum_{|\omega_k|\le\Omega} c_k \delta_{\omega_k} \right]\Bigg|_x = \sum_{|\omega_k|\le\Omega} c_k \, e^{i2\pi\omega_k x} \,.$$

3. [ZTF98, pp. 520–521] (*Deconvolution*) Suppose S is a system given by

$$S[f_I] = h * f_I \,,$$

but with h and $H = \mathcal{F}[h]$ being unknown. Since

$$f_O = \mathcal{F}^{-1}[HF_I] \qquad \text{where } F_I = \mathcal{F}[f_I] \,,$$

both h and H can be determined as follows:

- Find the output $f_O = S[f_I]$ for some known input f_I for which $F_I = \mathcal{F}[f_I]$ is never zero.
- Compute

$$H = \frac{F_O}{F_I} \qquad \text{where } F_O = \mathcal{F}[f_O] \,.$$

- Compute $h = \mathcal{F}^{-1}[H]$.

Similarly, an input f_I can be reconstructed from an output f_O by

$$f_I = \mathcal{F}^{-1}\left[\frac{F_O}{H}\right]$$

provided the transfer function H is known and is never zero.

58.3 The Discrete Theory

Definitions:

In all the following, $N \in \mathbb{N}$, and the indexing of any "N items" (including rows and columns of matrices) will run from 0 to $N - 1$.

An Nth **order sequence** is an ordered list of N complex numbers,

$$(c_0, c_1, c_2, \ldots, c_{N-1}).$$

Such a sequence will often be written as the column vector

$$\mathbf{c} = [c_0, c_1, c_2, \ldots, c_{N-1}]^T,$$

and the kth component of the sequence will be denoted by either c_k or $[\mathbf{c}]_k$. In addition, any such sequence will be viewed as part of an infinite repeating sequence

$$(\ldots, c_{-1}, c_0, c_1, c_2, \ldots, c_{N-1}, c_N, \ldots)$$

in which $c_{N+k} = c_k$ for all $k \in \mathbb{Z}$.

Let $\mathbf{c}$ be an Nth order sequence. The (Nth order) **discrete Fourier transform (DFT)** $\widehat{\mathcal{F}}_N\mathbf{c}$ and the (Nth order) **inverse discrete Fourier transform (inverse DFT)** $\widehat{\mathcal{F}}_N^{-1}\mathbf{c}$ of $\mathbf{c}$ are the two Nth order sequences given by

$$[\widehat{\mathcal{F}}_N\mathbf{c}]_n = \frac{1}{\sqrt{N}}\sum_{k=0}^{N-1} c_k\, e^{-i2\pi kn/N} \quad \text{and} \quad [\widehat{\mathcal{F}}_N^{-1}\mathbf{c}]_k = \frac{1}{\sqrt{N}}\sum_{n=0}^{N-1} c_n\, e^{i2\pi nk/N}.$$

Additionally, the terms "discrete Fourier transform" and "inverse discrete Fourier transform" can refer to the processes for computing these sequences from $\mathbf{c}$.

Given two Nth order sequences $\mathbf{a}$ and $\mathbf{b}$, the corresponding product $\mathbf{ab}$ and convolution $\mathbf{a} * \mathbf{b}$ are the Nth order sequences given by

$$[\mathbf{ab}]_k = a_k b_k \quad \text{and} \quad [\mathbf{a} * \mathbf{b}]_k = \frac{1}{\sqrt{N}}\sum_{j=0}^{N-1} a_{k-j} b_j\,.$$

Facts:

All the following facts can be found in [BH95, Chap. 2, 3] or [How01, Chap. 38, 39].

1. Warning: Slight variations of the above summation formulas, e.g.,

$$\sum_{k=0}^{N-1} c_k\, e^{-i2\pi kn/N} \quad \text{and} \quad \frac{1}{N}\sum_{n=0}^{N-1} c_n\, e^{i2\pi nk/N},$$

are also often used to define, respectively, $\widehat{\mathcal{F}}_N\mathbf{c}$ and $\widehat{\mathcal{F}}_N^{-1}\mathbf{c}$ (or even $\widehat{\mathcal{F}}_N^{-1}\mathbf{c}$ and $\widehat{\mathcal{F}}_N\mathbf{c}$).
2. $\widehat{\mathcal{F}}_N$ and $\widehat{\mathcal{F}}_N^{-1}$ are one-to-one linear transformations from $\mathbb{C}^N$ onto $\mathbb{C}^N$. They preserve the standard inner product and are inverses of each other.
3. Letting $w = e^{-i2\pi/N}$, the matrices (with respect to the standard basis for $\mathbb{C}^N$) for the Nth order discrete Fourier transforms (also denoted by $\widehat{\mathcal{F}}_N$ and $\widehat{\mathcal{F}}_N^{-1}$) are

$$\widehat{\mathcal{F}}_N = \frac{1}{\sqrt{N}}\begin{bmatrix} 1 & 1 & 1 & 1 & \cdots & 1 \\ 1 & w^{1\cdot 1} & w^{1\cdot 2} & w^{1\cdot 3} & \cdots & w^{1(N-1)} \\ 1 & w^{2\cdot 1} & w^{2\cdot 2} & w^{2\cdot 3} & \cdots & w^{2(N-1)} \\ 1 & w^{3\cdot 1} & w^{3\cdot 2} & w^{3\cdot 3} & \cdots & w^{3(N-1)} \\ \vdots & \vdots & \vdots & \vdots & \ddots & \vdots \\ 1 & w^{(N-1)1} & w^{(N-1)2} & w^{(N-1)3} & \cdots & w^{(N-1)(N-1)} \end{bmatrix}$$

and

$$\widehat{\mathcal{F}}_N^{-1} = \frac{1}{\sqrt{N}} \begin{bmatrix} 1 & 1 & 1 & 1 & \cdots & 1 \\ 1 & w^{-1\cdot 1} & w^{-1\cdot 2} & w^{-1\cdot 3} & \cdots & w^{-1(N-1)} \\ 1 & w^{-2\cdot 1} & w^{-2\cdot 2} & w^{-2\cdot 3} & \cdots & w^{-2(N-1)} \\ 1 & w^{-3\cdot 1} & w^{-3\cdot 2} & w^{-3\cdot 3} & \cdots & w^{-3(N-1)} \\ \vdots & \vdots & \vdots & \vdots & \ddots & \vdots \\ 1 & w^{-(N-1)1} & w^{-(N-1)2} & w^{-(N-1)3} & \cdots & w^{-(N-1)(N-1)} \end{bmatrix}.$$

4. $\widehat{\mathcal{F}}_N$ and $\widehat{\mathcal{F}}_N^{-1}$ are symmetric matrices, and are complex conjugates of each other (i.e., $\left[\widehat{\mathcal{F}}_N\right]^* = \widehat{\mathcal{F}}_N^{-1}$).
5. (*Convolution identities*) Let $\mathbf{v}$, $\mathbf{V}$, $\mathbf{w}$, and $\mathbf{W}$ be Nth order sequences with $\mathbf{V} = \widehat{\mathcal{F}}_N\mathbf{v}$ and $\mathbf{W} = \widehat{\mathcal{F}}_N\mathbf{w}$. Then

$$\widehat{\mathcal{F}}_N[\mathbf{vw}] = \mathbf{V} * \mathbf{W} \qquad \text{and} \qquad \widehat{\mathcal{F}}_N[\mathbf{v} * \mathbf{w}] = \mathbf{VW}.$$

Examples:

1. The matrices for the four lowest order DFTs are $\widehat{\mathcal{F}}_1 = [1]$,

$$\widehat{\mathcal{F}}_2 = \frac{1}{\sqrt{2}} \begin{bmatrix} 1 & 1 \\ 1 & -1 \end{bmatrix}, \quad \widehat{\mathcal{F}}_3 = \frac{1}{2\sqrt{3}} \begin{bmatrix} 2 & 2 & 2 \\ 2 & -1-\sqrt{3} & -1+\sqrt{3} \\ 2 & -1+\sqrt{3} & -1-\sqrt{3} \end{bmatrix},$$

and

$$\widehat{\mathcal{F}}_4 = \frac{1}{2} \begin{bmatrix} 1 & 1 & 1 & 1 \\ 1 & -i & 1 & i \\ 1 & -1 & 1 & -1 \\ 1 & i & -1 & -i \end{bmatrix}.$$

2. The DFT of $\mathbf{v} = [1, 2, 3, 4]^T$ is $\mathbf{V} = [V_0, V_1, V_2, V_3]^T$ where

$$\begin{bmatrix} V_0 \\ V_1 \\ V_2 \\ V_3 \end{bmatrix} = \frac{1}{2} \begin{bmatrix} 1 & 1 & 1 & 1 \\ 1 & -i & 1 & i \\ 1 & -1 & 1 & -1 \\ 1 & i & -1 & -i \end{bmatrix} \begin{bmatrix} 1 \\ 2 \\ 3 \\ 4 \end{bmatrix} = \begin{bmatrix} 5 \\ 2+i \\ -1 \\ -1-i \end{bmatrix}.$$

Applications:

1. [CG00, p. 153] Assume

$$f(z) = \sum_{n=0}^{2M} f_n z^n$$

is the product of two known Mth order polynomials

$$g(z) = \sum_{k=0}^{M} g_k z^k \qquad \text{and} \qquad h(z) = \sum_{k=0}^{M} h_k z^k.$$

The coefficients of f form a sequence $\mathbf{f} = [f_0, f_1, f_2, \ldots, f_{2M}]^T$ of order $N = 2M + 1$. These coefficients can be computed from the coefficients of g and h using either of the following approaches:

- Let $\mathbf{g} = [g_0, g_1, g_2, \ldots, g_{2M}]^T$ and $\mathbf{h} = [h_0, h_1, h_2, \ldots, h_{2M}]^T$ be the Nth order sequences in which g_k and h_k are the corresponding coefficients of the polynomials g and h when $k \leq M$ and are zero when $M < k \leq N$. The coefficients of f can then be computed by

$$f_k = \sum_{j=0}^{2M} g_{k-j} h_j = \sqrt{N}[\mathbf{g} * \mathbf{h}]_k \qquad \text{for } k = 0, 1, 2, \ldots, 2M.$$

- Let $\mathbf{F} = [F_0, F_1, F_2, \ldots, F_{2M}]^T$ be the Nth order DFT of $\mathbf{f}$, and note that, for $n = 0, 1, 2, \ldots, N-1$,

$$F_n = \frac{1}{\sqrt{N}} \sum_{k=0}^{N-1} f_k \, e^{-i2\pi kn/N} = \frac{1}{\sqrt{N}} f(w^n) = \frac{1}{\sqrt{N}} h(w^n) g(w^n)$$

where $w = e^{-i2\pi/N}$. Thus, the coefficients of f can be computed by first using the formulas for g and h to compute

$$F_n = \frac{1}{\sqrt{N}} h(w^n) g(w^n) \qquad \text{for } n = 0, 1, 2, \ldots, N-1,$$

and then taking the inverse DFT of $\mathbf{F} = [F_0, F_1, F_2, \ldots, F_{N-1}]^T$.

2. [BH95, p. 247] Consider finding the solution $\mathbf{v} = [v_0, v_1, v_2, \ldots, v_{N-1}]^T$ to the difference equation

$$\alpha v_{k-1} + \beta v_k + \alpha v_{k+1} = f_k \qquad \text{for } n = 0, 1, 2, \ldots, N-1$$

where α and β are constants and $\mathbf{f} = [f_0, f_1, f_2, \ldots, f_{N-1}]^T$ is a known Nth order sequence. If the boundary conditions are periodic (i.e., $v_0 = v_N$ and $v_{-1} = v_{N-1}$), then setting

$$\mathbf{v} = \widehat{\mathcal{F}}_N^{-1}\mathbf{V} \qquad \text{and} \qquad \mathbf{f} = \widehat{\mathcal{F}}_N^{-1}\mathbf{F}$$

yields the sequence $\mathbf{v}$ that satisfies the periodic boundary conditions, and whose DFT, $\mathbf{V}$, satisfies

$$\sum_{n=0}^{N-1} C_n \, e^{i2\pi nk/N} = 0 \qquad \text{for } k = 0, 1, 2, \ldots, N-1$$

where

$$C_n = \left[2\alpha \cos\left(\frac{2\pi n}{N}\right) + \beta\right] V_n - F_n .$$

From this, it follows that the solution $\mathbf{v}$ to the original difference equation is the DFT of the sequence $[V_0, V_1, V_2, \ldots, V_{N-1}]^T$ given by

$$V_n = \frac{F_n}{2\alpha \cos\left(\frac{2\pi n}{N}\right) + \beta},$$

provided the denominator never vanishes.

3. A difference equation of the form just considered with periodic boundary conditions arises when considering the steady-state temperature distribution on a uniform ring containing heat sources and sinks. In this case, the temperature v, as a function of angular position θ, is modeled by the

one-dimensional Poisson's equation

$$\frac{d^2 v}{d\theta^2} = f(\theta),$$

where $f(\theta)$ describes the heat source/sink density at angular position θ. The discrete analog of this equation is

$$v_{k-1} - 2v_k + v_{k+1} = f_k \qquad \text{for } n = 0, 1, 2, \ldots, N-1$$

where

$$\mathbf{v} = [v_0, v_1, v_2, \ldots, v_{N-1}]^T \qquad \text{and} \qquad \mathbf{f} = [f_0, f_1, f_2, \ldots, f_{N-1}]^T$$

describe, respectively, the temperatures at N evenly space positions around the ring, and the net sources and sinks of thermal energy about these positions. Setting

$$\mathbf{v} = \widehat{\mathcal{F}}_N^{-1}\mathbf{V} \qquad \text{and} \qquad \mathbf{f} = \widehat{\mathcal{F}}_N^{-1}\mathbf{F}$$

and applying the formulas given above yields

$$\left[2\cos\left(\frac{2\pi n}{N}\right) - 2\right] V_n = F_n \qquad \text{for } n = 0, 1, 2, \ldots, N-1.$$

The coefficient on the left is nonzero when $n \neq 0$. Thus,

$$V_n = \frac{F_n}{2\cos\left(\frac{2\pi n}{N}\right) - 2} \qquad \text{for } n = 1, 2, \ldots, N-1.$$

However, for $n = 0$,

$$0 \cdot V_0 = F_0 = \frac{1}{\sqrt{N}} \sum_{k=0}^{N-1} f_k \, e^{-i2\pi k \cdot 0/N} = \frac{1}{\sqrt{N}} \sum_{k=0}^{N-1} f_k \,,$$

pointing out that, for an equilibrium temperature distribution to exist, the net applied heat energy, $\sum_{k=0}^{N-1} f_k$, must be zero. Assuming this, the last equation then implies that V_0 is arbitrary, which, since

$$V_0 = \frac{1}{\sqrt{N}} \sum_{k=0}^{N-1} v_k \, e^{-i2\pi k \cdot 0/N} = \frac{1}{\sqrt{N}} \sum_{k=0}^{N-1} v_k$$

means that the given conditions are not sufficient to determine the average temperature throughout the ring.

58.4 Relating the Functional and Discrete Theories

Definitions:

Let f be a (continuous) function on a finite interval $[0, L]$. For any $N \in \mathbb{N}$ and $\Delta x > 0$ satisfying $N\Delta x \leq L$, the corresponding (**Nth order**) **sampling** (with **spacing** Δx) is the sequence

$$\mathbf{f} = [f_0, f_1, f_2, \ldots, f_{N-1}]^T \qquad \text{where } f_k = f(k\Delta x).$$

Corresponding to this is the **scaled sampling**

$$\widehat{\mathbf{f}} = [\widehat{f}_0, \widehat{f}_1, \widehat{f}_2, \ldots, \widehat{f}_{N-1}]^T$$

and the **discrete approximation**

$$\widehat{f} = \sum_{k=-\infty}^{\infty} \widehat{f}_k \delta_{k\Delta x}$$

where, in both,

$$\widehat{f}_k = \begin{cases} f_k \Delta x & \text{for } k = 0, 1, 2, \ldots, N-1 \\ \widehat{f}_{k+N} & \text{in general} \end{cases}.$$

Facts:

1. [How01, pp. 713–715] Let $\widehat{f}$ be the discrete approximation of a continuous function f based on an Nth order sampling with spacing Δx. Then $\widehat{f}$ is a periodic array of delta functions with spacing Δx and index period N that approximates f over the interval $(0, N\Delta x)$. In particular, for any other function ψ which is continuous on $[0, N\Delta x]$,
$$\int_0^{N\Delta x} \psi(x) f(x)\, dx \approx \int_0^{N\Delta x} \psi(x)\widehat{f}(x)\, dx = \sum_{k=0}^{N-1} \psi(k\Delta x)\widehat{f}_k .$$
Attempting to approximate f with $\widehat{f}$ outside the interval $(0, N\Delta x)$, however, cannot be justified unless f is also periodic with period $N\Delta x$.
2. [How01, chap. 38] Let
$$\mathbf{f} = [f_0, f_1, f_2, \ldots, f_{N-1}]^T \quad \text{and} \quad \mathbf{F} = [F_0, F_1, F_2, \ldots, F_{N-1}]^T$$
be two Nth order sequences, and let
$$f = \sum_{k=-\infty}^{\infty} f_k \delta_{k\Delta x} \quad \text{and} \quad F = \sum_{n=-\infty}^{\infty} F_n \delta_{n\Delta\omega}$$
be two corresponding periodic arrays of delta functions with index period N, and with spacings Δx and $\Delta\omega$ satisfying $\Delta x \Delta\omega = 1/N$. Then
$$F = \mathcal{F}[f] \iff \mathbf{F} = \Delta\omega\sqrt{N}\,\widehat{\mathcal{F}}_N \mathbf{f}.$$
In particular, if $\Delta\omega = \Delta x = N^{-1/2}$, then
$$F = \mathcal{F}[f] \iff \mathbf{F} = \widehat{\mathcal{F}}_N \mathbf{f}.$$
3. [How01, pp. 719–723] Suppose f is a continuous, piecewise smooth function that vanishes outside the finite interval $(0, L)$. Let $F = \mathcal{F}[f]$, and let $\widehat{\mathbf{f}}$ be the Nth order scaled sampling of f with spacing Δx chosen so that $L = (N-1)\Delta x$. Then, for each $n \in \mathbb{Z}$,
$$F(n\Delta\omega) \approx \sqrt{N}\,[\widehat{\mathcal{F}}_N \widehat{\mathbf{f}}]_n$$
where $\Delta\omega = (N\Delta x)^{-1}$. The error in this approximation is bounded by
$$(\max|f'| + 2\pi|n|\Delta\omega \max|f|)\frac{L^2 N}{2(N-1)^2} .$$
In practice, this bound may significantly overestimate the actual error.

4. [BH95, pp. 181–188] If f is a continuous, piecewise smooth, periodic function with period p and Fourier series $\sum_{n\in\mathbb{Z}} c_n e^{i2\pi\omega_n x}$, then

$$c_n \approx \frac{1}{\sqrt{N}}[\widehat{\mathcal{F}}_N \mathbf{f}]_n \qquad \text{for } -\frac{N}{2} < n \leq \frac{N}{2}$$

where $\mathbf{f}$ is the Nth order sampling of f over $[0, p]$ with spacing $\Delta x = p/N$. Moreover, if f is m-times differentiable on $\mathbb{R}$ and $f^{(m)}$ is piecewise continuous and piecewise monotone for some $m \in \mathbb{N}$, then the error in this approximation is bounded by

$$\frac{\beta}{N^{m+1}}$$

for some positive constant β independent of n and N.

Examples:

1. Let

$$f = \sum_{k=-\infty}^{\infty} f_k \delta_{k/3}$$

be the periodic array with index period $N = 4$ and coefficients

$$f_0 = 0, \quad f_1 = 1, \quad f_2 = 2, \quad \text{and} \quad f_3 = 1\,.$$

The Fourier transform of f is then the periodic array

$$F = \sum_{n=-\infty}^{\infty} F_n \delta_{n\Delta\omega}$$

with index period 4, spacing $\Delta\omega = 3/4$, and coefficients given by

$$\begin{bmatrix} F_0 \\ F_1 \\ F_2 \\ F_3 \end{bmatrix} = \Delta\omega\sqrt{4}\,\widehat{\mathcal{F}}_4 \mathbf{f} = \frac{3}{4}\begin{bmatrix} 1 & 1 & 1 & 1 \\ 1 & -i & 1 & i \\ 1 & -1 & 1 & -1 \\ 1 & i & -1 & -i \end{bmatrix}\begin{bmatrix} 0 \\ 1 \\ 2 \\ 1 \end{bmatrix} = \frac{1}{2}\begin{bmatrix} 6 \\ 3 \\ 0 \\ -3 \end{bmatrix}.$$

2. The 8th order sampling with spacing $\Delta x = 1/6$ of

$$f(x) = \begin{cases} x & \text{for } 0 \leq x \leq 1/2 \\ 1 - x & \text{for } 1/2 \leq x \leq 1 \\ 0 & \text{otherwise} \end{cases}$$

is $\mathbf{f} = [f_0, f_1, f_2, \ldots, f_7]^T$ with

$$f_k = f\left(\frac{k}{6}\right) = \begin{cases} k/6 & \text{for } k = 0, 1, 2, 3 \\ 1 - k/6 & \text{for } k = 4, 5, 6 \\ 0 & \text{for } k = 7 \end{cases}.$$

That is,

$$\mathbf{f} = \left[0, \frac{1}{6}, \frac{2}{6}, \frac{3}{6}, \frac{2}{6}, \frac{1}{6}, 0, 0\right]^T.$$

Multiplying this by the spacing $\Delta x = 1/6$ yields the scaled sampling

$$\widehat{\mathbf{f}} = \left[0, \frac{1}{36}, \frac{2}{36}, \frac{3}{36}, \frac{2}{36}, \frac{1}{36}, 0, 0\right]^T .$$

Letting $F = \mathcal{F}[f]$, $\Delta\omega = (N\Delta x)^{-1} = 3/4$, and $w = e^{-i2\pi/8}$,

$$\begin{aligned} F(n\Delta\omega) &\approx \sqrt{8}\,[\widehat{\mathcal{F}_8}\widehat{\mathbf{f}}]_n \\ &= \frac{1}{36}\left[0w^{0n} + 1w^{1n} + 2w^{2n} + 3w^{3n} + 2w^{4n} + 1w^{5n} + 0w^{6n} + 0w^{7n}\right] \\ &= \frac{1}{36}w^{3n}\left[w^{-2n} + 2w^{-n} + 3 + 2w^{n} + w^{2n}\right], \end{aligned}$$

which simplifies to

$$F(n\Delta\omega) \approx \sqrt{8}\,[\widehat{\mathcal{F}_8}\widehat{\mathbf{f}}]_n = \frac{1}{36}e^{-i3n\pi/4}\left[3 + 2\cos\left(\frac{n\pi}{2}\right) + 4\cos\left(\frac{n\pi}{4}\right)\right].$$

The error bound on this approximation is

$$E_n = (\max|f'| + 2\pi|n|\Delta\omega\max|f|)\frac{L^2 N}{2(N-1)^2} = \frac{4 + 3\pi|n|}{36}.$$

In fact, the Fourier transform of f is easily found to be

$$F(\omega) = e^{-i\pi\omega}\left[\frac{\sin(\pi\omega/2)}{\pi\omega}\right]^2.$$

So

$$F(n\Delta\omega) = F\left(\frac{3n}{4}\right) = e^{-i3n\pi/4}\left[\frac{4\sin(3n\pi/8)}{3n\pi}\right]^2,$$

and the actual error in the approximation is

$$\begin{aligned} \varepsilon_n &= \left|F(n\Delta\omega) - \sqrt{8}\,[\widehat{\mathcal{F}_8}\widehat{\mathbf{f}}]_n\right| \\ &= \frac{1}{36}\left|64\left[\frac{\sin(n\pi/8)}{n\pi}\right]^2 - 3 - 2\cos\left(\frac{n\pi}{2}\right) - 4\cos\left(\frac{n\pi}{4}\right)\right|. \end{aligned}$$

For $n = 0$ through 4, the actual and approximate values of $F(n\Delta\omega)$ (to five decimal places) are

$$\begin{aligned} &F(0\Delta\omega) = 0.25000 && \sqrt{8}\,[\widehat{\mathcal{F}_8}\widehat{\mathbf{f}}]_0 = 0.25000 \\ &F(1\Delta\omega) = -0.10872(1+i) && \sqrt{8}\,[\widehat{\mathcal{F}_8}\widehat{\mathbf{f}}]_1 = -0.11448(1+i) \\ &F(2\Delta\omega) = 0.02252i && \sqrt{8}\,[\widehat{\mathcal{F}_8}\widehat{\mathbf{f}}]_2 = 0.02778i \\ &F(3\Delta\omega) = 0.00207(1-i) && \sqrt{8}\,[\widehat{\mathcal{F}_8}\widehat{\mathbf{f}}]_3 = 0.00337(1-i) \\ &F(4\Delta\omega) = -0.01126 && \sqrt{8}\,[\widehat{\mathcal{F}_8}\widehat{\mathbf{f}}]_3 = -0.02778. \end{aligned}$$

The corresponding error bounds and actual errors (to five decimal places) are

$$\begin{aligned} E_0 &= 0.11111 & \varepsilon_0 &= 0.00000 \\ E_1 &= 0.37291 & \varepsilon_1 &= 0.00815 \\ E_2 &= 0.63471 & \varepsilon_2 &= 0.00526 \\ E_3 &= 0.89651 & \varepsilon_3 &= 0.00183 \\ E_4 &= 1.15831 & \varepsilon_4 &= 0.01652. \end{aligned}$$

Applications:

1. Let f be a continuous, piecewise smooth function to be sampled that is nonzero only in the interval $(0, 1)$ and which satisfies

$$|f(x)| \leq \frac{1}{4} \qquad \text{and} \qquad |f'(x)| \leq 1 \qquad \forall x \in \mathbb{R}.$$

Consider the problem of approximating $F(n\Delta\omega)$ where $F = \mathcal{F}[f]$, $\Delta\omega = 1/2$, and n is any integer from 0 to 10. For each $N \in \mathbb{N}$, set

$$\Delta x = \frac{1}{N\Delta\omega} = \frac{2}{N} \qquad \text{and} \qquad L = (N-1)\Delta x = \frac{2(N-1)}{N},$$

and let $\widehat{\mathbf{f}} = [\widehat{f}_0, \widehat{f}_1, \widehat{f}_2, \ldots, \widehat{f}_{N-1}]^T$ be the corresponding scaled sampling of f,

$$\widehat{f}_k = f(k\Delta x)\Delta x = f\left(\frac{2k}{N}\right) \cdot \frac{2}{N}.$$

Then, for $n = 0, 1, 2, \ldots, 10$,

$$F(n\Delta\omega) \approx \sqrt{N}\,[\widehat{\mathcal{F}}_N\widehat{\mathbf{f}}]_n = \frac{2}{N}\sum_{k=0}^{N-1} f\left(\frac{2k}{N}\right) e^{-i2\pi kn/N}$$

with an error bound of

$$(\max|f'| + 2\pi|n|\Delta\omega \max|f|)\frac{L^2 N}{2(N-1)^2} = \frac{4+10\pi}{2N} < \frac{18}{N}.$$

2. [CG00, p. 157] (*Deconvolution*) Assume $g = f * h$ where g and h are known by Nth order samplings $\mathbf{g}$ and $\mathbf{h}$, each with spacing Δx. The deconvolution formula for finding f,

$$f = \mathcal{F}^{-1}[V] \qquad \text{where } V = \frac{G}{H} = \frac{\mathcal{F}[g]}{\mathcal{F}[h]},$$

is approximated by

$$f(k\Delta x) \approx \sqrt{N}\left[\widehat{\mathcal{F}}_N^{-1}\widehat{\mathbf{V}}\right]_k$$

where, letting $\Delta\omega = (N\Delta x)^{-1}$, $\widehat{\mathbf{V}}$ is the sequence given by

$$\widehat{V}_n = \Delta\omega \frac{\sqrt{N}\left[\widehat{\mathcal{F}}_N[\widehat{\mathbf{g}}]\right]_n}{\sqrt{N}\left[\widehat{\mathcal{F}}_N[\widehat{\mathbf{h}}]\right]_n} \qquad \text{for } n = 0, 1, 2, \ldots, N-1.$$

This reduces to

$$f(k\Delta x) \approx \frac{1}{\Delta x\sqrt{N}}\left[\widehat{\mathcal{F}}_N^{-1}\mathbf{V}\right]_k$$

where, letting $\mathbf{G} = \widehat{\mathcal{F}}_N\mathbf{g}$ and $\mathbf{H} = \widehat{\mathcal{F}}_N\mathbf{h}$,

$$V_n = \frac{G_n}{H_n} \qquad \text{for } n = 0, 1, 2, \ldots, N-1 .$$

This requires, of course, that each H_n be nonzero.

58.5 The Fast Fourier Transform

A **fast Fourier transform** (**FFT**) is any of a number of algorithms for computing DFTs in which the number of arithmetic computations is greatly reduced through clever use of symmetries and cyclic structures in the DFT matrices. The FFT described here is the standard *radix 2* FFT for computing the Nth order "alternative" discrete Fourier transform $\mathcal{D}_N\mathbf{v}$ given by

$$[\mathcal{D}_N\mathbf{v}]_n = \sqrt{N}[\widehat{\mathcal{F}}_N\mathbf{v}]_n = \sum_{k=0}^{N-1} v_k\, e^{-i2\pi nk/N} .$$

For additional details on implementing a radix 2 FFT and descriptions of other FFTs, see [BH95, chap. 10], [CG00], or [Wal96].

Algorithms:

Two algorithms for computing the alternative DFT

$$\mathbf{V} = [V_0, V_1, V_2, \ldots, V_{N-1}]^T$$

of an Nth order sequence

$$\mathbf{v} = [v_0, v_1, v_2, \ldots, v_{N-1}]^T$$

are given. The first is the "first level" radix 2 FFT illustrating the basic concepts. The second is the more complete radix 2 FFT.

1. (*First level radix 2 FFT*) This requires $N = 2M$ for some $M \in \mathbb{N}$:
   - First, split $\mathbf{v}$ into two Mth order sequences $\mathbf{v}^O$ and $\mathbf{v}^E$ composed, respectively, of the even-indexed and odd-indexed elements of $\mathbf{v}$,

     $$\mathbf{v}^E = [v_0^E, v_1^E, v_2^E, \ldots, v_{M-1}^E]^T = [v_0, v_2, v_4, \ldots, v_{2M-2}]^T$$

     and

     $$\mathbf{v}^O = [v_0^O, v_1^O, v_2^O, \ldots, v_{M-1}^O]^T = [v_1, v_3, v_5, \ldots, v_{2M-1}]^T .$$

   - Then compute the Mth order alternative DFTs

     $$[V_0^E, V_1^E, V_2^E, \ldots, V_{M-1}^E]^T = \mathbf{V}^E = \mathcal{D}_N\mathbf{v}^E$$

     and

     $$[V_0^O, V_1^O, V_2^O, \ldots, V_{M-1}^O]^T = \mathbf{V}^O = \mathcal{D}_N\mathbf{v}^O .$$

- Then construct the Nth order sequence $\mathbf{V} = [V_0, V_1, V_2, \ldots, V_{N-1}]^T$ from $\mathbf{V}^E$ and $\mathbf{V}^O$ via the **butterfly relations**

$$V_n = \begin{cases} V_n^E + w^n V_n^O & \text{for } n = 0, 1, 2, \ldots, M-1 \\ V_{n-M}^E - w^{n-M} V_{n-M}^O & \text{for } n = M, M+1, M+2, \ldots, N-1 \end{cases}$$

where $w = e^{-i2\pi/N}$.

2. (*Full radix 2 FFT*) This requires $N = 2^P$ for some $P \in \mathbb{N}$. To simplify the description, let

$$\sigma_k \in \begin{cases} \{0\} & \text{if } k = 0 \\ \{E, O\} & \text{if } k = 1, 2, 3, \ldots, P \end{cases}.$$

- Recursively, split $\mathbf{v}$ into the sequences in the set

$$\{\mathbf{v}^{\sigma_0\sigma_1\sigma_2\cdots\sigma_K} : K = 0, 1, 2, \ldots, P\}$$

where $\mathbf{v}^{\sigma_0} = \mathbf{v}$ and, letting $\Sigma = \sigma_0\sigma_1\sigma_2\cdots\sigma_K$ and $M = 2^{P-K-1}$,

$$\mathbf{v}^{\Sigma E} = \left[v_0^{\Sigma E}, v_1^{\Sigma E}, v_2^{\Sigma E}, \ldots, v_{M-1}^{\Sigma E}\right]^T = \left[v_0^{\Sigma}, v_2^{\Sigma}, v_4^{\Sigma}, \ldots, v_{2M-2}^{\Sigma}\right]^T$$

and

$$\mathbf{v}^{\Sigma O} = \left[v_0^{\Sigma O}, v_1^{\Sigma O}, v_2^{\Sigma O}, \ldots, v_{M-1}^{\Sigma O}\right]^T = \left[v_1^{\Sigma}, v_3^{\Sigma}, v_5^{\Sigma}, \ldots, v_{2M-1}^{\Sigma}\right]^T.$$

- For each of the 2^P first order sequences in $\{\mathbf{v}^{\sigma_0\sigma_1\sigma_2\cdots\sigma_P} = [v^{\sigma_1\sigma_2\cdots\sigma_P}]\}$, set

$$\mathbf{V}^{\sigma_1\sigma_2\cdots\sigma_P} = [V^{\sigma_1\sigma_2\cdots\sigma_P}] = [v^{\sigma_1\sigma_2\cdots\sigma_P}] = \mathbf{v}^{\sigma_1\sigma_2\cdots\sigma_P}$$

- Set $\mathbf{V} = \mathbf{V}^{\sigma_0}$ where

$$\{\mathbf{V}^{\sigma_0\sigma_1\sigma_2\cdots\sigma_K} : K = 0, 1, 2, \ldots, P\}$$

is the set of sequences recursively constructed via the **butterfly relations**

$$V_n^{\Sigma} = \begin{cases} V_n^{\Sigma E} + w^n V_n^{\Sigma O} & \text{for } n = 0, 1, 2, \ldots, M-1 \\ V_{n-M}^{\Sigma E} - w^{n-M} V_{n-M}^{\Sigma O} & \text{for } n = M, M+1, M+2, \ldots, 2M-1 \end{cases}$$

where $\Sigma = \sigma_0\sigma_1\sigma_2\cdots\sigma_K$, $M = 2^{P-K-1}$, and $w = e^{(-i2\pi/N)2^K}$.

Facts:

All the following facts except those with specific reference can be found in [BH95], [CG00], or [Wal96].

1. Let $\mathbf{V}$ be the sequence computed from an Nth order sequence $\mathbf{v}$ by either of the two algorithms above. Then $\mathbf{V} = \mathcal{D}_N\mathbf{v}$. Multiplying $\mathbf{V}$ by $N^{-1/2}$ yields $\widehat{\mathcal{F}}_N\mathbf{v}$.
2. For corresponding "fast" algorithms for computing the inverse DFT of $\mathbf{v}$, replace $w = e^{-i2\pi/N}$ and $w = e^{(-i2\pi/N)2^K}$, respectively, with $w = e^{i2\pi/N}$ and $w = e^{(i2\pi/N)2^K}$ in the above algorithms.
3. [BH95, p. 393] Assume $N = 2^P \gg 1$ for some $P \in \mathbb{N}$. Using just the defining formulas, the computation of an Nth order DFT requires approximately $2N^2$ arithmetic operations ($(N-1)^2$ multiplications and $N(N-1)$ additions). Using the above described radix 2 FFT allows this same DFT to be computed with approximately $(3N/2)\log_2 N$ arithmetic operations ($N \log_2 N$ additions and $(N/2)\log_2 N$ multiplications, plus another N multiplications by $N^{-1/2}$ if it is desired to convert

the alternative DFT to the one previously described). Thus, the ratio of the number of arithmetic operations used to compute an Nth order DFT by the two methods is

$$\frac{\text{operations if using the FFT}}{\text{operations if not using the FFT}} \approx \frac{3\log_2 N}{4N}.$$

4. Assume $N = 2^P \gg 1$ for some $P \in \mathbb{N}$. Simply using the basic formulas, the computation of the discrete convolution of two sequences of order N,

$$[\mathbf{v} * \mathbf{w}]_k = \frac{1}{\sqrt{N}} \sum_{j=0}^{N-1} v_{k-j} w_j,$$

requires approximately $2N^2$ arithmetic operations ($(N+1)N$ multiplies and $(N-1)N$ additions). On the other hand, this convolution can be computed using the convolution identity

$$\mathbf{v} * \mathbf{w} = \widehat{\mathcal{F}}_N^{-1}[\mathbf{VW}] \qquad \text{where} \quad \mathbf{V} = \widehat{\mathcal{F}}_N \mathbf{v} \quad \text{and} \quad \mathbf{W} = \widehat{\mathcal{F}}_N \mathbf{w}.$$

Employing this identity along with the radix 2 FFT to compute the three DFTs involved requires approximately

- $3 \times (3N/2)\log_2 N$ arithmetic operations for the DFTs,
- Plus N multiplications for the product $\mathbf{VW}$,

for a total of approximately $(9N/2)\log_2 N$ arithmetic operations. The ratio of operations required to compute this convolution by the two methods is

$$\frac{\text{operations if using identities and the FFT}}{\text{operations if not using identities and the FFT}} \approx \frac{9\log_2 N}{4N}.$$

Examples:

1. Consider finding the DFT of

$$\mathbf{v} = [v_0, v_1, v_2, v_3]^T = [1, 2, 3, 4]^T$$

using the full radix 2 FFT. The order here is $N = 4$ (so $P = 2$). Recursively splitting $\mathbf{v}$ into the even and odd index subsequences yields

$$\mathbf{v}^E = [1, 3]^T \qquad \text{and} \qquad \mathbf{v}^O = [2, 4]^T$$

and then

$$\mathbf{v}^{EE} = [1], \quad \mathbf{v}^{EO} = [3], \quad \mathbf{v}^{OE} = [2], \quad \text{and} \quad \mathbf{v}^{OO} = [4].$$

(The $\sigma_0 = 0$ superscript is being suppressed.) Setting $\mathbf{V}^{\sigma_1\sigma_2} = \mathbf{v}^{\sigma_1\sigma_2}$ yields

$$\mathbf{V}^{EE} = [1], \quad \mathbf{V}^{EO} = [3], \quad \mathbf{V}^{OE} = [2], \quad \text{and} \quad \mathbf{V}^{OO} = [4].$$

Applying the first round of butterfly relations (in which $K = 1$, $M = 2^{P-K-1} = 1$ and $w = e^{(-i2\pi/4)2^1} = -1$),

$$\begin{aligned} V_0^E &= V_0^{EE} + w^0 V_0^{EO} = 1 + (-1)^0 3 = 4 \\ V_1^E &= V_0^{EE} - w^{1-1} V_0^{EO} = 1 - (-1)^0 3 = -2 \\ V_0^O &= V_0^{OE} + w^0 V_0^{OO} = 2 + (-1)^0 4 = 6 \\ V_1^O &= V_0^{OE} - w^{1-1} V_0^{OO} = 2 - (-1)^0 4 = -2, \end{aligned}$$

yields the second order sequences

$$\mathbf{V}^E = [4, -2]^T \qquad \text{and} \qquad \mathbf{V}^O = [6, -2]^T.$$

Applying the second (and last) round of butterfly relations (in which $K = 0$, $M = 2^{P-K-1} = 2$ and $w = e^{(-i2\pi/4)2^0} = -i$),

$$\begin{aligned} V_0 &= V_0^E + w^0 V_0^O = 4 + (-i)^0 6 = 10 \\ V_1 &= V_1^E + w^1 V_1^O = -2 + (-i)^1(-2) = -2 + 2i \\ V_2 &= V_{2-2}^E - w^{2-2} V_{2-2}^O = 4 - (-i)^0 6 = -2 \\ V_3 &= V_{3-2}^E - w^{3-2} V_{3-2}^O = -2 - (-i)^1(-2) = -2 - 2i, \end{aligned}$$

yields the alternative DFT

$$\mathbf{V} = \mathcal{D}\mathbf{v} = [\, 10, \, -2 + 2i\,, \, -2, \, -2 - 2i \,]^T.$$

2. Consider the computation of an Nth order DFT. When $N = 1024 = 2^{10}$, the ratio of operations required to compute this DFT with and without using the FFT is

$$\frac{\text{operations if using the FFT}}{\text{operations if not using the FFT}} \approx \frac{3 \cdot 10}{4 \cdot 2^{10}} \approx 7.3 \times 10^{-3}.$$

When $N = 11{,}048{,}576 = 2^{20}$, this ratio is

$$\frac{\text{operations if using the FFT}}{\text{operations if not using the FFT}} \approx \frac{3 \cdot 20}{4 \cdot 2^{20}} \approx 1.4 \times 10^{-5}.$$

3. Consider computing the convolution of two Nth order sequences. When $N = 1024 = 2^{10}$, the ratio of operations required to compute this convolution with and without using the FFT and identities is

$$\frac{\text{operations if using identities and the FFT}}{\text{operations if not using identities and the FFT}} \approx \frac{9 \cdot 10}{4 \cdot 2^{10}} \approx 2.2 \times 10^{-2}.$$

When $N = 11{,}048{,}576 = 2^{20}$, this ratio is

$$\frac{\text{operations if using identities and the FFT}}{\text{operations if not using identities and the FFT}} \approx \frac{9 \cdot 20}{4 \cdot 2^{20}} \approx 4.3 \times 10^{-5}.$$

References

[BH95] W. Briggs and V. Henson. *The DFT: An Owner's Manual for the Discrete Fourier Transform.* SIAM, Philadelphia, 1995.

[BC01] J. Brown and R. Churchill. *Fourier Series and Boundary Value Problems, 6th ed.*, McGraw-Hill, New York, 2001.

[CG00] E. Chu and A. George. *Inside the FFT Black Box.* CRC Press, Boca Raton, FL, 2000.

[Col88] D. Colton. *Partial Differential Equations: An Introduction.* Random House, New York, 1988.

[How01] K. Howell. *Principles of Fourier Analysis.* Chapman & Hall/CRC, Boca Raton, FL, 2001.

[Str94] R. Strichartz. *A Guide to Distribution Theory and Fourier Transforms.* CRC Press, Boca Raton, FL, 1994.

[Wal96] J. Walker. *Fast Fourier Transforms, 2nd ed.*, CRC Press, Boca Raton, FL, 1996.

[ZTF98] R. Ziemer, W. Trantor, and D. Fannin. *Signals and Systems: Continuous and Discrete, 4th ed.*, Prentice Hall, Upper Saddle River, NJ, 1998.

Applications to Physical and Biological Sciences

59

Linear Algebra and Mathematical Physics

Lorenzo Sadun
The University of Texas at Austin

59.1 Introduction

Linear algebra appears throughout physics. Linear differential equations, both ordinary and partial, appear through classical and quantum physics. Even where the equations are nonlinear, linear approximations are extremely powerful.

Two big ideas underpin linear analysis in physics. The first is the *Superposition Principle.* Suppose we have a linear problem where we need to compute the output for an arbitrary input. If there is a solution to the problem with input I_1 and output O_1, a solution with input I_2 and output O_2, etc., then the response to the input $c_1 I_1 + \cdots + c_k I_k$ is $c_1 O_1 + \cdots c_k O_k$. It is, therefore, enough to solve our problem for a limited set of inputs I_k, as long as an arbitrary input can be written as a linear combination of these special cases.

The second big idea is the *Decoupling Principle.* If a system of coupled differential equations (or difference equations) involves a diagonalizable square matrix A, then it is helpful to pick new coordinates $\mathbf{y} = [\mathbf{x}]_{\mathcal{B}}$, where $\mathcal{B}$ is a basis of eigenvectors of A. Rewriting our equations in terms of the y variables, we discover that the evolution of each variable y_k depends only on y_k, and not on the other variables, and that the form of the equation depends only on the kth eigenvalue of A. We can then solve our equations, one variable at a time, to get $\mathbf{y}$ as a function of time and, hence, get $\mathbf{x}$ as a function of time. (When A is not diagonalizable, one uses a basis for which $[A]_{\mathcal{B}}$ is in Jordan canonical form. The resulting equations for $\mathbf{y}$ are not completely decoupled, but are still relatively easy.)

Thanks to Newton's Law, $F = ma$, much of classical physics is expressed in terms of systems of second order ordinary differential equations. If the force is a linear function of position, the resulting equations are linear, and the special solutions that come from eigenvectors of the force matrix are called *normal modes of oscillation*. For nonlinear problems near equilibrium, the force can always be expanded in a Taylor series, and for small oscillations the leading (linear) term is dominant. Solutions to realistic nonlinear problems, such as small oscillations of a pendulum, are then closely approximated by solutions to linear problems.

Linear field equations also permeate classical physics. Maxwell's equations, which govern electromagnetism, are linear. There are an infinite number of degrees of freedom, namely the value of the field at each point, but the Superposition Principle and the Decoupling Principle still apply. We use a continuous basis of possible inputs, namely Dirac δ functions, and the resulting outputs are called Green's functions. The response to an arbitrary input is then the convolution of the input and the relevant Green's function.

Nonrelativistic quantum mechanics is governed by Schrödinger's equation, which is also linear. Much of quantum mechanics reduces to diagonalizing the Hamiltonian operator and applying the Decoupling Principle.

Symmetry plays a big role in quantum mechanics. Both vectors and operators decompose into representations of the rotation groups $SO(3)$ and $SU(2)$. The irreducible representations are finite-dimensional, so the study of rotations (and angular momentum) often reduces to a study of finite matrices.

59.2 Normal Modes of Oscillation

Suppose we have two blocks, each with mass m, attached to three springs, as in Figure 59.1, with the spring constants as shown, and let $x_i(t)$ be the displacement of the ith block from equilibrium at time t. It is easy to see that if $x_1(0) = x_2(0)$, and if $\dot{x}_1(0) = \dot{x}_2(0)$, then $x_1(t) = x_2(t)$ for all time. The middle spring never gets stretched, and the two blocks oscillate, in phase, with angular frequency $\omega_1 = \sqrt{k_1/m}$. If $x_2(0) = -x_1(0)$ and $\dot{x}_2(0) = -\dot{x}_1(0)$, then by symmetry $x_2(t) = -x_1(t)$ for all time, and each block oscillates with angular frequency $\omega_2 = \sqrt{(k_1 + 2k_2)/m}$. (This example is worked out in detail below.) These two solutions, with $x_1(t) = \pm x_2(t)$, are called *normal modes of oscillation*. Remarkably, every solution to the equations of motion is a linear combination of these two normal modes.

Definitions:

Suppose we have an arrangement of blocks, all of the same mass m, and springs with varying spring constants. Let $x_1(t), \ldots, x_n(t)$ denote the locations of the blocks, relative to equilibrium, and $\mathbf{x} = [x_1, \ldots, x_n]^T$. For any function $f(t)$, let $\dot{f}(t) = df/dt$. The **kinetic energy** is $T = m\sum_k \dot{x}_k^2/2$. The **potential energy** is

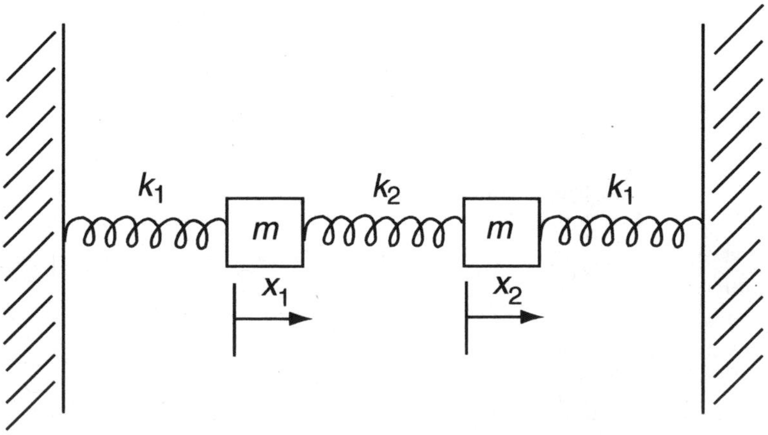

FIGURE 59.1 Coupled oscillators.

$V(\mathbf{x}) = \sum_{ij} a_{ij}x_i x_j/2$, where $A = (a_{ij})$ is a symmetric matrix. The **equations of motion** are

$$m\frac{d^2\mathbf{x}}{dt^2} = -A\mathbf{x}.$$

Let $\mathcal{B} = \{\mathbf{z}_1, \ldots, \mathbf{z}_n\}$ be a basis of eigenvectors of A, and let $\mathbf{y}(t) = [\mathbf{x}(t)]_{\mathcal{B}}$ be the coordinates of $\mathbf{x}(t)$ in this basis.

Facts:

(See Chapter 13 of [Mar70], Chapter 6 of [Gol80], and Chapter 5 of [Sad01].)

1. A is diagonalizable, and the eigenvalues of A are all real.
2. The eigenvectors can be chosen orthonormal with respect to the standard inner product: $\langle \mathbf{z}_i, \mathbf{z}_j\rangle = \delta_{ij}$.
3. The initial conditions for $\mathbf{y}(t)$ can be computed using the inner product: $y_k(0) = \langle \mathbf{z}_k, \mathbf{x}(0)\rangle$, $\dot{y}_k(0) = \langle \mathbf{z}_k, \dot{\mathbf{x}}(0)\rangle$.
4. In terms of the $\mathbf{y}$ variables, the equations of motion reduce to $md^2y_k/dt^2 = -\lambda_k y_k$, where λ_k is the eigenvalue corresponding to the eigenvector $\mathbf{z}_k$.
5. The solution to this equation depends on the sign of λ_k. If $\lambda_k > 0$, set $\omega_k = \sqrt{\lambda_k/m}$. We then have

$$y_k(t) = y_k(0)\cos(\omega_k t) + \frac{\dot{y}_k(0)}{\omega_k}\sin(\omega_k t).$$

   If $\lambda_k < 0$, set $\kappa_k = \sqrt{-\lambda_k/m}$ and we have

$$y_k(t) = y_k(0)\cosh(\kappa_k t) + \frac{\dot{y}_k(0)}{\kappa_k}\sinh(\kappa_k t).$$

   Finally, if $\lambda_k = 0$, then

$$y_k(t) = y_k(0) + \dot{y}_k(0)t.$$

6. If the system has translational symmetry, then there is a $\lambda = 0$ mode describing uniform motion of the system.
7. If the system has rotational symmetry, then there is a $\lambda = 0$ mode describing uniform rotation.
8. All solutions of the equations of motion are of the form $\mathbf{x}(t) = \sum y_k(t)\mathbf{z}_k$, where for each nonzero λ_k, $y_k(t)$ is of the form given in Fact 5.

Examples:

1. In the block-and-spring example above, the kinetic energy is $m(\dot{x}_1^2 + \dot{x}_2^2)/2$, while the potential energy is $(k_1x_1^2 + k_2(x_1 - x_2)^2 + k_1x_2^2)/2 = \langle \mathbf{x}, A\mathbf{x}\rangle/2$, where $A = \begin{pmatrix} k_1 + k_2 & -k_2 \\ -k_2 & k_1 + k_2 \end{pmatrix}$. The eigenvalues of A are k_1 and $k_1 + 2k_2$, with normalized eigenvectors $(\sqrt{2}/2, \sqrt{2}/2)^T$ and $(\sqrt{2}/2, -\sqrt{2}/2)^T$. Both eigenvalues are positive, so we have oscillations with angular frequencies $\omega_1 = \sqrt{k_1/m}$ and $\omega_2 = \sqrt{(k_1 + 2k_2)/m}$. Suppose we start by pushing the first block to the right and letting go. That is, suppose $\mathbf{x}(0) = (1, 0)^T$ and $\dot{\mathbf{x}}(0) = (0, 0)^T$. From the initial data we compute

$$y_1(0) = \frac{\sqrt{2}}{2}(x_1(0) + x_2(0)) = \sqrt{2}/2$$
$$y_2(0) = \frac{\sqrt{2}}{2}(x_1(0) - x_2(0)) = \sqrt{2}/2$$
$$\dot{y}_1(0) = \frac{\sqrt{2}}{2}(\dot{x}_1(0) + \dot{x}_2(0)) = 0$$
$$\dot{y}_1(0) = \frac{\sqrt{2}}{2}(\dot{x}_1(0) - \dot{x}_2(0)) = 0$$
$$y_1(t) = y_1(0)\cos(\omega_1 t) + \frac{\dot{y}_1(0)}{\omega_1}\sin(\omega_1 t) = \sqrt{2}\cos(\omega_1 t)/2$$
$$y_2(t) = y_2(0)\cos(\omega_2 t) + \frac{\dot{y}_2(0)}{\omega_2}\sin(\omega_2 t) = \sqrt{2}\cos(\omega_2 t)/2$$
$$\mathbf{x}(t) = y_1(t)\mathbf{z}_1 + y_2(t)\mathbf{z}_2 = \frac{1}{2}\begin{pmatrix} \cos(\omega_1 t) + \cos(\omega_2 t) \\ \cos(\omega_1 t) - \cos(\omega_2 t) \end{pmatrix}.$$

2. LC circuits obey the same equations as blocks and springs, with the capacitances playing the role of spring constants and the inductances playing the role of mass, and with the current around each loop playing the role of x_i.
3. Small oscillations: A particle in an arbitrary potential $V(\mathbf{x})$, or a system of identical-mass particles in an arbitrary n-body potential, follows the equation $md^2\mathbf{x}/dt^2 = -\nabla V(\mathbf{x})$. If $\mathbf{x} = \mathbf{x}_0$ is a critical point of the potential, so $\nabla V(\mathbf{x}_0) = 0$, then we expand $V(\mathbf{x})$ around $\mathbf{x} = \mathbf{x}_0$ in a Taylor series:

$$V(\mathbf{x}) = V(\mathbf{x}_0) + \frac{1}{2}\sum_{ij} a_{ij}(\mathbf{x} - \mathbf{x}_0)_i(\mathbf{x} - \mathbf{x}_0)_j + O(|\mathbf{x} - \mathbf{x}_0|^3),$$

where $a_{ij} = \left.\frac{\partial^2 V}{\partial x_i \partial x_j}\right|_{\mathbf{x}=\mathbf{x}_0}$, so $\nabla V(\mathbf{x}) = A(\mathbf{x} - \mathbf{x}_0) + O(|\mathbf{x} - \mathbf{x}_0|^2)$, and our displacement $\mathbf{x} - \mathbf{x}_0$ from equilibrium is governed by the approximate equation $md^2(\mathbf{x} - \mathbf{x}_0)/dt^2 = -A(\mathbf{x} - \mathbf{x}_0)$.

For example, a pendulum of mass m and length ℓ has quadratic kinetic energy $m\ell^2\dot{\theta}^2/2$ and nonlinear potential energy $mg\ell(1 - \cos(\theta))$. For θ small, this potential energy is approximated by $mg\ell\theta^2/2$, and the equations of motion are approximated by $\ell d^2\theta/dt^2 = -g\theta$, and yields oscillations of angular frequency $\sqrt{g/\ell}$. The same ideas apply to motion of a pendulum near the top of the circle: $\theta = \pi$. Then $V(\theta) \approx mg\ell(2 - (\theta - \pi)^2/2)$, and our equations of motion are approximately $\ell d^2(\theta - \pi)/dt^2 = +g(\theta - \pi)$. The deviation of θ from the unstable equilibrium grows as $e^{\kappa t}$, with $\kappa = \sqrt{g/\ell}$, until $\theta - \pi$ is large enough that our quadratic approximation for $V(\theta)$ is no longer valid.

Finally, one can consider two pendula, near their stable equilibria, attached by a weak spring. The resulting equations are almost identical to those of the coupled springs of Figure 59.1.
4. Central force motion (see Chapter 3 of [Gol80] or Chapter 8 of [Mar70]): In systems with symmetry, it is often possible to use conserved quantities to integrate out some of the variables, obtaining reduced equations for the remaining variables. For instance, if an object is moving in a central force (e.g., a planet around a star or a classical electron around the nucleus), conservation of angular momentum allows us to integrate out the angular variables and get an equation for the distance r. The radius then oscillates in a pseudopotential $V(r)$, obtained by adding a $1/r^2$ centrifugal term to the true potential $V_0(r)$. Orbits that are almost circular are described by small oscillations of the variable r around the minimum of the pseudopotential $V(r)$, and the frequency of oscillation is $\sqrt{V''(r_0)/m}$, where the pseudopotential has a minimum at $r = r_0$. When the true potential is a $1/r$ attraction (as with gravitation and electromagnetism), these oscillations have the same period as the orbital motion itself. Planets traverse elliptical orbits, with the sun at a focus, and the nearest approach to the sun (the perihelion) occurs at the same point each year. When the true potential is an r^2 attraction (simple harmonic motion), the radial oscillations occur with frequency twice that of the orbit. The motion is elliptical with the center of force at the center of the orbit, and there are

two perihelia per cycle. For almost any other kind of force, the radial oscillations and the rotation are incommensurate, the orbit is not a closed curve, and the perihelion precesses.

59.3 Lagrangian Mechanics

In the previous section, we assumed that all the particles had the same mass or, equivalently, that the kinetic energy was proportional to the squared norm of the velocity. Here we relax this assumption and we also allow generalized coordinates.

Definitions:

The **Lagrangian function** is $L(\mathbf{q}, \dot{\mathbf{q}}) = T - V$, where T is the kinetic energy and V is the potential energy. One can express the Lagrangian in terms of arbitrary generalized coordinates $\mathbf{q}$ and their derivatives $\dot{\mathbf{q}}$. The **kinetic energy** is typically quadratic in the velocity: $T = \langle \dot{\mathbf{q}}, B(\mathbf{q})\dot{\mathbf{q}}\rangle/2$, where the symmetric "mass matrix" B may depend on the coordinates $\mathbf{q}$, but not on the velocities $\dot{\mathbf{q}}$. The **potential energy** V depends only on the coordinates $\mathbf{q}$ (and not on $\dot{\mathbf{q}}$), but may be nonlinear.

If $\mathbf{q}_0$ is a critical point of V, we consider motion with $\mathbf{q}$ close to $\mathbf{q}_0$ and $\dot{\mathbf{q}}$ small.

Facts:

(See Chapter 7 of [Mar70] or Chapters 2 and 6 of [Gol80].)

1. The Euler–Lagrange equations

$$\frac{d}{dt}\left(\frac{\partial L}{\partial \dot{q}_k}\right) = \frac{\partial L}{\partial q_k}$$

reduce to the approximate equations of motion

$$B\frac{d^2(\mathbf{q} - \mathbf{q}_0)}{dt^2} = -A(\mathbf{q} - \mathbf{q}_0),$$

where $a_{ij} = \left.\frac{\partial^2 V}{\partial q_i \partial q_j}\right|_{\mathbf{q}=\mathbf{q}_0}$, essentially as before, and the mass matrix B is evaluated at $\mathbf{q} = \mathbf{q}_0$. Instead of looking for eigenvalues and eigenvectors of A, we look for numbers λ_k and vectors $\mathbf{z}_k$ such that $A\mathbf{z}_k = \lambda_k B\mathbf{z}_k$. (See Chapter 43.) We then let $\mathbf{y} = [\mathbf{q} - \mathbf{q}_0]_B$.

2. The matrices A and B are symmetric, and the eigenvalues of B are all positive.
3. The numbers λ_k are the roots of the polynomial $\det(xB - A)$. When B is the identity matrix, these reduce to the eigenvalues of A.
4. One can find a basis of solutions $\mathbf{z}_k$ to $A\mathbf{z}_k = \lambda_k\mathbf{z}_k$, with λ_k real. The numbers λ_k are the eigenvalues of $B^{-1}A$, or equivalently of the symmetric matrix $B^{-1/2}AB^{-1/2}$, which explains why the λ_k's are real.
5. The eigenvectors can be chosen orthonormal with respect to an inner product involving B. (See Chapter 5.) That is, if $\langle \mathbf{u}, \mathbf{v}\rangle_B = \mathbf{u}^T B\mathbf{v}$, then $\langle \mathbf{z}_i, \mathbf{z}_j\rangle_B = \delta_{ij}$.
6. The initial conditions for $\mathbf{y}(t)$ can be computed using the modified inner product of the previous fact: $y_k(0) = \langle \mathbf{z}_k, \mathbf{q}(0) - \mathbf{q}_0(0)\rangle_B$, $\dot{y}_k(0) = \langle \mathbf{z}_k, \dot{\mathbf{q}}(0)\rangle_B$.
7. In terms of the $\mathbf{y}$ variables, the approximate equations of motion reduce to the decoupled equations $d^2y_k/dt^2 = -\lambda_k y_k$.
8. The solution to these equations depends on the sign of λ_k. If $\lambda_k > 0$, set $\omega_k = \sqrt{\lambda_k}$; if $\lambda_k < 0$, set $\kappa_k = \sqrt{-\lambda_k}$. With values of ω_k or κ_k, these the solutions take the same form as in Fact 5 of section 59.2.
9. If the system is symmetric under the action of a continuous group, then there is a $\lambda = 0$ mode for each generator of this group.

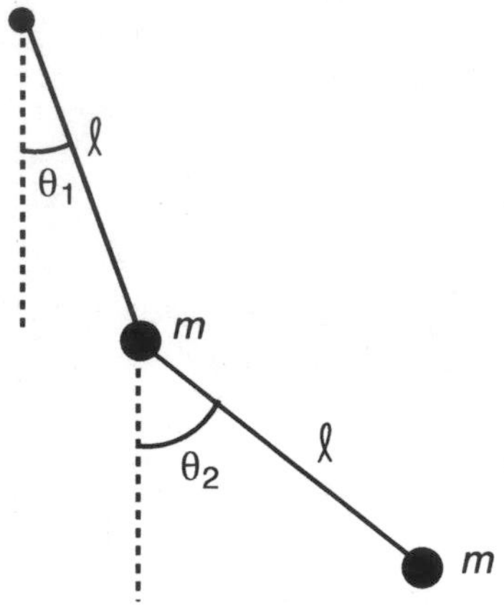

FIGURE 59.2 A double pendulum.

10. All solutions of the approximate equations of motion are of the form $\mathbf{q}(t) = \mathbf{q}_0 + \sum y_k(t)\mathbf{z}_k$, where for each nonzero λ_k, $y_k(t)$ is of the form given by Fact 8 above (and Fact 5 of Section 59.2).

Examples:

1. Consider the double pendulum of Figure 59.2, where each ball has mass m and each rod has length ℓ. For large motions, this system is famously chaotic, but for small oscillations it is simple. The two coordinates are the angles θ_1 and θ_2, and the potential energy of the system is $mg\ell(3 - 2\cos(\theta_1) - \cos(\theta_2)) \approx mg\ell(\theta_1^2 + \theta_2^2/2)$, so $A = mg\ell \begin{pmatrix} 2 & 0 \\ 0 & 1 \end{pmatrix}$. The kinetic energy is $\frac{m\ell^2}{2}\left(\dot\theta_1^2 + (\sin(\theta_1)\dot\theta_1 + \sin(\theta_2)\dot\theta_2)^2 + (\cos(\theta_1)\dot\theta_1 + \cos(\theta_2)\dot\theta_2)^2\right)$. For small values of θ_1 and θ_2, this is approximately $\frac{m\ell^2}{2}(2\dot\theta_1^2 + \dot\theta_2^2 + 2\dot\theta_1\dot\theta_2)$, so $B = m\ell^2 \begin{pmatrix} 2 & 1 \\ 1 & 1 \end{pmatrix}$. $\det(xB - A) = m^2\ell^4(x^2 - 4(g/\ell)x + 2g^2/\ell^2)$, with roots $\lambda_1 = (g/\ell)(2 + \sqrt{2})$ and $\lambda_2 = (g/\ell)(2 - \sqrt{2})$, and with $\mathbf{z}_i = c_i(1, \mp\sqrt{2})^T$, $i = 1, 2$, where c_i are normalization constants. The two normal modes are as follows: There is a fast mode, with $\omega_1 = \sqrt{(g/\ell)(2 + \sqrt{2})} \approx 1.8478\sqrt{g/\ell}$, with the two pendula swinging in opposite directions, and with the bottom pendulum swinging $\sqrt{2}$ more than the top; there is a slow mode, with $\omega_2\sqrt{(g/\ell)(2 - \sqrt{2})} \approx 0.7654\sqrt{g/\ell}$, with the two pendula swinging in the same direction, and with the bottom pendulum swinging $\sqrt{2}$ more than the top.

59.4 Schrödinger's Equation

In quantum mechanics, the evolution of a particle of mass m, moving in a time-dependent potential $V(x, t)$, is described by Schrödinger's equation,

$$i\hbar \frac{\partial \psi(x,t)}{\partial t} = -\frac{\hbar^2}{2m}\nabla^2\psi(x,t) + V(x,t)\psi(x,t),$$

where $\hbar$ is Planck's constant divided by 2π, and the square of the complex wavefunction $\psi(x, t)$ describes the probability of finding a particle at position x at time t. Space and time are not treated on equal footing. We consider the wavefunction ψ to be a square-integrable function of x that evolves in time.

Definitions:

Let $\mathcal{H} = L^2(\mathbb{R}^n)$ be the Hilbert space of square-integrable functions on $\mathbb{R}^n$ with "inner product"

$$\langle \phi | \psi \rangle = \int_{\mathbb{R}^n} \overline{\phi(x)} \psi(x) d^n x.$$

Note that this inner product is linear in the *second* factor and conjugate-linear in the first factor. Although mathematicians usually choose their complex inner products $\langle u, v \rangle$ to be linear in u and conjugate-linear in v, among physicists the convention, and notation, is invariably that of the above equation. The bracket of ϕ and ψ can be viewed as a pairing of two pieces, the "bra" $\langle \phi |$ and the "ket" $| \psi \rangle$. The ket $| \psi \rangle$ is a vector in $\mathcal{H}$, while $\langle \phi |$ is a map from $\mathcal{H}$ to the complex numbers, namely, "take the inner product of ϕ with an input vector."

The Hermitian adjoint of an operator A, denoted A^*, is the unique operator such that $\langle A^* \phi | \psi \rangle = \langle \phi | A \psi \rangle$, for all vectors ϕ, ψ. An operator A is called **Hermitian**, or **self-adjoint**, if $A^* = A$, and **unitary** if $A^* = A^{-1}$.

The **commutator** of two operators A and B, denoted $[A, B]$, is the difference $AB - BA$. A and B are said to **commute** if $AB = BA$.

The **expectation value** of an operator A in the state $|\psi\rangle$ is the statistical average of repeated measurements of A in the state $|\psi\rangle$, and is denoted $\langle A \rangle$, with the dependence on $|\psi\rangle$ implicit. The **uncertainty in** A, denoted ΔA, is the root mean squared variation in measurements of A.

The **generalized eigenvalues** of an operator A are points in the spectrum of A, and the **generalized eigenvectors** are formal solutions to $A|\psi\rangle = \lambda |\psi\rangle$. These may not be true eigenvalues and eigenvectors if ψ is not square integrable. (See Facts 11 to 15 below.) This use of the term "generalized eigenvector," which is standard in physics, has *nothing* to do with the same term in matrix theory (where it signifies vectors $\mathbf{v}$ for which $(A - \lambda I)^k \mathbf{v} = 0$ for some positive integer k).

Facts:

(See Chapter 6 of [Sch68] or Chapters 5 and 7 of [Mes00].)

1. The Schrödinger equation can be recast as an ordinary differential equation with values in $\mathcal{H}$:
$$i\hbar \frac{d|\psi\rangle}{dt} = H(t)|\psi\rangle,$$
where $H = -\frac{\hbar^2}{2m}\nabla^2 + V$ is the Hamiltonian operator.
2. Physically measurable quantities, also called *observables*, are represented by Hermitian operators. It is easy to see that the position operator $(X\psi)(x) = x\psi(x)$, the momentum operator $(P\psi)(x) = -i\hbar\nabla\psi(x)$, and the Hamiltonian $H = P^2/2m + V$ are all self-adjoint.
3. If an observable a is represented by the operator A, then the possible values of a measurement of a are the (generalized) eigenvalues of A.
4. Two Hermitian operators A, B can be simultaneously diagonalized if and only if they commute.
5. Suppose the state of the system is described by the vector $|\psi\rangle = \sum_n c_n |\phi_n\rangle$, where $\sum |c_n|^2 = 1$ and each $|\phi_n\rangle$ is a normalized eigenvector of A with eigenvalue λ_n. Then the probability of a measurement of a yielding the value λ_n is $|c_n|^2$.
6. If $|\psi\rangle$ is as in the previous Fact, then the expectation value of A is $\langle A \rangle = \sum_n \lambda_n |c_n|^2$.
7. The uncertainty of A satisfies $(\Delta A)^2 = \langle A^2 \rangle - \langle A \rangle^2$.
8. If A, B, and C are Hermitian operators with $[A, B] = iC$, then $\Delta A \Delta B \geq |\langle C \rangle|/2$.
9. In particular, $XP - PX = i\hbar$, so $\Delta X \Delta P \geq \hbar/2$. This is *Heisenberg's uncertainty principle.*
10. If the Hamiltonian operator does not depend on time, then energy is conserved. In fact, if $|\psi(0)\rangle = \sum c_n |\phi_n\rangle$, where $H|\phi_n\rangle = E_n|\phi_n\rangle$, then $|\psi(t)\rangle = \sum c_n e^{-iE_n t/\hbar}|\phi_n\rangle$. (Eigenvalues of the Hamiltonian are usually denoted E_n, for energy.) Solving the Schrödinger equation is tantamount to diagonalizing the Hamiltonian and using a basis of eigenvectors.

11. Operators may have continuous spectrum, in which case the generalized eigenvectors are not square-integrable. In particular, e^{ikx} is a generalized eigenvector for $P = -i\hbar d/dx$ with generalized eigenvalue $\hbar k$, and the Dirac delta function $\delta(x - a)$ is a generalized eigenvector for X with eigenvalue a.
12. Let $|A, \alpha\rangle$ be a generalized eigenvector of the operator A with generalized eigenvalue α. If A has continuous spectrum, then the decomposition of a state $|\psi\rangle$ involves integrating over eigenvalues instead of summing: $|\psi\rangle = \int f(\alpha)|A, \alpha\rangle d\alpha$. The generalized eigenstates are usually normalized so that $\langle\psi|\psi\rangle = \int |f(\alpha)|^2 d\alpha$. Equivalently, $\langle A, \alpha|A, \beta\rangle = \delta(\alpha - \beta)$.
13. For continuous spectra, $|f(\alpha)|^2$ is not the probability of a measurement of A yielding the value α. Rather, $|f(\alpha)|^2$ is a probability *density*, and the probability of a measurement yielding a value between a and b is $\int_a^b |f(\alpha)|^2 d\alpha$.
14. The two most common expansions in terms of generalized eigenvalues are for the position operator and the momentum operator: $|\psi\rangle = \int \psi(x)|X, x\rangle dx = \int \hat{\psi}(k)|P, \hbar k\rangle dk$. The coefficients $\hat{\psi}(k)$ of $|\psi\rangle$ in the momentum basis are the Fourier transform of the coefficients $\psi(x)$ in the position basis. From this perspective, the Fourier transform is just a change-of-basis.
15. An operator may have both discrete and continuous spectrum, in which case eigenfunction expansions involve summing over discrete eigenvalues and integrating over continuous eigenvalues. For example, for the Hamiltonian of a hydrogen atom, there are discrete negative eigenvalues that describe bound states, and a continuum of positive eigenvalues that describe ionized hydrogen, with the electron having broken free of the nucleus.

Examples:

1. *The one dimensional harmonic oscillator.* We have seen that a classical harmonic oscillator with potential energy $kx^2/2$ has frequency $\omega = \sqrt{k/m}$, so we write the Hamiltonian of a quantum mechanical harmonic oscillator as

$$H = \frac{P^2}{2m} + \frac{kX^2}{2} = \frac{P^2 + m^2\omega^2 X^2}{2m}.$$

We will compute the eigenvalues and eigenvectors of this Hamiltonian.

We define a *lowering operator*

$$a = \frac{P - im\omega X}{\sqrt{2m\hbar\omega}}.$$

The Hermitian conjugate of a is the *raising operator*

$$a^* = \frac{P + im\omega X}{\sqrt{2m\hbar\omega}}.$$

Note that a and a^* do not commute. Rather, $[a, a^*] = 1$. In terms of a and a^*, the Hamiltonian takes the form

$$H = \hbar\omega(a^*a + aa^*)/2 = \frac{\hbar\omega}{2}(2a^*a + 1) = \frac{\hbar\omega}{2}(2aa^* - 1).$$

Note that a^*a is positive-definite, since $\langle\psi|a^*a\psi\rangle = \langle a\psi|a\psi\rangle \geq 0$, so the eigenvalues of energy must all be at least $\hbar\omega/2$.

Since $Ha = a(H - \hbar\omega)$, the operator a serves to lower the energy of a state by $\hbar\omega$. If $|\phi\rangle$ is an eigenvector of H with eigenvalue E, then $Ha|\phi\rangle = a(H - \hbar\omega)|\phi\rangle = a(E - \hbar\omega)|\phi\rangle = (E - \hbar\omega)a|\phi\rangle$, so $a|\phi\rangle$ is either the zero vector or a state with energy $E - \hbar\omega$. Since we cannot reduce the energy below $\hbar\omega/2$, by applying a repeatedly (say, $n \geq 0$ times), we must eventually get a vector $|\phi_0\rangle$ for which $a|\phi_0\rangle = 0$. But then $H|\phi_0\rangle = \frac{\hbar\omega}{2}(2a^*a + 1)|\phi_0\rangle = \frac{\hbar\omega}{2}|\phi_0\rangle$. Since it took n lowerings to get the energy to $\hbar\omega/2$, our original state $|\phi\rangle$ must have had energy $(n + \frac{1}{2})\hbar\omega$.

The notation $|n\rangle$ is often used for this nth excited state, so $a|n\rangle$ is a constant times $|n-1\rangle$. It remains to compute that constant. Normalizing $\langle n|n\rangle = 1$ and picking the phases such that $\langle n-1|an\rangle > 0$, we compute $\langle an|an\rangle = \langle n|a^*an\rangle = \frac{\langle n|Hn\rangle}{\hbar\omega} - \frac{1}{2} = n\langle n|n\rangle = n$, so $a|n\rangle = \sqrt{n}|n-1\rangle$. A similar calculation yields $a^*|n\rangle = \sqrt{n+1}|n+1\rangle$ and, hence, $|n\rangle = (a^*)^n|0\rangle/\sqrt{n!}$.

The state $|0\rangle$ is in the kernel of a. In a coordinate basis, where X is multiplication by x and $P = -i\hbar d/dx$, the equation $a|0\rangle = 0$ becomes a first-order differential equation

$$\frac{d\psi(x)}{dx} + \frac{m\omega x}{\hbar}\psi(x) = 0,$$

whose solution is the Gaussian $\psi(x) = \exp(-m\omega x^2/2\hbar)$ times a normalization constant. The nth state is obtained by applying the differential operater $\frac{d}{dx} - \frac{m\omega x}{\hbar}$ to the Gaussian n times. The result is an nth order polynomial in x times the Gaussian.

59.5 Angular Momentum and Representations of the Rotation Group

The same techniques that solved the harmonic oscillator also work to diagonalize the angular momentum operator.

Definitions:

Angular momentum is a vector: $\vec{L} = \vec{X} \times \vec{P}$, or in coordinates, $L_1 = X_2P_3 - X_3P_2$, $L_2 = X_3P_1 - X_1P_3$, $L_3 = X_1P_2 - X_2P_1$. Each L_i is a self-adjoint observable. We define $L^2 = L_1^2 + L_2^2 + L_3^2$, and define a **raising operator** $L_+ = L_1 + iL_2$ and a **lowering operator** $L_- = L_1 - iL_2$.

Facts:

(See Chapter 7 of [Sch68] or Chapter 13 of [Mes00].)

1. The three components of angular momentum do not commute. Rather,

$$[L_1, L_2] = i\hbar L_3, \qquad [L_2, L_3] = i\hbar L_1, \qquad [L_3, L_1] = i\hbar L_2.$$

   By the uncertainty principle, this means that only one component of the angular momentum can be known at a time.
2. L^2 is Hermitian, and each $[L_i, L^2] = 0$. It is possible to know both L^2 and L_3, and we consider simultaneous eigenstates $|\ell, m\rangle$ of L^2 and L_3, where ℓ labels the eigenvalue of L^2 and $\hbar m$ is the eigenvalue of L_3.
3. $[L^2, L_\pm] = 0$ and $[L_3, L_\pm] = \pm\hbar L_\pm$. This means that L_+ does not change the eigenvalue of L^2, but increases the eigenvalue of L_3 by $\hbar$. Likewise, L_- decreases the eigenvalue of L_3 by $\hbar$.
4. Since $L^2 - L_3^2 = L_1^2 + L_2^2 \geq 0$, there is a limit to how big m (or $-m$) can get. For each ℓ, there is a state $|\ell, m_{max}\rangle$ for which $L_+|\ell, m_{max}\rangle = 0$, and a state $|\ell, m_{min}\rangle$ for which $L_-|\ell, m_{min}\rangle = 0$. We set the label ℓ to be equal to m_{max}.
5. $L_-L_+ = L_1^2 + L_2^2 - \hbar L_3$ and $L_+L_- = L_1^2 + L_2^2 + \hbar L_3$, so we can write L^2 in terms of $L_\pm$ and L_3: $L^2 = L_-L_+ + L_3^2 + \hbar L_3 = L_+L_- + L_3^2 - \hbar L_3$.
6. The minimum value of m is $-\ell$. Since $2\ell = m_{max} - m_{min}$ is an integer, ℓ must be half of a nonnegative integer.
7. The states $|\ell, m\rangle$, with m ranging from $-\ell$ to ℓ, form a $(2\ell + 1)$ dimensional irreducible representation of the Lie algebra of $SO(3)$. We denote this representation V_ℓ, and call it the "spin-ℓ" representation.
8. In V_ℓ, we have $L^2|\ell, m\rangle = \hbar^2\ell(\ell+1)|\ell, m\rangle$, $L_3|\ell, m\rangle = m\hbar|\ell, m\rangle$, and $L_\pm|\ell, m\rangle = \hbar\sqrt{\ell(\ell+1) - m(m\pm 1)}|\ell, m\pm 1\rangle$.

9. If u is a unit vector, then a rotation by the angle θ about the u axis is implemented by the unitary operator $\exp(-i\theta L \cdot u/\hbar)$.
10. Since rotation by 2π equals the identity, representations of the Lie group $SO(3)$ satisfy the additional condition $\exp(2\pi i L_3) = 1$, which forces m (and, therefore, ℓ) to be an integer.
11. If one particle has angular momentum ℓ_1 and another has angular momentum ℓ_2, then the combined angular momentum can be any integer between $|\ell_1 - \ell_2|$ and $\ell_1 + \ell_2$. In terms of representations, $V_{\ell_1} \otimes V_{\ell_2} = \oplus_{\ell=|\ell_1-\ell_2|}^{\ell_1+\ell_2} V_\ell$.
12. The Lie group $SU(2)$ is the double cover of $SO(3)$, and has the same Lie algebra. The generators are usually denoted J rather than L, and the maximum value of m is denoted j rather than ℓ, but otherwise the computations are the same. J describes the total angular momentum of a particle, including spin, and j can be either an integer or a half-integer.
13. Particles with j integral are called bosons, while particles with j half-integral are called fermions.
14. If the Hamiltonian is rotationally symmetric, then angular momentum is conserved, and our energy eigenstates can be chosen to be eigenstates of J^2 and J_3.

59.6 Green's Functions

Expansions in a continuous basis of eigenfunctions are not limited to quantum mechanics. The Dirac δ is an eigenfunction of position, and any function can be written trivially as an integral over δ functions:

$$f(x) = \int f(y)\delta(x-y)dy = \int f(y)|X, y\rangle dy.$$

It, therefore, suffices to solve linear input–output problems in the case where the input is a δ-function located at an arbitrary point y. The resulting solution $G(y, x)$ is called a **Green's function** (or in some texts, **Green function**) for the problem, and the solution for an arbitrary input $f(x)$ is the convolution $\int f(y)G(y, x)dy$.

Facts:

(See [Jac75], especially Chapters 1 to 3, for many applications to electrostatics, and see Chapter 11 of [Sad01] for a general introduction to Green's functions.)

1. Green's functions are sometimes called *integral kernels*, especially in the mathematics literature, or *propagators* in quantum field theory. The term *propagator* is also sometimes used for the Fourier transform of a Green's function.
2. Linear partial differential equations appear throughout physics. Examples include Maxwell's equations, Laplace's equation, Schrödinger's equation, the heat equation, the wave equation, and the Dirac equation. Each equation generates its own Green's function.
3. Some boundary value problems involve Neumann boundary conditions, in which the normal derivative of a function (as opposed to the value of a function) is specified on S, and some problems involve mixed Neumann and Dirichlet conditions. The formalism for these cases is a simple modification of the Dirichlet formalism.
4. Two common techniques for computing Green's functions are Fourier transforms and the method of images.
5. Fourier transforms apply when the problem has translational symmetry, as in the heat equation example, above. We decompose a δ function as a linear combination of exponentials e^{ikx}, compute the response for each exponential, and re-sum.
6. The method of images is illustrated in Example 2, where the actual response $G(y, x)$ is a sum of two terms. The first is the response $G_0(y, x)$ to the actual charge at y, computed without boundary, and the second is the response to a mirror charge, located at a point outside D, and chosen so that the sum of the two terms is zero on S.

Examples:

1. *Electrostatics without boundaries.* The electrostatic potential $\phi(x)$ is governed by Poisson's equation:

$$\nabla^2\phi = -4\pi\rho(x),$$

where $\rho(x)$ is the charge density. Here, ρ is the input and ϕ is the output. Since the solution to $\nabla^2 G(y,x) = -4\pi\delta^3(x-y)$ is $G(y,x) = |x-y|^{-1}$, the potential due to a charge distribution $\rho(x)$ is $\phi(x) = \int d^3y\rho(y)/|x-y|$. (Note that, when we write $\nabla^2 G(y,x)$, we are taking the second derivative of $G(y,x)$ with respect to x. The variable y is just a parameter.)

2. *Electrostatics with boundary conditions.* Poisson's equation on a domain D with boundary S is more subtle, as we need to apply boundary conditions on S. Suppose that D is the exterior of a ball of radius R, and that we apply the homogeneous Dirichlet boundary condition $\phi = 0$ on S. (This corresponds to S being a grounded conducting sphere.) The function $G_0(y,x) = 1/|x-y|$ satisfies $\nabla^2 G_0(y,x) = -4\pi\delta^3(x-y)$, but does not satisfy the boundary condition. The function $G(y,x) = G_0(y,x) - \frac{R}{|y|}G_0(R^2y/|y|^2, x)$ satisfies $\nabla^2 G(y,x) = -4\pi\delta^3(x-y)$ and $G(y,x) = 0$ for $x \in S$.

   Nonzero boundary values can be considered part of the input. If we want to solve the equation $\nabla^2\phi = -4\pi\rho$ on D with boundary values $f(x)$ on S, then we have two different Green's functions to compute. For each $y \in D$, we compute $G_1(y,x)$, the solution to $\nabla^2 G_1(y,x) = -4\pi\delta^3(x-y)$ with boundary value zero on S. For each $z \in S$, we compute $G_2(z,x)$, the solution to $\nabla^2 G_2(z,x) = 0$ on D with boundary value $\delta^2(x-z)$ on S. Our solution to the entire problem is then $\phi(x) = \int_D d^3y G_1(y,x)\rho(y) + \int_S d^2z f(z) G_2(z,x)$.

3. *The heat kernel.* In $\mathbb{R}^2$, with variables x and t, let D be the region $t > 0$, so S is the x-axis. We look for solutions to the *heat equation*

$$\frac{\partial f}{\partial t} - \frac{\partial^2 f}{\partial x^2} = 0,$$

with boundary value $f(x,0) = f_0(x)$. Since $G(y,x,t) = \exp(-(x-y)^2/4t)/\sqrt{4\pi t}$ is a solution to (3) and approaches $\delta(x-y)$ as $t \to 0$, the solution to our problem is

$$f(x,t) = \int G(y,x,t) f_0(y) dy = \frac{1}{\sqrt{4\pi t}} \int \exp(-(x-y)^2/4t) f_0(y) dy.$$

References

[Gol80] Herbert Goldstein, *Classical Mechanics, 2nd ed.* Addison-Wesley, Reading, MA, 1980.

[Jac75] J.D. Jackson, *Classical Electrodynamics, 2nd ed.* John Wiley & Sons, New York, 1975.

[Mar70] Jerry B. Marion, *Classical Dynamics of Particles and Systems, 2nd ed.*, Academic Press, New York, 1970.

[Mes00] Albert Messiah, *Quantum Mechanics, Vols. 1 and 2*, Dover Publications, NY, Mineola, 2000.

[Sad01] Lorenzo Sadun, *Applied Linear Algebra: the Decoupling Principle.* Prentice Hall, Upper Saddle River, NJ, 2001.

[Sch68] Leonard Schiff, *Quantum Mechanics, 3rd ed.* McGraw-Hill, New York, 1968.

60

Linear Algebra in Biomolecular Modeling

Zhijun Wu
Iowa State University

60.1 Introduction

Biomolecular modeling is an active research area in computational biology. It studies the structures and functions of biomolecules by using computer modeling and simulation [Sch03]. Proteins are an important class of biomolecules. They are encoded in genes and produced in cells through genetic translation. Proteins are life supporting (or sometimes, destructing) ingredients (Figure 60.1) and are indispensable for almost all biological processes [Boy99]. In order to understand the diverse biological functions of proteins, the knowledge of the three-dimensional structures of proteins is essential. Several structure determination techniques have been used, including x-ray crystallography, nuclear magnetic resonance spectroscopy (NMR), and homology modeling. They all require intensive mathematical computing, ranging from data analysis to model building [Cre93].

As in all other types of scientific computing, linear algebra is one of the most powerful mathematical tools for biological computing. Here we review several subjects in biomolecular modeling, where linear algebra has played a major role, including mapping from distances to coordinates in NMR structure determination (Section 60.2), solving the Procrustes problem for structural comparison (Section 60.3), exploiting the structure of the Karle–Hauptman matrix in protein x-ray crystallography (Section 60.4), computing the fast and slow modes of protein motions (Section 60.5), and solving the flux balancing equations in metabolic network simulation (Section 60.6). The last subject actually involves the modeling of a large biological system, something beyond conventional biomolecular modeling, yet of increased research interests in computational systems biology [Kit99].

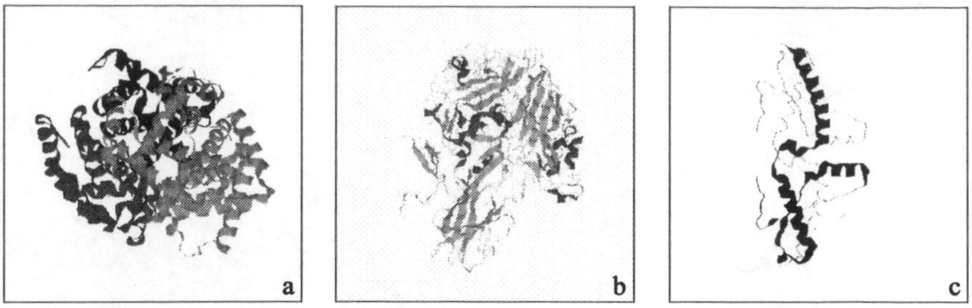

FIGURE 60.1 Example proteins: Humans have hundreds of thousands of different proteins (e.g., hemoglobin protein, 1BUW, in blood in 1a) and would not be able to maintain normal life even if short of a single type of protein. On the other hand, with the help of some proteins (e.g., protein, 2PLV, supporting poliovirus in 1b), viruses are able to grow, translate, integrate, and replicate, causing diseases. Some proteins themselves are toxic and even infectious such as the proteins in poisonous plants and in beef causing the Mad Cow Disease (e.g., prion protein, 1I4M-D, in human in 1c).

60.2 Mapping from Distances to Coordinates: NMR Protein Structure Determination

A fundamental problem in protein modeling is to find the three-dimensional structure of a protein and its relationship with the protein's biological function. One of the experimental techniques for structure determination is to use the nuclear magnetic resonance (NMR) to obtain some information on the distances for certain pairs of atoms in the protein and then find the coordinates of the atoms based on the obtained distance information. Mathematically, the second part of the work requires the solution of a so-called distance geometry problem, i.e., determine the coordinates for a set of points in a given topological space, given the distances for a subset of all pairs of points. We consider such a problem with the distances for all pairs of points assumed to be given.

Definitions:

The **coordinate vector** for atom i is a vector $\mathbf{x}_i = (x_{i,1}, x_{i,2}, x_{i,3})^T$, where $x_{i,1}$, $x_{i,2}$, and $x_{i,3}$ are the first, second, and third coordinates of atom i, respectively.

The **distance** between atoms i and j is defined as $d_{i,,j} = ||\mathbf{x}_i - \mathbf{x}_j||$, where $\mathbf{x}_i$ and $\mathbf{x}_j$ are coordinate vectors of atoms i and j, and $|| \cdot ||$ is the Euclidean norm.

The **coordinate matrix** for a protein is a matrix of coordinates denoted by $X = \{x_{i,j} : i = 1, \ldots, n, j = 1,2,3\}$, where n is the total number of atoms in the protein, and row i of X is the coordinate vector of atom i.

The **distance matrix** for a protein is a matrix of distances denoted by $D = \{d_{i,j} : i, j = 1, \ldots, n\}$, where $d_{i,j}$ is the distance between atoms i and j.

The problem of computing the coordinates of atoms (X) given a set of distances between pairs of atoms (D) is known as the **molecular distance geometry problem**.

Facts:

1. [Sax79] If the protein structure and, hence, X are known, D can immediately be computed from X°. Conversely, if D is known or even partially known, X can also be obtained from D, but the computation is not as straightforward. The latter is proved to be NP-complete for arbitrary sparse distance matrices.

2. [Blu53] Choose a reference system so that the origin is located at the last atom, or in other words, $x_n = (0, 0, 0)^T$. Let X° be a submatrix of X, $X^\circ = \{x_{i,j} : \; i = 1, \ldots, n-1, \; j = 1, 2, 3\}$, and D° be a matrix derived from D, $D^\circ = \{(d_{i,n}^2 - d_{i,j}^2 + d_{j,n}^2)/2 : \; i, j = 1, \ldots, n-1\}$. Then, matrix D° is maximum rank 3 and $X^\circ X^{\circ T} = D^\circ$.
3. [CH88] Let $D^\circ = U\Sigma U^T$ be the singular-value decomposition of D°, where U is an orthogonal matrix and Σ a diagonal matrix with the singular values of D° along the diagonal. If D° is a matrix of rank less than or equal to 3, the decomposition can be obtained with U being $(n-1) \times 3$ and Σ being 3×3, and $X^\circ = U\Sigma^{1/2}$ solves the equation $X^\circ X^{\circ T} = D^\circ$.

Algorithm 2: Computing Coordinates from Distances

Given an $n \times n$ distance matrix D,

1. Compute $D^\circ = \{(d_{i,n}^2 - d_{i,j}^2 + d_{j,n}^2)/2 : i, j = 1, \ldots, n-1\}$.
2. Decompose $D^\circ = U\Sigma U^T$ to obtain $X^\circ U[1 : n-1, 1 : 3]\Sigma^{1/2}[1 : 3, 1 : 3]$.
3. $X[1 : n-1, 1 : 3] = X^\circ[1 : n-1, 1 : 3]$, $X[n, 1 : 3] = [0, 0, 0]$.

Examples:

1. Given the distances among four atoms, D, determine the coordinates of the atoms, X, where

$$D = \begin{bmatrix} 0 & \sqrt{2} & \sqrt{2} & 1 \\ \sqrt{2} & 0 & \sqrt{2} & 1 \\ \sqrt{2} & \sqrt{2} & 0 & 1 \\ 1 & 1 & 1 & 0 \end{bmatrix}.$$

Following Algorithm 1,

$$D^\circ = \begin{bmatrix} 1 & 0 & 0 \\ 0 & 1 & 0 \\ 0 & 0 & 1 \end{bmatrix}.$$

Compute the singular value decomposition of D°. Obviously, $D^\circ = U\Sigma U^T$, with

$$U = \begin{bmatrix} 1 & 0 & 0 \\ 0 & 1 & 0 \\ 0 & 0 & 1 \end{bmatrix}, \qquad \acute{O} = \begin{bmatrix} 1 & 0 & 0 \\ 0 & 1 & 0 \\ 0 & 0 & 1 \end{bmatrix}.$$

Then,

$$X^\circ = \begin{bmatrix} 1 & 0 & 0 \\ 0 & 1 & 0 \\ 0 & 0 & 0 \end{bmatrix}$$

and

$$X = \begin{bmatrix} 1 & 0 & 0 \\ 0 & 1 & 0 \\ 0 & 0 & 1 \\ 0 & 0 & 0 \end{bmatrix}.$$

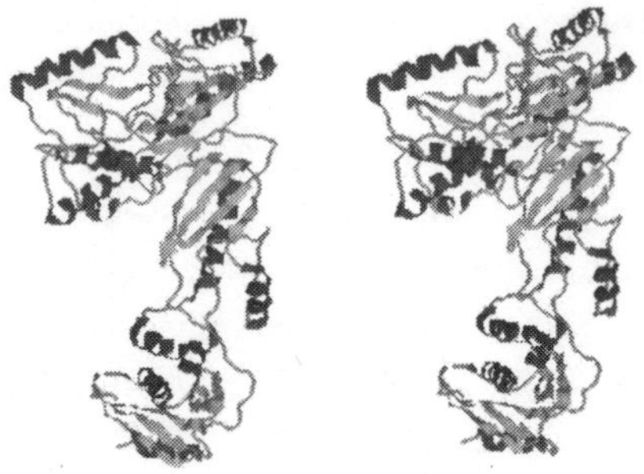

FIGURE 60.2 3D structures of protein 1HMV p66 subunit: The structure on the left was determined by x-ray crystallography, while on the right by solving a distance geometry problem given the distances for all the pairs of atoms. The RMSD for the two structures when compared on all the atoms is around 1.0e–04 Å. (Photo courtesy of Qunfeng Dong.)

2. Figure 60.2 shows two 3D structures of the p66 subunit of the HIV-1 retrotranscriptase (1HMV), one determined experimentally by x-ray crystallography [RGH95] and another computationally by solving a molecular distance geometry problem using the SVD method with the distance data generated from the known crystal structure. The RMSD (see description in section 60.3) for the two structures when compared on all the atoms is around 1.0e–04 Å, showing that the two structures are almost identical.

60.3 The Procrustes Problem for Protein Structure Comparison

The structural differences between two proteins can be measured by the differences in the coordinates of the atoms for all corresponding atom pairs. The comparison is often required for either structural validation or functional analysis. The calculation can be done by solving a special linear algebra problem called the Procrustes problem [GL89].

Definitions:

Let X and Y be two $n \times 3$ coordinate matrices for two lists of atoms in proteins A and B, respectively, where $\mathbf{x}_i = (x_{i,1}, x_{i,2}, x_{i,3})^T$ is the coordinate vector of the ith atom selected from protein A to be compared with $\mathbf{y}_i = (y_{i,1}, y_{i,2}, y_{i,3})^T$, the coordinate vector of the ith atom selected from protein B. Assume that X and Y have been translated so that their centers of geometry are located at the same position, say, at the origin. Then, the structural difference between the two proteins can be measured by using **the root-mean-square deviation** (RMSD) of the structures, $\mathrm{RMSD}(X, Y) = \min_Q \|X - YQ\|_F/\sqrt{n}$, where Q is a 3×3 rotation matrix and $QQ^T = I$, and $\|\cdot\|_F$ is the matrix Frobenius norm.

The RMSD is basically the smallest average coordinate errors of the structures for all possible rotations Q of structure Y to fit structure X. It is called **the Procrustes problem** for its analogy to the Greek story about cutting a person's legs to fit a fixed-sized iron bed. Note that X and Y may be the coordinate matrices for the same ($A = B$) or different ($A \neq B$) proteins and therefore, each pair of corresponding atoms do not have to be of the same type (when $A \neq B$). However, the number of atoms selected to compare must be the same from A and B (# rows of X = # rows of Y).

Facts:

1. Let A and B be two matrices. Suppose that A is similar to B, then $\mathrm{trace}(A) = \mathrm{trace}(B)$. In particular, $\mathrm{trace}(A) = \mathrm{trace}(V^T AV)$, for any orthogonal matrix V.

2. [GL89] Let $C = Y^T X$ and $C = U\Sigma V^T$ be the singular-value decomposition of C. Then, $Q = UV^T$ minimizes $||X - YQ||_F$.

Algorithm 2: Computing the RMSD of Two Protein Structures

1. Compute the geometric centers of X and Y:

$$\mathbf{x}_c[j] = \left(\sum_{i=1}^{n} X[i,j]\right)/n, \quad j = 1, 2, 3;$$

$$\mathbf{y}_c[j] = \left(\sum_{i=1}^{n} Y[i,j]\right)/n, \quad j = 1, 2, 3.$$

2. Translate X and Y to the origin:

$$X = X - \mathbf{1}_n \mathbf{x}_c^T, \quad Y = Y - \mathbf{1}_n \mathbf{y}_c^T, \quad \mathbf{1}_n = (1, \ldots, 1)^T \text{in} R^n.$$

3. Compute $C = Y^T X$ and $C = U\Sigma V^T$. Then,

$$Q = UV^T, \quad \text{RMSD}(X, Y) = ||X - YQ||_F/\sqrt{n}.$$

Examples:

1. Suppose that X and Y are given as the following.

$$X = \begin{bmatrix} -1 & -1 & -2 \\ -1 & -1 & 0 \\ -1 & 1 & -2 \\ -1 & 1 & 0 \\ 1 & -1 & -2 \\ 1 & -1 & 0 \\ 1 & 1 & -2 \\ 1 & 1 & 0 \end{bmatrix}$$

$$Y = \begin{bmatrix} 1 & 1 & 0 \\ 1 & -1 & 0 \\ -1 & 1 & 0 \\ -1 & -1 & 0 \\ 1 & 1 & 2 \\ 1 & -1 & 2 \\ -1 & 1 & 2 \\ -1 & -1 & 2 \end{bmatrix}.$$

Then $\mathbf{x}_c = (0, 0, -1)^T$ and $\mathbf{y}_c = (0, 0, 1)^T$. Following the Step 2 in Algorithm 2, X and $Y$are changed to

$$X = \begin{bmatrix} -1 & -1 & -1 \\ -1 & -1 & 1 \\ -1 & 1 & -1 \\ -1 & 1 & 1 \\ 1 & -1 & -1 \\ 1 & -1 & 1 \\ 1 & 1 & -1 \\ 1 & 1 & 1 \end{bmatrix}$$

$$Y = \begin{bmatrix} 1 & 1 & -1 \\ 1 & -1 & -1 \\ -1 & 1 & -1 \\ -1 & -1 & -1 \\ 1 & 1 & 1 \\ 1 & -1 & 1 \\ -1 & 1 & 1 \\ -1 & -1 & 1 \end{bmatrix}.$$

Let $C = Y^T X$. Then,

$$C = \begin{bmatrix} 0 & -8 & 0 \\ 0 & 0 & -8 \\ 8 & 0 & 0 \end{bmatrix}.$$

Compute the singular value decomposition of C to obtain $C = U \Sigma V^T$, with

$$U = \begin{bmatrix} 0 & 1 & 0 \\ 0 & 0 & -1 \\ -1 & 0 & 0 \end{bmatrix} \quad \text{and} \quad V = \begin{bmatrix} 0 & 1 & 0 \\ 0 & 0 & -1 \\ -1 & 0 & 0 \end{bmatrix}.$$

Then,

$$Q = UV^T = \begin{bmatrix} 0 & -1 & 0 \\ 0 & 0 & -1 \\ 1 & 0 & 0 \end{bmatrix}.$$

By calculating $\|X - YQ\|_F \,/\, \sqrt{n}$, we obtain RMSD $(X, Y) = 0$.

2. RMSD calculation has been widely used in structural computing. A straightforward application is for comparing and validating the structures obtained from different (x-ray crystallography, NMR, or homology modeling) sources for the same protein [Rho00]. Even from the same source, such as NMR, multiple structures are often obtained, and the average RSMD for the pairs of the multiple structures has been calculated as an indicator for the consistency and sometimes the flexibility of the structures [SNB03]. It has also been an important tool for structural classification, motif recognition, and structure prediction, where a large number of different proteins need to be aligned and compared [EJT00]. Figure 60.3 gives an example of using RMSD to compare NMR and x-ray crystal structures. Three structures of the second domain of the immunoglobulin-binding protein

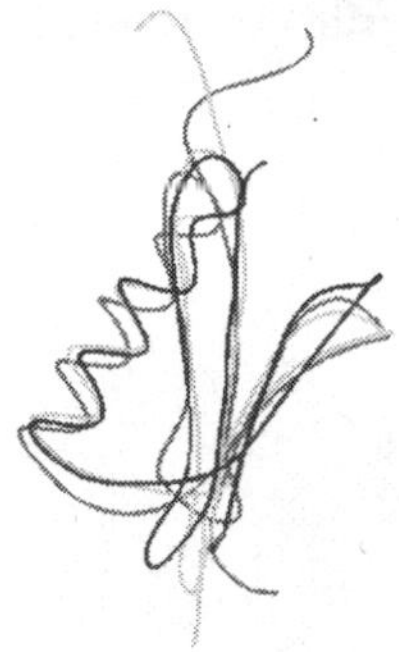

FIGURE 60.3 NMR and x-ray crystal structures of 2IGG: Two NMR structures of 2IGG are superposed to its x-ray crystal structure to find out which one is closer to the x-ray crystal structure (dark line). The RMSD values for the two NMR structures against the x-ray structure are 1.97 Å and 1.75 Å, respectively. (Photo courtesy of Feng Cui.)

(2IGG) [LDS92] are displayed in the figure. Two of them are NMR structures. They are compared using RMSD against the x-ray structure (dark line).

60.4 The Karle–Hauptman Matrix in X-Ray Crystallographic Computing

X-ray crystallography has been a major experimental tool for protein structure determination and is responsible for about 80% of 30,000 protein structures so far determined and deposited in the Protein Data Bank [BWF00]. The structure determination process involves crystallizing the protein, applying x-ray to the protein crystal to obtain x-ray diffractions, and using the diffraction data to deduce the electron density distribution of the crystal (Figure 60.4). Once the electron density distribution of the crystal is known, a 3D structure for the protein can be assigned [Dre94].

Definitions:

Define ρ: $R^3 \to R$ to be **the electron density distribution function** for a protein and F_H in complex space C to be the structure factor representing the diffraction spot specified by the integral triplet H [Dre94].

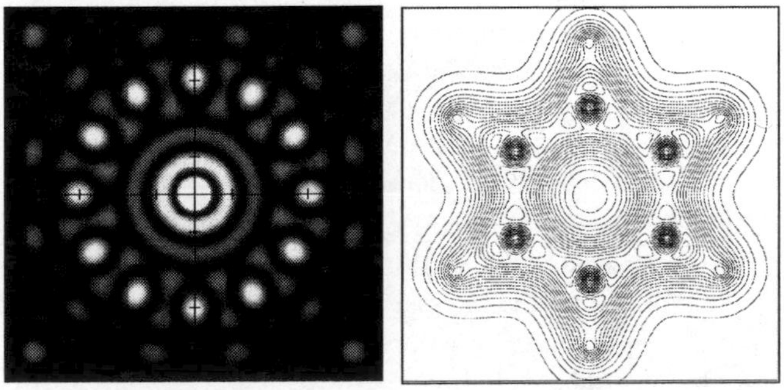

FIGURE 60.4 Example diffraction image and electron density map: The left one is the diffraction image of a 12-atom polygon generated by the program in [PNB01]. The right one is the electron density map of benzene generated by Stewart using program DENSITY in MOPAC [Ste02].

A Karle–Hauptman matrix for a set of structure factors $\{F_H : H = H_0, \ldots, H_{n-1}\}$ is defined as

$$K = \begin{bmatrix} F_{H_0} & F_{H_{n-1}} & \cdots & F_{H_1} \\ F_{H_1} & F_{H_0} & \cdots & F_{H_2} \\ \vdots & \vdots & \ddots & \vdots \\ F_{H_{n-1}} & F_{H_{n-2}} & \cdots & F_{H_0} \end{bmatrix} \quad [KH52].$$

Facts:

1. [Dre94] The electron density distribution function ρ can be expanded as a Fourier series with the structure factors F_H as the coefficients. In other words, F_H is a Fourier transform of ρ.

$$\rho(r) = \sum_{H \in Z^3} F_H \exp(-2\pi i H^T r),$$

$$F_H = \int_{R^3} \rho(r) \exp(2\pi i H^T r) dr.$$

2. [Bri84] If K is a Karle–Hauptman matrix, then the inverse of K is also a Karle–Hauptman matrix and can be formed directly as

$$K^{-1} = \begin{bmatrix} E_{H_0} & E_{H_{n-1}} & \cdots & E_{H_1} \\ E_{H_1} & E_{H_0} & \cdots & E_{H_2} \\ \vdots & \vdots & \ddots & \vdots \\ E_{H_{n-1}} & E_{H_{n-2}} & \cdots & E_{H_0} \end{bmatrix},$$

where

$$E_{H_j} = \int_{R^3} \rho^{-1}(r) \exp(2\pi i H_j^T r) dr, \quad j = 0, 1, \ldots, n-1.$$

3. [GL89] Suppose that we have a linear system $K\mathbf{x} = \boldsymbol{h}$, where K is an $n \times n$ Karle–Hauptman matrix and $\boldsymbol{h}$ an n-dimensional complex vector. If a conventional method, such as Gaussian Elimination, is used, the solution of the system usually takes $O(n^3)$ floating-point operations, which is expensive if n is larger than 1000 and if the solution is also required multiple times.
4. [Loa92], [WPT01] Since each element in the inverse matrix can be obtained by doing a Fourier transform for the inverse of ρ and only n distinct elements in the first column are required to form the whole matrix, the calculations can be done in $O(n\log n)$ floating-point operations by using the Fast Fourier Transform.
5. [TML97], [WPT01] The matrix K^{-1} as well as K has only n distinct elements listed repeatedly in the columns of the matrix with each column having the elements in the previous column circulated by one element from top to the bottom and then bottom to the top. This type of matrix is called the circulant matrix. According to the discrete convolution theory, if h is the Fourier transform of t, then $K^{-1}h$ can be computed by doing a Fourier transform for $\rho^{-1} \cdot t$, where t can be obtained through an inverse Fourier transform for h and the product $\cdot$ is applied component-wise. Therefore, the whole computation for the solution of $K\mathbf{x} = h$ can be done with at most $O(n\log n)$ floating-point operations.

Examples:

1. Let $\rho = [0.1250, 0.1250, 0.5000, 0.1250, 0.1250]$ be an electron density distribution. Then $\rho^{-1} = [8, 8, 2, 8, 8]$, and the Fourier transforms for ρ and ρ^{-1} are equal to

$$F = [0.2000, -0.0607 + 0.0441i, 0.0232\text{-}0.0713i, 0.0232 + 0.0713i, \text{-}0.0607\text{-}0.0441i] \text{ and}$$

$$F_{\text{inv}} = [6.8000, 0.9708 - 0.7053i, -0.3708 + 1.1413i, -0.3708 - 1.1413i, 0.9708 + 0.7053i],$$

respectively. And it is not hard to verify that the inverse of the Karle–Hauptman matrix $K(F)$ formed by using F is equal to the Karle–Hauptman matrix K_{inv} (F_{inv}) formed by using F_{inv}.

2. The Karle–Hauptman matrix is an important matrix in x-ray crystallography computing, named after two Nobel Laureates, chemist Jerold Karle and mathematician Herbert Hauptman, who received the Nobel Prize in chemistry in 1985 for their work on the phase problem in x-ray crystallography [HK53]. The Karle–Hauptman matrix is frequently used for computing the covariance of the structure factors [KH52] or the electron density distribution that maximizes the entropy of a crystal system [WPT01].

60.5 Calculation of Fast and Slow Modes of Protein Motions

In a reduced model for protein, a residue is represented by a point, in many cases, the position of the backbone atom C_α or the sidechain atom C_β in the residue, and a protein is considered as a sequence of such points connected with strings [HL92]. If the reduced model of a protein is known, a so-called contact map can be constructed to show how the residues in the protein interact with each other. The map is represented by a matrix with its i, j-entry equal to -1 if residues i and j are within, say 7Å distance, and 0 otherwise. The contact matrix can be used to compare different proteins. Similar contact patterns often imply structural or functional similarities between proteins [MJ85]. When a protein reaches its equilibrium state, the residues in contact can be considered as a set of masses connected with springs. A simple energy function can also be defined for the protein using the contact matrix.

Definitions:

Suppose that a protein has n residues with n coordinate vectors $\mathbf{x}_1, \ldots, \mathbf{x}_n$. **A contact matrix** Γ for the protein in its equilibrium state can be defined such that

$$\tilde{A}_{i,j} = \begin{cases} -1, & ||x_i - x_j|| \leq 7 \text{ Å} \\ 0, & \text{otherwise} \end{cases} \qquad i \neq j = 1, \ldots, n \qquad \text{[HBE97]}.$$

$$\tilde{A}_{i,i} = -\sum_{j=1}^{n} \tilde{A}_{i,j} \qquad i = 1, \ldots, n$$

A potential energy function E for a protein at its equilibrium state can be defined such that for any vector $\Delta\mathbf{x} = (\Delta\mathbf{x}_1, \ldots, \Delta\mathbf{x}_n)^T$ of the displacements of the residues from their equilibrium positions,

$$E(\Delta\mathbf{x}) = \frac{1}{2}\, k\, \Delta\mathbf{x}^T \tilde{A}\, \Delta\mathbf{x},$$

where k is a spring constant.

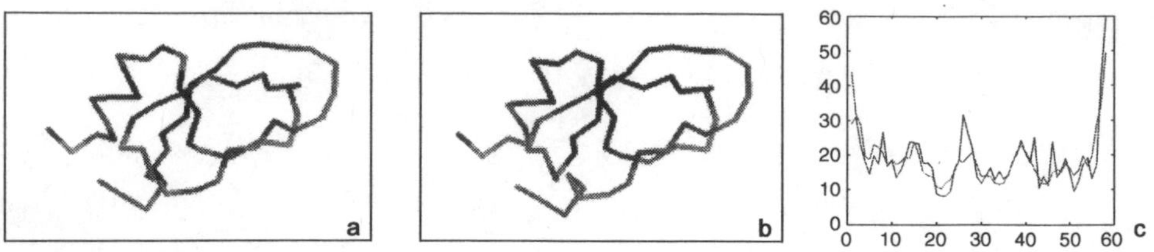

FIGURE 60.5 Mean-square fluctuations: The fluctuations for protein 2KNT based on the mean-square fluctuations calculated with GNM (a) and the B-factors determined by x-ray crystallography (b). The two sets of values show a high correlation (0.82) (c). (Photos courtesy of Di Wu.)

Facts:

1. [HBE97] Given a potential energy function E, the probability for a protein to have a displacement Δx at temperature T should be subject to the Boltzmann distribution,

$$p_T(\Delta \mathbf{x}) = \frac{1}{Z} \exp(-E(\Delta \mathbf{x})/k_B T) = \frac{1}{Z} \exp(-k\, \Delta \mathbf{x}^T \tilde{\mathbf{A}}\, \Delta \mathbf{x}/2k_B T),$$

where Z is the normalization factor and k_B the Boltzmann constant.

2. [HBE97] Let the singular-value decomposition of Γ be given as $\Gamma = U \Lambda U^T$. Then, the mean-square residue fluctuations of a protein at its equilibrium state can be estimated as

$$< \Delta \mathbf{x}_i, \Delta \mathbf{x}_i > \equiv \frac{1}{Z} \int_{R^{3n}} \Delta \mathbf{x}_i^T \Delta \mathbf{x}_i \exp(-E(\Delta \mathbf{x})/k_B T) d\Delta \mathbf{x} = \sum_{j=1}^{n} k_B T\, U_{i,j} \ddot{E}_{j,j}^{-1} U_{i,j}/k.$$

Examples:

1. The energy model defined above for a protein at its equilibrium state is called the Gaussian Network Model. The model can be used to find how the residues in the protein move around their equilibrium positions dynamically and in particular, to estimate the so-called mean-square fluctuations for the residues $<\Delta \mathbf{x}_i, \Delta \mathbf{x}_i>$, $i = 1, \ldots, n$. If the mean-square fluctuation is large, the residue is called hot, and otherwise, is cold, which often correlates with the experimentally detected average atomic fluctuation such as the B-factor in x-ray crystallography [Dre94] and the order parameter in NMR [Gun95]. In fact, the Gaussian Network Model is equivalent to the Normal Mode Analysis for predicting the mean-squares residue fluctuations of a protein, with the energy function defined for the residues instead of the atoms.
2. Figure 60.5 shows the mean-square fluctuations calculated using the Gaussian Network Model for the protein 2KNT and the comparison with the B-factors of the structure determined by x-ray crystallography. The two sets of values appear to be highly correlated. Based on the facts stated above, the calculation of the mean-squares fluctuations requires only a singular-value decomposition of the contact matrix for the protein.

60.6 Flux Balancing Equation in Metabolic Network Simulation

A metabolic system is maintained through constant reactions or interactions among a large number of biological and chemical compounds called metabolites [Fel97]. The reaction network describes the structure of a metabolic system and is key to the study of the metabolic function of the system. Figure 60.6 shows the reaction network for an example metabolic system of five metabolites given in [SLP00].

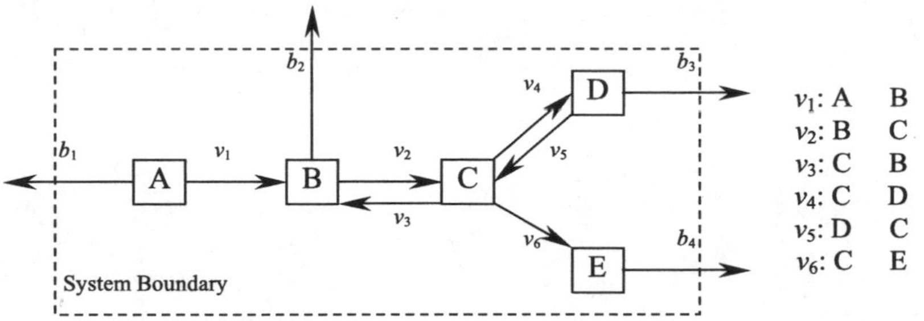

FIGURE 60.6 Example metabolic networks: A, B, C, D, E are metabolites; v_j, $j = 1, \ldots, 6$ are internal fluxes; b_j, $j = 1, \ldots, 4$ are external fluxes. Each flux v_j corresponds to an internal reaction.

Definitions:

Each metabolite has a concentration, which changes constantly. **The rate of the change** is proportional to the amount of the metabolite consumed or produced in all the reactions.

Let C_i be **the concentration of metabolite** i. Let v_j be **the chemical flux** in reaction j, i.e., the amount of metabolites produced in reaction j per mole. Then,

$$\frac{dC_i}{dt} = \sum_{j=1}^{n} s_{i,j}\, v_j,$$

where $s_{i,j}$ is the **stoichiometric coefficient of metabolite** i in reaction j, and $s_{i,j} = \pm k$, if $\pm k$ moles of metabolite i are produced (or consumed) in reaction j.

Let $C = (C_1, \ldots, C_m)^T$ be a vector of concentrations of m metabolites, and $v = (v_1, \ldots, v_n)^T$ a vector of fluxes of n reactions. Then, the equations can be written in a compact form,

$$\frac{dC}{dt} = S\, v,$$

where $S = \{s_{i,j} : i = 1, \ldots, m,\ j = 1, \ldots, n\}$ is called the **stoichiometry matrix**, and the equations are called **the reaction equations** [HS96].

The fluxes are functions of the concentrations and some other system parameters. Therefore, the above reaction equations are nonlinear equations of C. However, when the system reaches its equilibrium, $dC/dt = Sv = 0$, and the vector **v** becomes constant and is called a solution of **the steady-state flux equation** $Sv = 0$ [HS96].

The steady-state fluxes are important quantities for characterizing metabolic networks. They can be obtained by solving the steady-state flux equation $Sv = 0$. However, since the number of reactions is usually larger than the number of metabolites, the solution to the equation is not unique. Also, because the internal fluxes are nonnegative, the solution set forms a convex cone, called the **steady-state flux cone.**

Usually, a convex cone can be defined in terms of a set of extreme rays such that any vector in the cone can be expressed as a nonnegative linear combination of the extreme rays,

$$\text{cone}\,(S) = \{v = \sum_{i=1}^{l} w_i\, p_i, \quad w_i \geq 0\},$$

where $p = \{p_1, \ldots, p_l\}$ is a set of extreme rays. An **extreme ray** is a vector that cannot be expressed as a nonnegative linear combination of any other vectors in the cone.

A set of vectors is said to be systematically independent if none of them can be expressed as a nonnegative linear combination of others. Since the extreme rays can be used to express all the vectors in a convex cone, they are also called the generating vectors of the cone. For metabolic networks, they are called the **extreme pathways** [PPP02].

Facts:

1. [PPP02] A convex flux cone has a set of systematically independent generating vectors or extreme pathways. They are unique up to positive scalar multiplications. If the extreme pathways of the convex flux cone of a metabolic network are found, all the solutions for the steady-state flux equation can be generated by using the extreme pathways. In other words, the extreme pathways provide a unique description for the solution space of the steady-state flux equation, and can be used to characterize the whole steady-state capacity of the system.
2. [PPP02] Let P be a matrix with each column corresponding to an extreme pathway of a given metabolic network. Let Q be the binary form of P such that $Q_{i,j} = 1$ if $P_{i,j} \neq 0$ and $Q_{i,j} = 0$ otherwise. Then, the diagonal elements of $Q^T Q$ are equal to the lengths of the extreme pathways, while the diagonal elements of QQ^T show the numbers of extreme pathways the reactions participate in.

Examples:

1. Consider the example network in Figure 60.6 and let S be the stoichiometric matrix with 10 columns for the internal $(v_1, \ldots, v_6)$ as well as external $(b_1, \ldots, b_4)$ fluxes,

$$
\begin{array}{c}
\\
S =
\end{array}
\begin{array}{c}
\begin{array}{cccccccccc} v_1 & v_2 & v_3 & v_4 & v_5 & v_6 & b_1 & b_2 & b_3 & b_4 \end{array}\\
\begin{bmatrix}
-1 & 0 & 0 & 0 & 0 & 0 & -1 & 0 & 0 & 0\\
1 & -1 & 1 & 0 & 0 & 0 & 0 & -1 & 0 & 0\\
0 & 1 & -1 & -1 & 1 & -1 & 0 & 0 & 0 & 0\\
0 & 0 & 0 & 1 & -1 & 0 & 0 & 0 & -1 & 0\\
0 & 0 & 0 & 0 & 0 & 1 & 0 & 0 & 0 & -1
\end{bmatrix}
\end{array}
\begin{array}{l}
\leftarrow A\\ \leftarrow B\\ \leftarrow C.\\ \leftarrow D\\ \leftarrow E
\end{array}
$$

Then, by using an appropriate algorithm (such as the one given in [SLP00]), a matrix of 7 extreme pathways of the system can be found as follows:

$$
P^T =
\begin{array}{c}
\begin{array}{cccccccccc} v_1 & v_2 & v_3 & v_4 & v_5 & v_6 & b_1 & b_2 & b_3 & b_4 \end{array}\\
\begin{bmatrix}
1 & 0 & 0 & 0 & 0 & 0 & -1 & 1 & 0 & 0\\
0 & 1 & 1 & 0 & 0 & 0 & 0 & 0 & 0 & 0\\
0 & 1 & 0 & 1 & 0 & 0 & 0 & -1 & 1 & 0\\
0 & 1 & 0 & 0 & 0 & 1 & 0 & -1 & 0 & 1\\
0 & 0 & 1 & 0 & 1 & 0 & 0 & 1 & -1 & 0\\
0 & 0 & 0 & 1 & 1 & 0 & 0 & 0 & 0 & 0\\
0 & 0 & 0 & 0 & 1 & 1 & 0 & 0 & -1 & 1
\end{bmatrix}
\end{array}
\begin{array}{l}
\leftarrow p_1\\ \leftarrow p_2\\ \leftarrow p_3\\ \leftarrow p_4,\\ \leftarrow p_5\\ \leftarrow p_6\\ \leftarrow p_7
\end{array}
$$

where row i corresponds to extreme pathway $p_i, i = 1, \ldots, 7$.

2. By forming the binary form Q for P and computing

$$
Q^T Q =
\begin{bmatrix}
3 & 0 & 1 & 1 & 1 & 0 & 0\\
0 & 2 & 1 & 1 & 1 & 0 & 0\\
1 & 1 & 4 & 2 & 2 & 1 & 1\\
1 & 1 & 2 & 4 & 1 & 0 & 2\\
1 & 1 & 2 & 1 & 4 & 1 & 2\\
0 & 0 & 1 & 0 & 1 & 2 & 1\\
0 & 0 & 1 & 2 & 2 & 1 & 4
\end{bmatrix}
$$

and

$$QQ^T = \begin{bmatrix} 1 & 0 & 0 & 0 & 0 & 0 & 1 & 1 & 0 & 0 \\ 0 & 3 & 1 & 1 & 0 & 1 & 0 & 2 & 1 & 1 \\ 0 & 1 & 2 & 0 & 1 & 0 & 0 & 1 & 1 & 0 \\ 0 & 1 & 0 & 2 & 1 & 0 & 0 & 1 & 1 & 0 \\ 0 & 0 & 1 & 1 & 3 & 1 & 0 & 1 & 2 & 1 \\ 0 & 1 & 0 & 0 & 1 & 2 & 0 & 1 & 1 & 2 \\ 1 & 0 & 0 & 0 & 0 & 0 & 1 & 1 & 0 & 0 \\ 1 & 2 & 1 & 1 & 1 & 1 & 1 & 4 & 2 & 1 \\ 0 & 1 & 1 & 1 & 2 & 1 & 0 & 2 & 3 & 1 \\ 0 & 1 & 0 & 0 & 1 & 2 & 0 & 1 & 1 & 2 \end{bmatrix},$$

we obtain, from the diagonal elements of the matrices, the lengths of the pathways, p_1: 3, p_2: 2, p_3: 4, p_4: 4, p_5: 4, p_6: 2, p_7: 4, and the participations of the reactions in the extreme pathways, v_1: 1, v_2: 3, v_3: 2, v_4: 2, v_5: 3, v_6: 2, b_1: 1, b_2: 4, b_3: 3, b_4: 2. The off-diagonal elements of $Q^T Q$ show the numbers of common reactions in different extreme pathways. For example, $(Q^T Q)_{3,4} = 2$ means that p_3 and p_4 share two common reactions, and $(Q^T Q)_{3,5} = 1$ means that p_3 and p_5 have one common reaction. The off-diagonal elements of QQ^T show the numbers of extreme pathways in which different reactions participate. For example, $(QQ^T)_{2,3} = 1$ means that v_2 and v_3 participate in one extreme pathway together, and $(QQ^T)_{2,8} = 2$ means that v_2 and b_2 participate in two extreme pathways together.

60.7 Conclusion

In this chapter, we have reviewed several subjects in biomolecular modeling, where linear algebra has played a central role in related computations. The review has focused on simple showcases and demonstrated important applications of linear algebra in biomolecular modeling. The subjects discussed are of great research interest in computational biology and are related directly to the solutions of many critical but challenging computational problems in biology yet to be solved, including the general distance geometry problem in NMR, the phase problem in x-ray crystallography, the structural comparison problem in comparative modeling, molecular dynamics simulation, and biosystems modeling and optimization, which we have not elaborated in detail here, but the interested readers can further explore.

Linear algebra has also been used to support many basic algebraic calculations required for solving other types of mathematical problems in biomolecular modeling, such as the optimization problems in potential energy minimization [WS99], the initial value problem in molecular dynamics simulation [BKP88], and the boundary value problem in simulation of protein conformational transformation [EMO99]. They are usually straightforward or routine linear algebra calculations, so we have not covered them in this chapter, but they should be considered as equally important as the applications we have discussed.

Acknowledgments

I would like to thank my students, Feng Cui, Ajith Gunaratne, Kriti Mukhopadhyay, Rahul Ravindrudu, Andrew Severin, Matthew Studham, Peter Vedell, Di Wu, Eun-Mee YoonAnn, and Rich Wen Zhou, and my colleagues, Dr. Qunfeng Dong and Dr. Wonbin Young, who have read the paper carefully, helped produce some of the figures, and provided valuable comments and suggestions. The subjects discussed in the chapter have actually been the frequent topics of our research meetings. I would also like to thank the

Mathematical Biosciences Institute, Ohio State University, for providing support during my visit to the Institute in Spring 2005 where I completed the writing of this chapter.

References

[BWF00] H.M. Berman, J. Westbrook, Z. Feng, G. Gilliland, T.N. Bhat, H. Weissig, L.N. Shindyalov, and P.E. Bourne, The protein data bank, *Nuc. Acid. Res.* 28, 235–242, 2000.

[Blu53] L.M. Blumenthal, *Theory and Applications of Distance Geometry*, Clarendon Press, Oxford, U.K., 1953.

[Boy99] R. Boyer, *Concepts in Biochemistry*, Brooks/Cole Publishing, Belmont, CA, 1999.

[Bri84] G. Bricogne, Maximum entropy and the foundations of direct methods, *Acta Cryst.* A40, 410–445, 1984.

[BKP88] C.L. Brooks, III, M. Karplus, and B.M. Pettitt, *Proteins: A Theoretical Perspective of Dynamics, Structure, and Thermodynamics*, John Wiley & Sons, New York, 1988.

[CH88] G.M. Crippen and T.F. Havel, *Distance Geometry and Molecular Conformation*, John Wiley & Sons, New York, 1988.

[Cre93] T.E. Creighton, *Proteins: Structures and Molecular Properties*, Freeman and Company, New York, 1993.

[CJW05] F. Cui, R. Jernigan, and Z. Wu, Refinement of NMR-determined protein structures with database-derived distance constraints, *J. Bioinfo. Comp. Biol.* 3, 1315–1329, 2005.

[Dre94] J. Drenth, *Principles of Protein X-Ray Crystallography*, Springer-Verlag, Heidelberg, 1994.

[EJT00] I. Eidhammer, I. Jonassen, and W.R. Taylor, Structure comparison and structure patterns, *J. Comp. Biol.* 7, 685–716, 2000.

[EMO99] R. Elber, J. Meller, and R. Olender, Stochastic path approach to compute atomically detailed trajectories: application to the folding of C peptide, *J. Phys. Chem.* 103, 899–911, 1999.

[Fel97] D.A. Fell, Systems properties of metabolic networks, in *Proc. International Conference on Complex Systems*, 1997, 21–26.

[GL89] G. Golub and C. van Loan, *Matrix Computation*, Johns Hopkins University Press, Baltimore, MD, 1989.

[Gun95] H. Gunther, *NMR Spectroscopy: Basic Principles, Concepts, and Applications in Chemistry*, John Wiley & Sons, New York, 1995.

[HBE97] T. Haliloglu, I. Bahar, and B. Erman, Gaussian dynamics of folded proteins, *Phys. Rev. Lett.* 79, 3090–3093, 1997.

[HK53] H. Hauptman and J. Karle, *The Solution of the Phase Problem I: The Centrosymmetric Crystal*, Polycrystal Book Service, Handsville, AL, 1953.

[HS96] R. Heinrich and S. Schuster, *The Regulation of Cellular Systems*, Chapman and Hall, New York, 1996.

[HL92] D.A. Hinds and M. Levitt, A lattice model for protein structure prediction at low resolution, *Proc. Natl. Acad. Sci. U.S.A.*, 89, 2536–2540, 1992.

[KH52] J. Karle and H. Hauptman, The phases and magnitudes of the structure factors, *Acta Cryst.* 3, 181–187, 1952.

[Kit99] K. Kitano, Systems biology: towards system-level understanding of biological systems, in *Proc. 4th Hamamatsu International Symposium on Biology: Computational Biology*, Hamamatsu, Japan, 1999.

[LDS92] L.Y. Lian, J.P. Derrick, M.J. Sutcliffe, J.C. Yang, and G.C.K. Roberts, Determination of the solution structures of domain II and III of protein G from *Streptococcus* by ^{1}H nuclear magnetic resonance, *J. Mol. Biol.* 228, 1219–1234, 1992.

[Loa92] C. van Loan, *Computational Frameworks for the Fast Fourier Transform*, SIAM, Philadelphia, 1992.

[MJ85] S. Miyazawa and R.L. Jernigan, Estimation of effective inter-residue contact energies from protein crystal structures: quasi-chemical approximation, *Macromolecules* 18, 534–552, 1985.

[PPP02] J.A. Papin, N.D. Price, and B. Palsson, Extreme pathway lengths and reaction participation in genome-scale metabolic networks, *Genom. Res.* 12, 1889–1900, 2002.

[PNB01] T.H. Proffen, R.B. Neder, and S.J.L. Billinge, Teaching diffraction using computer simulations over the Internet, *J. Appl. Crys.*, 34, 767–770, 2001. (http://www.med.wayne.edu/biochem/~xray/education.html)

[Rho00] G. Rhodes, *Judging the Quality of Macromolecular Models*, Department of Chemistry, University of Southern Maine, 2000. (http://www.usm.maine.edu/~rhodes/ModQual/)

[RGH95] D.W. Rodgers, S.J. Gamblin, B.A. Harris, S. Ray, J.S. Culp, B. Hellmig, D.J. Woolf, C. Debouck, and S.D. Harrison, The structure of unliganded reverse transcriptase from the human immunodeficiency virus type 1, *Proc. Natl. Acad. Sci. U.S.A.* 92, 1222–1226, 1995.

[Sax79] J.B. Saxe, Embedability of weighted graphs in K-space is strongly NP-hard, in *Proc. 17th Allerton Conference in Communications, Control and Computing*, University of Illinois at Urbana-Champaign, 1979, 480–489.

[SLP00] C.H. Schilling, D. Letscher, and B. Palsson, Theory for the systemic definition of metabolic pathways and their use in interpreting metabolic function from a pathway-oriented perspective, *J. Theor. Biol.* 203, 229–248, 2000.

[Sch03] T. Schlick, *Molecular Modeling and Simulation: An Interdisciplinary Guide*, Springer, Heidelberg, 2003.

[SNB03] C.A.E.M. Spronk, S.B. Natuurs, A.M.J.J. Bonvin, E. Krieger, G.W. Vuister, and G. Vriend, The precision of NMR structure ensembles revisited, *J. Biomolecular NMR* 25, 225–234, 2003.

[Ste02] J.J.P. Stewart, *MOPAC Manual*, The Cache Group, Fujitsu America Inc., Sunnyvale, CA, 2002. (http://www.cachesoftware.com/mopac/Mopac2002manual/node383.html)

[TML97] R. Tolimieri, A. Myoung, and C. Lu, *Algorithms for Discrete Fourier Transform and Convolution*, Springer, Heidelberg, 1997.

[WS99] D.J. Wales and H.A. Scheraga, Global optimization of clusters, crystals, and biomolecules, *Science* 285, 1368–1372, 1999.

[WPT01] Z. Wu, G. Phillips, R. Tapia, and Y. Zhang, A fast Newton algorithm for entropy maximization in phase determination, *SIAM Rev.* 43, 623–642, 2001.

Applications to Computer Science

61

Coding Theory

Joachim Rosenthal
University of Zürich

Paul Weiner
Saint Mary's University of Minnesota

Sometimes errors occur when data is transmitted over a channel. A binary 0 may be corrupted to a 1, or vice versa. Error control coding adds redundancy to a transmitted message allowing for detection and correction of errors. For example, if 0 stands for "no" and 1 stands for "yes", a single bit error will completely change a message. If we repeat the message seven times, so 0000000 stands for "no" and 1111111 stands for "yes", then it would require at least 4 bit errors to change a message, assuming that a message is decoded to either 0000000 or 1111111 according to which bit is in the majority in the received string. Thus, this simple code can be used to correct up to 3 errors. This code also detects up to 6 errors, in the sense that if between 1 and 6 errors occur in transmission, the received string will not be a codeword, and the receiver will recognize that one or more errors have occurred. The extra bits provide *error protection*.

Let $\mathbb{F} = \mathbb{F}_q$ be the finite field with q elements. It is customary in the coding literature to write message blocks and vectors of the vector spaces $\mathbb{F}^k$ and $\mathbb{F}^n$ as row vectors, and we shall follow that practice in this chapter.

61.1 Basic Concepts

Definitions:

A **block code of length** n **over** $\mathbb{F}$ is a subset C of the vector space $\mathbb{F}^n$. Elements of C are **codewords**. Elements of the finite field $\mathbb{F}$ form the **alphabet**.

Message blocks of length k are strings of k symbols from the alphabet. Message blocks may also be identified with vectors in the vector space $\mathbb{F}^k$.

An **encoder** is a one-to-one mapping

$$\varphi : \mathbb{F}^k \longrightarrow \mathbb{F}^n$$

whose image is contained in the code C.

The process of applying φ to a k-digit message block to obtain a codeword is **encoding**.

A **noisy channel** is a channel in which transmission errors may occur.

A codeword $\mathbf{v} \in C \subseteq \mathbb{F}^n$ that has been transmitted may be received as $\mathbf{y} = \mathbf{v} + \mathbf{e}$, where $\mathbf{e} \in \mathbb{F}^n$ is an **error vector**.

Given a received word $\mathbf{y} = \mathbf{v} + \mathbf{e} \in \mathbb{F}^n$, the receiver estimates the corresponding codeword, thus, obtaining $\tilde{\mathbf{v}} \in C$. This estimation process is called **decoding**. If $\tilde{\mathbf{v}} = \mathbf{v}$, then correct decoding has occurred.

Decoding may be thought of as a mapping $\psi : \mathbb{F}^n \longrightarrow C$. The mapping ψ is a **decoder**. Generally a decoder ψ should have the property that if the received word is in fact a codeword, it is decoded to itself.

The **word error rate** for a particular decoder is the probability P_{err} that the decoder outputs a wrong codeword.

The process of encoding, transmission, and decoding is summarized by

$$\begin{array}{ccccccc} \mathbb{F}^k & \xrightarrow{\varphi} & \mathbb{F}^n & \xrightarrow{\text{n.t.}} & \mathbb{F}^n & \xrightarrow{\psi} & C \\ \mathbf{u} & \longmapsto & \mathbf{v} & \longmapsto & \mathbf{y} & \longmapsto & \tilde{\mathbf{v}}, \end{array}$$

where "n.t." stands for "noisy transmission."

For $\mathbf{v}, \mathbf{w} \in \mathbb{F}^n$ the **Hamming distance**, or simply the **distance**, between $\mathbf{v}$ and $\mathbf{w}$ is $\mathrm{d}(\mathbf{v}, \mathbf{w}) = |\{i \mid v_i \neq w_i\}|$.

The **distance of a code** $C \subseteq \mathbb{F}^n$ is defined by $\mathrm{d}(C) = \min\{\mathrm{d}(\mathbf{v}, \mathbf{w}) \mid \mathbf{v}, \mathbf{w} \in C \text{ and } \mathbf{v} \neq \mathbf{w}\}$.

The **Hamming weight** or **weight** of $\mathbf{v}$ is defined to be $\mathrm{wt}(\mathbf{v}) = \mathrm{d}(\mathbf{v}, \mathbf{0})$.

The **closed ball of radius** t **about the vector** $\mathbf{v} \in \mathbb{F}^n$ is the set

$$B(\mathbf{v}, t) = \{\mathbf{y} \in \mathbb{F}^n \mid \mathrm{d}(\mathbf{v}, \mathbf{y}) \leq t\}.$$

Nearest-neighbor decoding refers to decoding a received word $\mathbf{y}$ to a nearest codeword (with respect to Hamming distance).

A code $C \subseteq \mathbb{F}^n$ is called a **perfect code** if there exists a nonnegative integer t such that

1. $\mathbb{F}^n = \bigcup_{\mathbf{v} \in C} B(\mathbf{v}, t)$, and
2. For distinct $\mathbf{v}, \mathbf{w} \in C$, $B(\mathbf{v}, t) \cap B(\mathbf{w}, t) = \phi$.

That is, $\mathbb{F}^n$ is the disjoint union of closed t-balls about the codewords.

Facts:

1. The minimum number of symbol errors it would take for $\mathbf{v}$ to be changed to $\mathbf{w}$ is $\mathrm{d}(\mathbf{v}, \mathbf{w})$.
2. A nearest-neighbor decoder need not be unique as it is possible that some received words $\mathbf{y}$ might have two or more codewords at an equal minimum distance. The decoder may decode $\mathbf{y}$ to any of these nearest codewords.
3. [Hil86, p. 5] The Hamming distance is a metric on $\mathbb{F}^n$.
4. [Hil86, p. 7] If the distance of a code C is d, then C can detect up to $d - 1$ errors. It can correct up to $\left\lfloor \frac{d-1}{2} \right\rfloor$ errors.

Examples:

1. The binary message block 1011 can be identified with the vector $(1, 0, 1, 1) \in \mathbb{F}_2^4$.
2. The distance between the binary strings 0011 and 1110 is 3 since these two bit strings differ in three places.
3. We may picture nearest-neighbor decoding by taking a closed ball of expanding radius about the received word $\mathbf{y}$. As soon as the ball includes a codeword, that codeword is a nearest-neighbor to which we may decode $\mathbf{y}$.

61.2 Linear Block Codes

Definitions:

A **linear block code of length** n **and dimension** k over $\mathbb{F} = \mathbb{F}_q$ is a k-dimensional *linear* subspace, $\mathcal{C}$, of the vector space $\mathbb{F}^n$. One says $\mathcal{C}$ is an $[n, k]$-code over $\mathbb{F}$. If one wants to include the distance as a parameter, one says that $\mathcal{C}$ is an $[n, k, d]$-code over $\mathbb{F}$. A **binary linear block code (BLBC)** is a linear block code over the binary field $\mathbb{F}_2 = \{0, 1\}$.

Two linear codes over $\mathbb{F}_q$ are **equivalent** if one can be obtained from the other by steps of the following types: (i) permutation of the coordinates of the codewords, and (ii) multiplication of the field elements in one coordinate of the code by a nonzero scalar in $\mathbb{F}_q$.

The **rate** of a linear $[n, k]$-code is $\frac{k}{n}$.

Let $\mathcal{C}$ be a linear $[n, k]$-code over $\mathbb{F}$. A **generator matrix for** $\mathcal{C}$ is a matrix $G \in \mathbb{F}^{k \times n}$ whose k rows form a basis for $\mathcal{C}$. A generator matrix for $\mathcal{C}$ is also called an **encoder for** $\mathcal{C}$.

A generator matrix G for an $[n, k]$-code is a **systematic encoder** if G can be partitioned, perhaps after a column permutation, as $G = [I_k \mid A]$, where $A \in \mathbb{F}^{k \times (n-k)}$. Then $\mathbf{u} \in \mathbb{F}^k$ is encoded as $\mathbf{u}G = [\mathbf{u} \mid \mathbf{u}A]$ so that the first k symbols of the codeword $\mathbf{u}G$ are the message word $\mathbf{u}$, and the last $n - k$ symbols form the **parity check** symbols.

Let $\mathcal{C}$ be a linear $[n, k]$-code over $\mathbb{F}$. A **parity check matrix for** $\mathcal{C}$ is a matrix $H \in \mathbb{F}^{(n-k) \times n}$ such that $\mathcal{C} = \{\mathbf{v} \in \mathbb{F}^n \mid \mathbf{v}H^T = 0 \in \mathbb{F}^{(n-k)}\}$.

Let $\mathcal{C}$ be a linear $[n, k]$-code over $\mathbb{F}$ with generator matrix $G \in \mathbb{F}^{k \times n}$ and parity check matrix $H \in \mathbb{F}^{(n-k) \times n}$. Then H itself is the generator matrix of another code, denoted $\mathcal{C}^\perp$. $\mathcal{C}^\perp$ is the **dual code** of $\mathcal{C}$.

A linear block code, $\mathcal{C}$, is **self-dual** if $\mathcal{C} = \mathcal{C}^\perp$.

Let $\mathcal{C}$ be an $[n, k]$-code with parity check matrix H. For any vector $\mathbf{y} \in \mathbb{F}^n$, the **syndrome of y** is the vector $\mathbf{y}H^T \in \mathbb{F}^{(n-k)}$.

Facts:

1. [Hil86, p. 47] There are q^k codewords in an $[n, k]$-code over $\mathbb{F} = \mathbb{F}_q$.
2. The rate, $\frac{k}{n}$, of an $[n, k]$-code tells what fraction of transmitted symbols carry information. The complementary fraction, $\frac{n-k}{n}$, tells how much of the transmission is redundant (for error protection).
3. If G is a generator matrix for the linear $[n, k]$-code, $\mathcal{C}$, then $\mathcal{C} = \{\mathbf{u}G \mid \mathbf{u} \in \mathbb{F}^k\}$. Thus, the mapping $\varphi : \mathbb{F}^k \longrightarrow \mathbb{F}^n$ given by $\varphi(\mathbf{u}) = \mathbf{u}G$ is an encoder for $\mathcal{C}$.
4. [Hil86, p. 48] The distance of a linear code is equal to the minimum weight of all nonzero codewords.
5. Every $[n, k]$ linear block code has an $(n - k) \times n$ parity check matrix.
6. For a given $[n, k]$ linear code $\mathcal{C}$ with generator matrix G, an $(n - k) \times n$ matrix H is a parity check matrix for $\mathcal{C}$ if and only if the rows of H are a basis for the left kernel of G^T.
7. If H is a parity check matrix for the linear code $\mathcal{C}$, then $\mathcal{C}$ is the left kernel of H^T.
8. A generator matrix of $\mathcal{C}$ is a parity check matrix of $\mathcal{C}^\perp$.
9. [MS77, p. 33] Suppose $H \in \mathbb{F}^{(n-k) \times n}$ is a parity check matrix for an $[n, k]$-code $\mathcal{C}$ over $\mathbb{F}$. Then the distance of $\mathcal{C}$ is equal to the smallest number of columns of H for which a linear dependency may be found.
10. [Hil86, p. 70] If $G = [I_k \mid A]$ is a systematic encoder for the $[n, k]$-code $\mathcal{C}$, then $H = [-A^T \mid I_{n-k}]$ is a parity check matrix for $\mathcal{C}$.
11. [MS77, p. 22] *Shannon's Coding Theorem for the Binary Symmetric Channel.* Assume that the transmission channel has a fixed symbol error probability p that the symbol 1 is corrupted to a 0 or vice versa. Let

$$C := 1 + p \log_2 p + (1 - p) \log_2(1 - p).$$

Then for every $\epsilon > 0$ and every $R < C$ there exists, for any sufficiently large n, a binary linear $[n, k]$ code having transmission rate $k/n \geq R$ and word error probability $P_{\text{err}} \leq \epsilon$.

Examples:

1. The binary repetition code of length 7 is a $[7, 1]$ BLBC. A generator matrix is

$$G = \begin{bmatrix} 1 & 1 & 1 & 1 & 1 & 1 & 1 \end{bmatrix}.$$

This code consists of two codewords, namely 0000000 and 1111111. The rate of this code is $\frac{1}{7}$. For every 7 bits transmitted, there is only 1 bit of information, with the remaining 6 redundant bits providing error protection. The distance of this code is 7. A parity check matrix for this code is the 6×7 matrix

$$H = \begin{bmatrix} J_{6,1} & | & I_6 \end{bmatrix}.$$

2. *Binary even weight codes.* The binary even weight code of length n is a $[n, n-1]$-code that consists of all even weight words in $\mathbb{F}_2^n$. A generator matrix is given by

$$G = \begin{bmatrix} I_{n-1} & | & J_{(n-1),1} \end{bmatrix}.$$

This code detects any single error, but cannot correct any errors. A binary even weight code may also be constructed by taking all binary strings of length $n - 1$ and to each adding an nth parity check bit chosen so that the weight of the resulting string is even. The 8-bit ASCII code (American Standard Code for Information Interchange) is constructed in this manner.

Note that the binary even weight code of length n and the binary repetition code of length n are dual codes.

3. The ISBN (International Standard Book Number) code is a $[10, 9]$-code over $\mathbb{F}_{11}$ (the finite field $\{0, 1, 2, 3, 4, 5, 6, 7, 8, 9, X\}$ where field operations are performed modulo 11. X is the symbol for 10 in this field). This code may be defined by the parity check matrix

$$H = \begin{bmatrix} 1 & 2 & 3 & 4 & 5 & 6 & 7 & 8 & 9 & X \end{bmatrix}.$$

Thus, a string of 10 digits, $v_1v_2v_3v_4v_5v_6v_7v_8v_9v_{10}$, from $\mathbb{F}_{11}$ is a valid ISBN if and only if $\sum_{i=1}^{10} iv_i \equiv 0 \pmod{11}$. The ISBN code will detect any single error or any error where two digits are interchanged (transposition error). The first nine digits of an ISBN carry information about the book (such as publisher). The tenth digit is a check digit. Given the first 9 digits of an ISBN, the check digit may be computed by forming the sum $\sum_{i=1}^{9} iv_i$ and reducing modulo 11. The value thus obtained is v_{10} (if the value is 10, then $v_{10} = X$).

Note: Currently the ISBN system does not use X as any of the first 9 digits. However, mathematically there is no reason not to use X anywhere in an ISBN.

4. The binary linear $[8, 4]$ code with generator matrix

$$G = \begin{bmatrix} 1 & 0 & 1 & 1 & 0 & 0 & 0 & 1 \\ 0 & 1 & 0 & 1 & 1 & 0 & 0 & 1 \\ 0 & 0 & 1 & 0 & 1 & 1 & 0 & 1 \\ 0 & 0 & 0 & 1 & 0 & 1 & 1 & 1 \end{bmatrix}$$

is a self-dual code (note that over the binary field, $GG^T = 0$). This code is called the extended binary $[8, 4, 4]$ code. All pairs of different codewords are at distance 4 or 8.

61.3 Main Linear Coding Problem and Distance Bounds

Definitions.

Let $A_q(n,d)$ be the largest possible number so that there exists a block code $C \subseteq \mathbb{F}_q^n$ of distance $d = \text{d}(C)$ and cardinality $|C| = A_q(n,d)$. The **Main Coding Theory Problem** asks for the determination of $A_q(n,d)$ for different values of n, d, q.

Similarly, let $B_q(n,d)$ be the largest possible number so that there exists a linear $[n,k]$ block code C of distance $d = \text{d}(C)$ and cardinality $|C| = B_q(n,d)$. The **Main Linear Coding Problem** asks for the determination of $B_q(n,d)$ for different values of n, d, q.

An $[n,k,d]$ linear code is **maximum distance separable (MDS)** if $d = n - k + 1$.

Facts:

1. Finding a value of $B_q(n,d)$ is equivalent to finding the largest dimension, k, for which there is an $[n,k,d]$-code over $\mathbb{F}_q$. For such maximal k, we have $B_q(n,d) = q^k$.
2. $B_q(n,d) \leq A_q(n,d)$ for all n, q, d.
3. [Rom92, p. 170] $A_q(n,1) = B_q(n,1) = q^n$.
4. $B_q(n,2) = q^{n-1}$. Such a code is achieved by taking C to be a code with parity check matrix $J_{1,n}$.
5. [Rom92, p. 170] $A_q(n,n) = B_q(n,n) = q$.
6. [Rom92, p. 170] For $n \geq 2$, $A_q(n,d) \leq q\,A_q(n-1,d)$.
7. [Hil86, Theorem 14.5] $B_q(n,3) = q^{n-r}$, where r is the unique integer such that

$$(q^{r-1} - 1)/(q-1) < n \leq (q^{r-1} - 1)/(q-1).$$

8. [Rom92, Theorem 4.5.7] *The sphere-packing bound.* Suppose C is a block code of length n over $\mathbb{F}_q$ with distance d. Let $t = \left\lfloor \frac{d-1}{2} \right\rfloor$. Then

$$A_q(n,d)\left(1 + \binom{n}{1}(q-1) + \binom{n}{2}(q-1)^2 + \cdots \binom{n}{t}(q-1)^t\right) \leq q^n.$$

9. [Hil86, pp. 20–21] The code C is perfect if and only if equality holds in the sphere-packing bound.
10. [Rom92, Theorem 4.5.7] Suppose C is a block code of length n over $\mathbb{F}_q$ with distance d. Then

$$A_q(n,d) \leq q^{n-d+1}.$$

11. *The Singleton bound for linear block codes.* For any $[n,k,d]$ linear code over $\mathbb{F}_q$, we have

$$d \leq n - k + 1.$$

12. An $(n-k) \times n$ matrix H with entries in $\mathbb{F}$ is the parity check matrix of an MDS code if and only if any $n-k$ columns of H are linearly independent. Equivalently, H has the property that any full size minor of H is invertible.
13. [MS77, pp. 546–547] *The Griesmer bound.* Let $N(k,d)$ be the length of the shortest binary linear code of dimension k and distance d. Then one has

$$N(k,d) \geq \sum_{i=0}^{k-1} \left\lceil \frac{d}{2^i} \right\rceil.$$

The right-hand side in the above inequality is, of course, always at least $d + k - 1$.

14. [Rom92, Theorem 4.5.16] *The Plotkin bound.* If $d \geq \frac{qn-n}{q}$, then

$$A_q(n,d) \leq \frac{dq}{d - qn - n}.$$

15. [Hil86, pp. 91–92] *The Gilbert Varshamov bound.* Assume q is a prime power. For any n and d, if k is the largest integer satisfying

$$q^k < \frac{q^n}{\sum_{i=0}^{d-2} \binom{n-1}{i}(q-1)^i},$$

then $A_q(n,d) \geq q^k$.

Examples:

1. Consider the binary [7,4] code C defined by the parity check matrix

$$H = \begin{bmatrix} 0 & 0 & 0 & 1 & 1 & 1 & 1 \\ 0 & 1 & 1 & 0 & 0 & 1 & 1 \\ 1 & 0 & 1 & 0 & 1 & 0 & 1 \end{bmatrix}.$$

This code has $2^4 = 16$ elements. Since any two columns of H are linearly independent, but the first three columns are dependent, it follows that $\mathrm{d}(C) = 3$. So this code meets the sphere packing bound with $t = 1$. Observe that reading downward, the 7 columns of H form the numbers 1 through 7 in binary. This leads to a very nice nearest-neighbor decoding scheme that corrects any single error. If the received word is $\mathbf{y}$, the syndrome $\mathbf{s} = \mathbf{y}H^T$ consists of 3 bits. Interpret these 3 bits as a binary number, x. If x is 0, the received word is a codeword and we assume that no errors have occurred. If x is not 0, it is one of the numbers from 1 to 7. In $\mathbf{y}$ we change the bit in position x. This will give us a codeword (the unique codeword) at a distance of 1 from $\mathbf{y}$. We presume that this codeword is the intended transmission. For instance, if we receive the vector $\mathbf{y} = 1100100$, we compute the syndrome $\mathbf{s} = \mathbf{y}H^T = 110$. Interpreting 110 as a binary number, we obtain $x = 6$. We change the sixth bit of $\mathbf{y}$ and obtain the codeword 1100110. This codeword is at a Hamming distance 1 from $\mathbf{y}$, and we assume it was the intended transmission.
2. Assume $\alpha_1, \ldots, \alpha_n$ are pairwise different nonzero elements of a finite field $\mathbb{F}$. Let

$$H := \begin{bmatrix} \alpha_1 & \alpha_2 & \ldots & \alpha_n \\ \alpha_1^2 & \alpha_2^2 & \ldots & \alpha_n^2 \\ \vdots & & & \vdots \\ \alpha_1^{n-k} & \alpha_2^{n-k} & \ldots & \alpha_n^{n-k} \end{bmatrix}.$$

Then any full-size minor is the determinant of a Vandermonde matrix and, hence, nonzero. So H represents the parity check matrix of an MDS code.

61.4 Important Classes of Linear Codes 1

Definitions:

Let $\mathbb{F} = \mathbb{F}_q$.

Let r be a positive integer. Let $\mathbb{P}(\mathbb{F}^r)$ be the set of all 1-dimensional subspaces of $\mathbb{F}^r$. Choose a basis vector for each element of $\mathbb{P}(\mathbb{F}^r)$. Let H be a matrix whose columns are these basis vectors. The matrix H is the parity check matrix of a **Hamming code**, C.

A **cyclic code** is a linear code C for which every cyclic shift of a codeword is again a codeword.

Let $\mathbb{F}[x]$ be the polynomial ring in one indeterminate over $\mathbb{F}$. Let $\langle x^n - 1\rangle$ be the ideal generated by $x^n - 1$. The quotient ring $\mathbb{F}[x]/\langle x^n - 1\rangle$ is denoted R_n. The polynomials of R_n are identified with vectors in $\mathbb{F}^n$ via

$$a_0 + a_1x + a_2x^2 + \cdots a_{n-1}x^{n-1} \longmapsto [a_0 \quad a_1 \quad a_2 \quad \cdots \quad a_{n-1}].$$

Suppose $k < n$ are positive integers and $r = n - k$. A polynomial $g(x) = a_0 + a_1x + a_2x^2 + \cdots + a_rx^r$ dividing $x^n - 1 \in \mathbb{F}[x]$ is a **generator polynomial** of a cyclic $[n, k]$-code, C. The polynomial $h(x) = b_0 + b_1x + b_2x^2 + \cdots + b_kx^k$ chosen so that $g(x)h(x) = x^n - 1$ is the **parity check polynomial** of C.

Facts:

1. The ring R_n forms an n-dimensional vector space over $\mathbb{F}$. The monomials $\{1, x, x^2, \ldots, x^{n-1}\}$ form a basis. It further has an algebra structure over $\mathbb{F}$ (i.e., there is a well-defined multiplication). In fact, polynomials in R_n are multiplied as usual, but $x^n = 1$ in R_n, so exponents are reduced modulo n.
2. The cardinality of $\mathbb{P}(\mathbb{F}^r)$ is $n = (q^r - 1)/(q - 1)$. Thus, the parity check matrix, H, of a Hamming code is of size $r \times n$. The corresponding Hamming code C has q^k elements where $k = n - r$. Furthermore, the distance $d(C) = 3$. The Hamming code reaches the sphere packing bound and hence is a perfect code in all cases.
3. [Hil86, pp. 147–148] For cyclic codes, R_n has the structure of a principal ideal ring (any ideal is generated by a single polynomial). Thus, any cyclic code has a generator polynomial $g(x)$ where $g(x)$ is a divisor of $x^n - 1$. Moreover, every cyclic code has a parity check polynomial, namely $h(x) = (x^n - 1)/g(x)$.
4. [Hil86, Chap. 12] If C is a cyclic code with generator polynomial $g(x) = a_0 + a_1x + a_2x^2 + \cdots + a_rx^r$ and parity check polynomial $h(x) = b_0 + b_1x + b_2x^2 + \cdots + b_kx^k$, then the corresponding generator matrix, G, and parity check matrix, H, of C are given by

$$G = \begin{bmatrix} a_0 & a_1 & a_2 & \cdots & a_r & 0 & 0 & \cdots & 0 \\ 0 & a_0 & a_1 & a_2 & \cdots & a_r & 0 & \cdots & 0 \\ 0 & 0 & a_0 & a_1 & a_2 & \cdots & a_r & \cdots & 0 \\ & \ddots & \ddots & \ddots & \ddots & \ddots & \ddots & \ddots & \\ 0 & 0 & 0 & \cdots & a_0 & a_1 & a_2 & \cdots & a_r \end{bmatrix}$$

and

$$H = \begin{bmatrix} b_k & \cdots & b_2 & b_1 & b_0 & 0 & 0 & \cdots & 0 \\ 0 & b_k & \cdots & b_2 & b_1 & b_0 & 0 & \cdots & 0 \\ 0 & 0 & b_k & \cdots & b_2 & b_1 & b_0 & \cdots & 0 \\ & \ddots & \ddots & \ddots & \ddots & \ddots & \ddots & \ddots & \\ 0 & 0 & 0 & \cdots & b_k & \cdots & b_2 & b_1 & b_0 \end{bmatrix}.$$

Examples:

1. For $q = 5$ and $r = 2$, we get the $[6, 4, 3]$-Hamming code over $\mathbb{F}_5$ with parity check matrix

$$H = \begin{bmatrix} 1 & 0 & 1 & 1 & 1 & 1 \\ 0 & 1 & 1 & 2 & 3 & 4 \end{bmatrix}$$

and generator matrix

$$G = \begin{bmatrix} 4 & 4 & 1 & 0 & 0 & 0 \\ 4 & 3 & 0 & 1 & 0 & 0 \\ 4 & 2 & 0 & 0 & 1 & 0 \\ 4 & 1 & 0 & 0 & 0 & 1 \end{bmatrix}.$$

2. The binary repetition code of length n is a cyclic code with generator polynomial $1+x+\cdots+x^{n-1}$. The binary even weight code is a cyclic code with generator polynomial $1+x$.
3. Over $\mathbb{F}_2$, the polynomial $x^7-1 = x^7+1 = (1+x)(1+x+x^3)(1+x^2+x^3)$. Hence, $g(x) = 1+x+x^3$ is the generator polynomial of a binary $[7,4]$-cyclic code. The corresponding parity check polynomial is $h(x) = (1+x)(1+x^2+x^3) = 1+x+x^2+x^4$. So, the generator matrix G and parity check matrix H for this code are given by

$$G = \begin{bmatrix} 1 & 1 & 0 & 1 & 0 & 0 & 0 \\ 0 & 1 & 1 & 0 & 1 & 0 & 0 \\ 0 & 0 & 1 & 1 & 0 & 1 & 0 \\ 0 & 0 & 0 & 1 & 1 & 0 & 1 \end{bmatrix} \quad \text{and} \quad H = \begin{bmatrix} 1 & 0 & 1 & 1 & 1 & 0 & 0 \\ 0 & 1 & 0 & 1 & 1 & 1 & 0 \\ 0 & 0 & 1 & 0 & 1 & 1 & 1 \end{bmatrix}.$$

Noting that the columns of H represent all the nonzero elements of $\mathbb{F}_2^3$, we see that this code is an equivalent version of the Hamming $[7,4]$-perfect code.

61.5 Important Classes of Linear Codes 2

Definitions:

The **Golay codes** [Gol49] are two cyclic codes defined as follows.

The **binary Golay code,** G_{23}, is a $[23,12,7]$ binary cyclic code with generator polynomial $g(x) = \left(x^{11}+x^{10}+x^6+x^5+x^4+x^2+1\right)$.

The **ternary Golay code,** G_{11}, is an $[11,6,5]$ cyclic code over the ternary field $\mathbb{F}_3$ with generator polynomial $g(x) = \left(x^5+2x^3+x^2+2x+2\right)$.

Denote by $\mathbb{F}[x;k-1]$ the set of all polynomials of degree at most $k-1$ over $\mathbb{F}$. Assume $\mathbb{F}$ has size $|\mathbb{F}| > n$ and let α be a primitive of $\mathbb{F}$ (i.e., α is a generator of the cyclic group $\mathbb{F}^*$). Define the linear map (called the evaluation map) by

$$\begin{aligned} ev: \quad \mathbb{F}[x;k-1] & \longrightarrow \mathbb{F}^n \\ f & \longmapsto (f(\alpha),\ldots,f(\alpha^n)). \end{aligned}$$

The image of this map is a **Reed–Solomon code.**

Let $\mathbb{F} = \mathbb{F}_q$ be a fixed finite field and consider the polynomial $p(x) = x^n - 1 \in \mathbb{F}[x]$ and let the root w of $p(x)$ be a primitive nth root of unity. Let b, d be positive integers.

$$\mathcal{B}_q(n,d,b,w) := \{c(x) \in \mathbb{F}[x;k-1] \mid c(w^{b+i}) = 0 \text{ for } i = 0,\ldots,d-2\}$$

is called a **Bose–Chadhuri–Hocquenghem (BCH) code** having designed distance d. In the special case when $n = q^m - 1$, $\mathcal{B}_q(n,d,b,w)$ is a **primitive BCH code**; and if $b = 1$, $\mathcal{B}_q(n,d,b,w)$ is a **narrow sense BCH code.**

Facts:

1. Over the binary field $\mathbb{F}_2$,

$$x^{23}+1=(x^{11}+x^{10}+x^6+x^5+x^4+x^2+1)(x^{11}+x^9+x^7+x^6+x^5+x+1)(x+1).$$

   Therefore, the binary Golay code, G_{23}, has parity check polynomial $h(x)=(x^{11}+x^9+x^7+x^6+x^5+x+1)(x+1)$.
2. [Hil86, pp. 153–157] The binary Golay code G_{23} is a $[23,12,7]$ code having 4096 elements. G_{23} is a perfect code.
3. Over the ternary field $\mathbb{F}_3$,

$$x^{11}-1=(x^5+2x^3+x^2+2x+2)(x^5+x^4+2x^3+x^2+2)(x+2).$$

   Therefore, the ternary Golay code, G_{11}, has parity check polynomial $h(x)=(x^5+x^4+2x^3+x^2+2)(x+2)$.
4. [Hil86, pp. 157–159] The Golay code G_{11} is a ternary $[11,6,5]$ code having 729 elements. G_{11} is a perfect code.
5. [Hil86, Theorem 9.5] Here is a complete catalog of perfect linear codes up to equivalence:

   (a) The trivial perfect codes: All of $\mathbb{F}^n$, and the binary repetition codes of odd length.

   (b) The Hamming codes.

   (c) The Golay codes G_{23} and G_{11}.

6. Regarding Reed–Solomon codes, the linear map ev is one-to-one. Hence, the defined Reed–Solomon code is an $[n,k]$ code. By the fundamental theorem of algebra the minimum weight of a nonzero code word is $n-k+1$. It follows that a Reed–Solomon code is MDS. With regard to the natural basis $1,x,\ldots,x^{k-1}$ of $\mathbb{F}[x;k-1]$, this code has a generator matrix:

$$G:=\begin{bmatrix} 1 & 1 & \ldots & 1 \\ \alpha & \alpha^2 & \ldots & \alpha^n \\ \vdots & & & \vdots \\ \alpha^{k-1} & \alpha^{2(k-1)} & \ldots & \alpha^{n(k-1)} \end{bmatrix}.$$

7. Regarding BCH codes, it is immediate from the definition that $\mathcal{B}_q(n,d,b,w)$ has a parity check matrix of the form

$$H:=\begin{bmatrix} 1 & w^b & \ldots & w^{(n-1)b} \\ 1 & w^{b+1} & \ldots & w^{(n-1)(b+1)} \\ \vdots & & & \vdots \\ 1 & w^{b+d-2} & \ldots & w^{(n-1)(b+d-2)} \end{bmatrix}.$$

   We would like to stress that $\mathcal{B}_q(n,d,b,w)$ is the $\mathbb{F}_q$ kernel of H and that the entries of H are in general not in the base field $\mathbb{F}_q$. Assume $w\in\mathbb{F}_{q^s}$, where $\mathbb{F}_{q^s}$ is an extension field of $\mathbb{F}_q$. The $\mathbb{F}_{q^s}$ kernel of H is a Reed–Solomon code of distance d. As a consequence, we have that $\mathcal{B}_q(n,d,b,w)$ has distance at least d.
8. [Rom92, Chap. 8] The BCH code $\mathcal{B}_q(n,d,b,w)$ is a cyclic code. For $i=0,\ldots,d-2$, let $m_i(x)\in\mathbb{F}_q[x]$ be the minimal polynomial of w^{b+i} and let

$$g(x):=\operatorname{lcm}\{m_i(x)\mid i=0,\ldots,d-2\}\in\mathbb{F}_q[x].$$

   Then $g(x)$ is a generator of the cyclic code $\mathcal{B}_q(n,d,b,w)$ and its dimension is

$$\dim\mathcal{B}_q(n,d,b,w)=n-\deg g(x).$$

9. [Mas69], [FWRT94], [Sud97], [KV03] Reed–Solomon codes and BCH codes can be efficiently decoded up to half the designed distance using the Berlekamp–Massey algorithm [Mas69]. Extensions for this algorithm exist for algebraic geometric codes [FWRT94]. These algorithms have the major drawback that no **soft information** can be processed. This means that the decoder needs a vector in $\mathbb{F}^n$ as an input and it is not possible to process real numbers as they naturally occur in wireless communication. In 1997 Sudan [Sud97] achieved a major breakthrough when he came up with a polynomial time algorithm for list decoding of Reed–Solomon codes. Since that time the technique has been vastly improved and generalized and a broad overview is given in [Gur04] and [KV03].

Examples:

1. *Algebraic Geometric Codes.* In 1977, V.D. Goppa had the idea to vastly generalize the construction techniques of Reed–Solomon codes and BCH codes. Goppa codes became famous in 1982 when Tsfasman, Vladut, and Zink [TVZ82] constructed an infinite family of codes that exceeds the Gilbert Varshamov bound. (For further details, the reader is referred to the extensive literature on this subject [Gop81], [Gop88], [LG88], [TV91], and [Wal00].

   Consider the **projective plane** $\mathbb{P}^2_{\mathbb{F}}$ and let $V \subset \mathbb{P}^2_{\mathbb{F}}$ be a **smooth curve** of degree θ. This means there is an irreducible homogeneous polynomial $\varphi(x, y, z) \in \mathbb{F}[x, y, z]$ of degree θ with

$$V = \{(\alpha, \beta, \gamma) \in \mathbb{P}^2_{\mathbb{F}} \mid \varphi(\alpha, \beta, \gamma) = 0\}.$$

   The smoothness is characterized by the fact that there is no point on the curve where the partial derivatives vanish simultaneously. $\Gamma(V) := \mathbb{F}[x, y, z]/ < \varphi >$ is called the homogeneous coordinate ring. As $\Gamma(V)$ is a domain, there is a well-defined quotient field $k(V)$, which is called the **homogeneous function field.** (See e.g. [Ful89, p.91].)

   If $f \in k(V)$, then f has an **associated divisor**

$$(f) := \sum_{P \in V} \mathrm{ord}_P(f) P,$$

   where $\mathrm{ord}_P(f)$ is m if f has a pole of order m at P and $-m$ if f has a zero of order m at P.

   Let $D = P_1 + \cdots + P_n$ be a divisor on V of degree n. Define the vector space

$$L(D) := \{f \in k(V)^* \mid (f) + D \geq 0\} \cup \{0\}.$$

   Let G be a divisor on V having support disjoint from D. Consider the linear map:

$$\begin{array}{rccc} ev: & L(G) & \longrightarrow & \mathbb{F}^n \\ & f & \longmapsto & (f(P_1), \ldots, f(P_n)). \end{array}$$

   The image $\mathcal{C}(D, G)$ of the linear evaluation map ev is called an **algebraic geometric Goppa (AG) code.** Note that $\mathbb{F}[x; k-1]$ is a linear space of the form $L(G)$ and in this way Reed–Solomon codes are AG codes.

   Let $\mathcal{C}(D, G)$ be an algebraic geometric Goppa code defined from a curve of degree θ. If $\deg G < n$, then

$$\dim \mathcal{C}(D, G) = \dim L(G) - \dim L(G - D) \geq \deg G + 1 - \frac{1}{2}(\theta - 1)(\theta - 2)$$

   and for the distance one has the lower bound

$$d(\mathcal{C}(D, G)) \geq n - \deg G.$$

   Note that $\frac{1}{2}(\theta - 1)(\theta - 2)$ is the genus of the curve. One way to construct AG codes with good distance parameters is, therefore, to construct curves of low genus having many $\mathbb{F}_q$ rational points.

2. *Low density parity check codes.* A class of linear codes of tremendous practical importance was introduced by R.G. Gallager in his dissertation [Gal63]. These binary linear codes were called **low density parity check (LDPC) codes** by the author and are characterized by the property that the code can be written as the kernel of a very sparse matrix. For example, Gallager studied codes whose parity check matrix had three 1s in every column and six 1s in every row randomly chosen.

   LDPC codes come with the remarkable property that efficient decoding algorithms exist whose complexity increases linearly with the code length [RU01]. One of simplest algorithms is the **bit flipping algorithm.** For this, assume that H is a sparse parity check matrix, a code word $\mathbf{v}$ was sent, and a vector $\mathbf{y} = \mathbf{v} + \mathbf{e}$ was received. The algorithm then checks bit-by-bit if the weight of the syndrome vector $\mathbf{y}H^T$ increases or decreases if the bit under consideration is flipped. The bit flipping algorithm is one of the most simple type of message passing algorithms.

   LDPC codes were later generalized by R. M. Tanner [Tan81] who defined linear codes defined through graphs. A good overview to this active area of research can be found in [MR01]. The class of LDPC codes and codes on graphs became prominent again after the discovery of Turbo codes [BGT93].

61.6 Convolutional Codes

Aside from linear block codes, the codes most used in engineering practice are convolutional codes. For this, consider the situation where a large amount of data has to be encoded by a linear block code. Let $\mathbf{m}_0, \ldots, \mathbf{m}_\tau \in \mathbb{F}^k$ be $\tau + 1$ blocks of messages to be transmitted and assume G is the generator matrix of a linear $[n, k]$ code. The encoding scheme:

$$\mathbf{m_i} \longmapsto \mathbf{m_i}G = \mathbf{x_i}, \quad i = 0, \ldots, \tau$$

can be compactly written in the form

$$\mathbf{m}(z) \longmapsto \mathbf{m}(z)G = \mathbf{x}(z),$$

where

$$\mathbf{m}(z) := \sum_{i=0}^{\tau} \mathbf{m_i} z^i \in \mathbb{F}^k[z] \quad \text{and} \quad \mathbf{x}(z) := \sum_{i=0}^{\tau} \mathbf{x_i} z^i \in \mathbb{F}^n[z].$$

Elias [Eli55] had the original idea to replace in this encoding scheme the generator matrix G with a $k \times n$ matrix $G(z)$ whose entries are elements of the polynomial ring $\mathbb{F}[z]$.

Definitions:

Consider the polynomial ring $R := \mathbb{F}[z]$. A submodule $\mathcal{C} \subseteq R^n$ is called a **convolutional code**. The **rank** of the code $\mathcal{C}$ is the rank, k, of $\mathcal{C}$ considered as an R-module. $\mathcal{C}$ is an $[n, k]$-convolutional code. The **rate** of $\mathcal{C}$ is k/n.

A **generator matrix** for the rate k/n convolutional code $\mathcal{C}$ is a $k \times n$ polynomial matrix G with entries in R such that

$$\mathcal{C} = \{\mathbf{m}G = \mathbf{x} \mid \mathbf{m} \in R^k, \mathbf{x} \in R^n\}.$$

The **degree** of the convolutional code $\mathcal{C}$ with generator matrix G is defined to be the largest degree of any $k \times k$ minor of G.

Let $\mathbf{x}(z) = \sum_{i=0}^{\tau} \mathbf{x_i} z^i \in R^n$ be a code vector. Define the **weight of** $\mathbf{x}(z)$ by

$$\mathrm{wt}(\mathbf{x}(z)) := \sum_{i=0}^{\tau} \mathrm{wt}(\mathbf{x_i}).$$

The **free distance** of the convolutional code $\mathcal{C}$ is defined as

$$d_{\text{free}}(\mathcal{C}) := \min\{\text{wt}(\mathbf{x}(z)) \mid \mathbf{0} \neq \mathbf{x}(z) \in \mathcal{C}\}. \tag{61.1}$$

Facts:

1. [Hun74] R is a principal ideal domain and $\mathcal{C}$ is, therefore, a free R-module. As such, $\mathcal{C}$ has a well-defined rank k, and a $k \times n$ generator matrix G with entries in R, such that

$$\mathcal{C} = \{\mathbf{m}G = \mathbf{x} \mid \mathbf{m} \in R^k, \mathbf{x} \in R^n\}.$$

2. If G_1 and G_2 are both $k \times n$ generator matrices of the same convolutional code $\mathcal{C}$, then there exists a unimodular matrix $U \in Gl_k(R)$ such that $G_2 = UG_1$.
3. The degree of a convolutional code is well-defined (i.e., it does not depend on the particular generator matrix chosen). In the literature the degree is sometimes also called the **total memory** [LC83] or the **overall constraint length** [JZ99] or the **complexity** [Pir88].
4. The free distance of a convolutional code measures the smallest distance between any two different code words.
5. A major design problem in the area of convolutional codes is the construction of $[n, k]$ codes of degree δ whose free distance is maximal or near maximal. This question constitutes the generalization of the **main linear coding problem.**
6. If $\mathcal{C}$ is an $[n, k]$ convolutional code having $G(z)$ as a generator matrix, then there exists an $(n-k)\times n$ parity check matrix $H(z)$ with entries in R. $H(z)$ is characterized by the property that $G(z)H(z)^T = \mathbf{0}_{k\times k}$.
7. Convolutional codes of degree zero correspond to linear block codes.
8. Convolutional codes of small degree can be efficiently decoded using the Viterbi decoding algorithm [LC83, Chap. 11]. This algorithm has the distinct advantage that soft decision can be processed and due to this fact convolutional codes became the codes of choice for many wireless applications.
9. The free distance of an $[n, k]$ convolutional code of degree δ must satisfy the **generalized Singleton bound** [RS99]:

$$d_{\text{free}}(\mathcal{C}) \leq (n-k)\left(\left\lfloor \frac{\delta}{k} \right\rfloor + 1\right) + \delta + 1.$$

10. Convolutional codes achieving the generalized Singleton bound are called **MDS convolutional codes**. Such codes exist as soon as the field size is large enough [RS99].
11. The literature on convolutional codes is not consistent in terms of terminology and definitions. Classical articles are [MS67] and [For70]. The articles [McE98] and [Ros01] provide surveys on recent developments and further connections. Generalizations to multidimensional convolutional codes were considered and the reader will find further reading and references in [GLRW00].

    In contrast to the theory of linear block codes, there are not so many algebraic construction techniques of convolutional codes with a good designed distance and even fewer constructions of codes that can be efficiently decoded. The reader will find algebraic constructions of convolutional codes in [GLS04], [Jus75], [Pir88], [RSY96], [SGLR01], and in the bibliography of these articles.

Examples:

1. Consider the finite field $\mathbb{F}_5$. The generator matrix $G(z) = (z, z+1, z+2, z+3)$ defines an $[n, k] = [4, 1]$ convolutional code $\mathcal{C}$ of degree $\delta = 1$ and distance $d_{\text{free}}(\mathcal{C}) = 7$. The generalized Singleton bound for these parameters n, k, and δ is 8 and the defined code is, therefore, not an MDS convolutional code.

2. [RS99, Ex. 2.13]. Consider the finite field $\mathbb{F}_7$ and the generator matrix

$$G(z) = \begin{bmatrix} (z^2+1) & (3z^2+1) & (5z^2+1) \\ (z-1) & (z-2) & (2z-3) \end{bmatrix}.$$

Then $G(z)$ defines an $[n, k] = [3, 2]$ convolutional code C of degree $\delta = 3$ and distance $d_{\text{free}}(C) = 6$. The generalized Singleton bound for these parameters n, k, and δ is 6 and the defined code is therefore an MDS convolutional code.

References

[BGT93] C. Berrou, A. Glavieux, and P. Thitimajshima. Near Shannon limit error-correcting coding and decoding: Turbo-codes. In *Proc. of IEEE Int. Conference on Communication*, pp. 1064–1070, Geneva, Switzerland, May 1993.

[Bla03] R.E. Blahut. *Algebraic Codes for Data Transmission*. Cambridge University Press, Cambridge, 2003.

[Eli55] P. Elias. Coding for noisy channels. *IRE Conv. Rec.*, 4:37–46, 1955.

[For70] G.D. Forney, Jr. Convolutional codes I: Algebraic structure. *IEEE Trans. Inform. Theory*, IT-16(5):720–738, 1970.

[Ful89] W. Fulton. *Algebraic Curves*. Addison Wesley, Reading, MA, 1989. (Originally published by Benjamin/Cummings in 1969.)

[FWRT94] G.L. Feng, V.K. Wei, T.R.N. Rao, and K. Tzeng. Simplified understanding and efficient decoding of algebraic geometric codes. *IEEE Trans. Inform. Theory*, IT-40(4):981–1002, 1994.

[Gal63] R.G. Gallager. *Low-Density Parity Check Codes*. M.I.T. Press, Cambridge, MA, 1963. Number 21 in Research monograph series.

[GLRW00] H. Gluesing-Luerssen, J. Rosenthal, and P. A. Weiner. Duality between multidimensional convolutional codes and systems. In F. Colonius, U. Helmke, F. Wirth, and D. Prätzel-Wolters, Eds., *Advances in Mathematical Systems Theory, A Volume in Honor of Diederich Hinrichsen*, pp. 135–150. Birkhauser, Boston, 2000.

[GLS04] H. Gluesing-Luerssen and W. Schmale. On cyclic convolutional codes. *Acta Appl. Math*, 82:183–237, 2004.

[Gol49] M.J.E. Golay. Notes on digital coding. *Proc. IEEE*, 37:657, 1949.

[Gop81] V.D. Goppa. Codes on algebraic curves. *Soviet Math. Dolk.*, 24(1):170–172, 1981.

[Gop88] V.D. Goppa. *Geometry and Codes*. Kluwer Academic Publisher, Dordecht, The Netherlands, 1988.

[Gur04] V. Guruswami. *List Decoding of Error-Correcting Codes*, Vol. 3282 of *Lecture Notes in Computer Science*. Springer, Heidelberg, 2004.

[Hil86] R. Hill. *A First Course in Coding Theory*. Oxford Applied Mathematics and Computing Science Series. The Clarendon Press/Oxford University Press, New York, 1986.

[HP03] W.C. Huffman and V. Pless. *Fundamentals of Error-Correcting Codes*. Cambridge University Press, Cambridge, 2003.

[Hun74] T.W. Hungerford. *Algebra*. Graduate Texts in Mathematics. Springer, New York, 1974.

[Jus75] J. Justesen. An algebraic construction of rate $1/\nu$ convolutional codes. *IEEE Trans. Inform. Theory*, IT-21(1):577–580, 1975.

[JZ99] R. Johannesson and K.Sh. Zigangirov. *Fundamentals of Convolutional Coding*. IEEE Press, New York, 1999.

[KV03] R. Koetter and A. Vardy. Algebraic soft-decision decoding of Reed–Solomon codes. *IEEE Trans. Inform. Theory*, 49(11):2809–2825, 2003.

[LC83] S. Lin and D.J. Costello, Jr. *Error Control Coding: Fundamentals and Applications*. Prentice-Hall, Upper Saddle River, NJ, 1983.

[LG88] J.H. van Lint and G. van der Geer. *Introduction to Coding Theory and Algebraic Geometry*. Birkhäuser Verlag, Basel, 1988.

[Lin82] J.H. van Lint. *Introduction to Coding Theory*. Springer-Verlag, Berlin, New York, 1982.

[Mas69] J.L. Massey. Shift-register synthesis and BCH decoding. *IEEE Trans. Inform. Theory*, IT-15:122–127, 1969.

[McE98] R.J. McEliece. The algebraic theory of convolutional codes. In V. Pless and W.C. Huffman, Eds., *Handbook of Coding Theory*, Vol. 1, pp. 1065–1138. Elsevier Science Publishers, Amsterdam, The Netherlands, 1998.

[MR01] B. Marcus and J. Rosenthal, Eds. *Codes, Systems and Graphical Models*. IMA Vol. 123. Springer-Verlag, New York, 2001.

[MS67] J.L. Massey and M.K. Sain. Codes, automata, and continuous systems: explicit interconnections. *IEEE Trans. Automat. Contr.*, AC-12(6):644–650, 1967.

[MS77] F.J. MacWilliams and N.J.A. Sloane. *The Theory of Error-Correcting Codes*. North Holland, Amsterdam, 1977.

[PHB98] V.S. Pless, W.C. Huffman, and R.A. Brualdi, Eds. *Handbook of Coding Theory. Vol. I, II*. North-Holland, Amsterdam, 1998.

[Pir88] Ph. Piret. *Convolutional Codes, an Algebraic Approach*. MIT Press, Cambridge, MA, 1988.

[Rom92] S. Roman. *Coding and Information Theory*. Graduate Texts in Mathematics. Springer-Verlag, New York/Berlin, 1992.

[Ros01] J. Rosenthal. Connections between linear systems and convolutional codes. In B. Marcus and J. Rosenthal, Eds., *Codes, Systems and Graphical Models*, IMA Vol. 123, pp. 39–66. Springer-Verlag, Heidelberg, 2001.

[RS99] J. Rosenthal and R. Smarandache. Maximum distance separable convolutional codes. *Appl. Alg. Eng. Comm. Comp.*, 10(1):15–32, 1999.

[RSY96] J. Rosenthal, J.M. Schumacher, and E.V. York. On behaviors and convolutional codes. *IEEE Trans. Inform. Theory*, 42(6, part 1):1881–1891, 1996.

[RU01] T. Richardson and R. Urbanke. An introduction to the analysis of iterative coding systems. In B. Marcus and J. Rosenthal, Eds., *Codes, Systems and Graphical Models*, IMA Vol. 123, pp. 1–37. Springer-Verlag, Heidelberg, 2001.

[SGLR01] R. Smarandache, H. Gluesing-Luerssen, and J. Rosenthal. Constructions for MDS-convolutional codes. *IEEE Trans. Inform. Theory*, 47(5):2045–2049, 2001.

[Sud97] M. Sudan. Decoding of Reed–Solomon codes beyond the error-correction bound. *J. Complex.*, 13(1):180–193, 1997.

[Tan81] R.M. Tanner. A recursive approach to low complexity codes. *IEEE Trans. Inform. Theory*, 27(5):533–547, 1981.

[TV91] M.A. Tsfasman and S.G. Vlăduţ. *Algebraic Geometric Codes*. Mathematics and Its Application. Kluwer Academic Publishers, Dordecht, The Netherlands, 1991.

[TVZ82] M.A. Tsfasman, S.G. Vlăduţ, and Th. Zink. On Goppa codes which are better than the Varshamov–Gilbert bound. *Math. Nachr.*, 109:21–28, 1982.

[Wal00] J.L. Walker. *Codes and Curves*. American Mathematical Society, Providence, RI, 2000. IAS/Park City Mathematical Subseries.

Web Resources

1. http://www.win.tue.nl/˜aeb/voorlincod.html (A.E. Brouwer's Web site on bounds on the minimum distance of linear codes).
2. http://www.eng.tau.ac.il/˜litsyn/tableand/ (Simon Litsyn's table of nonlinear binary codes).
3. http://www.research.att.com/˜njas/codes/ (Neil J. A. Sloane's Web site of linear [and some non-linear] codes over various fields).

62

Quantum Computation

Zijian Diao
Ohio University Eastern

Modern computer science emerged when the eminent British mathematician Alan Turing invented the concept of Turing machine (TM) in 1936. Though very simple and primitive, TM serves as the universal model for all known physical computation devices. The principles of quantum mechanics, another revolutionary scientific discovery of the 20th century, had never been incorporated in the theory of computation until the early 1980s. P. Benioff first coined the concept of quantum Turing machine (QTM). Motivated by the problem that classical computers cannot simulate quantum systems efficiently, R. Feynman posed the quantum computer as the solution. The field of quantum computation was born.

Quantum computation mainly studies the construction and analysis of quantum algorithms that outperform the classical counterparts. In terms of computability, quantum computers and classical computers possess exactly the same computational power. But in terms of computational complexity, which measures the efficiency of computation, there are many examples confirming that quantum computers do solve certain problems faster. The two most significant ones are Shor's factorization algorithm and Grover's search algorithm, among other examples such as the Deutsch–Jozsa problem, the Bernstein–Vazirani problem, and Simon's problem.

Quantum computers share many common features of the classical computers. In a classical computer, information is encoded in binary states (for example, 0 denotes the low voltage state and 1 denotes the high voltage state), and processed by various logic gates. In a quantum computer, information is represented by the states of the microscopic quantum systems, called qubits, and manipulated by various quantum gates. A qubit could be a two-level atom in the excited/ground states, a photon with horizontal/vertical polarizations, or a spin-$\frac{1}{2}$ particle with up/down spins. The state of a qubit can be controlled via physical devices such as laser and microwave. The distinctions between quantum and classical computers originate from the special characteristic of quantum mechanics. In contrast to a classical system, a quantum system can exist in different states at the same time, an interesting phenomenon called *superposition*. Superposition enables quantum computers to process data in parallel. That is why a quantum computer can solve certain problems faster than a classical computer. But, when we measure a quantum system, it randomly collapses to one of the basis states. This indeterministic nature makes the design of efficient quantum algorithms highly nontrivial. Another distinctive feature of the quantum computer is that the operations performed by quantum gates must be unitary. This is the natural consequence of the unobserved quantum systems evolving according to the Schrödinger equation.

Throughout this chapter, we will use Dirac's bra-ket notation (see Section 59.4 for more information). In quantum mechanics, the state of a quantum system is described by a unit vector in a complex Hilbert space (a complete inner product vector space). Under Dirac's bra-ket notation, we use $|\psi\rangle$ ("ket") to denote a vector in the Hilbert space and $\langle\psi|$ ("bra") for its dual. The inner product of two vectors $|\psi\rangle$ and $|\phi\rangle$ is denoted $\langle\psi|\phi\rangle$. We also use $|\psi\rangle|\phi\rangle$ and $|\psi\phi\rangle$ interchangeably with the notation for the tensor product $|\psi\rangle \otimes |\phi\rangle$.

62.1 Basic Concepts

Definitions:

A **(classical) Turing machine** (TM) is an abstract computing device consisting of a finite control, a two-way infinite tape, and a read/write head that moves to the left or right on the tape. It can be described by a 6-tuple $(Q, A, B, \delta, q_0, q_a)$, where Q is a finite set of control states, A a finite alphabet, $B \in A$ the blank symbol, $q_0, q_a \in Q$ the initial and accepting states, and δ the transition function

$$\delta : Q \times A \to Q \times A \times \{L, R\}.$$

L and R stand for moving left and right, respectively.

A **quantum Turing machine** (QTM) is an abstract computing device consisting of a finite control, a two-way infinite tape, and a read/write head that moves to the left or right of the tape. It can be described by a 6-tuple $(Q, A, B, \delta, q_0, q_a)$, where Q is a finite set of control states, A a finite alphabet, $B \in A$ the blank symbol, $q_0, q_a \in Q$ the initial and accepting states, and δ the transition amplitude function

$$\delta : Q \times A \times Q \times A \times \{L, R\} \to \mathbb{C}.$$

The transition amplitude function satisfies

$$\sum_{(q_2,a_2,d)\in Q\times A\times\{L,R\}} |\delta(q_1, a_1, q_2, a_2, d)|^2 = 1,$$

for any $(q_1, a_1) \in Q \times A$. L and R stand for moving left and right, respectively.

A **quantum bit** (qubit) is a two-level quantum system, modeled by the two-dimensional Hilbert space H_2, with basis $\{|0\rangle, |1\rangle\}$. For example, for a spin-$\frac{1}{2}$ particle, $|0\rangle$ and $|1\rangle$ denote the spin-down and spin-up states, respectively. They can be mapped to the standard basis for H_2 as $|0\rangle = [1 \quad 0]^T$ and $|1\rangle = [0 \quad 1]^T$.

A **quantum register** of length n is an ordered system of n qubits, modeled by the 2^n-dimensional Hilbert space $H_{2^n} = H_2 \otimes H_2 \otimes \ldots \otimes H_2$ with basis $\{|00\ldots00\rangle, |00\ldots01\rangle, |00\ldots10\rangle, |00\ldots11\rangle, \ldots, |11\ldots11\rangle\}$. The basis states are ordered in lexicographic order. We may also write each basis state as $|i\rangle$ for $0 \leq i < 2^n$, interpreting the n-bit string of 0s and 1s as the binary representation of i.

An **one-bit quantum gate** is a unitary map $U : H_2 \to H_2$.

An **n-bit quantum gate** is a unitary map $U : H_2 \otimes H_2 \otimes \ldots \otimes H_2 \to H_2 \otimes H_2 \otimes \ldots \otimes H_2$.

A **quantum circuit** on n bits is a unitary map on H_{2^n}, which can be represented by a concatenation of a finite number of quantum gates.

Facts:

1. [BV93] Any function which is computable by a TM is computable by a QTM.
2. [Fey82] Any function that is computable by a QTM is computable by a TM.
3. [NC00, pp. 13] A general state of a qubit is a unit vector $a|0\rangle + b|1\rangle$, where $a, b \in \mathbb{C}$ and $|a|^2 + |b|^2 = 1$. a and b are the probability amplitudes of $|0\rangle$ and $|1\rangle$, respectively. A measurement of a qubit yields either $|0\rangle$ or $|1\rangle$, with probability $|a|^2$ or $|b|^2$, respectively.
4. [BB02] A general state of n-bit quantum register is a unit vector

$$\sum_{x=00\ldots0}^{11\ldots1} \psi_x |x\rangle,$$

where $\psi_x \in \mathbb{C}$ and $\sum_{x=00\ldots0}^{11\ldots1} |\psi_x|^2 = 1$. ψ_x is the probability amplitude of $|x\rangle$. A measurement of a quantum register yields $|x\rangle \in \{|00\ldots0\rangle, |00\ldots1\rangle, \ldots, |11\ldots1\rangle\}$, with probability $|\psi_x|^2$.

Examples:

This section contains a list of quantum gates that are frequently used. We provide the description of their effects on the basis states, matrix representations, and circuit diagrams. In the circuit diagrams, the horizontal lines stand for the qubits. When there is no gate on the line, no operation is done, which can be interpreted as the identity operation. When the diagram of a gate shows up on a horizontal line/lines, the corresponding gate operation is applied to the qubit/qubits, with the input coming from the left and the output (result) going out to the right. The entire circuit diagram is read from left to right.

1. NOT gate Λ_0: $\Lambda_0|0\rangle = |1\rangle$, $\Lambda_0|1\rangle = |0\rangle$, or $\Lambda_0 = \begin{bmatrix} 0 & 1 \\ 1 & 0 \end{bmatrix}$.

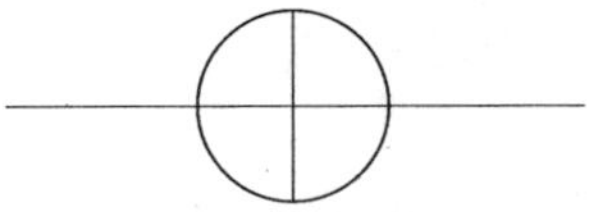

2. The Walsh–Hadamard gate H: $H|0\rangle = \frac{1}{\sqrt{2}}(|0\rangle + |1\rangle)$, $H|1\rangle = \frac{1}{\sqrt{2}}(|0\rangle - |1\rangle)$, or

$$H = \frac{1}{\sqrt{2}} \begin{bmatrix} 1 & 1 \\ 1 & -1 \end{bmatrix}.$$

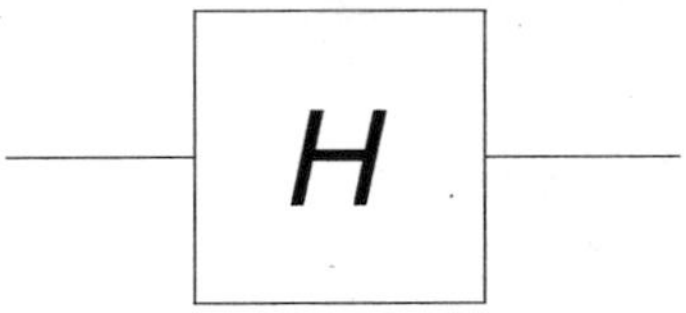

3. The x-rotation gate X_θ: $X_\theta|0\rangle = \cos\frac{\theta}{2}|0\rangle - i\sin\frac{\theta}{2}|1\rangle$, $X_\theta|1\rangle = -i\sin\frac{\theta}{2}|0\rangle + \cos\frac{\theta}{2}|1\rangle$, or

$$X_\theta = \begin{bmatrix} \cos\frac{\theta}{2} & -i\sin\frac{\theta}{2} \\ -i\sin\frac{\theta}{2} & \cos\frac{\theta}{2} \end{bmatrix}.$$

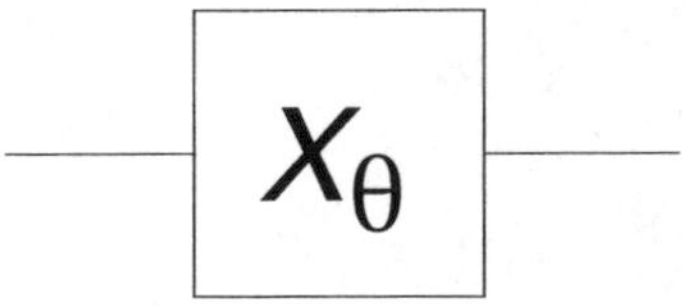

4. The y-rotation gate Y_θ: $Y_\theta|0\rangle = \cos\frac{\theta}{2}|0\rangle + \sin\frac{\theta}{2}|1\rangle$, $Y_\theta|1\rangle = -\sin\frac{\theta}{2}|0\rangle + \cos\frac{\theta}{2}|1\rangle$, or

$$Y_\theta = \begin{bmatrix} \cos\frac{\theta}{2} & -\sin\frac{\theta}{2} \\ \sin\frac{\theta}{2} & \cos\frac{\theta}{2} \end{bmatrix}.$$

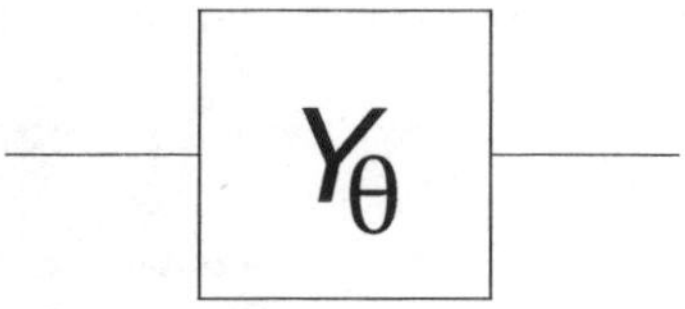

5. The z-rotation gate Z_θ: $Z_\theta|0\rangle = e^{-i\theta/2}|0\rangle$, $Z_\theta|1\rangle = e^{i\theta/2}|1\rangle$, or

$$Z_\theta = \begin{bmatrix} e^{-i\theta/2} & 0 \\ 0 & e^{i\theta/2} \end{bmatrix}.$$

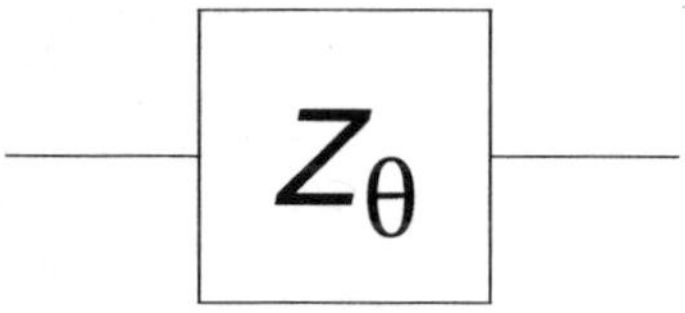

6. Controlled-NOT gate Λ_1: $\Lambda_1|00\rangle = |00\rangle$, $\Lambda_1|01\rangle = |01\rangle$, $\Lambda_1|10\rangle = |11\rangle$, $\Lambda_1|11\rangle = |10\rangle$, or,

$$\Lambda_1 = \begin{bmatrix} 1 & 0 & 0 & 0 \\ 0 & 1 & 0 & 0 \\ 0 & 0 & 0 & 1 \\ 0 & 0 & 1 & 0 \end{bmatrix}.$$

The first qubit acts as the control bit; the operation on the second qubit is controlled by it. If the control bit is $|0\rangle$, no operation is done on the second qubit. If the control bit is $|1\rangle$, the NOT gate is applied to the second qubit. There is no change on the control bit in either case. In the diagram for the Controlled-NOT gate, a black dot denotes the control bit and a vertical line signifies the control action.

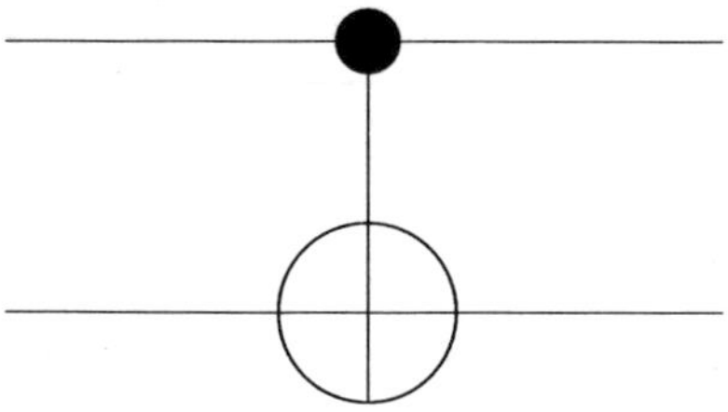

7. Two-bit Controlled-U gate $\Lambda_1(U)$, where U is any arbitrary one-bit unitary transform: $\Lambda_1(U)|00\rangle = |00\rangle$, $\Lambda_1(U)|01\rangle = |01\rangle$, $\Lambda_1(U)|10\rangle = |1\rangle U|0\rangle$, $\Lambda_1(U)|11\rangle = |1\rangle U|1\rangle$, or,

$$\Lambda_1(U) = \begin{bmatrix} 1 & 0 & 0 & 0 \\ 0 & 1 & 0 & 0 \\ 0 & 0 & \multicolumn{2}{c}{\multirow{2}{*}{U}} \\ 0 & 0 & & \end{bmatrix}.$$

The first qubit acts as the control bit; the operation on the second qubit is controlled by it. If the control bit is $|0\rangle$, no operation is done on the second qubit. If the control bit is $|1\rangle$, the one-bit gate U is applied to the second qubit. There is no change on the control bit in either case. In the diagram for the Controlled-U gate, a black dot denotes the control bit and a vertical line signifies the control action.

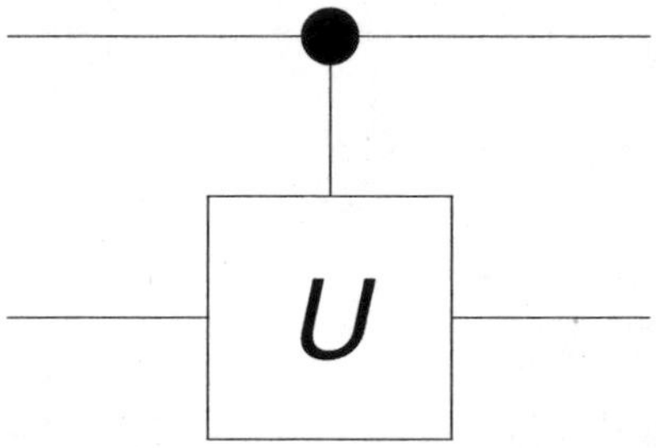

8. Function evaluation operator $U_f : H_{2^n} \otimes H_{2^m} \to H_{2^n} \otimes H_{2^m}$, where $f : \{0,1\}^n \to \{0,1\}^m$, $|x\rangle \in H_{2^n}$, and $|y\rangle \in H_{2^m}$, is given by

$$|x\rangle|y\rangle \to |x\rangle|y + f(x)\text{mod}2^m\rangle.$$

Two special cases of this operator will be used in the algorithms discussed later.

- $m = 1$, U_f is given by: $|x\rangle|y\rangle \to |x\rangle|y \oplus f(x)\rangle$, where $\oplus$ is the addition mod 2.

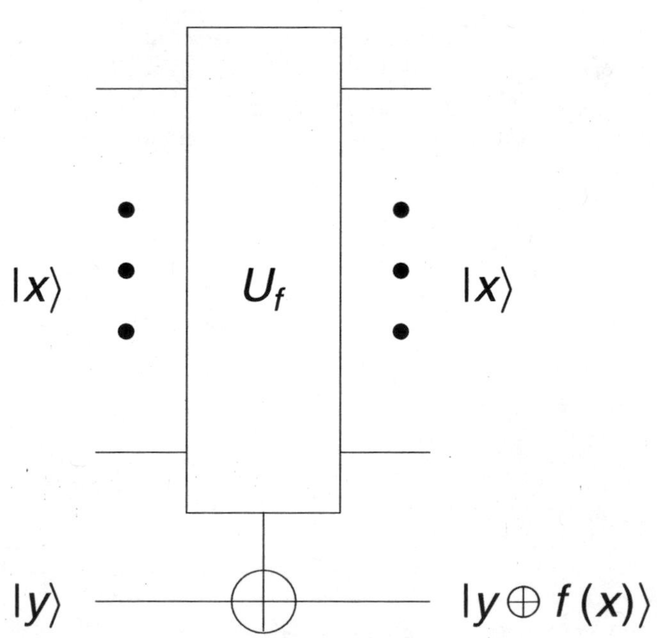

- $y = 0$, U_f is given by: $|x\rangle|0\rangle \to |x\rangle|f(x)\rangle$.

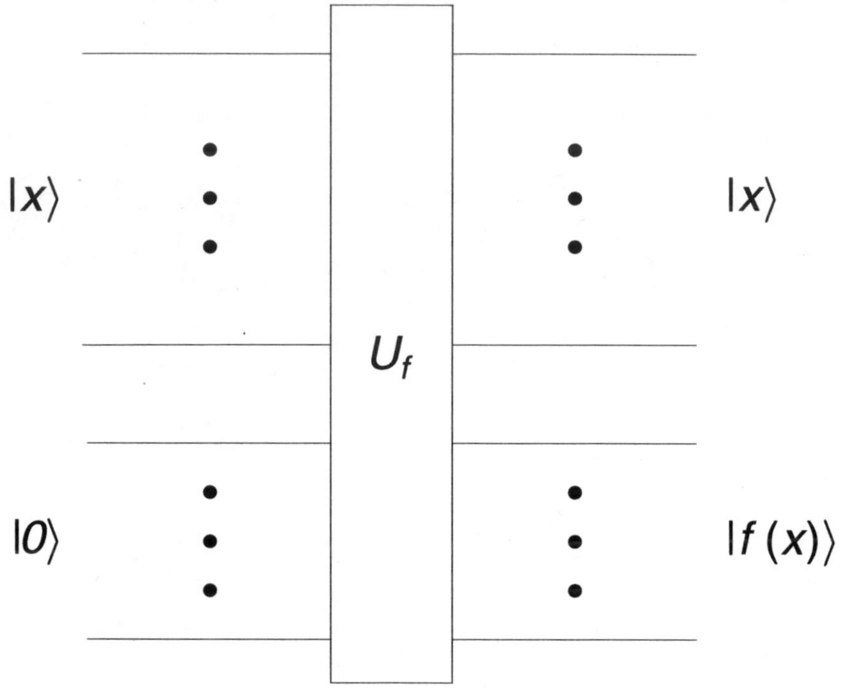

9. Quantum Fourier transform (QFT) (n-bit):

$$|x\rangle \to \frac{1}{2^{n-1}} \sum_{y=0}^{2^n-1} e^{2\pi i x y/2^n} |y\rangle,$$

where $x \in \{0, 1, \ldots, 2^n - 1\}$. This operator is a crucial building block of Shor's Factorization Algorithm. The following example manifests the power of QFT.

Example: Define a function $f : \{0, 1, 2, 3\} \to \{0, \frac{1}{\sqrt{2}}\}$ by $f(1) = f(3) = \frac{1}{\sqrt{2}}$ and $f(0) = f(2) = 0$. This function has period 2. Consider a 2-bit quantum system with state $\frac{1}{\sqrt{2}}(|1\rangle + |3\rangle)$, where the probability amplitudes of the basis states $|0\rangle$, $|1\rangle$, $|2\rangle$, and $|3\rangle$ are specified by the function values of f at 0, 1, 2, and 3. Apply QFT to this state.

$$\begin{aligned}
&\frac{1}{\sqrt{2}}(|1\rangle + |3\rangle) \\
&\to \frac{1}{2\sqrt{2}}(|0\rangle + i|1\rangle - |2\rangle - i|3\rangle) + \frac{1}{2\sqrt{2}}(|0\rangle - i|1\rangle - |2\rangle + i|3\rangle) \\
&= \frac{1}{\sqrt{2}}(|0\rangle - |2\rangle).
\end{aligned}$$

Measurement of the result yields 2, the period of f, with probability $\frac{1}{2}$. Thus, QFT provides a tool for period finding.

In the diagram for QFT, $x_{n-1}x_{n-2}\ldots x_2x_1x_0$ and $y_{n-1}y_{n-2}\ldots y_2y_1y_0$ are the binary representations of x and y, respectively.

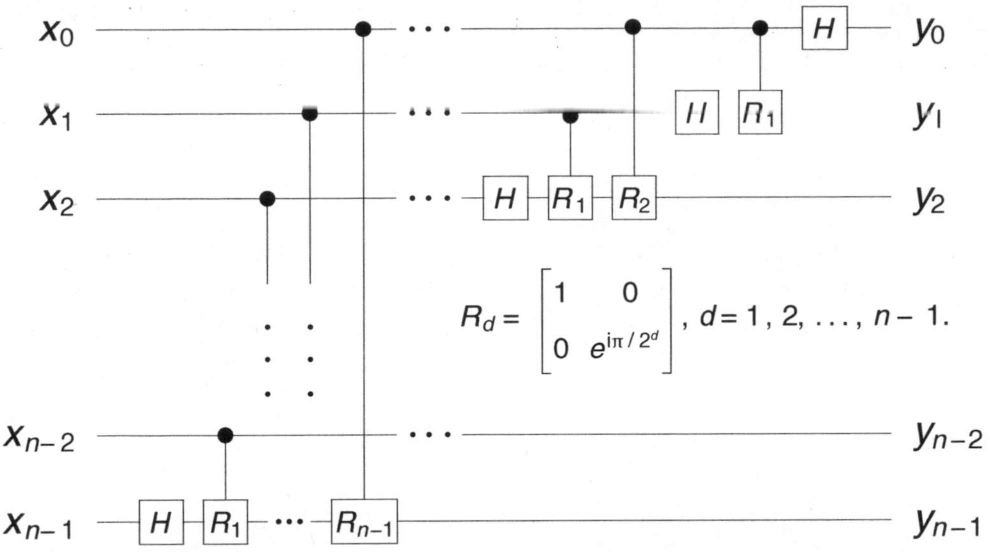

$$R_d = \begin{bmatrix} 1 & 0 \\ 0 & e^{i\pi/2^d} \end{bmatrix}, \quad d = 1, 2, \ldots, n-1.$$

62.2 Universal Quantum Gates

Definitions:

A 1-bit gate A is **special** if $\det(A) = 1$.

A 2-bit gate V is **primitive** if V is decomposable, i.e., there exist 1-bit gates S and T such that $V|xy\rangle = S|x\rangle \otimes T|y\rangle$, or $V|xy\rangle = S|y\rangle \otimes T|x\rangle$, for any state $|xy\rangle$.

A 2-bit gate V is **imprimitive** if it is not primitive.

A collection of quantum gates G is **universal** if, for each $n \in \mathbb{N}$, every n-bit quantum gate can be approximated with arbitrary accuracy by a circuit consisting of quantum gates in G.

A collection of quantum gates G is **exactly universal** if, for each $n \in \mathbb{N}$, every n-bit quantum gate can be obtained exactly by a circuit consisting of quantum gates in G.

Facts:

The following facts can be found in [BB02].

1. The collection of all 1-bit gates and any imprimitive 2-bit gate is universal.
2. The collection of all 1-bit gates and any imprimitive 2-bit gate is exactly universal.
3. The collection of all special 1-bit gates and any imprimitive 2-bit gate V with $\det(V)$ not being a root of unity is universal.

Examples:

1. The X_θ, Y_θ, and Z_θ gates are special.
2. The NOT gate and Walsh–Hadamard gate are not special.
3. The 2-bit gate $V = \frac{1}{\sqrt{2}} \begin{bmatrix} 1 & 0 & 1 & 0 \\ 0 & 1 & 0 & 1 \\ 1 & 0 & -1 & 0 \\ 0 & 1 & 0 & -1 \end{bmatrix}$ is primitive, $V|xy\rangle = H|x\rangle \otimes |y\rangle$.

4. The Controlled-NOT gate is imprimitive.
5. The collection of all 1-bit gates and Controlled-NOT gate is universal, and exactly universal.
6. The collection of all X_θ, Y_θ, and Z_θ gates and Controlled-phase gate $\Lambda_1(Q_\theta)$, where $Q_\theta = \begin{bmatrix} 1 & 0 \\ 0 & e^{i\theta} \end{bmatrix}$ and $e^{i\theta}$ is not a root of unity, is universal.

62.3 Deutsch's Problem

Definitions:

A **Boolean function** is a function with codomain $\{0, 1\}$.

A Boolean function $f : \{0, 1\} \to \{0, 1\}$ is **constant** if $f(0) = f(1)$.

A Boolean function $f : \{0, 1\} \to \{0, 1\}$ is **balanced** if $f(0) \neq f(1)$.

Problem: Deutsch

Given a Boolean function $f : \{0, 1\} \to \{0, 1\}$, determine whether it is constant or balanced.

Algorithm: Deutsch

1. Prepare a two-bit quantum register and initialize it to the state $|0\rangle|1\rangle$.
2. Apply the Walsh–Hadamard transform to both qubits.
3. Apply the function evaluation operator U_f

$$U_f : |x\rangle|y\rangle \to |x\rangle|y \oplus f(x)\rangle.$$

4. Apply the Walsh–Hadamard transform to the first qubit.
5. Measure the first qubit. If the outcome is $|0\rangle$, f is constant. If the outcome is $|1\rangle$, f is balanced.

Circuit Diagram: Deutsch

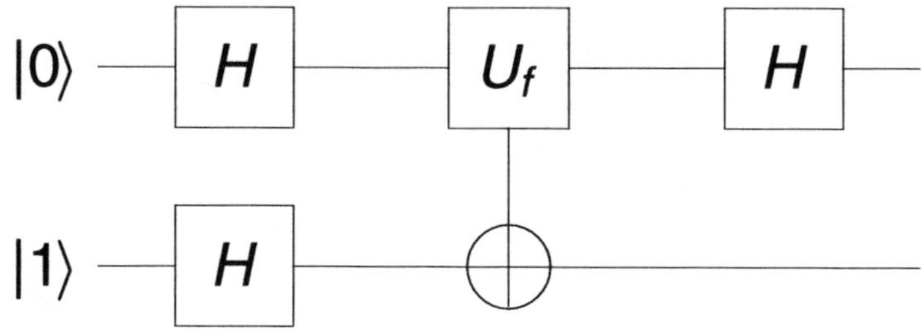

Facts:

The following facts can be found in [CEM98].

1. With the classical computer, we need two evaluations of f to determine whether f is constant or balanced.
2. With the quantum computer, we need only one evaluation of f to determine whether f is constant or balanced.

Examples:

1. Let $f(0) = 0$ and $f(1) = 1$. The following sequence of quantum states shows the result of computation utilizing Deutsch's Algorithm. We start from the initial state $|0\rangle|1\rangle$. Then,

$$\begin{aligned}
& |0\rangle|1\rangle \\
& \rightarrow \frac{|0\rangle + |1\rangle}{\sqrt{2}} \frac{|0\rangle - |1\rangle}{\sqrt{2}} = \frac{|0\rangle}{\sqrt{2}} \frac{|0\rangle - |1\rangle}{\sqrt{2}} + \frac{|1\rangle}{\sqrt{2}} \frac{|0\rangle - |1\rangle}{\sqrt{2}} \\
& \rightarrow \frac{|0\rangle}{\sqrt{2}} \frac{|0\rangle - |1\rangle}{\sqrt{2}} + \frac{|1\rangle}{\sqrt{2}} \frac{|1\rangle - |0\rangle}{\sqrt{2}} \\
& = \frac{|0\rangle}{\sqrt{2}} \frac{|0\rangle - |1\rangle}{\sqrt{2}} - \frac{|1\rangle}{\sqrt{2}} \frac{|0\rangle - |1\rangle}{\sqrt{2}} = \frac{|0\rangle - |1\rangle}{\sqrt{2}} \frac{|0\rangle - |1\rangle}{\sqrt{2}} \\
& \rightarrow |1\rangle \frac{|0\rangle - |1\rangle}{\sqrt{2}}.
\end{aligned}$$

The outcome of measuring the first qubit is $|1\rangle$, hence, f is balanced.

2. Let $f(0) = 0$ and $f(1) = 0$. The following sequence of quantum states shows the result of computation utilizing Deutsch's Algorithm. We start from the initial state $|0\rangle|1\rangle$. Then,

$$\begin{aligned}
& |0\rangle|1\rangle \\
& \rightarrow \frac{|0\rangle + |1\rangle}{\sqrt{2}} \frac{|0\rangle - |1\rangle}{\sqrt{2}} = \frac{|0\rangle}{\sqrt{2}} \frac{|0\rangle - |1\rangle}{\sqrt{2}} + \frac{|1\rangle}{\sqrt{2}} \frac{|0\rangle - |1\rangle}{\sqrt{2}} \\
& \rightarrow \frac{|0\rangle}{\sqrt{2}} \frac{|0\rangle - |1\rangle}{\sqrt{2}} + \frac{|1\rangle}{\sqrt{2}} \frac{|0\rangle - |1\rangle}{\sqrt{2}} = \frac{|0\rangle + |1\rangle}{\sqrt{2}} \frac{|0\rangle - |1\rangle}{\sqrt{2}} \\
& \rightarrow |0\rangle \frac{|0\rangle - |1\rangle}{\sqrt{2}}.
\end{aligned}$$

The outcome of measuring the first qubit is $|0\rangle$, hence, f is constant.

62.4 Deutsch–Jozsa Problem

Definitions:

A Boolean function $f : \{0,1\}^n \rightarrow \{0,1\}$ is **constant** if $f(x) = c$, $c \in \{0,1\}$, for all $x \in \{0,1\}^n$.

A Boolean function $f : \{0,1\}^n \rightarrow \{0,1\}$ is **balanced** if the function value is 1 (or 0) for exactly half of the input values, i.e., $\text{card}\left(f^{-1}(\{1\})\right) = \text{card}\left(f^{-1}(\{0\})\right)$.

Problem: Deutsch–Jozsa

Given a Boolean function $f : \{0,1\}^n \rightarrow \{0,1\}$, which is either constant or balanced, determine whether it is constant or balanced.

Algorithm: Deutsch–Jozsa

1. Prepare an $(n+1)$-bit quantum register and initialize it to the state $(|0\rangle)^n|1\rangle$.
2. Apply the Walsh–Hadamard transform to all the qubits.
3. Apply the function evaluation operator U_f:

$$U_f : |x\rangle|y\rangle \rightarrow |x\rangle|y \oplus f(x)\rangle.$$

4. Apply the Walsh–Hadamard transform to the first n qubits.
5. Measure the first n qubit. If the outcome is $|00\ldots0\rangle$, f is constant. If the outcome is not $|00\ldots0\rangle$, f is balanced.

Circuit Diagram: Deutsch–Jozsa

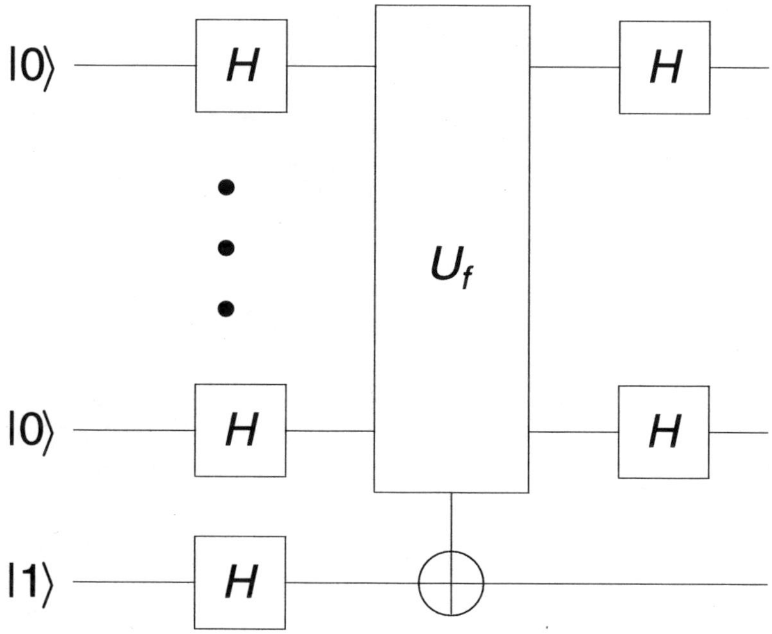

Facts:

The following facts can be found in [CEM98].

1. With the classical computer, we need at least $2^{n-1}+1$ evaluations of f to determine with certainty whether f is constant or balanced.
2. With the quantum computer, we need only one evaluation of f to determine with certainty whether f is constant or balanced.

Examples:

1. Let $n = 2$ and $f(00) = f(01) = f(10) = f(11) = 1$. The following sequence of quantum states shows the result of computation utilizing Deutsch–Jozsa's Algorithm. We start from the initial state $|00\rangle|1\rangle$. Then,

$$
\begin{aligned}
&|00\rangle|1\rangle \\
&\to \frac{|0\rangle+|1\rangle}{\sqrt{2}}\frac{|0\rangle+|1\rangle}{\sqrt{2}}\frac{|0\rangle-|1\rangle}{\sqrt{2}} = \frac{|00\rangle+|01\rangle+|10\rangle+|11\rangle}{2}\frac{|0\rangle-|1\rangle}{\sqrt{2}} \\
&= \frac{|00\rangle}{2}\frac{|0\rangle-|1\rangle}{\sqrt{2}} + \frac{|01\rangle}{2}\frac{|0\rangle-|1\rangle}{\sqrt{2}} + \frac{|10\rangle}{2}\frac{|0\rangle-|1\rangle}{\sqrt{2}} + \frac{|11\rangle}{2}\frac{|0\rangle-|1\rangle}{\sqrt{2}} \\
&\to \frac{|00\rangle}{2}\frac{|1\rangle-|0\rangle}{\sqrt{2}} + \frac{|01\rangle}{2}\frac{|1\rangle-|0\rangle}{\sqrt{2}} + \frac{|10\rangle}{2}\frac{|1\rangle-|0\rangle}{\sqrt{2}} + \frac{|11\rangle}{2}\frac{|1\rangle-|0\rangle}{\sqrt{2}} \\
&= -\frac{|00\rangle}{2}\frac{|0\rangle-|1\rangle}{\sqrt{2}} - \frac{|01\rangle}{2}\frac{|0\rangle-|1\rangle}{\sqrt{2}} - \frac{|10\rangle}{2}\frac{|0\rangle-|1\rangle}{\sqrt{2}} - \frac{|11\rangle}{2}\frac{|0\rangle-|1\rangle}{\sqrt{2}} \\
&= \frac{-|00\rangle-|01\rangle-|10\rangle-|11\rangle}{2}\frac{|0\rangle-|1\rangle}{\sqrt{2}} = -\frac{|0\rangle+|1\rangle}{\sqrt{2}}\frac{|0\rangle+|1\rangle}{\sqrt{2}}\frac{|0\rangle-|1\rangle}{\sqrt{2}} \\
&\to -|00\rangle\frac{|0\rangle-|1\rangle}{\sqrt{2}}.
\end{aligned}
$$

The outcome of measuring the first 2 qubit is $|00\rangle$, hence, f is constant.

2. Let $n = 2$, $f(00) = f(01) = 0$, and $f(10) = f(11) = 1$. The following sequence of quantum states shows the result of computation utilizing Deutsch–Jozsa's Algorithm. We start from the initial state $|00\rangle|1\rangle$. Then,

$$
\begin{aligned}
&|00\rangle|1\rangle \\
&\to \frac{|0\rangle+|1\rangle}{\sqrt{2}}\frac{|0\rangle+|1\rangle}{\sqrt{2}}\frac{|0\rangle-|1\rangle}{\sqrt{2}} = \frac{|00\rangle+|01\rangle+|10\rangle+|11\rangle}{2}\frac{|0\rangle-|1\rangle}{\sqrt{2}} \\
&= \frac{|00\rangle}{2}\frac{|0\rangle-|1\rangle}{\sqrt{2}} + \frac{|01\rangle}{2}\frac{|0\rangle-|1\rangle}{\sqrt{2}} + \frac{|10\rangle}{2}\frac{|0\rangle-|1\rangle}{\sqrt{2}} + \frac{|11\rangle}{2}\frac{|0\rangle-|1\rangle}{\sqrt{2}} \\
&\to \frac{|00\rangle}{2}\frac{|0\rangle-|1\rangle}{\sqrt{2}} + \frac{|01\rangle}{2}\frac{|0\rangle-|1\rangle}{\sqrt{2}} + \frac{|10\rangle}{2}\frac{|1\rangle-|0\rangle}{\sqrt{2}} + \frac{|11\rangle}{2}\frac{|1\rangle-|0\rangle}{\sqrt{2}} \\
&= \frac{|00\rangle}{2}\frac{|0\rangle-|1\rangle}{\sqrt{2}} + \frac{|01\rangle}{2}\frac{|0\rangle-|1\rangle}{\sqrt{2}} - \frac{|10\rangle}{2}\frac{|0\rangle-|1\rangle}{\sqrt{2}} - \frac{|11\rangle}{2}\frac{|0\rangle-|1\rangle}{\sqrt{2}} \\
&= \frac{|00\rangle+|01\rangle-|10\rangle-|11\rangle}{2}\frac{|0\rangle-|1\rangle}{\sqrt{2}} = \frac{|0\rangle-|1\rangle}{\sqrt{2}}\frac{|0\rangle+|1\rangle}{\sqrt{2}}\frac{|0\rangle-|1\rangle}{\sqrt{2}} \\
&\to |10\rangle\frac{|0\rangle-|1\rangle}{\sqrt{2}}.
\end{aligned}
$$

The outcome of measuring the first 2 qubit is $|10\rangle$, hence, f is balanced.

62.5 Bernstein–Vazirani Problem

Definitions:

Let $x = x_1x_2 \ldots x_n$ and $y = y_1y_2 \ldots y_n$ be two n-bit strings from $\{0, 1\}^n$. The **dot product** $x \cdot y$ is the mod 2 sum of the bitwise products:

$$x \cdot y = x_1y_1 \oplus x_2y_2 \oplus \ldots \oplus x_ny_n.$$

Problem: Bernstein–Vazirani
Given a Boolean function $f_a : \{0, 1\}^n \to \{0, 1\}$ defined by $f_a(x) = a \cdot x$, where a is an unknown n-bit string in $\{0, 1\}^n$, determine the value of a.

Algorithm: Bernstein–Vazirani

1. Prepare an $(n+1)$-bit quantum register and initialize it to the state $(|0\rangle)^n|1\rangle$.
2. Apply the Walsh–Hadamard transform to all the qubits.
3. Apply the function evaluation operator U_{f_a}:

$$U_{f_a} : |x\rangle|y\rangle \to |x\rangle|y \oplus f_a(x)\rangle.$$

4. Apply the Walsh–Hadamard transform to the first n qubits.
5. Measure the first n qubits. The outcome is $|a\rangle$.

Circuit Diagram: Bernstein–Vazirani

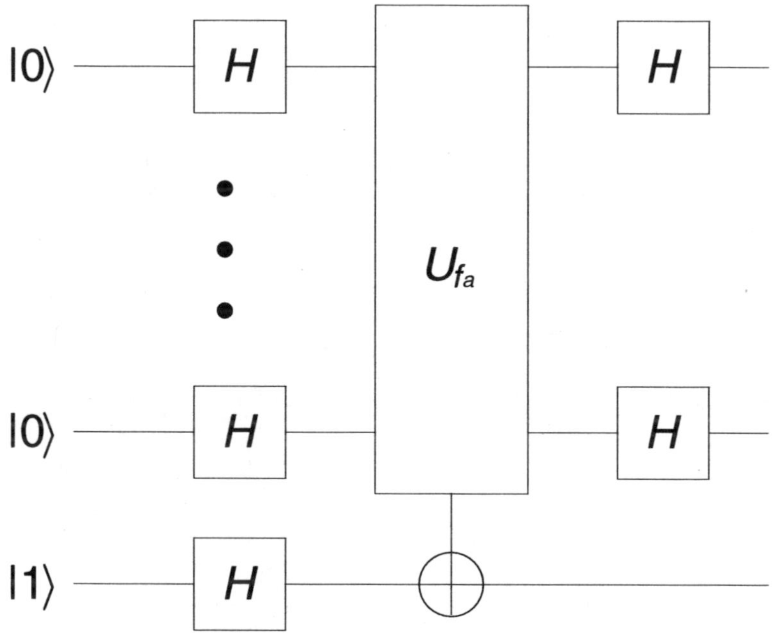

Facts:

The following facts can be found in [BV93].

1. With the classical computer, we need n evaluations of f_a to determine the value of a.
2. With the quantum computer, we need only one evaluation of f_a to determine the value of a.

Examples:

Let $a = 11$, a 2-bit string. The following sequence of quantum states shows the result of computation utilizing Bernstein–Vazirani's Algorithm. We start from the initial state $|00\rangle|1\rangle$. Then,

$$|00\rangle|1\rangle$$

$$\to \frac{|0\rangle+|1\rangle}{\sqrt{2}}\frac{|0\rangle+|1\rangle}{\sqrt{2}}\frac{|0\rangle-|1\rangle}{\sqrt{2}} = \frac{|00\rangle+|01\rangle+|10\rangle+|11\rangle}{2}\frac{|0\rangle-|1\rangle}{\sqrt{2}}$$

$$= \frac{|00\rangle}{2}\frac{|0\rangle-|1\rangle}{\sqrt{2}} + \frac{|01\rangle}{2}\frac{|0\rangle-|1\rangle}{\sqrt{2}} + \frac{|10\rangle}{2}\frac{|0\rangle-|1\rangle}{\sqrt{2}} + \frac{|11\rangle}{2}\frac{|0\rangle-|1\rangle}{\sqrt{2}}$$

$$\rightarrow \frac{|00\rangle}{2}\frac{|0\rangle - |1\rangle}{\sqrt{2}} + \frac{|01\rangle}{2}\frac{|1\rangle - |0\rangle}{\sqrt{2}} + \frac{|10\rangle}{2}\frac{|1\rangle - |0\rangle}{\sqrt{2}} + \frac{|11\rangle}{2}\frac{|0\rangle - |1\rangle}{\sqrt{2}}$$

$$= \frac{|00\rangle}{2}\frac{|0\rangle - |1\rangle}{\sqrt{2}} - \frac{|01\rangle}{2}\frac{|0\rangle - |1\rangle}{\sqrt{2}} - \frac{|10\rangle}{2}\frac{|0\rangle - |1\rangle}{\sqrt{2}} + \frac{|11\rangle}{2}\frac{|0\rangle - |1\rangle}{\sqrt{2}}$$

$$= \frac{|00\rangle - |01\rangle - |10\rangle + |11\rangle}{2}\frac{|0\rangle - |1\rangle}{\sqrt{2}} = \frac{|0\rangle - |1\rangle}{\sqrt{2}}\frac{|0\rangle - |1\rangle}{\sqrt{2}}\frac{|0\rangle - |1\rangle}{\sqrt{2}}$$

$$\rightarrow |11\rangle\frac{|0\rangle - |1\rangle}{\sqrt{2}}.$$

The outcome of measuring the first 2 qubits is $|11\rangle$, hence, $a = 11$.

62.6 Simon's Problem

Definitions:

A function $f : \{0, 1\}^n \rightarrow \{0, 1\}^n$ is **2–1** if for each $z \in \text{range}(f)$, there are exactly two distinct n-bit strings x and y such that $f(x) = f(y) = z$.

A function $f : \{0, 1\}^n \rightarrow \{0, 1\}^n$ has a **period** a if $f(x) = f(x \oplus a), \forall x \in \{0, 1\}^n$.

Problem: Simon

Given a function $f : \{0, 1\}^n \rightarrow \{0, 1\}^n$ which is 2-1 and has period a, determine the period a.

Algorithm: Simon

1. Repeat the following procedure for n times.
   (a) Prepare a $2n$-bit quantum register and initialize it to the state $(|0\rangle)^n(|0\rangle)^n$.
   (b) Apply the Walsh–Hadamard transform to the first n qubits.
   (c) Apply the function evaluation operator U_f:
   $$U_f : |x\rangle|y\rangle \rightarrow |x\rangle|y \oplus f(x)\rangle.$$
   (d) Apply the Walsh–Hadamard transform to the first n qubits.
   (e) Measure the first n qubits. Record the outcome $|x\rangle$.
2. With the n outcomes $x_1, x_2, \ldots, x_n$, solve the following system of linear equations:
$$\begin{cases} x_1 \cdot a = 0 \\ x_2 \cdot a = 0 \\ \vdots \\ x_n \cdot a = 0. \end{cases}$$

The solution a is the period of f.

Circuit Diagram: Simon

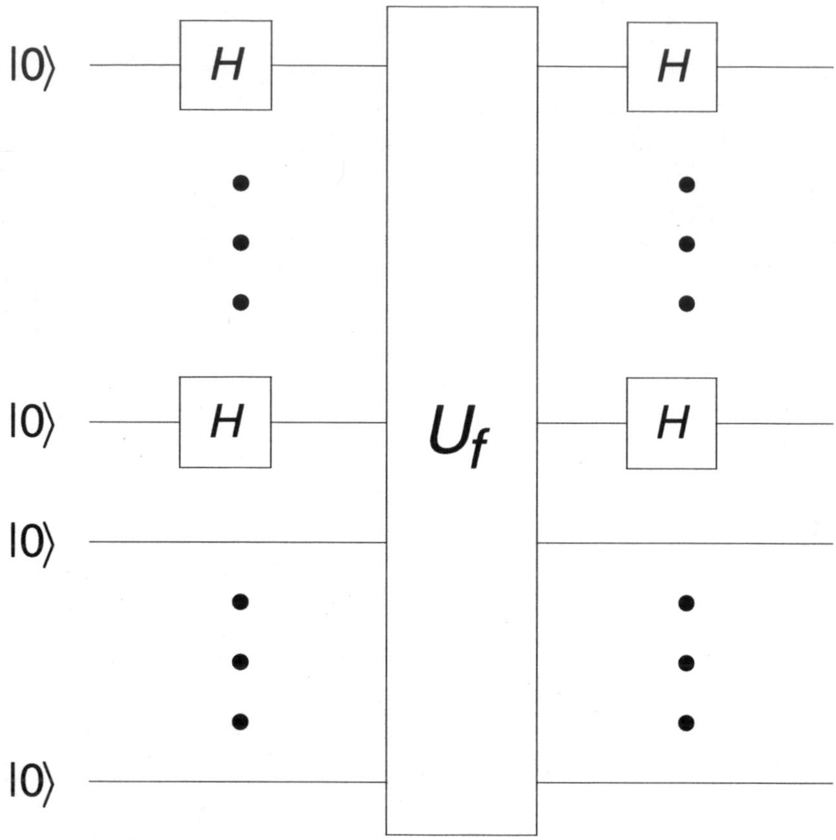

Facts:

The following facts can be found in [Sim94].

1. With the classical computer, we need exponentially many evaluations of f to determine the period a.
2. With the quantum computer, we need $O(n)$ evaluations (on average) of f to determine the period a.

Examples:

Let $f(00) = 01$, $f(01) = 11$, $f(10) = 01$, and $f(11) = 11$. The following sequence of quantum states shows the result of computation utilizing Simon's Algorithm. We start from the initial state $|00\rangle|00\rangle$. Then,

$$\begin{aligned}
&|00\rangle|00\rangle \\
&\to \frac{|0\rangle + |1\rangle}{\sqrt{2}}\frac{|0\rangle + |1\rangle}{\sqrt{2}}|00\rangle = \frac{|00\rangle + |01\rangle + |10\rangle + |11\rangle}{2}|00\rangle \\
&= \frac{|00\rangle|00\rangle}{2} + \frac{|01\rangle|00\rangle}{2} + \frac{|10\rangle|00\rangle}{2} + \frac{|11\rangle|00\rangle}{2} \\
&\to \frac{|00\rangle|01\rangle}{2} + \frac{|01\rangle|11\rangle}{2} + \frac{|10\rangle|01\rangle}{2} + \frac{|11\rangle|11\rangle}{2}
\end{aligned}$$

$$= \frac{|00\rangle + |10\rangle}{2}|01\rangle + \frac{|01\rangle + |11\rangle}{2}|11\rangle = \frac{|0\rangle + |1\rangle}{\sqrt{2}}\frac{|0\rangle}{\sqrt{2}}|01\rangle + \frac{|0\rangle + |1\rangle}{\sqrt{2}}\frac{|1\rangle}{\sqrt{2}}|11\rangle$$

$$\rightarrow |0\rangle\frac{|0\rangle + |1\rangle}{2}|01\rangle + |0\rangle\frac{|0\rangle - |1\rangle}{2}|11\rangle = \frac{|00\rangle + |01\rangle}{2}|01\rangle + \frac{|00\rangle - |01\rangle}{2}|11\rangle.$$

The outcome of measuring the first 2 qubits yields either $|00\rangle$ or $|01\rangle$. Suppose that we have run the computation above twice and obtained $|00\rangle$ and $|01\rangle$, respectively. We now have a system of linear equations:

$$\begin{cases} 00 \cdot a = 0 \\ 01 \cdot a = 0. \end{cases}$$

The solution is $a = 10$, the period of this function.

62.7 Grover's Search Algorithm

Problem: Grover

Given an unsorted database with N items, find a target item w. This problem can be formulated using an oracle function $f : \{0, 1, \ldots, N-1\} \rightarrow \{0, 1\}$, where

$$f(x) = \begin{cases} 0, & \text{if } x \neq w, \\ 1, & \text{if } x = w. \end{cases}$$

Given such an oracle function, find the w such that $f(w) = 1$.

Algorithm: Grover

Without loss of generality, let $N = 2^n$.

1. Prepare an $(n+1)$-bit quantum register and initialize it to the state $(|0\rangle)^n|1\rangle$.
2. Apply the Walsh–Hadamard transform to all the $n+1$ qubits.
3. Repeat the following procedure for about $\frac{\pi}{4}\sqrt{N}$ times. Cf. Figure 62.1.
   (a) Apply the function evaluation operator U_f (selective sign flipping operator):

   $$U_f : |x\rangle|y\rangle \rightarrow |x\rangle|y \oplus f(x)\rangle.$$

   This is equivalent to the unitary operator $\mathcal{I}_w = I - 2|w\rangle\langle w|$ on the first n qubits.

   (b) Apply unitary operator (inversion about the average operator) $\mathcal{I}_s = 2|s\rangle\langle s| - I$ on the first n qubits. Cf. Figure 62.2.
4. Measure the first n qubits. We obtain the search target with high probability.

Facts:

1. [Gro97] With the classical computer, on average, we need $O(N)$ oracle calls (evaluations of f) to find the search target.
2. [Gro97] With the quantum computer, on average, we need $O(\sqrt{N})$ oracle calls to find the search target.

Circuit Diagrams: Grover

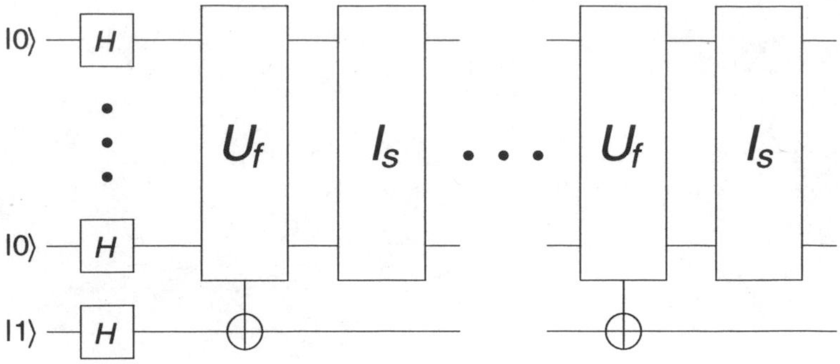

FIGURE 62.1 Grover's Algorithm.

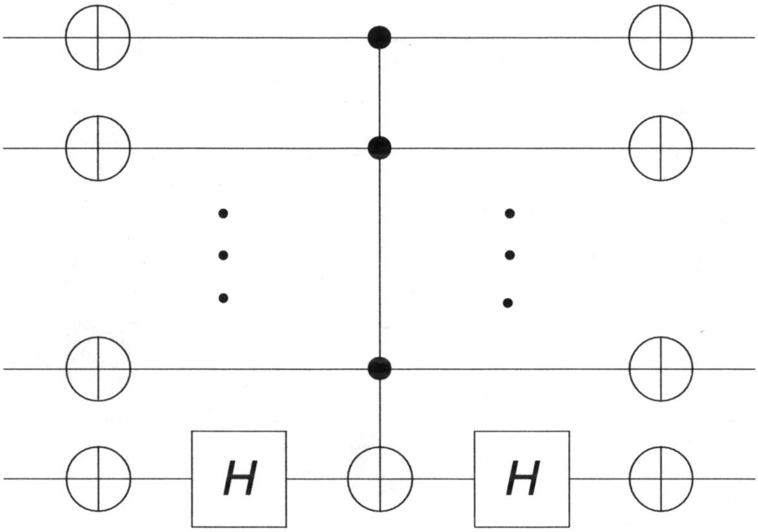

FIGURE 62.2 Inversion about the average operator $\mathcal{I}_s$.

3. [BBH98] When $N = 4$, using Grover's Algorithm, exactly one oracle call suffices to find the search target with certainty.

Examples:

Let $N = 2^2 = 4$ and Item 3 be the search target, which is encoded by the quantum state $|11\rangle$ (11 is the binary representation of 3). The following sequence of quantum states shows the result of computation utilizing Grover's Algorithm. We start from the initial state $|00\rangle|1\rangle$. Then,

$$\begin{aligned}
&|00\rangle|1\rangle \\
\to\ & \frac{|0\rangle + |1\rangle}{\sqrt{2}}\frac{|0\rangle + |1\rangle}{\sqrt{2}}\frac{|0\rangle - |1\rangle}{\sqrt{2}} = \frac{|00\rangle + |01\rangle + |10\rangle + |11\rangle}{2}\frac{|0\rangle - |1\rangle}{\sqrt{2}} \\
=\ & \frac{|00\rangle}{2}\frac{|0\rangle - |1\rangle}{\sqrt{2}} + \frac{|01\rangle}{2}\frac{|0\rangle - |1\rangle}{\sqrt{2}} + \frac{|10\rangle}{2}\frac{|0\rangle - |1\rangle}{\sqrt{2}} + \frac{|11\rangle}{2}\frac{|0\rangle - |1\rangle}{\sqrt{2}} \\
\to\ & \frac{|00\rangle}{2}\frac{|0\rangle - |1\rangle}{\sqrt{2}} + \frac{|01\rangle}{2}\frac{|0\rangle - |1\rangle}{\sqrt{2}} + \frac{|10\rangle}{2}\frac{|0\rangle - |1\rangle}{\sqrt{2}} + \frac{|11\rangle}{2}\frac{|1\rangle - |0\rangle}{\sqrt{2}}
\end{aligned}$$

$$= \frac{|00\rangle}{2}\frac{|0\rangle - |1\rangle}{\sqrt{2}} + \frac{|01\rangle}{2}\frac{|0\rangle - |1\rangle}{\sqrt{2}} + \frac{|10\rangle}{2}\frac{|0\rangle - |1\rangle}{\sqrt{2}} - \frac{|11\rangle}{2}\frac{|0\rangle - |1\rangle}{\sqrt{2}}$$

$$= \frac{|00\rangle + |01\rangle + |10\rangle - |11\rangle}{2}\frac{|0\rangle - |1\rangle}{\sqrt{2}}$$

$$\rightarrow |11\rangle\frac{|0\rangle - |1\rangle}{\sqrt{2}}.$$

The outcome of measuring the first 2 qubits yields $|11\rangle$, which is the search target $w = 3$.

Comments:

Grover's Algorithm was discovered by L.K. Grover of Bell Labs in 1996. This algorithm provides a quadratic speedup over classical algorithms. Although it is not exponentially fast (as Shor's Algorithm is), it has a wide range of applications. It could be used to accelerate any algorithms related to searching an unsorted database, including quantum database search, finding the solution of NP problems, finding the median and minimum of a data set, and breaking the Data Encryption Standard (DES) cryptography system.

62.8 Shor's Factorization Algorithm

Integer Factorization Problem:

Given a composite positive integer N, factor it into the product of its prime factors.

Algorithm: Shor

1. Choose a random number $a < N$; make sure that a and N are coprime. This can be done by using a random number generator and Euclidean algorithm on a classical computer.
2. Find the period T of function $f_{a,N}(x) = a^x \ mod \ N$. This step can be further expanded as follows:

   (a) Prepare two L-bit quantum registers in initial state

   $$\left(\frac{1}{\sqrt{2^L}}\sum_{x=0}^{2^L-1}|x\rangle\right)|0\rangle,$$

   where L is chosen such that $N^2 \leq 2^L < 2N^2$.

   (b) Apply the function evaluation operator $U_f : |x\rangle|0\rangle \rightarrow |x\rangle|f_{a,N}(x)\rangle$:

   $$\frac{1}{\sqrt{2^L}}\sum_{x=0}^{2^L-1}|x\rangle|0\rangle \rightarrow \frac{1}{\sqrt{2^L}}\sum_{x=0}^{2^L-1}|x\rangle|f_{a,N}(x)\rangle.$$

   (c) Apply QFT to the first register:

   $$\frac{1}{\sqrt{2^L}}\sum_{x=0}^{2^L-1}|x\rangle|f_{a,N}(x)\rangle \rightarrow \frac{1}{2^L}\sum_{y=0}^{2^L-1}\left(\sum_{x=0}^{2^L-1}e^{2\pi ixy/2^L}|y\rangle\right)|f(x)\rangle.$$

   (d) Make a measurement on the first register, obtaining y.

   (e) Find T from y via the continued fraction for $\frac{y}{2^L}$. This step might fail; in that case, repeat from 2a.
3. If T is odd, repeat from Step 1. If T is even and $N \mid (a^{T/2} + 1)$, repeat from Step 1. If T is even and $N \nmid (a^{T/2} + 1)$, compute $d = gcd(a^{T/2} - 1, N)$, which is a nontrivial factor of N.

Circuit Diagrams: Shor

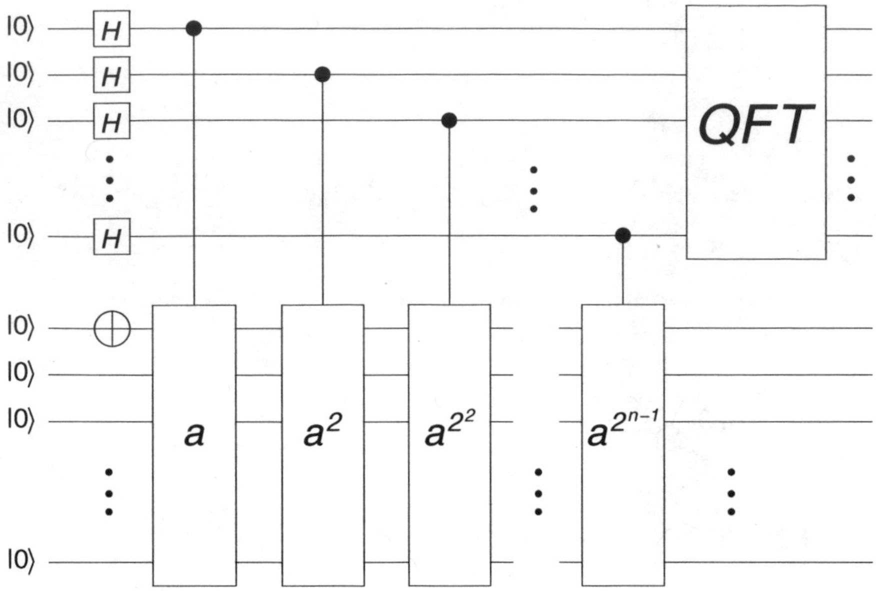

Facts:

The following facts can be found in [Sho94].

1. The integer factorization problem is classically *intractable*. The most efficient classical algorithm to date, a *number field sieve*, has a time complexity of $O(\exp(\log N)^{1/3}(\log\log N)^{2/3})$.
2. Shor's quantum factorization algorithm has a time complexity of $O((\log N)^2 \log\log N \log\log\log N)$. Hence, it is a polynomial time algorithm.

Examples:

Let $N = 15$. Choose $L = 8$ such that $N^2 = 225 < 2^L < 450 = 2N^2$. Choose a random integer $a = 2$, which is coprime with 15. Thus, $f_{a,N}(x) = 2^x \bmod 15$. The following sequence of quantum states shows the result of computation utilizing Shor's Algorithm:

$$|0\rangle|0\rangle \to \frac{1}{2^4}\sum_{x=0}^{2^8-1}|x\rangle|0\rangle$$

$$\to \frac{1}{2^4}\sum_{x=0}^{2^8-1}|x\rangle|f_{a,N}(x)\rangle$$

$$= \frac{1}{2^4}(|0\rangle|1\rangle + |1\rangle|2\rangle + |2\rangle|4\rangle + |3\rangle|8\rangle$$

$$+ |4\rangle|1\rangle + |5\rangle|2\rangle + |6\rangle|4\rangle + |7\rangle|8\rangle + \cdots + |2^8-2\rangle|4\rangle + |2^8-1\rangle|8\rangle)$$

$$\to \frac{1}{2^4}\sum_{x=0}^{2^8-1}\frac{1}{2^4}\sum_{y=0}^{2^8-1}\omega^{xy}|y\rangle|2^x \bmod 15\rangle$$

$$\to \frac{1}{2^8}\sum_{y=0}^{2^8-1}|y\rangle\sum_{x=0}^{2^8-1}\omega^{xy}|2^x \bmod 15\rangle,$$

where $\omega = e^{2\pi i/2^8}$. Suppose that the outcome of measuring the first n qubits is $|56\rangle$. We can compute the continued fraction of $\frac{56}{256}$ to be $[0, 4, \ldots]$. The second number 4 satisfies $2^4 = 1 \bmod 15$. So 4 is the period of $f_{a,N}$. $2^{4/2} - 1 = 3$ yields a factor of 15 and $15 = 3 \times 5$.

Comments:

Shor's Algorithm was discovered by P. Shor of AT&T Labs in 1994. It is the most important breakthrough in the research of quantum computation so far. It solves the integer factorization problem, an extremely hard problem for classical computers, in polynomial time. The security of the RSA cryptographic system, which is widely used nowadays over the Internet, is based on the difficulty of factoring large integers. Equipped with a quantum computer, one could easily break the RSA codes with Shor's Algorithm.

References

[BV93] E. Bernstein and U. Vazirani. Quantum complexity theory. *Proc. of the 25th Annual ACM Symposium on the Theory of Computing*, San Diego, CA, 11–20, 1993.

[BBH98] M. Boyer, G. Brassard, P. Hoyer, and A. Tapp. Tight bounds on quantum searching. *Fortsch. Phys.* 46:493–506, 1998.

[BB02] J. L. Brylinski and R. Brylinski. Universal quantum gates. *Mathematics of Quantum Computation* (R. Brylinski and G. Chen, Eds.). Chapman & Hall/CRC Press, Boca Raton, FL, 101–116, 2002.

[CEM98] R. Cleve, A. Ekert, C. Macchiavello, and M. Mosca. Quantum algorithms revisited. *Proc. R. Soc. London A*, 454:339–354, 1998.

[Fey82] R. Feynman. Simulating physics with computers. *Int. J. Theor. Phys.*, 21:467–488, 1982.

[Gro97] L.K. Grover. Quantum mechanics helps in searching for a needle in a haystack. *Phys. Rev. Lett.*, 78:325–328, 1997.

[NC00] M.A. Nielsen and I.L. Chuang. *Quantum Computation and Quantum Information.* Cambridge University Press, Cambridge, U.K., 2000.

[Sho94] P. Shor. Algorithms for quantum computation: Discrete logarithms and factoring. *Proc. of the 35th Annual IEEE Symposium on the Foundations of Computer Science*, Santa Fe, NM, 124–134, 1994.

[Sim94] D. Simon. On the power of quantum computation. *Proc. of the 35th Annual IEEE Symposium on Foundations of Computer Science*, Santa Fe, NM, 116–123, 1994.

63

Information Retrieval and Web Search

Amy N. Langville
The College of Charleston

Carl D. Meyer
North Carolina State University

Information retrieval is the process of searching within a document collection for information most relevant to a user's query. However, the type of document collection significantly affects the methods and algorithms used to process queries. In this chapter, we distinguish between two types of document collections: traditional and Web collections. Traditional information retrieval is search within small, controlled, nonlinked collections (e.g., a collection of medical or legal documents), whereas Web information retrieval is search within the world's largest and linked document collection. In spite of the proliferation of the Web, more traditional nonlinked collections still exist, and there is still a place for the older methods of information retrieval.

63.1 The Traditional Vector Space Method

Today most search systems that deal with traditional document collections use some form of the vector space method [SB83] developed by Gerard Salton in the early 1960s. Salton's method transforms textual data into numeric vectors and matrices and employs matrix analysis techniques to discover key features and connections in the document collection.

Definitions:

For a given collection of documents and for a dictionary of m terms, document i is represented by an $m \times 1$ **document vector** $\mathbf{d}_i$ whose jth element is the number of times term j appears in document i.

The **term-by-document matrix** is the $m \times n$ matrix

$$A = [\, \mathbf{d}_1\, \mathbf{d}_2\, \cdots\, \mathbf{d}_n \,]$$

whose columns are the document vectors.

Recall is a measure of performance that is defined to be

$$0 \leq \text{Recall} = \frac{\#\text{ relevant docs retrieved}}{\#\text{ relevant docs in collection}} \leq 1.$$

Precision is another measure of performance, defined to be

$$0 \leq \text{Precision} = \frac{\#\text{ relevant docs retrieved}}{\#\text{ docs retrieved}} \leq 1.$$

Query processing is the act of retrieving documents from the collection that are most related to a user's query, and the **query vector** $\mathbf{q}_{m\times 1}$ is the binary vector defined by

$$q_i = \begin{cases} 1 & \text{if term } i \text{ is present in the user's query,} \\ 0 & \text{otherwise.} \end{cases}$$

The **relevance** of document i to a query $\mathbf{q}$ is defined to be

$$\delta_i = cos\,\theta_i = \mathbf{q}^T\mathbf{d}_i / \|\mathbf{q}\|_2 \|\mathbf{d}_i\|_2.$$

For a selected tolerance τ, the **retrieved documents** that are returned to the user are the documents for which $\delta_i > \tau$.

Facts:

1. The term-by-document matrix A is sparse and nonnegative, but otherwise unstructured.
2. [BB05] In practice, weighting schemes other than raw frequency counts are used to construct the term-by-document matrix because weighted frequencies can improve performance.
3. [BB05] Query weighting may also be implemented in practice.
4. The tolerance τ is usually tuned to the specific nature of the underlying document collection.
5. Tuning can be accomplished with the technique of relevance feedback, which uses a revised query vector such as $\tilde{\mathbf{q}} = \delta_1\mathbf{d}_1 + \delta_3\mathbf{d}_3 + \delta_7\mathbf{d}_7$, where $\mathbf{d}_1$, $\mathbf{d}_3$, and $\mathbf{d}_7$ are the documents the user judges most relevant to a given query $\mathbf{q}$.
6. When the columns of A and $\mathbf{q}$ are normalized, as they usually are, the vector $\boldsymbol{\delta}^T = \mathbf{q}^T A$ provides the complete picture of how well each document in the collection matches the query.
7. The vector space model is efficient because A is usually very sparse, and $\mathbf{q}^T A$ can be executed in parallel, if necessary.
8. [BB05] Because of linguistic issues such as polysomes and synonyms, the vector space model provides only decent performance on query processing tasks.
9. The underlying basis for the vector space model is the standard basis $\mathbf{e}_1, \mathbf{e}_2, \ldots, \mathbf{e}_m$, and the orthogonality of this basis can impose an unrealistic independence among terms.
10. The vector space model is a good starting place, but variations have been developed that provide better performance.

Examples:

1. Consider a collection of seven documents and nine terms (taken from [BB05]). Terms not in the system's index are ignored. Suppose further that only the titles of each document are used for indexing. The indexed terms and titles of documents are shown below.

Terms	Documents
T1: Bab(y,ies,y's)	D1: *Infant* & *Toddler* First Aid
T2: Child(ren's)	D2: *Babies* and *Children's* Room (For Your *Home*)
T3: Guide	D3: *Child Safety* at *Home*
T4: Health	D4: Your *Baby's Health* and *Safety*: From *Infant* to *Toddler*
T5: Home	D5: *Baby Proofing* Basics
T6: Infant	D6: Your *Guide* to Easy Rust *Proofing*
T7: Proofing	D7: Beanie *Babies* Collector's *Guide*
T8: Safety	
T9: Toddler	

The indexed terms are italicized in the titles. Also, the stems [BB05] of the terms for baby (and its variants) and child (and its variants) are used to save storage and improve performance. The term-by-document matrix for this document collection is

$$A = \begin{bmatrix} 0 & 1 & 0 & 1 & 1 & 0 & 1 \\ 0 & 1 & 1 & 0 & 0 & 0 & 0 \\ 0 & 0 & 0 & 0 & 0 & 1 & 1 \\ 0 & 0 & 0 & 1 & 0 & 0 & 0 \\ 0 & 1 & 1 & 0 & 0 & 0 & 0 \\ 1 & 0 & 0 & 1 & 0 & 0 & 0 \\ 0 & 0 & 0 & 0 & 1 & 1 & 0 \\ 0 & 0 & 1 & 1 & 0 & 0 & 0 \\ 1 & 0 & 0 & 1 & 0 & 0 & 0 \end{bmatrix}.$$

For a query on *baby health*, the query vector is

$$\mathbf{q} = [1 \quad 0 \quad 0 \quad 1 \quad 0 \quad 0 \quad 0 \quad 0 \quad 0]^T.$$

To process the user's query, the cosines

$$\delta_i = cos\,\theta_i = \frac{\mathbf{q}^T\mathbf{d}_i}{\|\mathbf{q}\|_2\|\mathbf{d}_i\|_2}$$

are computed. The documents corresponding to the largest elements of $\boldsymbol{\delta}$ are most relevant to the user's query. For our example,

$$\boldsymbol{\delta} \approx [0 \quad 0.40824 \quad 0 \quad 0.63245 \quad 0.5 \quad 0 \quad 0.5],$$

so document vector 4 is scored most relevant to the query on *baby health*. To calculate the recall and precision scores, one needs to be working with a small, well-studied document collection. In this example, documents $\mathbf{d}_4$, $\mathbf{d}_1$, and $\mathbf{d}_3$ are the three documents in the collection relevant to baby health. Consequently, with $\tau = .1$, the recall score is 1/3 and the precision is 1/4.

63.2 Latent Semantic Indexing

In the 1990s, an improved information retrieval system replaced the vector space model. This system is called Latent Semantic Indexing (LSI) [Dum91] and was the product of Susan Dumais, then at Bell Labs. LSI simply creates a low rank approximation A_k to the term-by-document matrix A from the vector space model.

Facts:

1. [Mey00] If the term-by-document matrix $A_{m\times n}$ has the singular value decomposition $A = U\Sigma\ V^T = \sum_{i=1}^{r} \sigma_i \mathbf{u}_i \mathbf{v}_i^T$, $\sigma_1 \geq \sigma_2 \geq \cdots \geq \sigma_r > 0$, then A_k is created by truncating this expansion after k terms, where k is a user tunable parameter.
2. The recall and precision measures are generally used in conjunction with each other to evaluate performance.
3. A is replaced by $A_k = \sum_{i=1}^{k} \sigma_i \mathbf{u}_i \mathbf{v}_i^T$ in the query process so that if $\mathbf{q}$ and the columns of A_k have been normalized, then the angle vector is computed as $\boldsymbol{\delta}^T = \mathbf{q}^T A_k$.
4. The truncated SVD approximation to A is optimal in the sense that of all rank-k matrices, the truncated SVD A_k is the closest to A, and

$$\|A - A_k\|_F = \min_{rank(B)\leq k} \|A - B\|_F = \sqrt{\sigma_{k+1}^2 + \cdots + \sigma_r^2}.$$

5. This rank-k approximation reduces the so-called linguistic noise present in the term-by-document matrix and, thus, improves information retrieval performance.
6. [Dum91], [BB05], [BR99], [Ber01], [BDJ99] LSI is known to outperform the vector space model in terms of precision and recall.
7. [BR99], [Ber01], [BB05], [BF96], [BDJ99], [BO98], [Blo99], [BR01], [Dum91], [HB00], [JL00], [JB00], [LB97], [WB98], [ZBR01], [ZMS98] LSI and the truncated singular value decomposition dominated text mining research in the 1990s.
8. A serious drawback to LSI is that while it might appear at first glance that A_k should save storage over the original matrix A, this is often not the case, even when $k << r$. This is because A is generally very sparse, but the singular vectors $\mathbf{u}_i$ and $\mathbf{v}_i^T$ are almost always completely dense. In many cases, A_k requires more (sometimes much more) storage than A itself requires.
9. A significant problem with LSI is the fact that while A is a nonnegative matrix, the singular vectors are mixed in sign. This loss of important structure means that the truncated singular value decomposition provides no textual or semantic interpretation. Consider a particular document vector, say, column 1 of A. The truncated singular value decomposition represents document 1 as

$$A_1 = \begin{bmatrix} \vdots \\ \mathbf{u}_1 \\ \vdots \end{bmatrix} \sigma_1 v_{11} + \begin{bmatrix} \vdots \\ \mathbf{u}_2 \\ \vdots \end{bmatrix} \sigma_2 v_{12} + \cdots \begin{bmatrix} \vdots \\ \mathbf{u}_k \\ \vdots \end{bmatrix} \sigma_k v_{1k},$$

   so document 1 is a linear combination of the basis vectors $\mathbf{u}_i$ with the scalar $\sigma_i v_{1i}$ being a weight that represents the contribution of basis vector i in document 1. What we would really like to do is say that basis vector i is mostly concerned with some subset of the terms, but any such textual or semantic interpretation is difficult (or impossible) when SVD components are involved. Moreover, if there were textual or semantic interpretations, the orthogonality of the singular vectors would ensure that there is no overlap of terms in the topics in the basis vectors, which is highly unrealistic.
10. [Ber01], [ZMS98] It is usually a difficult problem to determine the most appropriate value of k for a given dataset because k must be large enough so that A_k can capture the essence of the document collection, but small enough to address storage and computational issues. Various heuristics have been developed to deal with this issue.

Examples:

1. Consider again the 9×7 term-by-document matrix used in section 63.1. The rank-4 approximation to this matrix is

$$A_4 = \begin{bmatrix} 0.020 & 1.048 & -0.034 & 0.996 & 0.975 & 0.027 & 0.975 \\ -0.154 & 0.883 & 1.067 & 0.078 & 0.027 & -0.033 & 0.027 \\ -0.012 & -0.019 & 0.013 & 0.004 & 0.509 & 0.990 & 0.509 \\ 0.395 & 0.058 & 0.020 & 0.756 & 0.091 & -0.087 & 0.091 \\ -0.154 & 0.883 & 1.067 & 0.078 & 0.027 & -0.033 & 0.027 \\ 0.723 & -0.144 & 0.068 & 1.152 & 0.004 & -0.012 & 0.004 \\ -0.012 & -0.019 & 0.013 & 0.004 & 0.509 & 0.990 & 0.509 \\ 0.443 & 0.334 & 0.810 & 0.776 & -0.074 & 0.091 & -0.074 \\ 0.723 & -0.144 & 0.068 & 1.152 & 0.004 & -0.012 & 0.004 \end{bmatrix}.$$

Notice that while A is sparse and nonnegative, A_4 is dense and mixed in sign. Of course, as k increases, A_k looks more and more like A. For a query on *baby health*, the angle vector is

$$\delta \approx [\,.244 \quad .466 \quad -.006 \quad .564 \quad .619 \quad -.030 \quad .619\,]^T.$$

Thus, the information retrieval system returns documents $\mathbf{d}_5$, $\mathbf{d}_7$, $\mathbf{d}_4$, $\mathbf{d}_2$, $\mathbf{d}_1$, in order from most to least relevant. As a result, the recall improves to 2/3, while the precision is 2/5. Adding another singular triplet and using the approximation matrix A_5 does not change the recall or precision measures, but does give a slightly different angle vector

$$\delta \approx [\,.244 \quad .466 \quad -.006 \quad .564 \quad .535 \quad -.030 \quad .535\,]^T,$$

which is better than the A_4 angle vector because the most relevant document, $\mathbf{d}_4$, *Your Baby's Health and Safety: From Infant to Toddler*, gets the highest score.

63.3 Nonnegative Matrix Factorizations

The lack of semantic interpretation due to the mixed signs in the singular vectors is a major obstacle in using LSI. To circumvent this problem, alternative low rank approximations that maintain the nonnegative structure of the original term-by-document matrix have been proposed [LS99], [LS00], [PT94], [PT97].

Facts:

1. If $A_{m\times n} \geq 0$ has rank r, then for a given $k < r$ the goal of a nonnegative matrix factorization (NMF) is to find the nearest rank-k approximation WH to A such that $W_{m\times k} \geq 0$ and $H_{k\times n} \geq 0$. Once determined, an NMF simply replaces the truncated singular value decomposition in any text mining task such as clustering documents, classifying documents, or processing queries on documents.
2. An NMF can be formulated as a constrained nonlinear least squares problem by first specifying k and then determining

$$\min \|A - WH\|_F^2 \quad \text{subject to} \quad W_{m\times k} \geq 0, \quad H_{k\times n} \geq 0.$$

The rank of the approximation (i.e., k) becomes the number of topics or clusters in a text mining application.

3. [LS99] The Lee and Seung algorithm to compute an NMF using MATLAB is as follows.

Algorithm 1: Lee–Seung NMF

```
W = abs(randn(m, k))                      % initialize with random dense matrix
H = abs(randn(k, n))                      % initialize with random dense matrix
for i = 1 : maxiter
H = H. * (W^T A)./(W^T WH + 10^-9)        % 10^-9 avoids division by 0
W = W. * (AH^T)./(WHH^T + 10^-9)
end
```

4. The objective function $\|A - WH\|_F^2$ in the Lee and Seung algorithm tends to tail off within 50 to 100 iterations. Faster algorithms exist, but the Lee and Seung algorithm is guaranteed to converge to a local minimum in a finite number of steps.
5. [Hoy02], [Hoy04], [SBP04] Other NMF algorithms contain a tunable sparsity parameter that produces any desired level of sparseness in W and H. The storage savings of the NMF over the truncated SVD are substantial.
6. Because $A_j \approx \sum_{i=1}^{k} W_i h_{ij}$, and because W and H are nonnegative, each column W_i can be viewed as a topic vector—if $w_{ij_1}, w_{ij_2}, \ldots, w_{ij_p}$ are the largest entries in W_i, then terms $j_1, j_2, \ldots, j_p$ dictate the topics that W_i represents. The entries h_{ij} measure the strength to which topic i appears in basis document j, and k is the number of topic vectors that one expects to see in a given set of documents.
7. The NMF has some disadvantages. Unlike the SVD, uniqueness and robust computations are missing in the NMF. There is no unique global minimum for the NMF (the defining constrained least squares problem is not convex in W and H), so algorithms can only guarantee convergence to a local minimum, and many do not even guarantee that.
8. Not only will different NMF algorithms produce different NMF factors, the same NMF algorithm, run with slightly different parameters, can produce very different NMF factors. For example, the results can be highly dependent on the initial values.

Examples:

1. When the term-by-document matrix of the MEDLINE dataset [Med03] is approximated with an NMF as described above with $k = 10$, the charts in Figure 63.1 show the highest weighted terms from four representative columns of W. For example, this makes it clear that W_1 represents heart related topics, while W_2 concerns blood issues, etc.

   When document 5 (column A_5) from MEDLINE is expressed as an approximate linear combination of $W_1, W_2, \ldots, W_{10}$ in order of the size of the entries of H_5, which are

$$h_{95} = .1646 > h_{65} = .0103 > h_{75} = .0045 > \cdots,$$

   we have that

$$A_5 \approx .1646\, W_9 + .0103\, W_6 + .0045\, W_7 + \cdots$$

$$= .1646 \begin{bmatrix} \text{fatty} \\ \text{glucose} \\ \text{acids} \\ \text{ffa} \\ \text{insulin} \\ \vdots \end{bmatrix} + .0103 \begin{bmatrix} \text{kidney} \\ \text{marrow} \\ \text{dna} \\ \text{cells} \\ \text{nephr.} \\ \vdots \end{bmatrix} + .0045 \begin{bmatrix} \text{hormone} \\ \text{growth} \\ \text{hgh} \\ \text{pituitary} \\ \text{mg} \\ \vdots \end{bmatrix} + \cdots.$$

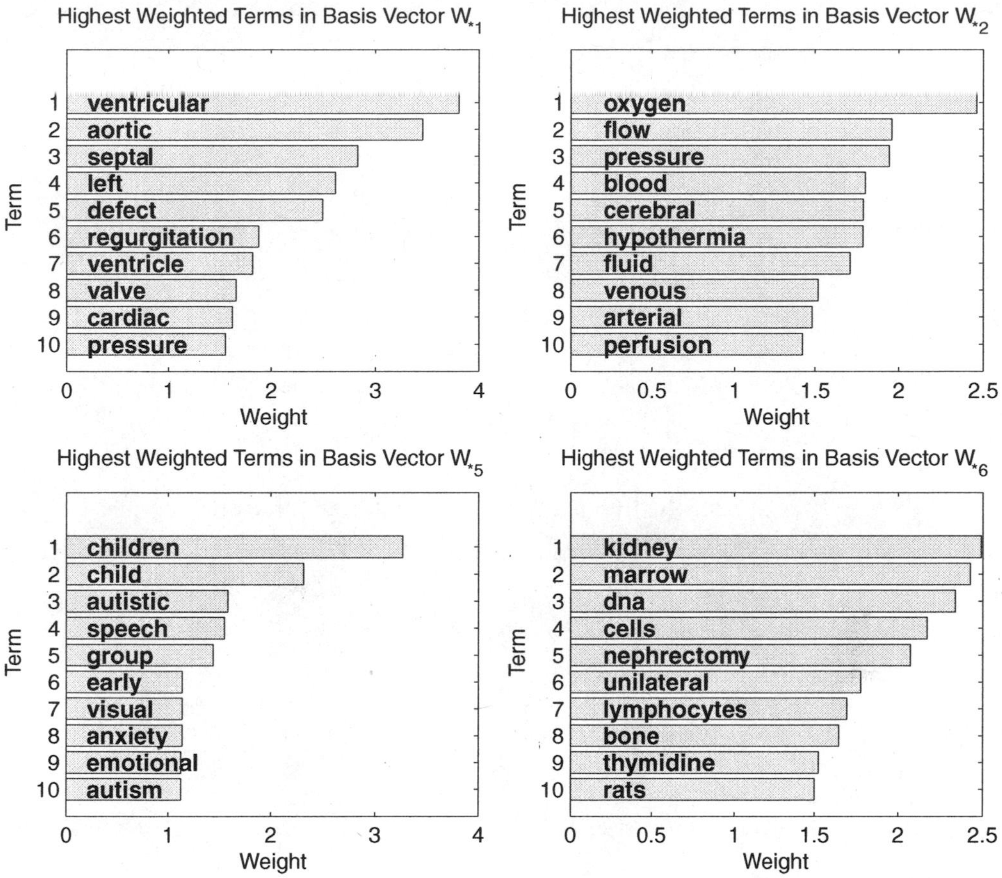

FIGURE 63.1 MEDLINE charts.

Therefore, document 5 is largely about terms contained in topic vector W_9 followed by topic vectors W_6 and W_7.

2. Consider the same 9×7 term-by-document matrix A from the example in Section 63.1. A rank-4 approximation $A_4 = W_{9\times 4}H_{4\times 7}$ that is produced by the Lee and Seung algorithm is

$$A_4 = \begin{bmatrix} 0.027 & 0.888 & 0.196 & 1.081 & 0.881 & 0.233 & 0.881 \\ 0 & 0.852 & 1.017 & 0.173 & 0.058 & 0.031 & 0.058 \\ 0.050 & 0.084 & 0.054 & 0.102 & 0.496 & 0.899 & 0.496 \\ 0.360 & 0.172 & 0.073 & 0.729 & 0.179 & 0.029 & 0.179 \\ 0 & 0.852 & 1.017 & 0.173 & 0.058 & 0.031 & 0.058 \\ 0.760 & 0.032 & 0.155 & 1.074 & 0.033 & 0.061 & 0.033 \\ 0.050 & 0.084 & 0.054 & 0.102 & 0.496 & 0.899 & 0.496 \\ 0.445 & 0.481 & 0.647 & 0.718 & 0.047 & 0.053 & 0.047 \\ 0.760 & 0.032 & 0.155 & 1.074 & 0.033 & 0.061 & 0.033 \end{bmatrix},$$

where

$$W_4 = \begin{bmatrix} 0.202 & 0.017 & 0.160 & 1.357 \\ 0 & 0 & 0.907 & 0.104 \\ 0.805 & 0 & 0 & 0.008 \\ 0 & 0.415 & 0 & 0.321 \\ 0 & 0 & 0.907 & 0.104 \\ 0 & 0.875 & 0 & 0.060 \\ 0.805 & 0 & 0 & 0.008 \\ 0 & 0.513 & 0.500 & 0.085 \\ 0 & 0.876 & 0 & 0.060 \end{bmatrix},$$

$$H_4 = \begin{bmatrix} 0.062 & 0.010 & 0.067 & 0.119 & 0.610 & 1.117 & 0.610 \\ 0.868 & 0 & 0.177 & 1.175 & 0 & 0.070 & 0 \\ 0 & 0.878 & 1.121 & 0.105 & 0 & 0.034 & 0 \\ 0 & 0.537 & 0 & 0.752 & 0.559 & 0 & 0.559 \end{bmatrix}.$$

Notice that both A and A_4 are nonnegative, and the sparsity of W and H makes the storage savings apparent. The error in this NMF approximation as measured by $\|A - WH\|_F^2$ is 1.56, while the error in the best rank-4 approximation from the truncated SVD is 1.42. In other words, the NMF approximation is not far from the optimal SVD approximation — this is frequently the case in practice in spite of the fact that W and H can vary with the initial conditions. For a query on *baby health*, the angle vector is

$$\delta \approx [\,.224 \quad .472 \quad .118 \quad .597 \quad .655 \quad .143 \quad .655\,]^T.$$

Thus, the information retrieval system that uses the nonnegative matrix factorization gives the same ranking as a system that uses the truncated singular value decomposition. However, the factors are sparse and nonnegative and can be interpreted.

63.4 Web Search

Only a few years ago, users of Web search engines were accustomed to waiting, for what would now seem to be an eternity, for search engines to return results to their queries. And when a search engine finally responded, the returned list was littered with links to information that was either irrelevant, unimportant, or downright useless. Frustration was compounded by the fact that useless links invariably appeared at or near the top of the list while useful links were deeply buried. Users had to sift through links a long way down in the list to have a chance of finding something satisfying, and being less than satisfied was not uncommon.

The reason for this is that the Web's information is not structured like information in the organized databases and document collections that generations of computer scientists had honed their techniques on. The Web is unique in the sense that it is *self organized*. That is, unlike traditional document collections that are accumulated, edited, and categorized by trained specialists, the Web has no standards, no reviewers, and no gatekeepers to police content, structure, and format. Information on the Web is volatile and heterogeneous — links and data are rapidly created, changed, and removed, and Web information exists in multiple formats, languages, and alphabets. And there is a multitude of different purposes for Web data, e.g., some Web pages are designed to inform while others try to sell, cheat, steal, or seduce. In addition, the Web's self organization opens the door for spammers, the nefarious people who want to illicitly commandeer your attention to sell or advertise something that you probably do not want. Web spammers are continually devising diabolical schemes to trick search engines into listing their (or their client's) Web pages near the top of the list of search results. They had an easy time of it when Web search relied on traditional IR methods based on semantic principles. Spammers could create Web pages that contained things such as miniscule or hidden text fonts, hidden text with white fonts on a white background, and misleading metatag descriptions designed to influence semantic based search engines. Finally, the enormous size of the Web, currently containing $O(10^9)$ pages, completely overwhelmed traditional IR techniques.

By 1997 it was clear that in nearly all respects the database and IR technology of the past was not well suited for Web search, so researchers set out to devise new approaches. Two big ideas emerged (almost simultaneously), and each capitalizes on the link structure of the Web to differentiate between relevant information and fluff. One approach, HITS (Hypertext Induced Topic Search), was introduced by Jon Kleinberg [Kle99], [LM06], and the other, which changed everything, is Google's *PageRank*™ that was developed by Sergey Brin and Larry Page [BP98], [BPM99], [LM06]. While variations of HITS and PageRank followed (e.g., Lempel's SALSA [LM00], [LM05], [LM06]), the basic idea of PageRank became the driving force, so the focus is on this concept.

Definitions:

Early in the game, search companies such as Yahoo!® employed students to surf the Web and record key information about the pages they visited. This quickly overwhelmed human capability, so today all Web search engines use **Web Crawlers,** which is software that continuously scours the Web for information to return to a central repository.

Web pages found by the robots are temporarily stored in entirety in a **page repository.** Pages remain in the repository until they are sent to an indexing module, where their vital information is stripped to create a compressed version of the page. Popular pages that are repeatedly requested by users are stored here longer, perhaps indefinitely.

The **indexing module** extracts only key words, key phrases, or other vital descriptors, and it creates a compressed description of the page that can be "indexed." Depending on the popularity of a page, the uncompressed version is either deleted or returned to the page repository.

There are three general kinds of **indices** that contain compressed information for each Web page. The **content index** contains information such as key words or phrases, titles, and anchor text, and this is stored in a compressed form using an **inverted file** structure, which is simply the electronic version of a book index, i.e., each morsel of information points to a list of pages containing it. Information regarding the hyperlink structure of a page is stored in compressed form in the **structure index**. The crawler module sometimes accesses the structure index to find uncrawled pages. Finally, there are **special-purpose indices** such as an image index and a pdf index. The crawler, page repository, indexing module, and indices, along with their corresponding data files, exist and operate independent of users and their queries.

The **query module** converts a user's natural language query into a language that the search engine can understand (usually numbers), and consults the various indices in order to answer the query. For example, the query module consults the content index and its inverted file to find which pages contain the query terms.

The **pertinent pages** are the pages that contain query terms. After pertinent pages have been identified, the query module passes control to the ranking module.

The **ranking module** takes the set of pertinent pages and ranks them according to some criterion, and this criterion is the heart of the search engine — it is the distinguishing characteristic that differentiates one search engine from another. The ranking criterion must somehow discern which Web pages best respond to a user's query, a daunting task because there might be millions of pertinent pages. Unless a search engine wants to play the part of a censor (which most do not), the user is given the opportunity of seeing a list of links to a large proportion of the pertinent pages, but with less useful links permuted downward.

PageRank is Google's patented ranking system, and some of the details surrounding PageRank are discussed below.

Facts:

1. Google assigns at least two scores to each Web page. The first is a popularity score and the second is a content score. Google blends these two scores to determine the final ranking of the results that are returned in response to a user's query.
2. [BP98] The rules used to give each pertinent page a content score are trade secrets, but they generally take into account things such as whether the query terms appear in the title or deep in the body of a Web page, the number of times query terms appear in a page, the proximity of multiple query words to one another, and the appearance of query terms in a page (e.g., headings in bold font score higher). The content of neighboring Web pages is also taken into account.
3. Google is known to employ over a hundred such metrics in this regard, but the details are proprietary. While these metrics are important, they are secondary to the the popularity score, which is the primary component of PageRank. The content score is used by Google only to temper the popularity score.

63.5 Google's PageRank

The popularity score is where the mathematics lies, so it is the focus of the remainder of this exposition. We will identify the term "PageRank" with just the mathematical component of Google's PageRank (the popularity score) with the understanding that PageRank may be tweaked by a content score to produce a final ranking.

Both PageRank and Google were conceived by Sergey Brin and Larry Page while they were computer science graduate students at Stanford University, and in 1998 they took a leave of absence to focus on their growing business. In a public presentation at the Seventh International World Wide Web conference (WWW98) in Brisbane, Australia, their paper "The PageRank citation ranking: Bringing order to the Web" [BPM99] made small ripples in the information science community that quickly turned into waves.

The original idea was that a *page is important if it is pointed to by other important pages.* That is, the importance of your page (its PageRank) is determined by summing the PageRanks of all pages that point to yours. Brin and Page also reasoned that when an important page points to several places, its weight (PageRank) should be distributed proportionately.

In other words, if YAHOO! points to 99 pages in addition to yours, then you should only get credit for 1/100 of YAHOO!'s PageRank. This is the intuitive motivation behind Google's PageRank concept, but significant modifications are required to turn this basic idea into something that works in practice.

For readers who want to know more, the book *Google's PageRank and Beyond: The Science of Search Engine Rankings* [LM06] (Princeton University Press, 2006) contains over 250 pages devoted to link analysis algorithms, along with other ranking schemes such as HITS and SALSA as well as additional background material, examples, code, and chapters dealing with more advanced issues in Web search ranking.

Definitions:

The **hyperlink matrix** is the matrix $H_{n\times n}$ that represents the link structure of the Web, and its entries are given by

$$h_{ij} = \begin{cases} 1/|O_i| & \text{if there is a link from page } i \text{ to page } j, \\ 0 & \text{otherwise,} \end{cases}$$

where $|O_i|$ is the number of outlinks from page i.

Suppose that there are n Web pages, and let $r_i(0)$ denote the initial rank of the ith page. If the ranks are successively modified by setting

$$r_i(k+1) = \sum_{j \in I_i} \frac{r_j(k)}{|O_j|}, \quad k = 1, 2, 3, \ldots,$$

where $r_i(k)$ is the rank of page i at iteration k and I_i is the set of pages pointing (linking) to page i, then the rankings after the kth iteration are

$$\mathbf{r}^T(k) = (r_1(k), r_2(k), \ldots, r_n(k)) = \mathbf{r}^T(0)H^k.$$

The **conceptual PageRank** of the ith Web page is defined to be

$$r_i = \lim_{k \to \infty} r_i(k),$$

provided that the limit exists. However, this definition is strictly an intuitive concept because the natural structure of the Web generally prohibits these limits from existing.

A **dangling node** is a Web page that contain no out links. Dangling nodes produce zero rows in the hyperlink matrix H, so even if $\lim_{k\to\infty} H^k$ exists, the limiting vector $\mathbf{r}^T = \lim_{k\to\infty} \mathbf{r}^T(k)$ would be dependent on the initial vector $\mathbf{r}^T(0)$, which is not good.

The **stochastic hyperlink matrix** is produced by perturbing the hyperlink matrix to be stochastic. In particular,

$$S = H + \mathbf{a}\mathbf{1}^T/n, \tag{63.1}$$

where $\mathbf{a}$ is the column in which

$$a_i = \begin{cases} 1 & \text{if page } i \text{ is a dangling node,} \\ 0 & \text{otherwise.} \end{cases}$$

S is a stochastic matrix that is identical to H except that zero rows in H are replaced by $\mathbf{1}^T/n$ ($\mathbf{1}$ is a vector of 1s and $n = O(10^9)$, so entries in $\mathbf{1}^T/n$ are pretty small). The effect is to eliminate dangling nodes. Any probability vector $\mathbf{p}^T > 0$ can be used in place of the uniform vector.

The **Google matrix** is defined to be the stochastic matrix

$$G = \alpha S + (1-\alpha)E, \tag{63.2}$$

where $E = \mathbf{1}\mathbf{v}^T$ in which $\mathbf{v}^T > 0$ can be any probability vector. Google originally set $\alpha = .85$ and $\mathbf{v}^T = \mathbf{1}^T/n$. The choice of α is discussed later in Facts 12, 13, and 14.

The **personalization vector** is the vector $\mathbf{v}^T$ in $E = \mathbf{1}\mathbf{v}^T$ in Equation 63.2. Manipulating $\mathbf{v}^T$ gives Google the flexibility to make adjustments to PageRanks as well as to personalize them (thus, the name "personalization vector") [HKJ03], [LM04a].

The **PageRank vector** is the left Perron vector (i.e., stationary distribution) $\boldsymbol{\pi}^T$ of the Google matrix G. In particular, $\boldsymbol{\pi}^T(I - G) = 0$, where $\boldsymbol{\pi}^T > 0$ and $\|\boldsymbol{\pi}^T\|_1 = 1$. The components of this vector constitute Google's popularity score of each Web page.

Facts:

1. [Mey00, Chap. 8] The Google matrix G is a primitive stochastic matrix, so the spectral radius $\rho(G) = 1$ is a simple eigenvalue, and 1 is the only eigenvalue on the unit circle.
2. The iteration defined by $\boldsymbol{\pi}^T(k+1) = \boldsymbol{\pi}^T(k)G$ converges, independent of the starting vector, to a unique stationary probability distribution $\boldsymbol{\pi}^T$, which is the PageRank vector.
3. The irreducible aperiodic Markov chain defined by $\boldsymbol{\pi}^T(k+1) = \boldsymbol{\pi}^T(k)G$ is a constrained random walk on the Web graph. The random walker can be characterized as a Web surfer who, at each step, randomly chooses a link from his current page to click on except that

(a) When a dangling node is encountered, the excursion is continued by jumping to another page selected at random (i.e., with probability $1/n$).

(b) At each step, the random Web surfer has a chance (with probability $1 - \alpha$) of becoming bored with following links from the current page, in which case the random Web surfer continues the excursion by jumping to page j with probability v_j.

4. The random walk defined by $\mathbf{r}^T(k+1) = \mathbf{r}^T(k)S$ will generally not converge because Web's graph structure is not strongly connected, which results in a reducible chain with many isolated ergodic classes.
5. The power method has been Google's computational method of choice for computing the PageRank vector. If formed explicitly, G is completely dense, and the size of n would make each power iteration extremely costly — billions of flops for each step. But writing the power method as

$$\boldsymbol{\pi}^T(k+1) = \boldsymbol{\pi}^T(k)G = \alpha\boldsymbol{\pi}^T(k)H + (\alpha\boldsymbol{\pi}^T(k)\mathbf{a})\mathbf{1}^T/n + (1-\alpha)\mathbf{v}^T$$

shows that only extremely sparse vector-matrix multiplications are required. Only the nonzero h_{ij}'s are needed, so G and S are neither formed nor stored.
6. When implemented as shown above, each power step requires only $nnz(H)$ flops, where $nnz(H)$ is the number of nonzeros in H, and, since the average number of nonzeros per column in H is significantly less than 10, we have $O(nnz(H)) \approx O(n)$. Furthermore, the inherent parallism is easily exploited, and the current iterate is the only vector stored at each step.
7. Because the Web has many disconnected components, the hyperlink matrix is highly reducible, and compensating for the dangling nodes to construct the stochastic matrix S does not significantly affect this.
8. [Mey00, p. 695–696] Since S is also reducible, S can be symmetrically permuted to have the form

$$S \sim \left[\begin{array}{cccc|cccc} S_{11} & S_{12} & \cdots & S_{1r} & S_{1,r+1} & S_{1,r+2} & \cdots & S_{1m} \\ 0 & S_{22} & \cdots & S_{2r} & S_{2,r+1} & S_{2,r+2} & \cdots & S_{2m} \\ \vdots & & \ddots & \vdots & \vdots & \vdots & \cdots & \vdots \\ 0 & 0 & \cdots & S_{rr} & S_{r,r+1} & S_{r,r+2} & \cdots & S_{rm} \\ \hline 0 & 0 & \cdots & 0 & S_{r+1,r+1} & 0 & \cdots & 0 \\ 0 & 0 & \cdots & 0 & 0 & S_{r+2,r+2} & \cdots & 0 \\ \vdots & \vdots & \cdots & \vdots & \vdots & \vdots & \ddots & \vdots \\ 0 & 0 & \cdots & 0 & 0 & 0 & \cdots & S_{mm} \end{array}\right],$$

where the following are true.

- For each $1 \leq i \leq r$, S_{ii} is either irreducible or $[0]_{1\times 1}$.
- For each $1 \leq i \leq r$, there exists some $j > i$ such that $S_{ij} \neq 0$.
- $\rho(S_{ii}) < 1$ for each $1 \leq i \leq r$.
- $S_{r+1,r+1}, S_{r+2,r+2}, \cdots, S_{m,m}$ are each stochastic and irreducible.
- 1 is an eigenvalue for S that is repeated exactly $m - r$ times.

9. The natural structure of the Web forces the algebraic multiplicity of the eigenvalue 1 to be large.
10. [LM04a][LM06][Mey00, Ex. 7.1.17, p. 502] If the eigenvalues of $S_{n\times n}$ are

$$\lambda(S) = \{\underbrace{1, 1, \ldots, 1}_{m-r}, \mu_{m-r+1}, \ldots, \mu_n\}, \quad 1 > |\mu_{m-r+1}| \geq \cdots \geq |\mu_n|,$$

then the eigenvalues of the Google matrix $G = \alpha S + (1 - \alpha)E$ are

$$\lambda(G) = \{1, \underbrace{\alpha, \ldots, \alpha}_{m-r-1}, (\alpha\mu_{m-r+1}), \ldots, (\alpha\mu_n)\}. \tag{63.3}$$

11. [Mey00] The asymptotic rate of convergence of any aperiodic Markov chain is governed by the magnitude of its largest subdominant eigenvalue. In particular, if the distinct eigenvalues λ_i of an aperiodic chain are ordered $\lambda_1 = 1 > |\lambda_2| \geq |\lambda_3| \geq \cdots \geq |\lambda_n|$, then the number of incorrect digits in each component of $\boldsymbol{\pi}^T(k)$ is eventually going to be reduced by about $-log_{10}|\lambda_2|$ digits per iteration.
12. Determining (or even estimating) $|\lambda_2|$ normally requires substantial effort, but Equation 63.3 says that $\lambda_2 = \alpha$ for the Google matrix G. This is an extremely happy accident because it means that Google's engineers can completely control the rate of convergence, regardless of the value of the personalization vector $\mathbf{v}^T$ in $E = \mathbf{1v}^T$.
13. At the last public disclosure Google was setting $\alpha = .85$, so, at this value, the asymptotic rate of convergence of the power method is $-log_{10}(.85) \approx .07$, which means that the power method will eventually take about 14 iterations for each significant place of accuracy that is required.
14. [LM04a] Even though they can control the rate of convergence with α, Google's engineers are forced to perform a delicate balancing act because while decreasing α increases the rate of convergence, decreasing α also lessens the effect of the true hyperlink structure of the Web, which is Google's primary mechanism for measuring Webpage importance. Increasing α more accurately reflects the Web's true link structure, but, along with slower convergence, sensitivity issues begin to surface in the sense that slightly different values for α near 1 can produce significantly different PageRanks.
15. [KHM03a] The power method can be substantially accelerated by a quadratic extrapolation technique similar to Aitken's Δ^2 method, and there is reason to believe that Google has adopted this procedure.
16. [KHM03b], [KHG04] Other improvements to the basic power method include a block algorithm and an adaptive algorithm that monitors the convergence of individual elements of the PageRank vector. As soon as components of the vector have converged to an acceptable tolerance, they are no longer computed. Convergence is faster because the work per iteration decreases as the process evolves.
17. [LM02], [LM04a], [LM04b], [LGZ03] Other methods partition H into two groups according to dangling nodes and nondangling nodes. The problem is then aggregated by lumping all of the dangling nodes into one super state to produce a problem that is significantly reduced in size — this is due to the fact that the dangling nodes account for about one fourth of the Web's nodes. The most exciting feature of all these algorithms is that they do not compete with one another. In fact, it is possible to combine some of these algorithms to achieve even greater performance.
18. The accuracy of PageRank computations is an important implementation issue, but we do not know the accuracy with which Google works. It seems that it must be at least high enough to differentiate between the often large list of ranked pages that Google usually returns, and since $\boldsymbol{\pi}^T$ is a probability vector containing $O(10^9)$ components, it is reasonable to expect that accuracy giving at least ten significant places is needed to distinguish among the elements.
19. A weakness of PageRank is "topic drift." The PageRank vector might be mathematically accurate, but this is of little consequence if the results point the user to sites that are off-topic. PageRank is a query-independent measure that is essentially determined by a popularity contest with everyone on Web having a vote, and this tends to skew the results in the direction of importance (measured by popularity) over relevance to the query. This means that PageRank may have trouble distinguishing between pages that are authoritative in general and pages that are authoritative more specifically to the query topic. It is believed that Google engineers devote much effort to mitigate this problem, and this is where the metrics that determine the content score might have an effect.

20. In spite of topic drift, Google's decision to measure importance by means of popularity over relevance turned out to be the key to Google's success and the source of its strength. The query-dependent measures employed by Google's predecessors were major stumbling blocks in maintaining query processing speed as the Web grew. PageRank is query-independent, so at query time only a quick lookup into an inverted file storage is required to determine pertinent pages, which are then returned in order by the precomputed PageRanks. The small compromise of topic drift in favor of processing speed won the day.
21. A huge advantage of PageRank is its virtual imperviousness to spamming (artificially gaming the system). High PageRank is achieved by having many inlinks from highly ranked pages, and while it may not be so hard for a page owner to have many of his cronies link to his page, it is difficult to generate a lot of inlinks from *important* sites that have a high rank.
22. [TM03] Another strength of PageRank concerns the flexibility of the "personalization" (or "intervention") vector $\mathbf{v}^T$ that Google is free to choose when defining the perturbation term $E = \mathbf{1}\mathbf{v}^T$. The choice of $\mathbf{v}^T$ affects neither mathematical nor computational aspects, but it does alter the rankings in a predictable manner. This can be a terrific advantage if Google wants to intervene to push a site's PageRank down or up, perhaps to punish a suspected "link farmer" or to reward a favored client. Google has claimed that it does not make a practice of rewarding favored clients, but it is known that Google is extremely vigilant and sensitive concerning people who try to manipulate PageRank, and such sites are punished. However, the outside world is not privy to the extent to which either the stick or carrot is employed.
23. [FLM$^+$04], [KH03], [KHG04], [KHM03a], [LM04a], [LM06], [LGZ03], [LM03], [NZJ01] In spite of the simplicity of the basic concepts, subtle issues such as personalization, practical implementations, sensitivity, and updating lurk just below the surface.
24. We have summarized only the mathematical component of Google's ranking system, but, as mentioned earlier, there are a hundred or more nonmathematical content metrics that are also considered when Google responds to a query. The results seen by a user are in fact PageRank tempered by these other metrics. While Google is secretive about most of its other metrics, it have made it clear that the other metrics are subservient to their mathematical PageRank scores.

Acknowledgment

This work was supported in part by the National Science Foundation under NSF grant CCR-0318575 (Carl D. Meyer).

References

[BB05] Michael W. Berry and Murray Browne. *Understanding Search Engines: Mathematical Modeling and Text Retrieval*. SIAM, Philadelphia, 2nd ed., 2005.

[BDJ99] Michael W. Berry, Zlatko Drmac, and Elizabeth R. Jessup. Matrices, vector spaces and information retrieval. *SIAM Rev.*, 41:335–362, 1999.

[Ber01] Michael W. Berry, Ed. *Computational Information Retrieval*. SIAM, Philadelphia, 2001.

[BF96] Michael W. Berry and R.D. Fierro. Low-rank orthogonal decompositions for information retrieval applications. *J. Num. Lin. Alg. Appl.*, 1(1):1–27, 1996.

[Blo99] Katarina Blom. *Information Retrieval Using the Singular Value Decomposition and Krylov Subspaces*. Ph.D. thesis, University of Chalmers, Götebong, Sweden, January 1999.

[BO98] Michael W. Berry and Gavin W. O'Brien. Using linear algebra for intelligent information retrieval. *SIAM Rev.*, 37:573–595, 1998.

[BP98] Sergey Brin and Lawrence Page. The anatomy of a large-scale hypertextual Web search engine. *Comp. Networks ISDN Syst.*, 33:107–117, 1998.

[BPM99] Sergey Brin, Lawrence Page, R. Motwami, and Terry Winograd. The PageRank citation ranking: bringing order to the Web. Technical Report 1999-0120, Computer Science Department, Stanford University, 1999.

[BR01] Katarina Blom and Axel Ruhe. Information retrieval using very short Krylov sequences. In *Computational Information Retrieval*, pp. 41–56, 2001.

[BR99] Ricardo Baeza-Yates and Berthier Ribeiro-Neto. *Modern Information Retrieval*. ACM Press, Pearson Education Limited, Essex, England, 1999.

[Dum91] Susan T. Dumais. Improving the retrieval of information from external sources. *Behav. Res. Meth., Instru. Comp.*, 23:229–236, 1991.

[FLM+04] Ayman Farahat, Thomas Lofaro, Joel C. Miller, Gregory Rae, and Lesley A. Ward. Existence and uniqueness of ranking vectors for linear link analysis. *SIAM J. Sci. Comp.*, 27(4): 1181–1201, 2006.

[GL96] Gene H. Golub and Charles F. Van Loan. *Matrix Computations*. Johns Hopkins University Press, Baltimore, 1996.

[HB00] M.K. Hughey and Michael W. Berry. Improved query matching using kd-trees, a latent semantic indexing enhancement. *Inform. Retri.*, 2:287–302, 2000.

[HKJ03] Taher H. Haveliwala, Sepandar D. Kamvar, and Glen Jeh. An analytical comparison of approaches to personalizing PageRank. Technical report, Stanford University, 2003.

[Hoy02] Patrik O. Hoyer. Non-negative sparse coding. In *Neural Networks for Signal Processing XII (Proc. IEEE Workshop on Neural Networks for Signal Processing)*, Martigny, Switzerland, pp. 557–565, 2002.

[Hoy04] Patrik O. Hoyer. Non-negative matrix factorization with sparseness constraints. *J. Mach. Learn. Res.*, 5:1457–1469, 2004.

[JB00] Eric P. Jiang and Michael W. Berry. Solving total least squares problems in information retrieval. *Lin. Alg. Appl.*, 316:137–156, 2000.

[JL00] Fan Jiang and Michael L. Littman. Approximate dimension equalization in vector-based information retrieval. In *The Seventeenth International Conference on Machine Learning*, Stanford University, pp. 423–430, 2000.

[KH03] Sepandar D. Kamvar and Taher H. Haveliwala. The condition number of the PageRank problem. Technical report, Stanford University, 2003.

[KHG04] Sepandar D. Kamvar, Taher H. Haveliwala, and Gene H. Golub. Adaptive methods for the computation of PageRank. *Lin. Alg. Appl.*, 386:51–65, 2004.

[KHM03a] Sepandar D. Kamvar, Taher H. Haveliwala, Christopher D. Manning, and Gene H. Golub. Extrapolation methods for accelerating PageRank computations. In *Twelfth International World Wide Web Conference*, New York, 2003. ACM Press.

[KHM03b] Sepandar D. Kamvar, Taher H. Haveliwala, Christopher D. Manning, and Gene H. Golub. Exploiting the block structure of the Web for computing PageRank. Technical Report 2003–17, Stanford University, 2003.

[Kle99] Jon Kleinberg. Authoritative sources in a hyperlinked environment. *J. ACM*, 46, 1999.

[LB97] Todd A. Letsche and Michael W. Berry. Large-scale information retrieval with LSI. *Inform. Comp. Sci.*, pp. 105–137, 1997.

[LGZ03] Chris Pan-Chi Lee, Gene H. Golub, and Stefanos A. Zenios. A fast two-stage algorithm for computing PageRank and its extensions. Technical Report SCCM-2003-15, Scientific Computation and Computational Mathematics, Stanford University, 2003.

[LM00] Ronny Lempel and Shlomo Moran. The stochastic approach for link-structure analysis (SALSA) and the TKC effect. In *The Ninth International World Wide Web Conference*, New York, 2000. ACM Press.

[LM02] Amy N. Langville and Carl D. Meyer. Updating the stationary vector of an irreducible Markov chain. Technical Report crsc02-tr33, North Carolina State, Mathematics Dept., CRSC, 2002.

[LM03] Ronny Lempel and Shlomo Moran. Rank-stability and rank-similarity of link-based web ranking algorithms in authority-connected graphs. In *Second Workshop on Algorithms and Models for the Web-Graph (WAW 2003)*, Budapest, Hungary, May 2003.

[LM04a] Amy N. Langville and Carl D. Meyer. Deeper inside PageRank. *Inter. Math. J.*, 1(3):335–400, 2005.

[LM04b] Amy N. Langville and Carl D. Meyer. A reordering for the PageRank problem. *SIAM J. Sci. Comp.*, 27(6):2112–2120, 2006.

[LM05] Amy N. Langville and Carl D. Meyer. A survey of eigenvector methods of web information retrieval. *SIAM Rev.*, 47(1):135–161, 2005.

[LM06] Amy N. Langville and Carl D. Meyer. *Google's PageRank and Beyond: The Science of Search Engine Rankings.* Princeton University Press, Princeton, NJ, 2006.

[LS99] Daniel D. Lee and H. Sebastian Seung. Learning the parts of objects by non-negative matrix factorization. *Nature*, 401:788–791, 1999.

[LS00] Daniel D. Lee and H. Sebastian Seung. Algorithms for the non-negative matrix factorization. *Adv. Neur. Inform. Proc.*,13:556–562, 2001.

[Med03] Medlars test collection, 2003. `http://www.cs.utk.edu/~lsi/`.

[Mey00] Carl D. Meyer. *Matrix Analysis and Applied Linear Algebra.* SIAM, Philadelphia, 2000.

[NZJ01] Andrew Y. Ng, Alice X. Zheng, and Michael I. Jordan. Link analysis, eigenvectors and stability. In *The Seventh International Joint Conference on Artificial Intelligence*, Seattle, WA, 2001.

[PT94] Pentti Paatero and U. Tapper. Positive matrix factorization: a non-negative factor model with optimal utilization of error estimates of data values. *Environmetrics*, 5:111–126, 1994.

[PT97] Pentti Paatero and U. Tapper. Least squares formulation of robust non-negative factor analysis. *Chemomet. Intell. Lab. Sys.*, 37:23–35, 1997.

[SB83] Gerard Salton and Chris Buckley. *Introduction to Modern Information Retrieval.* McGraw-Hill, New York, 1983.

[SBP04] Farial Shahnaz, Michael W. Berry, V. Paul Pauca, and Robert J. Plemmons. Document clustering using nonnegative matrix factorization. *J. Inform. Proc. Man.*, 42(2):373–386, 2006.

[TM03] Michael Totty and Mylene Mangalindan. As Google becomes web's gatekeeper, sites fight to get in. *Wall Street Journal*, CCXLI(39), 2003. February 26.

[WB98] Dian I. Witter and Michael W. Berry. Downdating the latent semantic indexing model for conceptual information retrieval. *Comp. J.*, 41(1):589–601, 1998.

[ZBR01] Xiaoyan Zhang, Michael W. Berry, and Padma Raghavan. Level search schemes for information filtering and retrieval. *Inform. Proc. Man.*, 37:313–334, 2001.

[ZMS98] Hongyuan Zha, Osni Marques, and Horst D. Simon. A subspace-based model for information retrieval with applications in latent semantic indexing. *Lect. Notes Comp. Sci.*, 1457:29–42, 1998.

64

Signal Processing

Michael Stewart
Georgia State University

A signal is a function of time, depending on either a continuous, real time variable or a discrete integer time variable. Signal processing is a collection of diverse mathematical, statistical, and computational techniques for transforming, extracting information from, and modeling of signals. Common signal processing techniques include filtering a signal to remove undesired frequency components, extracting a model that describes key features of a signal, estimation of the frequency components of a signal, prediction of the future behavior of a signal from past data, and finding the direction from which a signal arrives at an array of sensors.

Many modern methods for signal processing have deep connections to linear algebra. Rather than attempting to give a complete overview of such a diverse field, we survey a selection of contemporary methods for the processing of discrete time signals with deep connections to matrix theory and numerical linear algebra. These topics include linear prediction, Wiener filtering, spectral estimation, and direction of arrival estimation. Notable ommissions include classical methods of filter design, the fast Fourier transform (FFT), and filter structures. (See Chapter 58.)

64.1 Basic Concepts

We begin with basic definitions which are standard and can be found in [PM96].

Definitions:

A **signal** $\boldsymbol{x}$ is a real sequence with element k of the sequence for $k = 0, \pm 1, \pm 2, \ldots$ given by $x(k)$. The set of all signals is a vector space with the sum $\boldsymbol{w} = \boldsymbol{x} + \boldsymbol{y}$ defined by $w(k) = x(k) + y(k)$ and the scalar product $\boldsymbol{v} = a\boldsymbol{x}$ where a is a real scalar given by $v(k) = ax(k)$.

A **linear time-invariant system** is a mapping $L(\boldsymbol{x}) = \boldsymbol{y}$ of an input signal $\boldsymbol{x}$ to an output signal $\boldsymbol{y}$ that satisfies the following properties:

1. Linearity: If $L(\boldsymbol{x}_0) = \boldsymbol{y}_0$ and $L(\boldsymbol{x}_1) = \boldsymbol{y}_1$, then $L(a\boldsymbol{x}_0 + b\boldsymbol{y}_0) = a\boldsymbol{y}_0 + b\boldsymbol{y}_1$ for any $a, b \in \mathbb{R}$.
2. Time-invariance: If $L(\boldsymbol{x}) = \boldsymbol{y}$ and $\hat{\boldsymbol{x}}$ is a shifted version of $\boldsymbol{x}$ given by $\hat{x}(k) = x(k - k_0)$, then $L(\hat{\boldsymbol{x}}) = \hat{\boldsymbol{y}}$ where $\hat{y}(k) = y(k - k_0)$. That is, shifted inputs lead to correspondingly shifted outputs.

The **impulse response** of a linear time-invariant system is the output $\boldsymbol{h}$ that results from applying as input the unit impulse $\boldsymbol{\delta}$ where $\delta(0) = 1$ and $\delta(k) = 0$ for $k \neq 1$.

The **convolution** $\boldsymbol{y}$ of $\boldsymbol{h}$ and $\boldsymbol{x}$ is written $\boldsymbol{y} = \boldsymbol{h} * \boldsymbol{x}$ and is defined by the sum

$$y(k) = \sum_{j=-\infty}^{\infty} h(j)x(k-j).$$

A signal $\boldsymbol{x}$ is **causal** if $x(k) = 0$ for $k < 0$.

A linear time-invariant system is **causal** if every causal input gives a causal output. An equivalent definition is that a system is **causal** if its impulse response sequence is causal.

The z-**transform** of a signal or impulse response $\boldsymbol{x}$ is

$$X(z) = \sum_{k=-\infty}^{\infty} x(k)z^{-k}$$

where $X(z)$ is defined in the region of convergence of the sum.

The **transfer function** of a linear system with impulse response $\boldsymbol{h}$ is the z-transform of the impulse response

$$H(z) = \sum_{k=-\infty}^{\infty} h(k)z^{-k}.$$

The **discrete time Fourier transform** of $\boldsymbol{x}$ is the function of ω obtained by evaluating the z-transform on the unit circle

$$X(e^{i\omega}) = \sum_{k=-\infty}^{\infty} x(k)e^{-i\omega k}.$$

If $\boldsymbol{x}$ is an impulse response of a system, then $X(e^{i\omega})$ is the **frequency response** of the system. The discrete time Fourier transform is defined only if the region of convergence of the z-transform includes the unit circle.

A filter is **minimum phase** if the zeros of the transfer function are in the unit circle.

A **finite impulse response** (FIR) filter is a system that maps an input signal $\boldsymbol{x}$ to an output signal $\boldsymbol{y}$ through the sum

$$y(k) = \sum_{j=0}^{n-1} b(j)x(k-j).$$

A rational **infinite impulse response** (IIR) filter is a causal system that maps inputs $\boldsymbol{x}$ to outputs $\boldsymbol{y}$ via a relation of the form

$$y(k) = \sum_{j=0}^{n-1} b(j)x(k-j) - \sum_{j=1}^{m} a(j)y(k-j).$$

Causality is an essential part of the definition; without this assumption the above relation does not uniquely define a mapping from inputs to outputs.

A signal $\boldsymbol{x}$ has **finite energy** if

$$\sum_{k=-\infty}^{\infty} |x(k)|^2 < \infty.$$

A system mapping input $\boldsymbol{x}$ to output $\boldsymbol{y}$ is **stable** if $|x(k)| < B_x$ for $0 < B_x < \infty$ and for all k implies that there is $0 < B_y < \infty$ such that the output satisfies $|y(k)| < B_y$ for all k.

Facts:

1. The impulse response $\boldsymbol{h}$ uniquely determines a linear time-invariant system. The mapping from the input $\boldsymbol{x}$ to the output $\boldsymbol{y}$ is given by $\boldsymbol{y} = \boldsymbol{h} * \boldsymbol{x}$. Convolution is commutative so that if $\boldsymbol{y} = \boldsymbol{g} * (\boldsymbol{h} * \boldsymbol{x})$, then $\boldsymbol{y} = \boldsymbol{h} * (\boldsymbol{g} * \boldsymbol{x})$. Thus, the order in which two linear time-invariant filters are applied to a signal does not matter.
2. A causal system can be applied in real time in the sense that output $y(k)$ can be computed from the convolution $\boldsymbol{y} = \boldsymbol{h} * \boldsymbol{x}$ as soon as $x(k)$ is available.
3. A system defined by its impulse response $\boldsymbol{h}$ is stable if and only if

$$\sum_{k=-\infty}^{\infty} |h(k)| < \infty.$$

4. An expression for the z-transform of a signal (or an impulse response) does not uniquely determine the signal. It is also necessary to know the region of convergence. If $X(z)$ is the z-transform of $\boldsymbol{x}$, then the inverse transform is

$$x(k) = \frac{1}{2\pi i} \oint X(z) z^{k-1}\, dz$$

where the integral is taken over any contour enclosing the origin and in the region where $X(z)$ is analytic. A system is uniquely determined by its transfer function and the region of convergence of the transfer function.

5. An FIR filter has impulse response

$$h(k) = \begin{cases} b(k) & \text{for } 0 \le k \le n-1 \\ 0 & \text{otherwise.} \end{cases}$$

6. An FIR filter has transfer function

$$H(z) = b(0) + b(1)z^{-1} + \cdots + b(n-1)z^{-(n-1)}.$$

7. The impulse response $\boldsymbol{h}$ of a causal IIR filter is the unique solution to the recurrence

$$h(k) = \begin{cases} 0 & k \le 0 \\ b(k) - \sum_{j=1}^{m} a(j)h(k-j) & 0 \le k \le n-1 \\ -\sum_{j=1}^{m} a(j)h(k-j) & k \ge n \end{cases}.$$

This recurrence uniquely determines a causal impulse response $\boldsymbol{h}$ and, hence, uniquely determines the mapping from inputs to outputs.

8. A rational IIR filter has transfer function

$$H(z) = \frac{b(0) + b(1)z^{-1} + \cdots + b(n-1)z^{-(n-1)}}{1 + a(1)z^{-1} + \cdots + a(m)z^{-m}}.$$

9. A linear time-invariant system with input $\boldsymbol{x}$ and impulse response $\boldsymbol{h}$ with z-transform $X(z)$ and transfer function $H(z)$ has output $\boldsymbol{y}$ with z-transform

$$Y(z) = H(z)X(z).$$

10. A discrete version of the Paley–Wiener theorem states that $\boldsymbol{x}$ is causal and has finite energy if and only if the corresponding z-transform $X(z)$ is analytic in the exterior of the unit circle and

$$\sup_{1<r<\infty} \int_{-\pi}^{\pi} \left|X(re^{i\omega})\right|^2 d\omega < \infty.$$

If $|X(e^{i\omega})|$ is square integrable, then $|X(e^{i\omega})|$ is the Fourier transform of some causal system finite energy signal if and only if

$$\int_{-\pi}^{\pi} \left|\ln\left(|X(e^{i\omega})|\right)\right| d\omega < \infty.$$

11. For rational transfer functions $H(z)$ corresponds to a causal stable system if and only if the poles of $H(z)$ are in the unit circle. Thus, FIR filters are always stable and IIR filters are stable when the poles of the transfer function are in the unit circle.
12. A filter that is minimum phase has a causal stable inverse filter with transfer function $G(z) = 1/H(z)$. That is, filtering a signal $\boldsymbol{x}$ by a filter with z-transform $H(z)$ and then by a filter with z-transform $G(z)$ gives an output with z-transform $G(z)H(z)X(z) = X(z)$. The minimum phase property is of particular significance since the only filters that can be applied in real time and without excessive growth of errors due to noise are those that are causal and stable. In most circumstances only minimum phase filters are invertible in practice.

Examples:

1. The causal impulse response $\boldsymbol{h}$ with $h(k) = a^k$ for $k \geq 0$ and $h(k) = 0$ for $k < 0$ has z-transform $H(z) = 1/(1 - az^{-1})$ with region of convergence $|z| > |a|$. The anticausal impulse response $\hat{\boldsymbol{h}}$ with $\hat{h}(k) = -a^k$ for $k < 0$ and $\hat{h}(k) = 0$ for $k \geq 0$ has z-transform $H(z) = 1/(1 - az^{-1})$ with region of convergence $|z| < |a|$.
2. The causal impulse response $\boldsymbol{h}$ with

$$h(k) = \begin{cases} 2^{-k} + 3^{-k} & \text{for } k \geq 0 \\ 0 & k < 0 \end{cases}$$

has z-transform

$$H(z) = \frac{1}{1 - (2z)^{-1}} + \frac{1}{1 - (3z)^{-1}} = \frac{2 - \frac{5}{6}z^{-1}}{1 - \frac{5}{6}z^{-1} + \frac{1}{6}z^{-2}}$$

with region of convergence $|z| > 1/2$. The impulse response can be realized by a rational IIR filter of the form

$$y(k) = 2x(k) - \frac{5}{6}x(k-1) + \frac{5}{6}y(k-1) + \frac{1}{6}y(k-2).$$

The zeros of $z^2 - (5/6)z + (1/6)$ are $1/2$ and $1/3$ so that the system is stable.

64.2 Random Signals

Definitions of statistical terms can be found in Chapter 52.

Definitions:

A **random signal** or **random process** $\boldsymbol{x}$ is a sequence $x(k)$ of real random variables indexed by $k = 0, \pm 1, \pm 2, \ldots$.

A random signal $\boldsymbol{x}$ is **wide-sense stationary** (for brevity, we will use **stationary** to mean the same thing) if the mean $\mu_x = E[x(k)]$ and **autocorrelation sequence** $r_{xx}(k) = E[x(j)x(j-k)]$ do not depend on j. We assume that all random signals are wide-sense stationary. Two stationary random signals $\boldsymbol{x}$ and $\boldsymbol{y}$ are jointly stationary if the **cross correlation** sequence $r_{yx}(k) = E[y(j)x(j-k)]$ does not depend on j. When referring to two stationary signals, we always assume that the signals are jointly stationary as well.

A sequence $\mathbf{x}(k)$ of real random $n \times 1$ vectors indexed by $k = 0, \pm 1, \pm 2, \ldots$ is **stationary** if $E[\mathbf{x}(k)]$ and $E[\mathbf{x}(k)\mathbf{x}^T(k)]$ do not depend on k.

The **autocorrelation matrix** of a sequence of stationary random vectors $\mathbf{x}(k)$ is

$$R_{\mathbf{x}} = E\left[\mathbf{x}(k)\mathbf{x}^T(k)\right].$$

A zero mean stationary random signal $\boldsymbol{n}$ is a **white noise process** if it has autocorrelation sequence

$$r_{nn}(j) = E[n(k)n(k-j)] = \begin{cases} \sigma_n^2 & j = 0 \\ 0 & j \neq 0 \end{cases}$$

where σ_n^2 is the **variance** of $\boldsymbol{n}$.

A white noise driven **autoregressive process** (AR process) $\boldsymbol{x}$ is of the form

$$x(k) = n(k) - \sum_{j=1}^{n} a(j)x(k-j)$$

where $\boldsymbol{n}$ is a white noise process and the filtering of $\boldsymbol{n}$ to obtain $\boldsymbol{x}$ is causal.

Given a signal $\boldsymbol{x}$, we denote the z-transform of the autocorrelation sequence of $\boldsymbol{x}$ by $S_x(z)$. The **spectral density** of a stationary random signal $\boldsymbol{x}$ with autocorrelation sequence $r_{xx}(k)$ is the function of ω given by evaluating $S_x(z)$ on the unit circle

$$S_x(e^{i\omega}) = \sum_{j=-\infty}^{\infty} r_{xx}(j)e^{-ij\omega}.$$

A random signal with spectral density $S_x(e^{i\omega})$ is **bandlimited** if $S_x(e^{i\omega}) = 0$ on a subset of $[-\pi, \pi]$ of nonzero measure.

A random signal with spectral density $S_x(e^{i\omega})$ has **finite power** if

$$\int_{-\pi}^{\pi} S_x(e^{i\omega})\, d\omega < \infty.$$

The **spectral factorization**, when it exists, of a spectral density $S_x(z)$ is a factorization of the form

$$S_x(z) = L(z)\overline{L(\overline{z^{-1}})}$$

where $L(z)$ and $1/L(z)$ are both analytic for $|z| > 1$. Interpreted as transfer functions $L(z)$ and $1/L(z)$ both describe causal, stable systems. That is, $L(z)$ is causal and stable and causally and stably invertible.

Facts:

1. If $\mathbf{x}(k)$ is a sequence of vectors obtained by sampling a random sequence of vectors, then under suitable ergodicity assumptions

$$R_{\mathbf{x}} = \lim_{n\to\infty} \frac{1}{n} \sum_{k=0}^{n} \mathbf{x}(k)\mathbf{x}^T(k).$$

   In computational practice $R_{\mathbf{x}}$ is often estimated using a truncated sum in which n is finite.
2. The expectation $\langle x, y\rangle = E[xy]$ can be viewed as an inner product on a real Hilbert space of zero mean random variables with $E[|x|^2] < \infty$. This gives many results in optimal filtering a geometric interpretation in terms of orthogonalization and projection on the span of a set of random variables in a way that is perfectly analogous to least squares problems in any other Hilbert space. From this point of view, a white noise process $\boldsymbol{n}$ is a sequence of orthogonal random variables.

3. If a random signal $\boldsymbol{x}$ with autocorrelation sequence $\boldsymbol{r}_x$ is passed through a system with impulse response $\boldsymbol{h}$, then the output $\boldsymbol{y}$ has autocorrelation

$$\boldsymbol{r}_y = \boldsymbol{h} * \boldsymbol{r}_x * \tilde{\boldsymbol{h}}$$

where $\tilde{\boldsymbol{h}}$ is defined by $\tilde{h}(j) = h(-j)$.
4. The spectral density is nonnegative, $S_x(e^{i\omega}) \geq 0$.
5. The spectral density of white noise is $S_n(e^{i\omega}) = \sigma_n^2$.
6. [Pap85] If a random signal $\boldsymbol{x}$ with spectral density $S_x(z)$ passes through a filter with transfer function $H(z)$, then the output $\boldsymbol{y}$ is a stationary signal with

$$S_y(z) = S_x(z)H(z)\overline{H(\overline{z^{-1}})} \quad \text{or} \quad S_y(e^{i\omega}) = S_x(e^{i\omega})|H(e^{i\omega})|^2.$$

7. If $L(z)$ is the spectral factor of $S_x(z)$, then on the unit circle we have $S_x(e^{iw}) = |L(e^{iw})|^2$.
8. [SK01], [GS84] The spectral factorization exists if and only if the signal has finite power and satisfies

$$\int_{-\pi}^{\pi} \ln\left(S_x(e^{i\omega})\right) d\omega > -\infty.$$

If $S_x(e^{i\omega})$ is zero on a set of nonzero measure, then the condition is not satisfied. Thus, bandlimited signals do not have a spectral factorization.
9. If a spectral factor $L(z)$ exists for a signal $\boldsymbol{x}$ with spectral density $S_x(e^{i\omega})$ and

$$L(z) = \sum_{k=0}^{\infty} l(k)z^{-k} \quad \text{and} \quad M(z) = \frac{1}{L(z)} = \sum_{k=0}^{\infty} m(k)z^{-k},$$

then the filtered signal $\boldsymbol{n} = \boldsymbol{m} * \boldsymbol{x}$ given by

$$n(k) = \sum_{j=0}^{\infty} m(j)x(k-j)$$

is a white noise process with $S_n(e^{i\omega}) = 1$. The original signal can be obtained by causally and stably filtering $\boldsymbol{n}$ with $\boldsymbol{l}$:

$$x(k) = \sum_{j=0}^{\infty} l(j)n(k-j).$$

The signal $\boldsymbol{n}$ is the **innovations** process for $\boldsymbol{x}$ and the representation of $\boldsymbol{x}$ through filtering $\boldsymbol{n}$ by $\boldsymbol{l}$ is the **innovations representation** for $\boldsymbol{x}$. It is a model of $\boldsymbol{x}$ as a filter driven by white noise. The innovations representation is an orthogonalization of the sequence of random variables $\boldsymbol{x}$. The innovations representation for x is analogous to the QR factorization of a matrix.

Examples:

1. For a white noise process $\boldsymbol{n}$ with variance $\sigma_n^2 = 1$, define the autoregressive process $\boldsymbol{x}$ by

$$x(k) = n(k) + ax(k-1)$$

where $|a| < 1$. The impulse response and transfer function of the IIR filter generating $\boldsymbol{x}$ are

$$h(k) = \begin{cases} 0 & k < 0 \\ a^k & k \geq 0 \end{cases}, \quad \text{and} \quad H(z) = \frac{1}{1 - az^{-1}}.$$

The autocorrelation sequence for $\boldsymbol{x}$ is the sequence $\boldsymbol{r}_{xx}$ given by

$$r_{xx}(k) = \frac{1}{1 - a^2} a^{|k|}.$$

The autocorrelation sequence has z-transform

$$S_x(z) = \frac{1}{(1 - az^{-1})(1 - az)}$$

so that the spectral density is

$$S_x(e^{i\omega}) = \frac{1}{1 + a^2 - 2a\cos(\omega)}.$$

The spectral factor is

$$L(z) = \frac{1}{1 - az^{-1}}.$$

64.3 Linear Prediction

Definitions:

The **linear prediction** problem is to find an estimate $\hat{\boldsymbol{x}}$ of a stationary random signal $\boldsymbol{x}$ as a linear combination

$$\hat{x}(k) = \sum_{j=1}^{\infty} a(j)x(k - j),$$

with the $a(j)$ chosen to minimize the **mean square prediction error** $E_\infty = E[|e(k)|^2]$ where $e(k) = \hat{x}(k) - x(k)$ is the **prediction error**. Thus, the goal is to find a current estimate $\hat{x}(k)$ of $x(k)$ from the past values $x(k-1), x(k-2), \ldots$.

A random signal is **predictable** if $E_\infty = 0$. A signal that is not predictable is **regular.**

The **order** n **linear prediction problem** is to find an estimate of a stationary random signal $\boldsymbol{x}$ with a signal $\hat{\boldsymbol{x}}$ of the form $\hat{x}(k) = \sum_{j=1}^{n} a_n(j)x(k-j)$ so as to minimize the mean square error $E_n = E[|e(k)|^2]$ where $e(k) = \hat{x}(k) - x(k)$.

A random signal is **order** n **predictable** if $E_n = 0$.

The **prediction error filter** and the **order** n **prediction error filter** are the filters with transfer functions

$$E(z) = 1 - \sum_{j=1}^{\infty} a(j)z^{-j}$$

and

$$E_n(z) = 1 - \sum_{j=1}^{n} a_n(j)z^{-j}.$$

The **Wiener–Hopf equations** for the optimal prediction coefficients $a(j)$ are

$$r_{xx}(l) = \sum_{j=1}^{\infty} a(j)r_{xx}(l - j)$$

for $l \geq 1$.

Given a random signal $x(k)$ with autocorrelation $r_{xx}(k)$, the optimal prediction coefficients $a^{(n)}(j)$ and mean square prediction error E_n that satisfy the **Yule–Walker equations** are

$$\begin{bmatrix} r_{xx}(0) & r_{xx}(1) & \cdots & r_{xx}(n) \\ r_{xx}(1) & r_{xx}(0) & \ddots & \vdots \\ \vdots & \ddots & \ddots & r_{xx}(1) \\ r_{xx}(n) & \cdots & r_{xx}(1) & r_{xx}(0) \end{bmatrix} \begin{bmatrix} 1 \\ -a_n(1) \\ \vdots \\ -a_n(n) \end{bmatrix} = \begin{bmatrix} E_n \\ 0 \\ \vdots \\ 0 \end{bmatrix}$$

or $T_n \mathbf{a}_n = E_n \mathbf{e}_1$.

The **Levinson–Durbin algorithm** for the computation solution of the order n linear prediction problem is

$$\mathbf{a}_n = \begin{bmatrix} \mathbf{a}_{n-1} \\ 0 \end{bmatrix} + \gamma_n \begin{bmatrix} 0 \\ R\mathbf{a}_{n-1} \end{bmatrix}$$

where R is the permutation that reverses the order of the elements of $\mathbf{a}_{n-1}$. The quantity γ_n is computed from the relation

$$\gamma_n = -\frac{r_{xx}(n) - \sum_{j=1}^{n-1} a_{n-1}(j) r_{xx}(n-j)}{E_{n-1}}$$

and the mean square prediction error is computed from $E_n = E_{n-1}(1 - |\gamma_n|^2)$. The algorithm starts with $E_0 = r_{xx}(0)$ and $\mathbf{a}_0 = 1$. If $x(k)$ is order n predictable, then the process terminates with the order n predictor and $E_n = 0$.

The parameters γ_k are known as **reflection coefficients, partial correlation coefficients** or **Schur parameters.** The **complementary parameters** are $\delta_k = \sqrt{1 - |\gamma_k|^2}$.

Facts:

1. [Pap85] The Wold decomposition theorem: Every stationary random signal can be represented as the sum of a regular and a predictable signal

$$x(k) = x_r(k) + x_p(k).$$

   The two components are orthogonal with $E[x_r(k)x_p(k)] = 0$.
2. [Pap85] The Wiener–Hopf equations can be interpreted as stating that the prediction error $\hat{x}(k) - x(k)$ is orthogonal to $x(k-l)$ for $l \geq 1$ in the inner product $\langle x, y \rangle = E[xy]$.
3. The Toeplitz matrix T_n in the Yule–Walker equations is positive semidefinite.
4. The Levinson–Durbin algorithm is an $O(n^2)$ order recursive method for solving the linear prediction problem. In most cases the numerical results from applying the algorithm are comparable to the numerically stable Cholesky factorization procedure [Cyb80]. An alternative fast method, the Schur algorithm, computes a Cholesky factorization and can be proven to be numerically stable [BBH95].
5. The following are equivalent:

   (a) T_{n-1} is positive definite and T_n is positive semidefinite.

   (b) $|\gamma_j| < 1$ for $1 \leq j \leq n-1$ and $|\gamma_n| = 1$.

   (c) $E_j > 0$ for $1 \leq j \leq n-1$ and $E_n = 0$.

   (d) The polynomial $z^n E_n(z)$ has all zeros on the unit circle and each polynomial $z^k E_k(z)$ for $k = 1, 2, \ldots, n-1$ has all zeros strictly inside the unit circle.
6. [Pap85] The prediction errors are monotonic: $E_k \geq E_{k+1}$ and $E_k \to E_\infty \geq 0$ as $k \to \infty$.

7. [SK01] Let

$$a_k(z) = 1 - a_k(1)z - a_k(2)z^2 - \cdots - a_k(k)z^k$$

where $a_k(j)$ are the prediction coefficients for a signal $\boldsymbol{x}$ with spectral density $S_x(z)$. If $S_x(z) = L(z)\overline{L(\overline{z^{-1}})}$ is the spectral factorization of $S_x(z)$, then

$$L(z) = \lim_{k\to\infty} \frac{E_k^{1/2}}{r_{xx}(0)} \frac{1}{\overline{a_k(\overline{z^{-1}})}}.$$

A survey, including other representations of the spectral factor and algorithms for computing it, is given in [SK01].

8. [GS84] A finite power random signal $\boldsymbol{x}$ is predictable (i.e., $E_\infty = 0$) if and only if

$$\int_{-\pi}^{\pi} \ln(S_x(e^{i\omega}))\,d\omega > -\infty.$$

If the signal is regular, then we have the Kolmogorov–Szegö error formula

$$E_\infty = \exp\left(\int_{-\pi}^{\pi} \ln S(e^{i\omega})\,d\omega\right) > 0.$$

Examples:

1. Consider the autoregressive process $\boldsymbol{x}$ given by

$$x(k) = n(k) + ax(k-1)$$

where $\boldsymbol{n}$ is a white noise process with $\sigma_n^2 = 1$. The solution to the order 1 linear prediction problem is given by the Yule–Walker system

$$\begin{bmatrix} \frac{1}{1-a^2} & \frac{a}{1-a^2} \\ \frac{a}{1-a^2} & \frac{1}{1-a^2} \end{bmatrix} \begin{bmatrix} 1 \\ -a_1(1) \end{bmatrix} = \begin{bmatrix} E_1 \\ 0 \end{bmatrix}.$$

The Levinson–Durbin algorithm starts with $E_0 = r_{xx}(0) = 1/(1-a^2)$ and $\mathbf{a}_0 = 1$. Thus,

$$\gamma_1 = -\frac{r_{xx}(1)}{E_0} = -\frac{a/(1-a^2)}{1/(1-a^2)} = -a$$

so that

$$\begin{bmatrix} 1 \\ -a_1(1) \end{bmatrix} = \mathbf{a}_1 = \begin{bmatrix} \mathbf{a}_0 \\ 0 \end{bmatrix} + \gamma_1 \begin{bmatrix} 0 \\ \mathbf{a}_0 \end{bmatrix} = \begin{bmatrix} 1 \\ -a \end{bmatrix}.$$

Thus, $a_1(1) = a$ and the optimal linear prediction of $x(k)$ from $x(k-1)$ is given by

$$\hat{x}(k) = ax(k-1).$$

The prediction error is given by

$$E_1 = E[\hat{x}(k) - x(k)] = E[n(k)] = 1.$$

64.4 Wiener Filtering

Linear optimal filtering of stationary signals was introduced by Wiener [Wie49] and Kolmogorov [Kol41]. The material here is covered in books on adaptive filtering [Hay91] and linear estimation [KSH00].

Definitions:

The general **discrete Wiener filtering problem** is the following: Given stationary signals $\boldsymbol{x}$ and $\boldsymbol{y}$ find a filter with impulse response $\boldsymbol{w}$ such that

$$E_w = E\left[\left|y(k) - \sum_{j=-\infty}^{\infty} w(j)x(k-j)\right|^2\right]$$

is minimal. The filter with impulse response $\boldsymbol{w}$ minimizing E_w is the **Wiener filter**. The goal is to approximate a desired signal $\boldsymbol{y}$ by filtering the signal $\boldsymbol{x}$. Equivalently, we seek the best approximation to $y(k)$ in the span of the $x(j)$ for $-\infty < j < \infty$. Note that we leave open the possibility that the Wiener filter is not causal.

Depending on specific relations that may be imposed on $\boldsymbol{y}$ and $\boldsymbol{x}$, the Wiener filtering problem specializes in several important ways: The **Wiener prediction problem** for which $y(k) = x(k+l)$ where $l \geq 1$ and $\boldsymbol{w}$ is assumed causal; the **Wiener smoothing problem** for which $\boldsymbol{y}$ is arbitrary and $\boldsymbol{x}$ is a noise corrupted version of $\boldsymbol{y}$ given by $\boldsymbol{x} = \boldsymbol{y} + \boldsymbol{n}$ where $\boldsymbol{n}$ is the noise; and the **Wiener deconvolution problem** for which we have arbitrary desired signals $\boldsymbol{y}$ and $\boldsymbol{x}$ obtained from $\boldsymbol{y}$ by convolution and the addition of noise, $\boldsymbol{x} = \boldsymbol{h} * \boldsymbol{y} + \boldsymbol{n}$. For the Wiener prediction problem, the goal is to predict future values of $\boldsymbol{x}$ from past values. For the Wiener smoothing problem, the goal is to recover an approximation to a signal $\boldsymbol{y}$ from a noise corrupted version of the signal $\boldsymbol{x}$. For the Wiener deconvolution, the goal is to invert a filter in the presence of noise to obtain the signal $\boldsymbol{y}$ from the filtered and noise corrupted version $\boldsymbol{x}$.

The **FIR Wiener filtering problem** is to choose an FIR filter with impulse response $w(k)$ such that

$$E\left[\left|y(k) - \sum_{j=0}^{n-1} w(j)x(k-j)\right|^2\right]$$

is minimal.

The **causal part** of a signal $x(k)$ is

$$[x(k)]_+ = \begin{cases} x(k) & \text{if } k \geq 0 \\ 0 & \text{if } k < 0 \end{cases} \qquad \text{or} \qquad [X(z)]_+ = \sum_{k=0}^{\infty} x(k)z^{-k}.$$

Facts:

1. The stationarity assumptions ensure that the coefficients $w(j)$ that are used to estimate $y(k)$ from the signal $\boldsymbol{x}$ do not depend on k. That is, the coefficients $w(j)$ that minimize E_w also minimize

$$E\left[\left|y(k-k_0) - \sum_{j=0}^{n-1} w(j)x(k-k_0-j)\right|^2\right]$$

so that the Wiener filter is a linear time invariant system.

2. Let $S_x(z) = L(z)\overline{L(\overline{z^{-1}})}$,

$$r_{yx}(k) = E\left[y(j)x(j-k)\right], \qquad \text{and} \qquad S_{yx}(e^{i\omega}) = \sum_{k=-\infty}^{\infty} r_{yx}(k)e^{ik\omega}.$$

If we do not require causality, then the Wiener filter has z-transform

$$W(z) = \frac{S_{yx}(z)}{S_{xx}(z)},$$

If we do impose causality, then the Wiener filter is

$$\left[\frac{S_{yx}(z)}{L^*(z^{-*})}\right]_+ \frac{1}{L(z)}.$$

3. As seen above the Wiener filter depends on the cross correlations $r_{yx}(k)$, but not on knowledge of the elements of the unkown sequence $y(k)$. Thus, it is possible to estimate the signal $\boldsymbol{y}$ by filtering $\boldsymbol{x}$ without direct knowledge of $\boldsymbol{y}$. If the autocorrelation sequences $r_{xx}(k)$ and $r_{nn}(k)$ are available (or can be estimated), then $r_{yx}(k)$ can often be computed in a straightforward way. For the Wiener prediction problem, $r_{yx}(k) = r_{xx}(l+k)$. If the noise $\boldsymbol{n}$ and signal $\boldsymbol{y}$ are uncorrelated so that $r_{ny}(k) = 0$ for all k, then for the Wiener smoothing problem $\boldsymbol{r}_{yx} = \boldsymbol{r}_{xx} - \boldsymbol{r}_{nn}$. For the Wiener deconvolution problem, if $\boldsymbol{g}$ is the inverse filter of $\boldsymbol{h}$ and if $\boldsymbol{y}$ and $\boldsymbol{n}$ are uncorrelated, then $\boldsymbol{r}_{yx} = \boldsymbol{g} * (\boldsymbol{r}_{xx} - \boldsymbol{r}_{nn})$.
4. The coefficients for an FIR Wiener filter satisfy the Wiener–Hopf equation

$$\begin{bmatrix} r_{xx}(0) & r_{xx}(-1) & \cdots & r_{xx}(-m+1) \\ r_{xx}(1) & r_{xx}(0) & \ddots & \vdots \\ \vdots & \ddots & \ddots & r_{xx}(-1) \\ r_{xx}(m-1) & \cdots & r_{xx}(1) & r_{xx}(0) \end{bmatrix} \begin{bmatrix} w(0) \\ w(1) \\ \vdots \\ w(m-1) \end{bmatrix} = \begin{bmatrix} r_{yx}(0) \\ r_{yx}(1) \\ \vdots \\ r_{yx}(m-1) \end{bmatrix}.$$

As with linear prediction, this system of equations can be solved using fast algorithms for Toeplitz systems.

Examples:

1. We consider the first order FIR Wiener prediction problem of predicting $y(k) = x(k+2)$ from the autoregressive process $\boldsymbol{x}$ with $x(k) = n(k) + ax(k-1)$. Thus, we seek $w(0)$ such that

$$E\left[|x(k+2) - w(0)x(k)|^2\right]$$

is minimal. Since

$$r_{xx}(0) = \frac{1}{1-a^2}$$

and

$$r_{yx}(0) = r_{xx}(2) = \frac{a^2}{1-a^2}$$

the Wiener–Hopf equation for $w(0)$ is

$$\frac{1}{1-a^2} w(0) = \frac{a^2}{1-a^2}.$$

64.5 Adaptive Filtering

Definitions:

An FIR **adaptive filter** is a filter with coefficients that are updated to match the possibly changing statistics of the input $\boldsymbol{x}$ and the desired output $\boldsymbol{y}$. The output is

$$\hat{y}(k) = \sum_{j=0}^{m-1} w_k(j)x(k-j)$$

where the $w_k(j)$ depend on k. The goal is similar to that of the Wiener filter, that is, to approximate the sequence $\boldsymbol{y}$. The difference is that instead of computing a fixed set of coefficients $w(j)$ from knowledge of the statistics of stationary signals $\boldsymbol{x}$ and $\boldsymbol{y}$, the coefficients are allowed to vary with k and are computed from actual samples of the sequences $\boldsymbol{x}$ and $\boldsymbol{y}$.

A **recursive least squares (RLS)** adaptive FIR filter is a rule for updating the coefficients $w_k(j)$. The coefficients are chosen to minimize

$$\min_{\mathbf{w}_k} \left\| D_{\rho,k} \begin{bmatrix} x(l) & x(l-1) & \cdots & x(l-m+1) \\ x(l+1) & x(l) & \cdots & x(l-m+2) \\ \vdots & \ddots & \ddots & \vdots \\ x(k) & x(k-1) & \cdots & x(k-m+1) \end{bmatrix} \begin{bmatrix} w_k(0) \\ w_k(1) \\ \vdots \\ w_k(m-1) \end{bmatrix} - D_{\rho,k} \begin{bmatrix} y(l) \\ y(l+1) \\ \vdots \\ y(k) \end{bmatrix} \right\|_2$$

where $k \geq l+m-1$ and $D_{\rho,k} = \mathrm{diag}(\rho^{k-l}, \rho^{k-l-1}, \ldots \rho^0)$ with $0 < \rho \leq 1$. Equivalently, the vector $\mathbf{w}_k$ is chosen to solve $\min_{\mathbf{w}_k} \|D_{\rho,k} T_{l,k}\mathbf{w}_k - D_{\rho,k}\mathbf{y}_k\|$. The parameter ρ is chosen to decrease the influence of older data. This is of use when the signals $\boldsymbol{x}$ and $\boldsymbol{y}$ are only approximately stationary, i.e., their statistical properties are slowly time varying.

A signal $\boldsymbol{x}$ is **persistently exciting of order m** if there exist $c_1 > 0$, $c_2 \geq c_1$, and j such that

$$c_1 I \leq \lambda_p \left(\frac{1}{k-l+1} T_{l,k}^{\mathrm{T}} T_{l,k} \right) \leq c_2 I$$

for each eigenvalue λ_p of $T_{l,k}^{\mathrm{T}} T_{l,k}/(k-l+1)$ and for all $k \geq j$.

Facts:

1. The vector $\mathbf{w}_k$ can be updated using using the following recursive formulas:

$$\mathbf{w}_k = \mathbf{w}_{k-1} + \mathbf{g}_k \left(y(k) - \mathbf{w}_{k-1}^{\mathrm{T}} \mathbf{x}_k \right)$$

where

$$\mathbf{x}_k = \begin{bmatrix} x(k) \\ \vdots \\ x(k-m+1) \end{bmatrix},$$

$$\mathbf{g}_k = \frac{\rho^{-2} P_{k-1}\mathbf{x}_k}{1 + \rho^{-2}\mathbf{x}_k^{\mathrm{T}} P_{k-1}\mathbf{x}_k},$$

with

$$P_k = \rho^{-2} P_{k-1} - \rho^{-2}\mathbf{g}_k\mathbf{x}_k^{\mathrm{T}} P_{k-1}.$$

With initialization $P_k = (T_{l,k}^{\mathrm{T}} D_{\rho,k}^2 T_{l,k})^{-1}$ and $\mathbf{w}_k = P_k T_{l,k}^{\mathrm{T}} D_{\rho,k}^2 \mathbf{y}_k$, the recurrences compute the solution $\mathbf{w}_k$ to the recursive least squares problem.

2. [GV96] The weight vector $\mathbf{w}_k$ can also be updated as k increases with an $O(n^2)$ cost using a standard QR updating algorithm. Such algorithms have good numerical properties.
3. If $\boldsymbol{x}$ and $\boldsymbol{y}$ are stationary, if $\rho = 1$, and if $\boldsymbol{x}$ is persistently exciting, then $\mathbf{w}_k$ converges to the Wiener filter as $k \to \infty$. Under these assumptions the effect of initializing P_k incorrectly becomes a negligible as $k \to \infty$. Thus, convergence does not depend on the correct initialization of $P_k = T_{l,k}^{\mathrm{T}} T_{l,k}$. For simplicity, it is common to choose an initial P_k that is a multiple of the identity.
4. If $\boldsymbol{x}$ and $\boldsymbol{y}$ are not stationary, then ρ is typically chosen to be less than one in order to allow the filter to adapt to changing statistics. If the signals really are stationary, then choice of $\rho < 1$ sacrifices convergence of the filter.
5. There are a variety of algorithms that exploit the Toeplitz structure of the least squares problem, reducing the cost per update of $\mathbf{w}_k$ to $O(n)$ operations [CK84]. An algorithm that behaves well in finite precision is analyzed in [Reg93].

Examples:

1. Consider applying a recursive least squares adaptive filter to signals $\boldsymbol{x}$ and $\boldsymbol{y}$ with

$$x(1) = 1, x(2) = 2, x(3) = 3, x(4) = 4$$

and

$$y(2) = 5, y(3) = 6, y(4) = 8.$$

The vector of filter coefficients $\mathbf{w}_3$ satisfies

$$\begin{bmatrix} 2 & 1 \\ 3 & 1 \end{bmatrix} \mathbf{w}_3 = \begin{bmatrix} 5 \\ 6 \end{bmatrix},$$

so that

$$\mathbf{w}_3 = \begin{bmatrix} 4 \\ -3 \end{bmatrix}.$$

We also have

$$P_3 = \left(\begin{bmatrix} 2 & 1 \\ 3 & 1 \end{bmatrix}^{\mathrm{T}} \begin{bmatrix} 2 & 1 \\ 3 & 1 \end{bmatrix} \right)^{-1} = \begin{bmatrix} 13 & 8 \\ 8 & 5 \end{bmatrix},$$

so that

$$\mathbf{g}_4 = \frac{P_3 \mathbf{x}_4}{1 + \mathbf{x}_4^{\mathrm{T}} P_3 \mathbf{x}_4} = \frac{1}{6} \begin{bmatrix} -4 \\ 7 \end{bmatrix}.$$

Thus, the next coefficient vector $\mathbf{w}_4$ is

$$\mathbf{w}_4 = \mathbf{w}_3 + \mathbf{g}_3 \left(y(4) - \mathbf{w}_3^{\mathrm{T}} \mathbf{x}_4 \right) = \begin{bmatrix} 10/3 \\ -11/6 \end{bmatrix}.$$

The process can be continued by updating P_4, $\mathbf{g}_5$, and $\mathbf{w}_5$.

64.6 Spectral Estimation

Definitions:

The **deterministic spectral estimation** problem is to estimate the spectral density $S_x(e^{i\omega})$ from a finite number of correlations $r_{xx}(k)$ for $0 \leq k \leq n$. Note that this definition is not taken to imply that the signal $\boldsymbol{x}$ is deterministic, but that the given data are correlations rather than samples of the random signals.

The **stochastic spectral estimation** problem is to estimate $S(e^{i\omega})$ from samples of the random signal $x(k)$ for $0 \leq k \leq n$.

The problem of estimating **sinusoids in noise** is to find the frequencies ω_j of

$$\hat{x}(k) = \alpha_1 \cos(\omega_1 k + \eta_1) + \alpha_2 \cos(\omega_2 k + \eta_2) + \cdots + \alpha_p \cos(\omega_p k + \eta_p) + w(k) = x(k) + n(k)$$

where $\boldsymbol{n}$ is zero mean white noise with variance σ_n^2 and $\boldsymbol{x}$ is the exact sinusoidal signal with random frequencies $\omega_j \in [0, 2\pi)$ and phases η_j. The frequencies and phases are assumed to be uncorrelated with the noise $\boldsymbol{n}$. The available data can be the samples of the signal $\boldsymbol{x}$ or the autocorrelation sequence $r_{xx}(j)$.

Facts:

1. [PM96] A traditional method for the deterministic spectral estimation problem is to assume that any $r_{xx}(j)$ not given are equal to zero. When additional information is available in the form of a signal model, it is possible to do better by extending the autocorrelation sequence in a way that is consistent with the model.
2. [Pap85] The method of maximum entropy: The sequence $r_{xx}(k)$ is extended so as to maximize the entropy rate

$$\int_{-\pi}^{\pi} \ln S(e^{i\omega})\, d\omega,$$

where it is assumed that the random signal $\boldsymbol{x}$ is a sequence of normal random variables. It can be shown that the spectrum that achieves this is the spectrum of an autoregressive signal

$$S(e^{i\omega}) = \frac{E_n}{\left|1 - \sum_{j=1}^{n} a_n(j) e^{-i\omega j}\right|^2},$$

where the $a_n(j)$ are the coefficients obtained from the Yule–Walker equations.

3. [Pis73] Pisarenko's method: Given a signal comprised of sinusoids in noise, if $\sigma_n = 0$, then the signal sequence $\boldsymbol{x}$ is order p predictable and the signal and its autocorrelation sequence $\boldsymbol{r}_{xx}$ satisfy

$$x(k) = a(1)x(k-1) + a(2)x(k-2) + \cdots + a(p)x(k-p),$$
$$r_{xx}(k) = a(1)r(k-1) + a(2)r_{xx}(k-2) + \cdots + a(p)r_{xx}(k-p),$$

where $\mathbf{a}_p = \begin{bmatrix} 1 & -a_p(1) & \cdots & -a_p(p) \end{bmatrix}^{\mathrm{T}}$ is a null vector of the $(p+1) \times (p+1)$ autocorrelation matrix $T = [t_{ij}] = [r_{xx}(i-j)]$. This is the Yule–Walker system with $n = p$ and $E_p = 0$. When $\sigma_n \neq 0$, we have $\hat{T} = T + \sigma_n^2 I$ where $\hat{T}$ is the autocorrelation matrix for $\hat{x}(k)$. The value of σ_n^2 is the smallest eigenvalue of $\hat{T}$. Given knowledge of σ_n, T can be computed from $T = \hat{T} - \sigma_n^2 I$. The prediction coefficients can then be computed from the Yule–Walker equation $T\mathbf{a}_p = 0$. The zeros of the prediction error filter

$$z^p E_p(z) = z^p \left(1 - \sum_{j=1}^{p} a(j) z^{-j}\right)$$

all lie on the unit circle at $e^{i\omega_j}$ for $j = 1, 2, \ldots p$, giving the frequencies of the sinusoids.

4. [AGR87] The zeros of a polynomial are often sensitive to small errors in the coefficients. Computing such zeros accurately can be difficult. As an alternative the zeros of $z^p E_p(z)$ can be found as the eigenvalues of a unitary Hessenberg matrix. Let γ_j and δ_j be the Schur parameters and complementary parameters for the autocorrelation matrix R. Then $|\gamma_j| < 1$ for $1 \leq j \leq p-1$ and $|\gamma_p| = 1$. The matrix

$$H = G_1(\gamma_1)G_2(\gamma_2)\cdots G_p(\gamma_p)$$

with

$$G_j(\gamma_j) = I_{j-1} \oplus \begin{bmatrix} -\gamma_j & \delta_j \\ \delta_j & \overline{\gamma}_j \end{bmatrix} \oplus I_{p-j-1}$$

is unitary and upper Hessenberg. An explicit formula for the elements of H is

$$H = \begin{bmatrix} -\gamma_1 & -\delta_1\gamma_2 & -\delta_1\delta_2\gamma_3 & \cdots & -\delta_1\cdots\delta_{p-1}\gamma_p \\ \delta_1 & -\overline{\gamma}_1\gamma_2 & -\overline{\gamma}_1\delta_2\gamma_3 & \cdots & -\overline{\gamma}_1\delta_2\cdots\delta_{p-1}\gamma_p \\ & \delta_2 & -\overline{\gamma}_2\gamma_3 & & \vdots \\ & & \ddots & \ddots & \vdots \\ & & & \delta_{p-1} & -\overline{\gamma}_{p-1}\gamma_p \end{bmatrix}.$$

The eigenvalues of H are the zeros of $z^p E_p(z)$. The eigenvalues of a $p \times p$ unitary Hessenberg matrix can be found with $O(p^2)$ operations. An algorithm exploiting connections with Szegö polynomials was given in [Gra86] and stabilized in [Gra99]. A different, stable algorithm involving matrix pencils was given in [BE91].

Examples:

1. Consider a stationary random signal with known autocorrelations $r_{xx}(0) = 1$ and $r_{xx}(1) = a < 1$. The other autocorrelations are assumed to be unknown. The Yule–Walker system

$$\begin{bmatrix} 1 & a \\ a & 1 \end{bmatrix} \begin{bmatrix} 1 \\ -a_1(1) \end{bmatrix} = \begin{bmatrix} E_1 \\ 0 \end{bmatrix}$$

has solution $a_1(1) = a$ and $E_1 = 1 - a^2$. Thus, the maximum entropy estimate of the spectral density $S_x(e^{i\omega})$ is

$$S(e^{i\omega}) = \frac{1-a^2}{|1 - ae^{-i\omega}|}.$$

64.7 Direction of Arrival Estimation

Definitions:

The **direction of arrival estimation** problem can be described as follows. We assume that n plane waves $s_j(k)$, $1 \leq j \leq n$ arrive at an array of m sensors from angles θ_j resulting in output $x_l(k)$, $1 \leq l \leq m$ from the sensors. We assume that the sensors lie in a plane and that it is necessary to estimate only a single angle of arrival for each signal. It is assumed that each sensor output is corrupted by noise $n_l(k)$, $1 \leq l \leq m$.

The sensors experience different delays and attenuations of the signal $s_j(k)$ depending on their position and the direction of arrival. If the signals are **narrow-band signals**

$$s_j(k) = \text{Re}\left(u_j(k)e^{i(\omega_0 k + v_j(k))}\right)$$

for slowly varying $u_j(k)$ and $v_j(k)$, then a signal delayed by l is approximately

$$s_j(k-l) = \text{Re}\left(u_j(k-l)e^{i(\omega_0 k + v_j(k-l))}e^{-i\omega_0 l}\right) \approx \text{Re}\left(u_j(k)e^{i(\omega_0 k + v_j(k))}e^{-i\omega_0 l}\right).$$

Thus, both delay and attenuation of the signals can be described by multiplication by a complex scalar dependent on θ_j. The observed signals are, therefore,

$$\mathbf{x}(k) = \sum_{j=1}^{n} \mathbf{a}(\theta_j)s_j(k) + \mathbf{n}(k),$$

where

$$\mathbf{x}(k) = \begin{bmatrix} x_1(k) \\ \vdots \\ x_m(k) \end{bmatrix}$$

and $\mathbf{a}(\theta)$ is the complex **array response vector**. In matrix notation the **signal model** is

$$\mathbf{x}(k) = A\mathbf{s}(k) + \mathbf{n}(k),$$

where

$$A = \begin{bmatrix} \mathbf{a}(\theta_1) & \mathbf{a}(\theta_2) & \cdots & \mathbf{a}(\theta_n) \end{bmatrix}.$$

The problem is to estimate the angles θ_j from a sequence $\mathbf{x}(k)$ of sensor output vectors. We assume that the components of the noise vector $\mathbf{n}(k)$ are uncorrelated, stationary, and zero mean. A more complete description of the problem is given in [RK89].

The **array manifold** is the set of all $\mathbf{a}(\theta)$ for θ in $[-\pi, \pi]$.

The **signal subspace** is the range space of the matrix A.

The **noise subspace** is the orthogonal complement of the signal subspace.

We define the spatial correlation matrices

$$R_{\mathbf{x}} = E[\mathbf{x}\mathbf{x}^*], \qquad R_{\mathbf{s}} = E[\mathbf{s}\mathbf{s}^*], \qquad \text{and} \qquad R_{\mathbf{n}} = E[\mathbf{n}\mathbf{n}^*]$$

where we assume R_s is positive definite. These definitions apply directly when dealing with stationary random signals. In the deterministic case they can be replaced with time averages.

Facts:

1. If the signal and noise are uncorrelated, then

$$R_{\mathbf{x}} = AR_{\mathbf{s}}A^* + R_{\mathbf{n}}.$$

2. The signal model implies that except for the effect of noise the observation vectors $\mathbf{x}(k)$ lie in the signal subspace. Thus, in the absence of noise, any n linearly independent observation vectors $\mathbf{x}(k)$ will give a basis for the signal subspace. The directions θ_j are determined by the signal subspace so long as the mapping from the set of angles θ_j to the signal subspace is invertible. We assume that the sensor array is designed to ensure invertibility.

3. The signal subspace can be estimated by solving a generalized eigenvalue problem. The n eigenvectors associated with the n largest generalized eigenvalues of the pencil

$$R_x - \lambda R_n$$

are the basis for the estimated signal subspace. This method can be used to obtain matrices X_s and X_n such that $X_s X_s^*$ is a projection for the signal subspace and $X_n X_n^*$ is a projection for the noise subspace.

4. [Sch79] In order to obtain the angles θ_j the **MUSIC algorithm** searches the array manifold to find angles which are close to the signal subspace in the sense that

$$M(\theta) = \frac{\mathbf{a}^*(\theta)\mathbf{a}(\theta)}{\mathbf{a}^*(\theta) X_n X_n^* \mathbf{a}(\theta)}$$

is large. The values of θ for which $M(\theta)$ has a peak are taken as the estimates of the angles θ_j, $1 \leq j \leq n$. The basic approach of the MUSIC algorithm can be used for a variety of parameter estimation problems, including the problem of finding sinusoids in noise.

5. [RK89] The **ESPRIT** algorithm imposes a structure on the array of sensors. It is assumed that the array is divided into two identical subarrays displaced along an $\mathbb{R}^2$ vector $\mathbf{\Delta}$. Let $x_j(k)$ be the output of sensor j in the first subarray and $y_j(k)$ be the output of the corresponding sensor in the second subarray. The displacement $\mathbf{\Delta}$ causes a delay in the signal $x_j(k)$ relative to $y_j(k)$. If the signals arriving at the array are again narrow band signals centered at frequency ω_0, then the signal model is

$$\mathbf{x}(k) = \sum_{j=1}^{n} s_j(k)\mathbf{a}(\theta_j) + \mathbf{n}_x(k),$$

$$\mathbf{y}(k) = \sum_{j=1}^{n} s_j(k)e^{i\omega_0 \|\Delta\|_2 \sin(\theta_j)/c}\mathbf{a}(\theta_j) + \mathbf{n}_y(k),$$

where θ_k is the angle of arrival relative to the subarray displacement vector. The combined ESPRIT signal model including both subarrays is

$$\begin{bmatrix} \mathbf{x}(k) \\ \mathbf{y}(k) \end{bmatrix} = \begin{bmatrix} A \\ A\Phi \end{bmatrix} \mathbf{s}(k) + \begin{bmatrix} \mathbf{n}_x(k) \\ \mathbf{n}_y(k) \end{bmatrix},$$

where the diagonal matrix $\Phi = \text{diag}(e^{i\omega_0 \|\Delta\|_2 \sin(\theta_1)/c}, \ldots, e^{i\omega_0 \|\Delta\|_2 \sin(\theta_n)/c})$ characterizes the delay between the subarrays.

6. To compute the θ_j using the ESPRIT algorithm, we start with the signal subspace associated with the ESPRIT signal model estimated as before using a generalized eigenvalue problem. We assume that a $2m \times n$ matrix S with columns spanning the signal subspace

$$S = \begin{bmatrix} S_x \\ S_y \end{bmatrix}$$

has been computed and compute the $2n \times 2n$ matrix

$$V = \begin{bmatrix} V_{11} & V_{12} \\ V_{21} & V_{22} \end{bmatrix}$$

of right singular vectors of $\begin{bmatrix} S_x & S_y \end{bmatrix}$. Here, V_{11} and V_{22} are $n \times n$. The ESPRIT estimates of the values $e^{i\omega_0\|\Delta\|_2 \sin(\theta_k)/c}$ are the eigenvalues λ_j of $-V_{12}V_{22}^{-1}$ so that the estimate $\hat{\theta}_j$ of θ_j is

$$\hat{\theta}_j = \sin^{-1}(c \arg(\lambda_j)/(\|\Delta\|_2\omega_0))$$

for $j = 1, 2, \ldots, n$.

Examples:

1. We consider a single signal arriving at an array of two sensors with array response vector

$$\mathbf{a}(\theta) = \begin{bmatrix} e^{-i(\pi/2-\theta)} \\ 1 \end{bmatrix}.$$

If observations of the sensor outputs and solution of the generalized eigenvalue problem $R_x - \lambda R_n$ suggest that

$$X_s X_s^* = \begin{bmatrix} i/\sqrt{3} \\ 2/\sqrt{3} \end{bmatrix} \begin{bmatrix} -i/\sqrt{3} & 2/\sqrt{3} \end{bmatrix}$$

is the projection for the signal subspace, then the projection for the noise subspace is

$$X_n X_n^{\mathrm{T}} = \begin{bmatrix} -2i/\sqrt{3} \\ 1/\sqrt{3} \end{bmatrix} \begin{bmatrix} 2i/\sqrt{3} & 1/\sqrt{3} \end{bmatrix}.$$

In applying the MUSIC algorithm, we have

$$M(\theta) = \frac{2}{3|2e^{i\theta} + 1|^2}.$$

This function has a maximum of $2/3$ at $\theta = \pi$. Thus, the MUSIC estimate of the angle from which the signal arrives is $\theta = \pi$.

References

[AGR87] G. Ammar, W. Gragg, and L. Reichel, Determination of Pisarenko frequency estimates as eigenvalues of an orthogonal matrix, in *Advanced Algorithms and Architectures for Signal.* Processing II, Proc. SPIE, Vol. 826, *Int. Soc. Opt. Eng.*, pp. 143–145, 1987.

[BBH95] A. Bojanczyk, R.P. Brent, F.R. De Hoog, and D.R. Sweet, On the stability of the Bareiss and related Toeplitz factorization algorithms, *SIAM J. Matrix Anal. Appl.*, 16: 40–58, 1995.

[BE91] A. Bunse-Gerstner and L. Elsner, Schur parameter pencils for the solution of the unitary eigenproblem, *Lin. Alg. Appl.*, 154–156:741–778, 1991.

[CK84] J. Cioffi and T. Kailath, Fast, recursive least-squares filters for adaptive filtering, *IEEE Trans. Acoust., Speech, Sig. Proc.*, 32: 304–337, 1984.

[Cyb80] G. Cybenko, The numerical stability of the Levinson–Durbin Algorithm for Toeplitz systems of equations, *SIAM J. Sci. Stat. Comp.*, 1:303–310, 1980.

[GV96] G.H. Golub and C.F. Van Loan, *Matrix Computations*, 3rd ed., Johns Hopkins University Press, Baltimore, 1996.

[Gra86] W.B. Gragg, The QR algorithm for unitary Hessenberg matrices, *J. Comp. Appl. Math.*, 16:1–8, 1986.

[Gra93] W.B. Gragg, Positive definite Toeplitz matrices, the Arnoldi process for isometric operators, and Gaussian quadrature on the unit circle, *J. Comp. Appl. Math.*, 46:183–198, 1993.

[Gra99] W.B. Gragg, Stabilization of the UHQR algorithm, In Z. Chen, Y. Li, C. Micchelli, and Y. Xu, Eds., *Advances in Computational Mathematics*, pp. 139–154. Marcel Dekker, New York, 1999.

[GS84] U. Grenander and G. Szegö, *Toeplitz Forms and Their Applications.* 2nd ed., Chelsea, London, 1984.

[Hay91] S. Haykin, *Adaptive Filter Theory*, 2nd ed., Prentice Hall, Upper Saddle River, NJ, 1991.

[KSH00] T. Kailath, A. Sayed, and B. Hassibi, *Linear Estimation*, Prentice Hall, Upper Saddle River, NJ, 2000.

[Kol41] A.N. Kolmogorov, Interpolation and extrapolation of stationary random sequences, *Izv. Akad. Nauk SSSR Ser. Mat.* 5:3–14, 1941, (in Russian).

[Pap85] A. Papoulis, Levinson's algorithm Wold's decomposition, and spectral estimation, *SIAM Rev.*, 27:405–441, 1985.

[Pis73] V.F. Pisarenko, The retrieval of harmonics from a covariance function, *Geophys. J. Roy. Astron. Soc.*, 33:347–366, 1973.

[PM96] J.G. Proakis and D. Manolakis, *Digital Signal Processing: Principles, Algorithms and Applications*, Prentice Hall, Upper Saddle River, NJ, 1996.

[Reg93] P. Regalia, Numerical stability properties of a QR-based fast least-squares algorithm, *IEEE Trans. Sig. Proc.*, 41:2096–2109, 1993.

[RK89] R. Roy and T. Kailath, ESPRIT — Estimation of signal parameters via rotational invariance techniques, *IEEE Trans. Acoust., Speech, Sign. Proc.*, 37:984–995, 1989.

[Sch79] R. Schmidt, Multiple emitter location and signal parameter estimation, *IEEE Trans. Anten. Prop.*, 34:276–280, 1979.

[SK01] A.H. Sayed and T. Kailath, A survey of spectral factorization methods, *Num. Lin. Alg. Appl.*, 08:467–496, 2001.

[Wie49] N. Wiener, *Extrapolation, Interpolation, and Smoothing of Stationary Time Series with Engineering Applications*, MIT Press, Cambridge, MA, 1949.

Applications to Geometry

65

Geometry

Mark Hunacek
Iowa State University, Ames

Many topics taught in an introductory linear algebra course are often motivated by reference to elementary geometry. The geometry is treated as something the student is already familiar with, and reference to it is made to give the student a better understanding of algebraic concepts that may be considered, at least on first acquaintance, to be rather abstract. What is interesting and important is that the process can be reversed: Assuming the linear algebra as known, geometric concepts can be defined and developed on a rigorous basis. The use of vector methods in geometry often provides new ways of looking at old problems and also helps demonstrate deep and beautiful connections between algebra and geometry.

This chapter begins by discussing affine geometry, which can be defined, very roughly, as Euclidean geometry without any mention of measurement. Thus, affine theorems concern such things as incidence and parallelism. An affine space is defined in terms of an action of a vector space V on a set X, pursuant to which a vector $\mathbf{v}$ acts on a point A of X by sending it to another element $\mathbf{v}(A)$ of X, and two points A and B of X define a unique vector $\overrightarrow{AB}$, which acts on A by sending it to B. It might be helpful for the reader to think of $\overrightarrow{AB}$ as an arrow starting at point A and ending at point B. This vector then acts on an arbitrary point C of X by placing the starting point of the arrow at C and letting the end point of the arrow be $\mathbf{v}(C)$.

By introducing an inner product on a real affine space, Euclidean geometry can be defined. Actually, a very interesting mathematical theory can be developed by considering, more generally, an arbitrary bilinear form, but in the interest of simplicity, only (positive definite) real inner products are considered in this chapter. The general theory is discussed in Chapter III of [Art57] and Chapter 2 of [ST91].

Finally, projective geometry is considered. Here the points of the geometry are not vectors but one-dimensional subspaces and the resulting incidence properties are different than for Euclidean geometry. Projective geometry plays an important role in (among other things) the mathematical theories of elliptic curves and algebraic geometry, but these topics are far beyond the scope of this chapter.

Throughout this chapter, F is a field and V is a finite dimensional vector space over F.

65.1 Affine Spaces

Definitions:

An **affine n-space** is an ordered pair (X, V) where X is a nonempty set and V is an n-dimensional vector space over a field F, which acts on X in the following way: if $\mathbf{v} \in V$ and $A \in X$ then there is an element $\mathbf{v}(A) \in X$, and

1. If $\mathbf{u}, \mathbf{v} \in V$ and $A \in X$, then $(\mathbf{u}+\mathbf{v})(A) = \mathbf{u}(\mathbf{v}(A))$, and
2. For any two points $A, B \in X$ there exists a unique vector $\mathbf{v} \in V$ such that $\mathbf{v}(A) = B$. This vector $\mathbf{v}$ is denoted $\overrightarrow{AB}$.

The elements of X are called the **points** of the affine space (X, V), and sometimes, when no confusion will result, one speaks of "the affine space X."

A **real affine space** is an affine space (X, V) where V is a vector space over the field $\mathbb{R}$ of real numbers.

If (X, V) is an affine n-space, W is a (vector) subspace of V of dimension $m \leq n$, and A is a fixed point in X, then the **affine subspace determined by** A **and** W, denoted $S(A, W)$, is the set of all points $\mathbf{w}(A)$ as $\mathbf{w}$ ranges over W. A subset of X is called an affine subspace of dimension m if it is of the form $S(A, W)$ for some A in X and subspace W of dimension m (cf. Fact 2, below).

A one-dimensional affine subspace of X is called a **line**; a two-dimensional affine subspace, a **plane**.

The vector subspace W is called the **direction space** of the affine subspace $S(A, W)$.

Two affine subspaces of the same dimension are **parallel** if they have the same direction space. More generally, two affine subspaces (of possibly different dimension) are **parallel** if the direction space of one is a subspace of the direction space of another.

If C is a point in X, the map $P \rightarrow \overrightarrow{CP}$ is a bijection between X and V. By identifying a point of X with its image in V under this bijection, addition and scalar multiplication in X can be defined so that it becomes a vector space isomorphic to V, called the **tangent space** at C and denoted $X(C)$.

The elements of a set $\{X_1, \ldots, X_d\}$ of d points in an affine space are **in general position**, or **affine-independent**, if they are not contained in any affine subspace of dimension less than or equal to $d-2$.

Three points that are affine-independent are called **noncollinear**.

If A and B are distinct points in a real affine space, the set of points P such that $\overrightarrow{AP} = t\overrightarrow{AB}$ for some t, $0 \leq t \leq 1$, is called the **line segment** from A to B and is denoted $[A, B]$. The point P which corresponds to $t = 1/2$ is the **midpoint** of the line segment.

If A, B, and C are three points in general position in a real affine space, then the set of points P such that $\overrightarrow{AP} = t\overrightarrow{AB} + u\overrightarrow{AC}$, where u and t are nonnegative real numbers whose sum does not exceed 1, is called the **triangle** with vertices A, B, and C.

A subset Y of a real affine space is **convex** if whenever A and B are in Y, every point on the line segment $[A, B]$ is also in Y.

If (X, V) is an affine n-space and $n \geq 2$, then a bijection $T: X \rightarrow X$ is **semiaffine** if the image under T of any d-dimensional affine subspace is also a d-dimensional affine subspace, and is **affine** if it is semiaffine and satisfies the following additional condition: If A and B are distinct points and C and D are distinct points, and $\overrightarrow{AB} = k\overrightarrow{CD}$ for some nonzero scalar k, then $\overrightarrow{T(A)T(B)} = k\overrightarrow{T(C)T(D)}$.

If (X, V) is an affine n-space and $\mathbf{v} \in V$, the translation T_v is the map T_v: $X \rightarrow X$ defined by $T_v(A) = \mathbf{v}(\mathbf{A})$.

Facts: For proofs, see [ST71].

1. If A and B are points in an affine space, then
   (a) $\overrightarrow{AB} = 0$ if and only if $A = B$.
   (b) $\overrightarrow{AB} = -\overrightarrow{BA}$.
   (c) $\overrightarrow{AB} + \overrightarrow{BC} = \overrightarrow{AC}$.
2. If $S = S(A, W)$ is an affine subspace of the affine space (X, V), then $W = \{\overrightarrow{AB}: A, B \in S\}$. In particular, W is uniquely determined by S.
3. The affine subspaces $S(A, W)$ and $S(B, W)$ are equal if and only if $\overrightarrow{AB} \in W$.
4. The intersection of two affine subspaces is either empty or is an affine subspace. Specifically, if the point P is in both $S(A, U)$ and $S(B, W)$, then $S(A, U) \cap S(B, W) = S(P, U \cap W)$.

5. (Generalized Euclidean Parallel Postulate) If $S = S(A, W)$ is a d-dimensional affine subspace of an affine space X, and P is any point of X, then there exists a unique d-dimensional affine subspace of X, namely $S(P, W)$, that is parallel to S and contains the point P.
6. Given any two distinct points A and B in an affine space, there is a unique line containing them both.
7. The intersection of two convex subsets of a real affine space is convex (possibly empty). In addition, every affine subspace of a real affine space is convex.
8. The set of all semiaffine transformations of an affine n-space (X, V) forms a group under the operation of function composition. The set of all affine transformations of the affine space is a subgroup of this group.
9. A semiaffine transformation of an affine space of dimension at least two maps parallel subspaces to parallel subspaces, i.e, if S and S' are parallel subspaces of the affine space X and T is a semiaffine transformation, then $T(S)$ and $T(S')$ are parallel.
10. If (X, V) is an affine space of dimension at least two and V is a vector space over a field that, like the field of real numbers, has no nontrivial automorphisms, then any semiaffine transformation is an affine transformation.
11. A translation of an affine n-space, where n is at least two, is an affine transformation of that space.
12. If $\{A_0, \ldots, A_n\}$ and $\{B_0, \ldots, B_n\}$ are two sets of points in general position in an affine n-space X, then there is a unique affine transformation T such that $T(A_i) = B_i$ for each $i = 0, 1, \ldots, n$.
13. If C is an arbitrary but fixed point of X and T is an affine transformation of X, then T can be realized as the composition of a mapping that is a nonsingular linear transformation on the tangent space $X(C)$, followed by a translation of X.

Examples:

1. If V is any vector space over a field F, then by taking $X = V$ and $\mathbf{v}(A) = \mathbf{v} + A$ we obtain an affine space, denoted $\mathcal{A}(V)$. For two points (i.e., vectors) A and B, $\overrightarrow{AB} = B - A$. The affine subspaces of $\mathcal{A}(V)$ are precisely the additive cosets of the vector subspaces. When $V = F^n$, the vector space of column n-tuples of elements of a field F, the resulting affine space is often denoted $A^n(F)$.
2. Any affine subspace $S = \mathbf{v} + W$ of the affine space $\mathcal{A}(V)$ is an affine space (S, W), with $\mathbf{u}(\mathbf{v} + W) = (\mathbf{u} + \mathbf{v}) + W$. It can also be an affine space with a different (but isomorphic) vector space, however. For example, in the affine space $A^2(F)$, let S be the set of all points $\begin{bmatrix} x \\ y \end{bmatrix}$ which satisfy $x + y = 1$, so (S,W) is an affine space with $W = \left\{ \begin{bmatrix} u \\ -u \end{bmatrix} : u \in F \right\}$. But S is also an affine space with associated vector space F, with the action of F on W given by $u \begin{bmatrix} x \\ y \end{bmatrix} = \begin{bmatrix} x + u \\ y - u \end{bmatrix}$ for $u \in F$.
3. In the affine space $\mathcal{A}(V)$, the tangent space $V(\mathbf{0})$ is identical, not just as a set, but also with regard to the vector space operations, to the vector space V. Fact 13 therefore implies that any affine transformation of this affine space is of the form $T(\mathbf{v}) = L(\mathbf{v}) + \mathbf{b}$, where L is a nonsingular linear transformation from V to itself and $\mathbf{b}$ is a fixed vector.
4. The vectors $\{v_0, \ldots, v_n\}$ in $\mathcal{A}(V)$ are affine-independent if and only if the vectors $\{v_1 - v_0, \ldots, v_n - v_0\}$ are linearly independent in the vector space V. Thus, for example, if $\{\mathbf{u}, \mathbf{v}, \mathbf{w}\}$ and $\{\mathbf{x}, \mathbf{y}, \mathbf{z}\}$ are two sets of vectors in general position in the affine space $\mathcal{A}(V)$, where V is two-dimensional, then to find the unique affine transformation mapping $\mathbf{u}$ to $\mathbf{x}$, $\mathbf{v}$ to $\mathbf{y}$, and $\mathbf{w}$ to $\mathbf{z}$, we first use translation by $-\mathbf{u}$ to map $\{\mathbf{u}, \mathbf{v}, \mathbf{w}\}$ to $\{\mathbf{0}, \mathbf{v} - \mathbf{u}, \mathbf{w} - \mathbf{u}\}$, then use the (unique) linear transformation that maps the linearly independent vectors $\mathbf{v} - \mathbf{u}$ and $\mathbf{w} - \mathbf{u}$ to $\mathbf{y} - \mathbf{x}$ and $\mathbf{z} - \mathbf{x}$, respectively, and finally use translation by $\mathbf{x}$.

5. As a specific illustration of the idea discussed in the previous example, consider $A^2(\mathbb{R})$. Let $\mathbf{u} = \begin{bmatrix}1\\1\end{bmatrix}$, $\mathbf{v} = \begin{bmatrix}2\\1\end{bmatrix}$, $\mathbf{w} = \begin{bmatrix}1\\2\end{bmatrix}$, $\mathbf{x} = \begin{bmatrix}1\\4\end{bmatrix}$, $\mathbf{y} = \begin{bmatrix}4\\7\end{bmatrix}$, and $\mathbf{z} = \begin{bmatrix}6\\12\end{bmatrix}$. If T denotes translation by $\begin{bmatrix}-1\\-1\end{bmatrix}$, then T maps **u**, **v**, and **w**, respectively, to $\mathbf{u}' = \begin{bmatrix}0\\0\end{bmatrix}$, $\mathbf{v}' = \begin{bmatrix}1\\0\end{bmatrix}$, and $\mathbf{w}' = \begin{bmatrix}0\\1\end{bmatrix}$. The linear transformation L that maps $\mathbf{v}'$ and $\mathbf{w}'$ to **y-x** and **z-x** is given by the matrix $\begin{bmatrix}3 & 5\\3 & 8\end{bmatrix}$. Thus, the composite transformation LT, which in coordinates is given by $\begin{bmatrix}x\\y\end{bmatrix} \to \begin{bmatrix}3x+5y-8\\3x+8y-11\end{bmatrix}$, maps **u**, **v**, and **w** to $\begin{bmatrix}0\\0\end{bmatrix}, \begin{bmatrix}3\\3\end{bmatrix}, \begin{bmatrix}5\\8\end{bmatrix}$. Composing this composite map with translation by $\begin{bmatrix}1\\4\end{bmatrix}$ gives an affine map which maps **u**, **v**, **w** to **x**, **y**, **z**. In coordinates, this affine map is given by $\begin{bmatrix}x\\y\end{bmatrix} \to \begin{bmatrix}3x+5y-7\\3x+8y-7\end{bmatrix}$, which can be viewed as $T'L$, where T' is translation by $\begin{bmatrix}-7\\-7\end{bmatrix}$.
6. If $\{2\mathbf{u}, 2\mathbf{v}, 2\mathbf{w}\}$ is a set of three vectors in general position in the affine plane $A^2(\mathbb{R})$, then we can think of these vectors as the vertices of a triangle in the Cartesian plane. (The coefficients appear simply to make subsequent calculations less messy.) The midpoint of the line containing 2**u** and 2**v** is **u** + **v**, and similarly for the other sides of the triangle. Thus, the median of the triangle emanating from vertex 2**w** is the line containing 2**w** and **u** + **v**, which is the coset 2**w** + Span(**u** + **v** − 2**w**). It is easy to verify that $(1/3)(2\mathbf{u} + 2\mathbf{v} + 2\mathbf{w})$ is on this line. Similar calculations show that this point is on the other two medians as well. This gives an algebraic proof of the familiar result from high school geometry that the medians of a triangle are concurrent.
7. Let F be a field that has a nonidentity automorphism f, such as the complex numbers with complex conjugation. In the affine plane $A^2(F)$, the mapping T: $\begin{bmatrix}x\\y\end{bmatrix} \to \begin{bmatrix}f(x)\\f(y)\end{bmatrix}$ is a semiaffine transformation that is not affine. To verify that it is not affine, let a be an element of F that is not fixed by f, and consider the vectors $\mathbf{u} = \begin{bmatrix}0\\0\end{bmatrix}$, $\mathbf{v} = \begin{bmatrix}0\\a\end{bmatrix}$, $\mathbf{w} = \begin{bmatrix}0\\0\end{bmatrix}$, and $\mathbf{z} = \begin{bmatrix}0\\1\end{bmatrix}$. It is then easily verified that $\overrightarrow{\mathbf{uv}} = a(\overrightarrow{\mathbf{wz}})$ but $\overrightarrow{T(\mathbf{u})T(\mathbf{v})} = f(a)\overrightarrow{T(\mathbf{w})T(\mathbf{z})}$.

65.2 Euclidean Spaces

Definitions:

Euclidean n-space is an affine n-space (X, V) where V is a real inner product space. When $n = 2$, Euclidean n-space is called a **Euclidean plane**.

The **distance** between two points A and B in the Euclidean n-space (X, V) is $||\overrightarrow{AB}||$, where the norm is taken pursuant to the inner product in V. This distance is denoted $d(A, B)$.

Two lines in a Euclidean plane are **orthogonal** if the vectors that span their respective direction spaces are orthogonal with respect to the inner product of V.

If A and B are distinct points in a Euclidean plane, the **perpendicular bisector** of line segment $[A, B]$ is the line passing through the midpoint of this segment whose direction space is the orthogonal complement of the space spanned by the vector $\overrightarrow{AB}$.

An **isometry** or **rigid motion** of a Euclidean n-space (X, V) is a bijection T of X that preserves distances: $d(A, B) = d(T(A), T(B))$, for all A and B in X.

The **linear transformation associated with** T is the mapping T':$V \to V$ defined by $T'(\overrightarrow{AB}) = \overrightarrow{T(A)T(B)}$ (cf. Fact 4, below).

The isometry T is **direct** if T' has positive determinant, **indirect** if T' has negative determinant.

The remaining definitions in this section apply to the Euclidean plane $A^2(\mathbb{R})$ with the ordinary dot product as the inner product. (Recall from the preceding section that this is the affine plane obtained by taking $X = V = \mathbb{R}^2$.)

A 2 × 2 **rotation matrix** is a matrix of the form $R_\theta = \begin{bmatrix} \cos\theta & -\sin\theta \\ \sin\theta & \cos\theta \end{bmatrix}$.

A 2 × 2 **reflection matrix** is a matrix of the form $S_\theta = \begin{bmatrix} \cos\theta & \sin\theta \\ \sin\theta & -\cos\theta \end{bmatrix}$.

If C is a fixed point in $\mathbb{R}^2$, a **rotation about point** $\boldsymbol{C}$ is a mapping from $\mathbb{R}^2$ to itself of the form $T(X) = R_\theta(X - C) + C$, for some 2 × 2 rotation matrix R_θ.

If l is a line, the **reflection through** l is the mapping from $A^2(\mathbb{R})$ to itself which sends every point on l to itself, and which sends a point C not on l to the unique point C' with the property that l is the perpendicular bisector of the line segment $[C, C']$.

A **glide reflection** is a reflection through a line l followed by translation by a nonzero vector that spans the direction space of l.

Facts: For proofs, see [Roe93] and [Ree83].

1. The distance function d on a Euclidean n-space is a metric on the set X. Specifically, this means that if A, B, and C are arbitrary points of X, then
   (a) $d(A, B)$ is nonnegative, and equal to 0 if and only if $A = B$.
   (b) $d(A, B) = d(B, A)$.
   (c) $d(A, B) \leq d(A,C) + d(C, B)$.
2. The point C is in the line segment $[A, B]$ if and only if $d(A, C) + d(C, B) = d(A, B)$.
3. (Pythagorean Theorem) If A, B, and C are three noncollinear points in a Euclidean n-space and $\overrightarrow{AB}$ is orthogonal to $\overrightarrow{BC}$ then $d(A, B)^2 + d(B, C)^2 = d(A, C)^2$.
4. An isometry is an affine mapping.
5. The associated linear transformation of an isometry is a well-defined orthogonal linear transformation.
6. In the Euclidean space $A^n(\mathbb{R})$, when an isometry is written as $T(\mathbf{v}) = \mathbf{L}(\mathbf{v}) + \mathbf{b}$ for a nonsingular linear transformation L and vector $\mathbf{b}$, the associated linear transformation T' is simply L. (Such an expression for an isometry T is always possible.)
7. Any translation in a Euclidean n-space is an isometry of that space. The associated linear transformation of a translation is the identity map.
8. The set of all isometries of a Euclidean space (X, V) is a group (under the operation of function composition).
9. The map that associates to every isometry T its associated linear transformation T' is a homomorphism from the isometry group onto the group of all orthogonal linear transformations of V, the kernel of which is the set of all translations.
10. If l is a line in the affine space $A^2(\mathbb{R})$, then reflection through l is an isometry.
11. If l is a line that passes through the origin in $A^2(\mathbb{R})$, then reflection through l is a linear transformation given by a reflection matrix. Conversely, the linear transformation defined by multiplication of a vector by a (fixed) reflection matrix S_θ is a reflection through the line passing through the origin with direction space spanned by $\begin{bmatrix} \cos(\theta/2) \\ \sin(\theta/2) \end{bmatrix}$.
12. A rotation is an isometry.
13. In $A^2(\mathbb{R})$, a rotation about the origin $\mathbf{0}$ is a linear transformation given by a rotation matrix.
14. In $A^2(\mathbb{R})$, the point C is the unique fixed point of a nonidentity rotation about C. Any isometry with a unique fixed point is a rotation.
15. Any isometry of $A^2(\mathbb{R})$ can be written as the product of three or fewer reflections.

16. Every isometry of $A^2(\mathbb{R})$ is either a translation, rotation, reflection, or glide reflection.
17. In $A^2(\mathbb{R})$, the product of a reflection with itself is the identity mapping. The product of two distinct reflections is a translation if the reflections are through parallel lines; if the reflections are through two lines intersecting at a point, the product of the two reflections is a rotation about that point. The product of three reflections is either a reflection or glide reflection.
18. The translations and rotations are the direct isometries of $A^2(\mathbb{R})$ and the reflections and glide reflections are the indirect ones.
19. Two triangles ABC and $A'B'C'$ in $A^2(\mathbb{R})$ are congruent in the sense of high school geometry (i.e., corresponding sides and angles are equal) if and only if there is an isometry of $A^2(\mathbb{R})$, which maps A to A', B to B', and C to C'.

Examples:

1. In $A^2(\mathbb{R})$, the line segment between $\begin{bmatrix}1\\1\end{bmatrix}$ and $\begin{bmatrix}3\\3\end{bmatrix}$ lies on a line with direction space spanned by $\begin{bmatrix}1\\1\end{bmatrix}$; the span of $\begin{bmatrix}-1\\1\end{bmatrix}$ is the orthogonal complement of this direction space. The midpoint of this line segment is $\begin{bmatrix}2\\2\end{bmatrix}$. Therefore, the perpendicular bisector of this segment is the line l given by $\begin{bmatrix}2\\2\end{bmatrix} + \text{Span}\left(\begin{bmatrix}-1\\1\end{bmatrix}\right)$.
2. In $A^2(\mathbb{R})$ the map $\begin{bmatrix}x\\y\end{bmatrix} \to \begin{bmatrix}-y\\-x\end{bmatrix}$ has matrix $\begin{bmatrix}0 & -1\\-1 & 0\end{bmatrix}$, which is a reflection matrix. To determine the line through which this mapping is a reflection, note that its eigenvalues are 1 and -1. An eigenvector for the eigenvalue 1 is $\begin{bmatrix}1\\-1\end{bmatrix}$. Thus, the line spanned by $\begin{bmatrix}1\\-1\end{bmatrix}$ is the line through which this mapping reflects. Alternatively, $\begin{bmatrix}0 & -1\\-1 & 0\end{bmatrix} = S_{3\pi/2}$, so by Fact 11, the line of reflection is spanned by $\begin{bmatrix}\cos(3\pi/4)\\\sin(3\pi/4)\end{bmatrix} = \begin{bmatrix}-1/\sqrt{2}\\1/\sqrt{2}\end{bmatrix}$.
3. The mapping $\begin{bmatrix}x\\y\end{bmatrix} \to \begin{bmatrix}-y\\x\end{bmatrix}$ has matrix $\begin{bmatrix}0 & -1\\1 & 0\end{bmatrix}$, which is a rotation matrix (corresponding to $\theta = \pi/2$). The mapping T: $\begin{bmatrix}x\\y\end{bmatrix} \to \begin{bmatrix}-y+1\\x\end{bmatrix}$, which is a direct isometry and not a translation, must correspond to a rotation also, through a point other than the origin. To determine the center of this rotation, we compute the unique fixed point to be $x = 1/2 = y$. Thus, T is a rotation around the point $\begin{bmatrix}1/2\\1/2\end{bmatrix}$.

65.3 Projective Spaces

Definitions:

If V is an $(n+1)$-dimensional vector space over a field F, then the **n-dimensional projective space based on** V, denoted $\mathcal{P}(V)$, is the set of all subspaces of V. The one-dimensional subspaces of V are the **points** of $\mathcal{P}(V)$; the two-dimensional subspaces of V are the **lines** of $\mathcal{P}(V)$. More generally, the **k-dimensional projective subspaces** are the $(k+1)$-dimensional subspaces of V.

When V has dimension 3, $\mathcal{P}(V)$ is called the **projective plane** based on V.

A point Span(**v**) **lies on** a projective subspace if it is a subset of that projective subspace.

Relative to a fixed ordered basis of V, any nonzero element of V can be identified with an $(n+1)$-tuple of elements of F. Thus, by selecting a spanning vector of any point of $\mathcal{P}(V)$, that point can be identified with an $(n+1)$-tuple of elements of F, where not all of the components are zero and where two such $(n+1)$-tuples are identified if one is a nonzero scalar multiple of the other. Under this identification, the $(n+1)$-tuple is denoted $[a_1 : a_2 : \ldots : a_{n+1}]$ and called the **homogeneous coordinates** of the point. (In many geometry books, homogenous coordinates are denoted $[a_1, a_2, \ldots, a_{n+1}]$, but here that notation risks confusion with a $1 \times n$ matrix.)

When $\mathcal{P}(V)$ is a projective space of dimension n, with ordered basis $\mathcal{B}$ for V, then for any r-dimensional subspace W of V there is an $(n+1-r)$-dimensional subspace of the dual space V^* consisting of those linear functionals of V that vanish on W (cf. Section 3.8). In particular, if $\mathcal{P}(V)$ is a projective plane, then points (respectively, lines) of $\mathcal{P}(V)$ correspond to lines (respectively, points) of $\mathcal{P}(V^*)$. Since, relative to the dual basis of V^*, any point can be given homogenous coordinates, any line of $\mathcal{P}(V)$ can be given the homogenous coordinates of its annihilator; these are **homogenous line coordinates.**

The **dual** of a statement concerning points and lines in a projective plane is the statement obtained by interchanging the terms "point" and "line."

If $\{A, B, C\}$ and $\{A', B', C'\}$ are two sets of noncollinear points in a projective plane $\mathcal{P}(V)$, then the triangles ABC and $A'B'C'$ are **perspective from the point** P if and only if the lines containing A and A', B and B', and C and C' all pass through the point P.

The triangles ABC and $A'B'C'$ are **perspective from the line** l if the points of intersection of the corresponding sides of the triangles all lie on the line l.

If T is a nonsingular linear transformation of V, then the mapping $\hat{T}$ from $\mathcal{P}(V)$ to itself, defined by mapping Span($\{\mathbf{v}_1, \ldots, \mathbf{v}_m\}$) to Span$\{T(\mathbf{v}_1), \ldots, T(\mathbf{v}_m)\}$, is called the **projective transformation** (or **projectivity**) **determined by T.**

A **collineation** of $\mathcal{P}(V)$ is a bijective mapping from $\mathcal{P}(V)$ onto itself which maps subspaces to subspaces of the same dimension and which preserves set inclusion.

Facts: For proofs see [Kap69] or specific references.

1. In a projective plane $\mathcal{P}(V)$, points and lines satisfy the following incidence properties:

   (a) Two distinct points A and B lie on a unique line, denoted AB.

   (b) Two distinct lines meet in a unique point.

   (c) There are four distinct points, no three of which are collinear.

2. In the projective plane $\mathcal{P}(V)$, the dual of any theorem involving the incidence of points and lines is also a theorem. (In the preceding Fact, for example, statement (b) is the dual of statement (a), and vice versa.)
3. [Art57] Any projective transformation of the projective space $\mathcal{P}(V)$ is a collineation of $\mathcal{P}(V)$. If F is a field that does not have any nontrivial automorphisms (such as, for example, the field of real numbers), then any collineation of the projective space $\mathcal{P}(V)$ is a projective transformation.
4. In the projective plane $\mathcal{P}(V)$, if a point P has homogenous coordinates $[a{:}b{:}c]$ and a line l has homogenous coordinates $[x{:}y{:}z]$, then P lies on l if and only if $ax + by + cz = 0$. Thus, to find the homogenous line coordinates of the line containing the points $[a{:}b{:}c]$ and $[d{:}e{:}f]$, we can use the formula for vector cross product to find a vector $\begin{bmatrix} x \\ y \\ z \end{bmatrix}$, which is orthogonal to the two vectors $\begin{bmatrix} a \\ b \\ c \end{bmatrix}$ and $\begin{bmatrix} d \\ e \\ f \end{bmatrix}$ under the usual dot product. For an arbitrary field F, this formula yields the homogenous line coordinates $[x{:}\ y{:}\ z]$ of the line containing the two points.

5. (Desargues' Theorem and its converse) Two triangles in a projective plane $\mathcal{P}(V)$ are perspective from a point if and only if they are perspective from a line.
6. (Pappus' Theorem) Let A, B, C and A', B', C' be two triples of distinct collinear points in a projective plane. Let P, Q, and R denote, respectively, the points of intersection of the lines AB' and $A'B$, AC' and $C'A$, and BC' and $B'C$. Then P, Q, and R are collinear.
7. Let A, B, C, and D be four distinct points in a projective plane $\mathcal{P}(V)$ with the property that no three are collinear. Let A', B', C', and D' be four other distinct points with this property. Then there is a unique projective transformation which maps A to A', B to B', C to C', and D to D'.

Examples:

1. If V is a two-dimensional vector space over the field of complex numbers, then the projective line $\mathcal{P}(V)$ can be realized as the set of all homogeneous coordinates $[x\colon y]$ where x and y are complex numbers, not both zero. By identifying $[x\colon y]$ with the complex number x/y if y is nonzero, and with ∞ if $y = 0$, we can identify $\mathcal{P}(V)$ with the extended complex plane studied in courses on complex analysis. Note also that the complex-linear transformation of V given by a nonsingular matrix $\begin{bmatrix} a & b \\ c & d \end{bmatrix}$ represents the mapping of the extended plane given by $z \mapsto \dfrac{az+b}{cz+d}$; these are the linear fractional transformations studied in such a course.
2. The smallest projective space of dimension greater than 1 is obtained by letting V be a three-dimensional vector space over the two-element field Z_2. In homogenous coordinates, every element of $\mathcal{P}(V)$ is represented by a triple $[a\colon b\colon c]$ where each entry is either 0 or 1 and not all entries are 0. Thus, there are seven points in this projective plane, and by duality there are seven lines as well. It is easy to verify that each line of this projective plane consists of three points, and each point is contained on three lines. For example, the line containing the points [1:1:1] and [1:0:0], obtained by taking the span of these vectors, contains these points, the point with homogenous coordinates [0:1:1], and no other points (i.e., no other nonzero vector spans a one-dimensional subspace of the span of these two points).
3. If $V$is a three-dimensional vector space over the field of real numbers, then, thinking of the points of $\mathcal{P}(V)$ in terms of homogenous coordinates relative to a given fixed ordered basis of V, the points [0:1:0] and [0:0:1] lie on a unique line, which is the span of these vectors (i.e., the two-dimensional subspace of V consisting of all points with homogenous coordinates with first component zero). In the dual space of V, relative to the dual basis of the given ordered basis of V, this subspace corresponds to the subspace of all linear functionals of V that annihilate all such vectors. This is clearly the span of the first vector in the ordered dual basis. Thus, the homogenous line coordinates for the line containing the two given points is [1:0:0]. This can also be obtained as a cross product.
4. To illustrate Pappus' Theorem, let six points be given as follows in the real projective plane: $A = [1:0:0]$, $B = [0:1:0]$, $C = [1:1:0]$, $A' = [1:1:2]$, $B' = [1:2:2]$, $C' = [0:0:1]$. It was observed above that the line BC' consists of all points with homogenous coordinates with first component zero. It is easy to see that the line containing C and B' is the set of triples of real numbers of the form $\begin{bmatrix} a+b \\ a+b \\ a \end{bmatrix}$. The intersection of these two lines is obtained by letting $a = -b$, and, therefore, is the point with homogenous coordinates [0:0:1]. By similar calculations, the intersection of the lines AB' and $A'B$ is seen to be the point with homogenous coordinates [1:2:2] and the intersection of the lines CA' and $C'A$ is the point with homogenous coordinates [0:0:1]. Since the vectors $\begin{bmatrix} 0 \\ 0 \\ 1 \end{bmatrix}$, $\begin{bmatrix} 1 \\ 2 \\ 2 \end{bmatrix}$, and $\begin{bmatrix} 0 \\ 0 \\ 1 \end{bmatrix}$ are linearly dependent, the one-dimensional subspaces spanned by these vectors are collinear points, as required by Pappus' Theorem.

5. The three incidence relations described in Fact 1 are usually taken as the defining relations for an axiomatic definition of projective plane, in which "point," "line," and "incidence" are taken as undefined notions. It is not the case that any projective plane defined axiomatically in this way is of the form $\mathcal{P}(V)$ for some vector space V over a field F. As an example, if we interpret "point" to mean the set of all ordered pairs of real numbers and "line" to mean any horizontal Euclidean line, vertical Euclidean line, Euclidean line of negative slope, or broken Euclidean line of positive slope m above the x-axis and slope $2m$ below the x-axis, we obtain a geometric system that satisfies the standard incidence relations for points and lines, and the Euclidean parallel postulate. We can convert this example into a projective plane by the addition of "ideal points": To every line we add a new point, with the same point being added to two lines if and only if they are parallel. We also add a new line consisting of the ideal points, and no other points. The augmented geometric system satisfies the axioms of a projective plane, but neither Desargues' Theorem nor Pappus' Theorem holds in this plane. (This example is called the *Moulton Plane.*)
6. The addition of ideal points, as in the preceding example, can be carried out in any affine plane $A^2(F)$. An affine point $\begin{bmatrix} a \\ b \end{bmatrix}$ can be identified with the projective point with homogenous coordinates $[a{:}b{:}1]$ and the ideal points are those with homogenous coordinates with a zero in the third component.
7. In a projective n-space $\mathcal{P}(V)$, let H be a subspace of V of dimension n. Starting from an ordered basis for H, obtain an ordered basis for V by adding a single vector as the last element of the basis. Then, in homogenous coordinates, every point of $\mathcal{P}(V)$ that is not in H has nonzero last component. By associating the point with homogenous coordinates $[a_1 : \ldots : a_{n+1}]$ with the n-tuple $\begin{bmatrix} \frac{a_1}{a_{n+1}} \\ \vdots \\ \frac{a_n}{a_{n+1}} \end{bmatrix}$ we can think of the set of points that are not in H as the affine n-space $A^n(F)$.
8. Let $\mathcal{P}(V)$ be a projective plane, where V is a vector space over a field F that has a nonidentity automorphism f. The mapping $[a : b : c] \mapsto [f(a) : f(b) : f(c)]$ is a collineation. Since this mapping fixes the points $[0{:}0{:}1]$, $[0{:}1{:}0]$, $[1{:}0{:}0]$, and $[1{:}1{:}1]$ and is not the identity, it cannot, by Fact 7, be a projective transformation.
9. In the real projective plane, consider the six points $A = [1{:}2{:}1]$, $B = [2{:}0{:}1]$, $C = [5{:}5{:}1]$, $A' = [2{:}4{:}1]$, $B' = [4{:}0{:}1]$, and $C' = [10{:}10{:}1]$. Identifying the projective point with homogenous coordinates $[a{:}b{:}1]$ with the ordinary Euclidean point $\begin{bmatrix} a \\ b \end{bmatrix}$, as in Example 6 above, it can be seen by a simple diagram that the triangles ABC and $A'B'C'$ are perspective from the origin, since all lines AA', BB', and CC' pass through this point. Thus, in the real projective plane these triangles are perspective from the point $[0{:}0{:}1]$. Desargues' Theorem, therefore, asserts that the points of intersection of the lines AB and $A'B'$, AC and $A'C'$, BC and $B'C'$ will lie on a line. A simple calculation shows that these three points, in homogenous coordinates, have third component zero, so these points are indeed collinear. This corresponds to the fact that as Euclidean lines there are no points of intersection — the corresponding sides of the triangles are parallel in pairs. The points with third component zero are the "ideal points" that have been added to Euclidean geometry to form the real projective plane.

References

[Art57] E. Artin. *Geometric Algebra.* Wiley-Interscience, New York, 1957.

[Kap69] I. Kaplansky. *Linear Algebra and Geometry: A Second Course.* Allyn and Bacon, Inc., Boston, 1969.

[Ree83] E. Rees. *Notes on Geometry.* Springer-Verlag, Berlin, 1983.

[Roe93] J. Roe. *Elementary Geometry.* Oxford University Press, Oxford, U.K., 1993.

[Ser93] E. Sernesi. *Linear Algebra: A Geometric Approach.* Chapman and Hall, London, 1993.

[ST71] E. Snapper and R. Troyer. *Metric Affine Geometry.* Academic Press, New York, 1971.

66

Some Applications of Matrices and Graphs in Euclidean Geometry

Miroslav Fiedler
Academy of Sciences of the Czech Republic

This chapter presents some facts and examples illustrating the interplay between matrix theory, graph theory, and n-dimensional Euclidean geometry. The main objects of Euclidean geometry are, of course, the points. The simplest way of introducing them is to identify points with the endpoints of vectors; formally, one introduces an artificial point, the origin **O**, and the points are sums of this origin with any vector. A more general treatment of Euclidean n-space can be found in Chapter 65.

66.1 Euclidean Point Space

Definitions:

An **arithmetic Euclidean vector space** is the real inner product space $\mathbb{R}^n$ with the standard inner product $\langle \mathbf{x}, \mathbf{y} \rangle = \mathbf{y}^T \mathbf{x}$.

The **arithmetic point Euclidean** n**-space** E^n based on the vector space $\mathbb{R}^n$ has as **points** the column $n+1$-tuples with last coordinate 1, e.g., $\mathbf{C} = [c_1, c_2, \ldots, c_n, 1]^T$, and as **vectors** the column $n+1$-tuples with last coordinate 0, e.g., $\mathbf{v} = [v_1, v_2, \ldots, v_n, 0]^T$.

The origin **O** is the point $[0, \ldots, 0, 1]^T$; the numbers $c_1, \ldots, c_n$ are **coordinates** of the point **C**.

Algebraic operations are defined with points: If $\mathbf{A}_1, \mathbf{A}_2, \ldots, \mathbf{A}_m$ are points, $a_1, a_2, \ldots, a_m$ real numbers, then the symbol

$$a_1\mathbf{A}_1 + a_2\mathbf{A}_2 + \cdots + a_m\mathbf{A}_m$$

means a **point** if and only if $\sum a_k = 1$, and a **vector** if and only if $\sum a_k = 0$. In no other case is this symbol defined.

The **Euclidean distance**, or **distance**, of two points **A** and **B** is the **length** of the vector $\mathbf{A} - \mathbf{B}$, i.e., for $\mathbf{A} = [a_1, a_2, \ldots, a_n, 1]^T$ and $\mathbf{B} = [b_1, b_2, \ldots, b_n, 1]^T$,

$$\|\mathbf{A} - \mathbf{B}\| = \sqrt{(a_1 - b_1)^2 + \cdots + (a_n - b_n)^2}.$$

Let $\mathcal{S} = \{\mathbf{A}_1, \mathbf{A}_2, \ldots, \mathbf{A}_m\}$ be a set of points in E^n. The point

$$\mathbf{C} = a_1\mathbf{A}_1 + a_2\mathbf{A}_2 + \cdots + a_m\mathbf{A}_m \quad \text{and} \quad \sum_{k=1}^{m} a_k = 1,$$

is a **linear combination** of $\mathcal{S}$. The set of all linear combinations of $\mathcal{S}$ is the **linear hull** of $\mathcal{S}$, denoted by $\mathcal{L}(\mathcal{S})$.

A point **C** is **linearly dependent** on $\mathcal{S}$ if $\mathbf{C} \in \mathcal{L}(\mathcal{S})$. The set $\mathcal{S}$ is called **linearly dependent** if there is a point in $\mathcal{S}$ that is linearly dependent on the remaining points. Otherwise, the set $\mathcal{S}$ is called **linearly independent.**

Linear hulls of systems of points are called **linear subspaces** of E^n.

A linear subspace M has **dimension** k if the maximum number of linearly independent points in M is $k + 1$.

Linear subspaces of dimension 1 in E^n are called **lines.**

Linear subspaces of dimension $n - 1$ in E^n (i.e., of **codimension** 1) are called **hyperplanes.**

If $\widehat{\alpha}$ is a hyperplane in E^n defined by the equation $\alpha_1 x_1 + \alpha_2 x_2 + \cdots + \alpha_n x_n + \alpha_0 = 0$ (cf. Fact 10), the vector $\mathbf{u} = [\alpha_1, \ldots, \alpha_n, 0]^T$ is a **normal vector** to $\widehat{\alpha}$.

The **distance** of point **C** to a hyperplane $\widehat{\alpha}$ is the minimum of all distances of the point **C** from the points in the hyperplane $\widehat{\alpha}$.

Two hyperplanes are called **parallel** if their normal vectors are proportional, i.e., one is a scalar multiple of the other. They are called **orthogonal** (or **perpendicular**) if their normal vectors are orthogonal.

Let $\mathcal{S} = \{\mathbf{A}_1, \mathbf{A}_2, \ldots, \mathbf{A}_m\}$ be a set of points in E^n. Then the set of all points of the form

$$a_1\mathbf{A}_1 + a_2\mathbf{A}_2 + \cdots + a_m\mathbf{A}_m,$$

for which all the coefficients a_i are nonnegative and $\sum a_k = 1$, is called the **convex hull** of $\mathcal{S}$; we denote it by $\mathcal{C}(\mathcal{S})$.

The convex hull of two distinct points **A**, **B** is called the **segment** and is denoted by $\overline{\mathbf{AB}}$.

The point $\frac{1}{2}\mathbf{A} + \frac{1}{2}\mathbf{B}$ is the **midpoint** of the segment $\overline{\mathbf{AB}}$.

A set $\mathcal{S} \in E^n$ is **convex** if S has the property that with any two points **A** and **B** in $\mathcal{S}$, all points of the segment $\overline{\mathbf{AB}}$ are in $\mathcal{S}$.

A hyperplane $\widehat{\alpha}$ with normal vector $[\alpha_1, \alpha_2, \ldots, \alpha_n]$ determines two **open halfspaces**; one is the set of all points $\mathbf{X} = [x_1, x_2, \ldots, x_n, 1]^T$ satisfying

$$\alpha_1 x_1 + \alpha_2 x_2 + \cdots + \alpha_n x_n + \alpha_0 > 0,$$

while the other is the set of all points satisfying the reverse inequality.

In addition, one can speak about **closed halfspace** if in the inequality, equality is also admitted.

Facts:

These facts follow from facts in Chapter 1 and Chapter 65.

1. Points and vectors in arithmetic Euclidean n-space satisfy the following:

   (E1.) The sum of a point and a vector is a point.

   (E2.) The difference of two points is a vector.

   (E3.) $(\mathbf{C} + \mathbf{v}) + \mathbf{w} = \mathbf{C} + (\mathbf{v} + \mathbf{w})$, where **C** is a point and **v**, **w** are vectors.

2. Arithmetic Euclidean n-space, with vectors acting on points by addition, is a Euclidean n-space as defined in Chapter 65.
3. A linear hull is an affine subspace as defined in Chapter 65.
4. The set $\mathcal{S} = \{\mathbf{A}_1, \mathbf{A}_2, \ldots, \mathbf{A}_m\}$ is linearly independent if and only if $\lambda_1 = \lambda_2 = \cdots = \lambda_m = 0$ are the only numbers $\lambda_1, \ldots, \lambda_m$ for which the zero vector $\mathbf{0}$ satisfies $\mathbf{0} = \lambda_1\mathbf{A}_1 + \lambda_2\mathbf{A}_2 + \cdots + \lambda_m\mathbf{A}_m$ and $\sum_{k=1}^{m} \lambda_k = 0$.
5. The set $\mathcal{S} = \{\mathbf{A}_1, \mathbf{A}_2, \ldots, \mathbf{A}_m\}$ is linearly independent if and only if every point $\mathbf{P} \in \mathcal{L}(\mathcal{S})$ has a unique expression as

$$\mathbf{P} = \lambda_1\mathbf{A}_1 + \lambda_2\mathbf{A}_2 + \cdots + \lambda_m\mathbf{A}_m \text{ and } \sum_{k=1}^{m} \lambda_k = 1.$$

6. A linearly independent set in E^n contains at most $n+1$ points. A linearly independent set with $n+1$ linearly independent points in E^n exists.
7. Any linearly independent set has the property that each of its nonempty subsets is linearly independent as well.
8. Let the set $\mathcal{S}$ consist of m points $\mathbf{A}, \mathbf{B}, \ldots, \mathbf{G}$ defined by

$$\mathbf{A} = [a_1, a_2, \ldots, a_n, 1]^T, \mathbf{B} = [b_1, b_2, \ldots, b_n, 1]^T, \ldots, \mathbf{G} = [g_1, g_2, \ldots, g_n, 1]^T.$$

Then $\mathcal{S}$ is linearly independent if and only if the $m \times (n+1)$ matrix

$$\begin{bmatrix} a_1 & a_2 & \ldots & a_n & 1 \\ b_1 & b_2 & \ldots & b_n & 1 \\ \ldots & \ldots & \ldots & \ldots & \ldots \\ g_1 & g_2 & \ldots & g_n & 1 \end{bmatrix}$$

has rank m.

9. Let $\mathbf{A} = [a_1, a_2, \ldots, a_n, 1]^T$, $\mathbf{B} = [b_1, b_2, \ldots, b_n, 1]^T$, $\ldots$, $\mathbf{F} = [f_1, f_2, \ldots, f_n, 1]^T$ be n linearly independent points in E^n. Then the linear hull of these points consists of all points $\mathbf{X} = [x_1, x_2, \ldots, x_n, 1]^T$ in E^n which satisfy the equation

$$\det \begin{bmatrix} x_1 & x_2 & \ldots & x_n & 1 \\ a_1 & a_2 & \ldots & a_n & 1 \\ b_1 & b_2 & \ldots & b_n & 1 \\ \ldots & \ldots & \ldots & \ldots & \ldots \\ f_1 & f_2 & \ldots & f_n & 1 \end{bmatrix} = 0.$$

10. A hyperplane can be characterized as the set of points $\mathbf{X} = [x_1, x_2, \ldots, x_n, 1]^T$ such that the coordinates satisfy one linear equation of the form

$$\alpha_1 x_1 + \alpha_2 x_2 + \cdots + \alpha_n x_n + \alpha_0 = 0,$$

in which not all of the numbers $\alpha_1, \alpha_2, \ldots, \alpha_n$ are equal to zero.
11. We can generalize linear independence as well as the equation of the hyperplane in Fact 9 by including cases when some of the points (but not all of them) are replaced by vectors. We simply put into the corresponding row the $(n+1)$-tuple $a_1, a_2, \ldots, a_n, 0$ instead that for the point.
12. Two parallel but distinct hyperplanes have no point in common. Conversely, if two hyperplanes in E^n, $n \geq 2$, have no point in common, then they are parallel.
13. Hyperplanes are parallel if and only if they are parallel as affine subspaces as defined in Chapter 65.
14. The definitions of segment and midpoint are equivalent to the definitions of these terms in Chapter 65 for arithmetic Euclidean n-space.

15. The distance of a point $\mathbf{C} = [c_1, c_2, \ldots, c_n, 1]^T$ from a hyperplane $\widehat{\alpha}$ given by equation $\alpha_1 x_1 + \alpha_2 x_2 + \cdots + \alpha_n x_n + \alpha_0 = 0$ is

$$\frac{|\alpha_1 c_1 + \alpha_2 c_2 + \cdots + \alpha_n c_n + \alpha_0|}{\sqrt{\alpha_1^2 + \alpha_2^2 + \cdots + \alpha_n^2}}.$$

16. The distance of a point $\mathbf{C} = [c_1, c_2, \ldots, c_n, 1]^T$ from a hyperplane $\widehat{\alpha}$ given by equation $\alpha_1 x_1 + \alpha_2 x_2 + \cdots + \alpha_n x_n + \alpha_0 = 0$ is the distance of $\mathbf{C}$ from the point $\mathbf{F} = \mathbf{C} - \gamma \mathbf{u}$, where $\mathbf{u} = [\alpha_1, \alpha_2, \ldots, \alpha_n, 0]^T$ is the normal vector to $\widehat{\alpha}$ and

$$\gamma = \frac{\alpha_1 c_1 + \alpha_2 c_2 + \cdots + \alpha_n c_n + \alpha_0}{\alpha_1^2 + \alpha_2^2 + \cdots + \alpha_n^2}.$$

Examples:

1. In E^3, the points $\mathbf{A}_1 = [2, -1, 2, 1]^T$, $\mathbf{A}_2 = [1, 1, 3, 1]^T$, and $\mathbf{A}_3 = [0, 0, 1, 1]^T$ are linearly independent since the rank of the matrix

$$\begin{bmatrix} 2 & -1 & 2 & 1 \\ 1 & 1 & 3 & 1 \\ 0 & 0 & 1 & 1 \end{bmatrix}$$

is 3.

2. The point $\mathbf{A}_4 = [2, 0, 3, 1]^T$ is linearly dependent on the points $\mathbf{A}_1, \mathbf{A}_2$, and $\mathbf{A}_3$ from Example 1, since $\mathbf{A}_4 = \frac{2}{3}\mathbf{A}_1 + \frac{2}{3}\mathbf{A}_2 - \frac{1}{3}\mathbf{A}_3$.
3. The equation of the hyperplane (which is a plane) determined by the points in Example 1 is

$$0 = \det \begin{bmatrix} x_1 & x_2 & x_3 & 1 \\ 2 & -1 & 2 & 1 \\ 1 & 1 & 3 & 1 \\ 0 & 0 & 1 & 1 \end{bmatrix} = -3x_1 - 3x_2 + 3x_3 - 3.$$

4. The triangle with vertices $\mathbf{A}_1, \mathbf{A}_2, \mathbf{A}_3$ from Example 1 is the convex hull of the three points $\mathbf{A}_1, \mathbf{A}_2$, and $\mathbf{A}_3$.
5. Let $n \geq 3$ be an integer. In the Euclidean n-space E^n, define points

$$\mathbf{F}_k = [\underbrace{n-k, n-k, \ldots, n-k}_{k\text{–times}}, \underbrace{-k, -k, \ldots, -k}_{(n-k)\text{–times}}, 1], \quad k = 1, \ldots, n;$$

$$\mathbf{F}_0 = [n-1, n-2, \ldots, 0, 1]^T, \text{and}$$

$$\mathbf{C} = \left[\frac{1}{2}(n-1), \frac{1}{2}(n-3), \frac{1}{2}(n-5), \ldots, -\frac{1}{2}(n-3), -\frac{1}{2}(n-1), 1\right]^T.$$

Observe first that the points $\mathbf{C}, \mathbf{F}_1, \mathbf{F}_2, \ldots, \mathbf{F}_n$ are linearly dependent since $\mathbf{C} = \frac{1}{n}\mathbf{F}_1 + \frac{1}{n}\mathbf{F}_2 + \cdots + \frac{1}{n}\mathbf{F}_n$. On the other hand, the points $\mathbf{F}_0, \mathbf{F}_1, \ldots, \mathbf{F}_n$ form a linearly independent set by Fact 8 since the determinant (of an $(n+1) \times (n+1)$ matrix)

$$\det \begin{bmatrix} n-1 & n-2 & n-3 & n-4 & \ldots & 0 & 1 \\ n-1 & -1 & -1 & -1 & \ldots & -1 & 1 \\ n-2 & n-2 & -2 & -2 & \ldots & -2 & 1 \\ \ldots & \ldots & \ldots & \ldots & \ldots & \ldots & \ldots \\ 2 & 2 & 2 & 2 & \ldots & -(n-2) & 1 \\ 1 & 1 & 1 & 1 & \ldots & -(n-1) & 1 \\ 0 & 0 & 0 & 0 & \ldots & 0 & 1 \end{bmatrix}$$

is different from zero: Subtracting the $\frac{1}{n}$-multiple of the sum of the second till last row, from the first row, the first row is a $\frac{1}{2}(n-1)$-multiple of the row of all ones. Factoring out this number $\frac{1}{2}(n-1)$ from the determinant, subtracting the resulting first row from the nth row, the 2-multiple of the first row from the $(n-1)$st row, etc., till the $(n-1)$-multiple of the first row from the second row, we obtain the determinant of an upper triangular matrix with all diagonal entries different from zero. The value of the determinant is then also easily determined.

The Euclidean distances between the points $\mathbf{F}_i$ and $\mathbf{F}_{i+1}$, $i = 1, \ldots, n-1$, as well as between $\mathbf{F}_n$ and $\mathbf{F}_i$, are all equal, equal to $\sqrt{n(n-1)}$. The point $\mathbf{C}$ has same distances from all points $\mathbf{F}_i$, $i = 1, 2, \ldots, n$, equal to $\sqrt{\frac{1}{12}(n-1)n(n+1)}$.

All the points $\mathbf{F}_i$, $i = 1, 2, \ldots, n$, as well as the point $\mathbf{C}_i$, are points of the hyperplane H with equation $\sum_{i=1}^{n} x_i = 0$.

The vector $\mathbf{F}_0 - \mathbf{C}$ is the $\frac{1}{2}(n-1)$-multiple of the vector $\mathbf{u} = [1, 1, \ldots, 1]^T$ (which is throughout denoted as $\mathbf{1}$). This vector is at the same time the normal vector to the hyperplane H. It follows that the distance of the point $\mathbf{F}_0$ from H is equal to the length of the vector $\mathbf{F}_0 - \mathbf{C}$, which is $\frac{1}{2}(n-1)\sqrt{n}$. The same result can be obtained using Fact 15.

The hyperplane with equation

$$x_1 + x_2 + \cdots + x_n - 1 = 0$$

is parallel to H; the hyperplane

$$x_1 - x_2 = 0$$

is orthogonal to H since the vector $[1, -1, 0, \ldots, 0, 0]^T$ is orthogonal to the vector $\mathbf{u}$.

66.2 Gram Matrices

Definitions:

The **Gram matrix** $G(S)$ of an ordered system $S = (\mathbf{a}_1, \mathbf{a}_2, \ldots, \mathbf{a}_m)$ of vectors in the Euclidean vector n-space $\mathbb{R}^n$ is the $m \times m$ matrix $G(S) = G(\mathbf{a}_1, \mathbf{a}_2, \ldots, \mathbf{a}_m) = [\langle \mathbf{a}_i, \mathbf{a}_j \rangle]$. (See also Section 8.1.).

Let $\mathbf{O}$, $\mathbf{A}_1$, $\mathbf{A}_2$,..., $\mathbf{A}_k$, $k \geq 2$, be linearly independent points in E^n, and let $\mathbf{u}_1 = \mathbf{A}_1 - \mathbf{O}, \mathbf{u}_2 = \mathbf{A}_2 - \mathbf{O}, \ldots, \mathbf{u}_k = \mathbf{A}_k - \mathbf{O}$, be the corresponding vectors. We call the set of all points of the form $\mathbf{O} + \sum_{i=1}^{k} a_k \mathbf{u}_k$, where the numbers a_i satisfy $0 \leq a_i \leq 1$, $i = 1, \ldots, k$, the **parallelepiped** spanned by the vectors $\mathbf{u}_i$. For $k = 2$, we speak about the **parallelogram** spanned by $\mathbf{u}_1$ and $\mathbf{u}_2$.

If $\mathbf{u}_1, \mathbf{u}_2, \ldots, \mathbf{u}_n$ is a basis of an n-dimensional arithmetic Euclidean vector space and $\mathbf{v}_1, \mathbf{v}_2, \ldots, \mathbf{v}_n$ is a set of vectors such that the inner product of $\mathbf{u}_i$ and $\mathbf{v}_j$ is the Kronecker delta δ_{ij}, then this pair of ordered sets is a **biorthogonal pair of bases**.

Facts:

Facts for which no specific reference is given follow from facts in Chapter 1.

1. The Gram matrix is always a positive semidefinite matrix. Its rank is equal to the dimension of the Euclidean space of smallest dimension which contains all the vectors of the system.
2. Every positive semidefinite matrix is a Gram matrix of some system of vectors S in some Euclidean space.
3. Every linear relationship among the vectors in S is reflected in the same linear relationship among the rows of $G(S)$, and conversely.
4. The k-dimensional volume of the parallelepiped spanned by the vectors

   $\mathbf{u}_1, \mathbf{u}_2, \ldots, \mathbf{u}_k$ is

$$\sqrt{\det G(\mathbf{u}_1, \mathbf{u}_2, \ldots, \mathbf{u}_k)}.$$

5. To every basis $\mathbf{u}_1, \mathbf{u}_2, \ldots, \mathbf{u}_n$ of an n-dimensional Euclidean vector space there exists a set of vectors $\mathbf{v}_1, \mathbf{v}_2, \ldots, \mathbf{v}_n$ such that this pair is a biorthogonal pair of bases. The set of the $\mathbf{v}_i$s is also a basis and is uniquely determined.
6. If both bases in the biorthogonal pair coincide, the common basis is orthonormal, and an orthonormal basis forms a biorthogonal pair with itself.
7. The Gram matrices $G(\mathbf{u}_1, \mathbf{u}_2, \ldots, \mathbf{u}_n)$ and $G(\mathbf{v}_1, \mathbf{v}_2, \ldots, \mathbf{v}_n)$ of a pair of biorthogonal bases are inverse to each other:

$$G(\mathbf{u}_1, \mathbf{u}_2, \ldots, \mathbf{u}_n)G(\mathbf{v}_1, \mathbf{v}_2, \ldots, \mathbf{v}_n) = I.$$

8. [Fie64] Let $A = [a_{ij}]$ be a positive semidefinite matrix with row sums zero. Then

$$2 \max_i \sqrt{a_{ii}} \leq \sum_i \sqrt{a_{ii}}.$$

9. [Fie61b] If A is positive definite, then the matrix $A \circ A^{-1} - I$ is positive semidefinite and its row sums are equal to zero.
10. If $A = [a_{ij}]$ is positive definite and $A^{-1} = [\alpha_{ij}]$, then

$$a_{ii}\alpha_{ii} \geq 1 \quad \text{for all } i$$

and

$$2 \max_i \sqrt{a_{ii}\alpha_{ii} - 1} \leq \sum_i \sqrt{a_{ii}\alpha_{ii} - 1}.$$

11. [Fie64] Let $A = [a_{ik}]$ be a positive definite matrix, and let $A^{-1} = [\alpha_{ik}]$. Then the diagonal entries of A and A^{-1} satisfy the first condition in Fact 10 and

$$2 \max_i (\sqrt{a_{ii}\alpha_{ii}} - 1) \leq \sum_i (\sqrt{a_{ii}\alpha_{ii}} - 1).$$

Conversely, if some n-tuples of positive numbers a_{ii} and α_{ii} satisfy these conditions, then there exists a positive definite $n \times n$ matrix A with diagonal entries a_{ii} such that the diagonal entries of A^{-1} are α_{ii}.
12. [Fie64] Let the vectors $\mathbf{u}_1, \mathbf{u}_2, \ldots, \mathbf{u}_n, \mathbf{v}_1, \mathbf{v}_2, \ldots, \mathbf{v}_n$ form a pair of biorthogonal bases in a Euclidean n-space E. Then

$$\|\mathbf{u}_i\| \|\mathbf{v}_i\| \geq 1, \quad i = 1, \ldots, n,$$

$$2 \max_i (\|\mathbf{u}_i\| \|\mathbf{v}_i\| - 1) \leq \sum_i (\|\mathbf{u}_i\| \|\mathbf{v}_i\| - 1).$$

Conversely, if nonnegative numbers $\alpha_1, \alpha_2, \ldots, \alpha_n, \beta_1, \beta_2, \ldots, \beta_n$ satisfy

$$\alpha_i \beta_i \geq 1, \quad i = 1, \ldots, n,$$

$$2 \max_i (\alpha_i \beta_i - 1) \leq \sum_i (\alpha_i \beta_i - 1),$$

then there exists in E^n a pair of biorthogonal bases $\mathbf{u}_i, \mathbf{v}_j$, such that

$$\|\mathbf{u}_i\| = \alpha_i, \|\mathbf{v}_i\| = \beta_i, \quad i = 1, \ldots, n.$$

13. [Fie64] Let $A = [a_{ij}]$ be an $n \times n$ positive definite matrix, $n \geq 2$, and let $A^{-1} = [\alpha_{ij}]$. Then the following are equivalent:

(a) $\sqrt{a_{nn}\alpha_{nn}} - 1 = \sum_{i=1}^{n-1}(\sqrt{a_{ii}\alpha_{ii}} - 1)$.

(b) $\frac{a_{ij}}{\sqrt{a_{ii}}\sqrt{a_{jj}}} = \frac{\alpha_{ij}}{\sqrt{\alpha_{ii}}\sqrt{\alpha_{jj}}}$, $i, j = 1, \ldots, n-1$, and

$\frac{a_{in}}{\sqrt{a_{ii}}\sqrt{a_{nn}}} = -\frac{\alpha_{in}}{\sqrt{\alpha_{ii}}\sqrt{\alpha_{nn}}}$, $i = 1, \ldots, n-1$.

(c) A is diagonally similar to

$$C = \begin{bmatrix} I_1 + \omega \mathbf{c}\mathbf{c}^T & \mathbf{c} \\ \mathbf{c}^T & 1 + \omega \mathbf{c}^T\mathbf{c} \end{bmatrix},$$

where $\mathbf{c}$ is a real vector with $n-1$ coordinates and

$$\omega = \frac{\sqrt{1 + \mathbf{c}^T\mathbf{c}} - 1}{\mathbf{c}^T\mathbf{c}}$$

if $\mathbf{c} \neq \mathbf{0}$; if $\mathbf{c} = \mathbf{0}$, $\omega = 0$.

14. To realize a positive definite $n \times n$ matrix C as a Gram matrix of some n vectors, say $\mathbf{a}_1, \mathbf{a}_2, \ldots, \mathbf{a}_n$, it suffices to find a nonsingular matrix A such that $C = AA^T$ and to use the entries in the kth row of A as coordinates of the vector $\mathbf{a}_k$. Such matrix A can be found, e.g., by the Gram–Schmidt process (cf. Section 5.5).

Examples:

1. The vectors $\mathbf{u}_1 = [1, 3]^T$, $\mathbf{u}_2 = [2, -1]^T$ in $\mathbb{R}^2$ are linearly independent and form a basis in $\mathbb{R}^2$. The Gram matrix $G(\mathbf{u}_1, \mathbf{u}_2) = \begin{bmatrix} 10 & -1 \\ -1 & 5 \end{bmatrix}$ is nonsingular.
2. To find the pair of vectors $\mathbf{v}_1, \mathbf{v}_2$ that form a biorthogonal pair with the vectors $\mathbf{u}_1, \mathbf{u}_2$ in Example 1, observe that $\mathbf{v}_1$ should satisfy $\langle \mathbf{u}_1, \mathbf{v}_1 \rangle = 1$ and $\langle \mathbf{u}_2, \mathbf{v}_1 \rangle = 0$. Set $\mathbf{v}_1 = [x_1, x_2]^T$; thus $x_1 + 3x_2 = 1$, $2x_1 - x_2 = 0$, i.e., $\mathbf{v}_1 = [\frac{1}{7}, \frac{2}{7}]^T$. Analogously, $\mathbf{v}_2 = [\frac{3}{7}, -\frac{1}{7}]^T$. The Gram matrix is $G(\mathbf{v}_1, \mathbf{v}_2) = \begin{bmatrix} \frac{5}{49} & \frac{1}{49} \\ \frac{1}{49} & \frac{10}{49} \end{bmatrix}$. It is easily verified that $G(\mathbf{v}_1, \mathbf{v}_2)$ is the inverse of $G(\mathbf{u}_1, \mathbf{u}_2)$, as stated in Fact 7.
3. The pairs of vectors ($\mathbf{e}_1 = [1, 0]^T$, $\mathbf{e}_2 = [0, 1]^T$) and ($\mathbf{u}_1 = [\frac{3}{5}, \frac{4}{5}]^T$, $\mathbf{u}_2 = [\frac{-4}{5}, \frac{3}{5}]^T$) are orthonormal bases for $\mathbb{R}^2$, so $G(\mathbf{e}_1, \mathbf{e}_2) = I$ and $G(\mathbf{u}_1, \mathbf{u}_2) = I$ are inverses, but $(\mathbf{e}_1, \mathbf{e}_2)$ and $(\mathbf{u}_1, \mathbf{u}_2)$ are not a biorthogonal pair of bases.

66.3 General Theory of Euclidean Simplexes

An n-simplex in E^n is a generalization of the triangle in the plane and the tetrahedron in the three-dimensional space. Just as not every triplet of positive numbers can serve as lengths of three sides of a triangle (they have to satisfy the strict triangle inequality), we can ask about analogous conditions for the simplex.

Definitions:

An n-**simplex** is the convex hull of $n+1$ linearly independent points in E^n.

The points are called **vertices** of the simplex. The convex hull of a subset of the set of vertices is called a **face** of the simplex.

If the subset of vertices has $k+1$ elements, the face has **dimension** k. One-dimensional faces are called **edges**.

If $\mathbf{A}_1, \mathbf{A}_2, \ldots, \mathbf{A}_{n+1}$ are the vertices of an n-simplex Σ, we write $\Sigma = \{\mathbf{A}_1, \mathbf{A}_2, \ldots, \mathbf{A}_{n+1}\}$. The $(n-1)$-dimensional face opposite $\mathbf{A}_i$ is denoted by ω_i. A vector $\mathbf{n}$ is an **outer normal** to ω_i if $\mathbf{n}$ is a normal to ω_i and for $\mathbf{C} \in \omega_i$, $\mathbf{C} + \mathbf{n}$ is not in the same halfspace as $\mathbf{A}_i$.

The **dihedral interior angle** between ω_i and ω_j, $i \neq j$ (opposite the edge $\mathbf{A}_i\mathbf{A}_j$) is $\pi -$(the angle between an outer normal to ω_i and an outer normal to ω_j), and is denoted by ϕ_{ij}.

The **barycentric coordinates** with respect to the simplex $\Sigma = \{\mathbf{A}_1, \mathbf{A}_2, \ldots, \mathbf{A}_{n+1}\}$ of a point $\mathbf{P}$ in $\mathcal{L}(\mathbf{A}_1, \mathbf{A}_2, \ldots, \mathbf{A}_{n+1})$ are the unique numbers λ_i such that $\mathbf{P} = \lambda_1\mathbf{A}_1 + \lambda_2\mathbf{A}_2 + \cdots + \lambda_{n+1}\mathbf{A}_{n+1}$ and $\sum_{k=1}^{n+1} \lambda_k = 1$ (cf. Fact 5 of section 66.1). Strictly speaking, these numbers are the **inhomogeneous barycentric coordinates**.

The **homogeneous barycentric coordinates** of $\lambda_1\mathbf{A}_1 + \lambda_2\mathbf{A}_2 + \cdots + \lambda_{n+1}\mathbf{A}_{n+1}$ are the numbers λ_i (not all 0). The expression $\lambda_1\mathbf{A}_1 + \lambda_2\mathbf{A}_2 + \cdots + \lambda_{n+1}\mathbf{A}_{n+1}$ means a **proper point** if $\sum_{k=1}^{n+1} \lambda_k \neq 0$ (namely the point with inhomogeneous barycentric coordinates $\rho\lambda_1, \ldots, \rho\lambda_{n+1}$ with $\rho = (\sum_{k=1}^{n+1} \lambda_k)^{-1}$), or an **improper point** (determined by the (nonzero) vector $\lambda_1\mathbf{A}_1 + \lambda_2\mathbf{A}_2 + \cdots + \lambda_{n+1}\mathbf{A}_{n+1}$) if $\sum_{k=1}^{n+1} \lambda_k = 0$. These coordinates can be viewed as in projective n-dimensional geometry (see Chapter 65). The set of improper points is, thus, characterized by the equation $\sum_{k=1}^{n+1} \lambda_k = 0$ in homogeneous barycentric coordinates, and is called the **improper hyperplane**.

The **circumcenter** of a simplex is the point that is equidistant from the all vertices of the simplex. The set of all points at that common distance from the circumcenter is the **circumscribed hypersphere**.

The $(n+1) \times (n+1)$ matrix $M = [m_{ij}]$ with m_{ij} equal to the square of the distance between the ith and jth vertex of an n-simplex Σ is called the **Menger matrix** of Σ.

The **Gramian** of an n-simplex $\Sigma = \{\mathbf{A}_1, \mathbf{A}_2, \ldots, \mathbf{A}_{n+1}\}$ is defined as the Gram matrix Q of $n+1$ vectors $\mathbf{n}_1, \mathbf{n}_2, \ldots, \mathbf{n}_{n+1}$ determined as follows: The first n of them form the biorthogonal pair with the n vectors $\mathbf{u}_1 = \mathbf{A}_1 - \mathbf{A}_{n+1}, \mathbf{u}_2 = \mathbf{A}_2 - \mathbf{A}_{n+1}, \ldots, \mathbf{u}_n = \mathbf{A}_n - \mathbf{A}_{n+1}$, the remaining $\mathbf{n}_{n+1}$ is defined by $\mathbf{n}_{n+1} = -\sum_{i=1}^{n} \mathbf{n}_i$.

Facts:

1. There are $\binom{n+1}{2}$ edges $\{\mathbf{A}_i, \mathbf{A}_j\}$, $i \neq j$, in an n-simplex.
2. A 2-simplex is a segment and its circumcenter is its midpoint. The circumcenter of an n-simplex is the intersection of the lines ℓ_i, where ℓ_i is the line normal to face ω_i and through the circumcenter of ω_i.
3. [Blu53] (Menger, Schoenberg, etc.) The numbers m_{ij}, $i, j = 1, \ldots, n+1$, can serve as squares of the lengths of edges (i.e., of distances between vertices) of an n-simplex if and only if $m_{ii} = 0$ for all i, and
$$\sum_{i,j} m_{ij}x_ix_j < 0, \quad \text{whenever} \quad \sum_i x_i = 0.$$
4. (Consequence of Fact 3) The matrix
$$M_0 = \begin{bmatrix} 0 & \mathbf{1}^T \\ \mathbf{1} & M \end{bmatrix},$$
where $\mathbf{1}$ is the column vector of all ones and $M = [m_{ij}]$ is a Menger matrix, is elliptic, i.e., has one eigenvalue positive and the remaining negative.
5. The Menger matrix is the matrix of the quadratic form $\sum_{i,j} m_{ij}x_ix_j$. If $x_1, \ldots, x_{n+1}$ are considered as homogeneous barycentric coordinates of the point $\mathbf{X} = [x_1, \ldots, x_{n+1}]$, then the equation
$$\sum_{i,j} m_{ij}x_ix_j = 0$$
is the equation of the circumscribed hypersphere of the n-simplex Σ. (Thus, the condition in Fact 3 can be interpreted that for all improper points, the value of the quadratic form $\sum_{i,j} m_{ij}x_ix_j$ is strictly negative.)
6. The n-dimensional volume V of an n-simplex Σ satisfies
$$V^2 = \frac{(-1)^{n-1}}{2^n(n!)^2} \det M_0,$$
where M_0 is the matrix in Fact 4 using the Menger matrix M of Σ.

7. The volume V_s of the n-simplex with vertices in $\mathbf{O}, \mathbf{A}_1, \ldots, \mathbf{A}_n$ is equal to $V_s = \frac{1}{n!} V_p$, where V_p is the volume of the parallelepiped defined in Fact 4 of section 66.2 by n vectors $\mathbf{A}_i - \mathbf{O}$. Thus, $V_s = \frac{1}{n!}\sqrt{\det G(\mathbf{u}_1, \ldots, \mathbf{u}_n)}$, where $G(\ldots)$ means the Gram matrix and $\mathbf{u}_k = \mathbf{A}_k - \mathbf{O}$.
8. Let $Q_0 = \begin{bmatrix} q_{00} & \mathbf{q}_0^T \\ \mathbf{q}_0 & Q \end{bmatrix}$, where the matrix Q is the Gramian of the n-simplex Σ, the numbers in the column vector $\mathbf{q}_0$ are (-2)-multiples of the inhomogeneous barycentric coordinates of the circumcenter of Σ, and $\frac{1}{2}\sqrt{q_{00}}$ is the radius of the circumsphere of Σ. Then $M_0 Q_0 = -2I$.
9. The vectors $\mathbf{n}_1, \mathbf{n}_2, \ldots, \mathbf{n}_{n+1}$ from the definition of the Gramian are the vectors of outer normals of the simplex, in some sense normalized.
10. The Gramian of every n-simplex is an $(n+1) \times (n+1)$ positive semidefinite matrix of rank n with row sums equal to zero. Conversely, every such matrix determines uniquely (apart from the position in the space) an n-simplex in E^n whose Gramian this matrix is.
11. Let Σ_S be a face of an n-simplex Σ determined by the vertices with index set $S \subset \{1, \ldots, n+1\}$. Let Q be the Gramian of Σ. Then the Gramian of Σ_S is the Schur complement $Q/Q(S)$, where $Q(S)$ denotes the principal submatrix of Q corresponding to indices in the complement of S.

Examples:

1. Let us find the Menger matrix and the Gramian in the case of the segment $\{\mathbf{A}_1, \mathbf{A}_2\}$ of length d (i.e., a 1-simplex). The Menger matrix is $\begin{bmatrix} 0 & d^2 \\ d^2 & 0 \end{bmatrix}$. If $\mathbf{u}_1 = \mathbf{A}_1 - \mathbf{A}_2$, then $\mathbf{n}_1 = \frac{1}{d^2}\mathbf{u}_1$ since $\langle \mathbf{u}_1, \mathbf{n}_1 \rangle$ has to be 1. Thus, the Gramian is the Gram matrix $G(\mathbf{n}_1, -\mathbf{n}_1)$, i.e., $\begin{bmatrix} \frac{1}{d^2} & -\frac{1}{d^2} \\ -\frac{1}{d^2} & \frac{1}{d^2} \end{bmatrix}$. Indeed,
$$\begin{bmatrix} 0 & 1 & 1 \\ 1 & 0 & d^2 \\ 1 & d^2 & 0 \end{bmatrix} \begin{bmatrix} d^2 & -1 & -1 \\ -1 & \frac{1}{d^2} & -\frac{1}{d^2} \\ -1 & -\frac{1}{d^2} & \frac{1}{d^2} \end{bmatrix} = -2I_3,$$ as asserted in Fact 8.
2. Consider the simplex $\Sigma = \{\mathbf{A}_1 = [0, 0, 1]^T,\ \mathbf{A}_2 = [1, 0, 1]^T,\ \mathbf{A}_3 = [1, 2, 1]^T\}$ in E^2. We show how Fact 6 and Fact 7 can be used to find the volume V of this simplex.

   To use Fact 6, compute the squares of the distances between the vertices to obtain the Menger matrix: $M = \begin{bmatrix} 0 & 1 & 5 \\ 1 & 0 & 4 \\ 5 & 4 & 0 \end{bmatrix}$, so $M_0 = \begin{bmatrix} 0 & 1 & 1 & 1 \\ 1 & 0 & 1 & 5 \\ 1 & 1 & 0 & 4 \\ 1 & 5 & 4 & 0 \end{bmatrix}$. Since $\det M_0 = -16$, $V = 1$ by Fact 6.

   To use Fact 7, compute the vectors $\mathbf{u}_1 = [1, 0, 0]^T$ and $\mathbf{u}_2 = [1, 2, 0]^T$, so $G(\mathbf{u}_1, \mathbf{u}_2) = \begin{bmatrix} 1 & 1 \\ 1 & 5 \end{bmatrix}$. By Fact 7, $V = \frac{1}{2!}\sqrt{\det G(\mathbf{u}_1, \mathbf{u}_2)} = \frac{1}{2}\sqrt{4} = 1$.

   In this example, the volume can be found directly by finding the area of the triangle.
3. Consider the simplex in Example 2. Let us compute the Gramian to illustrate Fact 8 for this simplex. Since $\mathbf{u}_1 = \mathbf{A}_1 - \mathbf{A}_3 = [-1, -2, 0]^T$, $\mathbf{u}_2 = \mathbf{A}_2 - \mathbf{A}_3 = [0, -2, 0]^T$, the vectors $\mathbf{n}_1, \mathbf{n}_2$ forming a biorthogonal pair with $\mathbf{u}_1, \mathbf{u}_2$ are (as in Example 2 of Section 66.2) $\mathbf{n}_1 = [-1, 0, 0]^T$, $\mathbf{n}_2 = [1, -\frac{1}{2}, 0]^T$. Thus, $\mathbf{n}_3 = [0, \frac{1}{2}, 0]^T$ and $G(\mathbf{n}_1, \mathbf{n}_2, \mathbf{n}_3) = \begin{bmatrix} 1 & -1 & 0 \\ -1 & \frac{5}{4} & -\frac{1}{4} \\ 0 & -\frac{1}{4} & \frac{1}{4} \end{bmatrix}$.

   The line ℓ_1 is the set of points of the form $[x, 1, 1]^T$, and line ℓ_3 is the set of points of the form $[\frac{1}{2}, y, 1]^T$, so the circumcenter of the simplex is $\mathbf{c} = [\frac{1}{2}, 1, 1]^T$. The (inhomogeneous) barycentric coordinates b_1, b_2, b_3 of $\mathbf{c}$ can be found by solving $b_1\mathbf{A}_1 + b_2\mathbf{A}_2 + b_3\mathbf{A}_3 = \mathbf{c}$, $b_1 + b_2 + b_3 = 1$

to obtain $b_1 = \frac{1}{2}, b_2 = 0, b_3 = \frac{1}{2}$. The radius of the circumsphere is $\frac{\sqrt{5}}{2}$, so $q_{00} = 5$. Thus,

$$Q_0 = \begin{bmatrix} 5 & -1 & 0 & -1 \\ -1 & 1 & -1 & 0 \\ 0 & -1 & \frac{5}{4} & -\frac{1}{4} \\ -1 & 0 & -\frac{1}{4} & \frac{1}{4} \end{bmatrix},$$ which is indeed equal to the matrix $-2M_0^{-1}$.

4. We can use either Example 1 or Fact 11 and Example 3 to find the Gramian of Σ_S for $\{2, 3\}$ (again with Σ defined in Example 2). By Example 1, the Gramian is $\begin{bmatrix} \frac{1}{d^2} & -\frac{1}{d^2} \\ -\frac{1}{d^2} & \frac{1}{d^2} \end{bmatrix} = \begin{bmatrix} \frac{1}{4} & -\frac{1}{4} \\ -\frac{1}{4} & \frac{1}{4} \end{bmatrix}$.

66.4 Special Simplexes

Definitions:

Let us color an edge $\{\mathbf{A}_i, \mathbf{A}_j\}$ of an n-simplex with vertices $\mathbf{A}_1, \ldots, \mathbf{A}_{n+1}$, $n \geq 2$, in: **red**, if the opposite interior angle ϕ_{ij} is acute; **blue**, if the opposite interior angle ϕ_{ij} is obtuse; it will stay **uncolored**, if the opposite interior angle ϕ_{ij} is right.

The edge $\{\mathbf{A}_i, \mathbf{A}_k\}$ is called **colored** if it is colored red or blue. The assignment of red and blue colors is a **coloring** of the simplex.

The **graph** of the n-simplex is the graph with vertices $1, 2, \ldots, n+1$ and edges $\{i, k\}$, $i \neq k$, for which $\{\mathbf{A}_i, \mathbf{A}_k\}$ is colored. The **colored graph** of the simplex is the graph of the simplex colored in red and blue in the same way as the corresponding edges of the simplex.

An n-simplex is called **hyperacute** if each of its dihedral interior angles ϕ_{ij} is either acute or right.

A hyperacute n-simplex is called **totally hyperacute** if its circumcenter is either an interior point of the simplex, or an interior point of one of its faces.

Let $\mathbf{C}$ denote the circumcenter of the n-simplex with vertices $\mathbf{A}_1, \ldots, \mathbf{A}_{n+1}$. We extend the coloring of the simplex as follows: We assign the segment $\overline{\mathbf{CA}_k}$ the **red** color if the point $\mathbf{C}$ is in the same open halfspace determined by the face ω_k as the vertex $\mathbf{A}_k$, and the **blue** color, if $\mathbf{C}$ is in the opposite open halfspace. We do not assign to $\overline{\mathbf{CA}_k}$ any color if $\mathbf{C}$ belongs to ω_k. This is the **extended coloring** of the simplex.

In the same way, we speak about the **extended graph** and **extended colored graph** of the simplex, adding to $n+1$ vertices another vertex $n+2$ which corresponds to the circumcenter.

A **right simplex** (cf. [Fie57]) is an n-simplex which has exactly n acute interior angles and all the remaining $\binom{n}{2}$ interior angles right.

In a right simplex, the edges opposite the acute angles are called **legs**. The subgraph of legs and their endpoints is called **tree of legs** (see Fact 5).

A right simplex whose tree of legs is a path is called a **Schlaefli simplex**.

The face of a right simplex spanned by all pendent vertices (i.e., vertices of degree one) of the tree of legs is called the **hypotenuse** of the simplex.

A **net** on a simplex is a subset of the set of edges (not necessarily connected) such that every vertex of the simplex belongs to some edge in the net. A net is **metric** if for each edge in the net its length is given.

A **box** in a Euclidean n-space is a parallelepiped all of whose edges at some vertex (and then at all vertices) are mutually perpendicular.

Facts (cf. [Fie57]):

1. A simplex is hyperacute if and only if it has no blue edge in its coloring.
2. The set of red edges connects all the vertices of the simplex.
3. If we color the edges of an n-simplex by the two colors red and blue in such a way that the red edges connect all vertices, then there exists such deformation of the simplex that opposite red edges there are acute, opposite blue edges obtuse, and opposite uncolored edges right interior angles.

4. Every n-simplex has at least n acute interior angles.
5. There exist right n-simplexes. The red edges span a tree containing all the vertices of the simplex.
6. The legs of a right simplex are mutually perpendicular.
7. The tree of legs of a right n-simplex can be completed to a (n-dimensional) box; its center of symmetry is thus the circumcenter of the simplex. Conversely, every right n-simplex can be obtained by choosing among the edges of some box a subset of n edges with the property that any two are perpendicular and together form a connected set. These are then the legs of the simplex.
8. Let $G_T = (N, E, W)$ be a tree with the numbered vertex set $N = \{1, 2, \ldots, n+1\}$ and edge set E; let every edge $\{i, j\} \in E$ be assigned a positive number w_{ij}. Construct the matrices $Q_0 = [q_{rs}]$, $r, s = 0, 1, \ldots, n+1$, and $M = [m_{ij}]$, $i, j = 1, \ldots, n+1$, as follows:

$$q_{00} = \sum_{i,j \in E} \frac{1}{w_{ij}}, \quad q_{0i} = q_{i0} = d_i - 2, \quad i \in N,$$

where d_i is the degree of the vertex i in G_T,

$$q_{ij} = -w_{ij} \text{ if } \{i, j\} \in E, \quad q_{ii} = \sum_{j,(i,j) \in E} w_{ij}, \quad q_{ij} = 0 \text{ otherwise;}$$

$$m_{ii} = 0, \quad i \in N, \quad m_{ij} = w_{ik_1}^{-1} + w_{k_1 k_2}^{-1} + \cdots + w_{k_r j}^{-1},$$

where $i, j \in N, i \neq j$, and $i, k_1, k_2, \ldots, k_r, j$ is the (unique) path in G_T from i to j.

Then the matrices Q_0 and $M_0 = \begin{bmatrix} 0 & \mathbf{1}^T \\ \mathbf{1} & M \end{bmatrix}$ satisfy

$$M_0 Q_0 = -2I_{n+2}$$

and they are matrices corresponding to a right n-simplex whose tree of legs is G_T.
9. The inhomogeneous barycentric coordinates of the circumcenter of a right n-simplex are $q_{0i} = 1 - \frac{1}{2}d_i$, where d_i is the degree of the vertex i in the tree of legs.
10. The hypotenuse of a Schlaefli simplex is the longest edge; the midpoint of the hypotenuse is the circumcenter of the simplex.
11. Every face of a Schlaefli simplex is also a Schlaefli simplex.
12. The Schlaefli simplex is the only simplex all 2-dimensional faces of which are right triangles.
13. An n-simplex Σ is a Schlaefli simplex if and only if there exist real distinct numbers $c_1, \ldots, c_{n+1}$ such that the Menger matrix $M = [m_{ij}]$ of Σ has the form $m_{ij} = |c_i - c_j|$. If this holds for $c_1 > c_2 > \cdots > c_{n+1}$, then, in the usual notation, the edges $\{\mathbf{A}_k, \mathbf{A}_{k+1}\}$, $k = 1, \ldots, n$, are the legs of Σ.
14. Let Σ be a hyperacute n-simplex. Then the Gramian Q of Σ is a singular M-matrix of rank n, with annihilating vector $\mathbf{1}$.
15. Every face of a hyperacute simplex Σ is also a hyperacute simplex. The distribution of acute and right angles in the face is completely determined by the distribution of acute and right angles in the simplex as follows: If the face is determined by vertices with indices in S, the edge $\{\mathbf{A}_i, \mathbf{A}_j\}$ in the face will be red if and only if one can proceed from the vertex $\mathbf{A}_i$ to the vertex $\mathbf{A}_j$ along a path in the set of red edges of Σ which does not contain any vertex with index in S (except $\mathbf{A}_i$ and $\mathbf{A}_j$).
16. ([Fie61]). In the extended colored graph of an n-simplex, $n \geq 2$, the red part has vertex connectivity at least two.
17. ([Fie61]). In the definition of the extended graph and extended coloring, the vertex $n+2$ has no privileged position in the following sense: Let G be the extended colored graph of some n-simplex Σ_1, with vertex $n+2$ corresponding to the circumcenter of Σ_1. If $k \in \{1, 2, \ldots, n+1\}$, then there exists another n-simplex Σ_2 whose extended colored graph is G and such that k is the vertex corresponding to the circumcenter of Σ_2.

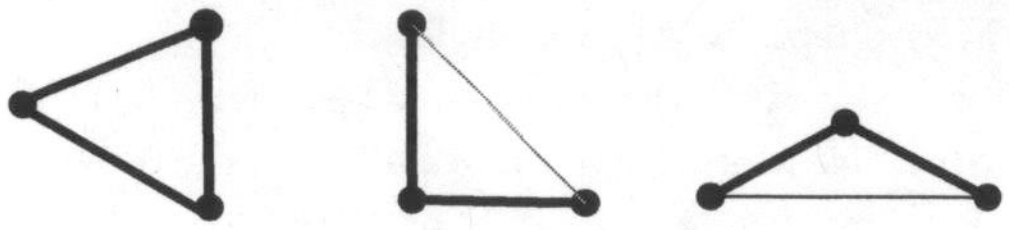

FIGURE 66.1 Coloring of triangles.

18. Every face (of dimension at least two) of a totally hyperacute simplex is also a totally hyperacute simplex. The extended coloring of the face is by the extended coloring of the simplex uniquely determined; it is obtained in the same way as in the usual coloring of a hyperacute simplex in Fact 15.
19. Let Σ be a totally hyperacute n-simplex. Then the extended graph of Σ is either a cycle, and then Σ is a Schlaefli simplex, or it has vertex connectivity at least three.
20. ([Fie61]) Let a metric net N on a simplex be given. There exists a simplex of maximum volume with this net if and only if the net N is connected, i.e., if it is possible to pass from any vertex of the net to any other vertex using edges in the net only. In addition, every simplex with this maximum volume has the property that every interior angle opposite an unspecified edge of the simplex (i.e., not belonging to N) is right.

Examples:

1. Figure 66.1 shows examples of three colored triangles. In these diagrams, the heaviest line represents red, the ordinary line represents blue, and the light gray line represents white.
2. Figure 66.2 shows examples of the two types of right tetrahedra. The dashed lines indicate the box described in Fact 7. The tetrahedron on the right is a Schlaefli simplex.
3. Figure 66.3 shows examples of extended graphs of triangles in Figure 66.1.
4. We apply the result in Fact 20 to so-called **cyclic simplexes** (cf. [Fie61]), which are maximum volume simplexes in the case that the metric net is a cycle. We call a simplex **regularly cyclic** if all edges in this net have the same length.

   The Gramian of the regularly cyclic n-simplex is a multiple of the matrix

$$\begin{bmatrix} 2 & -1 & 0 & \cdots & 0 & -1 \\ -1 & 2 & -1 & \cdots & 0 & 0 \\ \cdots & \cdots & \cdots & \cdots & \cdots & \cdots \\ 0 & 0 & 0 & \cdots & 2 & -1 \\ -1 & 0 & 0 & \cdots & -1 & 2 \end{bmatrix}$$

   with $n+1$ rows and columns, if the net is formed by the edges $\{\mathbf{A}_k, \mathbf{A}_{k+1}\}$, $k = 1, \ldots, n$, and $\{\mathbf{A}_{n+1}, \mathbf{A}_1\}$.

   The corresponding Menger matrix is then proportional to the matrix with entries $m_{ik} = |i-k|(n+1-|i-k|)$. It is possible to show that any two of the edges in the cycle span the same angle.

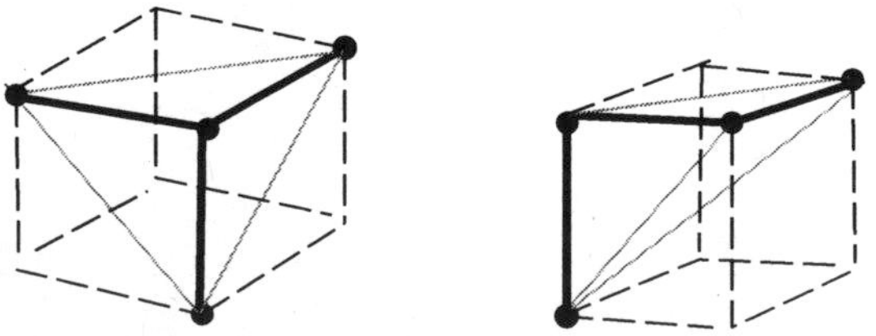

FIGURE 66.2 The two types of right tetrahedra.

FIGURE 66.3 Extended graphs of triangles.

It is immediate that the regularly cyclic 2-simplex is the equilateral triangle. The regularly cyclic 3-simplex is the tetrahedron, which is obtained from a square by parallelly lifting one diagonal in the perpendicular direction to the plane so that the distance of the two new diagonals equals half of the length of each diagonal.

Thus, the volume of every n-simplex with vertices $\mathbf{A}_1, \mathbf{A}_2, \ldots, \mathbf{A}_{n+1}$ in a Euclidean n-space for which all the edges $\{\mathbf{A}_1, \mathbf{A}_2\}, \{\mathbf{A}_2, \mathbf{A}_3\}, \ldots, \{\mathbf{A}_n, \mathbf{A}_{n+1}\}, \{\mathbf{A}_{n+1}, \mathbf{A}_1\}$ have length one, does not exceed $\frac{1}{n!}\sqrt{\frac{(n+1)^{n-1}}{n^n}}$.

66.5 An Application to Resistive Electrical Networks

We conclude with applications of matrices and graphs in another, maybe surprising, field.

Definitions:

A **resistive electrical network** is a network consisting of a finite number of nodes, say $1, 2, \ldots, n$, some pairs of which are directly connected by a conductor of some resistance. The assumptions are that the whole network is connected, and that there are no other electrical elements than resistors. As usual, **conductivity** of the conductor between two nodes is the reciprocal of the resistance of that conductor. If there is no direct connection between a pair of nodes, we set the conductivity of that "conductor" as zero.

A resistive electrical network of which just some of the nodes, **outlets**, are accessible, is called a **black-box**. The matrix of mutual resistances between the outlets of a black-box is called a **black-box matrix**.

The **problem** is: Characterize the set of all possible black-box $n \times n$ matrices.

Facts ([Moo68], [Fie78]):

1. Resistances of conductors in series add and conductivities of conductors in parallel add. If conductors having resistances R_1, R_2 are placed in series (see left illustration in Figure 66.4) the resistance R between the nodes is $R_1 + R_2$. If conductors having resistances R_1, R_2 are placed in parallel (see right illustration in Figure 66.4) the resistance R between the nodes satisfies $\frac{1}{R} = \frac{1}{R_1} + \frac{1}{R_2}$.
2. The 3×3 black-box matrices are all matrices

$$\begin{bmatrix} 0 & r_{12} & r_{13} \\ r_{12} & 0 & r_{23} \\ r_{13} & r_{23} & 0 \end{bmatrix}$$

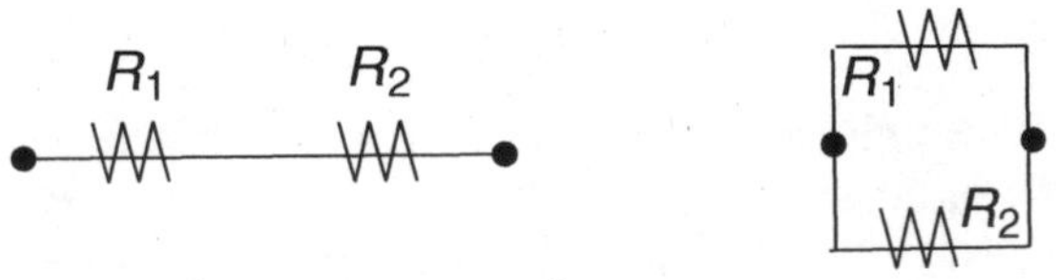

FIGURE 66.4 Resistors in series and parallel.

in which the numbers r_{12}, r_{13}, and r_{23} are nonnegative, at least two different from zero, and fulfill the nonstrict triangle inequality.

3. The $n \times n$ black-box matrices are exactly Menger matrices of hyperacute $(n-1)$-simplexes.
4. If a black-box matrix is given, then the corresponding network can be realized as such for which the conductivities between pairs of outlets are equal to the negatively taken corresponding entries of the Gramian of the simplex whose Menger matrix the given matrix is.
5. ([Fie78]) Let the given black-box B with n outlets correspond to an $(n-1)$-simplex Σ. Let S be a nonvoid proper subset of the set of outlets of B. Join all outlets in S by shortcuts. The resulting device can be considered as a new black-box B_S, which has just one outlet instead those outlets in S and, of course, all the remaining outlets. We construct the simplex corresponding to B_S and its black-box matrix as follows.

   If L is the linear space in the corresponding E^{n-1} determined by the vertices corresponding to S, project orthogonally all the remaining vertices as well as L itself on (some) orthogonal complement $L^{\perp}$ to L, thus obtaining a new simplex. The Menger matrix M_S of this simplex will be the black-box matrix of B_S. An algebraic construction is to pass from the Menger matrix of Σ, to the Gramian Q of Σ using Fact 8 of section 66.3; in this Gramian Q, add together all the rows corresponding to S into a single row and all the columns corresponding to S into a single column. The resulting symmetric matrix $\hat{Q}$ is again a singular M-matrix with row-sums zero. The simplex whose Gramian is $\hat{Q}$ is then that whose Menger matrix is B_S.
6. Let S_1 and S_2 be two disjoint nonempty subsets of the set of outlets of a black-box, and let Σ be the simplex in the geometric model. Join all outlets in S_1 with a source of potential zero, and all outlets in S_2 with the source of potential one. What will be the distribution of potentials in the remaining outlets using the geometric interpretation?

   The answer is: Let L_1, L_2, respectively, be linear spaces determined by vertices of Σ corresponding to S_1, S_2, respectively. There exists a unique pair of parallel hyperplanes H_1 and H_2, such that H_1 contains L_1 and H_2 contains L_2, and the distance between H_1 and H_2 is maximal among all such pairs of parallel hyperplanes. Then the square of the distance of the hyperplanes H_1 and H_2 measures the resistance between S_1 and S_2 (similarly like the square of the distance between two vertices of Σ measures the resistance between the corresponding outlets), and if V_0 is a vertex corresponding to an outlet S_0, then the potential in S_0 is obtained by linear interpolation corresponding to the position of the hyperplane H_0 containing V_0 and parallel to H_1 with respect to the hyperplanes H_1 and H_2. Let us remark that the whole simplex Σ is in the layer between H_1 and H_2, thanks to the property of hyperacuteness of Σ.

Examples:

1. Suppose the three nodes 1, 2, and 3 are connected as follows: The nodes 1 and 2 are connected by a conductor of resistance 18 ohms, the nodes 1 and 3 by conductor of resistance 12 ohms, and the nodes 2 and 3 by conductor of resistance 6 ohms. We compute the resistances for the black-box with the three outlets 1, 2, and 3. For r_{12}, the total resistance between outlets 1 and 2, note that there are two parallel paths between 1 and 2: the direct path with resistance 18 ohms, and the path through node 3 with resistance 6 ohms + 12 ohms = 18 ohms. Thus $\frac{1}{r_{12}} = \frac{1}{18} + \frac{1}{18} = \frac{1}{9}$, so $r_{12} =$ 9 ohms. Similar computations show that the resistance between outlets 1 and 3 is 8 ohms and the resistance between outlets 2 and 3 is 5 ohms. The black-box matrix is $\begin{bmatrix} 0 & 9 & 8 \\ 9 & 0 & 5 \\ 8 & 5 & 0 \end{bmatrix}$.
2. Let us remark that the properties of the outlets of a black-box do not depend on the way the conductors and resistors in the box are set. Here is another way the black-box-matrix in Example 1 could be obtained: There are no direct connections between nodes 1, 2, and 3, but there is a fourth node 4, which is connected with node 1 by a conductor of resistance 6 ohms, with node 2 by a conductor of resistance 3 ohms, and with node 3 by a conductor of resistance 2 ohms. Since

resistors in series add, this produces the same black-box matrix. If we then connect the outlets 1 and 2 by a short-circuit (conductor of no resistance), in both cases the resulting resistance between the new outlet $\{1, 2\}$ and outlet 3 will be the same, equal to 4 ohms.

References

[Blu53] L.M. Blumenthal, *Theory and Applications of Distance Geometry.* Oxford, Clarendon Press, 1953.

[Fie57] M. Fiedler, Über qualitative Winkeleigenschaften der Simplexe. *Czechoslovak Math. J.* 7(82):463–478, 1957.

[Fie61] M. Fiedler, Über die qualitative Lage des Mittelpunktes der umgeschriebenen Hyperkugel im n-simplex. *CMUC* 2,1:3–51, 1961.

[Fie61a] M. Fiedler, Über zyklische n-Simplexe und konjugierte Raumvielecke. *CMUC* 2,2:3 – 26, 1961.

[Fie61b] M. Fiedler, Über eine Ungleichung für positiv definite Matrizen. *Math. Nachrichten* 23:197–199, 1961.

[Fie64] M. Fiedler, Relations between the diagonal elements of two mutually inverse positive definite matrices. *Czech. Math. J.* 14(89):39 – 51, 1964.

[Fie78] M. Fiedler, Aggregation in graphs. In: Combinatorics. (A. Hajnal, Vera T. Sós, eds.) *Coll. Math. Soc. J. Bolyai*, 18. North-Holland, Amsterdam, 1978.

[Moo68] D.J.H. Moore, *A Geometric Theory for Electrical Networks.* Ph.D. Thesis, Monash University, Australia, 1968.

Applications to Algebra

67

Matrix Groups

Peter J. Cameron
Queen Mary, University of London

The topics of this chapter and the next (on group representations) are closely related. Here we consider some particular groups that arise most naturally as matrix groups or quotients of them, and special properties of matrix groups that are not shared by arbitrary groups. In representation theory, we consider what we learn about a group by considering all its homomorphisms to matrix groups. In this chapter we discuss properties of specific matrix groups, especially the general linear group (consisting of all invertible matrices of given size over a given field) and the related "classical groups." Most group theoretic terminology is standard and can be found in any textbook or in the Preliminaries in the Front Matter of the book.

67.1 Introduction

Definitions:

The **general linear group** $\mathrm{GL}(n, F)$ is the group consisting of all invertible $n \times n$ matrices over the field F.

A **matrix group** is a a subgroup of $\mathrm{GL}(n, F)$ for some natural number n and field F.

If V is a vector space of dimension n over F, the **group of invertible linear operators** on V is denoted by $\mathrm{GL}(V)$.

A **linear group** of **degree** n is a subgroup of $\mathrm{GL}(V)$ (where $\dim V = n$). (A subgroup of $\mathrm{GL}(n, F)$ is a linear group of degree n, since an $n \times n$ matrix can be viewed as a linear operator acting on F^n by matrix multiplication.)

A linear group $G \leq \mathrm{GL}(V)$ is said to be **reducible** if there is a G-invariant subspace U of V other than $\{0\}$ and V, and is **irreducible** otherwise. If $G \leq \mathrm{GL}(V)$ is irreducible, then V is called G-**irreducible**.

A linear group $G \leq \mathrm{GL}(V)$ is said to be **decomposable** if V is the direct sum of two nonzero G-invariant subspaces, and is **indecomposable** otherwise.

If V can be expressed as the direct sum of G-irreducible subspaces, then G is **completely reducible**. (An irreducible group is completely reducible.)

If the matrix group $G \leq \mathrm{GL}(n, F)$ is irreducible regarded as a subgroup of $\mathrm{GL}(n, K)$ for any algebraic extension K of F, we say that G is **absolutely irreducible**.

A linear group of degree n is **unipotent** if all its elements have n eigenvalues equal to 1.

Let X and Y be group-theoretic properties.

A group G is **locally X** if every finite subset of G is contained in a subgroup with property X.

Facts:

For these facts and general background reading see [Dix71], [Sup76], and [Weh73].

1. If V is a vector space of dimension n over F, then $\mathrm{GL}(V)$ is isomorphic to $\mathrm{GL}(n, F)$.
2. Every finite group is isomorphic to a matrix group.
3. Basic facts from linear algebra about similarity of matrices can be interpreted as statements about conjugacy classes in $\mathrm{GL}(n, F)$. For example:
   - Two nonsingular matrices are conjugate in $\mathrm{GL}(n, F)$ if and only if they have the same invariant factors.
   - If F is algebraically closed, then two nonsingular matrices are conjugate in $\mathrm{GL}(n, F)$ if and only if they have the same Jordan canonical form.
   - Two real symmetric matrices are conjugate in $\mathrm{GL}(n, \mathbb{R})$ if and only if they have the same rank and signature.

   (See Chapter 6 for more information on the Jordan canonical form and invariant factors and Chapter 12 for more information on signature.)
4. A matrix group G of degree n is reducible if and only if there exists a nonsingular matrix $M \in F^{n\times n}$ and k with $1 \leq k \leq n-1$ such that for all $A \in G$, $M^{-1}AM$ is of the form $\begin{bmatrix} B_{11} & B_{12} \\ \mathbf{0} & B_{22} \end{bmatrix}$, where $B_{11} \in F^{k\times k}$, $B_{22} \in F^{(n-k)\times(n-k)}$.
5. (See Chapter 68) The image of a representation of a group is a linear group. The image of a matrix representation of a group is a matrix group. We apply descriptions of the linear group to the representation: If $\rho : G \to \mathrm{GL}(V)$ is a representation, and $\rho(G)$ is irreducible, indecomposable, absolutely irreducible, etc., then we say that the representation ρ is irreducible, etc.
6. If every finitely generated subgroup of a group G is isomorphic to a linear group of degree n over a field F (of arbitrary characteristic), then G is isomorphic to a linear group of degree n.
7. Any free group is linear of degree 2 in every characteristic. More generally, a free product of linear groups is linear.
8. (Maschke's Theorem) Let G be a finite linear group over F, and suppose that the characteristic of F is either zero or coprime to $|G|$. If G is reducible, then it is decomposable.
9. A locally finite linear group in characteristic zero is completely reducible.
10. (Clifford's Theorem) Let G be an irreducible linear group on a vector space V of dimension n, and let N be a normal subgroup of G. Then V is a direct sum of minimal N-spaces $W_1, \ldots, W_d$ permuted transitively by G. In particular, d divides n, the group N is completely reducible, and the linear groups induced on W_i by N are all isomorphic.
11. A normal (or even a subnormal) subgroup of a completely reducible linear group is completely reducible.
12. A unipotent matrix group is conjugate (in the general linear group) to the group of upper unit triangular matrices.
13. A linear group G on V has a unipotent normal subgroup U such that G/U is isomorphic to a completely reducible linear group on V; the subgroup U is a nilpotent group of class at most $n-1$, where $n = \dim(V)$.
14. (Mal'cev) If every finitely generated subgroup of the linear group G is completely reducible, then G is completely reducible.
15. Let G be a linear group on F^n, where F is algebraically closed. Then G is irreducible if and only if the elements of G span the space $F^{n\times n}$ of all $n \times n$ matrices over F.

Examples:

1. The matrix group

$$G = \left\{ \begin{bmatrix} 1 & 0 \\ 0 & 1 \end{bmatrix}, \begin{bmatrix} -1 & 0 \\ 0 & -1 \end{bmatrix}, \begin{bmatrix} \frac{1}{3} & \frac{2}{3} \\ \frac{4}{3} & -\frac{1}{3} \end{bmatrix}, \begin{bmatrix} -\frac{1}{3} & -\frac{2}{3} \\ -\frac{4}{3} & \frac{1}{3} \end{bmatrix} \right\}$$

is decomposable and completely reducible: the subspaces of $\mathbb{R}^2$ spanned by the vectors $[1,1]^T$ and $[-1,2]^T$ are G-invariant. That is, with $M = \begin{bmatrix} 1 & -1 \\ 1 & 2 \end{bmatrix}$, for any $A \in G$, $M^{-1}AM$ is a diagonal matrix.

2. The matrix group

$$\left\{ \begin{bmatrix} 1 & a \\ 0 & 1 \end{bmatrix} : a \in \mathbb{R} \right\}$$

is reducible, but neither decomposable nor completely reducible.

3. The matrix group

$$\left\{ \begin{bmatrix} \cos x & \sin x \\ -\sin x & \cos x \end{bmatrix} : x \in \mathbb{R} \right\}$$

of real rotations is irreducible over $\mathbb{R}$ but not over $\mathbb{C}$: The subspace spanned by $[1,i]^T$ is invariant. The matrices in this group span the 2-dimensional subspace of $\mathbb{R}^{2\times 2}$ consisting of matrices $A = [a_{ij}]$ satisfying the equations $a_{11} = a_{22}$ and $a_{12} + a_{21} = 0$.

4. A group is locally finite if and only if every finitely generated subgroup is finite.

67.2 The General and Special Linear Groups

Definitions:

The **special linear group** $\mathrm{SL}(n, F)$ is the subgroup of $\mathrm{GL}(n, F)$ consisting of matrices of determinant 1. If V is a vector space of dimension n over F, the group of invertible linear operators on V having determinant 1 is denoted by $\mathrm{SL}(V)$.

The **special linear group** $\mathrm{SL}(n, F)$ is the subgroup of $\mathrm{GL}(n, F)$ consisting of matrices of determinant 1.

The **projective general linear group** and **projective special linear group** $\mathrm{PGL}(n, F)$ and $\mathrm{PSL}(n, F)$ are the quotients of $\mathrm{GL}(n, F)$ and $\mathrm{SL}(n, F)$ by their normal subgroups Z and $Z \cap \mathrm{SL}(n, F)$, respectively, where Z is the group of nonzero scalar matrices.

Notation: If $F = \mathrm{GF}(q)$ is the finite field of order q, then $\mathrm{GL}(n, F)$ is denoted $\mathrm{GL}(n,q)$, $\mathrm{SL}(n, F)$ is denoted $\mathrm{SL}(n,q)$, etc.

A **transvection** is a linear operator T on V with all eigenvalues equal to 1 and satisfying $\mathrm{rank}(T-I) = 1$.

Facts:

For these facts and general background reading see [HO89], [Tay92], or [Kra02].

1. The special linear group $\mathrm{SL}(n, F)$ is a normal subgroup of the general linear group $\mathrm{GL}(n, F)$.
2. The order of $\mathrm{GL}(n,q)$ is equal to the number of ordered bases of $\mathrm{GF}(q)^n$, namely

$$|\,\mathrm{GL}(n,q)| = \prod_{i=0}^{n-1}(q^n - q^i) = q^{n(n-1)/2}\prod_{i=0}^{n-1}(q^{n-i} - 1).$$

3. A transvection has determinant 1, and so lies in $\mathrm{SL}(V)$.
4. A transvection on F^n has the form $I + \mathbf{v}\mathbf{w}^T$ for some $\mathbf{v}, \mathbf{w} \in F^n$ and $\mathbf{w}^T\mathbf{v} = 0$, and anything in this form is a transvection.
5. A transvection T on V has the form $T : x \mapsto \mathbf{x} + f(\mathbf{x})\mathbf{v}$, where $\mathbf{v} \in V$, $f \in V^*$, and $f(\mathbf{v}) = 0$, and anything in this form is a transvection.
6. The group $\mathrm{SL}(n, F)$ is generated by transvections, for any $n \geq 2$ and any field F.

7. The group $\mathrm{PSL}(n, F)$ is simple for all $n \geq 2$ and all fields F, except for the two cases $\mathrm{PSL}(2,2)$ and $\mathrm{PSL}(2,3)$. The groups $\mathrm{PSL}(2,2)$ and $\mathrm{PSL}(2,3)$ are isomorphic to the symmetric group on 3 letters and the alternating group on 4 letters, respectively. The groups $\mathrm{PSL}(2,4)$ and $\mathrm{PSL}(2,5)$ are both isomorphic to the alternating group on 5 letters.

Examples:

1. $\mathrm{GL}(2,2) = \mathrm{SL}(2,2) = \mathrm{PSL}(2,2) =$

$$\left\{\begin{bmatrix}1&0\\0&1\end{bmatrix},\begin{bmatrix}1&1\\0&1\end{bmatrix},\begin{bmatrix}1&0\\1&1\end{bmatrix},\begin{bmatrix}0&1\\1&0\end{bmatrix},\begin{bmatrix}0&1\\1&0\end{bmatrix},\begin{bmatrix}0&1\\1&0\end{bmatrix}\right\}.$$

2. $\mathrm{SL}(2,3) =$

$$\left\{\begin{bmatrix}1&0\\0&1\end{bmatrix},\begin{bmatrix}1&1\\0&1\end{bmatrix},\begin{bmatrix}1&2\\0&1\end{bmatrix},\begin{bmatrix}1&0\\1&1\end{bmatrix},\begin{bmatrix}1&0\\2&1\end{bmatrix},\begin{bmatrix}2&0\\0&2\end{bmatrix},\begin{bmatrix}2&1\\0&2\end{bmatrix},\begin{bmatrix}2&2\\0&2\end{bmatrix},\right.$$
$$\left.\begin{bmatrix}2&0\\1&2\end{bmatrix},\begin{bmatrix}2&0\\2&2\end{bmatrix},\begin{bmatrix}1&1\\1&2\end{bmatrix},\begin{bmatrix}1&2\\2&2\end{bmatrix},\begin{bmatrix}2&1\\1&1\end{bmatrix},\begin{bmatrix}2&2\\2&1\end{bmatrix},\begin{bmatrix}0&1\\2&0\end{bmatrix},\begin{bmatrix}0&2\\1&0\end{bmatrix}\right\}.$$

3.

$$T = \begin{bmatrix}-13&-10&-2\\14&11&2\\28&20&5\end{bmatrix}$$

is a transvection. $T = I + \mathbf{v}\mathbf{w}^T$ with $\mathbf{v} = [-2,2,4]^T$, and $\mathbf{w} = [7,5,1]^T$.

67.3 The BN Structure of the General Linear Group

Definitions:

A BN-**pair** (or **Tits system**) is an ordered quadruple (G, B, N, S) where

- G is a group generated by subgroups B and N.
- $T := B \cap N$ is normal in N.
- S is a subset of $W := N/T$ and S generates W.
- The elements of S are all of order 2.
- If $\rho, \sigma \in N$ and $\rho T \in S$, then $\rho B \sigma \subseteq B\sigma B \cup B\rho\sigma B$.
- If $\rho T \in S$, then $\rho B \rho \neq B$.

If (G, B, N, S) is a BN-pair, the subgroups B and $W = N/T$ are known as the **Borel subgroup** and **Weyl group** of G.

A **parabolic subgroup** of G (relative to a given BN-pair) is a subgroup of the form $P_I = \langle B, s_i : i \in I\rangle$ for some subset I of $\{1, \dots, |S|\}$.

Facts:

For these facts and general background reading see [HO89], [Tay92], or [Kra02].

1. The general linear group $\mathrm{GL}(n, F)$ with $n \geq 2$ has the following Tits system:
   - B is the group of upper-triangular matrices in G.
   - U the group of unit upper-triangular matrices (with diagonal entries 1).

- T the group of diagonal matrices.
- N is the group of matrices having a unique nonzero element in each row or column.
- $S = \{s_i\ ;\ i = 1, \ldots, n-1\}$, where $s_i = P_i T$ and P_i is the reflection which interchanges the ith and $(i+1)$st standard basis vectors (P_i is obtained from the identity matrix by interchanging rows i and $i+1$).
- N is the normalizer of T in $\mathrm{GL}(n, F)$.
- $B = UT$.
- $B \cap N = T$.
- N/T is isomorphic to the symmetric group S_n.

2. If G has a BN-pair, any subgroup of G containing B is a parabolic subgroup.
3. In $\mathrm{GL}(n, F)$, there are 2^{n-1} parabolic subgroups for the BN-pair in Fact 1, hence, there are 2^{n-1} subgroups of $\mathrm{GL}(n, F)$ containing the subgroup B of upper-triangular matrices.
4. More generally, with respect to any basis of V there is a BN-structure. The terms *Borel subgroup* and *parabolic subgroup* are used to refer to the subgroups defined with respect to an arbitrary basis. All the Borel subgroups of $\mathrm{GL}(V)$ are conjugate. The maximal parabolic subgroups are precisely the maximal reducible subgroups.

Examples:

1. The maximal parabolic subgroups of $\mathrm{GL}(n, F)$ with the BN-pair in Fact 1 are those for which $I = \{1, \ldots, n-1\} \setminus \{k\}$ for some k; it is easy to see that in this case P_I is the stabilizer of the subspace spanned by the first k basis vectors. This subgroup consists of all matrices with block form as in 67.1, Fact 4.

67.4 Classical Groups

The classical groups form several important families of linear groups. We give a brief description here, and refer to the books [HO89], [Tay92], or the article [Kra02] for more details. For information on bilinear, sesquilinear, and quadratic forms, see Chapter 12.

Definitions:

A φ-sesquilinear form B is **φ-Hermitian** if $B(\mathbf{v}, \mathbf{w}) = \varphi(B(\mathbf{w}, \mathbf{v}))$ for all $\mathbf{v}, \mathbf{w} \in V$. In the case where $F = \mathbb{C}$ and φ is conjugation, a φ-Hermitian form is called a **Hermitian** form.

A **formed space** is a finite dimensional vector space carrying a nondegenerate φ-Hermitian, symmetric, or alternating form B.

A **classical group** over a formed space V is the subgroup of $\mathrm{GL}(V)$ consisting of the linear operators that preserve the form. We distinguish three types of classical groups:

1. **Orthogonal group:** Preserving a nondegenerate symmetric bilinear form B.
2. **Symplectic group:** Preserving a nondegenerate alternating bilinear form B.
3. **Unitary group:** Preserving a nondegenerate σ-Hermitian form B, with $\sigma \neq 1$.

We denote a classical subgroup of $\mathrm{GL}(V)$ by $\mathrm{O}(V)$, $\mathrm{Sp}(V)$, or $\mathrm{U}(V)$ depending on type. If necessary, we add extra notation to specify which particular form is being used. If $V = F^n$, we also write $\mathrm{O}(n, F)$, $\mathrm{Sp}(n, F)$, or $\mathrm{U}(n, F)$.

The **Witt index** of a formed space V is the dimension of the largest subspace on which the form is identically zero. The Witt index of the corresponding classical group is the Witt index of the formed space.

An **isometry** between subspaces of a formed space is a linear transformation preserving the value of the form.

A representation of a group G over the complex numbers is said to be **unitary** if its image is contained in the unitary group.

Facts:

For these facts and for general background reading see [HO89], [Tay92], or [Kra02].

1. The only automorphism of $\mathbb{R}$ is the identity, so any sesquilinear form on a real vector space is bilinear and any φ-Hermitian form is symmetric. A real formed space has a symmetric or alternating bilinear form as its form. The classical subgroups of a real formed space are orthogonal or symplectic.
2. The only automorphisms of $\mathbb{C}$ that preserve the reals are the identity and complex conjugation. Any φ-Hermitian form such that φ preserves the reals is a Hermitian form.
3. Classification of classical groups up to conjugacy in $\mathrm{GL}(n, F)$ is equivalent to classification of forms of the appropriate type up to the natural action of the general linear group together with scalar multiplication. Often this is a very difficult problem; the next fact gives a few cases where the classification is more straightforward.
4. (a) A nondegenerate alternating form on $V = F^n$ exists if and only if n is even, and all such forms are equivalent. So there is a unique conjugacy class of symplectic groups in $\mathrm{GL}(n, F)$ if n is even (with Witt index $n/2$), and none if n is odd.
   (b) Let $F = \mathrm{GF}(q)$. Then, up to conjugacy, $\mathrm{GL}(n, q)$ contains one conjugacy class of unitary subgroups (with Witt index $\lfloor n/2 \rfloor$), one class of orthogonal subgroups if n is odd (with Witt index $(n-1)/2$), and two classes if n is even (with Witt indices $n/2$ and $n/2 - 1$).
   (c) A nondegenerate symmetric bilinear form on $\mathbb{R}^n$ is determined up to the action of $\mathrm{GL}(n, \mathbb{R})$ by its signature. Its Witt index is $\min\{s, t\}$, where s and t are the numbers of positive and negative eigenvalues. So there are $\lfloor n/2 \rfloor + 1$ conjugacy classes of orthogonal subgroups of $\mathrm{GL}(n, \mathbb{R})$, with Witt indices $0, 1, \ldots, \lfloor n/2 \rfloor$.
5. (Witt's Lemma) Suppose that U_1 and U_2 are subspaces of the formed space V, and $h : U_1 \to U_2$ is an isometry. Then there is an isometry g of V that extends h.
6. From Witt's Lemma it is possible to write down formulas for the orders of the classical groups over finite fields similar to the formula in Fact 2 general linear group.
7. The analogues of Facts 6 and 7 in section 67.2 hold for the classical groups with nonzero Witt index. However, the situation is more complicated. Any symplectic transformation has determinant 1, so $\mathrm{Sp}(2r, F) \leq \mathrm{SL}(2r, F)$. Moreover, $\mathrm{Sp}(2r, F)$ is generated by symplectic transvections (those preserving the alternating form) for $r \geq 2$, except for $\mathrm{Sp}(4, 2)$. Similarly, the **special unitary group** $\mathrm{SU}(n, F)$ (the intersection of $\mathrm{U}(n, F)$ with $\mathrm{SL}(n, F)$) with positive Witt index is generated by unitary transvections (those preserving the Hermitian form), except for $\mathrm{SU}(3, 2)$. Results for orthogonal groups are more difficult. See [HO89] for more information.
8. Like the general linear groups, the classical groups contain BN-pairs (configurations of subgroups satisfying conditions like those in the previous section). The difference is that the Weyl group $W = N/H$ is not the symmetric group, but one of the other types of Coxeter group (finite groups generated by reflections).
9. Although this treatment of classical groups has been as far as possible independent of fields, for most of mathematics, the classical groups over the real and complex numbers are the most important, and among these, the real orthogonal and complex unitary groups preserving positive definite forms most important; see [Wey39].
10. The theory can be extended to classical groups over rings. This has important connections with algebraic K-theory. The book [HO89] gives details.
11. Every representation of a finite group is equivalent to a unitary representation.

Examples:

1. The function B given by $B((x_1, y_1), (x_2, y_2)) = x_1 y_2 - x_2 y_1$ is an alternating bilinear form on F^2. Any matrix with determinant 1 will preserve this form. So $Sp(2, F) = SL(2, F)$.
2. The symmetric group S_6 acts on F^6, where F is the field with two elements. It preserves the 1-dimensional subspace U spanned by (1, 1, 1, 1, 1, 1), as well as the 5-dimensional subspace consisting of vectors with coordinate sum zero. The usual dot product on F^6 is alternating when

restricted to W, and its radical is U, so it induces a symplectic form on W/U. Thus S_6 is a subgroup of the symplectic group $Sp(4, 2)$. Since both groups have order 720, we see that $Sp(4, 2) = S_6$.

Acknowledgment

I am grateful to Professor B. A. F. Wehrfritz for helpful comments on this chapter.

References

[HO89] A. Hahn and T. O'Meara, *The Classical Groups and K-Theory*, Springer-Verlag, Berlin, 1989.

[Kra02] L. Kramer, Buildings and classical groups, in *Tits Buildings and the Model Theory of Groups* (Ed. K. Tent), London Math. Soc. Lecture Notes **291**, Cambridge University Press, Cambridge, 2002.

[Tay92] D. E. Taylor, *The Geometry of the Classical Groups*, Heldermann Verlag, Berlin, 1992.

[Wey39] H. Weyl, *The Classical Groups*, Princeton University Press, Princeton, NJ, 1939 (reprint 1997).

[Dix71] J. D. Dixon, *The Structure of Linear Groups*. Van Nostrand Reinhold, London, 1971.

[Sup76] D. A. Suprunenko, *Matrix Groups*. Amer. Math. Soc. Transl. 45, American Mathematical Society, Providence, RI, 1976.

[Weh73] B. A. F. Wehrfritz, *Infinite Linear Groups*. Ergebnisse der Matematik und ihrer Grenzgebiete, 76, Springer-Verlag, New York-Heidelberg, 1973.

68

Group Representations

Randall R. Holmes
Auburn University

T. Y. Tam
Auburn University

Representation theory is the study of the various ways a given group can be mapped into a general linear group. This information has proven to be effective at providing insight into the structure of the given group as well as the objects on which the group acts. Most notable is the central contribution made by representation theory to the complete classification of finite simple groups [Gor94]. (See also Fact 3 of Section 68.1 and Fact 5 of Section 68.6.)

Representations of finite groups can be defined over an arbitrary field and such have been studied extensively. Here, however, we discuss only the most widely used, classical theory of representations over the field of complex numbers (many results of which fail to hold over other fields).

68.1 Basic Concepts

Throughout, G denotes a finite group, e denotes its identity element, and V denotes a finite dimensional complex vector space.

Definitions:

The **general linear group** of a vector space V is the group $GL(V)$ of linear isomorphisms of V onto itself with operation given by function composition.

A (**linear**) **representation** of the finite group G (over the complex field $\mathbb{C}$) is a homomorphism $\rho = \rho_V : G \to GL(V)$, where V is a finite dimensional vector space over $\mathbb{C}$.

The **degree** of a representation ρ_V is the dimension of the vector space V.

Two representations $\rho : G \to GL(V)$ and $\rho' : G \to GL(V')$ are **equivalent** (or **isomorphic**) if there exists a linear isomorphism $\tau : V \to V'$ such that $\tau \circ \rho(s) = \rho'(s) \circ \tau$ for all $s \in G$.

Given a representation ρ_V of G, a subspace W of V is G-**stable** (or G-**invariant**) if $\rho_V(s)(W) \subseteq W$ for all $s \in G$.

If ρ_V is a representation of G and W is a G-stable subspace of V, then the induced maps $\rho_W : G \to GL(W)$ and $\rho_{V/W} : G \to GL(V/W)$ are the corresponding **subrepresentation** and **quotient representation**, respectively.

A representation ρ_V of G with $V \neq \{0\}$ is **irreducible** if V and $\{0\}$ are the only G-stable subspaces of V; otherwise, ρ_V is **reducible**.

The **kernel** of a representation ρ_V of G is the set of all $s \in G$ for which $\rho_V(s) = 1_V$.

A representation of G is **faithful** if its kernel consists of the identity element alone.

An **action** of G on a set X is a function $G \times X \to X$, $(s, x) \mapsto sx$, satisfying

- $(st)x = s(tx)$ for all $s, t \in G$ and $x \in X$,
- $ex = x$ for all $x \in X$.

A $\mathbb{C}G$-**module** is a finite-dimensional vector space V over $\mathbb{C}$ together with an action $(s, v) \mapsto sv$ of G on V that is linear in the variable v, meaning

- $s(v + w) = sv + sw$ for all $s \in G$ and $v, w \in V$,
- $s(\alpha v) = \alpha(sv)$ for all $s \in G$, $v \in V$ and $\alpha \in \mathbb{C}$.

(See Fact 6 below.)

Facts:

The following facts can be found in [Isa94, pp. 4–10] or [Ser77, pp. 3–13, 47].

1. If $\rho = \rho_V$ is a representation of G, then
   - $\rho(e) = 1_V$,
   - $\rho(st) = \rho(s)\rho(t)$ for all $s, t \in G$,
   - $\rho(s^{-1}) = \rho(s)^{-1}$ for all $s \in G$.
2. A representation of G of degree one is a group homomorphism from G into the group $\mathbb{C}^\times$ of nonzero complex numbers under multiplication (identifying $\mathbb{C}^\times$ with $GL(\mathbb{C})$). Every representation of degree one is irreducible.
3. The group G is abelian if and only if every irreducible representation of G is of degree one.
4. *Maschke's Theorem*: If ρ_V is a representation of G and W is a G-stable subspace of V, then there exists a G-stable vector space complement of W in V.
5. *Schur's Lemma*: Let $\rho : G \to GL(V)$ and $\rho' : G \to GL(V')$ be two irreducible representations of G and let $f : V \to V'$ be a linear map satisfying $f \circ \rho(s) = \rho'(s) \circ f$ for all $s \in G$.
   - If ρ' is not equivalent to ρ, then f is the zero map.
   - If $V' = V$ and $\rho' = \rho$, then f is a scalar multiple of the identity map: $f = \alpha 1_V$ for some $\alpha \in \mathbb{C}$.
6. If $\rho = \rho_V$ is a representation of G, then V becomes a $\mathbb{C}G$-module with action given by $sv = \rho(s)(v)$ $(s \in G, v \in V)$. Conversely, if V is a $\mathbb{C}G$-module, then $\rho_V(s)(v) = sv$ defines a representation $\rho_V : G \to GL(V)$ (called the **representation of** G **afforded by** V). The study of representations of the finite group G is the same as the study of $\mathbb{C}G$-modules.
7. The vector space $\mathbb{C}G$ over $\mathbb{C}$ with basis G is a ring (the **group ring** of G over $\mathbb{C}$) with multiplication obtained by linearly extending the operation in G to arbitrary products. If V is a $\mathbb{C}G$-module and the action of G on V is extended linearly to a map $\mathbb{C}G \times V \to V$, then V becomes a (left unitary) $\mathbb{C}G$-module in the ring theoretic sense, that is, V satisfies the usual vector space axioms (see Section 1.1) with the scalar field replaced by the ring $\mathbb{C}G$.

Examples:

See also examples in the next section.

1. Let $n \in \mathbf{N}$ and let $\omega \in \mathbb{C}$ be an nth root of unity (meaning $\omega^n = 1$). Then the map $\rho : \mathbf{Z}_n \to \mathbb{C}^\times$ given by $\rho(m) = \omega^m$ is a representation of degree one of the group $\mathbf{Z}_n$ of integers modulo n. It is irreducible.
2. *Regular representation*: Let $V = \mathbb{C}G$ be the complex vector space with basis G. For each $s \in G$ there is a unique linear map $\rho(s) : V \to V$ satisfying $\rho(s)(t) = st$ for all $t \in G$. Then $\rho : G \to GL(V)$ is a representation of G called the **(left) regular representation**. If $|G| > 1$, then the regular representation is reducible (see Example 3 of Section 68.5) .

3. *Permutation representation*: Let X be a finite set, let $(s, x) \mapsto sx$ be an action of G on X, and let V be the complex vector space with basis X. For each $s \in G$ there is a unique linear map $\rho(s) : V \to V$ satisfying $\rho(s)(x) = sx$ for all $x \in X$. Then $\rho : G \to GL(V)$ is a representation of G called a **permutation representation**. The regular representation of G (Example 2) is the permutation representation corresponding to the action of G on itself given by left multiplication.
4. The representation of G of degree 1 given by $\rho(s) = 1 \in \mathbb{C}^\times$ for all $s \in G$ is the **trivial representation.**
5. *Direct sum*: If V and W are $\mathbb{C}G$-modules, then the $\mathbb{C}$-vector space direct sum $V \oplus W$ is a $\mathbb{C}G$-module with action given by $s(v, w) = (sv, sw)$ $(s \in G, v \in V, w \in W)$.
6. *Tensor product*: If V_1 is a $\mathbb{C}G_1$-module and V_2 is a $\mathbb{C}G_2$-module, then the $\mathbb{C}$-vector space tensor product $V_1 \otimes V_2$ is a $\mathbb{C}(G_1 \times G_2)$-module with action given by $(s_1, s_2)(v_1 \otimes v_2) = (s_1v_1) \otimes (s_2v_2)$ $(s_i \in G_i, v_i \in V_i)$. If both groups G_1 and G_2 equal the same group G, then $V_1 \otimes V_2$ is a $\mathbb{C}G$-module with action given by $s(v_1 \otimes v_2) = (sv_1) \otimes (sv_2)$ $(s \in G, v_i \in V_i)$.
7. *Contragredient*: If V is a $\mathbb{C}G$-module, then the $\mathbb{C}$-vector space dual V^* is a $\mathbb{C}G$-module (called the **contragredient** of V) with action given by $(sf)(v) = f(s^{-1}v)$ $(s \in G, f \in V^*, v \in V)$.

68.2 Matrix Representations

Throughout, G denotes a finite group, e denotes its identity element, and V denotes a finite dimensional complex vector space.

Definitions:

A **matrix representation** of G of **degree** n (over the field $\mathbb{C}$) is a homomorphism $R : G \to GL_n(\mathbb{C})$, where $GL_n(\mathbb{C})$ is the group of nonsingular $n \times n$ matrices over the field $\mathbb{C}$. (For the relationship between representations and matrix representations, see the facts below.)

The **empty matrix** is a 0×0 matrix having no entries. The trace of the empty matrix is 0. $GL_0(\mathbb{C})$ is the trivial group whose only element is the empty matrix.

Two matrix representations R and R' are **equivalent** (or **isomorphic**) if they have the same degree, say n, and there exists a nonsingular $n \times n$ matrix P such that $R'(s) = PR(s)P^{-1}$ for all $s \in G$.

A matrix representation of G is **reducible** if it is equivalent to a matrix representation R having the property that for each $s \in G$, the matrix $R(s)$ has the block form

$$R(s) = \begin{bmatrix} X(s) & Z(s) \\ 0 & Y(s) \end{bmatrix}$$

(block sizes independent of s).

A matrix representation is **irreducible** if it has nonzero degree and it is not reducible.

The **kernel** of a matrix representation R of G of degree n is the set of all $s \in G$ for which $R(s) = I_n$.

A matrix representation of G is **faithful** if its kernel consists of the identity element alone.

Facts:

The following facts can be found in [Isa94, pp. 10–11, 32] or [Ser77, pp. 11–14].

1. If R is a matrix representation of G, then

   - $R(e) = I$,
   - $R(st) = R(s)R(t)$ for all $s, t \in G$,
   - $R(s^{-1}) = R(s)^{-1}$ for all $s \in G$.

2. If $\rho = \rho_V$ is a representation of G of degree n and $\mathcal{B}$ is an ordered basis for V, then $R_{\rho,\mathcal{B}}(s) = [\rho(s)]_{\mathcal{B}}$ defines a matrix representation $R_{\rho,\mathcal{B}} : G \to GL_n(\mathbb{C})$ called the **matrix representation of G afforded by the representation ρ** (or **by the $\mathbb{C}G$-module V**) **with respect to the basis $\mathcal{B}$**. Conversely, if R is a matrix representation of G of degree n and $V = \mathbb{C}^n$, then $\rho(s)(v) = R(s)v$ $(s \in G, v \in V)$ defines a representation ρ of G and $R = R_{\rho,\mathcal{B}}$, where $\mathcal{B}$ is the standard ordered basis of V.
3. If R and R' are matrix representations afforded by representations ρ and ρ', respectively, then R and R' are equivalent if and only if ρ and ρ' are equivalent. In particular, two matrix representations that are afforded by the same representation are equivalent regardless of the chosen bases.
4. If $\rho = \rho_V$ is a representation of G and W is a G-stable subspace of V and a basis for W is extended to a basis $\mathcal{B}$ of V, then for each $s \in G$ the matrix $R_{\rho,\mathcal{B}}(s)$ is of block form

$$R_{\rho,\mathcal{B}}(s) = \begin{bmatrix} X(s) & Z(s) \\ 0 & Y(s) \end{bmatrix},$$

where X and Y are the matrix representations afforded by ρ_W (with respect to the given basis) and $\rho_{V/W}$ (with respect to the induced basis), respectively.
5. If the matrix representation R of G is afforded by a representation ρ, then R is irreducible if and only if ρ is irreducible.
6. The group G is Abelian if and only if every irreducible matrix representation of G is of degree one .
7. *Maschke's Theorem* (*for matrix representations*): If R is a matrix representation of G and for each $s \in G$ the matrix $R(s)$ is of block form

$$R(s) = \begin{bmatrix} X(s) & Z(s) \\ 0 & Y(s) \end{bmatrix}$$

(block sizes independent of s), then R is equivalent to the matrix representation R' given by

$$R'(s) = \begin{bmatrix} X(s) & 0 \\ 0 & Y(s) \end{bmatrix}$$

$(s \in G)$.
8. *Schur relations*: Let R and R' be irreducible matrix representations of G of degrees n and n', respectively. For $1 \le i, j \le n$ and $1 \le i', j' \le n'$ define functions $r_{ij}, r'_{i'j'} : G \to \mathbb{C}$ by $R(s) = [r_{ij}(s)]$, $R'(s) = [r'_{i'j'}(s)]$ $(s \in G)$.
   - If R' is not equivalent to R, then for all $1 \le i, j \le n$ and $1 \le i', j' \le n'$

$$\sum_{s\in G} r_{ij}(s^{-1}) r'_{i'j'}(s) = 0.$$

   - For all $1 \le i, j, k, l \le n$

$$\sum_{s\in G} r_{ij}(s^{-1}) r_{kl}(s) = \begin{cases} |G|/n & \text{if } i = l \text{ and } j = k \\ 0 & \text{otherwise.} \end{cases}$$

Examples:

1. An example of a degree two matrix representation of the symmetric group S_3 is given by

$$R(e) = \begin{bmatrix} 1 & 0 \\ 0 & 1 \end{bmatrix}, \quad R(12) = \begin{bmatrix} 0 & 1 \\ 1 & 0 \end{bmatrix}, \quad R(23) = \begin{bmatrix} -1 & -1 \\ 0 & 1 \end{bmatrix},$$

$$R(13) = \begin{bmatrix} 1 & 0 \\ -1 & -1 \end{bmatrix}, \quad R(123) = \begin{bmatrix} 0 & 1 \\ -1 & -1 \end{bmatrix}, \quad R(132) = \begin{bmatrix} -1 & -1 \\ 1 & 0 \end{bmatrix}.$$

(Cf. Example 2 of section 68.4.)

2. The matrix representation R of the additive group $\mathbf{Z}_3$ of integers modulo 3 afforded by the regular representation with respect to the basis $\mathbf{Z}_3 = \{0, 1, 2\}$ (ordered as indicated) is given by

$$R(0) = \begin{bmatrix} 1 & 0 & 0 \\ 0 & 1 & 0 \\ 0 & 0 & 1 \end{bmatrix}, \quad R(1) = \begin{bmatrix} 0 & 0 & 1 \\ 1 & 0 & 0 \\ 0 & 1 & 0 \end{bmatrix}, \quad R(2) = \begin{bmatrix} 0 & 1 & 0 \\ 0 & 0 & 1 \\ 1 & 0 & 0 \end{bmatrix}.$$

3. Let $\rho : G \to GL(V)$ and $\rho' : G \to GL(V')$ be two representations of G, let $\mathcal{B}$ and $\mathcal{B}'$ be bases of V and V', respectively, and let $R = R_{\rho,\mathcal{B}}$ and $R' = R_{\rho',\mathcal{B}'}$ be the afforded matrix representations.

   - The matrix representation afforded by the direct sum $V \oplus V'$ with respect to the basis $\{(b, 0), (0, b') \mid b \in \mathcal{B}, b' \in \mathcal{B}'\}$ is given by $s \mapsto R(s) \oplus R'(s)$ (direct sum of matrices).
   - The matrix representation afforded by the tensor product $V \otimes V'$ with respect to the basis $\{b \otimes b' \mid b \in \mathcal{B}, b' \in \mathcal{B}'\}$ is given by $s \mapsto R(s) \otimes R'(s)$ (Kronecker product of matrices).
   - The matrix representation afforded by the contragredient V^* with respect to the dual basis of $\mathcal{B}$ is given by $s \mapsto (R(s)^{-1})^T$ (inverse transpose of matrix).

68.3 Characters

Throughout, G denotes a finite group, e denotes its identity element, and V denotes a finite dimensional complex vector space.

Definitions:

The **character of** G **afforded by a matrix representation** R of G is the function $\chi : G \to \mathbb{C}$ defined by $\chi(s) = \text{tr } R(s)$.

The **character of** G **afforded by a representation** $\rho = \rho_V$ of G is the character afforded by the corresponding matrix representation $R_{\rho,\mathcal{B}}$, where $\mathcal{B}$ is a basis for V.

The **character of** G **afforded by a** $\mathbb{C}G$**-module** V is the character afforded by the corresponding representation ρ_V.

An **irreducible character** is a character afforded by an irreducible representation.

The **degree** of a character χ of G is the number $\chi(e)$.

A **linear character** is a character of degree one.

If χ_1 and χ_2 are two characters of G, their **sum** is defined by $(\chi_1 + \chi_2)(s) = \chi_1(s) + \chi_2(s)$ and their **product** is defined by $(\chi_1\chi_2)(s) = \chi_1(s)\chi_2(s)$ $(s \in G)$.

If χ is a character of G, its **complex conjugate** is defined by $\overline{\chi}(s) = \overline{\chi(s)}$ $(s \in G)$, where $\overline{\chi(s)}$ denotes the conjugate of the complex number $\chi(s)$.

The **kernel** of a character χ of G is the set $\{s \in G \mid \chi(s) = \chi(e)\}$.

A character χ of G is **faithful** if its kernel consists of the identity element alone.

The **principal character** of G is the character 1_G satisfying $1_G(s) = 1$ for all $s \in G$.

The **zero character** of G is the character 0_G satisfying $0_G(s) = 0$ for all $s \in G$.

If χ and ψ are two characters of G, then ψ is called a **constituent** of χ if $\chi = \psi + \psi'$ with ψ' a character (possibly zero) of G.

Facts:

The following facts can be found in [Isa94, pp. 14–23, 38–40, 59] or [Ser77, pp. 10–19, 27, 52].

1. The principal character of G is the character afforded by the representation of G that maps every $s \in G$ to $[1] \in GL_1(\mathbb{C})$. The zero character of G is the character afforded by the representation of G that maps every $s \in G$ to the empty matrix in $GL_0(\mathbb{C})$.

2. The degree of a character of G equals the dimension of V, where ρ_V is a representation affording the character.
3. If χ is a character of G, then $\chi(s^{-1}) = \overline{\chi(s)}$ and $\chi(t^{-1}st) = \chi(s)$ for all $s, t \in G$.
4. Two characters of G are equal if and only if representations affording them are equivalent.
5. The number of distinct irreducible characters of G is the same as the number of conjugacy classes of G.
6. Every character χ of G can be expressed in the form $\chi = \sum_{\varphi \in Irr(G)} m_\varphi \varphi$, where $Irr(G)$ denotes the set of irreducible characters of G and where each m_φ is a nonnegative integer (called the **multiplicity** of φ as a constituent of χ).
7. A nonzero character of G is irreducible if and only if it is not the sum of two nonzero characters of G.
8. The kernel of a character equals the kernel of a representation affording the character.
9. The degree of an irreducible character of G divides the order of G.
10. A character of G is linear if and only if it is a homomorphism from G into the multiplicative group of nonzero complex numbers under multiplication.
11. The group G is abelian if and only if every irreducible character of G is linear.
12. The sum of the squares of the irreducible character degrees equals the order of G.
13. *Irreducible characters of direct products*: Let G_1 and G_2 be finite groups. Denoting by $Irr(G)$ the set of irreducible characters of the group G, we have $Irr(G_1 \times G_2) = Irr(G_1) \times Irr(G_2)$, where an element (χ_1, χ_2) of the Cartesian product on the right is viewed as a function on the direct product $G_1 \times G_2$ via $(\chi_1, \chi_2)(s_1, s_2) = \chi_1(s_1)\chi_2(s_2)$.
14. *Burnside's Vanishing Theorem*: If χ is a nonlinear irreducible character of G, then $\chi(s) = 0$ for some $s \in G$.

Examples:

See also examples in the next section.

1. If V_1 and V_2 are $\mathbb{C}G$-modules and χ_1 and χ_2, respectively, are the characters of G they afford, then the direct sum $V_1 \oplus V_2$ affords the sum $\chi_1 + \chi_2$ and the tensor product $V_1 \otimes V_2$ affords the product $\chi_1\chi_2$.
2. If V is a $\mathbb{C}G$-module and χ is the character it affords, then the contragredient V^* affords the complex conjugate character $\overline{\chi}$.
3. Let X be a finite set on which an action of G is given and let ρ be the corresponding permutation representation of G (see Example 3 of Section 68.1). If χ is the character afforded by ρ, then for each $s \in G$, $\chi(s)$ is the number of ones on the main diagonal of the permutation matrix $[\rho(s)]_X$, which is the same as the number of fixed points of X under the action of s: $\chi(s) = |\{x \in X \mid sx = x\}|$. The matrix representation of $\mathbf{Z}_3$ given in Example 2 of Section 68.2 is afforded by a permutation representation, namely, the regular representation; it affords the character χ given by $\chi(0) = 3$, $\chi(1) = 0$, $\chi(2) = 0$ in accordance with the statement above.

68.4 Orthogonality Relations and Character Table

Throughout, G denotes a finite group, e denotes its identity element, and V denotes a finite dimensional complex vector space.

Definitions:

A function $f : G \to \mathbb{C}$ is called a **class function** if it is constant on the conjugacy classes of G, that is, if $f(t^{-1}st) = f(s)$ for all $s, t \in G$.

The **inner product** of two functions f and g from G to $\mathbb{C}$ is the complex number

$$(f, g)_G = \frac{1}{|G|} \sum_{s \in G} f(s)\overline{g(s)}.$$

The **character table** of the group G is the square array with entry in the ith row and jth column equal to the complex number $\chi_i(c_j)$, where $Irr(G) = \{\chi_1, \ldots, \chi_k\}$ is the set of distinct irreducible characters of G and $\{c_1, \ldots, c_k\}$ is a set consisting of exactly one element from each conjugacy class of G.

Facts:

The following facts can be found in [Isa94, pp. 14–21, 30] or [Ser77, pp.10–19].

1. Each character of G is a class function.
2. *First Orthogonality Relation*: If φ and ψ are two irreducible characters of G, then

$$(\phi, \psi)_G = \frac{1}{|G|} \sum_{s \in G} \varphi(s)\overline{\psi(s)} = \begin{cases} 1 & \text{if } \varphi = \psi \\ 0 & \text{if } \varphi \neq \psi. \end{cases}$$

3. *Second Orthogonality Relation*: If s and t are two elements of G, then

$$\sum_{\chi \in Irr(G)} \chi(s)\overline{\chi(t)} = \begin{cases} |G|/c(s) & \text{if } t \text{ is conjugate to } s \\ 0 & \text{if } t \text{ is not conjugate to } s, \end{cases}$$

   where $c(s)$ denotes the number of elements in the conjugacy class of s.
4. *Generalized Orthogonality Relation*: If φ and ψ are two irreducible characters of G and t is an element of G, then

$$\frac{1}{|G|} \sum_{s \in G} \varphi(st)\overline{\psi(s)} = \begin{cases} \varphi(t)/\varphi(e) & \text{if } \varphi = \psi \\ 0 & \text{if } \varphi \neq \psi. \end{cases}$$

   This generalizes the First Orthogonality Relation (Fact 2).
5. The set of complex-valued functions on G is a complex inner product space with inner product as defined above. The set of class functions on G is a subspace.
6. A character χ of G is irreducible if and only if $(\chi, \chi)_G = 1$.
7. The set $Irr(G)$ of irreducible characters of G is an orthonormal basis for the inner product space of class functions on G.
8. If the character χ of G is expressed as a sum of irreducible characters (see Fact 6 of Section 68.3), then the number of times the irreducible character φ appears as a summand is $(\chi, \varphi)_G$. In particular, $\varphi \in Irr(G)$ is a constituent of χ if and only if $(\chi, \varphi)_G \neq 0$.
9. Isomorphic groups have identical character tables (up to a reordering of rows and columns). The converse of this statement does not hold since, for example, the dihedral group and the quaternion group (both of order eight) have the same character table, yet they are not isomorphic.

Examples:

1. The character table of the group $\mathbf{Z}_4$ of integers modulo four is

	0	1	2	3
χ_0	1	1	1	1
χ_1	1	i	-1	$-i$
χ_2	1	-1	1	-1
χ_3	1	$-i$	-1	i

2. The character table of the symmetric group S_3 is

	(1)	(12)	(123)
χ_0	1	1	1
χ_1	1	-1	1
χ_2	2	0	-1

Note that χ_2 is the character afforded by the matrix representation of S_3 given in Example 1 of Section 68.2.
3. The character table of the symmetric group S_4 is

	(1)	(12)	(12)(34)	(123)	(1234)
χ_0	1	1	1	1	1
χ_1	1	-1	1	1	-1
χ_2	2	0	2	-1	0
χ_3	3	1	-1	0	-1
χ_4	3	-1	-1	0	1

4. The character table of the alternating group A_4 is

	(1)	(12)(34)	(123)	(132)
χ_0	1	1	1	1
χ_1	1	1	ω	ω^2
χ_2	1	1	ω^2	ω
χ_3	3	-1	0	0

where $\omega = e^{2\pi i/3} = -\frac{1}{2} + i\frac{\sqrt{3}}{2}$.
5. Let ρ_V be a representation of G and for each irreducible character φ of G put

$$T_\varphi = \frac{\varphi(e)}{|G|} \sum_{s \in G} \varphi(s^{-1}) \rho_V(s) : V \to V.$$

Then the Generalized Orthogonality Relation (Fact 4) shows that

$$T_\varphi T_\psi = \begin{cases} T_\varphi & \text{if } \varphi = \psi \\ 0 & \text{if } \varphi \neq \psi \end{cases}$$

and

$$\sum_{\varphi \in Irr(G)} T_\varphi = 1_V,$$

where 1_V denotes the identity operator on V. Moreover, $V = \bigoplus_{\varphi \in Irr(G)} T_\varphi(V)$ (internal direct sum).

68.5 Restriction and Induction of Characters

Throughout, G denotes a finite group, e denotes its identity element, and V denotes a finite dimensional complex vector space.

Definitions:

If χ is a character of G and H is a subgroup of G, then the **restriction** of χ to H is the character χ_H of H obtained by restricting the domain of χ.

A character φ of a subgroup H of G is **extendible** to G if $\varphi = \chi_H$ for some character χ of G.

If φ is a character of a subgroup H of G, then the **induced character** from H to G is the character φ^G of G given by the formula

$$\varphi^G(s) = \frac{1}{|H|}\sum_{t\in G}\varphi^\circ(t^{-1}st),$$

where φ° is defined by $\varphi^\circ(x) = \varphi(x)$ if $x \in H$ and $\varphi^\circ(x) = 0$ if $x \notin H$.

If φ is a character of a subgroup H of G and s is an element of G, then the **conjugate character** of φ by s is the character φ^s of $H^s = s^{-1}Hs$ given by $\varphi^s(h^s) = \varphi(h)$ $(h \in H)$, where $h^s = s^{-1}hs$.

Facts:

The following facts can be found in [Isa94, pp. 62–63, 73–79] or [Ser77, pp. 55–58].

1. The restricted character defined above is indeed a character: χ_H is afforded by the restriction to H of a representation affording χ.
2. The induced character defined above is indeed a character: Let V be a $\mathbb{C}H$-module affording φ and put $V^G = \bigoplus_{t\in T} V_t$, where $G = \bigcup_{t\in T} tH$ (disjoint union) and $V_t = V$ for each t. Then V^G is a $\mathbb{C}G$-module that affords φ^G, where the action is given as follows: For $s \in G$ and $v \in V_t$, sv is the element hv of $V_{t'}$, where $st = t'h$ $(t' \in T, h \in H)$.
3. The conjugate character defined above is indeed a character: φ^s is afforded by the representation of H^s obtained by composing the homomorphism $H^s \to H$, $h^s \mapsto h$ with a representation affording φ.
4. *Additivity of restriction*: Let H be a subgroup of G. If χ and χ' are characters of G, then $(\chi+\chi')_H = \chi_H + \chi'_H$.
5. *Additivity of induction*: Let H be a subgroup of G. If φ and φ' are characters of H, then $(\varphi+\varphi')^G = \varphi^G + \varphi'^G$.
6. *Transitivity of induction*: Let H and K be subgroups of G with $H \subseteq K$. If φ is a character of H, then $(\varphi^K)^G = \varphi^G$.
7. *Degree of induced character*: If H is a subgroup of G and φ is a character of H, then the degree of the induced character φ^G equals the product of the index of H in G and the degree of φ: $\varphi^G(e) = [G : H]\varphi(e)$.
8. Let χ be a character of G and let H be a subgroup of G. If the restriction χ_H is irreducible, then so is χ. The converse of this statement does not hold. In fact, if H is the trivial subgroup then $\chi_H = \chi(e)1_H$, so any nonlinear irreducible character (e.g., χ_2 in Example 2 of section 68.4) provides a counterexample.
9. Let H be a subgroup of G and let φ be a character of H. If the induced character φ^G is irreducible, then so is φ. The converse of this statement does not hold (see Example 3).
10. Let H be a subgroup of G. If φ is an irreducible character of H, then there exists an irreducible character χ of G such that φ is a constituent of χ_H.
11. *Frobenius Reciprocity*: If χ is a character of G and φ is a character of a subgroup H of G, then $(\varphi^G, \chi)_G = (\varphi, \chi_H)_H$.
12. If χ is a character of G and φ is a character of a subgroup H of G, then $(\varphi\chi_H)^G = \varphi^G\chi$.
13. *Mackey's Subgroup Theorem*: If H and K are subgroups of G and φ is a character of H, then $(\varphi^G)_K = \sum_{t\in T}(\varphi^t_{H^t\cap K})^K$, where T is a set of representatives for the (H,K)-double cosets in G (so that $G = \dot{\bigcup}_{t\in T} HtK$, a disjoint union).
14. If φ is a character of a normal subgroup N of G, then for each $s \in G$, the conjugate φ^s is a character of N. Moreover, $\varphi^s(n) = \varphi(sns^{-1})$ $(n \in N)$.
15. *Clifford's Theorem*: Let N be a normal subgroup of G, let χ be an irreducible character of G, and let φ be an irreducible constituent of χ_N. Then $\chi_N = m\sum_{i=1}^h \varphi_i$, where $\varphi_1, \ldots, \varphi_h$ are the distinct conjugates of φ under the action of G and $m = (\chi_N, \varphi)_N$.

Examples:

1. Given a subgroup H of G, the induced character $(1_H)^G$ equals the permutation character corresponding to the action of G on the set of left cosets of H in G given by $s(tH) = (st)H$ $(s, t \in G)$.
2. The induced character $(1_{\{e\}})^G$ equals the permutation character corresponding to the action of G on itself given by left multiplication. It is the character of the (left) regular representation of G. This character satisfies

$$(1_{\{e\}})^G(s) = \begin{cases} |G| & \text{if } s = e \\ 0 & \text{if } s \neq e. \end{cases}$$

3. As an illustration of Frobenius Reciprocity (Fact 11), we have $((1_{\{e\}})^G, \chi)_G = (1_{\{e\}}, \chi_{\{e\}})_{\{e\}} = \chi(e)$ for any irreducible character χ of G. Hence, $(1_{\{e\}})^G = \sum_{\chi \in Irr(G)} \chi(e)\chi$ (cf. Fact 8 of section 68.4), that is, in the character of the regular representation (see Example 2), each irreducible character appears as a constituent with multiplicity equal to its degree.

68.6 Representations of the Symmetric Group

Definitions:

Given a natural number n, a tuple $\alpha = [\alpha_1, \ldots, \alpha_h]$ of nonnegative integers is a **(proper) partition** of n (written $\alpha \vdash n$) provided

- $\alpha_i \geq \alpha_{i+1}$ for all $1 \leq i < h$,
- $\sum_{i=1}^{h} \alpha_i = n$.

The **conjugate partition** of a partition $\alpha \vdash n$ is the partition $\alpha' \vdash n$ with ith component α_i' equal to the number of indices j for which $\alpha_j \geq i$. This partition is also called the **partition associated with** α.

Given two partitions $\alpha = [\alpha_1, \ldots, \alpha_h]$ and $\beta = [\beta_1, \ldots, \beta_k]$ of n, α **majorizes** (or **dominates**) β if

$$\sum_{i=1}^{j} \alpha_i \geq \sum_{i=1}^{j} \beta_i$$

for each $1 \leq j \leq h$. This is expressed by writing $\alpha \succeq \beta$ (or $\beta \preceq \alpha$).

The **Young subgroup** of the symmetric group S_n corresponding to a partition $\alpha = [\alpha_1, \ldots, \alpha_h]$ of n is the internal direct product $S_\alpha = S_{A_1} \times \cdots \times S_{A_h}$, where S_{A_i} is the subgroup of S_n consisting of those permutations that fix every integer *not* in the set

$$A_i = \left\{ 1 \leq k \leq n \,\middle|\, \sum_{j=1}^{i-1} \alpha_j < k \leq \sum_{j=1}^{i} \alpha_j \right\}$$

(an empty sum being interpreted as zero).

The **alternating character** of the symmetric group S_n is the character ϵ_n given by

$$\epsilon_n(\sigma) = \begin{cases} 1 & \text{if } \sigma \text{ is even,} \\ -1 & \text{if } \sigma \text{ is odd.} \end{cases}$$

Let G be a subgroup of S_n and let χ be a character of G. The **generalized matrix function** $d_\chi : \mathbb{C}^{n \times n} \to \mathbb{C}$ is defined by

$$d_\chi(A) = \sum_{s \in G} \chi(s) \prod_{j=1}^{n} a_{js(j)}.$$

When $G = S_n$ and χ is irreducible, d_χ is called an **immanant**.

Facts:

The following facts can be found in [JK81, pp. 15, 35–37] or [Mer97, pp. 99–103, 214].

1. If χ is a character of the symmetric group S_n, then $\chi(\sigma)$ is an integer for each $\sigma \in S_n$.
2. *Irreducible character associated with a partition*: Given a partition α of n, there is a unique irreducible character χ_α that is a constituent of both the induced character $(1_{S_\alpha})^{S_n}$ and the induced character $((\epsilon_n)_{S_{\alpha'}})^{S_n}$. The map $\alpha \mapsto \chi_\alpha$ defines a bijection from the set of partitions of n to the set $Irr(S_n)$ of irreducible characters of S_n.
3. If α and β are partitions of n, then the irreducible character χ_α is a constituent of the induced character $(1_{S_\beta})^{S_n}$ if and only if α majorizes β.
4. If α is a partition of n, then $\chi_{\alpha'} = \epsilon_n \chi_\alpha$.
5. *Schur's inequality*: Let χ be an irreducible character of a subgroup G of S_n. For any positive semidefinite matrix $A \in \mathbb{C}^{n\times n}$, $d_\chi(A)/\chi(e) \geq det\ A$.

Examples:

1. $\alpha = [5, 3^2, 2, 1^3]$ (meaning $[5, 3, 3, 2, 1, 1, 1]$) is a partition of 16. Its conjugate is $\alpha' = [7, 4, 3, 1, 1]$.
2. $\chi_{[n]} = 1_{S_n}$ and $\chi_{[1^n]} = \epsilon_n$.
3. In the notation of Example 3 of section 68.4, we have $\chi_0 = \chi_{[4]}$, $\chi_1 = \chi_{[1^4]}$, $\chi_2 = \chi_{[2^2]}$, $\chi_3 = \chi_{[3,1]}$, and $\chi_4 = \chi_{[2,1^2]}$.
4. According to Fact 4, a partition α of n is self-conjugate (meaning $\alpha' = \alpha$) if and only if $\chi_\alpha(\sigma) = 0$ for every odd permutation $\sigma \in S_n$.
5. As an illustration of Fact 3, we have
$$\left(1_{S_{[2,1^2]}}\right)^{S_4} = \chi_{[4]} + \chi_{[3,1]} + \chi_{[3,1]} + \chi_{[2^2]} + \chi_{[2,1^2]}.$$
The irreducible constituents of the induced character $(1_{S_{[2,1^2]}})^{S_4}$ are the terms on the right-hand side of the equation. Note that $[4]$, $[3, 1]$, $[2^2]$, and $[2, 1^2]$ are precisely the partitions of 4 that majorize $[2, 1^2]$ in accordance with the fact.
6. When $G = S_n$ and $\chi = \epsilon_n$ (the alternating character), $d_\chi(A)$ is the determinant of $A \in \mathbb{C}^{n\times n}$.
7. When $G = S_n$ and $\chi = 1_G$ (the principal character), $d_\chi(A) = \sum_{s\in G} \prod_{j=1}^n a_{js(j)}$ is called the **permanent** of $A \in \mathbb{C}^{n\times n}$, denoted $per\ A$.
8. The following open problem is known as the **Permanental Dominance** (or **Permanent-on-Top**) **Conjecture**: Let χ be an irreducible character of a subgroup G of S_n. For any positive semidefinite matrix $A \in \mathbb{C}^{n\times n}$, $per\ A \geq d_\chi(A)/\chi(e)$ (cf. Fact 5 and Examples 6 and 7).

References

[Gor94] D. Gorenstein. *The Classification of the Finite Simple Groups.* American Mathematical Society, Providence, RI 1994.

[Isa94] I.M. Isaacs. *Character Theory of Finite Groups.* Academic Press, New York, 1976. Reprinted, Dover Publications, Inc., Mineola, NY, 1994.

[JK81] G. James and A. Kerber. *The Representation Theory of the Symmetric Group.* Encyclopedia of Mathematics and Its Applications **16**. Addison-Wesley Publishing Company, Reading, MA, 1981.

[Mer97] R. Merris. *Multilinear Algebra.* Gordan and Breach Science Publishers, Amsterdam, 1997.

[Ser77] J.P. Serre. *Linear Representations of Finite Groups.* Springer-Verlag, New York, 1977.

69

Nonassociative Algebras

Murray R. Bremner
University of Saskatchewan

Lúcia I. Murakami
Universidade de São Paulo

Ivan P. Shestakov
Universidade de São Paulo and Sobolev Institute of Mathematics

One of the earliest surveys on nonassociative algebras is the article by Shirshov [Shi58] that introduced the phrase "rings that are nearly associative." The first book in the English language devoted to a systematic study of nonassociative algebras is Schafer [Sch66]. A comprehensive exposition of the work of the Russian school is Zhevlakov, Slinko, Shestakov, and Shirshov [ZSS82]. A collection of open research problems in algebra, including many problems on nonassociative algebra, is the *Dniester Notebook* [FKS93]; the survey article by Kuzmin and Shestakov [KS95] is from the same period. Three books on Jordan algebras that contain substantial material on general nonassociative algebras are Braun and Koecher [BK66], Jacobson [Jac68], and McCrimmon [McC04]. Recent research appears in the *Proceedings of the International Conferences on Nonassociative Algebra and Its Applications* [Gon94], [CGG00], [SSS06]. The present chapter provides very limited information on Lie algebras, since they are the subject of Chapter 70. The last section (Section 69.9) presents three applications of computational linear algebra to the study of polynomial identities for nonassociative algebras: Pseudorandom vectors in a nonassociative algebra, the expansion matrix for a nonassociative operation, and the representation theory of the symmetric group.

69.1 Introduction

Definitions:

An **algebra** is a vector space A over a field F together with a bilinear **multiplication** $(x, y) \mapsto xy$ from $A \times A$ to A; that is, **distributivity** holds for all $a, b \in F$ and all $x, y, z \in A$:

$$(ax + by)z = a(xz) + b(yz), \qquad x(ay + bz) = a(xy) + b(xz).$$

The **dimension** of an algebra A is its dimension as a vector space.

An algebra A is **finite dimensional** if A is a finite dimensional vector space.

The **structure constants** of a finite dimensional algebra A over F with basis $\{x_1, \ldots, x_n\}$ are the scalars $c_{ij}^k \in F$ $(i, j, k = 1, \ldots, n)$ defined by:

$$x_i x_j = \sum_{k=1}^{n} c_{ij}^k x_k.$$

An algebra A is **unital** if there exists an element $1 \in A$ for which

$$1x = x1 = x \quad \text{for all } x \in A.$$

An **involution** of the algebra A is a linear mapping $j \colon A \to A$ satisfying

$$j(j(x)) = x \quad \text{and} \quad j(xy) = j(y)j(x) \qquad \text{for all } x, y \in A.$$

An algebra A is a **division algebra** if for every $x, y \in A$ with $x \neq 0$ the equations $xv = y$ and $wx = y$ are solvable in A.

The **associator** in an algebra is the trilinear function

$$(x, y, z) = (xy)z - x(yz).$$

An algebra A is **associative** if the associator vanishes identically:

$$(x, y, z) = 0 \quad \text{for all } x, y, z \in A.$$

An algebra is **nonassociative** if the above identity is not necessarily satisfied.

An algebra A is **alternative** if it satisfies the **right and left alternative identities**

$$(y, x, x) = 0 \quad \text{and} \quad (x, x, y) = 0 \qquad \text{for all } x, y \in A.$$

An algebra A is **anticommutative** if it satisfies the identity

$$x^2 = 0 \quad \text{for all } x \in A.$$

(This implies that $xy = -yx$, and the converse holds in characteristic $\neq 2$.)

The **Jacobian** in an anticommutative algebra is defined by

$$J(x, y, z) = (xy)z + (yz)x + (zx)y.$$

A **Lie algebra** is an anticommutative algebra satisfying the **Jacobi identity**

$$J(x, y, z) = 0 \quad \text{for all } x, y, z \in A.$$

A **Malcev algebra** is an anticommutative algebra satisfying the identity

$$J(x, y, xz) = J(x, y, z)x \quad \text{for all } x, y, z \in A.$$

The **commutator** in an algebra A is the bilinear function

$$[x, y] = xy - yx.$$

The **minus algebra** A^- of an algebra A is the algebra with the same underlying vector space as A but with $[x, y]$ as the multiplication.

An algebra A is **commutative** if it satisfies the identity

$$xy = yx \quad \text{for all } x, y \in A.$$

A **Jordan algebra** is a commutative algebra satisfying the **Jordan identity**

$$(x^2, y, x) = 0 \quad \text{for all } x, y \in A.$$

The **Jordan product** (or **anticommutator**) in an algebra A is the bilinear function

$$x * y = xy + yx.$$

(The notation $x \circ y$ is also common.)

The **plus algebra** A^+ of an algebra A over a field F of characteristic $\neq 2$ is the algebra with the same underlying vector space as A but with $x \cdot y = \frac{1}{2}(x * y)$ as the multiplication.

A Jordan algebra is called **special** if it is isomorphic to a subalgebra of A^+ for some associative algebra A; otherwise it is called **exceptional.**

Given two algebras A and B over a field F, a **homomorphism** from A to B is a linear mapping $f: A \to B$ that satisfies $f(xy) = f(x)f(y)$ for all $x, y \in A$.

An **isomorphism** is a homomorphism that is a linear isomorphism of vector spaces.

Let A be an algebra. Given two subsets $B, C \subseteq A$ we write BC for the **subspace spanned by the products** yz where $y \in B, z \in C$.

A **subalgebra** of A is a subspace B satisfying $BB \subseteq B$.

The **subalgebra generated by a set** $S \subseteq A$ is the smallest subalgebra of A containing S.

A (two-sided) **ideal** of an algebra A is a subalgebra B satisfying $AB + BA \subseteq B$.

Given two algebras A and B over the field F, the **(external) direct sum** of A and B is the vector space direct sum $A \oplus B$ with the multiplication

$$(w, x)(y, z) = (wy, xz) \quad \text{for all } w, y \in A \quad \text{and all } x, z \in B.$$

Given an algebra A with two ideals B and C, we say that A is the **(internal) direct sum** of B and C if $A = B \oplus C$ (direct sum of subspaces).

Facts: ([Shi58], [Sch66], [ZSS82], [KS95])

1. Every finite dimensional associative algebra over a field F is isomorphic to a subalgebra of a matrix algebra $F^{n\times n}$ for some n.
2. The algebra A^- is always anticommutative. If A is associative, then A^- is a Lie algebra.
3. (Poincaré–Birkhoff–Witt Theorem or PBW Theorem) Every Lie algebra is isomorphic to a subalgebra of A^- for some associative algebra A.
4. The algebra A^+ is always commutative. If A is associative, then A^+ is a Jordan algebra. (See Example 2 in Section 69.9.) If A is alternative, then A^+ is a Jordan algebra.
5. The analogue of the PBW theorem for Jordan algebras is false: Not every Jordan algebra is special. (See Example 4 below.)
6. Every associative algebra is alternative.
7. (Artin's Theorem) An algebra is alternative if and only if every subalgebra generated by two elements is associative.
8. Every Lie algebra is a Malcev algebra.
9. Every Malcev algebra generated by two elements is a Lie algebra.
10. If A is an alternative algebra, then A^- is a Malcev algebra. (See Example 3 in Section 69.9.)
11. In an external direct sum of algebras, the summands are ideals.

Examples:

1. Associativity is satisfied when the elements of the algebra are mappings of a set into itself with the composition of mappings taken as multiplication. Such is the multiplication in the algebra End V, the algebra of linear operators on the vector space V. Every associative algebra is isomorphic to a subalgebra of the algebra End V, for some V. Thus, the condition of associativity of multiplication

characterizes the **algebras of linear operators.** (Note that End V is also denoted $L(V, V)$ elsewhere in this book, but End V is the standard notation in the study of algebras.)

2. **Cayley–Dickson doubling process.** Let A be a unital algebra over F with an involution $x \mapsto \overline{x}$ satisfying

$$x + \overline{x},\ x\overline{x} \in F \quad \text{for all } x \in A. \tag{69.1}$$

Let $a \in F, a \neq 0$. The algebra (A, a) is defined as follows: The underlying vector space is $A \oplus A$, addition and scalar multiplication are defined by the vector space formulas

$$(x_1, x_2) + (y_1, y_2) = (x_1 + y_1, x_2 + y_2), \quad c(x_1, x_2) = (cx_1, cx_2) \qquad \text{for all } c \in F, \tag{69.2}$$

and multiplication is defined by the formula

$$(x_1, x_2)(y_1, y_2) = (\,x_1 y_1 + a y_2 \overline{x_2},\ \overline{x_1} y_2 + y_1 x_2\,). \tag{69.3}$$

This algebra has an involution defined by

$$\overline{(x_1, x_2)} = (\overline{x_1}, -x_2). \tag{69.4}$$

In particular, starting with a field F of characteristic $\neq 2$, we obtain the following examples:

(a) The algebra $\mathbb{C}(a) = (F, a)$ is commutative and associative. If the polynomial $x^2 + a$ is irreducible over F, then $\mathbb{C}(a)$ is a field, otherwise $\mathbb{C}(a) \cong F \oplus F$ (algebra direct sum).

(b) The algebra $\mathbb{H}(a, b) = (\mathbb{C}(a), b)$ is an algebra of **generalized quaternions**, which is associative but not commutative.

(c) The algebra $\mathbb{O}(a, b, c) = (\mathbb{H}(a, b), c)$ is an algebra of **generalized octonions** or a **Cayley–Dickson algebra**, which is alternative but not associative. (See Example 1 in Section 69.9.)

The algebras of generalized quaternions and octonions may also be defined over a field of characteristic 2 (see [Sch66], [ZSS82]).

3. **Real division algebras** [EHH91, Part B]. In the previous example, taking F to be the field $\mathbb{R}$ of real numbers and $a = b = c = -1$, we obtain the field $\mathbb{C}$ of **complex numbers**, the associative division algebra $\mathbb{H}$ of **quaternions**, and the alternative division algebra $\mathbb{O}$ of **octonions** (also known as the **Cayley numbers**). Real division algebras exist only in dimensions 1, 2, 4, and 8, but there are many other examples: The algebras $\mathbb{C}$, $\mathbb{H}$, and $\mathbb{O}$ with the multiplication $x \cdot y = \overline{x}\,\overline{y}$ are still division algebras, but they are not alternative and they are not unital.
4. **The Albert algebra.** Let $\mathbb{O}$ be the octonions and let $M_3(\mathbb{O})$ be the algebra of 3×3 matrices over $\mathbb{O}$ with involution induced by the involution of $\mathbb{O}$, that is $(a_{ij}) \mapsto (\overline{a_{ji}})$. The subalgebra $H_3(\mathbb{O})$ of Hermitian matrices in $M_3(\mathbb{O})^+$ is an exceptional Jordan algebra, the Albert algebra: There is no associative algebra A such that $H_3(\mathbb{O})$ is isomorphic to a subalgebra of A^+.

69.2 General Properties

Definitions:

Given an algebra A and an ideal I, the **quotient algebra** A/I is the quotient space A/I with multiplication defined by $(x + I)(y + I) = xy + I$ for all $x, y \in A$.

The algebra A is **simple** if $AA \neq \{0\}$ and A has no ideals apart from $\{0\}$ and A.

The algebra A is **semisimple** if it is the direct sum of simple algebras. (The definition of semisimple that is used in the theory of Lie algebras is different; see Chapter 70.)

Set $A^1 = A^{(1)} = A$, and then by induction define

$$A^{n+1} = \sum_{i+j=n+1} A^i A^j \quad \text{and} \quad A^{(n+1)} = A^{(n)} A^{(n)} \qquad \text{for } n \geq 1.$$

The algebra A is **nilpotent** if $A^n = \{0\}$ for some n and **solvable** if $A^{(s)} = \{0\}$ for some s. The smallest natural number n (respectively s) with this property is the **nilpotency index** (respectively, **solvability index**) of A.

An element $x \in A$ is **nilpotent** if the subalgebra it generates is nilpotent.

A **nil algebra** (respectively **nil ideal**) is an algebra (respectively ideal) in which every element is nilpotent.

An algebra is **power associative** if every element generates an associative subalgebra.

An **idempotent** is an element $e \neq 0$ of an algebra A satisfying $e^2 = e$. Two idempotents e, f are **orthogonal** if $ef = fe = 0$.

For an algebra A over a field F, the **degree** of A is defined to be the maximal number of mutually orthogonal idempotents in the scalar extension $\overline{F} \otimes_F A$, where $\overline{F}$ is the algebraic closure of F.

The **associator ideal** $D(A)$ of the algebra A is the ideal generated by all the associators. The **associative center** or **nucleus** $N(A)$ of A is defined by

$$N(A) = \{\, x \in A \mid (x, A, A) = (A, x, A) = (A, A, x) = \{0\}\,\}.$$

The **center** $Z(A)$ of the algebra A is defined by

$$Z(A) = \{\, x \in N(A) \mid [x, A] = \{0\}\,\}.$$

The **right and left multiplication operators** by an element $x \in A$ are defined by

$$R_x\colon y \mapsto yx, \qquad L_x\colon y \mapsto xy.$$

The **multiplication algebra** of the algebra A is the subalgebra $M(A)$ of the associative algebra End A (of endomorphisms of the vector space A) generated by all R_x and L_x for $x \in A$.

The **right multiplication algebra** of the algebra A is the subalgebra $R(A)$ of End A generated by all R_x for $x \in A$.

The **centroid** $C(A)$ of the algebra A is the centralizer of the multiplication algebra $M(A)$ in the algebra End A; that is,

$$C(A) = \{\, T \in \text{End}\, A \mid TR_x = R_x T = R_{Tx}, \quad TL_x = L_x T = L_{Tx}, \qquad \text{for any } x \in A\,\}.$$

An algebra A over a field F is **central** if $C(A) = F$.

The **unital hull** $A^\sharp$ of an algebra A over a field F is defined as follows: If A is unital, then $A^\sharp = A$; and when A has no unit, we set $A^\sharp = A \oplus F$ (vector space direct sum) and define multiplication by assuming that A is a subalgebra and the unit of F is the unit of $A^\sharp$.

Let $\mathcal{M}$ be a class of algebras closed under homomorphic images. A subclass $\mathcal{R}$ of $\mathcal{M}$ is said to be **radical** if

1. $\mathcal{R}$ is closed under homomorphic images.
2. For each $A \in \mathcal{M}$ there is an ideal $\mathcal{R}(A)$ of A such that $\mathcal{R}(A) \in \mathcal{R}$ and $\mathcal{R}(A)$ contains every ideal of A contained in $\mathcal{R}$.
3. $\mathcal{R}(A/\mathcal{R}(A)) = \{0\}$.

In this case, we call the ideal $\mathcal{R}(A)$ the $\mathcal{R}$-**radical** of A. The algebra A is said to be $\mathcal{R}$-**semisimple** if $\mathcal{R}(A) = \{0\}$.

If the subclass Nil of nil-algebras is radical in the class $\mathcal{M}$, then the corresponding ideal Nil A, for $A \in \mathcal{M}$, is called the **nil radical** of A. In this case, the algebra A is called **nil-semisimple** if Nil $A = \{0\}$. By definition, Nil A contains all two-sided nil-ideals of A, and the quotient algebra $A/\text{Nil}\, A$ is nil-semisimple, that is, $\text{Nil}\,(A/\text{Nil}\, A) = \{0\}$.

If the subclass *Nilp* of nilpotent algebras (or the subclass *Solv* of solvable algebras) is radical in the class $\mathcal{M}$, then the corresponding ideal Nilp A (respectively Solv A), for $A \in \mathcal{M}$, is called the **nilpotent radical** (respectively the **solvable radical**) of A.

For an algebra A over F, an A-**bimodule** is a vector space M over F with bilinear mappings

$$A \times M \to M,\ (x,m) \mapsto xm \quad \text{and} \quad M \times A \to M,\ (m,x) \mapsto mx.$$

The **split null extension** $E(A, M)$ of A by M is the algebra over F with underlying vector space $A \oplus M$ and multiplication

$$(x+m)(y+n) = xy + (xn + my) \quad \text{for all } x, y \in A,\ m, n \in M.$$

For an algebra A, the **regular bimodule** Reg(A) is the underlying vector space of A considered as an A-bimodule, interpreting mx and xm as multiplication in A.

If M is an A-bimodule, then the mappings

$$\rho(x)\colon m \mapsto mx, \qquad \lambda(x)\colon m \mapsto xm,$$

are linear operators on M, and the mappings

$$x \mapsto \rho(x), \qquad x \mapsto \lambda(x),$$

are linear mappings from A to the algebra $\text{End}_F\, M$. The pair (λ, ρ) is called the **birepresentation** of A associated with the bimodule M.

The notions of **sub-bimodule, homomorphism of bimodules, irreducible bimodule**, and **faithful birepresentation** are defined in the natural way. The sub-bimodules of a regular A-bimodule are exactly the two-sided ideals of A.

Facts: ([Sch66], [Jac68], [ZSS82])

1. If A is a simple algebra, then $AA = A$.
2. (Isomorphism theorems)
   (a) If $f\colon A \to B$ is a homomorphism of algebras over the field F, then $A/\ker(F) \cong \text{im}(f) \subseteq B$.
   (b) If B_1 and B_2 are ideals of the algebra A with $B_2 \subseteq B_1$, then $(A/B_2)/(B_1/B_2) \cong A/B_1$.
   (c) If S is a subalgebra of A and B is an ideal of A, then $B \cap S$ is an ideal of S and $(B + S)/B \cong S/(B \cap S)$.
3. The algebra A is nilpotent of index n if and only if any product of n elements (with any arrangement of parentheses) equals zero, and if there exists a nonzero product of $n - 1$ elements.
4. Every nilpotent algebra is solvable; the converse is not generally true. (See Example 1 below.)
5. In any algebra A, the sum of two solvable ideals is again a solvable ideal. If A is finite-dimensional, then A contains a unique maximal solvable ideal Solv A, and the quotient algebra A/Solv A does not contain nonzero solvable ideals. In other words, the subclass *Solv* of solvable algebras is radical in the class of all finite dimensional algebras.
6. An algebra A is associative if and only if $D(A) = \{0\}$, if and only if $N(A) = A$.
7. Every solvable associative algebra is nilpotent.
8. The subclass *Nilp* of nilpotent algebras is radical in the class of all finite dimensional associative algebras.
9. A finite dimensional associative algebra A is semisimple if and only if Nilp $A = \{0\}$.
10. The previous two facts imply that every finite dimensional associative algebra A contains a unique maximal nilpotent ideal N such that the quotient algebra A/N is isomorphic to a direct sum of simple algebras.

11. Over an algebraically closed field F, every finite dimensional simple associative algebra is isomorphic to the algebra $F^{n\times n}$ of $n \times n$ matrices over F, for some $n \geq 1$.
12. The subclass $Nilp$ is not radical in the class of finite dimensional Lie algebras. (See Example 1 below.)
13. Over a field of characteristic zero, an algebra is power associative if and only if

$$x^2x = xx^2 \quad \text{and} \quad (x^2x)x = x^2x^2 \qquad \text{for all } x.$$

14. Every power associative algebra A contains a unique maximal nil ideal Nil A, and the quotient algebra $A/\text{Nil } A$ is nil-semisimple, that is, it does not contain nonzero nil ideals. In other words, the subclass Nil of nil-algebras is radical in the class of all power associative algebras.
15. For a finite dimensional alternative or Jordan algebra A we have Nil $A =$ Solv A.
16. For finite dimensional commutative power associative algebras, the question of the equality of the nil and solvable radicals is still open, and is known as *Albert's problem.* An equivalent question is: Are there any simple finite dimensional commutative power associative nil algebras?
17. Every nil-semisimple finite dimensional commutative power associative algebra over a field of characteristic $\neq 2, 3, 5$ has a unit element and decomposes into a direct sum of simple algebras. Every such simple algebra is either a Jordan algebra or a certain algebra of degree 2 over a field of positive characteristic.
18. Direct expansion shows that these two identities are valid in every algebra:

$$x(y,z,w) + (x,y,z)w = (xy,z,w) - (x,yz,w) + (x,y,zw),$$
$$[xy,z] - x[y,z] - [x,z]y = (x,y,z) - (x,z,y) + (z,x,y).$$

From these it follows that the associative center and the center are subalgebras, and

$$D(A) = (A,A,A) + (A,A,A)A = (A,A,A) + A(A,A,A).$$

19. If $z \in Z(A)$, then for any $x \in A$ we have

$$R_zR_x = R_xR_z = R_{zx} = R_{xz}.$$

20. If A is unital, then its centroid $C(A)$ is isomorphic to its center $Z(A)$. If A is simple, then $C(A)$ is a field which contains the base field F.
21. Let A be a finite dimensional algebra with multiplication algebra $M(A)$. Then
    (a) A is nilpotent if and only if $M(A)$ is nilpotent.
    (b) If A is semisimple, then so is $M(A)$.
    (c) If A is simple, then so is $M(A)$, and $M(A) \cong \text{End}_{C(A)} A$.
22. An algebra A is simple if and only if the bimodule Reg(A) is irreducible.
23. If A is an alternative algebra (respectively a Jordan algebra), then its unital hull $A^\sharp$ is also alternative (respectively Jordan).

Examples:

1. Let A be algebra with basis x, y, and multiplication given by $x^2 = y^2 = 0$ and $xy = -yx = y$. Then A is a Lie algebra and $A^{(2)} = \{0\}$ but $A^n \neq \{0\}$ for any $n \geq 1$. Thus, A is solvable but not nilpotent.
2. Let A be an algebra over a field F with basis x_1, x_2, y, z and the following nonzero products of basis elements:

$$yx_1 = ax_1y = x_2, \quad zx_2 = ax_2z = x_1,$$

where $0 \neq a \in F$. Then $I_1 = Fx_1 + Fx_2 + Fy$ and $I_2 = Fx_1 + Fx_2 + Fz$ are different maximal nilpotent ideals in A. By choosing $a = 1$ or $a = -1$ we obtain a commutative or anticommutative algebra A.

3. In general, in a nonassociative algebra, a power of an element is not uniquely determined. In the previous example, for the element $w = x_1 + x_2 + y + z$ we have

$$w^2w^2 = 0 \quad \text{but} \quad w(ww^2) = (1+a)(x_1 + x_2).$$

4. Let $A_1, \ldots, A_n$ be simple algebras over a field F with bases

$$\{v_i^1 \mid i \in I_1\}, \ \ldots, \ \{v_i^n \mid i \in I_n\}.$$

Consider the algebra $A = Fe \oplus A_1 \oplus \cdots \oplus A_n$ (vector space direct sum) with multiplication defined by the following conditions:

(a) The A_i are subalgebras of A.

(b) $A_iA_j = \{0\}$ for $i \neq j$.

(c) $ev_i^j = v_i^j e = e$ for all i, j.

(d) $e^2 = e$.

Then $I = Fe$ is the unique minimal ideal in A, and $I^2 = I$. In particular, Solv $A = \{0\}$, but A is not semisimple (compare with the Lie algebra case).

5. **Suttles' example.** (Notices AMS 19 (1972) A-566) Let A be a commutative algebra over a field F of characteristic $\neq 2$, with basis x_i $(1 \leq i \leq 5)$ and the following multiplication table (all other products are zero):

$$x_1x_2 = x_2x_4 = -x_1x_5 = x_3, \quad x_1x_3 = x_4, \quad x_2x_3 = x_5.$$

Then A is a solvable power associative nil algebra that is not nilpotent. Moreover, Nil A = Solv A = A, and Nilp A does not exist (if F is infinite then A has infinitely many maximal nilpotent ideals).

69.3 Composition Algebras

Definitions:

A **composition algebra** is an algebra A with unit 1 over a field F of characteristic $\neq 2$ together with a **norm** $n(x)$ (a nondegenerate quadratic form on the vector space A) that **admits composition** in the sense that

$$n(xy) = n(x)n(y) \quad \text{for all } x \in A.$$

A **quadratic algebra** A over a field F is a unital algebra in which every $x \in A$ satisfies the condition $x^2 \in \text{Span}(x, 1)$. In other words, every subalgebra of A generated by a single element has dimension ≤ 2.

A composition algebra A is **split** if it contains zero-divisors, that is, if $xy = 0$ for some nonzero $x, y \in A$.

Facts: ([Sch66], [Jac68], [ZSS82], [Bae02])

1. Every composition algebra A is alternative and quadratic. Moreover, every element $x \in A$ satisfies the equation

$$x^2 - t(x)x + n(x) = 0,$$

where $t(x)$ is a linear form on A (the **trace**) and $n(x)$ is the original quadratic form on A (the norm).

2. For a composition algebra A the following conditions are equivalent:
   (a) A is split;
   (b) $n(x) = 0$ for some nonzero $x \in A$;
   (c) A contains an idempotent $e \neq 1$.
3. Let A be a unital algebra over a field F with an involution $x \mapsto \overline{x}$ satisfying Equation 69.1 from Example 2 of Section 69.1. The Cayley–Dickson doubling process gives the algebra (A, a) defined by Equations 69.2 to 69.4. It is clear that A is isomorphically embedded into (A, a) and that $\dim(A, a) = 2 \dim A$. For $v = (0, 1)$, we have $v^2 = a$ and $(A, a) = A \oplus Av$. For any $y = y_1 + y_2 v \in (A, a)$, we have $\overline{y} = \overline{y_1} - y_2 v$.
4. In a composition algebra A, the mapping $x \mapsto \overline{x} = t(x) - x$ is an involution of A fixing the elements of the field $F = F\,1$. Conversely, if A is an alternative algebra with unit 1 and involution $x \mapsto \overline{x}$ satisfying Equation 69.1 from Example 2 of Section 69.1, then $x\overline{x} \in F$ and the quadratic form $n(x) = x\overline{x}$ satisfies $n(xy) = n(x)n(y)$.
5. The mapping $y \mapsto \overline{y}$ is an involution of (A, a) extending the involution $x \mapsto \overline{x}$ of A. Moreover, $y + \overline{y}$ and $y\overline{y}$ are in F for every $y \in (A, a)$. If the quadratic form $n(x) = x\overline{x}$ is nondegenerate on A, then the quadratic form $n(y) = y\overline{y}$ is nondegenerate on (A, a), and the form $n(y)$ admits composition on (A, a) if and only if A is associative.
6. Every composition algebra over a field F of characteristic $\neq 2$ is isomorphic to F or to one of the algebras of types 2a–2c obtained from F by the Cayley–Dickson process as in Example 2 of Section 69.1.
7. Every split composition algebra over a field F is isomorphic to one of the algebras $F \oplus F$, $M_2(F)$, Zorn(F) described in Examples 2 to 4 below.
8. Every finite dimensional composition algebra without zero divisors is a division algebra, and so every composition algebra is either split or a division algebra.
9. Every composition algebra of dimension > 1 over an algebraically closed field is split, and so every composition algebra over an algebraically closed field F is isomorphic to one of the algebras F, $F \oplus F$, $M_2(F)$, Zorn(F).

Examples:

1. The fields of real numbers $\mathbb{R}$ and complex numbers $\mathbb{C}$, the quaternions $\mathbb{H}$, and the octonions $\mathbb{O}$, are real composition algebras with the Euclidean norm $n(x) = x\overline{x}$. The first three are associative; the algebra $\mathbb{O}$ provides us with the first and most important example of a nonassociative alternative algebra.
2. Let F be a field and let $A = F \oplus F$ be the direct sum of two copies of the field with the exchange involution $\overline{(a, b)} = (b, a)$, the trace $t((a, b)) = a + b$, and the norm $n((a, b)) = ab$. Then A is a two-dimensional split composition algebra.
3. Let $A = M_2(F)$ be the algebra of 2×2 matrices over F with the symplectic involution

$$x = \begin{bmatrix} a & b \\ c & d \end{bmatrix} \longmapsto \overline{x} = \begin{bmatrix} d & -b \\ -c & a \end{bmatrix},$$

   the matrix trace $t(x) = a + d$, and the determinant norm $n(x) = ad - bc$. Then A is a 4-dimensional split composition algebra.
4. An eight-dimensional split composition algebra is the **Zorn vector-matrix algebra** (or the **Cayley–Dickson matrix algebra**), obtained by taking $A = \text{Zorn}(F)$, which consists of all 2×2 block matrices with scalars on the diagonal and 3×1 column vectors off the diagonal

$$\text{Zorn}(F) = \left\{ x = \begin{bmatrix} a & \mathbf{u} \\ \mathbf{v} & b \end{bmatrix} \;\middle|\; a, b \in F,\ \mathbf{u}, \mathbf{v} \in F^3 \right\},$$

with norm, involution, and product

$$n(x) = ab - (\mathbf{u}, \mathbf{v}), \qquad \bar{x} = \begin{bmatrix} b & -\mathbf{u} \\ -\mathbf{v} & a \end{bmatrix},$$

$$x_1 x_2 = \begin{bmatrix} a_1 a_2 + (\mathbf{u}_1, \mathbf{v}_2) & a_1 \mathbf{u}_2 + \mathbf{u}_1 b_2 - \mathbf{v}_1 \times \mathbf{v}_2 \\ \mathbf{v}_1 a_2 + b_1 \mathbf{v}_2 + \mathbf{u}_1 \times \mathbf{u}_2 & b_1 b_2 + (\mathbf{v}_1, \mathbf{u}_2) \end{bmatrix};$$

the scalar and vector products are defined for $\mathbf{u} = [u_1, u_2, u_3]^T$ and $\mathbf{v} = [v_1, v_2, v_3]^T$ by

$$(\mathbf{u}, \mathbf{v}) = u_1 v_1 + u_2 v_2 + u_3 v_3, \qquad \mathbf{u} \times \mathbf{v} = [u_2 v_3 - u_3 v_2, u_3 v_1 - u_1 v_3, u_1 v_2 - u_2 v_1].$$

Applications:

1. If we write the equation $n(x)n(y) = n(xy)$ in terms of the coefficients of the algebra elements x, y with respect to an orthogonal basis for each of the composition algebras $\mathbb{R}, \mathbb{C}, \mathbb{H}, \mathbb{O}$ given in Example 3 of Section 69.1, then we obtain an identity expressing the multiplicativity of a quadratic form:

$$\left(x_1^2 + \cdots + x_k^2\right)\left(y_1^2 + \cdots + y_k^2\right) = z_1^2 + \cdots + z_k^2.$$

Here the z_i are bilinear functions in the x_i and y_i: To be precise, z_i is the coefficient of the ith basis vector in the product of the elements $x = (x_1, \ldots, x_k)$ and $y = (y_1, \ldots, y_k)$. By Hurwitz' theorem, such a *k-square identity* exists only for $k = 1, 2, 4, 8$.

69.4 Alternative Algebras

Definitions:

A **left alternative algebra** is one satisfying the identity $(x, x, y) = 0$.

A **right alternative algebra** is one satisfying the identity $(y, x, x) = 0$.

A **flexible algebra** is one satisfying the identity $(x, y, x) = 0$.

An **alternative algebra** is one satisfying all three identities (any two imply the third).

The **Moufang identities** play an important role in the theory of alternative algebras:

$$(xy \cdot z)y = x(yzy) \qquad \textbf{right Moufang identity}$$

$$(yzy)x = y(z \cdot yx) \qquad \textbf{left Moufang identity}$$

$$(xy)(zx) = x(yz)x \qquad \textbf{central Moufang identity}.$$

(The terms yzy and $x(yz)x$ are well-defined by the flexible identity.)

An **alternative bimodule** over an alternative algebra A is an A-bimodule M for which the split null extension $E(A, M)$ is alternative.

Let A be an alternative algebra, let M be an alternative A-bimodule, and let (λ, ρ) be the associated birepresentation of A. The algebra A **acts nilpotently** on M if the subalgebra of End M, which is generated by the elements $\lambda(x)$, $\rho(x)$ for all $x \in A$, is nilpotent. If $y \in A$, then y **acts nilpotently** on M if the elements $\lambda(y)$, $\rho(y)$ generate a nilpotent subalgebra of End M.

A finite dimensional alternative algebra A over a field F is **separable** if the algebra $A_K = K \otimes_F A$ is nil-semisimple for any extension K of the field F.

Facts: ([Sch66], [ZSS82]) Additional facts about alternative algebras are given in Section 69.1 and facts about right alternative algebras are given in Section 69.6:

1. Every commutative or anticommutative algebra is flexible.
2. Substituting $x + z$ for x in the left alternative identity, and using distributivity, we obtain

$$(x, z, y) + (z, x, y) = 0.$$

   This is the **linearization on** x of the left alternative identity. Linearizing the right alternative identity in the same way, we get

$$(y, x, z) + (y, z, x) = 0.$$

   From the last two identities, it follows that in any alternative algebra A the associator (x, y, z) is a skew-symmetric (alternating) function of the arguments x, y, z.
3. Every alternative algebra is power associative (Corollary of Artin's Theorem, Fact 7, Section 69.1). In particular, the nil radical Nil A exists in the class of alternative algebras.
4. Every alternative algebra satisfies the three Moufang identities and the identities

$$(x, y, yz) = (x, y, z)y, \qquad (x, y, zy) = y(x, y, z).$$

5. A bimodule M over an alternative algebra A is alternative if and only if the following relations hold in the split null extension $E(A, M)$:

$$(x, m, x) = 0 \quad \text{and} \quad (x, m, y) = (m, y, x) = (y, x, m) \quad \text{for all } x, y \in A \quad \text{and} \quad m \in M.$$

6. It follows from the definition of alternative bimodule and the Moufang identities that

$$[\rho(x), \lambda(x)] = 0, \qquad \rho(x^k) = (\rho(x))^k \quad \text{for } k \geq 1, \qquad \lambda(x^k) = (\lambda(x))^k \quad \text{for } k \geq 1.$$

   This implies that any nilpotent element of an alternative algebra acts nilpotently on any bimodule.
7. If every element of an alternative algebra A acts nilpotently on a finite dimensional alternative A-bimodule M, then A acts nilpotently on M.
8. A nilpotent algebra A acts nilpotently on the A-bimodule M if and only if the algebra $E(A, M)$ is nilpotent.
9. In a finite dimensional alternative algebra A, every nil subalgebra is nilpotent. In particular, the nil radical Nil A is nilpotent.
10. The subclass *Nilp* of nilpotent algebras is radical in the class of all finite dimensional alternative algebras. For any finite dimensional alternative algebra A, we have

$$\text{Nil } A = \text{Solv } A = \text{Nilp } A.$$

11. Let A be a finite dimensional alternative algebra. The quotient algebra $A/\text{Nil } A$ is semisimple, that is, it decomposes into a direct sum of simple algebras. Every finite dimensional nil-semisimple alternative algebra is isomorphic to a direct sum of simple algebras, where every simple algebra is either a matrix algebra over a skew-field or a Cayley–Dickson algebra over its center.
12. Let A be a finite dimensional alternative algebra over a field F. If the quotient algebra $A/\text{Nil } A$ is separable over F, then there exists a subalgebra B of A such that B is isomorphic to $A/\text{Nil } A$ and $A = B \oplus \text{Nil } A$ (vector space direct sum).
13. Every alternative bimodule over a separable alternative algebra is completely reducible (as in the case of associative algebras).
14. Let A be a finite dimensional alternative algebra, let M be a faithful irreducible A-bimodule, and let (λ, ρ) be the associated birepresentation of A. Either M is an associative bimodule over A (which must then be associative), or one of the following holds:

(a) The algebra A is an algebra of generalized quaternions, λ is a (right) associative irreducible representation of A, and $\rho(x) = \lambda(\overline{x})$ for every $x \in A$.

(b) The algebra $A = \mathbb{O}$ is a Cayley–Dickson algebra and M is isomorphic to Reg($\mathbb{O}$).

15. Every simple alternative algebra (of any dimension) is either associative or is isomorphic to a Cayley–Dickson algebra over its center.

69.5 Jordan Algebras

In this section, we assume that the base field F has characteristic $\neq 2$.

Definitions:

A **Jordan algebra** is a commutative nonassociative algebra satisfying the **Jordan identity**

$$(x^2 y)x = x^2(yx).$$

The **linearization on** x of the Jordan identity is

$$2((xz)y)x + (x^2 y)z = 2(xz)(xy) + x^2(yz).$$

A Jordan algebra J is **special** if it is isomorphic to a subalgebra of the algebra A^+ for some associative algebra A; otherwise, it is **exceptional.**

Facts: ([BK66], [Jac68], [ZSS82], [McC04]) Additional facts about Jordan algebras are given in Section 69.1, and facts about noncommutative Jordan algebras are given in Section 69.6:

1. (Zelmanov's Simple Theorem) Every simple Jordan algebra (of any dimension) is isomorphic to one of the following: (a) an algebra of a bilinear form, (b) an algebra of Hermitian type, (c) an Albert algebra. For definitions see Examples 3, 4, and 5 below.
2. Let J be a Jordan algebra. Consider the regular birepresentation $x \mapsto L_x, x \mapsto R_x$ of the algebra J. Commutativity and the Jordan identity imply that for all $x, y \in J$ we have

$$L_x = R_x, \qquad [R_x, R_{x^2}] = 0, \qquad R_{x^2 y} - R_y R_{x^2} + 2R_x R_y R_x - 2R_x R_{yx} = 0.$$

   Linearizing the last equation on x we see that for all $x, y, z \in J$ we have

$$R_{(xz)y} - R_y R_{xz} + R_x R_y R_z + R_z R_y R_x - R_x R_{yz} - R_z R_{yx} = 0.$$

3. For every $k \geq 1$, the operator R_{x^k} belongs to the subalgebra $A \subseteq \text{End}\, J$ generated by R_x and R_{x^2}. Since A is commutative, we have $[R_{x^k}, R_{x^\ell}] = 0$ for all $k, \ell \geq 1$, which can be written as $(x^k, J, x^\ell) = \{0\}$.
4. It follows from the previous fact that every Jordan algebra is power associative and the radical Nil J is defined.
5. Let J be a finite dimensional Jordan algebra. As for alternative algebras, we have

$$\text{Nil}\, J = \text{Solv}\, J = \text{Nilp}\, J,$$

   that is, the radical Nil J is nilpotent. The quotient algebra $J/\text{Nil}\, J$ is semisimple, that is, isomorphic to a direct sum of simple algebras. If the quotient algebra $J/\text{Nil}\, J$ is separable over F, then there exists a subalgebra B of J such that B is isomorphic to $J/\text{Nil}\, J$ and $J = B \oplus \text{Nil}\, J$ (vector space direct sum).

6. If a Jordan algebra J contains an idempotent e, the operator R_e satisfies the equation $R_e(2R_e - 1)(R_e - 1) = 0$, and the algebra J has the following analogue of the Pierce decomposition from the theory of associative algebras:

$$J = J_1 \oplus J_{1/2} \oplus J_0, \quad \text{where} \quad J_i = J_i(e) = \{x \in J \mid xe = ix\}.$$

For $i, j = 0, 1$ $(i \neq j)$, we have the inclusions

$$J_i^2 \subseteq J_i, \quad J_i J_{1/2} \subseteq J_{1/2}, \quad J_i J_j = \{0\}, \quad J_{1/2}^2 \subseteq J_1 + J_2.$$

More generally, if J has unit $1 = \sum_{i=1}^n e_i$ where e_i are orthogonal idempotents, then

$$J = \bigoplus_{i \leq j} J_{ij}, \quad \text{where} \quad J_{ii} = J_1(e_i), \qquad J_{ij} = J_{1/2}(e_i) \cap J_{1/2}(e_j) \qquad \text{for } i \neq j,$$

and the components J_{ij} are multiplied according to the rules

$$J_{ii}^2 \subseteq J_{ii}, \quad J_{ij}J_{ii} \subseteq J_{ij}, \quad J_{ij}^2 \subseteq J_{ii} + J_{jj}, \quad J_{ii}J_{jj} = \{0\} \quad \text{for distinct } i, j;$$
$$J_{ij}J_{jk} \subseteq J_{ik}, \quad J_{ij}J_{kk} = \{0\}, \quad J_{ij}J_{k\ell} = \{0\} \quad \text{for distinct } i, j, k, \ell.$$

7. Every Jordan algebra that contains >3 strongly connected orthogonal idempotents is special. (Orthogonal idempotents e_1, e_2 are **strongly connected** if there exists an element $u_{12} \in J_{12}$ for which $u_{12}^2 = e_1 + e_2$.)
8. (Coordinatization Theorem) Let J be a Jordan algebra with unit $1 = \sum_{i=1}^n e_i$ $(n \geq 3)$, where the e_i are mutually strongly connected orthogonal idempotents. Then J is isomorphic to the Jordan algebra $H_n(D)$ of Hermitian $n \times n$ matrices over an alternative algebra D (which is associative for $n > 3$) with involution $*$ such that $H(D, *) \subseteq N(D)$, where $N(D)$ is the associative center of D.
9. Every Jordan bimodule over a separable Jordan algebra is completely reducible, and the structure of irreducible bimodules is known.

Examples:

1. **The algebra A^+.** If A is an associative algebra, then the algebra A^+ is a Jordan algebra. Every subspace J of A closed with respect to the operation $x \cdot y = \frac{1}{2}(xy + yx)$ is a subalgebra of the algebra A^+ and every special Jordan algebra J is (up to isomorphism) of this type. The subalgebra of A generated by J is called the **associative enveloping algebra** of J. Properties of the algebras A and A^+ are closely related: A is simple (respectively nilpotent) if and only if A^+ is simple (respectively nilpotent).
2. The algebra A^+ may be a Jordan algebra for nonassociative A; for instance, if A is a right alternative (in particular, alternative) algebra, then A^+ is a special Jordan algebra.
3. **The algebra of a bilinear form.** Let X be a vector space of dimension > 1 over F, with a symmetric nondegenerate bilinear form $f(x, y)$. Consider the vector space direct sum $J(X, f) = F \oplus X$, and define on it a multiplication by assuming that the unit element $1 \in F$ is the unit element of $J(X, f)$ and by setting $xy = f(x, y)1$ for any $x, y \in X$. Then $J(X, f)$ is a simple special Jordan algebra; its associative enveloping algebra is the Clifford algebra $C(X, f)$ of the bilinear form f. When $F = \mathbb{R}$ and $f(x, y)$ is the ordinary dot product on X, the algebra $J(X, f)$ is called a **spin-factor**.
4. **Algebras of Hermitian type.** Let A be an associative algebra with involution $*$. The subspace $H(A, *) = \{x \in A \mid x^* = x\}$ of $*$-symmetric elements is closed with respect to the Jordan multiplication $x \cdot y$ and, therefore, is a special Jordan algebra. For example, let D be an associative composition algebra with involution $x \mapsto \overline{x}$ and let $D^{n\times n}$ be the algebra of $n \times n$ matrices over D. Then the mapping $S\colon (x_{ij}) \mapsto (\overline{x_{ji}})$ is an involution of $D^{n\times n}$ and the set of D-Hermitian matrices $H_n(D) = H(D^{n\times n}, S)$ is a special Jordan algebra. If A is $*$-simple (if it contains no proper ideal I with $I^* \subseteq I$), then $H(A, *)$ is simple. In particular, all the algebras $H_n(D)$ are simple. Every

algebra A^+ is isomorphic to the algebra $H(B, *)$ where $B = A \oplus A^{\text{opp}}$ (algebra direct sum) and $(x_1, x_2)^* = (x_2, x_1)$.

5. **Albert algebras.** If $D = \mathbb{O}$ is a Cayley–Dickson algebra, then the algebra $H_n(\mathbb{O})$ of Hermitian matrices over D is a Jordan algebra only for $n \leq 3$. For $n = 1, 2$ the algebras are isomorphic to algebras of bilinear forms and are, therefore, special. The algebra $H_3(\mathbb{O})$ is exceptional (not special). An algebra J is called an Albert algebra if $K \otimes_F J \cong H_3(\mathbb{O})$ for some extension K of the field F. Every Albert algebra is simple, exceptional, and has dimension 27 over its center.

69.6 Power Associative Algebras, Noncommutative Jordan Algebras, and Right Alternative Algebras

A natural generalization of Jordan algebras is the class of algebras that satisfy the Jordan identity but which are not necessarily commutative. If the algebra has a unit element, then the Jordan identity easily implies the flexible identity. The right alternative algebras have been the most studied among the power associative algebras that do not satisfy the flexible identity.

As in the previous section, we assume that F is a field of characteristic $\neq 2$.

Definitions:

A **noncommutative Jordan algebra** is an algebra satisfying the flexible and Jordan identities. In this definition the Jordan identity may be replaced by any of the identities

$$x^2(xy) = x(x^2y), \qquad (yx)x^2 = (yx^2)x, \qquad (xy)x^2 = (x^2y)x.$$

A subspace V of an algebra A is **right nilpotent** if $V^{\langle n \rangle} = \{0\}$ for some $n \geq 1$, where $V^{\langle 1 \rangle} = V$ and $V^{\langle n+1 \rangle} = V^{\langle n \rangle} V$.

Facts: ([Sch66], [ZSS82], [KS95])

1. Let A be a finite dimensional power associative algebra with a bilinear symmetric form (x, y) satisfying the following conditions:

   (a) $(xy, z) = (x, yz)$ for all $x, y, z \in A$.

   (b) $(e, e) \neq 0$ for every idempotent $e \in A$.

   (c) $(x, y) = 0$ if the product xy is nilpotent.

   Then $\text{Nil}\, A = \text{Nil}\, A^+ = \{x \in A \,|\, (x, A) = \{0\}\}$, and if F has characteristic $\neq 2, 3, 5$, then the quotient algebra $A/\text{Nil}\, A$ is a noncommutative Jordan algebra.

2. Let A be a finite dimensional nil-semisimple flexible power associative algebra over an infinite field of characteristic $\neq 2, 3$. Then A has a unit element and is a direct sum of simple algebras, each of which is either a noncommutative Jordan algebra or (in the case of positive characteristic) an algebra of degree 2.

3. The structure of arbitrary finite dimensional nil-semisimple power associative algebras is still unclear. In particular, it is not known whether they are semisimple. It is known that in this case new simple algebras arise even in characteristic zero.

4. An algebra A is a noncommutative Jordan algebra if and only if it is flexible and the corresponding plus-algebra A^+ is a Jordan algebra.

5. Let A be a noncommutative Jordan algebra. For $x \in A$, the operators R_x, L_x, L_{x^2} generate a commutative subalgebra in the multiplication algebra $M(A)$, containing all the operators R_{x^k} and L_{x^k} for $k \geq 1$.

6. Every noncommutative Jordan algebra is power associative.
7. Let A be a finite dimensional nil-semisimple noncommutative Jordan algebra over F. Then A has a unit element and is a direct sum of simple algebras. If F has characteristic 0, then every simple summand is either a (commutative) Jordan algebra, a quasi-associative algebra (see Example 3 below), or a quadratic flexible algebra. In the case of positive characteristic, there are more examples of simple noncommutative Jordan algebras.
8. Unlike alternative and Jordan algebras, an analogue of the Wedderburn Principal Theorem on splitting of the nil radical does not hold in general for noncommutative Jordan algebras.
9. Every quasi-associative algebra (see Example 3 below) is a noncommutative Jordan algebra.
10. Every flexible quadratic algebra is a noncommutative Jordan algebra.
11. The right multiplication operators in every right alternative algebra A satisfy

$$R_{x^2} = R_x^2, \qquad R_{x \cdot y} = R_x \cdot R_y,$$

where $x \cdot y = \frac{1}{2}(xy + yx)$ is the multiplication in the algebra A^+. (Recall that, in this section, we assume that the characteristic of the base field is $\neq 2$.)
12. If A is a right alternative algebra (respectively, noncommutative Jordan algebra), then its unital hull $A^\sharp$ is also right alternative (respectively, noncommutative Jordan).
13. If A is a right alternative algebra, then the mapping $x \mapsto R_x$ is a homomorphism of the algebra A^+ into the special Jordan algebra $R(A)^+$. If A has a unit element, then this mapping is injective. For every right alternative algebra A, the algebra A^+ is embedded into the algebra $R(A^\sharp)^+$ and, hence, is a special Jordan algebra.
14. Every right alternative algebra A satisfies the identity

$$R_{x^k} R_{x^\ell} = R_{x^{k+\ell}} \quad \text{for any } x \in A \quad \text{and} \quad k, \ell \geq 1.$$

Therefore, A is power associative and the nil radical Nil A is defined.
15. Let A be an arbitrary right alternative algebra. Then the quotient algebra $A/\text{Nil } A$ is alternative. In particular, every right alternative algebra without nilpotent elements is alternative.
16. Every simple right alternative algebra that is not a nil algebra is alternative (and, hence, either associative or a Cayley–Dickson algebra). The nonnil restriction is essential: There exists a nonalternative simple right alternative nil algebra.
17. A finite dimensional right alternative nil algebra is right nilpotent and, therefore, solvable, but such an algebra can be nonnilpotent. In particular, the subclass *Nilp* is not radical in the class of finite dimensional right alternative algebras.

Examples:

1. The Suttles algebra (Example 5 in Section 69.2) is a power associative algebra that is not a noncommutative Jordan algebra. For another example see Example 5 below.
2. The class of noncommutative Jordan algebras contains, apart from Jordan algebras, all alternative algebras (and, thus, all associative algebras) and all anticommutative algebras.
3. **Quasi-associative algebras**. Let A be an algebra over a field F and let $a \in F, a \neq \frac{1}{2}$. Define a new multiplication on A as follows:

$$x \cdot_a y = axy + (1 - a)yx,$$

and denote the resulting algebra by A^a. The passage from A to A^a is reversible: $A = (A^a)^b$ for $b = a/(2a - 1)$. Properties of A and A^a are closely related: The ideals (respectively subalgebras) of A are those of A^a; the algebra A^a is nilpotent (respectively solvable, simple) if and only if the same holds for A. If A is associative, then A^a is a noncommutative Jordan algebra; furthermore, if the identity $[[x, y], z] = 0$ does not hold in A, then A^a is not associative. In particular, if A is a simple noncommutative associative algebra, then A^a is an example of a simple nonassociative noncommutative Jordan algebra. The algebras of the form A^a for an associative algebra A are **split** quasi-associative algebras. More generally, an algebra A is quasi-associative if it has a scalar extension, which is a split quasi-associative algebra.

4. **Generalized Cayley–Dickson algebras.** For $a_1, \ldots, a_n \in F - \{0\}$ let

$$A(a_1) = (F, a_1), \quad \ldots, \quad A(a_1, \ldots, a_n) = (A(a_1, \ldots, a_{n-1}), a_n),$$

be the algebras obtained from F by successive application of the Cayley–Dickson process (Example 2 of section 69.1). Then $A(a_1, \ldots, a_n)$ is a central simple quadratic noncommutative Jordan algebra of dimension 2^n.

5. Let V be a vector space of dimension $2n$ over a field F with a nondegenerate skew-symmetric bilinear form (x, y). On the vector space direct sum $A = F \oplus V$ define a multiplication (as for the Jordan algebra of bilinear form) by letting the unit element 1 of F be the unit of A and by setting $xy = (x, y)1$ for any $x, y \in V$. Then A is a simple quadratic algebra (and, hence, it is power associative), but A is not flexible and, thus, is not a noncommutative Jordan algebra.

69.7 Malcev Algebras

Some of the theory of Malcev algebras generalizes the theory of Lie algebras. For information about Lie algebras, the reader is advised to consult Chapter 70.

Definitions:

A **Malcev algebra** is an anticommutative algebra satisfying the identity

$$J(x, y, xz) = J(x, y, z)x, \quad \text{where} \quad J(x, y, z) = (xy)z + (yz)x + (zx)y.$$

In a **left-normalized product** we omit the parenthesis, for example,

$$xyzx = ((xy)z)x, \qquad yzxx = ((yz)x)x.$$

A **representation** of a Malcev algebra A is a linear mapping $\rho: A \to \text{End } V$ satisfying the following identity for all $x, y, z \in A$:

$$\rho(xy \cdot z) = \rho(x)\rho(y)\rho(z) - \rho(z)\rho(x)\rho(y) + \rho(y)\rho(zx) - \rho(yz)\rho(x).$$

We call V a **Malcev module** for A. The anticommutativity of A implies that the notion of a Malcev module is equivalent to that of Malcev bimodule; we set $xv = -vx$ for all $x \in A, v \in V$.

The **Killing form** $K(x, y)$ on a Malcev algebra A is defined (as for a Lie algebra) by

$$K(x, y) = \text{trace}(R_x R_y).$$

Facts: ([KS95], [She00], [PS04], [SZ06])

1. After expanding the Jacobians, the Malcev identity takes the form

$$xyzx + yzxx + zxxy = xy \cdot xz,$$

(using our convention on left-normalized products). If F has characteristic $\neq 2$, the Malcev identity is equivalent to the more symmetric identity

$$xyzt + yztx + ztxy + txyz = xz \cdot yt.$$

2. Any two elements in a Malcev algebra generate a Lie subalgebra.

3. The structure theory of finite dimensional Malcev algebras repeats the main features of the corresponding theory for Lie algebras. For any alternative algebra A, the minus algebra A^- is a Malcev algebra. Let $\mathbb{O} = \mathbb{O}(a,b,c)$ be a Cayley–Dickson algebra over a field F of characteristic $\neq 2$. Then $\mathbb{O} = F \oplus M$ (vector space direct sum), where $M = \{x \in \mathbb{O} \mid t(x) = 0\}$. The subspace M is a subalgebra (in fact, an ideal) of the Malcev algebra $\mathbb{O}^-$, and $M \cong \mathbb{O}^-/F$ (in fact, $\mathbb{O}^-$ is the Malcev algebra direct sum of the ideals F and M). The Malcev algebra $M = M(a,b,c)$ is central simple and has dimension 7 over F; if F has characteristic $\neq 3$, then M is not a Lie algebra.
4. Every central simple Malcev algebra of characteristic $\neq 2$ is either a Lie algebra or an algebra $M(a,b,c)$. There are no non-Lie simple Malcev algebras in characteristic 3.
5. Two Malcev algebras $M(a,b,c)$, $M(a',b',c')$ are isomorphic if and only if the corresponding Cayley–Dickson algebras $\mathbb{O}(a,b,c)$, $\mathbb{O}(a',b',c')$ are isomorphic.
6. Let A be a finite dimensional Malcev algebra over a field F of characteristic 0 and let Solv A be the solvable radical of A. The algebra A is semisimple (it decomposes into a direct sum of simple algebras) if and only if Solv $A = \{0\}$ (in fact, this is often used as the definition of "semisimple" for Malcev algebras, following the terminology for Lie algebras). If the quotient algebra $A/\text{Solv } A$ is separable, then A contains a subalgebra $B \cong A/\text{Solv } A$ and $A = B \oplus \text{Solv } A$ (vector space direct sum).
7. The Killing form $K(x,y)$ is symmetric and associative:

$$K(x,y) = K(y,x), \qquad K(xy,z) = K(x,yz).$$

   The algebra A is semisimple if and only if the form $K(x,y)$ is nondegenerate. For the solvable radical we have Solv $A = \{x \in A \mid K(x, A^2) = \{0\}\}$. In particular, A is solvable if and only if $K(A, A^2) = \{0\}$.
8. If all the operators $\rho(x)$ for $x \in A$ are nilpotent, then they generate a nilpotent subalgebra in End V (A acts nilpotently on V). If the representation ρ is almost faithful (that is, ker ρ does not contain nonzero ideals of A), then A is nilpotent.
9. Every representation of a semisimple Malcev algebra A is completely reducible.
10. If A is a Malcev algebra and V is an A-bimodule, then V is a Malcev module for A if and only if the split null extension $E(A,V)$ is a Malcev algebra.
11. Let A be a Malcev algebra, and let V be a faithful irreducible A-module. Then the algebra A is simple, and either V is a Lie module over A (which must then be a Lie algebra) or one of the following holds:

    (a) $A \cong M(a,b,c)$ and V is a regular A-module.

    (b) A is isomorphic to the Lie algebra $sl(2,F)$ with dim $V = 2$ and $\rho(x) = x^*$, where x^* is the matrix adjoint to $x \in A \subseteq M_2(F)$. (Here the matrix adjoint is defined by the equation $xx^* = x^*x = \det(x)I$, where I is the identity matrix.)

12. The **speciality problem** for Malcev algebras is still open: Is every Malcev algebra embeddable into the algebra A^- for some alternative algebra A? This is the generalization of the Poincaré–Birkhoff–Witt theorem for Malcev algebras.

69.8 Akivis and Sabinin Algebras

The theory of Akivis and Sabinin algebras generalizes the theory of Lie algebras and their universal enveloping algebras. For information about Lie algebras, the reader is advised to consult Chapter 70.

Definitions:

An **Akivis algebra** is a vector space A over a field F, together with an anticommutative bilinear operation $A \times A \to A$ denoted $[x,y]$, and a trilinear operation $A \times A \times A \to A$ denoted (x,y,z), satisfying the

Akivis identity for all $x, y, z \in A$:

$$[[x, y], z] + [[y, z], x] + [[z, x], y] = (x, y, z) + (y, z, x) + (z, x, y) - (x, z, y) - (y, x, z) - (z, y, x).$$

A **Sabinin algebra** is a vector space A over a field F, together with multilinear operations

$$\langle x_1, \ldots, x_m; y, z\rangle \qquad (m \geq 0),$$

satisfying these identities:

$$\langle x_1, \ldots, x_m; y, y\rangle = 0, \tag{69.5}$$

$$\langle x_1, \ldots, x_r, u, v, x_{r+1}, \ldots, x_m; y, z\rangle - \langle x_1, \ldots, x_r, v, u, x_{r+1}, \ldots, x_m; y, z\rangle$$
$$+ \sum_{k=0}^{r} \sum_{s} \langle x_{s(1)}, \ldots, x_{s(k)}, \langle x_{s(k+1)}, \ldots, x_{s(r)}; u, v\rangle, x_{r+1}, \ldots, x_m; y, z\rangle = 0, \tag{69.6}$$

$$K_{u,v,w}\left[\langle x_1, \ldots, x_r, u; v, w\rangle + \sum_{k=0}^{r} \sum_{s} \langle x_{s(1)}, \ldots, x_{s(k)}; \langle x_{s(k+1)}, \ldots, x_{s(r)}; v, w\rangle, u\rangle\right] = 0, \tag{69.7}$$

where s is a $(k, r-k)$-shuffle (a permutation of $1, \ldots, r$ satisfying $s(1) < \cdots < s(k)$ and $s(k+1) < \cdots < s(r)$) and the operator $K_{u,v,w}$ denotes the sum over all cyclic permutations. (See Fact 9 below for an alternative formulation of this definition.)

An algebra $\mathcal{F}_{\mathcal{M}}[X]$ from a class $\mathcal{M}$, with a set of generators X, is called the **free algebra** in $\mathcal{M}$ with the set X of free generators, if any mapping of X into an arbitrary algebra $A \in \mathcal{M}$ extends uniquely to a homomorphism of $\mathcal{F}_{\mathcal{M}}[X]$ to A.

Let I be any subset of $\mathcal{F}_{\mathcal{M}}[X]$. The T-**ideal** in $\mathcal{F}_{\mathcal{M}}[X]$ determined by I, denoted by $T = T(I, X)$, is the smallest ideal of $\mathcal{F}_{\mathcal{M}}[X]$ containing all elements of the form $f(x_1, \ldots, x_n)$ for all $f \in I$ and all $x_1, \ldots, x_n \in \mathcal{F}_{\mathcal{M}}[X]$.

Facts: ([HS90], [She99], [SU02], [GH03], [Per05], [BHP05], [BDE05])

1. Free algebras may be constructed as follows. Let S be a set of generating elements and let Ω be a set of operation symbols. Let $r: \Omega \to \mathbb{N}$ (the nonnegative integers) be the arity function, that is, $\omega \in \Omega$ will represent an n-ary operation for $n = r(\omega)$. The set $W(S, \Omega)$ of nonassociative Ω-words on the set S is defined inductively as follows:

   (a) $S \subseteq W(S, \Omega)$.

   (b) If $\omega \in \Omega$ and $x_1, \ldots, x_n \in W(S, \Omega)$ where $n = r(\omega)$, then $\omega(x_1, \ldots, x_n) \in W(S, \Omega)$.

   Let F be a field and let $F(S, \Omega)$ be the vector space over F with basis $W(S, \Omega)$. For each $\omega \in \Omega$ we define an n-ary operation with $n = r(\omega)$ on $F(S, \Omega)$, denoted by the same symbol ω, as follows: Given any basis elements $x_1, \ldots, x_n \in W(S, \Omega)$, we set the value of ω on the arguments $x_1, \ldots, x_n$ equal to the nonassociative word $\omega(x_1, \ldots, x_n)$, and extend linearly to all of $F(S, \Omega)$. The algebra $F(S, \Omega)$ is the free Ω-algebra on the generating set S over the field F with the operations $\omega \in \Omega$.
2. The quotient algebra $F(S, \Omega)/T(I, S, \Omega)$ is the free $\mathcal{M}$-algebra for the class $\mathcal{M} = \mathcal{M}(I)$ of Ω-algebras defined by the set of identities I.
3. Every subalgebra of a free Akivis algebra is again free.
4. Every Akivis algebra is isomorphic to a subalgebra of *Akivis*(A) for some nonassociative algebra A. This generalizes the Poincaré–Birkhoff–Witt theorem for Lie algebras. The free nonassociative algebra is the universal enveloping algebra of the free Akivis algebra. (See Example 1 below.)
5. The free nonassociative algebra with generating set X has a natural structure of a (nonassociative) Hopf algebra, generalizing the Hopf algebra structure on the free associative algebra. The **Akivis elements** (the elements of the subalgebra generated by X using the commutator and associator) are properly contained in the **primitive elements** (the elements satisfying $\Delta(x) = x \otimes 1 + 1 \otimes x$ where Δ is the co-multiplication). The Akivis elements and the primitive elements have a natural

structure of an Akivis algebra. The primitive elements have the additional structure of a Sabinin algebra.

6. The Witt dimension formula for free Lie algebras (the primitive elements in the free associative algebra) has a generalization to the primitive elements in the free nonassociative algebra.
7. Sabinin algebras are a nonassociative generalization of Lie algebras in the following sense: The tangent space at the identity of any local analytic loop (without associativity assumptions) has a natural structure of a Sabinin algebra, and the classical correspondence between Lie groups and Lie algebras generalizes to this case.
8. Every Sabinin algebra arises as the subalgebra of primitive elements in some nonassociative Hopf algebra.
9. Another (equivalent) way to define Sabinin algebras, which exploits the Hopf algebra structure, is as follows. Let A be a vector space and let $T(A)$ be the tensor algebra of A. We write $\Delta\colon T(A) \to T(A) \otimes T(A)$ for the co-multiplication on $T(A)$: the algebra homomorphism that extends the diagonal mapping $\Delta\colon u \mapsto 1 \otimes u + u \otimes 1$ for $u \in A$. We will use the Sweedler notation and write $\Delta(x) = \sum x_{(1)} \otimes x_{(2)}$ for any $x \in T(A)$. Then A is a Sabinin algebra if it is equipped with a trilinear mapping

$$T(A) \otimes A \otimes A \to A, \qquad x \otimes y \otimes z \mapsto \langle x; y, z\rangle, \qquad \text{for } x \in T(A) \text{ and } y, z \in A,$$

satisfying the identities

$$\langle x; y, y\rangle = 0, \tag{69.8}$$

$$\langle x \otimes u \otimes v \otimes x'; y, z\rangle - \langle x \otimes v \otimes u \otimes x'; y, z\rangle + \sum \langle x_{(1)} \otimes \langle x_{(2)}; u, v\rangle \otimes x'; y, z\rangle = 0, \tag{69.9}$$

$$K_{u,v,w}\left[\langle x \otimes u; v, w\rangle + \sum \langle x_{(1)}; \langle x_{(2)}; v, w\rangle, u\rangle\right] = 0, \tag{69.10}$$

where $x, x' \in T(A)$ and $u, v, w, y, z \in A$. Identities (8 to 10) exploit the Sweedler notation to express identities (5 to 7) in a more compact form.

Examples:

1. Any nonassociative algebra A becomes an Akivis algebra *Akivis*(A) if we define $[x, y]$ and (x, y, z) to be the commutator $xy - yx$ and the associator $(xy)z - x(yz)$. If A is an associative algebra, then the trilinear operation of *Akivis*(A) is identically zero; in this case the Akivis identity reduces to the Jacobi identity, and so *Akivis*(A) is a Lie algebra. If A is an alternative algebra, then the alternating property of the associator shows that the right side of the Akivis identity reduces to $6(x, y, z)$.
2. Every Lie algebra is an Akivis algebra with the identically zero trilinear operation. Every Malcev algebra (over a field of characteristic $\neq 2, 3$) is an Akivis algebra with the trilinear operation equal to $\frac{1}{6}J(x, y, z)$.
3. Every Akivis algebra A is a Sabinin algebra if we define

$$\langle a, b\rangle = -[a, b], \qquad \langle x; a, b\rangle = (x, b, a) - (x, a, b), \qquad \langle x_1, \ldots, x_m; a, b\rangle = 0 \ (m > 1),$$

for all $a, b, x, x_i \in A$.

4. Let L be a Lie algebra with a subalgebra $H \subseteq L$ and a subspace $V \subseteq L$ for which $L = H \oplus V$. We write $P_V\colon L \to V$ for the projection onto V with respect to this decomposition of L. We define an operation

$$\langle -, -; -\rangle\colon T(V) \otimes V \otimes V \to V$$

by (using the Sweedler notation again)

$$\{x \otimes a \otimes b\} + \sum \{x_{(1)} \otimes \langle x_{(2)}; a, b\rangle\} = 0,$$

where for $x = x_1 \otimes \cdots \otimes x_n \in T(V)$ we write

$$\{x\} = P_V([x_1, [\ldots, [x_{n-1}, x_n]] \cdots]).$$

Then the vector space V together with the operation $\langle -, -; -\rangle$ is a Sabinin algebra, and every Sabinin algebra can be obtained in this way.

69.9 Computational Methods

For homogeneous multilinear polynomial identities of degree n, the number of associative monomials is $n!$ and the number of association types is C_n (the Catalan number); hence, the number of nonassociative monomials grows superexponentially:

$$n! \cdot \frac{1}{n}\binom{2n-2}{n-1} = \frac{(2n-2)!}{(n-1)!} > n^{n-1}.$$

One way to reduce the size of the computations is to apply the theory of superalgebras [Vau98]. Another technique is to decompose the space of multilinear identities into irreducible representations of the symmetric group S_n. The application of the representation theory of the symmetric group to the theory of polynomial identities for algebras was initiated in 1950 by Malcev and Specht. The computational implementation of these techniques was pioneered by Hentzel in the 1970s [Hen77]; for detailed discussions of recent applications see [HP97] and [BH04]. Another approach has been implemented in the Albert system [Jac03]. In this section, we present three small examples ($n \leq 4$) of computational techniques in nonassociative algebra.

Examples:

1. *The identities of degree 3 satisfied by the division algebra of real octonions.*
   There are 12 distinct multilinear monomials of degree 3 for a nonassociative algebra:

   $$(xy)z,\ (xz)y,\ (yx)z,\ (yz)x,\ (zx)y,\ (zy)x,\ x(yz),\ x(zy),\ y(xz),\ y(zx),\ z(xy),\ z(yx).$$

   We create a matrix of size 8×12 and initialize it to zero; the columns correspond to the nonassociative monomials. We use a pseudorandom number generator to produce three octonions x, y, z represented as vectors with respect to the standard basis $1, i, j, k, \ell, m, n, p$. We store the evaluation of monomial j in column j of the matrix. For example, generating random integers from the set $\{-1, 0, 1\}$ using the base 3 expansion of $1/\sqrt{2}$ gives

   $$x = [1, -1, 0, -1, -1, 1, 0, 0], \quad y = [-1, 1, 1, 1, 0, 0, 1, 0], \quad z = [-1, 1, 1, 1, 0, 0, 0, -1].$$

   Evaluation of the monomials gives the matrix in Table 69.1; its reduced row echelon form appears in Table 69.2. The nullspace contains the identities satisfied by the octonion algebra: the span of the rows of the matrix in Table 69.3. These rows represent the linearizations of the right alternative identity (row 1), the left alternative identity (row 2), and the flexible identity (row 5), together with the assocyclic identities $(x, y, z) = (y, z, x)$ and $(x, y, z) = (z, x, y)$ (rows 3 and 4).
2. *The identities of degree 4 satisfied by the Jordan product $x * y = xy + yx$ in every associative algebra over a field of characteristic 0.*

TABLE 69.1 The octonion evaluation matrix

−3	−9	−9	−3	−3	−9	−3	−9	−9	−3	−3	−9
1	−5	−3	5	−1	−1	−3	−1	1	1	−5	3
1	1	3	5	1	3	1	1	1	3	3	5
2	2	−4	−2	0	−2	2	2	−4	−2	0	−2
1	1	1	3	−9	3	5	−3	−3	7	−5	−1
−3	−3	3	−1	5	−1	−3	−3	3	−1	5	−1
−10	−2	0	4	2	4	−8	−4	−2	6	4	2
0	0	0	6	−2	−2	−2	2	2	4	−4	0

TABLE 69.2 The reduced row echelon form of the octonion evaluation matrix

1	0	0	0	0	0	0	1	1	−1	−1	1
0	1	0	0	0	0	0	1	0	0	0	0
0	0	1	0	0	0	0	0	1	0	0	0
0	0	0	1	0	0	0	0	0	1	0	0
0	0	0	0	1	0	0	0	0	0	1	0
0	0	0	0	0	1	0	0	0	0	0	1
0	0	0	0	0	0	1	−1	−1	1	1	−1

TABLE 69.3 A basis for the nullspace of the octonion evaluation matrix

−1	−1	0	0	0	0	1	1	0	0	0	0
−1	0	−1	0	0	0	1	0	1	0	0	0
1	0	0	−1	0	0	−1	0	0	1	0	0
1	0	0	0	−1	0	−1	0	0	0	1	0
−1	0	0	0	0	−1	1	0	0	0	0	1

The operation $x*y$ satisfies commutativity in degree 2, and there are no new identities of degree 3, so we consider degree 4. There are 15 distinct multilinear monomials for a commutative nonassociative operation, 12 for association type $((--)-)-$ and 3 for association type $(--)(--)$:

$$
\begin{gathered}
((w*x)*y)*z,\ ((w*x)*z)*y,\ ((w*y)*x)*z,\ ((w*y)*z)*x,\ ((w*z)*x)*y,\\
((w*z)*y)*x,\ ((x*y)*w)*z,\ ((x*y)*z)*w,\ ((x*z)*w)*y,\ ((x*z)*y)*w,\\
((y*z)*x)*w,\ ((y*z)*w)*x,\ (w*x)*(y*z),\ (w*y)*(x*z),\ (w*z)*(x*y).
\end{gathered}
$$

When each of these monomials is expanded in terms of the associative product, there are 24 possible terms, namely the permutations of w, x, y, z in lexicographical order: $wxyz, \ldots, zyxw$. We construct a 24×15 matrix in which the i, j entry is the coefficient of the ith associative monomial in the expansion of the jth commutative monomial (see Table 69.4). The nontrivial identities of degree 4 satisfied by $x * y$ correspond to the nonzero vectors in the nullspace. The reduced row echelon form appears in Table 69.5. The rank is 11 and, so, the nullspace has dimension 4. A basis for the nullspace consists of the rows of Table 69.6. The first row represents the linearization of the Jordan identity; this is the only identity that involves monomials of both association types. (This proves that the plus algebra A^+ of any associative algebra A is a Jordan algebra.) The Jordan identity implies the identities in the other three rows, which are permuted forms of the identity

$$
w*(x*(yz)) - x*(w*(yz)) = (w*(x*y))*z - (x*(w*y))*z + y*(w*(x*z)) - y*(x*(w*z));
$$

that is, the commutator of multiplication operators is a derivation.

TABLE 69.4 The Jordan expansion matrix in degree 4

$$
\begin{bmatrix}
1 & 0 & 0 & 0 & 0 & 0 & 1 & 1 & 0 & 0 & 0 & 1 & 1 & 0 & 0 \\
0 & 1 & 0 & 0 & 0 & 0 & 0 & 0 & 1 & 1 & 0 & 1 & 1 & 0 & 0 \\
0 & 0 & 1 & 0 & 0 & 0 & 1 & 1 & 0 & 1 & 0 & 0 & 0 & 1 & 0 \\
0 & 0 & 0 & 1 & 0 & 0 & 0 & 0 & 0 & 1 & 1 & 1 & 0 & 1 & 0 \\
0 & 0 & 0 & 0 & 1 & 0 & 0 & 1 & 1 & 1 & 0 & 0 & 0 & 0 & 1 \\
0 & 0 & 0 & 0 & 0 & 1 & 0 & 1 & 0 & 0 & 1 & 1 & 0 & 0 & 1 \\
1 & 0 & 1 & 1 & 0 & 0 & 0 & 0 & 0 & 0 & 1 & 0 & 1 & 0 & 0 \\
0 & 1 & 0 & 0 & 1 & 1 & 0 & 0 & 0 & 0 & 1 & 0 & 1 & 0 & 0 \\
0 & 0 & 1 & 1 & 0 & 1 & 1 & 0 & 0 & 0 & 0 & 0 & 0 & 0 & 1 \\
0 & 0 & 0 & 0 & 0 & 1 & 0 & 1 & 0 & 0 & 1 & 1 & 0 & 0 & 1 \\
0 & 0 & 0 & 1 & 1 & 1 & 0 & 0 & 1 & 0 & 0 & 0 & 0 & 1 & 0 \\
0 & 0 & 0 & 1 & 0 & 0 & 0 & 0 & 0 & 1 & 1 & 1 & 0 & 1 & 0 \\
1 & 1 & 1 & 0 & 0 & 0 & 0 & 0 & 1 & 0 & 0 & 0 & 0 & 1 & 0 \\
0 & 0 & 0 & 1 & 1 & 1 & 0 & 0 & 1 & 0 & 0 & 0 & 0 & 1 & 0 \\
1 & 1 & 0 & 0 & 1 & 0 & 1 & 0 & 0 & 0 & 0 & 0 & 0 & 0 & 1 \\
0 & 0 & 0 & 0 & 1 & 0 & 0 & 1 & 1 & 1 & 0 & 0 & 0 & 0 & 1 \\
0 & 1 & 0 & 0 & 1 & 1 & 0 & 0 & 0 & 0 & 1 & 0 & 1 & 0 & 0 \\
0 & 1 & 0 & 0 & 0 & 0 & 0 & 0 & 1 & 1 & 0 & 1 & 1 & 0 & 0 \\
1 & 1 & 0 & 0 & 1 & 0 & 1 & 0 & 0 & 0 & 0 & 0 & 0 & 0 & 1 \\
0 & 0 & 1 & 1 & 0 & 1 & 1 & 0 & 0 & 0 & 0 & 0 & 0 & 0 & 1 \\
1 & 1 & 1 & 0 & 0 & 0 & 0 & 0 & 1 & 0 & 0 & 0 & 0 & 1 & 0 \\
0 & 0 & 1 & 0 & 0 & 0 & 1 & 1 & 0 & 1 & 0 & 0 & 0 & 1 & 0 \\
1 & 0 & 1 & 1 & 0 & 0 & 0 & 0 & 0 & 0 & 1 & 0 & 1 & 0 & 0 \\
1 & 0 & 0 & 0 & 0 & 0 & 1 & 1 & 0 & 0 & 0 & 1 & 1 & 0 & 0
\end{bmatrix}
$$

TABLE 69.5 The reduced row echelon form of the Jordan expansion matrix

$$
\begin{bmatrix}
1 & 0 & 0 & 0 & 0 & 0 & 0 & 0 & 0 & -1 & 0 & 0 & 0 & 0 & 0 \\
0 & 1 & 0 & 0 & 0 & 0 & 0 & 0 & 0 & 1 & 1 & 1 & 0 & 0 & 1 \\
0 & 0 & 1 & 0 & 0 & 0 & 0 & 0 & 0 & 0 & 0 & -1 & 0 & 0 & 0 \\
0 & 0 & 0 & 1 & 0 & 0 & 0 & 0 & 0 & 1 & 1 & 1 & 0 & 0 & 1 \\
0 & 0 & 0 & 0 & 1 & 0 & 0 & 0 & 0 & 0 & 0 & -1 & 0 & 0 & 0 \\
0 & 0 & 0 & 0 & 0 & 1 & 0 & 0 & 0 & -1 & 0 & 0 & 0 & 0 & 0 \\
0 & 0 & 0 & 0 & 0 & 0 & 1 & 0 & 0 & 0 & -1 & 0 & 0 & 0 & 0 \\
0 & 0 & 0 & 0 & 0 & 0 & 0 & 1 & 0 & 1 & 1 & 1 & 0 & 0 & 1 \\
0 & 0 & 0 & 0 & 0 & 0 & 0 & 0 & 1 & 0 & -1 & 0 & 0 & 0 & 0 \\
0 & 0 & 0 & 0 & 0 & 0 & 0 & 0 & 0 & 0 & 0 & 0 & 1 & 0 & -1 \\
0 & 0 & 0 & 0 & 0 & 0 & 0 & 0 & 0 & 0 & 0 & 0 & 0 & 1 & -1
\end{bmatrix}
$$

TABLE 69.6 A basis for the nullspace of the Jordan expansion matrix

$$
\begin{bmatrix}
0 & -1 & 0 & -1 & 0 & 0 & 0 & -1 & 0 & 0 & 0 & 0 & 1 & 1 & 1 \\
0 & -1 & 0 & -1 & 0 & 0 & 1 & -1 & 1 & 0 & 1 & 0 & 0 & 0 & 0 \\
0 & -1 & 1 & -1 & 1 & 0 & 0 & -1 & 0 & 0 & 0 & 1 & 0 & 0 & 0 \\
1 & -1 & 0 & -1 & 0 & 1 & 0 & -1 & 0 & 1 & 0 & 0 & 0 & 0 & 0
\end{bmatrix}
$$

3. *The identities of degree 4 satisfied by the commutator $[x, y] = xy - yx$ in every alternative algebra over a field of characteristic zero.*

   The group algebra $\mathbb{Q}S_n$ decomposes as a direct sum of full matrix algebras of size $d_\lambda \times d_\lambda$ where the index λ runs over all partitions λ of the integer n; here d_λ is the dimension of the irreducible representation of S_n corresponding to the partition λ. We choose the "natural representation" to fix a particular decomposition. For each λ there is a projection p_λ from $\mathbb{Q}S_n$ onto the matrix algebra of size $d_\lambda \times d_\lambda$. In the case $n = 4$ the partitions and the dimensions of the corresponding irreducible

TABLE 69.7 Partitions of 4 and irreducible representations of S_4

λ	4	31	22	211	1111
d_λ	1	3	2	3	1

representation S_4 are given in Table 69.7. For a nonassociative operation in degree 4 there are 5 association types:

$$((--)-)-, \quad (-(--))-, \quad (--)(--), \quad -((--)-), \quad -(-(--)),$$

and so any nonassociative identity can be represented as an element of the direct sum of 5 copies of $\mathbb{Q}S_n$: Given a partition λ, the nonassociative identity projects via p_λ to a matrix of size $d_\lambda \times 5d_\lambda$. For an anticommutative operation in degree 4 there are 2 association types:

$$[[[-,-],-],-], \quad [[-,-],[-,-]],$$

and so any anticommutative identity projects via p_λ to a matrix of size $d_\lambda \times 2d_\lambda$. The linearizations of the left and right alternative identities are

$$L(x,y,z) = (xy)z - x(yz) + (yx)z - y(xz), \qquad R(x,y,z) = (xy)z - x(yz) + (xz)y - x(zy).$$

Each of these can be "lifted" to degree 4 in five ways; for $L(a,b,c)$ we have

$$wL(x,y,z), \quad L(xw,y,z), \quad L(x,yw,z), \quad L(x,y,zw), \quad L(x,y,z)w;$$

and similarly for $R(x,y,z)$. Altogether we have 10 lifted alternative identities that project via p_λ to a matrix of size $10d_\lambda \times 5d_\lambda$. Using the commutator to expand the two anticommutative association types gives

$$\begin{aligned}
[[[x,y],z],w] &= ((xy)z)w - ((yx)z)w - (z(xy))w + (z(yx))w \\
&\quad - w((xy)z) + w((yx)z) + w(z(xy)) - w(z(yx)), \\
[[x,y],[z,w]] &= (xy)(zw) - (yx)(zw) - (xy)(wz) + (yx)(wz) \\
&\quad - (zw)(xy) + (zw)(yx) + (wz)(xy) - (wz)(yx).
\end{aligned}$$

Given a partition λ we can store these two relations in a matrix of size $2d_\lambda \times 7d_\lambda$: We use all 7 association types, store the right sides of the relations in the first 5 types, and $-I$ (I is the identity matrix) in type 6 (respectively 7) for the first (respectively second) expansion. For each partition λ, all of this data can be stored in a matrix A_λ of size $12d_\lambda \times 7d_\lambda$, which is schematically displayed in Table 69.8. We compute the reduced row echelon form of A_λ: Let i be the largest number for which rows $1 - i$ of RREF(A_λ) have a nonzero entry in the first 5 association types. Then the remaining rows of RREF(A_λ) have only zero entries in the first 5 types; if one of these

TABLE 69.8 The matrix of Malcev identities for partition λ

$$A_\lambda = \left[\begin{array}{c|cc}
 & 0 & 0 \\
\text{Lifted alternative identities} & \vdots & \vdots \\
 & 0 & 0 \\
\hline
\text{Expansion of } [[[x,y],z],w] & -I & 0 \\
\text{Expansion of } [[x,y],[z,w]] & 0 & -I
\end{array}\right]$$

TABLE 69.9 The lifted and expansion identities for partition $\lambda = 22$

0	0	0	0	0	0	−2	−2	2	2	0	0	0	0
0	0	0	0	0	0	1	1	−1	−1	0	0	0	0
0	1	1	0	0	−1	−1	0	0	0	0	0	0	0
−1	−1	0	1	1	1	0	−1	0	0	0	0	0	0
0	1	1	0	0	−1	−1	0	0	0	0	0	0	0
1	0	−1	−1	−1	0	1	1	0	0	0	0	0	0
0	0	0	0	2	0	0	0	−2	0	0	0	0	0
0	0	0	0	−1	0	0	0	1	0	0	0	0	0
2	0	−2	0	0	0	0	0	0	0	0	0	0	0
−1	0	1	0	0	0	0	0	0	0	0	0	0	0
0	0	0	0	0	0	−1	0	1	0	0	0	0	0
0	0	0	0	0	0	−1	0	1	0	0	0	0	0
−1	0	0	0	1	0	0	0	0	0	0	0	0	0
−1	0	0	0	1	0	0	0	0	0	0	0	0	0
0	0	1	0	0	1	−1	0	0	−1	0	0	0	0
0	0	−1	−1	1	0	1	1	−1	0	0	0	0	0
0	0	−1	−1	1	0	1	1	−1	0	0	0	0	0
0	0	1	0	0	1	−1	0	0	−1	0	0	0	0
1	1	−1	−1	0	0	0	0	0	0	0	0	0	0
1	1	−1	−1	0	0	0	0	0	0	0	0	0	0
0	0	0	0	0	0	0	0	0	0	−1	0	0	0
1	2	−1	1	0	0	1	−1	−1	−2	0	−1	0	0
0	0	0	0	0	0	0	0	0	0	0	0	−1	0
0	0	0	0	0	0	0	0	0	0	0	0	0	−1

TABLE 69.10 The reduced row echelon form of the lifted and expansion identities

1	0	0	0	0	0	0	0	−1	0	0	0	0	0
0	1	0	0	0	0	0	0	0	−1	0	0	0	0
0	0	1	0	0	0	0	0	−1	0	0	0	0	0
0	0	0	1	0	0	0	0	0	−1	0	0	0	0
0	0	0	0	1	0	0	0	−1	0	0	0	0	0
0	0	0	0	0	1	0	0	0	−1	0	0	0	0
0	0	0	0	0	0	1	0	−1	0	0	0	0	0
0	0	0	0	0	0	0	1	0	−1	0	0	0	0
0	0	0	0	0	0	0	0	0	0	1	0	0	0
0	0	0	0	0	0	0	0	0	0	0	1	0	0
0	0	0	0	0	0	0	0	0	0	0	0	1	0
0	0	0	0	0	0	0	0	0	0	0	0	0	1

TABLE 69.11 The anticommutative identities for partition $\lambda = 22$

0	0	0	0	0	0	0	0	0	0	2	0	0	0
0	0	0	0	0	0	0	0	0	0	−1	0	0	0
0	0	0	0	0	0	0	0	0	0	0	0	2	0
0	0	0	0	0	0	0	0	0	0	0	0	−1	0
0	0	0	0	0	0	0	0	0	0	0	0	2	0
0	0	0	0	0	0	0	0	0	0	0	0	0	2

TABLE 69.12 The reduced row echelon form of the anticommutative identities

0	0	0	0	0	0	0	0	0	0	1	0	0	0
0	0	0	0	0	0	0	0	0	0	0	0	1	0
0	0	0	0	0	0	0	0	0	0	0	0	0	1

rows contains nonzero entries in the last 2 types, this row represents an identity that is satisfied by the commutator in every alternative algebra. However, such an identity may be a consequence of the obvious anticommutative identities:

$$[[[x, y], z], w] + [[[y, x], z], w] = 0,$$
$$[[x, y], [z, w]] + [[y, x], [z, w]] = 0, \qquad [[x, y], [z, w]] + [[z, w], [x, y]] = 0.$$

These identities are represented by a matrix of size $3d_\lambda \times 7d_\lambda$ in which the first $5d_\lambda$ columns are zero. We need to determine if any of the rows $i + 1$ to $12d_\lambda$ of $\text{RREF}(A_\lambda)$ do not lie in the row space of the matrix of anticommutative identities. If such a row exists, it represents a nontrivial identity satisfied by the commutator in every alternative algebra. For example, consider the partition $\lambda = 22$ ($d_\lambda = 2$). The 24×14 matrix A_λ for this partition appears in Table 69.9, and its reduced row echelon form appears in Table 69.10. The 6×14 matrix representing the anticommutative identities for this partition appears in Table 69.11, and its reduced row echelon form appears in Table 69.12. Comparing the last four rows of Table 69.10 with Table 69.12, we see that there is one new identity for $\lambda = 22$ represented by the third-last row of Table 69.10. Similar computations for the other partitions show that there is one nontrivial identity for partition $\lambda = 211$ and no nontrivial identities for the other partitions. The two identities from partitions 22 and 211 are the irreducible components of the Malcev identity: The submodule generated by the linearization of the Malcev identity (in the S_4-module of all multilinear anticommutative polynomials of degree 4) is the direct sum of two irreducible submodules corresponding to these two partitions.

Acknowledgment

The authors thank Irvin Hentzel (Iowa State University) for helpful comments on an earlier version of this chapter.

References

[Bae02] J.C. Baez, The octonions, *Bull. Am. Math. Soc.* 39, 2 (2002), 145–205.

[BDE05] P. Benito, C. Draper, and A. Elduque, Lie-Yamaguti algebras related to g_2, *J. Pure Appl. Alg.* 202, 1–3 (2005), 22–54.

[BK66] H. Braun and M. Koecher, *Jordan-Algebren* [German], Springer-Verlag, Berlin, New York, 1966.

[BH04] M.R. Bremner and I.R. Hentzel, Invariant nonassociative algebra structures on irreducible representations of simple Lie algebras, *Exp. Math.* 13, 2 (2004), 231–256.

[BHP05] M.R. Bremner, I.R. Hentzel, and L.A. Peresi, Dimension formulas for the free nonassociative algebra, *Comm. Alg.* 33, 11 (2005), 4063–4081.

[CGG00] R. Costa, A. Grishkov, H. Guzzo, Jr., and L.A. Peresi (Eds.), *Nonassociative Algebra and its Applications, Proceedings of the Fourth International Conference* (São Paulo, Brazil, 19–25 July 1998), Marcel Dekker, New York, 2000.

[EHH91] H.D. Ebbinghaus, H. Hermes, F. Hirzebruch, M. Koecher, K. Mainzer, J. Neukirch, A. Prestel, and R. Remmert, *Numbers*, translated from the 2nd 1988 German ed. by H.L.S. Orde, Springer-Verlag, New York, 1991.

[FKS93] V.T. Filippov, V.K. Kharchenko, and I.P. Shestakov (Eds.), *The Dniester Notebook: Unsolved Problems in the Theory of Rings and Modules*, 4th ed. [Russian], Novosibirsk, 1993; English translation in [SSS06].

[GH03] L. Gerritzen and R. Holtkamp, Hopf co-addition for free magma algebras and the non-associative Hausdorff series, *J. Alg.* 265, 1 (2003), 264–284.

[Gon94] S. González (Ed.), *Non-Associative Algebra and its Applications, Proceedings of the Third International Conference* (Oviedo, Spain, 12–17 July 1993), Kluwer, Dordrecht, 1994.

[Hen77] I.R. Hentzel, Processing identities by group representation, pp. 13–40 in *Computers in Nonassociative Rings and Algebras (Special Session, 82nd Annual Meeting of the American Mathematical Society, San Antonio, Texas, 1976)*, Academic Press, New York, 1977.

[HP97] I.R. Hentzel and L.A. Peresi, Identities of Cayley–Dickson algebras, *J. Alg.* 188, 1 (1997) 292–309.

[HS90] K.H. Hofmann and K. Strambach, Topological and analytic loops, pp. 205–262 in *Quasigroups and Loops: Theory and Applications*, O. Chein, H.O. Pflugfelder, and J.D.H. Smith, Eds., Heldermann Verlag, Berlin, 1990.

[Jac03] D.P. Jacobs, Building nonassociative algebras with Albert, pp. 346–348 in *Computer Algebra Handbook: Foundations, Applications, Systems*, J. Grabmeier, E. Kaltofen, and V. Weispfennig, Eds., Springer-Verlag, Berlin, 2003.

[Jac68] N. Jacobson, *Structure and representations of Jordan algebras*, American Mathematical Society, Providence, 1968.

[KS95] E.N. Kuzmin and I.P. Shestakov, Nonassociative structures, pp. 197–280 in *Encyclopaedia of Mathematical Sciences 57, Algebra VI*, A.I. Kostrikin and I.R. Shafarevich, Eds., Springer-Verlag, Berlin, 1995.

[McC04] K. McCrimmon, *A Taste of Jordan Algebras*, Springer-Verlag, New York, 2004.

[Per05] J.M. Pérez-Izquierdo, Algebras, hyperalgebras, nonassociative bialgebras and loops, Advances in Mathematics, in press, corrected proof available online 6 May 2006.

[PS04] J.M. Pérez-Izquierdo and I.P. Shestakov, An envelope for Malcev algebras, *J. Alg.* 272, 1 (2004) 379–393.

[SSS06] L. Sabinin, L. Sbitneva, and I.P. Shestakov (Eds.), Non-associative algebra and its applications, *Proceedings of the Fifth International Conference* (Oaxtepec, Mexico, 27 July to 2 August, 2003), Chapman & Hall/CRC, Boca Raton, 2006.

[Sch66] R.D. Schafer, *An Introduction to Nonassociative Algebras*, corrected reprint of the 1966 original, Dover Publications, New York, 1995.

[She99] I.P. Shestakov, Every Akivis algebra is linear, *Geometriae Dedicata* 77, 2 (1999), 215–223.

[She00] I.P. Shestakov, Speciality problem for Malcev algebras and Poisson Malcev algebras, pp. 365–371 in [CGG00].

[SU02] I.P. Shestakov and U.U. Umirbaev, Free Akivis algebras, primitive elements, and hyperalgebras, *J. Alg.* 250, 2 (2002), 533–548.

[SZ06] I.P. Shestakov and Natalia Zhukavets, Speciality of Malcev superalgebras on one odd generator, *J. Alg.* 301, 2 (2006), 587–600.

[Shi58] A.I. Shirshov, Some problems in the theory of rings that are nearly associative [Russian], *Uspekhi Matematicheskikh Nauk* 13, 6 (1958), 3–20; English translation, in [SSS06].

[Vau98] M. Vaughan-Lee, Superalgebras and dimensions of algebras, *Int. J. Alg. Comp.* 8, 1 (1998), 97–125.

[ZSS82] K.A. Zhevlakov, A.M. Slinko, I.P. Shestakov, and A.I. Shirshov, *Rings That are Nearly Associative*, translated from the Russian by Harry F. Smith, Academic Press, New York, 1982.

70

Lie Algebras

Robert Wilson
Rutgers University

A Lie algebra is a (nonassociative) algebra satisfying $x^2 = 0$ for all elements x of the algebra (which implies anticommutativity) and the Jacobi identity. Lie algebras arise naturally as (vector) subspaces of associative algebras closed under the commutator operation $[a, b] = ab - ba$. The finite-dimensional simple Lie algebras over algebraically closed fields of characteristic zero occur in many applications. This chapter outlines the structure, classification, and representation theory of these algebras. We also give examples of other types of algebras, e.g., one class of infinite-dimensional simple algebras and one class of finite-dimensional simple Lie algebras over fields of prime characteristic. Section 70.1 is devoted to general definitions about Lie algebras. Section 70.2 discusses semisimple and simple algebras. This section includes the classification of finite-dimensional simple Lie algebras over algebraically closed fields of characteristic zero. Section 70.3 discusses module theory and includes the classification of finite-dimensional irreducible modules for the aforementioned algebras as well as the explicit construction of some of these modules. Section 70.4 discusses graded algebras and modules and uses this formalism to present results on dimensions of irreducible modules.

70.1 Basic Concepts

Unless specified otherwise, F denotes an arbitrary field. All vector spaces and algebras are over F. The reader is referred to Chapter 69 for definitions of many basic algebra terms.

Definitions:

Let A be an algebra over a field F. An **automorphism** is an algebra isomorphism of A to itself. The set of all automorphisms of A is denoted $Aut(A)$.

A linear transformation $D : A \to A$ is a **derivation** of A if $D(ab) = D(a)b + aD(b)$ for all $a, b \in A$. The set of all derivations of A is denoted $Der(A)$.

Let A be an associative algebra and X be a subset of A. Then the smallest ideal of A containing X is denoted by $< X >$ and called the **ideal generated by** X.

Let V be a vector space over a field F. Let $V^{\otimes n}$ denote the tensor product of n copies of V and set $T(V) = F1 + \sum_{n=1}^{\infty} V^{\otimes n}$. Define a linear map $T(V) \otimes T(V) \to T(V), u \otimes v \mapsto uv$, by $1u = u1 = u$ for all $u \in T(V)$ and

$$(v_1 \otimes \ldots \otimes v_m)(u_1 \otimes \ldots \otimes u_n) = v_1 \otimes \ldots \otimes v_m \otimes u_1 \otimes \ldots \otimes u_n$$

whenever $m, n \geq 1, v_1, \ldots, v_m, u_1, \ldots, u_n \in V$. $T(V)$ is the **tensor algebra** on the vector space V. This algebra is defined, and denoted $\bigotimes V$, in Section 13.9.

An algebra L over a field F with product $[\ ,\ ] : L \times L \to L, (a,b) \mapsto [a,b]$ is a **Lie algebra** if it satisfies both

$$[a,a] = 0$$

and

$$[a,[b,c]] + [b,[c,a]] + [c,[a,b]] = 0 \qquad \textbf{(Jacobi identity)}$$

for all $a, b, c \in L$. The first condition implies

$$[a,b] = -[b,a] \qquad \textbf{(anticommutativity)}$$

and is equivalent to anticommutativity if the characteristic of $F \neq 2$.

A Lie algebra L is **abelian** if $[a,b] = 0$ for all $a, b \in L$.

If A is an algebra, the vector space A together with the product $[\ ,\ ] : A \times A \to A$ defined by $[a,b] = ab - ba$ is an algebra denoted by A^-.

Let L be a Lie algebra. Let I denote the ideal in $T(L)$ (the tensor algebra on the vector space L) generated by $\{a \otimes b - b \otimes a - [a,b] | a, b \in L\}$. The quotient algebra $T(V)/I$ is called the **universal enveloping algebra** of L and is denoted by $U(L)$.

Let V be a vector space with basis X. Define a map $\iota : X \to T(V)^-$ by $\iota : x \mapsto x \in V^{\otimes 1} \subseteq F1 + \sum_{n=1}^{\infty} V^{\otimes n} = T(V)^-$. Let $Fr(X)$ be the Lie subalgebra of $T(V)^-$ generated by $\iota(X)$. $Fr(X)$ is called the **free Lie algebra generated by** X.

Let V be a vector space and let I be the ideal in $T(V)$ generated by $\{a \otimes b - b \otimes a | a, b \in V\}$. The quotient $T(V)/I$ is called the **symmetric algebra** on V and denoted by $S(V)$. The image of $V^{\otimes n}$ in $S(V)$ is denoted by $S^n(V)$. An equivalent construction of this algebra (as a subalgebra, denoted $\bigvee V$, of $T(V)$) is given in Section 13.9.

Let V be a vector space and let I be the ideal in $T(V)$ generated by $\{a \otimes a | a \in V\}$. The quotient $T(V)/I$ is called the **exterior algebra** on V and denoted by $\Lambda(V)$. The image of $a_1 \otimes \ldots \otimes a_l$ is denoted by $a_1 \wedge \ldots \wedge a_l$ and the image of $V^{\otimes n}$ in $\Lambda(V)$ is denoted by $\Lambda^n(V)$. An equivalent construction of this algebra (as a subalgebra, denoted $\bigwedge V$, of $T(V)$) is given in Section 13.9.

Let $End\ V$ denote the vector space of all linear transformations from V to V (also denoted $L(V, V)$ elsewhere in this book). Let L be a Lie algebra. If $a \in A$, the map $ad : L \to End\ L$ defined by $ad(a) : b \mapsto [a,b]$ for all $b \in L$ is called the **adjoint map**.

Let F be a field of prime characteristic p and let L be a Lie algebra over F. If for every $a \in L$ there is some element $a^{[p]} \in L$ such that $(ad(a))^p = ad(a^{[p]})$, then L is called a **p-Lie algebra**.

Facts:

The following facts (except those with a specific reference) can be found in [Jac62, Chap. 5].

1. [Jac62, p. 6] If A is an associative algebra, then A^- is a Lie algebra. If F is a field of prime characteristic p, then A^- is a p-Lie algebra.
2. Let A be an algebra over F. Then $Der(A)$ is a Lie algebra. If F is a field of prime characteristic p, then $Der(A)$ is a p-Lie algebra.
3. Let V be a vector space. The tensor algebra $T(V)$ has the structure of an associative algebra and $T(V)^-$ is a Lie algebra.
4. Let L be a Lie algebra. A subspace $I \subseteq L$ is an *ideal* of L if $[a,b] \in I$ whenever $a \in A, b \in I$. The quotient space L/I with the product $[a + I, b + I] = [a,b] + I$ is a Lie algebra.
5. *Universal property of* $U(L)$: Let L be a Lie algebra. Define a map $\iota : L \to U(L)$ by $\iota : a \mapsto a \in L^{\otimes 1} \subseteq F1 + \sum_{n=1}^{\infty} L^{\otimes n} = T(L) \to U(L)$. Then ι is a (Lie algebra) homomorphism of L into

$U(L)^-$. If A is any associative algebra with unit 1 and ϕ is a homomorphism of L into A^-, then there is a unique homomorphism $\psi : U(L) \to A$ such that $\psi(1) = 1$ and $\phi = \psi\iota$.

6. *Poincaré–Birkhoff–Witt Theorem:* Let L be a Lie algebra with ordered basis $\{l_i | i \in I\}$. Then $\{l_{i_1} \dots l_{i_k} | k \geq 0, i_1 \leq \dots \leq i_k\}$ is a basis for $U(L)$. Consequently, $\iota : L \to U(L)$ is injective.
7. *Universal property of the free Lie algebra:* If L is any Lie algebra and if $\phi : X \to L$ is any map, there is a unique homomorphism of Lie algebras $\psi : Fr(X) \to L$ such that $\phi = \psi\iota$.
8. *Universal property of* $S(V)$: Let V be a vector space. Define a map $\iota : V \to S(V)$ by $\iota : a \mapsto a \in V^{\otimes 1} \subseteq F1 + \sum_{n=1}^{\infty} V^{\otimes n} \to S(V)$. Let A be an commutative associative algebra and $\phi : V \to A$ be a linear map. Then there is a unique algebra homomorphism $\psi : S(V) \to A$ such that $\phi = \psi\iota$.
9. *Structure of* $S(V)$: Let V be a vector space with ordered basis $B = \{b_i | i \in I\}$. Then $S(V)$ has basis $\{b_{i_1} \dots b_{i_l} | l \geq 0, i_1 \leq \dots \leq i_l\}$. Consequently, $S(V)$ is isomorphic to the algebra of polynomials on B.
10. [Lam01, p. 12] *Structure of* $\Lambda(V)$: Let V be a vector space with ordered basis $B = \{b_i | i \in I\}$. Then $\Lambda(V)$ has basis $\{b_{i_1} \wedge \dots \wedge b_{i_l} | l \geq 0, i_1 < \dots < i_l\}$. Consequently, if V has finite dimension l, then $\dim \Lambda(V) = 2^l$.

Examples:

1. $(End\ V)^-$ is a Lie algebra, denoted $gl(V)$. Similarly, $(F^{n\times n})^-$ is a Lie algebra, denoted $gl(n, F)$. These algebras are isomorphic.
2. $sl(n, F) = \{x \in gl(n, F) | tr(x) = 0\}$ is a Lie subalgebra, and in fact, a Lie ideal of $gl(n, F)$.
3. Let L be a vector space with basis $\{e, f, h\}$. Defining $[e, f] = h, [h, e] = 2e$, and $[h, f] = -2f$ gives L the structure of a Lie algebra. The linear map $L \to sl(2, F)$ defined by $e \mapsto \begin{bmatrix} 0 & 1 \\ 0 & 0 \end{bmatrix}, f \mapsto \begin{bmatrix} 0 & 0 \\ 1 & 0 \end{bmatrix}, h \mapsto \begin{bmatrix} 1 & 0 \\ 0 & -1 \end{bmatrix}$ is an isomorphism of Lie algebras.

70.2 Semisimple and Simple Algebras

Definitions:

Let L be a Lie algebra. The subspace spanned by all products $[a, b], a, b \in L$, is a subalgebra of L. It is called the **derived algebra** of L and is denoted by $L^{(1)}$.

Let L be a Lie algebra. For $n \geq 2$, $L^{(n)}$ is defined to be $(L^{(n-1)})^{(1)}$ and is called the **nth-derived algebra** of L.

A Lie algebra L is **solvable** if $L^{(n)} = \{0\}$ for some $n \geq 1$.

Let L be a Lie algebra. The sum of all the solvable ideals of L is the **radical** of L and denoted $Rad(L)$. (In Section 69.2, $Rad(L)$ is called the *solvable radical* and denoted Solv L.)

A Lie algebra L is **semisimple** if $Rad(L) = \{0\}$. N.B. This is standard terminology in the study of Lie algebras, but does *not* always coincide with the definition of semisimple given for nonassociative algebras in Section 69.2; cf. Fact 5 and Example 3.

A Lie algebra L is **simple** if L contains no nonzero proper ideals and $L^{(1)} \neq \{0\}$. (The second condition excludes the one-dimensional algebra.)

Let A, B be Lie algebras. Then the vector space $A \oplus B$ can be given the structure of a Lie algebra, called the **direct sum** of A and B and also denoted $A \oplus B$, by setting $[a_1 + b_1, a_2 + b_2] = [a_1, a_2] + [b_1, b_2]$ for $a_1, a_2 \in A, b_1, b_2 \in B$.

Let L be a finite-dimensional Lie algebra. The **Killing form,** κ_L, is the symmetric bilinear form on L defined by $\kappa_L(a, b) = tr((ad(a))(ad(b)))$.

For V a finite dimensional vector space over an algebraically closed field F and $x \in End\ V$, x is **semisimple** if the minimum polynomial of x has no repeated roots.

Let L be a Lie algebra over an algebraically closed field F and $x \in L$. x is **ad-nilpotent** if $ad(x)$ is a nilpotent linear transformation of L and x is **ad-semisimple** if $ad(x)$ is a semisimple linear transformation of L.

Let L be a Lie algebra. A subalgebra $T \subseteq L$ is a **torus** if every element of T is ad-semisimple.

Let T be a torus in L and let $\alpha \in T^*$, the dual of T. Define L_α, the **α-root space** of L by

$$L_\alpha = \{x \in L \mid [t, x] = \alpha(t)x \; \forall \, t \in T\}.$$

A vector space E over $\mathbb{R}$ with an inner product (i.e., a positive definite symmetric bilinear form)$\langle .,. \rangle$ is a **Euclidean space.**

Let E be a Euclidean space.

For $0 \neq x \in E$ and $y \in E$ set $< y, x > = \frac{2\langle y,x \rangle}{\langle x,x \rangle}$.

For $0 \neq x \in E$ define σ_x, the **reflection** in the hyperplane orthogonal to x, by

$$\sigma_x(y) = y - < y, x > x$$

for all $y \in E$.

A finite subset $R \subseteq E$ that spans E and does not contain 0 is a **root system** in E if the following three conditions are satisfied:

- If $x \in R$, $a \in \mathbb{R}$, and $ax \in R$, then $a = \pm 1$.
- If $x \in R$, then $\sigma_x R = R$
- If $x, y \in R$, then $< x, y > \in \mathbb{Z}$.

Let R be a root system in E. The **rank** of R is dim E.

A root system R is **decomposable** if $R = R_1 \cup R_2$ with $\emptyset \neq R_1, R_2 \subset R$ and $(R_1, R_2) = \{0\}$. If R is not decomposable, it is **indecomposable.**

Let R be a root system in E. The subgroup of *End* E generated by $\{\sigma_\alpha \mid \alpha \in R\}$ is called the **Weyl group** of R.

Let R be a root system in E. A subset $B \subseteq R$ is a **base** for R if B is a basis for E and every $x \in R$ may be written

$$x = \sum_{b \in B} k_b b$$

where all $k_b \geq 0$ or all $k_b \leq 0$.

Let R be a root system in E with base B and let $\alpha \in R$. α is a **positive** root if $\alpha = \sum_{b \in B} k_b b$ where all $k_b \geq 0$. Denote the set of positive roots by R^+.

Let $B = \{b_1, \ldots, b_l\}$ be a base for a root system R. The matrix $\left[< b_i, b_j >\right]$, is called the **Cartan matrix** of R with respect to B.

Facts:

Most of the following facts (except those with a specific reference) can be found in [Hum72, pp. 35–65].

1. The **radical**, $Rad(L)$, is a solvable ideal of L.
2. For V a finite dimensional vector space over an algebraically closed field F, $x \in End\ V$ is semisimple if and only if x is similar to a diagonal matrix.
3. [Jac62, p. 69] *Cartan's Criterion for Semisimplicity:* Let L be a finite-dimensional Lie algebra over a field of characteristic 0. Then L is semisimple if and only if κ_L is nondegenerate.
4. [Jac62, p. 74] Let L be a finite-dimensional semisimple Lie algebra over a field of characteristic zero and let $D \in Der(L)$. Then $D = ad(a)$ for some $a \in L$.
5. [Jac62, p. 71] Let L be a finite-dimensional semisimple Lie algebra over a field of characteristic 0. Then L is a direct sum of simple ideals.

6. Let L be a finite-dimensional Lie algebra over an algebraically closed field of characteristic 0. Any torus of L is abelian.
7. Let L be a finite-dimensional semisimple Lie algebra over an algebraically closed field of characteristic 0. Let T be a maximal torus of dimension l in L and $\Delta = \{\alpha \in T^* \mid \alpha \neq 0, L_\alpha \neq \{0\}\}$. Then:

   - $\dim L_\alpha = 1$ for all $\alpha \in \Delta$.
   - $L = T \oplus \sum_{\alpha \in \Delta} L_\alpha$.
   - For $\alpha \in \Delta$ there exists $t_\alpha \in T$ such that $\alpha(t) = \kappa_L(t_\alpha, t)$ for all $t \in T$. Then, defining $(\alpha, \beta) = \kappa_L(t_\alpha, t_\beta)$ gives the $\mathbb{R}$ span of Δ the structure of a Euclidean space of dimension l and Δ is a root system of rank l in this space.
   - Δ is indecomposable if and only if L is simple.
   - Δ spans T^* and so each $w \in W$ acts on T^*.

8. Let R be a root system in E with Weyl group W. Then W is finite and R has a base. Furthermore, if B_1, B_2 are two bases for R, then $B_1 = w(B_2)$ for some $w \in W$. Consequently, when B_1 and B_2 are appropriately ordered, the Cartan matrix of R with respect to B_1 is the same as the Cartan matrix of R with respect to B_2. Thus, we may refer to the Cartan matrix of R.
9. Let L be a semisimple Lie algebra over a field of characteristic 0, let T_1, T_2 be maximal tori in L, and let Δ_1, Δ_2 be the corresponding root systems. Then there is an automorphism $\phi \in Aut(L)$ such that $\phi(T_1) = T_2$. Consequently, when bases for Δ_1 and Δ_2 are appropriately ordered, the Cartan matrix for Δ_1 is the same as the Cartan matrix for Δ_2. Thus, we may refer to the Cartan matrix of L.
10. Let L_1, L_2 be semisimple Lie algebras over an algebraically closed field of characteristic 0. If the Cartan matrices of L_1 and L_2 coincide, then L_1 and L_2 are isomorphic.
11. Let $M = [m_{i,j}]$ be the $l \times l$ Cartan matrix of an indecomposable root system, with base appropriately ordered. Then the diagonal entries of M are all 2 and one of the following occurs.

   - M is of type A_l for some $l \geq 1$: $m_{i,j} = -1$ if $|i - j| = 1$; $m_{i,j} = 0$ if $|i - j| > 1$.
   - M is of type B_l for some $l \geq 3$: $m_{l-1,l} = -2$; $m_{i,j} = -1$ if $|i - j| = 1$ and $(i, j) \neq (l-1, l)$; $m_{i,j} = 0$ if $|i - j| > 1$.
   - M is of type C_l for some $l \geq 2$: $m_{l,l-1} = -2$; $m_{i,j} = -1$ if $|i - j| = 1$ and $(i, j) \neq (l, l-1)$; $m_{i,j} = 0$ if $|i - j| > 1$.
   - M is of type D_l for some $l \geq 4$: $m_{i,j} = -1$ if $|i - j| = 1$ and $(i, j) \neq (l-1, l), (l, l-1)$ or if $(i, j) = (l-2, l), (l, l-2)$; $m_{i,j} = 0$ if $|i - j| > 1$ and $(i, j) \neq (l-2, l), (l, l-2)$ or if $(i, j) = (l-1, l), (l, l-1)$.
   - M is of type $E_l, l = 6, 7, 8$: $m_{i,j} = -1$ if $|i-j| = 1$ and $i, j \neq 2$ or if $(i, j) = (1,3)(3,1)(2,4), (4,2)$; $m_{i,j} = 0$ if $|i - j| > 1, (i, j) \neq (1,3)(3,1)(2,4), (4,2)$ or if $|i - j| = 1$ and $i = 2$ or $j = 2$.
   - M is of type F_4: $M = \begin{bmatrix} 2 & -1 & 0 & 0 \\ -1 & 2 & -2 & 0 \\ 0 & -1 & 2 & -1 \\ 0 & 0 & -1 & 2 \end{bmatrix}$.
   - M is of type G_2: $M = \begin{bmatrix} 2 & -1 \\ -3 & 2 \end{bmatrix}$.

12. Let L be a finite-dimensional simple Lie algebra over an algebraically closed field of characteristic zero. Then L is determined up to isomorphism by its root system with respect to any maximal torus. The Cartan matrix of the root system is of type $A_l, l \geq 1$; $B_l, l \geq 3$; $C_l, l \geq 2$; $D_l, l \geq 4$; E_6, E_7, E_8, F_4, or G_2.

Examples:

1. The set $\mathbb{R}^3$ with $[\mathbf{u}, \mathbf{v}] = \mathbf{u} \times \mathbf{v}$ (vector cross product) is a three dimensional simple Lie algebra.
2. The set of all upper triangular complex $n \times n$ matrices is a solvable subalgebra of $gl(n, \mathbb{C})$.
3. [Pol69, p. 72] Let $p > 3$ be a prime. The only ideals of $gl(p, \mathbb{Z}_p)$ are $0 \subset scal(p, \mathbb{Z}_p) \subset sl(p, \mathbb{Z}_p) \subset gl(p, \mathbb{Z}_p)$, where $scal(p, \mathbb{Z}_p)$ is the set of scalar matrices. Thus, the only ideals of $L = gl(p, \mathbb{Z}_p)/scal(p, \mathbb{Z}_p)$ are L, $S = sl(p, \mathbb{Z}_p)/scal(p, \mathbb{Z}_p)$ and $\{0\}$. By considering $D = diag(0, 1, 2, \ldots, p-1)$, we see that $[sl(n, \mathbb{Z}_p), sl(n, \mathbb{Z}_p)] = sl(n, \mathbb{Z}_p)$, so $[S, S] = S$. Thus, S is not solvable and $Rad(L) = 0$. But L cannot be the sum of simple ideals.
4. The following are Cartan matrices of type A_3, B_3, C_3 respectively:

$$\begin{bmatrix} 2 & -1 & 0 \\ -1 & 2 & -1 \\ 0 & -1 & 2 \end{bmatrix}, \quad \begin{bmatrix} 2 & -1 & 0 \\ -1 & 2 & -2 \\ 0 & -1 & 2 \end{bmatrix}, \quad \begin{bmatrix} 2 & -1 & 0 \\ -1 & 2 & -1 \\ 0 & -2 & 2 \end{bmatrix}.$$

5. If $n > 1$ and if n is not a multiple of the characteristic of F, let $L = sl(n, F)$. Then L is a simple Lie algebra of dimension $n^2 - 1$. If T denotes the set of diagonal matrices in $sl(n, F)$ and if $\epsilon_i \in T^*$ is defined by $\epsilon_i(diag(d_1, \ldots, d_n)) = d_i$, then T is a maximal torus in $sl(n, F)$, Δ, the set of roots of $sl(n, F)$ with respect to T, is $\{\epsilon_i - \epsilon_j | i \neq j\}$, and the root space $L_{\epsilon_i - \epsilon_j} = F E_{i,j}$. In addition, $\{\epsilon_i - \epsilon_{i+1} | 1 \leq i \leq n-1\}$ is a base for Δ and so the Cartan matrix of $sl(n, F)$ is of type A_{n-1}.
6. Let $(.,.)$ be the symmetric bilinear form on F^{2l+1} with matrix $\begin{bmatrix} 1 & 0 & 0 \\ 0 & 0 & I_l \\ 0 & I_l & 0 \end{bmatrix}$. Then $L = \{x \in gl(2l+1, F) | (xu, v) = -(u, xv) \forall u, v \in F^{2l+1}\}$ is a Lie subalgebra of $gl(2l+1, F)$, denoted by $o(2l+1, F)$. If the characteristic of F is not 2, it is a simple algebra of dimension $2l^2 + l$. Let T denote the set of diagonal matrices in $o(2l+1, F)$. If $\epsilon_i \in T^*$ is defined by $\epsilon_i(diag(d_1, \ldots, d_{2l+1})) = d_i$, and if $\nu_i = \epsilon_{i+1}$ for $1 \leq i \leq l$, then T is a maximal torus in $o(2l+1, F)$; Δ, the set of roots of $o(2l+1, F)$ with respect to T, is $\{\pm\nu_i | 1 \leq i \leq l\} \cup \{\pm\nu_i \pm \nu_j | 1 \leq i \neq j \leq l\}$, and, for $1 \leq i \neq j \leq l$, $L_{\nu_i} = F(E_{1,l+i+1} - E_{i+1,1})$, $L_{-\nu_i} = F(E_{1,i+1} - E_{l+i+1,1})$, $L_{\nu_i - \nu_j} = F(E_{i+1,j+1} - E_{l+j+1,l+i+1})$, $L_{\nu_i+\nu_j} = F(E_{i+1,l+j+1} - E_{j+1,l+i+1})$, $L_{-\nu_i-\nu_j} = F(E_{l+i+1,j+1} - E_{l+j+1,i+1})$. In addition, $\{\nu_i - \nu_{i+1} | 1 \leq i \leq l-1\} \cup \{\nu_l\}$ is a base for Δ, and so the Cartan matrix of $o(2l+1, F)$ is of type B_l.
7. Let $(.,.)$ be the skew-symmetric bilinear form on F^{2l} with matrix $\begin{bmatrix} 0 & I_l \\ -I_l & 0 \end{bmatrix}$. Then $L = \{x \in gl(2l, F) | (xu, v) = -(u, xv) \; \forall \, u, v \in F^{2l}\}$ is a Lie subalgebra of $gl(2l, F)$, denoted by $sp(2l, F)$. If the characteristic of F is not 2, it is a simple algebra of dimension $2l^2 + l$. Let T denote the set of diagonal matrices in $sp(2l, F)$. If $\epsilon_i \in T^*$ is defined by $\epsilon_i(diag(d_1, \ldots, d_{2l})) = d_i$, and if $\mu_i = \epsilon_i$ for $1 \leq i \leq l$, then T is a maximal torus in $sp(2l, F)$, Δ, the set of roots of $sp(2l, F)$ with respect to T, is $\{\pm 2\mu_i | 1 \leq i \leq l\} \cup \{\pm\mu_i \pm \mu_j | i \neq j\}$, and, for $1 \leq i \leq l$, $L_{2\mu_i} = F E_{i,l+i}$, $L_{-2\mu_i} = F E_{l+i,i}$, $L_{\mu_i - \mu_j} = F(E_{i,j} - E_{l+j,l+i})$, $L_{\mu_i+\mu_j} = F(E_{i,l+j} + E_{j,l+i})$, $L_{-\mu_i-\mu_j} = F(E_{l+i,j} + E_{l+j,i})$. In addition, $\{\mu_i - \mu_{i+1} | 1 \leq i \leq l-1\} \cup \{2\mu_l\}$ is a base for Δ, and so the Cartan matrix of $sp(2l, F)$ is of type C_l.
8. Let $(.,.)$ be the symmetric bilinear form on F^{2l} with matrix $\begin{bmatrix} 0 & I_l \\ I_l & 0 \end{bmatrix}$. Then $L = \{x \in gl(2l, F) | (xu, v) = -(u, xv) \; \forall \; u, v \in F^{2l}\}$ is a Lie subalgebra of $gl(2l, F)$, denoted by $o(2l, F)$. If the characteristic of F is not 2, it is a simple algebra of dimension $2l^2 - l$. Let T denote the set of diagonal matrices in $o(2l, F)$. If $\epsilon_i \in T^*$ is defined by $\epsilon_i(diag(d_1, \ldots, d_{2l})) = d_i$, and if $\nu_i = \epsilon_i$ for $1 \leq i \leq l$, then T is a maximal torus in $o(2l, F)$, Δ, the set of roots of $o(2l, F)$ with respect to T, is $\{\pm\nu_i \pm \nu_j | 1 \leq i \neq j \leq l\}$, and, for $1 \leq i \neq j \leq l$, $L_{\nu_i - \nu_j} = F(E_{i,j} - E_{l+j,l+i})$, $L_{\nu_i+\nu_j} = F(E_{i,l+j} - E_{j,l+i})$, $L_{-\nu_i-\nu_j} = F(E_{l+i,j} - E_{l+j,i})$. In addition, $\{\nu_i - \nu_{i+1} | 1 \leq i \leq l-1\} \cup \{\nu_{l-1} + \nu_l\}$ is a base for Δ, and so the Cartan matrix of $o(2l, F)$ is of type D_l.

9. Let V be a vector space of dimension $n \geq 1$ over a field of characteristic 0. Let $W(n) = Der(S(V))$. Then $W(n)$ is an infinite-dimensional simple Lie algebra.
10. Let F be a field of characteristic $p > 0$. Let V be a vector space of dimension $n \geq 1$ with basis $\{x_1, \ldots, x_n\}$. Let I denote the ideal $\langle x_1^p, \ldots, x_n^p \rangle \subseteq S(V)$, $B(n : \mathbf{1})$ denote $S(V)/I$, and $W(n : \mathbf{1}) = Der(B(n : \mathbf{1}))$. Then $W(n : \mathbf{1})$ is a p-Lie algebra of dimension np^n. It is a simple Lie algebra unless $p = 2$ and $n = 1$.

70.3 Modules

Definitions:

Let A be an associative algebra and V be a vector space over F. A **representation** of A on V is a homomorphism $\phi : A \to End\ V$.

Let L be a Lie algebra and V be a vector space over F. A **representation** of L on V is a homomorphism $\phi : L \to gl(V)$.

Let B be an associative algebra or a Lie algebra. A representation $\phi : B \to gl(V)$ is **reducible** if there is some nonzero proper subspace $W \subset V$ such that $\phi(x)(W) \subseteq W$ for all $x \in B$. If ϕ is not reducible, it is **irreducible**.

Let L be a Lie algebra and M be a vector space. M is an **L-module** if there is a linear map $L \otimes M \to M, a \otimes m \mapsto am$ such that $[a, b]m = a(bm) - b(am)$ for all $a, b \in L, m \in M$.

Let M be an L-module. A subspace $N \subseteq M$ is a **submodule** of M if $LN \subseteq N$.

Let M be an L-module. M is **reducible** if M contains a nonzero proper submodule. If M is not reducible it is **irreducible**. If M is a direct sum of irreducible submodules, it is **completely reducible**.

Let M be an L-module and X be a subset of M. The submodule of M **generated by** X is the smallest submodule of M containing X.

Let L be a Lie algebra and M, N be L-modules. A linear transformation $\phi : M \to N$ is a **homomorphism** of L-modules if $\phi(xm) = x\phi(m)$ for all $x \in L, m \in M$. The set of all L-module homomorphisms from M to N is denoted $Hom(M, N)$.

Let L be a finite-dimensional semisimple Lie algebra over an algebraically closed field of characteristic 0. Let T be a maximal torus in L and M be an L-module. For $\lambda \in T^*$ define M_λ, the **λ-weight space** of M, to be $\{m \in M | tm = \lambda(t)m \forall t \in T\}$.

Let L be a finite-dimensional simple Lie algebra over an algebraically closed field of characteristic 0. Let T be a maximal torus, Δ the corresponding root system, B a base for Δ, and Δ^+ the corresponding set of positive roots. Let M be an L-module and $\lambda \in T^*$. An element $0 \neq m \in M_\lambda$ is a **highest weight vector of weight** λ if $L_\alpha m = 0$ for all $\alpha \in \Delta^+$.

Facts:

Unless specified otherwise, V denotes a vector space over a field F.

The following facts may be found in [Hum, Sect. 6].

1. Let L be a Lie algebra and $\phi : L \to gl(V)$ be a representation of L on V. Then V may be given the structure of an L-module by setting $xv = \phi(x)(v)$ for all $x \in L, v \in V$. Conversely, if M is an L-module, then the map $\phi : L \to gl(M)$ defined by $\phi(x)(m) = xm$ is a representation of L on M. A representation ϕ is irreducible if and only if the corresponding module is.
2. Let ϕ be a representation of a Lie algebra L on V. Then, by the universal property of the universal enveloping algebra, ϕ extends to a representation of $U(L)$ on V. Conversely, every representation of $U(L)$ on V restricts to a representation of L on V. A representation ϕ of $U(L)$ on V is irreducible if and only if its restriction to L is.
3. Let L be a Lie algebra, M be an L-module, and $N \subseteq M$ be a submodule. Then the quotient space M/N may be given the structure of an L-module by setting $x(m + N) = xm + N$ for all $x \in L, m \in M$.

4. Let L be a Lie algebra and M, N be L-modules. Then the vector space $M \oplus N$ may be given the structure of an L-module by setting $x(m+n) = xm + xn$ for all $x \in L, m \in M, n \in N$.
5. Let L be a Lie algebra and M, N be L-modules. Then the vector space $M \otimes N$ may be given the structure of an L-module by setting $x(m \otimes n) = xm \otimes n + m \otimes xn$ for all $x \in L, m \in M, n \in N$.
6. Let L be a Lie algebra and M, N be L-modules. Then $Hom(M, N)$ may be given the structure of an L-module by setting $(x\phi)(m) = x\phi(m) - \phi(xm)$ for all $x \in L, m \in M$.
7. Let L be a Lie algebra and V be an L-module. Then $T(V)$ is an L-module and the ideals occurring in the definitions of $S(V)$ and $\Lambda(V)$ are submodules. Hence, $S(V)$ and $\Lambda(V)$ are L-modules. Furthermore, each $S^n(V)$ is a submodule of $S(V)$ and each $\Lambda^n(V)$ is a submodule of $\Lambda(V)$.
8. [Jac62, p. 79] *Weyl's Theorem:* Let L be a finite-dimensional semisimple Lie algebra over a field of characteristic zero and let M be a finite-dimensional L-module. Then M is completely reducible.
9. [Hum72, pp. 107–114] Let L be a finite-dimensional semisimple Lie algebra over an algebraically closed field of characteristic 0 and M be a finite-dimensional L-module. Let T be a maximal torus in L, Δ be the corresponding root system, $B = \{\alpha_1, \ldots, \alpha_l\}$ a base for Δ, and Δ^+ be the corresponding set of positive roots. Then:

   - $M = \sum_{\lambda \in T^*} M_\lambda$.
   - M contains a highest weight vector of weight λ for some $\lambda \in T^*$ and setting $h_i = \frac{2t_{\alpha_i}}{(\alpha_i, \alpha_i)}$ for $1 \leq i \leq l$, we have $\lambda(h_i) \geq 0, \lambda(h_i) \in \mathbb{Z}$ for $1 \leq i \leq l$.
   - If M is irreducible and m_1, m_2 are highest weight vectors corresponding to $\lambda_1, \lambda_2 \in T^*$, then $\lambda_1 = \lambda_2$ and $Fm_1 = Fm_2$.
   - If M, N are irreducible finite-dimensional L-modules containing highest weight vectors corresponding to the same $\lambda \in T^*$, then M and N are isomorphic.
   - Let $\lambda \in T^*$ satisfy $\lambda(h_i) \in \mathbb{Z}, \lambda(h_i) \geq 0$ for $1 \leq i \leq l$. Then there exists a finite-dimensional L-module with highest weight λ.

Examples:

1. Let V be a vector space of dimension $n > 1$. Then V is a $gl(V)$-module and, hence, a $gl(n, F)$-module. Therefore, for each $k > 0$, $S^k(V)$ and $\Lambda^k(V)$ are modules for $gl(V)$ and, thus, modules for any subalgebra of $gl(n, F)$.
2. Let V be a vector space with basis $\{x, y\}$. Let $\{e, f, h\}$ be a basis for $sl(2, F)$ with $[e, f] = h, [h, e] = 2e, [h, f] = -2f$. The linear map $sl(2, F) \to Der(F[x, y])$ defined by $e \mapsto \frac{x\partial}{\partial y}, f \mapsto \frac{y\partial}{\partial x}, h \mapsto \frac{x\partial}{\partial x} - \frac{y\partial}{\partial y}$ is an isomorphism of $sl(2, F)$ into $Der(F[x, y])$. Consequently, $F[x, y] = S(V)$ is an $sl(2, F)$-module and each $S^n(V)$ is an $sl(2, F)$ submodule. $S^n(V)$ has basis $\{x^n, x^{n-1}y, \ldots, y^n\}$ and so is an $(n+1)$-dimensional $sl(2, F)$-module. It is irreducible.

70.4 Graded Algebras and Modules

Definitions:

Let V be a vector space and A be an additive abelian group. For each $\alpha \in A$, let V_α be a subspace of V. If $V = \oplus_{\alpha \in A} V_\alpha$, then V is an A-**graded vector space.**

Let B be an algebra and an A-graded vector space. B is an A-**graded algebra** if $B_\alpha B_\beta \subseteq B_{\alpha+\beta}$ for all $\alpha, \beta \in A$.

Let B be an A-graded associative algebra or an A-graded Lie algebra. Let M be a B-module and an A-graded vector space. M is an A-**graded module** for B if $B_\alpha M_\beta \subseteq M_{\alpha+\beta}$ for all $\alpha, \beta \in A$.

Let V be an A-graded vector space. V has **graded dimension** if $\dim(V_\alpha) < \infty$ for all $\alpha \in A$. In this case, we define the **graded dimension** of V to be the formal sum

$$gr\ dim(V) = \sum_{\alpha \in A} \dim(V_\alpha) t^\alpha.$$

The graded dimension of V is sometimes called the **character** of V.

Facts:

1. [FLM88, Sect. 1.10] Let V, W be A-graded vector spaces with graded dimensions. Then $V \oplus W$ is a graded vector space with graded dimension and $gr\ dim(V \oplus W) = gr\ dim(V) + gr\ dim(W)$. If W is a subspace of V and $W_\alpha \subseteq V_\alpha$ for all $\alpha \in A$, then the quotient space V/W has graded dimension and $gr\ dim(V/W) = gr\ dim(V) - gr\ dim(W)$. If $\{(\alpha, \beta) \in A \times A | V_\alpha, W_\beta \neq \{0\}, \alpha + \beta = \gamma\}$ is finite for all $\gamma \in A$, then $V \otimes W$ is a graded vector space with graded dimension, where $(V \otimes W)_\gamma = \sum_{\alpha+\beta=\gamma} V_\alpha \otimes W_\beta$ and $gr\ dim(V \otimes W) = (gr\ dim(V))(gr\ dim(W))$ (where we set $t^\alpha t^\beta = t^{\alpha+\beta}$).
2. [Bou72, p. 36] Let V be a vector space with basis X. Setting $Fr(X)_i = Fr(X) \cap V^{\otimes i}$ gives $Fr(X)$ the structure of a graded Lie algebra. If $|X| = l$ is finite, $Fr(X)$ has graded dimension and $gr\ dim(Fr(X)) = \sum_{n>0} \frac{1}{n}(\sum_{d|n} \mu(d) l^{\frac{n}{d}}) t^n$, where μ is the Möbius function (i.e., $\mu(p_1 \ldots p_r) = (-1)^r$ if $p_1, \ldots, p_r$ are distinct primes and $\mu(n) = 0$ if $p^2 | n$ for some prime p).
3. [Jac62, Sect. VIII.3] Let L be a finite-dimensional semisimple Lie algebra over an algebraically closed field of characteristic 0, T be a maximal torus in L, Δ be the corresponding root system, W be the Weyl group, Δ^+ be the set of positive roots with respect to some base, and M be a finite-dimensional L-module. Then:

   - Let $\mathbb{Z}\Delta$ denote the additive subgroup of T^* generated by Δ. Then the root space decomposition $L = T + \sum_{\alpha \in \Delta} L_\alpha$ gives L the structure of a $\mathbb{Z}\Delta$-graded Lie algebra. Here $L_0 = T$.
   - The weight space decomposition $M = \sum_{\alpha \in T^*} M_\alpha$ gives M the structure of a graded module with graded dimension.
   - *Weyl character formula:* Assume M is irreducible with highest weight λ. Let $\delta = \frac{1}{2} \sum_{\alpha \in \Delta^+} \alpha$. Then

$$gr\ dim(M) = \left(\sum_{w \in W} (\det(w) t^{w(\lambda+\delta)}) \right) \Big/ \left(\sum_{w \in W} (\det(w) t^{w\delta}) \right).$$

Examples:

1. Setting $T(V)_i = V^{\otimes i}$ gives $T(V)$ the structure of a $\mathbb{Z}$-graded algebra. If V has finite dimension l, then $T(V)$ has graded dimension and $gr\ dim(T(V)) = (1 - lt)^{-1}$.
2. Setting $S(V)_i = S^i(V)$ gives $S(V)$ the structure of a $\mathbb{Z}$-graded algebra. If V has finite dimension l, then $S(V)$ has graded dimension and $gr\ dim(S(V)) = (1 - t)^{-l}$.
3. Setting $\Lambda(V)_i = \Lambda^i(V)$ gives $\Lambda(V)$ the structure of a $\mathbb{Z}$-graded algebra. If V has finite dimension l, then $\Lambda(V)$ has graded dimension and $gr\ dim(\Lambda(V)) = (1 + t)^l$.
4. Let L be a finite-dimensional semisimple Lie algebra over an algebraically closed field of characteristic 0, T be a maximal torus in L, Δ be the corresponding root system, and $B = \{\alpha_1, \ldots, \alpha_l\}$ be a base. For $1 \leq i \leq l$ define λ_i by $\lambda_i(\frac{2t_{\alpha_j}}{(\alpha_j, \alpha_j)}) = \delta_{i,j}$. Then:

   - If $L = sl(V)$, where V is an $l+1$-dimensional vector space and the base for Δ is as described in Example 5 of section 70.2, then $\Lambda^i(V)$ is the irreducible $sl(V)$-module of highest weight λ_i for $1 \leq i \leq l$.

- If $L = o(V)$, where V is a $2l + 1$-dimensional vector space and the base for Δ is as described in Example 6 of section 70.2, then $\Lambda^i(V)$ is the irreducible $o(V)$-module of highest weight λ_i for $1 \leq i \leq l - 1$.
- If $L = o(V)$, where V is a $2l$-dimensional vector space and the base for Δ is as described in Example 8 of section 70.2, then $\Lambda^i(V)$ is the irreducible $o(V)$-module of highest weight λ_i for $1 \leq i \leq l - 2$.

References

[Bou72] N. Bourbaki, *Groupes et algebres de Lie, Chapitres 2 et 3*. Hermann, Paris, 1972.

[FLM88] I. Frenkel, J. Lepowsky, and A. Meurman, *Vertex Operator Algebas and the Monster*. Academic Press, New York, 1988.

[Hum72] J. Humphreys, *Introduction to Lie Algebras and Representation Theory*. Third printing, revised: Springer-Verlag, New York, 1980.

[Jac62] N. Jacobson, *Lie Algebras*. Reprint: Dover Publications, New York, 1979.

[Lam01] T. Lam, *A First Course in Noncommutative Rings*. Springer-Verlag, New York, 1980.

[Pol69] R.D. Pollack, *Introduction to Lie Algebras*. Queen's Papers in Pure and Applied Mathematics–No. 23, Queen's University, Kingston, Ontario, 1969.

V

Computational Software

Interactive Software for Linear Algebra

Packages of Subroutines for Linear Algebra

Interactive Software for Linear Algebra

71

MATLAB®

Steven J. Leon
University of Massachusetts Dartmouth

MATLAB® is generally recognized as the leading software for scientific computation. It was originally developed in the 1970s by Cleve Moler as an interactive Matrix Laboratory with matrix routines based on the algorithms in the LINPACK and EISPACK software libraries. In the original 1978 version everything in MATLAB was done with matrices, even the graphics. MATLAB has continued to grow and expand from the classic 1978 Fortran version to the current version, MATLAB 7, which was released in May 2004. Each new release has included significant improvements. The graphics capabilities were greatly enhanced with the introduction of Handle Graphics and Graphical User Interfaces in version 4 (1992). A sparse matrix package was also included in version 4. Over the years, dozens of toolboxes (application libraries of specialized MATLAB files) have been added in areas such as signal processing, statistics, optimization, symbolic math, splines, and image processing. MATLAB's matrix computations are now based on the LAPACK and BLAS software libraries. MATLAB is widely used in linear algebra courses. Books such as [Leo06], [LHF03], and [HZ04] have extensive sets of MATLAB exercises and projects for the standard first course in linear algebra. The book by D.J. Higham and N.J. Higham [HH03] provides a comprehensive guide to all the basic features of MATLAB.

71.1 Matrices, Submatrices, and Multidimensional Arrays

Facts:

1. The basic data elements that MATLAB uses are matrices. Matrices can be entered into a MATLAB session using square brackets.
2. If A and B are matrices with the same number of rows, then one can append the columns of B to the matrix A and form an augmented matrix C by setting `C = [A, B]`. If E is a matrix with the same number of columns as A, then one can append the rows of E to A and form an augmented matrix F by setting `F = [A; E]`.

3. Row vectors of evenly spaced entries can be generated using MATLAB's : operator.
4. A submatrix of a matrix A is specified by A(u, v) where u and v are vectors that specify the row and column indices of the submatrix.
5. MATLAB arrays can have more than two dimensions.

Commands:

1. The number of rows and columns of a matrix can be determined using MATLAB's size command.
2. The command length(x) can be used to determine the number of entries in a vector x. The length command is equivalent to the command max(size(x)).
3. MATLAB's cat command can be used to concatenate two or more matrices along a single dimension and it can also be used to create multidimensional arrays. If two matrices A and B have the same number of columns, then the command cat(1, A, B) produces the same matrix as the command [A; B]. If the matrices B and C have the same number of rows, the command cat(2, B, C) produces the same matrix as the command [B, C]. The cat command can be used with more than two matrices as arguments. For example, the command cat(2, A, B, C, D) generates the same matrix as the command [A, B, C, D]. If A1, A2, . . . , Ak are $m \times n$ matrices, the command S = cat(3, A1, A2, . . . , Ak) will generate an $m \times n \times k$ array S.

Examples:

1. The matrix

$$A = \begin{bmatrix} 1 & 2 & 4 & 3 \\ 3 & 5 & 7 & 2 \\ 2 & 4 & 1 & 1 \end{bmatrix}$$

is generated using the command

```
A = [1  2  4  3
     3  5  7  2
     2  4  1  1].
```

Alternatively, one could generate the matrix on a single line using semicolons to designate the ends of the rows

```
A = [ 1 2 4 3; 3 5 7 2; 2 4 1 1 ].
```

The command size(A) will return the vector (3, 4) as the answer. The command size(A,1) will return the value 3, the number of rows of A, and the command size(A,2) will return the value 4, the number of columns of A.

2. The command

```
x = 3 : 7
```

will generate the vector $\mathbf{x} = (3, 4, 5, 6, 7)$. To change the step size to $\frac{1}{2}$, set

```
z = 3 : 0.5 : 7.
```

This will generate the vector $\mathbf{z} = (3, 3.5, 4, 4.5, 5, 5.5, 6, 6.5, 7)$. The MATLAB commands length(x) and length(z) will generate the answers 5 and 9, respectively.

3. If A is the matrix in Example 1, then the command A(2,:) will generate the second row vector of A and the command A(:,3) will generate the third column vector of A. The submatrix of elements that are in the first two rows and last two columns is given by A(1:2, 3:4). Actually there is no need to use adjacent rows or columns. The command C = A([1 3], [1 3 4]) will generate the submatrix

$$C = \begin{bmatrix} 1 & 4 & 3 \\ 2 & 1 & 1 \end{bmatrix}.$$

4. The command

$$\mathsf{S} = \mathsf{cat}(3,\ [\,5\ 1\ 2;\ 3\ 2\ 1\,],\ [\,1\ 2\ 3;\ 4\ 5\ 6\,])$$

will produce a $2 \times 3 \times 2$ array S with

$$\mathsf{S}(:,:,1) = \begin{matrix} 5 & 1 & 2 \\ 3 & 2 & 1 \end{matrix}$$

and

$$\mathsf{S}(:,:,2) = \begin{matrix} 1 & 2 & 3 \\ 4 & 5 & 6. \end{matrix}$$

71.2 Matrix Arithmetic

The six basic MATLAB operators for doing matrix arithmetic are: $+$, $-$, $*$, ^, \, and /. The matrix left and right divide operators, \ and /, are described in Section 71.5. These same operators are also used for doing scalar arithmetic.

Facts:

1. If A and B are matrices with the same dimensions, then their sum and difference are computed using the commands: A + B and A − B.
2. If B and C are matrices and the multiplication BC is possible, then the product $E = BC$ is computed using the command

$$\mathsf{E} = \mathsf{B} * \mathsf{C}.$$

3. The kth power of a square matrix A is computed with the command A^k.
4. Scalars can be either real or complex numbers. A complex number such as $3 + 4i$ is entered in MATLAB as 3+4i. It can also be entered as 3+4∗sqrt(−1) or by using the command complex(3, 4). If i is used as a variable and assigned a value, say $i = 5$, then MATLAB will assign the expression 3 + 4 ∗ i the value 23; however, the expression 3 + 4i will still represent the complex number $3 + 4i$. In the case that i is used as a variable and assigned a numerical value, one should be careful to enter a complex number of the form $a + i$ (where a real) as a + 1i.
5. MATLAB will perform arithmetic operations element-wise when the operator is preceded by a period in the MATLAB command.
6. The conjugate transpose of a matrix B is computed using the command B′. If the matrix B is real, then B′ will be equal to the transpose of B. If B has complex entries, then one can take its transpose without conjugating using the command B′.

Commands:

1. The inverse of a nonsingular matrix C is computed using the command inv(C).
2. The determinant of a square matrix A is computed using the command det(A).

Examples:

1. If

$$A = \begin{bmatrix} 1 & 2 \\ 3 & 4 \end{bmatrix} \quad \text{and} \quad B = \begin{bmatrix} 5 & 1 \\ 2 & 3 \end{bmatrix}$$

the commands A ∗ B and A ∗ B will generate the matrices

$$\begin{bmatrix} 9 & 7 \\ 23 & 15 \end{bmatrix}, \quad \begin{bmatrix} 5 & 2 \\ 6 & 12 \end{bmatrix}.$$

71.3 Built-In MATLAB Functions

The inv and det commands are examples of built-in MATLAB functions. Both functions have a single input and a single output. Thus, the command d = det(A) has the matrix A as its input argument and the scalar d as its output argument. A MATLAB function may have many input and output arguments. When a command of the form

$$[A_1, \ldots, A_k] = \text{fname}(B_1, \ldots, B_n) \tag{71.1}$$

is used to call a function fname with input arguments $B_1, \ldots, B_n$, MATLAB will execute the function routine and return the values of the output arguments $A_1, \ldots, A_k$.

Facts:

1. The number of allowable input and output arguments for a MATLAB function is defined by a function statement in the MATLAB file that defines the function. (See section 71.8.) The function may require some or all of its input arguments. A MATLAB command of the form (71.1) may be used with j output arguments where $0 \leq j \leq k$. The MATLAB help facility describes the various input and output options for each of the MATLAB commands.

Examples:

1. The MATLAB function pi is used to generate the number π. This function is used with no input arguments.
2. The MATLAB function kron has two input arguments. If A and B are matrices, then the command kron(A,B) computes the Kronecker product of A and B. Thus, if A = [1, 2; 3, 4] and B = [1, 1; 1, 1], then the command K = kron(A, B) produces the matrix

```
K =

     1     1     2     2
     1     1     2     2
     3     3     4     4
     3     3     4     4
```

and the command L = kron(B, A) produces the matrix

```
L =

     1     2     1     2
     3     4     3     4
     1     2     1     2
     3     4     3     4.
```

3. One can compute the QZ factorization (see section 71.6) for the generalized eigenvalue problem using a command

$$[\mathsf{E}, \mathsf{F}, \mathsf{Q}, \mathsf{Z}] = \mathsf{qz}(\mathsf{A}, \mathsf{B})$$

with two input arguments and four outputs. The input arguments are square matrices A and B and the outputs are quasitriangular matrices E and F and unitary matrices Q and Z such that

$$QAZ = E \quad \text{and} \quad QBZ = F.$$

The command

$$[\mathsf{E}, \mathsf{F}, \mathsf{Q}, \mathsf{Z}, \mathsf{V}, \mathsf{W}] = \mathsf{qz}(\mathsf{A}, \mathsf{B})$$

will also compute matrices V and W of generalized eigenvectors.

71.4 Special Matrices

The ELMAT directory of MATLAB contains a collection of MATLAB functions for generating special types of matrices.

Commands:

1. The following table lists commands for generating various types of special matrices.

Matrix	Command Syntax	Description
eye	eye(n)	Identity matrix
ones	ones(n) or ones(m, n)	Matrix whose entries are all equal to 1
zeros	zeros(n) or zeros(m, n)	Matrix whose entries are all equal to 0
rand	rand(n) or rand(m, n)	Random matrix
compan	compan(p)	Companion matrix
hadamard	hadamard(n)	Hadamard matrix
gallery	gallery(matname, p1, p2, ...)	Large collection of special test matrices
hankel	hankel(c) or hankel(c,r)	Hankel matrix
hilb	hilb(n)	Hilbert matrix
invhilb	invhilb(n)	Inverse Hilbert matrix
magic	magic(n)	Magic square
pascal	pascal(n)	Pascal matrix
rosser	rosser	Test matrix for eigenvalue solvers
toeplitz	toeplitz(c) or toeplitz(c,r)	Toeplitz matrix
vander	vander(x)	Vandermonde matrix
wilkinson	wilkinson(n)	Wilkinson's eigenvalue test matrix

2. The command gallery can be used to access a large collection of test matrices developed by N. J. Higham. Enter help gallery to obtain a list of all classes of gallery test matrices.

Examples:

1. The command rand(n) will generate an $n \times n$ matrix whose entries are random numbers that are uniformly distributed in the interval $(0, 1)$. The command may be used with two input arguments

to generate nonsquare matrices. For example, the command rand(3,2) will generate a random 3×2 matrix. The command rand when used by itself with no input arguments will generate a single random number between 0 and 1.

2. The command

$$\mathsf{A} = [\,\mathsf{eye}(2),\ \mathsf{ones}(2,3);\ \mathsf{zeros}(2),\ 2 * \mathsf{ones}(2,3)\,]$$

will generate the matrix

```
A =
        1   0   1   1   1
        0   1   1   1   1
        0   0   2   2   2
        0   0   2   2   2.
```

3. The command toeplitz(c) will generate a symmetric toeplitz matrix whose first column is the vector c. Thus, the command

$$\mathsf{toeplitz}([1;\ 2;\ 3])$$

will generate

```
T =
        1   2   3
        2   1   2
        3   2   1.
```

Note that in this case, since the toeplitz command was used with no output argument, the computed value of the command toeplitz(c) was assigned to the temporary variable ans. Further computations may end up overwriting the value of ans. To keep the matrix for further use in the MATLAB session, it is advisable to include an output argument in the calling statement.

For a nonsymmetric Toeplitz matrix it is necessary to include a second input argument r to define the first row of the matrix. If $\mathsf{r}(1) \neq \mathsf{c}(1)$, the value of $c(1)$ is used for the main diagonal. Thus, commands

$$\mathsf{c} = [1;\ 2;\ 3], \quad \mathsf{r} = [9,\ 5,\ 7], \quad \mathsf{T} = \mathsf{toeplitz}(\mathsf{c}, \mathsf{r})$$

will generate

```
T =
        1   5   7
        2   1   5
        3   2   1.
```

The Toeplitz matrix generated is stored using the variable T.

4. One of the classes of gallery test matrices is circulant matrices. These are generated using the MATLAB function circul. To see how to use this function, enter the command

help private\circul.

The help information will tell you that the circul function requires an input vector v and that the command

C = gallery('circul', v)

will generate a circulant matrix whose first row is v. Thus, the command

$$\mathsf{C} = \mathsf{gallery}('\mathsf{circul}', [4, 5, 6])$$

will generate the matrix

$$\mathsf{C} = \begin{matrix} 4 & 5 & 6 \\ 6 & 4 & 5 \\ 5 & 6 & 4. \end{matrix}$$

71.5 Linear Systems and Least Squares

The simplest way to solve a linear system in MATLAB is to use the matrix left divide operator.

Facts:

1. The symbol \ represents MATLAB's matrix left divide operator. One can compute the solution to a linear system $A\mathbf{x} = \mathbf{b}$ by setting

$$\mathsf{x} = \mathsf{A} \backslash \mathsf{b}.$$

If A is an $n \times n$ matrix, then MATLAB will compute the solution using Gaussian elimination with partial pivoting. A warning message is given when the matrix is badly scaled or nearly singular. If the coefficient matrix is nonsquare, then MATLAB will return a least squares solution to the system that is essentially equivalent to computing $A^{\dagger}\mathbf{b}$ (where $A^{\dagger}$ denotes the pseudoinverse of A). In this case, MATLAB determines the numerical rank of the coefficient matrix using a QR decomposition and gives a warning when the matrix is rank deficient.

If A is an $m \times n$ matrix and B is $m \times k$, then the command

$$\mathsf{C} = \mathsf{A} \backslash \mathsf{B}$$

will produce an $n \times k$ matrix whose column vectors satisfy

$$\mathbf{c}_j = A \backslash \mathbf{b}_j \quad j = 1, \ldots, k.$$

2. The symbol / represents MATLAB's matrix right divide operator. It is defined by

$$\mathsf{B/A} = (\mathsf{A}' \backslash \mathsf{B}')'.$$

In the case that A is nonsingular, the computation B/A is essentially the same as computing BA^{-1}, however, the computation is carried out without actually computing A^{-1}.

Commands:

The following table lists some of the main MATLAB commands that are useful for linear systems.

Function	Command Syntax	Description
rref	U = rref(A)	Reduced row echelon form of a matrix
lu	[L , U] = lu(A)	LU factorization
linsolve	x = linsolve(A , b , opts)	Efficient solver for structured linear systems
chol	R = chol(A)	Cholesky factorization of a matrix
norm	p = norm(X)	Norm of a matrix or a vector
null	U = null(A)	Basis for the null space of a matrix
null	R = null(A, 'r')	Basis for null space rational form
orth	Q = orth(A)	Orthonormal basis for the column space of a matrix
rank	r = rank(A)	Numerical rank of a matrix
cond	c = cond(A)	2-norm condition number for solving linear systems
rcond	c = rcond(A)	Reciprocal of approximate 1-norm condition number
qr	[Q , R] = qr(A)	QR factorization
svd	s = svd(A)	Singular values of a matrix
svd	[U , S , V] = svd(A)	Singular value decomposition
pinv	B = pinv(A)	Pseudoinverse of a matrix

Examples:

1. The null command can be used to produce an orthonormal basis for the nullspace of a matrix. It can also be used to produce a "rational" nullspace basis obtained from the reduced row echelon form of the matrix. If

$$A = \begin{bmatrix} 1 & 1 & 1 & -1 \\ 1 & 1 & 1 & -1 \\ 1 & 1 & 1 & 1 \end{bmatrix},$$

then the command U = null(A) will produce the matrix

```
U =

   -0.8165   -0.0000
    0.4082    0.7071
    0.4082   -0.7071
   -0.0000    0.0000
```

where the entries of U are shown in MATLAB's format short (with four-digit mantissas). The column vectors of U form an orthonormal basis for the nullspace of A. The command R = null(A, 'r') will produce a matrix R whose columns form a simple basis for the nullspace.

```
R =

   -1   -1
    1    0
    0    1
    0    0.
```

2. MATLAB defines the numerical rank of a matrix to the number of singular values of the matrix that are greater than

$$\text{max(size(A))} * \text{norm(A)} * \text{eps}$$

where eps has the value 2^{-52}, which is a measure of the precision used in MATLAB computations. Let H be the 12×12 Hilbert matrix. The singular values of H can be computed using the command s = svd(H). The smallest singular values are s(11) $\approx 2.65 \times 10^{-14}$ and s(12) $\approx 10^{-16}$. Since the value of eps is approximately 2.22×10^{-16}, the computed value of rank(H) will be the numerical rank 11, even though the exact rank of H is 12. The computed value of cond(H) is approximately 1.8×10^{16} and the computed value of rcond(H) is approximately 2.6×10^{-17}.

71.6 Eigenvalues and Eigenvectors

MATLAB's eig function can be used to compute both the eigenvalues and eigenvectors of a matrix.

Commands:

1. The eig command. Given a square matrix A, the command e = eig(A) will generate a column vector e whose entries are the eigenvalues of A. The command [X, D] = eig(A) will generate a matrix X whose column vectors are the eigenvectors of A and a diagonal matrix D whose diagonal entries are the eigenvalues of A.
2. The eigshow command. MATLAB's eigshow utility provides a visual demonstration of eigenvalues and eigenvectors of 2×2 matrices. The utility is invoked by the command eigshow(A). The input argument A must be a 2×2 matrix. The command can also be used with no input argument, in which case MATLAB will take [1 3; 4 2]/4 as the default 2×2 matrix. The eigshow utility shows how the image $A\mathbf{x}$ changes as we rotate a unit vector $\mathbf{x}$ around a circle. This rotation is carried out manually using a mouse. If A has real eigenvalues, then we can observe the eigenvectors of the matrix when the vectors $\mathbf{x}$ and $A\mathbf{x}$ are in the same or opposite directions.
3. The command J = jordan(A) can be used to compute the Jordan canonical form of a matrix A. This command will only give accurate results if the entries of A are exactly represented, i.e., the entries must be integers or ratios of small integers. The command [X, J] = jordan(A) will also compute the similarity matrix X so that $A = XJX^{-1}$.
4. The following table lists some additional MATLAB functions that are useful for eigenvalue related problems.

Function	Command Syntax	Description
poly	p = poly(A)	Characteristic polynomial of a matrix
hess	H= hess(A) or [U , H] = hess(A)	Hessenberg form
schur	T= schur(A) or [U , T] = schur(A)	Schur decomposition
qz	[E,F,Q,Z]=qz(A,B)	QZ factorization for generalized eigenvalues
condeig	s = condeig(A)	Condition numbers for the eigenvalues of A
expm	E = expm(A)	Matrix exponential

71.7 Sparse Matrices

A matrix is *sparse* if most of its entries are zero. MATLAB has a special data structure for handling sparse matrices. This structure stores the nonzero entries of a sparse matrix together with their row and column indices.

Commands:

1. The command sparse is used to generate sparse matrices. When used with a single input argument the command

$$S = \text{sparse}(A)$$

will convert an ordinary MATLAB matrix A into a matrix S having the sparse data structure. More generally, a command of the form

$$\mathsf{S} = \mathsf{sparse}(\mathsf{i}, \mathsf{j}, \mathsf{s}, \mathsf{m}, \mathsf{n}, \mathsf{nzmax})$$

will generate an $m \times n$ sparse matrix S whose nonzero entries are the entries of the vector s. The row and column indices of the nonzero entries are given by the vectors i and j. The last input argument nzmax specifies the total amount of space allocated for nonzero entries. If the allocation argument is omitted, by default MATLAB will set it to equal the value of length(s).

2. MATLAB's spy command can be used to plot the sparsity pattern of a matrix. In these plots the matrix is represented by a rectangular box with dots corresponding to the positions of its nonzero entries.
3. The MATLAB directory SPARFUN contains a large collection of MATLAB functions for working with sparse matrices. The general sparse linear algebra functions are given in the following table.

MATLAB Function	Description
eigs	A few eigenvalues, using ARPACK
svds	A few singular values, using eigs
luinc	Incomplete LU factorization
cholinc	Incomplete Cholesky factorization
normest	Estimate the matrix 2-norm
condest	1-norm condition number estimate
sprank	Structural rank

All of these functions require a sparse matrix as an input argument. All have one basic output argument except in the case of luinc, where the basic output consists of the L and U factors.

4. The SPARFUN directory also includes a collection of routines for the iterative solution of sparse linear systems.

MATLAB Function	Description
pcg	Preconditioned Conjugate Gradients Method
bicg	BiConjugate Gradients Method
bicgstab	BiConjugate Gradients Stabilized Method
cgs	Conjugate Gradients Squared Method
gmres	Generalized Minimum Residual Method
lsqr	Conjugate Gradients on the Normal Equations
minres	Minimum Residual Method
qmr	Quasi-Minimal Residual Method
symmlq	Symmetric LQ Method

If A is a sparse coefficient matrix and B is a matrix of right-hand sides, then one can solve the equation $AX = B$ using a command of the form $\mathsf{X} = \mathsf{fname}(\mathsf{A}, \mathsf{B})$, where fname is one of the iterative solver functions in the table.

Examples:

1. The command

$$\mathsf{S} = \mathsf{sparse}([25, 37, 8], [211, 15, 92], [4.5, 3.2, 5.7], 200, 300)$$

will generate a 200×300 sparse matrix S whose only nonzero entries are

$$s_{25,211} = 4.5, \quad s_{37,15} = 3.2, \quad s_{8,92} = 5.7.$$

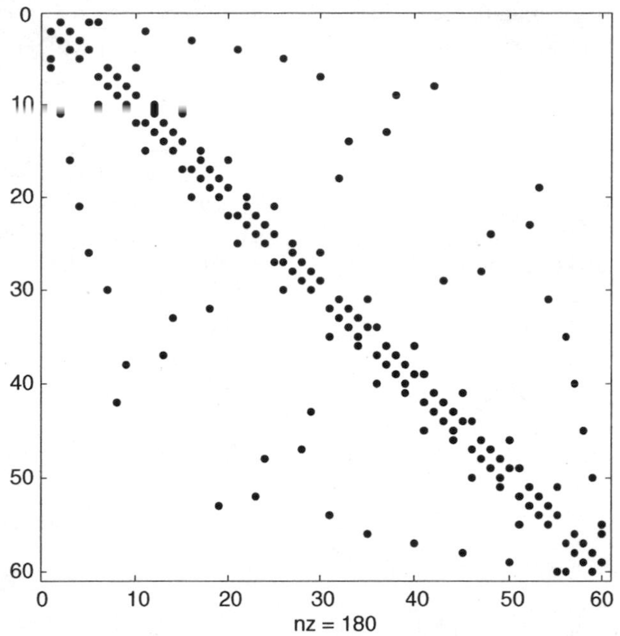

FIGURE 71.1 Spy(B).

2. The command B = bucky will generate the 60×60 sparse adjacency matrix B of the connectivity graph of the Buckminster Fuller geodesic dome and the command spy(B) will generate the spy plot shown in Figure 71.1.

71.8 Programming

MATLAB has built in all of the main structures one would expect from a high-level computer language. The user can extend MATLAB by adding on programs and new functions.

Facts:

1. MATLAB programs are called *M-files* and should be saved with a .m extension.
2. MATLAB programs may be in the form of *script* files that list a series of commands to be executed when the file is called in a MATLAB session, or they can be in the form of MATLAB functions.
3. MATLAB programs frequently include for loops, while loops, and if statements.
4. A function file must start with a function statement of the form

   function [oarg1, . . . , oargk] = fname(iarg1, . . . , iargj)

   where fname is the name of the function, iarg1,. . . ,iargj are its input arguments, and oarg1,. . . ,oargk are the output arguments. In calling a MATLAB function, it is not necessary to use all of the input and output allowed for in the general syntax of the command. In fact, MATLAB functions are commonly used with no output arguments whatsoever.
5. One can construct simple functions interactively in a MATLAB session using MATLAB's inline command. A simple function such as $f(t) = t^2 + 4$ can be described by the character array (or string)

"$t^2 + 4$." The inline command will transform the string into a function for use in the current MATLAB session. Inline functions are particularly useful for creating functions that are used as input arguments for other MATLAB functions. An inline function is not saved as an m-file and consequently is lost when the MATLAB session is ended.

6. One can use the same MATLAB command with varying amounts of input and output arguments. MATLAB keeps track of the number of input and output arguments included in the call statement using the functions nargin (the number of input arguments) and nargout (the number of output arguments). These commands are used inside the body of a MATLAB function to tailor the computations and output to the specifications of the calling statement.
7. MATLAB has six relational operators that are used for comparisons of scalars or elementwise comparisons of arrays. These operators are:

Relational Operators	
<	less than
<=	less than or equal
>	greater than
>=	greater than or equal
==	equal
~=	not equal

8. There are three logical operators as shown in the following table:

Logical Operators	
&	AND
|	OR
~	NOT

These logical operators regard any nonzero scalar as corresponding to TRUE and 0 as corresponding to FALSE. The operator & corresponds to the logical AND. If a and b are scalars, the expression $a\&b$ will equal 1 if a and b are both nonzero (TRUE) and 0 otherwise. The operator | corresponds to the logical OR. The expression $a|b$ will have the value 0 if a and b are both 0 and otherwise it will be equal to 1. The operator $\sim$ corresponds to the logical NOT. For a scalar a, the expression $\sim a$ takes on the value 1 (TRUE) if $a = 0$ (FALSE) and the value 0 (FALSE) if $a \neq 0$ (TRUE). For matrices these operators are applied element-wise. Thus, if A and B are both $m \times n$ matrices, then $A\&B$ is a matrix of zeros and ones whose (i, j) entry is $a(i, j)\&b(i, j)$.

Examples:

1. Given two $m \times n$ matrices A and B, the command C = A < B will generate an $m \times n$ matrix consisting of zeros and ones. The (i, j) entry will be equal to 1 if and only if $a_{ij} < b_{ij}$. If

$$A = \begin{bmatrix} -1 & 1 & 0 \\ 4 & -2 & 5 \\ 1 & -3 & -2 \end{bmatrix},$$

then command $A >= 0$ will generate

```
ans =

     0  1  1
     1  0  1
     1  0  0.
```

2. If

$$A = \begin{bmatrix} 3 & 0 \\ 0 & 2 \end{bmatrix} \quad \text{and} \quad B = \begin{bmatrix} 0 & 2 \\ 0 & 3 \end{bmatrix},$$

then

$$A\&B = \begin{bmatrix} 0 & 0 \\ 0 & 1 \end{bmatrix}, \quad A|B = \begin{bmatrix} 1 & 1 \\ 0 & 1 \end{bmatrix}, \quad \sim A = \begin{bmatrix} 0 & 1 \\ 1 & 0 \end{bmatrix}.$$

3. To construct the function $g(t) = 3\cos t - 2\sin t$ interactively set

```
g = inline('3 * cos(t) - 2 * sin(t)').
```

If one then enters g(0) on the command line, MATLAB will return the answer 3. The command ezplot(g) will produce a plot of the graph of $g(t)$. (See Section 71.9 for more information on producing graphics.)

4. If the numerical nullity of a matrix is defined to be the number of columns of the matrix minus the numerical rank of the matrix, then one can create a file *numnull.m* to compute the numerical nullity of a matrix. This can be done using the following lines of code.

```
function k = numnull(A)
% The command numnull(A) computes the numerical nullity of A
[m,n] = size(A);
k = n - rank(A);
```

The line beginning with the % is a comment that is not executed. It will be displayed when the command help numnull is executed. The semicolons suppress the printouts of the individual computations that are performed in the function program.

5. The following is an example of a MATLAB function to compute the circle that gives the best least squares fit to a collection of points in the plane.

```
function [center,radius,e] = circfit(x,y,w)
% The command [center,radius] = circfit(x,y) generates
% the center and radius of the circle that gives the
% best least squares fit to the data points specified
% by the input vectors x and y. If a third input
% argument is specified then the circle and data
% points will be plotted. Specify a third output
% argument to get an error vector showing how much
% each point deviates from the circle.
if size(x,1) == 1 & size(y,1) == 1
      x = x'; y = y';
end
A = [2*x, 2*y, ones(size(x))];
b = x.^2 + y.^2;
c = A\b;
center = c(1:2)';
radius = sqrt(c(3) + c(1)^2 + c(2)^2);
```

```
if nargin > 2
        t = 0:0.1:6.3;
        u = c(1) + radius * cos(t);
        v = c(2) + radius * sin(t);
        plot(x,y,'x',u,v)
        axis('equal')
end
if nargout == 3
        e = sqrt((x - c(1)).^2 + (y - c(2)).^2) - radius;
end
```

The command plot(x,y,'x',u,v) is used to plot the original (x, y) data as discrete points in the plane, with each point designated by an "x," and to also, on the same axis system, plot the (u, v) data points as a continuous curve. The following section explains MATLAB plot commands in greater detail.

71.9 Graphics

MATLAB graphics utilities allow the user to do simple two- and three-dimensional plots as well as more sophisticated graphical displays.

Facts:

1. MATLAB incorporates an objected-oriented graphics system called *Handle Graphics*. This system allows the user to modify and add on to existing figures and is useful in producing computer animations.
2. MATLAB's graphics capabilities include digital imaging tools. MATLAB images may be indexed or true color.
3. An indexed image requires two matrices, a $k \times 3$ colormap matrix whose rows are triples of numbers that specify red, green, blue intensities, and an $m \times n$ image matrix whose entries assign a colormap triple to each pixel of the image.
4. A true color image is one derived from an $m \times n \times 3$ array, which specifies the red, green, blue triplets for each pixel of the image.

Commands:

1. The plot command is used for simple plots of x-y data sets. Given a set of (x_i, y_i) data points, the command plot(x,y) plots the data points and by default sequentially draws line segments to connect the points. A third input argument may be used to specify a color (the default color for plots is black) or to specify a different form of plot such as discrete points or dashed line segments.
2. The ezplot command is used for plots of functions. The command ezplot(f) plots the function $f(x)$ on the default interval $(-2\pi, 2\pi)$ and the command ezplot(f,[a,b]) plots the function over the interval [a,b].
3. The commands plot3 and ezplot3 are used for three-dimensional plots.
4. The command meshgrid is used to generate an xy-grid for surface and contour plots. Specifically the command [X, Y] = meshgrid(u, v) transforms the domain specified by vectors u and v into arrays X and Y that can be used for the evaluation of functions of two variables and 3-D surface plots. The rows of the output array X are copies of the vector u and the columns of the output array Y are copies of the vector v. The command can be used with only one input argument in which case meshgrid(u) will produce the same arrays as the command meshgrid(u,u).

5. The mesh command is used to produce wire frame surface plots and the command surf produces a solid surface plot. If [X,Y] = meshgrid(u,v) and $Z(i, j) = f(u_i, v_j)$, then the command mesh(X,Y,Z) will produce a wire frame plot of the function $z = f(x, y)$ over the domain specified by the vectors u and v. Similarly the command surf(X,Y,Z) will generate a surface plot over the domain.
6. The MATLAB functions contour and ezcontour produce contour plots for functions of two variables.
7. The command meshc is used to graph both the mesh surface and the contour plot in the same graphics window. Similarly the command surfc will produce a surf plot with a contour graph appended.
8. Given an array C whose entries are all real, the command image(C) will produce a two-dimensional image representation of the array. Each entry of C will correspond to a small patch of the image. The image array C may be either $m \times n$ or $m \times n \times 3$. If C is an $m \times n$ matrix, then the colors assigned to each patch are determined by MATLAB's current colormap. If C is $m \times n \times 3$, a true color array, then no color map is used. In this case the entries of C(:,:,1) determine the red intensities of the image, the entries of C(:,:,2) determine green intensities, and the elements of C(:,:,3) define the blue intensities.
9. The colormap command is used to specify the current colormap for image plots of $m \times n$ arrays.
10. The imread command is used to translate a standard graphics file, such as a gif, jpeg, or tiff file, into a true color array. The command can also be used with two output arguments to determine an indexed image representation of the graphics file.

Examples:

1. The graph of the function $f(x) = \cos(x) + \sin^2(x)$ on the interval $(-2\pi, 2\pi)$ can be generated in MATLAB using the following commands:

```
x = -6.3 : 0.1 : 6.3;
y = cos(x) + sin(x).^2;
plot(x,y)
```

The graph can also be generated using the ezplot command. (See Figure 71.2.)

```
f = inline('cos(x) + sin(x).^2')
ezplot(f)
```

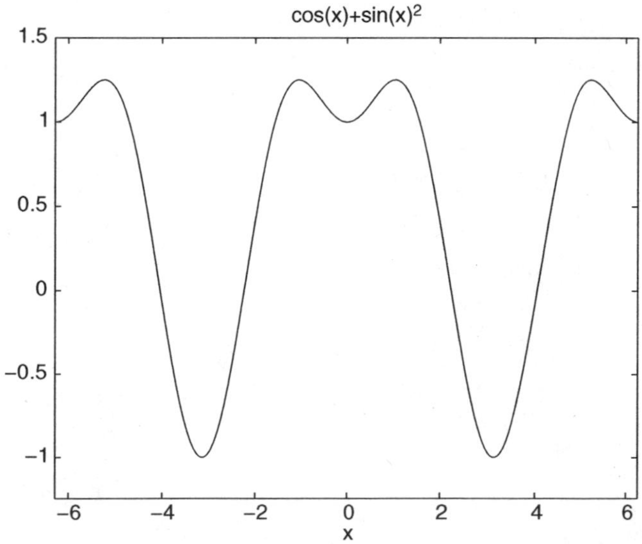

FIGURE 71.2

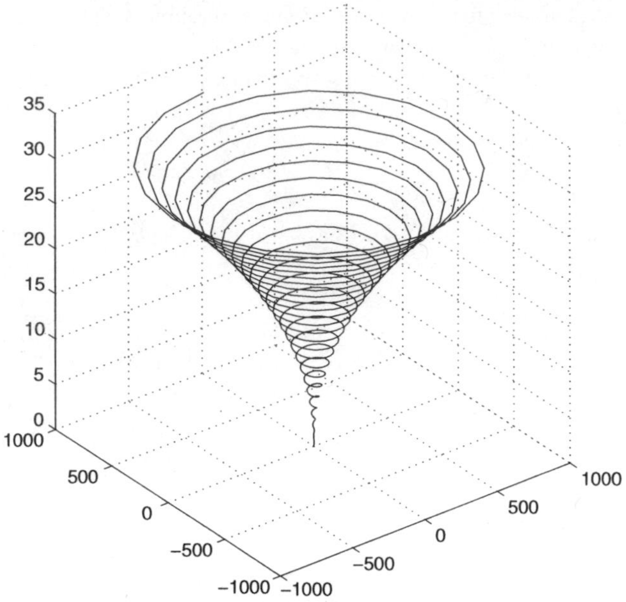

FIGURE 71.3

2. We can generate a three-dimensional plot using the following commands:

```
t = 0:pi/50:10*pi;
plot3(t.^2.*sin(5*t), t.^2.*cos(5*t), t)
grid on
axis square
```

These commands generate the plot shown in Figure 71.3.

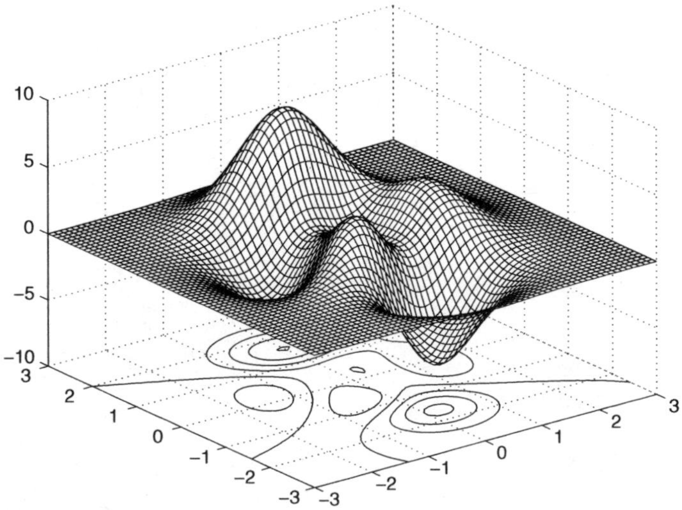

FIGURE 71.4

3. MATLAB's peaks function is a function of two variables obtained by translating and scaling Gaussian distributions. The commands

```
[X,Y] = meshgrid(−3:0.1:3);
Z = peaks(X,Y);
meshc(X,Y,Z);
```

generate the mesh and contour plots of the peaks function. (See Figure 71.4.)

71.10 Symbolic Mathematics in MATLAB

MATLAB's Symbolic Toolbox is based upon the Maple kernel from the software package produced by Waterloo Maple, Inc. The toolbox allows users to do various types of symbolic computations using the MATLAB interface and standard MATLAB commands. All symbolic computations in MATLAB are performed by the Maple kernel. For details of how symbolic linear algebra computations such as matrix inverses and eigenvalues are carried out see Chapter 72.

Facts:

1. MATLAB's symbolic toolbox allows the user to define a new data type, a symbolic object. The user can create symbolic variables and symbolic matrices (arrays containing symbolic variables).
2. The standard matrix operations $+, -, *, \hat{}, '$ all work for symbolic matrices and also for combinations of symbolic and numeric matrices. To add a symbolic matrix and a numeric matrix, MATLAB first transforms the numeric matrix into a symbolic object and then performs the addition using the Maple kernel. The result will be a symbolic matrix. In general if the matrix operation involves at least one symbolic matrix, then the result will be a symbolic matrix.
3. Standard MATLAB commands such as det, inv, eig, null, trace, rref, rank, and sum work for symbolic matrices.
4. Not all of the MATLAB matrix commands work for symbolic matrices. Commands such as norm and orth do not work and none of the standard matrix factorizations such as LU or QR work.
5. MATLAB's symbolic toolbox supports variable precision floating arithmetic, which is carried out within the Maple kernel.

Commands:

1. The sym command can be used to transform any MATLAB data structure into a symbolic object. If the input argument is a string, the result is a symbolic number or variable. If the input argument is a numeric scalar or matrix, the result is a symbolic representation of the given numeric values.
2. The syms command allows the user to create multiple symbolic variables with a single command.
3. The command subs is used to substitute for variables in a symbolic expression.
4. The command colspace is used to find a basis for the column space of a symbolic matrix.
5. The commands ezplot, ezplot3, and ezsurf are used to plot symbolic functions of one or two variables.
6. The command vpa(A,d) evaluates the matrix A using variable precision floating point arithmetic with d decimal digits of accuracy. The default value of d is 32, so if the second input argument is omitted, the matrix will be evaluated with 32 digits of accuracy.

Examples:

1. The command

```
t = sym('t')
```

transforms the string 't' into a symbolic variable t. Once the symbolic variable has been defined, one can then perform symbolic operations. For example, the command

```
factor(t^2 - 4)
```

will result in the answer

```
(t - 2) * (t + 2).
```

2. The command

```
syms a b c
```

creates the symbolic variables a, b, and c. If we then set

```
A = [ a, b, c; b, c, a; c, a, b ]
```

the result will be the symbolic matrix

```
A =
        [ a,  b,  c ]
        [ c,  a,  b ]
        [ b,  c,  a ].
```

Note that for a symbolic matrix the MATLAB output is in the form of a matrix of row vectors with each row vector enclosed by square brackets.

3. Let A be the matrix defined in the previous example. We can add the 3×3 Hilbert matrix to A using the command B = A + hilb(3). The result is the symbolic matrix

```
B =
        [ a + 1,    b + 1/2,  c + 1/3 ]
        [ c + 1/4,  a + 1/5,  b + 1/6 ]
        [ b + 1/7,  c + 1/8,  a + 1/9 ].
```

To substitute 2 for a in the matrix A we set

```
A = subs(A, a, 2).
```

The matrix A then becomes

```
A =
        [ 2,  b,  c ]
        [ c,  2,  b ]
        [ b,  c,  2 ].
```

Multiple substitutions are also possible. To replace b by b + 1 and c by 5, one need only set

```
A = subs(A, [ b, c ], [ b + 1, 5 ]).
```

4. If a is declared to be a symbolic variable, the command

```
A = [ 1 2 1; 2 4 2; 0 0 a ]
```

will produce the symbolic matrix

$$\begin{aligned} \mathsf{A} \ = \ & \\ & [\,1,\ \ 2,\ \ 1\,] \\ & [\,2,\ \ 4,\ \ 6\,] \\ & [\,0,\ \ 0,\ \ a\,]. \end{aligned}$$

The eigenvalues 0, 5, and a are computed using the command eig(A). The command

$$[\mathsf{X}, \mathsf{D}] = \mathsf{eig}(\mathsf{A})$$

generates a symbolic matrix of eigenvectors

$$\begin{aligned} \mathsf{X} \ = \ & \\ & [\,-2,\ \ 1/2 * (a+8)/(-2+3*a),\ \ 1\,] \\ & [\ \ 1,\ \ 1,\ \ 2\,] \\ & [\ \ 0,\ \ 1/2*a*(a-5)/(-2+3*a),\ \ 0\,] \end{aligned}$$

and the diagonal matrix

$$\begin{aligned} \mathsf{D} \ = \ & \\ & [\,0,\ \ 0,\ \ 0\,] \\ & [\,0,\ \ a,\ \ 0\,] \\ & [\,0,\ \ 0,\ \ 5\,]. \end{aligned}$$

When $a = 0$ the matrix A will be defective. One can substitute 0 for a in the matrix of eigenvectors using the command

$$\mathsf{X} = \mathsf{subs}(\mathsf{X}, \mathsf{a}, 0).$$

This produces the numeric matrix

$$\begin{aligned} \mathsf{X} \ = \ & \\ & -2\ \ -2\ \ \ 1 \\ & \ \ \ 1\ \ \ \ \ \ 1\ \ \ 2 \\ & \ \ \ 0\ \ \ \ \ \ 0\ \ \ 0. \end{aligned}$$

5. If we set z = exp(1), then MATLAB will compute an approximation to e that is accurate to 16 decimal digits. The command vpa(z) will produce a 32 digit representation of z, but only the first 16 digits will be accurate approximations to the digits of e. To compute e more accurately one should apply the vpa function to the symbolic expression 'exp(1)'. The command z = vpa('exp(1)') produces an answer z = 2.7182818284590452353602874713527, which is accurate to 32 digits.

71.11 Graphical User Interfaces

A graphical user interface (GUI) is a user interface whose components are graphical objects such as pushbuttons, radio buttons, text fields, sliders, checkboxes, and menus. These interfaces allow users to perform sophisticated computations and plots by simply typing numbers into boxes, clicking on buttons, or by moving slidebars.

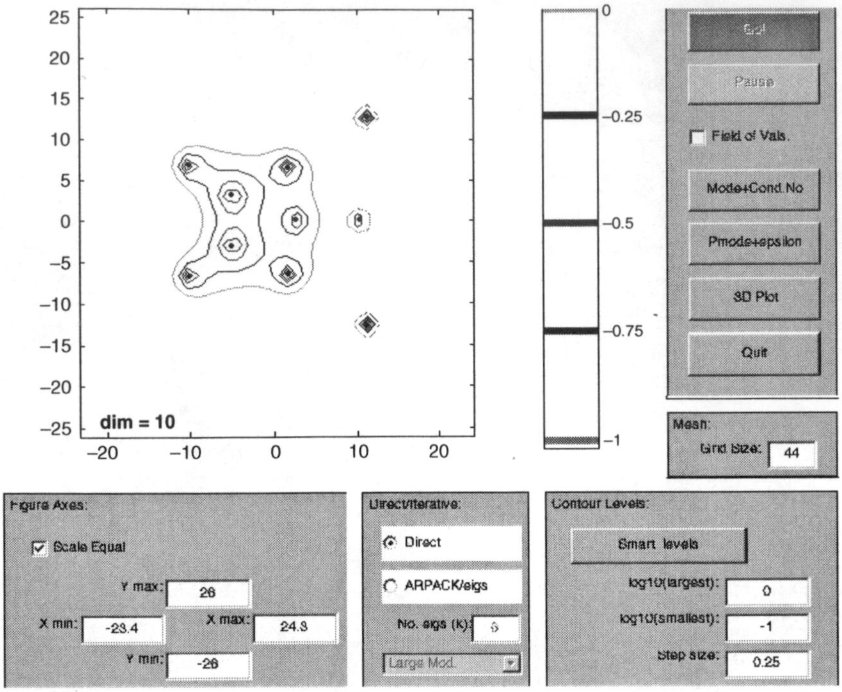

FIGURE 71.5 Eigtool GUI.

Commands:

1. The command guide opens up the MATLAB GUI Design Environment. This environment is essentially a GUI containing tools to facilitate the creation of new GUIs.

Examples:

1. Thomas G. Wright of Oxford University has developed a MATLAB GUI, eigtool, for computing eigenvalues, pseudospectra, and related quantities for nonsymmetric matrices, both dense and sparse. It allows the user to graphically visualize the pseudospectra and field of values of a matrix with just the click of a button.

   The *epsilon–pseudospectrum* of a square matrix A is defined by

$$\Lambda_\epsilon(A) = \{z \in C \mid z \in \sigma(A + E) \text{ for some } E \text{ with } \|E\| \leq \epsilon\}. \tag{71.2}$$

   In Figure 71.5 the eigtool GUI is used to plot the epsilon–pseudospectra of a 10×10 matrix for $\epsilon = 10^{-k/4}$, $k = 0, 1, 2, 3, 4$.

   For further information, see Chapter 16 and also references [Tre99] and [WT01].

2. The NSF-sponsored ATLAST Project has developed a large collection of MATLAB exercises, projects, and M-files for use in elementary linear algebra classes. (See [LHF03].) The ATLAST M-file collection contains a number of programs that make use of MATLAB's graphical user interface features to present user friendly tools for visualizing linear algebra. One example is the ATLAST cogame utility where students play a game to find linear combinations of two given vectors with the objective of obtaining a third vector that terminates at a given target point in the plane. Students can play the game at any one of four levels or play it competitively by selecting the two person game option. (See Figure 71.6.) At each step of the game a player must enter a pair of coordinates. MATLAB then plots the corresponding linear combination as a directed line segment. The game terminates when

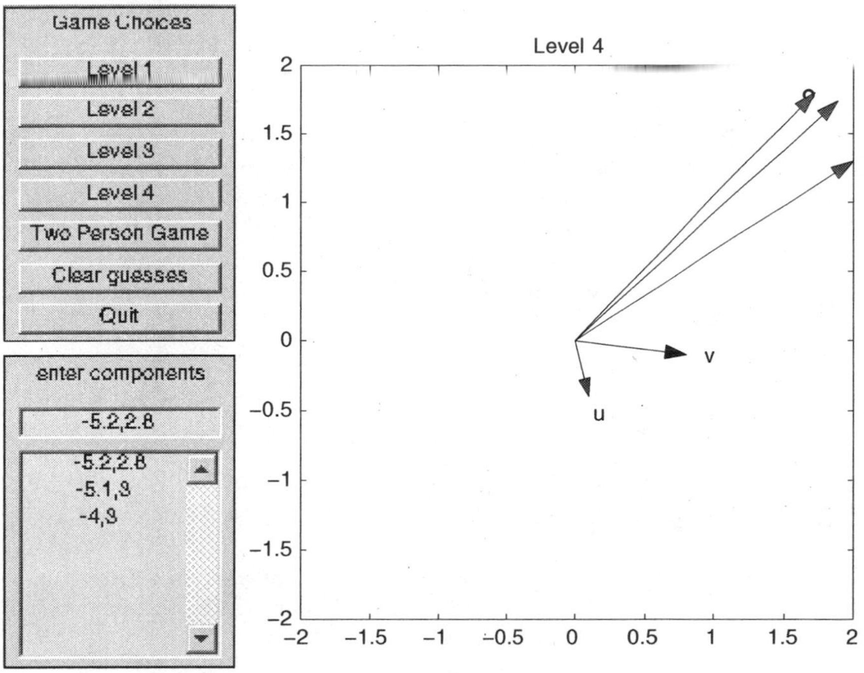

FIGURE 71.6 ATLAST Coordinate Game.

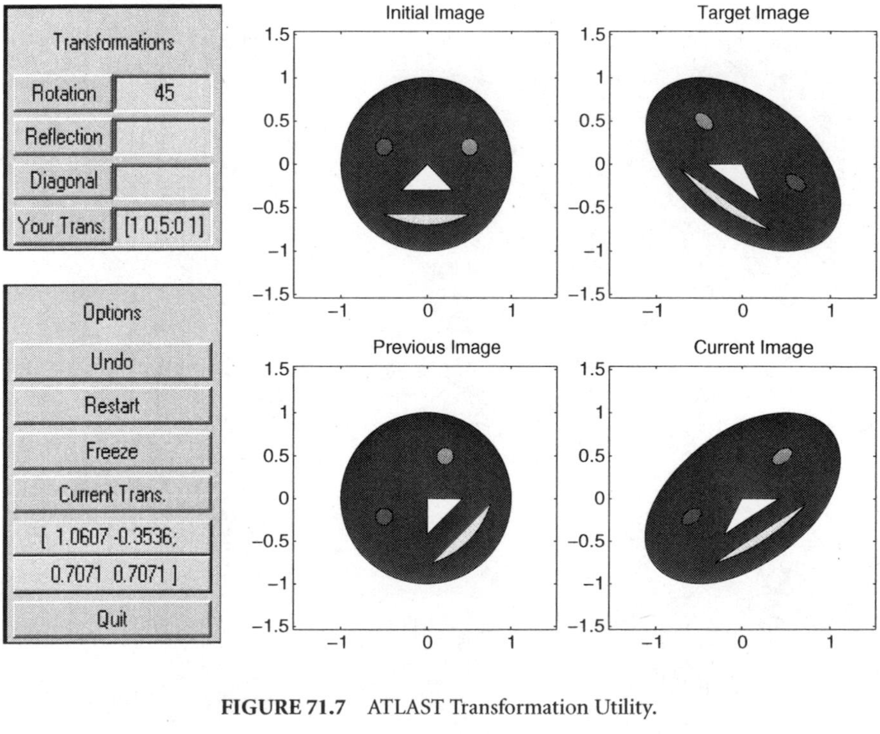

FIGURE 71.7 ATLAST Transformation Utility.

the tip of the plotted line segment lies in the small target circle. A running list of the coordinates entered in the game is displayed in the lower box to the left of the figure. The cogame GUI is useful for teaching lessons on the span of vectors in R^2 and for teaching about different bases for R^2.

3. The ATLAST transform GUI helps students to visualize the effect of linear transformations on figures in the plane. With this utility students choose an image from a list of figures and then apply various transformations to the image. Each time a transformation is applied, the resulting image is shown in the current image window. The user can then click on the *current transformation* button to see the matrix representation of the transformation that maps the original image into the current image. In Figure 71.7 two transformations were applied to an initial image. First a 45° rotation was applied. Next a transformation matrix $[1, 0; 0.5, 1]$ was entered into the "Your Transformation" text field and the corresponding transformation was applied to the lower left image with the result being displayed in the Current Image window on the lower right. To transform the Current Image into the Target Image directly above it, one would need to apply a reflection transformation.

References

[HH03] D.J. Higham and N.J. Higham. *MATLAB Guide, 2nd ed.* Philadelphia, PA.: SIAM, 2003.

[HZ04] D.R. Hill and David E. Zitarelli. *Linear Algebra Labs with MATLAB, 3rd ed.* Upper Saddle River, NJ: Prentice Hall, 2004.

[Leo06] S.J. Leon. *Linear Algebra with Applications, 7th ed.* Upper Saddle River, NJ: Prentice Hall, 2006.

[LHF03] S.J. Leon, E. Herman, and R. Faulkenberry. *ATLAST Computer Exercises for Linear Algebra, 2nd ed.* Upper Saddle River, NJ: Prentice Hall, 2003.

[Tre99] N.J. Trefethen. Computation of pseudospectra. *Acta Numerica*, 8, 247–295, 1999.

[WT01] T.G. Wright and N.J. Trefethen. Large-scale computation of pseudospectra using ARPACK and eigs. *SIAM J. Sci. Comp.*, 23(2):591–605, 2001.

72

Linear Algebra in Maple®

David J. Jeffrey
The University of Western Ontario

Robert M. Corless
The University of Western Ontario

72.1 Introduction

Maple® is a general purpose computational system, which combines symbolic computation with exact and approximate (floating-point) numerical computation and offers a comprehensive suite of scientific graphics as well. The main library of functions is written in the Maple programming language, a rich language designed to allow easy access to advanced mathematical algorithms. A special feature of Maple is user access to the source code for the library, including the ability to trace Maple's execution and see its internal workings; only the parts of Maple that are compiled, for example, the kernel, cannot be traced. Another feature is that users can link to LAPACK library routines transparently, and thereby benefit from fast and reliable floating-point computation. The development of Maple started in the early 80s, and the company **Maplesoft** was founded in 1988. A strategic partnership with NAG Inc. in 2000 brought highly efficient numerical routines to Maple, including LAPACK.

There are two linear algebra packages in Maple: `LinearAlgebra` and `linalg`. The `linalg` package is older and considered obsolete; it was replaced by `LinearAlgebra` in MAPLE 6. Here we describe only the `LinearAlgebra` package. The reader should be careful when reading other reference books, or the Maple help pages, to check whether reference is made to `vector, matrix, array` (notice the lower-case initial letter), which means that the older package is being discussed, or to `Vector, Matrix, Array` (with an upper-case initial letter), which means that the newer package is being discussed.

Facts:

1. Maple commands are typed after a prompt symbol, which by default is "greater than" (>). In examples below, keyboard input is simulated by prefixing the actual command typed with the prompt symbol.
2. In the examples below, some of the commands are too long to fit on one line. In such cases, the Maple continuation character backslash (\ ) is used to break the command across a line.
3. Maple commands are terminated by either semicolon (;) or colon (:). Before Maple 10, a terminator was required, but in the Maple 10 GUI it can be replaced by a carriage return. The semicolon terminator allows the output of a command to be displayed, while the colon suppresses the display (but the command still executes).
4. To access the commands described below, load the `LinearAlgebra` package by typing the command (after the prompt, as shown)
   `> with( LinearAlgebra );`
   If the package is not loaded, then either a typed command will not be recognized, or a different command with the same name will be used.
5. The results of a command can be assigned to one or more variables. Thus,
   `> a := 1 ;`
   assigns the value 1 to the variable a, while
   `> (a,b,c) := 1,2,3 ;`
   assigns a the value 1, b the value 2 and c the value 3. *Caution*: The operator colon-equals (:=) is assignment, while the operator equals (=) defines an equation with a left-hand side and a right-hand side.
6. A sequence of expressions separated by commas is an expression sequence in Maple, and some commands return expression sequences, which can be assigned as above.
7. Ranges in Maple are generally defined using a pair of periods (..). The rules for the ranges of subscripts are given below.

72.2 Vectors

Facts:

1. In Maple, vectors are not just lists of elements. Maple separates the idea of the mathematical object `Vector` from the data object `Array` (see Section 72.4).
2. A Maple `Vector` can be converted to an `Array`, and an `Array` of appropriate shape can be converted to a `Vector`, but they cannot be used interchangeably in commands. See the help file for `convert` to find out about other conversions.
3. Maple distinguishes between column vectors, the default, and row vectors. The two types of vectors behave differently, and are not merely presentational alternatives.

Commands:

1. Generation of vectors:

   - `Vector(` $[x_1, x_2, \ldots]$ `)` Construct a column vector by listing its elements. The length of the list specifies the dimension.
   - `Vector[column](` $[x_1, x_2, \ldots]$ `)` Explicitly declare the column attribute.
   - `Vector[row](` $[x_1, x_2, \ldots]$ `)` Construct a row vector by initializing its elements from a list.
   - $<v_1, v_2, \ldots>$ Construct a column vector with elements v_1, v_2, etc. An element can be another column vector.

- $<v_1|v_2|\ldots >$ Construct a row vector with elements v_1, v_2, etc. An element can be another row vector. A useful mnemonic is that the vertical bars remind us of the column dividers in a table.
- `Vector( n, k->f(k) )`. Construct an n-dimensional vector using a function $f(k)$ to define the elements. $f(k)$ is evaluated sequentially for k from 1 to n. The notation `k->f(k)` is Maple syntax for a univariate function.
- `Vector( n, fill=v )` An n-dimensional vector with every element **v**.
- `Vector( n, symbol=v )` An n-dimensional vector containing symbolic components v_k.
- `map( x->f(x), V )` Construct a new vector by applying function $f(x)$ to each element of the vector named `V`. *Caution*: the command is `map` *not* `Map`.

2. Operations and functions:

- `v[i]` Element i of vector **v**. The result is a scalar. *Caution*: A symbolic reference `v[i]` is typeset as v_i on output in a Maple worksheet.
- `v[p..q]` Vector consisting of elements v_i, $p \leq i \leq q$. The result is a `Vector`, even for the case `v[p..p]`. Either of p or q can be negative, meaning that the location is found by counting backwards from the end of the vector, with -1 being the last element.
- `u+v, u-v` Add or subtract `Vectors` **u**, **v**.
- `a*v` Multiply vector **v** by scalar a. Notice the operator is "asterisk" (*).
- `u . v`, `DotProduct( u, v )` The inner product of `Vectors` **u** and **v**. See examples for complex conjugation rules. Notice the operator is "period" (.) not "asterisk" (*) because inner product is not commutative over the field of complex numbers.
- `Transpose( v ), v^%T` Change a column vector into a row vector, or *vice versa*. Complex elements are not conjugated.
- `HermitianTranspose( v ), v^%H` Transpose with complex conjugation.
- `OuterProductMatrix( u, v )` The outer product of `Vectors` **u** and **v** (ignoring the row/column attribute).
- `CrossProduct( u, v ), u &x v` The vector product, or cross product, of three-dimensional vectors **u**, **v**.
- `Norm( v, 2 )` The 2-norm or Euclidean norm of vector **v**. Notice that the second argument, namely the 2, is necessary, because `Norm( v )` defaults to the infinity norm, which is different from the default in many textbooks and software packages.
- `Norm( v, p )` The p-norm of **v**, namely $(\sum_{i=1}^{n} |v_i|^p)^{(1/p)}$.

Examples:

In this section, the imaginary unit is the Maple default I. That is, $\sqrt{-1} = I$. In the matrix section, we show how this can be changed. To save space, we shall mostly use row vectors in the examples.

1. Generate vectors. The same vector created different ways.

```
> Vector[row]([0,3,8]): <0|3|8>: Transpose(<0,3,8>): Vector[row]
(3,i->i^2-1);
```

$$[0, 3, 8]$$

2. Selecting elements.

```
> V:=<a|b|c|d|e|f>: V1 := V[2 .. 4]; V2:=V[-4 .. -1];
V3:=V[-4 .. 4];
```

$$V1 := [b, c, d], \quad V2 := [c, d, e, f], \quad V3 := [c, d]$$

3. A Gram–Schmidt exercise.

```
> u1 := <3|0|4>: u2 := <2|1|1>: w1n := u1/Norm( u1, 2 );
```

$$w1n := [3/5, 0, 4/5]$$

```
> w2 := u2 - (u2 . w1n)*w1n; w2n := w2/Norm( w2, 2 );
```

$$w2n := \left[\frac{2\sqrt{2}}{5}, \frac{\sqrt{2}}{2}, -\frac{3\sqrt{2}}{10}\right]$$

4. Vectors with complex elements. Define column vectors $\mathbf{u_c}$,$\mathbf{v_c}$ and row vectors $\mathbf{u_r}$,$\mathbf{v_r}$.

```
> uc := <1 + I,2>: ur := Transpose( uc ): vc := <5,2 - 3*I>:
vr := Transpose( vc ):
```

The inner product of column vectors conjugates the first vector in the product, and the inner product of row vectors conjugates the second.

```
> inner1 := uc . vc; inner2 := ur . vr;
            inner1 := 9-11 I  , inner2 := 9+11 I
```

Maple computes the product of two similar vectors, i.e., both rows or both columns, as a true mathematical inner product, since that is the only definition possible; in contrast, if the user mixes row and column vectors, then Maple does not conjugate:

```
> but := ur . vc;
```

$$but := 9 - I$$

Caution: The use of a period (.) with complex row and column vectors together differs from the use of a period (.) with complex $1 \times m$ and $m \times 1$ matrices. In case of doubt, use matrices and conjugate explicitly where desired.

72.3 Matrices

Facts:

1. One-column matrices and vectors are not interchangeable in Maple.
2. Matrices and two-dimensional arrays are not interchangeable in Maple.

Commands:

1. Generation of Matrices.

- `Matrix( [[a,b,...],[c,d,...],...] )` Construct a matrix row-by-row, using a list of lists.
- `<< a|b|...>,<c|d|...>,...>` Construct a matrix row-by-row using vectors. Notice that the rows are specified by row vectors, requiring the | notation.
- `<< a,b,...>|< c,d,...>|...>` Construct a matrix column-by-column using vectors. Notice that each vector is a column, and the columns are joined using | , the column operator.

  Caution: Both variants of the `<< ... >>` constructor are meant for interactive use, not programmatic use. They are slow, especially for large matrices.

- `Matrix( n, m, (i,j)->f(i,j) )` Construct a matrix $n \times m$ using a function $f(i, j)$ to define the elements. $f(i, j)$ is evaluated sequentially for i from 1 to n and j from 1 to m. The notation `(i,j)->f(i,j)` is Maple syntax for a bivariate function $f(i, j)$.
- `Matrix( n, m, fill=a )` An $n \times m$ matrix with each element equal to a.
- `Matrix( n, m, symbol=a )` An $n \times m$ matrix containing subscripted entries a_{ij}.
- `map( x->f(x), M )` A matrix obtained by applying $f(x)$ to each element of M.

  [*Caution*: the command is `map` *not* `Map`.]
- << A|B>, < C|D>> Construct a partitioned or block matrix from matrices A, B, C, D. Note that < A|B > will be formed by adjoining columns; the block < C|D > will be placed below < A|B >. The Maple syntax is similar to a common textbook notation for partitioned matrices.

2. Operations and functions

- `M[i,j]` Element i, j of matrix M. The result is a scalar.
- `M[1..-1,k]` Column k of `Matrix` M. The result is a `Vector`.
- `M[k,1..-1]` Row k of `Matrix` M. The result is a row `Vector`.
- `M[p..q,r..s]` Matrix consisting of submatrix m_{ij}, $p \leq i \leq q, r \leq j \leq s$. In HANDBOOK notation, $M[\{p, \ldots, q\}, \{r, \ldots, s\}]$.
- `Transpose( M )`, `M^%T` Transpose matrix M, without taking the complex conjugate of the elements.
- `HermitianTranspose( M )`, `M^%H` Transpose matrix M, taking the complex conjugate of elements.
- `A±B` Add/subtract compatible matrices or vectors A, B.
- `A.B` Product of compatible matrices or vectors A, B. The examples below detail the ways in which Maple interprets products, since there are differences between Maple and other software packages.
- `MatrixInverse( A )`, `A^(-1)` Inverse of matrix A.
- `Determinant( A )` Determinant of matrix A.
- `Norm( A, 2 )` The (subordinate) 2-norm of matrix A, namely $\max_{\|u\|_2=1} \|Au\|_2$ where the norm in the definition is the vector 2-norm.

  Cautions:

  (a) Notice that the second argument, i.e., 2, is necessary because `Norm( A )` defaults to the infinity norm, which is different from the default in many textbooks and software packages.

  (b) Notice also that this is the largest *singular value* of A, and is usually different from the Frobenius norm $\|A\|_F$, accessed by `Norm( A, Frobenius )`, which is the Euclidean norm of the vector of elements of the matrix A.

  (c) Unless A has floating-point entries, this norm will not usually be computable explicitly, and it may be expensive even to try.
- `Norm( A, p )` The (subordinate) matrix p-norm of A, for integers $p >= 1$ or for p being the symbol `infinity`, which is the default value.

Examples:

1. A matrix product.

```
> A := <<1|-2|3>,<0|1|1>>; B := Matrix(3, 2, symbol=b); C := A .
B;
```

$$A := \begin{bmatrix} 1 & -2 & 3 \\ 0 & 1 & 1 \end{bmatrix}, \qquad B := \begin{bmatrix} b_{11} & b_{12} \\ b_{21} & b_{22} \\ b_{31} & b_{32} \end{bmatrix},$$

$$C := \begin{bmatrix} b_{11} - 2b_{21} + 3b_{31} & b_{12} - 2b_{22} + 3b_{32} \\ b_{21} + b_{31} & b_{22} + b_{32} \end{bmatrix}.$$

2. A Gram–Schmidt calculation revisited.

If u_1, u_2 are $m \times 1$ column matrices, then the Gram–Schmidt process is often written in textbooks as

$$w_2 = u_2 - \frac{u_2^T u_1}{u_1^T u_1} u_1.$$

Notice, however, that $u_2^T u_1$ and $u_1^T u_1$ are strictly 1×1 matrices. Textbooks often skip over the conversion of $u_2^T u_1$ from a 1×1 matrix to a scalar. Maple, in contrast, does not convert automatically.

Transcribing the printed formula into Maple will cause an error. Here is the way to do it, reusing the earlier numerical data.

```
> u1 := <<3,0,4>>; u2 := <<2,1,1>>; r := u2^%T . u1;
 s := u1^%T . u1;
```

$$u1 := \begin{bmatrix} 3 \\ 0 \\ 4 \end{bmatrix}, \quad u2 := \begin{bmatrix} 2 \\ 1 \\ 1 \end{bmatrix}, \quad r := [10], \quad s := [25].$$

Notice the brackets in the values of `r` and `s` because they are matrices. Since `r[1,1]` and `s[1,1]` are scalars, we write

```
> w2 := u2 - r[1,1]/s[1,1]*u1;
```

and reobtain the result from Example 3 in Section 72.2. Alternatively, $u1$ and $u2$ can be converted to `Vectors` first and then used to form a proper scalar inner product.

```
> r := u2[1..-1,1] . u1[1..-1,1]; s := u1[1..-1,1] . u1[1..-1,1];
w2 := u2-r/s*u1;
```

$$r := 10, \quad s := 25, \quad w2 := \begin{bmatrix} 4/5 \\ 1 \\ -3/5 \end{bmatrix}.$$

3. Vector–Matrix and Matrix–Vector products.

Many textbooks equate a column vector and a one-column matrix, but this is not generally so in Maple. Thus

```
> b := <1,2>; B := <<1,2>>; C := <<4|5|6>>;
```

$$b := \begin{bmatrix} 1 \\ 2 \end{bmatrix} B := \begin{bmatrix} 1 \\ 2 \end{bmatrix} C := \begin{bmatrix} 4 & 5 & 6 \end{bmatrix}.$$

Only the product `B . C` is defined, and the product `b . C` causes an error.

```
> B . C
```

$$\begin{bmatrix} 4 & 5 & 6 \\ 8 & 10 & 12 \end{bmatrix}.$$

The rules for mixed products are

```
Vector[row]( n ) . Matrix( n, m )        =   Vector[row]( m )
Matrix( n, m ) . Vector[column]( m )     =   Vector[column]( n )
```

The combinations `Vector(n). Matrix(1, m)` and `Matrix(m, 1). Vector[row](n)` cause errors. If users do not want this level of rigor, then the easiest thing to do is to use only the `Matrix` declaration.

4. Working with matrices containing complex elements.
First, notation: In linear algebra, I is commonly used for the identity matrix. This corresponds to the `eye` function in MATLAB. However, by default, Maple uses I for the imaginary unit, as seen in section 72.2. We can, however, use I for an identity matrix by changing the imaginary unit to something else, say `_i`.
```
> interface( imaginaryunit=_i):
```
As the saying goes: An `_i` for an I and an I for an `eye`.
Now we can calculate eigenvalues using notation similar to introductory textbooks.
```
> A := <<1,2>|<-2,1>>; I := IdentityMatrix( 2 );
p := Determinant ( x*I-A );
```

$$A := \begin{bmatrix} 1 & -2 \\ 2 & 1 \end{bmatrix}, \quad I := \begin{bmatrix} 1 & 0 \\ 0 & 1 \end{bmatrix}, \quad p := x^2 - 2x + 5.$$

Solving $p = 0$, we obtain eigenvalues $1 + 2i, 1 - 2i$. With the above setting of `imaginaryunit`, Maple will print these values as `1+2 _i, 1-2 _i`, but we have translated back to standard mathematical i, where $i^2 = -1$.

5. Moore–Penrose inverse. Consider `M := Matrix(3,2,[[1,1],[a,a^2],[a^2,a]]);`, a 3×2 matrix containing a symbolic parameter a. We compute its Moore–Penrose pseudoinverse and a proviso guaranteeing correctness by the command `> (Mi, p):= MatrixInverse\ (M, method=pseudo, output=[inverse, proviso]);` which assigns the 2×3 pseudoinverse to `Mi` and an expression, which if nonzero guarantees that `Mi` is the correct (unique) Moore–Penrose pseudoinverse of M. Here we have

$$Mi := \begin{bmatrix} \left(2 + 2a^3 + a^2 + a^4\right)^{-1} & -\dfrac{a^3 + a^2 + 1}{a\,(a^5 + a^4 - a^3 - a^2 + 2a - 2)} & \dfrac{a^4 + a^3 + 1}{a\,(a^5 + a^4 - a^3 - a^2 + 2a - 2)} \\ \left(2 + 2a^3 + a^2 + a^4\right)^{-1} & \dfrac{a^4 + a^3 + 1}{a\,(a^5 + a^4 - a^3 - a^2 + 2a - 2)} & -\dfrac{a^3 + a^2 + 1}{a\,(a^5 + a^4 - a^3 - a^2 + 2a - 2)} \end{bmatrix}$$

and $p = a^2 - a$. Thus, if $a \neq 0$ and $a \neq 1$, the computed pseudoinverse is correct. By separate computations we find that the pseudoinverse of $M|_{a=0}$ is

$$\begin{bmatrix} 1/2 & 0 & 0 \\ 1/2 & 0 & 0 \end{bmatrix}$$

and that the pseudoinverse of $M|_{a=1}$ is

$$\begin{bmatrix} 1/6 & 1/6 & 1/6 \\ 1/6 & 1/6 & 1/6 \end{bmatrix}$$

and *moreover that these are not special cases of the generic answer returned previously*. In a certain sense this is obvious: the Moore–Penrose inverse is *discontinuous*, even for square matrices (consider $(A - \lambda I)^{-1}$, for example, as $\lambda \to$ an eigenvalue of A).

72.4 Arrays

Before describing Maple's `Array` structure, it is useful to say why Maple distinguishes between an `Array` and a `Vector` or `Matrix`, when other books and software systems do not. In linear algebra, two different types of operations are performed with vectors or matrices. The first type is described in Sections 72.2 and 72.3, and comprises operations derived from the mathematical structure of vector spaces. The other type comprises operations that treat vectors or matrices as data arrays; they manipulate the individual elements directly. As an example, consider dividing the elements of `Array` $[1, 3, 5]$ by the elements of $[7, 11, 13]$ to obtain $[1/7, 3/11, 5/13]$.

The distinction between the operations can be made in two places: in the name of the operation or the name of the object. In other words we can overload the data objects or overload the operators. Systems such as MATLAB choose to leave the data object unchanged, and define separate operators. Thus, in MATLAB the statements $[1, 3, 5]/[7, 11, 13]$ and $[1, 3, 5]./[7, 11, 13]$ are different because of the operators. In contrast, Maple chooses to make the distinction in the data object, as will now be described.

Facts:

1. The Maple `Array` is a general data structure akin to arrays in other programming languages.
2. An array can have up to 63 indices and each index can lie in any integer range.
3. The description here only addresses the overlap between Maple `Array` and `Vector`.

Caution: A Maple `Array` might look the same as a vector or matrix when printed.

Commands:

1. Generation of arrays.
   - `Array(`$[x_1, x_2, \ldots]$`)` Construct an array by listing its elements.
   - `Array( m..n )` Declare an empty 1-dimensional array indexed from m to n.
   - `Array( v )` Use an existing `Vector` to generate an array.
   - `convert( v, Array )` Convert a `Vector v` into an `Array`. Similarly, a `Matrix` can be converted to an `Array`. See the help file for `rtable_options` for advanced methods to convert efficiently, in-place.
2. Operations (memory/stack limitations may restrict operations).
   - `a ^ n` Raise each element of a to power n.
   - `a * b, a + b, a - b` Multiply (add, subtract) elements of b by (to, from) elements of a.
   - `a / b` Divide elements of a by elements of b. Division by zero will produce `undefined` or `infinity` (or exceptions can be caught by user-set traps; see the help file for `Numeric_Events`).

Examples:

1. Array arithmetic.

```
> simplify( (Array([25,9,4])*Array(1..3,x->x^2-1 )/Array(<5,3,2>\
))^(1/2));
```

$$[0, 3, 4]$$

2. Getting `Vectors` and `Arrays` to do the same thing.

```
> Transpose( map(x->x*x,<1,2,3> )) - convert(Array( [1,2,3] )^2,\
Vector);
```

$$[0, 0, 0]$$

72.5 Equation Solving and Matrix Factoring

Cautions:

1. If a matrix contains exact numerical entries, typically integers or rationals, then the material studied in introductory textbooks transfers to a computer algebra system without special considerations. However, if a matrix contains symbolic entries, then the fact that computations are completed without the user seeing the intermediate steps can lead to unexpected results.
2. Some of the most popular matrix functions are discontinuous when applied to matrices containing symbolic entries. Examples are given below.
3. Some algorithms taught to educate students about the concepts of linear algebra often turn out to be ill-advised in practice: computing the characteristic polynomial and then solving it to find eigenvalues, for example; using Gaussian elimination without pivoting on a matrix containing floating-point entries, for another.

Commands:

1. `LinearSolve( A, B )` The vector or matrix X satisfying $AX = B$.
2. `BackwardSubstitute( A, B ), ForwardSubstitute( A, B )` The vector or matrix X satisfying $AX = B$ when A is upper or lower triangular (echelon) form, respectively.
3. `ReducedRowEchelonForm( A ).` The reduced row-echelon form (RREF) of the matrix A. For matrices with symbolic entries, see the examples below for recommended usage.
4. `Rank( A )` The rank of the matrix A. *Caution*: If A has floating-point entries, see the section below on Numerical Linear Algebra. On the other hand, if A contains symbolic entries, then the rank may change discontinuously and the generic answer returned by `Rank` may be incorrect for some specializations of the parameters.
5. `NullSpace( A )` The nullspace (kernel) of the matrix A. *Caution*: If A has floating-point entries, see the section below on Numerical Linear Algebra. Again on the other hand, if A contains symbolic entries, the nullspace may change discontinuously and the generic answer returned by `NullSpace` may be incorrect for some specializations of the parameters.
6. `( P, L, U, R ) := LUDecomposition( A, method='RREF' )` The $PLUR$, or Turing, factors of the matrix A. See examples for usage.
7. `( P, L, U ) := LUDecomposition( A )` The PLU factors of a matrix A, when the RREF R is not needed. This is usually the case for a Turing factoring where R is guaranteed (or known *a priori*) to be I, the identity matrix, for all values of the parameters.
8. `( Q, R ) := QRDecomposition( A, fullspan )` The QR factors of the matrix A. The option `fullspan` ensures that Q is square.
9. `SingularValues( A )` See Section 72.8, Numerical Linear Algebra.
10. `ConditionNumber( A )` See Section 72.8, Numerical Linear Algebra.

Examples:

1. Need for Turing factoring.

   One of the strengths of Maple is computation with symbolic quantities. When standard linear algebra methods are applied to matrices containing symbolic entries, the user must be aware of new mathematical features that can arise. The main feature is the discontinuity of standard matrix functions, such as the reduced row-echelon form and the rank, both of which can be discontinuous. For example, the matrix

$$B = A - \lambda I = \begin{bmatrix} 7-\lambda & 4 \\ 6 & 2-\lambda \end{bmatrix}$$

has the reduced row-echelon form

$$\text{ReducedRowEchelonForm}(B) = \begin{cases} \begin{bmatrix} 1 & 0 \\ 0 & 1 \end{bmatrix} & \lambda \neq -1, 10 \\ \begin{bmatrix} 1 & -4/3 \\ 0 & 0 \end{bmatrix} & \lambda = 10, \\ \begin{bmatrix} 1 & 1/2 \\ 0 & 0 \end{bmatrix} & \lambda = -1. \end{cases}$$

Notice that the function is discontinuous precisely at the interesting values of λ. Computer algebra systems in general, and Maple in particular, return "generic" results. Thus, in Maple, we have

```
> B := << 7-x | 4 >, < 6 | 2-x >>;
```

$$B := \begin{bmatrix} 7-x & 4 \\ 6 & 2-x \end{bmatrix},$$

```
> ReducedRowEchelonForm( B )
```

$$\begin{bmatrix} 1 & 0 \\ 0 & 1 \end{bmatrix}.$$

This difficulty is discussed at length in [CJ92] and [CJ97]. The recommended solution is to use Turing factoring (generalized PLU decomposition) to obtain the reduced row-echelon form with provisos. Thus, for example,

```
> A := <<1|-2|3|sin(x)>,<1|4*cos(x)|3|3*sin(x)>,<-1|3|cos(x)-3|\
cos(x)>>;
```

$$A := \begin{bmatrix} 1 & -2 & 3 & \sin x \\ 1 & 4\cos x & 3 & 3\sin x \\ -1 & 3 & \cos x - 3 & \cos x \end{bmatrix}.$$

```
> ( P, L, U, R ) := LUDecomposition( A, method='RREF' ):
```

The generic reduced row-echelon form is then given by

$$R = \begin{bmatrix} 1 & 0 & 0 & (2\sin x\cos x - 3\sin x - 6\cos x - 3)/(2\cos x + 1) \\ 0 & 1 & 0 & \sin x/(2\cos x + 1) \\ 0 & 0 & 1 & (2\cos x + 1 + 2\sin x)/(2\cos x + 1) \end{bmatrix}$$

This shows a visible failure when $2\cos x + 1 = 0$, but the other discontinuity is invisible, and requires the U factor from the Turing $(PLUR)$ factors,

$$U = \begin{bmatrix} 1 & -2 & 3 \\ 0 & 4\cos x + 2 & 0 \\ 0 & 0 & \cos x \end{bmatrix}$$

to see that the case $\cos x = 0$ also causes failure. In both cases (meaning the cases $2\cos x + 1 = 0$ and $\cos x = 0$), the RREF must be recomputed to obtain the singular cases correctly.

2. QR factoring.
   Maple does not offer column pivoting, so in pathological cases the factoring may not be unique, and will vary between software systems. For example,
   `> A := <<0,0>|<5,12>>: QRDecomposition( A, fullspan )`

$$\begin{bmatrix} 5/13 & 12/13 \\ 12/13 & -5/13 \end{bmatrix}, \quad \begin{bmatrix} 0 & 13 \\ 0 & 0 \end{bmatrix}.$$

72.6 Eigenvalues and Eigenvectors

Facts:

1. In exact arithmetic, explicit expressions are not possible in general for the eigenvalues of a matrix of dimension 5 or higher.
2. When it has to, Maple represents polynomial roots (and, hence, eigenvalues) *implicitly* by the `RootOf` construct. Expressions containing `RootOf`s can be simplified and evaluated numerically.

Commands:

1. `Eigenvalues( A )` The eigenvalues of matrix A.
2. `Eigenvectors( A )` The eigenvalues and corresponding eigenvectors of A.
3. `CharacteristicPolynomial( A, 'x' )` The characteristic polynomial of A expressed using the variable x.
4. `JordanForm( A )` The Jordan form of the matrix A.

Examples:

1. Simple eigensystem computation.
   `> Eigenvectors( <<7,6>|<4,2>> );`

$$\begin{bmatrix} -1 \\ 10 \end{bmatrix}, \begin{bmatrix} -1/2 & 4/3 \\ 1 & 1 \end{bmatrix}.$$

   So the eigenvalues are -1 and 10 with the corresponding eigenvectors $[-1/2, 1]^T$ and $[4/3, 1]^T$.
2. A defective matrix.
   If the matrix is defective, then by convention the matrix of "eigenvectors" returned by Maple contains one or more columns of zeros.
   `> Eigenvectors( <<1,0>|<1,1>> );`

$$\begin{bmatrix} 1 \\ 1 \end{bmatrix}, \begin{bmatrix} 1 & 0 \\ 0 & 0 \end{bmatrix}.$$

3. Larger systems.
   For larger matrices, the eigenvectors will use the Maple `RootOf` construction,

$$A := \begin{bmatrix} 3 & 1 & 7 & 1 \\ 5 & 6 & -3 & 5 \\ 3 & -1 & -1 & 0 \\ -1 & 5 & 1 & 5 \end{bmatrix}.$$

   `> ( L, V ) := Eigenvectors( A ):` The colon suppresses printing. The vector of eigenvalues is returned as

```
> L;
```

$$\begin{bmatrix} \texttt{RootOf}\left(\_Z^4 - 13\,\_Z^3 - 4\,\_Z^2 + 319\,\_Z - 386, \texttt{index} = 1\right) \\ \texttt{RootOf}\left(\_Z^4 - 13\,\_Z^3 - 4\,\_Z^2 + 319\,\_Z - 386, \texttt{index} = 2\right) \\ \texttt{RootOf}\left(\_Z^4 - 13\,\_Z^3 - 4\,\_Z^2 + 319\,\_Z - 386, \texttt{index} = 3\right) \\ \texttt{RootOf}\left(\_Z^4 - 13\,\_Z^3 - 4\,\_Z^2 + 319\,\_Z - 386, \texttt{index} = 4\right) \end{bmatrix}.$$

This, of course, simply reflects the characteristic polynomial:

```
> CharacteristicPolynomial( A, 'x' );
```

$$x^4 - 13x^3 - 4x^2 + 319x - 386$$

The `Eigenvalues` command solves a 4th degree characteristic polynomial explicitly in terms of radicals unless the option `implicit` is used.

4. Jordan form. *Caution*: As with the reduced row-echelon form, the Jordan form of a matrix containing symbolic elements can be discontinuous. For example, given

$$A = \begin{bmatrix} 1 & t \\ 0 & 1 \end{bmatrix},$$

```
> ( J, Q ) := JordanForm( A, output=['J','Q'] );
```

$$J, Q := \begin{bmatrix} 1 & 1 \\ 0 & 1 \end{bmatrix}, \begin{bmatrix} t & 0 \\ 0 & 1 \end{bmatrix}$$

with $A = QJQ^{-1}$. Note that Q is invertible precisely when $t \neq 0$. This gives a proviso on the correctness of the result: J will be the Jordan form of A only for $t \neq 0$, which we see is the generic case returned by Maple.

Caution: Exact computation has its limitations, even without symbolic entries. If we ask for the Jordan form of the matrix

$$B = \begin{bmatrix} -1 & 4 & -1 & -14 & 20 & -8 \\ 4 & -5 & -63 & 203 & -217 & 78 \\ -1 & -63 & 403 & -893 & 834 & -280 \\ -14 & 203 & -893 & 1703 & -1469 & 470 \\ 20 & -217 & 834 & -1469 & 1204 & -372 \\ -8 & 78 & -280 & 470 & -372 & 112 \end{bmatrix},$$

a relatively modest 6×6 matrix with a triple eigenvalue 0, then the transformation matrix Q as produced by Maple has entries over 35,000 characters long. Some scheme of compression or large expression management is thereby mandated.

72.7 Linear Algebra with Modular Arithmetic in Maple

There is a subpackage, `LinearAlgebra[Modular]`, designed for programmatic use, that offers access to modular arithmetic with matrices and vectors.

Facts:

1. The subpackage can be loaded by issuing the command
`> with( LinearAlgebra[Modular] );` which gives access to the commands

```
[AddMultiple, Adjoint, BackwardSubstitute, Basis, Characteristic
Polynomial, ChineseRemainder, Copy, Create, Determinant, Fill,
```

`ForwardSubstitute, Identity, Inverse, LUApply, LUDecomposition, LinIntSolve, MatBasis, MatGcd, Mod, Multiply, Permute, Random, Rank, RankProfile, RowEchelonTransform, RowReduce, Swap, Transpose, ZigZag]`

2. Arithmetic can be done modulo a prime p or, in some cases, a composite modulus m.
3. The relevant matrix and vector datatypes are `integer[4]`, `integer[8]`, `integer[]`, and `float[8]`. Use of the correct datatype can improve efficiency.

Examples:

```
> p := 13;
> A := Mod( p, Matrix([[1,2,3],[4,5,6],[7,8,-9]]), integer[4] );
```

$$\begin{bmatrix} 1 & 2 & 3 \\ 4 & 5 & 6 \\ 7 & 8 & 4 \end{bmatrix},$$

```
> Mod( p, MatrixInverse( A ), integer[4] );
```

$$\begin{bmatrix} 12 & 8 & 5 \\ 0 & 11 & 3 \\ 5 & 3 & 5 \end{bmatrix}$$

Cautions:

1. This is not to be confused with the `mod` utilities, which together with the inert `Inverse` command, can also be used to calculate inverses in a modular way.
2. One must always specify the datatype in `Modular` commands, or a cryptic error message will be generated.

72.8 Numerical Linear Algebra in Maple

The above sections have covered the use of Maple for exact computations of the types met during a standard first course on linear algebra. However, in addition to exact computation, Maple offers a variety of floating-point numerical linear algebra support.

Facts:

1. Maple can compute with either "hardware floats" or "software floats."
2. A hardware float is IEEE double precision, with a mantissa of (approximately) 15 decimal digits.
3. A software float has a mantissa whose length is set by the Maple variable `Digits`.

Cautions:

1. If an integer is typed with a decimal point, then Maple treats it as a software float.
2. Software floats are significantly slower that hardware floats, even for the same precision.

Commands:

1. `Matrix( n, m, datatype=float[8] )` An $n \times m$ matrix of hardware floats (initialization data not shown). The elements must be real numbers. The 8 refers to the number of bytes used to store the floating point real number.
2. `Matrix( n, m, datatype=complex(float[8] ))` An $n \times m$ matrix of hardware floats, including complex hardware floats. A complex hardware float takes two 8-byte storage locations.

3. `Matrix( n, m, datatype=sfloat )` An $n \times m$ matrix of software floats. The entries must be real and the precision is determined by the value of `Digits`.
4. `Matrix( n, m, datatype=complex(sfloat) )` As before with complex software floats.
5. `Matrix( n, m, shape=symmetric )` A matrix declared to be symmetric. Maple can take advantage of shape declarations such as this.

Examples:

1. A characteristic surprise.
   When asked to compute the characteristic polynomial of a floating-point matrix, Maple first computes eigenvalues (by a good numerical method) and then presents the characteristic polynomial in factored form, with good approximate roots. Thus,
   ```
   > CharacteristicPolynomial( Matrix( 2, 2, [[666,667],[665,666]],\
         datatype=float[8]), 'x' );
   ```
   $$(x - 1331.99924924882612 - 0.0\,i)\,(x - 0.000750751173882235889 - 0.0\,i).$$
   Notice the signed zero in the imaginary part; though the roots in this case are real, approximate computation of the eigenvalues of a nonsymmetric matrix takes place over the complex numbers. (n.b.: The output above has been edited for clarity.)
2. Symmetric matrices.
   If Maple knows that a matrix is symmetric, then it uses appropriate routines. Without the symmetric declaration, the calculation is
   ```
   > Eigenvalues( Matrix( 2, 2, [[1,3],[3,4]], datatype=float[8]) );
   ```
   $$\begin{bmatrix} -0.854101966249684707 + 0.0\,i \\ 5.85410196624968470 + 0.0\,i \end{bmatrix}.$$
   With the declaration, the computation is
   ```
   > Eigenvalues( Matrix( 2, 2, [[1,3],[3,4]], shape=symmetric,
    datatype=float[8]) );
   ```
   $$\begin{bmatrix} -0.854101966249684818 \\ 5.85410196624968470 \end{bmatrix}.$$
   Cautions: Use of the `shape=symmetric` declaration will force Maple to treat the matrix as being symmetric, even if it is not.
3. Printing of hardware floats.
   Maple prints hardware floating-point data as 18-digit numbers. This does not imply that all 18 digits are correct; Maple prints the hardware floats this way so that a cycle of converting from binary to decimal and back to binary will return to exactly the starting binary floating-point number. Notice in the previous example that the last 3 digits differ between the two function calls. In fact, neither set of digits is correct, as a calculation in software floats with higher precision shows:
   ```
   > Digits := 30: Eigenvalues( Matrix( 2, 2, [[1,3],[3,4]], \
        datatype=sfloat, shape=symmetric ) );
   ```
   $$\begin{bmatrix} -0.854101966249684544613760503093 \\ 5.85410196624968454461376050310 \end{bmatrix}.$$
4. NullSpace. Consider
   ```
   > B := Matrix( 2, 2, [[666,667],[665,666]] ): A := Transpose(B).B;
   ```

We make a floating-point version of A by

```
> Af := Matrix( A, datatype=float[8] );
```

and then take the `NullSpace` of both `A` and `Af`. The nullspace of `A` is correctly returned as the empty set — A is not singular (in fact, its determinant is 1). The nullspace of `Af` is correctly returned as

$$\left\{ \begin{bmatrix} -0.707637442412755612 \\ 0.706575721416702662 \end{bmatrix} \right\}.$$

The answers are different — quite different — even though the matrices differ only in datatype. The surprising thing is that both answers are correct: Maple is doing the right thing in each case. See [Cor93] for a detailed explanation, but note that `Af` times the vector in the reported nullspace is about $3.17 \cdot 10^{-13}$ times the 2-norm of `Af`.

5. Approximate Jordan form (not!).[1]
   As noted previously, the Jordan form is discontinuous as a function of the entries of the matrix. This means that rounding errors may cause the computed Jordan form of a matrix with floating-point entries to be incorrect, and for this reason Maple refuses to compute the Jordan form of such a matrix.
6. Conditioning of eigenvalues.
   To explore Maple's facilities for the conditioning of the unsymmetric matrix eigenproblem, consider the matrix "gallery(3)" from MATLAB.

```
> A := Matrix( [[-149,-50,-154], [537,180,546], [-27,-9,-25]] ):
```

The Maple command `EigenConditionNumbers` computes estimates for the reciprocal condition numbers of each eigenvalue, and estimates for the reciprocal condition numbers of the computed eigenvectors as well. At this time, there are no built-in facilities for the computation of the sensitivity of arbitrary invariant subspaces.

```
> E,V,rconds,rvecconds := EigenConditionNumbers( A, output= \
['values', 'vectors','conditionvalues','conditionvectors'] ):
> seq( 1/rconds[i], i=1..3 );
                        417.6482708, 349.7497543, 117.2018824
```

A separate computation using the definition of the condition number of an eigentriplet $(\mathbf{y}^*, \lambda, \mathbf{x})$ (see Chapter 15) as

$$C_\lambda = \frac{\|\mathbf{y}^*\|_\infty \|\mathbf{x}\|_\infty}{|\mathbf{y}^* \cdot \mathbf{x}|}$$

gives exact condition numbers (in the infinity norm) for the eigenvalues 1, 2, 3 as 399, 252, and 147. We see that the estimates produced by `EigenConditionNumbers` are of the right order of magnitude.

72.9 Canonical Forms

There are several canonical forms in Maple: Jordan form, Smith form, Hermite form, and Frobenius form, to name a few. In this section, we talk only about the Smith form (defined in Chapter 6.5 and Chapter 23.2).

[1] An out-of-date, humorous reference.

Commands:

1. `SmithForm( B, output=['S','U','V'] )` Smith form of B.

Examples:

The Smith form of

$$B = \begin{bmatrix} 0 & -4y^2 & 4y & -1 \\ -4y^2 & 4y & -1 & 0 \\ 4y & -1 & 4\left(2y^2-2\right)y & 4\left(y^2-1\right)^2 y^2 - 2y^2 + 2 \\ -1 & 0 & 4\left(y^2-1\right)^2 y^2 - 2y^2 + 2 & -4\left(y^2-1\right)^2 y \end{bmatrix}$$

is

$$S = \begin{bmatrix} 1 & 0 & 0 & 0 \\ 0 & 1 & 0 & 0 \\ 0 & 0 & 1/4\left(2y^2-1\right)^2 & 0 \\ 0 & 0 & 0 & (1/64)\left(2y^2-1\right)^6 \end{bmatrix}.$$

Maple also returns two unimodular (over the domain $\mathbb{Q}[y]$) matrices **u** and **v** for which $A = U.S.V$.

72.10 Structured Matrices

Facts:

1. Computer algebra systems are particularly useful for computations with structured matrices.
2. User-defined structures may be programmed using *index functions*. See the help pages for details.
3. Examples of built-in structures include *symmetric*, *skew-symmetric*, *Hermitian*, *Vandermonde*, and *Circulant* matrices.

Examples:

Generalized Companion Matrices. Maple can deal with several kinds of generalized companion matrices. A generalized companion matrix[2] *pencil* of a polynomial $p(x)$ is a pair of matrices C_0, C_1 such that $\det(xC_1 - C_0) = 0$ precisely when $p(x) = 0$. Usually, in fact, $\det(xC_1 - C_0) = p(x)$, though in some definitions proportionality is all that is needed. In the case $C_1 = I$, the identity matrix, we have $C_0 = C(p(x))$ is *the companion matrix* of $p(x)$. MATLAB's `roots` function computes roots of polynomials by first computing the eigenvalues of the companion matrix, a venerable procedure only recently proved stable.

The generalizations allow direct use of alternative polynomial bases, such as the Chebyshev polynomials, Lagrange polynomials, Bernstein (Bézier) polynomials, and many more. Further, the generalizations allow the construction of generalized companion matrix pencils for *matrix polynomials*, allowing one to easily solve *nonlinear eigenvalue problems*.

We give three examples below.

If $p := 3 + 2x + x^2$, then `CompanionMatrix( p, x )` produces "the" (standard) companion matrix (also called Frobenius form companion matrix):

$$\begin{bmatrix} 0 & -3 \\ 1 & -2 \end{bmatrix}$$

[2] Sometimes known as "colleague" or "comrade" matrices, an unfortunate terminology that inhibits keyword search.

and it is easy to see that $\det(tI - C) = p(t)$. If instead

$$p := B_0^3(x) + 2B_1^3(x) + 3B_2^3(x) + 4B_3^3(x)$$

where $B_k^n(x) = \binom{n}{k}(1-x)^{n-k}(x+1)^k$ is the kth Bernstein (Bézier) polynomial of degree n on the interval $-1 \le x \le 1$, then `CompanionMatrix( p, x )` produces the pencil (note that this is not in Frobenius form)

$$C_0 = \begin{bmatrix} -3/2 & 0 & -4 \\ 1/2 & -1/2 & -8 \\ 0 & 1/2 & -\frac{52}{3} \end{bmatrix}$$

$$C_1 = \begin{bmatrix} 3/2 & 0 & -4 \\ 1/2 & 1/2 & -8 \\ 0 & 1/2 & -\frac{20}{3} \end{bmatrix}$$

(from a formula by Jonsson & Vavasis [JV05] and independently by J. Winkler [Win04]), and we have $p(x) = \det(xC_1 - C_0)$. Note that the program does not change the basis of the polynomial $p(x)$ of Equation (72.9) to the monomial basis (it turns out that $p(x) = 20 + 12x$ in the monomial basis, in this case: note that C_1 is singular). It is well-known that changing polynomial bases can be ill-conditioned, and this is why the routine avoids making the change.

Next, if we choose nodes $[-1, -1/3, 1/3, 1]$ and look at the degree 3 polynomial taking the values $[1, -1, 1, -1]$ on these four nodes, then `CompanionMatrix( values, nodes )` gives C_0 and C_1 where C_1 is the 5×5 identity matrix with the $(5, 5)$ entry replaced by 0, and

$$C_0 = \begin{bmatrix} 1 & 0 & 0 & 0 & -1 \\ 0 & 1/3 & 0 & 0 & 1 \\ 0 & 0 & -1/3 & 0 & -1 \\ 0 & 0 & 0 & -1 & 1 \\ -\frac{9}{16} & \frac{27}{16} & -\frac{27}{16} & \frac{9}{16} & 0 \end{bmatrix}.$$

We have that $\det(tC_1 - C_0)$ is of degree 3 (in spite of these being 5×5 matrices), and that this polynomial takes on the desired values ± 1 at the nodes. Therefore, the finite eigenvalues of this pencil are the roots of the given polynomial. See [CWO4] and [Cor04] for example, for more information.

Finally, consider the nonlinear eigenvalue problem below: find the values of x such that the matrix C with $C_{ij} = T_0(x)/(i+j+1) + T_1(x)/(i+j+2) + T_2(x)/(i+j+3)$ is singular. Here $T_k(x)$ means the kth Chebyshev polynomial, $T_k(x) = \cos(k \cos^{-1}(x))$. We issue the command

```
> ( C0, C1 ) := CompanionMatrix( C, x );
```

from which we find

$$C0 = \begin{bmatrix} 0 & 0 & 0 & -2/15 & -1/12 & -\frac{2}{35} \\ 0 & 0 & 0 & -1/12 & -\frac{2}{35} & -1/24 \\ 0 & 0 & 0 & -\frac{2}{35} & -1/24 & -\frac{2}{63} \\ 1 & 0 & 0 & -1/4 & -1/5 & -1/6 \\ 0 & 1 & 0 & -1/5 & -1/6 & -1/7 \\ 0 & 0 & 1 & -1/6 & -1/7 & -1/8 \end{bmatrix}$$

and

$$C1 = \begin{bmatrix} 1 & 0 & 0 & 0 & 0 & 0 \\ 0 & 1 & 0 & 0 & 0 & 0 \\ 0 & 0 & 1 & 0 & 0 & 0 \\ 0 & 0 & 0 & 2/5 & 1/3 & 2/7 \\ 0 & 0 & 0 & 1/3 & 2/7 & 1/4 \\ 0 & 0 & 0 & 2/7 & 1/4 & 2/9 \end{bmatrix}.$$

This uses a formula from [Goo61], extended to matrix polynomials. The six generalized eigenvalues of this pencil include, for example, one near to $-0.6854 + 1.909i$. Substituting this eigenvalue in for x in C yields a three-by-three matrix with ratio of smallest to largest singular values $\sigma_3/\sigma_1 \approx 1.7 \cdot 10^{-15}$. This is effectively singular and, thus, we have found the solutions to the nonlinear eigenvalue problem. Again note that the generalized companion matrix is not in Frobenius standard form, and that this process works for a variety of bases, including the Lagrange basis.

Circulant matrices and Vandermonde matrices.

```
> A := Matrix( 3, 3, shape=Circulant[a,b,c] );
```

$$\begin{bmatrix} a & b & c \\ c & a & b \\ b & c & a \end{bmatrix},$$

```
> F3 := Matrix( 3,3,shape=Vandermonde[[1,exp(2*Pi*I/3),exp(4*Pi*I/3)]]
);
```

$$\begin{bmatrix} 1 & 1 & 1 \\ 1 & -1/2 + 1/2\,i\sqrt{3} & \left(-1/2 + 1/2\,i\sqrt{3}\right)^2 \\ 1 & -1/2 - 1/2\,i\sqrt{3} & \left(-1/2 - 1/2\,i\sqrt{3}\right)^2 \end{bmatrix}.$$

It is easy to see that the `F3` matrix diagonalizes the circulant matrix `A`.

Toeplitz and Hankel matrices. These can be constructed by calling `ToeplitzMatrix` and `HankelMatrix`, or by direct use of the `shape` option of the `Matrix` constructor.

```
> T := ToeplitzMatrix( [a,b,c,d,e,f,g] );
> T := Matrix( 4,4,shape=Toeplitz[false,Vector(7,[a,b,c,d,e,f,g])] );
```

both yield a matrix that looks like

$$\begin{bmatrix} d & c & b & a \\ e & d & c & b \\ f & e & d & c \\ g & f & e & d \end{bmatrix},$$

though in the second case only 7 storage locations are used, whereas 16 are used in the first. This economy may be useful for larger matrices. The shape constructor for Toeplitz also takes a Boolean argument `true`, meaning symmetric.

Both Hankel and Toeplitz matrices may be specified with an indexed symbol for the entries:

`> H := Matrix( 4, 4, shape=Hankel[a] );` yields

$$\begin{bmatrix} a_1 & a_2 & a_3 & a_4 \\ a_2 & a_3 & a_4 & a_5 \\ a_3 & a_4 & a_5 & a_6 \\ a_4 & a_5 & a_6 & a_7 \end{bmatrix}.$$

72.11 Functions of Matrices

The exponential of the matrix A is computed in the `MatrixExponential` command of Maple by *polynomial interpolation* (see Chapter 11.1) of the exponential at each of the eigenvalues of A, including multiplicities. In an exact computation context, this method is not so "dubious" [Lab97]. This approach is also used by the general `MatrixFunction` command.

Examples:

```
> A := Matrix( 3, 3,  [[-7,-4,-3],[10,6,4],[6,3,3]] ):
> MatrixExponential( A );
```

$$\begin{bmatrix} 6-7e^1 & 3-4e^1 & 2-3e^1 \\ 10e^1-6 & -3+6e^1 & -2+4e^1 \\ 6e^1-6 & -3+3e^1 & -2+3e^1 \end{bmatrix} \tag{72.1}$$

Now a square root: `> MatrixFunction( A, sqrt(x), x ):`

$$\begin{bmatrix} -6 & -7/2 & -5/2 \\ 8 & 5 & 3 \\ 6 & 3 & 3 \end{bmatrix} \tag{72.2}$$

Another matrix square root example, for a matrix close to one that has no square root:

```
> A := Matrix( 2, 2, [[epsilon^2, 1], [0, delta^2] ] ):
> S := MatrixFunction( A, sqrt(x), x ):
> simplify( S ) assuming positive;
```

$$\begin{bmatrix} \epsilon & \dfrac{1}{\epsilon+\delta} \\ 0 & \delta \end{bmatrix} \tag{72.3}$$

If ϵ and δ both approach zero, we see that the square root has an entry that approaches infinity. Calling `MatrixFunction` on the above matrix with $\epsilon = \delta = 0$ yields an error message, `Matrix function x^(1/2) is not defined for this Matrix`, which is correct.

Now for the matrix logarithm.

```
> Pascal := Matrix( 4, 4, (i,j)->binomial(j-1,i-1) );
```

$$\begin{bmatrix} 1 & 1 & 1 & 1 \\ 0 & 1 & 2 & 3 \\ 0 & 0 & 1 & 3 \\ 0 & 0 & 0 & 1 \end{bmatrix} \tag{72.4}$$

```
> MatrixFunction( Pascal, log(x), x );
```

$$\begin{bmatrix} 0 & 1 & 0 & 0 \\ 0 & 0 & 2 & 0 \\ 0 & 0 & 0 & 3 \\ 0 & 0 & 0 & 0 \end{bmatrix} \tag{72.5}$$

Now a function not covered in Chapter 11, instead of redoing the sine and cosine examples:

```
> A := Matrix( 2, 2, [[-1/5, 1], [0, -1/5]] ):
```

```
> W := MatrixFunction( A, LambertW(-1,x), x );
```

$$W := \begin{bmatrix} \text{LambertW}(-1,-1/5) & -5\,\dfrac{\text{LambertW}\,(-1,-1/5)}{1+\text{LambertW}\,(-1,-1/5)} \\ 0 & \text{LambertW}\,(-1,-1/5) \end{bmatrix} \tag{72.6}$$

```
> evalf( W );
```

$$\begin{bmatrix} -2.542641358 & -8.241194055 \\ 0.0 & -2.542641358 \end{bmatrix} \tag{72.7}$$

That matrix satisfies $W \exp(W) = A$, and is a *primary matrix function*. (See [CGH+96] for more details about the Lambert W function.)

Now the matrix sign function (cf. Chapter 11.6). Consider

```
> Pascal2 := Matrix( 4, 4, (i,j)->(-1)^(i-1)*binomial(j-1,i-1) );
```

$$\begin{bmatrix} 1 & 1 & 1 & 1 \\ 0 & -1 & -2 & -3 \\ 0 & 0 & 1 & 3 \\ 0 & 0 & 0 & -1 \end{bmatrix}. \tag{72.8}$$

Then we compute the matrix sign function of this matrix by
`> S := MatrixFunction( Pascal2, csgn(z), z ):` which turns out to be the same matrix (`Pascal2`).

Note: The complex "sign" function we use here is not the usual complex sign function for scalars

$$\text{signum}(re^{i\theta}) := e^{i\theta},$$

but rather (as desired for the definition of the matrix sign function)

$$\text{csgn}(z) = \begin{cases} 1 & \text{if } \text{Re}(z) > 0 \\ -1 & \text{if } \text{Re}(z) < 0\,. \\ \text{signum}(\text{Im}(z)) & \text{if } \text{Re}(z) = 0 \end{cases}$$

This has the side effect of making the function defined even when the input matrix has purely imaginary eigenvalues. The signum and `csgn` of 0 are both 0, by default, but can be specified differently if desired.
Cautions:

1. Further, it is not the `sign` function in MAPLE, which is a different function entirely: That function (`sign`) returns the sign of the leading coefficient of the polynomial input to `sign`.
2. (In General) This general approach to computing matrix functions can be slow for large exact or symbolic matrices (because manipulation of symbolic representations of the eigenvalues using `RootOf`, typically encountered for $n \geq 5$, can be expensive), and on the other hand can be unstable for floating-point matrices, as is well known, especially those with nearly multiple eigenvalues. However, for small or for structured matrices this approach can be very useful and can give insight.

72.12 Matrix Stability

As defined in Chapter 19, a matrix is (negative) *stable* if all its eigenvalues are in the left half plane (in this section, "stable" means "negative stable"). In Maple, one may test this by direct computation of the eigenvalues (if the entries of the matrix are numeric) and this is likely faster and more accurate than any purely rational operation based test such as the Hurwitz criterion. If, however, the matrix contains symbolic entries, then one usually wishes to know for what values of the parameters the matrix is stable. We may obtain conditions on these parameters by using the `Hurwitz` command of the `PolynomialTools` package on the characteristic polynomial.

Examples:

Negative of `gallery(3)` from MATLAB.

```
> A := -Matrix( [[-149,-50,-154], [537,180,546], [-27,-9,-25]] ):
> E := Matrix( [[130, -390, 0], [43, -129, 0], [133,-399,0]] ):
> AtE := A - t*E;
```

$$\begin{bmatrix} 149 - 130\,t & 50 + 390\,t & 154 \\ -537 - 43\,t & -180 + 129\,t & -546 \\ 27 - 133\,t & 9 + 399\,t & 25 \end{bmatrix} \tag{72.9}$$

For which t is that matrix stable?

```
> p := CharacteristicPolynomial( AtE, lambda );
> PolynomialTools[Hurwitz]( p, lambda, 's', 'g' );
```

This command returns "FAIL," meaning that it cannot tell whether p is stable or not; this is only to be expected as t has not yet been specified. However, according to the documentation, all coefficients of λ returned in `s` must be positive, in order for p to be stable. The coefficients returned are

$$\left[\frac{\lambda}{6+t}, 1/4\,\frac{(6+t)^2\,\lambda}{15 + 433453\,t + 123128\,t^2}, \frac{\left(60 + 1733812\,t + 492512\,t^2\right)\lambda}{(6+t)\,(6 + 1221271\,t)}\right] \tag{72.10}$$

and analysis (not given here) shows that these are all positive if and only if $t > -6/1221271$.

Acknowledgements

Many people have contributed to linear algebra in Maple, for many years. Dave Hare and David Linder deserve particular credit, especially for the LinearAlgebra package and its connections to CLAPACK, and have also greatly helped our understanding of the best way to use this package. We are grateful to Dave Linder and to Jürgen Gerhard for comments on early drafts of this chapter.

References

[CGH+96] Robert M. Corless, Gaston H. Gonnet, D.E.G. Hare, David J. Jeffrey, and Donald E. Knuth. On the Lambert *W* function. *Adv. Comp. Math.*, 5:329–359, 1996.

[CJ92] R.M. Corless and D.J. Jeffrey. Well... it isn't quite that simple. *SIGSAM Bull.*, 26(3):2–6, 1992.

[CJ97] R.M. Corless and D.J. Jeffrey. The Turing factorization of a rectangular matrix. *SIGSAM Bull.*, 31(3):20–28, 1997.

[Cor93] Robert M. Corless. Six, lies, and calculators. *Am. Math. Month.*, 100(4):344–350, April 1993.

[Cor04] Robert M. Corless. Generalized companion matrices in the Lagrange basis. In Laureano Gonzalez-Vega and Tomas Recio, Eds., *Proceedings EACA*, pp. 317–322, June 2004.

[CW04] Robert M. Corless and Stephen M. Watt. Bernstein bases are optimal, but, sometimes, Lagrange bases are better. In *Proceedings of SYNASC*, pp. 141–153. Mirton Press, September 2004.

[Goo61] I.J. Good. The colleague matrix, a Chebyshev analogue of the companion matrix. *Q. J. Math.*, 12:61–68, 1961.

[JV05] Gudbjorn F. Jónsson and Steven Vavasis. Solving polynomials with small leading coefficients. *Siam J. Matrix Anal. Appl.*, 26(2):400–414, 2005.

[Lab97] George Labahn, Personal Communication, 1997.

[Win04] Joab R. Winkler. The transformation of the companion matrix resultant between the power and Bernstein polynomial bases. *Appl. Num. Math.*, 48(1):113–126, 2004.

73

Mathematica

Heikki Ruskeepää

University of Turku

73.1 Introduction

About *Mathematica*®

Mathematica is a comprehensive software system for doing symbolic and numerical calculations, creating graphics and animations, writing programs, and preparing documents.

The heart of *Mathematica* is its broad collection of tools for symbolic and exact mathematics, but numerical methods also form an essential part of the system. *Mathematica* is known for its high-quality graphics, and the system is also a powerful programming language, supporting both traditional procedural techniques and functional and rule-based programming. In addition, *Mathematica* is an environment for preparing high-quality documents.

The first version of *Mathematica* was released in 1988. The current version, version 5, was released in 2003. *Mathematica* now contains over 4000 commands and is one of the largest single application programs ever developed. *Mathematica* is a product of Wolfram Research, Inc. The founder, president, and CEO of Wolfram Research is Stephen Wolfram.

Mathematica contains two main parts — the **kernel** and the **front end**. The kernel does the computations. For example, the implementation of the `Integrate` command comprises about 500 pages of *Mathematica* code and 600 pages of C code. The front end is a user interface that takes care of the communication between the user and the kernel. In addition, there are **packages** that supplement the kernel; packages have to be loaded as needed.

The most common type of user interface is based on interactive documents known as **notebooks**. *Mathematica* is often used like an advanced calculator for a moment's need, in which case a notebook is simply an interface to write the commands and read the results. However, often a notebook grows to be a useful document that you will save or print or use in a presentation.

Notebooks consist of **cells**. Each cell is indicated with a **cell bracket** at the right side of the notebook. Cells are grouped in various ways so as to form a hierarchical structure. For example, each input is in a cell, each output is in a cell, and each input–output pair forms a higher-level cell.

About this Chapter

Within the limited space of this chapter, we cover the essentials of doing linear algebra calculations with *Mathematica*. The chapter covers many of the topics of the handbook done via *Mathematica*, and the reader should consult the relevant section for further information.

Most commands are demonstrated in the Examples sections, but note that the examples are very simple. Indeed, the aim of these examples is only to show how the commands are used and what kind of result we get in simple cases. We do not demonstrate the full power and every feature of the commands.

Commands relating to packages are frequently mentioned in this chapter, but they are often not fully explained or demonstrated. *Mathematica* has advanced technology for sparse matrices, but we only briefly mention them in Section 73.3; for a detailed coverage, we refer to [WR03]. The basic principle is that all calculations with sparse matrices work as for usual matrices. [WR03] also considers performance and efficiency questions.

The Appendix contains a short introduction to the use of *Mathematica*. There we also refer to some books, documents, and Help Browser material where you can find further information about linear algebra with *Mathematica*.

This chapter was written with *Mathematica* 5.2. However, many users of *Mathematica* have earlier versions. To help these users, we have denoted by (Mma 5.0) and (Mma 5.1) the features of *Mathematica* that are new in versions 5.0 and 5.1, respectively. In the Appendix we list ways to do some calculations with earlier versions.

As in *Mathematica* notebooks, in this chapter *Mathematica* commands and their arguments are in boldface while outputs are in a plain font. *Mathematica* normally shows the output below the input, but here, in order to save space, we mostly show the output next to the input.

Matrices are traditionally denoted by capital letters like A or M. However, in *Mathematica* we have the general advice that all user-defined names should begin with lower-case letters so that they do not conflict with the built-in names, which always begin with an upper-case letter. We follow this advice and mainly use the lower-case letter **`m`** for matrices. In principle, we could use most upper-case letters, but note that the letters **`C`**, **`D`**, **`E`**, **`I`**, **`N`**, and **`O`** are reserved names in *Mathematica* and they cannot be used as user-defined names.

Because the **`` LinearAlgebra`MatrixManipulation` ``** package appears quite frequently in this chapter, we abbreviate it to **`LAMM`**.

73.2 Vectors

Commands:

Vectors in *Mathematica* are lists. In Section 73.3, we will see that matrices are lists of lists, each sublist being a row of the matrix. Note that *Mathematica* does not distinguish between row and column vectors. Indeed, when we compute with vectors and matrices, in most cases it is evident for *Mathematica* how an expression has to be calculated. Only in some rare cases do we need to be careful and write an expression in such a way that *Mathematica* understands the expression in the correct way. One of these cases is the multiplication of a column by a row vector; this has to be done with an outer product (see **`Outer`** in item 4 below).

1. Vectors in *Mathematica*:
   - **`{a, b, c, ...}`** A vector with elements **`a`**, **`b`**, **`c`**, ...
   - **`MatrixForm[v]`** Display vector **`v`** in a column form.

- **Length[v]** The number of elements of vector **v**.
- **VectorQ[v]** Test whether **v** is a vector.

2. Generation of vectors:
   - **Range[n_1]** Create the vector $(1, 2, \ldots, n_1)$. With **Range[n_0, n_1]** we get the vector $(n_0, n_0 + 1, \ldots, n_1)$ and with **Range[n_0, n_1, d]** the vector $(n_0, n_0 + d, n_0 + 2d, \ldots, n_1)$.
   - **Table[expr, {i, n_1}]** Create a vector by giving **i** the values $1, 2, \ldots, n_1$ in **expr**. If the iteration specification is **{i, n_0, n_1}**, then **i** gets the values $n_0, n_0 + 1, \ldots, n_1$, and for **{i, n_0, n_1, d}**, the values of **i** are between n_0 and n_1 in steps of **d**. For **{n_1}**, simply n_1 copies of **expr** are taken.
   - **Array[f, n_1]** Create the n_1 vector **(f[1], ..., f[n_1])**. With **Array[f, n_1, n_0]** we get the n_1 vector **(f[n_0], ..., f[n_0 + n_1 − 1])**.
3. Calculating with vectors:
   - **a v** Multiply vector **v** with scalar **a**.
   - **u + v** Add two vectors.
   - **u v** Multiply the corresponding elements of two vectors and form a vector from the products (there is a space between **u** and **v**); for an inner product, write **u.v**.
   - **u/v** Divide the corresponding elements of two vectors.
   - **v^p** Calculate the **p**th power of each element of a vector.
   - **a^v** Generate a vector by calculating the powers of scalar **a** that are given in vector **v**.
4. Products of vectors:
   - **u.v** The inner product of two vectors of the same size.
   - **Outer[Times, u, v]** The outer product (a matrix) of vectors **u** and **v**.
   - **Cross[u, v]** The cross product of two vectors.
5. Norms and sums of vectors:
   - **Norm[v]** (Mma 5.0) The 2-norm (or Euclidean norm) of a vector.
   - **Norm[v, p]** (Mma 5.0) The **p**-norm of a vector (**p** is a number in $[1, \infty)$ or ∞).
   - **Total[v]** (Mma 5.0) The sum of the elements of a vector.
   - **Apply[Times, v]** The product of the elements of a vector.
6. Manipulation of vectors:
   - **v[[i]]** Take element **i** (output is the corresponding scalar).
   - **v[[i]] = a** Change the value of element **i** into scalar **a** (output is **a**).
   - **v[[{i, j, ...}]]** Take elements **i**, **j**, ... (output is the corresponding vector).
   - **First[v]**, **Last[v]** Take the first/last element (output is the corresponding scalar).
   - **Rest[v]**, **Most[v]** Drop the first/last element (output is the corresponding vector).
   - **Take[v, n]**, **Take[v, -n]**, **Take[v, {n_1, n_2}]** Take the first **n** elements / the last **n** elements / elements $n_1, \ldots, n_2$ (output is the corresponding vector).
   - **Drop[v, n]**, **Drop[v, -n]**, **Drop[v, {n_1, n_2}]** Drop the first **n** elements / the last **n** elements / elements $n_1, \ldots, n_2$ (output is the corresponding vector).
   - **Prepend[v, a]**, **Append[v, a]** Insert element **a** at the beginning/end of a vector (output is the corresponding vector).
   - **Join[u, v, ...]** Join the given vectors into one vector (output is the corresponding vector).

7. In the **LinearAlgebra`Orthogonalization`** package:
   - **GramSchmidt[{u, v, ...}]** Generate an orthonormal set from the given vectors.
   - **Projection[u, v]** Calculate the orthogonal projection of **u** onto **v**.
8. In the **Geometry`Rotations`** package: rotations of vectors.

Examples:

1. Vectors in *Mathematica*.

```
v = {4, 2, 3}   {4, 2, 3}
MatrixForm[v]   (4
                 2
                 3)
Length[v]   3
VectorQ[v]   True
```

2. Generation of vectors. **Range** is nice for forming lists of integers or reals:

```
Range[10]   {1, 2, 3, 4, 5, 6, 7, 8, 9, 10}
Range[0, 10, 2]   {0, 2, 4, 6, 8, 10}
Range[1.5, 2, 0.1]   {1.5, 1.6, 1.7, 1.8, 1.9, 2.}
```

Table is one of the most useful commands in *Mathematica*:

```
Table[Random[], {3}]   {0.454447, 0.705133, 0.226419}
Table[Random[Integer, {1, 6}], {5}]   {2, 3, 6, 3, 4}
Table[x^i, {i, 5}]   {x, x^2, x^3, x^4, x^5}
Table[x[i], {i, 5}]   {x[1], x[2], x[3], x[4], x[5]}
Table[x_i, {i, 5}]   {x_1, x_2, x_3, x_4, x_5}
```

Array is nice for forming lists of indexed variables:

```
Array[x, 5]   {x[1], x[2], x[3], x[4], x[5]}
Array[x, 5, 0]   {x[0], x[1], x[2], x[3], x[4]}
```

3. Calculating with vectors. The arithmetic operations of multiplying a vector with a scalar and adding two vectors work in *Mathematica* as expected. However, note that *Mathematica* also does other types of arithmetic operations with vectors — multiplication, division, and powers. All arithmetic operations are done in an element-by-element way. For example, **u v** and **u/v** form a vector from the products or quotients of the corresponding elements of **u** and **v**. This is a useful property in many calculations, but remember to use **u.v** (which is the same as **Dot[u, v]**) for an inner product.

```
u = {a, b, c}; v = {4, 2, 3};
{10 v, u + v, u v}
{{40, 20, 30}, {4 + a, 2 + b, 3 + c}, {4 a, 2 b, 3 c}}
{1/v, u/v, u^2}   {{1/4, 1/2, 1/3}, {a/4, b/2, c/3}, {a^2, b^2, c^2}}
```

Functions of vectors are also calculated elementwise:

```
Log[v]   {Log[4], Log[2], Log[3]}
```

4. Products of vectors. With **u** and **v** as in Example 3, we calculate an inner product, an outer product, and a cross product:

```
v.u   4 a + 2 b + 3 c
Outer[Times, v, u]
{{4 a, 4 b, 4 c}, {2 a, 2 b, 2 c}, {3 a, 3 b, 3 c}}
```

MatrixForm[%] $\begin{pmatrix} 4a & 4b & 4c \\ 2a & 2b & 2c \\ 3a & 3b & 3c \end{pmatrix}$

Cross[v, u] $\{-3b + 2c,\ 3a - 4c,\ -2a + 4b\}$

5. Norms and sums of vectors. The default vector norm is the 2-norm:

   Norm[u] $\sqrt{\text{Abs}[a]^2 + \text{Abs}[b]^2 + \text{Abs}[c]^2}$

   Norm[u, 2] $\sqrt{\text{Abs}[a]^2 + \text{Abs}[b]^2 + \text{Abs}[c]^2}$

   Norm[u, 1] Abs[a] + Abs[b] + Abs[c]

   Norm[u, ∞] Max[Abs[a], Abs[b], Abs[c]]

   Total[u] a + b + c

   Apply[Times, u] a b c

Applications:

1. (Plotting of vectors) A package [WR99, p. 133] defines the graphic primitive **Arrow** which can be used to plot vectors. As an example, we compute the orthogonal projection of a vector onto another vector and show the three vectors (for graphics primitives like **Arrow**, **Line**, and **Text**, see [Rus04, pp. 132–146]):

```
<< LinearAlgebra`Orthogonalization`
<< Graphics`Arrow`
u = {5, 1}; v = {2, 4};
pr = Projection[v, u]   {35/13, 7/13}
Show[Graphics[{Arrow[{0, 0}, u], Arrow[{0, 0}, v],
     Arrow[{0, 0}, pr],Line[{v, pr}], Text["u", u, {-1, 0}],
     Text["v", v, {-1, -0.4}]}], Axes → True,
     AspectRatio → Automatic, PlotRange → All];
```

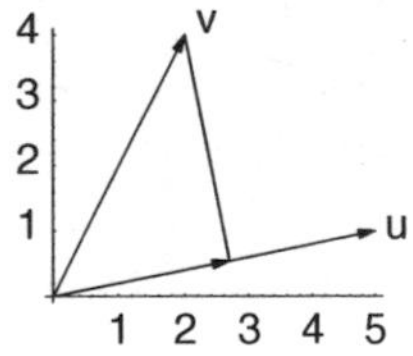

73.3 Basics of Matrices

Commands:

Matrices in *Mathematica* are lists of lists, each sublist being a row of the matrix. Although the traditional symbols for matrices are capital letters like A or M, we use the lower-case letter **m** for most matrices because, in *Mathematica*, all user-defined names should begin with a lower-case letter.

Here we only briefy mention sparse matrices; a detailed exposition can be found in [WR03]; this document is the item Built-in Functions ▷ Advanced Documentation ▷ Linear Algebra in the Help Browser of *Mathematica*.

1. Matrices in *Mathematica*:

   - **{{a, b, ...}, {c, d, ...}, ...}** A matrix with rows (**a**, **b**, ...), (**c**, **d**, ...), ...
   - **MatrixForm[m]** Display matrix **m** in a two-dimensional form.
   - **Length[m]** The number of rows of matrix **m**.
   - **Dimensions[m]** The number of rows and columns of matrix **m**.
   - **MatrixQ[m]** Test whether **m** is a matrix.
   - The menu command Input ▷ Create Table/Matrix/Palette generates an empty matrix into which the elements can be written.

2. Generation of matrices:

   - **IdentityMatrix[n]** An **n** × **n** identity matrix.
   - **DiagonalMatrix[{a, b, ...}]** A diagonal matrix with diagonal elements **a**, **b**, ...
   - **HilbertMatrix**, **HankelMatrix**, **ZeroMatrix**, **UpperDiagonalMatrix**, **LowerDiagonalMatrix**, **TridiagonalMatrix** (in the **LAMM** package).
   - **Table[expr, {i, m1}, {j, n1}]** Create a matrix by giving, in **expr**, **i** the values 1, 2, ... , **m1** and, for each **i**, **j** the values 1, ... , **n1**. Other forms of an iteration specification are {**m1**}, {**i, m0, m1**}, and {**i, m0, m1, d**} (see item 2 in Section 73.2).
   - **Array[f, {m1, n1}]** Create an **m1** × **n1** matrix with elements **f[i, j]**.
   - **Array[f, {m1, n1}, {m0, n0}]** Create an **m1** × **n1** matrix using starting values **m0** and **n0** for the indices (the default values of **m0** and **n0** are 1).

3. Sparse matrices:

   - **SparseArray[rules]** (Mma 5.0) Create a vector or matrix by taking nonzero elements from **rules** and setting other elements to zero.
   - **Normal[s]** (Mma 5.0) Show a sparse vector or matrix **s** in the usual list form.
   - **ArrayRules[m]** (Mma 5.0) Show all nonzero elements of a vector or matrix as rules.

4. Arithmetic of matrices (see also Section 73.4):

   - **a m** Multiply matrix **m** by scalar **a**.
   - **m + n** Add two matrices.
   - **m n** Multiply *the corresponding elements* of two matrices (giving the Hadamard product of the matrices; there is a space between **m** and **n**); for a proper matrix product, write **m.n** (see item 2 in Section 73.4).
   - **m^p** Calculate the **p**th power of *each element* of a matrix; for a proper matrix power, write **MatrixPower[m, p]** (see item 3 in Section 73.4).
   - **m^-1** Calculate the reciprocal of *each element* of a matrix; for a proper matrix inverse, write **Inverse[m]** (see item 5 in Section 73.4).

5. Sums and trace of matrices:

   - **`Total[m]`** (Mma 5.0) The vector of column sums (i.e., the sums of the elements of each column).
   - **`Total[Transpose[m]]`** or **`Map[Total, m]`** (Mma 5.0) The vector of row sums.
   - **`Total[Flatten[m]]`** or **`Total[m, 2]`** (Mma 5.0) The sum of all elements.
   - **`Tr[m]`** Trace.
   - **`Tr[m, List]`** List of diagonal elements.

6. Plots of matrices:

   - **`ArrayPlot[m]`** (Mma 5.1) Plot a matrix by showing zero values as white squares, the maximum absolute value as black squares, and values between as gray squares.
   - **`ArrayPlot[m, ColorRules → {0 → White, _→ Black}]`** (Mma 5.1) Show all nonzero values as black squares (also **`MatrixPlot[m]`** (Mma 5.0) from the **`LAMM`** package).

Examples:

1. Matrices in *Mathematica*. Matrices are lists of rows:

```
m = {{5, 2, 3}, {4, 0, 2}}   {{5, 2, 3}, {4, 0, 2}}
```

`MatrixForm[m]` $\begin{pmatrix} 5 & 2 & 3 \\ 4 & 0 & 2 \end{pmatrix}$

The number of rows and the size of the matrix are

```
Length[m]   2
Dimensions[m]   {2, 3}
```

With the menu command Input ▷ Create Table/Matrix/Palette, we get a matrix with empty slots that can then be filled with the help of the Tab key:

$\begin{pmatrix} \square & \square & \square \\ \square & \square & \square \end{pmatrix}$

With the **`TableSpacing`** option of **`MatrixForm`** we can control the space between rows and columns (the default value of the option is 1):

`MatrixForm[m, TableSpacing → {.5, .5}]` $\begin{pmatrix} 5 & 2 & 3 \\ 4 & 0 & 2 \end{pmatrix}$

Note that *Mathematica* will not do any calculations with a matrix that is in a matrix form! For example, writing

`m = {{5, 2, 3}, {4, 0, 2}} // MatrixForm` $\begin{pmatrix} 5 & 2 & 3 \\ 4 & 0 & 2 \end{pmatrix}$

defines **`m`** to be the matrix form of the given matrix and *Mathematica* does not do any calculations with this **`m`**. Write instead

`MatrixForm[m = {{5, 2, 3}, {4, 0, 2}}]` $\begin{pmatrix} 5 & 2 & 3 \\ 4 & 0 & 2 \end{pmatrix}$

or **`(m = {{5, 2, 3}, {4, 0, 2}}) // MatrixForm`** or do as we did above, define the matrix with **`m = {{5, 2, 3}, {4, 0, 2}}`** and then display the matrix with **`MatrixForm[m]`**.

2. Generation of matrices.

```
IdentityMatrix[3]  {{1, 0, 0}, {0, 1, 0}, {0, 0, 1}}
DiagonalMatrix[{3, 1, 2}]  {{3, 0, 0}, {0, 1, 0}, {0, 0, 2}}
Table[Random[], {2}, {2}]
{{0.544884, 0.397472}, {0.308083, 0.191191}}
Table[1/(i + j - 1), {i, 3}, {j, 3}]
```

$$\left\{\left\{1, \frac{1}{2}, \frac{1}{3}\right\}, \left\{\frac{1}{2}, \frac{1}{3}, \frac{1}{4}\right\}, \left\{\frac{1}{3}, \frac{1}{4}, \frac{1}{5}\right\}\right\}$$

If, **Which**, and **Switch** are often useful in connection with **Table**. In the following examples, we form the same tridiagonal matrix in three ways (we show the result only for the first example):

```
Table[If[i == j, 13, If[i == j - 1, 11,
  If[i == j + 1, 12, 0]]], {i, 4}, {j, 4}] // MatrixForm
```

$$\begin{pmatrix} 13 & 11 & 0 & 0 \\ 12 & 13 & 11 & 0 \\ 0 & 12 & 13 & 11 \\ 0 & 0 & 12 & 13 \end{pmatrix}$$

```
Table[Which[i == j, 13, i == j - 1, 11, i == j + 1, 12, True,
  0], {i, 4}, {j, 4}];
Table[Switch[i - j, -1, 11, 0, 13, 1, 12, _, 0],
  {i, 4}, {j, 4}];
```

Here are matrices with general elements:

```
Table[f[i, j], {i, 2}, {j, 2}]
{{f[1, 1], f[1, 2]}, f[2, 1], f[2, 2]}}
```

Table[f$_{\mathbf{i,j}}$**, {i, 2}, {j, 2}]** $\{\{f_{1,1}, f_{1,2}\}, \{f_{2,1}, f_{2,2}\}\}$

```
Array[f, {2, 2}]  {{f[1, 1], f[1, 2]}, {f[2, 1], f[2, 2]}}
```

3. Sparse matrices. With **SparseArray** we can input a matrix by specifying only the nonzero elements. With **Normal** we can show a sparse matrix as a usual matrix. One way to specify the rules is to write them separately:

```
SparseArray[{{1, 1} → 3, {2, 3} → 2, {3, 2} → 4}]
SparseArray[<3>, {3, 3}]
% // Normal  {{3, 0, 0}, {0, 0, 2}, {0, 4, 0}}
```

Another way is to gather the left- and right-hand sides of the rules together into lists:

```
SparseArray[{{1, 1}, {2, 3}, {3, 2}} → {3, 2, 4}];
```

Still another way is to give the rules in the form of index patterns. As an example, we generate the same tridiagonal matrix as we generated in Example 2 with **Table**:

```
SparseArray[{{i_, i_} → 13, {i_, j_} /; i - j == -1 → 11,
  {i_, j_} /; i - j == 1 → 12}, {4, 4}];
```

4. Arithmetic of matrices. The arithmetic operations of multiplying a matrix with a scalar and adding two matrices work in *Mathematica* as expected. However, note that *Mathematica* also does other types of arithmetic operations with matrices, especially multiplication and powers. All arithmetic operations are done in an element-by-element way. For example, **m n** forms a matrix from the products of the corresponding elements of **m** and **n**. This is a useful property in some calculations, but remember to use **m.n** (which is the same as **Dot[m, n]**) for an inner product, **MatrixPower[m, p]** for a matrix power, and **Inverse[m]** for a matrix inverse (see Section 73.4).

```
m = {{2, 5}, {3, 1}};
n = {{a, b}, {c, d}};
```

```
3 n   {{3 a, 3 b}, {3 c, 3 d}}
m + n   {{2 + a, 5 + b}, {3 + c, 1 + d}}
m n   {{2 a, 5 b}, {3 c, d}}
n^2   {{a^2, b^2}, {c^2, d^2}}
n^-1   {{1/a, 1/b}, {1/c, 1/d}}
```

5. Sums and trace of matrices. With **m** as in Example 4, here are the column sums, row sums, the sum of all elements, and trace:

```
{Total[m], Total[Transpose[m]], Total[Flatten[m]], Tr[m]}
{{5, 6}, {7, 4}, 11, 3}
```

6. Plots of matrices.

```
ArrayPlot[{{5, 2, 3}, {4, 0, 2}}];
m = SparseArray[{{i_, j_} /; Mod[i - j, 10] == 0 → 1}, {30, 30}];
ArrayPlot[m];
```

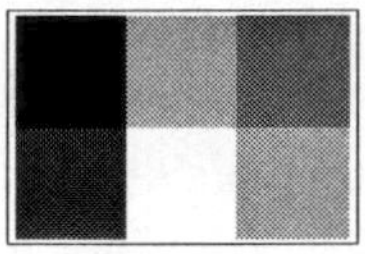

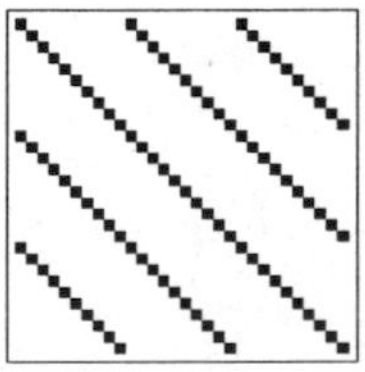

73.4 Matrix Algebra

Commands:

1. Transpose:
   - **Transpose[m]** Transpose.
   - **ConjugateTranspose[m]** (Mma 5.1) Conjugate transpose.
2. Product:
   - **m.n** Product of two matrices.
   - **m.v** Product of a matrix and a (column) vector.
   - **v.m** Product of a (row) vector and a matrix.
3. Power and matrix exponential:
   - **MatrixPower[m, p]** **p**th power of a square matrix.
   - **MatrixExp[m]** Matrix exponential $e^{\mathbf{m}} = \sum_{i=0}^{\infty} \frac{1}{i!}\mathbf{m}^i$ of a square matrix.

4. Determinant and minors:
   - **`Det[m]`** Determinant of a square matrix.
   - **`Minors[m]`** Minors of a square matrix.
   - **`Minors[m, k]`** **`k`**th minors.
5. Inverse and pseudo-inverse:
   - **`Inverse[m]`** Inverse of a square matrix.
   - **`PseudoInverse[m]`** Pseudo-inverse (of a possibly rectangular matrix).
6. Norms and condition numbers:
   - **`Norm[m]`** (Mma 5.0) The 2-norm of a numerical matrix.
   - **`Norm[m, p]`** (Mma 5.0) The **`p`**-norm of a matrix (**`p`** is a number in $[1, \infty)$ or ∞ or **`Frobenius`**).
   - **`InverseMatrixNorm[m, p]`** The **`p`**-norm of the inverse (in the **`LAMM`** package).
   - **`MatrixConditionNumber[m, p]`** The **`p`**-norm matrix condition number (in the **`LAMM`** package).
7. Nullspace, rank, and row reduction:
   - **`NullSpace[m]`** Basis vectors of the null space (kernel).
   - **`MatrixRank[m]`** (Mma 5.0) Rank.
   - **`RowReduce[m]`** Do Gauss–Jordan elimination to produce the reduced row echelon form.

Examples:

1. Transpose.

```
m = {{2 + 3 I, 1 - 2 I}, {3 - 4 I, 4 + I}};
MatrixForm[m]
```

$$\begin{pmatrix} 2+3i & 1-2i \\ 3-4i & 4+i \end{pmatrix}$$

```
MatrixForm[Transpose[m]]
```

$$\begin{pmatrix} 2+3i & 3-4i \\ 1-2i & 4+i \end{pmatrix}$$

```
MatrixForm[ConjugateTranspose[m]]
```

$$\begin{pmatrix} 2-3i & 3+4i \\ 1+2i & 4-i \end{pmatrix}$$

2. Product. Consider the following matrices and vectors:

```
m = {{1, 2, 3}, {4, 5, 6}}; n = {{p, q}, {r, s}, {t, u}};
v = {1, 2}; w = {1, 2, 3};
Map[MatrixForm, {m, n, v, w}]
```

$$\left\{ \begin{pmatrix} 1 & 2 & 3 \\ 4 & 5 & 6 \end{pmatrix}, \begin{pmatrix} p & q \\ r & s \\ t & u \end{pmatrix}, \begin{pmatrix} 1 \\ 2 \end{pmatrix}, \begin{pmatrix} 1 \\ 2 \\ 3 \end{pmatrix} \right\}$$

Here are some products:

```
m.n  {{p + 2r + 3t, q + 2s + 3u}, {4p + 5r + 6t, 4q + 5s + 6u}}
n.v  {p + 2q, r + 2s, t + 2u}
w.n  {p + 2r + 3t, q + 2s + 3u}
```

3. Power and matrix exponential. Powers can be calculated with **`MatrixPower`** or by using the dot repeatedly:

```
k = {{2, 3}, {-2, 1}};
MatrixPower[k, 4]  {{-50, -63}, {42, -29}}
k.k.k.k  {{-50, -63}, {42, -29}}
MatrixExp[k // N]  {{-2.6658, 3.79592}, {-2.53061, -3.93111}}
```

4. Determinant and minors. First, we consider a nonsingular matrix:

```
m = {{6, 2, 3}, {2, 7, 5}, {4, 2, 5}};
Det[m]  98
Minors[m]  {{38, 24, -11}, {4, 18, 4}, {-24, -10, 25}}
```

Here is a singular matrix:

```
n = {{7, 7, 2, 7}, {2, 5, 0, 7}, {8, 7, 8, 3}, {5, 6, 4, 5}};
Det[n]  0
```

5. Inverse and pseudo-inverse. First, we consider the nonsingular matrix **m** from Example 4:

i = Inverse[m] $\left\{\left\{\frac{25}{98}, -\frac{2}{49}, -\frac{11}{98}\right\}, \left\{\frac{5}{49}, \frac{9}{49}, -\frac{12}{49}\right\}, \left\{-\frac{12}{49}, -\frac{2}{49}, \frac{19}{49}\right\}\right\}$

```
m.i-IdentityMatrix[3]  {{0, 0, 0}, {0, 0, 0}, {0, 0, 0}}
```

Then, we consider the singular matrix **n**. The inverse does not exist, but the Moore–Penrose pseudo-inverse does exist:

```
Inverse[n]
Inverse::sing :
 Matrix {{7, 7, 2, 7}, {2, 5, 0, 7}, {8, 7, 8, 3}, {5, 6, 4, 5}} is
     singular. More...
Inverse[{{7, 7, 2, 7}, {2, 5, 0, 7}, {8, 7, 8, 3},
     {5, 6, 4, 5}}]
ps = PseudoInverse[n // N]
{{0.324527, -0.235646, -0.0236215, -0.129634},
 {-0.054885, 0.0700676, 0.020876, 0.0454718},
 {-0.256434, 0.102381, 0.121511, 0.111946},
 {-0.0535186, 0.136327, -0.031972, 0.0521774}}
```

Next, we check the four characterizing properties of a pseudo-inverse:

```
{n.ps.n - n, ps.n.ps - ps, ConjugateTranspose[n.ps] - n.ps,
   ConjugateTranspose[ps.n] - ps.n} // Chop
{{{0, 0, 0, 0}, {0, 0, 0, 0}, {0, 0, 0, 0}, {0, 0, 0, 0}},
 {{0, 0, 0, 0}, {0, 0, 0, 0}, {0, 0, 0, 0}, {0, 0, 0, 0}},
 {{0, 0, 0, 0}, {0, 0, 0, 0}, {0, 0, 0, 0}, {0, 0, 0, 0}},
 {{0, 0, 0, 0}, {0, 0, 0, 0}, {0, 0, 0, 0}, {0, 0, 0, 0}}}
```

For the option **Tolerance** of **PseudoInverse**, see **SingularValueList** in section 73.7.

6. Norms and condition numbers. The default matrix norm is the 2-norm (the largest singular value):

Norm[{{3, 1}, {4, 2}}] $\sqrt{15 + \sqrt{221}}$

MatrixForm[m = {{p, q}, {r, s}}] $\begin{pmatrix} p & q \\ r & s \end{pmatrix}$

Norm[m, Frobenius] $\sqrt{\text{Abs}[p]^2 + \text{Abs}[q]^2 + \text{Abs}[r]^2 + \text{Abs}[s]^2}$

```
Norm[m, 1]  Max[Abs[p] + Abs[r], Abs[q] + Abs[s]]
Norm[m, ∞]  Max[Abs[p] + Abs[q], Abs[r] + Abs[s]]
```

7. Nullspace, rank, and row reduction. The null space (or kernel) of a matrix M consists of vectors v for which $Mv = 0$. **NullSpace** gives a list of basis vectors for the null space. The matrix multiplied with any linear combination of these basis vectors gives a zero vector. For a nonsingular matrix M, the null space is empty (*Mathematica* gives the empty list {}), that is, there is no nonzero vector v such that $Mv = 0$. The null space of the following matrix contains two basis vectors:

```
p = {{7, 4, 8, 0}, {7, 4, 8, 0}, {1, 8, 3, 3}};
ns = NullSpace[p]   {{12, -21, 0, 52}, {-4, -1, 4, 0}}
```

The matrix **p** multiplied with any linear combination of the basis vectors gives a null vector:

```
p.(a ns[[1]] + b ns[[2]]) // Simplify   {0, 0, 0}
```

Matrix **p** is rank deficient:

```
MatrixRank[p]   2
```

The number of basis vectors of the null space plus the rank is the number of columns:

```
Length[NullSpace[p]] + MatrixRank[p]   4
```

The reduced row echelon form of a nonsingular square matrix is the identity matrix. Here is the reduced row echelon form of **p**:

RowReduce[p] // MatrixForm
$$\begin{pmatrix} 1 & 0 & 1 & -\frac{3}{13} \\ 0 & 1 & \frac{1}{4} & \frac{21}{52} \\ 0 & 0 & 0 & 0 \end{pmatrix}$$

The number of nonzero rows in the reduced row echelon form is the rank. **RowReduce** is also considered in item 6 of Section 73.9.

Applications:

1. (Prediction of a Markov chain) We classify days to be either wet or dry and assume that weather follows a Markov chain having the states wet and dry [Rus04, pp. 730–735]. From the weather statistics of Snoqualmie Falls in Western Washington, we have estimated the transition probability matrix to be $P = \begin{pmatrix} 0.834 & 0.166 \\ 0.398 & 0.602 \end{pmatrix}$. This means that if today is a wet day, then tomorrow will be a wet day with probability 0.834 and a dry day with probability 0.166, and if today is a dry day, tomorrow will be a wet day with probability 0.398 and a dry day with probability 0.602. (See Application 1 of Section 4.3, or Chapter 54 for more information on Markov chains.)

   If $\mu(n)$ is a row vector containing the probabilities of the states for day n, then $\mu(n) = \mu(0)P^n$. Assume that today is Monday and we have a dry day, then $\mu(0) = (0, 1)$. We predict the weather for Friday:

   ```
   p = {{0.834, 0.166}, {0.398, 0.602}};
   {0, 1}.MatrixPower[p, 4]   {0.680173, 0.319827}
   ```

   Thus, Friday is wet with probability 0.68 and dry with probability 0.32.

2. (Permanent) (See Chapter 31 for the definition and more information about permanents.) A permanent of a square matrix M with n rows can be computed as the coefficient of the product $x_1x_2\cdots x_n$ in the product of the components of the vector Mv, where $v = (x_1, \ldots, x_n)$. So, a permanent can be calculated with the following program:

   ```
   permanent[m_]:= With[{v = Array[x, Length[m]]},
     Coefficient[Apply[Times, m.v], Apply[Times, v]]]
   m = {{6, 2, 3}, {2, 7, 5}, {4, 2, 5}};
   permanent[m]   426
   ```

73.5 Manipulation of Matrices

Commands:

1. Taking and resetting elements:
   - **m[[i, j]]** Take element (**i**, **j**) (output is the corresponding scalar).
   - **m[[i, j]] = a** Change the value of element (**i**, **j**) into scalar **a** (output is **a**).
2. Taking, dropping, and inserting rows:
   - **m[[i]]** Take row **i** (output is the corresponding vector).
   - **m[[{i_1, i_2, ...}]]** Take the given rows (output is the corresponding matrix).
   - **First[m]**, **Last[m]** Take the first/last row (output is the corresponding vector).
   - **Rest[m]**, **Most[m]** Drop the first/last row (output is the corresponding matrix).
   - **Take[m, n]**, **Take[m, -n]**, **Take[m, {n_1, n_2}]** Take the first **n** rows / the last **n** rows / rows $n_1, \ldots, n_2$ (output is the corresponding matrix).
   - **Drop[m, n]**, **Drop[m, -n]**, **Drop[m, {n_1, n_2}]** Drop the first **n** rows / the last **n** rows / rows $n_1, \ldots, n_2$ (output is the corresponding matrix).
   - **Prepend[m, r]**, **Append[m, r]** Insert row **r** at the beginning/end of a matrix (output is the corresponding matrix).
   - **Insert[m, r, i]** Insert row **r** between rows **i** − 1 and **i** (so, **r** becomes row **i**) (output is the corresponding matrix).
3. Taking and dropping columns:
   - **m[[All, j]]** Take column **j** (output is the corresponding vector).
   - **m[[All, {j_1, j_2, ...}]]** Take the given columns (output is the corresponding matrix).
   - **Take[m, All, n]**, **Drop[m, {}, n]** Take/drop the first **n** columns (output is the corresponding matrix).
4. Taking submatrices:
   - **m[[{i_1, i_2, ...}, {j_1, j_2, ...}]]** Take the elements having the given indices.
   - **Take[m, {i_1, i_2}, {j_1, j_2}]]** Take the elements having row indices between i_1 and i_2 and column indices between j_1 and j_2.
5. Combining and extending matrices:
   - **Join[m_1, m_2, ...]** Put matrices m_i one below the other.
   - **MapThread[Join, {m_1, m_2, ...}]** Put matrices m_i side by side; another way: **Transpose[Join[Transpose[m_1], Transpose[m_2], ...]]**.
   - **PadRight[m, {n_1, n_2}]** Extend **m** with zeros to an $n_1 \times n_2$ matrix.
   - **PadRight[m, {n_1, n_2}, a]** Extend **m** with replicates of matrix **a** to an $n_1 \times n_2$ matrix.
6. In the **LAMM** package:
   - Taking parts of matrices: **TakeRows**, **TakeColumns**, **TakeMatrix**, **SubMatrix**.
   - Combining matrices: **AppendColumns**, **AppendRows**, **BlockMatrix**.
7. Permuting rows and columns:
   - **m[[{i_1, ..., i_n}]]** Permute the rows into the given order.
   - **RotateLeft[m]**, **RotateRight[m]** Move the first/last row to be the last/first row.
   - **RotateLeft[m, {0, 1}]**, **RotateRight[m, {0, 1}]** Move the first/last column to be the last/first column.

8. Flattening and partitioning:
   - **Flatten[m]** Flatten out a matrix into a vector by concatenating the rows.
   - **Partition[v, k]** Partition a vector into a matrix having **k** columns and as many rows as become complete.

73.6 Eigenvalues

Commands:

1. Eigenvalues and eigenvectors:
   - **CharacteristicPolynomial[m, x]** Characteristic polynomial of a square matrix [note that *Mathematica* defines the characteristic polynomial of a matrix M to be $\det(M - xI)$ while the handbook uses the definition $\det(xI - M)$].
   - **Eigenvalues[m]** Eigenvalues of a square matrix, in order of decreasing absolute value (repeated eigenvalues appear with their appropriate multiplicity).
   - **Eigenvectors[m]** Eigenvectors of a square matrix, in order of decreasing absolute value of their eigenvalues (eigenvectors are not normalized).
   - **Eigensystem[m]** Eigenvalues and eigenvectors of a square matrix.
2. Options:
   - **Eigenvalues**, **Eigenvectors**, and **Eigensystem** have the options **Cubics** (Mma 5.0) and **Quartics** (Mma 5.0) with default value **False**; if you want explicit radicals for all cubics and quartics, set **Cubics → True** and **Quartics → True**.
3. Largest eigenvalues and generalized eigenvalues:
   - **Eigenvalues[m, k]** (Mma 5.0) Calculate the **k** largest eigenvalues; similarly for **Eigenvectors** and **Eigensystem**.
   - **Eigenvalues[{m, a}]** (Mma 5.0) Calculate the generalized eigenvalues of matrix **m** with respect to matrix **a**; similarly for **Eigenvectors** and **Eigensystem**.

Examples:

1. Eigenvalues. First, we calculate the characteristic polynomial of **m** in two ways:

   ```
   m = {{6, 2, 3}, {2, 7, 5}, {4, 2, 5}};
   chp = CharacteristicPolynomial[m, x]    98 - 81 x + 18 x^2 - x^3
   Det[m - x IdentityMatrix[3]]    98 - 81 x + 18 x^2 - x^3
   ```

   Then, we calculate the eigenvalues in two ways:

   lam = Eigenvalues[m] $\{8+\sqrt{15},\ 8-\sqrt{15},\ 2\}$
   Solve[chp == 0] $\{\{x \to 2\}, \{x \to 8-\sqrt{15}\}, \{x \to 8+\sqrt{15}\}\}$

   Here are the eigenvectors:

   vec = Eigenvectors[m]
   $$\left\{\left\{1,\ -\frac{9}{2}+\frac{1}{2}(8+\sqrt{15}),\ 1\right\},\ \left\{1,\ -\frac{9}{2}+\frac{1}{2}(8-\sqrt{15}),\ 1\right\},\ \left\{-5,\ -14,\ 16\right\}\right\}$$

   An eigenvalue λ of a matrix M and the corresponding eigenvector x should satisfy $Mx = \lambda x$. We check this for the first eigenvalue and then for all the eigenvalues:

```
m.vec[[1]] – lam[[1]] vec[[1]] // Simplify {0, 0, 0}
m.Transpose[vec] – Transpose[vec].DiagonalMatrix[lam] //
Simplify
{{0, 0, 0}, {0, 0, 0}, {0, 0, 0}}
```

With **Eigensystem** we can compute both the eigenvalues and the eigenvectors. The result is a list where the first element is a list of eigenvalues and the second element a list of the corresponding eigenvectors:

```
{lam, vec} = Eigensystem[m]
```

$\{\{8+\sqrt{15},\ 8-\sqrt{15},\ 2\},$
$\{\{1,\ -\frac{9}{2}+\frac{1}{2}(8+\sqrt{15}),\ 1\},\ \{1,\ -\frac{9}{2}+\frac{1}{2}(8-\sqrt{15}),\ 1\},$
$\{-5,\ -14,\ 16\}\}\}$

2. Options. For more complex eigenvalues, we may get only representations of the values as some root objects, but with **N** we get numerical values and with the **Cubics** and **Quartics** options we may get explicit eigenvalues. Note that often we only need the decimal values of the eigenvalues (not their exact expressions), and then we can input the matrix **m // N**.

```
n = {{7, 7, 2, 7}, {2, 5, 0, 7}, {8, 7, 8, 3}, {5, 6, 4, 5}};
Eigenvalues[m]
{Root[-171 + 112 #1 – 25 #1^2 + #1^3 &, 1],
 Root[-171 + 112 #1 – 25 #1^2 + #1^3 &, 3],
 Root[-171 + 112 #1 – 25 #1^2 + #1^3 &, 2], 0}
% // N {19.7731, 2.61345 + 1.34834 i, 2.61345 – 1.34834 i, 0.}
Eigenvalues[n // N]
{19.7731, 2.61345 + 1.34834 i, 2.61345 – 1.34834 i,
   -2.85924 × 10^-16}
Eigenvalues[n, Cubics → True][[1]]
```

$$\frac{25}{3}+\frac{1}{3}\left(\frac{10667}{2}-\frac{261\sqrt{253}}{2}\right)^{1/3}+\frac{1}{3}\left(\frac{1}{2}(10667+261\sqrt{253})\right)^{1/3}$$

Applications:

1. (Positive and negative definite matrices) A symmetric (or Hermitian) matrix is positive [negative] definite if and only if all of the eigenvalues are positive [negative]. (See Section 8.4 for more information about positive definite matrices.) With this result we can easily test the definiteness of a matrix (note that *Mathematica* does not have a special command for this test). For example, the following symmetric matrix is positive definite:

```
m = {{2, 0, 1}, {0, 4, 2}, {1, 2, 3}};
Eigenvalues[m] // N {5.66908, 2.47602, 0.854897}
```

2. (A sufficient condition for a minimum or maximum) A sufficient condition for a local minimum [maximum] point of a function is that the Hessian is positive [negative] definite [BSS93, p. 134]. Calculate the stationary points of a function (for a program for classical optimization with constraints, see [Rus04, pp. 543–547]):

```
f = x^3 + x^2 y^2 + 2 x y – 3 x – y
   -3 x + x^3 – y + 2 x y + x^2 y^2
grad = {D[f, x], D[f, y]}
   {-3 + 3 x^2 + 2 y + 2 x y^2, -1 + 2 x + 2 x^2y}
stat = Solve[grad == 0, {x, y}] // N
{{y → 1.22959, x → -1.16294},
 {y → 0.956992, x→ 0.369407}, {y → -0.497601, x → 1.07442},
 {y → -0.844491 – 2.24741 i, x → -0.140442 – 0.584262 i},
 {y → -0.844491 + 2.24741 i, x → -0.140442 + 0.584262 i}}
```

Pick the real points (see Section 73.5, the second command in item 2 of Commands):

```
stat2 = stat[[{1, 2, 3}]]
{{y → 1.22959, x → -1.16294},
 {y → 0.956992, x → 0.369407}, {y → -0.497601, x → 1.07442}}
```

Calculate the Hessian and its eigenvalues at the real stationary points:

```
hes = {{D[f, x, x], D[f, x, y]}, {D[f, y, x], D[f, y, y]}}
{{6 x + 2 y^2, 2 + 4 x y}, {2 + 4 x y, 2 x^2}}
Map[Eigenvalues[hes /. #] &, stat2]
{{-5.61664, 4.36764}, {6.06166, -1.74063}, {6.94587, 2.30462}}
```

Thus, the first two points are saddle points while the third point is a point of minimum. Note that the gradient and the Hessian can in *Mathematica* 5.1 also be calculated as follows:

```
grad = D[f, {{x, y}, 1}]
      {-3 + 3 x^2 + 2 y + 2 x y^2, -1 + 2 x + 2 x^2 y}
hes = D[f, {{x, y}, 2}]
      {{6 x + 2 y^2, 2 + 4 x y}, {2 + 4 x y, 2 x^2}}
```

3. (Eigenvector–eigenvalue decomposition) Let diag(λ) be a diagonal matrix whose diagonal elements are the eigenvalues of a matrix M, and let P be a matrix whose columns are the eigenvectors of M. If P is nonsingular, we have the decomposition $M = P\text{diag}(\lambda)P^{-1}$ or diagonalization $P^{-1}MP = \text{diag}(\lambda)$. The former equality shows that M and diag(λ) are similar matrices, so that, for example, they have the same determinant, trace, and rank. (See Section 4.3 for more information about diagonalization.) As an example of the eigenvector–eigenvalue decomposition, first compute the eigenvalues and eigenvectors:

```
m = {{6, 2, 3}, {2, 7, 5}, {4, 2, 5}};
{lam, vec} = Eigensystem[m]
{{8 + √15, 8 - √15, 2},
 {{1, -9/2 + 1/2 (8 + √15), 1}, {1, -9/2 + 1/2 (8 - √15), 1},
   {-5, -14, 16}}}
```

Then, define the diagonal matrix of eigenvalues and the matrix whose columns are the eigenvectors:

```
diag = DiagonalMatrix[lam]
      {{8 + √15, 0, 0}, {0, 8 - √15, 0}, {0, 0, 2}}
p = Transpose[vec]
{{1, 1, -5}, {-9/2 + 1/2 (8 + √15), -9/2 + 1/2 (8 - √15), -14},
  {1, 1, 16}}
```

The following computation verifies the decomposition:

```
p.diag.Inverse[p] - m // Simplify
   {{0, 0, 0}, {0, 0, 0}, {0, 0, 0}}
```

73.7 Singular Values

Commands:

1. Singular values:

   - **`SingularValueList[m]`** (Mma 5.0) Nonzero singular values of a (possibly rectangular) numerical matrix with at least one entry that has a decimal point; the singular values are given in order of decreasing value (repeated singular values appear with their appropriate multiplicity).

2. Singular value decomposition:

   - **SingularValueDecomposition[m]** (Mma 5.0) Singular value decomposition of a (possibly rectangular) numerical matrix with at least one entry that has a decimal point. For matrix M, the decomposition is $M = UWV^*$ where U and V are orthonormal and W is diagonal with singular values as the diagonal elements (* means conjugate transpose). Output is $\{U, W, V\}$.

3. An option:

   - The default value **Automatic** of the option **Tolerance** (Mma 5.0) keeps only singular values larger than 100×10^{-p}, where p is **Precision[m]**. **Tolerance → t** keeps only singular values that are at least **t** times the largest singular value. **Tolerance → 0** returns all singular values.

4. Largest singular values and generalized singular values:

   - **SingularValueList[m, k]** (Mma 5.0) Calculate the **k** largest singular values.
   - **SingularValueList[{m, a}]** (Mma 5.0) Calculate the generalized singular values of **m** with respect to **a**.

Examples:

1. Singular values. The singular values of a matrix M are the square roots of the eigenvalues of the matrix $M^* M$:

```
m = {{6, 2, 3}, {2, 7, 5}, {4, 2, 5}};
SingularValueList[m // N]    {12.1048, 4.75064, 1.70418}
Sqrt[Eigenvalues[ConjugateTranspose[m].m]] // N
{12.1048, 4.75064, 1.70418}
```

   Only nonzero singular values are returned by **SingularValueList**:

```
n = {{7, 7, 2, 7}, {2, 5, 0, 7}, {8, 7, 8, 3}, {5, 6, 4, 5}};
SingularValueList[n // N]    {21.5225, 7.09914, 1.84001}
```

   By giving **Tolerance** the value 0, we get all singular values:

```
SingularValueList[n // N, Tolerance → 0]
{21.5225, 7.09914, 1.84001, 5.46569 × 10^-17}
```

2. Singular value decomposition. Here is an example of this decomposition:

```
Map[MatrixForm, {u, w, v} = SingularValueDecomposition[m // N]]
```

$$\left\{\begin{pmatrix} -0.516443 & 0.629389 & -0.580652 \\ -0.672542 & -0.717852 & -0.179933 \\ -0.53007 & 0.297588 & 0.79402 \end{pmatrix}, \begin{pmatrix} 12.1048 & 0. & 0. \\ 0. & 4.75064 & 0. \\ 0. & 0. & 1.70418 \end{pmatrix}, \begin{pmatrix} -0.542264 & 0.743264 & -0.391801 \\ -0.561826 & -0.667491 & -0.488678 \\ -0.624741 & -0.0448685 & 0.779542 \end{pmatrix}\right\}$$

The following verifies the decomposition:

```
  m - u.w.ConjugateTranspose[v] // Chop
   {{0, 0, 0}, {0, 0, 0}, {0, 0, 0}}
u.ConjugateTranspose[u] // Chop
   {{1., 0, 0},{0, 1., 0}, {0, 0, 1.}}
v.ConjugateTranspose[v] // Chop
   {{1., 0, 0}, {0, 1., 0}, {0, 0, 1.}}
```

Applications:

1. (Condition numbers) Relative to a matrix norm $||M||$, the condition number of an invertible matrix M is $||M||||M^{-1}||$. (See Chapter 37 for more information about condition numbers.) A condition number is ≥ 1; a large condition number indicates sensitivity to round-off errors. The 2-norm condition number can be shown to be the maximum singular value divided by the minimum singular value:

```
cond[m] := With[{s = SingularValueList[m // N,
  Tolerance → 0]}, First[s]/Last[s]]
```

For example, consider the condition numbers of Hilbert matrices:

```
h[n] := Table[1/(i + j - 1), {i, n}, {j, n}]
Table[cond[h[i]], {i, 1, 6}]
{1., 19.2815, 524.057, 15513.7, 476607., 1.49511 × 10^7}
```

The condition numbers grow rapidly and they are large even for small Hilbert matrices. This indicates numerical difficulties associated with Hilbert matrices. Note that the **LAMM** package defines both **HilbertMatrix** and **MatrixConditionNumber**.

73.8 Decompositions

Commands:

1. Decompositions into triangular matrices:
   - **LUDecomposition[m]** PLU decomposition of a square matrix. For matrix M, the decomposition is $PM = LU$, where L is unit lower triangular (with ones on the diagonal), U is upper triangular, and P is a permutation matrix. Output is $\{K, p, c\}$, where matrix K contains both L and U, p is a permutation vector, that is, a list specifying rows used for pivoting, and c is the L^{∞} condition number of M (however, for exact matrices with no decimal points, c is 1).
   - **CholeskyDecomposition[m]** (Mma 5.0) Cholesky decomposition of a Hermitian, positive definite matrix. For matrix M, the decomposition is $M = U^*U$, where U is upper triangular. Output is U.
2. Orthogonal decompositions:
   - **SingularValueDecomposition[m]** (Mma 5.0) Singular value decomposition of a (possibly rectangular) numerical matrix with at least one entry that has a decimal point. For matrix M, the decomposition is $M = UWV^*$ where U and V are orthonormal and W is diagonal with singular values as the diagonal elements (* means conjugate transpose). Output is $\{U, W, V\}$. See item 2 in Section 73.7.
   - **QRDecomposition[m]** QR decomposition of a numerical matrix. For matrix M, the decomposition is $M = Q^*R$, where Q is orthonormal and R is upper triangular. Output is $\{Q, R\}$.
3. Decompositions related to eigenvalue problems:
   - **JordanDecomposition[m]** Jordan decomposition of a square matrix. For matrix M, the decomposition is $M = SJS^{-1}$, where S is a similarity matrix and J is the Jordan canonical form of M. Output is $\{S, J\}$..

- **`SchurDecomposition[m]`** Schur decomposition of a square, numerical matrix with at least one entry that has a decimal point. For matrix M, the decomposition is $M = QTQ^*$, where Q is orthonormal and T is block upper triangular. Output is $\{Q, T\}$.
- **`HessenbergDecomposition[m]`** (Mma 5.1) Hessenberg decomposition of a square, numerical matrix with at least one entry that has a decimal point. For matrix M, the decomposition is $M = PHP^*$, where P is an orthonormal and H a Hessenberg matrix. Output is $\{P, H\}$.
- **`PolarDecomposition[m]`** Polar decomposition of a numerical matrix. For matrix M, the decomposition is $M = US$, where U is orthonormal and S is positive definite. Output is $\{U, S\}$. (In the **`LAMM`** package.)

Examples:

1. Decompositions into triangular matrices. As an example, we consider the PLU decomposition:

```
m = {{2., 7, 5}, {6, 2, 3}, {4, 2, 5}};
Map[MatrixForm, {lu, perm, cond} = LUDecomposition[m]]
```

$$\left\{\begin{pmatrix} 6. & 0.333333 & 0.666667 \\ 2. & 6.33333 & 0.105263 \\ 3. & 4. & 2.57895 \end{pmatrix}, \begin{pmatrix} 2 \\ 1 \\ 3 \end{pmatrix}, 9.42857\right\}$$

The matrix **`lu`** contains both the L and the U matrix, **`perm`** is a permutation vector specifying rows used for pivoting, and **`cond`** is the L^∞ condition number of **`m`**. With a package we can extract the L and U matrices from the **`lu`** matrix:

```
<< LinearAlgebra`MatrixManipulation`
Map[MatrixForm, {l, u} = LUMatrices[lu]]
```

$$\left\{\begin{pmatrix} 1. & 0. & 0. \\ 0.333333 & 1. & 0. \\ 0.666667 & 0.105263 & 1. \end{pmatrix}, \begin{pmatrix} 6. & 2. & 3. \\ 0 & 6.33333 & 4. \\ 0 & 0 & 2.57895 \end{pmatrix}\right\}$$

The following calculation verifies the PLU decomposition (see Section 73.5, the first command in item 7 of Commands):

```
m1 = m[[perm]]   {{6, 2, 3}, {2., 7, 5}, {4, 2, 5}}
m1 - l.u   {{0., 0., 0.}, {0., 0., 0.}, {0., 0., 0.}}
```

We can also verify the decomposition with the corresponding permutation matrix:

```
permMatr = IdentityMatrix[3][[perm]]
  {{0, 1, 0}, {1, 0, 0}, {0, 0, 1}}
m2 = permMatr.m   {{6., 2., 3.}, {2., 7., 5.}, {4., 2., 5.}}
m2 - l.u   {{0., 0., 0.}, {0., 0., 0.}, {0., 0., 0.}}
```

Once the PLU decomposition of **`m`** is ready, we can solve a system of linear equations with coefficient matrix **`m`**. Define **`b`**, the constant vector, and use then **`LUBackSubstitution`**:

```
b = {2, 4, 1};
LUBackSubstitution[{lu, perm, cond}, b]
  {0.826531, 0.530612, -0.673469}
```

However, the PLU decomposition and the back substitution are automatically done with **`LinearSolve`**:

```
LinearSolve[m, b]   {0.826531, 0.530612, -0.673469}
```

73.9 Linear Systems

Commands:

1. Solving equations written explicitly:
   - **lhs == rhs** An equation with left-hand side **lhs** and right-hand side **rhs** (after you have written the two equal signs ==, *Mathematica* replaces them with a corresponding single symbol).
   - **eqns** = {**eqn**$_1$**, eqn**$_2$**, ...**} A system of equations with the name **eqns**.
   - **Solve[eqns]** Solve the list of equations **eqns** for the symbols therein.
   - **Solve[eqns, vars]** Solve the list of equations **eqns** for the variables in list **vars**.
   - **Reduce[eqns, vars]** Give a full analysis of the solution.
2. Solving equations defined by the coefficient matrix and the constant vector:
   - **LinearSolve[m, v]** Solve the linear system **m.vars == v**.
   - **Solve[m.vars == v]** Solve the linear system **m.vars == v**.
   - If it is possible that the system has more than one solution, it is better to use **Solve**, as **LinearSolve** only gives one solution (with no warning).
3. Conversion between the two forms of linear equations:
   - **eqns = Thread[m.vars == v]** Generate explicit equations from a coefficient matrix **m** and a constant vector **v** by using the list of variables **vars**.
   - {**m, v**} = **LinearEquationsToMatrices[eqns, vars]** Give the coefficient matrix **m** and the constant vector **v** of given linear equations **eqns** containing variables **vars** (in the **LAMM** package).
4. Solving several systems with the same coefficient matrix:
   - **f = LinearSolve[m]** (Mma 5.0) Give a function **f** for which **f[v]** solves the system **m.x == v**.
   - **LUBackSubstitution[lu, v]** Solve the system **m.x == v**; here **lu = LUDecomposition[m]**.
5. Using the inverse and pseudo-inverse:
   - **Inverse[m].v** Give the solution of the equations **m.x == v** (this method is not recommended).
   - **PseudoInverse[m].v** Give the least-squares solution of the equations **m.x == v**.
6. Special methods:
   - **RowReduce[a]** Do Gauss–Jordan elimination to the augmented matrix **a** to produce the reduced row echelon form, where the last column is the solution of the system.
   - **TridiagonalSolve[sub, main, super, rhs]** Solve a tridiagonal system (in the **LinearAlgebra`Tridiagonal`** package).
7. Eliminating variables:
   - **Solve[eqns, vars, elims]** Solve **eqns** for **vars**, eliminating **elims**.
   - **Eliminate[eqns, elims]** Eliminate **elims** from **eqns**.

Examples:

1. Solving equations written explicitly. If the system does not have any symbols other than the variables for which the system has to be solved, we need not declare the variables in **Solve**:

```
sol = Solve[{2 x + 5 y + z == 2, 3 x + 4 y + 3 z == 2,
    6 x + y + 2 z == 5}]
```
$\left\{\left\{x \to \frac{47}{49},\ y \to \frac{5}{49},\ z \to -\frac{3}{7}\right\}\right\}$

```
sol // N  {{x → 0.959184, y → 0.102041, z→ -0.428571}}
```

Here is an example where we declare the variables:

```
eqns = {2 x + 5 y + z == 2, 3 x + 4 y + 3 z == 2,
    6 x + y + 2 z == 5};
vars = {x, y, z};
sol = Solve[eqns, vars]
```
$\left\{\left\{x \to \frac{47}{49},\ y \to \frac{5}{49},\ z \to -\frac{3}{7}\right\}\right\}$

Solve gives the solution in the form of transformation rules. If we want a list of the values of the variables, we can write

```
vars /. sol[[1]]
```
$\left\{\frac{47}{49},\ \frac{5}{49},\ -\frac{3}{7}\right\}$

To check the solution, write

```
eqns /. sol   {{True, True, True}}
```

(Note that for an inexact solution the test may yield False for some equations due to round-off errors.) If a solution does not exist, we get an empty list:

```
Solve[{x + 2 y == 2, 2 x + 4 y == 1}]   {}
```

If an infinite number of solutions exist, we get all of them (with a message):

```
Solve[{x + 2 y == 2, 2 x + 4 y == 4}]   {{x → 2 - 2 y}}
Solve::svars : Equations may not give solutions for all solve
   variables. More...
```

This means that **x** has the value **2 − 2y**, while **y** can be arbitrary. The message tells that the equations may not determine the values of all of the variables (this is, indeed, the case in our example because the solution does not give the value of **y**).

The system may contain symbols:

```
eqns2 = {2 x + 5 y == 3, 3 x + d  y == 2};
Solve[eqns2, {x, y}]
```
$\left\{\left\{x \to -\frac{10 - 3\,d}{-15 + 2\,d},\ y \to -\frac{5}{-15 + 2\,d}\right\}\right\}$

Note that this is a generic solution that is valid for most values of the parameter **d**. To get a full analysis of the equations, use **Reduce**. The result is a logical statement where, for example, **&&** means logical AND and || means logical OR:

```
Reduce[eqns2, {x, y}]
```
$-15 + 2\,d \neq 0 \quad \&\& \quad x == \frac{-10 + 3\,d}{-15 + 2\,d} \quad \&\& \quad y == \frac{1}{5}\,(3 - 2\,x)$

2. Solving equations defined by the coefficient matrix and the constant vector. We continue with the same system that we considered in Example 1:

```
m = {{2, 5, 1}, {3, 4, 3}, {6, 1, 2}}; v = {2, 2, 5};
sol = LinearSolve[m, v]
```
$\left\{\frac{47}{49},\ \frac{5}{49},\ -\frac{3}{7}\right\}$

```
sol // N  {0.959184, 0.102041, -0.428571}
```

We can check the solution:

```
m.sol - v   {0, 0, 0}
```

We can also use **Solve** if we form an equation with variables:

```
Solve[m.{x, y, z} == v]
```
$\left\{\left\{x \to \frac{47}{49},\ y \to \frac{5}{49},\ z \to -\frac{3}{7}\right\}\right\}$

If a solution does not exist, **LinearSolve** gives a message:

```
LinearSolve[{{1, 2}, {2, 4}}, {2, 1}]
```

```
LinearSolve::nosol : Linear equation encountered which has no
   solution. More...
LinearSolve[{{1, 2}, {2, 4}}, {2, 1}]
```

If an infinite number of solutions exist, we are given one of them (with no warning):

```
LinearSolve[{{1, 2}, {2, 4}}, {2, 4}]   {2, 0}
```

The system may contain symbols:

```
LinearSolve[{{2, 5}, {3, d}}, {3, 2}]
```

$$\left\{\frac{-10 + 3\,d}{-15 + 2\,d},\ -\frac{5}{-15 + 2\,d}\right\}$$

3. Conversion between the two forms of linear equations. Consider again the familiar system:

```
eqns = {2 x + 5 y + z == 2, 3 x + 4 y + 3 z == 2,
     6 x + y + 2 z == 5};
```

The corresponding coefficient matrix and constant vector are

```
<< LinearAlgebra`MatrixManipulation`
{m, v} = LinearEquationsToMatrices[eqns, {x, y, z}]
{{{2, 5, 1}, {3, 4, 3}, {6, 1, 2}}, {2, 2, 5}}
```

If we know **m** and **v**, we can obtain explicit equations as follows (here we use indexed variables):

```
eqns = Thread[m.Table[x_i, {i, 3}] == v]
```

$$\{2\,x_1 + 5\,x_2 + x_3 == 2,\ 3\,x_1 + 4\,x_2 + 3\,x_3 == 2,\ 6\,x_1 + x_2 + 2\,x_3 == 5\}$$

4. Solving several systems with the same coefficient matrix. Suppose that we have one coefficient matrix **m**, but we would like to solve three systems by using the constant vectors **v1**, **v2**, and **v3**:

```
m = {{2, 5, 1}, {3, 4, 3}, {6, 1, 2}};
v1 = {2, 2, 5}; v2 = {1, 4, 3}; v3 = {4, 2, 6};
```

First, we create a function with **LinearSolve**:

```
f = LinearSolve[m]   LinearSolveFunction[{3, 3}, "<>"]
```

Then, we give the three constant vectors as arguments:

```
{f[v1], f[v2], f[v3]}
```

$$\left\{\left\{\frac{47}{49}, \frac{5}{49}, -\frac{3}{7}\right\}, \left\{\frac{2}{49}, -\frac{5}{49}, \frac{10}{7}\right\}, \left\{\frac{68}{49}, \frac{26}{49}, -\frac{10}{7}\right\}\right\}$$

In this way we solve three systems, but the PLU decomposition is done only once, thus saving computing time. We could also use **LUDecomposition**:

```
lu = LUDecomposition[m];
Map[LUBackSubstitution[lu, #] &, {v1, v2, v3}]
```

$$\left\{\left\{\frac{47}{49}, \frac{5}{49}, -\frac{3}{7}\right\}, \left\{\frac{2}{49}, -\frac{5}{49}, \frac{10}{7}\right\}, \left\{\frac{68}{49}, \frac{26}{49}, -\frac{10}{7}\right\}\right\}$$

5. Using the inverse and pseudo-inverse. The inverse of the coefficient matrix can, in principle, be used to solve a linear system:

```
m = {{2, 5, 1}, {3, 4, 3}, {6, 1, 2}}; v = {2, 2, 5};
Inverse[m].v
```

$$\left\{\frac{47}{49}, \frac{5}{49}, -\frac{3}{7}\right\}$$

However, this method should not be used, since inverting a matrix is a more demanding task than solving a linear system with Gaussian elimination. If the coefficient matrix is singular or nonsquare, we can, however, use a pseudo-inverse to calculate a least-squares solution. (See Sections 5.7, 5.8, and 5.9 for definitions and more information on the pseudo-inverse and least squares solution.) If the system is $Mx = v$, the least-squares solution minimizes the norm $\|Mx - v\|_2$. As an example, consider a system with coefficient matix **p** and constant vector **c**:

```
p = {{7, 4, 8, 0}, {7, 4, 8, 0}, {1, 8, 3, 3}}; c = {4, 1, 2};
```

The equations do not have any solutions in the usual sense, but a least-squares solution can be calculated:

```
PseudoInverse[p].c      {671/6738, 565/3369, 1907/13476, 201/4492}
```

6. Special methods. **RowReduce** can be used to solve a system of equations. First we form the augmented matrix by appending the constant vector into the rows of the coefficient matrix:

```
m = {{2, 5, 1}, {3, 4, 3}, {6, 1, 2}}; v = {2, 2, 5};
a = Transpose[Append[Transpose[m], v]]
{{2, 5, 1, 2}, {3, 4, 3, 2}, {6, 1, 2, 5}}
```

Then, we do Gauss–Jordan elimination to produce the reduced row echelon form; the solution is the last column:

```
g = RowReduce[a]
    {{1, 0, 0, 47/49}, {0, 1, 0, 5/49}, {0, 0, 1, -3/7}}
sol = g[[All, 4]]      {47/49, 5/49, -3/7}
```

7. Eliminating variables. The following equations contain the parameter **d**, which we would like to eliminate:

```
eqns = {x + y - z == d, x - 2 y + z == -d,
   -x + y + z == 2 d}; vars = {x, y, z};
Solve[eqns, vars, {d}]    {{x → 3 z/5, y → 6 z/5}}
Solve::svars : Equations may not give solutions for all solve
   variables. More...
```

Alternatively, we can use **Eliminate**. The result is then expressed as a logical expression:

```
Eliminate[eqns, {d}]    5 x == 3 z  &&  5 y == 6 z
```

Applications:

1. (Stationary Distribution of a Markov chain) The stationary distribution π (a row vector) of a Markov chain with transition probability matrix P is obtained by using the equations $\pi P = \pi$ and $\sum \pi_i = 1$. One of the equations of the linear system $\pi P = \pi$ can be dropped. We continue Application 1 of Section 73.4:

```
p = {{0.834, 0.166}, {0.398, 0.602}}; st = {π1, π2};
eqns = Append[Most[Thread[st.p == st]], Total[st] == 1]
{0.834 π1 + 0.398 π2 == π1, π1 + π2 == 1}
Solve[eqns]    {{π1 → 0.705674, π2 → 0.294326}}
```

So, in the long run, about 71% of days are wet and 29% dry.

73.10 Linear Programming

Commands:

1. Solving linear optimization problems written explicitly:

- **Minimize[{obj, cons}, vars]** (Mma 5.0) Give the global minimum of **obj** subject to constraints **cons** with respect to variables **vars**.
- **Maximize[{obj, cons}, vars]** (Mma 5.0) Give the global maximum of **obj** subject to constraints **cons** with respect to variables **vars**.

2. Solving linear optimization problems defined by vectors and matrices:
   - **LinearProgramming[c, a, b]** Minimize **c.x** subject to **a.x ≥ b** and **x ≥ 0**.
   - **LinearProgramming[c, a, {{b1, s1}, {b2, s2}, ...}]** The ith constraint is **ai.x ≤ bi**, **ai.x == bi**, or **ai.x ≥ bi** according to whether **si** is $< 0, = 0$, or > 0.

Examples:

1. Solving linear optimization problems written explicitly.

```
obj = 2 x + 3 y;
cons = {x + y ≥ 4, x + 3 y ≥ 5, x ≥ 0, y ≥ 0};
vars = {x, y};
  Minimize[{obj, cons}, vars]
```

$\left\{\frac{17}{2}, \left\{x \to \frac{7}{2}, y \to \frac{1}{2}\right\}\right\}$

Here the minimum value of the objective function is 17/2. Next, we solve an integer problem. The constraint **vars ∈ Integers** or **Element[vars, Integers]** restricts the variables in **vars** to be integer-valued.

```
Minimize[{obj, cons, vars ∈ Integers}, vars]
{9, {x → 3, y → 1}}
```

2. Solving linear optimization problems defined by vectors and matrices. We solve the first problem in Example 1:

```
c = {2, 3}; a = {{1, 1}, {1, 3}}; b = {4, 5};
sol = LinearProgramming[c, a, b]
```

$\left\{\frac{7}{2}, \frac{1}{2}\right\}$

```
c.sol
```

$\frac{17}{2}$

Applications:

1. (A transportation problem) (See Chapter 50 for definitions and more information about linear programming.) Consider the following transportation problem (cf. [Rus04, pp. 524–526]). Supplies of plants 1 and 2, demands of cities 1, 2, and 3, and costs between the plants and the cities are as follows:

```
supply = {40, 60}; demand = {35, 40, 25};
costs = {{4, 9, 6}, {5, 3, 7}};
```

For example, to transport one unit from plant 1 to city 3 costs \$6. The problem is to minimize the total cost of transportation. Let $x_{i,j}$ be the amount transported from plant i to city j:

```
vars = Table[x_{i,j}, {i, 2}, {j, 3}]
  {{x_{1,1}, x_{1,2}, x_{1,3}}, {x_{2,1}, x_{2,2}, x_{2,3}}}
```

The objective function is

```
obj = Flatten[costs].Flatten[vars]
4 x_{1,1} + 9 x_{1,2} + 6 x_{1,3} + 5 x_{2,1} + 3 x_{2,2} + 7 x_{2,3}
```

The supply, demand, and nonnegativity constraints are as follows:

```
sup = Thread[Map[Total, vars] ≤ supply]
{x_{1,1} + x_{1,2} + x_{1,3} ≤ 40, x_{2,1} + x_{2,2} + x_{2,3} ≤ 60}
dem = Thread[Total[vars] ≥ demand]
{x_{1,1} + x_{2,1} ≥ 35, x_{1,2} + x_{2,2} ≥ 40, x_{1,3} + x_{2,3} ≥ 25}
non = Thread[Flatten[vars] ≥ 0]
{x_{1,1} ≥ 0, x_{1,2} ≥ 0, x_{1,3} ≥ 0, x_{2,1} ≥ 0, x_{2,2} ≥ 0, x_{2,3} ≥ 0}
```

Then, we solve the problem:

```
sol = Minimize[{obj, sup, dem, non}, Flatten[vars]]
{430, {x_{1,1} → 15, x_{1,2} → 0, x_{1,3} → 25, x_{2,1} → 20,
      x_{2,2} → 40, x_{2,3} → 0}}
```

Appendix

Introduction to Mathematica

1. Executing commands:
   - To do a calculation with *Mathematica*, write the command into the notebook document (possibly with the help of a palette) and then execute it by pressing the Enter key in the numeric keypad or by holding down the Shift key and then pressing the Return key.
2. Arithmetic:
   - For addition, subtraction, division, and power, use **+**, **-**, **/**, and **^**.
   - For multiplication, usually the space key is used, but the ***** key can also be used. For example, **a b** or **a*b** is a product but **ab** is a single variable.
3. Useful techniques:
   - To give a name for a result, use **=**; for example, **a = 17^2**.
   - To clear the value of a name, use **=.**; for example, **a =.**.
   - To refer to the last result, use **%**; for example, **% + 18^2**.
   - If an input is ended with a semicolon (**a = 17^2;**), the input is processed (after pressing, say, Enter), but the result is not shown in the notebook.
   - Input and output labels of the form In[*n*] and Out[*n*] can be turned off with the menu command Kernel ▷ Show In/Out Names.
4. Important conventions:
   - All built-in names start with a capital letter.
   - The argument of functions and commands are given within square brackets **[ ]**.
   - Parentheses **( )** are only used for grouping terms in expressions.
5. Constants and mathematical functions:
   - **Pi** (3.141...), **E** (2.718...), **I** ($\sqrt{-1}$), **Infinity**.
   - **Sqrt[z]** or **z^(1/2)**, **Exp[z]** or **E^z**, **Log[z]** (the natural logarithm), **Log[b, z]** (a logarithm to base **b**), **Abs[z]**, **Sin[z]**, **ArcSin[z]**, **n!**, **Binomial[n, m]**.
6. Lists:
   - A list is formed with curly braces { }; an example: **v = {4, 7, 5}**. A list is simply an ordered collection of elements. Lists can be nested, for example **m = {{4, 7, 5}, {2, 6, 3}}**.
   - Parts of lists (and of more general expressions) can be picked up with double square brackets **[[ ]]**. For example, **m[[2]]** gives {2, 6, 3} and **m[[2, 1]]** gives 2.
7. Useful commands:
   - A decimal value can be obtained using **N[expr]**.
   - A decimal value to **n** digit precision can be obtained using **N[expr, n]**.
   - Numbers near to zero can be set to 0 with **Chop[expr]**.
   - An expression can be simplified with **Simplify[expr]**.
   - An expression can be expanded out with **Expand[expr]**.
   - All commands with a single argument can also be applied with **//**; e.g., **expr // N** means **N[expr]**.

8. Replacements:

   - The syntax **expr /. x → a** is used to replace **x** in **expr** with **a**. Here, **x → a** is a transformation rule and **/.** is the replacement operator. The arrow → can be written by typing the two characters **->** with no space in between; *Mathematica* transforms **->** to a genuine arrow.

9. Mapping and threading:

   - With **Map** we can map the elements of a list with a function, that is, with **Map** we can calculate the value of a function at points given in a list. For example, **Map[f[#] &, {a, b, c}]** gives {f[a], f[b], f[c]} (a function like **f[#] &** is called a pure function). If **f** is a built-in function with one argument, the name of the function suffices. For example, if we want to calculate the 2-norms of the rows of a matrix **m**, we can write **Map[Norm[#] &, m]**, but also simply **Map[Norm, m]**.
   - With **Thread** we can apply an operation to corresponding parts of two lists. For example, **Thread[{x + y, x − y} == {2, 5}]** gives {x + y == 2, x − y == 5}.

10. Functions and programs:

    - The syntax **f[x_] := expr** is used to define functions. For example, **f[x_, y_] := x + Sin[y]**.
    - In programs, we often use **Module** to define local variables or **With** to define local constants. The syntax is **f[x_] := Module[**{local variables}**,** body**]** (similarly for **With**).

11. Loading packages:

    - A package is loaded with **<<**. For example, **<<LinearAlgebra`MatrixManipulation`**.

12. Using saved notebooks:

    - When you open a saved notebook, *Mathematica* will not automatically process the inputs anew. So, if the notebook contains the definition **a = 17^2** and you also want to use this definition in the new *Mathematica* session, you have to process the input anew; simply put the cursor anywhere in the input and press Enter or Shift-Return.

Getting Help

1. Books, documents, and internet addresses:

   - [Wol03]: A complete description of *Mathematica* 5; for linear algebra, see Sections 1.5.7 (Solving Equations), 1.8.3 (Vectors and Matrices), and 3.7 (Linear Algebra).
   - [WR99]: A description of the packages of *Mathematica* 4.
   - [WR03]: Linear algebra with *Mathematica* 5; a long Help Browser document.
   - [Sza00]: Linear algebra with *Mathematica* 3.
   - [Rus04]: A general introduction to *Mathematica* 5.
   - http://library.wolfram.com/infocenter/BySubject/Mathematics/Algebra/LinearAlgebra/: about 100 documents relating to linear algebra with *Mathematica*.

2. Using the Help Browser:

   - To open the Help Browser, select the menu command Help ▷ Help Browser.
   - In the Help Browser, write a command into the input field and press the Go-button, or choose from the lists of topics.

- In a notebook, write a command like **Eigenvalues** and then press the F1 key (Windows) or Help key (Macintosh) to read the corresponding Help Browser material.
- In a notebook, execute a command like **?Eigenvalues** to get a short description of a command and a link to the more complete information in the Help Browser.

3. Useful material in the Help Browser:

- The *Mathematica* Book ▷ A Practical Introduction to *Mathematica* ▷ Symbolic Mathematics ▷ Solving Equations: Section 1.5.7 of [Wol03].
- The *Mathematica* Book ▷ A Practical Introduction to *Mathematica* ▷ Lists ▷ Vectors and Matrices: Section 1.8.3 of [Wol03].
- The *Mathematica* Book ▷ Advanced Mathematics in *Mathematica* ▷ Linear Algebra: Section 3.7 of [Wol03].
- Built-in Functions ▷ Advanced Documentation ▷ Linear Algebra: [WR03].
- Built-in Functions ▷ Lists and Matrices ▷ Vector Operations: A list of commands related to vectors.
- Built-in Functions ▷ Lists and Matrices ▷ Matrix Operations: A list of commands related to matrices.
- Add-ons and Links ▷ Standard Packages ▷ Linear Algebra: Descriptions of linear algebra packages.

For Users of Earlier Versions

Some commands presented in this chapter are new in *Mathematica* 5.0 or 5.1. Below we have collected information about how to do similar calculations with earlier versions (mainly with 4.2).

- **ArrayPlot[m]**: replace with **ListDensityPlot[Reverse[Abs[m]], ColorFunction → (GrayLevel[1 – #] &), AspectRatio → Automatic, Mesh → False, FrameTicks → False]**.
- **CholeskyDecomposition**: replace with a command of the same name from the **LinearAlgebra`Cholesky`** package.
- **ConjugateTranspose[m]**: replace with **Conjugate[Transpose[m]]**.
- **HessenbergDecomposition**: replace with **Developer`HessenbergDecomposition**.
- **MatrixRank[m]**: replace with **Dimensions[m][[2]] – Length[NullSpace[m]]**.
- **Maximize[{f, cons}, vars]**: replace with **ConstrainedMax[f, cons, vars]**.
- **Minimize[{f, cons}, vars]**: replace with **ConstrainedMin[f, cons, vars]**.
- **Most[v]**: replace with **Drop[v, -1]**.
- **Norm**: Replace with **VectorNorm** or **MatrixNorm** from the **LAMM** package (note: the default is the ∞-norm).
- **SingularValueDecomposition**: replace with **SingularValues**.
- **SingularValueList[m]**: replace with **SingularValues[m][[2]]**.
- **Total**: Replace with **Apply**. For vectors use **Apply[Plus, v]**. For column sums of matrices, use **Apply[Plus, m]**; for row sums, use **Apply[Plus, Transpose[m]]**; and for the sum of all elements, use **Apply[Plus, Flatten[m]]**.

References

[BSS93] M.S. Bazaraa, H.D. Sherali, and C.M. Shetty. *Nonlinear Programming: Theory and Algorithms, 2nd ed.* John Wiley & Sons, New York, 1993.

[Rus04] H. Ruskeepää. *Mathematica Navigator: Mathematics, Statistics, and Graphics, 2nd ed.* Elsevier Academic Press, Burlington, MA, 2004.

[Sza00] F. Szabo. *Linear Algebra: An Introduction Using Mathematica.* Academic Press, San Diego, 2000.

[Wol03] S. Wolfram. *The Mathematica Book, 5th ed.* Wolfram Media, Champaign, IL, 2003.

[WR99] Wolfram Research. *Mathematica 4 Standard Add-on Packages.* Wolfram Media, Champaign, IL, 1999.

[WR03] Wolfram Research. *Linear Algebra in Mathematica.* In the Help Browser, 2003.

Packages of Subroutines for Linear Algebra

74

BLAS

Jack Dongarra
University of Tennessee and Oakridge National Laboratory

Victor Eijkhout
University of Tennessee

Julien Langou
University of Tennessee

74.1 Introduction

A significant amount of execution time in complicated linear algebraic programs is known to be spent in a few low-level operations. Consequently, reducing the overall execution time of an application program leads to the problem of optimizing these low-level operations. Such low-level optimization are highly machine-dependent and are a matter for specialists. A separation has, therefore, been made by the computational linear algebra community: On the one hand highly efficient, machine-dependent building blocks of linear algebra, the BLAS, (basic linear algebra subprograms) are provided on a given computational platform. On the other hand, linear algebra programs attempt to make the most use of those computational blocks in order to have as good performance as possible across a wide range of platforms.

The first major concerted effort to achieve agreement on the specification of a set of linear algebra kernels resulted in the Level-1 BLAS [LHK79]. The Level-1 BLAS specification and implementation are the result of a collaborative project in 1973 through 1977. The Level-1 BLAS were extensively and successfully exploited by LINPACK [DBM79], a software package for the solution of dense and banded linear equations and linear least squares problems.

With the advent of vector machines, hierarchical memory machines, and shared memory parallel machines, specifications for the Level-2 and -3 BLAS [DDD90a], [DDD90b], [DDH88a], [DDH88b], concerned with matrix–vector and matrix–matrix operations, respectively, were drawn up in 1984 through 1986 and 1987 to 1988. These specifications made it possible to construct new software to utilize the memory hierarchy of modern computers more effectively. In particular, the Level-3 BLAS allowed the construction of software based upon block-partitioned algorithms, typified by the linear algebra software package LAPACK (see Chapter 75).

In this chapter, we present information about the BLAS. We begin with basic definition, then describe the calling sequences, report a number of facts, and finally conclude with a few examples.

Definitions:

Two-dimensional column-major format. A two-dimensional array is said to be in column-major format when its entries are stored by column. This is the way the Fortran language stores two-dimensional

arrays. For example, the declaration `double precision A(2,2)` stores the entries of the 2-by-2 two-dimensional array in the one-dimensional order: `A(1,1)`, `A(2,1)`, `A(1,2)`, `A(2,2)`.
Row-major format. Storage of the entries by row. This is the way C stores two-dimensional arrays.
Packed format. The packed format is relevant for symmetric, Hermitian, or triangular matrices. Half of the matrix (either the upper part or the lower part depending on an `UPLO` parameter) is stored in a one-dimensional array. For example, if `UPLO='L'`, the 3-by-3 symmetric matrix `A` is stored as `A(1,1)`, `A(2,1)`, `A(3,1)`, `A(2,2)`, `A(3,2)`, `A(3,3))`.
Leading dimension. In column-major format, the leading dimension is the first dimension of a two-dimensional array (as opposed to the dimension of the matrix stored in that array). The leading dimension of an array is necessary to access its elements. For example, if `LDA` is the leading dimension of `A`, and `A` is stored in the two-dimensional column-major format, $A_{i,j}$ is stored in position $i + (j - 1) * LDA$. The leading dimension of an m-by-n matrix is often m. It enables convenient abstraction to access submatrices (see examples). An m-by-n matrix should have `LDA` $\geq m$.
Increment. The increment of a vector is the number of storage elements from one element to the next. If the increment of a vector is 1, this means that the vector is stored contiguously in memory; an increment of 2 corresponds to using every other array element. The increment is useful to manipulate rows of a matrix stored in column-major format (see Example 3).
BLAS vector description. A vector description in BLAS is defined by three quantities — a vector length, an array or a starting element within an array, and the increment. This gives `(n,X,1)` for a vector of size `n` starting at index `X` with increment of `1` .
BLAS description of a general matrix. A general matrix is described in BLAS by four quantities: a `TRANS` label (`'N'`, `'T'`, or `'H'`), its dimensions `m` and `n`, an array or a starting element within an array, and a leading dimension. This gives `(TRANS,M,N,A,LDA)` if one wants to operate on the `TRANS` of an `M`-by-`N` matrix starting in `A(1,1)` and stored in an array of leading dimension `LDA`.
Prefixes The first letter of the name of a BLAS routine indicates the Fortran type on which it operates.

S - REAL	C - COMPLEX
D - DOUBLE PRECISION	Z - COMPLEX*16

Prefixes for Level-2 and Level-3 BLAS
Matrix types:

GE - GEneral	GB - General Band	
SY - SYmmetric	SB - Sym. Band	SP - Sym. Packed
HE - HErmitian	HB - Herm. Band	HP - Herm. Packed
TR - TRiangular	TB - Triang. Band	TP - Triang. Packed

Level-2 and Level-3 BLAS Options

TRANS	= '**N**o transpose', '**T**ranspose', '**C**onjugate transpose'
UPLO	= '**U**pper triangular', '**L**ower triangular'
DIAG	= '**N**on-unit triangular', '**U**nit triangular'
SIDE	= '**L**eft', '**R**ight' on the right)

Calls:

1. Below is the calling sequence for the `GEMV` matrix-vector multiply subroutine. The `GEMV` routine performs the mathematical operation

$$y \longleftarrow \alpha Ax + \beta y, \quad \text{or} \quad y \longleftarrow \alpha A^T x + \beta y.$$

For double precision, the calling sequence looks like

```
      SUBROUTINE DGEMV  ( TRANS, M, N, ALPHA, A, LDA, X, INCX,
     $                    BETA, Y, INCY )
```

The types of the variables are as follows:

```
      DOUBLE PRECISION   ALPHA, BETA
      INTEGER            INCX, INCY, LDA, M, N
      CHARACTER*1        TRANS
      DOUBLE PRECISION   A( LDA, * ), X( * ), Y( * )
```

The meaning of the variables is as follows:

`TRANS` : Specifies the operation to be performed as
`TRANS='N'` $y \leftarrow \alpha Ax + \beta y$,
`TRANS='T'` $y \leftarrow \alpha A^T x + \beta y$.

`M` : Specifies the number of rows of the matrix A.

`N` : Specifies the number of columns of the matrix A.

`ALPHA`: Specifies the scalar α.

`A` : Points to the first entry of the two-dimensional array that stores the elements of A in column major format.

`LDA` : Specifies the leading dimension of A.

`X` : Points to the incremented array that contains the vector x.

`INCX` : Specifies the increment for the elements of `X`.

`BETA` : Specifies the scalar β.

`Y` : Points to the incremented array that contains the vector y.

`INCY` : Specifies the increment for the elements of `Y`.

All variables are left unchanged after completion of the routines, except `Y`.

2. The following table is the quick reference to the BLAS.

Facts:

1. The BLAS represent fairly simple linear algebra kernels that are easily coded in a few lines. One could also download reference source codes and compile them (`http://www.netlib.org/blas`). However, this strategy is unlikely to give good performance, no matter the level of sophistication of the compiler. The recommended way to obtain an efficient BLAS is to use vendor BLAS libraries, or to install the ATLAS [WPD01] package. Figure 74.1 illustrates the fact that ATLAS BLAS clearly outperform the reference implementation for the matrix–matrix multiply by a factor of five on a modern architecture.
2. To a great extent, the user community has embraced the BLAS, not only for performance reasons, but also because developing software around a core of common routines like the BLAS is good software engineering practice. Highly efficient, machine-specific implementations of the BLAS are available for most modern high-performance computers. To obtain an up-to-date list of available optimized BLAS, see the BLAS FAQ at `http://www.netlib.org/blas`.
3. Level-1 BLAS operates on $\mathcal{O}(n)$ data and performs $\mathcal{O}(n)$ operations. Level-2 BLAS operates on $\mathcal{O}(n^2)$ data and performs $\mathcal{O}(n^2)$ operations. Level-3 BLAS operates on $\mathcal{O}(n^2)$ data and performs $\mathcal{O}(n^3)$ operations. Therefore, the ratio operation/data is $\mathcal{O}(1)$ for Level-1 BLAS and Level-2 BLAS, but $\mathcal{O}(n)$ for Level-3 BLAS. On modern architecture, where memory access is particularly slow compared to computation time, Level-3 BLAS exploit this $\mathcal{O}(n)$ ratio to mask the memory access bottleneck (latency and bandwidth). Figure 74.1 illustrates the fact that Level-3 BLAS on a modern

Level-1 BLAS

Calling sequence (dim, scalar, vector, vector, scalars, 5-element array)	Operation	prefixes
`SUBROUTINE xROTG ( A, B, C, S )`	Generate plane rotation	S, D
`SUBROUTINE xROTMG( D1, D2, A, B, PARAM )`	Generate modified plane rotation	S, D
`SUBROUTINE xROT ( N, X, INCX, Y, INCY, C, S )`	Apply plane rotation	S, D
`SUBROUTINE xROTM ( N, X, INCX, Y, INCY, PARAM )`	Apply modified plane rotation	S, D
`SUBROUTINE xSWAP ( N, X, INCX, Y, INCY )`	$x \leftrightarrow y$	S, D, C, Z
`SUBROUTINE xSCAL ( N, ALPHA, X, INCX )`	$x \leftarrow \alpha x$	S, D, C, Z, CS, ZD
`SUBROUTINE xCOPY ( N, X, INCX, Y, INCY )`	$y \leftarrow x$	S, D, C, Z
`SUBROUTINE xAXPY ( N, ALPHA, X, INCX, Y, INCY )`	$y \leftarrow \alpha x + y$	S, D, C, Z
`FUNCTION xDOT ( N, X, INCX, Y, INCY )`	$dot \leftarrow x^T y$	S, D, DS
`FUNCTION xDOTU ( N, X, INCX, Y, INCY )`	$dot \leftarrow x^T y$	C, Z
`FUNCTION xDOTC ( N, X, INCX, Y, INCY )`	$dot \leftarrow x^H y$	C, Z
`FUNCTION xxDOT ( N, X, INCX, Y, INCY )`	$dot \leftarrow \alpha + x^T y$	SDS
`FUNCTION xNRM2 ( N, X, INCX )`	$nrm2 \leftarrow \|\|x\|\|_2$	S, D, SC, DZ
`FUNCTION xASUM ( N, X, INCX )`	$asum \leftarrow \|\|re(x)\|\|_1 + \|\|im(x)\|\|_1$	S, D, SC, DZ
`FUNCTION IxAMAX( N, X, INCX )`	$amax \leftarrow 1^{st} k \|re(x_k)\| + \|im(x_k)\| = max(\|re(x_i)\| + \|im(x_i)\|)$	S, D, C, Z

Level-2 BLAS

Calling sequence (options, dim, b-width, scalar, matrix, vector, scalar, vector)	Operation	prefixes
`xGEMV ( TRANS, M, N, ALPHA, A, LDA, X, INCX, BETA, Y, INCY )`	$y \leftarrow \alpha Ax + \beta y, y \leftarrow \alpha A^T x + \beta y, y \leftarrow \alpha A^H x + \beta y, A - m \times n$	S, D, C, Z
`xGBMV ( TRANS, M, N, KL, KU, ALPHA, A, LDA, X, INCX, BETA, Y, INCY )`	$y \leftarrow \alpha Ax + \beta y, y \leftarrow \alpha A^T x + \beta y, y \leftarrow \alpha A^H x + \beta y, A - m \times n$	S, D, C, Z
`xHEMV ( UPLO, N, ALPHA, A, LDA, X, INCX, BETA, Y, INCY )`	$y \leftarrow \alpha Ax + \beta y$	C, Z
`xHBMV ( UPLO, N, K, ALPHA, A, LDA, X, INCX, BETA, Y, INCY )`	$y \leftarrow \alpha Ax + \beta y$	C, Z
`xHPMV ( UPLO, N, ALPHA, AP, X, INCX, BETA, Y, INCY )`	$y \leftarrow \alpha Ax + \beta y$	C, Z
`xSYMV ( UPLO, N, ALPHA, A, LDA, X, INCX, BETA, Y, INCY )`	$y \leftarrow \alpha Ax + \beta y$	S, D
`xSBMV ( UPLO, N, K, ALPHA, A, LDA, X, INCX, BETA, Y, INCY )`	$y \leftarrow \alpha Ax + \beta y$	S, D
`xSPMV ( UPLO, N, ALPHA, AP, X, INCX, BETA, Y, INCY )`	$y \leftarrow \alpha Ax + \beta y$	S, D
`xTRMV ( UPLO, TRANS, DIAG, N, A, LDA, X, INCX )`	$x \leftarrow Ax, x \leftarrow A^T x, x \leftarrow A^H x$	S, D, C, Z
`xTBMV ( UPLO, TRANS, DIAG, N, K, A, LDA, X, INCX )`	$x \leftarrow Ax, x \leftarrow A^T x, x \leftarrow A^H x$	S, D, C, Z
`xTPMV ( UPLO, TRANS, DIAG, N, AP, X, INCX )`	$x \leftarrow Ax, x \leftarrow A^T x, x \leftarrow A^H x$	S, D, C, Z
`xTRSV ( UPLO, TRANS, DIAG, N, A, LDA, X, INCX )`	$x \leftarrow A^{-1}x, x \leftarrow A^{-T}x, x \leftarrow A^{-H}x$	S, D, C, Z
`xTBSV ( UPLO, TRANS, DIAG, N, K, A, LDA, X, INCX )`	$x \leftarrow A^{-1}x, x \leftarrow A^{-T}x, x \leftarrow A^{-H}x$	S, D, C, Z
`xTPSV ( UPLO, TRANS, DIAG, N, AP, X, INCX )`	$x \leftarrow A^{-1}x, x \leftarrow A^{-T}x, x \leftarrow A^{-H}x$	S, D, C, Z
options, dim, scalar, vector, vector, matrix		

xGER (		M, N, ALPHA, X, INCX, Y, INCY, A, LDA)	$A \leftarrow \alpha xy^T + A, A - m \times n$	S, D
xGERU (		M, N, ALPHA, X, INCX, Y, INCY, A, LDA)	$A \leftarrow \alpha xy^T + A, A - m \times n$	C, Z
xGERC (		M, N, ALPHA, X, INCX, Y, INCY, A, LDA)	$A \leftarrow \alpha xy^H + A, A - m \times n$	C, Z
xHER (	UPLO,	N, ALPHA, X, INCX, A, LDA)	$A \leftarrow \alpha xx^H + A$	C, Z
xHPR (	UPLO,	N, ALPHA, X, INCX, AP)	$A \leftarrow \alpha xx^H + A$	C, Z
xHER2 (	UPLO,	N, ALPHA, X, INCX, Y, INCY, A, LDA)	$A \leftarrow \alpha xy^H + y(\alpha x)^H + A$	C, Z
xHPR2 (	UPLO,	N, ALPHA, X, INCX, Y, INCY, AP)	$A \leftarrow \alpha xy^H + y(\alpha x)^H + A$	C, Z
xSYR (	UPLO,	N, ALPHA, X, INCX, A, LDA)	$A \leftarrow \alpha xx^T + A$	S, D
xSPR (	UPLO,	N, ALPHA, X, INCX, AP)	$A \leftarrow \alpha xx^T + A$	S, D
xSYR2 (	UPLO,	N, ALPHA, X, INCX, Y, INCY, A, LDA)	$A \leftarrow \alpha xy^T + \alpha yx^T + A$	S, D
xSPR2 (	UPLO,	N, ALPHA, X, INCX, Y, INCY, AP)	$A \leftarrow \alpha xy^T + \alpha yx^T + A$	S, D

Level-3 BLAS

	options	dim	scalar	matrix	matrix	scalar	matrix		
xGEMM (	TRANSA, TRANSB,	M, N, K,	ALPHA,	A, LDA,	B, LDB,	BETA,	C, LDC)	$C \leftarrow \alpha op(A)op(B) + \beta C$, $op(X) = X, X^T, X^H, C - m \times n$	S, D, C, Z
xSYMM (	SIDE, UPLO,	M, N,	ALPHA,	A, LDA,	B, LDB,	BETA,	C, LDC)	$C \leftarrow \alpha AB + \beta C, C \leftarrow \alpha BA + \beta C$, $C - m \times n, A = A^T$	S, D, C, Z
xHEMM (	SIDE, UPLO,	M, N,	ALPHA,	A, LDA,	B, LDB,	BETA,	C, LDC)	$C \leftarrow \alpha AB + \beta C, C \leftarrow \alpha BA + \beta C$, $C - m \times n, A = A^H$	C, Z
xSYRK (	UPLO, TRANS,	N, K,	ALPHA,	A, LDA,		BETA,	C, LDC)	$C \leftarrow \alpha AA^T + \beta C, C \leftarrow \alpha A^T A + \beta C$, $C - n \times n$	S, D, C, Z
xHERK (	UPLO, TRANS,	N, K,	ALPHA,	A, LDA,		BETA,	C, LDC)	$C \leftarrow \alpha AA^H + \beta C, C \leftarrow \alpha A^H A + \beta C$, $C - n \times n$	C, Z
xSYR2K(	UPLO, TRANS,	N, K,	ALPHA,	A, LDA,	B, LDB,	BETA,	C, LDC)	$C \leftarrow \alpha AB^T + \bar{\alpha} BA^T + \beta C$, $C \leftarrow \alpha A^T B + \bar{\alpha} B^T A + \beta C$, $C - n \times n$	S, D, C, Z
xHER2K(	UPLO, TRANS,	N, K,	ALPHA,	A, LDA,	B, LDB,	BETA,	C, LDC)	$C \leftarrow \alpha AB^H + \bar{\alpha} BA^H + \beta C$, $C \leftarrow \alpha A^H B + \bar{\alpha} B^H A + \beta C$, $C - n \times n$	C, Z
xTRMM (	SIDE, UPLO, TRANSA, DIAG,	M, N,	ALPHA,	A, LDA,	B, LDB)			$B \leftarrow \alpha op(A)B; B \leftarrow \alpha Bop(A)$, $op(A) = A, A^T, A^H, B - m \times n$	S, D, C, Z
xTRSM (	SIDE, UPLO, TRANSA, DIAG,	M, N,	ALPHA,	A, LDA,	B, LDB)			$B \leftarrow \alpha op(A^{-1})B, B \leftarrow \alpha Bop(A^{-1})$, $op(A) = A, A^T, A^H, B - m \times n$	S, D, C, Z

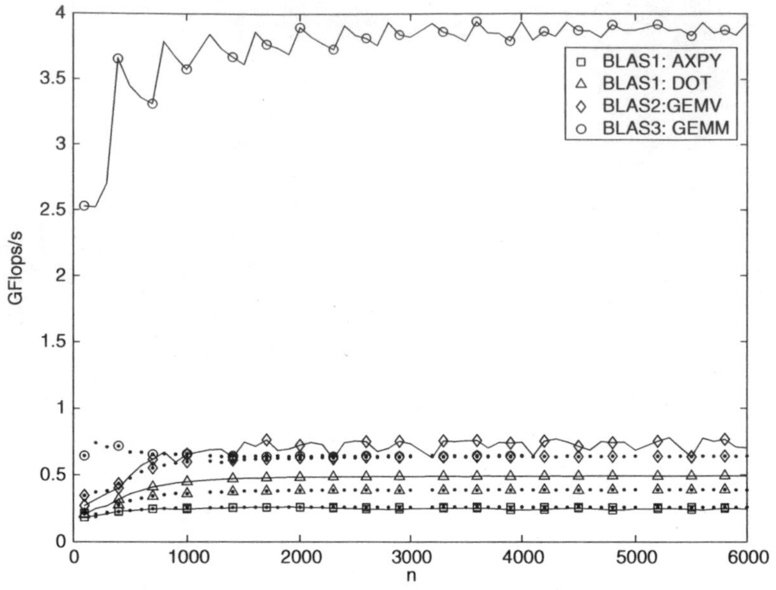

FIGURE 74.1 Performance in GFlops of four BLAS routines for two different BLAS libraries on an Intel Xeon CPU running at 3.20 GHz. The first BLAS library represents an optimized BLAS library (here, ATLAS v3.7.8), its four performance curves are given with a solid line on the graph. The second library represents straightforward implementation (here reference BLAS from netlib); its four performance curves are given with a dotted line on the graph. This graph illustrates two facts: An optimized Level-3 BLAS is roughly five times faster than Level-1 or Level-2 BLAS and it is also roughly five times faster than a reference Level-3 BLAS implementation. To give an idea of the actual time, multiplying two 6000-by-6000 matrices on this machine will take about 2 minutes using the ATLAS BLAS while it will take about 10 minutes using the reference BLAS implementation.

architecture performs 5 times more floating-point operations per second than a Level-2 or Level-1 BLAS routine. Most of the linear algebra libraries try to make as much as possible use of Level-3 BLAS.

4. Most of shared memory computers have a multithreaded BLAS library. By programming a sequential code and linking with the multithreaded BLAS library, the application will use all the computing units of the shared memory system without any explicit parallelism in the user application code.
5. Although here we present only calling sequences from Fortran, it is possible to call the BLAS from C. The major problem for C users is the Fortran interface used by the BLAS. The data layout (the BLAS interface assumes column-major format) and the passage of parameters by reference (as opposed to values) has to be done carefully. See Example 2 for an illustration. Nowadays most BLAS distributions provide a C-interface to the BLAS. This solves these two issues and, thus, we highly recommend its use.
6. The Level-1, Level-2, and Level-3 BLAS are now extended by a new standard with more functionality and better software standard; see [BDD02], [Don02]. For example, the C-interface to the BLAS (see Fact 5) is included in this new BLAS.

Examples:

In all of these examples, A is an m-by-n matrix and it is stored in an `M`-by-`N` array starting at position `A`, B is an n-by-n matrix and it is stored in an `N`-by-`N` array starting at position `B`, C is an m-by-m matrix and it is stored in an `M`-by-`M` array starting at position `C`, X is a vector of size n and it is stored in an `N` array starting at position `X`, and Y is a vector of size m and it is stored in an `M` array starting at position `Y`. All the two-dimensional arrays are in column-major format, all the one-dimensional arrays have increment one. We assume that m and n are both greater than 11 in Example 3.

1. To perform the operation $y \longleftarrow Ax$ the BLAS calling sequence is

```
CALL DGEMV ( 'N', M, N, 1.0D0, A, M, X, 1, 0.0D0, Y, 1 )
```

2. To perform the operation $x \longleftarrow \alpha A^T y + \beta x$ the BLAS calling sequence is

```
CALL DGEMV ( 'T', M, N, ALPHA, A, M, Y, 1, BETA, X, 1 )
```

From C, this would give

```
IONE = 1;
dgemv ( "T", &M, &N, &ALPHA, A, &M, Y, &IONE, &BETA, X, &IONE );
```

3. To perform the operation
$y(2:10) \longleftarrow 2A(3:11,4:11) * B(4,3:10)^T - 3y(2:10)$;
the BLAS calling sequence is

```
CALL DGEMV ( 'N', 9, 8, 2.0D0, A(3,4), M, B(4,3), N, -3.0D0,
Y(2), 1 )
```

(Note the use of `LDA` to operate on the submatrix of A and the use of `INCX` to operate on a row of B.)

4. LAPACK (see Chapter 75) is a library based on the BLAS. Opening its Fortran files enables one to gain a good understanding of how to use BLAS routines. For example, a good way to start is to have a look at `dgetrf.f`, which performs a right-looking LU factorization by making calls to `DTRSM` and `DGEMM`.

References

[BDD02] S. Blackford, J. Demmel, J.J. Dongarra, I. Duff, S. Hammarling, G. Henry, M. Heroux, L. Kaufman, A. Lumsdaine, A. Petitet, R. Pozo, K. Remington, and R.C. Whaley. An updated set of basic linear algebra subprograms (BLAS). *ACM Trans. Math. Softw.*, 28(2):135–151, 2002.

[Don02] J.J. Dongarra. Basic linear algebra subprograms technical forum standard. *Int. J. High Perform. Appl. Supercomp.*, 16(1–2):1–199, 2002.

[DBM79] J.J. Dongarra, J.R. Bunch, C.B. Moler, and G.W. Stewart. *LINPACK Users' Guide.* SIAM, Philadelphia, 1979.

[DDD90a] J.J. Dongarra, J. Du Croz, I.S. Duff, and S. Hammarling. A set of Level 3 basic linear algebra subprograms. *ACM Trans. Math. Softw.*, 16:1–17, 1990.

[DDD90b] J.J. Dongarra, J. Du Croz, I.S. Duff, and S. Hammarling. Algorithm 679: a set of Level 3 basic linear algebra subprograms. *ACM Trans. Math. Softw.*, 16:18–28, 1990.

[DDH88a] J.J. Dongarra, J. Du Croz, S. Hammarling, and R.J. Hanson. An extended set of FORTRAN basic linear algebra subprograms. *ACM Trans. Math. Softw.*, 14:1–17, 1988.

[DDH88b] J.J. Dongarra, J. Du Croz, S. Hammarling, and R.J. Hanson. Algorithm 656: an extended set of FORTRAN basic linear algebra subprograms. *ACM Trans. Math. Softw.*, 14:18–32, 1988.

[LHK79] C.L. Lawson, R.J. Hanson, D. Kincaid, and F. Krogh. Basic linear algebra subprograms for FORTRAN usage. *ACM Trans. Math. Softw.*, 5:308–323, 1979.

[WPD01] R.C. Whaley, A. Petitet, and J.J. Dongarra. Automated empirical optimizations of software and the ATLAS project. *Parallel Comp.*, 27:3–35, 2001.

75

LAPACK

Zhaojun Bai
University of California/Davis

James Demmel
University of California/Berkeley

Jack Dongarra
University of Tennessee

Julien Langou
University of Tennessee

Jenny Wang
University of Caifornia/Davis

75.1 Introduction

LAPACK (linear algebra package) is an open source library of programs for solving the most commonly occurring numerical linear algebra problems [LUG99]. Original codes of LAPACK are written in Fortran 77. Complete documentation as well as source codes are available online at the Netlib repository [LAP]. LAPACK provides driver routines for solving complete problems such as linear equations, linear least squares problems, eigenvalue problems, and singular value problems. Each driver routine calls a sequence of computational routines, each of which performs a distinct computational task. In addition, LAPACK provides comprehensive error bounds for most computed quantities. LAPACK is designed to be portable for sequential and shared memory machines with deep memory hierarchies, in which most performance issues could be reduced to providing optimized versions of the Basic Linear Algebra Subroutines (BLAS). (See Chapter 74).

There have been a number of extensions of LAPACK. LAPACK95 is a Fortran 95 interface to the Fortran 77 LAPACK [LAP95]. CLAPACK and JLAPACK libraries are built using the Fortran to C (f2c) and Fortran to Java (f2j) conversion utilities, respectively [CLA], [JLA]. LAPACK++ is implemented in C++ and includes a subset of the features in LAPACK with emphasis on solving linear systems with nonsymmetric matrices, symmetric positive definite systems, and linear least squares systems [LA+]. ScaLAPACK is a portable implementation of some of the core routines in LAPACK for parallel distributed computing [Sca]. ScaLAPACK is designed for distributed memory machines with very powerful homogeneous sequential processors and with homogeneous network interconnections.

The purpose of this chapter is to acquaint the reader with 10 essential numerical linear algebra problems and LAPACK's way of solving those problems. The reader may find it helpful to consult Chapter 74, where some of the terms used here are defined. The following table summarizes these problems and sections that are treated in version 3.0 of LAPACK.

Type of Problem	Acronyms	Section
Linear system of equations	SV	75.2
Linear least squares problems	LLS	75.3
Linear equality-constrained least squares problem	LSE	75.4
General linear model problem	GLM	75.5
Symmetric eigenproblems	SEP	75.6
Nonsymmetric eigenproblems	NEP	75.7
Singular value decomposition	SVD	75.8
Generalized symmetric definite eigenproblems	GSEP	75.9
Generalized nonsymmetric eigenproblems	GNEP	75.10
Generalized (or quotient) singular value decomposition	GSVD (QSVD)	75.11

Sections have been subdivided into the following headings: (1) Definition: defines the problem, (2) Background: discusses the background of the problem and references to the related sections in this handbook, (3) Driver Routines: describes different types of driver routines available that solve the same problem, (4) Example: specifies the calling sequence for a driver routine that solves the problem followed by numerical examples.

All LAPACK routines are available in four data types, as indicated by the initial letter "x" of each subroutine name: x = "S" means real single precision, x = "D", real double precision, x = "C", complex single precision, and x = "Z", complex*16 or double complex precision. In single precision (and complex single precision), the computations are performed with a unit roundoff of 5.96×10^{-8}. In double precision (and complex double precision) the computations are performed with a unit roundoff of 1.11×10^{-16}.

All matrices are assumed to be stored in column-major format. The software can also handle submatrices of matrices, even though these submatrices may not be stored in consecutive memory locations. For example, to specify the 10–by–10 submatrix lying in rows and columns 11 through 20 of a 30–by–30 matrix A, one must specify

- `A(11, 11)`, the upper left corner of the submatrix
- `30` = Leading dimension of `A` in its declaration (often denoted `LDA` in calling sequences)
- `10` = Number of rows of the submatrix (often denoted `M`, can be at most `LDA`)
- `10` = Number of columns of submatrix (often denoted `N`)

All matrix arguments require these 4 parameters (some subroutines may have fewer inputs if, for example, the submatrix is assumed square so that M = N). (See Chapter 74, for more details.)

Most of the LAPACK routines require the users to provide them a workspace (`WORK`) and its dimension (`LWORK`). The *optimal* workspace dimension refers to the workspace dimension, which enables the code to have the best performance on the targeted machine. The computation of the optimal workspace dimension is often complex so that most of LAPACK routines have the ability to compute it. If a LAPACK routine is called with `LWORK=-1`, then a workspace query is assumed. The routine only calculates the optimal size of the `WORK` array and returns this value as the first entry of the `WORK` array. If a larger workspace is provided, the extra part is not used, so that the code runs at the optimal performance. A *minimal* workspace dimension is provided in the document of routines. If a routine is called with a workspace dimension smaller than the minimal workspace dimension, the computation cannot be performed.

75.2 Linear System of Equations

Definitions:

The problem of linear equations is to compute a solution X of the system of linear equations

$$AX = B, \tag{75.1}$$

where A is an n–by–n matrix and X and B are n–by–m matrices.

Backgrounds:

The theoretical and algorithmic background of the solution of linear equations is discussed extensively in Chapter 37 through Chapter 41, especially Chapter 38.

Driver Routines:

There are two types of driver routines for solving the systems of linear equations — simple driver and expert driver. The expert driver solves the system (Equation 75.1), allows A be replaced by A^T or A^*; and provides error bounds, condition number estimate, scaling, and can refine the solution. Each of these types of drivers has different implementations that take advantage of the special properties or storage schemes of the matrix A, as listed in the following table.

	Routine Names	
Data Structure (Matrix Storage Scheme)	Simple Driver	Expert Driver
General dense	xGESV	xGESVX
General band	xGBSV	xGBSVX
General tridiagonal	xGTSV	xGTSVX
Symmetric/Hermitian positive definite	xPOSV	xPOSVX
Symmetric/Hermitian positive definite (packed storage)	xPPSV	xPPSVX
Banded symmetric positive definite	xPBSV	xPBSVX
Tridiagonal symmetric positive definite	xPTSV	xPTSVX
Symmetric/Hermitian indefinite	xSYSV/xHESV	xSYSVX/xHESVX
Symmetric/Hermitian indefinite (packed storage)	xSPSV/xHPSV	xSPSVX/xHPSVX
Complex symmetric	CSYSV/ZSYSV	CSYSVX/ZSYSVX

The prefixes GE (for general dense), GB (for general band), etc., have standard meanings for all the BLAS and LAPACK routines.

Examples:

Let us show how to use the simple driver routine SGESV to solve a general linear system of equations. SGESV computes the solution of a real linear Equation 75.1 in single precision by first computing the LU decomposition with row partial pivoting of the coefficient matrix A, followed by the back and forward substitutions. SGESV has the following calling sequence:

```
CALL SGESV( N, NRHS, A, LDA, IPIV, B, LDB, INFO )
```

Input to SGESV:

`N`: The number of linear equations, i.e., the order of A. $\mathtt{N} \geq 0$.

`NRHS`: The number of right-hand sides, i.e., the number of columns of B. $\mathtt{NRHS} \geq 0$.

`A, LDA`: The N–by–N coefficient matrix A and the leading dimension of the array `A`. $\mathtt{LDA} \geq \max(1, \mathtt{N})$.

`B, LDB`: The N–by–NRHS matrix B and the leading dimension of the array `B`. $\mathtt{LDB} \geq \max(1, \mathtt{N})$.

Output from SGESV:

`A`: The factors L and U from factorization $A = PLU$; the unit diagonal elements of L are not stored.

`IPIV`: The pivot indices that define the permutation matrix P; row i of the matrix was interchanged with row $\mathtt{IPIV}(i)$.

`B`: If $\mathtt{INFO} = 0$, the N–by–NRHS solution X.

`INFO`: $= 0$, successful exit. If $\mathtt{INFO} = -j$, the jth argument had an illegal value. If $\mathtt{INFO} = j$, $U(j, j)$ is exactly zero. The factorization has been completed, but the factor U is singular, so the solution could not be computed.

Consider a 4–by–4 linear system of Equation (75.1), where

$$A = \begin{bmatrix} 5 & 7 & 6 & 5 \\ 7 & 10 & 8 & 7 \\ 6 & 8 & 10 & 9 \\ 5 & 7 & 9 & 10 \end{bmatrix} \quad \text{and} \quad B = \begin{bmatrix} 23 \\ 32 \\ 33 \\ 31 \end{bmatrix}.$$

The exact solution is $\mathbf{x} = \begin{bmatrix} 1 & 1 & 1 & 1 \end{bmatrix}^T$. Upon calling SGESV, the program successfully exits with `INFO` $= 0$ and the solution X of (75.1) resides in the array `B`

$$X = \begin{bmatrix} 0.9999998 \\ 1.0000004 \\ 0.9999998 \\ 1.0000001 \end{bmatrix}.$$

Since SGESV performs the computation in single precision arithmetic, it is normal to have an error of the order of 10^{-6} in the solution X. By reading the lower diagonal entries in the array `A` and filling the diagonal entries with ones, we recover the lower unit triangular matrix L of the LU factorization with row partial pivoting of A as follows:

$$L = \begin{bmatrix} 1 & 0 & 0 & 0 \\ 0.8571429 & 1 & 0 & 0 \\ 0.7142857 & 0.2500000 & 1 & 0 \\ 0.7142857 & 0.2500000 & -0.2000000 & 1 \end{bmatrix}.$$

The upper triangular matrix U is recovered by reading the diagonal and upper diagonal entries in `A`. That is:

$$U = \begin{bmatrix} 7.0000000 & 10.0000000 & 8.0000000 & 7.0000000 \\ 0 & -0.5714293 & 3.1428566 & 2.9999995 \\ 0 & 0 & 2.5000000 & 4.2500000 \\ 0 & 0 & 0 & 0.1000000 \end{bmatrix}.$$

Finally, the permutation matrix P is the identity matrix that exchanges its ith row with row `IPIV`(i), for $i = n, \dots, 1$. Since

$$\texttt{IPIV} = \begin{bmatrix} 2 & 3 & 4 & 4 \end{bmatrix},$$

we have

$$P = \begin{bmatrix} 0 & 0 & 0 & 1 \\ 1 & 0 & 0 & 0 \\ 0 & 1 & 0 & 0 \\ 0 & 0 & 1 & 0 \end{bmatrix}.$$

75.3 Linear Least Squares Problems

Definitions:

The linear least squares (LLS) problem is to find

$$\min_{\mathbf{x}} \|\mathbf{b} - A\mathbf{x}\|_2, \tag{75.2}$$

where A is an m–by–n matrix and $\mathbf{b}$ is an m element vector.

Backgrounds:

The most usual case is $m \geq n$ and $\text{rank}(A) = n$. In this case, the solution to the LLS problem (75.2) is unique, and the problem is also referred to as finding a least squares solution to an overdetermined system of linear equations. When $m < n$ and $\text{rank}(A) = m$, there are an infinite number of solutions $\mathbf{x}$ that exactly satisfy $\mathbf{b} - A\mathbf{x} = 0$. In this case, it is often useful to find the unique solution $\mathbf{x}$ that minimizes $\|\mathbf{x}\|_2$, and the problem is referred to as finding a *minimum norm solution* to an underdetermined system of linear equations. (See Chapter 5.8 and Chapter 39 for more information on linear least squares problems.)

Driver Routines:

There are four types of driver routines that solve the LLS problem (75.2) and also allow A be replaced by A^*. In the general case when $\text{rank}(A) < \min(m, n)$, we seek the *minimum norm least squares* solution $\mathbf{x}$ that minimizes both $\|\mathbf{x}\|_2$ and $\|\mathbf{b} - A\mathbf{x}\|_2$. The types of driver routines are categorized by the methods used to solve the LLS problem, as shown in the following table.

Type of Matrix	Algorithm	Routine Names
General dense	QR or LQ factorization	xGELS
General dense	Complete orthogonal factorization	xGELSY
General dense	SVD	xGELSS
General dense	Divide-and-conquer SVD	xGELSD

xGELSD is significantly faster than xGELSS, but it demands somewhat more workspace depending on the matrix dimensions. Among these routines, only xGELS requires A to be full rank while xGELSY, xGELSS, and xGELSD allow less than full rank.

Note that all driver routines allow several right-hand side vectors $\mathbf{b}$ and corresponding solutions $\mathbf{x}$ to be handled in a single call, storing these vectors as columns of matrices B and X, respectively. However, the LLS problem (75.2) is solved for each right-hand side independently; that is not the same as finding a matrix X which minimizes $\|B - AX\|_2$.

Examples:

Let us show how to use the simple driver routine SGELS to solve the LLS problem (75.2). SGELS computes the QR decomposition of the matrix A, updates the vector $\mathbf{b}$, and then computes the solution $\mathbf{x}$ by back substitution. SGELS has the following calling sequence:

```
CALL SGELS( TRANS, M, N, NRHS, A, LDA, B, LDB, WORK, LWORK, INFO )
```

Input to `SGELS`:

`TRANS`: = `'N'` or `'T'`: solves the LLS with A or A^T.

`M, N`: The numbers of rows and columns of the matrix A. `M` $\geq$ `0` and `N` $\geq$ `0`.

`M, NRHS`: The number of rows and columns of the matrices B and X. `NRHS` $\geq$ `0`.

`A, LDA`: The `M`–by–`N` matrix A and the leading dimension of the array `A`, `LDA` $\geq \max(1, \texttt{M})$.

`B, LDB`: The matrix B and the leading dimension of the array `B`, `LDB` $\geq \max(1, \texttt{M}, \texttt{N})$.
If `TRANS` = `'N'`, then B is `M`–by–`NRHS`. If `TRANS` = `'T'`, then B is `N`–by–`NRHS`.

`WORK, LWORK`: The workspace array and its dimension. `LWORK` $\geq \min(\texttt{M}, \texttt{N}) + \max(1, \texttt{M}, \texttt{N}, \texttt{NRHS})$.
If `LWORK` $= -1$, then a workspace query is assumed; the routine only calculates the optimal size of the `WORK` array, and returns this value as the first entry of the `WORK` array.

Output from `SGELS`:

`B`: It is overwritten by the solution vectors, stored columnwise.

- If `TRANS` = `'N'` and `M` $\geq$ `N`, rows 1 to `N` of `B` contain the solution vectors of the LLS problem $\min_{\mathbf{x}} \|\mathbf{b} - A\mathbf{x}\|_2$; the residual sum of squares in each column is given by the sum of squares of elements `N` + `1` to `M` in that column;

- If TRANS = 'N' and M < N, rows 1 to N of B contain the minimum norm solution vectors of the underdetermined system $AX = B$;
- If TRANS = 'T' and M ≥ N, rows 1 to M of B contain the minimum norm solution vectors of the underdetermined system $A^T X = B$;
- If TRANS = 'T' and M < N, rows 1 to M of B contain the solution vectors of the LLS problem $\min_{\mathbf{x}} \|\mathbf{b} - A^T\mathbf{x}\|_2$; the residual sum of squares for the solution in each column is given by the sum of the squares of elements M+1 to N in that column.

WORK: If INFO = 0, WORK(1) returns the optimal LWORK.

INFO: INFO = 0 if successful exit. If INFO = $-j$, the jth input argument had an illegal value.

Consider an LLS problem (75.2) with a 6–by–5 matrix A and a 6–by–1 matrix $\mathbf{b}$:

$$A = \begin{bmatrix} -74 & 80 & 18 & -11 & -4 \\ 14 & -69 & 21 & 28 & 0 \\ 66 & -72 & -5 & 7 & 1 \\ -12 & 66 & -30 & -23 & 3 \\ 3 & 8 & -7 & -4 & 1 \\ 4 & -12 & 4 & 4 & 0 \end{bmatrix} \quad \text{and} \quad \mathbf{b} = \begin{bmatrix} 51 \\ -61 \\ -56 \\ 69 \\ 10 \\ -12 \end{bmatrix}.$$

The exact solution of the LLS problem is $\mathbf{x} = \begin{bmatrix} 1 & 2 & -1 & 3 & -4 \end{bmatrix}^T$ with residual $\|\mathbf{b} - A\mathbf{x}\|_2 = 0$. Upon calling SGELS, the first 5 elements of B are overwritten by the solution vector $\mathbf{x}$:

$$\mathbf{x} = \begin{bmatrix} 1.0000176 \\ 2.0000196 \\ -0.9999972 \\ 3.0000386 \\ -4.0000405 \end{bmatrix},$$

while the sixth element of B contains the residual sum of squares 0.0000021. With M = 6, N = 5, NRHS = 1, LWORK has been set to 11. For such a small matrix, the minimal workspace is also the optimal workspace.

75.4 The Linear Equality-Constrained Least Squares Problem

Definitions:

The linear equality-constrained least squares (LSE) problem is

$$\min_{\mathbf{x}} \|\mathbf{c} - A\mathbf{x}\|_2 \quad \text{subject to} \quad B\mathbf{x} = \mathbf{d}, \tag{75.3}$$

where A is an m–by–n matrix and B is a p–by–n matrix, $\mathbf{c}$ is an m-vector, and $\mathbf{d}$ is a p-vector, with $p \leq n \leq m + p$.

Backgrounds:

Under the assumptions that B has full row rank p and the matrix $\begin{bmatrix} A \\ B \end{bmatrix}$ has full column rank n, the LSE problem (75.3) has a unique solution $\mathbf{x}$.

Driver Routines:

The driver routine for solving the LSE is xGGLSE, which uses a generalized QR factorization of the matrices A and B.

Examples:

Let us show how to use the driver routine SGGLSE to solve the LSE problem (75.3). SGGLSE first computes a generalized QR decomposition of A and B, and then computes the solution by back substitution. SGGLSE has the following calling sequence:

```
CALL SGGLSE( M, N, P, A, LDA, B, LDB, C, D, X, WORK, LWORK, INFO )
```

Input to SGGLSE:

- `M, P`: The numbers of rows of the matrices A and B, respectively. $\mathtt{M} \geq 0$ and $\mathtt{P} \geq 0$.
- `N`: The number of columns of the matrices A and B. $\mathtt{N} \geq 0$. Note that $0 \leq \mathtt{P} \leq \mathtt{N} \leq \mathtt{M+P}$.
- `A, LDA`: The `M`–by–`N` matrix A and the leading dimension of the array `A`. $\mathtt{LDA} \geq \max(\mathtt{1,M})$.
- `B, LDB`: The `P`–by–`N` matrix B and the leading dimension of the array `B`. $\mathtt{LDB} \geq \max(\mathtt{1,P})$.
- `C, D`: The right-hand side vectors for the least squares part, and the constrained equation part of the LSE, respectively.
- `WORK, LWORK`: The workspace array and its dimension. $\mathtt{LWORK} \geq \max(\mathtt{1}, \mathtt{M} + \mathtt{N} + \mathtt{P})$.
  If $\mathtt{LWORK} = -1$, then a workspace query is assumed; the routine only calculates the optimal size of the `WORK` array, and returns this value as the first entry of the `WORK` array.

Output from SGGLSE:

- `C`: The residual sum of squares for the solution is given by the sum of squares of elements `N-P+1` to `M` of vector `C`.
- `X`: The solution of the LSE problem.
- `WORK`: If `INFO = 0`, `WORK(1)` returns the optimal `LWORK`.
- `INFO`: $=0$ if successful exit. If $\mathtt{INFO} = -j$, the jth argument had an illegal value.

Let us demonstrate the use of SGGLSE to solve the LSE problem (75.3), where

$$A = \begin{bmatrix} 1 & 1 & 1 \\ 1 & 3 & 1 \\ 1 & -1 & 1 \\ 1 & 1 & 1 \end{bmatrix}, \quad B = \begin{bmatrix} 1 & 1 & 1 \\ 1 & 1 & -1 \end{bmatrix}, \quad \mathbf{c} = \begin{bmatrix} 1 \\ 2 \\ 3 \\ 4 \end{bmatrix}, \quad \mathbf{d} = \begin{bmatrix} 7 \\ 4 \end{bmatrix}.$$

The unique exact solution is $\mathbf{x} = \frac{1}{8}[46 \quad -2 \quad 12]^T$. Upon calling SGGLSE with this input data and $\mathtt{M} = 4, \mathtt{N} = 3, \mathtt{P} = 2, \mathtt{LWORK} = 9$, an approximate solution of the LSE problem is returned in `X`:

$$X = [5.7500000 \quad -0.2500001 \quad 1.4999994]^T.$$

The array `C` is overwritten by the residual sum of squares for the solution:

$$\mathtt{C} = [4.2426405 \quad 8.9999981 \quad 2.1064947 \quad 0.2503501]^T.$$

75.5 A General Linear Model Problem

Definitions:

The general linear model (GLM) problem is

$$\min_{\mathbf{x},\mathbf{y}} \|\mathbf{y}\|_2 \quad \text{subject to} \quad \mathbf{d} = A\mathbf{x} + B\mathbf{y}, \tag{75.4}$$

where A is an n–by–m matrix, B is an n–by–p matrix, and $\mathbf{d}$ is a n-vector, with $m \leq n \leq m + p$.

Backgrounds:

When $B = I$, the problem reduces to an ordinary linear least squares problem (75.2). When B is square and nonsingular, the GLM problem is equivalent to the *weighted linear least squares problem*:

$$\min_{\mathbf{x}} \|B^{-1}(\mathbf{d} - A\mathbf{x})\|_2.$$

Note that the GLM is equivalent to the LSE problem

$$\min_{\mathbf{x},\mathbf{y}} \left\| 0 - \begin{bmatrix} 0 & I \end{bmatrix} \begin{bmatrix} \mathbf{x} \\ \mathbf{y} \end{bmatrix} \right\|_2 \quad \text{subject to} \quad \begin{bmatrix} A & B \end{bmatrix} \begin{bmatrix} \mathbf{x} \\ \mathbf{y} \end{bmatrix} = \mathbf{d}.$$

Therefore, the GLM problem has a unique solution of the matrix $\begin{bmatrix} 0 & I \\ A & B \end{bmatrix}$ and has full column rank $m + p$.

Driver Routines:

The driver routine for solving the GLM problem (75.4) is xGGGLM, which uses a generalized QR factorization of the matrices A and B.

Examples:

Let us show how to use the driver routine SGGGLM to solve the GLM problem (75.4). SGGGLM computes a generalized QR decomposition of the matrices A and B, and then computes the solution by back substitution. SGGGLM has the following calling sequence:

```
CALL SGGGLM( N, M, P, A, LDA, B, LDB, D, X, Y, WORK, LWORK, INFO )
```

Input to SGGGLM:

`N`: The number of rows of the matrices A and B. `N` $\geq$ `0`.

`M, P`: The number of columns of the matrices A and B, respectively. `0` $\leq$ `M` $\leq$ `N` and `P` $\geq$ `N-M`.

`A, LDA`: The `N`–by–`M` matrix A and the leading dimension of the array `A`. `LDA` $\geq$ max(`1,N`).

`B, LDB`: The `N`–by–`P` matrix B and the leading dimension of the array `B`. `LDB` $\geq$ max(`1,N`).

`D`: The left-hand side of the GLM equation.

`WORK, LWORK`: The workspace array and its dimension. `LWORK` $\geq$ max(`1,N+M+P`).
If `LWORK = -1`, then a workspace query is assumed; the routine only calculates the optimal size of the `WORK` array, and returns this value as the first entry of the `WORK` array.

Output from SGGGLM:

`X, Y`: Solution vectors.

`WORK`: If `INFO = 0`, `WORK(1)` returns the optimal `LWORK`.

`INFO`: `INFO = 0` if successful exit. If `INFO` $= -j$, the jth argument had an illegal value.

Let us demonstrate the use of SGGGLM for solving the GLM problem (75.4), where

$$A = \begin{bmatrix} 1 & 2 & 1 & 4 \\ -1 & 1 & 1 & 1 \\ -1 & -2 & -1 & 1 \\ -1 & 2 & -1 & -1 \\ 1 & 1 & 1 & 2 \end{bmatrix}, \quad B = \begin{bmatrix} 1 & 2 & 2 \\ -1 & 1 & -2 \\ 3 & 1 & 6 \\ 2 & -2 & 4 \\ 1 & -1 & 2 \end{bmatrix}, \quad \mathbf{d} = \begin{bmatrix} 7.99 \\ 0.98 \\ -2.98 \\ 3.04 \\ 4.02 \end{bmatrix}.$$

Upon calling SGGGLM with this input data, $\texttt{N} = \texttt{5}, \texttt{M} = \texttt{4}, \texttt{P} = \texttt{3}, \texttt{LWORK} = \texttt{12}$, the program successfully exits and returns the following solution vectors:

$$X = [1.002950 \quad 2.001435 \quad -0.987797 \quad 0.009080]^T$$

and

$$Y = [0.003435 \quad -0.004417 \quad 0.006871]^T.$$

75.6 Symmetric Eigenproblems

Definitions:

The symmetric eigenvalue problem (SEP) is to find the eigenvalues, λ, and corresponding eigenvectors, $\mathbf{x} \neq 0$, such that

$$A\mathbf{x} = \lambda\mathbf{x}, \tag{75.5}$$

where A is real and symmetric. If A is complex and Hermitian, i.e., $A^* = A$, then it is referred to as the *Hermitian eigenvalue problem.*

Backgrounds:

When all eigenvalues and eigenvectors have been computed, we write

$$A = X\Lambda X^*, \tag{75.6}$$

where Λ is a diagonal matrix whose diagonal elements are real and are the eigenvalues, and X is an orthogonal (or unitary) matrix whose columns are the eigenvectors. This is the classical *spectral decomposition* of A. The theoretical and algorithmic backgrounds is of the solution of the symmetric eigenvalue problem are discussed in Chapter 42.

Driver Routines:

There are four types of driver routines for solving the SEP (75.5) and each has its own variations that take advantage of the special structure or storage of the matrix A, as summarized in the following table.

Types of Matrix (Storage Scheme)	Routine Names			
	Simple Driver	Divide-and-Conquer	Expert Driver	RRR Driver
General symmetric	xSYEV	xSYEVD	xSYEVX	xSYEVR
General symmetric (packed storage)	xSPEV	xSPEVD	xSPEVX	–
Band matrix	xSBEV	xSBEVD	xSBEVX	–
Tridiagonal matrix	xSTEV	xSTEVD	xSTEVX	xSTEVR

The simple driver computes all eigenvalues and (optionally) eigenvectors. The expert driver computes all or a selected subset of the eigenvalues and (optionally) eigenvectors. The divide-and-conquer driver has the same functionality as, yet outperforms, the simple driver, but it requires more workspace. The relative robust representation (RRR) driver computes all or a subset of the eigenvalues and (optionally) the eigenvectors. The last one is generally faster than any other types of driver routines and uses the least amount of workspace.

Examples:

Let us show how to use the simple driver SSYEV to solve the SEP (75.5) by computing the spectral decomposition (75.6). SSYEV first reduces A to a tridiagonal form, and then uses the implicit QL or QR algorithm to compute eigenvalues and optionally eigenvectors. SSYEV has the following calling sequence:

```
CALL SSYEV( JOBZ, UPLO, N, A, LDA, W, WORK, LWORK, INFO )
```

Input to `SSYEV`:

`JOBZ:` = `'N'`, compute eigenvalues only;
= `'V'`, compute eigenvalues and eigenvectors.

`UPLO:` = `'U'`, the upper triangle of A is stored in the array **A**; if `UPLO = 'L'`, the lower triangle of A is stored.

`N:` The order of the matrix A. `N` $\geq$ `0`.

`A, LDA:` The symmetric matrix A and the leading dimension of the array `A`. `LDA` $\geq$ max(`1,N`).

`WORK, LWORK:` The workspace array and its dimension. **LWORK** $\geq \max(1, 3 * \mathrm{N} - 1)$.
If **LWORK** $= -1$, then a workspace query is assumed; the routine only calculates the optimal size of the `WORK` array, and returns this value as the first entry of the `WORK` array.

Output from `SSYEV`:

`A:` The orthonormal eigenvectors X, if `JOBZ = 'V'`.

`W:` The eigenvalues λ in ascending order.

`WORK:` If `INFO = 0`, `WORK(1)` returns the optimal `LWORK`.

`INFO:` = 0 if successful exit. If **INFO** $= -j$, the jth input argument had an illegal value. If **INFO** $= j$, the j off-diagonal elements of an intermediate tridiagonal form did not converge to zero.

Let us demonstrate the use of SSYEV to solve the SEP (75.5), where

$$A = \begin{bmatrix} 5 & 4 & 1 & 1 \\ 4 & 5 & 1 & 1 \\ 1 & 1 & 4 & 2 \\ 1 & 1 & 2 & 4 \end{bmatrix}.$$

The exact eigenvalues are 1, 2, 5, and 10. Upon calling SSYEV with the matrix A and `N` $= 4$, `LWORK` $= 3 *$ `N` $- 1 = 11$, `A` is overwritten by its orthonormal eigenvectors X.

$$X = \begin{bmatrix} 0.7071068 & -0.0000003 & 0.3162279 & 0.6324555 \\ 0.7071068 & 0.0000001 & 0.3162278 & 0.6324555 \\ 0.0000002 & 0.7071069 & -0.6324553 & 0.3162278 \\ -0.0000001 & -0.7071066 & -0.6324556 & 0.3162278 \end{bmatrix}.$$

The eigenvalues that correspond to the eigenvectors in each column of `X` are returned in `W`:

$$\texttt{W} = [0.9999996 \quad 1.9999999 \quad 4.9999995 \quad 10.0000000].$$

75.7 Nonsymmetric Eigenproblems

As is customary in numerical linear algebra, in this section the term *left eigenvector* of A means a (column) vector $\mathbf{y}$ such that $y^* A = \lambda \mathbf{y}^*$. This is contrary to the definition in Section 4.3, under which y^* would be called a left eigenvector.

Definitions:

The nonsymmetric eigenvalue problem (NEP) is to find the eigenvalues, λ, and corresponding (right) eigenvectors, $\mathbf{x} \neq 0$, such that

$$A\mathbf{x} = \lambda \mathbf{x} \tag{75.7}$$

and, perhaps, the left eigenvectors, $\mathbf{y} \neq 0$, satisfying

$$\mathbf{y}^* A = \lambda \mathbf{y}^*. \tag{75.8}$$

Backgrounds:

The problem is solved by computing the *Schur decomposition* of A, defined in the real case as

$$A = ZTZ^T,$$

where Z is an orthogonal matrix and T is an upper quasi-triangular matrix with 1–by–1 and 2–by–2 diagonal blocks, the 2–by–2 blocks corresponding to complex conjugate pairs of eigenvalues of A. In the complex case, the Schur decomposition is

$$A = ZTZ^*,$$

where Z is unitary and T is a complex upper triangular matrix.

The columns of Z are called the *Schur vectors*. For each k $(1 \leq k \leq n)$, the first k columns of Z form an orthonormal basis for the invariant subspace corresponding to the first k eigenvalues on the diagonal of T. It is possible to order the Schur factorization so that any desired set of k eigenvalues occupies the k leading positions on the diagonal of T. The theoretical and algorithmic background of the solution of the nonsymmetric eigenvalue problem is discussed in Chapter 43.

Driver Routines:

Both the simple driver xGEEV and expert driver xGEEVX are provided. The simple driver computes all the eigenvalues of A and (optionally) the right or left eigenvectors (or both). The expert driver performs the same task as the simple driver plus the additional feature that it balances the matrix to try to improve the conditioning of the eigenvalues and eigenvectors, and it computes the condition numbers for the eigenvalues or eigenvectors (or both).

Examples:

Let us show how to use the simple driver SGEEV to solve the NEP (75.7). SGEEV first reduces A to an upper Hessenberg form (a Hessenberg matrix is a matrix where all entries below the first lower subdiagonal are zeros), and then uses the implicit QR algorithm to compute the Schur decomposition, and finally computes eigenvectors of the upper quasi-triangular matrix. SGEEV has the following calling sequence:

```
  CALL SGEEV( JOBVL, JOBVR, N, A, LDA, WR, WI, VL, LDVL, VR, LDVR, WORK,
LWORK, INFO )
```

Input to SGEEV:

> JOBVL, JOBVR: = 'V', the left and/or right eigenvectors are computed;
>
> = 'N', the left and/or right eigenvectors are not computed.
>
> N: The order of the matrix A. N $\geq$ 0.
>
> A, LDA: The matrix A and the leading dimension of the array A. LDA $\geq \max(1, N)$.
>
> LDVL, LDVR: The leading dimensions of the arrays VL and VR if the left and right eigenvectors are computed. LDVL, LDVR $\geq$ N.
>
> WORK, LWORK: The workspace array and its dimension. LWORK $\geq \max(1, 3 * N)$. If eigenvectors are computed, LWORK $\geq 4 * N$. For good performance, LWORK must generally be larger.
>
> If LWORK $= -1$, then a workspace query is assumed; the routine only calculates the optimal size of the WORK array, and returns this value as the first entry of the WORK array.

Output from SGEEV:

> WR, WI: The real and imaginary parts of the computed eigenvalues. Complex conjugate pairs of eigenvalues appear consecutively with the eigenvalue having the positive imaginary part first.
>
> VL: If the jth eigenvalue λ_j is real, then the jth left eigenvector $\mathbf{y}_j$ is stored in VL$(:, j)$. If the jth and $(j+1)$-st eigenvalues λ_j and λ_{j+1} form a complex conjugate pair, then VL$(:, j) + i \cdot$ VL$(:, j+1)$ and VL$(:, j) - i \cdot$ VL$(:, j+1)$ are the corresponding left eigenvectors $\mathbf{y}_j$ and $\mathbf{y}_{j+1}$.
>
> VR: If the jth eigenvalue λ_j is real, then the jth right eigenvector $\mathbf{x}_j$ is stored in VR$(:, j)$. If the jth and $(j+1)$-st eigenvalues λ_j and λ_{j+1} form a complex conjugate pair, then VR$(:, j) + i \cdot$ VR$(:, j+1)$ and VR$(:, j) - i \cdot$ VR$(:, j+1)$ are the corresponding right eigenvectors $\mathbf{x}_j$ and $\mathbf{x}_{j+1}$.
>
> WORK: If INFO = 0, WORK(1) returns the optimal LWORK.
>
> INFO: = 0 if successful exit. If INFO $= -j$, the jth argument had an illegal value. If INFO $= j$, the QR algorithm failed to compute all the eigenvalues, and no eigenvectors have been computed; elements $j+1$: N of WR and WI contain eigenvalues, which have converged.

Let us demonstrate the use of SGEEV for solving the NEP (75.7), where

$$A = \begin{bmatrix} 4 & -5 & 0 & 3 \\ 0 & 4 & -3 & -5 \\ 5 & -3 & 4 & 0 \\ 3 & 0 & 5 & 4 \end{bmatrix}.$$

The exact eigenvalues are $12, 1 + i \cdot 5, 1 - i \cdot 5$, and 2. Upon calling SGEEV with this matrix and N $= 4$, LWORK $= 4 * N = 16$, each eigenvalue, λ_j, is retrieved by combining the jth entry in WR and WI.

such that $\lambda_j = \texttt{WR}(j) + i \cdot \texttt{WI}(j)$. If $\texttt{WI}(j)$ is 0, then the jth eigenvalue is real. For this example, we have

$$\lambda_1 = 12.0000000$$
$$\lambda_2 = 1.000000 + i \cdot 5.0000005$$
$$\lambda_3 = 1.000000 - i \cdot 5.0000005$$
$$\lambda_4 = 1.9999999.$$

The left eigenvectors are stored in VL. Since the first and fourth eigenvalues are real, their eigenvectors are the corresponding columns in VL, that is, $\mathbf{y}_1 = \texttt{VL}(:,1)$ and $\mathbf{y}_4 = \texttt{VL}(:,4)$. Since the second and third eigenvalues form a complex conjugate pair, the second eigenvector, $\mathbf{y}_2 = \texttt{VL}(:,2) + i \cdot \texttt{VL}(:,3)$ and the third eigenvector, $\mathbf{y}_3 = \texttt{VL}(:,2) - i \cdot \texttt{VL}(:,3)$. If we place all the eigenvectors in a matrix Y where $Y = [\mathbf{y}_1, \mathbf{y}_2, \mathbf{y}_3, \mathbf{y}_4]$, we have

$$Y = \begin{bmatrix} -0.5000001 & 0.0000003 - i \cdot 0.4999999 & 0.0000003 + i \cdot 0.4999999 & 0.5000000 \\ 0.4999999 & -0.5000002 & -0.5000002 & 0.5000001 \\ -0.5000000 & -0.5000000 - i \cdot 0.0000002 & -0.5000000 + i \cdot 0.0000002 & -0.4999999 \\ -0.5000001 & -0.0000003 + i \cdot 0.5000000 & -0.0000003 - i \cdot 0.5000000 & 0.5000001 \end{bmatrix}.$$

The right eigenvectors $\mathbf{x}_j$ can be recovered from VR in the way similar to the left eigenvectors. The right eigenvector matrix X is

$$X = \begin{bmatrix} -0.5000000 & 0.5000002 & 0.5000002 & 0.5000001 \\ 0.4999999 & -0.0000001 - i \cdot 0.5000000 & -0.0000001 + i \cdot 0.5000000 & 0.5000000 \\ -0.5000000 & -0.0000001 - i \cdot 0.4999999 & -0.0000001 + i \cdot 0.4999999 & -0.5000000 \\ -0.5000001 & -0.5000001 & -0.5000001 & 0.5000000 \end{bmatrix}.$$

75.8 Singular Value Decomposition

Definitions:

The singular value decomposition (SVD) of an m–by–n matrix A is

$$A = U\Sigma V^T \quad (A = U\Sigma V^* \quad \text{in the complex case}), \tag{75.9}$$

where U and V are orthogonal (unitary) and Σ is an m–by–n diagonal matrix with real diagonal elements, σ_j, such that

$$\sigma_1 \geq \sigma_2 \geq \ldots \geq \sigma_{\min(m,n)} \geq 0.$$

The σ_j are the singular values of A and the first $\min(m, n)$ columns of U and V are the left and right singular vectors of A.

Backgrounds:

The singular values σ_j and the corresponding left singular vectors $\mathbf{u}_j$ and right singular vectors $\mathbf{v}_j$ satisfy

$$A\mathbf{v}_j = \sigma_j \mathbf{u}_j \quad \text{and} \quad A^T\mathbf{u}_j = \sigma_j \mathbf{v}_j \quad (\text{or} \quad A^*\mathbf{u}_j = \sigma_j \mathbf{v}_j \quad \text{in complex case}),$$

where $\mathbf{u}_j$ and $\mathbf{v}_j$ are the jth columns of U and V, respectively. (See Chapter 17 and Chapter 45 for more information on singular value decompositions.)

Driver Routines:

Two types of driver routines are provided in LAPACK. The simple driver xGESVD computes, all the singular values and (optionally) left and/or right singular vectors. The divide and conquer driver xGESDD has the same functionality as the simple driver except that it is much faster for larger matrices, but uses more workspace.

Examples:

Let us show how to use the simple driver SGESVD to compute the SVD (75.9). SGESVD first reduces A to a bidiagonal form, and then uses an implicit QR-type algorithm to compute singular values and optionally singular vectors. SGESVD has the following calling sequence:

```
  CALL SGESVD( JOBU, JOBVT, M, N, A, LDA, S, U, LDU, VT, LDVT, WORK,
LWORK, INFO )
```

Input to SGESVD:

JOBU: Specifies options for computing all or part of the left singular vectors U:

= 'A', all M columns of U are returned in the array U:

= 'S', the first min(M, N) columns of U are returned;

= 'O', the first min(M, N) columns of U are overwritten on the array A;

= 'N', no left singular vectors are computed. *Note that* JOBVT *and* JOBU *cannot both be* 'O'.

JOBVT: Specifies options for computing all or part of the right singular vectors V^T:

= 'A', all N rows of V^T are returned in the array VT;

= 'S', the first min(M, N) rows of V^T are returned;

= 'O', the first min(M, N) rows of V^T are overwritten on the array A;

= 'N', no right singular vectors are computed.

M, N: The number of rows and columns of the matrix A. M, N $\geq$ 0.

A, LDA: The M–by–N matrix A and the leading dimension of the array A. LDA $\geq$ max(1, M).

LDU, LDVT: The leading dimension of the arrays U and VT. **LDU, LDVT** $\geq$ **1**;

If JOBU = 'S' or 'A', **LDU** $\geq$ **M**.

If JOBVT = 'A', **LDVT** $\geq$ **N**; If JOBVT = 'S', **LDVT** $\geq$ min(**M**, **N**).

WORK, LWORK: The workspace array and its dimension. **LWORK** $\geq$ max(3 min(**M**, **N**) + max(**M**, **N**), 5 min(**M**, **N**)).

If LWORK = -1, then a workspace query is assumed; the routine only calculates the optimal size of the WORK array and returns this value as the first entry of the WORK array.

Output from SGESVD:

A: If JOBU = 'O', A is overwritten with the first min(M, N) columns of U (the left singular vectors, stored columnwise);

If JOBVT = 'O', A is overwritten with the first min(M, N) rows of V^T (the right singular vectors, stored rowwise);

S: Singular values of A, sorted so that S(i) $\geq$ S($i+1$).

U: If JOBU = 'A', U contains M–by–M orthogonal matrix U. If JOBU = 'S', U contains the first min(M, N) columns of U. If JOBU = 'N' or 'O', U is not referenced.

`VT`: If `JOBVT = 'A'`, `VT` contains right N–by–N orthogonal matrix V^T. If `JOBVT = 'S'`, `VT` contains the first min(`M`,`N`) rows of $\mathtt{V}^T$ (the right singular vectors stored rowwise). If `JOBVT = 'N'` or `'O'`, `VT` is not referenced.

`WORK`: If `INFO = 0`, `WORK(1)` returns the optimal LWORK.

`INFO`: $=0$ if successful exit. If **INFO** $= -j$, the jth argument had an illegal value. If **INFO** >0, the QR-type algorithm (subroutine SBDSQR) did not converge. `INFO` specifies how many superdiagonals of an intermediate bidiagonal form B did not converge to zero. `WORK`(2:min(`M`,`N`)) contains the unconverged superdiagonal elements of an upper bidiagonal matrix B whose diagonal is in `S` (not necessarily sorted). B satisfies $A = UBV^T$, so it has the same singular values as A, and singular vectors related by U and V^T.

Let us show the numerical results of SGESVD in computing the SVD by an 8–by–5 matrix A as follows:

$$A = \begin{bmatrix} 22 & 10 & 2 & 3 & 7 \\ 14 & 7 & 10 & 0 & 8 \\ -1 & 13 & -1 & -11 & 3 \\ -3 & -2 & 13 & -2 & 4 \\ 9 & 8 & 1 & -2 & 4 \\ 9 & 1 & -7 & 5 & -1 \\ 2 & -6 & 6 & 5 & 1 \\ 4 & 5 & 0 & -2 & 2 \end{bmatrix}.$$

The exact singular singular values are $\sqrt{1248}, 20, \sqrt{384}, 0, 0$. The rank of the matrix A is 3. Upon calling SGESVD with $\mathtt{M} = 8, \mathtt{N} = 5, \mathtt{LWORK} = 25$, the computed singular values of A are returned in `S`:

$$\mathtt{S} = [35.3270454 \quad 20.0000038 \quad 19.5959187 \quad 0.0000007 \quad 0.0000004].$$

The columns in `U` contain the left singular vectors U of A:

$$U = \begin{bmatrix} \text{-7.0711e-001} & 1.5812e-001 & -1.7678e-001 & 2.4818e-001 & -4.0289e-001 & -3.2305e-001 & -3.3272e-001 & -6.9129e-002 \\ \text{-5.3033e-001} & 1.5811e-001 & 3.5355e-001 & -6.2416e-001 & 2.5591e-001 & -3.9178e-002 & 3.0548e-001 & -1.3725e-001 \\ \text{-1.7678e-001} & -7.9057e-001 & 1.7677e-001 & 3.0146e-001 & 1.9636e-001 & -3.1852e-001 & 2.3590e-001 & -1.6112e-001 \\ 0 & 1.5811e-001 & 7.0711e-001 & 2.9410e-001 & 3.1907e-001 & 4.7643e-002 & -5.2856e-001 & 7.1055e-002 \\ \text{-3.5355e-001} & -1.5811e-001 & -1.0000e-006 & 2.3966e-001 & -7.8607e-002 & 8.7800e-001 & 1.0987e-001 & -5.8528e-002 \\ \text{-1.7678e-001} & 1.5812e-001 & -5.3033e-001 & 1.7018e-001 & 7.9071e-001 & -7.0484e-003 & -9.0913e-002 & -8.3220e-004 \\ 0 & 4.7434e-001 & 1.7678e-001 & 5.2915e-001 & -1.5210e-002 & -1.3789e-001 & 6.6193e-001 & 7.9763e-002 \\ \text{-1.7678e-001} & -1.5811e-001 & -1.0000e-006 & -7.1202e-002 & 1.3965e-002 & -2.0712e-002 & 4.9676e-002 & 9.6726e-001 \end{bmatrix}.$$

The rows in `VT` contain the right singular vectors V^T of A:

$$V^T = \begin{bmatrix} -8.0064e-001 & -4.8038e-001 & -1.6013e-001 & 0 & -3.2026e-001 \\ 3.1623e-001 & -6.3246e-001 & 3.1622e-001 & 6.3246e-001 & -1.8000e-006 \\ -2.8867e-001 & -3.9000e-006 & 8.6603e-001 & -2.8867e-001 & 2.8868e-001 \\ -4.0970e-001 & 3.4253e-001 & -1.2426e-001 & 6.0951e-001 & 5.7260e-001 \\ 8.8224e-002 & -5.0190e-001 & -3.3003e-001 & -3.8100e-001 & 6.9730e-001 \end{bmatrix}.$$

75.9 Generalized Symmetric Definite Eigenproblems

Definitions:

The generalized symmetric definite eigenvalue problem (GSEP) is to find the eigenvalues, λ, and corresponding eigenvectors, $\mathbf{x} \neq 0$, such that

$$A\mathbf{x} = \lambda B\mathbf{x} \quad \text{(type 1)} \tag{75.10}$$

or

$$AB\mathbf{x} = \lambda\mathbf{x} \quad \text{(type 2)} \tag{75.11}$$

or

$$BA\mathbf{x} = \lambda\mathbf{x} \quad \text{(type 3)} \tag{75.12}$$

where A and B are symmetric or Hermitian and B is positive definite.

Backgrounds:

For all these problems the eigenvalues λ are real. The matrix Z of the computed eigenvectors satisfies $Z^*AZ = \Lambda$ (problem types 1 and 3) or $Z^{-1}AZ^{-*} = I$ (problem type 2), where Λ is a diagonal matrix with the eigenvalues on the diagonal. Z also satisfies $Z^*BZ = I$ (problem types 1 and 2) or $Z^*B^{-1}Z = I$ (problem type 3). These results are consequences of spectral theory for symmetric matrices. For example, the GSEP type 1 can be rearranged as

$$B^{-\frac{1}{2}}AB^{-\frac{1}{2}}\mathbf{y} = \lambda\mathbf{y},$$

where $\mathbf{y} = B^{\frac{1}{2}}\mathbf{x}$.

Driver Routines:

There are three types of driver routines for solving the GSEP, and each has variations that take advantage of the special structure or storage of the matrices A and B, as shown in the following table:

Types of Matrix	Routine Names		
(Storage Scheme)	Simple Driver	Divide-and-Conquer	Expert Driver
General dense	xSYGV/xHEGV	xSYGVD/xHEGVD	xSYGVX/xHEGVX
General dense (packed storage)	xSPGV/xHPGV	xSPGVD/xHPGVD	xSPGVX/xHPGVX
Band matrix	xSBGV/xHBGV	xSBBVD/xHBGVD	xSBGVX/xHBGVX

The simple driver computes all the eigenvalues and (optionally) the eigenvectors. The expert driver computes all or a selected subset of the eigenvalues and (optionally) eigenvectors. The divide-and-conquer driver solves the same problem as the simple driver. It is much faster than the simple driver, but uses more workspace.

Examples:

Let us show how to use the simple driver SSYGV to compute the GSEPs (75.10), (75.11), and (75.12). SSGYV first reduces each of these problems to a standard symmetric eigenvalue problem, using a Cholesky decomposition of B, and then computes eigenvalues and eigenvectors of the standard symmetric eigenvalue problem by an implicit QR-type algorithm. SSYGV has the following calling sequence:

```
CALL SSYGV( ITYPE, JOBZ, UPLO, N, A, LDA, B, LDB, W, WORK, LWORK, INFO )
```

Input to `SSYGV`:

`ITYPE`: Specifies the problem type to be solved:

`JOBZ`: = `'N'`, compute eigenvalues only;
= `'V'`, compute eigenvalues and eigenvectors.

`UPLO`: = `'U'`, the upper triangles of A and B are stored;
= `'L'`, the lower triangles of A and B are stored.

`N`: The order of the matrices A and B. `N` $\geq$ `0`.

`A, LDA`: The symmetric matrix A and the leading dimension of the array `A`. `LDA` $\geq$ `max(1,N)`.

`B`: The symmetric positive definite matrix B and the leading dimension of the array `B`. `LDB` $>$ `max(1,N)`

`WORK, LWORK`: The workspace array and its length. `LWORK` $\geq$ `max(1,3 * N - 1)`.
If `LWORK` $=$ `-1`, then a workspace query is assumed; the routine only calculates the optimal size of the `WORK` array, and returns this value as the first entry of the `WORK` array.

Output from `SSYGV`:

`A`: Contains the normalized eigenvector matrix Z if requested.

`B`: If `INFO` $\leq$ `N`, the part of B containing the matrix is overwritten by the triangular factor U or L from the Cholesky factorization $B = U^T U$ or $B = LL^T$.

`W`: The eigenvalues in ascending order.

`WORK`: If `INFO` $= 0$, `WORK(1)` returns the optimal `LWORK`.

`INFO`: $= 0$, then successful exit. If `INFO` $= -j$, then the jth argument had an illegal value. If `INFO` > 0, then SSYGV returned an error code:

- `INFO` $\leq$ `N`: if `INFO` $= j$, the algorithm failed to converge;
- `INFO` $>$ `N`: if `INFO` $=$ `N` $+ j$, for $1 \leq j \leq$ `N`, then the leading minor of order j of B is not positive definite. The factorization of B could not be completed and no eigenvalues or eigenvectors were computed.

Let us show the use of SSYGV to solve the type 1 GSEP (75.10) for the following 5–by–5 matrices A and B:

$$A = \begin{bmatrix} 10 & 2 & 3 & 1 & 1 \\ 2 & 12 & 1 & 2 & 1 \\ 3 & 1 & 11 & 1 & -1 \\ 1 & 2 & 1 & 9 & 1 \\ 1 & 1 & -1 & 1 & 15 \end{bmatrix} \quad \text{and} \quad B = \begin{bmatrix} 12 & 1 & -1 & 2 & 1 \\ 1 & 14 & 1 & -1 & 1 \\ -1 & 1 & 16 & -1 & 1 \\ 2 & -1 & -1 & 12 & -1 \\ 1 & 1 & 1 & -1 & 11 \end{bmatrix}.$$

Upon calling SSYGV with `N` $= 5$, `LWORK` $= 3 *$ `N` $- 1 = 14$, `A` is overwritten by the eigenvector matrix Z:

$$Z = \begin{bmatrix} -0.1345906 & 0.0829197 & -0.1917100 & 0.1420120 & -0.0763867 \\ 0.0612948 & 0.1531484 & 0.1589912 & 0.1424200 & 0.0170980 \\ 0.1579026 & -0.1186037 & -0.0748390 & 0.1209976 & -0.0666645 \\ -0.1094658 & -0.1828130 & 0.1374690 & 0.1255310 & 0.0860480 \\ 0.0414730 & 0.0035617 & -0.0889779 & 0.0076922 & 0.2894334 \end{bmatrix}.$$

The corresponding eigenvalues are returned in `W`:

$$\texttt{W} = \begin{bmatrix} 0.4327872 & 0.6636626 & 0.9438588 & 1.1092844 & 1.4923532 \end{bmatrix}.$$

75.10 Generalized Nonsymmetric Eigenproblems

Definitions:

The generalized nonsymmetric eigenvalue problem (GNEP) is to find the eigenvalues, λ, and corresponding (right) eigenvectors, $\mathbf{x} \neq 0$, such that

$$A\mathbf{x} = \lambda B\mathbf{x} \tag{75.13}$$

and optionally, the corresponding left eigenvectors $\mathbf{y} \neq 0$, such that

$$\mathbf{y}^* A = \lambda \mathbf{y}^* B, \tag{75.14}$$

where A and B are n–by–n matrices. In this section the terms *right eigenvector* and *left eigenvector* are used as just defined.

Backgrounds:

Sometimes an equivalent notation is used to refer to the GNEP of the pair (A, B). The GNEP can be solved via the generalized Schur decomposition of the pair (A, B), defined in the real case as

$$A = QSZ^T, \quad B = QTZ^T,$$

where Q and Z are orthogonal matrices, T is upper triangular, and S is an upper quasi-triangular matrix with 1-by-1 and 2-by-2 diagonal blocks, the 2-by-2 blocks corresponding to complex conjugate pairs of eigenvalues. In the complex case, the generalized Schur decomposition is

$$A = QSZ^*, \quad B = QTZ^*,$$

where Q and Z are unitary and S and T are both upper triangular. The columns of Q and Z are called left and right generalized Schur vectors and span pairs of deflating subspaces of A and B. Deflating subspaces are a generalization of invariant subspaces: For each k, $1 \leq k \leq n$, the first k columns of Z span a right deflating subspace mapped by both A and B into a left deflating subspace spanned by the first k columns of Q. It is possible to order the generalized Schur form so that any desired subset of k eigenvalues occupies the k leading position on the diagonal of (S, T). (See Chapter 43 and Chapter 15 for more information on generalized eigenvalue problems.)

Driver Routines:

Both the simple and expert drivers are provided in LAPACK. The simple driver xGGEV computes all eigenvalues of the pair (A, B), and optionally the left and/or right eigenvectors. The expert driver xGGEVX performs the same task as the simple driver routines; in addition, it also balances the matrix pair to try to improve the conditioning of the eigenvalues and eigenvectors, and computes the condition numbers for the eigenvalues and/or left and right eigenvectors.

Examples:

Let us show how to use the simple driver SGGEV to solve the GNEPs (75.13) and (75.14). SGGEV first reduces the pair (A, B) to generalized upper Hessenberg form (H, R), where H is upper Hessenberg (zero below the first lower subdiagonal) and R is upper triangular. Then SGGEV computes the generalized Schur form (S, T) of the generalized upper Hessenberg form (H, R), using an QZ algorithm. The eigenvalues are computed from the diagonals of (S, T). Finally, SGGEV computes left and/or right eigenvectors if requested. SGGEV has the following calling sequence:

```
  CALL SGGEV( JOBVL, JOBVR, N, A, LDA, B, LDB, ALPHAR, ALPHAI, BETA, VL,
LDVL, VR, LDVR, WORK, LWORK, INFO )
```

Input to SGGEV:

`JOBVL, JOBVR:` = `'N'`, do not compute the left and/or right eigenvectors;

= `'V'`, compute the left and/or right eigenvectors.

`N:` The order of the matrices A and B. `N` $\geq$ `0`.

`A, LDA:` The matrix A and the leading dimension of the array `A`. `LDA` $\geq$ max(`1,N`).

`B, LDB:` The matrix B and the leading dimension of the array `B`. `LDB` $\geq$ max(`1,N`).

LDVL, LDVR: The leading dimensions of the eigenvector matrices VL and VR. **LDVL, LDVR** $\geq$ **1**. If eigenvectors are required, then **LDVL, LDVR** $\geq$ **N**.

WORK, LWORK: The workspace array and its length. **LWORK** $\geq \max(1, 8 * \mathtt{N})$. For good performance, LWORK must generally be larger.

If **LWORK** $= -1$, then a workspace query is assumed; the routine only calculates the optimal size of WORK, and returns this value in WORK(1) on return.

Output from SGGEV:

ALPHAR, ALPHAI, BETA: $(\mathbf{ALPHAR}(j) + i \cdot \mathbf{ALPHAI}(j))/\mathbf{BETA}(j)$ for $j = 1, 2, \ldots, \mathrm{N}$, are the generalized eigenvalues. If ALPHAI(j) is zero, then the jth eigenvalue is real; if positive, then the jth and $(j+1)$-st eigenvalues are a complex conjugate pair, with ALPHAI($j+1$) negative.

VL: If JOBVL = 'V', the left eigenvectors $\mathbf{y}_j$ are stored in the columns of VL, in the same order as their corresponding eigenvalues. If the jth eigenvalue is real, then $\mathbf{y}_j = \mathbf{VL}(:, j)$, the jth column of VL. If the jth and $(j+1)$th eigenvalues form a complex conjugate pair, then $\mathbf{y}_j = \mathbf{VL}(:, j) + i \cdot \mathbf{VL}(:, j+1)$ and $\mathbf{y}_{j+1} = \mathbf{VL}(:, j) - i \cdot \mathbf{VL}(:, j+1)$.

VR: If JOBVR = 'V', the right eigenvectors $\mathbf{x}_j$ are stored one after another in the columns of VR, in the same order as their eigenvalues. If the jth eigenvalue is real, then $\mathbf{x}_j = \mathbf{VR}(:, j)$, the jth column of VR. If the jth and $(j+1)$th eigenvalues form a complex conjugate pair, then $\mathbf{x}_j = \mathbf{VR}(:, j) + i \cdot \mathbf{VR}(:, j+1)$ and $\mathbf{x}_{j+1} = \mathbf{VR}(:, j) - i \cdot \mathbf{VR}(:, j+1)$.

WORK: If **INFO** $= 0$, **WORK(1)** returns the optimal LWORK.

INFO: INFO = 0 if successful exit. If **INFO** $= -j$, the jth argument had an illegal value. If INFO = 1, . . . , N, then the QZ iteration failed. No eigenvectors have been calculated, but ALPHAR(j), ALPHAI(j), and BETA(j) should be correct for j = **INFO** + **1**, . . . , **N**. If INFO = N+1, then other than QZ iteration failed in SHGEQZ. If INFO = N+2, then error return from STGEVC.

Note that the quotients **ALPHAR**(j)/**BETA**(j) and **ALPHAI**(j)/**BETA**(j) may easily over- or underflow, and **BETA**(j) may even be zero. Thus, the user should avoid naively computing the ratio. However, ALPHAR and ALPHAI will be always less than and usually comparable to $\|A\|$ in magnitude, and BETA always less than and usually comparable to $\|B\|$.

Let us demonstrate the use of SGGEV in solving the GNEP of the following 6–by–6 matrices A and B:

$$A = \begin{bmatrix} 50 & -60 & 50 & -27 & 6 & 6 \\ 38 & -28 & 27 & -17 & 5 & 5 \\ 27 & -17 & 27 & -17 & 5 & 5 \\ 27 & -28 & 38 & -17 & 5 & 5 \\ 27 & -28 & 27 & -17 & 16 & 5 \\ 27 & -28 & 27 & -17 & 5 & 16 \end{bmatrix} \quad \text{and} \quad B = \begin{bmatrix} 16 & 5 & 5 & 5 & -6 & 5 \\ 5 & 16 & 5 & 5 & -6 & 5 \\ 5 & 5 & 16 & 5 & -6 & 5 \\ 5 & 5 & 5 & 16 & -6 & 5 \\ 5 & 5 & 5 & 5 & -6 & 16 \\ 6 & 6 & 6 & 6 & -5 & 6 \end{bmatrix}.$$

The exact eigenvalues are $\frac{1}{2} + i \cdot \frac{\sqrt{3}}{2}, \frac{1}{2} + i \cdot \frac{\sqrt{3}}{2}, \frac{1}{2} - i \cdot \frac{\sqrt{3}}{2}, \frac{1}{2} - i \cdot \frac{\sqrt{3}}{2}, \infty, \infty$. Upon calling SGGEV with **N** $= 6$, **LWORK** $= 48$, on exit, arrays ALPHAR, ALPHAI, and BETA are

$$\mathbf{ALPHAR} = \begin{bmatrix} -25.7687130 & 6.5193458 & 5.8156629 & 5.8464251 & 5.5058141 & 11.2021322 \end{bmatrix},$$

$$\mathbf{ALPHAI} = \begin{bmatrix} 0.0000000 & 11.2832556 & -10.0653677 & 10.1340599 & -9.5436525 & 0.0000000 \end{bmatrix},$$

$$\mathbf{BETA} = \begin{bmatrix} 0.0000000 & 13.0169611 & 11.6119413 & 11.7124090 & 11.0300474 & 0.0000000 \end{bmatrix}.$$

Therefore, there are two infinite eigenvalues corresponding to $\texttt{BETA}(1) = \texttt{BETA}(6) = 0$ and four finite eigenvalues $\lambda_j = (\texttt{ALPHAR}(j) + i \cdot \texttt{ALPHAI}(j))/\texttt{BETA}(j)$ for $j = 2, 3, 4, 5$.

$$(\texttt{ALPHAR}(2:5) + i \cdot \texttt{ALPHAI}(2:5))/\texttt{BETA}(2:5) = \begin{bmatrix} 0.50083 + i \cdot 0.86681 \\ 0.50083 - i \cdot 0.86681 \\ 0.49917 + i \cdot 0.86524 \\ 0.49917 - i \cdot 0.86524 \end{bmatrix}.$$

The left eigenvectors $\mathbf{y}_j$ are stored in VL. Since $\texttt{ALPHAI}(1) = \texttt{ALPHAI}(6) = 0$, $\mathbf{y}_1 = \texttt{VL}(:, 1)$ and $\mathbf{y}_6 = \texttt{VL}(:, 6)$. The second and third eigenvalues form a complex conjugate pair, the $\mathbf{y}_2 = \texttt{VL}(:, 2) + i \cdot \texttt{VL}(:, 3)$ and $\mathbf{y}_3 = \texttt{VL}(:, 2) - i \cdot \texttt{VL}(:, 3)$. Similarly, $\mathbf{y}_4 = \texttt{VL}(:, 4) + i \cdot \texttt{VL}(:, 5)$ and $\mathbf{y}_5 = \texttt{VL}(:, 4) + i \cdot \texttt{VL}(:, 5)$. If we place all the left eigenvectors in a matrix Y, where $Y = [\mathbf{y}_1, \mathbf{y}_2, \mathbf{y}_3, \mathbf{y}_4, \mathbf{y}_5, \mathbf{y}_6]$, we have

$$Y = \begin{bmatrix} -0.1666666 & 0.2632965 + i \cdot 0.3214956 & 0.2632965 - i \cdot 0.3214956 & -0.4613968 + i \cdot 0.1902102 & -0.4613968 - i \cdot 0.1902102 & 0.1666667 \\ -0.1666666 & -0.2834885 - i \cdot 0.7165115 & -0.2834885 + i \cdot 0.7165115 & 0.9231794 - i \cdot 0.0765849 & 0.9231794 + i \cdot 0.0765849 & 0.1666667 \\ -0.1666666 & 0.1623165 + i \cdot 0.7526108 & 0.1623165 - i \cdot 0.7526108 & -0.9240005 - i \cdot 0.0759995 & -0.9240005 + i \cdot 0.0759995 & 0.1666667 \\ -0.1666666 & 0.0396326 - i \cdot 0.4130635 & 0.0396326 + i \cdot 0.4130635 & 0.4619284 + i \cdot 0.1907958 & 0.4619284 - i \cdot 0.1907958 & 0.1666666 \\ -0.1666671 & -0.0605860 + i \cdot 0.0184893 & -0.0605860 - i \cdot 0.0184893 & 0.0000969 - i \cdot 0.0761408 & 0.0000969 + i \cdot 0.0761408 & 0.1666666 \\ 1.0000000 & -0.0605855 + i \cdot 0.0184900 & -0.0605855 - i \cdot 0.0184900 & 0.0000959 - i \cdot 0.0761405 & 0.0000959 + i \cdot 0.0761405 & -1.0000000 \end{bmatrix}.$$

The right eigenvectors can be recovered from VR in a way similar to the left eigenvectors. If we place all the right eigenvectors in a matrix X, where $X = [\mathbf{x}_1, \mathbf{x}_2, \mathbf{x}_3, \mathbf{x}_4, \mathbf{x}_5, \mathbf{x}_6]$, we have

$$U = \begin{bmatrix} 0.1666672 & -0.2039835 - i \cdot 0.5848466 & -0.2039835 + i \cdot 0.5848466 & 0.5722237 - i \cdot 0.0237538 & 0.5722237 + i \cdot 0.0237538 & 0.1666672 \\ 0.1666664 & -0.7090308 - i \cdot 0.2908980 & -0.7090308 + i \cdot 0.2908980 & 0.4485306 - i \cdot 0.5514694 & 0.4485306 + i \cdot 0.5514694 & 0.1666664 \\ 0.1666666 & -0.7071815 + i \cdot 0.2928185 & -0.7071815 - i \cdot 0.2928185 & -0.0709520 - i \cdot 0.7082051 & -0.0709520 + i \cdot 0.7082051 & 0.1666666 \\ 0.1666666 & -0.2013957 + i \cdot 0.5829236 & -0.2013957 - i \cdot 0.5829236 & -0.4667411 - i \cdot 0.3361499 & -0.4667411 + i \cdot 0.3361499 & 0.1666666 \\ 1.0000000 & -0.2023994 + i \cdot 0.0000001 & -0.2023994 - i \cdot 0.0000001 & 0.0536732 - i \cdot 0.1799536 & 0.0536732 + i \cdot 0.1799536 & 1.0000000 \\ 0.1666666 & -0.2023991 - i \cdot 0.0000002 & -0.2023991 + i \cdot 0.0000002 & 0.0536734 - i \cdot 0.1799532 & 0.0536734 + i \cdot 0.1799532 & 0.1666664 \end{bmatrix}.$$

75.11 Generalized Singular Value Decomposition

Definitions:

The generalized (or quotient) singular value decomposition (GSVD or QSVD) of an m–by–n matrix A and a p–by–n matrix B is given by the pair of factorizations

$$A = U\Sigma_1 \begin{bmatrix} 0 & R \end{bmatrix} Q^T \quad \text{and} \quad B = V\Sigma_2 \begin{bmatrix} 0 & R \end{bmatrix} Q^T. \tag{75.15}$$

The matrices in these factorizations have the following properties:

- U is m–by–m, V is p–by–p, Q is n–by–n, and all three matrices are orthogonal. If A and B are complex, these matrices are unitary instead of orthogonal, and Q^T should be replaced by Q^* in the pair of factorizations.
- R is r–by–r, upper triangular and nonsingular. $\begin{bmatrix} 0 & R \end{bmatrix}$ is r–by–n (in other words, the 0 is an r–by–$(n-r)$ zero matrix). The integer r is the rank of $\begin{bmatrix} A \\ B \end{bmatrix}$.
- Σ_1 is m–by–r and Σ_2 is p–by–r. Both are real, nonnegative, and diagonal, satisfying $\Sigma_1^T\Sigma_1 + \Sigma_2^T\Sigma_2 = I$. Write $\Sigma_1^T\Sigma_1 = \text{diag}(\alpha_1^2, \ldots, \alpha_r^2)$ and $\Sigma_2^T\Sigma_2 = \text{diag}(\beta_1^2, \ldots, \beta_r^2)$. The ratios α_j/β_j for $j = 1, 2, \ldots, r$ are called the *generalized singular values.*

Σ_1 and Σ_2 have the following detailed structures, depending on whether $m - r \geq 0$ or $m - r < 0$.

- In the first case, when $m - r \geq 0$,

$$\Sigma_1 = \begin{matrix} & \begin{matrix} k & \ell \end{matrix} \\ \begin{matrix} k \\ \ell \\ m-k-\ell \end{matrix} & \begin{pmatrix} I & 0 \\ 0 & C \\ 0 & 0 \end{pmatrix} \end{matrix} \quad \text{and} \quad \Sigma_2 = \begin{matrix} & \begin{matrix} k & \ell \end{matrix} \\ \begin{matrix} \ell \\ p-\ell \end{matrix} & \begin{pmatrix} 0 & S \\ 0 & 0 \end{pmatrix} \end{matrix} . \qquad (75.16)$$

Here $k + \ell = r$, and ℓ is the rank of B. C and S are diagonal matrices satisfying $C^2 + S^2 = I$, and S is nonsingular. Let c_j and s_j be the diagonal entries of C and S, respectively. Then we have $\alpha_1 = \cdots = \alpha_k = 1$, $\alpha_{k+j} = c_j$ for $j = 1, \ldots, \ell$, $\beta_1 = \cdots = \beta_k = 0$, and $\beta_{k+j} = s_j$ for $j = 1, \ldots, \ell$. Thus, the first k generalized singular values $\alpha_1/\beta_1, \ldots, \alpha_k/\beta_k$ are infinite and the remaining ℓ generalized singular values are finite.
- In the second case, when $m - r < 0$,

$$\Sigma_1 = \begin{matrix} & \begin{matrix} k & m-k & k+\ell-m \end{matrix} \\ \begin{matrix} k \\ m-k \end{matrix} & \begin{pmatrix} I & 0 & 0 \\ 0 & C & 0 \end{pmatrix} \end{matrix} \quad \text{and} \quad \Sigma_2 = \begin{matrix} & \begin{matrix} k & m-k & k+\ell-m \end{matrix} \\ \begin{matrix} m-k \\ k+\ell-m \\ p-\ell \end{matrix} & \begin{pmatrix} 0 & S & 0 \\ 0 & 0 & I \\ 0 & 0 & 0 \end{pmatrix} \end{matrix} . \qquad (75.17)$$

Again, $k + \ell = r$, and ℓ is the rank of B. C and S are diagonal matrices satisfying $C^2 + S^2 = I$, and S is nonsingular. Let c_j and s_j be the diagonal entries of C and S, respectively. Then we have $\alpha_1 = \cdots = \alpha_k = 1$, $\alpha_{k+j} = c_j$ for $j = 1, \ldots, m-k$, $\alpha_{m+1} = \cdots = \alpha_r = 0$, and $\beta_1 = \cdots = \beta_k = 0$, $\beta_{k+j} = s_j$ for $j = 1, \ldots, m-k$, $\beta_{m+1} = \cdots = \beta_r = 1$. Thus, the first k generalized singular values $\alpha_1/\beta_1, \ldots, \alpha_k/\beta_k$ are infinite, and the remaining ℓ generalized singular values are finite.

Backgrounds:

Here are some important special cases of the QSVD. First, when B is square and nonsingular, then $r = n$ and the QSVD of A and B is equivalent to the SVD of AB^{-1}, where the singular values of AB^{-1} are equal to the generalized singular values of A and B:

$$AB^{-1} = (U\Sigma_1 RQ^T)(V\Sigma_2 RQ^T)^{-1} = U(\Sigma_1\Sigma_2^{-1})V^T .$$

Second, if the columns of $\begin{bmatrix} A^T & B^T \end{bmatrix}^T$ are orthonormal, then $r = n$, $R = I$, and the QSVD of A and B is equivalent to the CS (Cosine-Sine) decomposition of $\begin{bmatrix} A^T & B^T \end{bmatrix}^T$:

$$\begin{bmatrix} A \\ B \end{bmatrix} = \begin{bmatrix} U & 0 \\ 0 & V \end{bmatrix} \begin{bmatrix} \Sigma_1 \\ \Sigma_2 \end{bmatrix} Q^T.$$

Third, the generalized eigenvalues and eigenvectors of the pencil $A^T A - \lambda B^T B$ can be expressed in terms of the QSVD of A and B, namely,

$$X^T A^T A X = \begin{bmatrix} 0 & 0 \\ 0 & \Sigma_1^T \Sigma_1 \end{bmatrix} \quad \text{and} \quad X^T B^T B X = \begin{bmatrix} 0 & 0 \\ 0 & \Sigma_2^T \Sigma_2 \end{bmatrix},$$

where $X = Q \begin{bmatrix} I & 0 \\ 0 & R^{-1} \end{bmatrix}$. Therefore, the columns of X are the eigenvectors of $A^T A - \lambda B^T B$, and the "nontrivial" eigenvalues are the squares of the generalized singular values. "Trivial" eigenvalues are those corresponding to the leading $n - r$ columns of X, which span the common null space of $A^T A$ and $B^T B$.

The "trivial eigenvalues" are not well defined.[1] (See Chapter 15 for more information on generalized singular value problems.)

Driver Routines:

The driver routine xGGSVD computes the GSVD (75.15) of A and B.

Examples:

Let us show how to use the driver routine SGGSVD to compute the QSVD (75.15). SGGSVD first reduces the matrices A and B to a pair of triangular matrices, and then use a Jacobi-like method to compute the QSVD of the triangular pair. SGGSVD has the following calling sequence:

```
CALL SGGSVD( JOBU, JOBV, JOBQ, M, N, P, K, L, A, LDA, B, LDB, ALPHA,
BETA, U, LDU, V, LDV, Q, LDQ, WORK, IWORK, INFO )
```

Input to SGGSVD:

- `JOBU, JOBV, JOBQ:` ,= `'U'`, orthogonal matrices U, V and Q are computed;
  = `'N'`, these orthogonal matrices are not computed.
- `M, N, P:` The number of rows or columns of the matrices A and B as defined in (15)
- `A, LDA:` The M–by–N matrix A and the leading dimension of the array A. LDA $\geq$ max(1, M).
- `B, LDB:` The P–by–N matrix B and the leading dimension of the array B. LDB $\geq$ max(1, P).
- `LDU, LDV, LDQ:` The leading dimension of the arrays U, V, and Q if the orthogonal matrices U, V, and Q are computed, LDU $\geq$ max(1, M), LDV $\geq$ max(1, P), LDQ $\geq$ max(1, N).
- `WORK:` The real workspace array, dimension max(3N, M, P) + N.
- `IWORK:` The integer workspace array, dimension N.

Output from SGGSVD:

- `K, L:` The dimension of the subblocks described in the definition of GSVD. K + L is the effective numerical rank of the matrix $\begin{bmatrix} A^T & B^T \end{bmatrix}^T$.
- `A:` The entire triangular matrix R is stored in `A(1:K+L,N-K-L+1:N)` if $m - r \geq 0$. Otherwise, the subblock $R(1:m, 1:k+\ell)$ of R are stored in `A(1:M,N-K-L+1:N)`.
- `B:` The subblock $R(m+1:k+\ell, m+1:k+\ell)$ of R are stored in `B(M-K+1:L,N+M-K-L+1:N)` if $m - r < 0$.
- `ALPHA, BETA:` The generalized singular value pairs;
  `ALPHA(1:K) = 1` and `BETA(1:K) = 0`.
  - If `M-K-L` $\geq$ 0, then ALPHA(K + 1 : K + L) = C and BETA(K + 1 : K + L) = S.
  - If `M-K-L` $<$ 0, then

    ALPHA(K + 1 : M) = C and ALPHA(M + 1 : K + L) = 0,

    BETA(K + 1 : M) = S and BETA(M + 1 : K + L) = 1;

  And `ALPHA(K+L+1:N) = 0`, `BETA(K+L+1:N) = 0`.
- `U, V, Q:` Contains computed orthogonal matrices U, V, and Q if requested.

[1] If we tried to compute the trivial eigenvalues in the same way as the nontrivial ones, that is by taking ratios of the leading $n - r$ diagonal entries of $X^T A^T AX$ and $X^T B^T BX$, we would get 0/0.

`INFO`: `INFO` = 0 if successful exit. If `INFO` $= -j$, then the jth argument had an illegal value. If `INFO` = 1, the Jacobi-type procedure failed to converge.

Let us demonstrate the use of SGGSVD in computing the QSVD of the following 6–by–5 matrices A and B:

$$A = \begin{bmatrix} 1 & 2 & 3 & 1 & 5 \\ 0 & 3 & 2 & 0 & 2 \\ 1 & 0 & 2 & 1 & 0 \\ 0 & 2 & 3 & 0 & -1 \\ 1 & 0 & 2 & 1 & 1 \\ 0 & 2 & 1 & 0 & 1 \end{bmatrix} \quad \text{and} \quad B = \begin{bmatrix} 1 & -2 & 2 & 1 & 1 \\ 0 & 3 & 0 & 0 & 0 \\ 1 & -2 & 2 & 1 & 1 \\ 0 & 2 & 0 & 0 & 0 \\ 2 & -4 & 4 & 2 & 2 \\ 1 & 3 & 2 & 1 & 1 \end{bmatrix}.$$

Upon calling SGGSVD with $\mathtt{M} = 6, \mathtt{P} = 6, \mathtt{N} = 5, \mathtt{LWORK} = 20$, we have $\mathtt{K} = 2$ and $\mathtt{L} = 2$. The QSVD (75.15) of A and B falls in the first case (75.16) since $\mathtt{M} - \mathtt{K} - \mathtt{L} = 6 - 2 - 2 = 2 > 0$. The arrays `ALPHA` and `BETA` are

$$\mathbf{ALPHA} = \begin{bmatrix} 1.0000000 & 1.0000000 & 0.1537885 & 0.5788464 & 0.0000000 \end{bmatrix},$$

$$\mathbf{BETA} = \begin{bmatrix} 0.0000000 & 0.0000000 & 0.9881038 & 0.8154366 & 0.0000000 \end{bmatrix}.$$

Hence, Σ_1 and Σ_2 have the structure as described in (75.16), namely,

$$\Sigma_1 = \begin{bmatrix} 1 & 0 & 0 & 0 \\ 0 & 1 & 0 & 0 \\ 0 & 0 & 0.1537885 & 0 \\ 0 & 0 & 0 & 0.5788464 \\ 0 & 0 & 0 & 0 \\ 0 & 0 & 0 & 0 \end{bmatrix} \quad \text{and} \quad \Sigma_2 = \begin{bmatrix} 0 & 0 & 0.9881038 & 0 \\ 0 & 0 & 0 & 0.8154366 \\ 0 & 0 & 0 & 0 \\ 0 & 0 & 0 & 0 \\ 0 & 0 & 0 & 0 \\ 0 & 0 & 0 & 0 \end{bmatrix}.$$

The first two generalized singular values are infinite, $\alpha_1/\beta_1 = \alpha_2/\beta_2 = \infty$, and the remaining two generalized singular values are finite, $\alpha_3/\beta_3 = 0.15564$ and $\alpha_4/\beta_4 = 0.70986$.

Furthermore, the array `A(1:4,2:5)` contains the 4–by–4 upper triangular matrix R as defined in (75.15):

$$R = \begin{bmatrix} 3.6016991 & -1.7135643 & -0.2843603 & 1.8104467 \\ 0 & -2.6087811 & -4.2943931 & 5.1107349 \\ 0 & 0 & 6.9692163 & 3.5063875 \\ 0 & 0 & 0 & 7.3144341 \end{bmatrix}.$$

The orthogonal matrices U, V, and Q are returned in the arrays `U`, `V`, and `Q`, respectively:

$$U = \begin{bmatrix} -0.6770154 & -0.4872811 & -0.4034495 & -0.2450049 & -0.2151961 & 0.1873468 \\ -0.0947438 & -0.5723576 & 0.4163284 & 0.1218751 & 0.0785425 & -0.6848933 \\ 0.2098812 & 0.0670342 & 0.2612190 & -0.7393155 & -0.5670457 & -0.1228532 \\ 0.6974092 & -0.5903998 & -0.3678919 & 0.0010751 & -0.0196356 & 0.1712235 \\ 0.0000000 & 0.0000001 & -0.0735656 & -0.6152450 & 0.7822418 & -0.0644937 \\ -0.0473719 & -0.2861788 & 0.6744684 & -0.0019711 & 0.1170180 & 0.6687696 \end{bmatrix},$$

$$V = \begin{bmatrix} -0.3017521 & -0.2581125 & 0.9018297 & -0.0002676 & -0.1695592 & -0.0166328 \\ 0.4354534 & -0.2679386 & 0.1028928 & 0.0704557 & 0.2595005 & -0.8097517 \\ -0.3017520 & -0.2581124 & -0.1784097 & -0.8828155 & -0.0002829 & -0.1764375 \\ 0.2903022 & -0.1786257 & -0.1298870 & -0.0008522 & -0.9259184 & -0.0980879 \\ -0.6035041 & -0.5162248 & -0.3568195 & 0.4625078 & -0.0224125 & -0.1660080 \\ 0.4240036 & -0.7046767 & -0.0097810 & -0.0419325 & 0.2146671 & 0.5250862 \end{bmatrix},$$

$$Q = \begin{bmatrix} -0.7071068 & -0.2073452 & -0.5604916 & -0.0112638 & -0.3777966 \\ 0.0000000 & 0.0000000 & 0.0000000 & 0.9995558 & -0.0298012 \\ 0.0000001 & 0.5853096 & 0.2932303 & -0.0225276 & -0.7555932 \\ 0.7071067 & -0.2073452 & -0.5604916 & -0.0112638 & -0.3777965 \\ -0.0000001 & -0.7559289 & 0.5345224 & -0.0112638 & -0.3777965 \end{bmatrix}.$$

References

[CLA] http://www.netlib.org/clapack/.

[JLA] http://www.netlib.org/java/f2j/.

[LUG99] E. Anderson, Z. Bai, C. Bischof, S. Blackford, J. Demmel, J. Dongarra, J. Du Croz, A. Greenbaum, S. Hammarling, A. McKenney, and D. Sorensen, *LAPACK Users' Guide*, 3rd ed., SIAM, Philadelphia, 1999.

[LAP] http://www.netlib.org/lapack/.

[LAP95] http://www.netlib.org/lapack95/.

[LA+] http://www.netlib.org/lapack++/.

[Sca] http://www.netlib.org/scalapack/.

76

Use of ARPACK and EIGS

D. C. Sorensen
Rice University

ARPACK is a library of Fortran77 subroutines designed to compute a selected subset of the eigenvalues of a large matrix. It is based upon a limited storage scheme called the implicitly restarted Arnoldi method (IRAM) [Sor92]. This software can solve largescale non-Hermitian or Hermitian (standard and generalized) eigenvalue problems. The IRA method is described in Chapter 44, Implicitly Restarted Arnoldi Method.

This chapter describes the design and performance features of the eigenvalue software ARPACK and gives a brief discussion of usage. More detailed descriptions are available in the papers [Sor92] and [Sor02] and in the ARPACK Users' Guide [LSY98].

The design goals were robustness, efficiency, and portability. Two very important principles that have helped to achieve these goals are modularity and independence from specific vendor supplied communication and performance libraries.

In this chapter, multiplication of a vector $\mathbf{x}$ by a scalar λ is denoted by $\mathbf{x}\lambda$ so that the eigenvector–eigenvalue relation is $A\mathbf{x} = \mathbf{x}\lambda$. This convention provides for direct generalizations to the more general invariant subspace relations $AX = XH$, where X is an $n \times k$ matrix and H is a $k \times k$ matrix with $k < n$.

76.1 The ARPACK Software

The ARPACK software has been used on a wide range of applications. P_ARPACK is a parallel extension to the ARPACK library and is targeted for distributed memory message passing systems. Both packages are freely available and can be downloaded at `http://www.caam.rice.edu/software/ARPACK/`.

Features (of ARPACK and P_ARPACK):

1. A reverse communication interface.
2. Computes k eigenvalues that satisfy a user-specified criterion, such as largest real part, largest absolute value, etc.
3. A fixed predetermined storage requirement of $n \cdot \mathcal{O}(k) + \mathcal{O}(k^2)$ words.
4. Driver routines are included as templates for implementing various spectral transformations to enhance convergence and to solve the generalized eigenvalue problem, or the SVD problem.
5. Special consideration is given to the generalized problem $A\mathbf{x} = M\mathbf{x}\lambda$ for singular or ill-conditioned symmetric positive semidefinite M.
6. A Schur basis of dimension k that is numerically orthogonal to working precision is always computed. These are also eigenvectors in the Hermitian case. In the non-Hermitian case eigenvectors are available on request. Eigenvalues are computed to a user specified accuracy.

76.2 Reverse Communication

Reverse communication is a means to overcome certain restrictions in the Fortran language; with reverse communication, control is returned to the calling program when interaction with the matrix is required. This is a convenient interface for experienced users. However, it may be more challenging for inexperienced users. It has proven to be extremely useful for interfacing with large application codes.

This interface avoids having to express a matrix-vector product through a subroutine with a fixed calling sequence or to provide a sparse matrix with a specific data structure. The user is free to choose any convenient data structure for the matrix representation.

Examples:

1. A typical use of this interface is illustrated as follows:

```
10   continue
     call snaupd (ido, bmat, n, which,..., workd,..., info)
     if (ido .eq. newprod) then
        call matvec ('A', n, workd(ipntr(1)), workd(ipntr(2)))
     else
        return
     endif
     go to 10
%
```

This shows a code segment of the routine the user must write to set up the reverse communication call to the top level ARPACK routine `snaupd` to solve a nonsymmetric eigenvalue problem in single precision. With reverse communication, control is returned to the calling program when interaction with the matrix A is required. The action requested of the calling program is specified by the reverse communication parameter `ido`, which is set in the call to `snaupd`. In this case, there are two possible requests indicated by `ido`. One action is to multiply the vector held in the array `workd` beginning at location `ipntr(1)` by A and then place the result in the array `workd` beginning at location `ipntr(2)`. The other action is to halt the iteration due to successful convergence or due to an error.

When the parameter `ido` indicates a new matrix vector product is required, a call is made to a subroutine `matvec` in this example. However, it is only necessary to supply the action of the matrix on the specified vector and put the result in the designated location. No specified data structure is imposed on A and if a subroutine is used, no particular calling sequence is specified. Because of this, reverse communication is very flexible and even provides a convenient way to use ARPACK interfaced with code written in another language, such as C or C++.

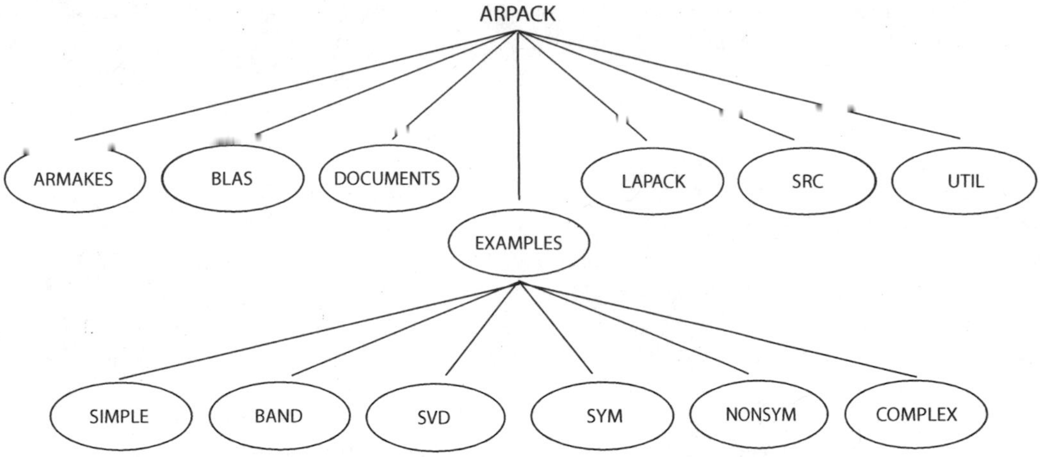

FIGURE 76.1 The ARPACK directory structure.

76.3 Directory Structure and Contents

Once the ARPACK software has been downloaded and unbundled, a directory structure will have been created as pictured in Figure 76.1.

Subdirectories:

1. The `ARMAKES` subdirectory contains sample files with machine specific information needed during the building of the ARPACK library.
2. The `BLAS` and `LAPACK` subdirectories contain the necessary codes from those libraries.
3. The `DOCUMENTS` subdirectory contains files that have example templates showing how to invoke the different computational modes offered by ARPACK.
4. Example driver programs illustrating all of the computational modes, data types, and precisions may be found in the `EXAMPLES` directory.
5. Programs for banded, complex, nonsymmetric, symmetric eigenvalue problems, and singular value decomposition may be found in the directories `BAND, COMPLEX, NONSYM, SYM, SVD`.
6. The `README` files in each subdirectory provide further information.
7. The `SRC` subdirectory contains all the ARPACK source codes.
8. The `UTIL` subdirectory contains the various utility routines needed for printing results and timing the execution of the ARPACK subroutines.

76.4 Naming Conventions, Precisions, and Types

1. ARPACK has two interface routines that must be invoked by the user. They are `__aupd` that implements the IRAM and `__eupd` to post process the results of `__aupd`.
2. The user may request an orthogonal basis for a selected invariant subspace or eigenvectors corresponding to selected eigenvalues with `__eupd`. If a spectral transformation is used, `__eupd` automatically transforms the computed eigenvalues of the shift-invert operator to those of the original problem.
3. Both `__aupd` and `__eupd` are available for several combinations of problem type (symmetric and nonsymmetric), data type (real, complex), and precision (single, double). The first letter `(s,d,c,z)` denotes precision and data type. The second letter denotes whether the problem is symmetric (`s`) or nonsymmetric (`n`).

4. Thus, dnaupd is the routine to use if the problem is a double precision nonsymmetric (standard or generalized) problem and dneupd is the post-processing routine to use in conjunction with dnaupd to recover eigenvalues and eigenvectors of the original problem upon convergence. For complex matrices, one should use _naupd and _neupd with the first letter either c or z regardless of whether the problem is Hermitian or non-Hermitian.

76.5 Getting Started

Perhaps the easiest way to rapidly become acquainted with the possible uses of ARPACK is to run the example driver routines that have been supplied for each of the computational modes. These may be used as templates and adapted to solve specific problems. To get started, it is recommended that the user execute driver routines from the SIMPLE subdirectory.

The dssimp driver implements the reverse communication interface to the routine dsaupd that will compute a few eigenvalues and eigenvectors of a symmetric matrix. It illustrates the simplest case and has exactly the same structure as shown previously except that the top level routine is dsaupd instead of snaupd. The full call issued by dssimp is as follows.

```
call dsaupd ( ido, bmat, n, which, nev, tol, resid,
&                 ncv, v, ldv, iparam, ipntr, workd,
&                 workl, lworkl, info )
```

This dssimp driver is intended to serve as template to enable a user to create a program to use dsaupd on a specific problem in the simplest computational mode. All of the driver programs in the various EXAMPLES subdirectories are intended to be used as templates. They all follow the same principle, but the usage is slightly more complicated.

The only thing that must be supplied in order to use this routine on your problem is to change the array dimensions and to supply a means to compute the matrix-vector product

$$\mathbf{w} \leftarrow A\mathbf{v}$$

on request from dsaupd. The selection of which eigenvalues to compute may be altered by changing the parameter which.

Once usage of dsaupd in the simplest mode is understood, it will be easier to explore the other available options such as solving generalized eigenvalue problems using a shift-invert computational mode.

If the computation is successful, dsaupd indicates that convergence has taken place through the parameter ido. Then various steps may be taken to recover the results in a useful form. This is done through the subroutine dseupd as illustrated below.

```
call dseupd(rvec, howmny, select, d, v, ldv, sigma, bmat,
&               n, which, nev, tol, resid, ncv, v, ldv,
&               iparam, ipntr, workd, workl, lworkl, ierr)
```

Eigenvalues are returned in the first column of the two-dimensional array d and the corresponding eigenvectors are returned in the first NCONV (=IPARAM(5)) columns of the two-dimensional array v if requested. Otherwise, an orthogonal basis for the invariant subspace corresponding to the eigenvalues in d is returned in v.

The input parameters that must be specified are

- The logical variable rvec = .true. if eigenvectors are requested,
  .false. if only eigenvalues are desired.
- The character*1 parameter howmny that specifies how many eigenvectors are desired.
  howmny = 'A': compute nev eigenvectors;
  howmny = 'S': compute some of the eigenvectors,
  specified by the logical array select.

- `sigma` should contain the value of the shift used if `iparam(7) = 3,4,5`. It is not referenced if `iparam(7) = 1 or 2`.

When requested, the eigenvectors returned by `dseupd` are normalized to have unit length with respect to the M semi-inner product that was used. Thus, if $M = I$, they will have unit length in the standard 2-norm. In general, a computed eigenvector $\mathbf{x}$ will satisfy $1 = \mathbf{x}^T M\mathbf{x}$.

76.6 Setting Up the Problem

To set up the problem, the user needs to specify the number of eigenvalues to compute which eigenvalues are of interest, the number of basis vectors to use, and whether or not the problem is standard or generalized. These items are controlled by the following parameters.

Parameters for the top-level ARPACK routines:

`ido` — Reverse communication flag.
`nev` — The number of requested eigenvalues to compute.
`ncv` — The number of Arnoldi basis vectors to use through the course of the computation.
`bmat` — Indicates whether the problem is standard `bmat = 'I'` or generalized (`bmat = 'G'`).
`which` — Specifies which eigenvalues of A are to be computed.
`tol` — Specifies the relative accuracy to which eigenvalues are to be computed.
`iparam` — Specifies the computational mode, number of IRAM iterations, the implicit shift strategy, and outputs various informational parameters upon completion of IRAM.

The value of `ncv` must be at least **nev** + 1. The options available for `which` include `'LA'` and `'SA'` for the algebraically largest and smallest eigenvalues, `'LM'` and `'SM'` for the eigenvalues of largest or smallest magnitude, and `'BE'` for the simultaneous computation of the eigenvalues at both ends of the spectrum. For a given problem, some of these options may converge more rapidly than others due to the approximation properties of the IRAM as well as the distribution of the eigenvalues of A. Convergence behavior can be quite different for various settings of the `which` parameter. For example, if the matrix is indefinite then setting `which` = `'SM'` will require interior eigenvalues to be computed and the Arnoldi/Lanczos process may require many steps before these are resolved.

For a given `ncv`, the computational work required is proportional to $\mathbf{n} \cdot \mathbf{ncv}^2$ FLOPS. Setting `nev` and `ncv` for optimal performance is very much problem dependent. If possible, it is best to avoid setting `nev` in a way that will split clusters of eigenvalues. For example, if the the five smallest eigenvalues are positive and on the order of 10^{-4} and the sixth smallest eigenvalue is on the order of 10^{-1}, then it is probably better to ask for $\mathbf{nev} = 5$ than for $\mathbf{nev} = 3$, even if the three smallest are the only ones of interest.

Setting the optimal value of `ncv` relative to `nev` is not completely understood. As with the choice of `which`, it depends upon the underlying approximation properties of the IRAM as well as the distribution of the eigenvalues of A. As a rule of thumb, $\mathbf{ncv} \geq 2 \cdot \mathbf{nev}$ is reasonable. There are tradeoffs due to the cost of the user supplied matrix-vector products and the cost of the implicit restart mechanism and the cost of maintaining the orthogonality of the Arnoldi vectors. If the user supplied matrix-vector product is relatively cheap, then a smaller value of `ncv` may lead to more user matrix-vector products, but an overall decrease in computation time.

Storage Declarations:

The program is set up so that the setting of the three parameters `maxn, maxnev, maxncv` will automatically declare all of the work space needed to run `dsaupd` on a given problem.

The declarations allow a problem size of `N` $\leq$ `maxn`, computation of `nev` $\leq$ `maxnev` eigenvalues, and using at most `ncv` $\leq$ `maxncv` Arnoldi basis vectors during the IRAM. The user may override the

```
Double precision
&                    v(ldv,maxncv), workl(maxncv*(maxncv+8)),
&                    workd(3*maxn), d(maxncv,2), resid(maxn),
```

FIGURE 76.2 Storage declarations needed for ARPACK subroutine `dsaupd`.

default settings used for the example problem by modifying `maxn`, `maxnev`, and `maxncv` in the following parameter statement in the `dssimp` code.

```
integer      maxn, maxnev, maxncv, ldv
parameter   (maxn=256, maxnev=10, maxncv=25, ldv=maxn)
```

These parameters are used in the code segment listed in Figure 76.2 for declaring all of the output and work arrays needed by the ARPACK subroutines `dsaupd` and `dseupd`. These will set the storage values in ARPACK arrays.

Stopping Criterion:

The stopping criterion is determined by the user through specification of the parameter `tol`. The default value for `tol` is machine precision ϵ_M. There are several things to consider when setting this parameter. In absence of all other considerations, one should expect a computed eigenvalue λ_c to roughly satisfy

$$|\lambda_c - \lambda_t| \leq \texttt{tol}\|A\|_2,$$

where λ_t is the eigenvalue of A nearest to λ_c. Typically, decreasing the value of `tol` will increase the work required to satisfy the stopping criterion. However, setting `tol` too large may cause eigenvalues to be missed when they are multiple or very tightly clustered. Typically, a fairly small setting of `tol` and a reasonably large setting of `ncv` is required to avoid missing multiple eigenvalues. However, some care must be taken. It is possible to set `tol` so small that convergence never occurs. There may be additional complications when the matrix A is nonnormal or when the eigenvalues of interest are clustered near the origin.

Initial Parameter Settings:

The reverse communication flag is denoted by `ido`. This parameter must be initially set to `0` to signify the first call to `dsaupd`. Various algorithmic modes may be selected through the settings of the entries in the integer array `iparam`. The most important of these is the value of `iparam(7)`, which specifies the computational mode to use.

Setting the Starting Vector:

The parameter `info` should be set to `0` on the initial call to `dsaupd` unless the user wants to supply the starting vector that initializes the IRAM. Normally, this default is a reasonable choice. However, if this eigenvalue calculation is one of a sequence of closely related problems, then convergence may be accelerated if a suitable starting vector is specified. Typical choices in this situation might be to use the final value of the starting vector from the previous eigenvalue calculation (that vector will already be in the first column of `V`) or to construct a starting vector by taking a linear combination of the computed eigenvectors from the previously converged eigenvalue calculation. If the starting vector is to be supplied, then it should be placed in the array `resid` and `info` should be set to `1` on entry to `dsaupd`. On completion, the parameter `info` may contain the value `0` indicating the iteration was successful or it may contain a nonzero value indicating an error or a warning condition. The meaning of a nonzero value returned in `info` may be found in the header comments of the subroutine `dsaupd`.

Trace Debugging Capability:

ARPACK provides a means to trace the progress of the computation as it proceeds. Various levels of output may be specified from no output (`level = 0`) to voluminous (`level = 3`). A detailed description of trace debugging may be found in [LSY98].

76.7 General Use of ARPACK

The Shift and Invert Spectral Transformation Mode:

The most general problem that may be solved with ARPACK is to compute a few selected eigenvalues and corresponding eigenvectors for

$$A\mathbf{x} = M\mathbf{x}\lambda,$$

where A and M are real or complex $n \times n$ matrices.

The shift and invert spectral transformation is used to enhance convergence to a desired portion of the spectrum. If $(\mathbf{x}, \lambda)$ is an eigen-pair for $(\mathbf{A}, \mathbf{M})$ and $\sigma \neq \lambda$, then

$$(A - \sigma M)^{-1} M\mathbf{x} = \mathbf{x}\nu \quad \text{where} \quad \nu = \frac{1}{\lambda - \sigma},$$

where we are requiring that $A - \sigma M$ is nonsingular. Here it is possible for A or M to be singular, but they cannot have a nonzero null vector in common. This transformation is effective for finding eigenvalues near σ since the `nev` eigenvalues ν_j of $C \equiv (A - \sigma M)^{-1} M$ that are largest in magnitude correspond to the `nev` eigenvalues λ_j of the original problem that are nearest to the shift σ in absolute value. As discussed in Chapter 44, these transformed eigenvalues of largest magnitude are precisely the eigenvalues that are easy to compute with a Krylov method. Once they are found, they may be transformed back to eigenvalues of the original problem.

***M* Is Hermitian Positive Definite:**

If M is Hermitian positive definite and well conditioned ($\|M\| \cdot \|M^{-1}\|$ is of modest size), then computing the Cholesky factorization $M = LL^*$ and converting $A\mathbf{x} = M\mathbf{x}\lambda$ into

$$(L^{-1} A L^{-*})\mathbf{y} = \mathbf{y}\lambda, \quad \text{where} \quad L^*\mathbf{x} = \mathbf{y}$$

provides a transformation to a standard eigenvalue problem. In this case, a request for a matrix vector product would be satisfied with the following three steps:

1. Solve $L^*\mathbf{z} = \mathbf{v}$ for $\mathbf{z}$.
2. Matrix-vector multiply $\mathbf{z} \leftarrow A\mathbf{z}$.
3. Solve $L\mathbf{w} = \mathbf{z}$ for $\mathbf{w}$.

Upon convergence, a computed eigenvector $\mathbf{y}$ for $(L^{-1} A L^{-*})$ is converted to an eigenvector $\mathbf{x}$ of the original problem by solving the the triangular system $L^*\mathbf{x} = \mathbf{y}$. This transformation is most appropriate when A is Hermitian, M is Hermitian positive definite, and extremal eigenvalues are sought. This is because $L^{-1} A L^{-*}$ will be Hermitian when A is the same.

If A is Hermitian positive definite and the smallest eigenvalues are sought, then it would be best to reverse the roles of A and M in the above description and ask for the largest algebraic eigenvalues or those of largest magnitude. Upon convergence, a computed eigenvalue $\hat{\lambda}$ would then be converted to an eigenvalue of the original problem by the relation $\lambda \leftarrow 1/\hat{\lambda}$.

***M* Is NOT Hermitian Positive Semidefinite:**

If neither A nor M is Hermitian positive semidefinite, then a direct transformation to standard form is required. One simple way to obtain a direct transformation to a standard eigenvalue problem $C\mathbf{x} = \mathbf{x}\lambda$

is to multiply on the left by M^{-1}, which results in $C = M^{-1}A$. Of course, one should not perform this transformation explicitly since it will most likely convert a sparse problem into a dense one. If possible, one should obtain a direct factorization of M and when a matrix-vector product involving C is called for, it may be accomplished with the following two steps:

1. Matrix-vector multiply $\mathbf{z} \leftarrow A\mathbf{v}$.
2. Solve: $M\mathbf{w} = \mathbf{z}$.

Several problem dependent issues may modify this strategy. If M is singular or if one is interested in eigenvalues near a point σ, then a user may choose to work with $C \equiv (A - \sigma M)^{-1}M$ but without using the M-inner products discussed previously. In this case the user will have to transform the converged eigenvalues ν_j of C to eigenvalues λ_j of the original problem.

76.8 Using the Computational Modes

An extensive set of computational modes has been constructed to implement all of the shift-invert options mentioned previously. The problem set up is similar for all of the available computational modes. A detailed description of the reverse communication loop for the various modes (Shift-Invert for a Real Nonsymmetric Generalized Problem) is available in the users' guide [LSY98]. To use any of the modes listed below, the user is strongly urged to modify one of the driver routine **templates** that reside in the `EXAMPLES` directory.

When to Use a Spectral Transformation:

The first thing to decide is whether the problem will require a spectral transformation. If the problem is generalized ($M \neq I$), then a spectral transformation will be required. Such a transformation will most likely be needed for a standard problem if the desired eigenvalues are in the interior of the spectrum or if they are clustered at the desired part of the spectrum. Once this decision has been made and `OP` has been specified, an efficient means to implement the action of the operator `OP` on a vector must be devised. The expense of applying `OP` to a vector will of course have direct impact on performance.

Shift-invert spectral transformations may be implemented with or without the use of a weighted M-inner product. The relation between the eigenvalues of `OP` and the eigenvalues of the original problem must also be understood in order to make the appropriate specification of `which` in order to recover eigenvalues of interest for the original problem. The user must specify the number of eigenvalues to compute, which eigenvalues are of interest, the number of basis vectors to use, and whether or not the problem is standard or generalized.

Computational Modes for Real Nonsymmetric Problems:

The following subroutines are used to solve nonsymmetric generalized eigenvalue problems in real arithmetic. These routines are appropriate when A is a general nonsymmetric matrix and M is symmetric and positive semidefinite. The reverse communication interface routine for the nonsymmetric double precision eigenvalue problem is `dnaupd`. The routine is called as shown below. The specification of which `nev` eigenvalues is controlled by the `character*2` argument `which`. The most commonly used options are listed below. There are templates available as indicated for each of these.

```
  call dnaupd (ido, bmat, n, which, nev, tol, resid, ncv, v,
 &             ldv, iparam, ipntr, workd, workl, lworkl, info)
```

There are three different shift-invert modes for nonsymmetric eigenvalue problems. These modes are specified by setting the parameter entry `iparam(7) = mode` where `mode = 1,2,3, or 4`.

In the following list, the specification of `OP` and B are given for the various modes. Also, the `iparam(7)` and `bmat` settings are listed along with the name of the sample driver for the given mode. Sample drivers for the following modes may be found in the `EXAMPLES/NONSYM` subdirectory.

1. Regular mode (`iparam(7) = 1, bmat = 'I'`).
   Use driver `dndrv1`.
   (a) Solve $A\mathbf{x} = \mathbf{x}\lambda$ in regular mode.
   (b) OP $= A$ and $B = I$.
2. Shift-invert mode (`iparam(7) = 3, bmat = 'I'`).
   Use driver `dndrv2` with `sigma` a real shift.
   (a) Solve $A\mathbf{x} = \mathbf{x}\lambda$ in shift-invert mode.
   (b) OP $= (A - \sigma I)^{-1}$ and $B = I$.
3. Regular inverse mode (`iparam(7) = 2, bmat = 'G'`).
   Use driver `dndrv3`.
   (a) Solve $A\mathbf{x} = M\mathbf{x}\lambda$ in regular inverse mode.
   (b) OP $= M^{-1}A$ and $B = M$.
4. Shift-invert mode (`iparam(7) = 3, bmat = 'G'`).
   Use driver `dndrv4` with `sigma` a real shift.
   (a) Solve $A\mathbf{x} = M\mathbf{x}\lambda$ in shift-invert mode.
   (b) OP $= (A - \sigma M)^{-1}M$ and $B = M$.

76.9 MATLAB's® EIGS

MATLAB has adopted ARPACK for the computation of a selected subset of the eigenvalues of a large (sparse) matrix. The MATLAB function for this is called `eigs` and it is a MATLAB driver to a mex-file compilation of ARPACK. In fact, a user can directly reference this mex-file as `arpackc`. Exactly the same sort of interfaces discussed above can be written in MATLAB to drive `arpackc`.

However, it is far more convenient just to use the provided `eigs` function. The command `D = eigs(A)` returns a vector of the 6 eigenvalues of `A` of largest magnitude, while `[V,D] = eigs(A)` returns eigenvectors in `V` and the corresponding (6 largest magnitude) eigenvalues on the diagonal of the matrix `D`.

The command `eigs(A,M)` solves the generalized eigenvalue problem `A*V = M*V*D`. Here M must be symmetric (or Hermitian) positive definite and the same size as A. If `M` is nonsymmetric, this can also be handled directly, but the user will have to write and pass a function that will compute the action `w <- inv(A - sigma*M)* M*v` as needed. Of course, a sparse direct factorization of `(A - sigma*M)` should be computed at the outset and reused to accomplish this action.

Other capabilities are easily utilized through `eigs`. The commands `eigs(A,k)` and `eigs(A,M,k)` return the `k` largest magnitude eigenvalues.

The commands `eigs(A,k,sigma)` and `eigs(A,M,k,sigma)` return `k` eigenvalues based on `sigma`, where `sigma` is either a scalar or one of the following strings: `'LM'` or `'SM'` - Largest or Smallest Magnitude

```
For real symmetric problems, SIGMA may also be:
   'LA' or 'SA' - Largest or Smallest Algebraic
   'BE' - Both Ends, one more from high end if K
           is odd
For nonsymmetric and complex problems, SIGMA
may also be:
   'LR' or 'SR' - Largest or Smallest Real part
   'LI' or 'SI' - Largest or Smallest Imaginary
    part
```

If `sigma` is a scalar including 0, `eigs` finds the eigenvalues in the complex plane closest to `sigma`. The required sparse matrix factorizations are done automatically when a scalar is passed.

More control and flexibility can be obtained by using the commands `eigs(A,M,sigma,opts)` and `eigs(A,M,k,sigma,opts)` where `opts` is a `struct`. The fields of the `struct opts` will specify the following options:

```
opts.issym: symmetry of A or A-SIGMA*B
     represented by AFUN [{0} | 1]

opts.isreal: complexity of A or A-SIGMA*B
     represented by AFUN [0 | {1}]

opts.tol: convergence: Ritz estimate
     residual <= tol*NORM(A) [scalar | {eps}]

opts.maxit: maximum number of
     iterations [integer | {300}]

opts.p: number of Lanczos vectors:
     K+1<p<=N [integer | {2K}]

opts.v0: starting vector
     [N-by-1 vector | {randomly generated by ARPACK}]

opts.disp: diagnostic information display
     level [0 | {1} | 2]

opts.cholB: B is actually its Cholesky
     factor CHOL(B) [{0} | 1]

opts.permB: sparse B is actually
     CHOL(B(permB,permB)) [permB | {1:N}]

opts(AFUN,n) accepts the function AFUN
     instead of a matrix
```

In the above list, the items in square brackets denote the possible settings and the item given in curly brackets $\{\cdot\}$ is the default setting. For example, to set the convergence tolerance to 10^{-3}, one should give the command `opts.tol = .001`.

The function passing mechanism can be used to define a customized spectral transformation as in the example with a nonsymmetric `M`. Arguments may be passed to `AFUN(X,P1,P2,...)` by invoking the command `eigs (AFUN,n,k,sigma,opts,P1,P2,...)`.

Thus, all of the functionality of ARPACK is available through `eigs` and, in many cases, it is much easier to use. Moreover, the sparse matrix direct factorizations and reorderings available in the `sparfun` suite are easily used to implement various desirable spectral transformations. It is highly advisable to run sample problems with the characteristics of a desired very large problem in order to get an idea of the best spectral transformation to use and to get an indication of the expected convergence behavior.

References

[LSY98] R. Lehoucq, D.C. Sorensen, and C. Yang, *ARPACK Users Guide: Solution of Large Scale Eigenvalue Problems with Implicitly Restarted Arnoldi Methods*, SIAM Publications, Philadelphia, (1998). (Software available at: http://www.caam.rice.edu/software/ARPACK).

[Sor92] D.C. Sorensen, Implicit application of polynomial filters in a k-step Arnoldi method, *SIAM J. Matrix Anal. Applic.*, 13: 357–385 (1992).

[Sor02] D.C. Sorensen, Numerical methods for large eigenvalue problems, *Acta Numerica*, **11**, 519–584, (2002).

77

Summary of Software for Linear Algebra Freely Available on the Web

Jack Dongarra
University of Tennessee and Oakridge National Laboratory

Victor Eijkhout
University of Tennessee

Julien Langou
University of Tennessee

Table 77.1 to Table 77.5 present a list of freely available software for the solution of linear algebra problems. The interest is in software for high-performance computers that is available in "open source" form on the Web for solving problems in numerical linear algebra, specifically dense, sparse direct and iterative systems, and sparse iterative eigenvalue problems.

Additional pointers to software can be found at:
`www.nhse.org/rib/repositories/nhse/catalog/#Numerical_Programs_and_Routines.`

A survey of Iterative Linear System Solver Packages can be found at:
`www.netlib.org/utk/papers/iterative-survey.`

TABLE 77.1 Support Routines for Numerical Linear Algebra

		Type		Language			Mode			
Package	Support	Real	Complex	f77	c	c++	Seq	Dist	Dense	Sparse
ATLAS	yes	X	X	X	X		X		X	
BLAS	yes	X	X	X	X		X		X	
FLAME	yes	X	X	X	X		X		X	
LINALG	?									
MTL	yes	X				X	X		X	
NETMAT	yes	X				X	X		X	
NIST S-BLAS	yes	X	X	X	X		X			X
PSBLAS	yes	X	X	X	X		X	M		X
SparseLib++	yes	X	X		X	X	X			X
uBLAS	yes	X	X		X	X	X		X	

TABLE 77.2 Available Software for Dense Matrix

		Type		Language			Mode	
Package	Support	Real	Complex	f77	c	c++	Seq	Dist
LAPACK	**yes**	X	X	X	X		X	
LAPACK95	**yes**	X	X	**95**			X	
NAPACK	**yes**	X		X				
PLAPACK	**yes**	X	X	X	X			**M**
PRISM	**yes**	X		X			X	**M**
ScaLAPACK	**yes**	X	X	X	X			**M/P**

TABLE 77.3 Sparse Direct Solvers

		Type		Language			Mode			
Package	Support	Real	Complex	f77	c	c++	Seq	Dist	SPD	Gen
DSCPACK	**yes**	X			X		X	**M**	X	
HSL	**yes**	X	X	X			X		X	X
MFACT	**yes**	X			X		X		X	
MUMPS	**yes**	X	X	X	X		X	**M**	X	X
PSPASES	**yes**	X		X	X			**M**	X	
SPARSE	**?**	X	X		X		X		X	X
SPOOLES	**?**	X	X		X		X	**M**		X
SuperLU	**yes**	X	X	X	X		X	**M**		X
TAUCS	**yes**	X	X		X		X		X	X
UMFPACK	**yes**	X	X		X		X			X
Y12M	**?**	X		X			X		X	X

TABLE 77.4 Sparse Eigenvalue Solvers

		Type		Language			Mode			
Package	Support	Real	Complex	f77	c	c++	Seq	Dist	Sym	Gen
(B/H)LZPACK	**yes**	X	X	X			X	**M/P**	X	X
HYPRE	**yes**	X			X		X	**M**	X	
QMRPACK	**?**	X	X	X			X		X	X
LASO	**?**	X		X			X		X	
P_ARPACK	**yes**	X	X	X			X	**M/P**		X
PLANSO	**yes**	X		X			X	**M**	X	
SLEPc	**yes**	X	X		X		X	**M**	X	X
SPAM	**yes**	X		**90**			X		X	
TRLAN	**yes**	X		X			X	**M**	X	

TABLE 77.5 Sparse Iterative Solvers

		Type		Language			Mode		Precond.		Iterative Solvers	
Package	Support	Real	Comp.	f77	c	c++	Seq	Dist	SPD	Gen	SPD	Gen
AZTEC	no	X			X		X	M	X	X	X	X
BILUM	yes	X		X			X	M		X		X
BlockSolve95	?	X		X	X	X		M	X	X	X	X
BPKIT	yes	X		X	X	X	X	M	X	X		
CERFACS	yes	X	X	X			X	M			X	X
HYPRE	yes	X		X	X		X	M	X	X	X	X
IML++	?	X		X	X	X	X				X	X
ITL	yes	X				X	X				X	X
ITPACK	?	X		X			X				X	X
LASPack	yes	X			X		X				X	X
LSQR	yes	X		X	X		X					X
pARMS	yes	X		X	X		X	M			X	X
PARPRE	yes	X			X			M	X	X		
PETSc	yes	X	X	X	X		X	M	X	X	X	X
P-SparsLIB	yes	X		X				M		X		X
PSBLAS	yes	X	X	f90	X		X	M	X	X	X	X
QMRPACK	?	X	X	X			X				X	X
SLAP	?	X		X						X		
SPAI	yes	X			X		X	M			X	X
SPLIB	?	X		X			X				X	X
SPOOLES	?	X	X		X		X	M	X	X	X	X
SYMMLQ	yes	X	X	X			X				X	X
TAUCS	yes	X	X		X		X		X	X	X	X
Templates	yes	X		X	X		X				X	X
Trilinos	yes	X				X	X	M	X	X	X	X

Glossary

This glossary covers most of the terms defined in Chapters 1 to 49. It does not cover some terminology used in a single chapter (including most of the terminology that is specific to a particular application (Chapters 50 to 70)), nor does it cover most of the terminology used in computer software (Chapters 71 to 77). When two sections are listed, both define the term; the first listed chapter is the first instance, the second is the primary chapter dealing with the topic. Definitions in this glossary may not give all details; the reader is advised to consult the chapter/section following the term and detinition).

A

abelian (group G): A commutative group, i.e., $ab = ba$ for all $a, b \in G$. **Preliminaries**
absolute (matrix norm): As a vector norm, each member of the family is absolute. **37.3**
absolute (vector norm $\|\cdot\|$): For all vectors $\mathbf{x}$, $\|\,|\mathbf{x}|\,\| = \|\mathbf{x}\|$. **37.1**
absolute algebraic connectivity (of simple graph $G = (V, E)$): $\max \alpha(G_w)$ where the maximum is taken over all nonnegative weights w of the edges of G such that $\sum_{e \in E} w(e) = |E|$. **36.5**
absolute error (in approximation $\hat{\mathbf{z}}$ to $\mathbf{z}$): $\|\mathbf{z} - \hat{\mathbf{z}}\|$ **37.4**
absolute value (of complex number $a + bi$): $\sqrt{a^2 + b^2}$. **Preliminaries**
absolute value (of real matrix A): The nonnegative matrix obtained by taking element-wise absolute values of A's entries. **9.1**
access: Vertex u has access to vertex v in a digraph Γ if there is a walk in Γ from u to v; also applied to sets of vertices. **9.1, 29.5**
access equivalence class (of a digraph): An equivalence class under the equivalence relation of access. **9.1, 29.5**
access equivalence class (of a matrix): An access equivalence class of its digraph. **9.1**
access equivalent: Two vertices in a digraph that have access to each other. **9.1, 29.5**
active branch (at a Type I characteristic vertex of a tree): For some Fiedler vector $\mathbf{y}$ the entries in $\mathbf{y}$ corresponding to the vertices in the branch are nonzero. **36.3**
acyclic (square matrix A): The graph of A has no cycles. **19.3**
additive (map $\phi : F^{m\times n} \to F^{m\times n}$): $\phi(A + B) = \phi(A) + \phi(B)$, for all $A, B \in F^{m\times n}$. **22.4**
additive coset (of subspace W): A subset of vectors the form $\mathbf{v} + W = \{\mathbf{v} + \mathbf{w} : \mathbf{w} \in W\}$. **2.3**
additive D-stable (real square matrix A): $A + D$ is stable for every nonnegative diagonal matrix D. **19.4**
additive inverse eigenvalue problem (AIEP): Given $A \in \mathbb{C}^{n\times n}$ and $\lambda_1, \ldots, \lambda_n \in \mathbb{C}$ find a diagonal matrix $D \in \mathbb{C}^{n\times n}$ such that $\sigma(A + D) = \{\lambda_1, \ldots, \lambda_n\}$. **20.9**
additive preserver: An additive map preserving a certain property. **22.4**
adjacency matrix (of a digraph Γ of order n): The $n \times n$ 0, 1-matrix whose i, jth entry is 1 if there is an arc from the ith vertex to the jth vertex and 0 otherwise. **29.2**
adjacency matrix (of a graph G of order n): The symmetric $n \times n$ matrix whose i, jth entry is equal to the number of edges between the ith vertex and the jth vertex. **28.3**
adjacent (matrices A, B): $\operatorname{rank}(A - B) = 1$. **22.4**
adjacent (vertices u and v in a graph): There exists an edge with endpoints u and v. **28.1**
adjoint (of a matrix): See *Hermitian adjoint.*
adjoint (of a linear operator T on an inner product space V): $\langle T(\mathbf{u}), \mathbf{v}\rangle = \langle \mathbf{u}, T^*(\mathbf{v})\rangle$ for all $\mathbf{u}, \mathbf{v} \in V$. **5.3**
adjugate: The transpose of the matrix of cofactors. **4.2**
affine parameterized inverse eigenvalue problem: See *Section* **20.9.**
AIEP: See *additive inverse eigenvalue problem.*

algebra (associative): A vector space A over a field F with a bilinear **multiplication** $(\mathbf{x}, \mathbf{y}) \mapsto \mathbf{xy}$ from $A \times A$ to A satisfying $(\mathbf{xy})\mathbf{z} = \mathbf{x}(\mathbf{yz})$. **Preliminaries, 69.1**
algebra (nonassociative): See *Section* **69.1**.
algebraic connectivity (of simple a graph): The second least Laplacian eigenvalue. **28.4, 36.1**
algebraic multigrid preconditioner: A preconditioner that uses principles similar to those used for PDE problems on grids, when the "grid" for the problem is unknown or nonexistent and only the matrix is available. **41.4**
algebraic multiplicity (of an eigenvalue): The number of times the eigenvalue occurs as a root in the characteristic polynomial of the matrix. **4.3**
algorithm: A precise set of instructions to perform a task. **37.7**
allows: If P is a property referring to a real matrix, then a sign pattern A allows P if some real matrix in $Q(A)$ has property P. **33.1**
allows a properly signed nest ($n \times n$ sign pattern A): There exists $B \in Q(A)$ and a permutation matrix P such that $\text{sgn}(\det(P^T B P[\{1, \dots, k\}])) = (-1)^k$ for $k = 1, \dots, n$. **33.4**
Alt: the map $\text{Alt}(\mathbf{v}_1 \otimes \cdots \otimes \mathbf{v}_k) = \frac{1}{m!} \sum_{\pi \in S_m} \text{sgn}(\pi) \mathbf{v}_{\pi(1)} \otimes \cdots \otimes \mathbf{v}_{\pi(k)}$. **13.6**
alt multiplication: $(\mathbf{v}_1 \wedge \cdots \wedge \mathbf{v}_p) \wedge (\mathbf{v}_{p+1} \wedge \cdots \wedge \mathbf{v}_{p+q}) = \mathbf{v}_1 \wedge \cdots \wedge \mathbf{v}_{p+q}$. **13.7**
alternate path to a single arc (in a digraph): A path of length greater than 1 between vertices i and j such that the arc (i, j) is in the digraph. **35.7**
alternating (bilinear form f): $f(\mathbf{v}, \mathbf{v}) = 0$ for all $\mathbf{v} \in V$. **12.3**
alternating ($n \times n$ matrix A): $a_{ii} = 0$, $i = 1, 2, \dots, n$ and $a_{ji} = -a_{ij}$, $1 \leq i, j \leq n$. **12.3**
alternating (multilinear form): See *antisymmetric.*
alternating algebra: See *Grassman algebra.*
alternating product: See *exterior product.*
alternating space: See *Grassman space.*
alternator: See *Alt.*
angle (between two nonzero vectors $\mathbf{u}$ and $\mathbf{v}$ in a real inner product space): The real number θ, $0 \leq \theta \leq \pi$, such that $\langle \mathbf{u}, \mathbf{v} \rangle = \|\mathbf{u}\| \|\mathbf{v}\| \cos \theta$. **5.1**
annihilator: The set of linear functionals that annihilate every vector in the given set. **3.8**
antidiagonal (of $n \times n$ matrix A): The elements $a_{i,k-i}$, $i = 1, \dots, k-1$ with $2 \leq k \leq n+1$. **48.1**
anti-identity matrix: The $n \times n$ matrix with ones along the main antidiagonal, i.e., $a_{i,n+1-i} = 1$, $i = 1, \dots, n$ and zeros elsewhere. **48.1**
antisymmetric (bilinear form f): $f(\mathbf{u}, \mathbf{v}) = -f(\mathbf{v}, \mathbf{u})$ for all $\mathbf{u}, \mathbf{v} \in V$. **12.3**
antisymmetric (multilinear form): A multilinear form that is an antisymmetric map. **13.4**
antisymmetric (map $\psi \in L^m(V; U)$): $\psi(\mathbf{v}_{\pi(1)}, \dots, \mathbf{v}_{\pi(m)}) = \text{sgn}(\pi) \psi(\mathbf{v}_1, \dots, \mathbf{v}_m)$ for all permutations π. **13.4**
aperiodic (matrix): An irreducible nonnegative matrix of period 1. **9.2**
appending G_2 at vertex v of G_1: Constructing a simple graph from $G_1 \cup G_2$ by adding an edge between v and a vertex of G_2. **36.2**
Arbitrary Precision Approximating (APA) algorithms: See *Section* **47.4.**
arc: An ordered pair of vertices (part of a directed graph). **29.1**
argument (of a complex number): θ in the form $re^{i\theta}$ with $0 \leq \theta < 2\pi$. **Preliminaries**
Arnoldi factorization (k-step): $AV_k = V_k H_k + \mathbf{f}_k \mathbf{e}_k^*$ where $V_k \in \mathbb{C}^{n \times k}$ has orthonormal columns, $V_k^* \mathbf{f}_k = 0$ and $H_k = V_k^* A V_k$ is a $k \times k$ upper Hessenberg matrix with nonnegative subdiagonal elements. **41.3, 44.2**
associated undirected graph (of a digraph $\Gamma = (V, E)$): The undirected graph with vertex set V, having an edge between vertices u and v if and only if at least one of the arcs (u, v) and (v, u) is in Γ. **29.1**
associates (in a domain): a, b are associates if if $a|b$ and $b|a$. **23.1**
association scheme: A set of graphs $G_0, \dots, G_d$ on a common vertex set satisfying certain axioms. **28.6**
associative algebra: See *algebra (associative).*
asymmetric (digraph $\Gamma = (V, E)$): $(i, j) \in E$ implies $(j, i) \notin E$ for all distinct $i, j \in V$. **35.1**
augmented matrix (of a system of linear equations): Matrix obtained by adjoining the constant column to the coefficient matrix. **1.4**

B

backward error (for $\hat{f}(\mathbf{x})$): A vector $\mathbf{e} \in \mathbb{R}^n$ of smallest norm for which $f(\mathbf{x}+\mathbf{e}) = \hat{f}(\mathbf{x})$. **37.8**
backward stable (algorithm): The backward relative error $\mathbf{e}$ exists and is small for all valid input data $\mathbf{x}$ despite rounding and truncation errors in the algorithm. **37.8**
badly conditioned: See *ill-conditioned.*
balanced ((0, 1)-design matrix W): Every row of $m \times n$ matrix W has exactly $(n+1)/2$ ones if is n odd; exactly $n/2$ ones or exactly $(n+2)/2$ ones if n even. **32.4**
balanced (real or sign pattern vector): It is a zero vector or it has both a positive entry and a negative entry. **33.3**
balanced column signing (of matrix A): A column signing of A in which all the rows are balanced. **33.3**
balanced row signing (of matrix A): A row signing of A in which all the columns are balanced. **33.3**
banded (family of Toeplitz matrices with symbol a): There exists some $m \geq 0$ such that $a_{\pm k} = 0$ for all $k \geq m$. **16.2**
barely L-matrix: An L-matrix that is not an L-matrix if any column is deleted. **33.3**
bases: Plural of *basis.*
basic class (of square nonnegative matrix P): An access equivalence class B of P with $\rho(P[B]) = \rho$. **9.3**
basic reduced digraph (of square nonnegative matrix P): The digraph whose vertex-set is the set of basic classes of P and whose arcs are the pairs (B, B') of distinct basic classes of P for which there exists a simple walk from B to B' in the reduced digraph of P. **9.3**
basic solution (to a least squares problem): A least squares solution with at least $n - \text{rank}A$ zero components. **39.1**
basic variable: A variable in a system of linear equations whose value is determined by the values of the free variables. **1.4**
basis: A set of vectors that is linearly independent and spans the vector space. **2.2**
BD: See *Bezout domain.*
Bezout domain (BD): A GCCD $\mathbb{D}$ such that for any two elements $a, b \in \mathbb{D}$, $(a, b) = pa + qb$, for some $p, q \in \mathbb{D}$. **23.1**
biacyclic (matrix): A matrix whose sparsity pattern has an acyclic bipartite graph. **46.3**
biadjacency matrix (of bipartite graph G): A matrix with rows indexed by one of the bipartition sets and columns indexed by the other, with the value of the entry being the number of edges between the vertices (or weight of the edge between the vertices). **30.1**
BiCG algorithm: Iterative method for solving a linear system using non-Hermitian Lanczos algorithm; the approximate solution $\mathbf{x}_k$ is chosen so that the residual $\mathbf{r}_k$ is orthogonal to Span$\{\hat{\mathbf{r}}_0, A^*\hat{\mathbf{r}}_0, \ldots, A^{*k-1}\hat{\mathbf{r}}_0\}$. **41.3**
BiCGSTAB algorithm: Iterative method for solving a linear system using non-Hermitian Lanczos algorithm with improved stability. See *Section* **41.3.**
biclique (of a graph): A subgraph that is a complete bipartite graph. **30.3**
biclique cover (of graph $G = (V, E)$): A collection of bicliques of G such that each edge of E is in at least one biclique. **30.3**
biclique cover number (of graph G): The smallest k such that there is a cover of G by k bicliques. **30.3**
biclique partition (of $G = (V, E)$): A collection of bicliques of G whose edges partition E. **30.3**
biclique partition number (of graph G): The smallest k such that there is a partition of G into k bicliques. **30.3**
bidual space: The dual of the dual space. **3.8**
big-oh: function f is $O(g)$ if beyond a certain point $|f|$ is bounded by a multiple of $|g|$. **Preliminaries**
bigraph (of the $m \times n$ matrix A): The simple graph with vertex set $U \cup V$ where $U = \{1, 2, \ldots, m\}$ and $V = \{1', 2', \ldots, n'\}$, and edge set $\{\{i, j'\} : a_{ij} \neq 0\}$. **30.2**
bilinear form (on vector space V over field F): A map f from $V \times V$ into F that satisfies $f(a\mathbf{u}_1 + b\mathbf{u}_2, \mathbf{v}) = af(\mathbf{u}_1, \mathbf{v}) + bf(\mathbf{u}_2, \mathbf{v})$ and $f(\mathbf{u}, a\mathbf{v}_1 + b\mathbf{v}_2) = af(\mathbf{u}, \mathbf{v}_1) + bf(\mathbf{u}, \mathbf{v}_2)$. **12.1**
bilinear noncommutative algorithms: See *Section* **47.2.**
binary matrix: A (0,1)-matrix, i.e., a matrix in which each entry is either 0 or 1. **31.3**

biorthogonal (sets of vectors $\{\mathbf{v}_1, \ldots, \mathbf{v}_k\}$ and $\{\mathbf{w}_1, \ldots, \mathbf{w}_k\}$ in an inner product space): $\langle \mathbf{v_i}, \mathbf{w_j} \rangle = 0$ if $i \neq j$. **41.3**
bipartite (graph G): The vertices of G can be partitioned into disjoint subsets U and V such that each edge of G has the form $\{u, v\}$ where $u \in U$ and $v \in V$. **28.1, 30.1**
bipartite fill-graph of $n \times n$ matrix A with each diagonal entry nonzero: The simple bipartite graph with vertex set $\{1, 2, \ldots, n\} \cup \{1', 2', \ldots, n'\}$ with an edge joining i and j' if and only if there exists a path from i to j in the directed graph, $\Gamma(A)$, of A each of whose intermediate vertices has label less than $\min\{i, j\}$. **30.2**
bipartite graph (of sparsity pattern $\mathcal{S}$): The graph with vertices partitioned into row vertices $r_1, \ldots, r_m$ and column vertices $c_1, \ldots, c_n$, where r_k and r_l are connected if and only if $(k, l) \in \mathcal{S}$. **46.3**
bipartite sign pattern: A sign pattern whose digraph is bipartite. **33.5**
bipartition: The sets into which the vertices of a bipartite graph are partitioned. **30.1**
block (in a 2-design): A subset B_i of X (see *2-design*). **32.2**
block (of a graph or digraph): A maximal nonseparable sub(di)graph. **35.1**
block (of a matrix): An entry of a block matrix, which is a submatrix of the original matrix. **10.1**
block-clique (graph or digraph): Every block is a clique. **35.1**
block diagonal (for a particular block matrix structure): A square block matrix A with all off-diagonal blocks 0. **10.2**
block matrix: A matrix that is partitioned into submatrices with the row indices and column indices partitioned into consecutive subsets sequentially. **10.1**
block lower triangular (matrix A): A^T is block upper triangular. **10.2**
block–Toeplitz matrix: A block matrix $A = [A_{ij}]$ such that $A_{i+k,i} = A^{(k)}$, **48.1**
block–Toeplitz–Toeplitz–block (BTTB) matrices: a block–Toeplitz matrix in which the blocks $A^{(k)}$ are themselves Toeplitz matrices. **48.1**
block upper triangular (for a particular block matrix structure): A block matrix having every subdiagonal block 0. **10.2**
(0,1)-Boolean algebra: $\{0, 1\}$, with $0+0 = 0, 0+1 = 1 = 1+0, 1+1 = 1, 0*1 = 0 = 1*0, 0*0 = 0$, and $1*1 = 1$. **30.3**
Boolean matrix: A matrix whose entries belong to the Boolean algebra. **30.3**
Boolean rank (of $m \times n$ Boolean matrix A): The minimum k such that there exists an $m \times k$ Boolean matrix B and a $k \times n$ Boolean matrix C such that $A = BC$. **30.3**
Bose–Mesner algebra: The algebra generated by the adjacency matrices of the graphs of an association scheme. **28.6**
bottleneck matrix (of a branch of tree T at vertex v): The inverse of the principal submatrix of the Laplacian matrix corresponding to the vertices of that branch. **36.3**
boundary (of a subset S of $\mathbb{R}$ or $\mathbb{C}$): The intersection of the closure of S and the closure of the complement of S. **Preliminaries**
branch (of tree T at vertex v): A component of $T - v$. **34.1**
Bunch–Parlett factorization (of symmetric real matrix H): The factorization $P^T H P = L B L^T$, where P is a permutation matrix, B is a block–diagonal matrix with diagonal blocks of sizes 1×1 or 2×2, and L is a full column rank unit lower triangular matrix, where the diagonal blocks in L which correspond to 2×2 blocks in B are 2×2 identity matrices. **15.5**
Businger–Golub pivoting: a particular pivoting strategy for QR-factorization. **46.2**

C

canonical angles between the column spaces of $X, Y \in \mathbb{C}^{n\times k}$: $\theta_i = \arccos \sigma_i$, where $\{\sigma_i\}_{i=1}^k$ are the singular values of $(Y^*Y)^{-1/2}Y^*X(X^*X)^{-1/2}$ **15.1** (equivalent to principal angles **17.1**)
canonical angle matrix: $\operatorname{diag}(\theta_1, \theta_2, \ldots, \theta_k)$, where $\theta_1, \theta_2, \ldots, \theta_k$ are the canonical angles. **15.1**
canonical correlations (between subspaces X and Y of $\mathbb{C}^r$): Cosines of principal (canonical) angles **17.7**
Cauchy matrix: Given vectors $\mathbf{x} = [x_1, \ldots, x_n]^T$ and $\mathbf{y} = [y_1, \ldots, y_n]^T$, the Cauchy matrix $C(\mathbf{x}, \mathbf{y})$ has i, j-entry equal to $\frac{1}{x_i + y_j}$. **48.1**

center See *central vertex.*
central (real matrix B): The zero vector is in the convex hull of the columns of B. **33.11**
central vertex (of a generalized star): a vertex such that each of its neighbours are pendant vertices of their branches, and each branch is a path. **34.1**
CG: See *conjugate gradient algorithm.*
CGS algorithm: Iterative method for solving a linear system using non-Hermitian Lanczos algorithm. See *Section* **41.3.**
change-of-basis matrix (from $\mathcal{B}$ to $\mathcal{C}$): The matrix consisting of the coordinate vectors with respect to basis $\mathcal{C}$ of the basis vectors in $\mathcal{B}$. **2.6**
characteristic equation (of the pencil $A - xB$): $\det(xB - A) = 0$. **43.1**
characteristic polynomial (of matrix A): $\det(xI - A)$. **4.3**
characteristic polynomial (of the pencil $A - xB$): $\det(xB - A)$. **43.1**
characteristic polynomial (of a graph): The characteristic polynomial of its adjacency matrix. **28.3**
characteristic vector (of a subset of m vertices of a graph): The m-vector whose ith entry is 1 if $i \in X$, and 0 otherwise. **30.3**
characteristic vertex (of tree T): See *Section* **36.3.**
Cholesky decomposition (of positive-definite matrix A): $A = GG^*$ with $G \in \mathbb{C}^{n\times n}$ lower triangular and having positive diagonal entries. **38.5**
Cholesky factorization: See *Cholesky decomposition.*
chord (of a cycle in a graph): An edge joining two nonconsecutive vertices on the cycle. **30.1**
chordal bipartite graph: Every cycle of length 6 or more has a chord. **30.1**
chordal graph: Every cycle of length 4 or more has a chord. **30.1**
1-chordal: See *block-clique.*
chromatic index (of graph G): The smallest number of classes in a partition of edges of G so that no two edges in the same class meet. **27.6**
chromatic number (of graph G): The smallest number of color classes of any vertex coloring of G (not defined if G has loops). **28.5**
circulant matrix: A Toeplitz matrix in which every row is obtained by a single cyclic shift of the previous row. **20.3, 48.1**
clique (of a digraph): Every vertex has a loop and for any two distinct vertices u, v, both arcs $(u, v), (v, u)$ are present. **35.1**
clique (of a graph that allows loops but not multiple edges): Every vertex has a loop and for any two distinct vertices u, v, the edge $\{u, v\}$ is present. **35.1**
clique (of a simple graph): A subgraph isomorphic to a complete graph. **28.5**
clique number (of graph G): The largest order of a clique in G. **28.5**
closed cone: A cone that is a closed subset of the vector space. **8.5, 26.1**
closed under matrix direct sums (class of matrices X): Whenever $A_1, A_2, \ldots, A_k$ are X-matrices, then $A_1 \oplus A_2 \oplus \cdots \oplus A_k$ is an X-matrix. **35.1**
closed under permutation similarity (class of matrices X): Whenever A is an X-matrix and P is a permutation matrix, then P^TAP is an X-matrix. **35.1**
closed under taking principal submatrices: See *hereditary.*
closed walk (in a digraph): A walk in which the first vertex equals the last vertex. **54.2**
4-cockade: A bipartite graph created by adding 4 cycles through edge identification to a 4-cycle. **30.2**
cocktail party graph: The graph obtained by deleting a disjoint edges from the complete graph K_{2a}. **28.2**
coclique (of a graph): An induced subgraph with no edges. **28.5**
co-cover (of a (0, 1)-matrix): A collection of 1s such that each line of A contains at least one of the 1s. **27.1**
co-index (of square nonnegative matrix P): $\max\{\nu_P(\lambda) : \lambda \in \sigma(P), |\lambda| = \rho \text{ and } \lambda \neq \rho\}$. **9.3**
codomain (of $T\colon V \to W$): The vector space W. **3.1**
coefficient matrix: The matrix of coefficients of a system of linear equations. **1.4**
coefficients (of a linear equation): The scalars that occur multiplied by variables. **1.4**
i, j**-cofactor**: $(-1)^{i+j}$ times the i, j-minor. **4.1**

Colin de Verdière parameter: A parameter of a simple graph. **28.5**
color class: A coclique in a vertex coloring partition of the vertices of a graph. **28.5**
column: The entries of a matrix lying in a vertical line in the matrix. **1.2**
column equivalent (matrices $A, B \in \mathbb{D}^{m\times n}$): $B = AP$ for some $\mathbb{D}$-invertible P. **23.2**
column signing (of (real or sign pattern) matrix A): AD where D is a signing. **33.3**
column space: See *range.*
column-stochastic (matrix): A square nonnegative matrix having all column sums equal to 1. **9.4**
column sum vector (of a matrix): The vector of column sums. **27.4**
combinatorially orthogonal (sign pattern vectors $\mathbf{x}, \mathbf{y}$): $|\{i \,:\, x_i y_i \neq 0\}| \neq 1$. **33.10**
combinatorially symmetric (partial matrix B): b_{ij} specified implies b_{ji} specified. **35.1**
combinatorially symmetric sign pattern A: $a_{ij} \neq 0$ if and only if $a_{ji} \neq 0$. **33.5**
communicate: See *access equivalent.*
commute: Matrices or linear operators A, B such that $AB = BA$. **1.2, 3.2**
companion matrix (of a monic polynomial): Square matrix with ones on the subdiagonal and last column consisting of negatives of the polynomial coefficients **4.3, 6.4**
comparison matrix (of real square matrix A): Matrix having i, j-entry $-|a_{ij}|$ for $i \neq j$ and i, i-entry $|a_{ii}|$. **19.5**
compatible (generalized sign patterns): The intersection of their qualitative classes is nonempty. **33.1**
complement (of set X in universe S): The elements of S not in X. **Preliminaries**
complement (of binary matrix M): $J - M$ (where J is the all 1s matrix). **31.3**
complement (of simple graph $G = (V, E)$): The simple graph having vertex set V and as edges all unordered pairs from V that are not in E. **28.1**
complement (orthogonal): See *orthogonal complement.*
complete (bipartite graph): A simple bipartite graph with bipartition $\{U, V\}$ such that each $\{u, v\}$ is an edge (for all $u \in U, v \in V$). **28.1, 30.1**
complete (simple graph): The edge set consists of all unordered pairs of distinct vertices. **28.1**
complete orthonormal set: An orthonormal set of vectors whose orthogonal complement is 0. **5.2**
completed max-plus semiring: The set $\mathbb{R} \cup \{\pm\infty\}$ equipped with the addition $(a, b) \mapsto \max(a, b)$ and the multiplication $(a, b) \mapsto a + b$, with the convention that $-\infty + (+\infty) = +\infty + (-\infty) = -\infty$. **25.1**
completely positive (matrix A): $A = CC^T$ for some nonnegative $n \times m$ matrix C. **35.4**
completely reducible (matrix A): There is a permutation matrix P such that $PAP^T = A_1 \oplus A_2 \oplus \cdots \oplus A_k$ where $A_1, A_2, \ldots, A_k$ are irreducible and $k \geq 2$. **27.3**
completion (of a partial matrix): A choice of values for the unspecified entries. **35.1**
X-**completion property** (where X is a type of matrix): Digraph (respectively, graph) G has this property if every partial X-matrix B such that $\mathcal{D}(B) = G$ (respectively, $\mathcal{G}(B) = G$) can be completed to an X-matrix. **35.1**
complex conjugate (of $a + bi \in \mathbb{C}$): $a - bi$. **Preliminaries**
complex sign pattern matrix: A matrix of the form $A = A_1 + iA_2$ for sign patterns A_1 and A_2. **33.8**
complex vector space: A vector space over the field of complex numbers. **1.1**
component-wise relative backward error of the linear system: See *Section* **38.1.**
component-wise relative distance (between matrices H and $\tilde{H}$): $\text{reldist}(H, \tilde{H}) = \max_{i,j} \dfrac{|h_{ij} - \tilde{h}_{ij}|}{|h_{ij}|}$, where $0/0 = 0$. **15.4**
composite cycle (in sign pattern A): A product of disjoint simple cycles. **33.1**
$k \times k$-**compound matrix**: A matrix formed from the $k \times k$ minors of a given matrix. **4.2**
condensation digraph: See *reduced digraph.*
condition number (of $\mathbf{z}$ with respect to problem P):
$\text{cond}_P(\mathbf{z}) = \lim_{\varepsilon\to 0} \sup \left\{ \left(\frac{\|P(\mathbf{z}+\delta\mathbf{z}) - P(\mathbf{z})\|}{\|P(\mathbf{z})\|} \right) \left(\frac{\|\mathbf{z}\|}{\|\delta\mathbf{z}\|} \right) \,\middle|\, \|\delta\mathbf{z}\| \leq \varepsilon \right\}$. **37.4**
condition number (of an eigenvalue): See *individual condition number.*
condition number (of matrix A for linear systems): $\kappa_\nu(A) = \|A^{-1}\|_\nu \, \|A\|_\nu$. **37.5, 38.1**

condition number of the linear system: $A\hat{\mathbf{x}} = \mathbf{b}$: $\kappa(A, \hat{\mathbf{x}}) = \|A^{-1}\| \frac{\|\mathbf{b}\|}{\|\hat{\mathbf{x}}\|}$. **38.1**
condition number (of least squares problem $A\mathbf{x} = \mathbf{b}$): $\kappa(A) = \sigma_1/\sigma_p$, where rank $A = p$. **39.6**
condition numbers for polar factors ($A = UP$) in the Frobenius norm: $\text{cond}_F(X) = \lim_{\delta \to 0} \sup_{\|\Delta A\|_F \leq \delta} \frac{\|\Delta X\|_F}{\delta}$, for $X = P$ or U. **15.3**
conductance: A parameter of a simple graph. **28.5**
cone: A subset K of a real or complex vector space such that for each $\mathbf{x}, \mathbf{y} \in K, c \geq 0$, $\mathbf{x} + \mathbf{y} \in K$ and $c\mathbf{x} \in K$; in Section 26.1 cone is used to mean proper cone. **8.5, 26.1**
conformable: See *conformal.*
conformal (partitions of matrices A, B): Partitions of block matrices that allow multiplication via the block structure. **10.1**
congruent (square matrices A, B over field F): There is an invertible matrix C such that $B = C^T AC$. **12.1**
***congruent** (square matrices A, B over $\mathbb{C}$): There is an invertible matrix C such that $B = C^* AC$. **8.3**
φ-congruent (square matrices A, B over field F with automorphism φ): There is an invertible matrix C such that $B = C^T A\varphi(C)$. **12.4**
conjugate: See *complex conjugate.*
conjugate gradient (CG) algorithm (to solve $H\mathbf{x} = \mathbf{b}$ for preconditioned Hermitian positive definite matrix H): At each step k, the approximation $\mathbf{x}_k$ of the form $\mathbf{x}_k \in \mathbf{x}_0 + \text{Span}\{\mathbf{r}_0, H\mathbf{r}_0, \ldots, H^{k-1}\mathbf{r}_0\}$ for which the $\sqrt{H}$-norm of the error, $\|\mathbf{e}_k\|_{\sqrt{H}} \equiv \langle \mathbf{e}_k, H\mathbf{e}_k \rangle^{1/2}$, is minimal. **41.2**
conjugate partition (of s sequence of positive integers $(u_1, u_2, \ldots, u_n)$): The ith element of the conjugate partition is the number of js such that $u_j \geq i$. **Preliminaries**
connected (graph): A graph with nonempty vertex set such that there exists a walk between any two distinct vertices. **28.1**
connected (digraph): A digraph whose associated undirected graph is connected. **29.1**
connected component (of a graph): A connected (induced) subgraph not properly contained in a connected subgraph. **28.1**
consecutive ones property (of a (0, 1)-matrix A): There exists a permutation matrix P such that PA is a Petrie matrix. **30.2**
consistent: A system of linear equations that has one or more solutions. **1.4**
k-consistent sign pattern A: Every matrix $B \in Q(A)$ has exactly k real eigenvalues. **33.5**
constant (of a linear equation): The scalar not multiplied by a variable. **1.4**
constant vector: The vector of constants of a system of linear equations. **1.4**
contraction (matrix): A matrix $A \in \mathbb{C}^{n \times n}$ such that $\|A\|_2 \leq 1$. **18.6**
contraction (of edge e in graph $G = (V, E)$): The operation that merges the endpoints of e in V, and deletes e from E. **28.2**
convergent (square nonnegative matrix P): $\lim_{m \to \infty} P^m = 0$. **9.3**
convergent regular splitting: Square real matrix A has a convergent regular splitting if A has a representation $A = M - N$ such that $N \geq 0$, M invertible with $M^{-1} \geq 0$ and $M^{-1}N$ is convergent. **9.5**
converges geometrically to a with (geometric) rate β (complex sequence $\{a_m\}_{m=0,1,\ldots}$): $\{\frac{a_m - a}{\gamma^m} : m = 0, 1, \ldots\}$ is bounded for each $\beta < \gamma < 1$. **9.1**
convex (set of vectors): Closed under convex combinations. **Preliminaries**
convex combination (of vectors $\mathbf{v}_1, \mathbf{v}_2, \ldots, \mathbf{v}_k$): A vector of the form $a_1\mathbf{v}_1 + a_2\mathbf{v}_2 + \cdots + a_k\mathbf{v}_k$ with a_i nonnegative and $\sum a_i = 1$ (vector space is real or complex). **Preliminaries**
convex cone: See *cone.*
convex function: A function $f : S \to \mathbb{R}$ where S is a convex set and for all $a \in \mathbb{R}$ such that $0 < a < 1$ and $\mathbf{x}, \mathbf{y} \in S$, $f(a\mathbf{x} + (1 - a)\mathbf{y}) \leq af(\mathbf{x}) + (1 - a)f(\mathbf{y})$. **Preliminaries**
convex hull (of a set of vectors): The set of all convex combinations of the vectors. **Preliminaries**
convex polytope: The convex hull of a finite set of vectors in $\mathbb{R}^n$. **Preliminaries**
coordinate (part of a vector): In the vector space F^n, one of the entries of a vector. **1.1**
coordinate mapping: The function that maps a vector to its coordinate vector. **2.6**
coordinate vector: The vector of coordinates of a vector relative to an ordered basis. **2.6**

coordinates (of a vector relative to a basis): The scalars that occur when the vector is expressed as a linear combination of the basis vectors. **2.6**
copositive (matrix A): $\mathbf{x}^T A\mathbf{x} \geq 0$ for all $\mathbf{x} \geq 0$. **35.5**
coprime (elements in a domain): 1 is a greatest common divisor of the elements. **23.1**
corner minor: The determinant of a submatrix in the upper right or lower left corner of a matrix. **21.3**
correlation matrix: A positive semidefinite matrix in which every main diagonal entry is 1. **8.4**
cosine (of a complex square matrix A): The matrix defined by the cosine power series, $\cos(A) = I - \frac{A^2}{2!} + \cdots + \frac{(-1)^k}{(2k)!}A^{2k} + \cdots$. **11.5**
cospectral (graphs): Having the same spectrum. **28.3**
cover (of a $(0,1)$-matrix): A collection of lines that contain all the 1s of the matrix. **27.1**
curve segment (from $\mathbf{x}$ to $\mathbf{y}$): The range of a continuous map ϕ from $[0,1]$ to $\mathbb{R}^n$ with $\phi(0) = \mathbf{x}$ and $\phi(1) = \mathbf{y}$. **28.2**
cut-vertex (of connected simple graph G): A vertex v of G such that $G - v$ is disconnected. **36.2**
(k-)**cycle** (in a graph or digraph): A walk (of length k) with all vertices distinct except the first vertex equals the last vertex. **28.1, 29.1**
cycle (permutation): A permutation τ that maps a subset to itself cyclically. **Preliminaries**
k-**cycle** (in sign pattern A): A formal product of the form $a_{i_1 i_2} a_{i_2 i_3} \ldots a_{i_k i_1}$, where each of the elements is nonzero and the index set $\{i_1, i_2, \ldots, i_k\}$ consists of distinct indices. **33.1**
n-**cycle matrix**: The $n \times n$ matrix with ones along the first subdiagonal and in the $1,n$-entry, and zeros elsewhere. **48.1**
n-**cycle pattern**: A sign pattern A where the digraph of A is an n-cycle. **33.5**
cycle-clique (digraph): The induced subdigraph of every cycle is a clique. **35.7**
cycle product: A walk product where the walk is a cycle. **29.3**
cyclic normal form (of matrix A): A matrix in specific form that is permutation similar to A. **29.7**
cyclically real (square ray pattern A): The product of every cycle in A is real. **33.8**

D

Δ(G): $\max(p - q)$ over all ways in which q vertices may be deleted from graph G leaving p paths. **34.2**
$\mathbb{D}$-invertible, $\mathbb{D}$-module, $\mathbb{D}$-submodule: Alphabetized under *invertible, module, submodule.*
D-optimal (design matrix W): $\det W^T W$ is maximal over all (± 1)- (or $(0,1)$-)matrices of the given size. **32.1**
damped least squares solution: The solution to the problem $(A^T A + \alpha I)\mathbf{x} = A^T\mathbf{b}$, where $\alpha > 0$. **39.8**
data fitting: See *Section* **39.2.**
data perturbations (for linear system $A\mathbf{x} = \mathbf{b}$): Perturbations of A and/or $\mathbf{b}$. **38.1**
decomposable tensor: A tensor of the form $\mathbf{v}_1 \otimes \cdots \otimes \mathbf{v}_m$. **13.2**
defective (matrix $A \in F^{n\times n}$): There is an eigenvalue of A (over algebraic closure of F) having geometric multiplicity less than its algebraic multiplicity. **4.3**
deflation: A process of reducing the size of the matrix whose eigenvalue decomposition is to be determined, given that one eigenvector is known. **42.1**
degenerate (bilinear form f on vector space V): Not nondegenerate, i.e., the rank of f less than $\dim V$, **12.1**; also applied to φ-sesquilinear form **12.4.**
degree (of polynomial $p(x_1, \ldots, x_n) \neq 0$): The largest integer m such that there is a term $cx_1^{m_1} \ldots x_n^{m_n}$ with $c \neq 0$ in p having degree m. **23.1**
degree (of term $cx_1^{m_1} \ldots x_n^{m_n}$ in polynomial $p(x_1, \ldots, x_n)$): The sum of the degrees of the x_i, i.e., $\Sigma_i m_i$.
degree (of a vertex v): The number of times that v occurs as an endpoint of an edge. **28.1**
deletion (of edge e from graph $G = (V, E)$): The operation that deletes e from E. **28.2**
deletion (of vertex v from graph $G = (V, E)$): The operation that deletes v from V and all edges with endpoint v from E. **28.2**
depth (of eigenvector $\mathbf{x}$ for eigenvalue λ of A): The natural number h such that $\mathbf{x} \in \text{range}(A - \lambda I)^h - \text{range}(A - \lambda I)^{h+1}$. **6.2**

derogatory (matrix $A \in F^{n\times n}$): Some eigenvalue of A (over algebraic closure of F) has geometric multiplicity greater than 1. **4.3**
2-design with parameters (v, k, λ): A collection of k-element subsets (blocks) B_i of a finite set X with $|X| = v$, such that each 2-element subset of X is contained in exactly λ blocks. **32.1**
(± 1)**-design matrix**: A matrix whose entries are 1 or -1. **32.1**
$(0, 1)$**-design matrix**: A matrix whose entries are 1 or 0. **32.1**
det: See *determinant.*
determinant (det): A scalar-valued function of a square matrix or linear operator, defined inductively, and also given by $\det(A) = \sum_{\sigma\in S_n} \operatorname{sgn}\sigma \prod_{i=1}^{n} a_{i\sigma(i)}$ **4.1**
determinantal region (of complex sign pattern or ray patern A): $S_A = \{\det(B) \;:\; B \in Q(A)\}$. **33.8**
determined by its spectrum (graph G): Every graph cospectral with G is isomorphic to G. **28.3**
diagonal (of a matrix): *The* diagonal means the *main diagonal,* i.e., the set of diagonal entries of a matrix, **1.2.** A diagonal is a collection of n nonzero entries in A no two on the same line. **27.2**
diagonal entry: An entry that lies in row k and column k for some k. **1.2**
diagonal matrix: A matrix all of whose off-diagonal entries are zero. **1.2, 10.2**
diagonal pattern: A square sign pattern all of whose off-diagonal entries are zero. **33.1**
diagonal product (of a matrix): The product of the entries on a diagonal of the matrix. **27.6**
diagonalizable (matrix $A \in F^{n\times n}$): There exist nonsingular matrix $B \in F^{n\times n}$ and diagonal matrix $D \in F^{n\times n}$ such that $A = BDB^{-1}$. **4.3**
diagonally dominant ($n \times n$ complex matrix A): $|a_{ii}| \ge \sum_{j=1, j\neq i}^{n} |a_{ij}|$ for $i = 1, \ldots, n$. **9.5**
diagonally scaled representation (of symmetric matrix H with positive diagonal entries): A factored representation $H = DAD$ with $D = \operatorname{diag}(\sqrt{h_{11}}, \ldots, \sqrt{h_{nn}})$, $a_{ij} = \dfrac{h_{ij}}{\sqrt{h_{ii}h_{jj}}}$. **15.4**
diagonally scaled totally unimodular (DSTU) (real matrix A): There exist diagonal matrices D_1, D_2 and totally unimodular Z such that $A = D_1 Z D_2$. **46.3**
diagonally stable: See *Lyapunov diagonally stable.*
diameter (of a connected graph): The largest distance that occurs between two vertices. **28.1**
digraph: A directed graph having at most one arc between each ordered pair of vertices (loops permitted). **29.1**
digraph (of $n \times n$ matrix A): The digraph having vertices $\{1, \ldots, n\}$ and having arc (i, j) exactly when $a_{ij} \neq 0$. **9.1, 29.2**
digraph (of $n \times n$ partial matrix B): The digraph having vertices $V = \{1, \ldots, n\}$ and for each $i, j \in V$, (i, j) is an arc exactly when b_{ij} is specified. **35.1**
dilation: Matrix $A \in \mathbb{C}^{n\times n}$ has a dilation $B \in \mathbb{C}^{m\times m}$ if there is $X \in \mathbb{C}^{m\times n}$ such that $X^*X = I_n$ and $X^*BX = A$. **18.6**
dim: See *dimension.*
dimension (dim) (of a vector space): The number of vectors in a basis for the vector space. **2.2**
dimension (dim) (of a convex polytope): The smallest dimension of an affine space containing it. **27.6**
direct sum (of matrices): A block diagonal matrix with these matrices as the diagonal blocks. **2.4, 10.2**
direct sum (of subspaces): $W_1 + \cdots + W_k$ is direct if for all $i = 1, \ldots, k$, $W_i \cap \sum_{j\neq i} W_j = \{\mathbf{0}\}$. **2.3**
direct sum (of vector spaces): See *external direct sum.*
directed bigraph (of the real $m \times n$ matrix A): The directed graph with vertices $1, 2, \ldots, m, 1', 2', \ldots, n'$, with arc (i, j') if and only if $a_{ij} > 0$, and with arc (j', i) if and only if $a_{ij} < 0$. **30.2**
directed edge: See *arc.*
directed graph: A finite nonempty set of vertices and a finite multiset of arcs, where each arc is an ordered pair of vertices. **29.1**
directed multigraph: A directed graph having more than one arc between at least one pair of vertices. **29.1**
disjoint (graphs): The vertex sets are disjoint sets. **28.1**
dispersion: See *Section* 21.3.
distance (between two vectors): The norm of their difference. **5.1**

distance (between two vertices in a graph): The length of a shortest path between the vertices (infinite if no path). **28.1**
distance of $W(A)$ to the origin: $\widetilde{w}(A) = \min\{|\mu| : \mu \in W(A)\}$. **18.1**
distance-regular graph: A connected graph with the property that for every pair of vertices u, v, the number of vertices that have distance i from u and distance j from v depends only on i, j, and the distance between u and v. **28.6**
distinguished eigenvalue (of nonnegative square matrix P): An eigenvalue of P that has a semipositive eigenvector. **9.1**
d **divides** a (in a domain $\mathbb{D}$): $a = db$ for some $b \in \mathbb{D}$. **23.1**
divisor (i.e., d is a divisor of a): See *divides.*
domain (ring): An integral domain, i.e., a commutative ring without zero divisors and containing identity 1. **23.1**
domain (of $T\colon V \to W$): The vector space V. **3.1**
dominant eigenvalue: λ_1 where $|\lambda_1| > |\lambda_2| \geq \cdots \geq |\lambda_n|$ are the eigenvalues. **42.1**
dot product: See *standard inner product.*
double echelon form: A special type of real matrix, see **21.1**
double generalized star: A tree constructed by joining the centers of two generalized stars by an edge. **34.7**
double path: A tree constructed by joining two degree two vertices of two paths by an edge. **34.7**
double precision: Typically, floating-point numbers have machine epsilon roughly 10^{-16} and precision roughly 16 decimal digits. **37.6**
doubly directed tree: A digraph Γ whose associated undirected graph is a tree and whenever (i, j) is an arc in Γ, (j, i) is also an arc in Γ. **29.1**
doubly nonnegative (matrix): A positive semidefinite matrix having every entry nonnegative. **35.4**
doubly stochastic (matrix): A square nonnegative matrix having all row and column sums equal to 1. **9.4**
downdating (QR factorization): See *Section* **39.7**.
DSTU: See *diagonally scaled totally unimodular.*
ΔTP: See *triangular totally positive.*
dual basis: A specific basis for the dual space defined from a given basis for the vector space. **3.8**
dual cone (of cone K): See *dual space of cone.*
dual norm (of vector norm $\|\cdot\|$): $\|\mathbf{y}\|^D = \max_{\mathbf{x}\neq 0} \frac{|\mathbf{y}^*\mathbf{x}|}{\|x\|}$. **37.1**
dual space: The vector space of linear functionals on a given vector space. **3.8**
dual space (of cone K): $K^* = \{\mathbf{y} \in V \mid \langle \mathbf{x}, \mathbf{y}\rangle \geq 0 \quad \forall\, \mathbf{x} \in K\}$ **8.5, 26.1**

E

ED: See *Euclidean domain.*
ED-RCF basis (for a linear operator): See *Section* **6.4.**
ED-RCF matrix (over F): A block diagonal matrix of the form $C(h_1^{m_1}) \oplus \cdots \oplus C(h_t^{m_t})$ where each $h_i(x)$ is a monic polynomial that is irreducible over F. **6.4**
EDD: See *elementary divisor domain.*
edge: A pair of vertices, part of a graph. **28.1**
edge cut: Let $G = (V, E)$ be a graph, and let $X, V \setminus X$ be a nontrivial partitioning of V. The set E_X of edges of G that have one end point in X and the other end point in $V \setminus X$ is an edge cut. **36.2**
efficacy (of a processor): See *speed.*
efficient: One algorithm is more efficient than another, if it accomplishes the same task with a lower cost of computation. **37.7**
eigenpair (of $A \in \mathbb{C}^{n\times n}$): A (eigenvalue, eigenvector) pair. **15.1**
eigenspace (of eigenvalue λ of A): $\ker(A - \lambda I)$. **4.3**
k-**eigenspace** (of matrix or linear operator A at λ): $\ker(A - \lambda I)^k$ **6.1**

eigentriplet (of $A \in \mathbb{C}^{n\times n}$): A vector-scalar-vector triplet $(\mathbf{y}, \lambda, \mathbf{x})$ such that $\mathbf{x} \neq 0$, $\mathbf{y} \neq 0$, and $A\mathbf{x} = \lambda\mathbf{x}$, $\mathbf{y}^* A = \lambda \mathbf{y}^*$. **15.1**
eigenvalue (of matrix or linear operator A): A scalar λ such that there exists a nonzero vector $\mathbf{x}$ such that $A\mathbf{x} = \lambda\mathbf{x}$. **4.3**
eigenvalue (of the pencil $A - xB$): $\lambda = \mu/\nu$ where $\nu A\mathbf{v} = \mu B\mathbf{v}$ and $\mathbf{v} \neq \mathbf{0}$ (if $\nu = 0, \lambda = \infty$). **43.1**
eigenvalue (of a graph): An eigenvalue of its adjacency matrix. **28.3**
eigenvalue decomposition (EVD) (of real symmetric matrix A): $A = U\Lambda U^T$, where $U^T U = UU^T = I_n$, $\Lambda = \text{diag}(\lambda_1, \ldots, \lambda_n)$ is a real diagonal matrix of eigenvalues, and the columns of U are eigenvectors. **42.1**
eigenvector (of matrix or linear operator A): A nonzero vector $\mathbf{x}$ such that there exists a scalar λ such that $A\mathbf{x} = \lambda\mathbf{x}$. **4.3**
eigenvector (of the pencil $A - xB$): Nonzero vector $\mathbf{v} \in \mathbb{C}^n$ such that there are scalars $\mu, \nu \in \mathbb{C}$, not both zero, such that $\nu A\mathbf{v} = \mu B\mathbf{v}$. **43.1**
elementary bidiagonal matrix: An $n \times n$ real matrix whose main diagonal entries are all equal to one, and there is at most one nonzero off-diagonal entry and this entry must occur on the super- or sub-diagonal. **21.2**
elementary divisor domain (EDD): A GCDD such that for any three elements $a, b, c \in \mathbb{D}$ there exists $p, q, x, y \in \mathbb{D}$ such that $(a, b, c) = (px)a + (py)b + (qy)c$. **23.1**
elementary divisors (of an ED-RCF matrix): The polynomials whose companion matrices are the diagonal blocks. **6.4**
elementary divisors (of matrix or linear operator A): The elementary divisors of $\text{RCF}_{ED}(A)$. **6.4**
elementary divisors rational canonical form (of matrix $A \in F^{n\times n}$): An ED-RCF matrix that is similar to A. **6.4**
elementary divisors rational canonical form (of a linear operator): See *Section* **6.4**.
elementary divisors rational canonical form matrix: See *ED-RCF matrix.*
elementary matrix: The result of doing one elementary row operation on the identity matrix. **1.5**
elementary column operation: Column analog of an elementary row operation. **6.5, 23.2**
elementary row operation (on a matrix over a domain): Add a multiple of one row to another, interchange two rows, or multiply a row by an invertible domain element. **6.5, 23.2**
elementary row operation (on a matrix over a field): Add a multiple of one row to another, interchange two rows, or multiply a row by a nonzero field element. **1.3**
elementary symmetric function: The sum of all products of n variables taken k at a time. **Preliminaries**
elimination graph: A graph, digraph or bigraph associated with the principal submatrix that remains to be factored during Gaussian elimination. **40.4**
elimination ordering (for the rows of a matrix): A bijection that specifies the order in which the rows of the matrix are eliminated during Gaussian elimination. **40.5**
elimination sequence: See *elimination ordering.*
embedding: A representation of a graph in $\mathbb{R}^n$ where edges intersect only at vertices. **28.2**
empty graph: A graph with no edges. **28.1**
endpoint (of an edge in a graph): One of the two vertices on the edge. **28.1**
energy norm: See *M-norm* (alphabetized under norm).
i, j-**entry**: The scalar in row i and column j in a matrix. **1.2**
entry sign symmetric P- **matrix** A: A P-matrix such that for all i, j, either $a_{ij}a_{ji} > 0$ or $a_{ij} = a_{ji} = 0$. **35.10**
entry sign symmetric P_0- **matrix** A: A P_0-matrix such that for all i, j, either $a_{ij}a_{ji} > 0$ or $a_{ij} = a_{ji} = 0$. **35.10**
entry sign symmetric $P_{0,1}$- **matrix** A: A $P_{0,1}$-matrix such that for all i, j, either $a_{ij}a_{ji} > 0$ or $a_{ij} = a_{ji} = 0$. **35.10**
entry weakly sign symmetric P- **matrix** A: A P-matrix such that for all i, j, $a_{ij}a_{ji} \geq 0$. **35.10**
entry weakly sign symmetric P_0- **matrix** A: A P_0-matrix such that for all i, j, $a_{ij}a_{ji} \geq 0$. **35.10**
entry weakly sign symmetric $P_{0,1}$- **matrix** A: A $P_{0,1}$-matrix such that for all i, j, $a_{ij}a_{ji} \geq 0$. **35.10**

envelope (of sparse matrix A): The set of indices of the elements between the first and last nonzero elements of every row. **40.5**
equivalence relation: A relation that is reflexive, symmetric, and transitive. **Preliminaries**
equivalent (bilinear forms f, g): There exists an ordered basis $\mathcal{B}$ such that the matrix of g relative to $\mathcal{B}$ is congruent to the matrix of f relative to $\mathcal{B}$. **12.1**
φ-equivalent (bilinear forms f, g): There exists an ordered basis $\mathcal{B}$ such that the matrix of g relative to $\mathcal{B}$ is φ-congruent to the matrix of f relative to $\mathcal{B}$. **12.4**
***equivalent** (bilinear forms f, g on a complex vector space): φ-equivalent with φ = complex conjugation. **12.5**
equivalent (matrices $A, B \in F^{m\times n}$): $B = QAP$ for some invertible matrices Q, P. **2.4**
equivalent (matrices $A, B \in \mathbb{D}^{m\times n}$): $B = QAP$ for some $\mathbb{D}$-invertible matrices Q, P. **23.2**
equivalent: Systems of linear equations that have the same solution set. **1.4**
ergodic class (of stochastic matrix P): A basic class of P. **9.4**
ergodic state: A state in an ergodic class. **9.4**
ergodicity coefficient (of complex matrix A): $\max\{|\lambda| : \lambda \in \sigma(P) \text{ and } |\lambda| \neq \rho(A)\}$. **9.1**
error at step k of an iterative method for solving $A\mathbf{x} = \mathbf{b}$: The difference between the true solution $A^{-1}\mathbf{b}$ and the approximate solution $\mathbf{x}_k$: $\mathbf{e}_k \equiv A^{-1}\mathbf{b} - \mathbf{x}_k$. **41.1**
Euclid's Algorithm: See *Section* **23.1.**
Euclidean distance (matrix A): There exist vectors $\mathbf{x}_1, \ldots, \mathbf{x}_n \in \mathbb{R}^d$ (for some $d \geq 1$) such that $a_{ij} = \|\mathbf{x}_i - \mathbf{x}_j\|_2$ for all $i, j = 1, \ldots, n$. **35.3**
Euclidean domain (ED): A domain $\mathbb{D}$ with a function $d : \mathbb{D}\backslash\{0\} \to \mathbb{Z}^+$ such that for all nonzero $a, b \in \mathbb{D}$, $d(a) \leq d(ab)$ and there exist $t, r \in \mathbb{D}$ such that $a = tb + r$, where either $r = 0$ or $d(r) < d(b)$. **23.1**
Eulidean norm (of a matrix): See *Frobenius norm.*
Eulidean norm (of a vector): See *2-norm.*
Euclidean space: Finite dimensional real vector space with standard inner product. **5.1**
EVD: See *eigenvalue decomposition.*
even (permutation): Can be written as the product of an even number of transpositions. **Preliminaries**
even (cycle in a sign pattern): The length of the simple or composite cycle is even. **33.1**
expanders (or family of expanders): An infinite family of graphs with constant degree and isoperimetric number bounded from below. **28.5**
explicit restarting (of the Arnoldi algorithm): A straightforward but inferior way to implement polynomial restarting by explicitly constructing the starting vector $\psi(A)\mathbf{v}$ by applying $\psi(A)$ through a sequence of matrix-vector products. **44.5**
exponent (of floating point number x): e in floating point number $x = \pm\left(\frac{m}{b^{p-1}}\right) b^e$. **37.6**
exponent (of a primitive digraph): Period; equivalently, the smallest value of k that works in the definition of primitivity. **29.6**
exponent (of a primitive matrix A): The exponent of the digraph of A. **29.6**
exponent of matrix multiplication complexity: The infimum of the real numbers w such that there exists an algorithm for multiplying $n \times n$ matrices with $O(n^w)$ arithmetic operations. **47.2**
exponential (of a square complex matrix A): The matrix defined by the exponential power series, $e^A = I + A + \frac{A^2}{2!} + \cdots + \frac{A^k}{k!} + \cdots$. **11.3**
extended precision: Floating point numbers that have greater than double precision. **37.6**
extension: Signing $D' = \text{diag}(d_1', d_2', \ldots, d_k')$ is an extension of signing D if $D' \neq D$ and $d_i' = d_i$ whenever $d_i \neq 0$. **33.3**
external direct sum (of vector spaces): The cartesian product of the vector spaces with component-wise operations. **2.3**
exterior power (of a vector space): See *Grassman space.*
exterior product (of vectors $\mathbf{v}_1, \ldots, \mathbf{v}_m$): $\mathbf{v}_1 \wedge \cdots \wedge \mathbf{v}_m = m!\text{Alt}(\mathbf{v}_1 \otimes \cdots \otimes \mathbf{v}_k)$. **13.6**
extreme point (of a closed convex set S): A point in S that is not a nontrivial convex combination of other points in S. **Preliminaries**
extreme vector $\mathbf{v} \in K$: Either $\mathbf{v}$ is the zero vector or $\mathbf{v}$ is nonzero and $\Phi(\mathbf{v}) = \{\lambda\mathbf{v} : \lambda \geq 0\}$. **26.1**

F

face F (of cone K): A subcone of K such that $\mathbf{v} \in F$, $\mathbf{v} \geq^K \mathbf{w} \geq^K 0 \Longrightarrow \mathbf{w} \in F$. **26.1**
face of K **generated by** $S \subseteq K$: The intersection of all faces of K containing S. **26.1**
family of Toeplitz matrices (defined by symbol $a = \sum_{k=-\infty}^{\infty} a_k z^k$): The set $\{T_n\}_{n\geq 1}$ where T_n has constants $a_{1-n}, \ldots, a_{n-1}$. **16.2**
Fiedler vector (of simple graph G): An eigenvector of the Laplacian of G corresponding to the algebraic connectivity. **36.1**
field: A nonempty set with two binary operations, addition and multiplication, satisfying commutativity, associativity, distributivity and existence of identities and inverses. **Preliminaries**
field of rational functions: The quotient field of the field of polynomials over F. **23.1**
field of values: See *numerical range.*
fill element (in a sparse matrix A): An element that is zero in A but becomes nonzero during Gaussian elimination. **40.3**
fill graph (of sparse matrix A): The (graph, digraph, or bigraph) of A, together with all the additional edges corresponding to the fill elements that occur during Gaussian elimination. **30.2, 40.5**
fill matrix (of Gaussian elimination on a matrix): The matrix $M_1+M_2+\cdots+M_{n-1}-(n-1)I+U$, where the M_i are Gauss transformations used in the elimination and U is the resulting upper triangular matrix. **40.3**
final (subset α of vertices): no vertex in α has access to a vertex not in α. **9.1**
finite dimensional (vector space V): V has a basis containing a finite number of vectors. **2.2**
finitely generated (ideal I of a domain $\mathbb{D}$): An ideal of $\mathbb{D}$ that is finitely generated as a $\mathbb{D}$-module. **23.1**
finitely generated (module $\mathbf{M}$ over a domain $\mathbb{D}$): There exist k elements (*generators*) $\mathbf{v}_1, \ldots, \mathbf{v}_k \in \mathbf{M}$ so that every $\mathbf{v} \in \mathbf{M}$ is a linear combination of $\mathbf{v}_1, \ldots, \mathbf{v}_k$, over $\mathbb{D}$. **23.3**
fixed space: the set of vectors fixed by a linear transformation. **3.6**
flip map: See *Section* **22.2.**
floating point addition, subtraction multiplication, division: Operations on floating point numbers, See *Section* **37.6.**
floating point number: A real number of the form $x = \pm\left(\frac{m}{b^{p-1}}\right) b^e$. **37.6**
floating point operation: One of floating point addition, subtraction, multiplication, division. **37.6**
flop: A floating point operation. **37.7**
FOM: See *full orthogonalization method.*
forest: A graph with no cycles. **28.1**
forward error (in $\hat{f}(\mathbf{x})$): $f(\mathbf{x}) - \hat{f}(\mathbf{x})$, the difference between the mathematically exact function evaluation and the perturbed function evaluation. **37.8**
forward stable (algorithm): The forward relative error is small for all valid input data $\mathbf{x}$ despite rounding and truncation errors in the algorithm. **37.8**
free variable: A variable x_i in the solution to a system of linear equations that is a free parameter. **1.4**
Frobenius norm (of $m \times n$ complex matrix A): $\|A\|_F = \sqrt{\sum_{i=1}^m \sum_{j=1}^n |a_{ij}|^2}$. **7.1, 37.3**
Frobenius normal form (of matrix A): A block upper triangular matrix with irreducible diagonal blocks that is permutation similar to A. **27.3**
full (cone): Has a nonempty interior. **8.5, 26.1**
fully indecomposable ((0, 1)-matrix): Not partly decomposable. **27.2**
full orthogonalization method (FOM): Generates, at each step k, the approximation $\mathbf{x}_k$ of the form $\mathbf{x}_k \in \mathbf{x}_0 + \text{Span}\{\mathbf{r}_0, A\mathbf{r}_0, \ldots, A^{k-1}\mathbf{r}_0\}$ for which the residual is orthogonal to the Krylov space. **41.3**
fundamental subspaces (of the least squares problem $A\mathbf{x} = \mathbf{b}$): range$A$, ker A^T, ker A and rangeA^T. **39.3**

G

Galerkin condition: The defining condition for a Ritz vector and Ritz value. **44.1**
Gauss multipliers: The nonzero entries in a Gauss vector. **38.3**
Gauss transformation: The matrix $M_k = I - \ell_k \mathbf{e}_k^T$, where ℓ_k is a Gauss vector. **38.3**

Gauss vector: A vector $\ell_k \in \mathbb{C}^n$ with the leading k entries equal to zero. **38.3**
Gauss–Jordan elimination: A process used to find the reduced row echelon form of a matrix. **1.3**
Gaussian elimination: A process used to find a row echelon form of a matrix. **1.3, 38.3**
g.c.d.: See *greatest common divisor.*
GCDD: See *greatest common divisor domain.*
general linear group (of order n over F): The multiplicative group consisting of all invertible $n \times n$ matrices over the field F. **67.1**
general solution: A formula that describes every vector in the solution set to a system of linear equations. **1.4**
generalized cycle (in a digraph): A disjoint union of one or more cycles such that every vertex lies on exactly one cycle. **29.1**
generalized cycle product: A walk product where the walk is a generalized cycle. **29.4**
generalized diagonal (of square matrix A): Entries of A corresponding to a generalized cycle in the digraph of A. **29.4**
generalized eigenspace (of matrix or linear operator A at λ): $\ker(A-\lambda I)^\nu$, where ν is the index of A at λ. **6.1**
generalized eigenvector (of a matrix or linear operator): A nonzero vector in the generalized eigenspace. **6.1**
generalized Laplacian (of simple graph G): A matrix M such that for $i \neq j$, $m_{ij} < 0$ if the ith and jth vertices are adjacent and $m_{ij} = 0$ otherwise (no restriction for $i = j$). **28.4**
generalized line graph: See *Section* **28.2**.
generalized minimal residual (GMRES): Algorithm generates, at each step k, the approximation $\mathbf{x}_k$ of the form $\mathbf{x}_k \in \mathbf{x}_0 + \text{Span}\{\mathbf{r}_0, A\mathbf{r}_0, \ldots, A^{k-1}\mathbf{r}_0\}$ for which the 2-norm of the residual is minimal. **41.3**
generalized sign pattern: A matrix whose entries are from the set $\{+, -, 0, \#\}$, where # indicates an ambiguous sum (the result of adding + with −). **33.1**
generalized star: A tree in which at most one vertex has degree greater than two. **34.6**
generators (of an ideal): See *finitely generated.*
generic matrix: A matrix whose nonzero elements are independent indeterminates over the field F. **30.2**
geometric multiplicity: The dimension of the eigenspace. **4.3**
Geršgorin discs (of $A \in \mathbb{C}^{n\times n}$): $\{z \in \mathbb{C} : |z - a_{ii}| \leq \sum_{j\neq i} |a_{ij}|\}$ for $i = 1, \ldots, n$. **14.2**
Givens matrix: See *Givens transformation.*
Givens rotation: See *Givens transformation.*
Givens transformation: An identity matrix modified so that the (i,i) and (j,j) entries are replaced by $c = \cos(\theta)$, the (i,j) entry is replaced by $s = e^{\imath\vartheta}\sin(\theta)$, and the (j,i) entry is replaced by $-\bar{s} = -e^{-\imath\vartheta}\sin(\theta)$. **38.4**
GMRES: See *generalized minimal residual, restarted GMRES algorithm.*
graded (associative algebra A): There exist $(A_k)_{k\in\mathbb{N}}$ vector subspaces of A such that $A = \bigoplus_{k\in\mathbb{N}} A_k$ and $A_i A_j \subseteq A_{i+j}$ for every $i, j \in \mathbb{N}$. **13.9**
graded matrix $A \in \mathbb{C}^{m\times n}$: A can be scaled as $A = GS$ (or $A = S^*GS$, depending on situation) such that G is "well-behaved" (i.e., $\kappa_2(G)$ is of modest magnitude), where S is a scaling matrix (often diagonal). **15.3**
Gram matrix (of vectors $\mathbf{v}_1, \ldots, \mathbf{v}_n$ in a complex inner product space): the matrix whose i, j-entry is $\langle \mathbf{v}_i, \mathbf{v}_j\rangle$. **8.1**
graph: A finite set of vertices and a finite multiset of edges, where each edge is an unordered pair of vertices (not necessarily distinct). **28.1**
graph (of $n \times n$ matrix A): The simple graph whose vertex set is $\{1, \ldots, n\}$ and having an edge between vertices i and j ($i \neq j$) if and only if $a_{ij} \neq 0$ or $a_{ji} \neq 0$. **19.3, 34.1**
graph (of a convex polytope $\mathcal{P}$): Vertices are the extreme points of $\mathcal{P}$ and edges are the pairs of extreme points of 1-dimensional faces of $\mathcal{P}$. **27.6**
graph (of $n \times n$ combinatorially symmetric partial matrix B): The graph having vertices $V = \{1, \ldots, n\}$ and for each $i, j \in V$, $\{i, j\}$ is an edge exactly when b_{ij} is specified (loops permitted, no multiple edges). **35.1**
graph of a combinatorially symmetric $n \times n$ sign pattern $A = [a_{ij}]$: The graph with vertex set $\{1, 2, \ldots, n\}$ where $\{i, j\}$ is an edge iff $a_{ij} \neq 0$ (note: the graph may have loops). **33.5**
Grassman algebra (on vector space V): $\bigwedge V = \bigoplus_{k\in\mathbb{N}} \left(\bigwedge^k V\right)$ with alt multiplication. **13.9**

Grassmann space (of vector space V): $\bigwedge^m V = \text{Alt}(\bigotimes^m V)$. **13.6**
greatest common divisor (g.c.d.) (of $a_1, \ldots, a_n \in \mathbb{D}$): $d \in \mathbb{D}$ such that $d|a_i$ for $i = 1, \ldots, n$, and if $d'|a_i, i = 1, \ldots, n$, then $d'|d$. **23.1**
greatest common divisor domain (GCDD): A domain in which any two elements have a g.c.d. **23.1**
greatest integer (of a real number x): the greatest integer less than or equal to x. **Preliminaries**
group: A set with one binary operation, satisfying associativity, existence of identity and inverses. **Preliminaries**
group inverse (of square complex matrix A): A matrix X satisfying $AXA = A$, $XAX = X$ and $AX = XA$. **9.1**

H

H**-matrix** (real square matrix A): The comparison matrix of A is an M-matrix. **19.5**
Hadamard matrix: A ± 1-matrix H_n with $H_n H_n^T = nI_n$. **32.2**
Hadamard product (of $A, B \in F^{n \times n}$): The $n \times n$ matrix whose i, j-entry is $a_{ij} b_{ij}$. **8.5**
Hall matrix: An $n \times n$ $(0,1)$-matrix that does not have a $p \times q$ zero submatrix for positive integers p, q with $p + q > n$. **27.2**
Hamilton cycle: A cycle in a graph that includes all vertices. **28.1**
Hamiltonian cycle: See *Hamilton cycle.*
Hankel matrix: A matrix with constant elements along its antidiagonals. **48.1**
height (of basic class B): The largest number of vertices on a simple walk that ends at B in the basic reduced digraph. **9.3**
hereditary (class of matrices X): Whenever A is an X-matrix and $\alpha \subseteq \{1, \ldots, n\}$, then $A[\alpha]$ is an X-matrix. **35.1**
Hermite normal form: A generalization of reduced row echelon form used in domains. **23.2**
Hermitian (linear operator on an inner product space): An operator that is equal to its adjoint. **5.3**
Hermitian (matrix): A real or complex matrix equal to its conjugate transpose. **1.2, 7.2, 8.1**
Hermitian adjoint (of a matrix): The conjugate transpose of a real or complex matrix. **1.2**
Hermitian form f (on complex vector space V): A sesquilinear form such that $f(\mathbf{v}, \mathbf{u}) = \overline{f(\mathbf{u}, \mathbf{v})}$ for all $\mathbf{u}, \mathbf{v} \in V$. **12.5**
Hoffman polynomial (of graph G): A polynomial $h(x)$ of minimum degree such that $h(A_G) = J$. **28.3**
Hölder norm: See *p-norm.*
homogeneous (digraph): Either symmetric or asymmetric. **35.7**
homogeneous (element of graded algebra $A = \bigoplus_{k \in \mathbb{N}} A_k$): An element of some A_k. **13.9**
homogeneous (pencil associated with $A(x) = A_0 + xA_1 \in \mathbb{D}[x]^{m \times n}$): $A(x_0, x_1) = x_0 A_0 + x_1 A_1 \in \mathbb{D}[x_0, x_1]^{m \times n}$. **23.4**
homogeneous (polynomial): All terms have the same degree. **23.1**
homogeneous (system of linear equations): A system in which all the constants are zero. **1.4**
Householder matrix: See *Householder transformation.*
Householder reflector: See *Householder transformation.*
Householder transformation (defined by $\mathbf{v} \in \mathbb{C}^n$): The matrix $I - \frac{2}{\|\mathbf{v}\|_2^2} \mathbf{v}\mathbf{v}^*$. **38.4**
Householder vector: The vector $\mathbf{v}$ used to define a Householder transformation. **38.4**
hyperbolic SVD decomposition (of matrix pair $(G, \mathcal{J})$): A decomposition of G, $G = W \Sigma B^{-1}$, where W is orthogonal, Σ is diagonal, and B is $\mathcal{J}$–orthogonal. **46.5**

I

IAP: See *inertially arbitrary pattern.*
ideal (of a domain $\mathbb{D}$): A nonempty subset I of $\mathbb{D}$ such that if $a, b \in I$ and $p, q \in \mathbb{D}$, then $pa + qb \in I$. **23.1**
idempotent: A matrix or linear transformation that squares to itself. **2.7, 3.6**
identity matrix: A diagonal matrix having all diagonal elements equal to one. **1.2**
identity pattern (of order n): The $n \times n$ diagonal pattern with + diagonal entries. **33.1**
identity transformation: A linear operator that maps each vector to itself. **3.1**

IEP: See *inverse eigenvalue problem.*
IF-RCF matrix: A block diagonal matrix of the form $C(a_1) \oplus \cdots \oplus C(a_s)$, where $a_i(x)$ divides $a_{i+1}(x)$ for $i = 1, \ldots, s-1$. **6.6**
ill-conditioned (input **z** for P): Some small relative perturbation of **z** causes a large relative perturbation of $P(\mathbf{z})$. **37.4**
image: See *range.*
imaginary part (of complex number $a + bi$): b. **Preliminaries**
immanant: A function f_λ from complex square matrices to complex numbers defined by the irreducible character of a partition λ of n, specifically, $f_\lambda(M) = \sum_{\sigma \in S_n} \chi_\lambda(\sigma) \prod_{i=1}^{n} m_{i\sigma(i)}$. **31.10**
implicit restarting (of the Arnoldi algorithm): Apply a sequence of implicitly shifted QR steps to an m-step Arnoldi or Lanczos factorization to obtain a truncated form of the implicitly shifted QR-iteration. **44.5**
imprimitive (digraph): Not primitive. **29.6**
imprimitive (matrix A): The digraph $\Gamma(A)$ is imprimitive. **29.6**
improper (divisor of a in a domain): An associate of a or a unit. **23.1**
incidence matrix (of a graph): A matrix with rows indexed by the vertices and columns indexed by the edges, having i, j entry 1 if vertex v_i is an endpoint of edge e_j and 0 otherwise. **28.4**
incidence matrix (of subsets $S_1, S_2, \ldots, S_m$ of a finite set $\{x_1, x_2, \ldots, x_n\}$): The $m \times n$ (0,1)-matrix $M = [m_{ij}]$ in which $m_{ij} = 1$ iff $x_j \in S_i$. **31.3**
($\pm$1)-incidence matrix (of a 2-design): A matrix W whose rows are indexed by the elements x_i of X and whose columns are indexed by the blocks B_j having $w_{ij} = -1$ if $x_i \in B_j$ and $w_{ij} = +1$ otherwise. **32.2**
incomplete Cholesky decomposition: A preconditioner for a Hermitian positive definite matrix A of the form $M = LL^*$, where L is a sparse lower triangular matrix. **41.4**
incomplete LU decomposition: A preconditioner for a general matrix A of the form $M = LU$, where L and U are sparse lower and upper triangular matrices. **41.4**
inconsistent: A system of linear equations that has no solution. **1.4**
indefinite (Hermitian matrix A): Neither A nor $-A$ is positive semidefinite. **8.4**
independent (subspaces W_i): $\mathbf{w}_1 + \cdots + \mathbf{w}_k = \mathbf{0}$ and $\mathbf{w}_i \in W_i, i = 1, \ldots, k$ implies $\mathbf{w}_i = \mathbf{0}$ for all $i = 1, \ldots, k$. **2.3**
independent set of vertices: See *coclique.*
index (of an eigenvalue): The smallest integer k such that the k-eigenspace equals the $k+1$-eigenspace. **6.1**
index (of square nonnegative matrix P): The index of the spectral radius of P. **9.3**
index (of $A \in \mathrm{H}_0$): See *Section* **23.3.**
index of imprimitivity (of an irreducible matrix A): The greatest common divisor of the lengths of all cycles in the digraph $\Gamma(A)$; same as period. **29.7**
index of primitivity: See *exponent.*
individual condition number for eigenvalue λ: $\frac{\|\mathbf{x}\|_2 \|\mathbf{y}\|_2}{|\mathbf{y}^* \mathbf{x}|}$ where $(\mathbf{y}, \lambda, \mathbf{x})$ is an eigentriplet. **15.1**
induced basis: The nonzero images in the quotient V/W of vectors in a basis that contains a basis for W. **2.3**
induced norm: The matrix norm induced by the family of vector norms $\|\cdot\|$ is $\|A\| = \max_{\mathbf{x} \neq 0} \frac{\|A\mathbf{x}\|}{\|\mathbf{x}\|}$. **37.3**
induced subdigraph (of $\Gamma = (V, E)$): A subdigraph $\Gamma = (V', E')$ containing all arcs from E with endpoints in V'. **29.1**
induced subgraph (of $G = (V, E)$): A subgraph $G = (V', E')$ containing all edges from E with endpoints in V'. **28.1**
inertia (of complex square matrix A): The ordered triple $\mathrm{in}(A) = (\pi(A), \nu(A), \delta(A))$ where $\pi(A)$ is the number of eigenvalues of A with positive real part, $\nu(A)$ is the number of eigenvalues of A with negative real part, and $\delta(A)$ is the number of eigenvalues of A on the imaginary axis. **8.3, 19.1**
inertia preserving (real square matrix A): The inertia of AD is equal to the inertia of D for every nonsingular real diagonal matrix D. **19.3**
inertially arbitrary pattern (IAP): A sign pattern A of size n such that every possible ordered triple (p, q, z) of nonnegative integers p, q, and z with $p + q + z = n$ can be achieved as the inertia of some $B \in Q(A)$. **33.6**
infinite dimensional (vector space): A vector space that is not finite dimensional. **2.2**

initial (minor): See *Section* **21.3**
injective: See *one-to-one.*
inner distribution: Vector parameter of an association scheme. **28.6**
inner product (on a vector space V over $F = \mathbb{R}$ or $\mathbb{C}$): A function $\langle\cdot,\cdot\rangle : V \times V \to F$ satisfying certain conditions. For $F = \mathbb{R}$, $\langle\cdot,\cdot\rangle$ is symmetric, bilinear and positive definite. **5.1**
inner product space: A real or complex vector space with an inner product. **5.1**
integral domain: A commutative ring without zero divisors and containing identity 1. **23.1**
iterative method (for solving a linear system $A\mathbf{x} = \mathbf{b}$): Any algorithm that starts with an initial guess $\mathbf{x}_0$ for the solution and successively modifies that guess in an attempt to obtain improved approximate solutions $\mathbf{x}_1, \mathbf{x}_2, \ldots$. **41.1**
internal direct sum: See *direct sum.*
intersection (of two graphs $G = (V, E)$ and $G' = (V', E')$): The graph with vertex set $V \cap V'$, and edge (multi)set $E \cap E'$. **28.1**
intersection numbers: Parameters of an association scheme. **28.6**
P**-invariant face** (of K-nonnegative matrix P): $PF \subseteq F$. **26.1**
invariant factors (of IF-RCF matrix $C(a_1) \oplus \cdots \oplus C(a_s)$): The polynomials $a_i(x), i = 1, \ldots s$. **6.6**
invariant factors (of matrix or linear operator $A \in F^{n\times n}$): The invariant factors of the invariant factors rational canonical form of A. **6.6**
invariant factors (of a matrix over a domain): See *Section* **23.1.**
invariant factors rational canonical form (of matrix $A \in F^{n\times n}$): The IF-RCF matrix similar to A. **6.6**
invariant factors rational canonical form (of a linear operator): See *Section* **6.6.**
invariant factors rational canonical form matrix: See *IF-RCF matrix.*
invariant polynomials: See *invariant factors.*
invariant subspace (of linear transformation $T : V \to W$): A subspace U of V such that for all $\mathbf{u} \in U$, $T(\mathbf{u}) \in U$. **3.6**
inverse (of square matrix A): A matrix B such that $AB = BA = I$. **1.5**
inverse (of linear transformation $T : V \to W$): A linear transformation $S : W \to V$ such that $TS = I_W$ and $ST = I_V$. **3.7**
inverse eigenvalue problem: The problem of constructing a matrix with prescribed structural and spectral constraints. **20.1**
inverse eigenvalue problem with prescribed entries: Construct a matrix with given eigenvalues subject to given entries in given positions. **20.1**
Inverse Eigenvalue Problem of tree T: Determine all possible spectra that occur among matrices in $\mathcal{S}(T)$. **34.5**
inverse iteration: The power method applied to the inverse of a shifted matrix. **42.1**
inverse M**-matrix**: An invertible matrix whose inverse is an M-matrix. **9.5**
inverse nonnegative: A square sign pattern that allows an entrywise nonnegative inverse. **33.7**
inverse-positive (real square matrix A): A is nonsingular and $A^{-1} \geq 0$. **9.5**
inverse positive (square sign pattern): Allows an entrywise positive inverse. **33.7**
invertible (matrix or linear transformation): Has an inverse. **1.5, 3.7**
$\mathbb{D}$**-invertible** (matrix in $\mathbb{D}^{n\times n}$): Has an inverse within $\mathbb{D}^{n\times n}$. **23.2**
irreducible (element a of a domain): a is not a unit and every divisor of a is an associate of a or a unit. **23.2**
irreducible (matrix): Not reducible. **9.2, 27.3**
K**-irreducible** (K-nonnegative matrix P): The only P-invariant faces are the trivial faces. **26.1**
irreducible components (of matrix A): The diagonal blocks in the Frobenius normal form of A. **27.3**
isolated vertex: A vertex of a graph having no incident edges. **28.1**
isomorphic (graphs $G = (V, E)$ and $G' = (V', E')$): There exist bijections $\phi : V \to V'$ and $\psi : E \to E'$, such that $v \in V$ is an endpoint of $e \in E$ if and only if $\phi(v)$ is an endpoint of $\psi(e)$. **28.1**
isomorphic (vector spaces): There is an isomorphism from one vector space onto the other. **3.7**
isomorphism (of vector spaces): An invertible linear transformation. **3.7**
isoperimetric number: A parameter of a simple graph. **28.5**

J

$\mathcal{J}$–orthogonal: Alphabetized under *orthogonal.*
Jacobi method (for computing the EVD of A): A sequence of matrices, $A_0 = A$, $A_{k+1} = G(i_k, j_k, c, s) A_k G(i_k, j_k, c, s)^T$, $k = 1, 2, \ldots$, where $G(i_k, j_k, c, s)$ is a Givens rotation matrix. **42.7**
Jacobi rotation: A Givens rotation used in the Jacobi method. **42.7**
join (of disjoint graphs $G = (V, E)$ and $G' = (V', E')$): The union of $G \cup G'$ and the complete bipartite graph with vertex set $V \cup V'$ and partition $\{V, V'\}$. **28.1**
join (of faces F, G of cone K): Face generated by $F \cup G$. **26.1**
Jordan basis (for matrix A): The ordered set of columns of C where $\text{JCF}(A) = C^{-1}AC$. **6.2**
Jordan basis (of a linear operator): See *Section* **6.2.**
Jordan block (of size k with eigenvalue λ): The $k \times k$ matrix having every diagonal entry equal to λ, every first superdiagonal entry equal to 1, and every other entry equal to 0. **6.2**
Jordan canonical form (of matrix A): A Jordan matrix that is similar to A. **6.2**
Jordan canonical form (of a linear operator T): ${}_{\mathcal{B}}[T]_{\mathcal{B}}$ that is a Jordan matrix. **6.2**
Jordan chain (above eigenvector $\mathbf{x}$ for eigenvalue λ of A): A sequence of vectors $\mathbf{x}_0 = \mathbf{x}, \mathbf{x}_1, \ldots, \mathbf{x}_h$ satisfying $\mathbf{x}_i = (A - \lambda I)\mathbf{x}_{i+1}$ for $i = 0, \ldots, h-1$ (where h is the depth of $\mathbf{x}$). **6.2**
Jordan invariants (of matrix or linear operator A): The set of distinct eigenvalues of A and for each eigenvalue λ, the number b_λ and sizes $p_1, \ldots, p_{b_\lambda}$ of the Jordan blocks with eigenvalue λ in a Jordan canonical form of A. **6.2**
Jordan matrix: A block diagonal matrix having Jordan blocks as the diagonal blocks. **6.2**

K

K-irreducible, K-nonnegative, K-positive, K-semipositive: Alphabetized under *irreducible, nonnegative, positive, semipositive.*
ker: See *kernel.*
kernel (ker) (of a linear transformation): The set of vectors mapped to zero by the transformation. **3.5**
kernel (ker) (of a matrix A): The set of solutions to $A\mathbf{x} = \mathbf{0}$. **2.4**
Kronecker product (of matrices A and B): The block matrix whose i, j block is $a_{ij}B$ **10.4**
Krylov matrix (of matrix A, vector $\mathbf{x}$, and positive integer k): $[\mathbf{x}, A\mathbf{x}, A^2\mathbf{x}, \cdots, A^{k-1}\mathbf{x}]$. **42.8**
Krylov space: A subspace of the form $\text{Span}\{\mathbf{q}, A\mathbf{q}, A^2\mathbf{q}, \ldots, A^{k-1}\mathbf{q}\}$, where A is an n by n matrix and $\mathbf{q}$ is an n-vector. **41.1, 44.1**
Ky Fan k norms (of $A \in \mathbb{C}^{m\times n}$): $\|A\|_{K,k} = \sum_{i=1}^{k} \sigma_i(A)$. **17.3**

L

L-matrix (sign pattern or real matrix A): For every $B \in Q(A)$, the rows of B are linearly independent. **33.3**
Lanczos algorithm (for Hermitian matrices): A short recurrence for constructing an orthonormal basis for a Krylov space. **41.2**
Laplacian: See *Laplacian matrix.*
Laplacian eigenvalues (of simple graph G): The eigenvalues of the Laplacian matrix of G. **28.4**
Laplacian matrix (of simple graph G): The matrix $D - \mathcal{A}_G$, where D is the diagonal matrix of vertex degrees and $\mathcal{A}_G$ is the adjacency matrix of G. **28.4**
Laplacian matrix (of a weighted graph): The matrix $L = [\ell_{ij}]$ such that $\ell_{ij} = 0$ if vertices i and j are distinct and not adjacent, $\ell_{ij} = -w(e)$ if $e = \{i, j\}$ is an edge, and $\ell_{i,i} = \sum w(e)$, where the sum is taken over all edges e incident with vertex i. **36.4**
largest size of a zero submatrix: The maximum of $r + s$ such that the matrix has an $r \times s$ (possibly vacuous) zero submatrix. **27.1**
LDU factorization: A factorization of a matrix as the product of a unit lower triangular matrix, a diagonal matrix, and a unit upper triangular matrix. **1.6**
leading principal minor: The determinant of a leading principal submatrix. **4.2**

leading principal submatrix: A principal submatrix lying in rows and columns 1 to k. **1.2**
leading entry: The first nonzero entry in a row of a matrix in REF. **1.3**
least squares data fitting: See *Section* **39.2.**
least squares problem ($A\mathbf{x} = \mathbf{b}$): Find a vector $\mathbf{x} \in \mathbb{R}^n$ such that $\|\mathbf{b} - A\mathbf{x}\|_2$ is minimized. **5.8, 39.1**
least squares solution: A solution to a least squares problem. **5.8, 39.1**
left singular space: See *singular spaces.*
left singular vector: See *singular vectors.*
left eigenvector (of matrix A): A nonzero row vector $\mathbf{y}$ such that there exists a scalar λ such that $\mathbf{y}A = \lambda\mathbf{y}$. **4.3**
length (of a walk, path, cycle in a graph or digraph): The number of edges (arcs) in the walk. **28.1, 29.1**
length (of a permutation cycle): The number of elements in the cycle. **Preliminaries**
length (of a vector): See *norm.*
$(C,0)$-**limit** (of complex sequence $\{a_m\}_{m=0,1,\dots}$): Ordinary limit $\lim_{m\to\infty} a_m$. **9.1**
$(C,1)$-**limit** (of complex sequence $\{a_m\}_{m=0,1,\dots}$): $\lim_{m\to\infty} m^{-1}\sum_{s=0}^{m-1} a_s$. **9.1**
(C,k)-**limit** (of complex sequence $\{a_m\}_{m=0,1,\dots}$): Defined inductively from $(C,k-1)$-limit. See *Section* **9.1.**
line (of a matrix): A row or column. **27.1**
line graph (of simple graph G): The simple graph that has as vertices the edges of G, and vertices are adjacent if the corresponding edges of G have an endpoint in common. **28.2**
linear combination: A (finite) sum of scalar multiples of vectors. **2.1**
linear equation: An equation of the form $a_1x_1 + \cdots + a_nx_n = b$. **1.4**
linear form: See *linear functional.*
linear functional: A linear transformation from a vector space to the field. **3.8**
m-**linear map**: See *multilinear map.*
linear mapping: See *linear transformation.*
linear operator: A linear transformation from a vector space to itself. **3.1**
linear preserver (of function f**)**: A linear operator $\phi : \mathcal{V} \to \mathcal{V}$ such that $f(\phi(A)) = f(A)$ for every $A \in \mathcal{V}$, where $\mathcal{V}$ is a subspace of $F^{m\times n}$. **22.1**
linear preserver problem (corresponding to the property P): The problem of characterizing all linear (bijective) maps on a subspace of matrices satisfying P. **22.2**
linear transformation: A function from a vector space to a vector space that preserves addition and scalar multiplication. **3.1**
linear transformation associated to matrix A: The transformation that multiplies each vector in F^n by A. **3.3**
linearly dependent (set of vectors): There is a linear combination of the vectors with at least one nonzero coefficient, that is equal to the zero vector. **2.1**
linearly independent (set of vectors): Not linearly dependent. **2.1**
linked: Two disjoint Jordan curves in $\mathbb{R}^3$ such that there is no topological 2-sphere in $\mathbb{R}^3$ separating them. **28.2**
linklessly embeddable: There is an embedding of graph G in $\mathbb{R}^3$ such that no two disjoint cycles of G are linked. **28.2**
little-oh: Function f is $o(g)$ if the limit of $\frac{|f|}{|g|}$ is 0. **Preliminaries**
local similarity: A map $\phi : F^{n\times n} \to F^{n\times n}$ such that for every $A \in F^{n\times n}$ there exists an invertible $R_A \in F^{n\times n}$ such that $\phi(A) = R_A A R_A^{-1}$. **22.4**
Loewner (partial) ordering (on Hermitian matrices A, B): $A \succ B$ if $A - B$ is positive definite and $A \succeq B$ if $A - B$ is positive semidefinite. **8.5**
logarithm (of a square complex matrix A): Any matrix B such that $e^B = A$. **11.4**
look-ahead strategy: A technique to overcome breakdown in some algorithms. **41.3**
loop: An edge (arc) of a graph (directed graph) having both endpoints equal. **29.1**
Lovász parameter: A parameter of a simple graph. **28.5**
lower Hessenberg (matrix A): A^T is upper Hessenberg. **10.2**
lower shift matrix: A square matrix with ones on the first subdiagonal and zero elsewhere. **48.1**

lower triangular matrix: A matrix such that every entry having row number less than column number is zero. **1.2, 10.2**
Löwner (partial) ordering: See *Loewner (partial) ordering*
LU **factorization**: A factorization of a matrix as the product of a unit lower triangular matrix and an upper triangular matrix. **1.6, 38.3**
Lyapunov diagonally semistable (real square matrix A): There exists a positive diagonal matrix D such that $AD + DA^T$ is positive semidefinite. **19.5**
Lyapunov diagonally stable (real square matrix A): There exists a positive diagonal matrix D such that $AD + DA^T$ is positive definite. **19.5**
Lyapunov scaling factor (of real square matrix A): A positive diagonal matrix D such that $AD + DA^T$ is positive (semi)definite. **19.5**

M

M-**matrix** (real square matrix A): A can be written as $A = sI - P$ with P a nonnegative matrix and s a scalar satisfying $s > \rho(P)$. **9.5**
M_0-**matrix:** (real square matrix A): A can be written as $A = sI - P$ with P a nonnegative matrix and s a scalar satisfying $s \geq \rho(P)$. **9.5**
machine epsilon: The distance between the number one and the next larger floating point number. **37.6**
main angles (of a graph): The cosines of the angles between the eigenspaces of the adjacency matrix and the all-ones vector. **28.3**
main diagonal (of a matrix): The set of diagonal entries of a matrix. **1.2**
majorizes: Weakly majorizes and the sums of all entries are equal. **Preliminaries**
(k-)**matching** (of a graph): A set of k mutually disjoint edges. **30.1**
matrices: Plural of *matrix*.
matrix: a rectangular array of elements of a field. **1.2**
Or if specifically stated, a rectangular array of elements of a domain, ring, or algebra. The following standard matrix terms are applied to the matrices over a domain $\mathbb{D}$ viewed as in the quotient field of $\mathbb{D}$: determinant, minor, rank, *Section* **23.2**. The following standard matrix terms are applied to matrices over a Bezout domain within the domain: kernel, range, system of linear equations, coefficient matrix, augmented matrix, *Section* **23.3**
matrix condition number: See *condition number (of matrix A for linear systems).*
matrix cosine: See *cosine.*
matrix direct sum: A block diagonal matrix. **2.4, 10.2**
matrix exponential: See *exponential.*
matrix function: A function of a square complex matrix; can be defined using power series, Jordan Canonical Form, polynomial interpretation, Cauchy integral, etc. **11.1**
matrix logarithm: See *logarithm.*
matrix of a transformation T (with respect to bases $\mathcal{B}$ and $\mathcal{C}$): The matrix consisting of the coordinate vectors with respect to $\mathcal{C}$ of the images under T of the vectors in $\mathcal{B}$. **3.3**
matrix norm $\|\cdot\|$: A family of real-valued functions defined on $m \times n$ real or complex matrices for all positive integers m and n, such that for all matrices A and B (where A and B are compatible for the given operation) and all scalars, (1) $\|A\| \geq 0$, and $\|A\| = 0$ implies $A = 0$; (2) $\|\alpha A\| = |\alpha|\|A\|$; (3) $\|A + B\| \leq \|A\| + \|B\|$; (4) $\|AB\| \leq \|A\|\,\|B\|$. **37.3**
matrix pencil: See *pencil.*
matrix product: The result of multiplying two matrices. **1.2**
matrix representing f **relative to basis** $\mathcal{B} = (\mathbf{w_1}, \mathbf{w_2}, \ldots, \mathbf{w_n})$: The matrix whose i, j-entry is $f(\mathbf{w_i}, \mathbf{w_j})$, Section **12.1**; also applied to φ-sesquilinear form. **12.4**
matrix sine: See *sine.*
matrix sign function: See *sign (of a matrix).*
matrix square root: See *square root, principal square root.*

matrix–vector product: The result of multiplying a matrix and a vector. **1.2**
maximal (ideal I of a domain $\mathbb{D}$): The only ideals that contain I are I and $\mathbb{D}$, and $I \neq \mathbb{D}$. **23.1**
maximal (sign-nonsingular-sign pattern matrix): If a zero entry is set nonzero, then the resulting pattern is not SNS. **33.2**
maximal rank (of sign-pattern A): $\max\{\text{rank}B : B \in Q(A)\}$. **33.6**
maximum column sum norm (of matrix $A \in \mathbb{C}^{m\times n}$): $\|A\|_1 = \max_{1\le j\le n} \sum_{i=1}^{m} |a_{ij}|$. **37.3**
maximum multiplicity (of simple graph G): The maximum multiplicity of any eigenvalue among matrices in $\mathcal{S}(G)$. **34.2**
maximum row sum norm (of matrix $A \in \mathbb{C}^{m\times n}$): $\|A\|_\infty = \max_{1\le i\le m} \sum_{j=1}^{n} |a_{ij}|$. **37.3**
max-plus semiring: The set $\mathbb{R} \cup \{-\infty\}$, equipped with the addition $(a, b) \mapsto \max(a, b)$ and the multiplication $(a, b) \mapsto a + b$. The identity element for the addition, zero, is $-\infty$, and the identity element for the multiplication, unit, is 0. **25.1**
measure of relative separation (of positve real numbers a, b): $|\sqrt{a/b} - \sqrt{b/a}|$. **17.4**
meet (of faces F, G of cone K): $F \cap G$ **26.1**
metric: A distance function. **Preliminaries**
MIEP: See *multiplicative inverse eigenvalue problem.*
minimal (matrix norm $\|\cdot\|$): For any matrix norm $\|\cdot\|_\nu$, $\|A\|_\nu \le \|A\|$ for all A implies $\|\cdot\|_\nu = \|\cdot\|$. **37.3**
minimal polynomial (of matrix A): The unique monic polynomial of least degree for which $q_A(A) = 0$. **4.3**
minimal rank (of sign pattern A): $\min\{\text{rank}B : B \in Q(A)\}$. **33.6**
minimal residual (MINRES) algorithm (to solve $H\mathbf{x} = \mathbf{b}$ with H preconditioned Hermitian): At each step k, the approximation $\mathbf{x}_k$ is of the form $\mathbf{x}_k \in \mathbf{x}_0 + \text{Span}\{\mathbf{r}_0, H\mathbf{r}_0, \ldots, H^{k-1}\mathbf{r}_0\}$ for which the 2-norm of the residual, $\|\mathbf{r}_k\|_2$, is minimal. **41.2**
minimal sign-central: A sign-central pattern that is not sign-central if any column of A is deleted. **33.11**
minimally connected (digraph): Strongly connected and the deletion of any arc produces a subdigraph that is not strongly connected. **29.8**
minimally potentially stable (sign pattern A): A is a potentially stable, irreducible sign pattern such that replacing any nonzero entry by zero results in a pattern that is not potentially stable. **33.4**
minimum co-cover (of a (0, 1)-matrix): A co-cover with the smallest number of 1s. **27.1**
minimum cover (of a (0, 1)-matrix): A cover with the smallest number of lines. **27.1**
minimum-norm least squares solution: The least squares solution of minimum Euclidean norm. **39.1**
minimum rank (of simple graph G): The minimum of rank$A$ among matrices $A \in \mathcal{S}(G)$. **34.2**
minor (of a graph G): Any graph that can be obtained from G by a sequence of edge deletions, vertex deletions, and contractions. **28.2**
minor (of a matrix): The determinant of a submatrix; for the i, j-minor, the submatrix is obtained by deleting row i and column j. **4.1**
MINRES: See *minimal residual.*
Möbius function: $\mu : \mathbb{N} \mapsto \{-1, 0, 1\}$ is defined by $\mu(1) = 1$, $\mu(m) = (-1)^e$ if m is a product of e distinct primes, and $\mu(m) = 0$ otherwise. **20.5**
$\mathbb{D}$-module: A generalization of vector space over a field; an additive group with a (scalar) multiplication by elements $a \in \mathbb{D}$ that satisfies the standard distribution properties and $1\mathbf{v} = \mathbf{v}$ for all $\mathbf{v} \in \mathbb{D}$. The following standard terms are also applied to modules over a Bezout domain—linear combination, basis (may not exist), standard basis for $\mathbb{D}^n$, dimension (if there is a basis). **23.3**
monic (polynomial $p(x)$): The coefficient of the highest power of x in $p(x)$ is 1. **23.1**
monotone (real vector $\mathbf{v} = [v_i]$): $v_1 \ge v_2 \ge \cdots \ge v_n$. **27.4**
monotone (vector norm $\|\cdot\|$): For all $\mathbf{x}, \mathbf{y}$, $|\mathbf{x}| \le |\mathbf{y}|$ implies $\|\mathbf{x}\| \le \|\mathbf{y}\|$. **37.1**
Moore–Penrose pseudo-inverse (of matrix $A \in \mathbb{C}^{m\times n}$): A matrix $A^\dagger \in \mathbb{C}^{n\times m}$ satisfying: $AA^\dagger A = A$; $A^\dagger AA^\dagger = A^\dagger$; $(AA^\dagger)^* = AA^\dagger$; $(A^\dagger A)^* = A^\dagger A$. **5.7**
Moore–Penrose inverse (of a sign pattern): Analog of the Moore–Penrose inverse of a real matrix. **33.7**
multibigraph (of a nonnegative integer matrix A): Like the bigraph of A, except that edge $\{i, j'\}$ has multiplicity a_{ij}. **30.2**

multigrid preconditioner: A preconditioner designed for problems arising from partial differential equations discretized on grids. **41.4**
multilinear form: A multilinear map into the field F. **13.1**
multilinear map (**m-linear map**): A map φ from $V_1 \times \cdots \times V_m$ into U that is linear on each coordinate. **13.1**
multiple (eigenvalue λ of real symmetric matrix B): $\alpha_B(\lambda) > 1$. **34.1**
multiplicative (map $\phi : F^{n\times n} \to F^{n\times n}$): $\phi(AB) = \phi(A)\phi(B)$, for all $A, B \in F^{n\times n}$. **22.4**
multiplicative D-stable (real square matrix A): DA is stable for every positive diagonal matrix D. **19.3**
multiplicative inverse eigenvalue problem (MIEP): Given $B \in \mathbb{C}^{n\times n}$ and $\lambda_1, \ldots, \lambda_n \in \mathbb{C}$ find a diagonal matrix $D \in \mathbb{C}^{n\times n}$ such that $\sigma(BD) = \{\lambda_1, \ldots, \lambda_n\}$. **20.9**
multiplicative perturbation (of $A \in \mathbb{C}^{m\times n}$): $D_L^* A D_R$ for some $D_L \in \mathbb{C}^{m\times m}$, $D_R \in \mathbb{C}^{n\times n}$. **15.3**
multiplicative preserver: A multiplicative map preserving a certain property. **22.4**
multiplicity (of an eigenvalue): See *algebraic multiplicity, geometric multiplicity.*
multiplicity (of a singular value): The dimension of the right or left singular space. **45.1**
multiset: An unordered list of elements that allows repetition. **Preliminaries**

N

NaN: Result of an operation that is not a number in a standard-conforming floating point system. **37.6**
nearly decomposable ((0, 1)-matrix): Fully indecomposable and each matrix obtained by replacing a 1 with a 0 is partly decomposable. **27.2**
nearly reducible (matrix A): Irreducible and each matrix obtained from A by replacing a nonzero entry with a zero is reducible. **27.3**
nearly sign-central: A sign pattern that is not sign-central, but can be augmented to a sign-central pattern by adjoining a column. **33.11**
nearly sign nonsingular (NSNS) (sign pattern): A square pattern having at least two nonzero terms in the expansion of its determinant, with precisely one nonzero term having opposite sign to the others. **33.2**
negative definite (real symmetric or Hermitian bilinear form f): $-f$ is positive definite. **12.2, 12.5**
negative semidefinite (real symmetric or Hermitian bilinear form f): $-f$ is positive semidefinite. **12.2, 12.5**
negative set of edges α (in a signed bipartite graph): sgn(α) is -1. **30.1**
negative stable (complex polynomial): Its roots lie in the open left half plane. **19.2**
negative stable (complex square matrix): Its eigenvalues lie in the open left half plane. **19.2**
neighbor (of vertex v): Any vertex adjacent to v. **28.1**
NIEP: See *nonnegative inverse eigenvalue problem.*
nilpotent: A matrix or linear transformation that can be multiplied by itself a finite number of times to obtain the zero matrix or transformation. **2.7, 3.6**
node: See *vertex.*
noncommutative algorithms: Algorithms that do not use commutativity of multiplication. **47.2**
nondefective (matrix $A \in F^{n\times n}$): For each eigenvalue of A (over algebraic closure of F), the geometric multiplicity equals the algebraic multiplicity. **4.3**
nondegenerate (bilinear form f on vector space V): The rank of f is equal to dim V, *Section* **12.1**; also applied to φ-sesquilinear form. **12.4**
nonderogatory (matrix $A \in F^{n\times n}$): The geometric multiplicity of each eigenvalue of A (over algebraic closure of F) is 1. **4.3**
nondifferentiable (boundary point μ of convex set S): There is more than one support line of S passing through μ. **18.2**
non-Hermitian Lanczos algorithm See *Section* **41.3**
nonnegative (real matrix A): All of A's elements are nonnegative. **9.1**
nonnegative (sign pattern): All of its entries are nonnegative. **33.7**
K-**nonnegative** (matrix $A \in \mathbb{R}^{n\times n}$): $AK \subseteq K$. **26.1**
K-**nonnegative** (vector $\mathbf{v} \in \mathbb{R}^n$): $\mathbf{v} \in K$. **26.1**
nonnegative inverse eigenvalue problem: Construct a nonnegative matrix with given eigenvalues. **20.3**

nonnegative integer rank (of nonnegative integer matrix A): The minimum k such that there exists an $m \times k$ nonnegative integer matrix B and a $k \times n$ nonnegative integer matrix C with $A = BC$. **30.3**
nonnegative P-matrix: A P-matrix in which every entry is nonnegative. **35.9**
nonnegative P_0-matrix: A P_0-matrix in which every entry is nonnegative. **35.9**
nonnegative $P_{0,1}$-matrix: A $P_{0,1}$-matrix in which every entry is nonnegative. **35.9**
nonnegative stable (complex square matrix A): The real part of each eigenvalue of A is nonnegative. **9.5**
nonprimary matrix function: Function of a matrix defined using the Jordan Canonical Form using different branches for f and its derivatives for two Jordan blocks for the same eigenvalue. **11.1**
nonseparable (graph or digraph): Connected and does not have a cut-vertex. **35.1**
nonsingular (linear transformation): A linear transformation whose kernel is zero. **3.7**
nonsingular (matrix): An invertible matrix. **1.5**
norm (of a vector $\mathbf{v}$, in an inner product space): $\sqrt{\langle \mathbf{v}, \mathbf{v} \rangle}$. **5.1**
norm: See *vector norm, matrix norm* or *specific norm.*
1-norm (of vector $\mathbf{x} \in \mathbb{C}^n$): $\|\mathbf{x}\|_1 = |x_1| + |x_2| + \cdots + |x_n|$. **37.1**
1-norm (of matrix $A \in \mathbb{C}^{m \times n}$): See *maximum column sum norm.*
2-norm (of vector $\mathbf{x} \in \mathbb{C}^n$): $\|\mathbf{x}\|_2 = \sqrt{|x_1|^2 + |x_2|^2 + \cdots + |x_n|^2}$. **37.1**
2-norm (of matrix $A \in \mathbb{C}^{m \times n}$): see *maximum column sum norm.*
∞-norm (of vector $\mathbf{x} \in \mathbb{C}^n$): $\|\mathbf{x}\|_\infty = \max_{1 \le i \le n} |x_i|$. **37.1**
∞-norm (of matrix A): See *maximum row sum norm.*
M-norm (where $\|\cdot\|$ is a vector norm and M is a nonsingular matrix): $\|\mathbf{x}\|_M \equiv \|M\mathbf{x}\|$. **37.1**
$\mathcal{M}$-norm (matrix norm): Norm induced by a family of M_n-norms for $\mathcal{M} = \{M_n : n \ge 1\}$ a family of nonsingular $n \times n$ matrices. **37.3**
p-norm (of matrix $A \in \mathbb{C}^{m \times n}$, with $p \ge 1$): The matrix norm induced by the (vector) p-norm. **37.1**
p-norm (of vector $\mathbf{x} \in \mathbb{C}^n$, with $p \ge 1$): $\|\mathbf{x}\|_p = (|x_1|^p + \cdots + |x_n|^p)^{\frac{1}{p}}$. **37.1**
normal (complex square matrix or linear operator): Commutes with its Hermitian adjoint. **7.2**
normal equations (for the least squares problem $A\mathbf{x} = \mathbf{b}$): The system $A^* A\mathbf{x} = A^*\mathbf{b}$. **5.8**
normalized (representation $\pm(m/b^{p-1})b^e$ of a floating point number): $b^{p-1} \le m < b^p$. **37.6**
normalized immanant: Function defined by the irreducible character of a partition λ of n, specifically, $f_\lambda(M) = \frac{1}{\chi_\lambda(\varepsilon)} \sum_{\sigma \in S_n} \chi_\lambda(\sigma) \prod_{i=1}^n m_{i\sigma(i)}$. **31.10**
normalized scaling (of rectangular matrix A): A scaling DAE of A with $\det(D) = \det(E) = 1$. **9.6**
NSNS: See *nearly sign nonsingular* .
null graph: A graph with no vertices. **28.1**
null space: See *kernel.*
nullity: The dimension of the kernel. **2.4, 3.5**
numerical radius (of $A \in \mathbb{C}^{n \times n}$): $w(A) = \max\{|\mu| : \mu \in W(A)\}$. **18.1**
numerical range (of $n \times n$ complex matrix A): $W(A) = \{\mathbf{v}^* A\mathbf{v} | \mathbf{v}^*\mathbf{v} = 1, \mathbf{v} \in \mathbb{C}^n\}$. **7.1, 18.1**
numerical rank (of matrix A, with respect to the threshold τ): $\min\{\text{rank}(A + E) : \|E\|_2 \le \tau\}$. **39.9**
numerically orthogonal (vectors $\mathbf{x}, \mathbf{y}$): $|\mathbf{x}^T\mathbf{y}| \le \varepsilon \|\mathbf{x}\|_2 \|\mathbf{y}\|_2$. **46.1**
numerically orthogonal matrix: Each pair of its columns are numerically orthogonal. **46.1**
numerically stable (algorithm): Produces results that are roughly as accurate as the errors in the input data allow. **37.8**
numerically unstable (algorithm): Allows rounding and truncation errors to produce results that are substantially less accurate than the errors in the input data allow. **37.8**

O

odd (permutation): Can be written as the product of an odd number of transpositions. **Preliminaries**
odd (cycle in a sign pattern): The length of the simple or composite cycle is odd. **33.1**
off-diagonal entry: An entry in a matrix that is not a diagonal entry. **1.2**
off-norm (of A): The Frobenius norm of the matrix consisting of all off-diagonal elements of A and a zero diagonal. **42.7**

oh: See *big-oh, little-oh.*
one-to-one (linear transformation T): $\mathbf{v}_1 \neq \mathbf{v}_2$ implies $T(\mathbf{v}_1) \neq T(\mathbf{v}_2)$. **3.5**
onto (linear transformation): The codomain equals the range. **3.5**
open left half plane: $\{z \in \mathbb{C} : \mathrm{re}(z) < 0\}$. **Preliminaries**
open right half plane: $\{z \in \mathbb{C} : \mathrm{re}(z) > 0\}$. **Preliminaries**
open sector (from ray $e^{i\alpha}$ to ray $e^{i\beta}$): The set of rays $\{re^{i\theta} \; : r > 0, \alpha < \theta < \beta\}$. **33.8**
operator norm: See *induced norm.*
order (of a graph): The number of vertices in the graph. **28.1**
order (of a square matrix): See *size.*
ordered multiplicity list (of real symmetric matrix B): $(m_1, \ldots, m_r)$ where the distinct eigenvalues of B are $\breve{\beta}_1 < \cdots < \breve{\beta}_r$ with multiplicities $m_1, \ldots, m_r$. **34.5**
oriented (vertex-edge) incidence matrix (of graph G): A matrix obtained from the incidence matrix of G by replacing a 1 in each column by a -1. **28.4**
orthogonal (vectors): Two vectors whose inner product is 0. **5.2**
orthogonal (set of vectors): Any two distinct vectors in the set are orthogonal. **5.2**
$\mathcal{J}$**–orthogonal** (real matrix B) $B^T \mathcal{J} B = \mathcal{J}$ (where $\mathcal{J}$ is a nonsingular real symmetric matrix). **46.5**
orthogonal basis: A basis that is orthogonal as a set. **5.2**
orthogonal complement (of subset S): Subspace of vectors orthogonal to every vector in S. **5.2**
orthogonal iteration: Method for computing the EVD of real symmetric A: given starting $n \times p$ matrix X_0 with orthonormal columns, the sequence of matrices $Y_{k+1} = AX_k$, $Y_{k+1} = X_{k+1}R_{k+1}$ $k = 0, 1, 2, \ldots$, where $X_{k+1}R_{k+1}$ is the reduced QR factorization of Y_{k+1}. **42.1**
orthogonal matrix: A real or complex matrix Q such that $Q^T Q = I$. **5.2, 7.1**
orthogonal projection: Projection onto a subspace along its orthogonal complement. **5.4**
orthogonal with respect to symmetric bilinear form f (vectors $\mathbf{v}, \mathbf{u}$): $f(\mathbf{u}, \mathbf{v}) = 0$. **12.2**
orthonormal (set of vectors): An orthogonal set of unit vectors. **5.2**
orthonormal basis: A basis that is orthonormal as a set. **5.2**
oscillatory (real square matrix A): A is totally nonnegative and A^k is totally positive for some integer $k \geq 1$. **21.1**
outerplanar (graph G): There is an embedding of G into $\mathbb{R}^2$ with the vertices on the unit circle and the edges contained in the unit disc. **28.2**
overflow: $|x|$ equals or exceeds a threshold at or near the largest floating point number. **37.6**

P

P**-invariant face**: Alphabetized under *invariant.*
P**-matrix**: every principal minor is positive. **19.2**
P_0**-matrix**: Every principal minor is nonnegative. **35.8**
P_0^+**-matrix**: Every principal minor is nonnegative and at least one principal minor of each order is positive. **19.2**
$P_{0,1}$**-matrix**: Every principal minor is nonnegative and every diagonal element is positive. **35.8**
pairwise orthogonal (orthogonal projections P and Q): $PQ = QP = 0$. **7.2**
Parter vertex: See *Parter–Wiener vertex.*
Parter–Wiener vertex (for eigenvalue λ of $n \times n$ Hermitian matrix A): j such that $\alpha_{A(j)}(\lambda) = \alpha_A(\lambda) + 1$. **34.1**
partial completely positive matrix B: Every fully specified principal submatrix of B is a completely positive matrix, whenever b_{ij} is specified, then so is b_{ji} and $b_{ji} = b_{ij}$, and all specified off-diagonal entries are nonnegative. **35.4**
partial copositive matrix B: Every fully specified principal submatrix of B is a copositive matrix and whenever b_{ij} is specified, then so is b_{ji} and $b_{ji} = b_{ij}$. **35.5**
partial doubly nonnegative matrix B: Every fully specified principal submatrix of B is a doubly nonnegative matrix matrix, whenever b_{ij} is specified, then so is b_{ji} and $b_{ji} = b_{ij}$, and all specified off-diagonal entries are nonnegative. **35.4**

partial entry sign symmetric P-matrix B: Every fully specified principal submatrix of B is an entry sign symmetric P-matrix and if both b_{ij} and b_{ji} are specified, then $b_{ij}b_{ji} > 0$ or $b_{ij} = b_{ji} = 0$. **35.10**

partial entry sign symmetric P_0-matrix B: Every fully specified principal submatrix of B is an entry sign symmetric P_0-matrix and if both b_{ij} and b_{ji} are specified, then $b_{ij}b_{ji} > 0$ or $b_{ij} = b_{ji} = 0$. **35.10**

partial entry sign symmetric $P_{0,1}$-matrix B: Every fully specified principal submatrix of B is an entry sign symmetric $P_{0,1}$-matrix and if both b_{ij} and b_{ji} are specified, then $b_{ij}b_{ji} > 0$ or $b_{ij} = b_{ji} = 0$. **35.10**

partial entry weakly sign symmetric P-matrix B: Every fully specified principal submatrix of B is an entry weakly sign symmetric P-matrix and if both b_{ij} and b_{ji} are specified, then $b_{ij}b_{ji} \geq 0$. **35.10**

partial entry weakly sign symmetric P_0-matrix B: Every fully specified principal submatrix of B is an entry weakly sign symmetric P_0-matrix and if both b_{ij} and b_{ji} are specified, then $b_{ij}b_{ji} \geq 0$. **35.10**

partial entry weakly sign symmetric $P_{0,1}$-matrix B: Every fully specified principal submatrix of B is an entry weakly sign symmetric $P_{0,1}$-matrix and if both b_{ij} and b_{ji} are specified, then $b_{ij}b_{ji} \geq 0$. **35.10**

partial Euclidean distance matrix B: Every diagonal entry is specified and equal to 0, every fully specified principal submatrix of B is a Euclidean distance matrix, and whenever b_{ij} is specified, then so is b_{ji} and $b_{ji} = b_{ij}$. **35.3**

partial inverse M-matrix B: Every fully specified principal submatrix of B is an inverse M-matrix and every specified entry of B is nonnegative. **35.7**

partial M-matrix B: Every fully specified principal submatrix of B is an M-matrix and every specified off-diagonal entry of B is nonpositive. **35.6**

partial M_0-matrix B: Every fully specified principal submatrix of B is an M_0-matrix and every specified off-diagonal entry of B is nonpositive. **35.6**

partial matrix: A square array in which some entries are specified and others are not. **35.1**

partial nonnegative P-matrix: Every fully specified principal submatrix is a nonnegative P-matrix and all specified entries are nonnegative. **35.9**

partial nonnegative P_0-matrix: Every fully specified principal submatrix is a nonnegative P_0-matrix and all specified entries are nonnegative. **35.9**

partial nonnegative $P_{0,1}$-matrix: Every fully specified principal submatrix is a nonnegative $P_{0.1}$-matrix and all specified entries are nonnegative. **35.9**

partial P-matrix: Every fully specified principal submatrix is a P-matrix. **35.8**

partial P_0-matrix: Every fully specified principal submatrix is a P_0-matrix. **35.8**

partial $P_{0,1}$-matrix: Every fully specified principal submatrix is a $P_{0,1}$-matrix. **35.8**

partial positive P-matrix: Every fully specified principal submatrix is a positive P-matrix and all specified entries are positive. **35.9**

partial semidefinite ordering: See *Loewner ordering*.

partial strictly copositive matrix B: Every fully specified principal submatrix of B is a strictly copositive matrix and whenever b_{ij} is specified, then so is b_{ji} and $b_{ji} = b_{ij}$. **35.5**

partitioned (matrix): A matrix partitioned into submatrices by partitions of the row and column indices. **10.1**

partly decomposable ($n \times n$ $(0, 1)$-matrix): Has a $p \times q$ zero submatrix for positive integers p, q with $p + q = n$. **27.2**

path (in a graph): A walk with all vertices distinct. **28.1**

path (in sign pattern A): A formal product of the form $\gamma = a_{i_1 i_2} a_{i_2 i_3} \dots a_{i_k i_{k+1}}$, where each of the elements is nonzero and the index set $\{i_1, i_2, \dots, i_{k+1}\}$ consists of distinct indices. **33.1**

path-clique (digraph): The induced subdigraph of every alternate path to a single arc is a clique. **35.7**

path-connected (subset S of complex numbers): There exists a path in S (i.e., a continuous function from $[0,1]$ to S) from any point in S to any other point in S. **Preliminaries**

path cover number (of simple graph G): The minimum number of vertex disjoint induced paths of G that cover all vertices of G. **34.2**

pattern (of matrix A): The $(0, 1)$-matrix obtained from A by replacing each nonzero entry with a 1. **27.1**

pattern block triangular form (partial matrix B): The adjacency matrix of $\mathcal{D}(B)$ is in block triangular form. **35.1**

PEIEP: See *inverse eigenvalue problem with prescribed entries.*
pencil (defined by $A, B \in \mathbb{C}^{n\times n}$): $A - xB$, or pair (A, B). **43.1**
pencil (in $\mathbb{D}[x]^{m\times n}$): A matrix $A(x) = A_0 + xA_1$, $A_0, A_1 \in \mathbb{D}^{m\times n}$. **23.4**
perfect elimination ordering: An elimination ordering that does not produce any fill elements during Gaussian elimination. **30.2, 40.5**
perfect matching: A matching M in which each vertex of G is in one edge of M. **30.1**
perfectly-well determined to high relative accuracy (singular values of A): Changing an arbitrary nonzero entry $a_{k\ell}$ to $\theta a_{k\ell}$, with arbitrary $\theta \neq 0$ causes a relative perturbation in σ_i bounded by $|\theta|$ and $1/|\theta|$. **46.3**
period (of access equivalence class J of reducible nonnegative P): The period of the irreducible matrix $P[J]$. **9.3**
period (of irreducible nonnegative matrix P): Greatest common divisor of the lengths of the cycles of the digraph of P. **9.2**
period (of reducible nonnegative matrix P): The least common multiple of the periods of its basic classes. **9.3**
permanent (of $n \times n$ matrix A): $\text{per}(A) = \sum_{\sigma\in S_n} \prod_{i=1}^{n} a_{i\sigma(i)}$ (also defined for rectangular matrices). **31.1**
permutation: A 1-1 onto function from a set to itself. **Preliminaries**
permutation equivalent (matrices A, B): There exist permutation matrices P and Q such that $B = PAQ$. **31.1**
permutation invariant (vector norm $\|\cdot\|$): $\|P\mathbf{x}\| = \|\mathbf{x}\|$ for all $\mathbf{x}$ and all permutation matrices P. **37.1**
permutation invariant absolute norm: s function $g : \mathbb{R}^n \to \mathbb{R}_0^+$ that is a norm, and $g(x_1, \dots, x_n) = g(|x_1|, \dots, |x_n|)$ and $g(x) = g(Px)$ for all $\mathbf{x} \in \mathbb{R}^n$ and all permutation matrices $P \in \mathbb{R}^{n\times n}$. **17.3**
permutation matrix: A matrix whose rows are a rearrangement of the rows of the identity matrix. **1.2**
permutation pattern: A square sign pattern matrix with entries 0 and +, where the entry + occurs precisely once in each row and in each column. **33.1**
permutation similar (square matrices A, B): There exists a permutation matrix P such that $B = P^{-1}AP$ $(= P^TAP)$. **27.2**
permutational equivalence (of sign pattern A): A product of the form P_1AP_2, where P_1 and P_2 are permutation patterns. **33.1**
permutational similarity (of sign pattern A) is a product of the form P^TAP, where P is a permutation pattern. **33.1**
Perron branch (at vertex v of a tree): The Perron value of the corresponding bottleneck matrix is maximal among all branches at v. **36.3**
Perron value (of square nonnegative matrix P): the spectral radius of P. **9.1**
perturbation (of matrix A): $A + \Delta A$ **15.1**
perturbation (of scalar β): $\beta + \Delta\beta$ **15.1**
perturbation (of vector $\mathbf{v}$): $\mathbf{v} + \Delta\mathbf{v}$ **15.1**
Petrie matrix: A $(0, 1)$-matrix with the 1s in each of its columns occurring in consecutive rows. **30.2**
PIEP: Affine parameterized inverse eigenvalue problem. **20.8**
pinching (of a matrix): Defined recursively. See *Section* **17.4.**
pivot: An entry of a matrix in a pivot position. **1.3**
pivot column: A column of a matrix that contains a pivot position. **1.3**
pivot position: A position in a matrix in row echelon form that contains a leading entry. **1.3**
pivot row: A row of a matrix that contains a pivot position. **1.3**
planar (graph G): There an embedding of G into $\mathbb{R}^2$. **28.2**
$PLDU$ **factorization**: A factorization of a row permutation of a given matrix as the product of a unit lower triangular matrix, diagonal matrix, and a unit upper triangular matrix. **1.6**
PLU **factorization**: A factorization of a row permutation of a given matrix as the product of a unit lower triangular matrix and an upper triangular matrix. **1.6, 38.3**
PO: See *potentially orthogonal.*
pointed (cone K): $K \cap -K = \{0\}$. **8.5, 26.1**

polar decomposition (of matrix $A \in \mathbb{C}^{m\times n}$ with $m \geq n$): A factorization $A = UP$ where $P \in \mathbb{C}^{n\times n}$ is positive semidefinite and $U \in \mathbb{C}^{m\times n}$ satisfies $U^*U = I_n$. **17.1**
polar form: See *polar decomposition.*
polynomial restarting (of the Arnoldi algorithm): Use $\psi(A)\mathbf{v}$ instead of vector $\mathbf{v}$ when restarting. **44.4**
polytope: See *convex polytope*
positionally symmetric: See *combinatorially symmetric.*
positive (linear map $\phi : \mathbb{C}^{n\times n} \to \mathbb{C}^{m\times m}$): $\phi(A)$ is positive semidefinite whenever A is positive semidefinite. **18.7**
positive (real matrix A): All of A's elements are positive. **9.1**
K**-positive** (matrix $A \in \mathbb{R}^{n\times n}$): $AK \subseteq \text{int } K$. **26.1**
K**-positive** (vector $\mathbf{v} \in \mathbb{R}^n$): $\mathbf{v} \in \text{int } K$. **26.1**
positive definite (matrix): An $n \times n$ Hermitian matrix satisfying $\mathbf{x}^* A\mathbf{x} > 0$ for all nonzero $\mathbf{x} \in \mathbb{C}^n$. **5.1, 8.4**
positive definite (real symmetric or Hermitian bilinear form f): $f(\mathbf{v}, \mathbf{v}) > 0$ for all nonzero $\mathbf{v} \in V$. **12.2, 12.5**
positive P**-matrix**: A P-matrix in which every entry is positive. **35.9**
positive semidefinite (function $f : \mathbb{R} \to \mathbb{C}$): For each $n \in \mathbb{N}$ and all $x_1, x_2, \ldots, x_n \in \mathbb{R}$, the $n \times n$ matrix $[f(x_i - x_j)]$ is positive semidefinite. **8.5**
positive semidefinite (matrix): An $n \times n$ Hermitian matrix A satisfying $\mathbf{x}^* A\mathbf{x} \geq 0$ for all $\mathbf{x} \in \mathbb{C}^n$. **8.4**
positive semidefinite (real symmetric or Hermitian bilinear form f): $f(\mathbf{v}, \mathbf{v}) \geq 0$ for all $\mathbf{v} \in V$. **12.2, 12.5**
positive semistable (complex square matrix): See *nonnegative stable*
positive set α **of edges** (in a signed bipartite graph): $\text{sgn}(\alpha)$ is $+1$. **30.1**
positive stable (complex polynomial): Its roots lie in the open right half plane. **19.2**
positive stable (complex square matrix): Its eigenvalues lie in the open right half plane. **9.5, 19.2**
k**-potent** (square sign pattern or ray pattern A): k is the smallest positive integer such that $A = A^{k+1}$. **33.9**
potentially orthogonal (PO) (sign pattern): Allows an orthogonal matrix. **33.10**
potentially stable (square sign pattern A): Allows stability, i.e., some matrix $B \in Q(A)$ is stable. **33.4**
Powell–Reid's complete (row and column) pivoting: A particular pivoting strategy for QR-factorization. **46.2**
power method: Method for computing the EVD of A: given starting vector $\mathbf{x}_0$, compute the sequences $\nu_k = \mathbf{x}_k^T A\mathbf{x}_k$, $\mathbf{x}_{k+1} = A\mathbf{x}_k / \|A\mathbf{x}_k\|$, $k = 0, 1, 2, \ldots$, until convergence. **42.1**
powerful (square sign or ray pattern A): All the powers $A^1, A^2, A^3, \ldots$, are unambiguously defined. **33.9**
precision: The number of digits, i.e., p in $x = \pm\left(\frac{m}{b^{p-1}}\right) b^e$. **37.6**
preconditioner: A matrix M designed to improve the performance of an iterative method for solving the linear system $A\mathbf{x} = \mathbf{b}$. **41.1**
preserves: Linear operator T preserves bilinear form f if $f(T\mathbf{u}, T\mathbf{v}) = f(\mathbf{u}, \mathbf{v})$ for all $\mathbf{u}, \mathbf{v}$, **12.1**; Also applied to φ-sesquilinear form. **12.4**
preserves: Linear operator ϕ preserves subset of matrices $\mathcal{M}$ if $\phi(\mathcal{M}) \subseteq \mathcal{M}$. **22.1**
preserves: Linear operator ϕ preserves relation $\sim$ on matrix subspace $\mathcal{V}$ if $\phi(A) \sim \phi(B)$ whenever $A \sim B$, $A, B \in \mathcal{V}$. **22.1**
primary decomposition (of a nonconstant monic polynomial $q(x)$ over F): A factorization $q(x) = (h_1(x))^{m_1} \cdots (h_r(x))^{m_r}$, where the $h_i(x)$, $i = 1, \ldots, r$ are distinct monic irreducible polynomials over F. **6.4**
primary factors: The factors in a primary decomposition. **6.4**
primary matrix function: Function of a matrix defined using the Jordan Canonical Form and using the same branch for f and its derivatives for each Jordan block for the same eigenvalue. **11.1**
prime (in domain $\mathbb{D}$): A nonzero, nonunit element p such that for any $a, b \in \mathbb{D}$, $p|ab$ implies $p|a$ or $p|b$. **23.1**
prime (ideal I of a domain): $ab \in I$ implies that either a or b is in I. **23.1**
primitive (digraph): There is a positive integer k such that for every pair of vertices u and v, there is a walk of length k from u to v. **29.6**
primitive (polynomial $p(x) = \sum_{i=0}^{m} a_i x^{m-i} \in \mathbb{Z}[x], a_0 \neq 0, m \geq 1$): 1 is a g.c.d. of $a_0, \ldots, a_m$. **23.1**
primitive matrix: A square nonnegative matrix whose digraph is primitive. **29.6**

principal angles (between subspaces X and Y of $\mathbb{C}^r$): See *Section* **17.7** (equivalent to *canonical angles, Section* **15.1.**)
principal ideal: An ideal generated by one element (in a domain). **23.1**
principal ideal domain domain (PID) $\mathbb{D}$: Any ideal of $\mathbb{D}$ is principal. **23.1**
principal minor: The determinant of a principal submatrix. **4.2**
principal logarithm (of complex square matrix that has no real eigenvalues ≤ 0): The logarithm with all eigenvalues in the strip $\{ z : -\pi < \text{im}(z) < \pi \}$. **11.4**
principal square root (of a complex square matrix that has no real eigenvalues ≤ 0): The square root with each eigenvalue having real part > 0. **11.2**
principal submatrix: A submatrix lying in the same rows as columns. **1.2, 10.1**
principal submatrix at a distinguished eigenvalue λ (of square nonnegative matrix P): The principal submatrix of P corresponding to a set of vertices of $\Gamma(P)$ having no access to a vertex of an access equivalence class C that satisfies $\rho(P[C]) > \lambda$. **9.3**
principal vectors (between subspaces X and Y of $\mathbb{C}^r$): See *Section* **17.7.**
product (of simple graphs): See (*strong*) *product.*
profile (of sparse matrix A): The number of elements in the envelope of A. **40.5**
projection (of a vector onto V_1 along V_2, assuming $V = V_1 \oplus V_2$): The (unique) part of the vector in V_1; also, the linear transformation that maps a vector to its projection. **3.6**
proper cone K (in a finite-dimensional real vector space V): A convex cone that is closed, pointed, and full. **26.2**
proper subdigraph (of a digraph $\Gamma = (V, E)$): A subdigraph $\Gamma' = (V', E')$ such that V' is a proper subset of V or E' is a proper subset of E. **29.1**
properly signed nest: See *allows a properly signed nest.*
Property C: An $n \times n$ M-matrix A satisfies property C if there exists a representation of A of the form $A = sI - P$ with $s > 0$, $P \geq 0$ and $\frac{P}{s}$ semiconvergent. **9.4**
Property L: A property that complex square matrices A, B have if their eigenvalues $\alpha_k, \beta_k, (k = 1, \cdots, n)$ may be ordered in such a way that the eigenvalues of $xA + yB$ are given by $x\alpha_k + y\beta_k$ for all complex numbers x and y. **7.2, 24.3**
PSD square root (of a positive semidefinite matrix): The square root with each eigenvalue ≥ 0 (same as principal square root for a positive definite matrix). **8.3**
pseudo-code: An informal computer program used for writing algorithms. **37.7**
ε-**pseudoeigenvalue** (of matrix $A \in \mathbb{C}^{n\times n}$): A number $z \in \mathbb{C}$ such that there exists a nonzero vector $\mathbf{v} \in \mathbb{C}^n$ such that $\|A\mathbf{v} - z\mathbf{v}\| < \varepsilon\|\mathbf{v}\|$. **16.1**
ε-**pseudoeigenvector** (of matrix $A \in \mathbb{C}^{n\times n}$): A nonzero vector $\mathbf{v} \in \mathbb{C}^n$ such that there exists $z \in \mathbb{C}$ such that $\|A\mathbf{v} - z\mathbf{v}\| < \varepsilon\|\mathbf{v}\|$. **16.1**
ε-**pseudospectral abscissa** (of a complex square matrix): The rightmost extent of the ε-pseudospectrum. **16.3**
ε-**pseudospectral radius** (of a complex square matrix): The maximum magnitude of the ε-pseudospectrum. **16.3**
ε-**pseudospectrum** (of complex square matrix A): The set $\{z \in \mathbb{C} : z \in \sigma(A + E)$ for some $E \in \mathbb{C}^{n\times n}$ with $\|E\| < \varepsilon\}$. **16.1**
ε-**pseudospectrum of a matrix pencil**: See *Section* **16.5.**
ε-**pseudospectrum of a matrix polynomial**: See *Section* **16.5.**
ε-**pseudospectrum of a rectangular matrix**: See *Section* **16.5.**

Q

QMR algorithm: Iterative method for solving a linear system using non-Hermitian Lanczos algorithm. **41.3**
QR factorization (of matrix A): $A = QR$ where Q is unitary and R is upper triangular. **5.5**
QR factorization with column pivoting (of $A \in \mathbb{R}^{m\times n}$): The factorization $A\Pi = Q \begin{bmatrix} R \\ 0 \end{bmatrix}$, where Π is a permutation matrix, Q is orthogonal, and R is $n \times n$ upper triangular. **46.2**

QR iteration: Method for computing the EVD of A: starting from the matrix $A_0 = A$, the sequence of matrices $A_k = Q_k R_k$, $A_{k+1} = R_k Q_k$, $k = 0, 1, 2, \ldots$ where $Q_k R_k$ is the QR factorization of A_k. **42.1**
quadrangular (bipartite graph): Simple and each pair of vertices with a common neighbor lie on a cycle of length 4. **30.1**
quadratic form (corresponding to symmetric bilinear form f): The map $g : V \to F$ defined by $g(\mathbf{v}) = f(\mathbf{v}, \mathbf{v})$, $\mathbf{v} \in V$. **12.2**
qualitative class (of complex sign pattern $A = A_1 + iA_2$): $Q(A) = \{B_1 + iB_2 : B_1 \in Q(A_1)$ and $B_2 \in Q(A_2)\}$. **33.8**
qualitative class (of real matrix B): The qualitative class of $\operatorname{sgn}(B)$. **33.1**
qualitative class (of sign pattern A): The set of all real matrices B with $\operatorname{sgn}(B) = A$. **33.1**
quotient (of vector space V by subspace W): The set of additive cosets of W with operations $(\boldsymbol{v}_1 + W) + (\boldsymbol{v}_2 + W) = (\boldsymbol{v}_1 + \boldsymbol{v}_2) + W$ and $c(\boldsymbol{v} + W) = (c\boldsymbol{v}) + W$ for $c \in F$. **2.3**
quotient field (of an integral domain): The set of equivalence classes of all quotients $\frac{a}{b}$, $b \neq 0$, constructed the same way $\mathbb{Q}$ is constructed from $\mathbb{Z}$. **23.1**

R

(R, S)-**standard map**, (R, c, f)-**standard map**: Alphabetized under *standard*.
radix: The exponentiation base b in floating point number $x = \pm\left(\frac{m}{b^{p-1}}\right) b^e$. **37.6**
range (of a linear transformation $T : V \to W$): $\{T(\mathbf{v}) : \mathbf{v} \in V\}$. **3.4**
range (of a matrix): The span of the columns. **2.4**
rank (of bilinear form f): The rank of a matrix representing f relative to an ordered basis. **12.1.** Also applied to φ-sesquilinear form. **12.4**
rank (of a matrix or linear transformation): The dimension of the range; for a matrix, equals the number of leading entries in the reduced row echelon form of the matrix. **1.3, 2.4, 3.5**
rank (of a tensor $\mathbf{z}$): The rank of $\mathbf{z}$ is k if $\mathbf{z}$ is the sum of k decomposable tensors, but it cannot be written as sum of l decomposable tensors for any l less than k. **13.3**
rank revealing (QR factorization (with pivoting) of A): Small singular values of A are revealed by correspondingly small diagonal entries of R. **46.2**
rank revealing decomposition (of real matrix A): a two-sided orthogonal decomposition of the form $A = U\widehat{R}V^T = U\begin{bmatrix} R \\ 0 \end{bmatrix}V^T$, where U and V are orthogonal, and R is upper triangular and reveals the (numerical) rank of A in the size of its diagonal elements. **39.9**
rational canonical form: See *invariant factors rational canonical form* or *elementary divisors rational canonical form.*
rational function: A quotient of polynomials. **23.1**
ray nonsingular (ray pattern A): The Hadamard product $X \circ A$ is nonsingular for every entry-wise positive $n \times n$ matrix X. **33.8**
ray pattern: A matrix each of whose entries is either 0 or a ray in the complex plane represented by $e^{i\theta}$. **33.8**
ray pattern class (of ray pattern A): $Q(A) = \{B : b_{pq} = 0$ iff $a_{pq} = 0$, and otherwise $\arg b_{pq} = \arg a_{pq}\}$. **33.8**
Rayleigh quotient: $\dfrac{\mathbf{x}^* A\mathbf{x}}{\mathbf{x}^*\mathbf{x}}$, where A is Hermitian and $\mathbf{x}$ is a nonzero vector. **8.2**
real generalized eigenspace: See *Section* **6.3.**
real generalized eigenvector: See *Section* **6.3.**
real-Jordan block (of size $2k$ with eigenvalue $\alpha + \beta i$): The $2k \times 2k$ matrix having k copies of $M_2(\alpha, \beta) = \begin{bmatrix} \alpha & \beta \\ -\beta & \alpha \end{bmatrix}$ on the (block matrix) diagonal, $k - 1$ copies of $I_2 = \begin{bmatrix} 1 & 0 \\ 0 & 1 \end{bmatrix}$ on the first (block matrix) superdiagonal, and copies of $0_2 = \begin{bmatrix} 0 & 0 \\ 0 & 0 \end{bmatrix}$ everywhere else. **6.3**

real-Jordan canonical form (of real square matrix A): A real-Jordan matrix that is similar to A. **6.3**
real-Jordan matrix: A block diagonal real matrix having diagonal blocks that are Jordan blocks or real-Jordan blocks. **6.3**
real part (of complex number $a + bi$): a. **Preliminaries**
real structured ε-pseudospectrum (of real square matrix A): See *Section* **16.5.**
real vector space: A vector space over the field of real numbers. **1.1**
reduced digraph (of digraph Γ): A digraph whose vertices are the access equivalence classes of Γ (or $\{1, \ldots, k\}$ where k is the number of access equivalence classes) and having an arc from the ith vertex to the jth precisely when the ith access class has access to the jth access class. **9.1, 29.5**
reduced QR factorization (of matrix $A \in \mathbb{C}^{m\times n}, m \geq n$): $A = \hat{Q}\hat{R}$ where columns of $\hat{Q}$ are orthonormal and R is upper triangular. **5.5**
reduced row echelon form (RREF): A matrix is in RREF if it is in REF, the leading entry in any nonzero row is 1, and all other entries in a column containing a leading entry are 0; matrix B is the RREF of matrix A if B is in RREF and A and B are row equivalent. **1.3**
reduced singular value decomposition (reduced SVD) (of complex matrix A): $A = \hat{U}\hat{\Sigma}\hat{V}^*$, $\hat{\Sigma} = \mathrm{diag}(\sigma_1, \sigma_2, \ldots, \sigma_r) \in \mathbb{R}^{r\times r}$, where $\sigma_1 \geq \sigma_2 \geq \ldots \geq \sigma_r > 0$ and the columns of $\hat{U} \in \mathbb{C}^{m\times r}$ and the columns of $\hat{V} \in \mathbb{C}^{n\times r}$ are both orthonormal. **5.6, 45.1**
reducible (square matrix A): There is a permutation matrix P such that $PAP^T = \begin{bmatrix} B & C \\ \mathbf{0} & D \end{bmatrix}$, where B, D are square. **9.1, 27.3**
reducing eigenvalue (λ of $A \in \mathbb{C}^{n\times n}$): A is unitarily similar to $[\lambda] \oplus A_2$. **18.2**
REF: See *row echelon form.*
(k-)**regular** (graph): Every vertex has the same degree k. **28.2**
regular ($m \times n$ $(0, 1)$-design matrix W): W is balanced and $W^T W = t(I + J)$ for some integer t. **32.4**
regular (pencil $A - xB$): There exists a $\lambda \in \mathbb{C}$ such that $A - \lambda B$ is nonsingular. **43.1**
regular (pencil $A_0 + xA_1 \in \mathbb{D}[x]^{n\times n}$): $\det(A_0 + xA_1)$ is not the zero polynomial, or equivalently, there exists a $\lambda \in \mathbb{D}$ such that $A_0 + \lambda A_1$ is nonsingular. **23.4**
relative backward error of the linear system: See *Section* **38.1.**
relative condition number: See *condition number.*
relative error (in approximation $\hat{\mathbf{z}}$ to $\mathbf{z}$): $\|\mathbf{z} - \hat{\mathbf{z}}\|/\|\mathbf{z}\|$. **37.4**
requires: If P is a property referring to a real matrix, then a sign pattern A requires P if every real matrix in $Q(A)$ has property P. **33.1**
requires unique inertia (sign pattern A): $\mathrm{in}(B_1) = \mathrm{in}(B_2)$ for all symmetric matrices $B_1, B_2 \in Q(A)$. **33.6**
residual vector (of $\tilde{\mathbf{x}}$ when solving the linear system $A\mathbf{x} = \mathbf{b}$): The vector $\mathbf{r} = \mathbf{b} - A\tilde{\mathbf{x}}$. **38.3**
residual vector (at step k of an iterative method for solving $A\mathbf{x} = \mathbf{b}$): The vector $\mathbf{r}_k = \mathbf{b} - A\mathbf{x}_k$, where $\mathbf{x}_k$ is the approximate solution generated at step k. **41.1**
resolvent (of matrix $A \in \mathbb{C}^{n\times n}$ at a point $z \notin \sigma(A)$): The matrix $(zI - A)^{-1}$. **16.1**
restarted GMRES algorithm, GMRES(j): Restart GMRES every j steps, using the latest iterate as the initial guess for the next GMRES cycle. **41.3**
right singular vector: See *singular vectors.*
right singular space: See *singular spaces.*
ring automorphism of $F^{n\times n}$ induced by f: The map $\phi : F^{n\times n} \to F^{n\times n}$ defined by $\phi([a_{ij}]) = [f(a_{ij})]$. **22.4**
Ritz pair: $(\theta, \mathbf{v})$ where θ is the Ritz value for Ritz vector $\mathbf{v}$. **43.3**
Ritz value: The scalar θ for a Ritz vector. **43.3**
Ritz vector (of matrix of A from subspace $\mathcal{S}$ of $\mathbb{C}^n$): There is a $\theta \in \mathbb{C}$ such that $A\mathbf{v} - \theta\mathbf{v} \perp \mathcal{S}$. **43.3**
rook polynomials: Generating functions for the rook numbers defined using permanents. **31.8**
rounding error: In one floating point arithmetic operation, the difference between the exact arithmetic operation and the floating point arithmetic operation; in more extensive calculations, refers to the cumulative effect of the rounding errors in the individual floating point operations. **37.6**

rounding mode: Maps $x \in \mathbb{R}$ to a floating point number $\mathrm{fl}(x)$; default rounding mode in standard-conforming arithmetic is round-to-nearest, ties-to-even. **37.6**
Routh-Hurwitz matrix: See *Section* **19.2**.
row (of a matrix): The entries of a matrix lying in a horizontal line in the matrix. **1.2**
row echelon form (REF): A matrix is in REF if every zero row is below all nonzero rows and for two nonzero rows, the leading entry in the upper row is to the left of the leading entry of the lower row; matrix B is a REF of of matrix A if B is in REF and A and B are row equivalent. **1.3**
row equivalent: Matrix A is row equivalent to matrix B if B can be obtained from A by a sequence of elementary row operations (equivalently, $B = QA$ for some invertible matrix Q). **1.3**
row equivalent (matrices $A, B \in \mathbb{D}^{m\times n}$): $B = QA$ for some $\mathbb{D}$-invertible matrix Q. **23.2**
row operation: See *elementary row operation.*
row signing (of (real or sign pattern) matrix A): DA where D is a signing. **33.3**
row space: The span of the rows of a matrix. **2.4**
row-stochastic (matrix): A square nonnegative matrix having all row sums equal to 1. **9.4**
row sum vector (of a matrix): The vector of row sums. **27.4**
RRD (of real matrix A): A decomposition $A = XDY^T$ with D a diagonal matrix and X and Y full column rank, well-conditioned matrices. **46.3**
RREF: See *reduced row echelon form.*

S

S**-matrix**: An $n \times (n+1)$ matrix B such that it is an S^*-matrix and the kernel of every matrix in $Q(B)$ contains a vector all of whose coordinates are positive. **33.3**
S^***-matrix**: An $n \times (n+1)$ matrix B such that each of the $n+1$ matrices obtained by deleting a column of B is an SNS matrix. **33.3**
SAP: See *spectrally arbitrary pattern.*
scalar: An element of a field. **1.1**
scalar matrix (transformation): A scalar multiple of the identity matrix (transformation). **1.2, 3.2**
scaling (of a real matrix A): A matrix of the form $D_1 A D_2$ where D_1 and D_2 are diagonal matrices with positive diagonal entries. **9.6, 27.6**
Schatten-p norm (of $A \in \mathbb{C}^{m\times n}$): $\|A\|_{S,p} = \left(\sum_{i=1}^{q} \sigma_i^p(A)\right)^{1/p}$. **17.3**
Schur complement: A matrix defined from a partitioned matrix. **4.2, 10.3**
Schur product: See *Hadamard product.*
score vector: The row sum vector of a tournament matrix. **27.5**
SDR: See *system of distinct representatives.*
Seidel matrix (of simple graph G): The matrix $J - I - 2\mathcal{A}_G$. **28.4**
Seidel switching: An operation on simple graphs. **28.4**
self-adjoint: See *Hermitian.*
self-inverse sign pattern: An S^2NS-pattern A such that $B^{-1} \in Q(A)$ for every matrix $B \in Q(A)$. **33.2**
semiconvergent (square nonnegative matrix P): $\lim_{m\to\infty} P^m$ exists. **9.3**
semipositive (real matrix A): A is nonnegative and some element is positive. **9.1**
K**-semipositive** (matrix $A \in \mathbb{R}^{n\times n}$): A is K-nonnegative and $A \neq 0$. **26.1**
K**-semipositive** (vector $\mathbf{v} \in \mathbb{R}^n$): $\mathbf{v} \in K$ and $\mathbf{v} \neq 0$. **26.1**
semisimple: An eigenvalue having algebraic multiplicity equal to its geometric multiplicity. **4.3**
semistable matrix: Either positive semistable (i.e., nonnegative stable) or negative semistable, depending on section.
separation (between two square matrices A_1 and A_2): $\inf_{\|X\|_2=1} \|XA_1 - A_2X\|_2$. **15.1**
φ**-sesquilinear form** (on vector space V over field F with automorphism φ): A map f from $V \times V$ into F that satisfies $f(a\mathbf{u}_1 + b\mathbf{u}_2, \mathbf{v}) = af(\mathbf{u}_1, \mathbf{v}) + bf(\mathbf{u}_2, \mathbf{v})$ and $f(\mathbf{u}, a\mathbf{v}_1 + b\mathbf{v}_2) = \varphi(a)f(\mathbf{u}, \mathbf{v}_1) + \varphi(b)f(\mathbf{u}, \mathbf{v}_2)$. **12.4**
sesquilinear form (on a complex vector space): A φ-sesquilinear form where φ is complex conjugation. **12**

set of Hermitian matrices associated with graph G: $\mathcal{H}(G) = \{B \in \mathcal{H}_n | \mathcal{G}(B) = G\}$. **34.1**
set of symmetric matrices associated with graph G: $\mathcal{S}(G) = \{B \in \mathcal{S}_n | \mathcal{G}(B) = G\}$. **34.1**
sgn: See *sign.*
Shannon capacity: A parameter of a simple graph. **28.5**
shift: The scalar μ in a shifted matrix $A - \mu I$. **42.1**
shifted matrix (of matrix A): The matrix $A - \mu I$, where μ is the shift. **42.1**
shifted QR iteration: QR iteration of the shifted matrix. **42.1**
sign (denoted sign, of a complex number): A nonzero complex number; sign(0) = 1, otherwise the complex number with the same argument having absolute value 1. **Preliminaries**
sign (denoted sign, of a complex square matrix): A function defined from the Jordan Canonical Form of the matrix. See *Section* **11.6.**
sign (denoted sgn, of a permutation): 1, if the permutation is even and -1, if the permutation is odd. **Preliminaries**
sign (denoted sgn, of a real number): $+, 0, -$ according as the number is $> 0, = 0, < 0$. **Preliminaries**
sign (denoted sgn, of a set α of edges in a signed bipartite graph): The product of the weights of the edges in α. **30.1**
sign (of a simple cycle in a sign pattern A): The product of the entries in the cycle. **33.1**
sign-central: A sign pattern matrix that requires centrality. **33.11**
sign nonsingular (SNS) (square sign pattern A): Every matrix $B \in Q(A)$ is nonsingular. **33.2**
sign pattern: A matrix whose entries are in $\{+, 0, -\}$. **33.1**
sign pattern class: See *qualitative class.*
sign pattern matrix: See *sign pattern.*
sign pattern of real matrix B: The sign pattern whose entries are the signs of the corresponding entries in B. **33.1**
sign potentially orthogonal SPO (square sign pattern): Does not have a zero row or zero column and every pair of rows and every pair of columns allows orthogonality. **33.10**
sign semistable (square sign pattern A): Every matrix $B \in Q(A)$ is semistable **33.4**
sign singular (square sign pattern A): Every matrix $B \in Q(A)$ is singular. **33.2**
sign-solvable (system of linear equations $Ax = b$): For each $\tilde{A} \in Q(A)$ and for each $\tilde{b} \in Q(b)$, the system $\tilde{A}x = \tilde{b}$ is consistent and $\{\tilde{x} \ : \text{ there exist } \tilde{A} \in Q(A) \text{ and } \tilde{b} \in Q(b) \text{ with } \tilde{A}\tilde{x} = \tilde{b}\}$ is entirely contained in one qualitative class. **33.3**
sign stable (square sign pattern A): Requires stability, i.e., every matrix $B \in Q(A)$ is stable. **33.4**
sign symmetric (square matrix A): $\det A[\alpha, \beta] \det A[\beta, \alpha] \geq 0$, $\forall \alpha, \beta \subseteq \{1, \ldots, n\}$, $|\alpha| = |\beta|$. **19.2**
signature (of a real symmetric or Hermitian bilinear form): The signature of a matrix representing the form relative to some basis. **12.2, 12.5**
signature (of real symmetric or Hermitian matrix A): The number of positive eigenvalues minus the number of negative eigenvalues. **12.2, 12.5**
signature (of a composite cycle γ in sign pattern A): $(-)^{\sum_{i=1}^{m}(l_i - 1)}$ where i_i are the lengths of the simple cycles in γ. **33.1**
signature matrix: A ± 1 diagonal matrix. **32.2**
signature pattern: A diagonal sign pattern matrix, each of whose diagonal entries is $+$ or $-$. **33.1**
signature similarity (of the square sign pattern A): A product of the form SAS, where S is a signature pattern. **33.1**
signed bigraph (of real matrix A): The weighted graph obtained from the bigraph of A weighting the edge $\{i, j'\}$ by $+1$ if $a_{ij} > 0$, and by -1 if $a_{ij} < 0$. **30.2**
signed bipartite graph: A weighted bipartite graph with weights $X = \{-1, 1\}$. **30.1**
signed 4-cockade: A 4-cockade whose edges are weighted by ± 1 in such a way that every 4-cycle is negative. **30.2**
signed digraph (of an $n \times n$ sign pattern A): The digraph with vertex set $\{1, 2, \ldots, n\}$ where (i, j) is an arc (bearing a_{ij} as its sign) iff $a_{ij} \neq 0$. **33.1**
significand: The (base b) integer m in floating point number $x = \pm \left(\frac{m}{b^{p-1}}\right) b^e$. **37.6**

signing (of order k): A nonzero $(0, 1, -1)$- or $(0, +, -)$-diagonal matrix of order k. **33.3**
signless Laplacian matrix (of graph G): The matrix $D + \mathcal{A}_G$, where D is the diagonal matrix of vertex degrees and $\mathcal{A}_G$ is the adjacency matrix of G. **28.4**
similar (to matrix A): Any matrix of the form $C^{-1}AC$. **2.4**
similarity scaling (of square matrix A): A scaling DAD^{-1} of A. **9.6**
simple (cycle) (in a digraph): The induced subdigraph of the vertices in the cycle is the cycle. **35.1**
simple cycle (in sign pattern A): See *k-cycle (in sign pattern A)*.
simple (digraph): A digraph with no loops (a digraph does not have multiple arcs). **29.1**
simple (eigenvalue): An eigenvalue having algebraic multiplicity 1. **4.3**
simple (graph): A graph with no loops where each edge has multiplicity at most one. **28.1**
simple (walk) (in a digraph): A walk in which all vertices, except possibly the first and last, are distinct. **29.1**
simple row operation: A generalization of an elementary row operation used in domains. **23.2**
sine (of a complex square matrix A): The matrix defined by the sine power series, $\sin(A) = A - \frac{A^3}{3!} + \cdots + \frac{(-1)^k}{(2k+1)!}A^{2k+1} + \cdots$. **13.5**
single precision: Typically, floating point numbers have machine epsilon roughly 10^{-7} and precision roughly 7 decimal digits. **37.6**
singular (matrix or linear operator): Not nonsingular. **1.5, 3.7**
singular (pencil): A pencil that is not regular. **23.4, 43.1**
singular space (right or left) (of complex matrix A): The subspace spanned by the right (or by the left) singular vectors of A. **45.1**
singular-triplet (of $A \in \mathbb{C}^{m\times n}$): A (left singular vector, singular value, right singular vector) triplet. **15.2**
singular value decomposition (SVD) (of matrix $A \in \mathbb{C}^{m\times n}$): $A = U\Sigma V^*$, $\Sigma = \text{diag}(\sigma_1, \sigma_2, \ldots, \sigma_p) \in \mathbb{R}^{m\times n}$, $p = \min\{m, n\}$, where $\sigma_1 \geq \sigma_2 \geq \ldots \geq \sigma_p \geq 0$ and both $U = [\mathbf{u}_1, \mathbf{u}_2, \ldots, \mathbf{u}_m] \in \mathbb{C}^{m\times m}$ and $V = [\mathbf{v}_1, \mathbf{v}_2, \ldots, \mathbf{v}_n] \in \mathbb{C}^{n\times n}$ are unitary. Also forms with other dimensions. **5.6, 45.1, 17.1**
singular value vector (of complex matrix A): The vector of singular values of A in nonincreasing order. **17.1**
singular values (of a complex matrix): The diagonal entries σ_i of Σ in a singular value decomposition. **5.6, 45.1, 17.1**
singular vectors (of a complex matrix): Left: the columns of U, and right: the columns of V, both in a singular value decomposition. **5.6, 45.1**
size (of matrix A): $m \times n$, where A is an $m \times n$ matrix; also m if $n = m$. **1.2**
Skeel condition number of the linear system $A\hat{\mathbf{x}} = \mathbf{b}$: $\text{cond}(A, \hat{\mathbf{x}}) = \frac{\|\,|A^{-1}|\,|A|\,|\hat{\mathbf{x}}|\,\|_\infty}{\|\hat{\mathbf{x}}\|_\infty}$. **38.1**
Skeel matrix condition number (of matrix A): $\text{cond}(A) = \|\,|A^{-1}|\,|A|\,\|_\infty$. **38.1**
skew-Hermitian (matrix): A real or complex matrix equal to the negative of its conjugate transpose. **1.2, 7.2**
skew-symmetric (matrix): A matrix equal to the negative of its transpose. **1.2**
Smith invariant factors (of $A \in F^{n\times n}$): The nonconstant polynomials on the diagonal of the Smith normal form of $xI - A$. **6.5**
Smith normal form (of $M \in F[x]^{n\times n}$): The Smith normal matrix obtained from M by elementary row and column operations. **6.5**
Smith normal form (over GCD domain $\mathbb{D}_g$): Matrix $B \in \mathbb{D}_g^{m\times n}$ is in Smith normal form if $B = \text{diag}(b_1, \ldots, b_r, 0, \ldots, 0)$, $b_i \neq 0$ for $i = 1, \ldots, r$ and $b_{i-1} | b_i$ for $i = 2, \ldots, r$. **23.2**
Smith normal matrix (in $F[x]^{n\times n}$): A diagonal matrix $\text{diag}(1, \ldots, 1, a_1(x), \ldots, a_s(x), 0, \ldots, 0)$, where the $a_i(x)$ are monic nonconstant polynomials such that $a_i(x)$ divides $a_{i+1}(x)$ for $i = 1, \ldots, s - 1$. **6.5**
SNS: See *sign nonsingular*.
S^{2}NS: See *strong sign nonsingular*.
solution: An ordered tuple of scalars that when assigned to the variables satisfies a system of linear equations. **1.4**
solution set: The set of all solutions to a system of linear equations. **1.4**
span, Span (noun): The set of all linear combinations of the vectors in a set. **2.1**
span, spans (verb): A set $S \subseteq V$ spans V if $V = \text{Span}(S)$. **2.1**
spanning subgraph (of a connected graph (V, E)): A connected subgraph (V', E') with $V' = V$. **28.1**

spanning tree (of a connected graph): A spanning subgraph that is a tree. **28.1**
sparse (matrix A): A matrix for which substantial savings in either operations or storage can be achieved when the zero elements of A are exploited during the application of Gaussian elimination to A. **40.2**
sparse approximate inverse (of matrix A): A sparse matrix M^{-1} constructed to approximate A^{-1}. **41.1**
sparsity pattern: See *sparsity structure.*
sparsity structure (of matrix A): The set of indices of nonzero elements of A. **40.3**
special linear group (of order n over F): The subgroup of the general linear group consisting of all matrices that have determinant 1. **67.1**
spectral absolute value (of complex matrix A): $(A^*A)^{1/2}$. **17.1**
spectral norm (of complex matrix A): $\|A\|_2 = \sqrt{\rho(A^*A)} = \max\{\|A\mathbf{v}\|_2 : \|\mathbf{v}\|_2 = 1\}$ = the largest singular value of A. **7.1, 37.3**
spectral radius (of $A \in \mathbb{C}^{n\times n}$): $\max\{|\lambda| : \lambda$ an eigenvalue of $A\}$. **4.3**
spectrally arbitrary pattern (SAP): A sign pattern A of size n such that every monic real polynomial of degree n can be achieved as the characteristic polynomial of some matrix $B \in Q(A)$. **33.6**
spectrum (of a matrix): The multiset of all eigenvalues of a matrix, each eigenvalue appearing with its algebraic multiplicity. **4.3**
spectrum (of a graph): The spectrum of its adjacency matrix. **28.3**
speed (of a processor): The number of floating-point operations per second (flops). **42.9**
splitting: See *preconditioner.*
SPO: See *sign potentially orthogonal.*
square matrix: The number of columns is the same as the number of rows. **1.2**
square root (of a square complex matrix A): A matrix B such that $B^2 = A$. **11.2** (See also *principal square root, PSD square root*)
stable (algorithm): See *numerically stable.*
stable matrix: Either positive stable or negative stable, depending on section; positive in *Section* 9.5 and *Section* 19.2.
standard basis (for F^n or $F^{m\times n}$): The set of vectors or matrices having one entry equal to 1 and all other entries equal to 0. **2.1**
standard-conforming (floating point system): conforming to IEEE standard. **37.6**
standard inner product (in $\mathbb{R}^n$ (or $\mathbb{C}^n$)): $\langle\mathbf{u},\mathbf{v}\rangle = \mathbf{u}^T\mathbf{v}$ ($\langle\mathbf{u},\mathbf{v}\rangle = \mathbf{v}^*\mathbf{u}$). **5.1**
standard linear preserver problem: A linear preserver problem where the preservers take one of the standard forms. **22.2**
(R, c, f)-standard map: For $A \in F^{n\times n}$, $\phi(A) = cRAR^{-1} + f(A)I$, or $\phi(A) = cRA^TR^{-1} + f(A)I$. R invertible, c nonzero, f a linear functional. **22.2**
(R, S)-standard map: For $A \in F^{m\times n}$, $\phi(A) = RAS$, or $m = n$ and $\phi(A) = RA^TS$. R and S must be invertible and may be required satisfy some additional assumptions. **22.2**
standard matrix of transformation $T : F^n \to F^m$: The matrix of T with respect to the standard bases for F^n, F^m. **3.3**
star on n vertices: a tree in which there is a vertex of degree $n - 1$. **34.6**
state (of $n \times n$ stochastic matrix P): An index $i \in \{1, \ldots, n\}$. **9.4**
stationary distribution (of stochastic matrix P): Nonnegative vector π that satisfies $\pi^T\mathbf{1} = 1$ and $\pi^TP = \pi^T$. **9.4**
k-step Arnoldi factorization: See *Arnoldi factorization.*
stochastic (square nonnegative matrix P): Same as row-stochastic; the sum of the entries in each row is 1. **9.4**
stopping (matrix): A transient substochastic matrix. **9.4**
strict column signing (of (real or sign pattern) matrix A): AD where D is a strict signing. **33.3**
strict row signing (of (real or sign pattern) matrix A): DA where D is a signing. **33.3**
strict signing: A signing where all the diagonal entries are nonzero. **33.3**
strictly block lower triangular (matrix A): A^T is strictly block upper triangular. **10.3**

strictly block upper triangular (matrix): A block upper triangular matrix in which all the diagonal blocks are 0. **10.3**
strictly copositive (matrix A): $\mathbf{x}^T A\mathbf{x} > 0$ for all $\mathbf{x} \geq 0$ and $\mathbf{x} \neq 0$. **35.5**
strictly diagonally dominant ($n \times n$ complex matrix A): $|a_{ii}| > \sum_{j=1, j\neq i}^{n} |a_{ij}|$ for $i = 1, \ldots, n$. **9.5**
strictly equivalent (pencils $A - xB$ and $C - xD$): If there exist nonsingular matrices S_1 and S_2 such that $C - \lambda D = S_1(A - \lambda B)S_2$ for all $\lambda \in \mathbb{C}$. **43.1**
strictly equivalent (pencils $A(x), C(x) \in \mathbb{D}[x]^{m\times n}$): $C(x) = QA(x)P$ for some $\mathbb{D}$-invertible matrices P, Q. **23.4**
strictly lower triangular (matrix A): A^T is strictly upper triangular **10.2**
strictly upper triangular (matrix): A matrix such that every entry having row number greater than or equal to column number is zero. **10.2**
strictly unitarily equivalent (pencils $A - xB$ and $C - xD$): strictly equivalent pencils in which S_1, S_2 can be taken unitary. **43.1**
Strong Arnold Hypothesis: Satisfied by a real symmetric matrix M provided there does not exist a real symmetric nonzero matrix X such that $MX = \mathbf{0}$, $M \circ X = \mathbf{0}$, $I \circ X = \mathbf{0}$. **28.5**
strong combinatorial invariant (of a matrix): A quantity or property that does not change when the rows and/or columns are permuted. **27.1**
strong Parter vertex: a Parter–Wiener vertex j for an eigenvalue λ of an $A \in \mathcal{H}(G)$ such that λ occurs as an eigenvalue of at least three direct summands of $A(j)$. **34.1**
(strong) product (of simple graphs $G = (V, E)$, $G' = (V', E')$): The simple graph with vertex set $V \times V'$, where two distinct vertices are adjacent whenever in both coordinate places the vertices are adjacent or equal in the corresponding graph. **28.1**
strong sign nonsingular ($\mathbf{S^2NS}$) (square sign pattern A): A is an SNS-pattern such that the matrix B^{-1} is in the same sign pattern class for all $B \in Q(A)$. **33.2**
strongly connected: A digraph whose vertices all lie in a single access equivalence class. **9.1, 29.5**
strongly connected components (of a digraph): The subdigraphs induced by the access equivalence classes. **29.5**
strongly inertia preserving (real square matrix A): The inertia of AD is equal to the inertia of D for every real diagonal matrix D. **19.5**
strongly nonsingular (square matrix): All its principal submatrices are nonsingular. **47.6**
ϕ **strongly preserves** (subset of matrices $\mathcal{M}$): $\phi(\mathcal{M}) = \mathcal{M}$. **22.1**
ϕ **strongly preserves** (relation $\sim$ on matrix subspace $\mathcal{V}$): For every pair $A, B \in \mathcal{V}$, we have $\phi(A) \sim \phi(B)$ if and only if $A \sim B$. **22.1**
strongly regular: A simple graph with parameters (n, k, λ, μ) that has n vertices, is k-regular with $1 \leq k \leq n - 2$, every two adjacent vertices have exactly λ common neighbors, and every two distinct nonadjacent vertices have exactly μ common neighbors. **28.2**
strongly stable: See *additive D-stable.*
subdigraph (of a digraph $\Gamma = (V, E)$): A digraph $\Gamma' = (V', E')$ with $V' \subseteq V$ and $E' \subseteq E$. **29.1**
subgraph (of a graph (V, E)): A graph $G' = (V', E')$ with $V' \subseteq V$ and $E' \subseteq E$. **28.1**
submatrix: A matrix lying in certain rows and columns of a given matrix. **1.2, 10.1**
$\mathbb{D}$**-submodule:** A nonempty set $\mathbf{N} \subseteq \mathbf{M}$ that is closed under addition and multiplication by ($\mathbb{D}$) scalars. **23.3**
submultiplicative (norm $\|\cdot\|$ on $\mathbb{C}^{n\times n}$): A vector norm satisfying $\|AB\| \leq \|A\|\|B\|$ for all $A, B \in \mathbb{C}^{n\times n}$ (satisfies the conditions of a matrix norm except not required to be part of a family). **18.4**
subordinate (matrix norm): See *induced norm.*
(k-th)-subpermanent sum (of matrix A): The sum of the permanents of all order k submatrices of A. **31.7**
subpattern (of a sign pattern): A sign pattern obtained by replacing some (possibly none) of the nonzero entries with 0. **33.1**
subspace: Subset of a vector space V that is itself a vector space under the operations of V. **1.1**

subspace iteration: See *orthogonal iteration.*
substochastic (square nonnegative matrix P): The sum of the entries in each row is ≤ 1. **9.4**
subtractive cancellation of significant digits: Occurs in floating point sums when the relative error in the rounding-error-corrupted approximate sum is substantially greater than the relative error in the summands. **37.6**
sum (of subspaces): $W_1 + \cdots + W_k = \{\mathbf{w}_1 + \cdots + \mathbf{w}_k \ : \mathbf{w}_i \in W_i\}$. **2.3**
sum norm: See *1-norm.*
sup-norm: See ∞*-norm.*
support (of rectangular matrix A or vector): The set of indices ij with $a_{ij} \neq 0$. **9.6**
support line (of convex set S): A line ℓ that intersects ∂S such that S lies entirely within one of the closed half-planes determined by ℓ. **18.2**
G **supports rank decompositions**: Each matrix $A \in \mathcal{M}(G)$ is the sum of rank(A) matrices in $\mathcal{M}(G)$ each having rank 1 (also a variant for signed bipartite graphs). **30.3**
surjective: See *onto.*
SVD: See *singular value decomposition.*
switching equivalent: A simple graph that can be obtained from another simple graph by a Seidel switching. **28.4**
Sym: The map $\text{Sym}(\mathbf{v}_1 \otimes \cdots \otimes \mathbf{v}_k) = \frac{1}{m!}\sum_{\pi \in S_m} \mathbf{v}_{\pi(1)} \otimes \cdots \otimes \mathbf{v}_{\pi(k)}$. **13.6**
sym multiplication: $(\mathbf{v}_1 \vee \cdots \vee \mathbf{v}_p) \vee (\mathbf{v}_{p+1} \vee \cdots \vee \mathbf{v}_{p+q}) = \mathbf{v}_1 \vee \cdots \vee \mathbf{v}_{p+q}$. **13.7**
symbol (of a family of Toeplitz matrices): The function $a(z) = \sum a_k z^k$ where the a_k are the constants of the Toeplitz matrices. **16.2**
symbol curve (of a Toeplitz family): The image of the complex unit circle under the symbol. **16.2**
symmetric (bilinear form f): $f(\mathbf{u}, \mathbf{v}) = f(\mathbf{v}, \mathbf{u})$ for all $\mathbf{u}, \mathbf{v} \in V$. **12.2**
symmetric (digraph $\Gamma = (V, E)$): $(i, j) \in E$ implies $(j, i) \in E$ for all $i, j \in V$. **35.1**
symmetric (form): A multilinear form that is a symmetric map. **13.4**
symmetric (map $\psi \in L^m(V; U)$): $\psi(\mathbf{v}_{\pi(1)}, \ldots, \mathbf{v}_{\pi(m)}) = \psi(\mathbf{v}_1, \ldots, \mathbf{v}_m)$, for all permutations π. **13.4**
symmetric (matrix): A matrix equal to its transpose. **1.2**
symmetric algebra (on vector space V): $\bigvee V = \bigoplus_{k \in \mathbb{N}} \left(\bigvee^k V\right)$ with sym multiplication. **13.9**
symmetric inertia set (of a symmetric sign pattern A): $\text{in}(A) = \{\ \text{in}(B) : \ B = B^T \in Q(A)\ \}$; **33.6**
symmetric matrices associated with graph G: See *set of symmetric matrices associated with graph G.*
symmetric maximal rank (of symmetric sign pattern A): $\max\{\text{rank}B : \ B^T = B \text{ and } B \in Q(A)\}$. **33.6**
symmetric minimal rank (of symmetric sign pattern A): $\min\{\text{rank}B : \ B^T = B \text{ and } B \in Q(A)\}$. **33.6**
symmetric power (of a vector space): See *symmetric space.*
symmetric product (of vectors $\mathbf{v}_1, \ldots, \mathbf{v}_m$): $\mathbf{v}_1 \vee \cdots \vee \mathbf{v}_m = m!\text{Sym}(\mathbf{v}_1 \otimes \cdots \otimes \mathbf{v}_k)$. **13.6**
symmetric rank revealing decomposition (SRRD) (of symmetric real matrix H): A decomposition $H = XDX^T$, where D is diagonal and X is a full column rank well-conditioned matrix. **46.5**
symmetric space (of vector space V): $\bigvee^m V = \text{Sym}(\bigotimes^m V)$. **13.6**
symmetric scaling (of a matrix): A scaling where $D_1 = D_2$. **9.6, 27.6**
system of distinct representatives (SDR): For the finite sets $S_1, S_2, \ldots, S_n$, a choice of $x_1, x_2, \ldots, x_n$ with the properties that $x_i \in S_i$ for each i and $x_i \neq x_j$ whenever $i \neq j$. **31.3**
system of linear equations: A set of one or more linear equations in the same variables. **1.4**

T

tensor: An element of a tensor product. **13.2**
tensor algebra (on vector space V): $\bigotimes V = \bigoplus_{k \in \mathbb{N}} \left(\bigotimes^k V\right)$. **13.9**
tensor multiplication: $(\mathbf{v}_1 \otimes \cdots \otimes \mathbf{v}_p) \otimes (\mathbf{v}_{p+1} \otimes \cdots \otimes \mathbf{v}_{p+q}) = \mathbf{v}_1 \otimes \cdots \otimes \mathbf{v}_{p+q}$. **13.7**
tensor power: A tensor product of copies of one vector space. **13.2**
tensor product (of matrices): See *Kronecker product.*

tensor product (of vector spaces $V_1, \ldots, V_m$): a multilinear map satisfying a universal factorization property, See *Section* **13.2.**
term rank (of a (0, 1)-matrix A): The largest size of a collection of 1s of A with no two 1s in the same line. **27.1**
term rank (of sign pattern A): The maximum number of nonzero entries of A no two of which are in the same row or same column. **33.6**
tight sign-central: A sign-central pattern A for which the Hadamard product of any two columns contains a negative element. **33.11**
TN: See *totally nonnegative.*
Toeplitz (matrix): A matrix whose entries are constant along each sub- and super- diagonal, i.e., the value of the i, j-entry is the constant a_k whenever $i - j = k$. **16.2, 48.1**
Toeplitz-block matrix: A block matrix $A = [A_{ij}]$ where each block A_{ij} is an $n \times n$ Toeplitz matrix. **48.1**
Toeplitz inverse eigenvalue problem (ToIEP): Given $\lambda_1, \ldots, \lambda_n \in \mathbb{R}$, find $\mathbf{c} = [c_1, \ldots, c_n]^T \in \mathbb{R}^n$ such that $\left[t_{ij}\right]_{i,j=1}^{n}$ with $t_{ij} = c_{|i-j|+1}$ has spectrum $\{\lambda_1, \ldots, \lambda_n\}$. **20.9**
ToIEP: See *Toeplitz inverse eigenvalue problem.*
total least squares problem: See *Section* **39.1.**
total signed compound (TSC) (sign patterns): Every square submatrix of every matrix $A \in Q(S)$ is either sign nonsingular or sign singular. **46.3**
total support: An $n \times n$ (0, 1)-matrix A has total support if $A \neq \mathbf{0}$ and each 1 of A is on a diagonal of A. **27.2**
totally nonnegative (real matrix A): Every minor is nonnegative. **21.1**
totally positive (real matrix A): Every minor is positive. **21.1**
totally unimodular (real matrix A): All minors of A are from $\{-1, 0, 1\}$). **46.3**
tournament: Digraph of a tournament matrix. **27.5**
tournament matrix: A (0,1)-matrix with 0 diagonal and exactly one of a_{ij}, a_{ji} equal to 1 for all $i \neq j$. **27.5**
TP: See *totally positive.*
trace: The sum of all the diagonal entries of the matrix. **1.2**
trace-minimal (graph G in a family $\mathcal{F}$ of graphs): The trace-sequence of the adjacency matrix of G is least in lexicographic order among all graphs in $\mathcal{F}$. **32.7**
trace norm (of $A \in \mathbb{C}^{m \times n}$): The sum of the singular values of A. **17.3**
trace-sequence (of $n \times n$ matrix A): $(\text{tr}(A), \text{tr}(A^2), \cdots, \text{tr}(A^n))$. **32.7**
transient class (of a stochastic matrix P): an access equivalence class of P that is not ergodic. **9.4**
transient matrix: See convergent.
transient state: A state in a transient class. **9.4**
transition matrix: See *change-of-basis matrix.*
transitive tournament matrix A: $a_{ij} = a_{jk} = 1$ implies $a_{ik} = 1$. **27.5**
transpose (of a linear transformation): A specific linear transformation from the dual of the codomain to the dual of the domain. **3.8**
transpose (of $m \times n$ matrix A): The $n \times m$ matrix $B = [b_{ij}]$ where $b_{ij} = a_{ji}$. **1.2**
transposition: A 2-cycle. **Preliminaries**
tree (digraph): A digraph whose associated graph is a tree. **29.1**
tree (graph): A connected graph with no cycles. **28.1**
tree sign pattern (t.s.p.): A combinatorially symmetric sign pattern matrix whose graph (suppressing loops) is a tree. **33.5**
triangular (matrix): Upper or lower triangular. **10.2**
triangular property (of class of matrices X): Whenever A is a block triangular matrix and every diagonal block is an X-matrix, then A is an X matrix. **35.1**
triangular system: A linear system $T\mathbf{x} = \mathbf{b}$ where T is a triangular matrix. **38.2**
triangular totally positive (Δ**TP**): A triangular matrix all of whose nontrivial minors are positive. **21.2**
tridiagonal matrix: A square matrix A such that $a_{ij} = 0$ if $|i - j| > 1$.
trivial face (F of cone K): $F = \{0\}$ or $F = K$. **26.1**
trivial linear combination: A linear combination in which all the scalar coefficients are 0 (or over the empty set). **2.1**

truncation error: The error made by replacing an infinite process by a finite process. **37.6**
TSC: See *total signed compound.*
t.s.p. See *tree sign pattern.*
two-sided Lanczos algorithm: See *Section* **41.3**.
type I (tree): Has exactly one characteristic vertex. **36.3**
type II (tree): Has two characteristic vertices. **36.3**

U

UFD: See *unique factorization domain.*
u.i.: See *unitarily invariant.*
underflow: $\mathrm{fl}(x) = 0$ for $x \neq 0$. **37.6**
unicyclic: A graph containing precisely one cycle. **36.2**
unimodular (matrix over a domain): See $\mathbb{D}$-*invertible.*
union (of two graphs $G = (V, E)$ and $G' = (V', E')$): The graph with vertex set $V \cup V'$, and edge (multi)set $E \cup E'$. **28.1**
unique factorization domain (UFD): Any nonzero, nonunit element a can be factored as a product of irreducible elements $a = p_1 \cdots p_r$, and this factorization is unique within order and unit factors. **23.1**
unisigned (real or sign pattern vector): Not balanced. **33.3**
unit (in a domain): An element a such that a divides 1. **23.1**
unit lower triangular matrix: A lower triangular matrix such that all diagonal entries are equal to one.
unit round: $u = \inf\{\delta > 0 \mid \mathrm{fl}(1+\delta) > 1\}$. **37.6**
unit upper triangular matrix: An upper triangular matrix such that all diagonal entries are equal to one. **1.2**
unit vector: A vector of length 1. **5.1**
unital (linear map $\phi : \mathbb{C}^{n\times n} \to \mathbb{C}^{m\times m}$): $\phi(I_n) = I_m$. **18.7**
unitarily equivalent: See *unitarily similar.*
unitarily invariant (vector norm $\|\cdot\|$ on $\mathbb{C}^{m\times n}$): $\|A\| = \|UAV\|$ for any unitary $U \in \mathbb{C}^{m\times m}$ and $V \in \mathbb{C}^{n\times n}$ and any $A \in \mathbb{C}^{m\times n}$. **17.3**
unitarily similar (matrices A and B): There exists a unitary matrix U such that $B = U^*AU$. **7.1**
unitarily similarity invariant (vector norm $\|\cdot\|$ on $\mathbb{C}^{n\times n}$): $\|U^*AU\| = \|A\|$ for all $A \in \mathbb{C}^{n\times n}$ and unitary $U, V \in \mathbb{C}^{n\times n}$. **18.4**
unitary matrix: A matrix U such that $U^*U = I$. **5.2, 7.1**
unknown vector: The vector of variables of a system of linear equations. **1.4**
unreduced upper Hessenberg matrix: An upper Hessenberg matrix A such that $a_{j+1,j} \neq 0$ for $j = 1, \ldots, n-1$. **43.2**
unstable (algorithm): See *numerically unstable.*
updating (QR factorization): See *Section* **39.7**.
upper Hessenberg (matrix A): $a_{ij} = 0$, for all $i \geq j+2$, $1 \leq i, j \leq n$. **10.2, 43.2**
upper triangular (pencil $A - xB$): Both A and B are upper triangular. **43.1**
upper triangular matrix: A matrix such that every entry having row number greater than column number is zero. **1.2, 10.2**

V

(v, k, λ)-**design**: See *2-design.*
valency: See *degree.*
Vandermonde matrix: A matrix having each row equal to successive powers of a number. **48.1**
variables (of a linear equation): The unknowns. **1.4**
vec-function (of matrix A): The vector formed by stacking the columns of A on top of each other in their natural order. **10.4**
vector: An element of a vector space. **1.1**

vector norm: A real-valued function $\|\cdot\|$ on $\mathbb{R}^n$ or $\mathbb{C}^n$ such that for all vectors $\mathbf{x}$, $\mathbf{y}$ and all scalars α: (1) $\|\mathbf{x}\| \geq 0$, and $\|\mathbf{x}\| = 0$ implies $\mathbf{0}$; (2) $\|\alpha\mathbf{x}\| = |\alpha|\|\mathbf{x}\|$; (3) $\|\mathbf{x}+\mathbf{y}\| \leq \|\mathbf{x}\| + \|\mathbf{y}\|$. **37.1**
vector seminorm: A real-valued function $\|\cdot\|$ on $\mathbb{R}^n$ or $\mathbb{C}^n$ such that for all vectors $\mathbf{x}$, $\mathbf{y}$ and all scalars α: (1) $\|\mathbf{x}\| \geq 0$; (2) $\|\alpha\mathbf{x}\| = |\alpha|\|\mathbf{x}\|$; (3) $\|\mathbf{x}+\mathbf{y}\| \leq \|\mathbf{x}\| + \|\mathbf{y}\|$. **37.2**
vector space (over field F): A nonempty set V with two operations, addition and scalar multiplication, such that V is an abelian group under addition, scalar multiplication distributes over addition, scalar multiplication is associative, and the multiplicative identity of F acts as the identity on V. **1.1**
vertex: An element of the vertex set of a graph or digraph **28.1, 29.1**
vertex coloring: A partition of the vertex set of a graph into cocliques. **28.5**
vertex-edge incidence matrix: See *incidence matrix*. **30.1**
vertex independence number: The largest order of a coclique in G. **28.5**
vertices: Plural of *vertex*.
Volterra–Lyapunov stable: See *Lyapunov diagonally stable*.

W

walk (in a digraph): A sequence of arcs $(v_0, v_1), (v_1, v_2), \ldots, (v_{k-1}, v_k)$ (the vertices need not be distinct). **29.1**
walk (in a graph): An alternating sequence $(v_{i_0}, e_{i_1}, v_{i_1}, e_{i_2}, \ldots, e_{i_\ell}, v_{i_\ell})$ of vertices and edges, not necessarily distinct, such that $v_{i_{j-1}}$ and v_{i_j} are endpoints of e_{i_j} for $j = 1, \ldots, \ell$. **28.1**
walk product: $\prod_{j=1}^{k} a_{v_{j-1},v_j}$ where A is a square matrix and $(v_0, v_1), (v_1, v_2), \ldots, (v_{k-1}, v_k)$ is a walk in the digraph of A. **29.3**
walk-regular (graph): For every vertex v the number of walks from v to v of length ℓ, depends only on ℓ (not on v). **28.2**
weak combinatorial invariant (of a matrix): A quantity or property that does not change when the rows and columns are simultaneously permuted (by the same permutation). **27.1**
weakly majorizes: Real sequence α weakly majorizes β if for all k, the sum of the first k entries of α in decreasing order is $\geq$ the sum of the first k entries of β in decreasing order. **Preliminaries**
weakly numerically stable (algorithm): The magnitude of the forward error is roughly the same as the magnitude of the error induced by perturbing the data by small multiple of the unit round. **37.8**
weakly sign symmetric (square matrix A): $\det A[\alpha,\beta] \det A[\beta,\alpha] \geq 0$, $\forall \alpha, \beta \subseteq \{1, \ldots, n\}$, $|\alpha| = |\beta| = |\alpha \cap \beta| + 1$. **19.2**
weakly unitarily invariant: See *unitarily similarity invariant*.
weighted bigraph (of matrix A): The bigraph of A with the weight of $\{i, j'\} = a_{ij}$. **30.2**
weighted bipartite graph: A simple bipartite graph with a weight function on the edges. **30.1**
weighted digraph: A digraph with a weight function on the arcs. **29.1**
weight function: A function from the edges or arcs of a graph. bipartite graph, or digraph to $\mathbb{R}^+$ **36.4, 30.1, 29.1**
weighted graph: A simple graph with a weight function. **36.4**
weighted least squares problem: See *Section* **39.1.**
Wiener vertex: See *Parter–Wiener vertex*.
well-conditioned (input $\mathbf{z}$ for P): Small relative perturbations of $\mathbf{z}$ cause small relative perturbations of $P(\mathbf{z})$. **37.4**
Wilkinson's shift: The shift μ is the eigenvalue of the bottom right 2×2 submatrix of T which is closer to $t_{n,n}$. **42.3**

Z

***Z*-matrix** (square real matrix): All off-diagonal elements are nonpositive. **9.5, 19.2**
zero completion (of a partial matrix): Obtained by setting all unspecified (off-diagonal) entries to 0 (partial matrix must have all diagonal entries specified). **35.6**

zero divisor (in a ring R): A nonzero element $a \in R$ such that there exists a nonzero $b \in R$ with $ab = 0$ or $ba = 0$. **Preliminaries**
zero-free diagonal (property of matrix A): All the diagonal elements of A are nonzero. **40.3**
zero line (in a matrix): A line of all zeros. **27.1**
zero matrix: A matrix with all entries equal to zero. **1.2**
(zero) pattern (of a matrix): See *pattern.*
zero pattern (of sign pattern A): The $(0, +)$-pattern obtained by replacing each $-$ entry in A by a $+$. **33.2**
zero transformation: A linear transformation that maps every vector to zero. **3.1**

Notation Index

This notation index covers most of the terms defined in Chapters 1 to 49. It does not cover some terminology used in a single section (including most of the terminology that is specific to a particular application (Chapters 50 to 70)), nor does it cover most of the terminology used in computer software (Chapters 71 to 77).

Notation is in "alphabetical" order. If you are looking for something done to a matrix, like the transpose, look under A. If you are looking for something done to a field, look under F. If you are looking for something done to a linear transformation, look under T. If you are looking for something done to a vector or vector space, look under V.

The meaning of a symbol depends on what it is applied to; e.g., $\rho(A)$, where A is a matrix, is the spectral radius of A, whereas $\rho(G)$, where G is a group, is a representation of the group.

Warning: Tilde and hat, as in $\tilde{A}$, $\hat{Q}$, frequently have the meanings *perturbation* and *reduced*, respectively, but are also redefined in some chapters to mean other things.

For most symbols, the section where the symbol is introduced is listed at the end of the definition.

$\mathbf{0}_{mn}$	the zero matrix (in $F^{m\times n}$), can be shortened to $\mathbf{0}$. **1.2**
$\mathbf{1}_n$	all 1s vector in F^n can be shortened to $\mathbf{1}$.
$A = [a_{ij}]$	matrix A and its elements. **1.2**
A^T	transpose of matrix A. **1.2**
$\overline{A}$	complex conjugate of matrix A. **1.2**
A^*	Hermitian adjoint (conjugate-transpose) of matrix A. **1.2**
A^{-1}	inverse of square matrix A. **1.5**
$A^\dagger$	Moore–Penrose pseudo-inverse of matrix A. **5.7**
$A^\#$	group inverse of square matrix A. **9.1**
$A[\alpha,\beta]$	submatrix of A with row indices in α and column indices in β. **1.2, 10.1**
$A[\alpha]$	$= A[\alpha,\alpha]$, principal submatrix, also denoted $A[1,2,3]$ for $A[\{1,2,3\}]$. **1.2, 10.1**
$A(\alpha,\beta)$	submatrix of A with row indices not in α and column indices not in β. **1.2, 10.1**
$A(\alpha)$	$= A(\alpha,\alpha)$, principal submatrix with row and column indices not in α. **1.2, 10.1**
$A_{i:k,j:l}$	submatrix $A[\{i,i+1,\dots,k\},\{j,j+1,\dots,l\}]$, analogously $A_{i,j:l}$, $A_{i:k,j}$.
$(A)_{ij}$	i,j-entry of A. **1.2**
A_{ij}	i,j-block in a block matrix of A. **10.1**
$A/A[\alpha]$	Schur complement of $A[\alpha]$. **4.2, 10.3**
$\lvert A\rvert$	matrix having as entries the absolute values of the entries of matrix A. **9.1**
$\lvert A\rvert_{pd}$	spectral absolute value $(A^*A)^{1/2}$. **17.1**
$\lvert A\rvert$	zero pattern of sign pattern A. **33.2**
$\sqrt{A}$	a square root of a square complex matrix A. **11.2**
$A^{1/2}$	the principal square root of a square complex matrix A. **11.2**
$A^{1/2}$	the positive semidefinite square root of a positive semidefinite matrix A. **8.3**
$\widetilde{A}$	a perturbation $A+\Delta A$ of matrix A. **15.1** (also other meanings)
$\lVert A\rVert_k$	operator norm of matrix A (induced by $\lVert\mathbf{v}\rVert_k$). **37.3**
$\lVert A\rVert_F$	Frobenius (Euclidean) norm of matrix A. **7.1, 37.3**
$\lVert A\rVert_{K,k}$	Ky Fan k norm. **17.3**
$\lVert A\rVert_{S,p}$	Schatten-p norm. **17.3**
$\lVert A\rVert_{tr}$	trace norm (*not* same as Frobenius norm). **17.3**

$\|A\|_{UI}$	unitarily invariant matrix norm. **17.3**
$A > 0$	matrix A is positive. **9.1**
$A \geq 0$	matrix A is nonnegative. **9.1**
$A \gneq 0$	matrix A is semipositive. **9.1**
$A > B$	$A - B$ is positive.
$A \geq B$	$A - B$ is nonnegative.
$A \gneq B$	$A - B$ is semipositive.
$A \geq^K 0$	A is K-nonnegative. **26.1**
$A >^K 0$	A is K-positive. **26.1**
$A \gneq^K 0$	A is K-semipositive. **26.1**
$A \geq^K B$	$A - B$ is K-nonnegative. **26.1**
$A >^K B$	$A - B$ is K-positive. **26.1**
$A \gneq^K B$	$A - B$ is K-semipositive. **26.1**
$A \succ 0$	matrix A is positive definite. **8.4**
$A \succeq 0$	matrix A is positive semidefinite. **8.4**
$A \succ B$	$A - B$ is positive definite. **8.4**
$A \succeq B$	$A - B$ is positive semidefinite. **8.4**
$\hat{A} \preceq A$	sign pattern $\hat{A}$ is a subpattern of sign pattern A. **33.1**
$A \stackrel{c}{\sim} B$	A and B are $*$-congruent. **8.3**
$A \sim_c B$	A and B are column equivalent (in a domain). **23.2**
$A \sim_r B$	A and B are row equivalent (in a domain). **33.2**
$A \sim B$	A and B are equivalent (in a domain). **23.2**
$[A\|\mathbf{b}]$	augmented matrix. **1.4**
$A + B$	sum of matrices A and B. **1.2**
AB	matrix product of matrices A and B. **1.2**
$A \oplus B$	direct sum (block diagonal matrix) of matrices A and B. **2.4, 10.2**
$A \circ B$	Hadamard product of A and B. **8.5**
$A \otimes B$	Kronecker product of A and B. **10.4**
$\mathcal{A}_G$	adjacency matrix of graph G, can be shortened to $\mathcal{A}$. **28.3**
$\mathcal{A}_\Gamma$	adjacency matrix of digraph Γ, can be shortened to $\mathcal{A}$. **29.2**
$\mathcal{A}(R, S)$	the class of all $(0, 1)$-matrices with row sum vector R and column sum vector S. **27.4**
$A^m(V; U)$	subset of $L^m(V; U)$ consisting of the antisymmetric maps. **13.5**
$a\|b$	a divides b (in a domain). **23.1**
$a \equiv b$	a, b are associates (in a domain). **23.1**
$\{\{a\}\}$	the set of associates (equivalence class) of a (in a domain). **23.1**
$(a_1, \ldots, a_n)$	any g.c.d. of $a_1, \ldots, a_n$ (in a domain). **23.1**
$\{\{(a_1, \ldots, a_n)\}\}$	the equivalence class of all g.c.d.s of $a_1, \ldots, a_n$ (in a domain). **23.1**
α^c	set complement of α in $1, \ldots, n$. **Preliminaries**
$\alpha^\downarrow$	permutation of real number sequence α, with entries in nonincreasing order. **Preliminaries**
$\alpha^\uparrow$	permutation of real number sequence α, with entries in nondecreasing order. **Preliminaries**
$\alpha \succeq \beta$	real number sequence α majorizes β. **Preliminaries**
$\alpha \succeq_w \beta$	natural number sequence α weakly majorizes β. **Preliminaries**
$\alpha_\varepsilon(A)$	ε-pseudospectral abscissa of complex matrix A. **16.3**
$\alpha(G)$	algebraic connectivity of graph G. **36.1**
$\hat{\alpha}(G)$	absolute algebraic connectivity of G. **36.1**
$\alpha(\lambda)$ or $\alpha_A(\lambda)$	algebraic multiplicity of eigenvalue λ (of A). **4.3**
$\alpha(m, n)$	$\max\{\det W^T W \| W \in \{\pm 1\}^{m\times n}\}$. **32.1**
$\alpha(n)$	$= \alpha(n, n)$. **32.1**

$\frac{\partial f}{\partial x}$	the partial derivative of F with respect to x.
$\partial(S)$	the boundary of a set S of real or complex numbers. **Preliminaries**
$\partial(V')$	number of edges of a graph having one endpoint inside vertex subset V' and other outside. **28.5**
δ_{ij}	Kronecker delta, i.e., 1 if $i = j$ and 0 otherwise.
$\delta(A)$	the zero part of the inertia of complex square matrix A. **8.3, 19.1**
$\delta_G(v)$	degree of vertex v in graph G, also denoted $\delta(v)$. **28.1**
Δ_n^k	the set of $n \times n$ matrices of nonnegative integers with each row and column sum $= k$. **31.4**
$d(\mathbf{v}, \mathbf{u})$	distance from vector $\mathbf{v}$ to vector $\mathbf{u}$ in a normed vector space. **5.1**
$d_G(v, u)$	distance from vertex v to vertex u in graph G, also denoted $d(v, u)$. **28.1**
$d(T)$	diameter of tree T, same as diam(T). **34.3**
$\deg(p)$	degree of polynomial $p(x)$. **23.1**
$\det A$	determinant of matrix A, also denoted $\det(A)$. **4.1**
$\operatorname{diag}(d_1, \ldots, d_n)$	$n \times n$ diagonal matrix with listed diagonal entries. **1.2**
$\operatorname{diam}(G)$	diameter of graph. **28.1**
$\dim \mathcal{P}$	dimension of convex polytope $\mathcal{P}$, also denoted $\dim(\mathcal{P})$. **27.6**
$\dim V$	dimension of vector space V, also denoted $\dim(V)$. **2.2**
ϵ	machine precision (i.e., machine epsilon). **37.6**
$\mathbf{e}_i$	ith standard basis vector (1 in ith coordinate, 0s elsewhere). **2.1**
E_{ij}	standard basis matrix (1 in i, jth coordinate, 0s elsewhere). **2.1**
$\mathcal{E}_n$ or $\mathcal{E}$	standard basis $\mathbf{e}_1, \ldots, \mathbf{e}_n$ for F^n. **2.1**
ε_n	the identity in the symmetric group S_n. **Preliminaries**
E_λ or $E_\lambda(A)$	the eigenspace for eigenvalue λ (of A). **4.3**
$E(A, \alpha + \beta i)$	the real generalized eigenspace of real matrix A for eigenvalue $\alpha + \beta i$. **6.3**
$E_k(\mu)$	a particular lower elementary bidiagonal matrix. **21.2**
e^A	exponential function of matrix A. **11.3**
$\eta(G)$	a graph parameter involving minimum rank. **28.5**
$\eta_p(A, \mathbf{b}; \tilde{\mathbf{x}})$	relative backward error. **38.1**
End V	algebra of linear operators on vector space V. **69.1**
$\exp(A)$	exponential function of matrix A, same as e^A. **11.3**
$f \mid g$	polynomial $f(x)$ divides $g(x)$ in F[x]. **20.2**
$\mathbb{F}_q$	finite field with q elements. **61**
$F^{m \times n}$	$m \times n$ matrices over the field F. **1.2**
F^n	F-vector space of column n-tuples of elements of F. **1.1**
$F[x]$	polynomials over F. **1.1**
$F[\mathbf{x}] = F[x_1, \ldots, x_n]$	ring of polynomials $p(\mathbf{x}) = p(x_1, \ldots, x_n)$ with coefficients in F. **23.1**
$F(\mathbf{x})$	field of rational functions in $x_1, \ldots, x_n$ over F. **23.1**
$F[x; n]$	polynomials of degree $\leq n$ over F.
$\mathcal{F}(A)$	all doubly stochastic matrices that have 0's at least wherever the (0,1)-matrix A has 0s. **27.6**
$F \vee G$	the join $\Phi(F \cup G)$ (where (F, G are cone faces). **26.1**
$F \wedge G$	the meet $F \cap G$ (where (F, G are cone faces). **26.1**
$\Phi(G)$	conductance (isoperimetric number) of a graph. **28.5**
$\Phi(S)$	face generated by S. **26.1**
$\Phi(\mathbf{x})$	face generated by $\{\mathbf{x}\}$. **26.1**
fix T	fixed space of linear transformation T. **3.6**
G	group. **Preliminaries**
G or $G = (V_G, E_G)$	graph = (vertices, edges), also denoted $G = (V, E)$. **28.1**
Γ or $\Gamma = (V_\Gamma, E_\Gamma)$	digraph = (vertices, arcs), also denoted $\Gamma = (V, E)$. **29.1**
$\overline{G}$	graph complement. **28.1**

$\ker T$	kernel of linear transformation T, also denoted $\ker(T)$. **3.5**
λ	eigenvalue of a matrix or transformation (can also use other Greek). **4.3**
Λ	diagonal matrix of eigenvalues. **15.1**
Λ_n^k	the set of $n \times n$ binary matrices in which each row and column sum is k. **31.3**
L_G	the Laplacian matrix of graph G. **28.4**
$L(G_w)$	Laplacian matrix of weighted graph G_w. **36.4**
$\|L_G\|$	the signless Laplacian matrix of graph G. **28.4**
$L(G)$	the line graph of graph G. **28.2**
$L(G; a_1, \dots, a_n)$	a generalized line graph. **28.2**
$\ell_k(x)$	Laguerre polynomial. **31.8**
$L(V, W)$	the set of all linear transformations of V into W. **3.1**
$L(V_1, \dots, V_m; U)$	the set of all multilinear maps from $V_1 \times \cdots \times V_m$ into U with operations. **13.1**
$L^m(V; U)$	$L(V_1, \dots, V_m; U)$ with $V_i = V$ for $i = 1, \dots, m$. **13.5**
$\operatorname{lcm}(a_1, \dots, a_n)$	least common multiple of $a_1, \dots, a_n$.
$\lim_{m\to\infty} a_m = a\ (C, 0)$	$\lim_{m\to\infty} a_m = a$. **9.1**
$\lim_{m\to\infty} a_m = a\ (C, k)$	(C, k) limit. **9.1**
$\log(A)$	principal logarithm of matrix A. **11.4**
$M(A)$	the comparison matrix of A. **19.5**
$m(G)$	the maximum rank deficiency of graph G. **34.2**
μ_i	ith eigenvalue of the Laplacian matrix of a graph (nondecreasing order). **28.4**
$\mu(G)$	Colin de Verdière parameter of graph G. **28.5**
$M(G)$	maximum (eigenvalue) multiplicity of graph G. **34.2**
$\mathcal{M}(G)$	set of matrices A such that bigraph of A is a subgraph of G. **30.3**
$\mathcal{M}(\Omega)$	the quotient field of $\mathrm{H}(\Omega)$. **23.1**
$\mathcal{M}_\zeta$	$= \mathcal{M}(\{\zeta\})$. **23.1**
$\mathrm{MR}(A)$	maximal rank of sign pattern A. **33.6**
$\mathrm{mr}(A)$	minimal rank of sign pattern A. **33.6**
$\mathrm{mr}(G)$	minimum rank of graph G. **34.2**
$\mathbb{N}$	natural numbers, i.e., $\{0, 1, 2, \dots\}$.
$\langle n \rangle$	the set of integers $\{1, 2, \dots, n\}$. **9.1**
$N_\lambda^k(A)$	the kth generalized eigenspace of square matrix A. **6.1**
$N_\lambda^\nu(A)$	the generalized eigenspace of matrix A, can be abbreviated $N_\lambda(A)$. **6.1**
$\mathcal{N}(T)$	the minimum number of distinct eigenvalues of a matrix in $\mathcal{S}(T)$ where T is a tree. **34.3**
$\nu(A)$	the negative part of the inertia of complex square matrix A. **8.3, 19.1**
$\nu_A(\lambda)$	the index of matrix A at eigenvalue λ. **6.1**
ν_P	the index of P, i.e. $\nu_P(\rho)$. **9.3**
$\bar{\nu}_P$	the co-index of P. **9.3**
N_G	the (vertex-edge) incidence matrix of graph G. **28.4**
$nnz(A)$	the number of nonzero entries in sparse matrix A. **40.2**
$\mathrm{null}\ A$	nullity of a matrix, also $\mathrm{null}(A)$. **2.4**
$\mathrm{null}\ T$	nullity of a linear transformation or of a matrix, also $\mathrm{null}(T)$. **3.5**
$O(f)$	big-oh of function f. **Preliminaries**
$o(f)$	little-oh of function f. **Preliminaries**
Ω_n	set of $n \times n$ doubly stochastic matrices. **9.4**
$\omega(G)$	clique number of graph G. **28.4**
P_n	path on n vertices. **28.1**
$P[\lambda]$	principal submatrix at a distinguished eigenvalue. **9.3**
$P(G)$	path cover number of graph G. **34.2**
$P(n, d)$	tree constructed in a specific way. **36.3**
$\pi(A)$	the positive part of the inertia of complex square matrix A. **8.3, 19.1**

$\sigma(G)$	spectrum of graph G. **28.3**
$\sigma_\varepsilon(A)$	ε-pseudospectrum of complex square matrix A. **16.1**
$\sigma_\varepsilon^{\mathbb{R}}(A)$	real structured ε-pseudospectrum of real square matrix A. **16.5**
$\sigma_\varepsilon(A, B)$	ε-pseudospectrum of the matrix pencil $A - xB$. **16.5**
$\sigma_\varepsilon(A, B, C)$	spectral value set of the matrix triplet A, B, C. **16.5**
$\sigma_\varepsilon(P)$	ε-pseudospectrum of matrix polynomial P. **16.5**
$\lvert S\rvert$	cardinality of set S.
$S_k(A)$	sum of all principal minors of size k of matrix A. **4.2**
$\mathcal{S}(G)$	the set of real symmetric matrices A such that $G(A) = G$. **34.1**
$S^m(V; U)$	subset of $L^m(V; U)$ consisting of the symmetric maps. **13.5**
$S_k(\alpha_1, \ldots, \alpha_n)$	kth elementary symmetric function of $\alpha_i, i = 1, \ldots, n$. **Preliminaries**
$S(u)$	$\sum_{k\in V}(d(u,k))^2$ where $G = (V, E)$ is a simple graph. **36.5**
$\operatorname{sep}(A_1, A_2)$	separation between matrices A_1, A_2. **15.1**
$\operatorname{sgn}(\pi)$	sign of the permutation π. **Preliminaries**
$\operatorname{sgn}(a)$	sign of the real number a as used in sign patterns, one of $+, 0, -$. **Preliminaries**
$\operatorname{sgn}(A)$	sign pattern matrix of the real matrix A, with entries in $+, 0, -$. **33.1**
$\operatorname{sgn}(\alpha)$	sign of a set α of edges in a signed bipartite graph. **30.1**
$\operatorname{sign}(A)$	matrix sign function of the complex square matrix A. **11.6**
$\operatorname{sign}(z)$	sign of the complex number z as used in numerical linear algebra, always nonzero. **Preliminaries**
$\operatorname{SL}(n, F)$	special linear group of order n over F. **67.1**
$\operatorname{SMR}(A)$	symmetric maximal rank of sign pattern A. **33.6**
$\operatorname{smr}(A)$	symmetric minimal rank of sign pattern A. **33.6**
$\operatorname{Span}(S)$	span of the set S of vectors. **2.1**
$\operatorname{Struct}(A)$	the sparsity structure (support) of A. **9.6, 40.3**
$\operatorname{sv}(A)$	vector of singular values of complex matrix A. **15.2, 17.1**
$\operatorname{sv}_{\text{ext}}(A)$	extended vector of singular values of complex matrix A. **15.2**
$\operatorname{Sym}(\mathbf{v}_1 \otimes \cdots \otimes \mathbf{v}_k)$	$= \frac{1}{m!}\sum_{\pi\in S_m} \mathbf{v}_{\pi(1)} \otimes \cdots \otimes \mathbf{v}_{\pi(k)}$. **16.6**
T	linear transformation T. **3.1**
T^{-1}	inverse of linear transformation T. **3.7**
T^T	transpose of linear transformation T. **3.8**
${}_{\mathcal{B}'}[T]_{\mathcal{B}}$	matrix of T with respect to bases $\mathcal{B}$ (input) and $\mathcal{B}'$ (output). **3.3**
$[T]_{\mathcal{B}}$	matrix of T with respect to bases $\mathcal{B}$ and $\mathcal{B}$ (same as ${}_{\mathcal{B}}[T]_{\mathcal{B}}$). **3.3**
$[T]$	for $T : F^n \to F^m$, matrix of T with respect to the standard bases, $[T] = {}_{\mathcal{E}_m}[T]_{\mathcal{E}_n}$). **3.3**
T_A	linear transformation associated to matrix A, $T_A(\mathbf{v}) = A\mathbf{v}$. **3.3**
T_r^n	triples of subsets of $\{1, \ldots, n\}$ of cardinality r. **17.6**
$T(k, l, d)$	tree constructed from P_d by appending k and l isolated vertices to the ends. **36.3**
$\tau(A)$	the ergodicity coefficient of matrix A. **9.1**
$\mathcal{T}(R)$	the class of all tournament matrices with score vector R. **27.5**
$\vartheta(G)$	Lovász parameter of graph G. **28.5**
$\Theta(G)$	Shannon capacity of graph G. **28.5**
$\Theta(X, Y)$	canonical angle matrix between X and Y. **15.1**
$\operatorname{tr} A$	trace of matrix A, also denoted $\operatorname{tr}(A)$. **1.2**
$\mathcal{U}(T)$	the minimum number of simple eigenvalues of a matrix in $\mathcal{S}(T)$ where T is a tree. **34.4**
$U\Sigma V^*$	singular value decomposition of a real or complex matrix. **5.6**
$\hat{U}\hat{\Sigma}\hat{V}^*$	reduced singular value decomposition of a real or complex matrix. **5.6**
V	vector space V. **1.1**
V^*	dual space of V. **3.8**

Index

B

C

D

E

F

G

H

I

J

K

L

M

N

O

P

Q

R

S

T

U

V

W

X

Y

Z